13th Global Congress on Manufacturing and Management (GCMM 2016)

MATEC Web of Conferences Volume 100 (2017)

Zhengzhou, China
28 - 30 November 2016

Part 1 of 2

Editors:

Leo Zhao
Anthony Xavior
Jeffrey Cai
Linda You

ISBN: 978-1-5108-3748-5

Printed from e-media with permission by:

Curran Associates, Inc.
57 Morehouse Lane
Red Hook, NY 12571

Some format issues inherent in the e-media version may also appear in this print version.

Printed by Curran Associates, Inc. (2017)

For additional information, please contact EDP Sciences – Web of Conferences at the address below.

EDP Sciences – Web of Conferences
17, Avenue du Hoggar
Parc d'Activité de Courtabœuf
BP 112
F-91944 Les Ulis Cedex A
France

Phone: +33 (0) 1 69 18 75 75
Fax: +33 (0) 1 69 28 84 91

contact-edps@webofconferences.org

Additional copies of this publication are available from:

Curran Associates, Inc.
57 Morehouse Lane
Red Hook, NY 12571 USA
Phone: 845-758-0400
Fax: 845-758-2633
Email: curran@proceedings.com
Web: www.proceedings.com

TABLE OF CONTENTS

PART 1

PART 2

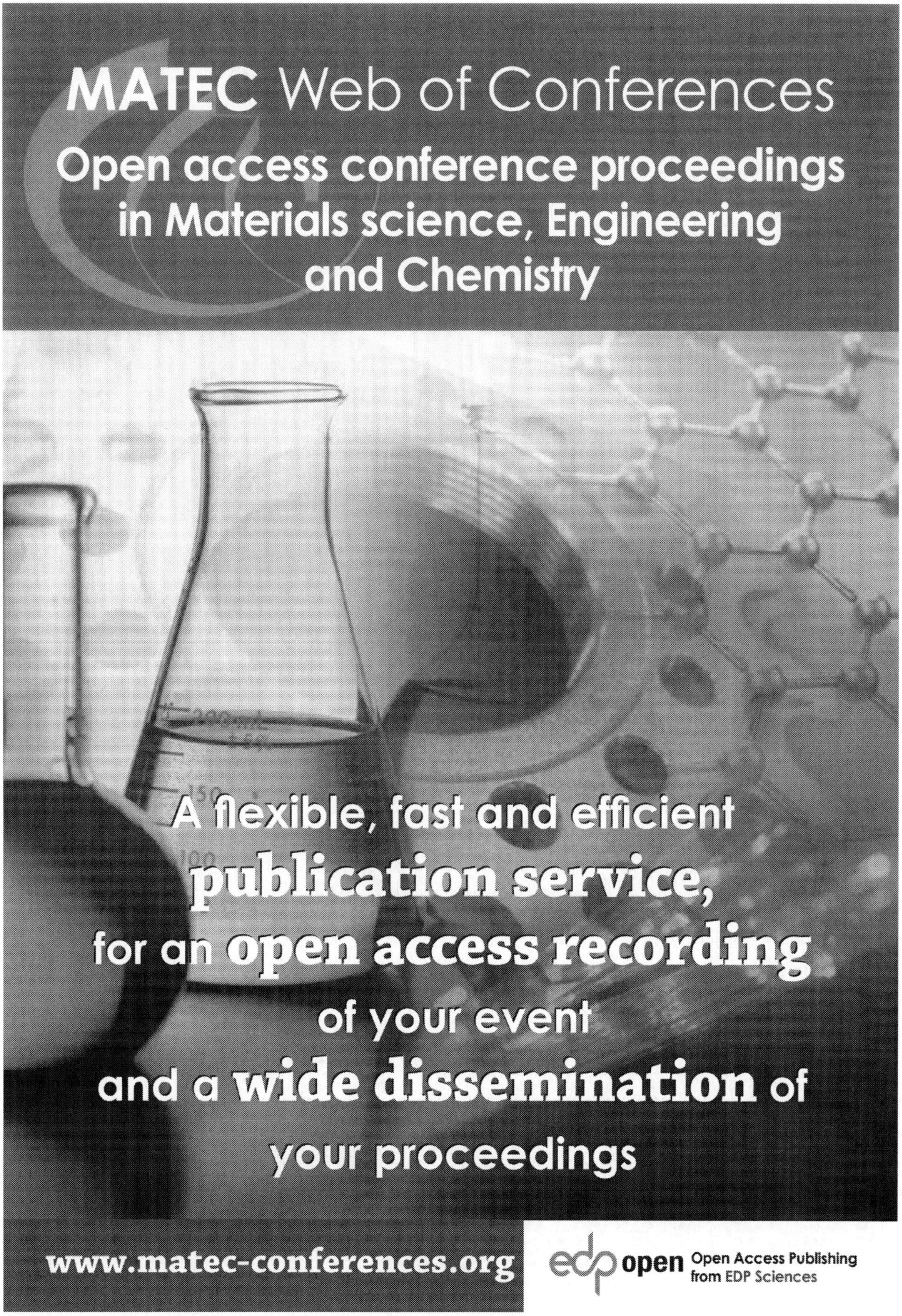
MATEC Web of Conferences
Open access conference proceedings in Materials science, Engineering and Chemistry
A flexible, fast and efficient publication service, for an open access recording of your event and a wide dissemination of your proceedings
www.matec-conferences.org
edp open
Open Access Publishing from EDP Sciences

ISSN (electronic edition): 2261-236X

Extending Driving Vision Based on Image Mosaic Technique

Deng Chen[1,*], Yanduo Zhang[1], Wei Wei[1], Xiaolin Li[1], Huabing Zhou[1], Rui Zhu[1], Tao Lu[1], Xun Li[1], Haijiao Xu[2] and Li Peng[3]

[1] Hubei Provincial Key Laboratory of Intelligent Robot, Wuhan Institute of Technology, Wuhan, China
[2] Department of Computer&Science, Guangdong University of Education, Guangzhou, China
[3] Teaching Guidance Center, Hubei Radio & TV University, Wuhan, China
Corresponding Email: dchen@wit.edu.cn

Abstract. Car cameras have been used extensively to assist driving by make driving visible. However, due to the limitation of the Angle of View (AoV), the dead zone still exists, which is a primary origin of car accidents. In this paper, we introduce a system to extend the vision of drivers to 360 degrees. Our system consists of four wide-angle cameras, which are mounted at different sides of a car. Although the AoV of each camera is within 180 degrees, relying on the image mosaic technique, our system can seamlessly integrate 4-channel videos into a panorama video. The panorama video enable drivers to observe everywhere around a car as far as three meters from a top view. We performed experiments in a laboratory environment. Preliminary results show that our system can eliminate vision dead zone completely. Additionally, the real-time performance of our system can satisfy requirements for practical use.

1 Introduction

In order to reduce car accidents, many electronic devices have been introduced to assist driving, such as parking sensors and Rear-View Car Cameras (RVCCs) [1]. RVCCs facilitate driving enormously by making it visible. However, the Angle of View (AoV) of existing RVCC is less than 180 degrees, that is, there exists dead zone outside our view, which is a primary origin of traffic accidents. To resolve the problem, we design a vehicle panorama system, which can extend the vision of drivers to 360 degrees around a car.

Our system comprises four wide-angle cameras, each of which has an AoV near to 180 degrees. The cameras are mounted at different sides around a car. Obviously, the cameras can capture 4-channel videos around a car. Via switching among the videos, drivers can observe everywhere around a car. However, switching videos may cause distraction from driving. Additionally, the experience is bad to observe four distinct videos, especially when videos overlapped with each other. In order to mitigate the problem, we seamlessly integrate the 4-channel videos into a panorama video. The image processing algorithms used by our system are camera calibration, image distortion correction and image mosaic. The output of our system includes a panorama video and a side video. The panorama video enables drivers to observe the environment around a car as far as three meters from a top view. The side video can be switched among the four cameras automatically. In order to investigate the effect of our system, we conduct experiments based on a mimical car and good results are achieved.

The contributions of this paper are:

• Hardware design of a driving assist system based on DM6437, which can extend driving vision to 360 degrees around a car;

• Image processing algorithms used by our system;

• The investigation of the effect of our system.

The rest of this paper is organized as follows: Section 2 presents the general view of our system. Section 3 introduces our image mosaic technique. Section 4 reports our preliminary experimental results. Section 5 concludes this paper.

2 General view of our system

The hardware of our system mainly consists of the following components: CCD Cameras, TVP5158, DM6437, DDR2 RAM, NOR Flash, Video Output Connector and CH376. Its functioning is illustrated in Fig. **1**. As we can see, it captures video images through four wide-angle cameras mounted at different sides of a car. The video decoder

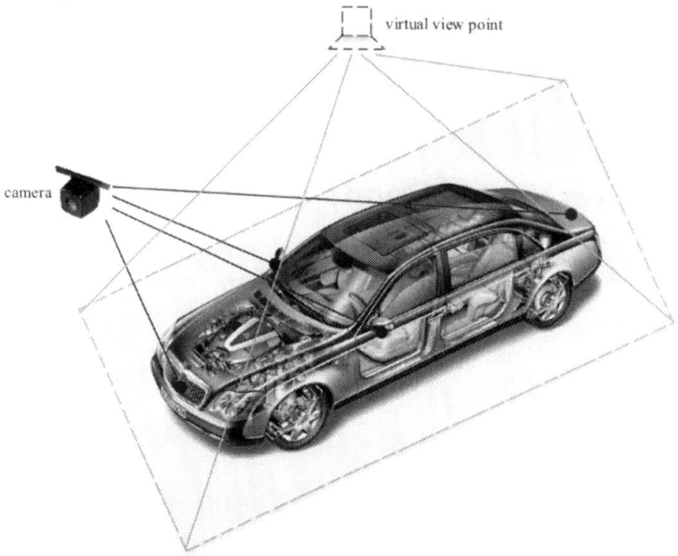

Fig. 1. The functioning of our system.

TVP5158 digitizes and decodes all popular base-band analog video formats into digital video output [2]. The main processor DM6437 [3-7] takes the digital videos as input and processes images based on OpenCV [8]. The storage of our system comprises a 16M NOR Flash and a 128M DDR2 RAM. The former is used to save programs and configuration information. The latter is used to cache videos. Additionally, relying on the file management controller CH376, our system can upgrade software through USB flash drives or SD cards.

Based on the above design, our system can provide drivers with a panorama image around a car from a virtual view point as illustrated in Fig. **1**. The virtual view point is above the centre of a car. Its height can be assign a value between zero and ten meters. By raising the virtual view point, we can broaden the panorama view to a maximum extent of three meters around a car.

3 Image processing

Our system processes images based on the open sourced computer vision library OpenCV. In this section, we elaborate the key image processing algorithms used by our system, inclusive of camera calibration [9,10], image distortion correction [11, 12] and image mosaic [13, 14].

3.1 Camera calibration

Camera calibration is a necessary step for many image processing tasks, such as 3D computer vision and image distortion correction. Through camera calibration, we can achieve the following metric information about a camera from 2D images:

- The extrinsic parameters $(\mathbf{R}, \mathbf{t})$ of a camera, where **R** is a 3 × 3 rotation matrix, **t** is a 3D translation vector;
- The intrinsic matrix **A** of a camera, which is given by

$$\mathbf{A} = \begin{bmatrix} \alpha & \gamma & u_0 \\ 0 & \beta & v_0 \\ 0 & 0 & 1 \end{bmatrix}$$

with (u_0, v_0) the coordinates of the principle point, α and β the scale factors in image u and v axes, and γ the parameter describing the skewness of the two image axes;

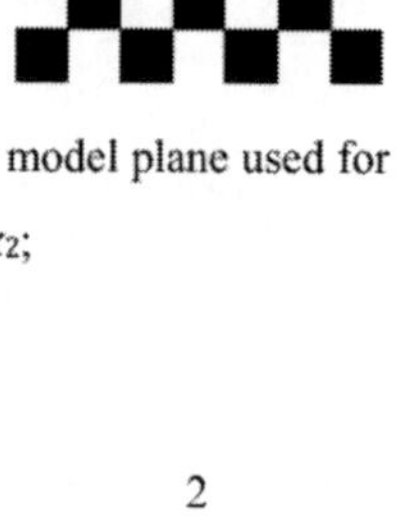

Fig. 2. The model plane used for calibration.

- The coefficients of radial distortion k_1 and k_2;

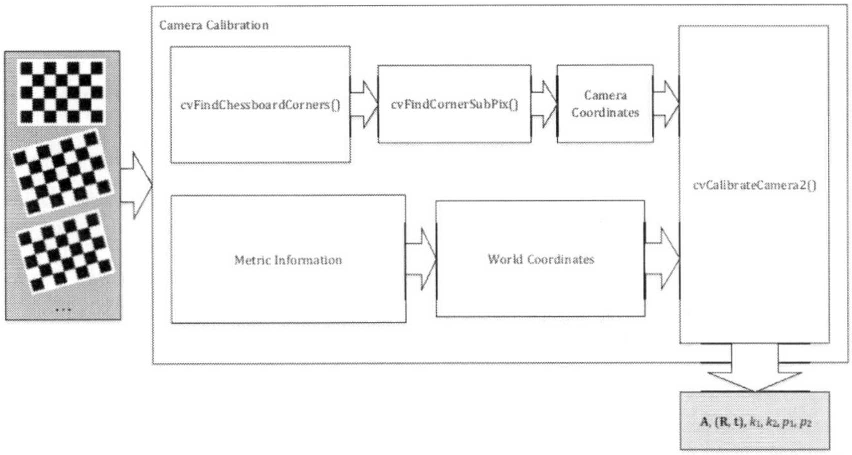

Fig. 3. The working principle for camera calibration.

• The coefficients of tangential distortion p_1 and p_2.

The extrinsic parameters and intrinsic matrix relate the world coordinate system to the camera coordinate system. Let $\tilde{m}$ and $\tilde{M}$ be a 2D and a 3D point respectively (homogeneous coordinates). The relationship between a 3D point $\tilde{M}$ and its image projection $\tilde{m}$ is given by

$$s\tilde{\mathrm{m}} = \mathbf{A}[\mathbf{R} \quad \mathbf{t}]\tilde{\mathrm{M}},$$

where s is an arbitrary scale factor. The distortion coefficients are essential for the image distortion correction.

We calibrate cameras using the method proposed by Zhang [10], which only requires the camera to observe a planar pattern shown at a few different orientations. As shown in Fig. **2**, our system utilizes a chessboard for calibration. It contains a pattern of 7 × 5 squares, so there are 24 internal corners (i.e., points where the black squares touch each other.) The size of the chessboard is 12.9 × 9.2 centimeters and that of each square is 1.9 × 1.9 centimeters. We print the chessboard with a high-quality printer and then put it on a plastic board. Before calibrating a camera, we hold the plastic board by hand and place it under the camera as far as its image fills in the whole screen. Our system takes eight gray images of the chessboard as input of the calibration algorithm (the gray image can be obtained by extracting the Y channel of an image in format of YCbCr-422.) Since the hand will inevitably shake, the eight images must be under different orientations.

The working principle of the calibration algorithm is shown in Fig. **3**, which mainly consists of three OpenCV functions, i.e., cvFindChessboardCorners(), cvFindCornerSubPix() and cvCalibrateCamera2(). The first function attempts to determine whether the input image is a view of the chessboard pattern and locate the internal chessboard corners. A non-zero value is returned if all of the corners are found and ordered. Otherwise, it returns zero. What should be noted is that the coordinates detected by cvFindChessboardCorners() are approximate. To determine their position more accurately, we should subsequently call the function cvFindCornerSubPix() which can refine the results obtained by cvFindChessboardCorners() and achieve the sub-pixel location of corners. After that, we compute the world coordinates of the corners based on the metric information of the chessboard. Finally, we take the camera coordinates achieved by cvFindCornerSubPix() and the computed world coordinates as input and call the function cvCalibrateCamera2() to estimate the intrinsic camera parameters, distortion coefficients and extrinsic parameters for each of the chessboard images.

3.2 Image distortion correction

As shown in Fig. **4**, serious distortion exists in the images captured by our system. To facilitate image mosaic, we should correct them. We perform the image distortion correction based on the OpenCV function cvUndistort2(). Let's assume that G' is the corrected image of G. cvUndistort2() corrects image G by mapping each pixel of G' to a pixel in G. The mapping is calculated based on the intrinsic matrix **A** and distortion coefficients k_1, k_2, p_1 and p_2, which are achieved by the above camera calibration. The corrected image output by cvUndistort2() has the same size and type as the input image. What should be noted is that the function cvUndistort2() only works with gray images.

Fig. 4. An example image captured by our system.

3.3 Image mosaic

Our system performs the task of image mosaic based on a two-phase approach. The first phase is carried out only once, in which we calculate and save the image projection parameters into the Flash. The second phase is executed in each frame, which integrates the 4-channel images into a whole based on the projection parameters achieved in the first phase. Below we will elaborate our approach in detail.

As we can see from Fig. **5** (a), Ow-XwYwZw is the world coordinate system. Its origin is located on the ground, at the center of the car. We assume that there exists a virtual camera, which is h meters high above the center of the car. Its camera coordinate system is Oc-XcYcZc. The goal of our image mosaic technique is to project images captured by four cameras into the image I_v, which is output by the virtual camera. Let I_1, I_2, I_3, I_4 be the corrected images captured by the front, rear, left and right cameras respectively. In order to project these images to I_v, we should achieve the homography matrixes $\mathbf{H}_i$, i = 1...4, which are between I_i and I_v, respectively. As shown in Fig. **5** (a), we compute the homography matrixes based on four chessboards, which are the same as that used for camera calibration, except that they have different sizes. The size of the chessboards used for image mosaic depends on the installation height of cameras. To prepare for image mosaic, we lay all chessboards on the ground, around the car. The front and rear ones are in the middle of the car. The left and right ones are l_1 and l_2 meters off the front of the car. What should be noted is that, the chessboards should be placed in a position where their images fill in the screen as much as possible. Obviously, based on the above metric information and the size of the car, we can calculate the world coordinates of all chessboard corners conveniently. Assuming P_w to be the set of corner points (world coordinates) of the front chessboard, our method of computing the homography matrix proceeds as follows:

1) Projecting Pw to the set of image points Pv from the perspective of the virtual camera. This task can be accomplished with the OpenCV function cvProjectPoints2(), which takes Pw and the parameters of the virtual camera as input.
2) Finding the set of corner points P1 from image I1 captured by the front camera. According to the introduction of camera calibration, this task can be carried out by using OpenCV functions cvFindChessboardCorners() and cvFindCornerSubPix().
3) Calculating the homography matrix H1 based on Pv and P1 achieved in step 1) and 2) respectively. We perform this task leveraging the OpenCV function cvFindHomography().
4) Computing the inverse matrix $\boldsymbol{H}_1^{-1}$ of H1 by using the OpenCV function cvInvert().

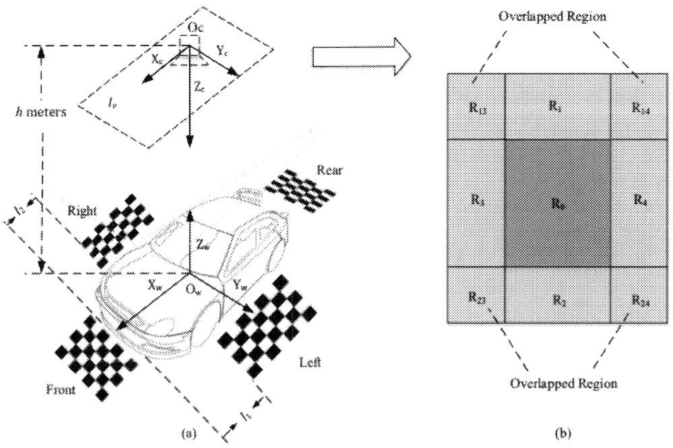

Fig. 5. The functioning of our image mosaic technique.

After that, we construct image I_v based on the following equation:

$$\mathrm{clr}(I_v,P)=\begin{cases}\mathrm{clr}(I_{car},SP) & if\ P\in R_0\\ \mathrm{clr}(I_1,\mathrm{H}_1^{-1}P) & if\ P\in R_1\\ \mathrm{clr}(I_2,\mathrm{H}_2^{-1}P) & if\ P\in R_2\\ \mathrm{clr}(I_3,\mathrm{H}_3^{-1}P) & if\ P\in R_3\\ \mathrm{clr}(I_4,\mathrm{H}_4^{-1}P) & if\ P\in R_4\\ \mathrm{mrg}\left(\mathrm{clr}(I_1,\mathrm{H}_1^{-1}P),\mathrm{clr}(I_3,\mathrm{H}_3^{-1}P)\right) & if\ P\in R_{13}\\ \mathrm{mrg}\left(\mathrm{clr}(I_1,\mathrm{H}_1^{-1}P),\mathrm{clr}(I_4,\mathrm{H}_4^{-1}P)\right) & if\ P\in R_{14}\\ \mathrm{mrg}\left(\mathrm{clr}(I_2,\mathrm{H}_2^{-1}P),\mathrm{clr}(I_3,\mathrm{H}_3^{-1}P)\right) & if\ P\in R_{23}\\ \mathrm{mrg}\left(\mathrm{clr}(I_2,\mathrm{H}_2^{-1}P),\mathrm{clr}(I_4,\mathrm{H}_4^{-1}P)\right) & if\ P\in R_{24}\end{cases} \quad (1)$$

where P is a point (homogeneous coordinates) in image I_v. Given an image I and a point p in I, clr(I, p) represents the color of pixel p in I. I_{car} denotes a car image, which is in the format of RGB565. S is a 3D vector (sx, sy, 1), where sx and sy are the horizontal and vertical scale factors between I_{car} and I_v respectively. R_0~R_4, R_{13}, R_{14}, R_{23} and R_{24} are different regions of image I_v, which are shown in Fig. **5** (b). According to the composition of pixels, we can divide these regions into three categories: 1) Region R_0 is the projection of the car, which will be filled with a car image. Its size can be determined according to the projection of car corners. 2) Regions R_1, R_2, R_3 and R_4 comprise pixels coming from I_1, I_2, I_3 and I_4, respectively. 3) Regions R_{13}, R_{14}, R_{23} and R_{24} are overlapped regions, that is, a pixel of the regions may be the fusion of two pixels originating from different cameras. For example, a pixel of R_{13} may be a mixture of two pixels coming from images I_1 and I_3 respectively. For an overlapped pixel, we compute its color based on the mrg function, which takes two source pixels as input and outputs the mixed pixel. The mrg function is crucial for our system. Taking a pixel $(x,y)\in R_{13}$ for example, we utilize the following mrg function in this work:

$$\mathrm{mrg}(x,y)=w_1(x_1,y_1)I_1(x_1,y_1)+w_3(x_3,y_3)I_3(x_3,y_3) \quad (2)$$

where $w_i(x_i,y_i)$, $i\in\{1,3\}$ is the weight function. Its form is given as follows:

$$w_i(x_i,y_i)=\frac{k_i(x_i,y_i)}{\sum_i k_i(x_i,y_i)}$$

$$k_i(x_i,y_i)=\left(1-\left|\frac{x_i}{width(I_i)}-\frac{1}{2}\right|\right)\times\left(1-\left|\frac{y_i}{height(I_i)}-\frac{1}{2}\right|\right)$$

Additionally, we have

$$(x_i \quad y_i \quad 1)^T=\mathrm{H}_i^{-1}(x \quad y \quad 1)^T$$

According to [15], the image mosaic technique based on the above weight function should work well.

In order to improve the real-time performance of our system, for each pixel (x, y) in image I_v, we save the following information into the Flash in the first phase: 1) pixel (x_1, y_1) in the first source image; 2) pixel (x_2, y_2) in the second source image; 3) the weight of the first source pixel; and 4) the weight of the second source pixel. If a pixel has only one source pixel (e.g. pixels in regions R_0~R_4), we invalidate the other source pixel by setting its weight to zero. Based on the results of the first phase, we can generate the panorama image I_v from (2) directly in each frame. Our method is efficient, because the first phase is performed only once. In each frame, we carry out only two multiplications and one addition for each pixel.

4 Experiments

In order to investigate the effect of our system, we performed experiments based on a mimical car, which is a cardboard box of the size of 1.4 × 1.2 meters. The chessboards used for image mosaic have the same size of an A4 paper. We mounted the cameras on the four sides of the cardboard box and launched our system. The effect of our system is shown in Fig. **6**. As we can see, the panorama video is displayed on the left side, from which we can observe everywhere around a car from a top view. The side video is displayed on the right side, which can be switched automatically among the four cameras.

We also investigated the real-time performance of our system and found that our system achieved 25 frames per second on average, which can satisfy the requirements for practical use.

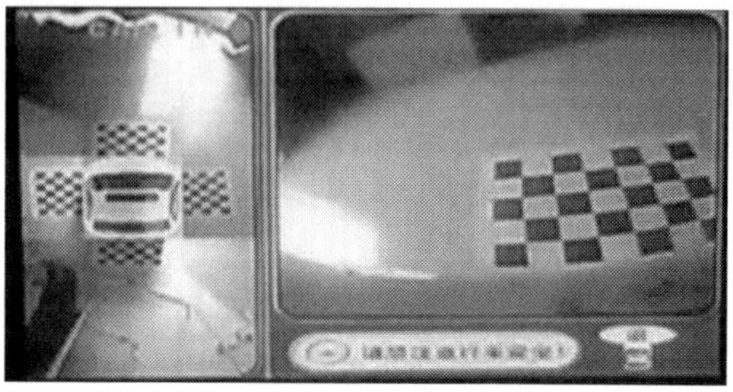

Fig. 6. The effect of our system.

5 Conclusions

In this paper, we gave the design of a vehicle panorama system. Our system leverages the techniques of camera calibration, image distortion correction and image mosaic to provide drivers with a panorama view of a car. The panorama view is able to extend the vision of drivers to 360 degrees, which provides a strong assurance for driving cars safely. In this work, we introduced design and implementation details of our system. We believe it is a good reference for the development of similar systems.

Acknowledgements

This work was supported by the Youths Science Foundation of Wuhan Institute of Technology (No. k201622), Surveying and Mapping Geographic Information Public Welfare Scientific Research Special Industry (No. 201412014), National Natural Science Foundation of China (No. 41501505, 61502355 and 61502354), Nature Science Foundation of Hubei Province (No. 2014CFB779) and the Key Program of Higher Education Institutions of Henan Province (No. 17A520040).

References

1. J. Ju, G. Han, M. Lin, et al., Design of a vehicle panorama reversing system based on embedded system, in *Chinese Automation Congress (CAC)*, Wuhan, China, pp. 1528-1532 (2015)
2. Texas Instruments, Data Manual for TVP5158, TVP5157 and TVP5156 (2011)
3. L. Kang, Z. Zhao, G. Xie, The hardware design of dual-mode wireless video surveillance system based on DM6437, in *2nd International Conference on Networks Security, Wireless Communications and Trusted Computing*, pp. 546-549 (2010)
4. D. Chen, Y. Zhang, W. Wei, et al., Design of a panorama parking system based on DM6437, in *Proceedings of the 9th International Congress on Image and Signal Processing, BioMedical Engineering and Informatics* (to be published)
5. B. Li, W. Zhang, Z. Liu, et al., Development and implement of the IP camera, in *2011 International Conference on Mechatronic Science, Electric Engineering and Computer*, Jilin, China, pp. 1961-1964 (2011)
6. Texas Instruments, User's Guide for TMS320DM643x DMP Video Processing Back End (VPBE) (2007)
7. Texas Instruments, User's Guide for TMS320DM643x DMP Video Processing Front End (VPFE) (2007)
8. OpenCV, http://opencv.org (2016)
9. Q. Chen, H. Wu, S. Higashino, et al, Camera calibration by recovering projected centers of circle pairs, in *ACM SIGGRAPH 2016 Posters*, Anaheim, California, pp. 1-2 (2016)
10. Z. Zhang, A flexible new technique for camera calibration, IEEE Transactions on Pattern Analysis and Machine Intelligence **22**, 11, pp. 1330-1334 (2000)
11. M.S. Brown, D. Tsoi, Correcting common distortions in camera-imaged library materials, in *Proceedings of the 3rd ACM/IEEE-CS joint conference on Digital libraries*, Houston, Texas, pp. 367-368 (2003)
12. S. Lee, J. Lee, J. Choi, Lens distortion correction using a checkerboard pattern, in *Proceedings of the 7th ACM SIGGRAPH International Conference on Virtual-Reality Continuum and Its Applications in Industry*, Singapore, pp. 1-2 (2008)
13. X. Shao, C. Xu, J.H. Lim, Image mosaics base on homogeneous coordinates, in *2002 Pan-Sydney Workshop on Visualisation*, Sydney, Australia, pp. 93-96 (2002)
14. R. Szeliski, H. Shum, Creating full view panoramic image mosaics and environment maps, in *Proceedings of the 24th Annual Conference on Computer Graphics and Interactive Techniques*, pp. 251-258 (1997)
15. Q. Ruan, Shi shi shi pin pin jie xi tong guan jian ji shu yan jiu, M. Chs thesis, Huazhong University of Science and Technology, Wuhan, China (2011)

Parameters Designing of Slide Mode Variable Structure Controller of Bus Voltage of DC Microgrid Based on Proportion Switching Function

Zhenchuan Sun[1,2], Jiasheng Zhang[1] and Zhen Guo[1,3]

[1]College of Information and Control Engineering, China University of Petroleum(East China), Qingdao, Shandong, China
[2] School of Mechanical and Electronic Engineering, Zaozhuang University, Zaozhuang, Shandong, China
[3] College of Automation Engineering, Qingdao University of Technology, Qingdao, Shandong, China
Corresponding Email: author@e-mail.org

Abstract. Constant value control of the DC-bus voltage is a essential problem of the control system of the DC microgrids. DC-DC converters are applied in parallel to realize the transform of energy from the distributed generations (DGs) to the DC-bus. Droop control methods are applied to the DC-bus voltage while PI controllers are used in controlling the duty ratios of the converters. This method may bring out the slow response speed of the system accompanied by the large ripple of the voltage. The slide mode variable structure control can speed up the response and reduce the ripple of the voltage as well. In the traditional slide mode control based on the proportion switching function, the denominator of the transfer function of the controlled plant is a second-order characteristic polynomial without the constant term. The denominators of the transfer functions of the buck DC-DC converters contain the constant terms. The designing of the parameters of the slide mode control based on the proportion switching function is analyzed based on mathematics deductions. Simulation results show that the selected parameters can not only speed up the response of the system but also greatly reduce the ripple of the voltage.

1 Introduction

To solve the problem caused by energy voltage and envi-ronmental pollution, more and more attentions are payed to the distributed power generation technology.There are many advantages in the DGs such as lesspollutions, higher energy efficiency, and flexible loca-tions. Compared with the centralized power generations,it save the operating costs with reducing the line losses. With the development of the power electronics and themodern control theory, the microgrid emerges as anorganic whole combined with DGs, loads, energystorage devices and control devices. It may be operatedby the grid-connected mode and the island mode. ACmicrogrids are the main form of the microgrids.DC-buses are applied to transform the energy from theDGs to the loads in the DC microgrids.The reliabilityand controllability is improved as the control of thefrequency and the phase position is undesired in the DCmicrogrid [1,2].

Constant value control of the DC-bus voltage is aessential problem of the control system of the DC microgrid [3]. The droop control method is applied in the traditional control method of the dc-bus voltage [4].The duty ratios of the DC-DC converters are controlled by the PI controllers which make the dc-bus voltage constant [5]. With the droop controllers and the PI controllers applied to the system, the response time may be long and the ripple of the dc-bus voltage may be large. In the dynamic process of the control system, the slide mode variable structure controller can vary on purpose according to the current state of the system which make the state move along the predetermined path of the slide state. The fast response may reduce the ripple of the dc-bus voltage and show the strong robustness [6].

By analyzing the parameters designing of the slide mode variable structure controller based on the proportion switching function, using the transform function of the buck DC-DC converters with small-signal equivalent model, this paper makes a lot of theoretical calculations as the selecting of the parameters of the controller. With the phase plane as the analytical tool, the motion characteristics and stability is analyzed while the method proposed in this paper is tested by Matlab/Simulink. The simulation results show that the selected parameters can speed up the response and reduce the ripple of the dc-bus voltage. The robustness is present in the proposed method of the designing of the parameters.

2 The circuit of the DC microgrid control system

The circuit of the DC microgrid control system is shown as Figure 1.

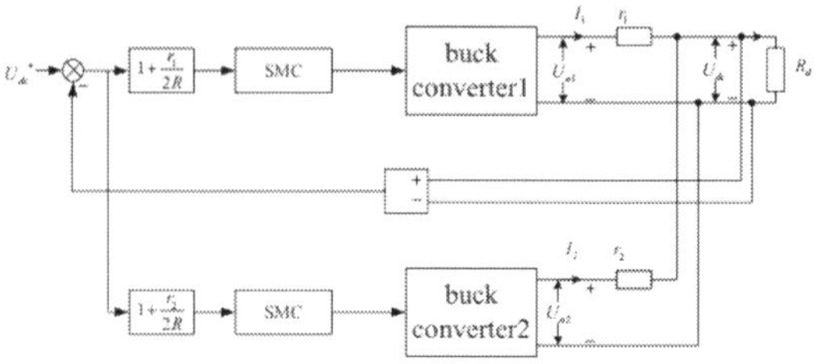

Fig. 1. The circuit of the DC microgrid control system.

In the DC microgrid control system, two buck DC-DC converters are applied to realize the transform of the energy. The values of the inductances, capacitances and resistances are the same while the input voltages of them are not, so as to simplify the analysis. The modified droop control strategy is used to realize the current-sharing control of the two converters [7,8]. The SMC is the slide mode variable structure controller based on the proportion switching function.The transfer function of the buck converter is approximate using small-signal equivalent model. The impedances of the cables from the convertes to the load is supposed as the pure resistances. The load is a equivalent pure resistor load [9,10].

3 Parameters designing of the SMC of the DC microgrid system

The transfer function of the buck converter of the small-signal equivalent model is as formula (1).

$$G(s)\big|_{U_i=const} = \frac{U_o(s)}{d(s)} = \frac{U_i}{LCs^2+\frac{L}{R}s+1} = \frac{\frac{U_i}{LC}}{s^2+\frac{1}{RC}s+\frac{1}{LC}} \tag{1}$$

While U_i is the input voltage of the buck converter. It is not the same as the transfer function of the SMC based on the proportion switching function in form because of its denominator is a second-order characteristics polynomial with a constant term. The form of common transfer function is $\frac{b}{s^2+as}$.

Suppose that $G(s)=\frac{U_o(s)}{d(s)}=\frac{x_1(s)}{u(s)}$,formula (1) can be rewrited as the state space model as formula (2).

$$\dot{x} = Ax + Bu$$
$$\begin{bmatrix}\dot{x}_1\\ \dot{x}_2\end{bmatrix} = \begin{bmatrix}0 & 1\\ -\frac{1}{LC} & -\frac{1}{RC}\end{bmatrix}\begin{bmatrix}x_1\\ x_2\end{bmatrix} + \begin{bmatrix}0\\ \frac{U_i}{LC}\end{bmatrix}u \tag{2}$$

While $\dot{x} = \begin{bmatrix}\dot{x}_1\\ \dot{x}_1\end{bmatrix}$.

According to the control method of the SMC based on the propotion swithching function, the position command signals of the two SMC controllers are the output voltages of the two converters. That is to say

$$\begin{cases} e = r - x_1 = U_{oref} - x_1 \\ \dot{e} = \dot{r} - \dot{x}_1 = -x_2 \end{cases} \tag{3}$$

From formula (2) and (3),we can get

$$-\begin{bmatrix}\dot{e}\\ \ddot{e}\end{bmatrix} = \begin{bmatrix}0 & 1\\ -\frac{1}{LC} & -\frac{1}{RC}\end{bmatrix}\begin{bmatrix}U_{oref}-e\\ -\dot{e}\end{bmatrix} + \begin{bmatrix}0\\ \frac{U_i}{LC}\end{bmatrix}u \tag{4}$$

While $u = (\alpha|e| + \beta|\dot{e}|)\operatorname{sgn}(s)$, $s = ce + \dot{e}$. This paper focuses on the designing of the parameters α, β and c.

In order to simplify the problem, we make following assumptions:

(1) We suppose that $\alpha > 0$, $\beta > 0$, and $c < 0$.

(2) Only the locus of the slide mode from the initial point to the switching surface is considered

(3) The two SMCs of the two converters are designed on the same theory.

That is to say, the only situation that should be researched is that $e>0$, $\dot{e}<0$ and $s>0$. This situation shows that the actual value is gradually approaching the setting value.

If $s>0$,then $\mathrm{sgn}(s)=1$, $u=\alpha e-\beta\dot{e}$.

$$-\begin{bmatrix}\dot{e}\\ \ddot{e}\end{bmatrix}=\begin{bmatrix}0 & 1\\ -\frac{1}{LC} & -\frac{1}{RC}\end{bmatrix}\begin{bmatrix}U_{oref}-e\\ -\dot{e}\end{bmatrix}+\begin{bmatrix}0\\ \frac{U_i}{LC}\end{bmatrix}\begin{bmatrix}\alpha & -\beta\end{bmatrix}\begin{bmatrix}e\\ \dot{e}\end{bmatrix} \tag{5}$$

$$\begin{bmatrix}\dot{e}\\ \ddot{e}\end{bmatrix}=\begin{bmatrix}0 & 1\\ -\frac{1+U_i\alpha}{LC} & \frac{U_i\beta}{LC}-\frac{1}{RC}\end{bmatrix}\begin{bmatrix}e\\ \dot{e}\end{bmatrix}+\begin{bmatrix}0\\ \frac{U_{oref}}{LC}\end{bmatrix} \tag{6}$$

According to the reaching condition of the slide mode, we get

$$s\dot{s}<0 \tag{7}$$

So, $\dot{s}<0$.

Using formula (6) we can get

$$\begin{aligned}\dot{s}&=c\dot{e}+\ddot{e}=\begin{bmatrix}c & 1\end{bmatrix}\begin{bmatrix}\dot{e}\\ \ddot{e}\end{bmatrix}\\ &=\begin{bmatrix}c & 1\end{bmatrix}\left\{\begin{bmatrix}0 & 1\\ -\frac{1+U_i\alpha}{LC} & \frac{U_i\beta}{LC}-\frac{1}{RC}\end{bmatrix}\begin{bmatrix}e\\ \dot{e}\end{bmatrix}+\begin{bmatrix}0\\ \frac{U_{oref}}{LC}\end{bmatrix}\right\}\\ &=\frac{U_{oref}}{LC}-\frac{1+U_i\alpha}{LC}e+\left(\frac{U_i\beta}{LC}+c-\frac{1}{RC}\right)\dot{e}\\ &=\frac{U_{oref}}{LC}-\frac{1+U_i\alpha}{LC}e+\frac{U_i\beta R+RLCc-L}{RLC}\dot{e}<0\end{aligned} \tag{8}$$

If $\frac{U_i\beta R+RLCc-L}{RLC}>0$, then the last line of formula (8) can be calculated as

$$\dot{e}<\frac{(1+U_i\alpha)R}{U_i\beta R+RLCc-L}e-\frac{RU_{oref}}{U_i\beta R+RLCc-L} \tag{9}$$

Since $1+U_i\alpha>0$, so formula (9) can be expressed as the straight line l_2 in Figure 2 which phase plane plot with e as its x axes and $\dot{e}$ as its y axes.

As shown in Figure 2, the x-coordinate of A is $\frac{U_{oref}}{1+U_i\alpha}$, while the y-coordinate of B is $-\frac{RU_{oref}}{U_i\beta R+RLCc-L}$.

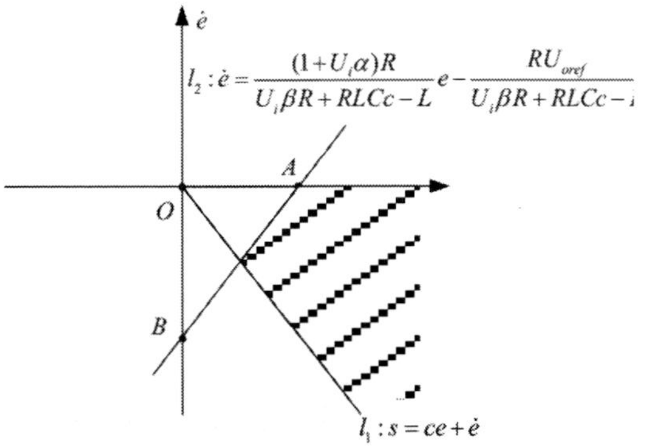

Fig. 2. The phase plane plot of the SMC.

According to the plane geometry,inequality systems

$$\begin{cases} s = ce + \dot{e} > 0 \\ \dot{s} = \frac{U_{oref}}{LC} - \frac{1+U_i\alpha}{LC}e + \frac{U_i\beta R + RLCc - L}{RLC}\dot{e} < 0 \end{cases} \tag{10}$$

is expressed by the shaded area in Figure 2.That is to say, the locus of the slide mode from the initial point to the line l_1 should be within the shaded area in Figure 2.

So,we can get inequality constrains as follows:

$$\begin{cases} \frac{U_{oref}}{1+U_i a} < U_{oref} \\ U_i\beta R + RLCc - L > 0 \end{cases} \tag{11}$$

Formula (11) can be calculated as

$$\begin{cases} \alpha > 0 \\ \beta > \frac{L - RLCc}{U_i R} \\ \frac{L - U_i\beta R}{RLC} < c < 0 \end{cases} \tag{12}$$

If $\frac{U_i\beta R + RLCc - L}{RLC} < 0$, then the last line of formula (8) can be calculated as

$$\dot{e} > \frac{(1+U_i\alpha)R}{U_i\beta R + RLCc - L}e - \frac{RU_{oref}}{U_i\beta R + RLCc - L} \tag{13}$$

As the analysis mentioned above, formula (13) can be expressed as the straight line l_3 in Figure 3.

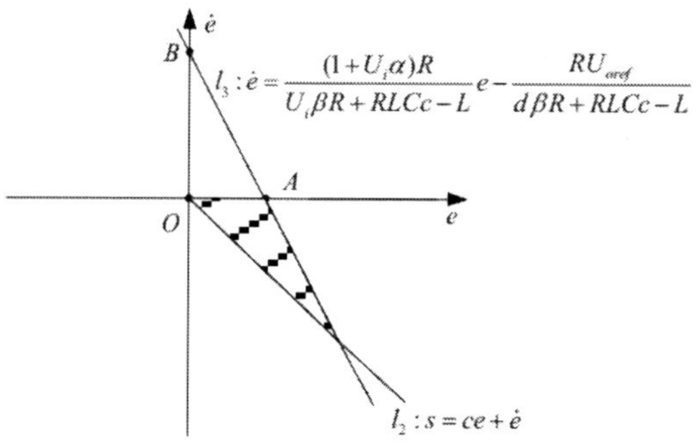

Fig. 3. The phase plane plot of the SMC.

Suppose the locus of the slide mode lies in the shaded area, we can get

$$\frac{U_{oref}}{1+U_i a} > U_{oref}, 0 < 1 + U_i\alpha < 1 \tag{14}$$

Formula (14) is in conflict with the initial conditions including that $\alpha > 0$ .So the suppose that $\frac{U_i\beta R + RLCc - L}{RLC} > 0$ is false.

From the above, the parameters should satisfy the following inequality constrains from the initial point to the switching surface is formula (12).

4 Simulation Results and Analysis

The circuit parameters of the two buck converters are as follows: $R = 0.6\Omega$, $L = 6\times10^{-5}$ H, $C = 4\times10^{-3}$ F, $f_s = 4\times10^4$ Hz. The input voltages of the two buck converters are U_{i1} and U_{i2}. $U_{i1} = 200$ V,

$U_{i2}=400\text{V}$. The initial duty cycles of the two buck converters are d_1 and d_2. $d_1=56.25\%$, $d_2=31.25\%$. The reference of the dc-bus voltage changes with time as follows:

$$t:\ 0\sim 0.1s, U_{dcref}=100\text{V},$$
$$0.1s\sim 0.2s, U_{dcref}=120\text{V},$$
$$0.2\sim 0.3s, U_{dcref}=110\text{V},$$
$$0.3\sim 0.4s, U_{dcref}=130\text{V}.$$

The parameters α, β and c are selected by the formula (12) are as follows:

$$\begin{cases} \alpha>0 \\ \beta>\dfrac{6\times10^{-5}-1.44\times10^{-7}c}{1.2\times10^{2}}>\dfrac{6\times10^{-5}-1.44\times10^{-7}c}{2.4\times10^{2}} \\ \dfrac{6\times10^{-5}-2.4\times10^{2}\beta}{1.44\times10^{-7}}<\dfrac{6\times10^{-5}-1.2\times10^{2}\beta}{1.44\times10^{-7}}<c<0 \end{cases} \tag{15}$$

As the formula (15) mentioned above, we take $\alpha=500$, $\beta=10$, $c=-50$ as an example.

The simulation results based on the theory above are figures as follows:

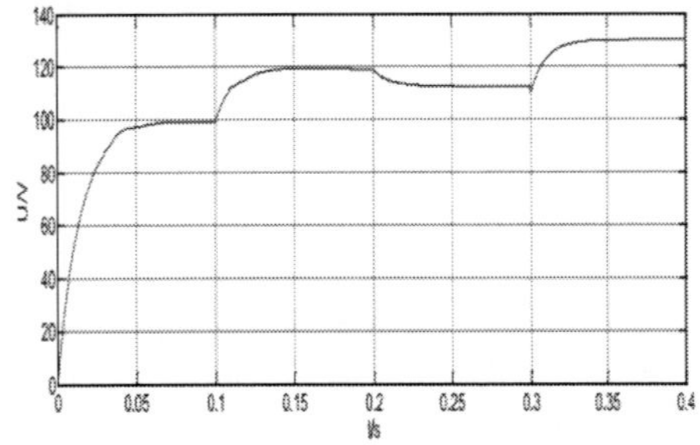

Fig. 4. The dc-bus voltage of the DC microgrid based on SMC.

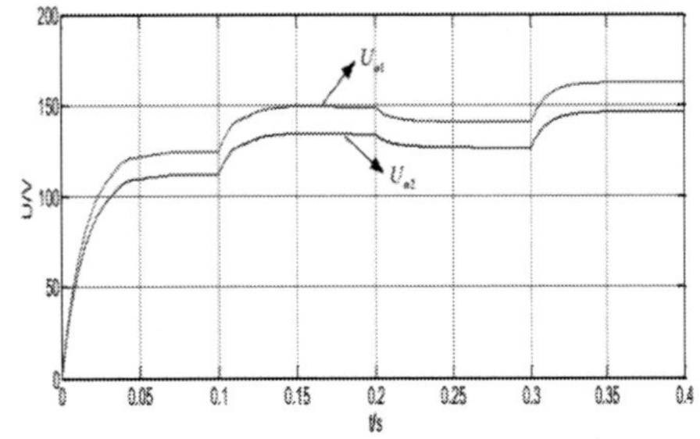

Fig. 5. The output voltages of the two buck converters based on SMC.

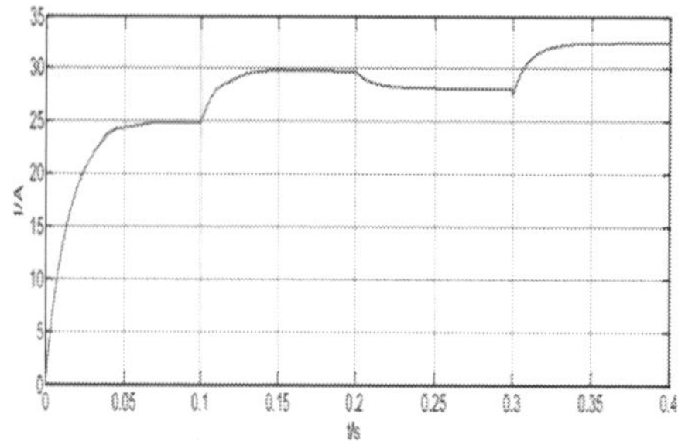

Fig. 6. The output currents of the two buck converters based on SMC.

By constract, the simulation results based on the PI controllers are shown as follows. The parameters of the two PI controllers are $k_{P_1} = 1.25$, $k_{I_1} = 1$; $k_{P_1} = 1.5$, $k_{I_1} = 1$ [11]. Based on that the response time of PI controllers is longer than it of SMC.The reference of the dc-bus voltage is reset as follows:

$$t:\ 0 \sim 5s, U_{dcref} = 100\text{ V},$$

$$5s \sim 10s, U_{dcref} = 120\text{ V},$$

$$10s \sim 15s, U_{dcref} = 110\text{ V},$$

$$10s \sim 15s, U_{dcref} = 130\text{ V}.$$

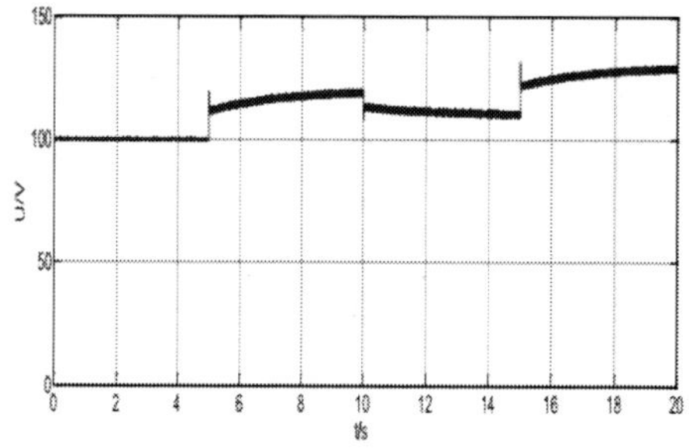

Fig. 7. The dc-bus voltage of the DC microgrid based on PI controller.

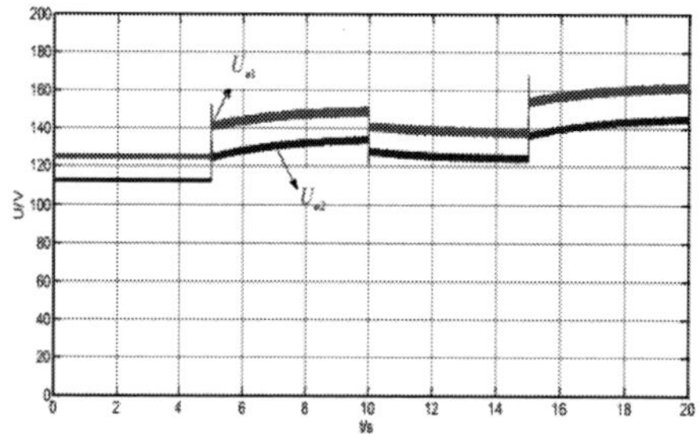

Fig. 8. The out voltages of the two buck converters based on PI controller.

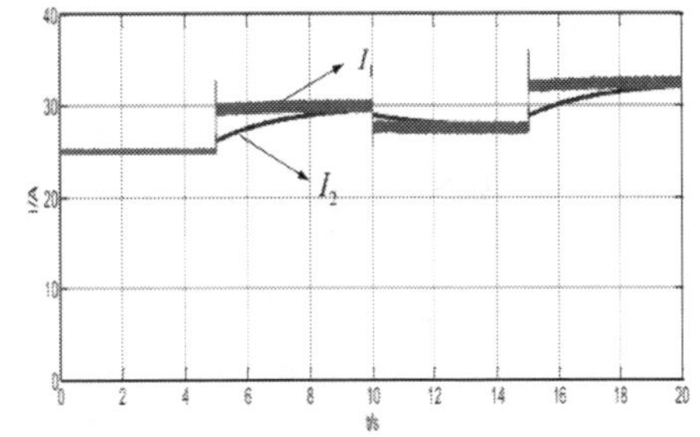

Fig. 9. The output currents of the two buck converters based on PI controller.

From Figure 4 to Figure 6 we can come to the conclusion that the actual dc-bus can track its four different set values within 0.1s based on the SMC.The output currents of the two buck dc-dc converters are identical.

From Figure 7 to Figure 9 we can come to the conclusion that the actual dc-bus can also track its four different set values based on the PI controllers. But the response time is 5s. The output currents of the two buck dc-dc converters are the same after 5s. Besides, from Figure 4 to Figure 9, we can get that the ripples of the voltages and the currents based on the PI controllers are larger than them based on the SMC.

The simulation results show that the SMC including reasonable parameters is superior to the PI control. The method of the parameter designing may also be applied to other controlled plant which transfer function has the second-order characteristic polynomial as its denominator including constant term.

5 Conclusion

Based on the mathematical derivations of the theory of the proportion switching function of the SMC, a new parameter designing of the SMC has been present in this paper. Taking the buck dc-dc converters of the DC microgrids as an example, using the transfer function between and duty cycle and the output voltage based on the small-signal model, this paper has provided us many valuable conclusions. The simulation results show that the SMC including the reasonable parameters is superior to the PI control. Furthermore, the research of the state which has reached the switching surface along with the change of the parameters of the SMC based on the proportion switching function will be dealt with in future.

Acknowledgement

First of all, I would like to extend my sincere gratitude to my supervisor, Zhang Jiasheng, for his instructive advice and useful suggestions on my thesis. I am deeply grateful of his help in the completion of this thesis. I am also deeply indebted to all the other tutors and teachers for their direct and indirect help to me. Finally, I am indebted to my mother and my wife for their continuous support and encouragement.

References

1. J.H.Zhang, W.Huang. Operation control and protection technology of Microgrid. 1-5(2009)
2. C.S.Wang. Analysis and simulation theory of Microgrid. Science Press, Beijing(2013)
3. W.M.Wu,Y.B.He,P.Geng. Transactions of China Electrotechnical Society , **Vol.27,No.1**, pp. 98-113, (2012)
4. F.Gao, Y.Gu, S.Bozhko, G.Asher, P. Wheeler.2014 16th European Conference on Power Electronics and Applications(2014)
5. D.H.Xu. Modeling and control of power electronics system. 43(2005)
6. J.K.Liu. MATLAB simulation for sliding mode control. 25-30(2005)
7. P. Chaudhari, M. Khadse, V. Jadhav, P. Patil, K. Daware, M.Rajwade, S.Sharma.2015 IEEE International Conference on Signal Processing,Informatics, Communication and Energy Systems,(2015)
8. L. Zhang, T.J. Wu,Y. Xing,K. Sun,J.M. Gurrero. IEEE Applied Power Electronics Conference and Exposition,1926-1932(2011)
9. S.Augustine ,M. K. Mishra, N.Lakshminarasamma. IEEE Transactions on Sustainable Energy. **Vol.6,No.1**, pp. 132-141(2015)
10. L. Zhang, T.J. Wu, Y. Xing, K. Sun, J.M. Gurrero. IEEE Applied Power Electronics Conference and Exposition, 1926-1932(2011)
11. J.K. Liu. MATLAB simulation for advanced PID control(3rd edition),1-8(2011)

Analysis on Dynamic Transmission Accuracy for RV Reducer

Fengshou Zhang[1], Pengfei Li[2,*], Peng Zhu[2], Xingxing Yang[2] and Wenhe Jiang[3]

[1]School of Medical Technology and Engineering, Henan University of Science and Technology, Luoyang, China
[2]School of Mechatronics Engineering, Henan University of Science and Technology, Luoyang, China
[3]School of Mechatronics Engineering, China University of Mining and Technology, Xuzhou, China
Corresponding Email:18238832971@163.com

Abstract. By taking rotate vector (RV) reducer as the research object, the factors affecting the transmission accuracy are studied, including the machining errors of the main parts, assembly errors, clearance, micro-displacement, gear mesh stiffness and damping, bearing stiffness. Based on Newton second law, the transmission error mathematical model of RV reducer is set up. Then, the RV reducer transmission error curve is achieved by solving the mathematical model using the Runge-Kutta methods under the combined action of various error factors. Through the analysis of RV reducer transmission test, it can be found that there are similar variation trend and frequency components compared the theoretical research and experimental result. The presented method is useful to the research on dynamic transmission accuracy of RV reducer, and also applies to research the transmission accuracy of other cycloid drive systems.

1 Introduction

Rotate vector (RV) reducer is a high-performance precision transmission system, developed from the cycloidal pinwheel transmission. Currently, the research of high-performance precision transmission component and system is one of the key technologies of the mechanical engineering science, the analysis of RV reducer transmission precision has become an important research topic. To get higher transmission precision of RV reducer, the transmission error should be strictly controlled.

Domestic and foreign scholars have done lots of in-depth researches on the precision of general gear transmission, while the study of transmission accuracy for cycloidal-pin wheel mechanism and RV reducer is insufficient. In foreign countries, the United States Hughes companies Blanche and the Japanese Professor Teruaki Hidaka had studied it [1-3]. In China, the research on driving accuracy of cycloid system started relatively late and mostly confined geometry or static methods [4-7]. To improve the accuracy and stability of system output, it is necessary to analyse and research the dynamic transmission error on RV reducer, thus to reveal its dynamic behaviors influence on transmission precision in nature. This paper applies the method of nonlinear to study the transmission accuracy of RV reducer.

2 The transmission error mathematical model of RV reducer

2.1 System equivalent model

The structure of RV reducer is complex and the parts are more. Based on lumped mass method and dynamic substructure method, the equivalent model of RV reducer is established. The lumped mass method is also called mass-spring method, that is to say, the parts of large mass, big inertia and small elasticity can be regarded as the mass concentration of particles and rigid bodies, and the parts of little mass, high elasticity and large damping can be regarded as elastic elements or damping elements, their mass can be neglected or equivalent to the concentrated mass, rigid bodies are coupled by elastic elements or damping elements, the spring forces and damping forces are rigid body inter-atomic forces. The dynamic substructure method is to decompose the complex structure into some simple substructures, and then analyse the model of each substructure, according to the relationship between the each substructure, the analysis models of substructure are integrated into the analysis model of the whole structure, the whole process that is the analysis process of the whole to the local and then to the whole.

The RV-40E reducer is mainly composed of a sun gear, two planetary gears, two crank shafts, two cycloids, pin gears, planetary frame and the pin wheel housing. The equivalent model [2] is shown in Figure1. Taking the fixed pin gear distribution circle center as the origin to establish static coordinate system (X,Y), regarding the J-th cycloid centroid O_j as origin and the eccentric direction of cycloid as η_j direction to establish the dynamic coordinate system(η_j,ξ_j), Φ_i are the relative position of the planetary gears(i=1,2, i is the number of planetary gears), Ψ_j are the relative position of cycloid centroids(j=1,2, j is the number of cycloids).

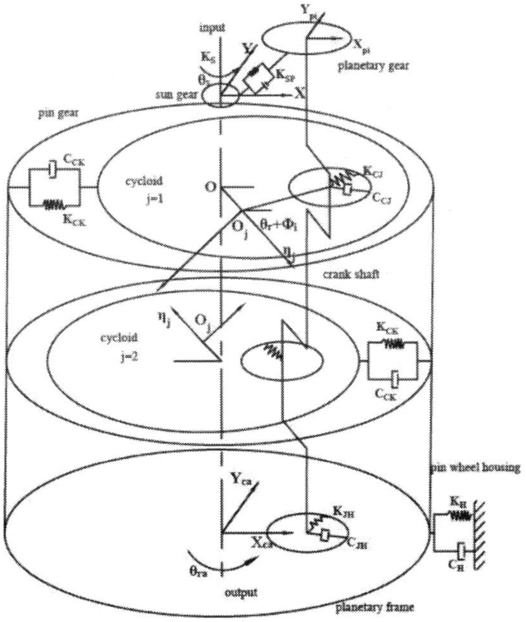

Fig. 1. Dynamic equivalent model of RV reducer.

In the equivalent model, the main parts of reducer are replaced by lumped mass, the bearing stiffness, and the gear meshing stiffness and damping are replaced by the spring element and damping element, such as, KS is the supporting stiffness of gear shaft, The sun gear and planetary gear meshing stiffness is KSP, CSP is the meshing damping, KCJ is the bearing stiffness between the crank shaft cam and cycloid shaft hole, KJH is the bearing stiffness between the crank shaft and planet carrier shaft hole, KH is the bearing stiffness between the planet carrier and pin gear housing, The cycloid and pin gear meshing stiffness is KCK, CCK is the meshing damping, The stiffness and damping values are calculated by using the related formulas.

When a small torque acts on the output shaft of reducer, because of machining errors, assembly errors, clearances and deformations, the parts' centroid positionand angle will deviation from the theoretical position, the tiny linear displacement and angular displacement are referred to as micro displacement. Reflected in the equivalent model. Xs, Ys are micro displacement of the center gear, Xpi, Ypi and θpi-θp (the planetary gear and the crank shaft are regarded as an integral whole) are micro displacement of the planetary gear and crank shaft, ηsj, θj-θr and θoj-θp are micro displacement of the cycloid, Xc, Yc and θra-θr are micro displacement of the planet carrier. θs is the theoretical angle of sun gear, θp is the theoretical rotation angle of planetary gear, θpi is the actual rotation angle of planetary gear, θr is the theoretical angle of planetary carrier, θra is the actual angle of planetary carrier, θj is the actual rotation angle of cycloid, θoj is the revolution angle of cycloid .

2.2 Description of the various errors of parts

When analyzing the impact of error factors on RV reducer transmission accuracy, the following error factors are considered: the machining eccentric error of sun gear is (E_s, β_s), E_s is the size of error, β_s is the direction of error. Other eccentric or assembly error representation methods are similar to this. The assembly error of sun gear is (E_{sa},β_{sa}), the machining eccentric error of Planetary gear is (E_{pi},β_{pi}), the eccentric error of crank shaft eccentric cam is (E_{dji},β_{dji}), the eccentric error of planet carrier crank shaft hole is (E_{cji},β_{cji}), the eccentric error of crank shaft hole is (E_{hi},β_{hi}), the assembly error of planet carrier is (E_c,β_c), the profile deviation of pin gear is E_{Rk}, the tooth pitch accumulated error is E_{ARk}, the profile deviation of cycloid is P_{jk}, the tooth pitch accumulated error is AP_{jk}, the amount of equidistant modification is Δr_{rp}, the amount of moved distance modification is Δr_p, δ_{cj} is the bearing clearance at cycloid crank shaft hole, δ_{jh} is the bearing clearance at planet carrier crank shaft hole, δ_h is the bearing clearance between planet carrier and pin gear housing, e_k is the sum of pin wheel radius error and pin gear arc-groove radius error.

2.3 The mathematical model of RV reducer dynamic transmission accuracy

In the equivalent model, the spring force and damper force are the interaction force between the parts, that are the product of spring stiffness coefficient and the equivalent error, the product of damping coefficient and the equivalent error first derivative, namely $F = K\delta$ and $P = C\dot{\delta}$. Analyzing the force of the main parts in the ideal position, based on the newton second law, the mathematical model of dynamic transmission accuracy is established.

2.3.1 The dynamic equation of center gear

$$\begin{cases} m_s\ddot{X}_s + F_{SX} + \sum_{i=1}^{2}(F_{SPi} + P_{SPi})\sin(\theta_r + \phi_i - \alpha) = 0 \\ m_s\ddot{Y}_s + F_{SY} - \sum_{i=1}^{2}(F_{SPi} + P_{SPi})\cos(\theta_r + \phi_i - \alpha) = 0 \\ J_s\ddot{\theta}_s + R_s\sum_{i=1}^{2}(F_{SPi} + P_{SPi}) - T_0 = 0 \end{cases} \tag{1}$$

Where, m_s is the mass of center gear, J_s is the moment of center gear, R_s is the radius of base circle of center gear, T0 is the input shaft torque.

2.3.2 The dynamic equation of planet carrier

$$\begin{cases} m_c\ddot{X}_c + \sum_{i=1}^{2}F_{JHX} - F_{HX} = 0 \\ m_c\ddot{Y}_c + \sum_{i=1}^{2}F_{JHY} - F_{HY} = 0 \\ J_c\ddot{\theta}_r + R_{sp}\sum_{i=1}^{2}F_{JHX}\sin(\theta_r + \phi_i) - R_{sp}\sum_{i=1}^{2}F_{JHY}\cos(\theta_r + \phi_i) + T = 0 \end{cases} \tag{2}$$

Where, mc is the mass of the planet carrier, Jc is the moment of inertia of the planet carrier, T is the output shaft torque.

2.3.3 The dynamic equation of planetary gear (include the crank shaft)

$$\begin{cases} m_{cp}[\ddot{X}_{pi} - R_{sp}w_r^2\cos(\theta_r + \phi_i) - R_{sp}\ddot{\theta}_{ra}\sin(\theta_r + \phi_i)] - (F_{SPi} + P_{SPi})\sin(\theta_r + \phi_i - \alpha) + \sum_{j=1}^{2}F_{CJX} + F_{JHX} = 0 \\ m_{cp}[\ddot{Y}_{pi} - R_{sp}w_r^2\sin(\theta_r + \phi_i) + R_{sp}\ddot{\theta}_{ra}\cos(\theta_r + \phi_i)] + (F_{SPi} + P_{SPi})\cos(\theta_r + \phi_i - \alpha) + \sum_{j=1}^{2}F_{CJY} + F_{JHY} = 0 \\ J_{cp}\ddot{\theta}_{pi} + R_p(F_{SPi} + P_{SPi}) - e\sum_{j=1}^{2}F_{CJX}\sin(\theta_p + \psi_j) - e\sum_{j=1}^{2}F_{CJY}\cos(\theta_p + \psi_j) = 0 \end{cases} \tag{3}$$

Where, m_{cp} is the sum of planetary gear mass and crank shaft mass, J_{cp} is the moment of inertia of planetary gear and crank shaft, w_r is the theoretical angular velocity of planet carrier, Rsp is the center distance of center gear and planetary gear.

2.3.4 The dynamic equation of cycloid

$$\begin{cases} m_b[\ddot{\eta}_{jx}\cos(\theta_p + \psi_j) - ew_p^2\cos(\theta_p + \psi_j) - e\ddot{\theta}_{oj}\sin(\theta_p + \psi_j)] - \sum_{i=1}^{2}F_{CJX} - \sum_{k=m}^{n}(F_{CK} + P_{CK})\cos(\alpha_{jk} + \theta_p + \psi_j) = 0 \\ m_b[-\ddot{\eta}_{js}\sin(\theta_p + \psi_j) + ew_p^2\sin(\theta_p + \psi_j) - e\ddot{\theta}_{oj}\cos(\theta_p + \psi_j)] + \sum_{i=1}^{2}F_{CJY} + \sum_{k=m}^{n}(F_{CK} + P_{CK})\sin(\alpha_{jk} + \theta_p + \psi_j) = 0 \\ J_o\ddot{\theta}_j - R_d\sum_{k=m}^{n}(F_{CK} + P_{CK})\sin\alpha_{jk} + R_{oh}\sum_{i=1}^{2}F_{CJX}\sin(\theta_r + \phi_i) - R_{oh}\sum_{i=1}^{2}F_{CJY}\cos(\theta_r + \phi_i) = 0 \end{cases} \tag{4}$$

Where, mb is the mass of cycloid, J0 is the moment of cycloid, wp is the theoretical rotation angular velocity of the planetary gear.

The dynamic equations can be converted into the general matrix form:

$$M\ddot{X} + C\dot{X} + KX = F \tag{5}$$

Where, **X** is the displacement vector of equivalent model, $\mathbf{X}^T=[X_s,Y_s,\theta_s,X_{p1},Y_{p1},\theta_{p1},X_{p2},Y_{p2},\theta_{p2},\eta_{s1},\theta_1,\theta_{o1},\eta_{s2},\theta_2,\theta_{o2},X_c,Y_c,\theta_{ra}]$, **M** is the mass matrix, **K** is the stiffness matrix, **C** is the damp matrix, **F** is the force vector.

By solving the differential equations, the actual angle of the output shaft can be obtained, if the transmission ratio is i, the transmission error is $\Delta\theta_{ra}=\theta_{ra}-\theta_s/i$.

3 Solving the equations and experimental confirmation

Taking the RV-40E reducer as the research object, the transmission error dynamic model of RV reducer is established. The solid part models of RV reducer are built, the quality and the moment of inertia of each part is solved. The damping and stiffness of involute planetary gear is calculated by using the theory of elasticity, iterativing the meshing damping and stiffness between cycloid gear which has been modified and the pin gear by using the computer programming, the equivalent stiffness of the bearings are calculated by Hertz formula. The basic parameters and correlated errors are shown in Table 1 to Table 4. the errors' unit is (μm,°), due to the errors of sun gear and planetary gear have little impact on the transmission accuracy, thus can be neglected [3].

Table 1. The basic structure parameters of RV-40E reducer.

Planetary gear drive		Cycloidal pin-gear drive	
sun gear tooth number	12	Cycloid tooth number	39
planetary gear tooth number	36	pin gear tooth number	40
modulus	1.5	the radius of pinwheel distribution circle/mm	64.05
meshing pressure angle	20	pin wheel radius/mm	3
transmission ratio	121	Eccentricity/mm	1.3

Table 2. The manufacturing error between the crank shaft and cycloid.

error location	crank shaft cam eccentric error		cycloid crank shaft hole eccentric error	
	crank shaft 1	crank shaft 2	crank shaft 1	crank shaft 2
cycloid 1	(3.5,30)	(3,195)	(3.5,40)	(3.5,215)
cycloid 2	(4,140°)	(4.5,85)	(3,130).	(3,130)

Table 3. Planetary carrier crank shaft hole eccentric error and assembling error.

planetary carrier crank shaft hole eccentric error		planetary carrier assembling error
crank shaft 1	crank shaft 2	(4,30)
(3.5,195)	(3,75)	

Table 4. The correlated errors of cycloid gear and pin gear.

error types	cycloid	pin gear
profile error	8	5
total cumulative pitch error	6	8

The basic parameters and correlated errors are substituted into dynamic equations, and the equations are solved by using the Runge-Kutta method, the actual transmission error is measured by means of test platform. The numerical simulation and experimental result are respectively shown in Figure2 and Figure3.

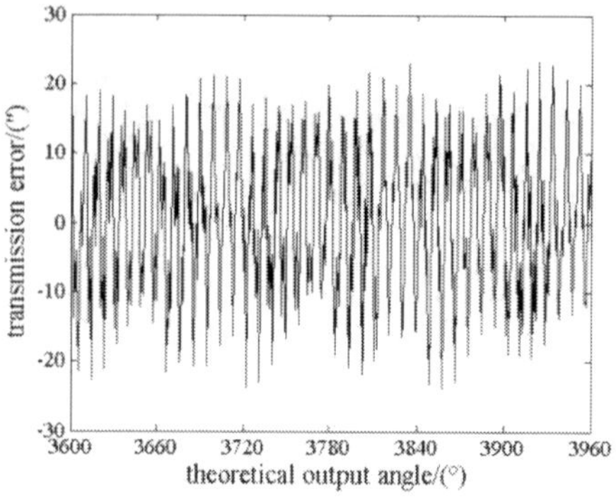

(a) The theoretical transmission error curve.

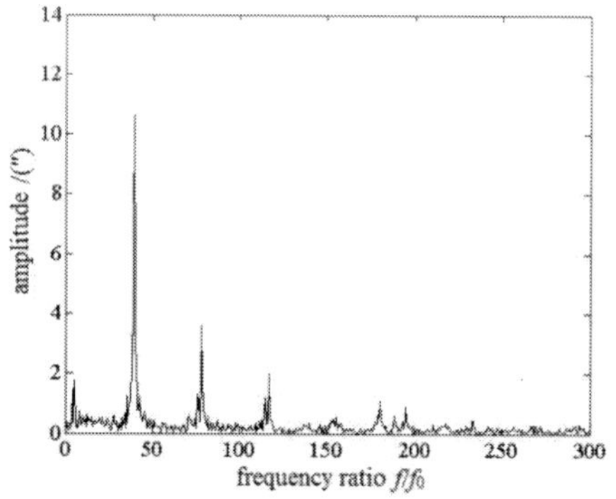

(b) The spectrum characteristic

Fig. 2. The theoretical results of RV reducer transmission error.

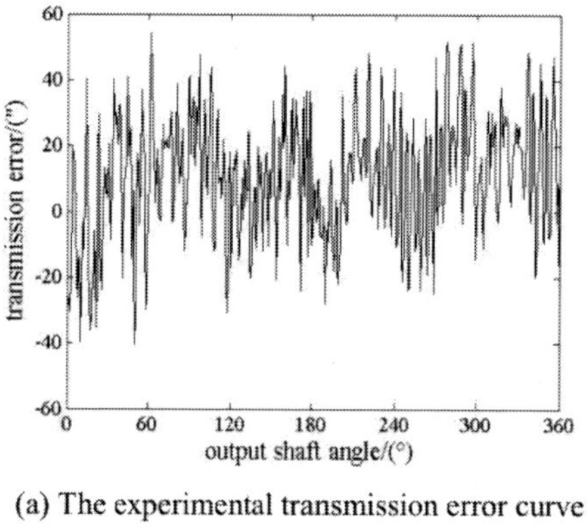

(a) The experimental transmission error curve

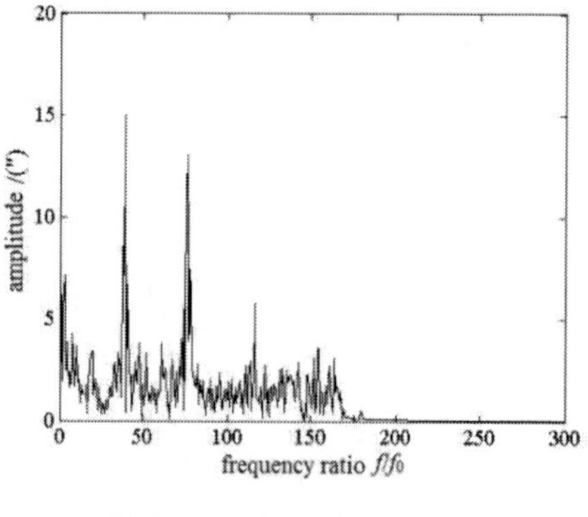

(b) The spectrum characteristic

Fig. 3. The experimental results of RV reducer transmission error.

To explore the spectral characteristics of RV reducer transmission error, the spectrum can be obtained by Fourier transform, as shown in Fig 2(b) and Fig 3(b), f is the frequency of output error, f_0 is the frequency of output shaft rotation, The frequency ratio f/f_0 is the number of output error changes per revolution. As shown in Fig 2 and Fig 3, the theoretical result and experimental result have similar trend and frequency components, Therefore, the mathematical model established in this paper can correctly reveal the variation of RV reducer transmission error.

In addition, by simulating the different error parameters and comparing the effection of each error factor on the accuracy of RV reducer. It can be obtained that the correlated errors of the involute gear in one-stage gearing have little influence on the transmission accuracy of RV reducer. In the two stage of the transmission part, crank shaft cam eccentric error, cycloid crank shaft hole eccentric error, cycloid profile error and total cumulative pitch error, pin gear profile error and total cumulative pitch error have a great influence on transmission accuracy, which depend on the error size, the action direction and the combination mode. So the error factors in secondary transmission part should be taken seriously when a RV reducer is designed and produced, by controlling the greater impact factors, the transmission accuracy can be effectively improved.

4 Conclusion

In comprehensive consideration of the main parts machining errors, assembly errors, clearance, micro-displacement, gear meshing stiffness and damping, bearing stiffness, the dynamic transmission accuracy mathematical model of RV reducer is established according to Newton second law and is solved by using Runge-Kutta methods. The theoretical research and test analysis show that the results of numerical simulation agree with that of experiment. The method provides a theoretical basis for further research and improvement of cycloid pin gear transmission system.

Acknowledgement

Research is supported by Scientific and Technological Project of Henan Province (162102210049).

References

1. Yang D. C. H & Blanche J. G. Design and application guidelines for cycloid drives with machining tolerances. *Mechanism and Machinery Theory*, **25(5)**, pp. 487-501,(1990).
2. Teruaki Hidaka, Takeshi Ishida & Hongyou Wang. Rotational transmission error of K-H-V-type planetary gear with cycloid gears (1st Report, Analytical Method of the Rotational Transmission Error). Japan Society of Mechanical Engineers. **60(570)**, pp. 645-653, (1994).
3. Teruaki Hidaka, Takeshi Ishida & Hongyou Wang. Rotational transmission error of K-H-V-type planetary gear with cycloid gears (2nd Report, Effects of Manufacturing and Assembly Errors on Rotational Transmission Error). *Japan Society of Mechanical Engineers*. **60(578)**, pp. 3510-3517, (1994).
4. Han Lin-Shan, Wu Lan-Ying & Shen Yun-Wen. Sensitivity analysis of the 2K-V reducer transmission precision. *Mechanical Science and Technology*, **25(10)**, pp. 1366-1369,(2010).
5. Yang Yu-Hu, Zhang Jie Xu Li-Xin. The transmission accuracy analysis of RV reducer. *Journal of Tianjin University*, **46(7)**, pp. 623-628, (2013).
6. Zhu Bin, Qin Wei, Liu Jing, et al. Simulation and Analysis of Dynamical Transmission Precision of 2K-V Cycloid Pin Gear Reducer Based on Multi-body System Dynamics. Advanced Materials Research: Advanced Design Technology,**3(08)**: 2205-2210.(2012).
7. Zhang Jin, Sun YongGuo. Backlash Analysis of RV Reducer Based on Error Factor Sensitivity and Monte-carlo Simulation. Science and Engineering Research Support Society, **7(2)**: 283-291,(2014).

Analysis of Wheel Hub Motor Drive Application in Electric Vehicles

Yuechao Sun*, Man Li and Cong Liao

Mechanical and Electrical Engineering Department, Lingnan Normal University, Zhanjiang 524048 , China
Corresponding Email: yuechaosun@sina.com

Abstract. Based on the comparative analysis of the performance characteristics of centralized and distributed drive electric vehicles, we found that the wheel hub motor drive mode of the electric vehicles with distributed drive have compact structure, high utilization ratio of interior vehicle space, lower center of vehicle gravity, good driving stability, easy intelligent control and many other advantages, hence in line with the new requirements for the development of drive performance of electric vehicles, and distributed drive will be the ultimate mode of electric vehicles in the future.

Keywords. Electric Vehicle, Drive Mode, Wheel Hub Motor, Development Analysis.

1 Foreword

Compared with conventional vehicles, electric vehicles have the advantages of high efficiency of energy conversion, low noise, zero emission, etc., and the load-carrying property and wide range speed control characteristics of motors can remove the mechanical devices such as clutch and gearbox, simplifying the structure and facilitating maintenance [1, 2]. Driven by the dual pressures of energy and environment nowadays, the world's major automobile producing countries are developing electric vehicle industry with unprecedented efforts. Electric vehicles are creating a new pattern of the automobile industry, which will sure lead the main direction of the automobile industry development.

As the core component of an electric vehicle, the quality of driving motor has a great influence on the power, economical efficiency and safety of the electric vehicle. However, different from other industrial motors, the motor drive system of vehicle drive motor is not only affected by the size of vehicle structure, but shall also meet the operating conditions under complex conditions [3-6]. Therefore, in addition to the requirements of high efficiency, low mass, high power density, small size, good reliability and low cost of the drive motor, it shall also adapt to the frequent start, stop, climbing, acceleration and deceleration conditions of vehicles, which require a wide range of speed and high overload factor of automotive drive motor in order to meet the performance requirements of high torque under low speed or climbing or low torque under high speed [7, 8].

2 Drive Mode

According to the different installation locations of drive motor in electric vehicles, there are single-motor centralized drive and multi-motor distributed drive modes. Distributed drive is further divided into wheel side motor drive and wheel hub motor drive.

2.1 Centralized drive

Centralized drive is close to the conventional vehicle structure, in which the internal combustion engine is replaced by a motor, and the motor torque is transmitted to the drive wheel through the power train to drive the vehicle. It can be achieved by making a slight change based on the conventional vehicle structure, which has mature, safe and reliable operation technology, but has the shortcomings of small interior size, heavy mass, low transmission efficiency and complex control.

DOI: 10.1051/matecconf/201710001004

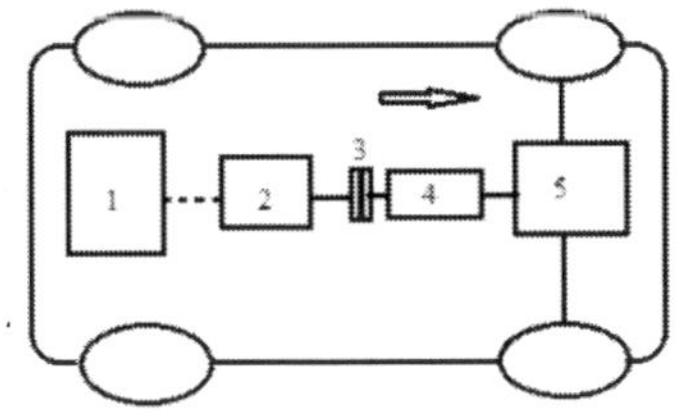

1- Power battery,2- Drive motor,3- Clutch,4- Transmission,5- Differential

a Mechanical transmission with a clutch.

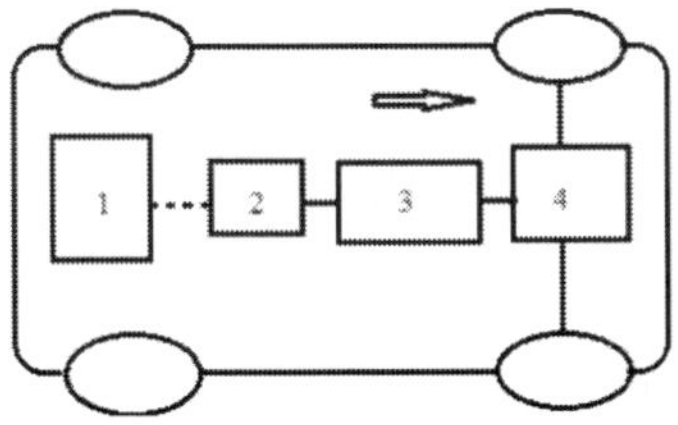

1- Power battery,2- Drive motor,3- Fixed speed ratio reducer 4- Differential

b Mechanical transmission without a clutch.

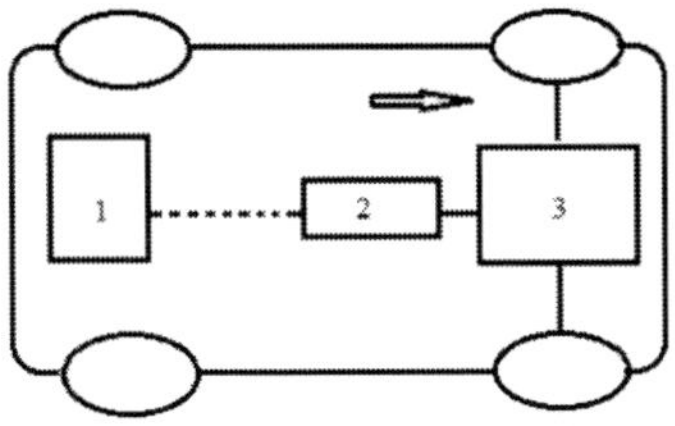

1-Power battery,2-Drive motor,3-Transmission and Differential

c Electromechanical integrated transmission.

Fig.1. Centralized drive transmission.

There are 3 transmission modes of centralized drive, see Figure [9]. Fig. 1 a shows the transmission mode with a clutch, the transmission of pure electric vehicle generally has 2-3 gear shifts, the clutch plays the role of breaking off power and reducing shift shock while shifting. Fig. 1b eliminates the clutch and connects the motor to the fixed speed ratio reducer via the transmission shaft, reducing the power train mass and the transmission volume, thus increasing the space inside the vehicle. Fig. 1c integrates the motor, reducer and differential, drives the corresponding side wheel through the left and right axle shafts, respectively, and is compact in layout, mainly for small cars.

2.2 Distributed drive

Distributed drive transmits power to the corresponding wheel by integrating multiple motors in the vicinity of the wheel or rim, greatly shortening the transmission chain. According to the motor position and different transmission, it can be divided into wheel side motor drive and wheel hub motor drive.

2.2.1 Wheel side motor drive

Wheel side motor drive installs the drive motor on the side of the drive wheel of the subframe, and drives the corresponding side wheel with or without the reducer. The drive mode with a reducer connects the motor to the fixed speed ratio reducer and drives the corresponding side wheel through the axle shaft.

As shown in Figure 2 [9,10], the two motors can drive the corresponding side wheel separately through the corresponding side reducer. The motor speed can be controlled independently, and the speed differential on the left and right axle shafts can be realized by an electronic differential.

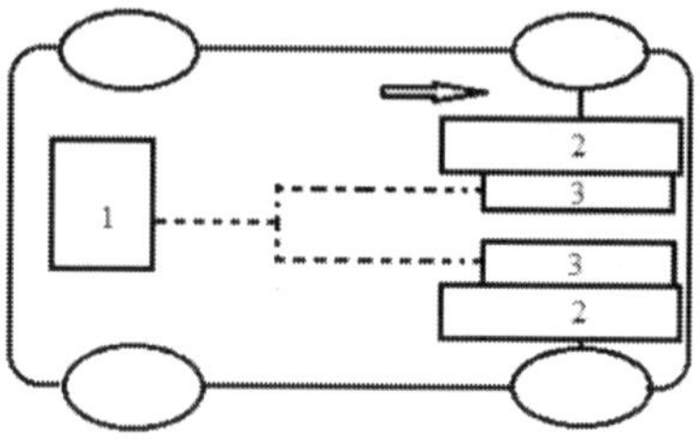

1-Power battery,2- Fixed speed ratio reducer,3- Drive motor
a Layout of wheel rim motor drive.

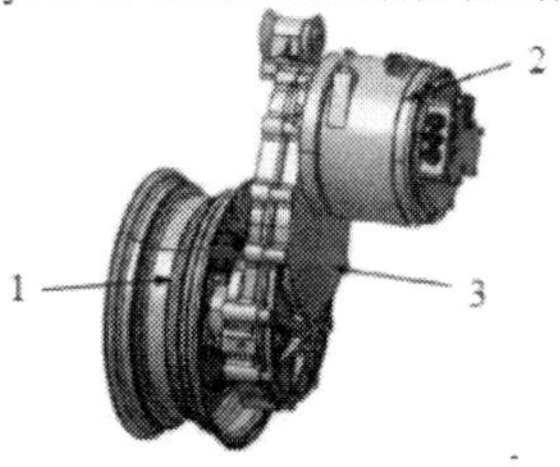

1- Rim,2- Power battery,3- Reducer
b Schematic diagram of wheel rim motor drive assembly.

Fig. 2. Wheel side motor drive.

The vehicle transmission chain and transmission space of wheel side motor drive are further reduced, making the mechanical structure of the chassis simple, reducing the vehicle mass, achieving more reasonable layout, and improving the transmission efficiency. Audi R8e-tron pure electric sports car is driven by four asynchronous motors, with 100 km acceleration of only 4.8s.

2.2.2 Wheel hub motor drive

Wheel hub motor drive is the most advanced electric vehicle driving technology, which installs two, four or more motors in the wheel, direct driving wheels, commonly known as electric wheels, which is especially suitable for pure electric vehicles.

a Conventional vehicle drive. b Wheel hub electric vehicle drive.

Fig. 3. Comparison of the drive of conventional vehicles and wheel hub electric vehicles.

Figure3 shows a comparison of the chassis of conventional vehicles and vehicles driven by a wheel hub motor, the wheel hub motor drive completely eliminates the power train components such as clutch, transmission, differential and axle shafts, making the chassis structure simple, improving the transmission efficiency, reducing the vehicle mass, achieving more reasonable layout, and facilitating the realization of intelligent chassis and electrification control. Based on whether there is a retarding mechanism, wheel hub motor drive can be divided into direct drive and deceleration drive (deceleration drive electric wheel) [11].

The external rotor of the motor is mechanically connected with the wheel hub and has no deceleration structure, which is a direct drive mode, see Fig. 4a [9]. Motor speed is generally 1500r/min or so, due to the lack of retarding mechanism, driving structure is compact with high transmission efficiency. But large current is required at the start, climbing or under large loads, which could easily damage the battery and permanent magnets. In order to ensure a large start torque and good dynamic property, the requirements of the motor are high, and low-speed external rotor permanent magnet synchronous motor is generally adopted.

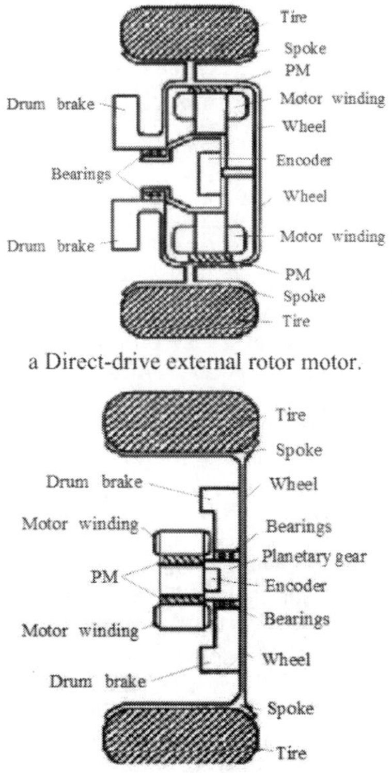

a Direct-drive external rotor motor.

b Deceleration-driven inner rotor motor.

Fig. 4. Layout of wheel hub motor drive.

Deceleration drive is to installed a fixed speed ratio reducer between the motor and the wheel, playing the role of deceleration and torque-up. The deceleration device is generally a high reduction ratio planetary gear mechanism, see Fig. 4b [9]. The working speed of the motor is about 10000r/min, usually using high-speed inner rotor permanent magnet synchronous motor, the drive wheel hub drives the vehicle after deceleration and torque-up of motor output power. The motor is small, light and with high specific power of high-speed running. After deceleration and torque-up, the vehicle has good climbing ability, and it can guarantee the vehicle has a large steady torque when running at low speed. However, the structure is relatively complex, and the increase of unsprung mass affects the ride comfort and handling stability.

3 Drive Performance and Development Analysis

3.1 Drive performance analysis

Table 1 shows the performance comparison of the electric motors in different drive modes, and shows the transmission efficiency, economical efficiency and handling stability. Table 1 and the above drive modes show that: centralized drive vehicles are similar to the conventional internal combustion engine vehicles, the latter have many power train components, high cost, low transmission efficiency, complex control and other shortcomings.

Table 1. Performance comparison of electric vehicles' drive modes.

Performance parameters	Centrali zed	Distribute d wheel rim motor drive	Distribute d wheel hub motor
Transmission chain length	Long	Short	The shortest
The interior space utilization	General	High	High
Body centroid height	General	Low	Low
Transmission efficiency	Medium	Highe r	High

Economical efficiency	Better	Good	Good
Kinetic controllability	General	Good	Good
Energy recovery (regenerative braking)	Yes	Yes	Yes
Noise	Small	Small	Small
Wheel drive torque fluctuation	Small	Big	Big
Wading performance	Good	General	Genera l

Compared with centralized drive, wheel rim motor drive mode has short transmission chain, good economical efficiency, high interior space utilization, small braking energy feedback loss, but the transmission efficiency is not as high as wheel hub motor drive.

Wheel hub motor drive completely abolishes the power train components, integrates the motor, suspension system and brake system in the rim, achieving compact vehicle structure, lower center of gravity and improved driving stability. The wheel hub motor is connected to the power battery and the controller using a wire harness, as shown in Fig. 3b. This allows for more flexibility in the layout of the interior space and saving more space for occupants while improving ride comfort. Meanwhile, the movement of each wheel is independent from each other, with no rigid connection, the wheel driving force and braking force can be quickly optimized, assigned precisely and arbitrarily by computer and motor control systems as per the driving state, easy to implement by-wire steering, ABS, TCS, ESP and other functions, achieving flexible motor steering, reliable kinetic performance, and good handling stability. In addition, wheel hub drive can also achieve electrical braking, electromechanical composite braking and braking energy feedback, with low energy consumption and high braking energy recovery, energy conversion efficiency can reach 90% [12].

Although the vehicles with wheel hub motor drive have outstanding performance, the wheel hub motor, brake system and even suspension system are integrated in the wheel, resulting in significant increase of vehicle unsprung mass and rotary inertia of wheel rotating parts, thereby increasing the vibration amplitude along vertical direction of vertical, and even affecting the tire adhesion, which is not conducive to vehicle control and could also reduce ride comfort.

3.2 Development analysis

Wheel hub motor drive has unparalleled advantages compared with other drive modes, although there is yet no mature wheel hub drive automotive product applications, as a new advanced drive mode, wheel hub drive is a focus in the current domestic and international electric vehicles research. Many scholars at home and abroad have done a lot of research on the shortcomings of wheel hub motor drive, and put forward a series of corresponding improvement measures. For example: Nagaya et al [13] controlled vertical vibration negative effect caused by unsprung mass by making a vibration absorber using motor mass. Johansen, Yang et al [14-15] reduced the unsprung mass by converting the motor stator mass into the sprung mass through a special motor design. Zhao Yan'e et al [16] converted the wheel hub motor stator mass from unsprung mass to sprung mass by setting a vibration reduction mechanism in parallel with the suspension system. Luo Hong et al [17] proposed to take the entire motor mass as the sprung mass, namely to take the motor mass as the vibration absorber mass in the power driven system that directly drives wheel hub motor, with no additional mass. Shi Tianze et al [18-20] developed a special suspension and steering system that applies to wheel hub motor vehicles, which comprises a double-arm front suspension, torsion beam rear suspension and mechanical steering system.

In view of the problems of wheel hub motor drive, researchers are exploring, improving and developing from various aspects, including the development of high-torque wheel hub motor, the integration and control of intelligent chassis, the impact of non-spring mass on vehicle performance and other technologies. We believe that in the near future, researchers will explore a reasonable new technology to solve the current problems, give full play to the advantages of wheel hub motor drive, making it the ultimate drive mode of electric vehicles.

4 Conclusion

Electric vehicles that adopt wheel hub motor drive mode have compact structure, high utilization ratio of interior vehicle space, lower center of vehicle gravity, good driving stability, easy intelligent control and many other advantages, and have unique advantages in maintenance costs, safety, high-torque drive and other aspects compared with other drive modes. Therefore it is in line with the requirements for the development of small electric vehicle drive motor, high power density, high reliability, etc., which will sure be the ultimate drive mode of electric vehicles in the future.

References

1. Chai Haibo, Yan Zhiguo, Kuang Mingwei et al. Development status of electric vehicle drive motors. Small & Special Electrical Machines, 2013, 41(4): 52-57
2. Zhou Fengjun. Research on integrated control strategy of distributed drive electric vehicles based on two-tier structure. Beijing: Beijing Institute of Technology, 2014.
3. Luo Longfei, Yuan Shouhua. Application overview of electric vehicle drive motor and its control system. Automobile Applied Technology, 2014, 4: 5-7
4. Liu Jinfeng, Zhang Xueyi, Hu Jianlong. Development prospect of electric vehicle drive motor. Agricultural Equipment & Vehicle Engineering, 2012, 50(10): 35-38
5. Santiago J, Bernhoff H, Ekergrrd B, et al. Electrical motor drivelines in commercial all-electric vehicles: a review. IEEE Trans Veh Technol, 2012,61(2):475-484.
6. Liu Gang. Study on driving and control system of switched reluctance motor in electric drive vehicles. Beijing: University of Science and Technology Beijing, 2015.
7. Wang Chengyuan, Xia Jiakuan, Sun Yibiao. Modern motor control technology. Beijing: China Machine Press, 2010.
8. Dr.-Ing.Klaus Webers, Dr.-Ing. Rainer Busch, Fort Integrated Vehicle Dynamics Control Realization. Research & Advanced Engineering. 2003.
9. Xu Wei, Zhang Ruoping. Selection of transmission scheme for electric vehicles. Auto Engineer, 2014, 8: 54-56
10. Liu Hao, Zhong Zaijing, Jing Hui, et al. Dynamic characteristics simulation of wheel rim motor drive system in distributed drive electric vehicles. Automotive Engineering, 2014, 36(5): 597-607
11. He Ren, Zhang Ruijun. Research and development of wheel hub motor drive technology. Journal of Chongqing University of Technology (Natural Science), 2015, 29(7): 10-18
12. Hori Y. Future Vehicle driven by Electricity and Control-Research on 4 Wheel Motored 'UOT March II', 2004.
13. Nagaya G， Wakao Y， Abe A． Development of an in-wheel drive with advanced dynamic-damper mechanism.JSAE Review,2003,24(4) : 477-481.
14. Johansen P R， Pattersib D， O' keefe C． The use of an axial flux permanent magnet in-wheel direct drive in an electric bicycle． Renewable Energy， 2001， 22(1 /2 /3) : 151 — 157.
15. Yang Y P， Luh Y P， Cheng C H． Design and control of axial flux brushless DC wheel motors for electric vehi cles-part I: multi-objective optimal design and analysis． IEEE Transaction on Magnetics， 2004， 40(4) : 1873-1882.
16. Zhao Yan'e, Zhang Jianwu, Han Xu. Design and research of dynamic vibration reduction mechanism of electric vehicles independently driven by wheel hub motors. Mechanical Science and Technology for Aerospace Engineering, 2008, 27(3): 395-398.
17. Luo Hong, Xie Dan, Ma Ying. Analysis and design of power transmission mechanism of dynamic absorber power wheels. Mechanical Science and Technology for Aerospace Engineering, 2014, 33(2): 249-253.
18. Shi Tianze. Design and performance matching of wheel hub motor drive electric vehicle suspension and steering system. Changchun: Jilin University, 2015.
19. Tan Di. Dynamic characteristics and structural optimization of wheel hub motor drive system with built-in suspension. Guangzhou: South China University of Technology, 2013.
20. Wang R, Chen Y, Feng D, et al. Development and performance characterization of an electric ground vehicle with independently actuated in-wheel motors. Journal of Power Sources, 2011,196(8):3962-3971.

A Fully Automatic Fresh Apple Juicer: Peeling, Coring, Slicing and Juicing

Fuwen Hu [1,a], Li Li[1,b], Yancheng Liu [1,c] and Tadeusz Mikolajczyk[2,d]

[1]School of Mechanical and Material Engineering, North China University of Technology, Beijing 100144, China
[2] Department of Production Engineering, University of Science and Technology, Bydgoszcz 85-796, Poland
[a]hfw@ncut.edu.cn,[b]442578001@qq.com,[c]lyc19740828123@qq.com,[d]tami@utp.edu.pl

Abstract. With the fresh apple juice as an example, a fully automatic and intelligent juicer prototype was built via the integrated application of servo positioning modules, human-machine interface, image vision sensor system and 3D printing. All steps including peeling, coring, slicing and juicing were achieved automatically. The challenging technical problems about the identification and orientation of apple core, and adaptive peeling were settled creatively. The trial operation results illustrated that the fresh apple juice can be produced without manual intervention and the system has potential application in the crowded sites, such as mall, school, restaurant and hospital.
Keywords: Fresh apple juice, machine vision, removal of apple core, servo positioning.

1 Introduction

Generally, fresh fruit juice is full of nutrition and active to people health. Meanwhile, the process of squeezing juice is almost transparent and timely without any additive and decaying fruits. Therefore, fresh fruit juice has become the flavour of drink market. However unlike sodas and concentrated juice, the fresh juice is unsuitable for batch production due to the short shelf life and demanding procedure. Production of fresh fruit juice first need to select perfect fruits and usually also need to peel and core, namely "just squeeze -just sale-just drink". Hence the production of fresh juice belongs to individualized on-demanding production. Currently, at the level of domestic or personal juicer, manual operation is essential within loading, peeling, coring, slicing and such operations. The fully automatic fresh apple juicer still is a gap in the market.

Certainly, there are several kinds of fresh fruit juicer in the market, such as orange juicer. But it does not peel the orange before juicing. Thereby this may lead to doubts and hesitations: what about the surface clearness, what about the wax residues, etc. For instance, the apple juicer in the market may achieve automatic peeling, but the coring automation is not implemented. Zeng et al [1] presented a principle solution using the spinning copying cutter to peel the apple. Bai et al [2] put forward an automatic system including the processes of conveying, location of apple core, peeling, and coring.

Briefly, there are two core difficult issues when developing a fully automatic fresh apple juicer. First, it is the adaptive peeling of apple. The nutrition or detriment of apple peel is always a highly controversial topic. It is better to peel for ensuring the flavour and safety. However, there are no two identical apples in the world. The apples have irregular shapes, dents of stalk and calyx. These factors lead to big difficulties for complete peeling. And the second problem is automatic coring problem. Since the apple core contains little cyanide which does not benefit our health, it is better to core when juicing. However, how to identify the position and orientation of the core and to separate it from flesh is another problem for fresh juicing [3]. To develop a fully automatic fresh apple juicer, the two key issues mentioned above must be settled.

In this paper, we put forward a fully automatic juicing concept at the level of domestic or commercial environment rather than the industrial environment. And we have developed a fully automatic fresh apple juicer prototype in laboratory environment based on a creative solution using machine vision system.

2 Solution presentation

Figure 1 has shown the main configuration of the presented system as two generation prototypes. The basic working procedure includes seven steps: loading apples using robot, cutting the end, identification of core, wedging, peeling, slicing and coring, squeezing.

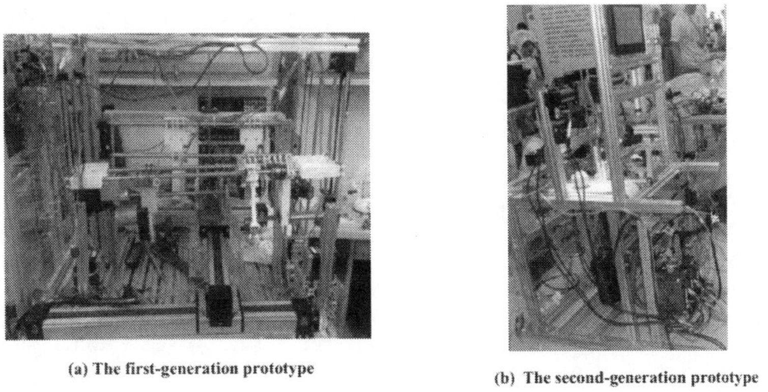

Fig. 1. Structure of the system

Two generation of automatic fresh apple juicer system are made. The first generation of the system (Figure 1a), making a bit rough, is in the exploratory stage to find reasonable workflow and some problems. In the first juicer system, the problems are that the force inserting the apple is underestimated leading to system instability as well as motor power shortage. The core position of the apple is not accurate without utilizing the vision system. In the second generation juicer system (Figure 1b), more powerful motors and profiles are introduced to the system. Vision systems are adopted to identify the accurate position of the apple core. The full automatic fresh apple juicer system' s performs using procedure presented on Figure 2 as the following:

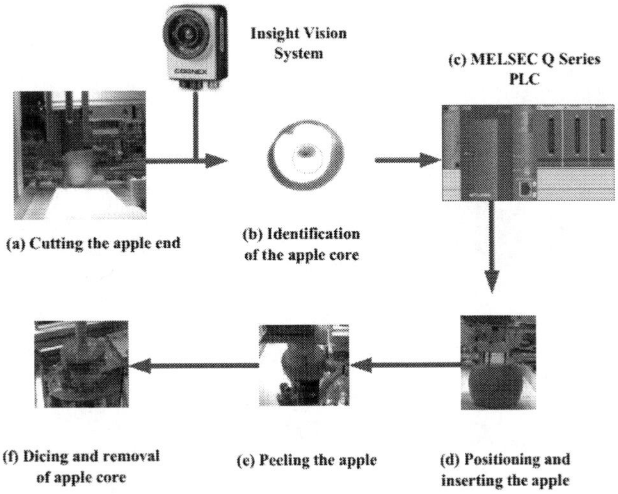

Fig. 2. Automatic fresh apple juicer work procedure

- *Loading*: The industrial robot picks the apple from the storing box and places it onto the conveyors. Here, the basic assumption is that apples in storing box have been washed.
- *End cutting*: When the apple was transmitted to the position at the center of the clamp device, it would be held on. Then under the drive of the linear slide module, the spinning cutter would cut off the ride of the apple end, as shown in the Figure 2. The flesh of apple end would come out, and its colour is in total contrast to the black apple base. This can lay the foundations for the identification of apple core in next process. Meanwhile, this work is equivalent to removing the peel on the apple base. The peel on the apple base is usually difficult to be removed when adaptive peeling.
- *Identification of apple core*: We identified the position of apple core using the machine vision camera. As illustrated in Figure 2, the core is viewed as "spot" and then is identified by the vision tool of spot detection. First in the vision area, the camera sensor roughly determines the core position. Second, in the obtained area, the black spot would be obtained exactly。 After two times detection, the apple core can be exactly confirmed and its position coordinates (x_d, y_d) or offsets from the reference axis can be calculated.
- *Wedging:* The COGNEX vision system can be connected directly on the Ethernet port on the Q-series PLC using Ethernet based on the seamless message protocol (SLMP) or MELSEC communication (MC) protocol.

Consequently, the apple core position parameters would be transmitted from vision camera to the programmable logic controller (PLC). PLC would drive the line slide module and bring the wedging pins to the target position. In the Z direction, the wedging pins plate moves down driven by screw motor and plugs into the flesh (Figure 3).

- *Adaptive peeling:* After the insertion of wedging pins plate into the flesh, the apple would move to the peeling position with the drive of linear slide module. Then the wedging pins plate start rotating and moves down along Z direction. Though the apple has irregular shape, the spring at the peeling cutter bar can make it close with the peeling cutter. Hence, the adaptive peeling of apple can be achieved as shown in Figure 2. Maybe this mechanical method of peeling has lower efficiency than the lye corrosion method [4]. But it can avoid steeping and washing processes. Therefore mechanical peeling method has priority for the personal commercial juicer.
- *Slicing and coring:* After peeling, the apple the apple would further move to the below the pushing plate. The pushing plate driven by screw motor would push the apple into the slicing and coring cutter, and the core is removed and the flesh is cut into slices (Figure 2). The core slides into the hopper of the juicer through the slide way, and the removed core drops into the collector through soft pipe.
- *Juicing:* Here, the domestic juicer is directly used to squeeze the apple slices without further improvement.

Through the sequential process above, the system first solved two key issues: adaptive peeling and coring. This realized the possibility of fully automatic juicing without manual intervention. Particularly, the process of cutting end made the identification of apple core very easy via the machine vision, and also removed the peel on the apple base. It is shoot two hawks with one arrow. Thirdly, the system used servo positioning technology and ensured the precision requirement of every shift position.

3 System establishment

Next we discussed the integration and connection strategies of the whole system. Since the PLC, touch screen, motion module, servo drive, servomotor and industrial robot are chosen from Mitsubishi Electric Corporation, they naturally have good compatibility and easy communication interfaces. Therefore, we mainly described the servo positioning system configuration, as well as the communication between the CONGEX camera and the Q series PLC.

The COGNEX vision system can be connected directly on the Ethernet port on the Q-series PLC using Ethernet based on the seamless message protocol (SLMP) or MELSEC communication (MC) protocol [5,6]. Thereby, the SLMP/MC protocol communication enables COGNEX vision system to communicate directly with Mitsubishi PLCs without requiring ladder logic. COGNEX vision can read and write many different data types via SLMP/MC protocol, including tool results, data strings and tolerances. The following diagram (Figure 3) illustrates the communication method between In-Sight and Q-Series PLCs using MC Protocol.

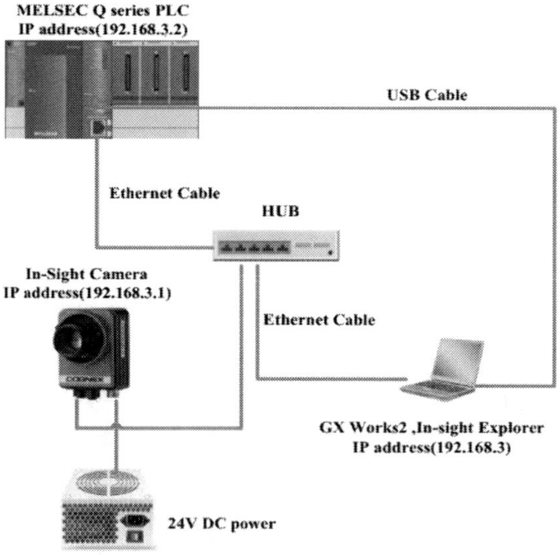

Fig. 3. Connection configuration for the vision system

4 Conclusion

In the present work, we put forward the fully automatic juicing concept at the level of domestic or commercial environment rather than the industrial environment. And we have developed a fully automatic fresh apple juicer in laboratory environment. The system synthetically used the technologies of image sensor, servo positioning, PLC as well as additive manufacturing. The whole processes of peeling, coring, slicing and juicing can be achieved automatically without manual intervention. This potential application spots of the system embraced the crowded sites, such as mall, school, restaurant and hospital. Obviously, this system prototype had many aspects that can be improved, for instance, industrial design, operation reliability etc. Further research is required to completely build a full product type and more simple system.

Acknowledgements

This work was financially supported by the Scientific Research Project of Beijing Educational Committee (No.XN088).

References

1. S.Q. Zeng, C.Z. Zhang, J.N. Zhang, Packaging and Food Machinery, **25** (2009)
2. X.N. Bai, H.F. Chen. J: Sci. & Tech, **50** (2009)
3. A.M. Lefcourt, P. Narayanan, U.Tasch, et al., Bio. Eng, **64** (2009)
4. Z.CH. Zhang, Fine Chem, **20** (1999)
5. T. Mikolajczyk, Applied Mechanics and Materials, **282** (2013)
6. T. Mikolajczyk, P. Wasiak, Applied Mechanics & Materials, **186** (2012)

A Survey of Vehicle Routing Problem

Wujun Cao[1] and Wenshui Yang[2,*]

[1]School of Management Engineering, Zhengzhou University, Zhengzhou 450001, China
[2]School of Management Engineering, Zhengzhou University, Zhengzhou 450001, China
Corresponding Email: yangwenshui2009@163.com

Abstract. In recent years, vehicle routing problem (VRP) has become an important content in logistics management research, and has been widely used in transportation system, logistics distribution system and express delivery system. In this paper, we discuss the classification of VRP, and summarize the common constraints of VRP, model algorithm and the main research results in recent years. Finally, we analyzes the future of VRP, and it is considered that the intelligent vehicle routing problem and intelligent heuristic algorithm will be an important field of future research.

1 Introduction

Vehicle Routing Problem was first proposed by Dantzig and Ramsar 1959, and is mainly used to solve the transport route optimization problem: Atlanta's refinery problem. The subject quickly attracted the attention of experts and scholars, such as operations research, management, computer, graph theory, and proved to be widely used in transportation system, logistics distribution system and express delivery system. After several decades of development, the vehicle routing problem has become an important part of logistics management research, and is classified as the general term for such a type of problem: by a number of vehicles from one or more warehouses to multiple geographic On the distribution of customers, how to arrange the vehicle and its route to the total distribution costs can be minimized. Then, in theory, the VRP problem is defined as: organizing a series of loading and unloading points, as well as the corresponding traffic line organization, so that vehicles can be ordered through them. That is, to achieve the objectives and solve certain problems (such as shortest distance, minimum cost, time limit) under certain constraints (goods demand, delivery, delivery time, vehicle capacity constraints, travel restrictions, time constraints, etc.) May be less, the number of vehicles used as little as possible, etc.). From this simple definition, it is easy to see that TSP is a special case of VRP problem, then VRP problem is also a NP problem [1,2]. Shown in Figure 1:

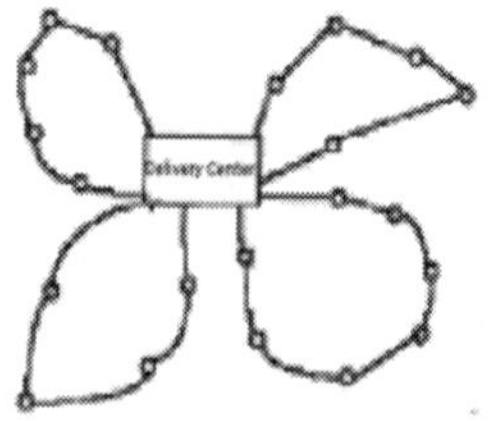

Fig. 1. VPR Schematic diagram.

At present, the enthusiasm of experts and scholars of the VRP research is still high, and according to different perspectives, the classification of VRP:

①According to the number of distribution centers: single distribution center and multi-distribution center problem;

②According to the type of vehicle: single-vehicle type and multi-vehicle type problem;

③According to the characteristics of the task: pure send (take) cargo problems and loading and unloading mixing problem;

④According to whether the time constraints: no time window problem and time window problem;

⑤By vehicle loading: And the problem of non-full load;

⑥According to the optimization of the number of goals: a single objective and multi-objective problem;

⑦Vehicle and vehicle by the ownership of the points: the vehicle open problem and vehicle closure problems;

⑧By mastering the information of certainty: Sexual VRP and non-deterministic VRP problems;

As can be seen from these classifications, solutions to the VRP problem are varied, each category will be mapped a corresponding models and algorithms for solving the natural process is varied. But back to the original point of view to build the model of VRP, VRP problem can be seen as a variant or in combination with integer programming model of graph theory model or other models among which we can find some rules, to achieve the purpose of solving.

2 Constraints of vehicle routing problem

As mentioned earlier, VRP problems have a variety of classification methods, for different classification methods also have different corresponding models and algorithms. However, no matter how complex the model of the algorithm, common constraints in the discussion of VRP problem is the same capacity constraints [3]:

①Capacity constraints: regardless of which vehicle, it should be less than the total path of the vehicle load capacity. The vehicle routing problem with capacity constraints is brought forward.

②Priority constraints: leads to the priority of the vehicle to constrain the path problem.

③Vehicle constraints: leads to multi-vehicle vehicle routing problem.

④Time window constraints: including hard time window, soft time window constraints. The vehicle routing problem with time window (including hard time window and soft time window) is brought forward.

⑤Compatibility constraints: leads to compatibility constraints of the vehicle routing problem.

⑥Random demand: leads to a random demand for the vehicle routing problem.

⑦Open: leads to open the path of the vehicle problem.

⑧Multi-transport center: leads to a multi-transport vehicle routing problem.

⑨Return transport: leads to a return route with the vehicle routing problem.

⑩The last time period: leads to the vehicle with the last time period issue.

⑪Vehicle speed changes with time: with time, leads to the vehicle speed with time changes in vehicle routing problem.

3 Model and algorithm of vehicle routing problem

To solve the VRP problem in different classes, you need to find the relationship between the VRPs of different classifications. A simple VRP problem can be solved by building an integer programming model, or by constructing a graph theory model or other related models. Complex VRP problems are only a combination of these models [4].

After analysis, we can extract the following three types from these complex models:①vehicle-based model; ② logistics-based model; ③multi-coverage model.

After the VRP problem model is set up, the most important step is to solve the model. In recent years, scholars at home and abroad have also proposed many algorithms for obtaining the optimal or suboptimal solution to the different VRP problems. In general, it can be divided into precise algorithms, heuristic algorithms and sub-heuristic algorithms.

3.1 Accurate algorithm

In general, it is more efficient to use exact algorithms to compute small-scale VRP problems, and the optimal solution can be solved in a certain time. However, with the increase in the scale of logistics and distribution, it is difficult in a short time to find a satisfactory solution. At present, scholars at home and abroad are not many of the precise algorithm research, including: branch and bound method, cutting plane method, dynamic programming method, network flow algorithm.

(1) Branch and bound method. Branch and bound method is a commonly used, simple and pure integer algorithm for integer programming, mixed integer programming problem. When using this algorithm to solve, we must first set two variables:

A-The problem of Maximizing the integer programming;

B-Linear Programming Problem without Integer Constraints;

If the optimal solution of B does not satisfy the integer condition of A, then the optimal objective function value of B can be regarded as an upper bound of the optimal objective function value Z * of A, denoted as Z; then the objective function of any feasible solution of A The value can be used as a lower bound of Z *, denoted as Z. Then we use the adjacency integer of B's non-integer solution as the additional condition to get two branches or two subproblems. In the feasible region of the two subproblems, we can find all the satisfactory solutions of the original integer programming problem. Finally, after continuous branching, and gradually reduce the Z, Z increased, the final will be able to obtain Z *.

(2) Cutting plane method. The cutting plane method was first proposed to solve the integer integer programming problem. When solving this problem, it should be noted that no matter what condition (which is often set as the relaxation problem), it is first determined whether the solution is an integer optimal solution, and if so, the operation is stopped. If not, add a new constraint, the purpose is to not meet the conditions before the non-integer solution cut off. After a time of cutting, all the integer solution retained. So that in the end will form an integer point coordinates, that is, the original integer optimal solution.

(3) Dynamic programming method. Dynamic programming method is a kind of optimization algorithm commonly used in system analysis to solve multi-stage decision process problem. This method is to ensure the dynamic nature of the planning process, that is, at every stage to achieve system optimization. Any complex problem can be divided, and can be classified after segmentation, each category in the planning problem in the position occupied is not the same, we need to do is to ensure that these different categories occupy the position of the most optimization. And then through a step by step recursive operation, so that each stage can achieve the best decision-making and choice, so that the final can achieve global optimal.

(4) Network flow algorithm. Network flow algorithm is an algorithm to solve small-scale VRP problem, because the network model of the algorithm vertex number and the number of sides, will significantly affect the network flow algorithm time efficiency. Therefore, in the construction of VRP network model, to find the network need to adjust the largest arc; then, by adjusting the arc flow and potential, making the need to adjust the arc flow reduction until the network flow is zero.

3.2 Heuristic algorithm

Heuristic algorithm has been used to solve the VRP research, now talked about more: saving method, two-stage method, scanning method.

(1) Saving method. The conservation law was proposed by Clark and Wright in 1964 to solve the problem of VRPs where the number of vehicles is not fixed. The basic idea of the saving algorithm is that the path is sorted according to the size of the saving value, and the corresponding two customer points are discharged into the path in the order of the capacity of the vehicle and in the order of big to small until all the customers can be discharged. Into the path. Solomon in 1983, the law will be applied to solve the problem with the time window constraints VRP. At this time, to consider the savings in the two customers into the path of the larger, but also consider the time window restrictions. The advantage of the saving algorithm is that it has a faster calculation speed, especially for small-scale distribution optimization, the optimization accuracy is very close to the optimal solution, and can be obtained in a short time. The disadvantage is that with the increase in the size of customers, optimization results will be some deviation.

(2) Two-stage method. Two-stage method is mainly used to solve the shortcomings of the construction algorithm, it can be said that the two-stage heuristic algorithm for an improvement. In the first stage, the algorithm is used to construct a heuristic algorithm, and then a feasible solution can be obtained. In the second stage, Insert method, 2 - Exchange, 2-swap, 2-OPT, 3-OPT and other methods to improve the technology to ensure the feasibility of the solution, through the adjustment of the point to achieve the goal of optimization results. The main goal of this algorithm is to replace the existing solution of the objective function with the feasible solutions generated in each stage, and to get closer to the optimal solution step by step, and finally to optimize the objective function. Therefore, using the two-stage method, the interactive optimization technique can be used to effectively combine the subjective active action with the VRP problem to solve the problem quickly and easily.

(3) Scanning method. The scanning method was proposed by Gillette and Miller. The method is proposed to solve the problem of vehicle routing problem. In the use of scanning method, the first point to be divided about the location of a demand point, where the polar coordinates are often used; and then set any point corresponding to the point of zero polar coordinates, that is, as the starting point, on this basis , The required service range of clockwise or counterclockwise to divide the division of the restriction is the car capacity; Finally, the use of Lin and Kernighan exchange of previously set demand points sorted, and ultimately to the establishment of the vehicle row Process line. Scanning method is not complicated to use, but the need to ensure the accuracy of program editing.

3.3 Heuristic algorithm

The heuristic algorithm, also known as the intelligent heuristic algorithm, uses the heuristic algorithm to replace the decision criterion of the precise algorithm, so it can effectively narrow the search space. Since the 1990s, artificial intelligence has shown some powerful functions in the field of production, especially when solving combinatorial optimization problems, its intelligence is particularly prominent. At present, it is common to introduce artificial intelligence into the solution of VRP problem, and construct an intelligent heuristic algorithm. Commonly include: genetic algorithm, simulated annealing algorithm, tabu search algorithm, ant colony algorithm [5].

(1) Genetic algorithm. The earliest use of genetic algorithms for VRP research is J. Lawrence, and also designed to effectively solve the problem with the time window of the VRP. As the traditional genetic algorithm is a large-scale, coarse-grained optimization algorithm. Therefore, Barnier combines it with the constraint satisfaction problem, and uses the genetic algorithm to deal with the subdomain of the parameter which satisfies the constraint satisfaction problem (the fitness of the gene can be obtained by calculating the solution of CSP), which can reduce the search space and reduce the constraint satisfaction The objective function of the problem and the complexity of the genetic algorithm constraints. Some scholars use genetic algorithm to ensure the global search, with 3-OPT algorithm to enhance the local search capabilities, to solve the VRP joint algorithm. This hybrid algorithm can now be used to solve large-scale VRP problems (about 199 customers).

(2) Simulated annealing algorithm. The simulated annealing algorithm is a stochastic optimization algorithm based on the Monte-Carlo iterative method. The basic idea is mainly derived from the solid annealing principle. The solid is first heated to a sufficiently high temperature and allowed to cool slowly. Because the particles in the solid state become disordered with the temperature when they are heated, then the internal energy increases, but after the slow cooling, the particles gradually become orderly state, in which each temperature Are maintained in equilibrium, so that the reference state at room temperature, the internal energy will be reduced to a minimum. According to the Metropolis criterion, the probability of the particle tends to equilibrium at temperature T is: $e(-\Delta E / (kT))$, where E is the internal energy at temperature T, ΔE is the change, and k is Boltzmann's constant. With solid annealing simulation combination optimization problem, the internal energy E can be modeled as the objective function value f, the temperature T evolved into the control parameter t, that is, the simulated annealing algorithm to solve the combinatorial optimization problem: the initial solution i and the control parameter initial value t At the end of the algorithm, the current solution is the approximated optimal solution, which is based on the Monte Carlo Iteration (iterative method), which is based on Monte Carlo Iteration A Heuristic Stochastic Search Process for Solving. The annealing process is governed by a cooling

schedule consisting of the initial value of the control parameter t and its attenuation factor Δt, the number of iterations L for each t value, and the stopping condition S. To sum up, the simulated annealing algorithm can effectively avoid the optimization of the serial structure which is local minimally and eventually tends to the global optimum by giving the search process a time-varying and finally to zero probability jumping.

(3) Tabu search algorithm. The first tabu search algorithm proposed by Glover, mainly to solve the problem of global optimization step by step. When tabu search algorithm is used, it always starts with an initial feasible solution, after a series of specific search test behavior, and finally chooses the behavior that can change the most value of a particular objective function value. In order to avoid the dilemma of local optimal solution, tabu search algorithm uses a flexible "memory" technology, the process has been recorded and selected in order to guide the direction of the next search, saving time, Improve efficiency.

(4) Ant colony algorithm. Ant colony algorithm is a natural phenomenon of the algorithm, has a strong anti-biochemical capacity. The algorithm was first proposed in 1992, inspiration is to observe the ants in the path of food to find inspiration. Ant colony in the process of looking for food, there is a particularly important volatile substances, called pheromones, ant can also be said to find a variety of food behavior, are under the guidance of this pheromone. Ants in the ant can be released according to the companion pheromone concentration, to determine their own forward direction will be. You know, each ant will walk through a road when the release of pheromones, it can be said that when the concentration of a particular location pheromone exceptionally high, it should be the region where the highest concentration of food. We refer to the problem of the path, that is, the optimal objective function value, the optimal path. Of course, when using this algorithm will have some drawbacks, because the ant colony and ant colony is still relatively scattered before, a single ant colony to find the optimal food concentration may be due to geographical constraints, and mistakenly believe that the local food Is the most, quoted into our equation solving process, that is, we may fall into the local optimal dilemma.

But on the whole, the advantage of ant colony algorithm is greater than the disadvantage, in some of the requirements of the accuracy of the situation is not high, you can quickly solve the problem. Of course, for some complex problems, you can set some restrictions, optimize some parameters, the structure of some of the necessary processes, but also to achieve good results. In recent years, in solving the problem of parameter optimization design of PID controller, the results of ant colony algorithm design and the result of genetic algorithm design are simulated and compared. The results show that the ant colony optimization algorithm has shown a new Effectiveness and application value.

4 Research prospects

Logistics distribution cost is one of the main sources of logistics costs, and logistics and distribution costs to a large extent by the arrangement of vehicles and routes to decide. This paper first introduces the classification of vehicle routing problem, and then summarizes the main research results of VRP algorithm in recent years according to the classification of VRP algorithm [6,7]. The research shows that, with the expansion of the scale of logistics distribution, the research of exact algorithms has little significance. In the future for a long time, construct the approximate algorithm, especially the intelligent heuristic algorithm will be the important content of VRP algorithm research. The rapid development of social economy, consumer demand for quality of service is also progress, logistics, as a producer services industry, the importance of this area should be increased. VRP problem can be said to be the key link in the process of logistics system optimization, more can not exist slackness psychology. Faced with such a fierce market competition, we should also pay attention to the competition among logistics companies can say who can make full use of existing information technology and Internet technology, who will be able to rapid development. Therefore, in the future, the alliance vehicle scheduling problem between enterprises will also be a new application research field of VRP problem.

References

1. Shi Chun-yan, Huang Hui. Vehicle Routing Problem: Research Summary and Prospect. Logistics Technology, 2014 (12): 75-77.
2. Chen Wen-lan, DAI Shu-gui. Review on the algorithm of vehicle routing problem.Journal of Chuzhou University, 2007 (3): 19-25.
3. Liu Yun-Zhong, XUAN Hui-Yu. Review of Models and Algorithms for Vehicle Routing Problem. Chinese Journal of Management, 2005 (1): 124-130.
4. HAN Juan-juan, LI Yong-xian. Review of dynamic vehicle routing problem. Green Science and Technology, 2015 (5): 285-288.
5. Fang Jincheng, Zhang Qi-shan. VRP routing algorithm for vehicle routing problem. Journal of Shenyang Institute of Engineering, 2006 (4): 357-360.
6. Li Xiang-yong, Tian Peng. Vehicle routing problem model and algorithm research. Shanghai Jiaotong University Antai College of Economics and Management, 2007 (9).
7. Xie Binglei. Multi-loop optimization strategy for stochastic vehicle routing problem .Systems Engineering -Theory & Practice, 2007 (9): 167-171.

Numerical Investigation of the Leakage and Rotordynamic Characteristic for a Short Staggered Labyrinth Seal

Minglei Zhong

Department of Thermal Engineering, Tsinghua University, Beijing 100084, PR China
Corresponding author: lxs78@139.com

Abstract. Short staggered labyrinth seals are widely used in high pressure cylinder shroud seals of steam turbines. They are good at restricting the leakage flow but do not respond well to rotordynamics and often lead to turbomachine instabilities. A numerical investigation into the leakage and rotordynamic characteristics of a short staggered labyrinth seal is carried out in this paper, concerning effects of pressure ratio, inlet preswirl, rotational speed, seal clearance and whirling frequency, using both the whirling rotor method and the multiple frequencies transient simulation method. The numerical model is verified by experimental data. The result shows that the staggered labyrinth seals have a similar but not identical leakage and rotordynamic performance compared with that of teeth-on-stator see-through labyrinth seals. The cross-coupled stiffness can be reduced by decrease inlet swirl with considering the factor of rotational speed, which can significantly benefit the stability of rotor system.

1 Introduction

Labyrinth seals are commonplace in turbomachinery to reduce leakage through rotor–stator clearances from high pressure regions to low pressure regions. This is a necessity to enhance the thermodynamic performance of modern turbomachinery, since the leakage flows are one of the main sources of losses. Along with leakage performance, the importance of rotordynamic assessment for noncontact seals is well recognized. Many researchers have shown that labyrinth seals used in high-speed, high-power-density applications can provoke nonsynchronous self-excited vibrations and often lead to turbomachine instabilities.

In order to estimate the instability of rotor systems, researchers have developed several bulk-flow models to calculate the rotordynamic coefficients of labyrinth seals, such as Childs [1, 2] et al. The primary advantage of these bulk-flow models is that they can predict the rotordynamic coefficients with efficient computational cost. However, these models require the empirical correction, such as friction factors that may change for varying applications and geometry, to have the analysis predictions match the experimental results.

Recently computational fluid dynamics (CFD) has obtained great development such as the schemes [3-5], large eddy simulation [6, 7] and detached eddy simulation [8, 9], and the capability of CFD has been shown for simulation flows of small clearances such as bearings [10, 11] and seals [12-14]. Therefore, CFD programs and software are now gradually displacing the bulk-flow approach in predicting the rotordynamic coefficients of annular gas seals. Compared to the experimental measurements and bulk-flow methods, the CFD method can offer a great deal of insight for understanding the fluid dynamics of the annular gas seal flows and developing new seal design. Currently, the CFD analysis for the rotordynamic characteristics of the labyrinth seal have been carried out by many authors. Huang researched rotordynamic characteristics [12], coupled fluid–solid vibration [14], and developed a non-linear model [13] of seals by CFD. Moore [15] studied two labyrinth seals using a specially developed CFD code named SCISEAL. Hirano [16] applied CFD technique to a compressor eye seal leakage and rotordynamic simulation with commercial CFD program CFX-TASCflow. Pugachev [17, 18] used CFD software ANSYS CFX, studied the rotordanymic performance of a short staggered labyrinth seal, and compared the results to experimental data.

In practice, the short staggered labyrinth seals are widely used in high pressure cylinder shroud seals of steam turbine, since the physical dimensions are limited. In the earlier research, most work are concentrated on leakage and rotordynamic performance of see-through labyrinth seals, while few work is about staggered labyrinth seal. Therefore, a detailed study of the short staggered labyrinth seals is quite essential and significant. In this paper, a short staggered labyrinth seal is studied to find out the influence of pressure ratio, inlet swirl, rotating speed, seal clearance and whirling frequency to leakage and rotordynamic performance.

2 Numerical model

2.1 Linear rotordynamic model

The influence of the seal on rotordynamics is described by the seal's stiffness K, and damping C coefficients. For small motion about a centered position (x-axis), these coefficients appear when linearizing the aerodynamic force which arises due to the squeezed air film in the seal:

$$-\begin{pmatrix} F_y \\ F_z \end{pmatrix} = \begin{bmatrix} K & k \\ -k & K \end{bmatrix} \begin{pmatrix} y \\ z \end{pmatrix} + \begin{bmatrix} C & c \\ -c & C \end{bmatrix} \begin{pmatrix} \dot{y} \\ \dot{z} \end{pmatrix} \tag{1}$$

Assuming a circular whirl orbit around the seal center and a small shaft eccentricity e, as shown in Figure 1, the linear rotordynamic model of Eq. (1) can be written in a new coordinate system aligned with vectors normal and tangential to the shaft surface as

$$F_r / e = -K - c \cdot \Omega \tag{2a}$$

$$F_t / e = k - C \cdot \Omega \tag{2b}$$

According to Eq. (2), knowing aerodynamic forces acting in the seal at different operating conditions one can determine rotordynamic coefficients. The cross-coupled stiffness k and direct damping C terms are of particular interest. The cross-coupled stiffness is governed by the tangential force F_t which contributes to the destabilizing of the rotor system. The direct damping reflects the capability to suppress severe vibrations. A negative value of the direct stiffness also contributes to instability. In this instance the radial force F_r is acting in the direction of the rotor excursion and tries to increase the rotor eccentricity.

However, information on the stiffness of the whole rotor system is necessary to judge the effect of the seal direct stiffness on rotordynamics. The assessment of the stability performance of a seal can be evaluated by the whirl-frequency ratio, as defined in Eq. (3). It is the ratio between the destabilizing force due to the cross-coupled stiffness k and stabilizing force due to the direct damping C. A lower Ω_w, which means a lower cross-coupled stiffness coefficient or a bigger direct damping coefficient, is expected to obtain better stability of the rotor system.

$$\Omega_w = k / (\omega C) \tag{3}$$

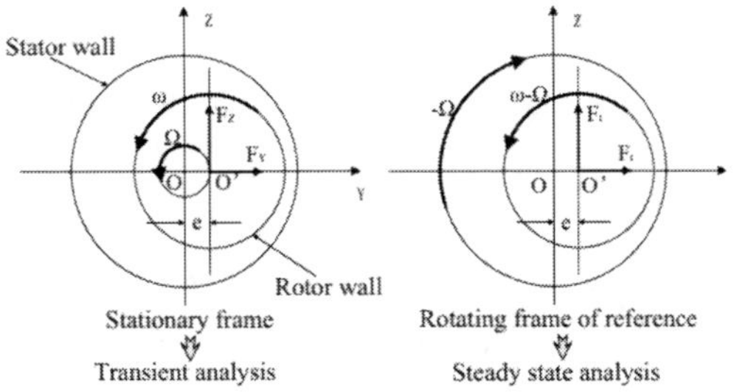

Fig. 1. Motion Model of Seal-Whirling Rotor in Stationary and Rotating Frame.

2.2 Identification of rotordynamic coefficients

There are two methods used in this paper to calculate rotordynamic coefficients which are appropriate for the CFD analysis. They are described as below.

The whirling rotor method is based on CFD calculations in a rotating frame, as shown in Figure 1(b). the rotor is supposed to whirl with a circular orbit and small amplitude around the seal centerline. The dynamic behavior of the seal is modeled by steady-state calculations. Rotational speed in the rotating frame of reference is the difference between the speed in the stationary frame of reference (ω for rotor and zero for stator) and the rotor whirl speed:

$$\omega^r = \omega - \Omega \tag{4}$$

The procedure of calculating four rotordynamic coefficients involves at least two CFD solutions at different rotor whirl speeds. The rotordynamic coefficients are determined from the linear interpolation of the force components that are functions of the whirl speed.

The transient simulation method is the most computationally expensive technique to determine rotordynamic coefficients, but it can simulate any user defined complex movements. Li [19] has presented a 3D transient CFD method based on the multi-frequency elliptical whirling orbit model. In the present paper, a revised model based on this multi-frequency approach is developed. To obtain the rotordynamic coefficients, the first step in the present numerical analysis is to develop the equations of whirling motion for the rotor surface. Here a y-direction sine curve to the rotor movement is assigned, as described below:

$$y = \sum_i A_i \sin(\Omega_i t), z = 0 \tag{5}$$

During the calculation, the computational grid is deformed at every time step according to the new position of the shaft. The capability of mesh deformation is provided in ANSYS CFX software. In this approach, the shaft surface rotates around its center with a rotational speed ω, and the stator wall keeps still.

For the multi-frequency model in the present paper, the response forces (F_y , F_z), rotor motions (y , z), and rotor velocity ($\dot{y}$, $\dot{z}$) in Eq. (1) are multi-frequency signals that contain the same frequency components, but different initial phases. Therefore, to obtain the rotordynamic coefficients for each frequency component, it is necessary to transform Eq. (1) into the frequency domain expression given by Eq. (6) using the FFT.

$$-F_y^{FFT} = (K + j\Omega C)y^{FFT} + (k + j\Omega c)z^{FFT} \tag{6a}$$

$$-F_z^{FFT} = -(k + j\Omega c)y^{FFT} + (K + j\Omega C)z^{FFT} \tag{6b}$$

where $j = \sqrt{-1}$, (y^{FFT} , z^{FFT}) and (F_y^{FFT} , F_z^{FFT}) are the frequency domain components of the relative displacement and the response forces after the transformation. Therefore, introducing the y-direction sine curve rotor movement assumption, Eq. (6) can be written into a new form:

$$-\begin{pmatrix} F_y^{FFT} \\ F_z^{FFT} \end{pmatrix} = y^{FFT} \begin{pmatrix} K + j\Omega C \\ -(k + j\Omega c) \end{pmatrix} \tag{7}$$

Once the force impedances are obtained through the transient solutions in this paper, rotordynamic coefficients are calculated by solving Eq. (7) and separating the right-side array into real and imaginary parts, as defined in Eqs.(8)-(11).

$$K = -\mathrm{Re}\left(F_y^{FFT} / y^{FFT}\right) \tag{8}$$

$$k = \mathrm{Re}\left(F_z^{FFT} / y^{FFT}\right) \tag{9}$$

$$C = -\mathrm{Im}\left(F_y^{FFT} / y^{FFT}\right) / \Omega \tag{10}$$

$$c = \mathrm{Im}\left(F_z^{FFT} / y^{FFT}\right) / \Omega \tag{11}$$

2.3 Numerical model

In this paper, a short staggered three-tooth-on-stator labyrinth seal model is utilized as the computational model. The seal geometrical parameters and operational conditions in this work are based on the experimental and computational results published by Pugachev [17, 18].

The geometry of the seal typical for blade tip and shroud seal locations, is shown in Figure 2. The shaft band height is 3mm and the length is 6mm. The tooth width is 1mm, with a 0.3mm width at the tip, and the tooth taper is 75 deg.

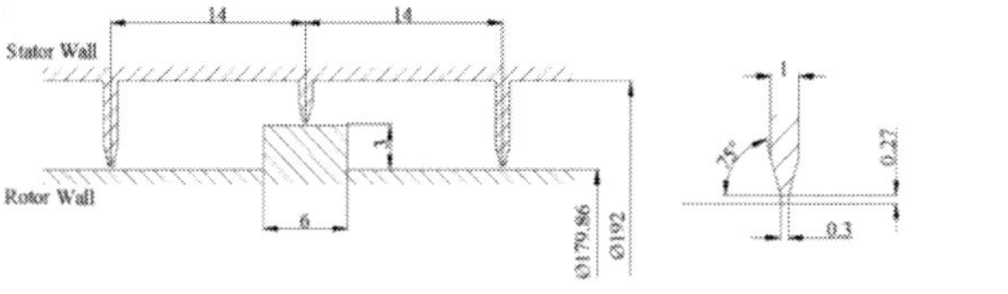

Fig. 2. Geometry of the Three-tooth-on-stator Staggered Labyrinth Seal.

A full eccentric 3D CFD model of the seal is constructed in the commercial CFD code ANSYS CFX. Three-dimensional structural grids are generated using ANSYS ICEM CFD. Figure 3 shows a cross-sectional plane illustrating distribution of the computational grid. A compressible Reynolds-averaged Navier–Stokes (RANS) analysis with the total energy heat transfer model is considered. Turbulence is simulated using the two-equation k-ε model, with a scalable wall function. Ideal gas law is used with the assumption of a constant dynamic viscosity. A stable pressure condition is set at the inlet, with specified velocity direction, keeping a constant inlet swirl velocity of 145 m/s. A static pressure condition is used at the outlet. Two methods are used here to attain the rotordynamic coefficients, as described in section 2.2. The shaft rotate at 750 rpm, and whirls around x-axis at -750 rpm, 0 rpm, 750 rpm when using the whirling rotor method. Multiple frequencies from 0-150 Hz of sinusoid shaft displacement is assigned when using the transient simulation method.

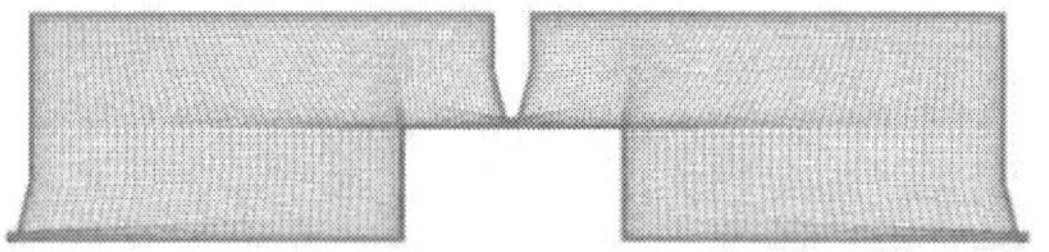

Fig. 3. 2D Cross-sectional Distribution of the Computational Grid.

3 Verification of numerical model

Figure 4 is a illustration of the predicted pressure and velocity distribution in the seal cavities. The sharp pressure drop mainly happens at the tip of seal tooth, which is the primary cause for restriction of leakage flows. The rotor band disturbs heavily the flow especially in the second sealing cavity which results in developing of secondary vortices. The kinetic energy of the leakage flow is converted into dynamic pressure at locations where the flow hits the wall. This leads to considerable changes in pressure in radial and axial directions within a cavity.

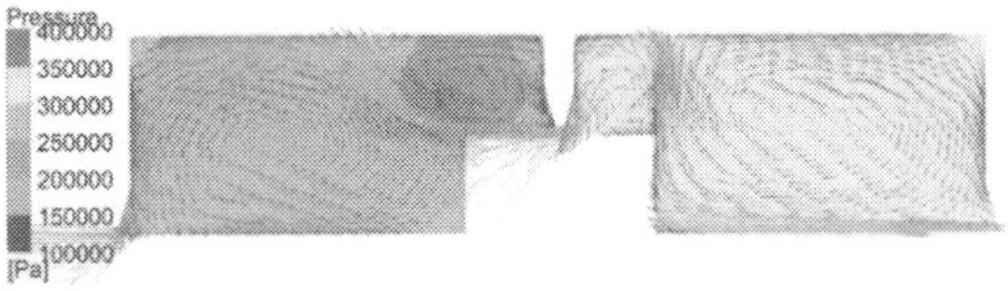

Fig.4. Predicted pressure and velocity distribution in two sealing cavities.

The theoretical rotordynamic coefficients attained by the whirling rotor method is presented in Figure 5, in comparison with experimental result published by Pugachev [17]. Here the linear correlations Pugachev used in his article are followed. That is, there is a linear relationship between direct stiffness with the leakage, and the cross-coupled stiffness with the inflow swirl force(product of leakage and inlet swirl velocity), and the direct and cross-coupled damping with the averaged density. A good conformity between the theoretical and experimental results shown in Figure 5 is a well confirmation of the numerical model.

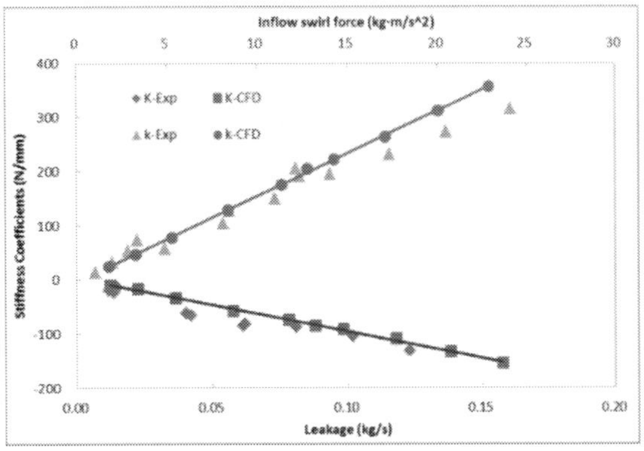

Fig. 5 Experimental and theoretical stiffness coefficients for the short staggered labyrinth seal (direct stiffness versus leakage, cross-coupled stiffness versus inflow swirl force).

4 Results and discussion

For see-through labyrinth seals, earlier researches have done lots of experimental and numerical work about the effect of pressure ratio, inlet swirl, rotational speed, seal clearance, and whirling frequency on the leakage and rotordynamic performance. Childs has made prominent contributions and summarized his achievements in his book [20]. Researchers have put forward several useful method [21-25] to solve nonsynchronous self-excited vibrations, such as anti-preswirl method.

In this section a study of the leakage and rotordynamic performance of a short staggered labyrinth seal is carried out, trying to figure out whether the principles of see-through labyrinth seals can be applied to short staggered labyrinth seals.

4.1 Effects of inlet swirl

To investigate the effect of inlet swirl on the leakage and rotordynamic characteristics of short staggered labyrinth seal, the whirling rotor method is performed with inlet circular velocity varying from -150 m/s to 150 m/s, for two inlet pressures 0.4 MPa and 0.15 MPa.

Figure 6-8 illustrates correlation of the leakage flow and rotordynamic performance with inlet swirl. The flow rate decreases as the absolute inlet circular velocity increases, while absolute value of direct stiffness and direct damping increase as the inlet swirl getting stronger. The cross-coupled stiffness and cross-coupled damping show an approximately linear relationship with inlet swirl.

This correlation with inlet swirl is consistent with that of see-through labyrinth published by Childs. Hence inlet swirl control is a straightforward way to attain a little cross-coupled stiffness, and to improve rotor system stability.

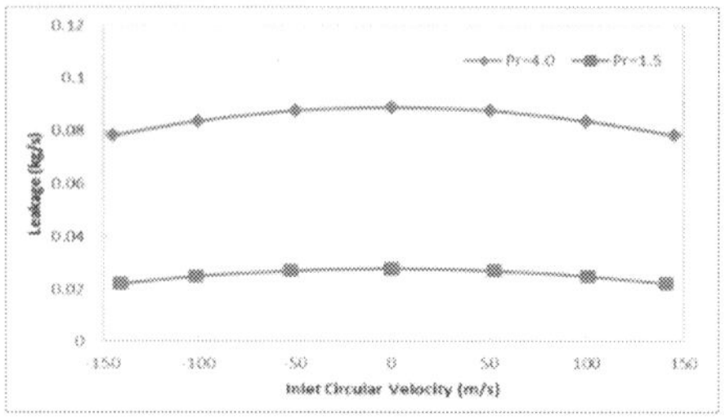

Fig.6. Leakages for inlet circular velocities.

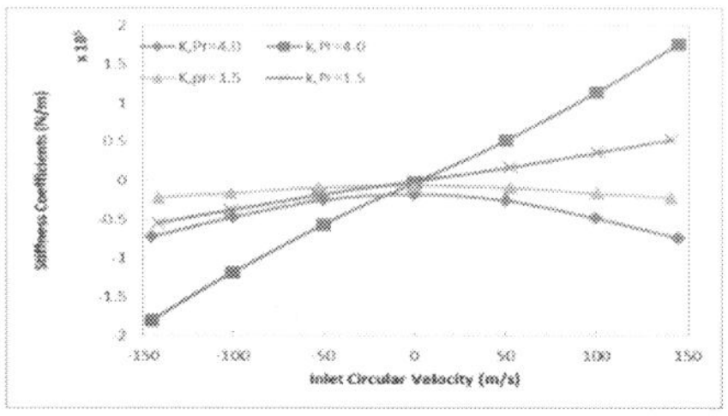

Fig.7. Stiffness coefficients for inlet circular velocities.

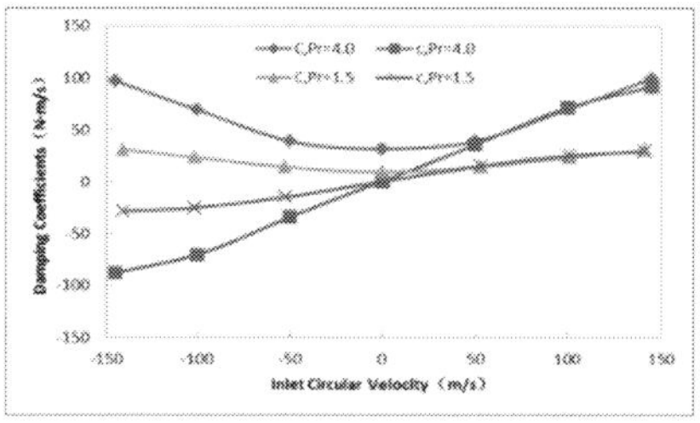

Fig.8. Damping coefficients for inlet circular velocities.

4.2 Effects of seal clearance

In order to explore the relationship between seal clearances and seal performance, a variety of cases with different seal clearances and pressure ratios are studied, using the whirling rotor method. A constant inlet circular velocity,145 m/s, is guaranteed.

Figure 9 illustrates the leakage performance of the short staggered labyrinth seal as a function of the seal clearance. As expected, the leakage increases as the seal clearance increases. Results of different pressure ratio show a similar trend.

With a seal clearance increase, the cross-coupled stiffness decreases and direct stiffness increases, while the absolute value of direct stiffness decreases, as shown in Figure10. A slight drop of direct damping happens as the seal clearance increases, as shown in Figure11.

These principles are in conformity with the rotordynamic performance of teeth-on-rotor see-through labyrinth. A bigger seal clearance will result in better stability, but will cause a severe leakage. The optimized seal clearance happens when the contradictory compromises.

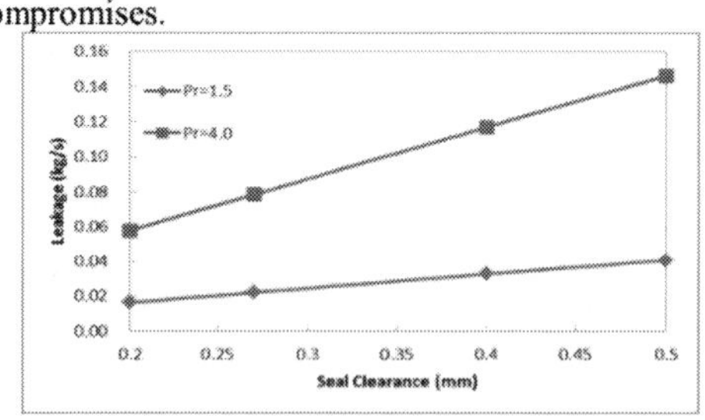

Fig.9. Leakages for seal clearances.

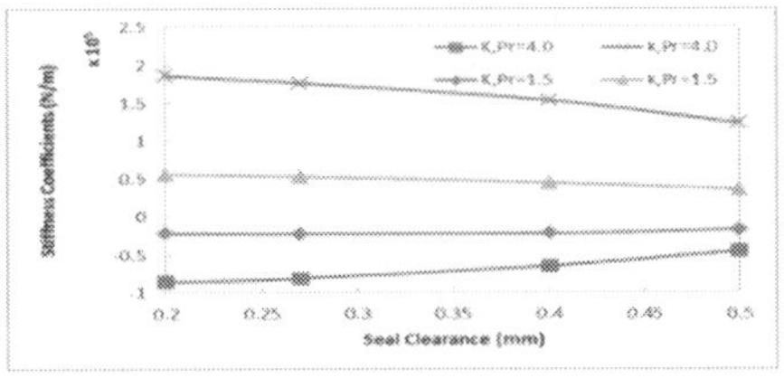

Fig.10. Stiffness coefficients for seal clearances.

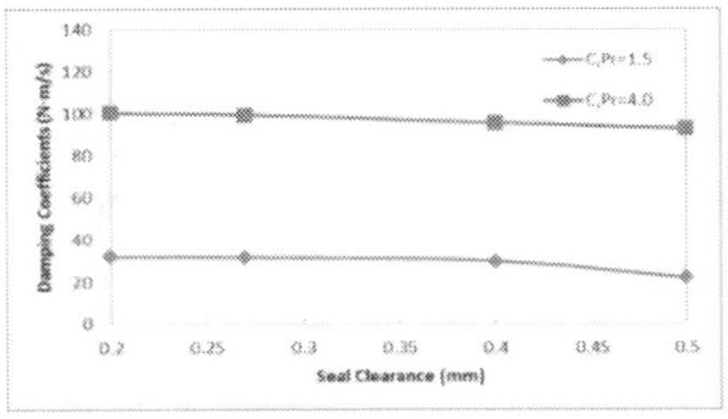

Fig.11. Direct damping coefficients for seal clearances.

4.3 Effects of rotational speed

Figure 12-14 illustrate dependence of the leakage flow rate and rotordynamic coefficients versus rotational speed from 5000 rpm to 35000 rpm. There is no obvious trend of leakage as rotational speed increases. The direct and cross-coupled stiffness decrease with the increasing rotational speed, while the absolute value of direct stiffness increase because of its negative value. A slight increase of direct damping is observed as the rotational speed increases.

These effects of rotational speed on short staggered labyrinth seal is in consistent with that of see-through labyrinth seal. Especially, the cross-coupled stiffness is obviously effected by the rotational speed.

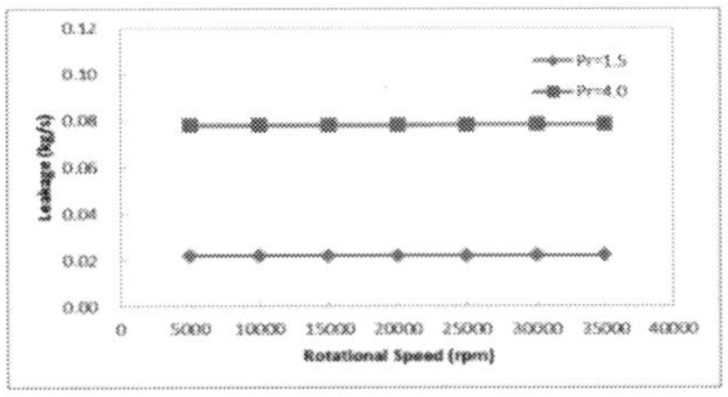

Fig.12. Leakages for rotational speeds.

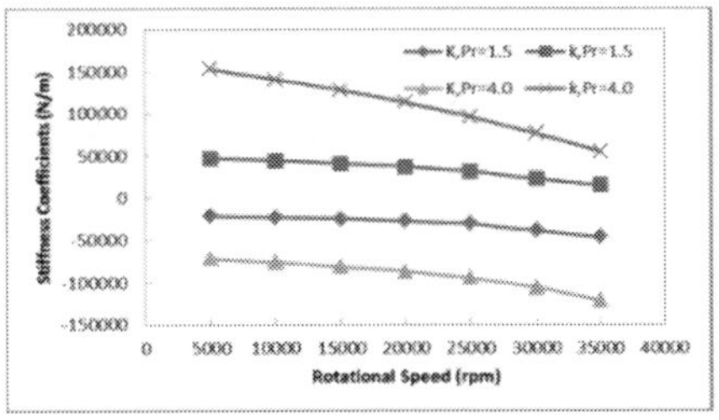

Fig.13. Stiffness coefficients for rotational speeds.

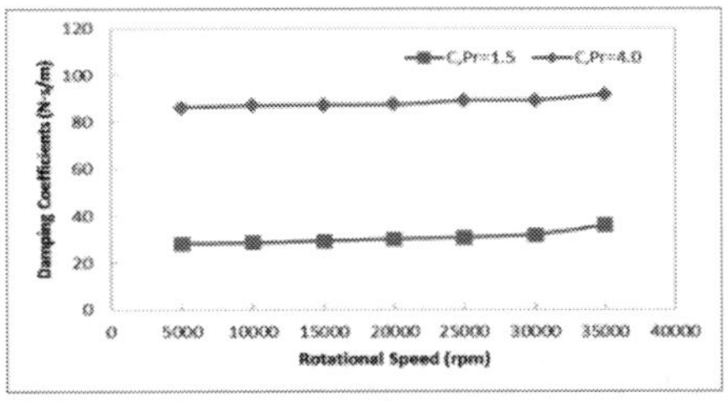

Fig.14. Direct damping coefficients for rotational speeds.

4.4 Whirling frequency dependency

A multi-frequency transient simulation method is used in this section to investigate frequency dependence of rotordynamic coefficients, since it is convenient to attain several rotordynamic coefficients of different whirling frequencies in a single calculation. Prior studies on see-through labyrinth seal have shown that damping coefficients are independent, while direct stiffness coefficient is frequency dependent and cross-coupled stiffness shows frequency independence out to a frequency of 150 Hz. Other types of seals, such as honeycomb seal, hole-pattern seal, and pocket seal, are known as frequency dependent.

As for short staggered labyrinth seal, Figure 15 shows a frequency independence of cross-coupled and direct stiffness out of a frequency of 50 Hz. Direct and cross-coupled damping seem to be frequency independent, as shown in Figure 16. This phenomenon is similar but not identical to that of see-through labyrinth seal.

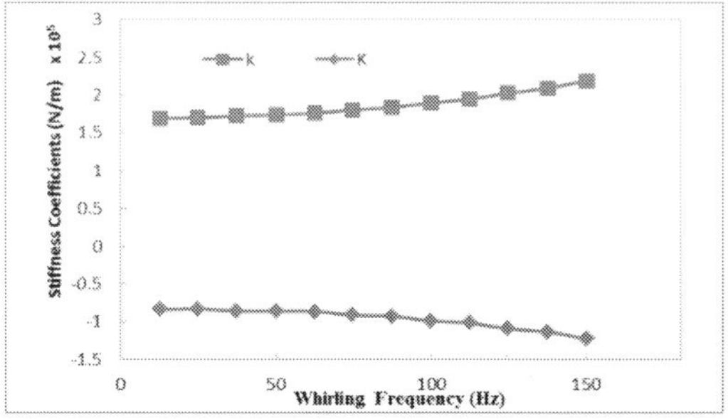

Fig.15. Stiffness coefficient for whirling frequency.

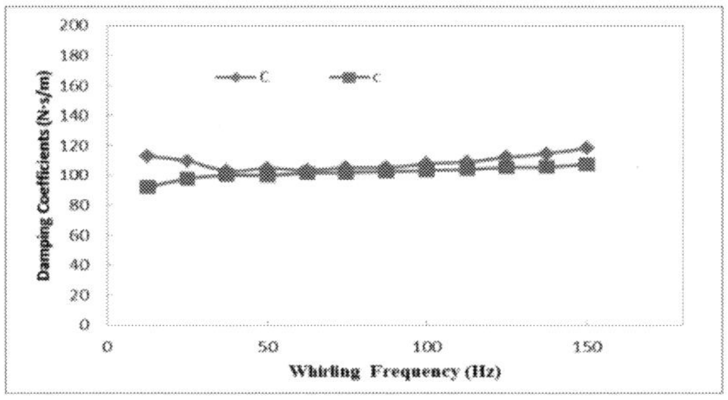

Fig.16. Damping coefficient for whirling frequency.

5 Conclusions

A short staggered labyrinth seal is studied in the present paper using two numerical approaches, including the whirling rotor method and the multi-frequency transient simulation method. Verification of these methods is made by comparison with experimental data from Pugachev [17]. A detailed research about the effects of pressure ratio, inlet swirl, rotational speed, seal clearance, and whirling frequency on the leakage and rotordynamic performance is carried out.

(1) Inlet swirl has a great effect on leakage and rotordynamic performance of the short staggered labyrinth seal. The cross-coupled stiffness can be controlled by restricting the inlet circular velocity.

(2) The seal clearance has an obvious influence on leakage and stiffness performance. A bigger seal clearance will result in better stability and severe leakage. Hence compromise is required.

(3) An increase of rotational speed will cause decrease of direct and cross-coupled stiffness.

(4) Cross-coupled and direct stiffness are frequency independent out of a frequency of 50 Hz. Direct and cross-coupled damping seems to be frequency independent.

The result shows that the performance of short staggered labyrinth is similar but not identical to that of see-through labyrinth. Since no evident trend of direct damping is observed in the research, the best way to improve rotor stability lies in controlling the cross-coupled stiffness. A lower inlet circular velocity is expected to restrict cross-coupled stiffness and avoid rotor system instability. Because the rotational speed has obvious effect on the cross-coupled stiffness, it should be an important factor for choosing appropriate value of inlet circular velocity.

References

1. D. Childs, J. Scharrer. ASME Journal of Vibration, Acoustics, Stress, and Reliability in Design, 110, 11 (1988).
2. D. Childs, J. Scharrer. ASME Journal of Vibration, Acoustics, Stress, and Reliability in Design, 110, 7 (1988).
3. X. Li. International Journal of Computational Fluid Dynamics, 28, 10 (2014).

4. X. Li, C. Gu, Computers & Fluids, 86, 15 (2013).
5. X. Li, C. Gu, Journal of Computational Physics, 229, 13 (2010).
6. X. Li, J. Xu, C. Gu, Science in China - Series G: Physics Mechanics and Astronomy, 51, 11 (2008).
7. X. Li., X. Li. International Journal of Computational Fluid Dynamics, 30, 10 (2016).
8. C. Gu, F. Feng, X. Li, ASME Paper, No. GT2008-50203 (2008).
9. C. Gu, M. Chen, X. Li, ASME Paper, No. GT2008-50040 (2008).
10. X. Li, Y. Song, Z. Hao, C. Gu, Journal of Tribology - Transactions of the ASME, 134, 7 (2012).
11. Y. Song, X. Ren, C. Gu, X. Li, Journal of Tribology - Transactions of the ASME, 137, 9 (2015).
12. D. Huang, X. Li, Proceedings of the Institution of Mechanical Engineers - Part A: Journal of Power and Energy, 218, 7 (2004).
13. D. Huang, X. Li, Proceedings of the Institution of Mechanical Engineers - Part A: Journal of Power and Energy, 218, 7 (2004).
14. D. Huang, X. Li, Proceedings of the Institution of Mechanical Engineers - Part A: Journal of Power and Energy, 218, 11 (2004).
15. J. Moore, ASME Journal of Vibration and Acoustics, 125, 7 (2003).
16. T. Hirano, Z. Guo, R. Kirk, ASME Journal of Engineering for Gas Turbines and Power, 127, 7 (2005).
17. A. Pugachev, M. Deckner, ASME Paper, No. GT2010-22058 (2010).
18. A. Pugachev, U. Kleinhans, M. Gaszner, ASME Journal of Engineering for Gas Turbines and Power, 134, 10 (2012).
19. Z. Li, J. Li, X. Yan, Journal of Vibration and Acoustics, 135, 9 (2013).
20. D. Childs, Turbomachinery rotordynamics: phenomena, modeling, and analysis (John Wiley & Sons Press, New York, USA, 1993).
21. X. Li, C. Gu, J. Xu, Computers & Fluids, 38, 8 (2009).
22. X. Li, C. Gu, Journal of Computational Physics, 227, 16 (2008).
23. C. Gu, W. Hao, X. Ji, X. Li, Energy, 112, 15 (2016).
24. J. Song, C. Gu, X. Li, Applied Thermal Engineering, 110, 9 (2017).
25. X. Ren, C. Gu, International Journal for Numerical Methods in Fluids, DOI: 10.1002/fld.4351.

 DOI: 10.1051/matecconf/201710001008

Problems and Countermeasures of Aviation Industry Development in Henan Province

Shaoxia Zhao *

Administrative Engineering College, Zhengzhou University, Zhengzhou, China

Corresponding Email: 1358804759@qq.com

Abstract. With the rapid development of China's aviation industry, the province is also accelerating the pace of development of the aviation industry, but in the high-speed development also encountered some problems.This paper starts from the current situation of the aviation industry in Xinzheng airport pilot area of Henan Province, and then briefly introduces the development of aviation industry in Henan Province. On this basis, it points out that the high - end talent shortage in the development of Henan's aviation industry and the effect of enterprises' agglomeration have not been formed. The countermeasures are put forward to perfect the traffic system, build the green aviation city, strengthen the government guidance, esteem the innovation, improve the technical content of the products and enhance the industrial competitiveness. Finally, based on the current development of the aviation industry, the future development prospects of aviation industry in our province are forecasted.

1 The development status of aviation industry in Henan province

In Henan, there are 13 aerospace and spacecraft manufacturing enterprises. In addition, there is a related university: Zhengzhou Institute of Aviation Industry Management, and a test area: Zhengzhou Airport Economic Comprehensive Experimental Zone which is also the focus of the aviation industry in Henan Province. Xinzheng Airport, as the air portal of Henan Province and one of the important domestic airports, is located in the test area. Xinzheng airport is responsible for most air transport tasks of Henan Province, and the transport sector accounted for more than 96% of the province's civil aviation transport. In addition, as the international airport of Zhengzhou, Xinzheng International Airport has many international flights to different countries, and these countries include Russia, Singapore, Japan, South Korea and so on. Here, there are now 32 passenger airlines, Passengers can easily fly to 73 cities at home and abroad through 116 routes [1]. Zhengzhou airport development status is as follows:

(1) Zhengzhou airport transport capacity continues to increase, passenger traffic steadily increased year by year. With the "Zhengzhou Airport Comprehensive Experimental Zone Planning" issued, Zhengzhou began a comprehensive construction phase of being an aviation hub. There are 31 airlines, including China Southern Airlines Henan Branch, operate at Zhengzhou Xinzheng International Airport. They have opened 82 routes, more than 1,400 flights per week. And these routes can reach 52 cities and regions.

(2) Rapid development of air transport and aviation logistics.

The construction of the comprehensive bonded zone in Zhengzhou aviation port area greatly improved the speed of goods transportation and saved the transportation time. Now with the large-scale promotion of e-commerce, online shopping has become a very popular way of shopping. The province's bonded port logistics center is also under way cross-border e-commerce services experimental work. If the pilot is successful, combined with the advantages of the Central Plains is located, Zhengzhou is expected to become the country's online shopping distribution center , which will further promote the rapid development of the aviation industry[2].

(3) The introduction of aviation-related industries, making the function of the airport area is increasingly perfect.

Now, the Zhengzhou airport area already has a comprehensive bonded area, export processing zones, bonded logistics center, container center station and other large projects [3]. In addition, it has attracted many electronic precision manufacturing, bio-medicine, optical materials and other industries to come. These high-end manufacturing and modern service industry will promote the prosperity of the industry in our province, at the same time, will also lead to the rapid development of aviation industry to achieve industry and industry rely on each other, and ultimately achieve common development [4].

2 The problems in the development of aviation industry in Henan province

The research of China's aviation industry started relatively late. In recent years, although in China's aviation technology and strength have been greatly improved, but compared with the developed countries there is still room for further improvement. Traditionally, Henan is an agricultural province. At the same time, because it is located in the middle of China, Henan Province has become a transportation hub. With China's comprehensive strength, the upgrading of international status, as well as industrial, high-end manufacturing, transportation development, the national and provincial governments have begun to focus on the development of aviation industry. However, due to various factors, there are still many problems the province's aviation industry in development of the province's aviation industry.

2.1 Aviation industry has not yet formed industrial agglomeration

From the above analysis of the development of aviation industry in Henan Province can be seen, there are only 13 spacecraft manufacturing enterprises, a related university, two research institutes in Henan Province [5]. But, the spacecraft manufacturing enterprises are small in size and scattered in layout, and the links between enterprises and enterprises are not much contact, and the products are not the core products of spaceflight, the competitiveness is not strong, the technology content is low and the branding is weak. And the research shows that there is a positive correlation between industrial agglomeration and competitiveness, if the industry is more dispersed, it is difficult to enhance industrial competitiveness, so the competitiveness of aviation industry in our province is not strong, in part because industrial agglomeration is not formed.

2.2 The lack of high-end aviation industry talent

There are dozens of colleges and universities in Henan Province. But, there is only one aviation university, which is undoubtedly an important constraint for the development of aviation industry in terms of my province. And Henan has not yet introduced policies conducive to the introduction of talent.

2.3 The weak international cooperation and insufficient transport capacity of Zhengzhou Airport

Zhengzhou International Airport is only with Japan, Korea, Singapore, Russia and other countries have flights, and other countries have not yet opened routes, from this we can see that its international cooperation is not strong. In addition, the Zhengzhou International Airport traffic exceeded 10 million in 2011. Compared with Xi'an, Wuhan, Chengdu and other regions, Zhengzhou is not only in time but also in the number fall behind of them. In addition, due to geographical and economic factors, the development of Henan's aviation industry is far behind the speed of the coastal areas, Henan aviation industry development foundation is weak, operational capacity and management level remain to be improved.

2.4 The gradually increasing competitive pressure in the central region

The province is building a pilot area of the airport, at the same time, the surrounding provinces and cities are also active in the airport planning and development. For example, Hubei is building Wuhan Tianhe Airport as the core of the Wuhan Airport; Hunan, Xi'an and other places are also actively develop their aviation industry, which undoubtedly has brought great pressure competition to the development of aviation industry in Henan Province.

2.5 The environmental impact

The environmental impact of the aviation industry mainly includes the following: (1) airport aircraft noise and equipment noise[6]; (2) combustion gas, wastewater, industrial wastewater; (3) the transport, production, storage of various types of hazardous chemical substances hazardous materials;(4) waste and garbage, produced by the surrounding industrial, will pollute the surrounding living environment; (5) the construction of large-scale aviation industry will cause soil erosion and impact on the ecological environment. As we all know, the pilot area of airport development belongs to regional linkage development. The impact of this regional development on the environment, not only from the scope of the impact, but also from the extent of the impact is much larger than a single enterprise, and with the continued activity, the impact will continue. Obviously, this departure from the green economy, low-carbon economy, and sustainable development policies, and long-term will restrict the development of the aviation industry [7].

2.6 Innovation system and innovation ability need to be improved

Since the establishment of the airport pilot area, the province's aviation industry has achieved initial results. The development of the aviation industry in our province cannot just copy the development model of other regions, and should have our own characteristics, innovation. In the initial stage of development of the aviation industry, Aviation industry has a strong technical dependence, and the pattern is probably the same. However, in the development stage, the province's aviation industry should take the road of collaborative innovation, and form the core competitiveness with regional characteristics.

3 Responding measures to the development of aviation industry

Henan Province's aviation industry is still in the stage of development and exploration. In the development process of a new industry will encounter many problems, only by daring to find problems, face these problems, and find the response measures to correct it, can make an industry more reasonable and perfect. The following countermeasures are put forward for the problems in the development of aviation industry of Henan Province:

3.1 The formation of industrial agglomeration improves competitiveness of aviation industry

Measures to enhance the competitiveness of the province's aviation industry are as follows: (1) the government should build an aviation research institute here, and increase investment in the aviation industry; (2) spacecraft production enterprises need to gather together, while strengthening the cooperation between our province aviation industry and advanced spacecraft enterprise;(3) the government should vigorously attracts foreign investment, and give those aviation industries who enter our province good policies to improve the construction of the airport pilot area system. Industrial agglomeration can facilitate the cooperation between enterprises, which not only can enhance the competitiveness of the aviation industry, promote the aviation industry well developed, but also can promote the local manufacturing industry, the development of the electronics industry, and promote the economic development of our province[8].

3.2 Training and introducing high-end technical personnel

Aviation industry is a high-end technology industry; the support of relevant high-tech talent is a necessary condition for its development. In the process of vigorously developing the aviation industry, Henan Province needs to pay attention to the cultivation and introduction of talents. And Zhengzhou Airport Economic Experimental Zone is in the construction stage, a large number of high-end talents are needed. The relevant government departments should also introduce relevant policies to attract and retain these senior personnel, so that they can devote themselves to the construction of the airport pilot area[2].

3.3 Making full use of geographical and transportation advantages

Henan is located in the Central of our country; here is the national transport hub. Henan Province is building an airport pilot area, which will open the door to the world. In recent years, Henan Province is focusing on building roads and the development of modern logistics. These measures to a large extent alleviate the competitive pressures brought by the surrounding areas. However, the Zhengzhou International Airport is located in Xinzheng, where is distance from Zhengzhou city, Zhengzhou Railway Station and high-speed rail stations, which gives passengers in the plane brought a lot of inconvenience. Therefore, a subway from the city to the airport should be built as soon as possible o facilitate passengers to travel.

3.4 Building green city, strengthen ecological construction, and protect the environment

In response to environmental pollution caused by the development of the aviation industry, the following measures can be taken: (1) rational planning and rational development. In the construction of the airport pilot area, government departments should lead the industrial layout, optimize the industrial structure, and actively respond to the national green economy, low-carbon development policy; (2) protecting the environment and providing reasonable supporting. In the process of development, the concept of environmental protection is not ignored. In the production process to reduce waste, to avoid unnecessary operations and steps to reduce industrial waste production and reduce machine noise; (3) green aviation city. Scientific planning urban service area, efficient use of land, creating green livable living environment, improving the transport system, all of them are service support for the development of the aviation industry;(4) ecological construction. The surrounding area of the test area needs to build more green landscape belts, so as to expand the green area and improve the greening rate.

3.5 Promoting cross-border e-commerce services

As mentioned above, our province is trying to pilot cross-border e-commerce services. With the prevalence of online shopping, the implementation of the pilot, is bound to greatly increase the capacity of our province's aviation, and promote the development of aviation industry in our province. Specific measures include the construction of e-commerce centers and cross-border network shopping items cluster evacuation center. E-commerce promotes the development of the real economy, r to achieve the physical store and online store interactive development.

3.6 Government should strengthen the guidance of the aviation industry and establish monitoring system

Compared with the developed countries, China's aviation industry is still relatively backward. Our province is facing a weak foundation for the development of the aviation industry. Substantial progress is inseparable from the strong support of government departments. The government needs give more help in the capital, personnel and business introduction policies[9]. In addition, the aviation industry is an industry, the airport pilot area is a large area, where will involve a lot of money. At the same time, the aviation industry is involved in many national technology secrets, in order to avoid fraud, technology leaks and other phenomena, the government should set up the corresponding supervision system to ensure that the airport experimental zone in the right path of development on the rapid, healthy and harmonious development.

4 Conclusions

In the context of vigorous development of aviation industry, Henan Province should seize this historical opportunity to develop aviation industry in Zhengzhou pilot area as the core. On the one hand, high-end technology industries related to aircraft manufacturing can be developed, including the design of large body components, the manufacture of aircraft parts, the manufacture of aeronautical standard parts and the manufacture of light aircraft. On the other hand, develop logistics, e-commerce, technical services, human resources, entertainment, tourism and other modern services as the corresponding facilities. These will contribute to the formation of Aviation Industry Park with the characteristics of Henan province, the good wishes of aviation industry strides forward and rapid development of our province's economic.

References

1. Zhang Yan. Zhengzhou airport economic comprehensive experimental zoneresearch on talent requirement and development. Contemporary Economics, **12:90-91**(2014).
2. Hao Ai-min. Research center of aeronautics & economics, Zhengzhou institute of aeronautical industry management. Journal of Kaifeng University, **01:24-28**(2014).
3. Guo Cheng. Study on the development of bonded logistics system in China. China Chuyun, **02:85-87**(2011).
4. Xia Hong. Competitiveness of Chinese aviation industry: An analysis based on diamond model. Journal of Zhongzhou University, **05:18-22**(2014).
5. He Shu-ping. The problems and countermeasures in the development of aviation industry.Journal of Zhengzhou Institute of Aeronautical Industry Management,2009(03).
6. Dong Xiao-lin, Yang Meng-yao. Relationship between industrial waste gas emission and control investment based on the pollution degree of investment index. Journal of Earth Science and Environment, **03:113-118**(2013).
7. Bai Shu-wei. Study on the development path of low - carbon economy in Henan province .Zhengzhou University, 2013.
8. Wang Win-fei. The route choice of china aviation industry's future development. Nanjing University of Aeronautics and Astronautics,2008.
9. Wang Xue-feng, Feng De-xian, Yang Yan-zhe, Du Jun, Zhang Hong. Construction of Zhengzhou airport economic comprehensive experimental zone and its spatial organization of growth triangle in Zhengzhou-Kaifeng-Xuchang. Areal Research and Development, 2013(06).

Research on high-end manufacturing and aviation economy development of Henan Province

Yongming Zhu, Youcun Shi and Minge Jia

Management and Engineering Department, Zhengzhou University, Zhengzhou, Hennan, China

Abstract. With the development and construction of zhengzhou airport, Henan province' seconomic transformation and upgrading, aviation economy has become an integral part.High-end manufacturing will be the future development direction of China's manufacturing industry, so the organic combination of high-end manufacturing and aviation economic development is imperative. This article mainly introduced the aviation economy, high-end manufacturing as the breakthrough point, using SWOT to study how to perfect and improve the high-end manufacturing to develop the aviation economy, advanced manufacturing and aviation economy to promote common progress.

1 Introduction

Aviation economy is a new economic form formed by aviation hub as the center to attract air transport industry, high-end manufacturing industry and modern service industry agglomeration development. Zhengzhou Airport Economy Zone is approved by the State Council as the first national strategy of the airport economic development and reform of the first area. Whether the area can develop smoothly and obtain experience, the air cargo industry will further promote the development of China's air cargo industry, especially to promote the rise of the Central Plains, the revitalization of Henan Province and the expansion of the central and western regions.

2 The Current Development of Aviation Economy in Henan Province

Henan Province, the most significant aviation economic development is the Zhengzhou airport port area. Zhengzhou Airport Economy Zone is China's first airport economic development zone, which is the Henan Zhengzhou City Xinzheng International Airport near the comprehensive bonded zone as the core of the aviation economy. It has far-reaching siginificance.

The relevant national development planning report Zhengzhou airport construction is the construction of Central Plains' economic zone top priority, as Henan to enhance the degree of opening up, to create the level of opening up the Central Plains three-dimensional handle. In the port area, positive innovation and development of new ideas provide mature experience.for the overall leap-forward development of Henan. At the same time, actively develop a reasonable experimental area planning, the construction of high standards around the infrastructure, build efficient service-oriented management mechanism, and focus on the experimental area of environmental protection, so that high-speed economic sustainable development. The initial development goal is to 2017,and the experimental area of the facilities, basic services, industrial chain basic shape wil be done; by 2025, the experimental zone will become competitive, the international influence of large, leading the rapid economic development of Henan Zhongyuan open upland .

3 The SWOT analysis of the economic development for Zhengzhou Airport Economy Zone

3.1 Strengths

3.1.1 Traffic and Location advantages

Comprehensive experimental area has a good location and traffic advantages, the location of Zhengzhou City, although located inland hinterland, but as an integrated transport hub in central China, has an important geographical advantage which can not be copied. Beijing-Hong Kong and Macao high-speed, even the Huo-speed and other national backbone highway in this convergence, even around the city high-speed, Zhengzhou Airport Expressway, Zheng Shao-speed high-speed road network, and other highways, plus national high-speed rail, 107,310 State Road cross the experimental area. With the Zhengzhou City to the airport expressway, light rail line 6 and the "four port" linkage avenue complered, the airport port and Zhengzhou city merged into one, high-speed rail, light rail and road and air traffic Way will achieve the "zero" transfer. This will create good airspace conditions for the economic development of the integrated experimental zone, connecting east-west and north-south routes, facilitating the development of intermodal transport, and providing a powerful guarantee for the development of airport logistics.

3.1.2 Infrastructure improving

Over the years, Henan adheres to the high standard of design, high-intensity inputs and high-performance management concept, and strive to create a comprehensive experimental area of the airport industry development new platform. At present, relatively complete infrastructure such as power supply, drainage, communication, heating, natural gas, cable TV and broadband network have been built in the experimental area. As well as the number of quality to meet the requirements of the schools, hospitals, banks, hotels and passenger terminals and other living facilities for large and medium-sized enterprises settled in the experimental area to provide adequate support.

3.2 Weaknesses

The industrial structure of the aviation harbor area is unreasonable. In the aviation economy, high-end manufacturing industry is neglected, such as lack of core technology, insufficient R & D investment, low degree of concentration, high-tech manufacturing industry development. There is no complete industrial chain, and the number of enterprises is small, lack of world-class brand, imperfections but other issues are still outstanding. So Zhengzhou Airport Economy Zone in the high-end manufacturing industry gathering and development still has a long road to go. In this case, it is necessary to study the present situation and prospects, development environment and development strategy of the high-end manufacturing industry in Zhengzhou Airport Economy Zone.

In August 2012, the strategic high-end manufacturing industry will be included in the scope of emerging industries in the "second five" strategic emerging industries in Henan Province Development Plan ". In December 2012, the State Council promulgated the "Central Plains Economic Zone Planning" and clearly put forward to speed up the process of new industrialization to develop advanced manufacturing actively cultivate equipment manufacturing, bio-medicine, electronic information, new energy and new strategic new industries. In March 2013, the State Council approved the "Zhengzhou Airport Economy Zone Development Plan" on the pillars of high-end manufacturing industry has made a specific description. The promulgation of these documents for the Henan Province to speed up industrial restructuring, promote industrial transformation and upgrading pointed out the direction. Development of high-end manufacturing industry will undoubtedly be a focus of our province's development, which is of great significance to change the transformation of economic growth mode, optimize the industrial structure and improves overall competitiveness. But there is no doubt that the development of high-end manufacturing industry in Henan has just started, low-end manufacturing industry still occupies the majority of the manufacturing sector, and the development of the airport port can meet the high-end industry on the speed of a series of stringent requirements. High-end manufacturing industry will promote the aviation economy and high-end manufacturing industry and common development.

3.3 Opportunities

Our province port has external market potential. Zhengzhou is located in the hinterland of the Central Plains, radiation surrounding the economic location and rich labor resources in Henan Province, the comprehensive experimental area in the industry to undertake the transfer of the advantage is very significant, so the market has great potential. In the Ministry of Industry and Information Technology released the "Industrial Transfer Guidance Catalog (2012)" clearly defined Zhongyuan Economic Zone as the focus of regional industrial transfer. In recent years, Zhengzhou has made remarkable progress in industrial transfer, and its economic aggregate and market space have been enlarged obviously. This has provided a good external opportunity for enhancing the competitiveness of the comprehensive experimental area.

3.4 Threats

The central region competing to develop the airport economy, Zhengzhou airport in attracting investment and other areas faces fierce competition.Such as Wu Han, in 2006, became the first comprehensive pilot city for air transport and then Chang Sha, He Fei have established the development of regional industrial development strategy of the regional economy. Brothers provinces competing to develop the airport industry, will inevitably bring investment and industrial competition is increasing.

In addition, the rapid development of other modes of transport for the aviation industry is also a huge impact. Henan Province is the national railway transportation hub city, extending in all directions of the rapid development of high-speed rail industry will undoubtedly have a certain impact on the airport economy. The major feature of air transport is speed, but in recent years the full coverage of high-speed rail network is gradually affecting the air transport and aviation economy. Convenient high-speed rail compressed part of the civil aviation transport market. Therefore, Zhengzhou airport economic development to find a breakthrough, to seek their own better development quickly.

4 Development strategy of high - end manufacturing industry and aviation economy

Comprehensive analysis of the impact of comprehensive experimental area of the airport industry development of various factors, highlighting the development of high-end manufacturing, in the external opportunities to seek development is the better Countermeasures. So, we choose Weaknesses and Opportunities as development strategy. As shown in the Table1.

Table 1. Development Strategy of Zhengzhou Airport Economy Zone.

Strengths (S)	Weaknesses (W)	Opportunities (O)	Threats(T)	Policy Selection (WO)
Traffic and Location advantages ,Infrastructure improving	unreasonable industrial structure	external market potential	Competitive	Political strategy, Government support, talent introduction

4.1. Government policy guidance

The government can directly give subsidies or the implementation of tax cuts, improving the government's high-end manufacturing financial guidance policies to increase subsidies to stimulate the development of high-end manufacturing. Opening land use policy, to attract high-end enterprises settled in order to promote high-end enterprise development of aviation economy. Government campaign and increasing publicity attract high-end manufacturing companies. What's more providing a platform to increase the network and other information technology advocacy is necessary.

4.2 Developing high-end manufacturing agglomeration and base park strategy

High-end industrial agglomeration-different enterprises in the same industry and the same region can share basic public facilities and technical personnel in the region and form industrial clusters. Successful manufacturing clusters have the characteristics of export-oriented and centralized. They are in the region in a high-end manufacturing sector significantly exceeded the average level of the host country, showing obvious characteristics, comparative advantages and industrial scale. Meanwhile, high-end manufacturing cluster also has the effect of capital accumulation, which can effectively promote regional economic development.

Developing high-end manufacturing industry needs together and economic scale, so it is necessary to make high-end manufacturing base and park concentration. Large high-end manufacturing enterprises need to have their own base, which is similar to the manufacturing industry in the "Auto City", "Space City." Smaller high-end manufacturing enterprises should enter the industrial park. Only in this way can we truly realize the high-end manufacturing industries and enterprises gathered, resulting in economies of scale, in order to save infrastructure costs, rational use of infrastructure and a variety of public service facilities.

4.3 Introducing leading high-end manufacturing

The region's high-end manufacturing may not be able to meet local demand and the construction of such enterprises need a long time, so to speed up the local development we can choose to introduce high-end manufacturing. Many high-end manufacturing industries are in developed areas. Our province has the advantage of labor resources, and the high labor costs in developed areas, so many companies are willing to transfer to the central region, the province can introduce high-end manufacturing to attract our needs. At the same time, the government also can introduce some preferential policies to attract excellent enterprises, with the introduction of preferential policies, large projects, highlighting the introduction of the manufacturing industry leader, not blind investment, and it can enhance the introduction of enterprise level and influence. The focus is to attract large multinational companies, Taiwan-funded enterprises to settle down in the experimental area, and identify the weak links in the manufacturing sector in Henan Province, to guide the flow of funds to the R & D and other sectors to form high-end manufacturing industry cluster effect.

4.4 Developing high-end services

Services and manufacturing in parallel, can not develop high-end manufacturing alone, so we should pay attention to high-end manufacturing and related services in the coordinated development, in order to achieve mutual melting and mutual promotion. Focus on the development in the production period can provide services directly to the service industry. What's more focusing on the development of high-end manufacturing industry chain in the technical services, research and opening up, information security, logistics and transportation and other related industries, can do a good job in the development of high-end manufacturing protection.

4.5 Introducing and training high-end technical personnel

High-end manufacturing industry requires a high quality of labor. Although Henan is a large population, but meeingt the needs of high-end manufacturing is still a big problem. Henan has developed traditional manufacturing, so its labor requirements are relatively low and the labor force may not be able to adapt to high-end manufacturing production. The labor force itself attracts the important advantages of foreign investment, but now something does not seem obvious, so training high-quality technical personnel is imminent. At present, there is a large gap in the number of high-quality skilled workers. Most of the laborers are migrant workers. Their cultural level is low, and it is difficult to upgrade their technical level in the short term. There is also a large number of college graduates who are graduated from colleges and universities. As a result, Adapt to the needs of production. Therefore, we should further strengthen the vocational and technical education, while training in the enterprise, the formation of a gang of a model and pass it down, so that more first-line technical staff to quickly get started.

4.6 Guide enterprises to increase investment

High-end manufacturing needs high invest to support its development of the industry, so the funding problem must be solved. It is essential to develop high-end manufacturing industry to integrate financial services with the development of financial service industry, expand its financing channels, and consider the capital chain of small and medium-sized enterprises in high-end manufacturing industry, and provide more financial support. Besides their own need to find funding channels, the national policy support is also very important. And continuously improve the manufacturing enterprise R & D innovation capacity for national science and technology special funds, and take the national special research project funding support, cultivate a number of technology research and development strength can lead the industry progress model enterprise.

5 Conclusion

In summary, the weak development of aviation industry in Henan Province is a high-end manufacturing industry, which can be a breakthrough of aviation development in our province. In the development of high-end manufacturing sector, there are many areas we need to improve.High-end manufacturing agglomeration can form a complete production chain, promote and serve the entire aviation economy. To the development of aviation economy in our province to bring more opportunities, while the rise of the central contribution to a force, is committed to building a more harmonious and prosperous society.

References

1. Hongbin Li. Bei Jing: Journal of Beijing University of Aeronautics and Astronautics. On the Concept and Development of Aviation Economy. **02** (2014)
2. Jilin Long, Guangcai Liu. China Transportation Review. Aviation Economy: New Forms of Future Development of China 's Aviation Industry. **09** (2013)
3. Zhenzhong Tian. Logistics Technology. Development Strategy of Aviation Logistics Based on SWOT Analysis in Zhengzhou Airport Economy Zone. **21:77-79** (2014)
4. Yang Liu. Industry Productivity Research. The Strategic Orientation of the Development of China 's High - end Manufacturing . **03** (2011)
5. Lina Gao, Ping Wei. Journal of Industrial Technological Economics. An Empirical Study on the Spatial Structure Change of China 's High - end Manufacturing Industry:**2003-2009. 01** (2012)
6. Daoxu Zhang. Economic Research Guide. Research on Development of High - end Manufacturing Industry in Zhengzhou Airport Economy Zone. **26** (2014)
7. Bianmin Kou. Statistics and Management. Discussion on Construction and Development of Zhengzhou Airport Economic Comprehensive Experiment. **04:99-100** (2016)
8. Zhifu Wei, Zhanjie Zhou. Public Finance Research. A Study on the Financial Policy to Support the Zhengzhou Airport Economic Comprehensive Experimental Zone. **01:66-69** (2014)

The Exploration on the Development Mode of Airport Industry

Yongming Zhu and Hao Zhang

School of Management and Engineering, Zheng Zhou University, Hen Nan, China

Abstract: Airport industry is a new platform of the global allocation of resources. It is indeed necessary that we need start with the basic theory. Also, we need classify the airport industry and discuss the type of airport industry and the development model. According to the international development experience, this article summarizes the development modes and the present situations of the airport industry. On this base, this paper summarized the successful experience, and induct some inspiration.

1 Introduction

The airport industry, also known as the "airport economy", relying on the aviation hubs, the airport attraction and radiation force with a specific industrial structure and spatial form so as to promote the capital, labor, technology, knowledge, management and other factors of production to the airport and its surrounding gathered, and has enormous influence as a new form of regional economic that consists of air transport industry, air transport services and aviation industry.

These industries briefly cover: air transport, air transport-related logistics, processing, high-tech industries, commercial and other services. The airport industry is a new economic model with regional economic system, which is based on the advantages of machine flow, logistics, information flow and capital flow. It is a new type of service model with modern service characteristics and the new economic characteristics of the new era of industry. The tremendous effects of air transport make production, technology, capital, trade, population aggregation in the airports adjacent to the airport and along the airport corridor which to form the multifunctional economic zone - airport economic zone.

From the domestic and foreign practices, the airport economic zone are mostly concentrated in the airport around 6-20 km range, or in the airport traffic corridor along the 15-minute drive to the airport as the core, to develop airport industry, airport and the formation of interrelated, Interdependence, mutual promotion of interaction.

According to domestic and international experience, the airport's investment efficiency ratio is 1: 8, in fact it is a low-input and high-output industries.

A lot of local governments in China have come to realize the importance of relying on the airport, and have begun planning and develop the "airport economic zone", and thus effectively stimulate the development of regional economic. For example, Heilongjiang Mohe Airport, which is a county-level airport, the cost of operating less than 20 million each year, and has about 60 airport's manager. It will provide revenue to Mohe's border trade, tourism, logistics and other areas after the completion of the airport. In terms of the benefit ratios of 15%, when the revenue from the airport is more than 3 billion, the profit generated from it will reach 45 million.

For example, the United States, as the world's largest aviation economy, the international airport association of north America released the "economic effects of commercial airports" (2010) analysis shows that: in terms of employment in 2010, the US civil airport and related fields to create 10.5 million or 7 percent of the US workforce (139 million), the airport is the second largest employer in the US after Wal-Mart, based on direct hiring. In terms of total annual salaries, the airport's total annual salary in 2010 was $ 365 billion, equivalent to the size of Michigan's economy. In terms of GDP contribution, civil airports create a GDP value of $ 1.2 trillion, which is higher than the total value of all goods and services produced in Mexico, Switzerland and South Korea, and the airport creates economic effects that account for more than 8 percent of US GDP.

2 The development model of airport industry

The emergence of airport industry is the the product of world's high-tech industries and the rapid development of global economic integration. The Shannon airport in Ireland began to try in 1959. As a trans-Atlantic transshipment station, Shannon airport set up the Shannon international airport free trade zone, relying on convenient transportation conditions and huge passenger flow and logistics. It attracted a lot of domestic and foreign capital and raw materials, and developed processing and export industries, which led to the rise of local economic and social.

Since the founding of Shannon international airport free trade area, the international industrial began a sustained practice and exploration on the industry layout of airport economy. the international community has gradually formed several airport industry development model after more than 10 years of practice.

3 The typical development model of foreign airport industry

3.1 Shannon International Airport Free Trade Zone, Ireland

The Shannon international airport free trade zone of Ireland is characterized by the airport industrial zone. Shannon international airport free trade zone is established in Ireland in 1959, using its advantages about the location of the only route across the Atlantic, mainly for transit through the aircraft to provide, refueling, maintenance and other aviation support services, while the use of foreign capital and raw materials export processing industry. With the continuous development of science and technology, Shannon free trade zone gradually tend to upgrade to the aviation industry, and introduced lots of high-tech industry like communications information, electronic product manufacturing, software development, etc. In addition, the national aeronautical research center of Ireland was also established.

3.2 The Memphis airport economic zone in United States

Memphis airport economic zone is dominated by aviation logistics industry; it is the common development of various industries. Memphis aviation logistics industry benefited from the Federal Express, FedEx, driven by the aviation logistics industry chain enterprises in the various areas around the airport gathered to form a complete aviation logistics industry chain, the air logistics industry to become Memphis-led industry. At the same time, the aviation logistics industry led to the development of many industries around the airport, such as auto parts, medical equipment, bio-medicine, electronic communications. FedEx maked Memphis air logistics industry prosperous, and aviation logistics industry further affected the development of other industries.

3.3 The Frankfurt airport logistics park, Germany

The main development characteristic of Frankfurt airport logistics park is business logistics. Frankfurt is located in the center of Germany, with Germany's largest air station and railway hub, Rhine - Meiyin airport is Europe's second largest airport, and also the important transit center in Europe. Relying on the global air transport network which is the center of the airport and all over the world, Frankfurt, give full play to the traditional advantages of trade center, to develop the financial industry and exhibition industry. In the vicinity of the airport, hundreds of logistics and transport companies stationed in Germany to manufacture large-scale machinery and equipment and key parts, cars and auto parts in the world can be promptly supply, which makes Frankfurt become Germany's major industrial and commercial, international exhibition, financial and transportation center, also known as the "Rhine River Manhattan."

3.4 Korea Incheon airport international aviation city

Korea Incheon airport is a multi-functional aviation city model. Incheon airport economic zone includes the airport area, international business district, free trade and tourism development zone, the Korean government is ready to develop the airport into a shipping, logistics, finance in 2011, and also introduced the Incheon airport airport development plan, high-tech in a body with a variety of industrial functions of the economic region. In the industry supporting, the South Korean government plans to introduce logistics, IT-based 64 green growth industries, products related to semiconductor and related components, network and communications equipment, computer software, medical and medical products by 2020. In the facilities, a large amount of land is reserved around Incheon airport for the construction of business and exhibition center, entertainment facilities, gaming hotels, shopping and leisure areas and other cultural experiences and recreational facilities.

3.5 Netherlands Schiphol airport economic zone

The characteristics of industry in Netherlands Schiphol airport economic zone is a variety of industries integrated development. Schiphol airport in Amsterdam, the Netherlands, continued to maintain the fourth passenger traffic in the European airports, the third place in cargo volume, is the base for air France, Lufthansa. From the initial simple airports to the development of a diversified and integrated aeronautical city, Schiphol Airport has created sustainable value for its stakeholders. Its excellent traffic conditions have attracted more than 200 international logistics operators to operate there. Many of the world's top 500 enterprises in Europe headquarters, close to the Schiphol Airport World Trade Center become the headquarters of many multinational trading company locations. At present, the Amsterdam airport economic zone covers industries such as flowers and vegetables, IT, aerospace, automotive, pharmaceutical, electronics, finance, entertainment and shopping, etc.

3.6 Industrial base in Bangalore, India

Bangalore, India industrial base is supported by high-tech industries. Bangalore has set up electronic city and international computer software technology park around the airport. It attracts many world-class elite technical talents and scientists, and actively develops the business of software design outsourcing. International famous companies such as Intel and Microsoft, IBM, Siemens, General Electric, HP, Compaq, Oerick, Acer have set up development centers and production bases in Bangalore, it has become the largest offshore software development outsourcing base of Silicon Valley now.

3.7 Dubai, United Arab Emirates airport

The main development industry of Dubai, the United Arab Emirates is the airport shopping tourism. Dubai Airport is the largest airport in the Middle East, Dubai air city in the industrial label is undoubtedly a unique tourist shopping, leisure and free trade zone. Duty-free shopping center can attract a large number of airport visitors, and is also the platform which can embody the values of business value directly. Dubai International airport duty free shop is the middle east and north Africa's largest duty-free goods retail store, the store's transit passenger spending of more than 30 US dollars per capita, ranking first in the world. Dubai's duty-free shopping area reached 9,000 square meters. In addition, it has gorgeous facilities, elegant environment, and world-renowned brands, and also keep pace with Europe and the United States in popular fashion Making it a truly world-class shopping airport.

4 The development status of China's airport industry zone

4.1 Beijing airport economic core area

The establishment of Beijing airport economic core area was approved by the Beijing Municipal Organization Committee in March 25, 2014. The core area is one of the six key high-end industrial functional areas in Beijing. The total planned area is about 178 square kilometers. It covers an area of about 56 square kilometers and consists of three functional areas, including the Beijing Tianzhu airport economic development zone, the former Beijing airport logistics base and the former Beijing Guomen Business District (including the Capital Airport of 27 square kilometers). This area collects the enterprises such as Air China, China Eastern Airlines, China Eastern Airlines, China Southern Airlines, Aviation Aviation, Aviation Materials, SF Express, and other strategic emerging industries such as the National Geographic Information Industry Park, and other industrial and financial enterprises such as Huaxia Fund and Kokai Innovation Capital, Yachang, airport advertising and other cultural and creative enterprises and the New China International Exhibition Center and other exhibition categories of enterprises, there are more than 2,000 enterprises in the core area, including Procter & Gamble, PetroChina and more than 50 world top 500 projects. In 2014, the core area achieved territorial tax totaled 8.25 billion yuan, an increase of 26%, accounting for 12% of Shunyi District, and local public revenue 2.4 billion, an increase of 17%, accounting for 22% of Shunyi District.

4.2 Tianjin airport economic zone

It is one of the important functional areas of Tianjin Binhai New Area, and also a national aviation high-tech industry base relies on the scientific research resources of China Civil Aviation University and the convenient conditions of Tianjin Binhai International Airpor, with the airport logistics processing zone as a medium. According to the industrial development plan, it will build civil aircraft integrated assembly base, aviation industry gathering base, national aviation industry clusters innovation base, airport air traffic control and general aviation equipment R&D and manufacturing base.

Tianjin airport industrial park sticks to the policy of "market traction, policy guidance, open cooperation, breakthroughs and supporting follow-up", and breakthroughs in institutional and institutional innovation, in line with local conditions and scientific planning, and actively develop aviation manufacturing, Aviation logistics, aviation education and training and aviation business service industry. It will become a new aviation industry base with the largest scale, perfect complement and internationalization in Asia, with obvious effect of industry agglomeration and rooting, brand effect and demonstration effect.

At present, the aviation industry of the airport industry has begun to take shape, attracting 17 aviation projects with a total investment of over 2 billion US dollars, including 10 foreign-funded projects and 7 domestic projects, including Airbus A320 series aircraft assembly line, AVIC helicopter industry The United States Goodrich aircraft nacelle, France Zuo Diyagao aviation equipment maintenance, Thales radar assembly, STTS aircraft painting services, Canada FTG aviation instrument panel, the German Lufthansa air cargo and HNA lease holding, the Xinhua logistics, etc. World-class aviation projects, involving aircraft assembly, research and development, spare parts manufacturing, aviation leasing, logistics and services and other fields.

4.3 Shanghai Hongqiao airport economic park

The planned Shanghai Hongqiao airport economic park covering an area of 5.14 square kilometers, adjacent to Hongqiao international airport. With the unique advantages of the Hongqiao international business district, the park is the largest Hongqiao integrated transportation hub in the world. it is also the largest transportation hub in the world. It has become the western core of the development of modern service industry in Shanghai Golden Corridor with the superiority of service and support of innovative environment, the advantages of clustering many well-known corporate headquarters, connecting the entire Yangtze River Delta region, the Yangtze River Basin region one of the most Vitality and radiation of the modern service industry cluster area.

The airport has initially formed three major industrial agglomeration of the situation, the formation of the information service industry-based modern service industry cluster situation: Ericsson, Staples, Synnex, Digital China, BenQ, Ctrip and other information industry leader More than 20 well-known headquarters enterprises at home and abroad to take root in the air, including the world's top 500 regions (the world's top 500 enterprises in the world), the United States, Headquarters of the 8; the formation of the aviation service industry, modern logistics industry agglomeration of the situation, FedEx, TNT, Jin Da International, Hong Lin International, Yangtze River Express, a large number of international freight logistics company has also brought together.

4.4 Zhengzhou airport economic comprehensive experimental zone

The State Council formally approved the "Zhengzhou airport economic development experimental zone development plan (2013-2025)" on March 7, 2013, Zhengzhou airport economic comprehensive experimental zone (hereinafter referred to as the Zhengzhou airport port) became the first national strategy and the first area of economic development of the airport. In 2014, the fixed asset investment in the region completed 40.1 billion yuan, an increase of 91.8%; GDP completed 41.3 billion yuan, an increase of 18%; above-scale industrial added value (the value-added industrial added value) Completed a total of 34.3 billion yuan, an increase of 21%; large-scale industrial main business income of 209 billion yuan, an increase of 17.7%; public budget revenue 2.12 billion yuan, an increase of 40.4%; total import and export of 37.92 billion US dollars, 58.3% of the province.

The passenger throughput is 15.8 million in 2014, an increase of 20.3%; cargo throughput of 370,000 tons, an increase of 44.9%; two growth rates are ranked the first large airport. The total number of routes to 185, navigation city 97. In particular, air cargo, new cargo routes 9, a total of 32, of which all international shipping routes 28, continue to lead the inland, initially built up a global coverage of air cargo network.

At present, the Zhengzhou airport comprehensive economic experimental zone is developing aviation logistics, aviation preference high-end manufacturing industry and modern service industry, and strive to build a modern global aviation metropolis with Unicom global, ecological livable, intelligent innovation.

5 The experience of foreign airport industry development

5.1 Integrated transport hub is the basic condition of the economic development of the airport

It is not difficult to find that extending in all directions of the multi-modal transport system is the basic condition of the construction of the airport through the comparative analysis of the international mainstream airport development experience. For example, Memphis is located on the Mississippi River, with the world's leading air cargo airport, five first-class rail companies, the fourth largest US inland port and seven US national and intercontinental highways, cargo containers can be reached in 10 hours by truck The United States local 2/3 of the region, two-hour range can cover all the major cities in the United States, convenient surface and air transport is more conducive to saving transportation costs. Frankfurt is located in Germany's busiest intersection of two highways, with Europe's largest railway station, highway, rail system and airport seamlessly to Frankfurt and Germany are closely linked.

5.2 Both soft and hard facilities are the necessary factors for the economic development of the airport

Perfect airport operation resources, rich air routes network, convenient passenger transit process, comfortable entertainment facilities are the key elements of building the large-scale international aviation hub indispensable prerequisite. For example, Incheon International Airport has an airport runway that utilizes the world's most advanced technology in airfield facilities. It can safely take off and land when sight visibility is less than 200 meters. In addition, the airport's automatic baggage handling system is extremely advanced, Within the passenger baggage handling transit is completed. Beijing Capital International Airport has a wealth of routes, the world's three major aviation hub as an important transit hub, Air China, China Southern Airlines, China Eastern Airlines, Hainan Airlines and other major domestic airlines are in its establishment of operating base. Changi Airport is complemented by a recreational facility centered on the "guest experience". It also features 5 different themed gardens, a science museum and a 4-storey slide.

5.3 Complete system of policy support system is necessary to ensure the economic development of the airport

In a region, it is necessary for the government to play a dominant role in developing the regional economy based on the advantage of airport transportation. For example, Dubai Air City is a typical investment policy to create a shopping paradise. First, Dubai to implement a more liberal visa-free policy, the implementation of the 96-hour transit visa-free foreigners, only this one on the increase of nearly 200 million US dollars in revenue; Second, the establishment of many preferential policies in Dubai free trade zone to attract investors, the most successful The free trade of capital, free import and re-export tariffs, exempt from personal income tax, no exchange control and other incentives; re-implementation of the concessions Of the tariff policy, the general tax rate of 5%, and another 421 items to the GCC countries are exempt from tariffs. South Korea Incheon Airport Airport City in the airport free trade zone to implement 7 free 3 half (7 years free, 3 and a half years) rental concessions and tariff reductions, lower import VAT, special import agricultural tax, fishery tax, And other tax policies to attract the strength of the logistics companies and enterprises stationed. Schiphol Airport encourages the development of air transport enterprises, the introduction of relevant tax incentives and economic subsidies, such as tax relief for the airport, reduction of airline take-off and landing fees, the implementation of the pilot personal income tax return, increase route subsidies and other policies.

6 Enlightenment from the development of airport industry at home and abroad

First, it has a comprehensive integrated transportation network. The comprehensive integrated transportation network is the basic support for the development of airport industry economic zone and the important factor to be taken into consideration when selecting the site. Only have a good integrated transportation network can ensure the flow of airport passenger and logistics. Developed air transport network is also the key element to ensure the enterprises transport products to the world in the shortest time and at the least cost.

Second, it has a high level of service and service efficiency. The biggest function of the airport is to provide fast and convenient services, so the competition between the airport is actually the efficiency. In addition, the airport's service capacity and service efficiency are also the important key elements of enterprise location.

Third, it has excellent research environment and comfortable living environment. Excellent research environment can provide technical and intellectual support and to attract such as product development, creative and other knowledge-intensive industries. As the senior white-collar workers in the port, they tend to live in workplace which is around the natural environment that requires a higher, comfortable living, the work environment has gradually become an important factor in the enterprise site selection.

Fourth, it developed the trade economy and producer services. The location advantages and strong connectivity of the hub airport give it the natural advantage of developing international trade. In the case of the world hub airport economic behavior of the evolution of the surrounding, the trade economy has become a through-airport economic development of the main line. The strong passenger and cargo distribution capabilities and global accessibility of hub airport also make it a best place for logistics, training, finance, consulting and other productive service industry gathering.

References

1. CAO Yun-chun. Airport economy - the economic growth space of speed. Beijing: Economic Science Press, 2009.
2. CAO Yun-chun, XI Yan-rong.Analysis of International Experiences of Development of Airport Economy and Reference to China. Market Modernization, 2009, (1).
3. ZHOU Shao-hua, WEI Hui-zhen. The main development mode of airport economy. China's National Conditions and National Strength, 2009, (11).
4. LIU Chun-ling.Analysis of International Experience of Industrial Development in Aviation Economic Zone. World Geography Research, 2014, 23 (4).
5. Liu Daoxing, Ren Xiaoli, etc. Advancing the construction of Zhengzhou Airport Economic Comprehensive Experimental Zone of a number of issues. Regional Economic Review, 2015 (02).

Questions and Countermeasures on Developing General Aviation Industry in China

Yongming Zhu and Yongchao Liu

School of Management Engineering, Zheng Zhou University, Henan, China

Abstract. China's general aviation industry has exposed many problems with the low-altitude airspace expanding openness. General aviation is an important part of national economy and defence forces, the development of general aviation has a practical and far-reaching significance. By analysing the China's current general aviation industry, combined with the importance of the development of general aviation, propose some countermeasures and suggestions for the development of China's general aviation industry.

1 Overview on general aviation

General Aviation refers to the use of civil aircraft to engage in civil aviation activities other than public air transport, including operational flight in industry, agriculture, forestry, fisheries and construction, as well as medical and health, emergency rescue, meteorological detection, Scientific experiments, education and training, cultural and sports aspects of flight activities. General aviation is an important component of modern integrated transportation system, and it is an industry with rich content and active development in civil aviation field. Years ago, due to domestic policy constrains, the development of China's general aviation industry was limited, thus the pace was slow. With the reform and implementation of China's low-altitude airspace management, the obstacles hindering the development of China's general aviation industry are gradually disappearing, and China's general aviation industry also ushered in its golden age of development.

General aviation has a flexible, fast and efficient, environmental adaptability and other characteristics. In the area of emergency relief, general aviation provides humanitarian relief to people facing natural disasters, hunger and war by means of cargo transport and refugee movements. General aviation can provide strong protection for environmental protection and comprehensive management, scientific research and other social public services. It can also provide effective supply in many aviation consumption fields like civil flight, recreation flight, sightseeing flight, aviation education and training [1]. At the same time, with the 2015 low-altitude airspace reform, the private jet market is also about to become the next growth point of promoting domestic demand. All in all, the level of comprehensive development of general aviation, can fully reflect the country's technological level and economic strength. General aviation industry chain as shown in Figure 1.

2 The Significance of Developing General Aviation

2.1 Improve the Traffic Conditions and Promote Opening to the Outside World

The western region is inland in China, and the terrain hinders the exchange of its property and culture with other regions. The western region covers an area of about 6.86 million square kilometers, accounting for 72% of the total area. The western region is rich in resources, mineral resources, land, water, and great potential for development, which is the formation of the western characteristics of the economy and an important basis for advantageous industries and favorable conditions.

However, for the development of western resources, relying solely on highways and railways is not enough. The railway network in the western region is thin, the road grade is low, the density is small, the majority of townships in the country that do not have access to roads are concentrated here, and the airport is small, it is inconvenient for large aircraft to take off and land. To carry out the development of the western region and expand the opening-up of the western region, there will certainly be a large flow of personnel and materials. However, the construction of highways and railways takes a long time, and the construction of high-speed railway is very difficult and costly, and it cannot achieve the expected purpose in short time. The flow of goods and personnel will be accelerated because of the development of general aviation in the western region.

2.2 Promote the logistics industry to upgrade

Parcel delivery is the most profitable part of the air transport industry. International courier industry in the world is an annual growth rate of 18%. China accounts for a large proportion of the volume of postal services, with an annual turnover of US $ 2.5 billion and an annual growth rate of 25%, However, due to restrictions on air transport conditions, Speed post business cannot fully reflect the timeliness. In addition, there is a clear market for the development of small, rapid, special transport and postal express services in the region, but it cannot be compared with developed countries due to the influence of mode and price of air transport, but such business is precisely the best opportunity for the general aviation market [3]. Air transport is characterized by the one hand, can greatly reduce the goods, parcels from the origin to the destination of the time, bringing business to enhance and higher customer satisfaction; the other hand, in the extensive use of air transport under the premise, land transport will be relatively reduced, to some extent alleviate the large and medium-sized urban transport problems. At the same time, the logistics speed, will in turn promote the development of e-commerce industry, thus changing the entire e-commerce industry and people's lifestyles, greatly enhance people's quality of life.

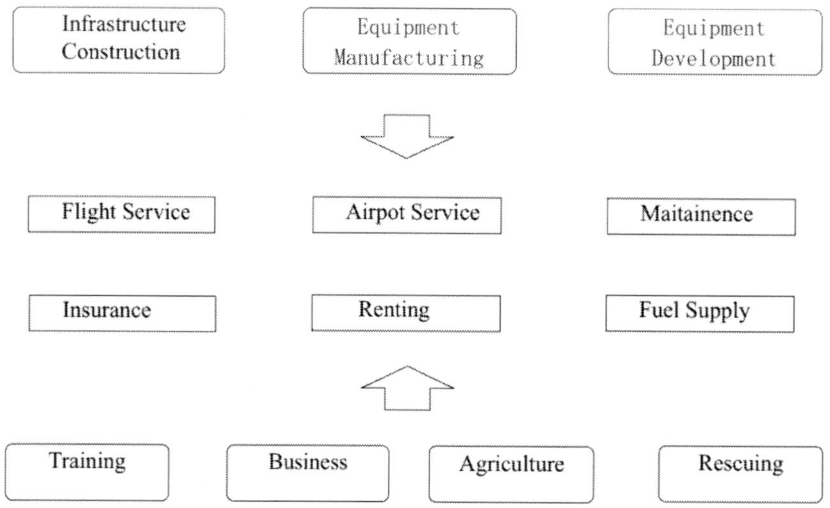

Fig.1. General aviation industry chain

2.3 Promote the development of aviation rescue enterprises

In the past, sudden international and social emergencies occurred in China, which were handled by the government to organize temporary institutions. In the case of neither experience nor preparation, rescue effect is seriously affected, in the face of the victims or their families, the international community censure, the government often bear unnecessary pressure. Rescue work in the West has a history of half a century, has been developed to be the 4th largest industry following banking, post and telecommunications, insurance. Adapt to this objective trend, through the international and commercial rescue services to promote the industrialization of China's rescue is an inevitable trend. The birth of the aviation rescue industry has become a requirement of the current situation for the vast territory and less developed traffic in western China.

2.4 Promote the development of private airc raft market

China's low-altitude air China's low-altitude airspace reform began in 2015, the national plan by 10 years will be all the country's airspace is divided into control, surveillance and reporting of three categories of low-altitude airspace for classification management, continue to promote low-altitude airspace reform will further stimulate the private aircraft marketspace reform began in 2015, the national plan by 10 years will be all the country's airspace is divided into control, surveillance and reporting of three categories of low-altitude airspace for classification management, continue to promote low-altitude airspace reform will further stimulate the development of private aircraft market. and the United States more than 0.2 million private aircraft compared to a huge difference. According to market research, the mainland of China has 87.5 million millionaires and 60 million billionaires in the rich, 1/6 of the rich plan to buy a private jet, so projections, China will have 155,800 potential customers. At the same time, because the private airplane market related

DOI: 10.1051/matecconf/201710001011

industry chain is relatively long, profit comes from the aircraft pre-purchase, consulting services to pilot training, and follow- up maintenance, operation and fuel consumption.

3 General Aviation Industry in China

China's general aviation industry was established in 1952, but the use of relatively was small after the formation, there was a large growth after reforming and opening. According to the CAAC Statistical Yearbook, from 1978 to 2002, the average annual operating time of general aviation is only about 40,000 flight hours. In 2007, it reached 100,000 flight hours, the highest in history, but mainly in agriculture and forestry operations. From the flight time, flight and flight sorties, the northeastern region of the general aviation industry in the country occupy the largest proportion. Active domestic general aircraft mainly Yun 5 / Yun 5B, transport 11 / transport 12 and other models, belonging to the 70s and

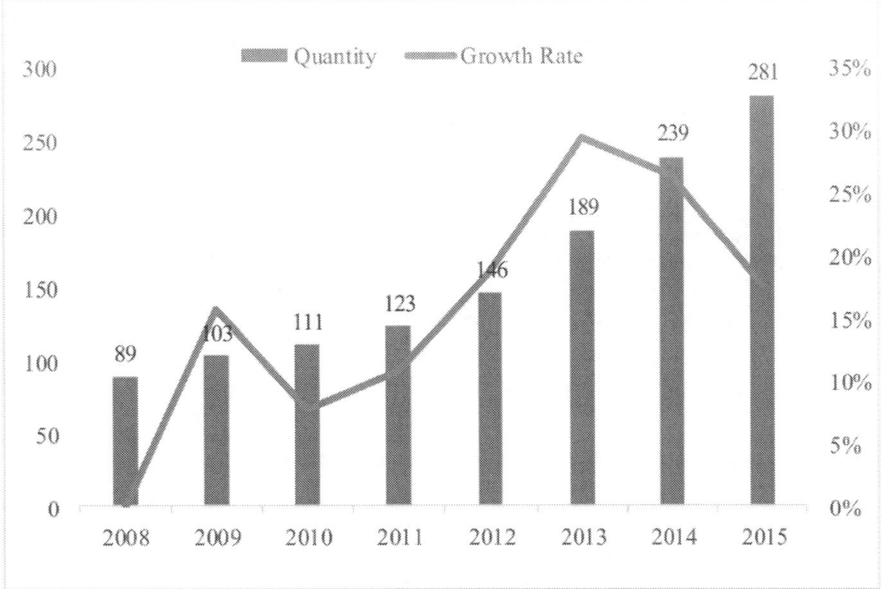

Fig.2. The number of general aviation enterprises in China(2008-2015)

80s products. According to the Civil Aviation Administration of China, "the Civil Aviation Industry Development Statistics Bulletin in 2015" shows that by the end of 2015, access to general aviation business license of 281 general aviation enterprises, among them, there are 72 in North China, 55 in Central South China, 56 in East China, 30 in Northeast China, 37 in Southwest China, 25 in Northwest China, 6 in Xinjiang, General aviation industry to complete the production of 779,300 hours of flight operations, an increase of 15.5% over the previous year; general aviation enterprises in the total number of airworthiness aircraft registered 1904, of which 508 teaching and training aircraft[2].

From 2016 to 2020, the national low-altitude resources will be fully developed, and gradually formed to supervision, supervision, reporting for the three types of airspace control mode. On this basis, will also build a low-altitude airspace regulatory standard system to achieve service standards and service forms of innovation. Based on the national investment support, to fully mobilize social forces to carry out the construction of aviation service sites, so as to effectively enhance the overall level of general aviation protection, and further improve the flight personnel training system, the establishment of a more standardized industry supervision and evaluation system to enhance general aviation management ability. 2008-2015 The number of general aviation enterprises in China is shown in Figure 2.

4 The Problems of Developing General Aviation in China

4.1 General aviation manufacturing industry development is weak

With the opening of low-altitude airspace and the increasing market demand, according to the forecast of 2020, China needs 10260 aircraft of various types of peers, including private aircraft, business and trainer demand is growing rapidly, the next 10 years, private and business market demand may reach 100,000. With the strengthening of comprehensive national strength and policy support, China's aviation industry is developing rapidly, but compared with the developed countries, we are still a lot behind, general aircraft and even in its infancy. By the end of 2015, 71.18% of China's general aviation aircraft was made by foreign manufacturers or joint ventures. At the same time, the development of various models imbalance, rotor aircraft accounted for less than 30%. The high cost of general aviation aircraft and high aviation materials tariffs, making the cost of general aviation operations high, seriously limited the development of China's general aviation.

4.2 Policies and regulations are imperfect

China's general aviation industry due to the late start, the lack of appropriate laws, regulations and standards system, which to some extent, seriously hampered the development of China's general aviation industry. Such as general aviation airworthiness, airport management regulations are still not perfect. The construction of private airports, private heliports and airports lacks the standards and regulations, and the construction of airports is often costly, which limits the construction of airports. At the same time, airport utilization is not high and airport charges are also raised. Increasing the operating costs of shipping companies. Administrative regulations, industry regulations have not yet formed a standard system, the public airport can arbitrarily refuse to accept general aviation aircraft and arbitrary charges, operation and management has yet to be standardized. General aviation management has no clear attribution. Therefore, the establishment of general aviation in line with China's general aviation requirements of the laws and regulations and standards system has become a top priority.

4.3 Lack of professional staff

General aviation industry will become China's new economic growth point but at the same time, also highlights the shortage of professionals in this area the problem. General aviation talent shortage is mainly reflected in two aspects: First, the low quality of personnel. China's general aviation institutions are mostly from veterans, the lack high-quality professionals. Second, a small number of talent. According to the People's Daily reported data, assuming that each aircraft is equipped with two pilots, by 2020 the demand for general aviation pilots will reach 10,000, the pilot gap of 8,000 people. At the same time, China's general aviation industry, recruitment of difficult flight personnel, and ground crew, air control and other personnel scarce, restricted the development of general aviation industry.

5 Suggestions on Promoting General Aviation Industry Development in China

5.1 Improve industrial policies and optimize the macro environment

In the process of general aviation development, the relevant state departments need to support the industry as soon as possible to develop the industry chain, and develop a new general aviation development in line with the policy. According to the relevant provisions of the international civil aviation, the definition of general aviation, and to determine its legal protection mechanism. The existing laws and regulations to adjust and improve the establishment of ordinary shipping and general aviation in the orderly convergence. The general aviation personnel management, safety management system planning, simplification of management processes, and to develop independent regulations [4].

5.2 To play the guiding role of the government

In the future development process, the Government should encourage and support the development of the aviation industry guide, give financial and technical support. Government should allow investors and enterprises to realize that participation in the development and construction of general aviation industry, not only the purchase of aircraft, the aviation-related business, but also can be around the general aviation industry and derived from the business. It can be the construction of airport facilities, which can be the operation and management of aviation operations. It can be engaged in air transportation, photography, sightseeing, emergency rescue, etc., or operate general airport, fixed maintenance base (FBO), land development and so on. At the same time, it may also be considered to participate in the training and assessment of aircraft sales leasing, financing, management and related pilots and management personnel.

In the process of reforming the administrative system, we should strengthen the strength of the government management organization and establish the organization specifications, personnel team and function compilation that are compatible with China's general aviation development.

5.3 Focus on the relevant personnel training

General aviation industry chain covering manufacturing, aircraft maintenance and maintenance, flight services and a series of core technologies or expertise required as a support project, which requires practitioners with a high degree of professional knowledge and skills. Due to the lack of core technology and aircraft manufacturing capacity is not high, resulting in a large number of imports to the status quo, the high price of aviation equipment makes the general business and cannot easily purchase aircraft in the general aviation industry, to a certain extent hindered the development of general aviation industry. Therefore, on the one hand, should be policy-oriented, focusing on the aircraft manufacturing industry in personnel training; the other hand, to further improve the flight service post identification mechanism, to make up the flight and flight services industry personnel gap, and constantly promote flight skills by specialization to the popularization.

5.4 Improve the security mechanism

Coordination of the State Air Traffic Control Committee and other relevant departments to speed up the reform of low-altitude pilot to improve the airspace use of the environment; in capital, equipment and projects to support the introduction of new technologies to promote the use of ADS-B system to improve command and support capabilities; to encourage social forces development FBO Model; reference to the developed countries' general aviation operations system, improve the general aviation fuel subsidies.

Conclusion and Prospect

China's general aviation industry started late, slow progress in supporting facilities, imperfect system measures, lack of open air resources, manufacturing capacity and technology is relatively backward, resulting in China's general aviation industry behind the status quo. However, the experience of developed countries tells us the role of general aviation in economic development and social welfare is growing. Constantly improve the system in today, the development of general aviation industry ushered in its golden age.

With the development of society and the progress of science and technology, factors restricting the development of general aviation industry will be gradually eliminated. The general aviation industry plays a vital role in the process of China's opening to the outside world. Under the guidance of the continuous improvement system, it will become another growth point of China's economy. In the near future, we will see a vibrant general aviation industry.

References

1. Li Chengzhi, Su Daoning. The General Aviation of China: Problems, Reasons and Solutions. Journal of Engineering Studies, **01**(2010)
2. Xu Xing. Analysis on Current Dvelopment Situation and Countermeasures of China's General Aviation. Journal of Xi'an Aeronautical University, (**06**)2014
3. Kang Yong, Zhou Jianmin. Analysis on Development Status, Trend and Countermeasure of General Aviation. Modern Navigation, (**05**)2012
4. Sun Qinglei. Status Quo and Development Countermeasures of China General Aviation Industry. Shandong Industrial Technology, (**17**)2015

Residual Deformation Control Design for Car Handbrake Operating Arm Assembly

Yuan Li[1] and Li Ma[2]

[1]School of Automotive Engineering, Wuhan Huaxia University of Technology, Wuhan, China
[2]School of Automotive Engineering, Wuhan University of Technology, Wuhan, China

Abstract. The aim of the paper is to control residual deformation of the handbrake arm assembly under lateral force in an effective way. Compared with the test method, nonlinear FEA method is higher in accuracy and fewer in cost of test specimen, which is an available means to control the residual deformation before producing.

1 Introduction

Handbrake is an important safety car assembly. Under lateral and longitudinal forces, handbrake will be in large displacement or unrecoverable residual deformation, or even occurs partial fracture, which seriously affect its performance. Foreign references specifiy acceptable residual deformation. The handbrake companies often use test method to get deformation datas, but it needs huge amount of samples, which make waste of materials and increase of costs. The paper provide another mothed of controlling residual deformation to decrease the cost and waste of resource.

2 Tensile test of handbrake operating arm

There is a quality standard to check the strength and stiffness of the handbrake arm assembly, which mentions that the lateral offset of the arm end should not exceed 30mm if the maximal load on the arm reaches 400N, and the residual offset should be less than 10mm when unloading.A lateral tensile test is designed to test whether the strength and stiffness is up to standard.

As the lateral clearance tester shown in Figure 1, the sample of the handbrake arm,which was fixed on the positioning fixture of tester, located at the same position with the actual condition, and the arm rotated to a certain angle according to the specification. The force sensor connected assistor fixed on the test bench with one end, another end of which jointed contacting device by nuts,and its signal were output on the digital screen. Electronic display vernier cliper were also fixed on the test bench,and parallel to the tester's surface. The load was applied on the screw assistor until the contacting device touched the arm end. When the depth ruler of the vernier caliper was moved to the position that just touched the contacting chip--the minimal data were shown on the caliper, reset the caliper and unload the lateral force.

Fig.1. Lateral clearance tester

The lateral force test started after pre-adjustment when illustrated above. First,when contacting chip contacted with the arm after screw rotating and the lateral force was 5N,the displacement on the caliper was recorded and all parts of the assembly were not loose. Unloading the lateral force and rotating the screw to make two parts contact again,the load was applied until the lateral force peaked 100N,and the caliper data was recorded. Applying load as the same steps as above until the lateral force reaches 400N, record the lateral displacement L_p (Y residual offset on the arm end) and unload the force.we also find that there was no loose phenomenon between different parts. Finally, changed the location of the loading device and the measuring equipment,and test again to get the negative side lateral force results.

3 FEA control design for the residual deformation of handbrake operating arm

Although the offset and residual offset in the lateral tensile conditions can be measured by test, there are several problems in the testing process and results. After the positive lateral force test, the residual deformation of the handbrake occurs. Using this assembly again to carry out the negative force test would result in error, because the state of the assembly with plastic deformation is different from which in the specification. To reduce the error in negative force test, it is necessary to use a new assembly that may inceasing testing cost and material consumption. To avoid this error, another method like FEA method can be used.

3.1 The geometry model and finite element meshing

Compared with the beam or shell model, 3D solid model is more available to detect the slight displacement and contacting deformation between parts. Therefore,it is better to use 3D solid model of the handbrake arm assembly for FEA(Finite Element Analysis).

Four-node solid element is a reasonable type to ensure high accuracy for the FEA model in finite computing time and scale. Because this FEA model is a nonlinear problem,including nonlinear of state and material, as contacting state changes in lateral tensile process and the material characteristic curve is not linear.Consequently, there are 4034000 solid elements and 10000 nodes meshed in FEA model. Partial enlarged figure of meshing model is shown in Figure 3.

Fig.2. 3D model of the assembly

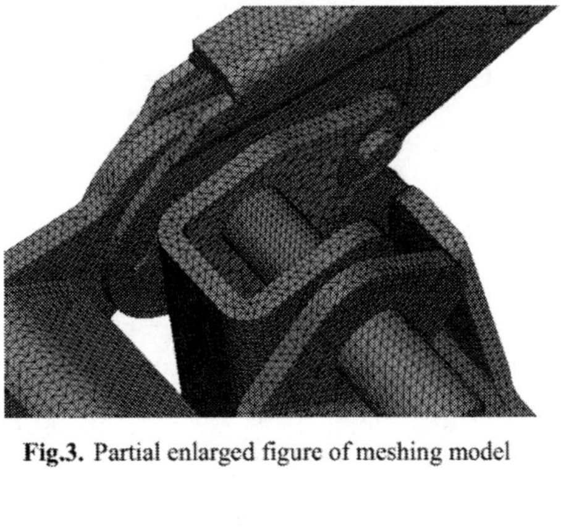

Fig.3. Partial enlarged figure of meshing model

3.2 Material model

Considering the specification focus on the residual deformation causing by the lateral tensile force, material nonlinear problem ought to be taken into consideration. The reason is that the residual plastic deformation would occur and the stress-plastic curve of material would be no more linear provided that the stress exceeded the material yield limit. In principle, nominal stress-nominal strain curves obtained by the material axial tensile test should transform to true stress-plastic strain curves. The relationship between nominal stress and true stress is

$$\sigma = \sigma_{nom}(1+\varepsilon_{nom}) \quad (1)$$

And the relationship between nominal strain and plastic strain is

$$\varepsilon_{mon} = \varepsilon_t - \sigma / E \quad (2)$$

In these formulas, σ -true stress, σ_{mon} -nominal stress, ε_{mon} -nominal strain, ε_{pl} -plastic strain, ε_t -true strain, E-elastic modulus

According to the formula (1) and (2), the true stress-plastic strain curve of main material of handbrake arm assembly is shown in Figure 4.

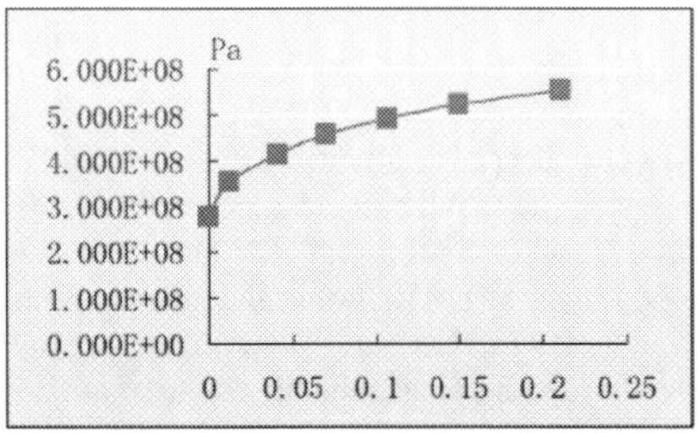

Fig.4. True stress-plastic strain curve of main material of handbrake arm assembly

3.3 Setting up the contact pairs

There are two main kinds of contacting pairs between parts of handbrake arm assembly. One is constant contact, which are riveting, welding and articulated relations. Another is variational contact . In FEA, contact is a complex state nonlinear issue. So it is necessary to set different relationships in relative kinds of contact to ensure the nonlinear calculation convergence within a controllable time.

The connection between the rivet and the manipulation arm, as well as little rivet and the cover is riveting; the connection between tooth base and the stationary seat, as well as between the pin shaft and the cover is welding;the connection between pin shaft and it mating parts is articulation.All these relationships can be set to constraint.

While another relationship varies with the increase of the tensile force, such as contact between the convex surface on the cover and the square limited convex surface on the manipulation arm. So sliding contacts are applied on these surfaces.

3.4 Definition of boundary condition

To simulate the screw fastening of the handbrake arm assembly, set the nodes near the areas of 3 bolt holes on the tooth plate and the base constraint.

According to the specification, the lateral load of 400N is applied on the end of the manipulator arm,which is decomposed into 4 points on the top of arm and cover to avoid local stress concentration caused by load on one single point.

4 Results contrast between FEA and test

Table 1. Results in FEA and test method (mm)

Test name	FEA results	
	Deformation	Residual deformation
Positive lateral force	25.78	7.122
Negative lateral force	29.04	9.346
Test name	Test result	

	Loading offset	Loaded offset
Positive lateral force	38.4	6.62
Negative lateral force	39.32	7.04

The results(Table 1) of offset and residual offset in positive/negative lateral force test are achieved by FEA and testing method. From Table 1,the offsets when loading in two conditions are 38.4mm and 39.32mm, both of which outnumber 30mm the specification required,showing that the loading offset does not meet the standard.And the residual offsets,which are 6.62mm and 7.04mm respectively, are under 10mm.

Comparing the results between the calculation and the tests in lateral force conditions, the displacement and residual deformation by the nonlinear FEA method are much close to the test results, which show that this residual deformation controlling method is effective.

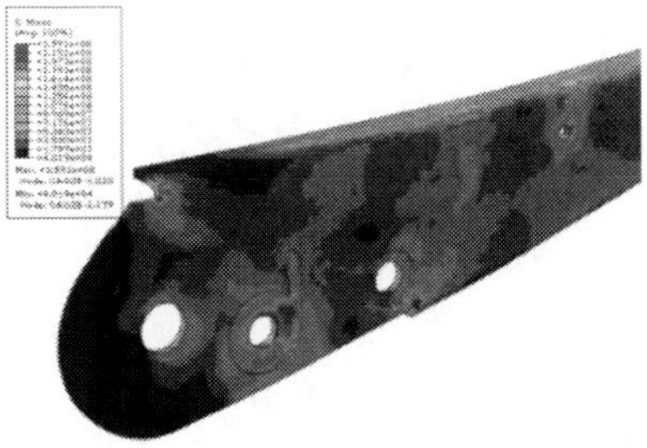

Fig.5. The arm residual stress nephogram

In addition, the lateral force test can measure the displacement and residual deformation on the end of the arm, but whether the assembly occurs fracture can be only achieved by macroscopic observation,while the internal surfaces of contacting pairs are hard to recognise. Nevertheless, not only the residual stress but also the failure position can be easily obtain via nonlinear FEA method. The residual stress nephogram ot the arm is shown in Figure 5, which demostrates that the residual stress in the shape corner in the upper area are 218.7MPa and the lower 359.2MPa,both of which are higher than the yield limit. Therefore replacing the corner area from sharp to smooth is a suggested way to reduce the partial residual stress.

5 Application of the new product design by nonlinear FEA method

Table 2. The results of W2 assembly in FEA and test method (mm)

Test name	FEA results	
	Deformation	Residual deformation
Positive lateral force	21.88	2.367
Negative lateral force	15.66	1.906
Test name	Test result	
	Loading offset	Loaded offset
Positive lateral force	20.48	3.07
Negative lateral force	15.57	4.85

This study of nonlinear FEA method applied to a new product W2 handbrake arm assembly. The test and FEA results of W2 are shown in Table2, showing the approaching of the results between these two methods and high precision in the FEA nonlinear method to control the residual deformation.

6 Conclusion

The nonlinear FEA method to control the residual deformation of the handbrake operation arm under the lateral force has achieved good results. If the residual deformation is controlled before test, it can effectively reduce low accuracy results in test method and waste of resources by trial and error. Modifing the product's design according to the nonlinear FEA results, the structure will be improved effectively with less time in designing, producing and testing.

References

1. J.Xiao.Transpo World,**6**,42(1998)
2. R.A.C.Slater.*Engineering plasticity*. (Thomson Press, 1977)
3. Delphi Corporation. Automotive Engineering,(2005)
4. Si.Xue.*Elastic-plastic mechanics*. (China buliding material press, 2005)
5. Y.Guo , W.Tao , Z.Zhuang. *Linear and nonlinear finite element*. (Mechanical Industry Press: Beijing, 2005)

The Research on Sensor Optimal Layout on Detection Device for Automobile Seat Horizontal Drive Mechanism

Hongxian Ye[1,2] and shixi Yang[1]

[1]College of Mechanical Engineering, Zhejiang University, Hangzhou, China
[2] School of Mechanical Engineering, Hangzhou Dianzi University, Hangzhou, China

Abstract. The type of the sensor and the optimal placement of the installation location of the sensor are important factors in vibration signal acquisition on automobile seat horizontal drive mechanism (HDM) for quality grading. According to the characteristics of HDM, the actual installation condition and the attenuation characteristic of the vibration signal transmission, the type of sensor was selected reasonably. By means of finite element analysis method, the best placement of the sensors to mount on the fixture was analyzed aim to maximize the information reflecting the fault and be sensitivity to the direction of the vibration. The finite element simulation and experimental results show that the acceleration sensor can effectively detect the fault characteristic signal, which provides a basis for further quality grading for HDM.

1 Introduction

Horizontal drive mechanism (HDM) is an adjustment mechanism that converts the rotary motion of the drive motor into the linear motion of automobile seat glides back and forth in order to achieve suitable position. Since the worm gear tooth surface bumps or residue burrs in the processing, a small number of HDM will produce repetitive tapping noise during operation. When the defective portion size is small, the vibration energy is weak, the inspectors cans not perceive directly even in a very quiet and semi-anechoic chamber. It makes the factory quality grading very difficult. Currently, most of the auto parts manufacturers determine the presence of percussion in HDM by means of the artificial methods using the sound signal. The company is trying on developing the online quality inspection system to solve the detection problem. Liu1 and Wang2 analyzed the sound signals of HDM with the anthropogenic breakdowns by Gabor time-frequency analysis and Hilbert envelope demodulation method to extract the percussive information. Although both of them obtained the defect feature information, they could not use directly in the detection systems because of two reasons: One is the sound signal detection needs higher environmental requirements. The other one is that the energy strength of the noise signal in the HDM with quality problems is weaker than the artificially generated breakdown's. To find the weak fault signal, the signal processing methods applied to analyze the collected signals to obtain frequency characteristics associated with the fault. Time-frequency signal processing methods such as wavelet denoising algorithm3-4and feature extraction methods5-7adopted to analyze the weak vibration signal for detecting weak source signals. Some scholars introduced nonlinear theory, such as the theory of chaos8-9, stochastic resonance10-12 to identify weak signal frequency characteristics. All these methods were based on the premise that the analyzed signal contains the weak source signal. For small closed driving device such as HDM, the energy generated by the vibration signal is very weak. Such weak signal can be collected effectively or not is a question.

Compared with the general gear box, the horizontal drive mechanism gear box is too small to install the sensor directly. The sensor has to be installed on the fixture. For small closed driving device, the efficient acquisition of the weak signal depends on the optimal layout of the vibration sensor on the horizontal driver.

2 HDM weak vibration signal detection device

The vibration sensor is mounted directly on the gear box for the general gearbox. The vibration signal on gear box is strong. Compared with the general gear box, the horizontal drive mechanism is too small to install the sensor. It is shown in Fig. 1. The impulse vibration signal in the horizontal driver is weak. It is easy to be interference by and submergence in the noise signal. The effective transmission of weak vibration signal from HDM to the fixture is the premise on the HDM quality grading.

The structure of fixture to detect vibration from HDM mounted on the side of the HDM is symmetrical. Fixture connected to the base plate fixedly with bolts. The original fixtures contact with HDM by driving force. The improved fixture connected with HDM via preload, the mechanism shows in Fig.2. During the gearbox operation, the vibration and impact generated by the worm and worm wheel travelled through the shaft and gearbox bearing to the surface of the gearbox. The vibration signal transmits to the jig. The sensor is installed on the jig.

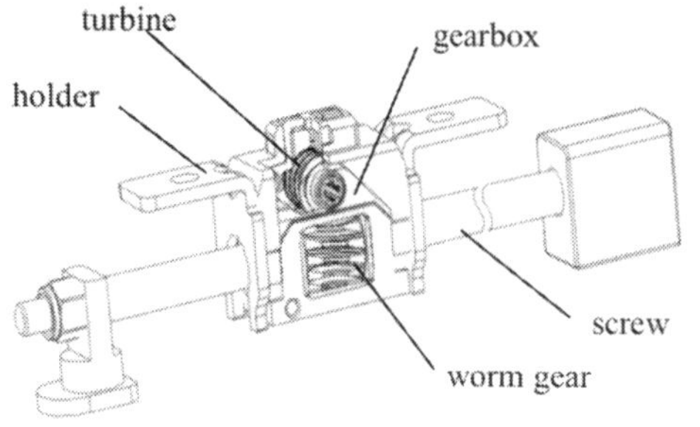

Fig.1. Structure of HDM

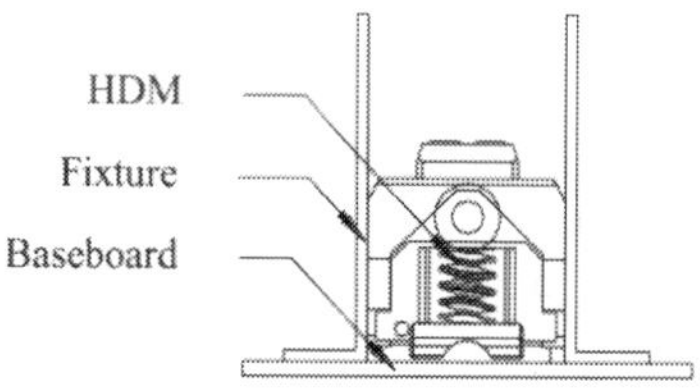

Fig.2. HDM Fixture Installation Diagram

The energy of the vibration signal generated by HDM is very weak. It is need to choice the type and location of the sensor reasonably.

3 Selection of Horizontal Driver Vibration Sensor

The sensors for the vibration signal acquisition have non-contact laser displacement sensors, force sensors, acceleration sensors. As the non-contact laser displacement sensor is too large to the horizontal drive mechanism, and has to be installed in a distance of 5-10cm from the HDM. The vibration caused by the operation horizontal drive mechanism has interference on the vibration signals collected by laser displacement sensor. It is improper to use the displacement sensor to acquire the vibration signal of the HDM.

Excitation force is unchanged in the process of transmission. The signal collected by the force sensor is the deformation on the vibration fixture. The strength of the vibration fixture is large, and the force generated by the HDM is very weak. The force is seriously attenuated from the HDM to the fixture. So it is not suitable to use the force sensor to collect the vibration signal.

The signal collected by acceleration sensor is the acceleration of the measurement point. The mechanical model of the fixture to detect weak vibration signal shows in Fig.3. The vibration generated by internal excitation force in HDM, transferred to the jig through the spring and damping.

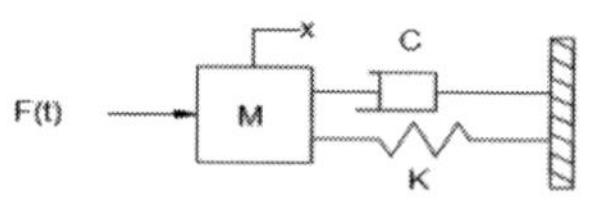

Fig.3. Vibration transferring model of the fixture

The differential equation of the system is shown in equation (1):

$$M\ddot{x}+c\dot{x}+kx=F_0\sin(\omega t) \quad (1)$$

where $F_0 \sin(\omega t)$ is the excitation vibration source generated by HDM which is transmitted to the gearbox surface. M is the equivalent mass of the fixture. c is equivalent damping between fixture and base plate. k is equivalent stiffness between fixture and base plate.

The acceleration of motion equation can be obtained from equation (1) as equation (2). And we can get the acceleration amplitude in equation (3).

$$\ddot{x} = \frac{\omega^2 F_0 / M}{\sqrt{\left(\omega_0^2 - \omega^2\right)^2 + 4\left(\frac{c}{2M}\right)^2 \omega^2}} \sin(\omega t + \varphi) \tag{2}$$

$$D = \frac{F_0}{\sqrt{M^2 (1 - (\frac{\omega_0}{\omega})^2)^2 + (\frac{c}{\omega})^2}} \tag{3}$$

From equation (3), we can see that the nearer the fault signal frequency is close to the natural frequency, the higher is the amplitude of the acceleration on the frequency corresponding to fault. The harmonic response of the fixture was analyzed as the bottom of the jig was fixed and the acceleration was set to 1 mm/s2. The acceleration response relative to the original fixture increased by about 1000 times in the vicinity of 100 Hz where fault signal frequency located.

In the transmission path, the acceleration signal can well reproduce the characteristics of the fault vibration signal. At the same time acceleration sensor is small and can be installed directly on the fixture. The acceleration sensor is selected to acquire the vibration signal.

4 Position optimization for acceleration sensor on HDM

The finite element mesh model of the HDM vibration detection fixture was shown in Fig. 4.

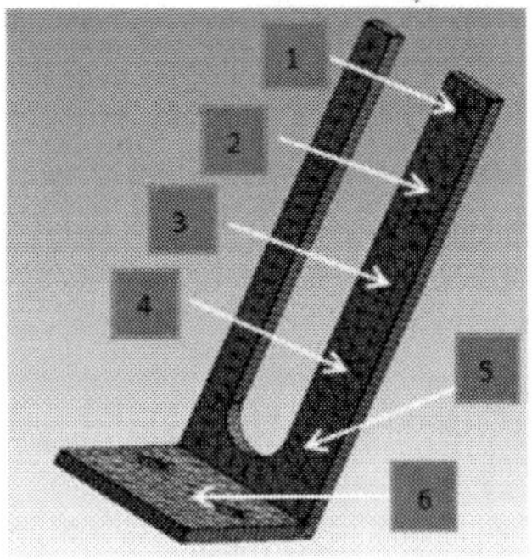

Fig.4. Diagram of the Node Location Corresponding to the Measuring Points of HDM Detection Fixture

The node 1 to 6 in the finite element model were select as the location of the sensor.The characteristic frequency of the vibration signal caused by the fault HDM is around 100Hz. The vibration response of the 6 test point in the detection fixture was analyzed within 700 Hz.

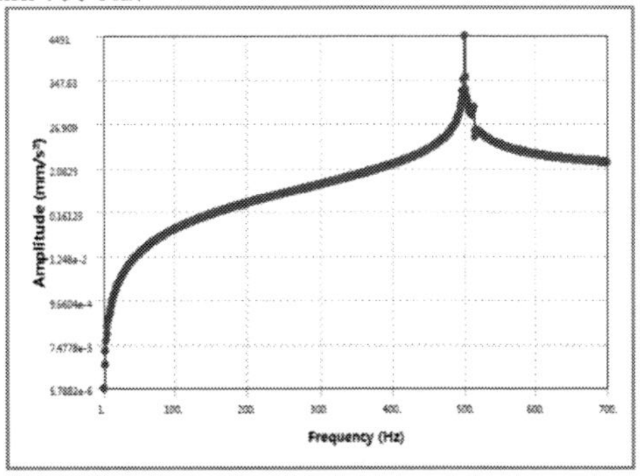

(a) Harmonic Response of Node 1

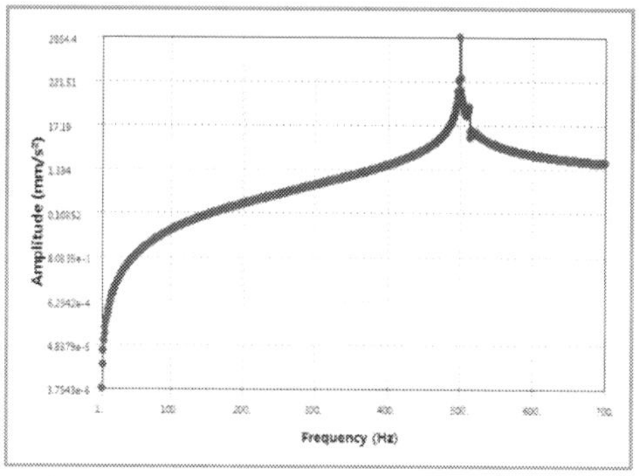

(b) Harmonic Response of Node 2

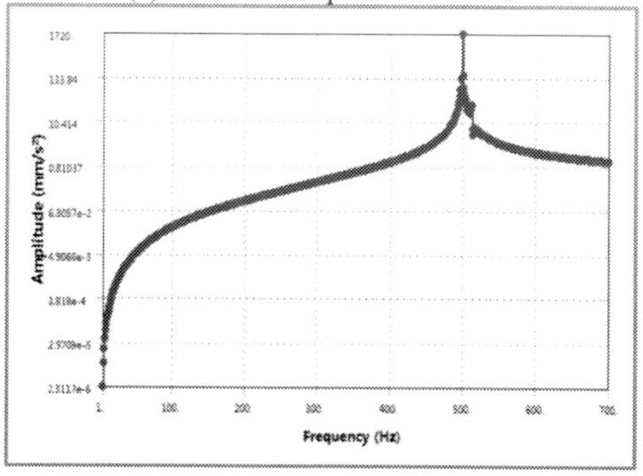

(c) Harmonic Response of Node 3

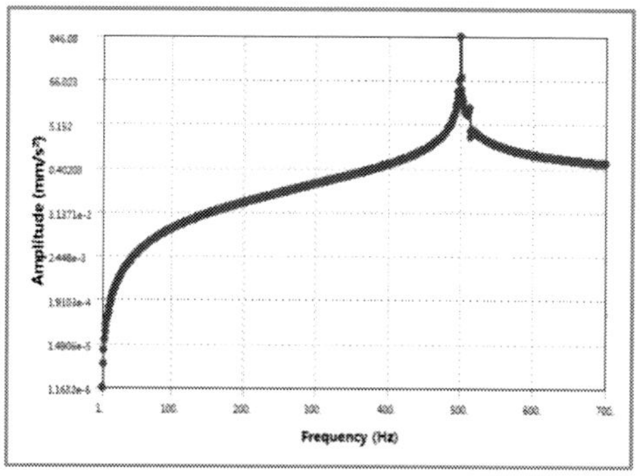

(d) Harmonic Response of Node 4

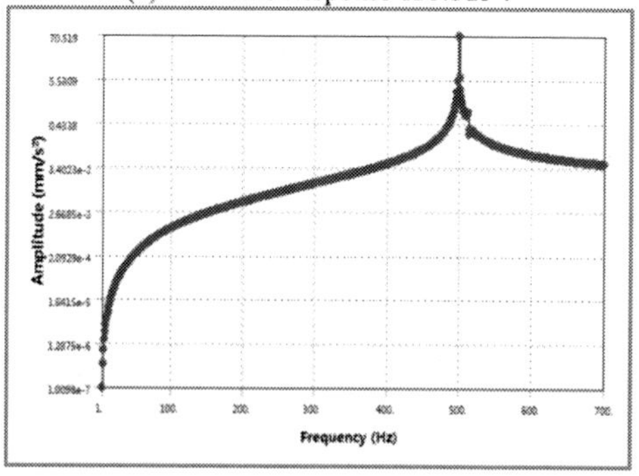

(e) Harmonic Response of Node 5

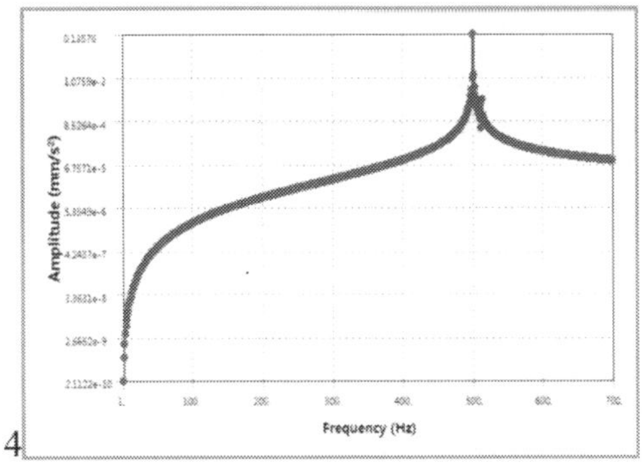

(f) Harmonic Response of Node 6

Fig.5. Harmonic Response of Testing Point

The harmonic response values at 100Hz of each node are listed as follows:
Node 1:0.1379; Node 2:9.7146e-2; Node3:4.3514e-2;
Node 4:2.9727e-3; Node 5:2.1794e-3; Node 6:5.1378e-6.
We can conclude that:

(1) Comparing the harmonic response values of node 1 with node 5, we can know that when the sensor is at the same direction with the vibration propagation direction, the harmonic response value of node 1 is the largest one. The optimal location to install the sensor is at node 1, which is sensitivity to signal for effective vibration signal acquisition.

(2) Comparing the harmonic response values of node 6 with node 5, we can conclude that it is difficult to collect the vibration signal effectively when the direction of sensor is different from the vibration propagation direction. The sensor should be installed in the same direction with the vibration propagation direction.

5 Experiment

The vibration signal of the horizontal driver is sampled by the sensor on the test point optimization of the detection fixture. The vibration signal of the normal and innormal HDM with worm wheel fault are shown in Fig. 6. Compare Fig.6(a) with Fig.6(b), we can directly determine the quality grade of HDM through the amplitude of the time-domain vibration signal.

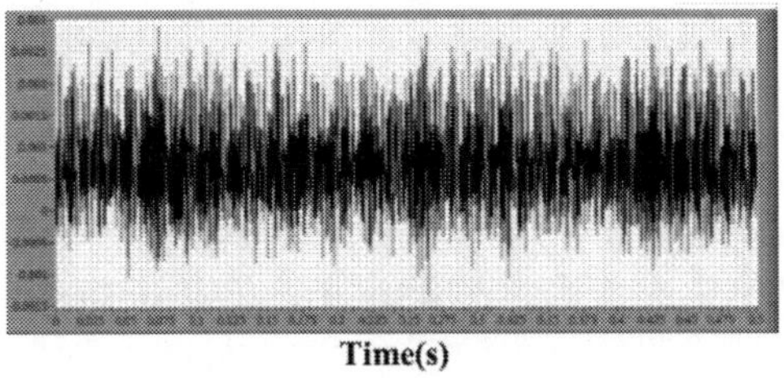

(a) Vibration Signal of Normal HDM

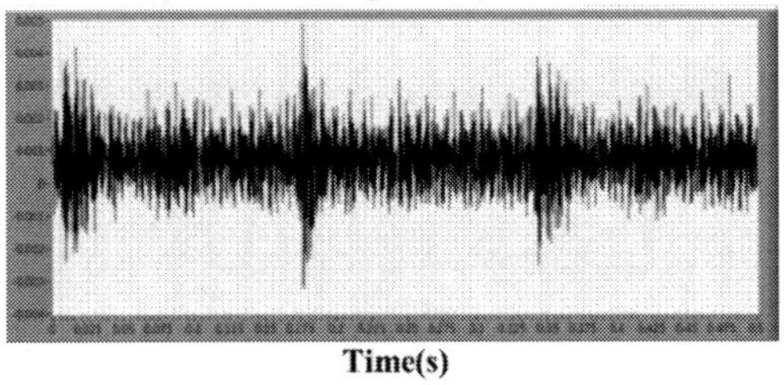

(b) Vibration Signal of Normal HDM

Fig.6. Vibration Signal of HDM with Worm Wheel Fault

6 Conclusions

The type of the sensor and the optimal placement of the installation location of the sensor are important factors in vibration signal acquisition. According to the characteristics of HDM, the acceleration sensor can effectively acquire the vibration signal including the fault information according to the actual installation condition and the attenuation characteristic of the vibration signal transmission. By means of finite element analysis method, we can optimize the best placement of the sensors with the goal of maximizing the information reflecting the fault and sensitivity to the direction of the vibration. On the base of installation conditions, the sensor shall be mounted on the top of the vibration detection fixture, in the same direction as the vibration propagation direction.

Acknowledgement

This research was supported by the National Natural Science Foundation of China (No. 50905050), Zhejiang Postdoctoral Science Foundation of China (No. BSH1402014), Zhejiang Province Science and Technology Project of China (No. 2016C31045). The authors would like to thank the reviewer for his valuable comments and suggestions.

References

1. L. J, LIU, K. J. YANG, F. LI: Knocking Noise Identification of Horizontal Driving Machine in Automobile Seat. Chin. Mech. Eng., **22**(21), 2760(2010).
2. Q. WANG, L. J. LIU, K. J. YANG: Sound Quality Evaluation of a Kind of Compact Horizontal Driver in Advanced Automobile Seats. Mech. Sci. Technol. Aerospace Eng., **31**(1), 167(2012).
3. L. P. WANG, X. F. LI, T. F. WU: Research of the Wavelet Denoising Applied to the Weak Signal Detection of the Magnetofluid Gyroscope. Chin. J. Sensors Actuators, **27**(10), 1355(2014).
4. Z. MING, Y. YU, J. D. ZHENG: Normalized Complex Teager Energy Operator Demodulation Method and its Application to Fault Diagnosis in a Rubbing Rotor System. MECH SYST SIGNAL PR., 50-51, 380(2015).
5. Y. QIN, Y. F. MAO, B. P. TANG: Vibration Signal Component Separation by Iteratively Using Basis Pursuit and its Application in Mechanical Fault Detection. J SOUND VIB, **332**(20), 5217(2013).
6. N. SARAVANAN, V. N. S. KUMAR, K. I. RAMACHANDRAN: Fault Diagnosis of Spur Bevel Gear Box Using Artificial Neural Network (ANN) and Proximal Support Vector Machine (PSVM). Appl. Soft. Comput. , 10, 344(2010).
7. G. L. HE, K. DING, H. B. LIN: Matching Pursuit Method for Coupling Modulation Signal Separation of Gearbox Vibration. Chin. J. Mech. Eng., **32**(1), 102(2016).
8. M. HANSARD, R. HORAUD, M. AMAT: Automatic Detection of Calibration Grids in Time-of-Flight Images. Comput. Vis. Image Und., **121**(4), 108(2014).
9. WILSON: Applying Algorithms for Machine Vision. Vision Syst. Design, **18**(6), 26(2013).
10. V. MARINCA, N. HERISANU: Explicit and Exact Solutions to Cubic Duffing and Double Well Duffing Equations. Math. Comput. Model. , (53), 604(2011).
11. J. F. WEBER, S. D. WALDMAN: Stochastic Resonance is a Method to Improve the Biosynthetic Response of Chondrocytes to Mechanical Stimulation. J. Orthop. Res., **34**(2), 231(2016).
12. J. WANG, Q. HE, F. KONG: An Improved Multiscale Noise Tuning of Stochastic Resonance for Identifying Multiple Transient Faults in Rolling Element Bearings. J. Sound Vib., **333**(26), 7401(2014).

An Identification Method of Main Harmonic Current Indicator Based On Active Power Direction Method

Junyi Zhao [1], Chaoying Yang [1], Zhiwei Xue[1] and Lin Yuan [2,*]

[1] State Grid Shanxi Electric Power Company, Taiyuan, Shanxi province, China
[2] School of Electrical Engineering and Information, Sichuan University, Chengdu, Sichuan province, China
Corresponding Email: 583599338@qq.com

Abstract. It is difficult to determine the harmonic current which flows through the line element in power grid, the paper proposed a new method for identifying the main harmonic current indicator at the Point of Common Coupling (PCC) based on the strict mathematic constraints. At the fundamental of the harmonic Norton equivalent circuit, the paper analysed that the harmonic active power direction method can be completely correct and valid in a certain interval. Combined the basic principles of the circuit and the strict inequality constraints, the paper discussed various fundamental characteristic of the harmonic impedance on both the system side and the user side, with the harmonic voltage and current measured amplitude and phase angle at the PCC. When the indicator of main harmonic current is the user side, the method can directly get the conclusion with the phase angle difference between harmonic voltage and current. When the indicator of main harmonic current is the system side, the method can get the conclusion combined the phase angle difference between harmonic voltage and current and harmonic impedance. Finally, the paper used the measured data of a variety of practical engineering scenarios to verify the correctness and effectiveness of the method.

1 Introduction

Harmonic current flowing through line components (such as transformers) is one of the main factors that cause adverse effects, such as increased losses and reduced operating efficiency. It is not conducive to the operation efficiency of the power grid, and makes the power supply sector bear huge economic losses [1-6]. At present, there are few literatures to study the mechanism of identifying the main harmonic current source from the system or the user side, but the engineering is always concerned [4, 5, 7]. Based on this, the paper deduced from the basic circuit law that can directly identify the main harmonic current source through the simple condition, but also provides the theoretical basis for further quantifying the harmonic current contributions from both sides. It's very important to theoretical and practical significance.

The core of the main harmonic current source identification is to compare the launch harmonic current contribution at PCC of the system side or the user side, that is, use the harmonic current indicator as the basic criteria [2, 8, 9]. Most of the relevant methods to estimate harmonic impedance of the system side or the user side firstly [10-15], and determine secondly. The main methods are fluctuation method [10-12], threshold voltage method [13], independent random vector covariance method [14] and fast independent component analysis method [15] and so on. The fluctuation method is used to estimate the harmonic impedance based on the measured harmonic voltage and current information at the PCC. According to the basic properties of the grid impedance, it can distinguish the estimation result is system-side or user-side. Independent random vector covariance method is used to estimate the harmonic impedance by using the basic properties that harmonic voltage and current are independent approximation in the power grid and the covariance equals zero. It weakens the effect of background harmonics to a certain extent. The threshold voltage method uses the estimated harmonic impedance value to identify. Fast independent component analysis is one of blind source separation methods. This method is applied to the estimation of harmonic impedance, which has a better effect than the previous. The main harmonic current source identification is the qualitative conclusion, but the above method to determine whether the system side or the user side are required to estimate harmonic impendent form the historical data, and cannot draw the conclusion directly only through the measured harmonic voltage and current information.

Harmonic active power direction method was proposed on the base of harmonic state estimation. It was widely used in the actual power grid because of simple and easy to use [2, 8, 16]. But the relevant literature reported that the method result was seriously affected by the harmonic source phase angle difference value between the system side and the user side. And the difference phase angle of the both harmonic source sides can vary arbitrarily in the whole circumference and cannot be measured directly, so the use of this method is limited [16]. However, there is no literature to study the wide range of the method which can be applied correctly.

On the basis of harmonic active power direction method, the paper deduced to the corresponding interval constraint identification method with various situations of harmonic impedance on the system side and the user side under the physical conditions of actual power grid. For the situation that the main harmonic current source is the user side, it can get identification result only depending on the harmonic voltage and current phase angle difference; and for the situation that the main harmonic current source is the system side, it can get result by using the harmonic voltage and current phase angle and the user-side impedance range which is estimated by relevant literature. Finally, the paper method is validated by using the measured data of several engineering scenarios. And the results show that the proposed method is effective and accurate.

2 Harmonic active power direc-tion method

Figure 1 is the Norton equivalent circuit model and it's the basis of active power direction method. In the figure, $\dot{I}_s$ and $\dot{I}_c$ respectively denote system-side and user-side harmonic source current; I_s and I_c respectively denote their own modulus value; Z_s and Z_c respectively denote system-side and user-side harmonic impedances; $\dot{U}_{pcc}$ and $\dot{I}_{pcc}$ respectively denote harmonic voltage and current value which measured at the PCC; U_{pcc} and I_{pcc} respectively denote the modulus value of the two.

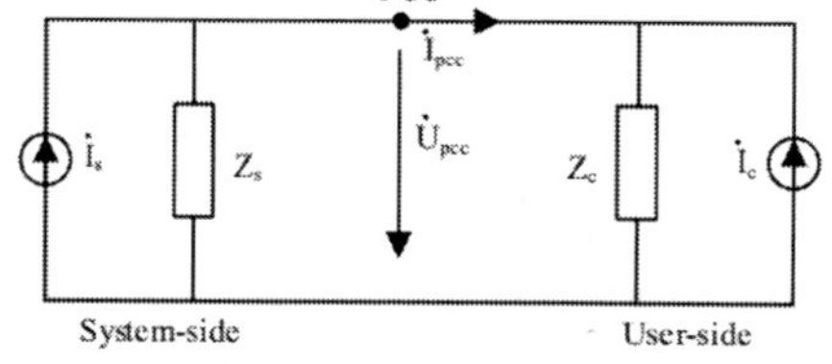

Fig. 1. The model of Norton equivalent circuit.

$$P=U_{pcc}I_{pcc}\cos\varphi \tag{1}$$

Harmonic active power direction method considered that when the harmonic active power flows from the system side to the user side, that is, P > 0, the system side's harmonic current contribution is bigger; when the harmonic active power flows from the user side to the system side, that is P < 0, the user side's harmonic current contribution is bigger [2, 16].

Equation (1) is the harmonic active power calculation formula and we can see the essence of this method. It though the positive or negative cosine value of the harmonic voltage and current phase angle difference φ to determine which side is the main harmonic current contribution.

2.1 Harmonic Current Indicators

According to the superposition theorem of the circuit, the PCC harmonic current $\dot{I}_{pcc}$ is the superposition of the harmonic current which contributed by the system side and the user side, as shown in formula (2).

$$\dot{I}_{pcc}=\dot{I}_{spcc}-\dot{I}_{cpcc}=\frac{Z_s\dot{I}_s-Z_c\dot{I}_c}{Z_s+Z_c} \tag{2}$$

Harmonic current phasor decomposition is shown in Figure 2, Isf and Icf respectively denote the projection value of $\dot{I}_{spcc}$ and $\dot{I}_{cpcc}$ in the $\dot{I}_{pcc}$ direction, they each ratio of I_{pcc} are the harmonic current indicator.

$$I_{sf}-I_{cf}=\frac{I_{spcc}^2-I_{cpcc}^2}{I_{pcc}}=\frac{(|Z_s|I_s)^2-(|Z_c|I_c)^2}{|Z_s+Z_c|^2 I_{pcc}} \tag{3}$$

With the help of the cosine theorem and formula(3), it can see the essence of harmonic current indicator is to compare the size of $|Z_s|I_s$ and $|Z_c|I_c$.

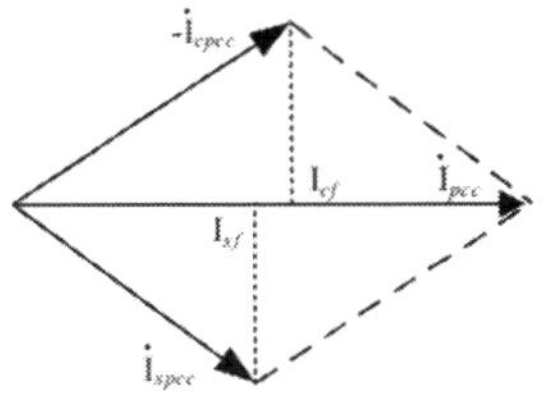

Fig. 2. Decomposition of current phasor.

2.2 Applicable Analysis of the Method

Literature [16] pointed out that the harmonic active power direction method is not entirely valid, there is a misjudgment. The harmonic voltage and current size and phase angle difference of actual power grid can be measured only at the PCC, and the phase angle difference can range from 0 ° to 360 °. The paper take the PCC harmonic voltage and current phase angle difference φ as the change amount, and the step length takes 1 °. The other parameters are shown in Tab1 and the experimental result is shown in Figure 3.

Table 1. The parameters of experiment.

Impedance characteristic	$\|U_{pcc}\|$/V	$\|I_{pcc}\|$/A	Z_s/Ω	Z_c/Ω
Both sides are inductive	15	1	1+j8	5+j25

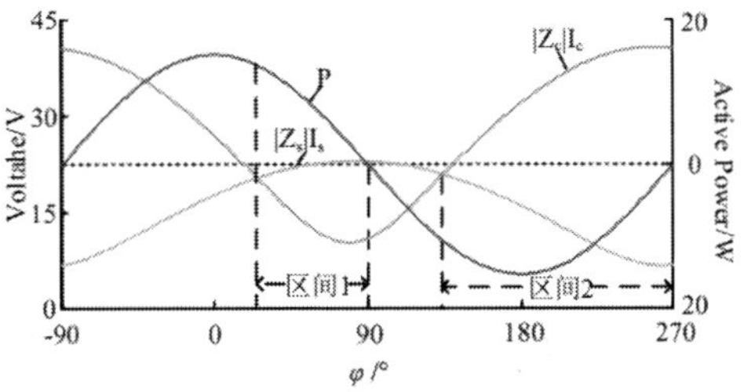

Fig. 3. The judgment conditions of active harmonic power direction method.

According to Figure 3, we can see that in the interval 1, $|Z_c|I_c < |Z_s|I_s$ and $P > 0$, in the interval 2, $|Z_c|I_c > |Z_s|I_s$ and $P < 0$, they are all in line with harmonic active power direction method. In other words, harmonic active power direction method is able to correctly identify which the harmonic current index side is greater in a certain interval. There is a certain range of application for it.

3 The identification method of main harmonic current indica-tor

3.1. Fundamental Theory

The purpose of the main harmonic current indicator identification is to compare the size of the two sides' harmonic current indicator. That is, to compare the size between $|Z_s|I_s$ and $|Z_c|I_c$. According to the basic circuit theory, it can be drawn from Figure 1:

$$\begin{cases} \dot{I}_s Z_s = \dot{U}_{pcc} + \dot{I}_{pcc} Z_s \\ \dot{I}_c Z_c = \dot{U}_{pcc} - \dot{I}_{pcc} Z_c \end{cases} \tag{4}$$

Both sides calculate the modulus values and the formula can be:

$$E_s^2 = U_{pcc}^2 + 2U_{pcc} I_{pcc} |Z_c| \frac{\cos(\varphi - \theta_s)}{k} + I_{pcc}^2 \frac{|Z_c|^2}{k^2} \tag{5}$$

$$E_c^2 = U_{pcc}^2 - 2U_{pcc} I_{pcc} |Z_c| \cos(\varphi - \theta_c) + I_{pcc}^2 |Z_c|^2 \tag{6}$$

Where,

$$E_s = |\dot{I}_s Z_s|, E_c = |\dot{I}_c Z_c|, k = |Z_c| / |Z_s| \tag{7}$$

θ_s is the system-side's harmonic impedance angle; θ_c is the user-side's harmonic impedance angle; k is the harmonic impedance modulus ratio between the user-side and the system-side, generally $k > 1$.

Because of the actual grid physical limit, the real part of the harmonic impedance must be positive, that is, harmonic impedance angle θ_s and θ_c range [-90 °, 90 °]. For theory, the impedance angle ranges from 0° to 90° is inductive, and it ranges from -90° to 0° is capacitive [10]. The components and loads in the power grid are mostly inductive. For the system side impedance, its characteristic is usually inductive and its ratio of the inductive part is greater than the resistive part, and the higher the voltage level is, the higher the ratio [17]. It's easy to know the impedance angle is in [45°, 90°] interval. For the user-side impedance, the relevant literature [18, 19] study gives out the load harmonic impedance model and its calculation method. Without considering the reactive power compensation devices or a large number of urban distribution network cable case, power factor of the distribution network load is less than or equal to 0.85 mostly[20]. For example this takes the load power factor as 0.85, the harmonic impedance angle is shown in Fig 4. It can be seen that the impedance angles of the harmonics are in the range of [45 °, 90 °] and the impedance is inductive, so the same conclusion can be obtained in the other cases where the power factor is less than 0.85. According to the relevant literature [21, 22], the resonance point of the city distribution network which uses cable, low-voltage capacitors and other reactive power compensation device in the large-scale must be appeared in the higher harmonics. That led to the character of the low-order harmonic impedance is capacitive, but in this case the capacitive reactance part is less than the resistive part, so the harmonic impedance angle is range in [-45 °, 0 °].

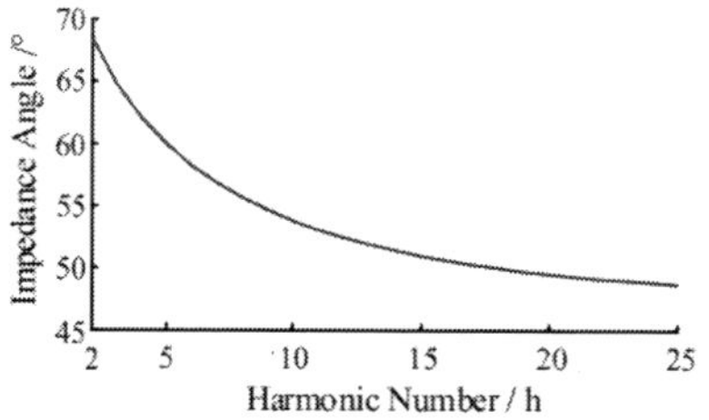

Fig. 4. Each harmonic impedance phase angle.

In summary, this paper considers the both sides' characteristics of harmonic impedance to meet the common engineering scene in the grid, but also makes the method in the course of simple. Therefore, in this paper, when the harmonic impedance is inductive, it is regarded as the impedance angle range in [45 °, 90 °]. When the harmonic impedance is capacitive, it is regarded as the impedance angle range in [45 °, 0 °].

3.2 The Identification Method: User-Side

Place the figure as close as possible after the point where it is first referenced in the text. If there is a large number of figures and tables it might be necessary to place some before their text citation. If a figure or table is too large to fit into one column, it can be centred across both columns at the top or the bottom of the page.

$$E_c^2 - E_s^2 = I_{pcc}^2 |Z_c|^2 (1-\frac{1}{k^2}) - 2I_{pcc}U_{pcc}|Z_c|[\frac{1}{k}\cos(\varphi-\theta_s)+\cos(\varphi-\theta_c)] \quad (8)$$

In order to get the user side identification method of main harmonic current indicators, the paper uses the formula (6) minus (5) and the result is formula(8). When the right part of the formula is greater than 0, then $E_c^2 > E_s^2$, that is, harmonic current indicator identification result is the user side. With the same and different characteristic of the harmonic impedance, we discuss the following four cases:

Both sides' characteristic are the same

The harmonic impedance characteristic of the system-side and the user-side are inductive, that is, θ_s and θ_c range in [45 °, 90 °], and the phase angle difference φ is between [180 °, 270 °]. We can easily get $\cos(\varphi-\theta_s) < 0$, $\cos(\varphi-\theta_c) < 0$, what's more $k > 1$ and $1-1/k^2 > 0$. So the right part of equation (8) is greater than 0, it meets the conclusion.

Similarly, both the system side and the user side of the harmonic impedance properties are capacitive, that is, θ_s and θ_c are between [-45 °, 0 °], the user side identification method of main harmonic current indicators is the phase angle φ ranging in the (90 °, 225 °] interval.

Both sides' characteristic are the different

The harmonic impedance characteristic of the system-side is inductive and the user-side's is capacitive, that is, θ_s ranges in [45 °, 90 °] and θ_c ranges in [-45 °, 0 °], and the phase angle difference φ is between [180 °, 225 °]. We can easily get $\cos(\varphi-\theta_s) < 0$, $\cos(\varphi-\theta_c) < 0$, what's more $k > 1$ and $1-1/k^2 > 0$. So the right part of equation (8) is greater than 0. So it's the identification method.

Similarly, the harmonic impedance characteristic of the system-side is capacitive and the user-side's is inductive, that is, θ_s ranges in [-45 °, 0 °] and θ_c ranges in [45 °, 90 °], the identification method is that the phase angle difference φ is between [180 °, 225 °].

In summary, based on the reference direction shown in Fig1 and the both sides' various cases of impedance characteristic, the identification method of the main harmonic current indicator for the user side is as shown in Table 2.

Table 2. The identification method of main harmonic current indicator for the user side.

impedance characteristic		the phase angle difference φ
Z_s	Z_c	
inductive	inductive	[180°, 270°)
inductive	capacitive	[180°, 225°]
capacitive	capacitive	(90°, 225°]
capacitive	inductive	[180°, 225°]

3.3 The Identification Method: System-Side

In order to get the system side identification method of main harmonic current indicators, the paper uses the formula (6) minus (5) and the result is formula(9).

$$E_s^2 - E_c^2 = I_{pcc}^2 |Z_c|^2 (\frac{1}{k^2} - 1) + \\ 2I_{pcc}U_{pcc}|Z_c|[\frac{1}{k}\cos(\varphi - \theta_s) + \cos(\varphi - \theta_c)] \quad (9)$$

When the right part of the formula is greater than 0, then $E_s^2 > E_c^2$, that is, harmonic current indicator identification result is the system side. With the same and different characteristic of the harmonic impedance, the paper discussed the following four cases:

Both sides' characteristic are the same

The harmonic impedance characteristic of the system-side and the user-side are inductive, that is, θ_s and θ_c range in [45 °, 90 °], and the phase angle difference φ is between [0 °, 90 °]. We can easily get $cos(\varphi-\theta_s) > 0$, $cos(\varphi-\theta_c) > 0$, what's more $k > 1$ and $1-1/k^2 < 0$. So it cannot draw the conclusion directly.

According to the cosine correlation knowledge, E_s^2 equals $E_{s\,min}^2$ when θ_s is under the condition of formula (10). Similarly, when θ_c is in the formula (11) conditions, E_c^2 equals $E_{c\,max}^2$. So we can get the conclusion.

$$\begin{cases} \theta_s = 90^\circ, \varphi \in [0^\circ, 67.5^\circ) \\ \theta_s = 45^\circ, \varphi \in [67.5^\circ, 90^\circ) \end{cases} \quad (10)$$

$$\begin{cases} \theta_c = 90^\circ, \varphi \in [0^\circ, 67.5^\circ) \\ \theta_c = 45^\circ, \varphi \in [67.5^\circ, 90^\circ) \end{cases} \quad (11)$$

This makes the equation(9) > 0 and combine with the formula (10) and (11), it's easy to get the flowing identification method addition condition.

$$\begin{cases} \cos(\varphi - 90^\circ) > \dfrac{I_{pcc}|Z_c|}{2U_{pcc}}, \varphi \in [0^\circ, 67.5^\circ) \\ \cos(\varphi - 45^\circ) > \dfrac{I_{pcc}|Z_c|}{2U_{pcc}}, \varphi \in [67.5^\circ, 90^\circ) \end{cases} \quad (12)$$

Similarly, both the system side and the user side of the harmonic impedance is all capacitive, that is, θ_s and θ_c are in the [-45 °, 0 °], the identification method of main harmonic current indicator is the phase angle difference φ range in (-90 °, -22.5 °] and satisfies the addition condition (13).

$$\cos\varphi > \frac{I_{pcc}|Z_c|}{2U_{pcc}} \quad (13)$$

Both sides' characteristic are the different

The harmonic impedance characteristic of the system-side is inductive and the user-side's is capacitive, that is, θ_s ranges in [45 °, 90 °] and θ_c ranges in [-45 °, 0 °], and the phase angle difference φ is between [0 °, 45 °]. We can easily know when, θ_s is 90°, E_s^2 equals $E_{s\,min}^2$ and when θ_c is -45°, E_c^2 equals $E_{c\,max}^2$. Make equation (9) greater than 0, we can get the additional constraint condition as follows.

$$\cos(\varphi + 45^\circ) > \frac{I_{pcc}|Z_c|}{2U_{pcc}} \quad (14)$$

Similarly, the harmonic impedance characteristic of the system-side is capacitive and the user-side's is inductive, that is, θ_s ranges in [-45 °, 0 °] and θ_c ranges in [45 °, 90 °], the identification method is that the phase angle difference φ is between [0 °, 45 °] and the addition condition is formula(15).

$$\cos(\varphi - 90^\circ) > \frac{I_{\text{pcc}}|Z_c|}{2U_{\text{pcc}}} \tag{15}$$

From the above analysis, the user-side harmonic impedance modulo is included in the additional condition that the identification method of main harmonic current indicator. And only when the fluctuation of the harmonic source on the user side is small, a better impedance estimation result can be obtained [10, 11, 13]. Therefore, this paper only uses the correlation method to estimate the range of the harmonic impedance of the user side for auxiliary judgment.

The voltage waveform of the medium-low voltage distribution network is usually a flat-top wave, so it can be equivalent to the user-side model by the harmonic Norton circuit. Related literature [14, 15, 23] is a new harmonic impedance calculation method proposed in recent years, first measured harmonic voltage and current phase angle difference φ meet the interval constraints, and the user side of the small harmonic fluctuations situation, can be more accurate calculation of the user side harmonic impedance mode value. At the same time, in order to make this method as simple as possible in the application, the method of literature [15] is used to calculate the harmonic impedance of the user side.

To sum up, according to the reference direction shown in Figure 1 and the both sides of the impedance properties of classification, the identification method of main harmonic current indicator for the system side of the interval constraints. It is summarized in Table 3 below.

Table 3. The identification method of main harmonic current indicator for the system side.

Impedance Characteristic		the phase angle difference φ	Addition Conditions
Zs	Zc		
inductive	inductive	[0°, 90°)	formula(12)
inductive	capacitive	[0°, 45°]	formula(14)
capacitive	capacitive	(-90°, -22.5°]	formula(13)
capacitive	inductive	[0°, 45°]	formula(15)

4 Case studies

4.1 Electric Arc Furnace Load Calculation

Figure 5 shows the measured data of the 3rd harmonic voltage and current of a 150kV bus (PCC) from an industrial electric arc furnace plant. There is about 1h downtime in this period, and the PCC point on both sides of this harmonic impedance Were inductive.

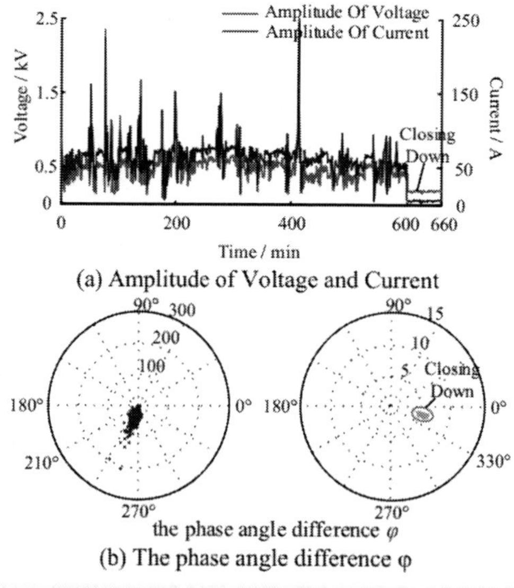

Fig. 5. The measured data of the 3rd harmonic voltage and current.

In the figure, the harmonic voltage and current amplitude fluctuate very much in 0 to 600 minutes, but the phase angle difference of harmonic voltage and current is almost entirely concentrated in the range of [210 °, 270 °]. According to the constraint in Table 2, the harmonic current contributed by PCC at the user side is larger and the main harmonic current source is the user side. As we all know, the use of thermistor rectifier load in the work of the arc furnace will launch a large number of harmonic current, especially in the low harmonic [6]. At the same time, the harmonic voltage and current amplitude of PCC point decrease sharply during the 600 to 660min downtime period, and the phase angle difference between PCC point and PCC point is in the range of [330 °, 0 °], not within the constraint condition of Table 2. And the contribution from the user side to the harmonic current index of the PCC point is greater; and the evaluation results in [11] show that the harmonic level emitted by the user side is greater. Therefore, the above analysis and results verify the correctness and effectiveness of this method.

4.2 Example Calculation of Urban Distribution Network

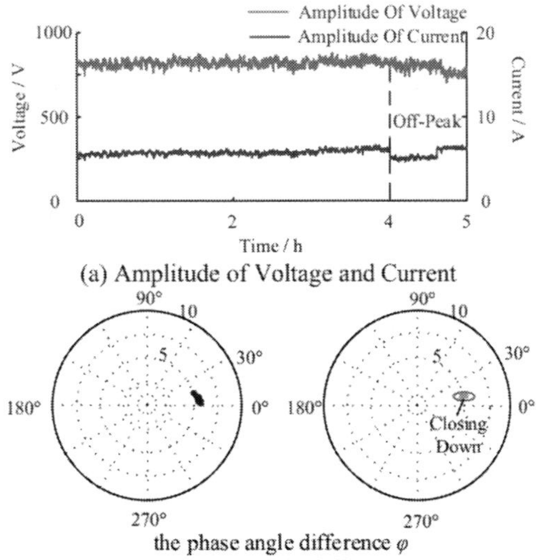

(b) The phase angle difference φ

Fig. 6. The measured data of the 11th harmonic voltage and current.

Table 4. The parameters of PCC.

$\|Z_c\|/\Omega$	$\|U_{pcc}\|/V$	$\|I_{pcc}\|/A$	$\varphi/(°)$
103.7	781	5.3	15.2

Figure 6 shows the measured data of the 11th harmonic voltage and current of the 66kV bus (PCC) of a city power grid near the DC drop point. The whole data interval is about 1h. PCC point known as the equivalent system-side harmonic impedance inductive nature of the city because of the large-scale use of underground power cables, and to ensure that the user power quality with a large number of capacitors and other devices, resulting in the user side of the sub-harmonic impedance nature of Capacitive [17].

It can be seen from the figure, 0 to 4h time period of the normal power consumption period, the harmonic voltage and current amplitude changes are small, which is the user side of the basic conditions of harmonic impedance estimation, while the figure shows the harmonic voltage and current The phase angle difference is concentrated in the [0 °, 30 °] interval. Using the constraints in Table 3, it can be seen that the current condition satisfies the range of the phase angle difference φ of the main harmonic current source as the system side, and the additional condition of Equation (14) is satisfied. Using the method of literature [15] to calculate the harmonic impedance of the user side, we can get PCC related parameters of 95% probability maximum value in this time period, as shown in Table 4. With the formula (14) Constraint, so you can determine the current situation, the main harmonic current source for the system side.

From 4h to 5h trough time can be seen, compared to the PCC normal point of time the second harmonic voltage and current amplitude are not significantly reduced, the harmonic voltage and current phase angle difference is still concentrated 0 ° to 30 ° interval. In [17], using Dig-silent to simulate the urban power network, about 70% of the 11th harmonic current measured by each branch bus near the DC drop point is contributed by the 500kV DC drop point, so it can be determined that the system side pair PCC's 11th harmonic current contribution is greater, the main harmonic current source for the system side. Therefore, the correctness and validity of the method are verified.

5 Conclusions

Based on the analysis of harmonics active power direction method the paper deduced the identification method of the main harmonic current indicator, and the following conclusions are drawn:

1) Analyzing and drawing the harmonic active power direction method can be completely accurate conclusions in a certain region. Combined with the basic principle of the circuit and the constraint condition of the strict inequality constraints, it can effectively solve the problem that the main source of the harmonic current flowing through the line components in practical engineering.

2) In the basic properties of the power grid, the paper considered the various cases of the two sides' harmonic characteristics to meet the most project usage scene. Finally, the correctness and effectiveness of the method are verified by using the measured data from several engineering scenarios.

The identification method of main harmonic current indicator is deduced from the mechanism, and it provides a theoretical basis for the use of harmonic current active power direction method and the quantitative division of power equipment caused by the loss of harmonic current. There is a great theoretical significance and engineering practical value.

References

1. Wilsun X, Yilu L. A method for determining customer and utility harmonic contributions at the point of common coupling. IEEE Transactions on Power Delivery, 2000, 15(2): 804-811.
2. Wilsun X, Xian L, Yilu L. An investigation on the validity of power-direction method for harmonic source determination. IEEE Transactions on Power Delivery, 2003, 18(1): 214-219.
3. Li C, Xu W, Tayjasanant T. A critical impedance based method for identifying harmonic sources. IEEE Transactions on Power Delivery, 2004, 19(2): 671-678.
4. LIU Hua, LIU Rui, FU Yangyang, et al. Calculation of Harmonic Loss in Converter Transformer Tank Shells Based on the Impedance Boundary Condition. High Voltage Engineering, 2015, 41(9): 3171-3176.
5. YE Jianxin, HUANG Jiaohao, ZHANG Luliang, et al. Analysis of harmonic harm to distribution transformers. Electric Power Automation Equipment, 2011, 31(7): 146-149.
6. YE Jianxin, HUANG Jiaohao, ZHANG Luliang, et al. Analysis of harmonic harm to distribution transformers. Electric Power Automation Equipment, 2011, 31(7): 146-149.
7. ZHANG Liangxian, CHEN Mosheng, PENG Zongren, et al. Study on the Harmonic Losses of UHV Converter Transformer Windings Subject to Non-sinusoidal Load Current. Proceedings of the CSEE, 2014, 34(15): 2452-2459.
8. ZHOU Lin, ZHANG Feng, LI Qiuhua, et al. Methods for Localizing the Harmonic Source in Power Distribution Network. High Voltage Engineering, 2007, 33(5): 103-108.
9. Hua Huichun, JIA Xiufang, CAO Dongsheng, et al. Harmonic Contribution Estimation Under Power Quality Data Interchange Format. Power System Technology, 2013, 37(11): 3110-3117.
10. Yang H, Pirotte P, Robert A. Harmonic emission levels of industrial loads statistical assessment. CIGRE, 1996, 96: 1936-1996.
11. HUI Jin, YANG Honggeng, YE Maoqing. Assessing Harmonic Emission Level Based on the Impedance Gathering Trend Discrimination, Proceedings of the CSEE, 2011, 31(10): 73-80.
12. CHE Quan, YANG Honggeng. ASSESSING THE HARMONIC EMISSION LEVEL BASED ON ROBUST REGRESSION METHOD, Proceedings of the CSEE, 2004, 24(4): 39-42.
13. SUN Yuanyuan, LI Peixin, YIN Zhiming. Harmonic Source Identification Based on Threshold Voltage Method . Automation of Electric Power Systems, 2015(23): 145-151.
14. HUI Jin, YANG Honggeng, LIN Shunfu, et al. Assessment Method of Harmonic Emission Level Based on Covariance Characteristic of Random Vectors. Automation of Electric Power Systems, 2009, 33(7): 27-31.
15. ZHAO Xi, YANG Honggeng. Method of Calculating System-side Harmonic Impedance Based on FastICA . Automation of Electric Power Systems, 2015, 39(23):139-144.
16. Xu W. Power direction method cannot be used for harmonic source detection// Power Engineering Society Summer Meeting. IEEE, 2000: 873-876 vol. 2.
17. LI Yong, XU Youping, XIAO Hua, et al. Simulation Study on Load Model Parameters for Stability Analysis of Central China Power Grid. Power System Technology, 2007, 31(5): 17-20.
18. Robert A, Deflandre T. Guide for assessing the network harmonic impedance. Electra, 1996 (167): 96-131.
19. Wu X, Sadullah S, Matthews B, et al. Nodal harmonic impedance derivation of AC network in PSS/E// 9th IET International Conference on Ac and Dc Power Transmission. IET, 2010: 02-02.
20. Zhao X, Yang H. A New Method to Calculate the Utility Harmonic Impedance Based on FastICA. IEEE Transactions on Power Delivery, 2016, 31(1):381-388.

21. CHEN Tiantian, HU Rong, Jin Zuyang, et al . Analysis Research on Harmonic Transmission Characteristics of Urban Power Network Adjacent to UHVDC Terminal Location. Power System Technology, 2015, 39(10)：3000-3004.
22. WANG Jianxun, LIU Huijin, LIU Chunyang. Identification of harmonic and interharmonic sources based on composite criterion . Electric Power Automation Equipment, 2013, 33(7)：63-69.
23. QiuSiyu, YANG Honggeng. Assessment Method of Harmonic Emission Level Based on the Improved Weighted Support Vector Machine Regression. Transactions of China Electrotechnical Society, 2016, 31(5)：85-90.

Study on Determination Method of Load Measurement Sample Length of the Loader Working Device

Yipin Wan, Jie Jia and Xuding Song

The Key Laboratory of Road Construction Technology and Equipment, Chang'an University, Xi'an 710064, China

Abstract. In order to determine the sample size in the load spectrum measure of the loader working device, the methods of the sample length determining were studied. The load test system of the loader working device was constructed and the original load time history data were collected. Considering the load characteristics and the confidence level of the population and the sample data, a new method was proposed and the minimum sample size required for the load test of the loader working device on ZL50 loader was 70 based on the measured load. The results of sample length of different methods were compared and analyzed. The new method presented in this paper was more reliable and credible than other methods. Methods and results to determine the test data is obtained, which provides a reference for the load test and life cycle spectrum compiling.

1 Introduction

In earthwork construction, working medium and working conditions are complicated and changeable; fatigue failure is the main failure mode of the loader working device [1]. The load time history is the comprehensive performance of the structure in the working environment, which could describe the strength of the structure and the size of the stress [2-3]. Load spectrum is an important foundation for the structural strength analysis and fatigue life reliability analysis of shaft parts. [4], the length of the load sample determines the load spectrum accuracy and load test costs [5]. In general, the larger the load sample capacity, the more accurate the load spectrum data and the fatigue life are obtained. However, too large samples brought about by the high cost of testing, the accuracy of the load spectrum is not obvious. Statistical characteristics of data and error are the theoretical basis for determining the length of the sample. In literature [6], the sample length is studied, and the influence of sample size on the model parameter estimation is defined, and the basic selection principle of the minimum and maximum sample length is given. The [7] analysis method to determine the error meet the required sample length from two aspects of amplitude and frequency, given power spectral density to determine the sample length of the theoretical calculation formula. The [8] in the index model is based on the coefficient of variation of sample length of random data were calculated; the literature [9] considering the statistical accuracy, the accuracy of approximate mean value formula for determining the length of the sample estimate; literature using [10-11] method to fit the trend line of the obtained load test sample length.

Minimum sample size is the key foundation for the testing and compilation of load spectrum. Taking the domestic 50 type loader as the research object, the load signal acquisition system of the loader working device is constructed. In order to test the data as the foundation, proposed a method to determine the actual load of the loading shaft parts based on the length of the sample, considering a variety of statistical characteristics of the actual load data, is given to meet the given error precision of the loading shaft parts minimum sample size, which provides a reference for the loader working device load test method and load spectrum compiling.

2 Methods for determining the number of load samples

It is the basic principle that the sample size is determined by the statistical characteristics of the sample data [12], and the minimum length of the sample is obtained within the error tolerance range. For shaft loaders parts load parameters, which are stochastic data obtained by a stable physical phenomenon, could be considered as having characteristics of stationary ergodic [13]. Commonly used sample length determination methods are spectral density method, approximate mean estimation method and trend line fitting method.

The spectral density method is based on the time of the load cycle and the approximate solution of the formula (1).

$$\varepsilon_1 \approx 1/\sqrt{B_e T} \tag{1}$$

Approximate mean value estimation method, the parameter load time history data are considered as random variables. From statistical knowledge, the overall mean and standard deviation could not be obtained. But we could use a certain length of the sample mean and standard deviation to replace the overall, at this time could be approximated by a mathematical model of the mean estimation method (2).

$$\varepsilon_2 = \sigma_x / \left(\sqrt{K} u_x\right) \approx s_x / \left(\sqrt{K} \bar{x}\right) \quad (2)$$

In the above formulas, ε_1 is the spectral density statistics error; B_e is the frequency band width; T is the sample time length under the statistical error. ε_2 is the statistical error of the approximate mean estimator; K is the number of samples under a given statistical precision; σ_x and u_x are the overall standard deviation and the mean, respectively; s_x and $\bar{x}$ are the sample standard deviation and the mean value, respectively.

In a given statistical accuracy and sample mean and standard deviation, we could calculate the statistical error of the sample size required to test the test. Trend line fitting method is based on a large number of experimental data, analysis of the parameters of discrete data. To increase the sample length and calculate the mean value of different samples, the mathematical model which is closest to the change trend of the parameters is selected to fit the changing process, so that the variation trend of these parameters with the length of the sample is obtained. When the length of the sample is enough, we could use the sample to replace the overall parameter, and then determine the length of the sample with the statistical precision. According to the trend of the finite sample capacity data, the data fit the exponential model or the hyperbolic model, and the length of the sample is the minimum sample size when the fitting curve is stable.

In general, the sample variance is used to replace the total variance. From the central limit theorem, we could know that the random variable $\bar{x} - u_x / \left(s_x / \sqrt{n}\right)$ obeys the t distribution. The probability relationship between the significant level and the number of samples from the population is shown in the formula (3).

$$P\left\{\left|\bar{x} - u_x\right| \le \left(s_x / \sqrt{n}\right) t_{\alpha/2}(n-1)\right\} = 1 - \alpha \quad (3)$$

The confidence interval of the sample mean value could be got at this time for a given confidence level of 1-α as shown in the formula (4).

$$\bar{x} - \left(s_x / \sqrt{n}\right) t_{\alpha/2}(n-1) \le u_x \le \bar{x} + \left(s_x / \sqrt{n}\right) t_{\alpha/2}(n-1) \quad (4)$$

$V_x = s_x / u_x$ as the coefficient of variation; $\varepsilon_3 = \left(u_x - \bar{x}\right) / u_x$ is the relative error of the overall mean and the sample mean, and the lower the relative error of the overall parameters of the estimated accuracy is higher. The relationship between the coefficient of variation and the relative error is shown in the formula (5).

$$\varepsilon_3 = 1 - V_x \cdot \bar{x} / s_x \quad (5)$$

At this time, the relationship between the relative error and the length of the sample could be obtained. The model is shown in the formula (6).

$$\varepsilon_3 \ge \left|\left(s_x / u_x \sqrt{n}\right) t_{\alpha/2}(n-1)\right| = \left|\left(V_x / \sqrt{n}\right) t_{\alpha/2}(n-1)\right| \quad (6)$$

Putting the formula (5) into the formula (6) could have the formula (7) as follows.

$$n \ge \left[\left(\left(1 - \varepsilon_3\right) / \varepsilon_3\right) \cdot s_x / \bar{x} \cdot t_{\alpha/2}(n-1)\right]^2 \quad (7)$$

For a given confidence level, according to the sample mean and standard deviation and the formula (7) could calculate the minimum sample size required for a given statistical error under. However, the sample size is unknown in the practical solution, and the $t_{\alpha/2}(n_1 - 1)$ is unknown at this time. Therefore, it is required to determine the size of the $t_{\alpha/2}(n_1 - 1)$ according to the initial value of the selected sample number. By the formula (7) a new sample length, calculated the number of samples for recycling, until by $t_{\alpha/2}(n_1 - 1)$ and formula (7) the number of samples to obtain consistent loop termination.

3 Load testing and data processing for loader working device

The loading shaft parts including load test and axle shaft torque and rotational speed and real time shift information. The stress characteristics of the working device of the loader are analyzed. The axial force of each pin is the basic data of the structural design and fatigue life analysis of the working device. Select the corresponding test parameters to be tested, build the load spectrum test system of work device as shown in Figure 1.

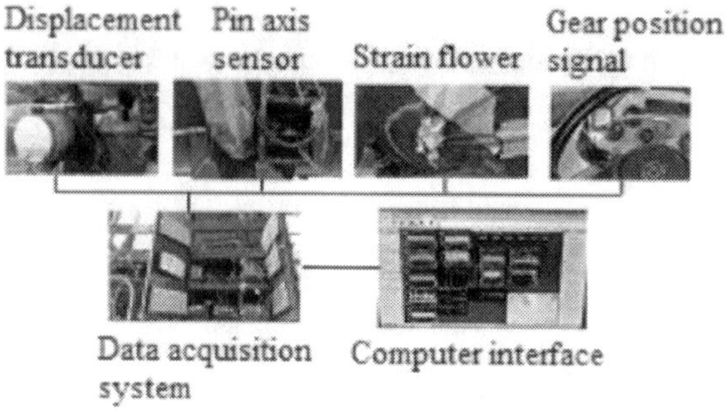

Fig. 1. The load spectrum testing system of the loader working device.

Select the bulk materials containing iron ore particles, with the highest shovel loading efficiency as the principle of shovel loading test. Select the driver with a wealth of driving experience, shovel loading route using commonly used L type line, material loading test as shown in Figure 2.

Fig. 2. The load spectrum testing system of the loader working device.

In the measured signal, the interference signal and the peak value of the interference signal caused by the surrounding environment and road surface are included. Preprocessing includes filtering, eliminating trend items, and abnormal peak points. The loading shaft parts load signal for low frequency signal, using 2Hz low pass filter. Trend elimination is implemented using Matlab programming. Anomaly points are proposed by using the gradient threshold method to recognize the singular point. After the pretreatment of loader front axle shaft of a torque load signal as shown in Figure 3.

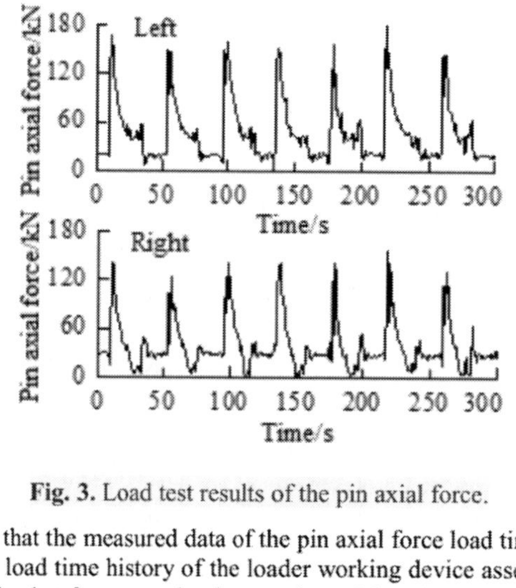

Fig. 3. Load test results of the pin axial force.

We could know from Figure 3 that the measured data of the pin axial force load time history curve shows significant seasonal changes could reflect the load time history of the loader working device assembly process in shovel real stress characteristics. The test data to 5 bucket for a sample, the test data of the 120 bucket, the mean value of the 24 sample data is shown in Table 1.

Table 1. Table of the mean changes in sample data.

Sample	Mean/kN	Sample	Mean/ kN

number		number	
1	51.89	13	48.69
2	50.48	14	48.90
3	50.94	15	48.73
4	50.24	16	48.70
5	49.11	17	48.49
6	49.41	18	48.38
7	48.89	19	48.28
8	48.64	20	48.59
9	48.93	21	48.91
10	48.52	22	48.92
11	48.25	23	48.72
12	48.20	24	48.83

4 Result analysis

The average time for a shovel loading test cycle is 45s, by the formula (1) could get the result of the spectral density method is about 16000s, that is, 350 buckets. By using the method of approximate mean value estimates, measured 120 bucket sample data calculation of the sample mean is 48.83kN, the sample standard deviation as 3.48kN, the statistical error of 0.05, by the formula (2) could get the required number of samples is rounded up to 17. Each sample contains 5 buckets, and the sample length is approximately 105.

Statistical analysis of the measured data by using the trend line fitting method. Describe the change trends of the mean of Table 1 with the increase of sample number, respectively, with the exponential model, polynomial model and S curve model to fit the trend of the data, the fitting trend line as shown in Figure 4

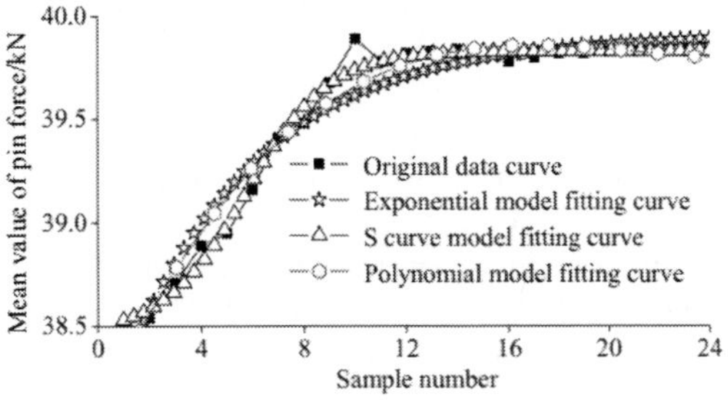

Fig. 4. Schematic diagram of the trend line fitting method.

Figure 4 shows the mean value of the sample, the exponential model of the fitting curve of the goodness of fit is 0.94, the fitting equation as shown in the formula (8).

$$M = -0.0009n^3 + 0.0471n^2 - 0.7620n + 52.36 \quad (8)$$

The goodness of fit curve of polynomial model is 0.95, and the fitting equation is as shown in formula (9).

$$M = 4.603e^{-n/3.057} + 48.57 \quad (9)$$

The goodness of fit of the S curve model is 0.92, and the fitting equation is as shown in the formula (10).

$$M = 48.61 - 3.863/\left(1 + e^{(n-3.042)/1.643}\right) \quad (10)$$

The required sample sizes of the three models are 17, 19 and 15, respectively. The sample length is 85, 95 and 75, respectively. With the curve fitting method, the results are quite different with the different models.

By using the method that sample length calculation based on measured load data proposed in this paper, the iterative calculation is realized by computer programming. For a given confidence level is 0.95, the statistical error is 0.05, the initial sample selection is 20. According to the formula (6) and the iterative procedure described in this paper, the output value of the sample is calculated in the range of 70~75. With the increase of the initial value of the sample length, the length of the output sample is gradually stabilized, and the sample length is 70.

From the above results, the trend line fitting method results in dependence on the selected model is too strong, the fitting results contain too many artificial factors, the length of the sample results of lack of credibility. The mean value and the standard deviation of the sample load are obtained by using the approximate mean value estimation method, and the result of the sample size is 105. The proposed estimation method based on the central limit theorem, consider not only the mean and standard deviation of measured load characteristics of sample loading, but also consider the overall mean probability falls on the sample mean interval is the confidence level, makes the results more reasonable. The 70 bucket sample size obtained by this method can be considered as the minimum sample size required to be tested, and the model parameters are estimated to be stable. The results of the new analysis method proposed by common methods and the minimum length of the sample shows that the load testing required is 70.

5 Conclusion

In this study, the method of determining the sample in the load test of the loader working device is studied, and the following conclusions are obtained.

According to the loading characteristics of the loader working device, the load test system was constructed. The load sample data of loading test reflect the real operation characteristics of loader working device, load sample length determination method to determine the basis. The method of determining the length of the load sample is analyzed. The sample length estimation method based on the measured load data is proposed. The method not only considers the characteristics of sample loading and standard deviation, but also considers the confidence level of the population mean and the sample mean, and the result is more reliable and credible than the other methods. The load test samples required minimum length is 70, and the research results provide an important reference spectrum and fatigue life prediction of the loader working device. The next step is to increase the amount of data, and further study the effect of sample size on the results of the load spectrum and the results of fatigue reliability analysis.

Acknowledgements

In this paper, the research was sponsored by the National Key Technology Research and Development Program of the Ministry of Science and Technology of China (Project No. 2015BAF07B02) and the Fundamental Research Funds for the Central Universities (No. 310825163407).

References

1. J. Samuelsson, Jonsson B, Barsoum B. Service fatigue design of complex welded construction equipment. Material-wiss. Werkstofftech **10**, 734-739 (2008).
2. H. Guo, W.H. Chen, X.Y. Fan. Research of enhancement coefficient of automobile reliability enhancement test on proving ground. Chinese Journal of Mechanical Engineering **10**, 102 (2004).
3. S. Murer, D. Leguillon. Static and fatigue failure of quasi-brittle materials at a V-notch using a dugdale model. European Journal of Mechanics A/Solids **8**, 109-118(2010).
4. J.X. Wang, J. Hu, X.T. Hou. Loader axle shaft finite element analysis based on load spectrum. Modern Manufacturing Engineering **5**, 1-3(2010).
5. C.H. Yan, Z.T. Gao. Design criterion of the flight takeoff-landing number of minimum no-loading measurement sampling. Acta Aeronautica Et Astronautica Sinica **20**, 514-517(1999).
6. P.X. Gao. Selection of sample length for parameter modeling. Journal of Vibration, Measurement and Diagnosis **4**, 5-10 (1991).
7. L.D. Shi. Finite life design and testing of machinery Seventh speak: the development process of the load spectrum. Construction Machinery **10**, 35-45 (1986).
8. L.D. Shi, J.G. Zhang, S.Q. Huang. Discussion on the problem of determining the length of the sample in the process of random data processing. Construction Machinery **4**, 16-21 (1993) .
9. Y.L. Zhang, W.N. Chu, C.X. Xu. Determination of load sample length for loader driveline. Construction Machinery and Equipment **6**, 16-20 (1994).
10. C.H. He, H. Zhou, J. Xiang. Study on the method of load spectrum sample size determination in vehicle road testing. Automobile Technology **11**, 14-17 (2003).
11. Y.N. Li, S.F. Lin, S.H. Liu. Application of regression analysis to sample size's error analysis. Journal of Jimei University **1**, 72-75 (2010).
12. D.L. Cheng, J.B. Adam, D.S. James. A Bayesian approach to sample size determination in mechanical engineering. Computational Statistics and Data Analysis **54**, 298-307 (2010).
13. Y.P. Wan, J. Jia, J. Liang. Structural strength analysis and experimental research for working device of loader. Journal of Mechanical Strength **4**, 772-776 (2016).

Topology Optimization in Electric Car Body Frame Based on Optistruct

Dongdong Ge [1] ,Liangrong Zhu [1], Dongji Xuan [2]

[1]Institute of vehicle, Zhejiang Industry Polytechnic College, Shaoxing 312000, China;
[2]College of Mechanical & Electrical Engineering, Wenzhou University, Wenzhou 325035, China
[a]gdd1988@126.com, [b]348607826@qq.com, [c]xuandongji@163.com

Abstract. In order to optimize the structure of the electric car body frame, the static analysis of the car frame were carried on. For the goal of the frame's weight minimum, OptiStruct software was used to topology optimization design. And the optimal material distribution program of the frame structure was got. Static strength before and after optimization was comprehensive compared through the stress, deformation. The results showed that the weight of frame after optimization was reduced by 18.96%,and the requirements of the strength and stiffness were ensured.
Keywords. Electric car, static analysis, OptiStruct, topology optimization.

1 Introduction

The goal of structure optimation is achieving the optimal performance of the structure, including objectives such as strength, stiffness and stability by using the least amount of material, the most simple process, low cost 1.Depending on the simulation experiment and the experience of the traditional optimization design method is difficult to solve the design problems such as plane, automobile2, ship, mold3 and other large-scale complex structure parts and bridge4,tunnel and other large engineering. Topology optimization method is becoming one of the key means to solve such problems 5.

The insufficient mileage is one of the major bottleneck factors which affect the promotion and application of the electric cars. This study based on the electric car body frame static analysis, the minimum body frame weight as the optimization goal,OptiStruct software was used to topology optimization design, get the optimal material distribution program of body frame structure.Finally HyperWorks and general numerical analysis software Radioss were used to static analysis of body frame before and after optimization, from the stress, deformation, weight and other aspects of the calculation results were compared.

2 Static analysis of the electric car body frame

2.1 The analysis model

In this paper, UG software was used to model one electric car (shown in figure 1), and body frame model (shown in figure 2) was isolated.The total mass of lower body frame part is 17.88 Kg, a total of 17 beams. Based on the finite element technology, using Hypermesh software to carry on finite element modeling of the body frame. Radioss software was used to make static analysis on the body frame under four kinds of typical working conditions. (horizontal bending condition, emergency braking condition, ultimate torsion condition, emergency turn condition)

The material of electric car body frame is Q345. Youngs's modulus, Poisson's ratio,density,thickness for the material of body frame is taken as 210 Gpa, 0.3, 7800 kg/m^3 and 2 mm.

DOI: 10.1051/matecconf/201710001016

Fig. 1. Electric car body model of UG

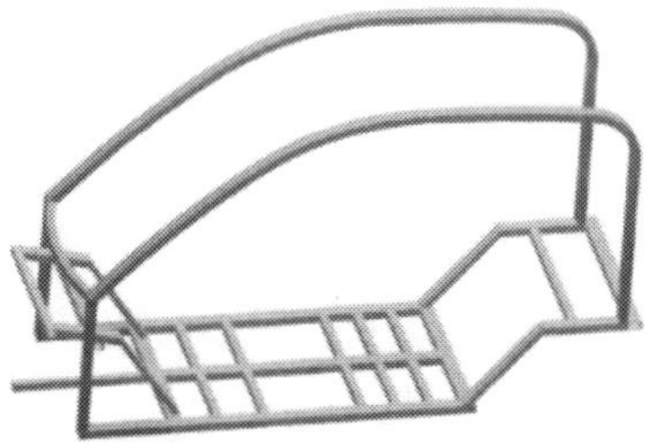

Fig. 2. Body frame model

2.2 Analysis of typical working condition

The electric car body frame material is Q345, and failure marks is plastic deformation.Make analysis of static strength and stiffness according to Von-Misses stress distribution and displacement deformation of the body frame. Referring to the literature6, 7and 8 on the treatment of boundary conditions, make static analysis on the body frame under four kinds of typical working conditions. By analyzing the stress and displacement response of the body frame conditions, improve unreasonable structure, and provide the basis for lightweight design body frame structure.

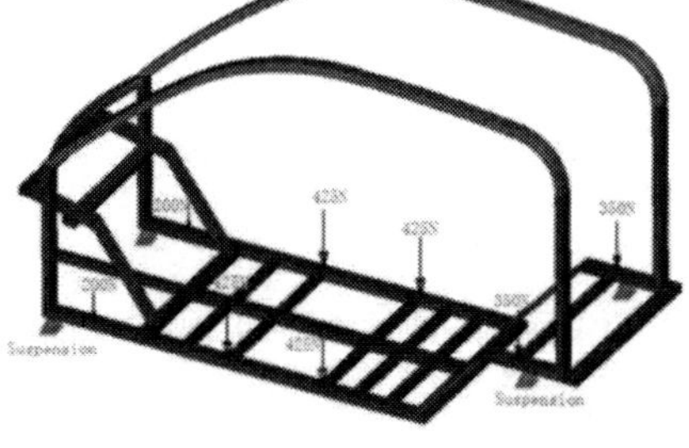

Fig. 3. The boundary conditions of electric car body frame

According the car model suspension, seat, power, occupant layout (Shown in figure 3) ,set the boundary conditions of the body frame. The total weight of driver and battery are 1700N, vertical load in the central body frame with 4 average concentration (425N).Weight of rear passenger is 700N, vertical load in the rear body frame with 2 average concentration (350N).Weight of front mechanism is 400N, vertical load in the front body frame with 2 average concentration (200N).The freedom restricted position of front and rear suspension is the car suspension lug actual installation location.

(1)Horizontal bending condition 678

Setting up forces on the body frame under the horizontal bending condition, DOF constraints suspension system.Constraint nodes at the left front suspension UX, UY, UZ freedom; Constraint nodes at the right front suspension UY, UZ freedom;Constraint nodes at the left rear suspension UX,UZ freedom; Constraint nodes at the right rear suspension UZ freedom,and release of other degrees of freedom.

(2)Emergency braking condition 67

The vertical load of this condition is the same as horizontal bending condition.Apply a 0.8g brake acceleration in the longitudinal direction of the model.Constraint nodes at the front suspension UX, UY, UZ freedom;Constraint nodes at the rear suspension UY, UZ freedom,and release of other degrees of freedom.

(3)Ultimate torsion condition 678

This condition is mainly to simulate the car round dangling or round up. Left and right wheels have height difference. The car body is subjected to greater torsion under this condition, and it is one of the most serious conditions of the car.

Left front wheel round dangling to simulate this working condition. Constraint nodes at the right front suspension UX, UY, UZ freedom; Constraint nodes at the rear suspension UZ freedom; Release all degrees of freedom at the left front suspension.

Figure 4 and 5 are the stress diagram, deformation displacement diagram under ultimate torsion condition.

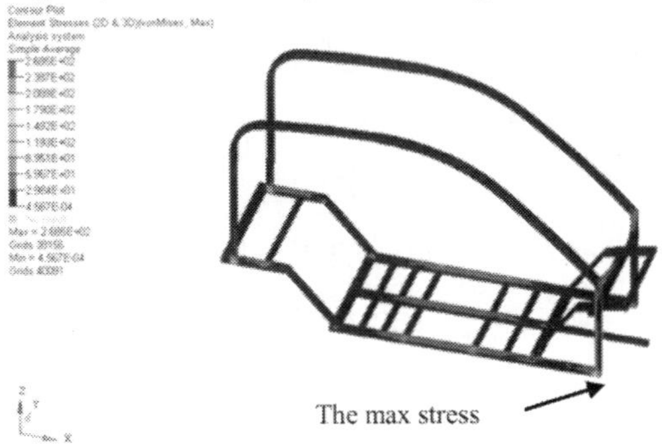

Fig. 4. Stress diagram of ultimate torsion condition

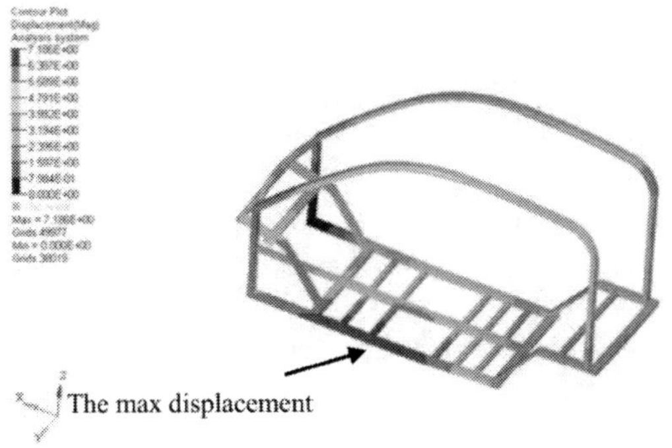

Fig. 5. Deformation of ultimate torsion condition

(4)Emergency turn condition 67

This condition is mainly to simulate the responsiveness of body frame at the lateral inertial force and vertical load when the car emergency turns.Apply a 0.4g lateral acceleration which simulates lateral load in the centroid of body frame.

Constraint nodes at the left front suspension UX,UY, UZ freedom; Constraint nodes at the right front suspension UY, UZ freedom; Constraint nodes at the rear suspension UY, UZ freedom, and release of other degrees of freedom.

Table 1 shows maximum stress and maximum deformation of the body frame in 4 typical working conditions.

Table 1. Comprehensive analysis of 4 typical working conditions

Condition	Max stress(MPa)	Max deformation (mm)
Horizontal bending	144	3.7
Emergency braking	268	7.2
Ultimate torsion	275	9.4
Emergency turn	147	4.5

3 Topology optimization in body frame

Based on the electric car body frame static analysis, the minimum body frame weight as the optimization goal, OptiStruct software was used to topology optimization design, get the optimal material distribution program of body frame structure. Finally HyperWorks and Radioss were used to static analysis of body frame before and after optimization.

3.1 Topology optimization

Figure 6 is topology optimization design process of an electric car body frame 8.

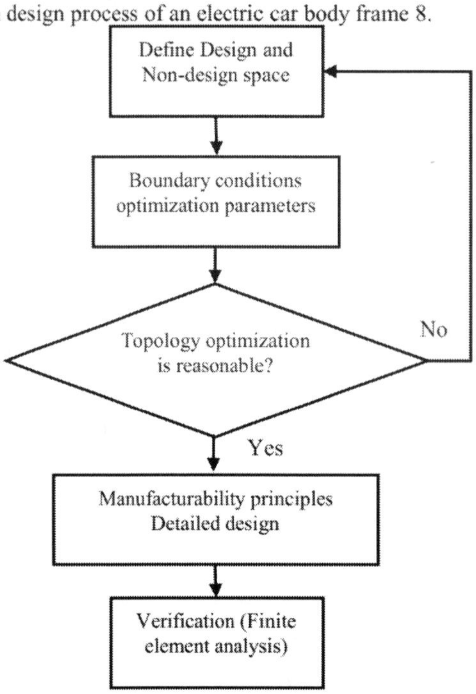

Fig. 6. Design process of topology optimization in body frame

Based on design process of topology optimization to define the design space and non-design space.Non-design space is some have been identified in the structure, which will not be changed in the design process.In Figure 7 (a) is non-design space. Figure 7 (b) is design space(part of the blue region in the model).

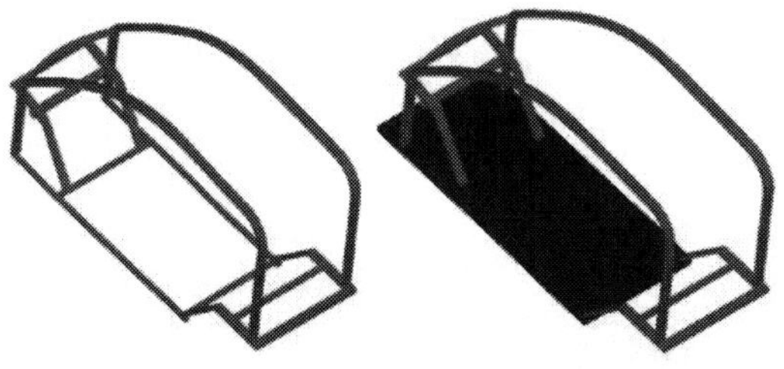

(a) Non-design space (b) Design space(Part of the blue area)

Fig. 7. The non-design space and design space of body frame

The car body is subjected to greater torsion under ultimate torsion condition, and it is one of the most serious conditions of the car.To meet the stiffness requirements of ultimate torsion conditions,we set the boundary conditions, optimization parameters and submit calculation, then get the topology optimization program.

Accordance with the previous ultimate torsion condition set boundary conditions.Optimization parameters include design variables, response function, design objective and design constraints.Element pseudo-density is design variables, the total mass and nodes(Position of arrow in Fig. 5) in the Z-direction displacement are the response functions.Minimum total mass of the model is design objective.The node maximum allowable displacement 7.2mm is design constraint.By constraining displacement of the body frame to ensure rigidity of the body frame.Optimization mathematical model is as follows.

$$\begin{cases} Min W(\beta_i) = \sum_i \beta_i \rho_i v_i \\ s.t. \begin{cases} d_j \le 7.2 \\ \beta_i \in [0,1] \end{cases} \end{cases} \quad (1)$$

W is the total mass of the model; β_i is design variable; ρ_i is element initial density of number i ; v_i is element volume of number i ; d_j is nodes in the Z-direction displacement.

This optimization process takes 23 iterations,and fig.8 shows topology result.After analysising this topology result, some nodes maximum deformation displacement of body frame reach 16 mm.So this topology optimization program has failed to fully meet the requirements of the initial stiffness of the body frame.

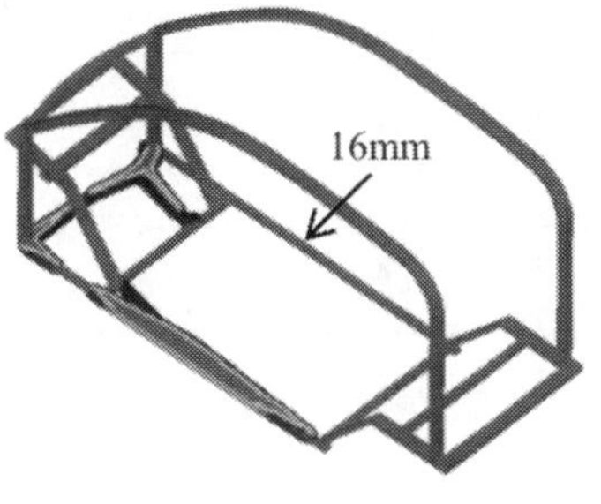

Fig. 8. Topology optimization of single-node constraint in body frame

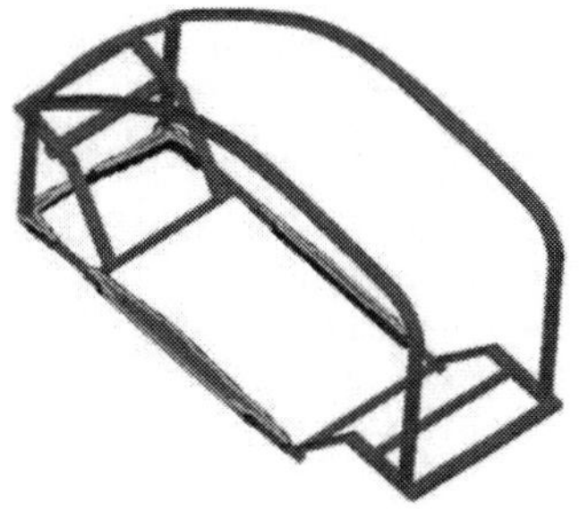

Fig. 9. Topology optimization of multi-nodes constraint in body frame

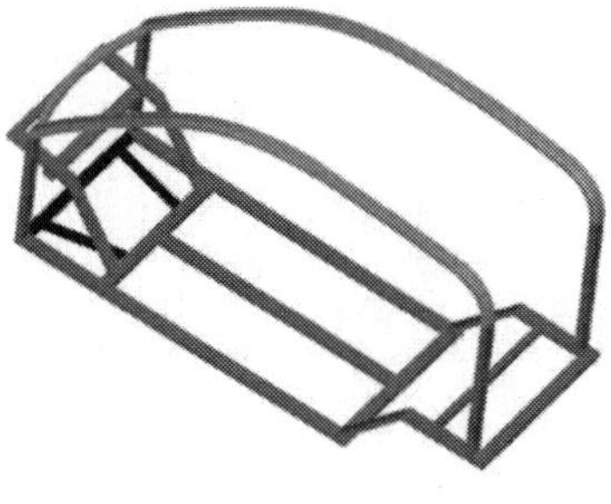

Fig. 10. Detailed design of body frame

According to the first results of the topology optimization, adjust optimization parameters. Element pseudo-density is design variables, the total mass and nodes(Positions of arrow in Fig.5 and Fig.8) in the Z-direction displacement are the response functions.Minimum total mass of the model is design objective.The nodes maximum allowable displacement 7.2 mm is design constraint.

The second optimization process takes 23 iterations,and fig.9 shows topology result.After analysising this topology result, all nodes maximum deformation displacement of body frame less than 7.2 mm.Topology optimization program meet the requirements of the initial stiffness of the body frame.

Based on two times topology optimization results about material distribution of body frame, and according to the principle of manufacturability detailed design the body frame.Figure 10 shows detailed design of body frame. The total mass of the lower part of the body frame is 14.49Kg, a total of 8 beams after optimization.

3.2 Finite Element Verification and contrast

HyperWorks and Radioss were used to static analysis of body frame before and after optimization under four kinds of typical working conditions(horizontal bending condition, emergency braking condition, ultimate torsion condition, emergency turn condition),from the stress, deformation, weight and other aspects of the calculation results were compared.

Table 2. Comparison of static strength stiffness before and after optimization

Condition		Max stress (MPa)	Max deformation (mm)
Horizontal bending	Before	144	3.7
	After	143	3.7
Emergency braking	Before	268	7.2
	After	279	7.4
Ultimate torsion	Before	275	9.4
	After	286	9.0
Emergency turn	Before	147	4.5
	After	144	4.6

The results showed that after optimizion the body frame meet the strength and stiffness requirements under four kinds of typical working conditions. The total mass of lower part of the body frame reduced from 17.88 KG to 14.49 KG,reduced 18.96%; The number of beams reduced from 17 to 8,and the length of the weld was ignificantly reduced.

Literature 8, 9 and 10 verify the effectiveness of above design process of topology optimization in body frame. The way which apply Optistruct software on the body frame to topology optimization design is an effective way to solve the problem of automobile lightweight.

4 Summary

This paper chooses the electric car body frame as the research object, using Radioss software on the body skeleton four typical conditions of static analysis.Radioss software was used to make static analysis on the body frame under four kinds of typical working conditions.OptiStruct software was used to topology optimization design, get the optimal material distribution program of body frame structure.Finally HyperWorks and Radioss were used to static analysis of body frame before and after optimization, from the stress, deformation, weight and other aspects of the calculation results were compared.The results showed that after optimizion the body frame meet the strength and stiffness requirements under four kinds of typical working conditions. The total mass of lower part of the body frame reduced from 17.88 KG to 14.49 KG, reduced 18.96%; The number of beams reduced from 17 to 8,and the length of the weld was ignificantly reduced.

References

1. H.S. Ho, B. Lui, X.H. Xing, Parametric Shape and Topology Optimization with Radial Basis Functions and Partition of Unity Method.American Institute of Physics, (2010)
2. H.S. Lee, A structural analysis and topology optimization on cylinder block of heavy duty diesel engine.International Journal of Modern Physic B., **24**, 15(2010)
3. Nilsson, F. Birath, Topology Optimization Of A Stamping Die.Materials Processing and Design, (2007)
4. F.Gerold, K.Beucke, F.Seible, M. Asce, Integrative Structural Design.Journal of Computing in Civil Engineering, **26**, 6(2012)
5. S. Amstutz, Analysis of a level set method for topology optimization.Optimization Methods & Software, **26**, 4(2011)
6. Y.B. Zhang, Full load bus body design analysis and research.Zhejiang University,(2008)
7. X.N. Zuo, Analysis and lightweight research of coach body frame.Harbin Institute of Technology,(2011)

8. X.Y.Xu, Y.J.Guo, Stiffness constrained topology optimization for body structure of electric low speed car. Machine Design and research, **26**, 2(2010)
9. W.J. Fan, Z.J. Fan, R.Y. Su, Research on Multi-objective Topology Optimization on Bus chassis frame.Journal of mechanical engineering, **19**, 12(2008).
10. W.J. Fan, Z.J. Fan, Muti-stiffness topology optimization of bus frame with multiple loading conditions.Automotive Engineering, **30**, 6(2008)

Optimization Design Of Setback Generator For Initiating Explosive Devices

Qiao Lu[a], Liming Li[b] and Guofu Yin[c]

State Key Laboratory of Applied physic-Chemistry Research, China

[a]1185256995@163.com, [b]biq213@163.com, [c]yinguofu521@126.com

Abstract. This paper describes an optimized design of setback generator based on the smart miniature initiator's demand for physical power, Through the Hopkinson bar impact experiments, we conducted a comparative study of multiple sets of high overload experiments and concluded that axial magnetized permanent magnets with the homopolar opposite and distribution winding can improve generator's power generation efficiency to certain degree. When axial magnetized permanent magnets with homopolar opposite and distribution winding are used to assemble the optimized setback generator, it can detonate ignition head consistently at 20000G acceleration.

Keywords: Initiating explosive devices; physical power source; setback generator.

1 Introduction

The complexity of the modern battlefield requires higher performance of weapon system, as an important component of the weapon system,the initiating explosive devices being developed toward to the intelligent, miniaturization direction[1]. The missile-borne power supply release electric energy as the acceptable energy for stimulating Initiating Explosive, currently most of the Initiating Explosive uses missile-borne battery as their Power supply. With the continuous development of Initiating Explosive, also spawned a variety of ways to provide power in addition to onboard battery power supply[2].

Physical power is being paid more attention because it has a lot of outstanding features such as non-destructive testing, reusability, long-term storage, withstand high and low temperatures, high security and others[3]. As an important physical power,setback generator plays an important role in Initiating Explosive applications. The setback generator is a pulsed power supply.When the device is under the influence of the recoil,the magnet in the device is doing axial linear motion relative to the coil. So that the magnetic flux through the coil is changed, which generate induced electromotive force and induced current in the coil. But due to the setback generation device using pulse power, which produces less energy. A cylindrical structure setback generator has been designed by Korea Agency For Defense Development, which have a diameter of 17.7mm, a height of 4.1mm, and a weight of 6.0g. This setback generators test results show that a voltage of 4.6V was charged on a capacitor of 100F within the charging time of 0.56ms under 30000g accelerations [3].

2 Setback Generator Ignition System Design and Theoretical Basis

This paper describes a diagram of missile-borne setback generator which provides power design. In the system, the setback generator firstly generates pulse voltage in the function of recoil, and then pulse voltage is stored in the

DOI: 10.1051/matecconf/201710001017

capacitor through the rectifier device and a voltage stabilizing device. Capacitor power can supply power for the control circuit. And explosive initiation energy can also be provided by it in the case of an open security. The whole system is controlled by the microcontroller.

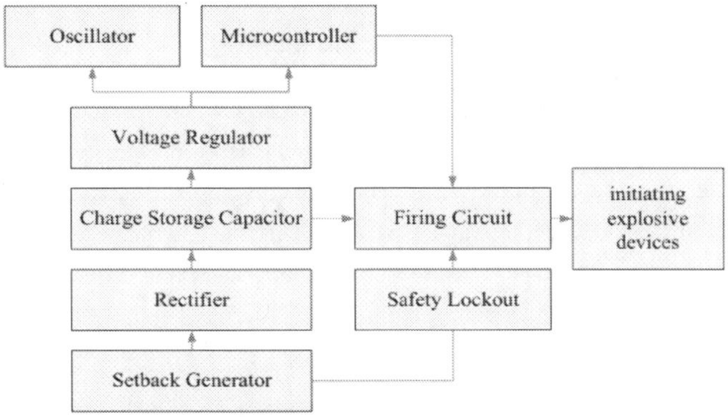

Fig. 1. Block diagram of power supply system

The setback generator is the key of the whole power supply system described above.When setback generator in working process, because the permanent magnet around the magnetic field distribution is non-uniform, and there is a relative motion between the coil and the magnet,through analysis of the Faraday's law of electromagnetic induction we can conclude that ,in this producing energy process, the electromotive force generate both induced electromotive force and motional electromotive force.In this paper, any one turn coil' s induce electromotive force generated by the expression:

$$E_l = E_i + E_m = -\int_A \frac{\partial \vec{B}}{\partial t} \cdot d\vec{s} + \oint_L (\vec{v} \times \vec{B}) \cdot d\vec{L} \tag{1}$$

For the N turns of the coil, the total induced electromotive force is generated in this process equals to the sum of each single coil induced electromotive force, expressed as:

$$E = \sum_{i=1}^{N} E_i \tag{2}$$

3 Experiment and Analyses

The disadvantage of low energy output has been restricting the development of setback generator, the experiments will be conducted to design new structure for the permanent magnet part and the coil part of the device, which aims to improve its performance. The reliability of initiating explosive device will be improved when it is integrated with optimized setback generator [5].

Hopkinson Pressure Bar (SHPB) is one of the most basic experimental apparatus in the material dynamics.It mainly measures the dynamic stress and strain relationship of the material,and it is widely used in the dynamic mechanical properties of the materials at home and abroad[4].

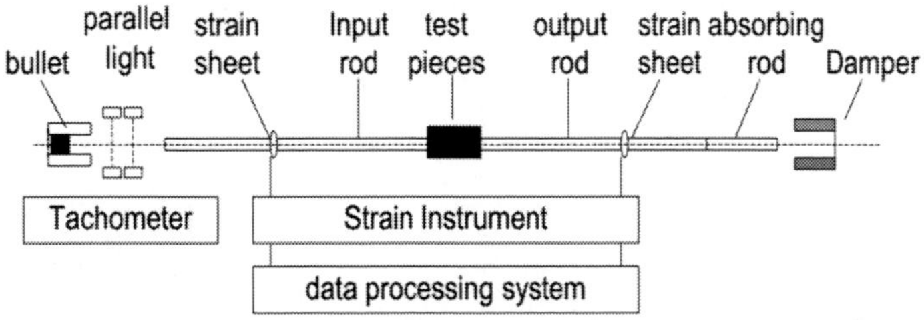

Fig. 2. Hopkinson bar system diagram

DOI: 10.1051/matecconf/201710001017

In this study we need to use the input rod of Hopkinson bar to directly hit impact on the setback generator in order to roughly simulate the ammunition launching process.When setback generator in the high acceleration of the environment, the permanent magnet and the coil in it will produce a high speed relative displacement and produce energy.The energy waveform is recorded and is used for comparison and analysis.

3.1 Comparative experiment of permanent magnet structure

Magnetization direction of the permanent magnets is generally divided into two types, they are radial and axial magnetization. This attribute has already been decided in the production process of permanent magnet. In addition to the magnetization direction, the topology of the permanent magnet is also an important factor affecting the output of setback generator. Based on their different topologies axially magnetized ,permanent magnets can be further divided into： a single length of the permanent magnet, two short permanent magnets of different poles opposite and two short permanent magnets with the homopolar opposite.

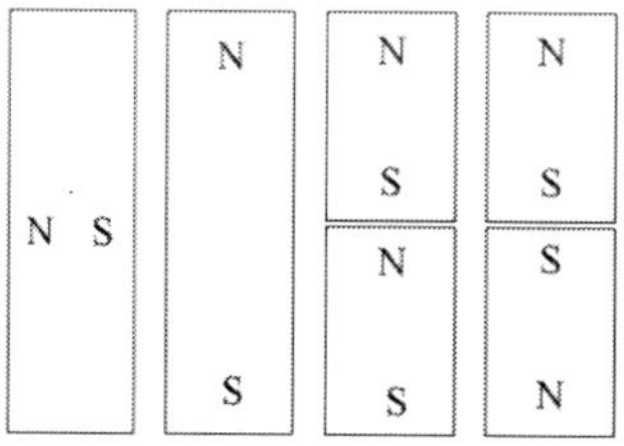

Fig. 3. Permanent magnet for testing

According to the different magnetizing direction and topology, we designed above four kinds of permanent magnet structures for comparison. We conducted the Hopkinson bar impact test without changing the other parameters. The output cable of the setback generator was connected with the Oscilloscope in order to collect no load voltage. And the data was compared and analyzed.

Experimental results and analysis: Based on the magnetization direction of the permanent magnets and the topology will be four groups of experiments were conducted. The voltage waveforms of the measured waveforms are shown in the following figure.

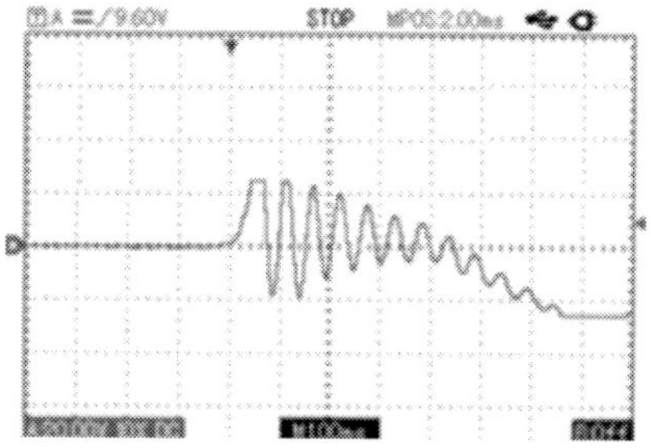

Fig. 4. Radial magnetization

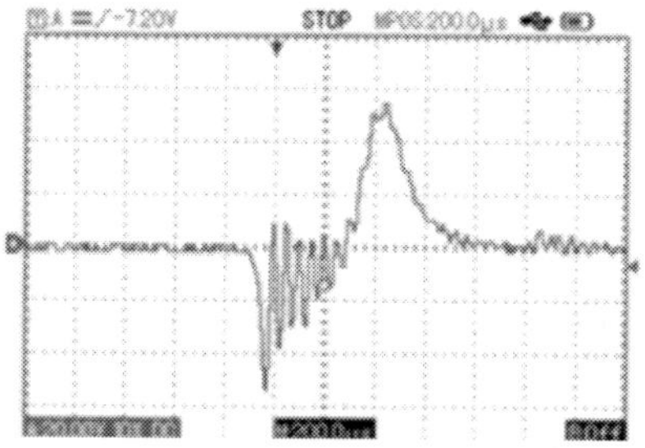

Fig. 5. Axial magnetization

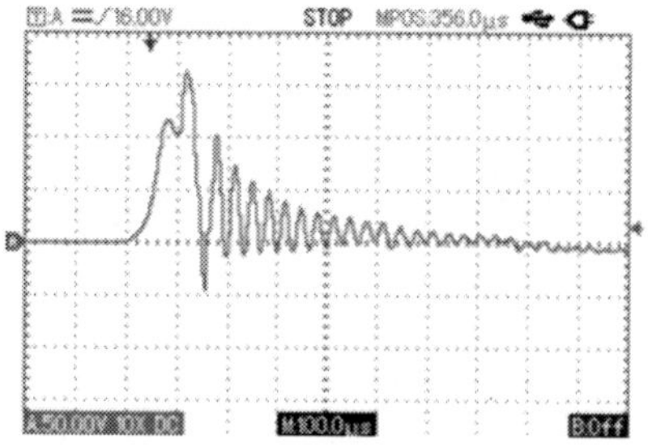

Fig. 6 Different polar opposite

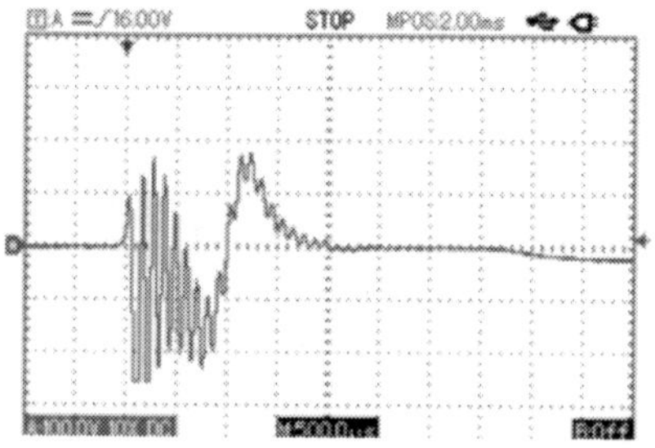

Fig. 7 Homo polar opposite

We made some conclusions through experiment:

(1)The setback generators which use the radial magnetized permanent magnet produce less energy, but longer discharge duration.

(2) Using axial magnetized permanent magnet setback generation compared to using a radially magnetized permanent magnet setback generation will produce more energy, but relatively short duration of the discharge.

(3) For axially magnetized permanent magnet, the setback generator which use magnets with the homopolar opposite have shorter discharge time, but the discharge waveform more tidy, complete and higher quality discharge.

3.2 comparative experiments of coil winding structure

Depending on the circumstances, coil winding method can be divided into the distribution winding method and non-distribution winding method. Multi slot and multi group coil mode is used in the distribution winding method, using series connection between each group. It found that using the distribution winding method can to some extent improve the generation efficiency of setback generator in the early simulation work, this conclusion is also required Hopkinson Bar experiments to validate it.

Fig. 8. Distribution winding and non-distribution winding

As shown in the above two kinds of different winding ways of the coil. In order to reduce the eddy current effect in power generation, the plastic is used as the skeleton of the coil. Non distribution winding height 16mm, diameter 10mm, with a diameter of 0.06mm enameled copper wire winding, winding turns 8000 turns, the resistance is 2.2KΩ.The distribution winding on the right using two smaller superposition of coil windings in series, each winding coil height 8mm, diameter 10mm, with a diameter of 0.06mm enamelled copper wire, is wound turns 4000 turns, the resistance is 1.1 KΩ.

In order to compare two winding method, the axial charge magnetism homopolar opposite permanent magnets was selected as the experiment permanent magnet. Respectively, we use distributed windings bad and non-distributed coil winding to conduct Hopkinson bar impact test. The results were compared and analyzed.

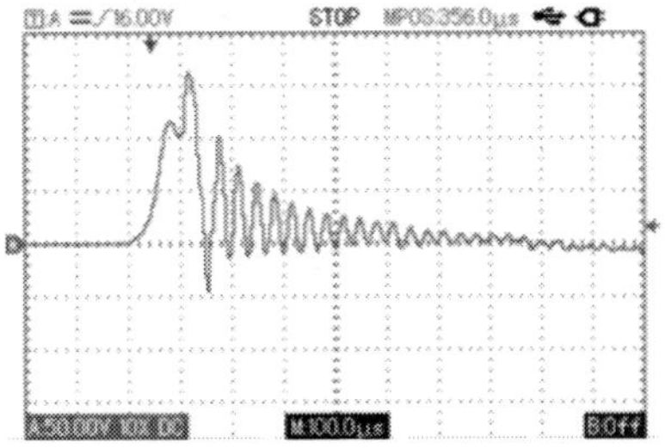

Fig. 9. Non distribution winding

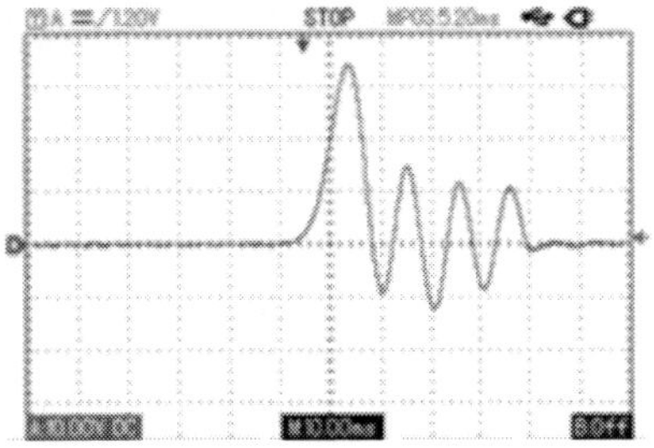

Fig. 10. Distribution winding

We made some conclusions through experiment:

(1) The setback generator with distributed windings will obviously produce a higher voltage, the energy conversion efficiency is higher.

(2) The discharge time of the setback generator using the distributed winding is longer than that of the setback generation using the non-distributed winding.

(3) The voltage waveform of the setback generator with the distribution winding is more complete. It shows that the distribution winding method can effectively eliminate the harmonic and improve the power quality.

3.3 setback generation device detonated igniter test

According to the permanent magnet experiment and coil experiment, we selected axial charge magnetism homopolar opposite permanent magnet and a distribution winding coil to integrate an optimized setback generator, and then connect the setback generator with follow-up circuit and ignition igniter test.

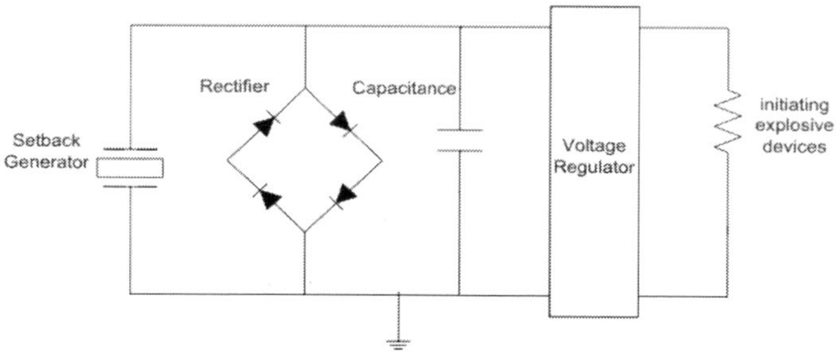

Fig.11. Ignition Experiment circuit diagram

Hopkinson bar produced an acceleration 20000g overload impact generating device. The energy generated flow into the capacitor through the bridge rectifier, and then detonated igniter after the regulator. A total of five trials were carried out, and all the ignition was successfully initiated.

4 Summary

Changing the structure of permanent magnet and the coil winding method can effectively improve production efficiency of the setback generating device. Most favorable factor for power generating efficiency is to have permanent magnet with axial magnetizing direction and homopolor opposite topological structure. As far as the choice of coil winding method is concerned, the most favorable factor is to have the distribution winding way. When axial magnetized permanent magnets with homopolor opposite and distribution winding are used and integrated to assemble the generator, it can detonate ignition head consistently under 20000G acceleration.

References

1. Amirtharajah, Chandrakasan, Sci. Forum, **22-23** (2011)
2. Sang-HeeYoon, submitted to Journal of Materials Research (2005)
3. Cheng Hok, Aw, Yong Lim Thoma, U.S. Patent 656, 471 (2003)
4. Hu Shi-sheng, Wang Dao-rong, Journal of Engineering Mechanics, (2001)
5. Zhao Ge-xiao, Chen Zhi-yuan, Journal of Engineering Mechanics, (2009)

Oil metal particles Detection Algorithm Based on Wavelet Transform

Wei Shang [a], Yanshan Wang [b], Meiju Zhang [c] and Defeng Liu[d]

AVIC Beijing Changcheng Aeronautic Measurement and Control Technology Research Institute, Beijing, China
[a]shangwei987654@163.com, [b]wangys@263.net, [c]zhangmj_bmc@163.com, [d]liudf@avic-aim.com

Abstract: In order to observe the real-time abrasion status of the aero-engine, we need to monitor the lubrication system online. As the aero-engine operating time and running state changes, the concentration, composition, size and other parameters of the metal debris can show different changes. They can be used as an important indicator to reflect the state of the aero-engine fault. However, due to the influence of electromagnetic, vibration disturbance and random noise signal introduced by the processing unit itself, the metal particles signal tend to comprise noise. Oil metal particles detection algorithm based on wavelet transform, utilizes the optimized localized nature in time domain and frequency domain of wavelet transform and the characteristics of multi-resolution analysis, combined with the signal characteristics in actual aero-engine condition to realize noise reduction and detection, while validating the algorithm using real experimental data. The result shows that noise can be effectively decreased and signal characteristics can be detected correctly.

Keywords: Sensor, metal debris, wavelet transform, variance iterative estimation, feature extraction.

1 Introduction

With the rapid development of sensor technology, oil metal particle on-line monitoring sensors have been widely used in aviation, marine power equipment and other online monitoring fields [1]. For example, in aero-engine lubrication system, the content of metal particles, as an important indicator, is often used as the basis for judging the status of the engine [2, 3]. Oil metal particles on-line monitoring sensor can not only detect size of metal particles but also distinguish ferromagnetic and non-ferromagnetic metal particles. The parameter of metal particles can also be used to analyze the health state of the aero-engine.

2 Oil particle detection algorithms

2.1 Structure of the oil metal particles on-line monitoring sensor

Oil metal particles on-line monitoring sensor comprises sensor probe, signal transmission cable and the signal processing unit, the overall connection shown in Figure 1.

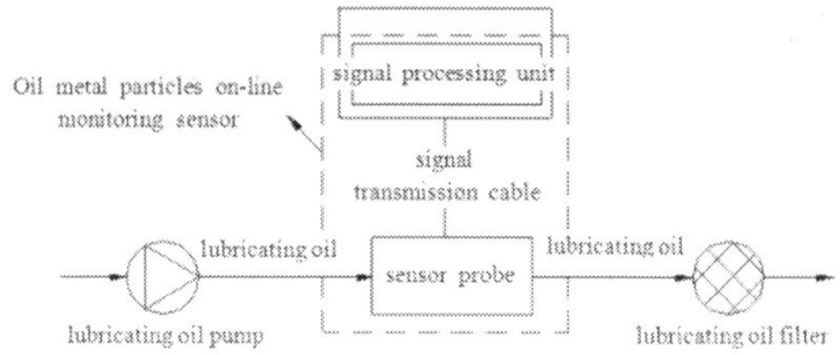

Fig.1. Oil metal particle on-line monitoring sensor overall connection

Sensor probe is installed in the engine oil pipe from filter oil booster to supply pipe, oil can pass through the sensor probe in order to detect metal particles, and signal transmission cable is installed between the sensor probe and the signal processing unit. Sensor probe induce particles signal mainly based on the principle of electromagnetic induction. When a metallic particle passes through the sensor, it generates a full period of sine-like waveform which is considered as particle signature(shown in Figure 6(a) or Figure7(a)). The amplitude of the sine-like waveform signal is proportional to the mass of ferromagnetic and the surface area of non-ferromagnetic, and the phase of them are opposite. Thus, we can obtain the size and the type of the debris by extracting amplitude, phase and frequency information.

However, on-line monitoring sensors are often exposed to vibration interferences during their operations, which will severely reduce the SNR[4] .Thus, it is difficult to estimate the real particle size due to the interference on the sensor. Since metal particle detection is an important aspect of sensor performance, achieving signal from the complex signals is particularly important. This calls for effective de-noising methods to extract the oil debris signature from noisy sensor output data. On the other hand, an ideal de-noising method can completely remove the noise component without any loss in the integrity of the target signal component. However, it is almost impossible to achieve such a goal in real applications. As described above, the oil metal particle monitoring sensors are often exposed to vibration interferences, and the signal is low-intensity and high-frequency wideband, so we can de-noising under the premise of ensuring the energy of signal.

Since oil metal particle on-line monitoring sensor signal is sparse, low frequency, sinusoidal and strong noise, we use a metal particle detection algorithm based on wavelet transform, make full use of the advantages of iterative least squares method and variance estimation algorithm for de-noising, and then complete signal detection according to the true characteristics of signal.

2.2 Algorithm overall Design

Depend on the result of signal processing unit design, the sensor signal after AD sampling will produce two route valid signal—the left and the right channel signal, they are respectively sensitive to ferromagnetic particle and non-sensitive ferromagnetic particle. When ferromagnetic particles pass through the sensor, the left and right channel can both generate cosine signal and the amplitude of left channel signal is larger, when the non-ferromagnetic particles pass through the sensor, left and right channel respectively generate cosine and sine signal and the amplitude of right channel signal is larger. The relationship between signal amplitude and phase can be as the basic aspect for signal detection and extraction.

Figure 2 shows the basic flow of data processing. Firstly, left and right channel respectively obtain the original data acquired by AD, then using the least square method to remove baseline wander, eliminating noise by wavelet noise reduction method, using nonlinear energy operator to test ferromagnetic and non-sensitive ferromagnetic particle, finally communicating the particle data up to host computer.

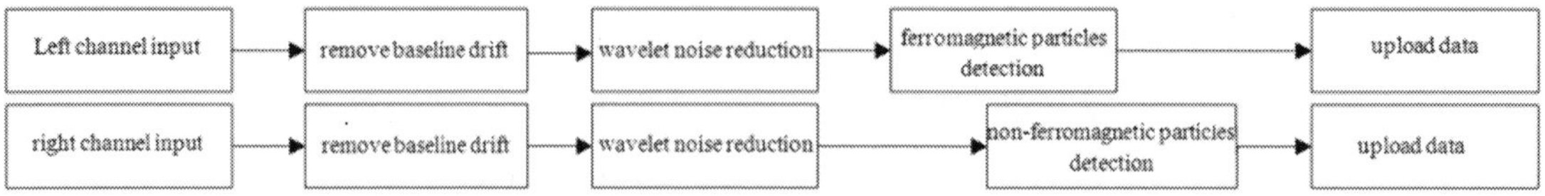

Fig.2. Signal handler Basic Flowchart

2.3 Wavelet Analysis

Figure 5(a) shows a ferromagnetic particle signature collected from oil metal particle on-line monitoring sensor in a vibration-free environment. If a particle signal is too weak, it may be masked by noise and cannot be detected directly. As such, we adopt a wavelet transform approach to detect particle footprint in a noisy environment. Wavelet transform analysis is a very effective tool for time-frequency analysis [5], the most prominent feature is that it has good localization properties and characteristics of multi-resolution analysis in the time and frequency domain. Multi-resolution analysis characteristic signal can be displayed in different levels, and its essence is to restore signal on a range of different levels of space. This restoration can decomposition a variety of signal containing different frequencies into the sub-signal with different frequency bands, which can be effectively used in many signal processing problems.

We can use two sets of filter coefficients $\{h_n\}$ and $\{g_n\}$ to decompose signal into rough section and detail section, $\{h_n\}$ is low-pass filter and $\{g_n\}$ is high-pass one. Rough section corresponds to the low-frequency portion of the signal, which represents the main outline of signal, detail part corresponds to the high frequency part of the signal. The two sections are the performance of signal by the wavelet transform in scale 1. For example, if the signal frequency is f_s, the rough part through $\{h_n\}$ will obtains the result which frequency is between 0 and f_s / 2, and detail part will obtains frequency between f_s / 2 and f_s. Then do the same operation on rough section and obtain the result in scale 2. And so on.

According to the sampling rate of sensor, the decomposition level of oil metal particle on-line monitoring sensor is 7, and according to the frequency range of signal, making the first five layers of detail coefficients set to 0 in order to remove the high frequency components.

Wavelet de-noising is based on Wavelet transform and it is frequently be used because of its outstanding advantages. Firstly, the original signal is decomposed into multiple scales by using wavelet transform, and then depends on different nature in wavelet coefficients of signal and noise, we can use corresponding roles in signal processing. The essence here is to reduce the coefficient generated by noise, as well as maximizing retaining real signal coefficients, and finally reconstruct the original signal based on wavelet coefficients. The result of the processing of noisy wavelet coefficients greatly affects the signal noise reduction effect, so it is an important part of the noise reduction process. This is described in detail in 2.5.

2.4 Remove the trend component

From above, the signal always includes low frequency trend component due to acquisition error , vibration interferences and other reasons. In normal circumstance, this trend component is usually negligible. However, when dealing with wavelet processing, due to the operational characteristics, low-frequency component is often amplified. It always causes a significant shift so we need take appropriate methods to remove these drift components.

The classical de-trending process such as least-squares scheme can often removes the trend effect by finding the best-fit line and then subtracts it from the original signal. Here, we use this method to remove the trend component. We assume a polynomial trend item and use least squares equation theory to make formula, then take matrix method to solve out the matrix coefficients and obtain trend term, finally original signal can be derived by subtracting the trend term useful signal[6].

Assuming the input data is X_{in} [N], the amount of data is N, use U_i to fit trend term

$$U_i = \sum_{i=0}^{N-1} b_k {X_{in}}^k(i). \tag{1}$$

Where is the polynomial coefficients. According to the principle of least squares error, the error between estimated value and real value is

$$E = \sum_{i=1}^{N} (X_{in}(i) - U_i)^2. \tag{2}$$

We solve out the minimum error in accordance with the least squares method and solve its partial derivative, then make it zero, sorting out the corresponding equation, using the matrix method for solving polynomial coefficients. In the sensor system described in the present, using a polynomial, obtained trend coefficient after solving, then coming to the mathematical expression

$$Y_{out}(i) = X_{in}(i) - \left(a + b * \left(i - \frac{N}{2}\right)\right). \quad (3)$$

$$a = \frac{\sum_0^{N-1} X_{in}(i)}{N}. \quad (4)$$

$$b = \frac{\sum_0^{N-1} X_{in}(i)(i-a)}{\sum_0^{N-1}(i-a)(i-a)}. \quad (5)$$

$Y_{out}(i)$ is the data item after de-trending.

2.5 Wavelet domain coefficient adjustment

The key of wavelet filtering is removing noise component in frequency domain. In the literature, some de-noising methods mainly focus on maintaining the integrity of the target signal component and remove only part of the noise component. These methods are called under-de-noising approaches which may lead to false alarm due to certain spurious noise components such as vibration remaining in the partially denoised data. On the contrary, another kind of methods primarily concentrates on the removal of noise component with less regard to the target signal integrity. In this case, part of the signal component could also be removed, leading to distortion or even loss of the particle signature. Considering of the characters of sensor, we point out a noise variance iterative estimation method which is carried out to removing large deviation from the mean weight, thus gathering signal energy.

2.5.1 Iterative noise variance estimation

Variance represents the deviant degree to average value, the part which has larger deviation is considered correlation-weak to valid signal and is not enough to reflect the signal characteristics. Here point out variance iterative estimation method, do the nonzero iterate calculation on wavelet decomposition coefficients, thus can be more accurate to estimate the noise variance. Here we can use threshold to measure the extent.

Firstly over all N wavelet coefficients s (n) at layer j (j=6 and 7)and rough wavelet coefficients separately with the M-length window and with 50% data overlap, then calculate all window variance. For example, in the detection system, N takes 2048, M takes 128, so the data is divided into 32 segments, each of 128 data, then calculate each variance, select the minimum v_{min}, using the following formula to calculation threshold

$$T = \sqrt{2 * v_{min} * \ln(N)}. \quad (6)$$

Then maintain the data which is between –T and T, thus constitute a new array and then do the same processing as described above, iterative 3 times, finally obtain the variance of the data retention.

The amplitude of valid signal is relatively concentrated through wavelet transform, so after removing the data outside the threshold, it can effectively remove large deviation from the overall level of the noise signal, the energy will be focused on preserving the data valid signal in order to achieve noise reduction purpose.

2.5.2 The improve coefficient adjustment method

There are three major wavelet signal extraction methods that are respectively based on modulus maxima, adjacent scale correlation and threshold shrinkage principles. Since the sensor output data is a mixture of debris signature, white noise and correlated vibration interference, it is very difficult to directly use modulus maxima or correlation-based methods. The threshold shrinkage in the wavelet domain is the most widely used due to its simplicity and effectiveness. There are two basic threshold shrinkage methods. There are hard threshold shrinkage and soft threshold shrinkage. Hard threshold adjustment can achieve a smaller mean square error after signal de-noising, but it has breakpoints, which leads to the shock point mutation after signal de-noising, when there are higher noise level, this phenomenon is particularly obvious. Soft threshold adjustment can achieve greater SNR after signal de-noising, but the wavelet coefficient after Soft threshold adjustment existent constant deviation, which will directly affect the degree of approximation of the reconstructed signal and the real signal, and it can bring inevitable errors to reconstructed signal.

In view of the advantages and disadvantages of threshold processing, here we use an improved threshold processing method. If $swad(i) > T$, then

$$swad(i) = \begin{cases} swad(i) - \frac{2*T}{1+e^{k*(swad^2(i)-T^2)}}, swad(i) \geq 0 \\ \frac{2*T}{1+e^{k*(swad^2(i)-T^2)}} - swad(i), swad(i) < 0 \end{cases}. \quad (7)$$

Else $swad(i) = 0$. T is the result in 2.5.1, $k = 1, e = 2.718$, swad(i)represents detail coefficients and rough wavelet coefficients at layer 6 and 7. This method can effectively improve the breakpoint effects which is induced by hard threshold adjustment method, at the same time, respect to the soft threshold adjustment, it can ensure better continuity and higher accuracy, thus can achieve better filtering purpose.

2.6 Particle detector

Here, particle detection method is based on the signal characteristics of oil particles, using nonlinear energy operator which can simultaneously reflect the instantaneous frequency and amplitude information of the signal. Output is proportional to the product of the amplitude and frequency of input signal. For signal sequence X (i)

$$\varphi(X(i)) = X(i) * X(i) - X(i-1) * X(i+1). \quad (8)$$

After nonlinear energy operator processing, signal output is larger at abrasive location, here we can set the threshold T to initially screened abrasive location. We get T according to the following formula

$$T = C * \frac{1}{N} * \sum_{n=1}^{N} \varphi(X(n)). \quad (9)$$

Here, C is the empirical constant, according to a large number of statistical results in this project, takes 2. Select the part which signal energy is above T to initially regard as the effective particle signal.

Then do further choose to the position which is obtained by non-linear energy operator, It includes: abrasive speed, symmetry and amplitude. Here the left channel is sensitive for ferromagnetic and it shows cosine phase which valley precedes peak in the post; the right channel is sensitive for non-ferromagnetic and it shows cosine phase which peak precedes valley in the post. By determining the particle morphology and symmetry to eliminate abnormal signal, then set the appropriate threshold based on the nature and magnitude, excluding the impact of mutations and small-amplitude interference signal.

3 Results and Discussion

In this section, we use oil particles online monitoring sensor signal detection algorithms mentioned above to test the actual data in a lab environment. The oil particles online monitoring sensor is shown in Figure 3.

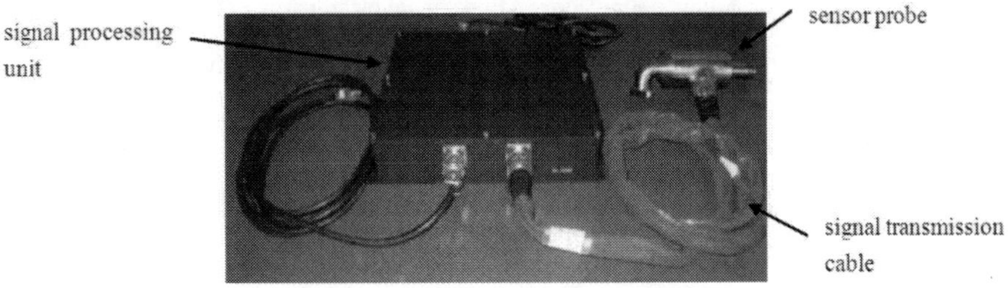

Fig.3. Oil particles online monitoring sensor

Take an 110μm ferromagnetic metal particle and a 508μm non-ferromagnetic metal particle which were embedded in a plastic sleeve (shows in Figure 4). In the test, in accordance with the requirements of sensor, respectively pump plastic sleeve through the sensor with the speed within 1 ~ 4m/s to ensure effective through of metal particles.

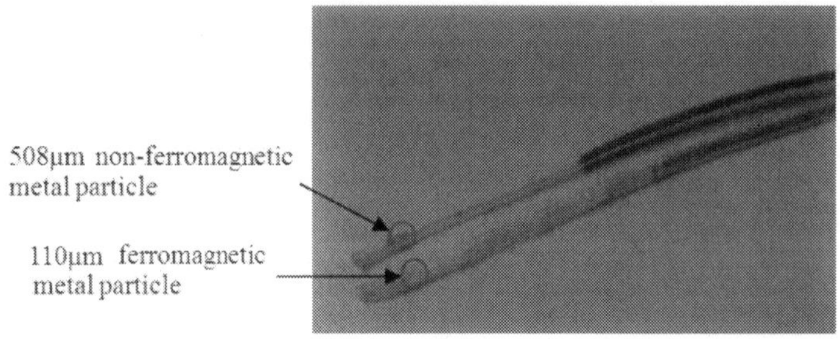

Fig.4. 110μm ferromagnetic metal particle and 508μm non-ferromagnetic metal particle

Connect the sensor system and give power, open PC software for testing. Smoke particles and observe original sensor signal and signal after processing, the result shows in Figure 5-6:

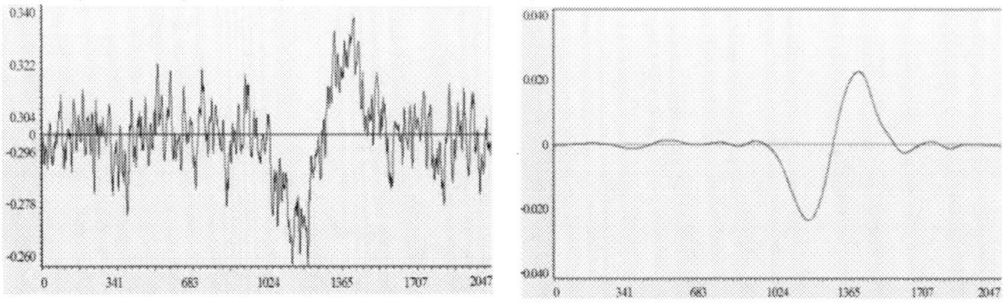

(a) (b)

Fig.5. 110μm Ferromagnetic metal particle

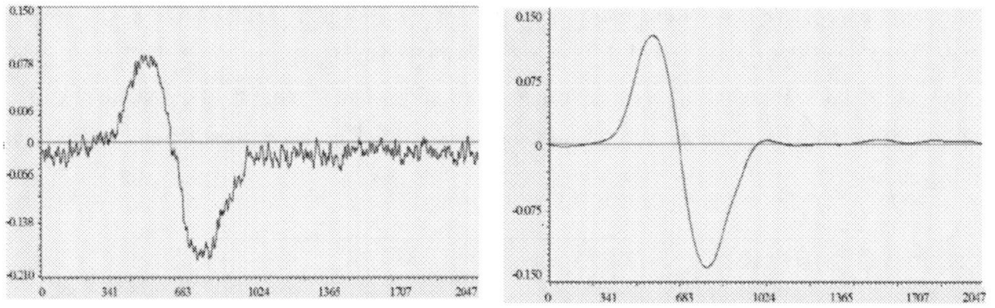

(a) (b)

Fig.6. 508μm Non-ferromagnetic metal particle

In Figure 5-6, (a) shows the raw data measured from the sensor, and (b) shows the data after disposing. As shown, the graphic above displays the original signal of metal particles, the graphic below shows the filtered signals after filtering, the indmax and indmin show the peak and valley position of waveform. As can be seen, the filtered signal has dropped baseline drift and has no noise. It can fully reproduce the original signal significantly, enhance signal to noise ratio, and can correctly point out the signal peak position.

4 Conclusion

A wavelet domain de-noising technique has been proposed to extract high quality debris signature for improving the measurement capability of the oil debris monitoring sensor. Metal particle detection algorithm has fully considered the impacts of both target and unwanted signals, it based on wavelet transform, using the least squares method to remove trend component, effectively avoiding to enlarge the low frequency component when doing wavelet transform. Other advantages of this method include the iterative variance estimation can effectively remove the remote components from the mean level of the signal. This method has been evaluated using sensor data, the results show the algorithm can effectively improve the oil debris monitoring accuracy in real applications where vibration interferences are inevitable.

So far, the application of the particle detection algorithm has successfully passed the test of the engine, and the present system has completed XX-engine two phases test and delivery.

References

1. J.L. Miller, D. Kitaljevch, In-line oil debris monitor for aircraft engine condition assessment, IEEE Aerospace Conference Proceedings, **6**(2000)
2. Q.B. He, X.X. Wang, Q. Zhou, Vibration sensor data denoising using a time-frequency manifold for machinery fault diagnosis, **14**(2014)
3. J. Wu, T.G. Liu, X. Tang, Qualitative ferrographic analysis method by quantitative parameter of wear debris characteristics, Industrial Lubrication and Tribology, **64**, 6(2012)
4. R.W. Kempster, D.B. George, Method and apparatus for detecting particles in a fluid having coils isolated from external vibrations, United States Patent, 5(1995)
5. Y. Zhang, S.R. Lee, H. Zhao, B. Wang, B. Liu, Research on wavelet analysis in fault signal processing. International Conference on Signal Processing proceedings, 1(2004)
6. S.C. Chan, H.C. Wu, K.M. Tsui, A new method for preliminary identification of gene regulatory networks from gene microarray cancer data using ridge partial least aquares with recursive feature elimination and novel brier and occurrence probability measures, IEEE Transactions on System, Man and Cybernetics, **42**, 6(2012)

Optimal Design of Hydraulic Fracturing for Deep Volcanic Reservoir in Zhungeer Basin

Hao Cheng [1], Bo Huang [2], Shijie Han [3,a] and Yanming Fu[3,a]

[1]Oil and Gas Development Company, PetroChina Xinjiang Oilfield Company, Karamay, China

[2]Engineering Technology Institute, PetroChina Xinjiang Oilfield Company, Karamay, China,

[3]State Key Laboratory of Oil-Gas Reservoir Geology & Exploitation, Southwest Petroleum University, Chengdu, China

[a]18583625735@163.com

Abstract. Deep volcanic reservoir in Xinjiang Zhungeer basin is buried over 4000m, the elastic modulus of reservoir rocks is high, hence it is difficult to initiate hydraulic fracture during fracturing operation; fracturing fluid filtration is serious and hard to be controlled due to the massive natural fractures in reservoir. Because of the serious heterogeneity along both horizontal and vertical direction, the optimum targets of design fracturing for different well layers are not the same. 4 out of 12 fracturing wells have not achieved the target amount of proppant. Ultimately, fracturing success rate is low and the production increase after fracturing shows a big difference among different wells. In order to improve the success rate and efficiency of deep volcanic reservoir fracturing, based on the analysis of the difficulties in previous fracturing operation, combined with the geological characteristics of reservoir, we optimized the key engineering parameters of deep volcanic rock fracturing, including perforation parameters, filtration parameters, fracture parameters and operation parameters. Results of fracturing design optimization bear an important guiding significance for improving the fracturing success rate and efficiency of deep volcanic reservoir in Zhungeer basin, Xinjiang province.

Keywords. Volcanic rock, facture fluid, filtration, natural fracture, hydraulic fracturing

1 Introduction

Deep volcanic reservoir is characterized by great burial depth and low matrix permeability. It is necessary to apply the hydraulic fracturing to attain an economic production [1]. Meanwhile, due to the volcanic reservoir develop a large number of natural fractures and dissolved pores, which has serious impact on the opening and expansion of hydraulic fractures, the existence and extension of multiple fractures limit the width of the opening hydraulic fracture [2-5], raised the risk of sand plug[6]. At the same time, the distribution of natural fractures complicates the fluid loss mechanism. Fluid filtration is mainly controlled by natural fracture, and the natural fractures network will change the propagating direction of hydraulic fractures, which will distort and intersect with each other, and finally turn into artificial fractures network [7-11]. Consequently, the opening width at hydraulic fracture tip would be severely restrained and the proppant at the fracture tip will form a bridge plug, lead to the failure of fracturing. Therefore, the corresponding fluid-loss-control measures should be taken to improve the success rate of fracturing [12]. According to the characteristics of volcanic rock reservoir, Vermani et al proposed the slug sand schedule, small diameter proppant, high viscosity fracturing fluid and other measures to improve the operation success rate of volcanic rocks fracturing [13]. Buried depth of volcanic reservoir leads to the high closure stress, and makes the proppant vulnerable under the long-term effect of the closure pressure [14], which may impair the hydraulic fracture conductivity. Additionally, the high stress can also lead to the conditions that the volcanic rocks cannot be cracked by existing equipment [15]. Besides the high stress, high temperature usually is an inevitable challenge, , for example, China's Daqing Xujiaweizi fault depression deep volcanic gas reservoir temperature ranged from 120 to 170℃[16], Temperature of deep volcanic reservoir in southern Song Liao basin even reach to 183℃[17], presenting a fundamental challenge to the selection of fracturing fluid.

Jinlong 2 reservoir is a typical deep fault block reservoirs, reservoir buried depth is about 4000m, the development of nearly north-south, east-west nearly two sets of faults, including three near NS faults: Jing201 west fault, Jinglong2 fault and Ke301 fault; seven near EW trending faults: Jing204 south fault, Jing213 south fault, Jing201 north fault, Jing207 south fault, Jing208 south fault, Jing214 south fault and Jing215 north fault. According core physical properties, average porosity of the reservoir is 10.82%, the average permeability is0.43mD, Based on seismic data prestack inversion to predict natural fractures development in northern; the degree of fracture development in southern is relatively weak; the natural fracture width is mainly distributed between the 0.01mm-0.19mm, which are mostly micro-fractures; fracture density is ranged from0.02m^{-1}-6.93 m^{-1}, the fracture dip angle is generally larger than 45°; the strike of natural fractures is East-West proximately. According to the indoor rock mechanics experiments, the elastic modulus of volcanic rock in Jinlong 2 reservoir is 30152MPa, poisson ratio is 0.186, the direction of maximum principal stress is 118°, with value of 82MPa, the minimum principal stress is 68MPa. In view of the geological characteristics of deep volcanic reservoir in the Zhungeer basin, the key fracturing parameters are optimized based on the analysis of the previous fracturing operation.

2 The characteristics of previous fractured well

Based on the fracturing operation of 12 wells in Jiamuhe groups, 4 wells did not complete the target amount of proppant, as is seen in Table 1:

(1) Success rate is low: Jing 218, 219, 203, 212 did not complete the target amount of proppant, which means that fracturing operation of 33.3% wells hasn't been completed thoroughly.

(2) Efficiency is low: 6 wells are dry layer or water layer after fracturing, which is half the total fractured wells, reflecting that fracturing parameters are required to be further optimized.

Combined with the geological characteristics of volcanic reservoir, the key features of fractured wells in early stage are:

(1) The fracturing effect of gas bearing block is high

Previous fracturing result show that fracturing effect better in two gas bearing fault blocks located in the high position of structure, such as Jing215 wells, Ke301 well(Jing214 fault block), Jing213 well and Jing201 wells(Jing201 fault block).

(2) The operational efficiency at edge and bottom of the reservoir is low

The early fracturing is inefficient for the Jing216 and the Jing203 wells located at the edge of the fault block; At the same time, the oil test results for the Jing215 well at 3940~3932m, Jing214 at 4210~4205.5m, Jing209 at 4350~4334m, and Jing209 at 4324~4302m are either dry layer or water layer.

(3) The well fracturing effect in natural fracture development zone is good

The fracturing efficiency and stimulation effect of wells that located in Jing208 fault, center part of the Jing202 fault and Jing209 fault where developed with natural fractures is preferable.

(4) The fault's influence on reconstruction is significant

Faults would not only dampen the fracturing effect, but also decrease the fracturing success rate, Jing212 well may be the typical unsuccessful well which is affected by the fault. Fault may change the circumferential stress distribution around wellbore, then alter the extension of fracture morphology and fracture mode, and finally directly affect the effectiveness of the fracturing design and operation.

(5) The success of fracturing is a prerequisite to fracturing effect

Table 1. Fracturing construction and post production data of deep volcanic reservoir.

Well	Well depth	Rate	Sand	Pad	Fluid	Sand ratio	Pressure	Production After fracturing
		m^3/min	m^3	m^3	m^4	%	MPa	t/d
JL2002	4275~4286	3.5	28	85.5	157.5	15.9	47-69	27.77
JL2006	3710~3725	2-3.5	16	90	105	13.3	56-60	0.00
Jing203	4268~4280	2.0-4	15	148	144	6.9	59-68	0.00
Jing 204	4205.5~4227	3.5-3.8	21	109.6	102.6	17.5	51-65	19.01
Jing 209	4334~4350	2.7-3.5	13	52.6	80.5	13.7	69-38	0.00
Jing 210	3158.5~3161.5	3.5	11	30.5	59.5	16.8	45-52	3.18
Jing 212	4364~4371	3.5	10	76	41	19.5	44-47	12.46
Jing 215	3866~3880	2-3.5	25	100	115.5	19.9	21-46	13.21
Jing 216	4390~4397	3.0-4	12	150	135	5.9	57-66	0.00
Jing 218	3892~3898	4.0-5	8	148	90	8	55-62	0.00
Jing 219	4245~4250	3.0	9.7	87.5	57.5	13.4	46-65	29.94
Jing 220	4278~4298	2.5-4	20	200	196	7.14	44-64	0.00

Figure 1 shows a comparison of the injection proppant, sand ratio and production of fractured wells. Among the 4 wells, JL2002, J203, J215 and J220 with injection proppant amount over $20m^3$, only J220 is ineffective with a relatively low sand ratio—7.14%. It suggests that a large amount of proppant being injected by efficient sand ratio would improve the efficiency of fracturing. For the 6 ineffective wells, which J203, J216, J218 and J220 four wells sand ratio is 10% or less (J203, J216 and J218 produced water in oil test, J220 is non-productive at all), inefficient sand ratio may directly result in ineffective fracturing. Therefore, both fracture conductivity and propped fracture length play an important role in a successful fracturing operation,

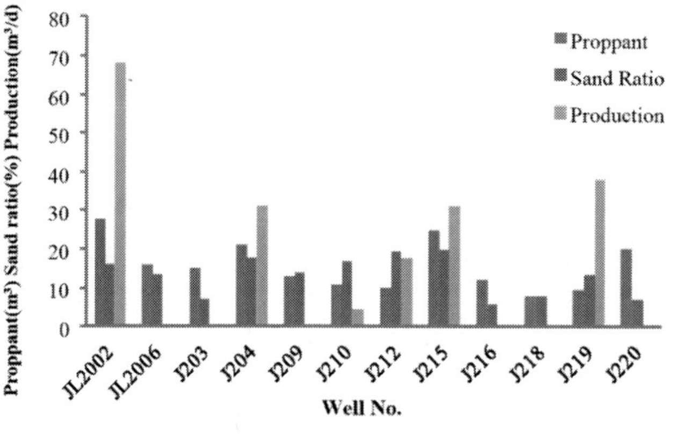

Fig. 1. Relationship among proppant amount, sand ratio and fractured well production

Previous fracturing practice and reservoir characteristics determine that the key factor to achieve the fracturing effect of volcanic reservoir is the sand concentration and the length of the hydraulic fracture, the fracturing filtration and the expansion of multi fractures needs be controlled to optimize fracturing effect. The primary guarantees for successful fracturing operation include avoiding excessive extension of multi fractures, as well as reducing the fracturing fluid filtration. The problem of the low success rate and low efficiency need to be solved by optimization design of fracturing parameters. It is difficult to control the expansion and extension of the multi fractures, which is essential part of the design for critical fracturing parameters.

3 Filtration control design

When the fracture width is larger than the diameter of proppant, proppant could move unrestrictedly in the fracture without forming effective blockage to control the fracture fluid loss. According to 3 times of particle diameter proppant access criterion, the fracture width should be greater than $2.5d_{prop}$. By using 70 / 140, 40 / 70, 20 / 40 mesh sand plug schedules to control fluid loss, the minimum fracture width to form a blockage is shown in Table 2.

Table 2. The minimum fracture width for different proppant to enter.

Proppant mesh	Proppant diameter (mm)	Minimum fracture width for different concentration proppant to enter (mm)		
		5%	10%	>25%
70/140	0.106~0.212	0.29	0.36	0.636
40/70	0.212~0.425	0.58	0.73	1.275
20/40	0.425~0.85	1.15	1.45	2.55

The low concentration proppant cannot control filtration dominated by fissured with over 0.36mm width, while the low concentration medium sand can form a bridge within 1.45mm fractures, but it is high permeability medium cannot reduce the filtration effectively. In most of the fractures with 0.4-1.5mm width, small diameter particles of 100 mesh cannot form an effective blockage, meanwhile, large diameter particles of 20/40 mesh can form a blockage, and its permeability is too high to control the fluid filtration. In order control the fluid loss in open fractures, a dualistic proppant scheme has been proposed— using mixed proppant with different diameters to fill fractures, large proppant for fracture bridging and small proppant for fluid loss control.

Because it is difficult for large proppant to enter the closed natural fractures, conventional filtrate reducer can be used to control the filtration, the commonly used filtration reducer contains silica and clay, but the defects of this kind of material is that when it is get into reservoir it might permanently block some pore channels. In order to address this problem, the JL-1 filter should be selected specifically. Table 3 is the experiment result of filtrate loss control by JL-1.When increase the dosage of JL-1 by 1%, the liquid filtrating time is delayed by 15min, and the wall building filtration coefficient was decreased by nearly 1/2; when increase the dosage of JL-1 by 2%, the wall building filtration coefficient decrease even more. It implies that JL-1 in fracturing fluid has a significant effect on reducing the filtration.

Table 3. JL-1 filtrate loss reducer.

Item	Total filtration (ml)	C (m/min$^{1/2}$)	Experimental phenomenon
Blank sample	12.5	8.5×10^{-4}	After pressure filtrate filter out soon
1%JL-1	7.2	4.6×10^{-4}	Filter out 15 min later
2%JL-1	6.0	3.5×10^{-4}	Filter out 30 min later

In order to analyze the effect of JL-1 on core permeability. Damage to core permeability in water based gel fracturing fluid is analyzed by using artificial core, result is shown in Table 4.

Table 4. The effect of JL-1 on core permeability.

JL1 concentration	Core quantity	K_g ($10^{-3}\mu m^3$)	K_{oil} ($10^{-3}\mu m^3$)	Total filtration	Damage rate

(%)				(ml)	(%)
0	3	9.7-25.6	16.18	3.62	16.8
2	4	10.1-25.2	16.64	2.36	15.2

As is seen from the above table, the damage effect of the gel fracturing fluid with 2% filtrate reducers concentration is equivalent to that without filtrate reducers. It shows that the impact of the filtrate reducers on formation damage is negligible.

During fracturing operation, the natural fracture opening width would be more than 1-3mm, 20/40 mesh and 100 mesh proppants are simultaneously used to control the fluid loss in multiple fractures. Based on the fluid-loss-control principle of naturally fractured reservoir, in addition to applying proppant slug in pre-pad fluid stage, employing dualistic proppant scheme is also an efficient method to reduce fluid loss in multiple fractures.

Table 5. Pre-pad schedule with proppant slugs for multi - fracture fluid loss control.

Equivalent fracture width (mm)	Total sand concentration (%)	Middle sand concentration %	Powder sand concentration %	Slug number	Fluid loss additive%
<0.5	3-6	0	3-6-9 (0.09-0.224mm)	3	1
0.5-1.0	4-10	2-4-6(0.45-0.9mm)	2-4(0.09-0.224mm)	3	1
1.0-3.0	**5-10**	**3-6(0.45-0.9mm)**	**3-6-9(0.09-0.224mm)**	**3**	**1.5**
>3.0	6-15	3-6-9(0.45-0.9mm)	3-6(0.224-0.45mm)	3	1.8

4 Optimal design of fracturing parameters

Perforation parameters. Perforating parameters have a direct impact on the generation and evolution of hydraulic fractures. In order to enhance the operation success rate in volcanic rock reservoir, the length and direction of the perforated section should be optimized. Since the hydraulic fractures initiate and propagate along maximum horizontal principal stress direction, and the maximum principal stress direction of target reservoir is NE113°~NE124°, the perforation azimuth should be controlled to determine the initiating point of hydraulic fracture, so that can not only avoid the hydraulic crack extension along other directions, and can guarantee the hydraulic fracture will not propagate deviously but in a straight line.

Too short perforation interval will result in high perforation friction during fracturing operation; too long perforation interval will lead to multiple fractures extension. The optimization of perforation length can learn from fracturing experience in domestic typical volcanic reservoir, taking the volcanic rock fracturing in DQ oilfield as an example, in order to limit the multiple fractures extension, perforation length is controlled in 5~8m, combined with the practice of previous experience, the perforation length should be controlled within the 10m.

Fracture parameters. Propped fracture length and fracture conductivity are the key parameters to influence fracturing effect. There is a strong positive correlation between the dose of the implanted proppant and the long crack length. The relationship between the implemented proppant amount and production after fracturing demonstrates that an adequate propped fracture length is indispensable for fracturing effect. Figure 2 shows that the higher proppant being implemented, the higher the production, therefore, increasing the propped fracture length will improve the effect of volcanic reservoir stimulation.

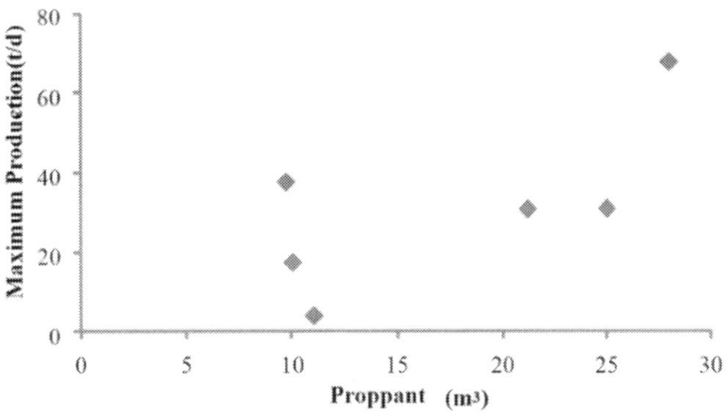

Fig. 2. The relationship between proppant amount and oil production after fracturing

There is a strong positive correlation between sand ratio and the conductivity of the propped fracture. The relationship between average sand ratio and production after fracturing demonstrates that an enough fracture conductivity is critical for fracturing effect. However, Figure 3 shows the relationship between the average sand ratio and well production is not obvious, so high fracture conductivity cannot reflect the fracturing effect. On the other hand, it implies that a long hydraulic fracture with appropriate conductivity could maximize the fracturing effect for low permeability volcanic reservoir.

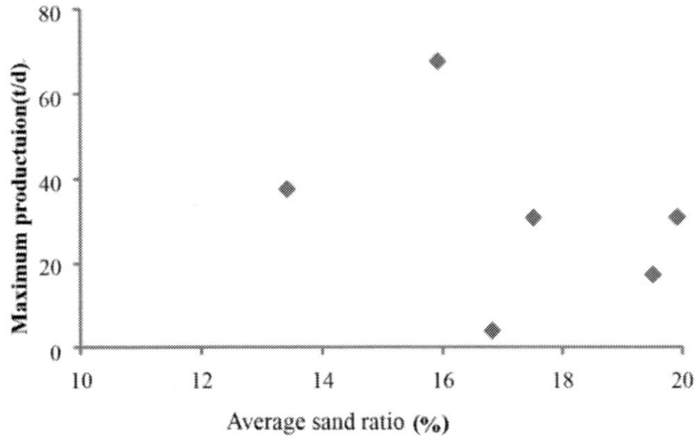

Fig. 3. Relationship between average sand ratio and well production.

Figure 4 shows the relationship between hydraulic fracture length and well production, and Fig. 5 shows the relationship between hydraulic fracture conductivity and well production. When the hydraulic fracture half-length exceeds 120m, the increasing length of hydraulic fracture will not contribute to raising well production, so the optimal hydraulic fracture in the reservoir is about 110-120m. Besides, when the conductivity of hydraulic fracture is larger than 25D.cm, the increasing of fracture conductivity dose little contribution to raising well production, so 20-25D·cm is the optimal conductivity to this reservoir. The analysis shows that appropriate conductivity is required for the long fracture in low permeability volcanic reservoir.

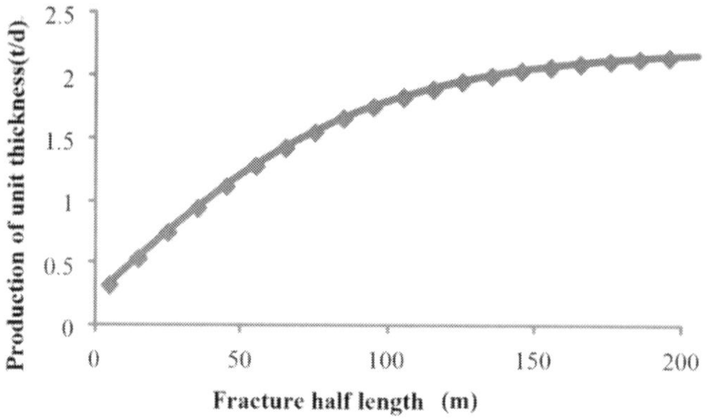

Fig. 4. The relationship between hydraulic fracture length and well production.

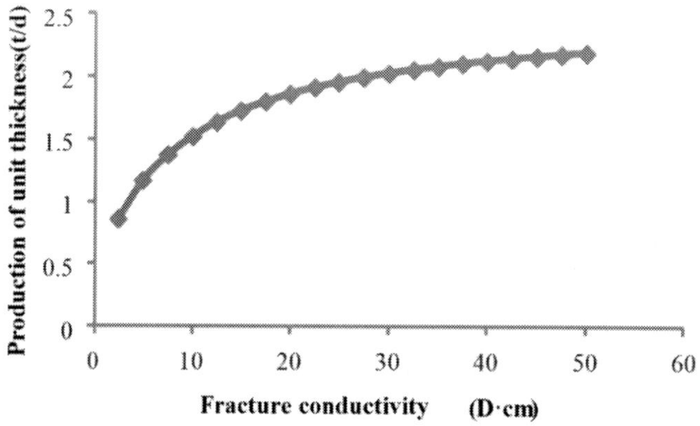

Fig. 5. The relationship between hydraulic fracture conductivity and well production.

5 Operation parameters

The operation parameters mainly include the pre-pad fluid ratio, sand ratio and pump rate.

Pre-pad fluid ratio. The optimization of the fracturing fluid volume is one of the key parameters in fracturing design. It is needed to reduce the amount of the pre-pad fluid to minimize the damage to reservoir. Volcanic reservoir is naturally fractured low permeability reservoir, fluid filtration is a serious problem, in order to reduce the damage of fracturing fluid to the reservoir, we should try to reduce the fluid into well, At the same time, to avoid sand settling to ensure the safety of fracturing, based on hydraulic fracturing fluid characteristics and considering the loss coefficient of reservoir filter is 1×10^{-3}m/min$^{0.5}$, when the volume of the liquid volume accounts for about 42% of the total fracturing fluid volume, effective ratio of propped fracture length could reach to 75%, which can meet the requirement of hydraulic fracturing. As a result of adding filtrate reducer in the pre-pad fluid can inhibit the fracturing fluid to flow into reservoir formation, The filtration of the reservoir can be analyzed by a mini fracturing test, so the pre-pad fluid amount could be optimized according to actual filtration, and the pre-pad fluid volume can be reduced to about 40% when the filtration coefficient is reduced to 8×10^{-4}m/min$^{0.5}$.

Sand ratio. Depending on the simulation of fracture conductivity, the optimal design of propped fracture can be realized when the fracture conductivity reaches 20D.cm. According to the software simulation analysis, assuming that the pre-pad fluid is completely lost at the end of operation, and the requirement of the fracture conductivity for different average sand ratio is shown in Table 6. 20D.cm fracture conductivity demands for 18% of the average sand ratio to achieve design goals, considering the existence of natural fracture in volcanic reservoir, the average sand ratio should be controlled in the range of 15~18% to reduce the risk of sand screen out.

Table 6. Minimum sand concentration/ratio for different fracture conductivities.

Fracture conductivity (D·cm)	Sand concentration (kg/m^2)	Average sand ratio (%)
15	4.24	13
20	5.27	18
25	5.91	27
30	6.88	37

Pump rate. The optimization of fracturing operation depends on many factors, and the high pump rate is beneficial to increase fracture width. And because of the increase of the injection speed and the decrease of the injection time, both sand settling time and fracturing fluid viscosity breaking are decreased consequently, therefrom, high pump rate could also improve the carrying capacity of fracturing fluid. On the other hand, the high pump rate will result in high net pressure, so that a large number of natural fractures will open, exacerbating the fracturing fluid filtration. From the above, the optimization of the pump rate should not only consider the fluid carrying capacity in volcanic reservoir, but also the control of fluid filtration.

Table 7. the relationship among pump rate, maximum sand ratio and natural fracture state.

Pump rate (m^3/min)	The maximum sand ratio (%)	Natural fracture state
1.0	17	Not activated
1.5	23	Not activated
2.0	30	Not activated
2.5	35	Not activated
3.0	39	Not activated
3.5	42	Shear failure
4.0	46	Shear failure

The critical maximum sand ratio and natural fracture state under different pump rates is shown in Table 7. To meet the 35% of the maximum design sand amount, pump rate at least need to reach 3.0m3/min, meanwhile, when pump rate is lower than 4.0m3/min, natural fractures cannot open, when pump rate is higher than 3.5m3/min shear failure will occur in natural fractures, the filtrate reducer can be used to control fluid filtration and avoid vast fluid loss through massive open natural fractures. From the above, pump rate should be chosen in the range of 3.0~4.0 m3/min.

6 Conclusion and cognition

(1) Based on the geological features of deep volcanic reservoir in the Zhungeer basin and the previous fracturing experience, the difficulties of fracturing in deep volcanic reservoir were analyzed, which is mainly reflected in the control of multiple fractures extension and fracturing fluid filtration, and the optimal design of key fracturing parameters.

(2) A comprehensive control methods based on dualistic proppant scheme was proposed to effectively control the fluid filtration in both open and closed fractures, satisfied the requirement of hydraulic fracturing in deep volcanic reservoir. . A pre-pad schedule with proppant slugs was designed to deal with fracturing fluid loss in multiple fractures. These method could dramatically increase the success rate of hydraulic fracturing in volcanic reservoir.

(3) Based on the requirement of controlling the multiple fractures propagation, the perforation parameters were optimized by numerical simulation method. The optimal hydraulic fracture length and conductivity were carried out by using numerical simulation method; Based on the requirement of fracturing operation, the ratio of the pre-pad fluid, sand ratio and pump rate were also optimized, and the results have an important guiding significance for improving the fracturing effect of deep volcanic reservoir.

References

1. Zhao Z, Liu D, Liu W, et al. Development and field application for integrated fracturing technology of volcanic reservoir. Canadian International Petroleum Conference. Petroleum Society of Canada, 2006.
2. Sun Yijian. Current status and trend of fracturing technology of volcanic reservoir fracturing technology. Henan chemical industry, 2010 (14): 56-58.
3. Warpinski N R, Teufel L W. Influence of geologic discontinuities on hydraulic fracture propagation (includes associated papers 17011 and 17074). Journal of Petroleum Technology, 1987, 39(02): 209-220.
4. Hainey B W, Weng X, Stoisits R F. Mitigation of multiple fractures from deviated wellbores//SPE Annual Technical Conference and Exhibition. Society of Petroleum Engineers, 1995.
5. Mishra S, Gupta A, Mathur M, et al. A new dimension added to the hydraulic fracture stimulation design of a volcanic reservoir based on microseismic monitoring. SPE-143524, SPE EUROPEC/EAGE Annual Conference and Exhibition, 2011.
6. Xu Tao, Zhao Dong, Cheng Fenglian, et al. Analysis showed multi fracture system volcano rock after lamination. Foreign Oilfield Engineering 2006 (6): 22-26.
7. Yuan S, Ran Q, Xu Z, et al. Reservoir Characterisation of Fractured Volcanic Gas Reservoir in Deep Zone//International Oil & Gas Conference and Exhibition in China. Society of Petroleum Engineers, 2006.
8. Meyer B R, Jacot R H. Implementation of fracture calibration equations for pressure dependent leakoff//SPE/AAPG Western Regional Meeting. Society of Petroleum Engineers, 2000.
9. He L, Yongping Z. The New Hydraulic Fracture Design Method and Good Performance in a Deep and Naturally Fractured Volcanic Gas Reservoir in China. Paper SPE, 2006, 100891: 3-6.
10. Weijers L, Griffin L G, Sugiyama H, et al. Hydraulic fracturing in a deep, naturally fractured volcanic rock in Japan-design considerations and execution results//SPE Asia Pacific Oil and Gas Conference and Exhibition. Society of Petroleum Engineers, 2002.
11. Weijers L, Griffin L G, Sugiyama H, et al. The first successful fracture treatment campaign conducted in Japan: Stimulation challenges in a deep, naturally fractured volcanic rock//SPE Annual Technical Conference and Exhibition. Society of Petroleum Engineers, 2002.
12. Sun Liangtian, Song Heping, Wang Peiyi, et al. Fluid loss reducing technique in the fracturing of Songnan volcanic rock reservoirs. Journal of Xi'an Shiyou University (Natural Science Edition) Vol. 25 No. 6 Nov. 2010.
13. Vermani S, Gupta A, Purusharthy N K, et al. Fracturing deep nonconventional volcanic reservoir: a case history Raageshwari gas field onshore India. SPE-132932, SPE Annual Technical Conference and Exhibition, 2010.
14. Yang Zhengzhou, Chen Mian, Xu Yun, et al. An experimental study of long-term flow conductivity of volcanic rock core plate. Natural Gas Industry volume 30, issue 10, pp.45-48, 10/25/2010.
15. Sun Yijian. Current status and trend of fracturing technology in volcanic rock reservoir [J]. Henan Chemical Industry, 2010 (14): 56-58.
16. Xie Zhaoyang, Zhang Hao, Han Song, et al. Research and application of fracturing fluid system in volcanic rock deep reservoir. Oilfield Chemistry, 2006, 22 (4): 310-312.
17. Zhang Yingan, Liu Guangyu, Zhou Xueping, et al. New CMG ultra-high temperature fracturing fluid used in the southern Songliao Basin deep volcano rock wells. Chinese Oil Exploration, 2009, 14 (4): 70-73.

Fracturing Fluid Leak-off for Deep Volcanic Rock in Zhungeer Basin: Mechanism and Control Method

Bo Huang [1], Hao Cheng [2], Yidong He [3,a], and Yanming Fu [3,a]

[1] Oil and Gas Development Company, PetroChina Xinjiang Oilfield Company, Karamay, China

[2] Engineering Technology Institute, PetroChina Xinjiang Oilfield Company, Karamay, China.

[3] State Key Laboratory of Oil-Gas Reservoir Geology & Exploitation, Southwest Petroleum University, Chengdu, China

[a]18583625735@163.com

Abstract. The deep volcanic reservoir in Zhungeer Basin is buried in over 4000m depth, which is characterized by complex lithology (breccia, andesite, basalt, etc.), high elastic modulus and massive natural fractures. During hydraulic fracturing, hydraulic fracture will propagate and natural fractures will be triggered by the increasing net pressure. However, the extension of fractures, especially natural fractures, would aggravate the leak-off effect of fracturing fluid, and consequently decrease the fracturing success rate. 4 out of 12 fracturing wells in the field have failed to add enough proppants due to fluid loss. In order to increase the success rate and efficiency of hydraulic fracturing for deep volcanic reservoir, based on theoretical and experimental method, the mechanism of fracturing fluid leak-off is deeply studied. We propose a dualistic proppant scheme and employ the fluid loss reducer to control the fluid leak-off in macro-fractures and micro-fractures respectively. The proposed technique remarkably improved the success rate in deep volcanic rock fracturing. It bears important theoretical value and practical significance to improve the hydraulic fracturing design for deep volcanic reservoir.

Keywords. Volcanic rock, fracturing fluid, leak-off, natural fracture, hydraulic fracturing

1 Introduction

According to the acoustic imaging logging data and core observation, the deep volcanic reservoir in Zhungeer basin, Xinjiang Province, is naturally developed with macro-frature, micro-fracture and induced fracture, which is triggered by temperature field variation during the drilling fluid encountered hot formation in the drilling process.

The most difference between hydraulic fracturing in naturally fractured volcanic reservoir and conventional homogeneous reservoir is the fracturing fluid leak-off mechanism [1-2]. In the hydraulic fracturing process for conventional homogeneous sandstone reservoir, the fracturing fluid leak-off is mainly controlled by matrix permeability, fracturing fluid viscosity and reservoir fluid compressibility, but for the naturally fractured volcanic, the leak-off is primarily dominated by natural fractures. The leak-off rate is dynamic, and related to net pressure in fracturing process [3], so it is hardly preestimated in hydraulic fracturing design. Due to natural fractures and the stress variation from far to near field, the fracture propagation path will curl, swerve, bifurcate and intertwined with each other, finally turn into a fracture network system. [4-6]. By the influence of high elastic modulus in the deep reservoir and the interaction of multiple propagating fractures, the fracture width is limited, leads to prematurely screen-out during fracturing operation. Thus in naturally fractured volcanic reservoir, fracturing fluid leak-off is hardly controlled because of the complexity of fracture propagation and natural fracture [7-8]. Pad fluid largely leaks off from extended major fracture into natural fractures, which cause proppant block up at the end of crack, and pressure rocket up, resulting in the failure of fracturing treatment. Thus for deep volcanic reservoir, controlling natural fracture leak-off in hydraulic fracturing design is crucial to improve the success rate of treatment.

2 Natural fracture width with leak-off effect

In the process of volcanic rock fracturing, the extended hydraulic fracture intersects with natural fracture, as shown in Figure 1, when the fluid pressure within the hydraulic fracture exceeds the normal stress imposed on the natural fracture surface, the natural fracture will be activated, and tensile failure occurs, then the fracturing fluid leak-off from hydraulic fracture into the natural fracture system will increase dramatically.

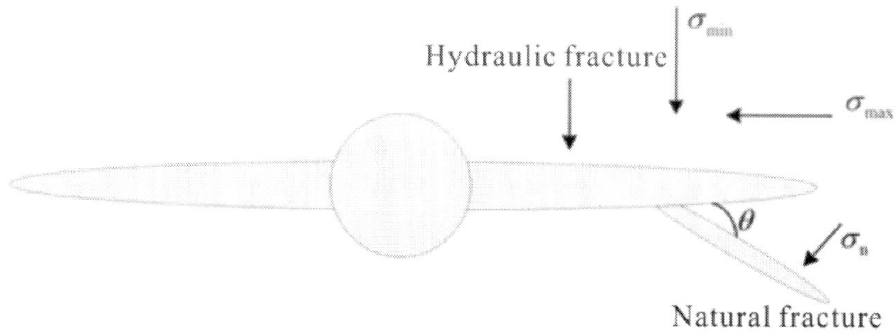

Fig. 1. Intersection graph between the hydraulic fracture and the natural fracture.

The dynamically width of natural fracture is a key factor to influence fracturing fluid loss, as well as a basic criterion to select fluid loss reducer. Based on the mutual relationship between the natural fracture direction and hydraulic fracture orientation controlled by in-situ stress, considering the hydraulic fracture will intersect with natural fracture in hydraulic fracturing process, given the fluid pressure in major fracture, the equation for natural fracture width can be written as [9]:

$$W(x) = \frac{2(1-\nu^2)h_{\rm f}(p(x)-\sigma_{\rm n}(x))}{E} \tag{1}$$

where $W(x)$ is the maximum width of fracture along extended direction, E is the rock Young's modulus, h_f is the fracture height, ν is the Poisson's ratio, $\sigma_n(x)$ is the normal stress of the fracture in the extended direction, $p(x)$ is the fluid pressure along fracture.

The fracture extension pressure can be expressed as the form of net pressure:

$$p(x) = p_{\rm net}(x) + \sigma_{\rm min} \tag{2}$$

where $p_{net}(x)$ is the net pressure in fracture, σ_{min} is the minimum horizontal principal stress.

The normal stress imposed on natural fractures surface is [10]:

$$\sigma_{\rm n} = \frac{\sigma_{\rm max}+\sigma_{\rm min}}{2} + \frac{\sigma_{\rm max}-\sigma_{\rm min}}{2}\cos 2(90^\circ - \theta) \tag{3}$$

where σ_{max} is the maximum horizontal principal stress, θ is the angle between the natural fracture and the horizontal maximum stress.

3 The selection of the fluid loss reducer in open natural fracture

For the natural fracture opened by hydraulic fracturing, when proppant diameter is less than natural fracture width, it can move in the fracture unrestrictedly, could not effectively form a blockage, so that fracturing fluid leak-off is out of control. According to the criterion of the triple proppant diameter, fracture width should be greater than 2.5 d_{prop}, when using slug that respectively consist of 70/140, 40/70, 20/40 sand to control fracturing fluid loss, the minimum fracture width that can prevent proppant from moving is as shown in Table 1

Table 1. The minimum width of facture that proppant could get through.

Proppant mesh	Proppant diameter mm	The minimum fracture width under different sand concentration (mm)		
		5%	10%	>25%
70/140	0.106-0.212	0.29	0.36	0.636
40/70	0.212-0.425	0.58	0.73	1.275
20/40	0.425-0.85	1.15	1.45	2.55

The low sand concentration (sand concentration<10%) totally can not control the leak-off dominated by natural fractures when fracture width is over 0.36mm, therefore, when a single type proppant is used alone, the effect that decreasing fracturing fluid leak-off is very limited. The low concentration medium sand can form a bridge when fracture width is less than 1.45mm, but it cannot reduce the leak-off, due to its high permeability characteristics. It can be known that, in the width of 0.4-1.5mm fracture, the minor proppant such as 100 mesh cannot form effective blockage. Although the large propping agent such as 20/40 mesh could form effective blockage, the natural fracture permeability which packed by proppant is high, fracturing fluid leak-off is out of control.

In order to control the leak-off through open natural fractures, we proposed a dualistic proppant scheme. Specifically, using the mixed propping agent in hydraulic fracturing, the large proppant are used to provide high conductivity, and the small propping agent aims at controlling fluid leak-off, as is shown in Figure 2. First of all, the temporary blocking agent with continuous particle size distribution is added according to fracture geometry, which can effectively block off different sizes of fractures and the gap between temporary blocking agents. Only from the reasonable particle size distribution, can we minimize the leak-off process.

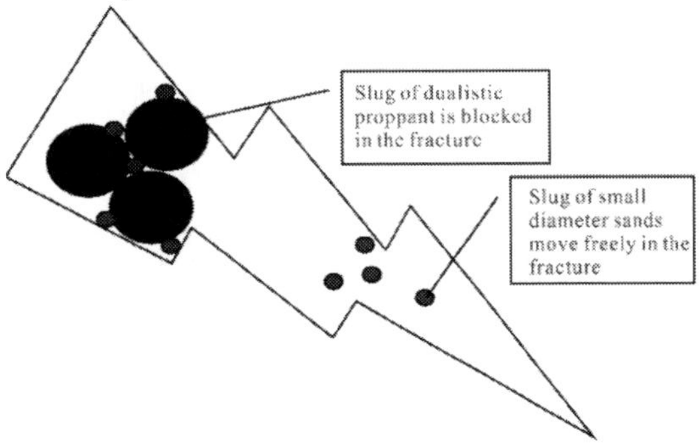

Fig. 2. Controlling fluid leak-off by dualistic proppant scheme.

When employ the dualistic proppant scheme to decrease the leak-off, the movement of small particles among large particles can influence fracture conductivity, and it also affects the stimulation effects, therefore, it is essential to optimize the diameter of small proppants.

Assuming the large particles are ideal spheres, which are closely packed in the fracture, as shown in figure 3. Four adjacent spheres form a cavity, and there is a smaller ball locates in this cavity, which can be considered to a throat formed by the ideal large ceramsite. The ball diameter is 0.41 times of the solid ball according to geometric relationship, for the 0.45-0.90mm (20/40 mesh) proppant in hydraulic fracturing, the distribution of diversion channel gap diameter is in the range of 0.18 to 0.36mm, According to the principle of bridging by 1/3 diameter, the channel allows small particles whose diameter are below 0.06mm to get through, the 0.06-0.36mm particles will block the gap, and large particles whose diameter over 0.36 mm will be excluded. So all particles whose diameter is under 0.06mm (240 mesh) may have the potential to affect the final conductivity of propped fracture. Therefore, when choosing large particles whose diameter are between 0.45 and 0.90mm（20/40 mesh）in hydraulic fracturing, small diameter particles can be selected in the range of 70 to 200 mesh.

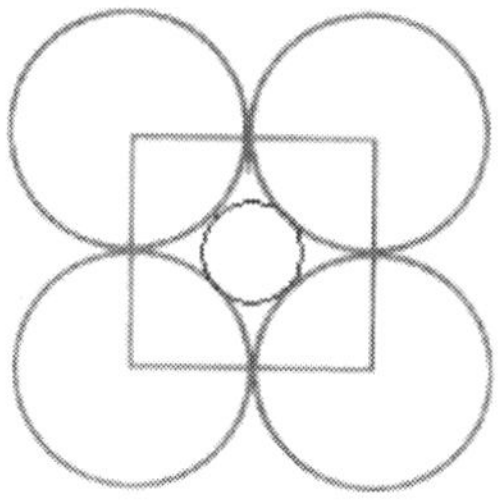

Fig. 3. The plane diagram of the gap volume among four accumulated particles.

When employ the dualistic proppant scheme to control the leak-off, the effect of sealing natural fractures can be achieved by adjusting the proportion of the two elements, through the indoor experiments, the sealing effect of two kinds of proppants for 20-40 mesh and 100 mesh ceramsite was compared. Assuming the sand concentration is 5kg/m^2, the experimental results are shown in Table 2.

Table 2. The permeability of mixed proppant for 20-40 mesh and 100 mesh.

Closure stress (MPa)	100% 100 mesh (D)	65%100 mesh 35%20-40 mesh (D)	35%100 mesh 65%20-40 mesh (D)	100% 20-40 mesh (D)
7	9.2	9.15	18.35	160.0
14	8.5	9.13	17.31	158.0
28	7.3	8.97	15.80	140.0
42	6.0	8.60	13.70	980.0
56	4.0	7.05	12.12	250.0

The experimental results show that fracture permeability varies with particle concentration ratio. Under the condition of the same closure stress, the smaller the proportion of small particles, the lower the permeability of the packed fracture, the better the control of leak-off.

4 The selection of fluid loss reducer for closed natural fractures

Conventional fluid loss reducer is often used to control leak-off in closed natural fractures, such as silica flour, clay, which result in the large proppant hardly enter natural fractures, the defect of this type material is that, once getting into the formation and pore channels, it will cause permanently damage to formation. To solve this problem, the JL-1 fluid loss reducer was selected, the JL-1 is a light yellow powder and can be dissolved in crude oil but water. The JL-1 prepared by the C9 component which generated in the process of pyrolysis petroleum, its fineness is over 95%, the softening point is above 70℃, and the dissolution is shown in table 3~4.

The results show that the JL-1 can be totally dissolved in kerosene but not dissolvable in water and acidizing fluid, and still maintain the characteristics of solid particles in water and acidizing fluid, ensured that no solid particles, which might block up the oil flow passage, be left over before production stage.

Table 3. The solubility of JL-1 fluid loss reducer.

Solvent	Acidizing fluid	Fresh water	Kerosene
Concentration of JL-1 （%）	2	2	2
Dissolution	Insoluble	Insoluble	Completely soluble

Table 4 is experiment result that the JL-1 decreases the leak-off under the condition of 90℃, 2MPa. When increasing the JL-1 by 1%, the time of producing liquid will be put off by 15 minutes, and the leak-off coefficient of building capacity decreased by nearly 1/2; when increasing the JL-1 by 2%, the leak-off coefficient will decrease more. It can be seen that, there is a significant effect on reducing fracturing fluid leak-off, when the JL-1 is added to the fracturing fluid. In the actual construction, the general design of fluid loss reducer ratio is 2%, you can get a better effect of controlling leak-off to meet the requirements of fracturing operation in micro-fractured formation.

Table 4. The effect of reducing leak-off with JL-1.

Item	Total filter loss (ml)	C ($m/min^{1/2}$)	Experimental phenomenon
Control sample	12.5	8.5×10^{-4}	Filtrate oozes out soon after pressurized
1% JL-1	7.2	4.6×10^{-4}	Filtrate oozes out in 15min after pressurized pressure
2% JL-1	6.0	3.5×10^{-4}	Filtrate oozes out in 30min after pressurized

In order to analyze the JL-1 effect on core permeability, we analyzed the damage of the JL-1 to core permeability in water based gel fracturing fluid by using the artificial core, which is shown in Table 5.

Table 5 Core permeability with different JL-1 concentration

JL-1 concentration (%)	Core quantity (piece)	K_g ($10^{-3}\mu m^3$)	K_{oil} ($10^{-3}\mu m^3$)	Total filter loss (ml)	Damage rate (%)
0	3	9.7-25.6	16.18	3.62	16.8
2	4	10.1-25.2	16.64	2.36	15.2

Table 5 shows that the damage rate caused by gel fracturing fluid in 2% concentration is equivalent to no using fluid loss reducer. It implies that formation damage rate caused by fluid loss reducer is low.

5 The treatment design for multiple fractures leak-off

The design of multi-fractures leak-off control technology for target volcanic reservoir is based on the following principles:

- Using the packed particles which match fracture width for temporary blockage;
- Employing the dualistic proppant scheme to substantially improve the effect of reducing fracturing fluid leak-off;
- Choosing the large size ratio of temporary plugging agent, the effect of reducing fracturing fluid leak-off will be better;
- The minimum particle size should meet the requirements of no relative movement in the space among large particles.

Considering the distribution of natural fractures width for target reservoir is in the range of 0.01~0.19mm, the net pressure is in the range of 5-6MPa, according to G function analysis. The angle between the natural fracture and the maximum principal stress is 23~35°, the average elastic modulus is 30152MPa, the maximum horizontal principal stress is 82MPa, and the minimum horizontal principal stress is 68MPa.

The calculation result of Equation (1)~(3) shows that: the natural fracture will occur tensile failure, and the open width of natural fractures is mainly 1 to 3mm, in the actual fracturing operation, the proppant system employing the dualistic proppant scheme, which consists of 20/40 mesh particles and less 100 mesh sand. According to the principle of fluid loss control technology for multi-fractured reservoir, the control of multi-fracture leak-off process should be carried out by using the following particle combination when using proppant slug to control multiple fractures leak-off in pad stage.

Table 6. Pad schedule for multi-fractures leak-off treatment.

Equivalent fracture width(mm)	Total sand concentration(%)	Medium sand concentration %	Small sand concentration %	Slug quantity	fluid loss reducer %
<0.5	3-6	0	3-6-9 (0.09-0.224mm)	3	1
0.5-1.0	4-10	2-4-6(0.45-0.9mm)	2-4(0.09-0.224mm)	3	1
1.0-3.0	5-10	3-6(0.45-0.9mm)	3-6-9(0.09-0.224mm)	3	1.5
>3.0	6-15	3-6-9(0.45-0.9mm)	3-6(0.224-0.45mm)	3	1.8

6 Conclusion

(1) Based on mechanism of natural fracture controlled fracturing fluid leak-off in the fracturing process of volcanic reservoir, a proposed method combining dualistic proppant scheme and fluid loss reducer could meet the requirements of reducing the leak-off in deep volcanic reservoir fracturing by decreasing the open fractures width and controlling the fluid loss in closed fractures;

(2) For the dualistic proppant scheme to control fluid leak-off in open fractures, the size and proportion of different particles are vital in design process, the selection of different proppant size has been demonstrated by theory, the experimental evaluation of different particle proportion effect on the leak-off is carried out, which provides theoretical basis for dualistic proppant scheme design;

(3) The experimental results for fluid loss reducer in closed fractures show that: the JL-1 has good oil solubility, the effect of reducing the leak-off is evident, and the damage rate to reservoir is low, which satisfies the requirements of reducing the fluid loss for deep volcanic reservoir;

(4) Based on the open behavior and extending behavior of natural fractures during fracturing operation in volcanic reservoir, the pad fluid schedule for multi-fractures leak-off treatment is optimally designed, which could remarkably improve the success rate in hydraulic fracturing for target reservoir.

References

1. Barree R D, Mukherjee H. Determination of pressure dependent leakoff and its effect on fracture geometry//SPE Annual Technical Conference and Exhibition, Society of Petroleum Engineers, 1996.
2. Yuan S, Ran Q, Xu Z, et al. Reservoir Characterisation of Fractured Volcanic Gas Reservoir in Deep Zone//International Oil & Gas Conference and Exhibition in China, Society of Petroleum Engineers, 2006.
3. Meyer B R, Jacot R H. Implementation of fracture calibration equations for pressure dependent leakoff//SPE/AAPG Western Regional Meeting, Society of Petroleum Engineers, 2000.
4. He L, Yongping Z. The New Hydraulic Fracture Design Method and Good Performance in a Deep and Naturally Fractured Volcanic Gas Reservoir in China// paper SPE, 2006, 100891.

5. Weijers L, Griffin L G, Sugiyama H, et al. Hydraulic fracturing in a deep, naturally fractured volcanic rock in Japan-design considerations and execution results//SPE Asia Pacific Oil and Gas Conference and Exhibition, Society of Petroleum Engineers, 2002.
6. Weijers L, Griffin L G, Sugiyama H, et al. The first successful fracture treatment campaign conducted in Japan: Stimulation challenges in a deep, naturally fractured volcanic rock//SPE Annual Technical Conference and Exhibition, Society of Petroleum Engineers, 2002.
7. CAO Baojun, LI Xiangfang, YAO Yuedong. Challenges and strategies in volcanic gas reservoir development. Natural Gas Industry, 2007, 27(8): 82-84.
8. WANG Xiaolei, LIU Hai, DA Yong, et al. Reservoir reconstruction technology of Kelameili volcanic gas reservoirs. Natural Gas Industry, 2010, 30(2): 63-65.
9. REN Lan, ZHAO Jinzhou, HU Yongquan, et al. Non-planar propagation characteristics of hydraulic fracture in naturally fractured formations. Journal of Central South University (Science and Technology), 2014, 45(1): 167-172.
10. ZHAO Jinzhou, REN Lan, HU Yongquan, et al. Numerical Simulation on Non-planar Propagation of Hydraulic Fracture in Naturally Fractured Formations. Journal of Southwest Petroleum University (Science & Technology Edition), 2012, 34(4): 174-180.

Research on CO_2 Quality Pipeline Transportation Based on Yanchang Oilfield CCUS

Yongchao Yang [1], Hongsheng Tao[1], Jinghua Yang[1], Qinghua Shang[1] and Shuangge Cao[2]

[1]Yanchang Oil Field co., LTD, Research Center of Exploration and Development, Shannxi Yan'an, China

[2]Xi'an Shiyou University, School of Mechanical Engineering, Shaanxi Xi'an 710065, China

740293171@qq.com

Abstract. CO_2 capture,utilization and storage (CCUS) is now recognized as an important technology in the global scope of CO_2 emission reduction, pipeline transportation is the main center to connect the capture point and the use storage point, the first issue to CO_2 pipeline transportation is to solve CO_2 source quality research. Yanchang Oilfield has the advantages of CCUS, its coal chemical capture of CO_2 contains different impurities. In the CO_2 pipeline transportation, the impurity content in CO_2 is based on its end use and the actual situation of pipeline. The impurities will affect the efficiency of CO_2-EOR, the choice of CO_2 state equation, the changes of CO_2 phase diagram and the capacity of pipeline transportation.

Keywords. CO_2 quality, pipeline transportation, CO_2-EOR

1 Introduction

CCUS (CO_2 capture, utilization and storage) is a new technology, which has the potential of large-scale CO_2 emission reduction, is expected to achieve low carbon utilization of fossil fuels, which is widely considered to be one of the important technologies to deal with global climate change and control greenhouse gas emissions [1]. CO_2-EOR is the most important part of CCUS, it refers to that CO_2 is separated from the industrial or energy production, then transported to some oilfields to enhanced oil recovery, at the same time, sequestrates some CO_2 [2].According to the geology evaluation, there are 10 billion tons of petroleum geological reserves are suitable for CO_2-EOR in China.If they all used for CO_2-EOR, it is expected to increase the recoverable reserves of 700 million tons to 1400 million tons, which provides a broad prospect for CO_2-EOR;the first problem to used CO_2-EOR to achieve industrial applications is to solve pipeline transportation, CO_2 from capture contains such as CH_4, H_2S, N_2 and other impurities, so CO_2 must need to be need to be purified, then transported by long distance pipeline to the injection point for EOR. According to the data, there are more than 65 million tons CO_2 for CO_2-EOR each year in America, it increases more than 15 million tons oil [3]. CO_2 pipeline construction and operation experience has been more than 40 years in America, which contains impurities CO_2 pipeline's experience is 20 years, however, CO_2 pipeline transportation in China has just started, building a large scale containing impurities CO_2 pipeline and ensuring safety and stable operation, the research of CO_2 quality is the first step.

2 The advantages of Yanchang Oilfield to carry out CCUS

Yanchang oilfield has coal, gas and oil resources in the same area, the natural conditions lays the foundation for the comprehensive development of the CCUS work. Domestic modern energy and chemical base has been formed, especially Coal-to-Oil and Coal-to-Olefins as the representative capture large numbers of high concentrations CO_2. There are more than 9.5 million tons CO_2 had emitted by coal chemical projects every year, of which about 60%, the concentration is above 85%. the use of CO_2 for EOR to reduce CO_2 emissions in Northern Shaanxi means that the coal utilization rate is higher than other areas and it's also the important module of coal chemical industry in Northern Shaanxi [4]. In recent years, there are a total of about 10 million tons of high pure CO_2 capacity in Northern Shaanxi, of which Yanchang oilfield has been formed a huge capture scale that is early 400000t CO_2, mid 700000 CO_2 and forward 4 million tons CO_2, this huge CO_2 capture is the prerequisite for the development of CO_2-EOR.

Yanchang oilfield is extra-low permeability reservoir, which has low oil recovery, CO2-EOR can efficiently enhance oil recovery, this can ensure the long-term stable production of the oilfield; water resources in North Shaanxi is deficient, CO2-EOR saves a large amount of water;according to the abroad experience of CO2 pipeline and the characteristics of CO2, the CO2 pipelines are laid away from the crowd, sparsely populated of northern shaanxi for CO2 pipeline provides a good geographical conditions.

The nature conditions of Yanchang oilfield provided a successful foundation for CCUS: (1) It makes CO2 emission reduction successful and the benefits from enhanced oil and gas recovery can make up for the cost of CCUS. (2) It saves a lot of water resources, these advantages are the key reason that America and China choose Yanchang oilfield as demonstration base of CCUS [5].

3 Requirements of CO_2-EOR for CO_2 Quality

The requirements of pipeline transportation for CO2 quality are considered its final use: enhance oil recovery. For EOR applications, CO2 flow must has a high purity, which generally provides the minimum content of CO2 is 95% and the average content of CO2 is about 98.45%~98.72% in foreign countries, which is to ensure that CO2 and oil are mixed in the suitable geothermal basin.different impurities in CO2 flow has different effects on CO2-EOR. H2S can increase the solubility of CO2 in the oil and reduce the mixed phase pressure, increase the water solubility in CO2, so H2S is favorable for CO2-EOR. Therefore, the restriction on H2S is based on security perspective, the recommended of H2S concentration is below 0.02% [6], but in the actual operation, the concentration is determined according to the actual operation situation eventually;CH4 can affect the minimum miscible pressure , and it is an important factor for affecting the mixing capacity, the volume concentration of CH4 is 2% in EOR applications; while N2 and other non-condensable species have a negative effect on the miscible capacity, density and viscosity of CO2 flow [7,8]. In conclusion, the impurities content in the CO2 flow is considered by having a minimal on EOR applications.

4 Requirements of Pipeline Transportation for CO_2 Quality

The pipeline system is very sensitive to the impurities in CO_2 flow, which will affect many parts of CO_2 pipeline transportation. The impurities will affect the phase change of the whole CO_2 flow, pipeline transport process, pipeline transport capacity, pipeline crack propagation and pipeline corrosion and protection, etc.

4.1 Requirements of pipeline transportation for water content

The impact of free water on the transport pipeline is mainly reflected in the following two aspects: (1) CO_2 will lead to the electrochemical reaction between carbon steel and water, then make the pipeline and equipment corrosive and corrosion rate is high; (2) The free water will form hydrates to block the pipes and even damage equipment. So, before entering the pipeline systems, CO_2 must be dehydrated.

Under super-critical conditions, the solubility of water in pure CO2 is 0.0026kg/m3 and it has been determined that the water content is less than 60% saturation (0.0015kg/m3), and carbon steel will not corrode. For super-critical CO2, along with the increase in pressure and temperature, the solubility of water in CO2 increased. In addition, the existence of free water also can cause hydrogen embrittlement or hydrogen brittleness, because the free water can cause the hydrogen atoms diffusing to the metal substrate to form hydrogen molecules, resulting in local internal pressure decreased the ductility and tensile strength of steel [9,10].

The formation of hydrate is not the main problem but it is a problem to be solved. The concerns about hydrates are mainly from the aspects of the flow protection, hydrate in the pipeline may cause the pipe blockage and flow content reduction and plug the pipeline system and accessories then causing accidents. In addition, if there is some water in CO2 flow, it is easy to generate hydrate in pipeline restart [11,12]. In order to avoid the occurrence of corrosion, the water content in current CO2 pipeline is far less than 60%, and the operation of these pipelines are safe, so in our country designing CO2 pipeline is also to meet this requirement. If in transit appear free water, there is no definite response, the International Petroleum Exchange gives the suggestion is that using ethylene glycol or corrosion inhibitor [13, 14].

4.2 Requirements of pipeline transportation for other impurities

The impurities affect the density, compressibility and viscosity of the whole CO_2 flow. These thermodynamic parameters affect the physical properties of the fluid and the pipeline transport properties, they can determined by the equation of state. The most popular used to predict the CO_2 phase diagram are: Peng-Robinson (PR) equation, Soave-Redlich-Kwong (SRK) equation, Benedict-Webb-Rubin-Starling (BWRS) equation and GERG-2008 equation.The differences of impurities content and kinds make the equation of state's more difficult, so the equation of state for CO_2 pipeline design has not yet reached a consensus, currently used for simulation of the impurities in the CO_2 state equation are based on the published experimental data available in the literature to evaluate, DNV RP J202 *Design and Operation of CO_2 Pipelines* recommended PR equation is used for CO_2 containing impurities. PR equation has higher accuracy, however, in the process of calculation, the equation of state needs use the experimental data to adjust and to assess the calculation of uncertainty.

Impurities affect the phase behavior of the whole CO_2 gas flow and then affect the process of pipeline transportation. Figure 1 show the phase diagram of CO_2, compared the different impurities on the critical point and phase curve of CO_2.

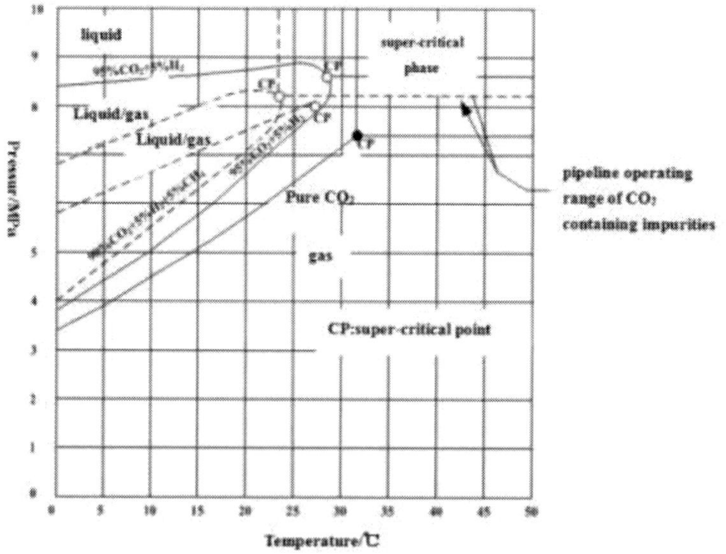

Fig. 1. CO_2 phase diagram

The critical point of pure CO2 is T=31.4, P=7.38MPa, which has a lower critical point. Because super-critical CO2 has good flow characteristics and transmission characteristics, it has high transmission efficiency in the process of conveying. So all foreign operation of CO2 pipelines are using super-critical CO2. Figure 1 shows that different kinds of impurities make CO2 phase diagram in a two phase region and changed CO2 critical point, this influences the operation scope and transportation way of pipelines. With the increasing of impurities, critical pressure increases and critical temperature decreases [15], this means compared with transporting pure CO2, CO2 containing impurities need higher pressure, so in the pipeline transportation, determining the types and contents of impurities in the CO2 flow is the first step for determining the operating conditions.

Impurities will also affect the transport capacity of the pipeline, it can consume additional compression, and with the increase of impurities, the pipeline transportation capacity will be reduced. It has been shown that [16, 17], for example, in the same transport conditions, compared with pure CO2, if the content of CH4 is 5% or 10%, the pipeline transport capacity decreased by 9.4% and 16%, if the content of N2 is 5% , the pipeline transport capacity decreased by 12.6%, which is due to the impurities reduce the volume of the pipeline transport capacity.

Impurities can also affect the design of pump and compressor, which is based on CO2 components such as NOX, H2S, CO, SOX, etc. The existence of such impurities increase the pipeline leakage risks; Impurities can affect the use of the pipeline infrastructure. Above all, the impurities in CO2 flow influence multiple aspects of the pipeline, so it must control the content of impurities reasonably and effectively during the process.

5 Conclusions and Recommendations

Compared with natural gas pipelines and other hazardous liquids pipelines, few accidents happened at the current CO_2 pipelines in America.Throughout the CO_2 pipeline technology, it can be seen that CO_2 pipeline transportation is not really difficult to overcome, so large-scale and long-distance CO_2 transmission pipeline will be fully realized in China.

(1) Yanchang oilfields as CCUS demonstration project implementation in China, CO_2 capture needs combine with CO_2-EOR, and explore the best way to capture for CO_2-EOR.

(2) It must be to do a good job of dehydration and removing impurities before CO_2 flow into the whole pipeline system and monitor the water content in the entrance to ensure the saturation limit of moisture content below 60%; Impurities also can affect the critical point of CO_2, pipeline transport process and transport properties, so it should be strictly controlled the content of impurities and monitored carefully.

(3) The research on CO_2 pipeline transportation needs to combine American's 40 years of experience in CO_2 pipeline transportation and China's experience in rich natural, gas oil pipelines, then develop a suitable standard for China's large-scale CO_2 transmission pipeline.

(4) Capture, transportation, storage, injection, separation, detection, monitoring of high pressure and large impurity CO_2, CO_2 anti-corrosion technology and equipment need to be tested research and optimized.

(5) With the development of experimental and industrial applications, the government and related enterprises should take into account the planning and construction of the CO_2 containing impurities transport pipeline in due time.

Acknowledgements

This paper is supported by Yanchang oilfield company science and technology project fund.
Project name: CO2 pipeline feasibility study of Yanchang oilfield. Project number: ycsy2015ky-B-02

References

1. HengweiLiu, Kelly Sims Gallagher. Preparing to ramp up large-scale CCS demonstrations: An engineering-economic assessment of CO_2 pipeline transportation in China. International Journal of Greenhouse Gas Control, 5(2011)798-804.

2. Qing Zhao, Yu-Xing Li. The influence of impurities on the transportation safety of an anthropogenic CO_2 pipeline .Process Safety and Environmental Protection, 92(2014)80-92.
3. Amin Ettehadtavakkol. CO_2-EOR and storage design optimization. International Journal of Greenhouse Gas Control, 25(2014)79-92.
4. Zhang Jian et al. Coal chemical industry CO_2 emissions and emissions reduction analysis . Coal chemical industry, 2008,6:8-12.
5. Xinhua. The United Nations Climate Change Statement by China and US. http://news.xinhuanet.com/politics/2015-09/26/c_1116685873.htm. 2015,9,26.
6. Bing Chen,Shuangge Cao.Key Techniques of Super-critical CO_2 Pipeline Transportation. International Journal of u- and e- Service, Science and Technology,7(2015)285-292.
7. Liu Yingbo et al. Enhanced oil recovery by CO_2 containing impurities. Foreign oilfield engineering.2010,9(26):20-24.
8. Det Norske Veritas. DNV-RP-J202; Design and Operation of CO_2 Pipelines. Hovik, Norway: DNV, 2010.
9. HENDRIKS C, HAGEDOORN S, WARMENHOVENH. Transportation of carbon dioxide and organizational issues of CCS in the Netherlands . Utrecht, the Netherland: Ecofys, 2007.
10. Li Yuxing, Liu Mengshi, Zhang Jian. Gas impurities affect the CO_2 pipeline system security. Natural gas industry.2014, 34(1):108-113.
11. =Rickard Svensson, etal.Transportation System for CO_2-application to Carbon Capture and Storage .Energy Conversion and Management, 45(2004)2343-2353.
12. HEGGUM G, WEYDAHL T,MO R et al.CO_2 Conditioning and transportation.Oxford,UK:Elsevier,2005.
13. BRATFOS H A,LEINUM B H,TORBERGSEN L E, et al.Challenge to the pipeline transportation of dense CO_2.Journal of Pipeline Engineering,2007,6(3):161-172.
14. Payam Zargarzadeh. Structural integrity of CO_2 transportation infrastructures.England: Cranfield University, 2013.
15. IPCC. Carbon dioxide capture and storage summary for policymakers and technical summary, ISBN 92-9169-119-4.Montreal; IPCC, 2005.
16. National Energy Technology Laboratory. CO_2 Impurity Design Parameters. US: Department of Energy, 2012.
17. Ayello F, EvansK, Thodla R.Effect of impurities on corrosion of steel in supercritical CO_2 . NACP Paper No 10193, 2010.

Study on the Water in Lignite by DSC

You Li[1], Shouyu Zhang[1], Menghao Zhao[1], Qing Mao[1], Yunlong Yao[1] and Jingning Yang

[1]School of Energy and Power Engineering, University of Shanghai for Science and Technology, Shanghai, China
Corresponding Email: zhangsy-guo@163.com

Abstract. As pointed out in several previous works, new drying technologies must be developed to ensure lignite remained competitive. In order to design and test them, it was needed to have a better understanding of the characters of water in lignite. Using differential scanning calorimetry (DSC), three types of the water storing in the lignite were determined from congelation character. The contents of the free water and bound water were experimentally measured from their heats of congelation, respectively. The results shows that the congelation temperature and the freezing heat per weight unit of the bound water vary in different types of lignite. It is deduced that the bound water in lignite can reflect the character of lignite. Besides, the total water content of free water and bound water is less than that measured by fully drying at 105°C. Therefore, it is deduced that there is a kind of water not frozen in the lignite, possibly because its molecular cluster is too small.
Keywords. Lignite, moisture, drying, differential scanning calorimetry

1 Introduction

The high moisture content of lignite in Eastern Inner Mongolia region reduces the efficiency of combustion, since in conventional processes part of the heat of combustion of the coal is wasted in removing this associated water. Therefore, the amount of carbon dioxide produced per MWh is higher for lignite than bituminous coal. New drying/dewatering technologies must be developed to ensure brown coal remains competitive [1], and in order to design and test them, it is necessary to have a better understanding of the fundamental physical and chemical characteristics of coal structure, particularly in relation to the forms of water in lignite.

Low temperature high resolution Differential Scanning Calorimetry (DSC) is a powerful tool for investigating water and solution behavior in porous materials [2]. It has been used for distinguishing the types of water in coal [1, 3, 4, 5]. When the temperature is changed, phase transitions occur, and the energy (enthalpy) released during this process can be accurately measured. Since the slope of the DSC curve, the freezing heat per weight unit, is easy to obtain, the weight of the water that undergoes a phase change can be calculated [6]. Norinaga et al. [3] observed two exothermic peaks when cooling wet coal samples at 8°C/min which were attributed to free or bulk water (large peak at -15°C) and to bound water (small peak at -47°C).

In this work, DSC was applied to characterize water in two kinds of lignite. The contents of different types of water in lignite were obtained.

2 Material and methods

Raw Samples. Two kinds of lignite, Datang-five-rooms (DT) lignite and Xilingol (XLGL) lignite, from China's Eastern Inner Mongolia region were chosen as the samples, whose datum of proximate and ultimate analyses were listed in table1, obtained by Institute of Coal Chemistry, Chinese Academy of Sciences.

Drying. The raw samples were smashed, and the lignite granules in size between 2mm and 3mm were selected. A portion of the samples were dried at 40°C for different time; the other part of the samples was dried at 40°C to constant mass and then at 105°C for different time. The drying samples were numbered and sealed for the follow-up analysis. The water content of the lignite was measured by heating at 105°C to constant mass, and the loss of mass as result of heating was taken as the water content.

Table 1. Proximate and ultimate analysis and calorific value of samples.

Samples	Proximate analysis [%]					Ultimate analysis [%]					$Q_{net,v,ad}$
	M_{ad}	A_{ad}	V_{ad}	FC_{ad}		C_{ad}	H_{ad}	O_{ad}	N_{ad}	$S_{t,ad}$	[MJ·kg^{-1}]
DT	5.56	22.56	31.12	40.76		53.46	3.22	13.76	0.89	0.56	15.50
XLGL	19.44	15.48	31.13	33.95		47.71	1.94	14.26	0.85	0.33	10.93

Cooling in DSC Analyzer. DSC analysis was performed on Netzsch DSC200 analyzer with liquid nitrogen refrigeration accessories. The drying samples were milled and then sieved to less than 0.2mm. Approximately 5 mg of coal sample was placed into the crucible each time. The crucible was cooled down from 15 to -150°C with a rate of 8°C/min, using nitrogen as carrier gas at a constant flow rate of 80 ml/min. The thermogram was recorded during the cooling process.

3 Results and discussion

Types of Freezable Water. The DSC curves of four representative samples were selected to compare with the curve of pure water. The curves in Fig.1 indicated the heat changing of samples in the cooling process, and the peaks indicated the exothermic process.

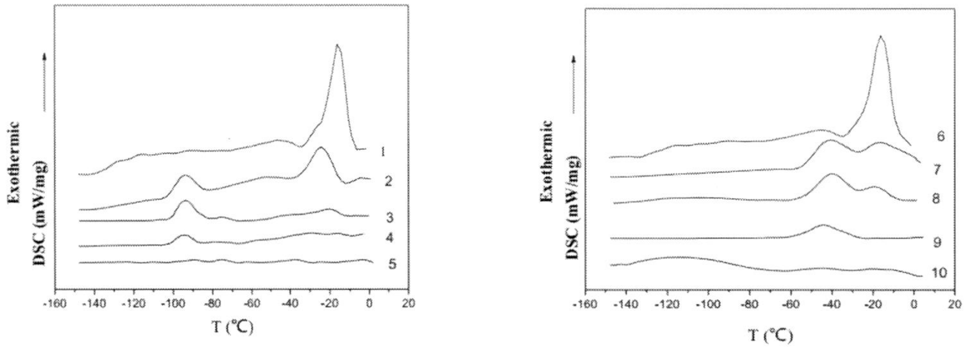

a - DT b - XLGL

1 and 6 - pure water; 2 and 7- raw lignite; 3 and 8- lignite dried at 40°C for 30 min;
4 and 9 - lignite dried at 105°C for 20 min; 5 and 10 - lignite fully dried at 105°C.

Fig. 1. DSC thermogram of lignite.

The DSC curves for the samples and the pure water under cooling were shown in Figure 1. The peaks appearing in the thermograms were the result of exothermic processes. These peaks were attributed to the freezing of the water absorbed in the coal because such peaks were never observed for the samples fully dried at 105°C.

According to the contrast between curve 1 and 2, the big peak of raw DT lignite was similar to that exhibited by pure water, which was attributed to free water. In addition, as previously noted by Norinaga et al [3], a small peak at -90°C was also shown by the curve of raw DT lignite in the Fig.1a, which was defined as bound water according to the study of Nakamura et al [7]. Contrasting curve 2, 3 and 4, it was shown that the bound water did not start to be separated out until the free water fully evaporated. The phenomenon indicated that it was harder for bound water in lignite to be separated out than bulk water in lignite because of the coal-water reaction [5].

Fig. 1b showed similar phenomenon to Fig. 1a. Differently, the peak of bound water in XLGL lignite appearing at -45°C, which indicated that the bound water in XLGL lignite was easier to separate out than that in DT lignite. Therefore, the coal-water interaction in XLGL lignite was weaker than that in DT lignite. To sum up, the bound water in lignite could represent the characteristics of water rather than the free water in lignite.

Content Measurement of Different Kinds of Water. The relationship between the freezing heat and moisture contents of the samples was shown in the Figure 2. As it was showed in Figure 2a, the decrease in freezing heat per unit reduction of water mass was 330.1 J/g of water in the range of 0.335 to 0.213g/g coal (db.), and it was in agreement with the congelation heat of pure water into ice, 334 J/g. It was proved again that the peaks at -15°C was ascribed to water which had little interaction with the coal.

In Figure 2a, for water contents ranging from 0.213 to 0.128 g/g coal (db.), ΔH decreased with a slope of 188 J/g of water, which was considered as bound water [7]. As it was analyzed previously, the congelation heat of capillary-condensed water was generally less than that of free water [8]. The content of the two kinds of water could be figured out based on enthalpy released and the slope during the cooling process.

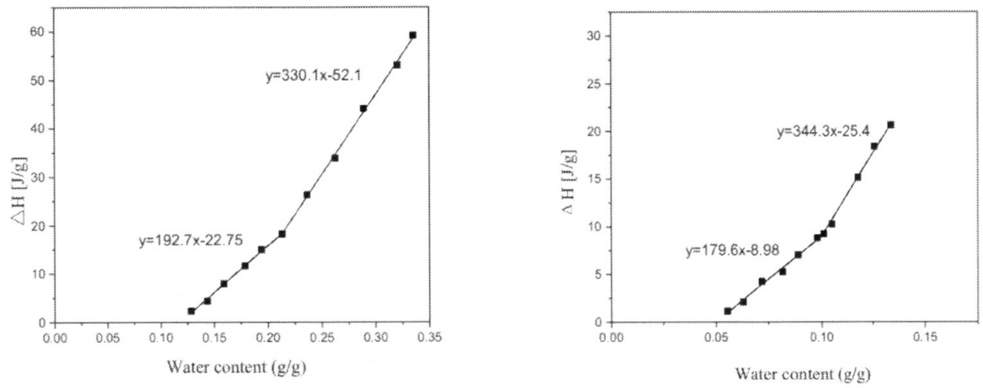

a – DT b - XLGL

Fig. 2. Heat released from lignite by freezing.

However, the total water content of free water and bound water was less than that obtained by fully drying at 105°C. Therefore, there was likely to be a kind of water that was not freezable in lignite, namely, non-freezable water. Sheng et al [9] reported that water became non-freezable when the molecular cluster was too small. The content of the non-freezable water could be calculated by difference. The similar phenomenon for XLGL lignite was shown in Fig.2b. In summary, the three kinds of moisture in two kinds of lignite were listed in Table 2, respectively.

Table 2. Results of DSC analysis.

Sample	Peak temperature [°C]		Congelation enthalpy[J/g]	Kinds of moisture [g/g coal (db.)]		
	Peak 1	Peak 2		Free water	Bound water	Non-freezable water
DT	-15	-90	330.1,192.7	0.1223 (37%)	0.0952 (28%)	0.118 (35%)
XLGL	-15	-42	344.3,179.6	0.036 (27%)	0.048 (36%)	0.05 (37%)

4 Conclusions

(1) There were two kinds of water contained in lignite, free water and bound water, which could be coagulated. In the DSC test environment, the coagulation temperature of free water was -15°C, and that of bound water in DT lignite was -90°C solidification, and that of bound water in XLGL lignite was -45°C. In addition to the water that could be condensed, lignite contained non-freezable water, which could be calculated by difference.

(2) The content of water in DT lignite included 0.14g/g coal (db.) as free water, 0.083g/g coal (db.) as bound water and 0.127g/g coal (db.) as non-freezable water; The content of water in XLGL lignite included 0.036g/g coal (db.) as free water, 0.048g/g coal (db.) as bound water and 0.05g/g coal (db.) as non-freezable water.

Acknowledgements

This research was supported by the National Science & Technology Pillar Program during the Twelfth Five-year Plan Period (Grant No.2012BAA04B01).

References

1. D.J. Allardice, L.M Clemow, G. Favas, W.R. Jackson, M. Marshall and R. Sakurovs: The Characterisation of Different Forms of Water in Low Rank Coals and Some Hydrothermally Dried Products. Fuel Vol. 82 (2003), p. 661-667.
2. L.G. Homshaw: High Resolution Heat Flow DSC: Application to Study of Phase Transitions, and Pore Size Distribution in Saturated Porous Materials. Journal of Thermal Analysis Vol. 19 (1980), p. 215-234.
3. N. Koyo, K. Haruo, J. Hayashi and T. Chiba: Classification of Water Sorbed in Coal on the Basis of Conflation Characteristics. Energy & Fuels. Vol. 12 (1998), p. 574-579.

4. F. Yi, C. A.L. Chaffee, M. Marshall and W.R. Jackson: Lignite–water Interactions Studied by Phase Transition—Differential Scanning Calorimetry. Fuel. Vol.84 (2005), p. 1557-1562.
5. T. Arash, J. Yu., H. Su, Y. Han, J. Lucas, H. Zheng and T. Wall: A Differential Scanning Calorimetric (DSC) Study on the Characteristics and Behavior of Water in Low-rank Coals. Fuel vol.135 (2014), p. 243-252.
6. G. Zhou, Q. Huang, B. Yu, F. Wang, Y. Chi and J. Yan: Study on the critical particle size of high moisture low rank coals drying in fluid-ized-bed reactor via low temperature Differential Scanning Calorimetry (DSC). Meitan Xuebao/*Journal of the China Coal Society* Vol.40 (2015), p. 185-189.
7. K. Nakamura, T. Hatakeyama and H. Hatakeyama: Relationship between hydrogen bonding and bound water in poly hydroxystyrene derivatives. Polymer Vol. 24 (1983), p. 871–876.
8. S.C. Mraw and D.F. O'Rourke: Water in Coal Pores: The Enthalpy of Fusion Reflects Pore Size Distribution. Journal of Colloid and Interface Science. Vol.89 (1982), p. 268-271.
9. P. Sheng, R.W. Cohen and J.R. Schrieffer: Melting Transition of Small Molecular Clusters. Journal of Physics C: Solid State Physics. Vol.14 (1981), p. 565-569.

The Mobile Application Social Mode of Manufacturing Enterprise under the Environment of Mobile Internet

Cen Zuo[1], Xumei Zhang[1] and Xingshan Wang[2]

[1]School of Economics and Business Administration, Chongqing University, Chongqing, 400044, China
[2]Inspur Group Limited Liability Company. Jinan, 25101, China

Abstract. The rapid development of the mobile internet, mobile communication technology and mobile terminal stimulate demand of mobile application in manufacturing enterprises, which makes the enterprise implement the mobile ERP. The implementation of mobile ERP changes the internal application management of manufacturing enterprises to social applications. We analyse the driving factors of implementing the mobile application social mode considering the complexity of organization relations and business processes of manufacturing enterprise. And then the mobile application social mode of the manufacturing enterprise is built based on mobile internet. Finally, we present the management strategies of the mobile application social network to provide references for the manufacturing enterprise.

Keywords. Mobile application social mode, enterprise social network (esn), mobile erp, manufacturing enterprise.

1 Introduction

With the development of mobile communication and internet technology, the mobile internet develops rapidly in recent years. Simultaneously, the 4G network technology and mobile terminals such as smartphones and tablets become more and more popular. These enable internet users to connect to the internet without space and time limit, and using mobile internet technology and terminal equipment is convenient for enterprises to operate more efficient. With the development of mobile internet, 4G network technology and mobile terminals, internet users can connect to the internet without space and time limit, which makes enterprises operate more efficient. At the same time, many international IT companies such as Microsoft, Oracle and SAP acquire enterprise social network (ESN) companies by merger & acquisition. Since then, ESN becomes an important trend and leads to enterprise to implement mobile social network Therefore, it is significant to research the enterprise mobile application social mode in the new environment.

ESN refers to apply mature social network to enterprise organization to realize the enterprise internal efficient communication and cooperation, and it is a subset of enterprise 2.0 [1].At present scholars mainly focus on the benefit of ESN such as promoting the performance of enterprise employees, improving the work process, enhancing enterprise knowledge management, reducing the cost and saving time. Kuegler [2] thinks social network can influence staffs' performance by improving the work process and employee motivation and it also have a positive effect on team cooperation. Matii [3] defines five functions of ESN from the perspective of knowledge management and studies the mechanism of how the five functions drive the value of ESN. Von Krogh [4] points out questions need to pay attention in the social network such as knowledge protection, the definition of enterprise boundary. Majchrzak [5] shows that social network has a positive effect on optimizing the enterprise process and teamwork by a case researching. By analysing the case of implementing technologies such as social network and web-based communities in a large multinational telecommunications company, Denyer [6] shows that though these technologies have the potential to bring significant business benefit, the experience of organizational actors suggest that their expectations are not met. Richter [7] thinks that different enterprise will get different benefits after deploying ESN. To solve the problem that the ERP system fails to meet user needs in informal organization processes, Greasley [8] puts forward a hybrid ERP system that combines the function of a traditional ERP with the capabilities of enterprise social software.

Current researches are mainly focus on traditional ESN, but there are few literatures that aim at the mode of mobile enterprise social network. Therefore, this paper intends to analyse the driving factors of the appearance of the mobile application social network, and then this paper puts forward the mobile application social mode in enterprise and its management strategies.

2 Driving factors of implementing mobile social network in manufacturing enterprise

2.1 The development of new technologies makes ERP mobile applications become reality

With the rapid development of network technology and the continuously upgrades of mobile terminal ,bandwidth and terminal equipment technology are no longer barriers to enter internet and the mobile terminal becomes an important way to surf the internet. The infrastructure of Mobile communications is improved gradually and WLAN's coverage area is continually expanding. According to the data released by China's Ministry of Industry and Information Technology [9], the amount of mobile phone users is 1.28 billion in January 2016. Because it is more convenient for netizens to use the mobile internet, the scale of mobile internet users keep growing and internet users' online equipment switch to the mobile terminal gradually. As of December 2015, the number of China's netizens reached 688 million, which included 620 million mobile internet users [10]. With the bigger screen of the mobile terminal and better experience on mobile applications, mobile terminal s become a main way to surf the internet, which lay the foundation for the development of enterprise mobile applications, and the trend of mobile office is further clear. ERP mobile applications do become reality and can be used in the enterprise's office, marketing, logistics and other fields.

2.2 The new technologies stimulate the demand of enterprise mobile application social network

In recent years, the mobile terminals such as smartphones and tablets are continued upgrading in hardware and software aspects. The mobile terminal is gradually switching the previous "equipment centered" into "people-centered" mode with humanized design and more powerful functions. Meanwhile, the mobile social network becomes a hot area in the mobile internet .According to the data released by Nielsen as of April 2014 [11], social media users took up 46% of total domestic users and more than 60% of mobile users used social applications via the mobile terminal. Moreover, after Oracle's acquisition of Virtue social marketing company in 2012, international IT companies such as Microsoft, IBM, and SAP have acquired enterprise social media companies. Since then, enterprise social network has become an important trend. Because of the large scale and complex business relationships of enterprises, staffs need process business information more immediate and high efficient to get rid of the complex organization structures, and the demand of mobile office increases sharply .So the enterprise mobile social network becomes necessary for enterprises to adapt to new environment.

2.3 Mobile ERP makes the enterprise mobile application social network develop fast

Since experienced more than 30 years from initial MRP to MRPII to ERP, ERP system upgrades and evolves in all aspects of the enterprise management. Nowadays, many enterprises pay attention to the research and development of ERP system and ERP is developing towards the mobile ERP. With the advent of mobile internet, tradition ERP cannot meet the need of staffs dealing with business affairs wherever and whenever in the fierce completion environment since the capability of quick response is important to enterprise to capture the market share. Meanwhile, enterprise staffs can deliver various views to communicate with others via various mobile terminal devices, which can break through traditional communication hierarchy and improve the efficiency greatly. So enterprise mobile social network can meet the demand of real-time information interaction in dealing with business affairs, and promote enterprise internal and external service value and competiveness in the mobile internet time.

2.4 The mobile social network promotes the business mode's innovation in enterprises

The traditional communication way in strict hierarchy leads to more time cost for enterprises in the new internet age. However, the implementation of ESN can make the enterprise internal and external communication get rid of traditional hierarchical structure. With the concept of enterprise 2.0, IT companies introduce enterprise mobile social software to guarantee smooth communication. SAP stream work mobile version developed by SAP can realize efficient collaborative decision-making; Oracle Fusion Middleware can establish the link among personnel, processes and applications based on enterprise2.0; IU social platform developed by UFIDA provides collaborative decision-making; Kingdee's K/3 WISE V13 can meet the demand of enterprise management for small and medium-sized enterprises in cloud era. Enterprise mobile social network can not only meet the needs of internal business social activities and integrate with scattered information including meetings, emails and so on, but also can overcome the disadvantages such as the complexity of enterprise hierarchy and the relatively closure in departments. Besides in each department levels, Enterprise mobile social network can speed up the information delivery based on swarm intelligence and public knowledge, then the enterprise can maximize knowledge sharing.

3 The mobile application social mode of manufacturing enterprise

The mobile application social network of manufacturing enterprise is based on mobile ERP, mobile internet and other basic technical supports. It realizes enterprise socializing by visiting ERP with the mobile terminal in mobile application social network. And the smooth and

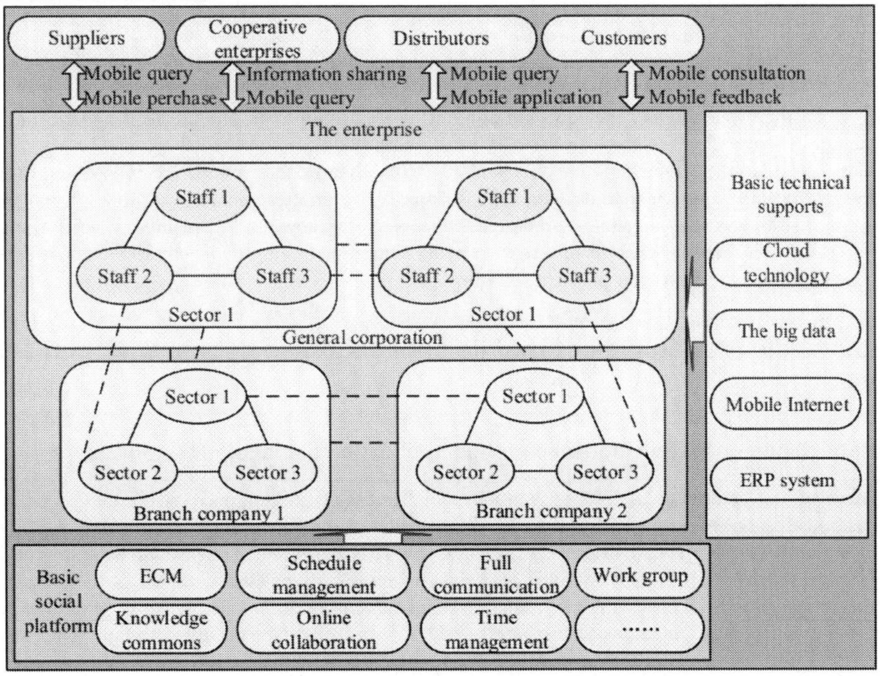

Fig. 1. The enterprise mobile application social mode.

friendly communication among departments stimulates employees' potential and increases work efficiency. The enterprise mobile application social mode includes internal social network, external social network, the basic social platform and basic technical supports, as shown in Figure 1.

The following analysis focus on the enterprise mobile application social mode in internal socializing and external socializing

3.1 The intra-enterprise mobile application social network

The intra-enterprise mobile application social network mainly can be divided into two levels ranging from staffs to departments, including the communication among staffs in the same department and the communication among different departments. There are many ways of collaboration and communication within enterprise. For examples, staffs create and manage the documents and track business workflow by building a team work space on the basic social platform; staffs can check the plan of project they participate in and contact with team members for online real-time collaboration according to the current project schedule; staffs can post documents and information allowing others' comments feedback, and realize enterprise content management and team knowledge sharing, which can remove information island. The intra-enterprise mobile application social network can achieve the collaboration in the work group while traditional ERP cannot. For example, it is difficult to define some virtual organizations crossing departments such as the strategic committee and the expert committee in traditional ERP, but the enterprise mobile application social network can define these by social groups.

Departments can benefit a lot from the enterprise mobile application social network. Staffs can do basic collaborative work such as sharing experience, transferring files and accessing to information through the social platform. Human resource department staffs use enterprise mobile social platform to simplify the application process; sales teams develop contacts and supervise sales channels via the mobile terminal; operation teams can track the supply chain information in details. In addition, chairman and general manager can query all real-time information in enterprise and communicate with all department managers and staffs to assist decision-making.

For group companies, besides the social network mentioned above, the intra-enterprise social network also includes the communication among different subsidiary corporations, the communication between Subsidiary Corporation and the general corporation. All levels of social communication need the support of the basic social platform and the basic technical supports.

3.2 The inter-enterprise mobile application social network

The enterprise mobile social network can also satisfy real-time social needs for the external suppliers, cooperative enterprises and customers. The enterprise staffs can communicate with customers instantly by literal statement and picture presentation on the mobile social platform when costumers have some doubts about products. Both can also join a video session to communicate directly, and enterprise staffs invite relevant product manager and staffs to help costumers if necessary. The customers can feed back and evaluate the products in enterprise mobile social network so that enterprises can provide better after-sales service.

When enterprise staffs have a project and need to sign a contract with suppliers, they can negotiate contracts through the enterprise social platform. Purchasing department staffs can set up a working group and invite product mangers join in to assist purchasing raw materials. Besides, the purchasing staffs can track procurement logistics information at all times via the mobile terminal.

Enterprises can establish communication mechanisms with distributors through the inter-enterprise mobile social network. Due to the uncertainty of market demand, distributors can apply replenishment through the enterprise social platform when distributors are in short supply caused by emergency, and distributors need track the logistics information at any time via the mobile terminal to reduce losses. Through the enterprise mobile social platform enterprises can communicate with their partners smoothly.

4 The management strategies of mobile application social network in manufacturing enterprise

4.1. To establish personalized enterprise mobile application social platform for users

The final users of the enterprise mobile application social network are internal staffs and external users such as suppliers, customers and cooperative enterprises. The main target of implementing the enterprise mobile social network is to meet the needs of all users to improve efficiency. So it is particularly important to distinguish the different demands and meet the demands of the two kinds of users. Generally, internal staffs' demands focus on the business, for instance, purchasing staffs need check the raw materials' logistics information and always keep in touch with suppliers, while the sales personnel need contact with customers and know the regional sales through the mobile social network at any time. Besides, external users' needs mainly focus on smooth communication to ensure the convenience, accuracy and timeliness of information interaction. For example, the suppliers can negotiate business and make sure the procurement details though the mobile application social network, and if they have any problems, they can consult the relevant staffs immediately.

4.2 To strength authority management and its alteration management

The clear role definition and content distribution are the key to ensure enterprise operate orderly within the enterprise mobile application social network, it is the vital problems of authority management and authority alteration that enterprises face during the implementation of the enterprise mobile application social network. Facing internal users and external users, the enterprise needs to define the enterprise's boundary by authority management and makes sure different kinds of users have different authority in the mobile social network. According to the different responsibilities of enterprise staffs, different staffs' authority makes the staff can only read and write information related to their business to prevent data leakage and illegal access. The external users' authority is mainly about the external exchange information with relevant department, for example, suppliers cannot access enterprise internal information but theirs related purchasing information such as the quantity and amount of material procurement, logistics, and history records, etc.

4.3 To pay attention to the security of the enterprise mobile application social network

While implementing the enterprise mobile application social network, the problems the enterprise faced are more complex than traditional ones in the mobile internet environment. So the enterprise needs to establish an effective information security system with some monitoring methods and safety precautions. To ensure the safety of core technologies and important customer information, the safety problems should be focused on all aspects including mobile terminal devices, information transmitting channel, system background, etc. Besides, the social network must be in charged with the firewall, antivirus program and other monitoring equipment. The staff enter into the enterprise mobile social network and do the daily business by the mobile terminal device, so when staffs access the enterprise internal system outside office area, it is necessary to ensure the mobile terminal devices are safe and the visitor is legal to prevent network wiretapping. In addition, staffs should separate the office application and their own applications to guarantee the office data security.

4.4 To promote staffs' corporate identity to adapt to the mobile social network

It is necessary for staffs to cultivate professional quality and improve the proficiency in the enterprise mobile application social network, which not only make staffs adapt to the mobile social environment and optimize business processes to create more value, but also strengthen corporate identity to ensure the information safety from inner enterprise. The staff who has professional ethics accomplishment and corporate identify is the first defence for enterprise information security, while the staff who has not is a threat to the enterprise mobile application social network and ERP system. Therefore, it is an essential task for the enterprise to improve staffs' professional quality and sense of security, and it is extremely important to guarantee the data security and the legal access to information in the security task when the enterprise deploy the mobile social network.

5 Conclusions

The rapid development of mobile internet requires enterprise to change internal management mode in response to new environment. The implementation of the mobile application social network can form a communication community for enterprises to communicate smoothly and improve the enterprise working environment. As the development of mobile terminal technology and enterprise social network and the ERP software's upgrade, the enterprise mobile application social mode becomes the new mobile office mode. The effective management strategies are the important guarantee for manufacturing enterprise to implement mobile application social networking.

Acknowledgment

The authors express sincere thanks to the National Science and Technology supporting Program of China for the financial support under the grant number No 2015BAF05B01, as well as the contributions from all partners of the mentioned project.

References

1. A. McAfee, Enterprise 2.0 new collaborative tools for your organization's toughtest challenges (Havard Business Rev. Press,Boston,2009)
2. M. Kuegler, S. Smolnik, G Kane. J. Strategic. Inf. Syst.24.2(2015),90-112.
3. M. Mäntymäki, K. Riemer, Int. J. Inform. Manage, 36, 6(2016), 1042-1052.
4. G. V. Krogh, J. Strategic. Inf. Syst, 21,2(2012), 154–164.
5. A. Majchrzak, L. P. Cooper, O. E. Neece, Manage. Sci, 50,2(2004), 174-188.
6. D. Denyer, E. Parry, P. Flowers, Long. Range. Plann. 44, 5–6 (2011), 375–396.
7. D. Richter, P. D. D. K. Riemer, J. V. Brocke, Bus. Inform. Syst, 53, 2 (2011), 89-103.
8. A .Greasley, Y .Wang, Comput. Ind, 82(2016),69-81.
9. China Internet Network Information Center. The thirty-eighth statistical report on the development of China Internet Network. Beijing: China Internet Network Information Center, 2016.08.03.
10. Ministry of Industry and Information Technology. Economic operation of the communications industry in December 2015. Beijing: Ministry of Industry and Information Technology, 2016.01.21.
11. Nielsen-CCData. The study on mobile social demand and behavior.Beijing: Nielsen-CCData, 2014

Management strategies of Bring Your Own Device

Peixuan Li and Liang Yang

School of Economics and Business Administration, Chongqing University, Chongqing, 400044, China
Corresponding Email: 357005895@qq.com

Abstract. The rapid development of mobile Internet and mobile terminals promote business office system from PC to mobile terminals gradually. Thus Bring Your Own Device (BYOD) has become one of the important development trends of enterprise office mode. We analyse the driving factors of implementing Bring Your Own Device, then point out some problems in the process of implementing Bring Your Own Device. Further, we propose the corresponding management strategies of Bring Your Own Device in order to provide references for enterprises to meet the need of mobile office.
Keywords. Bring your own device, management strategies, mobile internet.

1 Introduction

Internet and mobile communication are two major technology which grow rapidly and have great market potential today. Mobile Internet, the combination of them, is growing at an explosive speed and has brought great changes to various industries in recent years [1-2]. According to Cisco's 2016 global mobile data traffic forecast report release, the number of mobile users will reach 5.5 billion by 2020, accounting for 70 percent of the global population, and the number of smart mobile devices is expected to 72 percent of total global mobile devices, which is growing rapidly compared with 36 percent in 2015 [3]. This can be seen the advent of the era of mobile is unstoppable.

With the rapid development of mobile Internet and mobile intelligent terminals, employees' pursuit of new technology and personalization promotes enterprises' increasing demand for mobile applications, and enterprises' daily office work are gradually expanding from PC to mobile terminals. Bring Your Own Device (BYOD) as one of important mobile office modes has received the academia's extensive attention in recent years. At present, the literature mainly focus on BYOD's concept, development process, development status, demand level, the advantages and disadvantages, the risks and user experience and so on. Morrow B et al [4] think that Bring Your Own Device (BYOD) mainly refers to a policy that some enterprises allow employees to bring their own laptops, tablets, smart phones and other mobile devices to workplace, by which they can access enterprise's internal information and use enterprise's licensed applications. Zhang Jian et al [5] and Thomson G [6] show that the rise of Bring Your Own Device (BYOD) thanks to the transformation of IT consumption, and the development process of mobile office has mainly experienced three stages: offline mobile office, online mobile office, intelligent mobile office. Yuan Hong [7] thinks that the computing power of intelligent mobile terminals has far more than the demand of business, and there begin to appear a phenomenon of extensive use of personal devices for work in enterprises. In addition, Bring Your Own Device (BYOD) is also widely used in the field of education, and has become a new education information service mode in recent years [8]. For example, Song Yanjie [9] reports a one-year study on the project of "Bring Your Own Device (BYOD) for seamless science inquiry" in a primary school in Hong Kong, and finds that the students advanced their understanding of the anatomy of fish well beyond what was available in the textbook and they developed positive attitude toward seamless science inquiry supported by their own mobile devices. Kobus Martijn B. W. [10] investigates ownership and on-campus use of laptops, tablets, and smartphones, using survey information on Dutch university students, and it appears that the current high ownership rates of mobile IT devices by no means imply students' preference or support for university Bring Your Own Device (BYOD) strategies. What's more, Qian Yuming et al [11] analyze the three demand levels of enterprise business mobility, which in order are general office demand, enterprise application mobility demand, the scene perception of mobile devices and business smoothly handover with the scene demand. Chen Fen et al [12] and Singh N [13] analyze and compare the advantages and disadvantages of Bring Your Own Device (BYOD), then forecast the future development trend of Bring Your Own Device (BYOD) . Zhou Hua [14] and Miller K W [15] et al analyze the security risks of Bring Your Own Device (BYOD) , which provides an important reference for the security construction of mobile office. Tokuyoshi B [16] puts forward some suggestions of implementing Bring Your Own Device to enterprises from the perspective of user experience.

As we can see from the above, at present there is little literature that research management strategies of implementing Bring Your Own Device based on mobile Internet from the perspective of business management. In view of this, this paper bases on the analysis of the driving factors of implementing Bring Your Own Device, then points out some new problems in the process of implementing Bring Your Own Device. Further this paper proposes corresponding management strategies which can promote enterprises to implement Bring Your Own Device, so that we can provide references for enterprises to meet the arrival of the era of mobile office.

2 Driving factors of implementing BYOD under the environment of mobile internet

2.1 The rapid development of mobile Internet and mobile terminals make mobile office become reality

With the rapid development of mobile Internet, the characteristic of mobile terminals that can surf the Internet anytime and at anywhere makes it become an increasingly important part in our daily life. According to statistics shown by the Ministry of China, by December 2015 the traffic of accessing mobile Internet up to 4.187 billion G and was up 103 percent year on year [17]. Along with the government and enterprises vigorously constructing “smart city” and “wireless city”, public wireless network has been spread rapidly, and the proportion of Internet users who access the Internet through WiFi is as high as 91.8 percent [18]. The improvement of network environment and the rapid popularization of mobile Internet stimulate the demand for the application of mobile Internet and have shaped an new social life form. At the same time, the rapid development of mobile Internet and mobile terminals drives enterprises to implement mobile application solutions, and more and more employees want to freely choose their mobile intelligent devices to handle their work through mobile Internet. According to the survey data of Forrester, the majority of enterprises have begun to implement the plan of Bring Your Own Device, and 70 percent of enterprises will use some form of Bring Your Own Device program, and 62 percent of employees are already using smartphones to deal with their office affairs, and 56 percent of employees are using tablets to handle daily work. In the mobile Internet age, enterprises need more rapid and effective way of communication and way of handling business to enhance their competitiveness, and comprehensively implementing Bring Your Own Device is an inevitable trend.

2.2 The employees’ growing demand for mobile office prompts enterprises to implement BYOD

With the Promotion of mobile communication and mobile terminal in service quality, function design, user experience and other aspects, more and more employees want to handle their work through mobile Internet. Because some applications of intelligent terminals have simpler operation mode and superior user experience than PC, employees want to get rid of dependence oncable network and laptop by Bring Your Own Device mode. In this way, they can make full use of fragmented time to improve the utilization of time and their work efficiency. For employees, they sometimes need to handle some business on the way to and from work; for managers, they need to look, read and review internal documents on business trip. In addition, many employees need track the progress of project, submit collected information to the internal system timely, solve users’ question in the first time. So they need mobile office to help them free from the constraints of space and time, then easily deal with urgent business in hand at anytime and anyplace. The products about mobile office will not only help enterprises to implement Bring Your Own Device program to meet employee’s demands for mobile office, but also will make enterprises get more business opportunities.

2.3 Mobile office mode is becoming new competitiveness of enterprise

The emergence of mobile office bring enterprise office to another new level, and enterprises' demand for information is increasingly strong. It has become an important target for enterprises to improve the core competitiveness by mobile office. According to 2016 mobile office panorama report given by the Mobile Information Research Center, mobile office is developing at an explosive speed, and more than 93 percent of CIO indicates recognition of mobility, and the number of enterprises without planning in 1-3 years falls sharply from 82.8 percent in 2013 to 6.2 percent in 2015, and the enterprises that have already deployed or being deployed are more than 40 percent. Mobile office has become an important strategic plan of enterprises. The innovation of mobile office mode makes enterprises greatly accelerate the process of enterprise mobile informationization, which leads to the improvement of the executive power and competitiveness of enterprises. Mobile office is an assistant for managers and a platform for work and communication. With the aid of mobile office, managers can timely release business notices to let employees understand enterprise culture information. At the same time, employees can submit application to managers through the mobile office system to obtain timely approval, which makes enterprises’ work efficiency improved greatly. In addition, the application of mobile office, breaking the limitation of geographical space and organization, makes office action integrated. With the fierce of marke competition, mobile office has become a wise choice for enterprises to enhance the core competitiveness.

3 Problems in implementation of BYOD faced by enterprises

3.1 Enterprise’s application data information faces the risk of leakage

In the process of implementing Bring Your Own Device program, the influx of a large number of mobile devices has increased the amount of terminals that access enterprise. Portability of the mobile terminal and the randomness of accessing data information make enterprises' confidential data faces the risk of leakage. In general office environment, enterprises' data exist in the form of electronic documents, flowing in the office terminals and office applications, then the storage and transmission of the data are faced with different threats. The realization of mobile management of electronic documents is based on wireless network information transmission and cable network information transmission. In the office environment where information transmission entirely depends on network, no matter enterprise choose which kind of information transmission mode, the network always has many unsafe factors. In the mobile office system which involves information exchange and complex personnel communication, handling electronic documents often need information interaction. Because some electronic documents are confidential, it makes electronic documents information faces various risks brought by open network, such as computer Trojan, the invasion of virus, illegal users stealing and tamperingwith electronic documents, and so on. As a result, although Bring Your Own Device brings convenience to employees for work, it increases security risks of enterprise data information leakage.

3.2 The management and control boundary of mobile devices between enterprise and employees is not clear

Mobile terminal is an open platform, which can be installed software to meet the heterogenous needs of different users and be combined with related enterprise applications for mobile office. The first feature of Bring Your Own Devices is that employees can ues their own mobile terminal. In relevant legal category, the definition of ownership states a fact that employees have absolute ownership and control of the devices. But in the sense where employees can ues their own mobile terminal, the mobile device is a compromise between employees and enterprises. In terms of control, employees need to accept the control policy made by enterprises for security, such as strict entry authorization, device safety controls. Even enterprises and employees have reached a contract, but the devices belong to employees, so employees can control the devices as they like. Thus enterprises still need to resort to their control and management in a reasonable range. In addition, employees' personal information and enterprises' business information are mixed in the same device, while employees' main concern is their personal privacy data, so relevant monitor and management are needed in an appropriate range. This way can not only guarantee that enterprise's confidential information will not leak, but also ensure the security of user's personal privacy data. However, in the process of implementing Bring Your Own Devices program, the above mentioned management control boundary of mobile devices is not clear and most enterprises don't make strict and clear provisions in this aspect.

3.3 The employees' acceptability for BYOD is inconsistent

With the popularity of mobile Internet, more and more employees choose using Bring Your Own Device to work, but the popularity doesn't reach as high as we thought. In daily work, some employees don't use mobile intelligent terminals, such as older employees, most of them are used to the mobile devices easy to be operated, and they can not accept intelligent mobile office as fast as young people. Although mobile office is more convenient, the comprehensiveness and easy operation of desktop office system can not be denied, which is another reason why some employees can not accept Bring Your Own Device. Moreover, there exists a problem that employees' acceptability for Bring Your Own Device is inconsistent, which give rise to many difficulties when making related policies of Bring Your Own Device and implementing Bring Your Own Device , such as the scope of the executor is difficult to delineate, and employees' work is hard to perfect docking in the process of mobile officing. Therefore, while considering implementing Bring Your Own Device strategy, it should be thought carefully about employees' specific ideas and needs for mobile office. Then according to the enterprise's nature and employees' specific work, to gradually generating enterprise mobility strategies can promote mobile office and improve employees' work efficiency, which can become a new competitiveness of enterprises.

4 Management strategies of enterprises to implement BYOD

As shown in Figure 1, aiming at problems in implementation of Bring Your Own Devices faced by enterprises, next we will propose the corresponding management strategies of Bring Your Own Device in order to provide references for enterprises to meet the need of mobile office.

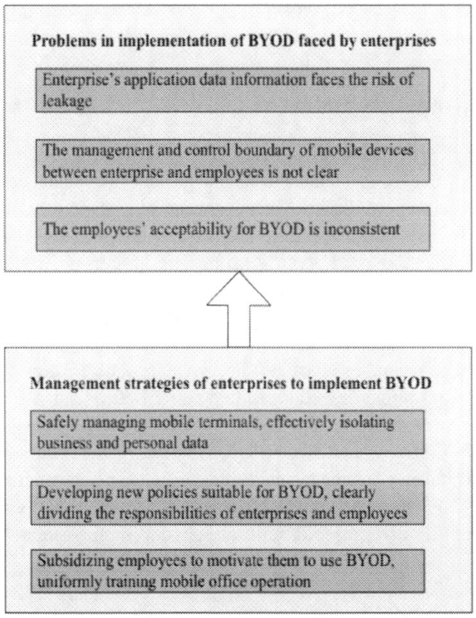

Fig. 1. Management strategies of Bring Your Own Device.

4.1 Safely managing mobile terminals, effectively isolating business and personal data

On the one hand, as a basis for judging terminals' current security state, IT management department should check whether terminals are installed anti-virus software and whether the anti-virus software is updated. And the policies which are used to ensure data security and make up for security vulnerabilities caused by employee's weak security consciousness, should include configurating and checking screen saver, checking the file sharing, checking account security and so on. On the other hand, enterprises should have full control of enterprise data information and can uniformly send the policies to each client through the BYOD server. Before the use of BYOD software, employees need to sign a confidentiality agreement to comply with the corporate policy and protect corporate data. Enterprises must ensure their data in mobile terminals have no risk to be leaked in the process of storing and transmitting. And for secure storage the data should be isolated in the sandbox. Enterprise data and applications have been isolated in the "security sandbox", and important documents and data are automatically encrypted. In this way the contents that users read and edit within the sandbox can not be copied to the outside, but also the outside data can not be copied into the sandbox. From the perspective of enterprises, this method not only ensures the safety of enterprise data, but also prevents the illegal business information (such as news, entertainment information), virus, Trojan and other illegal procedure from spreading and infecting the inner-enterprise. From the perspective of users, their personal data have been isolated outside of the "security sandbox", so they don't have to worry their personal data will flow into the enterprise Intranets and lead to the disclosure of personal privacy.

4.2 Developing new policies suitable for BYOD, clearly dividing the responsibilities of enterprises and employees

Before implementing Bring Your Own Devices strategy, enterprises should make clear, concise and strict rules for employees who use their own devices for work. Everyone who participates in the BYOD program has to sign the terms of use, and violators will not be allowed to use their own devices for work. Besides, enterprises and users should sign a privacy treaty, in which users can know what data BYOD software may involve, and obligations that enterprises and users need to follow. BYOD software should be able to allow users to decide whether the information in their terminals can be collected by the server, including GPS location, personal information and so on. In addition, it should be able to implement remote policy configuration management for mobile devices participating in the Bring Your Own Devices strategy, and the policies should be strictly divided into the policies that enterprises must follow and the policies that users can select. Users and managers could manage mobile terminals by the BYOD's server side, but their management authorities should not be the same due to their diffrenent types of job. The policies designed for managers are used for all mobile devices involved in BYOD, and the application range of these policies includes operating system types, roaming ways, encryption, the complexity of password settings and so on. Among the policies designed for users, the first are privacy policies, which includes GPS positioning, whether personal information needs to be checked, etc.. Followed by some service functions, such as remote locking, sending text messages to the terminal, removing password, clearing information and so on. When users act against these policies, it should promptly notify managers to warn the violator and record violations, and the BYOD function of serious violator should be disabled directly and relevant data should be cleared.

4.3 Subsidizing employees to motivate them to use BYOD, uniformly training mobile office operation

When employees bring their own devices to handle office business, there should have certain requirements for mobile devices, for example, it had better to choose reputable brand mobile terminals that carry intelligent operating system. But not all employees' mobile devices meet the above requirements, thus if enterprises want to implement Bring Your Own Device comprehensively, it will certainly increase employees' average purchase costs of mobile devices and extra Internet fees in the process of mobile officing. If condition permits, enterprises should give corresponding subsidies to employees according to their actual demand for mobile office to motivate them to bring their own devices. For example, the job nature of salesman determine that their demand for mobile office is large, then enterprises can bulky purchase and configure mobile office devices; for general demand for mobile office, such as managers who release notices and employees who submit application to manager to obtain approval timely through the flow system, enterprises can give employees appropriate Internet access subsidies. Except that, enterprises can also link employees' work performance with their subsidies, then develop a reasonable and scientific distribution system of subsidies to encourage the work passion of employees. In addition, there is part of employees who are not familiar with mobile office operation, then enterprises should set up a special training course to teach them how to operate relevant mobile office system, and make them adapt to the new office mode as soon as possible. Meanwhile it will make employees carry out cooperative work more smoothly in the process of mobile officing and facilitate unified management of enterprises.

5 Conclusion

Compared with traditional PC office mode, mobile office mode has significant change and innovation. Enterprises which adopt Bring Your Own Device program not only improve the satisfaction and work efficiency of employees, but also strengthen internal information communication and information exchange with external. With the rapid development of mobile Internet and mobile terminal, as well as the ERP software vendors do further researches and make breakthroughs in enterprise-class mobile applications, Bring Your Own Device will certainly be implemented by more enterprises, which will not only make enterprises experience more convenient and efficient office mode but also enhance enterprises' competitiveness.

Acknowledge

The authors acknowledge funding from the National Science and Technology supporting Program of China (grant number No. 2015BAF05B01), as well as the contributions from all partners of the mentioned project.

References

1. S. H. Lee, B. H. Chang. Int J Mob Commun, **3**, 11(2013), 279-298
2. G Wu, S. Talwar, K. Johnsson , et al. Commun Mag, **49**, 4(2011), 36-43
3. CISCO reports: the number of mobile users will reach 5.5 billion in 2020. http://network.chinabyte.com/279/13690279.shtml
4. B. Morrow, Net. Sec, **12**(2012), 5-8
5. J. Zhang, X. Liu, HLJ Sci Technol Info, (2013), 99
6. G. Thomson, Net Sec, **2**(2012), 5-8
7. H. Yuan, Chin New Technol New Prod, **22**, 1(2013), 22
8. L.Y. Li, Y.L. Zheng, Mod Dist Educ Res, **6**(2012), 71-76
9. Y. Song, Comput & Edu, **74**(2014), 50-60
10. M.B.W. Kobus, P Rietveld, JN.Van Ommeren, Comput & Edu, **68**(2013), 29-41
11. Y.M. Qian, Z.J. Dong, D Lu, W Wang, ZTE Technol, **6**(2013), 33-38+42
12. F. Chen, Z.W. Zheng, FJ Comput, **9**(2013), 95-96+111
13. N. Singh, J. Bus Manage & Soc Sci Rea, **3**, 1(2012), 1-12
14. H. Zhou, X.B. Zhang, L.L. Yang, W.M. Yang, J.H. Long, Silicon Vally, **22**(2014), 142-145
15. K.W. Miller, J.M. Voas, G.F. Hurlburt, IT Prof, **14**, 5(2012), 53-55
16. B. Tokuyoshi, Net Sec, 4(2013), 12-13

17. Ministry of industry and information, Economic operation of the communications industry in December 2015. 2016.01.21
18. China Internet Network Information Center. The thirty-seventh statistical report on the development of China Internet Network. 2016.01.2

Synchronization Time Division Multiplexing Bus Communication Method Using Serial Communication Interface

Xi Liu, Ruihai Dai and Bifu Qian*

State Grid Wenzhou Electric Power Supply Company, Zhejiang, Wenzhou, China
Corresponding Email: qianbifu@163.com

Abstract. It provided a synchronization time division multiplexing bus communication method using serial communication interface, in which input and output data lines of a host and several slaves being connected with the bus. Then the detailed communication mode was stated. The method could be applied in all MCU, meanwhile meets the requirements of high speed for the real time performance, and solves the problem of bus communication between internal modules of a device and realizes controllable communication real time performance, reduced hardware circuit complexity as well as enhanced universality and reliability.

1 Field of invention

At present, bus is commonly utilized as communication channels between the master board and the input or output board in automatic devices. On the input or output board MCU is provided for processing a variety of input or output signals, which communicates with the MCU on the master board. Consequently the timeliness and stability of automatic devices is dependent on architecture of the internal bus. However, normal serial or parallel communication manner could not meet requirements of real time communication under strong interference [1,2]. As a result, the manufacturers proceed to develop respective bus for their own devices so as to achieve the requirements of real time communication under strong interference. However, due to the fact that there is a wide variety of internal bus, such as parallel bus and serial bus, bus formed by FPGA and formed by CPLD, as well as formed by MCU of other types, bus with SPI interface and with SCI interface, as well as with CAN interface or other types of interface, it's very difficult to develop a bus with desirable simple architecture applicable to the communication between the master board and the input or output board using short and fixed length communication messages, thus making the requirements for real time communication hard to be achieved [3,4].

2 Summary of invention

The object of the present invention is to overcome above disadvantages in prior art. There is provided a synchronization time division multiplexing bus communication method, which utilizes just two physical connection lines to implement differential connection with high reliability and meet the requirements of high speed for real time communication on the basis of most ordinary MCUs, thereby solving the problem in bus communication between internal modules of a device and enabling for controllable communication real time performance, reduced hardware circuit complexity as well as enhanced universality and reliability [5,6].

The above objects of the present invention are achieved by the following technical solutions:

A synchronization time division multiplexing bus communication method using serial communication interface, including a host and plural slaves, the host and slaves all comprising SCI (Serial Communication Interface) and timer provided micro control unit (MCU), the input and output data lines of the host and slaves all being connected with the bus, characterized in that, the bus communication method is as follows: each time when the host sends a down-link data message required by respective slaves, the slaves receive the data via the SCI and proceed to send up-link data messages after waiting for a time interval, and each time when a previous slave completes to send the up-link data message, the next slave begins to send the up-link data message after waiting for an identical time interval, and so on, until the bus communication method is accomplished.

Preferably, the slaves send the up-link data messages in such a manner that after a slave receives data, determine whether the data is host data or not; if the data is not host data, the slave doesn't process the data and continue to receive data; and if the data is host data, the slave initiates the timer and implements data validation to the received host data; in the case of validation failure the slave turns off the timer and continues to receive data, and in the case of validation success, the slave proceeds to process data; when the timer interruption occurs, the slave sends data message via the SCI.

In order to enhance communication reliability and reduce bit error rate, in terms of bus physical architecture, the bus is converted into differential lines by a single-ended differential signal conversion chip, and the host and slaves are connected to the bus via the differential lines.

Preferably, the time interval mentioned above is a period of an Idle.

Preferably, the bus utilizes half-duplex communication mode.

By the above method, the bus physical architecture of the present invention could be simplified, in which real time bus communication with high reliability could be realized only using two differential lines, and the requirements of high speed for the real time performance could also be achieved. In the case of low external interference real time bus communication could be realized using even only one signal lines. The bus could be implemented using any kind of MCU provided with SCI and timer, therefore the demand for MCU hardware is low. For different slaves, respective baud rate and waiting time could be chosen according to actual situation so as to meet real time requirements. As synchronization time division multiplexing method is employed, the host could fill data memory according to the time interval, and the procedure for processing bus data of the slaves could be simplified. The message sent by the host could be used as beat of message transmission of the slaves. Each time when the host sends a message, time setting could be made to the slaves, thus ensuring the accuracy of transmission beat of the slaves. Variable period could be selected based on task amount of the host to eliminate loss of synchronism of the slaves. The present invention solves the problem of bus communication between internal modules of a device and realizes controllable communication real time performance, reduced hardware circuit complexity as well as enhanced universality and reliability.

3 Brief describe of the drawings

Figures and tables, as originals of good quality and well contrasted, are to be in their final form, ready for reproduction, pasted in the appropriate place in the text. Try to ensure that the size of the text in your figures is approximately the same size as the main text (10 point). Try to ensure that lines are no thinner than 0.25 point.
The present invention will be further described in detail below in connection with embodiment and accompanying drawings,

Figure 1 is a schematic diagram showing system configuration in an embodiment of the present invention

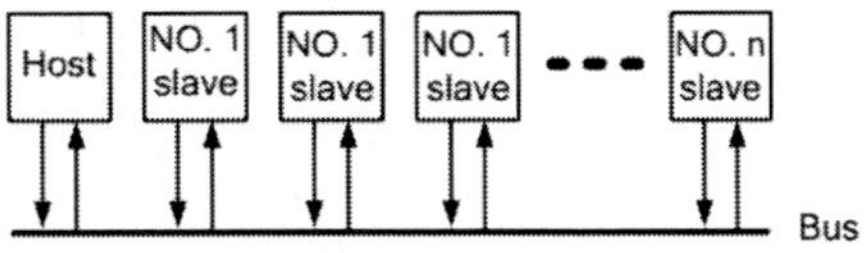

Fig. 1. System diagram.

Figure 2 is a schematic diagram showing format of a message frame with address bit in the embodiment of the present invention [7];

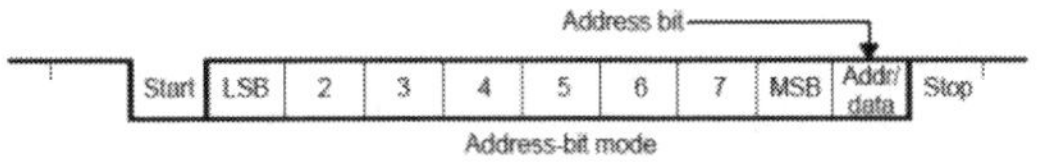

Fig. 2. The diagram of communication format.

Figure 3 is a schematic diagram showing communication procedure in the embodiment of the present invention;

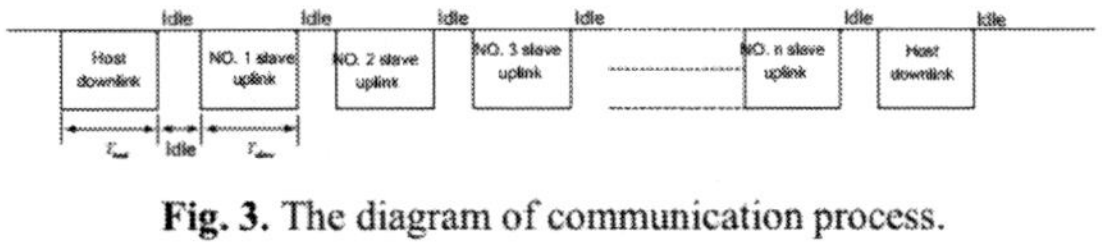

Fig. 3. The diagram of communication process.

Figure 4 is a flow chart showing a salve sending uplink data message in the embodiment of the present invention.

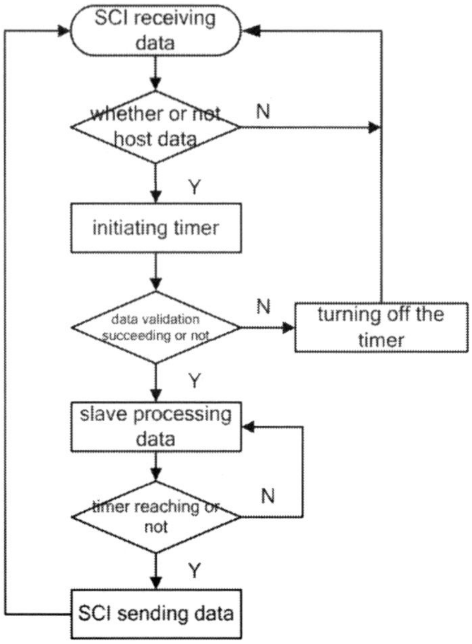

Fig. 4. The diagram of software process.

4 Describe of preferred embodiment

These and other aspects, features and advantages of which the invention is capable of will be apparent from the following description of embodiment of the present invention.

Referring to Figure1, there is provided synchronization time division multiplexing bus communication method using serial communication interface to solve the problem of bus communication between internal modules of a device. The bus utilizes half-duplex communication mode, which could realize real time bus communication with high reliability and meet the requirements of high speed for the real time performance. In the embodiment of the present invention, the input and output data lines of the host and multiple slaves are all connected with the bus. At any time only one equipment occupies bus resource to send data, and other equipment receives data. In order to enhance communication reliability and reduce bit error rate, in terms of bus physical architecture, the bus is converted into differential lines by a single-ended differential signal conversion chip, i.e. differential lines are utilized to send and receive data, and the host and slaves are connected to the bus via the differential lines.

Figure 2 is a schematic diagram showing frame format of a message with address bits in the embodiment of the present invention. As shown in Figure 2, there is a start bit, a data bit, an address bit and a stop bit. A message frame with address bit is employed for communication in the embodiment, which is accomplished as follows:

The communication loop is divided into up-link message and down-link message, wherein the down-link message has an address bit of "1" and is sent from the host to the slaves. In the manner of "one host and multiple slaves" and "the slaves receive what the host sends", the slaves pick up desired information respectively from the message received. The rate of the down-link message is treated as operation beat of the slaves.

The up-link message has an address bit of "0" and is sent from the slaves to the host in a time division multiplexing manner. When a slave receives a message with address bit of "1", it sends up-link data after a certain delay time period according to its own address number, thereby achieving multiplexing. When a slave receives a message with address bit of "0", it doesn't handle it.

In the embodiment the symbol "T_{host}" represents the time period of sending down-link message by the host, and the symbol "T_{slave}" represents the time period of sending up-link message by each slave, and the symbol "$Idle$" represents the waiting time between two times of sending.

Referring to Figure3, A synchronization time division multiplexing bus communication method using serial communication interface is provided in the embodiment, in which the host sends a down-link data message, the NO 1 slave receives the data via the serial communication interface, i.e. SCI and proceeds to send up-link data messages after waiting for a time interval equal to $Idle$; and after waiting for a time interval equal to $Idle$ again, the NO 2 slave proceeds to send up-link data messages, and so on. The NO N slave proceeds to send up-link data messages after waiting for a time interval equal to $Idle$, and then the host sends a down-link data message once more after waiting for a time interval equal to $Idle$, thereby the data communication being completed.

The slave opens the SCI and proceeds to receive the data message, then determines whether the address bit indicates host data or not. In the case of other slave data, it will not be handled. In the case of host data, the slave initiates the timer (the timing length is dependent on $Idle$, T_{slave} and N). Then the slave implements data validation to the received host data. In the case of validation failure the slave turns off the timer and continues to receive data, and in the case of validation success, the slave proceeds to process data. When the timer interruption occurs, the slave sends data message via the SCI.

The time for message sending is calculated as follows:

the sending time t_n of the Nth slave:

$$t_n = Idle + (Idle + T_{slave}) \cdot (n-1) \tag{1}$$

in which n is address number from 1 to N;

And one communication cycle T:

$$T = T_{host} + Idle + (Idle + T_{slave}) \cdot n \tag{2}$$

in which n is address number from 1 to N.

The one communication cycle T and the maximum number of slaves in system could be determined according to actual requirements of the system, and appropriate baud rate and waiting time could be chosen to accomplish data communication within predetermined period of time. For example, provided the communication cycle of the system is 500us, the maximum number of slaves is 10, the baud rate is set as 3.75mbps, and the down-link message of the host has 16 bytes while the up-link message of the slaves has 10 bytes, then the waiting time,

$$Idle = [500 - (16 \cdot 11/3.75) - 10 \cdot (10 \cdot 11/3.75)]/(10+1) \approx 14.5us \tag{3}$$

By setting above waiting time for the slaves, real time communication with cycle of 500us could be achieved.

In conclusion, the bus physical architecture of the present invention could be simplified, in which real time bus communication with high reliability could be realized only using two differential lines. In the case of low external interference real time bus communication could be realized 1using even only one signal lines. The bus could be implemented using any kind of MCU provided with SCI and timer, therefore the demand for MCU hardware is low. For different slaves, respective baud rate and waiting time could be chosen according to actual situation so as to meet real time requirements. As synchronization time division multiplexing method is employed, the host could fill data memory according to the time interval, and the procedure for processing bus data of the slaves could be simplified. The message sent by the host could be used as beat of message transmission of the slaves. Each time when the host sends a message, time setting could be made to the slaves, thus ensuring the accuracy of transmission beat of the slaves. Variable period could be selected based on task amount of the host to eliminate loss of synchronism of the slaves.

The internal bus of the present invention only uses two differential lines to implement differential connection with high reliability on the basis of most common MCU, meanwhile meets the requirements of high speed for the real time performance.

5 What is claimed is

1). The synchronization time division multiplexing bus communication method using serial communication interface, including a host and plural slaves, the host and slaves all comprising Serial Communication Interface and timer provided micro control unit, the input and output data lines of the host and slaves all being connected with the bus, characterized in that, the bus communication method is as follows: each time when the host sends a down-link data message required by respective slaves, the slaves receive the data via the SCI and proceed to send up-link data messages after waiting for a time interval, and each time when a previous slave completes to send the up-link data message, the next slave begins to send the up-link data message after waiting for an identical time interval, and so on, until the bus communication method is accomplished.

2). The synchronization time division multiplexing bus communication method using serial communication interface according to claim 1, characterized in that, the slaves send the up-link data messages in such a manner that after a slave receives data, determine whether the data is host data or not; if the data is not host data, the slave doesn't process the data and continue to receive data;

and if the data is host data, the slave initiates the timer and implements data validation to the received host data; in the case of validation failure the slave turns off the timer and continues to receive data, and in the case of validation success, the slave proceeds to process data; when the timer reaches, the slave sends data message via the SCI.

3). The synchronization time division multiplexing bus communication method using serial communication interface according to claims 1 or 2, characterized in that, in terms of bus physical architecture, the bus is converted into differential lines by a single-ended differential signal conversion chip, and the host and slaves are connected to the bus via the differential lines.

4). The synchronization time division multiplexing bus communication method using serial communication interface according to claims 1 or 2, characterized in that, the time interval mentioned above is a period of an Idle.

5). The synchronization time division multiplexing bus communication method using serial communication interface according to claims 1 or 2, characterized in that, the bus utilizes half-duplex communication mode.

References

1. Cao Hai' ou, Zheng Jianyong, Cai Yueming. Study of substation automation communication system based on CAN bus. Proceedings of the CSUEPSA. 2002, 14(6): 24-26. (in Chinese).
2. Liu Yan, Qin Wen. Desig n of communication network for integrated substation automation system based on CAN field-bus). Relay, 2007, 35(18): 54- 56. (in Chinese).
3. Zhu Huawei, Wu Ailian, Yang Rufeng, et al. Design of CAN field bus communication adapter for transformer station. Proceedings of the CSUEPSA. 2005, 17(5): 86- 89. (in Chinese).
4. Zhang Zhizhe, Li Xingyuan, Cheng Shijie. Structures functions and implementation of united information system for smart grid. Proceedings of the CSEE, 2010, 30(34): 1-7. (in Chinese).(in Chinese).
5. Lin Chuang, Shan Zhiguang, Ren Fengyuan. Quality of service for computer networks. Beijing: Tsinghua University Press, 2004: 174-177. (in Chinese).
6. Miao Xin, Zhang Kai, Tian Shiming, et al. Information communication system supporting smart grid. Power System Technology, 2009, 33 (17): 8-13. (in Chinese).
7. Dong Zhangzhu, Tang Ming, Li Ning. Network Management in Master Station of Electric Power Systems Dispatch. Power System Protection and Control, 2008, 36 (18) : 62-64. (in Chinese).

Under the Background of "Internet +" the Social Anomie and Governance Plan in Chinese Modernization Process

Yundong Jiang[1] and Feng Liu[2,*]

[1]School of Marxism Education, University of Electronic Science and Technology, Chengdu Sichuan, China
[2]Peking University, School of Marxism, Beijing, China
Corresponding Email: liufeng2010beijing@126.com

Abstract. During the reform and opening up in China, which appeared political anomie, values astray inevitably, and so on. To achieve the current benign social governance, the rule of law need to improve the Chinese characteristics of the socialist system, to promote the growth of the rule of law; which is need to improve socialism with Chinese characteristics system of moral standards to guide members of the community to actively Pursuits socialist core values; innovation and social governance mechanism, vigorously develop modern new social organizations; and which is need to explore the implementation of "Internet + supervision" mode, the innovation market regulatory approach, innovation in public service delivery.
Keywords. Reform and opening up, modernization, social anomie, social governance, "internet + supervision" mode.

1 Introduction

Since the reform and opening up, the basic process of social change can be attributed to institutional change and modernization in transition. The whole society to a market economy, mainly reflected "social resource allocation, resource ownership in the social distribution relations, social class structure, functional structure of society, social, political and public life, social spiritual and cultural life" [1] change. Big social changes and big changes for the modernization has brought a number of positive factors, but also for the social anomie increased incentives. "In the development of civilization, some temporary confusion difficult to avoid the growth of a civilization is in fact the order - disorder - order infinite process of negation of negation" [2] . For social anomie situation of China's modernization process of reform and opening up, the current need to absorb and draw on advanced social ideas and way of governance at home and abroad, to build legal, ethical, institutional mechanism, "four in one" social governance system to achieve benign social governance. Based on this understanding, the paper since the reform and opening up China's modernization drive "micro change" as the Object of social anomie in this period overall analysis. Based on this analysis, "Internet +" in the context of our response to the governance of social anomie.

2 The brief of social anomie and governance

Anomie means the loss of the rule, which is a detailed description of the old social values being questioned, or even destroy, not binding on the members of the society; and the new social norms, values have not yet fully accepted members of society, causing social behaviors of its members lack a clear norm-governed society into regulatory mechanisms failed state, order and social structure distortion. French sociologist Durkheim Duke use earliest ly of the concept of social anomie, used to refer to the phenomenon of the disintegration of social norms missing personal situation of chaos and social behavior. Laterly, the American sociologist Merton further that social anomie is a state that "When the means of social objectives set forth, and to achieve this goal and to society norms guiding inconsistent social anomie will appear" [3]. In short, the social anomie refers to the adjustment of norms of social life, the effectiveness of failure, causing a state of crisis. According to the content of social anomie involved, including social and political anomie anomie, anomie and cultural anomie and other areas of social life.

Facing the social anomie in social change, social governance gradually become a hot topic. Governance is essentially a process of the various parts and members of the community regulation, including the way the object of social governance and the like. The basic form of social governance was mainly two kinds of governance and the rule of law. In terms of content, social management system can be divided into non-institutional governance and management, governance and legal culture of governance. Governance and legal system of governance called "hard control", non-institutional governance and culture of governance called "soft governance."

For China's modernization inevitably emerging social anomie, in the "Internet +" in the background, it is necessary "hard control" and "soft governance" to effectively combine social governance body to mobilize the government, society, people take part among social governance, social benign operation.

3 The overall analysis of the process of modernization of social anomie since reform and opening

"Modernity breeds stability, but modernization process was the breeding of turmoil" [4]. For China's modernization process, the modern Kim Sung since the reform and opening up, China is in progressively abandon traditional to modern society among the major powers appeared to err corruption, political anomie, social crime and legal anomie, social values and so on.

3.1 Political anomie power corruption

The abuse of public power leads to political life in "privatization of power" and "power of capital" and anomie surge. In the context of power anomie, corruption will flourish, and thus exacerbated the alienation of public power. From the general characteristics of the process of modernization of the world's point of view, the transition from the traditional society to a modern society, must be accompanied by powers of anomie and the breeding of corruption. With our society "micro change", a "systemic corruption." Our power anomie mainly manifested in two aspects: Firstly, the cadre selection and appointment of cronyism, bureaucracy, dereliction of duty on the phenomenon of work style. Secondly, money, corruption, bribery, the amount and number of people involved in the growing phenomenon.

3.2 Social crime and legal anomie

Since the reform and opening up, the tide of commodity economy to break through all the barriers hinder the development of productive forces, the existence of material interests of legitimacy and rationality be recognized by the people, stimulate people's desire for sexual excesses, evoked people's motive for the crime. Especially, the failure of the legal norms and weakened social crime is increasing. On the situation in terms of drug trafficking, "1988-1989 Chinese public security organs seized from drug offenses in 4485; in 1991 seized a total of 8395 onwards, drug smugglers 18479; 1992 14701 cracked in drug offenses, confiscating 28,291 people involved arrest drug criminals 7025 " [5] .

3.3 Social values err situation

Since the reform and opening up, China's biggest achievement is the beginning of the human spirit free spirit is formed, the emergence of consciousness. "Consciousness and Spiritual Freedom is the presence of one of the two sides. People only exists as a subject it may be free, man is free only it may be subject" [6]. This consciousness continues awakening to social development has brought unlimited vigor and vitality, but also to the community brought considerable confusion, to individuals brought deep anxiety and loneliness. The values and norms reflected the value of the old challenges of the new specification by the values, social values rendered Multiple changeable situation. A variety of values, social norms conflict with each other, intertwined, constitute the most distinctive feature of reform and opening up the social value system of norms in China. Among them, the performance of the most fundamental values of the conflict is a conflict the new market economy values and the values of the old planned economy between.

4 Deal with the process of modernization of social anomie governance plan under the "Internet +" background

Availability for social anomie in China's modernization process since the reform and opening up, especially in the "Internet +" background, promoting healthy social governance, should proceed from the following aspects: Firstly, the need to absorb and draw on advanced social ideas and way of governance at home and abroad, improve the legal system of socialism Chinese characteristics, promote the growth of the rule of law; secondly, which is the need to improve the system of socialism with Chinese characteristics ethics to guide members of the socialist society actively Pursuits core values; thirdly, which is the need to improve social governance and innovation, vigorously develop modern new social organization; Fourthly, which explores the implementation of "Internet + supervision" mode, the innovation market regulatory approach, innovation in public service delivery.

4.1 Legal aspects: improving the Chinese legal system, socialist characteristics, and promote the growth of the rule of law

Performance of social transformation in the legal field, which is to realize the construction of the social order of law, the law controls social anomie, to achieve "from status to contract" law changes. In the practice of reform and opening process, the increased emphasis on law. Therefore, the current must actively improve the legal system of socialism Chinese characteristics, promote the growth of the rule of law. As the party's eighteen Fourth Plenary Session made to comprehensively promote the rule of law, the legal system of the construction of socialism with Chinese characteristics and the rule of law as the overall objective of the rule of law.

The current should be comprehensively deepen reforms as the basis, in respect of aspects to improve the socialist market economy, democracy, and ecological civilization construction process and other laws and regulations. These laws and regulations to protect the reform must reflect the party's leadership and the people are the masters. At the same time, improve the legal system is the premise, the key is to enhance the legal awareness of citizens and the rule of thinking, make learning the law, the use of law have become intrinsic qualities of each member of a society. As a government department set up to foster the rule of law should also be a platform for citizens, smooth channels of citizen participation in legal practice, advocacy and encouraging citizens to adhere to legal thinking to deal with the problem. Only in this way, a rule of law society will continue to grow up.

In addition, our government enriches and improves the socialist legal system and the rule of law norms, while social development, we should ethics education and legal system organically combine. So that legal norms play a fundamental role, and give fully the play to the guiding role of ethics for social governance level.

4.2 Moral dimensions: improving socialism with Chinese characteristics system of moral standards to guide members of the community to actively Pursuits socialist core values

Morality social governance plays an important regulatory role. We should focus on improving the current system of moral standards under the socialist market economy with Chinese characteristics, and this moral system and norms necessary to boldly absorb the beneficial ingredients include ideological capitalist countries, including, it should be fine ethical and traditional Chinese concept phase to undertake. Under the socialist market economy the most important thing is to carry forward the principles of collectivism, strengthen competition, cooperation and supervision moral education so that market players have a sense of responsibility and sense of honor, in the fierce competition consciously abide by ethics. Such code of ethics established by the Organic Coupling fully realized national, collective and individual interests. In moral practice, and resolutely opposed to selfish, Sungongfeisi, abuse of power, fraud behavior. The ultimate goal of moral education is to guide and regulate the activities of daily living members of the community, therefore, must be the people's ideals and beliefs, daily life and the pursuit of the value of life organically combined.

At the same time, the process of reform and opening up of society must pay attention to the current governance practices inspired by faith education level, positive Pursuits socialist core values to guide and regulate the behavior of members of society. Reform and opening up our first sight of social transformation is facing "survival Order" construct, thereby masking the "meaning order" of socialist modernization is about "economic modernization." Coupled with a long time, people above the emancipatory interest purport technology, scientific knowledge and critical knowledge of liberation outranked knowledge, resulting in power, money, status, influence on the life of the world erosion. Reform and opening up so that our material life has been greatly improved, but until around the 1990s, including the not yet formed a political elite, economic elite, cultural elite and the majority of the lower classes of society commonly accepted core values. In addition, the diversification of economic organization of diverse interests, so that the values of the people there multivariate changing situation, the objective requires citizens to accept the core values common to lead diverse marketplace of ideas.

The party's 18 report is based on the objective requirements of social development, we raised the need to actively cultivate and practice the socialist core values, in particular to make a prosperous, democratic, civilized, harmonious value at the national level on a quest to become every Chinese person part of the dream, so that freedom, equality, justice, rule of law, the value of the social dimension of the quest to become a part of every Chinese people's pursuit of life goals, make more patriotic, dedicated, honest, friendly and value pursuit individual level melt into each of China human growth and success inner beliefs. Only in this way, before the effective integration of diverse marketplace of ideas, to maximize the role of moral belief in social governance, laid the ideological foundation for the formation of real and lasting, harmonious social order.

4.3 Social institutional level: improving social governance and innovation, vigorously develop modern new social organization

To improve the current system of social governance, to adapt to the present stage of economic and social development situation in which the key is to build social governance mechanism of government as the core of the body. With the rapid development of the socialist market economy, the transformation of functions and roles of the government's increasingly high demands. In this regard, the Government should continue to decentralization, in order to service and rule of law-based government as the goal, to ensure that the government make a difference in social governance. This requires mechanisms to improve the performance evaluation of leading cadres, it is important to assess factors related to various parts of the social problems of governance and enhance the level of governance as the leading cadres appointment, promotion. In addition to this core government social governance body, the majority of the people should actively participate in community governance processes to strengthen the supervision of government action. The initiative requires the government members of the society to participate in social management to take measures to mobilize more need for government departments to improve the relevant policies, broaden the channels for participation, participation in social governance as members of the community to provide policy support and material basis.

Of course, for many of the growing grass-roots affairs, the government and people rely on participation, social governance practice cannot achieve the desired results. The current grassroots autonomous organizations emerged some degree of weakening, as a direct result of the weakening of the social fabric showing a "government - the people," the dual structure model. This requires increased between the government and the people of modern social organization that buffer zone, thereby reducing the difficulty of social governance. To this end, the positive development of social organizations, social organizations to stimulate vitality, great efforts to cultivate modern new organization; to "speed up the implementation of the government agency separate from social organizations to promote a clear mandate, autonomous law, play a role," [7] Thus contributing to the governance and oversight functions of modern society organizations to effectively play.

4.4 Operating mechanism level: exploring the implementation of "Internet + supervision" mode, the innovation market regulatory approach, innovation in public service delivery

Innovation market regulatory approach, in order to better play to the role of government to require strengthening a matter of ex post regulation. Firstly, Our government should deepen reform and commercial system, separate licenses to carry out a pilot's efforts to facilitate enterprise investment business and civic entrepreneurship employment, market access trial negative list system in some areas. Secondly, Our government should innovative regulatory mechanisms. To establish a horizontal edge, vertical in the end of the regulatory networks and scientific and effective regulatory mechanism, improve the information disclosure system, scientific inspection systems, retroactive liability management system and exception list, facilitate regulatory institutionalization, standardization and normalization. Thirdly, Our government should innovative forms of regulation. "Promote the comprehensive reform of administrative law enforcement, random inspection object randomly selected inspection of law enforcement personnel, timely publication of investigation results." [8] Fourthly, Our government should explore the implementation of "Internet + supervision" mode. Advancing the government sector, on open sharing, and interoperability between different levels of information resources, promote market supervision data open the whole society, to promote "smart" regulation. Fifthly, Our government should build credit as the core of the new regulatory mechanism. Comprehensively promote the social credit system, the establishment of a unified social credit system and the integrity of the code file system.

In addition, the public service function of our government need to further strengthen public service supply mode requires innovation, especially, " to mobilize the broad participation of social forces effectively, to build a wide range of public service delivery system, so as to provide more convenience for the people, fast and efficient Public Service" [9]. Specifically, to promote the marketization of public service, and promote the socialization of public services, to promote the equalization of public services, to promote the "Internet +-government." Particularly with regard to the "Internet + government" promotion, to speed up the e-government, convenience services, public resources trading, public opinion and other support services and platform construction, with big data, cloud computing technology integration of network resources, to break the "information island" to achieve data sharing between departments. To carry out online check, limit handling and transparent handling, improve service capacity and administrative efficiency.

References

1. Zhang Jun. Out of the Black Hole: A Critique of Contemporary Chinese anomie. China Economic Publishing House, 1(2000)
2. GAO Zhaoming. High social anomie theory Jiangsu People's Publishing House, 1(2000)
3. G • Duncan • Mitchell. New Dictionary of Sociology Shanghai Translation Publishing House, 12(1987)
4. Samuel • P • Huntington. Political Order in Changing Societies Joint Publishing, 38(1989)
5. Zhang Jun. Out of the Black Hole: A Critique of Contemporary Chinese anomie. China Economic Publishing House, 126(2000)
6. GAO Zhaoming. High social anomie theory Jiangsu People's Publishing House, 181-182(2000)
7. Chinese Communist Party Central Committee at its eighteenth plenary meeting of the Third Compilation People's Publishing House, 70(2013)
8. Li Keqiang. Government work report. - March 5, 2016 at the fourth meeting of the 12th National People's Congress on the People's Daily, 03-18-2016
9. Liu Gang.The current governance Promoting innovation focal point of the. SCHOOL OF CPC Tianjin Municipal Committee, 5(2016)

Study on Operational Availability Relational Model of RMS

Kai Cao[1,2], Zheng Huang [1,2] and Xiaowei Liu [1,2]

[1] College of Power Engineering, Naval University of Engineering, Wuhan, China
[2] Military Key Laboratory for Naval Ship Power Engineering, Naval University of Engineering, Wuhan, China
Email: kevin10001@163.com

Abstract. This paper is about a research of the relationship between equipment's reliability, maintainability, supportability and operational availability. It investigates the influencing factors of operational availability, analyzes the maintenance strategies of equipment and specifies distribution types. Based on these findings and Markow process, it then deduces computational model of operational availability. It also defines reliability, maintainability, supportability and other constraint condition and establishes optimizing equations to find feasible solution and optimal solution. At last, it tests correctness and feasibility of established model by case study. After knowing the requirement of operational availability and operational method provided here to define corresponding level of reliability, maintainability, supportability, it transforms operational-using requirement to technical design requirement in support for further research.

1 Introduction

Operational availability is one of the vital parameters in the development and application of warship equipment. The warship equipment needs to guarantee various missions carried out away from shore base, which requires equipment readiness to fully complete tasks assigned. Paying attention to the design of operational availability, as the key connotation in optimizing the operational effectiveness of warship equipment, will aid the concept transformation from solely stress on the performance to pursuing comprehensive mission effectiveness [1]. Operational availability is co-guaranteed by the design reliability, maintainability and supportability of the system; hence the process of designing the operational availability is also the process of confirming the reliability, maintainability and supportability of the equipment, converting operational requirements to technical design requirements.

Numerous researches had been taken on the evaluation of operational availability home and abroad, of which the common methods are classical model method, statistic analyzing method and analytical method.Sherbrooke divides the operational availability model into attainability availability and supply availability according to its definition, which is widely applied in the inventory field [2]. Diao et al established the calculation model based on the steady state availability with full consideration of the operational availability of three states of unmanned aircraft system—stored, combat and sealed [3]. Based on the statistical analysis of failure data, Liu et al raised a combined GM-RBF neural network model on grey theory [4]. Liu and Han et al discussed the application of BP neural network and RBF neural network on the evaluation of operational availability [5,6]. Zhang et al descry ibed a statistical evaluation method tightly related to the missile weapon system by analyzing the availability of different equipment using statistical method and connecting the system structure based on in situ data [7].Zhang et al studied the lower confidence limit of operational availability of single element product on a given distribution [8]. De Smidt-Destombes K S and Zhang et al established calculation model of the k/N warm standby reductant system's operational availability when the overhaul interval is set based on the Markov process [9,10]. Chan G K and Liu deduced a relation model among steady state availability and distribution, delayed maintenance time distribution, maintenance time distribution and preventative maintenance time distribution [11,12].Xu and Duan et al established the naval gun operational availability model [13,14].

It is known that great deal of results have been accumulated on the operational availability by related researchers. The classical model method is easy to calculate and suits to solve the inventory, however the results might be quite different from the simulation ones [15]. The statistic analyzing method requires relatively long period to gather data, moreover it cannot be applied to the equipment development stage [11]. In addition, the researches focused more on the already known maintenance strategy, system structure, and distribution types, and thus deduce the calculation models of the operational availability, but the researches on the other way around was quite few.

Due to the questions above, this article will investigate the factors influencing the operational availability, analyze the maintenance strategy, and determine the distribution types, and deduce the calculation model of operational availability based on Markov process; determine the reliability, maintainability, supportability and other limiting factors, establish the optimized equation set and its feasible solutions and optimal solution. Verify the validity and feasibility of the established model through case study.

2 Models of operational availability

The warship equipment needs to guarantee various missions carried out away from shore base, and the resources taken on the ship is limited while the support resources is hard to replenish, in order to guarantee the equipment readiness, a deep study on the operational availability is what it calls for to establish the models to calculate operational availability, and offer theoretical support to improve designation and optimize allocation of resources.

Before the models are established, some explanations and assumptions are made as follows:

(1) Two levels of spare parts supply system are considered, assume that all spare parts in local level can be fulfilled in the intermediate level warehouse.

(2) Assume that local level spare parts have 3 individuals, and when the third ran out, a request for the 4th is made to the base. If the equipment faces no absence of any failure before the spare parts' arrival, 3 parts will be supplied, and the extra ones will be sent back to the base; if the equipment faces any failure and caused delay, the failed one will be changed and 3 more supplied. The time from application to acquirement is t, which obeys index distribution with parameter.

(3) The longevity and maintenance time obeys index distribution with and parameter, and can be analyzed through Markov process.

(4) When the spare parts are used up and the requested part doesn't arrive but the equipment fails, thus is the logistice delay until the part arrives, mean logistic delay time is the average time of logistic delay.

(5) When any failure is noticed of the equipment, immediate maintenance should be made to its normal form. Only replacement of the equipment is made on the ship and the maintenance staff is enough at hand.

The working and local level storage conditions are expressed with E_{mn} .The subscript m stands for the working condition of the equipment,0 for normal working condition,1 for logistic delay condition, 2 for maintenance condition. The subscript n stands for local warehouse condition, 0 for the spare parts ran out and request to intermediate level warehouse for supply is already made, 1 for 1 spare part, 2 for 2 spare parts, 3 for 3spare parts, therefore 9 conditions are defined, E_{03} stands for normal working condition with 3 spare parts; E_{22} stands for under maintenance with 2 spare parts; E_{02} stands for normal working condition with 2 spare parts; E_{21} stands for under maintenance with 1 spare part; E_{01} stands for normal working condition with 1 spare part; E_{20} stands for under maintenance with 0 spare part; E_{00} stands for normal condition with 0 spare part; E_{10} stands for under maintenance with 0 spare part; E_{23} stands for under maintenance with 3 spare parts. We consider any condition that doesn't satisfy the assumptions doesn't exist.

The time of requesting for the spare part obeys index distribution with γ parameter, the final spare part's life span also obeys index distribution with λ ,parameter, therefore the supplied spare part can arrive before the last available element fails(event A) or after(event B).These two events should be discussed separately.

According to the assumption, when the last spare part is used, the application for more spare part is made, that is the last spare part come into working condition and the application is made coincide. If random variable X is used to express the elements' lifespan, Y to express the application time, then the possibility of event A happens is $P_A(X \geq Z)$,possibility of event B happens is $P_B(X \leq Z)$. Because the time of applying for spare part obeys index distribution with γ parameter, the longevity of the element obeys also index distribution with λ parameter, plus X and Y are independent, so the combined probability density function are as follows:

$$f(x,z) = \lambda \gamma e^{-\gamma z} e^{-\lambda x} \tag{1}$$

The solution of $P_B(X \leq Z)$

$$P_B(X \leq Z) = \iint_{D_{xz}} f(x,z)dxdz = \frac{\lambda}{\lambda+\gamma} \tag{2}$$

$$P_A(X \geq Z) = 1 - P_B(X \leq Z) = \frac{\gamma}{\lambda+\gamma} \tag{3}$$

D_{xz} means the range where X is less than or equal to Z.

Assume that failure happens at t_1, and the spare part arrives at t_2, then:

$$F(t_1) = 1 - e^{-\lambda t_1}, G(t_2) = 1 - e^{-\gamma t_2} \tag{4}$$

$$f(t_1) = \lambda e^{-\lambda t_1}, g(t_2) = \gamma e^{-\gamma t_2} \tag{5}$$

Due to the fact that the failures happens and the spare part arrives are two independent events, so the combined probability density function is :

$$h(t_1,t_2)=\lambda\gamma e^{-\gamma t_2}e^{-\lambda t_1} \tag{6}$$

If the spare part arrives after the last element fails, it will cause delay for $(t_2\text{-}t_1)$. Regardless of the delay time caused by other factors, the mean logistic delay time is:

$$MLDT=E(t_2-t_1) \tag{7}$$

According to the two-dimensional random variable function and its distribution, we can conclude that:

$$E(t_2-t_1)=\iint_{D_{t1t2}}(t_2-t_1)h(t_1,t_2)dt_1dt_2=\frac{\lambda}{\gamma(\lambda+\gamma)} \tag{8}$$

Among them D_{t1t2}represents a range $t_1\le t_2$.

In the following, we will take consideration of event A. Complementary element arrived before the last available element had broken. At this moment, the condition aggregate E_A is {1,2,3,4,5,6,7}, normal condition aggregate W_A is {1,3,5,7}, and breakdown condition aggregate F_A is {2,4,6}. The figure of condition transformation is reflected in Figure 1.

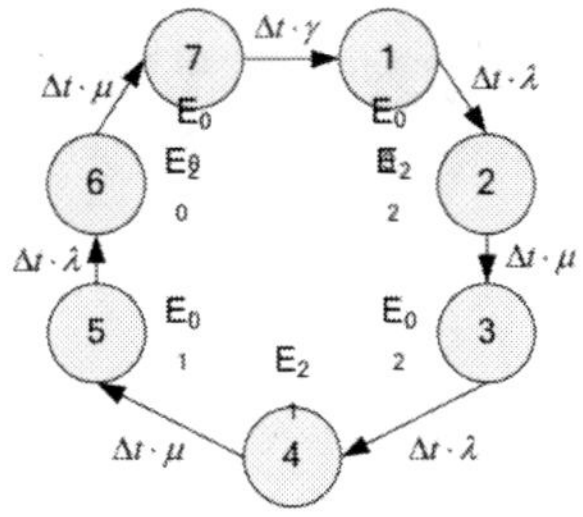

Fig. 1. Condition transformation of event A.

Condition transformation matrix $Q_A$can be showed in the following from condition transformation figure:

$$\mathbf{Q}_A=\begin{pmatrix} -\lambda & \lambda & & & & & \\ & -\mu & \mu & & & & \\ & & -\lambda & \lambda & & & \\ & & & -\mu & \mu & & \\ & & & & -\lambda & \lambda & \\ & & & & & -\mu & \mu \\ \gamma & & & & & & -\gamma \end{pmatrix} \tag{9}$$

According to bibliography [16] :

$$\left(P_1'(t),P_2'(t),\cdots,P_7'(t)\right)=\left(P_1(t),P_2(t),\cdots,P_7(t)\right)\mathbf{Q}_A \tag{10}$$

As for all i,j $\in$ E, $\lim_{t\to\infty}P_{ij}(t)=P_j \quad j=1,2,\cdots,7$.

As for any j $\in$ E, $\lim_{t\to\infty}P_j'(t)=0$

We can choose term limit value from both sides according to equation (10), and then obtain:

$$\left(P_1,P_2,\cdots,P_7\right)\mathbf{Q}_A=\left(0,0,\cdots,0\right) \tag{11}$$

Solve it to: $P_1=P_3=P_5,P_2=P_4=P_6=\frac{\lambda}{\mu}P_1,P_7=\frac{\lambda}{\gamma}P_1$

Due to $P_1+P_2+\cdots P_7=1$, we can make a conclusion:

$$P_1=\left(3+\frac{3\lambda}{\mu}+\frac{\lambda}{\gamma}\right)^{-1} \tag{12}$$

According to the definition of A_o, we know:

$$\begin{aligned} A_{oA} &= P_1 + P_3 + P_5 + P_7 \\ &= (3+\frac{\lambda}{\gamma})(3+\frac{3\lambda}{\mu}+\frac{\lambda}{\gamma})^{-1} \end{aligned} \tag{13}$$

When we take consideration of event B, complementary element arrived after the last available element had broken, and maintenance delay condition occurred before re-complementary element's arrival. At this moment, the condition aggregate E_B is {1,2,3,4,5,6,7,8,9}, normal condition aggregate W_B is {1,3,5,7}, and breakdown condition aggregate F_B is {2,4,6,8,9}. The figure of condition transformation is reflected in Figure 2.

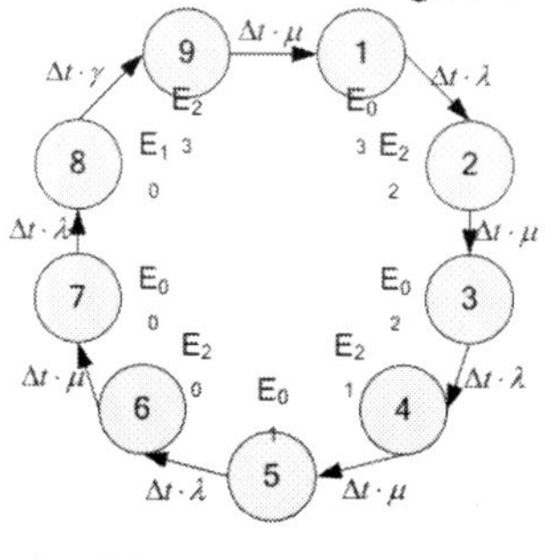

Fig. 2. Condition transformation of event.

Condition transformation matrix QBcan be showed in the following from condition transformation figure:

$$\mathbf{Q}_B = \begin{pmatrix} -\lambda & \lambda & & & & & & & \\ & -\mu & \mu & & & & & & \\ & & -\lambda & \lambda & & & & & \\ & & & -\mu & \mu & & & & \\ & & & & -\lambda & \lambda & & & \\ & & & & & -\mu & \mu & & \\ & & & & & & -\lambda & \lambda & \\ & & & & & & & -\gamma & \gamma \\ \mu & & & & & & & & -\mu \end{pmatrix} \tag{14}$$

According to bibliography 16 :

$$\left(P_1'(t), P_2'(t), \cdots, P_9'(t)\right) = \left(P_1(t), P_2(t), \cdots, P_9(t)\right)\mathbf{Q}_B \tag{15}$$

As for all i,j∈E, $\lim\limits_{t\to\infty} P_{ij}(t) = P_j \quad j = 1,2,\cdots,9$

As for any j∈E, $\lim\limits_{t\to\infty} P_j'(t) = 0.$

We can choose term limit value from both sides according to equation (15), and then obtain:

$$\left(P_1, P_2, \cdots, P_9\right)\mathbf{Q}_B = \left(0, 0, \cdots, 0\right) \tag{16}$$

Solve it to:

$$P_1 = P_3 = P_5 = P_7, P_8 = \frac{\lambda}{\gamma} P_1,$$

$$P_2 = P_4 = P_6 = P_9 = \frac{\lambda}{\mu} P_1 。$$

Due to $P_1 + P_2 + \cdots P_9 = 1$, we can make a conclusion:

$$P_1 = \left(4 + \frac{4\lambda}{\mu} + \frac{\lambda}{\gamma}\right)^{-1} \tag{17}$$

According to the definition of A_o, we know :

$$\begin{aligned} A_{oB} &= P_1 + P_3 + P_5 + P_7 \\ &= 4\cdot(4+\frac{4\lambda}{\mu}+\frac{\lambda}{\gamma})^{-1} \end{aligned} \tag{18}$$

Combining P_A and P_B, we can make a conclusion of systematic comprehensive A_o :

$$\begin{aligned} A_o &= P_A\cdot A_{oA} + P_B\cdot A_{oB} \\ &= \frac{\gamma}{\lambda+\gamma}\cdot(3+\frac{\lambda}{\gamma})(3+\frac{3\lambda}{\mu}+\frac{\lambda}{\gamma})^{-1}+\frac{4\lambda}{\lambda+\gamma}\cdot(4+\frac{4\lambda}{\mu}+\frac{\lambda}{\gamma})^{-1} \\ &= \frac{3\gamma+\lambda}{\lambda+\gamma}(3+\frac{3\lambda}{\mu}+\frac{\lambda}{\gamma})^{-1}+\frac{4\lambda}{\lambda+\gamma}\cdot(4+\frac{4\lambda}{\mu}+\frac{\lambda}{\gamma})^{-1} \end{aligned} \tag{19}$$

A_o, *MTBF*,*MTTR* and *MLDT* can be calculated according to known parameter λ, μ and γ in model (8)and (19). But thisset of parameters may not be the optimal parameters. In the process of equipment designation, we concern more about limit value of A_o and the feasible solution and optimal solution of design parameter limited by other term, thus we need to come up with an optimized method.

3 Ascertain limit term and establish optimal relational models

According to A_o, the larger *MTBF* is, the smaller *MTTR* is, the smaller *MLDT* is and the higher A_o is. But in fact, as the result of the limitation of technical level and other objective conditions, *MTBF*, *MTTR* and *MLDT* can only change in a range, so do parameter λ, μ and γ. Besides, maintenance delay time (event B) made by the drawn-out application of replacement part can be cut down by optimized allocation of maintenance system, so we can prescribe a limit that maintenance delay time takes up proportion in unusable time no more than limit value. Unusable time is made of average maintenance delay time and average repair time. Maintenance delay time taking up proportion in unusable time no more than limit value is equal to average repair time taking up proportion in unusable time more than limit value. Then regard available cost's limit value, range of λ, $\mu$and γ, the rate average repair time takes up in unusable time, average maintenance delay time as limit condition, and the maximal value of A_o as target function, to establish optimized relational models:

$$\begin{cases} \max A_o(\lambda,\mu,\gamma) = \\ \frac{3\gamma+\lambda}{\lambda+\gamma}(3+\frac{3\lambda}{\mu}+\frac{\lambda}{\gamma})^{-1}+\frac{4\lambda}{\lambda+\gamma}\cdot(4+\frac{4\lambda}{\mu}+\frac{\lambda}{\gamma})^{-1} \\ \text{s.t. } A_o \ge A_{oaccept} \\ \quad a \le E(t_1) = \frac{1}{\gamma} \le b \\ \quad c \le MTBF = \frac{1}{\lambda} \le d \\ \quad e \le MTTR = \frac{1}{\mu} \le f \\ \quad g \le MLDT = \frac{\lambda}{\gamma(\lambda+\gamma)} \le h \\ \quad \frac{MTTR}{MTTR+MLDT} \ge k \end{cases} \tag{20}$$

We can choose reciprocal from both sides of equation according to A_o equation, and then obtain 17:

$$MTTR + MLDT = (\frac{1}{A_o}-1)MTBF \tag{21}$$

so we can present the rate average repair time takes up in unusable time in the following:

$$\frac{MTTR}{MTTR+MLDT} = \frac{MTTR}{(\frac{1}{A_o}-1)MTBF} \tag{22}$$

4 Solve the equation set and analyze the case

There are three variables λ, $\mu$and γ in optimal relational model (20), which both has linear constraint and nonlinear constraint, belonging to multivariate nonlinear constraint optimized problem. According to optimal theory, we can obtain optimal solution which meets the limit condition and optimal value of corresponding parameters by solving optimal relational model through numerical value MATLAB 18.

We know that under the limitation of current technical level and other usage conditions, some equipment's *MTBF* is more than 500h and less than 1000h, *MTTR* is larger than 4h and less than 25h, and average time from the application of replacement parts to the arrival of them is more than 72h and less than 120h, as well as $A_{oaccept}$ is 0.96, lower limit value of *MLDT* is *g*, and the rate *MTTR* takes up in unusable time should be more than *k*,so the optimal relational model (20) can be transformed into (23).

$$\begin{cases} \max A_o(\lambda,\mu,\gamma)= \\ \dfrac{3\gamma+\lambda}{\lambda+\gamma}(3+\dfrac{3\lambda}{\mu}+\dfrac{\lambda}{\gamma})^{-1}+\dfrac{4\lambda}{\lambda+\gamma}\cdot(4+\dfrac{4\lambda}{\mu}+\dfrac{\lambda}{\gamma})^{-1} \\ \text{s.t. } A_o \geq 0.96 \\ \quad 72 \leq E(t)=\dfrac{1}{\gamma}\leq 120 \\ \quad 500 \leq MTBF=\dfrac{1}{\lambda}\leq 1000 \\ \quad 4 \leq MTTR=\dfrac{1}{\mu}\leq 25 \\ \quad g \leq MLDT=\dfrac{\lambda}{\gamma(\lambda+\gamma)} \\ \quad \dfrac{MTTR}{(\dfrac{1}{A_o}-1)MTBF}\geq k \end{cases} \tag{23}$$

In the relational model, λ, μ and γ are variables. After performing the procedure, we can obtain the maximal value and corresponding parameters of A_o. Then change the rate *k* which *MTTR* takes up in unusable time and lower limit value *g* of *MLDT*, and re-perform the procedure to obtain the maximal value and corresponding parameters of different A_o. Data are showed in the following table:

Table 1. The maximal value of A_o and corresponding parameters.

No.	*k*	*g*/h	*maxA_o*	*MTBF*/h	*MTTR*/h	*E(t)*/h	*MLDT*/h
1	0.8	5	0.9943	1000	4.6	73.4	5.02
2	0.85	5	0.9926	1000	6.4	73.4	5.02
3	0.8	6	0.9932	1000	5.5	81.1	6.08
4	0.85	6	0.9912	1000	7.6	81.1	6.08
5	0.8	7	0.9921	1000	6.4	87.4	7.03
6	0.85	7	0.9901	1000	8.5	87.4	7.03

When k is 0.85 and g is 6h, we can obtain relation figure showed in Figure 3 according to the relation between expected time $E(t)$ from replacement parts' application to their arrival and A_o.

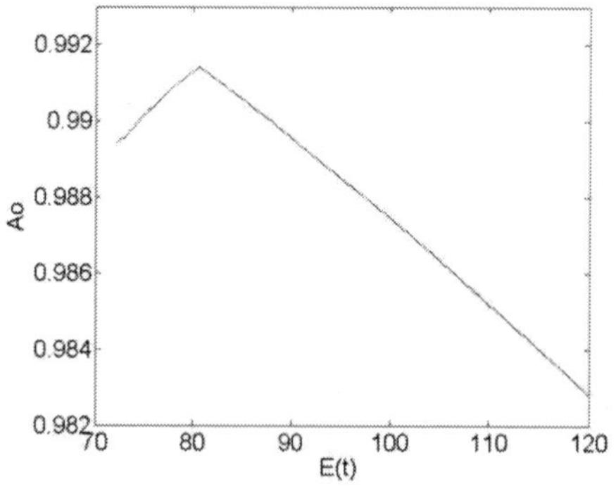

Fig. 3. The relation between *E(t)*and A_O.

After calculating, when *E(t)*is 81.1h, A_o arrives at the maximal value. At this moment, the three dimensional relational model of *MTTR*, *MTBF* and Ao can be showed in Figure 4.

From Table 1, we know that when k are same, the smaller g is, the larger the maximal value of operational availability is, and *MTTR*, *MLDT*and *E(t)* will gradually decrease. When g are same, the smaller kis, the larger the maximal value of operational availability is, and the optimal value of other parameters won't change, but only the optimal value of *MTTR* will decrease. The optimal value of MTBFwill choose the maximal value of the range all the time. Combining the relationship figure of *MTBF*, *MTTR* and A_o, as well as relationship figure of *E(t)*and A_o, we will analyze the phenomenon above.

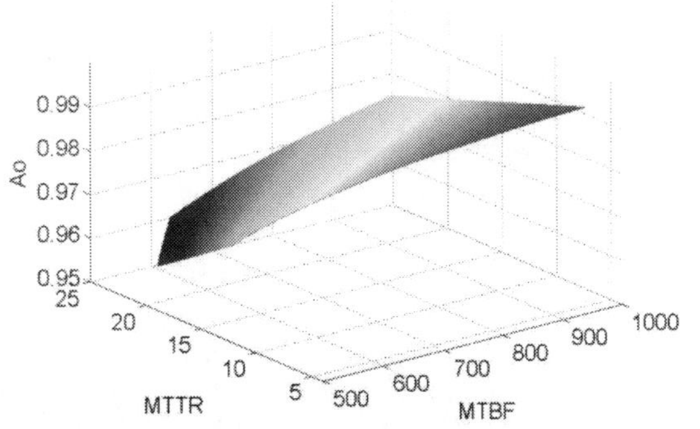

Fig. 4. The relation between *MTTR*, *MTBF* and A_o

According to formula (8), we know that *MLDT* is influenced by λ and γ, it is inversely proportional to MTBF, and it is in proportion to *E(t)*. When lower limit k that average repair time takes up proportion in unusable time are same, the lower limit value gof *MLDT* decreases and the constraint that MLDTputs on *MTBF* and *E(t)* also decreases. At this moment, *MTBF* will be bigger in its co-domain and *E(t)* will be smaller in its co-domain, as reflected in Figure 4 that the maximal value of A_o then will increase.

According to formula(22), we know that k is in proportion to *MTTR* and inversely proportional to *MLDT*. When g are same, and k decreases, the constraint that the lower limit that average repair time takes up proportion in unusable time puts on *MTTR* and *MLDT*also decreases. At this moment, *MTTR* will be smaller in its co-domain, as reflected in Figure 4 that the maximal value of A_o then will increase. Although the constraint that k puts on *MLDT* decreases, the constraint that g puts on *MLDT* doesn't change, so that the optimal value of other parameters won't change.

We can know from Figure(4) that when k is 0.85, g is 6h, and t is 81.1h, maxA_o that has concerns constraint conditions is the maximal value 0.9912 as *MTBF* is 1000h. When *MLDT* is inversely proportional to *MTBF*, and it is in proportion to *E(t)*, *MLDT*'s contour line can be showed in Figure(5). Under current constraint conditions, when g is between 5h and 7h, the minimal value of *MLDT* constrains the range of *MTBF*'s and *E(t)*'s feasible region to be pentagon area in the lower right corner above contour line gall the time. When A_o is the maximal value, *MTBF*can always be the maximal value 1000h in its feasible region. If taking costs, time limit and other constraints into consideration, *MTBF*'s value should be further restricted.

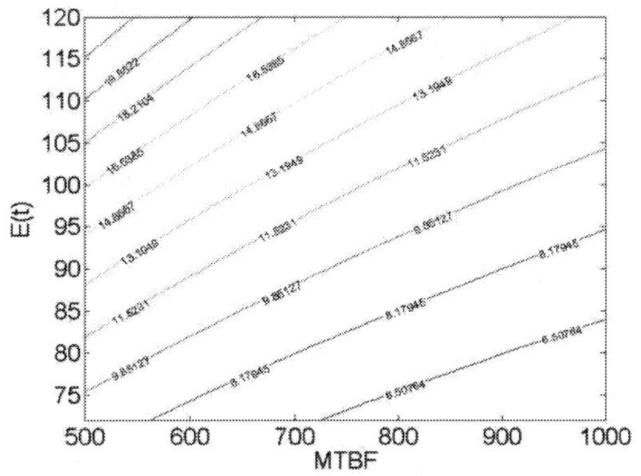

Fig. 5. Contour line of *MLDT*.

5 Conclusion

Operational availability is an integrity norm as same importance as other functional norms of equipment, and it is an insurance of equipment that it can finish tasks perfectly when needed. This paper analyzes the influencing factors of operational availability and has a research on the process of replacement parts' maintenance based on fully understanding of the importance of operational availability. And based on Markow process, it establishes relational models of operational availability and defines constraint condition and establishes optimizing equations to find feasible solution and optimal solution. At last, it tests correctness and feasibility of established model by case study. The arithmetic established here transforms operational-using requirement to technical design requirement and ascertains target of equipment's reliability, maintainability and supportability. It provides theoretical support and quantitative reference of parameter setting when testing equipment.

Although the arithmetic provided here can solve some physical problem, the optimal solution may have some distinction to reality without consideration of costs, limit time and other constraints. So it is the main problem we should solve in the future.

References

1. HJB 31.2-03, Management guidance of operational availability of naval ships' equipment.
2. Sherbrooke C C.Optimal inventory modeling of system: multi-echniques second edition. 371(2004)
3. W.Q. Diao, S.W. Lu, B. Zheng. Analysis of unmanned combat aircraft storage security and operational availability.Aircraft design , **35**, 48(2015)
4. W. Liu, T.K. Li, S.C. Yu. Study on appraisal method of missile weapon system's operational availability based on neural network of GM-RBF. Equipment environmental engineering, **10**, 108(2013)
5. Q.J. Liu, G.M. Chen, X.F. Liu, etc. An application of BP neural network in the prediction of missile weapon system's operational availability. Tactical missile technology,**1**, 26(2011)
6. Z. Han, T.X. Xu, J.K. Yang. Study on neural network of RBF in the prediction of missile weapon system's operational availability. Computers and modernization, **10**, 27(2012)
7. H.X. Zhang, C.H. Wang. Study on appraisal method of ground-to-air missile weapon system's operational availability. Modern defense technology, **43**,129(2015)
8. G.F. Zhang, H.P. Wu. Assessment of steady state availability of repairable system when maintenance delay time occurs. Journal of Zhejiang University, **33**, 0268(2006)
9. De Smidt-Destombes K S, van der Heijden M C, van Harten A. Availability of k-out-of-N systems under block replacement sharing limited spares and repair capacity. International Journal of Production Economics, **107**, 404(2007)
10. T. Zhang, J.J. Zhang, Guo B. Analysis of maintenance strategies of K/N system based on operational availability. Acta Astronautica, **30**, 395(2009)
11. F.S. Liu, W. Wu, Z.W. Shan, etc. Operational availability model of armored equipment based on Markov updating process. Journal of armored corps engineering college, **24**,15(2010)
12. Chan G K, Asgarpoor S. Optimum Maintenance Policy with Markov Processes. Electric Power Systems Research, **76**,452(2006)
13. T.G. Xu, Y. Wei.Operational availability modeling and simulation of shipborne gun based on spare parts inventory. Equipment environmental engineering, **7**,132(2010)
14. M.Q. Duan, J. Dai, J. Ma. Operational availability simulation of task oriented marine nuclear power plant. Traffic science and technology, **2**,160(2015)
15. Y. Wang, N.C Wang, L. Ma, ec.Computational method of multi-unit tadem system's operational availability concerned constraint conditions of spare parts. Aeronautical Journal, **36**,1195(2015)
16. J.H. Cao, K. Chen.An introduction to mathematics of reliability.(2005)
17. Y.G Zhao, C Qi, Y.L. Shang. Optimization of equipment's operational availability under the condition of two level protection. Science and technology of atomic energy, **47**,1202(2013)
18. C. Gong, Z.L Wang. Frequently-used algorithm assembly of MATLAB language.(2008)

MATEC Web of Conferences **100**, 02011 (2017) DOI: 10.1051/matecconf/201710002011
GCMM 2016

A Systematic Analysis of Functional Safety Certification Practices in Industrial Robot Software Development

Xie Tong[1,*] and Wu Lei[2]

[1]School of Software, Beijing Institute of Technology, Beijing, China
[2]Software Quality Engineering Research Centre, The Fifth Electronic Research Institute of MIIT, Guangzhou, China
Corresponding Email: vickyxt@163.com

Abstract. For decades, industry robotics have delivered on the promise of speed, efficiency and productivity. The last several years have seen a sharp resurgence in the orders of industrial robots in China, and the areas addressed within industrial robotics has extended into safety-critical domains. However, safety standards have not yet been implemented widely in academia and engineering applications, particularly in robot software development. This paper presents a systematic analysis of functional safety certification practices in software development for the safety-critical software of industrial robots, to identify the safety certification practices used for the development of industrial robots in China and how these practices comply with the safety standard requirements. Reviewing from Chinese academic papers, our research shows that safety standards are barely used in software development of industrial robot. The majority of the papers propose various solutions to achieve safety, but only about two thirds of the papers refer to non-standardized approaches that mainly address the systematic level rather than the software development level. In addition, our research shows that with the development of artificial intelligent, an emerging field is still on the quest for standardized and suitable approaches to develop safety-critical software.

1 Introduction

Industrial robot, as defined by ISO 8373 [1]: An automatically controlled, reprogrammable, multipurpose manipulator programmable in three or more axes, which may be either fixed in place or mobile for use in industrial automation applications. An automatically controlled, reprogrammable, multipurpose manipulator programmable in three or more axes, which may be either fixed in place or mobile for use in industrial automation applications. The main customer for industrial robots - the automotive industry - is changing and diminishing. There is a worldwide trend towards automation in the 'non-automotive industry'. Robot suppliers are offering increasingly tailored solutions to customers in China, as the "Made in China 2025" plan which focus on the development of fully-automated "smart" factories has a strong support from the government.

As the industrial robots becoming smarter, faster and cheaper, they're being called upon to do more. They're taking on more "human" capabilities and traits such as sensing, dexterity, memory and trainability. As a result, they're taking on more jobs - such as picking and packaging, testing or inspecting products, or assembling minute electronics. Also, a new generation of "collaborative" robots (Figure 1) ushers in an era of shepherding robots out of their cages and the robots work literally hand-in-hand with human workers who train them through physical demonstration.

Fig. 1. A new generation of "collaborative" robots by ABB (left) and Rethink Robotics (right).

The power and size of industrial robots means that they are capable of inflicting severe injury if programmed incorrectly or used in an unsafe manner. Due to the mass, high-speeds and the new "collaborative" behavior of industrial robots, it is always unsafe for a human to remain in the work area of the robot during automatic operation. The system can begin motion at unexpected times and a human will be unable to react quickly enough in many situations, even if prepared to do so. Thus, great care must be taken to make an industrial robot safe for human workers or human interaction.

Programming errors represent a serious safety consideration, particularly in large industrial robots. Those problem prompted the development of software safety standards, such as recommendations based on best practices (MISRA [2]), and formal standards(ISO 25119 [3] for agriculture and ISO 26262 for automotive [4]). For robotic systems, there are some researches analyze potential hazards [5,6] refer to ISO 13482 [7], which is aimed to achieve safety requirements not for industrial robots but for personal care robots. ISO 10218 [8] is conducted to recognize the particular hazards which are presented by industrial robots and industrial robot systems. Regarding the development of software for industrial robots, those related safety research can only be addressed by few standards indirectly. Moreover, those standards or studies gave the requirements only, but didn't provide the guidance of how to implement.

In summary, we present findings from a systematic review in which we collected and structured from the current body of knowledge regarding (software development) practices and standards applied to the development of safety-critical robotic software in China. Our research from reviewing academic papers written in Chinese, shows that safety standards are barely used in software development of industrial robot in domestic.

A systematic literature review is a means of identifying, evaluating and interpreting all available research relevant to a particular research question, or topic area, or phenomenon of interest. There are three existing guidelines for systematic reviews [9-11], but all these guidelines are intended to aid medical researchers. In particular, software engineering research has relatively little empirical research compared with the large quantities of research available on medical issues. This article aims to analysis the functional safety certification practices in industrial robot software development, following a guideline for systematic reviews which is appropriate for software engineering researchers proposed by Barbara Kichenham [12].

2 Review questions

We formulate the following research questions:

RQ1: what standards or certifications for industrial robot software development have been adopted in domestic? This research questions aims to investigate the current states of standards for software development.

RQ2: What practices are used for the development of safety-critical software for industrial robotic systems (including coding and testing)? This research question aims at understanding the practices that are used to develop robots in safety-critical contexts. The question addresses coding and testing practices, such as code generation or code reuse, unit testing or fault injection and so on.

RQ3: Which certification standards have provided feasible implementation guidelines for software for industrial robots? This research question aims to find out whether the current standards are suitable for implement and test safety-critical software for industrial robots.

3 Review methods

3.1 Data sources and search strategy

The process of performing a systematic review must be transparent and replicable. We constructed the search strings (Table1) based on ISO 10218-2011 [8].

Table 1. Overview of the final search queries.

	Search queries (searching scope: FT-full-text; AB-abstract)
S1	1. (FT="Robot" OR FT="Robots") AND (FT="Industry" OR FT="Industrial") AND (FT="Safety" OR FT="safe") AND (FT="Standard" OR FT="Standards" OR FT="ISO" OR FT="IEC" OR FT="GB") AND (FT="Software" OR FT="test" OR FT="testing") 2. (AB="Robot" OR AB="Robots") AND (AB="Industry" OR AB="Industrial") AND (AB="Safety" OR AB="safe") AND (AB="Standard" OR AB="Standards" OR AB="ISO" OR AB="IEC" OR AB="GB") AND (AB="Software" OR AB="test" OR AB="testing"

S2	1. (FT="Robot" OR FT="Robots") AND (FT="Industry" OR FT="Industrial") AND (FT="Safety" OR FT="safe") AND (FT="Standard" OR FT="Standards" OR FT="ISO" OR FT="IEC" OR FT="GB") 2. (AB="Robot" OR AB="Robots") AND (AB="Industry" OR AB="Industrial") AND (AB="Safety" OR AB="safe") AND (AB="Standard" OR AB="Standards" OR AB="ISO" OR AB="IEC" OR AB="GB")
S3	1. (FT="Robot" OR FT="Robots") AND (FT="Industry" OR FT="Industrial") AND (FT="Safety" OR FT="safe") AND (FT="Software" OR FT="test" OR FT="testing") 2. (AB="Robot" OR AB="Robots") AND (AB="Industry" OR AB="Industrial") AND (AB="Safety" OR AB="safe") AND (AB="Software" OR AB="test" OR AB="testing")
C1	(AB="medicin*" OR AB ="surgence*" OR AB ="health-care")
Final	(S1 OR S2 OR S3) AND NOT C1

In order to assure the accuracy and universality of the searching results, we used the following databases for querying (Table 2). The databases are the largest and continuously updated Chinese database in the world, including journals, doctoral dissertations and masters' theses full-text databases, which have a certain focus on software development.

Table 2. Databases used for searching.

Name of database	Search strategy for each database	Date of search	Years covered by search
China Academic Journals Full-text Database (CJFD)	(S1 or S2 or S3) and not C1	2016.5	1984-Now
China Doctoral Dissertations Full-text Database (CDFD)	(S1 or S2 or S3) and not C1	2016.5	1984-Now
China Masters' Theses Full-text Database (CMFD)	(S1 or S2 or S3) and not C1	2016.5	1984-Now

3.2 Study selection

Intended to identify these primary searches that provide direct evidence about our research questions, according to the questions from chapter 3, inclusion and exclusion criteria are defined as follow:

3.2.1 Inclusion criteria

• Title, keyword list or abstract make it explicit that the paper is related to safety in industrial robotics;

• The paper is on tools, procedures or development methods;
• The paper is in a journal, proceedings, magazine, doctoral dissertations or masters' theses;
• The paper describes a long term observation of the use of development methods in relation to safety-critical development;
• The paper surveys practitioners for the use of development methods;
• The paper is on tools implementing certain methods (infer information about method use), for development of safety-critical software;
• The paper is about important aspects for industrial networking, industrial control safety.

3.2.2 Exclusion criteria

• The paper is a proposal only;
• The paper is not within safety or industrial robotics;
• The paper occurred multiple times in the result set;
• The paper does not touch the domain of software engineering, computer science or robotics in general;
• The paper's full text is not available for download.

3.3 Data extraction

The objective of data extraction is to design data extraction forms to accurately record the information which are obtained from the primary studies by researchers. The data extraction forms (Table 3 & Table 4) were designed to collect all the information needed to address the review questions and the study quality criteria. Following the steps of conducting a systematic mapping study [13], we developed schemas to address RQ1, RQ2 and RQ3. Table 3 presents the classification schema that was used to classify the publications according to the practices used in the software development (RQ2). In particular, based on the different aspects of software development, we included software design, implement, and testing practices, such as formal specification, fault injection, simulation. In order to answer RQ3, we collected information about norms and standards used in safety-critical systems (Table 4).

Table 3. Development practices in software development (including software design, implement and testing).

Aspects	Practices
Architectural design	Notations for architectural design (Semi-formal notations, formal notations, etc.)
	Error detection at the software architectural level (Range checks, plausibility check, control flow monitoring, diverse software design, etc.)
	Error handling at the software architectural level (Static recovery mechanism, graceful degradation, independent parallel redundancy, correcting codes for data, etc.)
	Verification of the software architectural design (Walk-trough, inspection, simulation, prototype generation, formal verification, control flow analysis, data flow analysis, etc.)
Unit design and implementation	Notations for software unit design (Natural language, semi-formal notations, formal notations, etc.)
	Verification of software unit design and implementation (Walk-trough, inspection, semi-formal and formal verification, control flow analysis, data flow analysis, static code analysis, semantic code analysis, etc.)
testing	Methods for software testing (requirements-based test, interface test, fault injection test, resource usage test, back-to-back comparison test between

	model and code, etc.)
	Methods for deriving test cases for software testing (analysis of requirements, generation and analysis of equivalence classes, analysis of boundary values, error guessing, etc.)
	Structural coverage metrics at the testing level (statement coverage, branch coverage, modified condition/decision coverage)
Others	Misc (papers that either encompasses many of the above methods, or do not clearly define which method is used)
	Not in Software Development

Table 4. Standards used in safety-critical software development.

Standards	Description
ISO 10218	Robots and robotic devices Safety requirements for industrial robots (IDT by GB 11291-2013)
IEC 62061	Electrical safety of machinery—Functional safety of safety-related electrical electronic and programmable electronic control systems (IDT by GB 28526-2012)
IEC 61508	Functional Safety of Electrical/Electronic/Programmable Electronic Safety-related Systems (E/E/PE, or E/E/PES) (IDT by GB/T 20438-2007)
IEC 61131-6-2012	Programmable controllers -- Part 6: Functional safety (IDT by GB/T 15969-2007)
IEC 61511-2003	Functional safety - Safety instrumented systems for the process industry sector (GB/T21109-2007)
IEC 61499	Open standard for distributed control and automation
Guaranteeing safety	Not necessarily using a standard approach
Non-Standard Approach	When it is specifically mentioned that there are no standards available for the domain

3.4 Data synthesis & analyze

In this section, we present the results of the included primary studies using data synthesis and analyze. In Sect.4.4.1, we give an overview of the research. In Sect.4.4.2 – Sect.4.4.4, we answer the research questions.

3.4.1 General overview

An overview of results obtained from the different search steps are provided as Table 5. Step1, we used S1 or S2 or S3 as strings for full-text search, resulted in more than 92,000 hits. After filter the C1 strings, the result remained more than 50,000 hits. In order to assure the accuracy of the result, we change S1 or S2 or S3 not C1 as strings for abstract search (change the pefix to AB). After applying the different in-/exclusion criteria and peer review procedures, 109 papers were selected for further analyse.

Table 5. Overviews of the results obtained from the literature search.

Step \ Databases	CJFD	CDFD	CMFD	Total
Step1: Search (S1 or S2 or S3)	14956	10630	66873	92459
Step2: Filtering				
Not C1	12176	7387	33412	52975
Change FT (full text) to AB (abstract)	45	72	446	563
Final result set (after peer review)	28	36	45	**109**

According to the publication frequency over time (1984-2016), we classified the sources to illustrate the development of the considered domain over time, as Figure 2.

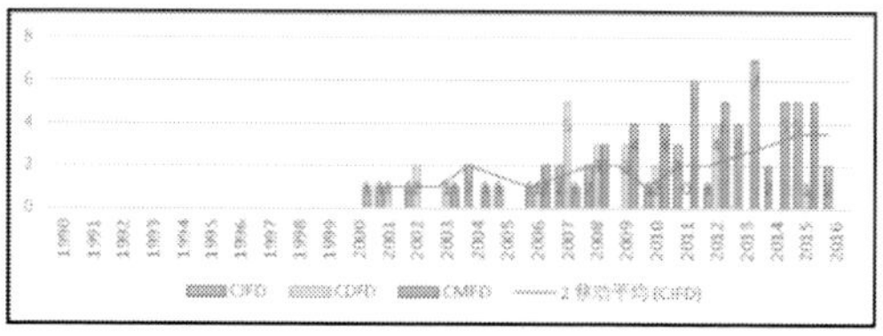

Fig. 2. Numbers of papers per year and distribution over the sources.

From Figure 2, we see the field of safe-critical software development of industrial robots which has been paid close attention since 2000, being a still emerging trend in China. Since then, we observed the majority of the published papers. From the data sources, we can see that there still have a big gap between the evaluation researches and the engineering practices, as three quarters of papers are written by graduate and doctoral students.

3.4.2 RQ1: what standards or certifications for industrial robot software development have been adopted in domestic?

Functionally safe systems are developed and validated against well-defined market-specific functional safety criteria. The umbrella standard for the majority of application-specific functional safety development in electronics is IEC 61508. It specifically addresses electrical, electronic and programmable electronic safety-related systems. Many market specific safety standards and guidelines have been derived from IEC61508. To present the standards or certifications for industrial robot software development in domestic, a systematic analysis is provided as Table6.

Table 6 illustrates the research- and standards type facets and shows the majority of the papers which still do not follow the safety standards in software development of industrial robots (68 out of 109). Table6 also illustrates that there are more pre-researches in colleges (81 out of 109). That is, the current publication body is focused on proposing fundamental researches to deal with the challenges coming along with developing safety-critical software for industrial robots. Nonetheless, table6 points to an emerging field.

3.4.3 RQ2: What practices are used for the development of safety-critical software for industrial robotic systems (including coding and testing)?

Integrating software onto a specific processor and hardware environment is one of the most challenging areas of a system to develop and assess demanding certification requirements. Table 7 shows that many different aspects are covered and that many different practices are addressed, especially verification of the software architecture design and software testing. However, 60 papers are classified into the others, meaning that there has not safety-related software development practices. Therefore, table7 indicates that, the safety-critical software development of industrial robots in domestic are not according to the prescribed standards or certifications. Nevertheless, there are some practices such as software simulation and software testing, which are implemented widely in software development of industrial robots.

3.4.4 RQ3: Which certification standards have provided feasible implementation guidelines for software for industrial robots?

This question aims at investigating whether current certification standards have provided feasible implementation guidelines for the software of industrial robots. Therefore, we collected the major standards addressing this topic for general safety-critical software development (Table4). According to 4.4.2, we found that a loose connection between software development practices and available standards. Only the IEC 61508 (general functional safety standard) was mentioned by those papers. However, the standard ISO 10218 (Robots and robotic devices—Safety requirements for industrial robots) was not referred at all! Thus, although software safety is becoming more critical due to the increasing number of industrial systems controlled by software, there are no feasible implementation guidelines of these standards to achieve safety requirement.

Table 6. Systematic analysis illustrating papers vs standards.

	ISO 10218	IEC 61508	IEC 61131-6-2012	IEC 61511-2003	IEC 61499	Guaranteeing safety	Non-Standard Approach
CJFD	0	4	1	0	0	9	14
CDFD	0	4	2	2	0	4	24
CMFD	0	7	1	0	0	7	30
Total	0	15	4	2	0	20	68

Table 7. Systematic analysis illustrating papers vs software development practices.

	Notations for architectural design	Error detection at the software architectural level	Error handling at the software architectural level	Verification of the software architectural design	Notations for software unit design	Verification of software unit design and implementation	Methods for software testing	Methods for deriving test cases for software testing	Others
CJFD	2	3	1	6	2	2	4	1	16
CDFD	3	3	5	7	4	4	5	2	16
CMFD	2	1	5	3	3	1	7	1	28
Total	7	7	11	17	11	9	16	3	60

4 Conclusion

Over recent years, there has been a drastic increase in the numbers of software usage in safety-related systems. Several safety-related systems such as industrial robot systems depend on software to manage the safety-related functionalities. The last several years have seen a sharp resurgence in the orders of industrial robots in China, and the areas addressed within industrial robotics has extended into safety-critical domains. The demand for software in industrial robots is expected to increase in the following years which turn out to be a challenge in the software safety and requirement engineering processes. This paper presents a systematic analysis of functional safety certification practices in software development for the safety-critical software of industrial robots, to identify the safety certification practices used for the development of industrial robots in China and how these practices comply with the safety standard requirements. Our research from reviewing Chinese academic papers, shows that safety standards are barely used in software development of industrial robot. The majority of the papers propose various solutions to achieve safety, but only about two thirds of the papers refer to non-standardized approaches that mainly address the systematic level rather than the software development level. In addition, our research shows that as the development of artificial intelligent, an emerging field is still on the quest for standardized and suitable approaches to develop safety-critical software.

The present study is a first step toward a deeper understanding of safety certification in industrial robot development. In future, we need to do further research on safety-related standards of industrial robots such as (IEC 61508, ISO 10128 etc.), to establish a feasible implementation guideline for software for industrial robots. Our end objective is not to meet safety standards or to pass the assessment test, but to actually deliver an industrial robot system that is functionally safe.

References

1. International Standard 8373:2012: Robots and Robotic Devices - Vocabulary. International Organization for Standardization (2014)
2. MISRA: MISRA-C Guidelines for the Use of the C Language in Critical Systems (2012)
3. International Standard ISO 25119-2010: Tractors and machinery for agriculture and forestry - safety-related parts of control systems, International Organization for Standardization (2010)
4. International Standard ISO 26262:2011: Road Vehicles Functional Safety, International Organization for Standardization (2011)
5. Dogramadzi, S., Giannaccini, M.E., Harper, C., Sobhani, M., Woodman, R., Choung, J.: Environmental hazard analysis - a variant of preliminary hazard analysis for autonomous mobile robots. J. Intell. Rob. Syst.**76**(1), 73–117 (2014)
6. International Standard ISO 13482:2014-Robots and robotic devices - Safety requirements for personal care robots. International Organization for Standardization (2014)
7. Jacobs, T., Virk, G.S.: ISO 13482 - the new safety standard for personal care robots. International Symposium on Robotics (ROBOTIK 2014), pp. 1–6. VDE-Verl (2014)
8. International Standard ISO 10218:2011- Robots and robotic devices—Safety requirements for industrial robots. International Organization for Standardization (2011)
9. Cochrane Handbook for Systematic Reviews of Interventions [EB/OL] / [2009-7]. http://www.cochrane-handbook.org/
10. Australian National Health and Medical Research Council. *How to Review the Evidence: Systematic Identification and Review of the Scientific Literature: Handbook Series on Preparing Clinical Practice Guidelines* (2000)
11. NHS Centre for Reviews and Dissemination Undertaking Systematic Reviews of Research on Effectiveness. CRD Guidelines for Those Carrying Out or Commissioning Reviews. University of York, York. (1996)
12. B.Kitchenham. Procedures for Performing Systematic Reviews. Keele University Technical Report TR/SE-0401 (1999)
13. K.Petersen , R.Feldt, S.Mujtaba, M.Mattsson: Systematic mapping studies in software engineering. In: International Conference on Evaluation and Assessment in Software Engineering, pp. 68–77. British Computer Society (2008)

Factor Analysis of Enterprise Internet+ Impetus Based on System Dynamics

Haiwang Cao[1], Chaogai Xue[2] and Xinfeng Wang[2]

[1] Department of Electronic and Communication Engineering, Zhengzhou Institute of Aeronautical Industry Management, Zhengzhou, Henan, 450015, China
[2]School of Management Engineering, Zhengzhou University, Zhengzhou, Henan, 450001, China

Abstract. Internet+ plays an important role in the transition and upgrading of enterprises in this era. To promote enterprise Internet+ implementation, this paper studies the influencing factors in enterprise Internet+ impetus. First, based on system dynamics, the influencing factors of enterprise Internet+ impetus are analyzed as well as their relationships. Then, a system dynamics model is established and Vensim PLE is utilized to simulate the model. Finally, some suggestions about the Internet+ implementation are given. The simulation results show that technological environment, government policy and information asymmetry all have different influence on the enterprise Internet+ impetus and different strategies can be taken to change their influence. This study is helpful for enterprises to understand the Internet+ impetus mechanism thus promote Internet+ implementation.

1 Introduction

With the government report of China in 2015 [1], Internet+ is proposed for the first time and become a national strategy of China. Under the background of national network strategy, cloud computing, big data, mobile internet, networking and other technologies provide a good technical support for enterprise Internet+ implementation. Internet+ pushes enterprises to fuse into the internet, and realizes integration of online and offline resource, as well as different enterprises and different industries [2]. However, facing new situation, enterprises confuse about what influences Internet+ implementation and how to promote Internet+ impetus. To this end, this study analyzes the factors influencing enterprise Internet+ impetus and their mutual relations, and tries to find the casual relationships of enterprise Internet+ impetus so as to provide references for the implementation of enterprise Internet+.

In recent years, there are studies about Internet+ in different aspects, which can be briefed as follows: (1) Concepts of Internet+. These studies are mainly about the concepts, structure and connotation of Internet+. For example, Ning[3] studies the connotation, background and content of Internet+. You [4] propose that Internet+ is more than informatization, and its essence is the online data. As for traditional enterprises, it is a process of transformation and upgrading of traditional industries. (2)Technologies of Internet+. These studies are mainly about the technologies in Internet+, such as cloud computing, big data, mobile Internet, networking and other technologies [5,6]. Sun et.al. [5] Discuss the general rules of the development of energy and information infrastructure. (3) Applications of Internet+ in different fields. These studies are mainly about the applications of Internet+ in different fields. For example, Zhao [7] discusses material information system railway engineering project under the background of Internet+. Though the studies on Internet+ provide good foundation, there is few study about factors influencing enterprise Internet+ impetus.

System dynamics was proposed by professor Forrster, W. Jay at the Massachusetts Institute of Technology, and is mainly used for the studies of relationships between complex system structure, functions and dynamic behaviour. Through dynamic simulation experiments, the system dynamic behaviour and trend can be studied under different input of factors. As to the studies of complex nonlinear systems, it has incomparable advantages, thus it has been widely used in almost all fields of society, economy, management, resource environment and so on [8-10]. However, there is no such relative reference in the factor analysis of enterprise Internet+ impetus.

From relevant literature, few scholars have studied the factors influencing enterprise Internet+ impetus based on system dynamics. Therefore, this paper further explores the relationships among the factors influencing Internet+ impetus, and tries to build system dynamic model of enterprise Internet+ impetus so as to provide decision making support for enterprises and relative agencies.

2 Factor analysis of enterprise Internet+ impetus

Based on system dynamic complexity science and feedback control theory, system dynamics is mainly used to study the dynamic behaviour of complex system to understand and solve system phenomena [9]. The implementation of Internet+ is a complex process with non-linear dynamic complex features. Therefore, system dynamics is suitable to identify the impetus elements of enterprise Internet+ implementation and build a system flow diagram or make feedback loop analysis. Based on previous studies, this paper divides the factors of enterprise Internet+ implementation into external and internal factors.

2.1 External Factors

The external factors of enterprise Internet+ impetus can be divided into the following ways:

(1)Government influence. From government work report in 2015 to Internet+ manufacturing in 2016, the era of enterprise Internet+ has come. From the national level, increasing policy support in tax, finance and land enable enterprises to implement Internet+. With the rapid development of internet and the rapid growth of information, relevant laws and regulations also need to regulate the network market, maintain information security, and solve the conflict between online and offline, so as to pave the way for Internet+ smoothly. Meanwhile, the introduction of technology, professional guidance, and coordination between online and offline are all things can be done by government.

(2)Competitor influence. When enterprises implement Internet+, they are facing more competitors. The Internet+ impetus of competitors affects their market competitiveness, thus influences the implementation of enterprises.

(3) Internet+ consciousness. In the era of innovation 2.0, the new ecosystem based on internet has updated as well as technologies. The essence of enterprise Internet+ is the fusion of internet and enterprises through the information communication technology. Procurement of raw materials, production, warehousing and sales et. al. need support of internet. Communication cannot do without internet, as well as internal management and daily information. Therefore, it is necessary for big data and cloud system to tap the potential resources and improve collaborative processing capabilities. The information communication technology has promoted the enterprise informatization, and enhanced the enterprise's Internet+ consciousness. To gain competition capacity, enterprises continue to improve consciousness of Internet+, and this is also an important external factor of Internet+ impetus.

2.2 Internal Factors

The internal factors of enterprise Internet+ impetus include the following aspects:

(1)Management mode. The rapid and timely information is the key to Internet+ era. Traditional linear functional organization structure cannot adapt to the information age, and the enterprise management mode is very important to Internet+ implementation.

(2)Information asymmetry. In the enterprise and customer communication, the asymmetry of information will lead to poor communication of internal enterprises as well as the communication with customers, thus results into the decline in market response ability. Thus, the degree of information asymmetry is one of the key internal factors influencing the enterprise Internet+ impetus.

(3)Innovation capacity of employee. In Internet+ era, the personnel training and management are very important for enterprises. If the employee have high level of innovation capacity, then they may have more Internet+ consciousness, thus affect the entire enterprise Internet+ development progress and implementation. Technical upgrading needs talents to operate the application, and new management mode also needs innovative personnel to coordinate. The cultivation of innovative and divergent thinking is an important power for Internet+ implementation.

(4)Innovation on product development. Product customization and intelligence are two new ways to improve the competitiveness of products in Internet+ era. Product research and development need data mining on customs, and then products can be promoted to customized development. In Internet+ environment, enterprises in the technological innovation has also increased the intelligent degree of products. Therefore, the enterprises' innovation on product development is an important factor influencing enterprise Internet+ implementation.

(5)Innovation on marketing. The innovation of marketing improves the competitiveness of products in Internet+ era. Diversification of marketing channels gave birth to the diversification of interactive channels, and solve the problem of poor customer communication in a certain degree. Thus, the enterprises' effort on market innovation determines the impetus of Internet+.

As discussed above, we have the following factors influencing the impetus of enterprise Internet+, as shown in Table1.

Table 1. Influencing factors of enterprise Internet+ impetus.

External factors		**Internal factors**	
Government strategies	National strategies	Management mode innovation	
	Regional strategies		
Technological environment	Development of communication technologies	Information asymmetry	Poor communication in internal enterprise
	Security of network technologies		Poor communication with customs
Competitor	Offline enterprises	Innovation capacity of employee	entrepreneur Internet+ consciousness
	Substitute enterprises		Employee Internet+ consciousness
	Online enterprises	Innovation on product development	Requirement of product design
			Requirement of product R&D
		Innovation on marketing	Marketing strategies
			Marketing channels

3 Internet+ impetus model based on system dynamics

3.1 Causal loop diagram of Internet+ impetus model

According to the factor analysis, both internal and external factors influence the enterprise Internet+ impetus. The external impetus of enterprise Internet+ is mainly government policies (including national policies and regional policies), the existing information and communication technology environment, and enterprise competition; internal impetus is mainly composed of business management, personnel innovation capacity, product innovation, marketing innovation etc.

The impetus mechanism of enterprise Internet+ is a combination of internal and external dynamic mechanism. Various elements of impetus compose causal relationships. In order to study the impetus mechanism of enterprise Internet+, it is necessary to use the viewpoints of system theory, and view enterprise Internet+ process as a dynamic, complex and open system, and there are dynamic relationships between elements. Based on above analysis, the causal relationship diagram of Internet+ impetus is shown in Figure 1.

There are several feedback loops in Figure 1, and entrepreneur Internet+ consciousness is taken as an example shown as follows:

entrepreneur Internet+ consciousness→management mode innovation→application of communication technology→employee skills→marketing innovation→product competition→market competition→entrepreneur Internet+ consciousness.

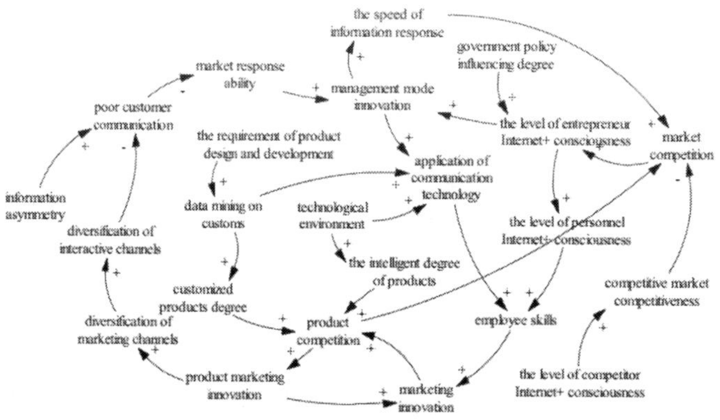

Fig. 1. Causal relationship diagram of Internet+ impetus.

3.2 Stock and flow diagram of Internet+ impetus model

The interaction of various factors in causality diagram of system dynamics model is actually a mathematical relation, and stock and flow diagram is specific to this mathematical relationship, shown in Figure 2. Based on the previous research and field investigation data of enterprise Internet+, the main parameters are set as follows:

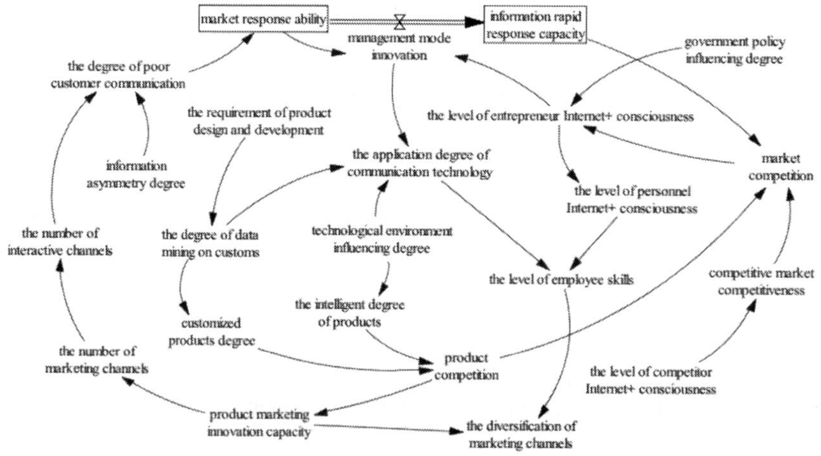

Fig.2. Stock and flow diagram of Internet+ impetus.

(1)Assume that the requirement of product design and development is 0.8, and the competitor Internet+ consciousness is 0.6.

(2)Assume that government policy influencing degree is a_1 and a_1=0.40, which is decided by the government concern degree and the development of enterprise Internet+, and it decreases with the improvement of Internet+.

(3)Assume that the promotion effect of technological environment is b_1 and b_1=0.45,which is decided by the maturity level and innovation level of technologies, and the more mature it is, the more obvious the impetus effect.

(4)Assume that information asymmetry is c_1 and c_1=0.35, which is determined by the information delivery speed and accuracy. It decrease with the improvement of information delivery speed and accuracy.

4 Simulation and analysis on Internet+ impetus model

When parameters of enterprise Internet+ impetus model are set, Vensim PLE is used to simulate the model. As Internet+ aims to solve the information asymmetry problem, improve the enterprise's ability to respond to information, and thus improve the ability of information processing. Therefore, government policy influencing degree, information asymmetry degree, technological environment are taken as the main factors to discuss their effect on enterprise Internet+ impetus.

The simulation results of the model is shown in Figure 3. Here, Curve 4 shows the simulation result of set parameters in Section 3.2. Curve 3 shows the simulation result when government policy influencing degree increases from 0.4 to 0.7. Curve 2 shows the simulation result when technological environment influencing degree increases from 0.45 to 0.75. Curve 1 shows the simulation result when information asymmetry degree increases from 0.35 to 0.7.

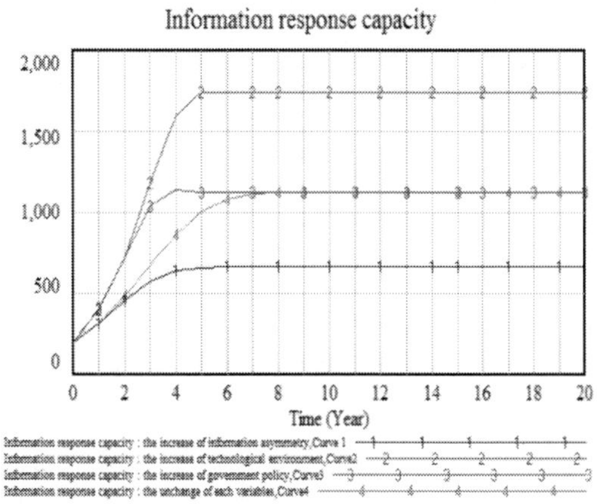

Fig.3. Different factor effect on enterprise Internet+ impetus.

From Figure 3, the following conclusions can be obtained:

(1) All curves are S - shaped curve. The slope is first increased and then decreased, and the overall trend is upward. Curve 2 reaches the maximum value when the saturation is reached, the maximum value of curve 1 is the smallest, the maximum value of curve 3 and the curve 4 are equal. The decrease order of the curves at saturation state follows: curve 2, curve 3, curve 4 and curve 1.

(2) Curve 3 shows that with the increase of government policy, the speed of information response capacity increases. In comparison with the curve 4, the shape of the curve is more steep, the time of reaching saturation is much earlier, and the saturation value of the two curves is same. It shows that government policy on information rapid response capacity is positive feedback, and increase in the government policy degree can enhance the ability to response to information, thus enhance enterprise Internet+ impetus.

(3) Curve 2 shows that with the increase of technological environment, the speed of information response capacity increases. Compared with curve 4, the shape of curve 2 is more steep, reaching the saturation state earlier. It shows that the increase of technological environment influence can increase the information response ability and can make it reach saturation earlier.

(4) Curve 1 shows that information asymmetry also influences information response ability. In comparison with curve 4, curve 1 reached saturation state earlier, and the saturation value of curve 1 is less than that of curve 4. It shows that the increase in information asymmetry degree can shorten the time to saturation but has less impact on the information response capacity.

5 Suggestions for decision makers

Factor analysis obtains the boundary of enterprise Internet+ impetus, and causal diagram and flow diagram realize qualitative and quantitative study of enterprise Internet+ impetus. According to the factor analysis of enterprise Internet+ impetus, we have a further understanding on enterprise Internet+ mechanism, thus the following strategies can be utilized to promote the impetus of enterprise Internet+:

(1)Government policies. In addition to the policies to promote enterprise Internet+ implementation, government can establish industrial parks and incubators, so as to format industrial belt, find potential enterprise, and promote the implementing of Internet+. At the same time, government can introduce new technologies, provide professional guidance to enterprises, and help enterprises to cultivate professional employees. Government can also be a matchmaker, and realize the enterprise integration and cooperation between enterprises and internet enterprise resources.

(2)Technological environment. Enterprises should grasp the technical environment provided by information and communication technologies. For product research and development (R&D) level, it is necessary to realize enterprise management mode reform to support personalized products. Enterprise should take strategies to improve their product design and R&D, so as to realize the customized products. That is the way to enterprise Internet+.

(3)Ways to solve information asymmetry. Information asymmetry can be reduced by the reform of management mode in enterprises, and the expansion of interactive channels which can reduce the interaction between enterprise and customer. Therefore, enterprises can reform their organization structure as flat one so as to adapt to the new Internet+ ecological system.

6 Conclusion

Internet+ has become an inevitable topic of this era, and enterprises are eager to implement it so as to increase their competition capacity. This paper discusses internal and external factors influencing the impetus of enterprise Internet+ through system dynamics. A model of enterprise Internet+ impetus is established, and simulated by Vensim PLE. The impact of government policy, technological environment and information asymmetry has been discussed in detail. According to the simulation results, we can obtain that government policy and technological environment degree play crucial roles in the process of enterprise Internet+. And information asymmetry can also influence enterprise Internet+ impetus, but its influence is more obvious at the beginning. This study is helpful for enterprise and government to make decisions on the key factors when implementing Internet+.

Acknowledgement

This work is supported by National Natural Science Foundation of China Grant No. #71371173, #70971119. Academic Technology Leaders of the Provincial Education Department of Henan Project under Grant number 2014GGJS-103. Technology Innovation Talent plan of the Provincial Education Department of Henan under Grant number 16HASTIT037.

References

1. http://www.gov.cn/zhengce/content/2015-07/04/content_10002.html.
2. Z Zhao. Crossover administration of internet plus: perspective of creative destruction. CIE. **10**(2015).
3. J J Ning. Background, connotation and content of Internet plus plan implementation. E-government. **06**(2015).

4. W Y You. "Internet plus"--driving the transformation and up grading of traditional industries. Statistical Theory and Practice,**07**(2015).
5. H B Sun, Q L Guo, S G Pan etc. Energy Internet: concept, architecture and frontier outlook. Automation of Electric Power Systems, **39**(2015).
6. M Xia, T Li etc. Closed-loop design evolution of engineering system using condition monitoring through internet of things and cloud computing. Computer Networks, **101**(2016).
7. L Q Zhao and J Y Liu. Research on the Internet+ construction of railway engineering project material information system. Journal of Railway Engineering Society, **03**(2016).
8. H Y Ren, S B Cheng and L C Huang. The strategy of emerging technology industrialization based on system dynamics. SRM. **34**(2013).
9. J H Liu, X J Qian and M Li. Study on the dynamic factors of technology innovation of software industry based on system dynamics. Journal of Changzhou University (SOCIAL SCIENCE EDITION),**04**(2015).
10. S Oh, H Baek and S Lee. Research on the relationship between information technology capabilities and sustainable competitive advantages of enterprises. International Journal of Plant Engineering and Management, **03**(2012).

A Large Group Decision Making Approach Based on TOPSIS Framework with Unknown Weights Information

Yupeng Li, Xiaozhen Lian*, Cheng Lu and Zhaotong Wang

Department of Industrial Engineering, School of Mines; Key Laboratory of Deep Coal Resource Mining, Ministry of Education of China; China University of Mining and Technology, Xuzhou, 221116, PR China
Corresponding Email: xzhen_l@163.com

Abstract. Large group decision making considering multiple attributes is imperative in many decision areas. The weights of the decision makers (DMs) is difficult to obtain for the large number of DMs. To cope with this issue, an integrated multiple-attributes large group decision making framework is proposed in this article. The fuzziness and hesitation of the linguistic decision variables are described by interval-valued intuitionistic fuzzy sets. The weights of the DMs are optimized by constructing a non-linear programming model, in which the original decision matrices are aggregated by using the interval-valued intuitionistic fuzzy weighted average operator. By solving the non-linear programming model with MATLAB®, the weights of the DMs and the fuzzy comprehensive decision matrix are determined. Then the weights of the criteria are calculated based on the information entropy theory. At last, the TOPSIS framework is employed to establish the decision process. The divergence between interval-valued intuitionistic fuzzy numbers is calculated by interval-valued intuitionistic fuzzy cross entropy. A real-world case study is constructed to elaborate the feasibility and effectiveness of the proposed methodology.

Keywords. Multiple-attributes large group decision making, optimization of weights of DMs, non-linear programming model, TOPSIS.

1 Introduction

Operational management of many areas are consisted by series of decision making issues. Multiple-attributes group decision making (MAGDM) is widely applied to solve these operational management problems such as supplier selection [1], product concept evaluation [2], and facility location planning [3]. Somehow, in actual decision-making process, MAGDM cannot be accurately analyzed due to the complicated requirement of social and technological trend and the limitations of knowledge and experience of decision-makers (DMs), thus the participation of large group of DMs is imperative. The MAGDM with large group DMs is defined as multiple attributes large group decision making (MALGDM).

Research on MALGDM can be classified into two categories: clustering-based methods and consensus reaching process-based methods. The principle of the clustering-based method is to decrease the number of decision group by clustering the DMs into some communities, and DMs in the same community is weighted by the same weight. By extending the analytic hierarchy process (AHP) into the Euclidean vector space (VAHP), Zadir [4] defined the similarity measure between DMs, based on which, the DMs in MALGDM is partitioned into homogeneous groups by a clustering procedure. Bolloju [5] identified DMs into subgroups based on similarities in preferences and aggregated preferences within the same subgroups by using AHP. Based on the claimed that, many DMs in large group decision making had similar opinion, Tapia-Rosero et al. [6] proposed a shape-similarity-based clustering approach to improve the weighting process of DMs in MALGDM. To simplify the weighting process of DMs, Liu et al. [7] proposed a classification method for DMs based on the partial binary tree DEA-DA cyclic classification model. Then, to determine the weights of DMs in the cluster and the cluster weights, Liu et al. [8] proposed a two-layer model including Minimized Variance Model and an Entropy Weight Model. The essence of clustering-based methods is to decrease the sample of DMs, the main drawback of which is that, much useful weight information for individual DM will be eliminated. The consensus reaching process-based methods are regarded as the iterative procedure aimed at seeking a high degree of agreement among DMs' preferences before making a group decision. Palomares et al. [9] proposed a consensus iteration model complementing with a visual analysis tool of the overall consensus reaching process based on self-organizing maps, which facilitates the monitoring of the process performance across the time. Aiming at the lower consensus and the urgency of large group emergency decision making, Xu et al. [10] proposed and investigated a dynamical consensus method based on an exit–delegation mechanism. Zhang et al. [11] presented a new consensus-based group decision making model that an aggregation of individual priority-based aggregation mechanism to employ flexible methods for deriving each DM's priority and to avoid information loss caused by unifying heterogeneous information. Comparing with the clustering-based methods, the consensus reaching process-based methods can incorporate the opinions of all DMs. Inspired by the principle of consensus-based methods, a non-linear programming model optimizing the weights of the DMs through an iterative procedures among individual decision matrices and comprehensive decision matrix is proposed in this article.

Fuzziness always exists in MALGDM, which is described by interval-valued intuitionistic fuzzy set (IVIFS) in this article owning to its advantages to address the fuzziness and hesitation of DMs [12-13]. For the MALGDM problem, the determination of the weights of criteria and the selection of decision framework should also be seriously considered. As it can utilize the original decision information of DMs to reduce the subjectivity of some other weight methods such as AHP [14-15] and Delphi [16], the information entropy theory is employed to calculate the weights of the criteria. The TOPSIS framework is addressed to establish the decision process as it is an ideal points-based framework concerning both positive ideal solution (PIS) negative ideal solution (NIS) [17]. The interval-valued intuitionistic fuzzy weighting average (IIFWA) operator [18] is used to aggregate individual decision matrices into the comprehensive decision matrix. The interval-valued intuitionistic fuzzy (IVIF) cross-entropy [19] is used to measure the divergence between IVIFSs.

The reminder of this article is organized as follows. Section 2 provides some definitions. Section 3 proposes the MALGDM methodology. In section 4, an example is provided to illustrate the effectiveness of the proposed method. Section5 summarizes the major contributes of this article.

2 Preliminaries

Definition 1[20] An IVIFS A in a finite set of the universe of discourse $X=\{x_1, x_2, \dots, x_m\}$ is: $A=\{\langle[u_A^L(x_i), u_A^U(x_i)], [v_A^L(x_i), v_A^U(x_i)]\rangle | x_i \in X\}$, and $\langle[u_A^L, u_A^U], [v_A^L, v_A^U]\rangle$ denotes the interval membership degree and non-membership degree of element x_i belonging to X, respectively. And $[u_A^L, u_A^U] \in [0,1]$, $[v_A^L, v_A^U] \in [0,1]$, $u_A^U+v_A^U \leq 1$. The hesitation degree $\pi(x_i)$ of x_i to X is: $\pi(x_i)=[\pi_A^L, \pi_A^U]=[1-u_A^U-v_A^U, 1-u_A^L-v_A^L]$.

Definition 2[18] For an IVIFS $\tilde{a}_k=([a_k, b_k], [c_k, d_k])$ $(k=1, 2,\dots, t)$, the IIFWA is defined as:

$$IIFWA_{\lambda_k}(\tilde{a}_1, \tilde{a}_2,\dots, \tilde{a}_k)=\left\{\left[1-\prod_{k=1}^{t}(1-a_k)^{\lambda_k}, 1-\prod_{k=1}^{t}(1-b_k)^{\lambda_k}\right], \left[\prod_{k=1}^{t}c_k^{\lambda_k}, \prod_{k=1}^{t}d_k^{\lambda_k}\right]\right\} \tag{1}$$

where $\boldsymbol{\lambda}=(\lambda_1, \lambda_2,\dots, \lambda_t)$is weight vector of DM and $\lambda_k \in [0, 1]$, $\sum_{k=1}^{t}\lambda_k=1$.

Definition 3[21] For two interval-valued intuitionistic fuzzy numbers (IVIFNs) $A=\{[u_A^L, u_A^U], [v_A^L, v_A^U]\}$, $B=\{[u_B^L, u_B^U], [v_B^L, v_B^U]\}$, the standard Euclidean distance between A and B is:

$$d(A, B)=\frac{1}{m}\sum_{i=1}^{m}\left(\frac{1}{4}\left[\begin{array}{c}|u_A^L-u_B^L|^2+|u_A^U-u_B^U|^2+|v_A^L-v_B^L|^2+|v_A^U-v_B^U|^2+\\ |\pi_A^L-\pi_B^L|^2+|\pi_A^U-\pi_B^U|^2\end{array}\right]\right)^{\frac{1}{2}} \tag{2}$$

where $\pi_A^L=1-u_A^U-v_A^U$, $\pi_A^U=1-u_A^L-v_A^L$, $\pi_B^L=1-u_B^U-v_B^U$, $\pi_B^U=1-u_B^L-v_B^L$.

Definition 4[22] The entropy of $A=\langle[u_A^L, u_A^U], [v_A^L, v_A^U]\rangle$ is defined as:

$$e(A)=1-d(A) \tag{3}$$

where $d(A)=\left|\frac{u_A^U-v_A^U}{2}\right|+\left|\frac{u_A^L-v_A^L}{2}\right|$. The information entropy of the jth criterion c_j is defined as:

$$E_j=\frac{1}{n}\sum_{i=1}^{n}e_{ij} \tag{4}$$

Then the weight of c_j is calculated by:(1)

$$\omega_j=\frac{1-E_j}{\sum_{j=1}^{m}(1-E_j)}=\frac{d_j}{\sum_{j=1}^{m}d_j} \tag{5}$$

Definition 5[17] A symmetric form of IVIF cross-entropy between IVIFNs A and B is defined as:

$$D^*(A, B) = D(A, B)+D(B, A) \tag{6}$$

where:

$$D(A, B)=\sum_{i=1}^{n}\left(\begin{array}{l}\frac{\underline{\mu}_A^i+\overline{\mu}_A^i+2-\underline{v}_A^i-\overline{v}_A^i}{4}ln\frac{2\left(\underline{\mu}_A^i+\overline{\mu}_A^i+2-\underline{v}_A^i-\overline{v}_A^i\right)}{\underline{\mu}_A^i+\overline{\mu}_A^i+2-\underline{v}_A^i-\overline{v}_A^i+\underline{\mu}_B^i+\overline{\mu}_B^i+2-\underline{v}_B^i-\overline{v}_B^i}+\\ \frac{\underline{v}_A^i+\overline{v}_A^i+2-\underline{\mu}_A^i-\overline{\mu}_A^i}{4}ln\frac{2\left(\underline{v}_A^i+\overline{v}_A^i+2-\underline{\mu}_A^i-\overline{\mu}_A^i\right)}{\underline{v}_A^i+\overline{v}_A^i+2-\underline{\mu}_A^i-\overline{\mu}_A^i+\underline{v}_B^i+\overline{v}_B^i+2-\underline{\mu}_B^i-\overline{\mu}_B^i}\end{array}\right) \tag{7}$$

3 Methodology

The mathematical description of the MALGDM can be defined as following. The alternative set is: $X=\{x_1, x_2,\dots, x_i,\dots, x_n\}$, $(i=1, 2, \dots, n)$, the evaluation criteria set is: $C=\{c_1, c_2,\dots, c_j,\dots, c_m\}$, $(j=1, 2, \dots, m)$, the weights of C is: $\Omega=\{\omega_1, \omega_2,\dots, \omega_j,\dots, \omega_m\}$, $(j=1, 2, \dots, m)$, and the DM set is: $E=\{e_1, e_2,\dots,e_k ,\dots, e_t\}$, $(k=1, 2, \dots, t)$, the weight of E is: $\Lambda=\{\lambda_1, \lambda_2,\dots, \lambda_k,\dots, \lambda_t\}$, $(k=1, 2, \dots, t)$. $\boldsymbol{\tilde{R}_k}=(\tilde{r}_{ijk})_{n\times m}$ is the interval-valued intuitionistic fuzzy matrix given by e_k, where $\tilde{r}_{ijk}=\langle[u_k^L(x_i), u_k^U(x_i)], [v_k^L(x_i), v_k^U(x_i)]\rangle$ is the ratings given by e_k with regard to c_j on x_i. Then the logic flow of the proposed methodology is demonstrated in Figure 1.

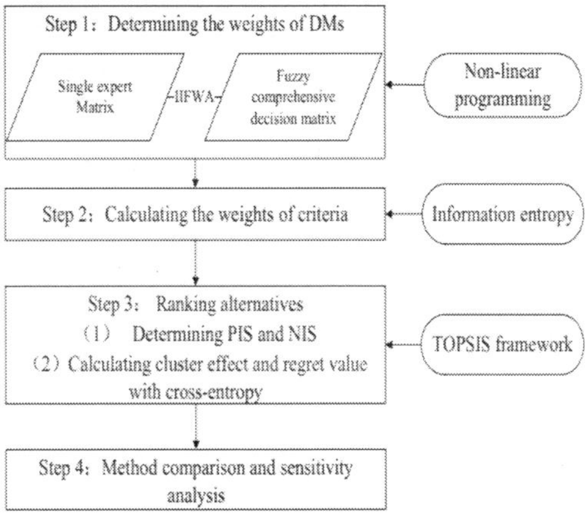

Fig 1. The logic flow of the proposed MGLGDM methodology.

The problem can be solved through three phases in detail.

Step 1: Calculating Λ by using non-linear programming optimization model.

The optimization model of Λ is constructed as following:

$$min\sum_{k=1}^{t} d(\tilde{\boldsymbol{R}}_k, \tilde{\boldsymbol{R}})$$
$$s.t.\begin{cases} 0\leq\lambda_k\leq 1 \\ \sum_{k=1}^{t}\lambda_k=1 \end{cases} \tag{8}$$

where $d(\tilde{\boldsymbol{R}}_k, \tilde{\boldsymbol{R}})$ is the Euclidean distance (Equation (2)) between $\tilde{\boldsymbol{R}}_k$ and the fuzzy comprehensive decision matrix $\tilde{\boldsymbol{R}}$:

$$d(\tilde{R}_k, \tilde{R})=\frac{1}{n}\sum_{i=1}^{n}\frac{1}{m}\sum_{j=1}^{m}\left\{\left(\frac{1}{4}\begin{bmatrix}|u_{ijk}^L-u_{ij}^L|^2+|u_{ijk}^U-u_{ij}^U|^2+\\ |v_{ijk}^L-v_{ij}^L|^2+|v_{ijk}^U-v_{ij}^U|^2+\\ |\pi_{ijk}^L-\pi_{ij}^L|^2+|\pi_{ijk}^U-\pi_{ij}^U|^2\end{bmatrix}\right)^{\frac{1}{2}}\right\} \tag{9}$$

By Equation (9), the objective function is transformed to: $min\sum_{k=1}^{t} d(\tilde{\boldsymbol{R}}_k, \tilde{\boldsymbol{R}})=$

$$min\sum_{k=1}^{t}\left(\frac{1}{n}\sum_{i=1}^{n}\left\{\frac{1}{m}\sum_{j=1}^{m}\left\{\frac{1}{4}\left(\begin{matrix}\left|u_{ijk}^L-\left[1-\prod_{k=1}^{t}(1-u_{ijk}^L)^{\lambda_k}\right]\right|^2+\\ \left|u_{ijk}^U-\left[1-\prod_{k=1}^{t}(1-u_{ijk}^U)^{\lambda_k}\right]\right|^2+\\ \left|v_{ijk}^L-\left(\prod_{k=1}^{t}v_{ijk}^{L\,\lambda_k}\right)\right|^2+\left|v_{ijk}^U-\left(\prod_{k=1}^{t}v_{ijk}^{U\,\lambda_k}\right)\right|^2+\\ \left|1-\prod_{k=1}^{t}(1-u_{ijk}^U)^{\lambda_k}+\prod_{k=1}^{t}v_{ijk}^{U\,\lambda_k}-u_{ijk}^U-v_{ijk}^U\right|^2+\\ \left|1-\prod_{k=1}^{t}(1-u_{ijk}^L)^{\lambda_k}+\prod_{k=1}^{t}v_{ijk}^{L\,\lambda_k}-u_{ijk}^L-v_{ijk}^L\right|^2\end{matrix}\right)^{\frac{1}{2}}\right\}\right\}\right)$$

$$s.t.\begin{cases}0\leq\lambda_k\leq1\\ \sum_{k=1}^{t}\lambda_k=1\end{cases} \tag{10}$$

Solve the model through MATLAB® to obtain the optimal solution of Λ. Then the fuzzy comprehensive decision matrix $\tilde{R}$ is acquired through IIFWA as:

$$R=(r_{ij})_{n\times m}=\begin{bmatrix}([u_{11}^L,u_{11}^U],[v_{11}^L,v_{11}^U]) & \cdots & ([u_{1m}^L,u_{1m}^U],[v_{1m}^L,v_{1m}^U])\\ \vdots & \ddots & \vdots\\ ([u_{n1}^L,u_{n1}^U],[v_{n1}^L,v_{n1}^U]) & \cdots & ([u_{nm}^L,u_{nm}^U],[v_{nm}^L,v_{nm}^U])\end{bmatrix}_{n\times m} \tag{11}$$

Step 2: Calculate the weight of criteria.

Calculating Ω based on Definition 4.

Step3: Alternative ranking.

(1) Determining the PIS and the NIS.

The comprehensive decision matrix R_i of x_i is denoted as:

$$R_i=\{r_{i1},r_{i2},\ldots,r_{im}\}=\begin{Bmatrix}\langle[u_{i1}^L,u_{i1}^U],[v_{i1}^L,v_{i1}^U]\rangle,\langle[u_{i2}^L,u_{i2}^U],[v_{i2}^L,v_{i2}^U]\rangle,\\ \ldots,\langle[u_{im}^L,u_{im}^U],[v_{im}^L,v_{im}^U]\rangle\end{Bmatrix}. \tag{12}$$

The PIS and the NIS are denoted in the form of IVIFS as, respectively:

$$PIS=\begin{Bmatrix}\langle\max_{i=1,2,\ldots,n}[u_{i1}^L,u_{i1}^U],\min_{i=1,2,\ldots,n}[v_{i1}^L,v_{i1}^U]\rangle,\langle\max_{i=1,2,\ldots,n}[u_{i2}^L,u_{i2}^U],\min_{i=1,2,\ldots,n}[v_{i2}^L,v_{i2}^U]\rangle,\\ \ldots,\langle\max_{i=1,2,\ldots,n}[u_{im}^L,u_{im}^U],\min_{i=1,2,\ldots,n}[v_{im}^L,v_{im}^U]\rangle\end{Bmatrix}$$

$$=\{\langle[u_{i1}^{L+},u_{i1}^{U+}],[v_{i1}^{L+},v_{i1}^{U+}]\rangle,\langle[u_{i2}^{L+},u_{i2}^{U+}],[v_{i2}^{L+},v_{i2}^{U+}]\rangle,\ldots,\langle[u_{im}^{L+},u_{im}^{U+}],[v_{im}^{L+},v_{im}^{U+}]\rangle\}. \tag{13}$$

$$NIS=\begin{Bmatrix}\langle\min_{i=1,2,\ldots,n}[u_{i1}^L,u_{i1}^U],\max_{i=1,2,\ldots,n}[v_{i1}^L,v_{i1}^U]\rangle,\langle\min_{i=1,2,\ldots,n}[u_{i2}^L,u_{i2}^U],\max_{i=1,2,\ldots,n}[v_{i2}^L,v_{i2}^U]\rangle,\\ \ldots,\langle\min_{i=1,2,\ldots,n}[u_{im}^L,u_{im}^U],\max_{i=1,2,\ldots,n}[v_{im}^L,v_{im}^U]\rangle\end{Bmatrix}$$

$$=\{\langle[u_{i1}^{L-},u_{i1}^{U-}],[v_{i1}^{L-},v_{i1}^{U-}]\rangle,\langle[u_{i2}^{L-},u_{i2}^{U-}],[v_{i2}^{L-},v_{i2}^{U-}]\rangle,\ldots,\langle[u_{im}^{L-},u_{im}^{U-}],[v_{im}^{L-},v_{im}^{U-}]\rangle\}. \tag{14}$$

(2) Calculate the weighted cross-entropy and relative closeness coefficient between each alternative and the PIS or NIS. $D^*(R_i, PIS)$ is the weighted cross-entropy between each alternative and the PIS. $D^*(R_i, NIS)$ is the weighted cross-entropy between each alternative and the NIS in the following. Then, the relative closeness degree (RCD) of each alternative is calculated, where$0\leq\theta\leq1$, and θ is the attention degree. Finally, the alternatives are ranked based on RCD_i (i=1, 2, …, n). The bigger the RCD_i is, the better the alternative x_i is.

$$RCD_i=\frac{\theta D^*(R_i,NIS)}{\theta D^*(R_i,PIS)+(1\text{-}\theta)D^*(R_i,PIS)} \tag{15}$$

$$D^*(R_i,PIS)=D(R_i,PIS)+D(PIS,R_i) \tag{16}$$

$$D(R_i,PIS)=\sum_{j=1}^{m}\begin{pmatrix}w_j\dfrac{u_{ij}^L+u_{ij}^U+2\text{-}v_{ij}^L\text{-}v_{ij}^U}{4}ln\dfrac{2(u_{ij}^L+u_{ij}^U+2\text{-}v_{ij}^L\text{-}v_{ij}^U)}{u_{ij}^L+u_{ij}^U\text{-}v_{ij}^L\text{-}v_{ij}^U+u_{ij}^{L+}+u_{ij}^{U+}\text{-}v_{ij}^{L+}\text{-}v_{ij}^{U+}+4}+\\ w_j\dfrac{v_{ij}^L+v_{ij}^U+2\text{-}u_{ij}^L\text{-}u_{ij}^U}{4}ln\dfrac{2(v_{ij}^L+v_{ij}^U+2\text{-}u_{ij}^L\text{-}u_{ij}^U)}{v_{ij}^L+v_{ij}^U\text{-}u_{ij}^L\text{-}u_{ij}^U+v_{ij}^{L+}+v_{ij}^{U+}\text{-}u_{ij}^{L+}\text{-}u_{ij}^{U+}+4}\end{pmatrix} \tag{17}$$

4 Case study and comparison analysis

4.1 Case study

Due to the decline of traditional manufacturing industry, automobile manufacturing enterprises are confronted with high cost and low margin pressure. Some enterprises even are confronted with the danger of survival in the fierce competition environment. Under this circumstance, a growing number of enterprises are preparing customer-oriented 5S (Sale, Spare-part, Service, Survey, Sustainability) service. There are four alternatives providing 5S service, denoted as x_1, x_2, x_3, x_4 according to their own industry and management characteristics. The criteria of four alternatives $C=\{c_1, c_2, c_3, c_4\}$ are: the ability of satisfying customer demand, the controlling ability of operating costs, the sustainable development ability, and the enhancing ability of production and sales. The initial interval-valued intuitionistic fuzzy decision matrices of fifteen DMs are listed in Table 1.

According to the proposed MALGDM methodology, the alternatives are ranked through three steps in the following:

Step 1: The optimized model of the weights of DMs is constructed by using Equation (8)~(10). The iteration number is set as 1000, the optimization target value is eventually obtained as 6.467 as shown in Figure 2. The weights of the DMs as shown in Table 2. The fuzzy comprehensive decision matrix $\tilde{\boldsymbol{R}}$ is obtained using Equation (1), and the PIS and the NIS are determined using Equation (13)~(14) as shown in Table 3.

Step 2: The weights of the criteria are obtained using Equation (4)~(5), the results are listed in Table 4.

Step 3: The cross-entropy and relative closeness degree of alternatives are calculated by using Equation (15)~(17), which are listed in Table 5.

According to the RCDs of each alternative, the alternatives are ranked as: $x_3 > x_2 > x_1 > x_4$.

Table 1. The initial fuzzy decision matrices.

		c_1	c_2	c_3	c_4
e_1	x_1	<[0.5,0.6],[0.2,0.3]>	<[0.4,0.5],[0.2,0.3]>	<[0.3,0.4],[0.3,0.4]>	<[0.4,0.5],[0.3,0.4]>
	x_2	<[0.3,0.4],[0.1,0.2]>	<[0.5,0.6],[0.3,0.4]>	<[0.6,0.7],[0.2,0.3]>	<[0.6,0.7],[0.1,0.2]>
	x_3	<[0.4,0.5],[0.2,0.3]>	<[0.3,0.4],[0.3,0.4]>	<[0.5,0.6],[0.2,0.3]>	<[0.7,0.8],[0.1,0.2]>
	x_4	<[0.6,0.7],[0.1,0.2]>	<[0.5,0.6],[0.1,0.2]>	<[0.4,0.5],[0.1,0.2]>	<[0.3,0.4],[0.2,0.3]>
......					
e_{15}	x_1	<[0.5,0.6],[0.2,0.3]	<[0.5,0.6],[0.2,0.3]	<[0.2,0.3],[0.4,0.5]	<[0.3,0.4],[0.4,0.5]

		>	>	>	>
	x_2	<[0.4,0.5], [0.1,0.2]>	<[0.8,0.9], [0,0.1]>	<[0.5,0.6], [0.3,0.4]>	<[0.7,0.8], [0.1,0.2]>
	x_3	<[0.6,0.7], [0.2,0.3]>	<[0.6,0.7], [0.2,0.3]>	<[0.3,0.4], [0.5,0.6]>	<[0.4,0.5], [0.2,0.3]>
	x_4	<[0.4,0.5], [0.4,0.5]>	<[0.4,0.5], [0.3,0.4]>	<[0.4,0.5], [0.3,0.4]>	<[0.3,0.4], [0.3,0.4]>

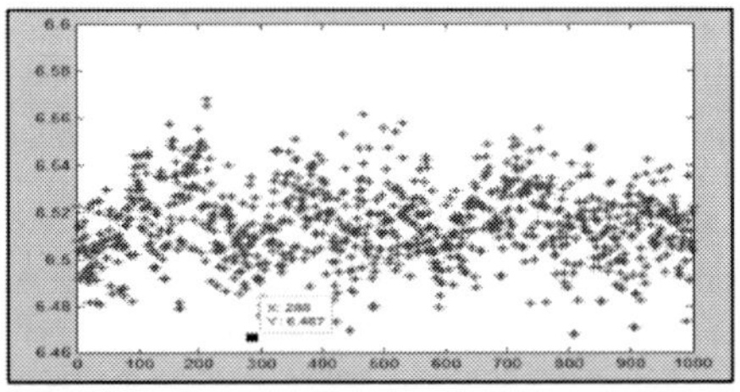

Fig. 2. The optimal value of the target function.

Table 2. The optimized results of weights of DMs.

e_k	λ_k	e_k	λ_k	e_k	λ_k
e_1	0.1245	e_6	0.0007	e_{11}	0.0126
e_2	0.0284	e_7	0.1080	e_{12}	0.1035
e_3	0.0702	e_8	0.0390	e_{13}	0.0711
e_4	0.1138	e_9	0.0310	e_{14}	0.0031
e_5	0.1393	e_{10}	0.1073	e_{15}	0.0474

Table 3. The comprehensive decision matrix and the PIS(NIS).

	c_1	c_2	c_3	c_4
x_1	<[0.543,0.63], [0.236,0.327]>	<[0.585,0.672], [0.0,0.308]>	<[0.538,0.626], [0.0,0.335]>	<[0.526,0.669], [0.0,0.289]>

x_2	<[0.585,0.696], [0.109,0.186]>	<[0.495,0.637], [0.0,0.344]>	<[0.496,0.583], [0.0,0.276]>	<[0.457,0.654], [0.166,0.293]>
x_3	<[0.501,0.578], [0.0,0.323]>	<[0.544,0.677], [0.0,0.277]>	<[0.513,0.604], [0.0,0.319]>	<[0.540,0.673], [0.0,0.282]>
x_4	<[0.586,0.667], [0.0,0.32]>	<[0.562,0.667], [0.0,0.289]>	<[0.472,0.564], [0.223,0.413]>	<[0.445,0.618], [0.0,0.348]>
PIS	<[0.586, 0.696], [0.0, 0.186]>	<[0.585,0.677], [0.0,0.276]>	<[0.539,0.626], [0.0, 0.277]>	<[0.540, 0.673], [0.0,0.282]>
NIS	<[0.501,0.578], [0.236,0.33]>	<[0.495,0.637], [0.,0.344]>	<[0.472,0.565], [0.223,0.413]>	<[0.445, 0.618], [0.166, 0.348]>

Table 4. The weights of the criteria.

	c_1	c_2	c_3	c_4
d_j	1.6472	1.81005	1.41685	1.81005
ω_j	0.255	0.279	0.218	0.248

Table 5. The cross-entropy and RCD of alternatives.

	Alternatives			
	x_1	x_2	x_3	x_4
$D^*(R_i, PIS)$	0.0047	0.0027	0.0022	0.0052
$D^*(R_i, NIS)$	0.0058	0.0067	0.0068	0.0041
RCD_i	0.4485	0.2828	0.2476	0.5614
Ranking order	3	2	1	4

4.2 Sensitivity analysis

The sensitivity of the proposed methodology is analyzed in this section by adjusting the original evaluation data [23]. For the highest importance degree, the hesitancy degree of c_2 in Table 4 is reduced to zero. The cross-entropy and the RCDs of each alternative are calculated by the proposed approach, the results are listed in Table 6. For comparison, the weighted Euclidean distance and RCDs of each alternative are calculated by using Equation (9), the results are listed in Table 7. The Radar Map of the three kinds of ranking results is shown in Figure 3 and finally the ideal alternative is x_3. The ranking order does not change compared to the original order which can indicate that the applicability of the proposed approach is acceptable. Yet, the ranking order obtained by the weighted Euclidean distance under TOPSIS framework changes, that is, the ranks of x_1 and x_2 are swapped, which indicates that the proposed methodology is effective.

Table 6. RCD of alternatives calculating by the proposed method.

	Alternative			
	x_1	x_2	x_3	x_4
$D^*(R_i, PIS)$	0.0047	0.0041	0.0027	0.0053
$D^*(R_i, NIS)$	0.0076	0.0068	0.0069	0.0048
RCD_i	0.3828	0.3791	0.2797	0.5237
Ranking order	3	2	1	4

Table 7. RCD of alternatives calculating by Euclidean distance.

	Alternative			
	x_1	x_2	x_3	x_4
$D(R_i, PIS)$	0.0025	0.0044	0.00053	0.0034
$D(R_i, NIS)$	0.0054	0.0020	0.0033	0.0032
RCD_i	0.3199	0.6872	0.1384	0.5177
Ranking order	2	4	1	3

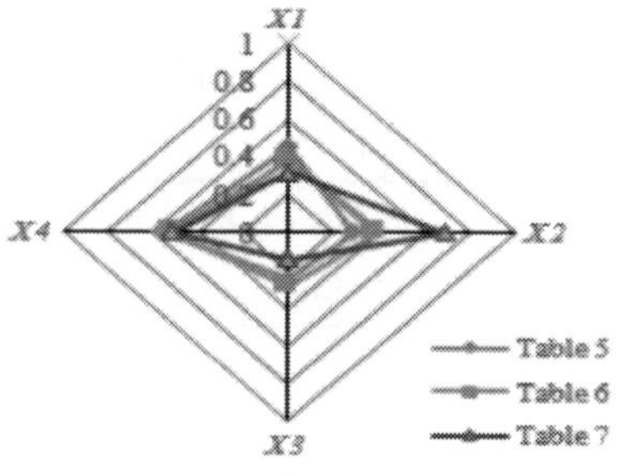

Fig 3. The three kinds of ranking result.

5 Conclusion

This paper reviews approaches dealing with MALGDM problem which is difficult to obtain the weights of the DMs and proposes an integrated MALGDM framework. The main contributions are as follows.

(1) First, the weights of the DMs are optimized by constructing a non-linear programming model, in which the original decision matrices are aggregated by using IIFWA operator. By solving the non-linear programming model with MATLAB®, the weights of the DMs and the comprehensive decision matrix are determined.

(2) Second, the IVIFS is introduced to describe the membership degree and non-membership degree of the DMs. The IVIF-cross entropy is employed to measure the divergence degree between IVIFNs and the information entropy is employed to determine weights of criteria under TOPSIS framework.

A selection problem of alternatives providing 5S service is implemented as the case study. There are still some issues need to be considered in the future. For example, the consensus analysis can be introduced to increase the consensus between $\tilde{\boldsymbol{R}}_k$ and $\tilde{\boldsymbol{R}}$.

Acknowledgments

The authors express sincere appreciation to the anonymous referees for their helpful comments to improve the quality of the paper.

This research is supported by the Fundamental Research Funds for the Central Universities (grant No. 2015QNA41), the National Natural Science Foundation, China (No. 51505480, 51475290), Natural Science Foundation of Jiangsu Province (No. BK20150197).

References

1. J. Qin, X. Liu, W. Pedrycz. A multiple attribute interval type-2 fuzzy group decision making and its application to supplier selection with extended LINMAP method . Soft Computing. 1-20 (2016).
2. D. Guha, D. Chakraborty. Fuzzy multi-attribute group decision making method to achieve consensus under the consideration of degrees of confidence of experts' opinions ☆. Computers & Industrial Engineering. **4,** 493-504 (2011).
3. J. Xiong, Y.W. Chen, K.W. Yang. A decision support model for multi-attribute group decision making using a multi-objective optimization approach. International Journal of Computational Intelligence Systems. **2**, 337-353 (2013).
4. S. Zahir. Clusters in a group: Decision making in the vector space formulation of the analytic hierarchy process. European Journal of Operational Research. **3,** 620-634 (1999).
5. N. Bolloju. Aggregation of analytic hierarchy process models based on similarities in decision makers' preferences ☆. European Journal of Operational Research. **3,** 499-508 (2001).
6. A. Tapia-Rosero, Bronselaer A, Tré G D. A method based on shape-similarity for detecting similar opinions in group decision-making. Information Sciences. **258,** 291-311 (2014).
7. B. Liu, Y Shen, X Chen. A partial binary tree DEA-DA cyclic classification model for decision makers in complex multi-attribute large-group interval-valued intuitionistic fuzzy decision-making problems. Information Fusion. **1,** 119-130 (2014).
8. B. Liu, Y. Shen, Y. Chen. A two-layer weight determination method for complex multi-attribute large-group decision-making experts in a linguistic environment. Information Fusion. **23,** 156-165 (2015).

9. I. Palomares, L. Martinez, F. Herrera. A Consensus Model to Detect and Manage Noncooperative Behaviors in Large-Scale Group Decision Making. IEEE Transactions on Fuzzy Systems. **3,** 516-530 (2014).

10. X.H. Xu, X.Y. Zhong, X.H. Chen. A dynamical consensus method based on exit–delegation mechanism for large group emergency decision making. Knowledge-Based Systems. **86,** 237-249 (2015).

11. F. Zhang, J. Ignatius, Y. Zhao. An improved consensus-based group decision making model with heterogeneous information. Applied Soft Computing. **35,** 850–863 (2015).

12. K. Atanassov, G. Gargov. Interval valued intuitionistic fuzzy sets. Fuzzy Sets & Systems. **3,** 343-349 (1989).

13. S.P. Wan, G.L. Xu, F. Wang. A new method for Atanassov's interval-valued intuitionistic fuzzy MAGDM with incomplete attribute weight information. Information Sciences. **316,** 329-347 (2015).

14. R. Al-Aomar. A combined AHP-entropy method for deriving subjective and objective criteria weights. International Journal of Industrial Engineering Theory Applications & Practice. **1,** 12-24 (2010).

15. Y. Yu, Y. Wu, N. Yu. Fuzzy comprehensive approach based on AHP and entropy combination weight for pipeline leak detection system performance evaluation// Systems Conference (SysCon), 2012 IEEE International. 1-6 (2012).

16. I.A. Curtis. Valuing ecosystem goods and services: a new approach using a surrogate market and the combination of a multiple criteria analysis and a Delphi panel to assign weights to the attributes. Ecological Economics. **3–4,** 163-194 (2004).

17. Y. Li, X. Chu, D. Chu. An integrated approach to evaluate module partition schemes of complex products and systems based on interval-valued intuitionistic fuzzy sets. International Journal of Computer Integrated Manufacturing **7,** 675-689 (2013).

18. K. Xu, J. Zhou, R. Gu. Approach for aggregating interval-valued intuitionistic fuzzy information and its application to reservoir operation. Expert Systems with Applications. **7,** 9032-9035 (2011).

19. H. Zhang, L. Yu. MADM method based on cross-entropy and extended TOPSIS with interval-valued intuitionistic fuzzy sets. Knowledge-Based Systems. **2,** 115-120 (2012).

20. K.T. ATANASSOV. Interval Valued Intuitionistic Fuzzy Sets. Intuitionistic Fuzzy Sets. Physica-Verlag HD. 343-349 1999.

21. L.C. Jang, W.J. Kim, T. Kim. A Note on Distances between Interval-Valued Intuitionistic Fuzzy Sets. International Journal of Fuzzy Logic & Intelligent Systems. **1,** 8-11 (2011).

22. S. Wang. A Novel Multi-attribute Allocation Method Based on Entropy Principle. Journal of Software Engineering. **1,** 16-20 (2012).

23. R. Simanaviciene, L. Ustinovichius. Sensitivity Analysis for Multiple Criteria Decision Making Methods: TOPSIS and SAW. Procedia - Social and Behavioral Sciences. **6,** 7743-7744 (2010).

Design of Flow Big Data System Based on Smart Pipeline Theory

Jianqing Zhang[1], Shuai Li[2,*] and Lilan Liu[3]

[1]Department of Mechatronic Engineering and Automation, Shanghai University, Shanghai, China
[2]Department of Mechatronic Engineering and Automation, Shanghai University, Shanghai, China
[3]Department of Mechatronic Engineering and Automation, Shanghai University, Shanghai, China
Corresponding Email: lsmichaelly@163.com

Abstract. As telecom operators to build intelligent pipe more and more, analysis and processing of big data technology to deal the huge amounts of data intelligent pipeline generated has become an inevitable trend. Intelligent pipe describes operational data, sales data; operator's pipe flow data make the value for e-commerce business form and business model in mobile e-business environment. Intelligent pipe is the third dimension of 3 D pipeline mobile electronic commerce system. Intelligent operation dimensions make the mobile e-business three-dimensional artifacts. This paper discusses the smart pipeline theory, smart pipeline flow big data system, their system framework and core technology.

Keywords. Smart pipeline, big data, e-commerce, framework design.

1 Smart pipeline theory

Smart pipeline [1] originating from the communications industry, the ultimate goal is to meet customer a variety of flexible ways of access through establishing the intelligent pipeline which can be controlled and the end-to-end difference. Meanwhile corresponding to different business forms and provide innovative business experience. Intelligent pipe is the third dimension theory of the pipeline system superimposed on a one-dimensional, two-dimensional pipe. If the superposition of 2D pipeline, mobile e-business system from a single bearing extensions to the plane, it be open from carriers and operators resources clouded. That 3D pipeline intelligent pipe [2] is for mobile electronic commerce further extension, extended to the three-dimensional model. 3D integration is the ultimate form of mobile electronic commerce theory model, the third dimension emphasized in the era of mobile Internet, the era of big data, operators' pipeline intelligent theory, pipeline intelligent technology and intelligent application of business model, business value. Intelligent pipe generates a lot of users surfing the Internet log files to provide date source for billed for billing and user behavior online query and management analysis. Analyzing business data effectively in the intelligent pipeline, and use the results of the analysis and processing for control and management of pipeline is essential for the successful operation of intelligent pipe to the realization of the rights of the operators. As shown in Figure 1.

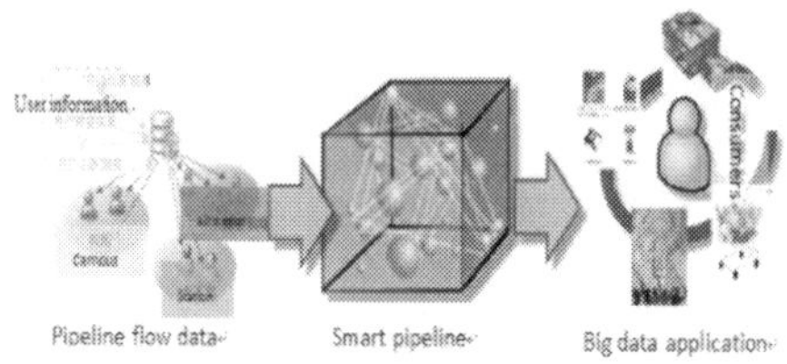

Fig.1. 3 D pipeline under the integration of intelligent mobile e-commerce ecosystem.

2 Big data systems of operating intelligent pipe flow TAS

Smart pipeline flow analysis support system TAS is the subsystem of Intelligent Mobile E-Commerce Cloud Platform (IMECCP). TAS mainly for operators in the network traffic data, traffic data collection and analysis, data mining applied in IMECCP platform.

2.1 Business mode

TAS platform business and application is mainly aimed at the target merchants, operation analysis means for merchants to provide depth, marketing intelligence (BI) capabilities. Its main business model has three categories:

(1) Operational analysis categories: TAS provide merchants basic operational data analysis function, operation analysis module provides free module and charging module, for daily general merchants sales, warehouse, order statistics report to provide free services, as for some merchants customized data report, customized report will be charged according to the needs of businesses.

(2) Smart marketing categories: Smart marketing business is a major core business model of TAS. Merchants can order intelligent marketing service, for example, obtaining merchant track customer behavior data ,smart marketing class can be divided into offline mode smart marketing and online marketing.

(3) Run the auction categories: Operation bidding class is to help merchants analysis industry information, TAS collecting data form various industries and send it to the data center, can provide Shared data within the scope of industry recognition, such as providing the same industry with many merchants business development and the consulting services about an overview of the operating data

2.2 TAS Platform application system

TAS platform application system mainly divided into three parts:

(1) Data gathering: Date gathering is based on the operator data platform to large flow data and gathering all kinds of data sources: Operators, mobile Internet traffic data and customer behavior data and so on. Such as Internet access to the user, use the terminal behavior and the content of the order package actual usage and data integration network quality, etc. Data gathering constructing data system as large operators + mobile Internet data. Completing the structured and unstructured data preprocessing and data storage function.

(2) Data driven：Data driven is the second largest system in TAS. Its role is to enable mobile e-business towards intelligent mobile e-commerce, data driven to gather the data for data processing, big data mining and data analysis, and the result used for driving the process of mobile e-commerce operation, marketing and service of each link. TAS can get the user data from baidu ranked, SEO index, etc. It can provide business decisions propaganda drainage on the strategy. In terms of customer behavior, TAS provides behavior data can drive merchants product operation optimization, such as purchasing, payment process, at the same time can also provide goods operation adjustment data basis. TAS provides segment traffic data, can make the merchants operating staff know the conversion rate of various business or goods.

(3) Data labels: It is the third largest TAS system. TAS after routinization and a large number of intelligent data processing. Customers, merchants and other attributes can be formed TAS data labels. On the one hand, it fixed-line

data-driven data processing results; On the one hand, it open data tag library can be used as a data model is a child module of TAS opened to the public. TAS tag library is characteristic of TAS formed after the big data processing collection classes. Merchants select existing tags for commercial applications by label adaptation. TAS label application is TAS data-driven upgrade version, on the one hand is data driven conceptual characteristic collection, classification, on the other hand is as TAS ability to open module.

3 TAS smart pipeline architecture design

TAS platform is mainly composed of data gathering, data driven and data labels. As shown in Figure 2. Data gathering is to realize the function of data acquisition deployment, data acquisition, data preprocessing, and data adapter; Data driven is business intelligence data mining, intelligent application rules, and other functions; Dateable is the data-driven model, the application and the ability to open the component library.

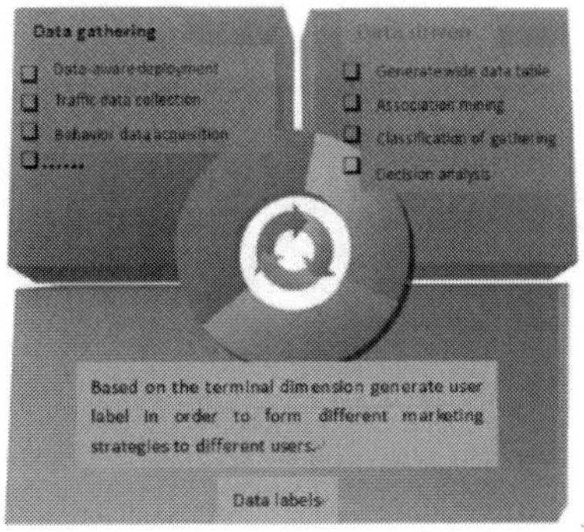

Fig.2. TAS system three big modules.

3.1 Based on the data gathering pipeline filter architecture

Pipeline filter [3] is one of the classic design of software architecture, in the pipe/filter style of software, architecture is composed of multiple hierarchy structural connections, each level component with one or a set of input and output, when data got into a single component (pipe), component begin internal processing and output data stream. To change the data input data stream, specifications and other processing becomes filtering, the component becomes the filter. Such a style of fitting is become pipe/filter. This is shown in Figure 3.

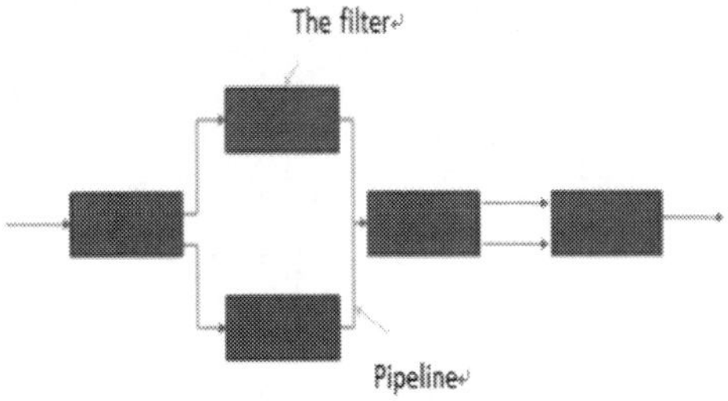

Fig.3. Pipe/filter design architecture.

Pipeline filter architecture design has the following characteristics:

(1) Pipeline filter has a good concealment, high cohesion and low coupling;

(2) Can support data gathering more data sources, the demand of the multi-layer data preprocessing;

(3) Have higher software reuse and reusability, similar to function of the filter can be reuse connection;

Data gathering architecture based on the pipeline filter architecture style as shown in Figure 4.

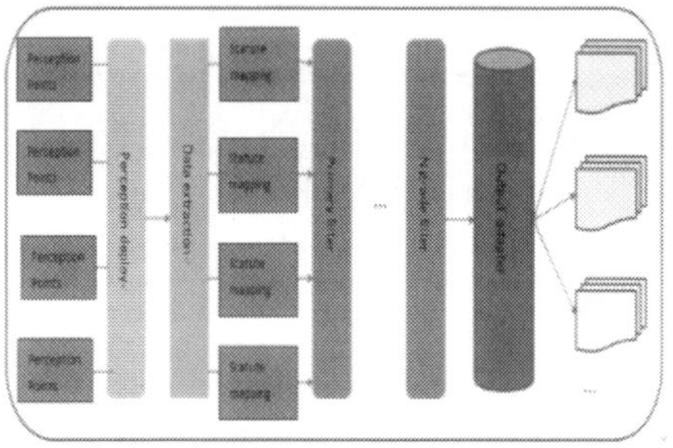

Fig.4. Data gathering architecture.

Perception of deployment: Requires data analysis is used to drive the aware of acquisition for deployment;

Data extraction: Is a set of integrated filter module, complete the basic data packet extraction, such as the data is divided into content, behavior, etc.;

Statute mapping: Refers to the rules on the application of the different data gathering data mapping, mainly according to the data gathering after mapping the drive application of management;

Filter layer: TAS data gathering data processing layer, the same data preprocessing function module at the same level is available for reuse and through the connection between layer and layer thinning data processing;

Output adapter: Output format is based on the outer data applications, data driven data requires;

3.2 Based on the data warehouse architecture and knowledge base

After complete the data gathering, different businesses have different business development needs. Therefore the function of the data driven goals and modules will be different, but the application of data driven commonness exists in the method. Thus cannot blindly to the expansion of data driven application modules. So the driver module for data of TAS, based on the warehouse and the knowledge base of architecture design style, as shown in Figure 5. Warehouse and knowledge base structure is to apply data driven and data-driven rules and unified, the warehouse storage component module has complete data driven capacity. For example, correlation algorithm, clustering, classification, linear regression and so on. It can categorize knowledge stored in the knowledge base and distinguish between the management of the private intelligence of the merchants. The knowledge base was shown as the blackboard.

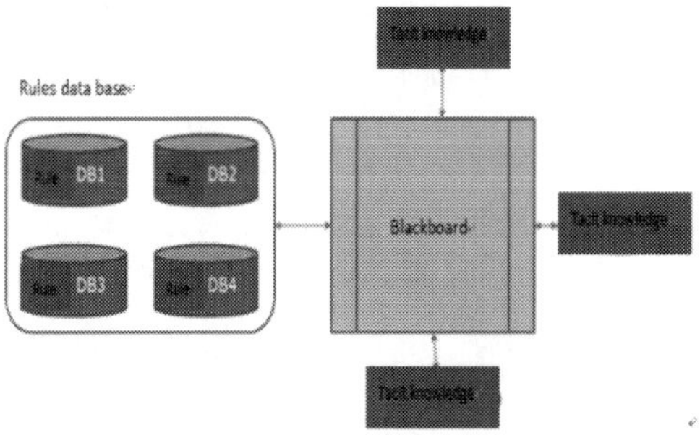

Fig.5. Warehouse and the knowledge base architecture.

Rule knowledge base is used to store different classification or different merchant's private business intelligence driving rules. The blackboard extracts data from data gathering of tacit knowledge, docking rules according to the corresponding application on the blackboard. The blackboard is a state of driver and application driver module, the data display and driver is on the blackboard to tacit knowledge and the corresponding knowledge base organization to solve the problem of specific business applications.

Warehouse and corresponding TAS data-driven knowledge base architecture features as follows:

(1) Users don't need to pay attention to the blackboard after the concrete application of transparent to the client, simple to use;

(2) Merchants can manage their own application rules, and can reuse from common rule knowledge base;

3.3 Data labels based on the Hadoop

Data labels are the curing and open ability of TAS platform module. Data label is to solve the problems of various merchants during the actual store operations, all kinds of data analysis and data intelligent marketing demand. Data labels can integrate, mining and manage the effective customer information.

Integration: Existing a lot of customer information resources stored separately in different systems, the need to form a unified view of customer information, raise the utilization ratio of information.

Mining: The current use of customer information with index is staying the level of analysis and report. The further analysis of customer information, deepen the characteristics analysis and customer demand is important.

Management: Daily operation of customer did not form the effective management of knowledge. Knowledge utilization is low, the need to strengthen the effective management of customer knowledge.

Data labels need to solve the problem of large data processing ability. Therefore in the project practice, this paper adopt the Hadoop [4] related architecture design technology. Hadoop contains two types of technology: One is the HDFS distributed file storage system; The second is graphs distributed technology. HDFS is used to solve the TAS to deal with many merchants, large data problem of data storage capacity, its technical architecture diagram as shown in Figure 6.

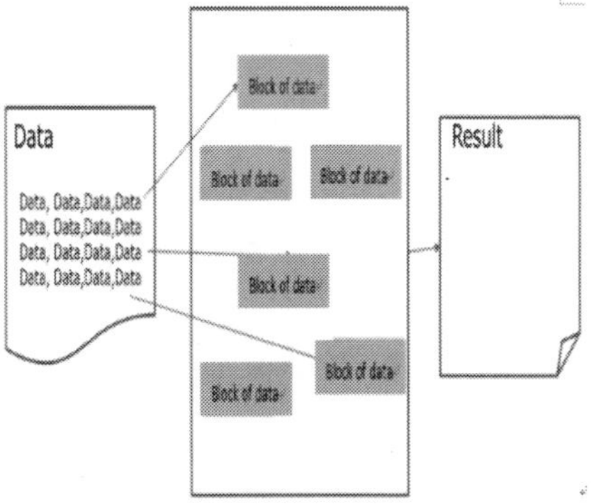

Fig.6. HDFS Schematic diagram.

HDFS file is composed of blocks. Each block has multiple copies, multiple copies in the HDFS architecture can be stored in a building on different computers in the system. The graphs can be split into multiple small pieces for processing shown as Figure7.

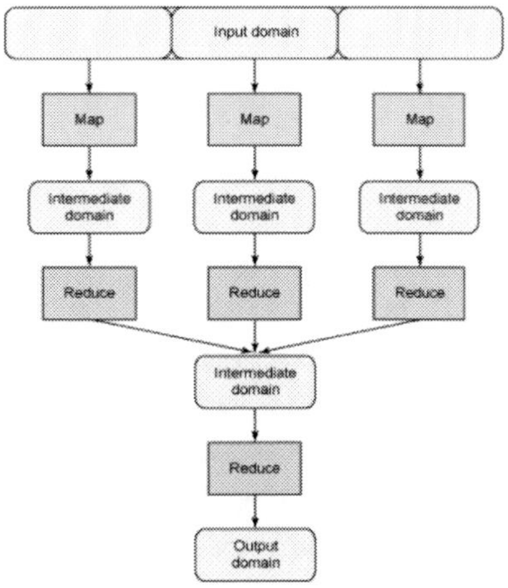

Fig.7. MapReduce Schematic diagram.

HDFS architecture in the aspect of application architecture is transparent for the outer applications. Applications need not care about the data location, as local calls for data calls.HDFS + graphs can solve data label under the TAS big data module, data processing ability, and TAS data label module technology architecture as shown in Figure 8.

4. The system key techniques in the big data

4.1Key techniques in data gathering

(1) Big data access technology

As can be seen from the operators of big data, the traditional billing domain data, how to file as a carrier; Signaling class data, the data stream as the carrier; Broadband Internet data, the data message as the carrier.

A. Grid collection: Grid acquisition [5] is suitable for gathering data in the file as a carrier. Based on the concept of worker bee collaboration, the traditional FTP / SFTP file collection technology was extended and realized the distributed collaborative concurrent acquisition. When data network generate batch files at a time, the traditional sampling need time sequence generated by file, it takes a long time; But grid sampling will split stay files into multiple division (divided by the data source ID, the information such as file name, start, end displacement), different node/application receive the respective division for concurrent collaborative collection. When the data network element at a time to generate large files, the traditional collection depends on the only data download channel, likely to cause incomplete data acquisition or acquisition multiple times, but grid acquisition will split the large file into multiple partitions, different nodes / Application to receive their own division for concurrent collaborative collection, stitching and verification when landing. Collection dispatching and breakpoint aimed at persistence management. A real-time synchronization of multi-node grid collection of calibration, guarantee grid to collect accurate and complete and at the same time security when fault occurs it can quickly move to the new node stability without gap operations

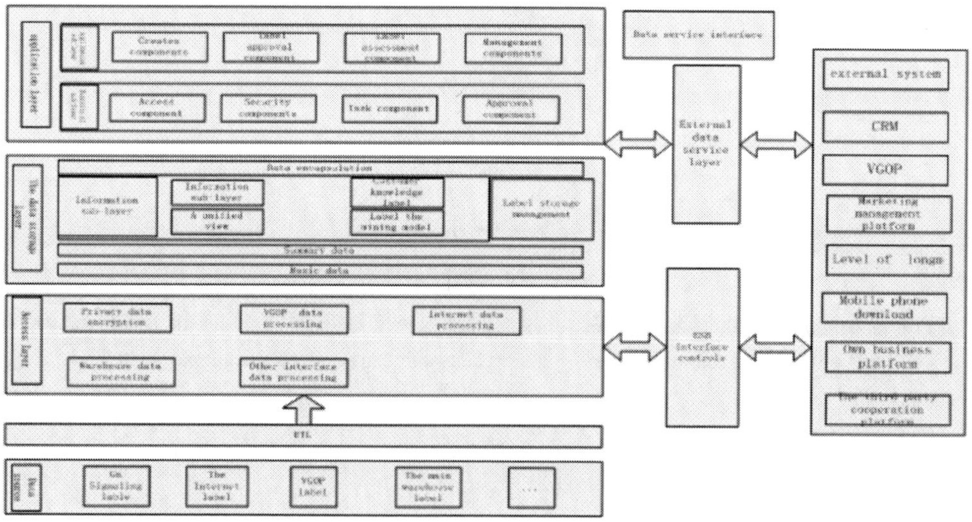

Fig.8.TAS data label module technology architecture.

b. Streaming Collection: Streaming Collection is suitable for aggregating signaling class data with high time efficiency. Streaming Collection need the large data center and data network element prescribed uniform flow interface, large data center send data subscription request to the data network element, waiting for the data network element returned data subscription response. The large data aggregation center analyzing and checking streaming message according to the message format, translating legitimate messages into internal format events, pushing to the data delivery module, and writing the message backup to the distributed file system.

c. Spectral collection: Spectral acquisition is suitable for aggregating broadband interconnect data from the core network. Spectral collection has the ability to capture data directly to ensure data comprehensiveness and timeliness; it protects the value of the data to maximize on the source. For example, the signaling location data information can be split and access from the signaling port A, and the preamble is used to collect the signaling A port signaling by the network access protocol converter. After the signaling data are analyzed and correlated, the synthesized position is updated, Called the message, the main called voice calls and other signaling CDR / TDR.

(2)Big data processing technology

As can be seen from the telecom operators of large data, the traditional billing domain data, there are billing, DPI file data, and also signaling, location information class message flow data. Billing domain and other types of data processing need to be processed in sequence and be calculated accurately with the high timeliness; Signaling, location information and other types of data processing need to quickly deal with the high timeliness. Hadoop, Storm, Spark is the most popular large-scale data processing tools, in the application of telecom operators in large data processing, the need to combine the characteristics of timing and timeliness to select the framework technology. Hadoop uses MapReduce distributed [6] computing framework, developed HDFS distributed file system according to GFS and developed HBase data storage system according to BigTable. Hadoop is often used for off-line complex large data processing.

MapReduce technology framework contains three levels of content: (1) Distributed file system; (2) Parallel programming model; (3) Parallel execution engine. The Google file system runs on a large cluster, which is built using cheap machines, and the data is stored in key / value mode. The entire file system uses metadata centralized management, data block distributed storage model, through data replication (at least 3 copies of each data.) to achieve a high degree of fault tolerance. Data is stored in chunks (64MB or 128MB for 1 block), it can be easily compressed on the data, saving storage space and transmission bandwidth. The MapReduce parallel programming model decomposes the computational process into two main phases, the Map phase and the Reduce phase. The Map function handles the Key / Value pair and generates a series of intermediate Key / Value pairs The Reduce function is used to merge all intermediate key-value pairs with the same Key value and calculate the final result.

4.2 The key algorithm of data mining

Data mining [7] also known as Knowledge Discovery in Database. It is from a large, incomplete, noisy, fuzzy and random big data to extract implicit in people is unknown in advance, has the potential value of the process of information and knowledge. Through association rule mining, can be implied in sea has potential value in the amount of data useful information.

Assuming assemble $\mathrm{I} = \{i_{1,}i_{2,}i_{3,} \dots i_{m,}\}$, i_k (k=1,2,…,m) represent the item. If $\mathrm{x} \in \mathrm{I}$, X is called item sets. When$|x| = k$, X is called K's item sets. Affair two-tuples $\mathrm{T} = (\mathrm{tid}, \mathrm{X})$, tid is the uniqueness identification called affair NO. Data assemble$\mathrm{D} = \{i_{1,}i_{2,}i_{3,} \dots . i_n\}$ consist of t_1, t_2… t_n. Association rules can be described as: like the formula A=>B in which $A \in I$, $B \in I$, 并且$A \cap B = \emptyset$.The support of item sets X is S which is the percentage of D contained X's number of transactions and All real number remembered as $\mathrm{s(x)} = \mathrm{p(x)} = \frac{\sup(x)}{|D|}$.The degree of confidence of item sets X is C which is the percentage of D contained $X \cup Y$ number of transactions and All real number remembered as $\mathrm{c(x)} = \mathrm{p}(X|Y) = \frac{\sup(X \cup Y)}{\sup(X)}$. Minimum support (minsup) and minimum confidence (minconf) are depended on the users. If sup(X)≥minsup, the item sets X is called cumbersome item sets in Which generated association rules all support and confidence in not less than minsup and minconf is called the strong association rules. Commonly used algorithm[8] for mining association rules has breadth-first algorithm、depth-first algorithm、dataset partition algorithm、sampling algorithm and parallel algorithm. Parallel algorithm is used at the same time to perform a collection of the process of the interaction and coordination to complete for a given problem solving. It has the higher efficiency.

Parallel FCM algorithm adopts the master/slave mode. A main process controls a number of process to calculate. In order to be kept in distributed parallel interconnect workstations environment mining fuzzy association rules. Improve the master/slave mode for single program multiple data (SPMD) model and divide the quantitative attributes. Each quantitative attribute values on the record as the initial data set. The PCM algorithm as follows:

FCM.1: In each process divided between the initial data set. Each process for n/s data, in which n is the total number of data records' to start the process.

$$\{x_1, x_2, \dots, x_n \,|\, x_{(\frac{n}{s})}, \dots, x_{\frac{2n}{s}} \Big| \dots \big| x_{(s-1)(n/s)}, \dots x_n\}$$

Process1 Process2 Process S

FCM2: Process by the root process (process id 0) initialization center $v_i, i = 1,2 \dots .. c$ and broadcast to all processes.

FCM3: Each process receives the $v_i, i = 1,2 \dots .. c$ and calculate the membership degree

$$u_{ik} = [\sum_{j=1}^{c}(||x_k - v_{io}||_A \backslash ||x_k - v_{jo}||_A)^{2(m-1)}]^{-1}$$

Each process j operation on its data subset $\left\{x_k, \mathrm{k} = \frac{(\mathrm{j}-1)\mathrm{n}}{\mathrm{s}} + 1, \dots j\frac{n}{s}\right\}$. Initializes is end.

FCM4: Each process according to the local data calculation.

$$\alpha_{ij} = \sum_{k=1}^{h}(u_{kj,t-1})^m X_r) \qquad \beta_{ij} = \sum_{k=1}^{h}(u_{k,t-1})^m)$$

h=n/s is the received data subset size for each process.

FCM5: Each process j send results $(\alpha_{ij}, \beta_{ij})$ to the root process, then the root process about computing v_{it} and broadcast to all processes.

$$v_{it} = \frac{\sum_{j=1}^{s}\alpha_{ij}}{\sum_{j=1}^{s}\beta_{ij}} \quad \mathrm{i} = 1,2 \dots . \mathrm{c}$$

FCM6: Each process receives class center value and calculates the membership degree. Each processes their operations on the data subset.

$$u_{ik\,t} = [\sum_{j=1}^{h}(||x_k - v_{it}||_A \backslash ||x_k - v_{jt}||_A)^{2(m-1)}]^{-1} \ \forall i, k$$

FCM7: Each process calculation error. Each process j only calculation error, and sent to a process.

$e_j = \sum_{k=1}^{h}\sum_{i=1}^{c}(u_{ik,t-1} - u_{k,t})^2$ For j=1, 2... S

Finally in the root process calculate summary error E_t.

$E_t = (\sum_{j=1}^{s} e_j)^{1/2}$.

If E_t.<ε, notify each process termination procedure.

For each quantitative attribute, the clustering center C and dividing matrix U. Each class represents a fuzzy set. And the elements in the matrix are divided into each values belonging to the degree of fuzzy sets. When using the Euclidean distance, the time complexity of parallel FCM algorithm is $\mathrm{o}\left(\frac{c^2\mathrm{pn}}{\mathrm{s}}\right)$.

5 Conclusion

This paper describes the concept of smart pipeline. In order to apply the network traffic data, traffic data collection and analysis, data mining applied in mobile e-commerce platform, it establish the system of intelligent pipe flow analysis support system TAS Reference. This paper mainly introduces the Business model, the platform body system, and framework design and core technology.

References

1. Huiling Zhao, Xianghui Xu. Connotation and characteristics of intelligent pipe. Telecom science,S1, 2011, pp.1-5

2. Anaraki FB. Elearning and mLearning at assumption university. 2011 International Conference on e-Education, Entertainment and e-Management (ICEEE), 2011, pp.31-34
3. Ping Xu, Jing Liu, Shengju Sang. Introduction to the software architecture style. Science and technology information. No.26, 2008, pp.42-43.
4. Shukui Hao. Introduction to Hadoop HDFS and MapReduce.Design technology of posts and telecommunications. No.7, 2012, pp.37-42.
5. Li,Guojie.The scientific value of big data research.Communications of CCF,2012,8(9):8-15
6. Jeffrey Dean Sanjay Ghemawat，Map R educe: Simplified Data Processing on Large Clusters． USA: San Frcmcisco，cakfcrnia． 2004:137 －150.
7. Han Jiawei , Kamb er M. The concept of data mining and technology.Beijing: Mechanical industry press, 2001: 149-176.
8. Cheung D W, Han J, Ng V T. A Fast Distr ibuted Algorithmfor Mining Assoeiation Rules //Proceedings of 1996 Inter-nat ional conference on Parallel and Distr ibuted InformationSystem. Miami Beaeh, F L: [s. n.] , 1996: 31-44.

Research on the Impact of Big Data on Logistics

Yaxing Wang[1,*], Lijie Feng[1], Heyu Chang[1] and Min Wu[1]

[1]School of management and engineering, Zhengzhou University, China
Corresponding Email:615871159@qq.com

Abstract. In the context of big data development, a large amount of data will appear at logistics enterprises, especially in the aspect of logistics, such as transportation, warehousing, distribution and so on. Based on the analysis of the characteristics of big data, this paper studies the impact of big data on the logistics and its action mechanism, and gives reasonable suggestions. Through building logistics data center by using the big data technology, some hidden value information behind the data will be digged out, in which the logistics enterprises can benefit from it.

1 Introduction

With the rapid popularization of the Internet and the Internet of things, information technology has brought the rapid growth of data scale, all walks of life will produce massive data in the process of operation. If you can not use the massive data effectively, the data will become a drag on the industry, into a data swamp. Technology application of big data is a kind of technical behavior by searching, calculation, analysis, storage, index, query complex structure, large quantity, changeable type and highly commercial value of the collection of data set. Big data applications bases on massive facts, we can use scientific methods to tap the potential value of data , make the mass of data into a new type of intellectual resources, prompte the traditional industry to change the way of thinking, use the data of this important factor in production better, create more benefits for society.

In this era of big data, there will be a lot of data in the logistics industry every day, especially the whole logistics, including transportation, warehousing, transportation, distribution, packaging and other links [1]. Each link to each other forms a complete information flow. If the logistics enterprises can not accurately process these data timely, then it will bring the logistics enterprise a data disaster and waste of resources. In recent years, a large number of domestic and foreign experts focus on the application of big data in the logistics industry. It is of great significance for enterprises to adapt to the changeable market environment and to deal with the fierce market competition. Also it will have a significant impact on the logistics industry's strategic decision-making, operational management, brand management, customer relationship management, service innovation and so on. This helps enterprises to optimize the allocation of logistics resources, accelerate the upgrading and transformation of the logistics industry, to meet the requirements of the information age [2]. In the context of large data, the logistics industry needs to seize the opportunity to improve the ability, use data resources, improve efficiency, in the market competition advantage.

2 Definition and characteristics of big data

Now, the concept of big data does not have a unified view. The contemporary society is an information and digital society.The information data is produced quickly, flows, in all kinds of information source, information platform, information channel, information terminal.In all,the data is filled with the whole world. The number of data is not only increasing in exponential form but also the structure of the data tends to be more complicated. This makes the big data has a different connotation compared with the previous data. McKinsey Global Research points out that, "Big data" is a large collection of data, greatly beyond the traditional database software tools in the ability to access, storage, management, analysis [3]. It is a way of decision making based on data driven [4],In a short period of time to quickly carry out data collection, storage, analysis, collation through the network platform. "Big data" is usually used to describe a large number of unstructured and semi-structured data created by a company. These data will cost too much time and money when they are downloaded to the relational database for analysis. Therefore, the strategic significance of "big data" technology does not lie in the master of the huge data information, and lies in the meaning of these data for professional processing, that is, from the "big data" mining of "small data".

At present, the more unified understanding is that: big data has four basic characteristics: huge data size, various data types , high data processing speed, low value of the data. These four characteristics make big data is different from the massive data". Huge data scale is the basic attribute of big data. Data size is large, on the one hand, because the data has a wide source, access to a wide range of channels. On the other hand, in order to make the description of the data closer to the original thing and thus lead to the proliferation of data. The reason for a wide variety of data mainly refers to the information age needs to pay attention to the dimensions and types of data have increased significantly. At the same time, focusing on the research and analysis of unstructured data which contains a large number of details of the information reserves space for personalization. Data processing is faster owing to the data generating speed, high speed of communication, but also for the timeliness of consideration. At the same time, we can explore the potential effective value, after eliminating the redundant data, invalid, no value data by fast processing. Low data value density is an important attribute of the unstructured data which is concerned by big data. Big data retained the original data, get all the details of things without abstraction and induction. Directly using the original data and all the data, you can analyze more information, but at the same time it also cited a large number of no sense and even the wrong information. Therefore, relatived to the specific application , the proportion of useful information of non structured data is very low, and the value density is also low.

Big data technology is a new technology, it brings opportunities to the logistics enterprises, but also brings challenges. Using big data technology scientifically and reasonably will play a positive role in promoting the management and decision-making, resource allocation and other aspects of logistics enterprises. If we compare big data to an industry, then the key to achieve profitability of this industry is to improve the data processing capacity and achieve value-added through the processing data. Therefore, in the background of big data, data information has become an important factor in production. This will change the way of management of logistics enterprises and innovation of the profit mode of logistics enterprises.

3 Big data helps to make scientific decision of logistics

The core of large data is finding and prediction. Through mining and analysis data, find the running law of history hidden under the data surface, and predict the future trend. Therefore, under the background of "big data", data information has become an important production factor, which will change the management mode of logistics enterprises and innovate the profit mode of logistics enterprises [5]. In the process of using large data assets, each step like acquisition, integration, governance, analysis, prediction, scientific decision-making to take precise action, plays an important role. Making decisions based on big data is a major highlight of the 2015 electricity suppliers logistics enterprises. Big data is variable, the algorithm is the rule, set what kind of rule will receive what kind of results. Take the "Rookie Network Technology Co., Ltd." (herein after referred to as the rookie network) as an example. Before the "double eleven" in 2015, this company analyzed the historical data of commercial tenant taking part in "double eleven" in recent four years which includes the location, previous orders and logistics of commercial tenant, and predicted the situation for 2015 according to the zone, and shared the predicting network sending and receiving information with information of each express enterprise. Each express logistics enterprise combined with three important nodes - warehousing, trunk and distribution of logistics system construction to deploy the corresponding logistics facilities according to these forecasts. More accurate data analysis and prediction make the courier company be more accurate and easy in response. The application of big data technology effectively reduces procurement, warehousing, internal logistics distribution share and finally expects to achieve zero stock, zero distance, zero freight [6].

4 Analysis of the impact of big data on the logistics industry

The strategic significance of big data technology lies in the professional processing of information and the value mining, and analysis of the long-term trends in the market and so on. With the gradual deepening of the application of large data in logistics industry, the data obtained in the future logistics industry is not only the industry's internal information, but also includes a large number of external information. Based on the data analysis, the logistics enterprises can tailor personalized and differentiated service for each customer with forecast.

Big data analysis can help logistics enterprises understand the industry development. At present, the logistics enterprises are faced with a highly competitive and rapidly changing market environment, many transportation load problems is in virtue of logistics enterprises lacking prediction of future market through data analysis, and they blind to increase transport capacity and storage area when they only see the immediate business growth. There will be a large number of surplus capacity and vacant warehouses when the market shrinks and the business volume declines, which will lead to the loss of logistics enterprises. Through the analysis of large data, logistics enterprises can make certain predictions on the future market and the behavior of competitors, timely adjust the development strategy and avoid blind asset investment, thus reduce losses.

Big data analysis can help logistics enterprises to enhance customer loyalty. For logistics enterprises, the analysis of customer's behavior habit can make their market promotion investment, supply chain investment and promotion investment returns be maximization. Through advanced statistical methods, the logistics enterprise can establish model through the historical record and analysis of users, predict their future behavior, then design a forward looking logistics service plan, therefore integrate resources and improve the level of cooperation with the customer to avoid the loss of customers. Logistics enterprises can not only mining value of existing users through large data, but also access new users efficiently through the data. Big data technology is changing the game rules in the market promotion with revolution. Through the promotion of information interaction, push service adjustment, price changes and market changes and other information to customers, and constantly meet the change of customer needs. In the context of the Internet, marketing will not be affected by time and place restrictions, and is no longer just a one-way flow of information. The great difference is that from contacting with customers and attracting customers to manage customers, launch promotions and finally to reach a sale, the entire marketing process can be realized only in the information interaction, and carry out precision marketing through the understanding of user behavior.

Big data analysis can improve the transparency and quality of service management of logistics industry. Big data analysis can make the logistics practitioners and the performance of the logistics organization be more transparent through the logistics information communication opening and information sharing, and indirectly promote the improvement of logistics service quality. According to the operation and performance data set by logistics service providers, the data can be analyzed and establish visualized flow chart and dashboard, and promote information transparency. There are world famous logistics enterprises are currently testing the instrument panel, see it as a part of the construction of active, transparent, open, collaborative companies. Opening logistics quality and performance data can help customers make more informed decisions, which will also help the logistics service provider to improve overall performance, finally enhance the competitiveness.

Big data analysis can optimize the profit mode of logistics enterprise. Through establishing the network platform of logistics industry and the community, the platform can produce a large number of valuable data, provide valuable data and summary the consumption records of logistics industry customers, and carry on the advanced analysis, and ultimately improve the decision-making ability of logistics demand side and logistics service provider. Data analysis of users on the platform is updated in real time to ensure that behavior prediction of users is always in line with the actual needs of users. At the same time, design some market strategy according to the behavior forecast dynamically, market expansion speed will depend on the developing speed of the big data acquisition and analysis in logistics industry. Establish a national customer database to provide accurate and timely logistics information consult, which will significantly improve the company's visibility and profitability.

5 Suggestions and measures

5.1 Paying enough attention to big data at the national level

The state should conform to the requirements in the era of big data, provide strong support to big data technology in terms of policy-making and resource input in the logistics industry, eliminate information barriers, concentrate forces to establish a healthy ecological environment of big data and attache great importance to big data theory and application as well as its safety. Furthermore, developing a long-term plan on a macro level and constructing a data sharing information platform.

5.2 Coordinating all kinds of logistics way and building a three-dimensional logistics network

The logistics industry should strengthen relation with various modes of transport, integrate transportation platform, coordinate different transport modes, adjust the network structure, strengthen the effective connection of various modes of transportation. Logistics enterprises should make full use of existing information infrastructure to integrate information resources and set up a modern logistics information processing platform in terms of railway, highway, aviation, shipping and and so on. In the aspect of construction of information platform, they also should pay attention to information system standardization and sharing in order to construct cross-platform, collect mechanism of multi-level information resources and create seamless information chain. Besides, the government needs to pay attention to traffic and logistics planning, improve the comprehensive transportation system and optimize logistics spatial layout.

5.3 Combing cloud logistics and big data and improving the logistics efficiency

In order to set up a good information platform, the logistics enterprises should speed up the logistics infrastructure standardization, informatization and intelligent hardware and software construction because the application of cloud logistics and big data is inseparable from the existing logistics information platform [7]. From a demand view, enterprises can have a complete data life cycle control through analyzing and integrating the data information which is corrected and extracted from the big data system. Furthermore, the complete logistics information platform can gather big data information and make both supplier and buyer to enhance mutual understanding. The supplier and buyer can get a full range of logistics services such as sharing information, making agreement and integrating logistics solution plan by using logistics management platform. Besides, combining cloud logistics and big data can also innovate logistics development mode and thus improve the logistics efficiency.

6 Conclusion

As a new technology, big data will bring both opportunity and challenge to logistics enterprises because its application transformed logistics enterprises from traditional logistics to data logistics and network logistics, and gradually developed into the cloud logistics. In the future, the application of mobile terminal logistics will be expanded and at that time, it will generate the larger data sets. So proper, efficient application of big data technology will play a positive role for logistics enterprises in terms of management and decision-making, customer relationship maintenance and resource allocation

Logistics enterprises will eventually become the data-driven enterprises, and it is a necessary condition for the further development of the enterprises in the era of big data. In the face of this opportunity, I call on all senior managers in logistics enterprises should give great attention to the big data, squarely face the problems of its application, properly apply big data to business management and bring greater benefits for the enterprise.

References

1. Tian Xue, Si Weipeng, Liu Yingying. The application of big data in the logistics enterprises. Electronic Commerce,2015(1):36-37.
2. He Jun. The impact of big data in enterprise management decision. Scientific and Technological Progress and Countermeasures, 2014(4):65-68.
3. James Manyika, Michael Chu. Big Data: The Next Frontier for Innovation, Competition and Productivity. McKinsey Quarterly, 2011(5):27-30.
4. Yan Xiaofeng, Zhang Dexin. Big data research. Computer Technology and Development, 2013(4):168-172.
5. Zi Wucheng. The Evolution of the Era of "Big Data" Enterprise Ecosystem and Construction. Social Science, 2013(12):55-62.
6. Xiang Meiying. Zero Inventory Management Analysis. China Market, 2009(45):27-28.
7. Han Zhaolong. Modern Logistics Industry Development in the Era of Big Data. Quality of Goods and Theory Research,2014.

Research of self-power Generation Scheduling Model Base on Multi-objective in Iron and Steel Enterprises

Xin Geng *, Honghai Niu, Jun Chen and Bing Li

Research and Development Center of NR Electric Co., Ltd., Nanjing 211102, China
Corresponding Email:genx@nrec.com

Abstract. With the targets of the minimum cost of power generation and the lowest rate of gas emission in iron and steel enterprises, a multi-objective self-power generation optimal scheduling model was built based on multi-objective particle swarm optimization the research of coupling relationship of gas and power. And by using the hierarchical decomposition method, the model was broken down into two parts: optimization of gas system and optimization of thermal and power system. The case analysis indicated that: the model could distribute the energy of gas and power reasonably, safely and efficiently when the production condition was changed, and improve the energy utilization efficiency.
Keywords. Iron and steel enterprises, multi-objective particle swarm optimization, cooperated optimization, self-power generation optimal scheduling.

1 Introduction

Iron and steel industry is an energy intensive industry, which accounts for more than 15% of energy consumption [1]. High energy consumption not only leads to increase cost of steel products, but also means more pollution and emissions. In actual production, considering the changes in production, maintenance plans and other factors, the amounts of gas generation and consumption are always fluctuating and the gas system is in an imbalance state, which leads the low utilization of gas. All the problems above are needed by the system optimize scheduling to resolve.

A large amount of by-product gas is generated in the production process of Iron and steel enterprises. The recycling utilization of the gas surplus in power generation of the steel industry is an important measure to develop recycling economy. Gas boiler, CCPP and other gas generating device are widely used in the iron and steel enterprises. At the same time, the optimal scheduling technology of surplus gas in the gas generating devices and the gas holder is much more concerned.

In recent years, many scholars have carried out extensive researches on steel enterprise energy optimization models. Literature [2] shows the relationship between the respective units of the gas pipeline network, which is modeled and solved by the method of partial least squares matrix. In Literature [3-5], a gas supply and demand forecasting model is established, which achieves the maximum integrated power efficiency and safety control of the gas holder.

Although the above algorithm theory is reasonable and simulation results are also very good, but in practice, they cannot meet the requirements of the real-time and effective gas scheduling. Meanwhile, the objective function of the models above is single, which cannot solve the problem of multi-objective, considering the minimum cost of power generation, diffusion and others. Therefore, a multi-objective self-power generation optimal scheduling model is proposed in this paper which based on multi-objective particle swarm optimization (MOPSO) the research of coupling relationship of gas and power.

According to the features of gas pipeline network system, combined with the actual scheduling situation, the by-product gas system is modeled and analyzed based on MOPSO rules and priorities. The results proved that this model is useful to reduce gas emission, achieve the balance of the entire gas system and have important theoretical and practical benefits.

2 Principle and algorithms

2.1 Mathematical description of PSO [6]

PSO is a population-based optimization approach. The basic idea behind the algorithm is to use a collection of particles to explore the fitness landscape of a particular problem. Each particle is a vector that describes a candidate solution, and can be evaluated along several quality dimensions. The algorithm is iterative, and at each iteration each particle moves through the fitness landscape according to its current fitness values as well as those of nearby particles, and the swarm as a whole.

Formal description of multi-objective optimization problem is shown as below:

$$\begin{aligned} &\min Y = f(X) = (f_1(X), f_2(X), ..., f_r(X)) \\ &g_j(X) \geq 0, (j = 1, 2, ..., r) \\ &h_i(X) = 0, (j = 1, 2, ..., r) \end{aligned} \tag{1}$$

Where X is the decision vector, Y is the target vector, $f(X)$ is the objective function, $g_j(X)$ and $h_i(X)$ are the restriction conditions.

The exact steps of the PSO algorithm for the single-objective case are as below:

(1) Initialize the swarm;

(2) For each particle in the swarm:

(a) Select leader

(b) Update velocity

(c) Update position

(3) Update global best;

(4) Repeat.

2.2 Mathematical description of MOPSO

Multi-Objective Particle Swarm Optimization (MOPSO) was proposed by Coello in 2004. It is a multi-objective version of PSO which incorporates the Pareto Envelope and grid making technique, similar to Pareto Envelope-based Selection Algorithm to handle the multi-objective optimization problems. Just like PSO, particle in MOPSO are sharing information and moving towards global best particles and their own personal (local) best memory. However, unlike PSO, there is more than one criterion to determine and define the best (global or local).

Most current researches are focused on this approach based on Pareto by Drawing on the successful experiences of previous evolutionary algorithm to solve multi-objective optimization problem. In the evolutionary process there are usually two groups, one is ‘population’: the basic evolution population and the other is the population ‘archive’ which is used to store the elite individuals in the evolution. First, the merits of individuals of the basic population are determined through evaluate, and then by determining the overall best individual and its location in basic population and using PSO formula to update, the next generation of basic groups is obtained. There are two operations in archive which are update and truncate, the former is used to update the archive of the individual which remains the best individual in evolution; the latter is used to remove the extra individuals when the number of actual individuals which enter the archive is greater than the capacity.

Typically MOPSO algorithm has a retention mechanism for elite solutions, the MOPSO algorithm steps are shown in Figure1.

Where Quality (pbest) is the selection of individual history best position; Quality (gbest) is the selection of the optimal particles.

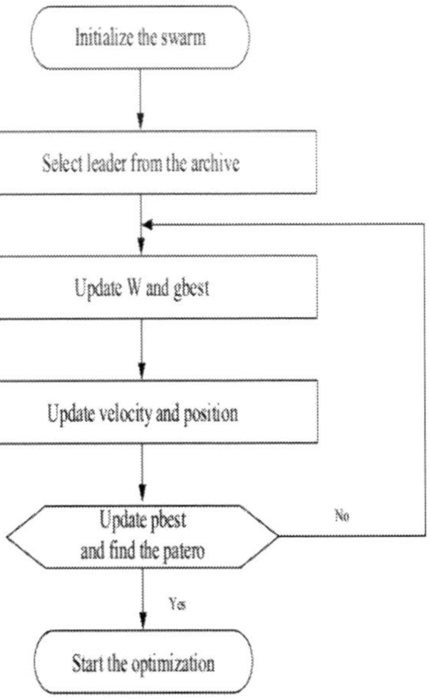

Fig. 1. MOPSO algorithm flow chart.

3 Modelling

Based on the above algorithm, with the target of the minimum cost of power generation and the lowest rate of gas emission in iron and steel enterprises, a multi-objective self-power generation optimal scheduling model was built based on MOPSO.

3.1 Problem description

The byproduct gas generated from iron making and steel making are supplied to gas users after pressurization. Surplus gas is sent to power plants' boilers for power generation. The small quantity of gas imbalance frequently aroused by fluctuation of working condition is absorbed by gas holder, when surplus gas is insufficient, the boiler will be switched into the model of coal firing or gas coal co-firing. Detail as Figure 2.

Generally, heating furnace and boiler have fuel interchangeability, that's to say, if allowed technically, such devices may use different fuels; when different fuels are used, the devices have differences in fuel consumption of unit output. Therefore, reasonable gas scheduling may lower the energy consumption of the system. Meanwhile, similar devices running in parallel have differences in conversion efficiency when manufacturing different energy resources, so self-power generation scheduling model optimization may be used to solve the following problems:

(1) In the precondition of knowing enterprises' production plan and equipment maintenance plan, we may reasonably allocate various fuels (blast furnace gas, converter gas, mixed gas, coal) to different energy consumption devices; in condition of meeting production demand, we shall minimize the loss of energies caused by the mixture of energy resources of different qualities, lower the energy diffusion of the system, and control the energy consumption of the whole enterprises' fuel system within the minimum scope;

(2) Buffer surplus gas between gas holder and power plant boiler reasonably, realize the optimum allocation of stem turbine's steam and electrical load according to the dynamic characteristics of steam turbine.

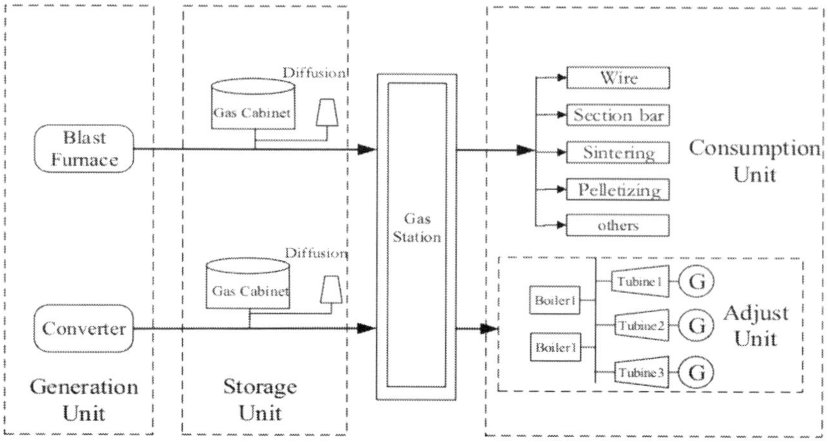

Fig. 2. Gas system flow chart.

3.2 Objective function

The objective function of steel and iron enterprises' multi-energy cooperated optimization model is as shown in Formula (1): the first equation indicates the consumption cost of various energy resources; the second equation indicates gas diffusion cost; the third equation indicates the minimum total costs.

$$\begin{aligned} \min A &= \min(\sum_{i=1}^{m+n}\sum_{k=1}^{4}\varphi_k(G_{i,k}+B_{i,k})) \\ \min B &= \min\sum_{k=1}^{3}\varphi_k V_k \\ \min C &= \min(\sum_{i=1}^{m+n}\sum_{k=1}^{4}\varphi_k(G_{i,k}+B_{i,k}))+\min\sum_{k=1}^{3}\varphi_k V_k \end{aligned} \tag{2}$$

Where $G_{i,k}$ is The fuel costs of energy resource k of equipment i, m^3/h or t/h, k=1 indicates blast furnace gas, k=2 indicates converter gas, k=3 indicates mixed gas, k=4 indicates coal; $B_{i,k}$ is The power conversion costs of energy resource k of equipment i, m^3/h or t/h; φ_k is the value coefficient of energy resource k ,kJ/kg or kJ/m^3; V_k is the diffusion of energy resource k ,m^3/h; m, n are the main working procedure has m sets of equipment and the power plant has n sets of machine.

3.3 Restriction conditions

(1)Restriction on fuel heat value:

$$Q_{i\_max} \geq Q_i \geq Q_{i\_min} \tag{3}$$

Where Q_i is The heat value of gas i ,kJ/m^3; Q_{i_max} is the maximum heat value that the production unit i can bear, kJ/m^3; Q_{i_min} is the minimum heat value that the production unit i can bear, kJ/m^3

(2)Restriction on energy demand of the production unit i:

$$\sum_{k=1}^{3}\varphi_{i,k}*\theta_k = q_i * p_i \tag{4}$$

Where p_j is the product output of the production unit i, t/h; q_i is the fuel consumption per unit output of the production unit i, kJ/t; θ_k is The standard coal coefficient of energy resource k, kJ/kg or kJ/m³; δ is the proportion of coke oven gas in mixed coal gas.

(3)Restriction on the unit capacity of different unit types:

$$U^z_{\min,y} \le U^z_{t,y} \le U^z_{\max,y} \tag{5}$$

Where $U^z_{\min,y}$ is the minimum unit capacity; $U^z_{\max,y}$ is the maximum gas proportioning; z is the unit type; y is the unit number.

(4)Restriction on unit ramp rate:

$$\left|U^z_{k+1,y} - U^z_{k,y}\right| \le v^z_y \tag{6}$$

Where v^z_y is the unit ramp rate, MW/s.

(5)Restriction on the capacity of gas holder:

$$S_{\min} \le S_{k,} \le S_{\max} \tag{7}$$

Where $S_{\min}$ is the minimum gas holder capacity, m³; $S_{\max}$ is the maximum gas holder capacity, m³.

The above restrictions are obtained by means of statistical analysis and regression analysis, which are used to solve the model based on MOPSO.

4 Examples and analysis

Use the above mentioned model to carry out multi-energy cooperated optimization with a steel and iron plant as example. The plant consists of five blast furnaces, five converter s, a BFG gas holder, and a LDG Gas holder, 8 gas pressurized stations ,respectively. There are four kinds of rolling plants and five boilers in generation system. Use the model mentioned in Section 1 to optimize the following two production conditions:

Working condition 1: Normal production; Working condition 2: Based on working condition 1, tubular production factory is stopped, the optimization results of gas system are as shown in Table1, and those of thermal power production system are as shown in Table1.

This model was found with rapid response and Accurate target characteristics. Table 1 show that volatility has been effectively controlled; thereby gas emission is avoided. Fluctuation caused by the remaining gases is stabilized by the cooperation between the gas holders and boilers. The method proposed in this study can be used to guide the rational allocation of byproduct gases and arrange the production in steel plants. Multi-energy cooperated optimization model may re-allocate the energy resources without energy diffusion, and meanwhile, thermal power generation system is also optimized.

Table 1. Comparison of the optimization results of gas system in different conditions.

Working uint	Working condition 1		Working condition 2	
	Blast furnace gas (m³/h)	Converter gas (m³/h)	Blast furnace gas (m³/h)	Converter gas (m³/h)
Sintering	14837	0	14837	0

Iron making	440638	0	440638	0
Steel making	4050	7772	4050	7772
Bar steel	66802	0	66802	0
Section steel	12990	0	0	0
Power plant	192958	15745	205968	15745
Diffusion	0	0	0	0

5 Conclusions

Comprehensively considering the coupling relationship between the production and consumption of coal gas, steam, and electric power, the steel and iron enterprises' multi-energy cooperated optimization model is established by MOPSO. This model is based on hierarchical theory and contains two objective functions which refer to gas system and thermal power system for optimization respectively. The result shows that, in condition of knowing production plan and maintenance plan, we may use the model to optimize and adjust energy production and distribution, avoid energy diffusion, and raise energy utilization rate. Meanwhile the scheduling model is helpful to enhance the security and stability of the production of the gas holder, get a good application results in practical applications, reduce gas emission and achieve balance of the entire gas system, which has important theoretical and practical benefits.

References

1. Kim J H, Yi H S, Han C H. A novel model for plant-wide multi-period optimization of byproduct gas supply system in the iron and steel making process, Trans IChemE, 2003, 81: 1015-1025.
2. ZHANG qi, CAI Jiu-ju, PANG Xing-lu, et al. Optimal distribution of by-product gases in iron and steel complex. Journal of Northeast University, 2011, 32(1): 98-101.
3. ZHANG qi, CAI Jiu-ju, DU Tao. Optimum utilization of gas system in incorporated business enterprises of iron and steel. Energy for Metallurgical Industry, 2005, 24(5): 9-16.
4. Su C L. Probabilistic load-flow computation using point estimate method. IEEE Transactions on Power Systems, 2005,20(4):1843-1851.
5. Coello C A C, Pulido G T, Lechuga M S. Handling multiple objectives with particle swarm optimization. IEEE Transactions on Evolutionary Computation, 2004,8(3):256-279.
6. Kim J H, Yi H S, Han C H. Plant-wide optimal byproduct gas distribution and holder level control in the iron and steel making process, Korean Journal of Chemical Engineering, 2003, 20(3): 429-435.

Man-hour Estimation Model based on Standard Operation Unit for Flexible Manufacturing System

Zhenggeng Ye[1,*], Jiachao Cui[1] and Fuli Zhou[2]

[1] The 713th Research Institute of China Shipbuilding Industry Corporation, Zhengzhou, China
[2] Department of Industrial Engineering, Chongqing University, Chongqing China
Corresponding Email: yezhenggeng@139.com

Abstract. In flexible manufacturing system, the estimation of man-hour is a difficult problem because of its production-diversity. To explore a more effective method, this paper tried to estimate man-hour from the perspective of operation`s character by establishing standard operation unit (SOU) in this paper. A method of parameterizing the SOU is proposed, and a new man-hour estimation model is established on the basis of SOU. At last, this paper verified the effectiveness of this method by the operation of large-scale welding parts.

1 Introduction

The diversity of productions that clients need promoted the enterprise transforming gradually from the standardization production to customized production[1, 2]. In this environment, the flexible manufacturing system is increasingly accepted and used in manufacturing factory which can achieve the producing of more than one production line at the same time[3]. With these advantages, the problems of complicated management, such as man-hour estimation and shop scheduling, also puzzled the enterprises. Man-hour is the basic data for enterprise to finish production management, but it is difficult to find a quick and effective way to estimate it for a long time because of its production-diversity. To solve this problem, XIAO Ying-jiao and Zhou Yu-jie put forward a new method to estimate man-hour based on the group technology and modular[4, 5]. They divided the productions into modules and established different kind of time modules based on the character of productions, which provides a new idea for the estimation of man-hour. In the following study, YANG Qing-qing divided the process time into three time modules: static time, flexible time and unique time, which is also on the basis of product`s structure[6]. This method is effective to estimate man-hour, but it also cannot quickly finishing the estimating when a large number of new structure products are putted into producing. In order to get rid of product structure`s constraints on man-hour estimation, it is necessary to explore the estimation method from the perspective of the operation.

To study virtual man`s action, LU Xiao-jun presented a virtual human`s action database on the basis of action unit level and motion element level, who began to study human behaviour from a hierarchical perspective[6, 7]. Later, to study virtual human's action models in virtual maintenance, CHANG Gao-xiang separated the process of maintenance into three courses and established three corresponding straticulate action modes[8]. He fully implemented studying operating behavior in a hierarchical perspective. With further research of hierarchical model, the study of Parameterized Action Representation (PAR) also made great progress. Norman I. Badler and Ramamani Bindiganavale presented the model of PAR by studying how to bridge the gap between natural language instructions and the virtual agents, and this opens up a path to parametric research on operation behaviour [9-11]. Based on these achievements, PAR is used into study of virtual maintenance. WANG Feng-chan presented the Maintenance Task Decomposition Model and Actions Database based on the PAR and Intelligent Virtual Maintenance Training System in Delmia environment[12]. All of these is the good foundation for the research of man-hour estimation in operation perspective.

2 Standard operation unit generation by work breakdown method

2.1 Work breakdown

In previous researches, operation is usually studied from the perspective of process. In Flexible Manufacturing System (FMS), one operation process will not comply with other products due to the difference of productions` structure and material. In production, operation is usually completed step by step and consisted of different discrete movement modules.

This paper decomposed the operation to modules, and summarized their analogous property and established the modular model of operation. This part will analyse the decomposing of operation based on Work Breakdown Structure.

Work Breakdown Structure (WBS) comes from project management, which is an effective method to decompose project. Through this method, work of project can be divided into more flexible and manageable units to improve management efficiency[13]. Based on this advantages, WBS has been used to study operation process of manufacturing. Chenqiang(2009) proposed the theory of Product-oriented Work Breakdown Structure(PWBS) by studying the application of WBS to Shipbuilding Industry, and decomposed shipbuilding-operation into different modules on the basis of Similarity Principle and Group Technology[14]. In this paper, the ideology of WBS is also used to study the operation process to acquire the standard operation unit of FMS.

Based on WBS, this paper builds a hierarchical model to describes the manufacturing process of products by three class: process, operation unit and therblings (Figure 1). In this model, process is a stage of a product`s manufacturing, which is defined as a continuous operation of a part or a group of parts by one operator or a group of operators on one device. Operation unit is the functional module of process. Different operation units can finish different function, and the goal of process will be achieved when they are combined together. Therbling is the basic element of operation and this paper will employs the therblings of MOD to analyse the operation unit. In addition, the operation method will be different in different manufacturing technology which will lead to the difference of therblings and operation unit, so the hierarchical model of product is not one and only. So this paper will study operation characteristics, summarize operation unit and establish the hierarchical model on the basis of the given manufacturing technology.

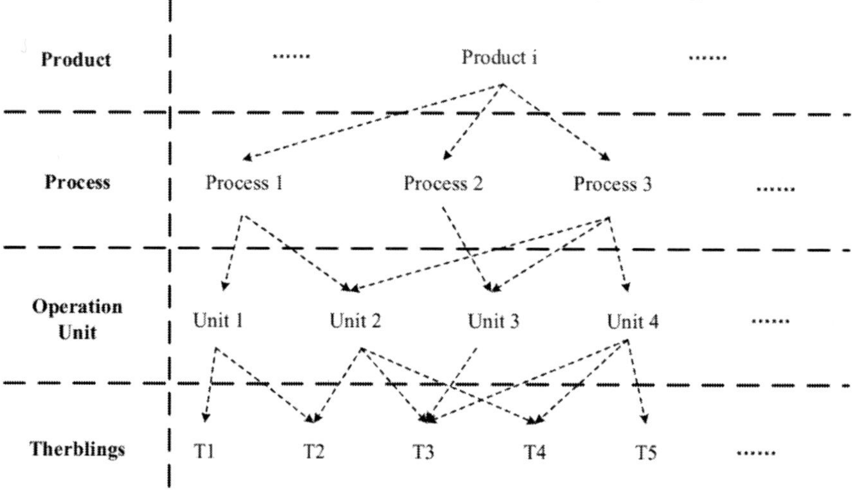

Fig. 1. Hierarchical Model of Manufacturing Process.

2.2 Character extraction of the operation unit

Operation unit is universal operation modular of operating process, which is extracted by work breakdown. This paper defined it as the portfolios and configurations of therblings that operators made to parts for a special operation goal. Generally, an operation unit may have the following properties (Figure 2):

• Including three key objects: operator, part and tool. Operator is the executant of the operation unit, part is the receptor.

• Purposiveness: the operation unit should have operation semantics and be finished for a special purpose of the process. In manufacturing, the purpose is the natural language instruction of process file. But therblings don`t have the operation semantics, which only represent the moving or turning of body.

• Connection: To achieve the special operation goal, these therblings must be connected. If the therblings of the operation unit are spaced, the goal will not be achieved.

• Order: therblings used in the operation unit obeys special order. Because the same sets of therblings which have different orders will be operated for different goal, they are different operation units. Similarly, the operation units will be also operated in order to finish the content of process.

• Time: the achievement of operation unit needs time, it is not instantaneous.

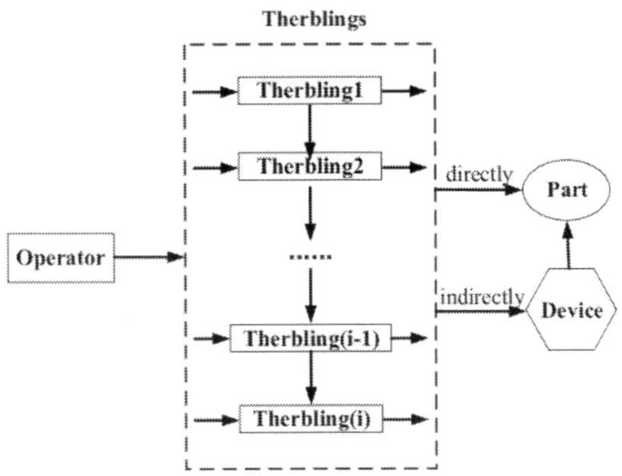

Fig. 2. Operation Unit Model.

2.3 Standard operation unit based on three-tuple model

Based on the above analysis, this paper proposed the concept of standard operation unit (SOU) to describe the universal character of operation unit. SOU is defined as the set of therglings that the operator finished for one purpose over a period of time, which can be used to describe the character of similar operations in the given manufacturing environment. Considering the natural language semantics of operations, this paper classified and defined the frequently-used operations in manufacturing environment, and built the SOU database. SOU can be classified into universal SOU that will be used by many kinds of process and special SOU that will be used by one kind of process. SOU database is an open database which can add new SOU according to the actual production situation. Some of the SOU is listed in Table 1.

Table 1. Classification and definition SOU Database.

Classification	Name	Definition
Universal SOU	Getting Parts	transport parts from storage area to operating area
	Measuring	check the size of parts
	Fixing	fasten parts securely in device
	...	...
Special SOU	Welding	join metal pieces or parts together by heating the surfaces
	Assembling	fit the component parts of a machine or other object together
	Turning	shape parts on a lathe
	...	...

To parameterize SOU, this paper builds a three-tuple model to describe it:

$$U = [O, D, Th]$$

Where:

O is the set of operators who maybe use this standard operation unit. For different kind of operators, their labour intensity is different, so they should be given different allowance rates for the same SOU.

D is the set of devices or tools used in the operation unit. If different devices or tools can be used for the same SOU, operators will need different degree of effort to finish it, so different hour coefficient should be assigned to different devices or tools.

Th is the set of basic therblings which used in the operation unit. This paper used the Therblings of MOD method to describe the operation and estimate the basic working hour.

For one SOU, the parameterized representation model can be established:

$$\text{SOU}\begin{cases}\text{Operators: } [\text{OPERATOR} \quad \text{RATE}]=\begin{bmatrix} O_1 & r_1 \\ \vdots & \vdots \\ O_m & r_m \end{bmatrix} \\ \text{Devices: } [\text{DEVICE} \quad \text{COEFFICIENT}]=\begin{bmatrix} D_1 & c_1 \\ \vdots & \vdots \\ D_n & c_n \end{bmatrix} \\ \text{Therblings: } [\text{THERBLING} \quad \text{TIME}]=\begin{bmatrix} Th_1 & t_1 \\ \vdots & \vdots \\ Th_k & t_k \end{bmatrix}\end{cases}$$

3 Man-hour estimation model

3.1 Hour calculation based on SOU

Based on above analysis, therblings are the basic elements of the SOU, which can be regarded as the combination of the different therblings portfolio. In order to identify the working hour of flexible operation for multi-specification and small-batch production in job shop, this paper proposed an estimation model integrating modularized hour unit and SOU portfolios. First, process of historical products was collected in this model and was decomposed into some operation units. Then the character of operation will be analyzed and the SOU was extracted which will constitute the SOU database. When new parts are designed, technologist analyzes their process, adopts the SOU from the database and makes sure the parameters of SOU. At last, the man-hour of SOU and process will be calculated on the basis them. Figure 3.

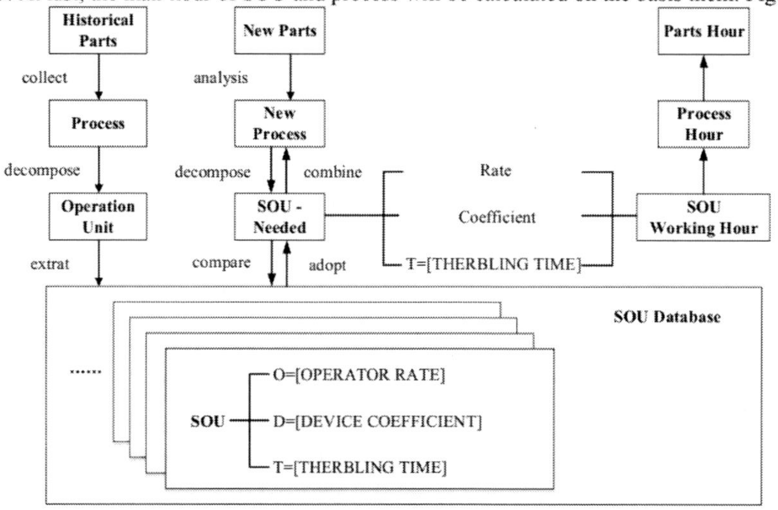

Fig. 3. Man-hour estimation process.

3.2 Man-hour estimation model by SOU portfolios

For one SOU, it can be parameterized as the following three matrices:

$$\text{OPERATOR: } [\text{OPERATOR} \quad \text{RATE}]=\begin{bmatrix} O_1 & r_1 \\ \vdots & \vdots \\ O_m & r_m \end{bmatrix}$$

$$\text{DEVICE: } [\text{DEVICE} \quad \text{COEFFICIENT}]=\begin{bmatrix} D_1 & c_1 \\ \vdots & \vdots \\ D_n & c_n \end{bmatrix}$$

$$\text{THERBLING: } [\text{THERBLING} \quad \text{TIME}]=\begin{bmatrix} Th_1 & t_1 \\ \vdots & \vdots \\ Th_k & t_k \end{bmatrix}$$

R is the set of allowance rates of different operations, and their values can be calculated based on different industry standards. C is the coefficients of different devices or tools, which represents the degree of difficulty to finish the SOU.

This paper defined the difficulty degree of device and tool`s operation in five class. The definition and value of coefficient is in table 2.

Table 2. Class of Device`s Difficulty Degree.

Class	Value	Definition
I	[0,0.01]	no device/Very easily
II	(0.01,0.03]	Easily
III	(0.03,0.06]	Slightly difficultly
IV	(0.06,0.08]	Difficultly
V	(0.08,0.09]	Very difficultly

O and D is the set of SOU's operators and devices. If operator i and device j is used in SOU, $O_i = 1$, D_j=1; else the value is 0. Assume that one operation only included operator i and device j, then the matrix O and D will be represented as:

$$O=\begin{bmatrix} O_1 \\ \vdots \\ O_{i-1} \\ O_i \\ O_{i+1} \\ \vdots \\ O_m \end{bmatrix}=\begin{bmatrix} 0 \\ \vdots \\ 0 \\ 1 \\ 0 \\ \vdots \\ 0 \end{bmatrix};\quad D=\begin{bmatrix} D_1 \\ \vdots \\ D_{j-1} \\ D_j \\ D_{j+1} \\ \vdots \\ D_n \end{bmatrix}=\begin{bmatrix} 0 \\ \vdots \\ 0 \\ 1 \\ 0 \\ \vdots \\ 0 \end{bmatrix}$$

Then the allowance rate r of the operator and the coefficient c of the device is:

$$r=O^T R$$

$$c=D^T C$$

Th is the set of therblings that the SOU uses and T is the set of their time. For one therbling i, Th_i is the frequency that it is used in the SOU and t_i is its time. So the basic time T_b of SOU is:

$$t_b=Th^T T$$

So the working hour of SOU can be estimated by the following formula:

$$time=(1+r)\times(1+c)\times t_b$$

$$=[1+(O^T R)]\times[1+(D^T C)]\times(Th^T T)$$

4 Case study

This paper takes the production of large-scale welding parts as the case to analyse the SOU in company G. The welding parts of company G usually have four process: splicing of components, welding, boring and drilling. This paper will estimate the working hour of spicing through analysing the SOU. The process of splicing can be decomposed into four SOUs: Getting parts, measuring, assembling, point-welding.

Getting parts: it is the operation to transport parts from storage area to operating area, and its parameterized representation in this manufacturing environment is

$$\text{OPERATOR: [OPERATOR RATE]}=\begin{bmatrix} \text{welder} & 0.2 \\ \text{boring machine operator} & 0.05 \\ \text{benchwork} & 0.1 \end{bmatrix}$$

$$\text{DEVICE: } [DEVICE \quad COEFFICIENT] = [\text{crane} \quad 0.02]$$

The therblings used in this SOU include 6 kinds (table 3), that is W5, M4, M3, G1, P1 and P0.so Th and T is:

Table 3. Therblings of Getting Parts.

Number	Movement	Therblings	Frequency
1	Walking	W5	5
2	Grabbing crane	M4G1	1
3	Hooking part	M4G1M3P1	1
4	Transporting	W5	5
5	Taking off hook	M4G1M3P0	1

6	Walking	W5	5
7	Placing crane	M3P0	1
8	Returning	W5	5

$$\text{THERBLING: [THERBLING TIME]}=\begin{bmatrix}15 & 0.645\\ 3 & 0.516\\ 3 & 0.387\\ 3 & 0.129\\ 1 & 0.129\\ 2 & 0\end{bmatrix}$$

For the splicing, the operation of getting parts includes the operator welder and device crane, so the O and D is:

$$O=\begin{bmatrix}O_1\\ O_2\\ O_3\end{bmatrix}=\begin{bmatrix}1\\ 0\\ 0\end{bmatrix};\ D=[D_1]=[1]$$

So the allowance rate r of the operator, the coefficient c of the device and the basic time respectively is:

$$r=O^TR=[1\ \ 0\ \ 0]\times\begin{bmatrix}0.2\\ 0.05\\ 0.1\end{bmatrix}=0.2$$

$$c=D^TC=[1]\times[0.02]=0.02$$

$$t_b=Th^TT=[15\ \ 3\ \ 3\ \ 3\ \ 1\ \ 2]\times\begin{bmatrix}0.645\\ 0.516\\ 0.387\\ 0.129\\ 0.129\\ 0\end{bmatrix}=4.473s$$

so the time of getting parts is:

$$time_1=(1+r)\times(1+c)\times t_b=[1+0.2]\times[1+0.02]\times4.473=5.47s$$

Similarly, the time of measuring, assembling and point-welding can be calculated:

$$time_2=2.84s$$
$$time_3=7.25s$$
$$time_4=4.53s$$

And the time of splicing one part is:

$$time=time_1+time_2+time_3+time_4=20.09s$$

5 Conclusions

Estimation of man-hour from the perspective of operation has a wide applicability. To establish this new man-hour model, this paper used the hierarchical model to analyse the operation process and built the parameterized operation representation model, which provides technology support to quickly obtaining working hour data in flexible manufacturing environment. As the technology continues to mature, it will also provide the technology support for the digital and intelligent manufacturing in the future.

References

1. D. C. SUN, H. B. SHI, W. I. SHANG, C. LIU, COMPUT. INTEGRA. MANUF. SYS. **5**, 10(2013).
2. J. J. BAI, Y. G. GONG, N.-S. WANG, D. D. TANG,COMPUT. INTEGRA. MANUF. SYS. **2**, 8(2009).
3. C. LI, S. LI, Y. P. FENG, G. RONG, J. SHANGHAI JIAOTONG U. **49**, 6(2015).
4. Y. J. Xiao, Y. L. Chen, Y. J. Zhou, Application Research of Computers. **28**, 4(2011).
5. Y. J. Zhou, Y. L. Chen, N. Cao, X. Z. Lu, WORLD SCI-TECH R&D. **34**,5(2012).
6. Q. Q. Yang, Y. L. Chen, X. Z. Lu, Application Research of Computers. 29, 5(2012).
7. X. J. Lu, Y. Li, H. G. He, Computer Simulation. **23**, 4(2006).
8. G. X. Chang, X. G. Xu, L. Zhang, JOURNAL OF ENGINEERING GRAPHICS. **32**, 5(2011).
9. N. Badler, R. Bindiganavale, J. Allbeck, W. Schuler, L. Zhao, S. J. Lee, H. Shin, M. Palmer, AAAI Spring Symposium, **20**,5(2000).
10. N. Badler, R. Bindiganavale, Embodied Conversational Agents, 9(2000).
11. R. Bindiganavale, University of Pennsylvania Scholarly Commons. 124(2000).
12. F. C. Wang, Y. Z. Sun, JOURNAL OF GRAPHICS. **34**, 7(2013).
13. Q. Chen, S. Haung, SHIPBUILDING OF CHINA. **44**, 13(2003).
14. A. Guide, PMBOK, 2013.

Study on the Value Creation of New B2B Platform—— Based On the Case of GongChang.com

Qian Wang*

Business school of Zhengzhou University, Zhengzhou, Henan Province 450001, China
Corresponding Email: author@e-mail.org

Abstract. Although B2B e-commerce platform has become an important driving force for economic development, its theoretical research is still at the producer-centric stage, which lags far behind the user-centric reality. A new generation of B2B e-commerce platform has four roles, including value driver, value diggers, value supplier and value providers and they create the value ecosystem synergistically. At the same time, it implements value creation mainly through the following three ways including "information symmetry + information mining", "network effects + open platform for integration" and "cooperative competition + incremental income ", and achieves a virtuous circle of the value ecosystem. The development of a new generation of B2B e-commerce platform is to achieve win-win situation and build industry ecosystem.

1 Why put forward this question

Currently, the research of the domestic and international scholars to the whole value creation including electronic commerce enterprise can be divided into three parts: the goods-dominant logic value creation theory, the service leading logic value theory together and the customer dominant logic theory [1]. Compared with the consumer goods marketing B2C(Business to Consumer), the B2B electronic commerce trades scale always occupies great majority of the whole electronic commerce development, and the B2B is the most worthy of study and the most promising mode of electronic commerce mode. But few researches focus to the value creation aspect of the B2B electronic commerce enterprise [2]. In 2016, Chinese "supply side reform" will become the new opportunity of the B2B e-commerce development, and there will be different B2B model through the integration and cooperation of upstream and downstream industry resources, from bargain service of B developing to depth service and even to the customized service stage. So, what is the value creation mechanism of the B2B e-commerce platform enterprise? This paper passes to set up the new theory about the value creation of B2B e-commerce platform and uses case of the GongChang.com to verify the rationality of the theoretical paradigm further.

2 Theoretical construction

The traditional B2B e-commerce platform takes producer as the center, mainly surrounding value provider to launch service, chasing a need homogenization of all purchase companies, which obviously already can't not apply customer's strategy ages of taking customer as the center at present [3] . A new era of B2B as consumer value drivers of demand is not homogeneous but various, B2B e-commerce platform should provide personalized service according to the demand condition. Therefore the traditional B2B e-commerce platform needs to transform into the new generation B2B e-commerce platform. On the one hand, this "new" R(need square) drive for main, in which every effort of the platform carries on around the purchase square ; On the other hand, it also has a community attribute in this kind of B2B the e-commerce platform, which carries out the value creation by creating the community ecosystem circle, namely B2B e-commerce platform must ask for help from the strength of industry value ecosystem and reconstruct new manufacturer's relation and make the whole value ecosystem together contribute to the value drivers. The key to community success lies in the sense of participation through the interaction of each value corpus of the community ecosystem [4]. Compared with the tradition of serving manufacturer, namely B2B e-commerce is the value promoter (principle of "taking manufacturer as the center"), the new generation B2B e-commerce has already changed into the principle of "taking consumer as the center" which launches around its consumption side--value drivers. At the same time, according to the principle of "taking consumer as the center", the new generation B2B electronic commerce platform business enterprise should serve around the value drivers and add the value drivers experience; For serving the value drivers better, it should make use of its own technique advantages to carry on the full data excavation, and this will appear a new role that is different from the traditional B2B electronic commerce platform value ecosystem. This paper defines it as the value digger. At the same time, in order to reveal better the value creation mechanism of value ecosystem of the new generation electronic commerce B2B platform business enterprise, this paper focuses on the each stakeholders of the value ecosystem: a new generation B2B e-commerce platform business enterprise supplier namely the value provider; a new generation of B2B e-commerce platform of enterprises demand namely value driver; value excavation of a new generation of B2B e-commerce platform of enterprises is mainly composed of industry experts and the community; a new generation of B2B e-commerce platform for enterprises as the integration/value reconstruction, connected with the value provider, value drivers and value mining. If a new generation of B2B e-commerce platform of enterprises can build its own ecosystem value and achieve a virtuous cycle, mass customization will be realized in manufacturing industry, boosting the early arrival of the new era of industrial civilization. From the analysis of the traditional B2B e-commerce platform of enterprises and its disadvantages of in-depth analysis, this chapter finally sets up the theory frame of a new generation B2B e-commerce platform value ecosystem to create value, and reveals mechanism of value creation.

In the process of the traditional B2B industrial products procurement, value drivers, value providers and e-commerce platform form a benign operation value of complex ecosystem.

Value drivers and value providers constitute their own community organizations to achieve the edge of the network effect in the community; and both enjoy a cross-border brought the network effect on the basis of and B2B platform integration. Same and across network effects make B2B e-commerce platform for enterprises build up the value of ecological system. A new generation of B2B e-commerce platform of enterprise uses the concepts of value digger innovatively. Through the specialized operation of the digger's value, it optimizes industrial products purchasing in the value of the ecosystem successfully.

In industrial products purchasing value of ecological system, each role has their value, and let their own into value a node in the ecological system. Value drivers acquire value provider's releasing information through electronic business platform. Value diggers provide professional services for value drivers, supervising and managing value provider. Finally value diggers clinch a deal set on both sides. For the aggregation of the product and the enterprise information, e-commerce platform's advantages can be embodied in comparison with search engines. In order to improve service quality and quantity, we need to focus on an industry, according to the characteristics of the industry to make services. But if we want to focus on an industry, we have to let the person who knows great industry rules do it. A platform is unable to put every niche business rules through, which is also the height of the information flow and cash flow can't reach. Here, technology cannot solve all problems; platform must have an experienced team which understands industry market segment and a certain category business rules. Only with this can we reach a degree on focusing on an industry that makes services according to the characteristics of the industry. This role is similar to industry experts, just like value diggers. Value diggers are a new generation of B2B e-commerce platform enterprise of traditional industry with rich experience of experts, teams, enterprises who make use of a new generation of B2B e-commerce platform to provide professional advice and guidance services for buyers to solve the value drivers' problems in production and corresponding problems from the source. They provide the best product recommendations and solutions. As a Value digger, their responsibilities include: the major function of the investment selection, mining, auditing, sign industry quality value provider; Information audit, the audit information of product and enterprise, meet the industry professional enterprise and product information; Business services, procurement consulting, recommend quality products, processing orders, etc.; Purchase mining, mining industry demand to provide value drivers with high value of the content. These responsibilities ensure value digger can service the value drivers better under a single industry. Value diggers' intervention can provide three aspects service for the classified providers: (1) Not only value diggers can provide professional advice and guidance services for value drivers also can directly improve the value provider inquiry products into a single conversion rate, which reverses the situation of "having the visitors, nothing" that value providers face in the promotion and provides effective help in the important link to promote transformation of value providers. ②The credit problems of traditional E-commerce are better solved with the value digger. ③Through those who expert in value digging, the new B2B plat can help the value driver with massive issues of business.

As to the stakeholders participate in value creation, we can see the roles in value ecosystem are both givers and beneficiaries. Not only should we concern how much profits stakeholders obtain, but also whether it shall create value with a sharing economy mindset of mutual creation and mutual sharing for each other when judging whether a value ecosystem is advanced. During the operation of platform, be equal to partners, providing them with good service and listen carefully to the views of them. Only by acting like that can your company achieve continuous self-innovation and exponential growth. And the new plat achieve a better shopping experience, through the introduction of the value digger, so does the service. Furthermore owing to the fact that the provider no longer to be dealers, the value driver can trade with manufacturers directly, then solving the problem of information asymmetry from the first time it created on traditional B2B plat. Also, the digger can help standardize the industry market, avoiding even eliminating unfair competition, malicious monopoly and other unjust activities. Thus what the providers can do to increase their market share is to improve the quality and service. Some capable value diggers can even handle the gray site problems effectively so that purchasing personnel can trade in a more open, mature and perfect market. In conventional membership fee model, providers enhance the exposure rate by paying a membership fee or buying extra advertisement, by which they get inquiries and achieve the transaction. The theoretical value creation framework of new B2B e-commerce plat is shown in Figure 1:

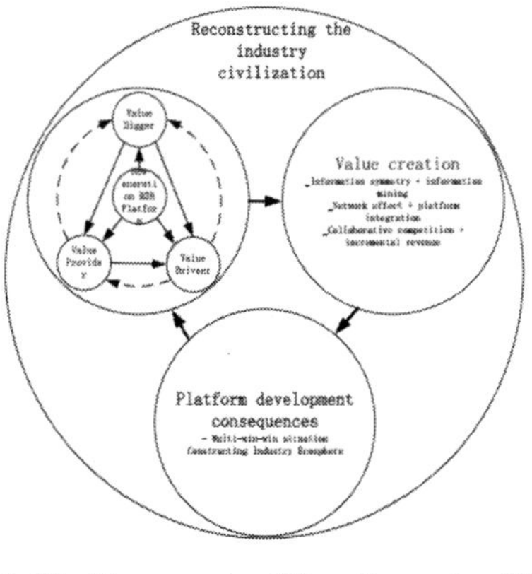

Fig. 1. The Framework of New Generation B2B Platform.

3 Case study of Gongchang.com

The factory (Chinese station: www.gongchang.com; international station: www.forbuyers.com) is a innovation mode of e-commerce platform designed by Zhengzhou Xizhi Information Technology Co., Ltd which mainly engaged in electronic commerce, trade services, information integration and data mining etc.. By providing quality professional purchasing services for buyers, providing enterprise website, promoting information dissemination, network advertising, online order tracking, transaction information management service e-commerce platform for suppliers, factory has successfully established a thriving online trading market, which has accumulated more than 700 registered buyers and more than 350 registered suppliers. Also, it has formatted 26 million enterprise database and 250 million product database, and then match transactions through online services, information technology services, software services and web promotion services and other value-added services revenue. The company has cooperated with the Google, Yandex and other international Internet giants to integrate the development of foreign trade e-commerce services. The main business of foreign trade e-commerce services to customers through the provision of foreign trade website, foreign trade e-commerce service training, search engine promotion hosting accounts and technical support, social media marketing and other services revenue. Gongchang.com to address the purchase of individual needs to provide solutions, the value of its ecosystem as shown in Figure 2:

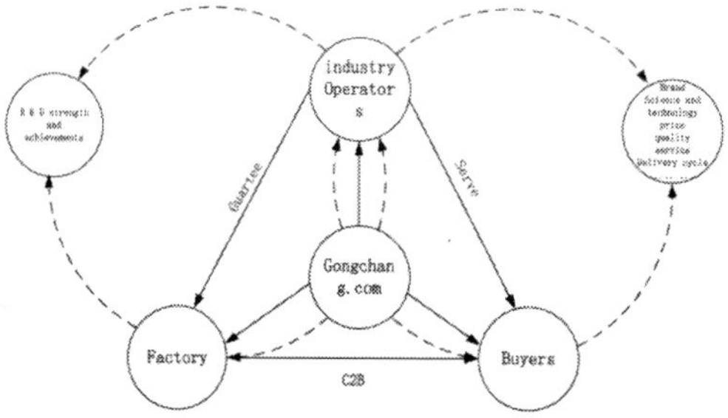

Fig.2. The Ecosystem of Gongchang.com.

From the Figure 2 we can see: the value of the world's factory network ecosystem consists of four main characters and three systems. The four protagonists are factory (Supplier), buyer, factory and industry operator. The network interconnection between enterprises value of the ecosystem is coming into being. For more industrial enterprises, they need to be able to provide an open interface and can fully dock with the purchasing manager third party platform, this is the value location of the factory.

It is the docking platform servicing purchasing managers, rather than the advertising platform of selling products, and traditional B2B electric business platform's basic function is to connect the buyer and the seller. From the platform point of view, what the traditional B2B electric business platform services are the sellers and it is not hard to find that its location is the supplier from the advertisements of Alibaba, hc360 and other enterprises on behalf of cross-border electrical business. Analysis from the perspective of traditional B2B electric business platform, what the Gongchang.com does is to connect factory and purchaser. So, there is three roles and the purchasing manager, factory value ecosystem and even the exchange of product is the front end of the industrial logic. Moreover, it is the most in need to explore and the factory's vision is to become" the first brand of global industrial products purchasing". Trade operators, the forth role, play an important role in the whole value ecosystem. Why there are operators? It is because industrial products purchasing is different from consumables purchasing. Everyone can purchase consumables and he can sell them after wholesaling and unfolding package, for example, selling foods, clothes and household articles. Industrial products(machine tool, crusher, food machine, engineering machine and so on) must have industrial background and it is no way to service upstream and downstream well, when you don't have industrial background, don't understand products parameter and what advantages every factory's product has, and don't know purchasing agent's specific requires to product. So, what the third party platform dose for the industrial products' platform service directly just stays on the information aspect and it never has ways to penetrate deeply to transaction aspect. If you don't penetrate deeply to the platform, your platform won't have value just like current platform, so it is necessary to penetrate deeply to the transaction service. Why there is trade operator? Because it is not possible for platform to train many teams who understanding industrial product. So, only relying on trade operator and needing the team having industry background and understanding products clearly to do it.

And it decides that industry products must be subdivided and vertical subdivide according to category, such as machine tool, mine machine, crusher, food machine, so each platform needs trade operator.

Different from the traditional B2B platform, the net of factory and industry operators select quality suppliers（manufacturing factory） worldwide strictly and then allow them free to enter, they obtain revenue through online matchmaking buyers and suppliers trading and then charge a certain percentage of service fee according to the turnover of suppliers. By providing quality service to purchasing manager and supplier, the net of factory has successfully established a thriving online trading market. The online market has more than 7 million registered buyers and more than 3 million 500 thousand registered suppliers, and the daily unique visitors (UV) reached to 1 million 500 thousand, forming 26 million enterprise library data and 250 million product library data which cover 31 industries, 339 categories and more than 2 thousand sub categories of the manufacturing industry under the" classification of national economic industry ",therefore become an e-commerce platform rich of professional data and a large number of customers. Based on the benign operation and innovation model called industry factory store, the company received huge amounts of data and resources of users. With the platform "net of factory "as the carrier, the company also provides enterprise users with information technology services (Dianjintai, Waixiaotong, search engine technology services), software services, web promotion services to get revenue. The benign cycle of the value of ecological system of the net of world factories shown in Figure 3: The value creation in the new era of industrial civilization is reflected in the personalized mass customization of humanistic economy. Personalized customization (DIY) is the abbreviation of Do It Yourself, originally refers to the whole process that users use raw materials and technology provided by planners to design, produce and enjoy the product. In the Internet era, personalized customization is the behavior applied to the level of information dissemination; audience participates in the productive process of information and even the products.

Internet makes innovation as the core power, never lack of new ideas. The individual needs of users promote individualized and personalized innovation of enterprises; innovation of enterprises lead to the individual needs of the market, personalized service has become the norm of the Internet. Through the analysis of users' behavior, Baidu , Google and other search engines launch personalized information recommendation and advertising, social media (such as micro-blog) also recommended interesting topics and characters to users. News APP such as "today's headlines" also played a slogan as "what you are concerned is the headline". Personalized customization is almost everywhere.

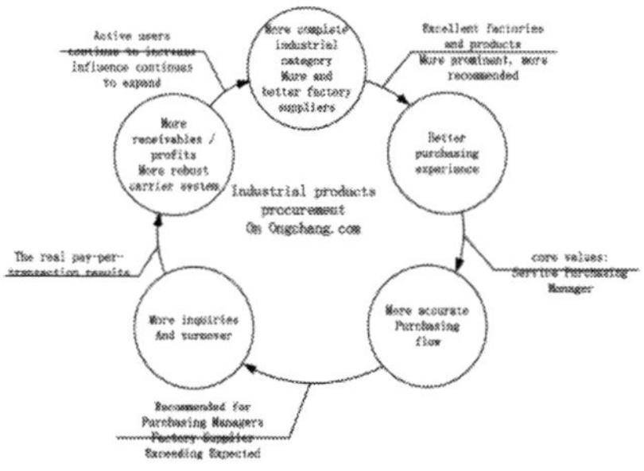

Fig. 3. The virtuous circle of the value network of Gongchang.com.

4 Further discussion

The mobile Internet era is different from the industrial economy era, even different from the PC Internet age. The original industrial economy era of the value creation theory has lost its theoretical guiding significance.

The community platform ecosystem which is an important goal of the value of ecosystem has become the important goal of internet Business model innovation.

Construct a model of a new generation of B2B e-commerce platform to create value the theory innovation of refined value reconstruction, value driven mining of the four roles and value provider, and predicts ecosystem community value;

a new generation of B2B e-commerce platform to the mode of operation, from the factory at present, operational status of industry experts such as the community is good, can be extended to other groups to the value of the ecosystem, so the development potential of the entire value ecosystem will be limitless, the future has come, community business model has been started to force, social ecosystem value of this new approach will enable mankind to create new industrial civilization, to achieve "people of the Economy".

The value of ecological system of the Gongchang.com via the lateral ecosystem and vertical value chain arranged in a crisscross pattern realize value creation, now from the information platform to the trading platform, but also to the future of the platform to integrate. The future of new ventures and traditional enterprise transformation to the value of the ecosystem consciousness, pay attention to the layout of ecological system. To provide practical reference for real enterprises, especially B2B enterprises. This paper can provide reference for other enterprises in the research of the Gongchang.com. Whether it is "Internet +" or "+ Internet" enterprises, should pay attention to the mobile Internet era of competition is no longer the competition between enterprises, not between industry and industry competition, the competition between the business model is not even, but between ecosystem competition. Set up this kind of ecosystem consciousness and practice ability to be in an invincible position in the competition.

References

1. Sun yanxia. A summary of research on enterprise value creation based on different perspectives.Nankai Economic Studies, 1:145—152,(2012).
2. Zhao jianhao.A comparative study of B2B e-commerce business model. Beijing: Capital University of Economics and Business,(2013).
3. Shapiro Aaron. User strategy: the ultimate solution to transition anxiety Beijing: CITIC Publishing House, (2014).
4. Tang xingtong.Detonated community: new mobile Internet era law 4C.Beijing: Machinery Industry Press, (2015).

Influence of Big Data on Manufacturing Industry and Strategies of Enterprises: A Literature Review

Zhao Zhao and Yifan Yang *

School of Economics and Management, Hebei University of Technology, Tianjin 300401, China
Corresponding Email: 330683898@qq.com

Abstract. Along with the rapid development of information technologies, such as cloud computing, mobile internet and internet of things, and the promotion of IT application, all kinds of data are generated and accumulated rapidly in various ways, big data era is coming, in which enterprises are faced with both opportunities and unprecedented challenges. Various processes, from decision making to operation and from designing to marketing, are being influenced by big data in manufacturing industry. This paper, according to the nature and features of big data, analyzes and extends a classical model of organizational change, Leavitt's model of organizational change, in order to explore the ways for enterprises to cope with challenges and seize chances of development in big data era. Then, using the extended Leavitt's model, opportunities and challenges derive from big data are combed, and approaches to making use of big data and coping with big data are generalized from five perspectives, including task, structure, people, technology and environment.

1 Introduction

"Big data", emerged at 2008, has profound influence on various fields such as commerce, consumers and society [1], even deeply changes them all [2]. All walks of life, like enterprises, governments and research institutes, have begun to explore the connotation of big data and the approaches to using it to optimize different activities one after another. For instance, many research institutes, including EMC, IBM, McKinsey and so on, reported their diverse understandings of "big data", many countries, such as America, England, France and Japan, also adopted different measures to make full use of it, so did China. Li Keqiang, the Chinese Premier, attended the China's big data industry summit and the development of Chinese e-commerce innovation summit, emphasizing that we should combine the technological innovation of big data and the craftsman spirit of traditional industries to rebuild the industry chain, supply chain and value chain, and indicating that the government will play a role in the process of promoting information sharing, strengthening information security and so on, in May 5, 2016.

According to the report from Cisco, mobile data traffic increased 4000 times over the past decade, and rose four billion times over the past 15 years. It is not only the data volume explodes, but the rate of data transfer also grows faster and faster, as well as the proportion of unstructured data becomes bigger and bigger. There is more and more various data around enterprises internal and external. At the same time, more and more companies regard it as their own data asset as they could create more opportunities, optimize allocation of resources, make real time decisions, and improve production and services, by using it. In addition, there are new types of business organizations, new data brokers, in the context of big data, they acquire benefits by circulating data [3]. Besides, the existence of big data impacts the competition mode between enterprises, which is to say that it is no longer focus on products but knowledge technologies [4]. Thus, there is an issue corporations must solve, how they take certain steps to seize the opportunities to gain competitive edge in the intense competitive environment, in the big data era. Being centred on the matter companies are facing with, for different purposes, a large number of researches are made by disparate scholars from diverse views. The aim of this paper is, basing on the Leavitt's model of organizational change and integrating literatures correlation with manufacturing enterprises in big data era, to comb the opportunities and challenges derived from big data, finally generalizes approaches to capitalizing on it.

Theoretical bases, the definition and features of big data and Leavitt's model of organization change, are introduced in the first part. The second part is the opportunities and challenges enterprises are faced with, while the third part is their coping strategies. Finally, there are a few conclusions.

2 Theoretical bases

2.1 The definitions and features of big data

Although the word, big data, has been received wide attention in recent years, there is no specific definition about it [5]. Some researchers define it in the technology field, while others consider it as data, which has certain characteristics. Big data is the accumulation of mass techniques that are related to processing a great deal of data [6]. Schroeck et al treat it as a technique that contains complex analysis tools, and enterprises would find as well as access to data and then analyze it rapidly to dig up its value by using them. However, big data has no longer been limited in the field of technology. A large portion of scholars think that it is a data asset of companies, which implies lots of value that can be used by them [5, 7]. Besides, Manyika et al deem it to a data set whose size go far beyond the abilities of getting, storing, managing and analyzing of traditional database software [3].

There is a definition, which has been widely applied, that big data is information asset whose volume is large, velocity is high, and forms are various [8]. It is the "3v" characteristic of big data, volume, velocity and variety, we can derive from the definition. Therein, volume refers to unceasing accumulation of data, its size expands constantly, velocity refers to its high speed of generating and timeliness, so it must be analyzed in time, while variety refers to its various forms, it can be classified into structural data, unstructured data and semi-structured data [5]. There are a few researchers to enrich the "3v" features to "4v" after Gartner's definition. Schroeck et al [6] added veracity into above three characteristics to emphasize on the uncertainty of certain types data such as weather, economy, the consuming intention of consumers and so on, while Li et al [9] added value into "3v" in order to explain the feature of huge value but low-density. So, big data is some kinds of data with above characteristics, as well as technique related to data analysis tools, because many researches defined it from diverse views in different fields.

The application value of big data has covered most fields, including education, communication, healthcare, manufacturing industry, financial industry, government, public utilities and so on [4, 10]. At the same time, it has permeated into different behaviors of enterprises, they can use it to make optimization. For instance, product bundling and pricing, reacting to market sentiment and consumer demand, product design, market positioning and obtaining new massage about dynamic competition could be improved or enhanced by using it [1, 11]. But the "big value" and "big impact" of big data are waiting to be further studied and explored.

2.2 Leavitt's model of organizational change

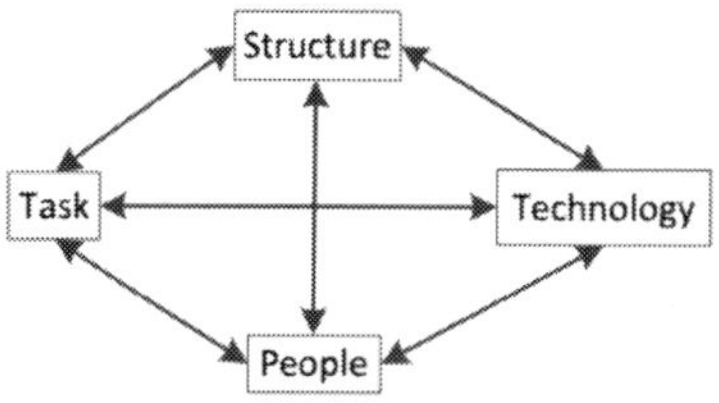

Fig. 1. Leavitt's model of organizational change.

In the Leavitt's model of organizational change, which has been wide applied and extended, Leavitt categorized organization into four variables, technology, people, task and structure, and considered that those four variables interacted with each other, if there were a certain variable changes, other one or more variables, including itself, would change as well [12]. Thus, there are a large number of researchers have a thought that it is necessary to keep a balance between them in order to make the organizations operate steadily and effectively [13-15].

Though Leavitt and other researchers, combining with certain specific issues, defined those four variables diversely in the process of model application, the essential connotation has been maintained. Technology refers to the techniques and mechanisms of solving issues directly [12, 13], like the platform of computer software [16], the storage and aging treatment of information and some technical norms [14, 17]. People accounts for everything related to individual or group inside of organizations, such as the power, responsibilities, abilities and attitude [12, 13], as well as the direct relationship between individuals [17]. While there are various specific definitions about task because of diverse research problems, it refers to the aims or missions of organizations or techniques, like the performance of production of products and services [12], the behavior of strategy implementation [13], the aim of a certain system or work supported by the system [14-17]. And structure mainly accounts for internal frameworks, no matter visible or invisible, of enterprises, such as business process, organizational departmentalization or systems of authority, information systems inside of organizations and size of it [12, 17] as well as the immaterial structure, like corporate culture [13]. This paper defines technology as data management and data techniques related to big data, people as the power, responsibilities, abilities and attitude of individuals, task as strategic management, operations management, marketing management and human resource management, and structure as the internal information systems, organizational structures, business process and corporate culture.

In addition, the Leavitt's model has been widely expanded in the process of application. Scott et al, the first one who added environment variable into this model, considered that environment interacted and interdepended with each other four variables, so did Lyytinen et al [15] and Mandt et al [18]. Although it is enough to categorize enterprise into above four variables or dimensions in the context of big data, factors outside of organizations, such as public policies, economy and environment of industry and technology, also influence them, which cannot be ignored. Thus, environment variable is added in this paper.

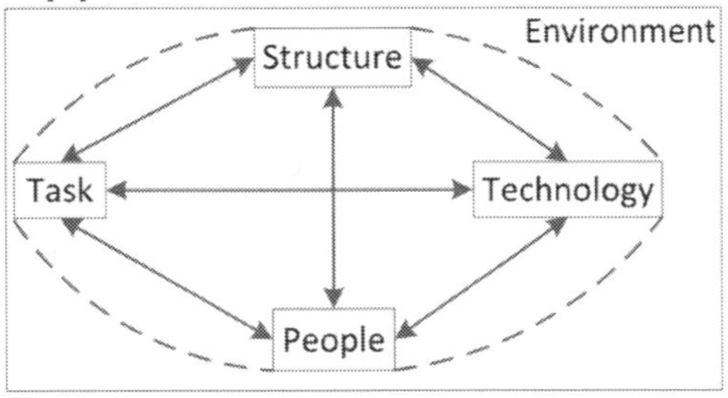

Fig. 2. Model in this paper [6].

Table 1. The definitions of model applied in this paper.

	Classifications of definitions	Examples
Technology	Data management [14]	Data privacy management, standards related to data management and so on.
	Data technology [14-16]	Techniques and infrastructures related to big data
People	Individual power [12,13]	Individual privacy
	Individual ability [14]	Ability of mastering multiple knowledge
	Individual attitude [14,17]	Support to big data application
Task	Strategic management [13]	Making sure competitive edge, generating strategy and so on
	Operations management [12]	Product design and manufacturing, specific policies and so on.
	Marketing management [12]	After-sales service
	Human resource management [12]	Recruitment, the vocational development planning and so on.
Structure	Internal information system [12, 13, 15]	Communication between individuals or groups
	Organizational structure [12, 13, 17]	Systems of power, distribution of employees and so on
	Business process [12-14]	Specific process about data management
	Corporate culture [13]	Culture of hiring and firing, sharing information atmosphere and so on
Environ-ment	Political system, economic, social and technological environment [13]	This paper focuses on technological environment such so techniques have not been applied.

The Leavitt's model has been applied in various research fields, like medicine [18], risk management [16], information system [15, 17, 19], information technology [13, 20], computer science [21], organizational design [20], performance evaluation [14] and so on, since it was created. Besides, there are three approaches to applying this model roughly. 1. Redefining variables according to the original model, from different perspectives. For instance, technology variable affects others [18, 19, 21] or itself, serves as dependent variable, is affected by others [17]. At the same time, some researchers apply the category of this model to categorizing diverse factors of different issues [16]. 2. Expanding model to do some researches, like discussing the relationship between different variables [13, 15]. 3. Using the thought of this model to state a certain perspective [22, 23]. Mainly, this paper applies the expanded Leavitt's model to integrating current living conditions and coping strategies of manufacturing corporations under the background of big data.

3 The opportunities and challenges in the big data era

In the big data era, there is more and more valuable and multiple-structure data inside or outside of enterprises. It is a focus all over the world that how to make full use of it availably [10]. However, it is well-known that the application of big data is not only an opportunity to developing and optimizing, but also a challenge for corporations.

Table 2. The opportunities and challenges in the big data era.

	Domains	Dimension	Explanation
Opportun-ities	Environ-ment	Technological environment	More and more data could be used by enterprises [1]
	Task	Strategic management	Market segmentation [7]
			Making decision more clearly [24, 25]
		Operations management	Optimizing production planning or obtain new business opportunities [26]
			Optimizing service or division of customers [3]
			Reducing the cost and flaws of product, or optimizing product design [3]
		Marketing management	Optimizing after-sales service [3]
		Human resource management	Improving the accuracy of online recruitment [27]
Challenges	Technology	Data management	Cannot apply and manage data effectively [28]
		Data technology	The types of data does not match techniques [29]
			Lacking of the abilities of data acquisition, analysis and processing [6, 26]
	Task	Human resource management	The lack of talent [1, 3]
	People	Individual power	Privacy security is threatened[31]
			Increased working pressure [31]
	Structure	Corporate culture	Stability is threatened [31]

3.1 The opportunities in the big data era

In the ever- changing technological environment of enterprises, there is more and more data and relevant techniques. Many behaviers of corporations, such as strategy, operation and human resource, would be optimized because of the application of big data, which is full of abundant value and waiting to be explored.

Big data techniques could provide a great deal of information, which has been processed already and abounds with a mass of commercial value, to enterprises, would help them to gain competition advantages as well[6]. Thus, they are able to observe the business environment more clearly, and then would find a way to become more accurately and quickly response to market. As a result, they could separate themselves from other competitors [7]. Moreover, the employ of big data could also render decision makers more clear-eyed, they are capable of understanding how to create, deliver and gain more value, when they are making either long-term or short-term decisions [24, 25].

Meanwhile, it is possible for enterprises to make use of relevant data to analyze in real-time, and make timely or near-timely decisions to enlarge the benefits [1, 6, 26]. 1. Using data about the market, they can comprehend the highly unstable market environment, get more accurate information of market demand, and then optimize allocation of resources and adjust production plans [27], or create new business opportunities [26]. 2. Using data about the consumers, corporations would focus on the features of products certain consumers pay more attention to, see the customers demand and forecast their consuming behaviours. Thereby, there will be more specific product positioning and higher level of services [3]. Otherwise, enterprises could divide the groups of consumers for marketing more efficaciously [32]. 3. Using data about the products, manufacturers are able to do quick and low-cost simulation experiments to test various designs, and chose both components and suppliers to calculate the relevant manufacturing costs. As a result, the production cost and development time could be saved. If there is any information about products flaws, productive process would be adjusted, and new design would be developed, then the product defects would be removed before production. Enterprises can also analyze data about the products, which were already sold, to improve their quality of after-sales service and level of marketing. Before the trouble occurs, it would be fixed up, like the preventive maintenance of aircrafts and elevators [3].

In addition, the analysis of big data may help companies to find the more desired employees and insure the success rate of online recruitment. They are capable to obtain the information of employees more comprehensively from the various social platforms, so the ones they need could be found more accurately.

In summary, enterprises could apply the big data and relevant techniques to gain market insight, guarantee the variation of demand [26], and change the operating strategy and business models to optimize the abilities of marketing, production, service and human resource [27]. But meanwhile, in the context of big data, many domains of enterprises are faced with a large number of challenges because of the features of big data, such as big size, multiple types, timeliness, uncertainty and so on.

3.2 The challenges in big data era

It is challenging for current enterprises to manage and analyse a mass of data, which is continuously and quickly generated and accumulated, unstructured or semi-structured, high timeliness or low certainty, not only in the technology domain, but also in the aspects of people, task, and structure domain.

3.2.1 Technology

It is challenging to the abilities of enterprises from grasping data to managing and analyzing it. Although there is a great deal of data generates and accumulates, it is a small part of it can be explored and applied, uncover the rest part is still beyond the abilities of companies [28]. According to IBM and Intel, the 90 percent of data is unstructured and unemployed. Meanwhile, maybe, the enterprises of downstream supply chain, such as distributors and retailers, have no intention to share the data they consider is full of competing values. Hence, it is hard for manufacturing firms to get the comprehensive information of consumers, and that is doubtless a barrier for them to create value of data [3]. Not only cannot enterprises obtain the whole data they need, but their abilities also cannot grasp, manage and analyze such high-volume and high-variety big data [6, 24, 26, 29, 30]. Furthermore, there may be semi-structured or unstructured data does not fit with the type their techniques can process [29].

The corporate abilities to manage and analyze data are needed to be enhanced, just as the thought of Lavalle et al [28] that the biggest obstacle enterprises are faced with is guaranteeing the data quality and eliminating the useless data analyses. The problems, how to manage the big-size and multiple-type data, and how to distinguish the valid and actual data from a mass of disturbed and invalid data in the process of analyzing, are impeding firms applying big data effectively.

3.2.2 Other domains

It is more and more important for enterprises to response to the consumer demand in time and accurately since the marketing environment, being impacted by big data, becomes more and more unstable, besides the position of consumers gradually rises in the supply chain [32]. There is an approach for them to confront the changeful business environment flexibly, using the techniques, which are relevant to big data, to seize lots of data generated in the market, then analyzing and integrating the information, which is rich in value. Yet when they are processing the data with wide resources and multiple structures, they would need to utilize the comprehensive knowledge of statistics, economics, computer science and so on synthetically. From that, there would be much higher requirements for the abilities of analysts, so it is a considerable problem that the lacking of analysts [1]. Just as the report of the US Bureau of Labor Statistics, the number of the skilled analytical talent in demand in the US would be 50%-60% higher than its expected supply by 2018. In other word, the enterprises are short of capacity of applying big data sufficiently [3].

In the people domain, the application of big data not only requires complete integrative abilities increasingly, but also threatens the individual privacies and securities. Analyzing a lot of data, which is gathered without one's attention and might derive from phone record, consumption history, GPS and so on, could make others understand the habits and hobbies of individuals in depth. Hence, big data has direct threat on personal privacy [31]. Concurrently, although managers could supervise the status of their employees further closely, by using the corporate internal information such as data about individual work, and intensify the work flow and staff working abilities, by applying the relevant techniques, there must be greater heavy pressure for employees to complete tasks faster [31]. In the meantime, Blasiak considers that paying more close attention to the staff would not only cause heavier pressure on them, but also change the hiring and firing institution, and the stability could be under threat.

Although there are still a lot of challenges in the face of the enterprises in the big data era, there will be abundant benefits to them because of investing in the techniques relevant big data [30, 33].

4 The coping strategies

In order to follow the wave of big data, a majority of enterprises have already explored the application and development of relative techniques and endeavored to achieve competitive edge by taking advantage of them flexibly. That needs firms to make efforts in all domains, including technology, people, task and structure, when the managers attach importance to data analyses in daily operation and make decision based on data [24, 28].

4.1 Technology

Enterprises, in the first place, ought to lay emphasis on the data management. They have to manage data effectually to be able to make some data reused. To achieve that, they need a large data warehouse to store and manage data, and have to affirm some certain issues, like the resource allocation, who has the responsibility to manage the data, what data and how long does it should be saved. Hence, there must be a series of criteria for the data scope and circulation to promote data to be obtained, delivered, compared and comprehended easily, as well as being reutilized [34]. However, it is worth nothing that the data management and storage are really not to treat the information as inventory in database, but to cognize, gather and analyze it continually and make it flow [29]. Beyond that, managers should pay more attention to the data privacy security management, if they cannot handle that, the stability of organization and the working conditions of individuals would be influenced. Just as the thought of Schroeck et al, it is one of the most fundamental details of data privacy management. Thus, it is necessary for companies to communicate about the privacy issue with employees, users and other collaborators for the purpose of making sure the scope and transparency of relevant information [3].

It is very hard for companies to create better performance by integrating information and making more complex analyses, because their capacities for data processing cannot fit with the large-scale and various-type data. So they need applying the advanced data techniques effectively to integrate and analyze the ever-increasing information [28]. Besides, it is the basic of effective technology application that being sufficiently appreciated by users. Therefore, techniques related to big data based on the exact sources must be easy to understand, by explaining the reason of each result [35]. In addition, guaranteeing the quality of data, like veracity, possesses a very high position in the process of data manipulation [4]. There is an approach to insuring the quality and the effective application of relevant techniques. Before making the formal analysis, data should be pre-processed to become suitable for being analysed [31].

Goes [4] deems that the data integration is the key of big data technology, managers and analysts must comprehensively utilize information from the whole corporate perspectives to optimize various steps [3, 6, 7]. 1. Integrating and analyzing the demand data, inventory data and market data to make more valid decision in the highly unstable marketing environment [27]. 2. Integrating the product development, design and historical records to establish a digital model of the whole manufacturing process, then optimize production [3].

4.2 People

Big data would help enterprises to response to market demand and variation rapidly, which request them to flexibly master multiple disciplines, including computer science, mathematics, statistics and so on, to explore and make full use of the value of data they obtained [3]. Meantime, the synthetic abilities of analysts must be promoted, which is to say when they are grasping the analysis techniques, they are supported to own other business skills, such as effective communication with decision-makers, to conduct more valid analyses [31].

When big data techniques are introduced in organizations, it is significant to receive the support from top managers. Because applying big data, as a range of new technologies, would involve changes in corporate many aspects, including information technology, enterprise culture, daily operation and so on. So the support of the top manager is needed [6]. Besides, the application of big data also requires managers to attach importance to data analysis, and make decisions based on information [24, 28]. There is another voice that transformational leadership is more suitable for firms in the big data era than servant leadership, because they are more adjusted to the dynamic environment [36].

4.3 Task

There should be a big data schema, which would confirm the anticipation, strategy and demand of the application of big data, and must set up the relationship between business requirements and information technology, at the enterprises level. The schema could generate a certain consensus, like how to use the big data to achieve their aims and create value, certainly that is conducive to the application of new techniques and creation of corporation value [6]. Accordingly, during the process of daily operation, relevant standards are needed to help decision-makers to make sure decision process, whether the decision must be made and when a specific decision is required. For instance, managers ought to make sure the scope of certain data, when it is out of the scope, they need to do some specific work to guarantee to the value of data [28].

Enterprises need to find a series of approaches such as production differentiation or changing the location of products and service to distinguishing themselves from other competitors in the fierce competition. They should analyze their advantages, then focus on the field that would bring the maximum value to them [6, 28], and by combining big data techniques and handling the dynamic information outside and inside of them, optimize allocation of resources to promote their competitive edge [27]. Moreover, Manyika et al recommend manufacturing corporations to make different strategies in diverse staged market. For example, in the emerging market, they ought to gain the advantages about relatively low labour cost, but in the developed market, they should reduce cost and innovate in products and service by taking advantage of big data.

Obtaining the whole market data, in order to apply big data to analyze availably, requires enterprises to establish certain policies involved with stimulating information sharing, to stimulate relevant cooperative partners share information to an extreme. At the same time, organizations also have to make a series of policies about data privacy to limit the usage of data and protect the benefits of stakeholders from threat [31].

For improving the abilities of enterprises to process data and apply relevant techniques, managers ought to take care of the professional and career development of analysts to enhance their comprehensive skills and advance the corporate abilities indirectly [6].

4.4 Structure

To release the value of big data sufficiently, enterprises ought to optimize their internal information systems. For instance, they could make information about product freely transmit between many departments, including marketing, research design and production department. Thereby, there is no any information isolated islands but a system inside of organizations [3]. Meanwhile, for managing data better, they need develop relevant work flow to define the responsibilities of data management [28].

With enterprises are paying more and more attention to the application of big data, the position of analysts inside of organizations should be changed too. For example, while there are analysts in internal consultant groups to help managers make decisions in the traditional firms, there are more and more analysts in product development groups, in big data era, to develop new products and product features in the manufacturing industry [28].

In addition, it is worth nothing that a large number of corporations need changing organization to make certain transformation as the groups, which have never shared data, would not share information simply because of the existence of information systems [3]. Enterprises could facilitate the application of big data, by changing culture to create the cultural atmosphere of using and sharing information [1]. Besides, according to the record, corporations with high benefit, which integrate data analysis into operation, combine analyses with making strategic direction more frequently than those with low benefit [28]. That might need them to re-establish their culture to promote employees to working on account of data more often [1].

Table 3. The coping strategies.

Approaches	Dimension	Explanations
Technology	Data management	Formulating relevant data standards [34]
		Grasping and analyzing data constantly [29]
		Communicating about data privacy and so on [3]
	Data technology	Data techniques must be understood sufficiently [35]
		Preprocessing the types of data [31]
		Integrating data [4,6,10,27]
People	Individual ability	Comprehensive talent [31]
	Individual attitude	Top managers must support to big data application [6]
		Transformational leadership is more suitable for big data firms [36]
Task	Strategic management	Making big data plan at strategic level [6]
		Optimizing allocation of resources [6, 27, 28]
		Different strategies to diverse markets [3]
		Establishing incentive policies about information sharing [3]
		Establishing policies related to information privacy security [31]
	Operations management	Establishing standards about the application of big data at operating level [29]
	Human resource management	Focusing on the development of internal analysts [6]
Structure	Internal information system	Circulating data among different departments [3]
	Organizational structure	Changing the position of analysts [29]
	Business process	Developing relevant work flow of data management [29]
	Corporate culture	Cultivating atmosphere to share information and so on [1, 28]

There is no entirely same technique applied in diverse firms. Similarly, the approaches they used to cope with the big data era need to be applied according to specific practical situations.

5 Conclusions

In the context of big data, enterprises are confronted with various obstacles, comprehensive abilities they have cannot obtain, analyze and process amount of diverse data effectively. At the same time, there is an urgent need for them to respond the unstable business environment timely. Different definitions of big data from different institutions and researchers are introduced. Then, combining the features of big data and the follow-up applications of Leavitt' model of organizational change, four internal dimensions, technology, people, task and structure, and an external dimension, environment, are defined under the background of big data. Manufacturing enterprises would optimize products and services, reduce the cost of production or obtain new business opportunities by using correlation technologies, which is more suitable for them, more effectively to make timely or near-timely decisions. Therefore, from the four internal domains, there are several solutions, which are for companies to enhance their competitive edges in the big data era business competition, are integrated.1. Develop data analysis techniques that could be made full sense of by users, and establish relevant rules and regulations to regularize the management and transfer of data. 2. Pay attention to train the comprehensive abilities of employees, and adjust the attitude of managers toward using big data and relative techniques. 3. Draw up a plan about the usage of big data to make staffs comprehend and use it more conjointly and effectively, and take the issues of privacy security into account. 4. Transmit data among different but relative departments to make information sharing more conveniently. However, it is worth nothing that there are various characteristics in diverse organizations, so as conditions. So, making relevant decisions, managers ought to focus on the strategies that are more suitable for their situations.

References

1. R.M. Chang, R.J. Kauffman, Y.O. Kwon, DSS, **63**,67-80(2014)
2. F. Frankel, R. Reid, Nature, **455**,30(2008)
3. J. Manyika, M. Chui, B. Brown, et al, McKinsey Global Institute(2011)
4. P.B. Goes, MIS Quarterly, **38**(2014)
5. P.M. Hartmann, M. Zaki, N. Feldmann, A. Neely, Cambridge Service Alliance, March 27(2014)
6. M. Schroeck, R. Shockley, J. Smart, D. Romero-Morales, P. Tufano, IBM(2012)
7. H. Demirkan, D. Delen, DSS, **55**,412-421(2013)
8. Gartner,http://www.gartner.com/it-glossary/big-data
9. G.J. Li, X.Q. Cheng, Bulletin of Chinese Academy of Sciences, **27**(2012)[In Chinese]
10. Z.B. Xu, Z.Y. Feng, X.H. Guo et al, Management world, **11**,158-163 (2014)[In Chinese]
11. H. Chen, R.H.L. Chiang, V.C. Store, MIS Quarterly, **36**,1165-1188(2012)
12. H. Leavitt, *Handbook of Organizations*. Rand Mc Nally, 1144-1170(1965)
13. M. Scott, J.F. Rockart, Interfaces, **14**,84-95(1983)
14. P. Kueng, A. Meier, T. Wettstein, Communications of the AIS, **7**,1-27 (2001)
15. K. Lyytinen, M. Newman, European Journal of IS, **17**, 589-613(2008)
16. K. Lyytinen, L. Mathiassen, J. Ropponen, Information Systems Research, **9**, 233-255(1998)
17. S. Petter, W. DeLone, E.R. McLean, Journal of MIS, **29**, 7-62(2013)
18. I. Mandt, A.M. Horn, A. Ekedahl et a, Research in Social & Administrative Pharmacy, **6**,6-17(2010)
19. R. Kishore, E.R. McLean, IEEE transactions on engineering management,**54**,756-775(2007)
20. H.C. Lucas, J. Baroudi, Journal of MIS, **10**, 9-23(1994)
21. J.N. Danziger, K.L. Kraemer, D.E. Dunkle, J.L. King, PAR, **53**, 161-169 (1993)
22. S. Biazzo, BPMJ, **8**,42-52(2002)
23. P.G.W. Keen, Communication of the ADM, **24**,24-33 (1981)
24. E. Brynjolfsson, L.M. Hitt, H.H. Kim, Ssrn Electronic Journal, 1-28(2011)
25. T. O'driscoll, Training, **51**,51(2014)
26. T.K. Das, P.M. Kumar, International Journal of Engineering & Technology, **5**, 153(2013)
27. D.Q. Chen, D.S. Preston, M. Swink, Journal of MIS, **32**,4-39(2015)
28. S. Lavalle, E. Lesser, R. Shockley et al, MIT Sloan Management Review, **52**,21-32(2011)
29. T.H. Davenport, P. Barth, R. Bean, MIT Sloan Management Review, **54**, 22-24(2012)
30. P. Tambe, Management Science, **60**,1452-1469(2014)
31. M. Halaweh, A.E. Massry, Journal of International Technology and Information Management, **24**,21-34(2015)
32. Y.B. Zhang, L.L. Ma, Economist, **8**, 44-51(2015)[In Chinese]
33. R.E. Bryant, R.H. Katz, E.D. Lazowska, http://www.cra.org/ccc/initiatives
34. C. Lynch, Nature, **455**, 28-29(2008)
35. A. Labrinidis, H.V. Jagadish, Proceedings of the Vldb Endowment, **5**, 2032-2033(2012)
36. L. Li, W.Z. Chen, Inquiry into Economic Issues, **8**, 26-29(2014) [In Chinese]

Design and Implementation of Mobile Blended Learning Model Based on WeChat Public Platform

Yanyan Han, Liyue Tian and Weijun Cheng

College of Information Engineering, Minzu University of China, Beijing, China

Abstract. Merging together the ideas of mobile learning, blended learning and flipped classroom, a Mobile Blended Learning Model (MBLM) is constructed. Based on WeChat Public Platform (WPP), MBLM can optimize the instructional process and improve the learning efficiency. A Mobile Blended Learning System (MBLS) is implemented by using MBLM, and it is constructed by both WPP and auxiliary learning system which based on Java Web. This system has reasonable designed function, easy operation, and beautiful interface, so it can effectively promote the popularization of MBLM.

1 Introduction

In 2012, the idea of "Internet+" was proposed, and brought revolutionary impact and challenge to the traditional educational idea [1]. Blended learning aims to find the correct balance between face-to-face and online teaching methodologies. This also integrates well with the idea of a "flipped classroom" [2]. The prevalence of mobile Internet has optimized the traditional online learning model, and learning will move to a larger extent out of the classroom [3]. Blended learning transformed from one with passive students who were recipients of knowledge, as is typical of traditional teaching models, to one with active students who constructed their own knowledge [4]. Since 2011, study of WeChat has presented an explosive growth. But now, few studies can apply a variety of functions of WeChat and WeChat Public Platform (WPP) to the whole instructional process.

Therefore, Mobile Blended Learning Model (MBLM) based on WPP is proposed. It integrates the idea of blended learning, mobile learning, and flipped classroom. Mobile Blended Learning System (MBLS) combines WPP and Java program. It can effectively promote the implementation of MBLM based on WPP and make it more convenient to use by learners.

2 MBLM based on WPP

The affordances of blended learning are now well understood, and its flexibility, ease of access, and the integration of sophisticated multimedia and technologies are high among the list of appeals [5]. Rong-huai Huang [6] et al. proposed "Design framework of Blended Learning Curriculum" in 2009. According to this framework, design of blended learning model should be divided into three stages: pre-phase analysis, design of activities and resources, and instructional evaluation design. In the third stage, the design of activities and resources can best embody the characteristics of blended learning. This design is guided by the basic theory of pedagogy, supported by the function of WPP, and took time as horizontal coordinate. We combine WPP and traditional classroom, optimize and design the MBLM based on WPP in two areas: learners and teachers. As showed in Figure 1, we will emphatically introduce the system design of online learning based on WPP as bellow.

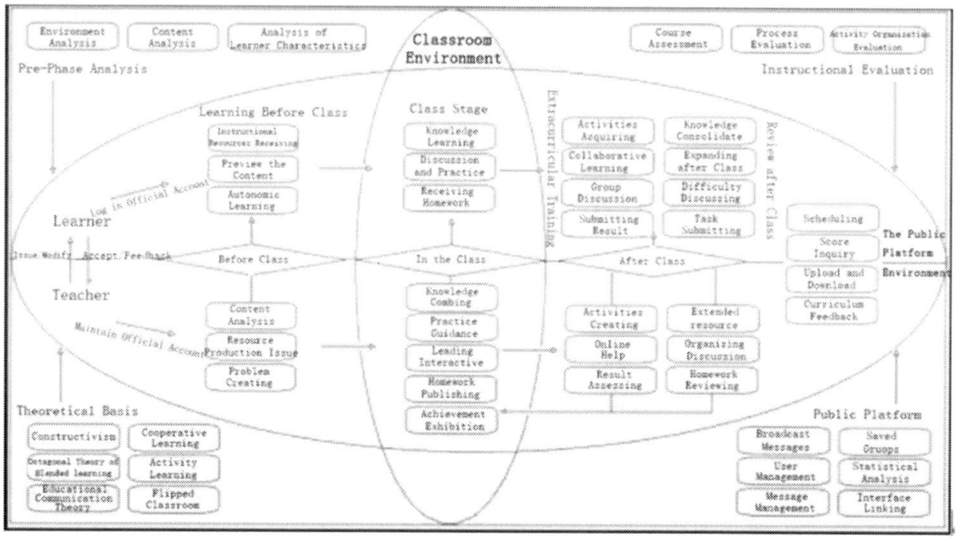

Fig. 1. MBLM Based on WPP Based on WeChat Official Platform.

There are a great many functions of WPP, such as interface, customize menus, management of Broadcast Groups, user management, information management, statistical analysis and so on. MBLM based on WPP uses the interface function of WPP. These functions assist the development of the whole blended learning.

The first stage is pre-class analysis. Before the curriculum, it is necessary to in-depth analyze the learning environment, learning content, and the characteristics of the learners, to determine whether the course is suitable for the development of MBLM.

Then, the next stage is Design of Activities and Resources. blended learning provides anytime anywhere access to rich learning materials and course content over the Internet[8]. Design of activities and resources can be divided into pre-class learning, instructional stage, after-class review, and extracurricular activities, training, implementation. Before class, the teacher releases new learning content framework, difficulty, provides relevant learning materials and instructional video and other multimedia courseware. In class, the teacher systematically summarizes the previewing knowledge and publishes the relevant information needed for class. After class, the teacher releases knowledge summaries, extracurricular activities, experiment and extracurricular learning materials, and collects the assignments and test papers to review corrections in the WPP. At the same time, the teacher will organize the discussion in the Saved Groups once more.

The instructional evaluation design is the last stage of the whole instructional process, mainly including curriculum evaluation, process evaluation, and activity organization evaluation, and they are respectively measured through the online testing, the situation of the discussion in the classroom, and activity experimental review.

3 Design of MBLS

The system is mainly based on WPP, and supplemented by the Java Web auxiliary system, combining with the Instant Messaging, Saved Groups, Moments, Official Account Security and other WeChat functions. The system aims realizes instant messaging, online testing, upload and download of related document, query results, course feedback, student management and statistical and other functions, then construct a multiple learning system based on WeChat.

Implementation of related functions fully relies on WPP and WeChat's basic functions, and the interface style and operational methods are consistent with WeChat's office. Auxiliary system requires rational design of the system, clear thinking, complete functions, and easier operation. Because the system is mainly used in mobile phone terminal, the interface design should be simple, and options should be clear and easy to operate. In addition, the use of support tools can improve the whole system. Official Account Security can meet the instant management function. Cloud notes and SkyDrive can accomplish online management and sharing of notes and documents. Analysis function of Public Platform can be used to analyze the data. WeChat Saved Groups' instant messaging feature can be used to discuss questions, as well as WeChat Moments can be shared promotion. Based on the above, the entire MBLS is constructed as showed in Figure 2.

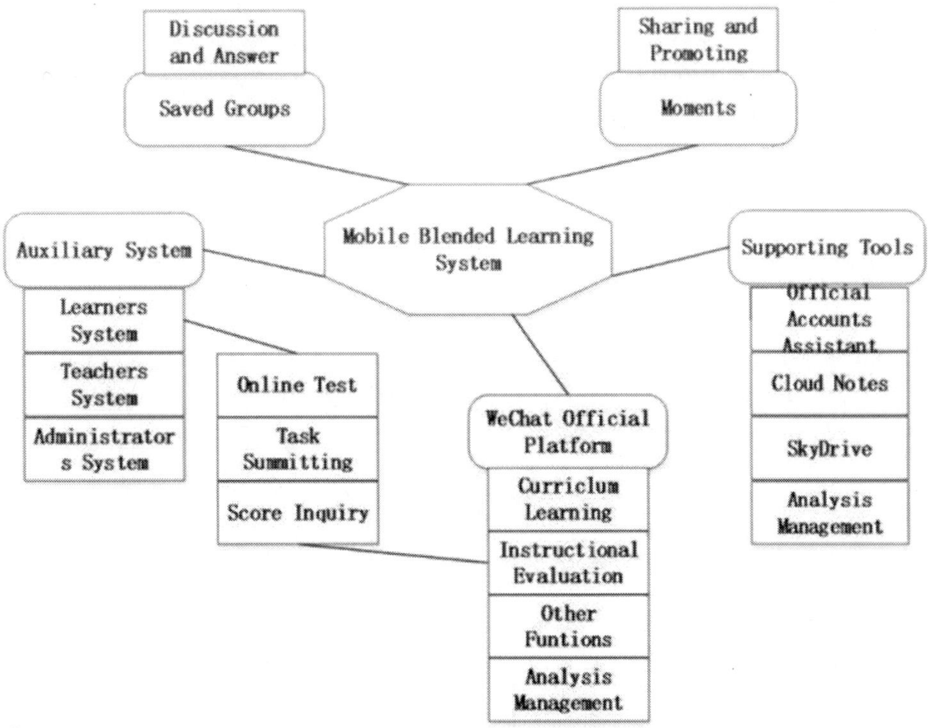

Fig. 2. Schematic diagram of MBLS.

3.1 WPP system

The system is divided into four modules: curriculum learning, instructional evaluation, analysis management, and other functions. In view of this, system functions decomposition diagrams can be designed as showed in Figure 3.

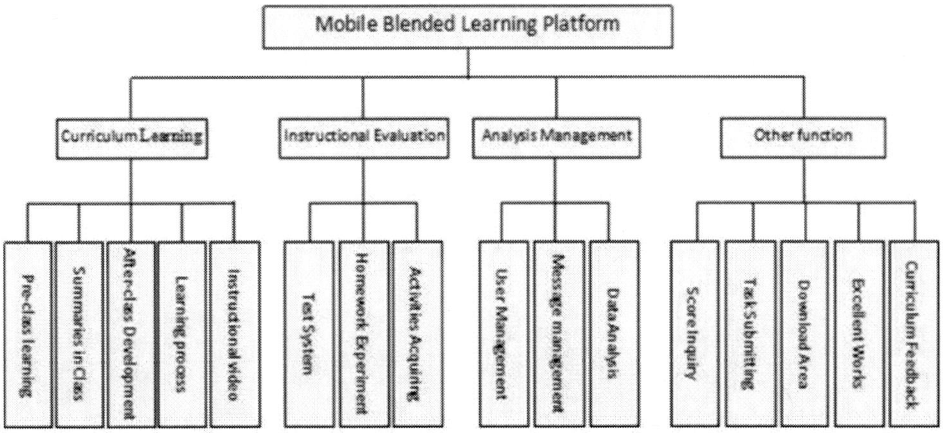

Fig. 3. Functional decomposition diagrams of WPP system.

Curriculum learning mainly consists of five modules: Pre-class Learning, summaries in class, After-class Development, Learning Progress, and Instructional Video. Through clicking on the submenus of WeChat platform, the teacher can present the learning resource needed to the learner, in image-text or video form.

By extending the classroom space to the online space for interactions, teacher noticed the importance of prompt feedback [7]. Instructional evaluation is accomplished through test, homework, experiment, activity and etc. Information is released by clicking on the submenu in image-text form. Because the function of test is unable to realize based on WPP, this design implements this function by clicking on the submenu to jump to an auxiliary system.

Through the functions of user management, message management and data analysis, teachers (or administrators) can realize the functions of message handling, instructional data analysis and student management.

The system sets up other auxiliary functions, to better serve the entire learning process. The functions of results submitting and score query are implemented through auxiliary system, and the functions of download and curriculum feedback can be achieved through clicking on the submenu.

3.2 Auxiliary system based on Java Web

This system implements some functions, such as online testing, upload and download of related document, performance and other related information query and other functions to enrich the whole MBLS through design of auxiliary system based on Java Web. This system users are divided into three categories: administrator, teacher and learner. The detailed functions of these three module is shown in Figure 4.

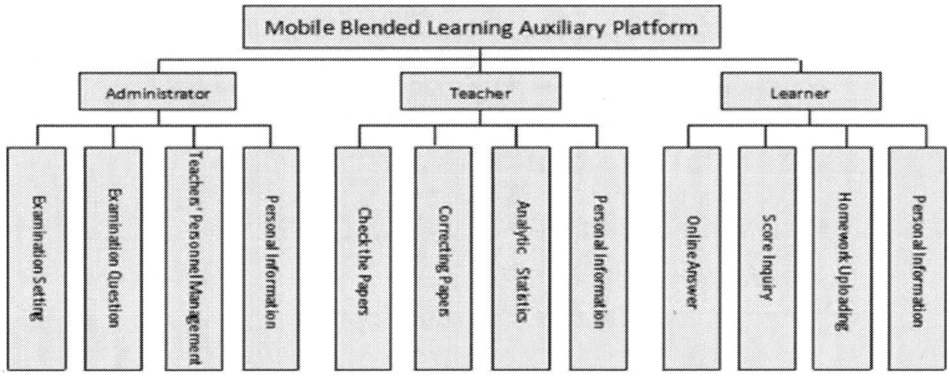

Fig. 4. Functional decomposition diagrams of Auxiliary System.

4 Implementation of MBLS

4.1 Implementation of WPP

You could register an Official Account and log in the official website of the WPP. Then you could see the homepage with managing and editing function, as shown in Figure 5.

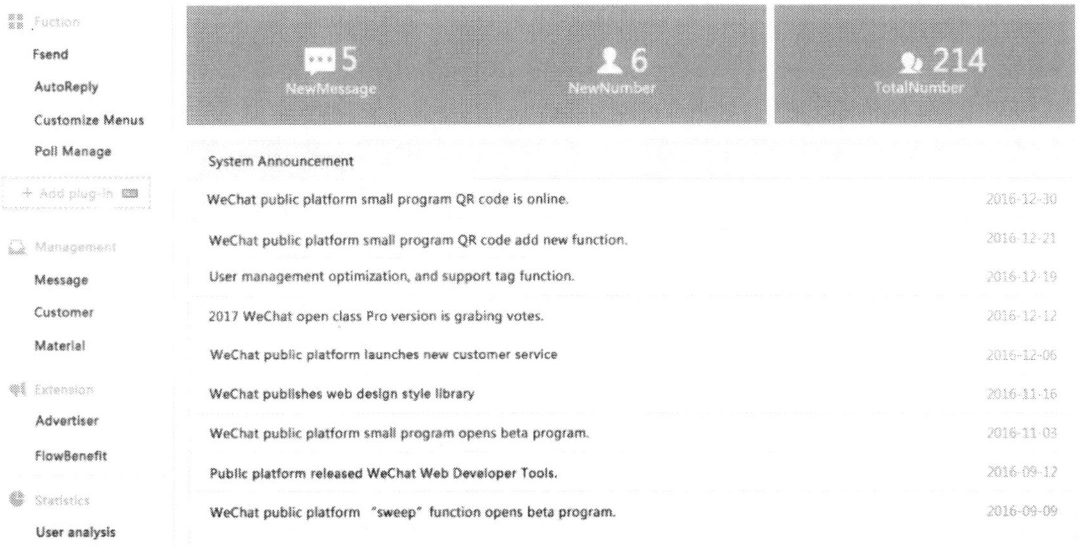

Fig. 5. Management editing interface of Official Account.

4.2 Implementation of auxiliary system

This system is a Java Web project based on B/C model, with SQLServer2005. The server uses Tomcat v8.0 to accomplish local test. And as the need for linking with WeChat Official Account, it is necessary to purchase the domain name, and lend the server and publish it on the line. For the system is not commercially available, and taking into account the problems related to cost, this test applies for free domain using peanut shell software, and uses the port mapping method to temporarily visit this system in the public net.

5 Conclusion

MBLM based on WPP optimizes the instructional process and improves the learning efficiency. MBLS uses WPP and its convenient third-party interface to implement MBLM. With enormous user groups, excellent instant messaging, rich present form of content, cross-platform, free and easy constructing and other advantages, WPP provides teachers with solutions which are universal, simple operation, strong extensibility. At the same time, with the open interface teachers can develop and extend functions to improve the feasibility of MBLM. But now, the design and implementation of MBLM based on WPP has not yet added the factors of subject. In the future research, we will further combine with the characteristics of different subjects, to implement MBLM based on WPP in different situations.

References

1. Y. Zhang. "Internet+ Education" idea and pattern analysis. China's higher EDU. RSRCH. **02**,70-73,(2016)(in Chinese)
2. C.E. Morton, S.N. Saleh, S.F. Smith et.al. Blended learning: how can we optimise undergraduate student engagement? Bmc Medical EDU. **16(1)**,(2016)
3. Sheng-quan Yu. From the knowledge transfer to cognitive construction, and situational awareness---Development and outlook of the three generations of mobile learning. China EDU TECH. **6**,(2007) (in Chinese)
4. E.M.T. Maza, M.T.G. Lozano, A.C.C. Alarcón et al. Blended learning supported by digital technology and competency-based medical education: a case study of the social medicine course at the Universidad de los Andes, Colombia. International JRNL. EDU. TECH. Higher EDU. **13(1)**,2016

5. NMC Horizon Report (2016 Higher Education Edition). (2016)
6. R.H. Huang, D. Ma, L.Q. Zheng. Based on blended learning theory of curriculum design. E-EDU. RSRCH. **(1)**:11-16,(2009) (in Chinese)
7. H. Çakir, B.A. Bichelmeyer. Effects of teacher professional characteristics on student achievement: an investigation in blended learning environment with standards-based curriculum. Interactive Learning ENVR. **24(1)**:1-13,(2013)
8. M. Lai, K.M. Lam, C.P. Lim. Design principles for the blend in blended learning: a collective case study. Teaching in Higher EDU. 1-14(2016)

MATEC Web of Conferences **100**, 02021 (2017) DOI: 10.1051/matecconf/201710002021
GCMM 2016

Research on Evaluation of Regional Logistics Capability under Big Data Background

Wenfeng Zhao [1,*] and Jiapeng Zhang [1]

[1] Management Engineering Department, Zhengzhou University, Zhengzhou, China
Corresponding Email: 15136836918@163.com

Abstract. Over the past few years, the emerging information technology has greatly promoted the development of e-commerce, and then the level of the logistics industry's ability to put forward higher requirements. At this stage, the development of regional logistics is not balanced, the problems of each region are not the same. In this paper, the regional logistics capability evaluation system is established by analytic hierarchy process (AHP). The paper analyzes and evaluates the regional logistics development in six typical provinces, and puts forward suggestions to promote the regional logistics capability, so as to promote the sound development of regional logistics.

1 Introduction

Now, a variety of new information technology emerging in the mobile computing, Internet of Things, cloud computing and other technical support, people in the information exchange and commodity trading is more convenient and efficient.

In the age of information explosion, the amount of data has been increasing rapidly. The redundancy and complexity of data are becoming more and more obvious. Traditional data mining methods and technologies are unable to cope with the large-scale data rapidly emerging in the field of technology. Level data processing and application, thus losing the auxiliary decision-making based on data analysis programs and capabilities [1].Large data comprehensive information and network based on the new business model and production model, is to redefine the social management and national strategic decision-making, business management decision-making, organizational business processes, personal decision-making process and way, the extensive use of large data will bring human society The Third Industrial Revolution. "Cannot be within a certain period of time using conventional software tools to capture, manage, process data collection" This sentence is Wikipedia on the definition of large data, the concept of large data, the industry does not have a unified view. The term "big data" first appeared in the 1980's, but the big data has only been used as another description in the definition of data mining technology. It is confined to the computer science disciplines and has not formed prominent research results with influence and representative significance. And research programs [2].When large data is seen as a resource, it cannot ignore the access to such resources, processing problems, application issues, each of which is an important management problem. At present, people are aware of large data perspective there is a difference, the main reason for this difference is that people analyze the background of large data and application of large data for different purposes [3].In recent years, China's rapid economic growth, regional logistics in the economy under the impetus of the rapid development, therefore, the development of regional logistics from the actual demand, how to effectively grasp the "big data" environment, opportunities and challenges, Data "to the traditional logistics management work brought about by the thinking system to the technical methods of the great changes, to seek a more scientific and effective way to enhance the level of logistics capacity in the region.[4]It is not only the foothold of this paper but also the practical problem that this paper is expected to solve in order to realize the decision-making innovation of regional logistics management in our country.

2 Basic concepts of regional logistics capability

Regional logistics in the academic and national standards so far has not yet formed a unified concept, which Dong Qianli that regional logistics and logistics control and supervision of the relevant operating system Haifeng emphasizes the regional logistics is the transport, Loading and unloading, handling, circulation processing, distribution and information processing and other basic activities of the logistics of organic synthesis.

The regional logistics described in this article is to maximize the integration of logistics resources, save logistics costs and improve economic efficiency, to provide efficient and rational organization and management of regional economic development, in the context of environmental, political, economic, military and other regional background. Technical support, so as to boost regional economic sustainable and rapid development of a system of logistics concepts.

3 Establishment of evaluation index system of regional logistics capability

Based on the statistical methods of the China Statistical Yearbook and considering the availability of data, this paper proposes four suggestions on regional logistics service capability B1, regional logistics talent contribution capacity B2, regional logistics fixed assets support capability B3 and regional logistics economic competitiveness B4. Logistics capacity of C1, C2, posts and telecommunications C3, the number of employees in the logistics industry C4, the number of employees in the logistics industry accounted for the tertiary industry, the number of employees in the logistics industry accounted for the first three indicators, The proportion of the number of employees C5, the logistics industry, investment in fixed assets C6, the fixed assets investment in the logistics industry, the proportion of total investment in fixed assets C7, at the end of the road mileage C8, C9 logistics industry, wholesale and retail trade added value C10, The proportion of the output value of the total output value of tertiary industry C11, the proportion of the total output value of logistics industry GDP.

4 Analytic hierarchy process (AHP)

4.1 Analytic hierarchy process

The analytic hierarchy process (AHP) is used to judge the importance degree of the two indexes by decomposing the complex problem into several layers and several factors.

The judgment matrix is established by calculating the maximum eigenvalue and the corresponding eigenvector of the judgment matrix. The weight of the importance of the factors at all levels, providing the basis for the choice of the best option [4].

4.2 The basic steps of AHP

4.2.1 Identify the problem

By clarifying the scope of the problem, the specific requirements put forward, the elements involved and the relationship between the elements, you can clear what to solve the problem, the need for intelligence information has to meet the demand.

4.2.2 Establish hierarchical hierarchical structure

Hierarchical structure can be divided into three levels, the first level is the target layer, the second layer is the standard layer, the third-level program layer, the different levels of the relationship between the elements with the connection.

4.2.3 Establish the judgment matrix

The evaluation scale of the judgment matrix is shown in Table 1:

Table 1. 1-9 scale table.

Scale	Meaning
1	means that the two factors have the same importance
3	one factor is slightly more important than the other
5	one factor is significantly more important than the other
7	one factor is more important than the other

9	one factor is more important than the other
2, 4, 6, 8	Intermediate value of the above adjacency judgment
reciprocal	Judgment bij of the reciprocal factor i is compared with j, and bij = 1 / bij

4.2.4Hierarchical single sorting and hierarchical total sorting

(1)The relative importance of the calculation. Then the maximum eigenvalue is obtained for the judgment matrix, and then the corresponding eigenvector W is obtained.

The component of W ($W_1, W_2, \ldots W_n$) is the weight coefficient.

(2)Consistency check. $\lambda_{max} = n$, The random consistency ratio C.R. = is used, C.R.= $\frac{C.I.}{R.I.}$ .and the consistency test is considered to be acceptable when C.R.≦0.10

The standard values of R.I. are shown in Table 2.

Table2. Average random consistency index.

Order	R.I.
3	0.52
4	0.89
5	1.12
6	1.26
7	1.36
8	1.41
9	1.46
10	1.49
11	1.52
12	1.54

4.2.5 Hierarchical total sorting

The total ranking of the hierarchy is calculated as shown in table3.

Table3. Hierarchical ranking calculation.

A / B	A_1 A_2 $\cdots$ A_m a_1 a_2 $\cdots$ a_m	B Hierarchical total sorting
B_1	b_1^1 b_1^2 $\cdots$ b_1^m	$\sum_{j=1}^{m} a_j b_1^j$
B_2	b_2^1 b_2^2 $\cdots$ b_2^m	$\sum_{j=1}^{m} a_j b_2^j$
$\vdots$	$\cdots$ $\cdots$ $\cdots$ $\cdots$	
B_n	b_n^1 b_n^2 $\cdots$ b_n^m	$\sum_{j=1}^{m} a_j b_n^j$

5 An empirical study on evaluation of regional logistics capability

5.1 Construction of the judgment matrix for consistency testing

This paper selects six provinces or municipalities of Beijing, Shanghai, Henan, Guangdong, Yunnan and Heilongjiang which are representative of our country as regional logistics capacity evaluation samples [5,6].

The relative importance judgment matrix A-B of the target layer and the criterion layer is constructed.

Table 4.A-B judgment matrix.

A	B1	B2	B3	B4	Wi	C.I.
B1	1	2	3	2	0.41	0.08<0.10
B2	0.50	1	3	3	0.32	
B3	0.33	0.33	1	2	0.15	
B4	0.5	0.33	0.50	1	0.12	

C.I. = 0.08, the average randomness consistency index R.I. = 0.89, the consistency ratio C.R. $=\frac{C.I.}{R.I.}= 0.09 \leq$ 0.10Through the consistency test. The relative importance judgment matrix-P of the criterion layer and the scheme layer is constructed, B_j-P,as shown in Table5 -8.

Table 5. B1-C judgment matrix.

B1	C1	C2	C3	Wi	C.I.
C1	1	3	3	0.59	0.05<0.10
C2	0.33	1	0.50	0.16	
C3	0.33	2	1	0.25	

Consistency test: C.I. = 0.08, mean random consistency index R.I. = 0.52, consistency ratio C.R. $=\frac{C.I.}{R.I.}=0.10\leq$ 0.10.Through the consistency test.

Table 6. B2 - Judgment Matrix.

B2	C4	C5	Wi	C.I.
C4	1	0.50	0.33	0.00<0.10
C5	2	1	0.67	

Consistency test: C.I. = 0.00, mean random consistency index R.I. = 0.52, consistency ratio C.R. $=\frac{C.I.}{R.I.}=0.00\leq$ 0.10.Through the consistency test.

Table 7. B3- judgment matrix.

B3	C6	C7	C8	Wi	C.I.
C6	1	2	2	0.49	0.05<0.10
C7	0.5	1	0.50	0.20	
C8	0.5	2	1	0.31	

Consistency test: C.I. = 0.05, mean random consistency index R.I. = 0.52, consistency ratio C.R. $=\frac{C.I.}{R.I.}=0.10\leq$ 0.10.Through the consistency test.

Table 8.B4-Judgment Matrix.

B4	C9	C11	C12	Wi	C.I.
C9	1	3	2	0.26	0.03<0.10
C10	2	4	5	0.51	
C11	0.33	1	0.50	0.09	
C12	0.5	2	1	0.14	

Consistency test: C.I. = 0.03, mean random consistency index R.I. = 0.52, consistency ratio C.R. $=\frac{C.I.}{R.I.}=0.03\leq$ 0.10.Through the consistency test.

5.2 Overall ranking

According to the ranking of the three levels of the overall ranking index weight from large to small order: freight, logistics, the number of practitioners in the proportion of the tertiary industry, the number of logistics practitioners, postal and telecommunications business, logistics industry, investment in fixed assets, freight turnover, wholesale and retail Value of logistics at the end of the year, total output value of logistics industry, investment in fixed assets in the logistics industry as a proportion of total investment, logistics industry in GDP and logistics industry in total output value of tertiary industry.

Among all the 12 factors, the first one is the freight volume, the weight of 0.24, is the primary factor affecting the regional logistics capacity. The second place is the number of logistics practitioners accounted for the proportion of tertiary industry practitioners, the weight of 0.21. The third is the number of logistics practitioners, the weight of 0.11. The fourth is the post and telecommunications business volume, its weight 0.10 [7,8].

Table 9. Hierarchical total sort.

Secondary indicators B	The third indicator C	Total weight
B1	Freight volume C1	0.24
	The rate of goods turnover C2	0.06
	Post and Telecommunications Services C3	0.10
B2	Number of logistics employees C4	0.11
	The proportion of logistics industry in the tertiary industry C5	0.21
B3	Logistics industry fixed asset investment C6	0.08
	Investment in Fixed Assets in the Logistics Industry as a Percent of Total Investment C7	0.03
	End of the year Highway mileage C8	0.05
B4	Added value of logistics industry C9	0.03
	Added value of wholesale and retail trade C10	0.06
	The added value of the logistics industry accounts for the proportion of tertiary industry added value C11	0.01
	Value - added of Logistics Industry as a Percentage of GDP C12	0.02

5.2.1 Instance-related data The final result and sum of the data after the data are normalized and multiplied by the weigh

The abbreviations in the table represent the cities respectively:
BJ:Beijing,SH:ShangHai,HN:HeNan,GD:GuangDong,YN:YunNan,HLJ:HeiLongJiang.

Table 10. The final result and the sum of each data multiplied by the weight.

City	BJ	SH	HN	GD	YN	HLJ
C1	0.06	0.05	0.13	0.24	0	0.03
C2	0.00	0.06	0.02	0.05	0	0.00
C3	0.00	0.02	0.02	0.10	0.02	0
C4	0.01	0.04	0	0.11	0.02	0.02
C5	0.12	0.21	0	0.04	0.16	0.18
C6	0.04	0	0.03	0.08	0.01	0.01
C7	0.01	0	0.01	0.03	0.00	0.00

C8	0.04	0	0.05	0.04	0.00	0.03
C9	0	0.01	0.02	0.03	0.01	0.00
C10	0	0.02	0.01	0.06	0.01	0.00
C11	0	0.00	0.01	0.00	0.00	0.00
C12	0	0.01	0.02	0.01	0.01	0.02
Sum	0.29	0.43	0.33	0.80	0.25	0.30

6 Conclusion and prospect

Based on AHP, this paper constructs the regional logistics capacity evaluation index data model, and draws the conclusion: in the six cities analyzed above, the regional logistics capacity The ranking order of the evaluation order is Guangdong, Shanghai, Henan, Heilongjiang, Beijing and Yunnan.

Coastal developed cities such as Shanghai and Guangdong should strengthen the publicity and construction of logistics standardization work, introduce advanced logistics infrastructure, optimize the allocation of resources, improve the management mode, enhance the competitiveness of enterprises from the system, and strive to reach the world leading level. How to make the developed city logistics management approach to adapt to the large data environment as soon as possible, and effective use of large data environment of unprecedented opportunities and challenges, is every science and technology workers inevitable and must be carefully considered.Inland areas such as Henan, Heilongjiang and other regions, regional logistics competitiveness in general, which is due to the less developed central and western regions, mainly relying on economic development, the second industry, logistics level is low, lack of regional logistics capacity. So to strengthen the region's logistics industry at all levels of reform efforts to change the mode of development and structure to tertiary industry development to drive the overall development. We should make use of the policies of central and western regions and other preferential policies to do the overall strategic planning of logistics, integrate the traditional "target-driven decision-making" and the "data-driven decision-making" concept and method in large data environment, and actively adapt to large data environment, Knowledge and perspective of the ability to form the core strengths of their respective regions in the fierce regional competition to occupy a favorable position.

References

1. Zhu Donghua,Wang Xuefeng,LiBing.Research on Management Method of Technological Innovation in Big Data Environment.Science of Science and Management of S. & T.,2013(04):172-180.
2. Xu Zongben,Feng Zhiyan,Guo Xunhua.Big Data - Driven Management and Decision.Management of the world,2014(11):158-163.
3. Yang Shanlin,Zhou Kaile.Management Problems in Big Data - Based on Big Data.Journal of Management Sciences,2015(5):1-8.
4. Nie Xiaoping.An Economic Analysis of Regional Logistics.Regional Economy,2007(9):214-222.
5. China Statistical Yearbook. Beijing: China Statistics Press, 2015.
6. Guo Jinyu, Zhang Zhongbin, Qing Yun. Research and application of analytic hierarchy process. Chinese Journal of Safety Science, 2008,18 (5).
7. LI Yong, WANG Yun, XU Guang-yin.Study on Comprehensive Evaluation of Regional Logistics Planning Journal of Henan Agricultural University, 2009 (02)
8. LiuBao,Xu Shubo,Zhao Huanchen.Analytic Hierarchy Process-Planning Decision Tools.Systems Engineering,1984(06)23-30.

The Transport Performance of Feedback Coupled Brownian Ratchets

LiMing Fan, MingTao Lv, RenZhong Huang and TianFu Gao *

College of Physical Science and Technology, Shenyang Normal University, Shenyang 110034, China
Corresponding Email: tianfugao@synu.edu.cn

Abstract. The directed transport properties of feedback coupled Brownian ratchets under the effect of external periodic force and load force are investigated. The influence of the coupling strength on the transport properties of coupled system is discussed in detail. It is found that the centre-of-mass mean velocity, the average diffusion coefficient and the energy conversion efficiency of the feedback system can achieve the maximum for the different optimal coupling strength, and the maximum move to right with the increase of the natural length. It implies that the optimal coupling strength and natural length can promote the directed transport and energy conversion efficiency of the feedback ratchets.

Keywords. Bioengineering, feedback-control, directed transport, energy conversion efficiency.

1 Introduction

Brownian motion in the environment of the thermal fluctuations is a long-study issue in nonequilibrium statistical physics. In recent years, the directed transport properties of Brownian ratchets attract the widespread attention of scholars [1, 2] Only when the systems meet the spatio-temporal symmetry-breaking feature can produce the directed transport. The Brownian ratchets satisfy this condition, and that can take advantage of noise as it converts random fluctuations into directed motion without external force [3-5] Some properties of directed transport can be obtained through this interesting phenomenon, such as the center-of-mass mean velocity, the energy conversion efficiency, etc. In addition, Brownian ratchets have the wide application in bioengineering and nanotechnology.

There are some work of the directed transport for Brownian ratchets [6, 7] The main concern of this paper is the directed transport properties of the feedback coupled Brownian ratchets under the external force and the effect of coupling. For discussing the transport properties of coupled ratchets in detail, the center-of-mass mean velocity, the average diffusion coefficient, the Pe number and the energy conversion efficiency of the coupled ratchets are investigated. It is interesting to find that there is an optimal coupling strength which can facilitate the directed transport and diffusion of the feedback system, and the energy conversion efficiency can reach a maximum as the coupling strength increasing. Meanwhile, these conclusions can provide some guidance for the enhancement of the efficiency in nano-machines.

2 Model

The coupled Brownian particles are acted on the asymmetric periodic ratchet potential $U(x_i)$, and then the particles are under the effect of asymmetric periodic force $F(t)$, load force λ, feedback control switch $\varepsilon(t)$, and the thermal noise. The dynamic behavior of particles can be described by the overdamped Langevin equation [8]:

$$\gamma\dot{x}_1(t)=\varepsilon(t)G(x_1(t))-k(x_1-x_2-l)-\lambda+F(t)+\xi_1(t) \quad (1)$$

$$\gamma\dot{x}_2(t)=\varepsilon(t)G(x_2(t))+k(x_1-x_2-l)-\lambda+F(t)+\xi_2(t) \quad (2)$$

The locations of the two coupled particles at the time t are represented by $x_1(t)$ and $x_2(t)$, γ indicates the friction coefficient. k is the elasticity coefficient and l is the natural length. λ denotes the load force. The external force acting on the particles is $F(t)$, and it can be expressed as follows,

$$F(t)=\begin{cases} A, & n\tau\le t<n\tau+\tau_1 \\ -A, & n\tau+\tau_1\le t<(n+1)\tau \end{cases}. \quad (3)$$

In Eq. (3), A is the amplitude of periodic external force, τ is the period of external force. In addition, $\xi_1(t)$ and $\xi_2(t)$ are Gaussian white noise of zero mean and $<\xi_i(t)\xi_j(t)>=2D\delta_{ij}\delta(t-t')$, D is the strength of noise. $G(x_i(t))$ in Eqs. (1) and (2) denotes the force acted on the particles by the ratchet potential, and it can be given by the relation

$$G(x_i(t))=-\frac{d}{dx_i}U(x_i), \quad (4)$$

where $U(x_i)$ indicates the external periodic potential, and can be expressed as follows,

$$U(x_i)=\begin{cases} \dfrac{dx_i}{c}, & x_i\le c, \\ \dfrac{d(1-x_i)}{1-c}, & c<x_i\le L, \end{cases} \quad (5)$$

d is the barrier height. L is the period of ratchet potential, and the ratchet period $L=1$.

In the Eqs. (1) and (2), we consider a controller $\varepsilon(t)$, which switches the potential on ($\varepsilon=1$) if the ensemble average of the force is positive or switches the potential off ($\varepsilon=0$) for otherwise. And the alternative controller can be expressed as follows

$$\varepsilon(t)=\Theta(g(t))=\begin{cases} 0, & g(t)\le 0, \\ 1, & g(t)>0, \end{cases} \quad (6)$$

where Θ is Heaviside function. In Eq. (6), $g(t)$ can be expressed as

$$g(t)=\frac{1}{2}(G(x_1(t))+G(x_2(t))). \quad (7)$$

So Eqs. (1) and (2) are the motion equations of the feedback coupled Brownian particles.

In order to obtain the transport properties of feedback ratchets, the center-of-mass mean velocity of the coupled particles is investigated, and it can be written as [9]

$$<V_{cm}>=\lim_{T\to\infty}\frac{1}{2T}\sum_{i=1}^{2}\int_0^T\dot{x}_i(t)dt, \quad (8)$$

which $<\cdot>$ indicates the ensemble average. Because of the diffusion of coupled Brownian particles at all times, we describe the fluctuation of particle displacement by introducing the average effective diffusion coefficient D_{eff} [10]

$$D_{eff}=\lim_{T\to\infty}\sum_{i=1}^{2}\frac{<x_i(t)^2>-<x_i(t)>^2}{2\cdot 2T}. \quad (9)$$

In addition, we also calculate the Pe number for describing the directed transport properties of the coupled Brownian particles [11]

$$Pe=\frac{|V|L}{D_{eff}}. \quad (10)$$

The Pe number reflects the ability of the directed transport of the feedback ratchets.

For discussing the properties of directed transport from the view of energy transform, the energy conversion efficiency of the coupled ratchets η is calculated [12]. As usual, the efficiency η is defined as the ratio of the useful work W which accomplished by the system in pumping coupled particles against the load force λ to the input of energy P_{in} from the external fluctuation. That is

$$\eta = \frac{W}{P_{in}}, \quad (11)$$

Where

$$W = \sum_{i=1}^{2} \lambda_i \cdot \langle V_i \rangle,$$

and,

$$P_{in} = \sum_{j=1}^{2} \frac{1}{\tau} \int_{x_j(n\tau)}^{x_j((n+1)\tau)} F_j(t) dx_j(t).$$

In this paper, the Eqs. (1) and (2) are numerically solved by using the stochastic Runge-Kutta algorithm. In order to obtain a stable ensemble average, we simulate 5×10^3 different trajectories, and each track evolves 50 periods with the time step $h = 10^{-3}$. All physical quantities are used the normalization processing, no special description, the parameters take $\gamma = 1.0$, $l = 0.4$, $L = 1.0$, $d = 2.5$, $\tau = 10.0$, $\tau_1 = 0.8$, $c = 0.96$, and $\lambda = 1.0$.

3 Results and discussion

Figure 1(a) shows the curves of the center-of-mass mean velocity $\langle V_{cm} \rangle$ varying with the coupling strength K for different natural lengths. We find that the too weak or strong coupling strength can not facilitate the directed transport. So there is an optimal value of the coupling strength which can make the enter-of-mass mean velocity achieve the maximum. Furthermore, the optimal coupling strength moves to right with the increase of the natural length, this indicates that the feedback ratchets need the larger coupling strength to make the directed transport of coupled system reach the maximum.

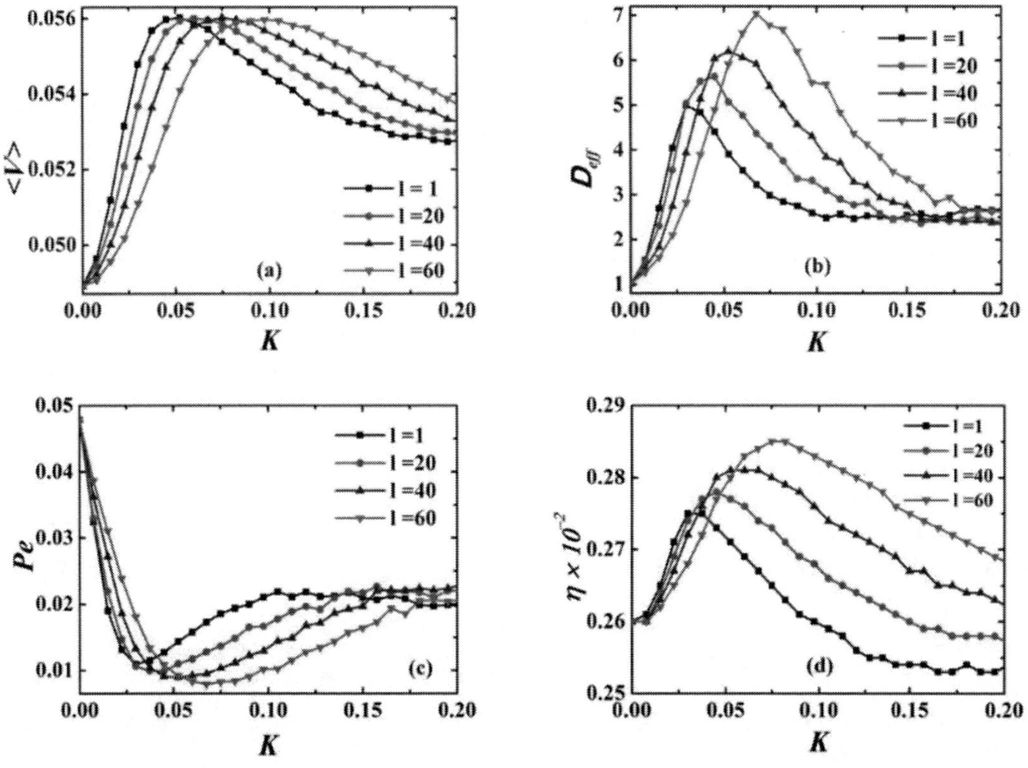

Fig. 1. The curves of (a) the center-of-mass mean velocity $\langle V_{cm} \rangle$, (b) the average effective diffusion coefficient D_{eff}, (c) the Pe number, and (d) the energy conversion efficiency η varying with the coupling strengths K.

In order to obtain the directed transport of Brownian particles in depth, the average diffusion coefficient D_{eff} varying with the coupling strength for different natural lengths is discussed. The result is shown in Figure 1(b). It can be found that the average diffusion coefficient D_{eff} can achieve the maximum with the increase of coupling strength, and the maximum of the coupling strength increases as the natural length increasing. In addition, we can clearly see that it is not constructive to promote the diffusion of the feedback ratchets when the coupling strength is too weak or strong. It can be explained that the weak coupling effect is not enough to cause the coupled particles diffusion. Moreover, the strong coupling effect restricts the particles and hinders the diffusion of the feedback ratchets. Furthermore, when the coupling strength is fixed, if $K = 0.075$, the feedback ratchets diffuse more easily with the increase of the natural length.

Figure 1(c) shows the relation between the Pe number and the coupling strength for different natural lengths. From Figure 1(a) and 1(b), it is easy to observe that the curves of the diffusion coefficient achieve the maximum when the center-of-mass mean velocity reaches the maximum. Due to the diffusion occupies the dominant in the competition between the directed transport and diffusion, so according to Eq. (10), the Pe number reaches the minimum for the same coupling strength, as shown in Figure 1(c).

For understanding the transport property of feedback ratchets from the perspective of energy utilization, the energy conversion efficiency is discussed. As shown in Figure 1(d), it is clearly found that the efficiency curves and the velocity curves have the similar results. The efficiency of the coupled ratchets is not large when the coupling strength is too weak or strong. Therefore, the optimal coupling strength can promote the efficiency of energy conversion to reach the maximum. From Figure 1(a) and 1(d), it is interesting to note that the coupled ratchets can produce larger transport when the particles use the input energy more efficiently. And the maximum of efficiency will move to right as coupling strength increasing. It implies that the feedback ratchets need larger coupling to make the energy conversion efficiency achieve the maximum with the increase of the natural length.

4 Conclusion

The directed transport properties of feedback coupled Brownian ratchets under the effect of external periodic asymmetric force and load force are investigated, and the impact of the coupling strength on the transport properties of coupled ratchets is discussed in detail. It is found that the center-of-mass mean velocity, the average diffusion coefficient, and the energy conversion efficiency can achieve the maximum as the coupling strength increasing for the different natural lengths. In addition, the Pe number reaches the minimum with the increase of coupling strength. Therefore, the optimal parameters are important to enhance the directed transport of feedback coupled ratchets. These conclusions can be applied in separation of particles and the improvement of the efficiency for biomolecular motors.

Acknowledgement

This work is supported by the National Natural Science Foundation of China (Grant No. 11347003) and the Excellent Talents Program of Shenyang Normal University (Grant No. 91400114005).

References

1. L. Angelani, A. Costanzo, R. Di Leonardo, EPL **96**, 68002 (2011).
2. P. K. Ghosh, V. R. Misko, F. Marchesoni, F. Nori, Phys. Rev. Lett. **110**, 268301 (2013).
3. J. E. Villegas, S. Savel'ev, F. Nori, E. M. Gonz´alez, J. V. Anguita, R. Garc´ıa, J. L. Vicent, Science **302**, 1188 (2003).

4. M. Salerno, Y. Zolotaryuk, Phys. Rev. E **65**, 056603 (2002).
5. S. Ooi, S. Savel'ev, M. B. Gaifullin, T. Mochiku, K. Hirata, F. Nori, Phys. Rev. Lett. **99**, 207003 (2007).
6. Ai B Q, Wu J C, J. Chem. Phys. **140**, 094103 (2014).
7. W. E. Uspal, M. N. Popescu, S. Dietrich, M. Tasinkevych, Soft Matter **11**, 434 (2015).
8. Li G, Tu Z C, Sci. China-Phys. Mech. Astron. **59**, 640501 (2016).
9. Zheng Z G, Cross M C, Hu G Phys. Rev. Lett. **89**, 154102 (2002).
10. Wang L F, Gao T F, Huang R Z, Zheng Y X, Acta. Phys. Sin. **62**, 070502 (2013).
11. Gao T F, Chen J C, J. Phys. A: Math. Theor. **42**, 065002 (2009).
12. Sumithra K, Sintes T, Physica A **297**, 112 (2001).

Selecting Visualization Alternative Based on Uncertain Theory

Ping Huang[1,*], Zhiying Lv[1,2] and Liwei Zheng[3]

[1]College of Management, Chengdu University of Information Technology, Chengdu 610103,China
[2]College of Mathematics, University of Electronic Science and Technology of China, Chengdu 610103,China
[3]College of Applied Mathematics, Chengdu University of Information Technology, Chengdu 610103,China
Corresponding Email: 229059014@qq.com

Abstract. Multiple attribute decision making (MADM) is an efficient way to solve complex systems, and has wide application. This research develops MCDM model based on uncertain theory, used for selecting a suitable visualization alternative for tourism. First, in order to achieve desirable decision making, a new concept is proposed, which is called the best and the worst reference uncertain linguistic variable as a datum uncertain linguistic variable. At the same time, a new method for ranking uncertain linguistic variable is also presented. Second, based on the preference order relation of attributes given by the experts, a new score function is introduced to get the weight vector of attributes. Finally, the evaluation system of tourism big data visualization alternatives is constructed and the order of those alternatives is acquired by the decision method.

Keywords. MADM, score function, reference uncertain linguistic variable, tourism big data visualization.

1 Introduction

Big data [1] has been risen as national strategic resources. In recent years, after many developed countries such as Europe, the United States, Japan and South Korea took large data as a national level strategy, China have brought the construction of large data into the national strategic choice. Large datawas written in the government work report in 2014. In January 2015, the National Tourism Administration in the "Guidance on the promotion of the development of intellectual tourism," said: "By 2020, large data mining and intelligent marketing capabilities will be improved significantly, mobile e-commerce, tourism data system analysis, artificial intelligence technology will be applied in the tourism industry more widely. And the main task of building tourism data has been clear many times in the views.

Tourism big data visualization [2] decision-making is an emerging field of the internet age. In the internet context, tourism big data has a large scale and has a high complexity. It is a huge challenge to search, analysis and understand the large number of unstructured data and semi structured data. The value of the rich information behind the hidden is reflected only through the collection, analysis, interpretation and expression. And visualization is the most effective way for people to understand easily the value of complex information data. Therefore, it is significant to study the decision making of tourism big data visualization. It can not only promote the fusion and innovation of data mining, analysis techniques and methods, computer graphics technology and decision theory and methods, but also have a transformative influence on thinking and methods of government departments, tourism enterprises and tourists. Therefore, it can provide a more rapid, effective and scientific decision-making protection.

In the process of decision, there are some difficulties for experts to express their preference degrees with crisp numerical values. So, it is another possible way to use linguistic labels [3], which represents qualitative aspects values. We will consider a finite and totally ordered label set $S = \{s_0, s_1, \cdots, s_{t-1}\}$ in the usual sense and with odd cardinality, where each label s_i represents a possible value for a linguistic real variable. The number of linguistic terms in the set S is called the cardinality of S. In the symbolic computation process, the discrete linguistic set S is extended to a continuous interval $\bar{S} = \{s_\alpha : \alpha \in [0,t]\}$. $\bar{S}$ is called an extended linguistic term set associated with S[4]. Let $s \in \bar{S}$, $I(s)$ is denoted as the position index of s and called the gradation of s in $\bar{S}$. For example, $I(s_\alpha) = \alpha$. The 2-tuple fuzzy linguistic representation model represents the linguistic information by means of a 2-tuple (s, α), where s is a linguistic label and the numerical value and α represents the value of the symbolic translation [4, 5].

In this paper, the focus is on the extension of discrete support model for MADM [6, 7] in which the experts express their opinions by means of uncertain linguistic setting in stead of precise numerical values. In order to get the weight vector, a new method by the definition of score function of attribute and the preference order of attribute presented by different experts is presented. Moreover, a new ranking method is presented to rank uncertain linguistic variable, which is based on the best and the worst reference uncertain linguistic variable.

2 Preliminaries

In this section, the basicconcepts and their extensions of the this paper are briefly introduced.

Definition 1 Let $\tilde{s} = [s_{a^-}, s_{a^+}]$, where $s_{a^-}, s_{a^+} \in \bar{S}$, s_{a^-} and s_{a^+} are lower and upper limits, respectively. Then $\tilde{s}$ is called an uncertain linguistic variable.

Definition 2 Let $\{\tilde{s}_{a_1}, \tilde{s}_{a_2}, \cdots, \tilde{s}_{a_n}\}$ be a set of uncertain linguistic variables, where $\tilde{s}_{a_i} = [s_{a_i^-}, s_{a_i^+}] \in \tilde{S}$. The associated weight vector is $\omega = (\omega_1, \omega_2, \cdots, \omega_n)$, where $\omega_i \geq 0$ and $\sum_{i=1}^{n} \omega_i = 1$. The uncertain linguistic weighted average (ULWA)operator comes from the extended linguistic representation model is defined as:

$$ULWA(\tilde{s}_{a_1}, \tilde{s}_{a_2}, \cdots, \tilde{s}_{a_n}) = \tilde{s}_a = [s_{a^-}, s_{a^+}] \tag{1}$$

where

$$\begin{aligned} a &= [a^-, a^+] \\ &= [\sum_{i=1}^{n} \omega_i I(s_{a_i^-}), \sum_{i=1}^{n} \omega_i I(s_{a_i^+})]. \\ &= [\sum_{i=1}^{n} \omega_i a_i^-, \sum_{i=1}^{n} \omega_i a_i^+] \end{aligned}$$

The deviation measures have been discussed between two uncertain linguistic terms. The following deviation measure for two uncertain linguistic variables is first introduced.

Definition 3 Let $\tilde{s}_{a_1} = [s_{a_1^-}, s_{a_1^+}]$ and $\tilde{s}_{a_2} = [s_{a_2^-}, s_{a_2^+}]$ be two uncertain linguistic variables. The deviation measure between $\tilde{s}_{a_1}$ and $\tilde{s}_{a_2}$ is defined as follows:

$$d(\tilde{s}_1,\tilde{s}_2)=\frac{1}{2U}(|I(s_{a_1^-})-I(s_{a_2^-})|^2+|I(s_{a_1^+})-I(s_{a_2^+})|^2)$$
$$=\frac{1}{2U}(|a_1^- - a_2^-|^2+|a_1^+ - a_2^+|^2) \tag{2}$$

where U is the number of linguistic terms in the set S .

Clearly, $0\le d(\tilde{s}_1,\tilde{s}_2)\le 1$; $d(\tilde{s}_1,\tilde{s}_2)=0$ if and only if $\tilde{s}_1=\tilde{s}_2$.

Definition 4 Suppose $s_{a_i}=[s_{a_i^-},s_{a_i^+}], i\in I$ is a group of linguistic interval valuables, the best and the worst referenceuncertain linguistic valuable as a datum uncertainlinguistic variables are $s_p=[s_{p^-},s_{p^+}], i\in I$ and $s_q=[s_{q^-},s_{q^+}], i\in I$, where

$$s_{p^+}=\max_{i\in I}\{s_{a_i^+}\},$$

$$s_{p^-}=\max_{i\in I}\{\{s_{a_i^-},s_{a_i^+}\}\setminus s_{p^+}\},$$

$$s_{q^-}=\min_{i\in I}\{s_{a_i^-}\},$$

$$s_{q^+}=\min_{i\in I}\{\{s_{a_i^-},s_{a_i^+}\}\setminus s_{q^-}\}.$$

Definition 5 Suppose $s_{a_i}=[s_{a_i^-},s_{a_i^+}], i\in I$ is a group of uncertain linguistic variables, the best and the worst referenceuncertain linguistic valuable as a datum uncertainlinguistic variables are $s_p=[s_{p^-},s_{p^+}], i\in I$ and $s_q=[s_{q^-},s_{q^+}], i\in I$, then the relative correlation coefficient of s_{a_i} is defined as:

$$\alpha_i=\frac{d(s_{a_i},s_q)-d(s_{a_i},s_p)}{d(s_p,s_q)} \tag{3}$$

Obviously,

(1) $-1\le\alpha_i\le 1,\quad i\in I.$

(2) If $\alpha_i>\alpha_j, \alpha_j>\alpha_k$, then $\alpha_i>\alpha_k, i,j,k\in I.$

For $\alpha_i>\alpha_j,\alpha_j>\alpha_k, i,j,k\in I$, then

$$\frac{d(s_{a_i},s_q)-d(s_{a_i},s_p)}{d(s_p,s_q)}$$
$$>\frac{d(s_{a_j},s_q)-d(s_{a_j},s_p)}{d(s_p,s_q)}$$
$$>\frac{d(s_{a_k},s_q)-d(s_{a_k},s_p)}{d(s_p,s_q)}$$

We get $\alpha_i > \alpha_k$, so can rank uncertain linguistic variables based on their relative correlation coefficient. **Theorem 1** Suppose $s_{a_i} = [s_{a_i^-}, s_{a_i^+}], i \in I$ is a group of linguistic interval valuable, the best and the worst referenceuncertain linguistic valuable as a datum uncertain linguistic variables are $s_p = [s_{p^-}, s_{p^+}], i \in I$ and

$$s_q = [s_{q^-}, s_{q^+}], i \in I \text{, then } s_p \geq s_{a_i} \geq s_q .$$

Proof. For $\forall k \in I$, according to the definition of s_p, we get $s_{a_k^+} \leq s_{p^+}$. So if $s_{a_k^+} \leq s_{p^+}$, suppose $s_{a_l^+} \leq s_{p^+}$, $l \in I$, then

$s_{p^-} = \max\limits_{\substack{i \in I \\ i \neq l}} \{\{s_{a_i^-}, s_{a_i^+}\} \setminus s_{p^+}\}$, so $s_{a_k^-} \leq s_{p^-}$.

If $s_{a_k^+} = s_{p^+}$, then $s_{p^-} = \max\limits_{\substack{i \in I \\ i \neq l}} \{\{s_{a_i^-}, s_{a_i^+}\} \setminus s_{a_k^+}\}$. Obviously, $s_{a_k^-} \leq s_{p^-}$. Overall, $s_{a_k} \leq s_p$.

so, for $\forall i \in I, s_{a_i} \leq s_p$. In the same way, $s_{a_i} \geq s_q$.

Definition 6 Suppose $C = \{c_1, c_2, \cdots, c_m\}$ is the set of attributes, the score function of attribute $c_i \in C$ is defined as:

$$s_i = \sum_{\substack{i,j=1 \\ i \neq j}}^{m} r_{ij} ,$$

where

$$r_{ij} = \begin{cases} 1 & c_i \succ c_j \\ 0.5 & c_i \approx c_j \quad c_i, c_j \in C . \\ 0 & c_i \prec c_j \end{cases} \tag{4}$$

3 The solution approach

Let $A = \{A_1, A_2, \cdots, A_n\}$ be the alternative set and $C = \{c_1, c_2, \cdots, c_m\}$ be the set of all attributes. Assume $\omega = (\omega_1, \omega_2, \cdots, \omega_m)^T$ is the weight vector of attributes, such that $\sum_{j=1}^{m} \omega_j = 1$, $\omega_j \geq 0$ and ω_j denotes the weight of attribute c_j. Let $E = \{e_1, e_2, \cdots, e_t\}$ be the set of experts who give the preference of attributes and $\lambda = (\lambda_1, \lambda_2, \cdots, \lambda_t)$ be the weight vector ofthe experts, where $\lambda_k \in (0,1)$, $\sum_{k=1}^{t} \lambda_k = 1$. Suppose $\widetilde{S} = (s_{a_{ij}})_{n \times m}$ is the uncertain linguistic decision matrix given by the experts, where $s_{a_{ij}} = [s_{a_{ij}^-}, s_{a_{ij}^+}]$ represents the performance of the alternative A_i with respect to the attribute c_j.

Now, we will give the process of decision based on the mentioned method before.

Step 1. Based on Eq .(4), compute the weight vector of attributes as:

$$\varpi_j = \frac{\sum_{k=1}^{t} \lambda_k s_j^k}{\sum_{j=1}^{m} \sum_{k=1}^{t} \lambda_k s_j^k} \quad (j = 1, 2, \cdots, m) \tag{5}$$

where s_j^k denote the score of c_j given by expert $e_k, k = 1, 2, \cdots, t$.

Step 2. Based on the decision matrix and Eq.(1), for every alternative A_i, compute the corresponding weighted decision value s_{a_i} as follows:

$$s_{a_i} = ULWA(s_{a_{i1}}, s_{a_{i2}}, \cdots, s_{a_{im}}) = [s_{a_i^-}, s_{a_i^+}] \tag{6}$$

Where

$$[a_i^-, a_i^+] = [\sum_{j=1}^{m} \omega_j I(s_{a_{ij}^-}), \sum_{j=1}^{m} \omega_j I(s_{a_{ij}^+})] = [\sum_{j=1}^{m} \omega_j a_{ij}^-, \sum_{j=1}^{m} \omega_j a_{ij}^+]$$

Step 3. For the set $\{s_{a_1}, s_{a_2}, \cdots, s_{a_n}\}$, decide the the best and the worst referenceuncertain linguistic valuable as a datum linguistic interval $s_p = [s_{p^-}, s_{p^+}]$ and $s_q = [s_{q^-}, s_{q^+}]$,based on definition 4,that is

$$s_{p^+} = \max_{i=1,2,\cdots n} \{s_{a_i^+}\},$$

$$s_{p^-} = \max_{i=1,2,\cdots n} \{\{s_{a_i^-}, s_{a_i^+}\} \setminus s_{p^+}\},$$

$$s_{q^-} = \min_{i=1,2,\cdots,n} \{s_{a_i^-}\},$$

$$s_{q^+} = \min_{i=1,2,\cdots,n} \{\{s_{a_i^-}, s_{a_i^+}\} \setminus s_{q^-}\}. \tag{7}$$

Step 4. Then, compute the relative correlation coefficient α_i of $s_{a_i}, (i = 1,2,\cdots,n)$ by Eq.(3) , that is

$$\alpha_i = \frac{d(s_{a_i}, s_q) - d(s_{a_i}, s_p)}{d(s_p, s_q)} \tag{8}$$

then the bigger the α_i , the better the alternative A_i , so the overall ranking is acquired.

4 The establishment of the evaluation

At present, the tourism big data visualization has gradually been widely used in industry. The characteristics of tourism big data visualization alternative can be describes by a group of factor attributes, and experts can evaluate it with these attributes. Based on the National Tourism Administration website and the latest research results, experts summarized the evaluation rules and related factors generally considered in the process of evaluating the tourism big data visualization alternative, then construct the evaluation system, which will be presented as follows:

Let $A = \{A_1, A_2, \cdots, A_5\}$ be the set of alternatives, where A_1 is statistic graph tool, A_2 is information graph tool, A_3 is map tool, A_4 is vectored graph tool, A_5 is statistic graph measured tool.

Let $C = \{c_1, c_2, c_3, c_4\}$ be the set of attributes, where c_1 is for the government, c_2 is for the enterprise, c_3 is for the tourism practitioners, c_4 is for the broad masses of the visitor.

Suppose there are four experts who give the preference order of attributes as follows:

$$c_4 \succ c_1 \approx c_3 \succ c_2$$

$$c_2 \succ c_1 \succ c_3 \succ c_4$$

$$c_4 \succ c_2 \succ c_1 \succ c_3$$

$$c_1 \succ c_2 \succ c_3 \succ c_4$$

After many years 'research analysis by many experts, we can construct the decision matrix as follows:

$$\tilde{S}=\begin{pmatrix}[s_7,s_9] & [s_7,s_9] & [s_5,s_7] & [s_5,s_6]\\ [s_4,s_6] & [s_3,s_4] & [s_6,s_8] & [s_7,s_9]\\ [s_5,s_7] & [s_5,s_6] & [s_5,s_7] & [s_7,s_9]\\ [s_4,s_6] & [s_7,s_8] & [s_3,s_4] & [s_3,s_5]\\ [s_5,s_6] & [s_5,s_7] & [s_7,s_8] & [s_6,s_7]\end{pmatrix}$$

In the following, the proposed method is applied to solve this problem and the computational procedure is summarized in the following.

Step1.Using Eq.(5), compute the weight vector of attribute:

$$\omega_1=0.333,$$
$$\omega_2=0.308,$$
$$\omega_3=0.112,$$
$$\omega_4=0.247$$

Step 2. Compute the weighted decision value by Eq.(6):

$$s_{a_1}=[(s_6,0.282),(s_8,0.035)],\ s_{a_2}=[(s_5,-0.343),(s_6,0.349)],$$
$$s_{a_3}=[(s_5,0.494),(s_7,0.186)],$$
$$s_{a_4}=[(s_5,-0.435),(s_6,0.145)],$$
$$s_{a_5}=[(s_5,0.471),(s_7,-0.221)]$$

Step 3. Decide the best and the worst referenceuncertain linguistic valuables by Eq.(7):

$$s_p=[(s_7,0.186),(s_8,0.035)],$$
$$s_q=[(s_5,-0.435),(s_5,-0.343)]$$

Step4.Compute the relative correlation coefficient α_i of $z_i,(i=1,2,\cdots,n)$ by Eq.(8):

$$\alpha_1=0.785,$$
$$\alpha_2=-0.348,$$
$$\alpha_3=0.201,$$
$$\alpha_4=-0.435,$$

Then, we can rank all the alternatives:

$$A_1\succ A_3\succ A_5\succ A_2\succ A_4$$

and thus the most desirable alternative is A_1.

5 Conclusion

There are many types of visualization alternatives faced by traveler, so it is hard to see which one is better. In order to deal with this problem, visualization alternative evaluations should be done to help people to decide on the better option for tourism. In this paper, a new multiple attribute decision method based on uncertain theory is given to evaluate visualization alternatives. Compared to traditional way of assessment, the new selection model proposed by this article is less liable to fault and more feasible to practice. This method can not only be used to solve visualization alternatives selection problem, but it can also be used to deal with select problem of many similar issues. In the practice application, people can not only adopt the most desirable alternative, but also combine several desirable alternative.In future research, the decision model presented in this paper will be extended according to the frequency of the application in different scenarios and group decision [3,8].

Acknowledgment

The National Natural Science Foundation Project of China (No. 61203285;No.61673285; No.11671284)

The Province Department of Science Soft Science Project in Sichuan (2016ZR0095)

The high level research team of the major projects division of Sichuan province(Sichuan letter [2015] no.17-5)
The Project of Chengdu University of InformationTechnology (CRFKYTZ201644)

References

1. F.F. Luo, The systematic review about the theory of big data and the development of practice, Business, 11(2016):219.
2. Sébastien Liment, Millian Poquet, Sophie Robert, A frameworkfor efficient dynamic interactive scientific visualization, Procedia Computer Science. **29** (2014):692-702.
3. Z.S. Xu, Uncertain linguistic aggregation operators based approach to multiple attribute group decision making under uncertain linguistic environment, Information Sciences, 168(2004): 171-184.
4. D.F. Li, Multiple attribute decision making method using extended linguistic variables, Int J Uncertain Fuzziness Known Based System17(2009): 793-806.
5. Herrera F, Marteinez L, A model based on linguistic 2-tuples for dealing with multigranular hierarchical linguistic contexts in multiple experts decision making. IEEE Transactions on SystemsManand cybernetics, 31(2001):227-234.
6. Z. Y. Lv, A fuzzy multiple attribute decision making methodbased on possibility degree, Journal of Intelligent & Fuzzy Systems. 31 (2016) 787–794.
7. D.F. Li, Multiple attribute decision making method using extended linguistic variables, Int J Uncertain Fuzziness Known Based System,17(2009):793-806.
8. J.P. Xu, Z.B. Wu, A discrete consensus support for multiple attribute group decision making, Knowledge-Based systems 24(2011): 1196-1202..

Design and Implementation of Online Booking System of University Sports Venues

Can Li*, Junjie Li, Hongxiang Cao and Zhan Meng

Management Engineering School, Zhengzhou University, Zhengzhou, China
Corresponding Email: 365083792@qq.com

Abstract. To solve the imbalanced use of university sports venues, the online booking system of university sports venues based on the table tennis hall of Zhengzhou University is designed, combining the experience of the existing online booking systems, the main methods and the major ideal of building these systems both at home and aboard. The basic functions of the system are designed according to the general requirement analysis, including user's registration and login, online venue booking, online payment, personal center, message board and database construction. The development of the system used Java programming language, JSP (a Web platform development technology), MySQL database processing technology, JDBC data access model, MyEclipse development platform and Tomcat server. The system has accomplished such functions as online booking, online payment and online message. The tests of the system are run in good conditions. The use of the system has made up for the current luck of sports venues management, solved the problems of the online booking function of university sports venues, improved the efficiency of the venues and meet the needs of efficient use of the venues. The system can be an experience of the management of the other college sports venues.

1 Introduction

College sports venues are not only an essential part of physical education, but also an important place for fitness activities of the faculties, students as well as nearly residents. At present, sports venues of Zhengzhou University has such problems as places idled, venues are fulled at the high time of movement, which greatly affects the practical efficiency of sports venues resources and increases the management burden to venues. With the rapid development of information technology, college sports venues realizing online booking has become a new trend.

2 Research status and significance

2.1 Research status

Many venues used the reservation machines to make a booking in Shanghai World Expo in 2010. The State Grid Pavilion and other venues also launched online booking systems [1]. Dai Xiaojing and other scholars designed a sports venue booking system based on .NET [2], which has realized the basic function of venues online booking. They also developed a booking and detection system of sports venues, based on the three-tier architecture theory of the Internet of things(IOT) [3], but it failed to make detailed demands analysis and function design. Wei Honglei's article about the network service management platform system of college sports venues involved in a module of reservation center system [4], which has realized the function of users to booking venues at anytime. Zhang Huai and other scholars also covers a booking module in the design of a stadium management system [5].

Overall, current literature related to this study are less. Moreover, the design schemes of reservation system and realization technology are various. The imbalanced use of college sports venues resources has not been completely solved now. So lots of researches about the online booking system of college sports venues are seriously needed.

2.2 Research significance

People can know the usage of every sports venues in Zhengzhou university online, and book any places they need, which can reduce the imbalanced problems that there are no usable places or all the places are being used fully, etc. To a certain extent it can reduce the complicated human affairs, optimize resources management and reduce the imbalanced use of venues.

3 Requirements analysis and research methods

3.1 Requirement analysis

Many college sports venues are opened for free. But because the usage of venues can't be known good, people often encounter that students of athletics are in the class or there is no place for sports when go to the venues directly. So a online booking system for college sports venues is necessary. The venue booking system should provide humanized interface and complete functions. The basic functions are as follows:

(1) Identity authentication of users: For faculty and students, work number and student Id can respectively be used as an account , which are based on our campus card. Moreover, For the nearly residents, they need register at first, and then input legitimate user name and password. Eventually, after the certification they can booking venues.

(2) Classification of administrator users and ordinary users: Users are divided into administrators users and ordinary users. The administrators have more authorities, such as the management of users, place orders, changing information of places and the bulletin-board.

(3) Real-time information display for venues: Visual graphics based on the actual geographical position of venues need to be built. Users can know the current usage of every place of every venues of our college and can know the situation of the venue booking in the next seven days.

(4) Venue booking and cancel: Users can booking venues in advance in the next week. In order to cater to the teaching time, a day can be divided into fixed period of time for the user to make a order. The user can cancel the order if it hasn't meet the time.

(5) Pay the cost for booking: To standardize the process of the user to make a booking and to reduce the damages due to users go back on their word, payment for booking can be realized by deducting the virtual money. The payment success, the booking success.

(6) Message board: When submit orders, users can leave a message to the administrator if they have more demands.

(7) Personal center module: In this place, users can view their order and balance, and they can modify their personal information as well as password.

(8) Database: It mainly includes the following table: users' information table, venues' information table, booking orders table, opening schedule of venues, etc.

3.2 Research methods

(1)Technical route:

The development of system use JSP (Web platform development technology) + JAVA (programming language) + MySQL (database) + JDBC (data access model) [6]+ Javascript (scripting language) as a technical support.

(2)Development tools and environment of the system:

1> Development technology:

Development technology of dynamic webpages: JSP + DIV + CSS (Javascript, JAVA, JDBC) [7]

2> System development environment:

Operating system: Windows XP Pro3

Software: MyEclipse (tools) [8, 9]

Tomcat6.0(server) [10]

MySQL 5.0 (database)

PhotoshopCS3 (be used to make pictures)

3> System running environment

The server side:

Operating system: Windows 2000 Server

Web server: Tomcat6.0

Web database: MySQL 5.0

Client:

Operating system: Windows 98 / xp and above

The browser version: IE6, IE7 ,IE8 browser, Internet explorer, firefox, Google, Opera, etc.

4 Design and implementation of the system

4.1 Design of the interface

This design is based on two design concept:

(1) Simple design concept:The purpose of development of this system is to provide convenience to users when they booking venues, and to avoid flashy, so the system' interface is simple but contain complete functions.

(2) Humanized design concept: In the process of design the system, questions are pondered from the perspective of users in order to provide better service for users.

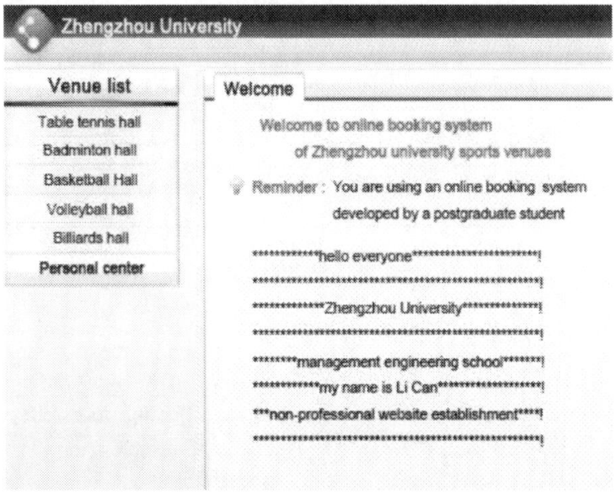

Fig. 1. The welcome screen.

4.2 Design and implementation of registered and login module

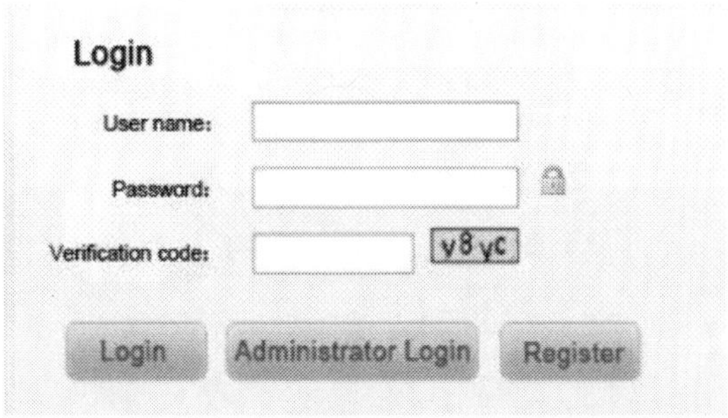

Fig. 2. User login.

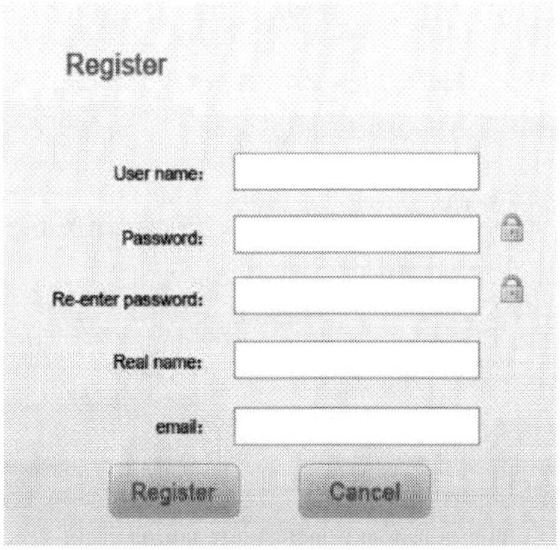

Fig. 3. User registration.

4.3 Design and implementation of the classification of users

Users are divided into two categories, they are administrator users and members, Their functions and permissions are as follows:

User classification and their functions	Members	Browse the real-time information of the fields
		Venue booking and cancel the order
		Check their own orders
		Change their personal information
	Administrator users	Change and delete the board information
		Change the real-time information of the fields
		Browse、change and delete all the booking orders
		Add or remove members

Fig. 4. Classification and functions of users.

4.4 Venue booking function

Venue booking is the core module of the system and the design of the venue booking process is also an important content of this study. Through a lot of references reading and investigation researches, the design of the process of venue booking functions are as follows:

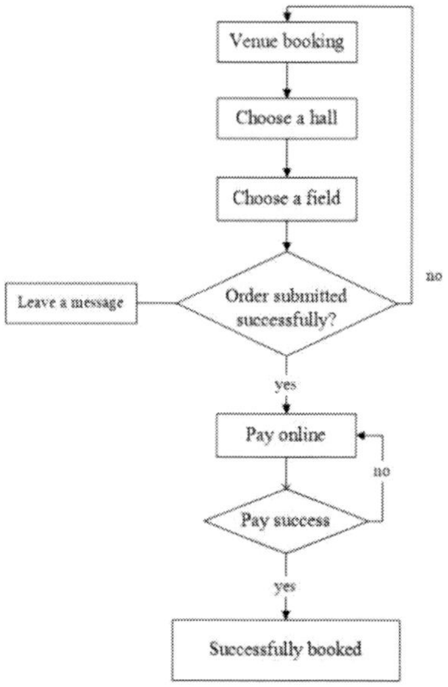

Fig. 5. Flow chart of venue booking.

Take table tennis hall as an example, the venues' distribution plan and its real-time information are shown to users in the form of a plane. The design is reasonable and seems to be clear at a glance. The user can booking venues within seven days from now on site, and the distribution of the time period is divided into 90 minutes a time period one day, They are 8:00-9:30,10:00-11:30 and other time periods in the afternoon. After a click, it can jump to the pay function page. If the payment is successful, the reservation can be achieved.

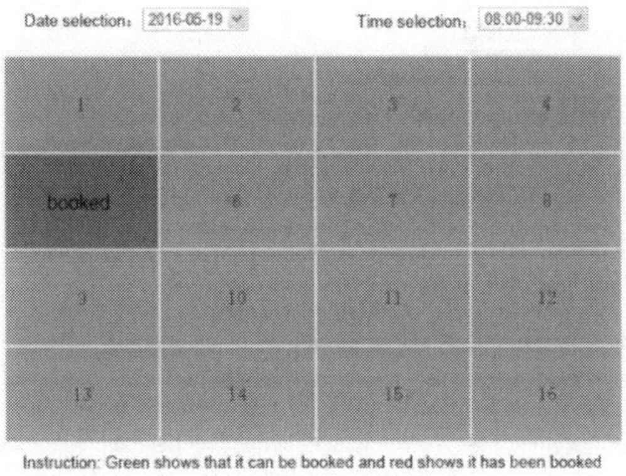

Fig. 6. Booking interface.

4.5 Design and implementation of payment function

The payment function is another difficulty of this study. To implement a online payment, the virtual account was introduced. Through deducting the user's account points at the time of booking site, the payment was achieved. When submitting an order, users will be asked if they do determine the venue booking. If they click yes, their accounts will be deducted, and then the venue booking is succeed. Users can view their orders or cancel them in a column called "my orders" in the personal center module.

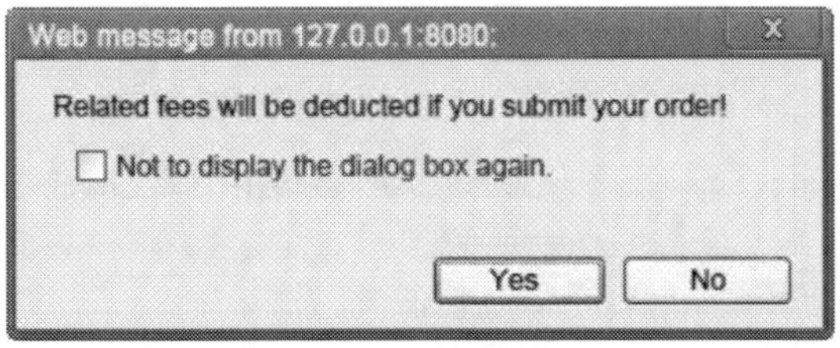

Fig. 7. Prompt of deduction.

4.6 Design and implementation of message board function

When submit orders, users can leave a message to the administrator if they have more demands, and then the administrator should meet users' demands as far as possible. Which can strengthen communication and exchange of information, making our services more thoughtful. Moreover, this is in accordance with the humanization design concept of the system.

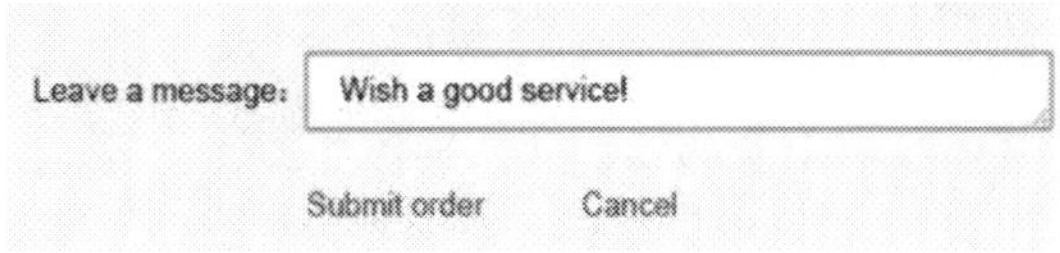

Fig. 8. Message board.

After success of submit orders:

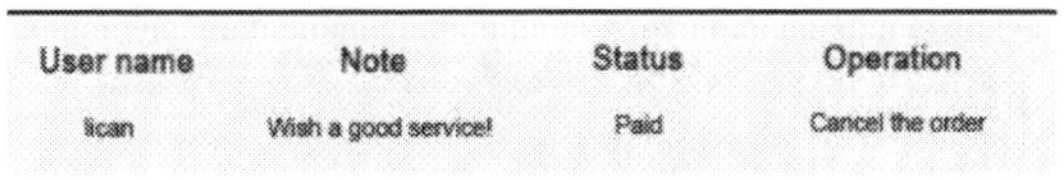

User name	Note	Status	Operation
lican	Wish a good service!	Paid	Cancel the order

Fig. 9. Display of message.

4.7 Design and implementation of personal center module

(1) Personal center module for ordinary users

It is mainly divided into two parts, they can view or cancel their order here. The second is a personal information module, the user can not only change or improve their personal information, but also change password, etc.

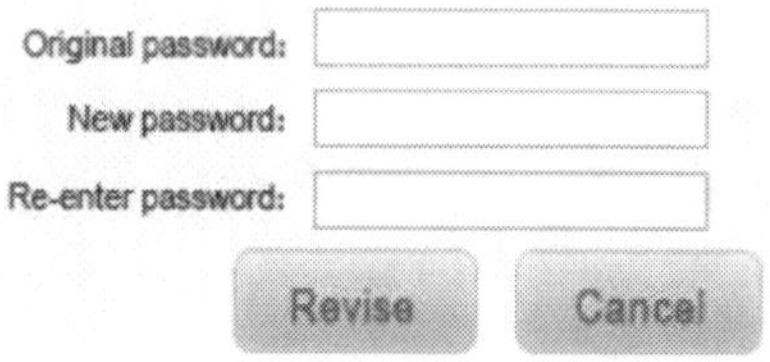

Fig. 10. Change password.

(2)Personal center module for administrator users

Besides with normal ordinary user' permissions, administrator uses have more functions which can mainly divided into the following three aspects: Member management, bulletin management and order management. Administrator users can implement a change of information of bulletin board, they can check all users' orders, change or delete ordinary users' information and recharge accounts for users, etc.

Fig. 11. Different permissions between ordinary users(left) and administrator users(right).

4.8 Construction of the database

The system mainly used a data access model named JDBC to access the data resources of database MySQL server. Several major tables are as follows:

seq	lasttime	content
1	2016-11-05 22:31:25	Volleyball pavilion is not opened to the public today.
2	2016-10-05 14:58:57	Here is Zhengzhou University.
3	2016-10-05 14:59:25	XXXXX........................
4	2016-10-05 15:13:37	There is a game in basketball hall at nine tonight.

Fig. 12. Bulletin board information table.

field	remark	status	user	id	date	time
1	1	paid	york	2015102608000930 1	2015-10-27	08:00-09:30
2	2	paid	york	201510270800093 02	2015-10-27	08:00-09:30
1		paid	lican	201511020800093 01	2015-11-02	08:00-09:30
1		paid	lican	201511050800093 01	2015-11-05	08:00-09:30
10	e33	paid	lican	2015110508000930 10	2015-11-05	08:00-09:30
1	6767	paid	york	201511080800093 01	2015-11-08	08:00-09:30
1	good	paid	york	201511081000113 01	2015-11-08	10:00-11:30

Fig. 13. Orders information table.

usertype	username	password	realname	sex	idcard	tel	email	createtime	money
guest	1	1	htx	boy			123@qq.com	2015-11-05 22:37:50	0
guest	[illegible]	1		girl			36508379	2015-11-04 16:09:0[illegible]	0
guest	qj	1		girl			365083792@qq.com	2015-11-04 16:12:0[illegible]	100
guest	qah	1		boy			365083792@qq.com	2015-11-06 13:48:12	0
guest	lican	2580		girl		15022369159	365083792@qq.com	2015-10-28 17:19:46	5
guest	mark	mark		boy	434324234 23	123123123 1	123123123123	2015-10-04 21:24:12	24
admin	york	2580		boy	434345243 24	15022369159	365083792@qq.com	2015-10-04 21:14:55	86
guest	lapei	lapei		girl			1461668626@qq.com	2015-11-03 16:10:5[illegible]	100

Fig. 14. Users information tabl.

5 Test and improvement of the system

5.1 Test of browser's compatibility

There are many different versions of browsers, so the browser's compatibility must be maintained to make sure the system are correctly displayed under different browsers. IE Tester can be used to test various versions of IE browsers' effects of the same system. Such browsers as IE6, IE7, IE8, firefox, Google and Opera are installed to test the system. Finally, the tests are all passed.

5.2 Test of operation of the system

Alpha and Beta are respectively used to make a internal and external tests. We find problems existed in the system and constantly improve it by summarzing test results.

6 Conclusion and prospect

The use of the system can made up for the current luck of sports venues management; It can solve the problems of the online booking function of university sports venues; It can improve the efficiency of the venues and meet the needs of efficient use of the venues. In this study, there are a lot of shortcomings. For instance, there is no limit to the number of reservation; It failed to set up a punishment and give a reasonable credentials into the venues. In the later study, the system needs completely improved. The order number, place number and the time may be feedbacked to users by moblephone messages in the future. The system has a great deal of portability.

The website can not only be bordered with web sites of school resources management, improving the management system, but also be extended to other sports venues of zhengzhou city or other colleges and universities, improving the use of stadiums resources.

References

1. W. Wang, W. Sun, http://news.xinhuanet.com/mrdx/2010-05/01/content_13451241.htm.
2. X. Dai, N. Zhang, Com. Mod. **11**, 123-126 (2012).
3. X. Dai, N. Zhang, Int. Thin. Tech. **7**, 69-71(2012).
4. H. Wei, Research and development of network service management platform of university sports facilities based on.NET Technology(Sichuan Normal University, 2011).
5. H. Zhang, The design and implementation of the management system of comprehensive sports facilities in Guizhou Normal University(Xiamen University, 2013).
6. C. Xia, J. Wuhan Eng Inst, **2**, 48-50(2013).
7. R.Yuan, DIV + CSS Layout Essence of the Case(Tsinghua University Press, 2011).
8. H. Kong, X. Lu, Web Java design and practical tutorial - based on the application development of Eclipse environment(China Machine Press, 2012).
9. J. Zhao, Design and implementation of mobile business hall charging system based on MYECLIPSE development tools(University of Electronic Science and Technology of China, 2012).
10. C. Feng, Y. Chao, Tomcat and Java Web development Technology Description(Tsinghua University Press, 2008)

Optimization of Multiple Traveling Salesman Problem Based on Simulated Annealing Genetic Algorithm

Mingji Xu[1,*], Sheng Li[1] and Jian Guo[1]

[1]School of Automation, Nanjing University of Science and Technology, Nanjing 210094, China
Corresponding Email: jahsonxu@hotmail.com

Abstract. It is very effective to solve the multi variable optimization problem by using hierarchical genetic algorithm. This thesis analyzes both advantages and disadvantages of hierarchical genetic algorithm and puts forward an improved simulated annealing genetic algorithm. The new algorithm is applied to solve the multiple traveling salesman problem, which can improve the performance of the solution. First, it improves the design of chromosomes hierarchical structure in terms of redundant hierarchical algorithm, and it suggests a suffix design of chromosomes; Second, concerning to some premature problems of genetic algorithm, it proposes a self-identify crossover operator and mutation; Third, when it comes to the problem of weak ability of local search of genetic algorithm, it stretches the fitness by mixing genetic algorithm with simulated annealing algorithm. Forth, it emulates the problems of N traveling salesmen and M cities so as to verify its feasibility. The simulation and calculation shows that this improved algorithm can be quickly converged to a best global solution, which means the algorithm is encouraging in practical uses.

1 Introduction

In recent years, there are many multiple Traveling Salesman Problems [1] in reality as society develops forward. For example, problems of multiple watering cart routing [2], urban medium-voltage distribution network planning [3], ventilation network, aircraft assembly sequence planning [4], hot rolling orders staffing issues in steel industry [5], etc.

Multiple Traveling Salesman Problem is evolved from Traveling Salesman Problem. This issue is involved in allocation between several tasks in several individuals [6]. Therefore, compared to pure path optimization in Traveling Salesman Problem, multiple Traveling Salesman Problem seems to be more complicated and challenging due to task allocation and dispatch.

From the points of MTSP description types, it can be divided into four cases [7]: M traveling salesmen set off at the same city and back to that city together; M traveling salesmen set off at N different cities and back to their own leaving city; M traveling salesmen set off at the same city but back to N different cities; M traveling salesmen set off at N different cities but back to the same city. While this paper studies and simulates the first description type.

Looking from MTSP solving ways, it can be divided into following four ways; MTSP's degradation to TSP; exact algorithm; heuristic algorithm and Meta-heuristic algorithm. The research on MTSP has a history of fifty years. At first exact algorithm was adopted to solve the problem, which used a lot Branch-and-bound algorithm [8]. While as Traveling Salesman increases and city enriches, it would spend more time on exact algorithm so it is replaced by heuristic algorithm. The representative of heuristic algorithm Lin-kernighan, but it would make space more complicated if MSTP is degraded to TSP. Genetic algorithm, ant colony algorithm, neural network are typical of Meta-heuristic algorithm. In recent decade, researches on Hierarchical Genetic Algorithm increases a lot. Different chromosome codings result in different solution space and it will have an impact on convergence and optimization of algorithm [9]. Compared to ordinary chromosome, hierarchical chromosome has a stronger binding capacity towards to information and their gene cluster can control to each other. Therefore, this kind of algorithm could solve multi-variables optimization problems that normal genetic algorithm can't deal with, but in terms of large dealing space and redundant solution, it will remain to be solved later [10].

This paper improves the chromosome coding of hierarchical structure, reduces solution space; designs self-adaptive genetic operator for crossover and mutation; modifies Simulated Annealing Algorithm and makes it apply in MTSP solution with hierarchical genetic algorithm.

2 Improved simulated annealing genetic algorithm

Necessary coding on chromosome structure can reduce solution space and enhance operation efficiency of algorithm [11]. The design of self-adaptive genetic operator for crossover and mutation could avoid premature convergence problem efficiently and protect population diversity. Improve simulated annealing genetic algorithm, blend it with genetic algorithm and stretch it in fitness can make algorithm get rid of partial optimization trap.

2.1 Suffix coding design of chromosome

The 'Node Accession' coding design of chromosome, like Figure 1, which could make multiple Traveling Salesman Problems degrade to single Traveling Salesman Problems [12]. Take the example of MTSP: m Traveling Salesman, n cities, this design may produce (n+m+1)! solutions, and other abundant solutions.

Hierarchical coding on chromosome, like Figure2, the upper parameter gene cluster is controlled by the bottom controlling gene cluster. Compared to ordinary chromosome, hierarchical chromosome has a stronger binding capacity which is beneficial to multi-parameter optimization problems. Also, taking the example of MTSP: m Traveling Salesman, n cities, there will be $n!m^n$ probable solutions and other abundant solutions [13].

The improvement of hierarchical coding on chromosome changes bottom controlling gene cluster into city quantities of each traveling salesman passed and it suffixes the upper parameter gene cluster, just like what is shown in Figure 3. This design sets the visiting sequence of traveling salesman [14], thus producing $n!C_{n-1}^{m-1}$ probable solutions and reduces a large number of abundant solutions.

8	4	2	9	1	6	10	7	3	5

Fig. 1. 'Node Accession' coding of three traveling Salesman and eight cities chromosome.

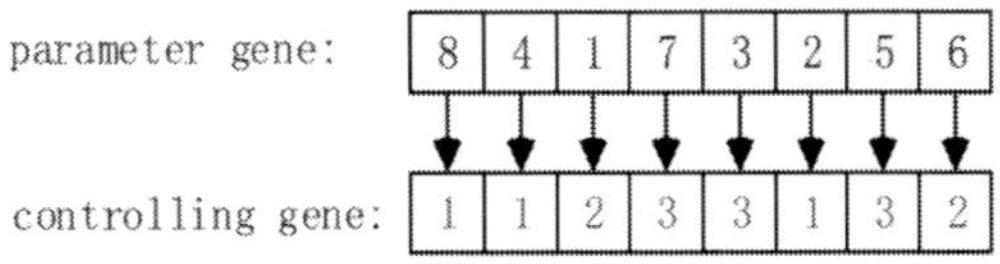

Fig. 2. The hierarchical coding of three traveling Salesman and eight cities chromosome.

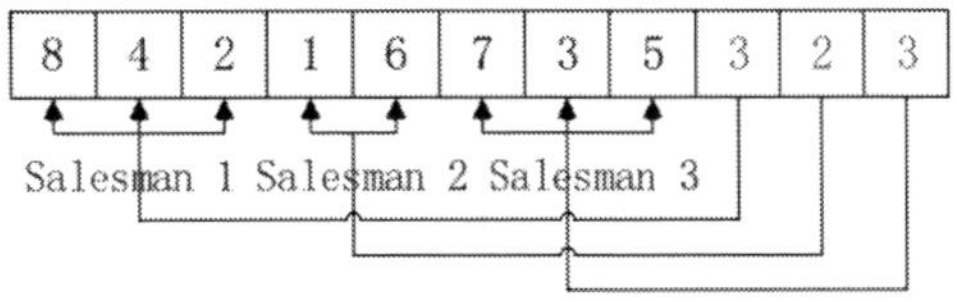

Fig. 3. The suffix coding of three traveling Salesman and eight cities chromosome.

2.2 Self-adaptive genetic operator design

There are many advantages of genetic algorithm while there is also a premature convergence problem, which means converging to partial optimization too early. After population going through several iterations, the individual will have a higher average adaption and it increases exponentially. Due to limited population, these excellent individuals will occupy a lot in population, while other individuals will disappear quickly. This trend is nearly in line with the principal Survival of the Fittest. If it continues developing without manual control then it will result in sole species and an increasing number of near neighbors of parent's population, thus, genetic operator will lost its own value, making algorithm fall into partial optimization. Whether premature convergence problem is obvious, depends on gene cluster, the crossover probability P_c and mutation probability P_m: If P_c is too small, the iterative number will increase and algorithm speed will slow down; whereas P_c is too big, then high-fitness individual will be destroyed quickly; if P_m is too small, it will be hard to produce new individual; whereas P_m is too big, then there will be irregular exhaustion for genetic algorithm. For premature convergence problem of genetic algorithm, it is necessary to design a self-adaptive genetic operator so as to keep population diversity [15].

Self-adaptive genetic operator consists of self-adaptive crossover probability operator and self-adaptive mutation probability operator. The crossover probability P_c has been improved like following formula:

$$P_c = \begin{cases} P_{c1} - \dfrac{(P_{c1} - P_{c2})(f' - f_{avg})}{f_{\max} - f_{avg}}, f' \geq f_{avg} \\ P_{c1}, f' < f_{avg} \end{cases} \quad (1)$$

As is shown above, $f_{\max}$ is the biggest fitness among the group, f_{avg} is the mean value of group fitness; f' is the bigger fitness between the two crossover individuals; P_{c1} =0.9; P_{c2} =0.6.

The self-adaptive improvement is made to the crossover probability, and the formula is as follows:

$$P_m = \begin{cases} P_{m1} - \dfrac{(P_{m1} - P_{m2})(f_{\max} - f)}{f_{\max} - f_{avg}}, f \geq f_{avg} \\ P_{m1}, f < f_{avg} \end{cases} \quad (2)$$

$f_{\max}$ is the biggest fitness among the group, f_{avg} is the mean value of group fitness; f is the fitness of mutation individuals; P_{m1} =0.1; P_{m2} =0.01.

2.3 Simulated annealing algorithm

The physical background of SA is the physical image and statistical properties during solid annealing process. SA accepts a worse solution than current solution at a certain probability so that the algorithm will jump out of the erroneous zone of local optimum. The procedures of SA can be summarized as follows:

(1) A solution is obtained under a certain temperature T (This is solution can be obtained through optimization of Genetic Algorithm);

(2) If this solution is better than the one under last certain temperature, then this solution will be accepted, otherwise follow (3) ;

(3) Calculate the probability under the temperature T:

$$P = e^{\frac{dE}{KT}} \quad (3)$$

It generates a random number a in interval [0,1]. If $a < P$, then the solution will be accepted, and turn to (1); otherwise give up this solution and turn to (1).

From formula (3) we can see the value of P is getting smaller and smaller with the rise of temperature T. With the proceeding of iterative (the temperature will go down with iterative number), the probability of accepting a bad solution decreases. This is rather similar to crystal's annealing crystallization in nature, hence, SA is realized.

2.4 Modified simulated annealing genetic algorithm

Genetic Algorithm has weaker local search ability. During the later period of Hierarchical Genetic Algorithm, the fitness converges, and less superior individuals are produced. However, Hybrid Simulated Annealing Algorithm can make it jump out of the erroneous zone of local optimum. Due to the compatibility of Genetic Algorithm [16], it is feasible to combine Simulated Annealing Algorithm and Hierarchical Genetic Algorithm to form the modified Simulated Annealing Genetic Algorithm. Now we are going to do the fitness stretch to Hybrid Algorithm:

$$f_i = \frac{e^{\frac{f_i}{T}}}{\sum_{i=1}^{M} e^{\frac{f_i}{T}}} \quad (4)$$

$$T = T_0(0.99^{g-1}) \quad (5)$$

f_i is the fitness of the i the individual; M is the size of population; g is the number of iterations; T is temperature; T_0 is initial temperature. The fitness stretch of Hybrid Simulated Annealing Algorithm has obvious effect. In the initial stage of Genetic Algorithm, individuals with consistent fitness had the similar probability of offspring generation; in the later period, because of significant decrease of temperature [17], the stretch effect got enhanced, so the fitness difference among different individuals is magnified which has made the excellent individuals more superior [18].

Here is the solution procedure of the modified Simulated Annealing Genetic Algorithm.

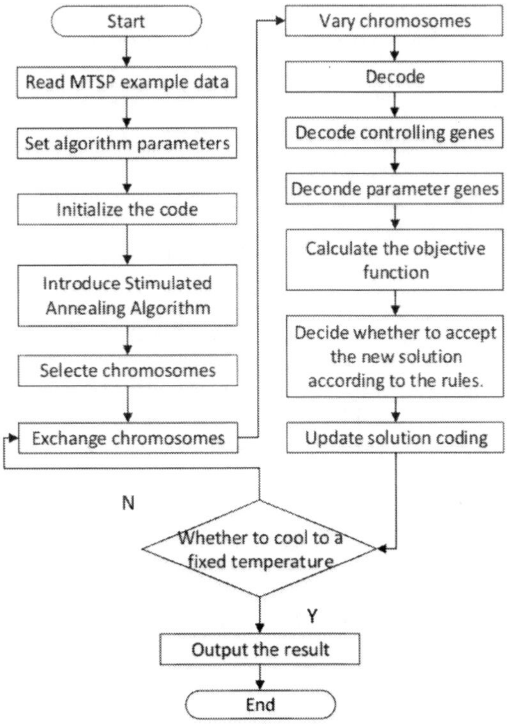

Fig. 4. The flow chart of modified Simulated Annealing Genetic Algorithm.

3 Simulation Results & Analysis

In order to compare modified Simulated Annealing Genetic Algorithm with Simulated Annealing Algorithm, 50 cities are given as simulation objects, and their coordinates in a plane are shown in Table 2. The simulation is divided into 8 groups. The numbers of traveling salesmen and cities in each group are shown in Table 1.

The first group of simulation takes the coordinates of the first 5 groups in Table 2, and uses Hierarchical Genetic Algorithm, Simulated Annealing Algorithm, and modified Simulated Annealing Genetic Algorithm to solve MTSP; each group increases gradually the numbers of traveling salesmen and cities, and the rest can be done in the same manner.

Table 1. Eight simulation variables.

Sequence Number	Number of Salesmans m	Number of Cities n
1	2	5
2	3	10
3	4	15
4	5	20
5	6	25
6	7	30
7	8	40
8	10	50

Table 2. Coordinate values of 50 cities.

City Number	City coordinate		City Number	City coordinate		City Number	City coordinate	
	X	Y		X	Y		X	Y
1	82	84	18	66	54	35	58	22
2	52	30	19	12	23	36	68	18

DOI: 10.1051/matecconf/201710002025

3	66	41	20	18	36	37	72	74
4	88	15	21	15	60	38	67	68
5	83	40	22	56	45	39	78	53
6	62	64	23	20	20	40	18	49
7	83	63	24	30	71	41	22	82
8	76	88	25	45	85	42	36	17
9	30	82	26	51	72	43	43	30
10	87	51	27	32	41	44	62	33
11	27	19	28	75	46	45	68	30
12	83	89	29	62	82	46	75	28
13	12	89	30	39	68	47	75	58
14	47	33	31	12	43	48	88	41
15	67	47	32	12	78	49	83	72
16	23	56	33	35	31	50	11	55
17	18	72	34	46	18			

3.1 Simulation results of 10 traveling salesmen and 15 cities

Three algorithms are used to simulate MTSP (m=10, n=50) and these are all the convergences and solutions.

The algorithm convergence of Hierarchical Genetic Algorithm is in Figure5, and the solution is in Figure6.

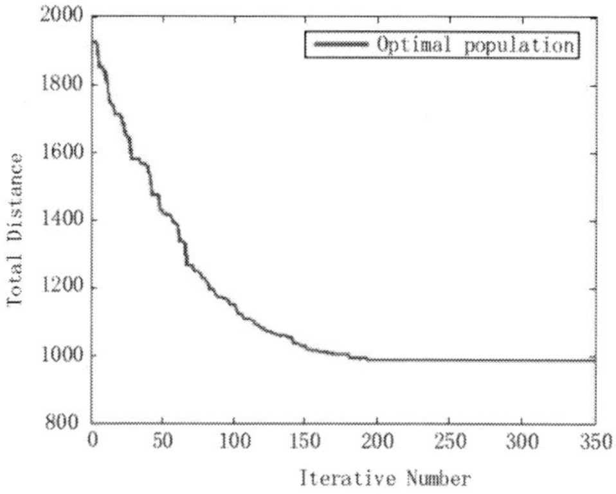

Fig. 5. The iteration curve of Hierarchical Genetic Algorithm optimum MTSP(m=10, n=50).

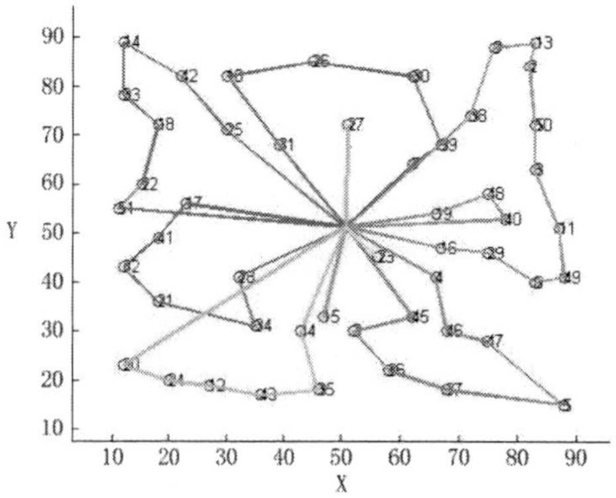

Fig. 6. The solution of Hierarchical Genetic Algorithm optimum MTSP(m=10，n=50).

The algorithm convergence of Simulated Annealing Genetic Algorithm is in Figure7, and the solution is in Figure8.

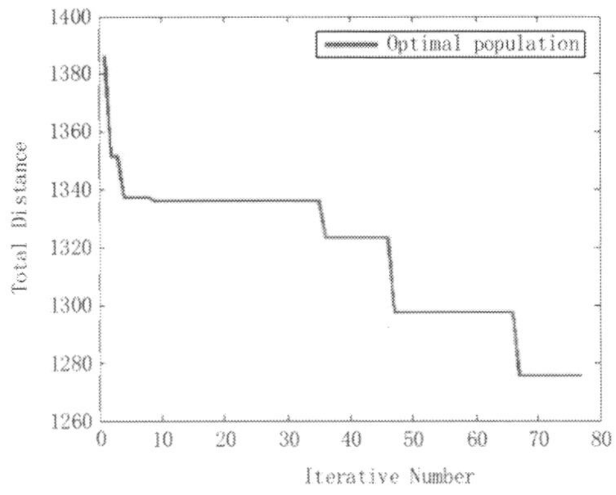

Fig. 7. the iteration curve of Simulated Annealing Genetic Algorithm optimum MTSP(m=10, n=50).

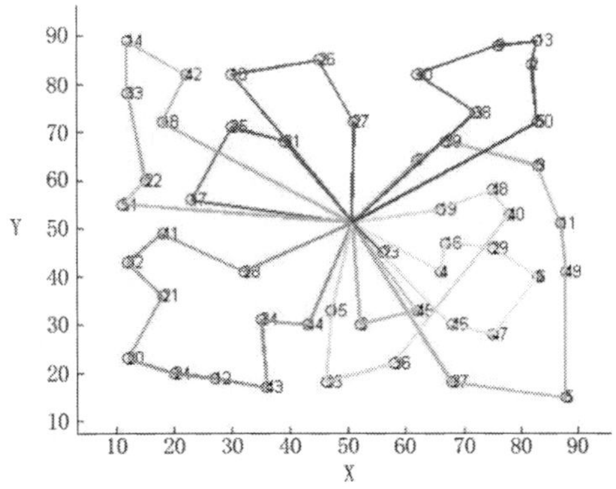

Fig. 8. The solution of Simulated Annealing Genetic Algorithm optimum MTSP (m=10, n=50).

The algorithm convergence of modified Simulated Annealing Genetic Algorithm is in Figure9, and the solution is in Figure 10.

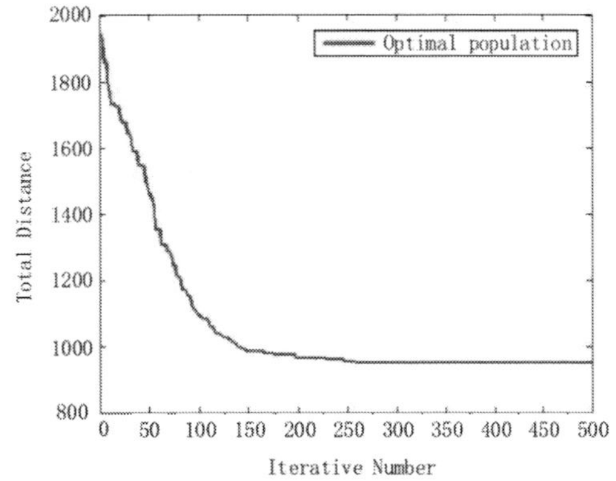

Fig. 9. the iteration curve of modified Simulated Annealing Genetic Algorithm optimum MTSP (m=10, n=50).

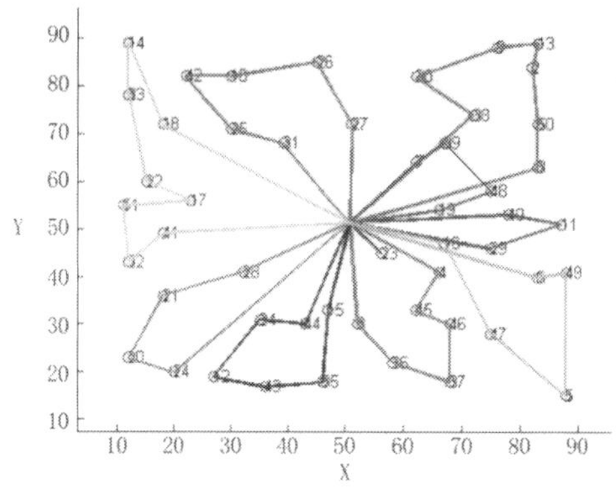

Fig. 10. The solution of modified Simulated Annealing Genetic Algorithm optimum MTSP (m=10, n=50).

3.2 Analysis of simulation results

By the comparison of the simulation results of the three algorithms in Table 3 and Figure11, it is clear that Hierarchical Genetic Algorithm is able to converge to a better solution, while Simulated Annealing Algorithm iterates faster. However, modified Simulated Annealing Genetic Algorithm can sooner achieve a better global optimum solution. The reasons are: the suffix structure design of chromosome reduced the space of the solution, the self-adaptive genetic operator and double crossover and mutation improved 'premature convergence problem'; the introduction of Simulated Annealing Algorithm stretched the fitness and enhanced the local search ability [19].

Table 3. The comparison of the simulation results of the three algorithms.

	m=2 n=5	m=3 n=10	m=4 n=15	m=5 n=20	m=6 n=25	m=7 n=30	m=8 n=40	m=10 n=50
Hierarchical Genetic Algorithm (HGA)	198	316	456	512	612	692	828	998
Simulated Annealing Algorithm (SA)	185	323	556	735	776	821	996	1270
Modified Simulated Annealing Genetic Algorithm (MSAGA)	193	299	413	474	576	668	778	956

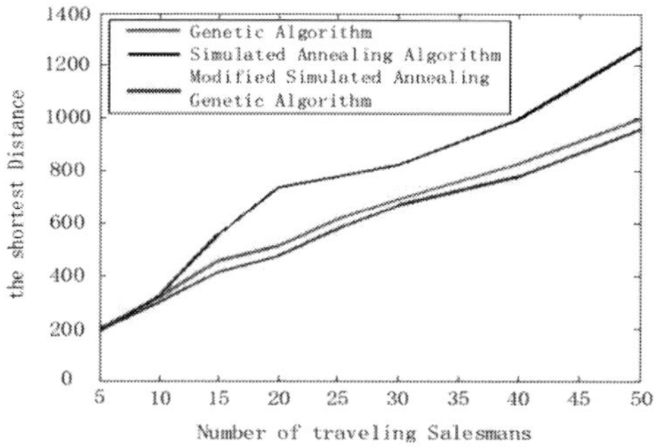

Fig. 11. The comparison of the simulation results of the three algorithms.

4 Conclusion

At present, there are numerous researches on MTSP. The situation of MTSP is complex, the solving of MTSP is difficult, and the research achievements of MTSP are inadequate [20].This thesis adopts a modified Simulated Annealing Genetic Algorithm to solve MTSP. The algorithm has following features:

(1) A suffix structure design of chromosome was put forward to reduce the space of the solution and eliminate most redundant solution;

(2) A self-adaptive genetic operator is devised for double crossover and mutation which can both remain population diversity and algorithm convergence;

(3) Simulated Annealing Algorithm and Hierarchical Genetic Algorithm are combined, and stretch the fitness, enhance the local search ability so as to achieve a better global optimum solution.

Acknowledgements

Finally, I would like to extend my sincere gratitude to my supervisor, Sheng Li, for his instructive advice and useful suggestions on my work. I am deeply grateful of his help in the completion of this work.

And, this work was supported by the National Natural Science Foundation of China (61673219), the Jiangsu Six Talents Peaks Project of Province (XNYQC-CXTD-001) and the Tianjin Major Projects of Science and Technology (15ZXZNGX00250).

References

1. Shim, V. A.,Tan, K. C.,Cheong, C. Y. et al.A Hybrid Estimation of Distribution Algorithm with Decomposition for Solving the Multiobjective Multiple Traveling Salesman Problem.IEEE transactions on systems, man and cybernetics, Part C. Applications and reviews: A publication of the IEEE Systems, Man, and Cybernetics Society,2012,42(5):682-691.
2. Daibo Liu,Mengshu Hou,Hong Qu et al.A simple model for the multiple traveling salesmen problem with single depot and multiple sink.COMPEL: The international journal for computation and mathematics in electrical and electronic engineering,2013,32(2):556-574.
3. Marco Casazza,Alberto Ceselli,Marc Nunkesser et al.Efficient algorithms for the double traveling salesman problem with multiple stacks.Computers & operations research,2012,39(5):1044-1053.
4. Cheang, B.,Gao, X.,Lim, A. et al.Multiple pickup and delivery traveling salesman problem with last-in-first-out loading and distance constraints.European Journal of Operational Research,2012,223(1):60-75.
5. Marco Casazza,Alberto Ceselli,Marc Nunkesser et al.Efficient algorithms for the double traveling salesman problem with multiple stacks.Computers & operations research,2012,39(5):1044-1053.
6. Carrabs, F.,Cerulli, R.,Speranza, M.G. et al.A branch-and-bound algorithm for the double travelling salesman problem with two stacks.Networks: An International Journal,2013,61(1):58-75.
7. Estevez-Fernandez A,Borm P,Hamers H et al.On the core of multiple longest traveling salesman games.European Journal of Operational Research,2006,174(3):1816-1827.
8. Elad Kivelevitch,Kelly Cohen,Manish Kumar et al.A Market-based Solution to the Multiple Traveling Salesmen Problem.Journal of Intelligent & Robotic Systems: Theory & Application,2013,72(1):21-40.
9. Angel Felipe,M. Teresa Ortuno,Gregorio Tirado et al.The double traveling salesman problem with multiple stacks: A variable neighborhood search approach.Computers & operations research,2009,36(11):2983-2993.
10. Qu H, Yi Z, Tang H J. A columnar competitive model for solving multi-traveling salesman problem . Chaos, Solitons and Fractals. Chaos, solitions & Fractals, 2007,31(4):1009-1019.
11. Laporte G, Nobert Y. A cutting planes algorithm for the m-salesmen problem . Journal of the Operational Research Society, 1980, 31:1017-1023.
12. Russell R A. An effective heuristic for the m-tour traveling salesman problem with some side conditions . Operations Research, 1977, 25(3):517-524.
13. Yonggang Peng,Xiaoping Luo,Wei Wei et al.A New Fuzzy Adaptive Simulated Annealing Genetic Algorithm and Its Convergence Analysis and Convergence Rate Estimation.International Journal of Control, Automation, and Systems,2014,12(3):670-679.
14. Hoseini, P.,Shayesteh, M.G..Efficient contrast enhancement of images using hybrid ant colony optimisation, genetic algorithm, and simulated annealing.Digital Signal Processing,2013,23(3):879-893.
15. Min Dai,Dunbing Tang,Adriana Giret et al.Energy-efficient scheduling for a flexible flow shop using an improved genetic-simulated annealing algorithm.Robotics and Computer Integrated Manufacturing: An International Journal of Manufacturing and Product and Process Development,2013,29(5):418-429.
16. Ying Xu,Rong Qu,Renfa Li et al.A simulated annealing based genetic local search algorithm for multi-objective multicast routing problems.Annals of operations research,2013,206(Jul.):527-555.
17. Symone Soares,Carlos Henggeler Antunes,Rui Araujo et al.Comparison of a genetic algorithm and simulated annealing for automatic neural network ensemble development.Neurocomputing,2013,121(Dec.9):498-511.
18. Xiaorong Xie*, Qirong Jiang, Yingduo.Damping multimodal subsynchronous resonance using a static var compensator controller optimized by genetic algorithm and simulated annealing.European transactions on electrical power,2012,22(8):1191-1204.
19. A.F. Crossland,D. Jones,N.S. Wade et al.Planning the location and rating of distributed energy storage in LV networks using a genetic algorithm with simulated annealing.International journal of electrical power and energy systems,2014,59(Jul.):103-110.
20. Meei-Yuh Ku,Michael H. Hu,Ming-Jaan Wang et al.Simulated annealing based parallel genetic algorithm for facility layout problem.International journal of production research,2011,49(6/8):1801-1812.

Logistics Distribution Center Location Optimizatio Model An Example Study

Lingling Li [1,*] and Jiapeng Zhang [1]

[1]Management Engineering Department, Zhengzhou University, Zhengzhou, China
Corresponding Email: sijiliuzhuan@163.com

Abstract. This paper analyzes and investigates the situation of the today's site of distribution centers .Then taking SF Express for an example ,In the light of minimize the total expense ,the location models for single distribution centers is established with the corresponding resolutions given .After that ,The location of the single distribution center of SF Express ,the starting point of which is Zhengzhou Railway Station ,can be determined with the help of distance-testing function of 360 map to conclude its coordinate and then determine its exact service spot .This paper will test the feasibility of above models by the chosen cases with the application of LINGO software.

1 Introduction

1.1 Research background

In recent years, with the rapid development of China's economy, the computer has become one of the necessary household appliances of every family.

Online shopping has become an indispensable part of people's life. "Internet plus" tide rise rapidly and various industries promote the transformation of enterprises .E-commerce has become an important strategic direction for the future development of enterprises. Distribution center as the most important infrastructure in the process of transport of goods,in the enterprise logistics network planning, the correct location decision is of great importance to reduce the logistics cost.

Distribution center work as a transit point for product transport in the entire logistics network system. It is of great significance to reduce the operating costs of logistics enterprises and improve the service level and efficiency of logistics enterprises. It also has important influence on obtaining enterprise competitive advantage, increasing enterprise efficiency and improving logistics distribution efficiency and customer satisfaction. Scholars at home and abroad for the research of logistics and distribution center are mostly theoretical research. Based on the theoretical research, this paper takes SF Express Company as an example to establish an example model for the cost data collection in a certain distribution area, obtaining the appropriate distribution center location address and applying the theory to the practical application.

1.2 Domestic and international logistics distribution center location study

In 1998, Barahona F,etc. established the warehouse location model by using mixed integer programming method which chose the sum of fixed costs, transportation costs, storage costs to be the evaluation criterias [1].

In 1999, Holmberg K studied the location problem on the condition of non-linear transportation cost, and solved the problem by branch-and-bound method [2]; In 2003, Zuo-Jun Max Shen studied the location of distribution center under the condition of joint inventory management, which is a relatively new management model. This model mainly researched the product transport system which the distribution of a supply point deliver a variety of products to multiple customers [3].In 2012, Zhu Rongrong, Hu David considering the characteristics of the cold chain logistics transport items easily degenerate , taking customer satisfaction of logistics services and total logistics costs as the evaluation index, established the cold chain logistics distribution center location model of multi-objective and then solved the research problem with the LINGO software, verifying the availability of this model in location cold chain logistics distribution center [4] ;In 2013, Hu Ziyu, using analytic hierarchy process and Delphi method, studied how to establish a scientific and rational logistics distribution center location model which was based on the theory of economies of scale [5].

This paper establishes a single distribution center location model with only one distribution center in a specific area, and finally taking SF Express Company as an example, the location of single distribution center is analyzed. Then we input the collected data into LINGO software to solve the model quickly. It proves the usability of the location model in practice, and has certain guiding significance for the construction of distribution centers in county and remote areas of China.

2 The establishment of a single distribution center location model

Single distribution center refers to a specific area where only build a distribution center, without regard to the distribution point from the supply point to the distribution center and other issues.

The model is relatively simple. We use the quantitative analysis method to get the optimal distribution center location which take the total cost of the minimum as the principle. First of all, in order to establish the model and make it easier to solve, we make the following assumptions:

(1)The fixed costs of construction and operation of distribution centers are known.

(2)Product distribution costs and distribution quantity is proportional and the cost of the unit product distribution is known.

(3)Distribution center unit product processing costs are known.

(4)The distribution cost is only related to the straight-line distance between the distribution center and the customer

(5)The distribution quantity and location of each customer is known.

(6)Only one product can be delivered and shipped once.

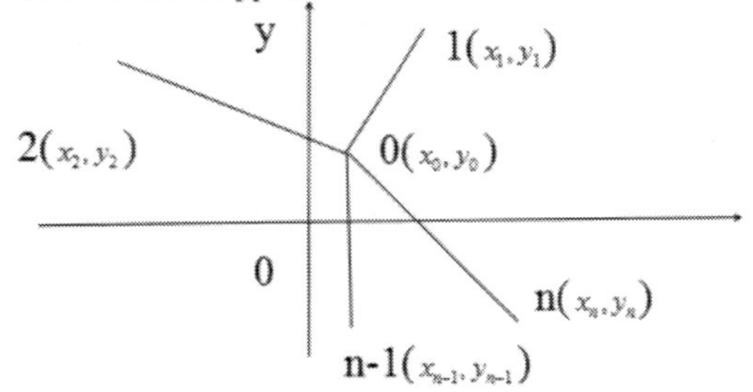

Fig.1. Location model diagram.

A distribution center distributes the goods to multiple distribution points.Our goal is to find the optimal distribution center location where the total cost can be minimized, with known the location of the distribution points.As shown in Figure 1, there is a distribution point with coordinates (x_i, y_i) (i =1,2,3...n)

Assume that the decision variables involved in the model and related parameters are as follows:

The relevant parameters are:

w represents distribution center construction, management and other fixed costs; k represents the total distribution quantity of the distribution center (that is the flow of the distribution center), k_i represents distribution quantity from the distribution center to the i distribution point; h_i represents from the distribution center to the first distribution point of the unit product unit cost of delivery; s represents distribution Center Processing Cost Coefficient; θ represents the economic benefit index of distribution center,can be taken as 1/2 ;

The relevant variables : d_i represents the straight-line distance between the distribution center and the distribution point i .

The total cost from the distribution center to the distribution point is:

$$TC = \sum_{i=1}^{n} h_i * k_i * \sqrt{(x_0 - x_i)^2 + (y_0 - y_i)^2} + s * k^{\theta} + w \tag{2-1}$$

$$d_i = \sqrt{(x_0 - x_i)^2 + (y_0 - y_i)^2} \tag{2-2}$$

$$\text{make : } \frac{\partial TC}{\partial_{x_0}} = \sum_{i=1}^{n} \frac{h_i k_i (x_0 - x_i)}{d_i} = 0 \tag{2-3}$$

$$\frac{\partial TC}{\partial_{y_0}} = \sum_{i=1}^{n} \frac{h_i k_i (y_0 - x_i)}{d_i} = 0 \tag{2-4}$$

Solution:

$$x_0^* = \frac{\sum_{i=1}^{n} \frac{h_i k_i x_i}{d_i}}{\sum_{i=1}^{n} \frac{h_i k_i}{d_i}} \tag{2-5}$$

$$y_0^* = \frac{\sum_{i=1}^{n} \frac{h_i k_i y_i}{d_i}}{\sum_{i=1}^{n} \frac{h_i k_i}{d_i}} \tag{2-6}$$

Due to the equations (3-5), (3-6) ,the right side of the equation d_i Still contain unknowns x_0, y_0 .So a calculation can not be calculated x_0^*, y_0^* .Generally,given a hypothetical value, and then using the iterative method calculates many times until the minimum total cost (x_0, y_0) is found .For the application of gravity method to solve the complex problems, we can input data into the LINGO software to get the answer which greatly reduce the cumbersome degree of calculation.

3 Example solving - taking SF express location study as an example

3.1 Data acquisition

Using 360 map searches SF Express near the Zhengzhou train station .

From left to right, we using A,B,C,D,E,F represent the delivery points in turn. Taking Zhengzhou Railway Station as the origin, using the 360 map ranging function, Respectively the coordinate of each point are A(-2.3,0.9) B(-2.1,-0.3) C(-1,-0.8) D(-0.2,1.1) E(1.3,-0.5) F(2,0) . The coordinate unit is km.We mark the points on the coordinates, As shown in Figure2.

By consulting a staff member of Zhengzhou SF Express distribution center, we learned that the fixed cost of a distribution center for construction and operation is about $w = 150000$ RMB. The distribution cost per km is approximately $h_i k_i = 2.5$ RMB.Because most of the courier vehicles are self-sufficiency, here we do not consider the distribution center to buy vehicle costs.We have to determine the best location through the coordinates of the distribution center, so we assume that these costs are known and fixed.

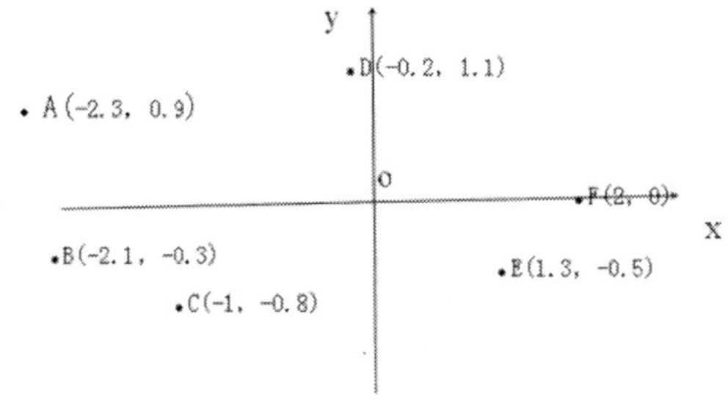

Fig.2. Example coordinate diagram.

3.2 Based on the study of a single distribution to solve the problem

Because the model involves so many (x_i, y_i) , the center of gravity method to solve this problem is more cumbersome.

Therefore, we use LINGO software to solve.LINGO software is from the United States. a company called Lingo System Inc. developed the system whose full name is "Linear Interactive and General Optimizer".It can not only be used to solve nonlinear programming problems quickly, but also can solve some linear and nonlinear equations.the function is very powerful and is used to solving optimization problems.

We do some adjustments for the single delivery center location model .The total monthly cost from the distribution center to the distribution point is:

$$\min = \sum_{i=1}^{n} 30 * h_i * k_i * \sqrt{(x_0 - x_i)^2 + (y_0 - y_i)^2} + s * k^{\theta} + w \tag{3-1}$$

$$\text{S.T: } d_i = \sqrt{(x_0 - x_i)^2 + (y_0 - y_i)^2} \quad (3\text{-}2)$$

$$s * k^{\theta} = 15000 \quad (3\text{-}3)$$

$$h_i k_i = 2.5 \quad (3\text{-}4)$$

$$w = 150000 \quad (3\text{-}5)$$

$$x_1 = \text{-}2.3 \quad y_1 = 0.9 \quad (3\text{-}6)$$

$$x_2 = -2.1 \quad y_2 = -0.3 \quad (3\text{-}7)$$

$$x_3 = \text{-}1 \quad y_3 = \text{-}0.8 \quad (3\text{-}8)$$

$$x_4 = \text{-}0.2 \quad y_4 = 1.1 \quad (3\text{-}9)$$

$$x_5 = 1.3 \quad y_5 = \text{-}0.5 \quad (3\text{-}10)$$

$$x_6 = 2 \quad y_6 = 0 \quad (3\text{-}11)$$

x_i Is an arbitrary value, and y_i is an arbitrary value.
The above data input LINGO software, the results are as follows:

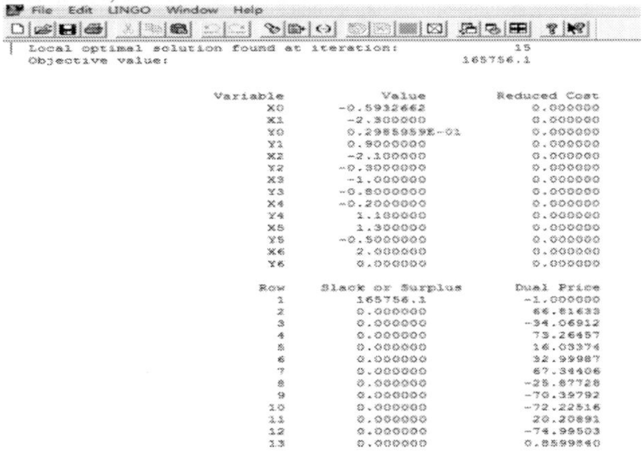

Variable	Value	Reduced Cost
X0	-0.5932662	0.000000
X1	-2.300000	0.000000
Y0	0.2985959E-01	0.000000
Y1	0.9000000	0.000000
X2	-2.100000	0.000000
Y2	-0.3000000	0.000000
X3	-1.000000	0.000000
Y3	-0.8000000	0.000000
X4	-0.2000000	0.000000
Y4	1.100000	0.000000
X5	1.300000	0.000000
Y5	-0.5000000	0.000000
X6	2.000000	0.000000
Y6	0.000000	0.000000

Row	Slack or Surplus	Dual Price
1	165756.1	-1.000000
2	0.000000	66.81633
3	0.000000	-34.06912
4	0.000000	73.26457
5	0.000000	16.03374
6	0.000000	32.99987
7	0.000000	67.34406
8	0.000000	-25.87728
9	0.000000	-70.39792
10	0.000000	-72.22516
11	0.000000	20.20891
12	0.000000	-74.99503
13	0.000000	0.8599840

Fig.3. Result output diagram.

$x_0 = \text{-}0.5932662 \approx \text{-}0.59$, $y_0 = 0.2985959\text{E-}01 \approx 0.03$ .Distribution center location as shown in Figure3, is shown in the middle of Zhengzhou Railway Station about 600 meters .If west.the price is right, you can rent the street facade room location used as SF Express distribution center.

4 Conclusion

This paper introduces the location of single distribution center in detail and established single distribution center location model,given the corresponding solution method.

With the help of 360 map,With the help of 360 map, we locate SF Express points near Zhengzhou Railway Station and construct a single distribution center to get the optimal location of distribution center.The innovation of this paper is combined the problem of distribution center location with the concrete examples of SF Express and combine the theory with reality further ,strengthening the applicability of the model.It is also of great significance to the establishment of distribution centers in county and remote areas of China.

References

1. Barahona F.,Jensen D.Plant Location with Minimum Inventory.Mathematical Programming,1998,83(1):101-111.
2. Holmberg K.Exact Solution Methods for Uncapacitated Location Problem with Convex Transportation Costs.European Journal of Operational Research,1999,114(1):127 -140.
3. Zuo-jun Max shen,Collette Coullard,Mark S.Daskin.Ajoint Location-inventory model.Transportation Science,2003,37(1):40-55.

4. Zhu Rongrong, Hu Dawei.Multi-objective optimization model of cold chain logistics distribution center selection. Logistics Technology, 2012,32 (01): 108-110.
5. .Hu Ziyu.Study on the Location of Logistics Distribution Center under Scale Economy .Commercial Times, 2013,32 (31): 29-30.

A Study on the Training of Logistics Management Talents in Colleges in the Big Data Age --Taking Luoyang area as an Example

Sijia Li[1,*] and Wenfeng Zhao[1]

[1]Management Engineering Department, Zhengzhou University, Zhengzhou, China
Corresponding Email: 598209309@qq.com

Abstract. Under the background of rapid development in era of Big Data, the logistics industry has become an important service industry to promote economic globalization. This paper explores how colleges and universities to cultivate talents of logistics management in order to better adapted the needs of society.This paper analyzes and summarizes the investigation based on reading relevant literature and using descriptive analysis after investigation. So as to provide some suggestions to the colleges which set up related professional courses of logistics management and the graduates who want to engage in the logistics industry.
Keywords. Big Data, talents of logistics management, training mode.

1 Introduction

1.1 Research background

At this stage of the world is in the era of information explosion.For such complex information, how to better integrate this information is the top priority to win advantage of the era of Big Data, and the logistics industry is one of the representative industries.The logistics industry has become a key to promote China's economic in the context of economic globalization. Therefore, the logistics industry plays a very important role in the development of China's Big Data age.

Data shows that Chinese modern logistics industry has shown a great upward trend, it has a strong market potential and optimistic about the development prospects of the industry. However, there is still a huge supply gap in the supply of professional logistics talent, which has become a short board to limited the development of Chinese logistics industry. Logistics professionals have been listed as one of 12 kinds of scarce talent in China, especially the demand for senior logistics management personnel a serious shortage [1]. However, due to logistics related professional teaching of our country started late, but also the lack of co-ordination arrangements and policies to support the guidance. So,on the whole has not yet entered the track of standardized development. At the same time, most of the community's awareness of the logistics industry still remain in the transport, storage, distribution and other individual aspects, and not have systematic understanding. In now advanced logistics technology of world is not limited to individual aspect, but the unitary plan of the whole process which from the manufacture of products to downstream sales, and which will not only involve the overall operation of the enterprise process but also the number of enterprises coordination and cooperation that related to upstream and downstream. Therefore, the study of logistics personnel's demand situation has become increasingly urgent.

1.2 Overview of the study

Chinese logistics industry has great potential and broad prospects, so the domestic scholars began to study after the start of logistics industry developed in China. Pan Wenjun in 2005 pointed out that our country in the logistics management personnel demand analysis on foreign logistics education and logistics personnel training education published research.This aspect of education is not mature, need to be more advanced with the times while strengthening the professional setting and the creation of this professional schools and their respective disciplines combined strengths [2].Wenhua, Wang Ying in 2007 based on the Shenzhen logistics enterprise study the logistics industry demand situation, they believe that with the development of China's logistics industry, logistics enterprises tend to compound-type, export-oriented and application-oriented talents,while a solid foreign language ability is equally important [3].Gan Hongyun aiming at the contradiction between the training of logistics talents and the demand of enterprises in our country explored the way to solve the problem of how to connect the talent of logistics industry with the demand of enterprises [4].Qu Qunzhen,Ma Jie,Tang Mengxue forecast of logistics talents demand in Shanghai from 2010 to 2015 by gray system, time series and combination forecasting method [5].Chen Xianyou, Chang Han and other scholars through the field survey, telephone interviews and e-mail and other survey methods of logistics enterprises in Wuhan conducted a demand for talent and logistics management professionals for the development of the recommendations [6].

Lim yu Pui, Dazmin Daud and other scholars using the method of hierarchical regression analysis the needs of logistics practice in the context of the Malaysian colleges and universities by the background of knowledge, skills and talent [7]. In his 2007 survey of CEOs of the top 10 third-party logistics companies,Robert Lieb gained insight into the current state of affairs and explored the future of third-party logistics in the Asia-Pacific region [8]. Matthias Sohn, Werner Sohn and other scholars to explore attracted situation of the logistics talent by logistics company's corporate social performance [9].

China is in the initial stage of logistics management personnel training.The first is that different enterprises have different understanding of logistics management talent.Followed by most of our enterprises are small and medium enterprises, logistics management does not attach importance and those enterprises think logistics management does not need higher education and professional accomplishment. Moreover, the training of logistics management professionals in colleges involves only the cultivation of theoretical knowledge, but the lack of practical training. The final college graduates on their own positioning is not accurate and do not attach importance to grass-roots part of the exercise.They want to become a senior management when they get the job. These have become the logistics professionals' important reasons for the contradiction between supply and demand in China today.

In this paper, based on the situation of Luoyang add up the demand for logistics professionals and the use of gray analysis to predict the next five years of logistics talent demand. In order to promote the establishment of logistics courses in colleges to ease China's logistics professionals supply and demand.

2 Research design

2.1 Sample design

This survey was conducted in Luoyang by sending a survey questionnaire online, and 70 valid questionnaires were retrieved. The contents of the survey include: the current situation of social demand for various logistics management personnel,logistics management personnel in the enterprise more engaged in the work of enterprises and the community that logistics professionals should have what professional quality, business logistics personnel basic requirements, On the logistics professionals should strengthen what aspects of education and other five elements.

2.2Descriptive statistical analysis

2.2.1 Current social demand for various logistics management personnel status

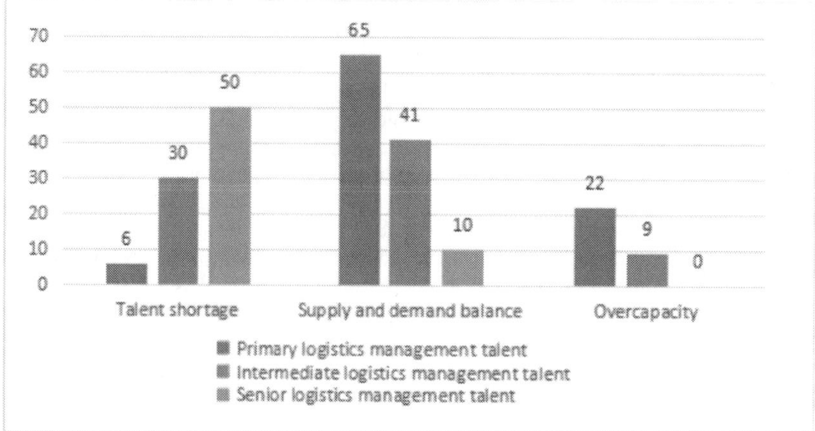

Fig. 1. The demand for logistics management personnel at all levels.

From Figure 1 can be seen that the basic logistics management personnel in the balance of supply and demand, even in some areas in the situation of excess; intermediate logistics management personnel in the three in the middle position, the basic situation is in short supply; and advanced logistics management personnel in the country is in a scarce situation which is serious shortage of talent.Although the major companies hired with high salary, but few candidates.So senior logistics management talent is still the most lacking talent at this stage.

2.2.2 logistics management personnel in the enterprise more engaged in the work

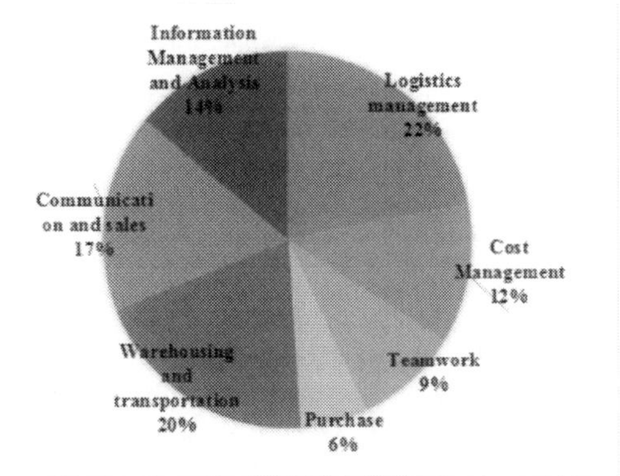

Fig. 2. Logistics management personnel engaged in more work.

Based on the survey of logistics management talented people in Luoyang, it can be seen from Figure 2 that 22% of the respondents believe that logistics management personnel in the work of more professional work; 20% of the 9% of the respondents believe that logistics management is the important factor in logistics management, and logistics management personnel in warehousing, transportation, distribution positions,and followed by communication between customers and sales, information management and analysis, and cost management.9% of the respondents believe that talent at work is generally a member of the team to join a project team responsible for logistics work; and fewer people think that logistics management personnel in the procurement position.

2.2.3 Enterprises and society that logistics management personnel should have what professional quality

Table1. Logistics management personnel should have the professional quality.

Professional knowledge and skills	total	proportion
Logistics management basics	56	80%
Logistics planning, operational management knowledge	42	60%
Development and Design of Logistics Information System	39	55.71%
Logistics management software operation	39	55.71%
Finance, accounting, cost management	39	55.71%
Management knowledge and skills	38	54.29%
Warehousing, inventory, distribution of knowledge and skills	37	52.86%
Supply chain management knowledge and skills	27	38.57%
Trade,marketing,business knowledge	26	37.14%
Foreign language ability	14	20%

As can be seen from Table 1, the proportion of the basic knowledge of logistics management with the highest proportion of 80%; logistics planning, operations management knowledge accounted for 60% of the proportion.As the rapid development of China's information technology, computer technology and widespread Application, logistics information system development and design and logistics management software operation from the beginning of the 21st century only in large companies and small and medium enterprises are not universal application of the state has become the country's large and small enterprises are almost all in the use.Therefore, the logistics information system development,design and logistics management software operation of the professional requirements in the third and fourth with the proportion was 55.7%. Sixth place is the financial, accounting, cost management, and the proportion is 55.7%.China's enterprises are aware of the logistics management in cost savings and enhance the profits of the huge role, so the logistics management personnel professional quality requirements are also stressed that the use of financial knowledge, or have financial aspects of the work experience.The seventh is management knowledge and skills, accounting for 54.29%.The eighth is the storage, inventory, distribution of knowledge and skills, accounting for 52.86%. Ninth is the supply chain management knowledge and skills, accounting for 38.57% Chain management is also the development trend of China's logistics management at present. Tenth is the trade, marketing, business knowledge, the proportion of 37.14%.While in the final is the foreign language ability, only 20%, which may be due to Luoyang City is located in the Central Plains region, but also is not the first-tier cities in China, so people do not know much about the ability of foreign languages. In Beijing, Shanghai, Guangzhou and other developed first-tier cities, logistics management talents should have the high foreign language level, and generally require a skilled foreign language can be used to read and write.

2.2.4 The basic requirements of logistics business personnel

Table2.enterprises of the basic requirements of logistics personnel.

Comprehensive quality	total	proportion
Good communication and communication skills	60	85.71%
Love and dedication, loyalty	58	82.86%
Hard-working	56	80%
Team work	56	80%
Pioneering and innovative ability	54	77.14%
Project planning, planning capabilities	51	72.86%
Self - learning ability	35	50%
Social practice ability	27	38.57%
Good physical fitness	22	31.43%
Excellent foreign language skills	11	15.71%

The basic requirements of logistics management personnel that is its overall quality requirements. From Table 2 can be seen, good communication and communication skills is the most important logistics management personnel comprehensive quality, the proportion of 85.71%. Good communication ability is the basic ability of every manager necessary. Ranked second in love and respect their jobs, loyalty to the enterprise, accounting for 82.86%,and showing that all companies are employees of the enterprise loyalty is very important.Third is hard-working and hard-working spirit and teamwork, both accounted for 80%, which can be seen in the work of individual hard work and teamwork with others ability to cooperate are very valued enterprises.The fourth is the pioneering and innovative ability, accounting for 77.14%. In today's society, innovation is the most important capacity.It is also important to plan, analyze and understand the tasks assigned by the higher authorities,accounting for 72.86%.The sixth is for the professional knowledge of self-learning ability, the proportion is 50%.Constantly in the work of the professional knowledge to update and learn is also a very important comprehensive quality management. Seventh is the social practice Capacity, accounting for 38.57%, and a manager of the social practice ability is very important, and managers are generally a few years of grass-roots work experience of personnel, so as to be familiar with their own need to manage the project or task exactly what to do,and clear which the experience in this work. Ranked in the latter two were good physical fitness and excellent foreign language level, respectively in 31.43% and 15.71%.

2.2.5 Colleges of logistics professionals should strengthen what aspects of education

Table 3. What kind of education should be strengthened in colleges.

Educational direction	total	proportion
Logistics management expertise	57	81.43%
Practical ability training	49	70%

Computer knowledge and ability	37	52.86%
Economic management knowledge	36	51.43%
Foreign language ability	11	15.71%

Colleges of logistics management personnel should strengthen what aspects of education survey, the results shown in Table 5. Ranked first is the logistics management expertise, accounting for 81.43%, because now China's logistics professional education theory is not keep pace with the pace of international development, the theory of some outdated and the international advanced logistics management expertise in the introduction and learning is imperative. The second is the practical ability of training, accounting for 70%. China's current emphasis on college education theory teaching, and practice with the actual derailment, resulting in employment difficulties of graduates, enterprises to recruit the right talent is also difficult to embarrassing situation.The third is the computer knowledge and ability, the proportion of 52.86%, the extensive use of computers makes computer knowledge has become an irreplaceable work in today's essential factor. The fourth is the knowledge of economic management, accounting for 51.43%. A certain degree of economic management knowledge and its application to the theory of logistics management in the current stage of economic and trade globalization can make enterprises more competitive. Fifth is the ability of foreign language 15.71%.In Luoyang, the more information shows that the foreign language ability does not value the situation.However this is not conducive to the development of logistics industry in Luoyang.

3 Empirical analysis

As the number of employees in the logistics management industry is scattered and difficult to collect, this paper intends to study the demand of logistics management talents in China in the next five years by calculating the proportion of employees in the logistics management industry in the tertiary industry. Use the data for the next few years to carry out a certain forecast and analysis for the demand for logistics personnel.

Since the number of employees in the tertiary industry is not a quantitative information,and it can not use the quantitative analysis method. So it can use the gray forecast analysis method to forecast it. This paper makes use of the number of employees in the tertiary industry from 2002 to 2014 in the China Statistical Yearbook [10].

Table 4. Number of Employed Persons in the Tertiary Industry in 1997-2014 (Unit: Ten Thousand)

year	1997	1998	1999	2000	2001	2002
people	18432	18860	19205	19823	20165	20958
year	2003	2004	2005	2006	2007	2008
people	21605	22725	23439	24143	24404	25087
year	2009	2010	2011	2012	2013	2014
people	25857	26332	27282	27690	29636	31364

3.1 Construction GM (1,1) model

From $\rho(\mathrm{k})=\frac{x^{(0)}(k)}{x^{(1)}(k-1)}$ we can see,when k>3, $\rho(k)<0.4$,So quasi-smooth conditions are met.

From $\sigma^{(1)}(k)=\frac{x^{(1)}(k)}{x^{(1)}(k-1)}$:when k>3, $\sigma^{(1)}(k)\in(1,1.4)$, $\delta=1.4-1=0.4$, $\delta\le 0.5$,So to meet the quasi-exponential law.

The quasi-exponential rule is obtained and the GM (1,1) model is established by using SPSS:

$$\frac{dx^{(1)}}{dt}-0.044x^{(1)}=96.504$$

3.2 GM (1,1) model test

Correlation test:

$$\gamma = \frac{1}{n}\sum_{k=1}^{n}\eta(k)$$

$$\eta(k) = \frac{\min|\varepsilon(k)| + \rho\max|\varepsilon(k)|}{|\varepsilon(k)| + \rho\max|\varepsilon(k)|}(k = 1,2,3,....,n)$$

From $\rho = 0.5$, $\gamma = 0.71 > 0.6$,meet the correlation test.

Relative error test $\Delta(k) = \frac{|\varepsilon(k)|}{x^{(0)}(k-1)} \times 100\%$,The relevant data into the test $\Delta(k)$ are less than 10%.

Posterior difference test: the standard deviation of the original sequence: $S_1 = \sqrt{\frac{\sum_{k=1}^{n}\left(x^{(0)}(k) - \overline{X}\right)^2}{n}}$

Standard deviation of residual series: $S_2 = \sqrt{\frac{\sum_{k=1}^{n}\left(\varepsilon(k) - \overline{\varepsilon}\right)^2}{n}}$

According to C, P value to determine the model accuracy, $C = \frac{S_1}{S_2}$ The relevant data into:

$S_1 = 133.31$, $S_2 = 14.38$,so C=0.13， P=0.99.So the accuracy of the above model is reliable.

According to the data published in the China Statistical Yearbook from 2002 to 2014, it can be seen that the employment of the logistics industry accounts for 2.75% of the total number of employees in the tertiary industry, and the demand is used to calculate the demand,shown in Table 5.

Table 5. Forecast value (Unit: ten thousand).

year / forecast	2015	2016	2017	2018	2019
Gray model prediction	32732	34228	35165	36697	38113
Logistics talent demand	900.13	941.27	967.04	1009.17	1048.11

3.3 Analysis of prediction results

As can be seen from Table 3-2, the demand for logistics talents in the next five years is rising. As the status of the logistics industry in China's economic activities continues to increase [11], the role of the logistics industry in the future is growing. The situation will continue to improve. Although the level of China's logistics industry is uneven and the hardware facilities are not perfect, the logistics industry in China is also developing rapidly. The market demand of logistics management personnel is increasing year by year. The popularity of electronic information technology, supply chain management wide dissemination and the country's logistics personnel to cultivate will accelerate China's logistics industry.

4 Conclusion and significance

Through the investigation and research, this paper from the five problems to summarizes the commonness and analyzes the reasons,and to finds out that the logistics-related enterprises pay more and more attention to the comprehensive quality of employees in the background of large data. Because Luoyang in the central plains, the degree of openness than the coastal and first-tier cities, so the level of foreign language requirements are not prominent. But in the context of China's great times, foreign language applications more widely, and has become an important indicator of comprehensive strength. Based on this strengthening foreign language skills for the mainland students to coastal cities and first-tier cities, and even overseas work is particularly important.

In background of Big Data, the senior logistics management talent is most scarce of the market demand. But because of the job candidates personal capacity, work experience, the overall planning and integration capacity is high, so the candidates are few. Due to the huge data information needs to be dealt with, the demand for logistics management talents in enterprises at this stage is more and more inclined to compound talents. Logistics management personnel in the complex talents such as proficient in the logistics management expertise at the same time familiar with economic and financial knowledge; proficient in logistics management expertise at the same time familiar with computer information technology knowledge; proficient in logistics management expertise at the same time familiar with legal knowledge and so on, these are the focus of the compound talents in the demand for talent in the modern enterprise increasingly sought-after.

In the logistics management professionals of the professional and comprehensive quality requirements can be seen that the higher the level of talent in their logistics expertise should be have the ability to request the higher, meanwhile on the computer level, information management, foreign language level and so on has a very high demand. Therefore, logistics students in order to adapt to the era of large data background in the four-year undergraduate study should also focus on strengthening their ability in these areas, in order to meet the needs of the community to improve their own quality.

In the logistics management personnel engaged in the investigation of the work point of view. Most are engaged in logistics management and related departments. At the context of China's large data logistics ,management talent in the future direction of development and development of view: more and more enterprises are willing to outsource their own logistics module, so professional third-party logistics business prospects are good, and these enterprises are also need more and more logistics management talents. With the rapid development of supply chain management in China, many large enterprises have begun to consider the transformation and upgrading for the establishment of a complete supply chain management system, and logistics management as a supply chain management is very important, so the future prospects are also optimistic, which also requires the future of logistics management personnel with knowledge of logistics management should also have knowledge of supply chain management.

This paper uses the gray forecast analysis method to establish the model, predicts the demand situation of the logistics talents in the next five years, and obtains that the demand situation of the logistics management talent is good in the next five years.The demand potential is also relatively large.

Due to the limitation of personal knowledge and time, there are still many shortcomings in the investigation and research on the demand of logistics management talents. The depth and breadth of the research are still to be expanded. Something need to in-depth study of the content: 1 more enterprises should be surveyed which will making the data more reliable. 2 it can be added such as time series forecasting methods to predict the number of the next few years, making the conclusion more credible and accurate.

References

1. Zou Min, Xie Ming, LI Ying. Discussion on the Development of Logistics Management Specialty and the Demand of Talents in Higher Vocational Colleges - The Investigation Report on the Status Quo of Hunan Logistics Enterprises and the Demand for Talents.Fourth Industrial Enterprise Logistics Forum Proceedings, 2007, 26 (8): 65-67.
2. Pan Wenjun.Gap_Status_Strategy - Foreign Logistics Education and Logistics Personnel Training in China.China Logistics & Purchasing, 2005 (14): 70-72.
3. Wen Hua, Wang Ying. From the needs of enterprises to see the logistics management professional bilingual teaching. Higher Education Exploration, 2007 (6): 152-155.
4. Gan Hongyun. Discussion on the Effective Docking between Logistics Talents Training and Enterprise Demand in China.Journal of Logistics Engineering and Management, 2009,31 (7): 139-140.
5. Qu Qunzhen, Ma Jie, Tang Mengxue. Shanghai 2010 ~ 2015 years of logistics talent demand forecast and countermeasures research modern management science, 2011 (10): 115-117.
6. Chen Xian-you, Chang Han, Li Jun, Wen Qiqi.The investigation and analysis of enterprises' demand for logistics management talents.Technology Start Monthly, 2013 (4): 45-47.
7. Lim Yu Pei, Dazmin Daud, Kholyn Ruran Jonathan. Perceptions of Logistics Students on Internship Programme: The Case of Private Higher Institution in Malaysia. Canadian Social Science,2012,8（4）:1-7.
8. Robert Lieb. The year 2007 survey: Provider CEO perspectives on the current status and future prospects of the third party logistics industry in the Asia-Pacific region. International Journal of Physical Distribution & Logistics Management,2008,38(6):220-224.
9. Matthias Sohn,Werner Sohn,Thorsten Klaas-Wissing,Bernhard Hirsch.The influence of corporate social performance on employer attractiveness in the transport and logistics industry. International Journal of Physical Distribution & Logistics Management,2015,45（5）:486-505.
10. China National Bureau of Statistics. China Statistical Yearbook .2006, 2007, 2008, 2009, 2010, 2011, 2012, 2013, 2014.
11. Li Guanyi.Electronic business logistics innovation and traditional logistics transformation under the Internet.Commercial Research,2016(4):187-192.

Intelligent Management System of Power Network Information Collection Under Big Data Storage

Yingying Qin[1,*], Hua Xuan[1] and Bailin Zhang[2]

[1] School of Management Engineering, Zhengzhou University, Zhengzhou 450001, China
[2] Inspur Electronic Information Industry Co. ,Ltd, Jinan 250101, China
Corresponding Email: 1484261055@qq.com

Abstract. With the development of economy and society, big data storage in enterprise management has become a problem that can't be ignored. How to manage and optimize the allocation of tasks better is an important factor in the sustainable development of an enterprise. Now the enterprise information intelligent management has become a hot spot of management mode and concept in the information age. It presents information to the business managers in a more efficient, lower cost, and global form. The system uses the SG-UAP development tools, which is based on Eclipse development environment, and suits for Windows operating system, with Oracle as database development platform, Tomcat network information service for application server. The system uses SOA service-oriented architecture, provides RESTful style service, and HTTP(S) as the communication protocol, and JSON as the data format. The system is divided into two parts, the front-end and the backs-end, achieved functions like user login, registration, password retrieving, enterprise internal personnel information management and internal data display and other functions.

1 Introduction

The demand of acquisition system is increasing, it is imperative to improve the data processing performance of the electric information acquisition system through leading technology. The amount of data generated by electricity is considerable, Therefore the data storage and mining analysis has become the most important one, including the selection of database and optimization of the SQL statement. These are precisely the direction of this research.

For the implementation plan of Henan electric power company "three years to enhance the quality of rural power", to support the carry out of Henan county company management, to deepen the application of measurement acquisition, and measurement acquisition related business city, county integration and other work, based on the principle of the establishment of the main station expansion and function upgrading of the electric information acquisition system issued by the state power grid, designed and realized the management system of Henan province electric information (including the County Department and the equipment, etc.). The upper controllers improved their work effectively, they can observe the power dynamics in real time, which is of practical significance.

Intelligent enterprise information management has become the main model in the information age, which is changing the way managers works. Firstly, the management of big data in enterprises has become a difficult problem in modern times, while using information management can find the required information more quickly, and improve work efficiency greatly. Secondly, which can make the management of enterprise data information organized, facilitate to operation the company. Thirdly, as diversified and personalized management is becoming more and more common in our daily fast pace of life, management software which is highly efficient and easy to operate is more popular.

Big data storage problem has been discussed by many scholars in related fields, and has achieved a lot of research results so far. For example, Literature [1] based on the analysis of cloud computing platform, a uniform hash storage algorithm was proposed for big data storage and fast access problem. Literature [2] introduced the big data management under the new storage technology. The information management system of power network was described in literature [3-5], among which, Literature [3] introduced smart grid information management system under the multi-agent technology. Literature [4] analyzed the complexity of smart grid information management system. Literature [5] studied the information share and control management under the smart grid information management system. Literature [6] introduced the features and functions of SQL Server database. Literature [7] introduced the system evaluation from traditional database to emergent big data. For a large number of enterprise data management issues, literature [8] studied the business intelligence data management. Literature [9] used J2EE, Oracle 9i design applications and upload to cloud under the enterprise ERP management system of the cloud computing platform. Based on the above research, this paper designed and developed power grid information management system to solve the effective management of big data, information diversity problems and so on.

The paper is organized as follows.The next section, section 2 introduces the technique used in this system. In section 3, we show the architecture design in detail. In section 4, we introduce the core modules, especially it's implementation.Finally,some conclusion are made in section 5.

2 Technical description

The development platform of this system is: operating system Microsoft Windows; programming language Java/Ajax; development tool UAP Studio 2.0.0; Database Oracle 10g; Java application server Tomcat 6.0, IDE Eclipse as the core of the enterprise software, to combine database, development programs and application servers quickly. The development owns a variety of languages, complete coding, debugging, testing and publishing, supports HTML, JavaScript, Sql, Spring, Hibernate, CSS, Ajax, JSON, jQuery technologies and so on. The platform uses two level technical service system, and the technical service includes technical consulting, training and examination, assistant design, design review, development direction, operation, maintenance support and so on. In the UAP, the Rest styles are interactive with the front and the back ends, Written in the URL address in the foreground form or the underlying code and the background to interact to realize system front desk and the background management individually.

The server framework as shown in Figure 1, the application system based on OSGI was embedded in the framework of application server. With the advantages of Web applications and advanced dynamic, modular characteristics, realized the high coupling between the application system modules. The platform was composed of a modular framework, which is the smallest unit that can run on the platform. Micro kernel inherited the thought and technology of SoTower platform, using Servlet bridge mode, the OSGI framework is embedded in the application container to run. The internal application of the platform and all the functions of the business were assigned to specific modules, which runs in the OSGI framework.

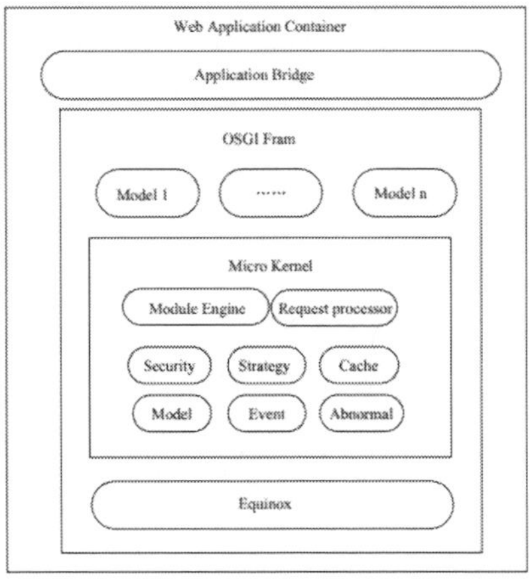

Fig.1. Server framework.

The design discarded the disadvantages of the previous logic module partition-cross store each function module file: business modules are easy to interfere with each other, reduce the efficiency of the development of business applications, which also has a certain influence on the stability of the product, SG-UAP platform micro kernel solved these problems.

The server was divided into three layers, control layer, logic layer and persistence layer. Control layer, also called interaction layer, the service interaction section is one of the most important parts of the core framework, which is responsible for receiving and processing requests from browser or system, providing users data transmission format, packet sending, URL handling of the distribution and other functions. Logic layer is generally considered as the service layer of SG-UAP project hierarchy, between with the interaction layer and the persistence layer, which generally dependent on the persistence layer not the interaction layer. The responsibility of the logic layer is to encapsulate the business logic. In SG-UAP, logic layer technology mainly refers to logic component, to a certain extent, simplified the use of the business development, web services, AOP and other functions. The use of the persistence layer is mainly reflected in the resource allocation and the call of the persistence method. By default, the platform configures a SessionFactory object and Hibernate file path in the file for the user automatically. Hibernate can perfectly be used on the platform, be good division of application levels, conduct transaction management and data access independently, improve the degree of polymerization between the modules greatly and reduce the coupling degree of the application objects.

ASP, AJAX, Oracle database, Annotation technology and Hibernate Technology were used in this system, AJAX technology integrated the development of CS, which can solve the problem more effectively. The most widely used database is Oracle and MySQL, Oracle suits large and medium-sized enterprises, while MySQL suits small and medium enterprises. As a large number of data requires the performance of the database very high, small database is difficult to meet the needs, so Oracle was chosen.

3 Architecture design

3.1 Functional analysis

The system platform uses SOA framework, SG-UAP software combined with Oracle database to achieve the general management operation of enterprise information, which contains the following function modules:

(1) User registration: the system needs to be registered for the first time, and needs to verify identity. The users are unique, the identity card number and the nickname of the registered person are unique.

(2) Retrieve password: Once the password error, users will not be able to enter the system, users needs to retrieve the password, and the nickname, ID number and other information is essential, must be consistent with the database data, otherwise the password will not be found.

(3) Validate logon: User login needs input name and password, considering the same name, so does not use the real name as the logon condition, and nickname as the login name, which requires the nickname must be unique. Click on the login button to verify the information.

(4) Department equipment information management: through a form, the department equipment information will be displayed to the manager dynamically. Managers query the associated information through the index, number and name. Different forms can be chosen to display the department equipment information, giving the users visual enjoyment.

(5) Staff information management: the information of staff is displayed on the left side of the page, through click the staff information node then display the corresponding information, which is convenient for the manager to manage.

(6) Enterprise basic information management: the enterprise information displays the related registration information preview to the users. Through this module, users can use histogram to view all of the registered capital of the enterprise and security authentication information simply.

(7) Qualification information management of state grid electric power enterprise: to record the information of enterprise registration certificate in the form of DataGrid.

(8) Security license information: to browse the safety license information of enterprises in the form of Grid Complex.

3.2 System structure diagram

The following figure 2 shows the design ideas of the state grid information acquisition system module, including the front-end and the back-end. The back-end module contains four parts: the persistent layer, the control layer, the logic layer and the UI layer. The first three sections are needed to interact with the database.

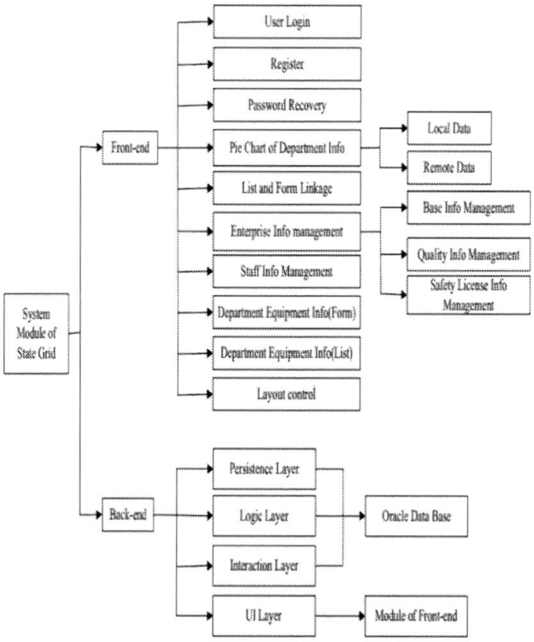

Fig.2. The diagram of system structure.

3.3 System use case diagram

The relationship between all modules of the system and the user is presented in the form of a use case diagram 3. Users can add, delete, edit and search the department information, equipment information, employee information and the State

Grid Electric Power Enterprise Information. The system provided corresponding interfaces for different users. If you do not have the authority to log in the system, you need to register the account. If you forget the password, you can apply for resetting the password. If you have an account and the authentication information is correct, you can re verify and login.

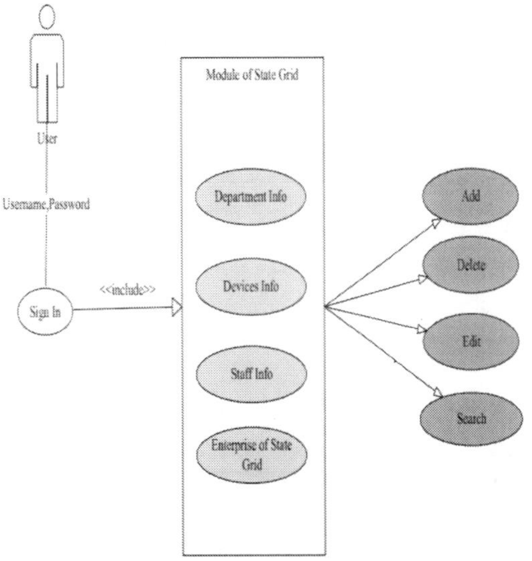

Fig.3. The diagram of system use case.

3.4 ER chart

The establishment of the primary key and the dependency relationship between the various data tables facilitates the index query, which can optimize the operation of the database. The relationship between tables is established by primary keys and dependencies, and the tables satisfy the criteria for implementing referential integrity, avoiding some incorrect operations, which can manage the data table effectively. The integral ER chart is shown as Figure 4.

4 Implementation of core modules

When users enter the homepage, they can manage enterprise information, for example, add / delete / edit / query / add attachments and other functions can be achieved, changing the previous operating location, the new, edit, delete and other operation can be integrated into each record, which can improve work efficiency.

Page layout as shown in Figure 5, HSplit as the main interface, dividing the page into two parts, displaying the basic information DataGrid in HSplit(0). In addition, creating a VSplit in HSplit(1), which divided the page into two parts vertically display, and VSplit(0) used to store chart information, setting the enterprise name and registered capital for the vertical and horizonta coordinates, the histogram shows the data information is stored in VSplit(1), the property can be set to display data in various forms.

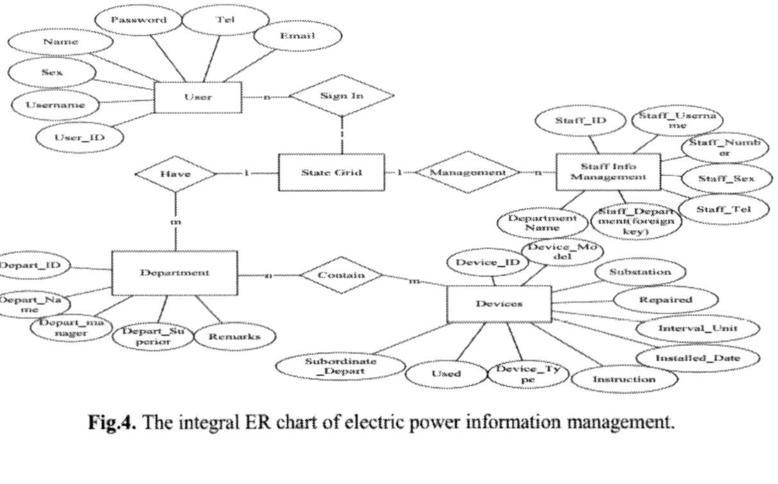

Fig.4. The integral ER chart of electric power information management.

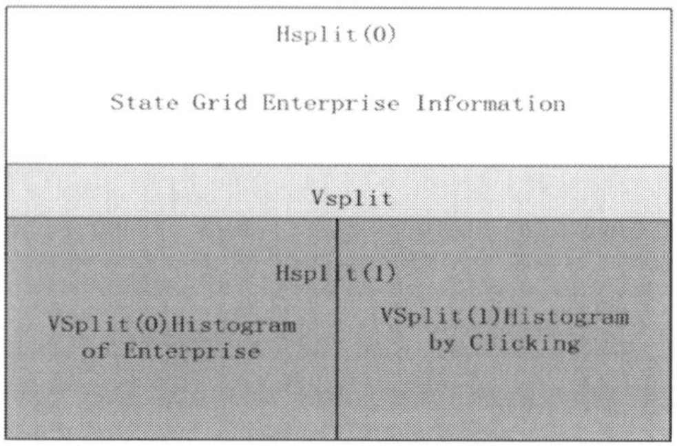

Fig. 5. The diagram of page layout.

5 Conclusion

In order to change the traditional enterprise management model, this paper provides a novel management plan, using software to manage the large enterprise data information, which can improve the efficiency of enterprise management greatly and save costs. For example, the existing power information acquisition system storage capacity, the data will not be updated in real time, so the operation time, reaction efficiency and access time should be improved. The use in the SQL statement is optimized to better save access to query time, do a great improvement of big data storage and access. This paper studied the development and implementation of this system, both in the visual and the operation gives users a new experience, which is simple and easy to operate. Further research may consider designing and managing more complex database, and web page dynamic display, more complex needs analysis and other factors.

References

1. SONG Y Q,ZHOU G L,ZHU Y L,LI L,WANG L W,WANG D W. Storage optimization and parallel processing of condition monitoring big data of transmission and transforming equipment based on cloud platform. Proceedings of the CSEE, ,35,255-267, (2015).
2. JIN P Q,HAO X J,YUE L H.A survey on storage architectures and core algorithms for big data management on new storages.Computer Engineering & Science,35,12-24,(2013).
3. LIU X M,TIAN H Y,QIN C. Research on smart grid information management system based on complex scientific management thinking and multi agent system technology. Power System Technology, 36,204-208,(2012).
4. LIU X M,JU J. Analysis of the complexity features for information management system of smart grid. Journal of North China Electric Power University (Social Sciences),2, 20-24, (2013).
5. CHENG S J, LI X Y,ZHANG Z Z. Entire-grid-area information-sharing and integrated applications in united information system for smart grid. Proceedings of the CSEE, 31, 8-31, (2011).
6. YAN X. Research of the characteristics and basic functions of the SQL server database. Value Engineering, 31,229-231, (2012).
7. JIN C Q,QIAN W N,ZHOU M Q,ZHOU AO Y. Benchmarking data management systems： from traditional database to emergent big data. Chinese Journal of Computers, 38,18-34, (2015).
8. SHEN Z P,XIE Y Q. Business Intelligence data management based on corporate information factory. Information Science, 31,102-106, (2013).
9. ZHAO L. Design and implementation of enterprise ERP system based on cloud computing platform. Electronic Design Engineering,24 ,47-49, (2016).

Forecasting Analysis of Shanghai Stock Index Based on ARIMA Model

Chenggang Li [1, 2], Bing Yang[1] and Min Li[1]

[1]Faculty of Finance, Guizhou University of Finance and Economics, Guiyang, 550025, China
[2]Faculty of Science, National University of Singapore, 117546, Singapore

Abstract. Prediction and analysis of the Shanghai Composite Index is conducive for investors to investing in the stock market, and providing investors with reference. This paper selects Shanghai Composite Index monthly closing price from Jan, 2005 to Oct, 2016 to construct ARIMA model. This paper carries on the forecast of the last three monthly closing price of Shanghai Stock Index that have occurred, and compared it with the actual value, which tests the accuracy and feasibility of the model in the short term Shanghai Stock Index forecast. At last, this paper uses the ARIMA model to forecast the Shanghai Composite Index closing price of the last two months in 2016.

1 Introduction

The stock market is an embodiment of the economic strength of the real industry. The Shanghai Composite Index is the embodiment of a country's overall entity economy level. In some ways, the stock price reflects a national economy strength. We can know the rise and fall of a country's real industry by observing the share price. The Shanghai Composite Index can reflect the most intuitive of the overall economy.

Therefore, the Shanghai Composite Index forecast has important significance for a country in latter regulation and control of the real economy, and the development of regional policy. But it is difficult to analyze and predict the Shanghai Composite Index by using the traditional structural method, and the result error is quite large. So choosing suitable *ARIMA* model to predict the Shanghai Composite Index can be satisfied with the results.

2 ARIMA(p,d,q) model

Setting X_t sequence is a random non-stationary time series. After the X sequence for D order difference, we can conclude:

$$Y_t = \nabla^d X_t \tag{1}$$

Among them, ∇ is the differential operation. d is the differential order [1]. After the difference, sequence Yt has the following structure model:

$$Y_t = \varphi_0 + \varphi_1 Y_{t-1} + \varphi_2 Y_{t-2} + \cdots + \varphi_p Y_{t-p} + \varepsilon_t - \theta_1 \varepsilon_{t-1} - \cdots - \theta_q \varepsilon_{t-q} \tag{2}$$

Integrated auto-regressive average moving is referred as the *ARIMA (p, d, q)* model. Among them, p is auto-regressive order. d is the moving average. $\{\varepsilon_t\}$ is the white noise sequence. φ_i indicates the auto regressive coefficient, and θ_i is the moving average coefficient.

The auto-regressive moving average model, which is first proposed by the American statistician G.E.P.Box and G.M.Jenkins in 1970, is widely used in time series data analysis [2]. It is a effect prediction method with high precision of short-term.

3.The application of the model in the forecasting analysis of Shanghai Stock Index

3.1 Data selection and processing

This paper selects Shanghai Composite Index monthly closing price from Jan, 2005 to Oct, 2016. All data are from the Shanghai Security Exchange [3]. All data processing and final data prediction are carried out using EViews7.2.

3.2 Stationarity test

We do the stationary test of Shanghai Composite Index monthly closing price by ADF test. The test results are shown in Table 1. From the test results, it can be considered that the original time series is not stationarity, but the first order differential sequence is stationarity.

Table 1. Stationarity test results.

Sequences	T statistics	P values	Conclusions
Original sequence	-2.0807	0.2528	Non-stationarity
First order Differential sequence	-6.5150	0.0000	Stationarity

3. 3 Parameters determination

There are three important parameters in the *ARIMA(p,d,q)* model, which are the auto-regressive order p, the difference order d and the moving average q. In the front, after one difference to determine the smooth of the sequence, we can get $d = 1$. Next, it is necessary to determine the value of parameter p and parameter q in order to determine the final model *ARIMA (p, d, q)*. By analyzing the auto-correlation and partial correlation graphs of time series $\{Y_t\}$, the value of p and q can be determined. The self-correlation and partial correlation test results of the sequence are shown in Figure 1.

From Figure 1, we can see that the partial correlation number is significantly truncated, while the self-correlation coefficient falls on the edge of 2 times standard deviation in the time lag of second and forth order [4]. In this case, it is very difficult to use the traditional Box-Jenkins method (auto-correlation partial auto-correlation function, residual variance, F test, the criterion function) to determine the model order. For this special case, this paper makes repeatedly a comparison of the parameters of the model to determine the significance of the parameters of different models.

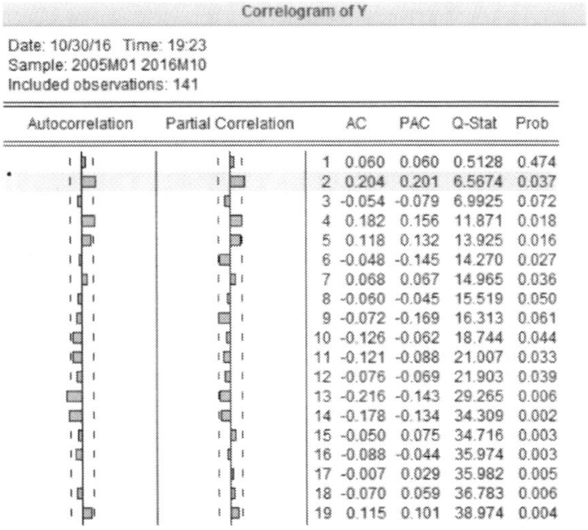

Fig. 1. Self-correlation and partial correlation of first order difference sequence.

For the sequence Y, we try several different models, such as *ARMA (2.2)*, *ARMA (4.4)*, etc. The results of t test (P value) of detecting the parameters of different models are used to select the best model.

According to the *ARMA (2,2)* model, the regression analysis results are shown Table 2.

Table 2. Regression analysis of *ARMA (2,2)*

Variable	Coefficient	Std.Error	t-Statistic	Prob
C	2795.319	390.4668	7.158915	0.0000
AR(1)	0.383696	0.201107	1.907921	0.0585
AR(2)	0.492393	0.194084	2.537005	0.0123
MA(1)	0.63379	0.194721	3.511645	0.0006
MA(2)	0.340521	0.08479	4.016054	0.0001
R-Squared	0.914277	Mean dependent var		2643.74
Adjust R-squared	0.911737	S.D.dependent var		937.99
SE of regression	278.6684	Akaike info criterion		14.132
Sum squared resid	10483574	Schwarz criterion		14.238
Log likelihood	-984.3088	Hannan-Quinn criter		14.175
F-statistic	359.9612	Durbin-Wast on stat		2.0367
Prob(F-statistic)	0.0000			

According to the *ARMA (4,4)* model, the regression analysis results are shown in Table 3.

Table 3. Regression analysis of ARMA (4,4).

Variable	Coefficient	Std.Error	t-Statistic	Prob
C	2720.817	46.94.33	57.96331	0.0000
AR(1)	0.618826	0.081399	7.602379	0.0000
AR(2)	0.779381	0.09541	8.168787	0.0000
AR(3)	0.303961	0.086531	3.513458	0.0006
AR(4)	-0.74622	0.066057	-11.29967	0.0000
MA(1)	0.367347	0.120799	3.040993	0.0029
MA(2)	-0.339289	0.043801	-7.746205	0.0000
MA(3)	-0.972002	0.046586	-20.86487	0.0000
MA(4)	-0.024311	0.115923	-0.209719	0.8342
R-Squared	0.925244	Mean dependent var		2664.049
Adjust R-squared	0.920607	S.D.dependent var		929.2741
SE of regression	261.8386	Akaike info criterion		14.03633
Sum squared resid	8844170	Schwarz criterion		14.22724
Log likelihood	-959.5066	Hannan-Quinn criter		14.11391
F-statistic	199.5754	Durbin-Wast on stat		2.000099
Prob(F-statistic)	0.0000			

Through the comparison of parameters significant t test results (P value) and the fitness of R-squared of the *ARMA (2.2)* and *ARMA (4.4)* model, it can be obtained, the t test is significant and R-squared is optimal in $p=4$ and $q=4$. The model *ARIMA (4, 1, 4)* is established.

3.4 ARIMA(p,d,q) model checking

After the completion of the model construction, it is necessary to carry out adaptive testing, to determine whether it is appropriate. For the model *ARIMA(p,d,q),* it is appropriate to determine whether the model is suitable by testing whether the residual sequence is white noise sequence. When the residual series is not a white noise sequence, it is shown that there is information contained in the residual error. The established model is certainly not the final model, and other parameters can not fully represent [5]. It is necessary to further estimate model. If the residual sequence is white noise, it is shown that the proposed model is suitable.

White noise test results (For space consideration, the results are not list) show that the auto-correlation coefficient of the residual sequence is in random interval, there is no residual serial correlation, and they are close to zero in each lag order auto-correlation and partial auto-correlation values. All of the Q-statistics is not significant, and the P values of lag of the Q-statistics are significantly greater than 0.05. So the residual sequence passes the white noise test. It is white noise sequence. So the *ARIMA(4,1,4)* model can be used for further analysis and prediction.

3.5 Model estimation

From the previous calculation, the final model is determined as the model *ARIMA(4,1,4)*. Through the Eviews7.2 software, we get the coefficients of the *ARIMA(4,1,4)* model. The results are shown in Table 4.

Table 4. Coefficient estimation of ARIMA(4,1,4) model

Variable	Coefficient	Std.Error	t-Statistic	Prob
C	2749.145	178.5409	15.39784	0.000
AR(4)	0.622969	0.079429	7.843075	0.000
MA(4)	0.373878	0.098223	3.806418	0.002

From the above table, the mathematical specification of the model *ARIMA(4,1,4)* is

$$Y_t = 2749.145 + 0.6229Y_{t-2} + \varepsilon_t + 0.3738\varepsilon_{t-2} \quad (3)$$

3.6 Forecasting analysis of Shanghai Composite Index

Firstly, we use the *ARIMA(4,1,4)* model to forecast the closing price of Shanghai Composite Index from July, 2016 to October, 2016. The relative errors between the predicted and actual values can be calculated to determine the feasibility and accuracy of the model [6]. The relative errors are calculated by the following formula:

The relative error = (actual value - prediction value / actual value)

The predicated value of the closing price of Shanghai Composite Index and the relative errors are shown in table 5 and Figure 2.

Table 5. Predicated value and error analysis.

Data	Predicative value	Actual value	Relative error	Average error
2016-08	3010.56	3085.49	-2.4%	2.4%
2016-09	306073	3004.70	1.8%	
2016-10	3005.10	3100.49	3.1%	

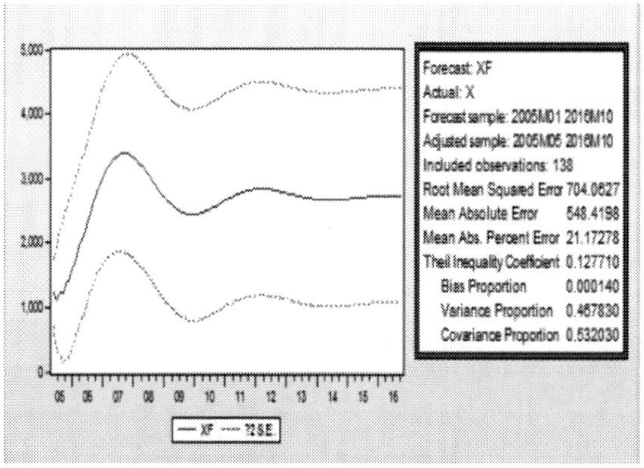

Fig. 2. Short term prediction.

Comparing the predicated value and the actual value, we can see that the average error is 2.4% less than 5%. Seen from the graph, the Theil inequality coefficient is 0.1277, which shows that the model has better prediction effect, and its decomposition table shows that the error ratio is close to zero, and the variance is small.

Secondly, we use the closing price of the Shanghai composite index from January, 2005 to October ,2016 to forecast the closing price of November, 2016. The main calculation method is the deferred method. We take the forecast closing price data of November, 2016 data back, combined the data from January 2005 to October, 2016 to forecast the closing price of December, 2016. The deferred method can get more accurate results [7]. The predicated closing price of the Shanghai Composite Index in November and December, 2016 are shown in Table 6.

Table 6. Predicted results.

Time	Nov,2016	Dec,2016
Predictive value	3273.52	3195.7

4 Conclusions

Changes of the Shanghai Stock Index are the focus of investors and investment institutions. The forecasting of the Shanghai Stock Index is the hot-spot and difficulty in the financial sector. In the varieties of the methods, the *ARIMA* model is one of the typical representative, and has its own unique advantages in the prediction of the Shanghai Stock Index. *ARIMA* model is good at capturing the linear characteristics of the data, and has quite ideal effect for the short-term prediction.

In this paper, the *ARIMA* model is used to analyze the Shanghai Composite Index in October, 2016 and predict the changes of the Shanghai Composite Index in November and December, 2016. The results show that in the last two months of 2016, the Shanghai Composite Index will have a small rise. This gives investors a strong basis for the future of the market judgment, and can let investors in the stock market reduce the loss impacting by the risk of future uncertainty.

References

1. Q. Wang. Time series analysis and application, Southwest Jiao- Tong University Press, (2008).
2. J. Yu. Econometrics software - the use of Eviews , University of International Business and Economics Press, (2006).
3. Y. Wang. Applied time series analysis ,China Renmin University Press, (2005).
4. B. Zhang, X. Liu. Dynamic ecological footprint simulation and prediction based on ARIMA Model: a Case study of Gansu Province, China, Acta Ecologica Sinica,31, (2011).
5. Z. Xiong. Research on RMB exchange rate forecasting model based on combining ARIMA with Neural Networks , The Journal of Quantitative & Technical Economics, 6, (2011).
6. S. Shao. Forecast research on ARIMA model for China's health insurance, Journal of Baoji University of Arts and Sciences. 3, (2016).
7. L. Xiao. Forecast of resident's consumption level based on seasonal ARIMA model, Statistics and decision making.8, (2016).

The Research on Influencing Factors of Medical Logistics Cost Based on ISM Model

Yunkai Zhai[1,2,3] and Yingchao Li[1,*]

[1]Management Engineering School, Zhengzhou University, Zhengzhou, People's Republic of China
[2]Henan Engineering Research Center of Digital Medicine, Zhengzhou, People's Republic of China
[3]The first Affiliate Hospital, Zhengzhou University, Zhengzhou, People's Republic of China
Corresponding Email: 995680589@qq.com

Abstract. The reason why medical logistics cost remains high is a system problem, this paper analyzes the system through the ISM model. The result presents that medical logistics cost factors have four levels of relationship, primary factor is the national policies, secondary factors are the talent construction and pharmaceutical enterprise scale, Intermediate factors are medical information management system and inventory cost, the key factors are transportation cost and distribution center location. Finally, according to the four levels of relationship, this paper put forward specific suggestions to reduce logistics cost.

1 Introduction

With the progress of science and application of online intelligence platform, social logistics cost gradually decreases, because of medical logistics has the characteristic of highly professional, high degree of complexity, strong timeliness and high security, the medical logistics cost still remains high. There is a lot of researches focus on reducing medical logistics cost, however, most of them are one-sided, the papers just analysis a factor and then put forward countermeasures to reduce medical logistics cost. The effect of this kind of research method is limited to reduce the cost of medical logistics.

Through reading literature, we find that the papers about reducing medical logistics cost focus on constructing online information sharing platform, improving the efficiency and establishing third party pharmaceutical logistics enterprises. However, there is no research on the relationships among the factors. Naturally ,there is no research to study the primary and secondary relation between factors of medical logistics cost. The main documents of reducing the medicine logistics cost are as follows.

Lin Xiaoxia [1](2014) points out the domestic medicine distribution centers have the problem of operation is not standardized and the degree of material flow is low; she believes that medical logistics distribution system can promote the distribution center's operation efficiency and improve the level of logistics, so as to reduce the medicine logistics cost. Zhang Luyue [2](2012) points out the composition of distribution costs (including the cost of distribution transport and sorting) and the influencing factors of distribution cost (market competition factors, distribution management factors and so on), she advocates to reduce the logistics cost by consolidation strategy and delay strategy. Wang Chuanlei [3](2015) points out the green distribution system of medical logistics. Based on the characteristics of the medical logistics, combining the operating system, recovery system and information system, Constructing the green distribution system of medical logistics, so as to reduce the cost of logistics, and ensure the development of medical logistics healthy and orderly. Wei Hua [4](2015) points out the construction of medical logistics information sharing platform has the positive effect to reduce the cost of medical logistics, and he proposes a modern medical logistics sharing information platform based on automatic recognition technology. These documents analysis and put forward recommendations just from one simple perspective.

This paper identifies the important factors through literature research and expert consultation and analysis the medical logistics cost factor from primary to secondary, then puts forward effective and reasonable suggestions.

2 The influencing factors of medical logistics cost

Through the analysis of the literature and Delphi method, this paper identify 7 main factors as follows.

2.1 Location of medical distribution center

The rationality of distribution center location has a direct impact on other factors in medical logistics, so it should be reasonable as far as possible. The location of medical distribution center has qualitative and quantitative method. Qualitative method has the analytic hierarchy process and the fuzzy clustering method. Quantitative methods include heuristic algorithm, gravity method, Baumol-Wolfe method and mixed 0-1 integer programming model combination method, simulation algorithm and so on.

2.2 Medical information management system

With the development and application of big data and other network technology, online medical information platform came into being, the influence of online medicine information platform on medical logistics cost is mainly reflected in the following two points.

2.2.1 Direct impact

The online medical information platform can integrate the medical chain enterprises with the requirements of hospital, so that the information can be shared on the platform, the drug demand can get the information about the supply through the platform so as to meet their needs. by this way ,they can reduce the cost of purchase time and transportation costs, at the same time, After receiving the order, the supplier can classify orders (drug types, distribution area, time, etc.), improving the efficiency of distribution.

2.2.2Indirect effect

The online medical information platform not only open for the medical chain enterprises and hospitals but also open to individuals, medical chain enterprises and hospitals have the ability to identify the authenticity of the drugs, however, it is difficult to identify the authenticity for the individual. Especially in the last few years, the report of the poison capsule event which lead to bad impression to the people, worse more, people may not buy drugs through online medical information platform , this may cause warehousing and logistics system can not play its due value and increase the cost of medical logistics indirectly.

2.3 Inventory cost

The effect of inventory is to adjust the balance between supply and demand, the high inventory increase the management costs and storage costs, the low inventory lead to short supply risk. Due to the storage of the medical products require certain temperature, humidity and other conditions, which leads to the high cost of storage.

2.4 The size of medical logistics enterprise

According to the *2013-2017 the research of China medical logistics industry and analysis report of investment strategy planning*, there are 13000 medical logistics enterprises in China, large enterprises accounted for only 20%, present the state of Many, small, scattered, chaotic; medical logistics enterprises in the top four accounts for only about 30% of the medical logistics market share, by contrast, the U.S. top three medical logistics enterprises reached more than 95% in the medical market share. If the size of medical logistics enterprises is too small, the standard, operation mode and distribution efficiency will affect the whole medical logistics cost.

2.5 Transportation cost

According to statistics, transport costs accounted for 40% of the overall logistics costs, such a high proportion makes the medical logistics enterprises recognize importance of reducing the transportation costs. However, due to the characteristics of medical products, in the process of transportation must keep the physical condition, such as temperature, humidity, etc. To meet the requirements, special transport facilities are required and transport personnel need professional operating skills.

2.6 Talents construction

Medical logistics is different from the general logistics, it is more professional and stronger, to achieve a certain standard, it requires the participation of professional talents who have professional skills and medical knowledge.

2.7 National policy

The national policy is also the influencing factor of medical logistics cost, because the policy can determine development of medical logistics, such as guiding the construction of the distribution center and encouraging the development of large medical logistics enterprises etc.. If the country does not pay attention to the medical logistics, then the development of medical enterprises will be in disarray, a series of problems may appear. For example, 570 million illegal vaccine case happened in 2016, if the Illegal vaccine can be seized in logistics chain, a series of vicious consequences can be avoided. Therefore, national policy-making, not only can reduce the cost of medical logistics but also ensure the medical market be more stable and healthy .

3 Systematic analysis of the influencing factors of medical logistics cost based on ISM model

Interpretive structure model [5,6](ISM model) is presented by American professor Warfield in 1973, the model based on qualitative analysis, its function is decomposing a complex system into subsystems (elements), then hierarchical processing of these elements, so as to show the relationship between the elements.

There are connections between the different influence factors of medical logistics cost, the connections constitute a system. Analyzing the system through ISM model can point out the way to reduce medical logistics cost.

3.1 Draw directed graph

Through the analysis of the literature and Delphi method, the relationship between the various factors as shown in Figure 1.

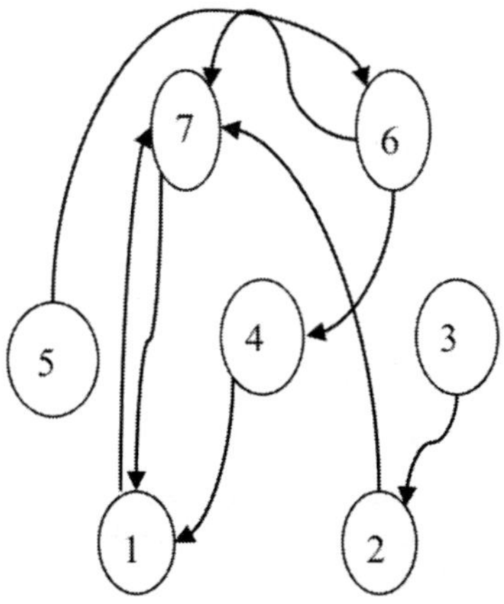

Fig. 1. the relationship between the various factors.

In Figure 1, 1 indicates the location of medical distribution center, 2 indicates the medical information management system, 3 indicates talents construction, 4 indicates inventory cost, 5 indicates the national policy, 6 indicates the size of medical logistics enterprise, 7 indicates the transportation cost.

3.2 Draw the relationship table

Based on the transfer of the relationship between the two elements, inferring the binary relation of the elements , and according to the directed graph, drawing the relationship table as shown in table1.

Table 1. The relationship table.

X	(A)			A	(A)	1
V				A	2	
				3		
V	A	(A)	4			
(V)	V	5				
V	6					
7						

3.3 Write out the reachable matrix

Write out the reachable matrix according to the relationship table.

$$M=\begin{array}{c}\\1\\2\\3\\4\\5\\6\\7\end{array}\begin{array}{c}\begin{array}{ccccccc}1&2&3&4&5&6&7\end{array}\\\left(\begin{array}{ccccccc}1&0&0&0&0&0&1\\1&1&0&0&0&0&1\\0&1&1&0&0&0&0\\1&0&0&1&0&0&1\\0&0&0&1&1&1&1\\1&0&0&1&0&1&1\\1&0&0&0&0&0&1\end{array}\right)\end{array}$$

Write out the reduced matrix according to the reachable matrix:

$$M'=\begin{array}{c}1\\2\\3\\4\\5\\6\end{array}\left(\begin{array}{cccccc}1&0&0&0&0&0\\1&1&0&0&0&0\\0&1&1&0&0&0\\1&0&0&1&0&0\\0&0&0&1&1&1\\1&0&0&1&0&1\end{array}\right)$$

Hierarchical processing:

$$M'(l)=\begin{array}{c}1\\4\\2\\3\\6\\5\end{array}\left(\begin{array}{cccccc}\boxed{1}&0&0&0&0&0\\1&1&0&0&0&1\\1&0&1&0&0&0\\0&0&1&1&0&0\\1&1&0&0&1&0\\0&1&0&0&1&\boxed{1}\end{array}\right)$$

After the Hierarchical processing, the elements are divided into four layers, the first layer is the factor 1, the second layer are factor 2 and factor 4, the third layer are factor 3 and factor 6, the forth layer is factor 5. Drawing the multi class directed graph based on hierarchical, as shown in Figure 2.

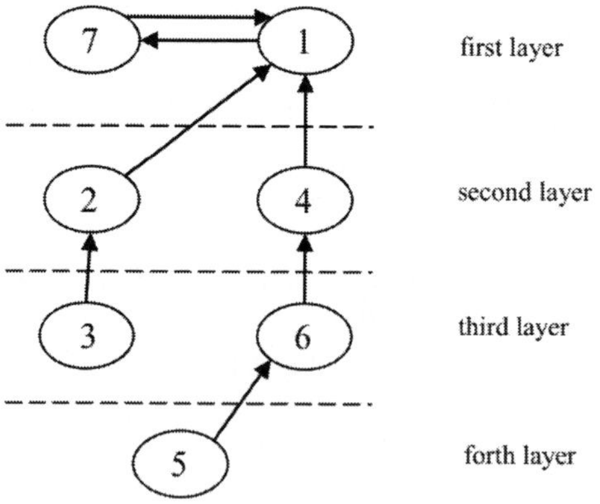

Fig .2. The multi class directed graph.

3.4 ISM model of influencing factors of medical logistics cost

According to the multi class directed graph, getting the ISM model of the influencing factors of the medical logistics cost as shown in Figure 3.

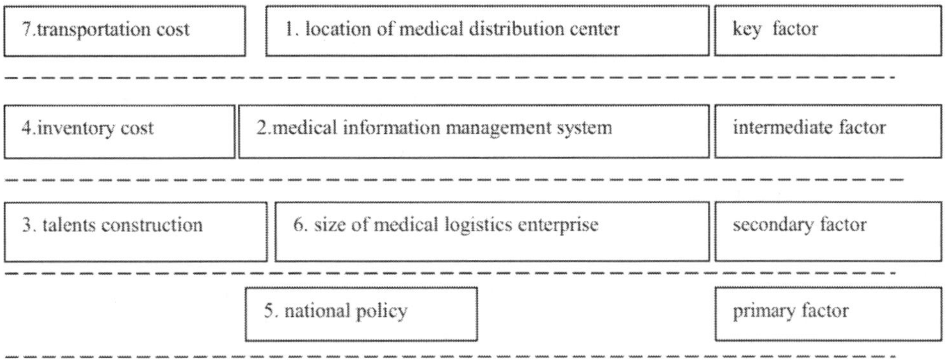

Fig. 3. ISM model of the influencing factors of the medical logistics cost.

According to Figure 3, medical logistics cost factors have four levels of relationship, primary factor is the national policies, secondary factors are the talent construction and pharmaceutical enterprise scale, Intermediate factors are medical information management system and inventory cost, the key factors are transportation cost and distribution center location.

4 The measures to reduce the medical logistics cost

4.1 Policy-making

The characteristics of medical logistics determine that it is different from the general logistics, in order to speed up the development of medical logistics, and reduce the medical logistics cost effectively, it is necessary to formulate a series of policies to reduce the medical logistics cost.

First, through administrative, legal and economic means to integrate medical market and optimize the allocation of resources, then promoting the merger of medical logistics enterprises, improving the share of large medical logistics enterprises in the medical market, so that laying a solid foundation for the standardization of medical logistics market. Secondly, formulating relevant policies to develop talents who good at medical logistics, what's more, it is necessary to put the talent construction on the strategic level.

The formulation of the national policy should not limited to these two aspects, it should be planned in overall picture , and perform it from the national level, The future of the medical logistics market is bound to have a new look, the medical logistics costs will also be controlled and gradually reduced.

4.2 Strengthen the construction of talent and expand the scale of pharmaceutical logistics enterprises

4.2.1 Strengthen the construction of talent

Talent is the basis for the development of any industry, leaving the talent building and talent supplement, the industry is like the air loft. The demand for talents of medical logistics is more unique , it requires of medical logistics staff is not only proficient in logistics knowledge but also good at medical knowledge , Inter-disciplinary talents are necessary to adapt to the development of medical logistics industry.

First of all, college should set up a courses that students can learn about logistics and medical , through this way , Cultivating students' comprehensive ability. secondly, promoting the students to practice in medical enterprises actively, to cultivate the student's practical ability so that laying a good foundation for joining the medical logistics industry.

4.2.2 Expand the scale of pharmaceutical logistics enterprises

In view of the present situation of the medical logistics enterprises in our country, which is many, small, scattered and chaotic, small medical logistics enterprises should be set up to form large medical logistics enterprises, or large medical logistics enterprises merge small medical logistics enterprises mergers. In this way, the present situation of the medical logistics enterprises can be solved , and medical logistics resources can be integrated. What's more, the medical distribution, transportation and storage can achieve scale and more manageable.

4.3 Promote the construction of medical information management system and reduce the Inventory costs

4.3.1 Promote the construction of medical information management system

The development of the Internet and big data provides good opportunity and technical support for the construction of medical information platform. Based on Internet and big data, constructing medical information platform that matches the medical characteristics, demand-side and supply-side can get the information through medical information platform. The medical information management system can also get rid of the information communication barriers and realize share of information, so as to achieve the goal of saving procurement time, improving the distribution efficiency, and reducing the transportation cost.

4.3.2 Reduce inventory cost

If the inventory cost of medical company is too high, then the cost will be added to the transportation and distribution, therefore reducing the cost of medical enterprises is particularly important. Due to the special requirements of the medicine storage, such as a certain temperature and humidity conditions, then it would spend much money to build warehouse, which result in high storage costs, so it requires the implementation of effective and reasonable inventory management methods to reduce the cost of management and warehousing. For example, VMI (Vendor Managed Inventory), CPFR (Collaborative Planning Forecasting and replenishment), RMI (Joint Inventory Management), etc., these method can help to reduce inventory cost, we can also combine these methods to create new way to reduce inventory cost.

4.4 Reduce the transportation cost and optimize distribution center allocation

4.4.1 Reduce transport costs

The reason why transport cost is so high is that too many transit times, distribution path is unreasonable, the vehicle load, therefore, we can reduce the cost of transport starting from the three aspects. First of all, Through the medical information platform, medical logistics enterprises can Load and distribute medicine uniformly. Secondly, at the end of the distribution stage, it is necessary to carry out a reasonable transport path planning, and ensure that there is no return in the process of transportation and distribution, so that improving the efficiency of transport. Finally, after the distribution, medical logistics enterprises should look for goods that suitable for function of vehicle, through this way, the cost due to no-load can be reduced.

4.4.2 Optimize distribution center allocation

Distribution center allocation is related to the distribution efficiency, how to measure distribution center allocation is essential to reduce the cost of medicine logistics. The allocation is an important and complex system engineering. Due to the characteristics of medical logistics, the medical logistics distribution center allocation is more complex and much more factors to consider. For example, Drug consumption in the region, traffic conditions, etc. So it is important to take all factors into account and never build the medical distribution center blindly that does not meet the needs of the region. Otherwise, the large scale will lead to high cost of construction, resulting in waste of resources, if the size is too small ,then it can't meet the needs of distribution.

5 Summary

This paper analyzes the medical logistics cost through the ISM model , clarifies the relationship between related factors, When looking for ways to reduce the cost of medical logistics, it is necessary to consider the relationship between the different factors, and analyze the effects of various factors on the medical logistics cost in the view of system.

References

1. Lin Xiaoxia. Analysis and design of medical logistics distribution system, *Fujian computer,* **30**,2(2014) 132-133.
2. Zhang Luyue. Study on the cost of drug logistics distribution, Business Culture **2**,2(2014) 193.
3. Chen Jiao, Wang Dongzi. Research on drug cloud distribution model based on Logistics 4.0 , Journal of Beihua University(Social Sciences) **16**,4(2015) 33-37.
4. Wei Hua. Construction of medical logistics sharing information platform, Logistics Technology **34**,7 (2004) 275-277.
5. Wang Yingluo. *Systems engineering.* Beijing:China Machine Press,(2008)220-235.
6. CHEN Senfa. *The theory and method of complex system modeling.* Nanjing: Southeast University Press,(2006)273-292.

The Discussion about International Agricultural Logistics and Distribution System of Henan Wanbang

Shumin Luo[1] and Hua Xuan [1]

[1]Management Engineering Department, ZhengZhou University, ZhengZhou, China

Abstract. With the development of economic globalization, The current development of China's agriculture has shifted from production areas to circulation areas , People on fresh produce a concern degree higher and higher, the key is to ensure agricultural products fresh degree is the logistic distribution . This article takes the distribution system He nan wan bang International Agricultural Logistics Park for the study, From the current situation, agricultural logistics perspective，Through Wan bang distribution status，Problems distribution system was investigated，Also it includes its logistics strategy, organizational design and distribution, distribution function, operations and process design, information systems and mechanical equipment analysis, business performance evaluation and so on. With related theory, To study the issue of logistics through knowledge learned，Put forward reasonable solutions，Wan bang optimized logistics and distribution system of agricultural products.

1 Introduction

From the point of the current situation in our country, lack of competitiveness of agricultural products, not only on the quality and appearance but also can be seen in terms of the transaction cost, loss, waste. To seek a more reasonable scientific agricultural products logistics distribution system, to promote the farmers' income, distribution of the improvement of service quality, reduce distribution costs, to solve the agricultural product circulation speed,there are great significance. In the rapid development of zhengzhou economy, the logistics industry of agricultural products will play an important support role for its development, agricultural products logistics industry is becoming more and more important, so the study of Chinese agricultural products logistics is imperative [1-3]. Among all nations based on the investigation and study of distribution system, starting from the distribution system of all the nations, and analyses its general situation, the analysis of the agricultural products logistics distribution system, existing problems.

2 Logistics distribution system analysis

2.1 Distribution of the organizational design and distribution functions

The operation of the Wanbang logistics distribution system are mainly composed of self-run logistics distribution, private home car transport is complementary. Of self-run logistics for all future in Henan, the main transportation forces are buying their cars. Information network and logistics management software which is called the logistics information system, its composition including people, processes and equipment, can provide the information for some functions including logistics managers perform the planning and implementation of control, it is not exist independently but part of the enterprise information system, is not only a subsystem of logistics system, is also a kind of interaction system [4]. Among all the distribution of organizational design including its logistics distribution center organization chart and personnel and job responsibilities. Distribution center structure diagram as shown in figure 1:

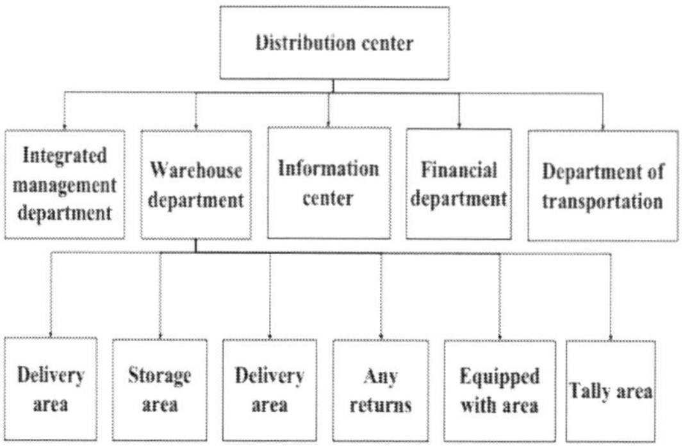

Fig. 1. Distribution center structure.

Distribution system, its distribution function mainly has several aspects: (1)Stock goods. From the aspects of distribution, the role is preparation and collection of the goods ,goods ordering, goods collection, purchasing goods and its interrelated settlement, such as quality inspection, etc. When costs are very high, in the very great degree will reduce distribution efficiency. Agricultural products, distribution center.(2)storage. The two forms of storage : store for future use and temporary storage. temporary storage is to distribution in the specific date, according to the provisions of sorting and distribution, do less work for mass storage on the reasonable freight yard. Because the whole storage efficiency is determined by the whole storage quantity, therefore, the portion of the temporary number will not influence the overall benefits of store, only cause certain influence to whether convenient work, so do not strict control the number of temporary storage. And the storage of Wanbang logistics configuration is mainly the temporary storage, this is according to its own cold storage of fresh and frozen goods storage depots and other facilities. (3)sorting and distribution. At the time of delivery the mutual competition between different distribution enterprises, and to improve their own economic benefits, sorting and distribution is an extension that must be carried out [5]. Sorting, distribution is one of the key elements for a distribution system level. sorting and distribution of Wanbang logistics mainly depends on manual work, its modernization level is not high. (4)Assembling. How to make the goods that the different users need to delivery together,for better use of resources through collocation, loading, this needs assembling; This can effectively reduce the delivery cost, to a large extent improve the level of service delivery. But Wanbang in many cases cannot achieve the payload of the vehicle. (5)Distribution transports.it is the regional transportation and the last transportation, its transportation facility are cars. Compared with the main transport has the following characteristics: short distance, small size, high cost. And it has a route selection problem. Due to the complexity of city traffic route condition, the big number of users, the characteristics of distribution transport: how to choose the best, let assemble and route fit effectively [6].

2.2 The process of logistics distribution center

The operations of Wanbang logistics distribution center include: Warehouse operation, Tally work, the customer and order management , EX-warehouse operation , Loading and unloading and handling, distribution operation. The process design as shown in figure 2:

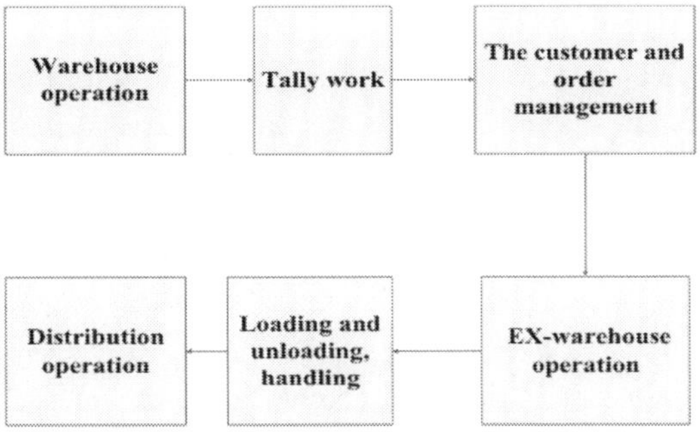

Fig. 2. Wangbang logistics distribution center process design.

3 SWOT analysis

Due to Wangbang logistics distribution system operation is a self-run logistics distribution, and their private car to do auxiliary work. SWOT analysis of this model of Wangbang : Superiority (S), weaknesses (W), opportunities (O), threats (T).such as follows:

(1)Superiority (S): save money, keep strong competitiveness and make four logistics(Macro and micro logistics and social logistics and logistics, international logistics and regional logistics, general logistics and special logistics)seems more tightly, because control agricultural products logistics, so as to improve the efficiency of the logistics operation, reduce capital takes up, save the expenses [7]. Its surrounding comprehensive traffic has formed the transportation network, has developed the highway transportation system and railway transportation system. the development of Wangbang with thick solid foundation. The highway of Zhengzhou to Minquan ,it is surrounded by a avenue, Kowloon, Hong Kong and Macao and neighboring Beijing highway, highway three thousand, north longhai railway lines [8].

(2)Weak (W):Because Wangbang 's distribution system increases the burden on its investment, making it less capable of resisting market risk. Because the initial investment is too large, the logistics of its own building has not yet reached a more appropriate level in the system management requirements, in the professional aspects of the high enough, difficult to manage and so on [9]. And because of the relatively inactive geographical location, the lack of professional agricultural logistics management personnel, logistics management personnel but also efforts, the private car market led to a certain degree of confusion.

(3)Opportunity (O): Adjacent to the Zhengzhou International Logistics Park under construction, the traffic advantage is being greatly strengthened, laying a solid foundation for its future expansion and development. The government attaches great importance to the logistics of agricultural products, the Ministry of Commerce, Henan Province, city and county leaders at all levels with the strong support of Henan Wanbang to accelerate the development of the central region has become the most influential large-scale distribution center at the national level, Key Functions of Agricultural Products Logistics Hub. With a large self-built, including vegetables, fruit trading area, cold storage facilities, and its strong competitiveness, will be the Central Plains region's largest agricultural products market [10].

(4)Threat (T): the market demand of modern agricultural products logistics service is insufficient, because there is contradiction between supply and demand, leading to supply greater than demand and short supply. Because the more accurate the information of supply and demand is relatively difficult, so make the agricultural products logistics links exist bigger risks. Because of China's entry into the WTO, foreign logistics enterprises also have threat to Wanbang logistics. Agricultural products logistics has a relatively high risk at the same time, because it is easy to corrupt, follow the seasonal change characteristics of agricultural products is "fresh" meaning and life place, but the short date of agricultural products with good quality and very easy to corrupt, and other characteristics, to a great extent, restricts the transport path as well as the trading time, so the preservation condition of circulation and the transport efficiency proposed higher specification, and customers need to improve the logistics service requirements, the need to work hard for future logistics.

4 comparative analysis with foreign

4.1 present situation of development and characteristics of the major foreign countries

The early country of the agricultural products logistics research is the United States, then other countries also gradually began to study. Logistics is towards the direction of further development in the 1980, the logistics theory gradually became mature. foreign logistics development of agricultural products, mainly has the following three models:

(1)Japan, South Korea, as the important representative of east Asian model.

They are in the wholesale market as the main channel, the buying and selling as an important means. Have the perfect logistics infrastructure, which included in the distribution center of packaging and processing facilities, warehouse at low temperature and normal temperature, but also engaged in the distribution of electronic commerce, form a complete set, ready to goods, etc. They supply of agricultural products logistics distribution has the following several characteristics: (1)the links in the circulation channels, cost is high; Its supply process is "product-wholesale-middle wholesale- retail - consumption" (2)through (1)the flow of process, it can be seen that the uneven distribution of profits; (3)although the uneven distribution of profits, but the circulation work standardization, the legal system more strengthen, speed and efficiency are high.

(2)France, the Netherlands and other countries as the main representative of the western European model.

Western mode of wholesale market circulation ratio relatively smaller than east Asian model, and a lot of wholesale market still adhere to this principle the public welfare, it is compared with the east Asian model. Mainly through the network to form agricultural products supply chain, including suppliers, manufacturers, wholesalers, retailers, and to the very many country in the world of consumer people to provide services through electronic trading of agricultural products. And they're in refrigeration industry development is very positive, has highly efficient cooling and freezing technology equipment.

The main characteristic of Western Europe pattern is : (1)To encourage the prenatal and postnatal related enterprises built in the countryside, and the production, processing and marketing in one level. (2)Built wholesale market about agricultural products, its nature is public welfare, it is relatively perfect, contemporary; (3)A relatively standardized product and manufacture of agricultural products.

(1)United States, Canada, and Australia is the main representative of the pattern of North America.

Direct marketing system development is better, and the main mode of agricultural products sales and direct sales. Information based development of agriculture is very good, they built information management system for their entire national crop variety resources, provide services to the entire country experts.

These countries agricultural products logistics distribution mainly has the following characteristics: (1)The origin and market are relatively concentrated. (2)Sales site are larger cities, wholesale market is located in big cities. (3)Circulation channels is shorter, less in the middle of the process, high efficiency.

4.2 Draw lessons from foreign agricultural products logistics mode

Three modes of foreign countries in the development of the agricultural products logistics is good with the government's macroeconomic regulation and control. Because the of government to promote the development of agricultural products logistics is very important. Government should also play a good role in macroeconomic control of the agricultural products logistics, it can from the following several aspects: formulate relevant laws and regulations to regulate some bad phenomena in China logistics market of agricultural products, slowly to control and improve the current non-standard phase at the present stage; Insist on open to other countries, to learn the modes and experience of the three countries, and make the necessary cooperation in the logistics of agricultural products; Our country can learn from the three modes, advanced agricultural logistics technology to our country, make our country more modern in the agricultural products logistics. Based on Wangbang analysis of the logistics system, which draw lessons from the above experience of foreign research, to improve its distribution facilities and establish a modern distribution center of logistics, it is a valuable asset for it.

5 The problems and solutions

5.1 Analysis on the Problem of Logistics Distribution System of Wanbang and the solutions

The Problem of Logistics Distribution System of Wanbang are as follows:(1)its own logistics and distribution system is not perfect(2)lack of professional logistics personnel, technology and facilities support(3)The coordination mechanism of departments, regulations and market access is not perfect.

The Solutions to the Problem of Wanbang 's Existence are as follows:

5.1.1 Establish and improve the sector, market access, regulatory aspects of the coordination mechanism

For the sector, regulations, market access and other aspects of the coordination mechanism is not perfect, Wanbang market sector should develop the market of modern agricultural logistics laws and regulations to unify the standards of the market, regulate market behavior on agricultural logistics, the establishment of quality standards and industry norms Agricultural products logistics services, but also the corresponding policy development, agricultural flow regulations, strengthen the management of the market, regulate some of the behavior of the market competition. Crack down on bad business bad behavior, but also a healthy and orderly market.

5.1.2 Professional agricultural products logistics distribution center establishment

In the distribution system for the production and sales from the role of communication and distribution logistics distribution center, which has the following characteristics：(1)fast response. On the upstream logistics and downstream logistics and distribution needs, it is faster and faster response. (2)functional integration. Just like integration between each channel of the logistics, for example in the logistics and other parts of the supply chain integration. (3)standardization of distribution operations. Improve the efficiency of logistics operations and its benefits, based on the operation of the logistics and distribution processes, so that relatively more complex operations tend to simplify (4)series of distribution services. In addition to door-to-door, you can expand the logistics of the following aspects: distribution consulting, inventory control, payment recovery. (5)distribution means modernization. If you want a larger scale, logistics and distribution technology, logistics equipment and management needs to be more modern, which can be through the use of advanced logistics technology, logistics equipment and management, logistics and distribution are the main support. (6)distribution organization network. Including logistics nodes, distribution centers and other network facilities, mainly refers to the logistics and distribution network system is complete and sound. (7)logistics and distribution management legalization. We must have sound laws and regulations and act according to law, according to the chapter, regardless of macro or micro.

And it has the following functions:(1)in the logistics cost control. (2)to achieve intensive inventory. (3)It is conducive to the realization of commercial separation. (4)It is conducive to grasp the sales information. (5)can promote product sales, improve customer service levels.

Through the above characteristics and role of logistics distribution center, and then for the Henan Wanbang on the status of agricultural logistics, as well as Wanbang Future Logistics development, the establishment of agricultural logistics and distribution center is a good choice. When a complete distribution center is established, the optimization of the transportation route can be solved by solving the shortest path problem according to the solution of operations research.

5.1.3 strengthen the logistics personnel training efforts ,take new logistics technology ,improve the utilization of facilities

Henan Wanbang can set up their own logistics personnel training institutions, their staff on a regular basis to other logistics companies to visit and learn to understand their positions in the company's entire system, learn modern information technology. In addition, with the pay and benefits to retain logistics personnel, let them play a greater role for the company.

Wanbang logistics vehicles, while equipped with GPS, bar code technology , EDI electronic data exchange and RF radio technology, but compared with other logistics companies still have a certain gap. In order to make its logistics and distribution system more better, cost more less, this technology is to be greatly improved.

Although Wangbang has its own cold storage, forklifts, warehouses and other equipment, but the utilization rate is very small, which can use "puerile" this idea, leased them to customers in need, rented them to the business park, not only Improve its utilization, but also reduce the waste of agricultural products.

6 Conclusion

With the development of economic globalization, the agricultural logistics industry will play an important supporting role in the economic development of our country on the basis of China's national conditions. The paper structure is clear, starting from the agricultural product logistics distribution system, Through the analysis of the general situation of enterprises, the problems of logistics and distribution of agricultural products, as well as the comparison of the same industry abroad, And through the knowledge learned by the author, put forward some suggestion to solve their problems, and also concluded that logistics and distribution system is very important for the development of logistics , It is an important factor for its success.

References

1. Liu Chunyan, Pan Feng.Study on the Logistics Distribution System of Fresh Agricultural Products under the Condition of Intelligent Logistics.Chinese Market, 10, (2013).
2. LUO Xuqin, TONG Xiaohu, YU Jun.Study on the Agricultural Products Logistics Distribution System Based on E-commerce. Electronic Commerce, 1, (2014).
3. Zhang Xue. Qingdao modern logistics and distribution system of agricultural products. Ocean University of China master's degree thesis printed version. 6, (2011).
4. Li Qing, Liu Binglu. JIT-based way of aquatic products cold chain logistics and distribution .Goods and quality of scientific theory . 2, (2011).
5. Liu Yangqing. Current situation of the development of Fresh agricultural product distribution model in E-commerce environment .Logistics engineering and management, 2, (2014).
6. Zhang Na. Liu Jian. Fresh agricultural products "last mile" distribution of Mianyang City.Trade, 21,(2014).
7. Liu Xudong. Analysis of JIT distribution model of aquatic products based on SWOT analysis. Logistics Engineering and Management, 6, (2012).
8. Xing Jiangbo. Fresh agricultural products logistics distribution network optimization of E-commerce environment. Dissertation,9,(2012).
9. GAO Xuedong, GU Shujuan, BAI Chen, WU Sen. Clustering Algorithm Considering the Structure of Logistics Distribution Network and Delivery Quantity Constraints.Systems Engineering Theory and Practice 1, (2012).
10. Wang Junjiao. Aquatic product distribution model optimization analysis of Dalian area. Shopping Center .22,(2014).

The Third Party Logistics Partner Selection of B2C E-Commerce Enterprise

Wei Xu[1] and Bing Li[1]

[1] School of Management Engineering, Zhengzhou University, Zhengzhou 450001, China

Abstract. In recent years, with the rapid development of Internet, e-commerce enterprises in China is at an alarming rate of sustainable development, the B2C e-commerce in the online shopping market accounted for a large proportion, with the rapid change of the market and customer needs to establish a flexible distribution service is one of the many e-commerce companies need to solve. Using self logistics capital occupancy bigger; the logistics alliance, with the risk of internal information disclosure; and on the third party logistics, not only capital and manpower can be effectively used, helps enterprises to strengthen the core business. Therefore, it is very important to choose the appropriate third party logistics providers to reduce operating costs and improve the market influence. This paper analyses the existing problems in the third party logistics, the establishment of a more complete index system, according to the actual situation of the suppliers, using the improved AHP method and the entropy coefficient method to get the comprehensive weight, finally get the approximation method to meet the business requirements of the logistics service providers use the ideal, for commercial enterprises to third party logistics service provider selection and puts forward some available models.

1 Introduction

Ai Rui Advisory data show that in 2015 China's online shopping market B2C transactions reached 1588.2 million, the proportion of the overall online shopping market transactions reached 45.8%, from the growth point of view, in 2015 China's shopping B2C market growth of 68.7%, visible, B2C market in the entire online shopping market and the rapid development of the dominant position [1]. Among them, the logistics is the guarantee of electronic commerce, the first, logistics guarantee production. Second, logistics service provider. At present, there are two forms of e-commerce logistics system, which are self built logistics and outsourcing logistics.

However, for most of the electricity supplier companies, a self built logistics system is very difficult, especially for emerging enterprises, only rely on their own strength is difficult to get through the logistics channel to occupy the market in the short term, in comparison, the third party logistics has a professional, all-round, multi-functional characteristics, can be better to adapt to the business enterprise distribution requirements, improve the level of comprehensive strength of the enterprise. However, there are more and more problems in the third party logistics service.

This requires enterprises to choose the best supplier for the problem of distribution service, reduce operating costs, to help enterprises to achieve the fastest, best service to meet the needs .

2 Research status

Based on the reference to many documents, through the evaluation system of supplier selection and supplier from two aspects proposed third party logistics supplier selection research in the field of more representative studies.

2.1 Research status of supplier evaluation system and method

Dickson is the founder of the research on supplier evaluation system, 23 supplier evaluation criteria given by him, and the evaluation index is divided into four grades, are extremely important, more important, important and less important, and pointed out that the quality factor is extremely important [2].

Johnson through the evaluation method of candidate enterprises, it is concluded that four important factors, delivery time, service, cost and quality are the choice of suppliers [3]. Ma Shihua, Lin Yong, the third party logistics service provider evaluation index system is divided into four parts, the quality system, enterprise performance, business structure and production capacity and business environment [4]. Liu Yiyong Sun Shudong, Ma Xuefen, Wu Xiuli put the development potential of the enterprise, logistics technology, equipment status and the service quality of the enterprise as the main factors of the evaluation index system [5] . Yan Xiuxia studied other third party logistics service provider selection of multi objective decision [6] .Zhao Xiaoyu,etc. which using fuzzy comprehensive evaluation method to evaluate the core enterprise, and ranking the importance of the core business, the design of the supply chain is the use of multi objective integer programming model [7] . Liu Gang put forward the supplier value - the supplier evaluation index system of supply risk, and use the entropy weight coefficient method to determine the index weight [8].

Chen Hangjuan, Lu Zhigang put forward the multi attribute decision making method of information entropy to determine the weight of each attribute [9].

At present, the choice of the third party logistics supplier is relatively mature in terms of evaluation system and selection method, but there are still some shortcomings. In the evaluation system, the former suppliers often pay more attention to some of the static indicators such as quality, price and ignore the customer's personalized and diversity requirements. On the choice of methods, qualitative methods such as Delphy. expert results will be influenced by subjective factors; quantitative methods such as purchasing cost method, although this method only needs to compare the supplier's quotation, and choosing a lower price supplier, looking simple but there is a big risk.

2.2 Establishment and research method of index system of this paper

Based on the index system, not only retains some of the traditional static indicators such as quality, price, these aspects such as the flexible delivery of the supplier, the prospects for the development of suppliers, the establishment of information capacity are also conclude to form more complete supplier evaluation system. In the research methods, this paper is not using a single research method, using the improved AHP and entropy coefficient method combined weight, and proposing TOPSIS comprehensive evaluation model.

This combination of qualitative and quantitative methods, both the advantages of a single method and to make up for the disadvantages of a single method, so that the research process is more abundant and comprehensive. However, each index weight is the key issue of the study, in order to solve this problem, this paper use the subjective and objective weighting method, proposing AHP and entropy coefficient method improved comprehensive weight, this method is effective to combine the two kinds of weight information, which fully use the objective information, and achieve the greatest possible of the decision-maker's subjective intention. Finally, the TOPSIS model is used to find the satisfactory solutions.

3 B2C e-commerce enterprise 3 party logistics supplier evaluation index system

3.1 B2C e-commerce enterprise development status quo

With the continuous development of network shopping market ,the proportion of B2C e-commerce market in the online shopping market is more and more comprehensive, in order to explain this question, this article illustrate these two aspects the B2C e-commerce market scale and market share in detail.

3.1.1 e-commerce market scale

In 2015 the network shopping market share of 22.9% in the entire e-commerce market segment.B2C market share trading volume in the online shopping market is a sharp increase in the recent six years survey,in 2011 B2C accounted for the first time exceeded 20%, accounting for 29.7% in 2012,in 2013 accounted for 35.1%,and in 2015 B2C market transactions accounted for 45.8% of the entire online shopping market. In just a few years, the B2C market in the entire online shopping market develops rapidly and gradually holds the dominant position.

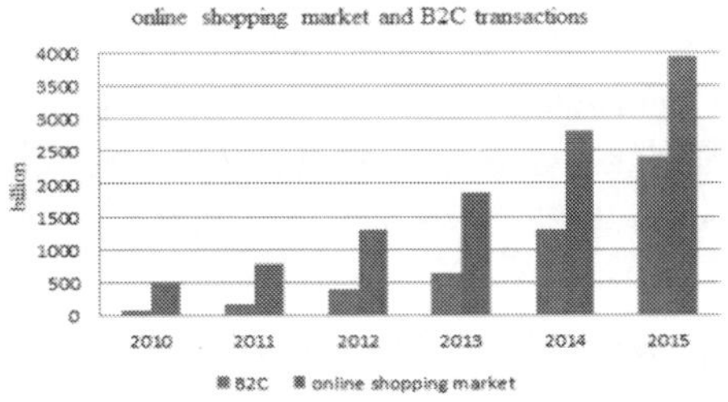

Fig. 1. Caption of the Figure 1. Below the figure.

3.1.2 Market share

According to the survey, with the B2C market demand, B2C shopping sites are also increasing, which are familiar with the Jing dong, such as T mall, suning.com plays an important role in the whole B2C market ,in 2015 the scale of the market share of B2C shopping website transactions in China see Figure 2.

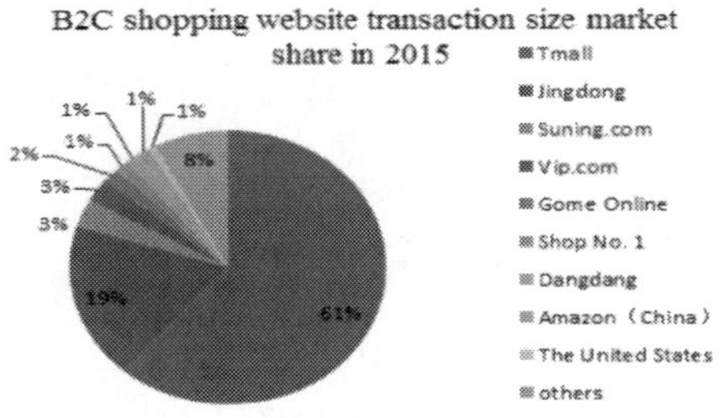

Fig. 2. Caption of the Figure 2. Below the figure.

From the above illustrations, it can see that the B2C market with the rapid pace of development gradually becomes the main driving force of the online shopping market.

3.2 Analysis in factors of supplier selection

Based on the above third party logistics development situation, analysing the factors that affect the supplier selection, among them, the third party logistics service provider is the focus of our research, this paper will analyze the following factors.

(1) third party logistics service quality

Third party logistics service quality is the most important factor that should be considered when selecting suppliers, the quality of service will directly affect the reputation of enterprises.

(2) the cost of third party logistics service

cost of logistics service is too high has been an important factor affecting domestic logistics development, enterprises in the selection of third party logistics, reduce the logistics cost of service must follow the principle of logistics service cost will directly determine the cost of enterprises so that this paper will treat the service cost as an important factor.

(3) service of the geographical Scope

The third party logistics provider selection should be consistent with the scope of business enterprise principle, for large multinational companies, its scope of business throughout the world, its sales is bound to be about local culture or influence, in order to further develop the market to help the choice of local suppliers.

(4) third party logistics flexibility

In the network economy environment, customers pay attention to the ability to adapt to rapid changes in the market and customer demand diversification so that logistics flexibility, time flexibility dynamic index is also equally important therefore, the flexible factors, the third party logistics is an indispensable factor.

(5) service provider information sharing ability

B2C enterprises and logistics providers should also realize information sharing, information ability of logistics providers will directly affect the B2C enterprise and its information exchange.

3.3 The establishment of index system of the third party logistics supplier

This article through to the domestic and foreign related literature retrieval and consulting experts and scholars, finishing induction, and ultimately determine the logistics distribution supplier selection index system, as shown in Figure 3.

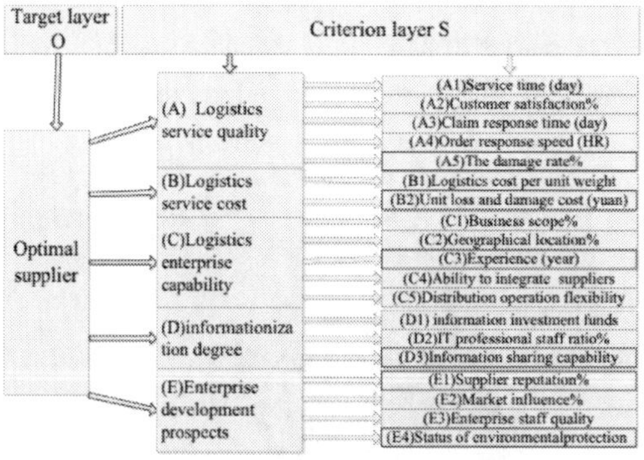

Fig. 3. Caption of the Figure 3. Below the figure.

4 Model establishment of evaluation and selection of supplier

4.1 Improved analytic hierarchy process to determine subjective weights

The traditional AHP method to construct the judgment matrix A, calculate feature vector, to obtain the maximum eigenvalue, but often in the next consistency check is especially cumbersome, because in practice, consistency is not satisfied, with most estimates to adjust the judgment matrix, it is possible to adjust one or not to meet the requirements, with a lot of blindness [10]. This article in order to avoid this kind of blindness, the use of the concept of the optimal transfer matrix, so that the judgment matrix nature to meet the consistency requirements. Steps are as follows.

constructing judgment matrix A

Construct judgment matrix to use of the double factors, determine the degree of importance, the rule layer S and sub rule layer S_{ij} to construct the judgment matrix in this paper respectively.

construct the anti symmetric matrix B

From above, A is Reciprocal matrix, that is $a_{ij} = 1/a_{ji}$, $B = \lg A$, that is $b_{ij} = \lg a_{ij} = \lg \frac{1}{a_{ij}} = -b_{ji}$,B is the structure of the anti symmetric matrix.

Calculation of the optimal transfer matrix C

If C is the transfer matrix, B is the anti symmetric matrix and $\sum_i^n \sum_j^n (c_{ij} - b_{ij})^2$ is the minimum, it can say that C is the optimal transfer matrix for B, $c_{ij} = \frac{1}{n} \sum_{k=1}^{n} (b_{ik} - b_{jk})$.

To calculate the quasi optimal consistent matrix of A*

From $A^*=10^C$,that is $a_{ij} = 10^{c_{ij}} = 10^{\frac{1}{n}\sum_{k=1}^{n}(b_{ik}-b_{jk})} = \frac{a_{ik}}{b_{jk}}$, by the definition of A is the reciprocal matrix, if $a_{ij} = \frac{a_{ij}}{b_{jk}}$,that A is consistent. From the above, A*as a quasi optimal consistent matrix

To determine the feature vector by extract method

A*will be matrix elements in line multiplication, n times of the product, do get normalized, $W=[W_1, W_2, W_3, , , W_n]^T$ as the feature vectors .

Using the above steps, it can get the supplier's final subjective weight vector $w_1=[w_{11},w_{21},w_{31},,,w_{n1}]^T$

4.2 improved entropy coefficient method to determine the objective weight

For the objective weight model more reasonable should be.the objective weight of the model obtained can not only reflect the information of the decision matrix and decision matrix; when the change, change degree of weight should be and the decision matrix changes. Therefore, in order to overcome the above problems, in this paper, based on the traditional entropy model, a flexible entropy coefficient model is established, and the specific steps are as follows.

(1) Structural decision matrix D

Order Y=(y_{ij}), beside, $y_{ij} = \frac{x_{ij}}{\sqrt{\sum_{i=1}^{m} x_{ij}^2}}$, standardized matrix normalization process is $D = (D_{ij})_{m \times n}$.

$$d_{ij} = \frac{y_{ij}}{y_j^{\max}}, j \in J_1 \tag{1}$$

$$d_{ij} = \frac{y_j^{\min}}{y_{ij}}, j \in J_2 \tag{2}$$

$$y_j^{\max} = \max(y_{1j}, y_{2j}, y_{3j},,, y_{mj}), j = 1,2,3,,,n \tag{3}$$

$$y_j^{\min} = \min(y_{1j}, y_{2j}, y_{3j},,, y_{mj}), j = 1,2,3,,,n \tag{4}$$

(2) Entropy coefficient model

For the normalized matrix D, the entropy of the first j attribute is defined as $h_j = \rho - E_j$, $E_j = -\sum_i^n d_{ij}\frac{\ln d_{ij}}{\ln n}$. Then the entropy coefficient model used to solve the objective weight is $\min = w^T kw, s.t. e^T w = 1, w \geq 0$. Lagrange number multiplication.

order $L = w^T kw - \lambda(e^T w - 1)$, it can get the objective weight $w = w_2 = \frac{k^{-1}e}{e^T k^{-1}e}$. From the above, the comprehensive weight is $w_j = \frac{w_{1j} \times w_{2j}}{\sum_{j=1}^n w_{1j} \times w_{2j}}$.

4.3 the TOPSIS method to build the choice of third party logistics service providers evaluation

TOPSIS model is a classical multi-objective decision-making method. Using the ideal approach, it should first construct the ideal and negative ideal solution, by calculating the ideal solution and negative solution of the Euclidean distance between each scheme and ideal scheme, and obtained the various positive and negative ideal solution close to the degree of closeness to the ideal for the supplier; and understanding recently and negative ideal solution the most distant scheme is the solution satisfactory solution. The specific steps are as follows.

(1) Weighted normalized matrix v

$$v = (v_{ij})_{m \times n} = (w_j y_{ij})_{m \times n} = \begin{bmatrix} v_{11} & v_{12} & \cdots & v_{1n} \\ v_{21} & v_{22} & \cdots & v_{2n} \\ \cdots & \cdots & \cdots & \cdots \\ v_{m1} & v_{m2} & \cdots & v_{mn} \end{bmatrix}$$

(2) determine the positive ideal solution v+ and negative ideal solution v-

$$v^+ = \{v_1^+, v_2^+, , , v_m^+\} = \left\{\max_{1\le i\le m} v_{ij} \middle| j \in J_1, \min_{1\le i\le m} v_{ij} \middle| j \in J_2, \right\}$$

$$v^- = \{v_1^-, v_2^-, , , v_m^-\} = \left\{\min_{1\le i\le m} v_{ij} \middle| j \in J_1, \max_{1\le i\le m} v_{ij} \middle| j \in J_2, \right\}$$

(3)Calculate the Euclidean distance and the relative closeness degree and order

The evaluation value of logistics supplier is the distance from the positive ideal solution and the negative ideal solution.

$$S_i^+ = \sqrt{\sum_j^n (v_{ij} - v_j^+)^2}, i = 1,2,,,m \quad S_i^- = \sqrt{\sum_j^n (v_{ij} - v_j^-)^2}, i = 1,2,,,m$$

The evaluation value of logistics supplier is the distance from the positive ideal solution and the negative ideal solution $C_i^* = \frac{S_i^-}{S_i^+ + S_i^-}, i = 1,2,,,m$

By the relative size of the progress of the paste to judge the merits of the supplier, close to a large degree of satisfaction is the best supplier.

5 case study analysis

5.1 Case introduction

There is a B2C e-commerce business clothing, food, daily necessities and other products, since its inception, in the principle of maximizing the interests of customers, quickly become one of the most influential B2C e-commerce business. The company originally only in the first tier cities such as Guangzhou, Shanghai and other implementation of door-to-door, but with the market and customer demand. The B2C e-commerce enterprises have opened the second tier cities of logistics and distribution services, however, the original logistics services have been unable to meet the needs of enterprises. After screening 5 logistics suppliers have become the candidate supplier, respectively A, B, C, D, E said, according to the above proposed evaluation index system and the logistics supplier performance in the previous situation, combined with the actual operating results of enterprises and related experts. The evaluation values of the 5 suppliers are as shown in Table 1.

Table 1. Candidate supplier index evaluation form.

evaluating indicator	**Distributor candidates**				
	A	B	C	D	E
A1	5	3	5	4	6
A2	65.6	70.5	68.9	78.9	58.3
A3	2	3	2	1	1
A4	24	6	18	12	24
A5	3.6	5.3	4.5	2.8	3.2
B1	26	18	29	15	20
B2	0.021	0.036	0.052	0.018	0.046
C1	80.2	78.8	85.6	76.3	89.1
C2	69.1	76.5	80.6	86.2	72.9
C3	12	8	6	10	15
C4	76.4	68.1	80.6	79.3	70.8
C5	72.8	86.5	88.6	78.2	63.4
D1	150	320	460	510	600
D2	16.8	20.1	25.6	19.7	22.4
D3	5	4	3	5	4
E1	82.4	78.6	80.7	79.8	83.9
E2	16.7	10.3	9.8	12.1	18.9
E3	81.2	85.8	69.6	78.5	86.7
E4	69.7	75.6	80.4	78.1	86.8

5.2 Supplier selection and comprehensive evaluation

According to the above evaluation model, the process is as follows.

5.2.1 Establish the initial decision matrix and standardization

Initial judgment matrix X

$$\begin{bmatrix} 5 & 65.5 & 2 & 24 & 3.6 & 26 & 0.021 & 80.2 & 69.1 & 12 \\ 3 & 70.5 & 3 & 6 & 5.3 & 18 & 0.036 & 78.8 & 76.5 & 8 \\ 5 & 68.9 & 2 & 18 & 4.5 & 29 & 0.052 & 85.6 & 80.6 & 6 \\ 4 & 78.9 & 1 & 12 & 2.8 & 15 & 0.018 & 76.3 & 86.2 & 10 \\ 6 & 58.3 & 1 & 24 & 3.2 & 20 & 0.046 & 89.1 & 72.9 & 15 \end{bmatrix}$$

$$\begin{bmatrix} 76.4 & 72.8 & 150 & 16.8 & 5 & 82.4 & 16.7 & 81.2 & 69.7 \\ 68.1 & 86.5 & 320 & 20.1 & 4 & 78.6 & 10.3 & 85.8 & 75.6 \\ 80.6 & 88.6 & 460 & 25.6 & 3 & 80.7 & 9.8 & 69.6 & 80.4 \\ 79.3 & 78.2 & 510 & 19.7 & 5 & 79.8 & 12.1 & 78.5 & 78.1 \\ 70.8 & 63.4 & 600 & 22.4 & 4 & 83.9 & 18.9 & 86.7 & 86.6 \end{bmatrix}$$

Normalized matrix Y

$$\begin{bmatrix} 0.475 & 0.427 & 0.459 & 0.590 & 0.404 & 0.524 & 0.253 & 0.437 & 0.400 & 0.503 \\ 0.285 & 0.458 & 0.688 & 0.147 & 0.595 & 0.362 & 0.434 & 0.429 & 0.443 & 0.335 \\ 0.475 & 0.448 & 0.459 & 0.442 & 0.505 & 0.584 & 0.627 & 0.466 & 0.252 & 0.479 \\ 0.380 & 0.513 & 0.229 & 0.295 & 0.314 & 0.302 & 0.217 & 0.415 & 0.499 & 0.419 \\ 0.569 & 0.379 & 0.229 & 0.590 & 0.359 & 0.403 & 0.554 & 0.485 & 0.422 & 0.629 \end{bmatrix}$$

$$\begin{bmatrix} 0.454 & 0.415 & 0.153 & 0.356 & 0.524 & 0.454 & 0.532 & 0.451 & 0.398 \\ 0.405 & 0.493 & 0.327 & 0.425 & 0.419 & 0.433 & 0.328 & 0.476 & 0.432 \\ 0.479 & 0.505 & 0.470 & 0.542 & 0.314 & 0.445 & 0.312 & 0.386 & 0.459 \\ 0.472 & 0.446 & 0.521 & 0.417 & 0.524 & 0.440 & 0.386 & 0.436 & 0.446 \\ 0.421 & 0.361 & 0.613 & 0.474 & 0.419 & 0.463 & 0.602 & 0.481 & 0.495 \end{bmatrix}$$

5.2.2 Determine the weight of indicators

According to the above requirements to improve the analytic hierarchy process for the subjective weight of the steps, the target layer S for the target layer O, the corresponding judgment matrix, anti symmetric matrix, the optimal transfer matrix and to be and consistent matrix are as follow.

Anti symmetric matrix B

O	S_1	S_2	S_3	S_4	S_5
S_1	0	0.477	0.602	0.778	0.699
S_2	−0.477	0	0.699	0.602	0.778
S_3	−0.602	−0.699	0	0.301	0.477
S_4	−0.778	−0.602	−0.301	0	0.602
S_5	−0.699	−0.778	−0.477	−0.602	0

Optimal matrix C

O	S_1	S_2	S_3	S_4	S_5
S_1	0	0.191	0.616	0.727	1.022
S_2	−0.191	0	0.425	0.536	0.832
S_3	−0.616	−0.425	0	0.111	0.407
S_4	−0.727	−0.536	−0.111	0	0.295
S_5	−1.022	−0.832	−0.407	−0.295	0

Fittingconsistent matrix A^*

O	S_1	S_2	S_3	S_4	S_5
S_1	1	1.552	4.130	5.333	10.52
S_2	0.644	1	2.661	3.436	6.792
S_3	0.242	0.376	1	1.291	2.551
S_4	0.186	0.291	0.775	1	1.972
S_5	0.095	0.147	0.392	0.507	1

Using the root method to deal with the quasi and consistent matrix can target on criterion layer relative weights for .

Subjective and Objective weight can be.

$$w_1 = [0.138,0.041,0.062,0.059,0.162,0.084,0.198,0.006,0.22,0.041,0.031,0.012,0.026,\\ 0.005,0.066,0.029,0.011,0.004,0.014]^T$$

$$w_2 = [0.032,0.025,0.123,0.234,0.033,0.035,0.078,0.027,0.029,0.046,0.026,0.032,0.106,\\ 0.028,0.026,0.031,0.11,0.024]^T$$

Comprehensive weight can be.

$$w = [0.073,0.017,0.123,0.231,0.088,0.049,0.252,0.003,0.11,0.031,0.014,0.007,0.043,\\ 0.002,0.028,0.003,0.018,0.005,0.002]^T$$

5.2.3 Calculate the weighted normalized matrix V

$$
V=\left[\begin{array}{cccccccccc}
0.035 & 0.008 & 0.057 & 0.136 & 0.036 & 0.036 & 0.026 & 0.064 & 0.002 & 0.503 \\
0.021 & 0.008 & 0.085 & 0.034 & 0.034 & 0.053 & 0.018 & 0.109 & 0.001 & 0.335 \\
0.035 & 0.008 & 0.057 & 0.102 & 0.102 & 0.045 & 0.029 & 0.158 & 0.002 & 0.479 \\
0.028 & 0.009 & 0.028 & 0.068 & 0.068 & 0.028 & 0.015 & 0.055 & 0.002 & 0.419 \\
0.042 & 0.007 & 0.028 & 0.136 & 0.136 & 0.032 & 0.020 & 0.140 & 0.002 & 0.629
\end{array}\right.
$$

$$
\left.\begin{array}{ccccccccc}
0.454 & 0.415 & 0.153 & 0.356 & 0.001 & 0.434 & 0.532 & 0.451 & 0.398 \\
0.405 & 0.493 & 0.327 & 0.425 & 0.001 & 0.433 & 0.328 & 0.476 & 0.432 \\
0.479 & 0.505 & 0.470 & 0.542 & 0.001 & 0.445 & 0.312 & 0.386 & 0.459 \\
0.472 & 0.446 & 0.521 & 0.417 & 0.001 & 0.440 & 0.386 & 0.436 & 0.446 \\
0.421 & 0.361 & 0.613 & 0.474 & 0.429 & 0.463 & 0.602 & 0.481 & 0.495
\end{array}\right]
$$

The positive and negative ideal solutions and the European distance between the candidate suppliers and the positive and negative ideal solutions are

$$V^{+}=(0.021,0.009,0.028,0.068,0.028,0.015,0.055,0.002,0.006,$$
$$0.019,0.007,0.004,0.026,0.001,0.015,0.002,0.011,0.003,0.001)$$

$$V^{-}=(0.042,0.007,0.085,0.136,0.053,0.029,0.158,0.001,0.005,$$
$$0.008,0.006,0.003,0.002,0.001,0.009,0.001,0.006,0.002,0.001)$$

$$S_1^{+}=0.0794,\ S_2^{+}=0.0905,\ S_3^{+}=0.116,\ S_4^{+}=0.0107,\ S_5^{+}=0.1109$$

$$S_1^{-}=0.1009,\ S_2^{-}=0.1163,\ S_3^{-}=0.0492,\ S_4^{-}=0.1415,\ S_5^{-}=0.0698$$

The closeness of each candidate supplier is calculated by the formula

$$C_1^{*}=0.560,\ C_2^{*}=0.788,\ C_3^{*}=0.298,\ C_4^{*}=0.93,\ C_5^{*}=0.086$$

From this, it can see that the supplier D close to the highest degree, is the best partner of B2C enterprise.

6 Concluding remarks

Aiming at the logistics service supplier business problems, establish the index system and determine the weight by using the improved AHP method and the entropy weight coefficient method, and then use the ideal approximation method to calculate the degree of closeness of each supplier, according to the order of suppliers close to the level, select the closest suppliers, is the best partner. Through the design and analysis of the case, it is proved that this method is feasible and effective for the third party logistics supplier selection provides a reference, but this article also has some shortcomings, such as the factor analysis, may not be able to cover all the factors may also change with the times, some new factors will be taken into account. I believe that research in this area will be more comprehensive, more and more mature.

References

1. Artis M, Krolzig H M, Toro J. The European Business Cycle// the ERM', International Journal of Finance and Economics. (1999).
2. Dickson G.W.An Analysis of Vendor Selection Systems and Decisions . Journal of Purchasing, **2**,6(1966).
3. Johnson M.V. Partner Selection in the Agile Environment, Creating the Agile Organization: Models, Metrics and Pilot, 4th Annual conference proceedings, **21**,6(1995) .
4. Ma Shihua, Lin Yong, Chen Zhixiang. Supply chain management (First Edition) . Beijing: China Machine Press, 2000.
5. Ma Xuefen, Liu Yiyong, Sun Shudong, et al. Evaluation and selection of the three party logistics enterprises under the environment of supply chain management . Computer engineering and applications, 14,3(2003).
6. Yan Xiuxia, Sun Linyan, Luo Bo. Multi objective decision making of third party logistics supplier selection . China mechanical engineering,2004.
7. Zhao Xiaoyu, Huang Xiaoyuan, Sun Wei. Supply chain design method for key suppliers and customers . system engineering,**23** ,6(2005).

8. Liu Gang. Classification and evaluation of suppliers based on entropy method [J]. Journal of systems management, **19**,6(2010).
9. Lu Zhigang, Chen Hangjuan. Fuzzy multiple attribute decision making supplier selection method basedon information entropy. application of computer systems, **18**,3(2012).
10. Liu H T, Wang W K. An integrated fuzzy approach for provider evaluation and selection in third-party logistics. **36**,3(2009).

A New Stochastic Geometry Model of Coexistence of Wireless Body Sensor Networks

Ming Yang[1], Ruixia Liu[1,2*], Yinglong Wang[1,2], Minglei Shu[1] and Yuanyuan Yang[1]

[1]Shandong Computer Science Center(National Supercomputer Center in Jinan), Shandong Provincial Key Laboratory of Computer Networks, Ji'nan 250014, China
[2]College of Information Science and Engineering, Shandong University of Science and Technology, Qingdao 266590, China
Corresponding Email: liurx@sdas.org

Abstract. Stochastic geometry, in particular Poission point process theory, has been widely used in the last decade to provide models and methods to analyze wireless networks. It is a branch of mathematics which deals with the study of random point processes. There are various models for point processes, typically based on but going beyond the classic homogeneous Poisson point process. Poisson point process cannot be used to model the spatial distribution of the simultaneously active transmitters. A novel framework has been presented for modeling the intensity of simultaneous active transmitters of a random carrier sense multiple access wireless sensor network. This thinning rule uses a second-neighbors distance-dependent method, which controls too many nodes deleted of points close together.

1 Introduction

WBSNs enable wireless communications between several miniaturized body sensors and a single coordinator worn on the human body. WBSNs offer many promising new applications in the area of remote health monitoring systems to measure specified physiological data and also provide location-based information. Since a WBSNs provide remote and continuous medical monitoring for patients without constraining their movements, it plays a crucial role in next generation healthcare applications [1-2].

WBSNs for health monitoring systems are required to meet stringent performance demands regarding the tradeoff between reliability, latency, and power efficiency. WBSNs feature limited range and bandwidth and they are prone to interference. These modes incur issues regarding the coexistence of multiple WBANs. There is often a group of BSNs together to happen in hospitals for patients and staff or for the elderly at nursing homes [3]. In many cases, several WBSNs coexist in a small area, resulting in very strong inter-BAN interference, which seriously disturbs intra-BAN communications [4].

In this paper, we use stochastic geometry to analyze the coexistence of IEEE802.15.4-based WBSNs Stochastic geometry is an established branch of mathematics that studies uncertainty in geometric structures. The density of effective points in model is a sufficient parameter to describe the network performance. However, The HCPP model still suffers from the underestimation problem. So we proposed a PCMP model to mitigate this problem.

We propose a new thinning method of CSMA network to mitigate the node intensity underestimation problem of HCPP model. We put forward a modified Poisson cluster and marked process (PCMP) to adapt to this network coexistence The PCMP model is generalized for CSMA protocol's network.

Define:

Contention domain: node y is in the contention domain of node x if the power received by x from y is above some detection threshold.

Neighbors: the neighbors of a node are the nodes in its contention domain.

Second Neighbors: the neighbors of a origin node's neighbors node.

The rest of the paper is organized as follows: Section 2 introduces the related work. Section 3 proposes and describes the system model in detail. In Section 4, thinning theory and stochastic geometry are used to investigate the CSMA-based wireless sensor networks coexistence. Section 5 illustrates and analyzes numerical results. We conclude our work in Section 6.

2 Related work

Since the Poisson Point Process (PPP) is highly tractable, it is frequently used to model a variety of networks, such as celluar networks, mobile ad hoc networks, cognitive radio networks and wireless sensor networks[5-6]. In [7], the authors assume the star topology network's transmission nodes to be uniformly distributed according to a PPP. this paper stands in the precise consideration of CSMA algorithms, considering the collision, the success transmit probability and interference in the same network. The matern hard core are often used to model concurrent transmitters in CSMA networks [8-9]. Hesham ElSawy et al. [10] adopt the Poisson cluster process to obtain the minimum required number of channels of the coexisting IEEE 802.15.4 networks, but the underestimation problem is not fully resoved.

3 System model

The carrier sense multiple access with collision avoidance (CSMA-CA) is widely employed in wireless networking due to its simplicity and performance efficiency [11] .

CSMA create exclusion zones to protect scheduled transmissions that nodes which are close by never transmit simultaneously. CSMA-type MACs essentially create a guard zone around the receiver [12]. The traditional homogeneous PPP doesn't suit to this coexisting IEEE 802.15.4 networks, so we proposed PCMP model.

We consider WBSNs, each contains one receiver(CN: the cord node) and N transmitters(SN: sensor node). In our proposed model, the CN is placed at center of the human body. Some SN are deployed on a human body, forming a star topology with the CN such as Figure 1. Each SN has equal power and computation capabilities.

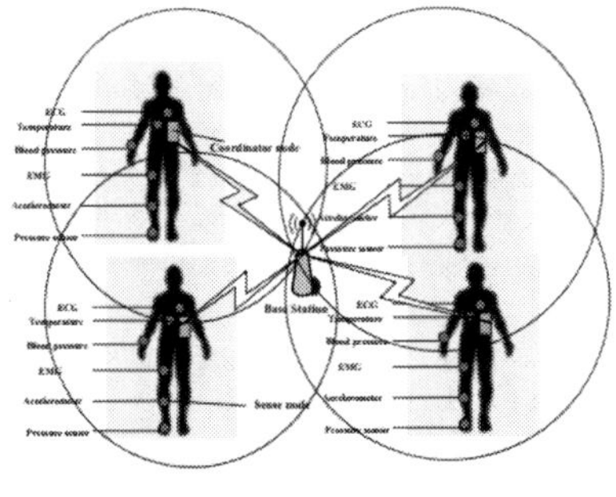

Fig. 1. System model.

The WBSN's locations are modeled as a stationary and isotropic Poisson cluster process (PCMP) coexisting in the R^2 Euclidean space. The cluster process consists of the cluster centers (CNs) and cluster members (SN). The CNs spatial distribution model is a PPP with intensity λ_{ppp} For the Poisson cluster process , each cluster member is uniformly distributed around the cluster center.

This PCMP is denoted by $\phi = \{x_i, m_i, p_i\}$, where

$\{x_i\}$ denote the locations of the transmit points, is always assumed Poisson with positive and finite intensity λ_{ppp} .

$\{m_i\}$ are marks, uniformly distributed on [0,1].

$\{p_i\}$ denote the transmit power by node i .

We denote target domain of area as M. Denoting by K the number of CNs in M. K is Poisson distributed with mean λ_{ppp} M and probability mass function (pmf). M is the standard Lebesgue measure of M. So the numbers of CN in the domain A is:

$$P(k = n) = \frac{(\lambda_{ppp} M)^k e^{-\lambda_{ppp} M}}{n!}$$

Define the contention domain:

$$N_x = \{(x_i, m_i, p_i) \in \phi : p_i h \,|\, |x_i - x_j| \,|^{-\alpha} \geq \beta\}$$

The medium access indicators $e_i = 1(\Delta x_i \in N_x, x_i < x_j)$

The lower the carrier-sensing threshold β the higher the sensitivity of each transmitter to other transmissions occurring in the spatial domain, which increases the distances between simultaneously active transmitters and decreases the mutual interference at the expense of decreasing the spatial frequency reuse [13].

4 Thinning method

To consider a realistic medium-access control (MAC) layer algorithm, successful signal reception is based on the protocol model, and nodes employ carrier-sense multiple access [14]. One of the best choices for such scenarios is to use pure carrier sense multiple access (CSMA), which only incorporates channel sensing. In the CSMA protocol, every communication entity first senses the ongoing transmission in the channel and then determines when to start transmitting [15]. A node which wants to access the shared wireless medium senses its occupation and refrains from transmitting if the channel is already locally occupied.

Throughout this paper, we will say that the transmitter x_i is older than the transmitter x_j (or equivalently x_j is younger than x_i) to denote that $m_i > m_j$ .The hard core point process (HCPP) accounts only for the points having the lowest mark in their contention domains [16].

The classical HCPP leads to the following scenario: node x_1 is not retained because of it detects its neighbors x_2; moreover node x_2 in turn is not retained because it detects its neighbor x_3 $(m_1 > m_2 > m_3)$, such as the nodes with marks 0.9,0.3,0.2 in Figure 2. Consequently, in the Matern CSMA model neither x_1 nor x_2 is retained. But if x_3 is not the neighbor of x_1, and if x_3 does not detect any of its neighbors then a more reasonable MAC would allow x_1 and x_3 to transmit simultaneously [17]. Such phenomena underestimate the density of the actual running network. Node is retained rules:

(1) $\{m_i < m_j \mid m_i, m_j \in B_{x_i}(\beta)\}$

(2) $\{m_j < m_d < m_l \mid m_d, m_l \in B_{x_j}(\beta) \setminus x_j\}$

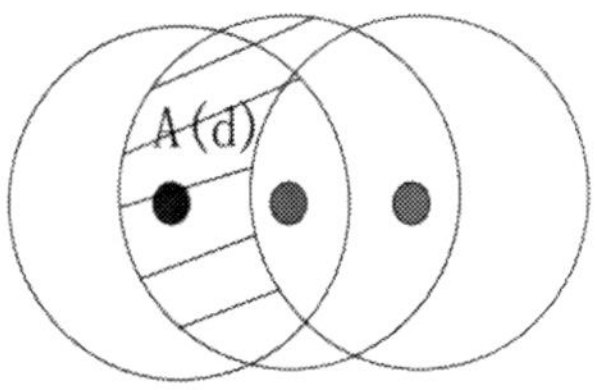

Fig. 2. The second neighbour node.

The shaded area:

$$A(\mathrm{d}) = \pi R^2 - 2R^2 \arccos(\frac{d}{2R}) + \frac{d}{2}\sqrt{4d^2 - R^2}$$

According to the distance properties of a PPP:

$$0 < d < R$$

$$E(\mathrm{d}) = \frac{1}{2\sqrt{\lambda_{ppp}}}$$

$$E(A(\mathrm{d})) = \pi R^2 - 2R^2 \arccos(\frac{E(d)}{2R}) + \frac{E(d)}{2}\sqrt{4E(d)^2 - R^2}$$

The probability of x_k is the lowest mark in the A(d):

$$P(A) = P\{m_k < m_u, m_u \in (B_{x_j}(\beta) - B_{x_i}(\beta)) \setminus x_j\}$$

The probability of the nodes retained:

$$
\begin{aligned}
p &= \sum_{n=0}^{\infty}\frac{1}{n+1}\frac{\Lambda_{B_{x_i}}^{n}e^{-\Lambda_{B_{x_i}}}}{n!}\sum_{m=0}^{\infty}\frac{1}{m+1}\frac{\Lambda_{A(d)}^{m}e^{-\Lambda_{A(d)}}}{m!}\\
&= \frac{1-e^{-F}}{F}\,\frac{1-e^{-H}}{H}\\
p_{PCMP} &= \frac{1-e^{-F}}{F}\left(1+\frac{1-e^{-H}}{H}\right)\\
\lambda_{PCMP} &= \lambda_{PPP}\,p_{PCMP}\\
&= \lambda_{PPP}\frac{1-e^{-F}}{F}\left(1+\frac{1-e^{-H}}{H}\right)
\end{aligned}
$$

5 Simulation results

We have distributed many transmission nodes in the 20m*20m area according to certain density. Set the contain domain R around the origin. If the closet node's distance is less than a certain value, the node with the smallest mark is retained.

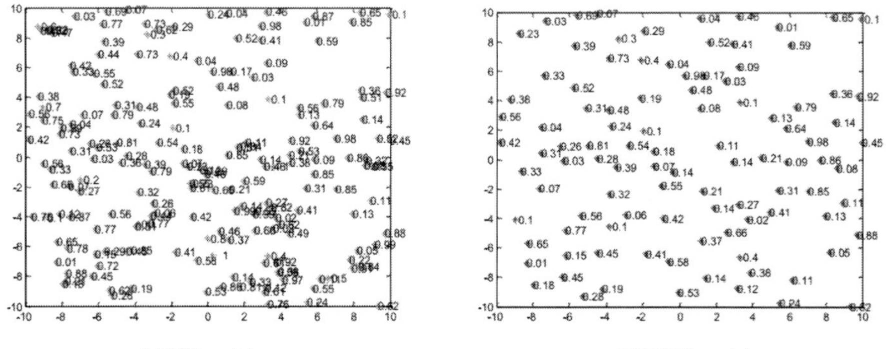

(a)PPP model. (b)PCMP model.

Fig. 3. PPP model and PCMP model.

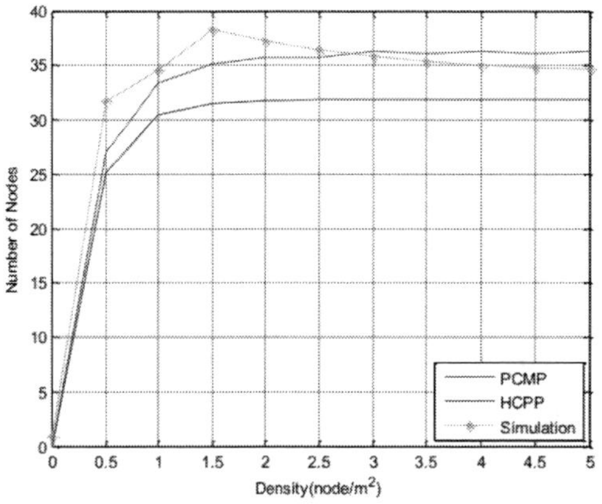

Fig. 4. The retained nodes of HCPP model and PCMP model.

Figure 3 compare the classical hard core point process and the PCMP via Monte Carlo simulations using Matlab to verify the accuracy of mode. We distributed many transmitter nodes in the 100 meters of area according to certain density. Set the contain domain is 2 m. If the closet node's distance less than 2, the node with the smallest mark is retained. The underestimation problem of HCPP model lead to the many of nodes is unselected, so in the Figure 4 the number of remaining nodes in PCMP model is much more than the HCPP mode. Their number of selected node is the same in the low density because the intersect probability of contain domain is very small.

6 Conclusion

PCMP is derived by applying a specific thinning rule to a homogeneous Poisson point process in R^2. In the following, the way of thinning a Poisson point process in order to obtain a moderate point process.The proposed model will be useful for capturing the intensity of simultaneously active nodes in a random CSMA wireless sensor networks.

APPENDIX A

$$p=\sum_{n=0}^{\infty}\frac{1}{n+1}p(|\mathrm{B}_{x_i}(\mathrm{R})\setminus \mathrm{x}_i|=\mathrm{n})\sum_{m=1}^{\infty}\frac{1}{m+1}p(|\mathrm{A(d)}|=\mathrm{m})$$
$$=(\frac{e^{-F}}{F}\sum_{n=0}^{\infty}\frac{F^{n+1}}{(\mathrm{n}+1)!})(\frac{e^{-H}}{H}\sum_{m=0}^{\infty}\frac{H^{m+1}}{(m+1)!})$$
$$=\frac{1-e^{-F}}{F}\frac{1-e^{-H}}{H}$$

Where $F=\wedge_{B_{x_i}}=\lambda_{ppp}\pi R^2$ is the mean number of nodes in B_{x_i}, $H=\wedge_{A(\mathrm{d})}=\lambda_{ppp}E(\mathrm{A(d)})$ E(A(d)) is the mean numbers of nodes in $E(\mathrm{A(d)})$.

References

1. G.Z. Yang, Body Sensor Networks, Springer, （2006）.
2. A. Pantelopoulos, N.G. Bourbakis, A survey on wearable sensor-based systems for health monitoring and prognosis, IEEE Transaction on System, Man, and Cybernetics **40**, (2010).
3. M. Deylami, E. Jovanov, Performance Analysis of Coexisting IEEE 802.15.4-Based Health Monitoring WBANs, 34th Annual International Conference of the IEEE EMBS, (2012).
4. W. Lusheng, C. Goursaud, N. Nikaein,etc, Cooperative Scheduling for Coexisting Body Area Networks, IEEE Trans.Wireless Commun., (2012).
5. F. Martelli, Modeling Query-Based Wireless CSMA NetworksThrough Stochastic Geometry, IEEE Trans. Veh. Technol., **63**, (2014).
6. B. Cho, K. Koufos, Bounding the Mean Interference in Matern Type II Hard-Core Wireless Networks, IEEE Wireless Commun. Letters, **2**, (2013).
7. A. Busson, G. Chelius, J. Gorce, Interference Modeling in CSMA Multi-Hop Wireless Networks. (2009).
8. E. Hesham, H. Ekram, C. Sergio, Spectrum-efficient multi-channel design for coexisting IEEE 802.15.4 networks:a stochastic geometry approach. IEEE J. Sel. Areas Commun, **13**,(2014).
9. F. Martelli, C. Buratti, R. Verdone, Modeling query-based wireless CSMA networks through stochastic geometry. IEEE Trans. Veh. Technol., **63**,(2014).
10. F. Baccelli , B. Blaszczyszyn . Stochastic geometry and wireless networks, Volume II: applications. Paris, Now Press, (2005).
11. E. Hesham, H. Ekram, A Modified Hard Core Point Process for Analysis of Random CSMA Wireless Networks in General Fading Environments, IEEE Trans.Wireless Commun., **61**, (2013).
12. S. Hamed, M. Pakravan, Analytical Modeling and Performance Analysis of Flooding in CSMA-Based Wireless Networks, IEEE Trans. Veh. Technol., **61**, (2013).
13. S. Chiu, D. Stoyan, Stochastic geometry and its applications. Third Edition. UK, (2013).
14. P. Yiyang, L. Yingchang, K. Chan, Energy-Efficient Design of Sequential Channel Sensing in Cognitive Radio Networks: Optimal Sensing Strategy, Power Allocation, and Sensing Order, IEEE J. Sel. Areas Commun., **29**, (2011).
15. J. Akhtman, L. Hanzo, Power versus bandwidth efficiency in wireless communications: the economic perspective. In:Proceeding of Proc. IEEE VTC-Fall09, (2009).
16. Y. Fengji, S. Yi, B. Sikdar, Scaling of Spatial Reuse and Saturation Throughput in a Class of MAC Protocols, IEEE Trans.Wireless Commun., **10**, (2007).
17. W. Xiaofei, T. Berger, Spatial channel reuse in wireless sensor networks, Wireless Netw., **14**, (2008).

Internet of Things: Application and Prospect

Ran Liu[1] and Jinfeng Wang[1]

[1]School of Management Engineering,Zhengzhou University,Zhengzhou Henan 450001,China

Abstract. Nowadays, Internet of Things (IOT) is growing rapidly, which is used to the extension of the Internet and the Web into the physical realm, by means of the widespread deployment of spatially distributed devices with embedded identification, sensing and actuation capabilities. Internet-of-Things envisions a future in which digital and physical entities can be linked, by means of appropriate information and communication technologies, to enable a whole new class of applications and services. In this paper, the origin and concept of the Internet of things are introduced, and then the potential application areas and the impact of the Internet of things are analyzed.
Keywords. Internet of Things, Application research.

1 Introduction

Now, about 2 billion people around the world use the Internet to browse web pages, send and receive messages, use social networking applications and or do the other activities. With the development of the Internet information technology, Internet of things (IOT) has become a new round of economic and technological development, one of the strategic high ground in the world. In the information industry field, whether engineers engaged in the information technology industry consulting and research of the company's or consumers of the information technology products, are optimistic about the prospects of the Internet of things. From Google to apple, Microsoft, more and more IOT related products appeared in the current industry, all of them investing a large amount of capital and manpower in the field of IOT to develop the consumers favorite products, hope to lead the development of the industry [1]. IOT as a global platform for physical entities and intelligent devices to communicate, calculate and coordinate, It is foreseeable that the deep integration of IOT technology and other industries can change the mode of economic development, achieve the real green growth and low-carbon economy. The IOT economy will lead the future of sustainable economic and social development.

The goal of this paper is to study the Internet of Things with a new angle, including the concept and application field of IOT. In fact, research topics related to IOT topics are still highly fragmented and largely focused on single application domains or single technologies. In addition, participation in the field of networking and communication science is still limited, although their contribution to the development of the field has great potential. The main contents of this paper are as follows: Firstly, introduced the main concepts of Internet of Things. Then the potential applications and impact areas of IOT are analyzed. Finally, the research contents of this paper are summarized.

2 Origin and definition

2.1 Origin

The concept of IOT in 1999, by the Massachusetts Institute of Technology Auto-ID Lab, the first idea of the IOT is using Radio Frequency Identification (RFID), encoding and Internet technology to establish a kind of global information sharing of the "Internet of Things". From its name can know that IOT are connected the Internet with things. There are two meanings: First, the core technology of things is still the Internet, is the extension and expansion of the Internet. Second, the client extends and expansions between the objects and objects, so as to carry out information exchange and communication. IOT through the intelligent perception, identification technology, pervasive computing and other communication-aware technology integrates into the Internet widely. Application characteristics is the core of IOT, user experience and innovation is the soul of the development of IOT [2]. In 2003, the United States, "Technology Review" propose the point that "sensor network technology will change the people's way of life in the future as the top ten technology." In 2005, the International Telecommunication Union (ITU) formally defined the concept of IOT, published "ITU Internet Report 2005: Internet of things", and describes the form of IOT, features, technologies, opportunities and challenges. With the development of science and technological progress, the concept of IOT is also expanding. The current IOT can achieve a comprehensive sense of things, identification and control, through the wireless connection to the network, intelligent reasoning and decision-making.

2.2 Definition

Sun Qibo [3] puts forward the concept of Internet of things from the definition origin, development, features. From a narrow sense, things are connected to multiple items of the network, and these items can be identified and managed. The broad sense of things can be considered to be a fusion of the physical world and information world, everything will be networked, digital. Achieve efficient information exchange between the environment and people, objects and objects, people and objects. the information technology will be into our society deep-dyed.

The Internet of things white paper, "the Internet of things is the extension of network's application and the Internet, it uses sensing technology and intelligent device for perception to sense the physical world. Calculating by Internet, process and knowledge mining to realize information exchange between people and objects, objects and objects, so as to seamless links to physics the real-time control, to accurate management and scientific decision-making [4]."

Zhou Hongbo [5] and other scholars believe that the concept of the definition refers to the terminal equipment and facilities everywhere, through a variety of wireless or wired, short or long distance to achieve interoperability ,application integration, and cloud computing operating mode based on network, Internet environment. Using appropriate security mechanisms to provide safe and controllable personalized location tracing, alarm linkage, dispatching plan management, remote control, security, remote maintenance, online upgrade, statistics, decision support, leadership desktop management and service functions, to achieve high efficient, energy saving, safety, environmental protection tube control, business integration."

Zhu Hongbo [6] that IOT is a name with the network will combine things together. IOT is currently define as a net that making use of radio frequency identification (RFID), infrared sensors, GPS, laser scanners and other information sensing equipment, and according to the agreement, the goods connect with the network to realize the exchange of information and communication, intelligent identification, positioning, tracking, monitoring and management of a network.

IOT use local network and Internet communication technology such as sensors, controllers, machines to realize communication between people and objects, objects and objects by a new contact links .So as to achieve information resources , Remote program control and intelligent control network [2] .

3 Application and prospect

IOT has a great potential for development in almost all areas, which mainly due to the Internet of things can be context-aware (for example, can collect information on natural parameters, medical parameters, or user habits), and provide tailored services. Regardless of the application area, the purpose of these applications is to improve the quality of people's everyday life and will have a profound impact on the economy and society. The application of IOT can be divided into three areas: industrial areas, smart city areas and health areas. Each domain is not separate but partially overlapping, because some applications are shared. For example, product tracking, usually in the industrial and health fields, can be used to monitor goods or food, but it can also be used to monitor the distribution of medicines.

Table 1. The main application areas of Internet of things.

Main Application Areas	Domain Segmentation	Examples
Industry	Logistics and Product Lifecycle Management	Inventory management, cargo transport
		Food preservation
	Agriculture and animal husbandry	Agricultural Production and Breeding
		Animal Disease Infection Control
	Industrial processes	Real - time vehicle diagnosis
		Baggage Management
Smart city	Smart Home / Building	Home Area Network (HAN)
		Multimedia management, power management
	Public Safety and Environmental Monitoring	Environmental monitoring
		Smart mobile devices
Health Care	Disease diagnosis and treatment	Real-time chiri-cal consmlfation
		Vital signs monitoring
	Independent live	Assisted persons with disabilities
		emergency rescue

Table 1 shows the breakdown of the above three application areas of the IOT. It is noteworthy that not all of the IOT applications have the same level of maturity. Some of these applications, especially for the user is the simplest and most intuitive application, is already part of our daily lives. But many are still in the experimental stage, because they need to work with different users for better cooperation. Finally, there are many applications in its infancy. The remainder of this section provides a detailed introduction to the various areas of application.

3.1 Industry

IOT can be applied to all industrial activities, including inter-firm commercial or financial transactions, organizations and other entities, mostly used in logistics, manufacturing, regulatory processes, services, banking, intermediaries and so on.

3.1.1 Logistics and product lifecycle management

A related example of industrial applications is logistics and supply chain management. Electronic tags attached to specific objects can be used to identify the type of material and commodity that can be clothing, furniture, equipment, food or liquids [7]. The use of electronic tags contributes to effectively manage warehouses and retail and reduce inventory through accurate knowledge of the current inventory, while reducing inventory uncertainties. The entire lifecycle of the goods can also be tracked [8], for example, a RFID reader installed at a manufacturing facility can monitor the production process, and the label can be traced back to the entire supply chain (e.g., packaging, transportation, warehousing). An advanced IOT system consisting of RFID equipment and smart shelf real-time tracking equipment can help to reduce material waste, thereby reducing costs and improving margins for retailers and manufacturers. For example, if the shelf keeps some free items, sales are reduced by approximately 8.3% [9]. If there is a correct estimate of what is needed, the shortfall and overproduction can be substantially reduced, which can be inferred from the data collected on the smart shelf. In addition, the real-time analysis of the sensor can determine the deterioration of the product, which is essential to food and liquid. For example, to ensure continuous monitoring of the temperature and humidity in a warehouse or cold store for fresh perishable goods such as fruit, vegetables, frozen food, the actuator can adjust the relevant parameters to best ensure the quality of the food.

3.1.2 Agriculture and animal husbandry

IOT contributes to agriculture and breeding. Actually, animal traceability provisions require continuous monitoring of animals and their migration in order to report any relevant events, such as disease, to the authorities in a timely and appropriate manner. The use of IOT identification systems (e.g., radio frequency identification, sensors) allows the identification and monitoring of animals and isolates infected animals from healthy animals [10], thereby avoiding the spread of infectious diseases. Advanced chips can store information about the physical condition of animals (e.g., quantity information, veterinary examination, disease infection, vaccination) or send information about the health of animals(e.g., temperature) to simplify animal health certificates, control trade and imports, and avoid possible fraud. By analyzing the data collected, the authorities can verify the actual number of livestock reported by the local breeder and provide subsidies. The use of advanced sensor systems to monitor and control agricultural production and feed (such as additives, pigments) is a further application of the Internet of Things.

3.1.3 Industrial processes

IOT can provide advanced solutions for the automotive industry. Real-time vehicle diagnosis is a key application, everything can be monitored by a specific sensor: tire pressure, motor data, fuel consumption, position, speed, distance from other vehicles, etc., and then report the sensed data to the central system [11]. The wireless identification technology attached to the vehicle component can record the history of a particular motor moving part and improve the assembly process by automatically finding missing parts. The application of IOT technology makes people and goods transportation system more advanced, such as ticket pricing, more secure baggage management based on automatic tracking and sorting. The industrial management system based on IOT technology can monitor the industrial plant, for example, to reduce the incidence of accidents, especially in high-risk areas (e.g., oil and gas plants), sensors connected to containers for the transport of dangerous goods may send different signals indicating the chemical composition of the cargo and the composition Of the highest level. In the case of critical situations (e.g., where a chemical composition is approaching the maximum allowed level in a particular geographical area), the sensor can automatically send an alarm to the control center.

3.2 The area of smart city

IOT will help to improve the environmental sustainability and the life quality of the people in our country. The key point is the energy and how to use the resources efficiently, and seeking intelligent solutions to enjoy personal life

3.2.1 Smart home / building

Integrating communication technology into building and housing, the future smart city building or home will be equipped with a variety of sensors and smart devices (e.g., broadband gateway, cell phone, computer, TV, surveillance camera and lamplight). Some application programs use the simplest functions in the IOT, for example, the program to ensure the safety (e.g., video surveillance and intrusion detection), plant management and maintenance (e.g., fault detection), service Automation (e.g., HVAC and lighting), and entertainment system (e.g., multimedia distribution at home). Other types of applications integrate smart grid, and optimize the domestic consumption situation at the same time [12]. For example, the home area network (HAN) allows the electrical appliances to interact with the intelligent instrument, and ensures the required performance while reduces the cost. It can also arrange a variety of home appliances with intelligent work to avoid the peak period. People can see the using situation of home appliances and other equipment through the smart phone anywhere and anytime, monitor the running state of the system intelligently, and control home appliances at a specified time to open, close or set running parameters to make the family environment more suitable for family members to live. For example, through the analysis of information flow, the system can understand the time to arrive home of a person, open the door and light, and fill the bathtub with water. Users can change or cancel the automatic operation at any time.

3.2.2 Public safety and environment monitoring

Public safety services include the maintenance of public order, the prevention and protection of citizens, and the protection of public and private property. Emergency management helps the society to prevent and respond to natural or man-made disasters, such as chemical leakage, water flood, conflagration, epidemic disease and power interruption. The IOT provides solutions to monitor and track these emergencies [13]. Collecting data from a fixed camera and personal equipment located within the city can implement advanced video surveillance and regional monitoring, and help the police in holding sports events, music performances, political meetings to maintain public order at the same time. Private and public building security (e.g., banks, stores) can be enhanced by using the sensor technology to trigger the alarm, emergency action can also be improved and strengthened. At present, the emergency system is short of accurate information about the scene of the emergency. Dedicated sensors and smart cameras, as well as global positioning system (GPS) providing real-time positioning and tracking of and wireless technology, can be used to outline the adumbration of an event, forecast the trend of events (e.g., the direction and speed of fire spread) so as to establish a dynamic emergency planning to coordinate the rescue operation.

3.3 Health care

IOT will play an important role in the development of intelligent services to improve people and social activities. IOT that involve citizens and communities in governance and government decision-making (for example, online political participation) to allow people to live independently or to maintain their social relationships, improve health and social care.

3.3.1 Disease diagnosis and treatment

The healthcare industries will be strongly influenced by IOT. Advanced sensing devices can monitor patient vital signs (e.g., body temperature, blood pressure, heart rate, and cholesterol levels) in real time and then transmit the acquired data to a doctor via a specific communication technique (e.g., Bluetooth, Wireless HART, ISA100) for diagnosis and monitoring of the patient's signs. Body area networks (BANs) are interconnected through wearable devices that allow remote monitoring of patient status outside the hospital [14]. Internet of things can also be used in the identification of drugs and medical equipment identification, for example, the use of smart labels will ensure that items are accurately tracked to prevent loss or theft, or items in a surgical procedure to be left in the patient (e.g., gauze or other small items). Smart tags are also important for the inventory management of medical devices.

3.3.2 Independent live

IOT has many advantages that can improve the quality of people's live. Wearable sensors if connect to personal computer devices and run specific applications can allow people to record their daily activities (walking distance, calorie expenditure, exercise conditions, etc.), providing many ways for them to improve their lifestyles and prevent health problems. IOT can be added through the sensor device to monitor the physical state of the elderly. By monitoring the sensor's real-time acquisition of physiological status information, the system will be able to set up a medical alert, advise on hospitalization. It is also possible to diagnose early Alzheimer's disease by observing abnormalities in normal life. The mobile assistant is also a basic service that allows the elderly to move safely in the city or to use public transport. Through the data collected by the personal mobile device (equipped with sensors such as position sensors, direction sensors, motion obstacle detection sensors, cameras) and the fixed sensor in the city, the mobile phone artificial intelligence system can perceive the surrounding environment and transmit the related information to the user through the synthesized speech. Similar applications can help visually impaired people improve their ability to move around the city [15].

4 Conclusion

IOT may represent the next leap forward in the information and communications technology field. The large-scale deployment of embedded applications increases the possibility of a real-world and virtual world to merge seamlessly, opening up new research and business directions.

This article provided an overview of the key issues related to the development of IOT technologies and services. The most relevant application fields have been presented. We do hope that this survey will be useful for researchers and practitioners in the field, helping them to understand the huge potential of IOT. Although we have outlined the main applications of Internet of things, but The challenges that IOT may be faced in practical applications of the future may not be covered in this paper, future research can proceed from this aspect.

References

1. J. Ma, X. Zhou, S. Li, Z. Lio, Connecting agriculture to the internet of things through sensor networks, in: Proceedings of Internet of Things (iThings/CPSCom), 2011, pp. 184–187.
2. Liu Shuncai. Discussion on the development of the Internet of things technology and its application of network security technology and application, 2016,04:93-94.
3. Sun Qibo, Liu Jie, Li Shan, Fan Chunxiao, Sun Juanjuan. The Internet of things: concept, architecture and key technology research. Journal of Beijing University of Posts and Telecommunications, 2010,03:1-9.
4. Zhou Jin. The basic theory and key technology of information fusion in the Internet of things. Jilin University, 2014.
5. Internet of things white paper (2011). China Public Security (integrated version), 2012, Z1:138-143.
6. Zhu Hongbo, Yang Longxiang,. Research on the technology and application of Internet of things. Journal of communication, 2010,11:2-9.
7. F. Villanueva, D. Villa, F. Moya, M. Santofimia, J. Lopez, Internet of things architecture for an RFID-based product tracking business model, in: Proceedings of 6th Innovative Mobile and Internet Services in Ubiquitous Computing (IMIS), 2012, pp. 811–816.
8. H. Cai, L.D. Xu, B. Xu, C. Xie, S. Qin, L. Jiang, IoT-based configurable information service platform for product lifecycle management, IEEE Trans. Indust. Inform. 10 (2) (2014) 1558–1567.
9. T.W. Gruen, D.S. Corsten, S. Bharadwaj, Retail Out-of-Stocks: A Worldwide Examination of Extent, Causes and Consumer Responses, Tech. rep, 2002.
10. A.S. Voulodimos, C.Z. Patrikakis, A.B. Sideridis, V.A. Ntafis, E.M. Xylouri, A complete farm management system based on animal identification using RFID technology, Comp. Electron. Agricult. 70 (2) (2010) 380–388.
11. P. Hank, S. Müller, O. Vermesan, J. Van Den Keybus, Automotive ethernet: in-vehicle networking and smart mobility, in: Proceedings of the Conference onDesign, Automation and Test in Europe (DATE'13), 2013, pp. 1735–1739.

12. J. Lu, T. Sookoor, V. Srinivasan, G. Gao, B. Holben, J. Stankovic, E. Field, K.Whitehouse, The smart thermostat: using occupancy sensors to save energyin homes, in: Proceedings of the 8th ACM Conference on EmbeddedNetworked Sensor Systems (SenSys '10), 2010, pp. 211–224.
13. E. Sun, X. Zhang, Z. Li, The internet of things (IOT) and cloud computing (CC)based tailings dam monitoring and pre-alarm system in mines, Safety Sci. 50(4) (2012) 811–815.
14. F. Delmastro, Pervasive communications in healthcare, Comp. Commun. 35(11) (2012) 1284–1295.
15. R. Manduchi, J. Coughlan, (computer) vision without sight, Commun. ACM 55(1) (2012) 96–104.

Applying TPM on Production Line of Huanwei

Yunkai Zhai [1,2,3] and Jing Zhang [2]

[1]Management Engineering School, Zhengzhou University, Zhengzhou, People's Republic of China
[2]Henan Engineering Research Center of Digital Medicine, Zhengzhou, People's Republic of China
[3]The first Affiliate Hospital, Zhengzhou University, Zhengzhou, People's Republic of China
Corresponding Email: 1174995417@qq.com

Abstract. As customers increasingly focus on the quality, cost, delivery time of the product, therefore the production line needs to improve production efficiency, improve product quality, which requires the introduction of a management system. The TPM (Total Productive maintenance) as a starting point to maintain the equipment, maintenance activities through full participation in the maintenance, increase productivity, reduce production costs.

The study based on the study and analysis of the TPM theory, based on my practice in the SMT2 production line of Victoria Electronics of Huanwei. Then it studied the promotion and implementation steps of TPM on SMT2 line. And it also introduced the improvement instance of SMT2 line. By contrasting the utilization of equipment on SMT2 line before and after the comprehensive of TPM, the necessity of implementing TPM is derived. For how to apply TPM on the production lines, and how to improve equipment comprehensive utilization, finish the order on time, provide high quality products to our customers, these are the main part of production line equipment management work. This article provides the reference for how to promote TPM on other production lines.

1 Introduction

1.1 research background and significance

Huanwei is an electronics manufacture to provide professional design services and solutions. It mainly for the domestic and international famous brands to provide the design and the material procurement of electronics, computers, phone and other five categories of electronic products.

There exists some problems in the production workshop SMT2 line in Huanwei.The cohesion is unreasonable,and production equipment has problems.So this article is based on SMT2 line, through the data that I collected.The data is what unreasonable about SMT2 line.Through the implying of TPM, we can increase the economic benefit.The application of TPM can raise the production efficiency, and to reduce production costs and to ensure the quality of the products, and to improve the utilization rate of the production workshop equipment.

1.2 research status

According to the fire brigade and the fire fighting equipment,and the operator's skill is at low status, Li Kailian, Yang yong [1] thought that carrying out "six big shots"(full equipment repair and maintenance construction) is urgent. Wei-hong wang, FanTingJiu [2] published the main points of medical equipment maintenance. It based on TPM management deals with the essentials of management of medical equipment maintenance, so as to promote medical equipment maintenance and management, and improve the utilization rate of medical equipment. Fan Yanjiang [3] mainly analyzed the eight generation of Beijing dongfang. He mainly studied the line array factory machinery and equipment management, and product quality, and production efficiency and cost and so on.

Song Guimin [4] pointed out the current pressure of equipment management, and he advocated the participation of the management of machinery and equipment.He also started from the actual situation of enterprises, and guided the operation of equipment field management practices. Elaine Aspinwall, Maged Elgharib [5] determined how to apply TPM in the large and medium-sized manufactures, and analysed why we introduced TPM process, and the results.

Yuan Shan [6] provided the basis to the research of TPM in the aspect of employee. Wang Chunsheng [7] provided scientific basis for the maintenance work. Zhu Haimin [8] imported TPM in the management of Jiangling foundry machinery and equipment. Adnan Hj.Bakri [9] further researched on integrated field of TPM and lean production.

2 The theory of TPM

TPM is total production maintenance.TPM emphasis on participation, such as through full participation, so that it can eliminate waste, and control the whole process, and improve the comprehensive utilization of equipment, and achieve the benefit maximization. TPM is a combination of methods, equipment and operators.Through improving operations, and equipment, we can change the thinking of the employees, and mainly to promote employee's ownership, so that improve the ability of the enterprise to adapt to the external environment.

TPM is an enterprise that improve the system through the improvement of equipment. Advanced machinery and equipment management is the most effective and simple way of increasing efficiency, and reducing the operating costs. TPM has certain coordination effect between each function in manufacturing enterprises. One of the most important is the effect between the two functions of maintain and production. The main goal of TPM is to improve the OEE of equipment. It likes to take measures to reduce or even eliminate the six big losses: the loss of construction, the loss of waste, the machine short outage losses because of machinery and equipment speed down operation, idle equipment, machinery and equipment, adjustment, setting downtime losses, machinery and equipment downtime time loss, etc.

3 The analysis of the problem of STM2 line

3.1 STM2 line

SMT2 line process is divided into:

Into the moistureproof ark → The moistureproof ark → Into the station→ the flap → SP2→SPI2→The patch (iron, connectors) → AOI2-F(furnace before test)→Reflow soldering → Collecting plate → AOI2 - B (furnace) → SMT2 overall check

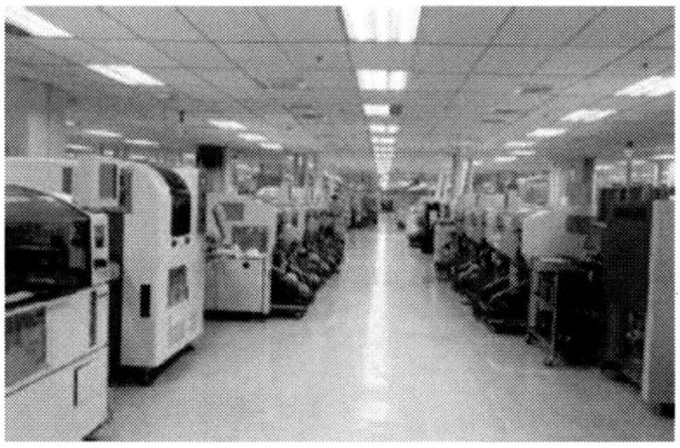

Fig.1. SMT2 line.

3.2 The problem of STM2 line

(1) We need to improve equipment management system, and improve the corresponding management mechanism

For SMT2 line, in order to ensure that equipment management work can be arranged on effectively for a long time, we need to establish and improve a series of related equipment management mechanism, and specify the responsibility of the management team and related staff. In apple watches front chip production process, there are many different kinds of chips and operating personnel, machinery and equipment transfers is frequent, and imperfect management mechanism, personnel strength is weak, the phenomenon is common on the production line.

(2) The disconnection between using and maintenance of machinery and equipment

On SMT2 line, even if the staff is assigned to the relatively fixed work, each staff has a corresponding machinery and equipment, but they did not implement machine equipment maintenance work of equipment maintenance personnel and staff. When there is serious problems, equipment maintenance personnel and operators have begun to pass the buck to each other.

(3) The maintenance of machinery and equipment lags, and the waste is serious

On the SMT2 line, daily maintenance measures can not get effective implementation. In the maintenance management of machinery and equipment, the concept of "maintenance" is already entrenched. In such aspects as material, manpower, financial resources, energy, causing unnecessary waste. A lot of equipment maintenance personnel, in order to save time and effort, the materials are useful on equipment repair not recycled, but its subjective to scrap processing.

3.3 The solution of the existing problem

(1) Improve the equipment management system

Training employees with TPM at all levels of production line; improving staff quality; making them more clearly about their own work; strengthening the management between leader and employees; ensuring the smooth implementation of equipment management system.

(2) The disconnect between using and maintenance of machinery and equipment

From the two aspects of operator and staff, improving the skill of operators and make them solve the problem of machine timely; and strengthening the sense of ownership of them. Professional maintenance employees should make regular maintenance strategy according to the collection of equipment fault information.

(3) The aintenance of machinery and equipment lags, and the waste is serious

Through the implementation of TPM, all related staff should participate in the machinery and equipment maintenance.When the problems occur, the operator can according to himself technology maintenance first; at the same time, he should inform the professional operators to the scene for further processing.

3.4 The effect on the SMT2 line

(1) The OEE calculation before implement TPM

To placement machine on SMT2 line as the research object to show that the utilization rate of machinery and equipment changes after implement TPM on SMT2 line.

Through the observation records, each shift work in 10.5 h, stop 30 min before the meeting. The production process downtime per class 4 times, and each time the fault processing time for 20 min. More refueling dish 10 times the average class, and each time for 5 min. SMT machine theory for each Carrier chips processing time of 0.25 min/Carrier, the actual processing cycle 0.5 min/Carrier. So daily processing chip 550 Carrier, there are 10 Carrier.

On the basis of the above data, before implement TPM SMT2 line, comprehensive utilization calculation to SMT machine are as follows:

Load time=630min-30min=600min

Start time=600min-80min-50min=470min

Rate of time to start=470/600=78.3%

Rate of speed to start=0.25/0.5=50%

Rate of net working =550*0.5/470=58.5%

Rate of working=50%*58.5%=29.2%

Rate of qualified products= （550-10） /550=98.2%

So,we can get that：

OEE=78.3%*29.2%*98.2%=22.5%

(2) The OEE calculation after implement TPM

Through to implement TPM in SMT2 line is for SMT machine, if the fault just reduce 1 time, fault handling time can reduce 2 min, and waste reduce 4 carrier, due to the shorter downtime, chip production increase of 580 carrier, then time statistics are as follows:

In various divisions work 10.5 h, and stop 30 min before the meeting, the production process downtime per class 3 times, each time the fault processing time for 18 min, more refueling dish 10 times the average class, each time for 5 min, the processing time of a placement machine for each Carrier chip 0.25 min/Carrier, the actual processing cycle 0.5 min/Carrier, processing chip Carrier 580 a day, six Carrier.

Load time=630min-30min=600min

Start time=600min-54min-50min=496min

Rate of time to start=496/600=82.7%

Rate of speed to start=0.25/0.5=50%

Rate of net working =580*0.5/496=58.5%

Rate of working=50%*58.5%=29.2%

Rate of qualified products= （580-6） /580=99.0%

So,we can get that：

OEE=82.7%*29.2%*99.0%=23.9%

(3) The OEE contrast before and after implementing TPM

The OEE for SMT2 line placement machine calculated above. Before and after implementing TPM, placement machine OEE improved about:

23.9%-22.5%=1.4%

Although the comprehensive utilization ratio of SMT machine only increased by 1.4%, for SMT2 line, has been greatly improved.

4 Summary

The purpose of this paper is the study of TPM on SMT2 line on the popularization and application. Application TPM on SMT2 line, not only can improve the comprehensive utilization of equipment, and improves the staff maintenance machinery and equipment of self-consciousness, also can improve the morale of the team members, and to improve the factory environment, and decrease the wasting of human resources. TPM is not just a policy, it also includes the production line on the organization structure change, full participation. Predictable, after the TPM really applied SMT2 line will have a lot of improvement.

References

1. Li Kailian, Yang yong. TPM full-time fire team in the enterprise management application. Journal of fire monthly, 2002, 28 (7) : 21-36.
2. Wang Wei-hong, FanTingJiu. Based on the TPM management of medical equipment maintenance points. China medical equipment, 2008, 11:88-89.
3. Fan Yanjiang. TPM in the application of liquid crystal display manufacturing boe research. Beijing: university of Chinese academy of sciences, 2014.
4. Song Guimin. Total productive maintenance equipment management new method. Journal of chemical engineering and equipment, 2012, 10 (5) : 46-58.
5. Elaine Aspinwall, Maged Elgharib. TPM implementation in large and medium size orginazations. Journal of Manufacturing Technology Management, 2013 (05) : 56-77.
6. Yuan Shan, photo, Wang Bulin. Total productive maintenance mode using the thinking of the unity of operating check. Science and technology of meishan, 2014 01:59-61.
7. Wang Chunsheng. Systemic power plant equipment maintenance mode and its preliminary application. North China electric power university, 2001.
8. Zhu Haimin. The application of TPM in jiangling foundry research. Nanjing: nanjing university of science and technology, 2006:41-65.
9. Adnan Hj.Bakri,Abdul Rahman Abdul Rahim,Noordin Mohd,Yusof Ramlli Ahmad. Boosting Lean Production via TPM.Procedia-Social and Behavioral Sciences, 2012, 21(05): 20-37.

DOI: 10.1051/matecconf/201710002036

O2O - Based Agricultural Products Supply Chain Process Integration Optimization Based on Internet +

Huijuan Li

Department, School of Management Engineering University, Zhengzhou University, China

Abstract. Traditional wholesale and retail, electricity supplier of agricultural products supply chain have many difficulties. The O2O supply chain of agricultural products of "Internet+", committed to the integration of online and offline advantage process, has become the main direction of the agricultural products supply chain transformation. Practice operation results show that O2O supply chain can effectively play the advantages of online and offline process integration, but its further development is still subject to the logistics, information flow of the dispersion, fracture and high cost. The integrated optimization of various regions and various enterprises and all sectors of the supply chain process is the key to optimize the process Internet plus era of agricultural products supply chain.

1 Introduction

In recent years, too many links in the agricultural supply chain,the lagging of network construction , the high operation costs, these problems and related strategies arouse much the concerns of all sectors of society.community. In promoting of Internet + boom, local governments and companies in the "Internet +" help,accelerate the transformation in the mode of production, the characteristics of agricultural products marketing and service mode.The "O2O" mode of the agricultural product supply chain is constructed through the models of "Internet + Production", "Internet + Logistics", "Internet + Information"and "Internet + Sales".

"O2O" model in"The Internet" environment , is committed to the integration of online and offline advantage process, has become the main direction of the agricultural products supply chain transformation. The actual operating results of the"O2O" model show that compared to the traditional model, a single production entity, a single flow-led "O2O" supply chain, although can effectively combine online, offline advantage process,but its further development depends on the dispersion, fracture, high cost of logistics and information flow.Integrated optimization of the various regions, the enterprises, the link of agricultural supply chain process, is the key of "O2O" type of agricultural products supply chain sustainable development in "Internet +" era.

2 Literature review and problems

2.1 Literature review

For the integration process research:

Bo Meilan [1](2011) believed that agricultural supply chain integration can affect the value chain of agricultural products, and promote the value-added agricultural products and value. The other part of the research on agribusiness supply chain integration is focused on the integration model. Li Jifang [2] (2007) argues that wholesale market-oriented circulation of agricultural products is not conducive to the value-added of agricultural products, should focus on the cultivation of core enterprises, through e-commerce means to achieve fresh agricultural product supply chain integration.Wang Puqing, Zhou Deyi etc. [3](2009) studied the organizational model of supply chain through field investigation, and concluded that the degree of agricultural product supply chain integration was directly proportional to the quality and safety level of agricultural products. Liu Chang [4](2011) studied the supply chain model of agricultural product supply chain in Japan, and proposed the supply chain model of "fresh produce-house-sale".Wang Ying [5](2013) through the study that the characteristics of agricultural products can be used as an important indicator of the choice of supply chain model. But in general, the re-use of "Internet +" advantage of "O2O" type of agricultural supply chain integration optimization research is still very small.

From the literature research, the domestic research started late, in recent years, agricultural supply chain research gradually increased, has become a research focus in the field of supply chain. Therefore, the supply chain optimization of agricultural products can not be ignored, the important problem to be solved, which is one of the main object of agricultural logistics research.

2.2 Statement of problem

By analyzing the supply chain optimization of agricultural products, it is found that the current scholars from different angles on China's agricultural supply chain and optimization of the parties to study and propose different solutions ideas and methods.

However, few scholars had put forward improvement in the supply chain of "O2O" type agricultural products. This paper analyzes the supply chain process of "O2O" type agricultural products under the background of "Internet +", and discusses the bottleneck problem of "O2O" type agricultural product supply chain optimization in order to solve the problem of "O2O" supply chain integration optimization.

3 "O2O" type agricultural product supply chain process

3.1 Agricultural production and processing enterprise-led supply chain process

Agricultural production and processing enterprises build and maintain their traditional channels, at the same time, actively through a variety of network platforms to sell their products and services and carry out a variety of marketing activities in the use of websites, microblogging and other platforms.They build the supply chain includes not only the traditional offline business processes, but also by the online transactions, online promotion business to promote the process.

3.2 Wholesale market-led supply chain processes

Since the new agricultural products wholesale market in Beijing in 2010 12 start the first electronic trading center - Beijing Xinfadi agricultural products electronic trading center,agricultural products wholesale markets across the country have self-built electronic trading center. In addition to self-built electronic trading center, the Beijing Xinfadi wholesale market also formed a strategic partnership with Jingdong Mall,and opened a new official flagship store in the Jingdong Mall.Its supply chain processes are mainly three types, the first category is the Beijing Xinfa electronic trading center as the core of the cash transaction process. The second category is the Jingdong new flagship store to the core of the process. The third category is the production process, logistics, wholesale, retail and other traditional processes.

3.3 "O2O" E-commerce-led agricultural supply chain process

3.3.1. Fresh "O2O" leading agricultural supply chain process

On the basis of summarizing the experience of the early online market failure, the fresh "O2O" continue to innovative business model, strengthen the characteristics of the trading network,and restructure supply chain processes.On the one hand, according to the development of the market constantly adjust their product supply process, on the other hand continue to expand its distribution system under the line, increase network publicity and promotion efforts.

3.3.2. Integrated electronic business platform-led supply chain processes

With the further development of the fresh "O2O", the local agricultural products online mall which has a wide range of services, large scale, many kinds of products, and various characteristics of "online and offline". Its operation mode mainly has two kinds of B2C and B2B2C.

COFCO I buy net is a typical representative of B2C in the food network. The website has made several changes to its supply chain for the past too narrow supply of goods is difficult to effectively meet the diverse needs of users,the third-party logistics with high logistics costs, long logistics, logistics, low quality of service, customer satisfaction and other issues.

The first is to reform the product supply process, changing the suppliers from the group production enterprises into all the quality of food suppliers. The second is to strengthen the procurement management, food quality management and logistics service management.The third is actively implement the "localization strategy" and "concentration strategy" to reduce logistics costs. The fourth is to combine self-logistics and outsourcing logistics to reduce the logistics cost.The fifth is to expand the marketing promotion, advertising, event organization, social media marketing and other promotional methods organic integration.The sixth is to increase advertising, sales agents and other processes to broaden the income channels. Specifically shown in Figure 1.

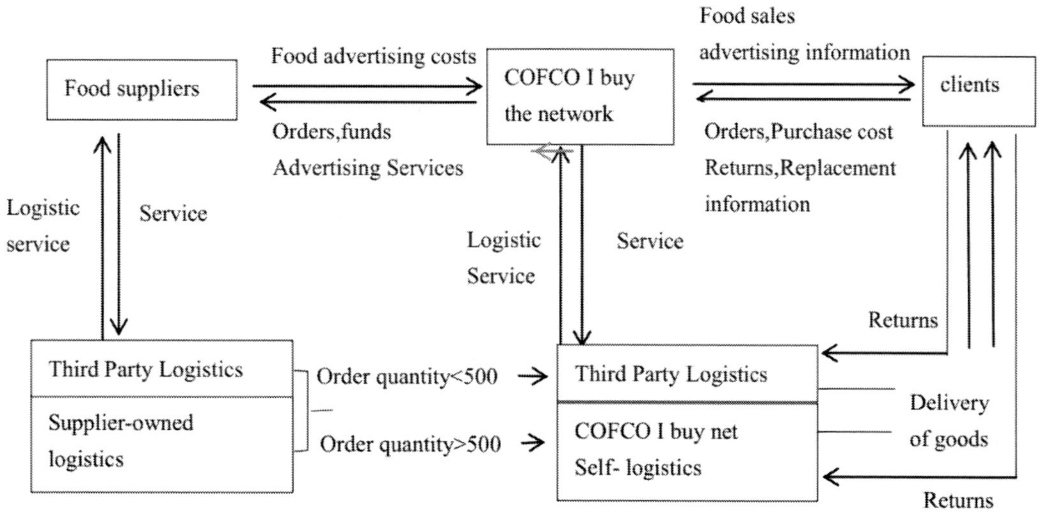

Fig. 1. Supply chain relationship and related processes of COFCO I buy the network.

China's geographical indications product store B2B2C model is a typical representative. Mall for geographical indications products production operators and terminal consumers, provide online trading services. In 2014, The operator of the mall ---Beijing Dadi Bo Electronic Commerce Co., Ltd. ("Lida") has intensified its construction in China Geographical Indications Industrial Park, Agricultural Production Base, High-end Community Store, PC Platform and Mobile Terminal Mall.Mall operators build a supply chain of agricultural products,which is a geographical indication products in China Mall as the core, into the geographical indication products manufacturers (suppliers), self-production base (park), third-party logistics service providers, self-line terminal service shops, advertising service providers , China Geographical Indications Culture Museum, customers (consumers).

The supply chain process centered by Chinese Geographical Indication Products Mall and the supply chain centered by I buy the network are differences. Chinese Geographical Indication Products Mall increase the business shop process, business certification process, business service processes and business shop management process.China's geographical indications product store to increase the business shop process, business certification process, business service processes and business shop management process; In addition, the two logistics service flow,and cash flow service process are different.

4 Bottleneck analysis on the optimization of supply chain process of "O2O" - type agricultural products

4.1 Production and processing enterprise-led "O2O" supply chain process integration is too low

Most of the production and processing enterprises dominated the "O2O" supply chain is still showing the logistics dispersion, fracture, high cost characteristics.Its online transactions need to pay high courier logistics costs, and even after several courier logistics to complete the product from the origin to the final consumer of the entire logistics process, resulting in express logistics costs are too high.In addition, the online model can not keep abreast of offline consumer demand, resulting in procurement, production, sales and logistics plan is still a great blindness. It is difficult to realize the reasonable transformation of the relationship and position of the online and offline.

4.2 Wholesale market - led "O2O" supply chain logistics is dispersed and its process integration is difficult

The integration of supply chain processes in this model is limited by many practical factors.First, the existing wholesale market facade is a private ownership, wholesale market management side only assume the responsibility and power of property management.Second, the wholesale market management side to establish online trading system is motivated by local, short-term interests.It is difficult to truly "online" servicing for the "offline".Third, the "online" and "offline" wholesale operations are decentralized operations.Integration gap of online transactions and offline transactions is the wholesale market-led "O2O" type of supply chain's common problem.More importantly,The independent wholesalers to carry out online and offline transactions simply can not explore the common distribution, the value of co-marketing.Fourth, the electronic trading center emphasizes online transactions, ignore the integration of associated logistics and information flow and other service processes.Therefore, the bottlenecks of the "O2O" model of the wholesale market development and optimization of the supply chain process are that the centralized, synchronous and coordinated integration of logistics and information flows before, during and after the wholesale transaction.

4.3 "O2O" -based supply chain-driven process lacks dynamic integration and dynamic adaptation

The development of small-scale "O2O" -based supply chain appears to be constrained by the lack of funds and technical and managerial capacity of core entities. In fact, it is constrained by the scale of logistics, information flow and business flow, high cost and consumer demand of the timely response.

The agricultural supply chain operations by the core of "I buy the network" show that the fast-growing B2C e-commerce's scale and all aspects of service capabilities can not quickly adjust and optimize its supply chain processes and supply chain capabilities with changes in consumer demand and competitor competition strategies.This is the main bottleneck in the development of the "O2O" supply chain, which is dominated by the B2C electricity supplier.

B2B2C model can not adjust its supply network, logistics network, offline transaction and service network according to the rapidly changing market.So adjustment and optimization of its supply chain process is its supply chain process is the bottleneck problem of the development of Direct store in B2B2C model.For the B2B2C electronic trading platform, the integration services of the flagship store, direct sales of the product supply, logistics and distribution and terminal, providing services of low cost, high response, integrated logistics, business flow and information flow for consumers are the key to development.

5 Integrated optimization of "O2O" agricultural product supply chain

By analyzing the bottleneck of "O2O" supply chain optimization process, it is found that all kinds of main supply chains are faced with the problem of high cost and low response speed caused by the decentralized operation of the logistics,faced the problem of information flow breakage, poor sharing and high sharing cost.These problems not only need to integrate online and offline Business flow, more importantly,need to integrate product supply, consumer demand, product supply and consumer demand for coordinated development to achieve product supply and consumer demand for collaborative development.Therefore, the integration optimization is the key to solve the "O2O" type of agricultural product supply chain process optimization bottleneck.

Integrated optimization of "O2O" type of agricultural products supply chain processes

Refers to the integration of local characteristics of agricultural production resources, agricultural logistics resources, agricultural information resources and agricultural products sales resources through the "Internet + production", "Internet + logistics", "Internet + information flow", "Internet + sales" mode and achieving "Online and offline" procurement process, "online and offline" promotion process, "online and offline" transactions (business flow), "online and offline" logistics and "online and offline" information flow and related after- Vertical integration, in order to reduce the supply chain operating costs, speed up the supply chain response speed, increase the dynamic adaptability of the supply chain.As the "O2O" model continues to evolve, the agri-food supply chain process will undergo the following integration optimization phases.

5.1 Decentralized independent optimization stage

In the initial stage of development, production, processing, wholesale, retail and other entities in the local link leveraging "Internet +" achieve business flow and information flow integration.

5.2 Each stage "O2O" collaborative optimization stage

In order to solve the problem of vertical coordination and integration of logistics and information flow, the production, processing, wholesale, retail and other main body come together to discuss the "O2O" mode of vertical integration of the whole supply chain, to realize the different online- , Information flow of collaborative integration.

The "O2O" supply chain, which cooperates with each link, can overcome the dilemma of logistics and information flow faced by scattered "O2O" type supply chain effectively.

As shown in Figure 2, the collaborative "O2O" model can achieve the initial supply of information and information on the final demand in the supply chain to deliver the seamless, integrated target. This model can shorten the information delivery time in the supply chain, eliminating the transmission of the "bullwhip effect."

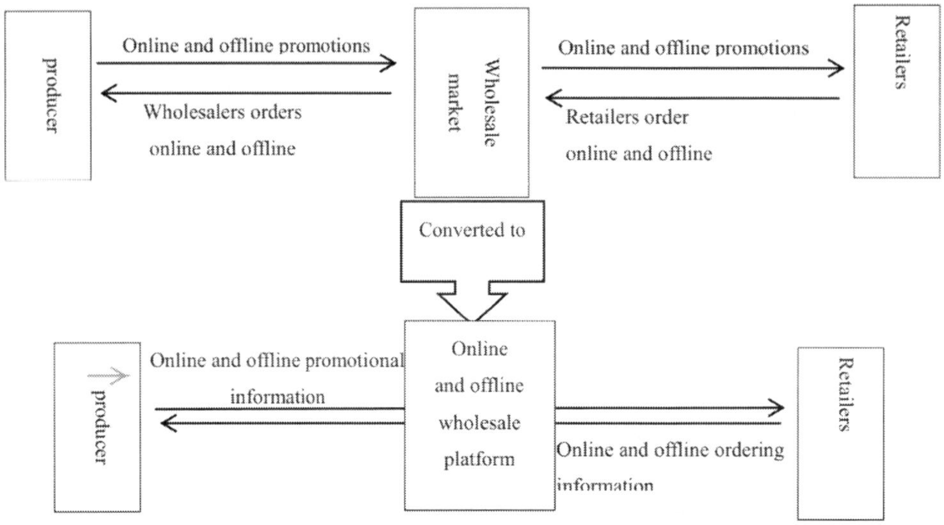

Fig.2. The supply chain information flow changes in the dispersion model.

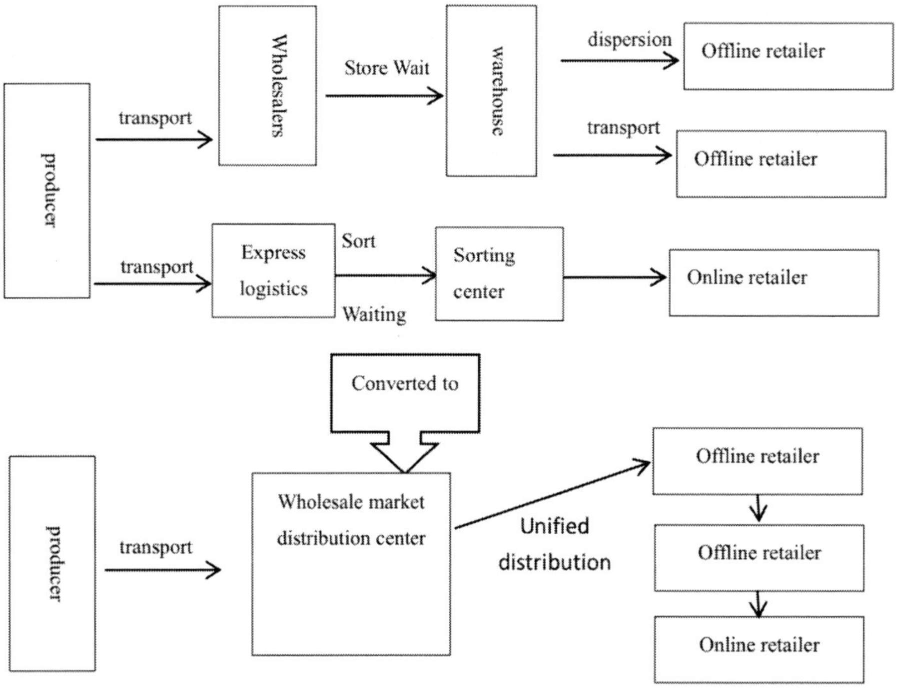

Fig.3. Supply chain logistics changes in collaborative model.

As shown in Figure 3, the collaborative "O2O" model can coordinate the logistics of the whole supply chain, carry out unified distribution, centralized storage, effectively alleviate the "O2O" mode of logistics small volume, high logistics costs, logistics and other issues.

5.3 Based on the "O2P" of localization, regional collaborative optimization stage

The "O2O" mode of coordinated optimization of the vertical flow of a single product will eventually develop into the stage of coordinated development of localized online and offline regional line on-line, that is, through the "O2P" model, the agricultural producers in the same area, Online and offline retailer collaboration phase.

As shown in Figure 4, the O2P model of ocalization and regional integration led by fourth-party logistics service providers will further achieve the transport and distribution of large-scale and intensive. Supply chain process integration capabilities will be further improved .

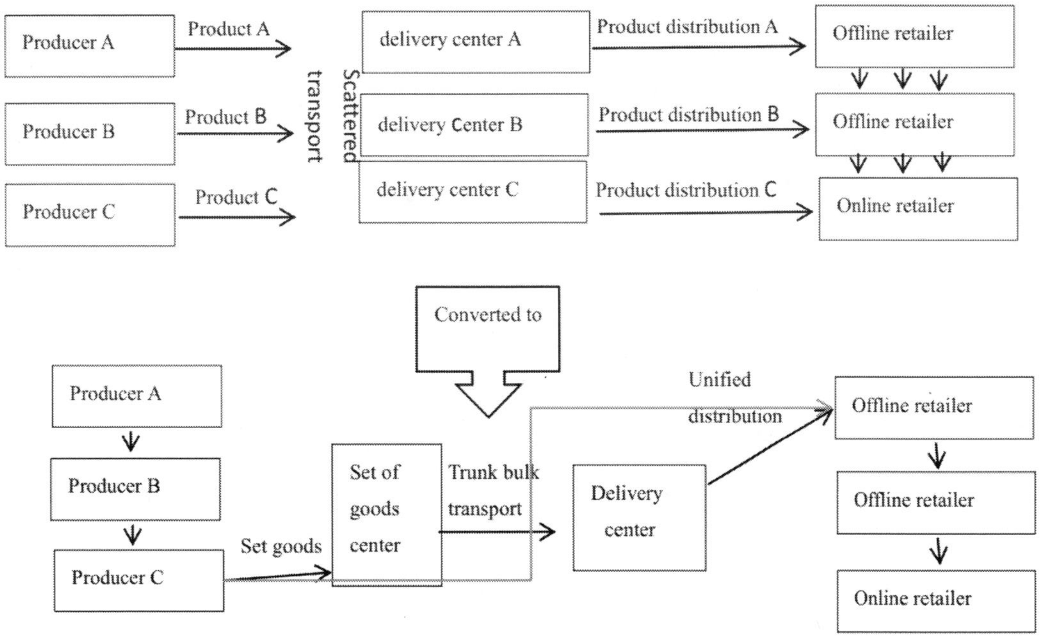

Fig.4. The logistics service chain changes in regional collaborative model.

The "O2P" model, which is led by 4PL, is expected to realize the diversification of the supply of the supply chain, the synergism of supply chain behavior, the synchronization of online and offline transactions, the scale and agility of logistics and distribution. As "O2O" type of agricultural products supply chain crack process optimization bottlenecks effective measures.

6 Concluding remarks

This paper through analysis the process of "O2O" agricultural supply chain and optimization bottlenecks in the context of "Internet +" ,comes to conclusion that integrated optimization of supply chain processes is the key of agricultural supply chain process optimization in "Internet +" era. The "O2P" model of Regional synergies led by 4PL is a effective model to crack "O2O" type of agricultural products supply chain optimization process bottlenecks ,also the "O2O" type of agricultural supply chain optimization process of the main direction.

References

1. Bu Meilan.Study on the Integration of Agricultural Product Supply Chain Based on Value Chain. [J] Journal of Chongqing University of Science and Technology.2011 (21): 79-81
2. Li Jifang.Study on Supply Chain Management of Fresh Agricultural Products in China [J]. China Circulation Economics, 2007,1: 17-19.
3. Wang Puqing, Zhou Deyi etc.Agricultural Product Supply Chain Organizational Model and Food Safety .[J] Journal of Agricultural Economics 2009(3): 8-12
4. Liu Chang,etc.The Change of the Supply Chain of Fresh Agricultural Products of JSA and Its Enlightenment to China[J]Modern Economic Research. 2011 (10): 88-92.

5. Wang Ying.Selection of Agricultural Product Supply Chain Organization Model Based on Agricultural Product Classification.[J]Journal of Business Economics.2013 (28):

The New Information Grey Direct Model and Its Use in The Forecasting of Energy Constitutes

Baohua Yang [1,*] and Jinshuai Zhao [2]

[1] Business school, Jiangsu Normal University, Xuzhou, China
[2] School of Computer Science and Technolo, Jiangsu Normal University, Xuzhou, China
Corresponding Email: mathyang@126.com

Abstract. To overcome the contradiction of the traditional grey first order accumulate with the new information priority principle, this paper establish the new grey fractional order accumulate, which is proposed to better reflect the priority of the new information. Based on this, This paper establish the fractional order grey direct model based on the principle of new information priority, which is proposed to improve the prediction precision. The results of practical numerical examples demonstrate that this new method has a good predication performance for small data set forecasting. This research also provides a new model for the less data forecasting.

1 Introduction

Time series prediction refers to the process by which the future values of a system is forecasted based on the information obtained from the past and current data points [1]. GM(1,1) was first proposed by Deng [2-3], which is one of the most important parts in grey system theory, for small sample forecasting. Because the grey prediction model needs little origin data, has simple calculate process and higher forecasting accuracy, it has been successfully applied in many disciplines [4–7].

The purpose of GM (1, 1) model is to work on system forecasting with poor, incomplete or uncertain messages. However, the existing GM (1,1) model cannot be used for accurate prediction for many actual systems, while the system behaviors are affected more or less by other relative factors and their characteristic values never follow the Grey exponential law completely [8]. So, A large number of studies on improved Grey model and applications have been reported, such as Li et al. [9], Xie and Liu [10], Yang et al. [11]. Those improved Grey models all integrated traditional GM (1,1) model and other modeling technology, but not including the research on accumulated generating operator of Grey model.

Although grey forecasting model had been widely adopted, its predicting performance still could to be improved [9-15]. However, most of improved methods do not emphasize the value of new data, which has been put forwarded in the grey system by Deng, and named the new information priority principle. Dang and Luo improved GM (1,1) using the nth item of $X(0)(n)$ as the starting condition of the Grey differential model to increase prediction precision [12–15]. Because of the new data represents the future law of development of small sample system [2], we should considered more priority of the nth data than the n-1th data in the model. Similarly, the 2th data is more prior than the 1th data.

Recently, fractional order [16,17] has been applied on the grey generating function, which uses fractional order accumulative generation to narrow the variation of the sequential data and then establish the prediction model. Wu [16] established the grey system model based fractional order accumulative generation by applying the fractional order to grey model. In fact, GM (1,1) model only can simulate homogeneous index sequence, it will has large error when the original data is non-homogenous exponential sequence. So, the grey direct grey model has been formulated to non-homogenous exponential sequence [18,19]. This paper studies the grey direct grey model based on the technology of fractional order accumulation generation and the principle of priority of new information theory, we establish the optimized model and solve the optimization accumulation order value, then give it in practical application.

The rest of the paper proceeds as follows, the next section presents an overview of the relevant literature on grey model. The third section provides a new fractional order grey direct model with the principle of new information priority of grey system. The fourth section proves the advantage of the grey model put forward in this paper over the traditional grey model by one real case in China. Finally, the paper concludes with some comments in last Section.

2 Review on GM(1,1) model and grey direct model

2.1 GM(1,1) model [3]

Suppose that the non-negative sequence of raw data is X(0)=(x(0)(1), x(0)(2),…, x(0)(n)), and its sequence of the first-order accumulative generation operator on X(0) is denoted as following,

X(1)=(x(1)(1), x(1)(2),…, x(1)(n)) (1)

In formula (1), $x^{(1)}(k)=\sum_{i=1}^{k}x^{(0)}(i)$,k=1,2,…,n

From the sequence X(1)=(x(1)(1), x(1)(2),…, x(1)(n)), we can derive the sequence of generated mean value of consecutive neighbors in the following,Z(1)=(z(1)(2),…, z(1)(n)); where

z(1)(k)=0.5 x(1)(k)+0.5x(1)(k-1),k=2,…,n.

Definition 1 For a non-negative sequence of raw data X(0); X(1)is a generated sequence with the application of the first-order accumulative generation operator on X(0); Z(1)is a sequence with the application of the generated mean value of consecutive neighbors operator on X(1), then the following equation

x(0)(k)+a z(1)(k)=b, k=2,…,n. (2)

is a grey differential equation, also called model GM(1,1).
And the equation

$$\frac{dx^{(1)}}{dt}+ax^{(1)}=b \tag{3}$$

is the whitened equation of GM(1,1). If we let

$$Y=\begin{bmatrix}x^{(0)}(2)\\x^{(0)}(3)\\\vdots\\x^{(0)}(n)\end{bmatrix},\quad B=\begin{bmatrix}-z^{(1)}(2) & 1\\-z^{(1)}(3) & 1\\\vdots & \\-z^{(1)}(n) & 1\end{bmatrix},$$

One can use formula (2) to make a least square estimation for parameters $\hat{a}=[a,b]^T$,

$$\hat{a}=(B^TB)^{-1}B^TY \tag{4}$$

After obtaining parameters a and b, use formula (5) to define time response formula and use formula (6) to define reducing value sequence.

$$\hat{x}^{(1)}(k+1)=(x^{(0)}(1)-\frac{b}{a})e^{-ak}+\frac{b}{a},\quad k=1,2,\cdots,n-1 \tag{5}$$

$$\hat{x}^{(0)}(k+1)=(1-e^{a})(x^{(0)}(1)-\frac{b}{a})e^{-ak},\quad k=1,2,\cdots,n-1 \tag{6}$$

Finally, one should make error test. The residual sum of squares is shown as formula (7).

$$s=\varepsilon^T\varepsilon=\sum_{k=1}^{m}(x^{(0)}(k)-\hat{x}^{(0)}(k))^2 \tag{7}$$

Professor Deng points out that the whitened equation and its corresponding time response function cannot be derived from the GM(1,1) directly, which are some approximate replacement.

2.2 Grey direct model [18-20]

Definition 2 Assume that a non-negative sequence X(0)=(x(0)(1), x(0)(2),…, x(0)(n)) is given, then

$$\hat{x}^{(0)}(k+1)=\beta_0+\beta_1\hat{x}^{(0)}(k) \tag{8}$$

is referred to as the original form of the auto-regressive GM(1,1) model (ARGM(1,1)),or grey direct model.
Theorem 1 Assumed that $X^{(0)}$ is defined by the Definition 2,

$$B=\begin{bmatrix}1 & x^{(0)}(1)\\1 & x^{(0)}(2)\\\vdots & \vdots\\1 & x^{(0)}(n-1)\end{bmatrix},\quad Y=\begin{bmatrix}x^{(0)}(2)\\x^{(0)}(3)\\\vdots\\x^{(0)}(n)\end{bmatrix},$$

Then the lease squares estimation values are

$$\hat{\beta}=(B^TB)^{-1}B^TY$$

Among that $\hat{\beta}=(\beta_0,\beta_1)^T$ is the parameters list of the formula (8).

Theorem 2 Assume that B、Y and $\hat{\beta}$ are defined by the Theorem 1, and let $\hat{x}^{(0)}(1)=x^{(0)}(1)$, then

$$\hat{x}^{(0)}(k+1)=[x^{(0)}(1)-\frac{\beta_0}{1-\beta_1}]\beta_1^k+\frac{\beta_0}{1-\beta_1} \tag{9}$$

Literature [18] pointed out that if the original data is monotonous movements bump geometry, then the application of the formula (8) as shown in the simulation data obtained with the same sequence of model geometry; moreover formula (8) model can fully simulate inhomogeneous index sequence.

Because grey auto-regressive method of GM(1,1) is formulated by using the original data rather than the accumulated generation data, it does not need to inverse accumulated generating. We can see that formula (6) is an exponential model, formula (9) is a non-homogenous exponential model.

3 Fractional order grey direct model with the new information priority

3.1 Fractional order accumulate with the new information priority

Fractional contains an "in between" thinking, more and more scholars recognized [16,22], The recent studies have shown that we can enhance the accuracy of gray forecast modeling by select the appropriate fractional order [16].

Definition 3 [16]. Assume that r=q/p and $\{x_j:j=1,\cdots,m\}$ which is a nonnegative sequence, are given, then

$$\sum_{j=1}^{m}{}^{(r)}x_j=\sum_{j=1}^{m}C_{m-j+r-1}^{m-j}x_j \tag{10}$$

is called the r order accumulated, and define

$$C_{r-1}^0=1,\ C_{m-j+r-1}^{m-j}=\frac{(m-j+r-1)(m-j+r-2)\cdots(r+1)r}{(m-j)!}.$$

Supposed that $X^{(0)}=(x^{(0)}(1), x^{(0)}(2),\dots, x^{(0)}(n))$ is non-negative sequence, application of gray system accumulate, we can get the r order accumulated generating sequence $X^{(r)}$ of $X^{(0)}$ is $X^{(r)}=(x^{(r)}(1), x^{(r)}(2),\dots, x^{(r)}(n))$, where $x^{(r)}(k)=\sum_{j=1}^{k}C_{k-j+r-1}^{k-j}x^{(0)}(j)$, k=1,…,n. Since $X^{(0)}$ is time series data, the acquired time from $x^{(0)}(1)$ to $x^{(0)}(n)$ is decreasing by the view of time, that is to say, $x^{(0)}(1)$is the earliest data and $x^{(0)}(n)$ is the latest data, so, $x^{(0)}(1)$ may be referred to as the oldest information and $x^{(0)}(n)$ for the newest message.

From the above r order accumulated generating function, we can see that every accumulate contains the oldest data $x^{(0)}(1)$, that is, we have used the oldest data n times, and as the latest information $x^{(0)}(n)$, is only used once in the $x^{(r)}(n)$ calculation, which is obviously contradicts with the principle of new information priority in grey system theory. In fact, $x^{(0)}(n)$, as the latest information, is the latest incarnation of the historical, and can be better reflect the internal development and trends of the system. The principle of new information priority is the key point of view about information applied in grey systems theory. That is, by applying additional weights on newer information, one can achieve a better effect from grey prediction. Therefore, the latest information should be paying more consideration when we establish the model and get model parameters. Based on this, we will provide a new cumulative method which can reflect the new information's effect in the following.

Definition 4 For r=q/p and non-negative sequence X(0)=(x(0)(1), x(0)(2),…, x(0)(n)), then, the r order cumulative X(r)=(x(r)(1), x(r)(2),…, x(r)(n))is called the fractional accumulated generating with the new information priority where

$$x^{(r)}(k)=\sum_{j=k}^{n}C_{n-j+r-1}^{n-j}x^{(0)}(j).$$

Fractional derivatives accumulate the whole history of the system in weighted form. The larger r of x(r)(k) is, the larger the weight of old data is; the smaller r of x(r)(k) is, the smaller the weight of old data is. Reducing r can reduce the weights of old data, which can put more emphasis on the newer data.

But from the definition 4, for any k, we can see that all new data have been contained in the $x^{(r)}(k)$, and the more new data, the more weight has been putted because of $C_{n-j+r-1}^{n-j}$ is accessioned with j's increases. So, the text of the accumulated sequence obtained this way is called the accumulation of new information accumulated priority sequence.

3.2 Fractional order grey direct model with the new information priority

Definition 5 Assume that a non-negative sequence $X^{(0)}=(x^{(0)}(1), x^{(0)}(2),\dots, x^{(0)}(n))$ is given, then

$$\hat{x}_{(0)}^{(r)}(k+1)=\beta_0\sum_{j=k}^{n-1}C_{n-1-j+r-1}^{n-1-j}+\beta_1\hat{x}_{(0)}^{(r)}(k) \tag{11}$$

is called r order grey direct model based on the new information priority, where

$x_{(0)}^{(r)}(k)=\sum_{j=k}^{n}C_{n-j+r-1}^{n-j}x^{(0)}(j)$, $k=1,2,\cdots,n-1$.

Formula (11) and formula (8) is the same in the form, in fact, they are different to each other, the difference lies in the way of data generated essentially. The innate character difference makes the new grey model can not only reflect the new information priority principle of gray system, but also has better accuracy than old, and the advantages of the new model will present in the example.In addition, if $r=1$, the new model reduces to the literature [20] given model, that is the model of formula (11) shown in more general.

Theorem 3 Assumed that $X^{(0)}$and $\hat{x}_{(0)}^{(r)}(k)$ are defined by the Definition 5,

$$B=\begin{bmatrix}\sum_{j=1}^{n-1}C_{n-1-j+r-1}^{n-1-j} & \sum_{i=1}^{n}C_{n-i+r-1}^{n-i}x^{(0)}(i)\\ \sum_{j=2}^{n-1}C_{n-1-j+r-1}^{n-1-j} & \sum_{i=2}^{n}C_{n-i+r-1}^{n-i}x^{(0)}(i)\\ \vdots & \vdots\\ \sum_{j=n-1}^{n-1}C_{n-1-j+r-1}^{n-1-j} & \sum_{i=n-1}^{n}C_{n-i+r-1}^{n-i}x^{(0)}(i)\end{bmatrix},\quad Y=\begin{bmatrix}\sum_{i=2}^{n}C_{n-i+r-1}^{n-i}x^{(0)}(i)\\ \sum_{i=3}^{n}C_{n-i+r-1}^{n-i}x^{(0)}(i)\\ \vdots\\ \sum_{i=n}^{n}C_{n-i+r-1}^{n-i}x^{(0)}(i)\end{bmatrix}$$

Then the lease squares estimation values are

$$\hat{\beta}=(B^TB)^{-1}B^TY \tag{12}$$

Among that $\hat{\beta}=(\beta_0,\beta_1)^T$ is the parameters lists of the formula (11).

Theorem 4 Assume that a non-negative sequence $X^{(0)}=(x^{(0)}(1), x^{(0)}(2),\ldots, x^{(0)}(n))$ is given. Actually, Let error is u ;then

$$x^{(0)}(k+1)=\beta_0+\beta_1x^{(0)}(k)+u_k,\ k=1,\ldots,n-1 \tag{13}$$

Via the treatment of new information priority, the mean of the error is $E(Au)$:

$$E(Au)=u_n+\frac{(n-2)C_{n-(n-1)+r-1}^{n-(n-1)}}{n-1}u_{n-1}+\cdots+\frac{2C_{n-3+r-1}^{n-3}}{n-1}u_3+\frac{C_{n-2+r-1}^{n-2}}{n-1}u_2;$$

Where

$$A=\begin{bmatrix}C_{n-2+r-1}^{n-2} & C_{n-3+r-1}^{n-3} & \cdots & C_{n-(n-1)+r-1}^{n-(n-1)} & C_{n-n+r-1}^{n-n}\\ 0 & C_{n-3+r-1}^{n-3} & \cdots & C_{n-(n-1)+r-1}^{n-(n-1)} & C_{n-n+r-1}^{n-n}\\ \vdots & \vdots & \vdots & \vdots & \vdots\\ 0 & \cdots & 0 & C_{n-(n-1)+r-1}^{n-(n-1)} & C_{n-n+r-1}^{n-n}\\ 0 & \cdots & 0 & 0 & C_{n-n+r-1}^{n-n}\end{bmatrix},\ u=\begin{bmatrix}u_1\\ \vdots\\ u_{n-2}\\ u_{n-1}\end{bmatrix}.$$

Proof: Let

$$Y_1=\begin{bmatrix}x^{(0)}(2)\\ \vdots\\ x^{(0)}(n-1)\\ x^{(0)}(n)\end{bmatrix},\ B_1=\begin{bmatrix}1 & x^{(0)}(1)\\ \vdots & \vdots\\ 1 & x^{(0)}(n-2)\\ 1 & x^{(0)}(n-1)\end{bmatrix}$$

Then via the treatment of new information priority, formula (13) becomes

$$AY=AB\begin{bmatrix}\beta_0\\ \beta_1\end{bmatrix}+A\mu$$

Obviously $E(Au)=u_{n-1}+\frac{(n-2)C_{n-(n-1)+r-1}^{n-(n-1)}}{n-1}u_{n-2}+\cdots+\frac{2C_{n-3+r-1}^{n-3}}{n-1}u_2+\frac{C_{n-2+r-1}^{n-2}}{n-1}u_1$ Because of

$$C_{n-i+r-1}^{n-i}=\frac{(n-i+r-1)(n-i+r-2)\cdots(r+1)r}{(n-i)!}\leq 1$$

So,

$$E(Au) = u_{n-1} + \frac{(n-2)C_{n-(n-1)+r-1}^{n-(n-1)}}{n-1}u_{n-2} + \cdots + \frac{2C_{n-3+r-1}^{n-3}}{n-1}u_2 + \frac{C_{n-2+r-1}^{n-2}}{n-1}u_1$$
$$\leq u_{n-1} + \frac{n-2}{n-1}u_{n-2} + \cdots + \frac{2}{n-1}u_2 + \frac{1}{n-1}u_1$$

That is the average error of the simulation sequence is more small when we consider the fractional order accumulate with the new information priority.

4 The determine to the fractional order parameter of grey direct model based on the new information priority

Based on the discussion above, we can see that the fractional cumulative model's average error is smaller, so the new model has better modeling accuracy. Of course, how to choose the accumulated order r is an important issue. A simple and natural way is establish an optimal model to get the cumulative order r, such as the average relative error of simulation data sequence to establish minimum target on the cumulative order optimization model. Therefore, we will establish the cumulative order optimization model, the goal is minimization the simulate sequence's average relative percentage error (*ARPE*).

$$\begin{aligned} &\min ARPE \\ &s.t\ \ ARPE = \frac{1}{n-1}\sum_{k=1}^{n-1} |\hat{x}^{(0)}(k+1) - x^{(0)}(k+1)| \\ &\hat{x}^{(0)}(k+1) = [x^{(0)}(1) - \frac{\beta_0}{1-\beta_1}]\beta_1^k + \frac{\beta_0}{1-\beta_1} \\ &r > 0 \end{aligned} \tag{14}$$

Obviously, the above optimization model (14) is a nonlinear optimization model; we will use the modern intelligent algorithms, such as genetic algorithms, simulated annealing algorithm, PSO et.al, to solve this model. So, in the next computational analysis, we will use the PSO algorithm to get the accumulated order r.

PSO(Particle Swarm ,PSO) is a kind bionic optimization algorithm, and the method comes from the simulation of motor behaviour of birds and fish [21].Assumed the objective is a n dimensional function, the size of particle population is m, the i-th particle's "best" value is p_i^{best}, and global "best" value is p_g^{best}, then the i-th particle modified its velocity and position according to the following formulas:

$$\begin{cases} v_{id} = w \cdot v_{id} + c_1 rand()(p_{id}^{best} - x_{id}) + c_2 Rand()(p_{gd}^{best} - x_{id}) \\ x_{id} = x_{id} + v_{id} \end{cases}$$

Where i=1,…,m, d=1,…,n, rand() and Rand() are two separately random function in the range[0,1], c_1 and c_1 are constants known as acceleration coefficients and c_1=c_1=2, w is the weigh value of the inertia.

Algorithm process of particle swarm optimization:

Step1: Initializing the population size of particles , every particle's velocity and position;

Step2: For every particle, according to the formula(3) to calculate $\hat{\beta} = (\beta_0, \beta_1)^T$,then according to the formula (4), $\hat{x}^{(0)}(k+1)$, k=1,2,…,n-1 are obtained;

Step3: Calculating every particle's fitness value and get p_i^{best} and p_g^{best};

Step4: Updating each particle's velocity v_i and position x_i according to the formula(6) ;

Step5: if the stop condition (the iteration number or the error precision satisfy the request) is satisfied, the iteration stop and output the p_g^{best} ,that is the optimization r; otherwise turn to step 2.

5 Example:the simulation of natural gas consumption in china

Energy has an influencing role in achieving economic and social progress. Forecasting energy Constitutes a vital part of energy policy of a country, especially for a developing country like China whose economy is in a stage of energy transition: from low efficiency solid fuels to oil, gas and electric power. A large number of studies on energy consumption forecasting using grey model and improved grey model have been reported, such as Wu et al [20] , Xu et al [23], Li et al [24].

In this section, the advantage of the fractional order grey direct model based on the new information priority over the other grey models is demonstrated by natural gas consumption in china. Average relative percentage error (*ARPE*) compares the real and forecasted values to evaluate the precision.

We consider an example from paper [23] which provides the sample data. Using the data from 1995 to 2006 (in-sample data) to construct fractional order grey direct model based on the new information priority:

$$x(k) = 24.2617 \times 1.268^k + 141.6864 \tag{15}$$

then predict the actual value from 2007 to 2008.

Actual values and the forecasting values of three compared models are presented in Table 1. As can be seen from Table 1, fractional order grey direct model based on the new information priority yielded the lowest *ARPE* compared with the other models across all the period from 1995-2008, which also illustrate that the new model is more suitable for natural gas consumption.

Table 1. The fitted values and ARPE of four grey models.

Year	Natural gas consumption (From China government webpage)			
	Actual value	model (15)	GMNI(1,1) [20]	reference [23]
1995	177.531	172.406	173.677	190.243
1996	188.049	180.655	181.593	183.658
1997	176.132	191.119	191.694	184.908
1998	218.699	204.395	204.583	193.994
1999	215.337	221.231	221.030	210.914
2000	244.811	242.592	242.015	235.669
2001	274.554	269.687	268.793	268.259
2002	292.180	303.058	302.961	308.685
2003	339.454	347.569	346.559	356.945
2004	397.285	402.968	402.191	413.040
2005	486.936	474.128	473.177	476.970
2006	561.051	562.129	563.756	548.735
ARPE (%)	-	3.12	3.092	4.406
2007	698.903	675.029	679.333	721.62
2008	807.857	818.246	826.811	925.69
ARPE (%)	-	2.35	2.573	10.75

6 Conclusion

Based on principle of priority of new information in the grey system theory, this paper establish a new information accumulates method by use of Fractional accumulate ideas; based on this, a new fractional order direct gray model with information priority has been presented, and given the model parameters with least squares estimation; and analyzed the simulation error of the new model. The results show that fractional order gray direct model with the new information priority can reduce simulation error; the application example also show that the new model has better prediction accuracy and prediction accuracy of analog data. Although the proposed model overcomes the defects of parameters estimation in traditional direct grey model, they may be superior to other modeling methods in some aspects. There are some potential drawbacks such as the selection of tradeoff parameter related to a least squares cost function. The performance of this model is more related to the selection of trade-off parameter. Thus, how to quickly and accurately select the model trade-off parameter should be further studied. Further, by calculating, we have found that the prediction error maybe worse when we use the accumulated order, which can be obtained by the minimize averaging the error between the raw data and simulation data. Therefore, in order to get the optimization cumulative order, what evaluation criteria as objective function is suitable is also an important issue for future research process.

Acknowledgement

This paper is supported by the National Nature Science Fund of China(No.71301064) and the Ministry of Education Humanities and Social Science Foundation of China (No.12YJC630262) .

References

1. E Kayacan, B Ulutas & O Kaynak Kaynak. Grey system theory-based models in time series prediction, *Expert Systems with Applications*, 33, 2(2010), 1784-1789.
2. J L Deng. Contral problems of grey system, *Systems& Contral Letters*, 1,5(1982), 288- 294.
3. Liu S F, Lin Y. *Grey systems Theory and Applications,* Berlin: Springer, 2010.

4. D C Li, C J Chang, C C Chen.et al. Forecasting short-term electricity consumption using the adaptive grey-based approach--An Asian case. Omega, 40,6(2012),767-773.
5. T S Chang, CY Ku, HP Fu. Grey theory analysis of online population and online game industry revenue in Taiwan. Technological Forecasting and Social Change, 80(2013),175-185.
6. M Evans. An alternative approach to estimating the parameters of a generalized grey Verhulst model: An application to steel intensity of use in the UK, Expert Systems with Applications, 41 (2014),1236–1244.
7. W Yan, C Y Yu, R Tu, et al. Fire detection model in Tibet based on Grey-fuzzy neural network algorithm. Expert Syst Appl, 38,8(2011),9580–9585.
8. 8.T Li. A research on the Grey prediction model GM(1,n). Appl Math Comput , 218,1(2012),4903-49018.
9. 9.D C Li, C J Chang, C C Chen. et al. Forecasting short-term electricity consumption using the adaptive Grey-based approach-an Asian case. Omega,40,6(2012),767-773.
10. N M Xie, S F Liu. Discrete Grey forecasting model and its optimization. App Math Model,33(2009),1173 – 86.
11. B H Yang, Z G Fang, K Zhang . Discrete GM(1,1) model based on sequence of stepwise ratio. Journal of Systems Engineering and Electronics, 34,4 (2012), 715-718.
12. T X Yao, S F Liu, N M Xie. On the properties of small sample of GM(1,1) model. App Math Model ,33(2009),1894 – 903.
13. Y G Dang, S F Liu, K J Chen. The GM models that be taken as initial value. Kybernetes,33,2(2004),247 – 54.
14. Y X Luo. Non-equidistant step by step optimum new information GM(1,1) and its application. Syst Eng Theor Pract,30,12(2010),2254 – 8.
15. Y H Wang, Y G Dang, Y Q Li, et al. An approach to increase prediction precision of GM(1,1) model based on optimization of the initial condition. Expert Syst Appl ,37,8 (2010),5640-5643.
16. L F Wu, S F Liu, L G Yao, et.al. Grey system model with the fractional order accumulation. Commun Nonlinear Sci Numer Simulat ,18(2013),1775-1785.
17. L F Wu, S F Liu, L G Yao. Discrete grey model based on fractional order accumulate, Systems Engineering--Theory & Practice,34,7(2014),1882-1887.
18. B Zeng, S F Liu. Direct modeling approach of DGM (1,1) with approximate non-homogeneous exponential sequence. Systems Engineering--Theory & Practice, 31,2(2014),297 -301.
19. Y N Wang. GM (1,1) direct modelling method and its characteristic. Systems Engineering--Theory & Practice, 8,1(1998):27 -31.
20. L FWu, S F Liu, H J Chen,et..al. Using a novel grey system model to forecast natural gas consumption in China, Mathematical Problems in Engineering,2015
21. J Kenned, R Eberhart. Particle swarm optimization, In: Proc IEEE Int Conf on Neural Networks,1995,1942-1946.
22. R Almeida, A B Malinowska. Generalized transversality conditions in fractional calculus of variations, Communications in Nonlinear Science and Numerical Simulation, 18,3(2013), 443-452.
23. G Xu, W G Wang. Forecasting China's natural gas consumption based on a combination model, Journal of Natural Gas Chemistry,19(2010):493-496.
24. J C Li, X C Dong, M Hook .et.al. Forecasting the growth of China's natural gas consumption, Energy, 36,3(2011),1380-1385.

New Construction of Fire Fighting Academies Under the View of Information Construction

Bin Ni

Public Security Fire Control Forces Command Academy in Kunming, Yunnan, China
Corresponding Email: 578659400@qq.com

Abstract. With the rapid development of economy, we have entered the era of knowledge-based economy. With the rapid reform and innovation of science and technology, people's production and life style has undergone tremendous changes, especially the more and more in-depth informatization, technicalization, networking and intelligentization of various trades and professions. In this paper, the author first explained the new changes brought by informatization to fire fighting academies at the different levels, and then combined with the actual construction of informatization of fire fighting academies, put forward feasible strategies to promote the management of informatization, providing a certain reference value for the development of informatization in fire fighting academies.
Keywords. Informatization, fire control construction, change, strategy, CIO.

1 New changes brought by informatization to fire fighting academies

1.1 Flat organizational structure

In the informatization era [1], the information interaction is more and more rapid and frequent, in order to ensure the high value of information use, improve the efficiency of information transmission, fire fighting academies are also developing toward the flat structure. Compared to the prior hierarchical bureaucratic management structure, the management scope of flattened organizational structure is widened, and the problem of duplication and multi-management is reduced, which greatly improves the efficiency of academies, information transmission and feedback, so that the academies can understand and master the general situation in a timely and comprehensive manner, strengthen the communication efficiency among departments, facilitate the integration of teaching resources, promote exchanges between various departments, and make the work more standardized and institutional.

1.2 Humanized management mode

Human resources have become one of the core resources of competition among different trades and occupations, so the concept of "people-oriented" has become more and more popular among the people. As the main resource of resource utilization [2], integration and innovation, human resources play an important role in the process of academy development. In fire fighting academies, the main human resources include three parts: administrative management layer, teachers and students. Therefore, when applying the "people-oriented" management concept, the management of fire fighting academies should be fully mobilized to meet the management needs, respect teachers, care students, and constantly stimulate potentials. The self-management of teachers and students shall be taken as the goal, give full play to their own initiatives to achieve the optimization of resource management, to achieve a people-oriented and self-administrative new trend.

1.3 Learning campus atmosphere

With the deepening of learning society, learning government and learning enterprise, the construction of learning campus has become the main trend of informatization, which is an inevitable requirement of social progress regarding academy development, so as to meet the requirements of the development of the times and train learning talents. As the education place of fire fighting professionals, fire fighting academies must establish the correct study orientation [3], create the learning atmosphere, construct a learning campus, and fully mobilize the value of mining all kinds of effective resources. Building a learning academy requires every teacher to strengthen self-learning and improvement, stimulate students' learning enthusiasm, render the atmosphere of academy learning, create a learning campus culture and atmosphere, and promote the academy reform and development, so that the academy will always keep vitality in the development process.

2 Implementation strategy of informatization management in fire fighting academies

2.1 The basis of academy informatization management--the establishment of the academy informatization management evaluation system

First, fire fighting academies shall actively promote and implement the establishment and promotion of the informatization management evaluation system as an important work in accordance with internationally recognized quality assurance system in the ISO9000 standards during construction of the system. In order to realize the whole-process development of academy management, fire fighting academies should actively introduce the "PDCA" standard and apply it in the whole process of informatization management to promote self-management and self-evaluation.

Second, the introduction of ISO9000 management standard system is the internationalized requirements of China's educational management. At presen, China's academies are still relatively backward, some of the more advanced academies already have their own management and evaluation system, but there are still many loopholes in the practice of the operation, management standards are not unified. Therefore, fire fighting academies should pay attention to the combination of their own situation in using and drawing lessons from the relevant management evaluation systems, refer to the school management and decision-making standards, take the mutual development of teachers and students as the management purpose, achieve proceduring academy management, inspire the enthusiasm of faculty and students, actively build a democratic, equal, transparent and scientific management system to promote orderly and sustained management, and to effectively guarantee the scientific management be strictly enforced and implemented.

Third, it is necessary to supervise the corresponding standards and systems in order to promote their continuous reform and innovation, and form a virtuous cycle of management of fire fighting academies. The ISO9000 management evaluation system of Fire fighting academies improves the normalization in running fire fighting academies, refines the management standards, making academy management more scientific and standardized, ultimately contributing to the perfection of academy management and promoting comprehensive and coordinated development of academies.

2.2 Perfect CIO--Accountability mechanism of academy management

CIO, namely the Chief Information Office, which is the school management responsibility system that we usually refer to, fire fighting academies can use computer and information management platform to manage the relevant information and help academy managers in analysis and decision-making. At present, management of fire fighting academies is not specific, informatization management is generally anchored in the network information center and can not grasp the full-range information, so the targeted allocation of existing teaching resources is difficult to be achieved.

On the one hand, a sound CIO should start from the following aspects: First, fire fighting academies should actively introduce high-quality and complex informatization management personnel, and constantly enrich academy management organization to improve the overall quality of the senior management team. Second, fire fighting academies shall conduct re-education and training of the relevant administrators of the existing school-related departments in accordance with the requirements of the learning-based campus, and constantly improve the overall quality of existing management workers. Third, it is necessary to make sure the posts and responsibilities of all departments and related personnel to be rationally distributed, to realize mutual supervision, clear power and responsibility, to establish and improve CIO responsible mechanism, and to improve the informatization of fire fighting academies management.

On the other hand, academies should make sure all managers in the CIO team are able to follow advanced educational philosophy, use technology-based means of education, with the help of computers, software and other equipment and platforms, in a comprehensive and accurate understanding of informatization, establish a sound fire fighting academies management informatization system, continue to guide academy management in the direction of technicalization and informatization, maximize the effectiveness of the existing funds, strive to maximize the value of academy management resources, help academies to develop appropriate reform strategies, improve the timeliness and accuracy of informatization management, so that managers of fire fighting academies can make rational allocation of school resources, make the decision of academy management more scientific, and keep close contact with various departments within the academy to improve the orderliness of informatization construction.

2.3 Teacher's knowledge management--virtual teaching research

The development of the knowledge economy era requires academies to adhere to the times and focus on the development and progress of the times, meanwhile, more knowledge growth and skill enhancement of teachers are required. Only comprehensive and good knowledge management can ensure that teachers' the invisible educational ideas and thoughts continue to reveal in the long-term education, the maximum play of the value of knowledge, and the best knowledge dissemination, so that knowledge can be deepened and continued.

In the current era of informatization, teachers of fire fighting academies can use the virtual platform to improve the rigid form of traditional teaching methods and avoid the singularity and boringness of the form. In the course of concept teaching, we can strengthen the use of multimedia, simulation system, network teaching platform, on the one hand, it will let students understand the actual situation of fire control, on the other hand, it can also provide students with the form of network to combine theory and practice, enhance the quality and level of network teaching. The use of informatization platform for teaching can also break the limitations of time and space in teaching, so that students can learn and practice anytime and anywhere, not only forming more effective knowledge management, but also driving students' learning initiative, and gradually forming a knowledge management mode.

2.4 An effective form of student management--virtual class management

The "invasion" of network, database, multimedia, and other information modes in campus is making students' management work different from the past, the formation of virtual classes is a new student management method through the application of network technology, which is of great help to the management of students in fire fighting academies.

Teachers can build network files for fire fighting academy students by establishing a virtual class through the website or the home page to form a complete student electronic archives database. Through this way, on the one hand allows students to learn knowledge in the network, have fire fighting practice, but also allows students to carry out more teaching activities on the network platform. In addition, teachers can also communicate with the parents effectively to help students form correct values and outlooks on life. The construction of a network virtual class can also track the future development of the students, arrange the data of the students' employment situation, summarize the problems encountered in the students' work, and provide better service and help for the students. Students can get better help inside or outside the campus, at work or in life, enabling their professional skills to get a greater degree of promotion.

3 Conclusion

School information management not only aims at teaching and management via information technology means, more importantly, it aims to establish a sound informatization system in line with the practical situation of fire fighting academies, to help fire fighting academies build an informatization basis, establish appropriate scientific management rules and processes, and improve the level of informatization management through feedback and improvement in the continuous practice.

References

1. J.G. Zhang, Reflections on Teaching Strategies of Rescue Course in Fire Fighting Academies (Career Horizon, 2014)
2. H.C. Yang, Network Security Technology & Application (2015)
3. W. Yan, The Guide of Science & Education (Early Journal), 06 (2012)

Improved PCA Method Based on RBF Neural Network for Multiple Response Parameters Optimization

Jianli Yu*, Xiaotian Pan and Hongqi Huang

Management Engineering Department, Zhengzhou University of Aeronautical, Zhengzhou, China
Corresponding Email: yjl837@163.com

Abstract. In the design of multiple response parameters optimization, weighted principal component analysis (weighted PCA) is used to build the relationship between the response variables and controllable factor model by linear regression. But in the complicated nonlinear production process, the fit of the linear regression model is not high that cannot satisfy the requirement of the parameter design model. This study proposed an improved weighted PCA based on RBF neural network prediction model. In this paper, RBF neural network was used to construct nonlinear prediction model of production process and to adjust the weighted PCA algorithm by adding the predict ability index of neural network model. In the design of multiple response parameters, this approach improve the effect of process parameters optimization. And applied this method to multiple response parameters optimization design of metallization polypropylene film capacitor thermal polymerization process, the results show that capacitance value and the loss tangent are all improved, and the effect of optimization parameters is achieve to satisfactory results.

1 Introduction

With the complexity of the production process and product quality requirements, it is need to consider multiple quality characteristics in the process of product optimization. The multiple quality parameters optimization method is aim to improve the products quality by process parameters optimization. So the multiple quality characteristics optimization design has a important role in the real process of quality continuous improvement.

The quality loss function and satisfaction function method have an extensive application in the multiple response optimization design, but it ignores the correlation between the response variables, which will affect the effect of optimization design [1]. So the problem of the correlation between the multiple quality characteristics caused more attention among the academics. Wang Jianjun, Ma Yizhong [2] used the partial least-squares regression model to solve the problem of the correlation between the multiple quality characteristics. Ali [3] proposed a new method of satisfaction function method and spherical rule model to solve the problem of correlation.

Liao [4] proposed weighted principal component analysis (PCA) which can combine the multiple response to the single response and eliminate the correlation between variables. In actual industrial production, PCA is applied to solve the multiple response parameters problem which have been achieved satisfactory results [5]. Zhang Yingdong [6] proposed an improved weighted principal component analysis(PCA) which can construct a mathematical model between the impact factor and multiple response variable. The traditional response surface method based on linear regression analysis to determine the the polynomial model of first-order or second-order. But the quality characteristics become more nonlinear and highly complicated in the modern advanced manufacturing progress, the traditional response surface based on linear model cannot meet the practical requirements [7]. However, artificial neural network can independently complete the nonlinear mapping from input space to the output space, it belong to nonlinear model which has the strong ability of prediction [8].

This paper proposed an improved PCA method based on radial basis function(RBF) neural network for multiple response parameters optimization. Firstly, construct a production process mapping relations between the factors and multiple response based on RBF neural network. Next, using PCA method eliminate the correlation between response variable through weighted multiple response variables into a single comprehensive response variable. Finally, it is priority to improve the response variable which have a strong prediction ability, and realize the overall effect of multiple response optimization. And applied this method for the metallized polypropylene film capacitor, which not only solving the parameters optimization problem of the polymerization temperature and polymerization time, but also realizing the whole optimization of the capacitor value and the loss tangent.

2 Improved PCA based on RBF neural network method

2.1. Radial basis function (RBF) neural network model

Radial Basis Function (RBF) is an efficient feedforward neural network [9]. Its approximation ability and generalization ability and learning speed is all superior to the BP neural network [10]. RBF neural network can arbitrary approach to the linear and nonlinear function, so there is no local optimum problem. Its structure is shown in figure 1.

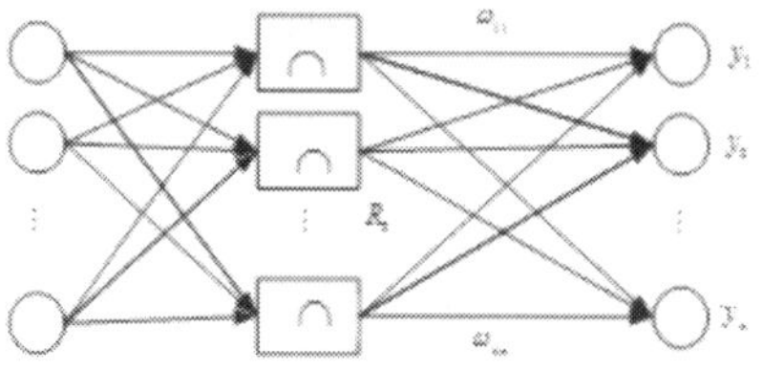

Fig. 1. The structure of RBF neural network.

In the complex nonlinear process, the fitting degree of the linear regression model is not high and the regression formula can not meet the forecast ability of the responses variable. Based on that, non-parametric model is established by using RBF neural network to forecast the process-response indicators, and the mean square error(MSE) as predictive ability evaluation index for RBF neural network model. The value of MSE is smaller, the RBF neural network model has a better prediction accuracy.

Assumes that the controllable factors variable is x_1, x_2,... x_i, which as the input vector of RBF neural network model. And assume the total number of the response variable is k. The first j response variable y_j as output vector of RBF neural network model. So the number of RBF model established is k.

Using the MATLAB computer software to construct and train the RBF neural network model. The statements as following

P=[x_1 x_2 ···x_i]'; 'x_1、x_2···x_i'is the controllable factors variable
T=[y_j]'; 'y_j'is the first j response variable
net=newrb(p, t, goal, spread); p is the input vector, t is the output vector
y=sim(net, p);
e=T-y; 'e' is the error vector between the expectations and forecast
MSE=mse(e); 'mse(e)' is the mean square error

2.2. Weighted principal component analysis

Weighted principal component analysis method can combine multiple indicators into a new set of unrelated indicators, and according to the variance contribution get a weighted sum, which can transfer the multiple response to the single response.

Assumes that the number of the response is p, and the response variable is Y1, Y2,... , Yp. According to the principal component analysis, the principal component is Z1, Z2,... ,Zk. The formula of Z1, Z2,... ,Zk is

$$Z_q = e_{q1}Y_1 + e_{q2}Y_2 + \cdots + e_{qp}Y_p \tag{1}$$

Zq is the principal component of first q. Yq is the response of the first q. "eq1,eq2,... ,eqp" is the coefficient of the first q principal component. And the formula of the variance contribution rate is

$$a_q = \frac{\lambda_q}{\sum_{q=1}^{k}\lambda_q} \tag{2}$$

λ_q is the eigenvalue of first q.

Using MINITAB software to make principal component analysis that can get component load matrix and principal components variance contribution rate. According to the variance contribution rates can calculate a weighted sum of the principal component. And the formula of the multiple response performance index(MPI) is

$$MPI = \sum_{q=1}^{k}\alpha_q Z_q \tag{3}$$

By comparing the values of MPI can convert multiple performance characteristics optimization problem into a single response optimization problem. And if the goal or target of the response is larger for better, so the MPI is larger, the level of quality characteristics is higher. If the target of the response is smaller for better, so the MPI is smaller, the level of quality characteristics is higher.

2.3. Improved weighted PCA method

For RBF neural network prediction model, the corresponding mean square error (MSE) is an effective evaluation index for network prediction ability. So the predictive ability index of first j. And the prediction ability of RBF neural network model is defined as

$$R_j = \frac{\sum_{i=1}^{p} mse_i - mse_j}{\sum_{i=1}^{p} mse_i}, r_j = \frac{R_j}{\sum_{i=1}^{p} R_i}, j = 1,2,\cdots,p \quad (4)$$

$$r_1+r_2+\cdots+r_p=1 \quad (5)$$

So the MSE is smaller, the predictive ability "r" of RBF neural network model is higher.

Combining the RBF neural network model and weighted principal component analysis method, the formula of predict the response (1) is revised as follows

$$Z_{q-new} = e_{q1}r_1Y_1 + e_{q2}r_2Y_2 + \cdots + e_{qp}r_pY_p \quad (6)$$

$eq_1, eq_2, \ldots, eq_p$ is the coefficient of the first q principal component, r_1, r_2,$\cdots$,r_p is prediction ability of RBF Neural network.

3 The parameter optimization design of metallized polypropylene film capacitors

The process of thermal polymerization is key to eliminating the air between the metallized film layers, which can improve the compactness of the capacitor and the stability of electrical performance. But if thermal polymerization is insufficient, which will drop down the capacity of capacitor and increase the loss tangent [11]. Therefore, it is necessary to optimization the two parameters of temperature and time in the process, which is expected to improve the capacitance value and loss tangent value.

In this paper, there is two responses in the optimization process of metallized polypropylene film capacitors. One of the response variables is the capacitance value(y1), the other response variable is loss tangent(y2). The target of capacitance value(y1) is 45.45μF, and the target of loss tangent(y2) is smaller for better. So the absolute value of the difference between the capacitance value(y1) and the target value(45.45μF) is converted to the smaller for better. And assume that the controllable factors variable of thermal polymerization temperature is x1, the polymerization time is x2. The process parameters and their level given in table 1.

Table 1. The process parameters and their level.

No	parameters	Level 1	Level 2	Level 3	Level 4	Level 5	Level 6	Level 7
1	Temperature	95	100	105	110	115	120	125
2	Time(h)	4	6	8	10	None	None	None

There are 28 groups of thermal polymerization experiments on metallized polypropylene film capacitor. In the level of (1, 1)、(1, 2)、(1, 3)、(1, 4)、(2, 1)、(2, 2)、(2, 3)、(2, 4)、(3, 1)、(3, 2)、(3, 3)、(3, 4) repeat the experiment 10 times respectively, other levels repeated the experiments 5 times respectively. The average of the two response variables in each experiment was calculated, so the experimental results are shown in table 2. Using the MINITAB to make the principal component analysis of response variables, and the eigenvalues and eigenvectors are shown in table 3.

Table 2. The experimental data and MPI.

Experi-ment No.	Factor variable		Response mean		Standardized mean		MPI_1 (first-order regression)	MPI_2 (second-order regression)	MPI_3 (RBF)
	x_1	x_2	y_1 (μF)	y_2 (×10^{-4})	Y_1	Y_2			
1	1	1	0.2051	3.8	0.534	1.081	0.3751	0.2901	0.3702
2	1	2	0.0331	3	-0.703	-1.390	-0.4939	-0.3817	-0.4874
3	1	3	0.0383	3.4	-0.666	-0.154	-0.4674	-0.3521	-0.4607
4	1	4	0.0101	3.6	-0.869	0.463	-0.6096	-0.4540	-0.6006
5	2	1	0.1268	3.6	-0.029	0.463	-0.0204	-0.0117	-0.0199
6	2	2	0.0336	3.1	-0.700	-1.081	-0.4913	-0.3773	-0.4847

7	2	3	0.0045	3.7	-0.909	0.772	-0.6378	-0.4728	-0.6283
8	2	4	0.0063	3.3	-0.896	-0.463	-0.6290	-0.4758	-0.6202
9	3	1	0.0668	3.9	-0.461	1.390	-0.3231	-0.2317	-0.3178
10	3	2	0.0144	3	-0.838	-1.390	-0.5883	-0.4525	-0.5805
11	3	3	0.0563	3.7	-0.536	0.772	-0.3763	-0.2764	-0.3705
12	3	4	0.06	3.5	-0.510	0.154	-0.3577	-0.2673	-0.3525
13	4	1	0.0262	3.4	-0.753	-0.154	-0.5284	-0.3979	-0.5209
14	4	2	0.097	3	-0.244	-1.390	-0.1712	-0.1394	-0.1695
15	4	3	0.0198	3.6	-0.799	0.463	-0.5607	-0.4173	-0.5524
16	4	4	0.0444	3.6	-0.622	0.463	-0.4364	-0.3240	-0.4299
17	5	1	0.1234	3.6	-0.054	0.463	-0.0375	-0.0245	-0.0368
18	5	2	0.0646	2.8	-0.477	-2.008	-0.3349	-0.2672	-0.3311
19	5	3	0.1074	3.6	-0.169	0.463	-0.1183	-0.0852	-0.1164
20	5	4	0.1316	3.8	0.005	1.081	0.0040	0.0115	0.0045
21	6	1	0.1644	3.8	0.241	1.081	0.1696	0.1358	0.1677
22	6	2	0.1638	3	0.237	-1.390	0.1661	0.1138	0.1630
23	6	3	0.2304	3.6	0.716	0.463	0.5027	0.3810	0.4957
24	6	4	0.1198	3.6	-0.080	0.463	-0.0557	-0.0382	-0.0547
25	7	1	0.367	3.8	1.699	1.081	1.1926	0.9038	1.1759
26	7	2	0.4234	2.8	2.104	-2.008	1.4767	1.0929	1.4545
27	7	3	0.506	3.6	2.698	0.463	1.8943	1.4257	1.8673
28	7	4	0.4194	3.4	2.075	-0.154	1.4569	1.0925	1.4359

Table 3. Principal Component Analysis.

Principal Component	Original eigenvalues			Eigenvectors	
	Eigenvalues	Variance ratio	Cumulative ratio	Y_1	Y_2
1	1.0432	0.522	0.522	0.707	0.707
2	0.9568	0.478	1.000	0.707	-0.707

3.1 The parameter optimization based on first-order regression prediction model

According to the experimental data in Table 1, using the MINITAB to establish the first-order regression model of influence factors and response variables, and calculating its predictive ability index by formula (7)

$$r_j = \frac{R_j^2}{\sum_{i=1}^{p} R_i^2} \tag{7}$$

the result is as follows

$Y_1 = -0.0406 + 0.0494x_1 - 0.0104x_2$

$R^2 = 53.1\%, r_1 = 0.993$

$Y_2 = 3.42 - 0.0027x_1 + 0.0171x_2$

$R^2 = 0.4\%, r_2 = 0.007$

Calculated by formula (6)

$Z_1 = 0.707 \times 0.993 \times Y_1 + 0.707 \times 0.007 \times Y_2$

$Z_2 = 0.707 \times 0.993 \times Y_1 - 0.707 \times 0.007 \times Y_2$

Calculated by formula (3)

$MPI_1 = 0.522Z_1 + 0.478Z_2 = 0.702Y_1 + 0.0002Y_2$

The main effect chart as shown in Figure 2.

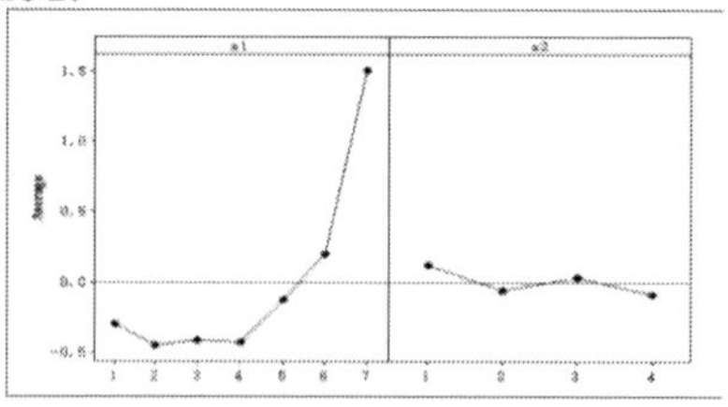

Fig. 2. Main effect chart.

Since the quality characteristic of y_1 and y_2 are all smaller for better, so the lower the value of MPI, the better effect of the response. According to figure 2, the minimum value of x_1 is second level, and x_2 is fourth level. So the result of parameter optimization design by first-order regression prediction model is (2, 4).

3.2 The parameter optimization based on second-order regression prediction model

According to the experimental data in Table 1, using the MINITAB to establish the second-order regression model of influence factors and response variables, and calculating its predictive ability index by formula (7), the result is as follows

$$y_1 = 0.35 - 0.155x_1 - 0.0695x_2 + 0.0113x_1x_2 + 0.022x_1x_2 + 0.0028x_2x_2$$

$$R^2 = 87.7\%, r_1 = 0.746$$

$$y_1 = 4.3 + 0.007x_1 - 0.867x_2 + 0.0068x_1x_2 - 0.0033x_1x_2 + 0.171x_2x_2$$

$$R^2 = 29.8\%, r_2 = 0.254$$

Calculated by formula (6)

$$Z_1 = 0.707 \times 0.746 \times Y_1 + 0.707 \times 0.254 \times Y_2$$
$$Z_2 = 0.707 \times 0.746 \times Y_1 - 0.707 \times 0.254 \times Y_2$$

Calculated by formula (3)

$$MPI_2 = 0.522Z_1 + 0.478Z_2 = 0.527Y_1 + 0.008Y_2$$

The main effect chart as shown in Figure 3.

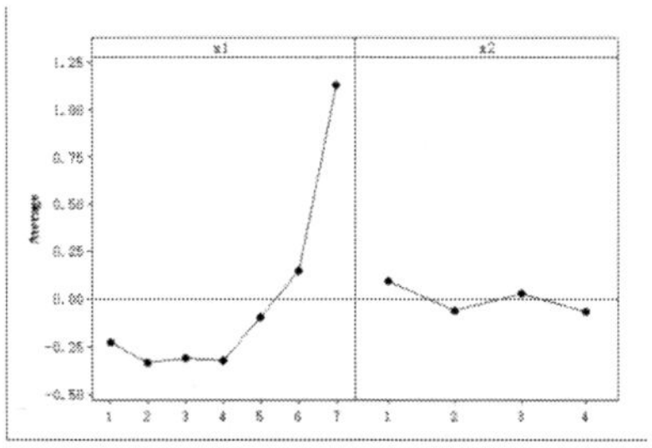

Fig. 3. Main effect chart.

So the result of parameter optimization design by second-order regression prediction model is (2, 4).

3.3 The parameter optimization Based on RBF neural network prediction model

Above that, the fitting degrees(R2) of the first-order and second-order regression formulas of polymerization process are respectively 53.1%, 0.4%and 87.7%, 29.8%, If the R2 less than 60%, it means that the prediction effect of first or second order parameter regression model is not ideal. So the RBF neural network is adopted to establish the nonlinear predictive model and optimize parameter.

Using the MATLAB to establish the RBF neural network model of influence factors and response variables. Two RBF neural network models are established with x_1 and x_2 as input variables and y_1 and y_2 as output variables respectively. When setting the goal of RBF neural network is 0 and spread is 8, the MSE of the RBF neural network model is smallest. The MSE of network training are respectively 1.3835*10-5 and 6.3297*10-4. It means that nonlinear prediction model based on RBF neural network is ideal. Calculating its predictive ability index by formula (4), the result is as follows

r_1=0.9786, r_2=0.0214

Calculated by formula(6)

$$Z_1 = 0.707 \times 0.9786 \times Y_1 + 0.707 \times 0.0214 \times Y_2$$
$$Z_2 = 0.707 \times 0.9786 \times Y_1 - 0.707 \times 0.0214 \times Y_2$$

Calculated by formula(3)

$$MPI_3 = 0.522Z_1 + 0.478Z_2 = 0.6919Y_1 + 0.0007Y_2$$

The main effect chart as shown in Figure 4.

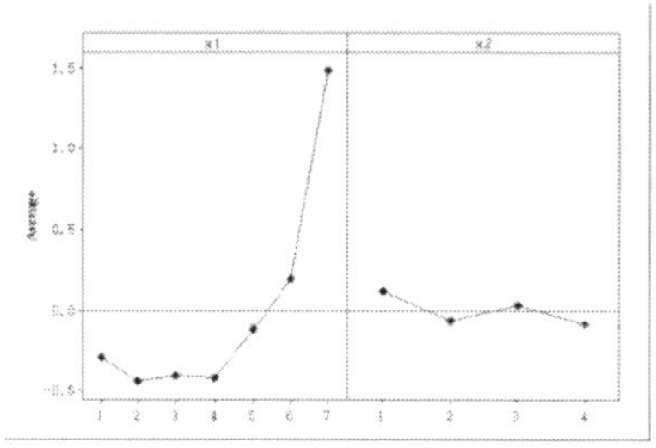

Fig. 4. Main effect chart.

The results of optimization parameters design based on RBF neural network prediction model is also (2, 4).

The results of optimization parameters design above the three model are all (2, 4). According to the experimental data in Table 1 and compare with the original process parameter (4, 3), the result of (2, 4) can be seen that the capacitance value and the loss tangent are all improved, and the effect of optimization parameters is achieve to ideal.

4 Conclusion

This paper proposed an improved weighted principal component analysis (PCA) method based on parametric and non-parametric prediction models. It is aim to solve the problem of multiple response process parameter optimization design, and this method is applied to optimize the multiple response process parameters in the thermal polymerization process. The results show that this method is effectively improving the two response quality characteristic.

(1)The method based on parametric and non-parametric prediction models to improved weighted principal component analysis, considers the predictive ability of multiple response models. It defined the evaluate the predictive ability index of each response model, and introduces it into the MPI model of the weighted principal component analysis method, which can improve the effectiveness of parameter optimization.

(2)This paper proposed an improved weighted principal component analysis (PCA) method based on RBF neural network prediction model. Because of the RBF neural network has a good generalization ability in nonlinear model. Therefore, in the complex nonlinear production process, it is an effective method to optimization the multiple response process parameter based on RBF neural network prediction model.

Acknowledgments

The research work was supported by National Natural Science Foundation of China under Grant No. U1404702, Aviation Science Fund under Grant No. 2014ZG55021, Nature Science Foundation of Henan Province under Grant No. 162102210083.

References

1. Wu F C. Optimization of correlated multiple quality characteristics using desirability function. Quality engineering, 17,1 (2005).
2. J Jun Wang, Y Zhong Ma, Y huan Zhai. Optimal Design of Correlated Multiple quality characteristics. Engineering Management, 25,2 (2011).
3. Ali S,·Mahdi B. A new desirability function-based method for correlated multiple response optimization. The International Journal of Advanced Manufacturing Technology, 76 (2015).
4. Liao H C. Multi-response optimization using weighted principal components. The International Journal of Advanced Manufacturing Technology, 27,7 (2006).
5. Shit J S, Tzen Y F, Yang J. B.. Principal components analysis for multiple quality characteristics optimization of mental inert gas welding aluminum foam plate. Materials and Design, 32, 3 (2011).
6. Y Dong Zhang, Z He. Application of Improved Principal Component Analysis Method to the Multi-Response Optimizatio. Modular Machine Tool & Automatic Manufacturing Technique, 11 (2012).
7. Q An Cui. Design of experiments and global modeling approach based on sub-domains significance for complicated relatonship process. Systems Engineering-Theory&Practice, 33, 9 (2013).
8. Ali N, Celik T M. Neural network based method for conversion of solar radiation data. Energy Conversion and Management, 67,(2013).
9. X Ming Yan, Z Zheng. Performance comparison between some nonlinear curves fitting with BP and RBF. Journal of Fujian Agriculture and Forestry University (Natural Science Edition), 40,6 (2011).
10. Afantitis A, Melagraki G, Makridima K. et al. Prediction of high weight polymers glass transition temperature using RBF neural networks. Journal of Molecular Structure Theochem, 716, 1(2005).

11. H Jie Fan. Design and optimization of capacitance decay program for metallized film capacitors. ELECTRONIC COMPONENTS AND MATERIALS, 30,9 (2011).

Research on the Development of E-commerce Model of Agricultural Products

Yaping Huo and Huiping Mu
Management Engineering Department, ZhengZhou University, ZhengZhou, China

Abstract. In recent years ,electronic commerce of agricultural products has developed rapidly.In the development process of electronic commerce of agricultural products,many classic cases and business models have emerged.Such as tootoo industrial commune model(product drive type),Original life model(marketing driven type),Suichang model(service platform type).The author analyzes and compares each model from 7 aspects,and summarizes their differences and links.Finally,The author puts forward suggestions on the development of electronic commerce of agricultural products in china.

1 Introduction

1.1 Background and significance of the study

Electronic commerce of agricultural products is a new way of selling and circulating agricultural products.In recent years ,electronic commerce of agricultural products has developed rapidly.According to Alibaba agricultural e-commerce white paper (2015),Sales of agricultural products reached 69.55 billion yuan on Ali platform in 2015.The number of sellers of agricultural products was 762,100 in 2014,and The number of sellers more than 900,000 in 2015,which accounted for 97.73% of retail platforms.[1]

In 2005, Fresh electricity business started to develop.Along with the development of e-commerce of agricultural products,there have been a variety of classic cases of e-commerce of agricultural products.Such as Futian network,tootoo industrial commune ,Original life,sfbest. Agricultural electricity suppliers is becoming more and more,and some electricity suppliers have begun to profit.Because farmers are not familiar with e-commerce,and the business of electronic business enterprise involves purchase, processing, sale, logistics and platform construction, operation and maintenance and other links,Agricultural Electronic Commerce face the following problems:(1) the slow development of electronic commerce in traditional enterprises.(2) the farmers only conduct online sales, ignoring the quality of agricultural products.(3) the government does not know how to support the development of e - commerce.

Through reading the references at home and abroad, the author finds out a phenomenon.There are many literatures about electronic commerce, but there is little research on the e-commerce of agricultural products.The reasearch on the classic cases of electronic commerce of agricultural products has have great significance:First, with the popularity of the Internet and the improvement of rural network coverage, farmers will be able to open the shop at home. However, due to the lack of knowledge of agricultural electricity suppliers, farmers do not know how to operate a good shop.So there are many shops, but the number of profitable shops is not much.How to help farmers to operate the shop is a problem that needs to be solved.Second,people need more and more kinds of agricultural products,and people require more and more high quality of food,so government regulation of agricultural products is becoming more and more difficult.How to reduce the difficulty of government regulation is an urgent need to be resolved. Third, the development of electronic commerce of agricultural products can increase farmers' income and create more jobs.

1.2 The research status at home and abroad

The case study of electronic commerce of agricultural products is relatively small at present.Zhao Ping, Luo Yi (2011) analyzed the service orientation, brand building and so on,and put forward five enlightenment to our country to carry out electronic commerce of agricultural products:First,the focus of the development of electronic commerce of agricultural products is to improve farmers' income and service consumers.Second,The situation of electronic commerce of agricultural products can be varied, But in the end, the whole process should be formed in the closed loop operation,which includes production, procurement to the transaction, distribution and other links.finally electronic commerce of agricultural products realizes the docking of the production, supply and sale of agricultural products.Third,the government should pay attention to regional development.Fourth,in order to carry out electronic commerce of agricultural products, Online retailers should first occupy a certain market, and then seek further development.Fifth,The integration of the supply chain should pay attention to the information sharing of each node.[2]Luo Yi (2012) analyzed the development of agricultural products e-commerce enterprises.He made five suggestions to the government.First,the government should affirm the prospect of the development of electronic commerce of agricultural products.Second, according to the actual situation,the government should selectively support the development of the electronic commerce enterprises of agricultural products.Third, the government should pay attention to the way to support and the key points of support.Fourth, the government should make a standard.Fifth, the government should strengthen its own management of the quality of agricultural products, and promote the formation of high quality and favorable price environment.[3]

1.3 Research ideas

In view of the current problems of e-commerce of agricultural products in China,The author analyzed the classic case of e-commerce of agricultural products,and sorted out the different patterns,then summarized the differences between the models and links.Finally, some suggestions were put forward from three different perspectives: farmer (or rural cooperatives), e-commerce enterprises and government.

2 Cases description

2.1 Tootoo industrial commune

Tootoo industrial commune officially launched,which is B2C online store and specializes in providing fresh products.In 2012, Tootoo industrial commune operating income increased by 181% over 2012, reaching 23,112,000 yuan,and gross profit is 4,749,000 yuan.[4]

Tootoo industrial commune controls the upstream side.Tootoo farm covers an area of 1050 acres in the Pinggu area of Beijing.From the choice of seed, fertilization, pest control, to the harvest of agricultural products, packaging,It is strictly in accordance with the standard implementation.It guarantees the quality of organic food in the whole production process.Ninetowns Group is the parent company of Tootoo industrial commune,which is a software company and provides inspection system software services to domestic export food enterprises.So Tootoo industrial commune can find outstanding foreign trade food export enterprises as their suppliers,and it can guarantee the distribution of all orders and master the price of the initiative.

Tootoo industrial commune establishes its own cold chain distribution.It has insisted on building its own logistics system from the beginning.The company set up a 4,000 square meters of logistics center,which integrates cold storage, freezing and processing into one body.It purchased a refrigerated truck.By using full cold chain distribution model,it achieves the goal of "fresh distribution everyday".

Tootoo industrial commune occupies the channel.In addition to their own Web site,It also opened its own shop on third-party platforms,such as Tmall, Jingdong, Kuba and so on.In addition to the initial market in Beijing,It still developed the market in Shanghai and Shenzhen and other first-tier cities.At first it took several categories to test and took the virtual inventory method to carry on the remote controll, and finally provide all products.

2.2 Originally living network

Originally living network caused a lot of attention at the beginning.On the one hand,It's Southern Newspaper that founded it,and It invested hundreds of millions of funds for warehousing and logistics distribution team. This is rare in the fresh electricity suppliers. [5]

Originally living network buys products in place of origin.Their buyers need to determine the topic through the opening of the topic,then began to go deep into the local area, and inspect the local specialty, and began sourcing products.Only choosing the best food in the season is the principle of buyers.

Originally living network integrated supply chain.In the process of purchasing of litchi, It made a comprehensive supply chain integration about Gaozhou litchi.Consumers believe there is no difference between Gaozhou litchi with good quality and Hainan litchi,so general manager Hu Haiqing proposed the local government to create a concept of origin which is similar to the "Gannan navel orange".

Originally living network promotes marketing through the media.Buyers should not only choose the high quality products, but also to write paperwork.Excellent marketing capabilities brought more partners.

2.3 Suichang model

Suichang model is considered to be China's first service platform driven agricultural e-commerce model.It is driven by a localized e-commerce integrated service provider.It drives the ecological development of county e-commerce and promotes local traditional industries. [6]

E-businessmen develop together.Suichang online shop association played a key role in the formation of e-businessmen cluster model.Association is not only the rural e-commerce public service platform, but also the company,and it is self-financing.[7]

Sales of agricultural products is its characteristics.Online trading of agricultural products include nuts and snacks, dried goods, tea, fresh fruits and vegetables.Now fresh fruits and vegetables gradually become the main products in the transaction.Local cooperatives have high-quality product resources, but lack of e-commerce experience and talent.On the agricultural products distribution platform set up by the association, the electricity businessmen achieve a lower cost of network sales,and it speeds up the transformation process of traditional enterprises.

Government support and investment in e-commerce.First, the government has built infrastructure,such as transportation, broadband and industrial parks.Second, the government has formulated policies.Beginning in 2011, each year 200 million in financial assistance is used for Suichang e-commerce development.

3 Analysis on the e-business model of agricultural products

3.1 Tootoo industrial commune model (product-driven type)

Tootoo industrial commune has its own farm.The author browsed its website,and found some information about the farm.The details are shown in table 1:

3.2 Original life model(marketing driven type)

The author browsed its website,and found some marketing activities.The details are shown in table 2:

3.3 Suichang model(service platform type)

The company provides members with the following services,As shown in Table 3.

3.4 Comparison of the models

According to the above description,The author compared the seven aspects of each model.The comparison results are shown in Table 4:

Table 1. The specific circumstances of the farm

Item	Detailed content
Farm profile	The farm is located in Ma Changying Town, Pinggu District, northeast of Beijing,and covers an area of 1050 mu.There are forest cultivation, organic cultivation, tourism and leisure, farming experience, science education, catering, etc.
Organic farming	From the purchase of seeds, seedlings, pest control to cultivation techniques,there are strict operation and control system.Farm refused to use pesticides and fertilizers, followed the natural growth of plants.
Organic farming	From the purchase of seeds, seedlings, pest control to cultivation techniques,there are strict operation and control system.Farm refused to use pesticides and fertilizers, followed the natural growth of plants.
free-rangeunder the woods	Poultry free to find food without the addition of any promoting agent.So livestock and poultry growth is relatively slow, resistance, immunity is very strong, no drug residues.here are standard operation and control system throughout
Organic certification	2009: through the domestic organic standards and the EU organic standard double certification; 2012: access to the state inspection and Quarantine Bureau for the record for the quality of vegetable base in Hong kong; 2013: won the title of excellent production base of Beijing food basket standardized the Beijing Municipal Council issued
Tracing system	It consists of an external and internal traceability system.Consumers to buy products can get a technology center certification card, scratch card can get a set of organic code, and then through the internal traceability system can get the product specific information. Thus ensuring the full traceability of the product.
Activity recruitment	Recruiting at WeChat.Experiencer can get free vegetables by providing screenshots to the staff.
Amusement	Organic farmer's tour,Family activities,Fun games,Health food,Organic market

Landlord plan	Experiencer spend money to contract a piece of land on the organic farm.Tootoo industrial commune is responsible for planting, fertilizing, weeding, pest and disease prevention, harvesting etc.When the crops are ripe, experiencer can get a part of the organic produce.
Live farm	Through the entire process monitoring and intelligent instrument long-distance observation and control terminal,Connecting the computer to observe the dynamics in the field.To understand the control of environmental data, environmental factors, etc.So that consumers really worry about food products.

Table 2. Marketing activities of originally life network

Item	Detailed content
Going home for dinner	Each theme activity will have dishes and recipes which matches the theme,and there are corresponding ingredients for consumers to buy.
Solar terms of food	They put the list of dishes corresponding to the Chinese traditional 21 solar terms out,and consumers can choose and learn the solar terms of health recipes and the corresponding ingredients.
the kitchen of Original life	The kitchen of Original life is a western restaurant with reservation service.It is mainly engaged in Italian, French and Japanese cuisine. Its ingredients are from the original life.

Table 3 .Services for Members

Item	Detailed content
Shop training	providing online shop training for businesses, students start their own businesses, internships, etc.
Supply sources	solving the problem of supply business and providing a variety of sources of supply channels.
Packaging services	Providing customized packaging items, good quality, low prices, thereby reducing the cost of packaging the shop.
Free training	Members can get free training.
Introducing qualif ied person	Recommending and introducing e-commerce qualified person to the member.
Solving Logistics	Providing members with integrated warehousing, logistics solutions.
Consulting service	Providing consulting and service to all kinds of problems encountered in the process of starting a business.
Policy interpretation	Integrating and interpreting of the various entrepreneurial policies.

Table 4. The difference and connection between the models

Item	Tootoo industrial commune	Original life	Suichang model
Whether it has its own farm(Or whether the product from its own production)	Yes	No	Yes
Whether there are overseas suppliers	Yes	Yes	No
Whether it has its own logistics system	Yes	Yes	No
Whether there is a professional food quality testing standards and institutions	Yes	Yes	No
Whether relying on third-party logistics	Yes	Yes	Yes
Whether it is operated by a specialized company or association	No	No	Yes
Whether it has access to government policy or financial support	No	No	Yes

4 Suggestions on the development of e-commerce of agricultural products in China

4.1 Recommendations to farmers (or agricultural cooperatives)

The peasant household should return to the land and give the shop to the specialized agency operation company.Farmers can not only participate in the basic free training, but also can entrust a professional e-commerce company operating shop.At present, there are many companies to undertake such a service.In this way, farmers can focus on the cultivation of agricultural products, after all, the quality of products is one of the key factors to improve sales.

Farmers should be rational use of chemical fertilizers, pesticides,and plant agricultural products in strict accordance with the quality and safety standards of agricultural products.Now people hope that the quality and safety of agricultural products is getting higher and higher,at the same time,people are becoming less sensitive to price.So farmers must comply with a series of standards in the process of the cultivation of agricultural products.

Farming should be arranged reasonably according to market information.On the one hand,Agricultural products sales channels have not been expanded,On the other hand,farmers lack of market information,Product supply exceeds demand.

4.2 Recommendations for e-commerce companies

E-commerce companies should strictly control the quality,and establish a set of agricultural products quality and safety system.Producers of agricultural products do not have or lack the technical and financial resources to develop a safety system,so e-commerce companies need to develop a security system.

If own logistics strength is not strong, do a good job within a small range.E-commerce enterprises should initially set the target market within a reasonable range,and do a good job within a small range,then slowly expand market.Tootoo industrial commune has its own logistics and distribution system.Its main target market is Beijing,so it not only reduces the damage rate of the product in transit,but also control the logistics costs of the downstream supply chain.

E-commerce companies should establish a customer relationship management system to improve consumer participation.The two main components of customer relationship management system are customer value and customer satisfaction.An important concept of customer value is the value of customer delivery,which refers to the difference between total customer value and total customer cost.The significance of maximizing the value of the transfer is that the customer will buy a product that provides the highest value of the transfer.Customer satisfaction refers to the consumer's sense of pleasure or disappointment.Consumers to participate in the company's daily activities.It not only increases their value of customer delivery and satisfaction, but also increases consumer loyalty.

4.3 Recommendations to the Government

In addition to supporting e-commerce of agricultural products in terms of policy and funding,government should also increase investment in infrastructure construction.Now the government has provided a healthy environment for the development of agricultural products.The government should increase the road traffic, broadband coverage and other aspects of investment,especially in remote areas.Governments increase investment in infrastructure,on the one hand,it is conducive to e-commerce business day-to-day business activities,on the other hand,it is conducive to cultivating the local people's online shopping habits,increasing sales and potential customers for e-commerce enterprises.

The government should guide farmers to cooperate with large-scale e-commerce platform.If there was a region with a variety of specialty products,the Government could cooperate with the relatively large electronic business platform for example, Taobao, Jingdong, No. 1 shop, etc.,then promote local specialties.

The government should do a good job on the quality and safety of agricultural products supervision.E-commerce companies establish a set of agricultural products quality and safety system.Farmers (or agricultural cooperatives) are strictly in accordance with the standards.The government regulates the quality of agricultural products."Three-pronged approach" can really guarantee the safety of agricultural products and consumer health.

References

1. Ali Research Institute.Ali agricultural e-commerce development white paper. 12(4), 2-4, 2015.
2. Zhao Ping, Luo Yi.Case Study and Enlightenment on the development of agricultural products e-commerce-taking Food Steward" and Freshdirect as examples.Business Economics andManagement, 19-23, 2011(07).
3. Luo Yi.Some Thoughts on the Development of e-commerce of Agricultural Products in China-Study on multiple cases, 110-116, 2012(09).

4. Cui Jing.Tootoo industrial commune:the control of whole supply chain.China 's Economy and Informa-tization, 56-57, 2013(14).
5. Zhi Ying.Those things about originally life network and chu orange-Interview with assistant to the President of Originally life network.China Advertising, 85-87, 2014(08).
6. Wu Wenmi.The way of Suichang mode of letting the agricultural products business platform landing. Innovation technology, 48-49, 2014(07).
7. Ali Research Institute.research on Suichang model. (30)10, 2013.

A Knowledge Modeling Method for Computer Graphics Design & Production Based on Ontology

Tong Chen[1], Beibei Fan[*1], Dongping Chen[2] and Cong Li[1]

[1]Shanghai University, School of Mechatronic Engineering and Automation, 200072.

[2]AVIC Commercial Aircraft Engine Co., Ltd, Shanghai 200241, China.

[*]Corresponding author: fanbeibei@shu.edu.cn

Abstract. As one of the most critical stages of CG (Computer Graphics) industry, CG design & production needs the support of professional knowledge and practice experience of multidisciplinary. With the outstanding performance in knowledge sharing, integration and reuse, knowledge modeling could increase greatly the efficiency, reduce the cost and avoid repeated error in CG design & production. However, knowledge modeling of CG design & production differs greatly from those of other fields. On the one hand, it is similar to physical product design, which involves great deal of tacit knowledge such as modeling skills, reasoning knowledge and so on. On the other hand, as film, CG design & production needs a lot of unstructured description information. The heterogeneity between physical product and film makes knowledge modeling more complicated. Thus a systematic knowledge modelling method based on Ontology is proposed to aid CG design & production in this paper. CG animation knowledge is capture and organized from viewpoint of three aspects: requirements and design and production. The knowledge are categorized into static and dynamic knowledge, and Ontology is adopted to construct a hierarchic model to organize the knowledge, so as to offer a uniform communication semantic foundations for designers from different fields. Based on animation script, the CG design task model is proposed to drive the organization and management of different knowledge involved in CG design & production. Finally, we apply this method in the knowledge modeling of naked-eye animation design and production to illustrate effectiveness of this method.

Key words. CG; Knowledge Modeling, Ontology, OWL, Static Knowledge, Dynamic Knowledge.

1 Introduction

CG (Computer Graphics) animation is a kind of film designed and produced mainly by computer. CG animation is an important media product, which is of great need in film industry, industrial products propaganda, education and some other fields. As one of the most critical stages of CG industry, CG design & production involves script design, 3D modeling, scene rendering, road planning and so on. Characterized by high density of knowledge and innovation, CG design & production needs the support of professional knowledge and practice experience of multidisciplinary. CG design & production is the process to transform the customer requirements into CG animation by related personnel (shown in Figure 1). A great deal of knowledge would be generated during this process, such as story, story board, scripts, modelling skills, parameters setting methods and so on.

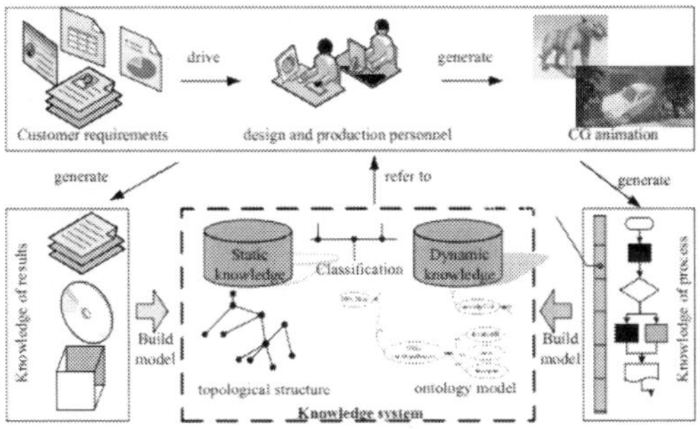

Figure 1. The framework of CG design & production process knowledge modeling

With the outstanding performance in knowledge sharing, integration and reuse, knowledge modeling could increase greatly the efficiency, reduce the cost and avoid repeated error in CG design & production. The Ontology technique is an important tool in knowledge engineering, which could support knowledge sharing, integrating and reusing on the level of semantic, and describe various kinds of phenomenon, facts and theoretical knowledge[1]. However, knowledge modeling of CG design & production differs greatly from those of other fields. On the one hand, it is similar to physical product design, which involves great deal of tacit knowledge such as modeling skills, reasoning knowledge and so on. On the other hand, as film, CG design & production needs a lot of unstructured description information. The heterogeneity between physical product and film makes knowledge modeling more complicated. Thus, this article aims to propose an integrating knowledge modelling method to aid CG design & production. The following part of this paper is as follows: related works on Ontology are presented in Part 2. CG design & production process is stated, and the knowledge generated during this process is analyzed in Part 3. The Ontology modelling process is illustrated in Part 4. A case of 3D film design & production is illustrated in Part 5. Part 6 concludes the whole paper.

2 Related works

Ontology is a philosophy terminology referred to the ultimate physical phenomenon which states philosophical theories. In the realm of knowledge engineering, there are no universal definition of Ontology. The most famous one is defined by Gruber-- "Ontology is a description of the concept of shared conceptualization"[2]. Ontology is studied and applied in many aspects, such as geographical area [3], biological and medical field[4-5], assembling process domain[6-7], service area[1] and so on.

The work of movie Ontology building is really great in progress. Some famous knowledge databases such as LinkedMdb, Freebase and so on, have been built on the open datacloud[8]. Wang[8] proposed a method to construct Bilingual Movie Knowledge Graph. However, these knowledge databases focus on film information about actors, audience, production companies and so on. Little knowledge about the design & production is involved.

Most Ontology researches about design & production were carried on tangible products[9-11]. Zhou et al. [9]proposed a modeling method of Diesel Ontology Model for Multidisciplinary Optimization Process, and solved the problems that sematic of information model was simple and information could not be inherited shared. However, as film, CG design & production needs a lot of unstructured description information. Thus the method of tangible products design & production knowledge building could not be used in CG directly.

3 CG design & production process analysis

CG design & production is the process to transform the customer requirements into CG animation by related personnel. A great deal of knowledge would be generated during this process. The detailed procedures are outlined in Figure 2. Story, story board, checking video, script and other kind of knowledge are generated in the design stage, while 3D model, camera layout, lighting layout, background materials are generated in the production stage. Besides, modelling skills, parameters setting methods and other tacit knowledge will be generated during the whole process. The above mentioned knowledge constitutes the knowledge system of CG design & production process.

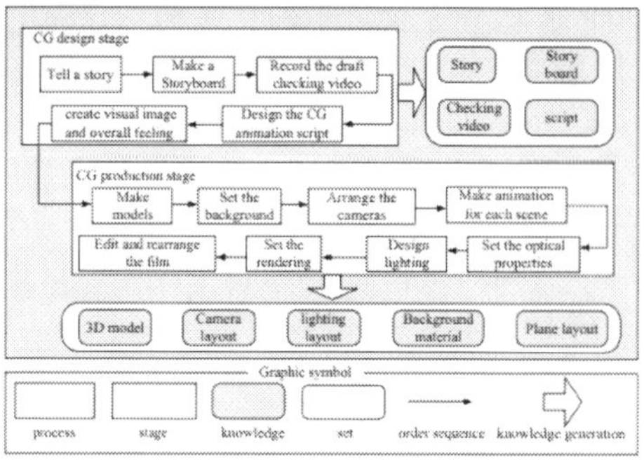

Figure 2. The knowledge generated within CG design & production process

In order to share information, the knowledge are classified into static knowledge and dynamic knowledge. Static knowledge could be referred to directly in other fields, while dynamic knowledge could be used in CG design & production process only. The upper part of Figure 3 shows the knowledge system classification of CG design & production process.

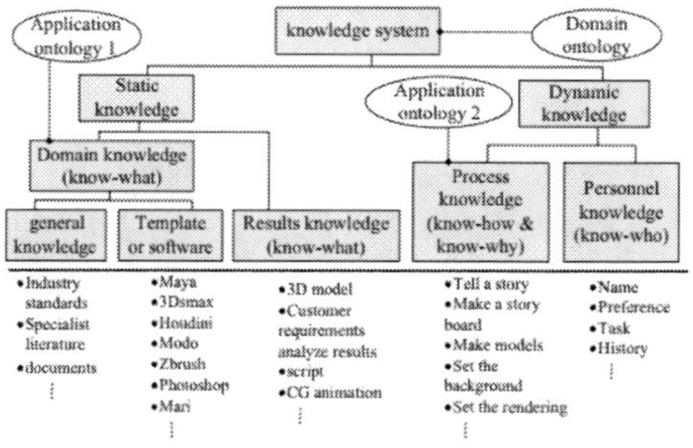

Figure 3. The knowledge system classification of CG design & production process

The main tasks of knowledge modeling in this article are to build the domain Ontology model and application Ontology model for each kind of knowledge appeared in Figure 2. The domain Ontology of CG design & production process is the base model for the classified knowledge. Due to limited space, only the domain Ontology model and two application ontologies (one in static knowledge sub-class and another in dynamic knowledge shown in Figure 3) are demonstrated in the following part.

4 Ontology modelling for CG design & production process

The domain Ontology of CG design & production process is the base model for the classified knowledge. Ontology model could be constructed by two methods – manual modeling and computer modeling. One the one hand, manual modeling is excellent in recognizing important concepts by experts due to the comprehensive property of human, while prone to omitting some details. On the other hand, computer modeling performed well in dealing with mass data, but can't summarize. An integrated method of Top-down mainly by manual and Bottom-up mainly by computer is adopted to construct Ontology model.

4.1 Domain Ontology

The domain Ontology of CG design & production process is denoted by a five-element model OCG={C, R, P, A, I}. C is the set containing all the concepts related in CG design & production process. Some main concepts and their hierarchical level in C is shown in Figure 3 (a). R represents the relationships among these concepts, containing parent-child relationship, whole-part relationship, object-attribute relationship and so on. The relationship is-a shown in Figure 3 (a) is a kind of parent-child relationship. As shown in Figure 3 (b), the relationship limitedBy means that CG design & production process is limited by CG animation laws. The relationship use means that CG design & production process will use some software, such as 3D modeling software, rendering software, texturing software and so on. The relationship use means that CG design & production process could produce CG animation. Besides, there are reciprocal relationships, transfer relationships, function relationships and symmetrical relationships. For example the relationship limitedBy and limited are reciprocal.

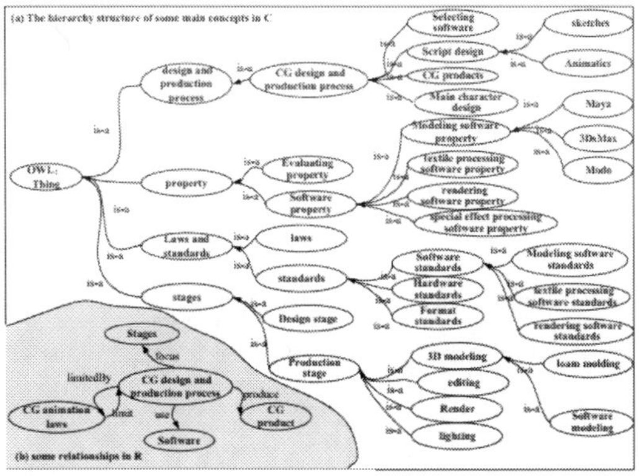

Figure 4. The domain Ontology of CG design & production process (part)

P represents the set of properties, used in depicting the concept. **A** is the Axioms defined on $\sum O_{CG}$ (where $\sum O_{CG}$=**C** ∪ **R**) depicting all the rules within this field. Axiom could put constraints on other concepts and properties. Take the concept *software rule* for example, it could be defined by the natural language as 'the software rule is used to select software'. This definition can be decomposed into two: (1) a software rule is a rule, and (2) this rule is used to select software. This concept could be expressed as follow axiom:

$$software\ rule = rule \cap (\forall rule.\ software\ selecting) \quad (1)$$

I represents the set1 of all the instances of **C**.

4.2 Application Ontology 1- domain knowledge

The application Ontology for domain knowledge is shown in Figure 5. Some main concepts and the hierarchy relationship among them are shown in Figure 5 (a). The Document_Feature Ontology represents the features of document in CG design & production process. According to their format, Document_Feature Ontology could be classified into several sub Ontology, such as usage feature, hierarchy feature, format feature, process feature, resource feature and so on. The Document Ontology represents knowledge carrier such as pictures, papers and so on. According to the classification of Document_Feature Ontology, document could be divided into corresponding classes. With the thining of document feature, document types are thinned further, and thus the hierarchy structure is formed just as shown in Figure 5 (a). The Ontology knowledge_Element could be divided into concept knowledge element, principle knowledge element, method knowledge element, fact knowledge element and so on. The properties of domain knowledge Ontology are shown in Figure 5 (b). The property hasFeature represents the features of domain knowledge.

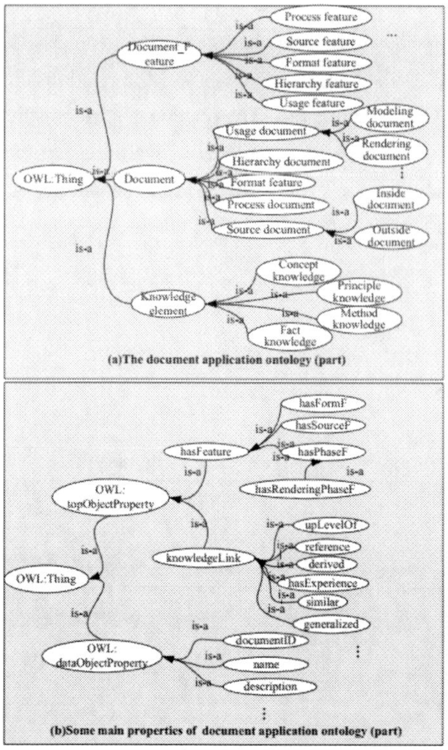

Figure 5. The application Ontology for domain knowledge (part)

4.3 Application Ontology 2- process knowledge

The process knowledge is the integration of all the interactive activities performed during the CG design & production process. The main ontologies are shown in Figure 6. The Ontology activity represents all the activities accord in the CG design & production process, which could be refined into *publicActivity* and *privateActivity*. The Ontology *publicActivity* signifies the activities occurred on the level of organizations, containing *taskPlanning*, *taskDecomposing*, *taskRealizing*, *taskResultReducing*. *requirementAnalysis* is the sub-class of taskPlanning. The Ontology *privateActivity* signifies the activities occurred on the level of individuals, containing *designProcessActivity* and *designRationalActivity*. *designRationalActivity* contains four thinking activities: convergent thinking activities, generative thinking activities, analytical thinking activities and synthesis thinking activities.

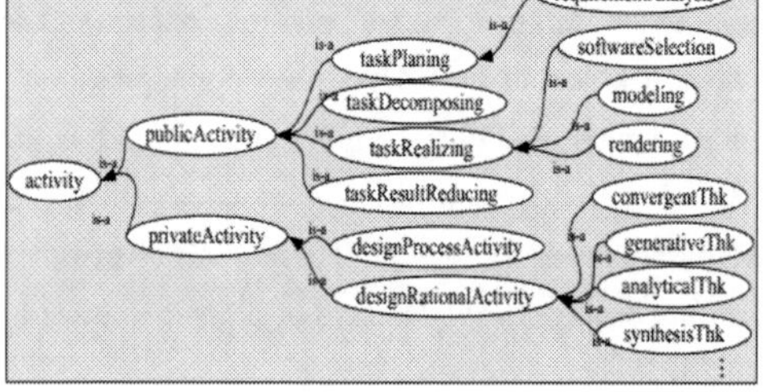

Figure 6. The process application Ontology (part)

5 Case study

This case is a naked-eye 3D advertisement for hand sanitizers, which comes from an IT company located in Shanghai. The naked-eye 3D film is an emerging CG product, which is regarded as the next generation of display technology with endless vitality. To get the best naked-eye 3D visual effect, design & production knowledge of each step should be reused and shared. Thus the proposed method was applied in the knowledge modeling of naked-eye 3D animation design and production to illustrate effectiveness of this method.

The procedures of naked-eye 3D film design & production include the following key steps: script design, 3D modeling, scene making, rendering, film production, synthesis, editing etc. The knowledge classification of naked-eye 3D animation design & production is shown in the lower part of Figure 3. Each kind of knowledge was modeling with Ontology technique. The activity of defining tasks could form links among the domain knowledge, design requirements (static knowledge) and the related tasks (dynamic knowledge). Thus the Ontology representation of the defining tasks activity is described in Figure 7 with OWL. Figure 7 (a) shows the integrated Ontology for defining tasks activity. The input and output of defining tasks activity Ontology are design target Ontology (Figure 7 (b)) and statement of task Ontology (Figure 7 (c)), respectively. The design target Ontology describes customer requirements in respects of pictures, voice, lens and so on. The statement of task Ontology is the formal statements of design task which are described in designer's language. During the defining tasks activity, domain knowledge Ontology (Figure 7 (d)) will be used. The final notes about Figure 7 are: "SPE****", such as "SPE2940", "SPE7293" etc., refers to relevant provisions. "TAB2098" refers to the task set. "RQT****", such as "RQT0712", "RQT0932" etc., refers to the design requirements of this design task. "YFSD2344" refer to the knowledge browsed in the process of task formulation.

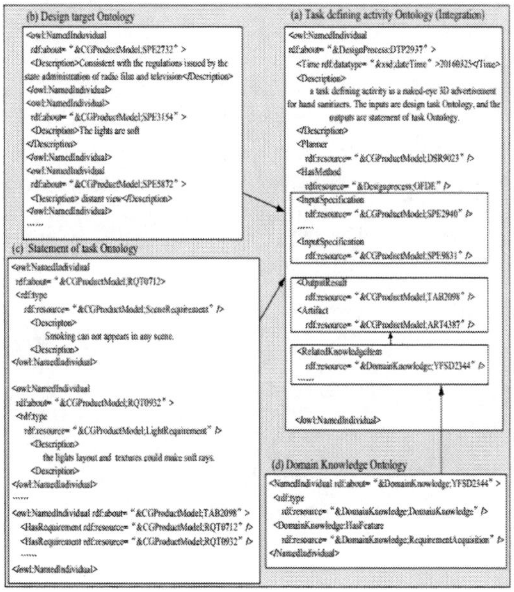

Figure 7. The OWL of the naked eye 3D film design & production Ontology (part)

6 Conclusion

This article proposed an integrating knowledge modelling method to aid CG design & production. CG animation knowledge is capture and organized from viewpoint of three aspects: requirements and design and production. The knowledge are categorized into static and dynamic knowledge, and Ontology is adopted to construct a hierarchic model to organize the knowledge, so as to offer a uniform communication semantic foundations for designers from different fields.

Based on animation script, the CG design task model is proposed to drive the organization and management of different knowledge involved in CG design & production.

Acknowledgements

This research was financially supported by National Natural Science Foundation of China (No.51405281) and Shanghai Science and Technology Innovation Action Plan (No.14511108303). The authors express sincere appreciation to the anonymous referees for their helpful comments to improve the quality of the paper.

References

1. Jin shen. Modelling and configuration of product extension services using an Ontology-based approach. Shanghai Jiao Tong University. 2012, Shanghai.
2. Gruber, T.R. A translation approach to portable ontology specifications. Knowledge Acquisition, 1993, 5(2): 199-220.
3. Jianjun Chen, Chenghu Zhou, Jinggui Wang. The research on geography Ontology. Geography frontier, 2006(03): 81-90.
4. FIORENTINI X, GAMBINO I, LIANG V, et al. An ontology for assembly representation. New York: Department of Commerce, 2007.
5. RAPHAEL B, KRIMA S, RACHURI S, et al. OntoSTEP : Enriching product model data using ontologies. Computer-Aided Design, 2012, 44(6): 575-590.
6. FIORENTINI X, GAMBINO I, LIANG V, et al. An ontology for assembly representation. New York: Department of Commerce, 2007.
7. MINNA L, GARCIA F, KALLELA T, et al. Product-process ontology for managing assembly specific knowledge between product design and assembly system simulation// Micro-Assembly Technologies and Applications. IFIP-International Federation for Information Processing 260. Springer US, 2008:99-108.
8. Weiwei Wang, Zhigang Wang, Liangming Pan, et al. Research on the construction of bilingual movie knowledge graph. Acta Scientiarum Naturalium Universitatis Pekinensis, 2016, 52 (1): 25-34.
9. Honggen Zhou, Mengying Wang, Xuwen Jing, et al. Modeling and application of diesel ontology model for multidisciplinary optimization. Computer Integrated Manufacturing Systems. 2014, 20 (10): 2348-2359.
10. Ma, S., Tian, L.Ontology-based semantic retrieval for mechanical design knowledge, International Journal of Computer Integrated Manufacturing, 2015, 28 (2):226-238.
11. Chen, Y., Zhao, M., Xie, Y., Zhang, Z. A new model of conceptual design based on Scientific Ontology and intentionality theory. Part II: The process model. (2015) Design Studies, 38, pp. 139-160.

Study on the Transformation and Upgrade of E-Commerce Application in Manufacturing Industry

Liang Zhao[1, 2] and Mengzhao Feng[1]

[1]Management Engineering School, Zhengzhou University, Zhengzhou, People's Republic of China
[2] E-commerce Department Director,Zhengzhou University, Zhengzhou, People's Republic of China

Abstract. Under the Internet thinking, manufacturing management mode is changed, it is not depend on the scale, but in the user as the center, more dependent on the reaction rate of user requirements. As the big data, cloud computing, Internet of things technology such as the development of technology, manufacturing is from automatic to intelligent evolution, on the user's personalized customization is possible. Using Internet thinking positive development manufacturing enterprise electronic commerce application mode of transformation and upgrading of manufacturing industry plays an important role. Promoting the continued ascent of manufacturing industry informatization level, promote the transformation of the manufacturing production mode, provide greater profit space for the export of products. In industrial 4.0 mode, the user is the core of the enterprise, C2B model will be the mainstream of future manufacturing e-commerce application mode. This paper to present the sorted out the basic model of manufacturing enterprise electronic commerce application, point out the existing problems, and the computing technology, communication technology, the constant development of storage technology, such as background, puts forward the corresponding countermeasure, points out that the C2B model is the development direction of manufacturing enterprise electronic commerce application mode.

1 Introduction

In recent years, China's e-commerce has entered a period of rapid development. According to the China Electronic Commerce Research Center report shows that in 2014, China's e-commerce market transactions amounted to 13.4 trillion, an increase of 31.4%. Among them, B2B e-commerce market transactions amounted to 10 trillion RMB, an increase of 21.9%. Network retail market transactions amounted to 2.82 trillion RMB, an increase of 49.7%. By 2015, China's e-commerce transaction volume is expected to more than 18 trillion RMB. During the two sessions last year, with the "Internet +" action plan proposed, mobile Internet, cloud computing, big data, Internet of things and the combination of the current manufacturing industry will become the future trend, further expanding the depth and breadth of industrialization and information fusion. China's e-commerce development will usher in a new historical opportunity. In the rapid development of e-commerce, the traditional manufacturing industry is facing the plight of resources and the environment, the growth rate has slowed. How to play the role of e-commerce, promote the transformation and upgrading of China's manufacturing industry is worthy of deep thinking [1].

2 Theories Review

2.1 E-commerce

E-commerce is using electronic ways and electronic tools to carry out the entire process of business activities [2]. E-commerce in general, using a variety of electronic tools to engage in business work or activities. Narrowly, E-commerce is defined as the main use of the Internet to engage in business activities. E-commerce is in the highly developed society of computing technology, communication technology, storage technology, with computer technology and business operations knowledge, the use of the Internet to carry out business activities.

B2B, B2C, C2C, C2B, O2O is the five major e-commerce business model. B2B mainly refers to enterprises and enterprises through the Internet to carry out business activities. Traditional business model in the procurement, sales and other business processes can be done through the B2B model. B2C model usually refers to the business-to-customer transactions. At present this model is used in online retail sales. C2C model is between customer-to-customer of transactions, this model generally need to rely on a third-party service platform to carry out business activities. C2B model is the recent rise of an application model, mainly refers to the transaction between customers and manufacturers, the biggest feature is that customers can customize the production of personalized. O2O model is an extension of e-commerce online activities, online diversion, offline experience, closer to the distance between customers and goods.

2.2 Industrial Internet

With the continuous development of e-commerce and enrichment of e-commerce applications, the penetration of e-commerce began to gradually extend from the consumer side, such as circulation enterprises, manufacturing enterprises, manufacturing enterprises, raw material supply enterprises and so on. Especially for manufacturing enterprises, it will take downstream market demand and upstream production more closely, forcing manufacturing transformation and upgrading that gives birth to the Chinese characteristics of the industry 4.0[3]. Industrial 4.0 was first proposed by the German Hannover Fair in 2013, whose purpose is to actively develop the next generation of manufacturing technology to promote the upgrading of modern manufacturing technology, and then to maintain Germany's leadership in the manufacturing field. With the development of artificial intelligence, big data, cloud computing, Internet of things technologies,Industry 4.0 shows up. Its core is intelligent manufacturing, to achieve the communications of material and material, people and objects, people and people. Intelligent machines can make decisions according to their own production situation. In the mode of intelligent manufacturing, production from the concentration to the decentralized, the scale effect is no longer the key factor in industrial production; products from the convergence to the personality change, the future will be fully in accordance with the wishes of individual production. Extreme cases will become automated, personalized one-piece manufacturing; users from partial participation to full participation, the users at the beginning and end of product manufacturing. They can interact with the production process, in the last part of the product can still make personalized products requirements. For customers, they can browse directly through the mobile terminal for related products, put forward personalized needs and personalized customization. Then the user's personalized requirements will be sent to manufacturing companies through the network. Manufacturing companies according to the order to purchase, to produce, to distribute arrangements, etc. All of these can be done automatically under the environment of the industry Internet. Industrial Internet for the manufacturing industry will bring a profound commercial revolution and industrial revolution [4].

3 Analysis on the Current Situation of E-commerce Service Platform for Manufacturing Industry

3.1 Manufacturing B2C Model Status

The current e-commerce model of the manufacturing industry is mainly B2C. Enterprises establish an official website to introduce products in detail, supplemented by the company's news, activities, recruitment and other information. Such as "Shanghai Volkswagen" of car manufacturing industry, "Changan Ford", "Yutong Bus" of passenger car industry and "Gree Mall" of home appliance industry. Gree Mall, for example, it's website provides five modules: product display, solutions, shopping processes, after-sales service, the strength of Gree. The products can be introduced to the user through the enterprise-related products, for customers encounter different problems, they can solve specific issues through specific solutions. Shopping process can be showed detailed. After-sales service eliminate the worries about after shopping. When users have problem, they can communicate with the online customer service in time.

3.2 Manufacturing B2B Model Status

At present, Alibaba, World Factory Network and Chinese Machinery Manufacturing Network are the better manufacturing e-commerce sites in Chnia[5]. Their main function is to provide a platform for communication between sellers and buyers. The sellers through the platform to publish product information and the buyers through the platform to browse the corresponding products. When the two sides reached a transaction intention, they can pay through the platform with the payment tool .

World Factory Network, one of the best manufacturing B2B websites, its biggest feature is to take into account other manufacturing industries, while it is free to the enterprises to provide the initial page and only for value-added services to charge the appropriate fees. Since the site has been on the line, there are already more than 500,000 manufacturers joined the site, which becomes a well-known, low-cost direct-to-business B2B website. What's more, it has a huge enterprise information database, such as the Global Enterprise Database, China Purchasing Information Database, the World's Factory Network Procurement Channels. In which you can easily find the information for the enterprise. At the same time the site also provides business schools and other resources for business owners, to change the traditional mode of thinking, re-use the Internet to rethink the business processes and how to use the Internet to maximize sales of enterprise products.

3.3 Manufacturing O2O Model Status

In 2013, the Dianping mobile side view more than 70%, the mobile end of the cumulative number of independent users more than 80 million; Meitua mobile side sales accounted for nearly 70%, exceeding the 50% target set at the beginning of the year[6]. Since 2013, all kinds of taxi software, car rental services for people to provide convenience, and integration will have far-reaching impact on the industry; all kinds of food services such as Yi Tao food, Taodiandian have a huge development and growth. Life services O2O field of public disclosure of investment and financing nearly

40. Into 2014, the mobile Internet era officially opened, O2O online and offline perfect integration of the most effective means. In such a combination, e-commerce will be a large number of integrated into the mainstream economy, "from there to nothing", and ultimately change people's lifestyles.

O2O model is one of the current well-established business models. In the home sector, Meilele is the most typical representative. It not only build its own official website platform, also build nearly 300 offline experience museum, provide goods and after-sales service [7]. Through the combination of online and offline, Meilele also has a traditional physical store and online virtual store double advantage. Erection of offline physical store, they can show the goods to customers, allowing users to personally experience the purchase of goods, while the store is also charged with the task of logistics and after-sales service, allowing users to purchase goods with greater confidence. Online sites play a role in the drainage, the use of the Internet low-cost marketing tools to expand the scope of the impact of the brand to attract more customers to browse the site.

3.4 Manufacturing C2B Model Status

The core idea of the C2B model is that customers drive the entire business [8]. Compared with traditional industrial high-volume, pipeline production, C2B model to the customers as the core, can accept the users' personalized needs, and then organize production and procurement. Through the first order of production, they can greatly reduce the production of blindness, reduce inventory and marketing costs, provide users with more intimate services, increase the viscosity of users.

In the household appliance industry, Haier Group and Tmall held household appliances customization activities. Customers can choose their own air-conditioning panels, refrigerator refrigeration ways. After the customers pay the deposit Haier will organize to produce, and delivery after 37 days. There are 65,000 orders in a month by this measure and 120 million RMB of sales. Customers, Haier and Tmall pre-sale platform tripartite win-win situation. Through the C2B, Haier dare to produce the products only a dozen users like. Customers' demand for long tail has been fully released. Nearly 10% of this activity comes from long tail demand, such as one of the seven-kilogram multi-drum washing machines, only from Xinjiang, Guangdong, Shanghai, Yunnan and other provinces of the 13 custom users. Without the Internet, Haier can not gather more than a dozen consumers prefer a product, can not predict the product sales, similar niche products often can not appear in the traditional sales channels. Haier through the C2B mode to receive customers' orders before the organization of production. Profit margins than the industry average increased by 0.5 percentage points [9]. By C2B mode, Haier can provide customers with only one order of the product, break the traditional model of regional sales forecast by the restrictions. Do not have to bear the bullwhip effect caused by the stock,achieve a supply chain without scale[10].

4 The Problems in the Development of Electronic Commerce in Manufacturing Industry

After rapid growth period in China, more and more companies recognize the use of Internet thinking to the enterprise competitive advantage. In particular, manufacturing industry in China is facing an important opportunity for transformation and upgrading. Many manufacturing enterprises have to strengthen the investment in e-commerce. However, due to the e-commerce develop not long in China, there are some outstanding problems in the process of rapid development.

4.1 The level of internal information to be improved

Manufacturing industry includes a very wide range, China's manufacturing e-commerce applications carry out mainly for the production of consumer-oriented products, such as clothing, household appliances and other industries. However, some traditional manufacturing industries also take the traditional sales methods, and the popularity of e-commerce to be further in-depth [11]. China's manufacturing industry is not strong, weak foundation of the low level of information technology. Such as customer relationship management system, supply chain management system, ERP system, the popularity and application level is still low. Developed countries in the manufacturing industry in the 1980s has been widely used within the enterprise computer network technology to improve the internal information exchange and operational efficiency. While China's current stage of the manufacturing industry is still not high degree of information [12]. But in China, most of the manufacturing industry with a serious shortage of investment on construction of information system, is still familiar with the application of information management model of the problem. In China, automobile industry is one of the high degree of automation production enterprises, but the car's sales model is still using the traditional methods. Because of the high price of the car, the customers for security reasons generally just browse commodity on the website and its functions, the low proportion of online payment of the customers. It needs innovation in the auto industry of electronic commerce application model, such as the use of O2O mode to achieve online and offline combination, the user's personalized customization and production systems connected to increase the user's desire to use the network to achieve online diversion and offline sales combined.

4.2 Manufacturing technology level behind

China's industry is still in the end of industry 2.0 and industrial 3.0 universal stage, the degree of automation of manufacturing processes to further significantly improved. In China, CAD technology coverage of machinery manufacturing industry only 5%[13]. With the manufacturing industry to become a new round of national competitiveness of the new high ground, Germany put forward the "Industrial 4.0 Development Strategic Plan", the United States developed the "Advanced Manufacturing Development Plan", through the development of big data, cloud computing and Internet of things to seize the manufacturing sector a new round of change opportunities. Through the adoption of new technologies to achieve material and material, people and objects, people and people connecting, to achieve from the automated manufacturing to intelligent manufacturing changes. Compared with foreign manufacturing technology, there is still a considerable gap. The integration of industrialization and informationization is the key to the development of e-commerce in manufacturing enterprises. However, the integration of the two needs a lot of manpower, material resources, financial resources and risks, so the manufacturing enterprises are seriously inadequate in investment in information technology and backward in manufacturing equipment. For most enterprises are still in the stand-alone automation, rigid automation stage. The majority of enterprise automation production level is still very low[14].

4.3 E-commerce development model to be improved

As enterprises develop e-commerce can eliminate the space and time barriers with customers, while users can grasp the huge amount of information. It is the enterprises through big data analysis that can grasp the changes in customers demand and make corresponding changes. B2B and B2C are the two main e-commerce application models of manufacturing enterprises. So many companies regardless of their own characteristics of the industry, blindly launched e-commerce projects, the results often counterproductive. With the development of technology, O2O model and C2B model are emerging, different companies should be based on their specific characteristics of the industry in which to choose the appropriate e-commerce model.

4.4 Macro environment to be optimized

Due to the e-commerce started late in China, it has been in a rough-style development. The relevant e-commerce laws and regulations are not perfect. What's more, enterprises and customers credit system has yet to be strengthened. Selling fake and shoddy products online have occurred. At the same time, the government departments of the e-commerce site monitoring capacity is not strong. In economic activities on the Internet is a lack of regulation means, such as a tax on online business activities[15].

5 Transformation and upgrading of manufacturing industry

5.1 Improve the informatization level of manufacturing enterprise

It is information economy era for 21 Century. The key to enhance the core competitiveness of enterprises is the realization of enterprise informationization. The enterprise informatization is the application of communication technology and computer technology in procurement, production, inventory, sales, customer service and other services sectors, to achieve the electronic operation process. But the information does not mean to buy equipment or use system. It is not the electronically instead of manual labor,but the transformation of business process and the production process, such as to realize the automation of production process, electronic financial management, accelerate the flow of information within the enterprise and between enterprises and enterprises. On one hand, it can grasp the latest information and to make a quick response to change. On the other hand, it can realize the sharing of information between various departments and enterprises break information barriers, which is conducive to the decision-making level of decision optimization. It can be said that the opening information construction is the extension of the enterprise management and organization management. At present the enterprise information systems of relatively simple common features include transaction process system, knowledge system experts, the office automation system, expert information system, management information system, advanced management information system, and information system with strong comprehensive such as ERP, SCM, CRM. Different information systems have different functions, and the enterprise level in it is not the same. Therefore, enterprise making the right choice should be based on their situation. The enterprise informatization is the foundation to develop the electronic commerce. Only if this, it is possible for external changes timely deliver to the enterprise decision-making, and make a change to enterprise production process.

The enterprise should increase their informatization construction investment to adapt to the needs of the development of electronic commerce. the enterprise is able to quickly respond to changes, strengthen the ability of enterprises to deal with the internal and external risks. Actively using big data, cloud computing, networking, wireless RF technology to remould technology, the production process, the transformation management flow. At the same time, it can strengthen the enterprise information construction level, the introduction of supply chain management,

management of customer relationship, enterprise resource planning, set up their own websites or the application of the third party service platform, achieving interoperability of Intranet and extranet, further appling collaborative commerce to manufacturing enterprises as soon as possible to join the torrent of e-commerce[16].

5.2 implementation of intelligent manufacturing

Our society has now entered the era from IT to DT, and will enter the IM era in the near future [17]. The technology of the enterprise should keep pace with the times. In the last century 80, 90 enterprises began to use the IT technology; the manufacturing enterprises should make DT technology; the enterprise must do IM technology well after10 years. With the development of artificial intelligence, networking, wireless RF technology, it can interoperability between objects and objects, and the production equipment can make the corresponding judgment according to the real-time, achieving intelligent production.

There is no doubt that the future direction of development of manufacturing enterprises is intelligent manufacturing. Artificial intelligence technology will be widely used in all aspects of the manufacturing process. Expert system can be used to project design, process design, production scheduling, fault diagnosis and other aspects and neural network calculation method of intelligent fuzzy control technology can also be used for product formulations, production scheduling and other aspects, so as to realize the intelligent manufacturing. For the complicated and uncertain problems in production process, it is especially suitable to use artificial intelligence technology. But it also needs a period of time for us to clearly realize the whole intelligentialize process of realizing the manufacturing enterprises in the production process. At present stage the intelligent production manufacturing enterprises realize is only partial intelligence, and it still has a long way to go for manufacturing enterprises intelligent production, and this trend is irreversible.

With the development of intelligent manufacturing technology, interoperability among things and things, people and objects, objects and objects into reality. Interconnection between people produces service internet, and interoperability between objects and objects forms industry internet. Through the links of service internet and industry internet, it will forms a comprehensive coverage of the network, to achieve full connectivity. The big data, cloud computing, networking technology were actively used in the process of upgrading of manufacturing enterprises, which forms the new value chain - knowledge of original research - development of new products - intelligent manufacturing - smart services on the Internet platform.

5.3 Actively explore the new model of the development of manufacturing electricity suppliers

In large quantities, scale, production process is fixed under traditional industrial production mode, which is based on the sales data in the past to predict the current enterprise product sales volume, and then determine the production scale. The manufacturing enterprise production decision as such exists blindness, which can't response timely to changes of consumers, leading to out of stock of some products, while the unsalable of some products, and out of stock of the market demand. It causes great waste of resources by manpower, material, and financial, and it is difficult to achieve maximum profit for the enterprise. But the demand pull type production is based on the electronic commerce mode of production. Enterprise organize production according to the user's orders, so that turn the production process from large quantities, assembly line production to small quantities, personalized customized production mode, and finally realize flexible production management[18].

In recent years, with the increase impact of Electronic Commerce on the garment industry, the competition between garment enterprises is increasing, leading to the positive thinking methods of change. Enterprises in the garment industry as an example, with the use of digital printing, CNC cutting, 3D body measurement instrument, enterprises can be customized according to the needs of individual users. Compared with the large-scale pipeline of the original mode of production, the production of the enterprise is more targeted after using C2B enterprise organization model of the production process, avoiding the blindness of production, saving a lot of manpower, material, and financial resources, improving the level of corporate earnings. It forms multi products species, small batch production mode.

With the continuous development of e-commerce, the dispersion of network users, personalized needs are constantly strengthened. The transformation from product sales end reverse promoting enterprise production mode, requires enterprise possesses stronger flexible production capability, and thus make changes of enterprise business processes and business model changes. In the continuous development of big data and cloud computing technology, the C2B model will be the future development direction of manufacturing enterprise e-commerce application mode.

5.4 Improve the external environment for the development of manufacturing industry

The development of the enterprise electronic commerce requires not only their own efforts, but also the government's actively creating a policy environment for the development of electronic commerce. Firstly, the government should further decentralization, reduce administrative examination and approval, issue support policy doing well to the development of electronic commerce, promoting the development of e-commerce in the manufacturing enterprises. Secondly, perfect the legislation of electronic commerce, and strength the supervision and actively guide of electronic transactions, security authentication, privacy protection, problems of tax laws and so on. Actively guide the electronic commerce enterprise innovation service pattern of electronic commerce, and realize the diversification of manufacturing enterprises in the development of electronic commerce. Put private capital into the range, and then into the field of electronic commerce, to increase the motive force of the development of electronic commerce. At the same time, the government should standardize the competition between enterprises, reduce the administrative intervention, and create a good market environment for the development of Chinese manufacturing enterprises of electronic commerce.

The government should strengthen the pilot work of electronic commerce, and play a leading role model. The successful experience of manufacturing enterprises is to carry out through a typical demonstration, expanding the scope of application of e-commerce in manufacturing enterprises. It should also need to strengthen the e-commerce development level of infrastructure construction such as speeding up the construction of "broadband Chinese", promote the popularization and application of the 4G network, and the pilot project of e-commerce integrated innovation. As to the key technologies of e-commerce development such as smart card, chip, independent research and development of safety management technology, realize the independent research and development and master the core technology. Through the leading role of the pilot to relevant technical standards. Encourage enterprises increase investment in enterprise informatization construction, promote the construction of manufacturing enterprise informatization and development of e-commerce applications, and then realize the transformation and upgrading of the application of electronic commerce in manufacturing enterprises[19].

6 Conclusion

Under the influence of e-commerce, from clothing, retail industry to the financial sector, from the agricultural industry to the manufacturing sector and even to the service industry are undergoing changes. In the use of Internet thinking to reshape business processes, and even change the business model to become a trend. Under the Internet thinking, the manufacturing business model is changing, not relying on economies of scale, but user-centric. It's more dependent on its response to user demand for speed. With the development of big data, cloud computing and Internet of Things technology, the manufacturing industry is evolving from automation to intelligence, and it is possible to customize the production of the users. The use of the Internet to actively develop the manufacturing enterprises of the e-commerce application model will be the manufacturing transformation and upgrading play an important role. In the industrial 4.0, users are the core of the enterprises, C2B model will be the mainstream of future manufacturing e-commerce application model.

References

1. Tang Guanghai.Research on the Path and Countermeasure of Transformation and Upgrading of Manufacturing Industry under Internet Thinking. Economic Outlook. 2015 (03): 22-23
2. Li Q. Introduction to E-commerce. Beijing: Higher Education Press, 2011: 13
3. Hu Jie.Development of China's Future Manufacturing Industry from the German Industrial "Industrial 4.0". Private Science and Technology. 2014 (12): 268
4. Luo Wen.Enlightenment of German Industrial Strategy 4.0 on China's Industrial Transformation and Upgrading.Journal of Industrial Economics and Management .2014 (7): 180
5. Zhang Shu. Intelligent manufacturing and future manufacturing. Modern Manufacturing, 2014 (1): 24-25
6. Fu Jun.Establishment and Balance of E-business Ecosystem in Enterprises. Business in Business, 2014, (15): 54-55
7. Liang Qiyu.How to Construct E-Business Ecosystem in Enterprises. Electronic Commerce .2015 (02): 221
8. Meng Cong.Application of Cloud Computing Technology in E-commerce Development. Electronic Commerce .2015 (04)
9. Han Qi. Realizing the Industry Catching up and Surpassing in Our Country .Industrial Weekly.2014 (09): 52-54
10. China Enterprise Confederation, Top 500 Manufacturing Enterprises in China, 2014: 5-9
11. Zhang Ping.Global Value Chain Division and the Growth of China's Manufacturing Industry. Beijing: Economic Management Press, 2014
12. Chen Aizhen.Constraints and Breakthroughs of China 's Equipment Manufacturing Industry under Global Competition - Analysis Based on Dual View of Value Chain and Industrial Chain .Beijing: Economic Science Press, 2012
13. Ding Yongjian.Evaluation of China's Manufacturing Industry Facing Global Industrial Value Chain. Beijing: Science Press, 2014
14. Jiang Wei, "The Upgrading of China's Manufacturing Industry in the Industrial Age of 4.0". China Development Review, 2015 (03): 37-38
15. An Lin.Considering German Industry 4.0 for Speeding up the Transformation and Upgrading of China's Manufacturing Industry .Manufacturing and Manufacturing. 2015 (03): 37

16. Bai Qing.Producer Services to promote the upgrading of the manufacturing sector analysis [J]. Finance and Economics Research. 2015 (04): 27-29
17. Wang Peng. Industrial 4.0 Manufacturing Power Dream [J]. Quality Exploration .2015 (02): 58-59
18. Zuehlke D. Smart Factory-Towards a factory-of things [J]. Annual reviews in Control， 2010(34):129-138
19. Michael GroB.Mobile shopping:a classification framework and literature reriew[J]. International Journal of Retail & Distribution Management,2015(3):221-223

An Optimizing Research about the Cloud Service Resource Based on the Requirements of the Users

Yanfang Kang *, Dapeng Li, Xuyang Li and Mingna Guo

State Grid Henan Economic Research, Zhengzhou, 45000, China
*Corresponding author e-mail: 154626491@qq.com

Abstract. Oriented the amounts of the service requests on the user's demand, it is possible to provide the service to the users with the lowest price, the most economical resource and the type of service closest to the users' requests. In order to provide the different resources to the users effectively, the optimization model of the management about the cloud services resources is set up. In the first place, under the guidance of the resources management center, it is best to match the most suitable cloud service resources for the user needs with the grey relational comprehensive evaluation, then according to the time, the price, the data workflow and the service type attribute value. Optimal resource deployment algorithm is established with the target of the best cloud service resources offered to the users. Finally, it is verified the validity and rationality of the method by the simulation.
Keywords. Requirements of the user, Cloud computing, Service resources, Grey correlation, Optimizing model, Management.

1 Introduction

With the rapid development of electronic technology and the Internet, there is a new innovative services based on the Internet environment - cloud computing [1]. This new approach provides a convenient, on-demand business service model [2].In order to meet user's growing requirements on huge amounts of data and the service processing, this model of the new information resource service provides a broad prospects in application. The cloud computing, with its five basic characteristics[3] the ability of rapid elasticity, the self-service on-demand, the broad network access, the shared resource pool, the measurable service, have received the unprecedented focus from all over the world. The problem of optimizing cloud resource is a core problem in cloud computing, which determines the efficiency of the tasks undertaken and the overall performance of the systems. The service users, according to their actual needs, orders the cloud services resources through the Internet from the cloud service providers. The cloud service providers, according to their users' requirements, schedule and optimize the cloud services resource distributed in different providers. So how to optimize the management service resources that satisfy the users' requirements is the main problem in this paper.

The rest of the paper is organized as follows. In the Section 3, we elaborate the cloud service model of the resources optimization management. In the Section 3, we first described the cloud service resources as quad group. We construct the associated correlation matrix about the service providers and data centers, and the associated correlation matrix about the demand and request service nodes. The users' requirements are matched to the cloud resources by the grey correlation theory in the Section 3, according to the four attributes about the time (T), the price (P), the data workflows(D) and the data service type value (Ts), we determine the optimal goal. The optimal constraints are considered in the Section 3.We introduce the optimal allocation algorithm after the mapping resources matched. In the Section 4, The effectiveness and superiority of the algorithm are analysis by the simulation experiments. We conclude in Section 5.

2 Related work

There is great difference of the information service model in the cloud computing between the traditional networks. The relationship between the users' requirements and the service resources will become more and more closely if they can access the service resources depend on the users' needs. The users' requirements gradually is various .In order to meet the users' needs, we should improve the complete efficiency of the tasks. In the paper, the users' requirements match the best cloud service resources by the method of the grey relational comprehensive evaluation, and then the optimal the service is provided to the users that according to the required target function. The cloud service resource management center as a medium, we set up the optimal management model to configure the cloud service resources.

3 Build optimal management model of the cloud service resources

3.1 Problem description

In cloud computing environment, the requirements of the users is complex and diverse. The demand of the cloud users for now mainly is the demand of information resources. The cloud users can purchase the necessary information resource according to their actual needs via the Internet. The cloud computing can offer the users various servers, personal computers, mobile phones. In order to meet the needs of the users' task better, to match the service resources conveniently, it is essential to classify the task requirements. The different fitness of the task will affect the optimal results of the allocated resources. The matching cost of the service resources would be increased if the divided fitness was too mint. The matching process of the resources would be complex if the fitness is too rough [4]. In the paper, we select the service time T_R, the service price P_R, the data workflow D_R, the attribute value of the service type (TS_R) to describe the cloud service resources. The task can be described as $S_R = < T_R, P_R, D_R, (TS_R) >$.

The service provided is not only quantity but also diversity. In the IT industry and the academia society, the cloud services can be divided into three categories, denoted [5, 6] the infrastructure as a service, the platform as a service [7, 8] [9] and the software as a service. The information resources allocated in different parts will be scheduled again and redeployed to build the cloud information resources by utilizing the cloud computing. The cloud service will be registered in the resource management center. In order to manage and match the cloud resource, the resources in the management center will be catalogued. Above analysis, according to the diversity characteristics of the cloud services, we should build the dynamic optimal model to manage the cloud resource and solve the optimal problem of the cloud resources, with the coordination of the cloud resource management center.

The optimal model can deploy the cloud service resource dynamically based on the users' requests. The request tasks are described into quad group as $S_{Ri} = < T_{Ri}, P_{Ri}, D_{Ri}, TS_{Ri} >$ i = 1, 2, 3,... M, in the center of the cloud service resource management. In order to match the cloud services resources, the requests tasks of $U_1, U_2, \ldots U_m$ is decomposed to different service node $N_1, N_2, \ldots$ and N_n, according to the service properties of the requests for time, the price, the data workflow and the attribute values of the service type. We build the matrix of the users - service nodes depend on the service properties. At the same time, the cloud service resources registered by the providers, $A_1, A_2, \ldots A_m$ in the resource management center are described as the quad group $Re_i = <T_{Si}, P_{Si}, D_{Si}, TS_{Si}>$.and then the resource is decomposed into the different data centers $D_1, D_2, \ldots, D_n$, according to the service properties of the time, the price, the service type, the data workflow and the attribute values of the service type. We build the matrix the services - data center. The users' request task is assigned to the management center based on the method of the grey relational comprehensive evaluation. The cloud services are matched by the map.The services is deployed optimally according to the intelligent optimal algorithm. Finally the optimal cloud services are allocated to the users. The specific process is shown in figure 1:

Above the figure 1, we can conclude the detailed explanation: matrix of the users - service nodes and matrix services - data center. They are defined as follows:

Definition 1: Matrix of service provider –property –data center

In matrix, x_{ij} denotes that the service resources in management center will be placed in the data center, by the time T_{ij}, the price P_{ij}, the data workflow D_{ij}, the attributes value of the service types Ts_{ij}. The i, j denotes the service resources provided by the Ai are placed in the data center DC_j.

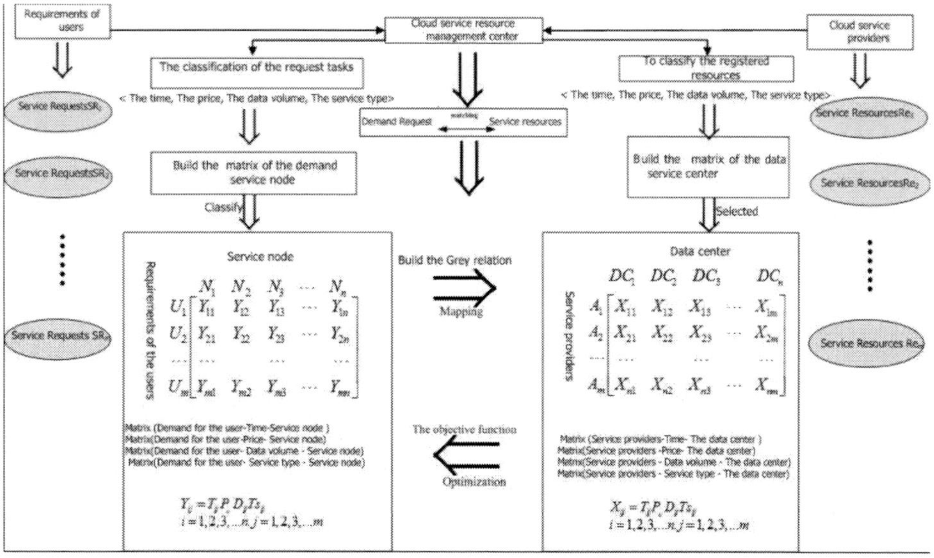

Fig. 1. The resources optimal management for the cloud service model.

Definition 2: Matrix of users –property –service node

In matrix, y_{ij} denotes that the users' requirements will be placed in the nodes by the time T_{ij}, the price P_{ij}, the data workflow D_{ij}, the attributes value of the service types Ts_{ij}. The i, j denotes the service requests of the users U_i are placed in the service node N_j.

3.2 Analysis about the optimal deployment of cloud service resources

Cloud service resources matched the cloud service requests will be deployed optimally. During the process, we should first determine the optimal goal. The objective functions represent the goal. They are established according to the time (T), the price (P), the data (D) workflow and the attributes value of the service types Ts (Ts).

The objective function 1: Execution time completed to satisfy the users' tasks service is minimum: MinT:

$$\text{Minimize obj-1}=\sum_{i=1}^{n}\sum_{j=1}^{m}T_{ij}\theta_{ij}+\sum_{i,i^*=1}^{n}\sum_{j,j^*=1}^{m}T(ij,i^*j^*)\theta_{ij}\theta_{i^*j^*} \quad (1)$$

The objective function 2: The price completed to satisfy the users' tasks service is lowest: MinP:

$$\text{Minimize obj-2}=\sum_{i=1}^{n}\sum_{j=1}^{m}P_{ij}\theta_{ij}+\sum_{i,i^*=1}^{n}\sum_{j,j^*=1}^{m}p(ij,i^*j^*)\theta_{ij}\theta_{i^*j^*} \quad (2)$$

The objective function 3: The service data workflow completed to satisfy the users' tasks service is least: MinD:

$$\text{Minimize obj-3}=\sum_{i=1}^{n}\sum_{j=1}^{m}D_{ij}\theta_{ij}+\sum_{i,i^*=1}^{n}\sum_{j,j^*=1}^{m}D(ij,i^*j^*)\theta_{ij}\theta_{i^*j^*} \quad (3)$$

The Objective function 4: The difference between the service attribute type values completed to satisfy the users' tasks service in a certain period of time is minimum: $MinSim(Ts_i,Ts_j)$

$$\text{Minimize obj-4}=\sum_{i=1}^{n}\sum_{j=1}^{m}Sim(Ts_i,Ts_j)\theta_{ij}+\sum_{i,i^*=1}^{n}\sum_{j,j^*=1}^{m}Sim[Sim(Ts_i,Ts_j)Sim(Ts_{i^*},Ts_{j^*})]\theta_{ij}\theta_{i^*j^*} \quad (4)$$

The resources in the data center DC_j or DC_{j^*} that provided by the provider A_i or A_i *is selected in the process of the optimal deployment

Otherwise,

$$\theta_{ij},\theta_{i^*j^*}=\begin{cases}1\\0\end{cases}$$

In the formula (1) $\sum_{i=1}^{n}\sum_{j=1}^{m}T_{ij}\theta_{ij}$ denotes the accumulated time generated by the map matching service resources, $\sum_{i,i^*=1}^{n}\sum_{j,j^*=1}^{m}T(ij,i^*j^*)\theta_{ij}\theta_{i^*j^*}$ denotes that the accumulated time generated is between the map matching service resources offered by the provider A_i from the data center DC_j and the provider A_i^* from the data center DC_i^*. In the formula (2) and (3) $\sum_{i=1}^{n}\sum_{j=1}^{m}P_{ij}\theta_{ij}$, $\sum_{i=1}^{n}\sum_{j=1}^{m}D_{ij}\theta_{ij}$ and $\sum_{i,i^*=1}^{n}\sum_{j,j^*=1}^{m}p(ij,i^*j^*)\theta_{ij}\theta_{i^*j^*}$, $\sum_{i,i^*=1}^{n}\sum_{j,j^*=1}^{m}D(ij,i^*j^*)\theta_{ij}\theta_{i^*j^*}$ have the same meaning as the above. In the formula (4) $\sum_{i=1}^{n}\sum_{j=1}^{m}Sim(Ts_i,Ts_j)\theta_{ij}$ denotes the accumulated attribute value required for the similar service types in the map matching process. $\sum_{i,i^*=1}^{n}\sum_{j,j^*=1}^{m}Sim[Sim(Ts_i,Ts_j)Sim(Ts_{i^*},Ts_{j^*})]\theta_{ij}\theta_{i^*j^*}$ denotes that the accumulated attribute value required for the map matching service resources is generated between the service resources offered by the provider Ai from the data center DCj and the provider Ai*from the data center DCi*.

Based on the above considerations, the target functions in the optimal process are always interconnected and conflicted. Sometimes maybe a target function has reached the optimal state ,the other objective function has been changed to the other optimal state. So the relative weight of each target function has to be distributed, the multi-objective optimal problem is converted into the single objective optimal problem [10], where

$$MinZ = w_1MinT + w_2MinP + w_3MinD + w_4MinSim(Ts_i,Ts_j) \quad (5)$$

In the formula (5) W_1 , W_2 , W_3 and W_4 denotes the weight coefficient that belong to the time, the price, the data workflow and the attributes value of the service types.

3.3 Optimal constraints

In the optimal process, we should consider the constraints from the following four aspects.

(1)In the whole executed task process, we should set the deadline; otherwise, the users can't wait for a long time.

$$\sum T \leq T_D \tag{6}$$

(2) In the whole executed task process, we should set the top price; otherwise, the users will give up the service, because of the price is too high.

$$\sum P \leq P_h \tag{7}$$

(3) In the whole executed task process, we should set the top line about the data flow; otherwise, the service process is overload, because of the excessive amount data.

$$\sum D \leq D_h \tag{8}$$

(4) In the whole executed task process, we should set the maximum of the service type attribute value. The sum of the service attribute value should no more than the maximum; otherwise, the unnecessary costs will be increased, because of the different gap of the service type is too large.

$$\sum Sim(Ts_i, Ts_j) \leq MaxSIM \tag{9}$$

3.4 Optimal algorithm of cloud service resources in deployment

In the cloud service resource management center, according to the correlation matrix established (the demand users - service node), we deploy the cloud service resources purposeful in the data center. First we encode the service nodes of the users' requirements. This paper adopts the indirect coding method. Each service node N_i denotes different users' requirement attributes. N_i=<T_R,P_R,D_R,TS_R>, where T_R is the time, P_R is the price, D_R is the data workflow, TS_R is the service attribute type value. Number of chromosome is the requirements. We should code the service resource that is taken by the tasks; the specific graph is as follows in figure 2:

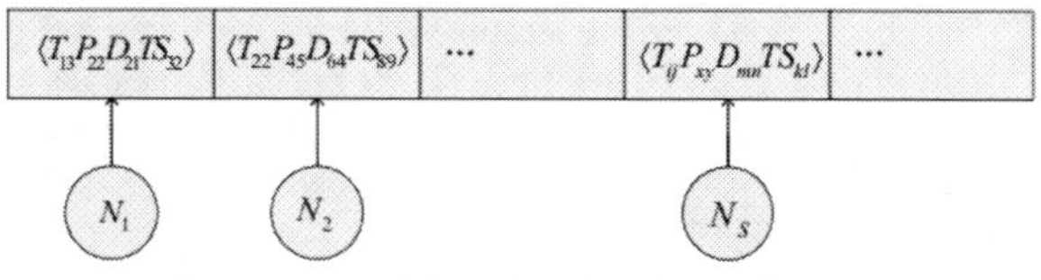

Fig. 2. Chromosome coding correspond the task node – the service provider- the data center.

In the figure 2, where T_{ij} is the execution time , P_{xy} is the pay price, D_{mn} is the service data workflow, TS_{kl} is the service attribute type value. They are all designed to complete the tasks in the service nodes N_i. In order to satisfy the requirements of the users, we set the parameters of the corresponding service nodes < T_{ij}, P_{xy}, D_{mn}, TS_{kl} >, where the execution time T_{ij} is stored in data center DC_j provided by the service provider A_i, where the pay price P_{xy} is stored in data center DC_y provided by the service provider A_x ,where the service data workflow D_{mn} is stored in data center DC_n provided by the service provider A_m, where the service attribute type value TS_{kl} is stored in data center DC_l provided by the service provider A_k. The main task of this paper is to deploy the cloud service resources optimally.

The optimal deployment problem of the cloud resources is to set the parameter problems < T_{ij}, P_{xy}, D_{mn}, TS_{kl} >in the service node. The parameters of the service nodes are processed by the cloud service resource management center. After the service registered by the cloud service providers in the cloud service resource management center, in order to schedule the resource conveniently, the cloud resource will be allocated in the different data centers with the form of the execution time, the paid price, the data workflow, the service attribute type value. This problem is converted to the optimal deployment about the cloud resource stored in the different data centers, registered by the different cloud service providers. In order to optimize the service resource allocation, it is necessary to use optimization algorithm to optimize the task.

4 Simulation

In this section, according to the simulation example, the effective of the optimal management model of the cloud services resources is analyzed. Hypothesis that at any time, the five users' requirements are received in the cloud service resource management center. They are the service requirement SR_1, the service requirement SR_2, the service requirement SR_3, the service requirement SR_4, the service requirement SR_5. First the five users' requirements will be set in the different service nodes, according to their different properties of the different users. Then according to the grey relation map, the highest correlation between the service requirements and the cloud service resources is to be found.

We hypothesis that there are three cloud service providers who offered the resources to meet the requirements of users. According to the above mentioned, the five kinds resources is selected to form the initial population respectably, according to the sub-objective functions. Hypothesis that the size of the initial population is 40, the length of the chromosome is 10. The optimal matched service resources offered to the users is selected by the optimal deployment algorithm, The information can be get in the table1 and table2.

When the resource is to be optimized, the coefficient of the weights are respectively that $w_1 = 0.42$, $w_2 = 0.21$, $w_3 = 0.19$, $w_4 = 0.18$, TD=45S, Ph=120￥/h, Dh=230Mb/h, MaxSim=0.55. We adopt the general genetic algorithm and the improved genetic algorithm as the above 2.5. Each algorithm is carried out experiments for 60 times. The results is MinZ = 21.86. The corresponding optimal resource group is seen as the table 3.

From the Figure 3, the global optimal solution can be obtained by the general genetic algorithm or the improved genetic algorithm in the paper. But in the searching process, the optimal solution fluctuates most by the general genetic algorithm (GA), the stability is poorer. However the optimal solution fluctuates slight by the improved genetic algorithm within 35 times. The wave number is 8. So the stability of the improved genetic algorithm is better than the general genetic algorithm. The improved GA during the global optimal process is more stability. Figure 4. shows the number of iterations, that the solution has reached the optimal results by the two kinds of genetic algorithm. The figure 4 illustrates that the average iterations number of the improved genetic algorithm is 51, the average iterations number is 157 by the general genetic algorithm. The average iterations number of the improved genetic algorithm is lower than that of the general genetic algorithm. Wave amplitude is less than that of the genetic algorithm. General genetic algorithm will fall into local optimal solution. Because their fitness functions set have the consistent standards. Search speed of the optimal solution is slower. However, the improved genetic algorithm has the unique method to set the fitness function, which can reach the optimal solution at a rapid speed.

5 Conclusion

The optimal deployment problem of the cloud service resources is the core of the cloud computing. At the same time the deployment problem is the basis that realizes the flexibility, the scalability, the resource sharing. It is the key that we choose the best cloud services provider to the users. It is related the basic interests about the cloud services providers and the cloud users. This paper constructs the optimal allocated management model about the cloud service resources. The model distributes the cloud service resources to the different service nodes by the cloud service resource management center, based on the diversity of the cloud users' requirements. The model takes advantage of the grey relational comprehensive evaluation to match the cloud service resources. It uses the improved genetic algorithm to optimize the matching cloud services, with the optimal goals of the time (T), the price (P), the data workflow (D) and the service type value (Ts). The optimal resource selected is offered to the users. Cloud service resource optimal deployment model and the solution algorithm proposed in the paper can provide the theoretical support and the research method for the optimized deployment management of the cloud computing resources.

Table 1. Distribution list of the requirments task nodes.

The task for Requirements	**SR_1**	**SR_2**	**SR_3**	**SR_4**	**SR_5**
	N_{11}	N_{21}	N_{31}	N_{41}	N_{51}
	N_{12}	N_{22}	N_{32}	---	N_{52}
The service nodes of the assigned tasks	N_{13}	---	N_{33}	N_{44}	---
	N_{14}	N_{24}	---	---	N_{54}
	---	N_{25}	N_{35}	N_{45}	N_{55}

Table 2. Service resource description matched for the user's requirements.

	Data Center	**Time (S)**	**Link time(S)**	**price(￥/h)**	**Link price(￥/h)**	**Data traffic(Mb/h)**	**Link flow(Mb/h)**	**Type attribute value**	**Link attribute values**
	DC_1	18	3	16	8	41	23	0.78	0.12
	DC_2	6	7	24	5	71	18	0.83	0.23
Service providers	DC_3	16	2	12	7	93	10	0.52	0.30
A_1	DC_4	20	5	19	4	46	15	0.90	0.25
	DC_5	7	4	31	10	52	19	0.64	0.28
	DC_1	13	1	30	12	80	12	0.85	0.31
	DC_2	8	6	18	9	66	17	0.82	0.15
Service providers	DC_3	10	4	24	6	58	16	0.7	0.24
A2	DC_4	15	3	15	20	35	25	0.73	0.19
	DC_5	17	7	18	15	48	27	0.59	0.27
	DC_1	11	8	34	17	78	11	0.75	0.18

	DC_2	13	6	20	11	67	18	0.68	0.16
Service providers	DC_3	14	3	22	13	50	14	0.55	0.34
A_3	DC_4	9	5	19	10	63	20	0.72	0.21
	DC_5	5	4	12	9	82	19	0.67	0.26

Table 3. Combination table of the optimal resource.

N_{11}: <(3,5)(1,5)(1,1)(2,3)>	N_{33}: <(1,3)(3,2)(2,3)(3,4)>
N_{12}: <(2,4)(3,5)(3,1)(3,2)>	N_{35}: <(2,4)(3,1)(3,4)(3,2)>
N_{13}: <(3,2)(3,4)(3,4)(3,5)>	N_{41}: <(2,1)(1,4)(3,5)(2,2)>
N_{14}: <(2,4)(3,2)(3,5)(1,5)>	N_{44}: <(2,2)(2,4)(3,5)(3,3)>
N_{21}: <(3,5)(3,2)(3,4)(1,5)>	N_{45}: <(1,4)(2,5)(3,4)(1,1)>
N_{22}: <(3,3)(2,5)(2,4)(3,4)>	N_{51}: <(3,2)(3,5)(3,4)(1,1)>
N_{24}: <(2,2)(3,2)(3,1)(1,2)>	N_{52}: <(1,5)(2,4)(2,3)(3,4)>
N_{25}: <(3,2)(1,5)(3,1)(2,2)>	N_{54}: <(2,2)(3,4)(1,3)(3,4)>
N_{31}: <(2,5)(1,4)(3,2)(3,1)>	N_{55}: <(2,4)(3,3)(2,1)(2,4)>
N_{32}: <(2,4)(3,2)(2,1)(3,1)>	

(note: N_i: < (a_1, b_1) (a_2, b_2) (a_3, b_3) (a_4, b_4) >.In the table, the (a_i, b_i), respectively, denote the task of the service nodes N_i is performed with the time, the price, the data traffic of the resource, the property values of the service types. The task node require to schedule the resource b_i in the data center provided by the service provider a_j.)

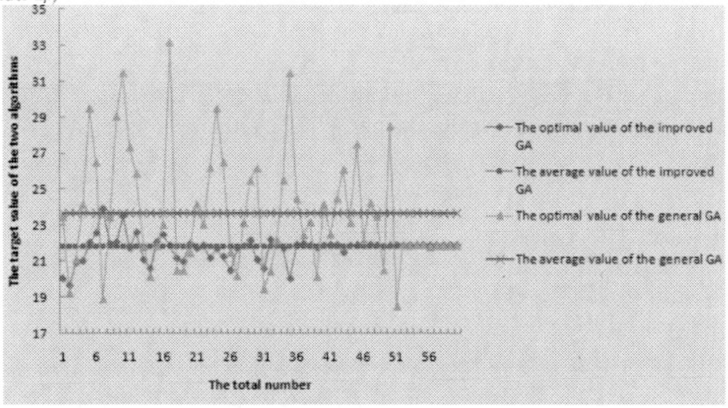

Fig. 3. Optimal target value of the two algorithm.

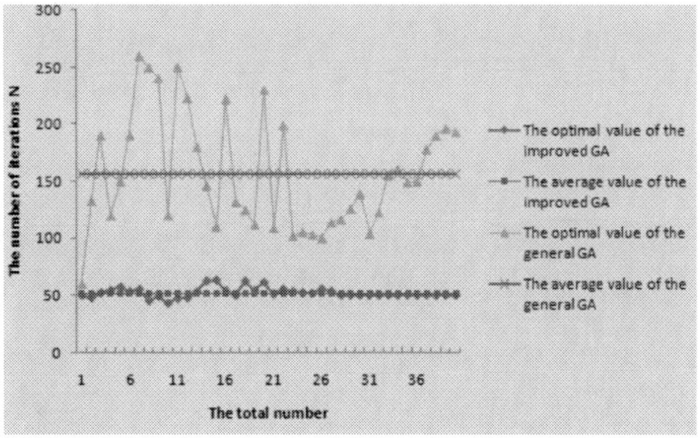

Fig. 4. Number of iterations on the optimal value of the two algorithms.

Acknowledgment

This work was financially supported by project that Research evaluation system on social benefits of major power transmission engineering

References

1. Wang J.J., Lv Z.H., Jie W.,et al., Analysis the development and application about the cloud computing technology, Journal of computer applications and design, vol.31,no.20,pp.4404-4409, (2014).
2. Han. W. The influence factors of the cloud computing services market competitive analysis, Journal of the managerial modernization, no. 2, pp. 25-27, (2015).
3. Lin W.W., Qi D.Y. The research review on the cloud computing resource scheduling, Journal of the computer science, vol. 39 ,no.10 ,pp.1-6, (2014).
4. Li B.M., Liu Q. Try on the match of the grid resources based on ontology, Journal of northwestern polytechnic university.vol.28,no.6, pp.608-611, (2015).
5. Ju J.H., Wang Y., Fu J.Q. Research on Key Technology in Saas, Intelligent Computing and Cognitive Information(ICICCI),International Conference on. pp.384-387, (2016).
6. Banerjee, Srikanth, Cukic. Log-Based Reliability Analysis of Software as Service(Saas), Software Reliability Engineering(ISSRE),IEEE 21st International Symposium on.pp.239-248, (2015).
7. Bernstein, Vidovic, Modi. A Cloud Paas for High Scale, Function, and Velocity Mobile Applications-With Reference Application as the Fully Connected Car, Systems and Networks Communications(ICSNC),Fifth International Conference on. pp.117-123, (2014).
8. Lv C.T., Li Q., Lei Z. Paas: A revolution for information technology platforms, Educational and Network Technology (ICENT),International Conference on.pp.346-349, (2014).
9. B.S. Lee, S.X.Yan, D.Ma etal. Aggregating IaaS Service.SRII Global Conference (SRII Global Conference).Annual. pp.335-338, (2015).
10. L.L. Hall. Evolution of telco-services utilizing Infrastructure as a Service(Iaas), Intelligencein Next Generation Networks(ICIN),15th International Conference on. pp.247-252, (2016).

DOI: 10.1051/matecconf/201710002044

Application of text mining in library book procurement

Ding Cong

Ding Cong Library, Jinan University, Guangzhou, China

Abstract. With the diversification of readers' preferences, the continuous growth of book price and the shortage fund for purchasing books, it becomes increasingly important to explore reads' real needs and spend limited money on those needed books. In this context, Chinese Word Segmentation technology combined with Chinese Library Classification are used to explore readers' reading preferences by using news, management and finance disciplines' borrowing data of Jinan University in the year of 2015. Results showed that readers showed remarkable preferences for books of a specific subcategory or theme. Meanwhile the hot borrowed books often gathered in a few core publishing house, with the core publishing house books accounting for 70%, 63% and 76% of the total sampling books for news, management and finance discipline respectively. All in all, this research is of great significance for the book procurement job of the high school library.

Keywords. Book procurement, Book borrow, Chinese Library Classification, Chinese word segmentation.

1 Introduction

The function of university library is to serve teaching and research of university. Book procurement play an important part in enriching the collection of resource and improving the level of teaching and research. In recent years, a series changes have taken place in the internal and external environment of the library, which include the diversification of readers' preferences, the growing variety of books ,the continuous growth of book prices and the shortage fund for purchasing books. Under such circumstances, how to use the limited funds to purchase readers' needed books become a tough problem placed in front of the majority library staff. In order to resolve the above problems, the library community have done a series of fruitful research in how to find a reasonable book procurement method. Summing up, the main research aspects of this task are as follows.

Firstly, book procurement principle obtained by setting mathematical models based on library collection data, circulation data and reader information data. For example, based on Bayesian network theory, Yu Yali and Zhou Qingsong[1-2] obtain the procurement proportion of social science, art and literature books by establishing a model using library collection data, readers' circulation data. Peng Jun[3] proposed a second-choice algorithm and corresponding decision-making formula for book procurement. There are many such studies. Most of them borrow some principles from mathematics or economics, such as particle swarm algorithm, genetic algorithm and neural network algorithm. These kinds of research process prominent scientific nature. But they are complex in modeling and data collecting and thus lack utility. And meantime, it is not objective in determine the weight of each factor because of lack of objective standard.

Secondly, principles based on core press and core authors. The first case is Google Scholar based method. The research process are roughly as follows. Firstly, core bibliography are acquired by searching Google Scholar database to get the citation situation of these books coming from publisher. Secondly, core authors and core press are obtained by analysis and screening these core bibliography. Mu Weiguo's[4] paper is such case. The second case is library collection data based method. Combining Bradfort's law and twenty-eight rule, core authors and core press are acquired by the statistics large number of library collection books. The papers of Wang Jingfen[5-6] are such case. The guiding significance of such research to book procurement are great. However, the acquisition of core press and core authors are often time-consuming because of large amount of labor statistics work.

Thirdly, principles based on Patron-Driven Acquisitions(PDA) mode. The realization of such research are achieved by using the book recommendation system and the OPAC online catalog. In the first place, the bibliography information provided by book sellers and publishers are imported into search system. And then certain trigger mechanism which assign a threshold value are set. If searching number of certain book item exceed the threshold value then book procurement order will be trigger automatically[7-9]. This method process great significance because it is based on users' real needs. However, as a passive means to acquire readers' needs, it requires readers' active participation which in itself is a problem.

Fourthly, principles based on data mining. Most of these studies are purely theoretical research and lack empirical analysis. The common mode of these studies are about how to collect data from library collection data, circulation data and reader information data, how to clean up, transform and integrate data and how to put forward corresponding data mining principles and methods using these data[10-13]. This kind of research mostly use the relevant data accumulated by library to explore readers' needs actively. In this respect, it has certain advantage over PDA mode and avoids the problem of readers' low participation.

Undoubtedly, text mining falls into the category of data mining. As a actively method for exploring readers' needs, the tough problem of Chinese Word Segmentation(CWS) must be solved first. The ICTCLAS CWS system of Chinese Academy of Sciences is adopted in this paper. The hierarchical Hidden Markov Model(HMM) is used in this system, which can realize the tagging of words, the recognition of new words and the recognition of names, and support both simplified and traditional CWS with 98.45% accuracy. And meanwhile the system provides API which can be used to realize personal CWS needs.

In this paper, text mining method is used to explore the borrowing data of news discipline, management discipline and financial discipline of Jinan University Library(hereinafter referred to as the library). The branch class of the three disciplines' book borrow number, disciplines' reading preferences and disciplines books' core press are explored deeply. The analysis results can be used to guide book procurement process, formulating reasonable book procurement strategy.

2 Data sources and research methods

Data sources. The data for text mining come from the library's annual borrow ranking data accumulated in the year of 2015. Each book ranking record contains title, author, call number, press, borrowing times and other relevant data.

Research methods. In order to illustrate data mining process and meantime simplify the discussion process. Only the borrow data of three key disciplines(news, management and finance) of Jinan university library are taken to explore the law reflected in these borrowed data. The discussion for other disciplines can be imitated. The research ideas of this paper are as follows.

Firstly, filtering data with the call number starting with G21(journalism), C93(management) and F8(finance) from the borrowing database. The matching number of records for each type are 409,252 and 1974 respectively.

Secondly, the borrowing distribution situation of the branch class of the three disciplines are analyzed. Take G21 as example, as the branch class of G21, the borrowing distribution of G212(news gathering and reporting) and G211(organization and management) will be examined in detail. The basis for dividing discipline sub class is Chinese Library Classification. During the analysis, the method of traversing borrowing data is used in program to judge the sub classification of each book.

Thirdly, book titles for each of the three discipline's borrowed books are collected to form text string respectively. And then Ansj java package implementing the ICTCLAS algorithm is used to execute the Chinese word segmentation process for the above three book title string. Finally, statistical results are artificially filtered and the synonymy words are merged to obtain the frequency distribution of the segmented term.

Fourthly, by using the method mentioned above, the information of the publishing house of these books are segmented by CLC and the frequency distribution of these publishing house are counted.

3 Data analysis

3.1 Analysis of Book borrowing distribution of the branch class of news, management, and finance discipline

According to Chinese Library Classification, further subdivision for news, management and finance disciplines are done. News disciplines can be subdivided to G210, G211,G212 and several other subcategories. Management discipline is divided into C93, C93-0, C931 and several other subcategories. Financial discipline can be divided into four subcategories of F81, F82,F83 and F84. And then book borrowing data for these subcategories are counted respectively. The results are shown in Figure 1.

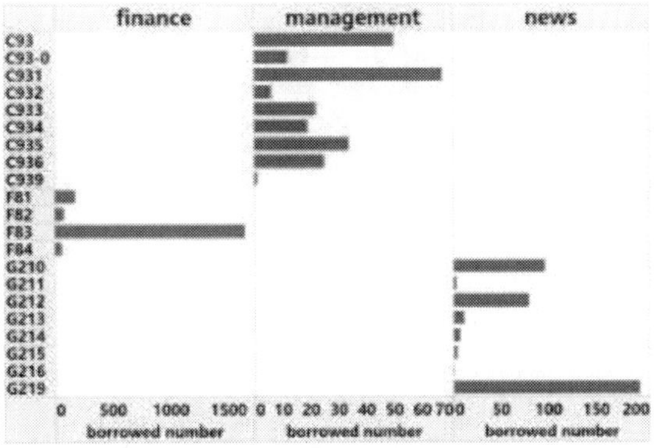

Fig. 1. Book Borrowing Distribution of the Subcategories of News, Management and Finance Disciplines

Note. G210-Journalism, G211-Organization and Management, G212-News Coverage, G213-Editing, G214-Journalists, G215-Newspaper publishing, G216-Various Newspaper, G219-Journalism in the World, C93-0-Management Theory and Methodology, C931-Management Technology and Methodology, C932-Consulting, C-933-Leadership, C93-Decision Science, C935-Management Planning and Control, C936-Management Histology, C939-Application Management, C93-Management, F81-Finance/State Finance, F82-Money, F83-Finance/Banking, F84-Insurance

As can be seen from Figure 1, the needs for book under the subcategories of news, management, finance disciplines have significance difference. Specifically, subcategories of Journalism in the World, Journalism and News Coverage rank the top three in the borrowing list. Whereas the borrowing number of management discipline are relative average with Management Technology and Methodology and Management ranking the top two list. Subcategory of Finance/Banking under finance discipline occupy an absolute advantage. The number of books borrowed under this class reached 1648, accounting for 83% in the total borrowed number. It is self-evidence that when formulating book procurement plans the above subcategories under the three disciplines should be focused. And meanwhile library collection distribution and subject development trend should also be considered to formulate a reasonable book procurement strategy.

3.2 Book title word frequency analysis for the borrowed books of news, management and finance discipline

Word frequency analysis is the concept of bibliometrics, which refers to the use of occurrences of key word capable of revealing the core content of literature to determine the hotspots and development trends of that field[14]. This paper uses this concept and apply it to analyze word frequency of book title of borrowed books. Word frequency of the book title of borrowed books are acquired according to research methods described above. And then for the three disciplines, word term with the occurrences above 5 times, 5 times and 10 times are filtered respectively and bubble chart is drawn into Fig 2 and Fig 3.

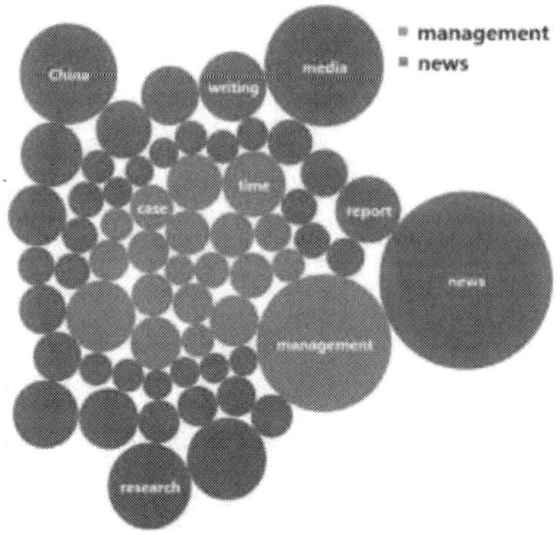

Fig. 2. Book titles' frequency distribution for news, management disciplines

Figure 2 and 3 use different colors to represent different disciplines with the size of the filled circle representing the occurrences times of that key word. It is possible to explore a reader's reading preference from these figure. For instance, the key word China appears in word segmentation results of the borrowed book title

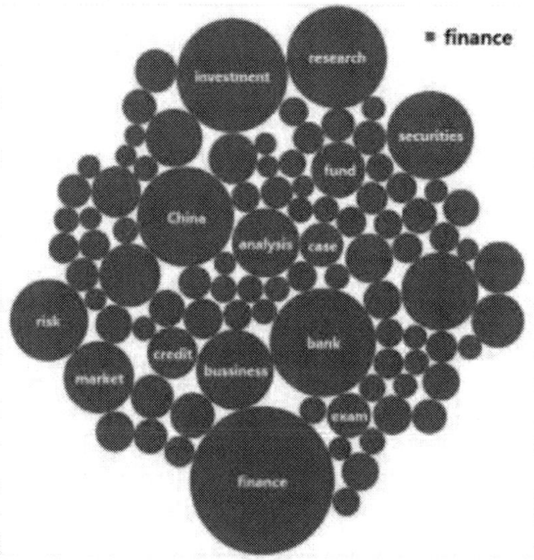

Fig. 3. Book titles' frequency distribution for finance discipline

of news discipline which may indicate that readers of news discipline probably show great interest in news event in China. Again, there are high frequency of practitioner, qualifications and exam in the word segmentation results of finance discipline. This may indicate that readers own great willingness for obtaining qualification certificate in finance field. Of course, in actual practice, sampling survey must be done to confirm those potential reading preference. Once those reading preference are confirmed, books of such kinds can be increased accordingly.

3.3 Word frequency analysis for press of borrowed books in news, management and finance disciplines

Unlike other research methods, this paper uses text mining strategy for determining core press. To avoid including books with low borrow times into the statistics analysis process, this paper choose only those book borrow data with the borrow times more than 6 times. After statistics, the number of book borrow records meeting the above criteria are 57, 38 and 297 respectively. And text miming results are shown in figure 4.

Fig. 4. Book publishers' frequency distribution for news, management and finance disciplines

Figure 4 represents different discipline with different color with the font size representing the occurrences times. After statistics, the press distribution of borrowed books of the three major disciplines are summarized as follows.

Firstly, there are 28 publishing houses in the journalism discipline, including NanFang Daily Publishing House, HuaXia Publishing House, Fudan University Press, Renmin University Press, Jinnan University Press, Xinhua Publishing House, Peking University Press, Tsinghua University Press, Higher Education Press, Social Sciences Literature Publishing House, China Petrochemical Press and etc. The mentioned 11 press have a total of 40 books on the list, accounting for 70% of the total sample.

Secondly, there are a total 20 publishing houses for the total 38 books sample of management discipline. Among them, Machinery Industry Press, Renmin University Press, CITIC Publishing House, People's Post Press, China Youth Publishing House and Electronics Industry Publishing House have a total of 24 books on the list, accounting for 63% of the total sample.

Thirdly, there are 70 press for the 297 books of finance discipline. Among them, Machinery Industry Press, Renmin University Press, CITIC Publishing House, China Financial Publishing House, China Financial and Economic Publishing House, China Economic Publishing House, Tsinghua University Press and other 10 press have a total 226 books on the list, accounting for 76% of the total sample. Notably, Machinery Industry Press, Renmin University Press and CITIC Press have 46, 56 and 24 books on the list respectively, occupying nearly one-third of the total sample.

4. Research conclusion and countermeasures

Based on the above analysis, the following conclusions can be drawn.

Firstly, it is possible to refine readers' demand based on the refinement borrowing data by Chinese Library Classification. In the process of making book purchase strategy, the borrow ratio for refined subcategories books can be referred. Meantime the actual library collection situation and development trend for a specific discipline should also be taking into consideration when making such book procurement strategy.

Secondly, readers' reading preferences can be explored from the word segmentation results of those borrowed books' title. These excavated information can be confirmed by way of sampling survey, which make the purpose of sampling survey greatly enhanced and is of great benefit for PDA-based book procurement.

Thirdly, publishing houses for borrowed book only distributed in a few of core publishing house. Books from these press usually possess high quality and are welcomed by readers. In the actual book procurement, it is a good practice to build on books from core press and at the same time supplement some high quality books from peripheral publishing house.

There are still several problems in this research. First of all, due to lack of reader information, there is an ambiguity in the direction of borrowing information which in in turn limits the use of reader information for accurate needs mining. Secondly, only single variable , such as call number, title or publishing house, not several variables combined, is used for data mining. This study can be enriched from the following aspects.

Firstly, acquiring reader information for borrowed books from OPAC and integrating those information to borrowing records. It is possible to utilize those information to analyze the reading preference of a specific reader or a group of reader(faculty or student) and formulate corresponding book procurement strategy aiming at those readers.

Secondly, integrating various data to explore readers' reading preferences. For example, according to the above analysis results, management books and finance books published by Machinery Industry Publishing House are welcomed. However, according to one's intuitive impression, Machinery Industry Press should be classified into engineering book publisher. Actually, if combined with author data, it is not difficult to find that a large part of the list books published by Machinery Industry House are English original translation. What management or finance readers really needs are the Chinese version for high level foreign original books.

References

1. Yu Yali. Application of Bayesian Network Theory in Library Book Procurement. Library and Information Service, 2015(7):126-127
2. Zhou Qingsong. Application of Bayesian Net in Book Procurement of Library. Yunan: Master of Yunan University, 2012:1-40
3. PengJun, Lumin, Ya Fayi et al. The Construction of Book Procurement Decision-making System in University Library Based on Quadratic Choice Algorithm. Practice Research, 2009, 32(9): 74-77
4. Mu Weiguo. Research on Book Purchasing Strategy Based on Core Bibliography and Popular Collection Analysis. Library Journal, 2015(8):44-50
5. Wang Jingfen, Gao Qi, Huang Jing et al. The Application of Core Publishing House in Purchasing Original Books. Library Journal, 2014(2): 64-66
6. Wang Jingfen, Gao Qi, Liang Weibo. The Application of the Core Authors in the Book Purchasing of Maritime. Library Journal, 2015(3):81-83
7. Jia Zhaoxia, Li Liwen. Acquisition Mode of Chinese Books in University Libraries Based on Purchasing Decision-making. Library Work and Study, 2014(4): 57-60
8. Hou Zhijiang, Hou Lingjuan. A New Method of Book Purchasing Based on OPAC Log User Behavior Analysis. Library Development, 2015(1):70-72
9. Tang Jisheng. A Review of Research on Patron-Driven Acquisitions(PDA) in China. Research on Library Science, 2015(2):22-28
10. Zhang Cunlu, Huang Peiqing, Wang Ziping. The Application of Data Mining in Book Procurement. Information Science, 2004, 22(5): 581-583
11. Song Yu. Research on Book Procurement Model Based on Data Mining. Research on Library Science, 2014(17): 53-55
12. Chi Chunjia, Mao Zhiyong. Research on Assistant Decision-making in Formulating University Book Purchasing Plan Based on Data Mining. Journal of Modern Information, 2009, 29(7): 108-110
13. Wu Jin, Mao Zhongxing. On the Book Purchasing Mode of University Libraries in the Background of Big Data. Journal of Changzhou University(Social Science Edition), 2014, 15(5): 133-135
14. Cun JieWang, Qian Qian. Analysis of Research Focus and Research Methods in the Field of Knowledge Management During the Past Decade. Information Science, 2014, 32(10): 157-157

The Online Bookstore

Yunkai Zhai[1,2,3] and Wei Lu[1*]

[1]Management Engineering School, Zhengzhou University, Zhengzhou, China
[2]Henan Engineering Research Center of Digital Medicine, China
[3]The first Affiliate Hospital, Zhengzhou University, Zhengzhou, China
[*]Corresponding author: author@e-mail.org

Abstract. Nowadays, the network plays an import role in people's life. In the process of the improvement of the people's living standard, people's demands of the life's quality and efficiency is more higher, the traditional bookstore's inconvenience gradually emerge, and the online bookstore has gradually be used in public. The online bookstore is a revolution of book industry. The traditional bookstores' operation time, address and space is limited, so the types of books and books to find received a degree of restriction. But the online bookstore broke the management mode of traditional bookstore, as long as you have a computer, you can buy the book anywhere, saving time and effort, shortening the time of book selection link effectively. The online bookstore system based on the principle of provides convenience and service to people.

1 Introduction

1.1 Research Background

The Internet has been favored by more and more people for its high efficiency and richness, and e-commerce has emerged[1]. The online bookstore is a form of e-commerce and book sales industry in one form, it has many advantages, such as: Bookstore size is relatively small, cost savings; transaction activities can be anytime, anywhere, improve service efficiency[2], The information is complete, more convenient retrieval [3], the new book information on the new, consumers can see in a timely manner, trading activities can be launched immediately, so online bookstore in today's era of development is extremely rapid. Online bookstore system is the main function of the trading platform for the site, consumers can connect to the Internet through the computer into the online bookstore and then check the book information, if you need to purchase should be registered landing, select their own books, submit orders and pay Operation to complete the entire book ordering process, to achieve online transactions.

1.2 The Purpose and Significance of The Project

With the online bookstore system, consumers do not need to blindly go to various places to find their own books, but only in a computer connected to the Internet log on online bookstore system[4], in the search box, type you want to find Of the book information retrieval, you can efficiently know whether the site has its own books, if you can online direct purchase, if not, you can change the home bookstore to continue to search or provide advice to the seller in order to supply, This greatly facilitates every consumer, saving time and labor.

The online bookstore system can not only reduce costs, save time, space, to bring convenience to everyone, but also to promote the development of the logistics industry, serve three purposes, mutual benefit. More importantly, in today's world, the increasingly close ties between countries, more frequent exchanges, the economy tends to globalization, which promote the future development of online bookstore system has some practical significance.

1.3 Technology and Development Environment

The system uses the JSP and Servlet technology, with C3P0 connection pool to connect to the development of MySQL database, the front interface design is to use Dreamweaver CS6 to achieve, and the page effects using the JavaScript language and its library function jQuery, enhanced User interaction with the interface experience[5-6].

2 Demand analysis

2.1 Analysis of Current Situation

The development of online bookstores in China has gone through four stages: the embryonic stage, the development stage, the mature stage, and the prosperous stage[7]. In recent years, the types of online bookstores are varied, including online bookstore, book sales online bookstore, information service online bookstore, etc. At the same time, online bookstore business model also has a variety of situations: BTOB mode, BTOC mode, BBC mode[7].

2.2 System User Analysis

The online bookstore's main users are divided into two categories, one is the front user, one is the background user. Front-end users are mainly customers who consume online bookstores. Front-end users can register, login, query, join shopping cart, place orders, submit orders, modify personal information, confirm receipt and add comments on the website. And add, modify or delete the book classification; add, modify or delete the book information, manage the order information and reply to the user's comments. The user is the administrator, the administrator to play the role of overall planning, master and control the front information and so on.

2.3 System function analysis

The online bookstore is opened in the network, not a physical store but a virtual store. In order to improve the efficiency of the query, the site can not only use a single condition query, but also can use multiple conditions to achieve accurate query, the query will be added after the shopping cart function, the query function, And then you can purchase to submit orders, generate orders and payment to complete the entire purchase process function, the user can submit orders before the order to modify, when the seller after the shipment can also confirm the receipt of the operation to complete the transaction , After the completion of the transaction the user can comment on the corresponding book, the other unregistered users can not purchase books or add to the shopping cart operation, the user must first register as a member of the bookstore, log in before the above operations. In addition to the foreground function, the need for background management, administrators can log in through the corresponding Web site background, after the classification of the book catalog and book information to add, modify and delete, but also the order information management, Such as the cancellation of unpaid or unpaid orders, the payment of orders for delivery operations, etc., can also reply to the user's comments, you can reply to each comment, and the client's book details page is displayed.

3 System feasibility analysis

3.1 Economic viability

An economic feasibility analysis is a criterion for determining the final market position of a website. Good economic feasibility analysis is helpful for project implementation and management.

In this system development process, the technology used, from the database to the development tools, and then to the server are free, so the cost of the development of the system is only invested in time and effort, therefore, the system is economical.

3.2 Operational feasibility

The system's user platform is for those interested in books of all types of personnel, the interface is simple, the use of B / S visual interface, the user simply enter the book you want to search or related information to query, great convenience The user's view, reflects the human side. The system is based on the B / S architecture of an application, to achieve the page layer, business logic layer and data storage layer separation, the system is more stable and flexible [14].

In summary: This system is completely feasible.

4 The system design

4.1 System function design

The main functions of the online bookstore system modules are divided into two categories, one for the consumer demand for books on the front of the functional modules, and the other is for the administrator of the background function modules.

Order management is an overlapping part of two functional modules. Throughout the system, the order of the system has five states, namely 1,2,3,4 and 5 state. 1 status is waiting for the payment status, indicating that the order has just been generated yet payment; 2 state is waiting for the delivery status, indicating that the order has been paid but the administrator has not shipped; 3 state is waiting for confirmation, that the administrator has But the consumer has not confirmed the receipt; 4 state is the transaction successful state, that the administrator shipped, the consumer has confirmed receipt, once the order for the 4 state, the status of the order can not be changed; The order is canceled and the order can be canceled in the background only if the orders are in the 1 and 2 states. In the background, the order is canceled.

4.1.1 The functions of the foreground module

（1）User Registration Login：Users can register as a member, the user can become a member after login, modify personal information, modify the password and exit.
（2）Category display：In the left navigation bar displays the primary classification name of the book and the secondary classification contained in the primary classification, and the corresponding secondary book information is under the secondary classification.
（3）Book display：Search for books by title or keyword, search for books by advanced query (by title, author, and publisher), search books by category, view book details, and more.
（4）Shopping Cart Management：Putting the book what you are interested in or want to buy to add to the shopping cart for later viewing, in the shopping cart can modify the number of books, delete the shopping cart book information can be deleted individually or batch delete.
（5）Order Management：Generate an order by submitting an order, review the details of an existing order, confirm receipt of the goods after the seller has shipped, and cancel the order in the correct order. After the transaction is successful, you can evaluate the book and display it under the Book Details page.

4.1.2 Background module function

（1）Classification management: View classification, add a classification of secondary classification, modify the primary classification of secondary classification, delete the primary classification of secondary classification.
（2）Book Management：Background administrators have the same functionality as foreground users in searching for books, but administrators can also control book information such as adding books, editing book information, and deleting book information.
（3）Order Management：View orders by order status, view order details, can cancel orders before they are shipped, and can be shipped after buyer payment.
（4）Comment Management：Administrators can see all the reviews of the book by the consumer, and can add a reply to any one of the reviews and display it under the Book Details page.

4.2 Logical Structure Design of System Database

According to the functional design of the system, by analyzing the system structure and requirements, we can know that the system should have 8 data tables.Users(T_user), a catagory of books (t_category), a book list (t_book), a shopping list (t_cartitem), an order form (t_order), an order entry table (t_orderitem), an administrator table (t_admin) and a comment table (T_comment), the difference between the eight tables, linked with each other, together constitute the system's database module. A partial data tables are given below:

Table 1. t_user.

number	name	type	status	Primary key	Foreign key
1	uid	char(32)	not null	Y	
2	loginname	varchar(50)	default null		
3	loginpass	varchar(50)	default null		
4	eamail	varchar（50）	default null		
5	address	varchar(80)	not null	5	address
6	name	varchar(20)	default null	6	name

7	tel	varchar(20)	not null	7	tel

Table 2. t_category.

number	**name**	**type**	**status**	**Primary key**	**Foreign key**
1	cid	varchar(32)	not null	Y	
2	cname	varchar(50)	unique		
3	pid	char(32)	default null		Y
4	desc	varchar(100)	default null		
5	orderBy	int(11)	auto_increment		

5 System detailed design and implementation

5.1 Build the development environment

（1）operating system：Windows 7.
（2） software：MyEclipse 10、Dreamweaver CS6、MySQL、Navicat Lite for MySQL.
（3）Server software：Tomcat 6.0.
（4）Browser：IE、360、Firefox、Google.

5.2 System modules to achieve the foreground

（1）Front page Index module
Index module includes three main parts, the use of embedded frame technology, using the traditional "factory" glyph web framework, the location were on, left, in between the modules linked to each other to form the home page.

Upper part of the realization: the upper contains and contains two parts, is also divided into upper and lower, the upper part of the user's menu is mainly a series of links, according to the user's login status display different information, the user is not logged in display "Welcome to the bookstore", "Home", "Member Login" and "registered members", and the user logs on display "Congratulations on your membership of the bookstore," "Member Name", "Home" Shopping Cart, My Orders, Modify Personal Information, and Exit options.

The left part of the realization of the left: the left part of the entire page, showing all the books classification, including the primary, secondary classification, when the user clicks on a classification will display the current classification of all the two sub-categories, Accurate search function, mainly through the title, author and publisher of these three options integrated search.

Central to achieve: Central is the main part of the site, book information, order information, shopping cart information and other core content will be displayed in the middle part. In the middle of the middle part of the default display database information in the database to display the list; the top of the default display is a picture of some recommended books for rotation, click on the picture will jump to the details of the corresponding book interface, and the following shows the default The list of all the books in the database is sorted by default in one field in the database, four books per line, and a total of 12 books per page.

(2) Comment module

When the transaction is successful, you can evaluate the operation, published after the success of the evaluation will be in the front of the corresponding book details page under the comments part of the show.

5.3 System Modules to Achieve the Background

（1）Background classification management module
Administrators can log on to the background system to manage the classification of books, can add, modify, delete the primary and secondary classification.

The code for adding a two-level classification is as follows:

```
public String addChildPre(HttpServletRequest req, HttpServletResponse resp)
        throws ServletException, IOException {
        String pid = req.getParameter("pid");
        List<Category> parents = categoryService.findParents();
        req.setAttribute("pid", pid);
        req.setAttribute("parents", parents);
        return "f:/adminjsps/admin/category/add2.jsp";
  }
  public String addChild(HttpServletRequest req, HttpServletResponse resp)
        throws ServletException, IOException {
        Category child = CommonUtils.toBean(req.getParameterMap(), Category.class);
        child.setCid(CommonUtils.uuid());
        String pid = req.getParameter("pid");
        Category parent = new Category();
        parent.setCid(pid);
        child.setParent(parent);
        categoryService.add(child);
        return findAll(req, resp);}
```

6 Summary

The system is a general hobby for reading and books on the needs of the consumer customer base, with a certain degree of practical online bookstore system. It is mainly different from the traditional physical bookstores, to overcome a series of physical bookstores limited variety, fixed location, limited space and narrow sales channels and other issues, for people to purchase the book has brought convenience.

Online bookstore system not only can easily find the information and purchase books, and the operating conditions are simple, user-friendly, to a large extent to solve real-life problems in the purchase of books, but the design of the system because of the completion time Of the restrictions, some of the functions of the system is not perfect, but the basic functions have been achieved.

References

1. Jie Liu.Design and Implementation of Online Bookstore Based on JSP and JavaBean Technology.Modern Information,2015,(12):42-47.
2. Lingxi Meng.Design of Online Bookstore System Based on B2C Model.Value engineering,2011,(36):101-102.
3. Yingqian Tan.Research on Evaluation Index System of Online Bookstore Based on University Students' Satisfaction.Journal of Sun Yat-sen University (Social Science Edition),2011,(4):174-184.
4. Yinghui Xu.Design and Realization of Online Used Book Sale System.Consumer Electronics,2012,(9):67-68.
5. Zhuanhong Chen,Huan Li.A typical application of jQuery in WEB query module.Electronic Technology,2013,(12):28-32.
6. Paz J R G. Ajax and jQuery.Beginning Asp Net Mvc,2013,(3):11-13.
7. Dan Deng.The Operating Mode and Development Prospect of China 's Online Bookstore.Entrepreneur World,2013,(4):110-111.

A Research on Development of Construction Industrialization Based on BIM Technology under the Background of Industry 4.0

Junjie Li[1*] and Hui Yang[1]
[1]School of Management Engineering, Zhengzhou University, Zhengzhou, Henan, China
[*]Corresponding author: junjie87@126.com

Abstract. "Industry 4.0" and "Made in China 2025", the future strategic plans of the manufacturing industry in Germany and China, are being noted and studied all over the world. In this wave, information and industrialization of production are Construction industry development direction in the future. Based on the introduction of the strategic objectives of "Made in china 2025", a new production pattern of construction industrialization, which is based on BIM technology, are put forward in this paper. Comparative analysis is used in the study of mechanism and development path of the production pattern. The dilemma of construction industrialization confronted at present is illustrated in this paper. Based on the background of implementation of "Made in China 2025", the countermeasures and suggestions of the new production pattern have been proposed.
Keyword. Industry 4.0, made in china 2025, construction industrialization, building information modelling, path research

1 Introduction

So far, man has experienced three great industrial revolution in history, in which every change has promoted the maximal productivity. At the end of the eighteenth Century, the "Steam Revolution" replaced human labor with steam power, which marked human beings have got into the industrial civilization from the agricultural civilization. At the beginning of twentieth Century, the "Electric Revolution" realized the scale production with electric power. Furthermore, the "Information Revolution" in 1970s improved the automation level of the manufacturing process. In early twenty-first Century, mankind will perform "Industry 4.0" era with intelligence as its core.

Concept of "Industry 4.0" was formally put forward at the Hannover Industrial Expo in April 2013.It is a common national strategy led by the German government, industry leaders, researchers, associations and trade unions, so as to improve the definition of industry, from the centralized production mode to the basic form of decentralized production and control. It is revolution about internet of things technology, manufacturing services and other technologies that integrate a virtual network with physical entities.

Since the "Industry 4.0" national strategy has been proposed by Germany, other countries put forward their own industrial development strategies, for instance, the US government puts forward the "Industrial Internet Project", France puts forward the strategy of "New Industrial France", the British government puts forward the "British Industrial 2050 Strategy", and Japan puts forward Manufacturing Revival Strategy. Under the background of economic new normality,

China's manufacturing industry is facing its own drawbacks and international environmental pressure. Premier Li Keqiang proposed "Made in China 2025" plan to reply the challenge of a new round of industrial revolution at the second session of the National People's Congress third conference in March 2015.

2 From "Made in China 2025" to the construction industrialization

The essence of "Made in China 2025" is the depth integration between Internet and manufacturing, namely realization of information in the process of manufacture. The fundamental purpose of this strategy is to realize the transformation and upgrading of manufacturing industry that is innovation driven, to promote the manufacturing industry from the low-end level of the global value chain to high-end level, and to use a resource-saving, environment-friendly, innovation driven sustainable development model instead of the original resource-depended, environmental-destructed, labor-intensive and capital-intensive traditional manufacturing industry development model[1]. "Made in China 2025" has always been regarded as China's "Industry 4.0". In June 13, 2016, the fourth round of consultations between Chinese government and German government jointly chaired by Prime Minister Li Keqiang and Prime Minister Angela Dorothea Merkel, the two governments reached a consensus to accelerate the docking of Industry 4.0 and China's industrial upgrading.

Although construction industry and manufacturing industry belong to two different categories, but the two are highly dependent each other. Construction industry can even be regarded as a special kind of manufacturing. As a pillar industry of Chinese national economy, construction industry plays an important role in achieving "Made in China 2025" strategic planning. There are a lot of conjunction points between the development idea of construction industrialization and "Made in China 2025" strategic planning.

Construction industrialization is a kind of the component-prefabricated, the construction-packaging, design-standardized, construction-assembled, management-scientific and operation-informational method, whose objective is reorganizing the construction industry in the way of industrialization so that energies are saved, environment is protected, the value of life cycle is used maximally and the sustainable development can be realized. In order to response to the strategic plan of "Made in China 2025", information has to be realized in construction industry. "Internet Plus" thought should be integrated into industrial production, new manufacture technology should be used to advance intelligent manufacturing of modern architecture, improve comprehensive level of integration, make the construction industry and manufacturing industry develop together, thus while promoting the reform of China's construction industry, grand strategic objectives of "Made in China 2025" can be achieved more swiftly.

3 Production mode of construction industrialization based on BIM technology

3.1 BIM and its characteristics

Building information modeling(BIM) is a concept that has risen rapidly in the field of construction engineering management. BIM is a technology that describes an engineering project consisting of intelligent facilities with their own data properties and parameter rules, in which each object's appearance and its internal components and features can be displayed in the form of three-dimensional figures.

BIM has five features: visualization, coordination, simulation, optimization, ability of plotting[2]. It can apply information technology to the whole life cycle of any project and enhance collaboration among the participants of the project. In addition, BIM also has a lot of advantages, as shown in Figure 1. The emergence and development of BIM technology has strongly influenced the construction industry. Now software and tools supporting BIM have came out one after another, more and more construction enterprises have entered the field of application of BIM to obtain benefits from the implementation of BIM technology.

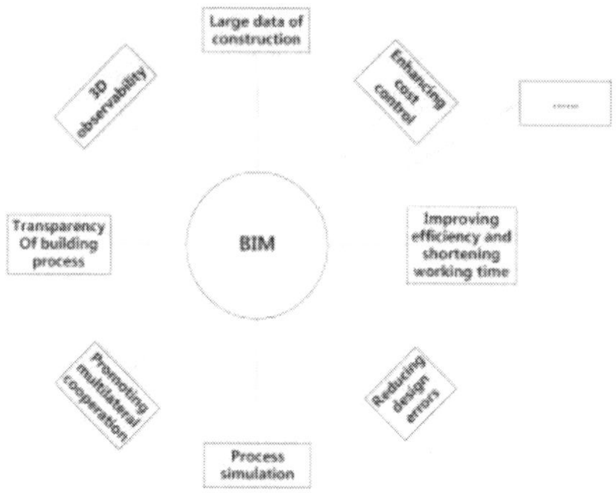

Fig. 1. Advantages of BIM

3.2 Mechanism analysis of production mode of construction industrialization based BIM technology

At present, BIM technology is applied to large-scale complex construction projects, but it has not been fully popularized to the design-standardized, construction-assembled construction projects that has huge amount of work. A mechanism of the new production pattern of construction industrialization is established and used in different phases of production of construction industrialization with BIM technology as the core, other engineering technology as auxiliary means.

3.2.1 Construction logistics management system based on BIM

In the pre construction preparation phase, BIM technology has some potentials in the material tracking. A real-time observation platform supported by lean thinking and practical experience to benefit supply chain management. It combines the BIM technology with the location based service (LBS). Implementing site management in a virtual construction site can effectively track and monitor workers, equipment and materials, and improve the supply chain management process. Evaluation of environmental impact by the carbon footprint of the supply chain, data and information on the safety and health of workers who is observed are contribute to understanding the situation of the construction site and making the corresponding decision.

3.2.2 Construction coordination system based on BIM

In the construction phase, BIM technology and augmented reality (AR) are combined to add digital information to physical components, thus BIM technology can be extended to the construction site and provide virtual information for project manager. Site activities will become much easier to manage, all the information needed for construction activities and model components can be simulated and retrieved in real environment[3]. Real-time cloud services of BIM platform will enable users to receive real-time feedback on their future design decisions and plans through mobile applications such as smart phone applications, etc.. In addition, integrated application of BIM technology and digital processing realizes efficient creation, refined management and accurate delivery of information of construction industrialization[4].

3.2.3 Facility management system based on BIM

In the operation phase, the facilities management can be carried out by BIM technology. When BIM technology and GIS are used in information extraction, conversion and download, geometric information can be separated from the related attributes to make the required data integrated effectively. BIM integrated facility management system will not only have a great impact on cost efficiency and cost savings, but also improve the value of assets. Moreover, photogrammetry and AR technology as well as BIM can be integrated together to improve the efficiency of facility management[5].

3.3 Implementation path of production mode of construction industrialization based BIM

The application of BIM technology is the core of the new pattern of construction industrialization. There are still many obstacles and problems in the development of China's construction industrialization, and the advantages of BIM technology, in order to use BIM technology in production mode of construction industrialization, several following aspects can be choose as "breakthroughs": construction simulation, component testing, component production, information exchange, etc.(As shown in Figure 2)

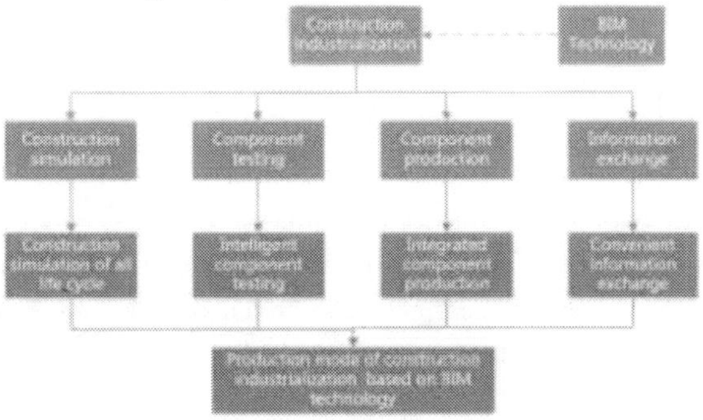

Fig. 2. System frame of development path of production mode of construction industrialization based on BIM

3.3.1 Carrying out the construction simulation in all life cycle

Different from the general product components, the volume and weight of buildings' components are generally quite large. It is difficult to transport in long distance[6]. The influences it causes compares to general engineering shows that more negative effects of engineering changes and secondary transportation result from materials produced need to be considered.

For builders, BIM technology is adopted in production of construction industrialization. On the basis of three dimensional visualization function, BIM adds time dimension to it and forms a 4D model. Builders use 4D model to virtually simulate the process of design, construction and operation, which helps them to determine the selection and use of prefabricated components, thus achieves accurate planning and proactive change management. In this way, they can not only ensure quality and safety, but also to reduce waste and improve the overall project performance.

For constructors, BIM technology is used to simulate the 4D model of each phase of projects to examine the appearance, function and situation of construction, and combined with the engineering cost and other information. So they can consider the feasibility and rationality of the project, and then decide whether to invest or not, in order to protect their own interests maximally. In addition, when using real-time cloud services of BIM platform, users are able to view and respond to any construction problems directly from their mobile devices, which is conducive to strengthening the communication between the two sides to prevent unnecessary delays or conflicts may occur in the construction site.

For the investigation unit, data and information of surveying and mapping on pre construction site are transferred to computer, and the data were compared with BIM. They need to calculated deviation for providing basis of further construction process, and keep the design and construction results consistent as far as possible.

3.3.2 Intelligent test mode of components

In the construction of construction industrialization, projects need a large number of various types materials or components. The situation that prefabricated components can not match often appear, and these conflicts are more complicated and difficult to solve than the problems that occur in concrete casting process. BIM technology can be a good solution to this problem.

BIM technology can be carried in conflict detection and construction analysis of pre construction phase. When projects use BIM software to collision check and build 3D model, the software automatically detects the collision position, and then provides the reasonable position and overlapping sequence of the components. It can quickly test component and record information in short time, and ensure installation can be completed successfully[7].

For the supplier of mechanical, electrical, plumbing (MEP), BIM model provides accurate digital information of service system for MEP contractors and projects. Integration and coordination of MEP system are supported by Autodesk and Solibri and other BIM software, solve that the physical installation conditions are normally not as ideal as the calculation model due to the varying quality of concrete structures. The issue is faced by early practitioners who fabricated system components such as ducts and pipes in factories[8].

3.3.3 Adopting integrated components production mode

For suppliers of prefabricated components, BIM model provides very valuable data to support the integration of manufacturer's design, manufacturing, construction, and maintenance process. The data in the BIM model makes the use of prefabricated components in the production process become very convenient, including material ordering, factory logistics, processing, packaging, inventory and transportation to the construction site etc., which accelerates suppliers and cooperation between Party A and Party B, thus saves time and costs.

With BIM and LBS, real-time observation platform supported by lean thinking and practical experience can support following four activities of prefabricated components in construction industrialization: components' design and manufacturing management activity, components transport activity, installation management activity and quality control activity. It not only makes the industrial construction project upgrade quality, but also saves costs and time.

3.3.4 Promoting convenient information exchange

In China, design units are independent organizations. They design for owner's requirements such as safety, beauty, economy and environmental protection. However, the situation of construction process, construction management are rarely considered. Design units are lack of communication with builders and supervisors results in design and construction are separated. In addition, all parties involved in construction project are stakeholders, they are facing the contradiction between interests and the overall interests of projects. Therefore information asymmetry between participants in the project also causes a lot of problems. If these problems can't be solved, the high quality and high efficiency of industrial and assembled construction way will be difficult to achieve.

The parametric model provided by BIM sayes all information at all life cycle of projects. BIM provides an information platform for engineering projects(As shown in Figure 3). Based on BIM and digital processing of integrated application, as well as the function of tracking of all life cycle of the project, working time, cost, responsibility, materials and suppliers, etc. of each phase of each part are clearly displayed. Participants can obtain information more easily and communicate more efficiently, and avoid conflicts and mistakes caused by poor communication and unclear responsibilities during the process.

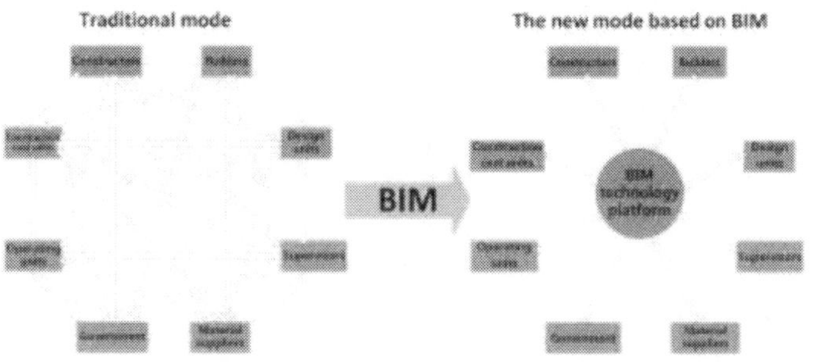

Fig. 3. Difference of information exchange between traditional mode and the new model based on BIM technology

Therefore, the integration of BIM and construction industrialization is utilizing advantages of BIM and construction industrialization, instead of simply combining the concepts. In this way, barriers in the implementation of construction industrialization is broken in China. It is promoting development of construction industrialization, and then enables China's new construction industry realize the goals of high efficiency economy, low carbon environmental protection and sustainable development.

4 Dilemmas of production mode of China's construction industrialization based BIM technology

Although BIM has many advantages in promoting construction industrialization, it still faces some dilemmas. This is mainly manifested in the following aspects.

4.1 The universal degree of construction industrialization and BIM is still low in China

Nowadays, the proportion of China's construction industrialization projects is still very low. According to a survey, merely 68% of the designers in China have only heard of BIM, a considerable proportion of designers have never heard, and the designers used BIM are less than 4%[9]. So, although the mode of construction industrialization based on BIM technology has absolute advantages, it still faces a tremendous challenge.

4.2 Research fund on BIM technology lacks

The time is very limited since BIM was introduced into China, degree of domestic research for BIM is small. At present, only several research institutions like China Construction Research Institute, several universities like Tongji University, HUST and several enterprises like Luban, Glodon are committing to the study of BIM applications. That's nowhere near enough. The biggest problem constraints research and development of BIM application is that government's attention and support is far from enough. Due to the insufficient research funds, popularity of research and development of BIM aren't able to come.

4.3 Enterprises lack a strong incentive to innovate

Due to the large investment in the early stage of BIM application, for most enterprises especially small and medium sized enterprises, more early they innovate, the return is greater while the investment is greater, which becomes an unfavorable factors blocking enterprise to innovate. Besides, the advantages of BIM is usually applied in all life cycle of building, in which there are a lot of stakeholders. Only few enterprises used BIM technologies can not give full play to its advantages, even it may occur mistakes because of the old and new technology cannot match successfully. State, society and construction industry have not advocated and promoted BIM application. Consequently, most of enterprises would rather sacrifice cost and schedule than pour their best to change their production pattern.

4.4 Lack of construction industrialization standard

There is the standard parameter of industrialized building at present, but the rules of mandatory implementation haven't established, therefore the standard of construction industrialization has not been unified. However, once BIM is applied in construction industrialization, it's more obvious that interface integration between prefabricated components, in particular, assembled concrete structures depend on the unified standard. Chinese government not only didn't establish relevant standards and provide a strong supervision, but also didn't coordinate and unify entire manufacturing industry, which can not maintain the norms and order of the production of prefabricated components.

4.5 Relevant laws and regulations are not perfect

In China, laws and regulations related BIM application are not perfect. Compared to traditional construction mode, the new production pattern has more uncertainty factors. It's possible that various disputes and disputes arise because the standardized contract templates, relevant laws and regulations haven't be provided by Chinese government. What's more, the government's protection for intellectual property rights is so lacking that it doesn't enable to fight against piracy software.

5 Countermeasures and suggestions on promoting the development of construction industrialization by using BIM technology

5.1 Seize opportunities, and take own advantages

Depends on "Made in China 2025" strategic plan, China should develop construction industrialization under the international background of "Industry 4.0". Construction industry is a special manufacturing industry, and China has one of the largest construction markets all over the world, China should seize opportunities to enable construction industry and real estate industry to become the world's manufacturing power. The goal of the Eighteenth Congress of the Communist Party of China that pushing forward the new urbanization steadily provides a comprehensive implementation platform. It could be used to build a great number of high-efficiency, and high-quality projects, which meets the requirements of the national planning, provides models for reform and accelerate the development of China's construction industrialization.

5.2 Attach importance to research and development of construction industrialization

At present, the trouble for scientific research institutions to study BIM and construction industrialization and enterprises to reform is that the preliminary study of BIM and construction industrialization need a large amount of investment. Thus China should increase investment in research and development of advanced technology for scientific research institutions and universities, and train some large enterprises as examples of innovation in the industry that bring a demonstration effect.

5.3 Improve laws and regulations and construction standards

To establish a framework for the standardization of construction, Chinese government must improve laws, regulations and construction standards. Avoiding chaos phenomenon without standard of implementing new production pattern, departments should improve the unified standard of construction industrialization and BIM application. In addition, guidelines for the implementation of production and contract templates should be provided, and laws and regulations should be improved. In order to strengthen the security, authenticity, reliability and the research of intellectual property rights of BIM technology. Solutions for problems and disputes should be provided to ensure that there must be laws to go by.

References

1. Yu Donghua, Hu Yanan, Lv Yinan. Research on the path of technological innovation and industrial choice of "Made In China" 2025 under the new industrial revolution. Tianjin Social Sciences, **4** (2015): 98-107.
2. Li Yong,Guan Changsheng. Information management model and strategy of engineering project based on BIM Technology.Journal of Engineering Management, **4** (2012): 17-21.
3. Park, X., Lee, C., Kwon, D. and Wang, X. (2013), "A framework for proactive construction defect management using BIM, augmented reality and ontology-based data collection template", Automation in Construction, Vol. **33**, pp. 61-71.
4. Anonymous.BIM+ leading the new direction of the development of BIM. Survey and Design of China, **10** (2015): 27-45.
5. Hou, L., Wang, X., Wang, J. and Truijens, M. (2014a), "An integration framework of advanced technologies for productivity improvement for LNG mega-projects", Journal of Information Technology in Construction, Vol. **19**, pp. 360-382.
6. He Yan, Chen Yan.The present and future of construction industrialization. Engineering Quality, **31.2** (2013): 1-8.
7. Liu Kading, Zhang Yongcheng, Chen Lijuan. Study on the collision analysis of metro station pipeline integrated installation based on BIM technology. Journal of Civil Engineering and Management, **1** (2015): 53-58.
8. Zhang J, Long Y, Lv S, et al. BIM-enabled Modular and Industrialized Construction in China. Procedia Engineering, 2016, **145**: 1456-1461.
9. Mu Xue.Research on the countermeasures of development of BIM in China's construction industry. Shandong Industrial Technology, **24** (2015): 95-96.

Flexible job-shop scheduling based on genetic algorithm and simulation validation

Erming Zhou[1,*], Jin Zhu[1] and Ling Deng[2]

[1]School of Transportation and Logistics, East China Jiaotong University, China
[2] School of Mechatronics and Vehicle Engineering, East China Jiaotong University, China
[*]Corresponding author: zhouermin@sina.com

Abstract. This paper selects flexible job-shop scheduling problem as the research object, and Constructs mathematical model aimed at minimizing the maximum makespan. Taking the transmission reverse gear production line of a transmission corporation as an example, genetic algorithm is applied for flexible job-shop scheduling problem to get the specific optimal scheduling results with MATLAB. DELMIA/QUEST based on 3D discrete event simulation is applied to construct the physical model of the production workshop. On the basis of the optimal scheduling results, the logical link of the physical model for the production workshop is established, besides, importing the appropriate process parameters to make virtual simulation on the production workshop. Finally, through analyzing the simulated results, it shows that the scheduling results are effective and reasonable.

1 Introduction

Flexible manufacturing system is a complex artificial system, which has the characteristics of complexity, randomness, multi-objective, multi constraint and so on. Production scheduling is a research hotspot of the manufacturing system. Because of its effective application in practical engineering, related research and control methods have been widely concerned, and have made great research achievements. Flexible Job shop scheduling problem is a typical NP hard problem, which is a class with multi-objective, multi variable and multi constrained combinatorial optimization problem. This problem is difficult to solve with exact algorithms within the available time, so, many scholars introduce and put forward a variety of heuristic algorithm to solve the flexible job shop scheduling problem. Among them, genetic algorithm has been widely used and developed for its good robustness and stability.

Many scholars made the research on the job shop scheduling problem based on genetic algorithm, Lei [1] proposed a effective collaborative evolutionary genetic algorithm and get the optimal solution to deal with the work order and machine assignment fusion. Toledo et al. [2] raised multi population genetic algorithm with migration strategy. When the algorithm continued to perform repeatedly, the objective function value did not changed, excellent population migration strategy would be started and performed until the algorithm satisfied the stop criterion. Teekeng et al. [3] put forward a kind of improved genetic algorithm which introduced a new mutation operator to help maintain the diversity of population and avoid premature convergence of the algorithm, and applied it to solve the flexible job shop scheduling problem successfully, Jalilvand-Nejad et al. [4] established the mixed integer programming model of circular flexible job shop scheduling, and solved the model based on the genetic algorithm to resolve the circular flexible job shop scheduling problem. Liu et al. [5] proposed crossbar collaborative multi-group genetic algorithm. It was difficult to obtain the optimal solution in multi-objective fuzzy flexible job shop scheduling by using common genetic algorithms. To solve this problem, based on the consideration of due date obeyed fuzzy time window distribution, a multi-objective fuzzy flexible job shop scheduling model was presented. It was aimed at maximizing customer satisfaction and minimizing completion time. Zhou et al. [6] created a dual clutch transmission scheduling model with DELMIA, and analyzed the collision and Gantt chart based on genetic algorithm to shorten the design period and reduce the design cost.

On the basis of previous studies, this paper constructs the mathematical model of flexible job shop scheduling problem. Taking the transmission reverse gear production line of a transmission corporation as an example, genetic algorithm is applied for flexible job-shop scheduling problem to obtain the optimal scheduling results. Finally, simulated model is made with DELMIA/QUEST to testify the scheduling results.

2 Descriptions and modeling for flexible job shop scheduling problem

2.1. Problem statement

Research on flexible job shop scheduling is the processing of n work-pieces to be processed' processing, each work-piece contains one or a plurality of processes, each process can be processed in a plurality of different processing machines, the processing time varies with the choice of processing machine. Under the premise of aiming at meeting process constraints and machine availability, the optimal processing machines are chosen for each process, and the optimal processing sequence and the start working time of the every process in every machine is decided, so that the processing time of a job is the shortest, the specific variables are defined as follows:

$N=\{1,2,\cdots,i,\cdots j,\cdots n\}$ Represents a set of artifacts, where n is the total number of artifacts.

$M=\{1,2,\cdots,k,\cdots h,\cdots m\}$ Represents a set of artifacts, where m is the total number of machines.

j_k Work-piece number (j_k=1,2,3,…,n).

i_e machine number (i_e=1,2,3,…,m).

h_i the total number of working procedures for the part i.

l the serial number of working procedures for the part i (i=1,2,3,…,h).

O_{ih} the procedure h of the part i.

t_{ik} the processing time of the part i on the machine k.

C_{ik} the makespan of the part i on the machine k.

$$a_{ihk} = \begin{cases} 1, \text{if the processing time of the part } i \text{ in the} \\ \quad \text{machine } h \text{ prior to the machine } k \\ 0, \text{others} \end{cases}$$

A a large positive number.

B a set of procedures for the constraint relations of before and after the process of the same kind of work-piece .

$$x_{iijk} = \begin{cases} 1, \text{the part } j \text{ prior to part } i \text{ is processed in the} \\ \quad \text{machine } k \\ 0, \text{others} \end{cases}$$

2.2 Mathematical model of flexible job shop scheduling problem

The maximum makespan is the most basic index to measure the scheduling scheme. In some degree, it reflects the production efficiency of the plant, aiming at minimizing the maximum make-span. Constructed objective functions are described as:

$$f = \min\{\max_{1\le k\le m}\{\max_{1\le i\le n}\{c_{ik}\}\}\} \quad (1)$$

In addition, the FJSP optimization process needs to satisfy the following constraints:

(1) Work-piece occupancy constraints: to ensure that every process of the work-piece can only be processed on a machine at the same time.

$$c_{jk} - c_{ik} + A(1 - x_{iijk}) \ge t_{jk}, i, j \in N; k \in M \quad (2)$$

(2) Process sequence constraints: to ensure that the processing sequence of every work piece to meet the requirements set in advance.

$$c_{jk} - c_{ih} \ge t_{ih}, (i, j) \in B \quad (3)$$

(3) Process time constraints : to ensure that the makespan of every process and the processing time is greater than or equal to zero.

$$t_{ik(h)}, c_{ik(h)} \ge 0 \quad (4)$$

3 Model solution for scheduling problem

The production workshop of the auto transmission reverse gear of some automobile transmission company in Jiang Xi province is set as an example. In the reverse gear production workshop, it produces a total of six parts which are machined in the 12 machines, and every work-piece contains 8 processes, process can choose machine table, processing schedule is shown in Table 1, table 2 .Genetic algorithm has the ability of doing the global parallel search, and focuses the main target on the high performance, so that it can improve the search efficiency and not easily fall into local minimum, thus, the approximate optimal solution of the model is solved by genetic algorithm in this paper.

Tab. 1 Production process of the optional machine

	J_1	J_2	J_3	J_4	J_5	J_6
O_{i1}	2	7	3	5	4	2
O_{i2}	4	[1,9]	[4,7]	4	10	8
O_{i3}	1	5	11	[3,6]	3	1
O_{i4}	[3,5]	2	4	8	[2,6]	7

O_{i5}	2	4	8	2	5	[2,3]
O_{i6}	3	10	6	[4,7]	2	[3,9]
O_{i7}	[6,9]	8	[1,2]	9	12	11
O_{i8}	7	[3,7]	5	12	7	[4,6]

Table 2. Production processes processing time

	J_1(s)	J_2(s)	J_3(s)	J_4(s)	J_5(s)	J_6(s)
O_{i1}	3	5	10	7	2	5
O_{i2}	6	[2,5]	[6,10]	5	3	11
O_{i3}	8	10	5	[7,9]	6	1
O_{i4}	[5,7]	6	3	6	[1,4]	4
O_{i5}	2	4	7	4	3	[4,6]
O_{i6}	7	3	2	[5,6]	7	[3,5]
O_{i7}	[4,9]	7	(3,5)	5	3	2
O_{i8}	3	[4,4]	4	7	10	[2,4]

3.1 Model solution based on genetic algorithm

Genetic algorithm is applied to solve flexible job-shop scheduling problem as shown in figure 1.Among them, the population initialization module initializes the initial solution of the population composition; the fitness value calculation module calculates the fitness value of chromosomes; the roulette wheel method is applied to select excellent individuals in the selection operation; it gets excellent individual by integer crossing method in the crossover operation; it gets excellent individual by integer variation method in the mutation operation [7-16].

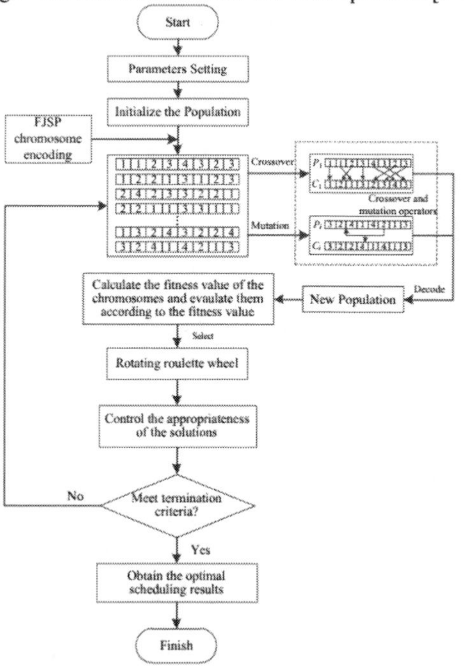

Fig. 1 Flow chart of genetic algorithm to solve FJSP

Step1: Setting parameters: input to the known data of the production workshop. In addition, maximum number of iterations, the size of the population, the crossover probability and mutation probability are also set.

Step2: Generating the initial population: initial population is randomly generated in the process of time operation, and there is no any rules, so the generation of initial population relies on computing ability and computational complexity of computer, the concrete production method is that the solution of the scheduling problem is expressed in the form of chromosomes with the chromosome coding, good coding style can generate feasible solutions more easily in subsequent genetic algorithm operation and improve execution efficiency. FJSP contains two problems of machine

 DOI: 10.1051/matecconf/201710002047

selection and process scheduling, integer coded MSOS combined with block coding is adopted to encode for the two sub problems. MSOS embraces machines selection, MS and operations sequencing, OS, the coding process is described in detail by taking the top three steps of the part 2 and the part 3 as an example. In the process of the machine selection coding, the process O22 of the part 2 have two machines to choose from, 1 indicates choosing the first machine, that is to say, it can be processed on the machine M1, and the rest can be done in the same manner. Process scheduling is to make use of the number of woke piece to encode directly, Emergence order of the work piece number represents the sequential processing between processes of the work piece. The coding process is shown in figure 2.

Step3: Crossover operation: crossover operator is the main method of generating new individuals, which determines the global search ability of genetic algorithm. The function of crossover operation is to give the original good genes to the next generation, and to generate new individuals with better genes. Usually the genetic operator we use is Point crossover, which is to cross as the probability of cross.

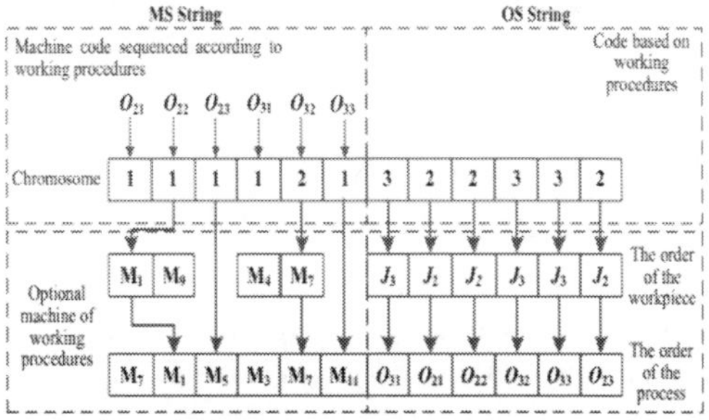

Fig. 2 FJSP chromosome encoding

Step4: Evaluation and Selection: evaluation and Selection is to choose the individuals with high fitness from the obtained fitness, and to eliminate the individuals with low fitness, namely, the better individuals continue to survive, which achieves survival of the fittest, to ensure that excellent genes can be passed on to the next generation.

Step5: Analyzing the termination conditions: Given that the genetic algorithm is a kind of repeated cycle selection, crossover and mutation algorithms, thus, when using genetic algorithms to calculate, termination conditions must be set. When setting the parameters, the final number of iterations should be set in accordance with the complexity of the problem to terminate the operation.

3.2 Model solution result

Writing the running code to solve the scheduling model with MATLAB, the basic parameter of algorithm is set to: population size pop is 50, maximum iteration parameters Ite for 50, crossover probability for 0.8, the mutation probability for 0.01, he shortest time of running program to a job done is 43s, the search process is shown in Figure 3. Figure 4 is the scheduling Gantt chart the optimal solution.

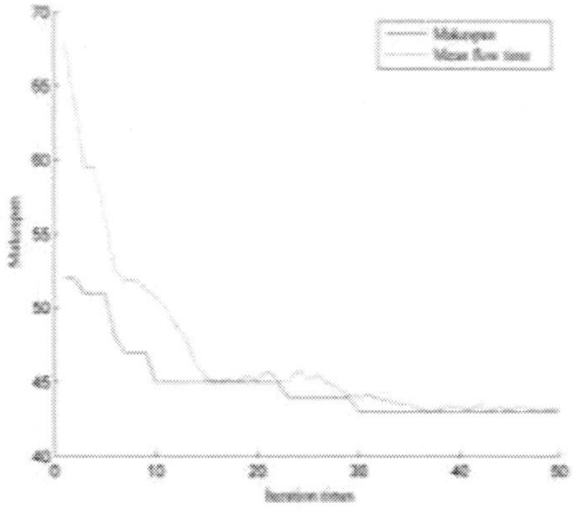

Fig.3 Genetic evolution curve

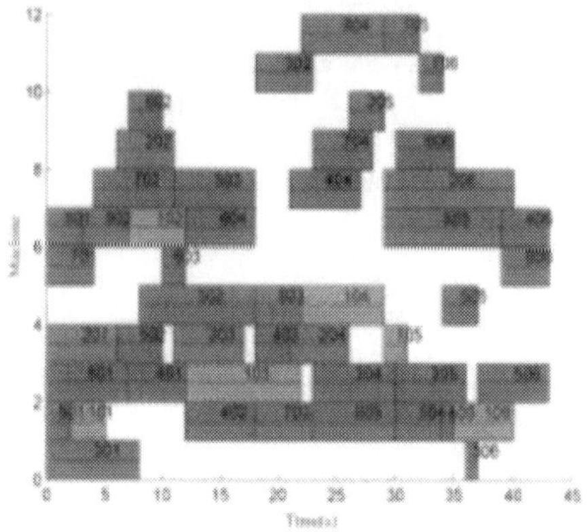

Fig.4 Scheduling Gantt chart

4 Simulation and verification of job shop scheduling based on DELMIA/QUEST

According to the actual situation of production system, simulation of production system is to establish a descriptive model of workshop with the computer software, simulation is made on the established model, which can not only evaluate the performance of the system, but also assist decision-making, besides, the optimization will be made. In the job shop scheduling problem, simulation verification of the scheduling schemes can predict the performance of the system under different scheduling strategies, and determine the reasonable and efficient production planning and scheduling scheme, consequently, the capacity of the production system will be made full use of, which improves the efficiency of the economy. Simulation of the production system is the simulation based on discrete event system, the process of simulation is shown in Figure 5. DELMIA/QUEST is a kind of simulation software, which is developed by the French Dassault company, and applied to the analysis of the process flow and workshop layout of discrete event. Because of its simple and easy-to-use visual interface and outstanding data analysis ability, it makes it become the preferred solution for the simulation and analysis of production process [17]. Generally, modeling and simulation with DELIMA/QUEST can be divided into four steps: the introduction of production line process layout, physical modeling and logical modeling, process design, result and analysis.

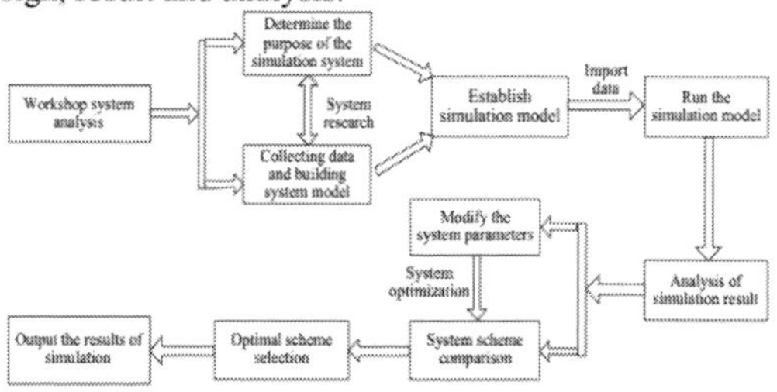

Fig.5 Flow chart of discrete system simulation

4.1 Production workshop simulation model

Process flow is a series of procedures from the initial raw materials to the final product in the process of production and processing, which is also the basis of the simulation of the production workshop. In practical engineering applications, the process layout is usually drawn with CAD software. Because Quest does not support common DWG format under CAD, at first, we need to transform DWG format of the process flow layout into the DXF format that Quest can identify, then directly importing.

Quest provides a powerful modeling function, where the modeling is the establishment of the simulation model, rather than the CAD model. The Quest model can be divided into two parts: the physical model and the logical model, CAD model is used to establish the real physical model, and the logical model controls simulation running.

The virtual physical model of production resources, including source, buffer, machine, sink, conveyor, automated guided vehicle (automated guided vehicles, namely AGV) and labor, Quest comes with a number of commonly used machinery and equipment, such as horizontal lathe, vertical drilling machine, fork-lift, it is convenient for modeling by defining the basic parameters. For initial models that quest does not have, CAD module is applied to set up the 3D geometry model of the equipment, importing the appropriate equipment parameter, ,according to the location of equipment in the system and the relationship between device and equipment, then modeling .Figure 6 is the modeling view for the human body model. The modeling view of the initial model of punch is shown as Fig. 7.

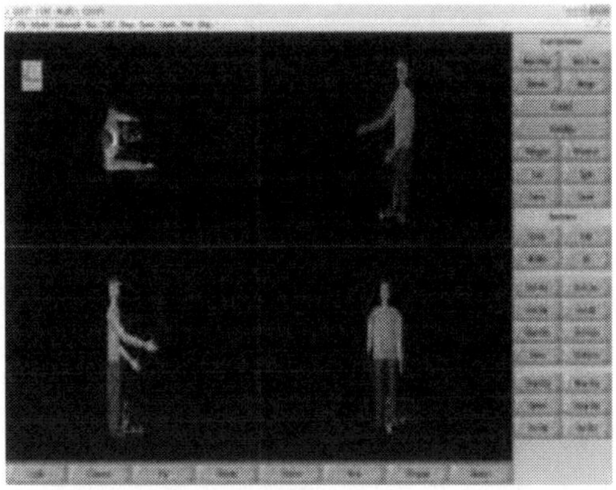

Fig.6 Mannequin modeling view

Logical model will be made for the established virtual physical model, establishing the relationships of links. Link (connection) is used to link the other elements in the Quest model. The overall flow of production system process is decided by establishing links and parts can only move between elements of link relations, once the link is established, so is the logical relation. The man-machine physical model and logical connection of a production line is shown in Figure 8, most of the elements have one input and one output link at least, in other words, the parts along the input link from a process into this element, and moves to a process along with the output link when the production and processing is finished. Thus each element has an input and output link, but the sources element and the reception element are two exceptions. This depends on their own properties, the function of sources is to create elements, and sinks is to destruct parts.

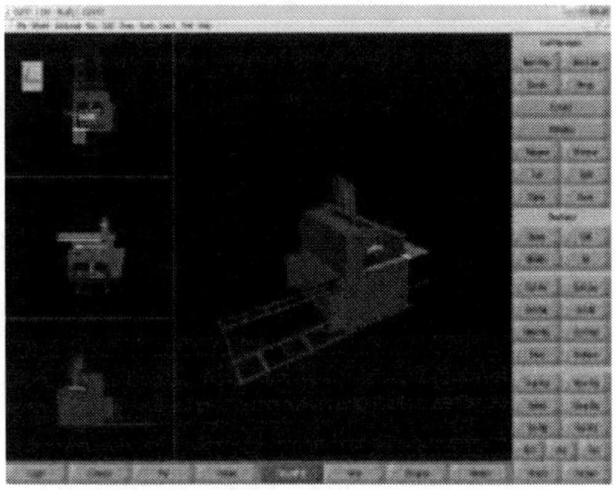

Fig.7 The initial model modeling view of punch

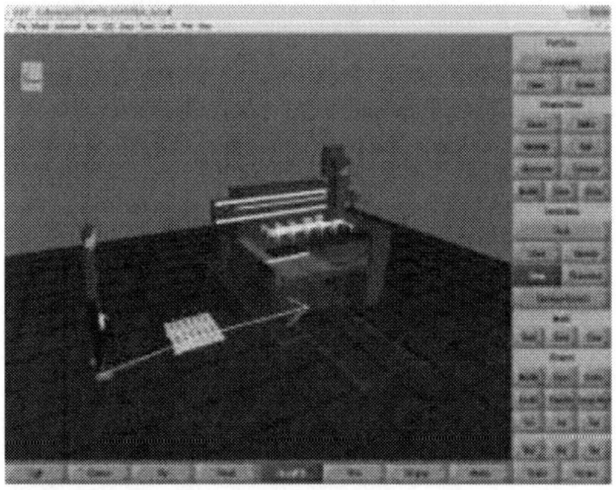

Fig.8 The physical model and logical connection of a production line

In Quest, selection and scheduling for the physical model is completed by controlling logical link. In fact, these links are a simulation program written by structured programming language (SCL). Quest provides a number of normative process and logic, in general, these normative process and logic are able to satisfy the demands of users. However, in some special production environment, the standard structured language cannot completely define the required process. Using SCL provided by the Quest to change or rewrite for the existing standard program, the logical link can be achieved in the special environment. This function enables the user to control the simulation model in any degree of their needs, and to use its two development function to achieve the simulation goal.

In order to make the simulation model more accurately reflect the running status of the actual production system, ensure the authenticity of the simulation results. The process parameters which are closely related to the production line are given to the established physical model, including the production rhythm, the transmission speed, the controlled logic and so on [18-20].

Simulation of the established model could dynamically display the utilization rate of the production equipment and human resources utilization in the form of pie chart and bar chart in the process of simulating. At the end of the simulation, the system will automatically generate a statistical table for the analysis of the entire model.

4.2 Simulation verification

Virtual simulation is made on the assembly workshop of the reverse gear. According to the previous scheduling results, this paper constructs a logical connection, running the simulation model. The scene view of the production operation simulation is shown in figure 9. In the process of simulation, we can clearly observe the entire process of production and assembly, at the same time, the simulation result can clearly reflect the utilization rate of equipment in the production and operation, and makes a true reflection of production transportation path, line, frequency and the operation process in the process of production and assembly, which can verify whether the results of job shop scheduling is reasonable. Through the simulation analysis, combined with the observations and the related report output, we can find that the equipment utilization rate is high in the process of simulation, and the cross roundabout phenomenon between production line rarely occurs, and the cost of the whole process is low. The simulation results show that the job shop scheduling scheme obtained by genetic algorithm is reasonable.

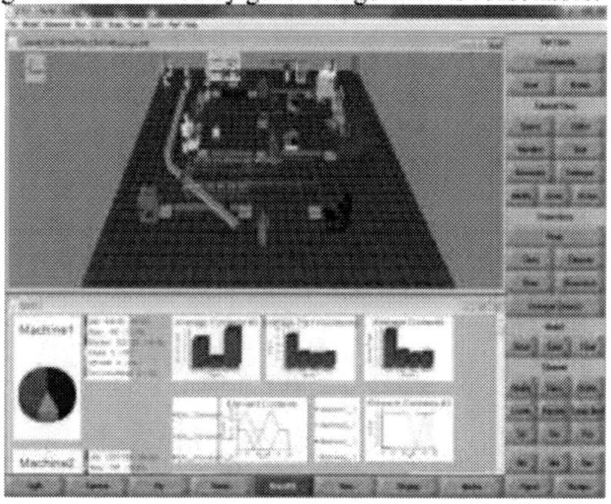

Fig.9 The physical model and logical connection of a production line

5 Conclusion

This paper constructs mathematical model of the job-shop scheduling problem aimed at minimizing the maximum makespan, and explains the meaning of each parameter in the mathematical model. Taking the transmission reverse gear production line of a transmission corporation as an example, genetic algorithm is applied for flexible job-shop scheduling problem to get the optimal scheduling results with MATLAB. Finally, simulation is made on the production workshop with the three-dimensional discrete-event simulation software DELMIA / QUEST. The simulation results show genetic algorithm in solving job shop scheduling problem is very effective and has a high application value.

Acknowledgement

The authors wish to acknowledge the support and motivation provided by Jiangxi Science Support Program (20151BBE 50053).

References

1. D. Lei. Co-evolutionary genetic algorithm for fuzzy flexible job shop scheduling. *Applied Soft Computing*, **12**,2237-2245(2012).

2. C.F.M. Toledo, P.M. França, R. Morabito, et al. Multi-population genetic algorithm to solve the synchronized and integrated two-level lot sizing and scheduling problem. *International Journal of Production Research*, **47**,3097-3119(2009).
3. W. Teekeng, A. Thammano. Modified Genetic Algorithm for Flexible Job-Shop Scheduling Problems. *Procedia Computer Science*, **12**,122-128(2012).
4. A. Jalilvand-Nejad, P. Fattahi. A mathematical model and genetic algorithm to cyclic flexible job shop scheduling problem. *Journal of Intelligent Manufacturing*, **26**,1085-1098(2013).
5. A.J. Liu, Y. Yang, Q.S. Xing, et al. Multi-population genetic algorithm in multiobjective fuzzy and flexible Job Shop scheduling. *Computer Integrated Manufacturing Systems*, **17**,1954-1961(2011).
6. E.M. Zhou, X.J. Peng , J.Q. Wang, et al. Visual Simulation of Dual Clutch Transmission Assmbly Process based on DELMIA. *Journal of Mechanical Transmission*, **37**,65-67(2013).
7. Q. Zhang, H. Manier, M.A. Manier. A genetic algorithm with tabu search procedure for flexible job shop scheduling with transportation constraints and bounded processing times. *Computers & Operations Research*, **39**,1713-1723(2012).
8. X. Li, L. Gao. An effective hybrid genetic algorithm and tabu search for flexible job shop scheduling problem. *International Journal of Production Economics*, **174**,93-110(2016).
9. R. Zhang, R. Chiong. Solving the energy-efficient job shop scheduling problem: A multi-objective genetic algorithm with enhanced local search for minimizing the total weighted tardiness and total energy consumption. *Journal of Cleaner Production*, **112**,3361-3375(2015).
10. J.C. Chen, C.C. Wu, C.W. Chen, et al. Flexible job shop scheduling with parallel machines using Genetic Algorithm and Grouping Genetic Algorithm. *Expert Systems with Applications*, **39**,10016-10021(2012).
11. N. Kundakcı, O. Kulak. Hybrid genetic algorithms for minimizing makespan in dynamic job shop scheduling problem. *Computers & Industrial Engineering*, **96**,31-51(2016).
12. C. Gutiérrez, I. García-Magariño. Modular design of a hybrid genetic algorithm for a flexible job - shop scheduling problem. *Knowledge-Based Systems*, **24**,102-112(2011).
13. J. Gao, M. Gen, L. Sun, et al. A hybrid of genetic algorithm and bottleneck shifting for multiobjective flexible job shop scheduling problems . *Computers & Industrial Engineering*, **53**,149-162(2007).
14. R.K. Phanden, A. Jain, R. Verma. A Genetic Algorithm-Based Approach for Flexible Job Shop Scheduling. *Applied Mechanics & Materials*, 2011, **7**,3930-3937(2011).
15. M. Zandieh, I. Mahdavi, A. Bagheri. Solving the Flexible Job-Shop Scheduling Problem by a Genetic Algorithm. *Journal of Applied Sciences*, **8**,4650-4655(2008).
16. J. Gao, L. Sun, M. Gen. A hybrid genetic and variable neighborhood descent algorithm for flexible job shop scheduling problems. *Computers & Operations Research*, **35**,2892-2907(2008).
17. Z.M. Bzymek, M. Nunez, M. Li, et al. Simulation of a Machining Sequence Using Delmia/Quest Software. *Computer-Aided Design and Applications*, **5**,401-411(2013).
18. E.M. Zhou, J. Zhu, G.Y. Wang. DELMIA/QUEST software-based digital process planning production line research. *Manufacturing Automation*, **37**,1-4(2015).
19. C. Guiling, L. Chuan, G. Jie, et al. Virtual human control technology for immersed virtual maintenance system in DELMIA environment: 2010 International Conference on Audio, *Language and Image Processing*(2010)
20. G. Pasca, I. Maniu. Synthesis of the design of flexible manufacturing system using Delmia/Quest software. *Annals of Daaam & Proceedings*, 695-697(2009).

A Big Data Decision-making Mechanism for Food Supply Chain

Guojun Ji[1] and KimHua Tan[2]

[1] Collaborative Innovation Center for Peaceful Development of Corss-Strait Relations; School of Management, Xiamen University, 361005 Fujian China
[2]Operations Management & Information Systems Division, Nottingham University Business School
*Corresponding author: jiking@xmu.edu.cn

Abstract. Many companies have captured and analyzed huge volumes of data to improve the decision mechanism of supply chain, this paper presents a big data harvest model that uses big data as inputs to make more informed decisions in the food supply chain. By introducing a method of Bayesian network, this paper integrates sample data and finds a cause-and-effect between data to predict market demand. Then the deduction graph model that translates foods demand into processes and divides processes into tasks and assets is presented, and an example of how big data in the food supply chain can be combined with Bayesian network and deduction graph model to guide production decision. Our conclusions indicate that the decision-making mechanism has vast potential by extracting value from big data.
Keywords. Big data, Bayesian network, deduction graph model, food supply chain.

1 Introduction

The data have now been woven into every sector of the global economy. Companies focus on capturing relevant information from multiple sources such as suppliers and customers made for a much clear and complete picture of the existing business process [1]. Big data analytics helps companies to identify new opportunities and requirements for new products and find ways of new services by integrating large amounts of trading information, real-time and historical information. Now, a complementary trend is under way. Information multiplies and is shared more widely around the world provides the basis for advance analysis of big data and enables us to find out new applications, such as the smartphone app that tells commuters when the next bus will arrive. This tendency carries profound significance for companies, governments, and individuals.

These developments have changed the operation management of the food supply chain beyond recognition. As companies capture, store, search, share and analyze huge volumes of data, radical customization and novel business models will be the new hallmarks of competition. Therefore, the application of big data in the food supply chain has been receiving increasing attention. Taylor & Fearne (2006) regard big data as the pre-requisites for the development of a more synchronised approach to demand and activity analysis for the food supply chain [2]. Tien (2012) pointed out that big data analytics is a key support technology to implement mass-customization in food production such as nano-modified and nano-additives [3]. Anica-popa (2012) indicates that data sharing in the food supply chain will improve the food quality and safety[4]. All the analyses mentioned above show that the importance of big data to the food supply chain can never be denied.

As a result, companies need not only skills but also new perspectives on how big data helps solve problems in the food supply chain due to the exploding data. In this paper, we propose a big data harvest model for the food supply chain. Our intent is to develop a decision-support tool that converts data into sights to make more informed strategic decisions. The paper is organized as follows. We first describe the value of big data in food network and develop a big data harvest model. The model is subsequently tested with an example. Then, we discuss and summarize our findings.

2 Value of Big Data in Food Network

Food supply chain is a changing system of organizations, people, activities, information, and resources engaged in production, processing, distribution and the disposal of food to move a product from farms to consumers [5]. Every year, $14 trillion of foods is produced, packaged, and sold in worldwide, through a series of transactions between suppliers, retailers and customers. It is estimated that $120 billion to nearly $150 billion in value per year could be achieved through the use of big data in food consumption. Big data has been applied to produce, package, sale, and use of food products (see Table 1, Source of Big Data). Suppliers track a large amount of useful data to understand better how customers evaluate the food and feed that information back into food design process. Retailers can make full use of big data to segment consumer types, carrying out precision marketing and cultivating the customers' loyalty. Also, customers have broader access to a massive amount of food information, making a more informed decision. For example, customers can be able to know the price of food ahead of time to decide which to buy.

What is more, it has been estimated that hundreds of billions in value per year could be enabled by the use of big data in food logistic. These data are captured by government, transportation operators, individuals, and third-party data providers. One of the largest potential benefits can be obtained by using big data to enhance the ability to deliver and adapt to customers in real time. Another benefit is that companies can optimize every process step from procurement to producing to marketing by uncovering new insights that are hidden within the data.

Table 1. Big Data in Food Network

Domains		Source of Big data	Value Creation Activities	Challenges
Food Supply Chain	Food consuming	• Retailers; • Suppliers; • Government; • Third-party data brokers; • Customers	• Improved product design and producing; • Efficient store operations; • More targeted marketing and sales; • Offering price transparency; • Considerate post-sales services	• Privacy concerns; • Lack of understanding ; • Technical challenges
	Food Logistics	• Government; • Transportation Operators; • Individuals; • Third-party data providers	• Technology investments and innovation; • Smarter and faster decision making; • Delivering the optimal experience for the customers;	• Cost and privacy concerns; • Technical challenges; • Extent and quality of the available data

Generally speaking, there are five main ways to leverage big data in food network that gain insights into opportunities and challenges and have implications on how organization will have to be designed, organized, and managed.

(1) Creating transparency: As big data in food network become more available across sectors, transparency of data drives transformation, increases productivity and leads to informed decision making.

(2) Enabling experimentation to identify anomalies, detect fraud and improve performance: Big data—much of it unstructured or machine-generated—needs to be collected, integrated and analyzed in real time to discover anomalies and fraud that help organizations improve operations and develop services.

(3) Micro-segmentation to customize actions: Big data make it possible to work through various streams of customer data to enable the definition of increasingly finer segments and take precise marketing to meet customers' needs.

(4) Replacing/supporting decision making and data analysing with automated algorithms: Big data analytics and visualization of automated algorithms allows organizations to find unknown patterns that occur in food network in a time-efficient and cost-effective manner.

Innovating new business models, products, and services: Using vast amounts of data provides new perspectives that can fuel innovation in food products and services, such as offering clues about how customers will behave.

3 Big data harvest model

Although the potential value in big data is tremendous, it is extremely hard for existing analytics to analyze high volume (and variety) of data in real time and produce useful information [6]. Although there are many data techniques might help managers to produce a lot of information, they are unfocused, and hence inefficient. So it is imperative to provide an analytical framework for structures and links various streams of data to create a coherent picture of a particular problem – so that a better insight into the issue been analyzed and could be gained.

Therefore, we propose a better analytic infrastructure to make use of the available big data to gain competitive advantages in food network management (please see Fig. 1). Firstly, we identify the products that could meet future markets from big data analytics; then, we translate products demand into processes and divide processes into tasks and assets; finally, we meet the market demand through chain coordination and continuous evaluation.

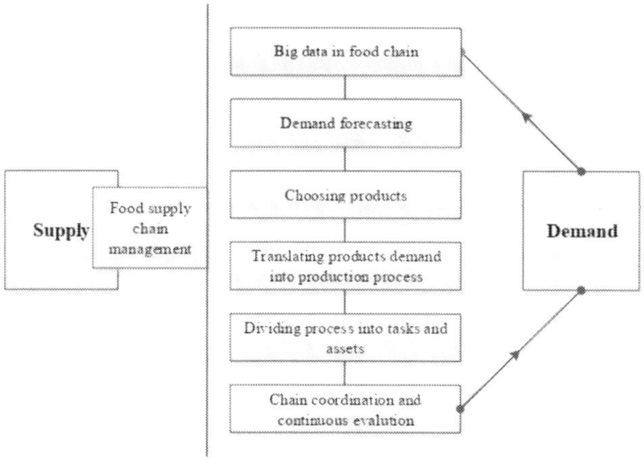

Fig. 1. Decision making based on big data in the food supply chain

For the first part, there are so many methods such as the Delphi method, time series analysis, regression analysis to predict the market demand for food. These methods mainly use historical data to forecast the market demand, but market demand depends on a variety of complex factors, including service quality, consumer groups and government policy. Moreover, these factors can be obtained from big data. If we adopt these factors into consideration, we can improve the precision of prediction to ensure product success. Because Bayesian networks can make effective use of all available data, diagnose what causes high preference and incorporate expert knowledge by representing the relationship among a set of variables [7,22], so that we use the Bayesian networks which link various streams of data in food chain to predict the market demand. Anderson et al. (2004) regard Bayesian network methodology as the implementation mechanism for causal modelling and build a Bayesian network model of customer service satisfaction [8]. Corney (2000) applies Bayesian networks to a typical food design problem and the results show that they are powerful tools to aid consumer preference modelling from a combination of data and expertise [9]. Further applications of Bayesian networks in food production include food security, food risk and consumer behaviours [10-12].

The structure of a Bayesian network is a directed graphical model in which nodes mean random variables of interest and directed arcs represent direct causal or influential relation between nodes [13]. Each node X has a probability distribution $P(X|\pi(X))$ which expresses the uncertainty of the interdependence of the variables, where $\pi(X)$ is the parent set of X ($\pi(X)=\varnothing$ if the node X has no parents). Therefore, together with the independence assumption, for a Bayesian network consisting of n nodes $(X_1, X_2, \cdots, X_n)$, we can factor out joint probability distribution:

$$P(X_1, X_2, \cdots\cdots, X_n) = \prod_{i=1}^{n} P(X_i \mid \pi(X_i)) \quad (1)$$

In particular, production decision will be provided by calculating and analyzing the Bayesian network which is set up based on the big data in the food supply chain. An analytical framework is presented in Fig. 2.

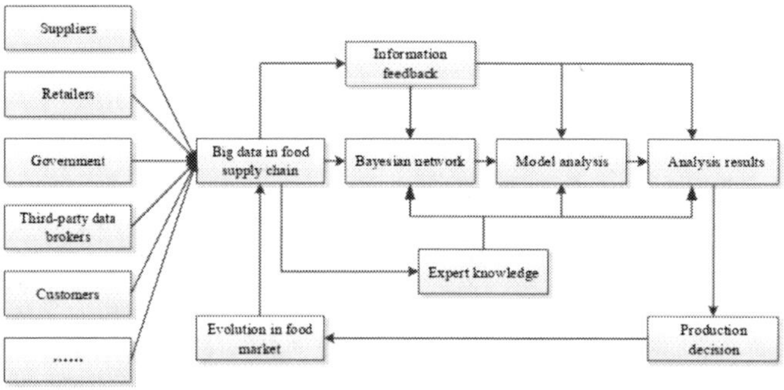

Fig. 2. An analytical framework based on big data in food supply chain

In the second part, to the products demand that the first part is analyzed, we probe into an analytic technique that translates products demand into processes and divides processes into tasks and assets.

Li (1994) proposes an analytic technique called deduction graph model that allows firms to incorporate their own competence sets with other firms [14]. It provides a sequence of optimized expanding process in a visual way by linking different competence sets from various sources [15]. Although this approach has not been adopted in big data analytic area, we have developed it and make it possible to provide the right analytic capabilities to help firms to produce a detailed process design to enhance food supply chain innovation.

4 Numerical Example

Our aim is to develop an optimization model to extract value from big data to improve food supply chain performance, which can also help incorporating capabilities and information (big data) of group decision makers to maximize big data benefits. The following sections describe the detailed application of the proposed analytic approach in a food company.

4.1. Construction of a Bayesian network

A food company is keen to explore how to make use of the value from big data to acquire potential value and enhance their supply chain performance. The company analyzes the market demand through the use of the Bayesian networks. A brief description of the steps is represented below.

The first step is sample selection. Big data in the food supply chain implies a large number of behaviour patterns related to consumers' preferences. In order to select an appropriate sample data from the big data in the food supply chain, the company, combined with prior knowledge from food market, identifies and describes the factors that affect the market demands under advice from experts and decision makers. These factors mainly include food attributes and the chemical and physical properties of the product related to these attributes [16]. Once these factors are determined, they will be the nodes $(X_1, X_2, \cdots, X_n)$ in the Bayesian network. Based on these factors, the company collects m representative consumers, where each consumer contains a value assignment for each factor.

The second step is to pre-processing the sample data. The values of factors need to be discrete by adopting the clustering algorithm or hierarchical category before modelling in order for the propagation and inference algorithms in the next couple of sections.

The third step is designed to build a Bayesian network. Building a Bayesian network includes two parts. One is to identify the network structure. The other is to determine the conditional probability table. A selection of search algorithms which can be used in learning of the Bayesian networks is shown in Table 2 [17].

Table 2. Search techniques of learning the network

Structure	Observability	Method
Known	Complete	Maximum-likelihood estimate
Known	Partial	Expectation maximization or a "greed" hill-climbing method, such as K2
Unknown	Complete	Search in model space
Unknown	Partial	Structural algorithm of expectation maximization or bound contraction

Figure 3 presents a part of a Bayesian Network for food preference, though it will be more complex in practice [12].And it is built by using the K2 algorithm [18]. Given the data D that the previous parts are processed, a Bayesian network is set up that maximizes $P(S^h \mid D)$.

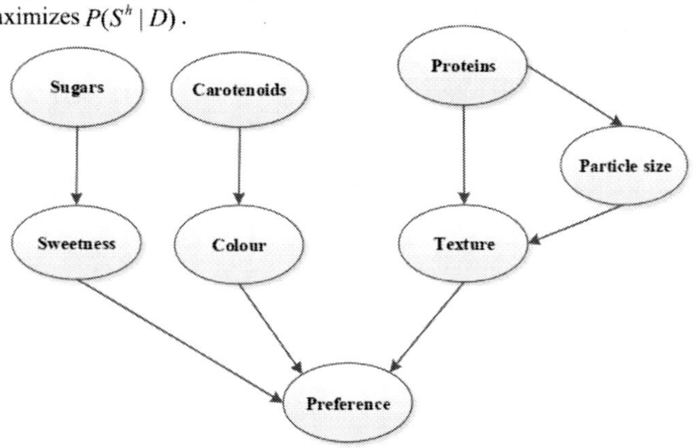

Fig. 3. Part of a Bayesian network for food preference

Based on the Bayesian theorem, then we have

$$P(S^h \mid D) = \frac{P(S^h)P(D \mid S^h)}{P(D)} \tag{2}$$

Where S^h represents an arbitrary network structure containing the variables in D. $P(D)$ is a constant independent of the network structure. $P(S^h)$ is the prior probability. In addition, under the assumptions of unconstrained multinomial distribution, the independence of parameters, Dirichlet prior distribution $Dir(\alpha)$ and data integrity, $P(D \mid S^h)$ is derived as

$$P(D \mid S^h) = \prod_{i=1}^{n} \prod_{j=1}^{q_i} \frac{\Gamma(\alpha_{ij})}{\Gamma(\alpha_{ij} + N_{ij})} \prod_{k=1}^{r_i} \frac{\Gamma(\alpha_{ijk} + N_{ijk})}{\Gamma(\alpha_{ijk})}, (i = 1, 2, \cdots n, j = 1, 2 \cdots, q_i; k = 1, 2, \cdots r_i) \tag{3}$$

Where a variable x_i has r_i possible value, π_i is the parent of x_i, N_{ijk} is the number of consumers in D when $x_i = v_{ik}$ and $\pi_i = \omega_{ij}$.

Once such the network structure is found, numerical conditional probability table should be determined. Let θ_{ijk} denote the conditional probability $P(x_i = v_{ik} \mid \pi_i = w_{ij})$. Let ξ denote the two assumptions: 1) there is no missing values; 2) θ_{ij} is independent and has Dirichlet prior distribution. Therefore, $P(\theta_{ij} | D, S^h, \xi) = Dir(\theta_{ij} \mid \alpha_{ij1} + N_{ij1}, \alpha_{ij2} + N_{ij2}, \cdots \alpha_{ijr_i} + N_{ijr_i})$
(4)

The fourth step is to forecast the market demand. A Bayesian network is a bi-direction inference method where inputs can predict the outputs and vice versa [19]. So given the values of the observed nodes, the company calculates the probability distribution of the target nodes to predict the demand for food or diagnose the likely causes of a perfect product. Then, the company identifies what kinds of food can satisfy most customers' preferences and have vast potential for future development.

Moreover, Bayesian network can be further updated to respond to the changing market demand. When new data are obtained, the company can continuously refine the Bayesian network by modifying some local part of it, so that the company is able to quickly change existing running processes to satisfy the customer requirements.

4.2 Deduction Graph Model

Specifically, the company identifies five different types of foods that will satisfy most customers' preferences through the Bayesian network analysis. The identified products are: A, B, C, D, E. The company also identified the features of the foods and the relevant production processes (raw materials, machines, skills and so on) needed to manufacture the five different foods i.e. a, b, c, d, e, f, g, h, i, j, k, l, m, and n, with each of a, b, c, d, e, f, g, h, i, j, k, l, m and n representing a unique required production process, respectively.

Specifically, different types of foods require different production processes to produce. Table 3 shows the needed production processes to make a specific product. For example, to produce C will require d, h, m and n.

Table 3. Different production processes required by products(**"** √**" means** ***required)***

	a	b	c	d	e	f	g	h	i	j	k	l	m	n
A						√	√							
B						√	√		√					
C				√				√					√	√
D		√	√							√				
E						√					√	√		

Having identified the required production processes for different products, both factory managers are asked to point out the existing production processes available in departments A and B. The existing production processes of department A (S_A) are identified as: c, d and e. Whereas, the existing production processes of department B (S_B) are: a, b and f.

A quick analysis shows that both departments A and B don't have all the required production processes to produce the five newly identified foods. Thus, to make foods that require new production processes, the departments should purchase the production processes from other departments or expand its existing production processes. The selling price for production processes in each department is estimated in Table 4. For example, the selling price for production process c in department A is 1 unit, and 1.5 units for production process f in department B.

Table 4. The selling price for each production process

	c	f
Department A	1	
Department B		1.5

Based on the selling price, the expanding cost for department A is shown in Table 5 (a), and for department B in Table 5 (b). The expanding cost for buying new production processes takes into account of the time, labour, energy, funds and so on. There are also compound nodes, such as d ^ e and a ^ b. In order to produce the new foods, the needed production processes will be obtained by learning from existing production processes or by purchasing from other departments directly.

Table 5 (a). Production process expanding cost for A (A owns competence c, d, and e)

	f	g	h	i	m	n
c	2	1.5		1		
d			2	1.5		
e	2.5				1	
f		1				2
g				1.5		
h						2.5
i		1				
m						

n	1		1.8			
d^e	1		1			1.5

Table 5 (b). Production processes expanding cost for B (B owns competence a, b, and f)

	c	j	k	l
a	1.8		2	
b			1.8	
c		0.8		
f	1.5	1		1
j	1			
k				1
l		1.5		
a^b	1		1.5	

Based on the above analysis, the two manufacturing departments should focus on different product families. From the production processes learning costs, we can figure that department A is more suitable to manufacture A, B and C, whereas, the department B should responsible for D and E producing. Table 4 shows the foods to be produced in departments A and B. In the Table 6, A, B and C are denoted as X1, X2, X3 respectively, whereas D and E are denoted as Y1 and Y2.

Table 6. Products in departments A and B

Department A	
X1	A
X2	B
X3	C
Department B	
Y1	D
Y2	E

The possible earning revenue for a different product mix is listed in Table 7. For instance, if department A makes food X1 and department B makes food Y1, the possible profit earned by A is 4.5 and the possible profit earned by B is 3. The assumption is that both departments are willing to collaborate. They are ready to communicate to achieve the entire maximum profit.

Table 7. Revenues for mixed product set

	Department A	Department B
X1, Y1	4.5	3
X1, Y2	5	3.5
X2, Y1	6.5	4
X2, Y2	6	3.5
X3, Y1	7	3
X3, Y2	8	4

4.2.1 The Competence Network

A production processes network can brightly depict the possible ways of expanding a production process to manufacture new foods [14]. The network developed in this case contains compound nodes and considers a cyclical situation. Fig. 4(a) shows the expanding process of department A to produce X1, X2, or X3, and Fig. 4(b) shows the expanding process of department B to produce Y1 or Y2 based on its current production processes a, b, and f.

Each node represents each production processes. The arc shows there is a connection between the two nodes, such as, a → c means production process c can be learned from production process a. As for d and m, there is no arc between these nodes, denoting that to learn d from m or to learn m from d is almost impossible. The number on the arc means the cost spent on obtaining the production processes. There are also compound nodes, such as d ^ e and a ^ b. The compound node can only be used when the decomposed nodes are obtained. In order to produce the new foods, the needed production processes will be obtained by learning from existing production processes or by purchasing from other departments directly. For example, in Figure 4(a), the production process f can be learned from production processes e, c, and d ^ e with the cost of 2.5, 2, and 1, respectively. But A also can purchase production process f from department B with the cost of 1.5. Also, e → f → g → i shows the learning sequence indicating that the learning process starts from e, learns f, then learns g, and then leans i from g. The final objective of the production process network is using optimization way to find the best sequence with the highest profit in food production.

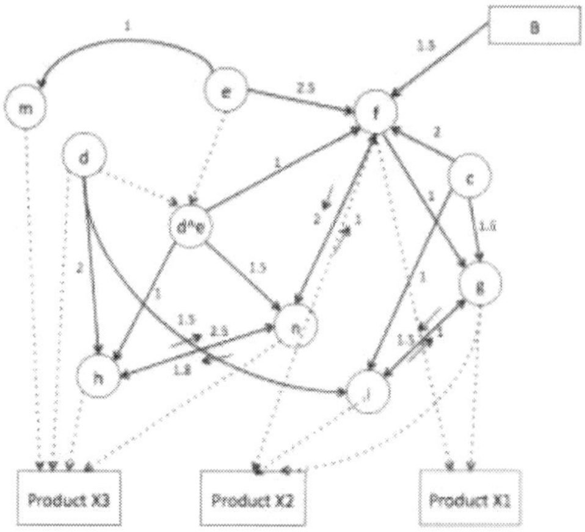

Fig. 4(a). Network of department A (existing production processes are c, d, and e)

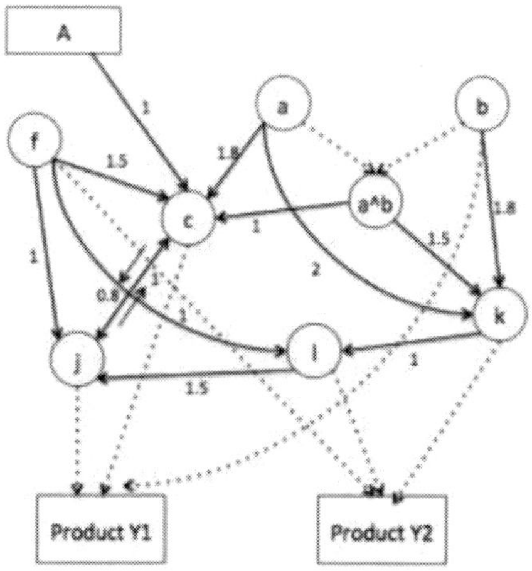

Fig. 4(b). Network of department B (existing production processes are a, b, and f)

4.2.2 Network flow approach

The results of the example problem can be formulated as the linear mixed 0-1 optimization model [14]. However, when the size of the problem is increasing, mixed 0-1 programming is not running quickly enough. Kim & Hooker (2002) indicate that a minimum-cost flow problem is already well suited for mixed 0-1 programming and can be solved better and faster with its advantage increasing with problem size [19]. So we translated the deduction graph model into a minimum-cost flow problem to find an optimal solution.

Let S is the set of department's existing production processes, T is the set of required production processes for products, I is the set of intermediate production processes. We define a directed graph $G=(V,E)$, $V=S\cup I\cup T$. Given a node i in $G=(V,E)$, $r(i,j)$ is the arc connecting j from i, and $w(i,j)$ is the corresponding cost of obtaining j from i.

To the graph $G=(V,E)$, we add a starting node s_0 and a terminal node t_0, then connect s_0 and $s\in S$, connect t_0 and $t\in T$ and get a new directed graph $G=(V',E')$. Let n is the value of $|T|$. So the capacity and the costs of edges are defined as

$$c(i,j)=\begin{cases}1, & if\ r(i,j)=r(t,t_0),\\ n, & otherwise,\end{cases}$$

$$w(i,j)=\begin{cases}0, & if\ r(i,j)=r(s_0,s)/r(t,t_0),\\ w(i,j), & otherwise.\end{cases}$$

And $f(i,j)$ denotes the flow of $r(i,j)$, then the minimum cost flow model is given by the following:

$$\min \sum_{(i,j)\in E'} w(i,j)f(i,j) \tag{5}$$

$$\text{s.t.} \sum_x f(i,x) - \sum_x f(x,i) = \begin{cases} n & \text{if } i = s_0, \\ -n & \text{if } i = t_0, \\ 0 & \text{otherwise.} \end{cases} \quad (6)$$

$$0 \le f(i,j) \le c(i,j) \quad (7)$$

There are many effective algorithms for solving the minimum cost flow problem [20]. After the optimal solution is found, the subgraph G' in which the flow of each arc is positive is the approximate solution of the problem. Based on the subgraph G', the company, combined with the earning revenue for a different product mix, selects an optimal combination of production strategies to obtain the largest benefits.

To solve the above problem of the food company by transforming into a minimum-cost flow problem [21], the solution listed in Table 8 is that department A chooses X3 and department B chooses Y2 respectively. Fig. 5(a) and 5(b) show the corresponding deduction graph. Take X3, for instance, the required production processes set for producing X3 are $\{d,m,n\}$ in which processes are all expanded by department A itself at the cost 3.5. The profit of producing X3 is

Profit(X3) = Revenue (X3|X3,Y2) - $\text{Cost}_A(\{c,d,e\},\{d,m,n\})$ =8-3.5=4.5

Table 8. Solutions of the example problem

	Department A	Department B
Selected product	X3	Y2
Required processes sets	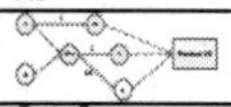	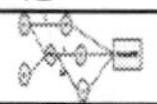
Processes developed by itself	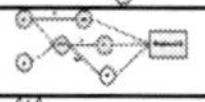	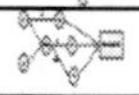
Processes purchased from other departments		
Total cost	3.5	2.5
Income	8	4
Profit	4.5	1.5

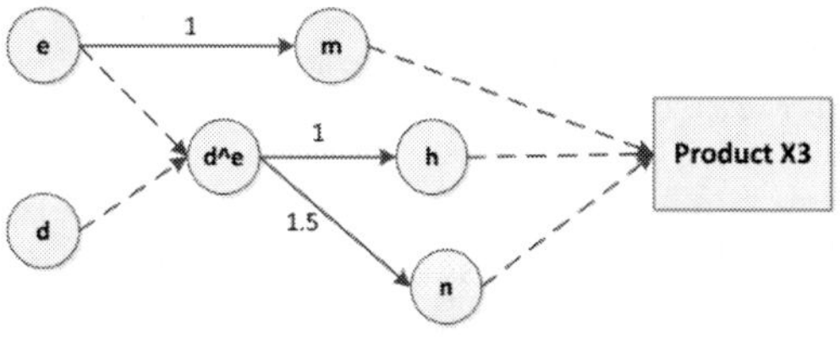

Fig. 5(a). Deduction of graph A

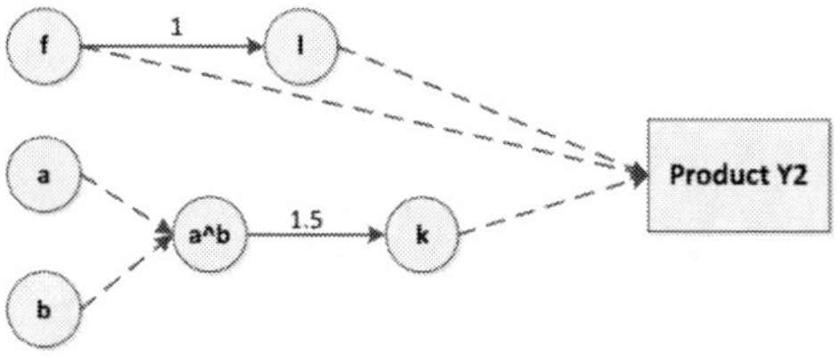

Fig.5(b). Deduction of graph B

Moreover, deduction graph model can be expanded to solve multi-level food quality problems, in which there are multi-levels of proficiency for the foods taken into deliberation. For example, the food X1 may have multi-quality levels designated by $X1_1$(Normal), $X1_2$(Good), $X1_3$(Excellent). Likewise, the food X2, X3, Y1 and Y2 may have multi-quality levels (i.e., $Y1_1$(Normal), $Y1_2$(Good), $Y1_3$(Excellent)). Each different level of food quality may lead to different results of reputation, intension of government's supervision and customer satisfaction. In this way, the proposed model can help us to select a feasible way, so that the expansion from the initial production processes to final products (five identified foods) can be reached at the lowest cost and the optimum proficiency of the food quality.

Furthermore, this model can be further developed to an optimization approach of incorporating information/skills/service/products (big data) of group decision makers to reap the entire maximum profit. In this way, it works well in cyclic situations and can be used in analyzing efficient information transmitting control of the food network.

4.3 Discussion

DOI: 10.1051/matecconf/201710002048

Big data analytics in food network makes it possible to discover needs and create value, which has implications on how organization will have to be designed, organized and managed. Hence, we develop a big data harvest model that links large amounts of data to create a coherent picture of a particular problem--having identified the products that can meet future markets from big data and then identified the required production processes to produce the products.

On one hand, comparing with other analytical approaches, Bayesian networks have a number of features that make them suitable for demand forecasting. The results indicate Bayesian networks are valuable tools for representing the relationship among a set of variables from a combination of big data and expertise in the food supply chain. Through Bayesian network analysis, the food company can build a Bayesian network for food preference, find the types and features that a food product must have in order to be preferred and decide what to produce.

On the other hand, once the company identifies the types of foods that can meet future markets, the next steps the company must translate products demand into production processes and divide processes into tasks and assets. We develop the deduction graph model and make it possible to provide the exact analytic capabilities to help firms to produce a detailed process design. The results indicate that the deduction graph model can effectively help the food company to select the product produced by each department and combine departments' respective production processes to make such products to maximize their profits. The results also indicate that network flow approach can be used to find the optimal solution of the deduction graph model with fast specialized algorithms. The optimal solution is that department A produces X3 and department B produces Y2, the corresponding profit, respectively, is 4.5 and 1.5.

5 Conclusions

In this paper, we propose a big data harvest model that converted data into sights to gain competitive advantages in food supply chain management. The purposes of this study are twofold. One of the goals is to use big data in the food supply chain as inputs to make production decisions. The other is to apply the deduction graph model to translate products demand into processes and divide processes into tasks and assets. First, using Bayesian network can integrate the prior information and sample information in the food supply chain and find a cause-and-effect relationship between data to effectively predict the market demand and direct food production.

Second, the results indicate a deduction graph model is capable of incorporating production processes of departments to realize the profit maximization. In order to find the optimal solution, the deduction graph model can be translated into a minimum-cost flow problem. We simply illustrate the application framework of using big data to make more informed production decisions in the food supply chain, however, it is necessary to provide technological support such as information-gathering techniques and Bayesian network inference techniques when the company plans and implements the application framework. What's more, the application of big data in other areas of the food supply chain should be addressed through further research.

Acknowledgement

This research was supported by the National Natural Science Foundation of China (NSFC) (Project Nos. 71571151 and 71371159), the National Planning Office of Philosophy and Social Science of China (Project No. 14AGL015).

References

1. Tien, J. M. (2013). Big data: Unleashing information. Journal of Systems Science and Systems Engineering, 22(2): 127-151
2. Taylor, D. H. & Fearne, A. (2006). Towards a framework for improvement in the management of demand in agri-food supply chains. Supply Chain Management: an International Journal, 11(5): 379-384
3. Tien, J. M. (2012). The next industrial revolution: integrated services and goods. Journal of Systems Science and Systems Engineering, 21(3): 257-296
4. Anica-popa I. (2012). Food traceability systems and information sharing in food supply chain. Management & Marketing, 7(4): 750-759
5. Yu, M. & Nagurney, A. (2013). Competitive food supply chain networks with application to fresh produce. European Journal of Operational Research,224(2): 273-282
6. Tien, J. M. & Goldschmidt-Clermont, P. J. (2009). Healthcare: A complex service system. Journal of Systems Science and Systems Engineering, 18(3): 257-282
7. Heckerman, D., Mamdani, A. & Wellman, M. P. (1995). Real-world applications of Bayesian networks. Communications of the ACM, 38(3): 24-26
8. Anderson, R. D., Mackoy, R. D., Thompson, V. B. & Harrell, G. (2004). A Bayesian Network Estimation of the Service - Profit Chain for Transport Service Satisfaction. Decision Sciences, 35(4): 665-689
9. Corney, D. (2000). Designing food with bayesian belief networks. In Evolutionary Design and Manufacture. Springer London

10. Stein, A. (2004). Bayesian networks and food security-an introduction. Frontis, 3: 107-116
11. Albert, I., Grenier, E., Denis, J. B. & Rousseau, J. (2008). Quantitative Risk Assessment from Farm to Fork and Beyond: A Global Bayesian Approach Concerning Food - Borne Diseases. Risk Analysis, 28(2): 557-571
12. Van Boekel, M. A. J. S. (2004). Bayesian solutions for food-science problems?. Frontis, 3: 17-27
13. Pearl, J. (1986). Fusion, propagation, and structuring in belief networks. Artificial Intelligence, 29(3): 241-288
14. Li, H. L. (1999). Incorporating competence sets of decision makers by deduction graphs. Operations Research, 47(2): 209-220
15. Li, H. L. & Yu, P. L. (1994). Optimal competence set expansion using deduction graphs. Journal of Optimization Theory and Applications, 80(1): 75-91
16. Wolters, C. J. & Van Gemert, L. J. (1989). Towards an integrated model of sensory attributes, instrumental data and consumer perception of tomatoes. Part I. Relation between consumer perception and sensory attributes. In Workshop on Measuring Consumer Perception of Internal Product Quality, 259: 91-106.
17. Bidyuk, P. I., Terent'Ev, A. N. & Gasanov, A. S. (2005). Construction and methods of learning of Bayesian networks. Cybernetics and Systems Analysis, 41(4): 587-598
18. Cooper, G. F. & Herskovits, E. (1992). A Bayesian method for the induction of probabilistic networks from data. Machine Learning, 9(4): 309-347
19. Lu, J., Bai, C. & Zhang, G. (2009). Cost-benefit factor analysis in e-services using bayesian networks. Expert Systems with Applications, 36(3): 4617-4625
20. Kim, H. J. & Hooker, J. N. (2002). Solving fixed-charge network flow problems with a hybrid optimization and constraint programming approach. Annals of Operations Research, 115(1): 95-124
21. Papadimitriou, C. H. & Steiglitz, K. (1998). Combinatorial optimization: algorithms and complexity. Courier Dover Publications
22. Jensen, F. V. (1996). An introduction to Bayesian networks. London: UCL press

Study on Effects of Incentive Factors on supply chain Performance in E-commerce Environment

Huiping Mu and Guopeng Yu

School of Management Engineering, Zhengzhou University, Zhengzhou 450001, China

Abstract. The past studies aimed at the effects which related incentive factors of supply chain mad-e on supply chain performance are mostly concentrated on the manufacturing sector,Few literature regards the e-commerce system leading enterprise as the core enterprise of supply chain. In the e-commerce environment, supply contract, trust, relationship commitment is regarded as a motivating factor in Inter-enterprise supply chain and the operating conditions with suppliers, financial situation of enterprises in two dimensions are used to measure the performance of the supply chain, and then construct the theoretical model assumptions .Using structural equation modeling empirical analysis,the results show that the supply contract, trust, relationship commitment have significant positive effects on suppliers operating performance and supply contracts have a significant positive impact on corporate financial performance; Effects of trust, relationship commitment and supplier operational performance on corporate financial performance are not significant.

Keywords. E-commerce system, incentive factor, Supply Chain Performance, Structural equat-ion model.

1 Introduction

In the supply chain management practice,In order to maximize its own interests and maintain the competitive advantage, Each node enterprise on supply chain have difficult to carry out effective cooperation[1]. So as to arouse the enthusiasm of each node enterprises on supply chain and improve supply chain performance, establish a reasonable incentive mechanism of supply chain is an effective way.

Supply Chain incentive mechanisms relevant factors in supply chain performance impact on the domestic and foreign scholars from different angles to answer. Wang xia Yang[2]and others believe that contract incentive, information sharing, coordination in the supply chain plays a vital role.Chen Shuzhen[3-4],who studied the impact of optimal supply chain promotion and compensation investment promotion for the supply chain contract design under dual channel mode, which proposed the optimal combination of supply chain incentive contract design. Some scholars have analyzed the causes of the supply chain, think through each node enterprise incentive analysis, a good supply chain information sharing incentives can significantly improve supply chain performance[5-7]..

However, the view is mostly based on manufacturing background, manufacturing enterprises as the core enterprise supply chain, but with the continuous development of e-commerce, enterprise supply chain management is also a result of e-commerce applications in various industries occurred very big changes, which increasingly reflected in e-commerce environment of the supply chain management[8-9].Ye Azhen[10]and others, who studied the problem of optimal incentives core enterprise e-commerce environment of the supplier's activities and information sharing activities that affect the core of the enterprise's own characteristics generate greater sharing of information, the greater the incentive coefficient, the greater the impact on supply chain performance. Liu Anwen[11]and others studied the incentives of supply chain e-commerce environment design from the agency's perspective, research shows that the core business of the agency cost earnings and asymmetric information in case of a great relationship. Shaik[12]and others, who analyzed the development trend of supply chain,they believe that the emergence of the internet and related technology changed the way of global business,and the sustainable development of supply chain management can not do without the e-commerce. But in terms of supply chain performance, Shkoukani[13]believed that supply chain management with the development of e-commerce is also changing after studied the changes in suppliers, customers and product flow, the relationship between supply chain integration and corporate performance.

Based on foregoing analysis,the paper orientation of core enterprise in supply chain management as the dominant enterprise in e-commerce system,selecting supply contract, trust and relationship commitment factors as supply chain incentive mechanism,and using vendor operating performance, the financial situation of the core business in two dimensions to measure the performance of the supply chain. Build a theoretical model and studid Supply chain incentive relevant factors on the supply chain performance.

2 Research hypothesis and model foundation

2.1 Supply contracts and supply chain performance

Building supply contract is to be able to develop a stable supply chain between enterprises and improve supply chain performance.Under reasonable conditions of supply contracts, e-commerce systems in the enterprise to respond quickly to market changes and meet customer needs, improve operational efficiency. Hou Yali[14]and others found a reasonable set about contract parameters of price subsidy can improve vendor enthusiasm, it have a significant impact on improving performance.Wang Xiayang[15]thinks that the reasonable design of supply contract has significant effect to improve supply chain performance.Thus, the paper propose the following hypothesis:

Hypothesis 1:Supply contracts have a significant positive impact on suppliers operating performance（H1）;

Hypothesis 2:Supply contracts have a significant positive impact on corporate financial performance（H2）.

2.2 Trust and supply chain performance

Trust is the precondition of cooperation, but also the key factor for win-win cooperation.Supply chain, the higher the trust level between each node enterprises of supply chain, the lower the transaction cost will be, thus improving the ability to respond quickly to the entire supply chain, and ultimately affect the overall performance of the supply chain.Ye Fei [16]and other studies have shown that trust each node in the supply chain between enterprises contribute to bilateral cooperation. Pan WenAn[17] and others found a significant relationship commitment through positive impact on organizational performance in the fierce competition in the market.Thus, the paper propose the following hypothesis:

Hypothesis 1: Trust has a significant positive impact on the operational performance of suppliers（H3）;

Hypothesis 2: Trust has a significant positive impact on corporate financial performance（H4）.

2.3 Relationship commitment and supply chain performance

Relationship commitment is considered to be the basis to establish long-term cooperative partnership.A high level of relationship commitment will enhance the trust between enterprises, reduce the incidence of opportunistic behavior, and have a significant impact on the share of each node enterprise between information and profit.Ye Fei [18]explored the relationship commitment on supply chain operational performance, we found that the relationship commitment between supply chain partners have significant positive impact on operational performance. However, some scholars have come to different conclusions.For example,Craig and Marianne found that the relationship commitment to supply chain nodes does not significantly affect the performance of cooperation[19].Based on the above study, we proposed the following hypothesis:

Hypothesis 1: Relationship commitment have significant positive impact on suppliers operating performance（H5）;

Hypothesis 1: Relationship commitment have significant positive impact on corporate financial performance（H6）.

2.4 Suppliers operating performance and corporate financial performance

Under e-commerce environment, the market competition becomes increasingly fierce, which requires companies to be more flexible, fast, high-quality product improvement, timely and accurate delivery of products to meet different consumer needs.So suppliers operating performance is considered to be in supply chain management decisive factor. Suppliers only timely and rapid response to the needs of downstream business supply chain, improve product quality, reduce inventory in order to reduce costs and improve ROI.Thus, the paper propose the following hypothesis:

Hypothesis 1: Suppliers operational performance has a significant positive impact on corporate financial performance（H7）.

Based on the above analysis and assumptions, we propose a conceptual model of this study, shown in Figure 1.

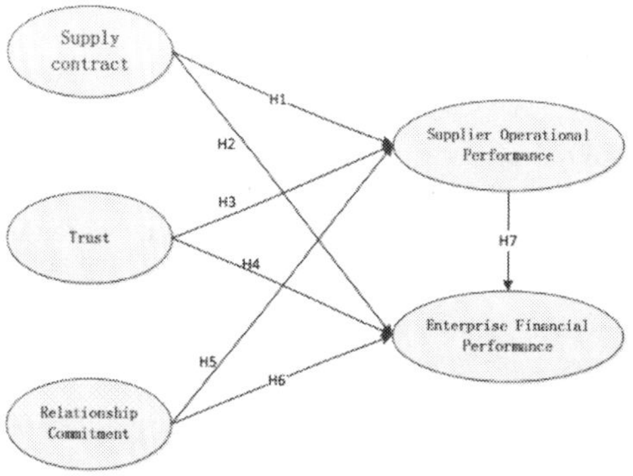

Fig1 .The conceptual model

3 Research design and analysis

3.1 Questionnaire Design and Data Collection

For the supply chain incentives and the study of supply chain performance , home and abroad have proven Q-item scale. While learn from home and abroad have been employed at the same scale, considered that we are ragard the e-commerce ecosystem of leading enterprise as a core enterprise in supply chain , paper is assumed to be the e-commerce ecosystem of leading enterprise and its main supplier of secondary supply chain,and survey involving materials, food, machinery, medicine, electronics and other 18 industries about e-commerce business in senior management, sales manager, in charge of logistics and supply chain operations and related personnel.Scoring systems and indicators using likert Scale for five, according to the actual situation review from "completely disagree" to "completely agree."

The main scope of the study is the e-commerce ecosystem of leading enterprise and its main suppliers in first line, second-tier cities , so there is no geographical restrictions.using a combination of email and the field survey, issued a total of 700 copies, of which e-mail 500 parts, 200 parts of paper questionnaires.Recycling 287 parts excluded for obvious errors and incomplete answer after 29 questionnaires were collected from 258 parts,which Effective recovery rate was 36.9%.

3.2 Correlation analysis

With SPSS21.0 to analyze three relevant factors about supply chain and supplier operational performance, financial performance. From Table 1, the correlation coefficient between 0.487-0.740, where the supply contract and corporate financial performance correlation weaker than other factors , while the most relevant between relationship commitment and supplier operational performance, and relations between the various other factors are significant.From the overall, good correlation.

Tab1. The correlation coefficient matrix

	Supply contract	Trust	Relationship Commitment	Supplier Operational Performance	Enterprise Financial Performance
Supply contract	1				
Trust	.646**	1			
Relationship Commitment	.606**	.680**	1		
Supplier Operational Performance	.708**	.686**	.740**	1	
Enterprise Financial Performance	.487**	.592**	.606**	.650**	1

Note:** At .01 level (bilateral) significant correlation.

3.3 Reliability and validity analysis

Reliability analysis, also known as reliability analysis, to measure the results of stability and consistency.In this paper, Cronbach ' s α coefficient to test the internal consistency of variables. α is more than 0.7 is generally considered feasible. Table 2 is made, Cronbach ' s α values in excess of 0.7, show that the measurement model has good reliability.

Questions of the questionnaire and evaluation are most used through empirical research results at home and abroad,thus ensuring content validity of the questionnaire. But the terms of construct validity, taking into account the characteristics of the contents of this paper, we consulted the relevant scholars, the questionnaire part of the structure, content was amended with SPSS21.0 factor analysis.From the table 2, load factor and AVE value of each Q-item is greater than 0.5, this shows that the inherent quality of the model is good,and measurement model has good construct validity.

Tab2. The reliability analysis and confirmatory factor analysis

Variable	Q-item	Means	Cronbach's α	Load factor	*AVE*
Supply contract	*SC1*	3.362	0.723	0.771	0.608
	SC2	3.576		0.775	
	SC3	3.543		0.551	
	SC4	3.610		0.550	
	SC5	3.562		0.951	
Trust	*Tr1*	3.600	0.845	0.789	0.623
	Tr2	3.648		0.723	
	Tr3	3.505		0.608	
	Tr4	3.662		0.498	
Relationship Commitment	*RC1*	3.552	0.844	0.736	0.701
	RC2	3.581		0.716	
	RC3	3.724		0.690	
Supplier Operational	*SOP1*	3.538		0.740	0.612

Performance	*SOP2*	3.595		0.680	
	SOP3	3.610	0.845	0.663	
	SOP4	3.595		0.564	
	SOP5	3.624		0.610	
Enterprise Financial Performance	*FP1*	3.591		0.614	
	FP2	3.505		0.705	
	FP3	3.586	0.861	0.790	0.663
	FP4	3.533		0.805	
	FP5	3.581		0.811	

4 Hypothetical model processing

4.1 model fit test

According Bogazzi and Yi's[20]point of view,we can find the basic goodness of fit, the overall model fit and internal structural adaptation which three test model whether is well suited for the actual data.

4.1.1 Basic adaptation degree

Basic adaptation of a standard model include four aspects: no negative measurement error, measurement error reaches a significant level, standard error control in the lowest range and the load factor must be between 0.5 to 0.95.Here in each latent variable measure of factor loadings are between 0.5 to 0.9, and the error variable is greater than 0, indicating that the theoretical model has a good basic adaptation degree.

4.1.2 The overall adaptation degree

Overall model adaptation degree is used to test the sample data and the degree of entire model.Generally it believed that the overall model adaptation degree measure include Absolute adaptation index, value-added adaptation index and Simple adaptation index.Overall modeladaptation degree in this paper are shown in Table 3, the overall adaptation in good condition.

Tab3. Model overall adaptation degree

	Adaptation index	**The model adaptation value**	**Acceptable values**
Absolute adaptation degree	Chi-square value X^2	292.546	The smaller the better
	RMR	0.032	<0.05
	RMSEA	0.041	<0.05
	GFI	0.957	>0.9
	AGFI	0.942	>0.9
Value-added adaptation degree	*NFI*	0.977	>0.9
	RFI	0.968	>0.9
	IFI	0.935	>0.9
	TLI	0.982	>0.9
	CFI	0.973	>0.9
Simple adaptation	*PNFI*	0.743	>0.5

degree	*PGFI*	0.649	>0.5
	PCFI	0.686	>0.5

4.1.3 Model of the internal structure of fit

Inner structure adaptation degree is used to evaluate the reliability and validity of each indicator variables, estimate significant degree of parameters .It is generally believed that the composite reliability of potential variable is greater than 0.6, the average variance extracted values of latent variables (AVE) is greater than 0.5, the project reliability of individual observed variables is greater than 0.5, the model can explain the internal structure adaptation degree is good.

4.2 hypothesis test

According to the model after AMOS 21.0 hypothesis proposed for analysist,Get the results shown in Table 4, hypothesis that H1, H2, H3, H5 is verified, the P value of H4, H6, H7 is not significant.

Tab4 .Model hypothesis test results

Hypothesis	Path Direction	Standard path coefficients	P value	Conclusion
H1	Supply contract→Supplier Operational Performance	0.496	0.026*	support
H2	Supply contract→Enterprise Financial Performance	0.812	0.032*	support
H3	Trust→Supplier Operational Performance	0.453	***	support
H4	Trust→Enterprise Financial Performance	0.157	0.076	support
H5	Relationship Commitment→Supplier Operational Performance	0.522	***	support
H6	Relationship Commitment→Enterprise Financial Performance	0.224	0.085	fefuse
H7	Supplier Operational Performance→Enterprise Financial Performance	0.539	0.078	fefuse

Note: The significant level * P <0.05, ** P <0.01, *** P <0.001

5 Discussion

Through theoretical model is verified by Table 4, we found that the supply contract to supplier operational performance and corporate financial performance has a significant impact. This means that H1 and H2 are supported.This article select revenue sharing, quality guaranteed, price subsidies, minimum order quantity commitment and information sharing as the parameters of the supply contract, fit the requirements change of supply chain management under the e-commerce environment , the resonance between each node enterprises in supply chain.This resonated with each node in the supply chain between enterprises.Thus the supply contract to improve the supply chain performance of the core businesses and suppliers have a significant impact.

Trust is the cornerstone of supply chain cooperation.Many scholars regard manufacturing enterprises as the core of supply chain enterprise , to study the effect of trust on supply chain performance is significant, and has been confirmed.The core business in supply chain is positioned as the dominant firms of e-commerce ecosystem , the results show, H3 was supported. This means that we can build trust mechanism between the supply chain node enterprises, enhance the relationship of mutual trust, promote the improvement of supplier performance, thus improving the performance of the entire supply chain.

Relationship commitment is a desire on the supply chain node enterprises to maintain cooperative relations.H5 is supported,So our supplier of core enterprise promises to maintain cooperative relations as an incentive,so as to maintain long-term stable relationship.

From Table 4, H4, H6, H7 have been rejected, the reason may be to improve the financial performance of corporate governance is the result of factors strategy development, market environment, partnerships and other joint action.

References

1. Ma Shihua,Lin Yong.Supply Chain Man-agement(3rd Edition).Beijing:Machin-ery Industry Press,2010.
2. Wang Xiayang.dynamic coordinate of c-ontract incentives, information sharing,supply chain.Management World,2005,(4):106-115.
3. Chen Shuzhen,Xiong zhongkai,LiangXi. Incentive Compensation Dual channel supply chain coordination contract design.ChinaManagement Science,2009,17(1):64-75.
4. Su Yuning,Chen Chuhung.Incentives in Supply Chain under Random Deman.Operations Research and Management,2010,14(1):150-154.
5. Zhang Xinfeng,Zhao Yan,Xu Guohua.Su-pply study on Chain Information Sharing Incentive Management. Engineering Management journal,2006,(2):123-125.
6. Lee H L ,So K C, Tang C. The Valueof Information Sharing in a Two-level Supply Chain. Management Science, 2000,46(5):626-643.
7. Gupta D,Weerawat W. Incentive Mechanisms and Supply Chain Design for Quick Response. Minneapolis: Univer -sity of Minneapolis,2010,12:135-146 .
8. Song Hua.supply chain management ch-anges in E-commerce Environment.Business Economics and Administration,2013,(12):4-7.
9. Tang Xuan,Wang Weifu.Study on electr-onic supply chain management system based on e-commerce. Technology management research,2006,(3):177-179.
10. Ye a zhen,Lin Yihua,Liu Xiaomin.Two stage supply chain incentive mechanismunder the e-commerce environment.C-hong qing University of Science and technolo-gy journal,2010,(21):85-88.
11. Liu Anwen,Wang YuePing.study on incentive mechanism of Supply Chain Principal agent Theory Electronic Commerce.chong qing university journal,2012,8(2):50-54.
12. Shaikh A,Rafiq M,Iyer R K.Exploring e-Business Trends with Supply Chain Management Perspective. International Journal of e-Education, 2014,4(3):58-64.
13. Shkoukani M, Alnagi E, Abulail R. Comparison between Upstream and Downstream Supply Chain Management and How they are affected by E-business. Oriental Journal of Computer Science and Technology, 2013,6(2):1-8.
14. Hou Yali,Zhou Dequn.Coordination of t-he three segments of the contract price subsidies Supply Chain. Systems engineering.2006,24(4):25-30.

15. Wang XiaYang.Nonlinear cost constraint-s,Contract Design and Supply Chain p-erformance.Modern management science,2008,(5):91-93.
16. Ye Fei,Xu Xuejun.On information sharing and operational performance of the supply chai partnership between the trust and relationship commitment.systems engineering theory and practice,2009,29(8):36-49.
17. Pan wenAn, Zhang Hong.Trust betweensupply chain partners, commitment on cooperative Performance.Psychologicalscience,2006,29(6):1502-1506.
18. Lu Hui, Ye Fei,Qiang Rui.Information on collaborative supply chain resource dependence, trust and relationship commitment.Industrial Engineering and Management, 2010,15(6):7-15.
19. Carig R,Marianne M.Social responsibility and supply chain relationships. Transpo -rtation Research Part E: Logistics and Transportation Review,2002,8(1):37 -2.
20. Bagazzi R P ,Yi Y Y.On the evaluationof struetural equation model. Academic of Marketing Science,1988,16:76-94.

Research on Multiple Particle Swarm Algorithm Based on Analysis of Scientific Materials

Hongwei Zhao[1*], Jingjing Zhao[1] and Abashe caleb tokan[1]

[1]School of Information Engineering, Shenyang University, Shenyang, China
Corresponding Email: zhw30@163.com

Abstract. This paper proposed an improved particle swarm optimization algorithm based on analysis of scientific materials. The core thesis of MPSO (Multiple Particle Swarm Algorithm) is to improve the single population PSO to interactive multi-swarms, which is used to settle the problem of being trapped into local minima during later iterations because it is lack of diversity. The simulation results show that the convergence rate is fast and the search performance is good, and it has achieved very good results.
Keywords. Swarm intelligence, particle swarm, scientific materials, bacterial foraging optimization

1 Introduction

In order to solve actual problems , a number of methods of algorithmic Swarm intelligence have been worked out,2 the most effective of which is Particle Swarm Optimization (PSO)[1]. Enlightened by the biological swarming's activities, such as school of fish, in flocks of birds, and even that of mankinds3,4. PSO is an optimization instrument, carried out to settle a variety of function optimization problems. Being as a skill to solve problems, the main advantage of PSO is its high speed of convergence, which exceeds that of Evolutionary Algorithms (EA) and the rest global optimization algorithms. However, if PSO is used to settle complicated problems may run into trouble as following: in the evolution of population, premature convergence to the local optimum in former generations may happen to all individuals, which could cause low population diversity of population together with adaptation stagnation in latter generations.[2,3]

The PSO algorithm first randomly initialized a group of particles, in each iteration, the particle is updated by following the two extremes: one is the optimal particle itself to find the solution, called individual best; another is the whole population found in the optimal solution, called global extreme value point, and the local PSO algorithm without the only one part of the population, adjacent particles, is the local extreme in the optimal solution. All the neighbors in the PSO algorithm, the position of the particle represents a potential solution to the optimization problem, the particle velocity indicates the direction and distance of flying particles, each particle at the extent of the position by the particle fitness function the value reflected. Particle velocity and position by tracking individual extreme and global extreme update, each iteration update once, to find the optimal solution or the maximum number of iterations is reached so far.[4,5]

We would analyzes this potential by evaluation of MPSO on both mathematical benchmark functions. In our experiments, there are five benchmark functions used to simulation results, which are compared to that of other optimization algorithms, the advantaged of the proposed algorithms would be clearly reported in this paper.

2 Basic flow of PSO algorithm

Particle swarm optimization algorithm is simple and easy to implement, compared with other evolutionary algorithms, the algorithm not only retains the global search Ability, but also to simple speed displacement operation model replaced the complex genetic operation, and has the specific memory function the according to the current search of dynamic adjustment of the search strategy, is a more efficient parallel search algorithm.[6]The basic particle swarm optimization algorithm is as follows:

The first step: to tell the truth within the scope of random initialization of the particle swarm, including random location and speed.

The second step: to calculate the fitness value of each particle.

The third step: for each particle, the fitness value and the best location fitness value are compared, if better, its value as the particle's best individual historical, personal historical best position is updated with the current position.

The fourth step: for each particle, its historical optimal fitness value and the community or neighborhood experience the best location of the fitness value for comparison, if better, then it as the current ah the best position.

The fifth step: according to the formula (1) and (2) to update the speed and position of the particles.

The sixth step: if the termination condition is reached, then the second step is to be reversed, and the termination condition is generally a good enough fitness value or to achieve a predetermined maximum iterative algebra. [7,8]

3 PSO algorithm related control parameters.

The selection of control parameters is very important, because it can directly affect the performance of the algorithm if the parameters are chosen properly.[9]

(1) Inertia weight

The inertia weight is an important parameter for the global search ability and local search ability of particle balance, from equation (3 and on the section of the velocity update in the interpretation of the inertia weight represents the size of particles retained the speed much of its state of motion. The degree of trust when the inertia weight is zero, the equivalent speed and update the particle velocity is independent of particle, the search space will be greatly reduced, only equivalent to the local search ability, and the particle is easy to flight to the same position, lead to local minima. When the inertia weight increases, which increases the current state of the particle retention, thus expanding the search space of particles, enhances the algorithm the global search ability, therefore, inertia weight changes and the choice of initial value is very important, choose appropriate or not directly related to the performance of the algorithm.[10]

(2) Accelerating factor,

Acceleration coefficient is adjusted parameters of their own experience and the experience of particle groups of particle trajectories. The update formula from the velocity of particle swarm algorithm (3.1) shows that the acceleration coefficient can balance the global search ability and local search ability of particle, when two values were increased, the local search capability of particle increases this, and know the inertia weight is just the opposite. At the same time, known as cognitive acceleration coefficient, represents the effect of particle velocity updating experience of its own, which is more dependent on their own experience of the particle is larger, vice versa; known as social acceleration, on behalf of the particle in the population social experience influence on speed, which is more dependent on the particle of social experience is, vice versa. Therefore, to adjust their own experience with the particle population social experience for speed, acceleration coefficient value is ten Important points.

(3) Population size M

The population size m, that the number of particles in a population. When the population size M is 1, namely only one particle in the search. At this time there is no groups between information sharing and cooperation to speak; when increasing the population size m, mutual collaboration cooperation of particle is more, the search algorithm can power will increase; when the population size is too large, it is clear, the search process of the algorithm is also more complex, the calculation time of the algorithm will increase a lot, so convergence speed will slow. Therefore, the size of the population size also need to according to the practical application of the selection of appropriate.[11]

(4) Maximum speed

Maximum speed is used to restrict the particle's velocity parameters, that define the maximum distance of particle in every iteration moves. When the maximum speed is too large, the moving distance of each particle will corresponding is too large, although the search speed of particles increases. Global search ability is enhanced, however, particles are easy to miss the optimal solution.[12,13]

4 The proposed multiple particle swarm optimization algorithm

MPSO modify the single population PSO to the interactive multiple swarms model by incorporating the behaviors of cooperation and competition in bacterial swarming. In the MPSO, respective individual moving in the range of solution space by 3 attractors:

(a) The particle previously best position.

(b) Best position of the particle in its own swarm.

(c) Best position of the particle in neighbor swarms.

By imitating cooperation and competition in bacterial swarming process, MPSO algorithm is proposed. Firstly, the section will introduce two kinds of operation, and then gives the specific implementation steps.[14]

In MPSO, the definition of cooperative operation is similar to standard PSO, each particle close to their optimal positions of its and population. The competitive operation indicates the worst position of particle and population in history. The update algorithm of particle velocity is as follows:

$$v_{id}^{k}(t) = \chi(v_{id}^{k}(t-1) + R_1 c_1 (W_{id} - x_{id}^{k}(t-1)) + R_2 c_2 (W_{gd} - x_{id}^{k}(t-1)) + R_3 c_3 (P_{gd}^{\theta} - x_{id}^{k}(t-1))) \quad (1)$$

$$x_{id}^{k}(t) = x_{id}^{k}(t-1) + v_{id}^{k}(t) \quad (2)$$

W_{id} is the worst position of population ,is the worst position of its history.is the optimal position of global.

5 Summary

In summary, MPSO modifies the single population PSO to interactive multi-swarms model based on analysis of scientific materials.[15]With the theoretical analysis, we've got the results of experiment showing that the MPSO reaches remarkably better results on all the test functions. In order to make a clear demonstration for the MPSO algorithm's performance, we compared it with optimization on several benchmark functions. Then comparison of experimental results shows that the MPSO algorithm is superior to the PSO in terms of accuracy and speed of convergence.

Acknowledgements

The authors are supported financially by the Natural Science Foundation of Liaoning Province (Project No.2013020011) and This work is supported in part by the International S&T Cooperation Program of China (ISTCP) under Grant 2011DFA91810-5 and Program for New Century Excellent Talents in University of Ministry of Education of China under Grant NCET-12-1012.

References

1. J. Kennedy and R. C. Eberhart, Swarm Intelligence, Morgan Kaufmann, San Francisco (2001).
2. Y. Xin. Improved firefly algorithm approach applied to chiller loading for energy conservation. Energy and Buildings, Vol **59** (2013),p.273-278.
3. M.H. Horng. Multilevel minimum cross entropy threshold selection based on the firefly algorithm. Expert Systems with Applications, Vol **38** (2011), p.14805-14811.
4. W. Peng, Image segmentation method based on firefly algorithm and maximum entropy method. Computer Engineering and Applications, Vol **12** (2014), p.115-119.
5. H. W. Tian, F. Xie, and J. M. Ni, Computer Technology and Development **21**, 22 (2011).
6. M. Clerc and J. Kennedy, IEEE Transactions on Evolutionary Computation **6**, 58 (2002).
7. Y. Liu, X. H. Wang, C. M. Xing, and S. Wang, Computer Technology and Development **21**, 19 (2011).
8. D. M. Li and H. H. Shi, A Hierarchical Load Balancing Scheduling Model Based on Rules Computer Science **30**, 16 (2003).
9. R. C. Eberchart and J. Kennedy, A new optimizer using par-ticle swarm theory, Proceeding of the 6th International Sympo-sium on Micromachine and Human Science, Nagoya, Japan (1995), p. 39–43.
10. Z. H. Zhang and X. J. Zhang, A load balancing mechanism based on ant population and complex network theory in open cloud com-puting federation, 2010 2nd International Conference on Industrial Mechatronics and Automation, Wuhan, China (2010), p. 240–243.
11. J. F. Schutte, Particle swarms in sizing and global optimization, Master's Thesis, University of Pretoria, Department of Mechanical Engineering (2002).
12. Y. Shi and R. C. Eberhart, A modified particle swarm optimizer, Proceedings of the IEEE International Conference on Evolution-ary Computation, Anchorage, Alaska, IEEE Press, Piscataway, USA (1998), p. 69–73.
13. Y. Shi and R. C. Eberhart, Parameter selection in particle swarm optimization, Evolutionary Programming VII, edited by V. W. Porto, N. Saravanan, D. Waagen, and A. E. Eiben, Lecture Notes in Com-puter Science, Springer, Berlin (1998), Vol. **1447**, p. 591–600.
14. Tao,X.M, Xu,J.and Yang,L.B. An Improved Hybrid Algorithm Based on Particle Swarm Optimization and K-means Algorithm. Journal of Electronics &Information Technology. **32**, 1,(2010),p.92-94.
15. Suresh,S. Sundararajan,N. Saratchandran, P. A sequential multi-category classifier using radial basis function networks, Neurocomputing **71**, (2008) p.1345–1358.

Laos Organization Name Using Cascaded Model Based on SVM and CRF

Shaopeng Duan[1,2], Lanjiang Zhou[1,2] and Feng Zhou[1,2]

[1]School of Information Engineering and Automation, Kunming University of Science and Technology, Kunming 650500, China
[2]The Key Laboratory of Intelligent Information Processing, Kunming University of Science and Technology, Kunming, Yunnan 650500, China

Abstract. According to the characteristics of Laos organization name, this paper proposes a two layer model based on conditional random field (CRF) and support vector machine (SVM) for Laos organization name recognition. A layer of model uses CRF to recognition simple organization name, and the result is used to support the decision of the second level. Based on the driving method, the second layer uses SVM and CRF to recognition the complicated organization name. Finally, the results of the two levels are combined, And by a subsequent treatment to correct results of low confidence recognition .The results show that this approach based on SVM and CRF is efficient in recognizing organization name through open test for real linguistics, and the recalling rate achieve 80.83%and the precision rate achieves 82.75%.

Keywords. organization name recognition, conditional random fields (CRF),support vector machine (SVM), cascaded model, laos.

1 Introduction

Named entity recognition (NER) is a basic requirement for the processing tasks of many natural languages, the recognition effect of NER has a direct impact on the deep processing of text information, and the recognition of organization name is one of the main tasks of NER. Compared with personal name and place name, Laos organization name has the characteristics of long and unstable, complex with large unlisted words, and nested structures etc. Hence the recognition is more difficult.

Based on existing material, there is few research on Laos entity name recognition, especially on the recognition of Laos organization name, therefore it has great research significance and value.

Rule-based approach were adopted for organization .In this paper, Laos organization names are divided into simple organization name and complicated organization name two categories. Simple organization name only has one word, like court, congress, republic,etc. Complicated9h organization name consists of mufti-letters, The complicated organization name of Laos is different from that of Chinese, which has the characteristic of post-modifier, feature words in the left boundary of the organization name. For example: the Confucius Institute in Laos. Therefore defined as S + P form, S is the feature words of organization name, like company , university etc and P is the rear word of organization name. Namely, Laos complicated organization name is composed of organization name feature word and one or more organization name rear word .me recognition in the early stage. Reference[1-2] builds a model of rules for university names, but the acquisition of those rules are mainly depend on specific area, which has become a bottleneck problem for this approach. Reference [3] adopted decision tree approach for NER, but the recognition accuracy is very low.

Reference [4-5] adopted HMM approach for NER, this model is based on a stringent independence assumption, but in fact most data could not be treated as a series of independent element. Reference [6] adopted SVM for NER. Reference [7-8] adopted CRF for organization name recognition, the result is more desirable, but still have room for improvements. Reference [9] established a role-tagging approach, but the inadequacy is the role set has a great impact on the recognition result, so repeated researches shall be done for the chosen of the right role set. Reference [10] combined machine learning and artificial knowledge for organization name recognition.

In this paper, Laos organization names are divided into simple organization name and complicated organization name two categories. Simple organization name only has one word, like court, congress, republic,etc. Complicate organization name consists of mufti-letters, The complicated organization name of Laos is different from that of Chinese, which has the characteristic of post-modifier, feature words in the left boundary of the organization name. For example: the Confucius Institute in Laos. Therefore defined as S + P form, S is the feature words of organization name, like company, university ,etc and P is the rear word of organization name. Namely, Laos complicated organization name is composed of organization name feature word and one or more organization name rear word.

2 Laos organization name using cascaded model based on svm and crf

2.1 The resources needed to recognition the laos organization name

Automatic extraction of each list organization name recognition required from the training corpus. The details are as follows:

(1)Feature word table D_f

Feature words refers to the organization name is characterized by significant words, such as "factory , University, company".
The organization name recognition of Laos is the first of left boundary, so the establishment of the list can be used as the trigger condition of organization name recognition.

(2)Rear word table D_b

Rear word refers to the words that in addition to the feature words of organization name, location names and common nouns larger proportion, but overall, the word is more complex, there is a strong randomness.
(3)The left and right boundary word table
The left boundary word is the first word of organization name, such as "representative ""admitted ";The right boundary word is the last word of organization name, such as "director" "host ".Different boundary word indicate different directions on the boundary of organization name. Therefore, .When statistical boundary word table, it is necessary to statistics the number of times as boundary word, and according to the number of times will be divided into different levels.
(4)Simple organization name table
Mainly used for simple organization name recognition, the words that in the vocabulary are considered to be the candidate words of simple organization name, for example: post office .

2.2 Laos organization name using cascaded model based on CRF and SVM

Laos organization name recognition model divided into two layers, the first layer use CRF to recognition simple organization name, and recognition results transmitted to the second layer; The second layer is based on the driven tagging method, which combines SVM and CRF to identify the complex organization name, that is to use the SVM to identify the name of the left boundary of the organization, the words which to be recognized as the left boundary word

DOI: 10.1051/matecconf/201710002051

backwards using CRF to rear marking. Then the recognition results of the two layer are combined. Figure 1 is an example of the organization name recognition is converted to a sequence annotation, figure 2 is the model structure.

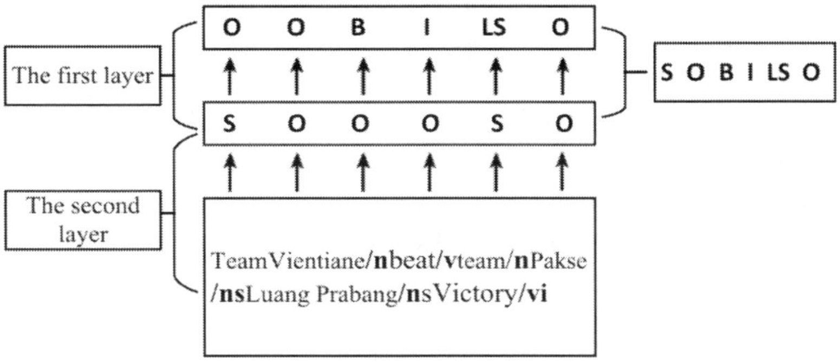

Fig.1 Recognition process of cascaded model

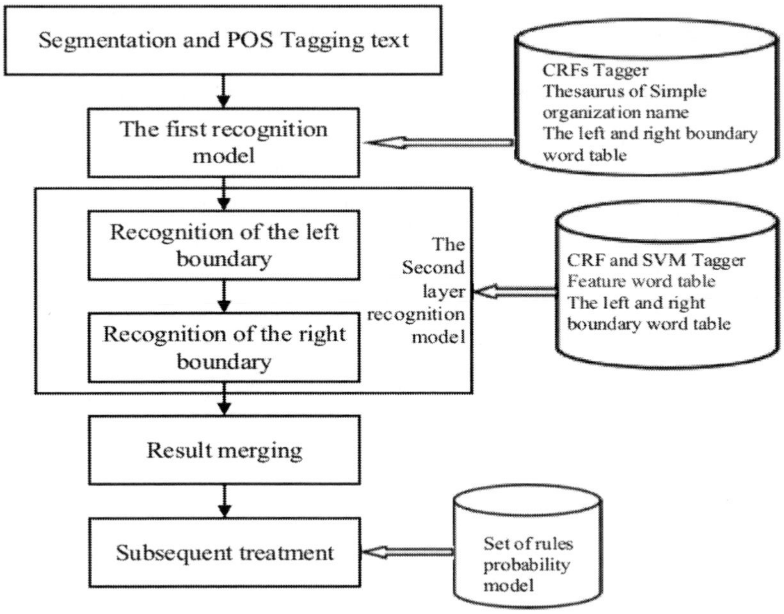

Fig.2 Architecture of Hybrid cascaded model

3 Svm and crf in combination with the identification of complex laos organization name

(1) SVM to determine the left boundary of the organization name

Left boundary is determined to be a two value classification problem, and SVM is an excellent two value classifier, so the SVM is used to determine the left boundary. To appear in the feature list of words were as the left border of the lookup, SVM was used to conduct screening determined whether indeed left word boundary of institutions . SVM also need to choose the appropriate features for a specific task, taking into account the two aspects of efficiency and recognition, select the part of speech and word form these two features. The definition of 11 dimensional vector format are as follows: $(S\ W(-2)\ P(-2)\ W(-1)\ P(-1)\ W(0)\ P(0)\ W(1)\ P(1)\ W(2)\ P(2))$,Where a is a class, in the left boundary of the task, $S=-1$ represents the word is not the left edge of the organization name, $S=+1$ represents the word is the organization name left boundary.

(2) CRF conduct rear label

After the left boundary is determined, use CRF to carry on the back label. Considering a smaller proportion of the organization name, use full labeled policy will cause a lot of waste of resources, decided to adopt the drive type annotation, namely the left boundary driven, only for the look-up annotation. Candidate words determine the rules as follows: assuming the longest organization name length is n, each determined a left boundary, the word directly labeled as "L", the words followed by N - 1 words become the organization name look-up, unless encounter punctuation, another left boundary or f the first of a line. Then tagging the word that Identified as candidate,in other words which are directly labeled as non organization name components .The use of this strategy, to a certain extent, reduces the training and tagging time, improve the recognition efficiency, and because of the reduction of redundant information, the recognition accuracy is also improved. The atomic features used here are also to be as follows in addition to the a used in the first layer.

This method is suitable for the complete identification of organization names, according to different materials need to make some adjustments in the way. If the text does not complete the organization name occupies a certain proportion, the use of two kinds of methods to identify, first in this paper, the second directly with CRF identification, then compare two recognition results, identification of different the confidence degree is higher as the final results.

3.1 Subsequent processing

Subsequent processing includes two parts, the first part for the construction of a probabilistic model, the recognition result confidence below the strings of a certain threshold calculating the confidence, a proper threshold is selected through experiments and reliability are higher than the threshold determination for organization name, or identified as non organization name. The Credibility T (org) of organization names including organization names feature words credibility T (S) and rear organization name word credibility T (P), is calculated as follows:

$$T(S)=\frac{\log(N_s+2)}{\sum_{y\in D_f}(N_y+2)}\ T(P)=\frac{\log(N_P+2)}{\sum_{y\in D_b}(N_y+2)}\ T(org)=\frac{C_n\times\sqrt[3]{n}\times(\sum_{i=1}^{n}T(F_i)+T(S))}{n+1}$$

Where N_s is the number of established institutions characteristics when S the feature word thesaurus; N_P is the number of the establishment of the former P of the rear table ; C_n is the adjustment factor, n is the number of rear word of the organization name. The second part for the construction of the rule model, mainly for identification of Incomplete organization name and concurrent mechanism, and the correction of recognition errors. Rules for example are as follows:

parallel relationship words (such as: and , versus , etc.) before and after the labeling should be consistent, inconsistent will mark the confidence degree is higher.

From the training corpus extraction framework of organization names, such as: (admitted to ,candidates to + Organization Name + (school, reading, work).And according to the number of times to streamline, Confidence is lower than a threshold of recognition and matching the results to determine the matching for organization name, or identified as non organization name.

4 Experiment analysis

In this paper, we use the data collected from the Lao language news website, We use word segmentation program to deal with the data,and in part by the Lao language experts and the Lao students to carry out a manual tagging, used as an experimental corpus .The remaining part is used as a test corpus. Test results taken three common evaluation indicators,that is the correct rate (P), recall (R) and a comprehensive index F value (F) to the evaluation result of organization name recognition.

Testing method is as follows:

Testing results were evaluated by correct rate, recall rate, and F-measure .

(1) correct rate(P): The P measures the numbers of correct named entities in the answer file over the total number of named entities in the answer file.

$$P = \frac{\text{The number of correct labeling of entities}}{\text{The total number of marked entities}} \times 100\%$$

(2) recall rate(R): The R measures the numbers of correct named entities in the answer file over the total number of named entities in the answer file.

$$R = \frac{\text{The number of correct labeling of entities}}{\text{The correct total number of entities}} \times 100\%$$

(3) F-measure(F): The F is a weighted combination of precision and recall.

$$F = \frac{2 \times P \times R}{P + R} \times 100\%$$

The experimental results of this method are shown in table 1.

Tab.1 Recognition result

Experiment	Correct rate(P)(%)	Recall rate (R)(%)	F-measure(F)(%)
Simple organization name recognition	83.62	85.19	84.40
Left boundary recognition	79.15	81.24	80.18
Complex organization name recognition	80.36	80.56	80.44
Global recognition	80.83	82.75	81.75

Experiments on complex organization names using different methods, the experiment is carried out with different methods, and the experimental results are compared as shown in Table 2.

Tab.2 Results of complicated organization name recognition

Recognition methods	Correct rate(P)(%)	Recall rate (R)(%)	F-measure(F)(%)
CRF	77.72	79.13	78.42
Driven tagging of SVM+CRF	80.83	82.75	81.75

As can be seen from the experimental results, the recognition effect of driven tagging of SVM+CRF is best, the training time is reduced due to the reduction of redundant information. But because the recognition of this paper is based on the correct word segmentation and part of speech tagging, the error of the word segmentation will decrease the recognition accuracy.

5 Conclusion

In this paper, a two layer model based on CRF and SVM is established to recognition the organization names,According to the different characteristics of the simple organization names and the complex organization names, at different levels, different methods are used to recognition.

Acknowledgements

Foundation item: China National Natural Science Foundation (61562049), (61662040).

References

1. Godeny, B., Rule Based Product Name Recognition and Disambiguation., Data Mining Workshops (ICDMW), 2012 IEEE 12th International Conference on, Brussels, 2012, pp: 858 - 860.
2. YanDanhui, BiYude., Identification of a rule-based Vietnamese named entity, Journal of Chinese Information Processing, 2014.9.
3. Sekine, S., Grishman, R., and Shinnou, H., A decision tree method for finding and classifying names in Japanese texts, in .6th Workshop on Very Large Corpora, Montreal, Canada, 1998.
4. Zheng Jiaheng, Zhang Hui, Automatic identification of Chinese organization names based on, Journal of Computer Applications, 2002 (11) .
5. Liu Qun, ZhangHuaping, YuHongkui, ChengXueqi., Based on cascaded hidden Markov model for Chinese lexical analysis, Journal of computer research and development., 2004 (08).
6. Xu-Dong Lin,Hong Peng,Bo Liu, Chinese Named Entity Recognition using Support Vector Machines., Machine Learning and Cybernetics, 2006 International Conference on, 2006, pp:4216- 4220.
7. Sisouvanh Vanthanavong, Choochart Haruechaiyasak., Lao Word Segmentation Based on Conditional Random Fields. Conference on Human Language Technology for Development, Alexandria, Egypt, 2011, pp: 2-5.
8. Su-Xiang Zhang, Guo-Yang Gao, Yin-Cheng Qi, and Wilks,Y, personal name and location name recognition based on conditional random fields(CRFs), Proceedings of the Eighth International Confreence on Machine Learning and Cybernetics, Baoding, 2009, pp:12-15.
9. Zhang Huaping, Liu Qun, Automatic Identification of Chinese Names Based Role Labeling, 2004, 27(1), pp:85-91.
10. Chanlekha, H., and Kawtrakul, A, Thai Named Entity Extraction by incorporating Maximum Entropy Model with Simple Heuristic Information, The Specialty Research Unit of Natural Language Processing and Intelligent Information System Technology, 2011, pp:720-836.

Chinese-Lao Bilingual Named Entity Alignment Research

Rui Han[1,2], Lanjiang Zhou[1,2*], Feng Zhou[1,2] and Jinpeng Zhang[1,2]

[1]Faculty of Information Engineering and Automation, Kunming University of Science and Technology, Kunming 650500, China
[2]Key Laboratory of Intelligent Information Processing, Kunming University of Science and Technology, Kunming 650500, China
*Corresponding author: 915090822@qq.com.

Abstract. Chinese-Lao bilingual NE alignment has a very important significance. Three entity alignment methods are proposed in this paper. Firstly, the paper proposes the similarity of bilingual entity fuzzy matching problem. Secondly, we use bilingual entity word sequence pattern similarity to propose Chinese entity model to match Lao entity method. Then we build a naive Bayes bilingual NE alignment model to align Chinese and Lao named entity in the comparable corpus, by mining knowledge information words of Chinese entities. In the end, the rules combine the advantages of the three methods are proposed to achieve the best results.

Keywords. Chinese, Lao, NE alignment, similarity, pattern matching, Naïve Bayes.

1 Introduction

Named entity refers to the entities which has a certain significance in the text [1], it is an important information in Natural Language Processing and an important support for information retrieval, automatic question answering, Machine Translation, topic discovery, and other information processing research [2]. For bilingual sequences, the alignment of bilingual entities aims at establishing the correspondence between the source language and the named entity in the target language and is an important task in the field of multi-language information processing such as Machine Translation, cross language information retrieval, etc. [3].

In recent years, the economic, political and cultural exchanges between China and Lao is deepening, which promote the development of Lao information processing technology, but also put forward higher requirements. But the research is still very weak in Lao languages.

Lao and Chinese Named Entities are quite similar, such as they both do not have special features like capitalization to help identify named entities. Moreover, in the sentence, there are not spaces to delimit the word. And the order of Subject, predicate and object is the same. Of course, it has its own characteristics. For example, if the personal name is the Lao local name, the first name is in front, the last name is in back, otherwise, the last name is in front, the first name is in back. In the sentence of Lao language, the adverbial is generally at last. The front of general location name of Lao has the special word to be distinguished. The personal name is represented as man if the front of name have Mr., the person name is represented as woman if the front of name have Mrs., etc. [4-5].

2 Selection of candidate equivalent entity

Candidate equivalent entity of Chinese named entity mainly were unknown words and identified similar entities in Chinese text in Lao parallel text. If the Lao parallel text is rare, consider comparable text with similarity more than 0.4. In the process of selecting candidate equivalent entities, we select the corpus first to find out the potential text set of candidate entities, then we consider the similar entities and unknown words recognized from the text set as a set of candidate equivalent entities.

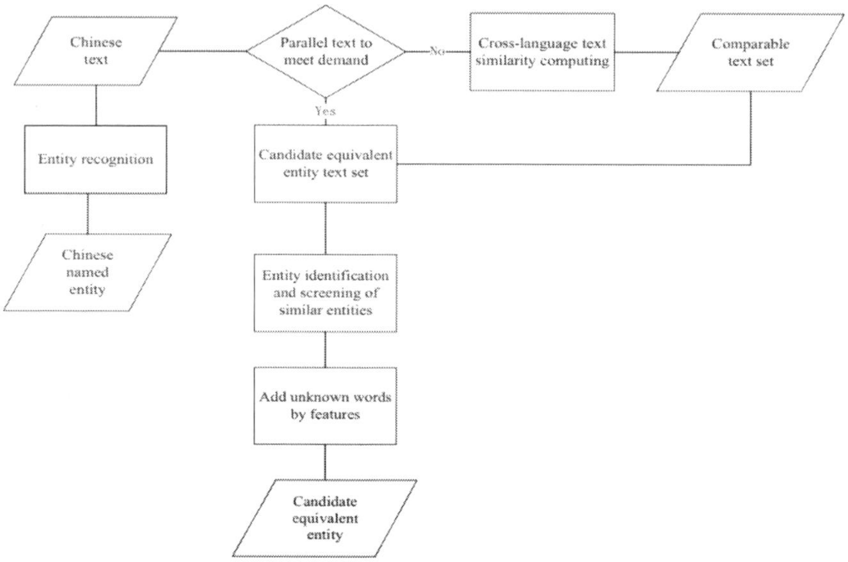

Fig.1. Flow chart of screening candidate equivalent entities

3 NE alignment based on similarity of bilingual entity fuzzy matching

To compare the similarity between Chinese named entity and the candidate entity selected by machine translation way, and translate the Lao candidate entity into Chinese by machine translation and compare the similarity between it and Chinese source entity. The higher similarity result between the two translation processes is the final similarity. If the similarity is greater than a certain threshold, we believe that the two are equivalent entities. Candidate entities are selected by making rules combined with unknown words. At present, Machine Translation has a poor translation of named entity, but some translation results can be used as reference. The correct part of the translation results can be referred to the entity screening and the similarity can be obtained by calculating the number of syllables and words accord with the feature of transliteration and paraphrase. The Lao candidate entity with the highest similarity and more than a certain threshold is the Lao equivalent entity.

The Chinese and Lao equivalent entities are determined by calculating their pronunciation consistency ratio, the specific calculation formula is as follows:

$$\text{similarty}(Ech|Ela) = \frac{\sum_{i=0,j=0}^{m,n} Count(sychi=sylaj)}{min(m,n)} \tag{1}$$

4 NE pattern recognition

Because of the differences between the Chinese and Lao pronunciation rules, the transliteration features are not fully applicable. Some Lao syllables have no corresponding pronunciation in Chinese which brings great obstacles to the named entity alignment based on transliteration.

Some entities do not fit in with the features of transliteration and paraphrase. For example, some people names often appear in the English, digital, abbreviated form such as Vill Wannarot. This paper argues that although there are differences between Chinese and Lao but in parallel corpora and comparable between Chinese and Lao entities in the corpus with the entity model should be the same or similar. We can get a high quality pattern that can meet the needs of people, place names matching by compiling statistics high frequency instance pattern through some common names, translating the pattern into Lao and adjusting the order according to Lao language habits. Lao words which fit in with the pattern are equivalent entities. For example, Huang Hua, served as China's foreign minister from 1976 to 1982. Translate the sentence into Lao and put the adverbial of time in the end of the sentence according to Lao language habits, generate a pattern. Compare this pattern with Lao text words sequence in the comparable corpus, and entity which fit in with the pattern is the equivalent Lao entity.

We can get a lot of Lao pattern by the manual adjustment pattern in Chinese. Lao entities mining through pattern have high accuracy in type judgement. For example, the most commonly used short pattern: [Mr. X says], the accuracy of matching Lao names in the corpus is 100%. If there is a need to align with Chinese entities, the pattern must contain sufficient context information. An equivalent pattern is shown as follows:

Chinese: Headmaster Yellen attends the meeting

Generate the following pattern:

Mr. X headmaster attend meeting

Headmaster can be replaced by similar name inspired feature words such as manager, doctor, etc.

5 NE alignment based on naïve bayes model

The work of the last chapter can achieve the alignment of entities in the simple sentences, but the matching effect of entities in some complex sentences is decreased. Therefore, this paper proposes statistical learning about the relevant knowledge information of the candidate entities, and Bayesian alignment through the knowledge information contained in the candidate words.

5.1 Knowledge information of names of people and place

A Chinese-Lao example sentence can be composed of a number of words. Consider words as data items and the assemblage of words is an item set. The occurrence frequency of an item set is the number of records that are included in the entire corpus, which is also called the degree of support of the item set. If the occurrence frequency of an item set is more than a certain frequency threshold value, it is the frequent item set. The set of all frequent set is denoted as L_k; its frequency is $\text{support—count}(L_k)$. The degree of support of this paper is set to 3.

Based on the study of this section, we recognize the Chinese named entities in Chinese-Lao corpus in a certain period. In the words sequence, nouns and entities that appears in a window of 8 words are knowledge information words related to the entity. Collect information words of the top three Co-occurrence degree with the entities and words whose degree of support is greater than 3, and form the two order frequent sets with entities.

5.2 NE alignment model based on Naïve Bayesian

Bayesian model is a kind of machine learning model, which is commonly used in the field of text classification and word sense disambiguation, and so on6. It uses the prior probability of the category and the distribution of words to calculate the probability of unknown text or word meaning to a certain class. According to the Bayesian model, the correct equivalence entities should be the largest in the probability of the occurrence of the given knowledge information words. We define the decision rule S as follows:

$$\text{if}\, \mathrm{P}(Ek|Cinf) > P(Elao|Cinf), Ek \neq e' then\ decide\ e$$

Cinf is the knowledge information word sequence, Ek is any candidate Lao equivalent entity variables of a Chinese entity, Elao is a right Lao equivalent entity and P is probability.

The P (Ek|Cinf) in formula is unknown and needs to be calculated by the following formula:

$$\mathrm{P}(Ek|Cinf) = \frac{P(Cinf|Ek)}{P(Cinf)} P(Ek) \tag{2}$$

P(Ek) is the prior probability of candidate entity Ek, $\frac{P(Cinf|Ek)}{P(Cinf)}$ is a correction factor in the knowledge information sequence, and the decision problem $\frac{1}{P(Cinf)}$ is a constant which does not affect the final decision result.

In order to calculate the conditional probability P(Cinf|Ek) need to use Naïve Bayesian Assumption. Naïve Bayesian Assumption assumes that each knowledge information words are independent. So P(Cinf|Ek) can be calculated by the following formula:

$$\text{Elao} = {}^{argmax}_{Ek}[\log P(Cinf|Ek) + \log P(Ek)] \quad (3)$$

$$P(Cinf|Ek) = P(\{K_i|K_i\ in\ Cinf\}|Ek) = \prod_{K_i\ in\ Cinf} P(K_i|Ek) \quad (4)$$

K_i is the information word in the sequence of knowledge information. Knowledge information words are independent with each other. The re definition of decision rules is as follows:

$$ifE_{lao} = {}^{\arg max}_{Ek}\left[\log \sum_{ki\ in\ Cknowledge} P(K_i|Ek) + \log P(Ek)\right] \quad (5)$$

$P(K_i|\text{Ek})$ and P(Ek) can be obtained from a Lao corpus using maximum likelihood estimation.

$$P(K_i|Ek) = \frac{C(Ek|K_i)}{C(Ek)} \quad (6)$$

$$P(Ek) = \frac{1}{k} \quad (7)$$

6 Rules combining three methods

There are advantages and disadvantages of these three ways of entity alignment. The similarity matching method is the fastest and there is a certain percentage of the correct rate but the matching rate of the entity is very low. The pattern matching method's accuracy is higher, but the recall rate depends on the quality of the existing model, the scale and the comparability of the comparable corpus. Bias alignment method has higher time complexity but can solve the equivalent entity that cannot be excavated by the former two methods. From the time complexity, the similarity fuzzy matching method is the lowest, the pattern matching method is higher, and the naive Bayes method is the highest. From the correct rate, the pattern matching method is the highest, and the similarity fuzzy matching is the lowest. So we use the following decision rules R2 mining equivalent entities:

Step 1. If the similarity between Chinese entity and the candidate Lao language entity exceeds the threshold value, then it is equivalent entity Ela1, otherwise the candidate result is null. And go to step 2.

Step 2. If the words sequence pattern matching between Chinese entity and the candidate Lao language entity is consistent, then it is equivalent entity Ela2, otherwise the candidate result is null. And go to step 3.

Step 3. Select the candidate entity with the maximum probability of the occurrence of the equivalent entity through the Bayesian formula and get the Lao equivalent entity Ela3, otherwise the candidate result is null. And go to step 4.

Step 4. Compare Ela1, Ela2 and Ela3. If the three are the same, we get the final equivalent entity Ela. If two of the three are the same, then take the entity same with the two. If none of them are the same, then take Ela2 as the final equivalent entity.

R2 combines the advantages of three equivalent entity mining alignment methods, and can dig out the Lao equivalent entities as much as possible and get the highest accurate rate.

7 Experiment and result

This paper sets up the bilingual corpus with Chinese-Lao parallel corpus and comparable corpus crawling from Internet. We have 214 Lao named entities and 214 Chinese entities corresponding to the Lao entities. In this paper, the effects of the three alignment methods are compared with the rule R2. The fuzzy similarity matching threshold is set to 0.4.

Table 1. Experimental result

	precision	recall	Fscore
Similarity matching	62.17%	35.42%	48.34%
Pattern matching	71.69%	21.84%	34.46%
Naive Bayesian	65.08%	67.69%	68.24%
R2	70.61%	73.93%	72.75%

The fuzzy similarity matching method can match equivalence entities of obvious characteristic of transliteration. Pattern matching method can match entities that does not have obvious characteristics of transliteration but the pattern sequence is close. Naïve Bayesian method can match entities that cannot be found by the first two methods, and mine the largest equivalent entities to achieve the highest recall rate. The combination of the three methods can get the most accurate entity alignment results.

8 Conclusions

Traditional word alignment approaches cannot come up with satisfactory results for named entity alignment. In this paper, we study Chinese-Lao entity common characteristic including translation similarity, pattern similarity and identical knowledge information. We propose three alignment methods based on similarity matching, pattern matching and Naïve Bayesian. Finally, the multiple matching rules are proposed by compromising the advantages of the three methods. Experiment results show the effect of the three methods each and the combination of the three methods and the method proposed in this paper obtained the ideal result under the present condition. The next step is to further improve the accuracy of bilingual NE alignment by making full use of more bilingual language features and some machine learning algorithm.

Acknowledgements

This paper is supported by National Nature Science Foundation (No.61662040, 61562049)

References

1. Chinchor, N.. MUC7 Named Entity Task Definition. Message Understanding Conference.1997.
2. M Sun, C Huang, H GAO. Automatic Identification of Chinese Names. Journal of Chinese Information Processing. Vol9, No.2, 1995.
3. Yu-Feng Chen, Cheng-Qing Zong, Keh-Yih SU. Joint Chinese-English Named Entity Recognition and Alignment. Chinese Journal of Computers. Vol.34, No.9, Sept. 2011.
4. Xin Zhao. A Survey on Named Entity Recognition, Disambiguation and Cross-Lingual Co-reference Resolution. Journal of Chinese Information Processing. Vol. 23, No. 2, Mar. 2009.
5. Yang, M., Zhou, L., Yu, Z., & Gao, S. ,2015. Lao Named Entity Recognition based on conditional random fields with simple heuristic information. International Conference on Fuzzy Systems and Knowledge Discovery. IEEE.
6. Da Wang, Kun Zhang. Application of Bayesian Model in Word Sense Disambiguation. Computer Era. No.7, 2009.

Method of Word Segmentation in Laos Based on Maximal Matching of Syllables

Wenjie Huo[1,2], Lanjiang Zhou[1,2], Feng Zhou[2] and Bei Yang[1,2]

[1]School of Information Engineering and Automation, Kunming University of Science and Technology, Kunming 650500, China

[2]The Key Laboratory of Intelligent Information Processing, Kunming University of Science and Technology, Kunming, Yunnan 650500, China

Abstract. Word segmentation is an important support of semantic analysis, Machine Translation, QA, knowledge mapping research work, mainly used in information retrieval, text processing, data processing and many other areas of Natural Language Processing. Therefore, the realization of word segmentation is a very meaningful work. The method of this paper is to segment the syllables of the text corpus of Lao language and complete the maximal matching of syllables and dictionaries. Then match the results of the word segmentation and the error dictionary, and correct some wrong words by the error dictionary. Finally, we use regular expressions to match the corresponding word strings in segmentation results and correct the wrong words by some artificially formulated rules of the alphabet, numbers, etc. in the Lao language. It can improve the efficiency and accuracy rate of Laos Word Segmentation.

Keywords: Word segmentation, the maximal matching, Syllable segmentation.

1. Introduction

Words are the smallest meaningful units in natural language. The goal of the word segmentation is to divide the sentence into words. Word segmentation is the most basic work in the Natural Language Processing. In information processing, word segmentation is the basis of many other work, the accuracy of the word segmentation directly affects the follow-up work. In the aspect of text similarity computation, text clustering, search engine and information retrieval can not be separated from the work of word segmentation. And these information processing work is also an indispensable part of this powerful Internet Age.

The Lao language is composed of syllables, and the syllables are composed of letters. Therefore, the word segmentation of Lao can not be regarded as a sequence of words and combine into words like Chinese. Considering the rules of language itself, we segment words from the aspect of syllable size. At present there are several commonly used methods of word segmentation: the method of word segmentation based on semantics[1] and the method of word segmentation based on dictionary[2]. The method of word segmentation based on dictionary is learning language knowledge[3,4] from large-scale corpus and getting a model of word segmentation, using this model to complete the word segmentation.

2. Word segmentation method based on maximal matching of syllables

2.1 Get a matching dictionary

The large-scale corpus on the Internet has played a very important role in the development of Natural Language Processing, which has a very good coverage. In the word segmentation method based on dictionary, dictionary resource is the most important condition to complete the word segmentation. We found the Lao language dictionary which contains 15768 commonly used words in total and a large number of English-Chinese dictionaries and Chinese-Laos dictionaries in the Lao language network and English Network. We use the English-Chinese dictionaries and Chinese-Laos dictionaries to intercept the part of the Lao language, and put these words into a file. Then use HashSet in the Java combined two files and remove duplicate. Finally we get the Lao language dictionary, containing 31719 words in total, and save words reversely according to the length of word.

2.2 The syllable rule of Lao language

The letter and the mark of Lao language can not appear in any placeholders. Table 1 is the position constraint of a word consisting of vowels, consonants, and tones. Among the *Xn*, *n* represents the input sequence of the letters in the word. Only *X* does not follow this order (*X* is typed between *X1* and *X2*). Tone is typed after the vowel *X0*, consonant *X1*, the main consonant *X* and vowel *X3*, *X4* (if *X0*, *X1* and *X2* are not empty, can not be empty).

Table 1. The syllable structure of Lao words.

		X5					
		X4					
X0	*X1*	*X*	*X6*	*X7*	*X8*	*X9*	*X10*
		X2					
		X3					

- *X0*: *X0* represents a vowel which always appears before the main consonant *X*;
- *X1*: *X1* represents a consonant which appears before the main consonant.
- *X*: *X* represents the consonant.
- *X2*: *X2* represents a consonant which is followed by the main consonant *X* and appears below the main consonant.
- *X3*: *X3* represents the vowel which appears below the main consonant.
- *X4*: *X4* represents the vowel which appears above the main consonant.
- *X5*: *X5* represents the tone which appears above the main consonant or vowel.
- *X6*: *X6* represents the vowel which appears behind the main consonant. Its role is that when there is no vowel in the syllable, it appears together with X_6.
- *X7*: *X7* appears behind a vowel. $X7_1$ represents the end of a syllable, and it does not appear together with the tone.
- *X8*: *X8* represents the consonant which can be alternated.
- *X9*: *X9* is the consonant that can be foreign language pronunciation, it appears together with $X10_3$ in syllables .
- *X10*: *X10* represents a mark and usually appears at the end of the syllable.

2.3 Syllable segmentation

The syllable segmentation algorithm of Lao language is defined as follows:

When we encounter punctuation and space which is not the letter of the Lao language in the input array, then record the boundaries of syllables.

Filtering the Lao language characters from the input array and removing the punctuation and space which is not the letter of the Lao language. And then reorder the character.

Then we reorder the letters which writing is not standardized and list the character which is marked appearing in all possible placeholders.

Steps of syllable segmentation:

Mark syllable boundary by input array, add up characters which are not Lao language, and store the Lao language text in the new array;

Reorder the letter which written order is not standardized;

Use rules to determine the syllable boundaries, and segment syllable;

Mark the position of placeholder of each character;

Put back the Lao language character to the original array.

2.4 Longest syllable matching

The word boundary and the syllable boundary of Lao Language are the same. We can use the matching based on syllable instead of the matching based on letter. It can avoid cutting the sentence into individual letters. The matching based on the longest syllable is matching the syllable sequence in the dictionary[5]. Given a syllable of the set *Sy*, and the word dictionary set *Dy*, the result of the longest syllable matching is the result of the word segmentation, denoted as *R*. In the matching process, firstly, the pointer *Bs* points to the beginning of the syllable which is defined and the pointer *Es* points to the last syllable. Then match words from the pointer *Bs* to *Es* in the dictionary. If the match is successful, the words are recorded in *R*. Then pointer *Bs* points to *Es*+1, and *Es* points to the end of next set of syllables. Else *Es*-1, until *Es*=*Bs*.

2.5 The acquisition and rule making of the error dictionary

In this paper, we do a statistical analysis on 37931 words in the Lao language word corpus. We select the common words of Laos which have high error frequency and always appear, and make a statistical classification. Then we can get a error dictionary of Lao language combine 176 words which case changes lower and the result does not produce word segmentation ambiguity with context.

The reason for this error is that there is no long word in the dictionary, which leads to the error of the segmentation. Rely on artificial to make rules in the process of word segmentation, and use these rules to deal with some special cases in the result of word segmentation such as numbers, English, and foreign words.

3. Experimental results and analysis

In the experiment, we use the common evaluation index to evaluate the segmentation effect of the word segmentation model, that is, the word segmentation accuracy. Usually use the following formula to calculate:

- R: Accuracy rate
- n: The number of words which is segmented correctly.
- N: The number of all the words.

$$A = (n \div N) \times 100\%$$

3.1 Word segmentation by different dictionary structure

The experimental data of this chapter are the dictionary with 15768 words and the extended dictionary with 31719 words. Then using the method of syllable segmentation to segment the syllable of the Lao language text which contains 30000 words and complete the longest matching based on syllables. The experimental results are as follows.

By extended dictionary we are correcting this type of error. Different results are shown in Table 2.

Table 2. Word segmentation results in different dictionary size

The number of words in the dictionary	Number of word segmentation	Accuracy of word segmentation

15768	30000	82.63%
31719	30000	86.09%

3.2 The influence of error dictionary on word segmentation

In this experiment, we segment the syllable of the Lao language text which contains 30000 words by the extended dictionary. Compare the results of the word segmentation with error dictionary and the results of the word segmentation without the error dictionary. The experimental results are as follows.

The influence of error dictionary on word segmentation is shown in Table 3·

Table 3. The influence of error dictionary on word segmentation

Word segmentation method	Number of word segmentation	Accuracy of word segmentation
without error dictionary	30000	83.03%
with error dictionary	30000	86.09%

3.3 The influence of rule correction on word segmentation

In this experiment, we segment the syllable of the Lao language text which contains 30000 words by the extended dictionary. Compare the results of the word segmentation with rule correction and the results of the word segmentation without rule correction. The experimental results are as follows.

The influence of rule correction on word segmentation is shown in Table 4.

Table 4. The influence of rule correction on word segmentation

Word segmentation method	Number of word segmentation	Accuracy of word segmentation
Word segmentation without rule correction	30000	82.66%
Word segmentation with rule correction	30000	86.09

4. Conclusions

In order to realize the Lao word segmentation, this paper combines the characteristics of Lao syllables. Using the structure rules of syllables, segment the syllable of the Lao language text and complete the maximum matching word segmentation. Finally, using the error dictionary and the rules improve the accuracy of the results through two-time correction. The experimental results show that the coverage of the dictionary directly affects the accuracy of word segmentation, the larger the scale, the higher the accuracy of the word segmentation. Through the error dictionary, correct of some wrong word segmentation effectively. At the same time, the error of some numbers, English letters, foreign words and so on are corrected by the rule correction, which can effectively optimize the segmentation results.

Foundation item: China National Natural Science Foundation (61562049), (61662040).

References

1. Quan-Shu Long, Zheng-Wen Zhao, Hua Tang. "Overview on Chinese Segmentation Algorithm". COMPUTER KNOWLEDGE AND TECHNOLOGY, 2009,5（10）:2605-2607.
2. Chang-Ning Huang. "The problem of Chinese word segmentation information processing". Applied Linguistics, 1997,6(1): 72-78.

3. De-Quan Zheng, Feng Yu, Kai-Tao Wang. "Ambiguity Word Segmentation Based on Two Chinese Characters Used as a Word in Chinese" Computer Engineering and Applications, 2003,39 (1): 17-19.
4. Qiao-Ming Zhu, Tao Wen, Pei-Feng Li, and Pei-De Qian. "Self-Adaptive Chinese Ambiguous Word Segmentation Method Based on Multi-Gram Library". MINI-MICRO SYSTEMS, 2006,27(8):1597-1600.
5. P.P issamay, et al .Syllabification of Lao Script for Line Breaking.Tech. Rep. of STEA, Lao PDR, 2004.

The Distribution of Words in Chinese and Laos Based on Cross Language Corpus

Yuquan Huang[1,a], Lanjiang Zhou[2,b], Feng Zhou[2,c] and Jinpeng Zhang[2,d]

[1]School of Information Engineering and Automation, Kunming University of Science and Technology, Kunming 650500, China
[2]Key Laboratory of Intelligent Information Processing, Kunming University of Science and Technology, Kunming 650500, China

Email: [a]120166881@qq.com, [b]915090822@qq.com, [c]zf158@sina.com, [d]939127870@qq.com

Abstract. Word representation is the basic research content of natural language processing .At present, distributed representation of monolingual words has shown satisfactory application effect in some Neural Probabilistic Language (NPL) research, while as for distributed representation of cross-lingual words, there is little research both at home and abroad. Aiming at this problem given distribution similarity of nouns and verbs in these two languages, we embed mutual translated words, synonyms, super-ordinates into Chinese corpus by the weakly supervised learning extension approach and other methods, thus Laos word distribution in cross-lingual environment of Chinese and Laos is learned. We applied the distributed representation of the cross-lingual words learned before to compute similarities of bilingual texts and classify the mixed text corpus of Chinese and Laos, Experimental results show that the proposal has a satisfactory effect on the two tasks.

Keywords. Weakly supervised learning extension, cross-lingual corpus, cross-lingual words distribution representations, neural probabilistic language model.

1 Introduction

Lexical representation is an important part of Natural Language Processing. [3] It is an important technical support for information retrieval, data mining, knowledge mapping and other research directions. The goal of vocabulary representation method based on statistical machine learning [6] is to learn the probability representation of word sequences from natural language texts. In the process of training, each word sequence is different from the word sequences in the training corpus. In the process of monolingual glossary space representation of a traditional but effective method is N meta grammar model , it through the learning of target words a short window information to predict the probability of the target vocabulary. The disadvantage of it is that it can not reflect the influence of the word's words on the sequence generation probability and the similarity of the distribution probability of the similar words. Neural probabilistic language model in monolingual environment to solve this problem well. Neural probabilistic language model by acquiring syntactic and semantic information of learning words represents the distribution of features from natural language text, the word sequence similarity is similar to the word distribution, validation of the word distribution can be well applied in POS tagging, named entity recognition, semantic role labeling and other natural language problems. Although the monolingual lexical distribution that has achieved good results, but in the cross language Natural Language Processing field research at home and abroad are scarce. There are two main methods at present: The first is the transfer of learning, the method will mark the learning information from one language to another language, so that fewer resources to obtain better processing results. The second approach is to translate the two languages into one or third languages, using a language to express cross language information.

The above methods have achieved certain results in dealing with cross language natural processing, but there are some disadvantages that cannot be transplanted, the algorithm process is complex, and the accuracy needs to be promoted. At present, the mainstream of the textual level analysis method only inspects the distribution characteristics of nouns and verbs. From this idea, in order to solve the above problems analysis of the distribution of Chinese, Lao noun, verb similarity, Lao nouns and verbs as nouns, verbs, Lao words into Chinese corpus, generating Chinese and Lao cross language lexical sequence corpus. Through the neural probabilistic language model to learn Laos nouns, verbs in cross language distribution in space. In this way in the cross linguistic data in learning the Chinese cross language word vector distribution, applied directly to the Laos text, learning resources and lack of resolve Laos corpus analysis of cross language text. The Chinese and Lao cross language text classification and text similarity based on the experiment, verify the Chinese and Lao cross language lexical distribution represents the effect.

2 Neural probabilistic language model

Neural probabilistic language model by Bengio Y[1,2] in 2003 first proposed joint probability function based on artificial neural network to learn the vocabulary of a language sequence. It solves the problem of sparse data dimensional and dictionary vector language of space effectively, and solve the distribution of N meta grammar model cannot solve the similar problems, and compared with the dictionary vectors and N meta grammar model can better represent the lexical distribution

Neural probabilistic language models are described as follows:

By a given word sequence: $\omega_1, \ldots, \omega_t$, $\omega_t \in$V,V represents all the lexical sets of the target language. V although very large but finite, the goal of the neural language model is to learn a good function to estimate the conditional probability of a word:

$$f(\omega_1, \omega_{t-1}, \ldots, \omega_{t-n+2}, \omega_{t-n+1}) = p(\omega_t | \omega_1^{t-1}) = p(\omega_t | \omega_{t-1}, \omega_{t-2}, \ldots, \omega_1) \qquad (1)$$

Among them, ω_t is the t word of the word sequence; The corresponding sub sequence written in $\omega_i^j = \omega_i, \omega_{i+1}, \ldots, \omega_{j-1}, \omega_j$; V represent for the thesaurus $|V|$ represent the size of the thesaurus. The joint probability of word sequences can be obtained by conditional probability.

Formula (1) includes two processes: Firstly, the mapping C is constructed to map any element in the set V to the feature vector of the word $C(i) \in R^d$ .The distribution of feature vector which represents the word association Thesaurus . d represents the dimension of feature vector. The free parameter matrix of $|V| * d$ is expressed in the experiment.

Then construct the probability function of word. We use the word feature vector C representation :Input context feature vector based on Corpus $(C(w_{t-n+1}) \ldots C(w_{t-1}))$ to prediction of probability distribution of the next word .The output of the g is a vector ,The i element of the vector is the estimated probability $p(w_t = i | w_1^{t-1})$,Calculated by the following way :

$$f(i, w_{t-1,}, \ldots, w_{t-n+2}, w_{t-n+1}) = g(i, C(w_{t-1}) \ldots C(w_{t-n+2}) C(w_{t-n+1})) \quad (2)$$

f from the above mapping C and g combination, These two maps are related to some parameters. The parameter of mapping C is the feature vector itself, which is represented as a $|V| * d$ matrix C . C line i is the feature vector of the word i . Function g can be realized by feedforward neural network or convolutional neural network. The formula (2) shows that the function of f through the context words to predict the word I word eventually transformed into a distribution function of g through the distribution of feature vectors to predict the i context of a word.

3 The distribution of linguistic words in Chinese and lao languages

The characteristics of the distribution of words in Chinese and Lao languages: Chinese and Lao language have a large degree of similarity, they have a lot in common with the syntax. For example, in the same sentence: The syntactic structure of Chinese is (+attribute)subject + (+adverbial)Predicate +(+attribute)object(+complement);The syntactic structure of Lao language is (time/adverbial)+subject(attribute)+Predicate. The main two kinds of sentences, SVO sequence is completely consistent .The main difference is reflected in the Lao adverbial must be placed in the center after the word, the Chinese attributive adverbial, must be placed in the center of the word before. In terms of the composition of the sentence, the main content of the sentence reflects the main content. Attributive, adverbial, complement is not essential components of leaves. The two main components of SVO language is completely consistent. Subject, predicate, object, corresponding to the noun and verb in part of speech, the main structure of the sentence is the same. The distribution of the word sequence of the two language nouns and verbs should be similar.

Our goal is to ignore the differences between Chinese and Lao language， Let Lao language nouns and verbs as Chinese nouns, verbs, learning their distribution in the Chinese language environment, so that the mature Chinese text analysis method can be directly applied in Lao language text.

Parallel corpus preprocessing: We have selected 1026 pairs of parallel sentences which are obtained from China Radio Station and have been manually corrected. Although the original text contains all the text information, but the current Natural Language Processing [3] technology cannot fully handle these text messages. [3] Therefore, it is necessary to preprocess the text. Traditional text preprocessing is mainly to remove the stop words. Since this method requires sequence distribution on word learning, so we did not remove stop words, but we will have nothing to do with Chinese and Lao language text symbols, meaningless digital removal, and some of the names into a unified symbol, avoid because of names caused by the change of word sequence distribution of learning effect, reduce noise interference.

Parallel corpus word alignment: We mainly use cross heuristic algorithm, by running from Chinese to Lao language and Lao language mapping from the map to the Chinese two directions to get to the word alignment. [4,5]

We only consider the alignment of words in two directions. By word alignment, we can obtain a corresponding cross language [5] translation word in the parallel corpus.

Example sentence:

(1)We are going to/0 play basketball/1 this afternoon/2

(2)ຕອນບ່າຍນີ້/2 ພວກເຮົາ/0 ຫຼິ້ນບ້ວງ/1

In the above of parallel sentences, the same suffix label of Chinese and Lao words indicate that they are aligned translation relations. Through $GIZA^{++}$ we can get the corresponding Chinese words in Lao words in Chinese instance. In Chinese examples, the Chinese word "we" will replace the ພວກເຮົາ generation for example as follows:

(3)ພວກເຮົາ are going to play basketball this afternoon

We place an example (1) and (3) into a neural probabilistic language model, get the distribution information of ພວກເຮົາ words in Chinese language environment. In the end, we can make the distribution of ພວກເຮົາ and Chinese words "we" are similar, in accordance with the law of linguistics, that is, the distribution of two words should be similar. We call this case (3) in the case of Chinese examples proper position is embedded into the corresponding Lao word for the derived instance. The distribution of the cross language vocabulary can be obtained by learning the Chinese language examples and the derivative examples in parallel sentences together as the learning materials.

Replace the Lao words and Chinese word similarity relation: Generally if between two words is the dictionary translation relations will assume they are justified, but in the more general case, play similar syntax semantic roles of the word have similar word sequence of joint probability distribution. Although Hot and cold are a antonym, but in many contexts in the sequence distribution is very similar. In the case of today's weather hot and today's weather cold, there is a relationship: $P(hot|today, weather) \approx P(cold|today, weather)$.

The massive Chinese and Lao language mixed corpus of weakly supervised learning process: We take the Chinese examples and examples of mixed learning derived instances can get a certain degree of Lao Chinese vocabulary. But by the parallel sentences required for manual correction in a larger extent, limited corpus, learning the Lao lexical distribution still can not fully reflect the distribution of it should be in the Chinese language environment. For example, we can learn to ພວກເຮົາ with the word "we" similar relations in the mixed case, but the ພວກເຮົາ and the "we" in Chinese, "I", "we" should be similar, and they also have some similarities such as pronouns. The neural probabilistic language model can obtain the best results only in the large scale corpus. As a result, we have added the above mixed case set to the search of the Chinese news corpus, which has been collected in a large scale. We first concentrated in the large-scale corpus of Lao nouns and verbs corresponding Chinese nouns and verbs are replaced. Then replace the generation of derivatives, while retaining the original corpus of Chinese examples. Such as Lao "ພວກເຮົາ" in addition to "we" can be replaced, you can also replace "I", "we". Through this process we get the first mixed data set. We learn synonyms through cross language learning corpus for learning, can obtain Lao nouns, verbs in the Chinese language environment more accurate sequence distribution of information. Through this process "ພວກເຮົາ" and "we", "I" have made a similar distribution. In the neural probabilistic language model, if there is a similar context semantic or syntactic structure, there can be a similar distribution.

We are on the Lao word corpus contains a weakly supervised learning extension:

(1) The similarity of Lao words and Chinese words are compared, if the similarity is greater than the threshold value, we put Chinese words into candidate replacement word set.

(2) The Lao word and word candidate replacement word set into English, query semantic relations between them in the hierarchy tree English, if they are similar words or directly on the meaning of a word can be directly replaced, generate new candidate derived instances

(3) Whether the candidate derived from the artificial judgment is in accordance with the knowledge logic, and the derivation is reasonable, otherwise it will remove the case.

(4) The selected examples of corpus derived added concentration, through the neural probabilistic models of language learning new vocabulary in Chinese and Lao language distribution, and jump to 1.

(5) Repeat the process from (1) to (4) until you can't learn a new Chinese word.

Through the propagation process of weakly supervised learning vocabulary learning focused on the corpus, we found that with the increase of the size of the corpus, the similarity between Chinese words and vocabulary will decline slightly, but the similarity between Lao vocabulary and Chinese vocabulary increased significantly, and the expected results of similarity of the synonymous relations between Lao vocabulary and Chinese vocabulary is closer, and close to the similarity between words in Chinese.

(6) Learning model

Back propagation algorithm for learning model parameters in neural probabilistic language model. [2] At present, there are many parameter improved learning algorithm for the back propagation algorithm. We choose the improved gradient descent algorithm to optimize the parameters of the model set. The method can dynamically adapt to the first order information, and the computation cost of the gradient descent is minimal.

4 Experiment and analysis

Text similarity computing method: We first used neural probabilistic language [2] on cross linguistic corpus study, get the cross linguistic distribution of Chinese and Lao lexical representation. Based on the experience, we set the vector dimension of each word to 200,The number of neural units in the hidden layer of neural probabilistic language model is 64 .The promise error is 0.001, and the training window is 5. The frequency is greater than or equal to 3 times the Chinese and Lao words only consider the process in the learning corpus. We have got a vocabulary learning based cross language distribution as text similarity computing. We have been the basis of learning Chinese and Lao language vocabulary distribution as text similarity computing.

We through the algorithm selected each document feature weight accounted for 5 of the feature words. The characteristic phrases for text t are $(\upsilon_{t1}, \upsilon_{t2}, ..., \upsilon_{t\varsigma})$, Weight as: $(\omega_{t1}, \omega_{t2}, ..., \omega_{t\varsigma})$, The features of the same text k are $(\upsilon_{k1}, \upsilon_{k2}, ..., \upsilon_{k\varsigma})$,The corresponding weights of characteristic words are $(\omega_{k1}, \omega_{k2}, ..., \omega_{k\varsigma})$.The similarity between the two texts through the text t each feature word and text K in each feature word of the word vector cosine similarity and the respective characteristics of the weight of the multiplication and accumulation, and then divided by the total number of 25. The cosine similarity between the words υ_{k1} and υ_{t1} is expressed as υ_{k1} & υ_{t1}. The formula for calculating the similarity of text:

$$t \,\&\, k = \frac{\sum_{i=1, j=1}^{\varsigma} \upsilon_{ti} \,\&\, \upsilon_{ki} * \omega_{ti} * \omega_{ki}}{2\varsigma} \tag{3}$$

Experimental results and analysis: We use Wikipedia on the Chinese and Thai text align text as the experimental text set, select the economic, political and technical sports, five kinds of Chinese and Thai parallel texts of 100 articles. The experiment consists of two parts. The first part: the Chinese and Lao parallel text similarity calculation; the second part: the text of Chinese and Lao mixed text concentrated randomly disrupted the order after the judgment in their classification in five categories. The similarity of Chinese and Lao text description of the cross language word distribution similarity between the two synonyms. Only the synonyms in the two texts are similar in the same vector space distribution, which can improve the similarity of the text.

The flow chart is as follows:

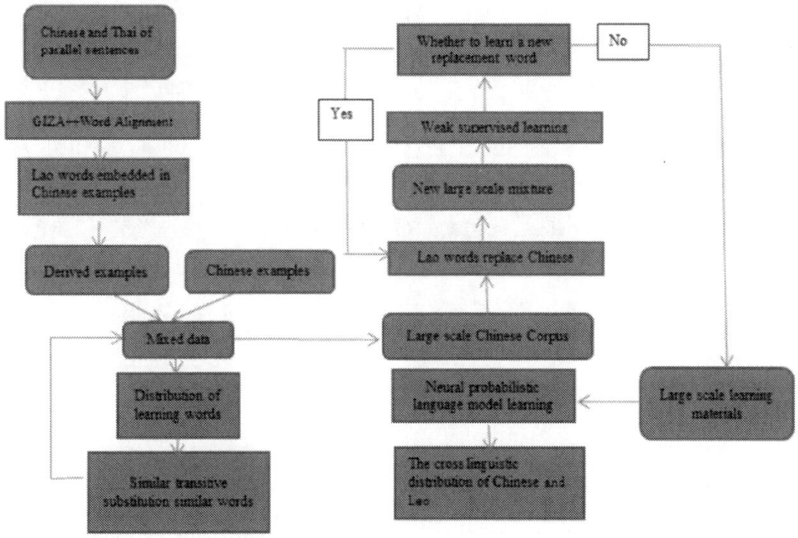

The experimental results show that: in the case of the same corpus size, the cross language word distribution in cross language text classification is better than two kinds of Machine Translation ways. For the expectation maximization algorithm considering the class information of the source language word translation as the biggest target language word translation probability, compared to cross language lexical similarity is the average similarity of all categories, higher accuracy, and the feature words combined with semi supervised adaptation can update the target language text classification, the best effect. Experiments show that the accuracy of a cross language lexical distribution, lexical meaning expression accuracy. Based on the cross language word distribution, the classification knowledge of the source language can be directly migrated to the target language, which has a certain effect. [1,2]

5 Conclusion

In this work, To solve the cross linguistic distribution of Chinese and Lao words, we ignoring the differences between the two languages. We will place Lao nouns and verbs into Chinese corpus to generate cross linguistic data. Extension of corpus by weakly supervised learning, and then through the neural probabilistic models of language learning Chinese vocabulary representation of cross linguistic distribution. [2,3] The application of mature text analysis method can be applied directly to the Lao text in Chinese, and the application method in the analysis of cross language text on the elimination process is relatively simple, not divergent very complex. Experiments have achieved certain results through text similarity and text categorization. The next step is to improve the neural probabilistic language model to improve the accuracy of cross language lexical distribution. And further study the influence of the distribution feature vector representation dimension of cross language vocabulary on the representation of cross language lexical distribution.

Acknowledgements

This paper is supported by National Nature Science foundation (No.61662040,61562049).

References

1. Bengio S,Bengio Y. Taking on the curse of dimensionality in joint distributions using neural networks. IEEE Transactions on Neural Networks,2000,11(3):550-557.
2. Bengio Y, Ducharme R, Vincent P, et al. A neural probabilistic language model.Journal of Machine Learning Research,2003,4(3):1137-1155
3. Collobert R,Weston J,Bottou L,et al.Natural language processing(almost)from scratch.Journal of Machine Learning Research,2011,12(1):2493-2537.
4. Zeman D,Resnik P. Cross-language parser adaptation between related languages//IJCNLP,2008:35-42.
5. Sogaard A. Data point selection for cross-language adaptation of dependency parsers//Proc of the 49th Annual Meeting of the Association for Computational linguistics: Human Language Technologies:Short Papers-Volume 2,2011:682-686.
6. Ando R K,Zhang T. A framework for learning predictive structures from multiple tasks and unlabeled data.Journal of Machine Learning Research,2005,(6):1817-1853.

Chinese and Thai Bilingual Topic Detection Online

Ziqiang Rang[1,2], Lanjiang Zhou[1,2*], Jinpeng Zhang[3], Yantuan Xian[1,2] and Zhengtao Yu[1,2]

[1]School of Information Engineering and Automation, Kunming University of Science and Technology, Kunming 650500, China

[2]The Key Laboratory of Intelligent Information Processing ,Kunming University of Science and Technology, Kunming, Yunnan 650500, China

[3]Information Management Center, Yunnan University Of Finance And Economics, Kunming, Yunnan 650500, China

Corresponding E-mail:915090822@qq.com

Abstract. Bilingual topic detection is a vital application of natural language processing in the Internet plus Era and trend of economic globalization. At present, the method of bilingual topic detection can't solve the problem of bilingual topic inconsistent distribution. Aiming at the shortcoming, this paper introduces a based on maximal clique method to find bilingual topic detection of Chinese and Thai feature words. First of all, extract the information of news with keywords of each Chinese and Thai documents through the TextRank algorithm. Next, disambiguate by means of the similarity combined with Chinese and Thai dictionary. Then, use credible association rules to cluster Chinese and Thai feature words, which generates maximal clique of bilingual topic. Finally, cluster similar maximal clique of topic to obtain the collection of final topic. According to the needs of users, the method can recommend a bilingual topic of different sizes. The test of Chinese and Thai news texts in January 2016 made good achievement. From the perspective of cross-language word clustering, the algorithm effectively solves the problem of inconsistency of bilingual topic distribution reasonably, and has the advantages of no need to estimate the number of topics and low time complexity, so it is suitable for the application of online discovery in bilingual topic.

Keywords. Chinese, Thai, maximal cliques, credible association rule, TextRank, bilingual topics detection.

1 Introduction

With the development of information technology, Internet has become an important channel for organizations or individuals to understand the trends of other countries. Now, Internet generates a massive amount of data every day, it is quite difficult that cover the useful news text, and there is a huge language barrier between Chinese and other languages. Therefore, it become a hot topic of public opinion monitoring, information competition, bilingual information retrieval that obtain the timely and effective foreign topic information in a massive amount of data. At present, in the field of cross-language topic detection, the mainly approaches are machine translation [1] or bilingual dictionary [2] and parallel corpus in the perspective of solving the language barrier. The method of based on machine translation or bilingual dictionary, it transforms cross-language texts into the texts of monolingual, then it analyzes topics of single language. BNN (2000) proposed a Chinese-English topic detection and tracking system based on Machine Translation [1]. Ying-Ju Chen et al. (2002) used word vector cosine similarity based on English-Chinese dictionary to calculate the multi-language related topic detection, and they compared similarity of topics with Chinese to English and English to Chinese and not translation etc [3]. The method of based on parallel corpus, such as the LDA of bilingual, it is obtained the topic distribution of cross-language [4,5]. Mimno D et al. (2009) presented the multi-language topic model based on the extension LDA, and it had topic detection effect slightly worse than the monolingual LDA in the condition of large language differences.

The above methods had achieved some effect. But they didn't solve the bilingual topic inconsistent problem of feature word distribution in the process of topic detection, and the number of topics must be predicted in advance. The language barrier in the cross-language topic detection increases the difficulty of accurately predicting the number of topic. Therefore, the above methods are less effective, and they are not suitable for the application of the topic detection online. In order to solve the above problems, this paper found a method of Chinese-Thai bilingual topic detection online base on maximal clique clustering. First of all, this method adopts TextRank algorithm [6] to extract keywords of Chinese-Thai news text, then combination with named entities and the first paragraph information of topic generated feature word sequence of news. Next, the use of bilingual dictionaries and similarity disambiguation algorithm obtain the co-occurrence correlation relationship of the Chinese-Thai feature word in the news corpus to construct the co-occurrence adjacency matrix of the Chinese-Thai feature word, then to obtain the set of initial maximum clique topic through the maximum clique mining algorithm. Finally, to cluster similar maximal clique of topic obtain the collection of final topic. This paper provides a description of the ordered bilingual news topics with different sizes. In this paper, the experiment by Chinese-Thai news corpus of January 2016, compared the different generation ways of maximum clique topic and the comparison results with other methods of bilingual topic detection. The experimental results show that the time efficiency of the method is the best, and the F value is 69.03%, which is achieved by the method of this paper.

2 The cross-language topic detection online process

2.1 Extraction news keywords and generating news information by TextRank algorithm

To express the same theme, there is a certain semantic relationships between words in the same text words. In this paper, it extracts Chinese-Thai news keywords in each news by TextRank algorithm. The essence of the algorithm is a sorting algorithm based on graph. It seen the Chinese-Thai vocabulary as vertices in the graph and reflect the relationship between them by undirected edges between the vertices.

According to basic theory of Sort algorithm based on graph, you can establish a connection between the words having a semantic relationship, then to calculate number of scores words according to mutual "vote" between the words recursive, each word of the score depends on the other word vote and fraction size of it, finally, select high-scoring words for the important words. And which does not have any words to connect words is an isolated point [6,11]. Mihalcea R and others have already proven that whether the graph have directions will not have much impact on keyword extraction. In this paper, I adopt undirected graph model. To calculate each word score by using iteration in the figure, the higher the score is, the more important to the word is. Finally Each Chinese-Thai News Documents obtained most important keywords before K corresponds. Candidate word score is calculated as shown in Equation 1:

$$WS(V_i) = (l-d) + d * \sum_{V_j \in Link(V_i)} \frac{w_{ji}}{\sum_{V_k \in Link(V_j)} w_{jk}} WS(V_j) \qquad (1)$$

Where $WS(V_i)$ is the vertex V_i (That word score of j), d is the ratio of two parts of the score distribution, $Link(V_i)$ is a cable of the point of the set of points.

2.2 Based on credit association rule mining the topic maximum clique of Chinese-Thai cross-language

The theoretical basis for mining topic maximum clique is the co-occurrence relationship of the related topic feature words. This paper addressed the statistical problem of the Chinese-Thai feature word's co-occurrence relationship by Chinese-Thai bilingual dictionary and the disambiguation algorithm of Chinese candidate translation similarity.

2.2.1 Chinese-Thai feature words similarity disambiguation algorithm

In this paper, it considers translation of Thai to Chinese. First of all, it makes part-of-speech tagging and entities realize in Thai text through the lexical analysis tool of the laboratory development [19]. Through the process of 2.1 to generate a news feature words sequence, then words of the words sequence find the Chinese translation words with part-of-speech based on bilingual dictionary. Because of the difference of Chinese and Thai language, there are some ambiguous words in Chinese-Thai test feature words. Through TextRank algorithm to find 5 unambiguous Thai nouns of the highest score vote for Thai translation words, and the highest score indicates that the five nouns are the most closely related to the translation word. To translate the five nouns into Chinese words, then to judge the similarity between the candidate word and the five Chinese words, the word of the highest similarity is the Chinese translation word. Such as, the Thai word of **นิวเคลียร์** has two meanings in the Thai-Chinese bilingual dictionaries, and the two meanings are atomic nucleus and cell nucleus [17]. But words if the five unambiguous Thai nouns translated into Chinese word for North Korea, country, Jeong-eun Kim, weapons and earthquake. It is obvious that the word distribution similarity between atomic nucleus and the five related words is higher than cell nucleus. So, it judges the word of Chinese translation as atomic nucleus.

In this paper, it calculated the word similarity through the Google word2vec [12] tool. The word2vec is a open-source tool for Google in 2013, it can be used to effectively characterize the real-value vector from word, then the similarity of vector space can be used to represent the semantic similarity of word. Using word2vec to train the news corpus of Sogou Laboratory. Each word is represented by 200-dimensional word vector distribution, and using cosine similarity to calculate the similarity between any two words.

2.2.2 The Chinese-Thai mapping dictionary of one-to-many

The differences of language culture between Chinese and Thai and the phenomena of logogram omission in news make some topic feature words between them be not the simple relationship of word alignment. There exists a phenomenon of verbal nominalization [13] and compound words formed by a combination of simple words [14] in Thai. For example, "express delivery" is a word in Chinese, while its meaning is "**จดหมาย**/piece, **ด่วน**/fast (this means something fast.)". Because some Chinese phrases correspond to a word of Thai, at the same time some Thai phrases may also correspond to a word of Chinese, this paper developed a special vocabulary of Chinese and Thai. The vocabulary contained some Chinese-Thai dictionaries which had segmentation of different sizes but the same meaning from Thai to Chinese. It was suitable for the translation of Chinese-Thai words in bilingual news. Some words are shown in Table 1.

Table 1. The one-to-many bilingual dictionary of Chinese-Thai.

Thai word	Chinese word
วัน day / เสาร six	Saturday
คน man / จีน China	Chinese
ตลาด bazaar / ขาย sell / ดอกไม flower	flower fair

คน man / โบราณ ancient times	the ancients
โทรศัพท์	call / phone
โทรกลับ	call back / phone

2.2.3 The semantic similarity translation of Chinese-Thai words

Because it is different of the description way and level of detail for the same event in news text of Chinese and Thai. It leads to the differences of news words in the two languages. Since the WordNet could better reflect the semantic relationship between words and the semantic information resource of the English WordNet was abundant, this paper used the synset_id correspondence of the WordNet synonym sets of different language versions to build the correspondence between the Chinese-Thai WordNet and the English WordNet, thus inquiring the synonymous and part-whole semantic relationship of the Chinese-Thai words on the English WordNet.

2.2.4 The construction of two-item credible set of Chinese-Thai feature words

The trusted association rule and the maximum clique algorithm [9] construct the two-item credible set and the trusted association rule according to the adjacency matrix. This paper did the processing of deduplication on the sequence set of Chinese-Thai feature words generated in section 2.1, and then the paper found the Chinese words corresponded to the Thai words according to the section 2.2.1-2.2.3, thus forming Chinese-Thai equivalent word pairs. If the unique word in Chinese or Thai didn't have a suitable translation item, the paper characterized it only in the form of Chinese or Thai, thus forming a characteristic equivalent vocabulary. The vocabulary was numbered sequentially. The paper regarded the adjacency matrix as A, the items of A as a_{ij}. Among them, i was the row index and j was the column index. They both started from 0. i and j both represented the id of a word in the Chinese-Thai equivalent vocabulary. The id was the characteristic equivalent word pairs of i, denoted $wordi$. If the part of Chinese or Thai in the equivalent word pairs appeared in the corpus, it was considered that the equivalent word appeared. The paper constructed the a_{ij} as follows:

(1)When $i=j$, if the $wordi$ appeared in the text of a document, a_{ii} + 1. If the $wordi$ appeared in the title of a document or the first sentence of the first paragraph, $a_{ij}+2$. Each document only counted once for a_{ii}, and finally the maximum bonus was taken. For example, the $wordi$ appeared twice in a document, once in the title and the other in the text, this document added 2 to a_{ii}, not repetitive;

(2)When $i \neq j$, if the $wordi$ and the $wordj$ never appeared in the same news, the a_{ij} was 0;

(3)When $i \neq j$, if the $wordi$ and the $wordj$ simultaneously appeared in the title of a news text, $a_{ij}+2$. If the $wordi$ and the $wordj$ appeared in the text or only one word appeared in the title or the first sentence of the first paragraph, while the other word appeared in the text, $a_{ij}+1$. Similarly, each document only added 1 to a_{ij} and the maximum bonus was also taken.

$$\begin{bmatrix} a_{11} & a_{12} & \cdots & a_{1n-1} & a_{1n} \\ a_{21} & \cdots & & & \\ \cdots & & a_{ij} & & \\ a_{n-11} & & & \cdots & \\ a_{n1} & a_{n2} & \cdots & a_{nn-1} & a_{nn} \end{bmatrix}$$

The paper constructed the adjacency matrix A through the above process. a_{ii} reflected the text frequency of the $wordi$ appearing in the Chinese-Thai news corpus. a_{ij} reflected the text frequency of the $wordi$ and the $wordj$ appearing simultaneously. The paper further constructed the two-item credible set of matrix C through the adjacency matrix A. The element c_{ij} of C reflected the co-occurrence relationship of the $wordi$ and the $wordj$. The initial definition of c_{ij} was calculated as follows:

$$C_{ij} = \frac{a_{ij}}{a_{ii} + a_{jj} - a_{ij} + 1} \tag{2}$$

Then the paper determined the c_{ij} was greater than the lowest confidence level or not. If the value of c_{ij} was greater than θ, there existed a trusted relationship between the $wordi$ and the $wordj$ and the c_{ij} was updated as 1, otherwise the c_{ij} was 0. Finally, the paper generated the two-item credible set of matrix C. The paper could find all the two-item feature word trusted relationship sets through the traversal on the matrix C.

2.2.5 Unearthing the maximum clique of topic feature words:

The paper regarded the two-item credible set matrix generated in the progress above as C_2. Since a topic needed more feature words to express, the paper needed to complete a credible set of items from k to $k+1$, that is, the matrix $C_k \rightarrow C_{k+1}$. This paper generated the k-item credible set through the following algorithm:

(1)The initial state was C_2, which was the two-item credible set obtained through calculation;

(2)This step required setting up a generating from C_k to C_{k+1}. Firstly, the paper arbitrarily selected two k-item sets. If the front $k-1$ items were completely the same and the last item of the two k-item sets appeared in the two-item credible set, the front $k-1$ items and the last item of the two k-item sets were regarded as the candidate set of C_{k+1}. Then the paper determined that all other items were credible or not. If they were credible, the paper generated C_{k+1} and marked all of its k-item subsets. When all the work was completed, if there were k-item sets which could not be trusted, they could not be merged.

2.3 The maximum clique topic generated the quasi maximum clique topic through clustering

This paper had a lot of topics through analyzing the topics generated in the above process. However, because the language culture between Chinese and Thai, the translation deviation of media and the public opinion of the two countries are different about the angle and the degree of emotional stance for the same topic, it leads to the different representations between Chinese and Thai for the same topic, that is, the topic distributions are inconsistent. For example, some Chinese-Thai news topics were shown in Table 2.

Table 2. The maximum clique news topic.

1.	North Korea / เกาหลีเหนือ, hydrogen bomb / ระเบิดไฮโดรเจน, test / การทดสอบ, success / ประสบความสำเร็จ, arrangement / รูปแบบ, security / การรักษาความปลอดภัย, America / ในสหรัฐอเมริกา, menace / การคุกคาม, Russia / รัสเซีย, Republic of Korea / เกาหลี, European Union / สหภาพยุโรป, expression / กล่าวว่า, condemn / ประณาม,sanction / การลงโทษ
2.	North Korea / เกาหลีเหนือ, hydrogen bomb / ระเบิดไฮโดรเจน, test / การทดสอบ China / จีน, Ministry of Foreign Affairs / กระทรวงการต่างประเทศ, statement / ประกาศ, oppose / การต่อต้าน, urge / เรียกร้องให้, denuclearization / โครงการอาวุธนิวเคลียร์, peninsula / คาบสมุทร, obligation / หน้าที่, strive / ความพยายาม
3.	Jeong-eun Kim / คิมจองอึน, North Korea / เกาหลีเหนือ, international / ระหว่างประเทศ, test / ทดสอบ, hydrogen bomb / ระเบิดไฮโดรเจน, celebrate / เฉลิมฉลอง, commend / ยกย่อง

This paper did clustering on the topics meeting certain similarity to create a quasi maximum clique. The topic similarity was calculated as follows:

$$similarity(topici, topicj) = \frac{wi \cap wj}{\min(wi, wj)} \quad (3)$$

Among them, $wi \cap wj$ was the number of synonymous and similar words between the two topics. If there was a pair of synonyms, $wi \cap wj + 1$. If there was a pair of similar words, such as China and the United States, opposition and condemnation, $wi \cap wj + 0.5$. The paper inquired the similarity of Chinese words mainly through the Word2vec. If the similarity of the two words exceeded 0.5, the two words were similar. $\min(wi, wj)$ was the number of words in the word-fewer document. The steps of merging maximum clique were as follows:

(1)Firstly, the paper loaded all maximum clique sets. Each group was placed into a word queue and the initial weight of each word was counted as 1. Then the data structure was placed into a list;

(4)The paper took the first group from the list, and calculated its similarity with other groups;

(3)If they were not similar, the paper placed the first group into the result set and returned back to the step 2 to continue;

(2)If the first group was similar with the m -th group after calculating through the formula 3, the paper did clustering statistics on the words of these two groups. The specific practices were as follows: The weight of the same word was added 1. Similar words were combined into a phrase, and then the weight was added 0.5. The weigh of the words neither different nor similar was changeless. The paper filtered the word sequence after statistics and abandoned the word whose weight was less than 1.5. The filtered phrase was added into the new group. Then the new group was placed into the m -th position of the list and the original group was replaced. Then the paper deleted the first group and returned back to the step 2 to continue;

(5)These steps ended until the list was empty.

The paper merged the bilingual topics whose distribution of feature words was inconsistent but satisfied the semantic relationship through the clustering process of maximum clique. The description purity of the quasi maximum clique topic was more higher after merger. It reduced the impact of noise word on the topic description and truthfully represented the bilingual content of the topic between Chinese and Thai.

2.4 Personalized recommendation of topics

Now, Internet will produce massive data everyday. But people often have no way to start, in the face of a various data. In this paper, it designed a method of Chinese-Thai bilingual topic detection online, and the method recommended the description of expectant maximum clique topic to the user after the merger of topics. If the user is interested in a topic, then the method pushes a maximum clique queue of the topic to the user. The method calculates the similarity between bilingual topics of the maximum clique queue and the news corpus texts, and the news text time of the highest similarity is defined as the topic build time. The method provides user with a topics set of the topic, sort it by topic build time. Personalized recommendation result of the North Korea H-bomb test as shown in Table 3.

Table 3. The personalized recommendation example for bilingual news.

quasi maximal group topic	language	before clustering topic
North Korea, hydrogen bomb, เกาหลีเหนือ[3], ระเบิดไฮโดรเจน[3],	Chinese:	1.(01/07/2016 08:34) North Korea, hydrogen bomb, test, success, arrangement, security, America, menace, Russia, Republic of Korea, European Union, expression, condemn, sanction
test,	Thai:	เกาหลีเหนือ, ระเบิดไฮโดรเจน, การทดสอบ,

การทดสอบ[3], { America, China, Russia,		ประสบความสำเร็จ, รูปแบบ, การรักษาความปลอดภัย, ในสหรัฐอเมริกา, การคุกคาม, รัสเซีย, เกาหลี, สหภาพยุโรป, กล่าวว่า, ประณาม, การลงโทษ
{ ในสหรัฐอเมริกา, จีน, รัสเซีย, Republic of Korea, European Union },	Chinese:	2.(01/08/2016 17:11) North Korea, hydrogen bomb, test, China, Ministry of Foreign Affairs, statement, oppose, urge, denuclearization, peninsula, obligation, strive
เกาหลี, สหภาพยุโรป}[3] , { expression, statement }, { กล่าวว่า, ประกาศ}[1.5] ,	Thai:	เกาหลีเหนือ, ระเบิดไฮโดรเจน, การทดสอบ, จีน, กระทรวงการต่างประเทศ, ประกาศ, การต่อต้าน, เรียกร้องให้, โครงการอาวุธนิวเคลียร์, คาบสมุทร, หน้าที่, ความพยายาม
{ condemn, oppose }, { ประณาม, การต่อต้าน}[1.5] ,	Chinese:	3. (01/12/2016 10:00) Jeong-eun Kim, North Korea, international, test, hydrogen bomb, celebrate, commend
{ celebrate, commend } { เฉลิมฉลอง, ยกย่อง}[1.5]	Thai:	คิมจองอึน, เกาหลีเหนือ, ระหว่างประเทศ, ทดสอบ, ระเบิดไฮโดรเจน, เฉลิมฉลอง, ยกย่อง

3 Experiments and results

In order to investigate the effect of a bilingual topic for online discovery, this paper has designed two sets of tasks.1. Through different keyword extraction methods to compare the effect of a maximum clique of bilingual topic for discovery; 2. Compare this method with other mainstream bilingual topic discovery methods. The experimental corpus has selected 161 bilingual topic news corpus from the internet between China and Thailand in Jan 2016, which contains 86 Chinese News and 75 Thai News. The experimental corpus involve 16 aspects of international, ASEAN, economic and trade, cultural and entertainment four topics.

Firstly, through different keyword extraction methods to compare the effect of a maximum clique of bilingual topic for discovery. The experiment are planed to compare the effects of generating the topic through three different ways: TextRank keyword extraction method, mainstream TF-IDF keyword extraction algorithm and not extracting keyword. TextRank keyword extraction method sets 5 co-occurrence window, d of the damping coefficient is 0.85, and confidence level of credible association rule is 0.22. The evaluation indicators include the precision rate of topic detection, recall rate and F-measure.

Table 4. Results of different extraction methods.

index / method	Precision rate	Recall rate	F-measure
not extracting keyword	67.09%	73.96 %	70.35%
TF-IDF	54.62%	62.37 %	58.23%
TextRank	65.54%	72.85 %	69.03%

Based on the Table 4, it is showed that the effect of TF-IDF keyword extraction algorithm is the worst but not extracting key words is the best. The reason of this result is that when we use TF-IDF in a text keyword extraction process, we need to consider the word frequency in the anti-external knowledge of other text. However, this will inevitably result in characteristic value deviations in the process of lexical translation between Chinese and Thai. This paper thinks that TF-IDF is not suitable to describe the task of a single text, it is suitable for the task of finding out the characteristic of text difference. However maximum clique method need to find out the commonness between the text from Chinese-Thai news, TextRank is more suitable for the discovery of a bilingual group than TF-IDF. Full text vocabulary without the extraction of keywords after removal of the stop words and generating the topic of the maximum clique contains more comprehensive information, so it's the best. But because it also contains some noise and translation errors, the final F value is only slightly better than TextRank. Considering that not extracting key words will increase the dimension of adjacency matrix and the number of translation disambiguation, which greatly increased time complexity. In this paper, we use TextRank keyword extraction method to generate topic.

In order to verify the effectiveness and performance of this experiment, Experiment 2 is based on machine Translation k-means [16] text clustering and bilingual LDA [4,5] to compare the method of this paper with the mainstream of bilingual topic. The text clustering and bilingual LDA methods need to point out the number of K values in advance, K value is set to 16. This paper select 621 on the Chinese encyclopedia Chinese-Thai parallel texts from China Radio International as the training data set of LDA model. $\alpha = 50/K$, $\beta = 0.01$. LDA method can only filter the topic distribution probability weight of more than 0.03 of document.

Table 5. Results of bilingual topic detection methods.

Index / Method	Precision rate	Recall rate	F-measure	Consuming time
Text Clustering (k-means)	51.89%	45.31%	48.37%	819.53s
Bilingual LDA	62.13%	69.20%	65.46%	118.72s
TextRank	65.54%	72.85%	69.03%	96.48s

We can find out the Table 5:

(1)The method of this paper is the highest value of F, then bilingual LDA , the last is text clustering. When we use the text clustering, we need to compare the similarity between the topic vector and the document vector and the document vector. Due to the feature vector clustering on text level need to cross, it inevitably increases the noise characteristics and translation errors.

(2)The effect of bilingual LDA bilingual topic discovery is not very ideal. The analysis considers that the training data set of LDA does not cover all the topic distribution and its ability to discover new topics is poor. Each of these topics has a fixed number of words, while the topic in the actual description of the number is not fixed. The alignment effect of the bilingual topic is entirely dependent on the quantity and quality of the training set of the LDA parallel corpus. Despite the massive collection of Chinese-Thai News parallel text training topic model can improve the F value of bilingual LDA, the Internet Chinese-Thai parallel corpus, especially news parallel corpus resources are scarce and we need to invest more in human and linguistic knowledge. The instability of quality parallel materials influence the effect of bilingual LDA.

(3)In time complexity, the method is the lowest. Owing to the extraction of the keywords of the news document, we can significantly reduce the number of word processing objects and do not need to carry out the convergence process of k-means and LDA text iteration. This method is most suitable for a bilingual online topic discovery application.

4 Conclusion

In this paper, we cluster the cross language words to generate a bilingual topic through maximal cluster. The method of this paper is a reasonable solution to the problem of the distribution of the topic in the bilingual environment and it is effective to solve the traditional topic detection methods to estimate the number of topics at the same time. The best results are obtained when the time complexity is lowest. This method provides a reasonable way of thinking for the application of bilingual topic on line. The next step is to consider the time factor into the topic discovery model and research the more reasonable way to solve a topic distribution of feature words inconsistent problem. We look forward to a better effect on the discovery of bilingual topics.

Acknowledgment

This paper is supported by China National Natural Science Foundation （61562049), （61662040）.

References

1. Leek, T., Jin, H., Sista, S., & Schwartz, R. (2000). The BBN crosslingual topic detection and tracking system. Tdt Evaluation System Summary Papers, 894--01.
2. Wactlar, H. D. (1999). New directions in video information extraction and summarization. Proceedings of Delos Workshop, 64(8), 229–232.
3. Y.J. Chen and H.H. Chen(2002). NLP and IR approaches to monolingual and multilingual link detection. Proceedings of the 19th international conference on Computational linguistics-Volume 1, pages 1–7.
4. Mimno, D., Wallach, H. M., Naradowsky, J., Smith, D. A., & Mccallum, A. (2009). Polylingual Topic Models. Conference on Empirical Methods in Natural Language Processing, EMNLP 2009, 6-7 August 2009, Singapore, A Meeting of Sigdat, A Special Interest Group of the ACL(Vol.2, pp.880--889).
5. Ni, X., Sun, J. T., Hu, J., & Chen, Z. (2009). Mining multilingual topics from wikipedia. Www, 1155-1156.
6. Mihalcea, R., & Tarau, P. (2004). TextRank: Bringing Order into Texts.Conference on Empirical Methods in Natural Language Processing, EMNLP 2004, A Meeting of Sigdat, A Special Interest Group of the Acl, Held in Conjunction with ACL 2004, 25-26 July 2004, Barcelona, Spain(pp.404-411).
7. Agrawal, R., Imieliński, Tomasz, & Swami, A. (1993). Mining association rules between sets of items in large databases. Acm Sigmod Record,22(2), 207-216.
8. Agrawal, R., & Srikant, R. (1994). Fast Algorithms for Mining Association Rules in Large Databases. International Conference on Very Large Data Bases (Vol.7, pp.487-499). Morgan Kaufmann Publishers Inc.
9. Bo, X., & Beijing. (2008). Credible association rule and its mining algorithm based on maximum clique. Journal of Software, 19(10), 2597-2610.
10. Chunying, L. I., Tang, Y., Tang, Z., Huang, Y., Yuan, C., & Zhao, J. (2015). Community detection model in large scale academic social networks. Journal of Computer Applications.
11. Jie Yang(2009). The research of keyword extraction technology in multi-document. Shenyang Institute of Aeronautical Engineering.
12. Mikolov, T., Sutskever, I., Chen, K., Corrado, G., & Dean, J. (2013). Distributed representations of words and phrases and their compositionfouality. Advances in Neural Information Processing Systems,26, 3111-3119.
13. Yiyuan Luo(2011). Based on Thai. World Publishing Corporation.
14. Xiaorui Pei(2001). New Thai grammar. Peking University Press.
15. Thoongsup, S., Robkop, K., Mokarat, C., Tan, S., Charoenporn, T., & Sornlertlamvanich, V., et al. (2009). Thai wordnet construction. 139-144.

16. Yuan, F., Zhou, Z., & Song, X. (2007). K-means clustering algorithm with meliorated initial center. Computer Engineering, 33(3), 65-66.
17. Chen, A. L., Tang, C. J., Tao, H. C., Yuan, C. A., & Xie, F. J. (2004). An improved algorithm based on maximum clique and fp-tree for mining association rules. Journal of Software, 15(8), 1198-1207.
18. ----"Thai Lexical Analysis Tool".

    </>.
19. ----"Thai-Chinese Dictionary".

    <http://www.thai-language.com/>.
20. Han, Z. M., Zhang H., Zhang M. (2015). Fast topic detection and evaluation towards massive short texts. Computer Application Research, 32(3), 717-722.

Investigate the Computer Information Network Security Technology and the Development Direction

Teng Ping

Liaoning Police College, Liaoning, 116036 China

Abstract. After China's accession to the WTO, the computer information network security technology of our country has a rapid development, bring many conveniences for the people's life and work, indirectly changing their daily life and working mode. For the whole development situation of our country, the development direction of informatization is the inevitable trend of development in our country, while the using process of computer information network security technology in the society still exist many problems, and the network security will be threatened to some extent. Therefore, people pay a higher attention to the security problems of computer information network technology.

Keywords. Computer, information network, security technology, development direction.

1 Introduction

As the technology of our country has already reached a certain height, the computer network resources become an important power to rapid developing of Chinese society. The development of computer network information technology has laid a solid foundation for our country's social development. It today's computer network information technology that has deep into the society in the economic, political, military, science and technology, daily life habits and work habits and other fields ,which transforms the people's daily life habits and way of life to a large extent, and realizes the socialization of computer development in the true sense. The development and change of the information age are very quick, so it often appears many problems inevitably. Consequently, we need continuously to develop, update and perfect computer network information security technology to solve these inevitable problems.

2 The computer information network security technology

There are many kinds of levels for computer information network security protection, which is being widely used and well-developed has the following kinds:

One is the firewall technology. Firewall technology refers to the separate by predefined strategy between intranet and extranet, is safety measures of implement mandatory access supervision on the Intranet and extranet communication. Its main contents are hiding the inside network structure, the promotion of security access and control to network, constraining the visiting etween internal network users and outside users, and preventing external users from using illegal means to attack the internal network, steal internal data and endanger the safety environment of internal network operating system. The way of checking the two or more network transmission data, link way, is carried out in accordance with certain security policies. Whether various network can communicate with each other is determined by testing results,. The firewall also can effectively monitor network running status, as long as some suspicious behavior appear, will conduct the early warning. It will automatically provide network testing on whether there areattacksand so on.Firewall uses the way of the separation of intranet and extranet can isolate the key website in the intranet site, in order to avoid the internal information leakage, reduce local focus, network security issues impact on the risks of the global network.

The second is data encryption technology. The data encryption information of computer information network security technology is to transcode and recode the clear data, transform the previous data into a cipher text data, then transform the cipher text back into clear data when applying. This is the data encryption technology. Its main purpose is to hide the information content and protest the information security. For the dynamic information, data encryption technology is a kind of protective measure. An encryption system including many types, such as expressly set, cipher, the key data collection and algorithm. On the specific, the algorithm includes some calculation formula, rules or procedures. We can say that for the confidentiality of data, data encryption technology has been a technology with high confidentiality, which is more used in the society, improving the confidentiality of data, preventing from being stolen and modified.

Three is intrusion detection technology. In computer information network security technology, intrusion detection technology can timely discover and report the technical means involving abnormal or unauthorized phenomenon in the computer system of. Under the premise that does not affect the network performance, real-time monitoring computer information network, including internet use behavior, packets, security logs, etc., indentifing movable which does not been wanted to happen in the network, protecting the network security problems caused by network attack and misoperation, to ensure the safety of the information resources in the system. Intrusion detection is known as the second exit after the firewall, that can be divided into three kinds of measurement methods, namely: the intrusion detection based on the network, intrusion detection based on computer hostand hybrid intrusion detection. The most widely used one is the intrusion detection based on network security.

Fourth, vulnerability scanning technology. Computer information network vulnerability refers to the safety defects of the using computer hardware, software and strategies, which will make the network system of attacker invasion into the internal network, with a very short time, easily, and access, steal, control network system resources. Therefore, the vulnerability scanning system can automatically detect the hardware, software, and strategy of the local or remote host, etc. Vulnerability scanner collect information through service ports of various network query, host, local host or a remote system scan , and use the shortest possible time to query each other bugs and security hidden danger existing in the network system, and help workers timely, accurately, and systematically grasp the network security situation, make system optimization, configuration, timely repair, update, improve the network system of loopholes and safe hidden trouble, remedy the holes of network security maximum, and eliminate safety hidden trouble.

Five is the visualization technology in network security. Visualization technology of the computer information network security is a reasonable complement and perfect for the firewall technology, vulnerability scanning technology, intrusion detection technology. The visualization technology of network security is a reference to human visual model and the ability of structure framework, to show people massive high abstract data in a network with presents, graphics, images, and other forms, to be used in the network communication real-time show special information to the people. Network security visualization also is the real reflection of the entire network running status currently we use. The visualization technology take the information network security risks exist in the operation on a humane way accurately to the user. Visualization technology can help network security analysis workers with the most intuitive way to fast, accurate and comprehensive information network of the bugs and security problems, analysis and grasp the status of the network operation fastly and accurately in the vast and complex high-dimensional data information, then quickly recognise the network abnormal or invasion, accurately judge the development direction of network security threat, make sure the computer information network security technology actively develop into intelligent direction.

3 The development direction of computer information network security technology

Computer information network technology constantly improve, at the same time, the number of the social hacker attacked on computer networks becomes more and more, their skills also are higher. Many hackers theft information by againsting software, or release some fraud information in computer network etc., that affect the property and information security of the consumer. As soon as possible in order to clear the security problem of computer information network, the computer information network security technology in the continuous development and innovation, and emerged many new thoughts, ideas, methods, technologies and products. The development direction of computer information network security technology in our country can be summarized as the followings: one is the computer information network security technology will become a kind of security management platform in the direction of development from a single product. The security management platform is centered on security policy and effectively combines the single product, build and implement a series of safety management system. Second, by the development of the passive form into active development. Active development form can effectively respond to cyber attacks, traps and other security issues, and can automatically recover, also can effectively protect itself. Three, to develop in the direction of visualization of computer information network , studying in a variety of visual structure represent data information, so as to realize the visibility, local and global of the computer network information, to achieve the ultimate goal of computer information network visualization man-machine exchange. Four is intranet network information security technology become more perfectly, can effectively and accurately prevent software for users to implement attacks, the error of safety factor. Five, the computer information network security technology development direction from static to dynamic development direction. We will change the past passive attack and invasion of the computer information network system, active prevent and avoid attack and invasion, timely response to safety concerns, auto repair loopholes, invasion of hackers can anti-tracking positioning network IP and other direction.

4 Conclusion

The rapid development of computer network technology makes the importance of safety in the computer information network is increasing high. With the rapid development of information age, computer information network technology also has a lot of flying. The computer has provided many conveniences for people's life, at the same time, also brought a lot of difficult to solve the problem of security. To ensure that improve and solve the computer information network security must take certain information network security technology, and continuously provide conduct in-depth research, development and development. There are still needs about the future development of the computer information network security technology, constantly promote people's understanding of computer information network security, what can ensure that make the correct use of the computer information network technology.

References

1. C. Xu: Information system engineering, Vol. 3 (2014).
2. W.B. Wu, Z.R. Yang: Journal of Hebei University, Vol. 38 (2013).
3. B. Yang: Technology Wind, Vol. 20 (2012).

A Generalized Mathematical Model for the Fracture Problem of the Suspended Highway

Ying Zhao and Ming Lu

Hebei University of Engineering, Handan, China
First author: Zhao Ying E-mail: 364408161@qq.com
Second author: Lu Ming E-mail: 469331499@qq.com

Abstract. In order to answer dangling fracture problems of highway, the suspended pavement equivalent for non - suspended pavement, through the special boundary conditions has been suspended highway stress field of expression, in accordance with the 3D fracture model of crack formation, and establish a vacant, a general mathematics model for fracture problems of highway and analysis in highway suspended segment weight and vehicle load limit of highway capacity of Pu For overturning road inPu is less than the force of carrying more than compared to the work and fruit Bridge Hydropower Station Road engineering examples to verify suspended highway should force field expressions for the correctness and applicability. The results show that: when the hanging ratio R 0.243177 limits of Pu design axle load 100kN. When the vertical crack in the vacant in the direction of length greater than 0.1, the ultimate bearing capacity is less than the design axle load 100kN; when the hanging ratio R is less than 0.5, the road to local fracture, the ultimate bearing capacity of suspended stress field expressions in solution; when the hanging ratio is greater than or equal to 0.5, the road does not reach the limit bearing capacity of the whole body; torque shear surface of the effect is far less than the bending moments on shear planes.

Keywords. Road engineering, fracture mechanics, generalized integral model, stress field expression.

1 Introduction

The problem of fracture and damage of suspended highway is related to the safety of driving and the progress of construction. It is a common concern of hydraulic engineering, road engineering and civil engineering field [1, 2]. Due to the coupling effect of the complex stress factor and the development effect of the crack port, it has brought many difficulties [3] to the research work.

In view of this, the domestic and foreign scholars have conducted a lot of research on the problem of the fracture and failure of the suspended highway. J L [4] based on the finite element theory to generalized the finite size of the plate with a crack and to obtain the numerical solution of the weight function, finally chalk up the stress intensity factor. Ozer H [5] based of experiment exploring the concrete pavement fracture mechanism under the action of stress in low, found that the transition layer between the surface layer and the base is the root cause of fracture, but Ozer H found that the transition layer and the classic smooth contact hypothesis conflict. X Chen [6] using the Fourier transform to convert the load on the suspended part of the highway pavement into the positive half wave load and then obtained the analytical expression of the deflection and the analytical expression of the stress. Li Chuang [7] Using the idea of semi analytical and generalized mathematical integral to solve water conservancy engineering problems, B Hill [8] and others based on Li Chuang's idea obtained the fatigue crack growth rate model . J Hebel [9] and others based on the finite element simulation method to explore the role of brittle materials under the load of the role of their own fracture damage and the degree of damage to the mathematical relationship with the load.

However, the literature is the use of westergarrd formula of boundary conditions, Seitl S [10] found that the popular westergarrd formula of boundary conditions is not suitable for the suspended highway fracture problem we should explore the new boundary conditions to answer problems of suspended highway. T P [11] Corrales and others found that the analytical method is more accurate than the numerical method, which can reflect the mathematical relationship between the crack and the fracture damage more directly and effectively. In view of this, this article from the Huaneng Lancang River Commission Ministry Gongguo bridge project, aiming at Gongguoqiao highway fracture cases used the classical Winkler foundation hypothesis, special boundary conditions, semi analytical solution and generalized integral mathematical model, studying on stress field distribution of suspended highway concrete, fracturing condition for suspended concrete ultimate load and the corresponding state , in order to provide reference for engineering project.

2 Derivation of stress field

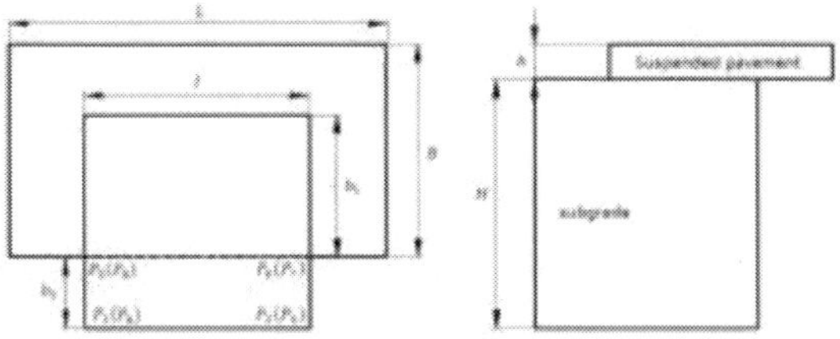

Fig.1. Top view (left) and left view (right) of the model of highway suspension.

Figure 1 shows the suspended highway model, P1 to P8 for the eight vertices of the suspended part. And the prescribed parameters b and R

$$b = b_1 + b_2 \tag{1}$$

Among them, b is the overall width of the highway pavement, b1 is the width of the non suspended part of the concrete road pavement, b2 is the width of the suspended part of the concrete road pavement.

$$R = \frac{A_{sus}}{A} \tag{2}$$

Among them, R is the suspension ratio, A is the total volume of the concrete highway pavement, Asus is the volume of the suspended part of the concrete road pavement.Figure1, in the mechanics studies literature was said elastic foundation, and highway engineering prefer subgrade pavement, in order to facilitate the presentation, do not distinguish between subgrade and roadbed, pavement and patch panels nor area in this paper.

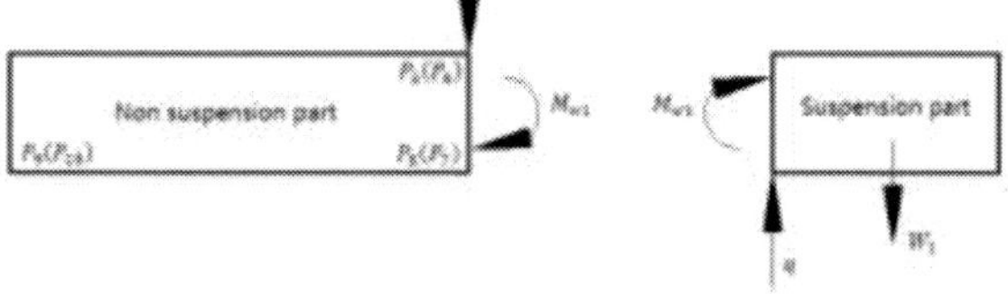

Fig.2. Pavement Stress Condition.

Figure 2 shows the stress situation of the suspended highway model, W_1 is the dead weight of the suspended road surface, M_{W1} is the moment of the dead weight of the suspended road ,generated in the P_8 P_5 P_6 P_7 section (shear surface), q for the gravity equivalent uniform load. So

$$W_1 = lb_2 h\gamma \tag{3}$$

$$q = \frac{W_1}{l} = b_2 h\gamma \tag{4}$$

$$M_{W_1} = \frac{W_1 b_2}{2} = \frac{1}{2} l b_2^2 h\gamma \tag{5}$$

Among them, the γ is the gradation of concrete.

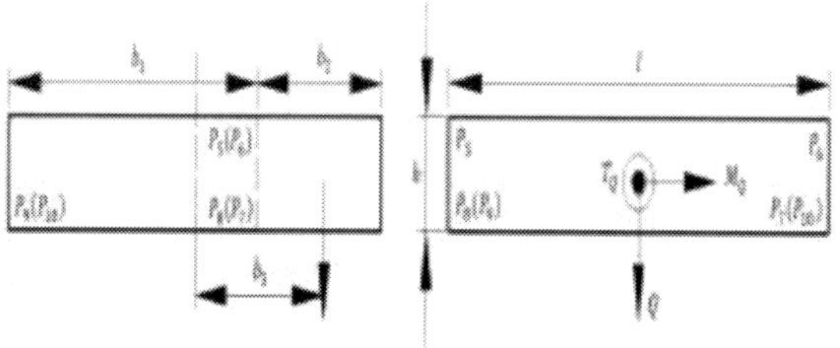

Fig.3. Traffic load condition.

In the same way, assuming that the vehicle load is Q, b_3 is the distance from the P_8 P_5 P_6 P_7 section (the shear surface), and the M_Q is the moment of the driving load, and T_Q is the torque generated by the driving load.

$$M_Q = Qb_3 \tag{6}$$

$$T_Q = \frac{1}{2} Ql \tag{7}$$

In the $\overrightarrow{p_8 p_9}$ direction is the positive direction of axis Y, in the $\overrightarrow{p_7 p_8}$ direction is the positive direction of axis X, in the $\overrightarrow{p_5 p_8}$ direction is the positive direction of axis X.Section P8 P5 P6 P7 to coordinate origin coordinate system. Because the Winkler elastic curved surface deflection equation can be expressed as:

$$D_g \nabla^4 \omega(x, y) + k\omega(x, y) = q(x, y) \tag{8}$$

Among them, ω(x, y) for the pavement deflection; q (x, y) for the uniform load; k for the foundation reaction modulus; D_g for the pavement bending stiffness, the expression for the:

$$D_g = \frac{Eh^3}{12(1-\mu^2)} \tag{9}$$

Among them, E is young's modulus; μ is Poisson's ratio. Type (8) is equivalent to the general solution of homogeneous equation:

$$\nabla^4 \omega(x, y) + \frac{k}{D_g} \omega(x, y) = 0 \tag{10}$$

On type decomposition:

$$\omega = \omega_1 + \omega_2 \tag{11}$$

Where ω_1 is deflections of the unit plate parallel to the Y direction, but has nothing to do with X; ω_2 is the actual deformation of the plate which is related to X and Y, the type can be substituted into the equation 8. Available

$$\begin{aligned}\omega_1 = & e^{\frac{-y}{\sqrt{2l}}}\left(A\cos\frac{y}{\sqrt{2l}} + B\sin\frac{y}{\sqrt{2l}}\right) \\ & + e^{\frac{-y}{\sqrt{2l}}}\left(C\cos\frac{y}{\sqrt{2l}} + D\sin\frac{y}{\sqrt{2l}}\right)\end{aligned} \tag{12}$$

Among them, A, B, C, D is the undetermined coefficients, according to the boundary conditions:

$$\omega_1\Big|_{y=b_1} = 0 \tag{13}$$

$$M_y = -D_g\left(\frac{\partial^2\omega}{\partial y^2} + \frac{\partial^2\omega}{\partial x^2}\right) \tag{14}$$

$$F_{sy} = -D_g\frac{\partial}{\partial y}\nabla^2\omega \tag{15}$$

Among them, M_y is bending moment perpendicular to the Y axis, F_{sy} is shear force perpendicular to the Y axis, simultaneous formula (12), type (13), type (14), type (15) can get undetermined coefficient A, B, C, D

$$\begin{aligned}A = & -\eta_1 + \eta_2 - \frac{\eta_2\lambda_1}{-\lambda_2 + 2\lambda_3 + \lambda_4} \\ & + \left(1 + \eta_2\lambda + \frac{\lambda_1 + 2\lambda_2 + \lambda_3 + 2\lambda_4}{-\lambda_2 + 2\lambda_3 + \lambda_4}\right)\end{aligned} \tag{16}$$

$$Q_1 = \frac{\eta_2\lambda_1(2\lambda_1 - \lambda_2 + \lambda_3)}{(-2\lambda_1 + \lambda_2 + \lambda_4)(-\lambda_2 + 2\lambda_3 + \lambda_4) - (\lambda_1 + 2\lambda_2 + \lambda_3 + 2\lambda_4)(2\lambda_1 - \lambda_2 + \lambda_3)} \tag{17}$$

$$B = \eta_1 - \frac{\eta_2\lambda_1(2\lambda_1 - \lambda_2 + \lambda_3)}{(-2\lambda_1 + \lambda_2 + \lambda_4)(-\lambda_2 + 2\lambda_3 + \lambda_4) - (\lambda_1 + 2\lambda_2 + \lambda_3 + 2\lambda_4)(2\lambda_1 - \lambda_2 + \lambda_3)} \tag{18}$$

$$C = -\frac{\eta_2\lambda_1}{-\lambda_2 + 2\lambda_3 + \lambda_4} + \frac{\eta_2\lambda_1(2\lambda_1 - \lambda_2 + \lambda_3)(\lambda_1 + 2\lambda_2 + \lambda_3 + 2\lambda_4)}{(-2\lambda_1 + \lambda_2 + \lambda_4)(-\lambda_2 + 2\lambda_3 + \lambda_4) - (\lambda_1 + 2\lambda_2 + \lambda_3 + 2\lambda_4)(2\lambda_1 - \lambda_2 + \lambda_3)} \tag{19}$$

$$D = \frac{-\eta_2\lambda_1(2\lambda_1 - \lambda_2 + \lambda_3)}{(-2\lambda_1 + \lambda_2 + \lambda_4)(-\lambda_2 + 2\lambda_3 + \lambda_4) - (\lambda_1 + 2\lambda_2 + \lambda_3 + 2\lambda_4)(2\lambda_1 - \lambda_2 + \lambda_3)} \tag{20}$$

$$\lambda_1 = e^{\frac{-y}{\sqrt{2l}}}\cos\frac{y}{\sqrt{2l}} \tag{21}$$

$$\lambda_2 = e^{\frac{-y}{\sqrt{2l}}}\sin\frac{y}{\sqrt{2l}} \tag{22}$$

$$\lambda_3 = e^{\frac{-y}{\sqrt{2l}}}\cos\frac{y}{\sqrt{2l}} \tag{23}$$

$$\lambda_4 = e^{\frac{-y}{\sqrt{2l}}} \sin\frac{y}{\sqrt{2l}} \tag{24}$$

$$\eta_1 = \frac{(M_{W_1} + M_Q)l^2}{D_g} \tag{25}$$

$$\eta_1 = \frac{(Q + W_1)l^2}{D_g} \tag{26}$$

By substitution type 12 from type 16 to type 26, get ω_1. Suppose:

$$\omega_2 = \sum_{m=2}^{+\infty} Y_m \cos\frac{m\pi}{\alpha}x \tag{27}$$

Among them, Y_m is strange function, and α is the spacing of Q. The type (27) substitution into the equation (12), available:

$$Y_m^{(4)} - 2Y_m''\left(\frac{m\pi}{\alpha}\right)^2 + Y_m\left[\left(\frac{m\pi}{\alpha}\right)^4 + \frac{k}{D_g}\right] = 0 \tag{28}$$

Because

$$M_{yx} = M_{xy} = -D_g(1-\mu)\frac{\partial^2\omega}{\partial x\partial y} \tag{29}$$

$$M_{yx}\Big|_{y=0} = Q \tag{30}$$

M_{xy} and M_{yx} is torque. The type (27), (28), (29), (30) substitution into the type (12), combined with Y_m is an odd function of this information, took the first item of the infinite series, available:

$$\omega = e^{\frac{-y}{\sqrt{21}}}\left(A\cos\frac{y}{\sqrt{21}} + B\sin\frac{y}{\sqrt{21}}\right) + e^{\frac{-y}{\sqrt{21}}}\left(C\cos\frac{y}{\sqrt{21}} + D\sin\frac{y}{\sqrt{21}}\right) + \frac{\alpha T_Q \cos\frac{m\pi}{\alpha} x \sin\gamma_m y(e^{\beta_m y} + e^{-\beta_m y})}{4\pi D_g(1-\mu)\gamma_m \sin\frac{m\pi}{\alpha}} \tag{31}$$

Among

$$m = 2 \tag{32}$$

$$\gamma_m = \sqrt{\frac{\sqrt{\left(\frac{m\pi}{\alpha}\right)^4 - \left(\frac{k}{D_g}\right)} - \left(\frac{m\pi}{\alpha}\right)^2}{2}} \tag{33}$$

$$\beta_m = \sqrt{\frac{\sqrt{\left(\frac{m\pi}{\alpha}\right)^4 - \left(\frac{k}{D_g}\right)} + \left(\frac{m\pi}{\alpha}\right)^2}{2}} \tag{34}$$

The analytical expression of the whole stress field can be got at this point:

$$M_x = -D_g\left(\frac{\partial^2\omega}{\partial x^2} + \mu\frac{\partial^2\omega}{\partial y^2}\right) \tag{35}$$

$$M_y = -D_g(\frac{\partial^2 \omega}{\partial y^2} + \mu \frac{\partial^2 \omega}{\partial x^2}) \tag{36}$$

$$M_{yx} = M_{xy} = -D_g(1-\mu)\frac{\partial^2 \omega}{\partial xy} \tag{37}$$

3 Generalized mathematical model of crack

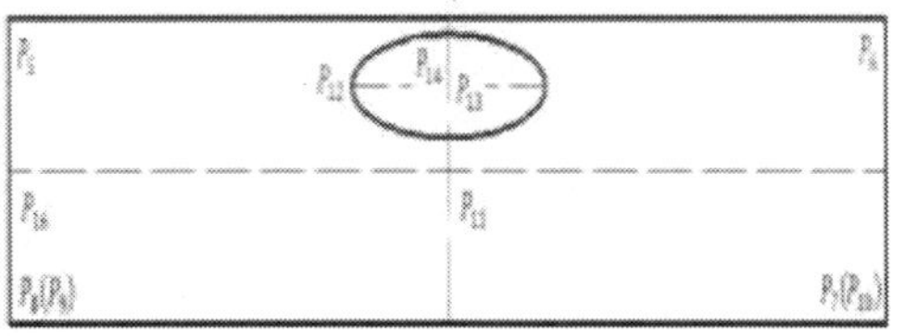

Fig.4. Pavement crack.

According to figure 4, elliptic P_{13} represents crack;P12 and P14 for the quadrant point of the ellipse; P15 is the midpoint of line P_5 P_6; P16 is the midpoint of the line segment P_8 P_5; P_{11} is the midpoint of section P_8 P_5 P_6P_7(shear plane), which is the length of the line $P_{14}P_{15}$

$$|P_{14}P_{15}| = h_1 \tag{38}$$

$$|P_{13}P_{12}| = a' \tag{39}$$

$$|P_{11}P_{11}| = e \tag{40}$$

$$|P_{13}P_{14}| = b' \tag{41}$$

iterature, stress intensity factor

$$\begin{cases} K_1 = M\sigma_y \dfrac{\sqrt{\pi b'}}{E(i)}\left[\left(\dfrac{b'}{a'}\right)\sin^2\theta + \cos^2\theta\right]^{\frac{1}{4}} \\ i^2 = 1 - \left(\dfrac{b'}{a'}\right)^2 \end{cases} \tag{42}$$

Among them, K_1 is the stress intensity factor; E (i) for the second kind of complete elliptic integral; θ is the circumferential angle of elliptic crack

$$\sigma_y = -\frac{Ez}{(1-\mu)^2}\left(\frac{\partial^2 \omega}{\partial y^2} + \mu \frac{\partial^2 \omega}{\partial x^2}\right) \tag{43}$$

The type (31) substitution into the equation (43), available

$$\sigma_y = -\frac{Ez}{(1-\mu)^2}\left\{\frac{A}{l^2}\sin\frac{y}{\sqrt{2}l}e^{\frac{-y}{\sqrt{2}l}} + \frac{1}{l^2}\left[\begin{matrix} D\left(\cos\dfrac{y}{\sqrt{2}l} + \sin\dfrac{y}{\sqrt{2}l}\right) \\ -C\sin\dfrac{y}{\sqrt{2}l} \end{matrix}\right] + Q_2 + Q_3\right\} \tag{44}$$

$$Q_2=\frac{\alpha T_Q\cos\frac{2\pi}{\alpha}x(\beta_2^2-\gamma_2^2)\cos\gamma_2 y(e^{\beta_2 y}+e^{-\beta_2 y})+2\gamma_2\beta_2\cos\gamma_2 y(e^{\beta_2 y}-e^{-\beta_2 y})}{4\pi D_g(1-\mu)\gamma_2\sin\frac{2\pi}{\alpha}x} \tag{45}$$

$$Q_3=\frac{-\pi\mu T_Q\sin\gamma_2 y(e^{\beta_2 y}+e^{-\beta_2 y})}{2\alpha D_g(1-\mu)\gamma_2}? \frac{4\sin^2\frac{2\pi}{\alpha}x\sin^2\frac{2\pi}{\alpha}x+\sin\frac{8\pi}{\alpha}x\cos\frac{2\pi}{\alpha}x}{\sin^4\frac{2\pi}{\alpha}x} \tag{46}$$

Assumed fracture toughness $K_{IC}=K_I$, according to the boundary conditions of y=0, simultaneous formula (7)、 formula (16) ~ (26)、 formula (44) ~ formula (46), the solution is obtained.

$$Q_{up}=\frac{W_1 l^2\lambda_1(2\lambda_1-\lambda_2+\lambda_3)}{(-2\lambda_1+\lambda_2+\lambda_4)(-\lambda_2+2\lambda_3+\lambda_4)-(\lambda_1+2\lambda_2+\lambda_3+2\lambda_4)(2\lambda_1-\lambda_2+\lambda_3)}-\frac{1-\mu^2 D_g E(i)K_{IG}}{MEz\sqrt{\pi b'}\left[\left(\frac{b'}{a'}\right)^2\sin^2\theta+\cos^2\theta\right]^{\frac{1}{4}}} \tag{47}$$

$$Q_{down}=\frac{-\lambda_1(2\lambda_1-\lambda_2+\lambda_3)}{(-2\lambda_1+\lambda_2+\lambda_4)(-\lambda_2+2\lambda_3+\lambda_4)-(\lambda_1+2\lambda_2+\lambda_3+2\lambda_4)(2\lambda_1-\lambda_2+\lambda_3)}+\frac{\alpha l(\beta_2^2-\gamma_2^2)\cot\frac{2\pi}{\alpha}x}{4\pi} \tag{48}$$

$$P_u=Q=\frac{Q_{up}}{Q_{down}} \tag{49}$$

Among them, Pu is the ultimate bearing capacity of highway pavement

4 Engineering example

Hydropower station is part of Huaneng Lancang River Hydropower Limited by Share Ltd, located in the town of Yunnan province. But the Yunnan province power fruit town summer rainfall along the mountain road, although the foundation treatment is good, But still by the rain, flash floods and other erosion, resulting in the suspension, which directly affects the people's lives and property safety, but also directly affect the work of the bridge hydropower station operation, construction safety. This article is entrusted by the project of gongguoqiao hydropower station , using Gongguoqiao observation data given by the Ministry of hydropower station project, combined with the theoretical derivation, explored the inner mechanism of the fracture problem, in order to provide reference for practical engineering

4.1 Ultimate bearing capacity

A set of data for the selection of Gongguoqiao provided: given the relevant size: l=5m, b=4m, h=0.3m; for crack, a' =0.0537m, b' =0.0068m, E (i) = 1.013304780919523, z=-0.145m, θ=90°; γ=25kN/m3, E=31GP, μ=0.15, K_{IC}=0.51kN/cm$^{1.5}$, M=1.1, k=33MPa/m, the design axle load is 100kN; Suspension degree R has a variety of vacant value, but 0.106, 0.132, 0.157, 0.198, 0.254, 0.259, 0.301, 0.378, 0.382, 0.400 in Gongguoqiao hydropower station is the most common. The ten set of values calculated by this method, and compared the measured data between fruit bridge project given by the Ministry, in order to verify the applicability of this method.

Defined error

$$\varepsilon = \frac{\sqrt{(value_1 - value_2)}}{value_1} \qquad (50)$$

Among them, $Value_2$ is the calculated value, and $value_1$ is the measured value.

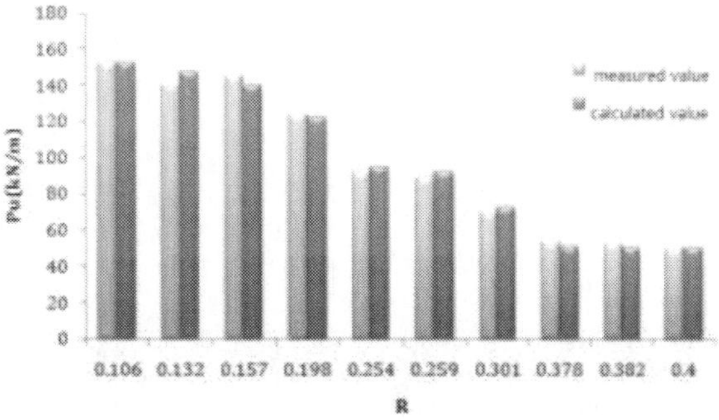

Fig.5. Comparison between calculated and measured values.

It can be seen in Figure 5, the calculated results are in agreement with the measured values, numerical value coincide basically: the maximum error is 0.0558842193609138, the minimum error is 0.0015798513235351, the average error is 0.03254999, indicating the applicability of this formula is excellent for practical engineering problems.

In addition, we can draw the following conclusions: first, with the change of the R, the ultimate bearing capacity of concrete pavement is reduced, and the numerical relationship between the two is as follows:

$$\begin{aligned} P_u &= 19.7111 + (149.733 + 578.026\cos R)\cot R \\ &+3.0797(0.948186 + R)^{\frac{17}{35}}\cot\sin R\csc(0.899379 + R^{\frac{1}{8}}) \\ &-14.7877R(0.948186 + R)^{\frac{17}{35}}\cot\sin R\csc(0.899379 + R^{\frac{1}{7}}) \\ &+191.282\csc R\csc(R^{\frac{5}{6}} + R)\csc\left|\sec R\right| - 170.909\sec R \\ &+10.2009R^{\frac{33}{14}}\sec R + R^{\frac{109}{35}}\csc(R^{\frac{1}{6}} + \left|R\right|) \\ &(0.608975\cos R - 0.00244245\tan R) - 85.3396R^{\frac{52}{35}}\sin R\tan R \end{aligned} \qquad (51)$$

Among them, $0<R<0.5$; secondly, when the hanging ratio of R is 0.3, ultimate bearing capacity is less than 100kN, the bearing capacity of the concrete at this time is less than that of the design axle load, may suddenly occur brittle fracture when driving, it will take a serious threat to traffic safety. According to equation (51) is calculated, when the hanging ratio of R is 0.243177, the ultimate bearing capacity is just equal to 100kN, if the suspension R is greater than 0.243177, the R is not allowed to travel; thirdly, according to equation (51) is calculated, when the hanging ratio R is greater than 0.5 and P_u is less than or equal to 0, When the floating ratio of R is greater than or equal to 0.5, the road is not broken, directly overturning

4.2 Crack size

Another set of data provided by the selection of Gongguoqiao: given the relevant size: l=5m, b=4m, h=0.3m; E=31GP, μ=0.15,γ=25kN/m3, K_{IC}=0.51kN/cm$^{1.5}$, M=1.1, k=33MPa/m, the design axle load is 100kN; suspended R is 0.198. Crack the values in Table 1; using the method of this paper to calculate the different values in Table 1and compared with the measured data, at the same time, the applicability of this method is verified by exploring the effect of crack size on the ultimate bearing capacity of Pu. (Table 1)

Table.1 Crack size.

a'/m	b'/m	θ	E(i)	z/m	calculated value /kN	measured value /kN	ε
0.050111	0.005017	90	1.00889	-0.152	129.153	127.17	0.015354
0.050111	0.010091	90	1.02917	-0.15324	129.328	135.261	0.045876
0.050111	0.015146	90	1.05731	-0.15393	129.481	129.183	0.002301
0.050111	0.020176	90	1.09193	-0.15467	129.612	128.678	0.007206
0.050111	0.025061	90	1.13152	-0.15525	129.721	135.252	0.042638
0.10035	0.005017	90	1.00264	-0.15567	120.471	118.357	0.017548
0.10035	0.010091	90	1.00896	-0.15668	120.647	126.082	0.045049
0.10035	0.015146	90	1.01794	-0.15756	120.8	125.452	0.03851
0.10035	0.020176	90	1.02909	-0.1582	120.931	122.037	0.009146
0.10035	0.025061	90	1.04177	-0.15833	121.039	125.448	0.036426
0.151123	0.005017	90	1.00128	-0.15858	98.3125	97.2247	0.011065
0.151123	0.010091	90	1.0044	-0.15991	98.4879	95.604	0.029282
0.151123	0.015146	90	1.00891	-0.15999	98.6408	96.0239	0.02653
0.151123	0.020176	90	1.01456	-0.16098	98.7720	101.262	0.02521
0.151123	0.025061	90	1.02104	-0.16124	98.8803	102.527	0.03688
0.201529	0.005017	90	1.00076	-0.16189	78.0032	76.7015	0.016688
0.201529	0.010091	90	1.00265	-0.16343	78.1786	77.4246	0.009645
0.201529	0.015146	90	1.00541	-0.16367	78.3315	80.351	0.025781
0.201529	0.020176	90	1.00889	-0.16376	78.4627	80.7021	0.028541

From Table 1 we can see that the calculated results are in agreement with the measured values, the formula deduced in this paper has excellent applicability to practical engineering problems.

In addition, the following conclusions can be drawn from table 1: first, a' or b' becomes larger, the ultimate bearing capacity of concrete pavement is reduced, the numerical relationship among the three is

$$\begin{aligned} P_u = & 216.302 - 3784.33a'^2 + 449912a'^5 - 1.34562\times10^6 a'^6 - 199.175b'^2 \\ & +23679.6b'^5 - 70821.9b^6 \\ & +24.2517Ina' + 1.2764Inb' \end{aligned} \tag{52}$$

Among them, 0< a' <0.25m, 0< b'<0.025m"; secondly, according to the formula (52) shows that the size of crack perpendicular to the direction of the impending greater influence on ultimate bearing capacity, the size of the crack parallel to the floating direction has little influence on the ultimate bearing capacity, a' and b' varied in the same range and the change of the same magnitude, resulting in the change a' of P_u caused by a is 19 times of the value of b' caused by P_u. Simultaneous formula (49) found that this is because the bending moment caused by the dead weight of the suspended part is relatively large.; thirdly, No matter what the value of b', when the long axle crack length is greater than 0.1M, the limit bearing capacity is less than 100kN, indicating that the bearing capacity of the concrete is far less than the design load, it may suddenly occur brittle fracture driving and take a serious threat to traffic safety, so, if the long axle crack length is greater than 0.1M, are not allowed to travel.

5 Conclusion

In this paper according to the Winkler assumption, the suspended pavement is equivalent to the non-suspended pavement, in accordance with the 3D fracture model of crack generation, through the establishment of special boundary conditions of suspended highway stress field and crack contains general ized mathematical model, so based on the principal Gongguoqiao hydropower station project department and provided the data and the generalized mathematical model to do comparison, results show that:

With the R larger than the left, the ultimate bearing capacity of concrete pavement is reduced, when the suspended ratio of R is greater than 0.243177, the ultimate bearing capacity is less than 100kN; this condition is not allowed to drive.

No matter what a' and b' became larger, the ultimate bearing capacity of concrete pavement is rapidly reduced. However a' and b' change the same amplitude, resulting in the change of value is 19 times the value of the change and moment for the shear plane is far greater than the role of torque for the shear surface.

No matter what the value, if the crack length and half shaft length greater than 0.1M, the ultimate bearing capacity is less than the design of axle load 100kN; it is not allowed to travel.

Under various conditions, the calculated results and measured results show the same trend. The more important is that the generalized mathematical model proposed in this paper is suitable for the fracture problem of the suspended concrete pavement.

References

1. ZHANG Di, FANG Lin, WANG Jian-wei. Strength and fatigue damage characteristics of cold recycled mixture of emulsified asphalt in curing period. Highway engineering,2016(1):236-241
2. MA Shi-bin, YU Qiu-fu, ZHANG Cai-li, et al. Study on the influence of pavement condition sampling inspection plan on the division of maintenance section. Highway engineering,2015,40(4):131-134
3. Page J L, Winston R J, Iii W F H. Soils beneath suspended pavements: An opportunity for storm water control and treatment. Ecological Engineering, 2015, 82:40-48
4. Ozer H, Al-Qadi I L, Lambros J, et al. Development of the fracture-based flexibility index for asphalt concrete cracking potential using modified semi-circle bending test parameters. Construction & Building Materials, 2016, 115:390-401
5. Chen X. Deflection and stress of hollow CRCP slab under concentrated vehicle load. Journal of Southeast University (English Edition), 2011
6. WANG Jun-qi, LI Chuang, DONG Ye. Semi analytical solution of Bishop method and its generalized mathematical model. Journal of hydraulic engineering and Construction Engineering, 2015(6):123-128. (In Chinese with English abstract).
7. Hill B, Oldham D, Behnia B, et al. Evaluation of low temperature viscoelastic properties and fracture behavior of bio-asphalt mixtures. International Journal of Pavement Engineering, 2016:1-8.
8. Kuruppu M D, Chong K P. Fracture toughness testing of brittle materials using semi-circular bend (SCB) specimen. Engineering Fracture Mechanics, 2012, 91:133-150..
9. Muralidhara S, Prasad B K R, Eskandari H, et al. Fracture process zone size and true fracture energy of concrete using acoustic emission. Construction & Building Materials, 2010, 24(4):479-486
10. Corrales T P, Friedemann K, Fuchs R, et al. Breaking Nano-Spaghetti - Bending and Fracture Tests of Nanofibers.. Langmuir the Acs Journal of Surfaces & Colloids, 2016, 32(5).
11. China Aviation Research Institute. Handbook of stress intensity factors. Science Press, 1993 (in Chinese with English abstract).

Application of Maintenance Simulate System of NC Machine Tools in Teaching

Xin He[1,a], Guosheng Xu [2,b], Ping He [3,c], Jiming Tan [4,d], Lifang Zhang[5,e] and Huili Zhang[6,f]

Tianjin vocational technology normal university, Tianjin 300222, China

Email:[a]3576554862@qq.com,[b]13820526206@163.com,[c]13110009352@126.com,[d]tanjiming69@163.com, [e]876898092@qq.com,[f]545278910@qq.com

Abstract. The project design is the foundation of training, which can insure the usefulness of human resource development system. The Training Project Maintenance Simulate System of NC Machine Tools is presented based on analyzed contradiction between supply and demand. This paper introduces several scheme of Maintenance Simulate System of NC Machine Tools.

Keywords. NC Machine Tool, maintenance, simulation system.

1 CNC machine maintenance personnel status quo

The numerical control technology is one of the important development directions of the machinery manufacturing industry today. It is the basic industry related to the national strategic position and comprehensive national strength, and its level is the core symbol of a country's manufacturing modernization.

China's CNC machine tool industry is lagging behind foreign advanced countries is an indisputable fact, about 100 million US dollars each year of imports and exports of millions of dollars highlights the seriousness of the problem. At present, many users to buy foreign CNC machine tools, the purpose is to ensure high-quality and guaranteed after-sales service, but the actual situation is often affordable, repair can not afford, foreign experts to spend a huge time. If the domestic CNC machine tools have high-quality maintenance services, this alone can make the majority of users to buy domestic CNC machine tools have incomparable advantages, at least in the mid-range CNC machine tool market can make CNC machine tools import and export ratio improved significantly, Which can save tens of millions of dollars of foreign exchange, and can make domestic CNC machine tool manufacturing industry access to valuable development funds to enable the industry to enter a virtuous circle. High-quality, low-cost maintenance personnel a steady stream of supply, will completely change the domestic CNC machine tool after-sales maintenance of poor image.

The Ministry of Education organized the relevant units to carry out the "NC personnel on the demand and numerical control education reform," the research task. The results are shown in Table 1.

The research results show that the utilization rate and the intact rate of NC equipment are generally low in our country, besides the insufficiency of production task due to the pervasive production task, the lack of maintenance power and programming ability are the main reasons for the underemployment, And 17.2% respectively. CNC equipment, the main reason for the failure rate of maintenance is insufficient, accounting for 38.4%. Thus, NC maintenance and NC programming and maintenance personnel is currently in urgent need of CNC talent.

Survey data also show that 59.16% of the CNC technology professionals for the technical secondary school and the following qualifications, 31.25% for the tertiary education, only 8.58% for undergraduate education, undergraduate or higher education accounted for only 1%. This situation is difficult to meet the technical needs of enterprises, so training a large number of undergraduate level of numerical control equipment maintenance personnel has been very urgent.

2 CNC maintenance personnel training training program

Today's CNC machine tool failure is completely not the traditional sense of machine failure, modern CNC machine tools using the latest power electronics technology, motor technology, test technology and computer technology and other technologies, so the operation of CNC machine tools, maintenance and management The new, completely different from the requirements of traditional machine tools.

At present, most of our CNC machine tool users are still using the master apprentice and short-term training in the way of training CNC machine tool maintenance personnel, and the proper use and maintenance requirements of the machine is not commensurate, and this is the low level of reliability of CNC machine tools One of the main reasons.

Today's CNC machine tools more to reflect the knowledge rather than experience, if the mechanical technology-based traditional machine tool use and maintenance can still be accumulated through the experience and short-term learning to make up, today's CNC machine tools across the machinery, (Sensors), electromagnetic interference and a lot of computer expertise, it is impossible to be short-term training to be quick. However, to establish a FANUC, SIEMENS and CNC systems such as NC numerical control machine tool repair training center, the school needs to invest about three to four million in education funding. This is difficult for most institutions, so it is necessary to develop a simulation system for CNC machine maintenance.

In the process of CNC machine tool training, most institutions have adopted a CNC machining simulation system. It can be a mountain of stone attack jade! Learning from successful experience, we can also develop a maintenance simulation system. In the training process, let the students operate the computer simulation system, in the grasp of the overall structure of CNC machine tools based on the familiar fault points and fault characteristics. And then let the students in the NC maintenance training on the experimental platform to master each CNC system and the spindle drive system, feed system, turret and limit switch electrical connections; Finally, let the students on the light machine for mechanical installation, electrical connection, Mechanical and electrical intermodulation, assembly of economic CNC lathes.

Students practice the simulation system on the computer for future training links have a very important role. CNC machine tools in the electrical cabinet in the 380 V power supply and 220 V switching power supply. If the students do not understand the wiring in advance is likely to be life-threatening, and may also damage the NC system and servo drives and other weak electrical equipment, so that the training schedule was interrupted, increase training equipment, capital investment.

Table 1. NC technical personnel status table.

Scope of investigation	NC talent sources			Educational background				Job type			
	From the school to recruit fresh students	Recruitment from the community	Self-cultivation operatives	Bachelor degree or above	Undergraduate	Specialist	Secondary and below	NC programming	System maintenance	Machine Operation	Above can be engaged
North-east area	71.10%	17.20%	11.70%	1.40%	17.80%	48.90%	31.90%	13.2%	16.7%	69.9%	0.2%
North China	60.20%	11.80%	28%	2%	8.50%	24.65%	65.45%	15.3%	10%	68.3%	4.9%
Central China	25%	10%	65%	0.60%	1.90%	26.20%	71.30%	16%	13%	67.0%	4.0%
North-west region	44.20%	13.65%	42.15%	0.80%	6.90%	33.60%	58.70%	6.7%	4.4%	77.9%	11%
Southwest Region	65.40%	16.90%	17.70%	1.60%	12.20%	29.10%	57.10%	12.0%	12.3%	74.1%	0
Huadong Region	29.30%	16.20%	54.50%	0.70%	6.70%	40.50%	52%	12.8%	19.5%	62.1%	5.6%
South China region	42.90%	22.30%	34.80%	0.80%	6.00%	15.70%	77.50%	9.9%	10.4%	71.6%	8.1%
Average	48.30%	15.44%	36.28%	1.0%	8.58%	31.25%	59.16%	12.6%	12.44%	70.13%	4.83%

3 CNC maintenance simulation system to achieve program

With reference to the successful experience of related fields at home and abroad, three feasible schemes are summarized:

(1) draws the numerical control processing simulation system development a set of numerical control maintenance simulation system.

The modeling of each part of CNC machine tool is made by ProE, then the friendly user interface is compiled with VC ++ and the model database and fault database are built with Access.

(2) draw lessons from Kingsoft and other software ideas to develop a set of CNC machine tool fault maintenance manual.

Using VC ++ to compile friendly user interface, using Access to build model database and fault database. Combined with data mining technology to write a fast intelligent search algorithm.

Data Mining (DM-Data Mining) is one of the most active research fields in recent years. It is a combination of artificial intelligence and database technology. It uses special algorithms to extract effective patterns from the data, Discover potential laws to extract useful information. It shows great vitality in the analysis of association rules, classification, prediction and clustering of massive data [2]. Therefore, the combination of data mining technology and CNC machine condition monitoring and fault diagnosis is conducive to break through the bottleneck of traditional condition monitoring and fault diagnosis system knowledge acquisition, so that the whole system development and application into a new stage of development.

(3) the use of interactive electronic manuals in the maintenance of weapons and equipment in the application of the development of interactive digital electronic maintenance manual [3-4].

IETM-Interactive Electronic Technical Manual (IETM-Interactive Electronic Technical Manual) is the development of technology of information network, multimedia computer, database, artificial intelligence expert system and electronic publishing. It meets the requirements of technical information management of mechanical equipment, realizes data interoperability and Highly shared technical information support system. It is able to equip the electronic equipment with a large amount of technical information and data, including technical specifications, maintenance manuals, commissioning manuals, external interface protocols, electrical schematics, cable connection diagrams, wiring tables, wiring diagrams, accessories and tools. Maintenance, maintenance manuals and other organically combined into one, based on electromagnetic media and a high degree of interaction with the ability of digital technology documents, so as to establish a paperless, intelligent, man-machine harmony and automated maintenance Environment, for technical staff to provide timely, appropriate and appropriate operational guidance and information support. The use of IETM technology can effectively improve the mechanical fault diagnosis and maintenance capabilities. Therefore, the IETM technology into the field of CNC machine maintenance, the establishment of a fully user-friendly and highly interactive electronic information maintenance environment for the maintenance of CNC machine tools to improve student ability, has a very important practical significance.

IETM includes product assurance data and product definition data. It not only emphasizes the interoperability and sharing of data, but also emphasizes the interaction between data and users. Internationally IETM is usually divided into five, five IETM, the most basic one is a (Class 1), the most advanced is five (Class 5). But as a complete IETM, it should have the following basic functions:

①IETM content includes product design, drawings, use, maintenance and other data.

② good access and interactive features. IETM provides a variety of information retrieval, search path, and can give users real-time response to the operation, accurate and timely help and guide users to operate.

③diversified display style. It can display technical information in various forms such as text, table, graphics, image, video, animation and virtual reality technology, and provide users with detailed, vivid and easy to understand technical information.

④ information real-time maintainability. For the technical staff to provide information update mechanism, so that technicians based on maintenance experience and equipment maintenance routine information, with the necessary Notes, update, enrich technical data and information.

⑤ with the computer information system of organic integration. IETM based on network and database can realize the integration with computer information system, and can easily interact with digital library, information center and technical data processing system in different places (or different places) to realize information sharing.

⑥ flexible and diverse storage media and application forms. It can be used in various types of electronic media, such as CD-ROM, hard disk, floppy disk, and tape, which can be used in desktop computers, laptops and even handheld computers, or in the form of web on the Internet.

To complete the basic functions, IETM should include at least three modules: product data database, expert system database, human-machine interface and interface program modules.

The product data base is the most important part of the system. It is based on the random electronic version of the drawing data. The contents of the electronic version of the drawing are: technical specifications, maintenance instructions, commissioning manual, external interface protocol, electrical schematics, cable connection Diagrams, wiring diagrams, wiring diagrams, accessories and tools summary table, list of components, photo albums, maintenance software and maintenance manuals, etc., is to support the use of operation and maintenance of important tools and resources. If there is no ready-made electronic version of the electronic version of the drawings, on the one hand through various channels to the development, production units to purchase, on the other hand, the organization of professional and technical personnel will be random paper data using computer keyboard or scanner input and computer processing, Paper manuals and document digitization. At the same time the organization of professional and technical personnel to carry out technical research, testing electronic equipment important nodes of the electronic technical parameters, the formation of graphics or tables to enrich the product information database content.

The expert system database is one of the important technical components of IETM. The contents of the expert system database are complex. The main contents include: specific failure phenomena, analysis of the phenomenon, possible causes and specific elimination steps. Through the analysis of the reasons for the failure to explain, and for different reasons are given troubleshooting methods, troubleshooting troubleshooting process may be difficult to answer. There are two ways to access it: browsing and finding. Browse is characterized by a list of all the symptoms in a table, followed by the current failure analysis. The user in the failure of the table and possible causes of failure to choose to find the corresponding symptoms and possible causes and solutions. As for the search, it is through the user input keywords automatically by the system to find fault and analyze the reasons and the exclusion method. The content of the fault diagnosis database will be enriched and renewed according to the change of the fault state of the NC machine tool, so that the fault diagnosis system has the memory ability. The knowledge update of the fault diagnosis system adopts the way of combining manual updating with system automatic updating.

The friendly man-machine interface and the effective management of the database are the necessary conditions for constructing IETM. Because IETM emphasizes "interactivity", it is necessary to provide users with a convenient access route. Users must use the efficient man-machine interface management program to link the data and data of the users interested in the database and show them to the users by computer multimedia technology.

4 Summary

In order to alleviate the pressure of lack of NC maintenance personnel, the author's teaching and research department opened the "CNC machine tool repair and testing" undergraduate professional, and a year ago applied for "CNC machine maintenance simulation system research and development" project. The author hopes that this article can serve as our country 's national conditions of NC maintenance personnel training program as soon as possible to play a role.

References

1. Lin Ying, Peng Huanang, Liu Qihong and so on. Research on Remote Fault Diagnosis and Maintenance Service System of NC Machine. Machine Tool & Hydraulics. 2005
2. Dai Yi. Thinking about Offering the Specialty of "NC Machine Fault Diagnosis and Maintenance". Journal of Tianjin Engineering Institute of Technology. 2005
3. Shu Yueqiang, Zhang Yunsheng and so on. Research on IETM Multimedia Production Based on Authorware. Equipment Manufacturing Technology. 2009
4. X.B.Li, Z.J.Yang, G.F.Li, H.W.Zhao, F.Chen, X.C.Zhu, Instantaneous Availability Modeling of CNC Machine Tools Based on Functional Data Analysis. Jilin Univer.J (Eng and Tech Edit). Vol42,No.5,1151-1156(2012)
5. P.Xu, W.D.Zhou, L.P.Li, Z.F.Ma, High - pressure Water - jet Collection and Analysis System Based on LabVIEW. Digi Petro and Chemic,No.6,44-46(2007)
6. Z.W.Sun, Design of Data Acquisition and Analysis System Based on LabVIEW. Elec and Mech Serv,No.1,16-18(2010)
7. S.J.Geng, J.Yu, Handling of Coarse Errors in Intelligent Measuring Systerm. Electri & Electro Educ.J.Vol27,No.3,258-261(2005)

Event Relation Recognition by Multi Part of Speech Association Distribution Characteristics

Chao Han, Feng Zhou *, Lanjiang Zhou and Xin Yan

Key Laboratory of Intelligent Information Processing, Kunming University of Science and Technology, Kunming, Yunnan, China

* Corresponding author: zf158@sina.com

Abstract. Event relation recognition, as one of natural language processing technologies, faces information stream of texts detecting event relation. By analyzing the influence of the words of different parts of speech on the relevance of events. And use the form of lexical chain to extract and store the relevant vocabulary between events, this paper propose an event relation recognization method based on lexical chain to detect latent semantic relation between events: whether events hold logical relation or not. Compared with the method based on dependency cue inference, the proposed method achieves 7.68% improvement.

Keywords. Multi part of speech distribution, lexical chain,event relation.

1 Introduction

Events refer to the objective facts that are involved in a number of roles and show some kind of behavior or state characteristics [1].

The occurrence of the event has the characteristics of objectivity, authenticity and so on. However, the occurrence of events is often not an isolated phenomenon. The occurrence of an event necessarily involves other events associated with it, such as the cause event, the result event, the concurrency event, and so on. The logical form of interdependence and association between events and their related events, called event relations. In the expression level of the text, the text information which expresses the event and the event relation contains a specific distribution rule. Therefore, it is great help to the realization of topic deduction and topic prediction for discrete events in large-scale information flows that the formation of a natural language analysis and information processing mechanism for automatic recognition and detection of event relations.

Event relation recognition is to achieve the shallow layer detection of the event logic relation, that is to judge whether there is a logical correlation between any events, and it is a kind of two element relation judgment. By parsing the text structure and semantic features, the results of "relevant" or "irrelevant" judgments are directly given to the text fragments (including phrases, clauses, sentences and paragraphs) describing the different natural events in the text.

The rest of the paper is organized as follows. Section 2 presents the related works. Section 3 analyzes the distribution characteristics of the multi part of speech association and construction of the lexical chain; Section 4 describes the method of Event Relationship Recognition Based on Lexical Chain in detail; Section 5 describes the experiment; Section 6 Conclusion.

2 Related work

Event detection is becoming a new research hotspot because of the increasing demand in automatic question answering, automatic summarization and event prediction. And the main mining methods in event logical relation recognition are divided into template matching and element analysis.

2.1 Template matching

One of the main methods of event relation detection is to use the pattern matching of event feature, according to the defined templates, event relation that matches with the template in the text is extracted. Chklovski [2] uses LSP (lexical syntactic pattern) to extract the resources with event relations, and the results are organized into a knowledge base called "VerbOcean". Chklovskii et al, extracted the event matching of six temporal relationships (similarity, strength, antonymy, enablement, happen, and before) using a manually collected LSP template. Manually defined event relation templates are often subject to a number of constraints, resulting in low recall rate of relational detection.

2.2 Element analysis

Most of the research on the event relation detection based on event elements inherited the distribution hypothesis of Harris [3]. The Harris hypothesis states that words in the same context have the same or similar meaning. Lin [4] proposed an unsupervised method, called DIRT algorithm, which combines the Harris distribution hypothesis with the idea of establishing dependency tree.

Ma Bin [5] through the analysis of event semantic dependency characteristics and events in the evolution of the semantic dependency of the law, proposed a semantic dependency clue based event correlation identification method. By locating the events in the text information flow, the method analyzes the syntactic dependency between events, and mining the reasoning clues (i.e. "dependent cues"), and then realizes the automatic identification of events.

3 Multiple part of speech semantic association distribution analysis and lexical chain

3.1 Multiple part of speech semantic association distribution analysis

The event can be thought of as consisting of a predicate and its accompanying six variables (who / whom, what, when, where, why, how), where predicates are usually verbs and the other six variables are nouns, Pronouns, adjectives and adverbs. In other words, the event is actually composed of a number of words with different parts of speech. It is found that even if the modified part (adjectives and adverbs) of the event is removed, The main information about the event can still be gotten. For example, a complete event: " Tom went to the big supermarket which is newly opened and close to his home early in the morning, and bought 3 pounds of fresh big crabs." After removing the modified part of the event: "Tom went to the supermarket to buy (3 pounds of) crabs." The main information of the event before and after processed is the same, the processed events are more concise in expression, which is mainly composed of verbs and nouns (and sometimes pronouns). It can be seen that in an event, verbs and nouns describe most of the information of the event, while adjectives and adverbs describe very little information about the event.

If there is a correlation between the main information of the two events, it is believed there is a correlation between the two events. This paper can convert the question of whether the two events are related to determine whether the words (verbs, nouns and pronouns) that describe the main information of the two events are related. Of course, a single word is not a complete expression of the main information of the event, so this paper determine the relevance of the two events is based on multiple words simultaneously related. So how many simultaneously related pair of words in the events are enough to determine that the two events are correlated? In order to explore this problem, this paper has carried on the related statistical experiment. Four press releases, 967 news events are selected out In Sohu, Sina and other news sites, after manual labeling and statistics, the statistical results shown in Table 1 and Table 2 .

Table 1. The probabilities of event-related in the same type related words.

Quantity / POS	1 pair	2 pair	3 pair
Nouns	26.77%	58.36%	83.54%
Verbs	61.34%	87.26%	
Pronouns	70.46%	83.85%	

Table 2. The probabilities of event-related in the multiple type related words.

Verbs & Nouns	Verbs & Pronouns	Pronouns & Nouns
93.24%	96.12%	89.67%

Table 1 shows that when there are one or more pairs of related words with the same part of speech in two events, the two events are the probability of the relevant event. And Table 2 shows that when there are multiple pairs of lexical correlations with different parts of speech in the two events, the two events are the probabilities of the relevant events. In the experiment, this paper find that the relevance of verbs and pronouns has a great impact on the relevance of events, while the relevance of nouns has little impact on the relevance of events. And the more the number of relevant vocabularies between events, the greater the probability of events being related events, while the probability of events being related events is generally high when there are multiple pairs of lexical correlations with different parts of speech between events. According to the feature of semantic relevance distribution, this paper proposes a method of event relation recognition based on the feature of multi part of speech association distribution. In terms of the processing of related words, this paper uses lexical chain to extract and store related vocabulary of event set.

3.2 Lexical chain

Halliday and Hasan [6] first proposed the concept of lexical chain in 1976. A lexical chain is a chain of words that are related to each other. There are many ways to construct a lexical chain. Morris and Hirst [7] first proposed a greedy algorithm to generate lexical chains, Barzilay and Elhadad [8] proposed a non-greedy algorithm model in 1997 to build a lexical chain. Silber and McCoy [9], Galleyz and McKeown [10] also proposed effective methods for constructing lexical chains. This paper argues that the lexical chain is a set of semantically related words which, using the method of semantic relevance is more suitable for the construction of lexical chain.

Semantic relevance of words reflects the degree of relevance between the semantic meaning of words. The computation method of semantic relatedness of words is divided into two categories, one is the machine learning method based on the corpus, the other is the method based on the semantic dictionary. This paper uses the computing method of the semantic relevance of Chinese words based on HowNet proposed by Li Gao [11] to calculate semantic relevance of words, this method takes into account various semantic relations among words.

3.3 Lexical chain construction algorithm

The construction of the lexical chain is based on the whole text. Firstly, the text is preprocessed, including text segmentation, low frequency word filtering, part of speech tagging and semantic annotation. Then after nouns which have been semantic annotated are processed, the semantic relevance between words is computed, the words whose semantic relevance value meets the specified criteria are taken as candidate words, and the synonyms are merged, then the words that contribute little significance to the meaning of literature are filtered out, Finally, semantically related words are aggregated together as a lexical chain. In order to reduce the workload of the automatic indexing system, some automatic indexing system after word segmentation, will first filter out those words with low frequency words, and then semantic annotation and part of speech tagging. The method in this paper chooses the sentences as the unit, and then merging these candidates to build a new lexical chain. The concrete construction of the lexical chain construction algorithm is as follows:

(1) Word segmentation, word frequency filtering, POS tagging and semantic tagging.

(2) Set semantic correlation threshold s.

(3) Based on sentence, calculate semantic correlation value between words in a sentence, adding the words whose correlation value is greater than s to the candidate words set H, the final result of H is $\{W_1, w_2,..., W_n\}$.

(4) Select w_1 as the first element of the initial lexical chain L from H and remove w_1 from H.

(5) From the remaining words of H, this paper selects the words whose semantic relevance is greater than s in L, add them to the lexical chain L, and delete these words from H at the same time. Repeat this step until no new words are added to the L, and L is a complete word chain.

(6) Repeat steps (4) and (5) to build other lexical chains until there is no word in H.

4 Event relation recognition by multiple part of speech semantic association distribution characteristics

The language is the carrier of the event, and the word is the smallest particle of language, that is to say, the event is actually composed of words. The lexical chain is a collection of semantically related vocabularies, so there must be more or less connections between events in the same lexical chain. These connections are partial, weak, and this paper cannot rely solely on this association to determine whether there is a relationship between the events. Many times, however, the same event is often associated with a number of lexical chains, and a lexical chain often involves multiple events, so the event set and the lexical chain will form a staggered relationship network. So as long as these parts, weak but staggered connection enough, this paper can use them to infer and even quantify the correlation between the events. This is the core idea of the event relation recognition method based on the feature of the multi part of speech semantic association distribution.

The flow chart of the event relation recognition method based on the feature of the multi part of speech semantic association distribution is shown in Figure 1. The method is divided into three parts: event preprocessing, lexical chain construction and event correlation calculation.

4.1 Event preprocessing

The main work of the event preprocessing is to carry out some necessary processing of the event set, to prepare for the following work. The processing flow of event preprocessing is given below.

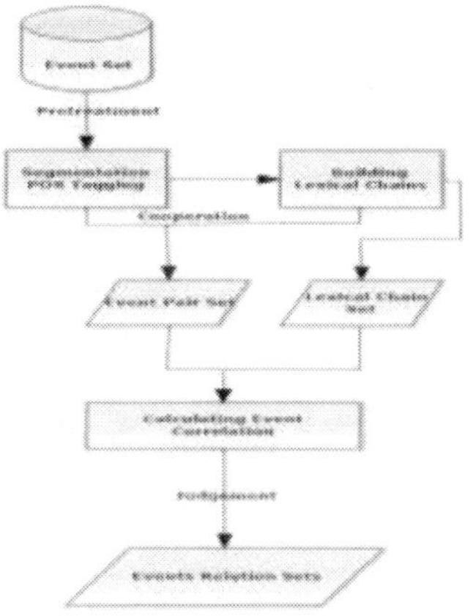

Fig.1. Flow chart of event relation recognition method.

(1) The events in the event set are numbered one by one, and the processing operations such as word segmentation and part-of-speech tagging are carried out.

(2) A new event expression is generated based on the result of the lexical chain construction in 5.2.. Such as e_1 (ChS_1), where e_1 represents the event with number 1 and ChS_1 represents the set of lexical chains associated with event 1.

(3) Event pairs set ECS is formed a by pairing the events in the lexical chain and excluding duplicated event pairs, the expression is ECS = $\{(e_1, e_2), (e_i, e_j),\ldots,(e_n, e_m)\}$.

4.2 Lexical chain construction

In this section, this paper builds the lexical chain with the algorithm introduced in 4.3 and the words extracted from the event set. Specific work is as follows:

(1) The lexical chain construction of the event set was carried out by the method of constructing the lexical chain introduced in 4.3.

(2) The lexical chain is processed and add the decision parameters v and p, the expression is ch_j (v_j, p_j). In which ch_j is the lexical chain whose number is j, the value of v_j is 0 or 1, which means that this lexical chain is (1) the verb chain or not (0); the value of p_j is also 0 or 1, which means that this lexical chain is (1) the pronoun chain or not (0),but v_j and p_j can not be 1 at the same time.

(3) Construct a lexical chain set ChS, whose expression is ChS = $\{ch_1, ch_2\ldots ch_j\ldots ch_n\}$.

4.3 Event correlation calculation

According to the first two parts of the work, to quantify the degree of correlation between the events, and then infer whether the event pairs are relevant, and finally output the relevant event pair set as the judgment result. This specific work is as follows:

(1) This paper gets the related lexical chain set of each event in the event pair from the event pair set ECS, and intersect them to get the common chain set ChRS. If there is an event pair who is composed of e_i and e_j, so ChRS = $ChS_i \cap ChS_j$.

(2) The degree of correlation between events can be expressed as the correlation factor θ, denoted as e_i e_j, $0\leq\theta\leq 2$, the greater the value of θ, the greater the correlation between events. θ is calculated as follows.

The number of shared lexical chains of the event pair and the characteristics of the shared lexical chain (v and p) have an effect on the value of θ. The parameters are weighted according to the influence of these factors. The formula is defined as (1).

Where α and β are weighting coefficients.

(3) Finally, the judgment of whether the events of the event pair are related is: If $\theta \geq X$, this paper believes that the event pair is the relevant event pair, otherwise the event pair is not relevant. Where X is a threshold set manually. The relevant event pair is then stored into the relevant event pair set ERS.

(4) Repeat steps (1), (2), (3) until all pairs of events in the event pair set are judged and processed.

5 Experimental analysis

5.1 Experimental data

In this experiment, this paper collected 10 news reports from Xinhua, Phoenix, Sina, Universal and aother global news website, and each news report contains 25 news events. Each news event is annotated ("Relevant" or "irrelevant"). Eventually, 1248 events "relationship pairs" are obtained, where the "event pair" with logical relationship (ie, "Relevant") is 346 pairs, accounting for the total "event pairs" 27.7%.

5.2 Evaluation criteria

In the event relation recognition task, the performance of the system depends mainly on the number of "event pairs" of the correctly identified association relationship. The general evaluation indexes in the field of text retrieval are used: Precision, Recall and F value.

5.3 Experimental results and analysis

In order to test and compare the effect of the event relation recognition model proposed in this paper, a system based on dependency cue [2] is chosen as the baseline system. First this paper adjusts the weighting coefficients in formula (1) to obtain the best result, and then apply the best value to the experiment. The weighting coefficients are: $\alpha = 0.4$, $\beta = 0.6$. Then this paper sets the decision threshold X to 1, that is, when $\theta \geq 1$ that the event pair is related event pair. The experimental results are shown in Table 3. The method proposed in this paper compared to the Baseline in the F value obtained 7.68% increase, while the P value and R value were also obtained 11.18% and 5.11% increase.

Table 3. Experimental results.

Method	P/%	R/%	F/%
Baseline	59.41	52.03	55.48
Our-method	70.59	57.14	63.16

Compared with Baseline, it can be seen that the event relation recognition method based on the feature of the multi part of speech semantic association has improved the precision and the recall rate. Which improve the precision is more significant, while the recall rate is not very significant increase, is still at a low level. The main reason is that the method of event relation recognition based on the characteristics of multiple parts of speech semantic association distribution is mainly based on the number and the part of speech of the shared lexical chain to determine whether there is a relationship between the events, and there are some related events whose number of shared lexical chain is small, and even some related events have not shared vocabulary chain. For example, the sentence: "It is estimated that during the Vietnam War, the United States dropped 2 million bombs to Laos, and 30% of the shells did not explode, bringing a huge hidden danger for the post - war Lao people." It contains event e1:" It is estimated that during the Vietnam War, the United States dropped 2 million bombs to Laos", event e2 :" and 30% of the shells did not explode" and event e3:" bringing a huge hidden danger for the post - war Lao people". Event e2 and event e3 are related events, but because there is no the shared lexical chain between event e2 and event e3, the system can not pair them as an event pair, and it is impossible to make a correct judgment.

6 Conclusion

This paper proposes an event relation recognition method based on the feature of the multi part of speech semantic association distribution, which is based on the task of event relation identification in the same topic. By using the relevance of the lexical chain, and then the shared lexical chain between the event pairs is extracted. According to the number and characteristics of these shared lexical chains, the correlation degree of event pairs is calculated, the recognition of the correlation between events is achieved. Experimental results show that the proposed method has higher P value, R value and F value compared with the event relation method based on event dependency cues. However, there are still some related events that can not be matched and judged (omitted and misjudged) by the method of this paper, and more event features and lexical chain features need to be extracted, and according to these features to carry out event matching and judgment.

Acknowledgment

Foundation item: China National Natural Science Foundation （61562049) ,(61462055).

References

1. X Yang, Y Hong, Y Chen, J Yao, Q Zhu. An Overview of Event Relation Detection System. JCIP 29, 4 (2015): 25-32
2. Chklovski, P Pantel. Global path-based refinement of noisy graphs applied to verb semantics//proceedings of the Joint Conference on Natural Language Processing, Jeju Island, Korea, (2005): 792—803
3. zS Harris. Mathematical Structure of Language. New York,(1968)
4. D Lin, P Pantel . Discovery of Inference Rules from Text// Proceedings of the 7th ACM SIGKDD,SanFrancisco,California,USA,(2001): 323-328
5. B MA ,Y Hong,X Yang,J Yao. Using Event Dependency Cue Inference to Recognize Event Relation. ASNUP 49,1 (2013): 109– 116
6. MAK Halliday,R Hasan. Cohesion in English.London, UK: Longman, (1976)
7. J Morris, G Hirst. Lexical cohesion computed by thesaural relations as an indicator of the structure of text. CL 17, 1 (1991): 21-48
8. R Barzilay, M Elhadad. Using Lexical Chains for Text Summarization. In Proceedings of the ACL Workshop on Intelligent Scalable Text Summari- zation. Madrid, Spain,(1997): 10-17
9. HG Silber, KF Mccoy. Efficiently Computed Lexical Chains as an Intermediate Representation for Automatic Text Summarization. CL 28, 4 (2006): 487-496
10. M Galley,K Mckeown. Improving Word Sense Disambiguation in Lexical Chaining. International Joint Conference on Ijcai,(2003): 1486-1488
11. G Li, Q Dai, Q Wei. A new approach to compute semantic relevance of Chinese words //Proceedings - 2010 International Conference on Artificial Intelligence and Education, ICAIE (2010): 610-613

Analysis and Application of River Surface Line in Hilly Area based on Hec-ras Model

Congshan Yang[1,2] ,Ming Lu[1] and Guifeng Li[3]

[1]College of hydroelectric engineering, Hebei University of Engineering, Handan 056001, China
[2] Handan water conservancy and Hydropower Survey and Design Research Institute, Handan 056001, China
[3]Handan water conservancy and Hydropower Construction Co., Ltd, Handan 056001, China

Abstract. For example—Cixian Fuyang River Regulation Project. Due to the character that Fuyang River is located in hilly areas of Cixian, we use the Hex-ras software to calculate the status of the river water surface line for the goal of determining the final treatment plan.We maintain the present situation of the river channel design as principle, select the most appropriate pushed water level and roughnessas the basic, and we combine the classification calculation of crossing structures of backwater and the encryption calculation section to get the more accurate result.We compare the water level elevation and the calculation of cross strait, analyze the design parameters, calculate repeated the water line section, analyze the rationality of the design plan, and then finally determine the applicability of Hex-rac software in the large continuous variation of cross section of embankment of river river surface line.

Keywords. Hilly area river, Water surface line calculation, hec-ras software, CSonstant flow, Roughness.

1 Introduction

In mountainous and hilly areas due to the river on both sides of the ground ladder structure, the embankment is not continuous, the river section or in a single section without beach land, or in a compound section with separating channel and beach, flood water surface width is not the law of the state, the method for determining the effective water line, is the priority among priorities of governance the river. The calculation methods of regular river surface line include graphic method and trial by trial method [1-7]. For the river with large cross-section change and steep slope in the hilly area, some unreasonable results may appear due to the above calculation. Wang Han, TianFamei et al., carry out the effective water surface line analysis of the gradually expanding river by improving the kinetic energy correction coefficient [8]. By establishing the variable step difference model, Qi Feng makes a high precision water line calculation for the river with large cross section,solves practical engineering problems [9], Lin Jinsong adopted the method of prediction correction to solve the problem of discontinuity calculation of water surface line under the complicated terrain conditions, such as positive slope and reversed slope, frequent expansion and sudden contraction [10].

2 Mathematical model and governing equation

The Hec-ras software is used to establish one-dimensional unsteady flow model, taking CixianFuyang River treatment project as an example, by changing the section parameters, analyze the change of water surface line and rules repeatedly, verify the reliability design. Hec-ras is the United States Army Corps of Engineers and one of the development of river simulation computer software, can be used to simulate constant calculation of unsteady flow, river flow and sediment transport calculation, steady flow, unsteady flow in river water line calculation [11]. Constant flow calculation principle based on one-dimensional energy equation, by section were calculated by direct method. In the calculation of constant water surface line, the formula is:

$$Z_2 + \frac{a_2 v_2^{\ 2}}{2g} = Z_1 + \frac{a_1 v_1^{\ 2}}{2g} + h_f + h_i$$

In the formula, Z_1、Z_2 is the downstream and upstream section of the water level elevation, a_1、a_2 is the downstream and upstream section of the velocity coefficient, v_1、v_2 is the downstream and upstream section of the velocity; g is gravity acceleration; and h_f、h_i for upstream and downstream section of the frictional head loss and the local water head loss [12].

3 Engineering survey

Cixian Fuyang started from Cixian Dongwushi Hydropower Station tailrace outlet, in Qinglan high-speed into the city of Handan within the length of Cixian River 26.1Km. The scope of governance for the Beijing Guangzhou railway upstream of the river 5km, calculate the range to the upper section of the water control 1km line. The purpose of governance is mainly through the channel dredging water serious place, does not meet the requirements of the high embankment of hilling, so as to meet the requirements of flood. In addition, the main building is three governance wading bridge, are the main factors causing backwater height, so the surface line calculation process need to increase the calculation of the three seat wading bridge.

4 Calculation parameters

4.1 Governance standards:

In 20 years the standard of flood control engineering calculation and analysis of river water line, according to the hydrological calculation, Cixian Fuyang River section of each hydrological results are shown in Table 1. The control section of the river is located in the upper reaches of interval Longwangmiao to Cixian railway bridge, so 20 years once the design flow using 174 m³/s.

Table 1. The peak flow results. m³/s

Section	10 years	20 years	50 years
Maximum reservoir discharge	50	51	450
The Dragon King Temple ditch after import	85	174	451
The railway near the peak flow	110	215	731

4.2 Roughness coefficient:

The roughness is the primary index to reflect river resistance, roughness selection directly affects the level of water flow rate, the size is an important factor affecting the water surface line calculation, in general, to determine the roughness of the need to repeatedly the natural river roughness should be solved according to the traffic data of the measured water level, or on the surface the measured line or flood investigation of water trace backstepping roughness [13]. According to the measured data, the water level of the corresponding pile number is compared with the measured water level through the selection of different roughness. The smaller the difference, the more reasonable the roughness selection is. This project due to the measured water level, so take this period of Fuyang governance adjacent segments roughness calculation, calculation of water surface profile of river cross section, channel roughness at the rate of 0.045, 0.06 of the beach roughness selection, after treatment of river cross section calculation of water surface profile, roughness of channel 0.03, the beach roughness coefficient unchanged.

4.3 Initial water level:

There is a range of influence of initial water level of the river to the river water level, general engineering design, the corresponding water level method or the highest water level method [14], when the channel is computed without measured data, by the control of downstream buildings (such as gates, weirs) to determine the control level, control of downstream buildings can be extended a distance,to the downstream river, the critical water depth as the downstream water depth control [15], because of the lower reaches of the river section has been governance, so the downstream water level control of the water level upstream has been calculated as the water surface line calculation.

4.4 Section selection:

At present, the main vector in flood period is calculated as the distance between two sections. Hilly area of river section mutation, dike is not continuous, in fact the flood and the main channel spacing is not uniform spacing [16], consider the main channel bend, using HEC-RAS software surface line calculation, the appropriate channel change area encryption section, can make the result more reasonable, the water surface line calculation process, using section spacing 50m.

4.5 Backwater height calculation of buildings across the river:

The present situation of the channel involving a total of three bridges, a highway bridge, two water bridges, the energy loss of buildings caused by wading river is divided into three parts in the software, the first part is the building occupancy wading flow section so that the upstream flow energy loss caused by energy loss; second part caused by contraction the water flow through the building after wading diffusion; third part for a variety of shape and arrangement of the building itself by wading [17-18]. The calculation of water diffuse bridge backwater, weir flow coefficient, entrance loss coefficient and wall roughness has a significant influence on the calculation results, the weir flow coefficient C is the result of the most sensitive parameters [19], therefore, choose a reasonable C value of backwater calculation is essential.

4.6 Computing method:

The establishment of system layer, draw the river curve from upstream to downstream and save the data, the result shown in Figure 1; Input the measured cross-section parameters in sequence and save it; add upstream section and downstream section fot every bridge, input current parameters of wading bridges, establish the boundary conditions, according to the classification of wading bridges, selectthe corresponding flow calculation method, according to the constant flow calculation process, input the flow and initial water level, then it can be calculated.

DOI: 10.1051/matecconf/201710002060

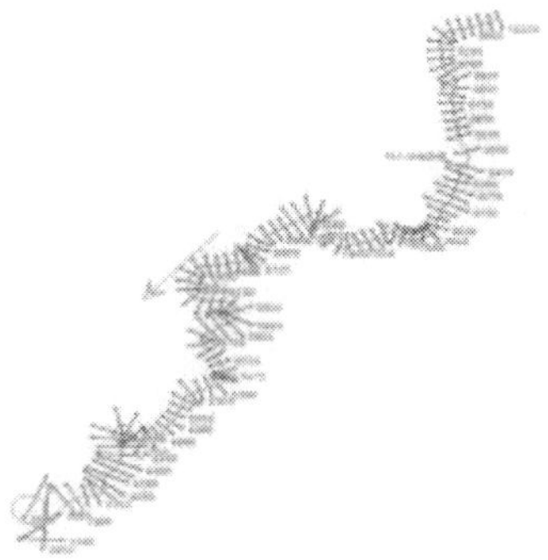

Fig.1. River Map.

5 Result analysis

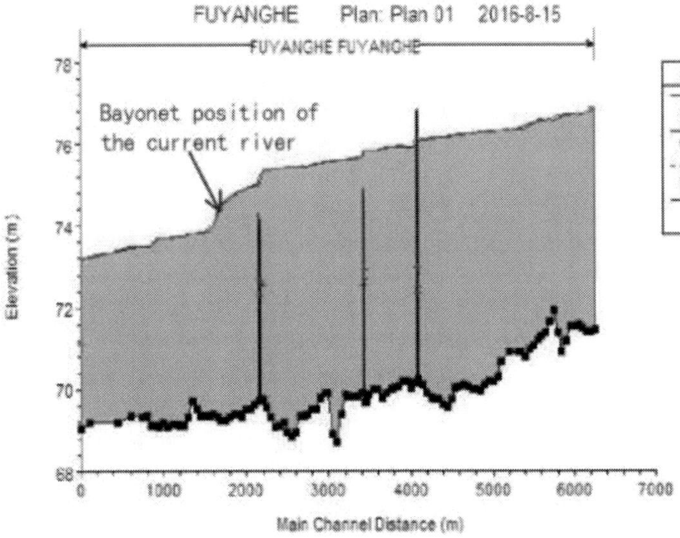

Fig.2. Longitudinal profile of current river surface line.

We can see from Figure 2, two bridges through the village have been completely submerged and the water level was raised, the highway bridge backwater height is small, Cross section of water of water bridge and cross river bridge can be seen in Figure 3 and Figure 4. In addition, in two km water suddenly increased by 90cm, the main reason is because of the serious 200m river siltation, narrow river channel, the bayonet, raise the water level directly determines the upper reaches of the river flood control capacity on both sides of the embankment, so it is necessary to compare the 20 year flood level and the status quo of estuary embankment elevation or the current channel,calculation result in Table 2.

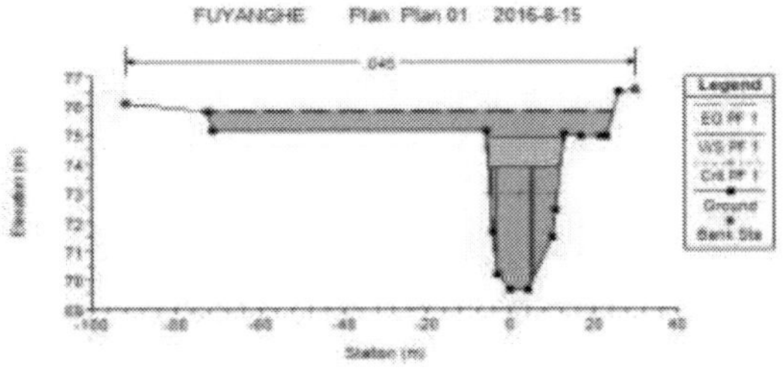

Fig.3. Cross flow section of a water bridge.

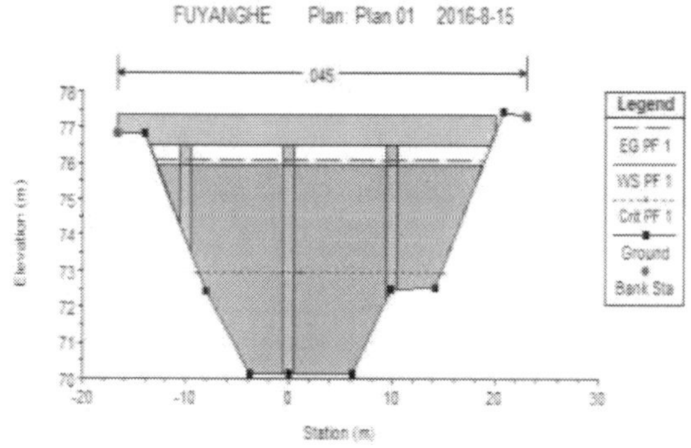

Fig.4. Cross flow section of Cross River Bridge.

Table 2. Current river level reference.

position	Flow (m³/s)	Thestatus quo of river bottom elevation (m)	Status flood level (m)	Status quo of left embankment elevation (m)	Status quo of right embankment elevation (m)	Current Speed (m/s)	Water surface width (m)
7+400	174	70.25	76.03	77.29	77.28	1.19	32.23
7+397	174	Cross River Bridge of Xiang Shuiliang village					
7+350	174	70.14	75.93	76.81	77.46	1.38	31.65
6+850	174	70	75.86	76.00	75.35	0.48	110.2
6+800	174	69.86	75.85	75.80	75.35	0.54	119.3
6+745	174	69.65	75.77	75.11	75.11	1.22	97.14
6+725	174			Water bridge of Li Zhuang village			
6+720	174	69.91	75.66	75.11	75.11	0.86	168.9
6+700	174	69.91	76.65	76.11	76.44	0.53	169.9
5+550	174	69.56	75.36	74.75	74.83	0.7	135
5+500	174	69.76	75.3	74.79	74.24	0.92	146.1
5+475	174			Water bridge of Bai Jiazhuagn village			
5+450	174	69.73	75.01	75.25	74.82	0.98	118.7
3+950	174	69.35	73.45	73.65	73.39	0.8	99.77
3+886	174	69.21	73.39	73.65	73.54	0.72	111.2
3+786	174	69.05	73.17	73.57	73.45	0.74	110.9

After the calculation of river governance ultra-high not less than 1m [20]. Can be seen from table 2, ultra high embankment are not almost intact to meet the requirements of flood control, embankment need to be filled; and surface width is large, most houses on both sides of the river will be flooded, considering the difficulty and investment immigration control, at the same time in order to keep the original river flow state, the initial design scheme is to design wide river close to the current width of 12m principle, longitudinal slope design close to the status of bottom slope is 0.00057,taking the slope of 1:2 on the narrow channel dredging cut slope, for the status quo is wider, the bottom section of the deeper, to maintain the status quo, not backfill. Design plan water surface profile in Figure 5, elevation contrast result in Table 3.

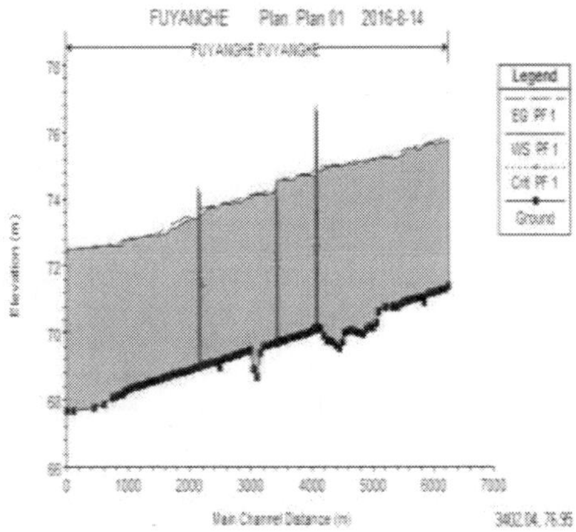

Fig.5. Longitudinal profile of river surface design.

Figure 5 shows that the river water is basically uniform trend, three bridges include only a water bridge, and the backwater height is obvious, but the backwater height does not affect the effect of flood control embankment upstream.The water level rising caused by the channel status of the bayonet region through dredging excavation, water level dropped to normal range. Table 3 shows, after dredging the river, flood situation is significantly reduced,the status of banks in addition to the local high position slightly lower, the vast majority of all meet the requirements of flood control,so We only need to add a heightened local bank; compared to the previous status of the speed increased, the calculation of scour depth is less than 1.5m, the maximum velocity at 1.99m/s, the dangerous position of the scour depth is less than 1m[20], so we just have to scour the dangerous position to take the masonry lining erosion prevention measures. The results show that the design scheme is reasonable, without debugging, on the contrary, we need to further debug the design parameters, repeatedly calculate the water line, so that the program in the construction and engineering benefits to achieve the best.

Table 3. Design river level reference.

position	Flow (m^3/s)	The designed bottom elevation (m)	Designed flood level (m)	Status quo of left embankment elevation (m)	Status quo of right embankment elevation (m)	Current Speed(m/s)	Water surface width (m)
7+400	174	70.12	74.79	77.29	77.28	1.64	32.73
7+397	174	Cross River Bridge of Xiang Shuiliang village					
7+350	174	70.14	74.67	76.81	77.46	1.82	30.14
6+850	174	69.79	74.56	76.00	75.35	0.75	103.9
6+800	174	69.76	74.54	75.80	75.35	0.82	88.14
6+745	174	69.7	74.42	75.11	75.11	1.63	33.71
6+725	174	Water bridge of Li Zhuang village					
6+720	174	69.73	74.14	75.11	75.11	1.99	30.84
6+700	174	69.7	74.22	76.11	75.44	1.08	115.2
5+550	174	69.04	73.62	74.75	74.83	1.45	62.06
5+500	174	69.01	73.53	74.79	74.24	1.75	33.1
5+475	174	Water bridge of Bai Jiazhuagn village					
5+450	174	68.98	73.43	75.25	74.82	1.83	34.3
3+950	174	67.85	72.59	73.65	73.39	0.46	374.2
3+886	174	67.77	72.5	73.65	73.54	0.51	262.2
3+786	174	67.65	72.45	73.57	73.45	0.81	174.6

6 Conclusion

(1)core work of the hilly area river is the calculation of water surface profile, which directly determines the size of the project, investment size, but because of the characteristics of ladder type hilly landform, the river section width disparity, embankment height fluctuation, it is difficult to determine the flow range, so the traditional method of water surface line calculation is the difficulty of river control in Hilly Area, Hec-ras software in the establishment of one dimensional unsteady flow model, through the simple steps will increase the river section, avoid the influence of frequent changes of water level caused by river section, the flow pattern changes in the river can be reflected in the calculation.
(2)According to the classification of water diffuse buildings and buildings across the river, the choice of formula of different simulation of wading structure calculation of backwater height, improve the accuracy of the design water level, the water line interface is reasonable, more important, which makes the design height of embankment is more accurate, to the requirements for flood control is more safe and reliable, fully applicable in the hilly area of discontinuous embankment, steep longitudinal slope, the river water dense buildings.

References

1. Liu Yang. Comparison of several surface line calculation method. Yellow River,2011,33(2):51-53 .(in Chinese).
2. Huang Yousheng. A new calculation method of water surface line.Development of water conservancy and Hydropower Science and technology, 2003, 23(3)： 31-33.(in Chinese) .
3. Zhang Jianmin ,Wang Yurong,Xu Weilin, et al.An iterative method for the calculation of constant gradient flow.Journal of hydraulic engineering,2005, 36(4):501-503. (in Chinese) .
4. Lei Fengze,Shi Xiangling.Analysis and analysis of water surface line calculation in design of embankment engineering.China water transport, 2007,5 （6） ： 83-84. (in Chinese) .
5. Wu Chigong.Hydraulics.Beijing:Higher Education Press,2008. (in Chinese) .
6. Gao Huanzhi,Qin Jianping,Li Ting, et al.Calculation of natural water surface of river channel in engineering design.Inner Mongolia Water Resources, 2014,4:32-33. (in Chinese) .
7. Song Tao.Study on influence of river channel change on water surface line.School of architecture and engineering, Nanchang University, 2013. (in Chinese).
8. Wang Han,Tian Famei,Lv Hongxing.Calculation and improvement of the water surface line of the gradual expansion of the mountainous area. Yellow River, 2007,29(5):66-67. (in Chinese) .
9. Qi Feng.Calculation of river water surface line based on variable step difference iterative method. Yangtze River, 2015,46(1):83-85. (in Chinese) .
10. Lin Jinsong,Ju Jiang,Zhang Kuandi, et al.Calculation of natural river surface line under complex terrain.Journal of Northwest Agriculture and Forestry University (Natural Science Edition),2010. (in Chinese) .
11. Ye Nan,Zhang Wei,Bai Zhongliang. Application of HEC-RAS software in small and medium river harnessing project in mountainous area .Computer application, 2013,3:70-73. (in Chinese) .
12. Liu Yang,Sun Xiaoying,Qang Junying, et al.Calculation of water surface line of natural river in mountainous area by HEC-RAS and SOBEK-RURAL software.Beijing Water, 2008,6:34-36. (in Chinese) .
13. Yao Zhiqiang,GU Hehe.MapInfo land use status thematic map digital production.Resource development and market, 2007,23(12) :1067－1069.(in Chinese) .
14. Chen Biao,Chen Changbai.Analysis and determination of the influence of water level on the water surface of the river.Science and Technology Forum, 2009:21-22. (in Chinese) .
15. Zhu Zhonglong.Determination of the water level in the calculation of the water surface line of natural river.Construction and Engineering, 2014,28(4):448-449. (in Chinese) .

16. Tian Fengjun,Ji Zhongguang,Liu Yuxia. Discussion on some problems in the calculation of the water surface line of rivers.Heilongjiang water conservancy science and technology, 1999,2:41-43. (in Chinese) .
17. Wang Xiaoyang,Deng Zanxin,Yu Weihou.Hec-ras model and its application in water resistance in the calculation of the high bridge.Hunan Hydro & Power,2008,3:31-32. (in Chinese) .
18. Liu Zhengfeng,He Runjin,Li Wenxiang.The application of HEC RAS model of continuous bridge backwater high resistance analysis.Water conservancy science and technology, 2011 , 3 :60-61. (in Chinese) .
19. Yang Yang,Han Fengxia,Liu Junmei, et al.Computation of water surface line of mountain river and culvert bridge backwater. Yellow River, 2013, 35(1):123-125. (in Chinese).
20. Code for design of embankment engineering, GB 50286-2013. (in Chinese) .

An Evaluation of Mobile with the Hierarchical Approach

Yi Cao

Depart. of Automobile Engineering, Zhengzhou University of Science & Technology, Zhengzhou, China

Abstract. The safety evaluation of MOBILE demonstrates the applicability and benefits of the hierarchical approach. Quantitative figures give a rough impression of the safety level. Additionally, relative changes in failure rates after modifications to components or the architecture of electronics can be observed. Working with a group of students showed that the hierarchical approach supports splitting of the complex vehicle design task into a number of smaller work packages that are easier to handle. At the same time, the system context is kept available and traceable for all developers.

Keywords. Evaluation, mobile, hierarchical approach.

1 Introduction

This paper evaluates the safety of the vehicle control function of MOBILE using the hierarchical approach introduced in the previous section. As MOBILE is a primarily student driven university project, some restrictions have to be regarded:

(1) Up to 20 students were working on parts of the vehicle in parallel. Each student works on a specialized field on a low hierarchical level. The students are assumed to be the experts for a specific field. The work on higher levels is mostly done by members of the scientific staff.

(2) Failure rates are not available for all parts of MOBILE. For these parts typical values were derived from literature. Failure rates of software under development is roughly estimated based on previous in-field experience.

(3) Aging effects are approximated by typical bath-tub curves given in literature. Thereby, one observes slightly higher failure rates of hardware components in early phases of the part's life time and a significant increase towards the end of the life time.

(4) Only the hierarchical layers "vehicle", "system"and "subsystem" have so far been taken into account. Some individual components of the underlying hierarchical layers are currently being investigated and results are fed back to the evaluation of MOBILE.

(5) Processes performed during development of MOBILE do not comply to the requirements imposed by ISO 26262.

(6) As the construction of MOBILE is not yet finished, only qualitative results are given that origin from quantitative but not yet complete data.

Still, the safety evaluation of MOBILE demonstrates the applicability and benefits of the hierarchical approach. Quantitative figures give a rough impression of the safety level. Additionally, relative changes in failure rates after modifications to components or the architecture of electronics can be observed. Working with a group of students showed that the hierarchical approach supports splitting of the complex vehicle design task into a number of smaller work packages that are easier to handle. At the same time, the system context is kept available and traceable for all developers.

2 Assumptions for the safety analysis of mobile

As indicated before, top level assumptions on the mode of operation of MOBILE are required to evaluate the functional safety of MOBILE: The mission time of MOBILE is limited to 30 min. After 30min, the lead-acid drive batteries are assumed to be emptied anyways or the test driver is expected to have a break. In case of a failure, the emergency operation interval that has to be guaranteed is set to 30s. This time span suffices to get MOBILE to a safe halt even if the failure occurred while driving at MOBILE stop speed of approx.160km/h(44m/s). It is assumed that only one independent fault has to be tolerated. For MOBILE, one "point in time" is defined as a 4ms time slot. This slot length is derived from the cycle time of the Flex Ray network in MOBILE that facilitates synchronization of all network nodes and precise triggering of the diagnostic algorithms. If two faults occur within a 4ms time slot, they are treated as a double fault at one point in time. A similar assumption for small diagnostic time intervals is, e.g., made by Sieglin.

3 Evaluation of complexity of the hierarchical approach

On "system layer" of MOBILE, eight units were defined: front and rear axle control system consisting of the according FTUs, user input control system (also embodied by the according FTU), two power supply systems, emergency off systems for front and rear drive motors and the stability control system [1]. Due to the design of MOBILE, these systems can be regarded as insusceptible to common cause failures—except loss of power. If cross couplings between the systems exist, the couplings are assumed to be irrelevant during the emergency operation interval of 30s, e.g., low voltage buffer batteries can compensate the loss of charging power due to failure of the high voltage system. In particular, this independence of the elements at "system level" led to the definition of the virtual systems as given and not to the classical system partitioning into braking, drive and steering system. For each of the chosen virtual systems, 2 to 9 generalized failure states, not including the "0k" / "no failure present" state, were defined. Resulting, 31 failure states have to be evaluated on vehicle layer. Thereby, the controllability of the vehicle has to be evaluated after occurrence of a given first and second system failure, summing up to 702 state transitions. Especially, for the second faults, several transitions need not be regarded as they do not furthermore impact the controllability of the vehicle [2].

Additionally, several transitions are identical for more than one system and thus only have to be considered once. For each failure scenario, the state of the vehicle is well defined as the generalized failure states are part of a first order Markov Chain. Thus, all relevant information is contained in the state descriptions and no knowledge about the failure history is needed. Given the knowledge about the effects of the system failures on vehicle dynamics, it takes the developer approximately an hour to go through all states and define the according consequences. Table8.5 shows a simplified classification for MOBILE after the first failure for each system [3]. Within the tool environment, the classification is done graphically based on Excel tables by color highlighting. For the evaluation of MOBILE on vehicle level, several experiments with a 1:5 scale vehicle were performed to estimate the effect of actuator or power supply failures on the controllability of the vehicle. Additionally research results of other groups were taken into account to fully exploit functional redundancies. Still, the classification at vehicle level is a challenging and not fully solved task from a scientific point of view but easy to handle formally, which allows the researcher to focus on his main tasks.

The failure states on "system layer" are derived from approximately 60 failure states on "sub system layer". In average, on system level approx. 100 state transitions have to be evaluated per system. Thereby, the behavior of the system for all "first faults" has to be considered. Additionally, selected "second faults" have to be investigated. Second faults that have to be regarded are identified automatically top down from "vehicle level". The number of state transitions that have to be investigated by the developer serves as an estimate for work load and complexity. If compared to "vehicle level" and depending on the individual system, the individual researcher on system level has to evaluate a similar amount of relevant combinations [4]. On lower levels (component and elementary) the number of total failure states furthermore increases but again can be handled due to the partitioning into virtual systems and allocation of tasks to local experts. Third party components can easily be integrated at any hierarchical level. Within the project MOBILE several such components exist (steering motors, drive motors, etc.). As mentioned, the evaluation process in the project MOBILE is supported by an Excel Sheet. Necessary calculations and the linking between hierarchical layers are automatically derived from "graphical" inputs of the user.

As the input tables are continuously being updated during the development process, the current state of the vehicle with regard to safety as well as the most critical components are known at any point in time. The generalized failure states including proper documentation support transparency and long time usability of the results of the safety analysis. These state descriptions also form the basis for discussions between experts in different fields and on different hierarchical levels. A further extension of the tool environment to automatically link graphical architecture descriptions or descriptions of state transitions with the inputs in the Excel environment would be useful. Currently, these steps are performed manually, which is acceptable for the scope and scale of the project. Summarized, the analysis results for MOBILE can serve as a well documented and tailored safety report and support continuous monitoring during development. The tailoring of the analysis by front loading knowledge on dependencies lowers work effort compared to other hierarchically structured approaches.

4 System monitoring and failure rates

This illustrates the failure rates of MOBILE at vehicle level over lifetime. Thereby, the failure rate was calculated using the approach detailed in Sect.8.3 for several points in time. The curvy form of the graph with high increase in failure rates towards the end of the vehicle lifetime results from the assumed aging of hardware parts. As introduced above, software parts that feature a high probability of failure are also taken into account—differently from the approach in ISO 26262. Of course, these failure rates are highly volatile, but are several orders of magnitude higher than the failure rates of the underlying hardware and thus have to be considered. Of course, software components are unconcerned by aging. Figure8.24b visualizes the huge benefit for failure compensation in the vehicle by considering interactions at "system" and "vehicle layer". E.g., the curve for "system layer" considers only cross-compensations between different systems up to "subsystem level" and so on. On higher layers, these cross-compensations are more and more due to functional redundancies. Thus, a highly flexible vehicle as MOBILE especially profits. Analogously, the efficiency of the diagnostic coverage over lifetime is automatically derived from the gathered data.

Tendencies show, that the proposed integrated safety concept relying on functional redundancies can increase functional safety while also maximizing the functional benefit from additional actuators and limiting system costs due to reduction in required hardware redundancy. Still, final results can only be provided after the vehicle has been completed, and further experiments can be conducted.

5 Conclusion

This contribution introduces a novel system architecture for an experimental drive by-wire vehicle with high functional integration and over-actuation. For this vehicle, a system architecture is derived top-down driven by according requirements. Especially, the top-down partitioning of the system can reveal novel structures also for series vehicles. Resulting, a system structure that exploits functional redundancies instead of hardware redundancies for safety purposes is presented. Exploiting functional redundancies necessitates a clearer definition of the safe state of the vehicle compared to typical part-oriented safe-state assumptions. For MOBILE, a model of the desired minimal vehicle dynamics is used. Consequently, control algorithms for vehicle dynamics play an important role in the proposed safety concept, and assessment of quality of these algorithms has to become more quantitative. To evaluate the safety of complex and integrated systems as proposed for MOBILE, a hierarchical approach to safety analysis is introduced. The approach complements already existing means for safety evaluation by taking a holistic view of the overall vehicle. It especially focuses on the targeted evaluation of highly integrated systems that provide functional redundancies. Therefore, virtual systems and generalized failure states support early reduction of the number of faults that have to be analyzed quantitatively. At the same time, system partitioning promotes allocation of work packages to developers that are best suitable. As given, the proposed approach features some restrictions and potential for further development. Especially, questions related to a development process in industry as intellectual property, responsibilities, or process management are not regarded in this contribution. Future work will focus on completion of the safety evaluation of

MOBILE. Starting from there, the further usage of the structured information on the system architecture for online self-representation of the vehicle and diagnostics will be investigated. Another important topic of future work will be the ongoing evaluation of control algorithms for vehicle dynamics to exploit functional redundancies between different types of actuators by coordinated control of remaining actuators. In parallel, analysis of critical components of the EE system will go on with regard to functional safety and failure rates.

References

1. Abele, A.: Design and realization of an integrated safety concept based on an architecture model with the given example for the serial development of a power train control unit used in electric driven vehicle. In: Hybrid and Electric Vehicles, pp. 481–525. Braunschweig (2012).
2. Adachi, M., Papadopoulos, Y., Sharvia, S., Parker, D., Tohdo, T.: An approach to optimization of fault tolerant architectures using HiP-HOPS. Softw. Pract. Experience 41(11),1303–1327(2011).
3. Anwar, S., Niu, W.: Analytical redundancy based predictive fault tolerant control of a steer-by-wire system using nonlinear observer. In: 2010 IEEE International Conference on Industrial Technology, pp. 477–482 (2010).
4. Baustein zukünftiger funktional strukturierter Domänenarchitektur im Fahrzeug. In: AUTOREG 2011, pp. 375–387. Baden-Baden (2011).

Fire Situations and Prevention Measures of residential building

Baixia Zhou

Public Security Fire Forces College, Kunming 650208, China.

Abstract. The proportion of residential building is the largest in all buildings. With the development of urbanization, residential building fires are at high momentum. The paper lists the residential building fires and the number of casualties' proportion to total fires from 2011 to 2014, analyzing the high incidence causes of the fires and casualties and putting forward suggestions and countermeasures including carrying out fire safety education to improve the fire safety awareness of residents, fulfilling responsibility to enhancing fire safety management capabilities, perfecting fire apparatus to develop fire safety environment and enhancing public awareness of fire safety, equipping evacuation equipments to promote response ability to deal with disasters etc.

Keywords. fire safety, residential building, the number of fires, fire casualties.

1 Foreword

Family is the most basic unit of the society, and family peace and harmony will bring about social harmony and stability. But the sudden fire accidents always endanger the residential buildings, affecting people's lives and property safety. Especially, with the rapid development of economy and society and the acceleration of urbanization in our country, there are more and more urban residential construction with growing volume and larger height. While some places do not attach importance to the fire safety of residential buildings, leading to prominent fire hazards. In addition, residents of our country generally have weak awareness of fire safety, so the residential building fire safety situation is not optimistic. In recent years, residential building fire has been at a high incidence, and the number of fire, personnel injury, death and other indicators of fire account for a rather high proportion to total fires. Table 1 shows the residential building fires number's proportion to total fires from 2011 to 2015 [1-3].

Table 1 The residential building fires number's proportion to total fires from 2011 to 2014

Year	Total number of fires	Number of residential building fires	Number of fires in education, office, public gathering places	Number of traffic fires	Number of industrial building fires
2011	125,000	48,500	12,000	13,000	26,000
2012	152,000	46,000	28,000	15,300	13,500
2013	388,000	117,000	26,100	35,300	20,400

2014	395,000	142,000	21,100	35,700	24,200

In the process of social and economic construction, especially in the process of urban construction and development, we do not pay attention to the fire safety of residential buildings, resulting in a low degree of standardization and scientification of fire prevention in residential buildings and frequent residential building fires. Not only the number of fires is large, but also the proportion of injuries and deaths caused by residential building fires are at a high level. Table 2 shows the proportion of residential building fire casualties to total fires in the past four years [1-3].

Table 2 The comparison of fire casualties of residential building fires and other building fires

Year	Residential building	Assembly occupancies	Production warehouse buildings	Others
2011	787 people, 71%	156 people, 14.1%	70 people, 6.3%	95 people, 8.6%
2012	732 people,71.2%	152 people,14.8%	83 people,8.1%	61 people,5.9%
2013	1429 people,67.6%	311 people,14.7%	262 people,12.4%	111 people,5.3%
2014	1244 people,70%	307 people, 17%	130 people,7%	108 people,6%

As can be seen from table 2, residential building fires, with a greater impact, are more likely to endanger the lives of people than other buildings. Therefore, residential building fire safety cannot be ignored.

2 The main reasons for the large number of fires and more residential building fire casualties in China

In order to effectively prevent and curb resident fires, create a safe and harmonious living environment, to ensure the safety of the lives and property of residents, it is necessary to analyze the main reasons for the large number of urban residential building fires and increasing resulting casualties in China.

2.1 Weak awareness of fire prevention of urban residents in China

Due to the lack of the national fire safety education and not widespread popularization of fire science popularization, insufficient awareness of the importance of urban residents on fire safety, people do not attach importance to or do not have the opportunity to learn fire safety knowledge, unable to use the fire-fighting equipment and do not know how to escape in case of fire5 [4] . Some residents still connect wires randomly, throw about cigarette butts and irregularly use fire, electricity and gas in daily life; some residents use a lot of flammable combustible materials in the process of home decoration, which increases the fire load; some residents store excessive hazardous chemicals at home; and some residents set off fireworks and firecrackers in the residential areas at any time, directly leading to the residential building fires. It will raise the proportion of casualties of residential building fires which are composed of vulnerable groups such as the elderly and children at home who have poor ability to dispose of the fire and escape.

2.2 Low level of fire safety management in urban residential areas in China

Some of urban residential quarters in the management mode of government, sub-district office, neighborhood committee, enterprise, etc. have weak awareness of the importance of fire prevention; some of residential quarters not only fail to include fire prevention publicity in the daily routine, even do not carry out fire control propaganda activities during the safety month and on 119 fire awareness day; some of residential quarters do not carry out fire safety management in accordance with the provisions of the fire law, fail to implement fire safety responsibility system or establish fire organizations as required, let alone extend the network of fire services organizations to the residents group, residential units and small business outlets so that the residential areas are lack of organized, scientific, normative fire safety management in the daily life. As for the residential quarters under the unified management of property management companies, since the property management fee does not cover the funds for daily operation, maintenance and update of fire facilities, fire facilities and equipment used are not regularly maintained, tested and replaced. In some communities, the security personnel are changed frequently, while the new personnel have not been trained in fire safety, do not know what is the automatic fire control facilities and unable to operate automatic fire facilities. They cannot quickly start the internal self fire protection facilities to control fire at the initial stage, which delays in fire fighting, causing a low flame spread to a fire hazard [5].

2.3 Bad situation of fire safety of high-rise residential buildings in China

With the rapid development of economic construction and the continuous increase of urban population, high-rise residential buildings are increasing. At present, China's public security fire control institutions attach more importance to monitor and manage public buildings instead of high-rise residential buildings. However, for high-rise residential buildings, the fire spreads fast inside the shaft, and it is difficult to evacuate the old and children left at home in case of fire. Besides, since complicated partition inside the residential buildings increases the difficulty of fire fighting and rescue, it is easy to cause major disasters. In recent years, many high-rise residential building fires occurred in China. At 15:23 on September 22, 2015, a fire broke out in the house on the 13th floor of a 33-storey residential building in Urban Garden on Shucheng Road, Luohu District, Shenzhen. The flame was as high as the height of 3 storeys and the fire quickly spread to some houses on the 15th, 16th and 17th floors. On November 15, 2010, a fire occurred in a 28-storey teacher's apartment at No. 728, Jing'an District, Shanghai due to irregular welding in heat preservation construction on outside wall, resulting in 58 deaths and 71 injuries. In recent years, the continuous occurrence of high-rise residential building fire is an alarm to the whole society.

2.4 Poor urban residential building fire infrastructure construction in China

In China, there were no strict requirements on the building fire protection infrastructure by related fire prevention specifications and regulations during their construction since some old urban residential buildings were built earlier, which results in the deficiency of fire fighting hardware facilities. Some of the old residential buildings are surrounded by newly developed markets with dead ends and narrow entry roads, which obstructs the fire fighting truck to rescue closely; some of the old residential buildings have no fire water and outdoor fire hydrants; some of the residential quarters occupy the fire separation and set up temporary places between buildings with combustible materials for community management; in some of the residential quarters, private cars are parked casually, blocking the fire fighting access and occupying safety evacuation site [6]. In other newly built urban residential quarters under the unified planning of the government with property and security management, the personnel are unable to use fire infrastructure according to codes.

3 Prevention countermeasures of frequent fires in residential buildings

It can be seen from fire data in Table 1 and 2 that the proportion of the number of residential building fires and resulting casualties is held at the highest level. Residential building fire not only is related to the safety of each family, but also it is easy to bring disaster the neighbors and cause the civil disputes, affecting social stability. How to do a good job of fire safety management of residential buildings to curb the frequent occurrence of fire in residential buildings is an important task in front of us.

3.1 Carrying out fire safety education to improve the fire safety awareness of residents

To enhance residents' awareness of fire safety is an effective way to prevent the occurrence of residential building fires. Neighborhood committee and property management and service departments should carry out rich and colorful fire control propaganda activities close to people's lives by dint of corridors and public places, such as setting up fire publicity windows inside the residential quarter, putting up fire warning signs in the building corridors to create a fire propaganda atmosphere. Some newly built residential areas in good condition can use modern facilities to propagandize, for example, use the unit LED screen to send fire safety knowledge, carry out fire control propaganda and education targeted at the vulnerable groups such as children and the elderly taking advantage of amusement room for children and activity room for the elderly. Meanwhile, the neighborhood committee and property management and security department should carry out targeted fire safety education activities in different holidays and seasons, like publicity of outdoor paper offerings burning on Tomb-sweeping Day, setting off fireworks before the Spring Festival, fire safety education to the youth during winter and summer vacation.

3.2 Fulfilling responsibility to enhancing fire safety management capabilities

Neighborhood committee and property security department should, according to the provisions of Fire Protection Law and Decree No. 61 of the Ministry of Public Security, clarify fire management responsibilities, set up fire safety organizations in the residential quarter, regularly inspect fire protection of electricity and gas using, regularly carry out fire-watch of fire-fighting equipment inside the unit, guide the residents in some of the residential quarters who block escape way by installing burglar mesh on the balcony and windows, dissuade those who park their bikes or stack waste furniture, packaging boxes on the platform and in the staircase or even clear up forcefully, mark out the areas for parking motor vehicles to prevent them from containing fire hydrants and blocking the fire passage, and strict manage residential buildings with commercial serving cubbies set on the ground floor according to relevant public building provisions [7].

3.3 Perfecting fire apparatus to develop fire safety environment

According to the requirements of Fire Law, fire cistern, fire pumps, outdoor hydrant system and other public fire-fighting installations should be set in the residential quarter; fire extinguishers, indoor fire hydrants and other fire extinguishing equipment should be outfitted in the units; sufficient fire separation, fire fighting access and safe evacuation site should be reserved between buildings. For some old residential quarters being short of fire-fighting equipment, the neighborhood committee should seek for the fund support of superior competent department and sub-district office to buy necessary fire hose, gun and fire extinguishers, etc. to improve fire safety situation [8]. For newly built residential quarters, fire infrastructure should be included in the construction of residential buildings, and fire passage should be planned at the same time to ensure that fire engines can put out the fire closely, and fire separation should be reserved to ensure that the flame will not be spread to other buildings. In the meanwhile, fire resistance rating and fire separation should be ensured to meet the specification.

3.4 Equipping evacuation equipments to promote response ability to deal with disasters

The newly revised *Code for Fire Protection Design of Buildings* was implemented in May 2015, adding the residential building fire protection design and explicitly stipulating fire resistance rating, fire separation, safety evacuation and evacuation range, etc. of the buildings. In addition to following normative design, new type of fire control products and technologies should be vigorously promoted in the newly built residential buildings, such as installing smoke type fire automatic detection alarm at home, gas leakage alarm and control system in the kitchen, simply-equipped automatic sprinkler system using the pressure of tap water and other facilities to automatically detect and put out the fire. Some families can also be equipped with fire extinguishers, household evacuation gas masks, fire blankets and other equipment [9]. With the development of the Internet of things technology in our country, a number of high-grade residential areas with the installation of digital security monitoring system can transmit the fire information to the safety monitoring data, helping property management company or security monitoring center find and deal with the fire instantly.

4 Conclusion

Residential building plays an important role in urban construction, and it is an important place for urban residents to live and engage in various activities. Frequent occurrence of residential building fires will seriously affect our people's lives. Therefore, it is necessary to strengthen the supervision of fire safety of residential buildings for timely detection and elimination of fire hazards; it is necessary to improve the residents' awareness of fire safety so as to prevent and reduce the occurrence of fire; it is necessary to perform a more rigorous residential building fire safety design in accordance with *Code for Design of Building Fire Protection* newly revised in 2015 to improve residential building fire resistance; it is necessary to perfect the construction of fire infrastructure in the residential quarters according to the new specification to improve the ability to control the initial fire, and to make efforts to promote the application of new fire alarm, fire fighting and escape equipment in the household so as to enhance the ability to respond to fire and ensure residents' personal and property security.

References

1. Yu Jianhua, China Fire Fighting Yearbook 20120, Beijing, China Personnel Publishing House, December 2012.
2. Yu Jianhua, China Fire Fighting Yearbook 20130, Beijing, China Personnel Publishing House, December 2013.
3. Yu Jianhua, China Fire Fighting Yearbook 20140, Kunming, Yunnan People's Publishing House, December 2014.
4. Chen Peng, Wang Da, Problems and Countermeasures of Fire Safety in Urban Residential Quarters, Fire Safety Journal, 2011 (1), 65-68.
5. Wang Yingjun, Strengthen Community Fire Prevention Work to Prevent Urban Residents Fire, Journal of Guangxi Teachers Education University (Philosophy and Social Science Edition), 2004 (6), 144-146.
6. Shi Haifeng, Main Problems and Countermeasures of Residents' Fire Prevention, Journal of Changzhou University (Social Science Edition), 2011 (4), 33-35.
7. Wang Li, Fire Safety Problems and Solutions in Residential Quarters, Jiangxi Building Materials, 2014 (20), 296-298.
8. Li Qinggong, Wu Dong, Xie Fei, Song Wenhua, Analysis on Residential Fire Hazard and Safety Protection Measures, Fire Science and Technology, 2009 (6), 457-460.
9. Wang Hui, Yuan Kexia, Study on Fire Safety Management of Residential Property, Science and Technology in Western China, 2014 (6), 93-95.

Software Sub-system in Loading Automatic Test System for the Measurement of Power Line Filters

Bo Yu [1*], Lin Fu[1] and Adam Barker[2]

[1] Chengdu Technology University Chengdu, 611730, China
[2] School of Computer Science, st andrews, Uk
Corresponding Email: 14392926@qq.com

Abstract. The loading automatic test system for measurement of power line filters are in urgent demand. So the software sub-system of the whole test system was proposed. Methods: structured the test system based on the virtual instrument framework, which consisted of lower and up computer and adopted the top down approach of design to perform the system and its modules, according to the measurement principle of the test system. Results: The software sub-system including human machine interface, data analysis and process software, expert system, communication software, control software in lower computer, etc. had been designed. Furthermore, it had been integrated into the entire test system. Conclusion: This sub-system provided a fiendly software platform for the whole test system, and had many advantages such as strong functions, high performances, low prices. It not only raises the test efficiency of EMI filters, but also renders some creativities.

Keywords. Loading, automatic test system, software sub-system, power line filter

1 Introduction

Power line filter plays an important role in each electrical and electric system for suppressing the EMI(Electromagnetic Interference) in order to meet the requirements of EMC(ElectroMagnetic Compatibility) Certification standards. The more strict rules of the standards are, the more extensive those filters are used, and the higher quantities of the filters have been demanded. How to make an efficient and accuracy test for the filters becomes an urgent job, which must not only design reliable measurement process but also implement the whole integrated test system to ensure the conformity of the electrical technical parameters and performance indexes of the filters. Those technical specifications include rated voltage/current, working voltage/current, withstand voltage, residual voltage, leakage current, insertion loss (IL), etc. Each specification has been ruled by certification standard(s) for its value or measuring method. GB7343-1987[1] rules the methods of measurement of suppression characteristics of passive radio interference filters and suppression components over the frequency range of 10khz to 30Mhz, and MIL-STD-220B[2] stipulates the methods for IL testing when filter is under full loading, while CISPR 17[3-4] fixes the approximate worst case impedance IL testing beyond the standard test method which uses a calibrated 50 Ω signal source and a 50 Ω receiver. The main problem is, faced with those specifications which use different test methods and different kinds of equipments, the whole test process had been separated to some individual measurement, and before now, each test of the index or indicator of the filters had been performed manually, in which there always some defaults been made, even catastrophic effect such as electric shock for operation staff, or physical damage for the equipments or devices under test. Thus there has so far seldom find the papers reporting and discussing automatically test for specifications of the filters except some researches for IL measurement only in which absolutely do not refer as loading test, not to mention that it integrate loading test to implement automatic measurement.

This paper addresses the software design in an actual automatic test system for main specifications stated above especially the loading IL automatically testing for the filter. The aims of software design in the test system are not only to overcome the shortcomings of traditional manually testing but also provide some advanced features and performances for the whole system and the test program. First, it sets up a friendly software platform to implement whole automatic specification test for each filter when it has been installed into the homemade test box which plays an important role in the testing configuration. Second, it performs an automatically test procedure according to the turns of index for DUT. Third, it realizes some intelligent function such as provides some redesign propositions for those unqualified DUT.

2 Framework and principle of the whole system

2.1 Framework

The whole test system was composed of an upper and lower computer architecture which based on the virtual instrument framework as shown in figure 1. Under the framework the hardware sub-system consisted of partly various commercial off the shelf(COTS) devices and equipments including PC, impedance tester, test voltage tester, etc., and partly homemade modules in which the key feature was that it integrated severally the signal generator and the receiver on a single card based on PCI bus plug in the PC to replace the vector network analyzer used in traditional manually test method, reduced the cost of the test system, and made the test more flexible accompanied with its more functions and no less performances. Note, the links among COTS equipments and homemade modules are totally standard interfaces, junctors, and shield cables.

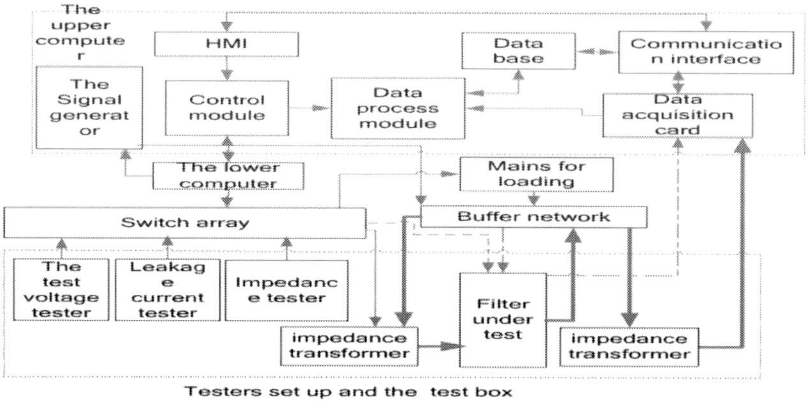

Fig. 1. Framework of the test system

2.2 Principle

As stated above, we must test multiple index for each filter, and the testing principles and equipments or instruments are different, so the test modes or methods are different respectively. Take IL test for instance, the principle of the system is: the operation command Issued by the upper computer, and the lower computer send the test signals which output by the signal generator to the filter under test in the test box via switch array, then the receiver collect the signals output by filter and convey them to the PC via communication interface. When the impedance between source and input port of filter and output vs. load are matching, signal follows the dotted line while the approximate worst case it follows the heavy line in figure 1.

As we know, the IL of EMI filter separated to two kinds, one is the common-mode (CM), and the other is the differential-mode (DM). Therefore the IL test configuration of hardware for EMI filters measurement must accordingly set up two forms. The CM set up as figure 2, while DM set up as figure 3. The flow chart of IL test shown as figure 4.

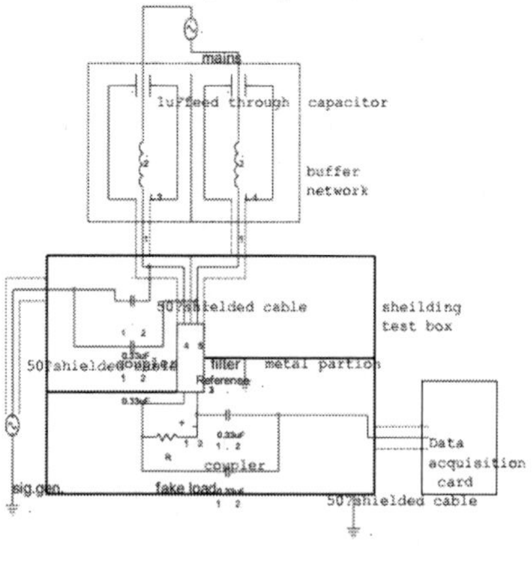

Fig.2. Hardware set up for IL of CM testing

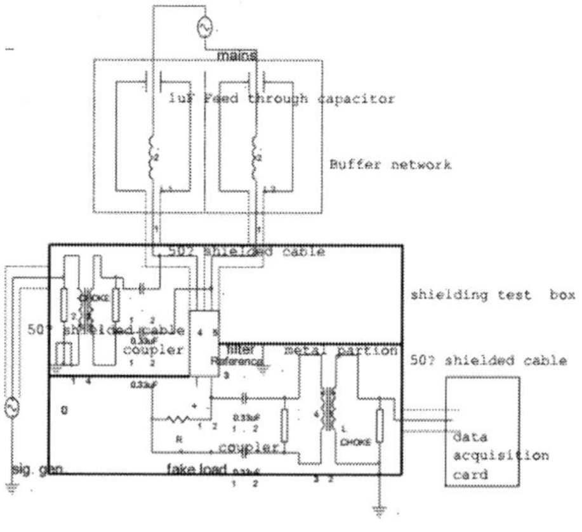

Fig.3. Hardware set up IL of DM testing

in which the automatic test process of IL is illustrated eventually, and the other index procedures such as insert loss, and test voltage, as well as leakage current are also including in the flowchart.

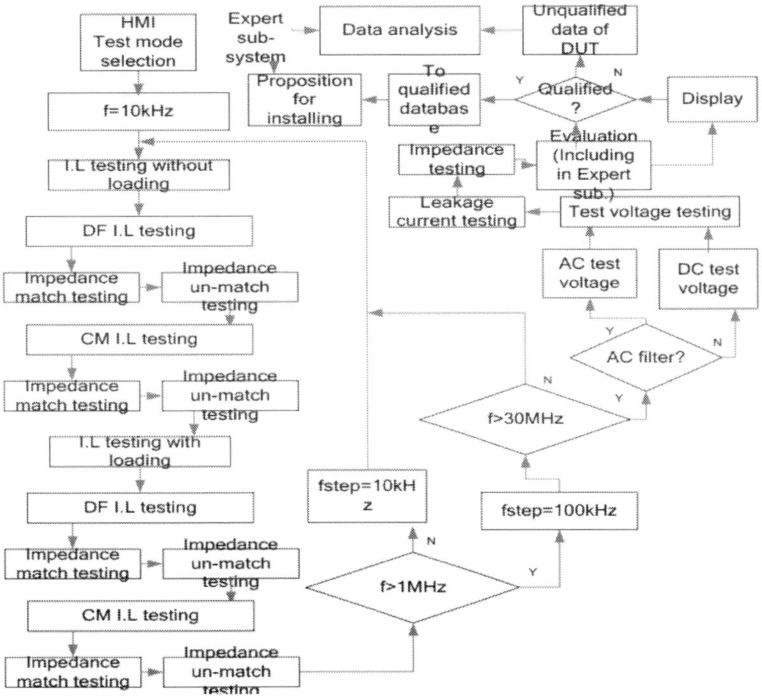

Fig. 4. Main index testing flowchart of the software

From figure 4, we note that there is an expert sub-system in the software, which is the one of our innovation points for the test system. Based on the expert sub-system, the test system can evaluate the quality of each DUT automatically, and further, it can diagnose the causes leading to invalidation for each unqualified DUT according to analysis module in the software sub-system.

3 Main intelligent modules of the software sub-system

The data sent from acquisition card which based on PCI bus plug in must have been analyzed using the intelligent modules integrated in the software. So the intelligent module plays an important role in the whole system, in which there consist with three parts, one is the I.L spectrum analysis module, second is the evaluation module, and third is the proposition module.

3.1 I.L spectrum analysis

The data analysis and process is the key component of the system, in which there are many methods including traditional time domain approaches and frequency domain approaches, and modern methods used to analyze and process the data from different input including output of data acquisition card, database, and internet. For the output of data acquisition card, there is a real time event, and for the data from database, there is an analysis and process for the third part data or the historical data. For the data from the internet, there is not only a remote measurement but also a real time event. According to the results of analysis module, the software can display time domain waves, spectrum plot.

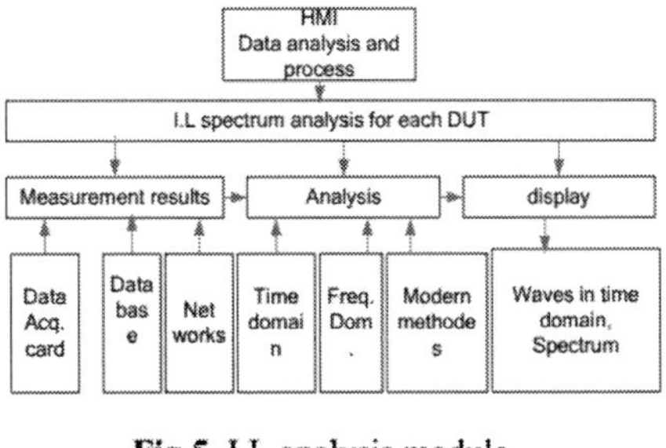

Fig.5. I.L analysis module

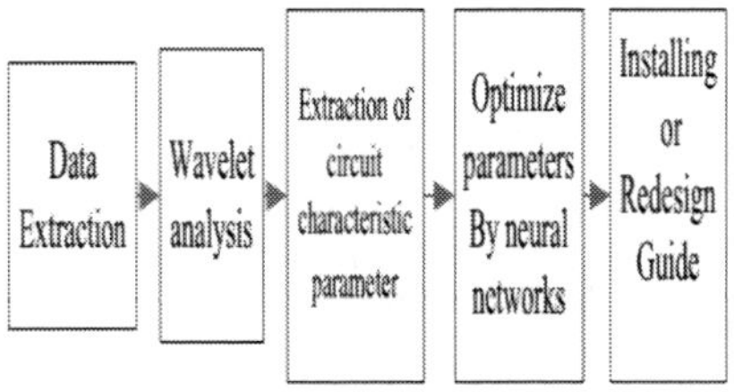

Fig. 6. Diagnosis module

3.2 Evaluation module

The evaluation module intellectually compares the data between test and analysis results and the database which keeps the records of good and standardized parameters for the DUT, furthermore, it makes a judgment that it is belong to unqualified or qualified ones for each DUT automatically. Finally, it stores the data into different database accordingly, noting that there is an intelligent discriminating between better DUT and good DUT in critical area, the data of later must send to the diagnosis module to obtain the installing guide In order to facilitate the efficiency of the DUT in the real system.

3.3 Diagnosis module

The diagnosis module shown as figure 6, it employs wavelet analysis and neural networks to analyze the DUTs which been judged to unqualified or in the critical area of good ones.

3.4 Software modules in the lower Computer

Software modules in the lower computer including MCU main module, up computer instruction handout module, switch array control module, etc. the MCU main module is the kernel of lower computer software, which operating the whole lower computer. The module of up computer instruction handout gives away the instructions from up c0puter through MCU main module to the modules implementing its function respectively. And, as stated above, to achieve all testing procedures, the whole system must employ some other instruments or equipments, so there must adopt multiple switches to change the test functions. Further, in order to deals with the approximate worst-case test method for testing the insertion loss of the EMI filters, the 0.1Ωand 100Ωand vice versa impedances usually used for this testing. Thus it must design the impedance transformer network which needs switches in proper order on or off according to the test process. The software of switch array control module is designed accordingly.

4 Results and conclusion

4.1 Results

The whole sub-system was designed perfectively, according to different testing function, the software had some different HMI. Noting that the software was designed for Chinese, so the language in the software is Chinese.

As shown in figure 7, there is a filter just under test been testing it I.L parameters. And as shown in figure 8, there is the data check HMI which indicated the output of testing.

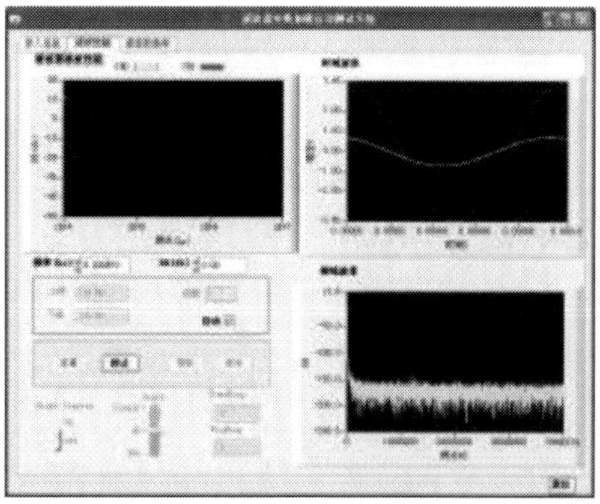

Fig.7. The HMI of I.L testing

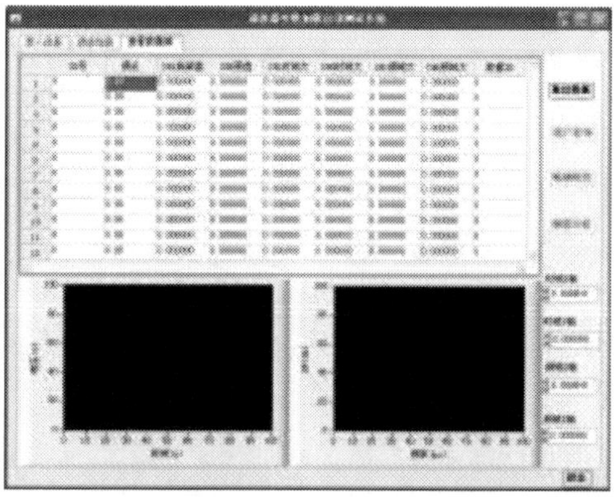

Fig.8. The data check function HMI

4.2 Conclusion

The software sub-system rendering a friendly and efficient platform to implement the automatic loading test for power line filters. Its main creativity is that it integrated the loading test and automatic test synchronously with its virtual instrument framework, and based on the framework, the strong power of the software in the test system can perform more intelligent. This sub-system had been granted with the utility model patent certificate for the whole system, which patent number is ZL 2014 20203665.8.

References

1. GB 7343-87, methods of measurement of suppression characteristics of passive radio interference filters and suppression components over the frequency range of 10khz to 30Mhz.
2. MIL-STD-220B, METHOD OF INSERTION LOSS MEASUREMENT
3. IEC CISPR 17, Methods of measurement of the suppression characteristics of passive radio interference filters and suppression components (1981)
4. Jiri DRINOVSKY, Zdenek KEJIK,etc. Insertion Loss Estimation of EMI Filters in Unmatched Input Output Impedance System. RADIOENGINEERING, 20, 1, APRIL (2011):295-298
5. Zhang Shuping, Zhao Yajun. Analysis of manual test mode for EMI filter of insert loss. Inner Mongolia Science Technology & Economy, No.16 (2008): 114-115
6. Ma Jianing, Song Bo. Automatic test system for insertion loss in EMI power filter. Intellectual Property Office of the people's Republic of China. CN203191451U. 2013.09
7. Ma Jianing, Song Bo. Automatic test system and methods for insertion loss in EMI power filter. Intellectual Property Office of the people's Republic of China.CN102944752A. 2013.02
8. Liu Bin, Li Gang, Jiang Quanxing. Design of Insertion Loss Testing System of EMI Power line Filter. Safty &EMC. No.2 (2005).
9. Wang Meng, Zhou Ping. Design of Automatic Test System for Insertion Loss in EMI Power Filter. Electronic quality, No.09 (2008):71-76
10. Ye Fen. A automatic method for EMI filter testing. Intellectual Property Office of the people's Republic of China.CN1657964A. 2005.05
11. Cao Quanming, Cai Dongxian. A automatic testing system for EMI filter. Intellectual Property Office of the people's Republic of China. CN101315404B. 2010.07.
12. LiuHao, Wei Lili, Zhong Xiaoyan. EMI power line filter automatic testing system based on virtual instrumentation. measurement techniques, No.2 (2008):10-13
13. Liu Hao, Ding Xue. Research of insert loss testing techniques for EMI power line filter. Electronic testing. No.7, (2007):51-53
14. Qiu Yang. A testing equipment of current loading for power line filters. Intellectual Property Office of the people's Republic of China.CN201607516U. 2010.10.
15. Zhang Jinkui. Development and Analysis of the test equipment of EMI filter with current applied. Master Thesis of Southeast University (2007).
16. Liu Bin. DESIGN OF LOAD TESTING SYSTEM OF EMI POWER LINE filter. Master Thesis of Southeast University (2007)
17. Liu Song. The integrating machine of filter reshaping automatic testing. Intellectual Property Office of the people's Republic of China. CN201985231U .2011.09
18. Qiu Yang. The design method and multiple function equipment of power line filter electrical performance. Intellectual Property Office of the people's Republic of China.CN101788631B. 2014.1
19. Tang Shiping,Yang Jingfa, Huang Jian. Research of testing method for insret loss feature of EMI filte. Saftye &EMC, No.2 (1999): 26-37

A Model of Trusted Connection Architecture

Xun Zhang [1] , Zhihao Wang [2], Yong Zhi [1] and Hui Yuan [1]

[1]Electric Power Research Institute, State Grid Gansu Electric Power Corporation, Lanzhou 730050, China

[2]Global Energy Interconnection Research Institute, Beijing 102209, China

Abstract. According to that traditional trusted network connection architecture (TNC) has limitations on dynamic network environment and the user behavior support, we develop TCA to propose a trusted connection architecture supporting behavior measurement (TCA-SBM), besides, the structure diagram of network architecture is given. Through introducing user behavior measure elements, TCA-SBM can conduct measurement on the whole network in time dimension periodically, and refine the measurement on network behavior in measure dimension to conduct fine-grained dynamic trusted measurement. As a result, TCA-SBM enhances the TCA's ability to adapt to the dynamic change of network and makes up the deficiency of trusted computing framework in the network connection.

Keywords. Behavior measurement, trusted network connection, fine-grained, adaptability

1 Introduction

At present, the trusted network connection architecture mainly focuses on the problem of network access, which is essentially a kind of network access control (NAC). NAC is a kind of computer network security method, which combines the technology of terminal security technology, user or system authentication and network security technology. Traditional network access control methods include: MAC address filtering, VLAN isolation, IEEE802.1Q authentication, access control list based on IP address and firewall control, host state inspection, isolation and recovery, etc. At present, the main network architectures include TCG trusted network connection (TNC), Cisco network admission control (NAC), Microsoft network access protection (NAP), and in our country the main network architecture is the trusted network connection architecture (TCA) with three entities and three levels [1-4].

TNC, NAC, NAP network connection all adopt the access control mechanism based on integrity measurement, on the time dimension, this access control focuses on the network connection establishment stage. With the increasing complexity of network applications, simply verifying the platform integrity in connection stage can't meet the requirements of credibility of the network system. The main limitations are as follows: lack of continuity integrity measurement mechanism; Lack of fine-grained trust metrics. Traditional trusted network connection control model has limitations, which result in that the trusted network connection mechanism can't ensure the credibility of the whole process in network connection, and can't adjust the trusted measurement to the dynamic change of network environment and the change of the user behavior, therefore, the application of trusted computing framework in real network system is restricted [5-15].

According to traditional trusted network connection architecture (TNC) has limitations on dynamic network environment and the user behavior support, this paper developed TCA to propose a trusted connection architecture supporting behavior measurement (TCA-SBM). Through introducing user behavior measure elements, TCA-SBM can conduct measurement on the whole network in time dimension periodically, and refine the measurement on network behavior in measure dimension to conduct fine-grained dynamic trusted measurement. As a result, TCA-SBM enhances the TCA's ability to adapt to the dynamic change of network and makes up the deficiency of trusted computing framework in the network connection [16-35].

2 The design of TCA-SBM network connection architecture

2.1 The Structure diagram of TCA-SBM

Here we give the structure diagram of TCA-SBM: Figure 1.

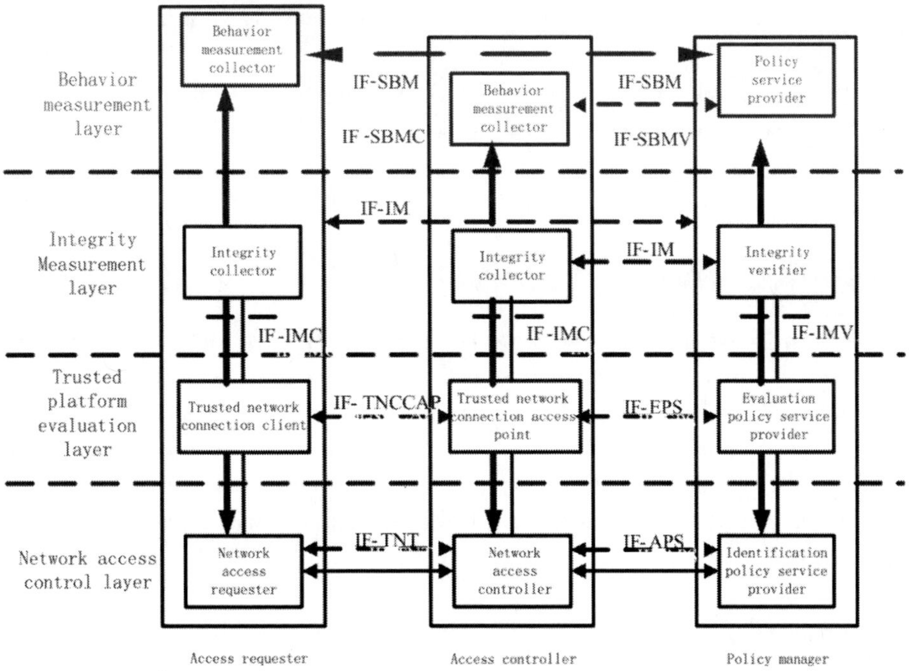

Fig.1. The architecture of TCA-SBM

2.2 The entity structure

2.2.1 Access requester

The main functions of access requester include: send request of accessing trusted network, and identify the user's identity with access controller; collect the platform integrity metric values of access requester and report to the access controller, complete the evaluation and report of trusted platform integrity between requester and controller; adopt the behavior report or behavior statement of requesters, and sent them to policy manager.

2.2.2 The access controller

The access controller is an entity that controls the requesters' access to a trusted network, whose functions mainly include: complete the user identity authentication, trusted platform evaluation and network behavior measurement between the access requesters; receive the integrity metric values of requesters and collect its own integrity metric values, and send the values to policy manager; adopt the behavior report or behavior statement of requesters, and sent them to policy manager; perform access control in case of the result generated during the processes such as user identity authentication, trusted platform evaluation and trusted behavior measurement.

This entity includes the following parts: network access controller, trusted network connection access point, integrity collector and Behavior measurement collector.

2.2.3 Policy manager

According to the different security requirements of trusted network environment, policy manager formulates the integrated security policy including user identity authentication, trusted platform evaluation and trusted behavior measurement, verify the validity of PIK certificates and platform integrity metric values from requesters and controllers, verify the results of trusted measurement for requesters' behavior, generate the results of user identity authentication, trusted platform evaluation and trusted behavior measurement performed by requesters and controllers; push network behavior statement, and verify the credibility of it.

This entity includes the following parts: evaluation policy service provider, identification policy service provider, integrity verifier and behavior policy service provider.

2.3 The architectural layers

2.3.1 Network access control layer

Network access control layer is the layer to realize the control in trusted network access, where network access requester, network access controller and identification policy service provider preform the protocol for verifying user identity, and realize the identification of user identity of requesters and controllers. As the trusted party in the protocol for verifying user identity, the service provider should verify whether the service provider can perform the verification for user identity, besides, it can perform the corresponding trusted network access control according to whether the results of user identity authentication, trusted platform evaluation and behavioral trusted measurement meet the integrated security policy.

2.3.2 Trusted platform evaluation layer

Trusted platform evaluation layer is the layer to perform the platform state evaluation, where trusted network connection client, trusted network connection access point and evaluation policy service provider perform trusted platform evaluation protocol, and realize the trusted platform evaluation for requesters and controllers. As the trusted party in the protocol for trusted platform evaluation, the evaluation policy service provider should verify the validity of PIK certificate of requesters and controllers, complete the platform integrity check of requesters and controllers by calling the integrity verifier form integrity measurement layer. The trusted network connection client and point generate the connection decision based on the corresponding trusted platform evaluation result, and send it to the requesters and controllers respectively.

2.3.3 Integrity measurement layer

Integrity measurement layer is the layer to perform the platform integrity measurement, where the platform integrity of requesters and controllers will be collected and verified. The integrity collector is to collect the integrity information in each component of platform, and the information include hardware factory information, software and hardware configuration, operating system certificate, browser certificate, third party application software certificate and so on, which the Integrity verifier should verify uniformly.

2.3.4 Behavior measurement layer

Behavior measurement layer is the layer to perform the behavior measurement for network visitor, collect the metric element information of behaviors, which try to access platform, such as create, delete, call, execution, access, upload or download and so on, determine whether the network visitor meet the network security policy according to the measurement results, besides, perform the corresponding access control for network visitor through network access control layer. Behavior measurement layer is the dynamic measure part during the trusted network connection, when having met the identity and integrity measurement policy and connected in trusted network, realizes the behavioral trusted measurement for network visitor during network connection, in addition, the metric elements and time accuracy of behavioral measurement can be self-configured according to behavior metric policy.

2.4 Functional components

2.4.1 Network access requester

Initiate an access request to access controller, perform user identity authentication protocol in company with network access controller and identification policy service provider, and realize the user identity authentication performed by requesters and controllers when they are in the network access control layer;

Forward the upper layer protocol data to the access controller and policy manager;

In case of the result of user identification the connection decision made by the identification policy service provider and the trusted network connection client respectively, control its own port, and realize the connection control of the access controller.

2.4.2 Network access controller

Activate user identity authentication protocol on the network access control layer, perform the user identity authentication protocol with requesters and service providers while realizing the two-way user authentication between them.

Forward the protocol data of the trusted platform evaluation layer and the behavior measurement layer to the access request and policy manager.

In case of the result of user identification, the result of platform state measurement and the result of behavior measurement, control its own port, and realize the access control of the access requester.

2.4.3 Identification policy service provider

Perform user identity authentication protocol in company with network access requester and network access controller, and as the trusted party in this protocol, realize the two-way user authentication between requesters and controllers.

2.4.4 Trusted network connection client

Request and receive the integrity values of measurement to the integrity collector through the interface IF-IMC, perform the trusted platform evaluation protocol with trusted network connection access point and evaluation policy service provider, let requesters and controllers realize the trusted platform evaluation.

In case of the trusted platform evaluation results generated by the service provider, make the connection decision and send it to the network access requester.

2.4.5 Trusted network connection access point

Activate user identity authentication protocol on the trusted platform evaluation layer, request and receive the integrity values of measurement to the integrity collector through the interface IF-IMC, perform the trusted platform evaluation protocol with trusted network connection client and evaluation policy service provider, realize the two-way trusted platform evaluation between requesters and controllers.

In case of the trusted platform evaluation results generated by the service provider, make the connection decision and send it to the network access requester.

2.4.6 Evaluation policy service provider

Perform the trusted platform evaluation protocol with trusted network connection access point and trusted network connection client, and as the trusted party in this protocol, realize the trusted platform evaluation between requesters and controllers.

The evaluation policy service provider verifies the validity of the PIK certificate of the access request and the access controller, when after confirming the validity of the PIK certificate, send to the integrality verifier e platform integrity metric values of requesters and controllers through the interface IF-IMV, and then receive the verified result returned by the verifier, finally generate a trusted platform evaluation result to send to the requesters and the controllers.

2.4.7 Integrity collector

Collect the platform integrity information of access requesters and access controllers with integrity services provided by trusted computing platform.

2.4.8 Integrity verifier

Check the platform integrity information of access requesters and access controllers with integrity reference values provided by platform components in integrity management.

2.4.9 Behavior measurement collector

Behavior measurement collector receives the request, report and declaration sent by requesters, and transmits them to the policy service provider. Behavior measurement collector bases on whether the metric result provided by the policy service provider meets the behavior security policy to determine whether the network connection port is connected to the network or disconnect from the network, which is depended on the network access controller.

2.4.10 Behavior policy service provider

The behavior policy service provider and behavior measurement collector carry out the behavior measurement protocol, and as the trusted party in this protocol, measure the behavior. What's more, the service provider can synthetically determine whether the requesters' behavior meet the security policy. Finally, the service provider has integrated network behavior metric functions including individual similarity and swarm similarity evaluation, direct trust evaluation and risk assessment.

3 The design of TCA-SBM network connection process

Trusted network measurement is divided into three levels: before the trusted network connection is established they need to perform the identity measurement on network access requesters; after that they will further perform the platform status measurement; when the two metrics are meet the security policy, the access controller and the access requester establish the trusted network connection. After the establishment of network connection, they perform the behavior measurement for visitors successfully accessing network. Behavioral measure layer is a dynamic measure layer where they perform the behavior measurement for visitors successfully accessing trusted network, configure the frequency of metric time according to behavior measurement policy, for the common safety requirements of system can adopt the snapshot, for all highly principal system can adopt the real-time behavior measurement.

The specific process and steps of the TCA-SBM network connection are as follows:

(0) Before the establishment of trusted network connection, client and access point must be respectively load their each IMC according to the platform specific binding function while evaluation policy service provider must load it its each IMV according to the platform specific binding function.

(1) The network access requester initiates an access request to the network access controller.

(2) Network access controller receives a request to access, and after that, performs user authentication protocol to realize two-way user authentication with network access requester and evaluation policy service provider, in which the service provider should not have to participate. If it participates in the user identity authentication protocol, the policy service provider acts as a trusted third party. In the protocol, the network access controller and the network access requester are also allowed to negotiate the session key.

If required to make the access decision immediately after the completion of user authentication, network access controller and network access requester generate the access decision respectively according to the user identity authentication results, and then perform the access control according to the generated access decision, otherwise go to the step (3).

(3) If network access requester requires the platform identification, network access requesters send the platform authentication request. If the network access controller requires the platform identification, network access controllers send the platform authentication request.

(4) A: Having received the platform authentication request messages, the network access controller starts the platform identification process, perform a round or multi round of platform authentication protocol to realize the authentication among platforms. If they did not receive the messages, they initiate a round of platform authentication protocol.

B: During the authentication process, client and access point interactive with their each IMC by IF-IMC respectively.

C: The evaluation policy service provider verifies the PIK certificate of the access controller and the access requester, and completes the integrity verification and evaluation by IF-IMC through calling its each IMV. The each IMV generate the component integrity evaluation results and send them to the policy service provider. Then, the providers generate the platform integrality evaluation results according to the evaluation strategy of component integrity evaluation results collected by evaluation protocol. Finally, they send the results, which include the PIK certificate validation results and platform integrity evaluation results, to the client and access point.

(5) When the access requester and access controller have completed the platform identification, they generate the access decision (allow, disallow, isolation), which is according to the PIK certificate validation results generated by the client and the access point and the platform integrality evaluation results, and then send the results to the client and the access point.

Network access requester and controller perform access controls according to the received access decision, realize the trusted network connect control, that is, they determine whether to connect this protected trusted network by the access decision and complete the measurement on identity and status in trusted network connection.

(6) After the establishment of the trusted network connection and before the behavior measurement, the trusted network connection client must bind to the specific platform to initialize the behavior measurement request.

(7) Behavior measurement collector carries out the measurement protocol by IF-SBM. Firstly, determine whether feasible network of the requester is same with access network on the attributes, if not, inform the controller to disconnect the network connection by control ports; if so, the performer and the requester negotiate a master key to negotiate key and establish a trusted pipeline to transmit confidential data.

(8) Behavior policy service provider conducts the policy evaluation for behavior metric reports, if it can meet the behavior policy, continue to keep the network connecting; otherwise, inform the network access control to

disconnect the network connection by control ports.

4 Conclusion

The scheme is mainly on the basis of TCA, integrates dynamic behavior measurement methods, ensure the credibility of the terminal identity and status through techniques of trusted computing platform and TCA integrity measurement mechanism, and ensure the credibility of the terminal behavior through the dynamic behavior measurement mechanism. This framework exploratively gathers measurement, report and declaration, verification mechanisms and so on, and combines the behavior measurement supporting the historical analysis. In theory, if the terminal does operations which do not meet the behavior security policy or malicious behavior and other events, it will be found by the behavior policy service provider, and the terminal will be isolated.

References

1. http://www.cisco.com/en/US/netsol/ns466/networking_solutions_package.
2. http://www.microsoft.com/technet/network/nap/napoverview.mspx
3. http://www.microsoft.com/technet/network/nap/naparch.mspx
4. https://www.trustedcomputinggroup.org/groups/network/
5. David Clark, Karen Sollins, John Wroclawski, NewArch Project: Future-Generation Internet Architecture [EB/OL]. http://www.isi.edu/newarch/iDOCS/final.finalreport.pdf
6. Neumann P.G., Principled assuredly trustworthy composable architectures [EB/OL].http://www.csl.sri.com/neumann/chats4.html
7. Rodriguez A, Egenhofer M, Determining semantic similarity among entity classes from different ontologies, IEEE Transactions on Knowledge and Data Engineering, 2003, 115, 2: 442-456
8. TCG Specification Trusted Network Connect -TNC Architecture for Interoperability Revision 1.1[EB/OL]. Trusted Computing Group, 2006.5, http://www.trustedcomputinggroup.org.
9. TCG Specification Trusted Network Connect IF-PEP: Protocol Binding for Radius Revision 0.7[EB/OL]. 2007.5, https://www.trustedcomputinggroup.org
10. TCG Specification Trusted Network Connect IF-T: Protocol Binding for Tunneled EAP Methods [EB/OL]. Revision 10, 2007.5, https://www.trustedcomputinggroup.org
11. TCG Specification Trusted Network Connect IF-TNCCS: TLV Binding Revision 10, 2008.1[EB/OL]. https://www.trustedcomputinggroup.org
12. TCG Specification Trusted Network Connect IF-IMC Revision 8 [EB/OL] .2007.2, https://www.trustedcomputinggroup.org
13. TCG Specification Trusted Network Connect IF-IMV Revision 8 [EB/OL] .2007.2, https://www.trustedcomputinggroup.org
14. TCG Specification Trusted Network Connect IF-M: TLV Binding Revision 30 [EB/OL]. 2008.1, https://www.trustedcomputinggroup.org
15. TCG Specification Trusted Network Connect IF-PTS Revision 1.0 [EB/OL]. 2006.11, https://www.trustedcomputinggroup.org
16. D.H.McKnight, N.L. Chervany, TheMeanings of Trust, Technical Report MISRC Working Paper Series. University of Minnesota, Management Information Systems Reseach Center, 1996: 96-102

17. Kini A. and Choobineh J. Trust in Electronic Commerce: Definition and Theoretical Considerations. In 31st Annual Hawaii International Conference on System Sciences, Hawaii,1998, http://ieeexplore.ieee.org/iel4/5217/14270/00655251.pdf
18. Beth T, Borcherding M , Klein B，Valuation of Trust in Open Network //Proceedings of the European Symposium on Research in Security. 1994: 152-157
19. Grandison, T, Sloman, M., A survey of trust in internet applications. IEEE Communications Surveys and Tutorials, 2000, 4, 4: 2-16
20. Audun Jøsang, Roslan Ismail, Colin Boyd. A Survey of Trust and Repuatition System for Online Service Provision. Decision Support System, 2007, 43, 2: 618-644
21. ITU-T Recommendation X.509, ISO/IEC 9594-8. Information Technology. Open systems interconnection. The Directory: Public-key and Attribute Certificate Frameworks. Draft ITU-T Rec. X.509, May, 2001: 22-44
22. Marsh, S. P., Formalizing Trust as a computational Concept. Doctoral Dissertion, University of Stirling, 1994: 7-33
23. Almenarez F, Matin A, Diaz D, Sanchez J. Developing a model for trust management in pervasive devices. In: Bob Werner，ed．Proc of the 3rd IEEE Int'l Workshop on Pervasive Computing and Communication Security. Washington, IEEE, 2006: 245-250
24. Jameel H, Hung LX, Kalim U, Asjjad A, Lee SY, Lee YK. A trust model for ubiquitous systems based on vectors of trust values. In Proc of the 7th IEEE Int'l Symp. on M ultimedia. Washington: IEEE Computer Society Press, 2005: 674-679
25. Sun Y, Yu W, Han Z, Liu KJR. Trust modeling and evaluation in ad hoc networks. In: Proc of the Global Telecommunications Conf, Globecom 2005. Washington: IEEE Computer Society Press, 2005: 1-10
26. Li DY, Meng HJ, Shi XM. Membership clouds and membership clouds generator. Journal of Computer Research and Development, 1995, 32, 6: 15-20
27. Sailer, X Zhang, T Jaeger, L vail Doom, Design and implementation of TCG-based integrity measureme architecture. In Proceedings of USENIX Security Symposium. Lake Tahoe, California, USA: ACM Press, Aug. 2004: 223-238
28. Haldar, D Chandra, M Franz, Semantic remote attestation a virtual machine directed approach to trusted computing. In Proc of the Third virtual Machine Research a Technology Symposium. San Jose, CA, USA: USENIX, 2004: 29-41
29. Gong B, Zhang J, Xiaolie Y E, et al. A Trusted Measurement Scheme Suitable for the Clients in the Trusted Network. Wireless Communication Over Zigbee for Automotive Inclination Measurement China Communications, 2014, 11(4):143-153.
30. Ning Z H, Shen C X, Zhao Y, et al. Trusted measurement model based on multitenant behaviors. Scientific World Journal, 2013, 2014:384967-384967.
31. Li R H, Zhu Z S, Li C, et al. Trusted measurement model of runtime process behavior based on turing// Machine Learning and Cybernetics (ICMLC), 2010 International Conference on. IEEE, 2010:2183-2187.
32. Liang P, Jiang W, Gong B, et al. A New Behavior Measurement for Cloud Computing. International Journal of Advancements in Computing Technology, 2012, 4(14):9-16.
33. Qiao A, Song R, Shen L. Study of the Network Behavior Measurement. Application Research of Computers, 2005.
34. Filva D A, Guerrero M J C, Forment M A. Google analytics for time behavior measurement in Moodle// 2014 9th Iberian Conference on Information Systems and Technologies (CISTI). 2014:1-6.
35. Guo Z H, University W, University W, et al. Research on Trusted Network Connection. Chinese Journal of Computers, 2010, 33(4):706-717.

The Empirical Analysis of Liaoning Equipment Manufacturing Industry Competitiveness

Lingyan Meng[1]* and Shijiang Xu[2]

[1]School of Economy and Law, University of Science and Technology, Liaoning, Anshan, Liaoning Province, China
[2] Liaoyang Vocational College of Technology, Liaoyang, Liaoning Province, China
Corresponding Email: mly7312@163.com

Abstract. This paper uses principal component analysis (PCA) method to effectively interpret the contribution of various factors by the idea of dimension reduction, a dimensionless processing on the original data and extraction of the representative variables in order to determine the weights of indicators, to identify meaningful factors which exert significant influence on the competitiveness of Liaoning Equipment Manufacturing Industry.

1 Introduction

Known as the "Armaments Department of China", Liaoning province is one of the old industrial bases of northeast and even the whole country, and equipment manufacturing industry (EMI) has already become the pillar of the rapid growth of regional economy and the important engines [1]. The EMI in Liaoning ranks the top ten in China and holds around 5% of the share in the whole country, and hence has obvious advantages of scale because of its size, production, gross assets, main business income and added value[2]. In division, industry of general purpose machinery in Liaoning has prominent advantage due to its 152 enterprises (including 131 large and medium-sized ones) ,72.79 billion yuan of total assets and 326.138 billion yuan of main business income in which Shenyang Machine Tool Co., Ltd and Shenyang Heavy Equipment Group play vanguard role in their special industry in China [3]. At the same time, major businesses such as AVIC Shenyang Aircraft Group, AVIC Liming Engine Group, Dalian Shipyard, Wafangdian Bearing Group enjoy strong influence on domestic and international market [4].

This paper uses principal component analysis (PCA) method to effectively interpret the contribution of various factors by the idea of dimension reduction, a dimensionless processing on the original data and extraction of the representative variables in order to determine the weights of indicators, to identify meaningful factor analysis instead of the original variables, and to boils down the complex multiple variables to a few unrelated principal component.

2 Choice of evaluation method

Use a two-column format, and set the spacing between the columns at 8 mm. Do not add any page numbers.

2.1 Basic model

Variables are respected as $X_1, X_2, \dots X_p$, linear combinations are expressed as $Z_1, Z_2, \dots Z_P$, X_{ij} shows the indicator data of number j belongs to unit i, and the covariance matrix is Σ, a shows the mean value of the random vector combination X. So, a linear combination of the original variables may be written as follows:

$$\begin{cases} Z_1=a_{11}X_1+a_{12}X_2+\cdots a_{1p}X_p \\ Z_2=a_{21}X_1+a_{22}X_2+\cdots a_{2p}X_p \\ \cdots\cdots \quad \cdots\cdots \quad \cdots\cdots \\ Z_p=a_{p1}X_1+a_{p2}X2+\cdots_{appXp} \end{cases} \quad (1)$$

2.2 Basic steps of principal component analysis

(1)Determine P as the number of index data, n as the estimated number of samples, and hence the number of the original data is $n*P$. we take the east 10 provinces (Hainan province is not included) as the research sample.

(2)Standardize the indicator data to insure the average value of each variable is 0, and the variance is 1.

(3)Establish a covariance matrix R to reflect and measure the affinity degree of standardized data. The bigger is R, the necessity of principal component analysis is higher.

(4)Determine the number of principal components according to the eigenvalues of covariance matrix R, the principal component contribution rate and cumulative contribution rate. Eigenvalue is the variance of each main component and used to measure its influence. Eigenvalue should bigger than numerical value *1* and the cumulative contribution rate should bigger than numerical value 85%. Given Principal component contribution rate as

$W_i = \lambda_i / \sum_1^m \lambda_i$, ($i = 1, 2, 3, 4 \cdots m$, m is the number of main components).

(5)Establish the initial factor loading matrix and then explain the main component.

(6) Calculate the comprehensive score function FJ (J is the number of samples), calculate the comprehensive values of regional competitiveness, and put them in descending order. The formula

$$F_j = W_1 F_{1J} + W_2 F_{2J} + \cdots W_m F_{mj} \quad (2)$$

3 Indexes and data selection

Firstly, the choice of indicator data must follow the principle of objectivity, availability and comparability [5]. Secondly, when estimate the comprehensiveness of industry, economic indicators, efficiency indicators, innovation and even government supporting indicators should be involved [5]. Specifically, this paper will choose 4 first-level indicators including industrial scale, market scale, profitability and scientific research activities, and choose 18 second-level indicators including assets per share, market share, product sales, asset-liability ratio, R&D input level, etc.(see table 1). Relative data are from 2015 statistical yearbooks of the eastern provinces, *2015 China Statistical Yearbook on Science and Technology*, and *2015 China Industry Economy Statistical Yearbook.*

Table 1. Indexes system of industry competitiveness evaluation.

	First-level indicators	Second-level indicators
exes System of Industry Competitiveness Evaluation	**Industrial scale**	Ratio of Enterprise number Assets per share Original value of fixed assets Net value of fixed assets Industry labor productivity Company sales revenue Tax paid levels
	Market scale	Market share Industrial relative specialization index Product sales
	Profitability	Contribution of total assets Asset-liability ratio profit rate of costs and expenses
	Scientific research activities	R&D input level Science and technology personnel number Enterprise scientific research activities Level of invention patents Ability of technical innovation

3.1 Industrial scale

Industry scale is important source and influencing factors of competitiveness and can be shown by 7 second-level factors including enterprise quantity, Per capita assets ratio, industry labor productivity, enterprise scale and tax paid level.

X_1: Ratio of enterprise number. It can be calculated as follow: *number of area formed EMI enterprises / number of national EMI enterprises*, which is used to measure the regional industry scale and that of industrial agglomeration. [7]The related data are from *2015 China Industry Economy Statistical Yearbook.*

X_2: Assets per share. It can be calculated as follow: *regional assets per person / assets per person in the East*, which can indicate and compare regional enterprises size and their strength, and has the effect that cannot be ignored to the regional industry competitiveness. Assets scale can be defined as *Net value of fixed assets + liquid assets*.

X_3: Fixed assets. It is defined as follow: *industrial total fixed assets in eastern areas / industrial fixed assets*. This index can be used to measure regional industry scale and the cluster status. The related data are from statistical yearbook 2015 of each province.

X_4: Net fixed assets. It is defined as follow: *fixed assets of regional industry / net fixed assets of industry in the east.* This index can measure both the regional industry scale, the cluster level, and can reflect the management efficiency and level of fixed assets. The related data are from statistical yearbook 2015 of each province.

X_5: Industry labor productivity. It can be showed as follow: *regional per capita output /per capita output of eastern area.* The related data are from *2015 China Industry Economy Statistical Yearbook.*

X_6: enterprise sales revenue. It can be defined as follow: *average sales revenue of regional enterprise / average sales revenue of the enterprise in eastern area.* This indicator can measure industry scale and regional enterprise scale of EMI. The related data are from statistical yearbook 2015 of each province.

X_7: Tax paid levels. It is defined as follow: *rate of regional tax paid / rate of tax paid in eastern area.* This indicator can reflect the scale of the enterprise, and can also reflect the level of enterprise capital operation. In general, the higher it is, the better is the enterprise's economic benefit. The related data are from statistical yearbook 2015 of each province.

3.2 Market scale

X_8: Market share. It is shown as follow: *regional industry sales revenue / industry sales revenue in the eastern area.* This index can indicate both the market scale of EMI and potential of EMI, and is a main factor that affects the core competitiveness of enterprises. The related data are from statistical yearbook 2015 of each province.

X_9: Industry specialization index. This index can be calculated as follow: regional EMI production value accounts for the proportion of industrial gross output value / EMI output value accounts for the proportion of gross industrial output value in the eastern area.[8] Industry specialization index can effectively reflect the comprehensive ability of regional EMI economic scale, benefit level, management level, etc. The related data are from 2015 China Industry Economy Statistical Yearbook.

X_{10}: Production sales rate. This index is defined as follow: regional sales of EMI / gross output of EMI in the eastern area. This index is a good indicator to reflect the economic benefits and market competitiveness of regional EMI.[9] The related data are from 2015 China Industry Economy Statistical Yearbook.

3.3 Industry profitability

X_{11}: Contribution rate of total assets. This index is defined as follow: *(profit + interest + tax) / total assets.* It reflects the assets quality, management capability and profitability of an enterprise. The related data are from *2015 China Industry Economy Statistical Yearbook.*

X_{12}: Asset-liability ratio. It is calculated as follow: *debt/total assets.* This index reflects the assets sufficiency of an enterprise, and at the same time it reflects the management capability and profitability of an enterprise. The related data are from *2015 China Industry Economy Statistical Yearbook.*

X_{13}: profit rate of costs and expenses. This index in defined as follow: *profit / cost and expense.* It is used to measure the economic benefits of enterprise capital operation and can reflect the enterprise operation and management ability. Usually, Total cost and expense can be calculated as: *sales cost + management fees + financial costs.* The related data are from *2015 China Industry Economy Statistical Yearbook.*

3.4 Ability of technical innovation

X_{14}: R&D input level. This index can be defined as follow: *proportion of R&D funds of regional industrial sales / R&D funds proportion of national industrial sales.* It can directly measure the *R&D* level, technical research and development ability and innovation ability of an enterprise. The related data are from *2015 China Statistical Yearbook on Science and Technology.*

X_{15}: Input level of science and technology personnel. It is defined as below: *personnel of regional science and technology / national science and technology personnel.* In general, input of scientific research personnel indicates the innovation vigour of regional industry, and at the same time, it well reflects the regional innovation environment and the science and technology policies of government. The related data are from *2015 China Statistical Yearbook on Science and Technology.*

X_{16}: Enterprise scientific research activities. This index is defined as below: *ratio of regional industrial enterprises that carry out R&D activities / ratio of national industrial that carry out R&D activities.* It indicates the overall strength of regional scientific and technological innovation. The related data are from *2015 China Statistical Yearbook on Science and Technology.*

X_{17}: Level of invention patents.it is defined as follow: *quantity of regional invention patents / quantity of national invention patents.* It can reflect the regional scientific and technological achievements and scientific research level, and has a certain influence on the regional industry level. The related data are from *2015 China Statistical Yearbook on Science and Technology.*

X_{18}: Ability of technical innovation which can be scribed as the proportion of funds for technology transfer, digestion and absorption. It can measure enterprise technology absorption and transformation ability, and also reflects the existing technical level and R&D capabilities. The related data are from *2015 China Statistical Yearbook on Science and Technology.*

4 Procedure of empirical analysis

4.1 Calculation of eigenvalue, variance contribution rate and cumulative variance proportion

After establishing competitiveness index evaluation system, this paper takes advantage of SPSS15.0 giving a principal component analysis by using 10 samples of 18 variables. 5 common factors are chosen in the analysis result, and the accumulative variance contribution rate (87.65%) meets the analysis conditions (>85%). Detailed analysis results are shown in table 2.

Table 2. Total variance explained.

	Initial Eigen values		**Extraction Sums of Squared Loadings**			
component	Eigenvalues	% of Variance	Cumulative %	Eigen values	% of Variance	Cumulative %
1	5.26	38.23	38.23	5.87	42.66	42.66
2	3.37	24.49	62.72	2.67	19.40	62.06
3	1.97	14.32	77.04	1.75	12.72	74.78
4	1.23	8.94	85.98	1.54	11.19	85.97

According to table 2, we replace the original 18 with 4 principal component variables (the rate has reached 85.97%), which is independent of each other. This processing course effectively avoided multicollinearity, and reflected the basic information of the original variables.

4.2 Original variable load matrix

After determining the eigenvalue, variance contribution rate and cumulative variance proportion, the original variable load matrix of 4 principle components are calculated and shown in table 3.

Table 3 The original variable load matrix.

Rotated Component Matrix				
	Component			
	Factor1	Factor2	Factor3	Factor4
X1	-0.448	0.007	-0.027	-0.001
X2	0.105	0.103	-0.213	-0.107
X3	0.554	0.47	0.017	0.005
X4	0.634	0.109	-0.009	-0.01
X5	0.706	0.107	0.005	0.005
X6	0.815	0.501	0.201	0.176
X7	0.457	0.27	0.09	0.001
X8	0.671	0.776	0.709	0.503
X9	0.289	-0.448	-0.448	-0.448
X10	0.105	0.105	0.105	0.105
X11	0.576	0.516	0.516	0.516
X12	-0.397	-0.024	-0.024	-0.024
X13	0.51	0.442	0.102	0.001
X14	0.771	0.403	0.103	0.009
X15	0.005	0.013	-0.003	-0.015
X16	-0.017	0.001	-0.011	-0.011
X17	-0.356	0.001	-0.002	-0.002
X18	0.516	0.317	0.516	0.417

Principal components taking place of the original 18 indicators, mainly reflects the basic information of the original variables. Based on rotating the original variable load matrix, and after 5 times Varimax orthogonal rotation, the rotated component matrix is determind and shown in table 4.

Table 4 Rotated component matrix

Rotated Component Matrix				
	Component			
	Factor1	Factor2	Factor3	Factor4
X1	-0.321	0.213	0.241	0.001
X2	0.437	0.772	0.152	0.007
X3	0.122	-0.101	0.001	-0.021
X4	0.476	0.709	0.019	-0.016
X5	0.122	0.101	-0.021	0.001
X6	0.851	0.372	0.013	-0.015
X7	0.103	0.115	-0.031	0.003
X8	0.785	0.401	-0.109	-0.021
X9	0.365	0.852	0.157	0.001
X10	0.837	0.201	0.007	-0.031
X11	0.525	0.103	-0.019	0.012
X12	0.103	-0.027	0.826	0.016
X13	0.279	-0.104	0.707	0.103
X14	0.124	-0.019	-0.015	0.007
X15	-0.301	-0.201	0.008	-0.061
X16	0.205	0.007	0.125	0.503
X17	0.101	0.021	-0.133	0.001
X18	0.351	0.106	0.004	0.839

From the table above, we can identify 4 principal components corresponding to each variable:Factor1: mainly determined by four variables, X_6, X_8, X_{10} and X_{11}, namely, the enterprise sales revenue, rate of market share, the original value of fixed assets, total assets. These 4 variables mainly reflect regional market size and market potential of EMI, can be seen as market performance factors.Factor2: mainly determined by two variables, X_2 and $X4$, namely, assets per share and net value of fixed assets. They can mainly measure regional EMI scale and its comprehensive strength, and can be seen as comprehensive strength factors.Factor3: mainly determined by X_{12} and $X13$, namely, asset-liability ratio, profit rate of costs and expenses. The two variables can indicate the enterprise management ability and market profitability, and can be seen as management factors.Factor4: mainly determined by variable X_{18}. The ability of technical absorbing and innovation reflects the technological level of the regional industry development, the regional investment level and innovation ability, and can be regarded as technical factors.

4.3 Construction of comprehensive evaluation model

By using the comprehensive evaluation model $F_j = W_1 F_{1J} + W_2 F_{2J} + \cdots W_n F_{mj}$, we can obtain the regional competitiveness shown in table 5.

Table 5 Four hierarchical model of EMI competitiveness evaluation

Destination layer	Score of main factor	Industry index layer
F_J ,comprehensive competitiveness Score of province(or city) J	W_1=0.4266	X_6, X_8, X_{10}, X_{11}
	W_2=0.1940	X_2, X_4
	W_3=0.1272	X_{12}, X_{13}
	W_4=0.1119	X_{16}, X_{18}

According to the data in table 5, scores of main factors in each region can be calculated and shown in formula (3):

$$
\begin{aligned}
F_{1J} &= \frac{0.851X_{6J} + 0.785X_{8J} + 0.837X_{10J} + 0.525X_{11J}}{0.851 + 0.785 + 0.837 + 0.525} \\
F_{2J} &= \frac{0.772X_{2J} + 0.709X_{4J}}{0.772 + 0.709} \\
F_{3J} &= \frac{0.826X_{12J} + 0.707X_{13J}}{0.826 + 0.707} \\
F_{4J} &= \frac{0.839X_{18J} + 0.503X_{16J}}{0.985 + 0.503}
\end{aligned} \tag{3}
$$

4.4 Comprehensive scores of EMI in different regions

Based on the analysis by using SPSS15.0, the EMI scores of 4 principal component in 10 provinces (or cities) can be obtained and shown in table 6.

Table 6 Factor score of each principal component and rank order

N	Area	PC1	R	PC2	R	PC3	R	PC4	R
1	LN	-0.96	10	-0.24	6	-1.21	9	1.57	1
2	BJ	0.03	6	0.11	5	0.03	6	1.03	2
3	TJ	-0.55	8	-1.35	10	1.02	1	0.85	3
4	HB	-0.56	9	-0.88	9	0.77	2	-0.56	8
5	SD	0.11	5	-0.17	7	0.73	3	-0.27	7
6	ZJ	0.19	4	-0.86	8	0.46	4	-0.56	6
7	SH	0.36	2	0.51	3	-0.85	7	0.2	4
8	JS	0.27	3	0.68	2	-1.07	8	0.16	5
9	FJ	-0.15	7	0.13	4	0.19	5	-1.45	10
10	GD	1.26	1	2.07	1	-1.25	10	-0.95	9

5 Conclusions of the empirical analysis

Firstly, Overall, EMI in Liaoning province has a large scale and some certain competitive power, and is in the row of medium place in the eastern region. This analysis result proves that EMI in Liaoning province has a good foundation, and its scale expansion has a certain influence on its competitiveness.

Secondly, after renovation, EMI in Liaoning province has a higher technical absorption and transformation ability. This analysis result proves that transformation of old industrial base in Liaoning province has obtained significant results. As for the old industrial base, it is a very good phenomenon and excellent developing trend.[10]

Thirdly, EMI in Liaoning province has poor assets quality and low capital management and operation efficiency due to heavy historical burden as an old industrial base which are far behind other provinces in the eastern region.

Fourthly, EMI in Liaoning province has weak competitiveness related to the immature of economic system in Liaoning province. As a result, growth of EMI in Liaoning province depends not on high product technology content and strong market operation ability, but mainly on scale expansion.

References

1. Jianfeng Cui. Brief Analysis of Industrial Clusters and Technical Innovation. SOCIAL SCIENCE FRONT, 2011(11): 252-257
2. Yanqing Zhang. technology innovation of small and medium-sized enterprisesand governmental behavior In industrial cluster .Enterprise Economy,2011(9):107-111
3. Shijie Li.Industry cluster development and the revitalization of old industrial areas : A study of Liaoning province . Beijing Shi : Zhongguo jing ji chu ban she, 2010
4. Guiping, Xiaqian. Enterprise strategic alliance stability analysis.Journal of Wuhan Science and Engineering University, 2007, 29(3): 151-152.

5. Lin Wenbao. Factors affecting the correlation between interactive mechanism of strategic alliance and technological knowledge transfer performance .Journal of High Technology Management Research, 2007(2): 139-155.
6. DAS T K, RAHMAN N. Determinants of partner opportunism in strategic alliances: a conceptual framework .Journal of Business and Psychology, 2010(1):55-74.
7. Bo Wu, Shenghua Jia. The cluster firms' network-based growth mechanics and models with the evolution of regional industry cluster. Hangzhou : Zhejiang da xue chu ban she, 2008
8. Mark Dodgson,David Gann; Ammon Salter.The management of technological innovation : strategy and practice . New York : Oxford University Press, 2008
9. Jici Wang.Industrial Clusters in China and Their Implications to Other Developing Countries. Areal Researchand Development, 2004(4):1-4
10. Paul R Krugman. Geography and trade. Cambridge, Mass:MIT Press, 1991

Analysis on Current Situation and Countermeasure of Domestic Electronic Commerce Logistics in the Internet Age—Based on Resource Dependence Theory

Jiapeng Zhang [1,*] and Sijia Li [2]

[1,2]Management Engineering Department, Zhengzhou University, Zhengzhou, China
Corresponding Email: author@e-mail.org

Abstract. This paper analyzes the status of electric business logistics in the current Internet era in China, and combines the SWOT analysis with AHP to do the empirical analysis, then puts forward the countermeasure that the electric business logistics resource should be shared based on the resource dependence theory. Through the empirical analysis, it is found that the disadvantages and opportunities of the logistics status are important in the Internet era.The resource sharing strategy based on the resource dependence theory is more scientific. The rational use of Internet technology in electric business logistics industry can achieve "sharing". It is of great significance for its balanced development, intelligent development and optimization and development.

1 Introduction

In rapid development of information technology society, the ability of mastering Internet technology and data information in enterprises usually determine their future development.

As a productive service industry, the essence of Logistics is to solve the contradiction between time and space. Electrical business logistics is different from the general logistics, it pays attention to timeliness, quality of service and goods information retrospective. Logistics is a large amount of data generated by the industry [1], Internet technology in the vast amount of data found in new knowledge and create new value. Data itself into information, and through the refining of information into a universal law, and ultimately become the ultimate profit.

E-commerce logistics is not only an important part of the implementation of e-commerce and a driving force for development, but also benefits from the depth of Internet innovation integration, which is an important pillar of e-commerce development. E-commerce drives new demands for logistics, with undergoing a transition from the IT era to the era of the DT era, the network will become the era of DT data of new energy [2], "Internet +" lead both opportunities and challenges the data revolution to e-commerce logistics he data revolution .

There is no doubt that the use of Internet technology is the importance of electricity business logistics, logistics services become more accurate through the professionalization of the data processing. Furthermore, electricity logistics problems are to be resolved. Yang Jimei put that China's logistics development lagged behind the emergence of logistics and distribution of e-commerce development but not in-depth; Xie Yurong argues that China's logistics starts late, lacks of support services; Shen Yan claims that other fresh produce of logistics and its distribution costs become higher , the distribution of agricultural products is very difficult to guarantee the quality. Based on the development of electronic business logistics, scholars also studied from different dimensions. SUN Ying and RONG Bing analyze the characteristics of reverse logistics under the e-commerce market, and put forward the suggestion of establishing effective reverse logistics system for the return logistics problem of electronic businessman. Jiang Minglin put forward the logistics cloud of agricultural product innovation and resource integration service collaborative management model; Meng and Guo Chunxia study the supply chain management based on the biopharmaceutical public logistics platform operating mode; Yang Ju-ping puts forward other customer-oriented multi-distribution program of the electricity business logistics "last mile" distribution model .

The research perspective above reflects the specific characteristics of the industrial transformation in the Internet era. Logistics in the era of large data under the external environment has undergone significant changes in value creation and business model has also undergone significant changes in the way[3].Resource dependence theory refers that the survival of organizations need to draw resources from the surrounding environment, interdependence with the surrounding environment and interaction, which can achieve their goals. Based on the theory of resource dependence, this paper puts forward the sharing of logistics resources and the realization of common distribution and common logistics for present situation and existing problems of electric business logistics. As today's logistics enterprises gradually realize their own network, most of the establishment on their own official website promote their businesses. Logistics and distribution design is relatively broad, rich in the form of logistics and distribution. However, in the distribution process, the construction of different logistics network is complex, which causes low efficiency of logistics and distribution of wasting resources very seriously. With the further development of the integration of the data resources, the interdependence and mutual influence between the logistics enterprises and other enterprises in the surrounding environment, the resource allocation has been fully enhanced to a certain extent.

2 Theoretical basis and status analysis

From the perspective of resource theory, the survival of the organization absorb resources from the surrounding environment and the interaction with the surrounding environment in order to achieve the purpose, "shared logistics" is a reasonable response to this opportunity to respond to the program.

In this paper, the sharing of logistics industry resources explains from two dimensions, including tangible resource sharing and intangible resource sharing. Physical resources, including logistics and transportation, logistics network, distribution personnel, distribution facilities, the sharing of these tangible resources can avoid duplication in the distribution process of wasting caused by social resources, while avoiding a lot of social funds, the logistics enterprises achieve the sharing of resources. This measure not only can improve the level of service delivery, but also can save excess funds and reduce the degree of road congestion and improve distribution of economic and social benefits. Intangible resources sharing can also be called information sharing. In the current Internet age, information covers a wide range of areas, including business data, buyer data, enterprise distribution of relevant personnel, network, transportation data information. With the logistics enterprise application of Internet technology, network information and data are constantly analyzed and excavated, sellers and logistics companies deal the information effectively, all the logistics resources in the Internet traction under the common services of e-commerce platform are helpful to the seller and the buyer., which reduce the cost and improve the level of service in the field of electrical distribution business.

The SWOT analysis and the AHP analysis method applied in this paper have been widely used in some project analysis and their respective development has been relatively mature. This paper provides a new way to use AHP and SWOT analysis by combining the two methods. The priority of SWOT element is calculated by AHP method, which makes the analysis result more reference, which can provide SWOT method reliability and accuracy. According to the status of the logistics, the advantages and disadvantages of the current logistics enterprises' own environment plus the opportunities and threats brought by the external environment are put forward, and the disadvantages of the logistics enterprises are analyzed and explained emphatically.

2.1 Strength

E-commerce logistics with Internet technology achieve the logistics, capital flow, business flow and information flow matching, integration and interactive development.

The development process with Internet technology make a breakthrough in traditional logistics innovation, format, including self-built logistics, third-party logistics, fourth party logistics, logistics and warehousing services surrounding services.At present there are Shunfeng, Shen Tong, the pass, Huitong, tact, rhyme, and other top 20 logistics enterprises, large and involved in the large number of areas. Therefore, the advantages are obvious: the vigorous development of electric power industry to the electricity business logistics market prospect is broad and abundant technical personnel. Only from the development of domestic current logistics enterprises, the industry is huge.

2.2 Opportunity

At present, during the transition from IT era to DT era, network data information will become the new energy in DT era [4].

The revolution of data led by Internet has brought opportunities and challenges to E-business logistics. Therefore, the development get the support of national policy and the extensive development and application of Internet technology, as well as the previously mentioned electronic commerce industry flourish.

2.3 Threat

With the existing opportunity, electricity and logistics industry are also facing some unavoidable threats, as more than 20 domestic logistics enterprises make the existence of this huge logistics industry increasingly competitive.

Along with competition at the same time, the scale of enterprise infrastructure will change in order to respond to the current competition actively for more development opportunities , causing that the threat of profit reduction.

2.4 Weakness

Based on the Internet technology era, in order to acquire better and more scientific countermeasures, this paper lists the existing problems on current status of electronic business logistics industry:

Unbalanced development of electricity logistics, mainly in the festive logistics imbalance and the development of urban and rural electricity logistics imbalance. The first is caused by peak holiday logistics imbalance. The peak of the electricity business holiday shopping, artificial festivals ,like the dual pairs of Taobao double eleven and Jingdong 618 ,and Spring Festival and other traditional result in holiday shopping spree and logistics congestion. The latter is the imbalance between the development of urban electricity supplier logistics and rural electricity supplier currently. The rural population is low in population density, the volume of electricity trader is not large and the logistics base is relatively backward.

The delivery of electronic business is slow and Internet technology is not intelligent enough [5], Express become "slow delivery" which has a increasingly high probability, actual level of comprehensive intelligence data is to be improved; the abilities of e-commerce enterprises in allocation of goods, coordination and other aspects determine the consumer experience for the logistics speed in a large extent. The current capacity of China's electricity providers in this area ,however, is also clearly to be improved. Now although online goods, businesses and the brand become more, the chance that express changes to "slow delivery" is getting higher. Although, people today in Internet industrial always concentrate on large data, cloud computing and other Internet technology, but in fact the comprehensive level of Internet technology has a long way. Despite China's large population base, broad market, transaction data is very rich, we has not been able to achieve the optimal use of data

Delivery service quality needs to be improved. Distribution services directly affect the end-user experience of online shopping. Poor logistics timeliness and delivery service attitude, and so on, are likely to affect the user's ultimate experience. Distribution service quality sometimes is poor mainly in: delivery timeliness is poor, not according to the agreed time to the goods; Second, the goods damage in the distribution process; Third, delivery staff attitude is poor, these issues have a direct impact on customer experience.

3 Empirical analysis

At present, there are twenty major logistics enterprises in China, such as SF, STM, Huitong, Huitong, Yuantong and so on ,which is large in size and involves a large number of regions .

So the research is based on SWOT analysis and AHP (Analytic Hierarchy Process), analyzing the current Logistics status qualitatively and quantitatively.

First, SWOT analysis, through the study of the logistics industry of internal and external environment, shows advantages weaknesses, opportunities and threats (Table 1).The factors in Table 1 qualitatively point out actual situation of the electricity supplier logistics, which shows opportunities and strengths brought by the enterprise's own resources and external environment as well as the weakness and the threat posed by the current environment [6].

Table 1. SWOT analysis of resource sharing in logistics.

Strength	Opportunity	Weakness	Threat
Market prospect S1	National policy support O1	Slow delivery speed W1	More intense competiti ons T1
Abundan t technical talent S2	Prosperity electricity business O2	Social resource allocation insufficient W2	Infrastru cture scale change T2
Huge industry S3	Internet technology using O3	Low service quality W3	Profit reduction T3

The SWOT group of four elements have the two pairs of comparison, and then pairs of SWOT group comparison.In the comparison process, taking into account a number of Internet technology experts and some of the company managers as well as the views of various departments, brainstorm the judgment matrix of each group is obtained (see Table 2, Table 3, Table 4 and Table 5).Then use the maximum eigenvalue method for each SWOT group do a single-level sequence, then list the total ranking of the hierarchy and finally calculate the results (see Table 6).

Table 2. Judgment matrix of strength group.

Strength group	S1	S2	S3
S1	1	1	1/3
S2	1	1	1/5
S3	3	5	1

Table 3. Judgment matrix of weakness group .

Weaknes s	W1	W2	W3
W2	1	1/4	2
W3	4	1	8
W3	1/2	1/8	1

Table 4. Judgment matrix of opportunity group .

Strength	O1	O2	O3
O1	1	4	1/3
O2	1/4	1	1/8
O3	3	8	1

Table 5. Judgment matrix of threat group.

Strength	T1	T2	T3
T1	1	5	3
T2	1/5	1	1/3
T3	1/3	3	1

We can see from Table 6 that the judgment matrix has satisfactory consistency, so the analysis result is desirable.

At the same time, opportunities and advantages of this project are in a relatively important position.

Through the combination of SWOT and AHP, this paper explains the current situation of electric business logistics with both qualitative and quantitative aspects, gives a more definite and measurable analysis result from all aspects. This result is the basis of coping scientific and systematic strategies in the current domestic situation of the electricity business logistics .

Table 6. Weights table.

SWOT	Each group priority	SWOT elements	C.R.	The priority of elements within each group	The total priority of the feature

S	0.1485	S1Market prospect	0	0.1867	0.0277
		S2 Abundant technical talent		0.1577	0.0234
		S3 Huge industry		0.6555	0.0973
W	0.4648	W1 Slow delivery speed	0.145	0.1818	0.0845
		W2 Social resource allocation insufficient		0.7273	0.3380
		W3 Low service quality		0.0910	0.0423
O	0.2970	O1 National policy support	0.0160	0.2572	0.0764
		O2 Prosperity electricity business		0.0738	0.0983
		O3 Internet technology using		0.6690	0.1987
T	0.0897	T1 More intense competitions	0.0194	0.6333	0.0568
		T2:Infrastructure scale change		0.1061	0.0095
		T3:Profit reduction		0.2604	0.0234

4 Conclusion and significance

The empirical results show that the inferiority group (0.4648) and the opportunity group (0.2970) have higher priority than the superiority group (0.1485) and the threat group (0.0897), the insufficient social resource allocation (W2) and technology development (O3) have the highest priority among its groups.

The theory of resource dependence emphasizes that under the normal environment, each organization depends on the surrounding environment and interaction between the various organizations, in order to better development.Based on the resource dependence theory and the current status of disadvantages and opportunities of logistics enterprises,this paper puts forward the strategy of resource sharing, which makes the electric business logistics enterprises better seize the opportunity and make up their own disadvantages.

In the era of Internet, logistics enterprises should seize the opportunities brought by external environment and adopt resource sharing strategy to make up their own disadvantages [7].Network of most logistics enterprises is sound, so resource sharing can play a certain role in the development of logistics enterprises. Applying resource sharing strategy to configure society resources, electrical and logistics enterprises can share tangible resource such as the network construction in various regions of the logistics enterprises , the integration of using social resources in the region in a certain extent, which make the sender efficiency improve, buyers and sellers benefit evenly. These all make the sharing of logistics resources play a positive role in promoting as to the development of imbalances. At the same time, the sharing of intangible resources, such as the integration of data and information in the entire logistics industry, run a great help. With the deepening of Internet technology, logistics information is constantly analyzed and excavated. Logistics itself is an industry which can generate large amount of data. It is a new way to professionally process and integrate the data and share information for the whole industry. The sharing of intangible resources or information sharing have a positive role in promoting the intelligent use of electrical logistics

Logistics industry in China started late [7], logistics construction lags behind the rapid development of electricity providers, so the current existence of some real business logistics problems remain unresolved. E-commerce and logistics should promote and coordinate each other to further realize the logistics sharing so that the logistics enterprises are able to complete the work efficiently, allocate social resources rationally and save social capital, these all make that the development of electric business logistics has far-reaching significance in future.

DOI: 10.1051/matecconf/201710002066

References

1. Xu Jiang-ping.Large-scale data power supply to promote the development of logistics. China Market,2014(40):22-23.
2. Li Guanyi.Electronic business logistics innovation and traditional logistics transformation under the Internet.Commercial Research,2016(4):187-192.
3. Zhao Jing."Internet +" era of the development of electronic business logistics problems and solutions. Network and informationization. 2015(5):191-211.
4. Wang Zhiling.From the era of large data to see the development of e-commerce logistics and distribution trends.Electronic Commerce .2015(4):27-28.
5. Yang Jimei. Analysis on the Current Situation and Trend of the Development of Electronic Commerce Logistics in China . Logistics Forum, 2014(4):1-2.
6. Hou Ying.E-commerce under the background of the development of China's logistics industry (Master's degree thesis) .2015.
7. Xie Yurong.Electrical and commercial logistics development model and trend analysis.Economic Logistics,2014(11)46-52.

Fatigue Analysis of Large-scale Wind turbine

Yongli Zhu[1,a] and David Lawson[2]

[1]Chongqing Vocational Institute of Engineering, Chongqing Engineering Polytechnic College, Chongqing, China; [a]E-mail: 1348125719@qq.com

[2]Northern Melbourne Institute of TAFE

Abstract. The paper does research on top flange fatigue damage of large-scale wind turbine generator. It establishes finite element model of top flange connection system with finite element analysis software MSC. Marc/Mentat, analyzes its fatigue strain, implements load simulation of flange fatigue working condition with Bladed software, acquires flange fatigue load spectrum with rain-flow counting method, finally, it realizes fatigue analysis of top flange with fatigue analysis software MSC. Fatigue and Palmgren-Miner linear cumulative damage theory. The analysis result indicates that its result provides new thinking for flange fatigue analysis of large-scale wind turbine generator, and possesses some practical engineering value.

Keywords. Large-scale wind turbine generator; top flange; fatigue load spectrum; fatigue analysis

1 Introduction

Top flange of wind turbine generator is a key component connecting tower and main frame, it bears various alternating load from the top of wind turbine generator including wind load, dead-weight of generator etc., and fatigue damage easily occurs, flange fatigue damage directly influences on performance of entire WTGS, therefore, fatigue analysis of top flange plays a significant role of WTGS structure design.

According to design requirements of homemade MW-level WTGS, top flange as one key component of WTGS can't present fatigue failure within 20 years life cycle. Currently, domestic and foreign wind power manufacturing enterprises firstly establish simulation model of components with common fatigue analysis method during design process of WTGS, then ensure fatigue life of component in combination of engineering experience, this requires design staffs not only to utilize simulation software effectively, but also possess rich and practical engineering experience, the advantages of certification authority embodies herein as well [1].

German Lloyd's Register of Shipping (Germanisher Lloyd, abbreviation as GL)is one of top world classification societies, which offers a series of certification service like inspection of materials, components, complete appliance and wind power farm [2].GL Wind Power Norm is one of international universal WTGS design, manufacturing and certification norm, which was formulated by Lloyd's Register of Shipping with its years of design and manufacturing experience in wind turbine generator field, it possesses highly reference value, and it is one of main reference standard for domestic wind power certification institutes, at present, most wind power farms require the products passing GL certification when purchasing WTGS [3].

The paper takes top flange of some MW-level wind power generator as example, and firstly establishes finite element model of top flange connection system with finite element analysis software MSC.Marc/Mentat, analyzes fatigue strain of top flange, corrects S-N curve line 1 of flange materials based on GL norm, realizes simulation of flange fatigue working condition with WTGS design software Bladed, and obtains flange fatigue load spectrum in connection with rain-flow counting method, at last, it analyzes flange fatigue damage with fatigue analysis software MSC.Fatigue and Palmgren-Miner linear cumulative damage law.

2 Analysis of finite element fatigue strain of top flange

2.1 Geometric model of top flange connection system

It establishes overall modelling for top flange connection structure with Pro/E software, makes simplified treatment for the parts playing unimportant role of top flange, such as main framework base and some accessory equipment etc., the simplified geometrical model shown as Figure 1:

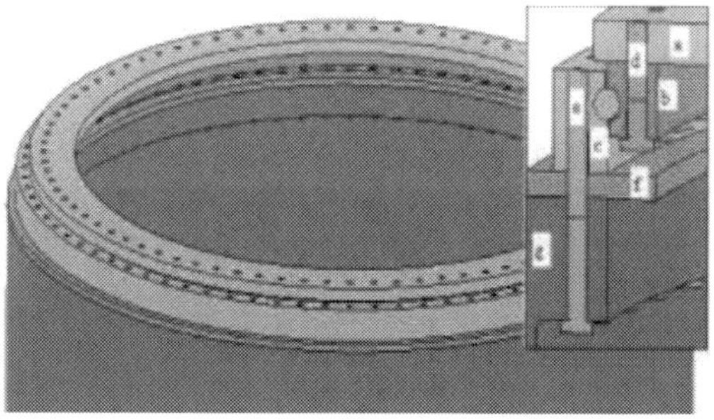

a- main frame, b-yaw bearing inner race, c-yaw bearing outer race, d-inner race bolt, e- outer race bolt, f-brake disc, g- top flange

Fig.1. Geometrical model of top flange system

2.2 Flange finite element model

The paper builds up top flange connection system in touch with finite element model by MSC.Marc/Mentat, model division adopts 8 nodes, hexahedron and No.7 unit, total units are 35,648, and total nodes are 389,788. Models include deformable contact bodies like main frame, yaw bearing, outer race screw, inner race screw, brake disc and top flange etc., materials of various component are made of low-alloy high-tensile structural steel, elasticity modulus quantity is 2.06×10^{5}MPa, Poisson's ratio is 0.3,density is 7.85×10^{-9} t/mm^3. Among these, flange material is forged part S355NL-Z25-0.

2.3 Computation of flange fatigue strain

When computing fatigue strain, it will respectively exert 6 limit loads(Fx, Fy, Fz, Mx, My, Mz), load setting shown as Table1:

Table 1. Limit load setting

Working condition	Mx(Nmm)	My(Nmm)	Mz(Nmm)	Fx(N)	Fy(N)	Fz(N)
Lc_Mx	5e9	0	0	0	0	0
Lc_My	0	-8e9	0	0	0	0
Lc_Mz	0	0	-7e9	0	0	0
Lc_Fx	0	0	0	-9e	0	0
Lc_Fy	0	0	0	0	7e5	0
Lc_Fz	0	0	0	0	0	-3e6

Strain distribution status of each working condition will be obtained from the analysis and computation of finite element analysis software MSC.Marc/Mentat, Limit fatigue strain of 6 limit loads are respectively 22.7Mpa, 16.58Mpa, 43.23Mpa, 90.04Mpa, 143.1Mpa, 58.73Mpa.

3 S-N curve line and its correction

S-N curve line indicates the relation between applied strain level and the strain experienced before fatigue failure of materials or strain circulation frequency [5]. S-N curve line is achieved from processing materials into standard component and making fatigue experiment. Although this method is reliable, a large amount of manpower and material resources will be consumed during design period, and test cycle is longer. GL norm regulates that it can utilize material parameters to imitate approximate S-N curve line replacing the obtained curve line [2] by experiment if fatigue experiment can't be made.

With log-log coordinate, S-N curve line is an ideal straight line [5].

Expression of S-N curve line in logarithmic coordinate system is:

$$\log S = \frac{1}{m}\log K - \frac{1}{m}\log N \tag{1}$$

(1)In Expression (1), m and K are parameters related to materials, loading method and strain ratio etc. [6]

Generally, S-N curve line in fatigue analysis gives survival probability of component as $P_U = 50\%$, on the basis of GL Norm, it can make components possess survival probability as $P_U > 97.7\%$ and guarantee non-fatigue failure within 20 years life time as the design when making fatigue analysis of components of wind turbine generator. Therefore, when S-N curve line requires the correction, reference strain amplitude within finite lifetime shall multiply reduction coefficient S_{PU}, consequently, there is:

$$\Delta\sigma_A^* = \Delta\sigma_A \cdot S_{PU} \tag{2}$$

(2) When survival probability is $P_U > 97.7\%$, $S_{PU} = 2/3$.

Considering some defects like chilling and cracks etc. in the flange forging process, it shall multiply forging defect coefficient S_t:

$$S_t = (t/25)^{-0.1} \tag{3}$$

(3) In Expression (3)[7], t is thickness impact coefficient.

Finally, reference strain amplitude of corrected S-N curve line within finite lifetime area is:

$$\Delta\sigma_A^* = 2 \cdot \Delta\sigma_A \cdot \frac{S_{PU} \cdot S_t}{\gamma_m} \tag{4}$$

(4)In Expression (4) [7], γ_m is partial security coefficient of material, γ_m takes 1.265.

When flange strain circulation frequency is more than limit strain circulation frequency, S-N curve line shall prolong with slope parameter $\frac{1}{1-2m}$.

In fatigue analysis software MSC. Fatigue, S-N curve line of flange material is corrected with PTMAT module.

4 Fatigue load spectrum

4.1 Fatigue load simulation

For fatigue life analysis, it shall firstly know load situation of component. The main source creating WTGS fatigue load is alternating load produced during the wind wheel running which is caused by mass eccentricity of wind wheel as well as alternating load caused by blade gravity when wind wheel turns.

Concerning analog simulation of WTGS fatigue load, it must firstly comprehend fatigue working condition of WTGS. According to GL Norm, WTGS has 6 large fatigue load working condition, including 46 fatigue load sub-working condition. For each fatigue working condition, it will compute fatigue load at the hub center, then make equivalent treatment to obtain fatigue load time sequence curve line of top flange.

4.2 Fatigue load spectrum

Fatigue load time sequence curve line records the process of fatigue load varying with time, which is a series of irregular data, and fatigue load spectrum [8] is acquired with certain amplitude statistical counting method for probability statistical disposition. Load spectrum presents like graph, digital form or matrix of occurrence frequency with size of random load. Based on regulation of GL Norm, fatigue analysis computation can make simplified fatigue analysis with equivalent constant amplitude fatigue load spectrum.

The paper adopts rain-flow counting method to treat data. At present, rain-flow counting method is the most popular calculation method in fatigue design and experiment, and it can comprehensively reflect entire process of immediate load, possess stricter mechanics base[9,10], and easily realize the programming in the computer.

In Bladed software, it will adopt rain-flow counting method to compute flange random load, and obtain a series of WTGS fatigue load as well as load circulation frequency, and acquire total circulation frequency of load within WTGS life cycle according to wind speed characteristics and required life time of WTGS design, then it can draw time-domain equivalent fatigue spectrum of wind turbine. There are 46 fatigue load sub-working conditions for top flange, and each sub-working condition includes 6 equivalent fatigue load spectrums, for example, equivalent load spectrum of sub-working condition DLC1.2ac shown as Figure 2. Regularization of equivalent fatigue load spectrum can achieve equivalent fatigue load spectrum.

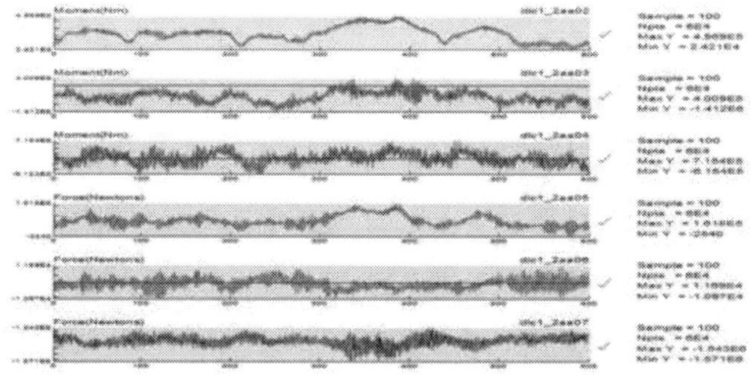

Fig.2. Working condition DLC1.2 ac time-domain equivalent load spectrum

5 Flange fatigue analysis

Flange finite element analysis model and flange fatigue strain result are imported into fatigue analysis software MSC. Fatigue, limit fatigue strain computed with limit load is treated as unitization, equivalent weight fatigue load spectrum under each working condition is connected with correspondent fatigue strain analysis result. Frequency is different under each working condition each year, therefore, it needs set yearly circulation indexes of each fatigue working condition. Moreover, S-N curve line shall be set, the paper utilizes the corrected S-N curve line in second section. Equivalent fatigue load spectrum setting is uploaded with sinusoid mode. Fatigue analysis results shown: it can see that maximum fatigue damage value of top flange is 0.0356 within one year, maximum value presents at the round corner connecting flange and yaw bearing, fatigue damage in other positions is very small, and it can be neglected to count. According to Palmgren-Miner cumulative damage theory, cumulative damage of wind power turbine within its life cycle can't surpass 1, or the component will lead to fatigue failure [11].Fatigue damage value of top flange is calculated as:

$$D_{20} = 20 \times 0.0356 = 0.712 < 1 \tag{5}$$

Therefore, fatigue damage will not take place on top flange within life cycle, and it will meet the requirements for 20 years of life cycle.

6 Conclusion

The paper takes top flange of some large-scale wind power turbine generator as an example, and discusses top flange fatigue damage analysis method of some wind power generator, the method constructs finite element model of top flange connection system with finite element analysis software MSC.Marc/Mentat, analyzes top flange fatigue strain, realizes load simulation of flange fatigue working condition with Bladed software, acquires flange fatigue load spectrum, finally achieves top flange fatigue analysis with fatigue analysis software MSC.Fatigu and Palmgren-Miner linear cumulative damage theory. The analysis method offers a new thinking way for flange fatigue analysis of large-scale wind power generator, and it possesses some practical engineering value.

References

1. H. Liu, WTGS System Dynamics Model and Key Component Optimization Study, Chongqing University, (2009)
2. Germanischer Lloyd Wind Energie GmbH. Guideline for the Certification of Wind Turbines.German: Germanischer Lloyd Wind Energie GmbH, (2010)
3. C. Y. Zhou, H. X. Zheng, H. Q. Luo, etc. Application and Samples of MSC.Fatigue Analysis, Beijing: Science Press, (2005)
4. G. J. Wang, Guidance of MSC. Fatigue Analysis Example, Beijing: China Machine Press, (2009)
5. B.N. Shi, MW-level WTGS Main Shaft Lifetime Analysis. Power Grid and Clean Energy 8(2010)
6. Y. Li, Ship Hull Fatigue Analysis Based on JBP Norm. Dalian University of Technology, (2006)
7. S.M. Li, Mechnical Fatigue and Reliability Design, Beijing: Science Press.
8. H.Q. Sheng, Research on Converter Fatigue Load Spectrum. Acta Mechanica Solida Sinica, 29(2008)
9. T.L. Li, S.J. Zou, D. Sun , etc. New Method of Airplane Fatigue Load Spectrum Analysis, Mechanical Science and Technology, 6(2006)
10. A.K. Khosrovanch, N.E. Dowling, Fatigue Loading History Reconstruction Based on the Rain-Flow Technique, International Journal of Fatigue, (1990).
11. A.M. Cai, D.T. Zhang, H.T. Dai, Wind Turbine Fatigue Analysis. 2009 Annual Conference of Clean and Efficient Coal-fired Power Generation Technology Collaboration Network, 10(2009)

Researched on Initiation Synchronization of Addressable Initiation System

Guofu Yin[1,a], Liming Li[1] and Qiao Lu[1,b]

[1]Shanxi Applied physic-Chemistry Research Institute, State Key Laboratory of Applied physic-Chemistry Research, Xi'an of China
[a]yinguofu521@126.com ,[b]chufengde01@163.com

Abstract. Addressable initiation system was established by bus topology structure and information coding which owns the ability to safely and reliably control up to 256 initiators on a signal bus according to the communication protocol which is a robust two-way command response type protocol. Also, the two-way digital communication was achieved between intelligent bus controller and smart initiator by multiplexed serial data bus. The synchronous performance of addressable initiation system was researched. The result shows that each smart initiator will performance the built-in logic control, and arm electronic safe switch after receiving the synchronous signal sent by intelligent bus controller. Then they discharged the electrical energy of the firing capacitor as approximately synchronous, and were initiated synchronously at 4±0.2μs. It indicates that this initiation system has favorable synchronization.

Keywords. Addressable initiation system, smart initiator, communication protocol, semiconductor bridge.

1 Introduction

Initiation system as the control center of weapons and ammunition system, responsible for safeguarding the warhead-war and wartime timely and reliable security arming and detonation operations, etc. With the developing of weapons systems toward miniaturization, low power, high accuracy, high damage, the addressing detonation initiation system is emerging which revolutionary transforms the initiation system from the typical centre firing control to the centralized control and distributed initiation [1].In paper, the addressable initiation system was established by bus topology structure and information coding. Also, the two-way communication was achieved between intelligent bus controller and smart initiator by multiplexed serial data bus. Intelligent bus controller can centralized control every smart initiators on bus and achieve multi-mode initiation of smart initiator. Also, the initiation synchronization, which is very important in weapons systems such as missile warhead, spacecraft solar panels unfold, solid rocket motor ignition, has been researched of the system.

2 Addressable Initiation System Design

The addressable initiation system was designed, and shown in Figure 1, which consists of intelligent bus controller, data bus, a number of smart initiators and communication protocols [2,3]. Intelligent bus controller can send identity address (ID) and instruction to all smart initiator on the initiation bus [4,5]. As each smart initiator received an ID, they will compare the ID with their own ID store in ROM. If the received ID of each smart initiator is entirely same with self ID, it will execute pre-set instruction, and release energy store in capacitance to fire the energy device such as semiconductor bridge (SCB), hot wire and bridge film which will result to fire charge sequence, and output strong explosion waves. If two ID is completely different, the smart initiator will stop all the action and wait for next ID and instruction which are sent by intelligent bus controller until two ID are completely match each other. The system is an electronically addressable, low power ordnance initiation system, consisting of a controller, smart initiators, and wire ham.

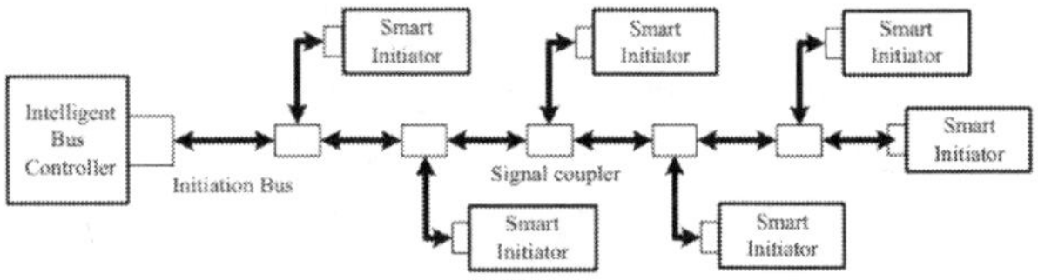

Fig. 1. Schematic of addressable initiation system

Intelligent bus controller as a initiation system central nervous system, centralized control the all action of the system. Shown in figure 2, it was designed including two energy regulation circuit, micro-control circuit, communication circuit and electronic Safe/Arm circuit. It receives instruction and energy from guidance and control unit, also control the instruction and the energy to reach every specified initiator. And then, it will receive the feedback information of the smart initiator, and continue to feedback the results of the implementation of the directive to the guidance and control unit.

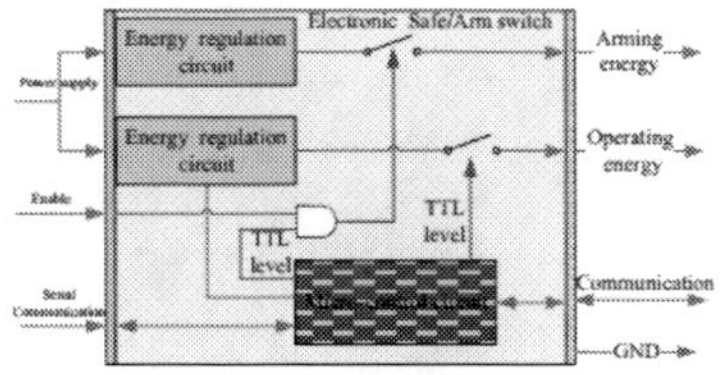

Fig.2. schematic of intelligent bus controller

Smart initiator was designed by using microelectronic integrated design technology. Shown in figure 3 and figure 4, some function circuits were integrated within it including anti-electromagnetic interference and (EMI) Static electricity (ESD)circuit, information identifying circuit, micro-control circuit, electronic safe/arm circuit, energy storage circuit, fire circuit and semiconductor bridge.

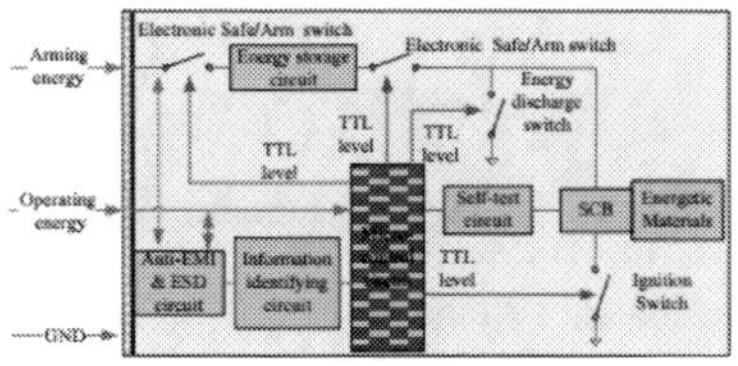

Fig.3. schematic of smart initiator

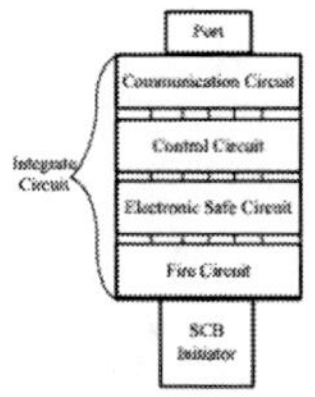

Fig.4. Integrate smart Initiator

Each smart initiator has a unique address code ID stored in micro-control circuit ROM. It can be ID to finish the safe/arm and fire function. Only receive Instruction signals consistent with their ID, smart initiator will carry out the initiation function. And other spurious signals will be rejected by the information identifying circuit in order to achieve a rapid addressing of the initiator. In addition, the anti-electromagnetic electronic circuit was designed in smart initiator, so that the information identifying circuit with anti-interference ability which can prevent accidental initiation caused the error identification. It will improve the safety and reliability of smart initiator. Static electronic is great harm for electrical explosion devices (EED), and must design antistatic electronic circuit in smart initiator. Due to the different characteristics of antistatic with each circuit, so two-stage anti-static circuit was designed. Information identifying circuit and logic control circuit is protected by semiconductor discharge transistors and TVS diodes. And then semiconductor bridge and electronic switch is protected by the CMOSFET which is sensitive to electrostatic. In addition , between the information identify circuit and logic control circuit designed the magnetic isolation circuit. Even in the case of a faulty initiator in system, it does not result in paralysis of the entire bus network.

3 Experiment and Analyses

The synchronization of initiation system is harsh demanding in the weapons systems such as missile warhead, spacecraft solar panels unfold, solid rocket motor ignition and other weapons systems. Therefore the synchronous performance of addressing detonation initiation system will be developed in this paper. Experiments using three-point addressable initiation system, three channel current loop use for testing the current characteristics of firing capacitor discharge, three channel oscilloscope probe use for test the voltage of across the semiconductor bridge chip, and three channel DET10A photo-detectors was be used to test the initiation output performance of SCB initiator. Experiment principle as shown, including the intelligent bus controller, data bus, signal couplers, three smart initiators, oscilloscope, current loop and photo-detectors. Each smart initiator integrates a 47μF firing capacitor and it charge voltage is 15V.

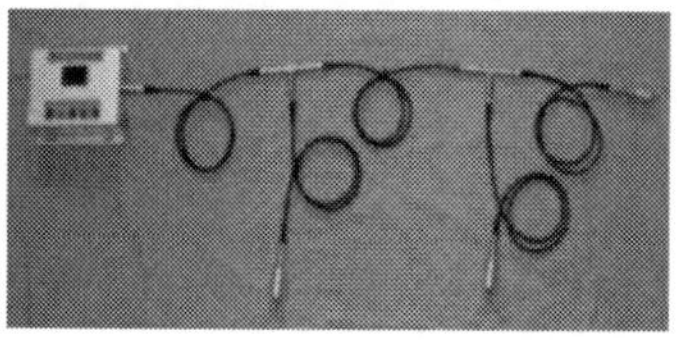

Fig.5. Three-point initiation system

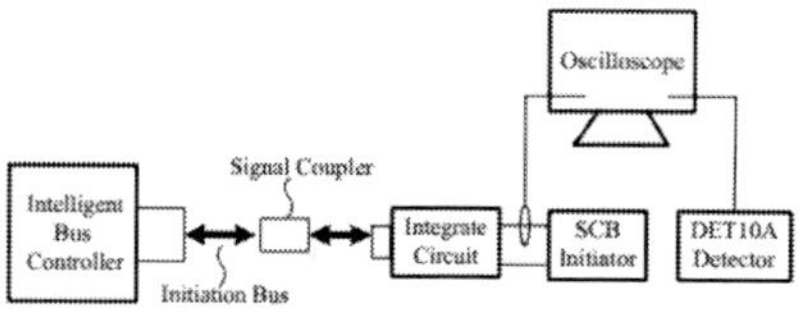

Fig.6. Schematics of firing test

(1) Synchronization of firing current and voltage

Firing current I reflect the changes of SCB resistor. As shown in Fig.7, firing current of three-point SCB energy element of smart initiator are gradually increased from 0, and reach a maximum 6±0.5A at 4±0.2μs, and at that time each smart initiator was initiated. It indicates that firing current of three-point initiators are approximately synchronous.

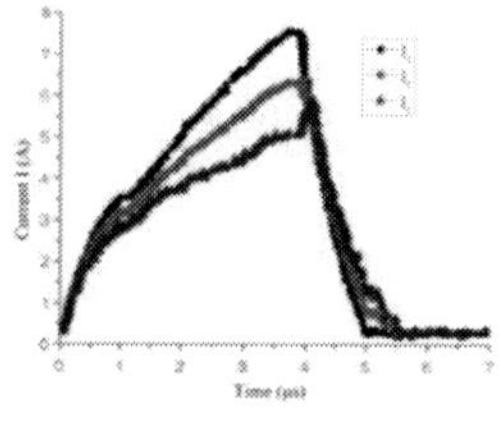

Fig.7. Graph of firing current characteristic

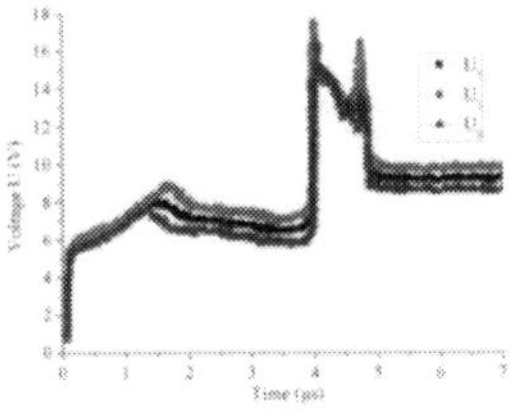

Fig.8. Graph of firing voltage characteristic

The voltage characteristic of across the SCB energy elements is shown in Fig.8. At the A-B stage, the SCB is heated by the passing current, and the voltage reaches its first peak at the point B. At the B-C stage, owing to ohmic heating, the silicon in SCB temperature increases, and solid silicon melts and transforms to liquid silicon. At the C-D, liquid silicon evaporates and transforms to gaseous silicon. At the D-E stage, the rapidly growing voltage synchronously reaches the maximum duration the 4±0.2μs, and initiates a discharge in gaseous silicon and plasma formation. It indicated the firing voltage of across the SCB of three-point smart initiators are approximatively synchronous.

(2)Output synchronization

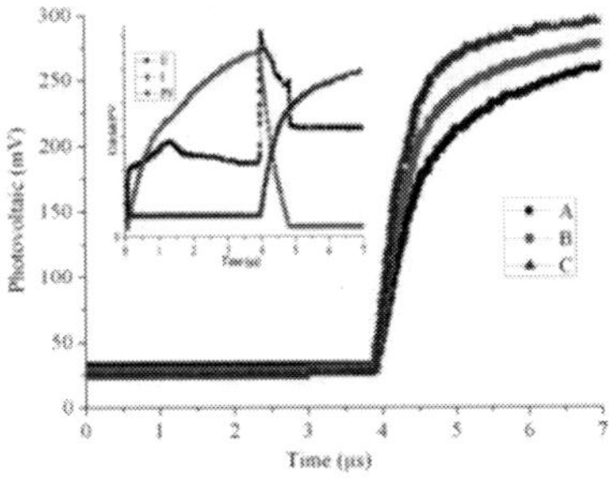

Fig.9. Graph of optical-eletronic detector output

DET10A photo-detectors were used to detect the initiation output of smart initiator. Fig.9 shows that the three-point smart initiators were initiated at 4±0.2μs when, and the initiation output ability of each smart initiator reached the maximum at 7μs. It indacates that the output of three-point smart initiators are approximately synchronous.

4 Summary

The synchronization of initiation system is harsh demanding in the weapons systems such as missile warhead, spacecraft solar panels unfold, solid rocket motor ignition and other weapons systems. Therefore the addressable initiation system was designed and this system has demonstrated the ability to arm and fire 256 initiators on a single bus . Also the synchronous performance of addressable initiation system was researched. The result shows that each smart initiator will performance the built-in logic control, and arm electronic safe switch after receiving the synchronous signal sent by intelligent bus controller. Then they discharged the electrical energy of the firing capacitor as approximatively synchronous, and were initiated synchronously at 4±0.2μs. It indicates that this initiation system has favorable synchronization.

References

1. David B. Novotney and Andrew Kochanek, Ensign-Bickford Aerospace & Defense Company, Simsbury, CT, 06070, Intelligent Initiation Systems for Divert and Attitude Control Applications, AIAA 2005- 4502
2. Steven D. Nelson, Redondo Beach, CA. Networked electronic ordnance system, US7752970B2, **13**(2010)

3. D.L. WANG, B.Y. GUO, S.K. NING, Design and implementation of explosive network of electronic detonators and communication protocol based on the RS-485, Joumal of Shanxi University of Technology (Natural Sicence Edition), **26**, 1(2010)
4. J.Y. WANG, L.M. LI, G.F. YIN, et al., Design on Intelligent Initiation System Based on RS-485 Bus, Initiators & Pyrotechnics,2 (2013)
5. T. LIU, Y.Z. FU, K.N. XIE, et al., Design and verification of new kind of integrated ordance initiation control system, Computer Science, **35**,4(2008)

Design of Bus Protocol Intelligent Initiation System Based On RS485

Liming Li [1, a], Qiao Lu [1,b] and Guofu Yin [1,c]

[1] State Key Laboratory of Applied physic-Chemistry Research, China

[a]biq213@163.com, [b] chufengde01@163.com, [c]yinguofu521@126.com

Abstract. In order to design an effective and reliable RS485 bus protocol based on RS485 bus, this paper introduces the structure and transmission mode of the command frame and the response frame, and also introduce four control measures and the communication in order to process quality of this system. The communication protocol is open, tolerant, reliable and fast, and can realize ignition more reliable and accurate in the intelligent initiation system.

Keywords. Intelligent initiation system; RS485 bus; communication protocol; pyrotechnics.

1 Introduction

The increasing demand of ordnance initiation system in the volume, weight, low energy consumption and integrated circuit with low energy input, high performance output led to this new way of detonating ordnance come out, which called addressable integrated intelligent initiation system (Intelligent detonation system). As shown in Figure 1, compared to the traditional intelligent ignition system ignition system: First, Time division multiplexing detonating control bus instead of multiple high current wire, in the guidance and control of the input signal module instruction function, all the detonators on the bus can be controlled by the bus controller, and the bus topology determines the changes in the number of wire does not increase with the initiation of nodes, which is very conducive to multi section industrial products group detonated ignition, greatly reducing the number of wires; Second, insurance and firing circuit are integrated into smart initiating device from the ignition controller, that is greatly reduces the system's volume and weight; Third, semiconductor bridge replaced the traditional hot bridge wire device, greatly reduces demand of fire energy , which can strengthen not firing performance of the system.

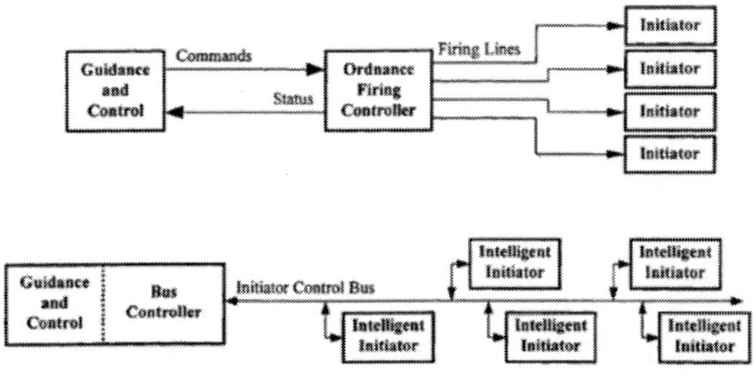

Fig. 1. Comparison between traditional and intelligent initiating system.

The "WizordTM" intelligent initiating system is a typical example of the system, which can be not only applied to the large launch vehicle and tactical missile,but also extended to the steering propulsion and attitude control system. As shown in Figure 2, it includes a bus controller, a number of intelligent initiating devices, a bus containing four transmission lines, and 124 initiating devices, which can relieve security or ignition. Figure 3 shows the the photo of the bus controller, multi node cable and the initiator [3-4].

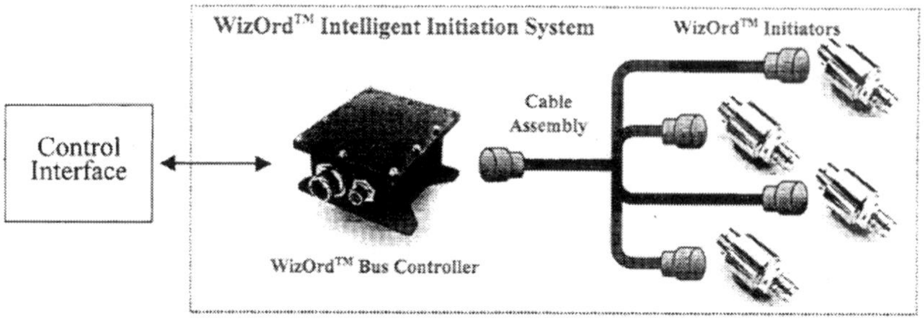

Fig. 2. "WizordTM" intelligent initiating system.

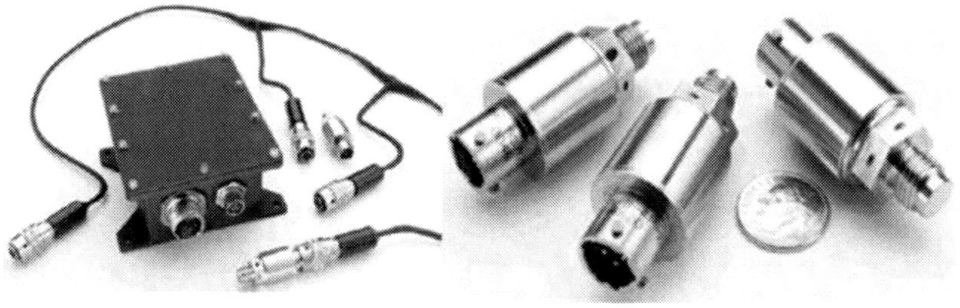

Fig. 3. "WizordTM" system hardware.

Intelligent detonation system, fundamentally changed the structure of initiation system, greatly optimizing the initiation system performance, reducing the cost of initiation system .Can be used for crew escape and carrier rocket, tactical missiles, aircraft ordnance initiation situations, realize multi-point synchronous or asynchronous accurate detonation or ignition [1,3].

2 Intelligent initiating system

According to the the study of intelligent initiation system at home and abroad, a kind of intelligent initiation system is designed,which can realize multi point synchronous or asynchronous accurate initiation or ignition. As shown in Figure 4, it includes a PC machine, a bus controller, a single chip controller, a plurality of smart initiating devices, and a multi connector shielded twisted pair cable. The whole system is divided into the following four parts:

(1) PC

PC simulation guidance and control module, send instructions to the intelligent initiator, and the implementation results of each instruction is displayed on the user interface,which is easy to understand the process of the system.

(2) Bus controller

The bus controller receives the energy and instructions from the PC, and through the control of the energy switch or the firing energy switch, the energy and the instruction is delivered to the designated initiator.

(3) Intelligent initiating device

In the intelligent initiating device, the system is integrated with the relieve security and ignition circuit, security and firing switch controlled by the received instruction, firing through capacitor discharge. In addition to the solution of the security and firing instructions, but also the implementation of state detection, write address and so on.

(4) Initiation control bus

The initiating control bus is composed of an operating power line, an insurance energy line, a communication line and a ground wire. Among them, the communication bus adopts RS485 standard, which can achieve high efficiency, accurate and reliable communication. It is still able to communicate with the transmission distance and the interference of the more serious, and meet the requirements of multi point initiation and small scale integration, which allows access to up to 256 initiator [5, 6].

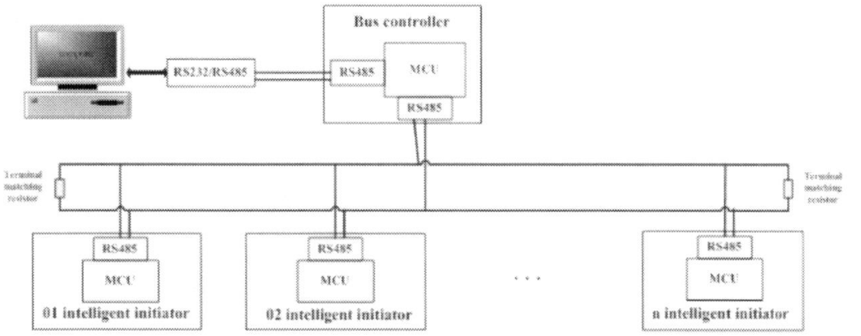

Fig.4. Schematic diagram of the intelligent initiating system constructed by RS485.

This is a master slave network that allows a master to communicate with one or more slave machine, the slave machine (initiator) can not send data, everything is controlled by the master (bus controller). Using the command / response mode, the bus controller forwards the instruction from the PC to the address of the initiator, and the initiator will perform the results to answer the bus controller in the way of the response frame.This way can ensure the system is reliable communication .Bus controller sends 2 types of requests to the initiator: (1)a single address mode: the master addresses a separate slave, which is received and processed from the machine, and returns an answer to the master. Every address from the machine must be unique. (2)broadcast mode: the master sends a request to all slave machines, every slave machine receive and process the request .After the provision delay time, each slave machine return the response to the master machine. By adopting the method of time response, it is effective to avoid the bus conflicts caused by multiple machines from the same time to the bus. All from the machine to share a broadcast address [7, 8].

3 RS485 bus

RS485 as the communication bus of intelligent initiating system, has the following advantages:

(1) According to the different input impedance of the receiver, the RS485 bus can be hung from 32 nodes, 64 nodes, 128 nodes, 256 nodes, is conducive to achieve more than one point of initiation.

(2) RS485 is used to transmit and receive the shield twisted pair, and the anti - mode interference capability is strong.

(3) RS485 bus structure is simple (half duplex 2 communication lines, full duplex 4 communication lines), the drive chip is small, and it is beneficial for the integration of the system.

(4) The sensitivity of the RS485 bus transceiver is high, and can detect the voltage of 200 mV, and the transmission signal can be recovered from the Km distance.

Due to the high reliability requirements of the initiating system, we must pay attention to the use of RS485 bus:

(1)In order to match the communication line impedance, RS485 bus terminal need to access two terminating resistance, the resistance is equal to the characteristic impedance of the cable transmission. The characteristic impedance of the twisted pair is generally 100Ω to 150Ω, which is related to the thickness of the insulation material and the diameter of the wire. The best matched impedance of the system should be measured by experiment.

(2)The maximum transmission rate of RS485 is 10Mb/s, the transmission distance is only 10m at this rate, the most distant transmission distance is about 1219 meters, the transmission rate is only 105b/s. The length of twisted pair is in inverse proportion to transmission rate, and the maximum transmission rate of 100m long twisted pair is 1Mb/s. In

 DOI: 10.1051/matecconf/201710003003

order to achieve the best transmission rate under different transmission distance, the limit transmission rate of each range can be measured by a large number of experiments.

(3)The length of the lead wire from the bus to each node should be as short as possible in order to minimize the impact of the reflected signal from the lead wire to the bus signal [5, 6].

4 Bus protocol design

RS485 bus only provides the electrical standard, in order to ensure the system communication is accurate and reliable , a clear provision have to be made on the data transmission mode, synchronization control mode (frame synchronization), error handling and other issues, in order to develop a general purpose of the intelligent initiation system of communication protocol.

In order to meet the need of multi point synchronization and different transmission distance of the intelligent initiating system, the communication protocol has two characteristics:

Baud rate selectable: the system is no longer limited to a particular baud rate, because the bandwidth resources can not be efficient used . According to different transmission distance, the best transmission rate is selected.

The broadcast instruction may be responsive: the slave machine can only perform the instruction but not to make any response. The time-sharing return mode, each slave machine can response, and if there is a slave machine dose not respond,master command retransmission, thus ensuring the synchronization and reliability of radio communication.

4.1 data frame structure design

Data frame is the basic unit of information transfer in the single chip computer system. According to the function, the data frame is divided into the command frame and the response frame, as shown in Figure 5.

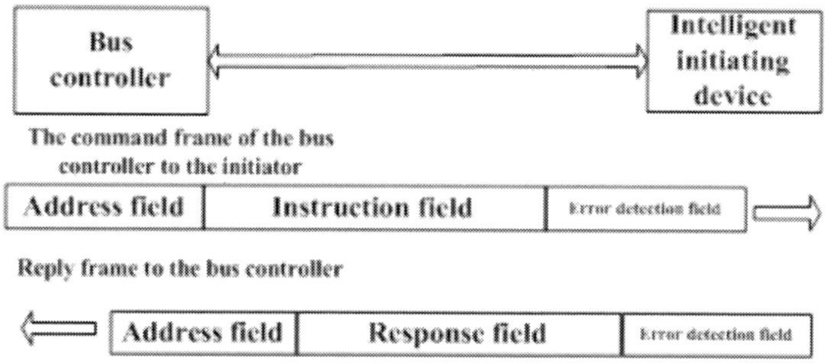

Fig. 5. Command frame and response frame.

The data frame structure is given below:

Table 1 Data frame structure.

Start byte	Address byte	Function byte	Data byte	Check byte
*	1 byte	1 byte	none, 1 byte, 2 byte	1 byte

- Start byte: the beginning of the identification of a frame of data, set to "*" characters, the value is 0x2A.
- Address byte: in the case of an intelligent initiating system, the address of the initiator is indicated. 0x00 - 0xEF, respectively, on behalf of the 0 - 254 initiator, the unique address, 0xFF as a general broadcast address; in the handheld detection and the encoder and the initiator, the byte is fixed to 0xFF.
- Function byte: the low 4 bits are the command code range. The effective command code is 0x07 - 0x01, which is 7 kinds of commands, such as address, state detection, system self check, single address charging, charging, single access, broadcast, etc..

- Data byte: additional information about the command code. Write address command requires 1 address byte, single access, broadcast firing command to 2 time delay time, other commands no data byte.
- Check and byte: binary arithmetic and the.
- Each byte of the transmission process contains 8 bit binary code, plus a start bit "0", and a stop bit "1", and a total of 10 bit [7-8].

4.2 measures to improve the quality of communication

(1)The frame structure is simple: the data frame only contains valid data and necessary expenses, which can improve the transmission rate, and reduce the bit error rate.

(2)Error control: Communication between the two parties using data and verification, frame check speed, simple process, although not error correction, but the RS485 communication reliability, bit error rate is very small, even if the parity check error code, just repeat once again is enough.

(3) Timeout control: the PC machine start the timeout control when it is waiting for the reply. If within the specified time has not received any response information, then repeat the command frame.

(4) Re sending control: if PC receives retransmission reply frame, receiving data is incorrect or no answer in time, repeat the command frame maximum retransmission 3 times [8-9].

4.3 system communication process

The main communication flow of intelligent initiation system is shown in figure 6.

1) System initialization: the power is applied to the bus controller, so that the the power supply and the initialization operation is completed.
2) System test:

- bus controller receiving from PC system self checking instructions, closed detection energy switch and forwarding self checking instructions broadcast to all detonator;
- When state detection of each detonator is completed, self results will returned to the bus controller in different time;
- The bus controller then uploaded to the PC machine, PC machine system self-test results display in the user interface.

3) Charging: after the system self-test is normal, PC machine will send charging instructions.

- Bus controller receives a charging instruction from PC, turn off the charging energy switch,and the charging instruction is transmitted to the initiator of the address;
- When the charge is completed, the initiator will return the charge state in time (single site mode) or the time (broadcast mode) to the bus controller;
- The bus controller then uploaded to the PC machine, PC machine will be charging status display in the user interface.

4) ignition: charging is completed, the PC machine will send fire instructions. According to the different firing instructions, it can achieve synchronous firing, asynchronous firing and delayed firing, etc.

- The bus controller will be forwarded to the detonator ignition instruction address match;
- When the time is up,automatic ignition detonator;
- If not normal initiation, initiation device closes automatically discharging switch, the excess energy is withdrawn to prevent the detonator charged, which can increase the security of the system.

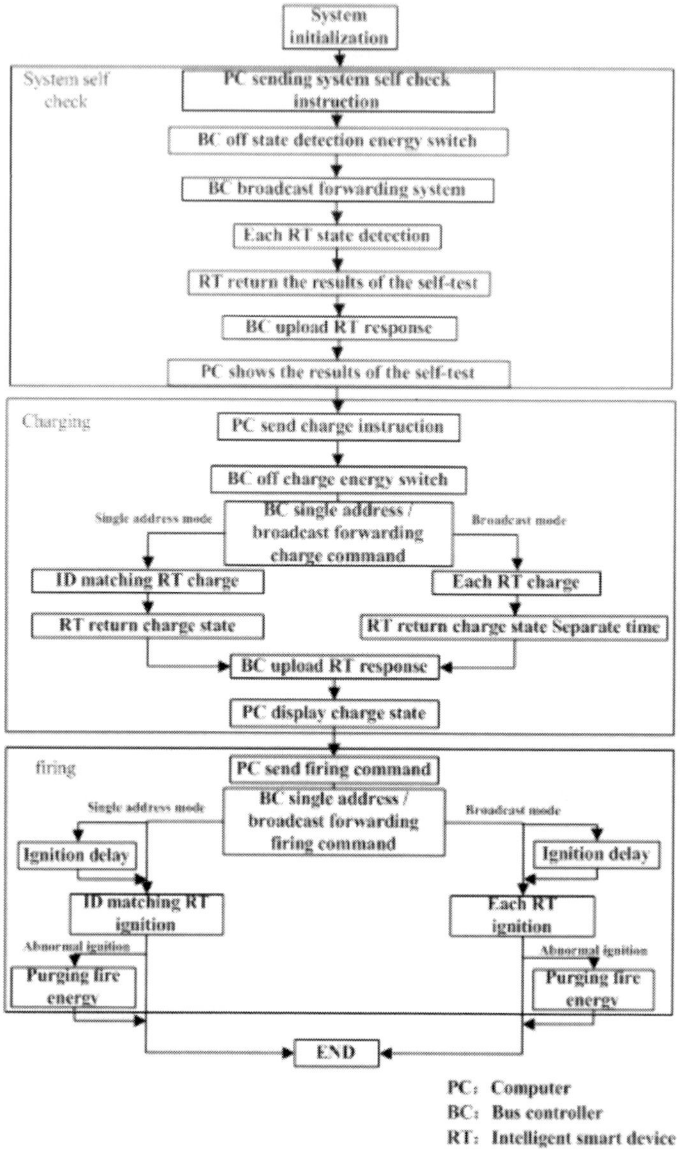

Fig. 6. Communication flow chart of intelligent initiating system.

5 Summary

In this paper, a communication protocol based on RS485 bus is designed, which can realize the accurate and reliable communication between the bus controller and multiple intelligent initiating devices, in order to realize intelligent initiating system more precise ignition in the environment where the interference is serious.

In addition to the use the RS485 bus which have strong anti-interference ability, open good, fault-tolerant ability of communication protocol is also the key to ensure data efficient transmission.RS485 bus protocol designed in this paper have a reasonable Frame structure ,perfect error control mechanism .The bus controller is controlled by a command / response mode to the control all the initiating devices in single address or broadcast way.In this way, the reliable communication of the intelligent initiating system is ensured in the physical layer and the data link layer.

References

1. Craig J. Boucher and David B. Novotney Ensign-Bickford Aerospace & Defense Company Simsbury. "Connecticut Performance evaluation of an addressable integrated ordnance system", AIAA 2001-3636.
2. David B. Novotney and Andrew Kochanek, Ensign-Bickford Aerospace & Defense Company, Simsbury, CT, 06070, "Intelligent Initiation Systems for Divert and Attitude Control Applications", AIAA 2005-4502.
3. United States Patent, US 2004/0020394 A1, "ORDNANCE FIRING SYSTEM" , Feb. 5, 2004.
4. United States Patent, US 2005/0150998 A1, "ORDNANCE SYSTEM WITH COMMON BUS, METHOD OF OPERATION AND AEROSPACE VEHICLE INCLUDING SAME" , Jul. 14, 2005.
5. Yang Jin-yan, Zheng Ying-qiang, Zhang Zhen-ren. 8051 single chip microcomputer data transmission interface technology and application examples. People's Posts and Telecommunications Press, 2005.
6. Wang Bao-li, Guo Bao-yi. The design and implementation of RS485 based electronic cap initiation network and communication protocol. Journal of Shaanxi University of Technology, 2010, 26(1):20-23.
7. Han Ke-hua, Ren Xi, Li Li-ming. Application design of RS485 bus in the long distance multi point ignition system. New special device application, 2008, 10(9): 31-34.
8. Liu Ting. Design and verification of a high performance ordnance initiation system bus. Shanghai Jiao Tong University, 2007.

Optimized Method for Knee Displacement Measurement in Vehicle Sled Crash Test

Hang Sun[1,a], Lei Lou[1,b], Zhixin Liu[1], Shicheng Li[1], Chao Chen [1] and Xing Pan[2,c]

[1]CATARC, China Automotive Technology and Research Center, Tianjin, China.
[2]North Carolina State University Computer Science Department, Raleigh, North Carolina, U.S.
[a]sunhang@catarc.ac.cn, [b]loulei@catarc.ac.cn, [c] xpan@ncsu.edu

Abstract. This paper provides an optimized method for measuring dummy's knee displacement in vehicle sled crash test. The proposed method utilizes completely new elements for measurement, which are acceleration and angular velocity of dummy's pelvis, as well as the rotational angle of its femur. Compared with the traditional measurement only using camera-based high-speed motion image analysis, the optimized one can not only maintain the measuring accuracy, but also avoid the disturbance caused by dummy movement, dashboard blocking and knee deformation during the crash. An experiment is made to verify the accuracy of the proposed method, which eliminates the strong dependence on single target tracing in traditional method. Moreover, it is very appropriate for calculating the penetration depth to the dashboard.
Keywords. Knee Mapping, vehicle sled test, motion image analysis, dashboard penetration

1 Introduction

During the vehicle frontal collision, knee injuries essentially come from the impact between the knee and the dashboard. The material and inner structure of the dashboard have a great influence on the injury level which is evaluated by femur compressive force and knee slider value [1]. Previous researches have shown that the injury level highly depends on the knee penetration against the dashboard and its severity will rise with the increasing depth of that penetration [2]. It was also a statement in Euro-NCAP SLED TEST PROCEDURE 2011 [3] that the manufacture has to provide the evidences for validating the knee mapping test result, which is an additional 20mm penetration inside the passenger dashboard [4]. At present, the issue for measuring the knee displacement is fully dependent on computer simulation, which can give out only a virtual result rather than an experimental one. However, the precise result is hard to be achieved in the sled crash experiment because the camera view on the knee could be easily blocked by a lot of matters, such as the dummy's arm and the dashboard. And the pasted tracing target may sometimes deviate from its original position because of strong impact force. Each of these unpredictable issues could disqualify the traditional single target-depended method from measuring the knee displacement correctly, especially for the measurement of penetration.

There is a practical way to calculate the displacement, which is an integral operation conducted twice on acceleration data [5]. Previous researches have made it practical in sled test to measure the displacement of the gravity center of the dummy's head [6], but the situation in measuring knee displacement is much more complicated because the acceleration sensor is hard to be mounted on the knee joint. However, the acceleration data from pelvis-mounted sensor could help to establish the knee displacement [3]. But experiments showed that the results calculated by pelvis acceleration were not very closed to the results using traditional target tracing method, unless the femur rotation around H-point had been taken into account as well. To obtain a precise knee displacement, femur rotaional angle is introduced in this paper, which can help to build a more reasonable way to obtain knee displacement.

2 Material and Methods

Traditional Method. In vehicle sled crash test, motion images captured by high-speed camera would be applied to a detailed analysis by engineers [7]. Target tracing measurement is a common analysis method, which is used to calculate the displacements of the parts under test by tracing quadrant symmetry targets [8] pasted on them. Here is the procedure in principle:

- Connecting several pre-determined points on the vehicle to establish a fundamental ruler for the following calculation.
- Establishing a proportion by the lengths of the same ruler in two dimensions, which are the length recorded in length units in the real space and in pixels in the scaled image.
- Establishing a proportion by the depths of the field of the ruler and the target under test. The proportion can reflect the relationship between the ruler length and the displacement of the target [7].

- According to the two proportions obtained, the displacement of the target can be calculated by its location change in pixels in different images.
- The velocity, acceleration and other related physical quantities are also available by introducing high-speed camera shutter frequency as a timing variable.

According to the procedure mentioned above, the dummy's knee displacement can be measured by following steps in vehicle sled test:

Step 1: Choosing two points (S_1 and S_2 in Fig. 1(a)) on the vehicle in the condition that the line segment connecting them should be parallel to the ground plane. Then, the line segment could be considered as a ruler which defines a coordinate system in both direction and length unit.

Step 2: Pasting quadrant symmetry target right on the knee joint which is under test (Target K).

Step 3: Measuring the depths of the field of ruler (d_1) and K (d_2) to the onboard camera lens.

Step 4: Recording the time (t_0) at the contact moment between the knee and the dashboard by high speed video analysis or contact detector.

Step 5: Using image analysis to plot a knee displacement curve vs. time as is shown in Fig. 1(b).

Step 6: The knee penetration could be calculated by $f(t_m) - f(t_0)$, in which the t_m is the time that the knee comes to its maximum penetration.

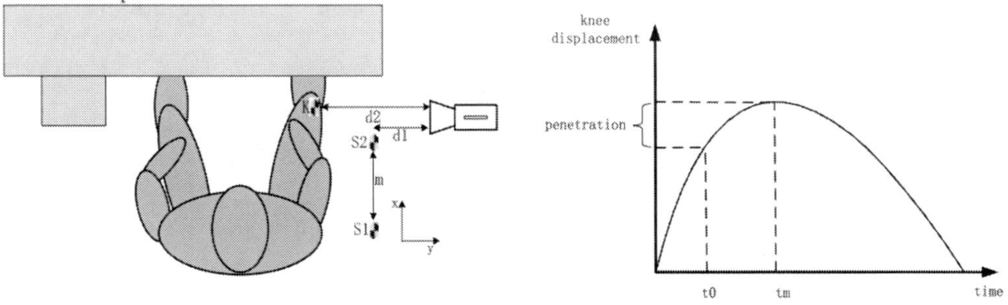

(a) testing setup (b) Profile of knee displacement

Fig.1. Knee penetration measured by high-speed motion image analysis.

Proposed Method. In vehicle sled crash test, acceleration and angular velocity sensors can be mounted in the dummy's pelvis in accordance with the position of its H-point. The displacement of H-point along X-axis can be calculated firstly projecting the acceleration to X-axis according to the pelvis angle and then integrating the acceleration to displacement. Dummy femur is a rigid body connecting pelvis with a hinge joint and its length could be considered unchangeable during the crash. So it is possible to use the displacement of pelvis to deduce knee displacement, and its penetration depth as well. But because of the counterforce coming from the impact of the knee against the dashboard, the femur would naturally perform a rotation around H-point, making the displacement of knee less than that of pelvis. And it also brings out the fact that pelvis itself is not enough to represent the knee movement, while an additional revision should be made upon the pelvis displacement to ensure a precise result of knee displacement. The method is proposed as follows:

The orientation of sensors mounted in the pelvis is the same as the pelvis angle. As is shown in Fig. 2, X_0 and Z_0 indicate the sensor coordinate system, which has an angle θ with the ground coordinate system.

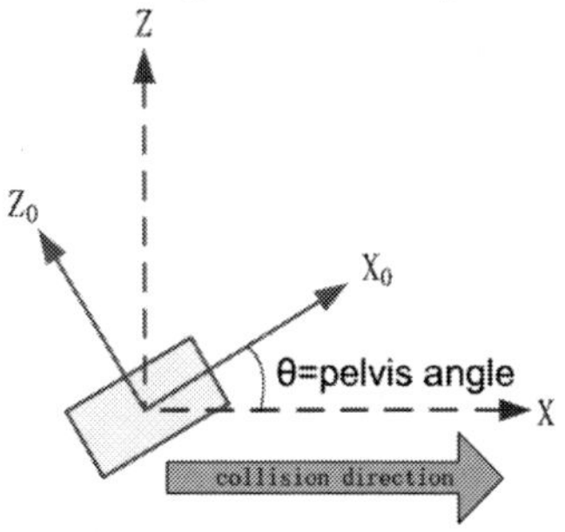

Fig. 2. The mount orientation of the pelvis acceleration sensor.

In order to calculate the displacement of pelvis, the pelvis acceleration should be projected to ground X-axis:

$$ACC_x = ACC_{X0} \cdot \cos\theta - ACC_{Z0} \cdot \sin\theta \tag{1}$$

The angle θ changes over time during the whole process of crash, which could be calculated by integration of angular velocities:

$$\theta = \int_0^t \omega(x)dx \tag{2}$$

Similarly, the displacements of pelvis and sled could be both calculated by integration of their accelerations, and the relative displacement could be figured out by a subtraction of their absolute displacements:

$$d_{pelvis} = \int_0^t \int_0^u ACC_{pelvis}(x)dxdu - \int_0^t \int_0^u ACC_{sled}(x)dxdu \qquad (3)$$

In practical test, the data captured by real-time data acquisition device are all discrete values at each sampling moment Δt (reciprocal of sample frequency), which turns the operation of integration into accumulation:

$$\theta_n = \sum_0^n \frac{\omega_n + \omega_{n+1}}{2} \cdot \Delta t \qquad (n=0, 1, 2\ldots) \qquad (4)$$

$$d_n = \sum_0^n \frac{\sum_0^n \frac{ACC_n + ACC_{n+1}}{2} \cdot \Delta t + \sum_0^{n+1} \frac{ACC_{n+1} + ACC_{n+2}}{2} \cdot \Delta t}{2} \cdot \Delta t - \sum_0^n \frac{\sum_0^n \frac{ACCsled_n + ACCsled_{n+1}}{2} \cdot \Delta t + \sum_0^{n+1} \frac{ACCsled_{n+1} + ACCsled_{n+2}}{2} \cdot \Delta t}{2} \cdot \Delta t \qquad (n=0, 1, 2\ldots) \qquad (5)$$

Figure 3 illustrates the knee penetration process. At the contact time(t_c), α is defined as the angle between X-axis and l_F(a line segment connecting joints of H-point and knee). And β is defined as the angle between X-axis and l_F at the time(t_{MAX}) that the knee reaches its maximum penetration.

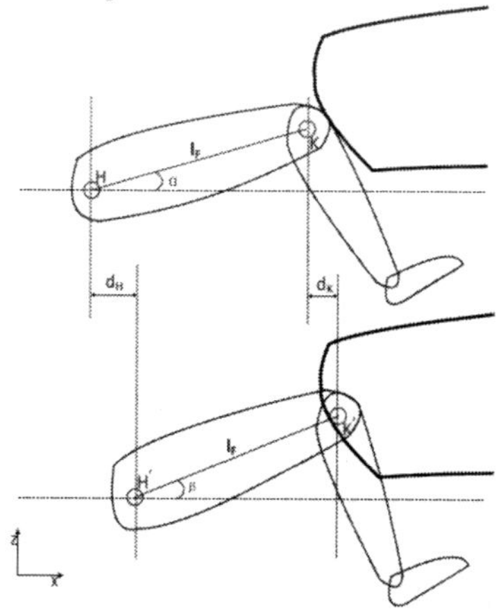

Fig. 3. Penetration measured by proposed method.

The penetration of knee is set for d_K, and the H-point displacement along X-axis is set for d_H, which is already worked out through Eq. 5. According to triangle relationship, the penetration can be represented as:

$$d_K = d_H - (\cos\alpha - \cos\beta) \cdot l_F \qquad (6)$$

3 Results

During the whole crash process, the measurement of knee displacement in penetration phase is much more difficult than that of knee displacement in other phases. A comparison is made between two methods provided above based on vehicle sled crash test and a thorough analysis is conducted upon the testing results. By use of accaleration sled facility, the test precisely simulated the crash pulse recorded in full scale vehical frontal collision. In order to achieve a significant depth of the penetraion for better comparison, a Hybrid 95th percentile dummy was uesd in the test with necessary instruments and high-speed cameras were set for recording. And the sample frequency was set to $f=10^4$ to ensure the integral precision. In addition, the arm and hand were all set on top of the femur rather than beside it and the target was pasted on the nearest point of the knee joint for a better camera view. Fig. 4 illustrates the knee displacement during the entire process of crash by the method of single target tracing. The displacement is 30.78mm at contact moment(40ms), and its maximum is 140.74mm at time of 101ms. So the penetration is 109.96mm that can be calculated by a subtraction operation.

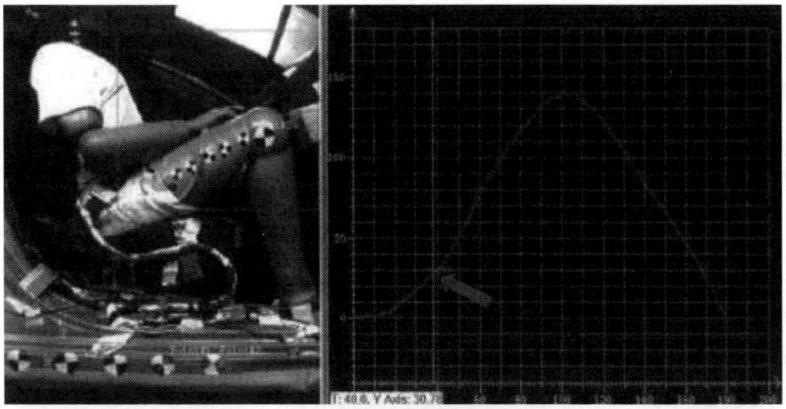

(a)t=40ms

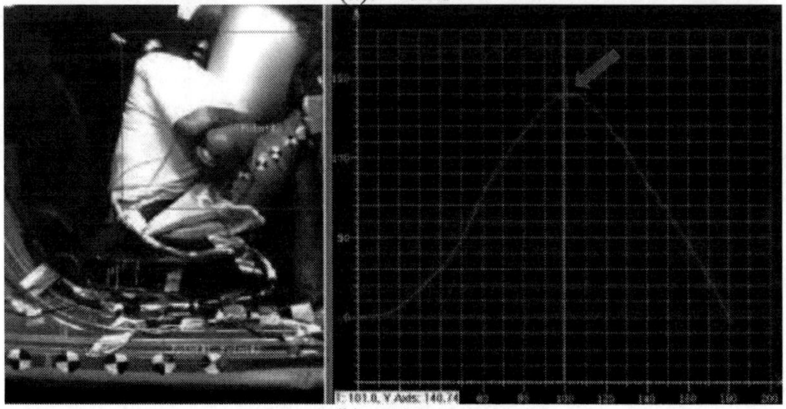

(b)t=101ms

Fig. 4. Knee displacement curve worked out by single target tracing.

Figure 5 illustrates the pelvis acceleration curve based on the data captured by the sensor with its own cordinate system X_0 and Z_0. The pelvis angle to the horizontal is calculated by integration of the angular velocity captured by angular sensor, which is illustrated in Figure 6 that indicates the changing trend of pelvis angle during the crash. According to the Eq. 1, the pelvis displacement along X-axis is shown in Figure 7(a), and the displacement curve of sled is shown in Figure 7(b) which is original in line with the same X-axis.

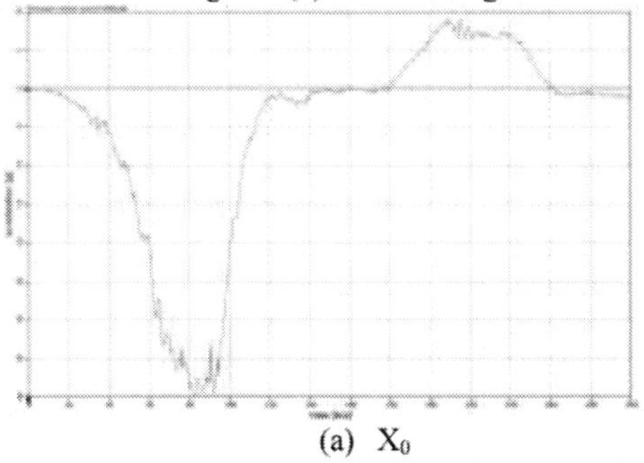

(a) X_0

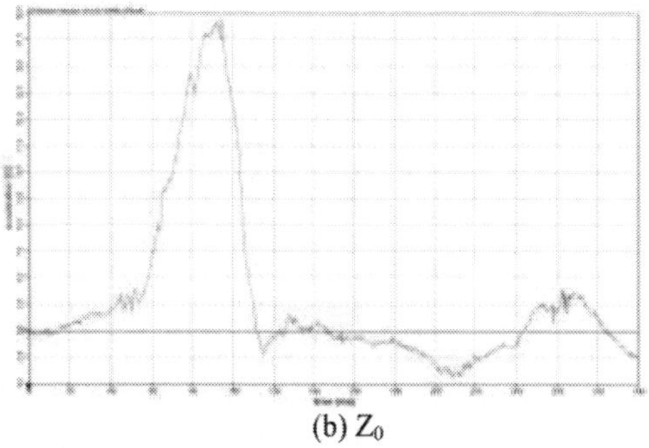

(b) Z_0

Fig. 5. Curves of original pelvis acceleration.

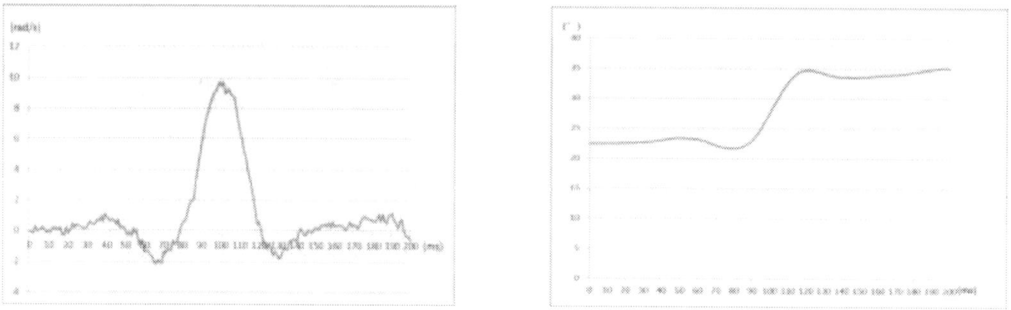

(a) Angular velocity (b) Angle degree

Fig. 6. Curves related to pelvis angle.

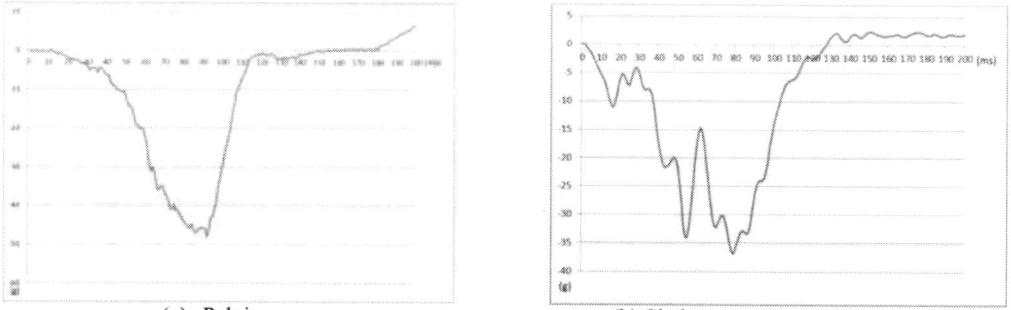

(a) Pelvis (b) Sled

Fig. 7. Curves of crash displacement

The relative dispalcement in Figure 8 can be calculated by Eq. 5. And the pelvis displacement is 132.19mm between the contact time(40ms) and the time(94ms) when pelvis reaches its peak point of displacement. Without considering femur angle change, there is a 22.23mm deviation to the previous result.

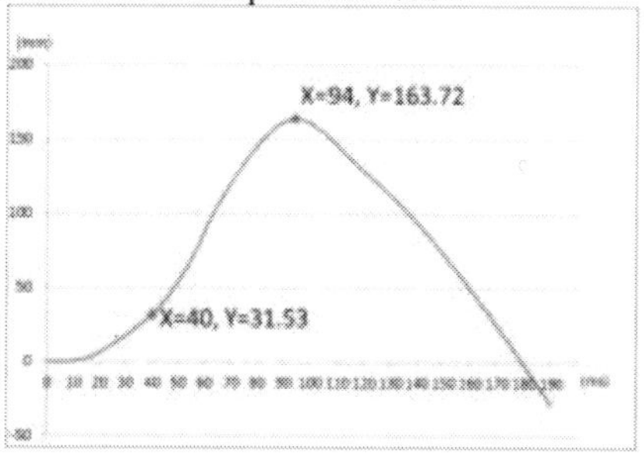

Fig. 8. Knee displacement curve worked out by proposed method

The femur angle changes from 27.19^0 to 34.18^0 shown in Figure 9. And the length between knee joint and H-point is 426.46mm [9], so the knee penetartion depth is 105.66mm calculated by Eq. 6.

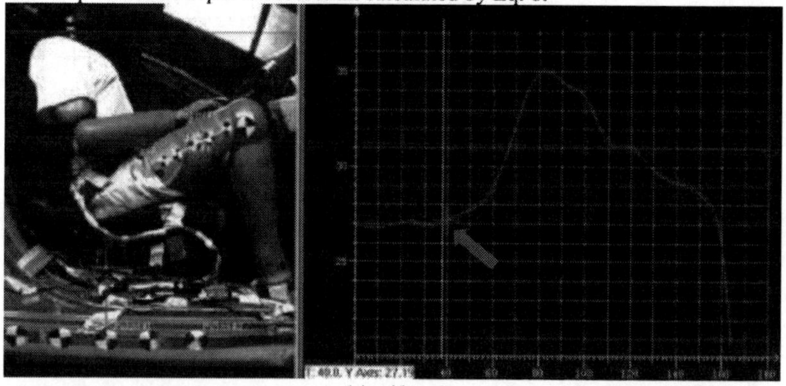

(a) 40ms

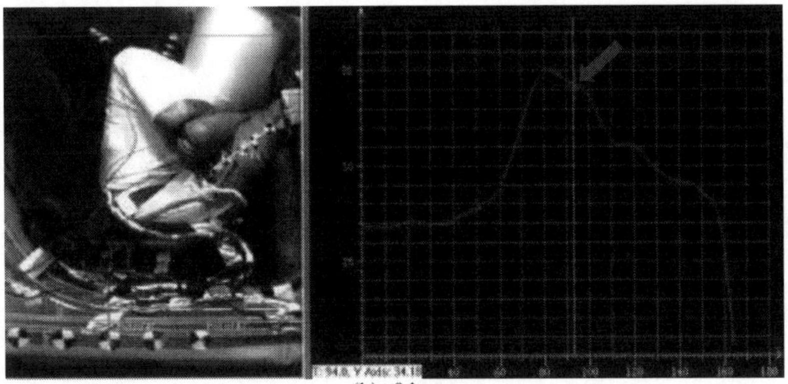

(b) 94ms

Fig. 9. Femur angle in two particular points of time.

4 Discussion

As shown in the results, the femur rotation establishes a correlation between the movement of pelvis and knee and revises the pelvis displacement 132.19mm to the knee displacement 105.66mm, which means a 26.53mm difference caused by femur rotation. Compared with the result of 109.96mm from single target tracing method, the proposed method has a difference of 4.3mm, 4% out of the traditional method, which means the results from both methods are basically identical. In addtion, because of the difference in rebound timing, there is a 7ms mismatch between the knee and pelvis reaching their maximum displacements respectively (Figure 4 & Figure 8). But it barely affects the final result because the knee is relatively stationary to the dashboard during that extremely short period.

5 Summary

Since the cost for vehicle collision test is significantly considerable, it would be a great risk if the knee displacement measuring is only dependent on single target, which can be easily blocked by several unexpected factors. The proposed method takes advantage of the pelvis acceleration and the femur angle which can be credibly obtained by connecting any two separate targets, which eliminates the dependence on single target. Additionally, since the result of proposed method is quite similar to that of traditional method, the proposed method can be considered as an alternative way to measure the knee displacement, especially effective for knee penetration measurement. In conclusion, the proposed method is applicable to knee impact test and would be very useful in validating knee mapping test.

References

1. Gokhale A V, Saravate V B, Chalipat S, et al: Femur and knee injury reduction by use of knee bolsters in frontal crashes. Papers. Automotive_Sector (2007).
2. Laurent Portier, Xavier Trosseille: Lower Leg injuries in Real-world Frontal Accidents. IRCOBI Conference (1993).
3. Sled Test Procedure for Assessing Knee Impact Areas (Euro-NCAP 2011-06).
4. Chen Tangpei, Lei Peng: The Study of Knee Mapping Sled Test. The 13th Conference of Automotive Safety Technology (2013).
5. Rong Tianping: The theory and analysis of displacement measuring by acceleration. Journal of Huazhong University of Science and Technology (2000).
6. Lei Peng, Chen Tangpei. A Method of Obtained Dummy Head Displacement by Acceleration and High Speed Camera during the sled test. The 16th International Conference of China Automotive Safety Technology (2013).
7. Li Shicheng, Lou Lei: Research on Parallax Correction Method for High-Speed Image Motion nalysis. Mechanical Research & Application (2014-01).
8. User Guide of TEMA (2008).
9. User's Manual for the Hybird III Large Male (2003).

Dynamic Characteristics of the Steam Accumulator Charging and Discharging

Peizhi Yang*, Xia Hu and Gang Liao

School of Energy Science and Engineering, Central South University, Changsha， Hunan province, 410083, P R China

*Corresponding author:yang_peizhi@mail.csu.edu.cn

Abstract. The method of numerical simulation is used to study the charging and discharging process of steam accumulator. By analyzing the charging and discharging thermal process of steam accumulator, establishing the mathematical model and using MATLAB software to simulate the solution, the dynamic characteristic curves of charging and discharging are obtained, and the charging and discharging conditions of the steam accumulator are defined. This research has an important significance in the design and application of the steam accumulator.

Keywords. Steam accumulator; Charging characteristics, Discharging characteristics, Numerical simulation.

1 Introduction

Energy crisis has become an important issue in the world, severe crisis has put forward a severe warning to mankind. Based on this energy situation, energy saving technology has been referred to an unprecedented height. Steam accumulator is a bridge which can connect steam generators and consumers smoothly, it can give supply and demand sides continuous and stable steam balance, so as to maintain the system running load balance, improve the combustion efficiency and save energy. With the continuous adjustment of China's energy policy and the continuous development of energy saving, steam accumulator will be paid more and more attention, and it will be widely used in industrial enterprises. Now steam accumulator has been widely used in iron and steel industry, power supply and daily necessities production (such as sugar refining, brewing, chemical, printing and dye works) and other industry, and it has achieved good economic benefits.

The dynamic and static characteristics of steam accumulator directly affect the economy and safety of the system; some scholars have carried on the related research. The operating characteristics of steam accumulator were studied by using numerical simulation method [1-5],a thermal test system of a small steam accumulator is established in Ref.[6], and the experimental study of the charging and discharging characteristics of the horizontal steam accumulator is carried out ,the dynamic process of the water space in the discharging process of the steam accumulator is analyzed by using the equilibrium model and Euler algorithm [7],a non-equilibrium numerical model is used to analyze the characteristics of the steam accumulator charging and discharging [8], and the effect of the inlet steam enthalpy on the change of the water content and the water level is studied. According to the analysis of the steam accumulator charging and discharging, the steam accumulator's mathematical model is established based on the assumption that the water and steam in the vessel in thermal equilibrium, using Matlab mathematical calculation software to write a program to simulate charging and discharging process and to analyze the simulation results of different control modes. The conclusion of this paper has certain significance for the design and application of the thermal storage system.

2. Establishment of mathematical model

The operation of steam accumulator includes two processes: charging and discharging, the process of charging, i.e. the saturated steam or superheated steam, which from the steam source, inflows into the container; the process of discharging, i.e. the pressure of saturated water in the steam accumulator drop till evaporate, and the released steam provide to the user. According to the thermodynamic characteristics of charging and discharging process, the mathematical model of the steam accumulator is established.

2.1. Basic hypothesis

Steam accumulator is divided into water space and steam space, In order to simplify the analysis, the following assumptions are given for the working process of steam accumulator:

(1)At any time, the state parameters of the water space and the steam space, which are the pressure of the steam space corresponding to the saturated liquid and saturated vapor , are uniform and in the state of the thermal equilibrium;

(2)Ignore the power of steam through the vessel wall and the heat dissipation to the environment;

(3) The steam, which released in the exothermic process, is saturated steam;

(4)The density of water is constant, and the density of water vapor is determined by the state equation of ideal gas P= ρRT.

2.2 .Mathematical model

The calculation and analysis of the steam accumulator charging or discharging ,the model used is shown in Fig.1.At a certain moment, the mass of saturated water in the heat accumulator is M_w. Specific enthalpy is h_w, the mass of saturated water vapor is M_s ,Specific enthalpy is h_s; The high temperature and high pressure steam mass flow, which is used in the process of charging, is m_1 ,specific enthalpy is h_1 , transient outlet steam mass flow during discharging process is m_2 ,specific enthalpy is h_2 ,the transient rate of evaporation of water inside the accumulator is m_3 ,specific enthalpy is h_3 .

According to the above model, the following balance equations are obtained:

Water space and steam space mass balance:

$$\frac{dM_w}{dt} = m_1 - m_3 \quad (1)$$

$$\frac{dM_s}{dt} = m_3 - m_2 \quad (2)$$

Water space and steam space energy balance:

$$\frac{d(M_w h_w)}{dt} = m_1 h_1 - m_3 h_3 \quad (3)$$

$$\frac{d(M_s h_s)}{dt} = m_3 h_3 - m_2 h_2 \quad (4)$$

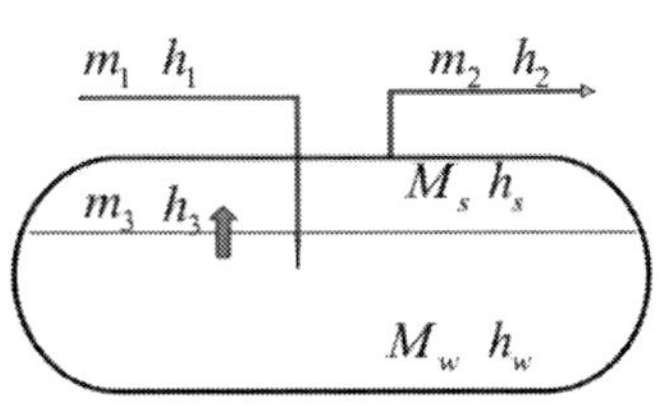

Fig.1. Schematic of steam accumulator

The calculation of charging and discharging process can use Eqs.(1) - (4),in the system that the steam accumulator parallel with steam source[9], the mass of charging process at the mathematical model is calculated as m_2=0(during discharging process, the mass is calculated as m_1=0).

3. Numerical solution

3.1. Characteristics of charging process of steam accumulator

According to the above mathematical model, the charging process of the steam accumulator is simulated by using MATLAB software [10].Heat steam from the steam source, which the temperature and pressure parameters have been determined, is the saturated steam or superheated steam, and its pressure is higher than the pressure in the vessel over the final charging state. Iterative calculation of time step is Δt, the volume of the steam accumulator is V, the thermal vapor pressure is p_0,temperature is T_0,specific enthalpy of h_0.In the Δt time, the mass flow rate of the hot steam is a constant, i.e. m_0, suppose the volume of water at a certain time is V_1, p_1 and T_1 are the pressure and temperature, respectively. h_{w1} and h_{s1} are the specific enthalpy of water and steam, respectively , ρ_{w1} and ρ_{s1} are the density of water and steam, respectively ; After Δt time ,the volume of water is V_2, pressure is p_2 and temperature is T_2, h_{w2} and h_{s2} are the specific enthalpy of water and steam , respectively , ρ_{w2} and ρ_{s2} are the density of water and steam, respectively.

After the steam accumulator charging Δt hours, the following equations are obtained.

Mass balance equation:

$$m_0\Delta t+\rho_{s1}(V-V_1)+\rho_{w1}V_1=\rho_{s2}(V-V_2)+\rho_{w2}V_2 \tag{5}$$

Energy balance equation:

$$m_0h_0\Delta t+\rho_{s1}h_{s1}(V-V_1)+\rho_{w1}h_{w1}V_1=\rho_{s2}h_{s2}(V-V_2)+\rho_{w2}h_{w2}V_2 \tag{6}$$

The enthalpy of saturated water and saturated steam are calculated according to the method suggested in Ref. [11], they are calculated as:

$$h\_water(T)=4.1868\times(1.001T-0.01) \tag{7}$$

$$\begin{aligned}h\_steam(T)=2.326\times((&0.071533T-13.08426)\times h\_water(T)\times 0.01\\&+(0.808956T+1075.18144))\end{aligned} \tag{8}$$

Specific enthalpy of saturated water and saturated steam in Eqs.(5)-(6) can be calculated according to the Eqs.(7)-(8)(the specific enthalpy is obtained from the data in the water and water vapor table fitting to get a pressure range of functional relation) ,the specific enthalpy of superheated steam is the dependent variable of pressure and temperature. In the equations, the density of water is a constant factor, it is determined by the ideal gas state equation. Therefore, the relevant thermodynamic parameters can be written as a function of pressure or temperature. The ratio of water volume to vessel volume(at the initial charging moment)and initial pressure are know, the hourly variation curves of pressure in the vessel under different conditions during the process of charging is obtained by simulation calculation.

The influence of different thermal steam parameters on the thermal characteristics of the steam accumulator is analyzed by using the above method of mathematical modelling approach in the process of charging, the following two factors are mainly considered: the steam pressure and steam flow.

3.1.1. Steam accumulator variable steam pressure and constant steam flow

Simulate the process of charging with saturated steam (the pressure is varied from 1MPa to 6MPa). The constant initial pressure of the steam accumulator of 600kPa is adopted, the steam flow in the process of charging is 3kg/s, maintain the heat steam flow constant, the hourly pressure change in the heat accumulator is shown in Fig.2, using different saturation pressure during the process of charging.

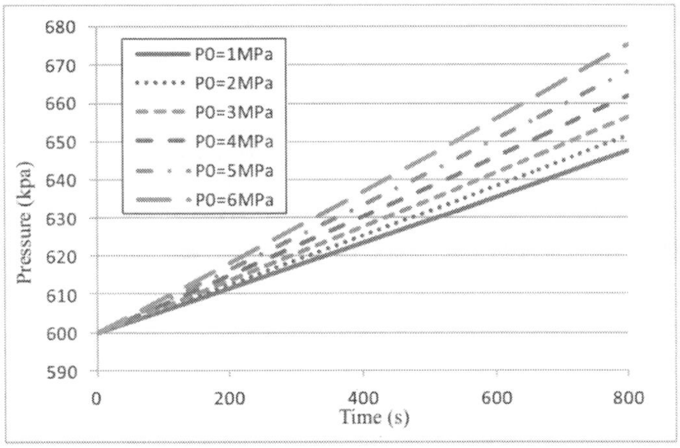

Fig.2. Calculated pressure change in the steam accumulator charged with the constant inlet steam flow

Heat steam enters the water space, and its condensation heat transfer to liquid saturated water, part of the condensation heat used to improve the parameters of the water itself, and the other part is used for the vaporization of water. The higher the saturated steam pressure, the more energy released during the process of condensation for saturated water with the same pressure, and speed up the evaporation of water and the heat storage of saturated water, therefore, the pressure inside the vessel is rising fast. With the charge of heat, energy stored in the vessel in the form of high pressure saturated water, the pressure in the vessel at the end of the charge must be lower than thermal steam pressure.

3.1.2. Steam accumulator variable steam flow and constant steam pressure

In actual operation process, after the steam source supply to the consumer, the remaining steam enters the steam accumulator for charging, due to the change of the consumer's demand, the steam inflow the accumulator will change too. The result of the steam accumulator operation (when simulating the linear change of steam flow during the process of charging) is shown in Fig. 3, the steam pressure during the process of charging is a constant (i.e.1Mpa), and the initial steam flow is 3kg/s. With the increase of the flow rate, the pressure inside the vessel is increased, so the schematic is a curve.

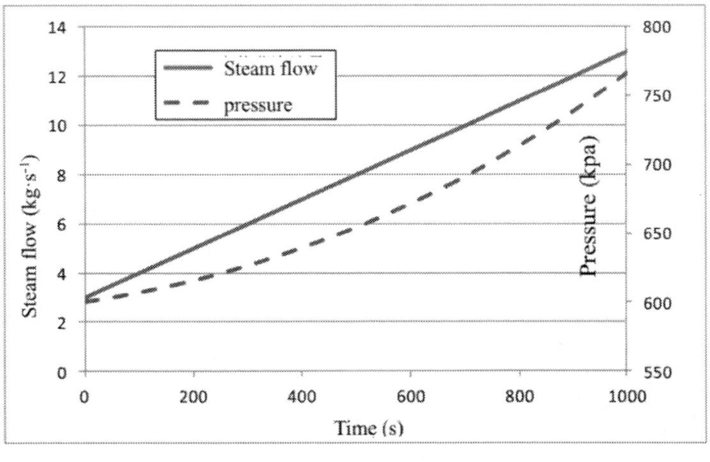

Fig.3. linear variation of steam flow during the process of charging

In the actual process of charging, the change of the steam source and the steam using side can affect the inlet steam parameters of the steam accumulator, the flow and pressure of the steam may change at the same time, and it is not possible to follow the simple linear change. Therefore, the pressure curve of the steam accumulator operation during the process of charging is usually irregular.

3.2. Characteristics of discharging process of steam accumulator

When the steam accumulator is discharged, the released steam come from the steam space in the current time, the steam parameter of the outlet is equal to the steam parameter of the steam space at the moment. The pressure during the process of discharging decreases ,so the steam parameters are changed hourly[12].The time step is $\Delta t'$,the volume of the steam accumulator is V ,suppose the volume of water at a time during the process of discharging is V_1', p_1' and T_1' are the pressure and temperature, respectively, h_{w1}' and h_{s1}' are the specific enthalpy of water and steam, respectively, ρ_{w1}' and ρ_{s1}' are the density of water and steam , respectively ; After $\Delta t'$ hours of discharging ,the volume of water is V_2' ,pressure and temperature are p_2' and T_2' , respectively, the specific enthalpy of water and steam are h_{w2}' and h_{s2}' , respectively, the density of water and steam are ρ_{w2}' and ρ_{s2}' , respectively .Because the time step is very short, it can be considered that the outlet steam flow is constant(i.e. m_0') in the process of $\Delta t'$ hours discharging, and the specific enthalpy of the outlet steam in this period is also a constant, i.e. $(h_{s1}' + h_{s2}')/2$.

The following equation is established after the steam accumulator discharge $\Delta t'$ time:

Mass balance equation:

$$\rho_{s1}'(V - V_1') + \rho_{w1}'V_1' = \rho_{s2}'(V - V_2') + \rho_{w2}'V_2' + m_0'\Delta t' \tag{9}$$

Energy balance equation:

$$\rho_{s1}'h_{s1}'(V - V_1') + \rho_{w1}'h_{w1}'V_1' = \rho_{s2}'h_{s2}'(V - V_2') + \rho_{w2}'h_{w2}'V_2' + m_0'\frac{h_{s1}' + h_{s2}'}{2}\Delta t' \tag{10}$$

In Eqs.(9) ~ (10),the temperature and density parameter of water and steam can be obtained by the change of the corresponding state pressure, the specific enthalpy of saturated water and saturated steam can be calculated according to Eqs.(7)-(8),the specific enthalpy of superheated steam is the dependent variable of pressure and temperature. The density of water is constant in the equations, the density of water vapor is determined by the ideal gas state equation. Therefore, the relevant thermodynamic parameters can be written as a function of pressure or temperature. The ratio of water volume to vessel volume at the initial discharging moment and initial pressure are know, the hourly variation curves of pressure in the vessel under different conditions during the process of discharging is obtained by simulation calculation.

The outlet of the accumulator is connected with the consumer, the process of discharging is controlled according to consumers' requirement, and it is often used in two modes: mass flow control and pressure control. The influence of different control modes on the thermal characteristics of the steam accumulator is analyzed by using the above method of mathematical modelling approach in the process of discharging (i.e. outlet flow control mode and pressure control mode).

3.2.1. Outlet flow control mode

Outlet flow control mode, i.e. controlling the change of the outlet steam flow, in this paper, simulated the linear variation of the outlet flow and calculated hourly variation curve of relative parameters in the vessel that based on regard m_0' as a function of the known time. The constant initial pressure of the charging process of 800kPa is adopted, control the steam flow rate from 3kg/s to start with time increasing linearly, then the pressure changing with time in the steam accumulator is shown in Fig. 4.With the increase of the outlet steam flow rate, the pressure decreasing rate of the steam accumulator is accelerated, the self - evaporation rate is also accelerated due to the pressure difference between saturated water and steam space, the hourly change of water level in the vessel is similar to pressure [13].

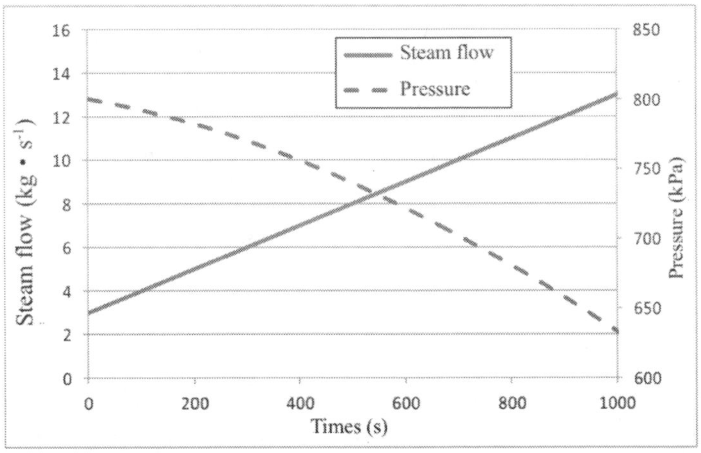

Fig.4. linear variation of outlet steam flow during the process of discharging

3.2.2. Pressure control mode

Pressure control mode, i.e. controlling the change of the outlet steam pressure of the accumulator, in this paper, simulated the operation of steam accumulator under the condition of simple linear variation of the outlet pressure, hourly variation curve of water quantity and outlet steam flow can be obtained by taking certain accumulator sliding pressure rate, and the total amount of steam released can be obtained by the flow integral. Simulation results of using the pressure control mode in the process of discharging are shown in Fig. 5. The decreasing rate of pressure over time is same, but the initial pressure in the range of 0.8MPa~1.3MPa has to take a different value (as shown in Fig.5 (a)), the outlet steam flow changing over time in the steam accumulator is shown in Fig. 5 (b).

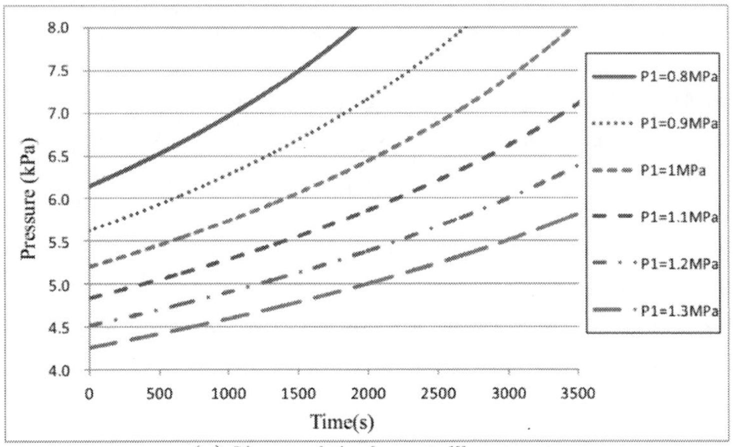

(a) Linear variation by controlling pressure

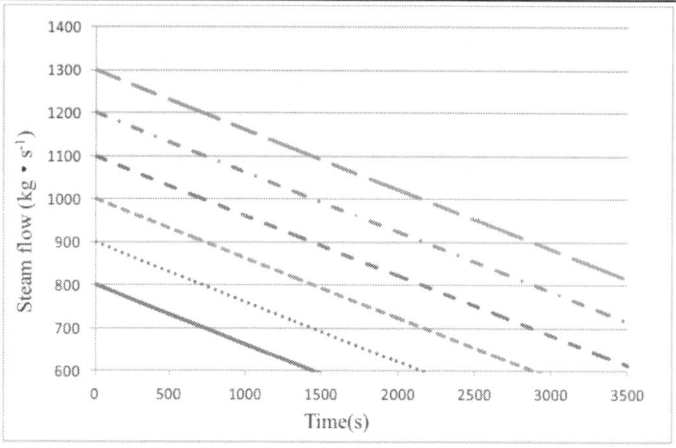

(b) Change of outlet steam flow in accumulator

Fig. 5 Linear variation of outlet steam pressure during the process of discharging

With the pressure in the vessel decreases, the outlet steam flow will gradually increasing and the rate of growth is accelerating. Analyzing the relationship between the specific enthalpy of saturated water and the corresponding saturation pressure can know that the slope of specific enthalpy change with pressure curve increases with the decrease of pressure, therefore, with the process of discharging, the energy released by the unit pressure decreases is gradually increasing and can make more saturated water evaporation, which can explain the phenomenon that the outlet steam flow increases with the pressure decreases[3].The higher the initial pressure (which is given), the smaller the outlet steam flow at the same time, this can also explained by the relationship curve between the specific enthalpy of saturated water and the corresponding saturation pressure . Therefore, the released energy of the steam accumulator is related to the initial pressure and the working pressure difference, The higher the initial release pressure, the less the amount of steam released by the same work differential pressure.

4. Conclusions

In this paper, established the mathematical model of the steam accumulator depending on analyzing thermodynamics of the steam accumulator charging and discharging and assuming the water and steam in the vessel are in thermal equilibrium, according to the mathematical model, the process of charging and discharging are simulated respectively and the dynamic characteristics of charging and discharging are obtained. Through the analysis of the simulation results, the following conclusions can be drawn:

(1)The pressure in the vessel is increased by time during accumulator charging;

(2) The regenerative rate of the steam accumulator is related to the steam flow and steam pressure, the greater the steam flow and the higher the pressure, the greater the regenerative rate of the accumulator, and the less time needed to achieve the same hot filling pressure.

(3) The steam pressure, temperature and other parameters of the steam accumulator, which are released during discharging, are decreased by time, but outlet steam flow rate increases;

(4) The released energy of the steam accumulator is related to the initial pressure and working pressure difference, the

References

1. E.S. Xu, W. Gao, F. Xu et al. Simulation of Dynamic Characteristics of Steam Accumulators in the Badaling Solar Power Tower Plant, Applied Chinese Journal of Electrical Engineering 32 (8) (2012) 112-117.
2. X. Li, T.Y. Zhu, X.Y. Xu et al. Simulation of thermal storage system for 1MW solar power tower plant, Applied Journal of solar energy 32 (5) (2011) 632-638.
3. C.L. Gong, X.F. Liu et al. Thermo-hydraulic Feature Analysis of Steam Accumulator Operation, Applied Journal of Wuhan automobile industry university 18 (3) (1996) 73-76.

4. X.F. Liu, C.L. Gong, S.J. Liu. Analytical calculation and experimental study on temperature rising process of steam accumulator, Applied Journal of solar energy 19 (1) (1998) 102-104.

5. Z.F. Zhao, L.J. Liu. Analysis on operation characteristics of steam accumulator, Applied Qinghai electric power 30 (4) (2011) 9-12.

6. Z.Q. Yang. Experimental study on the characteristics of steam accumulator charging and discharging, Applied Harbin Engineering University (2012).

7. Ismail M, Shah MH, Kamal MAA et al. Research Work on Steam Accumulator in Palm Oil Mill. European Journal of Scientific Research, 2009; 37(4): 628-640.

8. Stevanovic VD, Maslovaric B, Prica S. Dynamics of steam accumulation. Applied Thermal Engineering, 2012; 37: 73-79.

9. Shnaider DA, Divnich PN,Vakhromeev IE. Modeling the Dynamic Mode of Steam Accumulator. Automation and Remote Control, 2010; 71(9): 1994-1998.

10. X. Sun, L.M. Xu, Q. Wu. MATLAB7.0 basic tutorial, Beijing: Tsinghua University press (2005).

11. H.T. Le. Study on Modeling and Dynamic Simulation of Lithium Bromide Absorption Heat Pump, Master degree thesis of North China Electric Power University (2009).

12. Fengwu Bai, Chao Xu. Performance analysis of a two-stage thermal energy storage system using concrete and steam accumulator. Applied Thermal Engineering, 2011; 31: 2764-2771.

13. Hahne E, Chen Y. Numerical study of flow and heat transfer characteristics in hot water stores. Solar energy, 1998; 64(1-3): 9-18.

Degradation of 4-methylbenzylidene camphor (4-MBC) Using Fenton, UV Light Irradiation and Photo-Fenton

Chunmei Ji, Chen Wang *, Jing Sun and Yadong Yu

School of Environmental Science and Engineering, Qilu University of Technology
Key Laboratory of Cleaner Production and Industrial Waste Recycling and Resourcization in Universities of Shandong
Qilu University of Technology, Jinan, Shandong Province, China.
Corresponding Email: shanqing123@126.com

Abstract. As one of the most commonly employed UV-filters, 4-methylbenzylidene camphor (4-MBC) has been shown to accumulate in the environment and have endocrine disrupting activity. 4-MBC cannot be degraded completely by common methods in wastewater treatment plants. To prevent the environmental problems caused by 4-MBC from being more serious, finding effective method to degrade 4-MBC is essential. In this research, Fenton reaction and photo-Fenton was employed to degrade the 4-MBC. Orthogonal experimental design was employed to evaluate the influence of factors (Fe^{2+}, H_2O_2, pH and reaction time) on the degradation of 4-MBC. The degradation rate reached to 66.01% in Fenton process and 96.71% in photo-Fenton process. UV light irradiation was also employed to degrade 4-MBC. After being irradiated by mercury lamp (300W) 90 min, the concentration of the 4-MBC reduced 85.4%, but most of them translated to the isomeride. Compared with the Fenton, photo-Fenton and UV light irradiation processes, we deduces that photo-Fenton can get the better degradation efficiency of 4-MBC, and, Fenton reagents and light irradiation are synergistic.
Keywords. 4-MBC, fenton reaction, light degradation, photo-fenton.

1 Introduction

Pharmaceutical and Personal Care Products(PPCPs) include drugs, hormones, fragrances, cosmetics, eta, which have attracted much public and scientific attention due to widespread contamination in aquatic environment. Among them, sunscreen cosmetic products have been used for more than 80 years [1], and its essential components are UV-filters. The main concern used to focus merely on their utility and efficiency as UV-filters, but, only very recently, concern has been raised regarding their path and their endocrine disrupting activity in the environment. UV-filters enter the aquatic environment either directly via wash-off from skin during swimming and bathing activities, or indirectly via wastewater from wastewater treatment plants (WWTPs) [2]. Due to the high lipophilicity and relative stability against biotic degradation, UV-filters have been shown to accumulate in the food chain by bioaccumulation and potential biomagnification [3]. Studies have proved that some organic UV-filters exhibit endocrine disrupting activity [4, 5].

4-methylbenzylidene camphor (4-MBC) is an organic camphor derivative which has the ability to protect the skin against UV-B radiation, and it is one of the most commonly employed UV-filters [6]. However, it has been shown to accumulate in the environment. The concentration of 4-MBC can reach to 113±7ng·L-1 in the surface water around Majorca Island [7]. It also can be detected in the muscle tissue of fish from seven small Swiss Rivers, all receiving inputs from wastewater treatment plants [8]. And even in tap water, 4-MBC also exist and the concentration can reach to 35ng·L-1[9]. 4-MBC has also been proved to have endocrine disrupting activity. Researches demonstrate that it can affect the rats' development of reproductive organs [10], the number of embryos of Pantipodarum [11], and accelerate the cell proliferation in MCF-7 breast cancer cells [12].

Studies reveal that UV-filters cannot be degraded completely in WWTPs, and 4-MBC is the most prevalent UV filter detected in treated wastewater[13].To prevent the environmental problems caused by 4-MBC from being more serious, finding effective method to degrade 4-MBC is essential. Fenton and photo-fenton, which are the commonly used methods in degrading organic pollutants, were employed to degrade 4-MBC in the research. orthogonal experiments were employed to verify the interaction effects of factors responsible (Fe2+, H2O2, pH and reaction time) on the degradation efficiency of 4-MBC. UV light irradiation was also explored. The degradation eciency of Fenton, photo-Fenton and UV light irradiation were compared. The degradation efficiency was determined by HPLC (High Performance Liquid Chromatography). To the best of our knowledge, this is the first time todo the work.

2 Materials and Methods

2.1 Materials and Instruments

4-MBC was purchased from Brilliance Biochemical Company (Beijing, China) and the purity was higher than 99%. Acetonitrile (HPLC grade) was obtained from Damao Chemical Regent Factory (Tianjin, China).Analytically pure grade H_2O_2 (30%), $FeSO_4 \cdot 7H_2O$, NaOH used in the experiments were purchased from Damao Chemical Regent Factory(Tianjin,China) and H_2SO_4 was from Tieta (Laiyang, China). All reagents were used without further purification. High-purity water (18.2 MΩ cm) was used throughout all experiments, and it was prepared on an ELGA purelab flex system (ELGA, Veolia Water, Marlow, UK).

Photochemical reactor (BL-GHX) was manufactured by Bilon Instrument Company (shanghai, China).It was mainly comprised of light source, cylindrical reactor, magnetic stirrer and circulating cooling system. Mercury lamp was employed (300W) as light source and the lamp was vertically located in the center of the reactor. Orbital shaker (HV-4) used in the experiments was manufactured by guohua, (Changzhou, China). A Leici PHS-3C pH meter (Shanghai Leici Corporation, Shanghai, China) was used for pH measurements. The LC system comprised a Shimadzu SPD-20A UV detector, and a LC-20AT high-pressure solvent delivery pump with a 20 mL sample loop injector. The column used in our experiments was Anasil AQ-C18 (25 cm×4.6mm, 5μm).

2.2 Fenton process

In Fenton process, the vital factors are the initial concentration of the 4-MBC, the dosages of H_2O_2, pH value, amount of Fe^{2+}, reaction time. For the aim of the research is to assess the ability of Fenton in degrading 4-MBC, we fixed the initial concentration of 4-MBC and chose factors (Fe^{2+}, H_2O_2, pH and reaction time) as variables and evaluate their effect on degradation. Orthogonal experimental design was employed to optimize the reaction conditions of Fenton process. The Fenton reaction was carried out in a conical flask (200 mL). The initial concentration of 4-MBC aqueous was 10 mg/L and the volume of reaction solutions was 50 mL.The addition of H_2O_2 and $FeSO_4 \cdot 7H_2O$ were at beginning. The pH value, initial concentrations of H_2O_2 and $FeSO_4 \cdot 7H_2O$ were listed in Table 1. All the reaction solutions were well mixed and were vibrated by orbital shaker during the reaction. The whole Fenton process was working in darkness.

Table 1. Factors and levels of the Fenton's orthogonal experimental design.

Factors	A（H_2O_2 : Fe^{2+}*）	B（H_2O_2）	C（Time）	D（pH）
	5：1	0.1mL	15min	2
Levels	10：1	0.5mL	30min	3
	15：1	1.0mL	45min	4

* H_2O_2 : Fe^{2+} is the mole ratio.

2.3 UV light irradiation

In UV light irradiation process, the vital factors are reaction time, light source and the initial concentration of 4-MBC. For the feature of absorbing UV light, mercury lamp was employ as the light source. Same to the Fenton process, the concentration of 4-MBC was not thought about. Just reaction time was taken as the factor. Photochemical reactor was employed in UV light irradiation processes. Degradation was performed on 200 mL 4-MBC aqueous. During the reaction, about 3 mL samples were took out at 30 min, 45 min, 60 min, 75 min and 90 min respectively.

2.4 Photo-Fenton Process

The major factors affecting the photo-Fenton process are the initial concentration of the 4-MBC, the dosages of H_2O_2, pH value, amount of Fe^{2+}, reaction time and the light source [14]. According to Fenton and irradiation processes, we employed mercury lamp as the light source, fixed the initial concentration of 4-MBC and chose factors(Fe^{2+}, H_2O_2, pH and reaction time) as variables and evaluate their effect on degradation. Orthogonal experimental design was also employed to optimize the reaction conditions of photo-Fenton process. The factors and levels of the orthogonal experimental design were listed in Table 2. The addition of H_2O_2 and $FeSO_4 \cdot 7H_2O$ were at beginning. The whole photo-Fenton process was carried out in the photochemical reactor. The initial concentration of 4-MBC aqueous was 10 mg/L. Degradation of each group was performed on 200 mL 4-MBC aqueous.

Table 2. Factors and levels of the photo- Fenton's orthogonal experimental design.

Factors	A（H_2O_2 : Fe^{2+}*）	B（H_2O_2）	C（Time）	D（pH）
	5：1	0.1mL	30min	3
Levels	10：1	0.5mL	45min	5
	15：1	1.0mL	60min	7

* H_2O_2 : Fe^{2+} is the mole ratio.

2.5 Analytical procedures

The concentration of 4-MBC in solution was detected by HPLC (High Performance Liquid Chromatography). Before being injected into the LC system, the solution samples were passed through 0.45 μm Nylon, polyamide membranes filters. The column used in our experiments was Anasil AQ-C18 (25cm×4.6mm I.D. 5μm packing).The column temperature was maintained at 30℃. Water/acetonitrile (20/80%, v/v) was used for the isocratic elution of the analyte. Spectrum identification was performed at the wavelength of 300 nm [15].

3 Results

3.1 Fenton process

From Table 3, it could be saw that the highest degradation reached to 66.01%. It could be deduced from the range value (R) in table 3: RB>RA>RD>RC. It meant that the dosage of H_2O_2 had the greatest effect on the degradation of 4-MBC in Fenton process, then followed by the portion of H_2O_2 and Fe^{2+}, but the reaction time had the less effect. And the K value revealed that when the portion of H_2O_2 and Fe^{2+} was 1:5, the dosage of H_2O_2 was 1 mL, the reaction time was 30 min and the pH value was 3, the highest degradation could be reached. Fig. 1 illustrated the trend curves of different level vs different factor.

3.2 Irradiation process

When mercury lamp was used alone to degrading the 4-MBC, the degradation of 4-MBC at different time was tested (Fig.2). It revealed that as time growing, the concentration of 4-MBC was decreasing. After irradiating 90 min, the concentration decreased 85.4%.

Table 3. Orthogonal experimental design and results for Fenton reactions, C_0 (4-MBC) = 10 mg/L.

No.	**Factors**				**Degradation**
	A	**B**	**C**	**D**	**(%)**
1	1/5	0.1	15	2	32.14
2	1/5	0.5	30	3	65.25
3	1/5	1.0	45	4	64.51
4	1/10	0.1	30	4	34.70
5	1/10	0.5	45	2	40.52
6	1/10	1.0	15	3	66.01
7	1/15	0.1	45	3	18.01

8	1/15	0.5	15	4	29.37
9	1/15	1.0	30	2	46.04
K1	161.9	84.85	127.52	118.7	
K2	141.23	135.14	145.99	149.27	
K3	93.42	176.56	123.04	127.95	
R	68.48	91.97	22.95	30.57	

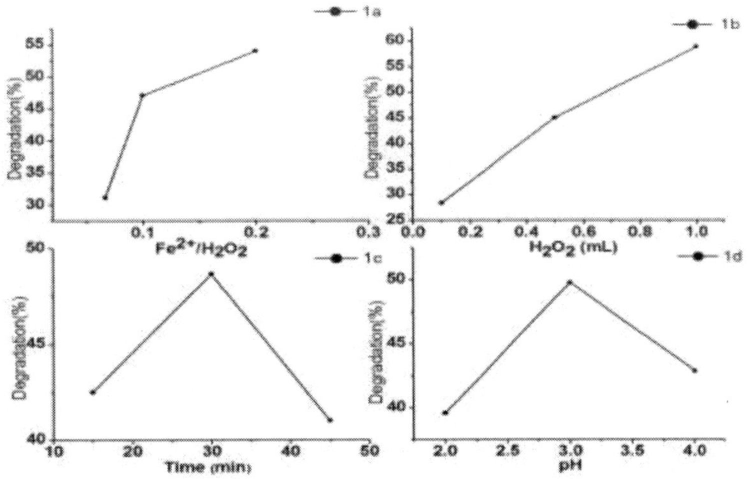

Fig.1. Trend curves of different level vs different factor of Fenton processes.

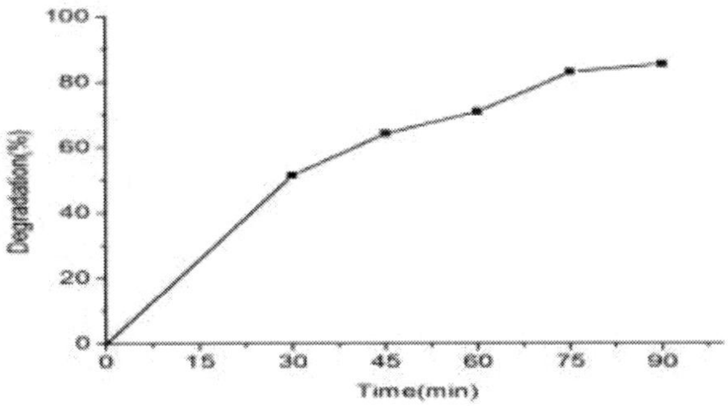

Fig.2. The effect of irradiation time on the degradation of 4-MBC. pH=7, C_0(4-MB) = 10mg/L.

3.3 Photo-Fenton process

The reaction condition and the degradation of each group were listed in Table 4. It could be deduced from the range value R in table 4: RB>RC>RD>RA. It meant that the dosage of H_2O_2 had the greatest effect on the degradation of 4-MBC in photo-Fenton process, then followed by the reaction time, but the portion of H_2O_2 and Fe^{2+} had the less effect. The degradation of 4-MBC varied between 76.77% and 96.71%. The K value revealed that when the portion of H_2O_2 and Fe^{2+} was 5: 1, the dosage of H_2O_2 was 0.1mL, the reaction time was 60 min and the pH value was 3, the highest degradation could be reached.

Table 4. Orthogonal experimental design and results for photo-Fenton reactions, C_0(4-MBC) = 10 mg/L.

No.	**Factors**				**Degradation**
	A	**B**	**C**	**D**	**(%)**
1	1/5	0.1	30	3	95.99
2	1/5	0.5	45	5	86.89
3	1/5	1.0	60	7	88.21
4	1/10	0.1	45	7	86.08
5	1/10	0.5	60	3	96.71
6	1/10	1.0	30	5	76.77
7	1/15	0.1	60	5	96.29
8	1/15	0.5	30	7	83.73
9	1/15	1.0	45	3	87.60
K1	271.09	278.36	256.49	280.3	
K2	259.56	267.33	260.57	259.95	
K3	267.62	252.58	281.21	258.02	
R	11.53	25.78	24.72	22.28	

4 Discussion

Figure 1a and Figure1b revealed that in Fenton process, along with the increase of the dosage of FeSO4 • 7H2O and H2O2, the degradation was increase. More FeSO4 • 7H2O and H2O2 were not employed, for the dilution effect caused by H2O2 and the economy issue. The degradation of 4-MBC varied between 18.01% and 66.01% in Fenton process. Even the highest degradation was lower than the photo-Fenton process's lowest degradation. It was known that the Fenton and photo-Fenton reactions are mostly based on the production of hydroxyl radical HO • in water. As it is classically described, hydroxyl radicals are produced in this technique by the decomposition of hydrogen peroxide when reacting with ferrous ions (reaction 1) [16]. Irradiation with sunlight, an artificial light source of wavelength 180-400nm, or even in the visible spectra [17] increases the rate of contaminant degradation mainly by stimulating the Fe3+ to Fe2+ reduction (reaction 2) [16]. Thus, the degradation of 4-MBC in photo-Fenton reaction is higher than Fenton reaction.

$$H_2O_2 + Fe^{2+} \rightarrow Fe^{3+} + HO^- + HO\cdot \quad (1)$$

$$Fe^{3+} + H_2O + h\upsilon \rightarrow Fe^{2+} + H^+ + HO\cdot \quad (2)$$

It was known that 4-MBC could exist as a (Z) or (E)-isomer due to the exocyclic carbon–carbon (styrene) double bond, and 4-MBC was shown to consist entirely of (E)- in commercial sunscreen products[18]. But when exposure to light, (E)-isomer is isomerized reversibly to (Z)-isomer [19]. Thus, although the degradation could reach to 85.41% after irradiating 90 min, we thought the light promoted the 4-MBC to translate to its isomeride, but did not decompose it.

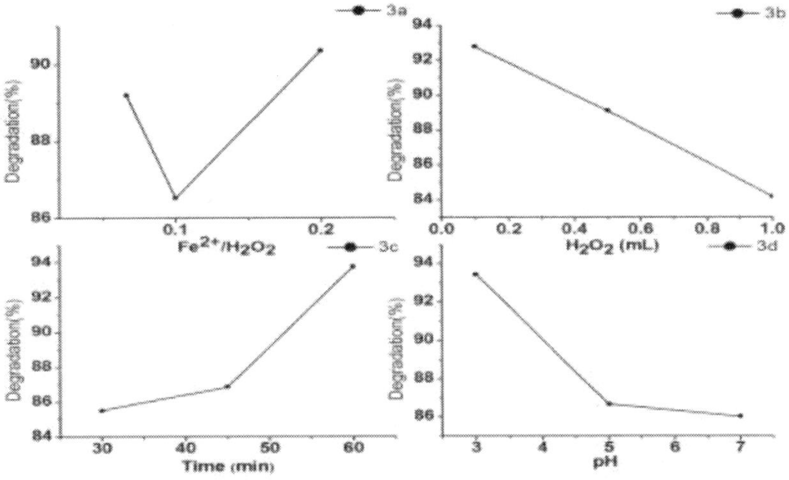

Fig.3. Trend curves of different level vs different factor of photo-Fenton processes.

Figure 3 illustrated the trend curves of different level vs different factors in photo-Fenton process. And Fig.3a revealed that the degradation rate of 4-MBC decreased firstly, and then increased. Comparing to Fig.1a, we speculated that the ferric ion weakened the irradiancy. The efficiency of Fenton reaction on degradation was better than the effect of irradiation when the concentration of ferric ion was high. Then the degradation decreased, because along with the decrease of the ferric ion, the efficiency of the Fenton reaction was also decreased and the irradiation process could not work well for the weakening of the irradiation by ferric ion. With the continuous decrease of the ferric iron, the degradation increased. We thought that irradiation make a significant contribution to the degradation in this stage. Fig.3b revealed that along with the increase of the H2O2, the degradation was decreased. It was because that the H2O2 also could consume the hydrogen radicals, which was achieved by reaction (3) [20]. Thus the degradation of 4-MBC would decrease when the amount of H2O2 increase. Fig.3c indicated that the degradation increased with the time going and Figure 3d indicated that the degradation decreased with the pH value increasing.

$$\cdot OH + H_2O_2 \rightarrow \cdot O_2^- + H_2O \tag{3}$$

5 Conclusion

When the initial concentration of 4-MBC was 10 mg/L, and the dosage of the Fenton regents (Fe^{2+},H_2O_2) were at the same level, degradation efficiency of 4-MBC in photo-Fenton process was better than Fenton process and the best degradation reached to 96.71% and 66.01% respectively. In UV light irradiation process, the concentration of 4-MBC decreased 85.4%. But most of them translated to the isomeride. Compared with the Fenton and photo-Fenton process, we deduced that Fenton reagents and light irradiation were synergistic.

Conflict of interest

The author confirms that this article content has no conflict of interest.

References

1. D.L. Giokas, A. Salvador, and A. Chisvert: UV filters: from sunscreens to human body and the environment. Trends in Analytical Chemistry. vol. 26(2007), p. 360–374.
2. F. Karl, Z. Armin, and R. Maja: Widespread occurrence of estrogenic UV-filters in aquatic ecosystems in Switzerland. Environmental Pollution. vol. 158(2010), p. 1817-1824.
3. M. S. Díaz-Cruz, and D. Barceló: Chemical analysis and ecotoxicological effects of organic UV-absorbing compounds in aquatic ecosystems. Trends in Analytical Chemistry. vol.28(2009), p. 708–717.

4. S. Durrer, K. Maerkel, M. Schlumpf, and W. Lichtensteiger: Estrogen target gene regulation and coactivator expression in rat uterus after developmental exposure to the ultraviolet filter 4-methylbenzylidene camphor. Endocrinology. vol. 146(2005), p. 2130-2139.
5. M. Schlumpf, S. Durrer, O. Faass, C. Ehnes, M. Fuetsch, C. Gaille, M. Henseler, L. Hofkamp, K. Maerkel, S. Reolon, B. Timms, J. A. F. Tresguerres, and W. Lichtensteiger: Developmental toxicity of UV filters and environmental exposure: a review. International Journal of Andrology. vol.31(2008), p. 144–151.
6. M. Badia-Fabregat, C. E. Rodríguez-Rodríguez, P. Gago- Ferrero, A. Olivares, B. Piña, M. S. Díaz-Cruz, T. Vicent, D. Barceló, and G. Caminal: Degradation of UV-filters in sewage sludge and 4-MBC in liquid medium by the ligninolytic fungus trametes versicolor. Journal of Environmental Management. vol. 104(2012), p. 114–120.
7. A.Tovar-Sánchez , D. Sánchez-Quiles, G.Basterretxea, J. L. Benedé, A. Chisvert, A. Salvador, I. Moreno-Garrido, and J. Blasco: Sunscreen products as emerging pollutants to coastal waters. PLOS ONE. vol. 8(2013), p. 1-8.
8. HR. Buser, M. E. Balmer, P. Schmid, and M. Kohler: Occurrence of UV-filters 4-methylbenzylidene camphor and octocrylene in fish from various Swiss rivers with inputs from wastewater treatment plants. Environmental Science & Technology. vol. 40(2006), p. 1427–1431.
9. M. S. Díaz-Cruz, P. Gago-Ferrero, M. Llorca, and D. Barceló: Analysis of UV filters in tap water and other clean waters in Spain. Analytical and Bioanalytical Chemistry. vol. 402(2012), p. 2325-2333.
10. M. Schlumpf, B. Cotton, M.Conscience, V. Haller, B. Steinmann, and W. Lichtensteiger: In vitro and in vivo estrogencity of UV screens. Environmental Health Perspectives. vol. 109(2001), p. 239-244.
11. C. SchmittM. Oetken, O. Dittberner, M. Wagner, and J. Oehlmann: Endocrine modulation and toxic effects of two commonly used UV screens on the aquatic invertebrates Potamopyrgus antipodarum and Lumbriculus variegates. Environmental Pollution. vol. 152(2008), p. 322-329.
12. A. Klann, G. Levy, I. Lutz, C. Müller, Werner Kloas, and JP. Hildebrandt: Estrogen-like effects of ultraviolet screen 3-(4-methylbenzylidene)-camphor (Eusolex 6300) on cell proliferation and gene induction in mammalian and amphibian cell. Environmental Research. vol.97(2005), p. 274–281.
13. M. E. Balmer, HR. Buser, M. D. Müller, and T. Poiger,"Occurrence of some organic UV-filters in wastewater, in surface waters, and in fish from Swiss lakes. Environmental Science & Technology. vol. 39(2005), p. 953–962.
14. A.S. Stasinakis: Use of selected advanced oxidation processes (AOPs) for wastewater tratement – a mini review. Global Nest. vol. 10(2008), p. 376-385.
15. V.A. Sakkas, P. Calza, M. Azharul Islam, C. Medana, C. Baiocchi, K. Panagiotou, and T. Albanis: TiO2/H2O2 mediated photocatalytic transformation of UV filter 4-methylbenzylidene camphor (4-MBC) in aqueous phase: statistical optimization and photoproduct analysis. Applied Catalysis B: Environmental, vol. 90(2009), p. 526–534.
16. C. Walling: Fenton's reagent revisited. Accounts of Chemical Research. vol. 8(1974), p. 125-131.
17. E. Oliveros, O. Legrini, M. Hohl, T. Müller, and A. M. Braun: Industrial waste water treatment: large scale development of a light enhanced Fenton reaction. Chemical Engineering and Processing. vol. 36(1997), p. 397–405.
18. HR. Buser, M.D. Müller, M. E. Balmer, T. Poiger, and I. J. Buerge: Stereoisomer composition of the chiral UV filter 4-methylbenzylidene camphor in environmental samples. Environmental Science & Technology. vol. 39(2005), p. 3013–3019.
19. R. Rodil, M. Moeder, R. Altenburger, and M. Schmitt-Jansen: Photostability and phytotoxicity of selected sunscreen agents and their degradation mixtures in water. Analytical and Bioanalytical Chemistry. vol. 395(2009), p. 1513-1524.
20. N. H. Ince, D. T. Gönenç: Treatability of a textile azo dye by UV/H2O2. Environmental Technology. vol. 18(1996), p. 179-185.

The Design and Implementation of the Remote Centralized-Monitoring System of Well-Control Equipment Based on RFID Technique

Bin Luo [1], Jiarui Liu [2], Junlan Li [3] and haibo Liang [1]

[1] School of Mechatronic Engineering, Southwest Petroleum University, Chengdu 610500, China
[2] Sichuan BOMCO Special Vehicle Manufacturing Co. Ltd, Guanghan 618300, China
[3] Southwest Branch, Engineering Design Co., CNPC, Chengdu 610413, China

Abstract. At present, in domestic for the management of well control equipment continue to the traditional way of nameplates identifies and paper-based registration, there are many issues like the separation of data information of device, easy lose, difficult query, confused management and many other problems, which will make the problem device into the well field, and then resulting in well control runaway drilling accident. To solve the above problems, this paper put forward to the integrated remote centralized-monitoring management mode of the well-control equipment. Taking the advantages of IOT technology, adopting the RFID technology, and combining with the remote transmission, this paper designs the remote centralized-monitoring system of well-control equipment based on RFID, which realizes the intelligent management of well-control equipment and meets the actual demand of the well-control equipment safe use and timely scheduling, and it has the ability of field application.

Keywords. Internet of Things; RFID, Well-control Equipment, Remote transmission, Remote Centralized-monitoring Management.

1 Introduction

Well control equipment is a safety measure of pressure control and safety drilling for oil and gas Wells. The well control equipment that is high quality, reliable, and meet the requirements of the target well, Not only can effectively implement well control, reduce drilling accidents, exploitation of oil and gas reached safety requirements, but also can ensure drilling quality and reduce non-work time.

However, since domestic well site geographically isolated and wide distributed, the use of well control equipment traditional nameplate, paper-based registration results in well control equipment trans-regional, intersect oral distribution, equipment and data separation, easy to lose and difficult inquiry. Even leads to well control equipment scheduling is difficult, and problem equipments easily flow into the well site, So that once the overflow, well kick occurs, you can not rely on well control equipment to achieve the purpose of preventing blowout, eventually leading to a vicious drilling accidents [1]. At present, as network technology has improved, network in the materials management, logistics, manufacturing and other fields has been widely used [2]. In other countries, a growing number of oil companies use network technology to achieve the construction of digital oilfield, intelligent field [3] - [6], and deal with the problems in the time of oilfield material using, perform modernization, such as centralized control and management of oil supplies. While the domestic oil industry, the researches and applications of network, mainly concentrate in the construction of wireless sensors and digital oil fields [7] - [9], management of oil supplies such as well control equipment, still remain in the traditional plate identification, paper registration management mode, unable to resolve problem of the well control equipment effective recognition and decentralized distribution fundamentally.

In this paper, well control equipment integrated remote centralized monitoring and management, the use of Networking technology, the use of RFID technology, combined with remote data transmission, Design and develop a well control equipment remote centralized control management system based on the network RFID. The system meets the actual demand of the safe use of well control equipment, real-time scheduling, changes the use of the traditional well control equipment nameplate, paper registration, and achieve that problem equipments can not flow into the well site, materials and equipment scheduling timely and well control equipment centralized monitoring and intelligent management. It is of great significance for drilling site providing reliable, reasonably well control equipments to ensure safe and high quality drilling, reduce the risk of accidents and increase economic efficiency.

2. Integrated remote centralized monitoring and well control equipment management

With the construction of the digital oil field, the networking technology in the oil industry has been widely used in development, but well control equipment still using the traditional nameplate, paper-based registration usage modes. Network could not be used for well control equipment remotely centralized control management to form a mature and perfect technology system, can not meet the actual demand on safe use of well control equipment and timely dispatch.

For the status of the management of well control equipment in oil industry, analyze the feasibility of networking well control equipment remote centralized control management from three aspects of the perception layer, transport layer, application layer, and the specific performance in the following areas:

(1) Use networking perception layer enables accurate identification of well control equipment, to avoid problems equipment into the well field.

(2) Use networking transport layer can provide effective interactive data transmission channels, to achieve timely and well control equipment scheduling.

(3) Use networking application layer enables centralized monitoring and well control equipment, digital and intelligent management.

Feasibility Analysis realized by the use of networking technology well control equipment remote centralized control management can be seen, through using RFID technology, combined with long-distance transmission can be realized from the well control equipment inspection and maintenance, inventory to a closed-loop drilling site centralized monitoring and management mode. In this integrated remote centralized monitoring and management mode, each well control equipment could be dated its source and ensure the reliability. Accordingly, to solve well control equipment identification, decentralized distribution, the device status is difficult to determine and so on, to prevent the problem device into the well site, timely dispatch equipment, centralized monitoring, remote centralized control management integrated intelligent management purposes, to provide protection for safe and efficient drilling.

3. The overall system design

According to well control equipment integrated remote centralized monitoring and management model, the well control equipment remote centralized control management system based on network consists of the perception layer (data acquisition system), the transport layer (Remote Data Transmission System) and application layer (well control equipment centralized control management system) constitute the overall design of the system shown in Figure 1.

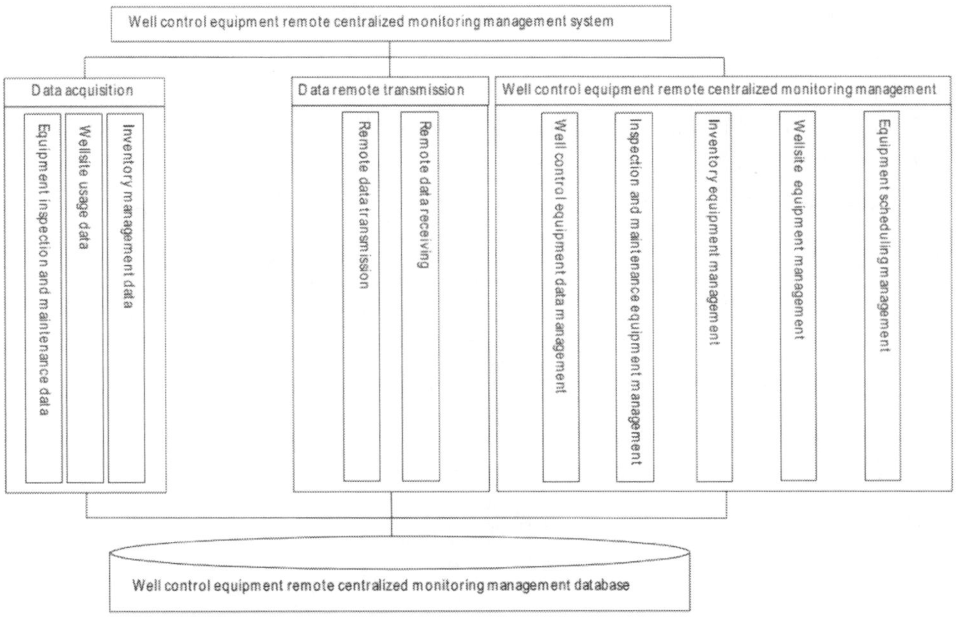

Fig.1 Overall design scheme

4. System features modular design

4.1 Perception layer

The main role of remote sensing layer of well control equipment centralized control management system is unique identification, automatic identification of well control equipment. Perception by reading and writing RFID tag data layer, it is possible to achieve the well control equipment to effectively identify and track, to achieve the purpose of providing high-quality, reliable and well control equipment to the well site. For questions of perception layer for high frequency (HF) and ultra high frequency (UHF) RFID data collection, transmission, storage and so on, through the serial communication, Windows Sockets network programming interface to achieve sensing layer data collection and transmission. The design block diagram of sensing layer in well control equipment remote centralized control management system is shown in Figure 2.

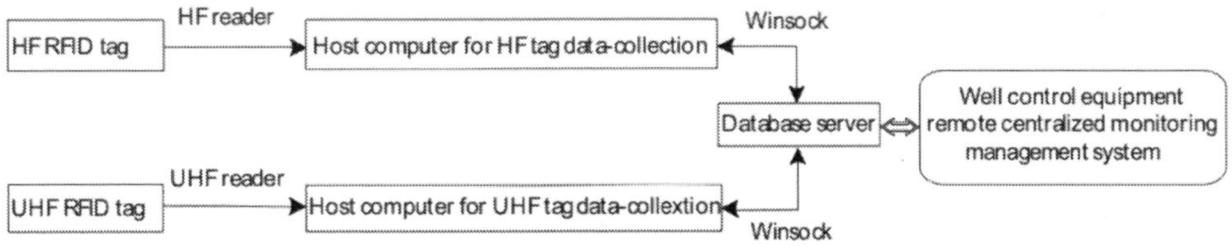

Fig.2 Block diagram of perception layer design

System perception layer use High Frequency (HF) and Ultra High Frequency (UHF) reader for the respective electronic tag to collect data or update storage information, Data collected or updated is sent to the management server to well control equipment by the host computer, and achieve the data and information of well control equipment management applications. Perception layer data acquisition and transmission program uses C # language, under the Visual Studio 2010 integrated development environment, it enables the development of data acquisition software systems and achieve data collection, and two-way communication between PC and the server.

4.2 Transport layer

The transport layer for well control equipment remote centralized control management system mainly communicates data between management center with well control equipments well site, In order to achieve the integration of centralized monitoring and management of well control equipment. According to the overall system design preferred remote communication, in use of CDMA / GPRS wireless communication network, design a suitable for well control equipment management center and well site to achieve interactive remote data transmission scheme, Transport Layer system design block diagram is shown in Figure 3.

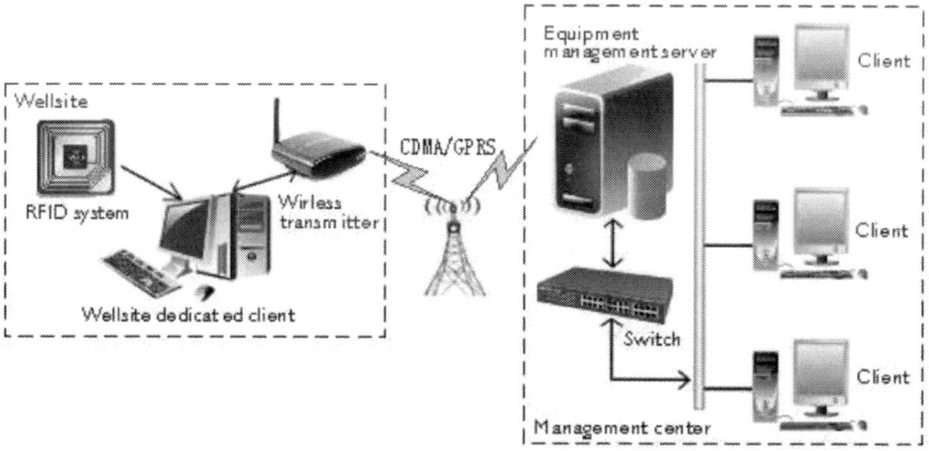

Fig.3 Block diagram of transport layer design

4.3 Application layer

Design of well control equipment remote centralized control management system application layer is based on a system of perception layer and transport layer. Application layer software system can combine data collected by layers sensing organic, to prevent the problem device flow into the well field, well control equipment are dispatched in time, integration of centralized monitoring, intelligent management. Application layer software system faces well control equipment management, between the various software functional subsystems, both independent, and can combine into one, to achieve the orderly management of well control equipment.

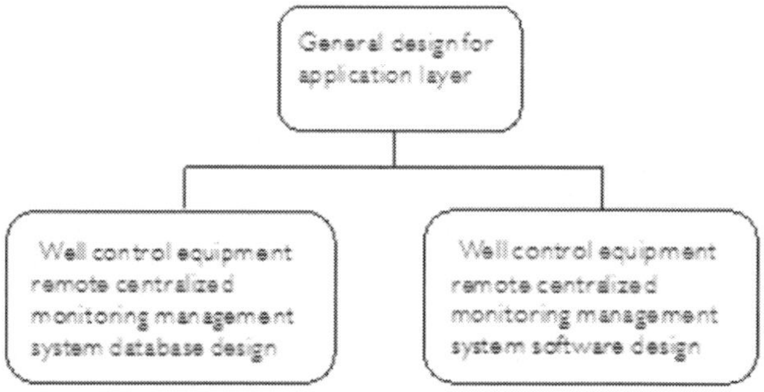

Fig.4 Block diagram of application layer design

5. The design of well control equipment remote centralized control management system software

Well control equipment remote centralized control management software is developed based on SQL Server 2008 database management platform and C # high-level programming language, mainly to complete, well control equipment basic information management, equipment inventory management, inspection and maintenance management, well site equipment management, system management, and other functions.

5.1 Basic information management subsystem

Well control equipment remote centralized control management software, basic information management subsystem, is used for setting well control equipment, well site, alarm reminders and other public categories data, in order to prevent the operator mistakenly fill. The subsystem consists of well control equipment basic data settings, setting the alarm to remind and other data settings of three functional modules, usually, it is operated by an administrator.

5.2 Inspection and maintenance management subsystem

Well control equipment inspection and maintenance management subsystem is used for checking for well control equipment, repairing and maintaining management. Well control equipment inspection management mainly finish safety inspection records and inquiries for well control equipment, housing and others; equipment maintenance management is used to restore faulty equipment functions management; well control equipment maintenance is maintaining equipments daily, such as lubrication, replacement of aging components, periodic inspections to ensure well control equipment status, and extend equipment life.

5.3 Equipment inventory management subsystem

Device inventory management subsystem is used to implement well control equipment in the library and out of storage management. Device inventory management files operate and manage all of inventory and equipment information, and it is able to view all equipment in the library, and print the related reports; in and out of storage management of device can operate well control equipment in and out of storage, Ordered to complete well control equipment inventory management, to ensure equipment inventory data is correct.

5.4 Well site equipment management subsystem

Well site equipment management subsystem has functions of well site equipment records management, query the device information, well site equipment statistical. Meanwhile, the subsystem has a well site equipment inspection management, the equipments transported to well site can be completed the inspection work as required, Avoid the problem equipments further be flew into the well field to be used; well site equipment management logs are in charge of the well field equipment inspection, everyday use, maintenance management by selecting notes category; well site equipment scheduling management is used to achieve the target device that wellsite remote desired, recover well site construction in time.

6 Conclusion

This article uses the advantage of network technology, and RFID technology, combined with remote data transmission, and puts forward to the integrated remote centralized-monitoring management mode of the well-control equipment. Design and develop a set of well control equipment remote centralized control management system collected of well control equipment data acquisition, remote transmission, centralized monitoring as one. After laboratory tests, the running result of system is stable, reliable, has a ability to field applications. Development and application of the system will change management of the traditional well control equipment nameplate paper-based registration, effectively prevent the problem device into the well field, and well control equipments meet the actual needs of safe use or real-time scheduling, and it can provide strong technical support for well control equipment digital intelligent management, security of well site drilling, reducing the risk of well control accidents, improving economic efficiency.

References

1. Wu J J. Monitoring and pre-warning method study on drilling engineering accidents. Mud Logging Engineering, 2006,3(1):53-55,70
2. Jiang Y B, You D Z, Ge Y J. Study on Popularization and Application of IOT in the Logistics Industry, Logistics Sci-Tech, 2012(10): 101-103.
3. van der Steen E, Knoppe R. BSP: An evolution from Smart Wells to Smart Fields. SPE100710, 2006.
4. L Ge, Pan Hu, Xiaohui Xie, Ze Hu, Qiang Zeng, Junbi Liao. Study on High Precision Swill-cooked Dirty Oil Detection System Based on Date Fusion Technology,Oxidation Communications,2016, book 1, 39,317-330
5. Shepard J S, Perkin G S, Denny LA. Electronic Identification of Drillstem and Other Components Used in Harsh Environments Proves Successful.SPE/IADC25774,Drilling Conference. Society of Petroleum Engineers, 1993.
6. L Ge, Ze Hu, Chen Ping et al. Research on Overflow Monitoring Mechanism Based on Downhole Microflow Detection，Mathematical Problems in Engineering，Volume 2014:1-6.
7. L Ge, Zhengyin Wang, Kui Deng, Qiang Zeng, Xi Wang, Xiaoshuang Chen, Junbi LIAO. Research on the Oil Life Estimation and Detection Method, Journal of the Balkan Tribological Association, 2015, book 4,21, 897-911.
8. L Ge, Qiang Zeng, Zhengyin Wang, Xiaohui Xie,Junbi Liao, Junlan Li. Research on Moisture Content Measuring Device of Edible Oil Based on Capacitance Detection Mechanism,Oxidation Communications,2016, book 1, 39,240-248
9. Wei Ming, Lu Yongjie WangHaiguo,et al.Practice of Internet of Things in Oilfield Enterprise. Journal of Chongqing University of Science and Technology(Social Sciences Edition),2013, 15(2): 154-158.

Under the Background of "New Normal", Research on Constructing the Productive and Technical Preparation System Based on Lean Thinking for Locomotives -Take M Company as an Example

Ming Ma*and Tao Lu

School of Economy and Management, Nanjing University of Science and Technology University, Nanjing, China
Corresponding Email: maming163@126.com

Abstract. This paper introduces the lean production theory and the application of lean production technology preparation system in related enterprises, put forward own views on a series of problems existing in the preparation of M company's production technology. Aiming at these problems, this paper combined with the practical application of lean thinking to the traditional production technology for system innovation and optimization, builds a clear M simulated production line concept, principles and framework, basically puts forward a complete set of methods and measures to construct the core technology of lean production system for part of the simulated production line.

1 Introduction

Under the background of "new normal", China's economy from high-speed growth to medium high speed, facing all kinds of users increasingly diversified, personalized, multi variety, small quantities of demand, production enterprise production technology preparation system flexibility is obviously insufficient, how to seize the opportunity of transformation and upgrading, become a new challenge in the face of various types of enterprises. M company is a locomotive manufacturing enterprise, in the long history of evolution and the process of production organization, has formed a set of relatively mature, fixed the new production of the locomotive production system. However, this set of system has been unable to fully adapt to the market demand for the production of new locomotives. Specifically, it mainly has the following four aspects of the problem: 1) production technology to prepare the elements is not comprehensive enough2) production technology preparation system flexibility is relatively short3) the preparation process of the production technology is poor4) the coordination of the functional departments in the preparation process is seriously insufficient.

In this paper, the use of Lean Thinking on the traditional production technology to prepare the system for innovation and optimization, clear M company based on lean thinking of the locomotive production technology preparation system construction strategy, methods and measures.

2 Literature review

The establishment of production technology preparation system based on lean thinking is the innovation practice of lean production mode, and its most important theoretical foundation lies in the research results of lean production. Lean production (Production Lean, referred to as LP) is an expert in the United States Massachusetts Institute of Technology Digital International Automotive Program Organization of Japan's TOYOTA on time production of JIT production mode of praise. James P. Womack and Daniel T. Jones increases lean production to a theoretical height [1].Art Smalley provides a good guide for the enterprise to use the balanced production method [2].Daniel T. Jones and James P. Womack promotes the lean supply to lean solutions, to achieve the value and wealth of the company and customers [3].M. E. Bayou and A.de Korin are based on previous studies put forward their own definition of lean production, the definition includes two dimensions of input and output in two aspects, namely the investment efficiency and effectiveness, tangible and intangible resources and cost, quality and service is the output of the final product. The research and practice of lean thinking in China is at the beginning of 1990s, Qi Ershi published a conference paper, gives the definition of lean production, and the grey theory is introduced to the analysis of the evaluation of the lean production area [4].Chen Lu in the "Journal of mechanical design and manufacturing engineering" published "the study on the production process optimization simulation of production line based on Simulation of rail vehicles, discusses the production line process design, process parameters and process route selection etc. [5].

3 Based on lean thinking of M company locomotive production technology preparation system composition strategy

3.1 The composition of locomotive production technology preparation system of M company based on Lean Thinking

The technology preparation system of locomotive production based on lean thinking consists of two parts, which are the simulation production line and the simulation distribution line, this paper only discusses the construction strategy of the simulation production line.

The simulation production line is the core component of the lean production technology preparation system, it refers to: Before the trial of the project node, recognition "three new"(New materials, new structures and new technologies),Through the "three off"(simulation, verification and coordination operations), to maximize the efficiency, economic and safety of the project trial production; To design and simulate the site work station, and to promote the preparation of the resources needed for the station before the start of construction; Through the simulation and verification of the trial production, reduce the abnormal production, shorten the trial cycle, so that mass production can enter the rhythm of production as soon as possible.

3.2 The construction strategy of simulation production line

The basic framework for the construction of the analog production line can be summarized as follows "Six stages, six contents, six elements".

The six stage refer to: Simulation planning, simulation preparation, simulation start, simulation deduction, simulation evaluation, simulation verification

The six contents refer to: The main contents of the construction of the analog production line, that is stage definition, division of labor, work content, input and output, operating points, the use of tools;

Six elements refer to: Six management factors of construction simulation production line, that is human, machine, material, method, ring, measurement

The division of responsibilities of the construction of each simulation production line is no longer detailed list.

3.2.1 Simulation planning

Simulation planning is the first stage of the whole production line construction project is simulated, in order to realize the simulation of production line construction and efficient operation of the target, the application of lean thinking and methods, process design, design and production of simulated production line implementation of the project. In the simulation planning stage, to do a good job in several aspects of the work: ①Fully understand customer needs ②The recognition of "three new" points③ Preparation of simulation plans

3.2.2 Simulation preparation

Simulation is prepared by the further decomposition and refinement of the implementation plan of the simulated production line, from the production, technology, management and other aspects, the simulation of the simulation of the production line to the specific point of the specific arrangements for the process, it is the most substantial working stage of the whole simulation production line. ①Clear six elements of the preparation requirements in order to stimulate the implementation of the preparation of simulation.② Define the task of each module.③ input and output of the simulation preparation stage are shown in table 1.

Table 1. The input and output of the simulation preparation stage.

Department name	input	output	Receiving unit
Process department	Simulation program	Analog line work network plan	Various departments of the project team
		Network plan map	
		Unit module task	
Production planning and management department	project plan	Pre-production plan	Various departments of the project team

3.2.3 Simulation start

The simulation starts according to the working network plan and the pre-production plan of the simulation production line, and the organization of each department is discussed and analyzed. In essence the simulation starts only as a landmark node of the whole simulation line. The input and output of the simulation start-up stage are shown in table 2.

Table 2. Simulation start stage of the input, output.

Department name	input	output	Receiving unit
Process department	Analog line work network plan; Unit module task	Analog start meeting minutes	Various departments of the project team

3.2.4 Simulation deduction

Simulation deduction is the simulation unit in accordance with the plan and responsible for the implementation of the scheme and the contents of six elements for each station on the spot inspection, through the passage of exercise, combing management ideas, found that the problem, the bottleneck and the implementation of the difficult process. Simulation deduction is a process of continuous implementation, verification, discovery and improvement, which is in line with the concept of the famous "Deming" management cycle in lean production, through the "PDCA" management cycle, so that the whole process of management level has been continuously improved. "PDCA" management cycle is shown in Figure 1.

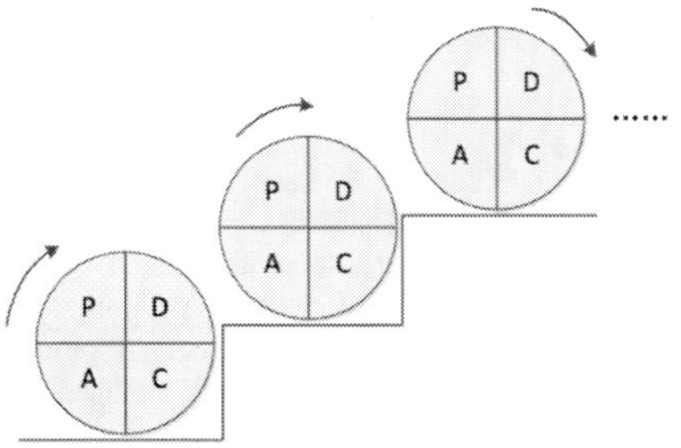

Fig. 1. "PDCA" management cycle diagram.

①Division of labor, that is, the Department of technology, manpower, equipment management, quality management, production planning and management department, procurement department, the Department of the ring, the logistics sector, manufacturing units and so on. ②The contents of the work, the planning and implementation of node, process control, project inspection and issue report etc.. ③Input and output content due to space constraints, is no longer listed.

3.2.5 Simulation evaluation

Simulation evaluation is to summarize the process of simulation, the risk assessment, analysis and rectification of all kinds of problems, and make the process of preparing for the simulation. In the simulation evaluation stage, the need to do a good job in two areas of focus.① Hold a meeting② Ready to start the "six elements"③ Input and output content due to space constraints, is no longer listed.

3.2.6 Simulation verification

Simulation verification is the base station file manufacturing unit formed in the pre-construction line simulation and evaluation report, for trial according to production plan, and the problem will produce abnormal production of timely feedback to the relevant departments, and in the process of rectification after curing into standard process documents.① Work content formulation② The input and output of the simulation verification stage are shown in Table 3.

Table 3. Simulation verification stage of the input and output.

Department name	input	output	Receiving unit
Production planning and management department	Project contract, Simulation program, Single start check	Trial production plan	Manufacturing unit

DOI: 10.1051/matecconf/201710003008

Manufacturing unit	Station file, Trial production plan	Field production line, Verification feedback table	
Process department	Verification feedback table, Problem improvement program	Validation report for analog line, Standard process file	Various departments of the project team
Design Department	Field production line, Before the new verification points	Problem improvement program, After verification three points	
Other departments	Station file, Verification feedback table	Problem to improve the program, the six elements of the standard to verify the standard document	

4 Conclusions

In this paper, a strong representative of the Q company as a case study, find that the existing production technology for system elements is not comprehensive, lack of flexibility and controllability during the functional departments of the poor, a serious shortage of coordination. On this basis, according to the relevant theory of enterprise production management and lean production, closely combined with the actual production and operation of Q company, puts forward the construction of Q company lean production technology of locomotive product preparation system based on strategy. Although, lean production technology of locomotive product preparation system although there are strong guidance on, but I have some limitations in the construction technology of lean production system for the theory of knowledge, there are a lot of work and details need to be sustained, in-depth study. Sincerely hope that there are more professional scholars and production management personnel in this field for more in-depth research, discussion and practice.

Acknowledgment

Nanjing University of Science and Technology, the Central University of basic scientific research business expenses special funds "Research on the main body position of the construction of scientific and technological innovation talents in Enterprises" (No. 30920140122009) ; Nanjing University of Science and Technology, the Central University of basic scientific research business expenses special funds "Research on efficiency evaluation of transportation industry based on network DEA" (No.30920130132009) ; Soft science research project of Jiangsu Province "Research on the reform of policy coordination management of Jiangsu science and technology innovation and Entrepreneurship Talent Policy" (BR2014099) .

References

1. Womack, J.P. & Jones, D.T. From lean production to the lean enterprise. *Harvard Business Review* 24, 38-46 (1996).
2. Smalley, A. Creating Level Pull. *Lean Enterprise Institute* (2004).
3. Jones, D.T. & Womack, J.P. Lean Consumption. *Harvard Business Review* 83, 58-68 (2005).
4. Tian Qing, Zhou Gang & Liu Haihong& Er Shi Qi. Research on logistics system of TOYOTA production mode. in China mechanical engineering society material handling branch annual meeting 20-23 (2000).
5. Chen Lu & Chu Weizhuo. Study on the optimization of the production process of the rail vehicle based on the simulated production line. Mechanical design and manufacturing engineering,61-66 (2015).

The Rigid-flexible Coupling Dynamics Analysis of the Powder Filling Quantity Control Mechanism Based on the ADAMS

Guihe Wang[1], Liu Ying[1], Na Cui[2], Yiqi Wang[3,*] and Yijing Zhang[1]

[1] Mechanical Engineering & Automation Department, Eastern Liaoning University, Dandong, China
[2] Faculty of Engineering Department, Tieling Normal College, Tieling, China
[3] International Department, Shenyang University, Shenyang, China
Corresponding Email: yiqiwang001@163.com

Abstract. Powder filling quantity control mechanism is an important part of medicine machine. The geometric model of the filling quantity control mechanism was established by Pro/E based on the NJP1200 capsule filling machine. By Lagrange multiplier method, the equation of movement characteristics was established. And also, the rigid-flexible coupling model of the mechanism was established by ADAMS, Based which, the simulation of displacement, velocity and acceleration curve were obtained. In the end, the change rule of CAM contact force was analyzed. It showed that ADAMS is an effective method in simulation the dynamic characteristics of mechanism.

1 Introduction

Automatic capsule filling machine is a key capsule preparation filling equipment. There are more than 7500 pharmaceutical enterprises in our country and 99% of the capsule filling machine belongs to the plunger type filling machine. Its advantage is easy to operate, safe and good sealing, but the powder filling quantity accuracy is greatly affected by the mechanical structure, powder liquidity and filling environment. At present, the drug companies are mostly using NJP1200 type, NJP2000 type and NJP3200 type capsule filling machine in our country. More advanced foreign capsule filling machine are represented G120N of Italy MG2, ZANAS125/40F of IMA company and GKF2000 of Germany BOSCH [1].

Capsule filling machine moves complex, including continuous movement, intermittent movement, linear movement, rotational movement. So the movement precision is high. In recent years, the related technology of the capsule filling machine design and manufacture has been mastered through the introduction, digestion and absorption from abroad in our country, but the powder filling quantity control technology remains to be further research and development. Dynamics research is doing based on ADAMS of the powder filling quantity adjusting mechanism to optimize the the structure and parameter of the related parts of filling quantity control, improving the design level of the fully automatic capsule filling machine in our country [2,3].

2 The working principle of capsule filling machine

Capsule filling machine is mainly composed of transmission mechanism, capsule delivery mechanism, capsule separation mechanism, lock capsule mechanism, module rotary and powder filling quantity control mechanism. Transmission mechanism completed the corresponding movement by gear motor through the chain corresponding CAM mechanism, as shown in Figure 1.

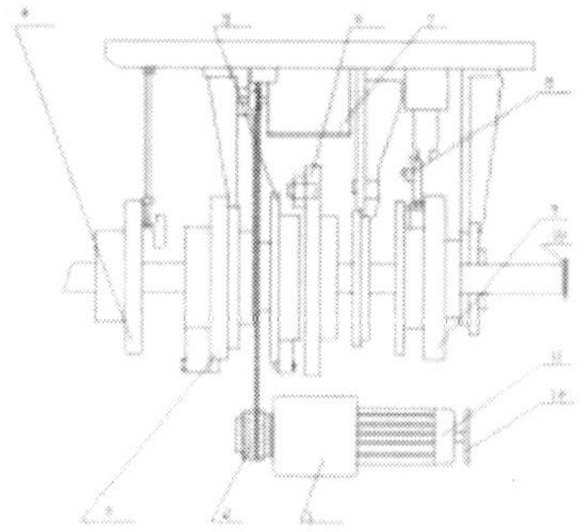

1-Reducer, 2- Safety clutch, 3,4,5,6,8,9- CAM, 7- Dividing box, 10- Dial, 11- Motor, 12- Motor handwheel

Fig. 1. Driving sketch of capsule filling machine.

The main shaft in transmission mechanism is installed six CAM mechanisms by holding linkage: CAM 3 completes lock capsule function; CAM 4 completes eliminating waste capsule function; CAM 5 completes combining capsule function; CAM 6 completes powder filling quantity control and filling function; CAM 8 completes separating capsule function and CAM 9 completes delivery capsule function. Intermittent mechanism includes ten locations and six location controlling the movement of the module and the powder filling quantity mechanism respectively. Through intermittent mechanism the module rotary movement and filling movement coordinate in order to complete capsule delivery, capsule separation, capsule filling, capsule elimination, capsule combination, capsule output and other functions in different location orderly to ensure that each location runs accurately and smoothly [4].

The powder filling movement is an important part of the movement of the capsule filling machine and can affect the powder filling quantity accuracy. Powder filling quantity control is completed by the filling CAM, linkage and so on, its working principle is shown in Figure 2.

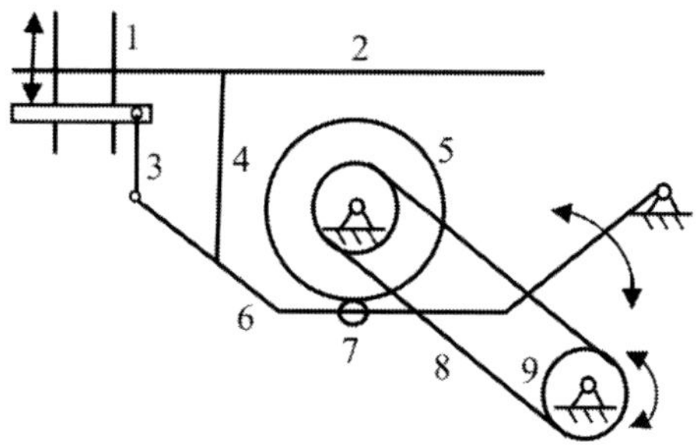

1-Push rod, 2Big board, 3-Buide rod, 4-Spring, 5-Filling CAM, 6-Connecting rod, 7-Roller, 8-Roller chain, 9-Sprocket

Fig. 2. Driving sketch of powder filling structure.

Gear motor drives the sprocket (9) and the rotation will be passed to the filling CAM (5) by the chain (8). The swinging movement of the connecting rod (6) is completed by the roller (7) on the connecting rod with the rotation of the filling CAM. Guide rod (3) and connecting rod (6) are linked together through joint bearing to make the guide rod (1) completing linear movement, namely filling movement.

3 The rigid-flexible coupling model of the filling quantity control mechanism

3.1 The motion equations

The dynamics model of powder filling quantity control mechanism was set up based on the theory of Lagrange equation [5], namely

$$q_i = [x, y, z, \psi, \theta, \varphi]_i^T$$

$$q = [q_1^T, \cdots, q_n^T]^T \tag{1}$$

The movement equation of the powder filling quantity mechanism was gotten combined with the generalized coordinates expression by Lagrange multiplier method, namely:

$$\frac{d}{dt} = \left[\frac{\partial T}{\partial \dot{q}}\right]^T - \left[\frac{\partial T}{\partial q}\right]^T + f_q^T + g_{\dot{q}}^T = Q \tag{2}$$

When the filling quantity mechanism is fully restrained: $f(q,t) = 0$; When the filling quantity mechanism is unfully restrained: $g(q,\dot{q},t) = 0$. Among them:

T—kinetic energy of the filling quantity mechanism;

Q—generalized coordinates array of the filling quantity mechanism;

q—the Laplace multiplier array corresponding with the fully restrained;

M—the Laplace multiplier array corresponding with the unfully restrained;

$\dot{q}$—generalized velocity array of the filling quantity mechanism.

3.2 The ADAMS model of the powder filling quantity control mechanism

Powder filling quantity control mechanism is composed of dose plate, filling rod, coppering, etc. Powder is compacted layer-by-layer and filled into the capsule body by 6 groups filling bar according to the demand of powder filling quantity. Spindle drives filling CAM rotation and CAM drives column through the linkage doing up and down reciprocating motion to realize the powder filling. The depth of the filling rod inset the dose plate can be adjusted by the knob to adjust to the volume of powder filling. Filling mechanism model was established based on Pro/E and imported into the ADAMS, as shown in Figure 3.

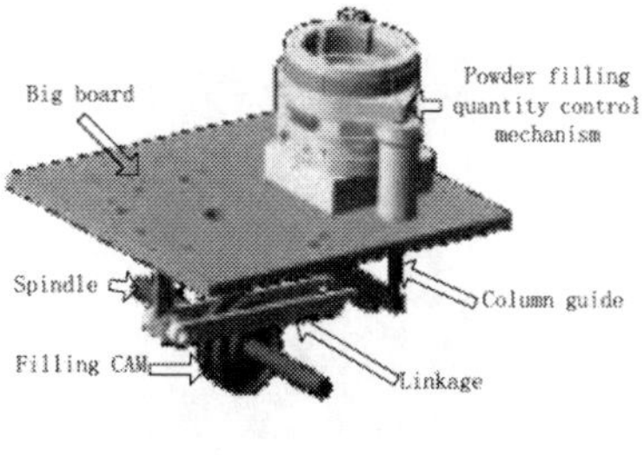

Fig. 3. Model of filling machine of ADAMS.

In order to reduce the constraints imposed in ADAMS, when rigid body is set in MECH /pro, the synchronous movement parts are set to a rigid body. Big board is set to be fixed joint; Connecting rod and big board are set to be revolute joint; Roller and CAM are imposed collision constraints, Roller and connecting rod are connected through the revolute joint.

3.3The rigid-flexible coupling model

Because there is no such thing as pure rigid body, the commonly used method is to simplify the flexible body for the rigid body in Mechanism Simulation for convenience. But the gap of the simulation results and the actual results is larger [6-8]. Because of the flexible body simulation computation time is long, therefore, in the simulation only the important components in the system will be treated as flexibility and other parts as the rigid body, namely the rigid-flexible coupled simulation analysis, the process as shown in Figure 4.

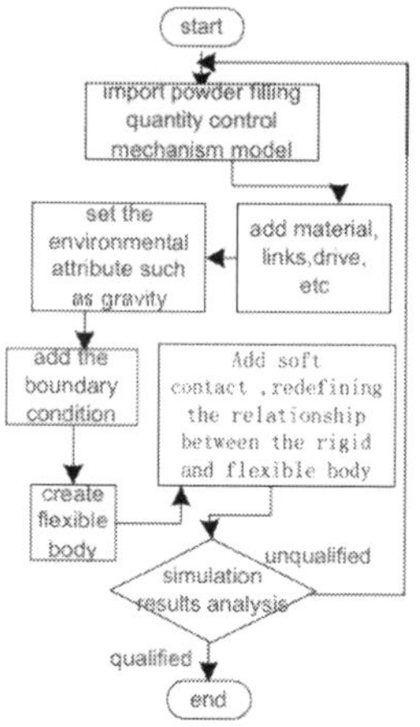

Fig. 4. Analysis chart of rigid-flexible coupling model in ADAMS.

The necessary simplification was done in the process of establishing the rigid-flexible coupling model of the powder filling quantity control mechanism in order to reduce the simulation computation. Two 30N forces were added at the top of the CAM rod to instead of the synthesis reaction of column guide on the CAM and corresponding environment variables, material properties, links and drive, etc. were set up. The rigid filling CAM was translated into flexible body using Adams/flex module and according to the actual working conditions of the CAM, the rigid-flexible coupling body was set up as shown in Figure 5.

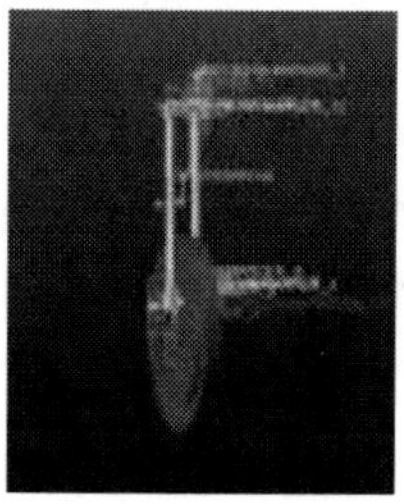

Fig. 5. Structuring of rigid-flexible coupling model.

4 The analysis of simulation results

Based on the established rigid-flexible coupling model and using the Adams Simulation function, kinematics and dynamics simulation was done by Adams/Postprocessor module. The movement simulation was done by setting the CAM speed at 90 ^{o} /s, end time of the simulation at 5 s and steps to 100. The displacement, velocity and acceleration curve along with the time change for the powder filling quantity control mechanism vertical movement are shown in Figure 6 (a), (b), (c) f change over time.

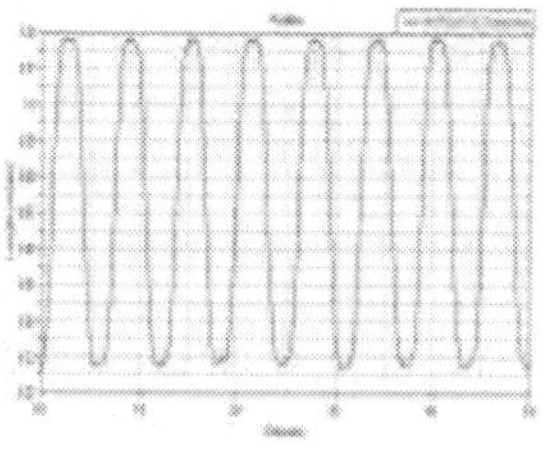

（a）displacement curve

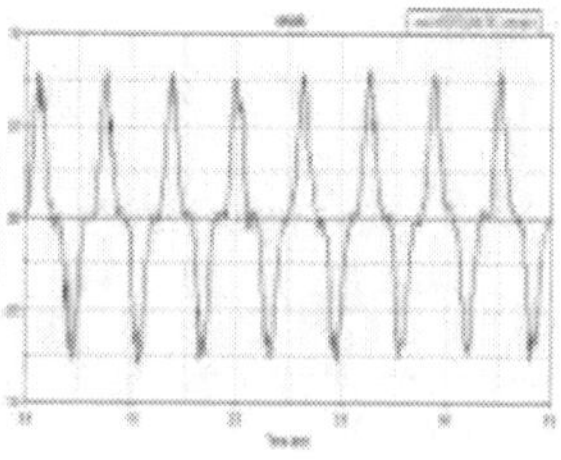

(b) speed curve

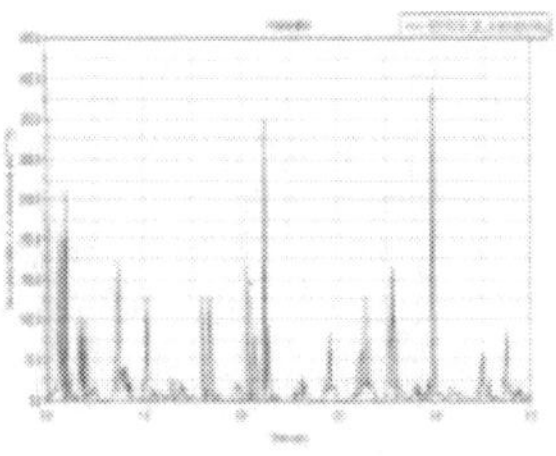

(c) acceleration curve

Fig. 6. Curve of follower.

From the filling quantity mechanism motion curve shown in Figure 6 (a), (b), (c), Comparison data between simulation and design data is shown in table 1.

Table1. Comparison data between simulation and design data.

project	simulation data	design data
displacement (m)	0.09	0.09
speed(m/s)	-0.7～0.79	-0.8～0.8
acceleration (m/s^2)	0～50	0～50

The simulation displacement is 0.09 m in agreement with the theoretical design data from table 1; Simulation speed is in the range of 0.7 ~ 0.79 m/s basic consistent with theoretical design parameters. Curve has pause phenomenon in zero and In this position the load regulation mechanism motion speed close to zero consistent with intermittent mechanism control; Simulation acceleration value is 0 ~ 50 m/s2 consistent with the theoretical design value, but the simulation curve has fluctuations related to the unsmooth CAM contour line surface. ADAMS simulation kinematics performance curve of powder filling quantity control mechanism and practical application motion characteristics of powder filling quantity control mechanism are basic identical with each other.

The surface contact force curve of filling CAM changing with time is shpwn in Figure 7 (a), (b), (c).

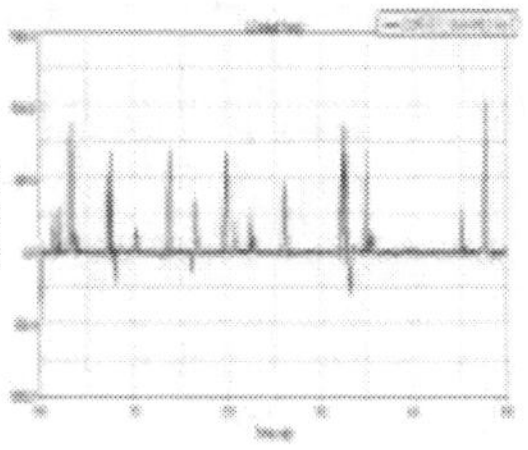

(a) X direction force curve

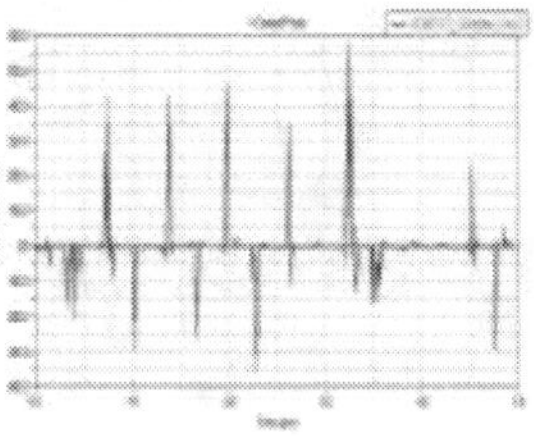

(b) Y direction force curve

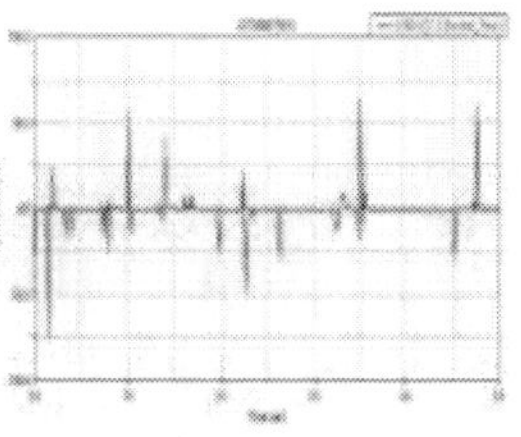

(c) Z direction force curve

Fig.7. Contact force curve of cam surface.

The maximum contact force in the X direction is 1006.8N appeared in 4.7s moment and minimum value is 501.07N appeared in the 0s moment from Figure 7 (a), (b), (c). The curve is instantaneous fluctuation and the average force is about 0N; The maximum contact force in the Y direction is 589.31 N appeared in 3.5s moment and minimum value is -354.61N appeared in the 2.3s moment .The curve is instantaneous fluctuation and the average force is about 0N; The maximum contact force in the Z direction is 65.68N appeared in 3.5s moment and minimum value is -749.65N appeared in the 0.1s moment .The curve is instantaneous fluctuation and the average force is about 0N; It is can be seen from the graph that although the contact force change is very frequent in simulation environment, the maximum forces value are all in a very small time scale and they are in line with the requirements of the CAM contact force by finding some information.

5 Summary

Capsule filling machine is a typical high speed, light load and complex movement equipment. It is a difficult point to research on powder filling quantity control. The rigid-flexible coupling model of the powder filling quantity control mechanism for NJP1200 type capsule filling machine was established based on Pro/E, ADAMS and the simulation analysis was also carried on. (1) The displacement, velocity and acceleration curve were obtained compared with the actual parameters. As a result, the simulation curve corresponds with the actual mechanism motion characteristics. (2) The contact force change regular was obtained based on the CAM mechanism contact force curves of the different direction. It is shown that the rigid-flexible coupling model based on ADAMS is an effective method of mechanism dynamic properties analysis.

Acknowledgement

This paper is supported by Excellent Talents Supporting Plan in Liaoning province(the first level, LR2013062); Public Welfare Projects in Liaoning province(2015004017); Industrial Research Project of Dandong city Technology Bureau (14207) and Eastern Liaoning University Scientific Research Fund Project(2015GG006).

References

1. H.Zh .Tang, Zh.Yao, Comparison Between Domestic and imported capsule filling machine,Mechanical and Electrical Information.l,137(2010).
2. G.H.Wang, Y.G. Zhang,T. B.Yu and W.Sh. Wang, Dynamics Simulation of Capsule Filling Machine Based Virtual Prototype, Applied Mechanics and Materials. 127(2012).
3. G.H. Wang, F.Xiao and W.Sh.Wang, Research on Capsule Filling Machine Based on Virtual Prototype, Advanced Materials Research. 411(2012).
4. G.H.Wang, Dynamic Simulation Rresearch of the Fully Automatic Capsule Filling Machine Based on ADAMS, Mechanical Design and Manufacturing. 6(2012).
5. W.H. Kim,T.W. Park, Study of optimization of the barrel cam in a paper-cup-forming machine, Journal of Mechanical Science and Technology.(2012).
6. Zh.Y. Wan, P.X. Li, X.H. Zhuang. Elastic Dynamics Analysis of CAM Mechanism Based on ADAMS, Journal of Dalian Jiaotong University. 1(2010).
7. D.Zhang, J.Zhang, W. Zhang, Research on Cylindrical CAM modeling method Based on Pro/E Relation, Mechanical Design and Manufacturing. 5(2012).
8. Q. Tang, CAM Mechanism Analysis and Dynamic Simulation Based on ADAMS, Journal of Anhui University of Science and Technology (natural science edition).4(2010).

Detection for Power line Inspection

Bing Han and Xiaoyu Wang

Hei Longjiang Electric Power Research Institute, Harbin, Hei Longjiang, China
Corresponding E-mail: han_bing@vip.sina.com

Abstract. Power line inspection is very important for electric company to keep good maintenance of power line infrastructure and ensure reliable electric power distribution. Research efforts focus on automating the inspection process by looking for strategies to satisfy all kinds of requirements. Following this direction, this paper proposes a learning approach for all kinds of detecting problems, where aggregate channel features are used to train the boost classifier. Adopting the sliding window paradigm, the electric tower, insulator and nest can be located very fast. The main advantage of this approach is its efficiency and accuracy for processing huge quantity of image data. Obtaining highly encouraging results shows that it is really a promising technique.

Keywords. Power line, inspection, aggregate channel features.

1 Introduction

Electric power companies invest significantly for power line inspection to ensure reliable electric power distribution. Now the common strategy is aerial inspection by manned helicopter equipped with multiple sensors such as visual, infrared and ultra-violet cameras etc. Data captured or recorded by expert crew with those above cameras are examined later manually to detect potential faults and damage on different power line components. This process is not only extremely time consuming, but also very expensive and prone to human error. With these problems in mind, power industry is actively seeking solutions to automate different aspects of power line inspection.

After data acquisition, the main task of power line inspection is fault identification where computer vision can contribute to help. It involves automatically detection and localization of electric devices such as wires, towers, insulators, conductors etc. The state of art has focused on tower detection. And the detection of tower can then be used to find various defects/faults of power line infrastructure.

This paper presents a learning based solution for all kinds of detection and localization. Including tower, insulator and nest on the tower. Therefore the background or other parts of image can be cropped quickly. Focus of Attention is concentrated on image region with what people care about. This process can also be used for images filtration. When millions of images were obtained by Unmanned Aerial Vehicles, Maybe lots of them don't include any power line devices. Those images which have no use for power line inspection should be deleted. Running this detection process can be used as preprocessing stage which can delete those images without ROI in them automatically.

Some researchers have also focused on detection and segmentation of electric towers in the images [1-5]. Several authors apply straight line segment extraction for tower detection or localization [1, 2, 4, 5]. Other authors then apply different segmentation approaches to extract the complete tower from the image: e.g. a template matching approach is used in [1]; graph-cut [6] based segmentation is used in [4]; a rule-based, as well as watershed segmentation [7] is used in [2]. Golightly and Jones [3] presented a different approach from the state of art where, instead of lines, the corners were considered the key identifying features of tower. They used a modified corner detector [8] to detect and track the tower tops. Although different approaches to tower detection and segmentation have reported promising results, Most of the results have been reported on just one type of tower. However, the electric towers are extremely diverse in shape, appearance and size. Therefore, most the state of the art results cannot be generalized to several different tower types.

To achieve the goal of complete autonomy, researchers must aim towards developing more general approaches. This paper following this direction considers tower detection as a learning problem. Sliding window paradigm is adopted for tower detection with boosted detector which is based on boosted decision trees computed over multiple feature channels such as color, gradient magnitude and histogram of gradient oration. Our solution for tower detection automatically can get top performing with very fast speed no matter the image size is large or small since multi-scale pyramid strategy is considered and implemented well. With the same method of tower detection, other power line devices can also be detected such as insulator. The nest which is one defect of power line can be detected perfectly too.

The rest of the paper is organized as follows: Section 2 states the problem addressed in this paper and describes several challenges which need to be addressed; Section 3 presents our approach to detection for power line inspection; The results are reported and discussed in Section 4; The final section is our conclusion and future research directions.

2 Problem statement

Currently, different projects are looking for automating either the acquisition process or the analysis process, or both, with the main objective of being able to detect and diagnose different defects of the power line infrastructure by using new sensors or by using new inspection platforms (e.g. robots; UAVs). In all these new possible approaches, computer vision plays an important role for automatically moving the camera in order to maintain the electric tower inside the filed of view of the camera, and for identifying and categorizing the different defects and failures of the power line infrastructure.

Nonetheless, computer vision is in fact a very challenging task for this technique. There are all kinds of complicated situation the visual system has to deal with such as viewpoint variation, illumination and background change. Because of the high variability of background, it is difficult to find a unique feature that can work in all the possible scenarios. Illumination changes also play an important role. That directly leads to power line segmentation from background algorithm not working for some low contrast images. Other problems such as constant viewpoint changes (e.g. especially when cameras are manually moved) and scale changes of the electric tower and its components add additional complexity to the idea of applying computer vision to solve this problem, in which, depending on the adopted strategy, could require a system that automatically defines which is the best frame to be used for detecting defects.

Currently, there is not a complete solution that satisfies the different requirements of automated power line inspection: simultaneously detect electric towers, check for defects, and also analyze security distances. Therefore, in terms of cost-benefits, it is important for energy companies to solve this problem and try to find a system that can deal with the different requirements of the inspections at high speed. In this paper we explore the electric tower detection problem applying a machine learning approach, using low quality images. Therefore, the system will help in reducing the maintenance cost of the electric system by being able to cope with one of the problems of increasing the vehicle speed (reducing the quality of the images).

3 Tower detection strategy

The objective of the proposed strategy is to determine the position of the ROI (Region Of Interest) in single images. Due to the difficulty of the task (e.g. wide variety of backgrounds), a learning-based approach is used. A boosted detector based on boosted decision trees is trained for tower localization and segmentation from background with very fast speed. Aggregate channel features are used to train boosted decision trees. Once the boosted decision trees have been trained, they can be applied to tower detection of power line inspection. In the following paragraphs the boosted decision trees and aggregate channel features are described.

Boosting is a simple yet powerful tool for classification and can model complex non-linear functions [9, 10]. The general idea is to train and combine a number of weak learners into a more powerful strong classifier. Decision trees are frequently used as the weak learner in conjunction with boosting, and in particular orthogonal decision trees, that is trees in which every split is a threshold on a single feature, are especially popular due to their speed and simplicity [11-13]. Decision trees with oblique splits can more effectively model data with correlated features as the topology of the resulting classifier can better match the natural topology of the data [14].

For training and detecting, the channel features are used. Given an input image, Aggregate Channel Features (ACF) computes several feature channels, where each channel is a per-pixel feature map such that output pixels are computed from corresponding patches of input pixels (thus preserving image layout). We use the same channels as [12]: normalized gradient magnitude (1 channel), histogram of oriented gradients (6 channels), and LUV color channels (3 channels), for a total of 10 channels. We down sample the channels by 2x and features are single pixel lookups in the aggregated channels. Thus, given a h×w detection window, there are h/2×w/2×10 candidate features (channel pixel lookups). We use RealBoost [10] with multiple rounds of bootstrapping to train and combine 2048 depth-3 decision trees over these features to distinguish object from background. Soft-cascades [14] and an efficient multiscale sliding-window approach are employed. Our implementation uses slightly altered parameters from [12] (RealBoost, deeper trees, and less down sampling); this increases model capacity and benefits our final approach.

The boosting algorithm for learning can be described as bellow:

- Given example images(x1,y1),...,(xn,yn) where yi=0,1 for negative and positive examples respectively.
- Initialize weights $w_{1,i} = \frac{1}{2m}, \frac{1}{2l}$ for $y_i = 0,1$ respectively, where m and l are the number of negatives and positives respectively.
- For $t = 1,\ldots,T$:

1. Normalize the weights, $w_{t,i} \leftarrow \frac{w_{t,i}}{\sum_{j=1}^{n} w_{t,j}}$
2. Select the best weak classifier with respect to the weighted error

$$\varepsilon_t = \min_{f,p,\theta} \sum_i w_i \left| h(x_i, f, p, \theta) - y_i \right|$$

3. Define $h_t(x) = h(x, f_t, p_t, \theta_t)$ where f_t, p_t and θ_t are the minimizers of ε_t
4. Update the weights: $w_{t+1,i} = w_{t,i} \beta_t^{1-e_i}$

Where $e_i = 0$ if example x_i is classified correctly, $e_i = 1$ otherwise, and $\beta_t = \frac{\varepsilon_t}{1-\varepsilon_t}$.

- The final strong classifier is :

$$C(x) = \begin{cases} 1 & \sum_{t=1}^{T} \alpha_t h_t(x) \geq \frac{1}{2} \sum_{t=1}^{T} \alpha_t \\ 0 & otherwise \end{cases}$$

Where $\alpha_t = \log \frac{1}{\beta_t}$

The algorithm described above [14] is used to select key weak classifiers from the set of possible weak classifiers. While the AdaBoost process is quite efficient, the set of weak classifier is extraordinarily large. Since there is one weak classifier for each distinct feature/threshold combination, there are effectively KN weak classifiers, where K is the number of features and N is the number of examples. In order to appreciate the dependency on N, suppose that the examples are sorted by a given feature value. With respect to the training process any two thresholds that lie between the same pair of sorted examples is equivalent. Therefore the total number of distinct thresholds is N.

The wrapper method can be used to learn a perceptron which utilizes M weak classifiers. The wrapper method also proceeds incrementally by adding one weak classifier to the perceptron in each round. The weak classifier added is the one which when added to the current set yields a perceptron with lowest error. Each round takes at least O(NKN); the time to enumerate all binary features and evaluate each example using that feature. This neglects the time to learn the perceptron weights. Even so, the final work to learn a 200 feature classifier would be something like O(MNKN).

The key advantage of AdaBoost over the wrapper method is the speed of learning. Using AdaBoost a feature classifier can be learned in O(MNK). In each round the entire dependence on previously selected features is efficiently and compactly encoded using the example weights. These weights can then be used to evaluate a given weak classifier in constant time.

We use AdaBoost for learning and training with ACF for feature computing. Further more we add multi-scale mechanism in our implementation. Given an image, ten channel features in original scale are computed as Figure 1.

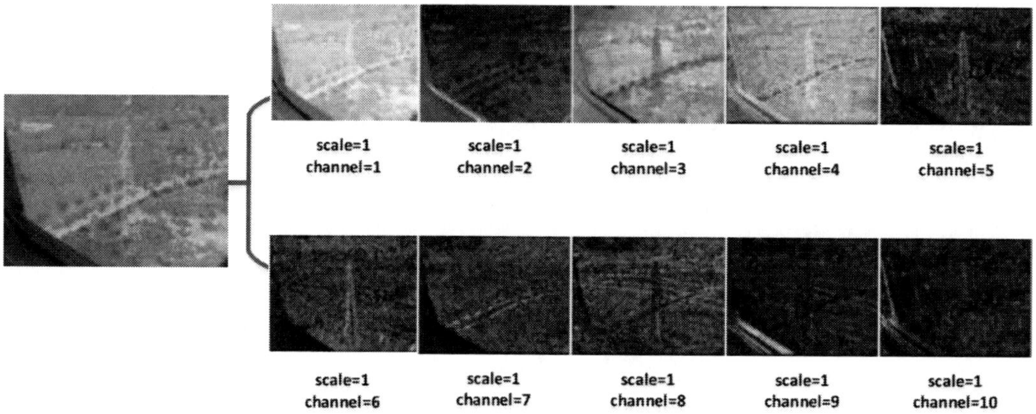

(a) Ten aggregate channel features of Tower.

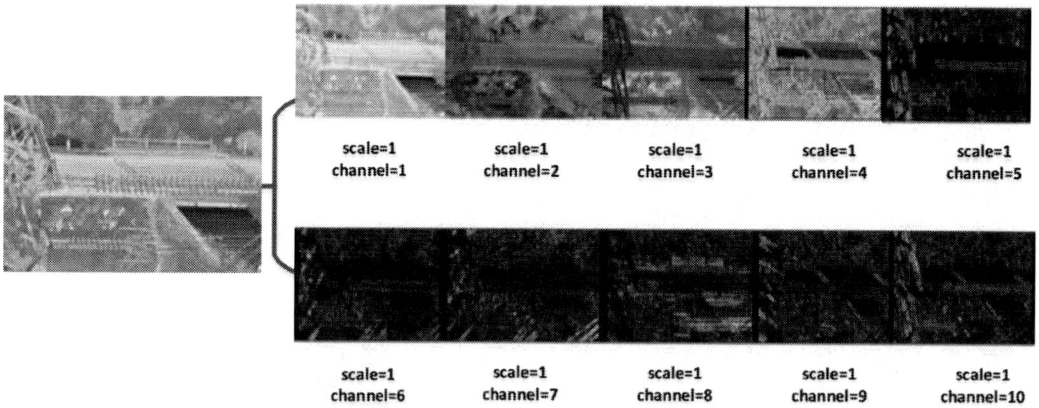

(b) Ten aggregate channel features of Insulator.

DOI: 10.1051/matecconf/201710003010

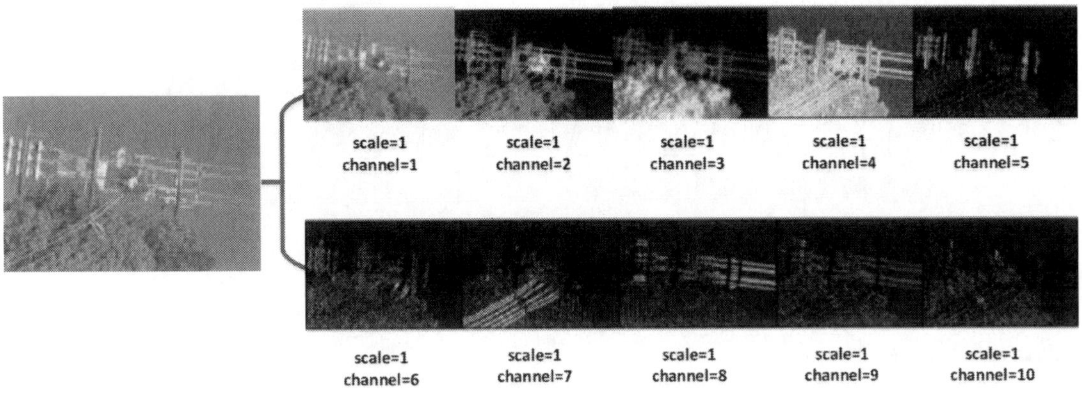

(c) Ten aggregate channel features of Nest.

Fig. 1. Ten aggregate channel features of ROI targets in an image.

4 Experiments and results

In order to train and evaluate the ACF boosted detector for detection, For each target, 1400 images have been divided into 3 sets: training, cross validation, and test set. 300 images (target and background) have been used for training, while 200 images are used for the cross validation and 200 for the test set. A total test error of 3.25% is attained. A false positive rate of 1.5% was achieved, which means that only 3 of the 200 background test images were incorrectly classified as target. On the other hand, we obtain a false negative rate of 2%, which indicates that 4 target images, out of 200 used for testing, were predicted as background. These results suggest that, although overall performance of the classifier is good, tower images get predicted as background more often than background images as target. See Figure 2.

(a)

(b)

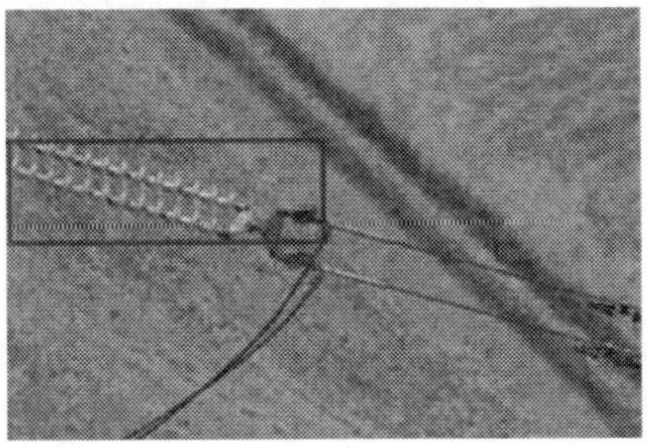

(c)

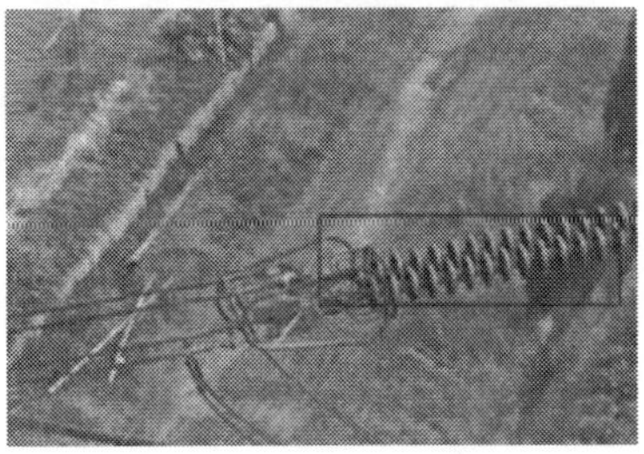

(d)

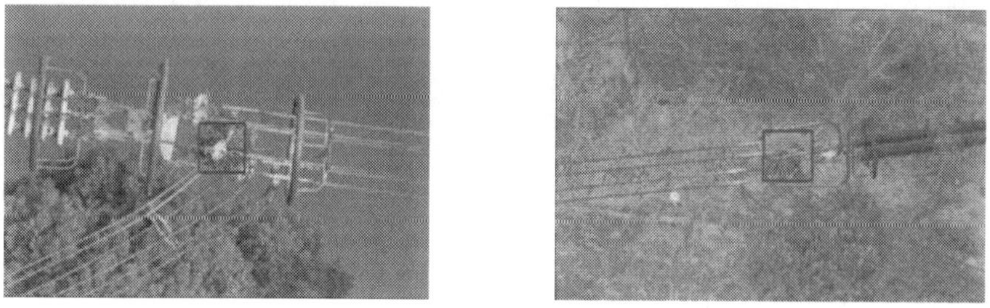

(e) (f)

Fig. 2. Targets detection results, (a)(b)for towers,(c)(d)for insulators and (e)(f) for nests.

5 Conclusions

Power line infrastructures are heterogeneous and complex, making automatic power line inspection a difficult problem. Therefore, there is a current considerable interest in this area of research. To achieve the goal of autonomous inspection, research efforts must aim towards developing general approaches that satisfy the several requirements: e.g. simultaneous detection of power lines and electric towers, check for defects in several power line components, analyze security distances, among others. The current paper is an effort in this direction, with emphasis on electric tower detection in aerial inspection data. We believe this is a key stage to be able to develop more complex tasks such as defects analysis, especially when the main source of information comes from poor quality images.

The key novelty of this paper is the investigation of a learning framework for providing a complete solution for detection which may be the fastest paradigm since we use boosted detector. Another main reason for the good performance is we use aggregate channel features including color, normalized gradient magnitude and histogram of gradients. However, such features may not be an ideal representation for all types of power line targets. This problem was more visible during the evaluation of the complete system, especially with higher number of misclassifications. Therefore, immediate future work is lined towards exploring other feature spaces. Finally, we hope to significantly enhance the results of power line targets detection.

References

1. C. Whitworth, A. Duller, D. Jones, and G. Earp, “Aerial video inspection of overhead power lines,” Power Engineering Journal, vol. 15, no. 1, pp. 25–32, 2001.

2. C. Sun, R. Jones, H. Talbot, X. Wu, K. Cheong, R. Beare, M. Buckley, and M. Berman, “Measuring the distance of vegetation from powerlines using stereo vision,” ISPRS journal of photogrammetry and remote sensing, vol. 60, no. 4, pp. 269–283, 2006.

3. I. Golightly and D. Jones, “Corner detection and matching for visual tracking during power line inspection,” Image and Vision Computing, vol. 21, no. 9, pp. 827–840, 2003.

4. W. Cheng and Z. Song, “Power pole detection based on graph cut,” in Image and Signal Processing, 2008. CISP’08. Congress on, vol. 3. IEEE, 2008, pp. 720–724.

5. J. Tilawat, N. Theera-Umpon, and S. Auephanwiriyakul, “Automatic detection of electricity pylons in aerial video sequences,” in Electronics and Information Engineering (ICEIE), 2010 International Conference On, vol. 1. IEEE, 2010, pp. 342–346.

6. Z. Wu and R. Leahy, “An optimal graph theoretic approach to data clustering: Theory and its application to image segmentation,” Pattern Analysis and Machine Intelligence, IEEE Transactions on, vol. 15, no. 11, pp. 1101–1113, 1993.

7. L. Vincent and P. Soille, “Watersheds in digital spaces: an efficient algorithm based on immersion simulations,” IEEE transactions on pattern analysis and machine intelligence, vol. 13, no. 6, pp. 583– 598, 1991.

8. J. Cooper, S. Venkatesh, and L. Kitchen, "Early jump-out corner detectors," Pattern Analysis and Machine Intelligence, IEEE Transactions on, vol. 15, no. 8, pp. 823–828, 1993.
9. R. E. Schapire, Y. Freund, P. Bartlett, and W. S. Lee. Boosting the margin: A new explanation for the effectiveness of voting methods. The Annals of Statistics, 1998.
10. J. Friedman, T. Hastie, and R. Tibshirani. Additive logistic regression: a statistical view of boosting. The Annals of Statistics, 38(2):337–374, 2000.
11. P. A. Viola and M. J. Jones. Robust real-time face detection. IJCV, 57(2):137–154, 2004.
12. P. Doll´ar, R. Appel, S. Belongie, and P. Perona. Fast feature pyramids for object detection. PAMI, 2014.
13. R. Benenson, M. Mathias, R. Timofte, and L. Van Gool. Pedestrian detection at 100 frames per second. In CVPR, 2012.
14. L. Bourdev and J. Brandt. Robust object detection via soft cascade. In CVPR, 2005.

Study on Measurement of Advanced Manufacturing: Case by China

Jinghuai She[1] and Xuedan Li[2, *]

[1] Department of Business Administration, Capital University of Economics and Business, Beijing, China
[2] Department of Business Administration, Capital University of Economics and Business, Beijing, China
Corresponding author: dxljyjy@163.com

Abstract. This article has built a system of China's Advanced Manufacturing measurement indicators. By applying the datum from 2004 to 2013, we estimate the level of development and current status of China's Advanced Manufacturing (AM), and evaluate the measurement results by establishing Hierarchical Linear Model (HLM). We confirmed that China's Advanced Manufacturing is in the rapid development trend. And due to the difference of initial conditions in Advanced Manufacturing development there is a greater imbalance. In contrast, a region with poor initial condition of has a relatively fast development speed.

1 Introduction

Advanced Manufacturing means using IT to propel industrialization, accelerate scientific development and improve industry with high-tech. Not only can advanced manufacturing improve the economic benefits and international competitiveness, but also reduce the consumption of resources, decrease emission of pollution and stick on the way of sustainable development. Judging China's AM development level accurately becomes an important indicator to test the AM and guide developing it, especially the developing status of different regions. This article is based on the background, combined with the industrialized classical theory, while constructing the index system of the AM in China, and applying the statistical datum from 2004 to 2013 to measure the 30 regions in China. Using hierarchical linear models (HLM) statistical methods to evaluate the results. This article has revealed the developing tendency of China's AM and the developing imbalance in those regions which are caused by the difference of initial conditions, especially those places in disadvantage with a higher speed to develop the industrialization.

2 Theory and hypothesis

2.1 Index system of intelligent manufacturing

The research of scholars majorly focused on the inevitability, path, content and characteristic of China's AM For example, Lu Zheng (2003)[1] , Jiang Xiaojuan (2002)[2], Qu Geping(1998)[3], Wei Liqun (2002)[4] , Hu Angang (2003) [5], Jin Bei (2004) [6], Wu Jinglian (2005) [7], Sun Xueguang[8] (2007) , etc. They were about how to handle those relations such as a AM and information, resource-environment relationship; also they confirmed how to take and how to realize the AM. Moreover, the study of the GI has promoted the relevant measurement. Such as Cao Jianhai and Li HaiJian(2003)[9], based on input-output method. Xie Delu[10] (2004), Yang Jie(2005)[11], based on characteristics of GI. Wu Xiaoqing(2007) [12], based on the GI connotation; Xie Chun (2011) [13] focuses on technology and resource environment. These studies were growing along from the speed and quality in single angle of industrialization..

AM revealed the interaction between information and industrialization, and AM is a process which takes the technology progressing as motive force and enhancing economic efficiency and international competitiveness as center. Thus the indicator system should reflect intensive and sustainable developing two factors. In all, based on the predecessors' theories and methods, we extract seven aspects which have the most typical connotation of AM as the sub-elements of the system and recompose them into 35 indicator elements and 50 measurable indicators to reveal the connotation and characteristics of AM.

3 Data and processing

This paper applied data from 2004 to 2013, all data are comes from the China statistical Yearbook, Industry statistics report, the science and technology statistical Yearbook, Environmental statistics Yearbook, The ministry of information and industry and other authoritative data. When evaluating the level of AM of China, the paper applied the 2004 – 2013 years' data, and standardization the raw data with dimensionless treatment, using a weighted method to calculate the index value of Advanced Manufacturing. Because there is no other means to deal with the original data, the index value represents of China's Advanced Manufacturing during this period, so called "development level"

In the empirical part of this article, the data is from 2002 to 2013. In the seven years, 30 provinces and cities try their best being industrialization which are different in resource endowments, economic development stage, history and geographical situation. We must select an appropriate method to deal with the longitudinal data so that we can analyze green trends as well as industrial growth differences between different provinces. The current method of researching variables growing trend using longitudinal data are Repeated Measures Analysis of Variance, Time Series Analysis and the Hierarchical Linear Model which is relatively late start, but developed rapidly.

4 Empirical assessment

Although a detailed analysis have been done from many Angle to China, contained from 2004 to 2013 Advanced Manufacturing stage of development, present development situation, the difference of Advanced Manufacturing level between provinces, the elements of green type industrialization constitution and so on, but the sample value just a symptom of development, the true discipline behind these sample points must through the establishment of the model can get real realization. For example, 30 provinces and cities of China in the process of the Advanced Manufacturing is different or not, whether the initial level of green type industrialization have influence to industrialization process, and how effective of green industrial factors to the Advanced Manufacturing process, all these can be solved only through the establishment of statistical model.

4.1 HLM

Through the preliminary analysis of development trend for the 30 provinces and cities, finding that the process of GI development all provinces and cities has a different in development rate and initial development level, that is to say, the intercept and slope is not same in each of the regression model. So, when using the provinces as sample group analysis the regression of the overall country, it is a set of random coefficient regression model, once using the time series model analysis the overall development trend, there will be heteroscedasticity and correlation problem, this will not be able to use OLS effectively estimates parameters, it can't directly use time series model dealing with layer structural data with individual differences. At the same time, because of the repeated measurement of variance analysis used to analysis overall development trends exist limitations, also cannot be used to analysis differences of individual development between regions. Therefore, this study used a green statistical method of data processing track which has been widely accepted - hierarchical linear models (HLM).

The so-called random coefficient regression model mainly set on the second floor of the linear development (growth) model. The second model is the equation about the first layer model intercept and slope, so the second model, without prediction variables, became a purely random effects equation. At this time, the linear development model has become a random coefficient regression model, also can be referred basic linear development model. The basic form is as follows:

The first layer model:

$$\text{GI level}=\beta_0+\beta_1(time)+r \quad (1)$$

The second layer model:

$$_0=\gamma_{00}+\mu_0 \quad (2)$$

$$_1=\gamma_{10}+\mu_1 \quad (3)$$

The equation is an unconditional model, for the second floor, there is no any variables. Although this model did not consider the factor influenced the level of industrialization, but you can use this model and the results analysis the initial level and individual variation, providing a reference for the next evaluation of the second layer model with variables.

In this model, the "time" encoding different variables, the results of model have a different interpretation. If the 2004 year make up for "0", the intercept of first layer model was the average level of Advanced Manufacturing of all provinces in 2004, the slope was units growth rate from 2004 to the next few years; if the 2013 series as "0 ", the intercept of first layer model was national average of Advanced Manufacturing on the 2013, the slope was unit growth rate after 2013. In this study, two different codes were calculated to last level and initial level to give an explanation.

In this study, the model used the HLM software which is specifically used for calculation of hierarchical linear model, developed by Stephen W. Raudenbush and others. We generated the original data using SPSS13.0 version, and calculated the final result using HLM6.08 version. In order to demonstrate the output in the form HLM6.08 software, using 2004 years as the initial level, we rearrange the output to the general table format according to the software, as shown in Table 1 and Table 2.

Table 1 shows, the estimated average intercept and average slope was 47.4 and 1.2 respectively. The results show that the initial implementation of the Advanced Manufacturing in 2004 the national average was 47.4 industrial units, which means that in 2004 the country's industrialization process has not come to the half, but the average annual growth rate was 1.2 percent after the implementation of the Advanced Manufacturing strategy since 2004, showing that the overall level of China's Advanced Manufacturing was a growing trend. Its "t" value is large enough, and P <0.001, so using these two factors is necessary to describe the growth of China's Advanced Manufacturing.

Table 1 estimated average intercept and slope were 54.7 and 1.2, as the model 2013 Numbers for 0, the intercept here was the average level of Advanced Manufacturing in 2013, that is to say in 2013, China's Advanced Manufacturing average level has reached 54.7, so only from the results of model, China's Advanced Manufacturing level from the average 47.4 in 2002 have been developed to 54.7. Because average slope is 1.2, it can expect China's Advanced Manufacturing annual growth rate is 1.2 since 2013, and hypothesis that two coefficients of testing results were significantly passing. This confirmed our hypothesis HI.

Table 1. GI model of linear growth (Unconditional, 2004).

Fixed effects	Coefficient	S.t error	t-ratio
Intercep β_0	47.371429	2.254750	21.010***
Slope β_1	1.214286	0.172327	7.046***
Random effects	**Variance**	**df**	**χ^2**	**Reliability**
Intercept u_0	154.6545 4	29	1465.78146***	0.980
Slope u_1	0.68150	29	111.30746***	0.739
The first layer of error r 6.72333				

$*P<0.05$, $**P<0.01$, $***P<0.001$

Table 2, For GI growth model, the variance of the estimated value for the two parameters β_0 and β_1 are 154.3 and 0.68.In the χ^2 test, the χ^2 -statistic of the two are 1462.2 and 111.3,the degree of freedom of the both two are 29,both the P values less than 0.001,significantly pass. Such results turn out that from 2004 the initial implementation of GI, 30 provinces and cities have existed significant differences, and from the value of variance, the level of GI in 2004 for China's 30 provinces and cities are very significant different. The slope of the variance shows from 2004 to 2013, seven years of development, the growth rate of 30 provinces and cities exists significant differences. 0.68 form the slope of the variance and standard deviation of 0.82, so for the specific area, if its growth rate is greater than the average value one standard deviation, it can reach 1.2+0.82=2.02percentage points of growth. Of course, for the slow-development region, it will develop by a speed of standard deviation below 1.2.

Table 2. GI model of linear growth (Unconditional, 2013).

Fixed effects	Coefficient	S.t error	t-ratio
Intercept β_0	54.657143	2.251987	24.271***
Slope β_1	1.214286	0.172327	7.046***
Random effects	**Variance**	**df**	**χ^2**	**Reliability**
Intercept u_0	154.2682 0	29	1462.19219***	0.980
Slope u_1	0.68150	29	111.30746***	0.739
The first layer of error r 6.72333				

4.2 The relations between initial level of regional GI and growth speed

According to literature, it can get related consistent estimates between initial conditions and real changes by use data repeatedly observed. Studying the relationship between the mentioned two above, generally determined by using the correlation between variance, the formula is :

$$\hat{\rho}(\beta_0,\beta_1) = \hat{\tau}_{01}/(\hat{\tau}_{00}+\hat{\tau}_{11})^{1/2} \tag{4}$$

Using equation (4) and HLM6.08 software, calculated the correlation coefficient is -0.202, as the negative correlation coefficient, the relationship for each region of the growth speed and the initial level of Advanced Manufacturing has a negative relationship, the low initial level of green industrial areas, the development has a faster speed.

4.3 The study on the influence of regional difference on GI

Variation of the process of Advanced Manufacturing, what percent are due to regional differences in the level of Advanced Manufacturing is an important issue which is worth focusing on. If regional differences influence on China's overall Advanced Manufacturing level can be reasonably measured, we can base the degree of difference and quantify the contradiction between regional differences in the development and Advanced Manufacturing strategy,. So we can decide how to choose between the two problems.

Hierarchical linear models have a linear cross-level concepts , Cross-level to be addressed is related to the total variance in the dependent variable is due to how many units of the second layer of the model caused by individual differences. The process of calculation is: firstly, the model for the first layer and the second layer of the model were created a pure random model, which can measure separately the random variation of the first and second floors. Among them, the first layer models' random variation is the total variation of dependent, which is caused by the second individual variation. Using the following formula:

$$\rho = \frac{\tau_{00}}{\tau_{00} + \sigma^2} \tag{5}$$

ρ stand for Cross-class correlation coefficient, τ_{00} is variance of individual variation, σ^2 is variance of total variation.

Table 3. The first layer and second layer stochastic model calculated results.

Random effects	**Variance**	**df**	χ^2	**Reliability**
The first layer model	146.91099	29	1821.10673	0.000
The second layer model	16.64127			

Calculated using the software HLM6.08 the first layer and second layer stochastic model results shown in Table 3, and then use (5) provides the calculation of inter-class correlation formula can calculate the cross-class correlation coefficient of the Advanced Manufacturing was 89.8%, the following formula:

$$\rho = \frac{\tau_{00}}{\tau_{00} + \sigma^2} = \frac{146.91099}{146.91099 + 16.64127} \times 100 = 89.8\% \tag{6}$$

Among the total variances of the development of Advanced Manufacturing, there are 89.8% are due to the 30 provinces and cities caused by the difference between the Advanced Manufacturing, which fully shows that the outstanding differences between the China region, while China also shows that green industrial development in narrowing the differences between regions is a very important issue.

4.4 China's GI initial level and the confidence level of the growth rate

As already described, reliability is used to describe the estimation accuracy. of model. Estimation accuracy of a model calculation for testing the validity of the model is very important because if the model parameters estimated from the variation of most of the errors caused by the model itself, then this model will be very unreliable, the use of such model for China's Advanced Manufacturing level of analysis, would be a mistake. Reliability is the variance divided by the estimated parameter variance, total variance equal to the parameter estimation error variance and the sum. With the formula is:

$$reliability(\beta_{0j}) = \tau_{00} / (\tau_{00} + \upsilon_{00}) \tag{7}$$

$$reliability(\beta_{1j}) = \tau_{11} / (\tau_{11} + \upsilon_{1j}) \tag{8}$$

Table 1 has been the reliability of the results list, the reliability of the intercept and slope of 0.980 and 0.739, respectively, belong to the relatively high credibility. The results show that both the intercept and the slope, the data have sufficient amount of information, each parameter can be as individual-level variables for modeling.

About the size of reliability, Raudenbush et al. (2004) described, usually, lower reliability does not prove that the statistical analysis carried out is not valid. Only when the reliability is very small, such as less than 0.1, research indicates that only the best will in the random coefficients as fixed factors, random items will be removed in order to facilitate subsequent model analysis.

4.5 The empirical results

Empirical results support our all previous assumptions, as follows:

·The average growth trajectory of China's Advanced Manufacturing from 2002 to 2013 shows sustained steady upward trend. The strategy implementation process of Advanced Manufacturing from 2004 to 2013, China's total Advanced Manufacturing level made a steady effects, Advanced Manufacturing index rise 54% in 2004 to 63% in 2013, while most of the provinces' Advanced Manufacturing level has considerable progress, although the albeit is different, the growth effect is obvious.

·There are large regional differences in China's Advanced Manufacturing. The results of the hierarchical model, the model intercept and slope variance test have passed a significant test, which means that the variance is significantly different from 0. From the perspective of the slope, the slope of the different provinces exist big differences, that is to say the green industrial growth in each region is very different..

·The initial level of industrialization and development rate within a certain range shows negative correlation. Pace of development of Advanced Manufacturing and regional initial level in a certain range exist negative correlation, the correlation coefficient is -0.202. In other words, the initial low level of green industrial areas, the growth rate slightly faster, but the initial high level of regional growth at a slower pace.

5 Conclusions

This article measured the development level and status of China's Advanced Manufacturing by building index system. And give evaluation, analysis and reasonable explanation on the development trend of China's Advanced Manufacturing and the effects that basic conditions will have on the Advanced Manufacturing of 30 areas by using hierarchical linear models (HLM) statistical methods. In our opinion, as regional differences have big influence on the development of Advanced Manufacturing, the government should plan the development policies of Advanced Manufacturing for different regions and promote the coordinated development; In addition, in the process of Advanced Manufacturing, because of different regional conditions ,we do not take the same road, according to the different regional resource endowments and historical conditions, we can follow our own pattern to achieve success.

Acknowledgement

The work described in this paper was supported by the National Social Science Foundation of China (No: 14BGL214) and the Science Foundation of Ministry of Education of China (No.13YJA630073).

References

1. Lv Z, Research on Chinese intelligent Manufacturing, Economy and Management Research, No. 2, 3-6 (2003).
2. XJ Jiang, Intelligent Manufacturing: the only way leading to XiaoKang life, RenMin.form., No.12, 4-6 (2002).
3. Qu GP, Chinese industrialization and environment protection, Chinese population, resources and environment, No. 2, 1-4 (1998).
4. Wei LQ, Stick to the way of intelligent Manufacturing, Econ.Times, No. 23, 17-20 (2002).
5. Hu AG, Wen J, China modernization pursuing in national regions: effect, characteristics, causes and its consequences, Guangxi. Univer. Nation. J, No. 1, 107-114 (2003).
6. Jin B, Several theoretical issues about intelligent Manufacturing, Chongqing. Tech. Busi. UIMversity Press (social science edition), Vol. 21, No. 3, 5-8 (2004).
7. Wu JL, Misunderstandings and distortions about intelligent Manufacturing, Busi. Week. J, No. 16, (2005).
8. XG Sun, Research on connotation and scientific advance of the new industrialization, China National Conditions and Strength, No. 10,18-20(2007).
9. JH Cao, HJ Li, The theory of new industrialization road, Chinese industrial economy, No.1,56-62(2003).
10. Xie DL, Li Q, Wang XM, Analysis of index system for intelligent Manufacturing and evaluation standards, Reform, No. 4, 30-38(2004).
11. Yang J, Luo ZH, Zhang CY, Research on evaluation index system of the new industrialization, Jilin. Univer. J (Social Sciences Edition), No. 3, 124-129(2005).
12. Wu XQ, Research on evaluation index system and model construction of new industrialization process in China, Chongqing University, (2007).
13. Xie C, Li J, The building and empirical analysis about evaluation index system of Chinese new characteristics industrialization, Syst. Engin. J, No. 3, 74-80(2011).

Research on the Influence Factors of Product Design in Complex Product Systems

Xiuhong Wang and Jing Suo*

School of Management Science and Engineering, Zhengzhou University of Aeronautics, Zhengzhou, China
Corresponding Email: 635013329@qq.com

Abstract. With the development of society and economy, CoPS play an increasingly important role in a country; it is obvious and urgent to research the product design in CoPS. Compared with the traditional ordinary products, this paper analyzes the requirements of CoPS product design and the main influence factors, and use a innovative mind to make the product design, in order to meet customer demand and reduce the risk of design, then put forward the influence factors in the design process. So, it can help enterprises improve product design process and provide favorable support to increase economic benefit.
Keywords. CoPS, product design, influence factors .

1 Introduction

Since the concept of CoPS generated, which scope is very broad, and the definition of CoPS is very fuzzy. Until the end of the 1990s, CoPS had a more clear understanding. Some scholars believed that CoPS usually need higher research costs, more complex technology, more specific requirements to customers, single product or system, or other facilities, and so on [1]. Related research scholars in domestic believed that CoPS is a kind of innovation activity, which has characteristics of high investment, high risk and high technique [2]. With the rapid development of science and technology, international competition is increasingly fierce, and the structure of CoPS is becoming more and more complicated, of cause products is also becoming more and more integrated and complicated. So, it is necessary to analyze the factor of product design in CoPS.

Chen Ming [3] pointed out that the complex product generally use high and new technology, and each subsystem have a more complex constraint relationship in various parts. Chen Jin and Tong Liang [4] study the management of CoPS innovation compared with the traditional management, from the high cost of custom, the subsystem structure, single production requirements to summarize the complexity of CoPS, and from five aspects such as product evolution characteristics to carry out innovation management. Ou Song [5] through the study of complex system structure, constructs a CoPS model of multiple views, and use this model to illustrate the concurrent engineering of product design.

Although many scholars do the in-depth research for CoPS, studying the product design this part of CoPS is very few. Therefore, research on the factors of product design help the enterprise have a good understanding in the products, then they will have a better grasp to the needs of society, in this way to recognize market requirement and meet customers satisfaction, at the same time, it can also avoid the waste of money. So, analyzing the factors of product design is imperative.

2 COPS and product characteristics

On the basis of some scholars' research, summarizing the characteristics of CoPS. From the production type, CoPS is custom, and it needs to negotiate with customer to determine the product requirements, then conduct customized production [6]. From the production quantity, which belongs to a single small batch production, because of the uncertainty of CoPS and inconsistency requirements, therefore it does not have scale production conditions [7]. From the product technique, which belongs to technology-intensive, its products cover many aspects of knowledge, and it also has complex function and interface [8].

CoPS product characteristics can be learned from analysis of CoPS, which mainly includes: complex product structure, diverse performance and stronger competitiveness this three aspects, from this perspective to grasp the factors analysis of product design is very effective and feasible.

3 CoPS product design requirements

After analysis of the product characteristics of CoPS, obviously, to study the product design is a activity that full of challenges and risks. It is influenced by many factors whether product design can achieve success or not, such as product performance, the degree of meeting customer demand, and so on. In order to reduce the risk of product design as much as possible, which need to analyze the requirements of CoPS product design, so this paper from the product performance and the degree of meeting customer demand this two aspects to discuss.

3.1 Excellent product performance

The requirements of CoPS product design must consider the product performance, this article from the product function, quality, efficiency, and manufacturing process requirements to consider the necessary factors in product design. Function is an important sign, the quality is customers should consider the core factor besides function, what's more, the final purpose of CoPS product design is to get benefit, and efficiency directly reflect the interests of enterprise, which should not be ignored. In order to make product more innovative and get more customers resources, enterprises must start from the manufacturing process to improve their product, which is a basic way to meet the requirements of product design.

3.2 Degree of meeting customer demand

The main purpose of design and trial-produce new products is to meet changing customer demand and obtain a better economic benefit. At the same time, design and trial-produce new products have to make profits as the premise. Only when there are society needs, they will have chance to get profits. What's more, enterprise usually need to consider the long period development, not just the current needs, so, they should make different measures to adapt society need in time, that is to say, through meeting the customer demand and greatly improve customer satisfaction, it can achieve the requirements of CoPS product design.

4 Analysis factors of product design

Through the above analysis, in CoPS product design factors analysis, this paper from the product hierarchy, the complexity of design techniques, the reliability of resource equipment and product innovation this four aspects to analyze the influence factors of product design.

4.1 Product hierarchy

The complexity of CoPS product is mainly embodied in the product has a hierarchical structure, the composition of complex products from low to high is divided into three levels, that is, component, system and integration. There are coupling relationship in each hierarchy, therefore, CoPS product own its inherent physical function, which reflect in product availability and functionality, and the function module of a product is usually composed of a multi-level hierarchical structure. At the same time, there is a kind of dynamic information interaction between each function module and parts, and the information feedback is also a multi-level hierarchical structure.

4.2 Complexity of design techniques

The stability of technique and difficult index are important index to measure the complexity of design techniques. Design techniques is a core factor of product design, which directly reflect the technical requirements of product design process, of cause, the higher products require, the more complex technology is. Technology stability refers to the deviation degree to its normal level design technology in the design phase, the difficult index refers to the biggest difficulty to meet the product design requirements. To ensure the product design is excellent, which need to guarantee a good stability and overcome the difficult index of the product.

4.3 Reliability of resources equipment

The input of resource and capital is enormous in CoPS [9], especially in product design, so, it is necessary to ensure the availability and reliability of resources, here mainly contain capital flow and system equipment. On the one hand, the capital is an important factor in enterprise, which is to guarantee a normal running of the system. Product design requires repeated trial, and the cycle time of CoPS product is very long and investment is large, so there are gaps about capital in the product design stage. Therefore, enterprise need to carry out reasonable budget and capital investment to guarantee product design stage can develop successfully. On the other hand, CoPS equipments usually have great complexity and integration, it needs various complex equipments work together in product design, especially design quality and innovation analysis are technical personnel through equipments running to accomplish. So, it is necessary to ensure the integrity equipment. In conclusion, to guarantee reliability of resources equipment is an indispensible factor for enterprise.

4.4 Product innovation

There is some difference between CoPS products and traditional products, CoPS will pay more attention to product innovation, actually, product innovation is the integration between technical knowledge and market demand, in this way to get a innovative product, through the effects of product for some customers, then continue to improve the product and further innovation, until to get the satisfied products. The specific innovation design process is shown in Figure 1.

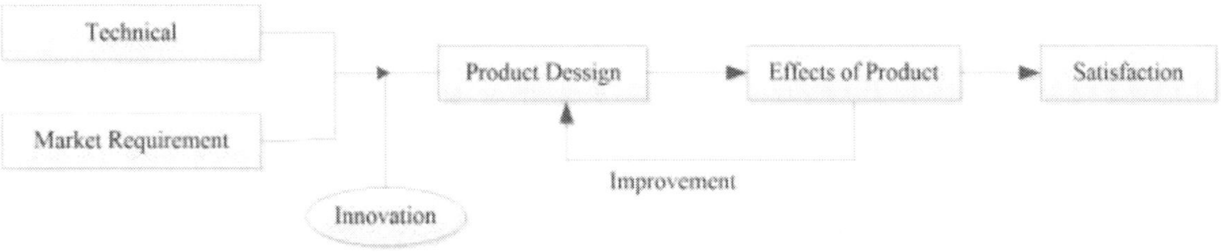

Fig. 1. Innovation design process.

5 Prospect of research

Compared with ordinary products, CoPS products design process is more complex, and capital inputs and product design cycle is greater than the ordinary products. Through the study of product design process, this paper analyzes the characteristics of CoPS product design factors in detail, and proposed four factors should be considered in the stage of product design, which is a very meaningful research. Through the concrete analysis of influence factors for CoPS, enterprise will get some useful advances to make a great product design, so it is very necessary to do this research. There is no doubt that there are still many issues deserve further research in the process of CoPS product design, such as the connection between the product design factors, how to realize the integration of product design process, and so on. Through in-depth study of these problems, enterprise can improve the success rate of product design, and make their product own a stronger competition in the market, so as to reduce the design risks and add economic benefits.

Acknowledgment

Thanks for the fund provided Humanities and social science project of Henan Provincial Department of Education（Grant No.2015-GH-039）and Key scientific research projects of Henan high school（Grant No.16A630063）

References

1. K.L. Hansen, H. Rush. Hotspots in Complex Product Systems: Emerging Issues in Innovation Management. Technovation, 1998, 18(9): 555- 561.
2. X.G. Gong, X. Song, Q.X. Zhu. the Cost Management of Technological Innovation System based on System Dynamics. Journal of Technology and Management,2009(4):128-131.
3. M. Chen, Y. Chen, B.S. Chen. the Research of Complex products in Technology Risk Management. Journal of Manufacturing Automation, 2005, 27(10): 6-16.
4. J. Chen, L. Tong, J.Z. Huang. Complex Product Systems Innovation to Traditional Management challenges. Journal of Science and Technology Management, 2004, 09: 47-51.
5. S. Ou, M.L. Zhong, D.Y. Qi, Q.J. Zhou. Multiple View Model of Complex Product Systems. Journal of Jinan University, 1998, 01: 150-156.
6. M. Hobday, T. Brady Rational vs. soft management in complex software: lessons from flight simulation. International Journal of Innovation Management, 1998, 2(1): 1- 43.
7. K. Hansen, H. Rush. Hotspots in complex product systems: Emerging issues in innovation management.Technovation, 1998, 18 (9): 555- 561.
8. Z.D. Chen, K.Y. Wang. the Complexity of Complex Product Systems to Study the Effect of Knowledge Management. Journal of Science and Technology Management, 2007, 05: 101-105.
9. J. Chen, L. Tong, Y. Gong. Research of Complex Product Systems Innovation Evaluation Index System. Journal of Management Research and Development, 2003, 04: 59-65.

Study on the Application of 6σ Method in Shoemaking Enterprise Quality Improvement

Ting Wang[1], Pengfei Ren[1,2,*] and Xin Wang[1]

[1]School of Management, Guizhou University, Guizhou, China
[2]Jinan Railway Bureau, Jinan, China
Corresponding Email: 953749935@qq.com

Abstract. With the development of science as well as technology and the influx of shoemaking companies abroad, our traditional shoemaking enterprises face greater pressure to survive, and product quality problems are in urgent need of improvement. Aiming at the quality problems of M shoemaking enterprise, DMAIC method is used to conduct the product quality improvement, and a mathematical statistics software Minitab is used to analyze the data. Moreover, the optimum combination of the key influencing factors is determined by orthogonal experiment design as well as a set of specific quality improvement plans are put forward. Finally, the production process capability of M enterprise is improved, the product bonding strength is enhanced, customer satisfaction is improved, and the thoughts of product quality improvement and reference are provided for domestic traditional shoemaking enterprises.

1 Introduction

6σ quality management research can be traced back to the end of the 20th century [1]. Since the beginning of 21 century,6σ quality management has been gradually introduced to China and now is growing with a booming tendency. At present, 6σ methods have been applied in manufacturing enterprises, construction enterprises site, the internet trading system, serving-oriented enterprises, retail industry, state organs, logistics and other fields [2-4].

China is one of the world's largest shoe-producing countries, and the continuous entrance of foreign shoemaking enterprises has brought the Chinese shoemaking enterprises opportunities and challenges [5], which promote domestic shoemaking enterprises to introduce the advanced theory of 6σ to face the increasingly fierce competition.

M corporation is a pretty prior domestic shoemaking enterprise and the production lines are traditional. Their customers reflected that there were some quality flaws of their products, and the most serious problem—glue problem accounts for more than 40% and in the face of increasingly fierce market competition, it is necessary to improve the problem of glue. Therefore, M corporation applies the DMAIC method in the production process to reduce product defective rate, improve quality, reduce bad costs, better meet the needs of customers, enhance core competitiveness and improve the quality of products, the company provides a reference for other domestic traditional production lines of manufacturing enterprises or shoe-making enterprises.

2 The specific application of DMAIC method in quality improvement

6σ is a complex system as well as a management change to the enterprise [6].The DMAIC approach is an abbreviation for the 6σ breakthrough improvement method and it is constituted by the first letter of Define, Measure, Analyze, Improve, and Control [7],it is a very rigorous process approach of process control and problem-solving of manufacturing. DMAIC method continuously improves the process by identifying the issues which are focused, establishing data collection methods, analyzing the causes of process variation, eliminating variation and providing improvement scheme, while monitoring the long-term process improvement. Stages of DMAIC form a closed-loop system where the former stage prepares for the latter stage, and the latter stage is the deep development of the previous stage.

From the production process of M company, it is known that the cause of the problem of glue including the defect of glue, unreasonable drying time after brushing and unqualified curing, but the cause of quality problems is not clear, in this case, M companies choose 6σ DMAIC method for the project research.

2.1 Definition phase

With the development of science & technology and economy, in recent years, as the old shoe company, M companies began to introduce advanced production equipment and production concepts, and continuously improve its production capacity and production quality. This article is a study of product quality improvement of M shoemaking enterprises and aims to help them improve the production process and product quality and finally win the market.

In this paper, customer's needs in many aspects are captured by "listening to the voice of the customer" and the "customer's voice" is transformed into the critical customer requirement (CCR), which converted into measurable items and indexes of output management [8].The percentage of M corporation's product quality problems which concluded from the customers' definition and collection of customers' needs are shown in Figure 1.

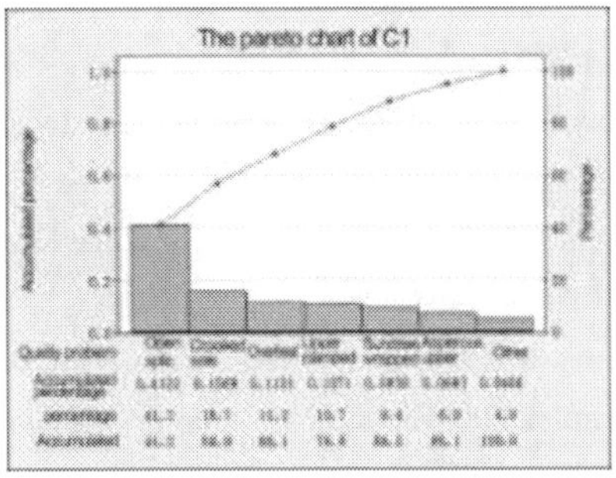

Fig. 1. The reflection of customer.

In this paper, through the analysis of "the customer's voice" and KANO, the product quality problems are identified as the key customer requirements, and after analyzing the Pareto which can reaction the quality problem from the customers, it is found that the glue problem of the product is the main problem that the customer complains a lot. So, this article will set improve the quality of the products problems as project object, and focus on improving the problem of glue.

2.2 Measuring phase

In the measurement phase, the existing measuring system—bonding degree measuring instrument is needed to analyze to ensure that the measured data is real and effective.

Ten products are chosed as samples to measure and record the data in this experiment. The other two inspectors measure 10 samples three times respectively, and the data is shown in Table 1.

Table 1. The results of the bonding degree measuring tools.

No.	Quality inspector 1	Quality inspector 2

	1(N/mm)	2(N/mm)	3(N/mm)	1(N/mm)	2(N/mm)	3(N/mm)
1	2.73	2.75	2.74	2.74	2.72	2.71
2	3.94	3.90	3.91	3.93	3.96	3.95
3	3.17	3.15	3.20	3.18	3.20	3.18
4	3.48	3.48	3.47	3.46	3.47	3.45
5	2.81	2.85	2.83	2.83	2.83	2.82
6	3.11	3.13	3.14	3.11	3.12	3.10
7	3.32	3.33	3.33	3.35	3.30	3.32
8	2.96	2.93	2.94	2.98	2.95	2.95
9	3.13	3.11	3.13	3.12	3.16	3.13
10	3.84	3.83	3.82	3.82	3.85	3.83

Using Minitab to process data [9], and then use it to analyze the accuracy of the machine, the results are shown in Figure2.

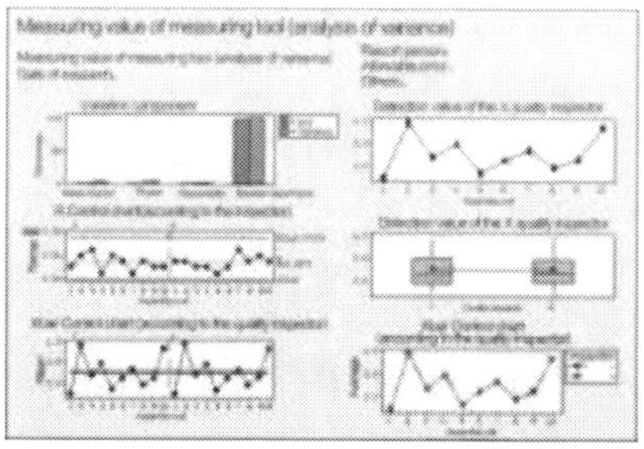

Fig. 2. Bonding degree measurement machine variance analysis report.

It can be seen from the Minitab results:

1) the variation of component between the measurement component is much bigger than the measuring tool R&R, the variation of component between the measurement component is 99.82%, and the combined measuring tool's is 0.18%,this means that the most parts of the total variation are caused by the difference of the measuring parts, so the variation caused by measuring system is very small.

2) There are a lot of differences between measurement components, and they are displayed with the broken line.

3) The difference between the operators is very small, and it is displayed with the straight line.

4) By X-bar control chart, most of the balls are falling outside the control line, this shows that the variation is mainly because of the difference between the measurement parts.

According to the automotive industry action group (AIAG) measurement system analysis manual 4th edition, it can use the following guidelines to determine whether the measurement system is acceptable: if the variation column (tolerance, process)of the total gage R&R contribute less than 10%, so the measurement system is acceptable; between 10% and 30%, whether the measurement system is acceptable or not depends on specific application, measuring equipment cost, maintenance cost and other factors; more than 30%, the measurement system cannot be acceptable, and the system should be improved. When the class number is less than 2,the measurement system is no value for the control process, because it is unable to distinguish the parts; when the number of class is 2,the data can be divided into 2 groups, such as high and low; when the number is 3,the data can be divided into 3 groups, like high, middle and low;5 or higher numbers shows the measurement system can be acceptable [10].

From the above Minitab processing results, it is said that the combine measuring tool R&R study variation (SV) values is 4.27%<10%,and the distinguishable class number is 25,so the measurement system is acceptable.

In addition, the process capability is analyzed in measuring phase. In order to know the process capability of M, each batch of this experiment took out only a sample, take 5 out each day, continuous for 7 days, and all the samples were measured with adhesion, get 35 sets of data. Using Minitab to analyze the collected data, get the results as shown in Figure 3.

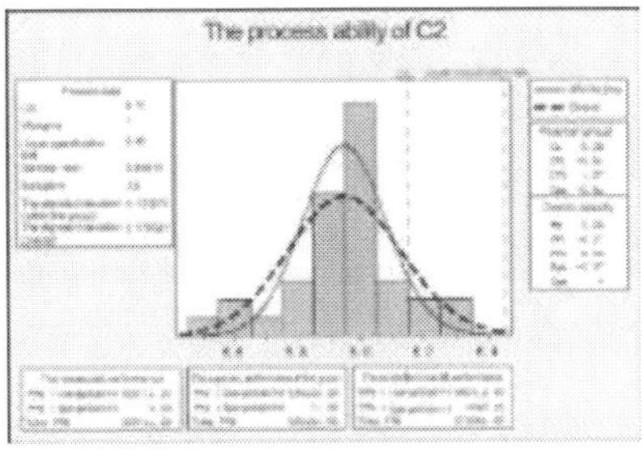

Fig. 3. Process capability analysis before improvement.

The results show that the data is normally distributed, and the Cpk = -0.51.The benchmark value is 1.33, so the process capability of M needs to be improved.

2.3 The analysis phase

According to Pareto diagram can get that the main quality problem of M company is the glue problem, in order to solve this problem, this paper uses cause and effect diagram, analyzes in people, machine, material, method, environment five aspects, gets the fishbone diagram as shown in Figure 4.

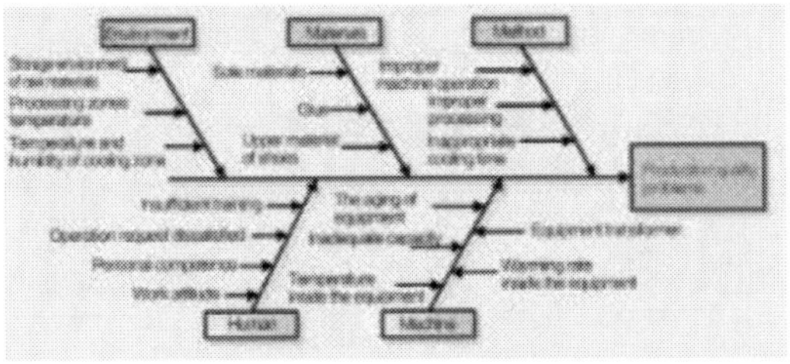

Fig. 4. Fishbone diagram.

Through the analysis of the key factors mentioned above and the cause analysis on the fishbone diagram, it can make sure that the main factor is in the XLQ—ф1700×400 horizontal vulcanizing tank. Because the main cause of the glue problem are raw materials, pressing and vulcanization process. The raw materials used by M company have been relatively fixed. There is no big difference, the glue used in production is purchased from the market, and after some processing, the glue can be used, so, the goal of improving is set as shoes pressing and vulcanization process. Shoes pressing begin after brushing glue, baotou and a series of preparations, put all the components of the shoe together with the glue, and pressure together. The vulcanization process occurs after all of the brush glue, baotou, pressing and a series of processing works, and put semi-finished shoes in vulcanizing pot to sulfide.

The study found that the main factors influencing the bonding process including whether workers operating specification and the cooling time length after pressing. The main factors influencing the sulfide concentration are listed as the following three points:1)the work pressure of the sulfide cans.2)the working temperature of the sulfide cans.3)the working ability of the equipment is insufficient. The dependent variable is set as X_1:temperature and fan speed are not changed, and change the tank pressure;X_2:tank pressure and fan speed are not changed, and the tank temperature is changed;X_3:the cooling time after pressing. Because the glue problems mainly reflects in the adhesiveness of the products, the adhesive degree of the product is set as Y, next will study the effect of the dependent variable X to Y.

2.4 Improvement phase

In improvement phase, the key influence factors of the optimal combination are found by designing experiments. According to the test, the relationship between cooling time, maximum temperature, maximum pressure and adhesion was obtained, and they are showed in Table2, 3, 4.

Table 2. Relationship between cooling time and adhesiveness.

Number	Cooling time (s)	Sampling number	adhesiveness (N/mm)
1	60	10	4.17
2	120	10	4.63
3	180	10	5.16
4	240	10	5.62
5	300	10	6.13
6	360	10	6.32
7	420	10	6.31
8	480	10	6.25
9	540	10	6.17
10	600	10	5.78

Table 3. Relationship between temperature and adhesiveness.

Number	highest temperature (C°)	Sampling number	adhesiveness (N/mm)
1	130	10	4.67
2	131	10	4.83
3	132	10	5.12
4	133	10	5.64
5	134	10	5.92
6	135	10	6.27
7	136	10	6.19
8	137	10	5.72
9	138	10	5.26
10	139	10	4.75

Table 4. Relationship between top pressure and adhesiveness.

Number	Top pressure (MPa)	Sampling number	adhesiveness (N/mm)
1	0.33	10	4.73
2	0.34	10	5.31
3	0.35	10	5.72
4	0.36	10	6.19
5	0.37	10	6.23
6	0.38	10	6.32
7	0.39	10	6.17
8	0.40	10	6.04
9	0.41	10	5.92
10	0.42	10	5.83

Through the table 2,table 3,table 4,it can be seen that cooling time after compression, the temperature of the vulcanizing tank and vulcanizing tank pressure can influence the adhesiveness of the products, therefore, it need to conduct the orthogonal experiment, get the optimal parameters of the three factors.

Next, conduct the orthogonal test on the drying time, temperature and pressure on three levels and three factors, first, assuming that the three parameters were independent of each other, there is no interaction, then orthogonal experiment factors level table as shown in Table 5.

Table 5. Factor level of orthogonal experiment.

Level	Factor		
	A. Cooling time (s)	B.Temperature (C°)	C. Pressure (MPa)
1	360	134	0.36
2	420	135	0.37
3	480	136	0.38

Orthogonal experiments needs to design orthogonal table, this experiment is 3 level and 3 factors, therefore, uses L9 (33) orthogonal table to arrange experiments, take 10 samples each experiment, measure their adhesiveness, and the average value is obtained as the experimental result Y, like Table 6 shows:

Table 6. Orthogonal experiment table.

No.	Factors			Results
	A	B	C	
1	1	1	1	5.76
2	1	2	2	6.13
3	1	3	3	6.04
4	2	1	2	6.23
5	2	2	3	6.41
6	2	3	1	5.97
7	3	1	3	6.13
8	3	2	1	5.91
9	3	3	2	5.86

Having a variance analysis in table 6 by Minitab, and Table 7 shows the result.

Table 7. Variance analysis of main factors improvement.

Source	SS	MS	F	P
A	0.1075	0.0537	1.51	0.294
B	0.0564	0.0282	0.64	0.560
C	0.1500	0.0750	2.63	0.151
Total	0.3139	0.1569		

Among them, SS stands for sum of squares (factors) between groups and the sum of squares in the group (error),MS stands for sum of squares divided by the number of degrees of freedom of the mean square, F stands for test statistics, P stands for significance level, and the P value is smaller, the influence is more significant. When the general P value is less than 0.05,it just means the factor has a significant effect. Table 4-12 is shows that factors C (vulcanizing tank top pressure) is the largest impact on adhesiveness, and the second is factor A (the cooling time after pressing), while the smallest one is factor B (the highest temperature in the curing tank).The main effect diagram of the average level of each factor is shown in Figure 5.

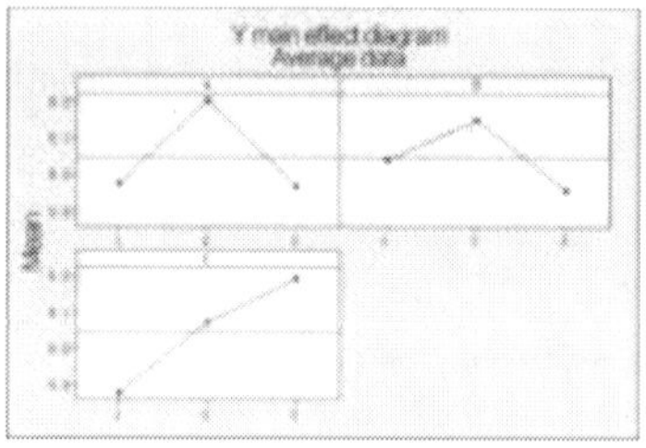

Fig. 5. The average level of factors effect diagram.

Then according to the Figure 5 "average level of factors effect chart" to determine the optimal parameter combination, the best combination by analyzing available for $A_2B_2C_3$ corresponding parameters: after brushing glue, the dry time will be 420s, the top temperature inside the vulcanizing tank will be 135℃,the top pressure inside the vulcanizing tank will be 0.38 MPa.

2.5 Control phase

After the implementation of several stages of DMAIC management method, the main factors affecting the glue problem is found, and the corresponding solution is proposed. And the specific factors were optimized to get the best combination of parameters. Then, as the implementation of the DMAIC, the project has entered the phase of control, monitoring the whole production process, feedback the implementation effect of DMAIC.

This paper has checked the products which were produced from the improved production process by sampling inspection. Also take 5 samples from different vulcanization tanks every day, take seven day in a row, and test the adhesiveness of each sample,35 groups of improved data were obtained.

The collected data were analyzed by Minitab software, get Xbar-R control chart, as shown in Figure 6.

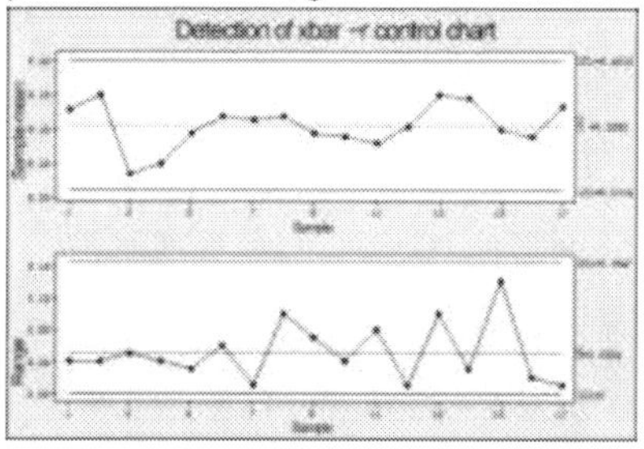

Fig. 6. Xbar -R control chart of control stage.

From this figure, $\overline{X}$ control chart's center line is at 6.3062,indicates that the process falls within the scope of the specification, and all of the points fall withinthecontrol limits, it shows that the process is stable. The center line of R control chart at 0.0504,also within the maximum allowable variation, so there is no greater variation in the process.

Process capability is obtained by the processing of Minitab, figure as shown in Figure 7.

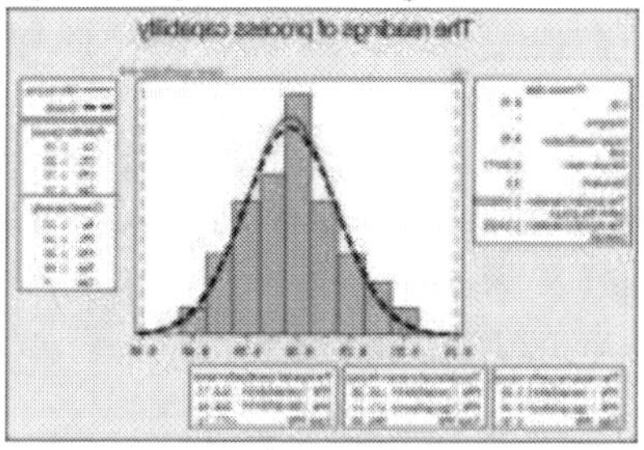

Fig. 7. The process capability of detection after improved.

Figure7 indicates that the improved sample average number is 6.31 N/mm,obviously higher than that of improved before:5.95 N/mm,and the improved Cpk is 1.10,higher than the process capability before: -0.51,so,the improvement is useful.

3 Conclusion

In this paper, the DMAIC improvement method of the Six Sigma theory is used in the traditional footwear enterprise product quality improvement project, found the major factors that impact the product quality problems, such as the drying time, curing tank temperature, pressure and so on, and using the orthogonal experiment method to determine the optimal parameter combination, a practical and effective improvement scheme is obtained. Successful implementation of DMAIC method, the process ability of process has been changed from -0.51 to 1.10,the adhesive strength of the product is increased from 5.95N/mm to 6.31N/mm, the customer's satisfaction has been improved, and enhance the competitiveness of the enterprises, implements the sustainable development of enterprises. At the same time also provides a reference for the development of similar enterprises to improve.

Acknowledgement

This research is supported by soft science research plan of Guizhou province (project number: 2016GZ67308) and the key projects of Guizhou province [2014]6004.

References

1. Jr. Blakeslee, J.A.. Qua. Pro. J.E **32**,77(1999).
2. XiangpingBo,HongLi..The.Theo.And.Prac.of.fin.and.eco, **31**,108(2010).
3. Yongdong Chen,. Tax Rese, **297**,86(2010).
4. Wei Yao, Jianzhun Liu.CHN.bus.and.mark. **03**,96(2014).
5. Weimin Chen. Chinese-foreign Footwear, **10**(2001).
6. Zhen He, Gang Yue, Lilin Wang. App.of.Sta.and Mana, **06**,1049(2007).
7. Quanli Wu, Jianan Qi. CHN.Sta.**06**,45(2005).
8. Li Shuang. Chem. Def. on Ships, 2005(1).
9. Zuocheng Wang. CHN. Sta, **01**,43(2006).
10. Qing Chang. CHN.Metr, 2007(1).

Research on the Innovation Elements In the Process of Technology Innovation

Yangguang Li [1], Junju Yue [1]and Min Wu [1]

[1] School of management and engineering, Zhengzhou University, Zhengzhou, China

Abstract. In order to improve the efficiency of innovation, the process of technology innovation has caused the attention of experts and scholars day by day. But in the process of technology innovation, the research on the extraction and classification of innovation elements needs to be further specific and systematic. Based on the full research on the innovation activities and innovation elements in many innovative fields, this paper elaborates the general process of the formation of innovation principle, and clusters innovation elements in many fields to form nine innovation dimensions, and then segments each innovation dimension into several sub dimensions of innovation. This paper aims to provide theoretical support for the research of innovative methods, and provide help for improving the quality and efficiency of innovation.

1 Introduction

With the advent of knowledge economy, innovation has become the only effective means to gain competitive advantage, while the improvement of innovation ability needs to be guided by scientific method. The role of the research on methods of technology innovation is to study the process of technology innovation, discover the law of technology innovation, and form a systematic and procedural innovation path, therefore improve the quality and efficiency of technology innovation[1, 2] .

The process of technology innovation is to rearrange and optimally design the elements of innovation system, information, process and production process and so on [3]. Analyzing the process of technology innovation will help to excavate the potential law of technology innovation. The research on the process of technology innovation, especially the research on the innovation elements in the process of technology innovation, has become the focus of the attention of experts and scholars day by day. Hu J uses the method of fuzzy cognitive map to identify the causal relationship between factors and constructs an evaluation framework of technological innovation to determine the important factors that influence the path dependence [4]. In the process of technology innovation, the system object should be analyzed structured and multi levels in order to better facilitate the innovation design [5, 6]. Some experts and scholars proposed that decomposing the process of technology innovation and studying on the innovation mechanism will help to dig the potential law of technology innovation [7, 8]. Meanwhile, existing research on the elements of technology innovation is also increasingly prominent, and the principle of innovation and invention obtained from the coupling of innovative elements and innovation law has been widely recognized [9, 10].

Above all, the existing research has fully analyzes the process of technology innovation, and promotes the development of the research on technology innovation. But most of the existing researches focus on the research of the principle of macro innovation in the process of technology innovation, and it is hardly to see the study on the extraction and classification of micro elements. Based on this, this paper sums up and extracts nine innovation dimensions in the process technology innovation through the systematic analysis on the innovation elements in the existing innovation areas based on the existing research, segments and defines the sub dimensions of each dimension. This paper aims to provide help for accelerating the process of technology innovation and offer methods and theoretical support for improving the quality and efficiency of technology innovation.

The chapters of this paper are arranged as follows. The second part studies the importance and necessity of the research on the elements of technology innovation. The third part studies nine innovation dimensions and sub dimensions induced and clustered after clustering the elements. The fourth part is the conclusion and prospect of this paper.

2 The importance and necessity of the research on the innovation elements in the process of technology innovation

After analyzing the process of technology innovation systematically, it is easy to find that the process of technology innovation can be summarized as the process of using the established principles of innovation to act on the elements of system innovation. The principle of innovation and invention are formed through combining different innovation laws and innovations elements. The forming process of the principle of innovation and invention is shown as the Figure 1.

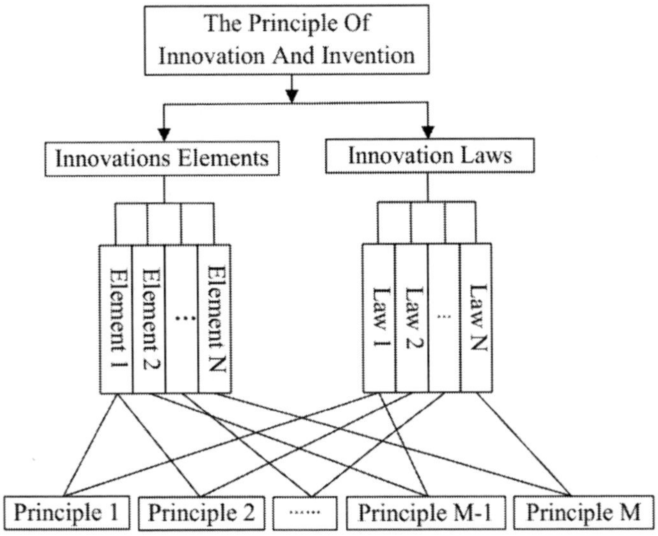

Fig. 1. Forming process of the principle of innovation and invention.

From Figure 1 we can learn that the selection of the innovation elements is the key link in the formation of the principle of innovation and invention. If we can quickly determine the innovation elements, and then use the existing innovation laws to act on the selected innovation elements, the principle of innovation and invention can be formed, and finally realize the technology innovation. Obviously, the way of forming the principle of innovation and invention can improve the efficiency and quality of technology innovation. Based on this, for the development of innovation, it is necessary and important to study the innovation elements in the process of technology innovation.

3 Research on the innovation elements

3.1 Dimension extraction of innovation elements

Based on the collecting and analyzing on the relevant knowledge and information of the ontology, environment and process of innovation[11-13], in the way of document retrieval and patent mining and through implementing the matrix and semantic analysis for the innovative knowledge, innovative mechanism and innovative principles in multiple innovation areas, this paper finds that innovation changes with the change of innovative elements, structure, system, environment, boundary conditions, time and space changes parameters. At the same time, the new model is constantly generated and transformed into a new form. The process of innovation is the process of the integration of key elements of, construction of network and structuring [14].

Therefore, in order to clarify the innovation elements in the process of technology innovation, and further guide the technology innovation, this paper constructs the knowledge map and patent technology base based on the existing theory knowledge of the innovation, systematically analyzes the innovative activities in many innovation areas, and clusters and analyzes the innovation elements involved in each field, finally condenses and summarize nine innovation dimensions which include space dimension, environment dimension, structure dimension, function dimension, mechanism dimension, material dimension, dynamic dimension, process dimension and human-machine relationship dimension. The process of extraction of the innovation dimensions is shown as Figure 2.

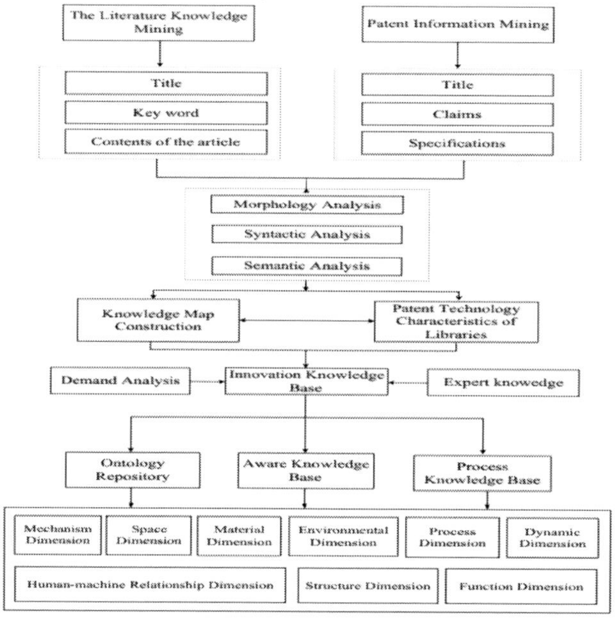

Fig. 2. Process of dimension extraction of the innovation elements.

3.2 Research on the sub dimensions of innovation

Innovative inventions in various fields are a complex system engineering involving a variety of innovation elements and methods, while innovative elements in innovation and invention of different areas are different. For example, in the innovation and invention of mechanical devices and transport areas, its innovation elements mainly include the space position and direction of the mechanical components, the optimization of structure relation, the way of function realization and the replacement and substitution of material and so on. While in the innovation and invention of biological medicine field, its innovation elements include biological and pharmaceutical materials, drug mechanism, environment relations and so on, and the elements involved are more specific, diversified.

Considering this, based on the above nine innovation dimensions, this paper comprehensively considers the technical features of several innovation fields, further distributes each innovation dimension into several sub dimensions in order to dig out the elements of technology innovation in the system quickly and improve the quality and efficiency of technology innovation through the established innovation dimensions and sub dimensions in the process of technology innovation of different areas. The name and definition of the sub dimensions of each dimension are shown in Table 1 to Table 9.

Table 1. Space dimension and its sub dimensions.

Innovation Dimensions	Sub Dimensions	Definition
Space Dimension	Direction	Includes not only the geographical location of the East, West, north and south or the left and right, up and down and other bearing elements. At the same time, it also includes the elements which include the operation direction, the heading direction, the direction of drilling, the relative direction and the absolute direction and so on.
	Location	All elements involving location relationship include the absolute location of the machine or the product and the relative location of the parts and so on, such as the job location of human, equipment and components and starting location of movement.
	Shape	It refers to the existence or manifestation of a particular thing or substance, such as rectangular, square, trapezoidal and so on.
	Volume	It refers to the amount of space occupied by the object or object being studied. For example, the volume of the wooden box is 3 cubic meters, etc.
	Weight	It refers to the magnitude of the gravity the object bears, while the weight differs with the difference of the earth's geographical position. For example, the weight of the host computer is 5Kg.
	Color	It refers to the color of an object, such as the red, yellow, blue and other colors.
	Morphology	It refers to the phase state of an object, such as the gas, solid, liquid, plasma state, neutron state, superconducting state, super fluid state, ultra solid state .

Table 2. Environment dimension and its sub dimensions.

Innovation Dimensions	Sub Dimensions	Definition
Environment Dimension	Environment Condition	It refers to the natural or induced stress conditions of physical environment which need to be considered when implement the environmental adaptability and environmental testing on the products and generate impact on product reliability, storage and transportation capacity, such as temperature, humidity, illumination, odor, ventilation and water flow conditions and other factors.
	Suspended Substance	It refers to the solid material suspended in a liquid, which is a generalized concept. It not only includes inorganic matter, organic matter and silt, clay, microorganism, etc, which are insoluble to water , but also includes relevant state parameters which include the state and concentration of the suspension of other solutions and solid materials and suspension.
	Field	It refers to a basic form of matter, employs energy and mass, and it can transmit the interaction among objects, such as electric field, magnetic field, gravitational field, atomic radiation, gravity field and so on.

Table 3. Structure dimension and its sub dimensions.

Innovation Dimensions	Sub Dimensions	Definition
Structure Dimension	Connection Relation	It means the connection of parts and components, including fastening, welding, bonding, hinge, bundling, traction, adsorption, lift, tolerance and fit relation, interference fit and other relations.
	Space and geometric relations	It refers to the relationship between the component and the component in the spatial direction. Spatial relations include the upper and lower, before and after, the left and right, internal and external and other relations. Geometric relations include grid, box, column beam, shear wall and slab reinforcement, etc.
	Stability And Reliability	Stability refers to the performance of the system which can be kept in a limited boundary or restored to the original equilibrium state after the system is disturbed. Reliability is the ability of the product to complete the specified function within the specified time under the specified conditions. Stability and reliability is a general term for the ability to resist interference and the function efficiency of the system.
	Life Cycle	It refers to all links of the product from the design stage to scrap recycling.

Table 4. Function dimension and its sub dimensions.

Innovation Dimensions	Sub Dimensions	Definition
Function Dimension	Movement	It refers to the movement of objects in space. The trajectory of its motion includes straight line, curve, rotation and irregular movement. And its state includes acceleration, uniform, slow and irregular movement.
	Production Efficiency	It refers to the ability to produce sound, light, electricity, heat, magnetic field, force, information and so on due to some kind of mechanism and principle.
	Operation Mode	It refers to the operation to accomplish a task which includes mechanical operations, chemical operations, biological operations, electromagnetic operations, optical operations, nuclear operations, artificial intelligence and other methods.

Table 5. Mechanism dimension and its sub dimension.

Innovation Dimensions	Sub Dimensions	Definition
Mechanism Dimension	Action Mechanism	It refers to that to realize a certain function, the internal working mode of the elements in a certain system structure and the elements connect and operate mutually to generate the rules and principles in a certain mechanism.

Table 6. Material dimension and its sub dimensions.

Innovation Dimensions	Sub Dimensions	Definition
Material Dimension	Material Element	It refers to the type and composition of elements that constitute the material.
	Material Properties	It refers to characteristics and properties of materials, which belong to the inherent nature of the material itself and include thermal properties, mechanical properties, electrical properties, magnetic properties, optical properties, chemical properties and other characteristics
	Phase State Of Materials	It refers to the state of the material and the most common is the solid, liquid, gas, etc.

Table 7. Dynamic dimension and its sub dimensions.

Innovation Dimensions	Sub Dimensions	Definition
Dynamic Dimension	Dynamic Source	It refers to the dynamic source of a machine or product, including electric power, coal power, fuel power, gas power, nuclear power, water power, wind power, tidal energy, animal power, human thermal power, compressed gas dynamic, compressed liquid dynamic, elastic energy and other dynamic sources.
	Transmission Mode	It refers to the transmission mode of energy in the process of power transmission of a machine or product, including mechanical transmission, electric transmission, fluid transmission, intermediary transmission, etc. It also includes work, heat transfer, radiation, etc.
	Energy Transfer Process	It refers to the connection mode between the structural components and the structural components of the machine or the product in the energy transfer process between. In the transmission mode of the dynamic system, the connection mode between the structural component and the structural component includes series, parallel and series-parallel connection and the like.
	Energy Storage Mode	It refers to the storage of energy and power in the energy transfer process of products and equipment, including chemical energy storage, energy storage of fluid compression, fluid energy storage, electromagnetic energy storage, material energy storage and the like.
	Intermediary And Media	It refers to the intermediary and media involved in the energy transfer of the product and equipment. Intermediary and medium is a kind of generalized definition, which includes not only the types, forms and properties of the intermediary and the medium, but also the interrelation between the intermediary and the medium and the object of action.

Table 8. Process dimension and its sub dimensions.

Innovation Dimensions	Sub Dimensions	Definition
Process Dimension	Time Sequence	It refers to the time related process and the order.
	Technology Flow	It refers to the technology related process and the order.

Table 9. Human-machine relationship dimension and its sub dimensions.

Innovation Dimensions	Sub Dimensions	Definition
Human-machine Relationship Dimension	human ability	It includes the basic size of human, human's ability to work, the limits of various organ function and influencing factors, etc.
	Human-computer Interaction	It refers to the scenario description of the role relationship between human and the machine with computer. And the human-machine interface is the form and method of human computer interaction.
	Physical Environment And Human	It refers to the environment in which people work, such as the coordination and design of the physical environment which includes the lighting conditions, noise, music, the color of work space.
	Security	It means that the machine or the product in the process of manufacture and using have no pollution to human body and other biological, and achieve certain safe using coefficient. To be friendly requires that the machine or product must be adapted to the user or operator.
	Environment Protection	It means that the machine or the product in the whole life cycle including manufacture, using and dumping has no influence on the environment, which Will not cause the deterioration of the environment and other conditions. It is different from the environment dimension.
	Portability	It means the convenience of the product or system to achieve mobility, portability and other function.

It should be pointed out that the nine innovation dimensions and their sub dimensions proposed in this paper are only the research results of the stage of study on the innovation elements in the process of technological innovation. With the emerging needs of the future and the emergence of new technology features, they can be made necessary changes.

4 Conclusions and prospect

This paper clusters and analyzes the innovation activities and innovation factors in each innovation field. It extracts and summarizes nine innovation dimensions in the process of technology innovation, and defines the sub dimensions in each innovation dimension. The research of innovation elements in the process of technology innovation is helpful to improve the quality and efficiency of the innovation. The determination of innovation dimension provides a new way to study the process of technology innovation, and provides help for accelerating the research of innovative methods.

With the continue emerge of new technology and new technologies, the innovation dimension and its sub dimensions proposed in this paper can be improved and enriched at any time. Inducing and constructing the proper innovation laws, combining with the innovation dimensions and studying the coupling mechanism to form a more scientific and complete innovation method system is the direction of the next research.

References

1. Huang H. Technology Innovation Method Research of Strategic Emerging Industry Based on the TRIZ Theory. Economy & Management, 2014.
2. Yeo W, Kim S, Park H, et al. A bibliometric method for measuring the degree of technological innovation. Technological Forecasting & Social Change, 2015, 95:págs. 152-162.

3. Wohlfeil F, Terzidis O. Critical Success Factors for the strategic management of radical technological innovation. International Ice Conference on Engineering, Technology and Innovation. IEEE, 2014:1-9.
4. Hu J, Zhang Y, Wang Y. Fuzzy cognitive research on influencing factors of technology innovation: In view of path dependence. Portland International Conference on Management of Engineering and Technology. 2015.
5. Xia G. Research about Relation of Working Ways and R&D Management Effected by Technology Innovation. Applied Mechanics & Materials, 2010, 40-41:881-886.
6. Harada T. Focusing device as innovation mechanism and cluster growth. Economics of Innovation & New Technology, 2014, 23(1):49-62.
7. Zhao J, Pablos P O D. Chinese firms' outward direct investment: Technological innovation mechanisms, organizational modes, and improving strategies. Human Factors & Ergonomics in Manufacturing & Service Industries, 2010, 20(2):149-160.
8. Mahoney R T. Product Development Partnerships: Case studies of a new mechanism for health technology innovation. Health Research Policy & Systems, 2011, 9(5):412-427.
9. Paredes-Frigolett H. A novel method for rule extraction in a knowledge-based innovation tutoring system. Knowledge-Based Systems, 2015, 92:183-199.
10. Iacono A, Seido Nagano M. Journal of Technology Management & Innovation. Journal of Technology Management & Innovation, 2014.
11. Kodama M. Boundaries Innovation and Knowledge Integration in the Japanese Firm. Long Range Planning, 2009, 42(4):463-494.
12. Hao J, Yan Y, Gong L, et al. Knowledge map-based method for domain knowledge browsing. Decision Support Systems, 2014, 61(1):106-114.
13. Kim Y G, Suh J H, Sang C P. Visualization of patent analysis for emerging technology. Expert Systems with Applications An International Journal, 2008, 34(3):1804-181.
14. Craig Mauelshagen, Mark Smith, Frank Schiller, David Denyer, Sophie Rocks, Simon Pollard. Effective risk governance for environmental policy making: A knowledge management perspective.Environmental Science & Policy,2014, 41: 23-32.

Research on De-icing Robot and Its Motion Characteristics Overhead Transmission Line

Zhuo Wang[1], Ming-zhi Cai[1], Tao Wang[2] and Bo Zhang[1,*]

[1]College of Mechanical and Electrical Engineering, Harbin Engineering University, Nantong Street 145,150001 Harbin, China
[2] School of MechanicalEngineering, Hebei University of Technology, 300401 Tianjin, China
Corresponding Email: jlwzh@163.com

Abstract. In recent years, most areas of our country are suffering transmission line ice-coating disasters caused by cold weather. Aiming at this phenomenon, this paper designs a robot that not only can avoid the various fittings on the line, but also can cross the tension tower of phase line to jumper line. According to the development status of the deicing robot at home and abroad, the specific environment of the transmission line, as well as the function of crossing obstacle and tension tower, the design specifications and structural scheme is presented. On the basis of the mechanical structure design, the motion characteristics of deicing robot is researched, and the virtual prototype of the robot is simulated based on ADAMS, and key research is the factors which affect the gravity center when the robot crosses obstacle and tower. With the analysis of factors mentioned, robot stability greatly improved.

1 Introduction

The ice on overhead transmission lines seriously affects the normal working of the grid and even brought a huge disaster [1]. The icing on the insulators and conductors ice cause flash trip conductor galloping, equipment damage, when the thickness of the ice cover significantly beyond the carrying capacity of the line, the ice accident such as tower down, lines break and the large blackouts will happen [2]. However, strategic assets such as transmission grids need to be operated in a safe, predictable, and reliable way. In light of this, in order to reduce the hazards of rain and snow on the transmission system and protect the safe operation of the power grid [3, 4], the design of effective de-icing tool to remove ice on the transmission cable, has important practical significance. Foreign scholars, such as Japan scholars, Institute de Recherché hydro-Québec (IREQ), as well as Portugal and Mexican scholars have done some research on the deicing robot, and a lot of domestic universities and institutes have also done the research [5-7].

2 Design requirements and overall scheme

2.1. Overhead transmission lines

The transmission line is mainly composed by the conductor, insulator strings, line fittings, tower, and guy, basic and grounding devices. There is also installed lightning conductor on the high overhead transmission lines. Insulator strings, high-altitude warning bell, and a variety of fittings such as suspension clamp, damping hammer and spacing rods, shown in the Figure1, these relative de-icing robot is a variety of obstacles. Therefore, the robot must have the capability to cross obstacles [8].

The strain section of the lines is mainly constituted by straight type tower and strain type tower. Straight type tower plays the role of supporting and fixing the cable without tension, while the strain type tower support cable with the tension between the wires and conductor gravity. The strain section of the lines is shown in the Figure2. Therefore, in order to continuously working between towers, the deicing robot must have the ability to cross towers [9].

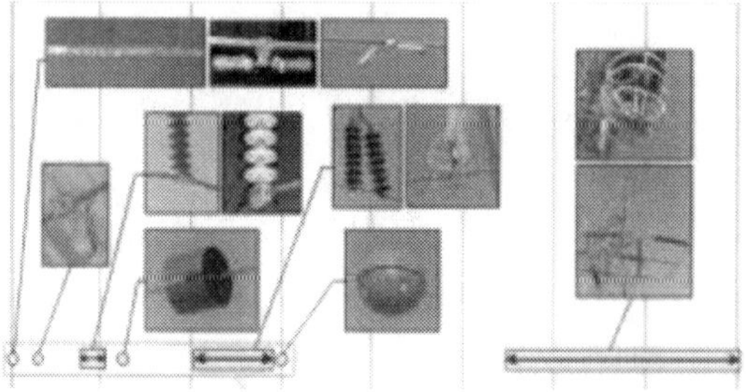

Fig. 1. Obstacles classified by length.

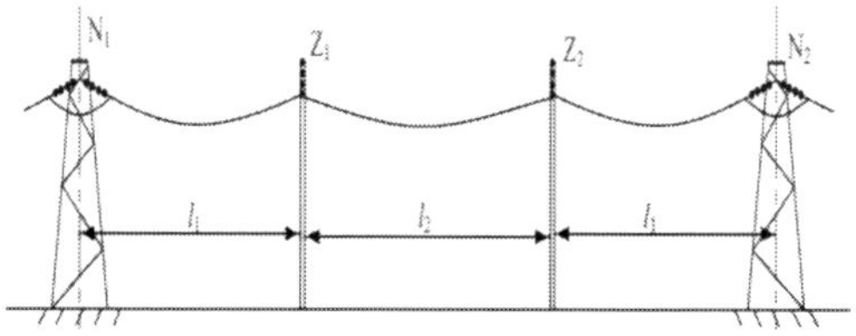

Fig. 2. A strain segment of the line.

2.2. Design constraints and technical specifications

The robot works on the long transmission lines at 15-30m high, where the environmental conditions are harsh. Firstly, to improve efficiency, the robot need to walk and deicing stable, and has the ability of cross the obstacles and towers. Meanwhile, in order to long adequate work, the robot is required light and small and low power consumption. And the environment and technical specifications of the robot are shown in the following Table 1.

Table 1. Design specifications.

Environmental specifications	
Voltage level	110kV
Conductor diameter	12mm-50mm
Maximum slope in span	30°
Maximum obstacle diameter	200mm
Maximum obstacle length	760mm
Maximum wind level	7
Robot specifications	
Weight	40kg
Traction force	500N
Inspection speed	1.0m/s
Deicing speed	3.0m/min
Battery life	5.0hours
Telecommunication signal range	5.0km

2.3. Mechanical structure schemes and working principle

The robot is composed by a pair of deicing mechanism, a pair of driving mechanism, a pair of clamping mechanism, a pair of support mechanisms, and a control box. Through the motor base of driving mechanism, the deicing mechanism and clamping mechanism are fixed on driving mechanism. The deicing mechanism is fixed in the front, used to get rid of the ice coating before the robot, and the mechanism is fixed in the post, when the robot crossing obstacles and brake. The motor base of driving mechanism is fixed on the top of the support mechanism, and two support mechanisms are fixed in two sliding arms of the control box, which can slide and pitch relative by each other.

Joints and mechanical structure scheme of the robot are shown in Figure3.

The working process of the robot includes avoid obstacles action, across obstacles action, drop back action, clamp and release cables action, and cross towers action. The following is specifically described.

(1)Avoid obstacles action. Through a pair of bevel gears, the arm motor drives the up-arm swing relative to down-arm along a certain angle, so as to make the driving mechanism swing away from cables.

(2)Across obstacles to action. Through a series of racks and pinions in control-box, box motor drive two arms slide relative, so as to make the driving mechanism pass obstacles.

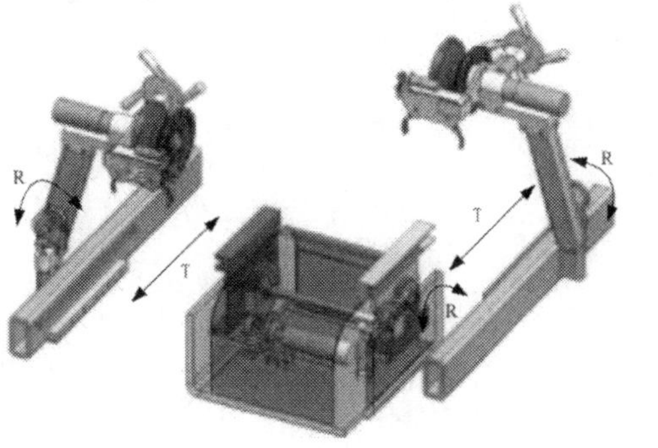

Fig. 3. Mechanism and movement joints of the deicing robot.

(3)Drop back action. Similar like the avoid obstacles action, the drop back action is the inverse process of avoiding obstacles action.

(4)Clamp and release cables action. In order, to ensure the stability of the robot, when one driving mechanism crosses obstacles, the clamping mechanism on another driving mechanism will clamp the cable. In this case, the motor in clamping mechanism will drive two clamping hands clamp the cable through the gears and turbine worms, so as to achieve clamp and release cables action.

(5)Across towers action. When the robot is walking to the tension tower end, it is required to cross the tower from the phase line to the leading line, and it has a certain angle between the phase line and the leading line. At this time, through the worms in the control box, the motor drives the arm rotates relative to the box, so as to achieve across towers action.

3 Motion characteristics

3.1. The deduction of connecting rod coordinates system and its motion equation

There're a total of five in deicing robot joints, respectively, T_1, T_2, R_1, R_2 and R_3, as showed in Figure 3. In Figure 4, the joints 2 and 6 for mobile vices, joints, 3, 5, and 7 for the rotation.

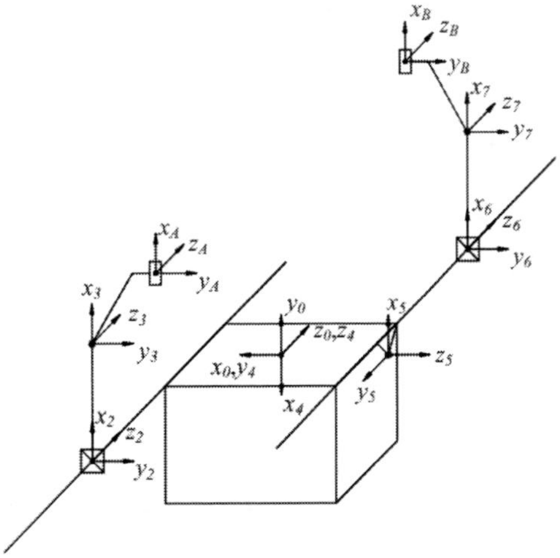

Fig. 4. B bracket each link coordinates system.

Set up between two connecting rods $i-1$ and i by two rotations and translation after the prescribed the coordinates of all connecting rod. This relationship is described by four homogeneous transformations. This relation A_i matrix is:

$$A_i = Rot(z,\theta_i)Trans(0,0,d_i)Trans(a_i,0,0)Rot(x,\alpha_i) \tag{1}$$

Open type is obtained:

$$A_i = \begin{bmatrix} c\theta_i & -s\theta_i c\alpha_{i-1} & s\theta_i s\alpha_{i-1} & a_{i-1}c\theta_i \\ s\theta_i & c\theta_i c\alpha_{i-1} & -c\theta_i s\alpha_{i-1} & a_{i-1}s\theta_i \\ 0 & s\alpha_{i-1} & c\alpha_{i-1} & d_i \\ 0 & 0 & 0 & 1 \end{bmatrix} \quad (2)$$

For prismatic joints, A matrix is:

$$A_i = \begin{bmatrix} c\theta_i & -s\theta_i c\alpha_{i-1} & s\theta_i s\alpha_{i-1} & 0 \\ s\theta_i & c\theta_i c\alpha_{i-1} & -c\theta_i s\alpha_{i-1} & 1 \\ 0 & s\alpha_{i-1} & c\alpha_{i-1} & d_i \\ 0 & 0 & 0 & 1 \end{bmatrix} \quad (3)$$

Connecting rod available transformation is given as follows:

$${}^{i-1}T_1 = \begin{bmatrix} c\theta_i & -s\theta_i & 0 & a_{i-1} \\ s\theta_i c\alpha_{i-1} & c\theta_i c\alpha_{i-1} & -s\alpha_{i-1} & -d_i s\alpha_{i-1} \\ s\theta_i s\alpha_{i-1} & c\theta_i s\alpha_{i-1} & c\alpha_{i-1} & d_i c\alpha_{i-1} \\ 0 & 0 & 0 & 1 \end{bmatrix} \quad (4)$$

Where s is sin, c is cos, θ_i is two connecting rod angle, α_{i-1} is connecting rod torsion angle, a_{i-1} is connecting rod base tangent length, d_i is two connecting rod distance.

According to the above principles, deicing robot A bracket connecting rod parameters is shown in Table 2.

Table 2. Deicing robot linkage parameters of A bracket

Connecting rod i	θ_i	α_{i-1}	a_{i-1}	d_i
1	0°	0°	0	d_1
2	-90°	0°	a_1	d_2 changeful
3	θ_3=0° changeful	0°	a_2	0

The coordinate system A observed changes in the matrix of T_A:

$$T_A = T_3 {}^3T_A = {}^0T_1 {}^1T_2 {}^2T_3 {}^3T_A = \begin{bmatrix} s_3 & c_3 & 0 & a_1 + Lc_3 + a_2 s_3 \\ -c_3 & s_3 & 0 & -a_2 - a_2 c_3 + Ls_3 \\ 0 & 0 & 1 & d_1 + d_2 \\ 0 & 0 & 0 & 1 \end{bmatrix} \quad (5)$$

In the same way, the robot B bracket each link coordinates system is established. Change the matrix of A relative to B ${}^BT_A$

$${}^BT_A = {}^BT_0 {}^0T_A = \left({}^0T_B\right)^{-1} \cdot {}^0T_A \quad (6)$$

The change of B relative to A matrix of the ${}^AT_B$

$${}^AT_B = {}^AT_0 {}^0T_B = \left({}^0T_A\right)^{-1} \cdot {}^0T_B \quad (7)$$

3.2. Cross obstacles process

Save the Solid works model of robot as para solid x_t file, and import it into ADAMS. Definitely the material or weight of all parts, and add constraints, drives, and forces to the robot model. Then set the contact impact parameters, verify the model, and take the simulation analysis.

Choose the Cartesian coordinate system in ADAMS model space, in which the z-axis is the forward direction of the robot, and the y-axis is the opposite direction of gravity. The clamping mechanism on arm B clamps cable before arm A awing. The rotation joint between up-arm and down-arm is the rotation center of the arm. Then the arm swings to side and the driving wheel out of cable, the curves are shown in Figure 5(a)-(d).

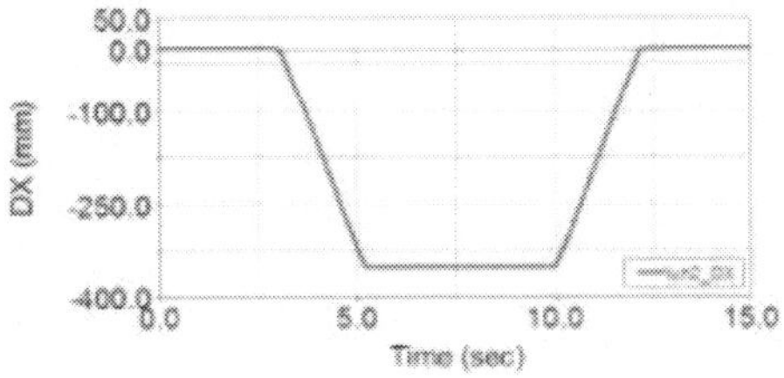

(a)

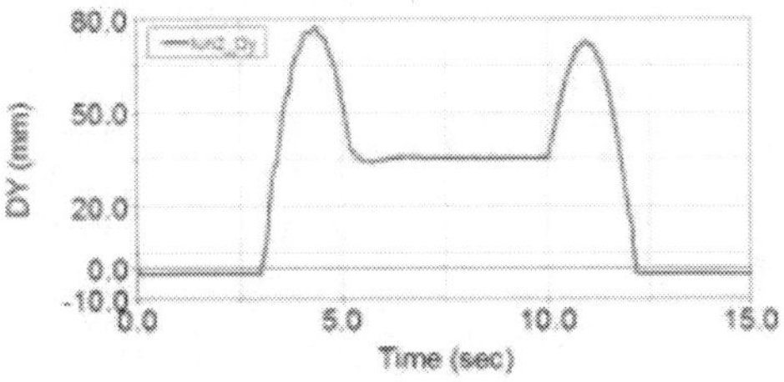

(b)

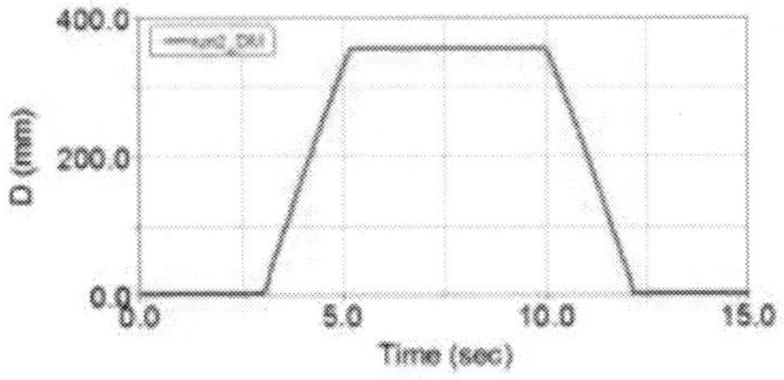

(c)

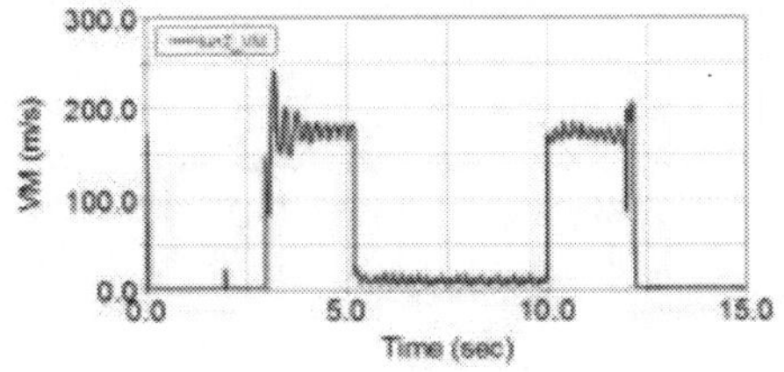

(d)

Fig. 5. The curve of the x-axis, y-axis, the displacement and velocity on driving wheel A.

The curve of the x-axis on driving wheel A is shown in Figure 5(a).The curve of the y-axis on driving wheel A is shown in Figure5 (b), the curve of the displacement on driving wheel A is shown in Figure5(c), the curve of the velocity on driving wheel A is shown in Figure5(d).

From the Figure5 (a), (b), (c), we can get that the clamp mechanism B clamps cable at 2 seconds, and begin to swing at 3 seconds, and swings to the maximum position at 5.2 seconds with the role of the positioning pins. As is shown in the Figure4 (d), the speed is vibration changes in the swing process, due to the change of robot gravity center caused by driving wheel swinging. Throughout the swing and drop back action, of the wheel speed has obvious mutation at 3 seconds, 5.2 seconds, 10 seconds and 12.2 seconds. The change of gravity center of the robot, caused by the swinging of the arm A, is shown specifically in Figure6(a),(b).

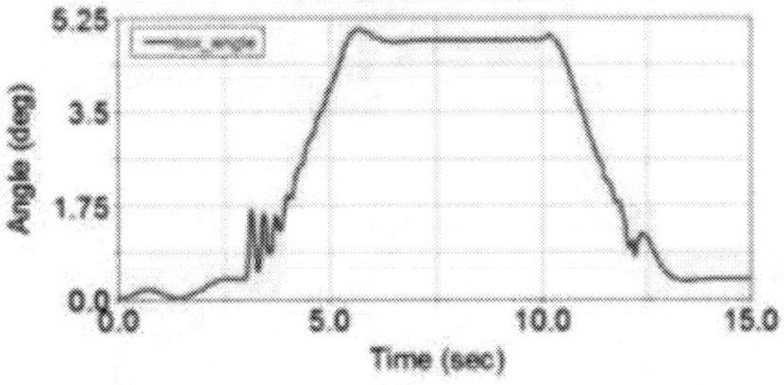

(a)

 DOI: 10.1051/matecconf/201710003015

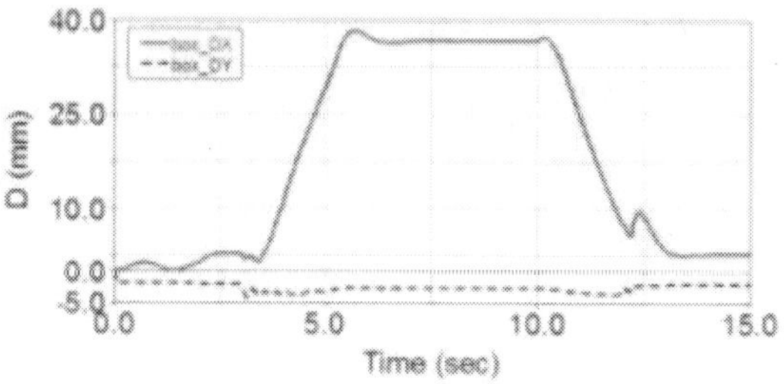

(b)

Fig. 6. The curve of eccentric angle α and the coordinates of the gravity center.

The curve of eccentric angle α and the coordinates of the gravity center is shown respectively in Figure 7. As showed in Figure7, eccentric angle α, the deflection angle of the robot relative to the central axis, is increasing gradually with driving wheel A swinging side, and reaches a maximum of 5.2 in 5.2 seconds. When the arm A starts to swing back to 10 seconds, the centroid of the entire robot gets back gradually again, so that the driving wheel A can drop back to the cable opportunely.

After the arm A swing side, two arms slides on the control box relative, and then the arm A drop back on the cable. During this crossing obstacles process, the curve of the gravity center of coordinates changed during crossing obstacle is shown in Figure7.

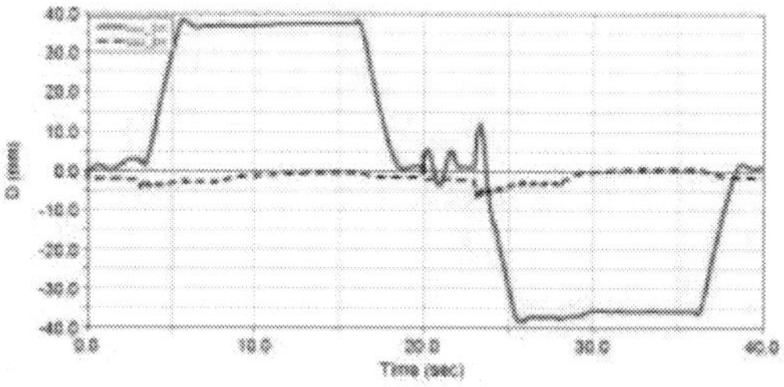

Fig. 7. The curve of the gravity center during crossing obstacle.

As is shown in Figure7, the centroid of robot changes little by its gravity before the arm A swinging. But when the arm A begins swinging to side in 3 seconds, the centroid m of the entire robot change significantly along the x-axis from 0 to 40mm, and gradually, the arm A get re-balanced at another side.

The mass center of the entire robot, when the arm A swing back again (t=18s), gradually returns to its original position, and then the driving wheel A down to the cable. In the process of the relative movement of the two sliding arms (5.2s, 18s) and (25s, 38s), the center of gravity is stable. Similarly, the gravity center swing with the swinging of the arm B. In the whole crossing obstacles process, the gravity center of robot changes very litter along the y-axis, in the range of 0-0.5mm.

The trajectory of driving wheel A， B and gravity center during crossing obstacle is shown in Figure8. As is described in Figure8, with the swing of the arm, the gravity center of robot offset to another side to make re-balance on the cable.

Fig. 8. The trajectory of driving wheel A, B.

3.3. Crossing towers process

When the robot has completed the de-icing task to the end of the cable, it needs across the tower. If it is straight tower, the robot can cross it through the crossing obstacles action, or it is strain type tower, the robot needs to transfer the path from phase line to the drainage line, so as to achieve continuous working between towers.

In the crossing towers process simulation, the gravity center of the robot also changed due to swinging of the arm, so that the robot can get to rebalance. The gravity center is offset along x-axis in the range of -40mm to 40mm, as is shown in Figure9.

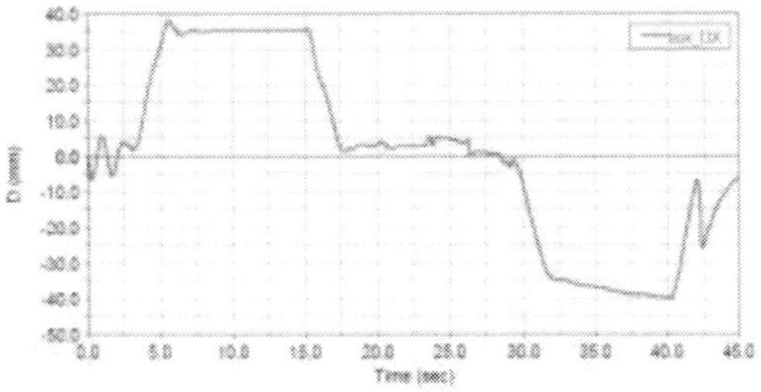

Fig. 9. The curve of gravity center during crossing tower.

Figure10 depicts the trajectory of two drive wheels and gravity center during crossing tension tower. With the swing of the arm, the gravity center of robot offset to another side to make re-balance on the cable.

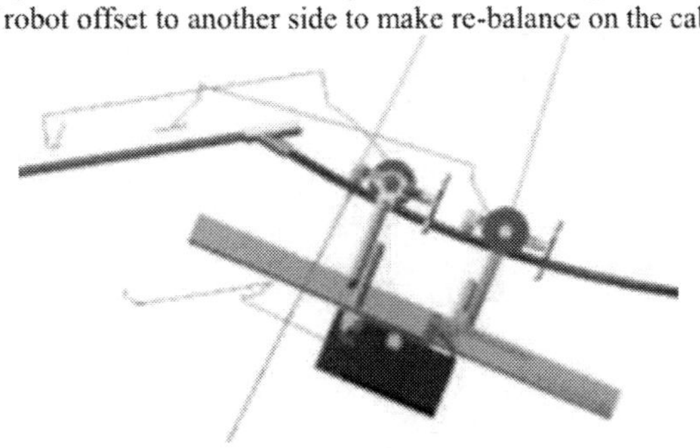

Fig. 10. The trajectory of two drive wheels.

3.4. Stability improvements

Through the crossing obstacles process and crossing towers process simulation, the position of rotation joint between the up-arm and the down-arm affect the driving wheels' swinging up action and drop back action, due to the wheel flange may collision with cable even cannot drop back.

The factors causes the gravity center offset are mainly constituted by following: 1) the mass ratio between the driving mechanism and the control box, 2) the swing range and swing speed of the driving wheel, 3) the clamping force of the clamping mechanism. Based on the points above, by adjusting the swing angle of the arm to 45 °, adjusting the final mass of the control box to 25kg in order to increase the mass ratio between the control box and the driving mechanism, while making the gravity center of the robot just below the cable; increased to intensify the coefficient of friction between the institutions and cable, increases the friction coefficient between the clamping mechanism and the cable.Figure11.

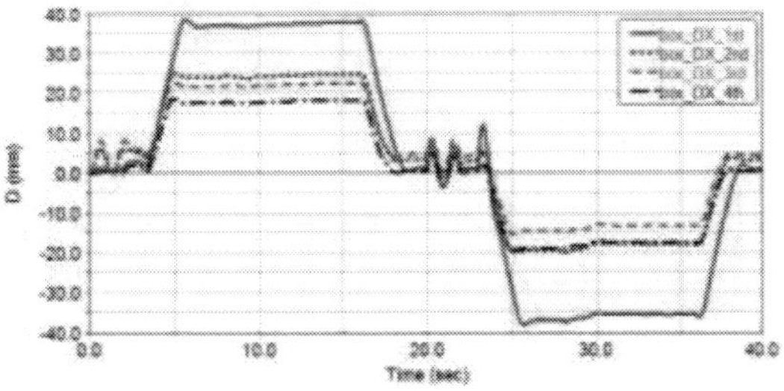

Fig. 11. The curve of the angle during crossing obstacle.

Figure12 depicts the curve of the eccentric angle and coordinate of the gravity center during crossing obstacle and tension tower.

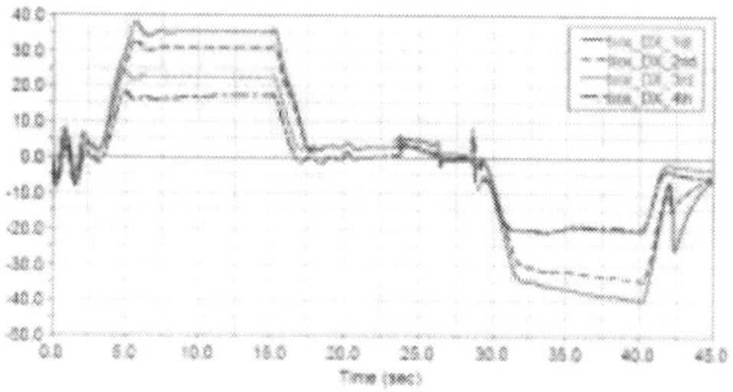

Fig. 12. The curve during crossing tension tower.

4 Conclusion

The coordinate system for the deicing robot was established at first, and deduced the kinematics equation. Then the virtual prototype model of deicing robot in kinematics simulation was analyzed. Deicing robot based on the simulation results can be obtained can smooth cable on the barrier; verify the correctness of the model. Factors influencing the robot center of gravity change was obtained by the simulation and it can be mainly composed of the following factors:

(1)The quality ratio between the driving mechanism and the control box;

(2) The location of the robot center of gravity;

(3) The driving mechanism of swing range and the size of the swing speed;

(4) The clamping force of clamping device. By adjusting the above factors, it can make the robot more stable cross obstacle or crossing tower, to improve the stability of the robot.

Acknowledgement

This material is based upon work supported by the National Science Foundation of China (Grant no. 51479043). The views expressed are authors alone.

References

1. K. Toussaint, N. Pouliot, S. Montambault. J Field Robot, **26**,5 (2009)
2. R. I. Egbert, R. L. Schrag, W. D. Bernhart, et al. IEEE T Power Deliver, **4**, 3 (1989)
3. K. Toussaint, N. Pouliot, S. Montambault. J Field Robot Wiley, **9** (2009)
4. J Michael, T. Moser, K. Bretter. IEEE SENS J, **35**,2 (2010)
5. T.G. Brown. Cold Regions Science and Technology, (2007)
6. H. Zangl, T. Bretterklieber, and G. Brasseur.IEEE T Instrum Meas, **58**,5 (2009)
7. S. Montambault, N. Pouliot. The HQ Line ROVer: contributing to innovation in transmission line maintenance, (2003)
8. M. J. Moser, T. Bretterklieber .IEEE T Ind Electron, **58**,7 (2011)
9. J. Sawada ,K. Kusumoto, Y. Maikawa , et al. IEEE T Power Deliver. **6**,1 (1991)

Creative Industries Development and Efficiency in China

Wentao Yu

Business School, Jiangsu Normal University, Jiangsu, China
Corresponding Email: zeroyu@mail.ustc.edu.cn

Abstract. China is becoming a leading player in the world market of creative goods and service. In this paper we first seek to explore the development and structure of urban creative industries in China. It indicates that China's city creative industries is fast increasing during the period of 2004-2008. Secondly, a Data Envelopment Analysis (DEA) model, based on the data of China's 36 capital and/or sub-provincial cities, is used to estimate the Technical Efficiency (TE) of urban creative industries, where the efficiency scores are decomposed into the Pure Technical Efficiency (PTE) and Scale Efficiency (SE). Due to poor input utilization and scale inefficiency, it shows most of cities in China are falling into the situation with low efficiency development on creative sectors. Finally, the paper provides some implications for policy makers on how to increase creative industries efficiency.

Keywords. Creative industries, technology efficiency, Data Envelopment Agreement (DEA).

1 Introduction

As our society moves into the so-called information age when the economic growth is greatly driven by innovation and creativity, the creative industries as one of most dynamic emerging sectors are growing rapidly and becoming a new engine of the world economic growth. UNCTAD (2010) shows that the world trade of creative goods and services continues its expansion in 2008, reaching $592 billion with an annual growth rate of 14 per cent during the period of 2002-2008, despite a decline of 12 per cent in global trade due to financial crisis. China has become a leading player in the world market for creative goods with an impressive 20 per cent exporting market share during the period from 2000 to 2008 (UNCTAD, 2010).

It is suggested creative industries is on the huge growth in the contemporary economy. Yusuf & Nabeshima (2005) highlight the economic center of gravity of the leading cities is changing from manufacturing to creative industries. To lead growth and own comparative advantage, cities need to evolve or remain as the centers of creative industries or service-providing hubs. The existing studies show the development of creative industries or the creative class plays an increasing role in driving and facilitating regional wealth and employment growth (Asheim & Hansen, 2009; Boschma & Fritsch, 2009; Rumpel et al., 2010; Cooke & De Propris, 2011). Creative industries rely on various factors, such as word of mouth, taste, culture, and popularity, in which individual choices are dominated by information feedback over social networks rather than innate preferences and price signals (Potts et al., 2008). Therefore, the creative industries should be explored by multidisciplinary approach (Pratt, 2013; Andres & Chapain, 2013; Kacerauskas, 2013), rather than only by case study or policy science perspective. Kacerauskas (2013) argues the analysis on creative and cultural industries should refer to philosophical, sociological and communication aspects. However, recent research lacks of paying attention to the growth of creative industries efficiency. As a result the efficiency score of the creative industries at the regional level or sector level is still under explored, particularly in China.

In this paper we attempt to respond to the existing unexplained questions and verify: how much efficiency scores does the creative industries achieve in China. Based on the data of China's 36 capital and/or sub-provincial cities, a data envelopment analysis (DEA) model is applied to assess technology efficiency of China's city creative industries (including pure technical efficiency and scale efficiency). The results show the technology efficiency score of urban creative industries is 58.3% during the period of 2002-2008 in China, including pure technical efficiency score of 66.3% and scale efficiency score of 89.3%. In general, due to poor input utilization and scale inefficiency, it shows most of cities in China are falling into the situation with low efficiency development on creative groups. This study not only can be helpful to understand problems hiding in China's creative industries despite of its fast growing but also provides a representative case for the world creative industries development from the biggest developing country in China.

The development structure of creative industries in China is depicted in the following section (Section 2). China's CIA at the city level is discussed in Section 3. China's city creative industries efficiency is estimated and analyzed in Section 4. The last section is the conclusions.

2 The development of creative industries in china

The participation in global markets for producing creative goods and services is extremely strong in the advanced economies. In recent years the rise of the creative industries are unprecedented in speed and scale in the developing economies, particularly in China. China is becoming a leading player in the world trade of creative goods, accounting for 20 per cent of world market. In 2008 China is turned out to be the largest trade surplus in creative goods with net exports of 79 billion, while the United States has the largest trade deficits in creative goods with net imports of 55 billion (UNCTAD, 2010).

Table 1 presents the development of China's creative industries compared to the general industry during the period of 2004-2008. The trend growth rate of the number of creative firm is calculated to be close to 20.11% per year while the growth rate of new firm in general industries appears to be 13.16% a year throughout the period of 2004-2008. The employment in the creative sectors grows 11.97% per year over 2004-2008, compared to 7.78% of employment growth for general industries. In 2008 the creative industries produced 4648.64 billion yuan in operating revenue with an average annual growth rate (AAGR) of 25.71% for 2004-2008, in contrast, the annual growth rate of operating revenue reach up to 34.05% in general industries.

Table 1. The development of creative industries in China.

Category	Industry	2004	2008	AAGR (%)
The number of firms	Creative industries (CI) (1000)	378.96	683.78	20.11
	General industries (GI) (1000)	3249.34	4959.67	13.16
	The share of CI in GI (%)	11.66	13.79	4.55
Employ-ment	Creative industries (million workers)	9.43	13.94	11.97
	General industries (million workers)	166.93	218.89	7.78
	The share of CI in GI (%)	5.65	6.37	3.19
Income	Creative industries (in billion of RMB)	2291.66	4648.64	25.71
	General industries (in billion of RMB)	44097.65	104167.2	34.05
	The share of CI in GI (%)	5.20	4.46	-3.53

Source: Author's calculations based on the China Economic Census Yearbook (2010).

In 2008 there are 683 781 creative firms registered in China, accounting for 13.79% in all registered firms, while the employment of creative industries is 13.94 million yuan, accounting for 6.37% of all employment in China. The revenue generated by creative firms is 4648.64 billion yuan, but accounting for 4.46% only of all industries revenue. It shows the proportion of business income of the creative industries in all industries revenue (i.e. 4.46%) is slightly less than the share of creative employment in total employment (i.e. 6.37%), but considerably less than the ratio of creative firm in all firms (i.e. 13.79%). It indicates that the average size of creative firms is slightly smaller than other firms, and the development of business incomes in creative industries is lagged behind general industries probably due to productivity shortfalls.

Data for the share of creative industries in general industries illustrated in Table 1 indicates that 13.79 per cent creative firms produce 6.37 per cent creative employments but only generate 4.46 per cent revenue incomes in all industries in 2008. Table 1 also shows a negative score (-3.53%) of the AAGR on operating revenue of creative industries during the period of 2004-2008, while the creative firms and employments are both highly increased by 4.55% and 3.19% respectively per year over the same period. In other words, it inputs the same creative firms and employments but outputs less incomes, implying there is serious inequality between the inputs and outputs of China's creative industries.

It is inferred that China's creative industries may fall into low efficiency by the observation that the share of creative industries revenue in total revenue is much less than the proportion of creative employment in total employment as revealed in Table 1. However, this argument may suffer from a suspicion because there is not sufficient empirical evidence to calculate how inefficient is the creative industries in China. To deal with this suspicion, empirically, the efficiency of creative industries by different sectors and regions will be estimated in Section 3.

3 China's city creative industries efficiency

Data envelopment analysis (DEA) is a mathematical programming approach to provide a relative efficiency assessment for a group of decision making units (DMU) by the ratio of inputs and outputs. Determining whether a DMU is efficient from the observed data is equivalent to testing whether the DMU is on the "frontier" of the production possibility set (Wei, 2000). In this paper, the variable returns to scale (VRS) model of DEA is used to calculate city efficiency of creative industries in China, where the VRS model allows breaking technical efficiency (TE) into two components: pure technical efficiency (PTE) and scale efficiency (SE). The PTE measures how a DMU utilizes the resources under exogenous environments, while the SE is the potential productivity gain from achieving optimal size of a firm (Wei, 2000).

Table 2. Summary statistics for the input and output data.

Inputs	**Unit**	**Mean**	**St. dv.**
Employment of creative industries	No. of workers	27571 1	353687
Total asset of creative industries	10 billion Yuan	492.40	995.19
Output	**Unit**	**Mean**	**St. dv.**
Operating revenue of creative industries	10 billion Yuan	93.30	169.28

Source: Author's calculations based on the China Economic Census Yearbook (2010).

The input and output variables selected should be best representative quality in estimating the efficiency. The DEA model, in this paper, includes two inputs and a single output. The output corresponds to the operating revenue of creative industries, while the input is measured by creative employment and total asset of creative industries. The data of inputs and output based on China's 36 capital and/or sub-capital cities all are from *China Economic Census Yearbook* (National Bureau of Statistic of China, 2010). Input and output variables have been identified in Table 2.

Table 3 presents the TE, PTE and SE scores of creative industries of top 10 and bottom 10 selected cities in 2008. On average, the technical efficiency score of China's creative industries is 58% in 2008, indicating 42% of efficiency has been lost. There are 15 cities (accounting for 42% of the sample) whose efficiency scores of creative industries have exceeded 60% (see Table 3). Nanchang and Urumchi being with TE scores of 100% are on the "frontier" of the production, suggesting these two cities have achieved the point of best relative efficiency in creative sectors, while Xining is found to be the least efficient relatively with TE scores of 27%. The TE score of top 10 cities are turned out to be more than 70%, in which 8 out of 10 cities, i.e. Urumchi, Nanchang, Beijing, Xi-an, Guangzhou, Shanghai, Chengdu and Lhasa are fully efficient on PTE score.

Table 3. Efficiency scores in urban China.

Top10 cities	Technical efficiency	Pure technical efficiency	Scale efficiency
Nanchang	1	1	1
Urumchi	1	1	1
Beijing	0.892	1	0.892
Changchun	0.874	0.968	0.902
Xi'an	0.842	1	0.842
Guangzhou	0.811	1	0.811
Shanghai	0.790	1	0.790
Chengdu	0.757	1	0.757
Jinan	0.752	0.88	0.854
Hohhot	0.749	0.763	0.981
Bottom 10 cities	**Technical efficiency**	**Pure technical efficiency**	**Scale efficiency**
Wuhan	0.451	0.487	0.926
Nanning	0.442	0.451	0.980
Harbin	0.439	0.439	0.999
Shenyang	0.423	0.428	0.989
Yinchuan	0.394	0.487	0.809
Guiyang	0.366	0.423	0.865
Tianjin	0.358	0.374	0.958
Lanzhou	0.302	0.326	0.927
Haikou	0.298	0.382	0.780
Xining	0.270	0.438	0.616
Mean	**0.583**	**0.663**	**0.893**

The mean score of scale efficiency is calculated as being 89.3% in creative industries, which is slightly higher than its TE scores (58.3%) and PTE scores (66.3%). In 2008, 30 out of 36 cities (approximately accounting for 83% of the sample) whose scale efficiency scores have exceeded 80%. However, 22 out of 36 cities (accounting for 61% of the sample) exhibit decreasing returns to scale on the production frontier of creative industries, implying the size of creative firm in those cities are engaged in production on a non-optimal scale. Alternatively, it may be a good way to increase their efficiency of creative industries by optimizing the size of creative firm appropriately.

If the cities aim to increase their technology efficiency of creative industries, both the PTE and SE should be paid attention to. Table 3 shows that some cities such as Lanzhou and Tianjin are calculated as low TE scores due to extremely PTE scores although they possess relatively high SE scores. In contrast, Zhengzhou is on the best efficient of scale, but its PTE score is only 48.2 per cent, leading to a relatively low technical efficiency.

Table 4. Efficiency scores estimated by creative sub-sectors

Creative sub-sectors	Technical efficiency	Pure technical efficiency	Scale efficiency
Software and computer services	0.689	0.758	0.911

Design	0.685	0.790	0.871
Performing arts and publishing	0.613	0.655	0.938
Leisure and Entertainment	0.572	0.666	0.858
Audio visuals	0.500	0.618	0.813
Creative R&D	0.396	0.474	0.879
Crafts and fashion	0.372	0.545	0.703
Advertising	0.301	0.406	0.830
Mean of creative sectors	0.583	0.663	0.893

Notes: Audio visuals include film, music, television and radio.

Table 4 presents the efficiency scores of creative industries by sub-sectors. It shows that China's software industry owns the highest technical efficiency scores (i.e. 68.9%), including PTE scores of 75.8% and SE scores of 91.1% in 2008. It is suggested the software industry has been growing rapidly and effectively over the last decades in China and boosts economic growth significantly (Yang et al., 2005), resulting in a higher technical efficiency. The TE scores of designing service (68.5%) and performing arts & publishing (61.3%) are found to be more than the average level (58.3%), while advertising is being on the least inefficient with TE score of 30.1% out of 8 creative sectors.

In summary, China's creative industries is experiencing tremendous growth over the world, but its productive efficiency and the ability to create business profit still have a big increasing room.

4 Conclusion

This paper, first explores the development and structure of China's creative industries, including the growth and development of the number of creative firm, creative employment and operating revenue. This exploration implies that China's creative industries is at the low efficiency level despite fast development. The DEA model, based on the data of China's 36 capital and/or sub-provincial cities, is used to investigate technology efficiency of China's city creative industries (including pure technical efficiency and scale efficiency). The empirical result shows the technology efficiency (particular in pure technology efficiency) of China's city creative industries is indeed relatively low. On average, the technical efficiency score of China's creative industries is estimated to be 58% in 2008. We suggest, on the one hand, the development of clustering of smaller creative firms is becoming an effective way to boost its efficiency, because most of cities in China now exhibit decreasing returns to scale on creative industries productivity. On the other hand, the pure technology efficiency of the creative groups should be much more increased by applying more new technologies and new business mode.

Acknowledgements

The author wants to acknowledge the support of the Natural Science Foundation of Jiangsu Province (No. BK20150234), and the Social Science Foundation of Jiangsu Province (No. 14EYC004).

References

1. UNCTAD, 2010, Creative Economy. Report 2008. UNDP, UNCTAD, Geneva New York. [R/OL]. http://unctad.org/en/Docs/ditctab20103_en.pdf. UNDP & UNCTAD, Geneva.
2. Yusuf S, Nabeshima K., 2005, Creative industries in east Asia. *Cities*, 22(2): 109-122.
3. Rumpel P, Slach O, Koutský J. 2010. Creative industries in spatial perspective in the old industrial Moravian-Silesian region. *E+M Ekonomie a Management*, 8(4): 30-46.
4. Asheim B T, Hansen H K. 2009. Knowledge Bases, Talents and Contexts: On the Usefulness of the Creative Class Approach in Sweden. *Economic Geography*, 85(4): 425-442.

5. Boschma R A, Fritsch M. 2009. Creative class and regional growth: empirical evidence from seven European countries. *Economic Geography*, 85(4): 391-423.

6. Cooke P, De Propris L. 2011. A policy agenda for EU smart growth: the role of creative and cultural industries. *Policy Studies*, 32(4): 365-375.

7. Potts J, Cunningham S. 2008. Four models of the creative industries. *International Journal of Cultural Policy*, 14(3): 233-247.

8. Pratt A C. 2004. Creative clusters: Towards the governance of the creative industries production system? *Media International Australia*, 112(1): 50–66.

9. Wei Q. 2000. Effective methods on evaluating relative efficiency (DEA). Chinese Science Bulletin, 45(17):1793-1807.

10. National Bureau of Statistics of China. 2009. China Statistical Yearbook (2010). China Statistics Press, Beijing.

11. Andres, L., Caroline C.2013. The integration of cultural and creative industries into local and regional development strategies in Birmingham and Marseille: towards an inclusive and collaborative governance? *Regional Studies*, 47(2): 161-182.

12. Kacerauskas, T. 2013. Creative and cultural industries: philosophical, sociological and communication aspects. *FILOSOFIJA-SOCIOLOGIJA*, 24(3): 112-120.

Research on Constructing a Model of Collaborative Innovation System for Equipment Manufacturing Industry

Baoyu Jia, Xiaonan Chen, Zhaoji Yu and Chuang Wei

School of Management,Shenyang University of Technology,Shenyang 110870,China
Corresponding author: shengchanjihua999@vip.sina.com

Abstract. In contrast to extant researches, this paper stats from the definition and analysis of manufacturing industry collaborative innovation system.Then,the author constructed a concept model of collaborative innovation system for equipment manufacturing industry.The model consists of a sub-system of cooperation education of government, enterprise, university and scientific research organization(CEGEUSRO),a sub-system of industry chain and a supporting platform system.On the foundation of the concept model,a system dynamic model was developed and simulation experiment was conducted.

1 Introduction

Liaoning intelligent manufacturing platform was forged to adapt to "Industry 4.0". However, there are still problems such as large but not strong and limited innovation ability. The contradiction between the surplus production capacity in lower or middle industries and the heavy dependence of the import of high-tech major equipment is not resolved [1].

2 Analysis of mechanism of actions among manufacturing industry collaborative subjects

On the basis of previous collaborative innovation models, this paper believes that the constructing subjects in a manufacturing industry collaborative innovation system involve the following seven enterprise or non-enterprise social organizations: manufacturing enterprises, government, universities or scientific research organizations, upstream and downstream enterprises in an industry chain, customer, financial institutions of science and technology, and intermediary service agencies [2].

As the real stimulator of market demands, customers play an important promoting role in the innovation of enterprises. Customers of manufacturing enterprises are characterized by centralization, complexity in purchase, and personalized demands. These distinctive features serve as the foundation of the collaboration between customers and manufacturing enterprise.

Technology collaboration and fund collaboration are the two main paths of collaboration between customer and manufacturing enterprises. The basis of collaboration of the technology collaboration path lies in customer's requirement for the improvements in techincal process of original products. Through technology development, enterprises continue upgrading their technologies in the constantly repeated process of "communication- improvement-communication " until new satisfactory products are produced for customers [3]. The basis of collaboration of the fund collaboration path lies in the financial guarantee provided to satisfy customers' demand for new products as well as enterprises' development of new products and expanded reproduction in the trading of products.

Knowledgeable and academic talented individuals gathering in universities and scientific research organizations greatly elevated the success rate and output rate of fundamental researches. Enterprises have advantages in the later stage of technology innovation, such as application research, research and development of products, and industrialization [4].

Enterprises' demand for technology is transferred to universities via the platform. Scientific research team in university develops new technologies that fit the demand of equipment manufacturing enterprise, consequently improve the competitiveness of enterprises. Meanwhile, talented individual output is another important collaboration path between universities and enterprises.

Intermediary service agencies serve as the adhesive among various innovation main bodies and the catalyst in the process innovation. They bridge the gaps in a collaborative innovation system [5].

After processing the technology, talented individual and industry development information gathered from the market [6], intermediary service agencies published them in digital form on intermediary service platform to satisfy enterprises' demand for specific innovation resource.

The financial system of science and technology is a service system aiming to combine science [7] and technology innovation with the capital chain and to stimulate the mutual promotion of the two [8].

Government financial departments and science and technology management department are the two major public financial main bodies of science and technology [9]. Through fiscal policy, government supports science and technology innovation of equipment manufacturing enterprises by direct science and technology appropriation [10]. Fiscal investment focuses on aiding the development of industry rather than gaining profits. Market financial main body of science and technology applies the principle of profit maximization to the investment in equipment manufacturing enterprises [11], therefore, it makes preferential investments. Whether in public financial system of science and technology or in the market financial system of science and technology, capital will be recycled in the process of collaboration [12]. The collaborative effect between the two parties will form a virtuous cycle and strengthen inner bodies in the system [13].

A complete industry chain contains processing of raw materials, production of intermediate products, product production, sales, and customers [14]. Every one of these five links can be considered as a node, and is made up of enterprises similar in nature. Studies showed that the equipment manufacturing industry chain is composed by upstream, middle stream and downstream enterprises [15]. Raw material manufacturers are the upstream enterprises. Middle ware manufacturers like component manufacturers and spare part manufacturers are middle stream enterprises. Module manufacturers and integrated manufacturers are downstream enterprises which ultimately serve customer [16].

When a customer makes a new demand for the technology of a product, based on features of the new product, downstream enterprises will submit improvement request to middle stream and upstream enterprises. Middle stream and upstream enterprises will upgrade the technology and integrate semi-finished products with downstream enterprise to produce the product the customer needed.

With the continuous development of modern market economy in China, the roles of government and market in resource allocation have changed. It has completed the transformation from "total planned economy" to "planned economy supplemented by market regulation" and finally to "market plays a decisive role". Government should admit the decisive role market mechanism plays in the development of equipment manufacturing industry, and take full advantage of overall planning to guide the sustained and healthy development of equipment manufacturing industry by taking active fiscal, currency, price and industry policies.

The collaborative paths between government and equipment manufacturing enterprise can be divided into direct path and indirect path. Direct collaboration refers to the direct collaborative effect between government and equipment manufacturing enterprises. Government promote the development of equipment manufacturing enterprises by formulating and implementing industrial policies. Indirect collaboration refers to the guarantee service provided by

DOI: 10.1051/matecconf/201710003017

government to ensure the rapid development of enterprises. It takes effect through optimizing supporting measures of the environment rather than direct collaboration with enterprises.

3 The construction of system dynamic model for collaborative innovation system for equipment manufacturing industry

3.1 System Boundary Determination

The system dynamic model for collaborative innovation system for equipment manufacturing industry this study constructed is composed of CEGEUSRO sub-system, supporting platform sub-system, and industry chain sub-system.

3.2 Basic Hypotheses

Hypothesis 1. The operation process of the system is not influenced by irresistible factors such as natural disasters or wars.

Hypothesis 2. The system operates in a continuous and gradual process during a period time from past to future.

Hypothesis 3. The intensity of collaboration between equipment manufacturing enterprise and university and scientific research organization is reflected by the proportion of enterprise investment into the R&D of university and scientific research organization.

Hypothesis 4. The number of patent application can fully reflect the innovation ability of equipment manufacturing enterprises. It can also be used to measure the overall operation effect of the collaborative innovation system for equipment manufacturing industry. It is the only representing variable of system output.

Hypothesis 5. The influence of upstream and downstream enterprises on innovation of equipment manufacturing enterprises is represented by "Income from New Product Sale".

Hypothesis 6. The service level of intermediary service agency is reflected by the information level of the environment an enterprise is in.

Hypothesis 7. University or scientific research organization pay back the corporate funding in R&D in form of paten transform.

3.3 The construction of causality loop chart and flow char of the model

Based on the analysis of internal components in the three sub-systems, this paper drafted the causality loop model of collaborative innovation system for equipment manufacturing industry, shown in Figure 1. The causality loop model is the integration of the causality loops of the three sub-systems. It vividly described the causality relationship among sub-systems and among components within the collaborative innovation system for equipment manufacturing industry.

To clearly illustrate the influence of different factor on the number of patent application of equipment manufacturing enterprise, this paper designed the following feedback circle to illustrate major feedback loops of the causality loop model. On the basis of previous causality loop model, this paper further put forward the flow chart model of the system to fully apprehend the internal structure and characteristics of dynamic behavior of the system.

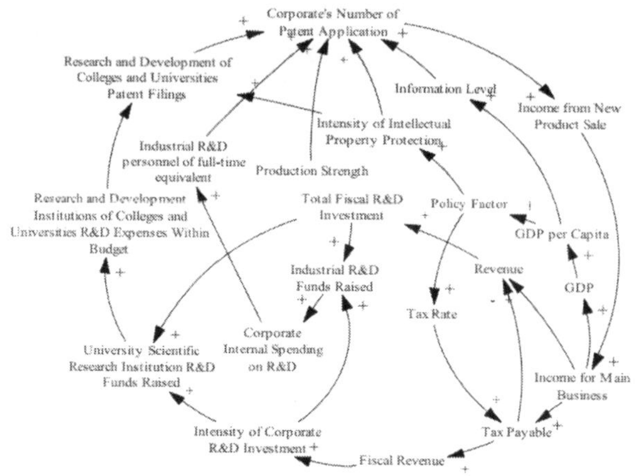

Fig.1. The causality loop model of collaborative innovation system for equipment manufacturing industry.

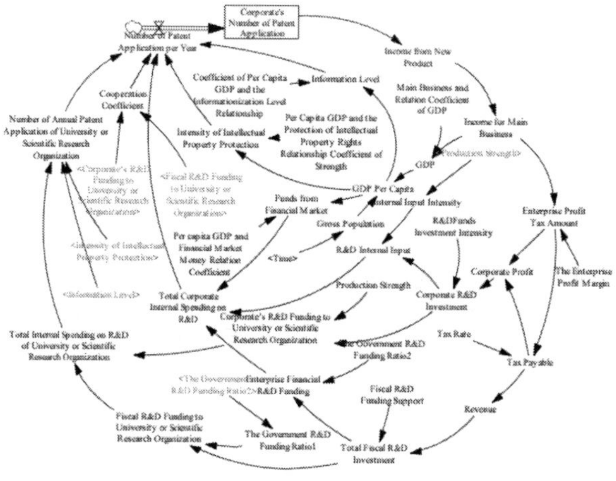

Fig.2. Flow chart model of collaborative innovation system for equipment manufacturing industry.

4 System Dynamic Simulation of the Collaborative Innovation System for Equipment Manufacturing Industry

4.1 The simulation of number of annual patent application of equipment manufacturing enterprise from 2014 to 2020

The aim of this experiment was to evaluate the general operation effect of the collaborative innovation system for equipment manufacturing industry when the innovation elements that all the collaborative main bodies invested in remained the same in future years. During the experiment process, no parameters in the dynamic model were altered, the model was run naturally to output the number of annual patent application from 2014 to 2020. The results are shown in Table 1.

Table 1. The increase rate of patent application from 2014 to 2020.

Year	2014	2015	2016	2017	2018	2019	2020
Number of Annual Patent Application	407614	488099	584971	701811	843098	1014480	1223156
Growth Rate	19.664%	19.745%	19.847%	19.974%	20.132%	20.328%	20.570%

The result shows that if the collaborative innovation system remain the same level, from the seven years between 2014 and 2020, the innovation ability of equipment manufacturing enterprise in China will steadfastly elevated with an increasing rate of 20.037%. This reflects a strong momentum for growth of the integral innovation ability of equipment manufacturing enterprise.

4.2 Simulation of Variable's Sensitivity

This simulation experiment was designed to determine sensitive factors that influenced the effect of collaborative innovation system for equipment manufacturing industry. The purpose was to identify the most influential and most sensitive variable to conduct further analysis to predict or estimate its degree of influence. The detection of the root of uncertainties would be the decision basis for effective countermeasures. Variables in this sensitivity analysis included funds from financial market, income from new product sale, information level, intensity of intellectual property protection, tax rate, strength of fiscal R&D funding support. The number of annual patent application was used as representing variable to measure the variables' sensitivity. To prepare for sensitivity analysis, judgment of positive or negative correlation between every variable and representing variables was carried out before sensitivity analysis.

(1) Analysis of the influence of variation of single variable on the number of patent application

The positive or negative influence of the variable on collaborative innovation system for equipment manufacturing industry was judged by measuring the variation of the representing variable when the amount of variation of a single variable was adjusted.

1) Analysis of the variation in corporate R&D investment

The intensity of corporate R&D investment was set to increase by 2%, 4%, 6%, 8%, and 10%.

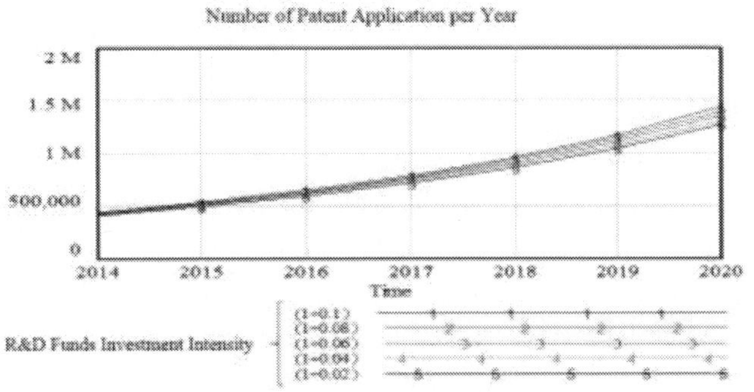

Fig.3. The influence of corporate R&D investment on number of patent application.

The result showed that, the number of patent application increased along with the increase of corporate R&D investment. This indicates that corporate R&D investment has a positive promoting effect on number of patent application. When the intensity of corporate R&D investment elevated, the money spent on R&D increased, and consequently, the corporate's number of patent application increased accordingly.

2) Analysis of the variation of total fiscal R&D investment

The strength of fiscal R&D funding support was set to increase by 2%, 4%, 6%, 8%, and 10%, the result of the output of number of annual patent application is shown in Table 2.

Table 2. The influence of total fiscal R&D investment on number of patent application.

Year	Funding Support (1+0.1)	Funding Support (1+0.08)	Funding Support (1+0.06)	Funding Support (1+0.04)	Funding Support (1+0.02)
2014	409215	408933	408652	408372	408093
2015	490412	489997	489584	489172	488762

2016	588287	587684	587083	586484	585887
2017	706545	705673	704804	703939	703077
2018	849836	848582	847333	846088	844848
2019	1024065	1022265	1020471	1018684	1016904
2020	1236798	1234214	1231640	1229076	1226524

The result showed that, the number of patent application increased along with the increase of total fiscal R&D investment. This indicates that total fiscal R&D investment has a positive promoting effect on number of patent application. The inflow of capital from government total fiscal R&D investment an enterprise received, increased corporate internal spending on R&D and resulted in the increase in the number of patent application.

3) Analysis of variation in tax revenue

The tax rate was set to decrease by 2%, 4%, 6%, 8%, and 10%, the result of the output of number of annual patent application is shown in Table 3.

Table 3. The influence of tax rate on number of patent application.

Year	Rate (1-0.1)	Rate (1-0.08)	Rate (1-0.06)	Rate (1-0.04)	Rate (1-0.02)
2014	406282	406574	406873	407180	407494
2015	485970	486423	486888	487365	487853
2016	581655	582345	583055	583783	584529
2017	696734	697774	698843	699940	701066
2018	835420	836973	838569	840208	841889
2019	1002970	1005274	1007643	1010076	1012572
2020	1206006	1209409	1212908	1216504	1220196

The result showed that, the number of patent application of equipment manufacturing enterprise was mildly influenced by decease of tax rate. The analysis of causality feedback loop indicated that on the one hand the reduced tax promoted the equipment manufacturing enterprise’ s total investment into R&D, on the other hand, it reduced government revenue. Since the government’ s R&D funding ratio of equipment manufacturing enterprise remained the same, as a result, the R&D funding equipment manufacturing enterprise received from the government was influenced. The simulation results told us that equipment manufacturing enterprise’ s total internal spending on R&D was affected, and ultimately influenced the output of patent application of equipment manufacturing enterprise. This indicated that though the adjustment of tax rate could increase corporate R&D investment, it was corporate R&D spending that directly influenced corporate’ s number of patent application. Besides, other factors also influenced corporate’ s total R&D spending, such as the amount of government funding, fund from financial market, etc.

4) Analysis of the variation in funds from financial market

The amount of funds from financial market was set to increase by 2 times, 4 times, 6 times, 8 times, and 10 times.The result showed that, the number of patent application increased together with the increase of financial market’ s investment. This indicates that fund from financial market has a positive promoting effect on number of patent application. Fund from financial market is an important capital guarantee of corporate’ s increase in investment, it can share corporate’ s innovation R&D risk.

5) Analysis of the variation in income from new product sale

The income from new product sale was set to increase by 2%, 4%, 6%, 8%, and 10%.

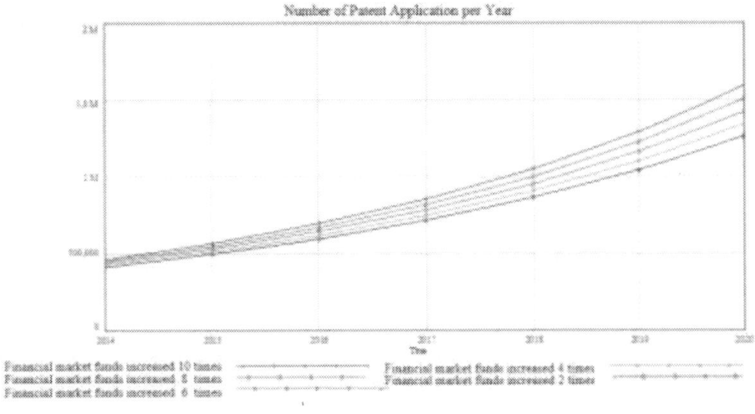

Fig.4. The influence of funds from financial market on number of patent application.

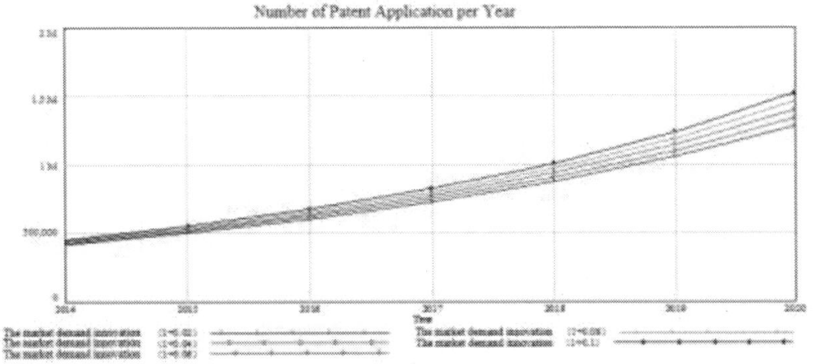

Fig.5. The influence of income from new product sale (customer demand for innovation) on number of patent application.

The result showed that, the number of patent application increased as income from new product increased. This indicates that income from new product sale has a positive promoting effect on number of patent application. Income from new product sale reflects customer' s demand for innovation. Corporate can gain a high additional profit above general product. As a result, the increase in profit encourages corporate' s enhancement of the investment in technology improvement to better satisfy customer' s demand for innovation, thus corporate' s innovation ability is constantly elevated.

6) Analysis of variation in information level

The information level was set to increase by 2%, 4%, 6%, 8%, and 10%, the result of the output of number of annual patent application is shown in Table 4.

Table 4. The influence of information level on number of patent application.

Year	Level (1+0.1)	Level (1+0.08)	Level (1+0.06)	Level (1+0.04)	Level (1+0.02)
2014	417027	415185	413342	411500	409657
2015	500452	498029	495607	493187	490769
2016	601081	597913	594751	591593	588441
2017	722719	718601	714492	710391	706300
2018	870143	864806	859485	854179	848888
2019	1049391	1042491	1035615	1028763	1021936
2020	1268184	1259270	1250393	1241553	1232750

The result showed that, the number of patent application increased in accordance with the improvement of information level. This indicates that information level has a positive promoting effect on number of patent application.

5 Conclusion

5.1 For Enterprises

5.1.1 Increase R&D Investment

Firstly, equipment manufacturing enterprise should cultivate the innovative consciousness of "Science and technology constitute a primary productive force" and grasp the concept that the corporate's core competitiveness is founded on technology R&D and new product R&D. They should fully realize that technology innovation is of vital importance to the development of an enterprise. Secondly, a special R&D fund should be set up in guarantee the sustainability of R&D investment. Lastly, under the guidance of government's macro industry policies, enterprises should actively cooperate with financial main body in the market to expand R&D financing resources.

5.1.2 Fastened the integration of enterprises and informatization

The result of simulation indicated that information level crucially promote the innovation ability of equipment manufacturing industry. At present, fastening the integration of informatization and industrialization has become a fundamental policy on equipment manufacturing industry in Liaoning Province. For an equipment manufacturing enterprise, on the one hand, investment in informatization should be increased so that complete informatization covering product R&D, production and sale can be realized by more advance information technologies. On the other hand, cultivation of talents in informatization should be improved to guide the integration between equipment manufacturing technology and information technology. Lessons can be drawn from overseas experience. There is plenty room to explore in the cooperation between universities and enterprises, university can cultivate talents for enterprises.

5.1.3 Promote cooperation with university and scientific research organization

The simulation result indicated that equipment manufacturing enterprises can enhance innovation efficiency by the cooperation with university and scientific research organization. Universities and scientific research organizations have advantages in original innovations. Whereas the advantages of equipment manufacturing enterprise lie in the application research and industrialization.

5.2 For Government

5.2.1 Establish a multi-source financing system

Government should explore more channels of financing for enterprises. The establishment of a multi-direction and multi-layer financing system will guarantee the sufficiency of funds to invest in the research and development of technology. Special attention should be paid to small and medium-size enterprises with stronger innovation vigor. These enterprises have remarkable innovation abilities. The guarantee of sufficient R&D funds is an important approach to improve innovation efficiency. In the meantime, government should promote the cooperation between enterprises and organization in financial market and other aspects of the society. Innovation risk fund of the industry should be created to share risk for enterprise innovation.

5.2.2 Provide enterprises with more complete intellectual property legal system

A more complete intellectual property protection system necessarily safeguards the fruit of corporate innovation. The result of simulation revealed that the intensity of intellectual property protection can extremely encourage corporate' s passion of innovation and produce a lasting effect on the number of corporate' s patent application. Government should create a serious atmosphere of being protected by laws in the society. Government should raise corporates' consciousness using law to protect themselves and teach them to keep their economic interests intact by using laws.

5.2.3 Build a favorable demand market for enterprises

The products produced by equipment manufacturing enterprise in our province have no advantages in the competition with products from developed countries because the technology contained in the products of is not high enough and the added value of products is low. This explains enterprises' incompetence Many enterprises, no matter how strong the intention of developing new product is, were forced to converted into accessory manufactures at the end of value chain. A favorable demand market built by government can promote the product sale made by domestic equipment manufacturing enterprises by means of lowing value-added tax. The market can support the export of innovative products from independent brands with core technology by the comprehensive application of export drawback. Besides government should take policy measures such as further increasing the proportion of purchase. As a result, the expanded demand market will promote the sale of products made by equipment manufacturing enterprises. In the meantime, the product and capital turnover rate will be improved so that the capital the corporate technology innovation needed will be guaranteed.

References

1. Michael T, Mark A.V, Jeen S L. Manufacturing technology and strategy formulation: keys to enhancing competitiveness and improving performance.Journal of Operations Management, 1999 (17): 411-428.
2. Gilbert B A. Knowledge spillovers and new venture performance: an empirical examination. Journal of Business Venturing, 2008, 23(4):405-422.
3. Hersh D. Manufacturing Matter.IEEE Engineering Management Review, 2003, 31(3):3-10.
4. Corbett L. Grappling with a Gusher Manufacturing`s Response to Business Success in Small and Medium Enterprises. Journal of Operations Management, 2002, 20(5):495-517.
5. Hyun-Sun R, Jung L. How Non-technological Innovation Reinforces the Effect of Technological Innovation on Firm Performance: An Empirical Study of Korean Manufacturing Industry. Communications in Computer and Information Science, 2015(10):176-182.
6. Patricia A. Eco-innovation on Manufacturing Industry: The Role of Sustainability on Innovation Processes.Greening of Industry Networks Studies, 2014(4):99-116.
7. Jose M B. Innovation in the Portuguese Manufacturing Industry: Analysis of a Longitudinal Company Panel. International Advances in Economic Research, 2005(11):301-314.
8. Marek V. Innovation behavior of firms in a small open economy: the case of the Czech manufacturing industry. Empirical, 2015(20):1-29.
9. Candace E, Thomas A. Turk. The evolution of trust in information technology alliances. Journal of High Technology Management Research, 1998(20):62-74.
10. Porter M E. The Competitive Strategy of Nations. New York: The Free Press,1990.
11. Hacklin F, Marxt C, Fahrni F. Strategic venture partner selection for collaborative innovation in production systems: A decision support system-based approach. International Journal of Production Economics, 2006, 104(1):100-112.

12. Islam N, Ozcan S. Nanotechnology Innovation System: An Empirical Analysis of the Emerging Actors and Collaborative Networks. IEEE Transactions on Engineering Management, 2013, 60(4):687-703.
13. Fang Q, Zheng J G. Structure characteristics of collaborative innovation system and its implementation paths—— Based on interpretative structural model. Economist, 2015.
14. Jian-Zhong X U, Zhao W F, Wang L J. Analysis of the Relationship among Agents of Collaborative Innovation System in Equipment Manufacturing Industry Based on Game Theory. China Soft Science, 2014.
15. Bambini M D, Coltri P P, Furtado A T, et al. Collaborative Innovation in Agrometeorology: Coordination Strategies to Develop a Monitoring IT System for Brazil. Journal of Technology Management & Innovation, 2014, 9(1):119-130.
16. Zhao Y, Ke L V. Construction and Implementation of Information Resources Allocation System for Collaborative Innovation. Information Science, 2013, 4(13):109-117.

The Internet of Things Technology Development and Application -- Zhengzhou Airport as An Example

Songlin Hu[1] and Bing Li[1]

[1]Management Engineering Department, Zhengzhou University, Zhengzhou, China
Corresponding author: 804541130@qq.com

Abstract. In order to understand the trend of the development of the Internet of things, and the role of development and construction in the future,this article introduces the developing situation of the Internet of things at home and abroad. Summarizes the architecture and key technologies of iot, this paper expounds the application of the Internet of things, the last development of the Internet of things industry of our country is prospected. Combining with the development of the Internet of things industry area, zhengzhou air port development makes us on the Internet of things has a comprehensive and accurate understanding and awareness.

1 Development status at home and abroad

Networking is an important part of the strategic emerging industries, is following the computer, Internet and mobile communication is a new round of revolution of information technology [1]. Information technology is promoting the further application in all walks of life in a new round of the wave of information technology. According to the definition of the International Telecommunication Union (1TU), the Internet of things mainly solves the problem of the thing to thing (T2T), human to thing (H2T), the human to human (H2H) [2]. The Internet is based on the Internet, through the radio frequency identification (RFII)) [3], infrared sensors, GPS, laser scanners and other information sensing equipment, according to the agreed protocol to anything connected with the Internet, information exchange and communication, to realize intelligent identification, positioning, tracking, monitoring and management of a network. The Internet of things is a new system of real time interaction between the virtual network and the real world [4-5]. It is the extension and expansion on the basis of the Internet, with the telecommunications network, radio and television network and Internet network [6-7].

1.1 The Internet of things in the developed countries and districts development

The United States attaches great importance to the strategic position of the Internet of things. In February 2009, Obama signed a $787 billion economic stimulus plan, including plans to invest $11 billion intelligent power grid, the wisdom medical $19 billion. In October 2009, the U.S. government approval to spend $3.4 billion m to establish intelligent power grid investment fund, the remaining $4.7 billion from the folk. In October 2009, the European commission in the form of a policy paper released iot strategy, based on the Internet to help the eu in leading the global intelligent infrastructure development, in addition to the ICT development plans to invest 400 million euros, launched more than 90 research and development projects to raise the level of intelligent network, the commission will also be in an annual increase of 200 million euros from 2011 to 2013 to further strengthen research and development efforts, at the same time fund earmarked 300 million euros. Support iot related public-private partnerships for short-term projects. South Korea in 2006 put forward a 10-year U - Kore8 strategy, Korea communications commission in September 2010 passed the "Internet of things the infrastructure construction of basic planning", established iot market as a new growth engine, determines the constructing iot infrastructure, development of the Internet service, developing the Internet of things technology, creating an environment of spread of the Internet of things, and other areas of the four, and 12 subjects. Japan, Australia, Singapore, France, Germany and other developed countries to speed up the deployment of the next generation of Internet of things network infrastructure.

1.2 Domestic development of Internet of things

China attaches great importance to the development of the Internet of things industry, the Internet of things as a strategic emerging industries to focus on and promote. In 2006 the State Council issued the "national long-term science and technology development plan (2006 ~ 2020)" on "important areas and priority themes", "major projects" and "advanced technology" has involved Internet content. In 2009, put forward 3 requirements of the development of the Internet of things in Wuxi Premier Wen Jiabao visits: (1) combine the sensing system in 3G and TD; (2) in the major national science and technology projects. To accelerate the development of sensor networks; (3) the establishment of the heart China sensing information as soon as possible, or "perception China center". Premier Wen Jiabao's important speech opened a new era in the development of China's Internet of things. 2010 two sessions of the government work report to foster strategic emerging industries to accelerate the development of the Internet of things applications. The local government subsequently period of time to carry out the construction of rapid development planning and the pilot project of the networking industry. Introduced a number of preferential policies to support industrial development; industry alliance, alliance enterprises have set up. In the standard setting, technical development and market environment has carried out a lot of work. In recent years, the development of the Internet of things Chinese the policy environment continues to improve, significantly accelerate the technological progress. In Wuxi set up a national sensing innovation demonstration zone as a sign, the Chinese Internet of things is accelerating the development of the local government of the Internet of things economic development showed a high enthusiasm and enthusiasm. Through the joint efforts of government and industry, China Internet industry rising strength, has made some achievements in the research of key technology, semiconductor, software and other fields. The domestic standardization work pace at present. China has made significant progress in wireless sensor network, communication technology, micro sensor, sensor, terminal G mobile station. Has the materials, technology, devices, network system to complete industrial chain.

2 Internet of things architecture and key technologies

Internet of things is a comprehensive technology of information, communication, sensor, automatic control and so on. The key technologies of sensing layer include sensor, RFID, near distance wireless communication, self organizing network, middleware, embedded system. Network layer key technologies include M2M, cognitive radio, heterogeneous integration, business capacity development, network environment awareness, etc. The application layer key technologies including software architecture, middleware, resource virtualization, data storage, data sharing, sea state, data mining, intelligent computing etc.. Internet of things also involves identification, analysis, security, network management and other important common technology.

2.1 Sensing layer key technologies

Sensing layer technology has a unified identification technology, RFID technology, sensor technology, sensor network technology, etc. Uniform identification technology to solve the problem of uniform identification of objects. RFID technology needs to solve the ultra high frequency and microwave band RFID chip, antenna, RFID middleware, tag anti collision algorithm, security authentication protocol and other issues. Sensor technology is the development of small size, low power consumption, self powered integrated central processing unit, micro processing system and other new types of chips. Since the sensor set foot technology mainly solve the multi-level network technology, wired and wireless mobile

2.2 Key technologies of network layer

Internet of things required to make full use of telecommunications networks, radio and television networks and other types of network. The Tao is actively promoting the integration of three networks, through the broadcast television network and Internet integration of telecommunications networks, development, realize the network interconnection, providing voice, data and broadcast television service for users. Triple play is the core of mechanism innovation and technological innovation, and secondly, the Internet of things need to carry the information of the sea, so new materials need to be used to research and development of new transmission network. Once again, only to achieve a variety of sensor networks, wide area of data sharing and multi shared, as well as the scale of the application, in order to really build it a valid Internet of things.

2.3 Application layer key technologies

Mainly includes the application processing center and system integration technology based on cloud computing. In the data networking and application of juvenile order asked quite nervous system networking throughout the network. It needs to solve data filtering, data aggregation and information transfer of 3 key issues. To achieve accurate interoperability, in order to meet the Internet of things in a hybrid network, the efficient operation of heterogeneous environment. Because there is a large amount of redundant information in the data collected by the perception layer, the data filtering technique and the data aggregation technique are very important. After filtering and aggregation processed data need to transfer technical support.

2.4 Safety problem

The network layer and service layer security is independent of each other, and a large part of the network security problem is special because the Internet is based on the existing mobile network integrated perception network and application platform to bring. There are three issues to consider. First, because the Internet of things in many places need wireless transmission, if there is no protection, it is easy to be disturbed and stolen. Two is due to the large number of nodes in the Internet of things. In order to cluster the existence of a large number of machine data will be sent to the network congestion. Three is the RFID system security issues. A RFID chip is designed to be bad and not protected. There are many ways to get the chip's structure and data. Therefore, in the study of Internet of things application, at the same time, to strengthen the research of Internet of things, the Internet of things to become an open, secure, trusted network.

3 The development of the Internet of things industry in Zhengzhou Airport

3.1 Development advantages

Zhengzhou air port is an important part of the Zhengzhou comprehensive experimentation area and Zhengzhou New District, is the first national strategy for the development of China's aviation port economic development zone. As an important part of the overall planning of Zhengzhou New District, the air port is the core growth zone of the Central Plains Economic Zone. It is also an important window and base in Henan province.

3.1.1 Outstanding traffic advantages

Zhengzhou airport is located in the hinterland, the spatial condition is better, the main city and 3/5 in 1.5 hours covering Chinese 2/3 population, Zhengzhou airport planning and construction of more than 4 runway, cargo throughput growth in China, large airport first week full cargo flights in the Midwest second major airports. The province's railway mileage 4822 km, with Zhengzhou as the center of the M shaped high-speed rail network layout is speeding up the construction, the highway traffic mileage of 5830 kilometers, all the county can be 20 minutes on a highway, a modern integrated transport system and air docking, multimodal transport of goods is becoming more and more perfect, the logistics cost and time cost advantage obviously thinning.

3.1.2 Things industry foundation is abundant

Zhengzhou airport networking industry has a good industrial base, driven by the convergence effect of Foxconn, there are more than and 100 industries, more than and 400 supporting enterprises located in its periphery. These enterprises mainly engaged in many fields of electronic information, communication network, transmission sensor and microelectronics, covering the Internet of things, and in the downstream industry chain, has great practical significance to the Zhengzhou air port to seize the future development of the networking industry of high-end areas.

3.1.3 Possess relevant advanced technology

At present, Zhengzhou air port has formed gas sensor, track monitoring, power sensor and other features of the networking industry chain, the emergence of Hanwei electronics, ANN, Wulian purple letter victory and a number of influential in the domestic enterprises, the networking industry development system initially formed. A number of electronic information, biological medicine, aviation transportation enterprises to speed up the gathering to the Zhengzhou airport, Zhengzhou airport with the enhancement of industrial agglomeration agglomeration effect, attracting Foxconn, the Russian air bridge, China Southern Airlines, SF express, aviation, electric Huadian Jin Yuan and other famous enterprises.

3.2 Existing problems

3.2.1 Understanding is not clear

Zhengzhou airport economic zone to promote urban modernization management and intelligent level of public service, using advanced information technology such as the Internet of things, to accelerate the "wireless city", "digital city", "wisdom city", "city" perception of construction. But due to consider their own conditions for application requirements and industrial reality, together with the Internet of things, cloud computing is the development of emerging industries such as law, stage, understanding unclear positioning concept lead to blind investment, speculation and repeated construction unreasonable phenomena occur frequently.

3.2.2 Industry chain is not perfect

At present, the Internet of things application or low levels of demand, is more of a stay in individual project pilot, is just a fragment application. At the same time, the industry application of parallel present a decentralized information island. Many enterprise application is a separate build your own business, communication is insufficient, can't form effective resource sharing. At the same time, the industry chain is relatively fragmented, the lack of dominant power, has not yet found a clear business model, and the Internet of things applications demand motivation is seriously insufficient, enterprise information alone obviously unable to support the whole industrial chain.

3.2.3 Industry achievement outflow phenomenon is serious

Zhengzhou air port iot industry and technological achievements is rich, but most of the industry results failed to form in the region economic benefit, but for other regions. This situation led to the zhengzhou air port iot industry although strong research and development, technological achievements more comprehensive results but less, the situation of application level is low, not conducive to the long-term development of the industry.

3.3 Suggestions for the development of the Internet of things

3.3.1 Increase research and development of industry standards

Zhengzhou air port should be independent innovation and merger fusion combined with industry standard to increase investment, and ensure that the networking industry related technology research and development status in the policy, actively seek independent intellectual property rights of the core technology, and actively participate in the work for the national standard of. On the basis of referring to national standards, the introduction of Zhengzhou air port networking industry standards as soon as possible, so that the development of enterprises to define the direction of the industry, the development and application of better working on networking technology, development of the networking industry regularly; encourage technological progress, how to set up a special project of national technology, breakthroughs in core technology research and development from the technical standard, independent to independent, so as to truly dominate the networking industry the right to speak.

3.3.2 Promote the integration of existing industries

Zhengzhou air port to exert the development of the Internet of things industry advantages and the existing industrial foundation, promote industrial integration. To strengthen the coordinated development of the overall coordination, the formation of resource sharing, collaborative work pattern and the various aspects of mutual support, mutual promotion of coordinated development effect. To strengthen the coordination of things related to planning, science and technology major projects, industrial projects such as the reasonable layout of things major demonstration and industrialization projects, strengthen the industrial chain and regional division of labor.

3.3.3 Government departments to actively guide and support

Establish and improve the beneficial iot application promotion, innovation motivation, orderly competition policy system, to promote to establish perfect information security and privacy protection laws and regulations. Encourage capital fair into the market access mechanism. To speed up the iot related standards, testing, certification and other public service platform construction, improve the supporting service system. Strengthen the protection of intellectual property rights, actively carry out Internet related technical analysis of the intellectual property review, accelerate the iot related patent layout.

4 Summary and outlook

The Internet of things technology application market is rapid growth, the world as the communication equipment, the deepening of management software and relevant technology, the Internet of things technology as a decline in the cost of related products, Internet business will gradually towards a comprehensive application. The Internet of things industry value brought by the dozens of times larger than the Internet, the huge economic interests will inevitably drive competition in the technology. Global technology powerhouse successively put forward the development strategy of the Internet of things, set off a new wave of Internet of things. The domestic famous universities and research institutions compete to itching, gaining momentum. Many provinces are also put forward the corresponding development strategies, and building demonstration project in succession. Looking to the future, the state and the government has clearly put forward the development of the Internet of things "experience China" the grand strategic objectives. It will next to the terminal standardization and intelligent development.

References

1. Zhang Hongzheng, Ma Teng. Design and practice of smart city information system. China CIO News, 2013 (4): 37-37.
2. Ran Qi. On the use of Internet of things and cloud computing to build a smart city information system. Chin New TeleCommuns, 2013 (19): 40-40.
3. Ye Tianchun. Development and construction of China's Internet of things industry. Internet of Things Technologies , 2012, 2 (12): 1-3.
4. Shi Lu. Principle of the wisdom of the city and in the city of our country in the development of the function and significance of. Forum on Science and Technology in China 2011 (5): 97-102.
5. Zhang Qun. Depth analysis of the Internet of things. communications business management, 2010 (1): 26-29.
6. Cao Yang, Wang Jianping. Study on the structure of Internet of things and its industrial chain. Technoeconomics & Management Research, 2013,02:98-101.
7. Wang Jianping, Cao Yang, the history of the software industry chain. As a philosophy of. Chin Soft Sci, 2011 (8): 27-32.

MATEC Web of Conferences **100**, 03019 (2017) DOI: 10.1051/matecconf/201710003019
GCMM 2016

The Development of Low Carbon Manufacturing Industry in Hainan Province

Xi Wang

Business Administration Department, Hainan College of Economics and Business, Haikou, Hainan Province, China
Corresponding author: cissywx2016@163.com

Abstract. As the main part of industrial production, the manufacturing industry promotes the economic growth of our country, but the subsequent large amount of energy consumption and greenhouse gas emissions have greatly undermined human's living environment. Therefore this paper aims at finding some solutions through the example of Hainan's low carbon manufacturing industry. It first analyzes the importance of low carbon economy for the development of the manufacturing industry, then describes the current status of Hainan's low carbon manufacturing industry, and finally puts forward five ways for the development of low carbon manufacturing in Hainan Province such as perfecting relevant laws and regulations, formulating supporting policies, setting up suitable industrial layout, improving the low-carbon technology innovation system, and establishing incentive mechanism of personnel in manufacturing enterprises. The third part is also the main innovation in this field.

1 The importance of low carbon economy to the development of manufacturing industry

Low carbon manufacturing industry is major breakthrough with low resource consumption and less environmental pollution [1]. The application of low carbon technology can increase productivity, reduce carbon emissions, realize the real clean production as well as improve the competitiveness of manufacturing industry.

1.1 Low carbon economy can enhance the scale and competitiveness of manufacturing

From the perspective of enterprise scale, manufacturing enterprises in China are scattered with small scale and low industrial concentration. Besides, the popularizing rate of energy-saving technology and related equipment is low. And, few ones have possessed their own brand and enter the world-top 500 large enterprises [2]. Low carbon technology can improve the comprehensive utilization of resources, optimize the allocation of resources as well as reduce enterprise cost, which contributes a lot to creating lower carbonation products, meeting consumer demand for low carbon products, adapting to the market demand in an effectively way. It is also beneficial for enhancing product competitiveness, and expanding the scale of industry. In addition, the application of low-carbon technology can phase out the large enterprises with relatively low-technical level, high-energy consumption as well as much pollution on environment. As a result, the enterprise with the core technology and innovation ability can acquire small businesses by means of relying on technology and innovation advantages, and thus quickly realize scale expansion, strengthen the intensive industry, and improve the efficiency of the industrial economy of scale.

1.2 Low carbon economy can promote the sustainable development of manufacturing industry

Low carbon economy means traditional ecological technology enterprises will gradually be replaced by more environmentally friendly low carbon technology, which can help enterprises realize green and ecological manufacturing, such as green design, green process planning, the selection of green materials and so on [3]. If manufacturing enterprises can make full use of low carbon technology for production, they will consume fewer resources, prevent environmental pollution, realize sustainable development, and comprehensively enhance their competitive advantages.

1.3 Low carbon economy can help develop new industries

Low carbon manufacturing is an important part of the development of emerging industries in china. The government has begun to advocate the application of low carbon technology in manufacturing enterprises, which accelerates the upgrading of manufacturing industry in our country, promotes the implementation of energy saving and emission reduction project, produces a great number of energy-efficient equipment and lead to the development of new industries [4]. What's more, the development of new industries will promote the steady growth of China's economy and help to realize sustainable development.

2 The development of low carbon manufacturing in Hainan

Since Hainan was listed in the second batch of low carbon pilot provinces by the state in 2012, Hainan has spared no effort to pursuing the development of low carbon manufacturing, and also achieving a lot of results.

2.1 Low carbon manufacturing output has grown steadily

During the "twelfth five-year" period, the low-carbon manufacturing industry of Hainan grew at the rate of 6.7% on every year, consumed only 2.2% of the industrial energy and contributed more than 25% of the industrial added value. Besides, it became the important support for the industrial optimization structure, transformation and upgrading of Hainan. Although the total profit of the low carbon manufacturing in Hainan province declined slightly from 2012 to 2015, their gross industrial output value, industrial added value, prime operating revenue increased to different extent, showing the low carbon manufacturing of Hainan province presented steady growth.

2.2 Low carbon manufacturing enterprises ascend innovation ability

By the end of 2015, there were 1800 low carbon manufacturing enterprises in Hainan province, including 200 scale enterprises, and they formed a larger scale of industry cluster in the fields of automobile manufacturing, biomedicine, new energy, new material and tropical agricultural products processing. Meanwhile, enterprises have improved innovation ability constantly, and more than half of the 169 high-tech enterprises in Hainan province belong to low carbon manufacturing enterprises. For example, Sanya ZTE Co., Ltd was the world's first enterprise to introduce the formal commercial H. 265 standard video terminal in 2015, and realized the software MCU and end products series software supporting cloud deployment; Hainan Litree Water Purification Technology Industry Co., Ltd contributed its technology to becoming shareholder of Russia Ecolos Group, building membrane technology water purification facilities with the total investment of over $100 million, as well as becoming the first high-tech enterprise in Hainan to be the shareholder of foreign factories by intellectual property rights.

2.3 The gradual appearance of green development

Low carbon manufacturing has made significant breakthrough in the field of saving energy and reducing consumption. In 2015, low carbon manufacturing of Hainan province consumed 1.9% of the total industrial energy, and contributed to more than 25% of the industry GDP. Besides, it made great contributions to the industrial saving energy and reducing consumption. Meanwhile, the oil and gas chemical industry, mineral processing and other traditional energy-intensive industries effectively controlled energy consumption, and the new comprehensive energy consumption of new products of some industry has reached the leading domestic level. In 2015, the industrial ten thousand Yuan GDP energy consumption in Hainan province fell to 2.331 tons of standard coal; pollutants such as chemical oxygen demand (COD) and sulfur dioxide (SO2) were brought under the control [5]; industrial structure is gradually optimized, and green, low carbon development trend have become increasingly obvious.

3 Development strategy of low carbon manufacturing industry in Hainan province

Hainan province should form a low carbon manufacturing industry with Hainan characteristics under the strategy of the Belt and Road, Made in China 2025, the development of the South China Sea. The followings are some suggestions.

3.1 Formulate relevant laws and regulations

At present, our country has not set up special laws and regulations for the development of low carbon manufacturing industry. Therefore, Hainan province should learn from other provinces and countries, formulate relevant laws and regulations to support the development of low carbon manufacturing industry, set out a standard system covering energy-saving production equipment, monitoring and management, energy management and auditing etc. [6]. For example, "Energy Conservation Regulations in Hainan Province" has made specific provisions on the construction of energy-saving industrial structure, which strengthens energy conservation supervision and improves energy-saving mechanism etc.. More similar regulations should be promulgated in the near future so as to provide a strong legal protection for the development of low carbon manufacturing industry in Hainan.

3.2 Formulate related policy and form the long-term mechanism of low carbon industry development

Formulate related policy and form the long-term mechanism of low carbon industry development. Due to the large risks and slow returns of low carbon industry, many enterprises are reluctant to invest. Thus, the government subsidies are required. Government departments should formulate the corresponding subsidy policy to encourage enterprises to conduct low carbon industry research and investment. For example, Special Fund Management Interim Measures for Low Carbon Industry Development of Hainan Province (hereinafter referred to as the Measures) stipulates that the provincial government should allocate special funds to support development of low carbon manufacturing every year, and give special funds in principle, in the way of loans with discounted interest, special subsidies, incentives, and equity investment. In addition, special funds are used for supporting the product R&D, technical innovation, brand building and talent incentive of the low carbon manufacturing industry enterprises in the fields of high-end equipment manufacturing (including new energy automobile, electrical and mechanical equipment, marine engineering equipment manufacturing and cruise yacht), green food processing, new energy, new materials and intelligent manufacture. In addition, the related policies, such as tax policy, price policy, and science and technology policy should be formulated as soon as possible, so as to optimize the financing environment of low carbon manufacturing, broaden the financing channels, and encourage all kinds of capital, and provide financing support for enterprise development from the aspects of innovation ability construction, technical transformation and industrialization, as well as enterprise merger and reorganization, which can eventually form long-term investment mechanism and provide policy support platform.

3.3 Set up suitable industrial layout

Hainan should focus on creating advantageous manufacturing industry of its own characteristics to maximize the benefits of limited resources. According to the "Made in China, 2025", Hainan will vigorously promote the development of pharmaceutical industry, automobile manufacturing, electronics manufacturing, marine equipment manufacturing, tourism manufacturing, new energy and materials, agricultural and sideline products processing, tropical marine biological industry, the integration of military and civilian industry and the oil and gas industry in the next ten years. And Hainan will also strive to set up a number of leading domestic manufacturing enterprises, brands and products as well as several industrial parks with complete functions and advanced facilities [7]. After determining the key areas, the government should actively improve the development of organizational leadership, policy environment, financing channels, investment-invitation and personnel structure to get out of a long-term development of the road.

3.4 Improve the low carbon technology innovation system

On the one hand, capital investment for low carbon technology innovation should be emphasized. Low carbon manufacturing enterprises should increase investment in new energy technology innovation to provide financial support for the application of new technologies [8]. On the other hand, low carbon technology innovation should be strengthened. The manufacturing industry can improve the technology innovation level by employing the advanced technical talents, cooperating with other enterprises and institutions, and introducing foreign advanced technology and so on. For instance, a number of enterprises reached a cooperation agreement, signed 15 cooperation projects such as new energy vehicles, marine equipment, civil military integration, 3D printing, sensors, seafood processing, etc. on a low-carbon manufacturing investment forum on July 29, 2016, in Hainan. These projects help promote technology innovation level of new energy vehicles and tropical agricultural products processing in Hainan Province, and lay a solid foundation for the development of the integration of military and civilian industry in Hainan.

3.5 Establish the incentive mechanism of personnel

As one of the new economic growth points in Hainan Province, Low carbon manufacturing industry needs a large number of excellent personnel who can help accelerate industrial development [9]. Therefore, the government should formulate relevant policies as soon as possible to attract high-level talents, while manufacturing enterprises should introduce and cultivate engineers and technicians in this field. Besides the talent introduction mechanism, in order to lay a solid foundation for the development and give full play to the initiative of high-level personnel, enterprises should also establish the incentive mechanism and make use of various supporting measures to motivate employees to improve their professional skills. For example, the manufacturer can gradually establish a full range of multi-channel human capital security system, which guarantees senior employees their housing, family employment, social insurance and other benefits. Finally, the enterprise should form a cultural atmosphere for innovation, encourage all staff try their best to learn new low carbon technology, which can lead to the development of low carbon manufacturing industry.

4 Conclusion

During the past five years, Hainan's low carbon manufacturing industry has witnessed its great development. However, there is still a long way to go to realize the optimization of the development. So the government departments and relevant enterprises shall work together to explore new ways and modes for further development from such aspects as regulations and policies, industrial layout, technological innovation and incentive system. In the next few years, Hainan's manufacturing industry will truly realize the rise of green low carbon.

References

1. Z.J. Liang, Create the Six Major Industries Around "Three Main Lines" Hainan daily (2015).
2. X.Y. Duan, Y.H. Zhang, The Low Carbon Road of Manufacturing Industry in China Environmental Protection **35** (2010).
3. R.S. Li, Research on the Core Competence of China's Manufacturing Eenterprises in Low Carbon Economy Ph D Thesis. Harbin University, **30** (2011).

4. H.J. Cao, H.C. Li, Y.B. Du, X.G. Li, The Research Status, Development Trends and Challenges of Low Carbon Manufacturing Aviation Manufacturing Technology, **28** (2012).
5. Z.J. Liang,The New Industry Road of Hainan Hainan daily (2013).
6. Chongqing Municipal People's Government Office on the Issuance of Chongqing City to Strengthen Energy Conservation Standardization Implementation Plan (2015).
7. Hainan's Manufacturing Industry Focus on 9 Areas, www.hainan.gov.cn.
8. J.B. Zhao, A Study on the Impact of Low Carbon Technology Innovation of Manufacturing Iindustry Master Thesis. Harbin University, **36** (2013).
9. Analysis Of the Introduction of Overseas High-level Personnel System in Hainan Province, http://hi.lss.gov.cn (2015).

Achievements of Theoretical Research on Chinese Technological Innovation

Hanxi Wang[1,*], Chengna Wan[2], Zude Zhou[1] and Shuhua Hu[3]

[1] Hubei Digital Manufacturing Key Laboratory, Wuhan University of Technology, Wuhan, China
[2] School of Arts and Law, Wuhan University of Technology, Wuhan, China
[3] Management College, Wuhan University of Technology, Wuhan, China
Corresponding author:dms2005@163.com

Abstract. Using the CNKI database as the sample source, we investigated the makeup of current research results on Chinese technological innovation, and found that the study of Chinese technological innovation consists of nine directions - underlying principles of innovation behavior, concepts of innovation cognition, methods of innovation, behavior of innovation subjects, structure of innovation mediators, innovation capacity, public service system of innovation, policy support system of innovation, and cultural environment of innovation, as well as several issues of innovation, including its behavior mechanism, operational mechanism, work mechanism, modes of implementation, modes of mathematical representation, modes of physical representation, modes of behavioral representation, modes of performance representation, modes of function realization, modes of structure realization, modes of workflow realization, modes of carrier realization, etc. Such a scattered makeup requires us to formulate the system of Chinese technological innovation achievements from the perspective of philosophy of science. This will be an important mission in the study of Chinese technological innovation.

1 Introduction

Although innovation was discussed on a large scale in China, the content of these discussions, mostly were conducted on the conception of innovation and on the purpose of innovation application. [1]

As science research, the behavioral mechanism, operational mechanism , working system and operational modes of innovation requires attention. The mathematic representative mode, physical representative modes, behavioral modes and performance representative modes of innovation need building up. The realizing modes of function, structure, procedure, embodiment of innovation need exploring [2]. From the perspective of retrieve CNKI database, the theoretical structure of China theoretical innovation research is reviewed in the paper.

2 Research on innovation behavioral principles

To construct innovation system on the base of innovation behavioral mechanism. This is the appropriate way for innovation to carry on from natural form to definite form. Therefore, through explorations on innovation operational rule which take as its targets the inspiring effects generated by innovation, life mechanism of innovation, working mechanism that innovation is implementing, to construct an innovation working paradigm to guide innovation development, is a key focus in China innovation theoretical research.

round the innovation mode in industrial system, Dr. Pan Xiongfeng [3], taking high-tech industry as his object, explored the life mechanism of innovation in high-tech industrial cluster, and depicted industrial cluster innovation effects by adopting innovation performance.

innovation behavioral mechanism is composed of mechanisms of impetus, matching, incentive, and operation etc. For impetus mechanism of innovation, Ji Yushan et al. [4] applied theories about social structure (i.e. evolutionary economics, technology) and organizational management in research of operation law of innovation system from intra-enterprise, regional and social network to the country at the top. And they also designed a technological innovation system with dynamic mechanism in it.For the innovation match mechanism, Zhao Dumin [5] proposed a conversion and matching mechanism for technological innovation achievements through the matching between technological innovator and entrepreneur, between technological characteristics and organizational form, and between technological entrepreneurship and supporting system.

For the innovation impetus mechanism, Dr. Guo Xiaolin [6]applied the theory of behavior sciences on the sharing and diffusion mechanism of industrial common technology; adopting motivation theories proposed incentive mechanism of industrial common technology including organizational incentive, financial incentive, spiritual incentive as well as rational combination and application of each single incentive according to the change of time condition.

Dr. Chen Fengdi [7] made a theoretical reconstruction for the operation mechanism of China technological innovation in accordance with operation mechanism of innovation system. She proposed operation mechanism of China technological innovation, composed of four parts including enterprise mechanism, government mechanism, market mechanism and social mechanism.

3 Discussion on the concept of innovation

Innovation is derived from discontent or dissatisfaction with present conditions, and the perception of this dissatisfaction is originated from the innovation subjects who are good at finding out problems, conflicts, and have the courage to challenge the present condition and oneself. The quality of the innovation subjects is depend on the their outlook on value conception, development conception, time conception. Therefore to expound the guiding function of innovation conception via value conception, development conception, and time conception is a fundamental issue in innovation theoretical research in China.

Dr. Lin jing [8] annotated the the possibility of science innovation through analysis of the existence significance of science innovation and demonstrated the reasons for its existence. And he also made an analysis on epistemological significance of science innovation and explained its epistemological contribution to philosophy of science and Marxist philosophy. What's more, he analyzed it on the basis of value and revealed the importance of science innovation to social progress and the all-around development of human beings.

Dr. Tian hui [9] discussed the currency significance of innovation. He proposed that innovative practice is the fundamental origin of epistemology, basic motive of epistemological development, and fundamental approach for verifying cognition; and also that cognition development is the history of innovative practice, and discussed the significance and value of innovative practice to the construction of contemporary epistemology.

Dr. Chen Li [10] discussed the time conception of innovation. He defined the scope of innovative practice from the perspective of philosophy according to the topic that innovation is core spirit of the time, and studied the structural problems of innovative practice (subject, object and intermediary), external environment problems (system, culture and resources) and mutual relationships between structure and environment. And he proposed that innovative practice is the prime impetus of human society progress, and inferred that as for society progress, practice of technological innovation is a basic impetus and practice of institutional innovation acts as an institutional impetus, while practice of intellectual innovation serves as a spiritual impetus.

4 Research on innovation approaches

True value of innovation research lies in making innovation a purposeful, active, automatic, orderly, normalized project activity. In order to establish a sustainable system of innovation project at human will, we need to make a deep investigation to the approaches to innovation. And only with normal innovative approaches to guide innovative project practice, can we avoid aimlessness and detours. Constructing innovative approach system is one of the key research in China innovation theory. Present researches mostly are focused on the explorations of innovation analysis approaches like its motivation analysis, risk analysis and organizational analysis.

For the motivation of innovation, Dr. Zhang Hongshi [11] constructed a motivation analysis framework of breakthrough innovation and identified a set of key motives. He found out the laws of breakthrough innovation through the interpretation of fitted curves, and set forth the breakthrough innovation mode 'Pan-Duality Organization' defined by six dimensions. And he also proved the effectiveness of 'Pan-Duality Organization' by comparison of statistic analysis and case analysis.

Around the innovation law exploration of organization structure, Dr. Liang Hongsong [12] built PAEI model in accordance with enterprise lifecycle phase judgment based on five-point scale,. And he also created an analytical model based on the organization and innovation mechanism of enterprise lifecycle.

round the exploration on innovation risk analytical approaches, Dr. Huang Yuansheng [13] constructed a selection method in technological innovation projects and an evaluation method of technological innovation capability based on the relationship between basic research and technological innovation. Meanwhile he created an identification method of technological innovation risks on the basis of the essence of technological innovation risks.

5 Research on the behaviors of innovation agents

Human beings are the subject of social practice, and also the agents of innovation practice. In order to inspire and achieve innovation, we need to make a research on the behavior of innovation subject inevitably. As a result, the research on the innovation behavior of innovation agents which takes traits, behavioral division, behavioral factors and leading capability of innovation subject as its parameters, is an important part of innovation theory research.

Innovative leadership is a key factor to innovation practice, so the behavior research of creative leaders is a very important aspect in the research of innovation subject. According to the motivation for innovation promotion of entrepreneurs, Dr. Guan Wei [14] described and analyzed the characters of technological innovation for entrepreneurs through four factors (variables) such as innovation motivation, innovation capability, innovation power and innovation decision, searched for their mutual relationship, established "an analysis model for technological innovation and enterprise scale based on control force" to make a research on dynamic relationship between technological innovation and enterprise scale in the transferring of small business to large enterprise.

Normal innovation behavior ensures stable innovation, so to construct innovation subjects' innovation behavioral norm is the ultimate goal of innovation subjects behavior. Dr. Li Zhiqiang [15] applied Principal-Agent theory in moral hazard analysis of the innovation activities. What's more, he constructed game models for describing entrepreneurs' innovation selections, and made use of institutional evolution theory analysis to propose an evolutionary model of entrepreneurs' innovation ability based on the absorption of tacit knowledge, and evolutionary logical model for institutional configuration of entrepreneurs' innovation activities.

6 Research on the structure of innovation medium

Innovation medium refers to the operation system established for people to implement innovation activities. The research on innovation medium is developed for the purpose of constructing a long-term, stable, creative, passionate and sustainable innovation operation project system. Generally, we need to construct the innovation operation project system in two aspects, one is project framework of innovation system including structural forms, construction paths, operation modes and tool platforms and the other, management framework of innovation system, including development strategy, behavior strategy, operational modes, and management systems. Structural framework of innovation operation project is a tool platform to achieve sustainable innovation, and also an experimental base for experimental innovation theory, applicable innovation theory and developing innovation theory, thus it is the top priority of innovation theory research.

Clear and stable innovation approach is a shortcut to implement innovation. Professor Hu Shuhua [16] created a theoretical system of product innovation management through a deep investigation of function and cost problems involved in the development and design of products under the support of Natural Science Foundation of China. His academic work, "Product Innovation and Management" published in 2000, set up a path pattern of product design and development intended for product innovation.

As the basis of standardization and engineering of innovation activity, innovation mode can provide innovation activity with behavioral paradigm. Dr. Wang Xueyuan [17] constructed a theoretical optimization framework for configuring system structure based on scientific and technological plan, in accordance with regional configuration problems of technological innovation resource. He proposed optimization rules for configuring system structure, and designed an optimization management model for configuring system structure. He also suggested a regional configuration mode of technological innovation resource and constructed a fractal evaluation model for the configuration effect of regional resources in technological innovation.

As the standard of innovation activity, innovation mode can make it follow established path, execute established regulations, use established tools and stand at established position according to project mode, in order to achieve innovation operation. However the generalization of innovation mode becomes the bottleneck for restraining its wide application. Dr. Zhao Zhenglong [18] analyzed the influence of network structure on innovation diffusion with relative advantages by constructing an innovation diffusion model based on complex social network.

Innovation mode is the high abstract for the behavior rules of innovation. It can show its value only when it is merged into specific innovation system, as a result, constructing innovation system based on innovation mode is an important topic which attracts great attention of academic circles. Dr. Liu Guoyan [19] proposed the framework, content and approaches of theoretical system for large innovative enterprises based on the two factors such as innovation and large enterprise

The building of innovation system doesn't represent launching real practice of innovation projects. The engineering of innovation not only needs innovation system, but also requires the development of innovation operation tools, the establishment of innovation operation procedures and the building of innovation operation workshop. As a result, related researches on the above three are indispensable directions of innovation research. Dr. He Shan [20] revealed the mechanism and laws of product innovation platform from the following four aspects functional combination, module distribution, level association and life cycle, and proposed three innovative behavior modes involving breakthrough, derivative and promotive product platforms. On this basis, he set forth a product innovation platform composed of management platform, information platform and technology platform.

As the basis of innovation management, innovation strategy is the strategic orientation for planning innovation activity, thus it is an important direction of innovation research. Dr. Wang Xueli [21] holds that, we can retrieve and identify innovation signals of internal and external environment from complex conditions through environmental scan and opportunity identification, then decide required extent and content of innovation from three aspects such as extant strategic evaluation, cooperative analysis of new and old strategies, as well as revolution cost of strategic innovation. Finally, we can set up new strategic location, create new business mode or construct new value network and the like, to realize effective strategic innovation.

7 Research on innovation capacity

Innovation capability is comprehensive subjective entity constructed by innovation subject with the support, subordination, and coordination from factors concerning innovation, and is also the premise and base to realize innovation's sustainable development, the important insurance of human civilization's sustainable progress as well. Therefore, to cultivate innovation capability of sustainable growth, is a key issue of China innovation theoretical research. From the present results of China innovation research, related China innovation capability research involves three aspects including existence forms, institutional structure and performance evaluation of innovation capability.

As for the theoretical origin of innovation capability, from the perspective of science philosophy, Dr. Lv Jianrong [22] clarified the history of innovation nation theory by historical analysis of intellectual economy, state innovation system, innovation state. He defined innovative nation by building a theoretical structure of innovative nation.

On the theory building of innovation capability, Dr. Jiang Hong [23] analyzed the application field of technological relevance and the functioning mechanism of industrial innovation and technology selection; put forward an industrial innovation model and technology selection theory model in view of technology relevance; built industry classification, industry selection model and technology selection and evaluation model based on the view of technology relevance.

On the cultivation and reproduction of innovation capabiliity, Dr. Zhang Zhongtang [24] discovered the law of enterprises' innovation system by defining the connotation of constant innovation system, put forward the functioning mechanism and structural characteristics of the leading force of enterprises' constant innovation system.

On the institutional structure of innovation capability, Dr. Chen Hong [25] found that the maintenance of stable core competitiveness depends on the knowledge innovation and knowledge management capacity of the organization; the internal knowledge innovation of the organization depends on the synergy of organization structure, organization system culture and internal knowledge network; the building of external knowledge network is a method to overcome the innovation paradox; to establish strategic knowledge network is an effective method to create stable competitive advantages.

Innovation performance is the core of innovation capability construction. So the evaluation of innovation performance is the hot issue of innovation capability theoretical research. There are four directions for the research on the evaluation of innovation capability performance, exploration on assessment mechanism, design of assessment mode, design of assessment index, design of assessment system.

On the design of performance evaluation mode of innovation capability, Dr. Sun Bing [26] built a comprehensive evaluation index system about the impetus of technological innovation of enterprises which include four levels and 30 basic indicators; he used Grid-Fuzzy Borda Number Analytical Method to give weight to each indicator in the evaluation index system, adopted principal component projection to determine the comprehensive evaluation result of impetus of technological innovation of enterprises, and hence established the comprehensive evaluation model of impetus of technological innovation of enterprises.

On the exploration on assessment of innovation capability performance, Dr. Li Qingdong [27] , brought forward the evolution mechanism of industrial innovation system and synergetic evolution theory of industrial innovation system, and built a performance evaluation model and evaluation index system of industrial innovation system.

Around the design of assessment index and assessment system, on the base of scientific and technologic innovation capability structure in colleges and universities, Wang Zhangbao et al . [28] built a quantification comprehensive evaluation index system of technological innovation capability in colleges and universities according to the framework of five principles, three modules, twelve evaluation items and thirty evaluation indicators and brought forward the indicator system for the performance evaluation of scientific research of colleges and universities.

8 Research on innovation service system

Innovation public service system is composed of at least four essential parts, that is, innovation achievement marker service (i.e. patent service and scientific publications), innovation technology transferring service (i.e. technology exchange market), innovation information sharing service which support innovation information dissemination and technological literature sharing, innovation knowledge popularization service. As a result, how to create a innovation public service framework intending to build innovation service public platform is regarded as a direction of innovation theoretical research.

Around innovation information transmission in information sharing service, Dr. Chen Xile [29]proposed three conditions of information dissemination in network including the information finiteness of innovation subject, the desirability of knowledge innovation activity and self-organized characteristics of technological information system, and created three mechanisms of information dissemination in network, including information selection, information coordination and information feedback.

Around the construction of technological information sharing service system, Hu Limei et al. [30] raised the problems about the construction of technological information service system, including service subject, network architecture and service platform.

On the resource configuration in innovation technology transmission service and innovation knowledge popularization service, Xu Jiang et al. [31] set forth a theoretical model for the optimization configuration of regional technology innovation service resources. They argued that the input ratio of public and private innovation service resources to different innovation service fields can have significant effect on the formation of regional technology innovation service capacity.

The ultimate goal of research on innovation public service system is the construction of innovation service platform. Zhang Qinglai et al. [32] designed an overall framework of knowledge service support system facing technological innovation, and defined the functions of each sub-system. They also proposed that on the basis of knowledge service supporting system, the knowledge service supporting hierarchy should be constructed through single-disciplinary knowledge portal, multi-disciplinary knowledge portal and even meta-knowledge portal, establishing the implement method in hierarchical construction.

9 Research on innovation supporting policy system

The realization of innovation needs innovation supporting policy system. The innovation supporting policy system includes innovation public policy system and government innovation management system. Which kind of innovation supporting system can stimulate innovation, and how to construct the system are the key directions of China innovation theory research.

Innovation policy means that the government, based on the acknowledgement and its demarcation of innovation value, adopts effective combination of functional macro-control methods, such as technology policy promotion, industry policy promotion, capital policy promotion and justice policy promotion. Around the synergetic promotion between technology and industry, and on the capital promotion, Wang Jingting et al. [33] analyzed the disadvantages of capital markets on each level in China from the point of supporting technological innovation, constructed multi-level capital market framework in support of technological innovation. They also made a research on the paths and policies for supporting technological innovation, involving capital markets of five levels, including main board market, growth enterprising market, agency share transfer system, property-rights exchange market and foreign capital market.

Around the government innovation management mode, Dr. Dongyou [34] discussed the relationship between technological innovation in colleges and universities and the government innovation management from the level of institutional level. He also explored government innovation management mode from two aspects: organization module and across-organization module, and designed policy adjusting tool for science innovation in colleges and universities, also developed a government macro-control evaluation index system, developed computer-aided information management system on the base of policy tool and policy match.

10 Discussion on cultural environment of innovation

Innovation must be supported by a wide and open cultural environment. China is an ancient nation with a history of 5000 years whose affluent cultural deposit is the base of innovation and hindrance as well.

Dr. Cui Kai [35] set up a dimensional model of enterprise innovation activities, composed of such enterprise dimensions as innovation directions, innovation behavior process, innovation risks and innovation environment for the relationship problems between innovation and culture. He explored the relationship between culture and innovation gradually with the dimensional model of enterprise innovation activities.

Dr. Cheng Liangbin [36] discussed the influence mechanism of culture on technological innovation through social capital, and illustrated the core status of cultural tradition as a basic part in Chinese technological innovation policies, and the temporal duality of cultural tradition to Chinese technological innovation policies. In the short run, cultural tradition can be regarded as a given constant, which determined our technological innovation policies; while in the long run, our technological innovation policies will have reaction to cultural tradition, making it change towards economy (non-economy but not against economy).

For the problems of relationship between social development and innovation, Dr. Liu Shiwen [37] set forth that innovation practice can be divided into three basic forms such as innovation practice in technology, institution, or knowledge, and any other non-basic forms such as innovation practice in education and art.. In his opinion, innovation practice is the natural impetus of social development.

For the purpose of constructing a humanity environment to support China innovation, Dr. Qiu Yongming [38] put forward a nation construction system which is composed of innovative nation construction soft system (innovative conception, innovative culture, innovative system , innovative intellectuals etc.)and innovative nation construction hard system (innovative system of technology, knowledge, technological and scientific state defense, region, and service system supported by science and technology).

11 Conclusion

The study of Chinese technological innovation consists of nine directions - underlying principles of innovation behavior, concepts of innovation cognition, methods of innovation, behavior of innovation subjects, structure of innovation mediators, innovation capacity, public service system of innovation, policy support system of innovation, and cultural environment of innovation, as well as several issues of innovation, including its behavior mechanism, operational mechanism, work mechanism, modes of implementation, modes of mathematical representation, modes of physical representation, modes of behavioral representation, modes of performance representation, modes of function realization, modes of structure realization, modes of workflow realization, modes of carrier realization, etc.

Formulation of the Chinese technological innovation achievements framework based on the results will be a highly difficult task. Without it, scientific integration of Chinese technological innovation achievements will be difficult to accomplish, and it will be difficult for China's scientific research capacity system to serve as guidance to the investment of China's technological innovations. Therefore, the formulation of the framework from the perspective of philosophy of science will be an important mission in the study of Chinese technological innovation.

References

1. Wang Hanxi, Wan Chengna. Wuhan University of Technology Press. 406-409(2014).
2. Wang Hanxi, Zhou Zude， Hu Shuhua. Hubei People Press. 799-805(2010).
3. Pan Xiongfeng. Dalian University of Technology. 2007.
4. Ji Yushan, Zhang Zhongyu. Technoeconomics &Management Research. **2.** 23-27 (2009).
5. Zhao Dumin. Liaoning Economy. 12. 36-37(2009).
6. Guo Xiaolin. Huazhong University of Science and Technology. 2006.
7. Chen Fengdi. Fujian Normal University. 2008.
8. Lin Jing. Jilin University. 2005.
9. Tian hui. Party School of the Central Committee of the Communist Party of China. 2008.
10. Chen Li. Party School of the Central Committee of the Communist Party of China. 2007.
11. Zhang Hongshi. Zhejiang University. 2005.
12. Liang Hongsong. Harbin Institute of Technology. 2008.
13. Huang Yuansheng. North China Electric Power University. 2005.
14. Guan Wei. Dongbei University of Finance and Economics. 2006.
15. Li Zhiqiang. Beijing Jiaotong University. 2008.
16. Hu Shuhua. Science Press. 2000.
17. Wang Xueyuan. Harbin University of Science and Technology. 2008.
18. Zhao Zhenglong. Shanghai Jiaotong University. 2008.
19. Liu Guoyan. Harbin Engineering University. 2008.
20. He Shan. Wuhan University of Technology. 2003.
21. Wang Xueli. Tianjin University. 2006.
22. Lv Jianrong. Xi'an: Xibei University. 2007.
23. Jiang Hong. Jilin University， 2008.
24. Zhang Zhongtang. Tianjin University. 2006.
25. Chen Hong. Tianjin University. 2005.
26. Sun Bing. Harbin Engineering University. 2003.
27. Li Qingdong. Jilin University. 2008.
28. Wang Zhangbao, Xu Congwei. Chinese Science & Technology Forum. **2**. 55-59(2005).
29. Chen Xile. Xiamen: Xiamen University. 2006.
30. Hu Limei, Mao Huifang, Shen Tongping. Journal of Library Science. **12**. 73-75(2009).
31. Xu Jiang, Shao Yunfei. Science & Technology Information(Academic Research). **14**. 3-6(2008).
32. Zhang Qinglai, Su Yun. Library and Information. **3**. 101-108(2009).
33. Wang Jingting, Wen Yue Chun. Journal of Shanghai Finance University. **1**. 22-26(2009).
34. Dong You. Hebei University of Technology. 2007.
35. Cui Kai. East China Normal University. 2002.
36. Cheng Liangbin. Huazhong University of Science and Technology. 2007.
37. Liu Shiwen. Party School of the Central Committee of the Communist Party of China. 2008.
38. Qiu Yongming. East China Normal University. 2010.

The Development Direction of Research on Chinese Technological Innovation

Hanxi Wang[1,*], Chengna Wan[2], Zude Zhou[1] and Shuhua Hu[3]

[1] Hubei Digital Manufacturing Key Laboratory, Wuhan University of Technology, Wuhan, China
[2] School of Arts and Law, Wuhan University of Technology, Wuhan, China
[3] Management College, Wuhan University of Technology, Wuhan, China

Abstract. The theoretical study of Chinese technological innovation assumes the perspective of prototype structure of scientific being, and consists of four directions of investigations: ontological study of innovation cognition, behavior study of innovation cognition, subjects of innovation cognition, and subjective material foundation of innovation cognition. The study aims to obtain technological innovation capacity through construction of an external technological innovation system. However, an external system would necessarily result in redundancy of systems and decreased operability. For the purpose of environmental friendliness and resource conservation, it is vital for our technological innovation to ignite the intrinsic innovation vitality of the system. Therefore the endogenous mechanism of innovation is the scientific direction to take for the theoretical study of technological innovation.

1 Introduction

A review of literature shows that the study of Chinese technological innovation consists of nine directions - underlying patterns of innovation behavior, concepts of innovation cognition, methods of innovation, behavior of innovation subjects, structure of innovation mediators, innovation capacity, public service system of innovation, policy support system of innovation, and cultural environment of innovation, as well as several issues of innovation, including its behavior mechanism, operational mechanism, work mechanism, modes of implementation, modes of mathematical representation, modes of physical representation, modes of behavioral representation, modes of performance representation, modes of function realization, modes of structure realization, modes of workflow realization, modes of carrier realization, etc.

When faced with such a multi-faceted system of achievements, we must carefully consider how to organize it into a theoretical system of technological innovation that can provide guidance to innovation projects. We intend to formulate a prototype structure of scientific being (PSSB) of technological innovation, and implement it in our investigation of Chinese technological innovation achievements and analysis of its future development.

2 Explore of technological innovation science being prototype framework

From the perspective of PSSB, China innovation theoretical research falls into four strata (see Figure 1).

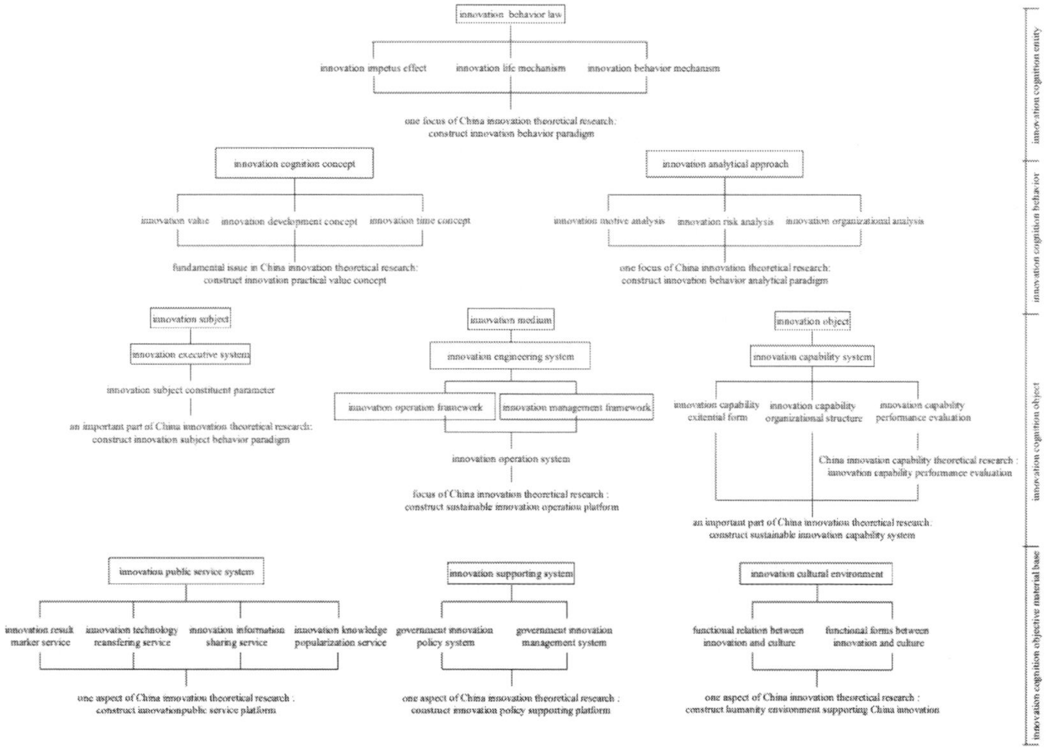

Fig. 1. The construct of technological innovation's PSSB.

2.1. Innovation cognition ontology

Around the law of innovation behavior, the impetus effect , life mechanism, behavior mechanism of innovation are explored. To construct innovation behavior paradigm becomes the focus of China innovation theoretical research.

2.2 Innovation cognition behavior

Consisting of innovation cognition concept and innovation analysis approach. Centered around innovation value, innovation development concept, and innovation time concept, the exploration and construction of innovation application value has become the fundamental problem in China innovation theoretical research. And taking innovation motive analysis, innovation risk analysis, innovation organization analysis as our objects, to explore innovation analysis approach and construct innovation behavior analysis paradigm becomes the focus of China innovation theoretical research.

2.3 Innovation cognition object

Consisting of innovation subjects, innovation medium, and innovation objects. Innovation subject refers to innovation implementing system. China scholars, based on the research of individual personality, behavior discrimination, behavior factors of innovation subject and leading ability of innovative leaders, gradually found the constituent parameters. So the exploration on innovation subject behavior paradigm on this base is an important part of China innovation theoretical research. Innovation medium refers to innovation engineering system, including innovation operation framework and management framework. Around innovation operation framework, china scholars discussed structuralforms of innovation organization, constructing paths of innovation behavior , innovation operational mode and innovation working platform. Around innovation management framework, China scholars discussed innovation developing strategy, behavior strategy, operation mode, and management system. Based on the innovation operation framework and innovation management framework, to construct a sustainable innovation operation platform composed of innovation operation tool, procedure and workshop is a heated issue in China innovation theoretical research. Innovation object refers to innovation capability system composed of forms of innovation capability, structure, performance evaluation. Around innovation capability forms, China scholars discussed innovation working forms ,organizational forms, representing forms, demonstrating forms. Around innovation capability organization structure China scholars discussed organizational constituents , structural function, transmission forms and constituent mode of innovation. Around innovation capability performance evaluation, China scholars discussed the assessment mechanism, mode, index, and system of innovation. To construct innovation capability performance evaluation system is also another heated issue of China innovation theoretical research.

2.4 Innovation cognition objective material base

Composed of innovation public service system, innovation supporting system and innovation cultural environment. Around innovation public service system, China scholars have discussed innovation result marker services, innovation technology transferring services, innovation information sharing services, and innovation knowledge popularization services. To construct innovation public service platform is one aspect of China innovation theoretical research. Around innovation supporting system, China scholars discussed government innovation policy system and government innovation management system. To construct innovation policy supporting platform is another aspect of China innovation theoretical research. Around innovation cultural environment, China scholars discussed the mutual relation and its forms between innovation and culture. To construct a humanity environment supporting China innovation is a third aspect of China innovation theoretical research.

3 Technological innovation operation framework

The major object of present China technological innovation theoretical research lies in constructing sustainable innovation operation platform. If innovation subject can be divided into innovation impetus subject, innovation idea subject and innovation implement subject, the thinking way of China technological innovation research and systematic research can be generalized as following: the innovation idea subject, guided by innovation development concept, based on the adjust of innovation policy supporting system, put forward innovative idea; innovation impetus subject, guided by innovation time concept, based on the support of innovation public service platform, initiated innovative activity; innovation executive subject, according to innovation impetus subject requirements and innovation subject innovative idea, guided by innovation value, based on innovation behavior paradigm and innovation behavior analysis paradigm, launched innovative activity on the sustainable innovation operation platform and construct innovation engineering system, hence obtained innovation capability system eventually. (see Figure 2)

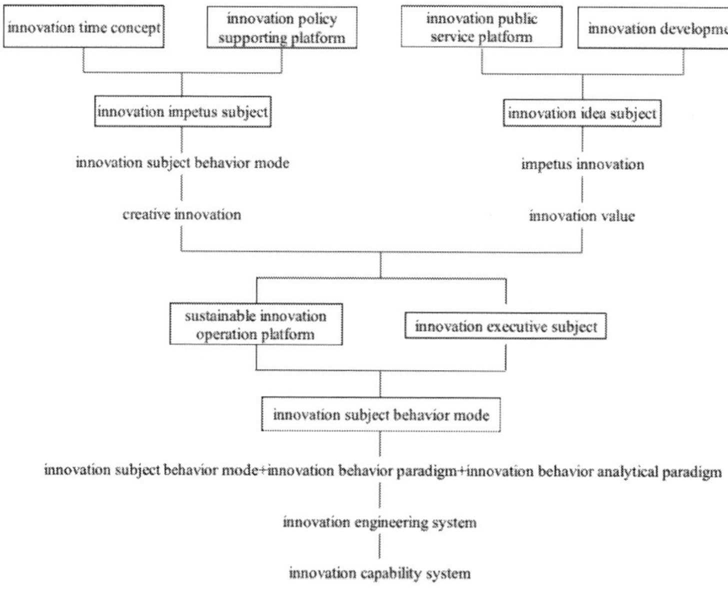

Fig.2. Thinking route of China technological innovation theoretical research and engineering application.

Innovation engineering system is attached existential scientific practice system (science research system, technology developing system, engineering operation system), productive practice system , or social practice system, and enables these practical system innovative function which is called exogenous innovation system.(see Figure 3)

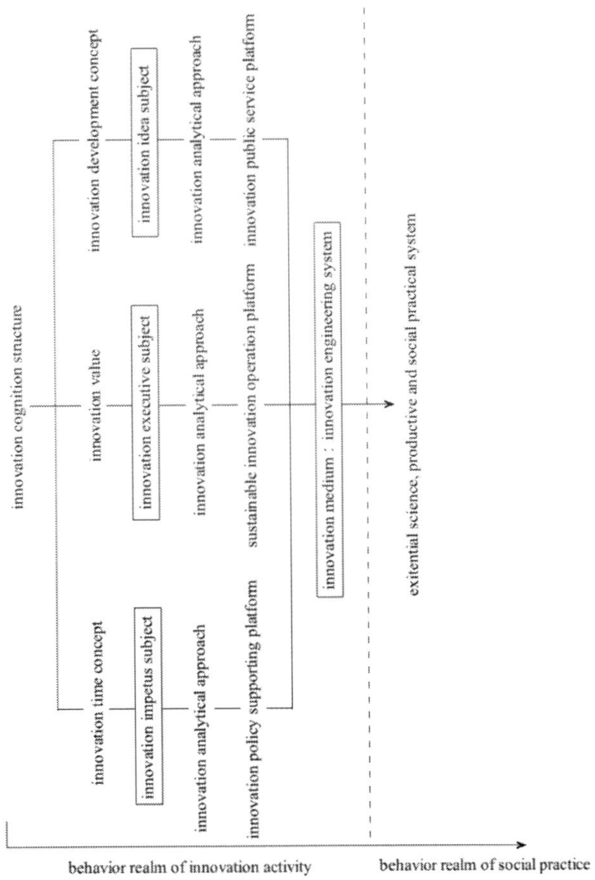

Fig. 3. Exogenous Technological Innovation System.

4 Discussion

From the perspective of Prototype Structure of Scientific Being(PSSB) [1], we can find that related theoretical achievements of technological innovation research are developed around innovation science being prototype framework, and constructed exogenous innovation practice engineering system, and reinforced the innovation capability of engineering system.

The above innovation operation framework seems to be a perfect function system. However, when observing it from the history of human civilization and the history of science, we can easily come to a conclusion: the history of human development itself is a perfect example for a history of innovations in civil activities, culture, science, technology, and society. As for this we can understand thoroughly from A History of Science by Dampier [2], History of Western Philosophy by Weber [3], Dictionary of History of Science by Bynum [4], Natural Philosophy by Hegel [5], Human Knowledge by Russell [6], A History of Science, Technology, and Philosophy in the Eighteenth Century by Wolf [7], Furthermore, a large number of historical facts prove that the scientific gains are increasing greatly on a geometric series scale and are generating massive achievements, which is specially verified in both Theoretical History of Science and Technology by Song Ziliang [8].

All the above facts sufficiently indicate that social practice system, or productive practice system or scientific practice system which in itself has powerful innovation capability, and creates abundant and colorful social civilization, production civilization and scientific civilization. Furthermore, our social practice system, productive practice system and scientific practice system have already presented the situation of accelerating innovation development, and we are living in a great knowledge-explosion era.

In light of this, the question we should raise is why we need to establish an artificial innovation engineering system out of the existing scientific practice system, productive practice system and social practice system? Have the original systems lost their innovation capability? Or are there any defects in the innovation functions of the original systems?

Among various practice systems, there must be numerous concrete individual systems in which the innovation functions have been restrained, shielded and obstructed for various reasons and accordingly results in the problems of losing or weakening of the innovation functions of practice systems, thus compelling us to address the issues of technological innovation research or rebuilding of technological innovation systems. According to the analysis from the perspective of control principle, the addition of auxiliary systems in the case the original systems are able to work (but with their functions possibly weakened) will certainly cause system redundancy and increase the control difficulties. Considering from friendly environment and resource saving, it may be an optimal choice to restore the original system functions.

So, we need to consider the following questions: how to restore the innovation functions of the original systems? How to stimulate the innovation functions of original systems? How to exploit the innovation functions of the original systems? How to magnify the innovation functions of the original systems? And how to perfect the innovation functions of the original systems.

Now we need not only to attach importance to the structure of and study on the exogenous innovation system, but also to the social practice system to look into the endogenous innovation functions of the system itself, including studies on the endogenous mechanism, operational mechanism, behavioral system and working model of system innovation, and on the reasons why endogenous innovation has been restrained, shielded and obstructed, and further including the researches on the mechanism to impel, revitalize, boost and extend the endogenous innovation functions of system.

In the end, the scope of theoretical research on technological innovation will expand from the single study and discussion on the exogenous innovation system to discussion on the endogenous mechanism and operation model of innovation, and there will be a wider field and space for the innovation research.

Acknowledgement

This paper is supported by Social Science Foundation of Hubei Province "Study of Evaluation Model for Scientific Innovation Activity Based on Endogenetic Mechanism of Scientific Innovation" (No.[2010]236).

References

1. Wang Hanxi. Wuhan University of Technology. 2010.
2. Dampier W C. Guangxi Normal University Press. 264-265(2001).
3. Alfred Weber(written). East China Normal University Press. 2007.
4. Bynum W. F., Brown E. J., Roy Porter(written). Hubei Secience And Technology Press. 1988.
5. Hegel(written). Liang Zhixue, Xue Hua, Qian Guanghua(translated). The Commercial Press. 1997.
6. Russell(written). Zhang Jinyan(translated). The Commercial Press. 2003.
7. Wolf(written), Zhou Changzhong, Miao Yishun, Mao Rongyun(translated. The Commercial Press. 1997.
8. Song Ziliang(compiled). Hubei Science & Technology Press. 1989.

The Application of Industry 4.0 in Customized Furniture Manufacturing Industry

Lin Wang[1,*], Jinfeng He[2] and Songjie Xu[2]

[1]School of Management Engineering, Zhengzhou University, Zhengzhou, China
[2] School of Management Engineering, Henan Institute of Engineering, Zhengzhou, China

Abstract. In the background of industrial 4.0, this paper analyzes the developmental road about customized furniture factory. Based on the concepts and features of industry 4.0, this paper analyzes the composition of the customized furniture factory, the main operation system and the operation process of the intelligent customized furniture factory. In order to achieve efficient and accurate production targets, intelligent customized furniture factory should be set up through the establishment of cyber physical system(CPS) to cover the intelligent network.

1 Introduction

At present, the traditional furniture manufacturing industry is facing with a challenge. It is becoming easier for people to obtain the necessary information. Consumers are looking for customized products and services. In order to meet people's living needs, Germany put forward the concept of "Industry 4.0". The Chinese government developed the concept of "China Made 2025" in 2015. At the same time, "Industrial Internet" is proposed by the United States in 2009. They all aim at transforming the traditional industries through the Internet [1]. Through the use of "Things of internet "and" business networking (service Internet)", traditional industries can achieve flexible production, zero inventory and Internet marketing. As a traditional manufacturing industry, furniture industry based on a solid industrial base. So it should learn the strategic thinking of industrial 4.0, and explore the "manufacturing + Internet" actively.

2 Industrial 4.0 concept and features

2.1 The concept of industry 4.0

Based on the CPS, industrial 4.0 can realize the timely communication, recognition between people by building up a highly flexible and intelligent manufacturing mode [2]. Industrial 4.0 is the fourth industrial revolution led by intelligent manufacturing. It aims at transforming the manufacturing industry into an intelligent one by the combination of information communication technology, cyberspace virtual system and cyber physical system (CPS).

2.2 Industrial 4.0 features

2.2.1 Interconnection

The interconnection between devices is the core content of "Industry 4.0". For example, interconnection between all kinds of equipment which have different functions is composed intelligent workshops and factories. Interconnection between equipment and products refers to that they can communicate by a special language. By reading the product information, equipment can achieve intelligent production. Cyber Physical System(CPS) can realize the integration of network world and physical world by connecting the single intelligent device with the internet. In this way, these intelligent devices can achieve self-adapting, self-diagnosing, self-repairing and remote assistance[3].

2.2.2 Integration

Through the physical information system ,"Industrial 4.0" can combine production system sensors, control systems, production equipment together to form an intelligent network. The goal of vertical integration is to realize seamless connection of product information, such as product design, manufacturing, logistics, transportation and maintenance.

2.2.3 Big Data

During the production process, the intelligent manufacturing equipment will produce a large number of production-related information and data. We need to collect the data information and give feedback to all aspects of production. In this way, we can achieve efficient high-quality operation process[4].

3 The Application of Industrial 4.0 in customized furniture factory

Furniture is closely related to people's life. But the increasingly cost, excess capacity and customized market all require a high degree of automation, intelligence, flexibility of the direction. The requirements mentioned above are in accordance with the requirements of the custom furniture factory.

3.1 Intelligent factory

Intelligent plant connect the intelligent physical devices by industrial internet intelligent physical devices. Intelligent devices can do computing, communication, precise control, remote assistance and autonomy[5]. In the future, firstly we need obtain the customer's customized information. Then the virtual factory which based on cloud based design and manufacturing will do the design and simulation. Finally, customer confirmed the product and the computer integrate and send the appropriate data to the appropriate equipment . Intelligent devices can read each other tags to achieve the information transfer function. Based on the information, intelligent devices can complete the manufacturing, testing and other activities.

The intelligent factory is the data-based enterprise. The daily operation of the intelligent factory mainly depends on following software and hardware: the industry thing networking, the industry network security, the industry big data, the cloud computation platform, the MES system, the virtual reality, RFID, 3D printing, machine vision, intelligent logistics and so on. The hardware includes industrial robots , data acquisition, industrial switches[6].

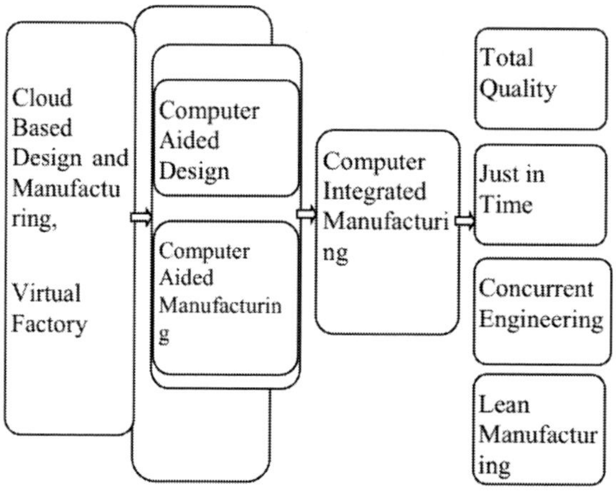

Fig. 1. Main operating activities of intelligent furniture factory

3.2 The main operating network system

The intelligent custom furniture factory uses information communication technology(ICT) and cyber physical system (CPS) to link related equipment, production lines, factories, related parties and products together. The entire intelligent network is divided into two parts .The information transmission in the custom-made furniture factory is called M2M. It transforms dialogue among human and machine. Factory exchange information through the Internet of Things which includes the factory's payment platform, intelligent logistics platform, and design platform to finish production. The process shows in Figure 2.

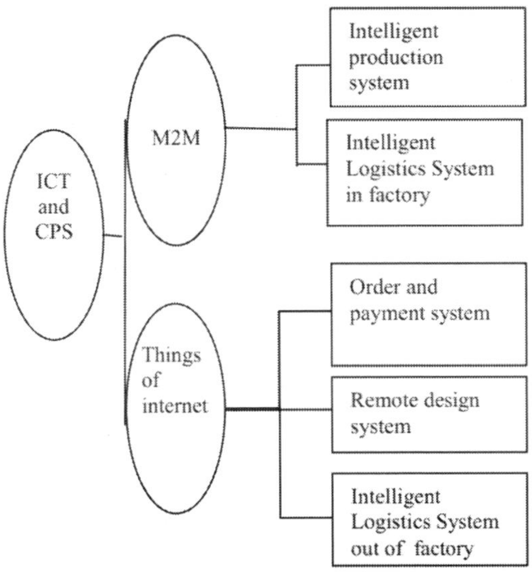

Fig2 Intelligent network and main operating system of intelligent furniture factory

3.3 Main operating system

Intelligent custom furniture factory is composed by an intelligent production line and network (financial enterprises, product design institute, production equipment factory, technical personnel training institutions, raw material suppliers, logistics companies, etc.). The intelligent operation system of the intelligent customized furniture factory can be divided into payment system, remote product development and design system, intelligent production system and intelligent logistics system.

1) Intelligent order and payment system. Through the system's intelligent ordering platform, the factory can accept the orders made by external factory. After the approval, the orders will be revised or negotiated according to order requirements.

2) Remote product development and design system.When receiving new products or customized product orders, they need to design products by computer simulation. The computer can set up three-dimensional model in accordance with the real work of the operation.

Then the product can be displayed on the screen.According to the views of customers, the designer will modify the model and print out new products to customers by the 3D printer.

3) Intelligent production system. Intelligent production system is the main system of intelligent machinery plant, involving the entire factory production processes, machinery, equipment and plant logistics management. In the intelligent factory, every part of the product blank, spare parts, tools, transport boxes, machines are equipped with sensors.

4) Intelligent logistics system. The intelligent logistics system is a modern logistics system which is composed by bar code, radio frequency identification technology, sensors, global positioning system and so on. Through the intelligent network formed by CPS, the logistics resources integrate and operate in the electronic commerce way. First of all, we collect the relevant data.Based on the data, we analyze and build the model. Then we establish intelligent logistics base database; Finally, we enhance the customer management and business process optimization.

3.4 Operation process

Through the cyber physical system(CPS) and the internet of things, the Semi-finished products are sent to the processing plant and workstation to arrange production process. Finally ,they send product to custom by modern logistics.

Intelligent machinery plant operating process shown in Figure3.

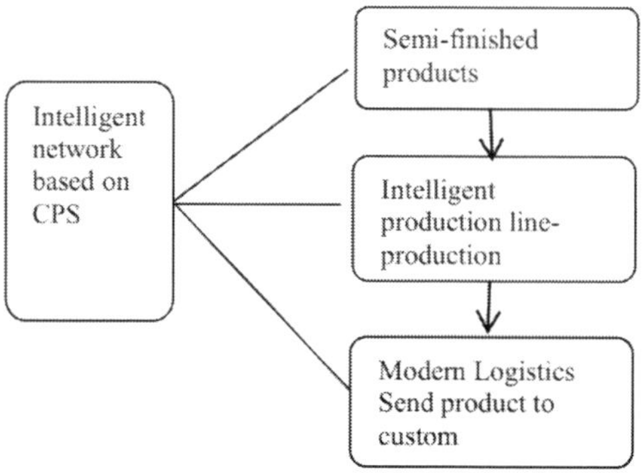

Figure 3 Operation flow of intelligent machinery plant

4 Conclusion

Because the furniture manufacturing industry is faced with many problems, such as low efficiency of resource use, great pressure of environment, high labor cost.The intelligent manufacturing is the inevitable trend of the future furniture manufacturing industry. "Industrial 4.0" has the advantages, such as high efficiency of resource allocation, quick response to market demand, low labor cost and logistics cost. In this way, we can apply "Industrial 4.0" into customized furniture manufacturing industry. With the poor foundation and high-tech backwardness, China's furniture manufacturing industry still face challenges.We should make full use of its huge customer market to catch up with the steps of industry automation, intelligence and green trend.

References

1. Zhihui, Wu. Manufacturing Model of Furniture Industry in Industry 4.0. China forest products industry, 2016, **43**(3): 7-7
2. Jian, Li. Made in China by 2025, the German Industrial 4.0, Industrial Internet Strategy of the United States and China Plastic Machinery. Plastics Technology and Equipment, 2015(21):3-6.
3. Zuehlke D. Smart Factory Towards a factory of things. Annual Reviews in control, 2010(34) :129-138
4. Hongmin, Dai. 4.0 Industrial and Intelligent Machinery Plant. Packaging engineering, 2016, **37**(19): 208-209
5. Jun, Yin. The application of industry 4.0 in aeronautical manufacturing industry. Manufacturing Automation, 2015, **37** (10): 5-5
6. Shu, Zhang. The Industry 4. 0 and Intelligent Manufacturing. Machine Design and Manufacturing Engineering, 2014, **43**(8): 4-4

The Evaluation Index System of Innovation-Driven Capability of Chinese Manufacturing Industry Based on Factor Analysis

Chuhao Zhao[1], Shanling Liao[2]and Jinfa Li[3]

[1]Chinese Academy of Governance, Beijing 100089, China

[2]Guangxi Normal University, Guangxi 541006, China

[3]School of Management Engineering, Zhengzhou University, Zhengzhou 450001, China

Abstract. In China's new normal, a strategy named "Made in China 2025" was proposed to speed up the manufacturing's transformation and upgrading . The hardship of China's manufacturing industry transformation and upgrading are the limit capability of independent innovation , the complexity and variability for competitive environment and the existence of institutional hurdles. At present, China is facing the powers of the industry from manufacturing to major, The paper will use factor analysis method to construct the evaluation index system of innovation-driven ability of Chinese manufacturing industry, and established evaluation model of innovation-driven ability of Chinese manufacturing industry. Based on the analysis of the model, the paper will propose the innovation driving strategy to speed up manufacturing transformation and upgrading.
Key words. Innovation-Driven, Manufacturing Industry, Factor Analysis.

1 Introduction

At present, China's manufacturing industry is featured by innovation ability insufficiency, low level of industrial structure, high resource consumption, low quality and efficiency. The popularization and the promotion of cloud computing, Internet of things, big data and intelligence engineering pose great pressures to China's manufacturing industry. The rapid development of information technology and mobile Internet technology promotes the change of industry. Moreover, with the China's economy coming into the new normal, the domestic manufacturing industry is faced with tremendous pressures from industrial transformation and upgrading, as well as the industrial competition caused by the reindustrialization of developed countries. Therefore, it is essential to implement the innovation driven strategy, which is a fundamental way to break through the bottleneck of China's economic development and solve the deep-seated contradictions and problems. The strategy is related to the future and destiny of the Chinese nation.
In 1982, Nelson and Winter published "the evolutionary theory of economic change", which made a ground-breaking contribution to apply the theory and method of evolutionary economics to study technology innovation. Based on the idea of evolution, Nalsonit concluded that innovation was a process of learning, searching, and (social) selection. In addition, Anderson and Tushman (1990) put forward the concept of "technology life cycle" to describe the speed of innovation relative to the technological development stage. Subsequently, Silverberg, Dosi and Orsenigo (1998) made a systematic analysis of the multidimensional structure and behaviour of variability within the technology innovation and diffusion system by building the dynamic evolution model of technology innovation. Moreover, Niosi and Banik (2005) analysed the evolution process of technological innovation in 90 biotechnology companies and demonstrated that the complex process of the evolution of biological technology innovation presented the three stage evolution law of "R & D organization innovation- enterprise innovation -regional technology innovation".
In summary, scholars at home and abroad have conducted relatively abundant researches about innovation driven models with diversified methods. However, the majority of the existing researches are carried out from the macroscopic aspect, with focus on the choice of technology innovation to enhance manufacturing industry, only taking the technology innovation as a countermeasure to promote competitiveness of manufacturing industries and enterprises. There is a lack of general rules in the innovation driven interior to study the development of manufacturing industry. Furthermore, the current domestic research is inadequate in the exploration of some quantitative law in the manufacturing technology innovation system.

On the basis of previous researches, this article took the problems faced by China in the innovation driven process as objectives and established the applicable evaluation index system for the innovation driven ability of manufacturing industry in China and developing countries. In addition, it tried to make the measurement scale and tools for the innovation driven ability of China's manufacturing industry. Moreover, using factor analysis to evaluate the innovation driven ability of China's manufacturing industry, a foundation has been laid for the further empirical research, so as to take the development countermeasures to improve the innovation driven ability of China's manufacturing industry.

2 Innovation-driven Strategy ——The Core Motive Force of the Transformation and Upgrading of China' s Manufacturing Industry

Currently, the Chinese government is appealing for the implementation of innovation-driven development strategy, emphasizing on using scientific and technological innovation to improve social productivity and comprehensive national strength, which places the scientific and technological innovation in the core position of national development. The innovation-driven strategy has become the major strategic choice in response to the China's economic growth shift and structural adjustment pains as well as the pre-stimulating preparation for the digestion period of the new era, in order to create a new China's economic upgrading version. Manufacturing industry is the foundation of comprehensive national strength, and an important strategic industry to maintain stable growth of national economy and solve employment problems. It is an important material carrier for China's construction of an innovation-oriented country and plays a pivotal role in China's economic development. Innovation-driven strategy is the necessity of for the expansion of the global manufacturing industry chain. It is the main engine to drive the transformation and upgrading of China's manufacturing industry and the key measure to realize the transition from "Made in China" to "Created in China".

3 The Predicament Innovation-Driven in the Transformation and Upgrading of Manufacturing Industry in China

In the new normal, the innovation driven transformation and upgrading of manufacturing industry in China is a necessary option to build the competitive advantage for manufacturing industry, which is conducive to consolidating and enhancing the international status of China's manufacturing industry. China's economy has begun to enter the new stage of speed shifting and economic restructuring. It not only means a slowdown in economic growth, but also suggests the change of economic growth momentum and the transformation of economic development. At present, China's manufacturing enterprises are in the "climbing plateau", so that the innovation-driven transformation and upgrading of manufacturing industry will face a series of restricting factors in the short term.

3.1 Lack of Independently Innovative Ability among Manufacturing Enterprises

The biggest problem faced by China's manufacturing industry is the lack of independent innovation ability and the development system combining original innovation and integrated innovation. Statistically, in large and medium-sized industrial enterprises, only 28.3% of them carry out researching activities and only 27.6% of them have established researching institutions, with researchers only accounting for 3.3% of total employees. It can be found that the engagement in research of enterprises is far from enough. At the same time, influenced by the conversion of original industry research institutes, the common technology research of industrial system is weakened, and the technologically sustained input to the innovation becomes inadequate, which further restricts the improvement of the manufacturing enterprises' ability of independent innovation.

Table 1. National Technical Innovation Model.

	Research And Development Model	Production-Education-Research Model Numbering
Silicon Valley	From basic research to applied research and development , mainly for original innovation	Represented by Stanford university research institutions is the model of close combination
Taiwan Hsinchu	Development model is priority to product application technology, integrated innovation and secondary innovation	Research institutes have a remarkable effect on innovation, especially the industrial technology research institute
Bangalore, India	Application software development	Indian institute of technology university and other research institutions, enterprises have built a close cooperate relationship
Zhonggu-Ancun	Mainly on basic research, at the	The university is the major power of

	same time paying attention to the transformation of achievements	technology innovation

3.2 Discontinuities Existing in the Support Power of Transformation and Upgrading of Manufacturing

Since China's reform and opening up, China's manufacturing industry has been dependent on the developed countries in technologies, and is taken as the product assembly line in a very long time. However, the advantages of "industrial chain" and "culture industry" consciousness have still not been accumulated and matured in the national economic and social development. In the past few decades, there were sufficient preferential policies, lands and labors, which provided a key content in comparative advantage for manufacturing industry in China to catch up with the global trend. But it is worth noting that the extensive development mode of manufacturing industry has not been fundamentally changed, and the pressure generated by population, resources and environment is more and more prominent.

3.3 The External Environment Manufacturing Industry Facing is Complicated

Since the outbreak of the 2008 international financial crisis, developed countries have implemented the strategy of "reindustrialization" and "returning of manufacturing industry", which brings many restrictions or domestic manufacturing industry on the market and technology. At the same time, attracted by the low cost, many industrialized countries transfer their mid-range products processing and manufacturing industries to those low and middle-income countries, which form relatively fierce competitions with China, leading up to the besieged position of demotic manufacturing industry. As entering the late stage of industrialization, market demand is declining, resulting in the highly excess of the traditional capacity. Therefore, the pressure from market is mainly focused on the demand of saturated low-end products.

4 Evaluation Method of Innovation Driving Capability of Chinese Manufacturing Industry

It is necessary to collect a large number of indicators (variables) in the study of China's manufacturing innovation driving ability evaluation, which is beneficial for extracting useful information. However, there may be some correlations among these variables; the extracted information will be repeated and invalid. Factor analysis is to maximize the preservation of the original information in the case, with a few factors to describe the original number of indicators or observations between the data relationship. Factor analysis can classify these closely related variables as a unit, reducing the number of factors to reflect the original data contained in the information.

4.1 Evaluation index

On the basis of the existing research, this paper constructs the evaluation index system of Chinese manufacturing innovation drive ability. The index system of this paper mainly revolves around the goal of innovation ability, which can be divided into five parts, namely innovation human input, innovation financial input, innovation main input, innovation technology output and innovation diffusion, and 23 sub-indexes.

Table 2. Evaluation Index System of Innovation Capability of Chinese Manufacturing Industry

targets (A)	**Main indicators (B)**	**Second indicators （C）**	
Innovation ability (A)	Manpower (B1)	The number of senior titles	C1
		R & D personnel	C2
		Number of returned overseas students	C3
		Scientific and technological activities personnel	C4
		College or above	C5
	Funding (B2)	R& d expenditure	C6
		Total government finance expenditure	C7
		financial science and technology funds	C8
		fixed investments	C9
		Investment in infrastructure	C10
	Innovation	Number of high - tech	C11

	system (B3)	enterprises	
		Number of incubation enterprises	C12
		Number of R & D Institutions	C13
		Number of Innovation Service Organizations	C14
	Output (B4)	Number of Patent Applications	C15
		Product sales revenue	C16
		Technical income	C17
		Total Labor Productivity	C18

4.2 Model Setup

With X_P representing the evaluation index of innovation capability of China's manufacturing industry. In each of the p economic variables, each variable contains m aspects of the evaluation, which are called the common factors of p economic variables. In the various economic variables, m common factor cannot explain the economic aspects of the variable known as a special factor. The model of factor analysis is:

$$X_1 = \mu_1 + \alpha_{11} f_1 + \alpha_{12} f_2 + \ldots + \alpha_{1m} f_m + \varepsilon_1 \quad (1)$$

$$X_2 = \mu_2 + \alpha_{21} f_1 + \alpha_{22} f_2 + \ldots + \alpha_{2m} f_m + \varepsilon_2 \quad (2)$$

$$\ldots \quad \ldots$$

$$X_p = \mu_p + \alpha_{p1} f_1 + \alpha_{p2} f_2 + \ldots + \alpha_{pm} f_m + \varepsilon_p \quad (p)$$

Where $\mu_1, \mu_2, \ldots \mu_p$ are the mean of $x_1, x_2, \ldots x_p$. $f_1, f_2, \ldots f_m$ are common factors; $\varepsilon_1, \varepsilon_2, \ldots \varepsilon_p$ are special factors. $f_1, f_2, \ldots f_m$ and $\varepsilon_1, \varepsilon_2, \ldots \varepsilon_p$ are unobservable random variables. $f_1, f_2, \ldots f_m$ in the expression for economic variables as common factors represent the correlation between the various economic variables of the components.
If there is not common factor (j=1,2, …, m) between at least two economic variables, these factors are special factors. Each special factor ε_i (i=1,2, …, p) exists only in the expression of the corresponding the i-th single economic variable, and it has only the effect on this economic variable. ∂_{ij} is called the load of the i-th variable on the j-th common factor. Namely, it is degree of explanation of the i-th economic variables on the j-th part of evaluation model.

5 Innovation-driven Strategy of China's Manufacturing Industry Transformation and Upgrading

At present, the situation is very complex with difficulties and opportunities intertwined. In order to implement innovation-driven development strategy, the country must think highly of strengthening the manufacturing innovation capacity to accelerate the process of China's manufacturing industry transformation and upgrading.

5.1 The emphasis to promote the transformation and upgrading of manufacturing is focused more on innovation.

The transformation and upgrading of China's manufacturing industry lie in enhancing the capability of independent innovation. The basic orientation of the transformation and upgrading of Chinese manufacturing industry is to create innovation chain, and build manufacturing innovation chain; it can be started from manufacturing chain to manufacturing innovation chain, and to manufacturing innovation method chain.
In the future, China will still need to take the intelligent manufacturing as the main direction, to accelerate the integration of industrialization and informatization. The implementation of the Internet + advanced manufacturing action plan will promote business model innovation and industry transformation and upgrading, and improve the national manufacturing innovation system as well.

5.2 Vigorously innovate manufacturing value chain

In the new era of global technological revolution and industrial transformation, if China's manufacturing industry wants to cross the "trap and high wall", the structural adjustment as well as transformation and upgrading must be conducted smoothly, which requires to innovate the manufacturing value chain, and improve the role of China's manufacturing enterprises in the global value chain. The manufacturing development in the future will change from the current mass customization stage to individual development stage. Moreover, the design innovation and technology research and development of manufacturing will become the focus of enterprise development. As a big manufacturing country, China is experiencing the transformation from "Made in China" to "China smart". The product and value can be improved through the combination of industrial design and industrial culture.

5.3 Create an institutional environment which is conducive to innovation

If "Made in China" wants to form a new advantage, China must participate in the international competition, so that Chinese enterprises can use extensively international resources, learn from each other, carry out research and development and compete for the market share in the open innovation, as well as further support the system innovation. Companies need to strengthen technological innovation, and strive to carry out the strategy of "mobile Internet +" to achieve interaction between products and culture carried by products with consumers.

5.4 Build Multi-dimensional Multi-level Manufacturing Industry Collaborative Innovation System

Innovation is the first driving force of development. In the new era of China's in-depth implementation of the innovation-driven development strategy, the development mode of manufacturing will be shifted from factor-driven to innovation-driven, with the industrial structure optimized and upgraded. In addition, it is inevitable to have mergers and acquisitions, and production concentration. At the same time, the role of emerging industries, services, small and micro enterprises are more prominent, and the small, intelligent and professional production will become a new feature of industrial organization. The country must become a driving force of innovation for the development, deepen reform and opening up, and accelerate the formation of a unified, transparent, orderly, and normative market environment.
Therefore, under the new trend of global technological development and the deep adjustment of industrial competition pattern, it is urgent to conduct the transformation and upgrading of manufacturing industry and the innovation-driven strategy has become the main driving force of this process. Therefore, the transformation and upgrading from government-led innovation to market-oriented enterprise independent innovation can be achieved, which is to promote China to be synonymous with manufacturer of quality from that of quantity.

5 Conclusion

Now China is facing a developmental environment to meet with profit and risk, chances and challenges, therefore, we suggest that China should take manufacturing innovation capacity as priority. First of all, put intelligent manufacturing as the main direction, and accelerate the integration of industrialization and informatization. Then, vigorously innovate manufacturing value chain and improving the role of China's manufacturing enterprises in the global value chain. The manufacturing development in the future will change from the current mass customization stage to individual development stage, and design innovation and technology research and development of manufacturing will become the focus of enterprise development. What's more, strengthen technological innovation, and strive to carry out the strategy of "mobile Internet +" to achieve interaction between products and culture carried by products with consumers. Finally, now that, the innovation-driven strategy is the main driving force of China's manufacturing transformation and upgrading, try to achieve conversion of transformation and upgrading from government-led innovation to market-oriented enterprise independent innovation to promote China one of quality from a manufacturer of quantity.

References

1. R. Nelson. Understanding Technical Change as Evolutionary Process. *Amsterdam: North-Holland,* (1987)
2. Anderson P, Tushman M. Technological discontinuities and dominant designs: A cyclical model of technological change. *Administrative Science Quarterly*, **35** (1990)
3. Silverberg G, Lehnert L. Long Waves and Evolutionary Chaos in a Simple Schumpeterian Model of Embodies Technical Change. *Structural Change and Economic Dynamics,* **9** (1993)
4. J.Niosi, M. Banik. The Evolution and Performance of Biotechnology Regional Systems of Innovation. Cambridge Journal of Economics,**29** (2005)
5. Hedberg, B. L. T.How Organizations Learn and Unlearn.In P.C.Nystrom and W.H.Starbuck. *Handbook of Organizational Design,* **11** (1981)
6. HousteinH D, NeuwirthE. Longwaves in World Industrial Production, Energy Consumption, Innovations, Inventions, and Patents and Their Identification by Spectral Analysis . *Technological Forecasting and Social Change,***22** (1982)
7. Kleinknecht A.Are there Schumpeterian Waves of Innovations, **14** (1990)
8. G • Dosi. Technological Paradigms and Technological Trajectories. *Research Policy,* **3** (1982)
9. W • B • Arthur.Competing Technologies,Increasing Returns,and Lock-in by Historical Events. *The Economic Journal,* **99** (1989)

10. G • Dosi.Perspectives on Evolutionary Theory. *Science and Public Policy,* **6** (1991)
11. Keun Lee, Chaisung Lim. Technological Regimes, catching-up and leap frogging: findings from the Korean Industries. *Research Policy,* **30** (2001)

Design and optimization of the large span dry-coal-shed latticed shell in Liyuan of Henan province

Wenfeng Du*, Chunyu Liu, Yun Sun and Qi Liu

Institute of Steel and Spatial Structures in School of Civil Engineering and Architecture, Henan University, Kaifeng 475004 China

*Corresponding author: dwf@henu.edu.cn

Abstract. The design and optimization about the large span dry-coal-shed latticed shell in Liyuan of Henan province were studied. On the basis of the structural scheme of double-layer cylindrical reticulated shell, the optimization scheme of the folding double-layer cylindrical reticulated shell was proposed. Through the analysis of a plurality of calculation models, the optimal geometric parameters were obtained after discussing the influence of different slopes of folding lines and shell thickness on the structural bearing capacity and the amount of steel. The research results show that in the case of the same amount of steel, the ultimate bearing capacity of the double-layer folding cylindrical reticulated shell whose folding line slope is 9% and the shell thickness is about 4.4m can be increased 27.3% compared with the original design scheme.

1 Introduction

With the expansion of the scale of power plants and environmental protection requirements, the application of dry-coal-shed in engineering has been rapidly popularized. Taking into account the needs of practical engineering, dry-coal-shed should have appropriate storage and work space, so the structure should meet the requirements which are large span, high altitude, wide coverage and good mechanical performance. Spatial latticed shell has become the priority structure of dry-coal-shed with the characteristics of large span, good integrated performance and less amount of steel, which has achieved good economic benefits. However, with the dry-coal-shed span requirements rising and the restrictions of steel quantity and safety, the commonly used form of latticed shell structure have already can't meet the demand of practical engineering. It is very necessary to design and optimize the original form of latticed shell structure. Combined with the practical engineering of the large span dry-coal-shed latticed shell in Liyuan of Henan province, the folding double-layer cylindrical reticulated shell was proposed on the basis of the structural scheme of double-layer cylindrical reticulated shell. In the case of the same amount of steel, the ultimate bearing capacity as the main target optimization function, folding line slope and shell thickness as optimization variables. By optimizing the structure, the optimal structure parameters are obtained.

2 Engineering situation

2.1 Basic situation and structural form selection

This project is large span dry-coal-shed latticed shell of 33000Nm3/h coke-oven gas LNG project in Liyuan of Henan province, the main structure is steel material. The structure of the horizontal span is 155.6m, which is currently the largest span of dry-coal-shed structure over the word. The longitudinal length is 365.82m, and the vector height is 47.3m. The height of the dry-coal-shed latticed shell is decided by the coal stacking height and the job requirement of the bucket wheel. In order to ensure the normal operation of the bucket wheel, overall dimension should meet the technological requirements and save space as soon as possible, and the structure as little as possible in contact with both sides of the coals. The dry-coal-shed latticed shell need to meet good mechanical properties, easy manufacture and installation, and the technical and economic indicators superior requirements.

The plane shape of the project is close to a rectangle. Structure shows multi-directional stress characters, planar structure lateral stiffness is weak. Based on the above consideration, the dry-coal-shed latticed shell preliminary adopts the double-layer cylindrical reticulated shell which combines skeletal structures simple structure and shell structure reasonable force characteristics. So the structure has the advantages of beautiful form, low steel consumption, better stiffness and simple force transmission form. The structure is a unidirectional force characteristic when the dry-coal-shed latticed shell length and span ratio is more than 1.5 times. Thus the project selection is square pyramid space grids, which can save the amount of steel and enhance the stiffness, so as to realize the purpose of better economic results.

According to the provisions of "Technical Specification for Space Frame Structures" (JGJ 7-2010) [6], the desirability thickness of the double-layer cylindrical reticulated shell is the width of 1/20 to 1/50, and the preliminary thickness is 4.83m. The both sides of bottom chord supported in 2m height of reinforced concrete columns, and the distance between supports spacing is 10.92m. The dry-coal-shed profile map and front view as shown in figure 1 and figure 2.

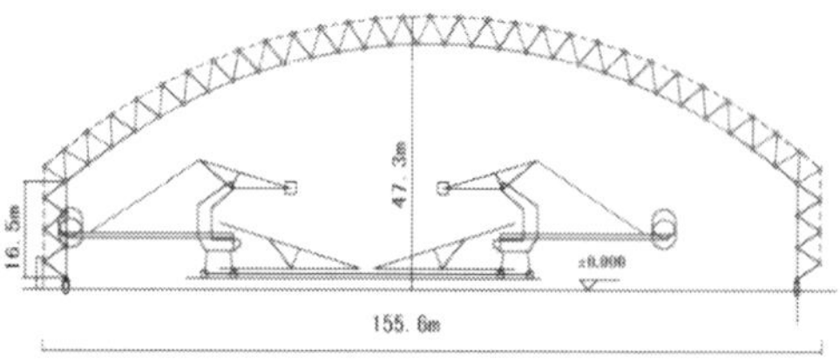

Fig. 1. The profile map of double-layer cylindrical reticulated shell in Liyuan

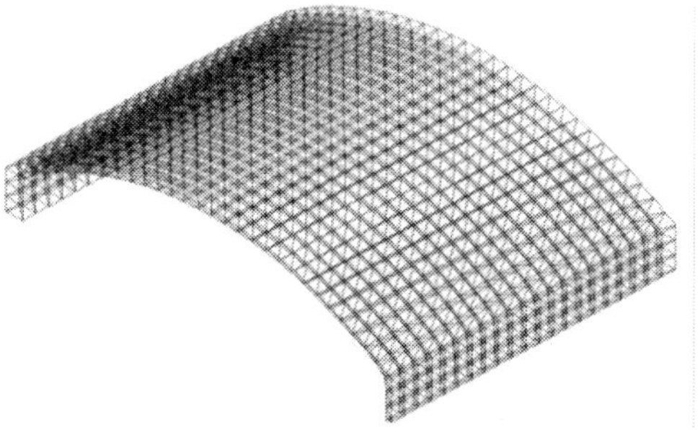

Fig. 2. The front view of double-layer cylindrical reticulated shell in Liyuan

2.2 Load and their combination

The dry-coal-shed latticed shell under the main loads include: (1) The constant load: The roof structure using single profiled steel sheet and purlin (excluding net weight), load is 0.25KN/m2; (2) The roof uniformly distributed live load: Taking into account the local snow load, roof live load is 0.5KN/m2 from live load and snow load maximum which apply in the first quarter node; (3) The wind load: The basic wind pressure is 0.55 KN/m2 (100 years return period), and the surface roughness is class B; (4)The earthquake: The seismic fortification intensity is 8 degrees, and the design of earthquake acceleration is 0.2g. The classification of design earthquake is the first group, the building site categories is class II, and the building seismic fortification type is class C; (5) The temperature difference: Local basic temperature from -8 to 36℃, so the temperature difference is ± 30 degrees when design the structure, requiring joining temperature from 6 to 22℃. Based on "Load code for the design of building structures" (50009-2012 GB) carry the load combination [7], and the load combination analysis is carried by MSTCAD, then calculate and analysis 19 kinds of working conditions combination. This paper takes the most unfavorable load combination to optimize: 1.2 static +1.4 live =1KN/m2.

2.3 Design parameters

(1) The material: The material of steel pipe, sealing plate and cone-shaped connectors all select the Q345B, and design strength is 315N/mm^2. High-strength bolt using 40Cr or 20MnTiB. Bolt sphere select 45 steel.

(2)The deflection limit value: Considering large deformation will affect the drainage and safety of the structure, so "Technical Specification for space Frame Structures" (JGJ7-2010) requirements structure deflection span ratio is L/250 (L is the span) and the displacement limit value is 0.6224m.

(3)The slenderness ratio: The lever limit slenderness ratios are determined by "Technical Specification for space Frame Structures" and "Code for seismic design of Buildings" (GB50011-2010) [8]. For compression member, the limit slenderness ratios is mainly to prevent too thin prone to bend, thereby affecting the ultimate capacity of compression bar, so limit slenderness ratio is 150; for tension member, the allowable slenderness ratio is mainly to ensure the lever in the production, transportation, installation and use of the process with a certain stiffness, so the limit value is 250; the bar where near the support may be affected by impact or supports uneven settlement, leading to the phenomenon of corrosion, bending, stress complex. The bar near the support is a sensitive area where affect the safety of the structure, so the limit slenderness ratios of the web member near the support are 150, which are the same as the limit slenderness ratios of the compression member.

(4)Stress ratio: The seismic code provisions the seismic combined internal force value of the key member in the 8 degree area should multiplied by the coefficient of 1.15. The general stress ratio limit is 0.9, so the control key bar stress ratio limit is 0.8.

2.4 Structure basic characteristics analysis

（1）Firstly, the structure is analysed and designed with MSTCAD, and the design stress is 283.5 N/mm2 which strength design value of 90%. The structure is applied load and full stress automatic optimization calculation is carried out, adjusting the software auto-configuration of the rods. The member section too small to be easy to produce the initial bending, taking into account the convenience of welds and anti-corrosion requirements. The wall thickness of circular steel tube should not be too thin, increasing the key parts of the member section properly, to strengthen the self-supporting performance of the shell. The final selection of the rod specifications has 10 species: Ø60.0×3.50、Ø89.0×4.00、Ø114.0×4.0、Ø140.0×6.0、Ø159.0×6.0、Ø180.0×12.0、Ø219.0×14.0、Ø219.0×16.0、Ø245.0×16.0、Ø299.0×22.0. The biggest bar in superior string layer edge arch part, inferior string layer maximum bar specifications is Ø219.0 × 14.0, web members selection bar specifications smaller. The steel consumption of the main structure (not counting node weight) is 809.36t. The overall displacement of the structure is shown in Figure 3, and the maximum displacement is 0.511m, to meet the specifications of the design requirements. The maximum stress is 279.8 N/mm2, which is less than control stress. The results of MSTCAD analysis show the strength and stiffness of the structure meet the requirements of the specification.

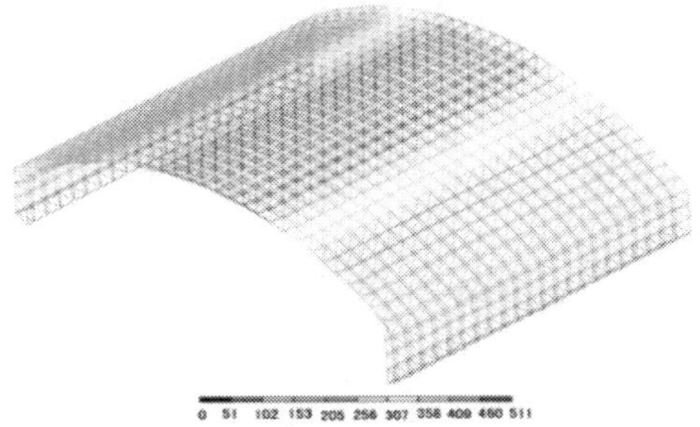

Fig. 3. The displacement diagram of Structure

(2) The structural model is input to the general finite element software ANSYS, carrying the eigenvalue buckling analysis [9], and obtain buckling mode and eigenvalues coefficient, as shown in Figure 4. As can be seen from the results of the analysis, the first order modal along the arch to the buckling of single-wave anti-symmetric, the first order eigenvalue coefficient is 10.459. In order to more accurately reflect the real structure of the large deformation, to carry the geometrically nonlinear analysis, as shown in Figure 5. The structure of the nonlinear buckling load is 1.339KN/m2. It is the weak link where structure across the region. Comparison of eigenvalue buckling analysis and geometric nonlinear analysis shows the buckling load of the latter is smaller than that of the former obviously, so the geometric nonlinear analysis of the structure is very necessary.

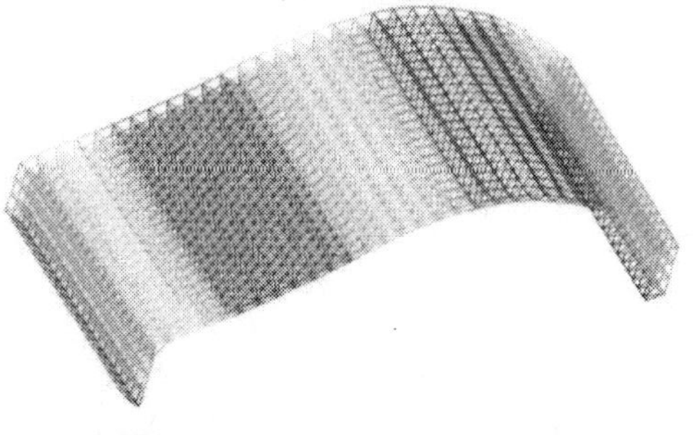

Fig. 4. The integral displacement of the first mode

(3) By checking the strength, rigidity and stability of the structure, we can obtain the following conclusions. Although they all meet the requirements, the structural geometric nonlinear buckling load is 1.339 times larger than the design load value. It is too small to cause the structure of the stability safety factor is small. The structure may be unstable failure if the structure is subjected to the load limit. Based on the above analysis, it is necessary to optimize the shell, achieve the purpose which the ultimate bearing capacity promotion in case of the same amount of steel.

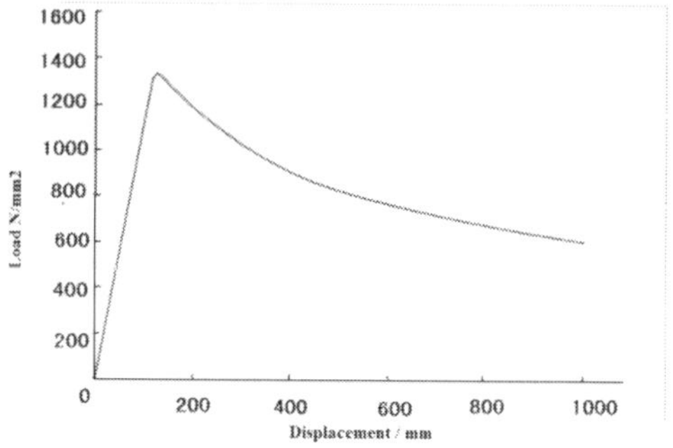

Fig. 5. Structural load displacement curve and failure mode

3 Optimizing Design

3.1 optimization method

Double-layer cylindrical reticulated shell structure by the action of "thin film" as the main force characteristic, most of the load are transmitted by the shell member axial force form, structure vertical stiffness is weak, and bending stiffness is small, greatly reduce the carrying capacity of the latticed shell structure. With the increase of the structure span, the safety defects of the double-layer cylindrical reticulated shell structure are more and more obvious, and the structure prone to damage.

Structural flexural rigidity is proportional to the third power of the thickness, folding the double-layer cylindrical reticulated shell plane, so that the structure is formed polygonal-shape reticulated shell along the longitudinal direction line, forming in transverse ridges and valleys. This method is equivalent to increasing the equivalent thickness of the structure, and the optimized structure is called as the folding double-layer cylindrical reticulated shell structure. The shape characteristics of the structure: the cross-section remains cylindrical form, but the form of a longitudinal cross section of a line, as shown in Figure 6.

In this paper, the flexural rigidity of the folding double-layer cylindrical reticulated shell structure is large, which can overcome the low of the vertical stiffness and the defects of the bearing capacity. In the optimization design, all folding line slope were optimization analysis firstly, and choose the optimal linear slope, then based on the influence of shell thickness on the structure performance, quadratic optimization of structure is carried, summary and analysis to determine the best structure selection.

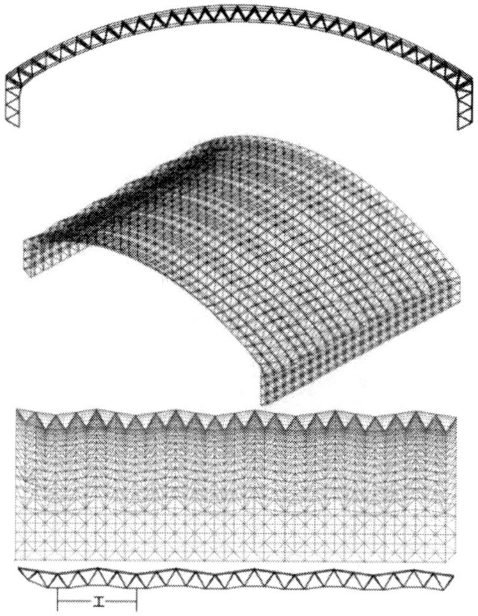

Fig. 6. (a) Cross section of the folding double-layer cylindrical reticulated shell; (b) The axis of the folding double-layer cylindrical reticulated shell; (c) the vertical elevation and section of the folding double-layer cylindrical reticulated shell

3.2 Optimization analysis of line slope

In the full stress design, changing the size of the rod size, to ensure that each calculation model is close to the amount of steel。According to the size of the h divide into six groups, respectively, 1m, 1.2m, 1.4m, 1.6m, 1.8 m, 2.0m. Geometric nonlinear analysis is carried out, and the displacement contour and the load displacement curve are observed and compared with the ultimate load bearing capacity.

Figure 7 is a geometric nonlinear analysis results which seven kinds of the folding double-layer cylindrical reticulated shell structure. From the load displacement curve, it can be seen that the structure has obvious geometric nonlinearity. Taking into account the different h values corresponding to the instability mode map is similar, so only the results of the geometric nonlinear analysis of the folding double-layer cylindrical reticulated shell structure in h=1.0m is listed, as shown in Figure 9.The results of seven kinds of the folding double-layer cylindrical reticulated shell structure with steel, maximum displacement and ultimate bearing capacity are summarized as shown in table 1.

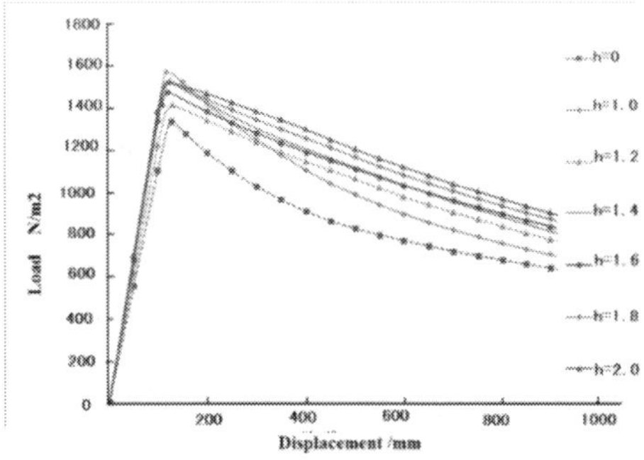

Fig.7. Different slopes corresponding structural load displacement curve

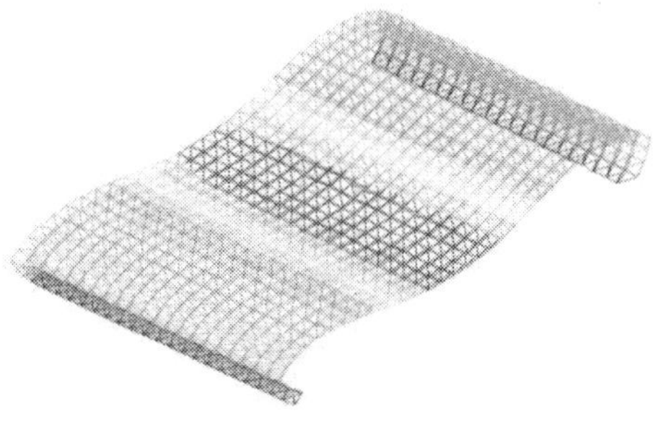

Fig.8. The structural instability mode figure when h=1.0m

Table 1. H1~H6 of different curves

h /m	The amount of steel /t	Maximum displacement /m	Ultimate bearing capacity KN/m^2
0	809.36	0.572	1.339
1.0	809.21	0.401	1.578
1.2	809	0.406	1.421
1.4	809.20	0.4	1.540
1.6	808.92	0.4	1.493
1.8	809.55	0.391	1.528
2.0	809.20	0.387	1.575

Comprehensive chart results can be obtained, the ultimate bearing capacity is the largest when h=1.0m (line slope is 9.16%), and the value is 1.578KN/m^2. Compared to the double-layer cylindrical reticulated shell in its bearing capacity increase 17.91%, at the same time the structure of the maximum stress is 262.4N/mm^2 and the maximum vertical displacement is 0.401m, the strength, stiffness and stability of the structure are improved significantly. The ultimate bearing capacity of the structure is the lowest bearing capacity of the structure with the change of the slope of the line when the h=1.2m, and the value is 1.42KN/m^2. However, the ultimate bearing capacity of the structure increased 5.97% compared to the original double-layer cylindrical reticulated shell, at the same time the structure of the maximum stress is 288.8 N/mm^2 and the maximum vertical displacement is 0.406m, the strength, stiffness and stability of the structure are improved significantly. The analysis results show the ultimate bearing capacity of the double-layer cylindrical reticulated shell structure can be significantly improved by changing the line slope, so this optimization method is feasible and effective. The appropriate line gradient is h=1.0m (the line slope is 9.16%).

3.3 Optimization analysis of shell thickness

On the basis of the optimization in the line slope, the shell what the h=1.0m and the shell thickness H=4.83m is selected to make the quadratic optimization. In the case of the same amount of steel, the shell thickness is an important factor to affect the bearing capacity. Use different shell thickness in this project, to analysis the impact of different shell thickness H on the structure bearing capacity. The results of different shell thickness with steel, maximum displacement and ultimate bearing capacity are summarized as shown in table 2. The influence of shell thickness on the bearing capacity can be directly seen through the different shell thickness H corresponding to bearing capacity curve, as shown in Figure 9.

Table 2. The data comparison of the different shell thickness

Shell thickness /m	The amount of steel /t	Maximum displacement /m	Ultimate bearing capacity KN/m^2
3.0	809.28	0.595	1.135
3.5	809.00	0.514	1.284
4.0	809.70	0.457	1.479
4.4	808.78	0.422	1.705
5.0	809.74	0.393	1.657
5.5	808.41	0.372	1.630
6.0	809.25	0.367	1.534

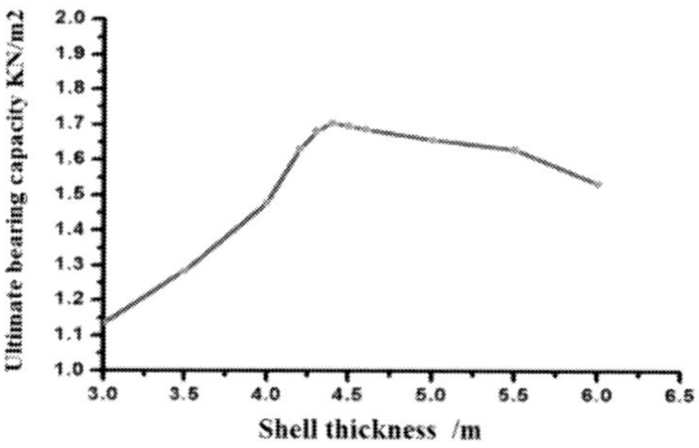

Fig.9. Different shell thickness corresponding bearing capacity curve

From table 2, figure 9 shows that shell thickness and ultimate bearing capacity has generally increased first and then decreased. The ultimate bearing capacity is the largest when the shell thickness H=4.4m, the value is 1.705KN/m2, bars maximum stress 248.5N/mm2, and the maximum vertical displacement of 0.512m. Compared to the structure of the front of the quadratic optimization, its strength and stability has significantly improved, so this optimization method is proved feasible and effective.

3.4. Optimization results

Through the structure optimization, then determine the structure type selection is folding double-layer cylindrical reticulated shell, the shell thickness is 4.4m and the line slope is 9% (h=1.0m). In the case of the same amount of steel, the ultimate bearing capacity of the shell what is optimized enhance 27.33% than the initial model, the maximum stress of the rod is reduced 11.19%, and the maximum vertical displacement of the structure can meet the requirements of the specification.

The final choice of rod specifications has nine species: Ø89.0×4.00、Ø114.0×4.0、Ø140.0×6.0、Ø159.0×6.0、Ø180.0×12.0、Ø219.0×14.0、Ø219.0×16.0、Ø245.0×16.0、Ø299.0×22.0. The biggest bar in superior string layer edge arch part, inferior string layer maximum bar specifications is Ø180.0×12.0mm, web members selection maximum bar specifications is Ø159.0×6.0mm, and the number of units is 6048. The steel consumption of the main structure is 808.78t.

4 Conclusion and suggestion

(1)The large span dry-coal-shed latticed shell in Liyuan use the optimized structural which the folding double-layer cylindrical reticulated shell has obvious advantages than double-layer cylindrical reticulated shell. The ultimate bearing capacity of the double-layer folding cylindrical reticulated shell can be increased 27.3% compared with the original design scheme, and increase economic efficiency.

(2)For the folding double-layer cylindrical reticulated shell, the strength, stiffness and stability are better when the line slope is 9% and the shell thickness is about 1/35 of span.

(3)In the construction of the folding double-layer cylindrical reticulated shell, along the longitudinal direction can take a fold line as the basic unit for installation, and finally connected with each fold line which between the bottom chord bar.

References

1. H. Deng, S.L. Dong, Shape optimization of spatial latticed shell structures, *Journal of Zhejiang University(Engineering Science)*, **33**,371(1999)
2. L.Y. Shan, H. Yan, Optimization analysis of large-span double-layer cylindrical latticed shell structure,**20**,47(1999)
3. Z.F. Liu, Z.T. Yan, Scheme selection and optimum design of single-layer reticulated domes, *Journal of Civil, Architectural & Environmental Engineering*,**29**,83(2007)
4. Y.M. Li, Y.G .Zhang, Section optimization of cable-supported latticed shell, *Industrial Construction*, **37**, 73(2007)
5. Y.Q. Qi, Optimization design of large-span spatial latticed, *Construction Technology*, **38**,393(2009)
6. Technical Specification for space Frame Structures: JGJ7—2010[S]. *Beijing: China Architecture & Building Press*,2010
7. Load code for the design of building structures:GB 50009—2012. Beijing: *China Architecture & Building Press*,2012
8. Code for seismic design of Buildings: GB50011—2010. Beijing: *China Architecture & Building Press*,2010
9. J.G. Zhang, Z.Z. Fan, S.Y. Liu, Optimization design for large span dry-coal-shed latticed shell in Xiangfan power plant, *Building Structure*, **39**,378(2015)
10. Y. G, H. Chen, H.F Yu, Structural design and analysis on the large- span dry coal shed in a thermo power station, *Building Structure*, **45**,49(2015)

Design of Power Cable UAV Intelligent Patrol System Based on Adaptive Kalman Filter Fuzzy PID Control

Siyu Chen[1], and *Wei* Chen[2*]

[1]School of Electric Power Engineering, Nanjing Institute of Technology, Jiangning District, Nanjing, 211167, China.

[2]Industrial Center, Nanjing Institute of Technology, Jiangning District, Nanjing, China

[*]Corresponding author chenwei-nj@163.com

Abstract. Patrol UAV has poor aerial posture stability and is largely affected by anthropic factors, which lead to some shortages such as low power cable tracking precision, captured image loss and inconvenient temperature measurement, etc. In order to solve these disadvantages, this article puts forward a power cable intelligent patrol system. The core innovation of the system is a 360°platform.This collects the position information of power cables by using far infrared sensors and carries out real-time all-direction adjustment of UAV lifting platform through the adaptive Kalman filter fuzzy PID control algorithm, so that the precise tracking of power cables is achieved. An intelligent patrol system is established to detect the faults more accurately, so that a high intelligence degree of power cable patrol system is realized.
Keywords. Adaptive Kalman filter; Fuzzy PID; Power cable; UAV; Intelligent patrol.

1 Introduction

Modern cities and enterprises all use the power cables for power transmission and distribution, and power transmission and distribution quality is affected by the running states of power cables. In order to improve power supply reliability and predict failure, cable temperature is a key indicator for fast judgment of fault points. Current power supply systems mostly adopt temperature patrol to realize fast troubleshooting, including overload, short circuit and aging, etc.

In traditional power line patrol process, patrol work is carried out by workers and patrol conditions are recorded in paper sheets, and then the patrol conditions will be input into the computer. Therefore, the patrol data is affected by human factors; meanwhile. Whether the workers patrol the whole line cannot be effectively managed, patrol quality cannot be ensured, and the safety conditions of power lines cannot be guaranteed, which leaves security risks. Then, what kind of effective measures can be taken becomes a difficult problem for the power sector. Therefore, the development of a temperature measurement system, which is safe and stable and has wide application, has a great significance to ensure the safe operation of power grid; this temperature measurement system can monitor the running temperatures of key points of the equipment in a real-time manner, grasp the running status, timely find out hidden faults, take corresponding measures and prevent further expanding of accidents[1].

Patrol UAV has poor aerial posture stability and is largely affected by anthropic factors, which lead to some shortages such as low power cable tracking precision, captured image loss and inconvenient temperature measurement, etc. In order to solve these disadvantages, this article puts forward a power cable intelligent patrol system developed based on adaptive Kalman filter fuzzy PID control algorithm. This system collects the position information of power cables by using far infrared sensors and carries out real-time all-direction adjustment of UAV lifting platform through the control algorithm, so that the precise tracking of power cables is achieve. This system uses STM32 as the control platform, uses GPS module to collect geographic information of power cables, uses temperature logger to collect temperature data of power cables and uses camera to capture image data; an intelligent patrol system is established to detect the faults more accurately, so that a high intelligence degree of power cable patrol system is realized. This system can directly display whether the power lines are patrolled and whether the patrol work has been completely finished, so as to meet the requirements of power line patrol in speed, efficiency and timeliness; it improves the power line patrol management to a new level and realizes the electronization, informationization and intelligence of patrol work, so as to guarantee the high efficiency, low failure rate and safe operation of power system[2].

2 System configuration

System configuration diagram is shown in figure 1:

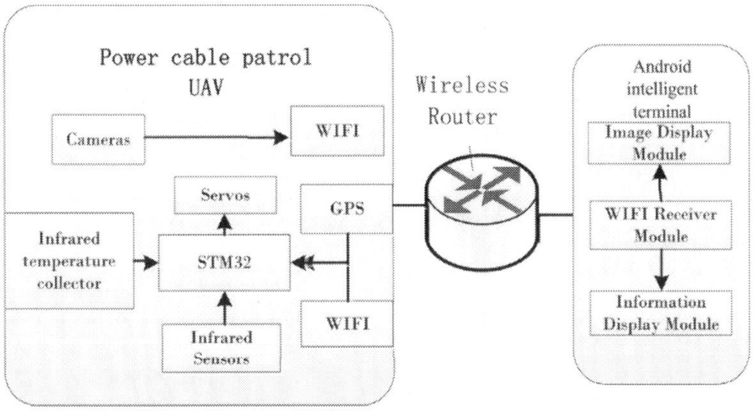

Figure 1. System configuration diagram

The whole system mainly consists of the power cable patrol UAV and an Andriod intelligent handheld terminal.

① Power cable patrol UAV: Patrol workers carry out rough patrols by means of remote control. The far infrared sensor on lifting platform acquires cable position information and controls the platform steering engine, so as to accurately capture power cables; temperature acquisition module on the platform is responsible for temperature collection; the camera acquires image information, realizing the monitoring on the waste, oil pollutant and other faults on power cables; the wireless router is used for data transmission and exchange.

② Android intelligent terminal: The intelligent Android mobile terminal, which is developed based on Android mobile platform and combines image and information display modules, can realize the integration of image transmission, temperature, geographic coordinates and time information, and it can monitor the patrol images and information in real-time manner.

3 Design of key structures of power cable patrol uav

Sensor Selection. Long-distance infrared sensor: As cable transmission will generate heat, the long-term allowable working temperature of power cable core is 70℃ ~ 90℃, and the temperature of insulation layer is 20℃ to 40℃ different from cable core. The temperature of external insulation layer is at least 30℃ according to the calculation. Considering the influence of environment temperature and the cable temperature is much higher than the environment temperature, the infrared wavelength of power cable is shorter than those of the surrounding environment. The measuring range of long-distance infrared sensor is 0 to 8 meters, meeting requirement of actual flying height[3].
Infrared temperature sensor: SMTIR99XX series are thermopile-based silicon infrared sensors which can measure temperatures in contactless ways; as infrared radiation characteristics are related to temperature, filters can be used to measure different temperature ranges. Mature semiconductor process can reduce the cost and miniaturize the overall dimension at the same time. In order to meet the requirements of some certain applications, the opening angle of infrared sensor can be designed to be less than 7°. Based on the Seebeck Effect, the temperature difference can be calculated according to the analog output from infrared sensor; SMT9902 is equipped with a built-in temperature sensor which can measure the temperature of its cold end, based on which the temperature of the object to be measured can be obtained through calculation. The working temperature of the sensor itself is -40 to 100℃. TO05 packaging is adopted for the sensor, and it has high precision, high sensitivity (110V/W), low impedance and a measuring temperature range from -100 to 2000℃.

Software Design. The software system flow chart of UAV patrol system is shown in Figure 2:

The main program of the temperature acquisition module is as shown in Figure 3:

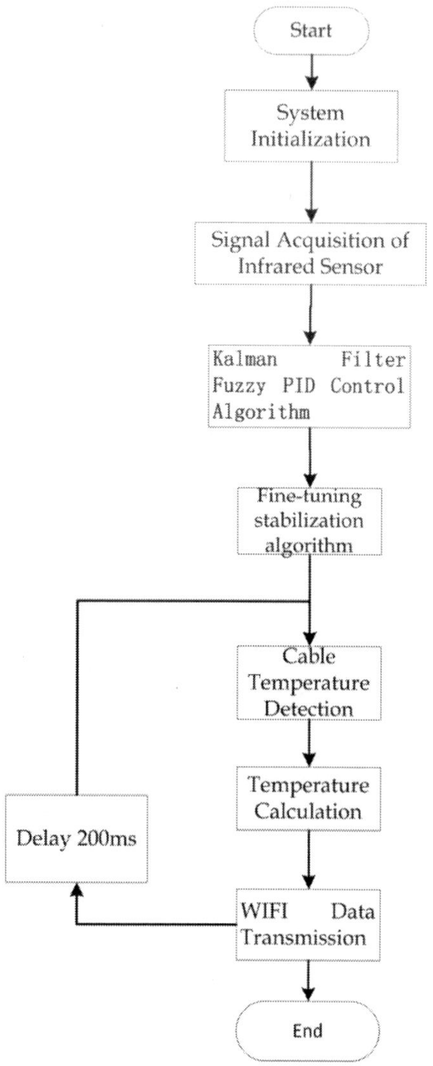

Figure 2: Software system flow chart

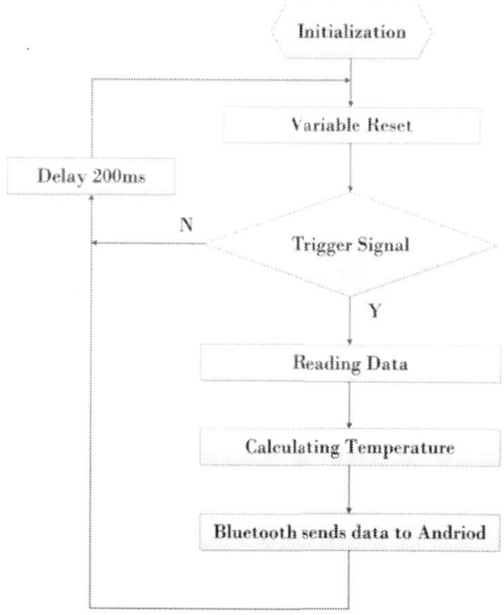

Figure 3: Flow chart of temperature acquisition modul

In the process of initialization, the definition of variables and the settings of ports and baud rate are completed. The program continuously detect whether there is a trigger signal, and only when it detects the trigger signal, will it drive SMT9902 to start temperature measurement and read temperature data. The temperature data in SMT9902 can be divided into high byte (DataH) and the low byte (DataL), and the actual Celsius temperature can be obtained after a certain amount of calculation on the two bytes, and the calculation formula is as follows:

$$Data = [(DataH \,\&\, 0X007F) << 8] + DataL$$

$$TempData = [(Data \times 0.02) - 0.01] - 273.15 \tag{1}$$

After obtaining the Celsius temperature, it will initiate the WIFI module transmission program, and as Bluetooth module are in series connection with Arduino, Bluetooth transmission program is the same like serial port output program[4].

 DOI: 10.1051/matecconf/201710003025

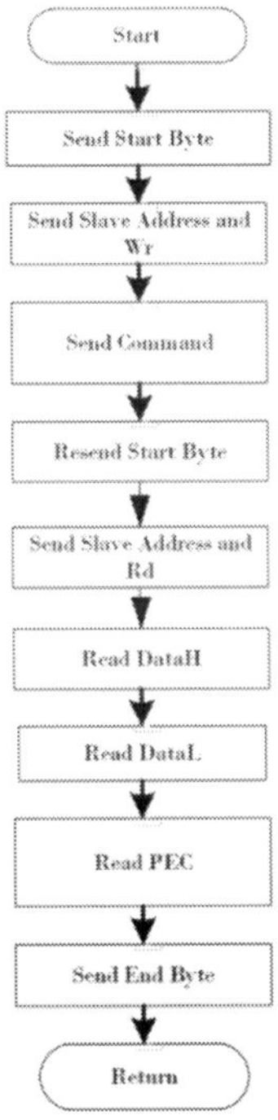

Figure 4. Data read flow chart

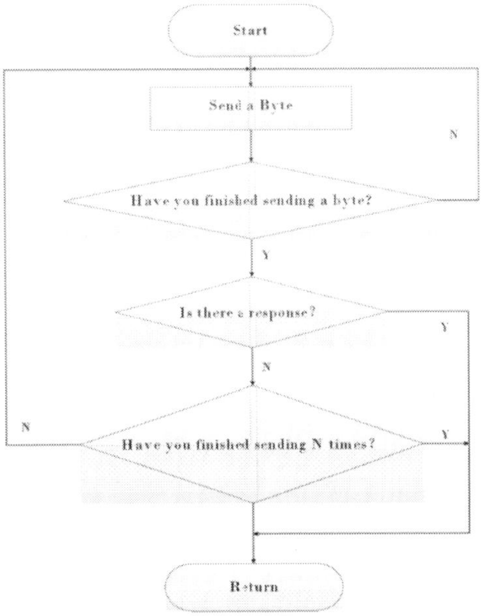

Figure 5. Operation process of each byte of data

Several SMT9902 can be used in one system, and different addresses are used to separate the components. Therefore, in a multiple SMT9902 system, each SMT9902 need to be allocated with an address, and the default address for components is 5AH. Data read procedures are shown in Figure 4. The data read from SMT9902 is in byte, and the operation procedures for each byte of data are shown in Figure 5. Each time only one byte (in bits, and 8 bits form one byte) is sent, then it will judge whether there is any response, and if there is a response, the next byte will be sent; if no, this byte will be sent repeatedly until there is a response, and if there is no response after repeating sending the same byte for a couple of times, the will come to an end[5].

4 Fuzzy pid control algorithm based on adaptive kalman filter

This article puts forward a fuzzy PID control algorithm that based on adaptive Kalman filter; carry out self-learning and self-tuning for PID parameters k_p, k_i and k_d, to find out the best combination, and meanwhile it can control interference noise and measurement noise to reach the filtering effect.
The control structure is as shown in Figure 6[6]:

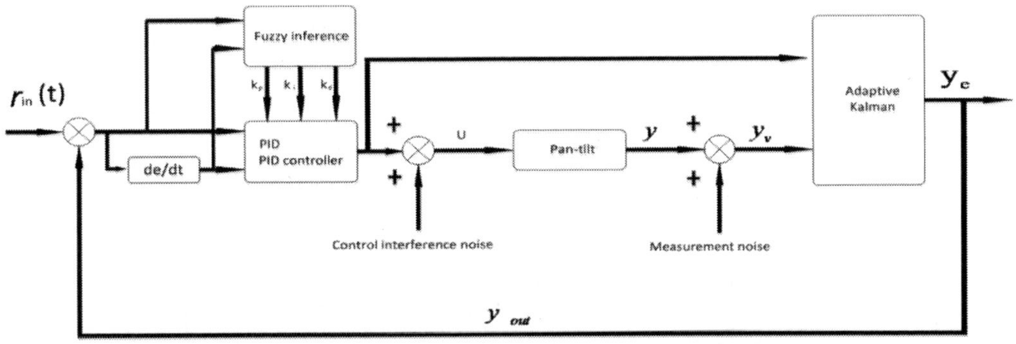

Figure 6: Structural diagram of fuzzy PID control algorithm based on adaptive Kalman filter

Fuzzy Adaptive Setting PID. The setting of PID parameters is a key problem in PID control. However, in actual applications, most industrial processes have different degrees of nonlinear, parameter time-varying and model uncertainty; therefore, the general PID control cannot realize the accurate control of such processes.

Fuzzy adaptive PID control adopts the basic theory and method of fuzzy mathematics, and uses fuzzy set to express the conditions and operations of the rules, and it stores these fuzzy control rules and relevant information (such as evaluation indicator and initial PID parameters, etc.) into the computer knowledge base, then the computer will automatically realize the best adjustment of PID parameters by means of fuzzy reasoning according to the actual response conditions of the control system.

PID controller is a linear controller, it makes up the control deviation according to this fixed value: $r(t)$ and actual output value $y(t)$.

$$e(t) = r(t) - y(t) \tag{2}$$

The continuous conditions of PID control law:

$$u(t) = k_p\left[e(t) + \frac{1}{T_I}\int_0^t e(t)dt + T_D\frac{de(t)}{dt}\right] \tag{3}$$

where, k_p is the scale factor; T_I is the integral time constant; T_D is derivative time constant

Discrete conditions:

PID parameters fuzzy self-setting functions by finding out the fuzzy relationship among three PID parameters

$$u(k) = k_p\left\{e(k) + \frac{1}{T_I}\sum_{j=0}^{k} Te(j) + T_D\frac{e(k)-e(k-1)}{T}\right\} \tag{4}$$

$$\Delta u(k) = k_p\big(e(k) - e(k-1)\big) + k_i e(k) + k_d\big(e(k) - 2e(k-1) + e(k-2)\big) \tag{5}$$

and e and e_c. It continuously detect e and e_c, and make online adjustment on the three parameters according to fuzzy control law to meet the different requirements of e and e_c on control parameters, so that the objectives can have better dynamic and static performances, as shown in Table 1.

Table 1: The relationship between PID control parameters with system performance index in time domain

Parameter name	Rising time	Overshoot	Transient time	Static error
k_p	Reduce	Increase	Minor change	Reduce
k_i	Reduce	Increase	Increase	Eliminate
k_d	Minor change	Reduce	Reduce	Minor change

In the quantitative field of input language variables: take 7 fuzzy subsets {NB NM PS PM PB NS Z}, which are Negative big, Negative middle, Negative small, Zero, Positive small, Positive middle and positive big, respectively.

Corresponding to the error e and error change rate e_c, 7 degrees are quantified as {-3, -2, -1, 0, 1, 2, 3}. Then, the membership grade of fuzzy subsets are determined, then design PID parameters according to membership grade assignment table of each fuzzy subset and fuzzy control model of each control parameter and based on fuzzy synthesized inference, then the fuzzy control tables for setting the three parameters k_p, k_i and k_d are obtained, as shown in Table 2, 3 and 4[7].

Table 2: Fuzzy rule table of k_p

ec \ e	NB	NM	NS	ZO	PS	PM	PB
NB	PB	PB	PM	PM	PS	ZO	ZO
NM	PB	PB	PM	PS	PS	ZO	NS
NS	PM	PM	PM	PS	ZO	NS	NS
ZO	PM	PM	PS	ZO	NS	NM	NM
PS	PS	PS	ZO	NS	NS	NM	NM

PM	PS	ZO	NS	NM	NM	NM	NB
PB	ZO	ZO	NM	NM	NM	NB	NB

Table 3: Fuzzy rule table of k_i

ec \ e	NB	NM	NS	ZO	PS	PM	PB
NB	NB	NB	NM	NM	NS	ZO	ZO
NM	NB	NB	NM	NS	NS	ZO	ZO
NS	NB	NM	NS	NS	ZO	PS	PS
ZO	NM	NM	NS	ZO	PS	PM	PM
PS	NM	NS	ZO	PS	PS	PM	PB
PM	ZO	ZO	PS	PS	PM	PB	PB
PB	ZO	ZO	PS	PM	PM	PB	PB

Table 4: Fuzzy rule table of k_d

ec \ e	NB	NM	NS	ZO	PS	PM	PB
NB	PS	NS	NB	NB	NB	NM	PS
NM	PS	NS	NB	NM	NM	NS	ZO
NS	ZO	NS	NM	NM	NS	NS	ZO
ZO	ZO	NS	NS	NS	NS	NS	ZO
PS	ZO	ZO	ZO	ZO	ZO	ZO	ZO
PM	PB	NS	PS	PS	PS	PS	PB
PB	PB	PM	PM	PM	PS	PS	PB

Adaptive Kalman Filtering. Because the lifting platform of UAV patrol system can be affected by various environmental factors, adaptive Kalman filter is used in this paper to suppress interference noise and measure the noise so as to improve the control accuracy.

The state equation and measurement equation of the system discrete model are: X(k+1)= Φ(k+1,k)X(k)+B(k)U(k)+W(k), y^v(k+1)=HX(k+1)+V（k), of which X (k) is the state matrix, y (k) is the measurement matrix, U (k) is the control output, Φ is the state transfer matrix, B is the input matrix, H is measurement transfer matrix, W (k) and V (k) respectively control the interference noise and measurement noise. The noise statistics are: E{W(k)}=q , E{W(k)W (j)}=Q(k) , δ_{kj} and $E\{V(k)V^T(j)\} = R(k)\delta_{kj}$, where, δ_{kj} is Keluonike Function: when $k = j$, =1, and when $k \neq j$, δ_{kj}=0.

Due to the randomness and uncertainty of disturbance, the variance and mean of noise cannot be determined precisely. In order to estimate the mean q (k) and variance Q(k) of control interference noise W(k) and measure the mean V(k) and variance R (k) of measurement noise V(k), this article will use the maximum posteriori noise estimator sage-husa.

Algorithm steps are as follows:

1) State estimation value in (k +1) moment;

$$\hat{X}(k+1\mid k)=\Phi(k+1\mid k)\hat{X}(k\mid k)+B(k)U(k)+\hat{q}(k); \quad (6)$$

2) Corresponding Covariance of(k+1 | k) : $P(k+1\mid k)=\Phi(k+1\mid k)P(k\mid k)\Phi^T(k+1\mid k)+Q(k)$; (7)

3) Kalman gain: $M(k+1)=P(k+1\mid k)H^T(k)\left[H(k)P(k+1\mid k)H^T(k)+R(k+1)\right]^{-1}$; (8)

4) New information in Kalman filter: $\varepsilon(k+1\mid k)=y^{\upsilon}(k+1)-H(k)\hat{X}(k+1\mid k)-\hat{r}(k)$;(9)

5) Optimal estimation value in (k+1) moment: $\hat{X}(k+1\mid k+1)=\hat{X}(k+1\mid k)+M(k+1)\varepsilon(k+1\mid k)$; (10)

6) $\hat{X}(k+1\mid k+1)$ corresponding covariance: $P(k+1\mid k+1)=\left[I-M(k+1)H(k)\right]P(k+1\mid k)$;(11)

7) Use noise estimator sage - husa to estimate ω(k), Q(k), υ(k) and R(k): $\hat{q}(k)=(1-d^k)\hat{q}(k-1)+d^k\left[X(k\mid k)-\Phi(k\mid k-1)X(k-1\mid k-1)-B(k)U(k)\right]$, (12)

$$Q(k)=(1-d^k)Q(k-1)+d^k\left[M(k)\varepsilon(k\mid k-1)\varepsilon^T(k\mid k-1)M^T(k)+P(k\mid k)-\Phi(k\mid k-1)P(k-1\mid k-1)\Phi^T(k\mid k-1)\right], \quad (13)$$

$$\hat{r}(k)=(1-d^k)\hat{r}(k-1)+d^k\left[y^{\upsilon}(k)-H(k)X(k\mid k-1)\right], \quad (14)$$

$$R(k)=(1-d^k)R(k)+d^k\left[\varepsilon(k\mid k-1)\varepsilon^T(k\mid k-1)-H(k)P(k\mid k-1)H^T(k)\right], \quad (15)$$

$d^k=(1-b)/(1-b^{k+1})$,$0<b<1$ is forgetting factor, which needs to be determined by experiment, 0.95 ~ 0.99 is preferable generally[8].

5 The system simulation results

The adaptive Kalman filter fuzzy PID control algorithm in this paper is applied to MATLAB2010 for simulation, with UAV platform frame as simulation model, open-loop transfer function is: $G(s)=16470/(S^2+1667S+3393)$; sampling time t = 1ms.

Figure 7 is the corresponding membership function:

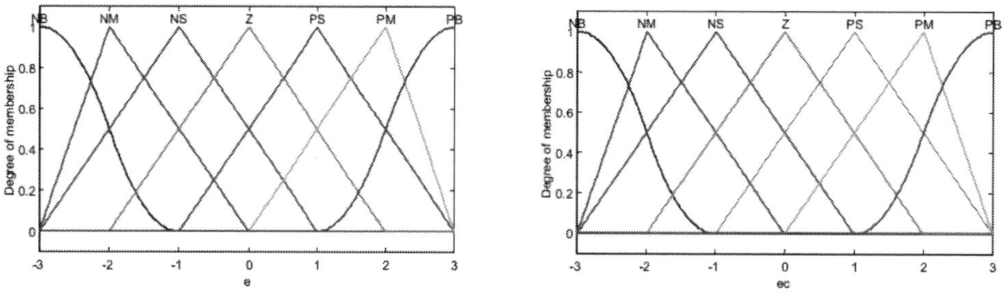

DOI: 10.1051/matecconf/201710003025

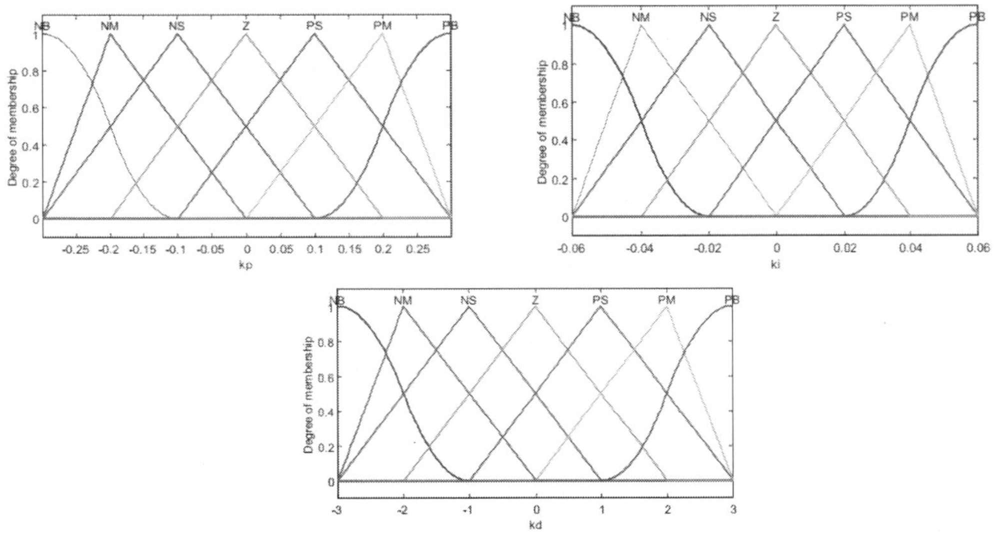

Figure 7: Membership function of fuzzy PID

Step signals are used as the simulation signal for the tests in this article. The comparison of image curves (as shown in figure 8)show that the effects of the performance of classic PID control and fuzzy PID control are inferior to that of fuzzy PID control that based on Adaptive Kalman Filter proposed in this article. Besides, Adaptive Kalman Filter Fuzzy PID control is faster in reaching the steady state and nearly no overshoot is required.

If white noise is added, its amplitude is 0.01 (as shown in figure 9), two algorithms are all interfered largely, fluctuating in the stable value. However by comparison, the algorithm used in this paper can quickly achieve the stable value, with less interference and no overshoot.

The sine wave simulation results show that:

①When the white noise with 0.01 amplitude is added, the classic PID and fuzzy PID all have large interference, but the algorithm in this paper can track signals fast and smoothly with no overshoot (as shown in figure10)

②when the white noise with 0.10 amplitude, the classic PID and fuzzy PID all have a big deviation and cannot satisfy tracking precision, but the algorithm in this paper still track precisely (as shown in figure11).

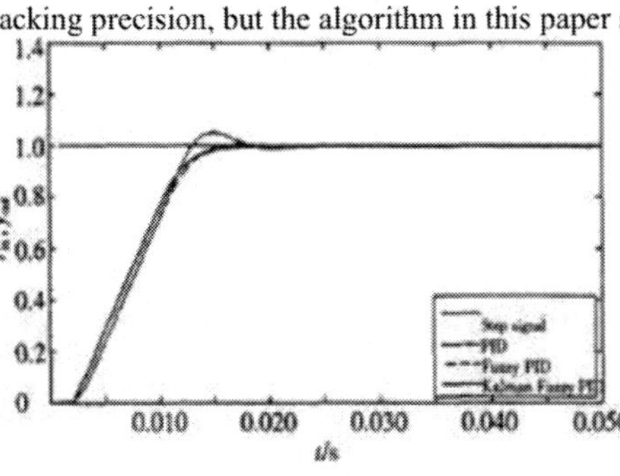

Figure 8:Simulation results of three algorithms for step signal without noise

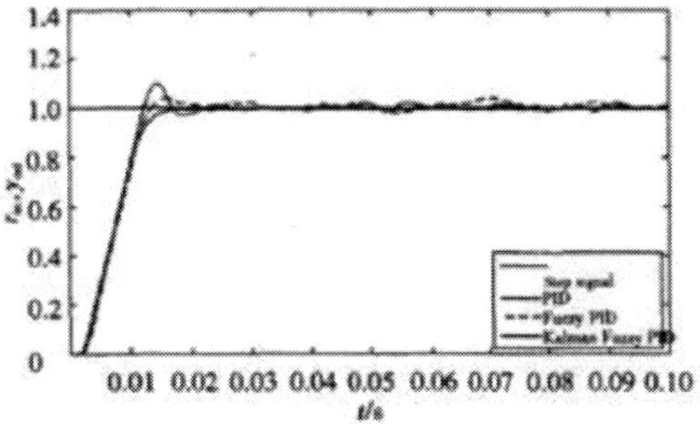

Figure9: Simulation results of three algorithms for step signal with magnitude 0. 01 white noise

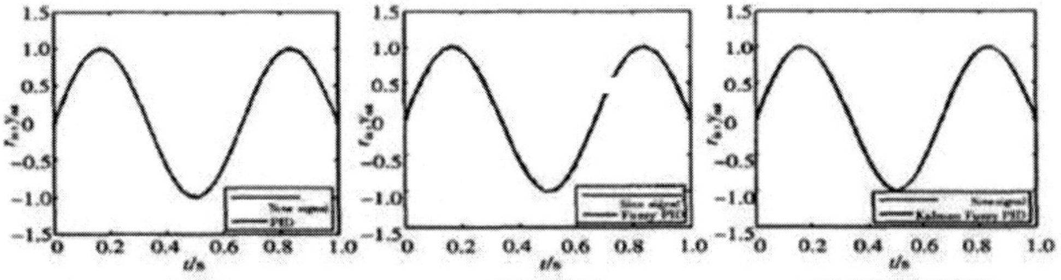

a PID b Fuzzy PID c Kalman Fuzzy PID

Figure 10: Simulation results of three algorithms for sine wave with magnitude 0. 01 white noise

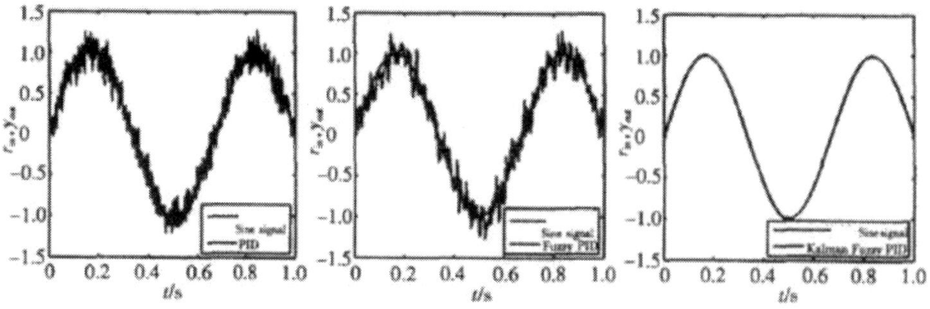

Figure 11: Simulation results of three algorithms for sine wave with magnitude 0. 1 white noise

6 Conclusion

In this article, an UAV power cable patrol system is designed, and a new algorithm of Adaptive Kalman Filter fuzzy PID control is put forward. Under the circumstance of unknown environment parameters and unstable UAV attitude, the algorithm uses infrared sensors to obtain the state of the UAV, so that the platform is controlled. In this way, the platform can precisely track the cable and make it always in the center of UAV, thus the inspection, image transmission and fault detection can be achieved. Theoretical analysis and simulation tests prove the feasibility of this kind of control algorithm. Subsequent research will be the whole set of system tests in practical operation.

Acknowledgements

The author is highly thankful for the financial support of Practice Innovation Training Projects for College Students in Jiangsu Province in 2015 (No. 201511276001Z), Support Projects of 2016 Challenge Cup Competition from University Students Science and Technology Innovation Fund of Nanjing Institute of Technology (No. TZ20160028) and Major Projects of Innovation Fund of Nanjing Institute of Technology (No. CKJA201606)

References

1. Zou Ling, Sun Yuqiang and Sun Qi: A Simulation Study of PID Control Based on Kalman filter. Control & Automation. Vol. 80-81(2007), p. 56
2. Xin Zhekui and Fang Yongchun: Modeling and Control for Tracking Ground Target Using an Unmanned Air Vehicle. Chinese High Techology Letters. Vol. 19(2009) , p. 398 – 403
3. Pragasen Pillay: Modeling, Simulation, and Analysis of Permanent-Magnet Drive. IEEE Transactions on Industry Applications. Vol.25(1989), p265~279
4. Liu Jinkun, in: *Advanced PID control and MATLAB simulation*/Electronic industry publishing house, Beijing(2003) , in press.
5. Guo Bingkun, Chen Shuili, Wu Yundong and Li Wei. Fuzzy PID control of three-axis stabilized platform of UAV. JournalofJimeiUniversity(Natural Science)Vol. 73 (2013), p. 309-312
6. Chen Shuili, Li Jianggong and Wang Xianggong, in: *Fuzzy set theory and its application*/ Science Press, Beijing(2005), in press.
7. Liu Jinkun, in: The 3rd edition of Advanced PID control and MATLAB simulation/ Electronic industry publishing house, Beijing(2011), in press.
8. Ma Jingbo and Yang Honggeng: The Application of Adaptive Kalman Filter in Electric Power System Short-term Load Forecasting. The Grid Technology. Vol. 29 (2005), p75-80

Building and Application of Cardiopulmonary Bypass Model in Rats

Kun Zhang[1], Rui Xue[2*] and Shuwei Shen[1]

[1] Department of Anesthesiology, Jingzhou Central Hospital,The Second Clinical Medical College, Yangtze University, Jingzhou, Hubei Province,434020,China.
[2] Department of Anesthesiology, Renmin Hospital, Hubei University of Medicine, shiyan,Hubei, China.
*Corresponding author: 7885909@qq.com

Abstract. To build a cardiopulmonary bypass model in rats, and research the feasibility. **Method** Cardiopulmonary bypass was built in 10 adult male SD rats of clean grade through intubation in jugular vein, caudal artery, and femoral artery, and bypass was sustained for 60 min at the flow rate of 100ml/(kg·min) to monitor heart rate, blood pressure, blood gas, and electrolyte. **Result** Puncture succeeded in all the 20 rats, and cardiopulmonary bypass was finished under given conditions. **Conclusion** The model has the following advantages, economical efficiency, simplicity, minimal invasion, cardiopulmonary bypass parameter setting similar to that of clinical trail, high rate of success. Thus, it is reliable for researching pathological and physiological changes after cardiopulmonary bypass and evaluating therapeutic strategy.

Keywords: Rat, anesthesia, cardiopulmonary bypass, building.

1 Experimental material and method

1.1 Material

10 healthy adult male SD rats of clean grade, body mass: 350~500g, (424±70) on average. Inhibited from drinking water within 6h before operation.

1.2 Test Equipment

Cardiopulmonary bypass rolling blood pump, with silicon pump with an internal diameter of 4mm; membrane oxygenator specially used for rats (surface area: 0.10m^2), ECG monitor; blood-gas analyzer and heat exchanger; temperature changing water bank [1].

1.3 Anesthesia and Monitor

A peritoneal injection of 0.2% pentobarbital sodium solution according to the standard of 50 mg/kg was given to rats for anesthesia, and then rats were fixed on rat fix table, monitored with an ECG monitor, and tested for rectal temperature. The body temperature of rats were sustained around 35. 5~ 36.0℃ with a temperature changing water bank [2]. A blood-gas analyzer was employed to test and analyze arterial blood gas, electrolyte, mixed venous oxygen saturation, and lactate levels. Tests were carried out before systemic circulation, 30min after systemic circulation, upon systemic circulation being finished, and 30min after the finish of systemic circulation.

2 Building of cardiopulmonary bypass

(1)The skin of left groin was cut about 0.8cm; left femoral artery was separated and ligated, a 24 G venous indwelling needle was stabbed through proximal end of ligation for fixation, and heparin solution was injected (half of 500U/kg), a monitor was connected via a transducer for continuous monitoring arterial blood pressure. Specimens can be taken via this bypass [3].

(2) The middle of on proximal end ventral side of cauda was cut about 0.8cm; a 20 G venous indwelling needle was stabbed in the same way for femoral artery for arterial perfusion.

(3) The 16 G venous indwelling needle clipped so that its frond end had multiple lateral pores, and the front end of the steel core was smoothened so as to avoid any injury to vessel wall at the time of puncture. The right neck was cut lengthways about 1cm for separating external jugular vein, of which the distal end was ligated, and the abovementioned venous indwelling needle was stabbed into the proximal end, so that the tip of puncture needle was close to the entrance of postcava of right atrium to serve as venous drainage tube.

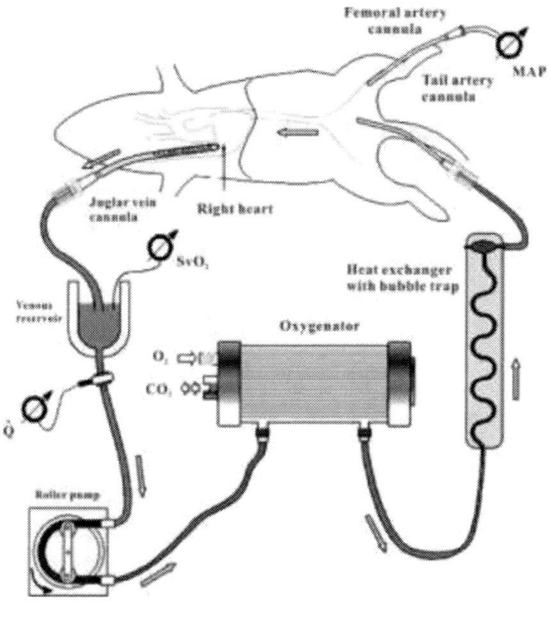

Cardiopulmonary Bypass Model

3 Result

Intubation tube and cardiopulmonary bypass in all rats succeeded. Please see Tab. 1 for haemodynamics and blood gas analysis and body temperatures before and after cardiopulmonary bypass. During bypass, the blood pressure of rats can be manually controlled via adjusting drainage and perfusion. The monitoring of blood pressure showed that early blood pressure during and after bypass was lower than that before bypass ($P<0.05$). Blood gas analysis showed that arterial partial pressure of oxygen and partial pressure of carbon dioxide roughly remained in the normal ranges. An acidosis trend was showed with the progress of bypass.

Tab. 1 Analysis and Comparison of Haemodynamics, Temperature and Blood Gas Before and After Cardiopulmonary Bypass ($\bar{x}$ ±s)

Time	MAP (mmhg)	HR	TEM	PO2	PCO2	PH	HCT
Before cardiopulmonary bypass	105.7±4.9	310.5±19.3	36.8±0.4	133.1±13.7	41.9±1.1	7.41±0.03	45.4±3.2
30min after cardiopulmonary bypass	75.9±9.8*	232.7±23.5*	29.4±0.7*	267.8±15.4*	36.9±0.9*	7.38±0.04*	30.6±2.8*
Upon the end of cardiopulmonary bypass	92.5±6.3*	307.5±19.9	37.2±0.5*	289.9±15.3*	39.2±1.3*	7.33±0.04*	28.7±1.6*
30min after the end of cardiopulmonary bypass	94.6±4.4*	347.5±25.6*	36.9±0.7	133.3±8.6	36.1±1.3*	7.34±0.05*	31.6±3.3*

*P < 0.05 between the four groups.

4 Discussion

The cardiopulmonary bypass model in rats we built has the following advantages, economical efficiency, simplicity, minimal invasion, cardiopulmonary bypass parameter setting similar to that of clinical trail, high rate of success, long duration, etc. Through preliminary studies, we found that these conditions would injure liver and kidney of rats to a certain extent, so can used for researching organ injury after cardiopulmonary bypass. Comparing with previous experimental models, reduction of trauma caused by factors other than cardiopulmonary bypass was emphasized in this model. Firstly, this model needs no thoracotomy, so that the integrity of thorax was maintained, sternum was not cut, and inflammatory response was reduced [4,5,6,7]. Secondly, autonomous respirations were retained, tracheal intubation or thracheotomy were not conducted for mechanical assisted ventilation. This is also significantly different from cases reported before. Since satisfactory anesthetic effect can be achieved through intraperitoneal injection of pentobarbital sodium, rats had no response to pain stimulation, operation had not significantly influenced blood pressure and heart rate, and no respiratory depression occurred. That these invasive manipulations were avoided is an important reason that the survival rate of out rats was high; besides, that thoracotomy or mechanical assisted ventilation that impact the body were not carried out will create conditions for researches to purely analyze the pathological physiological changes caused by cardiopulmonary bypass.

Comparing with cardiopulmonary bypass building via thoracotomy, the operation of cardiopulmonary bypass building via peripheral vessel intubation is simpler, because intubation is relatively easier. Besides, the latter is physiologically same with the former, but the thorax remains stable in the latter operation, good for the recovery of respiratory function after operation. Thus, in recent years, the cases of building of cardiopulmonary bypass in rats without thoracotomy were reported. The main intubation way is reverse perfusion through caudal artery. Blood flow varies with physiological conditions, and it is not suitable for cerebral perfusion. Therefore, we adopted internal carotid artery perfusion based on overseas experience, which conforms to the physiological conditions [8, 9, 10, 11, 12]. In most cases reported in China, cardiopulmonary bypass was realized under normal temperature, while DHLF and DHCA are needed for simulating physiological states in researches on cerebal protection. Thus, building DHLF or DHCA without thoracotomy is particularly important. Building cardiopulmonary bypass via peripheral vessel intubation can be used for researches on cardiopulmonary bypass of different types, such as CPB, DHLF, DHCA, etc. This experiment also verified this. Cardiopulmonary bypass in all rats succeeded, their vital signs, physical and chemical indicators during CPB remained in the acceptable ranges, and the rats all successfully got over cardiopulmonary bypass after the end of experiment. On this basis, I have reason to regard this model as a reliable platform for research on physiology of cardiopulmonary bypass.

CPB model in rats can be used for research on physiopathologic mechanism of system inflammatory response and organ function changes caused by CPB, and facilitates further discussions on strategy of organ protection in clinical perioperative period, worth further study and exploration.

References

1. Selnes OA, Goldsborough MA, Borowicz LM and Mckhann GM. Neurobehavioural sequelae of cardiopulmonary bypass. Lancet. (1999) 353: 1601-1606.
2. Joshi B, Brady K, Lee J, Easley B, Panigrahi R,Smielewski P, Czosnyka M and Hogue CW. Impaired autoregulation of cerebral blood flow during rewarming from hypothermic cardiopulmonary bypass and its potential association with stroke. Anesth Analg.(2010) 110: 321-328
3. Jeleazcov C, Saari TI, Ihmsen H, Schuttler J and Fechner J. Changes in total and unbound concentrations of sufentanil during target controlled infusion for cardiac surgery with cardiopulmonary bypass. Br. J.Anaesth. (2012) 109: 698-706.
4. Ranucci M, Isgrò G, Carlucci C, Torre T, Enginoli S and Frigiola A. Central venous oxygen saturation and blood lactate levels during cardiopulmonary bypass are associated with outcome after pediatric cardiac surgery. Crit. Care. (2010) 14: 1-10.
5. Townend WJ, Guy MJ, Pani MA, Martin B and Yates DW. Head injury outcome prediction in the emergency department: a role for protein S-100B? J. Neurol.Neurosurg. Psychiatry (2002) 73: 542-546.
6. Acharya UR, Joseph KP, Kannathal N, Lim CM and Suri JS. Heart rate variability: a review. Med. Bio. Eng.Comput. (2006) 44: 1031-1051
7. Wang Z, Leng Y, Tsai L-K, Leeds P and Chuang D-M.Valproic acid attenuates blood-brain barrier disruption in a rat model of transient focal cerebral ischemia: the roles of HDAC and MMP-9 inhibition. J. Cereb. Blood Flow. Metab. (2011) 31: 52-57.
8. Wolfler A, Salvo I, Sortino G, Bonati F and Izzo F.Epidural analgesia with ropivacaine and sufentanil is associated with transient fetal heart rate changes.Minerva. anestesiologica. (2010) 76: 340.
9. Li XM, Yang JM, Hu DH, Hou FQ, Zhao M, Zhu XH,Wang Y, Li JG, Hu P, Chen L, Qin LN and Gao TM.Contribution of downregulation of L-type calcium currents to delayed neuronal death in rat hippocampus after global cerebral ischemia and reperfusion. J.Neurosci. (2007) 27: 5249-5259.
10. Ranucci M, Isgrò G, Carlucci C, Torre T, Enginoli S and Frigiola A. Central venous oxygen saturation and blood lactate levels during cardiopulmonary bypass are associated with outcome after pediatric cardiac surgery. Crit. Care. (2010) 14: 1-10.
11. Arrowsmith JE, Grocott HP, Reves JG and Newman MF. Central nervous system complications of cardiac surgery. Br. J. Anaesth. (2000) 84: 378-393.
12. Schoknecht K and Shalev H. Blood-brain barrier dysfunction in brain diseases. Clin. Exper. Epileps.(2012) 53: 7-13.

Ultrafiltration Technology and Its Application in Cardiopulmonary Bypass

Kun Zhang, Lishen Wang* and Shuwei Shen

Department of Anesthesiology, Jingzhou Central Hospital, The Second Clinical Medical College, Yangtze University, Jingzhou, Hubei Province, 434020, China
Corresponding E-mail: 7885909@qq.com

Abstract. The application of cardiopulmonary bypass (CPB) technology makes it possible to safely conduct open-heart surgery. However, as an invasive adjuvant therapeutic measure, it cures and greatly injures a human body simultaneously. In CPB, hemodilution, the contact between blood and the surface of foreign matter and other various factors activate the stress response in the body and causes tissue edema, increase in total body water and systemic inflammatory response syndrome (SIRS); in severe cases, they may give rise to organ dysfunction. Ultrafiltration (UF) not only can remove the redundant moisture effectively from the body after CPB, concentrate blood cells and recover body fluid equilibrium, but also can clear away part of inflammatory medium, improve postoperative organ function and enhance the clinical effect after an open-heart surgery.

Keywords. Cardiopulmonary bypass.

1 Basic principle of ultrafiltration

The basic principle of ultrafiltration is to separate and filter out the moisture and soluble small molecular substances in blood from cellular constituents and plasma protein in the blood vessel. The driving force for filtration mainly relies on the transmembrane pressure. Generally, the allowable range of pressure difference on both sides of filter membrane is 100～500mmHg. The factors influencing the filtration effect are as follows: (1) transmembrane pressure (TMP): according to the Startling Law, the bigger the TMP is, the more liquid it will filter out; that the high limit of TMP is exceeded may lead to red cell rapture and even hemolysis; (2) blood flow volume: the low blood flow will result in that a large number of red cells pile up in hollow fiber, increase the possibility of hemolysis, lead to excessively fast blood flow, and prevent liquid from being filtered out in short time, so it is required to control the flow within 100～300ml/min; (3) thickness of membrane; (4) quantity and size of membrane apertures; (5) haematocrit (Hct); (6) temperature.

2 Filtrate ingredients

The filtrate, which is equivalent to the original urine of kidney, consists of K+, Na+, Cl-, UA, creatinine, glucose and most inflammatory media. Big molecular substances such as albumin, hemoglobin, fibrinogen and cellular constituent cannot penetrate through the filtering membrane, so they cannot be filtered out.

3 Types and characteristics of ultrafiltration method

3.1 Conventional ultrafiltration (CUF) is the earliest ultrafiltration method. When being used, the ultrafilter is in parallel connection with the CPB; its entry end is connected with the arterial line and is generally connected with the exit on the top of arterial micro sphere filter; its exit end is connected with the venous return passage. Establish the TMP with the vacuum aspiration (-150mmHg) and control the ultrafiltration flow with the additional pump. The ultrafiltration timing is generally the period from the beginning of rewarming to halt.

Characteristics: It is difficult for normal bypass to maintain the liquid level of oxygenator, so the moisture filtered out is limited; as for a patient with a short duration of operation, it is usually required to terminate the CPB before ultrafiltration.

3.2 Modified ultrafiltration (MUF) is the most common ultrafiltration method, which was created by Naik and other scholars [1] in the 1990s. The entry is connected with the arterial end close to the aortic cannulation with a Y-shaped tee; similarly, the exit end is connected with the venous return line; blood is transfused back to the right atrium; besides, a piezometer tube is equipped at the exit end to monitor the right atrial pressure; a pump is used to control the flow, which is generally within 100～150ml/min. The ultrafiltration timing is within 10～15min after the end of CPB.

Characteristics: In case of drop of blood pressure due to undercapacity during ultrafiltration, we can return the residual blood from the oxygenator to the patient through the aortic pump. As the technology is conducted after the CPB stops, the advantages of a certain degree of hemodilution in bypass can be shown; in particular, it is beneficial to patients with high cyanotic blood viscosity; however, once the CPB stops, the low Hct and hemodilution are harmful to infant patients, so it is required to correct them timely; the MUF has such an effect. Free from the influence of the liquid level inside the venous blood storage room and the duration of operation, it can eliminate moisture very thoroughly.

3.3 Zero balance ultrafiltration (ZBUF), which was created by Journois and other scholars [2] in 1996, is a way of ultrafiltration after modifying CUF, so the way to install its filter and the ultrafiltration time are the same with those of CUF. Zero balance, namely, means adding crystalloid solution with the same volume of liquid filtered into the venous blood storage room and filtering out inflammatory media through constant circulation.

Characteristics: It can get rid of inflammatory media effectively and reduce the concentration of inflammatory factors but cannot filter out the unnecessary moisture in vivo to concentrate blood. As shown in Journois's research, the combination of MUF and ZBUF technologies can remove inflammatory media, unnecessary moisture and concentrate blood in a better way. At present, the technology has been applied in many hospitals.

4 Function of ultrafiltration in CPB

4.1 Influence of ultrafiltration on water content and HCT in the body: It has been found through the research of Naik [1, 3] that MUF is superior to CUF in water drainage and generally can increase Hct to 40%. Hennein and other scholars [4] studied 38 patients with congenital heart disease in different groups and made comparative observation of CUF as well as the ability of the control group and the group using V-VMUF 3 to drain the body water. The result shows that only MUF can maximally drain body water, raise Hct to a relatively ideal degree and thus shorten patients' average hospitalization period.

4.2 Influence of ultrafiltration on hemorrhage and transfusion: through the large-sample study of Naik [1, 3], it is found that MUF can lower the postoperative blood loss and the dosage of banked blood and plasma; based on analysis, the causes are as follows: (1) hemoconcentration itself can reduce blood loss; (2) the concentration of blood coagulation factors including platelet increases; (3) some thrombolytic substances are filtered out. Friesen and other scholars [5] observed the changes in the coagulation system by conducting MUF for 20 infants whose weight was lower than 15kg. The result shows that MUF can enhance Hct significantly; both the fibrin and plasma total protein rise relatively significantly; there is no obvious change in platelet; the decrease in postoperative bleeding is because that the application of MUF can alleviate the disorder of children's coagulation system arising from pre-charge. Similarly, Draisma's result also shows that MUF can enhance the Hct level, reduce the dosage of banked blood, and significantly reduce the thoracic drainage fluid of postoperative patient. Koutlas et al [6] made the similar research between January 1995 and June 1996, showing that there are significant decreases in the dosage of banked blood, pleural effusion, pericardial drainage fluid and postoperative mortality of infant patients with MUF. In conclusion, it is certain that MUF can reduce hemorrhage and transfusion in cardiovascular surgery, so it provides us with a way to reduce the use of blood products.

4.3 Influence of ultrafiltration on hemodynamics: MUF, which improves patients' disorder of water metabolism arising from CPB, can help to improve intraoperative and postoperative patients' hemodynamics. Davies et al [7] monitored all preoperative and postoperative ultrasonic indicators of 21 patients through ICE. The result shows that the patients with MUF have their left ventricular function of contraction enhanced after CPB, have their cardiac compliance improved and use less vasoactive drugs than the control group. Chaturvedi [8] made an experimental observation of changes in the left ventricular function by grouping 22 children with congenital heart disease in the CPB process. The result shows that, as for the patients with the application of MUF, their Hct can be increased to 34% (21%~42%) and 40% respectively during and after operation; the proportion curve of end pressure myocardial contraction and capacity (Ees) increases by 58%; however, there is no significant change in the parameters of the control group. Schlunzen et al [9], Danish scientists, used MUF for 12min averagely for 134 infants with the average age of 5 months old, average weight of 5.3kg and congenital heart disease; the filtered moisture was 44ml/kg on average; the Hct increased from 28% to 40%; the mean arterial pressure was 56~74mmHg; the arterial partial pressure of oxygen was 30.8~34.1kPa; the heart rate decreased from 145 times/min to 136 times/min; however, there's no significant change in their parameters such as CVP and LAP.

4.4 Influence of ultrafiltration on inflammatory medium filtering: During CPB, the contact between blood and the surface of foreign matter results in the occurrence of numerous inflammatory responses, releases abundant inflammatory media, and finally causes injuries to tissue and organ. Ultrafiltration can filter out inflammatory media but the filtering effect of ultrafiltration on inflammatory media is greatly controversial at present.

Naik et al [10] have found through research that ultrafiltration can filter out some toxicants arising from CPB. The capillary electrophoretic analysis on ultrafiltrate shows that there are abundant inflammatory media in the ultrafiltrate, such as all sorts of cell factors, tumor necrosis factors and myocardial necrosis factors. Although Wang [11] and other scholars affirmed the existence of TNF-α and other inflammatory factors in the ultrafiltrate through measurement in their researches, the problem is that the concentration of inflammatory factors in plasma after application of MUF is the same with that before filtration or even increases; it is probably because inflammatory media cannot be filtered out after they are combined with the surface of protein or red cell. Most researches suggest that only ZBUF can filter out inflammatory media effectively. Zhu Deming et al [12] of Shanghai Xinhua Hospital in China compared the application of MUF and ZBUF in infants' CPB. The result shows that all forms of ultrafiltration can filter out inflammatory media but only ZBUF can reduce the plasma concentration of inflammatory media; besides, other forms of ultrafiltration can only achieve hemoconcentration.

5 Conclusion

The ultrafiltration technology is a method to effectively concentrate blood and remove excessive moisture and metabolites from the body. The application of ultrafiltration technology during or after CPB can attain a rather satisfactory balance liquid intake and output volume during operation, alleviate the cardiac load, enhance the plasma colloid osmotic pressure, accelerate moisture absorption among tissues, eliminate inflammatory media, alleviate systemic inflammatory response, and play an positive role in preventing clinical or subclinical pulmonary edema, cerebral injury and cardiac insufficiency etc. at the early stage after operation. However, we still cannot completely know the ultrafiltration action mechanism for CPB cardiac surgery; especially, its influence on cell factors is greatly controversial and further studies are needed. During MUF, the concentration of inflammatory media cannot remain at a rather low level, partly counteracting the effect of ZBUF. It is our new problem to figure out how to use the means of ultrafiltration effectively and how to give play to the ultrafiltration effect on alleviating inflammatory responses in a better way.

References

1. Naik SK, Knight A, Elliott MJ.A successful modification of ultrafiltration for cardiopulmonary bypass in children.Perfusion, 1991.6: 41.
2. Journois D, Israel-Biet D, Pouard P, et al.High volume, zero-balanced hemofiltration to reduce delayed inflammatory responce to cardiopulmonary bypass in children.Anesthesiology, 1996.86: 965-976.
3. Naik SK, Elliott MJ.Ultrafiltration and pediatric cardiopulmonary bypass.Perfusion, 1993.8（1）: 101-112.
4. Hennein HA, Kiziltepe U, Barst S, et al.Venous modified ult-rafiltration after cardiopulmonary bypass in children: a prospective randomized study.J Tborac Cardixasc Surg, 1999.117: 496-505.
5. Friesen RH, Campbell DN, Clarke DR, et al.Modified ultrafiltration attenuates dilutional coagulopathy in pediatric open heart operations.Ann Thorac Surg, 1997.64: 1787.
6. Koutlas TC, Gavnor JW, Nicolson SC, et al.Modified ultrafiltration reduces postoperative morbidity aftet cavopulmonary connection.Ann Thorac Surg, 1997.64（1）: 37-42.
7. Davies MJ, Nguyen K, Gaynor JW, et al.Modified ultrafiltration improves left ventricular systolic function in infants after cardiopulmonary bypass.J Thorac Cardiovasc Surg, 1998.115（2）: 361-369.
8. Chaturvedi RR, Shore DF, White PA, et al.Modified ultrafiltration improves global left ventricular systolic function after open-heart surgery in infants and children.Eur J Cardiothorac Surg, 1999.15（6）: 742-746.
9. Schlunzen L, Pedersen J, Hjortholm K, et al.Modified ultrafiltration in paediatric cardiac surgery.Perfusion, 1998.13（2）: 105-109.
10. Naik SK, Knight A, Elliott MJ.A prospective randomized study of a modified technique of ultrafiltration during pediatric open-heart surgery.Circulation, 1999.84（suppl 3）: 422.
11. Wang MJ, Chiu IS, Hsu CM, et al.Efficacy of ultrafiltration in removing inflammatory mediator during pediatric cardiac operations.Ann Thorac Surg, 1996.61: 651-666.
12. Ming ZD, Wei W, Hong C, et al.Blanced ultrafiltration, modified ultrafiltration, and balanced ultrafiltration with modified ultrafiltration in pediatric cardiopulmonary bypass.J Extra Corpor Technol, 2001.33（4）: 223.

Research on Error Compensation Technology for CNC Machining

Xianfeng Chen[1,a] and Weiming Zhang[1,b]

[1]School of Mechanical and Energy Engineering, Tongji University, 4800 Cao'an Road, Shanghai, China
[a]Email: xianfeng_chen@126.com, [b]Corresponding Email: iamt@tongji.edu.cn,

Abstract. Aiming at the thermal error problem in the procedure process of CNC machine, this study develops a new thermal error compensation technology. First, spindle and linear axis error are analyzed and measured. Then, a comprehensive error compensation model are proposed, which combined with spindle and linear axis. Finally, we use an example to illustrate the proposed method. Experimental results have shown that the proposed approach is able to reduce the influence of the thermal error on machining precision.

Keywords. Error Compensation, spindle and linear axis, multiple regression model.

1 Introduction

Recently, the CNC machine processing center plays a more and more important role in the manufacturing industry, it's important for companies to solve the problem about how to improve the machining efficiency and machining quality of the machine.

With the rapid development of CNC machining technology of China, the CNC machine accuracy has been greatly improved. With the unceasing enhancement of the CNC machine manufacturing technology and processing technology, there has got a solution for original machine tool error and machining error. However, it is important and difficult to eliminate the thermal error caused by temperature change and machine tool geometric error caused by the cutting load which affects the CNC machining accuracy [1].

In this study, we researched the measuring process and compensation mechanism of thermal error and geometric error. An error resource model and a new thermal error compensation technology are proposed.

2 Relate work

The research shows that thermal error is one of the main factors which influence the CNC machining accuracy. Temperature control method [2], thermal stability design [3], thermal error compensation [4] are the main three methods to eliminate the thermal error. However, the temperature control method is not easy to realize, because the temperature control mechanism is always lagging behind the temperature rise. The thermal stability design is difficult to analyze by the heat transfer process which is very complex and caused by the irregular structure of the machine tool. Thermal error compensation control technology is a kind of "soft" approach which do not need to change the machine parts, and can online real-time compensation, so it is has been widely accepted by the researchers. However, the Technology was blocked in the Advanced CNC control system, such as Siemens 840D and FANUC30i.

According to the structure characteristics of vertical CNC machining center, this study did a large amount of experiments under different operating conditions, proposed a thermal error prediction model to on-line predict the thermal error compensation and reduce the influence of the thermal error on machining precision.

3 Error compensation technology

3.1 Error analysis

3.1.1 Spindle error analysis and measurement

The spindle thermal error is caused by the thermal displacement between tool and the workbench, due to the effect of heat source of both inside and outside.

AS shown in the figure1, the spindle thermal displacement contains the axial thermal elongation dz (Z), the X axis thermal elongation dx (X) and the Y axis thermal elongation dy (Y).

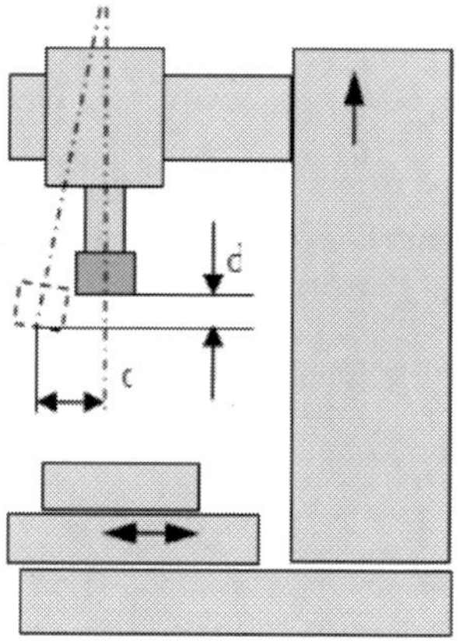

Fig.1. The spindle thermal displacement.

Five point method is adopted to measure the system thermal deformation parameter. Five point method which based on the theory of error separation and redundant information can minimize measurement results by reducing the reading error. And it widely used as a standard method for its high measurement accuracy and simple measurement.

AS shown in the figure2, five point method is adopted to improve the thermal error measurement. The environmental temperature measuring point (1), spindle temperature sensor (2), test mandrel (3), laser displacement sensor (4), bracket (5), the retaining bolt (6) are designed in different plane XY ,XZ, YZ.

3.1.2 Linear axis error analysis and measurement

The linear axis includes X axis, Y axis, and Z axes axis. The main heat source of linear axis are the heating of the motor, the rotating heat of the ball screw bearing seat, friction heat of screw and nut. The thermal error of the linear axis measurement step are, firstly, using laser interferometer measure the segmentation accuracy of linear axis of machine tool, then, let the linear axes empty run back and quick heating and warming. After a period of time, do the measurement positioning precision again, such a loop, and change the position precision under the different heat condition. The cycle operation and measurement signal of the linear axis X is shown as Figure 3.

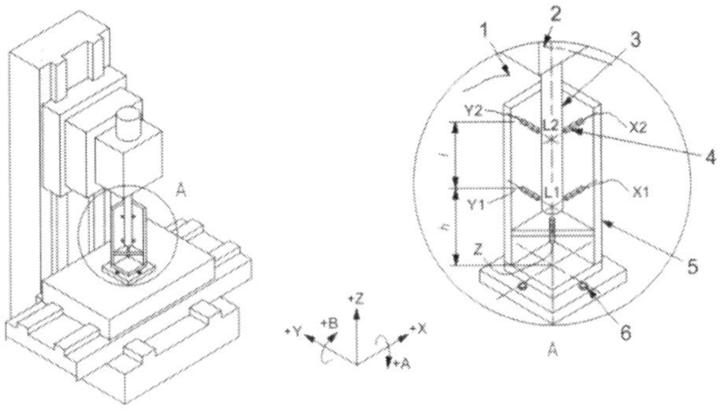

Fig. 2. Five point method.

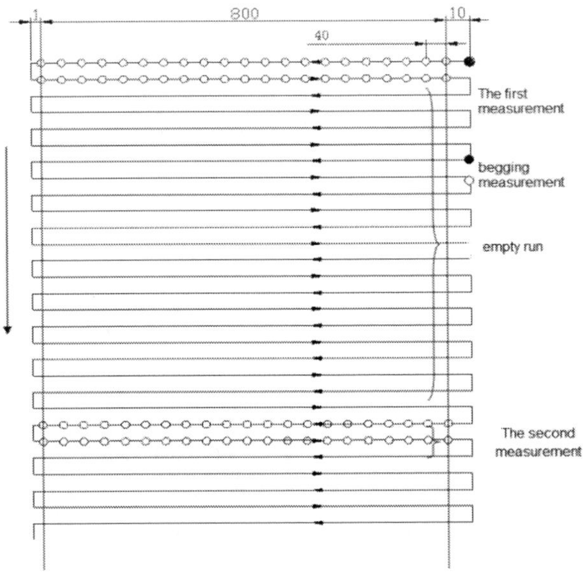

Fig. 3. The cycle operation and measurement signal of the linear axis X.

3.2 Model building

3.2.1 spindle thermal error modeling

Traditionally, the linear systems including univariate and multivariate are used to build engineering model. However, with the unceasing development of artificial intelligence, the neural network, genetic algorithm and fuzzy modeling technology are widely used in nonlinear systems.

In this study, a multiple regression model, which based on the spindle's thermal deformation mechanism of machine tool and the characteristics of the CNC control system, are proposed to analyze the spindle thermal error.

According to the analysis of information entropy clustering, and multivariate regression modeling method, the spindle thermal deflection deformation in the X direction, can be represented as formula 1:

$$Dx(T)= a0+a1*t1+a2*t2+a3*t3+a4*t4 \quad (1)$$

Here, ti（i=1，2，3，4）said the optimized heat sensitive temperature respectively, ai is the model coefficients.

When the each parameter coefficient is calculated from the temperature and deformation data, the error model can be established. As the same with X direction, the multiple regression model of spindle thermal deflection deformation in the Y and Z direction can be calculated.

3.2.2 The Linear axis thermal error modeling

The figure4 shows that the thermal error of linear axis X under T temperature, and it can be linear fitting. K0 is the error on the position P0 , Kx is the error on the position Px, and the error curve slope tanβ (temperature compensation coefficient) can be calculated in formula 2. Then the thermal error (Kx) of linear axis X under T temperature can be represented as formula 3.

$$\tan\beta(T) = (T - T_0) \cdot \frac{T\ K_{max}}{T_{max} - T_0} \tag{2}$$

$$Kx=K0(r)+\tan\beta(T)\cdot(Px-P0) \tag{3}$$

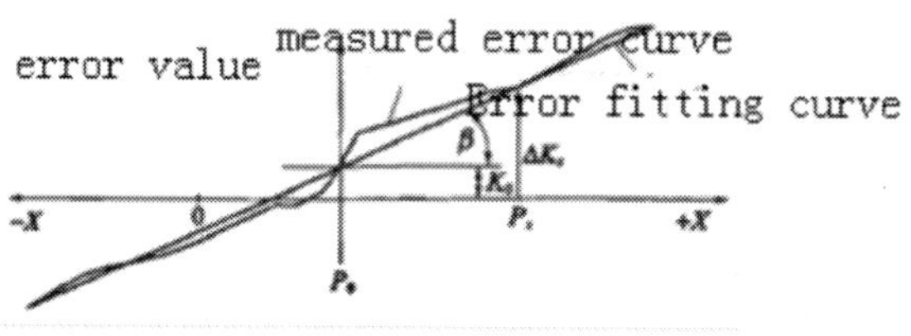

Fig. 4. The thermal error of linear axis X under T temperature.

3.2.3The comprehensive error compensation model

The error compensation should consider the relative error of spindle and linear axis. So the comprehensive error compensation model can be described as follows (take the X axis as the example) .

$$Dx= -(Dx(T)+\Delta Kx)= -（a0+a1*t1+a2*t2+a3*t3+a4*t4+ K0(r)+\tan\beta(T)\cdot(Px-P0)） \tag{4}$$

4 Case for error compensation technology

The research on error compensation can reduce the impact of machine tool movement and instability. Fig. 5 shows the effect of the abrupt change of error compensation on speed. The original interpolation output instruction is displayed on figure5 (a). When the machine is running at high acceleration, the error compensation is change from 0um to one 50um. It can be seen from the figure5 (b), the acceleration exceeds a predetermined allowable range if the output of the instruction signal occurs a jump.

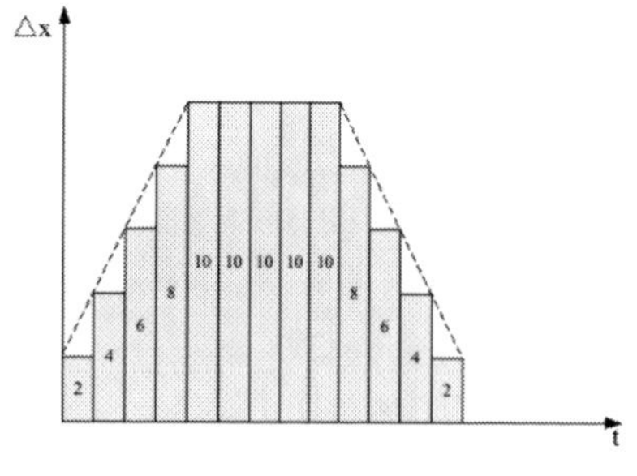

Fig. 5.(a) Instruction diagram after interpolation.

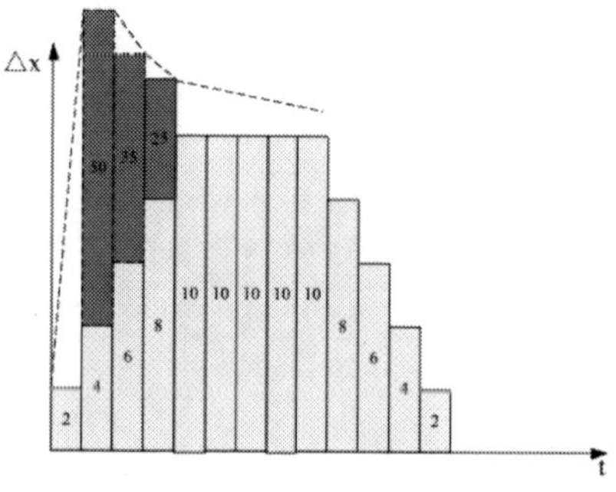

Fig. 5.(b) Instruction diagram after compensation.

5 Conclusion

We present an error compensation technology for CNC machine in this research. The experiment results have shown that the proposed method can reduce the influence of the thermal error on machining precision. However, improving the machining efficiency and reduce the error should be keep improving.

Acknowledgement

This work is supported by the Key Projects in the National Science & Technology Pillar Program during the Twelfth Five-year Plan Period (2012BAF01B02) "Key technologies development and application demonstration for high speed CNC machine tool of green manufacturing." The authors would like to thank all the members of the projects.

References

1. Miao, EM; Liu, Y; Liu, H; GAO, ZH; Li, W. Study on the effects of changes in temperature -sensitive points on thermal error compensation model for CNC machine tool. INT J MACH TOOL MANU 97:50-59(2015).
2. Guo, QJ ; Xu, RF ; Yang, TY; He, L ; Cheng, X; Li, ZY; Yang, JG . Application of GRAM and AFSACA-BPN to thermal error optimization modeling of CNC machine tools. INT J ADV MANUF TECH 83(5-8): 995-1002(2016).
3. Liu, K ; Liu, Y ; Sun, MJ ; Wu, YL ; Zhu, TJ. Comprehensive thermal compensation of the servo axes of CNC machine tools. INT J ADV MANUF TECH 85(9-12): 2715-2728(2016).
4. Jie, G; John, S. A ; Sheri, K; CNC machine tool work offset error compensation method. Journal of Manufacturing Systems 37(2):576-585.

Review of Cement Concrete Pavement of Noise Reduction Method

Hongliang Tao[1], Chen Chen[1], Ping Jiang[2,*] and Shuaihong Huang[2]
[1]Taffic Planning and Design Institute of Yunan Province, Kunming City, Yunnan Province,China
[2]Shaoxing University, Shaoxing City, Zhejiang Province, China
Corresponding E-mail: jiangping1119@163.com

Abstract. In recent years, with the rapid development of the economy, a sharp increase in car ownership in China, the rapid increase in cities traffic volume. Meanwhile, the effect on the normal life, work, study, rest and environment of residents along grows increasingly fierce both in extent and scope. Cement Concrete Pavement noise reduction measures seem more urgent. In this paper, introducing the current commonly used cement pavement noise reduction measures, and discussing the research and development direction of cement concrete pavement noise reduction measures.

Keywords. Traffic engineering, Cement concrete pavement, Noise causes, Noise reduction measures.

1 Introduction

Highway is the lifeline of the national economy, is the image of regional development, defense and stable basis, and only pavement keep long-term smooth condition, in order to ensure that the vehicle quickly and safely on the roads. According to the survey: noise has a negative impact on the human psychology and the body at the same time, particularly harmful to the nervous system and cardiovascular system; noise can damage children's brains, children who long-term in a noisy environment, their intellectual development than in the quiet environment for children approximately 20% less[1]. With the development of transportation construction, as of 2013, China's total mileage has more than 4 million km, cement concrete pavement with a mechanical strength, high load capacity, less prone to plastic deformation, etc., is one of the main basic form of road pavement. However, due to its stiffness, traffic noise is quite obvious. With more and more attention to environmental engineering, as well as comfort and road safety and environmental protection have become increasingly demanding. Cement concrete pavement noise pollution problem of environmental pollution has become most concerned of residents. Therefore, taking effective measures to reduce traffic noise pollution to maintain a good acoustic environment along the road has been a priority [2].

2 Cement concrete pavement noise generation mechanism

The noise of vehicles driving on the cement concrete pavement can be divided into: First, noise emitted by automobiles, mainly in the high-power motor vehicle engine noise for most, with its main class of vehicle, usually in the engine of the vehicle can be equipped with mufflers effectively reduce the noise. Second, car horn, mainly produce emergencies when the car in motion the process to remind drivers to produce, increased by about 8dB overall car tweeter automobiles can emit noise, general urban road approach can be used to reduce the noise control horn produced; Third, vibration of the interaction between tire and cement pavement produce the noise. This process of the generation of noise is very complicated, but mainly due to the air pump at high speed between the tire and the road surface and the vibration effect of the tire[3]. Tires at high speed rolling process, road to the vibration of the tyre tread and sidewall effect caused by noise, there are shock vibration noise, slip vibration noise, rehabilitation vibration noise and road vibration noise [4]. Generally, the proportional to the size of the noise and driving speed of the car. When all types of car speed exceeds 55km/h, tire and road noise generated by friction has become a major part of the road traffic noise. In addition, the strength of road traffic noise is also affected by the road line, slope and roadbed height and many other factors [2].

Reduce the noise generated by the interaction with the cement pavement concrete pavement and tire is the main method of noise reduction. Because this type of noise generation related with tire tread and road construction and road surface texture. Therefore, the study of the concrete pavement noise reduction measures along these two basic ideas that optimize the tire tread expand and optimize the structure of concrete pavement and road textures, this article will optimize the texture of the pavement structure and road noise reduction measures used to introduce the cement pavement.

3 Noise reduction measures pavement structure

3.1 Pavement Material Type

3.2.1Porous concrete pavement

The main reason is that porous concrete reduce noise is there is a wide gap in internal material structure. When sound waves enter the internal structure of the material, caused the air movement in the pores, and pore formation of solid body friction, due to the effects of viscosity and thermal conductivity, a portion of the sound energy into heat energy consumed; sound waves reflected by the rigid wall after the material returns to its surface, a portion of the acoustic wave back to the air permeable and partly reflecting back into the interior material, this process is repeated propagation of sound waves, is the energy dissipation constant conversion process, so the material on the "absorption" of the section sound energy [5].

3.2.2 Rubber Concrete Pavement

Rubber powder particles of cement-based composite materials is a new kind of impact resistant toughened functional materials developed in recent years, because of its existing large strain and stretch the limits of effects and impact resistance, low modulus, but also has excellent mechanical performance and good noise reduction, therefore, as the city noise insulation materials and impact-resistant bumper isolation member, it will have broad application prospects in municipal engineering neighbourhood.

The higher driving speed, the more significant of noise performance of rubber cement pavement. Rubber cement rubber particles uniformly distributed in the road pavement layer to form a multi pore characteristics, can weaken the sound waves produced between tire and road surface due to the "two sides effect", Simultaneously added such that the rubber particles in the surface acoustic wave transmitting and reflecting some of the noise at the time of the process through the rubber particles, sound waves into mechanical energy of vibration is weakened rubber particles, thus weakening the secondary noise transmission [6].

3.2.3 Exposed Aggregate Concrete Pavement

Exposed aggregate concrete pavement can be used in new construction, renovation or reconstruction of cement [7]. Exposed concrete pavement because of random convex surface of the aggregate, sound waves and pressure waves dissipate on their own in the space under the tread, 7dB lower than the old concrete pavement noise, and its surface is not easy to form a film of water, even in the rain can be maintain a high skid resistance. After treating the surface of exposed aggregate concrete pavement, rough road can form diffuse, eliminate or mitigate the diffuse reflection [1].

3.2.4 Asphalt Overlay Low Noise Pavement

Asphalt overlay low noise pavement to take effective measures against noise between the tire and the pavement, which a pavement structure type for the purpose of reducing noise. Porous asphalt pavement, or called permeable (or drainage) roads. It refers to the ordinary asphalt or cement concrete pavement or other paved pavement layer has a high porosity layer of asphalt mixture [8]. Its noise reduction mechanism is surface voids' sound absorption. In addition to absorb the noise of the engine and transmission parts radiation to the pavement outside, but also can absorb reflected noise road surface is reflected back through the car chassis road tire noise and other interfaces [9].

3.2.5 Diamond Polishing Concrete Pavement

Diamond grinding is a new technology of cement concrete pavement surface repair measure function refers to the use of fine-pitch diamond blade cutting abrasive concrete pavement, grinding away the surface of 4-6mm thickness, a certain vertical grain structure. The main purpose is to eliminate defects in the road surface, improving cement concrete pavement surface features and reduce road noise [7].

3.3 Road Surface Texture

3.3.1 Picking

Picking is instructed to produce shallow texture in concrete road surface to slow the pavement wear, thus reducing road noise. However, road surface texture produced by the technologies of nap is relatively, worn between tires and the road surface interaction will often reduce the feature of concrete pavement. Through the reasonable design improve the wear resistance of concrete pavement, to ensure that there is enough road surface features during the service period, so achieve the purpose of reducing road noise.

3.3.2 Groove

Groove is generated shallow channels on the surface of the concrete pavement, to increase the drainage capacity and the friction under the condition of slippery roads in varies weather conditions, but also has reduced road noise [7]. But the noise reduction effect on medium and heavy trucks of grooving roads.

3.3.3 Pressure Trough

Pressure trough, lateral pressure trough is the most common kind of concrete pavement surface texture forming method is a durable, high friction resistance and low-cost approach. Deep macro structure produced by lateral pressure trough can reduce the thickness of the film of water on rainy roads, reducing the potential threat of skidding in rainy wet road. But the noise generated by the interaction of the tire and the road surface on the lateral pressure trough was the largest.

4 Other noise reduction measures

4.1 Setting The Sound Barrier

Noise barrier mainly through the sound absorption and sound insulation to reduce noise, absorption is mainly by increasing the sound-absorbing material in the sound barrier to achieve, and insulation is mainly dependent on increasing the propagation distance of noise to achieve, in the process of through a few noise baffle of the sound barrier, the noise has formed a reflective, so as to continuously reduce its strength. So the noise reduction effect of sound barrier some extent depends on the height and length of the sound barrier [1].

4.2 Set Green Belt

Setting green belt refers to the land within the scope of the road for the green bar area. By using acoustic methods of setting green belt to protect the ecological environment, this noise reduction method is a long-term, sustainable improvement of ecological environment, the role of environmental protection is very prominent. Use of greening ecological protection measures to reduce noise is one of the effective and environmental protection measures, In the actual road construction, planting patterns of green belt and noise protective effect has a strong correlation.

5 Looking to cement pavement noise reduction measures

5.1 Noise Reduction Materials Research

All of the above concrete pavement noise reduction measures are based on hydraulic concrete materials, by changing the structure of the road to achieve the purpose of noise reduction[10]. With the rapid development of society, the use of cars is growing, more and more waste tires were produced, and now the repeated utilization of waste tires is very low, used tires, black pollution has become increasingly serious environmental problems. So based on the basic principle of exposed concrete pavement noise reduction, the rubber particles used in cement concrete pavement[11], not only can be clean utilization of scrap tire, save resources, also can improve the performance of the cement concrete.

5.2 Research and Development of Road Noise Test System

Since there is no test system to simulate actual road noise, and therefore cannot form a scientific research methods road noise, such as the use of the tire falling to simulate the impact of car tires on cement concrete pavement used by Hansen, Dong Yuming, And collecting the generated noise, to evaluate the noise reduction feature pavement materials [3]. Perfect noise test system can simulate different axle load and driving speed, and has the ability to obtain noise data between tire and road accurately.

6 Conclusion

The research for cement pavement noise reduction measures can create a better living and working environment for residents along, for now, noise reduction measures can be divided into the pavement structure and other noise reduction measures, Future noise reduction of cement pavement can around the development of new noise reduction materials, such as rubber aggregate concrete pavement and development of road noise test system.

References

1. Yang Xiang-tao: Analyse Cement concrete pavement noise reduction technology, Technology Guide Vol.2 (2012), p.397 .
2. Zhuo Yong-hong: Introduction to highway traffic noise and noise reduction measures,Highways & Transportation in Inner Mongolia Vol.4 (2013), p.17-18.
3. Han Sen, Dong yu-ming, Chen hai-feng: Noise Reduction Performance of Exposed-aggregate Cement Concrete Pavement, Journal of Traffic and Transportation Engineering Vol.3 (2005), p.32-34 .
4. Shan Yong-ti, Li Zhu-ming, Yuan Wei-ming, Zhu Wei-hong: Study on the Noise Reduction Effect by Monitoring of the Low-noise Pavement, Environmental Monitoring and Forewarning Vol.3(2011), p.45-50.
5. Jing Lu-bo, Tian Bo, Liu Ying, Niu Kai-min. Research on Noise Reduction Properties of Porous Cement Concrete Pavement, Highway Vol.7(2010), p.75-79.

6. He Zhuang-bin, Qin Feng: Mechanical Characteristics and Noise Reduction Performance of Rubberized Concrete, Highway Engineering Vol.5 (2013), p.128-131.

7. Yu Jie, Wang Jun-jie, Zhang Peng: Cement Concrete Pavement Noise Reduction Measures, North Traffic Vol.4 (2012), p.22-24.

8. Wang Tian-yang, Li Jing-yang: A Research on the Urban Noise Reduction Road Surface, Communications Science and Technology Heilongjiang Vol.8 (2011),p.50-52.

9. Wu Xin-feng: Types of Low Noise Pavement and their Applicability Analysis, Transpo World Vol.5-6 (2012), p.196-197.

10. Liu Ying, Tian Bo, Niu Kai-min: Skid-resistance and Denoising Properties of Cement Concrete Pavement with Different Surface Texture, Journal of Highway and Transportation Research and Development Vol. 29(2012), p.28-33.

11. Li Bo, Han Sen: Some Methods on Concrete Pavement Noise Reduction, Noise and Vibration Control Vol.3 (2008), p.127-130.

A Novel Assessment Method of Charging Station Planning Based on Fuzzy Matter Element Theory

Junyi Zhao[1], Chaoying Yang[1], Zhiwei Xue[1] and Yangyang Tan[2,*]

[1]State Grid Shanxi Electric Power Company, Taiyuan, China
[2]School of Electrical Engineering and Information, Sichuan University, Chengdu, China
Corresponding Email: tyy13683448505@136.com

Abstract. Scientific and rational planning of urban electric vehicles (EVs) charging station is an important prerequisite for large scale EVs interact with smart grid friendly. This article realizes the planning assessment of EV charging station based on fuzzy matter element theory. The features of urban EV charging station are analyzed, and the evaluation index system of alternative charging station is established. The paper applies fuzzy matter element analysis method to obtain the optimal fuzzy matter element sequence with alternative points, and as a reference sequence. The weights of alternative points corresponding evaluation index are obtained by entropy method. Then, the paper applies the gray correlation analysis to calculate the gray relational weighing degree of fuzzy matter element sequence of alternative points, and determine the EV charging station plan based on the size of gray relational weighing degree. Finally, the simulation results show that the proposed method is effective and feasible for EV charging station planning.
Keywords. Charging station planning, Fuzzy matter element theory, Entropy method, Gray correlation analysis method, Gray relational weighing degree.

1 Introduction

In recently years, with the improvement of China's social development and EV ownership increased year by year. The application of EV charging technology is also developing from the "small capacity and small scale" research to the "large capacity and large scale" [1, 2]. The development process of EV industry from the world's major countries point of view, EV industry support forces are mainly including government, enterprises, public institutions, research institutions and so on. According to the different participation characteristics of investment operators, charging station operating mode can be divided into government dominant mode, social enterprise dominant mode and user dominant mode. The social enterprise dominant mode is defined as a mode that EV manufacture or social enterprises investing and operating charging station.

The EV charging station planning problem is one of the important research topics in the large-scale application of EVs in cities. Many scholars have carried on massive researches upon the EV charging station planning. Reference [3] proposes an EV charging station planning model in local region considering the load requirements of distribution network and charging requirements of EV users. The model uses the analytic hierarchy process method to determine the weights of charging station alternative points. Reference [4] analyzes impact factors during planning the location of charging station and adopts some indicators that include load ratio, payback period of investment, driven distance and degree of comprehensive satisfaction to evaluate the economy of charging station plan. Reference [5] from the power grid indicators, traffic indicators, economic indicators and planning indicators point of view, basing on the cloud focus theory, the assessment method of charging station planning is conducted. Reference [6] takes into account the cost effectiveness and life cycle cost of EV charging station during charging station operation period. Estimate the capacity of charging station though the traffic flow information and establish the planning model of EV charging station. The objective function of model is maximizing the charging station operation net present value revenue and its constraints including traffic flow of traffic network, power quality of distribution network, economic of power grid and requirements of charging users. Reference [7] studies on the stochastic behavior of electric taxis, takes behavior characteristics of electric taxi and its demand distribution, structure and capacity of distribution network as constraint conditions. At the same time, takes the total cost of charging station construction, equipment maintenance, network loss and charging behavior time as objective function to establish a charging station planning model.

This paper analyzes many factors which influence the planning results of EV charging station and establish evaluation index system of charging station planning. The qualitative analysis and quantitative calculation of the candidate points were carried out and a multidimensional composite fuzzy matter element matrix is established. According to the evaluation method, the degree of membership matrix is obtained and the maximum membership value of each evaluation index is selected as the optimal fuzzy matter element sequence. The weighted gray correlation degree of each planning scheme is calculated by combining with entropy weight method and gray relational analysis method. The EV charging station planning program is selected according to the order of relevance degree. Finally, it is verified by simulation analysis and comparison with other methods.

2 Evaluation index of ev charging station

Charging station is an energy exchange center for EVs in urban cities and plays a very important role in meeting the needs of EV users. Scientific and reasonable charging station planning program will directly affect the entire city charging service network operating efficiency, and will also affect the urban distribution network and urban traffic safety.

2.1 Impact Factors Analysis EV Charging Station Planning

Many factors will affect the EV charging station planning program, that of can be summarized as the following aspects:

1)EV charging station should be built in the main roads or transportation hubs of urban city where the traffic flow is heavy and the consumption of electricity is also large.

2)EV charging station should have a convenient entrance and exit. It can facilitate EV charging the batteries, but also can avoid interfering with urban traffic.

3)The planning scheme of EV charging station not only reach a certain scale to meet the needs of urban charging users, but also control the cost of charging station in a reasonable range in order to achieve better ratio of input-output. It is the premise and basis that can attract more social investors.

4)Distribution of EV charging stations should avoid too concentrated to cause the waste of resources and ensure sufficient numbers of vehicle enter the charging station.

5)EV charging station can be regarded as a large load, charging facilities of charging station should avoid impacting on the grid when they are in operation. In enterprise dominant mode, the EV charging station can be seen as large users. First of all, the grid company will evaluate some indices when the EV charging station access the distribution network, such as harmonic pollution levels, load levels, voltage levels and loss of distribution network. Then, according to the results of the assessment, the grid company requires the EV charging station governing the harmonics until the harmonic meet the national standard. Therefore, charging station investors should focus on the degree of harmonic pollution.

2.2 Evaluation Index System for EV Charging Station Location

Considering the social benefit, the economic benefit of the investor and the economic benefit of the charger, the following assessment indices are established:

1) Design scale. Design scale refers to the total charge capacities of charging station, which according to the number of chargers generally.

2) Investment cost. Investment cost refers to the cost of the entire EV charging station, including the cost of land purchase, the cost of charging facilities and transformers purchase, the cost of equipment maintenance, the cost of power loss and the cost of personnel salaries. The total cost can be equivalent to a single charger investment cost by introducing a conversion factor, then, add the fixed investment cost of charging station in actual calculation. Consequantly, the investment cost is a function of the number of chargers.

3) Degree of convenience. Degree of convenience refers to total time of the traveling time and consuming time (including the waiting time and charging time) of EV. The traveling time and consuming time can be expected to get the expected value, the expected value of the sum of the two is the overall time.

4) Adjacent distance. Adjacent distance refers to the shortest distance between new charging station and adjacent charging station.

5) Total vehicle flow. The total vehicle flows refers to the total number of vehicles that passing through the EV charging station in a day.

6) Stop rate. Stop rate refers to the proportion of vehicles that entering the EV charging station in order to charge the batteries.

7) Charging capacity. Charging capacity refers to the single EV charging capacity when EV enters the charging station. Different types of EVs has different charging capacity, even the same type of EVs may also be different. EVs with similar charging capacity can be classified into a category by classification methods and the average amount of capacity is taken for each class of EVs.

8) Harmonic pollution degree. Harmonic pollution degree refers to the total harmonic distortion of the voltage at the point of common coupling when the charging facilities are in operation. This paper focus on harmonic pollution degree which is generated by charging station, whereas harmonic voltage that come from the high voltage and the other harmonic sources as background harmonic [8]. The harmonic pollution degree can be obtained by building simulation model PSCAD software [9].

3 Fuzzy matter element analysis

3.1 Multidimensional Complex Fuzzy Matter element Matrix

Matter element analysis was first proposed by Chinese scholar Cai Wen [10], is a new subject between mathematics and experimental analysis. The basic idea of analysis method is that using three elements "objects, features, values" to describe things, the orderly three elements called matter element, it can be expressed as $R=(M, C, V)$. Supposing that there are m comparison things, each thing has n features, the formula of n-dimensional complex fuzzy matter element matrix is:

$$R=\begin{bmatrix} & C_1 & \cdots & C_j & \cdots & C_n \\ M_1 & x_{11} & \cdots & x_{1j} & \cdots & x_{1n} \\ \vdots & \vdots & \ddots & \vdots & \ddots & \vdots \\ M_i & x_{i1} & \cdots & x_{ij} & \cdots & x_{in} \\ \vdots & \vdots & \ddots & \vdots & \ddots & \vdots \\ M_m & x_{m1} & \cdots & x_{mj} & \cdots & x_{mn} \end{bmatrix} \tag{1}$$

where, M_i is the ith evaluation object, C_j is the jth feature of an evaluation index, x_{ij} is the fuzzy value corresponds to the jth feature of the ith evaluation object.

3.2 Membership Matrix

According to the matter element theory, the membership degree of the fuzzy matter element matrix is obtained by the membership degree calculation method, and then, transform the complex fuzzy matter element matrix into the membership degree matrix. The fuzzy indices are generally divided into positive index, reverse index and fixed constant index. The membership degree calculation of the evaluation indices is also divided into three categories. Basing on the principle of merit, $\mu(x_{ij})$ represents the membership degree of the jth index of ith evaluation object.

*1)*Positive index of membership degree

According to the normalization formula of positive index, the value of index is larger, the index is better. The calculation formula is

$$\mu(x_{ij})=\frac{x_{ij}-\min\limits_{1\le i\le m} x_{ij}}{\max\limits_{1\le i\le m} x_{ij}-\min\limits_{1\le i\le m} x_{ij}} \tag{2}$$

*2)*Reverse index of membership degree

According to the normalization formula of reverse index, the value of index is smaller, the index is better. The calculation formula is

$$\mu(x_{ij})=\frac{\max\limits_{1\le i\le m} x_{ij}-x_{ij}}{\max\limits_{1\le i\le m} x_{ij}-\min\limits_{1\le i\le m} x_{ij}} \tag{3}$$

*3)*Fixed constant index of membership degree

The gap is smaller between fixed constant index and a certain constant, the index is better. The calculation formula is

$$\mu(x_{ij})=\frac{\min\limits_{1\le i\le m}(x_{ij},c_0)}{\max\limits_{1\le i\le m}(x_{ij},c_0)} \tag{4}$$

where , c_0 is constant.

According to the membership degree calculation of the above indices, the complex fuzzy matter element matrix R can be transformed into the membership degree matrix Q, which is expressed as follows:

$$Q=\begin{bmatrix} & C_1 & \cdots & C_j & \cdots & C_n \\ M_1 & \mu(x_{11}) & \cdots & \mu(x_{1j}) & \cdots & \mu(x_{1n}) \\ \vdots & \vdots & \ddots & \vdots & \ddots & \vdots \\ M_i & \mu(x_{i1}) & \cdots & \mu(x_{ij}) & \cdots & \mu(x_{in}) \\ \vdots & \vdots & \ddots & \vdots & \ddots & \vdots \\ M_m & \mu(x_{m1}) & \cdots & \mu(x_{mj}) & \cdots & \mu(x_{mn}) \end{bmatrix} \tag{5}$$

3.3 Optimal Fuzzy Matter element Matrix

According to the former calculation method of membership degree, regardless of the original index is a positive index or reverse index or a fixed constant index, the transformation of indices are positive indicators through the equation (2) ~ (4). The optimal fuzzy matter element matrix Q_0 of optimal evaluation object is constructed by selecting the maximum value of the membership degree of each index, the formula can be expressed as follows:

$$Q_0 = \begin{bmatrix} & C_1 & \cdots & C_j & \cdots & C_n \\ M_0 & \mu(x_{01}) & \cdots & \mu(x_{0j}) & \cdots & \mu(x_{0n}) \end{bmatrix} \tag{6}$$

where, $\mu(x_{0j}) = \max(\mu(x_{1j}),\ldots, \mu(x_{ij}), \ldots, \mu(x_{mj}))$.

4 Entropy weight method and gray relational analysis method

4.1 Entropy weight method

Entropy originated in thermodynamics, was used to express distribution uniformity of the energy in the space. Entropy theory [11] has been widely used in the field of probability theory, information theory. In order to reflect the importance of each evaluation index, different evaluation indices should be given different weight in the decision-making of EV charging station planning. Entropy weight method [12] is an objective weight method, widely used in the field of index empowerment. According to membership degree matrix, the specific calculation steps that using entropy weight method to calculate the weight is listed as follows:

1)Calculate the entropy of each index, the calculation formula is

$$H_j = -\frac{1}{\ln m}\sum_{i=1}^{m}\frac{\mu(x_{ij})+1}{\sum_{i=1}^{m}(\mu(x_{ij})+1)}\ln\left(\frac{\mu(x_{ij})+1}{\sum_{i=1}^{m}(\mu(x_{ij})+1)}\right) \tag{7}$$

2)The weight value of the index is calculated and normalized, the formula is

$$\omega_j = \frac{1-H_j}{\sum_{j=1}^{n}(1-H_j)} \tag{8}$$

4.2 Gray Relational Analysis

Gray relational analysis is a kind of evaluation method based on gray system theory [13-14]. The basic idea of gray relational analysis is that comprehensively evaluate the things and phenomena which affected by many factors from the holistic view.

According to membership matrix, fuzzy matter element sequences M_i and $M_i = (\mu(x_{i1}), \mu(x_{i2}),\ldots, \mu(x_{in}))$. According to the optimal fuzzy matter element matrix, an optimal fuzzy matter element sequence can be obtained as $M_0 = (\mu(x_{01}), \mu(x_{02}), \ldots, \mu(x_{0n}))$. In this paper, the sequence M_0 as a reference sequence, basing on the theory of gray relational analysis, the corresponding coefficient of element which comparing sequence M_i with sequence M_0 is calculated as follows:

$$\xi_{ij} = \frac{\min_i\left|\mu(x_{0j})-\mu(x_{ij})\right| + \beta\max_i\left|\mu(x_{0j})-\mu(x_{ij})\right|}{\left|\mu(x_{0j})-\mu(x_{ij})\right| + \beta\max_i\left|\mu(x_{0j})-\mu(x_{ij})\right|} \tag{9}$$

where, ξ_{ij} is correlation coefficient of the ith evaluation object corresponds to the jth evaluation index. β is resolution factor, $0 < \beta < 1$, the value of β is smaller, the greater the difference between the factors, the ability to distinguish is stronger, and value of β is usually taking 0.5.

The factors that influence the location decision of EV charging station are different in the comprehensive evaluation. Therefore, the influence factor weight coefficient is introduced, and then the weight gray incidence degree D_i of each fuzzy matter element sequence and the optimal fuzzy matter element sequence of the ith evaluation object is calculated, the calculation formula is

$$D_i = \sum_{j=1}^{n}\omega_j\xi_{ij} \tag{10}$$

The weight gray correlation degree is sorted by increase order. The value is larger, the more consideration will be given to the factors that influence the planning decision-making of EV charging station planning scheme, and the more it meets the actual planning requirements of EV charging station.

5 Case studies

In this section, case studies under different scenarios are conducted to demonstrate the proposed method.

A power grid company plans to build a new EV charging station around the beltway in a city. According to the actual market strategy and distribution network layout, at the same time, considering the traffic flows, site locations and surroundings, the power grid company proposing 4 sites as EV charging station candidates.

In terms of investment cost of EV charging station, assume that the investment cost is a linear function of the number of chargers and the unit price of one charger is 100,000$. After the actual inspection of candidates and comprehensive evaluation of the industry experts, the quantification value of each evaluation index of candidates can be listed in Table 1.

Table 1. Quantized values of evaluation index for EV charging station alternative points around the beltway.

Candidates	Design scale	Investment cost/$10^4$$	Convenience degree/min	Adjacent distance/km	Total vehicle flow	Stop rate /%	Charging capacity /KW·h	Harmonic pollution degree/%
1	12	180	38	4.0	6700	3.6	46	2.8
2	13	185	30	3.6	7400	2.9	52	3.1
3	11	160	40	2.8	6200	4.0	51	3.6
4	15	205	32	5.1	7900	2.8	47	2.9

The multidimensional complex fuzzy matter element matrix R can be expressed as follows.

$$R=\begin{bmatrix}12 & 180 & 38 & 4.0 & 0.67 & 3.6 & 46 & 2.8\\ 13 & 185 & 30 & 3.6 & 0.74 & 2.9 & 52 & 3.1\\ 11 & 160 & 40 & 2.8 & 0.62 & 4.0 & 51 & 3.6\\ 15 & 205 & 32 & 5.1 & 0.79 & 2.8 & 47 & 2.9\end{bmatrix}$$

The size of EV charging station must meet the public EV charging requirement, not the bigger the better, but estimating the total charging power according to the current charge needs, therefore, the EV charging station size index belonging fixed constant index and the numbers of charger defining as 13. The lower of the investment cost, the more attractive for investors. The shorter for EV charging users reaching the charging station, the more charging users choose the charging station for charge. The smaller for EV charging station produce harmonic pollution, the bigger of investment benefit. Thus, the investment cost index, the convenience degree index and the harmonic pollution degree index is explained above are belonging reverse index.

The farther away from the adjacent charging station, the more charging users will be attracted and charge the battery in this EV charging station. The bigger for traffic flow ahead of the charging station, stop rate and charging capacity of single EV, the bigger for investment benefit of EV charging station. As a result, the 3 indices above are belonging positive index.

According to the formula (2), (3) and (4), the transpose matrix of membership matrix Q^T is expressed as follows:

$$Q=\begin{bmatrix}0.923 & 0.556 & 0.200 & 0.522 & 0.294 & 0.667 & 0 & 1.000\\ 1.000 & 0.444 & 1.000 & 0.348 & 0.705 & 0.083 & 1.000 & 0.625\\ 0.846 & 1.000 & 0 & 0 & 0 & 1.000 & 0.833 & 0\\ 0.867 & 0 & 0.800 & 1.000 & 1.000 & 0 & 0.167 & 0.875\end{bmatrix}$$

The optimal fuzzy matter element matrix Q_0 is obtained by formula (6).

$$Q_0=\begin{bmatrix}1.0 & 1.0 & 1.0 & 1.0 & 1.0 & 1.0 & 1.0 & 1.0\end{bmatrix}$$

Firstly, the weight after normalization process and the correlation coefficient of each evaluation index is calculated by formula (7) to (10). Secondly, the degree of combined weighted gray relational of each candidate is calculated by formula (11). The calculation results are listed in Table 2.

Table 2. Gray relational weighing degree of alternative points.

i	1	2	3	4
D_i	0.5274	0.6399	0.5958	0.6302

According to Table 2, gray relational weighing degree of candidate 2 is larger than the other EV charging station candidates significantly. According to the evaluation indices are listed in Table 1, in terms of operator, they can reduce the investment cost by means of competitive negotiation, they can improve the stop rate by means of charging price concessions, and they can heighten the charging capacity of single EV by using the advanced charging technologies for chargers.

Basing on the deep analysis above, in this section, we choose candidate 2 for example, alter the indices including investment cost of charging station, stop rate and charging capacity of single EV and analyze how the index influence on gray relational weighing degree. The analysis results are described in Figure 1.

DOI: 10.1051/matecconf/201710003030

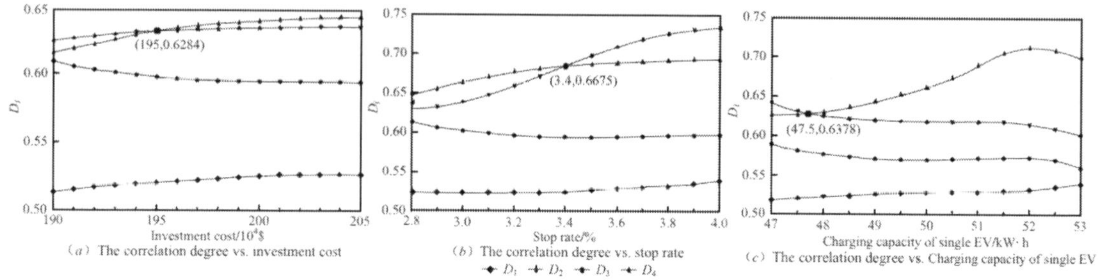

Fig.1. The relationship between gray relational weight degree of candidates with evaluation index.

According to the analysis results in Figure 1 (a), gray relational weighing degree of candidate 2 and 4 are reducing gradually by cutting down the investment cost of charging station. The gray relational weighing degree of candidate 2 is greater than candidate 4 when the investment cost of candidate 4 below 1,950,000$, at the same time, candidate 2 is regards as new charging station planning site. The analysis results in Figure 1 (b) show that investors improve the stop rate by adopting the effective incentive strategies or preferential measures can also achieve similar result, consequently, candidate 2 is replaced by candidate 4 when the stop rate is greater than 3.4%. Simultaneously, basing on the Figure 1 (c), the operators adopts the advanced charging technologies for charging, not only saving the charging time of EV, but also improving the charging capacity of single EV. We can see that candidate 2 is replaced by candidate 4 when the charging capacity of single EV is greater than 47.5kW·h.

It can be seen that the various evaluation indices of charging stations will have a greater impact on the final EV charging station planning scheme.

6 Conclusion

In this paper, an evaluation method of EV planning scheme based on fuzzy matter element theory is proposed for the existing charging station planning problem. It is found that the proposed method can find the final planning scheme conveniently by case study 1. By analyzing the impact of variable parameters such as investment cost, stop rate and charging capacity of single EV on the planning results, operators can adopt scientific and reasonable incentive measures to adjust the planning results and increase the investment benefit. Finally, the results of case study 2 demonstrate the feasibility and effectiveness of the proposed method by comparing it with other methods.

References

1. Fakhrazari, H. Vakilzadian, and F. F. Choobineh, "Optimal energy scheduling for a smart entity," IEEE Transaction on Smart Grid, vol. 5, pp. 2919-2928, 2014.
2. K. S. Reddy, L. K. Panwar, R. Kumar, and B. K. Panigrahi, "Distributed resource scheduling in smart grid with electric vehicle deployment using fireworks algorithm." Journal of Modern Power Systems and Clean Energy, vol. 4, pp. 1-12, 2016.
3. L. F. Kou, Z. F. Liu, and H. Zhou, "Modeling algorithm of charging station planning for regional electric vehicle." Modern Electric Power, vol. 27, pp. 44-48, 2010.
4. C. Zhang, and H. Teng, "Planning model and evaluation method for electric vehicle charging station." Proceedings of the CSU-EPSA, vol. 26, pp. 49-52, 2014.
5. Z. J. Peng, J. Lan, Y. Chen, Q. Zou, L. G. Jia, and J. Yang, "EV charging station sitting method based on cloud focus theory." Electric Power Construction, vol. 36, pp. 1-7, 2015.
6. X. Q. Huang, H. Yang, J. Chen, L. Jiang, and Y. J. Cao, "Optimal planning of electric vehicle charging stations based on life cycle cost and quantum genetic algorithm." Automation of Electric Power Systems, vol. 39, pp. 176-182, 2015.
7. S. Y. Ge, R. Li, J. Han, H. Liu, T. Li, and H. H. Lian, "Charging station planning considering probability behavior characteristic of electric taxi." Automation of Electric Power Systems, vol. 40, pp. 50-58, 2016.
8. X. Zhao, and H. G. Yang, " A new method to calculate the utility harmonic impedance based on FastICA." IEEE Transactions on Power Delivery, vol. 31, pp. 381-388, 2016.
9. N. Li, M. Huang, "Analysis on harmonics caused by connecting different types of electric vehicle chargers with power network." Power System Technology, vol. 35, pp. 170-174, 2011.
10. D. K. W. Ng, and W. Cai, "Treating non-compatibility problem from matter element analysis to extenics." Acm Sigice Bulletin, vol. 22, pp. 2-9, 1997.

11. R. Li, Z. Zhao, X. Zhou, and J. Palicot, "The prediction analysis of cellular radio access network traffic: from entropy theory to networking practice." IEEE Communications Magazine, vol. 52, pp. 234-240, 2014.
12. H. Deng, X. Sun, M. Liu, C. Ye, and X. Zhou, "Infrared small-target detection using multiscale gray difference weighted image entropy." IEEE Transactions on Aerospace and Electronic Systems, vol. 52, pp. 60-72, 2016.
13. J. Wang, J. Wang, T. Zhang, Y. Wu, "Probability estimation based on grey system theory for simulation evaluation." Journal of Systems Engineering and Electronics, vol. 27, pp. 871-877, 2016.
14. L. Chen, B. Tian, W. Lin, and B.Ji, "Analysis and prediction of the discharge characteristics of the lithium–ion battery based on the grey system theory." IET Power Electronics, vol. 8, pp. 2361-2369, 2015.

 DOI: 10.1051/matecconf/201710003031

The PLC-based Industrial Temperature Control System: Design and Implementation

Fanjie Wei

Chongqing College of Electronic Engineering, No. 76, Eastern Road, Daxuecheng, Shapingba District, Chongqing, 401331, China

Abstract. Targeting at the problem of slow response and low accuracy of the automatic temperature control system for material processing and boiler heating, a new design method is proposed to work with the PLC-based temperature control system, where the box temperature control may be achieved through the fan and the heating plate. The hardware design and software design of the system are analyzed in detail. In this paper, a combination of the traditional PID control and the more popular fuzzy control is taken as the control program to achieve the overall design of the control algorithm. Followed by the simulation in the MATLAB software, the designed system is highlighted by its the characteristics of impressive stability, precision and robustness.

Keywords. PLC, temperature control, PID algorithm.

1 Introduction

The modern sensing technology and control methods are undergoing continuous innovation, where the real-time temperature control is demanding higher accuracy and faster response more than ever. Temperature control is widely used in production and industrial control processes in all aspects. For example, in the iron and steel smelting process, iron and steel to be baked requires heat treatment in order to achieve their performance indicators; plastic qualitative process also needs to maintain a certain temperature range. The fact is that the temperature control system is a complex process object involving large inertia and pure delay with multi-variable and time-varying parameters. At present, the PID control methodology is adopted in most cases [1]. In this way, different PID parameters should be selected for different control objects, for which some practical experience is needed. As a language controller, the fuzzy PID control is to imitate the way of human thinking and experience to achieve its control process that can more closely reflect the best control behavior of the controller. With strong robustness and control stability, it can be applied to different control objects. The combination of fuzzy control and PLC, which is widely used in industrial control, is one of the hotspots in this research area. Therefore, this work involved the use of the PLC-based fuzzy PID control technology, by which the system temperature was set through the fan and the heating plate to control the box temperature.

2 System design

In this design, the temperature control system consists of hardware and software components.

2.1 System hardware

In the hardware part of the system, the acquisition module uses the temperature sensor to measure the measured object temperature, and the temperature signals are converted into electrical signals [2], which are then transmitted to the temperature transmitter, where the electrical signal is converted into a 4 ~ 20mA current signal, so that the module EM235 in the PLC expansion module can be facilitated as to the analog signal input. EM235 receives data, which will be sent directly to the PLC output control text display (display temperature) and the temperature control device [3] (heating &cooling device). The system block diagram is shown in Figure 1.

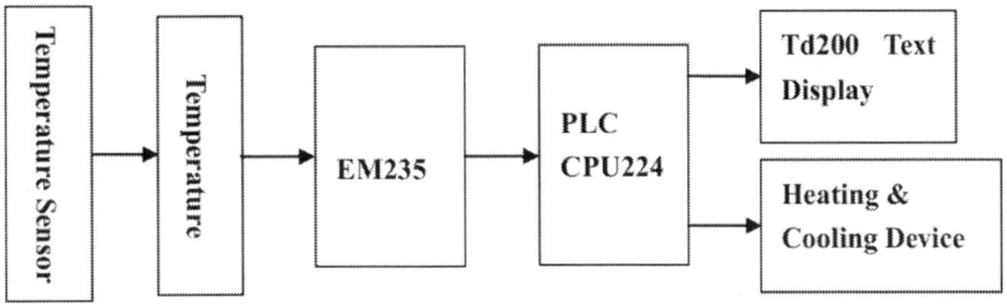

Fig.1. Block Diagram of Temperature Control System.

2.2 System software

The software of this system adopts STEP7 for PLC200, the popular programming software by Siemens, for software compilation [4], and the temperature controller device adopts fuzzy PID algorithm for temperature control, with the simulation to be implemented by MATLAB simulation software.

Figure 1 shows that the system consists of four modules, i.e., acquisition module, control module, display module and implementation module. The acquisition module includes PT100 temperature sensor and temperature transmitter. PT100 temperature sensor works with a temperature variable that can be converted into a standardized output signal. This instrument is mainly used for industrial process with measurement temperature and control parameters. The temperature transmitter is a signal conversion device, which is responsible for the signals collected by the temperature sensor to be converted to electrical signals of 4 ~ 20mA. This is quite convenient for PLC200's identification and collection of temperature signal. For control module, Siemens PLC200 is chosen as the core controller, playing the role of the completion of the temperature signal collection, signal processing and signal transmission. Display module using TD200 text display can be better compatible with the PLC to complete the data transmission. This display module displays the current temperature and the desired temperature. Implementation module works by using cold fan and heating resistor, through the PLC-controlled cold air fan and heating resistor to complete the instructions for the temperature rise or fall, and finally, the temperature reaches a constant value.

3 Fuzzy PID control technology

Fuzzy controller is a kind of new controller developed in recent years. The advantage of this device is that it does not require the precise mathematical model of the controlled object. Rather, the control decision table is organized according to the manual control rules, and then the size of the control variable is determined.

3.1 Fuzzy PID control: principle and rule

The CPU queries the corresponding fuzzy control table according to the system deviation (Deviation = Given Value - Feedback), and deviation change rate (Deviation Rate = Current Cycle Deviation - Upper Cycle Deviation) to get the setting values of Kp, Ki and Kd. Then PID operation is performed.

The formation of fuzzy control rules comes from the experience of the operators or experts with their control of knowledge and experience being made into a number of control decision tables. These rules can be expressed in natural language, only generally to be formalized. The current design of the fuzzy controller is basically used in the fuzzy control process, while the system and the deviation of the set value and its rate of change are taken as a fuzzy input.

This method can not only guarantee the stability of the system, but also reduce the overshoot and oscillation phenomenon.

3.2 Fuzzy PID control algorithm

PID control is very effective for the control of linear time-invariant systems, but not well controlled for nonlinear, time-varying complex systems and systems with unclear models. Remarkably, it is quite easy for the fuzzy controller to gain effectively control of the complex and the model unclear systems. Note that the fuzzy controller is lack of the integral link, so it is difficult to completely eliminate the static difference in the fuzzy control system. Furthermore, in the case of insufficient number of the variable classification, a small oscillation is often found near the equilibrium point. If we combine the two control methods, we can form a fuzzy PID controller with the advantages of both.

There are several methods that can be employed to combine fuzzy technology with the PID control algorithm to form a fuzzy PID controller: One is the use of fuzzy controller that allows the PID controller to get its online self-tuning PID parameters, making a fuzzy self-tuning parameter PID controller; another may be done in a large deviation range, where the proportional control is used, while the fuzzy control is adopted in the small deviation. These two control modes can be switched upon the pre-determined deviation threshold, which constitutes the FUZZY-P dual-mode segment controller. Similarly, multi-mode segmentation control algorithm may be designed to achieve sub-sections according to different conditions and requirements, with different modes of control.

3.3 Difference between fuzzy control and the traditional PID control

The distinction between the two may be understood as: PID control is time-based and fuzzy control is a process-based one. The structural difference between the two is shown in Figure 2.

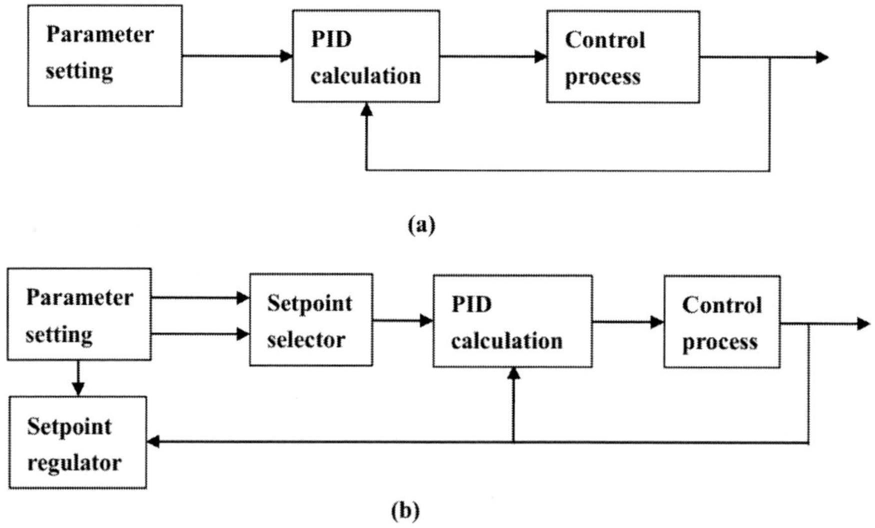

Fig.2. Difference Between Fuzzy Control and Traditional PID Control.

(a) Traditional PID Control System - Block Diagram (b) Fuzzy PID Control System - Block Diagram

The intelligence of the fuzzy controller, compared to the traditional PID controller, is reflected in an extra setpoint regulator and setpoint selector. The setpoint regulator simulates the control procedures and functions of an experienced operator. It focuses on the control process (in the range of PID control), the dynamic characteristics of the system and the non-linearity (including load changes), and the system is operated by inputting the assumed target value to the PID operator. As the fuzzy control significantly reduces the overshoot, it is capable of effectively improving the system response speed and temperature control accuracy, being easier to achieve computer control. For this reason, fuzzy control technology is used by many control systems. Concerning the control of heat treatment equipment, fuzzy control technology marks the most significant progress in the last decade, attracting more and more attention by the heat treatment sector. Fuzzy PID control is present in a series of control applications, including the resistance furnace temperature system, the raw material mixed system, beer fermentation monitoring system, the wind network throttle control system, magnetic levitation system, motor control system, cement cooling process control, air conditioning constant pressure water supply system, radar seeker servo system and so on. This is sufficient to illustrate the excellence of the fuzzy PID control algorithm.

4 Software design

According to the system design requirements, the software program flow is shown in Figure 3.

First of all, the parameters of the temperature control system undergoes wake-up initialization, mainly to set the control temperature and the PID initial value, including the value settings of PID gain, PID integral time, PID differential time and PID sampling time. Then, the ambient temperature is collected through the sensor in a range of 6400 ~ 320000, as the digital signal. As the fuzzy PID algorithm requires real-format temperature signal input, there is the need for A/D conversion of temperature signal prior to the PID algorithm process. The collected digital signal is converted into double integer signal, which is then transformed into a real figure. The actual temperature is calculated by the temperature calculation formula. The measured temperature is taken as the input signal for PID operation, and the output is ready for the control of the heating resistance and cold air fan.

In the design, the temperature control is based on the PID control algorithm. PID is the most commonly applied in industrial production, a control method being able to meet the need for high-precision measurement and control systems. Using PID algorithm to achieve the temperature control system can be more stable and reliable [5].

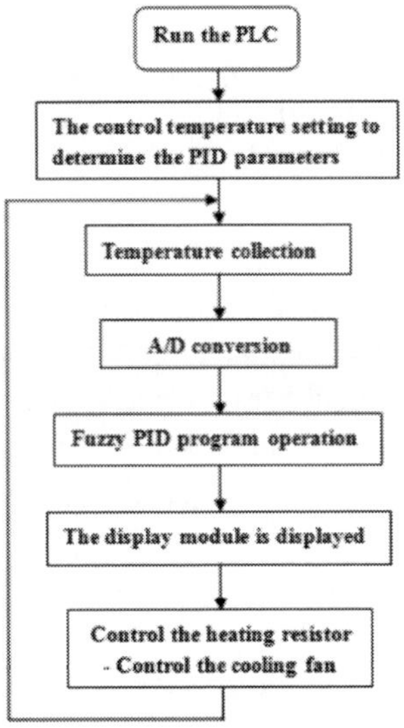

Fig.3. Program Flow.

5 Experiment and Simulation

Assume that the system has an open-loop transfer function as G (s) = 1/S (2*S + 5), and a fuzzy control system is established in Matlab, as shown in Figure 4. At this time, the amplifier Gain = -1000, Gain1 = 0.05, Gain2 = 0.01. Select the controlled object and its reference model, and we can get the simulation curve shown in Figure 5.

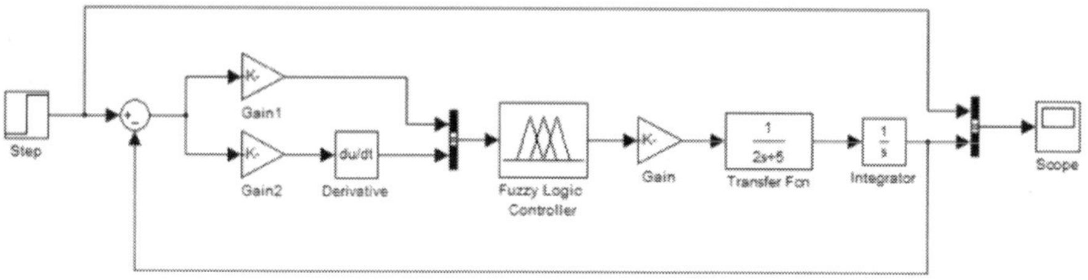

Fig.4. Block Diagram of Fuzzy Control System.

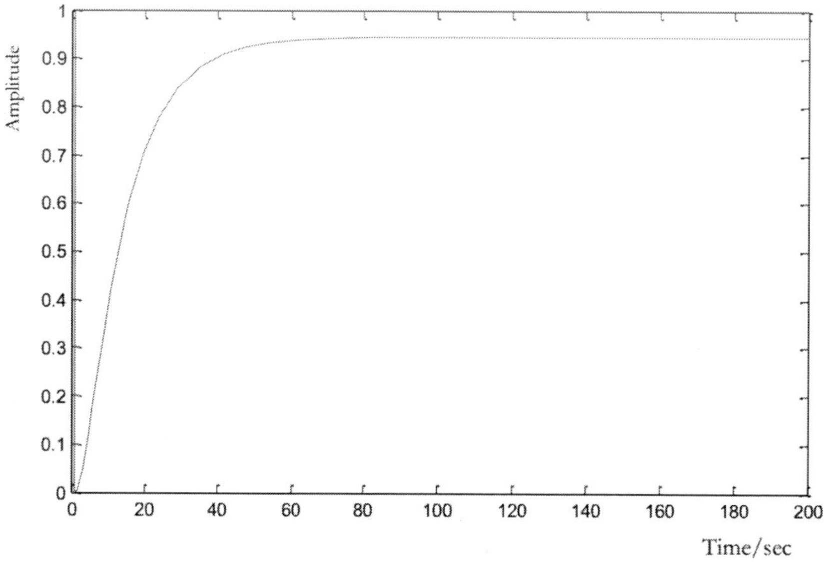

Fig.5. Simulation Results of Fuzzy Control System.

6 Conclusions

In this work, the temperature control system was available for real-time display of the temperature inside the box. By setting the temperature through the fan and the heating plate, PID control algorithm was introduced to control the temperature of the box to achieve the temperature control needs. With Siemens PLC200 as the controller, system control was quite impressive with high precision, stability and reliability, being not susceptible to outside interference. On-site commissioning revealed that box temperature could be successfully controlled by the system. This work is expected to bring about a good application prospect.

References

1. Fang Yanjun. Sun Jian. Intelligent Instrument Technology and its Application. Beijing: Chemical Industry Press .2003: 42-44.
2. Jiang Zhongliang, Chen Xiuyun. Measurement and Control of Temperature. Beijing: Tsinghua University Press, 2005: 26-27.
3. Chai Ruijuan. Chen Haixia. Siemens PLC Programming Technology and Engineering Applications. Machinery Industry Press .2006: 8-14
4. Tang Yifan. Electrical and Programmable Logic Controller Technology. Machinery Industry Press .2004

5. RenZhicheng. Application of Intelligent Digital Display Controller for Sensors and Transmitters. Beijing: China Electric Power Press .2007: 85-96

Influences of three-dimensional printing to product innovation design thinking

Ling Liu*, Xu Wang and Ping Guo

Beijing City, Daxing District Xinghua Street (two) No. 1, Beijing Institute of Graphic Communication, Beijing, China
*E-mail: liuling@bigc.edu.cn

Abstract. The increasing demand of individuation brings new challenges to traditional product design. This paper studies and analyzes three-dimensional print technology promotes the future development of product innovation design. The advantages of 3D printing equipment can help creative products enterprises shorten product manufacturing period and make them powerful in competition.
Keywords: three-dimensional printing, product design, innovative thinking.

1 Guidelines

Along with the enrichment of physical goods and the development oftechnology, more and more customers demand individuation service. Enterprises establish their customization system one by one to satisfy customers and win the competition. In the design process, more attention has been paid to user's individuation factors. In other words, the increasing demand of individuation brings new challenges to traditional product design. Facing with new customer demand, it is possible to realize product custom design through three-dimensional printing. So far, researches on custom design using three-dimensional printing have gained extensive attention. Researching on creative products custom design technology has profound theoretical significance and practical value of the establishment of custom design platform which meets customers' demand.

2 Characteristics of products innovation design thinking

According to the once shaping principle, 3D printer can manufacture complex structures which are hard to process or even cannot be processed under traditional methods. This pushes the limitations of traditional manufacturing. 3D printer shortens the distance between ideas and products by manufacturing three-dimensional objects as long as the designers have ideas and turn them into three-dimensional data model[1]. Yan Yongnian, professor and doctoral supervisor of Tsinghua University, mentioned that the biggest effect of three-dimensional printer is to train and inspire young men's intelligence when he gave a report in the lecture hall of Zhong tian institute of technology.

To meet the customers' changing demand, product manufacturers need to transform their old pattern which production promotes sales into a new way which requires manufacturers to arrange production according to customer requirement. Realizing product innovation design and personalized customization can help manufacturers reduce product cost, improve product quality, shorten time to market and provide quality service[2]. So far, customization pattern, which means implementing individual mass production to meet customer's individualized needs, has been applied widely. Facing with new customer demand, it is possible to realize product custom design through three-dimensional printing. This custom method is an interactive, authentic, seasonable method which enables customers to see custom outcome immediately and get satisfied product. Customers can also participate in the design at conceptual design phase.

3 Influences of three-dimensional printing to product innovation design thinking

Throughout the entire history, design has been playing an important role that makes life more efficient, advanced and interesting by applying technical innovation. Sometimes it plays the same role of art. Applying digital technology into producing also belongs to this traditional. For the same reason, 3D print has become a controversial topic in the design community. Some people regard it as an innovative technology as important as printing press and steam engine, while others consider that is too exaggerated. Hence it is the perfect time to evaluate the influence of three-dimensional printing to design and its future development.

Designers take more and more advantages of the progress of digital technology to manufacture products with complex geometric structure and elusive shape. As a result, an aesthetic pursuit of this kind of structure which as if comes from other worlds will become more and more common. Designers no longer pursue to control the final form of their product, they prefer to design a method which enables others personalize the production. The reasons come from aspects of style as well as function. Amputees can build a complete suitable artificial lime using 3D print and arthritics can adjust kitchen wares' hand shank for an easy use.

3D printer can not only build new objects but copy existing ones in case of lost and damage as well. Therefore, designers must consider the interchangeability of individual part when they design original items. Meanwhile, the growing requirement of maintenance brings a new challenge to the design of a new business model.

Firstly, conceptual model: at early stage of design flow, using 3D print technology structure models can examine the construction, shape and functions of an object. Any defects can be modified at the first time. After this, if it needs, the process of construction, examination and modification can be done again and again until the best conceptual model has been designed. Translate two dimensional design drawing into real objects can surely accelerate the development process of a product and reduce costs. In addition, three-dimensional objects can help designers make better decisions faster by presenting design preferably.

Secondly, function prototypes: designers can prove the rationality of their designs through manufacturing function prototypes. At the same time, they can test the performance and evaluate engineering strictly using three-dimensional objects. 3D print technology can shorten the time to market and maximize product performance.

Thirdly, finished products: 3D print technology has been a new fashion within the industry. The application fields are various including aviation enterprises, medical equipment manufacturers and small cars manufacturers. Using 3D print technology to replace traditional manufacturing process can save time and also reduce costs. What's more, 3D print technology breaks the limitations of traditional manufacturing process and enables designers to modify their design whenever there is a need. In this way, enterprises can exploit new markets of customized services or small amount application apparatus.

4 Three-dimensional print technology promotes the future development of product innovation design

4.1 Shorten the design period using 3D print technology

Digital production can not only change the relation between designers and users but also improve the cooperation between designers and producers. Ultimately, 3D printing equipment may spread every town, where local residents can acquire custom products or maintain existing products by themselves, just like their ancestors ask blacksmiths for forging products. Without question, enterprises which have technical experience have begun their preparation.

If this imagination comes true, designers no longer need to find producers who appreciate their designs. What they need to do is to make a prototype of their products and then sell them to consumers like entrepreneurs.

Producing products with complex shapes rapidly give a solution to the crazy ideas of designers. On the surface, 3D print can bring some changes temporarily. But if we connect it with some modern words like internet, cloud technology, open-source sharing

And big data, we can see a whole new connotation of the possibilities provided by 3D print. At this moment, 3D print becomes a key technology to subversion manufacturing industry. In aspect of production pattern, 3D can bring us changes of customization. Because of the affiliation dissociation between internet and manufacturing industry, clients are willing to design the product with manufacturers, which is feasible. Product manufacturing turns from patterned mass production into customization low-volume production, leading to some outcomes including the elimination of traditional industrial production line, the simplification of supply chain and even changes of world manufacturing economic distribution.

The longer a design cycle of one product is, the longer it takes to market, which means less potential profit of enterprise. Because of the importance to shorten time to market, enterprises have to shorten decision making time during the conceptual phase of a design. These decisions will decide most cost factors including material selection, manufacturing technology and designed life. 3D print technology can optimize design process and maximize potential profit for enterprises. There are many difficulties between an excellent idea and a successful product.When it comes to estimate an idea and decide whether it is worth to be invested, 3D print technology can help enterprises by shortening evaluation time and improving efficiency.

Design process which experienced prototype iterative optimization repeatedly can lower the risks of failure products effectively. Because that 3D printer can manufacture models which have excellent properties and can stand rigorous tests, it is much easier for designers to design. Models manufactured by 3D printer can help designers and engineers understand everything about potential products and decrease leftover problems.

4.2 Customize products using 3D print technology

The greatest advantage of 3D print is to expand the imaginary space of designers and develop their imagination and creativity[3]. Future equipment manufacturing industry will focus on personalized customization. 3D printer does not manufacture objects through incision or molds as traditional manufacturing machines. From physical objects by laying one by one expand the range of number concept in aspect of physics. 3D printer can manufacture cultural and creative products which demand accurate internal depressions or interlock shapes in physical world, which could develop huge design space. 3D print technology enables consumers customize products based on their requirements. Mass production will turn into mass customization. Current mass production method pursues to gain scale benefit through minimizing costs, detailing labor division and normalizing operation procedure. It cannot meet the customers' diversified demand any longer. The basic strategy of customization is to strengthen the standardization of inside structure of products in the manufacturing process and to increase the diversity of external structure which can be perceived by customers.

3D print will change our life tremendously in the future. This change, actually, has been taken its place for a long time. This revolution includes the invention of computers and the prosperity of internet. As said by Chris Anderson in *Makers: New Industrial Revolution*: In the process of manufacturing industry having a digital revolution, physical objects have become different designs on the screen which can be shared on line in forms of documents[4]. In the past several decades, factories and design studios experienced this transformation. Today, it happens on the clients' computers as well as in their basements. From retail industry to publishing industry, once an industry shift to digital, it will go through profound revolution. The biggest change is who is manufacturing rather than how to do it. As long as the process can be done by computer, everyone can participate in it. This is the revolution which manufacturing industry is experiencing.

5 Conclusion

No matter what revolution occurs to the manufacturing industry because of the combination between internet and 3D print technology, the essence of industrial design remains unchanged. It is to solve and coordinate the relationship between people and objects. Products bring enjoyments to people not only in physical level but also spiritual level. No matter using traditional artisan craftsmanship or 3D print technology, the essence of design is to bring people a beautiful life. This is the fundamental which would not change forever.

3D print technology provides a cost-effectiveness way to accomplish repeatedly design iteration. It can obtain immediate feedback information of product design on the critical initial phase of development process. 3D product technology can modify creative products immediately, which can help to reduce costs and shorten time to market. To enterprises which use 3D print technology in their design process, this gives them competitive advantages. Along with the decline of price, 3D printing equipment will enjoy a larger market share. The advantages of 3D printing equipment including speediness, precision and low-cost can help creative products enterprises shorten product manufacturing period and make them powerful in competition.

References

1. Y.G. Wang, X. Wang, *Three-dimensional Printing*, Beijing, Wuhan: Huazhong University of Science and Technology Press, (2013)
2. B. Evans, *Analysis 3D Printer: Science and Art of 3D Printer*, Beijing, China Machine Press, (2013)
3. H.Y. Wu, *3D Print: Three-dimensional Intelligent Digital Manufacturing*, Electronic Industry Press, (2014)
4. C. Anderson, *Makers: New Industrial Revolution*, Beijing, China Citic Press, (2012)

Real-time Solution via Dynamic Simulation for the Six Degree of Freedom Platform

Lizong Lin[1, a, *], Xiao Li[2,b], Xiaopeng Ni[3,c] and Gang Li[4,d]

[1, 2,3]Department of Mechanical Engineering, East China University of Science and Technology, Shanghai 200237, China
Email: [a,*]lzlin@ecust.edu.cn, [b]lixiao2016cv@163.com, [c]nixiaopeng1986@126.com, [d]1066332939@qq.com

Abstract. In order to provide a new way for dynamic simulation experiments, a real-time solution for the six degree of freedom platform was developed. The mathematical model of an improved Six-DOF Stewart platform was used to study positive solutions and inverse solutions. According to the parameters of different platforms, different function signals were selected to generate motion control data by using Visual C++ programming. Motion control card was embedded into industrial computer, data was sent automatically to the control card by the program when the platform ran. The output of the control card was analog voltage, and it was amplified to send to the proportional valve, then the flows of six hydraulic cylinders were controlled by the six proportional valves. So a closed-loop control CNC system was formed. The expected action could be realized by the platform. Experiments have proved that the method is simple, efficient and easy to operate. It can not only accompany the implementation moving of relevant actions in the 3D movie theater, but also provide the signal source for the road spectrum of simulated driving test of the automobile.

Keywords. Six DOF platform, positive solutions, inverse solution, visual C++, Googol NC Card.

1 Introduction

The 6-DOF motion platform is a very important experiment instrument, which is highly concerned about by most university and research institute in the world, for the simple and compact structure, high moving precision, stiffness, stability, ability of large load endurance [1]. For example, like the flight simulator [2], ship simulator [3], car -driving simulator [4], earthquake simulator, dynamic film and some kinds of device for entertainment. At the same time the platform is also provide a method which is engaged in different types of real-time driving operator. Stewart platform [5] is widely used here: it consists of two boards (up and down), six hydraulic cylinders, six couples of universal hinges (also up and down), the down board is fixed on the base, while the cylinders connect the up board and moving into the postures which we want by the movement of stretching and shrinking.

The precision and real-time property of the 6-DOF platform [6] is greatly dependent on the cylinders' movement, which is determined by the real-time position feedback and calculation. This calculation contains positive solutions and inverse solutions [7]. The positive solution is to get the posture of the platform by collecting the cylinders' amount of displacement, while the inverse solution is to calculate the cylinders' displacement by the posture feedback. In this paper, a mathematical model is made for the Stewart platform, and the solutions are calculated by the computer program. The 6-DOF platform is controlled by the numerical control system based on the PC, and Googol's GTS-800-PV-PCI-G motion controller is used, whose analog output is between -10V to 10V. The controller can handle almost eight axes. A hardware platform [8] of open CNC system is made by combining the computer and the Googol card (controller). In the whole system, the controller is the core, the hydraulic system is the drive unit and the cable encoder is the feedback sensing element. All those parts is made into a semi-closed loop. The implementation of the system is a mode of host computer and the slave computer. The host computer is used to charge the human-computer interaction and the data calculation. And the calculated data is put into the Googol card to realize the expected movement.

2 Platform structure

The 6-DOF platform is composed of the upper board, the down board and 6 hydraulic cylinders. The down board is stable and the upper board can move. Both boards are connected by the hydraulic cylinders. The structure of the platform is indicated in Figure 1. Suppose at first the two boards' center line is in the same vertical surface, and the stable coordinate system is named 'XYZ'. The new position of the board can be calculated by the coordinate transformation matrix [9] in kinematics after the upper board moved.

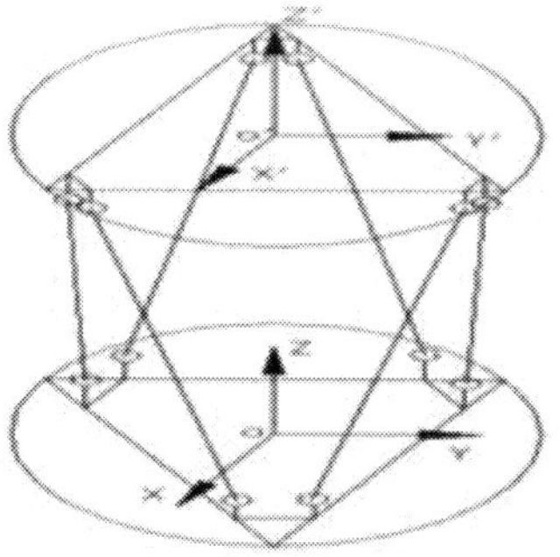

Fig.1. Structure Figure of 6-DOF Platform.

3 Hardware

The hardware of the 6-DOF platform mainly contains: the IPC, Googol motion controller, amplifier, encoder, hydraulic cylinders and motion platform. The hardware structure is showed in Figure 2. The pattern of setting communication between the host and the slave computer is used [10], because the IPC is very powerful machines. PC, as the host computer, which is in charge of the interaction between man and the system, and the real-time feedback [11] such as to show the status of the system and the code order input. The Googol NC card, working as the slave computer, can feedback the information of the movement control, which contains the start/stop order, the control of the direction and velocity, and I/O information. The proportional valve and the encoder are set to make up the closed-loop control system. The amplifier is to receive the analog signal released by the NC card, which is amplified and sent to the proportional valve to control the expenditure degree of the valve which constrains the liquid amount. The amplifier gets the feedback signal from the closed-loop control system, and then it sends the signal to the proportional valve to control the hydraulic cylinders.

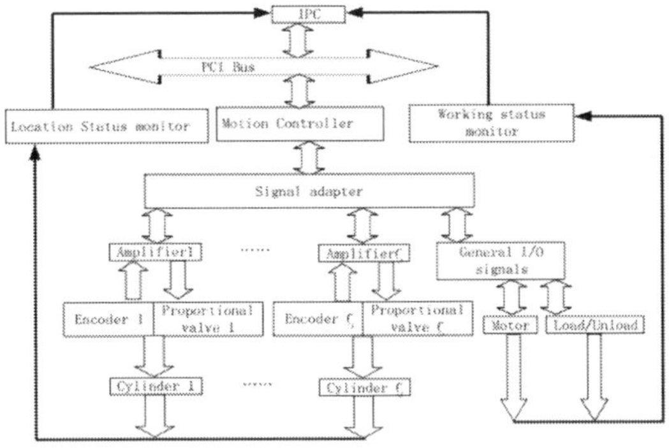

Fig.2. Composition Diagram of System Hardware.

4 The design of the software to reach the positive solution and the inverse solution

Considering the tough atmosphere in the factory (vibrations, dust, strong EMI, long running time), Yanhua IPC is selected. The software is based on Windows OS, which has the ability of handling more than one task at the same time, and win32 exe files is used to fit the upper software. The IPC software is made by Visual C++6.0, which is masked by many Windows API, called MFC (Microsoft Foundation Class) library [12]. The users can develop their own programs only by understanding the MFC library and Windows OS. The software is helped by App Wizard and Class Wizard, and set the exe files based on the dialog boxes. In this way, widely used functions is put into the interface to make it convenient to operate.

4.1 The Realize of the Inverse solution Research Program

According to the kinematics [13], the Inverse solution is defined as the cylinders' position parameters calculated from the board's position parameters. The relationship between the mathematics model and the closed solution is used to decide the length of the cylinders and get the displacement. This way is simpler than other methods. In the "CsixDofDlg" head files, the cylinder displacement matrix named "DeltaL" (double type) and the parameters of each board's displacement and angle are defined. The principle of Block diagram of anti-solution program [14] is showed in the following Figure 3.

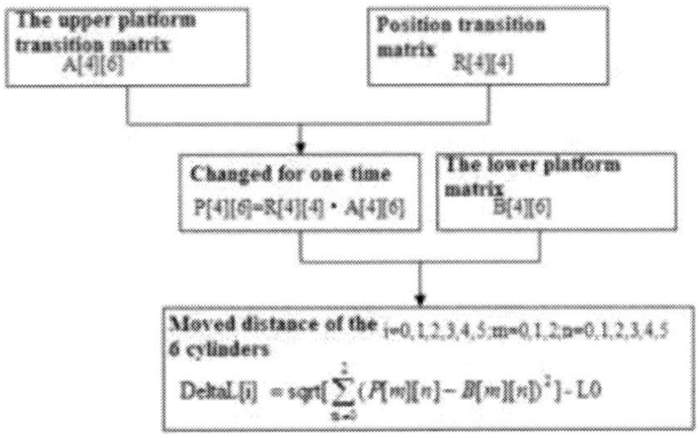

Fig.3. Block Diagram of Anti-solution Program.

The parameters of matrix A and matrix B, and L0 are from Access database, for the board and the cylinder length must be initiated before start. ADO object is used to define a combination and a set pointer: _ConnectionPtr, m_pConnection; _RecordsetPtr; m_pRecordset. At last the main interface of the inverse solution is showed in Figure 4.

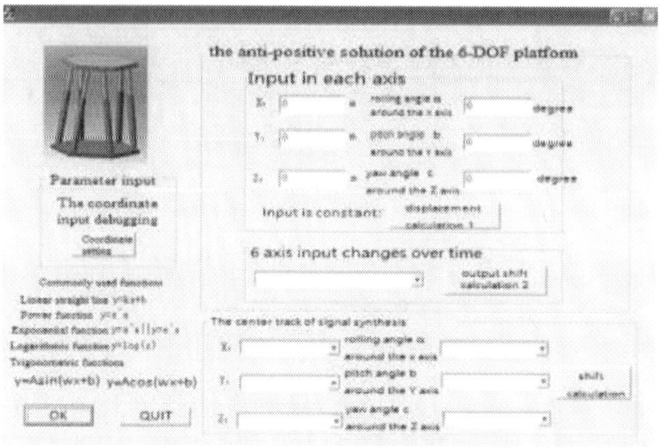

Fig.4. Interface of Platform Signal Generator.

The signal source can be superimposed in any way by using the higher mathematics function and it is easy to create many kinds of data. These data can supply many data for many simulation experiments. Among them the most widely used are the road spectrum and the 3D movie. The collection of road spectrum[15] is to get the wave which the road forces the vehicle to vibrate, and send them to the servo controller after filtered. The conclusion about the parameters of the road such as the flatness can be drawn by collecting the road information from the sensor set in car. With the help of other dynamics simulation software we can also check the flatness of the car, the fatigue of the parts in car, and thus work out the method to relief the car vibration. Collecting the data in real experiment is simple, but it feedbacks messes of data, and the time or road status are limited. However, the 6-DOF platform can realize any movement and movements, Vehicle motion based on the reproducibility of the measured data can also be based on the virtual experimental data vehicles to achieve a variety of road movement. Similarly when 3D movie is played, the platform also can generate the motion and make you feel you are in the movie. All these above are determined by our platform. And the signal source is the key to do these simulations. Appropriate signal source can help to achieve various analog devices motion control. In Figure 5, the signal of the upper board on the X direction is sin wave, while there is no signal on the other direction. The curves of displacement shown in Figure 5 is calculated by MATLAB.

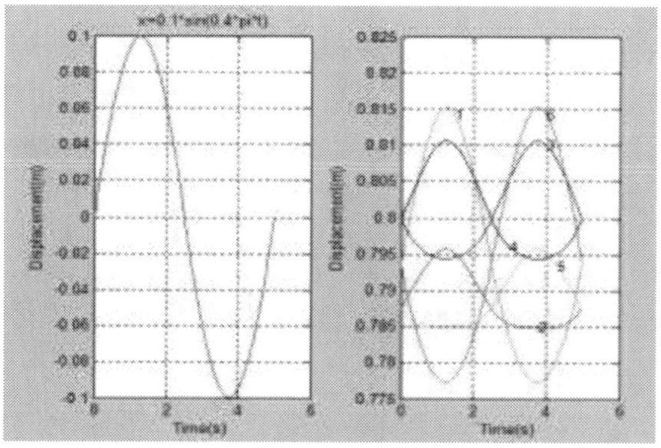

Fig.5. Displacement Curve of Each Cylinder.

4.2 The Research of the Positive Solution

According to the kinematics, the positive solution is defined as to get the boards' position parameters when given the cylinders' parameters. It is hard to get the positive solution, so numerical methods such as the Newton method with the anti-solving equations are often used. And the programming time is too long; the large platform is hard to move by the cylinders. In this paper, a small platform model is established shown in Figure 6. The two boards are connected with six displacement sensor [16] through hooker hinged. Data acquisition card collects the real-time data from the displacement sensor.

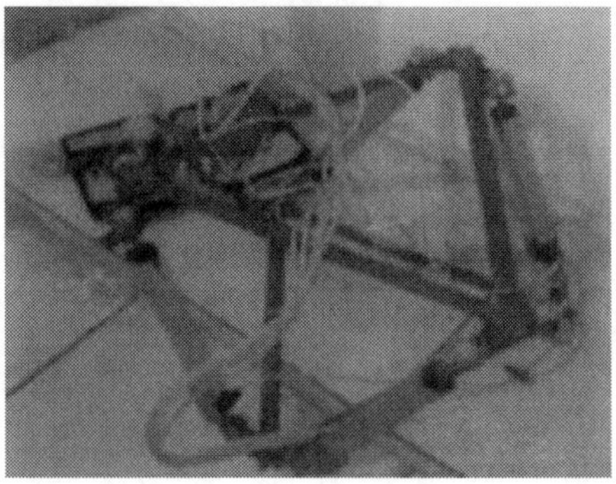

Fig.6. Platform Model.

From the Figure 7 we will know how it is operated. The collected data is stored in PC as txt files, and the first column is time (ms), the data was collected every 200ms, the second to the sixth column was set for displacement sensors whose maximum capacity is 1023, and these data is transferred and sent into ACCESS database as shown in Figure 8.

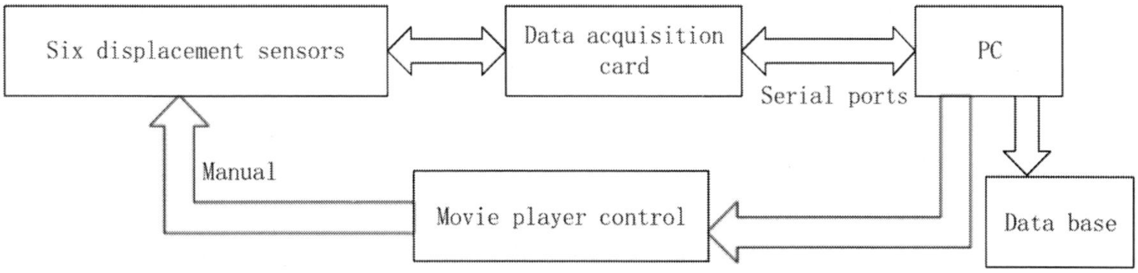

Fig.7. Positive Solutions for Data Acquisition.

0	0	0	0	0	0	0
200	0	0	0	0	0	0
400	0	0	0	0	0	0
600	0	0	0	0	0	0
800	0	0	0	0	0	0
1000	0	0	0	0	0	0
1200	0	0	0	0	0	0
1400	0	0	0	0	0	0
1600	0	0	0	0	0	0
1800	0	0	0	0	0	0
2000	0	0	0	0	0	0
2200	0	0	0	0	0	0
2400	0	0	0	0	0	0
2600	0	0	0	0	0	0
2800	0	0	0	0	0	0
3000	31	13	0	0	28	37
3200	55	73	22	20	31	103
3400	68	95	28	76	63	163
3600	90	95	39	104	99	188
3800	142	95	109	105	133	187
4000	177	95	180	104	180	140
4200	177	96	182	105	180	135
4400	177	101	152	137	180	160
4600	176	122	127	186	180	191
4800	180	123	133	188	180	165
5000	225	123	193	188	205	119
5200	236	123	214	188	224	111
5400	234	125	167	194	222	174

Fig.8. The Collected Data.

5 Conclusions

This paper is mainly about the research of the improved Stewart 6-DOF platform in the respect of the positive solution and the inverse solution. A method of getting the signals is proposed and proved on the Googol card system:

(1). IPC + Googol card works, and it can help to improve the research of the 6-DOF platform.

(2). Many kinds of signal sources can be used to mix them into some useful signals. To make the vehicle experiment more convenient, to improve the road spectrum testing, to let the audience of 3D films more comfortable.

(3). VC++ is a very useful and powerful software, and with the Googol card database, there are many functions in it and it does well to NC research.

References

1. C. Qian, J.Q. E, M. liu, Y.W. Deng, H. Zhu. Six degrees of freedom parallel Stewart platform vibration dynamic model analysis. Hunan Univ. J. (natural science edition). No.2, 36-42 (2016).
2. H. Wang, X.S. Chen. Simulating Six-DOF Motion Platform of Flight Simulator. J. Vol.31, No.5, 778-782 (2012)
3. Y.L. Wang, Y. Lu, J.M. Liang. 6 DOF motion platform location analysis and measurement control in Flight Simulator. Comput. Meas. Control. J. Vol13, No.11, 1243-1244, 1249 (2005).
4. Y.J. Pi, X.Y. Wang, X.H. Luo, X. Gu. Random Wave Spectrum Simulation on Six DOF Ship Motion Simulator. Vib. Testing. Diag. J. 2010, Vol.30, No.4, 75-378 (2010).
5. Q.H. Cao, X.X. Zhang. Car Driving Simulator Telepresence Research and Development on 6 DOF motion platform. Nanchang Inst. Tech. J. Vol.28, No.4, 8-11 (2009).
6. H. Zhao, S.Y. Zhang. Hydraulically Driven of Stewart platform and Trajectory Planning. Mach. Tool. Hydr. J. Vol.35, No.8, 137-139 (2007).
7. D.Y. Yang, L.M. Chen. Six DOF Hydraulic Platform Motion Simulation Based on SolidWorks. Mach. Tool. Hydr. J. Vol.36, No.9, 127-128 , 150 (2008).
8. B.F. Cai, C. Liao. Six DOF Stewart-type Platform for Positive and Negative Solution. Science Paper Online. J. http://www.paper.edu.cn: 1-6.
9. L.Z. Lin, G. Lan. G Code System Oriented Googol NC Card. Mold Tech. J. No.1: 5-8 (2009).
10. S.Y. Zhang, H. Zhao, J.W. Han. Real-time Control of 6 DOF Motion Platform Forward / Inverse Solution Algorithm Mach. Tool. Hydr. J. No.3, 133-135 (2010).
11. S.S. Zhu, H.P. Shen, Z. Wang, Y.F. Li, Z.H. Ma. Forward and Inverse Kinematics Solution for 4-3 Type 6-DOF Parallel Mechanism. Changzhou Univ. J. (natural science edition). Vol.25, No.4, 32-36 (2013).
12. J. Wang, Y. Y, S.F. Luo, Z.Y. Li. Development of the Stewart Platform Motion Control System Based on PMAC Controller. Mach. Tool. Hydr. J. Vol.40, No.9, 77-80 (2012) .
13. K. Song, R.N. Liu, W.M. Li. Book of Visual C++ Development. People's Posts and Telecommunications Press.. Beijing (2007).
14. L. Li. 6-DOF Parallel Platform Displacement and Control Method. Harbin Engineering University (2008).
15. Z. Q. Yao, X.N. Guo, X.B. Liu. Modeling and Simulation of 6 DOF motion platform inverse position. Fluid Power Trans. Control, No.6, 38-40 (2010).
16. R.P. Wang, X.Z. Jiao, G.L. Wang. Recognition Based on a Typical Road Surface of the Car Body Vertical Acceleration Spectrum. Auto. Eng. J. Vol.30, No.12, 1047-1051 (2008).

Intelligent Adjustment of Printhead Driving Waveform Parameters for 3D Electronic Printing

Na Lin[1, a], Shikai Jing[2,b], Hanning Chen[3,c] and Wenxin Liu[4,d]

[1]Beijing Shenzhou Aerospace Software Technology Co. Ltd., Beijing, China

[2]School of Mechanical Engineering, Beijing Institute of Technology, Beijing, China

[3] School of Computer Science and Software, Tianjin Polytechnic University, Tianjin, China

[4] Department of Electrical and Computer Engineering, Lehigh University, Bethlehem, USA.

[a]linn@htrdc.com, [b]jingshikai@126.com, [c]perfect_chn@hotmail.com, [d]wliu@lehigh.edu

Abstract. In practical applications of 3D electronic printing, a major challenge is to adjust the printhead for a high print resolution and accuracy. However, an exhausting manual selective process inevitably wastes a lot of time. Therefore, in this paper, we proposed a new intelligent adjustment method, which adopts artificial bee colony algorithm to optimize the printhead driving waveform parameters for getting the desired printhead state. Experimental results show that this method can quickly and accuracy find out the suitable combination of driving waveform parameters to meet the needs of applications.

Keywords. Intelligent adjustment, artificial bee colony algorithm, 3D Electronic Printing

1 Introduction

In recent years, drop-on-demand (DoD) inkjet printing has achieved a great success in 3D electronic printing [1]. This revolutionary technology can offer a high-resolution, high-speed, solution-thrift and high compatibility way of depositing picoliter droplets with diverse physical and chemical properties on various printing surfaces without contact [2]. Typically, flexible electronics, radio frequency identification (RFID) tags, solar cells, organic light emitting diode (OLED) are the popular applications of inkjet technology in 3D electronic printing engineering [3].

Nowadays, the developments of DoD printing are moving towards higher productivity and quality [4], requiring adjustable small droplet sizes fired at high jetting frequencies. Generally, the print quality delivered by an inkjet printhead depends on the properties of the jetted droplet, i.e., the droplet velocity, the jetting direction, and the droplet volume. However, in order to meet the challenging requirements of printable electronics fabrication, the conductive ink droplet properties have to be tightly controlled for higher inkjet performance [5].

To implement this issue, the theory field and engineering application field follow two different paths for high printing quality. For theory researchers, the analytical or numerical techniques are utilized to model the DoD piezoelectric inkjet printhead. These techniques can offer visualized information about the droplet formation process, such as the acoustic pressure wave traveling inside the channel and ink channel dynamics [6]. Unlike the theory researchers' enthusiasm towards the inkjet mechanism, practice engineering applications need to find an effective means to solve the problem of residual vibrations. A common method is to search an appropriate combination of the parameters of driving waveform for

DOI: 10.1051/matecconf/201710003034

the used ink with specific physical properties. Generally, these combinations of the parameters are obtained by exhaustive experiments and a wise guess [7]. However, the search process is extremely unbearable [8]. Especially, when the used ink is replaced, the exhausting search process must be re-established. Therefore, the goal of this work is to shorten the search process with an intelligent method for adjusting the parameters of printhead driving waveform.

In this work, we construct a droplet characteristic optimization system based on image detection method. And then, a canonical swarm intelligent algorithm – artificial bee colony algorithm – is adopted to optimize the combination of printhead driving waveform parameters. This method can quickly and accuracy finds out the suitable combination of driving waveform parameters to meet the needs of applications.

The rest of this paper is organized as follows: In Section 2, a droplet characteristic optimization system based on image detection method is present. Artificial bee colony algorithm (ABC) is introduced in Section 3. Experimental results are described in Section 4. Finally, concluding remarks are collected in Section 5.

2 Intelligent Adjustment of Printhead Driving Waveform Parameters

In this experiment, the investigated printhead is KM series IJ Head of Konica and the used ink is nano-silver ink provided by the LEED-PV Corporation, and the according parameters are listed in Table 1.

Table 1. Physical Parameters of Printhead and Ink

	Parameter	Unit	Value
Printhead	Nozzle radius, a_0	*um*	13.5
	Neck length, L	*um*	45
Ink	Free electrical capacitance, σ	*N/m*	$45*10^{-3}$
	Viscosity,u	*Kg/(m*s)*	15
	Density, ρ_0	*Kg/m*^{3}	$1.05*10^3$
	Sonic speed, C_0	*m/s*	1400

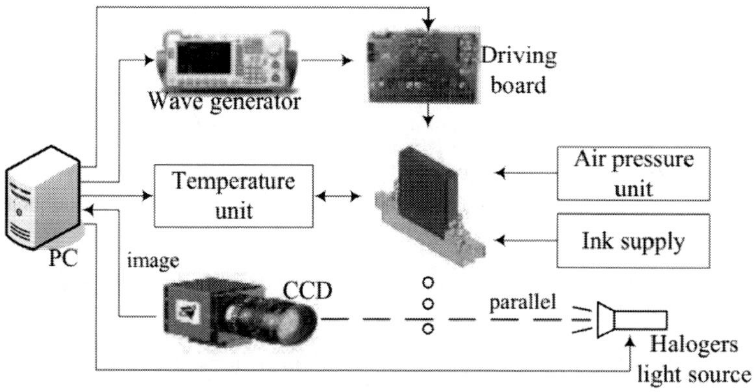

Fig. 1. Schematic of experimental setup.

A schematic overview of image detection system to observe thedroplet formation process is shown in Figure 1. The visualization system consists of a cold source of halogens and a CCD camera. The volume andvelocity of droplet and the profile of meniscus are measured through the images captured by CCD. The high-frequency driving waveform is programmed by arbitrary wave generator. In orderto repeatedly generate uniform droplets, ink supply unit connects air pressure unit, which offers suitable negative force to prevent the ink leaking from the printhead.

Here, we apply the ABC algorithm to search theappropriate combinations of the driving waveform parameters. A schematic diagram of the approach isillustrated in Figure 2.

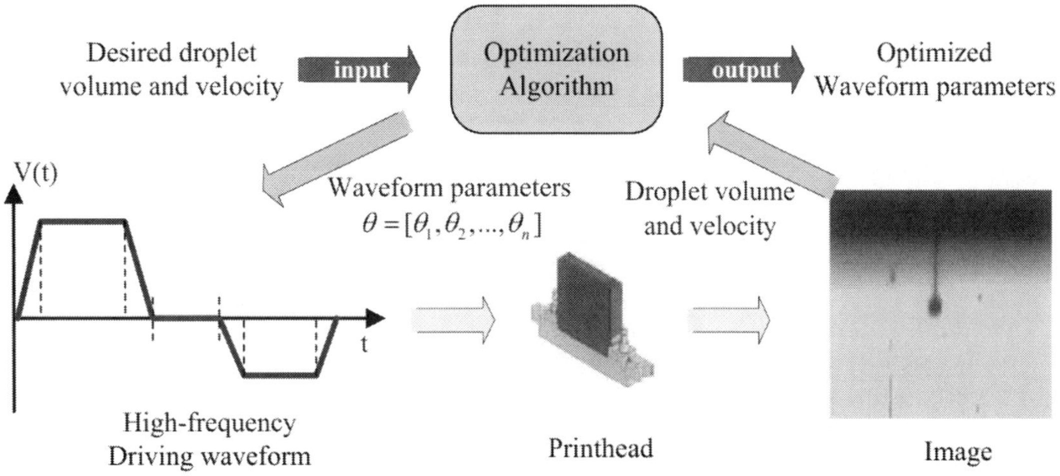

Fig.2. A schematic diagram of optimization framework for the appropriate combination of waveform parameters based on image analysis

The driving waveform input the PZT printhead is parameterized in a vector with several dimensions. The printhead works under the corresponding driving waveform and the pressure is generated in the ink channel. The pressure drives the ink out of the nozzle and the liquid filament is formed. The image detection system capture the image of droplet formation process, and the droplet volume and velocity are calculated. Through optimizing procedure with ABC algorithm, the optimal pulse parameters are obtained by minimizing the error betweenthe desired droplet volume (or velocity) and measured droplet volume (or velocity).

3 Artificial Bee Colony algorithm

The artificial bee colony (ABC) algorithm was firstly presented by Karaboga in 2005 [9], motivated by the intelligent foraging behavior of honeybees. The entire bee-colony is classified into three types of bees: employed bees, onlooker bees and scout bees. The employed bees explore the specificnectar (food) sources; meanwhile pass the food information to onlooker bees. Each food source corresponds to only one employed bee, which means the population size of employed bees is equal to the number of food sources. The onlooker bees are responsible forselecting good food sources, those with higher quality having a bigger chance to be chosen by onlookers. And then they further exploit the food near their selected food sources. If a food source found by the employed bee is exhausted, it will be replaced with a new one by a newly produced scout. The fundamental mathematic representations are listed as follows.

Step.1 Initialization phase

In the initialization phase, the ABC algorithm generates a randomly distributed initial food source positions of *SN* solutions, where *SN* denotes the size of employed bees or onlooker bees. Each solution x_i (i =1, 2,..., *SN*) is a *D*-dimensional vector. Here, *D* is the number of optimization parameters. And then evaluate each nectar amount $fitness_i$. In ABC model, nectar amount is the solution value of benchmark function or real-world problem.

Step.2 Employed bees' phase

In the employed bees' phase, the neighbor food source (candidate solution) can be generated from the old food source of each employed bee in its memory using the following expression:

$$v_{i,j} = x_{i,j} + \varphi(x_{i,j} - x_{k,j}) \tag{1}$$

where *k* is a randomly selected food source and must be different from *i*; *j* is a randomly chosen index; φ is a random number in range [-1,1].

Step.3 Onlooker bees' phase

In the onlooker bees' phase, one onlooker bee selects a food source depending on the probability value associated with that food source,P_i can be calculated as follows:

$$P_i = \frac{fitness_i}{\sum_{j=1}^{SN} fitness_j} \tag{2}$$

where $fitness_i$ is the fitness value of *j*th solution.

Table 2. Pseudocode of the canonical ABC algorithm

```
The canonical ABC algorithm
1: Initialization.
Initialize the food sources and evaluate the nectar amount (fitness) of food sources;
Send the employed bees to the current food source;
2:teration=0;
3: Do while(the termination conditions are   not   met)
4:        /*Employed Bees' Phase*/
   for (each employed bee)
find a new food source in its neighborhood following the Eq. (2);
Evaluate the fitnessof the new food source, Apply greedy selection;
end for
5: Calculate the probability P for each food source;
6:        /*Onlooker Bees'Phase*/
   for (each onlooker bee)
Send onlooker beesto food sources depending on P;
finda new food source in its neighborhood following the Eq. (2);
Evaluate the fitnessof the new food source, Apply greedy selection;
end for
7:      /*Scout Bees' Phase*/
    if (any employed bee becomes scout bee)
Send the scout bee to a randomlyproduced food source;
end if
8: Memorize the best solution achieved so far
Iteration = Iteration +1;
end while
9: Output the best solution achieved
```

Step.4 Scout bees' phase

In the scout bee phase, if a food source cannot be improved further through a predetermined number of cycles, then that food source is assumed tobe abandoned. The value of predetermined number of cycles is an important control parameter of ABC, which is called limit for abandonment. Then the food source abandoned by the bees is replaced with a new food source by the scouts. A new food source will be produced randomly in the search space usingtheequationbelow:

$$x_{i,j} = x_{\min,j} + rand[0,1](x_{\max,j} - x_{\min,j}) \tag{3}$$

where x^j_{min} and x^j_{max} are lower and upper bounds of parameter *j*, respectively.

The employed, onlooker and scout bees' phase will recycle until the termination condition is met. The pseudocode of the ABC algorithm is listed in Table 2.

4 Experimental results

In order to reflect the error between the optimized droplet characteristics and the desired droplet characteristics, the fitness function for calculating such error should be set first. The fitness function is defined as:

$$F = \min[w(1)(V - V_{\text{target}})^2 - w(2)(S - S_{\text{target}})^2] \quad (4)$$

where V_{target} is the desired droplet velocity, V is the optimized droplet velocity, S_{target} is the desired droplet volume, S is the optimized droplet volume, and $w(1)$ and $w(2)$ are the user-defined weights.

The desired droplet characteristics include droplet volumesmaller than 1.2×10^{-5} L and the droplet velocity largerthan 5 m/s. The weight is set as w=[0.2, 0.8].

In order to show the optimize accuracy, the convergence curve of artificial bee colony algorithm for optimizing the desired droplet characteristics is shown in Figure 3.

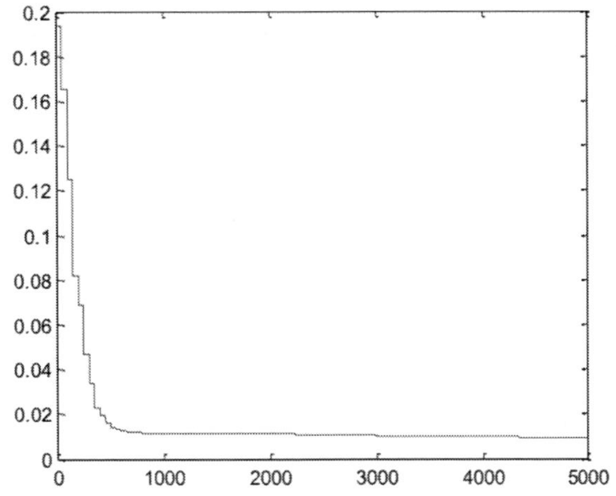

Fig.3. Convergence curve of optimization the desired droplet characteristics by artificial bee colony algorithm.

From Figure 3, it is obvious that the droplet characteristics (droplet velocity and droplet volume) gradually meet the desired values (V_{target} and S_{target}). The final optimized results is that the droplet velocity is 5.06 m/s, the droplet volume is 11.83 pL and the fitness value is 0.0087. The optimized results can satisfied the practical needs of 3D electronic printing.

5 Conclusion

In this paper, we proposed a new intelligent method to optimize the parameters of printhead driving waveform for getting the desired droplet characteristics. The optimization algorithm is a canonical artificial bee colony algorithm. Experimental results shown that the proposed adjustment method can quickly and accuracy find out the suitable combination of printhead driving waveform parameters. The method can effectively replace the traditional manual search process based on exhaustive experiments and a wise guess.

Acknowledgements

This research is partially supported by National Natural Science Foundation of China (Grand Nos. 51575158, and 61305028), and a self-planned task of CAS Key Laboratory of Networked Control System (Grand No. WLHKZ2015001).

References

1. M. Singh, H.M. Haverinen, Inkjet printing: Process and its applications. Advanced Materials, Vol.22 (2010), p. 673-685.
2. E. Macdonald, R. Salas, D. Espalin, M. Perez, E. Aguilera, D. Muse, R.B. Wicker, 3D printing for the rapid prototyping ofstructural electronics. IEEE Access, Vol. 2 (2014), p. 234-242.
3. H. Jaehyung, W. Alan, K. Antoine. Energetics of metal–organic interfaces: New experiments and assessment of the field, Materials Science and Engineering, Vol. 64(2009), p. 1-31.
4. J.C. Lioua, F. Tsenga. Multi-dimensional data registration CMOS/MEMS integrated inkjet printhead. Journal of Microelectro mechanical Systems, Vol.19, no.4 (2011), p.961-972.
5. J. Miettinen, K.Kaija, M. Mantysalo, P. Mansik kamaki, Molded substrates for inkjet printed modules. IEEE Transactions onComponents and Packaging Technologies, Vol.32 (2009), p. 293-301.
6. L. Sang, Y. Hong, F. Wang. Investigatin of viscosity effect ondroplet formation in T-shaped micro-channels by numerical andanalytical methods. Microfluidics and Nanofluidics, Vol.6 (2009), p. 6621-6635.
7. N.Reis, C. Ainsley, B. Derbyb, Ink-jet delivery of particlesuspensions by piezoelectric droplet ejectors. Journal of Applied Physics, Vol. 97 (2005), p. 094903-1-6.
8. K. Kwon, Waveform design methods for piezo inkjet dispensers based on measured meniscus motion. Journal of Microelectro mechanical Systems, Vol. 18, no.5 (2009), p. 1118-1125.
9. D. Karaboga, B. Basturk, On the performance of Artificial Bee Colony (ABC), Applied Soft Computing, Vol. 8, no.1 (2008), p. 687-697.

FEM-based Printhead Intelligent Adjusting Method for Printing Conduct Material

Xiaodan Liang[1a], Na Lin[2b], Hanning Chen[1c] and Wenxin Liu[3d]

[1]Tianjin Polytechnic University, Tianjin, 300387, China

[2]Beijing Shenzhou Aerospace Software Technology Co. Ltd., Beijing, China

[3] Department of Electrical and Computer Engineering, Lehigh University, Bethlehem, USA.

[a]lxdtjpu@163.com, [b]linn@htrdc.com, [c]perfect_chn@hotmail.com, [d]wliu@lehigh.edu

Abstract. Ink-jet printing circuit board has some advantage, such as non-contact manufacture, high manufacture accuracy, and low pollution and so on. In order to improve the and printing precision, the finite element technology is adopted to model the piezoelectric print heads, and a new bacteria foraging algorithm with a lifecycle strategy is proposed to optimize the parameters of driving waveforms for getting the desired droplet characteristics. Results of numerical simulation show such algorithm has a good performance. Additionally, the droplet jetting simulation results and measured results confirmed such method precisely gets the desired droplet characteristics.

Keywords. Finite Element Method (FEM), piezoelectric printhead, intelligent adjusting

1 Introduction

Ink-jet printing technology belongs to a kind of non-traditional machining process [1-2], which is also regarded as the increase-material manufacture, incremental, and rapid prototyping and so on. In recent 30 years, ink-jet printing technology is a collection of optical/mechanical/electrical, computer, numerical control and new materials in the integration of advanced manufacturing technology in global advanced manufacturing field. This technology deposits the powder, liquid, and flake on the substrate, which is different from the decimation method. This method greatly reduces the manufacturing complexity. Theoretically, as long as the design structure model on the computer, personnel can apply this technique to implement a fast manufacture without cutting tools, moulds and any complex design processes. The technique is particularly suitable for aerospace, weapons and equipment, in the field of biomedical, mold and other small batch and many asymmetric structure, surface and inner structure of empty parts (such as hollow aero engine blades, human bone restoration, conformal cooling channel) of rapid manufacturing, which conforms to the modern and the future development trend of manufacturing [3-4].

With the development of ink-jet printing technology, new problems also appear: (1) how to improve the efficiency of print? The efficiency of ink-jet printing technology manufacturing has a gap to the traditional manufacturing process (such as casting process) at mass production; (2) how to improve the stability of the printing? The stability of the ink jet printing technology is insufficient. In practice, it is difficult to guarantee the consistency of printing process, which causes printing product yield high enough; (3) how to improve the accuracy of print? Ink-jet printing precision greatly relies on the precision of the printhead. In terms of the present obtained industrial printhead, the adjustable range is very large, which has a lot of room to improve.

At present, in view of the above issues, in project, a kind of effective way is to optimize the choice of the working state of the printhead [5-8]: (1) improving print nozzle jet frequency and printing efficiency; (2) improving roundness of droplet ejected out of nozzle and enhancing the fault tolerance of printhead; (3) reducing the droplet volume ejected out of printhead and improving the print resolution. So it is need to control printhead to improve the print quality.

In this paper, we propose a new printhead intelligent adjusting method, which adopts the finite element method (FEM) to simulate the piezoelectric printhead jetting process. In order to improve the optimization efficiency, a new bacteria foraging algorithm with a lifecycle strategy is proposed to optimize the parameters of driving waveforms for getting the desired droplet characteristics.

The rest of this paper is organized as follows: In Section 2, the structure and working principle of piezoelectric printhead are introduced. The FEM-based intelligent adjustment system is described in Section 3. The bacteria foraging algorithm with a lifecycle strategy is proposed in Section 4. Section 5 demonstrates the simulation and experimental results. Finally, concluding remarks are collected in Section 6.

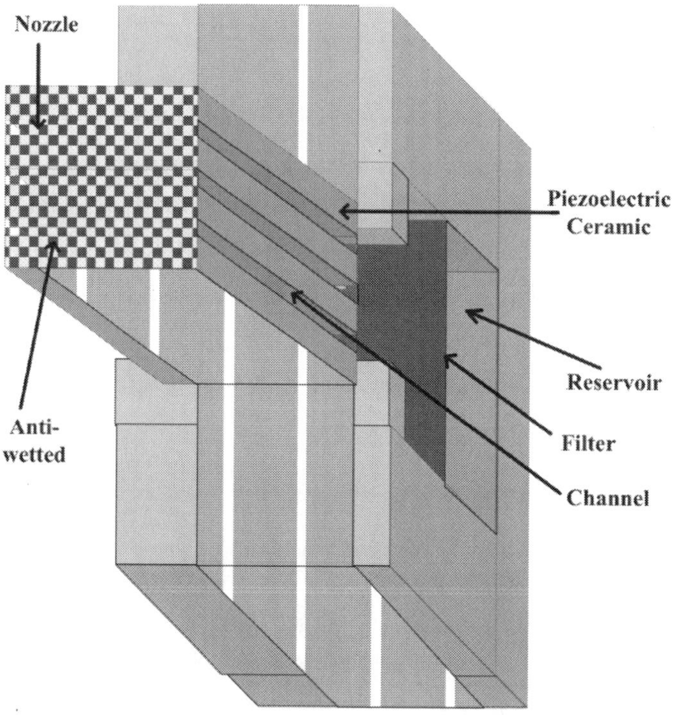

Fig. 1. Structure of Piezoelectric Printhead.

2 Piezoelectric Printhead

In this paper, the investigated printhead is KM series IJ Head of Konica. This is a typical piezoelectric (PZT) printhead. This kind of printhead has two arrays of nozzles in parallel. The structure of piezoelectric printhead is shown in Figure 1. Due to the need to spray printing ink mostly in high viscosity, it is necessary to heat the ink for decreasing ink viscosity. From Fig. 1, it is obvious that the nozzle has a reservoirwith a relatively large space; such design can effectiivly suppress the redundancy in ink-jet behavior after droplet jetting. The knitting filter can filter most of bubbles, particles pollution and large diameter particles to avoid them into ink channel, even to plug nozzle. Electric field on external electrode, making piezoelectric element deformation occurred. The change of the piezoelectric ceramic in ink channel generates pressure wave, and then pressure wave from the ink spreads to connect channel, through a narrow neck spreads to the spray hole, eventually forming droplets. Anti-wetted surface is to prevent the residual in the droplets on the surface of the nozzle clogging the nozzle.

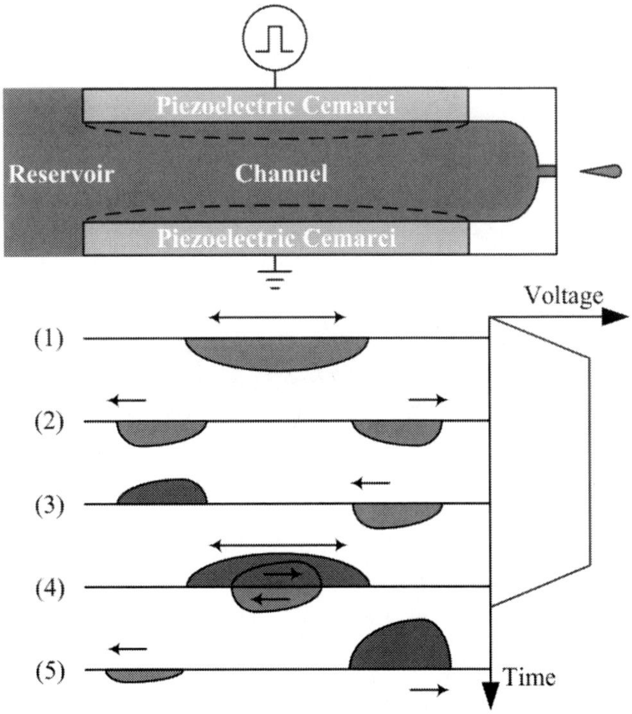

Fig. 2. Working principle of piezoelectric ink-jet printhead.

Figure 2 shows the cross-sectional schematic diagram of the ink channel. Ink channel length is greater than the diameter of the nozzle in the right hand side, and much smaller than the left side for the ink well geometry size. Figure 2 also shows a positive voltage applied to piezoelectric ceramics in trapezoidal wave when the pressure changes channel in the diagram.

3 FEM-based Intelligent Adjustment System

This research adopts the Finite Element Method (FEM) to simulate the working state of printheads. Finite element method is a kind of efficient and commonly used numerical method. Scientific computing, often need to solve all kinds of differential equation, and many general is hard to get the analytical solution of differential equation.Using the finite element method, the differential equation discretizes after programming with computer aided solution. At present, the widely used finite element analysis has been successfully applied to the fluid mechanics, electromagnetic mechanics, and structure mechanics calculation and so on. Using ANSYS and COMSOL finite element software for simulation, the preparatory design stage replaces the experimental testing for cost savings.

DOI: 10.1051/matecconf/201710003035

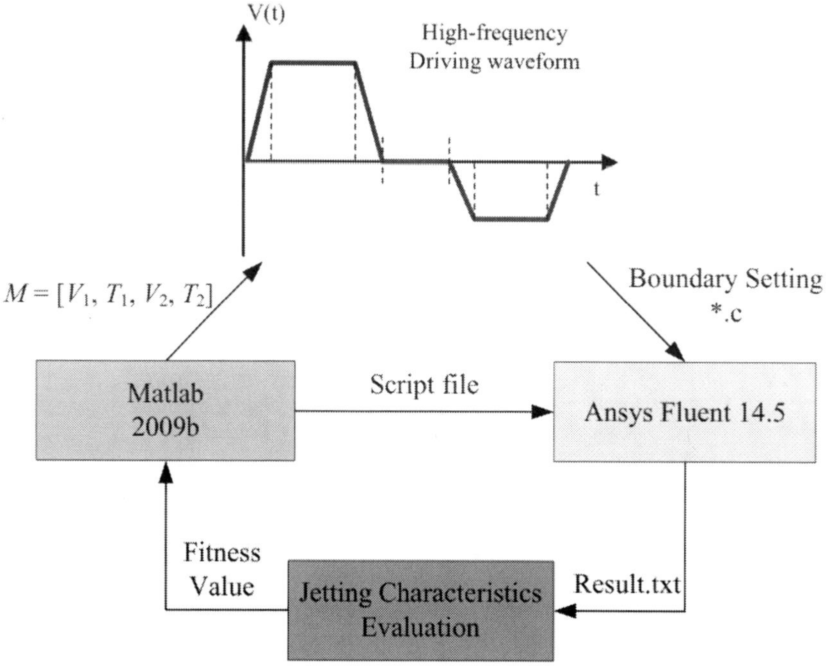

Fig.3. The simulation framework of jetting state.

This study sets a simulation framework of the printheand nozzle jetting state, as shown in Figure 3. Matlab script file generates the script file to drive FLUENT. The pressure generated by the intelligent optimization algorithm for FLUENT boundary configuration file is generated, too. The FLUENT software simulates the jetting process out of the nozzle according to the script files and configuration files, and generates the simulation results. According to preset evaluation standard, the fitness value is returned to the Matlab. According to the return of fitness value and the current waveform of driving parameter vector, intelligent algorithm generates the next generation of individuals, which drives the next cycle until the termination conditions meeting.

4 Bacteria Foraging Algorithm with a Lifecycle Strategy

The E. coli bacteria is one of the earliest bacteria which has been researched.Passino proposed the Bacterial ForagingOptimization (BFO) algorithm. The main mechanisms of BFO are illustrated as three steps: Chemotaxis, Reproduction and Eliminate-Dispersal. However, the original BFO has some shortages: (1) dispersal, reproduction, andelimination each happens; after a certain amount of chemotaxis operations. The appropriatetime andmethod for dispersal and reproduction are important. Otherwise, the stability of thepopulation may be destroyed. (2) The tumble angles in the chemotactic phase are generatedrandomly. As a result, the algorithm is more like a random searching algorithm except it willtry further in better directions. The bacteria swarm lacks interaction between individuals. Good information carried by those individuals in higher nutritional are as cannot be sharedwith and utilized by other bacteria. (3) The swim step length in the original BFO algorithm isa constant. In most cases, the bacterium will run one more step if the position is better than itslast position. If the swim step is large at the end stage (e.g., larger than the distance betweenits current position and the optimal point), it will skip the optimal point repeatedly. This will make the bacteria hard to converge to the optimal point.

To improve the optimization ability of BFO algorithms, for each bacterium, we will decide it to reproduce, die, or migrate by certainconditions in the bacteria's lifecycle.

We define the nutrient updating formula as Eq. 1. *Flast* represents the fitness ofthe bacterium's last position (for a minimum problem, fitness is larger when the functionvalue is smaller). In initialization stage, nutrients of all bacteria are zero.

$$N(i)=\begin{cases}N(i)+1 & if(F(i)>Flast)\\ N(i)-1 & if(F(i)<Flast)\end{cases} \quad (1)$$

In initialization stage, nutrients of all bacteria are zero. In the bacterium'schemotactic processes, if the new position is better than the last one, it is regarded that the bacterium will gain nutrient from the environment and the nutrient is added by one. Otherwise, it loses nutrient in the searching process and its nutrient is reduced by one.

5 Results

Lithium titanate solution (nanometer lithium titanate 35 g + alcohol 750 ml + ethylene glycol 250 ml) is choosen as test ink, its physical properties as shown in Table 1.

Table 1. Physical characteristic of test ink.

	Parameter	Unit	Value
Ink	Free electrical capacitance, σ	*N/m*	$45*10^{-3}$
	Viscosity,*u*	*Kg/(m*s)*	15
	Density, ρ_0	*Kg/m³*	$1.05*10^{3}$

At 50 ms after ejected droplet size, the speed of the front of droplet and the tail length as print nozzle are selected as the jet evaluationstates, and the evaluation function is as follows:

$$F=m_1*\frac{|V_S-V_G|}{V_G}+m_2*\frac{|S_S-S_G|}{V_G}+m_3*\frac{L}{D_{Nozzle}} \quad (2)$$

where V_G is the desired droplet velocity, V_S is the optimized droplet velocity, S_G is the desired droplet volume, S_S is the optimized droplet volume, L is the tail length, D_{nozzle} is the diameter of nozzle and m_1, m_2 and m_3 are the user-defined weights.

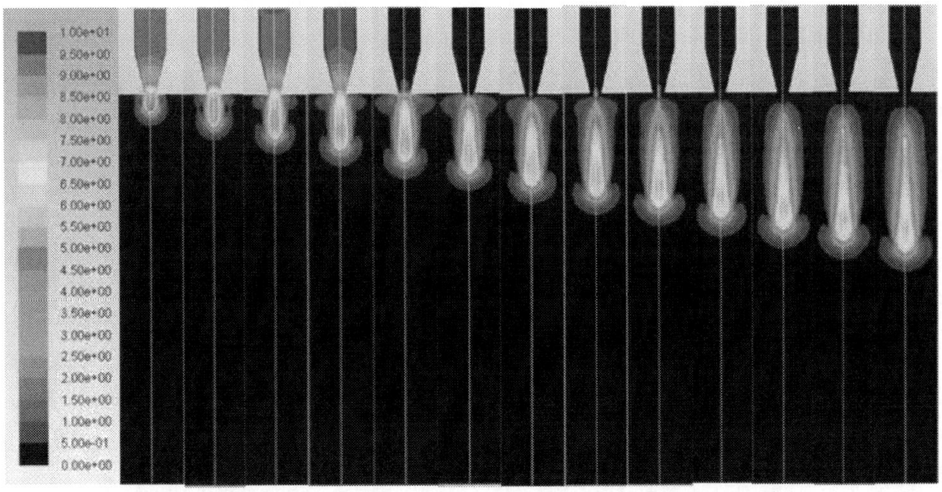

Fig. 4. Jetting characteristics optimized by the improved BFO

Through the simulation results, as shown in Fig. 4, it is obvious that LBCFO algorithm simulation waveform parameters get expected droplet size and velocity, minimum error, the much excellent results. LBCFO algorithm is very suitable for optimizing the piezoelectric printhead for printing conduct material.

6 Conclusion

On account of the research about information communication mode and life period searching approach, we propose a new algorithm: Life-cycle Bacterial Colony Foraging Optimization. In order to verify the performance of the proposed new type of bacterial swarm foraging algorithm for complex engineering optimization problems, the experiments of inkjet printing system in intelligent manufacturing field have been tested. The mode of printhead is structured in ANSYS software. The proposed algorithm got a small error towards target droplet velocity and volume.

Acknowledgement

This research is partially supported by National Natural Science Foundation of China (Grand Nos. 51575158, and 61305082), and a self-planned task of CAS Key Laboratory of Networked Control System (Grand No. WLHKZ2015001).

References

1. E. Macdonald, R. Salas, D. Espalin, M. Perez, E. Aguilera, D. Muse and R.B. Wicker, 3D Printing for the Rapid Prototyping of Structural Electronics, IEEE Access, **2** (2014), p. 234-242.
2. Information on https://en.wikipedia.org/wiki/3D_printing
3. G.K. Jan, J. Patrickand D.Y. Shin, in: *Inkjet-based Micro manufacturing*, Wiley (2012).
4. Information on https://en.wikipedia.org/wiki/Chuck_Hull
5. J. Sun, J, R Yang, K K Tan, J Y H Fuh and Y S Wong, Performance characterization of drop-on-demand micro-dispensing system with multi-printheads, Microsystem Technologies, , Vol. 16, no. 12 (2010), p. 2087-2097
6. A.K. Amol, X. Bombois, G. Scorletti, R. Babuška, S. Koekebakker, and W. Zeeuw, A Waveform Design Method for a PiezoInkjetPrinthead Based on Robust Feedforward Control, Journal of Microelectromechanical Systems, **21**, no. 6) (2012), p. 1365-1374.
7. A.L. Yarin, Drop Impact Dynamics: Splashing,Spreading, Receding, Bouncing, Annu. Rev. Fluid Mech, **38** (2006), p.159–92
8. B. Derby, Inkjet printing ceramics: From drops to solid, Journal of the European Ceramic Society, **31** (2010), p. 2543–2550.

An Ultraviolet-Visible (UV) Photometry System Based on the PDMS-based Microfluidic Chip

Changhua Xiang [1,a], Ning Yang [1,2,b], Rongbiao Zhang [1,c], Jianjiang Guo [1,d] and Hu Huang [3,e]

[1] School of Electrical and Information Engineering, Jiangsu University, Zhenjiang, PR China
[2] Institute of Agricultural Engineering, Jiangsu University, Zhenjiang, PR China
[3] Department of Biomedical Engineering, University of Minnesota, Minneapolis, United States
[a]xiangchanghua123@163.com, [b]yangning7410@163.com, [c]zrb@ujs.edu.cn, [d]jjguo18@foxmail.com,
[e]huhuang:hwangtiger@gmail.com

Abstract. In order to avoid a problem remains with the low accuracy and poor portability of the photometry system, the system based on the photometry method and microfluidic chip technology was built. As the characteristics of cheap, solid and good transmission, Polydimethylsiloxane (PDMS) was chosen as the material of the designed chip in the paper. To the designed UV photometry system, the light-emitting diode with the wavelength of 580m is chosen as the light source. The experimental result indicates that there is no significant deviation between the designed UV photometry system and the conventional immuneturbidimetric assay, the correlation coefficient is 0.95 obtained by adopting the linear regression analysis. The linearity of the designed UV photometry system based on the PDMS-based microfluidic chip has increased by 17.3% in comparison with the system based on the silicon-based microfluidic chip.
Keywords. Polydimethylsiloxane (PDMS), photometry, microfluidic chip, agglutination

1 Introduction

The photometry method is a qualitative and quantitative detection technology to substances, which adopts absorption spectra and emission spectra that are peculiar to substances [1]. It has some advantages (i.e., such as simple instruments, easy operation and rapid detection, etc.), which has a wide range of applications in science, engineering, agriculture, medicine and so on. The microfluidic chip technology is a platform that integrates one or several laboratory functions on a single chip of only millimeters to a few square centimeters to achieve automation and high-throughput screening [2]. The combination of the microfluidic chip technology and the photometry method has widely been applied in science, engineering, agriculture, medicine. Pinto etc. collected the samples of the natural water within 1 hour to detect the content of copper in water on the platform of the microfluidic chip by using the photometry method. The detection limit could reach up to 4.6μg/L and the average error is 5.32% [3]. Compare with other detection methods, photometry method based on the microfluidic chip technology can directly detect the samples without adding additional reagents, and it has the characteristics of high feasibility and simple operation. So the UV photometry system based on the microfluidic chip is built in the paper. However, the downside of the system is low metering accuracy. So it is necessary to build a portable photometric system with high precision.

For the designed microfluidic chip, there are many kinds of materials to fabricate the microfluidic chip including silicon, paper and PDMS. The process to prepare the microfluidic chip by the silicon is the ripest one [4]. However, the cost of silicon is relatively high which limits the application in the fabrication and application of microfluidic chip [5]. Paper is relatively cheap material compare to the silicon. However, paper is fragile which also limits the application too [6]. PDMS ($(C_2H_6OSi)_n$) has appeared in the last few years [7]. As the characteristics of cheap, solid and good transmission, PDMS is chosen as the material of the designed chip in the paper. Based on the designed chip, a portable photometric system with high precision is built.

2 The material of chip and corresponding fabrication method

There are many kinds of materials to fabricate the microfluidic chip like silicon, paper and PDMS. The silicon is the earliest in fabricating the microfluidic chip as the productions process of silicon is the most mature compare to paper and PDMS. However, there are also some disadvantages to the silicon for example the cost of silicon is relatively high which limits its application in lab-on-a-chip fabrication. Paper is relatively cheap compare to the silicon. However, the paper has a serious unfairness--the paper is fragile, which also limits the application in the field of the chip's fabrication. PDMS has appeared in the last few years, it has widely applied in fabricating the microfluidic chip as it has the characteristics of cheap, solid and good transmission. So, PDMS is regarded as the material to fabricate the microfluidic chip in the paper. Table 1 is the characters of three different materials.

Table 1. The characters of three different materials.

Character / Material	Cost	Transparency	Fragile article
Silicon	High	Bad	Yes
Paper	Low	Bad	Yes
PDMS	Low	Good	No

Considering the experimental condition and fabricating cost, the imprinting method is used to fabricate the microfluided chip, the fabrication processes are shown in Figure 1. During the processes, the imprinting method also needs a mould that should be made previously, such as silicon or quartz mould. The substrate and mould are pressed by the stamp technique. In other words, substrate and mould are bonded together by using appropriate force at an appropriate time. Then, the desired shape has been formed on the substrate. The final step is linkage. To protect substrate, the package is needed after all steps are finished. With the help of Suzhou Daya Electronics Co., LTD, the PDMS-based microfluidic chip is fabricated successfully.

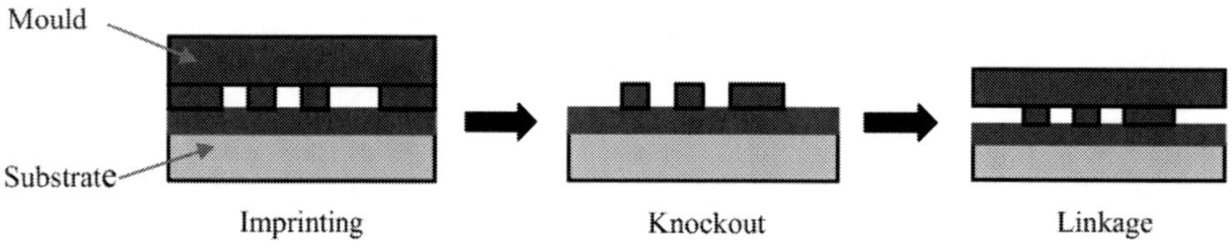

Fig.1. The production process of microfluidic chip.

3 The comprehensive system design

Design of the microfluidic chip. Figure 1 is the production process of the designed microfluidic chip. Figure 2 (a) is top view of the designed microfluidic chip and Fig.2 (b) is elevation view of the designed microfluidic chip. In the Figure 2 (a), *a* is the Y-shaped sampling zone, *b* is the mixing zone, *c* is the reaction zone and *d* is the photoelectric detection zone. The reagents are placed in the injection pumps to prepare for mixing and detecting. Then, reagents are injected into the microfluidic chip from the sampling zone *a* under the injection pumps. As reagents flow into the mixing *b*, the adequate mixing happens in the mixing zone *b*. In the paper, there is an immunization agglutination reaction between the sensitized latex reagent and the corresponding serum in the reaction zone *c*. Some time is needed before the chemical reaction has set off sufficiently. In the experiment, the visible agglutinators begin to appear as the chemical reaction goes on, and the corresponding photometric detection is needed at the photoelectric detection zone *d* to determine the extent of immunization agglutination reaction. The designed microfluidic chip is transparent as the material of the designed microfluidic chip is PDMS, so it is possible to determine extent of immunization agglutination reaction at the photoelectric detection zone *d* by using the transmission characteristic of light.

To the mixing zone *b*, the structure of inner rib is designed to achieve the intensive mixing between different liquids. Supposing that the titer of two different liquids is 1 and 0 respectively, the effect of reagents mixing could be simulated by the simulation software of COMSOL Multiphysics 5.1. The simulations reveal that the designed structure has a good mixing effect and the titer is close to 0.5 at the outlet. So, the intensive mixing could be achieved when liquids flow past the mixing zone *b*.

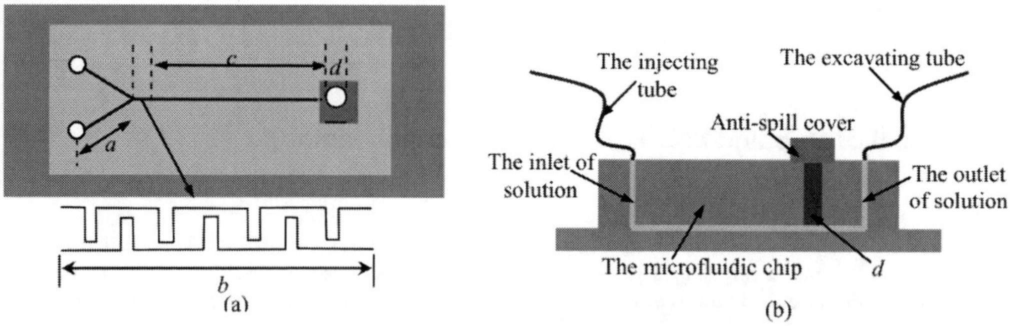

Fig.2. The schematic diagram of designed microfluidic chip.

(a) Top view of the designed microfluidic chip. (b) Elevation view of the designed microfluidic chip.

Optimal absorbance wavelength of the immune agglutination. The optimal wavelength is the wavelength that can reflects the optimal relationship between the solution concentration and light absorption, which is also called the resonant frequency light-wave. During the experiment, the light-emitting devices connect with the precision source—ocean optics HL-2000-HP-FHSA, the detection devices connect with ocean optics USB2000 spectrograph. The rheumatoid factor positive serums, with the titers of 20 IU/mL, 30 IU/mL, 40 IU/mL, 50 IU/mL, and 60 IU/mL, will have a reaction with the sensitized latex reagent respectively.

Agglutination reactions produce visible aggregates of antibody-antigen complexes when antibodies or antigens are conjugated to a carrier. Carriers used in agglutination methods could be artificial or biological. Table 2 is the absorption value of the designed UV photometry system varies according to the wavelength range from 500 nm to 600nm.

Table 2. The absorption value of the designed system varies according to the wavelength.

Absorbance (Abs) / Wavelength (nm) / Titer (IU/mL)	500	510	520	530	540	550	560	570	580	590	600
20	0.12	0.13	0.135	0.139	0.142	0.150	0.156	0.159	0.160	0.154	0.150
30	0.18	0.193	0.201	0.209	0.214	0.225	0.23	0.237	0.240	0.228	0.220
40	0.27	0.28	0.288	0.297	0.300	0.305	0.308	0.312	0.311	0.302	0.290
50	0.300	0.312	0.320	0.330	0.339	0.350	0.360	0.369	0.377	0.365	0.350
60	0.360	0.423	0.420	0.432	0.434	0.450	0.462	0.475	0.48	0.468	0.450

From the table 2, the result has revealed that the absorbance values increases with the detection wavelength between 500 nm and 580nm. The absorbance values decreases with the detection wavelength when the wavelength is greater than 580nm. So, the wavelength of 580nm is the optimal absorbance wavelength. When the detection wavelength is 580nm, the absorbance reaches maximum no matter what the titers of the positive serums are.

Design of detection circuit. The detection circuit is built by the designed microfluidic chip, the light-emitting diode and the photodiode. To enhance the sensitivity and accuracy of detection circuit, the detection circuit is built in a vertical way which is shown in Figure 3.

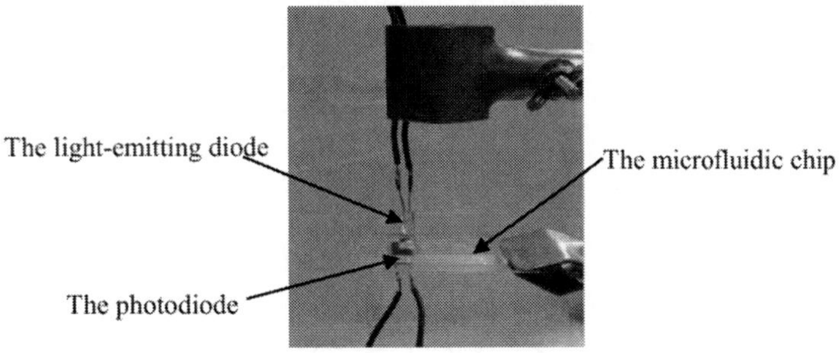

Fig.3. The physical picture of optical path detection circuit.

During the detecting, the light-emitting diode sends out the light of wavelength 580m, and the photodiode absorbs the light emitted by light-emitting diode. There is a lid over the photoelectric detection zone *c* for preventing the overflow of liquids, and the glass base under the chip is regard as base. The surface of PDMS substrate, anti-spill lid and glass base are dealt with by plasma oxidation method in advance. Then the PDMS substrate, anti-spill lid and glass base are recombined to enable the seam to be more solid and stable. As the thickness of chip is 5 mm, so the optical path length of 5 mm could be obtained.

Overall Experimental Platform Construction. Figure 4 is a picture of the designed photometry system. To ensure sufficient reaction of the agglutination reaction, the indoor temperature should be kept at 37℃. The microfluidic chip is placed between the light-emitting diode and the photodiode. The microfluidic chip connects with the sampling-drive pump (the velocity of pump is 0.098μm/step, and the accuracy of CVs is less than 1%) through the pipelines. The microfluidic chip is placed between the light-emitting diode and the photodiode. The microfluidic chip connects with the sampling-drive pump (the velocity of pump is 0.098μm/step, and the accuracy of CVs is less than 1%) through the pipelines.

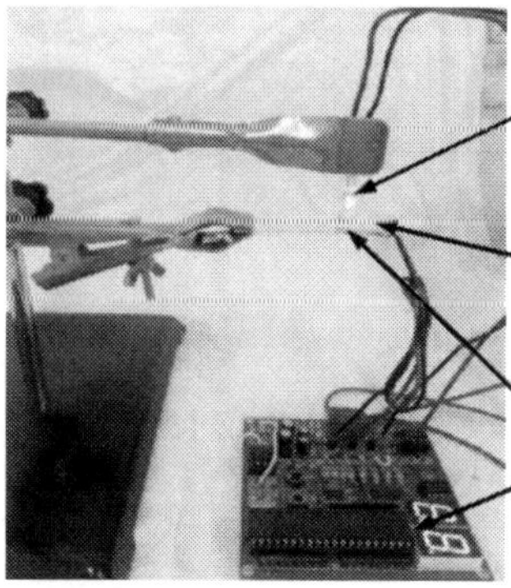

Fig.4. The designed UV photometry detection system based on microfluidic chip.

The sensitized latex reagent (pure and surface inactive polystyrene latex particles with the diameter of 50 nm, and it is provided by Shanghai Jiemen Bio-tech co.) is injected into one syringe, the positive serums sample with different titers of the rheumatoid factor is injected into the other syringe. The detection circuit connects to -10V and +5V power supplies, the designed system adopts the real time controlling MCS-51single-chip microprocessor as centre controller, and the result of absorbance detection could be displayed by the LED numerical code tube.

4 Experimental results

In the experiment, the wavelength of 580nm is chosen as the optimal wavelength, PDMS is chosen as the material of the chip and the microfluidic chip is fabricated by adopting the imprinting method. The PH value of the sensitized latex reagent is 7.5, the velocity of 0.08 mm/s is chosen as the injection rate and the experimental temperature of the photometry system is 37 ℃. After the basic parameters have been determined, the photometry detection method and the precise instrument are carried out. The titers of positive serums are 20 IU/mL, 25 IU/mL, 30 IU/mL, 35 IU/mL, 40 IU/mL, 45 IU/mL, 50 IU/mL, 55 IU/mL, and 60 IU/mL. From the Figure 5, we may draw a conclusion that the correlation coefficient is 0.95 obtained by adopting the linear regression analysis. So, there is no significant deviation between the designed UV photometry system and the conventional immuneturbidimetric assay. The linearity of photometry system based on the silicon-based microfluidic chip is 0.81 [8], so the linearity of photometry system based on the PDMS-based microfluidic chip has increased by 17.3% in comparison with the system based on the silicon-based microfluidic chip.

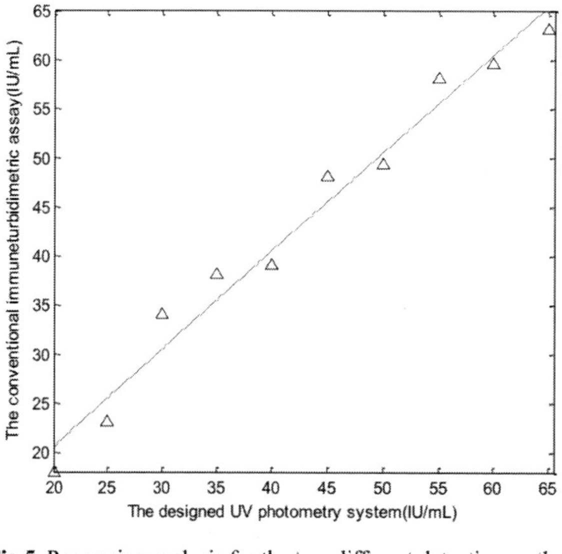

Fig.5. Regression analysis for the two different detection methods.

5 Conclusions

In the paper, a portable UV photometry system based on microfluidic chip is designed and the material of microfluidic chip is PDMS. The photometry system is a combination between the microfluidic chip technology and photometry detection technology. The microfluidic chip is fabricated by the imprinting method and the light-emitting diode which sends out

580m light is selected as one part of the detection circuit. The designed UV photometry system based on the microfluidic chip is built with the light-emitting diode, the photodiode, the detection circuit and other devices. Finally, the correlative analysis experiment between the photometry detection method and conventional precise instrument is carried out. The experimental result indicates that there is no significant deviation for the designed photometry system compares with the precise instrument, the correlation coefficient is 0.95 obtained by adopting the linear regression analysis. The linearity of the designed UV photometry system based on the PDMS-based microfluidic chip has increased by 17.3% in comparison with the system based on the silicon-based microfluidic chip. So, PDMS is a better material for fabricating the microfluidic chip and the designed UV photometry system has the advantage of portability.

Acknowledgements

Supported by the Priority Academic Program Development of Jiangsu Higher Education Institutions(PAPD), Natural Science Foundation of Jiangsu Province(BK20140550), China Postdoctoral Science Foundation (2015T80512), Changzhou Sci&Tech Program(CE20155054)and Research and Development Plan of Zhenjiang to Modern Agriculture(NY2015026).

References

1. M.I. Leong, S.D. Huang. Dispersive liquid-liquid microextraction method based on solidification of floating organic drop combined with gas chromatography with electron-capture or mass spectrometry detection. J. Chromatogr. A, Vol. 1211(2008), p. 8-12.
2. G.M. Whitesides. The origins and the future of microfluidics. Nature, Vol. 442(2006), p. 368-373.
3. J.J. Pinto, C. Moreno, M. García-Vargas. A very sensitive flow system for the direct determination of copper in natural waters based on spectrophotometric detection. Talanta, Vol. 64(2004), p.562-565.
4. R. Kangning, Z. Jianhua, W. Hongkai. Materials for Microfluidic Chip Fabrication. Accounts Chem. Res., Vol. 11(2013), p.2396-2406.
5. A. Mata, A.J. Fleischman, S. Roy. Characterization of Polydimethylsiloxane (PDMS) Properties for Biomedical Micro/Nanosystems. Biomed. Microdevice, Vol. 4(2006), P.281-293.
6. Q.H. He, C.C. Ma, X.Q. Hu, H.W. Chen. Method for fabrication of paper-based microfluidic devices by alkylsilane self-assembling and UV/O_3-patterning. Anal. Chem., Vol. 85(2013), P.1327- 1331.
7. M.A. Eddings, M.A. Johnson, B.K. Gale. Determining the optimal PDMS–PDMS bonding technique for microfluidic devices. J. Micromech. Microeng., Vol. 18(2008), P.1171-1185.
8. J.W. Lee, R. Investigation of thermal vibration correlation of silicon lattice atoms by ion scattering. Nucl. Instrum. Meth. B, Vol. 232(2005), p.285-289.

Electrokinetically Control Flow Mixing in Microfluidic Chip Utilizing Microelectrode Based on Indium Tin Oxide

Tianyuan Lv[1,a], Ning Yang [1,2,b], Zhiyu Zuo[1,c], Jianjiang Guo[2,d] and Hu Huang[3,e]

[1] Institute of Agricultural Engineering, Jiangsu University, Zhenjiang, PR China

[2] School of Electrical and Information Engineering, Jiangsu University, Zhenjiang, PR China

[3] Department of Biomedical Engineering, University of Minnesota, Minneapolis, United States

[a]18796086895@163.com, [b]yangning7410@163.com, [c] zuozy@ujs.edu.cn, [d] jjguo18@foxmail.com [e] huhuang:hwangtiger@gmail.com

Abstract. In order to blend the liquid in the microfluidic chip adequately, this paper presented a novel mixing scheme in which the species streams was mixed via the application of Lorenz chaotic system algorithm to three pair of parallel electrodes mounted on the surfaces of the mixing chamber to produces chaotic electric fields. To select the better microelectrode arrays, the paper also discussed the advantages of the indium tin oxide (ITO) materials and used ITO to design the microelectrode array. Finally, simulations of the mixing process under the control of the proposed chaotic electric fields are performed. Analysis result shows that the correlation coefficient between the largest Lyapunov exponents on Lorenz chaotic system and chaotic flow behavior is 0.72 which shows a better synchronism, and it also proved that ITO is a better material to fabricate the microelectrode arrays.
Keywords. Microfluidic chip, fluid chaotic mixing, chaotic electric fields, indium tin oxide.

1 Introduction

Efficient and rapid fluid mixing is a critical component in microfluidic applications because the mixing effect determines the speed and quality of chemical reaction. However, due to the small size of microfluidic devices, fluid flow is constrained to the low Reynolds number regime, and hence species mixing occurs primarily as a result of diffusion which always makes the required mixing length very long to ensure satisfactory mixing results. In consequence, a requirement exists for improved mixing schemes capable of accelerating the species mixing effect [1].

Kinds of micromixers called passive micromixers are used to generate chaotic advection effect by using specific channel geometry configurations to split or stretch the fluid. Some designs were investigated such as the 3D serpentine micromixer, the zigzag micromixer [1], the rapid three-dimensional passive rotation micromixer, and the staggered herringbone micromixer [2]. In contrast to passive micromixers, active micromixers use the external force to split or stretch the fluid. When the size of channel is reduced to the microscale, the ratio of surface to volume ratio becomes large [3]. Thus, forces such as the surface tension and electrokinetic force become the dominated force instead of the volume forces. This enables new techniques to develop active micromixers. Some external forces are due to ultrasonic vibration, magnetic force and electroosmosis [4]. Oddy et al. [5], developed an electrokinetic instability micromixer in which a rapid stirring effect was obtained via the application of external sinusoidally alternating electric fields. Biddiss et al. [6-7], presented a rapid mixer which mixes the fluids by controlling the distribution of the non-uniform zeta potential to generate vortices on the surface.

Chaos synchronization is an evaluation index to judge the effect of chaos control [8]. Therefore, the chaos synchronization also can be used to judge control effect of electric fields on microfluid in the microfluidic chip. This paper presents a novel microfluidic mixing scheme in which the species streams are mixed via the application of Lorenz chaotic system algorithm to three pairs of electrodes which is designed by ITO.

2 Materials manufacturing methods and physical models

The principle of ITO microelectrode array. The ITO films (In_2O_3) is transparent in the visible range (400~800nm), while the transmittance is above 90% and the reflectance in infrared region is more than 85%. Such a high visible light transmittance and low resistivity makes the ITO films become the typical electrical properties of transparent conductive. The electrical characteristics of ITO films are measured by the sheet resistance and the thickness of R. Sn and the formation of oxygen vacancy makes the carrier concentration of ITO films become very high(~$10^{20}cm^3$), while its resistivity is relatively low(~$10^4 \Omega$/cm). So it forms a highly degenerate N type semiconductor, which shows a kind of metal. For degenerate semiconductor, the carrier concentration basically remain static with the change of temperature, and the electrical properties of materials mainly depends on the migration rate. Therefore, the above characteristics show that the choice of ITO as the electrode material is more suitable than the metal electrode.

The manufacturing of ITO microelectrode array. Physical models Figure 1a illustrate the whole geometric configuration of the microfluidic chip. A Y-type channel is used to allow two different fluids to be injected from the two side inlets A, B and mixed in the main mixing chamber, which contains three pair of parallel electrodes mounted on the surfaces.

Microfluidic chip is produced by polydimethylsiloxane (PDMS) material. The indium tin oxide (ITO) conductive film is chosen as substrate. Ultrasonic cleaning is carried out by the deionized water of methylbenzene, acetone and alcohol. The water is baked in the temperature of 120℃. The negative photoresist ITO array electrode pattern can be obtained by 4 steps: whirl coating, prebake, exposure and developing. And then, the microelectrode array (20μm×20μm) is fabricated by ways of hardbaking, corrosion, removing of photoresist, scouring, stoving and so on. Figure 1b is the photo of microelectrode array. The overall microfluidic system can be obtained by the way of plasma bonding which is between ITO substrate with a microelectrode array and PDMS chip.

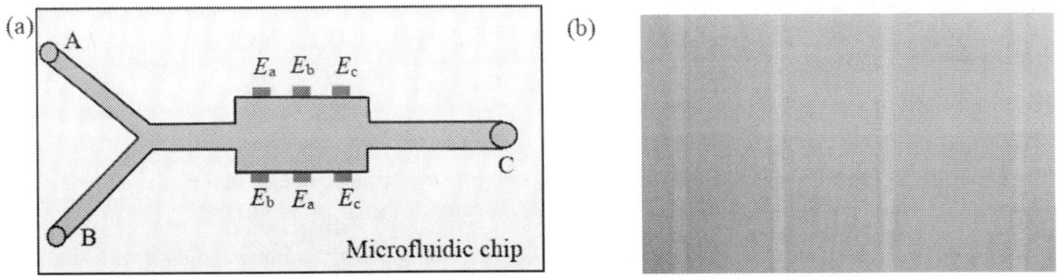

Fig. 1. The Y-shaped channel microfluidic chip chart.

(a) The whole geometric configuration of the microfluidic chip. (b) The Photo of microelectrode array.

TablesChaotic system algorithm. The object of this study is the chaotic mixing effect of mircrofluid perturbed by application of chaotic electrical fields provided by chaotic system algorithm. Therefore, in the eyes of Chaos anti-control theory, chaotic system can be regarded as control system and microfluid system can be regarded as controlled system. The model can be used to describe the chaotic phenomenon of fluid. Accordingly, we chose Lorenz chaotic system as the control system shown in the follow equations:

$$\frac{dx}{dt} = -c(x-y); \frac{dy}{dt} = ax - y; \frac{dz}{dt} = xy - bz \quad (1)$$

Where x, y and z are the oscillation signals applied to the three pair of parallel electrodes mounted on the surfaces of the mixing chamber; b and c are dimensionless parameters. a is the characteristic parameter of the Lorenz chaotic system. To ensure the Lorenz chaotic system exhibits chaotic behavior, the initial values are set at $x(0)=y(0)=z(0)=1$, with the parametric values c=10, a=28 and b=8/3. Under these settings, the signal x, $-y$ and z-a are applied to the electrodes E_a, E_b, and E_c which are shown in Figure1a. The phase trajectory of the control variables x, -y and z-a varying with time is shown in Figure 2a. Figure 2b shows the oscillatory trajectories of the control variables variables x, -y and z-a varying with time. The electric potentials φ_E of the three pair of parallel electrodes are βx, $-\beta y$ and $\beta(z\text{-a})$. Where β is the amplification factor, In this study, the amplification factor β=1.2. Therefore, the oscillatory intensityβx, $-\beta y$ and $\beta(z\text{-a})$ applied at electrodes E_a, E_b and E_c ranges between -28.9 and 26.2 V.

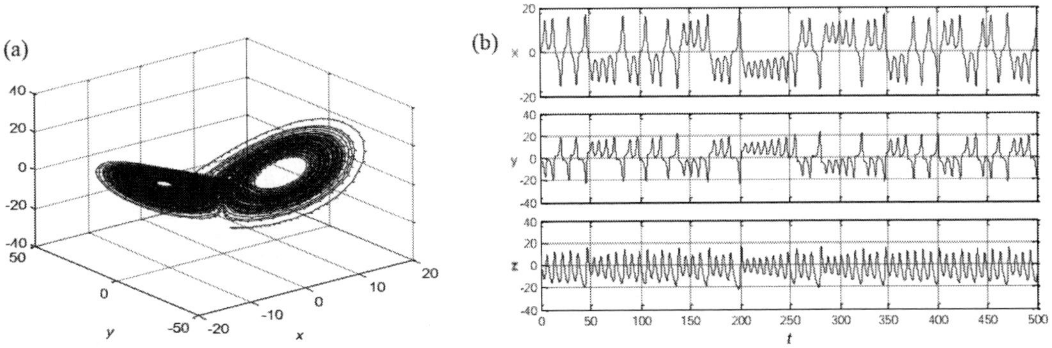

Fig. 2. Trajectories in the Lorenz system.

3 Results and discussion

COMSOL Multiphysics 4.3 was used to investigate the effect of the chaotic electric fields on the fluid flow characteristics within the mixing chamber. Figure 3 shows the initial distribution of the species concentration prior to the application of the chaotic electric fields. It is apparent that the two streams are essentially laminar flows which can be regarded as ordered microfluid system. Consequently, species mixing takes place primarily as a result of diffusion and initial mixing performance is poor.

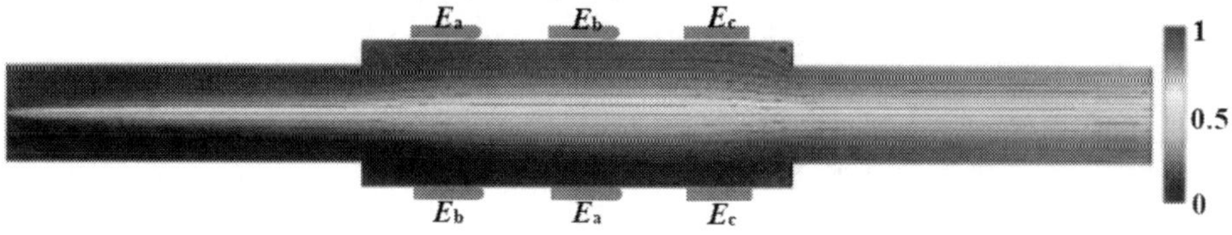

Fig. 3. The initial distribution of the species concentration prior to the application of the chaotic electric fields.

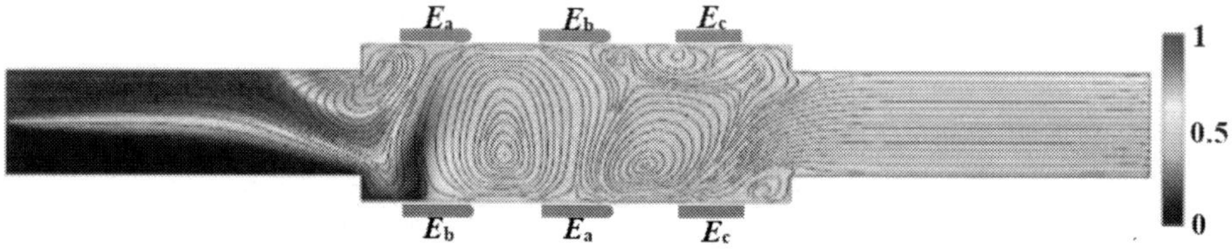

Fig. 4. The concentration and flow streamlines distribution within the mixing chamber with application of chaotic electrical field provided by Lorenz chaotic system at a=28 and t=300ms.

Figure 4 illustrates the concentration and flow streamlines distribution within the mixing chamber with application of chaotic electrical field provided by Lorenz chaotic system at a=28 and t=300ms. As shown, at this particular time instant, the electric potentials φ_E at electrodes E_a, E_b and E_c have a value of -15.2, 27.3 and 6.8 respectively. These circulations full of the mixing chamber make the microfluid generate a chaotic mixing effect.

The chaotic evaluation of control system. Since the control system is the known Lorenz chaotic system, we obtained the accurate Lyapunov exponent spectrum by using numerical iteration method to compute the differential equation. On the whole, we solved the approximate solution of the ordinary differential equations firstly. Sencondly, orthogonalization and reiteration were carried out to obtain the Lyapunov exponent at each time point.

We adjust the characteristic parameter "a" of the Lorenz chaotic system from 0 to 270. Figure 5 shows the largest lyapunov exponents vary with the characteristic parameter. When a<10, the largest lyapunov exponent is a negativevalue, which indicates that the Lorenz system is cyclical change. While, when a=32, the largest lyapunov exponent reaches a maximum, which indicates that the intensity of chaos reaches the maximum value. While, as a>200, the largest lyapunov exponent reduces to 0, which indicates that the Lorenz system return to cyclical change.

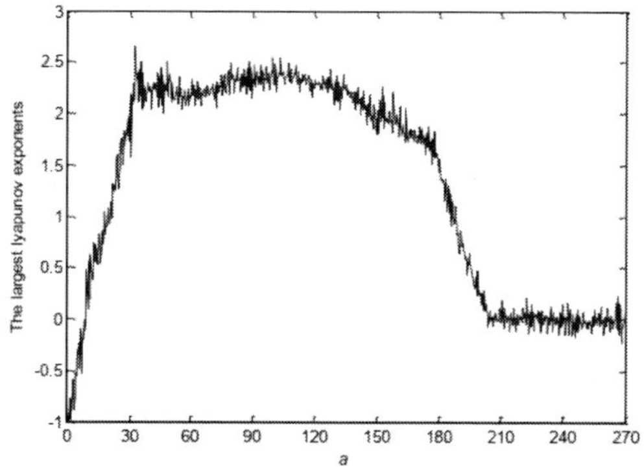

Fig. 5. The largest lyapunov exponents vary with the characteristic parameter.

The chaotic evaluation of controlled system. Because the controlled system is a microfluid system, the construction of the dynamic system by equation is difficult. Therefore, we have to reconstruct the series time of variables in gamma space and calculate the intensity of chaos by particle tracing method. The Particle Tracing Module in COMSOL Multiphysics version 4.3 provide us an environment for computing the trajectory of particles in the microfluidic mixer. Fig.6 shows the particle tracing at 300ms in the mixing chamber. Wolf et al., present algorithms that allow the estimation of Lyapunov exponents from a time series of particle tracing. Based on the wolf method, we can judge the state of system by largest lyapunov exponent (λ_1), second-largest lyapunov exponent (λ_2) and third-largest lyapunov exponent (λ_3). Though Wolf method provides the algorithms to estimate the Lyapunov exponents of system, the computation complexity is high and the result is susceptible to parameters. In fact, as long as the positive Lyapunov exponent exists, the microfuild system is a chaotic system this time. Therefore, there is no need to estimate the Lyapunov exponent spectrum from time series, just estimating largest lyapunov exponent is enough to judge the state of system.

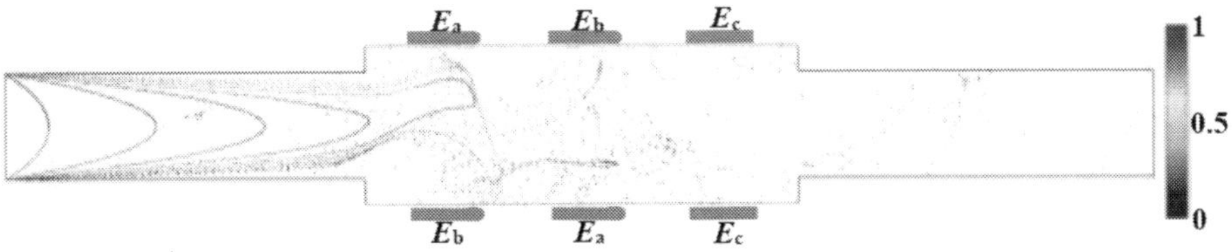

Fig. 6. The particle tracing at 300ms in the mixing chamber.

Chaotic synchronization. From the variation of the largest Lyapunov exponents in Figure 5 and Figure 6, It may be observed that the microfluidic system is synchronous with the Lorenz system. To figure this out, the relationship of the largest Lyapunov exponents between Lorenz system and microfluidic system are analyzed with regression analysis. Figure.7a shows the regression curve of the largest Lyapunov exponents between the system and controlled system. We commonly use the cross-correlation coefficient (also called the correlation coefficient) to describe the linear correlation between the data. R represents the correlation coefficient. The relationship between $|R|$ and correlation degree are shown in table 1.

The correlation coefficient R of the largest Lyapunov exponents between the two systems is 0.72, which indicates the chaos intensity of the two systems have a good synchronization, and these results also prove that ITO is a better material to fabricate the microelectrode arrays. However, from Figure 5 and Figure 7b, we can discover that when the characteristic parameter a of Lorenz system over 200, the largest Lyapunov exponents of Lorenz system fall to 0, while the largest Lyapunov exponents of microfluidic system is still greater than 0. Meantime, the Lorenz system is in a critical state, while the microfluidic system is in a chaotic state, though the intensity of chaos is small. The result indicates the state of the controlled system is not completely synchronous with the control system. However, When the characteristic parameter of chaotic electric fields is smaller than 200, the microfluidic system is well synchronous with the Lorenz system.

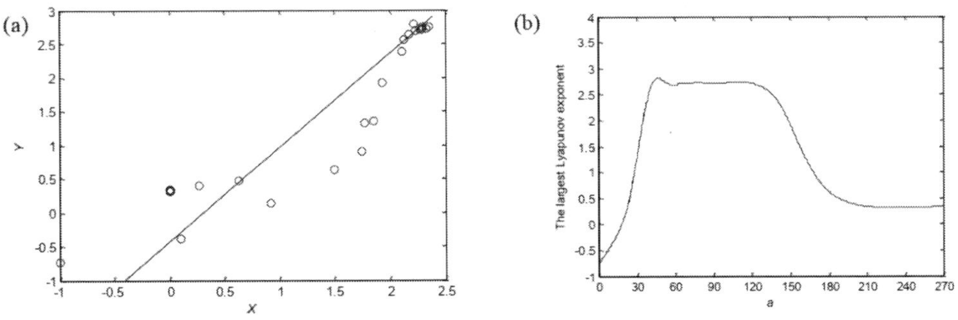

Fig.7. (a) The regression curve of the largest Lyapunov exponents between the system and controlled system. (b) The largest lyapunov exponents of the controlled microfluid system vary with characteristic parameter of the Lorenz chaotic control system.

Table 1. Relation between |R| and correlation degree.

Range of $\|R\|$	0.00-0.19	0.20-0.39	0.40-0.69	0.70-0.89	0.90-1.00
Correlation degree	Extreme low	Low	Moderate	High	Extreme high

4 Summary

This study has presented a novel microfluidic mixing scheme in which the species streams are mixed via the application of Lorenz chaotic system algorithm to three pair of parallel electrodes mounted on the surfaces of the mixing chamber to produces chaotic electric fields. And the microelectrode array is designed by the indium tin oxide (ITO) material. The advantages of ITO materials for the fabrication of microelectrode arrays are discussed. Simulations have been performed to investigate the control effect of the chaotic electric field on fluid chaotic characteristics. The simulation results have shown that the correlation coefficient between the largest Lyapunov exponents on Lorenz chaotic system and chaotic flow behavior is 0.72 which shows a better synchronism, and it also proved that ITO is a better material to fabricate the microelectrode arrays. When the characteristic parameter of Lorenz system is smaller than 200, the state of the chaotic flow system keep pace well with Lorenz system. This work is useful to the design of the mixing chamber via the application of chaotic electric fields to promote microfluid mixing in microfluidic chip.

Acknowledgements

Supported by the Priority Academic Program Development of Jiangsu Higher Education Institutions(PAPD), Natural Science Foundation of Jiangsu Province(BK20140550), China Postdoctoral Science Foundation (2015T80512), Changzhou Sci&Tech Program(CE20155054)and Research and Development Plan of Zhenjiang to Modern Agriculture(NY2015026).

References

1. S. J. Park, J. K. Ki2m, J. Park, S. Chung, C. Chung, K. Chang. Rapid three-dimensional passive rotation micromixer using the breakup process J, Micromech. Microeng, Vol. 12(2004), 6–14.
2. H. J. Song, Z. L. Cai, H. M. Noh, D. J. Bennett. Chaotic mixing in microchannels via low frequency switching transverse electroosmotic flow generated on integrated microelectrodes, Lab Chip, Vol. 10(2010), p. 734–740.
3. G. A. Mensing, T. M. Pearce, M. D. Graham, D. J. Beebe. An externally driven magnetic microstirrer, Phil. Trans. Roy. Soc, Vol. 362(2004), p. 1059–1068.
4. S. Qian, H. H. Bau. A chaotic electroosmotic stirrer. Anal, Chem. Vol. 74(2002), p. 3616–3625.
5. M. H. Oddy, J. G. Santiago, J. C. Mikkelsen. Electrokinetic instability micromixer, Anal, Chem., Vol. 73(2001), p.5822–5832.
6. E. Biddiss, D. Erickson, D. Li. Heterogeneous surface charge enhanced micromixing for electrokinetic flows, Anal. Chem,Vol. 76(2004), p.3208–3213.
7. H. J. Song, J. B. Dawn. Numerical study of enhancing the mixing effect in microchannels via transverse electroosmotic flow by placing electrodes on top and bottom of the channel, Microsyst. Technol, Vol. 17(2011), p.1427–1437.
8. S. Das, I. Pan, S. Das, A. Gupta. Master-slave chaos synchronization via optimal fractional order (PID mu)-D-lambda controller with bacterial foraging algorithm, Nonlinear Dynam, Vol. 69(2012), p.2193-2206.

Study on the Temporal and Spatial Effect of the Relief Pressure of Upper Protective Layer Mining

Chengcheng Li[1,2*]

[1] National Key Laboratory of Gas Disaster Detecting, Preventing and Emergency Controlling, Chongqing 400037, China
[2] Chongqing Research Institute of China Coal Technology & Engineering Group, Chongqing 400037, China
Corresponding Email:lccchn023@163.com

Abstract. In order to study the law of pressure relief of upper protective layer mining. Through theoretical analysis and numerical simulation method, studied the distance on the protection layer mining floor stress field evolution and distribution, found a protective layer on the longwall mining goaf bottom coal seam inclination direction is concave, and strike plane was "O" shape of relief ball shell, the pressure relief angle in the cut or end mining line near the corner of the middle bottom is minimum, the scope of protection of coal seam pressure relief effect than on both sides of the pressure relief effect, namely the scope of protection of inclined coal seam central drainage radius than the two larger; in the trend of protected layer pressure relief protection layer mining face lag distance based on this theory, research to guide the protected layer of gas drilling drainage is very important.
Keywords. Protective layer; pressure relief; temporal and spatial effect

1 Introduction

With the increase of mining depth, coal and gas outburst disaster is more and more serious. The mining protection layer is the most economical and reliable regional measure to prevent and control coal and gas outburst. But in some mining area in the delineation of protected within and in the vicinity of the prominent, protection layer mining process in gas drainage mining technique is not perfect, cannot meet the needs of the production safety, of different gas geological and mining conditions of protective layer mining technology has yet to be combined with the actual situation of the research and improve. Therefore, it is of great practical significance to study the temporal and spatial effects of the mining of upper protective layer.

2 Project overview

Experiment site is in Qidong Coal Mine mining area two of 71 coal seam, mining area three of 61 coal seam, and lower section of two upper protective layer working face were not mining. 71, 82 and 9 coal seam were the main section of the mining area two, 81 coal seam was locally recoverable coal seam. 61, 71, 82 coal seam were the main section of the mining area three, 63, 72, 9 coal seam were locally recoverable coal seam. The 7122 working face is in no danger zone. The 82 and 9 coal seams were serious outburst coal seams, the average coal thickness is 2.3m and 2.65m, the average dip angle is 12 degree and 16 degree, and the 61 coal seam has no danger. 7122 working face with a comprehensive mechanization to the long wall back type mining, Fully caving method. The basic situation of the two and three mining area test face (Table 1, table 2).

Table 1. Basic situation of working face.

working face	elevation (m)	wide (m)	lone(m)	The thickness of coal seam (m)	dip angle (°)
6136	-498.6~-584	158	908	1.93	13
7122	-460~-500	164	1090	1.75	13

Table 2. Coal seam spacing.

Mining area two	coal seam	63	71	81	82	9
	Layer spacing /m	13-36	11-25	7-11	6-21	
	Average interval /m	24	20	10	10	
Mining area three	coal seam	61	63	71	72	82	9
	Layer spacing /m	15-21	20-31	0-8	23-33	8-16	

DOI: 10.1051/matecconf/201710003038

	Average interval /m	19	22	3	27	11

3 Influence zoning law of upper protective layer mining

The evolution and distribution law of the bottom of upper protective layer surrounding rock stress field, fracture field and gas flow field were the important theoretical basis of coal and gas coordination mining for high gas outburst coal seam group, stress field evolution, distribution is the prerequisite and determinants of fracture field and gas flow field evolution and distribution, according to the relevant theoretical research, longwall inclined and gently inclined coal seam mining influence range of stress variation partitioning and zoning division (Figure 1).

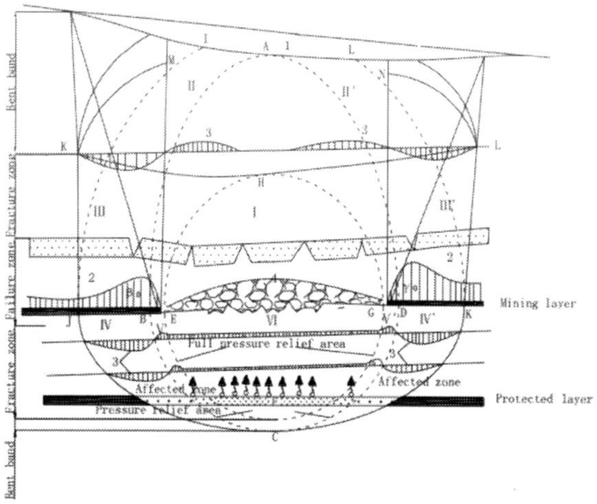

Fig.1. Schematic diagram of the influence of the internal stress in the mining area.

4 Numerical simulation analysis of protective layer mining

4.1. Evolution law of stress and displacement of bottom plate of upper protective layer

(1) Direction stress and displacement variation law

According to the actual conditions of the upper protective layer 7122 working face and its bottom coal seam, a 3-D numerical simulation model is established. To simulate 71 coal seam mining nowadays cover protected space of 82, 9 coal seam unloading pressure law more in line with the actual situation, the simulation distance protection 7122 working face, taking into account the overlying 61 coal seam working face mining pressure relief effect. Using ANSYS three-dimensional numerical simulation of protection layer mining protective layer was unloading pressure distribution and take protective layer of 71 coal seam calculation model to 100m in length and 7122 working face mining, the coal floor strata should stress and displacement changes(Figure 2).

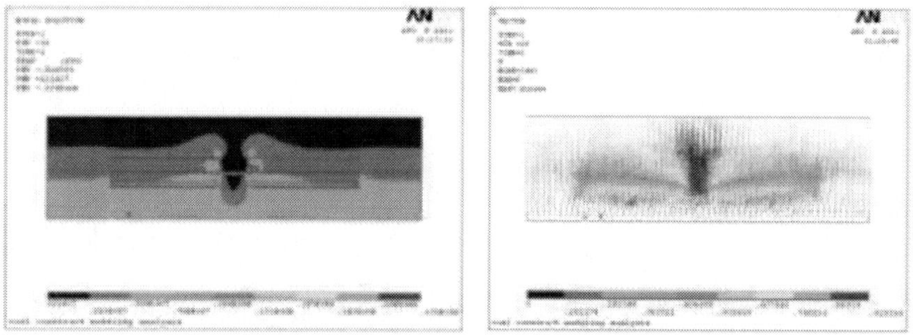

(a) Equivalent stress (b) Displacement

Fig.2. The evolution law of stress and displacement along the direction of the bottom coal seam.

From Figure 4 it can be seen that 7122 working face mining, bottom unloading pressure force and displacement isoline in elevation along the cut hole and end mining line exponentially (but curvature) to shrink, namely the facade unloading pressure isoline is concave pressure relief ring, plane O-shaped unloading pressure ring and the theory on shrinkage regularity; the roof unloading pressure force and displacement isoline and the bottom plate are similar to the distribution of the.

(2) The law of stress and displacement variation

Take the protective layer 71 coal seam calculation model to the length of 180m, 7122 working face mining, the stress and displacement of the floor coal seam (Figure 3).

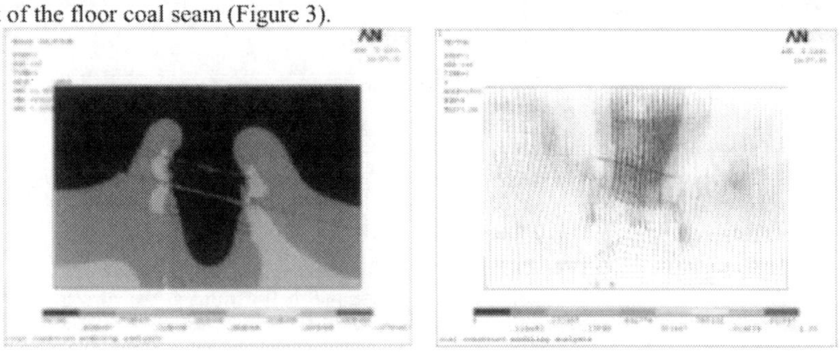

(a) Equivalent stress (b) Displacement

Fig.3. The evolution law of the stress and displacement of the coal seam in the bottom plate.

The stress field and displacement distribution of the bottom plate of the upper protective layer mining in the direction of the trend, the tendency of the stress field and the displacement distribution is a concave relief ring on the face of the face, and the equivalent line is "O". In order to provide theoretical support for the protective effect of the protective layer and the reasonable parameters of the gas drainage borehole layout.

5 Analysis on the pressure relief space of the protective layer in the middle and far distance

(1) According to the two mining area of Qidong mine protection layer mining working face 7122 spatial distribution characteristics and regularity of stress, lowering the pressure relief angle by 10% in the ground stress, in theory, the range of pressure relief gas drainage of protected layer is provided.

"O" ring close to the protective layer plane reaches the maximum value, and with the increase of the interlayer spacing of gradually decreasing trend, whether it is in the work surface or above the work surface below show the same rules. This is closely related to the evolution law of the pressure relief angle of the protective layer. When the distance protection layer face nearer, "O" ring relatively complete, but is relatively narrow, with the distance increase "O" ring the pressure relief effect gradually weakened, and its width it increased.

In order to reduce the equivalent of 10% of the equivalent line as a benchmark, the results obtained from the numerical simulation of the protective layer to remove the hysteresis of the coal mining face distance y and the distance between the X layer(Eq.1):

$$y = 16.7 \ln x - 37.2 \tag{1}$$

(2) Floor space pressure relief law

Floor strata due to stress lifted Poisson effect, near the floor strata pressure relief, level and vertical direction of fractured formation; mined out area at the top caving rock compaction, around goaf area there is still space, "O" - shaped pressure relief ring is formed, protect layer mining unloading pressure angle in the cut hole or terminal mining line near the corner is at a minimum.

7122 after the protection layer mining, the coal and rock mass compaction is more and more obvious, and the coal and rock mass around the mined out area is subjected to tensile action, resulting in a large number of cracks, and there is no compaction, the plane formed a "O" shaped ring.

7122 protective layer working face into the airways and return air lane, cuts the eye or terminal mining line near the corner corner should stress gradient is the largest, in the center of the work surface should be gradient force is relatively small, the corresponding upper and lower strata of the unloading pressure range is relatively small. Therefore, protective layer mining protected layer pressure relief protection angle in the plane is not the same everywhere, close to the protective layer working face of the central region is relatively large, close to the protection layer face cut eye and mining terminal line near the smaller. And production is protection layer face, return air lane must be arranged in a line, therefore, from the point of view of safety protective layer unloading pressure angle should be determined with protection layer face cut eye or terminal mining line near the corner of the minimum pressure relief angle as a benchmark.

6 Experimental investigation

Construction wear layer drilling in bottom pumping lane drilling field10, drainage pressure relief gas of 9 coal seam inclined top, to investigate the gas drainage volume before and after pressure relief. When the 10# drill field distance protection layer working face 40~60m, gas drainage and extraction in drilling field reduce, drainage volume in 0.01~0.02m^3/min, mainly by advance abutment pressure influence and protected 9 coal seam permeability reduced. Protection 7122 layer of work across the 10# investigation drilling field protective layer gradually been fully discharge pressure, gas drainage amount increases gradually, over 10# investigation drilling 30m near, gas drainage of it reached 0.28m^3/min, then gradually decay; 7122 working face to advance to the 160m and protected 9 coal seam bottom drainage roadway gas drainage stability in the volume of 0.02m^3/min (Figure 4).

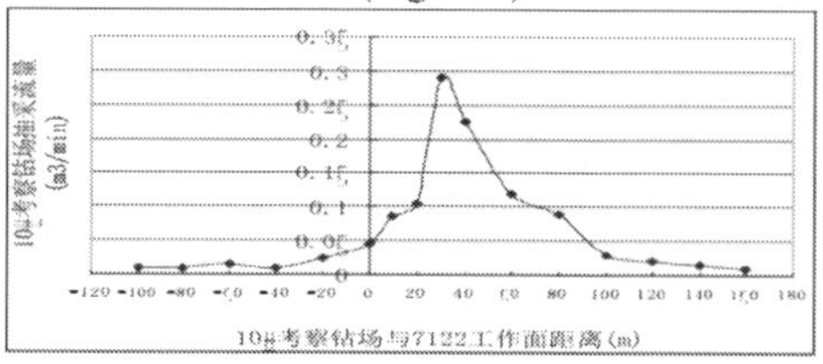

Fig. 4. Relationship between the gas drainage amount and the position of the 7122 working face.

Therefore, in medium or long distance protected floor rock roadway lag protection layer mining working face a certain distance construction crossing boreholes instant extraction to solve the traditional protected floor rock roadway wear layer drill gas drainage of the existence of drilling grouting hole, a clamping drilling, borehole outburst and dynamic pressure of mining damage caused by drilling leakage, coal hole hole collapse significant security risks.

7 Summary

Research on the theory and practice show that the protective layer of long wall mining at the bottom of the coal rock inclined direction concave, to the plane was "O" shape the discharging pressure spherical shell, protective layer and the protected layer spacing is larger, poorer protective effect, protection range is small, remote protective layer discharge pressure and gas flow lag protection layer mining working face certain distance of spatio-temporal relations.

Acknowledgements

This work was financially supported by the science and technology innovation fund of China coal science and Industry Group Co. Ltd. (2014QN002), (2015YBXM39).

References

1. Wang C.G,Zhang D.X. Research on Comprehensive Gas Control Technology of Deep Coal Mining. Coal Science and Technology, 2013(08):11-14 (in Chinese).
2. Wang Z.Q, Feng R.M. Research on mining technology of nether protective seam in inclined contiguous seams with continuous pressure-relief on outburst coal seam. Chinese Journal of Rock Mechanics and Engineering, 2013(S2):3795-3803(in Chinese).
3. Xue D.J, Zhou H.W. Evolution of mining-induced enhancement and distribution of gaspermeability in coal seam and surrounding rock. Journal of china coal society, 2013(06):930-935 (in Chinese).
4. Tian K.Y, Sun W.B. Determination of protection range for mining upper protective layers and its numerical simulation. Journal of Liaoning Technical University (Natural Science), 2013(01):7-13(in Chinese).
5. Liu Z,Li Z.H. Gas Control Technology of Coal Mining in UpperProtective Seam of Contiguous Outburst Seams Group. Coal Science and Technology, 2012(07):49-53 (in Chinese).
6. Shang Z.J, Cheng Y.P. Numerical Simulation of Overlying Rocks Changes with Underlying Protective Seam Mining.Safety in coal mines, 2010(03):5-9(in Chinese).
7. Liu M.J, Wang M. Study on Effect Evaluation of Protective Seam Mining. Coal Science and Technology, 2011(01):61-64(in Chinese).
8. Xue D.J, Zhou H. Mechanism of unloading-induced permeability increment of protected coal seam under mining. Chinese Journal of Geotechnical Engineering, 2012(10):1910-1916(in Chinese).

MATEC Web of Conferences **100**, 04001 (2017) DOI: 10.1051/matecconf/201710004001
GCMM 2016

A research of heat transfer enhancement of rectangular channel with dimples

Zhe Fu[a], Yingping Qian[b], Xizhi Zhou[c], Zhihong Yan[d] and Yang Jiao[e]

College of Mechanical Engineering of Hubei University of Technology, Hongshan District, Wuhan City, Hubei Province, PRC
[a]Email: daja_vu@163.com, [b] Email: 87467822@qq.com, [c] Email:51809621@qq.com, [d] Email: 871402560@qq.com, [e] Email: 2449558169@qq.com

Abstract: this thesis is aimed to do numerical simulations on rectangular channel with a new model of dimple in different condition of channel height, dimple depth, dimple distance by using CFD program and RNG k-ε model. It is concluded that, the existence of dimple has changed the velocity and direction of the fluid, as well as the original flow status. Dimple increased the heat exchange area of rectangular channel and broke the boundary layer, enhanced the flow rate of fluid, as a result of enhanced heat transfer performance.

Keywords: dimple, heat transfer enhancement, heat transfer performance, numerical simulation.

1 Introduction

In recently technology of heat transfer enhancement, in order to enhance the ability of heat exchange in fluid, most employers use the methods of roughen layer surface and extend the surface area. The general method is using horizontal or spiral fin, thread, groove and dimple to enhance heat transfer [1,2,3]. As for the research of heat transfer enhancement, not only need to take enhancement performance in to consideration but also have to consider the friction factors of fluid. The heat transfer enhancement of dimple breaks the boundary layer because of the sunken surface, and creates different sizes of vortex which accelerates the heat exchange in rectangular channel.

The factors of heat exchange and flow friction in dimple heat transfer enhancement are mainly affected by the shape, depth, distance, Re of dimple. Burgess [4] used air as medium and analyzed the effect of dimple depth, discovered that heat exchange factor η and friction factor are increased by the depth went up. Mahmood [5] used Infrared Thermometric Techniques to find out that leeward side of dimple has weaker heat exchange than windward side. Chen. J [6] has done the analysis of how the distance and depth and concentration affect heat transfer, ending up with found a best way to design heat transfer enhancement. Bunker [7] found that compared dimple to smooth channel, dimple has twice heat enhancement and 2~7 times stronger friction. Somin[8] has a research on channel height and dimple queue mode when Re between 30000~50000, found that heat transfer performance is good with high concentration dimple when the channel height kept constantly, and with narrow rectangular channel in the same dimple queue mode. The author analyzed that in different Re, the effect of channel height, dimple depth, dimple distance on heat transfer and friction in a new dimple and after that analyzed the integrated heat transfer performance under different individual working conditions.

2 Models and Judge Standard

2.1 Physical Model

The Figure1 below is the construction and geometry factors of dimple heat transfer rectangular channel. Channel height is H, dimple depth is H_d, dimple distance is S, this thesis has done numerical simulation of models: H=2mm、4mm、6mm、8mm; H_d =0.4mm、0.8mm、1.3mm、1.6mm; S=4mm、9mm、14mm、18mm, in Re= 2000, 4000, 7000, 10000, take rectangular dimple which length: width= 2:1, vertical distance: horizontal distance= 1:1.3.

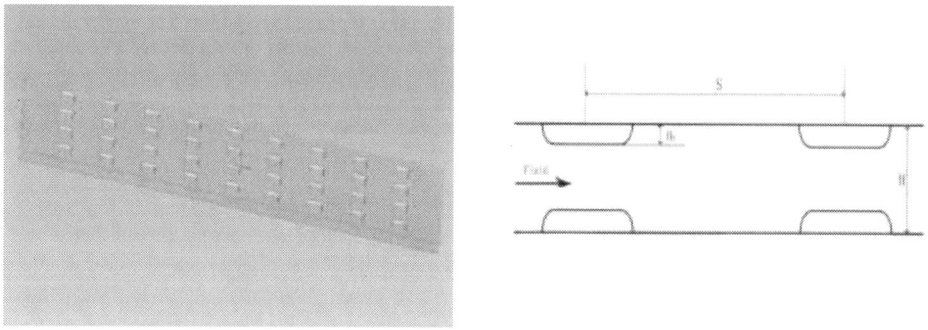

(a) Construction of dimple boa (b) Geometry factors of rectangular channel

Fig. 1. Physical model of rectangular channel.

2.2 Mathematic Model and Formula

Continuity equation, momentum equation, energy equation of fluid can be all presented as general equation:

$$\frac{\partial(\rho u\varphi)}{\partial x}+\frac{\partial(\rho v\varphi)}{\partial y}+\frac{\partial(\rho w\varphi)}{\partial z}=$$
$$\frac{\partial}{\partial x}\left(\Gamma_{\varphi,t}\frac{\partial\varphi}{\partial x}\right)+\frac{\partial}{\partial y}\left(\Gamma_{\varphi,t}\frac{\partial\varphi}{\partial y}\right)+\frac{\partial}{\partial z}\left(\Gamma_{\varphi,t}\frac{\partial\varphi}{\partial z}\right)+S_\varphi \tag{1}$$

The heat transfer performance, friction performance and integrated heat transfer performance of dimple board are presented as Nu, f, η. In the experiments, the Nu0 and f0 are calculated by numerical simulation. The Re, Nu, f, ηare defined as:

$$R_e=\frac{D_H\rho U}{\mu} \tag{2}$$

$$N_u=\frac{qD_H}{\left(T_{wall}-T_{ref}\right)\lambda} \tag{3}$$

$$f=\frac{2\Delta PD_H}{\rho U^2} \tag{4}$$

$$\eta=\frac{N_u/N_{u_0}}{\left(f/f_0\right)^{1/3}} \tag{5}$$

According to different Re convert into fluid mass Qm, and set Qm as initial value, both sides are meristic boundary, fluid in channel is air, fluid temperature is 600K, all calculation are carried out by SIMPLE and RNG k-ε. Momentum, Turbulent kinetic energy, Turbulent fluctuation dissipation rate and energy equation are second-order upwind scheme. The convergence precision of continuity equation is 10-6, ignore fouling resistance and gravity, phase transfer and reacting flow are not taken into consideration.

3 Results and Analysis

3.1 Numerical Simulation Results Analysis

It can be concluded that, although fluid has different working conditions in rectangular channel, their change trend are similar. Figure 2 shows the distribution of transverse section temperature field, it is obviously that the temperature in the exit of channel is lower than entry and the temperature surround dimples are lower especially close to dimples, it proves good heat transfer enhancement of dimples. In addition, the windward side of dimples has sparse temperature distribution, heat exchange has lower change and keeps a level, however, the leeward side of dimples has concentrated temperature distribution and has obvious decrease trend of temperature, higher heat exchange rate. With the increase of dimple depth, heat exchange rate is getting stronger. Figure 3 shows the transverse section velocity vector when fluid flows through the channel, we can see the flow rate and direction has changed when fluid pass through dimples, upstream velocity increases and divided when reach downstream air. Besides, adjacent heave force flow field overlying together thus enhances flow strength. It is similar to thread, heave can not only increase heat exchange area, but also can enhance turbulent fluctuation, cut off boundary layer so that enhance turbulent kinetic, in order to achieve good heat transfer enhancement.

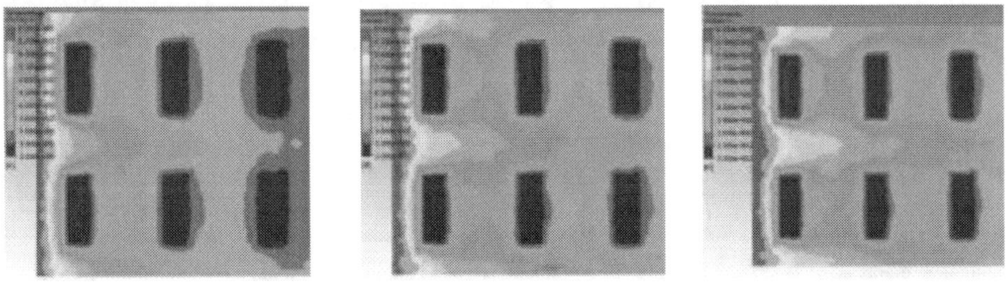

H_d=1.6 H_d=1.3 H_d=0.8

Fig. 2. Temperature areas of H_d=0.8~1.6

Fig. 3. Velocity vector of H_d=0.8

3.2 Data Analysis

(1) The effect of channel height on dimple heat transfer enhancement. The numerical simulation of model H=2mm, 4mm, 6mm, 8mm in Re= 2000, 4000, 7000, 10000, the results are shown in figure 4 below. It illustrates that, when Re increased, Nu increased and f decreased. When channel height decreased, Nu and f increased. It is because when Re is constant, narrow channel leads to higher flow velocity, turbulent increased thus heat transfer is enhanced. The figure 4

DOI: 10.1051/matecconf/201710004001

also shows that when Re increased, η decreased gradually, the curve of η intersected when Re= 4000, higher channel height leads to better heat transfer performance. In practical works, according to certain condition, channel height can be increased when Re is big and decreased channel height when Re is small so that heat exchange integrated performance can be optimized.

(2) The effect of dimple depth on dimple heat transfer enhancement. The numerical simulation of model Hd=0.4mm, 0.8mm, 1.3mm, 1.6mm in Re= 2000, 4000, 7000, 10000, the results are shown in figure 5 below. It illustrated that, heat transfer performance is better in dimple channel than in smooth channel, it is because dimple increased the heat exchange area and also breaks the boundary layer thus increased fluid kinetic and turbulent. When Re increased, Nu increased and f decreased, as a result of increasing Re caused turbulent enhanced heat transfer performance. When Re= 4100, with dimple depth increased, η increased. It is because in a small Re condition, the increase gradient of dimple heat transfer is bigger than friction. Above all, Hd= 0.8~1.3mm has a good heat transfer performance.

(3) The effect of dimple distance on dimple heat transfer enhancement. The numerical simulation of model S=4mm, 9mm, 14mm, 18mm in Re= 2000, 4000, 7000, 10000, the results are shown in figure 6 below. It illustrated that, with S decreased, Nu and f decreased, it is because when S went up, dimple concentration decreased, the effect to flow channel will be weaken. When Re increased, η decreased gradually, there was a threshold point of four different dimples when Re=5200. Dimple distance has negative effect on heat transfer performance when Re<5200 and good effect when S=9mm when Re>5200. It illustrated that when Re keeps a low value, the heat exchange caused by dimple concentration is stronger than friction, however in a high Re condition, terrible distribution of dimples may lead to bad result, so a good value of Re is necessary.

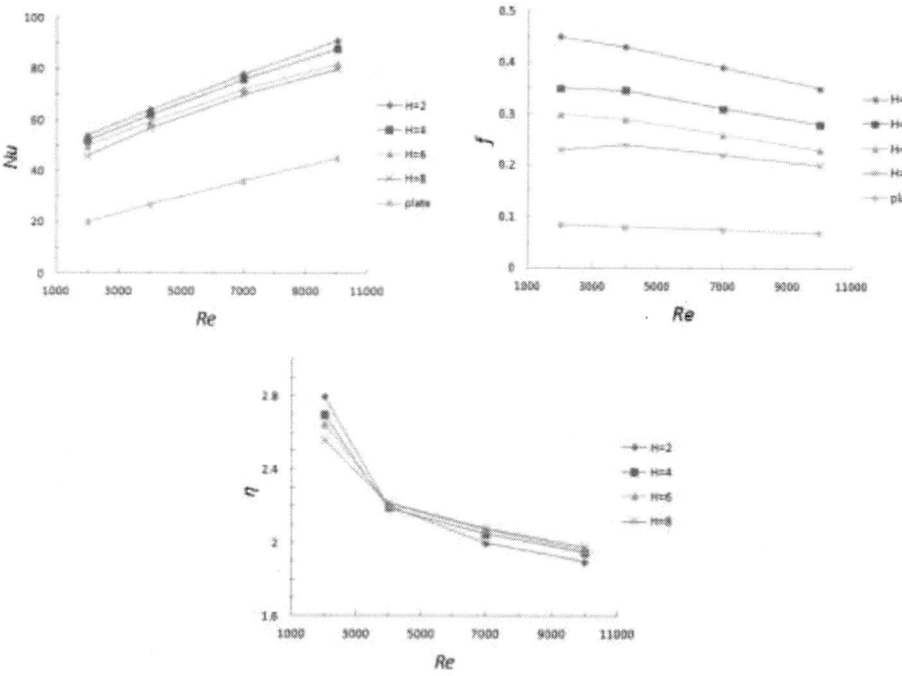

Fig. 4. The trend of *Nu*, *f*, η by changing Re in condition of different H.

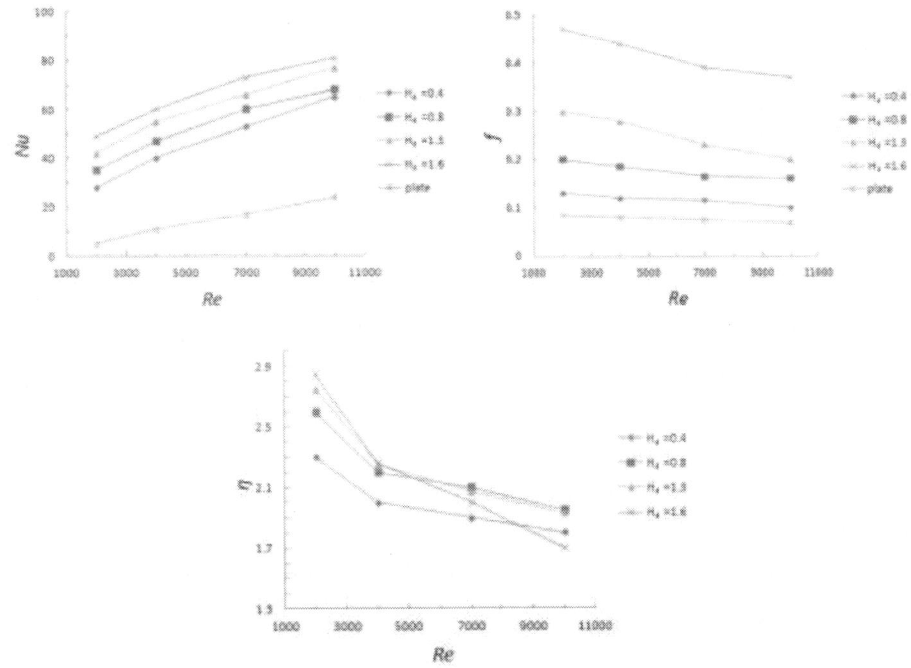

Fig. 5. The trend of *Nu*, *f*, η by changing Re in condition of different H_d.

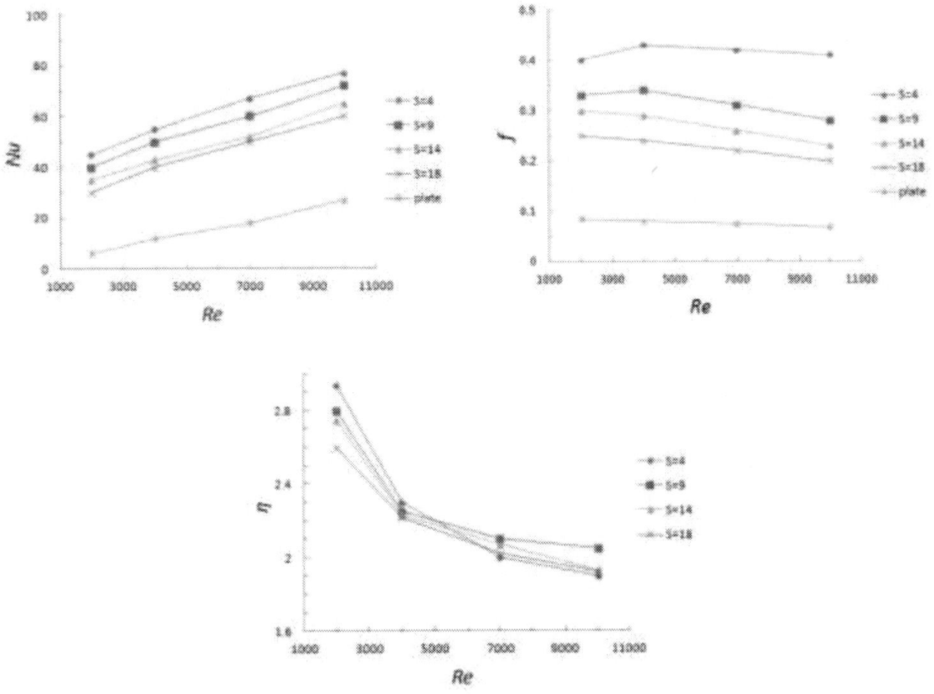

Fig. 6. The trend of *Nu*, *f*, η by changing Re in condition of different S.

4 Conclusion

According to the different numerical simulation analysis of heat transfer performance and friction in dimple rectangular channel under different geometry factors conditions, it is concluded that:

(1)The existence of dimple changed the flow velocity and direction of fluid and broke the internal status.
(2)The existence of dimple increased heat exchange area and broke boundary layer thus increased fluid kinetic and increased the ability of heat exchange.
(3)When Re is between 2000~10000, Nu increased following the increase of Re, η decreased when Re increased.
(4)When Re is between 2000~4100, ηdecreased when H increased, ηincreased when Hd increased. When Re is between 5200~10000, ηhas no obvious trend when other factors changed, however there is an optimized construction that has best heat transfer performance: H=8mm, H_d=0.8mm, S=9mm.

References

1. G.I. Mahmood, M.L. Hill, D.L. Nelson, et al. Local heat transfer and flow structure on and above a dimpled surface in a channel, ASME J Turbomach, **123**,1(2001)
2. V.S. Kesarev, A.P. Kozlov, Convective heat transfer in turbulized flow past a hemispherical Cavity, Heat Transfer Res, **25**, 2(1994)
3. V.I. Terekhcv, S.V. Kalinina, Y. M. Mshvidobadze, Flow structure and heat transfer on a surface with a unit hole depression, Russ J Eng Thermophys, 5(1999)
4. N. K. Burgess, P. M. Ligrani, Effects of dimple depth on nusselt numbers and friction factors for internal cooling in a channel, ASM E Paper GT, (2004)
5. G.I. Mahmood, M. L Hill, D.L. Nelson, et al., Local heat transfer and flow structure on and above a dimpled surface in a channel, ASM E J.of Turbomachinery, **123**,1(2001)
6. Juin Chen, M.S. Hans, G.D. Geoffrey, Heat Transfer Enhancement in Dimpled Tubes, Applied Thermal Engineering, **21** (2001)
7. R.S. Bunker, K.F. Donnellan, Heat transfer and friction factors for flows inside circular tubes with concavity surfaces//ASME Turbo Expo 2003, collocated with the 2003 International Joint Power Generation Conference. American Society of Mechanical Engineers, (2003)
8. S.M. Shin, K.S. Lee, S.D. Park, J.S. Kwak, Measurement of heat transfer coefficient in the dimpled channel: effects of dimple arrangement and channel height, Journal of Mechanical Science and Technology, (2009)

Study on Performance of Silicon Ignitiors based on Energetic Chip of Porous Silicon

Lan Liu [a], Yan Xue[b], Ruizhen Xie[c], Xiaoming Ren[d] and Kewei Li[e]

National Key Laboratory of Applied Physics and Chemistry, Shaanxi Applied Physics-Chemistry Research Institute, Xi'an, China,710061

[a]383131204@qq.com, [b]xueyan213@163.com, [c]xieruizhen@126.com, [d]rxm_2003@163.com@163.com, [e]994060346@qq.com

Abstract: In order to study the igniton properties of the silicon ignitior, firing sensitivity test with output charge B-KNO3 were carried out, and the igniton properties was studied as well. The result indicates that the energetic chip can ignite B-KNO3 which burns sufficiently. The 50% firing voltage was 21.3V, standard deviation was 1.87, the 99.9% point was 27.1V.

Keywords: porous silicon, energetic chip, silicon ignitior, electrochemical etching.

1 Introduction

As the modern weaponry developed toward miniaturization, intelligent, integrated, traditional Pyrotechnics can not meet the requirements[1,2]. Silicon ignitiors are based on the concept of MEMS manufacturing. The porous silicon chips have micronano-scale precision and the microscale planar structure which greatly reduced the volume and the weight of initiator[3,4]. The performance of porous silicon energetic chip can be adjust by different concentration of NaClO4/methanol solution[5,6].

Aiming at the need of integration development of pyrotechnics, the design and manufacture of silicon ignitior were researched, which include the whole design, the manufacture of porous silicon sequence and its performance. The design principle of initiators based on MEMS will support the development of mocro arms initiation, ignition and work.

2 Experiment

2.1 Reagent and apparatus

HF acid,analytical, XI'an chemical Reagent factory; Ethanol, Sichuan xilong Chemical Co.;Bno3,Xi'an213; P-type silicon, crystal-100,Resistivity of 0.02～0.03Ω·cm, thickness: 300±15μm, Shanghai Junhe electronic materials Co.; platinum electrodes, pureity 99.95%, diameter 70mm, Shanghai dingfu metal materials Co.; NaClO4, analytical, Tianjin fucheng chemical Reagent factory; current loop, and width :35MHz, measuring range: :0~500A, rise time :10ms.

2.2 Experiment equipment

The schematic diagram of ignition test system shown in Figure 1, which is made up of oscilloscope, optical probes, current loop and switch. The fire sensitivity tests of porous silicon ignitors were carried out and the range of voltage is from 10V to 50V.

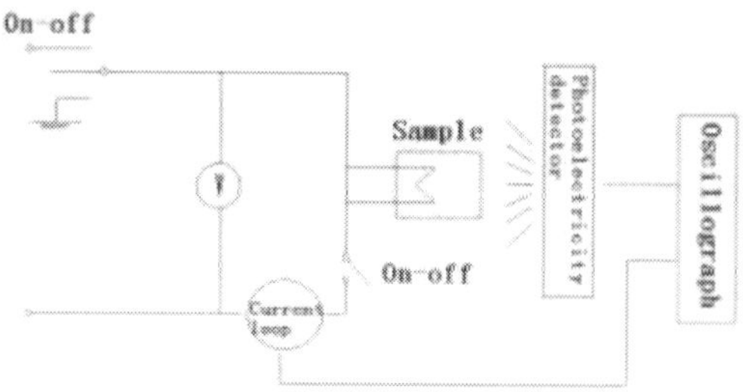

Fig.1. The test system of porous silicon energetic chip

2.3 Preparation of test sample

2.3.1 Design of porous silicon ignitor

Porous silicon ignitor was made of shell, output charge, energetic porous silicon chip, the energetic chip and electrode plug. The bridge film was produced through the process of sputtering, photolithography and wet etching. The micro energetic chip was made by in-situ charge method. The chip was stick to the groove in electrode plug. The sctech map showed in Figure 2.

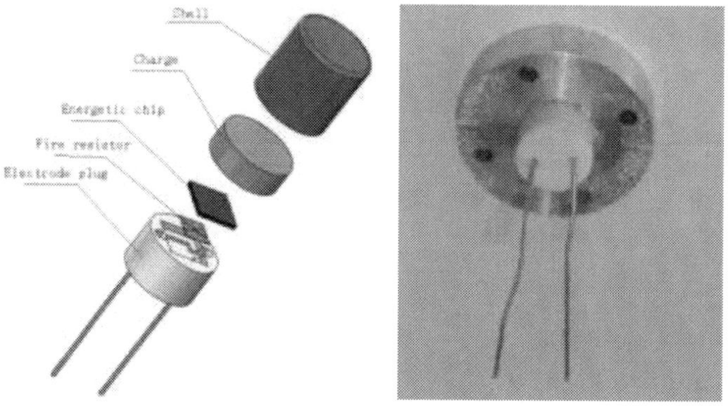

(a)Structure of silicon ignitior (b)Sample of silicon ignitior

Fig.2. Structure and sample of silicon ignitior and Silicon ignitior

2.3.2 Design of charge

The charge was 74mg B/KNO3 pellet, whose charge size was Φ4.2mm×3.8mm and density was 1.4g/cm3.

2.3.3 Preparation of porous silicon energetic chip

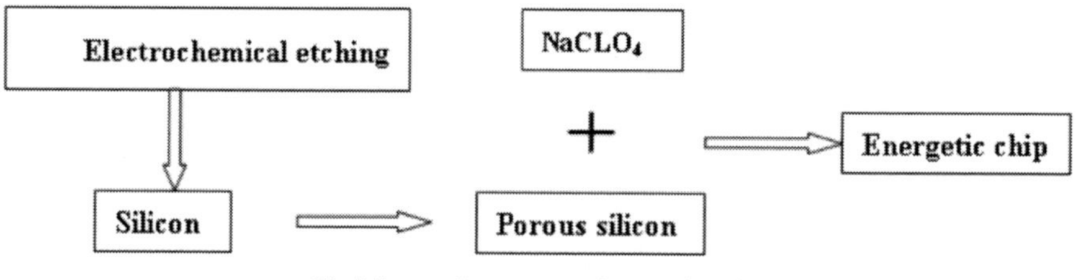

Fig.3 Preparation process of energetic chip

The principle of porous silicon energetic chip was showed in Figure 3. The silicon was prepared by electronchemical etching to form a sponge-like structure. The saturated solution of NaClO4 was added to the porous silicon to make the porous silicon energetic chip. The porous silicon was produced by electrochemical etching of silicon in HF/ethanol(HF/CH3CH2OH=1:1) electrolyte by five step constant current etching method.

Table 1. The parameter of Porous silicon Preparation

	Step 1	Step 2	Step 3	Step 4	Step 5
Current(A)	2.8	2.4	2.0	1.6	1.3
Etching time(min)	20	10	30	10	50

Placed porous silicon on the hot plate. Added saturated NaCLO3 to the porous silicon to made the porous silicon energetic chip. The temperature of hot plate controlled at 70°C~90°C and blowing hot air to the porous silicon 5~8second.Repeated dropping 3~5times. Carbinol volatilized, and oxidant aggregated in the porous silicon. The structural of porous silicon energetic chip shown in Figure 4.

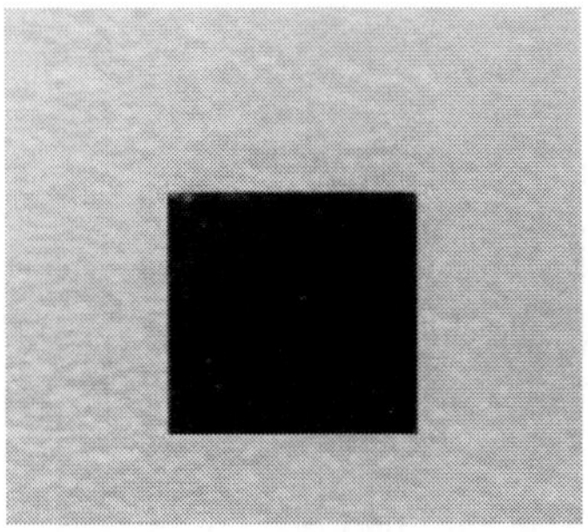

Fig.4. Image of porous silicon energetic chip

3 Results and discuss

The firing sensitivity was tested according to LangLie Method[7,8], firing current was less than 1.7A, firing voltage was less than 15V, ignition time was millisecond level, result can be seen from the table 2:

Table 2. Firing voltage of porous silicon energetic chip

number	sample siza	50% firing voltage	99.9%response point	standard deviation
1	12	21.3V	27.1 V	1.87
2	15	14.4V	20.8V	2.09
3	17	18.6V	17.9V	1.23

The firing sensitivity was tested according to LangLie Method, 50% firing voltage was less than 15V, test results wa shown in Figure 5. The image of porous silicon energetic chip after firing was shown in figure 6, BKNO3 effected completely shown in figure 6.

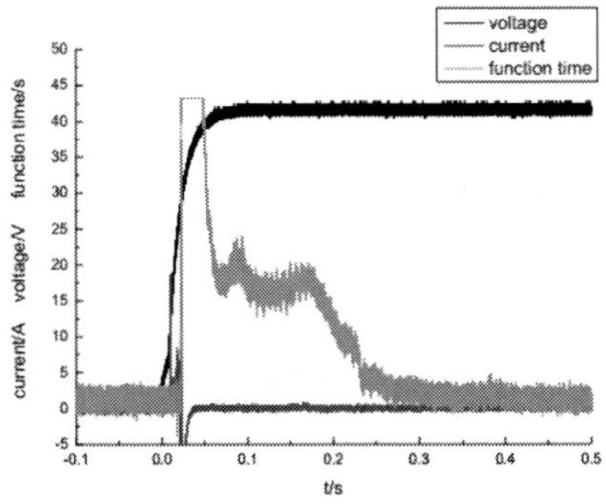

Fig.5. Curve of firing voltage、current and function time change

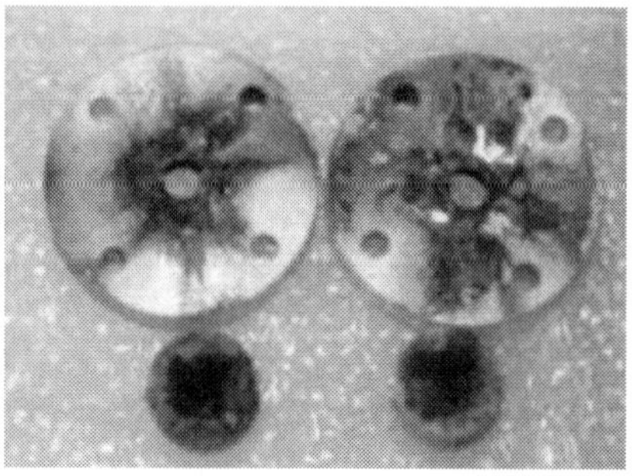

Fig.6. image of porous silicon energetic chip after firing of porous silicon energetic chip

4 Conclusion

Paper used electronchemical etching method to etched the silicon into a sponge-like structure.Added saturated solution of NaCLO3 to the porous silicon to made the porous silicon energetic chip. LangLie Method used to test the firing sensitivity of porous silicon energetic chip ignition, results indicated:

(1)Combined with the technics of MEMS, used electronchemical etching method to etched porous silicon energetic chip.Company with MEMS Ni/Cr-bridges and charge to facture the silicon ignitiors.

(2)The energetic chip can ignited B-KNO3 which burns sufficiently, text indicated porous silicon energetic chip can ignited charge and burning sufficiently.

(3)LangLie Method used to test the firing sensitivity of porous silicon energetic chip ignition, the 50% firing voltage 21.3V, standard deviation 1.87 the 99.9% point 27.1V.

References

1. Y. Xue, B. Lu, Properties of Energetic Materials Based on Nano-Porous Silicon. Chinese Journal of Energetic Materials, **18**,5(2010)
2. Churaman W., L. Currano, C. Becker, Initiation and reaction tuning of nanoporous energetic silicon. Journal of Physics and Chemistry of Solids, **71**, 2(2010)
3. Joachim Diener, Egon Gross, Dimitri Kovalev etal., Nanostructured Reactive Substance and Process for Producing the Same, US patent 0148569, (2003)
4. D. Clement etc, Highly explosive nanosilicon-based composite materials.phys.atat.sol, (a) 202, No.8 2005.
5. L. Liu, Y. Xue, etal., Influence of Electrochemical Etching Parameters on morphology of porous silicon, Mechanical Science and Engineering, **12**(2014)
6. S.X. Wang, R. Q. Shen, Y.H. Ye, Preparation and ProPerties of Porous Silicon Energe Chips, Chinese Journal of Energetic Materials, **18**, 5(2010)
7. R. Zhang, D.X. FU, etc. A Calculation Instrument for the Sensitivity Test of Initiating Explosive Device. Initiators&Pyrotechnics, **1**, 2(2009)
8. D.X. FU, R. Zhang. etc. Estimation and Its Applicability of D-Optimality-Based Sensitivity Test under Different Prior Information. Chinese Journal of Energetic Materials, **17**, 3(2009)

Study on Thermal Conversion of the Waste Oil Absorbin Activated Carbon

Qing Mao[1], Shouyu Zhang[1*], Menghao Zhao[1], Yunlong Yao[1], You Li[1] and Jingning Yang[1]

[1]School of Energy and Power Engineering, University of Shanghai for Science and Technology, Shanghai, China
Corresponding Email: zhangsy-guo@163.com

Abstract. The pyrolysis and combustion characteristics of a waste oil absorbing activated carbon were investigated using a thermogravimetric analyzer under heating rate of 10°C/min from 25°C to 1000°C. The activated carbon sample was used to adsorb waste oil. The results obtained, indicates that DTG curves of the pyrolysis process of the sample has three peaks at 106°C, 190°C and 249°C. The four peaks are observed in the DTG curves of the incineration process. The first peak at 110°C is caused by water emission. The second peak corresponds to the temperature of 190°C that is consistent with the second peak temperature of the pyrolysis process of the sample. The third peak at 240°C which is 9°C lower than the third peak temperature of the pyrolysis process. The second and third peaks maybe caused by the emission of the waste oil absorbed in the activated carbon. The fourth peak temperature is 516°C and the weight loss within the range of 400°C ~ 560°C is more obvious. The weight loss is caused by the residue char combustion.
Key words. Waste oil absorbing activated carbon, thermal conversion, pyrolysis, incineration.

1 Introduction

Activated carbon, because of the thermal treatment, which removes the moisture and the volatile matter contents of the biomass, the remaining solid char shows different properties than the parent biomass materials [1]. These changes in the properties usually lead to high reactivity, and hence, an alternative usage of char as an adsorbent material becomes possible [2]. Therefore, activated carbon is a specially treated carbon, with numerous small voids, high surface area, and it not only has a strong physical and chemical adsorption function, but also has the role of detoxification, so widely used [3-4]. However, when an activated carbon reached its saturation limit, it was simply discarded, generating a secondary source of pollution. Moreover, concern about environmental protection has increased over the years from a global viewpoint, and the spent activated carbon has to be considered a hazardous waste in itself, so it requires special treatment [5-6].

At present, the research on thermal decomposition of the spent activated carbon has been reported rarely. In this paper, the thermal decomposition characteristics of the waste oil absorbing activated carbon were studied by thermal conversion. The TG-DTG curves of the waste oil absorbing activated carbon at the atmosphere of air and nitrogen were researched, respectively. These experimental results helped to reveal the reaction process of the spent activated carbon decomposition.

2 Materials and methods

The materials used for the experiment was the spent activated carbon, which had been used to adsorb oil. In addition, the sample was ground and a particle size fraction of 0.3-0.45mm was used for the studies. What's more, the proximate analyses of the samples are listed in Table 1.

Thermal gravimetric method is a technique to measure the relationship between the mass and the temperature under the programed temperature. If a differential unit is attached to the analyzer, it can simultaneously be recorded the derivative thermogravimetric curve, namely the function of the mass change rate with temperature or time.

Pyrolysis and incineration was carried out with a thermogravimetric analysis (TA/TGA Q500, America). About 1mg of the sample was pyrolysed (or burned) under 40mL/min N2 (or air) flow at heating rate of 10°C /min from 25°C to 1000°C. The atmosphere was also maintained during the heating up and cooling-down intervals. This flow rate ensures a suitable atmosphere on the sample during the run, while the small amount of the sample and the slow heating rate ensure that the heat transfer limitations can be ignored. From these assays, the evolution with temperature of weight loss (TG) and the weight loss rate (DTG) for the samples studied were obtained.

Table 1. Proximate analyses of samples.

Name	M_d/%	A_d/%	Q_d/ （MJ/kg）

Value	21.8	6.9	12.21

3 Results and discussions

Pyrolysis Characteristics. Pyrolysis is one from of energy recovery process, which has the potential to generate char, oil and gas product [2]. Figure 1 shows the TG and DTG pyrolysis profiles of the sample [6].

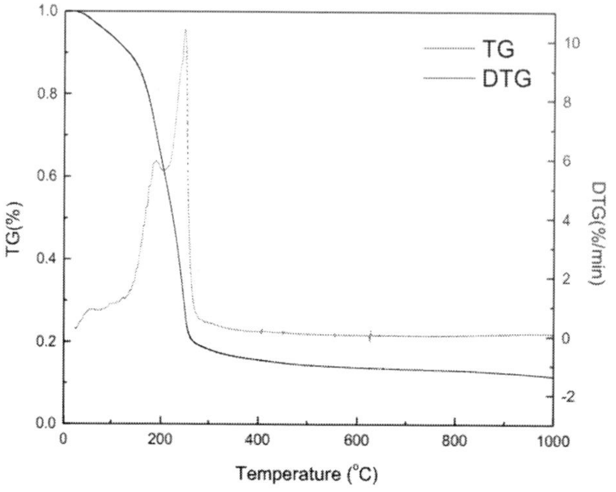

Fig. 1. TG and DTG pyrolysis profiles of the sample.

The TG-DTG curves showed that the decomposition proceeds through few well-defined steps with DTG peaks closely corresponding to the weight loss obtained. From Figure 1, the weight loss of the sample increased with temperature increasing (TG curve), and the whole process is divided into three distinct and successive steps can be deduced (DTG curve). Namely, the first weight loss step is from 25°C to 106°C, it is associated with water emission. It is well known that some of the water in the sample and the boiling point of the water is relatively low. What's more, the second step is from 106°C to 190°C, this step maybe can be considered as the emission of the waste oil absorbed in the activated carbon. Because the sample is the waste oil absorbing activated carbon. In addition, the third step, it is followed by a second step, between 190°C and 249°C, which corresponds to the decomposition of inorganic minerals.

Combustion Characteristics. The TG-DTG curves of the sample in air at heating rate of 10°C/min are shown in Figure 2.

The three peaks at 106°C, 190°C and 249°C in Fig.1 (DTG curve) can be observed. However, the four peaks were observed in Fig.2 (DTG curve). The DTG peaks closely corresponding to the weight changes observed on the TG curves. The first peak in the low temperature region is 110°C, in other words, the first weight loss step is from 25°C to 110°C, it is also related to water emission. Because some of the water in the sample and the boiling point of the water is relatively low. The second peak corresponds to the temperature of 190°C that is consistent with the second peak temperature of pyrolysis. The third peak at the temperature of 240°C which is 9°C lower than the third peak temperature of pyrolysis. Scilicet, the second and third weight loss maybe also associated with the emission of the waste oil absorbed in the activated carbon. Because the sample is the waste oil absorbing activated carbon. it is as same as the second weight loss step of the pyrolytic process. The fourth peak temperature is 516°C and the weight loss within the range of 400°C ~ 560°C is more obvious, and the reason of the weight loss is the residue char combustion [7-8].

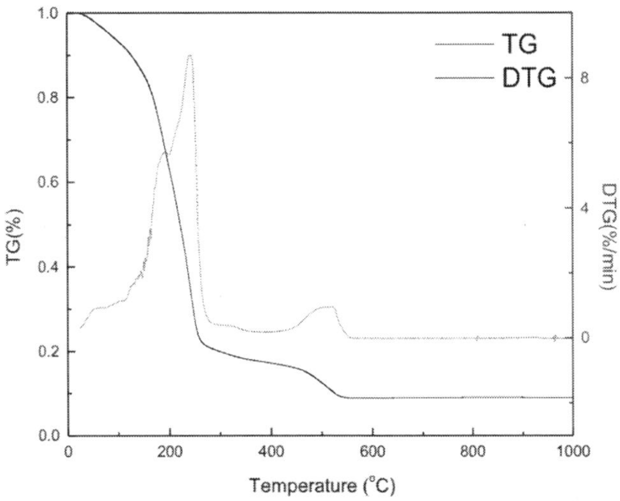

Fig. 2. TG and DTG Incineration profiles of the sample.

4 Conclusions

From the DTG profile of the waste oil absorbing activated carbon, the pyrolytic process is divided into three stages can be found, while the incineration process of the waste oil absorbing activated carbon is in four phases. For the incineration process, the first weight loss step is related to water emission. It is well known that some of the water in the sample and the boiling point of the water is relatively low. The second and third weight loss step maybe caused by the emission of the waste oil absorbed in the activated carbon. Because the sample is the waste oil absorbing activated carbon. The fourth weight loss step corresponds to the residue char combustion.

In another case, for the Pyrolytic process, the first weight loss step is also related to water emission. The second step maybe can be considered as the emission of the waste oil absorbed in the activated carbon. The third weight loss corresponds to the decomposition of inorganic minerals.

When an activated carbon reached its saturation limit, it was simply discarded, generating a secondary source of pollution. However, the spent activated carbon can be regarded as a kind of energy, especially the waste oil absorbing activated carbon. Therefore, this kind of energy should be greatly utilized in the future.

Acknowledgements

This research was supported by the National Science & Technology Pillar Program during the Twelfth Five-year Plan Period (Grant NO.2012BAA04B01).

References

1. IOANNIDOU O, ZABANIOTOU A: Agricultural residues as precursors carbon production: a review. Renewable and Sustainable Energy Reviews Vol 11(2007), 1966-2005.
2. Putun AE, Ozbay N, Onal EP, Putun E: Fixed-bed pyrolysis of cotton stalk for liquid and solid products. Fuel Process Technol Vol 86(2005), 1207–19.
3. R M Suzuki, A D Andrade, J C Sousa, et al: Preparation and characterization of activated carbon from rice bran. Bioresource Technology Vol 98(2007), 1985-1991.
4. PAUL T W, ANTON R: Reed pre-formed activated carbon matting derived from the pyrolysis of biomass natural fibre textile waste. J Anal Appl Pyrolysis Vol 70(2003), 563-577.
5. FOO K Y, HAMEED B H: Recent developments in the preparation and regeneration of activated carbon by microwaves. Advances in Colloid and Interface Science Vol 149(2009), 19-27.
6. ANIA CO, MENENDEZ J A, PARRA J B, et al: Pyrolysis of activated carbons exhausted with organic compounds, J. J Anal. Appl. Pyrolysis Vol 74(2005), 518-524.

7. Zhan D, Cong C J, Diakite K, et al: Kinetics of thermal decomposition of nickel oxalate dehydrate in air, J. Thermochimica Acta Vol 430(2005), 101-105.
8. Marisamy M, Tomoaki N, Kunio Y: Characteristics of co-combustion and kinetic study on hydrothermally treated municipal solid waste with different rank coals: A thermogravimetric analysis, J. Applied Energy 87(2010), 141-148.

Design of A Regional perimeter Optical fiber Security and Protection System

Guofan Chen and ShuangXin Wu

School of Mechanical Engineering, Anhui University of Science and Technology, Huainan, China

Abstract. The current commonly used widespread in electronic Security and Protection System Widespread the high cost of maintenance,Security distance short,Susceptible to electromagnetic interference.Aim at these problems, this paper based on source of access to separate The invasion of positioning signal and The invasion characteristic signal ,Design of a Regional perimeter Optical fiber Security and Protection System.The system can significantly improve the positioning accuracy, reduce the rate of false positives.

1 Introduction

At present, domestic Electronic security system include optical fiber probe, infrared detection, geomagnetism detection and closed-circuit television ,and the latter three ones due to the limitation of the power supply, they are usually security distance is short, are susceptible to electromagnetic interference, high maintenance cost disadvantage [1]. Because the distributed optical fiber security technology can overcome the above shortcomings and has had a rapid development in recent years, however, there are still some shortcomings in the current distributed optical fiber security technology [2]. Positioning accuracy is mostly outside the 20m, not suitable for the military and sensitive department of security requirements; to wind sways grass, animal intrusion events Characteristic can't accurate analysis, lead to high rate of false positives [3]. So it is necessary to research on optical fiber security invasion of precise positioning (0.5 meters) technology, reduce the intrusion event False alarm rate and research and development of its auxiliary equipment.

This system use the TDC (Time digital conversion technology) to realize the super short time difference (picosecond level) measurements, and solved the traditional measuring method in a short period of time is difficult to apply problem, and heightens the precision of Intrusion place location. When measuring time interval TDC measure the pulse signal converted into digital signals, don't need to do A/D conversion, Single-chip microcomputer connected to converter could have completed single chip microcomputer control of the converter and to deal with the results of the measurement [4]. TDC resolution can reach 65 ps, namely every measurement error is less than 0.98 cm.

2 System structure

The AT89C52 single chip microcomputer is the control core in this system,Mainly includes: the laser generator,optical coupler,four fiber cable,the photoelectric converter, electrical signal processing module,TDC, SCM and PC used as a upper computer,as shown in Figure 1.

2.1 System works

The signal source of the system come from the optical signals generated by the laser generator, After the laser beams was subjected to a series of Optical coupler action,then the photoelectric converter converts laser interference signals into electrical signals, the system uses four core optical fiber as the sensing device accept outside disturbance, The intrusion events generate signals on the optical cable,and input to the system for signal processing.then Electrical signal processing module output pulse signal. This time difference is converted to digital information by TDC,The disturbance signal emitted in both directions in a closed loop four core cable system, Thereby generating the pulse signal of a time difference T,

Through single chip microcomputer program for processing,And then the results transmitted to the upper computer display.

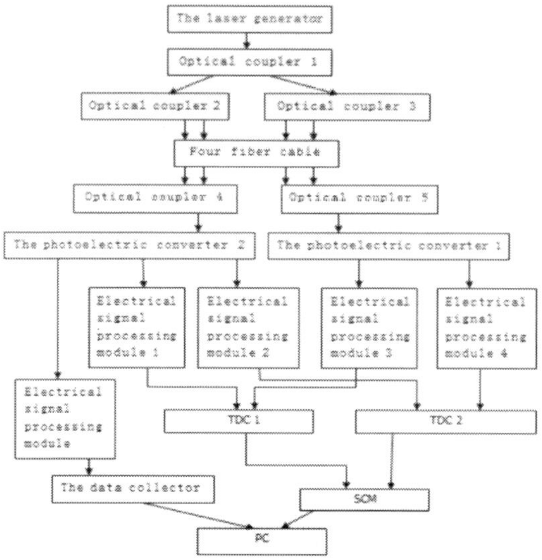

Fig. 1. System structure diagram.

2.2 System implementation process targeting

When the invasion of disturbance signal is stable , it is converted into stable electrical signals. The upper computer neither displays the invasion position nor gives an alarm. When the invasion occur , saltation light signals will be generated on intrusion location, and the volatility information would be obtained go through a series of signal processing process, and then the photoelectric conversion system and signal processing module can output the trigger pulse signal, lead to trigger the TDC work. Eventually,the computer calculates the location of the specific locations intrusion incident, and then Invasion of positioning is completed.

3 System program design

As soon as the system just starts, its initializes the interface of system and set parameters of a variety of hardware interfaces. Then enter the main loop program, the data can be analyzed, transmitted, analyzed and processed. continuous monitoring of the signature of four-core optical fiber, it continuous monitoring of the signature of four-core optical fiber, If there is an intrusion event occur , The system will respond immediately. The results are finally displayed on the upper computer. Accomplish Positioning function of the system.

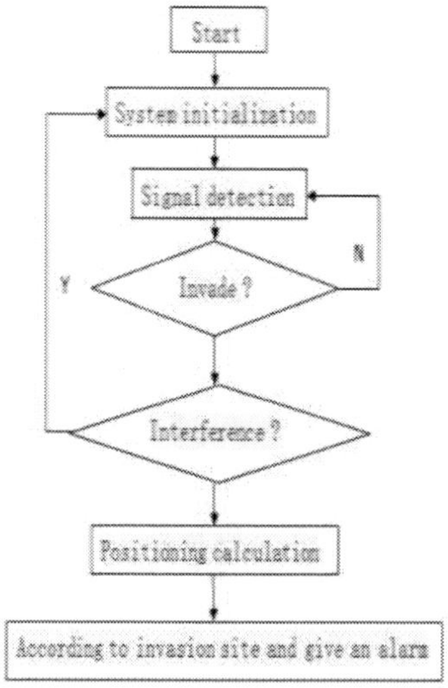

Fig. 2. Software program flowchart.

4 Conclusion

This region fiber perimeter security system adopts kernel 51 single-chip microcomputer as a core of Control Processor. the system uses the four-core cable device to obtain signal feature of intrusion incident. Invasion signal transforms the optical signals into electrical signals by the Optical coupler and Signal processing module. That pass through the electrical signal processing module and completed the signal processing by the TDC. Data acquisition card acquire system data and then transmits that to a upper computer. Finally the computer finish identify intrusion event properties.

The technical characteristics of this system includes innovation based on TDC invasion of the positioning system ,based on intrusion short time Fourier transform and wavelet analysis for Signal Feature Extraction and Intrusion alarm based on database technology.

References

1. Guangming Luo, Chunxi Zhang, etc. Distributed sensing and positioning technology fiber optic security system Piezoelectrics and Acoustooptics, 2009,31 (3): 327-333
2. Suiren Wan, Licheng Peng research extraction and classification algorithm security system features fiber optic signal Science & Technology Review, 2012,30 (36): 24-28
3. xing-qun zhao, Dongqing Xue, Distributed optical fiber vibration sensor signal data acquisition system Science and Technology News,2012,30(18):45-49
4. GuoFan Chen, Dawen Wang, Xiaoyan Wang, A regional fiber optic perimeter security systems. China Patent, 201220161011.2012-10-31.

Development of A Maintenance Device for Bus-bar PT Voltage Air Switch

Xiang Zhang*, Ying Chen, Gaoxiang Lin and Bifu Qian

Wenzhou Power Supply Company of SGCC, Wenzhou, Zhejiang 325000, China
Corresponding Email: xiangzju@126.com

Abstract. When PT breaks down, it takes long time of switching operation before maintenance, which seriously delays the restoration time. Based on the principle of multiple circuit, a live replacement maintenance device for PT voltage air switch is proposed. The following aspects are involved in the design of the device: the principle of device, component selection, device's assembly and operation process. Through functional test in simulation substation and on-site installation, it is proved that the failed air switch can be lively replaced by the device without switching operation, which greatly reduces the risk on power grid caused by such faulted air switch.
Keywords. Air switch, potential transformer, live replacement, switching operation.

1 Introduction

State Grid Corporation of China firmly considers building strong power grid as the development trend [1,2], so taking measures to protect the vulnerable points always remains a crux [3,4].Due to long-term operation in the outdoor terminal boxes, the voltage air switch of bus-bar potential transformer (PT) is prone to appear increasing internal resistance, serious voltage allocation problem and even single-phase or multi-phase interruption. These issues cause a voltage difference; therefore make the voltage acquisition inaccurate. Unbalanced voltage in high level will unlock the compound voltage lockout so that bus protection, transformer protection and line protection act incorrectly which leads to severe consequence [5]. For the maintenance department, the abnormity of the voltage air switch of bus-bar PT requires fast restoration. However, current maintenance procedure needs switching operation of dispatching department and operation department. Only after dropping out the PT can the maintenance of corresponding air switch begin. It is obvious that current time-taking process cannot meet the demand.

Live replacement can be an effective way to solve this problem, and until now seldom relevant reports have been seen. Literature [6] introduces a cabinet tool which can lively replace the DC fuse, but it isn't applicable to compact terminal array of PT. The essay presents a maintenance pedestal, with which maintenance personnel could replace the voltage air switch of bus-bar PT lively. Through the design of principle, device manufacture and system testing, this device complies fully with the requirements, and thus enhances the stability of power grid much further.

2 Design background

2.1 Potential transformer and corresponding air switch

Potential transformer is an essential element for linear voltage transition [7]. The secondary voltage can be used to calculate, monitor and count for load equipment such as protection equipment, measurement, control devices and metering devices. When the potential transformer voltage air switch in operation emerges short circuit, the strong short current will burn out the potential transformer [7].Therefore, one main air switch is setup in the potential transformer terminal boxes to protect the secondary circuit. Figure 1 shows the secondary circuit of 220kV bus-bar potential transformer.

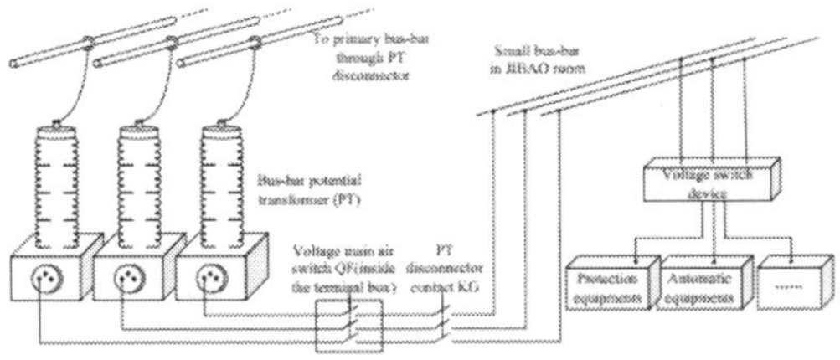

Fig. 1. Secondary circuit of 220kV bus-bar potential transformer.

2.2 Traditional maintenance mode

When the main air switch QF needs urgently replacing for abnormity, traditional maintenance mode has to conduct switching operation in order not to lose the secondary voltage that makes a difference in the reliability of protection devices. The switching operation includes transference of primary equipment and secondary equipment connected with PT to that in adjacent bus-bar, then the PT quits operation.

Take protection main air switch failure of PT in bus-bar I for example: in typical 220kV double bus-bar main wiring (refer to Figure 2), there are two approaches of switching operation. One is to transfer primary equipment from bus-bar I to bus-bar II so that bus-bar I and its PT quit operation. The other has to close bus-tie circuit breaker, then quit the PT in bus-bar I with the assistance of secondary voltage juxtaposition.

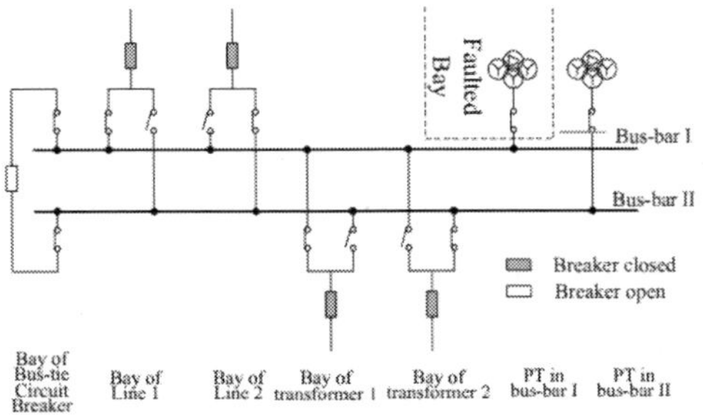

Fig. 2. Typical 220kV double bus-bar main wiring.

Table 1 compares the two switching operations, including equipment numbers, average operation time and risk rank of power grid. It is clear that the time-consuming procedure can't match the demand of timely recovery. Furthermore, it breaches building strong power grid because of the reduced reliability after operation [8].

Table 1. Comparison of the two switching operations.

No.	Mode One	Mode Two
Switch Operation Introduction	Operate primary devices from bus-bar I to bus-bar II and quit bus-bar I.	Close bus-tie circuit breaker, juxtapose the secondary voltage and quit the PT in bus-bar I.
Operation Times of Primary Devices	7	2
Operation Times of Secondary Devices	0	1
Operation Time	143 mins	71 mins
Main Wiring Mode after Operation Switch	Single-bus	Double-bus juxtaposition
Risk Grade of Power Grid and Audit Mode	Six, Audited by leaders of department in charge and approved by leaders of company	Eight, Approved by leaders of regulation and control center

3 Device designs and elements select

To decrease the impact on power grid that given by maintaining such air switch and to reduce the abnormal voltage time, this essay put forward a live replacement maintenance device for PT voltage air switch. Installing this device under original air switch makes it possible to isolate and replace the main air switch without outage of secondary voltage.

3.1 Device principle

The design is based on the principle that reserve protection circuit replaces the abnormal air switch circuit and parallel connecting. The two voltage circuit can realize seamless switching. Circuit one is used on normal operating while circuit two is activated during voltage air switch maintenance as reserve protection. The final maintenance device mainly consists of circuit switch, reserve protection and circuit monitoring modules. Figure 3 shows Schematic of the maintenance device and Table 2 clarifies components in the schematic.

3.1.1 Circuit switch module

Circuit switch module includes a 6-set and 2-level MBB rotary switch called K. MBB rotary switch remains a state that connects fore and back contacts when switching [9]. It enables the protection function of secondary circuit to seamlessly switch from normal operating (circuit one) to reserve protection (circuit two) without outage. Six sets of contacts realize the function that switches three-phase voltage in both sides of air switch synchronously. When rotary switch K is switched to reserve protection, the air switch in normal operating circuit is isolated. Therefore, maintenance personnel can replace the broken-down air switch in safety.

3.1.2 Reserve protection module

During isolation of original air switch, the protection function of voltage secondary circuit is provided by reserve protection module (R_A, R_B, R_C) .In the process of maintenance, the voltage circuit can be reliably switched off by reserve protection module when load secondary circuit occurs short circuit and grounding, which avoids short circuit in PT secondary side.

3.1.3 Circuit monitoring module

Circuit monitoring module consists of four sets of interphase indicator light, which monitor the voltage condition respectively. It provides direct operation feedback for reliability and safety of circuit switch. Considering reducing the operation power of circuit monitoring module and decreasing the ratio of PT capacity, the device chooses luminous diode of low nominal voltage with biggish circuit resistance.

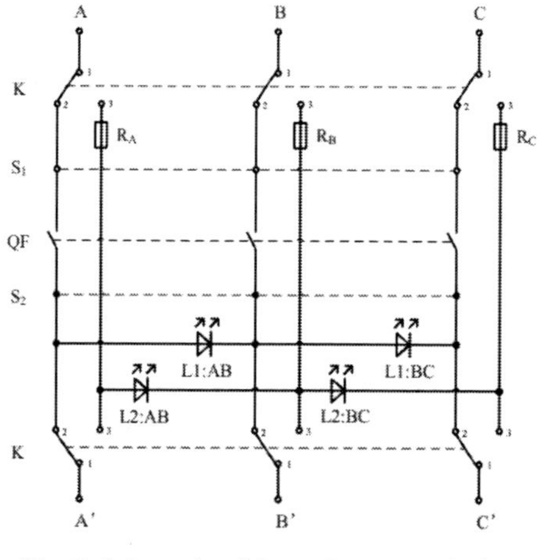

Fig. 3. Schematic of the maintenance device.

Table 2. Components in schematic.

No.	Symbol	Introduction
1	QF	Three-phase AC voltage air switch
2	A、B、C	Upper port of maintenance device, connected with three-phase voltage in secondary side of PT
3	A'、B'、C'	Lower port of maintenance device,

		connected with protection equipment, automatic equipment and so on
4	K	MBB rotary switch, used to switch voltage circuit
5	R_A、R_B、R_C	Overcurrent protection components of reserve circuit
6	S_1、S_2	Contacts connecting air switch QF and maintenance device
7	L1:AB 、 L1:BC、 L2:AB 、 L2:BC	Interphase circuit indicator lights

3.2 Components' selection

The physical device is based on schematic. In order to downsize the size, fuse, rotary switch and indicator light are intensively arranged. Three-phased voltage air switch is fixed on the right of the device with normative slot that makes it easier to clearing and wiring as is shown in Figure 4.

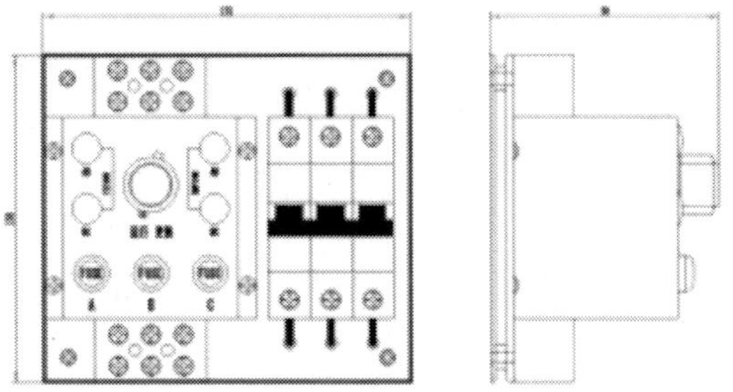

Fig. 4. Design drawing of device.

After confirming the final design drawing of device, correlation test and orthogonality analysis method are used to select component model with the aim of lowest operation power and resistance balance of three-phased voltage circuit. The indicator light in the normal operating circuit is selects as green while the reserve protection circuit red so that it is easily distinguished; The nominal current of ceramic tube fuse matches the trip setting current of voltage air switch. Components' selection is shown on Table 3.

Table 3. Components' selection.

Component	Type and specification	Quantity	Size/mm^3
Rotary fuse base	FH1-13	3	φ18*53.2
LED indicator light	AC12V 、 Red	2	φ10*35
LED indicator light	AC12V 、 Green	2	φ10*35
Rotary switch	XB2-BD	1	φ30*40
Terminal array	8003	2	40*26*26
Air switch	B6 three-phase AC voltage	1	90*53*70
Ceramic tube fuse	6A/250V	3	φ6*30
Metal film resistance	2W50kΩ	4	φ3*9

4 Device test and efficiency analysis

4.1 Function test

In the local simulation substation, the group imitates a construction of voltage secondary circuit to test the operation circumstance of circuits switch and air switch replacement.

When it happens one-phase or multi-phase abnormal interruption, the PT/TV breaking detection of protection devices sends out warning signal after judging time, and meanwhile quit the main protection function of distance protection element and direction element [10].

For the commonly used 220kV protection devices in Wenzhou, Table 4 shows time discrimination of PT/TV breaking in protection devices, from which we can know that CSC-103B developed by BEIJING SIFANG Automation CO.,LTD. and WXH-803A developed by XJ Group Corporation spend the shortest judging time as 1 second.

Table 4. Time discrimination of PT/TV breaking in protection devices.

Manufacturer	Protection type	Device type	Time /s
NARI Relay Protection Company	Line	RCS-931	1.25
	Transformer	RCS-978	10
	Bus	RCS-915	1.25
BEIJING SIFANG Automation CO.,LTD	Line	CSC-103B	1
	Transformer	CSC-326	10
	Bus	CSC-150	10
GUODIAN Nanjing Automation CO.,LTD	Line	PSL-603	1.25
	Transformer	PST-1202	1.25
	Bus	SG-B750	10
XJ Group Corporation	Line	WXH-803A	1
	Transformer	WBH-800	10
CYG SUNRI CO.,LTD.	Bus	BP-2B	9

Figure 5 illustrates the wiring of function test. The voltage sampling circuit of CSC-103B line protection device and maintenance device is constructed in simulation substation. HAOMAI 1600 test device can output simulative three-phase nominal voltage of PT secondary side. The result of fifty times air switch replacement tests shows the voltage sampling neither interrupting nor appearing PT/TV warning signal when the device conducts circuit switching. And under the condition of reserve protection circuit, the clearing and wiring of air switch won't influence the sampling.

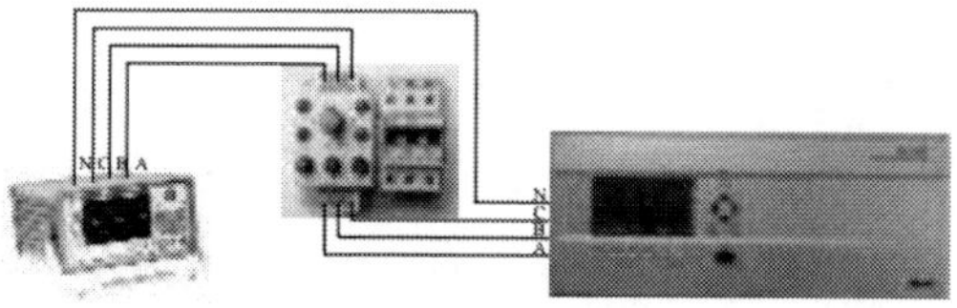

Fig. 5. Wiring of function test.

4.2 On-site installation and implementation efficiency

In the 220kV PT terminal box reconstruction project of one specific 220kV substation, the maintenance device is installed in the field (refer to Figure 6). The original three-phased AC air switch is replaced by maintenance device, and then placed on the maintenance device.

Fig. 6. On-site installation in 220kV PT terminal box.

Traditional maintenance mode needs switching operation to replace the abnormal air switch, involving nine sections of three departments(refer to Figure 7). With this device, maintenance can take operating personnel's switching operation and despatcher's operation compilation out so that the procedure cut down to five parts (refer to Figure 8); Working timespan decreases from 288.5 mins to 66.5 mins. What's more, operation personnel can take the task of air switch replacement with further implementation of operation-maintenance integration. Having no participation of maintenance department, the procedure can be reduced to less than 30 mins.

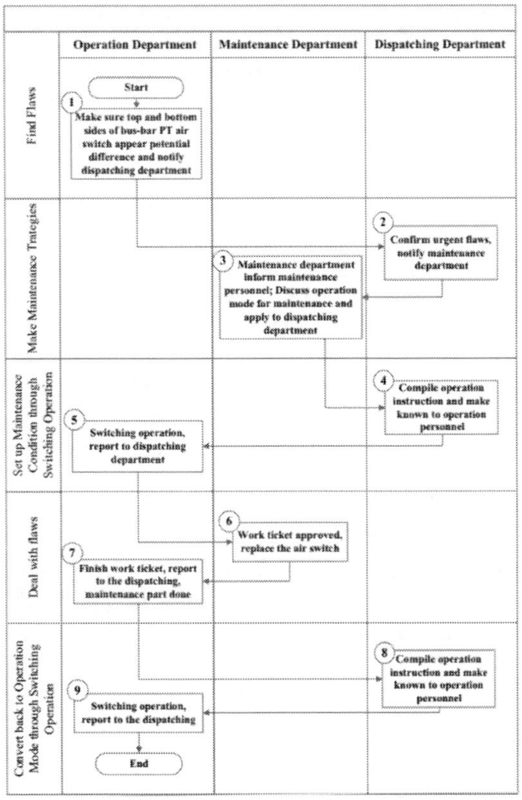

Fig. 7. Swim-lane Flowchart of replacement without maintenance device.

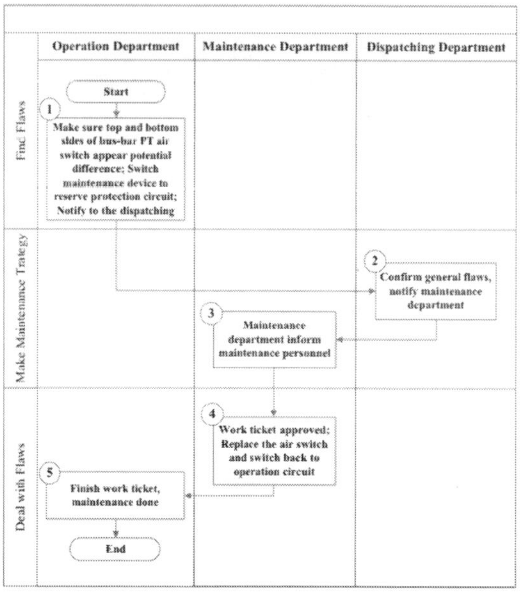

Fig. 8. Swim-lane Flowchart of replacement with maintenance device.

Operation personnel applies reserve circuit to temporarily substitute the faulted air switch, that helps protection voltage recovered at once and changes the grade of flaws from urgency to the general. Moreover, it cuts down the risk taken by switching operation and brings benefits to the stable operation of power grid.

5 Conclusion

This essay put forward a new maintenance device for bus-bar PT voltage air switch. Based on the principle of multiple circuit, it applies to replace air switch in different situations of switching operation. The device successfully simplifies work procedure from 9 sections into 5 sections, and the timespan reduces from 288.5 mins to 66.5 mins. It not only increases the efficiency but also tremendously reduces the risk taken by abnormal voltage for stable operation of power grid.

References

1. Liu Z Y. Smart grid technology. Beijing: China Electric Power Press,1-11(2010).
2. Zhang Dongxia, Yao Liangzhong;Ma Wenyuan. Development strategies of smart grid in china and abroad. *Proceedings of the CSEE* , **31**, 1-14(2013).
3. Liu Z Y. Electric power and energy in China. Beijing: China Electric Power Press, 18-44(2012).
4. Xie Kai, Liu Yongqi, Zhu Zhizhong, et al. The vision of future smart grid. *Electric Power*, **41**, 19-22(2008).
5. Gao Zhongde, Shu Zhihuai, Wang Delin, et al. State grid corporation of protection training materials. Beijing: China Electric Power Press, 411-416(2009).
6. Huang Wei, Zhang Jianghuan, Hu Junhua, et al. Substation DC system fuse Live Replacement. *Electric Engineering*, **43**, 51-52(2015).
7. Mao Jinqing, Zhao Zigang, Ma Jie, et al. Practical technical questions and answers of power system relay protection. Second edition. Beijing: China Electric Power Press, 125-133(2006).
8. Song Xinli, Tang Yong, Bu Guangquan, et al. Full dynamic simulation for the stability analysis of large power system. *Power System Technology*, **32**, 23-28(2008).
9. Wang Quantao, Yao Shujun, Han Minxiao, et al. Solid static transfer switching for duplex feeding. *Electric Power Automation Equipment*, **30**, 67-70(2010).
10. Yuan Zhian. Principle and application of loss-of-potential logic. *Electric Power Automation Equipment*, **29**,147-150(2009).

AUTHOR INDEX

AUTHOR INDEX

AUTHOR INDEX

AUTHOR INDEX

AUTHOR INDEX

AUTHOR INDEX

13th Global Congress on Manufacturing and Management (GCMM 2016)

MATEC Web of Conferences Volume 100 (2017)

Zhengzhou, China
28 - 30 November 2016

Part 2 of 2

Editors:

Leo Zhao
Jeffrey Cai
Anthony Xavior
Linda You

ISBN: 978-1-5108-3748-5

Printed from e-media with permission by:

Curran Associates, Inc.
57 Morehouse Lane
Red Hook, NY 12571

Some format issues inherent in the e-media version may also appear in this print version.

Printed by Curran Associates, Inc. (2017)

For additional information, please contact EDP Sciences – Web of Conferences at the address below.

EDP Sciences – Web of Conferences
17, Avenue du Hoggar
Parc d'Activité de Courtabœuf
BP 112
F-91944 Les Ulis Cedex A
France

Phone: +33 (0) 1 69 18 75 75
Fax: +33 (0) 1 69 28 84 91

contact-edps@webofconferences.org

Additional copies of this publication are available from:

Curran Associates, Inc.
57 Morehouse Lane
Red Hook, NY 12571 USA
Phone: 845-758-0400
Fax: 845-758-2633
Email: curran@proceedings.com
Web: www.proceedings.com

TABLE OF CONTENTS

PART 1

PART 2

Two Types of New Natural Materials for Fruit Vinegar in *Prunus* Plants

Han Zhao[1], Xiaoxing Zhou[2], Ying Luo[1], Yanli Huang[3], Tana Wuyun[1], Fangdong Li[1] and Gaopu Zhu[1,*]

[1]Non-timber Forestry Research and Development Center, Chinese Academy of Forestry, and China Paulownia Research and Development Center, Zhengzhou 450003, China
[2]Administrative Office, Chinese Academy of Forestry, Beijing 100091, China
[3]Henan Province Institute of Landscape Planning and Design Company Limited, Zhengzhou 450045, China
Corresponding Email: poog502@163.com

Abstract. To increase *Prunus armeniaca* × *P. sibirica* and *P. domestica* × *P. armeniaca* added value; three natural fruit vinegars were designed. The results showed the nutrition of *Prunus domestica* × *P. armeniaca cultivar* Fengweimeigui vinegar (T1) had high minerals and microelements, especially the Ca and Mg reached to the 150.00mg/L, 85.40 mg/L, respectively; the vinegar of *Prunus armeniaca* × *P. sibirica cultivar* Zhongren No.1 (T2) not only have rich Na (2800.00 mg/L), P (123.00 mg/L), but also have plentiful amino acid that content reached to 200.08 mg/L. However, the mixture vinegar (T3) with pulps from *Prunus domestica* × *P. armeniaca* and *Prunus armeniaca* × *P. sibirica* had the middle nutrient contents, but the property was balanced. We therefore conclude that solid fermentation is a suitable method to preserve nutrients and value-added for *Prunus* plants fruit, and three types vinegars are suitable for different age people, and the difference nutrient contents and typical characteristic indicate that three vinegars are competitive products in market.

1 Introduction

In general, for fruits, there are nourishing but short shelf life, for nuts, the pulps used discarded. For examples, fruits *Prunus domestica* × *P. armeniaca* cultivar Fengweimeigui is rich nutrient [1-2], but the shelf life only lasting 5-7 d; the pulps of nuts *Prunus armeniaca* × *P. sibirica cultivar* Zhongren No.1 have organic acid, Ca, K, Fe and Se [3] but the pulps is inedible because of sour and bitter. With the food structure development, human interest in the health benefits of functional foods is increasingly. Meanwhile, the farmers hope to add agricultural raw materials value. The both of them expectations are consistent and suitable for future economic trend. For these reasons, we developed the new, functional and special products will help stronger the body and economic recovery. Microbial fermentation have been efficient biocatalysts to make the functional foods, this is a suitable method to solve above problems [3-7]. However, fruits vinegar is not only used as a seasoning but also play important roles, such as antioxidant, digestive, lipid lowering effects, and regulations of blood-pressure, these results in the vinegar can resist or assist cure the imbalanced fat or excess energy intake harm [8-10]. In this paper, based on solid fermentation method to produce fruit vinegar and improve residuum utilization ratio have been explored.

2 Materials and methods

 DOI: 10.1051/matecconf/201710004006

2.1 Sample collection

Fully ripened and healthy (without diseases and insect pests) *Prunus domestica* × *P. armeniaca cultivar* Fengweimeigui and *Prunus armeniaca* × *P. sibirica cultivar* Zhongren No.1 fruits, whether big or small, were selected from garden of the Experimental Site of Non-timber Forestry Research and Development Center, Chinese Academy of Forestry, Yuanyang County, China, during June 2013 (day temperature 28±2℃, night temperature 25±2℃) (Figure 1). The three treatments were designed including T1, T2 and T3. The T1 and T2 are the pure *Prunus armeniaca* × *P. sibirica cultivar* Zhongren No.1, *Prunus domestica* × *P. armeniaca cultivar* Fengweimeigui, and the T3 is the mixture between *Prunus armeniaca* × *P. sibirica cultivar* Zhongren No.1 and *Prunus domestica* × *P. armeniaca cultivar* Fengweimeigui with weight is 1:1 (w/w) ratio because the pulps carbohydrate of *Prunus armeniaca* × *P. sibirica cultivar* Zhongren No.1 was lower, but the *Prunus domestica* × *P. armeniaca cultivar* Fengweimeigui was higher.

2.2 Fermentation process

The idea was solid fermentation. The total included seven stages (Figure 1). First, fruits were cleaned and dried. Second, the fruits were put a closed room and use ultraviolet disinfection with 8 h~12 h. Third, the fruits were crushed with machine and became pulp, meanwhile, the seeds was separated with pulp. Fourth, alcoholic fermentation stage including four steps (Figure 1). (a) Based on the preliminary experiment, the weight of saccharose with 2:5 (w/w) was added. (b) The *Saccharomyces cerevisiae* was added and the ratio is 0.2 g/100ml. (c) The pulps were stirred every 6 h interval until the saccharose were melted (about 3 times). (d) A stationary culture was conducted for 3 days at 35 ℃ in constant-temperature culture medium. The fifth stage, static acetic acid fermentation with three steps. (a) Acetic bacteria were added and the content reached to 1.0mg/100ml (w/v). (b) Fermentation at higher temperature at 33 ± 2℃ last 5 days. (c) Fermentation at constant lower temperature at 25±2℃ last 45 days. After fermentation, the filtration and sterilization was operation. Finally, the bottles were filled with vinegars. The procedure for making fruits vinegars is list in figure 1.

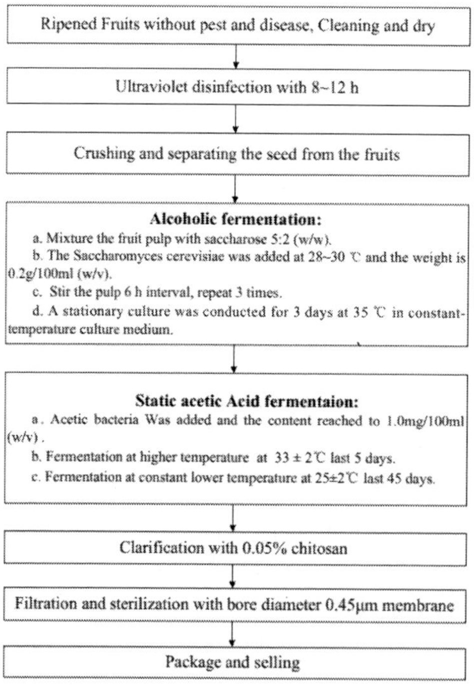

Fig. 1. Procedure for making fruit vinegars.

2.3 Composition and content analysis

The composition and content of amino acid, minerals and microelements, juice pH value and biochemical constituents (total sugar, total fat, total protein and ethanol) were analyzed based on GB/T 5009.124-2003 (China), GB/T 14924.12-2001 (China) and Ref. (Amerine et al., 1980; Zhu et al. 2015).

2.4 Data analysis

The experimental layout was a complete randomized plot with three replicates of five sampled per treatment. The data collect with Microsoft Office Excel 2013 software and the statistic analysis of variance analysis use Duncan's multiple range tests with DPS v6.05 software. Significance was accepted at $p \leq 0.05$.

3 Results and analysis

3.1 The content different of major components

Table 1. The proximate composition of main nutrition apricot vinegars (mg/L).

Treatment	Protein	Fat	Sugar	Fe	Mg	Na	Zn	Ca	P	pH
T1	0.12±0.02aA	0.01±0.00a	*138.00±1.70aA*	*3.30±0.16aA*	*85.40±2.90aA*	2600.00±14.84bB	*1.10±0.05aA*	*150.00±3.40aA*	63.70±2.73aA	*3.25±0.02a*
T2	*0.52±0.02cC*	0.02±0.01a	76.00±1.90cC	—	50.40±1.41cC	*2800.00±51.21aA*	0.64±0.04cB	108.00±2.75cB	*123.00±2.59bB*	3.14±0.05a
T3	0.18±0.01bB	0.02±0.00a	108.00±2.30bB	2.60±0.17bB	78.40±1.98bB	2600.00±6.08bB	0.74±0.06bB	143.00±2.43bA	68.20±2.56bB	3.19±0.01a

Note: values are means ± SD (n=5). Means in columns without letters in common differ significantly ($p<0.05$). ND = not detected.

In this study, all the treatments, the contents of protein and fat were lower, but the sugars 138.00 mg/L, 76.00 mg/L and 108.00 mg/L, respectively were higher and the different reached to the significantly (Table 1).

Table 2. The proximate content of essential amino acid in Fengweimeigui wine (mg/ L).

Treatment	Lys	Trp	Phe	Met	Thr	Ile	Leu	Val	Total
T1	1.8±0.04bB	—	1.84±0.03bB	0.16±0.01a	1.11±0.03b	0.82±0.03bB	0.65±0.05cC	1.86±0.09bB	8.24±0.78bB
T2	2.58±0.05aA	—	2.26±0.07aA	0.17±0.01a	—	1.78±0.05aA	2.08±0.05aA	3.36±0.06aA	12.23±0.55 aA
T3	1.52±0.03cC	—	1.56±0.06cC	0.16±0.00a	—	0.77±0.04bB	0.81±0.03bB	1.72±0.05cB	6.54±0.43cC

Note: values are means ± SD (n=5). Means in columns without letters in common differ significantly ($P<0.05$). ND = not detected.

Table 3. The proximate content of non-essential amino acid in vinegars (mg/ L).

Treatment	Asp	Ser	Gln	Gly	Ala	Cys	Tyr	His	Arg	Pro	Total
T1	1.32±0.05cC	1.83±0.06cC	1.10±0.06bB	0.36±0.04cC	4.40±0.40cC	1.03±0.00a	0.70±0.00bB	0.62±0.00bB	0.34±0.00bB	1.31±0.02cC	13.01±0.09cC
T2	44.56±3.51aA	9.48±0.09aA	4.54±0.57aA	1.94±0.08aA	29.64±1.10aA	1.14±0.00a	1.24±0.05aA	1.60±0.04aA	0.65±0.03aA	93.06±3.95aA	187.85±6.77aA

T3	10.02±0.08bB	2.61±0.13bB	1.18±0.04bB	0.56±0.06bB	7.12±0.10bB	1.02±0.01a	0.78±0.02bB	0.62±0.01bB	0.32±0.00bB	17.52±1.55bB	41.75±2.33bB

Note: values are means ± SD (n=5). Means in columns without letters in common differ significantly (P<0.05). ND = not detected.

In minerals and microelements part, T1 has the highest contents Fe (3.30 mg/L), Mg (85.40 mg/L), Zn (1.10 mg/L) and Ca (150.00 mg/L) shown that the vinegar included abundant mineral materials, and the characteristic constituent is the highest Mg and Ca contents (Table 1). Treatment T2 has the highest contents protein (0.52 mg/L), 2800.00 mg/L Na and 123.00 mg/L P, so the characteristic constituent are Na and P contents that these was difference than T1 (Table 1). However, Treatment T3 falls in between (Table 1).

3.2 The content different of essential amino acid

The total content of essential amino acid was lower among three treatments, and the amino acid Trp was not found in this paper (Table 2). The total content of essential amino acid from highest to lowest at different treatments was T2, T3 and T1. Treatment T1, have the only amino acid Thr (1.11 mg/L) and lowest total essential amino acid. Treatment T2 have the highest total essential amino acid (12.23 mg/L), and the components of amino acid Lys (2.58 mg/L), Phe (2.26 mg/L), Ile (1.78 mg/L), Leu (2.08 mg/L) and Val (3.36 mg/L) also has highest and the difference reached very significant Table 2). The T3 falls in the between except amino acid Leu and total essential amino acid (Table 2).

3.3 The content different of non-essential amino acid

The content tendency of total content of non-essential amino acid shown the same rule with essential amino acid, the contents from highest to lowest was T2>T3>T1 (Table 3), but the every constituents were significant difference. Treatment T2, not only have the highest total content but also have the highest constituents (44.56 mg/L Asp, 9.48 mg/L Ser, 4.54 mg/L Gln, 29.64 mg/L Ala, 1.24 mg/L Tyr and 0.65 mg/L Arg), especially in the content of amino acid Pro that reached to the 93.06 mg/L, the higher over 71, 5.3 times, respectively. And the next is the Asp, the difference reached to the 33.8, 4.5 times, respectively. So, the contents of amino acid Pro and Asp were the typical compositions in T2.

4 Conclusion and discussion

We have developed three new vinegar types. Vinegar prepared from *Prunus armeniaca × P. sibirica cultivar* Zhongren No.1 (T1) fruit pulp is a novel beverage rich minerals and microelements, especially in contents of Fe, Mg, Zn and Ca, so we putative that the content minerals Ca is the typical characteristic. Vinegar prepared from *Prunus domestica × P. armeniaca cultivar* Fengweimeigui (T2) fruit pule is a novel beverage rich amino acid (200.08 mg/L), especially in non-essential amino acid (187.85 mg/L), and this is the typical characteristic. But treatment T3, the contents of minerals, microelements and amino acids fall in between T1 and T2, the characteristic is mild and the nutrient is balanced.

In this study, the vinegars of proteins, fats and non-essential amino acid were lower than kernel-apricot vinegar, but the essential amino acid was significant higher, especially amino acid Lys, Phe, Met, Leu and Val, meanwhile, the amino acid Thr, Ile, Gln and Arg were only found in apricot vinegars [3], the results showed that the fermenting method was importance factor for nutritional ingredient separation from fruits pulp. Compared with apple vinegar, the essential amino acid was significant lower, but non-essential amino acid including the Ser, Ala and Pro were higher [11], especially the amino acid Pro (Table 3, 93.06 mg/L) was the over 180 times than apple vinegar (0.50 mg/L). Compared with persimmon vinegar, the amino acid was significant higher than persimmon vinegar [12], and the amino acid Cys was not found.

Acknowledgments

This work was supported by the Program of Research and Demonstration of Key Technology of High-efficient Production in Kernel-apricot and Almond under Grant No. 2013BAD14B02.

References

1. L.L. Lei, Study on the reproductive biology of *Prunus domestica* × *armeniaca* 'fengweimeigui', M.S. Thesis, Chinese Academy of Forestry, Beijing, China (2013).
2. G.P. Zhu, H. Zhao, X.X. Zhou, M.P. Liu, Y.L. Huang, T.N. Wuyun, F.D. Li, *Prunus domestica*×*P. armeniaca* cultivar Fengweimeigui: A New Natural Material for Fruit Wine, Adv. J. Food Sci. Technol. **10**(4), 277-280 (2016).
3. Y. Fan, Studies on the brewing techniques of apricot for kernel flesh vinegar and its health-care function, M.S. Thesis, Northwest A&F University, Yangling, China (2008).
4. S.M. Araújo, C.F. Silva, J.J.S. Moreira, N. Narain, R.R. Souza, Biotechnological process for obtaining new fermented products from cashew apple fruit by Saccharomyces cerevisiae strains, J. Ind. Microbiol. Biotechnol. **38**, 1161–1169 (2011).
5. T.S. Suryanarayanan, N. Thirunavukkarasu, M.B. Govindarajulu, V. Gopalan, Fungal endophytes: an untapped source of biocatalysts, Fungal Diversity **54**, 19–30 (2012).
6. Y. Jiang, S. Lin, L. Zhang, P. Yu, Upgrading the Fermentation Process of Zhejiang Rosy Vinegar by Purebred Microorganisms, Advances in Microbiology **3**, 297-301 (2013).
7. Z Nie, Y. Zheng, H. Du, S. Xie, M. Wang, Dynamics and diversity of microbial community succession in traditional fermentation of Shanxi aged vinegar, Food Microbiology **47**, 62-68 (2015).
8. J. Qui, C. Ren, J. Fan, Z. Li, Antioxidant activities of aged oat vinegar in vitro and in mouse serum and liver, Journal of the science and food agriculture **90**, 1951-1958 (2010).
9. World Health Organization, Obesity and overweight, Available from http://www.who.int/mediacentre/factsheets/fs311/en/ (2015).
10. C.H. Chou, C.W. Liu, D.J. Yang, Y.H. Samuel, Y.C. Wu, Chen, Amino acid, mineral, and polyphenolic profiles of black vinegar and its lipid lowering and antioxidant effects in vivo, Food Chemistry **168**, 63-69 (2015).
11. H.T. Zhang, X.L. Liu, Development and amino acid analysis of apple cider vinegar drink. Journal of Shandong Institute of Light Industry **24**(1), 41-44. (2010) (in Chinese).
12. X.L. Luo, Y. Yang, S.M. Lu, C.L. Zhang, The impact of fermentation modes on sensory and nutritional quality of persimmon vinegar, Food Science, Available from http://www.cnki.net/kcms/detail/11.2206.TS.20160108.1651.052.html (2016) (in Chinese) (to be published online).

The Influence of Chemical Composition on LNG Pool Vaporization

Zhidong Yu *

China Academy of Railway Sciences, Beijing, 100081, China
Corresponding Email: yygdupt_mmc@163.com

Abstract. A model is used to examine the influence of chemical composition on the vaporization rate of LNG during spreading. Calculations have been performed whereby the vaporization rate of the LNG mixtures has been compared to the vaporization of pure methane under the initial conditions. The detailed results indicate that the vaporization rate LNG mixture is different to that of pure methane. LNG as the liquid mixture gets rich in ethane and isobaric latent heat increases rapidly, leading to the rate vaporization of LNG decrease in compared to pure methane.

1 Introduction

The volume of liquefied natural gas (LNG) is only 1/600 that of its gaseous state, LNG allows more flexible means of transport and storage. In recent years demand for natural gas as a fuel has been on the increase. World-wide requirements necessitate long distance marine and tail transportation. Such an increase will bring renewed pressures on improved quantification of risk associated with LNG transport and loading operations, ensuring that safety precautions can be put in place [1] .

It is the most importance to understand and quantify the physical processes which take place in the spillage of LNG. In modeling the consequences of releases of LNG, the quantities of interest for the rate at which vapour is produced from the pool, the size of the pool, and the temperature of the vapour [2,3] .

2. Mathematic model analysis

2.1 Pool spreading model

We can describe the spreading of pool to give the surface area, which is an important variable in vaporization rate calculation. Different equations are used to calculate the pool radius depending upon whether the spill is on land or water, and whether it is an instantaneous, a continuous [4-6] .

The radius of a pool spreading on land is found by numerically integrating the spreading rate equation:

$$\frac{dr}{dt} = \sqrt{2g(h - h_{\min})} \quad (1)$$

Where r is the pool radius (m)
t is the time from the start of release (s)
g is the acceleration due to gravity (m/s^2)
h is the pool thickness (m)
h_{min} is a minimum pool thickness (m)

This assumes that the driving force for the spread is formed by the hydrostatic difference between the thickness of the current liquid layer and a minimum pool thickness characteristic for the substrate. This results in the rate of spreading decreasing as the pool approaches the minimum thickness.

The spreading of liquids on water is found by a three-regime model. Each regime is dominated by the balance of two main forces, the regimes occurring sequentially. The first regime to occur is the gravity-inertia regime when the gravity spreading and inertial forces are approximately equal. This is followed by the gravity-viscous regime when the gravity spreading and viscous drag forces are approximately equal. For non-volatile liquids which may spread to form thin slicks there may then be a third regime, the viscous-surface tension regime, where the viscous drag and surface tension forces are approximately equal. In practice, with the minimum thickness criterion implemented in the model the viscous-surface tension regime is rarely encountered.

The four forces considered in the three regimes are:
Gravity spreading force

$$F_G = M_{pool}(\frac{gh\Delta}{r}) \tag{2}$$

Inertial force

$$F_I = -M_{pool}\frac{d^2r}{dt^2} \cong M_{pool}\frac{r}{2t^2} \tag{3}$$

Viscous drag force

$$F_D = \pi r^2 \mu_w (\frac{r}{t})(\frac{\rho_w}{\mu_w t})^{\frac{1}{2}} \tag{4}$$

Surface tension force

$$F_S = 2\pi r\sigma \tag{5}$$

In the equations above $\Delta = (\rho_w - \rho_L) / \rho_w$ is a relative density factor, with ρ_w, ρ_L being the liquid and water densities (kg/m^3). Furthermore μ_w is the dynamic viscosity of the water (kg/m/s, and $\sigma = \sigma_{wa} - \sigma_{La} - \sigma_{Lw}$ is the net surface tension (N/m), with σ_{wa}, σ_{La}, σ_{Lw} being the surface tension of water-air, liquid-air, liquid-water, respectively.

2.2 instantaneous spills model

The radius is given as follows.
In the gravity-inertia regime when $0 < t < t_1$ by:

$$r(t) = K_{10}(\frac{g\Delta}{\rho_L})^{1/4} M_{pool}^{1/4}(0)t^{1/2} \tag{6}$$

In the gravity-viscous regime when $t_1 < t < t_2$ by:

$$r(t) = K_{20}\left(\frac{g\Delta}{\rho_L^2(\mu_w/\rho_w)^{1/2}}\right)^{1/6} M_{pool}^{1/3}(0)t^{1/4} \tag{7}$$

In the viscous-surface tension regime when $t > t_2$ by:

$$r(t) = (\frac{4\sigma^2}{\rho_w \mu_w})^{1/4} t^{3/4} \tag{8}$$

The transition from the gravity-inertia to the gravity-viscous regime occurs at the time:

$$t_1 = \left(\frac{K_{20}}{K_{10}}\right)^4 \left(\frac{\rho_w M_{pool}(0)}{\rho_L g \mu_w \Delta}\right)^{1/3} \tag{9}$$

and from the gravity-viscous to the viscous-surface tension regime at the time:

$$t_2 = (K_{20})^2 \left(\frac{M_{pool}^2(0) g \mu_w \Delta}{\rho_w \rho_L^2}\right)^{1/3} \left(\frac{\rho_w}{2\sigma}\right) \tag{10}$$

2.3 Continuous spills model

The case of a constant spillage S(t) = S is considered. The radius is given as follows.
In the gravity-inertia regime when $0 < t < t_3$ by:

$$r(t) = K_{11}\left(\frac{g\Delta}{\rho_L}\right)^{1/4} S^{1/4} t^{3/4} \tag{11}$$

In the gravity-viscous regime when $t > t_3$ by:

$$r(t) = K_{21}\left(\frac{g\Delta}{(\mu_w/\rho_w)^{1/2}\rho_L^2}\right)^{1/6} S^{1/3} t^{7/12} \tag{12}$$

The transition between these regimes occurs at the time:

$$t_3 = \left(\frac{K_{21}}{K_{11}}\right)^6 \left(\frac{\rho_w}{\rho_L g \Delta \mu_w} S\right)^{1/2} \tag{13}$$

For a continuous spill the viscous-surface tension phase occurs simultaneously with the gravity-viscous phase rather than following it in time, with the part of the slick that is spreading in the gravity-viscous phase supplying the mass for that part in the viscous-surface tension phase. No transition to the viscous-surface tension phase is made for continuous spills.

In the above equations K_{10}, K_{11}, K_{20}, K_{21} are dimensionless constants. conducted experiments specifically designed to test the validity of the above pool spread model and to determine the values for the constants K_{10}, K_{11}, K_{20} and K_{21}. their recommended vlues which have been implemented here are shown inTable 1.

Table 1. Constants in pool-spreading models.

constant	value
K_{10}	1.53
K_{20}	1.21
K_{11}	1.24
K_{21}	1.09

2.4 Pool vaporization model

The liquid spill model can describe the spreading and vaporization of pools of liquid spilt either on land or water. For spills on land the model takes into account heat conduction from the ground, ambient convection from the air, radiation and vapour diffusion. These are usually the main mechanisms for boiling and evaporation. Solution and possible reaction of the liquid in water are also included for spills on water, these being important for some chemicals. These are modeled numerically, maintaining mass and heat balances for both boiling and evaporating pools. This allows the pool temperature to vary as heat is either absorbed by the liquid or lost during evaporation [7,8].

At each time step in the numerical calculations the heat balance on the pool is evaluated and from this the vaporization mechanism, vaporization rate and pool temperature are determined.

Initially, the temperature T_{pool} of the pool is assumed to be equal to the temperature T_{spill} of the spilled liquid. The net rate of heat flow into the pool is calculated from

$$Q_{net} = Q_{cond} + Q_{conv} + Q_{rad} + Q_{sol} + Q_{spill} - Q_{evap} \tag{14}$$

Where Q_{cond} is the heat flow rate from conduction (W)
Q_{conv} is the heat flow rate from convection (W)
Q_{rad} is the heat flow rate from solar radiation (W)
Q_{sol} is the heat flow rate from dissolution (W)
Q_{spill} is the heat flow rate from spilled liquid (W)
Q_{evap} is the heat flow rate from evaporation (W)

2.5 Multi-component pool model

In order to account for real mixture behaviour, the model keeps track of the pool inventory at all times and employs established mixing rules to estimate overall pool properties. As in the single-component model two limiting cases in pool vaporization are distinguished: boiling and evaporation. The multi-component pool model is able to predict the evaporation rate and phase composition from both types of pool vaporization. Additionally, the model makes continuous checks for transitions between boiling and evaporation regimes, by performing a bubble point calculation at each step [9] .

The pool model described assumes ideal mixing of the components within the liquid pool components. The muti-component pool model is formulated from the following equations.

Spreading law:

For pools on land:

$$\frac{dr}{dt} = \sqrt{2g(h - h_{\min})} \tag{15}$$

For pools on water:

$$r = f\left(M_{spill}(0), S(t), t\right) \tag{16}$$

Mass balance per component i (i = 1,2,….,n; when n is the number of components):

$$\frac{dM_{pool}}{dt} = S_i(t) - E^i_{vap}(t) - E^i_{sol}(t) \tag{17}$$

Cumulative mass dissolved per component i (i = 1,2,….,n):

$$\frac{dM^i_{vap}}{dt} = E^i_{vap}(t) \tag{18}$$

Cumulative mass dissolved per component i (i = 1,2,….,n)(only for pools on water):

$$\frac{dM^i_{sol}}{dt} = E^i_{sol}(t) \tag{19}$$

Energy balance:

$$\frac{dT_{pool}}{dt}=\frac{\left\{Q_{in}+S(t)C_{pl}^{spill}(T_{spill})\left[T_{spill}-T_{pool}\right]\right\}}{C_{pl}^{pool}(T_{pool})M_{pool}}-\frac{\left\{E_{vap}(t)\sum_{i=1}^{n}Y_{vap}^{i}\Delta H_{vap}^{i}(T_{pool})-E_{sol}(t)\sum_{i=1}^{n}X_{sol}^{i}\Delta H_{sol}^{i}(T_{pool})\right\}}{C_{pl}^{pool}(T_{pool})M_{pool}} \quad (20)$$

where, M_{pool}^{i} is the mass in the pool of component i (kg) $S_i(t)$ is the mass spill rate of component i which is equal to 0 for Instantaneous spills (kg/s)

$E_{vap}^{i}(t)$ is the vaporization rate of component i (kg/s)

$E_{sol}^{i}(t)$ is the dissolution rate of component i only applicable or pools on water (kg/s)

Q_{in} sum of the heat inputs to the pool = sum of heat of conduction (J)

Q_{cond}, convection Q_{conv} and radiation Q_{rad} (J)

$C_{pL}^{spill}(T_{spill})$ is the specific liquid heat capacity at the temperatureand composition of the spill (J/kg/k)

$C_{pL}^{pool}(T_{pool})$ is the specific liquid heat capacity at the temperature and composition of the pool (J/kg/k)

$\Delta H_{vap}^{i}(T_{pool})$ is the heat of vaporization of component i at the temperatureof the pool (J/kg)

$\Delta H_{sol}^{i}(T_{pool})$ is the heat of dissolution of component i at the temperature of the pool only applicable for pools on water (J/kg)

Y_{vap}^{i} is the mass fraction of component i in the vaporization rate (fraction)

X_{sol}^{i} is the mass fraction of component i in the vaporization rate (fraction)

3. Simulating analysis

A number of simulations were performed, by means of the PHAST model outlined above, in order to ascertain how the rate of vaporization of LNG is different to that of pure methane under identical spill conditions. In each simulation run the initial LNG mixture is modeled as consisting of 90% methane and 10% ethane which can be considered as a reasonable representation of a typical LNG composition. The LNG mixture was assumed to be at -162 when released, the pressure was taken as atmospheric, while the water temperature is assumed to be 300 K throughout the simulation run. A number of simulations were run for LNG and methane released on land or water. No qualitative differences between methane and LNG were observed as a function of the mass released, and therefore all the results reported below are based on the release of 50m^3 tank of LNG and methane.

3.1 Case 1

Simulating LNG and methane spill on land. In the scenario it was assumed that calm atmospheric conditions prevail during the pool spreading and vaporization. LNG and methane volume are 50 cubic meters; Leak size is 100mm; Wind velocity is 1m/sec and the atmospheric stability is class G.

In the scenario, curve labeled LNG in the initial stages of the spill no marked difference is observed between the vaporization of pure methane. The rate of vaporization of LNG is higher than that of methane. The difference in the vaporization rate is approximate 8%, as illustrated in Figure 1 and Figure 2.This is primarily due to the unusual shape of the vapour-liquid phase envelope for the methane-ethane mixture. As the LNG liquid becomes ethane-rich, the total latent heat exhibits a rapid increase. This is due to a combination of the direct and indirect differential latent heat.

Therefore the influence of chemical composition on the rate of LNG vaporization is important. The increase of the latent heat leads to a substantial decrease of the rate of vaporization.

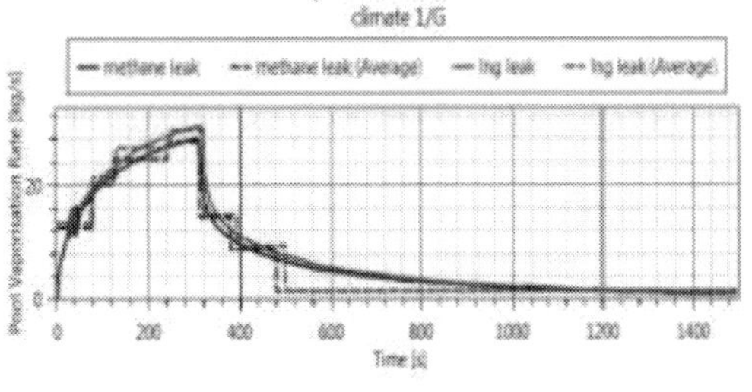

Fig. 1. Pool vaporization rate vs time on land.

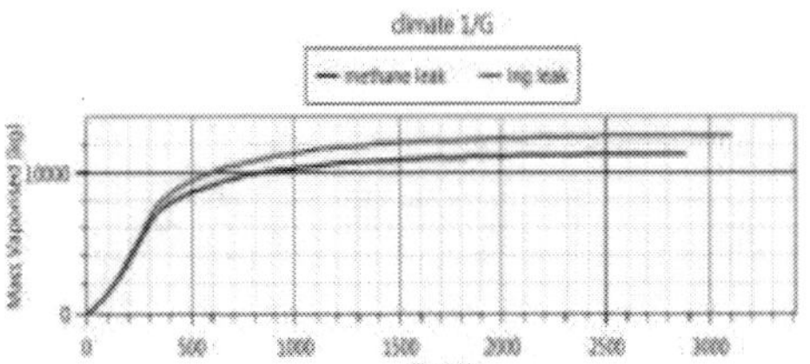

Fig. 2. Mass vaporized vs time on land.

3.2 Case 2

Simulating LNG and methane spill on water. In the scenario it was assumed that calm atmospheric conditions prevail during the pool spreading and vaporization. Wind velocity is 1m/sec and the atmospheric stability is class G.

In the scenario, the rate of vaporization of methane is smaller than that of LNG. The difference in the vaporization rate on water is bigger than the vaporization rate on land. This in turn leads to a partial decrease in the vaporization rate, since the amount of heat transferred from the water surface into the pool decreases due to a decrease in the temperature difference driving force. as illustrated in Figure 3 and Figure 4.

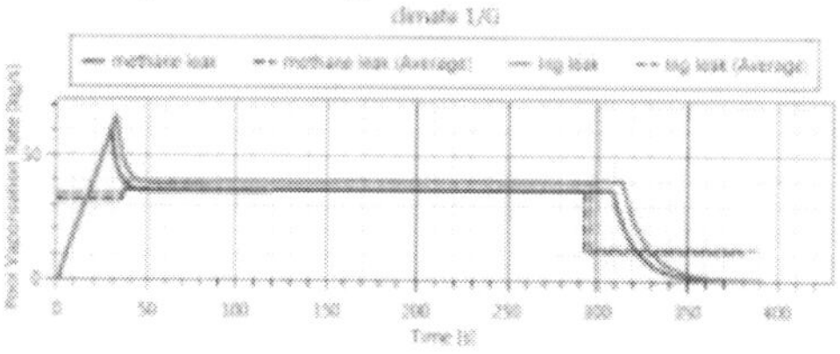

Fig. 3. Pool vaporization rate vs time on water.

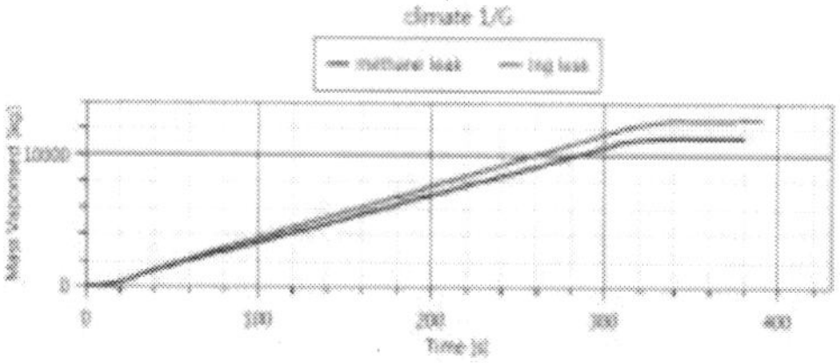

Fig. 4. Mass vaporized vs time on water.

4. Conclusions

The results indicate, the vaporization rate of LNG is different to that of pure methane. The difference can be attributed to the influence of chemical composition. For LNG gets rich in ethane, leading to a large decrease in the vaporization rate of LNG compared to pure methane. The overall results suggest that as the vaporization rate treating a pure methane spill results as an LNG spill in underestimation of approximate of 8% .

Acknowledgements

This Project is supported by the china railway corporation Science research Foundation of China, No.2016X007-F.

References

1. R.P. Koopman, D.L. Ermak: Lessons learned from LNG safety research. Journal of Hazardous Materials, 140, (2007), p.412-428.
2. M.J. Ivings, C.J. Lea, D.M. Webber: A protocol for the evaluation of LNG vapour dispersion models. Journal of Loss Prevention in the Process Industries, 26(1), (2013), p. 153-163.
3. T.A. Melton, J.B. Cornwell: LNG trench dispersion modeling using computational fluid dynamics. Journal of Loss Prevention in the Process Industries, 23(6), (2010), p. 762-767.
4. F. Gavelli, M.K. Chernovsky, E. Bullister: Modeling of LNG spills into trenches, Journal of Hazardous Materials, 180(1-3), (2010), p. 332-339.
5. H.A. Luketa: A review of large-scale LNG spills: experiments and modeling. Journal of Hazardous Materials, 132, (2006), p. 119-140.
6. J.A. Havens, T.O. Spicer: LNG vapor cloud exclusion zones for spills into impoundments. Process Safety Progress, 24(3), (2005), p. 181-186.
7. O. R. Hanna, F. Gavelli, M. Ichard: Validation of FLACS against experimental data sets from the model evaluation database for LNG vapor dispersion. Journal of Loss Prevention in the Process Industries, 23(6), (2010), p. 857-877.
8. J.A. Havens, T.O. Spicer: Wind tunnel studies of LNG vapor dispersion from impoundment, AIChE National Spring Meeting, Houston.
9. R. Bubblco, E. Salzano: Acoustic analysis of blast waves produced by rapid phase transition of LNG released on water, Safety Science, 47, (2009), p. 515-521.

Study on Dynamic Reconfiguration of Distribution Network Considering Distribution Generation

Xin Geng*, Honghai Niu, Jun Chen and Bing Li

Research and Development Center of NR Electric Co., Ltd., Nanjing 211102, China
Corresponding Email: genx@nrec.com

Abstract. A new dynamic reconfiguration method considering distribution generation (DG) was presented. The dynamic reconfiguration problem was solved from three aspects: firstly, the time interval partition of the entire scheduling period was optimized by load distribution variation index; secondly, a plurality of time intervals were optimized and reconfigured by using the multi-objective particle swarm optimization model of multi-period encoding; finally, the optimal time interval numbers was determined by gradually approaching the falling threshold of net loss. If the system contained DG, DG's output curve was determined by the analysis of its dynamic characteristics. The results of the test showed that the proposed method was effective to solve the dynamic reconfiguration with DG.

Keywords. Dynamic reconfiguration, distribution generation, multi-objective particle swarm optimization, distribution network reconfiguration.

1 Introduction

Distribution Network Reconfiguration contains static reconfiguration and dynamic reconfiguration. Static reconfiguration optimizes the network at a time section, and dynamic reconfiguration optimizes the network considering the time varying which has practical significance for application. With the development of Smart Gird requiring the high penetration of Distribution Generation, the research on the network reconfiguration with DG is essential and meaningful [1].

Dynamic reconfiguration of distribution network is applied to adjust the topology structure of distribution network dynamically on the premise of the switching times constraint according to the dynamic load variation over a period, so that the optimal running state of the distribution network can be achieved. The objective function of dynamic reconfiguration is different from the static reconfiguration. Static reconfiguration only take into account the optimization target of distribution network under a certain time cross section while the objective function of dynamic reconfiguration reveals the optimization goal over a period. At present, the minimum network loss in a scheduling cycle is usually treated as the objective function which is described as:

$$\min \quad \int_{t_b}^{t_e} P_{loss}(t)\cdot \mathrm{d}t \qquad (1)$$

where $P_{loss}(t)$ is the network loss at time t. In practice, the continuous load curve of the power distribution system is usually approximately replaced with trapezoidal curve to discretize the continuous variables. The discrete objective function can be written as

$$\min \sum_{t=1}^{T} P_{loss}(t) \cdot \Delta t \qquad (2)$$

where $P_{loss}(t)$ is the network loss, t represents an hour, T is the number of time segments. T equals 24 for a scheduling cycle of a day.

Considering that the frequent network topology transformation will greatly reduce the service life of the switchgear, the overall switching times limit in the time cycle should be added in addition to the traditional static constraints (including power flow constraints, branch capacity constraints, network topology constraints)

$$N_{total} \leq N_{total\max} \qquad (3)$$

where N_{total} is the switching times, $N_{total\max}$ is the maximum switching times limit.

According to the mathematical model of distribution network dynamic reconfiguration, the relation and difference between dynamic and static reconfiguration can be concluded as:

The time region of dynamic reconfiguration is a scheduling cycle while the time region of static reconfiguration is only a time section or a short time segment. The dynamic reconfiguration can be approximated as the composition of static reconfiguration problem and the static reconfiguration is a particular case of reconfiguration problem.

The decision variables of dynamic reconfiguration is the switch state throughout the scheduling cycle which is described as mixed integer optimization problem of discrete time systems. However, the number of control variables is constrained by the maximum switching times limit and the dynamic reconfiguration cannot be treated as the simple combination of static reconfiguration problems.

The solution method of dynamic reconfiguration is much more complicated comparing with static reconfiguration and excessive pursuit of the global optimal solution will greatly increase the complexity of the calculation. In this research, time interval segmentation method was applied to decouple the distribution network dynamic reconfiguration problem and time then the optimal number of time interval can be determined by the relation between network loss and the time interval number. The main steps are interpreted below:

(1)The whole scheduling cycle can be reasonably divided into some time intervals according to the variation of load distribution.

(2)After the partition of scheduling cycle, utilize multi-objective particle swarm optimization to code and reconfigure the time intervals.

(3)Determination of the optimal time interval number.

2 Principle and algorithms

2.1 Timer interval partition of scheduling cycle [2]

In some cases, load of various points might exhibit the same trend of variation during a period and the value change of a specific load point will not change the load distribution of the whole power distribution system. It is essential to standardize the load state at any time by

$$D_i(t) = \frac{L_i(t)}{L_{total}(t)} \tag{4}$$

where $L_i(t)$ is the load of point i at time t, $L_{total}(t)$ is the total load at time t, $D_i(t)$ is the standardized load value of point i at time t which represent the load proportion of point i at time t. After the standardization of all the points, the load variation of a single point can be calculated as

$$\Delta D_i(t_1,t_2) = |D_i(t_1) - D_i(t_2)| \tag{5}$$

where $D_i(t_1,t_2)$ represents the load variation of point i from time t_1 to time t_2, then the total variation of load distribution from time t_1 to time t_2, $\Delta D(t_1,t_2)$ can be calculated as

$$\Delta D(t_1,t_2) = \sum_{i=1}^{m} |D_i(t_1) - D_i(t_2)| \tag{6}$$

A high $\Delta D(t_1,t_2)$ value represents large variation of load distribution from time t_1 to time t_2. The standardized value is obviously reasonable for describing the change of load distribution and the load change will not be submerged by the variation of total load. In addition, a threshold of variation of load distribution ΔD_{max} is needed for the time interval partition.

Throughout the time interval partition, the threshold ΔD_{max} is first set as the initial value $\Delta D_{max.0}$, then the scheduling cycle is divided according to ΔD_{max}. Appropriate $\Delta D_{max.0}$ will greatly reduce iterations of the algorithm, which is related to the presupposed time interval number n. The initial value can be calculated $\Delta D_{max.0}$ from the largest variation of load distribution $\max \Delta D$ during a scheduling cycle:

$$\Delta D_{max,0} = \frac{\max \Delta D}{n} \tag{7}$$

If the time interval number doesn't equal the presupposed value n, ΔD_{max} will be adjusted to divide the scheduling cycle again until the time interval number equals n. Regarding the time interval number after i steps of iteration n_i, the threshold of variation of load distribution for the next step of iteration $\Delta D_{max,i+1}$ can be adjusted as:

$$\Delta D_{max,i+1} = \begin{cases} \Delta D_{max,i} + \varepsilon, & n_i > n \\ \Delta D_{max,i} - \varepsilon, & n_i < n \end{cases} \tag{8}$$

The time interval partition process is illustrated in Figure 1.

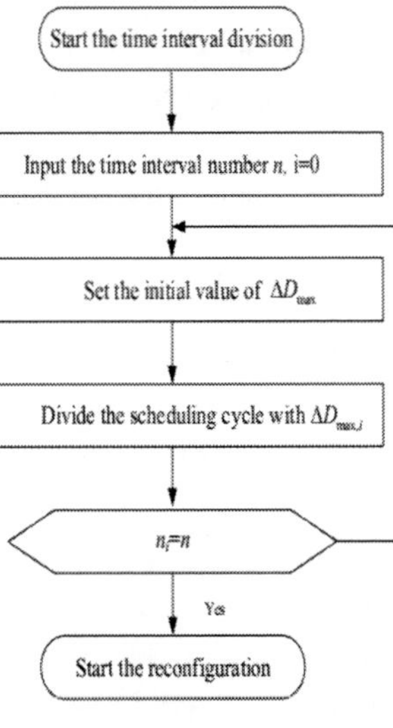

Fig. 1. The process of timer interval partition.

2.2 Dynamic reconfiguration of distribution network [3]

The mathematical model of dynamic reconfiguration requires the minimum network loss in a scheduling cycle and meets the maximum switching times constraint conditions. After the partition of scheduling cycle, the fitness function can be modified as:

(1)The minimal total network loss

$$\min f_1 = \sum_{i=1}^{n} P_{loss}(i) \cdot \Delta t(i) \tag{9}$$

where $P_{loss}(i)$、 $\Delta t(i)$ is the total network loss and segment length of time interval i, respectively.

(2)The minimal switching times

$$\min f_2 = N_{total} \tag{10}$$

where N_{total} represent the total switching times.

Regarding the multi-objective particle swarm optimization [4] (MOPSO) method, there are n substrings for the encoding of n time intervals and every substring represents the encoding form of the network topology. Figure 2 illustrates the encoding form for 3 timer intervals.

$$\begin{bmatrix} x_{1,1} & x_{1,2} & x_{1,3} & x_{1,4} & x_{1,5} \\ x_{2,1} & x_{2,2} & x_{2,3} & x_{2,4} & x_{2,5} \\ x_{3,1} & x_{3,2} & x_{3,3} & x_{3,4} & x_{3,5} \end{bmatrix} \begin{matrix} \text{Substring 1} \\ \text{Substring 2} \\ \text{Substring 3} \end{matrix}$$

Fig. 2. The encoding form for 3 timer intervals

Different time interval number will result in different dynamic reconfiguration strategies, however, the reconfiguration of every time interval is a good result and the network structure will not change a lot after one step of dynamic reconfiguration. Considering the local searching ability of multi-objective particle swarm optimization, the satisfactory result will be easily obtained near the particles if the optimal solution set and the particle optimal solution 'pbest' of the previous dynamic reconfiguration step are selected as the initial group of the next dynamic reconfiguration step. The method is interpreted below:

(1)The optimal solution set of the previous dynamic reconfiguration step and substrings of the 'pbest' are extracted and put into a feasible solution library, every substring represents a feasibly good network topology structure.

(2)According to the requirements of the current dynamic reconfiguration, substrings were randomly selected from the feasible solution library to make up the particle encoding form, and the initial particle group of the current dynamic reconfiguration will be built.

Among all the feasible strategies ($N_{total} \le N_{total\max}$), the optimal one exhibit the least network loss which is the desired reconfiguration strategy.

It can be concluded from experience that more reconfiguration number within a scheduling cycle brings about more reduction network loss. However, the network structure will not change a lot with reconfiguration number increases and the economic efficiency will reduce. In addition, more reconfiguration number means more switching times and the cost of switch operation, besides the network loss, should be taken into account. In practice, it is hard to precisely determine the maintenance and operating costs of switchgear. In order to find the optimal time interval number, the following criterion is introduced:

$$P_{loss}^{n} - P_{loss}^{n+1} \le \varepsilon_{loss} \tag{11}$$

where n is the time interval number, P_{loss}^{n} is the minimum network loss of n time intervals considering the switching times limit, ε_{loss} is the network loss reduction threshold when change the time interval number from n to $n+1$. The threshold ε_{loss} can be calculated as 1%~2% of the initial network loss according to experience. Equation (11) indicates that the time interval number will be increased from n to $n+1$ only if the minimum network loss reduction is beyond the threshold ε_{loss}. The process of dynamic reconfiguration is illustrated in Figure 3.

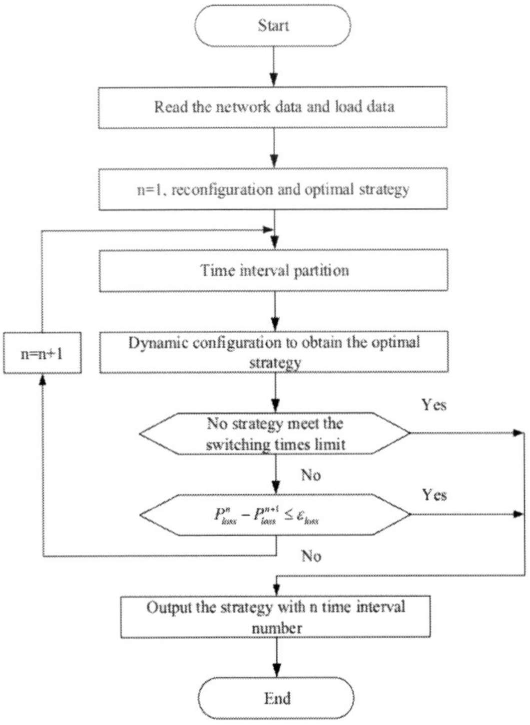

Fig. 3. The process of dynamic reconfiguration.

3 Examples and analysis

3.1 Example data

The measuring equipment is usually installed on the feeder in the substation exit, the total load variation of a feeder can be obtained and the load curve of each point is unavailable. The load curve estimation of each load point can be carried out according to the total load of the feeder and the historical data of the power station [5]. There are several types of load including resident load, industrial and commercial load.

It is assumed that resident, industrial and commercial load points are randomly distributed in the 69-node system and the standard node loads of the system is set as load peak value of each point. The sunlight intensity and wind speed data are collected from the one month historical data of a photovoltaic power station and wind power station, respectively.

For the sake of illustrating the influence of distribution generation on the distribution network dynamic reconfiguration, two different cases are discussed:

Case1: No distribution generation, only the three types of load (resident, industrial and commercial) are concerned.

Case2: Three photovoltaic power stations are connected to each of node 9, 60 and 69, three wind power stations are connected to each of node 18, 48 and 27.

3.2 Simulation results

Case1: No distribution generation, only the three types of load (resident, industrial and commercial) are concerned.

It is assumed that the maximum switching times $N_{total\max}$ equals 10, the network loss reduction threshold ε_{loss} is 15kWh. The dynamic configuration results according to the variation of load distribution and the value change of total load are shown in Table 1. The time interval partition based on the variation of load distribution is 0:00~6:00, 6:00~17:00, 17:00~24:00 while the partition result based on the value change of total load is 0:00~6:00, 6:00~22:00, 22:00~24:00. Dynamic configurations of the two kinds of time interval partition are carried out for comparison. According to the time interval partition based on the variation of load distribution, the total network loss caused by the dynamic configuration is 972.59kWh while the time interval partition based on the value change of total load creates 983.9kWh of dynamic configuration network loss. It can be concluded that the proposed variation of load distribution index provides a better description of the distribution network load change then contributes to a more reasonable time interval partition in comparison with the traditional index (value change of total load).

Table 1. The results of the two dynamic configuration method.

Time interval partition	Dynamic configuration strategy		Net-work loss
	Time interval	Open switching	kWh
Variation of load distribution	0:00~6:00	13-20,12-13,46-47,11-66,50-51	62.09
	6:00~17:00	13-20,14-15,44-45,11-66,50-51	759.07
	17:00~24:00	13-20,12-13,44-45,11-66,50-51	151.26
Total		6 switching actions	972.6
Value change of total load	0:00~6:00	13-20,12-13,46-47,11-66,50-51	6 2.09
	6:00~22:00	13-14,12-13,45-46,11-66,52-53	893.30
	22:00~24:00	13-20,12-13,47-48,11-66,50-51	28.03
Total		10 switching actions	983.4

Table 2 shows the network loss, loss reduction rate and switching times throughout the dynamic configuration of a scheduling cycle without switching times constraint considering different time interval numbers. When the time interval number increases from 1 to 3, the network loss falls sharply and the switching times are not less than 10. When the time interval number is beyond 4, the network loss of the system decreases slowly while the switching times increases sharply. It is suitable to divide one day into 2~4 time intervals [6].

Table 2. The results of dynamic configuration considering different time interval numbers.

Time interval number	Network loss after configuration (kWh)	Loss reduction rate (%)	Switching times
1	1125.7	47.84	0
2	988.55	54.20	4

3	972.59	54.93	8
4	970.41	55.03	12
5	969.56	55.07	16
6	968.34	55.13	22
7	965.92	55.25	24
8	964.71	55.30	26
9	963.34	55.36	28
10	963.34	55.36	28

Case 2: Three photovoltaic power stations are connected to each of node 9, 60 and 69, three wind power stations are connected to each of node 18, 48 and 27.

The maximum switching times $N_{total\max}$ is set as 10 and the network loss reduction threshold ε_{loss} is 15kWh. The results of the proposed dynamic configuration method are shown in Table 3.

Table 3. The dynamic configuration results of case 2.

Dynamic configuration strategy		Network loss
Time intervals	Open switching	kWh
0:00~14:00	13-20,13-14,8-9,9-10,50-51	515.01
14:00~24:00	13-20,12-13,47-48,11-66,50-51	250.12
Total	6 switching actions	778.13

The scheduling cycle is divided into two time intervals: 0:00~14:00, 14:00~24:00. The dynamic configuration results of case 2 is totally different from the results of case 1 which is attributed to the change of node load curve caused by the distributed generators. The network loss reduction is not very large regarding 3 time intervals and the unnecessary switching actions can be avoided if the cycle is divided into 2 intervals. Table 3 indicates that the proposed method provides effective strategies for the dynamic configuration of distribution network considering distribution generation [7].

4 Conclusions

The time interval partition method based on the variation of load distribution index is proposed. Multi-period encoding is applied to deal with the coordination among the switching states of different time intervals and the strategies for the dynamic configuration of distribution network is provided considering distribution generation. The simulation algorithms is carried out on the IEEE 69-node system sample and the proposed variation of load distribution index provides a more reasonable time interval partition in comparison with the traditional index (value change of total load). According to the dynamic configuration results considering different time interval numbers, the real economic value contributed by the increase of time interval number is limited.

References

1. Thomas A, Goran A, Lennart S. Distributed Generation: a Definitio. Electric Power System Research, 2001, 57(3) :195.204.
2. Lin W M, Cheng F S, Tsay M T. Distribution feeder reconfiguration with refined genetic algorithm. IEE Proceedings of Generation, Transmission and Distribution, 2000,147 (6):349-354.

DOI: 10.1051/matecconf/201710004008

3. Jonathan E Fieldsend, Sameer Singh. A multi –objective algorithm based upon particle swarm optimization: An efficient data structure and turbulence. Proc of the 2002 UK Workshop on Computational Intelligence. Manchester, 2002 : 37-44.
4. Coello C A C, Pulido G T, Lechuga M S. Handling multiple objectives with particle swarm optimization. IEEE Transactions on Evolutionary Computation, 2004,8 (3):256-279.
5. Su C L. Probabilistic load-flow computation using point estimate method. IEEE Transactions on Power Systems, 2005,20(4):1843-1851.
6. Hong H P. An efficient point estimate method for probabilistic analysis. Reliability Engineering & System Safety, 1998,59(3):261-267.
7. Charytoniuk W, Chen M S, Kotas P, et al. Demand forecasting in power distribution systems using nonparametric probability density estimation. IEEE Transactions on Power Systems, 1999, 14(4):1200-1206.

Research on Friction and Wear Performance of TiB_2 Based Self-lubricating Ceramic composites

Huagang Ding

School of Mechanical and Automotive Engineering, Qilu University of Technology, Jinan 250353, China
E-mail:dhggzyx@126.com

Abstract. TiB_2 based self-lubricating ceramic composites with the addition of different content of h-BN were prepared by hot pressing technique and their mechanical properties were tested. Friction and wear experiments are conducted on a MMW – 1A configuration control universal friction and wear testing machine. Friction and wear properties of the ceramic composites under different speed and load were studied. The results show that with the increase of the content of h-BN, the hardness, flexural strength and fracture toughness showed a trend of decrease. When load was fixed, with the increase of rotational speed, the friction coefficient rose and the wear rate were first decreased and then increased. When rotational speed was fixed, the friction coefficient and wear rate first decreased and then increased with the increase of load. In nitrogen, the friction coefficient and wear rate of material were higher than those in the air.

Keywords. Self-lubricating ceramic, mechanical properties, friction and wear, nitrogen atmosphere.

1 Introduction

Cause of friction is the relative motion between two surfaces contact each other ,and inevitably produced friction and wear [1].Ceramic tool because of its high wear resistance, heat resistance, oxidation resistance advantages, has been gradually expanded. However, because of its large friction coefficient and fragile, its ues of scope has been limited, so the study of friction and wear properties of ceramic tool materials is very important for development of ceramic cutting tool inspection role.

TiB_2 With high hardness, high melting point and good oxidation resistance, corrosion resistance and other characteristics, also has a good metal gloss [2]. h-BN has very good chemical stability, and it continue to be particularly stable at 2800degrees centigrade temperatures, has good electrical insulation and lubrication, it is very excellent high temperature lubricants [3]. Xu Xiuguo prepared TiB2/WC/h-BN ceramic cutting tools by hot-pressing sintering and experiments for cutting tool materials have good self-lubricating properties [4]. Cao Tongkun prepared Al2O3/TiC/CaF2(ATF) of ceramic materials by hot-press sintering process, used ring-type friction and wear tester for friction and wear properties of test, and analysised the formation processes and mechanisms of self-lubricating film [5].

In this thesis, self-lubricating ceramic material specimens was tested friction and wear, and studied the friction-wear properties and wear mechanisms, comparing different h-BN content wear resistance and wear mechanism of ceramic materials, and studied changes of friction coefficient and wear rate of materials under different loads and different speeds, then analysis the friction-wear properties and morphology.In addition, studied its friction and wear in nitrogen atmosphere compared with in the air, and its wear mechanism in a different atmosphere.

2 The experiment

2.1 Self-lubricating ceramic tool materials preparation

In TiB_2 based Self-lubricating ceramic material, h-BN contents are 0vol.% 、 5vol.% 、 10vol.% 、 15vol.% 、 20vol.% 、 25vol.% 、 30vol.% 、 35vol.% 、 40vol.% , Reinforced phase WC Content for 15vol.% ,Sintering additives Ni Content for 3vol.% , Mo Content for 2vol.% .

Different content TiB_2 powder, WC powder, h-BN powder, Ni powder, Mo powder into the amount of ethanol in the blend, ultrasonic dispersion, mixing, milling, drying and screening, followed by hot-pressing sintering,got different components of TiB2 based Self-lubricating ceramic material. The average size of TiB2 in experiment is 5~8 μ m, purity is 99.0%;WC particle size 1 μ m, purity is greater than 99.9%,h-BN average size of 1.5 μ m, the purity of 99.9%and additives Ni, and Mo purity respectively 99.5% and 99%.

2.2 Mechanical properties of self-lubricating ceramic material

By wire-cut edm machine cut TiB_2 matrix self-lubricating ceramic material, and rough grinding, fine grinding and polishing, tested hardness, bending strength and fracture toughness of samples.

By three-point bending method in measurement of bending strength of electronic universal testing machine, using digital measurement of Vickers hardness to test hardness; fracture toughness specimen is measured by indentation method. Results Figure 1 shows:

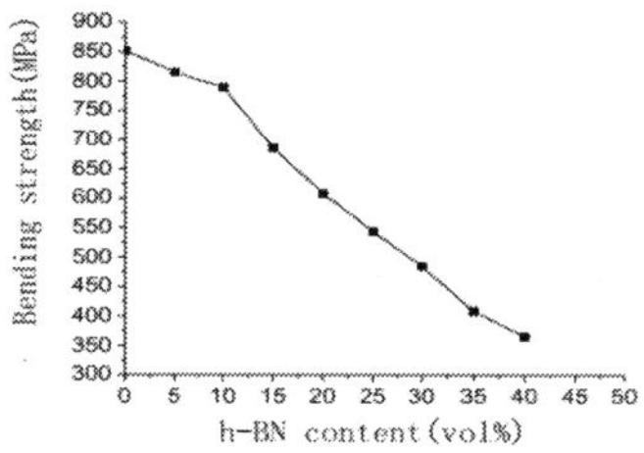

(a) Flexural strength h-BN Content changes.

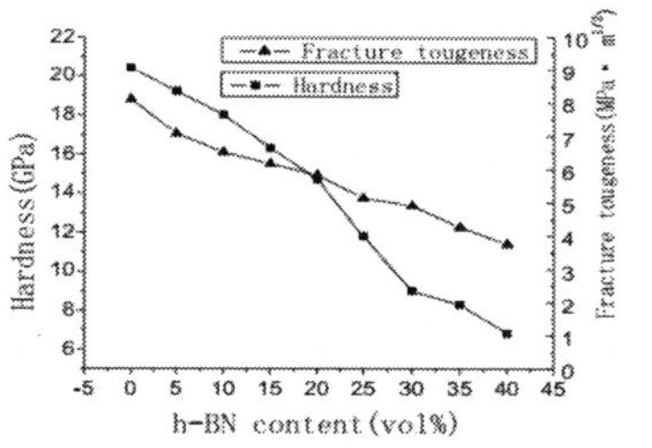

(b) Hardness, fracture toughness h-BN content changes.

Fig. 1. Mechanical properties h-BN Content changes.

Figure1 (a), The bending strength of materials decreases with h-BN Content increases, because h-BN Strength is very low, so, as the h-BN content increases, its bending strength is reduced. Figure 1(b), as the h-BN content increases, self-lubricating ceramic material hardness and fracture toughness decreases, this is because h-BN its hardness is very low, 2GPa, so with the h-BN content increasing, the hardness of the material reduce; the h-BN of plate-like structure in the sintering process will prevent the movement of particles and affect the density of the material, leading to the toughness of the material decreases.

2.3 Self-lubricating ceramic material for friction and wear Tester

MMW-1A Configuration control everything using pins on friction and wear testing machine - pairs form of friction and wear testing, taking into account the TiB2/WC/h-BN performance of a self-lubricating ceramic composite material, the inside diameter of ring for this test is Φ38mm Outside diameter is Φ54mm , Ring material 45# Steel, quenched hardness of 44 ~ 46HRC The surface roughness Ra 0.08μm .

Friction and wear test device schematic diagram diagram 2 Shown here:

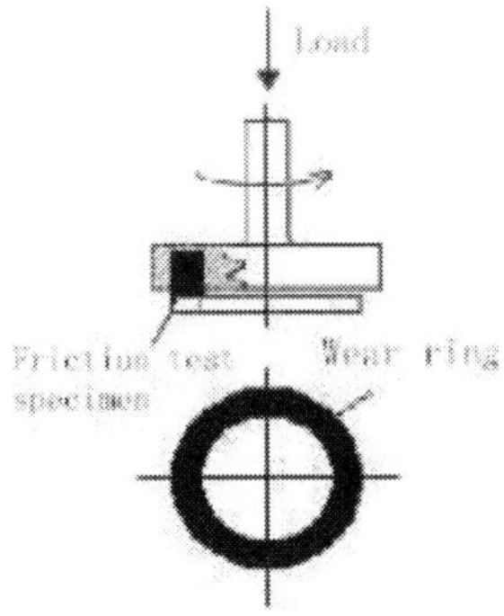

Fig. 2. Schematic diagram of friction and wear Tester.

Friction and wear test:

(1)In the case of the same rotating speed and load (v=200r/min 、 FN=30N) conducted friction and wear test in air and nitrogen For different content h-BN Samples;

(2)h-BN Content for 10vol.% to conduct friction and wear test under the same load different speed and the same speed different load;

(3)Records for all experimental coefficient of friction, weighing wear quality, calculate the rate of wear [6].

3 Results and discussion

3.1 Speed of the influence of friction coefficient and wear rate of materials

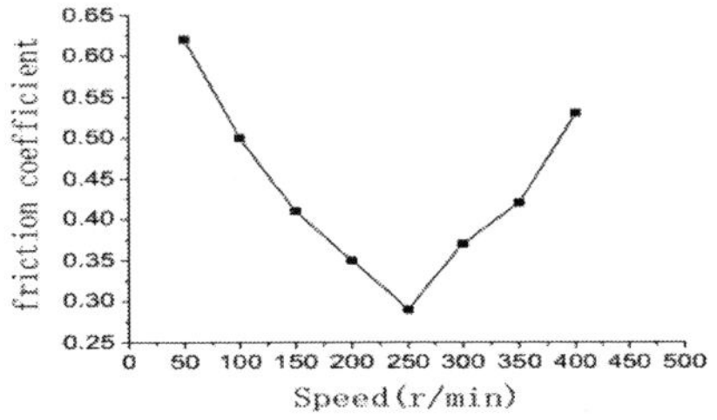

(a) Friction coefficient increases with the speed of change.

DOI: 10.1051/matecconf/201710004009

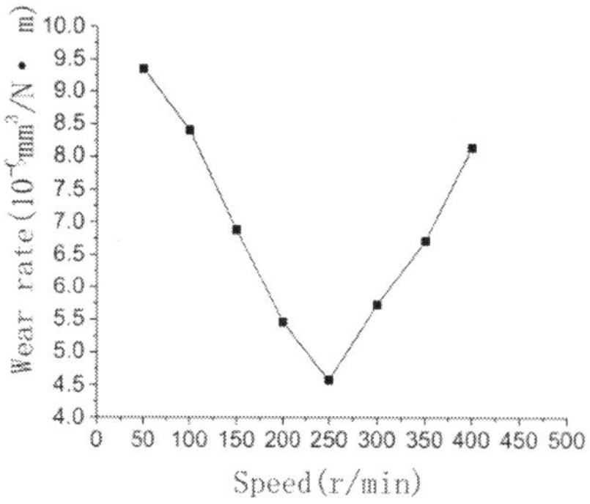

(b) Wear rate changes with speed.

Fig. 3. Coefficient of friction and wear rate and speed of change.

From Figure 3 (a) can directly see that when the load is constant, lower friction coefficient decreases with the speed increases ,and when the speed reaches 250r/min, friction coefficient increases. Because: with speed increasing, self-lubrication ceramic material and on mill deputy contact surface of friction speed gradually speed up, the temperature of contact surface rose, surface formed lubrication film, to reduced friction coefficient of material, when speed reached 250R/min, lubrication film was damaged seriously, the speed of forming lubrication film is lower than its wear, makes friction resistance increases, so friction coefficient increases with the speed up.

Figure 3 (b) can be seen , when speed increase from 50r/min to 250r/min, self-lubricating ceramic material wear rate decreases, when speed increase 250r/min to 400r/min, self-lubricating ceramic material's wear rate gradually reduced. Because when the speed is 50R/min, low speed, the temperature of contact surface is relatively low,h-BN separation more difficult from the ceramic material , cannot form a lubrication film, therefore it has a higher wear rate. As the speed increases (50R/min-250R/min), the grain of the surface damage, contact surface temperature increase, grain formed debris slowly, contact some of the debris at the contact surfaces form a surface layer, reduce the amount of wear of the material. In addition, with the speed continuing to increase,h-BN solid lubricant separated from surfaces more easily, form a lubrication film, which reduced the amount of wear. When the speed continues to increase (250R/min- 400r/min), the temperature of contact surface continued to rise, the damage of lubricating film is more serious, therefore, self-lubricating ceramic material wear rate also increased.

3.2 Effect of load on friction coefficient and wear rate

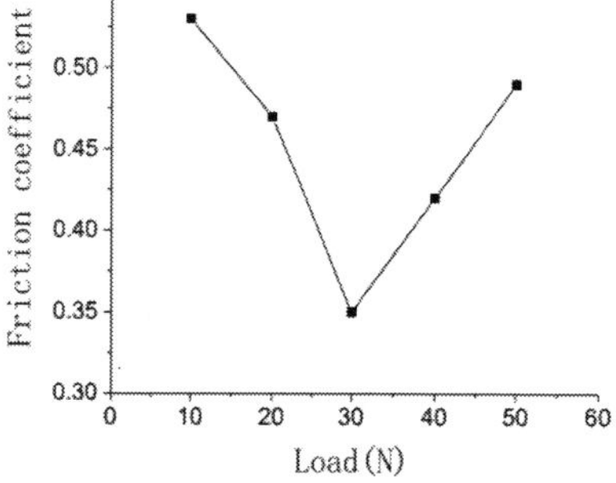

(a) Changes in friction coefficient as a function of load.

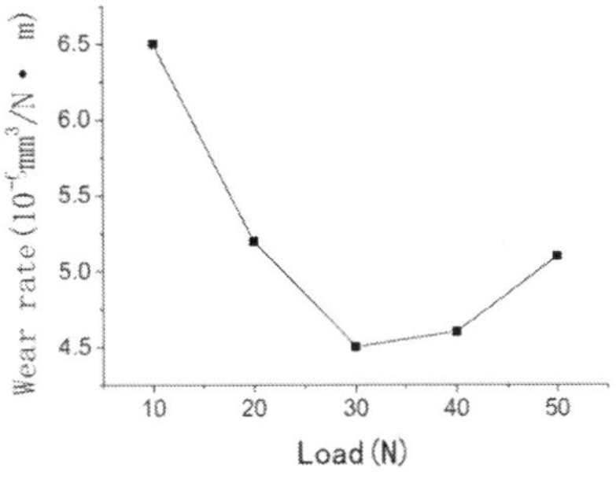

(b) Wear rate as the load changes.

Fig. 4. Trend of the coefficient of friction and wear rate as a function of load.

Figure 4 (a) can be seen, when the speed is constant, with the load increasing, the friction coefficient decreases and then increases. Because:

When the load is 10N, the stress of contact surface is relatively small, it is hard for h-BN to separate from the materials, causing friction of contact surface large and relatively high friction coefficient. As the load increases (10N-30N), the pressure between contact surface stress are larger, solid lubricant separate more easily, forming a lubricating film, and with the loss of material and material constant precipitation of solid lubricant, to repair the damaged film, so the friction is greatly reduced, the friction coefficient decreases. When the load continued increasing (30N-50N), the stress of contact surfaces also increases, lubricating film is damaged more seriously, self-lubricating materials wear increases, friction coefficient increases.

Figure 4 (b) can be seen , When the load from 10N 30N, self-lubricating ceramic material wear rate decreases, and when the load from 30N to 50N, self-lubricating film damaged more seriously, wear rate gradually reduced.

3.3 With nitrogen in the air h-BN Content on friction and wear properties of materials

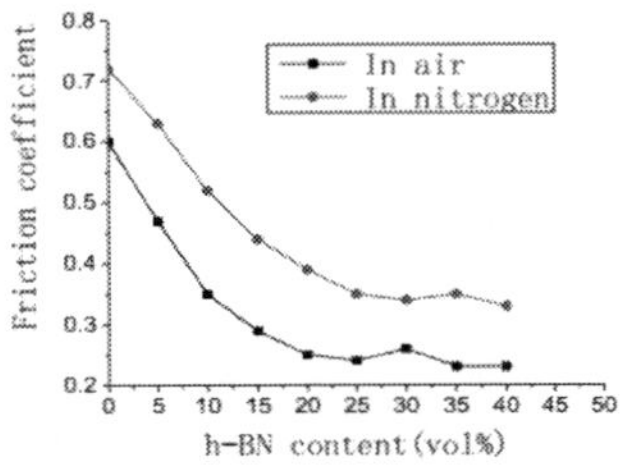

(a) Coefficient of friction h-BN Trends in content.

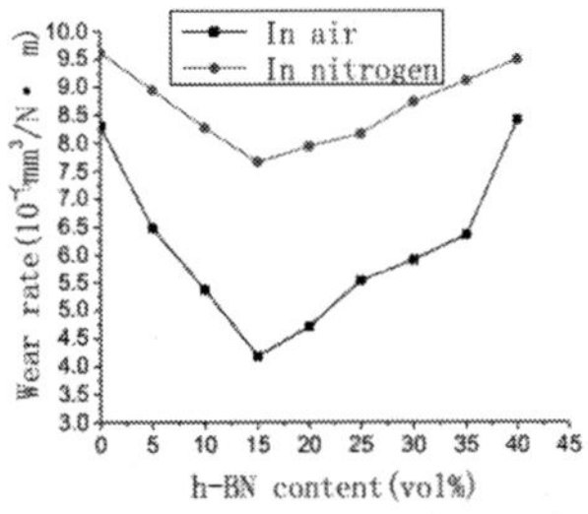

(b) Wear rate with h-BN content trends.

Fig. 5. Nitrogen and air friction coefficient and wear rate h-BN Trends in content.

Figure 5 (a), when the load is 30N, Speed 200r/min , with h-BN Content increases the friction coefficient of the material decreases, when increased to 20vol.%, friction coefficient remained stable in 0.25 .As the h-BN content increases, material produced by the friction of lubricating film has increased. In 20vol.%, self-lubricating film falling off is equal to its producing, so that the friction coefficient of the material remained stable, friction-reducing properties of the material remains stable. In nitrogen, self-lubricating ceramic material friction coefficient is higher than that in the air. Because, in the air,h-BN will be oxidized to B_2O_3,B_2O_3 is good self-lubricating film, it can reduce the friction coefficient of the material, while in nitrogen, material is prevented bing exposed to oxygen in case of friction,h-BN will not be oxidized to B_2O_3 film, just relying on h-BN self-lubricating characteristics of plate-like structure reducing friction, so the coefficient of friction is higher in nitrogen. Figure 6(a) is air friction surfaces in the EDS spectrum, shows that friction surface contains large amounts of oxygen in the air; in Figure 6(b) For nitrogen in the surface of EDS spectrum shows that the oxygen content is much lower than the oxygen content in the air. We can know that it did not form B_2O_3 film in nitrogen.

Figure 5 (b), as h-BN Contents of increasing, the wear rate of self-lubricating ceramic material first decreased and then increased, content 10vol.% reached the minimum. Because, joined h-BN, will increased its lubrication performance, improve the wear performance of the material; with continuing to add of h-BN, hardness, and bending strength and the fracture toughness of the material are reduced, load and speed on lubrication film of damage also became large, led to wear degree of material big, the wear rate of material increased. In nitrogen, the wear rate of materials was higher than in the air, because, in the air B2O3 film decreases the wear rate of materials, and in nitrogen, will not produce B2O3 film, material produced abrasion, resulting in increased wear rate. Figure 7(a,) were used in the air friction surfaces, surface scan photos and 3D graph shows it has formed good lubrication film on the surface. Figure 7(b) material surface friction in nitrogen from surface scans photos and 3D , surface has clear grooves, it is abrasion, surface quality is bad, so the wear rates are higher than in the air.

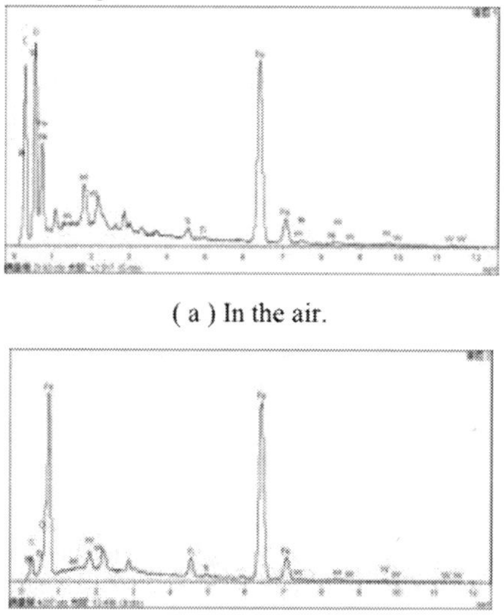

(a) In the air.

(b) Nitrogen.

Fig. 6. Materials on friction surface EDS Spectrum.

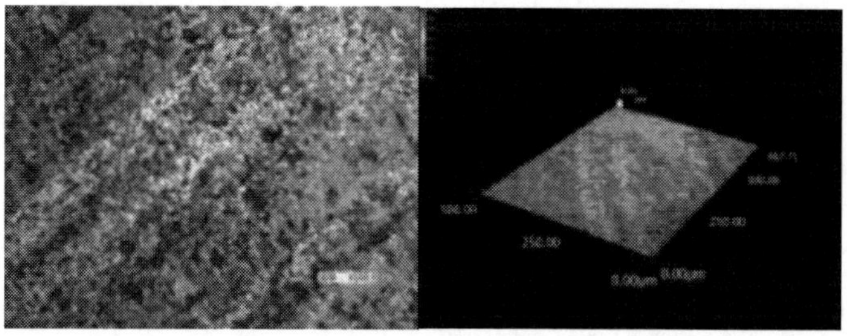

(a) In the air.

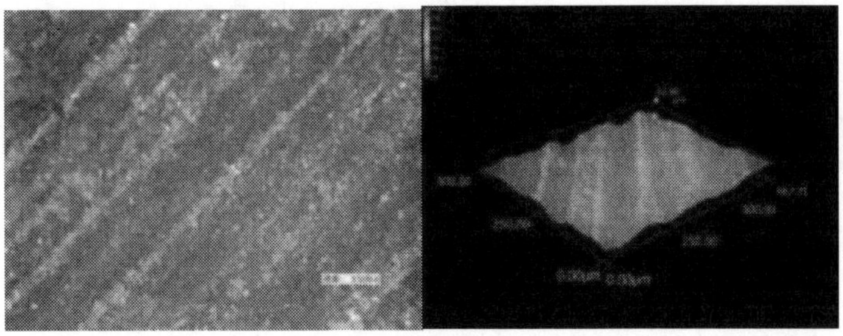

(b) Nitrogen.

Fig. 7. Super depth of field image of the friction surface.

4 Conclusion

(1)As h-BN Content increases, self-lubricating ceramic material hardness, flexural strength and fracture toughness are decrease;

(2)h-BN content for 10vol% material speed is 200r/min, as the load increases, friction coefficient first decreases and then increases, friction coefficient reached the minimum at the 30N, and the wear rate decreased at the first then increased, the lowest point is 30N;

(3)The material h-BN Content for 10vol%, when the load is 30N, as the speed increases the first decrease and then increase, wear rate first decreases and then increases, the lowest wear rate was found at 250r/min;

(4)Friction coefficient of lubricate ceramic materials decreases with h-BN content increases, when the content was increased to 20vol.%, stable about 0.25, wear rate lead reduced the lowest when the content is 15vol% Lowest rates of wear, then rise in nitrogen atmosphere, self-lubricating ceramic friction coefficient and wear rates are higher than in the air.

References

1. Wen Shizhu . Principles of Tribology. Beijing : Tsinghua University Press ,1990.
2. Seung-Ho Kim, Soo Wohn Lee. Wear and friction behavior of self-lubricating alumina–zirconia–fluoride composites fabricated by the PECS technique. Ceramics International, 2014, 40(40):779–790.
3. Tanemoto K,Kani T.Hot-pressed hexagonal boron nitride-aluminum nitride ceramic composites. Key Eng Mater,1995,108-110:85-96
4. Xu Xiuguo . Nano/Micro-composite-gradient development of a self-lubricating ceramic cutting tools and their cutting performance. Jinan: qilu University of technology 2013.
5. Cao Tongkun , Gao Wei . $Al_2O_3/TiC/CaF_2$ ceramic friction and wear and self-lubricating film formation mechanism. Materials engineering ,2009,(9):75-79.
6 Wu Guangyong . Gradient self-lubricating ceramic tool materials and its antifriction mechanism. Jinan : Cheeloo University of technology ,2012.

Study on Graphene Reinforced Copper Contact Material

Ruikun Song[1]

[1]School of Power and Mechanical Engineering, Wuhan University, Wuhan, 430072, Hubei, China.

Abstract. Cu is widely used to contact materials because of its excellent electrical conductivity and economical efficiency, but its high temperature strength is slightly insufficient. In this paper, the less layers of graphene oxide were dispersed in deionized water by ultrasonic treatment, then copper acetate aqueous solution was added to the graphene oxide suspension with mixing. Then add NaOH aqueous solution into the blend solution. With graphene as a substratum, Cu $(OH)_2$ precipitations were generated so that a molecular level dispersion can be achieved. The precipitations were isolated by filtering, rinsing and drying, and then, these powders were reduced at 400℃ under a hydrogen atmosphere to form the homogeneously dispersed Cu/graphene composite powders. Sinter powders by spark plasma sintered and we obtained Cu contact materials strengthened by graphene. Under the sintering pressure of 250 Mpa, the hardness of Cu/graphene composite was 171.4 HV, which was 4.3 times than that of annealed copper; the electrical conductivity of Cu/graphene only decreased 5% and can still meet the contact demand.

1 Introduction

Electrical contact is one of the most important elements in instruments and electric switches. It is mainly responsible for the switch-on and switch-off of load current and circuit [1]. Due to the fusion welding, oxidation, and the bridge can happen in the process of repeated disconnect the closure, electrical contact materials should meet these demands [2]: high electrical conductivity, high thermal conductivity, good plasticity and toughness, high strength and hardness, resistance to arc erosion, welding and small tendency of bridging.

Cu is widely used to contact materials because of its excellent electrical conductivity and economical efficiency, but the high temperature strength is slightly insufficient. Graphene has high electrical conductivity [3], high hardness and mechanical properties, it's hardness reaches 110-120Gpa, tensile strength reaches 125GPa, Young's modulus is about 1100GPa, thermal conductivity is about 5000J/(m·K·s), carrier mobility is 2×10^5 $cm^2/(V\cdot s)$. Graphene also has good chemical stability and physical structural stability, which provides the driving force for the development of composite materials. We believe that graphene is a very good nano-filler to strengthen copper contact [4].

2 Experimental method

2.1 Preparation of Cu/graphene composite

First, graphene oxide was weighed and added to deionized water. After ultrasonic treatment for 2 hours in ultrasonic cleaning machine, we got uniform and stable monolayer oxide graphene solution [5]. Then copper acetate solid was weighed to prepare $Cu(CH_3COO)_2$ solution. Heated the copper acetate solution to 80℃ and mixed it with the dispersed grapheme [6]. Then, added NaOH solution and kept stirring, a blue flocculent precipitate was formed in the solution.

Because the graphene was uniformly dispersed in the solution, it would be the nucleating agent and $Cu(OH)_2$ would be adsorbed onto the monolayer grapheme [7], thereby CuO/graphene composites were formed. After standing, the precipitate was filtered, washed and dried to obtain a dried CuO/graphene composite. The dried powder was placed in a planetary ball mill and grinded to obtain a fine CuO/graphene composite powder [8]. Finally, CuO was heated in a hydrogen atmosphere and reduced to Cu. So we can obtain Cu/graphene composite materials.

2.2 Preparation of graphene reinforced contact materials

After graphene composite powder qualified was by XRD analysis, we used SPS320MK II spark plasma sintering system to perform low temperature high pressure SPS sintering. SPS sintering mold was WC mold and Sintering process was that sintering pressure was 250MPa, heating rate was 50℃ /min, sintering temperature was 600℃, holding time was 5 min. Conventional SPS sintering molds are typically made of graphite, they can only withstand the pressure below 80 MPa. So we used a special WC mold for sintering.

3 Results and discussion

3.1 XRD scanning

Using XRD to scan prepared Cu/graphene composite powder, the results obtained are shown in Figure 1

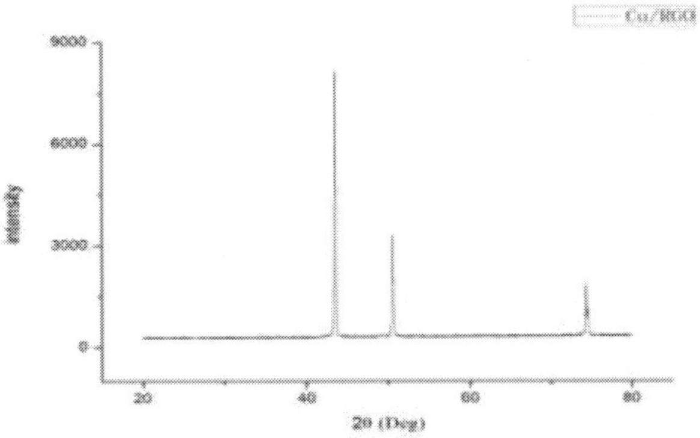

Fig.1. Results of XRD scanning of Cu/graphene composites.

Figure 1 shows that in the Cu/graphene composite powder, the characteristic peak of Cu is in full accordance with the PDF card result. The reduction process is sufficient.

By Debye-Scherrer formula ($D = \frac{K\lambda}{\beta\cos\theta}$) and using jade 6 software to draw the characteristic peak image of reduced copper, we calculated that D=79.5nm. That means the average thickness of the reduced Cu grains was 79.5nm, which shows in the Cu/graphene composite powder prepared by hydrogen reduction, Cu is nanometer.

3.2 Conductivity measurement

Conductivity was measured using an FQR-7501 eddy current conductivity meter. The conductivity is expressed in% IACS. The results are shown in Table. 1.

Table 1. Conductivity and Electrical Conductivity of Samples sintered at Different Pressure.

Sample / Conductivity (MS/m)	50 MPa	%IACS	250 MPa	%IACS
Reduced Cu	38	65.5	44.5	76.7
Cu/graphene composite	35	60.4	42	72.4

Table 1 shows that for reduced copper, the conductivity at 250Mpa reached 76.7%, increased by 17.1% than that of 50Mpa. For Cu/graphene composite, the conductivity at 250Mpa increased by 20% than 50Mpa. These results indicated that the increase of pressure has a positive effect on the improvement of conductivity. It is worth noting that at 50 MPa pressure, after adding graphene, the conductivity decreased by 7.9%. At 250MPa pressure, the conductivity only decreased by 5.6% and electrical conductivity also reached 75% IACS. At 20℃, the resistivity of pure copper is 1.7×10^-8Ω·m. The resistivity of samples prepared in this paper and that of pure copper are in the same order of magnitude.

After adding graphene, the conductivity of the sample decreased, the reason may be the following:

Effect of lattice distortion. In the actual crystal, impurities, defects, the structure of the interface between the grain is not complete will produce resistance and the addition of grapheme will cause lattice distortion defects which enhances the inhibition of electron motion, resulting in a decline in conductivity. But because of the high conductivity of graphene itself, the decline may be less than the other added components.

Effect of grain refinement. The finer the grain size, the more the number of grain boundaries, the stronger the resistance to electron motion, resulting in a decrease in conductivity.

3.3 Microhardness measurement

The samples obtained after SPS sintering were measured for microhardness with HXS-1000A. The results are shown in Table 2.

Table 2. Microhardness of Cu Samples sintered at Different Pressures.

Sample / Hardness (HV)	Pressure	
	50MPa	250MPa
Reduced Cu	130.6	137.5
Cu/graphene composite	163.7	171.4

Table 2 shows that at 50 MPa pressure, the hardness of the reduced copper reaches 3.3 times of the annealed copper. After adding graphene, the hardness reaches 4.1 times of annealed copper. The addition of graphene increased the hardness by 25.3%. At 250 MPa, the hardness of the reduced copper reaches 3.5 times of the annealed copper. After adding graphene, the hardness reaches 4.3 times of annealed copper. The addition of graphene increased the hardness by 24.7%. The increase in pressure has a certain effect on the hardness increase, they are 5.3% and 4.7%.But the effect of adding graphene on the hardness is obvious, reaching 25.3% and 24.7%.

The reasons are as follows. On one hand, as the strengthening phase, graphene has a very high hardness, the strengthening effect on the very soft Cu phase is obvious. On the other hand, increase in sintering pressure may reduce the voids and voids present within the sintered material, higher pressure results in a more compact material, thereby increasing the hardness.

3.4 SEM and EDS analysis

We conducted a SEM analysis of the Cu/graphene composites to observe the grain size of the samples and EDS surface scan of the Cu/graphene composites to analyze whether the dispersion of graphene was uniform. The results are shown in Figure 2 and Figure 3.

Fig.2. Cu/graphene composite SEM（50000X）scanning image.

Figure 2 shows Cu grain size is 500 nm-1μm, the Cu/graphene composites prepared in this paper are fine-grained materials.

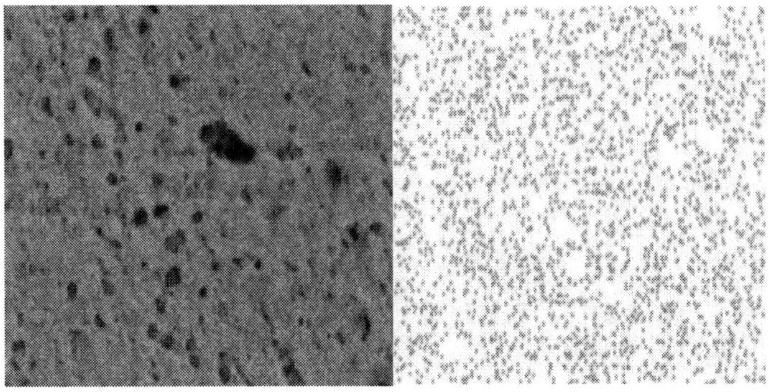

Fig.3. Micrograph of Cu/graphene composites and Carbon distribution state image.

In Figure 3, C is the graphene C is marked red. It can be seen that the distribution of graphene is uniform. Thus, we believe that the composite material is fine-grained material, graphene dispersion effect is good.

4 Conclusion

After XRD analysis, the prepared Cu/graphene composites have few impurity and high purity, Cu/graphene composites are the contact material of ultrafine grain.

The addition of graphene has a significant effect on the improvement of the hardness of copper. The conductivity of the Cu/graphene composites after graphene addition decreased only a little, still fully meet the electrical conductivity of the contact material requirements.

References

1. B.G. Qian, H.R. Geng, Z.Q. Guo. Mater Mech Eng, 28（3）, 7 (2004).
2. D.D. Zhang, L. Qin. Sci-Tech Information Development & Economy， 14（10）, 188 (2004).
3. A.K. Geim, K.S. Novoselov. Nat Mater, 6, 183 (2007).
4. T.W. Han, P.F. He, Y. Luo, X.Y. Zhang. Advances in Mechanics, 41（3）, 279 (2011).
5. S. Stankovich, D.A. Dikin, R.D Piner. Carbon, 45(7), 1558 (2007).
6. Z.M. Dang. Preparation and Characterization of Oxidized Graphene and Its Composites. 3 (2010).
7. Hwang J, Yoon T, S.H. JIN. Adv Mater, 2013, 25(46), 6724 (2013).
8. Y. Tu，Study on the Method of Obtaining Superfine Copper Oxide Powder by Interface Precipitation Method, 8 (2012).

Research of Multi-axis NC Machining Method of Cylindrical Cam Based on UG NX

Qianhua Liang

Electro-machinery Engineering Department, Chengdu Industrial Vocational Technical College, Sichuan, China
Corresponding author: 86755232@qq.com

Abstract. We have focused significant efforts on developing solutions for precision machining of cylindrical cams based on UG NX. A variety of processing method are put forward according to the digital model for a cylindrical cam which has been derived through parametric design, all of which are made a detailed comparison, analysis, research, elaborated. Simulation processing, post processing and NC program are carried out though optimized processing scheme. It will provide a reference for the numerical control programming of four-Coordinated axis.

1 Introduction

Cylindrical CAM is a cylinder on fluted artifacts, when moves, by higher pair contact follower can get according to the law of change of reciprocating motion [1]. Cylindrical CAM because of its small size, compact structure, high torque and rotational speed, etc. is widely used in the packaging, textile, food and pharmaceutical machinery automation [2, 3]. Because of cylindrical CAM profile curve is a closed space curve, with the traditional method is difficult to obtain accurate contour data so as to meet the needs of the CNC programming, it brings to the processing and manufacturing huge difficulties [4].

Using UG NX cylindrical CAM three-dimensional entity model is set up, and avoids the complex C language or other engineering software secondary development programming processing, make the CAM contour design is both accurate and simple [5, 6, 7]. At the same time, the use of UG processing module for CAM groove tool path is simulated, and through post-processing generate nc program can be used for machining center.

Through the technology analysis, the cylindrical CAM with four axis of A shaft is suitable for machining center. Generate four axis of cylindrical CAM slot method of tool path is very much, in this paper, by comparing, optimizing the reasonable solution.

2 Modeling

2.1 Parameter equation for cylindrical CAM

According to the principle of mechanical knowledge [8], cylindrical CAM contour curve parameter equation for (1), in which

x, y, z—curve on arbitrary point coordinates

s— follower motion

r_0—said base circle radius

j—said the CAM Angle

$$\begin{cases} x = r_0 \cos j \\ y = r_0 \sin j \quad (0^\circ \le j \le 360^\circ) \\ z = s(j) \end{cases} \tag{1}$$

2.2 Parameters

Stroke cylinder radius =30mm
follower h=20mm

curved groove b=12mm wide
groove depth =7mm
motion angle for rise travel =120°
farthest dwell angle =60°
motion angle for return travel =120°
nearest dwell angle =60°
both the rise travel and the return travel of the follower are follow simple harmonic motion.

2.3 CAM expression of UG

Design of CAM contour curve based on the above conditions, paragraphs U G expressions such as table 1. It is important to note:

(1) t, as UG system variables, $0 \leqslant t \leqslant 1$ [9].

(2) the working curve generated by rule curve respectively, and pasted to another file, each group of the curve and the graph object parameters (xt, yt, zt).

2.4 3D model

Generate the contour curve is shown in figure 1. Generate φ60 mm columns as shown in figure 2. To draw on in cylindrical center line of plane cross section sketch, drawing size to reflect slot width and groove depth, here to pay attention to the section size must be greater than the groove depth. With the method of differential scanning o generated CAM slot, pay attention to when scanning, cross section determination of locating method to φ60 mm for reference, the cylinder is direction to "plane normal" , get the required CAM as shown in figure 4.

Table 1. U G expression for paragraphs of contour curve.

Name	Formula	unit	note
a1	120	degree	motion angle for rise travel
a2	60	degree	farthest dwell angle
a3	120	degree	motion angle for return travel
a4	60	degree	nearest dwell angle
h	20	degree	Follower trip
j1	a1*t	degree	angle variable for rise
j2	a1+a2*t	degree	angle variable for farthest dwell
j3	a1+a2+a3*t	degree	angle variable for return
j4	a1+a2+a3+a4*t	degree	angle variable for nearest dwell
r0	30	mm	cylinder radius
s1	(h/2)*(1-cos(180*t))	mm	variable for rise
s2	h	mm	farthest trip
s3	(h/2)*(1+cos (180*t))	mm	variable for return
s4	0	mm	nearest trip

t	0		system variable (constant)
xt1	(r0+s1)*cos(j1)	mm	coordinate variables of x for rise
xt2	(r0+s2)*cos(j2)	mm	coordinate variables of x for farthest dwell
xt3	(r0+s3)*cos(j3)	mm	coordinate variables of x for return
xt4	(r0+s4)*cos(j4)	mm	coordinate variables of x for nearest dwell
yt1	(r0+s1)*sin(j1)	mm	coordinate variables of y for rise
yt2	(r0+s2)*sin(j2)	mm	coordinate variables of y for farthest dwell
yt3	(r0+s3)*sin(j3)	mm	coordinate variables of y for return
yt4	(r0+s4)*sin(j4)	mm	coordinate variables of y for nearest dwell
zt1	s1	mm	coordinate variables of z for rise
zt2	s2	mm	coordinate variables of z for farthest dwell
zt3	s3	mm	coordinate variables of z for return
zt4	s4	mm	coordinate variables of z for nearest dwell

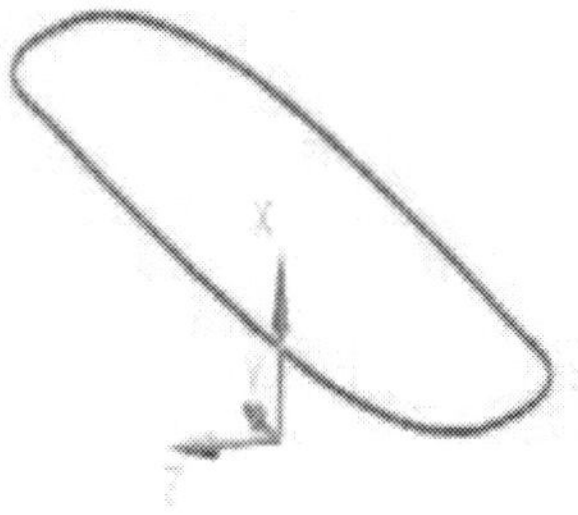

Fig. 1. Control curv.

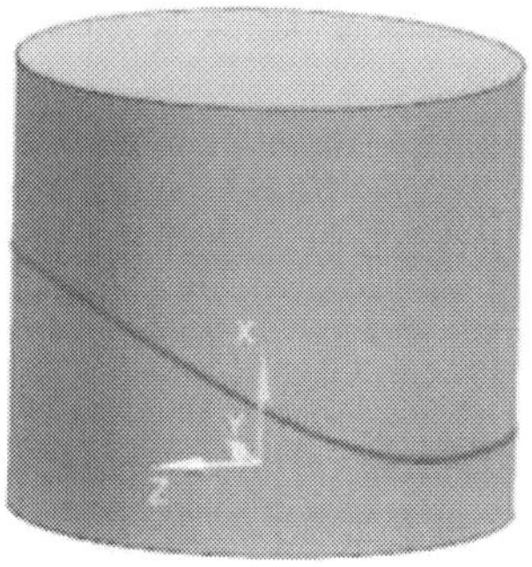

Fig. 2. Base circle cylinder.

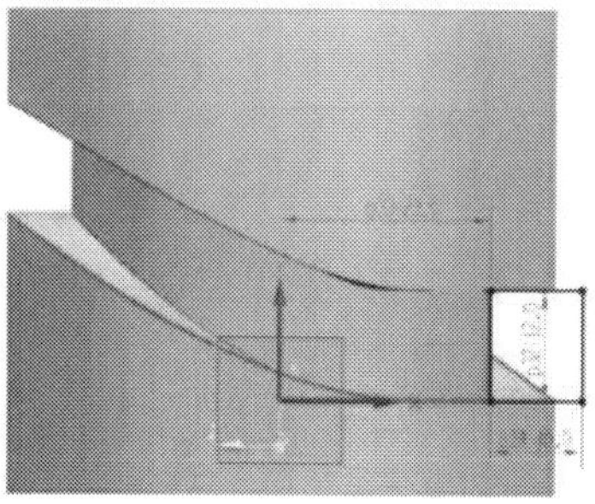

Fig. 3. Sketching of cam path.

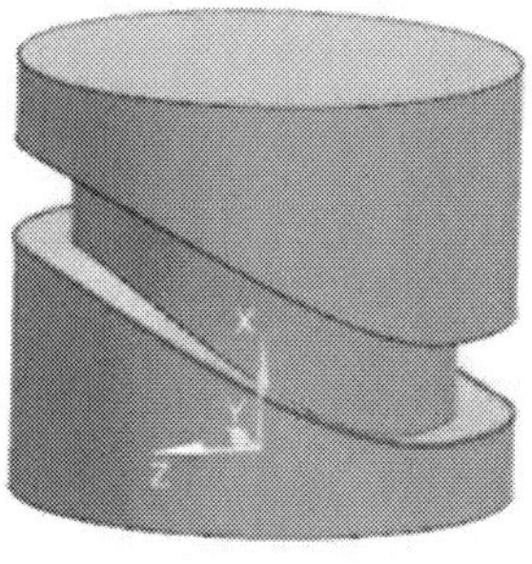

Fig. 4. Cylinder CAM.

3 CAM 4 axis machining of cylinder CAM

3.1 Machining of the Underside

Choose the first method. The need when using this method on the CAM groove bottom cylinder offset groove bottom CAM contour curve. Offset method is very simple, only can be produced to a command. Because processing is the bottom of the channel surface, thus can choose the bottom of the channel for cylinder part geometry. This method is safe and fast generated dao road. The rest of the way to do auxiliary plane and line, and reach cut but need to adjust the parameters. When machining CAM groove bottom also note that because use tool blanket out of the groove width, to generate tool path after must pay attention to check, especially when cutting, turning points in driving curve will cut, such as cutting, can move in non cutting parameters will be set as slotting cutting way. To generate tool path is shown in figure 5.

Table 2. Underside machining method comparison.

No.	drive method	Projection Vector	Shaft	note

1	Curve/ point	Shaft	Away from the straight line	Need to choose the appropriate tool by one cut
2	Curve/ point	Toward the straight line	4 axis, Perpendicular to the parts	Need to choose the appropriate tool by one cut
3	streamline	Toward the straight line	Away from the straight line	Need to make auxiliary plane according to the cutting tool
4	streamline	Toward the straight line	4 axis, Perpendicular to the parts	Need to make auxiliary plane according to the cutting tool

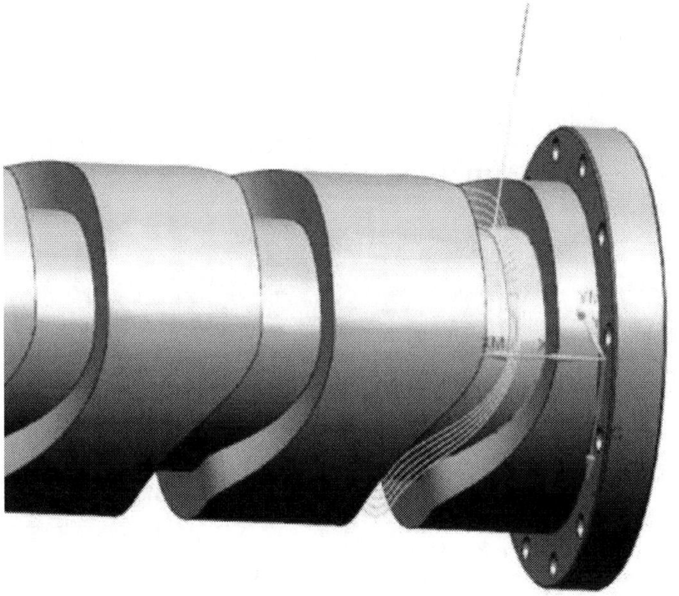

Fig. 5. Tool path of the Underside.

3.2 Machining of the side

Table 3. Side machining method comparison.

No.	drive method	Projection Vector	Shaft	note
1	Curve/point	shaft	Away from the straight line	more tool path by cutting parameters, can be offset
2	Curve/point	Toward the straight line	4 axis, Perpendicular to the parts	more Tool path by cutting parameters, can be offset

3	streamline	Perpendicular to the driver	Away from the straight line	Need to make auxiliary plane according to the cutting tool
4	surface	Perpendicular to the driver	Away from the straight line	Need to analyze surface UV direction
5	surface	Perpendicular to the driver	side edge driver	Need to analyze surface UV direction

First methods are preferred. When the method is adopted, the bottom cam profile curve of the bottom of the cam groove is needed, and the workpiece geometry can also be selected. This method is safe and fast. The remaining methods need to do the auxiliary surface and the line, and need to adjust the parameters to achieve but cut. In general, in the four axis machining, the cutter shaft is preferred to use far away from the straight line. Generate the tool path as shown in the following figure:

Fig. 6. Tool path of the side.

3.3 clean-up machining

Table 4. Clean-up machining method comparison.

No.	drive method	Projection Vector	Shaft	note
1	Curve/point	shaft	Away from the straight line	Need to choose the appropriate tool by one cut
2	Curve/point	Toward the straight line	4 axis, Perpendicular to the parts	Need to choose the appropriate tool by one cut
3	surface	Toward the driver	Away from the straight line	Need to modify the driving parameters

Choose the first method for the CAM groove side finishing machining, at the bottom of the finish machining by changes in rough machining method based on the corresponding parameters. The third method is commonly used, but need to choose the CAM groove profile as driving surfaces and components, projection vector selection toward the driver. If appear beat phenomenon in tool path generated by this method, the initial step length can be appropriately modified surface percentage and the end of the step length.

4 Program generation

On the basis of the tool path on UG own multi-axis reprocessing or post-processing can generate the corresponding NC system can use the processing program of. In machine cut using machine tools needed to view, after the clamping work piece can be generated by the program for processing simulation, observation program correctly or not [10]. After the

 DOI: 10.1051/matecconf/201710004011

confirmation process is correct, can through the network DNC, CF card and cable communication program into the corresponding machine tool can work out the parts [11].

5 Summary

Through the above examples can be seen that the UG software is very good to ensure the parts modeling accuracy, and provides a variety of simulation processing strategy, good maneuverability, generate NC program code and fast. At the same time, by contrast, found that for such four axis machining parts, drive method multi-select curve or surface drive, and try to use existing curve or surface parts, under the condition of the existing line or face can't meet to select new. If the tool path knife jumping, drive parameters can be appropriately adjusted to achieve the ideal effect. Away from the straight line option widely used knife shaft, linear choose cylindrical axis, fast accurate tool path.

References

1. Zheng wenwei, Wu. Edit. Mechanical principle. Beijing. Higher education press (1997).
2. machine design graphics manual. Chemical Industry Press (2004).
3. YuQing, rong-chang Liu etc. Spatial cylindrical CAM nc machining process design. Journal of mechanical design, 4 (2006) .
4. Yu-lin han. Full parameterized design of cylindrical CAM Model based on UG. Manufacturing technology and machine tool, 11 (2007).
5. Yang Ni, zheng-hao Ge, etc. Based on Pro/E design method for spatial cams and 3 D modeling. Journal of mechanical design, 12 (2007).
6. Huang Xinlin, ju-jiang Cao. The construction of a complex surface based on UG CAM contour surface. Journal of shaanxi university of science and technology, 11 (2004).
7. Ju-jiang Cao, Li Longgang etc. Based on UG NX6.0 globular indexing CAM 3 d entity modeling and simulation processing. mechanical design and manufacturing, 1 (2011).
8. Ya-hui Su., based on Pro/E of 4.0 cylindrical groove CAM parameterized design and simulation processing. Journal of coal mine machinery, 2 (2011).
9. Beijing trillion technology co., LTD. UG NX 8.5 bible. Beijing, China water conservancy and hydropower press (2013).
10. Beijing trillion technology co., LTD. UG NX 8.5 bible of Numerical control engineers. Beijing, China water conservancy and hydropower press (2013).
11. Geng GuoQing. Nc machine tool and computer serial port line connection and data transmission. Journal of machine tools and hydraulic, 14 (2012).

Study on the Load Sharing Capability of Closed-loop Gear Transmission in Braiding Machine

Xianqing Lei[1], Qi Meng[1,*] and Wensuo Ma[1]

[1]Henan University of Science & Technology , Luoyang, China
Corresponding author: 529566570@qq.com

Abstract. The static model is established based on the structural characteristics of closed loop gear transmission. The influence law of manufacturing and assembly error with load sharing coefficient was researched. The formula among transmission gear errors and load sharing coefficient is established. According to the formula of error coupling in the same angular frequency, transmission errors in closed loop gear transmission are coupled. The measure for reducing the transmission error is put forward. The study provides a theoretical basis for the design of high speed braiding machine.
Keywords. Braiding machine, closed-loop gears, load sharing coefficient, transmission error.

1 Introduction

The product of hose or cable has excellent property for tensile, bending, impact and wear by adding a layer or layers of mesh reinforcement layer. Mesh reinforcement layer is made by crossing the cable with each other [1-3]. Mesh reinforcement layer was often knitted by using knitting machine in order to realize cross cable with each other. The key part of the knitting machine is the mechanical drive system. Improving the transmission performance is part of the important ways to improve the speed and load of knitting machine. Currently, mechanical transmission system of the knitting machine is a closed-loop gear transmission at home and abroad. The closed-loop gear transmission is comprised of n gears. The number of gears is even, $n \geq 4$. Drive chain is made by n gears which are engaged with each two gears. If the closed-loop gear transmission exist transmission error, instantaneous ratio is changed. Asymmetric load distribution induces vibration and noise. The transmission stability of closed-loop gear transmission is reduced. Therefore, it is extremely important to study the relationship between transmission error and load.

The domestic and international scholars have done quite a lot of work on a transmission error and load sharing capability of closed-loop gear transmission. The transmission error is decreased by a modification gear profile , increasing manufacture and assembly accuracy [4,5]. But most academic researchers have focused on a pair of gears or open-loop gear transmission. Load sharing coefficient is put forward. The relationship between transmission error and load sharing coefficient is studied [6-8]. But most studies have focused on open-loop gear transmission like planetary gear transmission. There is less study on the load sharing capability of closed-loop gear transmission. Therefore, there has been a long lack of studies on the relationship between transmission and load sharing capability

The closed-loop gear transmission of knitting machine is investigated in this paper. According to the closed-loop gear transmission characteristic which is power split and converging, the static model of closed-loop gear transmission of knitting machine is established. The influence rule of manufacturing errors and assembly errors with closed-loop gear transmission is researched. According to the coupling formula of transmission error when the phase angle is the same, improvement measures are put forward to reduce the transmission error.

2 The working principle of closed-loop gear transmission

As figure 1 shows, power is input from input axis. Then the power is diverted to two power flows at the first gear. Each power flow transfers one-half of the load. The powers are confronted at the last gear.

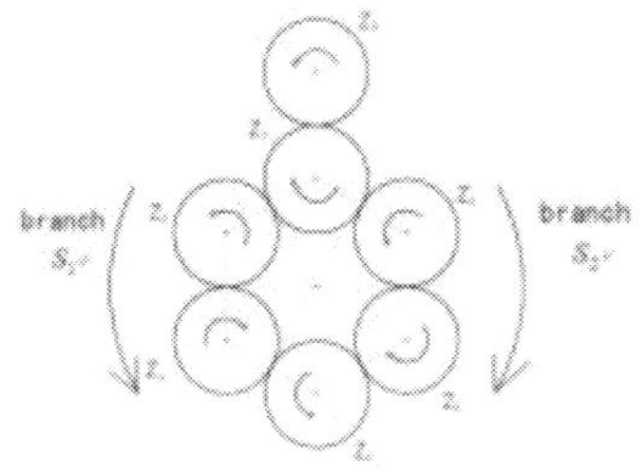

Fig.1 .Schematic diagram of closed-loop gear transmission.

3 Influence factors of transmission error

Gear error is part of the main reasons which induce the vibration. Therefore, the study of gear error is the most important thing. Gear error divided into two categories: and gears manufacture error and gear assembly error. The manufacture error mainly includes the gear eccentric error $F^{'}$ which leads to the asymmetric load distribution. The assembly error mainly includes the clearance e_1 between gear axle-hole and axle, the jumping value e_2 of the axis and the eccentric e_3 of bearing,which lead to the asymmetric load distribution.

4 Influence factors of transmission error

Equivalent mesh error is the projection of a manufacturing error and assembly error onto joggle line [9]. The direction of extending joggle line is positive, whereas its negative opposite. Taking manufacturing error E_i as an example, figure 2 shows the schematic of an equivalent mesh error which is the projection of eccentricity errors onto joggle line. E_i and A_i are manufacturing and assembly error. The direction of transmission is which is the direction of gear 1 to gear 2.Pressure angle is α . α_i and φ_i i are the phase angle of a manufacturing error and assembly error. β_{ij} is the angles between the direction of transmission and x axis when the direction of transmission is ij . γ_{ij} is the angles between the joggle line and the x axis when the direction of transmission is ij .

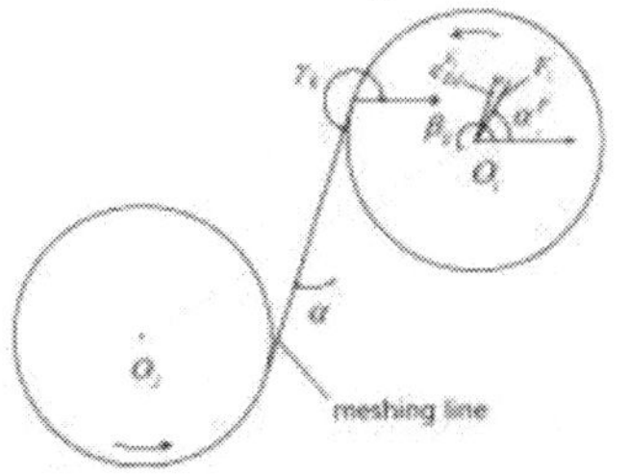

Fig.2. Schematic diagram of equivalent mesh error.

The equivalent mesh error of gear eccentric error $F^{'}$

$$e_{ij}^{F_i^{'}} = -F_i^{'} \cos\left(\gamma_{ij} - \alpha_i^{F^{'}} - \omega t\right) \tag{1}$$

The equivalent mesh error of clearance e_1 between gear axle-hole and axle

$$e_{ij}^{e_1} = -e_1 \cos\left(\gamma_{ij} - \alpha_i^{e_1} - \omega t\right) \tag{2}$$

The equivalent mesh error of jumping value e_2

$$e_{ij}^{e_2} = -e_2 \cos\left(\gamma_{ij} - \alpha_i^{e_2} - \omega t\right) \tag{3}$$

The equivalent mesh error of eccentric e_3 of bearing

$$e_{ij}^{e_3} = -e_3 \cos\left(\gamma_{ij} - \alpha_i^{e_3} - \omega t\right) \tag{4}$$

The equivalent mesh error of gear manufacturing error

$$e_{Eij} = e_{ij}^{F_i'} \tag{5}$$

The equivalent mesh error of gear assembly error

$$e_{Aij} = e_{ij}^{e_1} + e_{ij}^{e_2} + e_{ij}^{e_3} \tag{6}$$

e and ω are equivalent mesh error and gear speed.

5 The static model

The model of closed-loop gear transmission is as follow figure 3

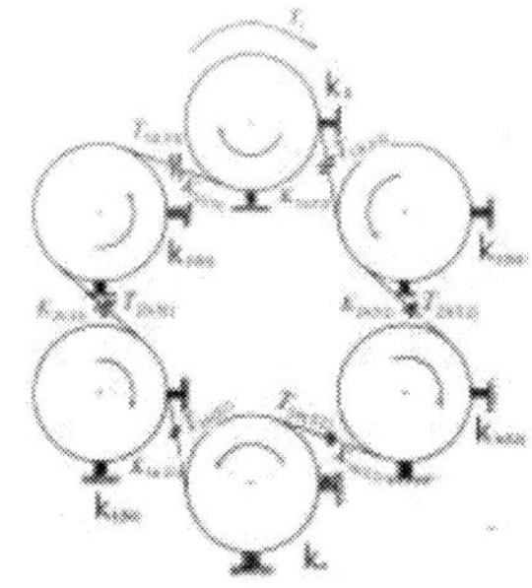

Fig.3. The model of closed-loop gear transmission.

The elastic deformation is represented by equivalent spring stiffness. k_i is the equivalent spring stiffness of gear supporting. K_{ij} is the meshing stiffness between gear i and gear j. Assuming the gear of output shaft doesn't rotation, gear Z_1 is applied a torque T . One of the gears which are Z_1 , $Z_{2(S1)}$ $Z_{2(S2)}$, $Z_{3(S1)}$, $Z_{3(S2)}$ and Z_4 first enter into meshing. Then the rest of gears has backlash. Finally all gears enter into meshing with the increasing torque T .

Considering the above analysis, the tooth surface load are set to $W_{ij(S1)}$ and $W_{ij(S2)}$

$$W_{ij(S1)} = K_{ij}\left(r_i\theta_i - r_j\theta_j - \Delta_{ij}\right) \tag{7}$$

$$W_{ij(S2)} = K_{ij}\left(r_i\theta_i - r_j\theta_j - \Delta_{ij}\right) \tag{8}$$

r_i and r_j in the formula are the radius of base circle of gear i and gear j. θ_i and θ_j are the angular displacement of gear i and gear j. Δ_{ij} is equivalent mesh error.

The formula can be derived from formulas (1)~(6) as

$$\Delta_{ij} = \sum e_{Eij} + \sum e_{Aij} \tag{9}$$

Due to the gear of output shaft doesn't rotation, the tooth surface load of output gear can be expressed as

$$W_{ij(S1)} = K_{ij}\left(r_i\theta_i - \Delta_{ij}\right) \tag{10}$$

$$W_{ij(S2)} = K_{ij}\left(r_i\theta_i - \Delta_{ij}\right) \tag{11}$$

$W_{ij(S1)}$ and $W_{ij(S2)}$ in the formula are the tooth surface load of branch S1 and branch S2.

The load sharing coefficient Ω_p is defined as the ratio of the actual load and the ideal load, which reflects the deviate degree of the actual load with the ideal load [8].

$$\Omega_p = \frac{W_{n\max}}{W_n} = \frac{W_n + \Delta W_n}{W_n} = 1 + \frac{\Delta W_n}{W_n} = 1 + \frac{K\sum \Delta_i}{T/r} \quad (12)$$

W_n is the ideal tooth surface load. ΔW_n is the error of tooth surface load.

6 The analysis of the static load

The figure 1, taking 6 gears as an example, shows the transmission principle. The power is input from input axis 0. Then the power is diverted to two power flow at the gear Z_1. Finally the powers are confluented at the gear Z_4.The power is transferred from 2 independent branches. The branch S_1 comprises $Z_{1(S1)}$ $Z_{2(S1)}$ $Z_{3(S1)}$ $Z_{4(S1)}$.The branch S_2 comprises $Z_{1(S2)}$ $Z_{2(S2)}$ $Z_{3(S2)}$ $Z_{4(S2)}$. The main parameters of closed-loop gear transmission with 6 gears are shown in table 1. The phase angle of error are shown in table 2.

Table 1. The main parameters of closed-loop gear transmission.

	Branch S_1	Branch S_2
Tooth number	60	60
Modulus/mm	5	5
Tooth width /mm	200	200
Pressure angle	20°	20°
Input speed (r/min)	6000	

Table 2. The phase angle of error.

	Gear	$\alpha_i^{F'}$	φ_i^{e1}	φ_i^{e2}	φ_i^{e3}
Branch S_1	$Z_{1(S1)}$	$25\pi/18$	$25\pi/18$	$25\pi/18$	$25\pi/18$
	$Z_{2(S1)}$	$15\pi/18$	$15\pi/18$	$15\pi/18$	$15\pi/18$
	$Z_{3(S1)}$	$3\pi/18$	$3\pi/18$	$3\pi/18$	$3\pi/18$
	$Z_{4(S1)}$	$25\pi/18$	$25\pi/18$	$25\pi/18$	$25\pi/18$
Branch $Z_{1(S1)}$	$Z_{1(S2)}$	$25\pi/18$	$25\pi/18$	$25\pi/18$	$25\pi/18$
	$Z_{2(S2)}$	$3\pi/18$	$3\pi/18$	$3\pi/18$	$3\pi/18$
	$Z_{3(S2)}$	$15\pi/18$	$15\pi/18$	$15\pi/18$	$15\pi/18$
	$Z_{4(S2)}$	$25\pi/18$	$25\pi/18$	$25\pi/18$	$7\pi/18$

The manufacturing error and assembly error are 50 μm. The change curve of load coefficient is shown in figure 4 when it is influenced by individual error.

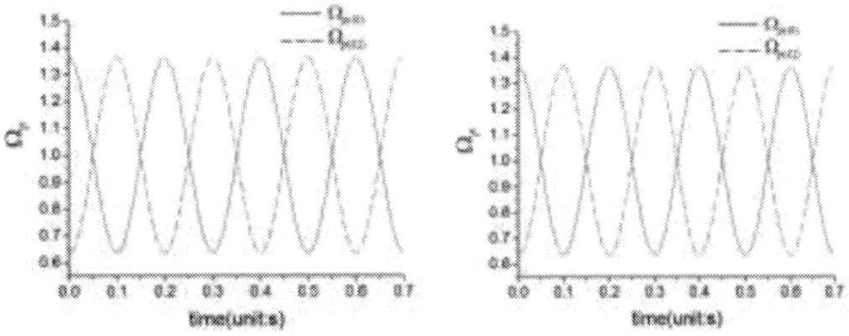

(a) Individual effect of F' (b) Individual effect of e_1

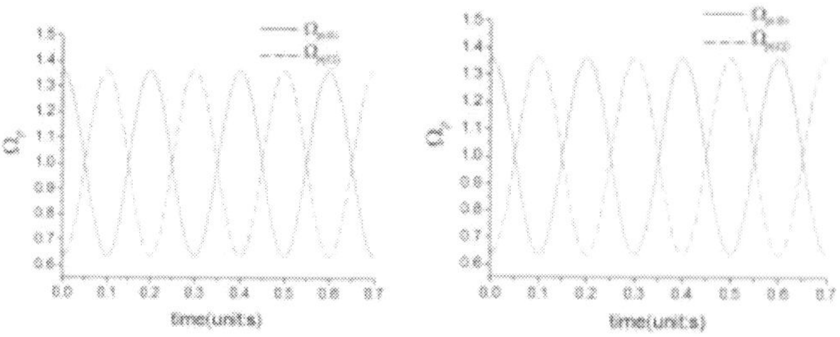

(c) Individual effect of e_2 (d) Individual effect of e_3

Fig.4. The change rule of load sharing coefficient.

The figure 4 shows the change curve of each error is same when the initial phase is the same. Actually, manufacturing error and assembly error are not equal. Each error randomly appears in the design standard. Manufacturing error and assembly error are shown in table 3. The change curve of load coefficient is shown in figure 5 when it is jointly influenced by all errors.

Table 3. The phase angle of error. μm

	Gear	F'	e_1	e_2	e_3
Branch S_1	$Z_{1(S1)}$	140	4	5	15
	$Z_{2(S1)}$	133	5	3	13
	$Z_{3(S1)}$	136	6	3	13
	$Z_{4(S1)}$	145	3	4	11
Branch S_2	$Z_{1(S2)}$	140	4	5	15
	$Z_{2(S2)}$	136	4	5	15
	$Z_{3(S2)}$	133	5	4	10
	$Z_{4(S2)}$	145	3	4	11

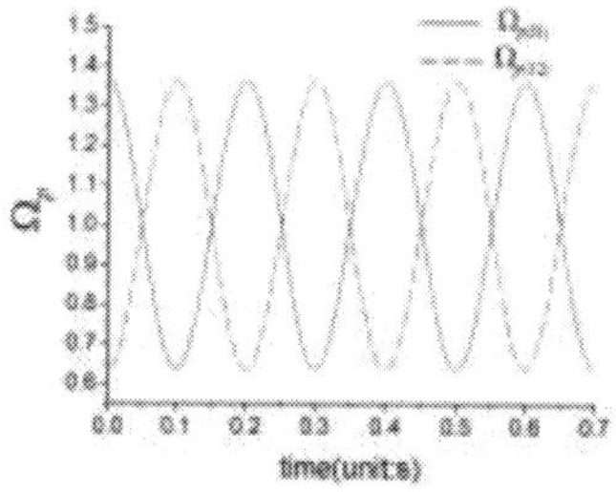

Fig.5. The change rule of load sharing coefficient.

The conclusion can be shown from the figure.

(1)The load sharing coefficient of each branch of closed-loop gear transmission show periodic changes.

(2)The maximum load sharing coefficient of branch S1 and branch S1 are 1.3625 and 1.3631.

(3)Through comparing the change rules of load sharing coefficient, the loading coefficient of two branches are not synchronized. Maximum instantaneous coefficients of each branch are also different.

7 Transmission error coupling

Due to the transmission features of closed-loop gear transmission, the calculation formula for the transmission error is derived. According to the principle of coupling errors, the error of closed-loop gear transmission is coupled, as showed in table 4. The change curve of load coefficient is shown in figure 6 when the error of closed-loop gear transmission is coupled.

Table 4. The phase angle of error.

	Gear	$\alpha_i^{F'}$	φ_i^{e1}	φ_i^{e2}	φ_i^{e3}
Branch S_1	$Z_{1(S1)}$	$7\pi/18$	$-11\pi/18$	$-11\pi/18$	$7\pi/18$
	$Z_{2(S1)}$	$15\pi/18$	$-3\pi/18$	$-3\pi/18$	$15\pi/18$
	$Z_{3(S1)}$	$3\pi/18$	$-15\pi/18$	$-15\pi/18$	$3\pi/18$
	$Z_{4(S1)}$	$25\pi/18$	$7\pi/18$	$7\pi/18$	$25\pi/18$
Branch S_2	$Z_{1(S2)}$	$7\pi/18$	$-11\pi/18$	$-11\pi/18$	$7\pi/18$
	$Z_{2(S2)}$	$3\pi/18$	$3\pi/18$	$3\pi/18$	$-15\pi/18$
	$Z_{3(S2)}$	$15\pi/18$	$15\pi/18$	$15\pi/18$	$-3\pi/18$
	$Z_{4(S2)}$	$25\pi/18$	$7\pi/18$	$7\pi/18$	$25\pi/18$

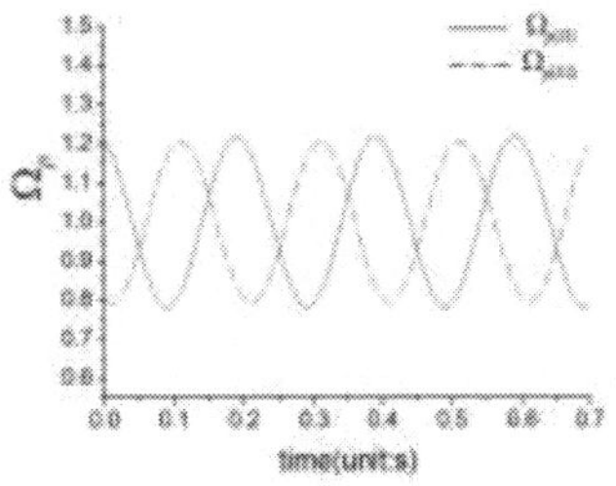

Fig.6. The change rule of load sharing coefficient.

The maximum load sharing coefficient of branch S1 and branch S1 after coupled are 1.2192 and 1.2073. The load sharing coefficient of closed-loop gear transmission has been reduced.

8 Conclusion

(1)The load sharing coefficient of closed-loop gear transmission have an obvious characteristic of periodic changes.

(2)According to the error accumulation, the load sharing coefficient under individual error effect is smaller than under common error effect

(3)The equivalent mesh error of every gear has the same effect on the load sharing coefficient

(4)According to the method of coupling, the load sharing coefficient of closed-loop gear transmission can be reduced. The transmission stability is improved.

References

1. P. Zhang, L. J. Gui, Z. J. Fan: The Experimental Study of Absorption Property for the Woven Fabric Composite Material Circular Tubes. Journal of Composite Materials, 2007, 01:146-150.
2. D. Q. Gu, L. Nie: Braiding Machine of Mesh Reinforcement Layer for the Plastic Pressure Hose. Journal of Textile Research, 2006, 01: 86-88.
3. Z. Zhao, Md. Hasab, Ikbal, W. Li: The development of the Knitting Machine and Knitting Process. Glass Fiber Reinforced Plastic/Composite Materials, 014, 10:90-95.
4. J. K. Jiang, Z. D. Fang, J. Z. Su: Optimal Designing of Helical Gear Tooth Surface Modification Based on the Minimum Transmission Error. Journal of Aerospace Power, 2013,07: 1637-1643.
5. L. D. Wang, S. Y. Ling, Y. Ma, X. D. Wang, Z. F. Lou: The Processing Method of Precision and Ultra-precision Cylindrical Gear. Optics and Precision Engineering, 2009,02: 321-326.
6. Krantz T L. A Method to Analyze and Optimize the Load Sharing Split Path Transmissions. San Diego: NASA Technical Memorandum107201, 1996: 1-10.

7. Ahmet. Load Sharing Characteristics of Planetary Transmission. Mech Mach Theory, 1994, 29(8): 1151-1161.
8. B.Li: Analyze the Method of Load Sharing Coefficient of Planetary Gear (MS., Nanjing University of Aeronautics and Astronautics, China 2005), pp.13-17.
9. Hidaka Teruaki, Yamamoto Nobuyuki and Takeshi Ishida: Japan Machinery Association proceedings (Tokyo, Japan, November 8-10, 1986). Vol. 52, pp.2200-2206.

 DOI: 10.1051/matecconf/201710004013

The Influence of Social Capital on Small Micro-manufacturing Venture in China

Shaoxia Zhao*and Yongming Zhu

Administrative Engineering College, Zhengzhou University, Zhengzhou, China

*Corresponding author: 1358804759@qq.com

Abstract. In recent years, because of the government's policy support and the scholars' attention, the small micro-manufacturing industry has been developing rapidly. However, due to small scale, limited financial resources, imperfect management and other constraints, small micro-manufacturing survival rate is relatively low. Scholars have been actively exploring the factors that affect micro-manufacturing entrepreneurship, and agreed that social capital has an important impact on micro-manufacturing industry. This paper analyzes the effects of social capital on the small and micro-manufacturing entrepreneurship from the perspective of structural capital, relational capital and cognitive capital by collecting relevant data, using questionnaire survey, structural equation method, and other empirical research. The results show that the three dimensions of social capital have positive effects on small and micro-manufacturing entrepreneurial behaviour. The results provide theoretical basis for small micro-manufacturing enterprises, and have a certain reference value for the follow-up study of small micro-manufacturing industry.

1 Introduction

Since the reform and opening up, China's small micro-manufacturing industry has gone through a process from small to large, from small to large. At present, the research on the micro-micro-manufacturing industry has aroused extensive attention in the academic circles. But so far, Small micro-manufacturing industry researches, more focused on the study of micro-manufacturing content, features, etc., are rarely involved in small micro-manufacturing entrepreneurial issues (Han Dongping, 2012 [1]). In the western academic circles, many scholars have studied the entrepreneurial behaviour of enterprises. Since the 1980s, social capital theory has been used in entrepreneurial research, and has been gradually introduced into the analysis of entrepreneurial behaviour (Stam, 2008 [2]). What is the role of social capital in entrepreneurship? To what extent? In what ways? None of these questions has been given a very rigorous conclusion. Therefore, the purpose of this paper is to study the entrepreneurial behaviour of small micro-manufacturing industry from the perspective of social capital, so as to reveal the impact of social capital on small and micro-manufacturing entrepreneurship.

2 Literature review

2.1 Social capital

The concept of "Corporate Social Capital" was first proposed in 1985 by the French scholar Pierre Bourdiet [3]. Social capital is increasingly concerned by many disciplines. Koka and Prescott (2000) [4] extended the study of social capital to the corporate level, making the academic community began to pay attention to corporate social capital [4]. At the same time, companies are beginning to realize the importance of the accumulation of social capital. More research has been done on the connotation and measurement of corporate social capital: China's scholar Qiu Zhongning (2000) divides the social capital of enterprises into horizontal, vertical and social relations, and points out that the enterprise can obtain the necessary resources by using these three links[5]; Westlund (2003) divides the social capital of an enterprise into internal and external social capital, in which the relationship between managers and employees is the internal social capital of an enterprise, and the resources related to the market and the environment are the external social capital of the firm[6]; Lin Yun (2011) divides the enterprise capital into cognitive dimension, relation dimension and structural dimension, and analyzes the influence of each dimension on enterprise innovation ability[7].

2.2 Small micro-manufacturing industry

Small micro-manufacturing industry refers to the number of employees is less than 10 people, a high degree of unity of property rights and management rights, products (services) single, small-scale enterprise organizations. Foreign theoretical circles have realized the prominent role of microenterprises in information technology. In order to facilitate research, the European Commission united the definition of standards in 19% of the year, with the number of employees ranging from 1 to 9 as very small (Li, 2012) [8].

China's theoretical circles of small micro-manufacturing research scattered in the study of small businesses. In China's current economic policy is not a small micro-manufacturing industry, the habit of small and micro enterprises as "self-employed." People only concerned about how large enterprises to enter the world's top 500 enterprises, "grasping big and small" from a certain perspective reflects the fate of the small micro-manufacturing disregard, so that people set up small micro-manufacturing power is inhibited, and further, Large-scale enterprises have lost their cooperation network composed of small micro-manufacturing and so on. Large-scale enterprises are in the cycle of self-service, and the returns of scale returns are not obvious (Zhou Ronghua, 2016) [9].

2.3 Research assumptions

China is in the economic transition period, social capital plays an important role in the development of enterprises. First, the enterprise can obtain the necessary information and resources through social capital. Corporate social capital can not only influence the transmission and application of market information in social networks, but also help enterprises to obtain important external environment information, and provide the basis for enterprise strategy formulation, and is an important source for enterprises to obtain specific resources or related policy support [11]. Second, corporate social capital helps enterprises and their network members to establish a trust mechanism to form a good environment for cooperation. Compared with the low degree of social capital accumulation, enterprises with rich social capital have higher credit level and can strengthen the relationship between enterprises and the stakeholders such as suppliers and sellers, so as to provide the necessary external environment support. Finally, corporate social capital can enhance the cohesion of enterprises and their own resources of creativity. Internal social capital can link different departments of enterprises, strengthen trust among members, and promote organizational goals to achieve agreement [12], for small micro-manufacturing venture to provide the necessary internal environmental support. Based on the above analysis, this paper proposes the following assumptions:

H1: The number of small-scale entrepreneurial social interactions positively affects entrepreneurial behaviour

H2: The social capital on the relationship dimension has a positive effect on small and micro manufacturing entrepreneurial behaviour

H3: The cognitive capital dimension has a positive effect on small and micro manufacturing entrepreneurial behaviour

3 Research and design

3.1 Research methods and data processing

In this study, a questionnaire survey was used, and the contents of the questionnaire were filled in by the founders of small and micro-manufacturing industries. This questionnaire is divided into two parts: small micro-manufacturing entrepreneur personal and small micro-manufacturing basic information, small micro-manufacturing entrepreneur social capital measurement of three dimensions.

The first is to investigate the basic situation of small and micro-manufacturing entrepreneurs and small and micro-manufacturing industry, namely: the name of the enterprise, the industry, the nature of the enterprise, the registered capital of enterprises, the number of employees and enterprises set up years. The social capital of small and small manufacturing entrepreneurs is divided into three dimensions: structural dimension, relationship dimension and cognitive dimension, followed by measurement of the four aspects of small and micro manufacturing entrepreneurial behaviour. The measurement design of this questionnaire is based on Vesa Puhakka (2006) [13] studied the measurement scale. Micro-manufacturing entrepreneurial behaviour is divided into four dimensions: access to business knowledge, competitive scrutiny, forward-looking search and innovative behaviour.

3.2 Variable selection

3.2.1 An interpreted variable

Micro-manufacturing entrepreneurial behaviour from four aspects to show, namely: access to business knowledge, competitive scrutiny, forward-looking search and innovative behaviour. First of all, access to business knowledge behaviour refers to the behaviour of the general social interaction and entrepreneur's personal relationship behaviour, the acquisition of the customer, technology, products, finance and scientific research information into the business knowledge behaviour. Second, competitive scrutiny includes the observation of market competition, the search for market vulnerabilities, and the formulation of competitive strategies and tactics. Third, forward-looking search refers to behaviour by understanding the external environment of the enterprise, predicting the future competitive situation of the enterprise, evaluating the behaviour of the enterprise resource situation, and observing and forecasting the development trend of the industry and enterprise. Fourth, innovation behaviour refers to the act of generating the innovative thinking and implementing the innovative practice, questioning the behavioural way and mode of thinking, and seeking to change the behaviour.

3.2.2 Explain the variables

At present, in a number of theories on social capital, Ghoshal (1998) gave a more complete measure. His theory represents social capital in three dimensions: structure, relationship and cognition.

The structural dimension of social capital measures the size of the entrepreneurial circle, that is, the number of social contacts between entrepreneurs and their social relations. The more social interaction, the greater the index of social capital structure dimensions. The relationship dimension of social capital measures the degree to which the relationship between entrepreneurs and their social relationships is close, that is, the frequency with which the entrepreneur interacts with his circle. The higher the frequency, the greater the degree of tightness, the greater the measure of the relationship dimension. The cognitive dimension of social capital measures the emotional loyalty between the entrepreneur and their relationship circle. The more the relationship circle provides, the more interdependent the team member in the entrepreneurial team. The more support, the larger the measure of cognitive dimension.

3.2.3 Control variables

There are many factors that affect the micro-manufacturing entrepreneurship. Taking into account the individual differences of small and small manufacturing entrepreneurs and their access to and use of social capital, this paper selects the age (*NL*), gender (XB), and educational level (*JY*) of the small micro-manufacturing entrepreneur as the control variable. The variables and definitions as shown in Table 1:

Table 1. Study variables and definitions

	The variable name	Variable symbol	Variable definition
Explained variable	Small micro manufacturing entrepreneurial behaviour	JH JJ JQ JC	Gaining business knowledge, Jcompetitive scrutiny, forward-looking search and Innovation behaviour
Explanatory variables	Social capital	SJ SG SR	Structural capital, relational capital, cognitive capital
Control variables	Entrepreneur Age	NL	The questionnaire was informed
	Entrepreneur gender	XB	The questionnaire was informed
	Entrepreneur education level	JY	The questionnaire was informed

4 Structural equation models

$$SJ=j_1S1+j_2SJ2+j_3SJ3+j_4SJ4+e_1 \quad (1)$$
$$SG=g_1SG5+g_2SG6+g_3SG7+e_2 \quad (2)$$
$$SR=r_1SR8+r_22SR9+e_3 \quad (3)$$
$$JH=h_1JH1+h_2JH2+h_3JH3+h_4JH4+e_4 \quad (4)$$
$$JJ=j_1JJ5+j_2JJ6+j_3JJ7+e_5 \quad (5)$$

$$JQ=q_1JQ8+q_2JQ9+q_3JQ10+e_6 \quad (6)$$
$$JC=c_1JC11+c_2JC2+e7 \quad (7)$$
$$M_1JH+m_2JJ+=3JQ+m_4JC=n_1SJ+n_2SG+n_3SR+n_4NL+n_5XB+n_6JY+e_9 \quad (8)$$

JH, JQ, JJ, JC are the four kinds of behaviours of small and micro-manufacturing entrepreneurial social capital structure, In the acquisition of business knowledge, competitive scrutiny, prospective search, and innovation behaviour represent potential variables. In addition, the years of establishment, the age of entrepreneurs, the degree of education (*JY*) are taken as the control variables.

5 Empirical analyses

5.1 Sample analysis

In order to ensure the effectiveness of the study, selected subjects must meet the following conditions: First, the entrepreneurs are the founder of business people who have achieved the process of enterprise entrepreneurs from scratch. Second, the enterprise must be a small micro-manufacturing, and survival time of 2 years. A total of 880 questionnaires were sent out and 743 were returned. Among them, the effective questionnaire is 650 copies. The effective recovery rate was 73.86%.

Using SPASS software, the reliability of the questionnaire was analysed, as shown in Table 2. Social capital and micro-manufacturing entrepreneurial behaviour of all dimensions of a value and the overall a value is higher than 0.7, in line with the requirements of the reliability requirements (a>0.6). The overall reliability and stability of the questionnaire were cantered. While the CITC values of all dimensions are greater than 0.5, indicating that the scale design project reliability is good, without removal.

Table 2. Reliability of social capital and entrepreneurial behaviour

Variables	Variable indicator	CITC	The value of each dimension （a）	Overall value （a）
Social capital	Structural dimension	0.629	0.841	0.850
	Capital dimension	0.686	0.865	
	Cognitive dimension	0.539	0.869	
Small and micro manufacturing Entrepreneurial behaviour	Get business knowledge	0.541	0.830	0.797
	Competitive review	0.556	0.804	
	Forward-looking search	0.510	0.780	
	Innovation behaviour	0.584	0.849	

5.2 Hypothesis testing

5.2.1 Test for hypothesis H1

As shown in rows (2), (3), (4) and (5) of Table 3, the path coefficients of social capital on the structural dimension are all significantly positive at 0.01 level, providing data on the assumption that H1 is established stand by. The small number of micro-manufacturing entrepreneurial social interaction positive impact on entrepreneurial behaviour, that is, micro-manufacturing to expand the number of social contacts, can get more information to help compete in the competency to review, make the right judgments, Forward-looking issues, the more conducive to innovation. It shows that the dimension of social capital structure has positive correlation with four kinds of behaviours, namely, business knowledge acquisition, competitive examination, forward-looking search and innovation in micro-micro-manufacturing entrepreneurial behaviour.

5.2.2 Hypothesis H2 test

As shown in lines (6), (7), (8) and (9) in Table 3, the path coefficients of the social capital on the relationship dimension are all significantly positive at 0.01 level, proving that the hypothesis H2 holds. Entrepreneurs social relations more closely, the more can be obtained from more channels can not get or get the first time the information, the future of the foreseeable than others grasp more accurate. Social relations, the degree of tightening, can also be used to help entrepreneurs to form innovative thinking, mutual awareness of the enhanced collaboration, from the intrinsic to enhance the quality of entrepreneurs. It shows that there is positive correlation between the dimensions of social capital and business knowledge, competitive examination, prospective search and innovation in small and micro manufacturing entrepreneurial behaviour.

5.2.3 Test for hypothetical H3

As shown in lines (10), (11), (12) and (13) in Table 3, the social capital in the cognitive capital dimension and the competitive scrutiny in the micro- and micro-manufacturing start- And there was a significant positive correlation between the level of entrepreneurial behaviour and the level of positive correlation in the level of obtaining business knowledge behaviour.

Table 3. The path coefficient table and the significance relation of the model

Variable relationship	Path coefficient	T value	The number of rows
SJ vs.JH	0.0436***	8.259	(2)
SJ vs.JJ	0.004***	4.808	(3)
SI vs.JQ	0.063***	10.826	(4)
SJ vs.JC	0.033***	6.847	(5)
SG vs.JH	0.090***	21.946	(6)
SG vs.JJ	0.068***	12.553	(7)
SG vs.JQ	0.069***	12.301	(8)
SG vs.JC	0.030***	6.169	(9)
SR vs.JH	0.099*	2.325	(10)
SR vs.JJ	0.030**	0.894	(11)
SR vs.JQ	0.068**	1.170	(12)
SR vs.JC	0.067***	8.112	(13)

Note: * indicates significant at the 0.1 level, ** indicates significant at the 0.05 level, *** indicates significant at the 0.01 level.

6 Conclusions and Enlightenment

Small micro-manufacturing industry is to provide new jobs, the main channel, but also the main platform for entrepreneurial growth. Therefore, the healthy development of small micro-manufacturing industry to maintain stable and rapid economic development is of great strategic significance. In this paper, the small micro-manufacturing entrepreneur target groups, the use of theoretical research and empirical research method of combining social capital into small micro-manufacturing entrepreneurial behaviour research frame. The results show that structural capital dimension and relational capital dimension of social capital have positive effects on small and micro manufacturing entrepreneurial behaviours. That is, social capital can promote micro-manufacturing entrepreneurial behaviour.

Therefore, in order to improve small and micro manufacturing entrepreneurial behaviour, on the one hand, entrepreneurs should pay attention to the development and accumulation of social capital, increasing personal social investment, enhance personal image and enhance personal social skills. First, small and micro-enterprise entrepreneurs need to expand social relations, access to peer or potential partners more opportunities for cooperation; Second, small and micro-enterprise entrepreneurs need to promote consensus with others; Third, small and micro enterprises entrepreneurs can join the entrepreneur associations or chambers of commerce to get to know more entrepreneurs and industry experts, to establish a diversified social network, which can obtain a richer quality of resources for the survival and development of enterprises. On the other hand, enterprises should constantly update their products and services according to market demand, continuously improve consumer satisfaction, grasp market opportunities, take the lead in using new methods, new technologies and new products, improve product productivity, and reduce production costs.

References

1. Han Dong-ping, Guo Mei-jia, Deng Xiu-li. The empirical research on the relationship between social capital and credit level of small and medium-sized enterprises. Accounting and Finance, 30: 99-101(2012).
2. W. Stam, T. Elfring, Entrepreneurial orientation and new venter performance: The moderating role of intra and extra industry social capital. Academy of Management Journal, 51(1):97-111(2008).
3. P. Bourdieu. Distinction: A social critique of the judgment of taste.Translated by Rich-and Nice, Roultedge Press, (1994).
4. B.R. Koka, J.E. Prescott. Strategic alliances as social capital: A multimensional view. Strategic Management Journal, 23(9):795-817(2000).
5. QIU Zhong-ning, CHEN Chuan-ming. Effects of internal social capital on organizational performance-based on the perspective of market knowledge transfer, Economic Management, 04: 41-50(2013).
6. H. Westlund. Implications of social capital for business in the knowledge economy: theoretical considerations. In International Forum on Economic Implication of social capital. Tokyo, 2003.
7. Lin Yun, Liu Wei, LI Sui-cheng. Empirical research on the impact of corporate social capital on technological innovation capability. Scientific Management, 01: 35-44(2011).
8. Li Wei, Xu Hui. Study on the relationship between social capital, market knowledge capability and business performance: An analysis from social networks .Soft Science, 10: 93-98(2012).
9. Zhou Ronghua, Zhang Minglin. Analysis on the regulatory effect of entrepreneurial orientation on social capital and small micro-manufacturing entrepreneurial performance. Business Times, 24: 108-109(2014).
10. P. Bourdueu. The forms of capital. In: Richardson G, Handbook of theory and research for the sociology of education[C]. Eestport, CT.: Greed wood Press,. 241-258(1985).
11. J. Liao, H. Welsch. Social capital and entrepreneurial growth aspiration: a comparison of technology — and non — technology based nascent entrepreneurs.Journal of High Technology Management Research: 149-170(2003).
12. D.M.D. Carolis, P. Saparito. Social capital, cognition and entrepreneurial opportunities: a theoretical framewor.Entrepreneur-shin Theory and Practice, 41-56(2006).
13. V. Puhakka. Effects of social capital on the opportunity recognition process.Journal of enterprising culture, 14(2):105-124(2006).
14. W. Tsai, S. Ghoshal. Social capital and value creation: The role of infra-firm networks.Academy of Management Journal, 41:464-476(1998).

Study on Stress Analysis and Welding Deformation of the Loader Boom

Zhengwen Yuan[1,2], Xi Lu[3], Yongneng Lu[1,2], Lei Xu[1,2] and Shumin Zhu[1,2]

[1] Jiangsu Xuzhou Construction Machinery Research Institute, Xuzhou, 221004, China
[2] State Key Laboratory of Intelligent Manufacturing of Advanced Construction Machinery, Xuzhou Construction Machinery Group, Xuzhou, 221004, China
[3] College of Mechanical Engineering, University of Shanghai for Science and Technology, Shanghai, 200093, China

Abstract. This paper takes the 5T loader boom as the research object. Firstly, the static analysis of the boom loader is carried out under the typical working conditions. The weak part of the welding seam is found, and then the welding deformation of the boom is analyzed for the weak welding part. Further, the deformation trends of the welding arm are studied. It is found that the weak part of the welding seam is at the welding arc of the boom plate and the beam. Besides, the deformation of the arm plate is about 30mm inward in the welding process due to the welding stress, and the residual stress of the boom. This work can provide theoretical and data support for the processing and deformation correction of the boom.
Keywords. loader boom, static analysis, welding deformation, stress.

1 Introduction

Wheel loader is a kind of main earthmoving machine used in construction, port, mine, railway, mining and other projects. It can carry out shoveling work on soil, stone and so on. It can also carry out material handling for trucks and other transportation vehicles. A loader completes the work mainly by its working device. It is a space multi-bar mechanism, mainly constructed by bucket, drawbar, swing arm, boom and fuel tank composition. Loader boom is the main load-bearing parts in the working device, mainly made by the plate welding process. During the process, a large weld deformation can occur during the welding of the loader boom plate. This will affect the loader assembly route. It will apply the additional assembly stresses that cause premature fatigue and fracture of the loader boom.

Many scholars have done the research works on the deformation of welded structure. To reduce the welding deformation of the loader tiger and get the optimal scheme of tiger shaped structure welding sequence, the numerical simulation of welding deformation of tiger-shaped structure for rear frame of loader was conducted using thermo elastic-plastic method [1]. As for butt-welding deformation of thick plates in large steel structure, the groove dimensions must be optimized and the rational welding sequence must be adopted [2]. Besides, the three-dimensional numerical simulation of welding process for aluminum alloy thin plate under the conditions of different clamping was also performed by recalls of thermal elastic-plastic FEM technology [3]. The welding deformation of high strength steel with T joint was studied based on inherent strain theory [4]. The reasons of the deformation is also analyzed aiming at the angle and distort deformation appearing in the course of welding in active arms of loading machine [5].

In this paper, the 5T loader boom is taken as the research object. Firstly, the static analysis of the boom loader is carried out under the typical working conditions. The weak part of the welding seam is found, and then the welding deformation of the boom is analyzed for the weak welding part. Further, the deformation trends of the welding arm are studied. This work can provide theoretical and data support for the processing and deformation correction of the boom.

2 Static analysis of loader boom

Figure 1 shows the structure of the loader working device. As shown in this figure, the loader working device consists of four parts: bucket, drawbar, swing arm and boom. When the loader carries out the shoveling operation, the arm cylinder can be stretched and compressed in order to transmit the force to the rod through the rocker arm. The drawbar can drive the bucket to swing around the hinge point of the boom. The boom cylinder can make the whole working device rotate around the front frame. So the materials can be upgraded. As long as the loads connected to the bucket at the three hinge points are calculated, the external workloads on the whole working device can be obtained.

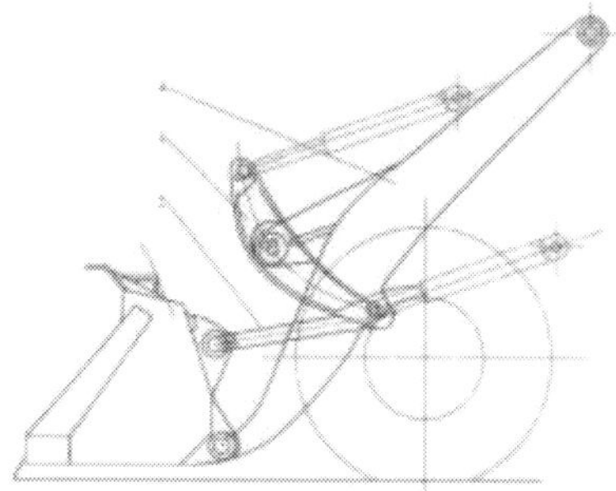

1 – bucket; 2 – drawbar; 3 - swing arm; 4 - boom

Fig. 1. structure of the loader working device.

The three-dimensional model is established by the use of three-dimensional modeling software. Under the precondition of not affecting the basic features and stress of the parts, the non-important features are simplified or neglected, such as smaller roundness, small holes and so on. Figure 2 shows the three-dimensional modeling results of the loader boom.

The three-dimensional solid model of loader boom was directly imported into the finite element software. The material was selected to be Q345. The density of the material Q345 is 7850 kg/m3, the Young's modulus was 206000 GPa and the Poisson's ratio was 0.28. The initial grid size was set to be 5mm. then the welding part of loader boom is carried out two grade refinement. The grade division results are shown in Figure 2.

As shown in Figure 2, after two grade refinement for the welding part, the grid of the finite element has been fully refined. Viewing the grid partition by the grid partition command, there are a total of 578,674 nodes and 358,652 units. The mesh quality option was used to check the meshing quality and the results can meet the requirement.

The holes of the boom are in hinged loading conditions under normal operating conditions. So when the constraints are applied, the degree constraint is applied to the hinge hole 3 to restrict the five freedom degrees of the hole, allowing it to rotate about the axis. The freedom degree of the hinge hole 2 is fully constrained. The boundary load is applied to the boom at the positions of the hinge holes 1, 4, and the boom constraint is shown in figure 2.

(a) Meshing result

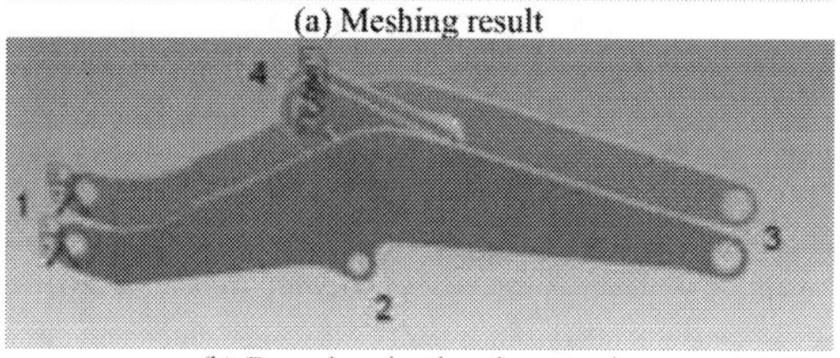

(b) Boundary load and constrain

Fig. 2. Three-dimensional modeling results of the loader boom.

Under the above boundary conditions, the deformation are shown in Figure 3-(a).

According to Figure 3-(a), it can be seen that the total deformation of the boom is in a state of symmetrical distribution. The largest deformation is on the left side, which is about 5.85mm and decreasing from left to right. The right side part had less deformation because there applying a hinge constraint at the hole. The deformation at the support arm is less than 1 mm. As shown in figure 3-(b), it is the deformation direction of the boom in working condition. The largest deformation occurs at the left part. The deformation direction is downward.

(a)Deformation analysis results of the loader boom

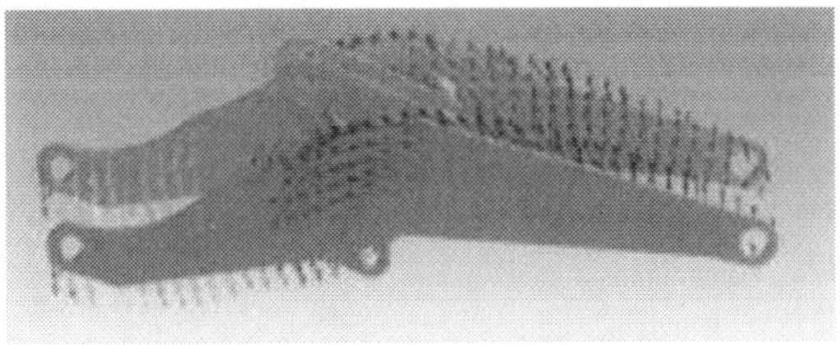

(b)Deformation direction of the boom

Fig. 3. FEM reults.

Based on the finite element simulation, the Von Mises Stress distribution is shown in Figure 4. It can be seen that the stress distribution on both left and right sides are the same. The upper and stress is larger than lower results on each side. The maximum stress of the upper side is about 225MPa, the middle part of the stress is about 20MPa, which is relatively small. The maximum stress of the bottom side is about 183MPa.

For the welded structure, the weld location in the four corners, that is the weld toe part, has relatively larger stress. The largest part occurs in the back corner, which is about 145MPa. Figure 5 shows the Von Mises Stress distribution in the weld part of the boom. As shown in Figure 6, it can be seen that the equivalent stress is larger at the toe of the back corner, and the maximum value is about 145 MPa. Figure 6 shows the cutaway view of the weld toe.

From the analysis above, the stress on the weld surface of the four corners is relatively large during the working process of the boom. Although the stress value is smaller than other locations of the boom plate, the heat affected zone exists at the weld part, the microstructure is complex and the welding process is easily led to be micro defects. So the fatigue life of this part is much lower than the base metal. So this position is the weak part of the boom loaded in working conditions. With the introduction of a certain residual stress and welding deformation caused by the assembly of stress, these conditions will certainly increase the risk of fatigue damage. The welding process is simulated as follows.

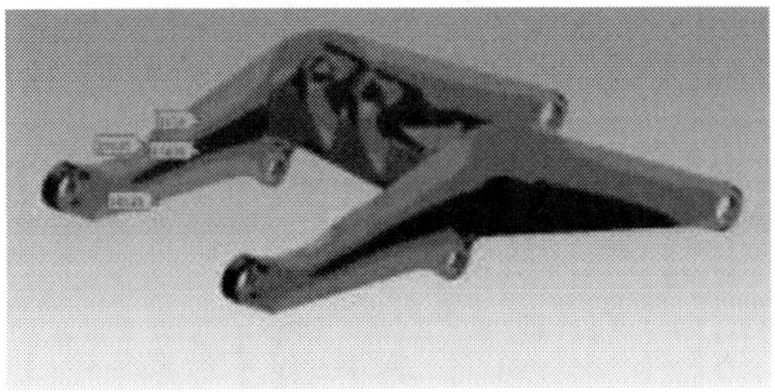

Fig. 4. Von Mises Stress distribution.

Fig. 5. Von Mises Stress distribution in the weld part of the boom.

Fig. 6. Cutaway view of the weld toe.

3 Simulation of the boom welding process

The physical model of the welding heat source is the prerequisite and condition for the analysis and numerical simulation of the welding heat flow behavior. The commonly used heat source models are the Gaussian surface heat source and the double ellipsoidal heat source model.

The Gaussian surface heat source is suitable for describing the surface welding heat and arc heat distribution on the workpiece surface:

$$q(r) = q_m \exp(-Kr^2) \tag{1}$$

In which is the heat flux at a distance of from the center of the heat source, is the maximum heat flux at the center of the heat source, is the thermal energy concentration factor.

As the total heat on the workpiece surface is equal to the effective power of the welding arc , so

$$\begin{aligned} Q &= \int_0^\infty q(r)2\pi r dr = q_m \int_0^\infty \exp(-Kr^2)2\pi r dr \\ &= \frac{q_m \pi}{-K} \int_0^\infty \exp(-Kr^2) d(-Kr^2) \end{aligned} \tag{2}$$

$$q_m = \frac{QK}{\pi} \tag{3}$$

The double ellipsoidal heat source model is widely used in the finite element calculation of the temperature field of various melting welding. Due to the effect of welding speed, the heating area in front of the arc is smaller than that behind the arc. The heated spots are not circular but elliptical. The heat flow distribution in the front of arc can be expression as follows:

$$q_f(x, y) = q_{mf} \exp(-Ax^2 - By^2) \tag{4}$$

In which is the Maximum heat flux, A and B is the elliptic distribution parameter.

The heat flow distribution at the back side of the arc can be expressed as follows:

$$q_r(x, y) = q_{mr} \exp(-A_1 x^2 - B_1 y^2) \tag{5}$$

In which is the Maximum heat flux, and is the elliptic distribution parameter.

So the total heat of the arc front region can be written:

$$\begin{aligned} Q_f &= 2\int_0^\infty \int_0^\infty q_{mf} \exp(-Ax^2 - By^2) dxdy \\ &= 2q_{mf} \int_0^\infty \exp(-Ax^2) dx \int_0^\infty \exp(-By^2) dy \end{aligned} \tag{6}$$

By the simulation of the boom welding process, it is found that the boom plate is deformed in the following two cases: the bending of the boom plate inward (see figure 7), the twisting deformation of the boom plate (see figure 8). between These two kinds of deformation, the first deformation is larger, which is about 30mm. the first deformation can be corrected by press straightening process or pre-deformation of the arm plate .the second deformation is relatively small and complex. So it can be ignored. The residual stress can reach 300MPa.

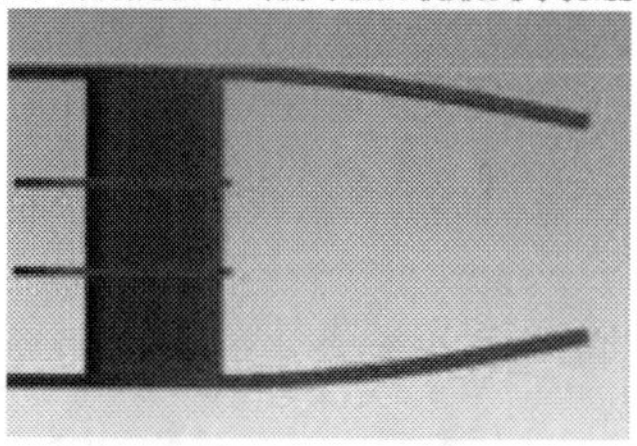

Fig. 7. Bending of the boom plate inward.

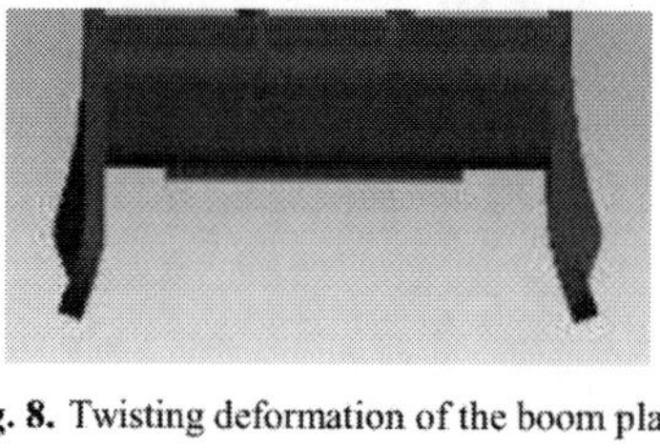

Fig. 8. Twisting deformation of the boom plate.

4 Conclusion

In this paper, the static analysis of the loader boom is carried out under the typical working conditions. It is found that the weak part of the welding seam is at the welding arc of the boom plate and the beam. Then for the weak part, the welding deformation analysis is carried out under the practical process conditions. The deformation and displacement of the boom are also studied. The results show that the deformation of the arm plate is about 30mm inward in the welding process due to the welding stress, and the residual stress of the boom. The residual stress can reach 300MPa. This study can provide theoretical and data support for the processing and deformation correction of the boom.

Acknowledgment

This work is supported by the National Natural Science Foundation of China (51175346), Shanghai automotive industry science and technology development foundation (1623), and the Key Scientific and Technological Projects of Shanghai (14521100500).

References

1. G.T. zhou, H. zhang, Z.W. Cheng, Numerical simulation analysis for welding sequence optimization of hinge structure of loader, Material science&technology, 24(3):62-67. (2016)
2. Z.B. Zeng, Z.Q. Shi, Y.J. Shi, Experimental Research on Butt-Welding Deformation of Thick Plates in Large Steel Structure, CHINA RAII WAY SCIENCE, 30(3):33-39. (2009)
3. J.Q.Zhang, H.Y. Zhao, Effects of Clamps on Welding Distortion in Thin Aluminum Alloy Plate, RARE METAL MATERIALS AND ENGINEERING, 38(z3):165-169. (2009)
4. D.Y. Huang, Numerical Modeling on Welding Deformation and Inherent Strain of High-strength Steel with T-joint, Hot Working Technology, 2013, 42(11):218-221,22. (2013)
5. X.G. Xu, Y.S. Bi, X. Wang, Control of Welding Deformation in the Active Arms Of Loading Machine, JOURNAL OF CHONGQING UNIVERSITY(NATURAL SCIENCE EDITION), 27(3):13-15. (2004)

An optimum method for latticed shells based on concept of the compression line

Liming Zhu, Qi Liu*, Peinan Gao, Yingyun Ma, Bingyang Guo and Wenfeng Du

Institute of Steel and Spatial Structures in School of Civil Engineering and Architecture, Henan University, Kaifeng 475004 China

*Corresponding author: dwf@henu.edu.cn

Abstract. To improve the load-carrying capacity latticed shells, an innovative optimum method based on the concept of compression line is proposed in this paper. The basic principle of this method is using the character without moment in compression line. The arc line which forms the contour line of sphere and latticed shell structures is substituted by compression line in this method. Then the latticed shell structures are in the state of compression, and the influence of bending stress reduces greatly. As a result, the load-carrying capacity of the latticed shell structure is increased. Through the geometrical nonlinear analysis of a sunflower-patterned single-layer latticed shell structure with a span of 48m, it is found that the load-carrying capacity of the single-layer latticed shell structure can be improved by 5.48%. Furthermore, the results of 84 structural analyses of single-layer or double-layer sphere and cylinder latticed shell structures show that the optimum method is right and effective. And especially, it is applicable to single-layer latticed shell structure with rise-span ratio 1/5 with the max improvement 6.4% of load-carrying capacity.

1 Introduction

Optimum design of the latticed shells can be divided into three levels. The first is the level of structure, which includes the macroscopical surface shape, the span, the rise, the thickness of the shells and so on. Deng Hua built a mathematical model to optimize the latticed shells to minimize the self-weight of the structure, and they discussed the accurate values for the rise and the thickness of the latticed shells [1]. The second is the level of components, which include the geometrical length of the member bar, the measurement of the grid and the design of the connecting joints. Shan Luyang set the measurement of the grid and the thickness of the latticed shell as the design variables, and the minimum of the cost as the objective function to optimize the structure [2]. The third is the level of section, which includes the optimization of the size of the section. Li Yongmei made on section optimizing under determined cable forces by means of improved optimization method of double objectives [3].

Based on the states of the research both in China and abroad, the main object of optimization is the rise-span ratio, the measurement of the grid, the thickness of the latticed shell, and the spherical or cylinder latticed shells structure are always adopted [4]-[10]. However, during the process of engineering practice, we noticed that the latticed shell structures are more likely to be in the full compressive stress states with no bending moment by substituting the arch line of the latticed shells into the compression line. From which it can increase the ultimate bearing capacity. Thus we proposed an optimum method for latticed shells based on concept of the compression line.

2 The principle of the method

The Peripheral hinges are used as the common constraint method of the latticed shells structure. If we take a profile of the single-layer latticed shells structure into consideration, it is easy to be found the stress characteristics of the structure are similar as two-hinged arch. As we know the selection of reasonable arch axis has a significant effect on the stress distribution and stress size of each section during the process of designing the two-hinged arch structure.

A major stress characteristic of single-layer latticed shells structure is the membrane action. Single-layer latticed shells structure has a powerful in-plane stiffness far outnumbering the out-plane stiffness. This structure uses the in-plane stress to resist the vertical load and the load-carrying capacity is controlled by the buckling strength. The single-layer latticed shell structures transmits most of the load as the axial forces of the members of latticed shells, while the sphere latticed shells structure remains some of the bending stress. Those bending stresses can sharply decrease the load-carrying capacity of the latticed shells structure. Therefore, the principle of axis selection decreases the moment created by the load as much as possible. The ideal axis is consistent with the compression line, which means all the sections in the latticed shell are under the axial compression with no bending moment and all the stress characteristics is the most reasonable at the same time.

The load acts on the latticed shell structures can simplify to the concentrated force of each intercept point. Under the action of the vertical load along gravity direction, the compression line is usually not the circular arch. In this situation, the compression line can be confirmed by the graphic method shown in Fig.1. Firstly, calculate the support reaction Rx and Ry; then assign the AB bar along the direction of the resultant force of Rx and Ry to make sure that bar BC has the same direction as the total force of Rx, Ry and F1. Then the rest can be done in the same method which means the bars are assigned along the direction of total force at nodes, in another word, the direction of the member bar coincides with the compression line, which can make the internal force transmit only as the axial force in each jointing point. Thus, we can replace the arc line with the compression line in the structure of latticed shells to contain the structure in a compressive stress state and increase the loading-capacity of the latticed shells structure eventually.

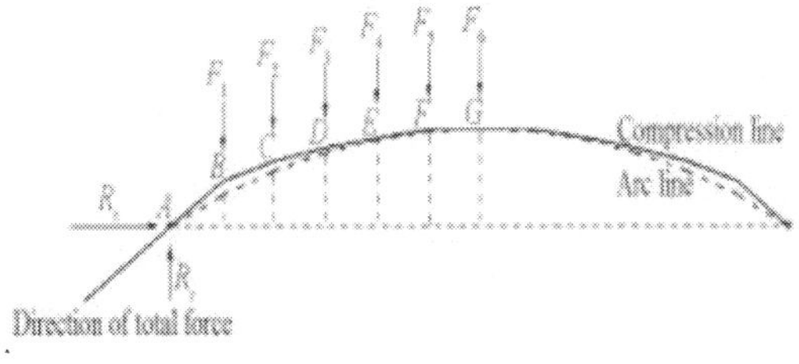

Fig. 1. Schematic compression line

3 Analysis of examples

In order to verify the Correctness of the optimum method, analysis is based on a sunflower-patterned single-layer latticed shell structure.

3.1. Model of structure

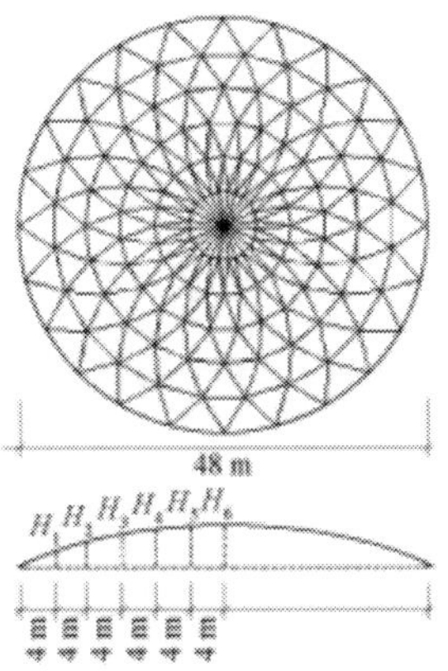

Fig. 2. Geometric model of latticed shell

Model is shown as the fig.2. The span of the structure is 48m, rise to span ratio is 1/6, the total number of structure bars is 408 and nodes is 145. Structure is clamped on supports, the steel bar is Q235 model, poison ratio is 0.3, yield strength is 215N/mm^2, elastic modulus is 206KN/mm^2, and the cross-section of pipes we used are 133mm×5mm. A general finite element analysis software ANSYS is used to analyse the structure and chose beam 4 as the element type of bar.

3.2 Ideas of analysis

At first, we set the model of the arc line, which forms the contour line of the sphere latticed shell structure. Then we set the compression line latticed shell structure model by the method indicated in fig.1. We also construct parabolic reticulated, sine line mesh and Ellipse line mesh latticed shell structure models to compare at the same time. Their difference is reflected in the data between H1~H6, as it is presented in table 1. We compare their ultimate carrying capacity through the geometrical nonlinear analysis.

Table 1. H_1~H_6 of different curves

Curves	H_1/m	H_2/m	H_3/m	H_4/m	H_5/m	H_6/m
Arc line	2.513	4.523	6.059	7.142	7.786	8.000
Parabolic	2.445	4.445	6.000	7.111	7.778	8.000

line						
Sine line	2.071	4.000	5.657	6.928	7.727	8.000
Ellipse line	4.422	4.422	6.928	7.543	7.888	8.000
Compress-ion line	2.552	4.600	6.146	7.152	7.667	8.000

3.3 The result of analysis

Fig.3 shows the result through the geometrical nonlinear analysis between five kinds of curves sunflower-patterned single-layer latticed shell structures. Ratio of current load level of design load as the load factor. We can learn curves of load displacement show highly geometrical nonlinearity from fig.3 and the load factors of five kinds of curves for single-layer latticed shell structures have a big difference. The highest load factor belongs to compression line latticed shell structure, as a number of 14.64, then we sorts others in the order as follows, arc line mesh latticed shell structure as a number of 13.88, parabolic line reticulated latticed shell structure as a number of 13.84, sine line mesh latticed shell structure as a number of 9.09, and ellipse line mesh latticed shell structure as a number of 4.37. From the above data we know that the shape of curves has a significant impact on carrying capacity. The load carrying capacity of the compression line latticed shell is increased by 5.48% compared with the circular arc line mesh latticed shell, but ellipse line mesh latticed shell structure just play the role as 31.48% of arc line mesh latticed shell.

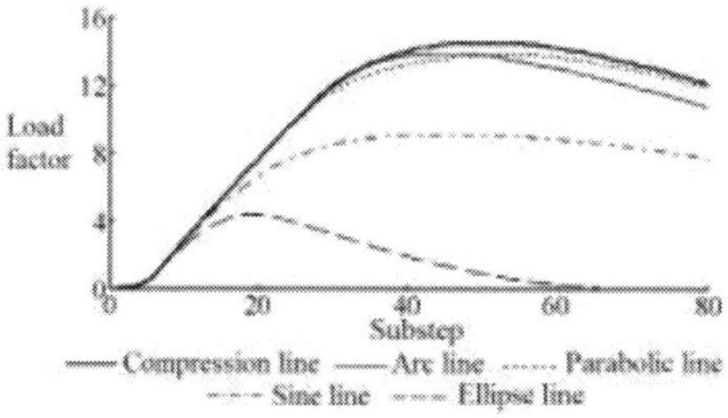

Fig. 3. Comparison of limited bearing capacity

Fig.4 shows the Nonlinear buckling geometrical model of five kinds of sunflower-patterned single-layer latticed shell structures. We can notice that the second hoop (start from the edge) of arc line mesh latticed shell structure collapsing, and inner hoop changes slightly. The feature of instability characteristics of parabolic reticulated, sine line mesh and arc line mesh latticed shell structure is nearly same. The buckling model of Ellipse line mesh latticed shell structure is the inner hoop collapsing, and outer hoop change merely. The buckling model of compression line latticed shell structure is the inner hoop has the same damage process with the outer hoop, the second hoop (start from the edge) collapsing and have large deformations at the inner hoop. We can find that arc line mesh, parabolic line reticulated and sine line mesh latticed shell structure broke down because of the instability of the outer hoop, while there are also have the potential carrying capacity at the inner hoop. The reason for Ellipse line mesh is the instability of the inner hoop, while there are also have the potential carrying capacity at the outer hoop. Due to the inner and outer hoop has the same damage process. Its potential carrying capacity plays an important role before they lose their stability. So the highest carrying capacity belongs to compression line mesh latticed shell.

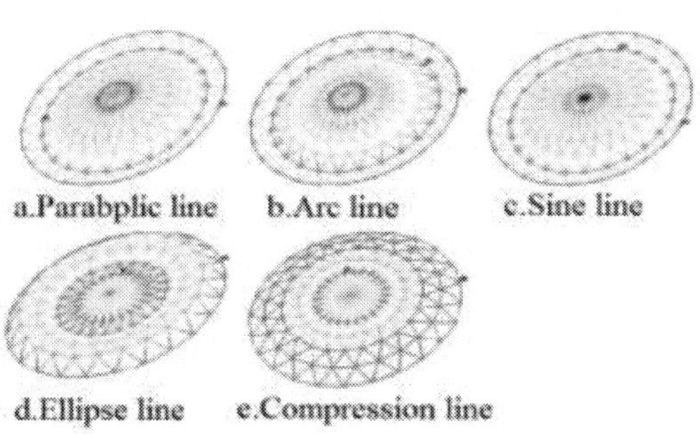

Fig. 4. Geometrical model of instability

3.4 The result influence of the rise to span ratio

Rise to span ratio do count much in mechanics performance of single-layer latticed shell structure. The former objects we analysed are flat reticulated latticed shell structure whose rise to span ratio are 1/6, and fig.5 provides the geometrical nonlinearity results from five kinds of curving latticed shell structures whose rise to span ratio equal 1/10, 1/8, 1/5, and 1/4.

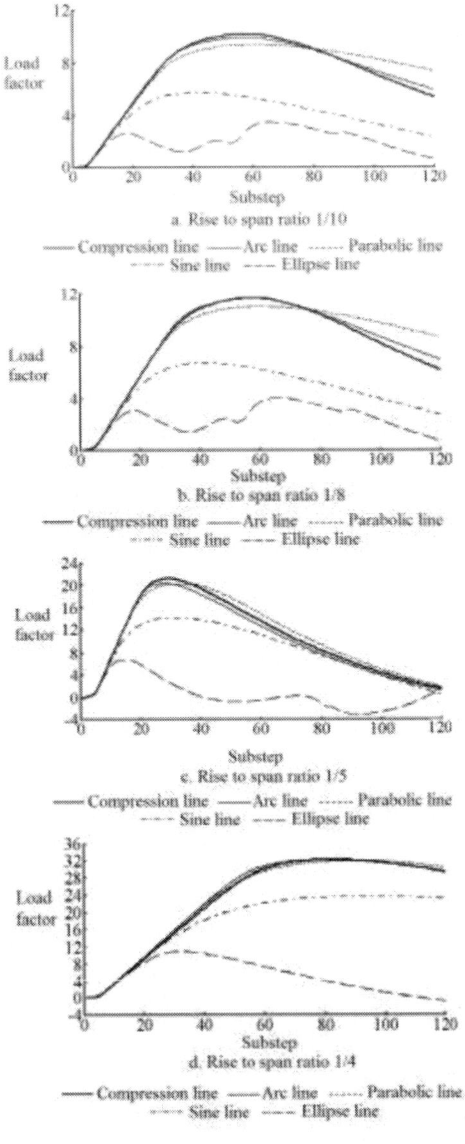

Fig. 5. Comparison of limited bearing capacity with different rise to span ratios

As we can learn from fig.5, due to vertical stiffness of the structure rising along with the number of rises to span ratio growing, the load factors of single-layer latticed shell structures increasing with. And load factor of compression line still occupied the peak. Arc line mesh, parabolic reticulated, sine line mesh, and ellipse line mesh latticed shell structure is as fallows.

We lay out 1/10, 1/8, 1/5, and 1/4 four kinds of rise to span ratios, by contrast. Load factor of compression line is improving 0.87%, 1.33%, 5.23%, and 4.53% than arc wire mesh latticed shell structure. It shows that carrying capacity of the compression line rising with a rise to span ratio growing at the primary period, then declining slightly. The carrying capacity goes through the roof when rise to span ratio equals 1/6. Carrying capacity increases merely when span ratio is small, but the amount of increase could be 5% when span ratio quite big.

4 Analysis of typical structure

To get a comprehensive understanding of the optimized effect by using this optimum method on all kinds spatial latticed shell structures, we also analysed the typical latticed shell structure which has wide application in engineering. The calculation results are presented in Table 2. Before analysing these structures, we build the model in spatial structure analysis software MATCAD firstly, then input into ANSYS to carry out geometrical nonlinear analysis.

Table 2. Optimum results of single-layer spherical latticed shell structures

Structure type and diagram	Span of structure	Rise to span ratio	Carrying capacity increase percentage
Kiewitt type 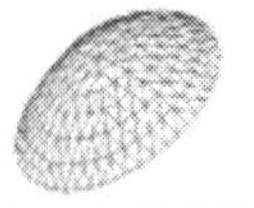	24	1/10	1.8
		1/5	4.1
	36	1/10	1.0
		1/5	4.2
	48	1/10	1.1
		1/5	4.5
Sunflower-patterned 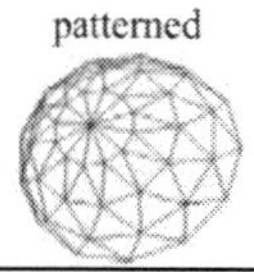	24	1/10	1.1
		1/5	5.1
	36	1/10	1.3
		1/5	5.6
	48	1/10	1.6
		1/5	5.8
Geiger type 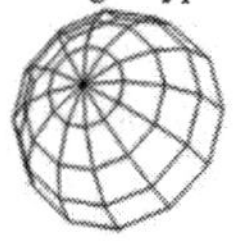	24	1/10	1.3
		1/5	4.5
	36	1/10	1.8
		1/5	6.0
	48	1/10	2.2
		1/5	6.4
Three-dimensional grids type 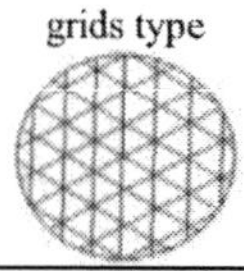	24	1/10	1.3
		1/5	5.0
	36	1/10	1.4
		1/5	5.3
	48	1/10	1.6
		1/5	5.7
Schwedler type 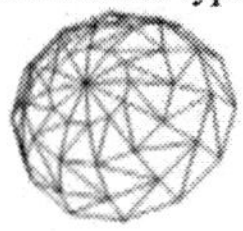	24	1/10	1.3
		1/5	4.6
	36	1/10	1.8
		1/5	5.2
	48	1/10	2.0
		1/5	5.5
Geodesic type 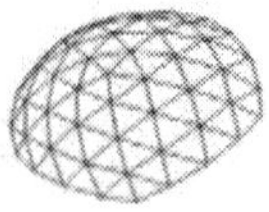	24	1/10	1.6
		1/5	5.4
	36	1/10	1.8
		1/5	5.6
	48	1/10	2.0
		1/5	5.8

As can be observed from the Table 2, after adopting the optimum method that is proposed in this paper, the carrying capacity of six single-layer latticed shell structures can be improved. The effectiveness of improvement is approximately 10% and the ribbed dome is seen as the most obvious structure. Along with the augment of the span of various latticed shell structure, the optimized effect increased gradually. For different rise to span ratio latticed shell structures, the smaller one is a little bit better, but not significant.

DOI: 10.1051/matecconf/201710004015

Table 3. Optimum results of single-layer cylinder latticed shell structures

Structure type and diagram	Span of structure	Rise to span ratio	Carrying capacity increase percentage
Single-diagonal type 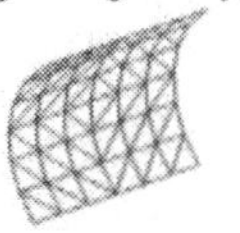	24	1/10	1.5
		1/5	5.3
	36	1/10	1.6
		1/5	5.5
	48	1/10	1.9
		1/5	5.6
Lamella type 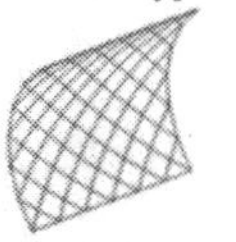	24	1/10	1.1
		1/5	5.1
	36	1/10	1.4
		1/5	5.5
	48	1/10	1.7
		1/5	5.9
S&P type 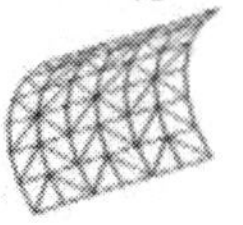	24	1/10	1.5
		1/5	5.5
	36	1/10	1.8
		1/5	5.9
	48	1/10	1.9
		1/5	5.6
Three-dimensional grids type 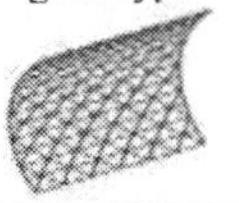	24	1/10	1.2
		1/5	5.0
	36	1/10	1.4
		1/5	5.5
	48	1/10	1.6
		1/5	5.7
Double-diagonal type 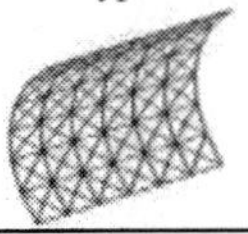	24	1/10	1.3
		1/5	5.2
	36	1/10	1.7
		1/5	5.8
	48	1/10	1.5
		1/5	6.0
"mi" type 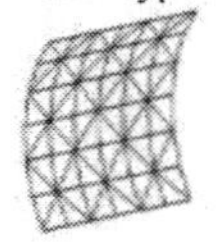	24	1/10	1.6
		1/5	5.4
	36	1/10	1.8
		1/5	5.6
	48	1/1	2.1
		1/5	6.1

Table 3 provides optimum calculation results of six single-layer cylinder latticed shell structures such as single-diagonal type, lamella type, S&P type, three-dimensional grids type, double-diagonal type, and "mi" type. As indicated in table 3, after adopting the optimum method proposed in the paper, the carrying capacity of six single-layer cylinder latticed shell improved too. The maximal improvement effect can reach about 6.4%, and union square type is the most obvious one. With the augment of span, the optimum effect increased a little. And the rule that a smaller rise-span ratio is fractional better than the bigger is the same as the rule of single-layer spherical shell structure.

The optimum results of double-layer latticed shell structure are shown in Table 4. Quadrangular pyramid system sphere and cylinder latticed shell structure which are frequently used were chose in the calculation. From the Table 4, the range of the carrying capacity of double-layer shell structure rarely improved after using the optimum method. The maximal effect of improvement is lower than 3%.

Table 4. Optimum results of double-layer latticed shell structures

Structure type and diagram	Span of structure	Rise to span ratio	Carrying capacity increase percentage
Sphere 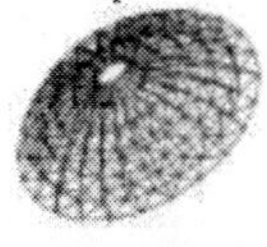	36	1/10	0.7
		1/5	1.1
	72	1/10	0.9
		1/5	1.3
	96	1/10	0.2
		1/5	2.0
Cylinder	36	1/10	1.3
		1/5	1.5

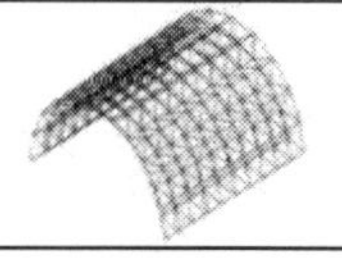	72	1/10	1.4
		1/5	1.4
	96	1/10	1.8
		1/5	2.2

5 Conclusions

1) Through the example analysis of the sunflower-patterned single-layer latticed shell structure with a span of 48m, it proved that the ultimate carrying capacity of the compression line latticed shell is the highest. The optimum method for latticed shells based on concept of compression line proposed in this paper is quite true.

2) Compared with the figures of H_1~H_6 of different curve forms, we can know that the dimension of several single-layer latticed shell structure is comparable with each other, but the carrying capacity reflects notable differences. From another perspective, this also can be an evidence to prove the single-layer latticed shell structure is defect-sensitive structure.

3) Summarizing 84calculation results of single-layer spherical latticed shell, single-layer cylinder latticed shell, double-layer spherical latticed shell and double-layer cylinder latticed shell. It is knowable that an optimum method for latticed shells based on concept of compression line has good effect on single-layer latticed shells with a rise-span ratio at about 1/5 and the ultimate carrying capacity can improve 6.4%. Yet the effect on double-layer latticed shells and single-layer latticed shells with rise-span ratio at about 1/10 is considered to be insignificant.

4) Carrying coefficients of latticed shells varies greatly when using different geometric curve types. The carrying capacity of the compression line, arch line and parabolic line is quite high. And the capacity of ellipse line is relatively low, only about 1/3 of arc line mesh latticed shell.

References

1. Hua D, Shilin D, "Shape optimization of spatial latticed shell structures" *Journal of Zhejiang University(Engineering Science)*, **4**,371, (1999)
2. Luyang S, Hui Y, "Optimization analysis of large-span double-layer cylindrical latticed shell structure" *Journal of Building Structures*, **6**,47, (1999)
3. Yongmei L, Yigang Z, Xuesong L, "Section optimization of cable-supported latticed shell" *Industrial Construction*, **4**,73, (2007)
4. Zongfa L, Zhitao Y, Yansheng W, Zuoliang L, "Scheme selection and optimum design of single-layer reticulated domes" *Journal of Civil,Architectural & Environmental* Engineering, **4**, 83, (2007)
5. Yueqin Q, "Optimization design of large-span spatial latticed" *Construction Technology*, **2**,393, (2009)
6. Kashani M, Croll J G A, "Non-linear buckling response of spherical space-domes" *Progress in shell and spatial structures, IASS*,(1989)
7. Chai S, Sun H C, "A relative difference quotient algorithm for discrete optimization" *Structural Optimization*, **1**, 46,(1966)
8. Dong W, Jing L, "Shape optimization of space trusses subject to frequency constralnts" *Engineering Mechanics*, **4**, 22, (2007)
9. Salajegheh E, Vanderplaats G N, "Optimum design of trusses with sizing and shape variables" *Structural Optimization*, **6**,79, (1993)
10. Rajah S D, "Sizing, shape, and topology design optimization of trusses using genetic algorithm" *Journal of Structural Engineering*, **10**, 1480, (1995)

Dynamical failure index of single layer Keiwitt latticed shell structures

Liming Zhu, Yun Sun*, Wenfeng Du and Zengguang Zhang

Institute of Steel and Spatial Structures in School of Civil Engineering and Architecture, Henan University, Kaifeng 475004 China

*Corresponding author: dwf@henu.edu.cn

Abstract. Based on the double control criterion of displacement and energy, a new index, dynamical damage index, for describing the dynamical damage extent in the single layer latticed shell structure is set up. The calculate method and change law of all kinds of consuming energy under the seismic action in the latticed shell structures are studied. Then the expression of dynamical damage index erected by combining the plastic consuming energy and maximum displacement is present. Moreover, the compare among the three index, dynamical damage index, displacement and plastic bar quantity, is completed by large numbers of parameter analysis in the single latticed shell structures. The results show that the discrepancy of analytical conclusion is distinct among the three index, dynamical damage index, displacement and plastic bar quantity to the latticed shell with local weakening stiffness, and the results of dynamical damage index are more reasonable. The dynamical failure index is in effect for the quantificational analysis of dynamical failure for single layer latticed shells.

1 Introduction

With the increase of external dynamic load, dynamic response of single layer latticed shell structure would develop. Structure will go through whole progress which the intact structure suffers from a dynamical failure and then collapses with a gradual increase of damage grade. At present, many of researches of single layer latticed shell structure describe the progress of structural dynamical failure and structural statue base on displacement index and plastic bar quantity index. However, according to practical analysis of latticed shell structure, displacement reflects the deformation of the structure but it is not able to distinguish structural damage accumulation process under cyclic load effect and plastic bar quantity reflects the grade of structural plasticity but it can't distinguish the deformation of latticed shell during damaging progress [1-3].

Initially, form the view of energy, this paper will analyse energy dissipation mechanism of single layer shell latticed. Subsequently, the paper will accord the dual-control guidelines of deformation and plastic cumulative energy dissipation to establish and apply dynamical failure index to analyse the extent of dynamical failure of single layer latticed shell structure. Ultimately, it will provide five cut-off values of dynamical failure index according to the degree of single layer latticed shell structure [4-7].

2 Mechanism of the structural energy dissipation

According to structural dynamics, differential equation system of motion that is considered the damping action of multi-degree-of-freedom system which is under seismic action [8-10]:

$$[M]\{\ddot{U}\}+[C]\{\dot{U}\}+[K]\{U\}=-[M]\{\ddot{U}_g\} \quad (1)$$

In the equation: *[M]*, *[K]* and *[C]* are mass matrix, stiffness matrix and damp matrix; *{U}* is the displacement of column vector with respect to ground; $\{\dot{U}\}$ is the relative velocity of column vector; $\{\ddot{U}\}$ is the relative acceleration the column vector; $\{\ddot{U}_g\}$ is the ground motion acceleration of column vector.

In equation (1), *[K]* and *{U}* are restoring vectors which can be denoted by *{F}*, so equation (1) can be transformed as:

$$[M]\{\ddot{U}\}+[C]\{\dot{U}\}+\{F\}=-[M]\{\ddot{U}_g\} \quad (2)$$

Multiply both sides of equation (2) by $\{\dot{U}\}$, and then do integral from zero time to t0 time, as a result:

$$[M]\int_0^{t_0}\{\ddot{U}\}\{\dot{U}\}dt+[C]\int_0^{t_0}\{\dot{U}\}\{\dot{U}\}dt+\int_0^{t_0}\{F\}\{\dot{U}\}dt = -\left[[M]\int_0^{t_0}\{\ddot{U}_g\}\{\dot{U}\}dt\right] \tag{3}$$

$$[M]\int_0^{t_0}\{\ddot{U}\}\{\dot{U}\}dt=\int_0^{t_0}[M]\{\ddot{U}\}d\{U\}=\frac{1}{2}\int_0^{t_0}[M]d\{\dot{U}\}^2 = \int_0^{t_0}d\frac{[M]\{\dot{U}\}^2}{2}=\int_0^{t_0}dE_d \tag{4}$$

$$[C]\int_0^{t_0}\{\dot{U}\}\{\dot{U}\}dt=\int_0^{t_0}[C]\{\dot{U}\}d\{U\} = \int_0^{t_0}\{F_c\}d\{U\}=\int_0^{t_0}dE_c \tag{5}$$

The Fc is damping force.

$$\int_0^{t_0}\{F\}\{\dot{U}\}dt=\int_0^{t_0}\{F\}d\{U\}=\int_0^{t_0}dE_y \tag{6}$$

$$-\left[[M]\int_0^{t_0}\{\ddot{U}_g\}\{\dot{U}\}dt\right]=-\int_0^{t_0}[M]\{\ddot{U}_g\}d\{U\} = -\int_0^{t_0}\{F_g\}d\{U\}=-\int_0^{t_0}dE_t \tag{7}$$

The F_{ig} is the impetus correspondent with the applied acceleration wave.

Therefore, the structure of the Eqs.(3) in turn indicates the structural kinetic energy (E_d), damping energy dissipation (E_c), total deformation energy (E_y) and total energy of structure absorption (E_t) from the zero moment to the t_0 moment.

When elastic-plastic analysis is conducted on the structure, the restoring force of elastic-plastic can be composed by elastic restring force and plastic yield force.

$$\{F\}=[K]\{U\}+\{B\} \tag{8}$$

The *{B}* is plastic yield force array. Thus,

$$\int_0^{t_0}\{F\}\{\dot{U}\}dt=\int_0^{t_0}\{F\}d\{U\} = \int_0^{t_0}[K]\{U\}d\{U\}+\int_0^{t_0}\{B\}d\{U\} = \int_0^{t_0}dE_e+\int_0^{t_0}dE_p \tag{9}$$

The total deformation energy of this structure is E_y, can be expressed as the summation of elastic strain energy is E_ε, and plastic cumulative hysteretic energy is E_p. It is shown:

$$E_y=E_e+E_p \tag{10}$$

Based on the principle of conservation of energy, the Eqs.(3) sets up during every moment of the whole process.

Therefore, plug Eqs.(10) into Eqs.(3), the equation of the total energy can be showed:

$$E_d+E_c+E_e+E_p=E_t \tag{11}$$

The above energy dissipation takes different proportions in different stages in the whole geological process.

The Fig.1. is relation curve of the total energy and each energy dissipation that the single-layer reticulated shell as the peak value 0.8g under the action of EL2-Centro(1940) 3d acceleration seismic wave after the above example in this paper.

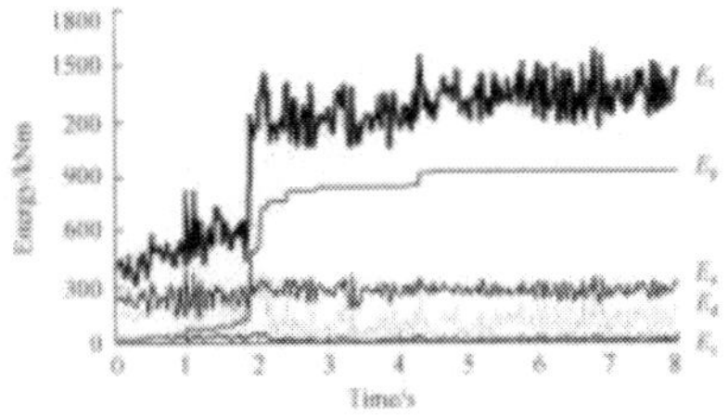

Fig. 1. Relationship of total energy and subentry

The Fig.1. illustrates that before one second, that seismic energy is shared by elastic deformation energy and damping energy dissipation structure kinetic energy, respectively 57.2%、5.2%、37.6%. After one second, the member bar goes into plasticity and produces plastic energy dissipation. With the time going by, the energy of plastic dissipation becomes more and then it takes the main part of energy dissipation, almost amounting to 64.1%.

3 Double-control of the criterion and dynamical failure index

From the mechanism of energy dissipation structure, for a moderate yield of net shell structure, the main energy consumption form is the plastic energy dissipation, E_p. Therefore, there is some evidence about damage index of the accumulated plastic energy dissipation structural. However, the energy usually reflects the destruction of the structure of overall situation not the whole details. For example, each energy value can be correspond more than one damage form, while it can't make out the specific form.

The other most important criteria in structural failure is the deformation, which shows structural maximum displacement response, but does not distinguish the structure under repeated load damage accumulation process. Therefore, from the two aspects, energy and deformation, the dynamic failure of the structure is expressed as a linear combination of the maximum deformation and the plastic cumulative consumption and can be expressed by dynamic damage index *D*. And the double control criteria of dynamic strength failure can be established, showing that the energy of maximum deformation and cumulative plastic deformation has an effect the influence of the structure damage. Try to ensure that lines are no thinner than 0.25 point.

$$D = \frac{\delta_M}{\delta_U} + \frac{\beta}{Q_y \delta_U} \int dE_p \tag{12}$$

The *D* as dynamic damage index is structural damage index; δ_M as structural maximum displacement response; δ_U as the limit displacement response under the static load; Q_y as the yield strength of material; β as the coefficient of structure system, usually 1.0 for the common structure; $\int dE_p$ as the cumulative plastic energy dissipation.

The first paragraph on the right of Eqs.(12) is the proportion of structural displacement response and ultimate displacement response, showing the development of displacement ductility of structure. The second paragraph is the proportion of structure of the plastic energy dissipation and limit of energy dissipation, showing the development of structure of plastic energy dissipation.

4 The comparison between dynamical failure index and the displacement index, the plastic bar quantity index

The $K_{8,6}$ type single layer lattice shell structure with span 40m as an example in this paper. Analyze the structure damage index becomes of law in the structural dynamic failure of the whole process, and compare it with displacement and plastic bar quantity.

4.1. Structure Model

The structural model as shows in Fig.2., it is a single layer latticed shell structure K8, 6. The structure span is 40 m, calculated five cases of the rise span ratios respectively 1/7, 1/6, 1/5, 1/4 and 1/3. The total number of structural bar quantities is 456, the total number of nodes is 169. The structure is clamped, the material is Q235 steel, the Young's modulus (E) is 206 kN/mm2, the Poisson's ratio is 0.3. Using the isotropic intensified Mises elastic-perfectly plastic material, the yield strength is 215 N/mm2, the Rayleigh damping and the ratio is 0.02. Select four kinds of round tube section 60×3.5mm, 75.5×3.75mm, 89×4mm, 114×4mm respectively.

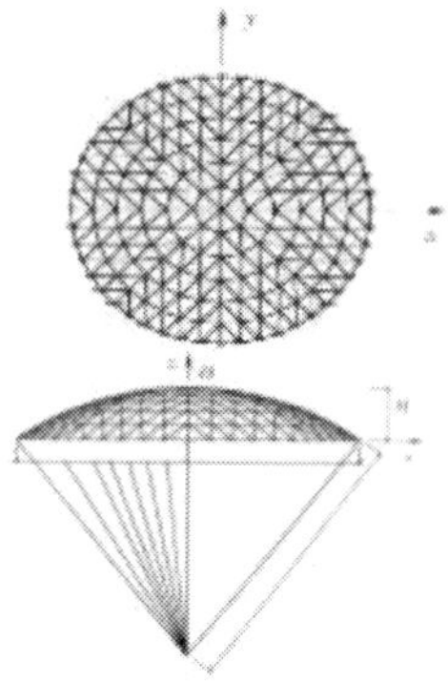

Fig. 2. Geometrical model of the latticed shell

Using the universal finite element software of ANSYS to analysis dynamic time-histories analysis, seismic load selected EL2-Centro (1940) three-dimensional acceleration speed of earthquake wave. Using Pipe20 beam elements, each section has eight output points, as shown in Fig.3. When there is an output point into the plastic state, recorded as 1P, there are two output points enter the plastic, recorded as 2P, and so on.

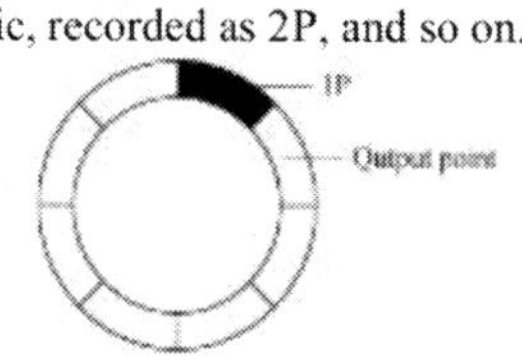

Fig. 3. Section of Pipe20 element

4.2. Calculative results

When span ratio structures takes 1/5, analyse the structure of elastic-plastic dynamic and time-historical, recording structural time displacement curve of the maximum displacement nodes under different peak of earthquake wave, as shown in Fig.4.

The Fig.4. shows that when the earthquake wave peak is less than 0.6g, structure remained stable vibration. When earthquake wave peaking at 0.7g, the maximum displacement node increases significantly, and there is a jump, after the jump, time displacement curve is in the new position to vibrate stably. Earthquake wave peak reaches 0.8g, there has a greater structural displacement jump, at the new location can remain steady vibration. Earthquake wave peak reaches 0.9g, the structure collapsed.

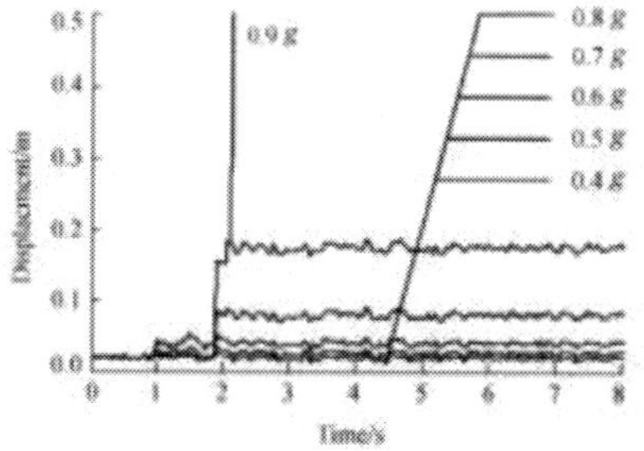

Fig. 4. Curve of time-history of maximum displacement node

The Fig.5 is a deformation of the structure and the bar distribution of plastic. The Table 1. shows the structural maximum displacement response, dynamic failure index, the number of plastic rod and a description of the extent of structural damage of calculated.

 DOI: 10.1051/matecconf/201710004016

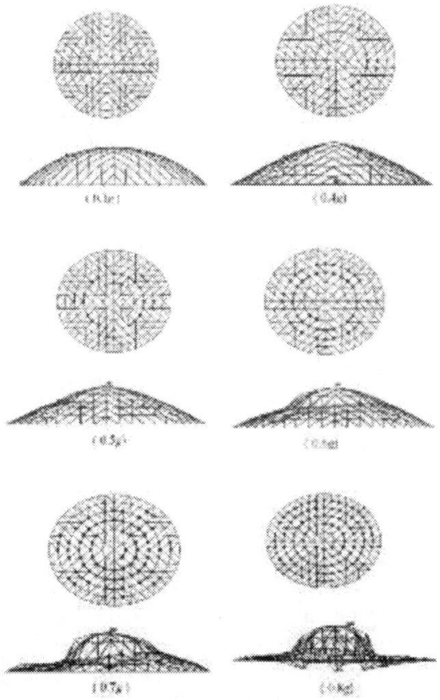

Fig. 5. Structural deformation and distribution of plastic bars (plastic bars are thick)

Table 1. Calculated results of indexes with different peak value of seismic wave

Peak value of earthquake wave	0.2g	0.3g	0.4g	0.5g	0.6g	0.7g	0.8g	0.9g
Max displacement/mm	25.43	26.30	27.27	33.04	49.56	103.55	392.06	∞
Dynamical failure index	0.098	0.101	0.115	0.14	0.282	0.724	1.945	∞
Number of bars of 1P	0	0	4	17	31	60	211	456
Number of bars of 8P	0	0	2	14	24	48	164	456
Grade of damage	intact	intact	intact	intact	slight damage	moderate damage	severe damage	collapse

From Fig.5 and Table 1, when the earthquake wave peak value reach 0.2g and 0.3g, there are no bars into the plastic, structure is perfectly elastic, the dynamic failure index in this range is below 0.1. When the earthquake wave peak value is 0.4g, the third ring portion of the bar member first to enter plastic, structure began in working condition with elastic-plastic, the maximum displacement nodes in the third ring and the diameter of the rib, the dynamical failure index is 0.115. When the earthquake wave peak at 0.5g and 0.6g, fourth and second rings began to enter the plastic, dynamical failure index are 0.14 and 0.282. The earthquake wave peak reaches 0.7g, into the plastic bars increased significantly, dynamical failure index increase to 0.724. When earthquake wave peak reaches 0.8g, into the plastic bars have been distributed to the ring bars, radial bars and diagonal bars, dynamical failure index is 1.945.

To further reflect the contact between dynamical failure index and development level of plastic structure, the maximum displacement of the structure, the number of bars into the plastic. Calculate the dynamical response of different rise span ratio and different earthquake waves peak value, results are shown in Table 2.

Table 2. Summary of calculated results of different indexes

Ratio of H to L	Peak value of earthquake wave	Dynamical failure index	Displacement /mm	Number of bars of 1P	Number of bars of 8P	Grade of damage
1/7	0.1g	0.275	46.47	30	22	intact
	0.2g	∞	∞	456	456	collapse
1/6	0.1g	0.095	23.23	0	0	intact
	0.2g	0.121	32.43	15	11	slight damage
	0.3g	0.327	66.95	49	38	slight damage
	0.4g	1.679	237.971	102	84	moderate damage
	0.5g	∞	∞	456	456	collapse
1/5	0.2g	0.098	25.43	0	0	intact
	0.3g	0.101	26.30	0	0	intact
	0.4g	0.115	27.27	4	2	slight damage

	0.5g	0.14	33.04	17	14	slight damage
	0.6g	0.282	49.56	31	24	slight damage
	0.7g	0.724	103.55	60	48	slight damage
	0.8g	1.945	392.06	211	164	moderate damage
	0.9g	∞	∞	456	456	collapse
1/4	0.2g	0.086	18.88	0	0	intact
	0.3g	0.090	19.79	0	0	intact
	0.4g	0.094	20.70	0	0	intact
	0.5g	0.098	21.61	0	0	intact
	0.6g	0.102	22.51	0	0	intact
	0.7g	0.112	23.45	4	2	slight damage
	0.8g	0.141	34.58	18	12	slight damage
	0.9g	0.203	36.08	29	24	slight damage
1/3	0.2g	0.066	14.47	0	0	intact
	0.3g	0.070	15.36	0	0	intact
	0.4g	0.074	16.26	0	0	intact
	0.5g	0.078	17.15	0	0	intact
	0.6g	0.082	18.05	0	0	intact
	0.7g	0.086	18.94	0	0	intact
	0.8g	0.090	19.84	0	0	intact
	0.9g	0.100	20.76	0	0	intact

Calculated from the data in Table 2. shows, many states experienced plastic evolving between shell structures to collapse entirely from the state in good condition. With the development of plastic, deepen structural damage, dynamical failure index increases, structure maximum displacement nodes increases, increasing the number of the bar members into plastic. When the dynamical failure index is less than 0.1, there is no structure in the bar into the plastic, structure is in elastic working condition, structurally in pretty good shape, this time corresponding to the maximum displacement response is about 1/1500 of structure spans. the basic dynamic failure index is within the range of 0.1 to 0.8, the structure has a small bar into the plastic, the plastic is not in-depth development, can consider there is a slight structural damage, then the corresponding to maximum displacement response is approximately 1/1500 to 1/400 of structure spans, the percentage of plastic bars is about 0% to 13%.

Dynamical failure index is basically in the range between 0.8 and 2.0, structure has a certain number of bars into the plastic, plastic level of development is moderate, that generates moderate structural damage, then the corresponding to maximum displacement response is approximately 1/400 to 1/100 of structure spans, the percentage of plastic bars is 13 % to 45%. Dynamical failure index exceeds 2.0, many bars of structure into the plastic state, plastic deformation development is also more serious, corresponding structure displacement are large, the structure can be considered at this time it has reached the ultimate of bearing capacity, generating dynamical strength failure.

Therefore, D is equal to 2 as the cut-off value of damage to the structure, the corresponding maximum displacement response and the percentage of 1P bars is about 1/100 and 45%, respectively. This is the same basically as the literature [1] proposed that use 1/100 of structure span (1P percentage is 42%) as limit of dynamic failure.

In order to contrast the result of dynamical failure index, maximum displacement and the analysis result of the proportion of the bars into the plastic. Fig.6. show the relational comparison chart respectively.

From Fig.6 dynamical failure index basically presents positive relationship with structural maximum displacement and the proportion of 1P bar; structural maximum displacement becomes larger with the increase of dynamical failure index. Analyzed the structure of the damage from failure index, structural maximum displacement, and the plastic bar, it also can get the same result.

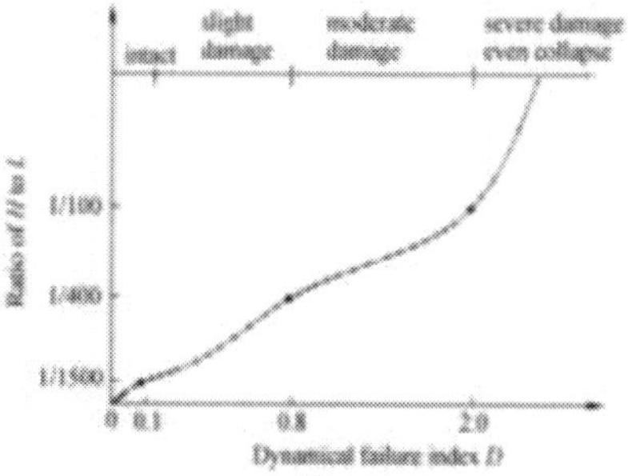

(a)

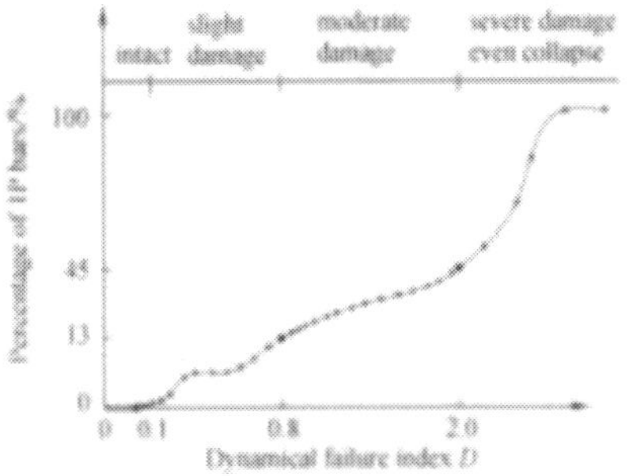

(b)

Fig. 6. (a) Relationship of dynamical damage index and maximum displacement; (b) Relationship of dynamical damage index and proportion of 1P plastic bars

While, the proportion of structural maximum displacement and plastic bar is from the development of displacement ductility and structure of plastic energy dissipation, while dynamical failure index is from the former two considerations. According to the method to calculate, when seismic wave peak is 0.5g, the structural maximum displacement is 375.12mm, nearly 1/100 of the whole span. From the point of structural maximum displacement, this structure has seriously been damaged. While, only changed the structure into the plastic bar section of this 4 root, from the point of proportion of plastic bar, the structure can be thought that it has been lightly damaged. Just as demonstrates in Fig.7. the structure in the maximum displacement has been damaged in some parts and it influences the formal use. If the four bars in damaged part are replaced, the whole structure will be fine. Therefore, it is not suitable to say that the structure has been seriously damaged or slightly damaged. The dynamical failure index is 0.82 from the point of dynamical failure index, in the secondary damage range

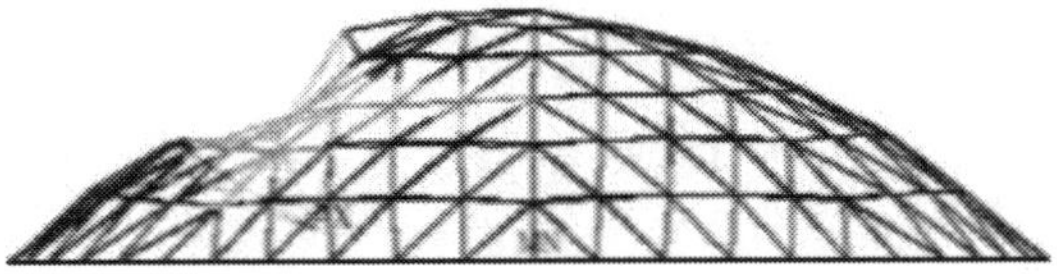

Fig. 7. Structural deformation after some bars are changed

5 Different damage levels of the corresponding to boundary index

According to the division of the habit of building structures, where the dynamical failure of single layer latticed shell structure is divided into five levels, they are largely intact, slight damage, moderate damage, serious damage, the collapse. According to the foregoing calculation results of the analysis, can correspond to given single layer Kiewit latticed shell structure of different control index extent of dynamical failure, as shown in Table 3.

Table 3. Division of dynamical damage grade and the corresponding indexes

Grade of damage	Intact	Slight damage	Moderate damage	Severe damage	Collapse
Dynamical failure index	$0 \leqslant D < 0.2$	$0.2 \leqslant D < 0.8$	$0.8 \leqslant D < 2.0$	$2.0 \leqslant D$	∞
Ratio of max displacement to span	$0 \leqslant U < 1/1500$	$1/1500 \leqslant U < 1/1400$	$1/400 \leqslant U < 1/100$	$1/100 \leqslant U$	∞
Percentage of plastic bars	P=0%	$0\% \leqslant P < 13\%$	$13\% \leqslant P < 45\%$	$45\% \leqslant P < 100\%$	P=100%
Description of phenomena	all elements are in elastic	few elements are plastic, displacement is minor, equipment are not damaged, cost for repair is few.	a few elements are plastic, displacement is big, some equipment are damaged, cost for repair is notable.	many elements are plastic, displacement is prodigious, equipment are devastated, cost for repair are more than that for build.	the structure is collapsed

6 Conclusion

1) Through the analysis of span of 40m of K8,6 single layer spherical latticed shell structure, it is appropriate to use dynamic failure index D = 2.0 as the dividing line when the structure is serious damaged, then the corresponding maximum displacement response and the 1P bar percentage are about 1/100 and 45%, respectively.

2) The dynamic failure index is also reflecting the impact of structural deformation and plastic cumulative energy to structural damage, it is favourable to comprehensive analyse the structure dynamical failure situation. For shell structure of uniform initial configuration, the result of analysis of dynamic failure index shows good consistency with responds of maximum displacement, the results of analysis of the percentage of plastic bars. But for the shell structure of a weak

local stiffness, there is a greater difference between the three, the analysis results of dynamical failure index are more moderate.

3) Structural damage state under dynamic loads cannot simply attributed two states, damage or no destruction. This article according to the division habit of building structures, gives largely intact, slight damage, moderate damage, serious damage, the collapse, five levels corresponding to dynamical failure index cut-off value, it is convenient to determine the dynamical damage extent of latticed shell structure.

References

1. S.Z. Shen, X.D. Zhi. Failure mechanism of reticular shells subjected to dynamic actions, *China Civil Engineering Journal* **38**,11(2005).
2. J.H. Ye, Z.Y. Shen. Experimental research on dynamical stability of single layer latticed shell structures, *Space Structure* **3**, 34(1997).
3. O.S. Kat, M.M. Shou, M. Mukaiyama. Study on Dynamic Behavior and Collapse Acceleration of single Layer Reticular Domes Subjected to Horizontal and Vertical Earthquake Motions, *Structural and Construction Engineering* **47**,89(1995)
4. F. Feng, S.Z. Shen. Study on the shaking table experiment and the vibration reducing experiment of single Layer reticulated shells, *Journal of H ARBIN University of Engineering and architecture* **33**,18(2000).
5. Z.X. Li, Z.Y. Shen, C.G. Deng. Judging criterion on nonlinear dynamical stability for truss structures, *Journal of Tong Ji University* **28**,148(2000).
6. W.F. Du, B.Q. Gao, S.L. Dong. New Judgment Method On Dynamic Instability of Truss Structures: Stress Rate Method, *Journal of Zhejiang University: Engineer science* **40**,506(2006).
7. Y.J. Park, A.H. Sang, Y.K. Wen. Mechanistic seismic damage model for reinforced concrete, *Journal of structural engineering*, **111**,722(1985).
8. X.Y. Yu, F. Fan, S.Z. Shen. Study on the aseismic performance of single layer reticulated domes J. *Xi'an Univ. of Arch. & Tech* (Natural Science Edition) **38**,442(2006).
9. J.F. Zhang, J.P. Hao. Research on the ultimate strength capacity of an arch steel frame using ANSYS J. *Xi'an Univ. of Arch. & Tech* (Natural Science Edition) **40**,71(2008).
10. YB9254295, Division standard grade of earthquake damaged industry buildings.

Quantitative Expression of Outlet Deviation Angle of Turbomachine Stator Based on Equivalent Moment of Momentum Principle

Wenjia Li [1,*], Anlin Wang [1], Wei Cheng [1], Xiaotian Li [1] and Jibin Han[2]

[1]School of Mechanical Engineering, Tongji University, 201804, Shanghai, China;
[2]Shantui Construction Machinery Co., Ltd, 272073, Shandong, China
Corresponding Email: pigeon_lwj_1989@126.com

Abstract. In order to express outlet deviation angle of turbomachine stator, outlet deviation angle dominantly expressed by fluid velocity distribution was derived based on equivalent moment of momentum under the condition of certain CFD simulation design parameters. On this basis, response surface function of outlet deviation angle was constructed by CFD simulation data of orthogonal experiment. The validity of the response surface function was proved by CFD simulation data of confirmatory models. An effective method is provided for calculating outlet deviation angle of turbomachine stator.

Keywords. Equivalent moment of momentum principle, outlet deviation angle, quantitative expression, turbomachine stator, computational fluid dynamic (CFD)

1 Introduction

The turbomachine is a series of power machines, which consists of several rotating impellers, and the energy transfers between impellers fixed on the shafts through the fluid. The turbomachine includes pump, fan, steam turbine, hydraulic torque converter and so on. The classical turbomachine design method is based on one-dimension flow theory, as the calculation speed is fast [1]. But the calculation precision is low, because the theory is based on the simplification of three-dimension flow to one-dimension flow, and it's hard to get accurate parameter values, such as the outlet deviation angle of stator. So, it's important to quantitatively research the outlet deviation angle of turbomachine stator.

The effect of the outlet deviation angle was mostly discussed in the papers [2-4]. Some papers discussed the reason that outlet deviation angle was produced and the phenomena of outlet deviation angle. For example, Willi [1] pointed out the reason that outlet deviation angle produced was the bending blade. Hayder [5] discussed the phenomena of wake formation. Padzillah [6] found the flow angle swings over a wide range in pulsating flow field. But the relationship between flow angle and design parameter was not studied in these papers. Nishi and Fukutomi [7] researched effect of outlet blade angle on outlet flow angle. But the effect of inlet blade angle and flow rate on outlet flow angle was not considered.

In this paper, equivalent moment of momentum principle was proposed to dominantly express outlet deviation angle of turbomachine stator by fluid velocity distribution. The stator of some type hydraulic torque converter was taken as a turbomachine stator, and the whole flow passage models of the stators with CFD simulation design parameters were built. The effects of the parameters on outlet deviation angle were analysed respectively by CFD simulation and dominantly expression of outlet deviation angle. Response surface function of outlet deviation angle was constructed by the effects of the parameters on outlet deviation angle and CFD simulation data of orthogonal experiment. The average deviation of the response surface function was proved 0.404 degree by CFD simulation data of confirmatory models.

2 Equivalent Moment of Momentum Principle

The classical turbomachine design method is based on one-dimension flow theory. According to one-dimension flow theory, a value is chosen as the outlet flow angle instead of the flow angles of all particles. But, the outlet flow angle and the flow angles of all particles must be equivalent. (The particle maybe distributes along stream wise direction or circumferential direction)

Among all of the physical quantities in turbomachine, the moment of every impeller is the macroscopic output physical quantity. The moment is determined by the moment of momentum [1]. So the moment of momentum is chosen as the equivalent principle of the outlet flow angle and the flow angles of all particles. It's called the equivalent moment of momentum principle in the paper.

The moment of momentum can be expressed by all the fluid particles as:

$$\boldsymbol{L} = \sum_{i=1}^{n} \boldsymbol{r}_i \times m_i \boldsymbol{v}_i \tag{1}$$

where L identifies moment of momentum; $\boldsymbol{r}$ identifies radius; m identifies mass; $\boldsymbol{v}$ identifies velocity vector; subscript i identifies particle i.

The direction of shaft is defined as direction z, then, the direction of moment of momentum is in direction z. The coordinate of fluid particle i is defined as (x_i, y_i, z_i), and the velocity component in each direction is v_{ix}, v_{iy}, v_{iz}, respectively. Then, it can be derived that:

$$|\boldsymbol{r}_i \times m_i \boldsymbol{v}_i| = m_i \left(x_i v_{iy} - y_i v_{ix} \right) \tag{2}$$

The moment of momentum can also be expressed by radius in the middle flow line, flow velocity, and total mass as:

$$\boldsymbol{L} = \boldsymbol{R} \times m\boldsymbol{v} \tag{3}$$

where $\boldsymbol{R}$ identifies radius vector in the middle flow line.

According to the definition of multiplication cross:

$$|\boldsymbol{R} \times m\boldsymbol{v}| = m|\boldsymbol{R}||\boldsymbol{v}|\sin\alpha \tag{4}$$

where the α is the angle between $\boldsymbol{R}$ and $\boldsymbol{v}$, as showed in figure 1.

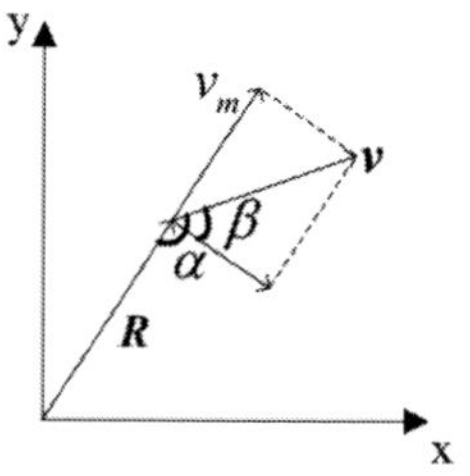

Fig. 1. Vector triangle of stator

According to one-dimension flow theory, vector triangle of stator can be showed as Fig. 1. Flow angle based on one-dimension flow theory can be expressed as

$$\sin\beta = \frac{|\boldsymbol{v}_m|}{|\boldsymbol{v}|} \tag{5}$$

where β identifies flow angle based on one-dimension flow theory; $\boldsymbol{v}_m$ identifies normal velocity component.

It can be derived in Fig.1 that the relationship between α and β can be expressed as

$$\alpha = \beta + \frac{\pi}{2} \tag{6}$$

According to one-dimension flow theory, normal velocity component can be expressed by flow rate and normal area as

$$|\boldsymbol{v}_m| = \frac{Q}{F_m} \tag{7}$$

where Q identifies flow rate; F_m identifies normal area.

For fluid, the mass in micro time t can be expressed by flow rate, density and time as

$$m_i = \rho_i Q_i t \tag{8}$$

where ρ identifies density.

The mass of all particles is equal to the total mass. So, it can be derived that

$$m = \sum_{i=1}^{n} m_i \tag{9}$$

The relationship among the flow angle, blade angle and deviation angle is

$$\Delta\beta = \beta - \beta_y \tag{10}$$

where β_y identifies blade angle; $\Delta\beta$ identifies deviation angle.

According to the equivalent moment of momentum principle, formula (1) is equal to formula (3). According to the formulas above, the deviation angle can be expressed as

$$\Delta\beta = \arctan\left(\frac{|\boldsymbol{R}| Q \sum_{i=1}^{n} \rho_i Q_i}{F_m \sum_{i=1}^{n} \rho_i Q_i \left(x_i v_{iy} - y_i v_{ix} \right)} \right) - \beta_y \tag{11}$$

3 Simulation Models for CFD

The main research contains of computational fluid dynamics (CFD) is to calculate the flow field with computer. The essence method of CFD is iterative calculation. A lot of studies have proved that the precision of CFD is high. A platform experiment of some type hydraulic torque converter has been done and compared to CFD simulation. As shown in figure

DOI: 10.1051/matecconf/201710004017

2, the average relative errors of torque ratio, efficiency and pump capacity are 0.86%, 0.47% and 1.02%. The reliability of CFD result is proved.

Besides the high precision, another characteristic of CFD is that almost all of the physical quantities can be observed. If CFD is used to calculate the flow field in turbomachine, the physical quantities in formula 11, such as normal area, flow rate, density, velocity and particle coordinate can be observed. So CFD is chosen as the research method.

When the turbomachine is under design, the blade angles of the impellers are the design variables. Circulating flow rate is a very important physical quantity in turbomachine and it has effect on the turbomachine design. Therefore, the simulation models with different inlet blade angles, different outlet blade angles and different flow rates need to be built for CFD.

The inlet impact angle of simulation model is zero, as the turbomachine is under design. Revolving spiral flow passage model was built to control the inlet impact angle.

The stator of some type hydraulic torque converter is taken as a turbomachine stator. The whole flow passage simulation models of the stator with different flow rates, inlet and outlet blade angles and zero inlet impact angles were built for CFD simulation. One of the models is shown in figure 3.

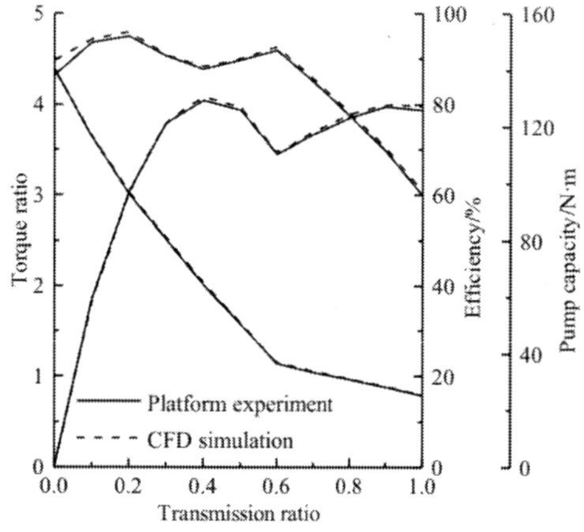

Fig. 2. Comparison of experiment and CFD simulation.

Fig. 3. One of the models for CFD simulation.

4 Effect of Design Parameter on Outlet Blade Angle

4.1 Effect of Inlet Blade Angle on Outlet Blade Angle

The method to analyse the effect of inlet blade angle on outlet deviation angle is to build some models of which the outlet blade angles and flow rates are the same, but the inlet blade angles are different. CFD simulation results of the models can be got and calculated to analyse the effect.

The inlet blade angle and outlet blade angle of the type hydraulic torque converter stator and the flow rate range are 90deg, 35deg and 0.043-0.149m3/s.

The design range of inlet blade angle is 20 degree [1]. So, four models need to be built to analyse the effect of inlet blade angle. According to table 1, the outlet blade angle and the flow rates of the models are 35 degree and 0.097m3/s. The inlet blade angles of the four models are 80 degree, 85 degree, 95 degree and 100 degree respectively.

According the parameters, four models can be built and simulation parameters for CFD can be set. Outlet deviation angles can be calculated by formula 11 and the physical quantities observed from simulation result. The corresponding relationship between inlet blade angle and outlet deviation angle is shown in figure 4.

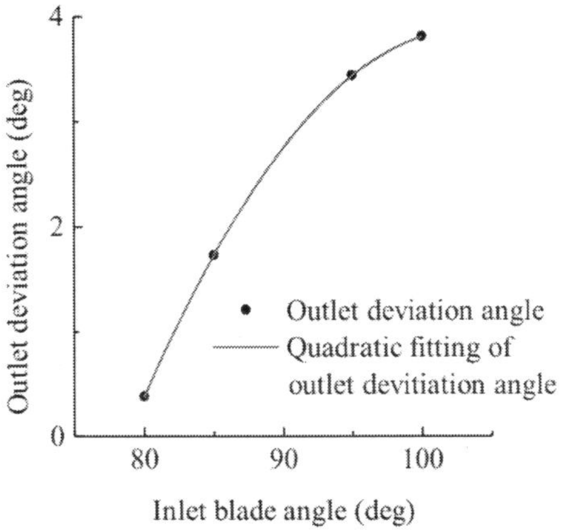

Fig. 4. Outlet deviation angle along inlet blade angle.

It can be seen that, the relationship between inlet blade angle and outlet deviation angle is almost quadric. Quadratic fitting is taken to fit the points. The Residual Sum of Squares (RSS) is 1.79×10-6. It can be concluded that the relationship between inlet blade angle and outlet deviation angle is almost quadric.

4.2 Effect of Outlet Blade Angle on Outlet Blade Angle

The method to analyse the effect of outlet blade angle on outlet deviation angle is similar to that of inlet blade angle. The design range of outlet blade angle is also 20 degree [1]. So, another five models need to be built to analyse the effect of outlet blade angle. According to table 1, the inlet blade angle and the flow rates of the models are 90 degree and 0.097m3/s. The outlet blade angles of the five models are 25 degree, 30 degree, 35 degree, 40 degree and 45 degree respectively.

According the parameters, five models can be built and simulation parameters for CFD can be set. The outlet deviation angles of stator models with different outlet blade angles can be got in the same way above. The corresponding relationship between outlet blade angle and outlet deviation angle is shown in figure 5.

It can be seen that, the relationship between outlet blade angle and outlet deviation angle is almost linear. Linear fitting is taken to fit the points. The RSS is 7.83×10-2. It can be concluded that the relationship between outlet blade angle and outlet deviation angle is almost linear.

4.3 Effect of Flow Rate on Outlet Blade Angle

The method to analyse the effect of flow rate on outlet deviation angle is similar to that of inlet blade angle. When the turbomachine works normally, the flow rate range is displayed in table 1. So, another five models need to be built to analyse the effect of flow rate. According to table 1, the inlet blade angle and the outlet blade angle of the models are 90 degree and 35 degree. The flow rates of the five models are 0.04 m3/s, 0.069m3/s, 0.097m3/s, 0.126m3/s and 0.154m3/s, respectively.

According the parameters, five models can be built and simulation parameters for CFD can be set. The outlet deviation angles of stator models with different flow rates can be got in the same way above. The relationship between flow rate and outlet deviation angle is shown in Fig. 6.

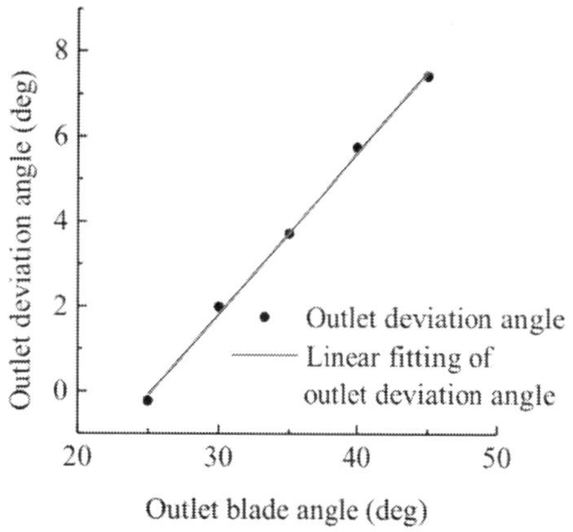

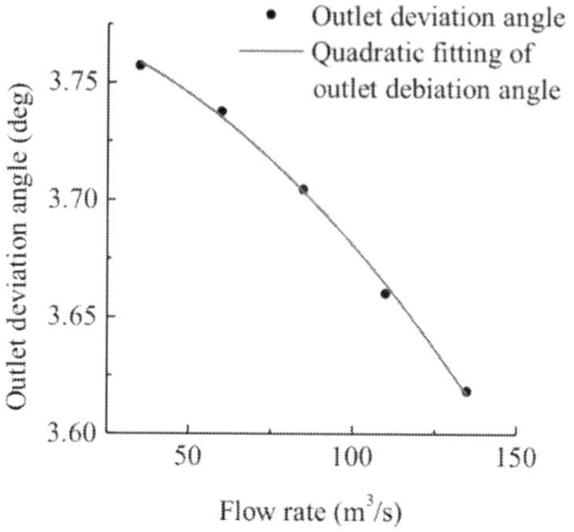

Fig. 5. Outlet deviation angle along outlet blade angle. **Fig. 6.** Outlet deviation angle along flow rate.

It can be seen that, the relationship between flow rate and outlet deviation angle is almost quadric. Quadratic fitting is taken to fit the points. The RSS is 3.01×10-5. It can be concluded that the relationship between flow rate and outlet deviation angle is almost quadric.

5 Response Surface Function

5.1 Orthogonal Experiment

According to the analyses above, the relationships between inlet blade angle, outlet blade angle, flow rate and outlet deviation angle are got. That means the response surface function form of outlet deviation angle is determined, but the coefficients of the function aren't obtained. It can be expressed as:

$$\begin{aligned}\Delta\beta = & c_1 + c_2\beta_{y1} + c_3\beta_{y2} + c_4Q + c_5\beta_{y1}^2 + c_6\beta_{y1}\beta_{y2} + c_7\beta_{y1}Q + c_8\beta_{y2}Q + c_9Q^2 \\ & + c_{10}\beta_{y1}^2\beta_{y2} + c_{11}\beta_{y1}^2Q + c_{12}\beta_{y1}\beta_{y2}Q + c_{13}\beta_{y1}Q^2 + c_{14}\beta_{y2}Q^2 \\ & + c_{15}\beta_{y1}^2\beta_{y2}Q + c_{16}\beta_{y1}^2\beta_{y2}^2 + c_{17}\beta_{y1}\beta_{y2}Q^2 + c_{18}\beta_{y1}^2\beta_{y2}Q^2\end{aligned} \quad (12)$$

where c identifies undetermined coefficients; subscript 1 and 2 identify inlet and outlet separately.

An equation can be determined by CFD simulation data of one model with a group of CFD simulation design parameter. If there are enough models for CFD simulation, there will be enough equations. When the quantity of equations is bigger than or equal to the quantity of coefficients, the least squares solution of equations can be got. The values of the least squares solution are the values of coefficients.

According to formula 12, the quantity of coefficients is 18. So, at least 18 equations are needed. That means at least 18 models with different groups of CFD simulation design parameter need to be built for CFD simulation.

According to formula 12, the quantity of variables is 3. For each variable, five values can be chosen. The values of the variables should distribute averagely. Values chosen by orthogonal experiment distribute averagely [8]. So, orthogonal experiment can be chosen to choose the values. Three factors and five levels orthogonal experiment is carried out. According to the orthogonal experiment table L25 (56) [8], 25 groups of parameters can be decided.

5.2 Response Surface Function

According to the 25 groups of parameters, 25 models can be built for CFD. Outlet deviation angels of the models can be calculated by formula 11 and fluid velocity distribution obtained from CFD simulation. According to CFD simulation design parameters and outlet deviation angles, coefficients can be got.

6 Verification

The accuracy of the response surface function needs to be verified to prove the function can reflect the real relationship among outlet deviation angle and design parameters.

In order to verify the accuracy of the response surface function, another seven models with different groups of design parameter are built. The parameters are random and in the design range.

CFD simulation of the seven models can be done, and outlet deviation angles can be calculated by formula 11 and fluid velocity distribution obtained from CFD simulation. Outlet deviation angles can also be calculated by the response surface function. The results are shown in figure 7.

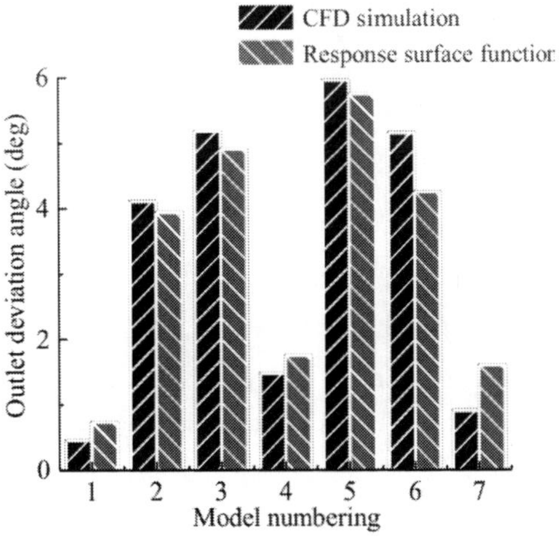

Fig.7. Contrast of outlet deviation angles calculated by CFD simulation and response surface function

It can be seen from figure 7 that the trends of the two kinds of results are the same. The average deviation of the two kinds of results is 0.404 degree.

The deviation of outlet deviation angle causes deviation of moment of momentum. The formula calculating moment of momentum with outlet deviation angle is derived by formula 3-6 and 10.

$$|\boldsymbol{L}| = m|\boldsymbol{R}|\frac{|\boldsymbol{v}_m|}{\tan\left(\Delta\beta + \beta_y\right)} \tag{13}$$

As introduced above, design range of outlet blade angle is between 25 degrees to 45 degrees [1]. It can be seen from Fig.7 that the average outlet deviation angle is about 3 degrees. So, the average error of moment of momentum caused by the 0.404 degree deviation of outlet deviation angle is 1.43%~1.71%.

That means the accuracy of formula quantitatively calculating outlet deviation angle in design range can satisfy the engineering demand.

7 Conclusion

In this paper, the equivalent moment of momentum principle was proposed according to the macroscopic output physical quantity of turbomachine. Outlet deviation angle dominantly expressed by fluid velocity distribution was derived based on the principle under the condition of certain CFD simulation parameter.

The stator of some type hydraulic torque converter was taken as a turbomachine stator. The effects of CFD simulation parameters on outlet deviation angle were analysed respectively by CFD simulation and dominantly expression of outlet deviation angle. Response surface function of outlet deviation angle was constructed by the effects of the parameters on outlet deviation angle and CFD simulation data of orthogonal experiment.

The average deviation of the response surface function was proved 0.404 degree by CFD simulation data of confirmatory models. The average error of moment of momentum in design range was 1.43%~1.71%. The accuracy of the response surface function could satisfy the engineering demand in design range.

Acknowledgements

Financial support for this investigation was received from the major science and technology achievements transformation projects of 2012.

References

1. Willi B. Stömungsmaschinen (Bere-chnung und Konstruktion). Vogel Press, West Germany (1980).
2. Kluß D, Wiedermann A, Stoff H. Effect of Wakes and Secondary Flow on Re-attachment of Turbine Exit Annular Diffuser Flow. Journal of Turbomachinery, **131**, 4 (2009):041012- 1-12.
3. Vassiliev V,Irmisch S,Abdel W S, et al. Impact of the Inflow Conditions on the Heavy-Duty Gas Turbine Exhaust Diffuser Performance. Journal of Turbomachinery, (2010): 1401-1412.
4. Tan M G, Guo B, Liu H L, et al. Investigation of radial force and hydraulic performance in a centrifugal pump with different guide vane outlet angle. **17**, 6 (2015): 3247-3260.
5. Hayder M M A. Wake Formation from a Pair of Circular Cylinders Traversing Between Small- and Large-Incidence Flow Regimes. Journal of Fluids Engineering, **134**, 1 (2012): 011204-1-12.
6. Padzillah M H, Rajoo S, Yang M, et al. Influence of Pulsating Flow Frequencies towards the Flow Angle Distributions of an Automotive Turbocharger Mixed-flowturbine. Energy Conversion and Management, **98**, 1 (2015):449-462.
7. Nishi Y, Fukutomi J. Effect of Blade Outlet Angle on Unsteady Hydrodynamic Force of Closed-Type Centrifugal Pump with Single Blade. International Journal of Rotating Machinery, (2014): 1-16.
8. Taguchi G, Yokoyama T. Experiment Design Method Textbook. Japanese Standards Association, Japan (1981).

Determination of Picogram Levels of Diacerein in a Pharmaceutical Formulation by Flow-Injection Chemiluminescence

Suya Gao, Binghong Ma, Pengfei Ma, Yun Zhang, Kaidi Huang and Chen Huang

School of Pharmacy, Xi'an Medical University, Xi'an 710021, China

Abstract. A simple, sensitive and inexpensive method for determination of diacerein by flow-injection chemiluminescence was proposed, based on the quenching effect of diacerein on the luminol-protein (bovine serum albumin, BSA) reaction. It was found that the decrement of CL intensity was linearly proportional to the logarithm of diacerein concentration ranging from 5.0 to 7.0×103 pg·mL-1 (r = 0.9968), with the limit of detection (LOD) of 1.0 pg·mL-1 (3σ). The proposed procedure was successfully applied to the determination of diacerein in pharmaceutical formulation, human saliva, and serum samples without interference from its potential impurities, with the recoveries ranging from 96.4% to 104.0% and the relative standard deviations (RSDs) less than 4.0% (n=6).
Keywords. diacerein, luminol, chemiluminescence, folw injection.

1 Introduction

Diacerein, named 1,8-diacetoxy-3-carboxyanthraquinone in chemical field, is a novel anti-inflammatory drug with pharmacological properties different from those of classical nonsteroidal anti-inflammatory drugs. Clinical studies have suggested that diacerein exerts a beneficial effect on the symptomatic treatment of osteoarthritis [1,2]. It is also known to inhibit the production of interleukin-1 and to prevent cartilage breakdown in a mouse granuloma model, and to slow the progression of cartilage lesions in a canine model of osteoarthritis [3,4]. Diacerein is known to be completely metabolized by animals and humans into rhein, an active metabolite of diacerein found in plasma and synovial fluid. Moreover, some researchers have related the use of diacerein or rhein to treat and prevent vascular diseases[5].

Several methods have been developed for determining diacerein such as high performance liquid chromatography, UV-vis spectrophotometry, flow-injection Chemiluminescence (FI-CL), etc [6-10]. Compared with other methods, FI-CL is an attractive detection method for determination because of the amazing advantages of ultra sensitivity, very low detection limit, rapidity, and wide linear working range that can be achieved by use of relatively simple instrumentation. Luminol as one of the highest quantum yield luminescent compounds, has been widely used in bioluminescent and chemiluminescent analysis. In recent years, the luminol-protein system has been widely used to determine some organic compounds in pharmaceutical preparations based on their enhancing effect on this CL reaction [11]. Bovine serum albumin (BSA) is one of protein, which has been commonly used as a model protein to study protein-drug interactions. To the best of our knowledge, no detailed investigation of diacerein by luminol-protein CL system has so far been reported. In this paper, it was found that the CL intensity from luminol-protein system could be significantly inhibited in the presence of diacerein, and the decrement of CL intensity was proportional to the logarithm of diacerein concentration in wide range of 5.0 to 7.0×103 pg·mL-1. Based on this observation, a simple, rapid and highly sensitive FI-CL method was established for the determination of diacerein in pharmaceutical formulations and biofluids with satisfactory results.

2 Experiment

Reagents. Luminol of 2.5×10-2 mol·L-1 was prepared by dissolving 0.44 g luminol (Sigma) in 100 mL of 0.01 mol·L-1 NaOH solution in a brown calibrated flask. The stock solution of BSA (Sigma, MW~68kDa) with the concentration of 5.0×10-9 mol·L-1 was prepared with purified water. All the reagents were of analytical grade and all solutions were prepared with doubly distilled water. The stock solution of diacerein (Xi'an Sino-Herb Bio-technology Co.Ltd., China) of 2.0 × 10-4 mg·mL-1 was prepared with HPLC-grade methanol up to 50mL in the ultrasonic generator for 60 min and diluted with water to prepare working solutions.

Apparatus and Procedures. The FI-CL apparatus consists of the sampling system, CL detector (photomultiplier tube), and the PC with a client system. Two peristaltic pumps (IFIS-C, Xi'an Ruike Electronic Science-Tech. Co. Ltd., China), equipped with an automatic sampler, were used to deliver all solutions at a flow rate of 2.0 mL·min-1. PTFE tubing (0.8mm i.d.) was used to connect all components in the flow system. Injection was made using a six-way

 DOI: 10.1051/matecconf/201710004018

injection valve equipped with a 100 μL sample loop. The flow cell was a 10.0 cm length of glass tubing (2.0mm i.d.) and was positioned in front of the detection window of the CR-150 photomutiplier tube (PMT, operated at -700V, Hamamatsu, Tokyo, Japan). The CL emission produced in the flow cell was detected by using a computerized BPCL ultra-weak luminescence analyzer (Institute of Biophysics, Acade-mia Sinica, Beijing, China). Data acquisition and treatment were performed with BPCL software. Four flow lines were inserted into the solutions of luminol, carrier (purified water), BSA and sample, respectively, and the solutions were propelled by peristaltic pumps. Luminol (100 μL) was quantitatively injected into the mixed solution of BSA and sample by a six-way valve, and then the mixture was delivered into the flow cell producing CL emission. The signal was displayed and integrated for a 0.1 s interval. The concentration of the diacerein was quantified by the decrement of CL intensity (ΔI = I0 −Is), where I0 and Is were CL signals in the absence and in the presence of diacerein, respectively.

Sample Preparation. Diacerein capsules (labelled amount 50mg·cap-1) purchased from local market, were disintegrated, weighed and finely powdered. Then the powder of average one capsule was accurately weighed and transferred into a 100 mL volumetric flask, and then diluted to volume with HPLC-grade methanol. Then, it was placed in an ultrasonic generator for 1 h and the resulting mixture was filtered. The filtrate was further diluted with water so that the final diacerein concentration was within the working range, and then analyzed according to the procedure described above. The saliva was collected from healthy volunteers, and the serum was supplied by the Shaanxi Province People's Hospital. To prepare the spiked samples, known quantities of standard solution of diacerein were spiked into 1.0mL of saliva or serum. After homogenization, the spiked samples with appropriate dilution were taken for the quantification.

3 Results and Discussion

Optimization of Luminol-BSA CL System. Under different concentrations of luminol solution from 1.0×10-8 mol·L-1 to 1.0×10-4 mol·L-1 and BSA solution from 1.0×10-11 mol·L-1 to 5.0×10-7 mol·L-1, the CL intensity was determined. It was found that the CL intensity arrived at the maximum with 1.0×10-5mol·L-1 lumimol and 5.0×10-9mol·L-1 BSA (shown in Fig.1a). NaOH was introduced into the luminol solution to improve the sensitivity of the system, and 0.01mol·L-1 NaOH was the optimal in the experiment (shown in Fig.1b). In addition, the influence of flow rate and the length of mixing tube on the CL intensity were also investigated in the ranges of 0.5~5mL·min-1 and 5.0~20.0cm. As a compromise between good precision and lower reagent consumption, the flow rate of 2.0mL·min-1 on each flow line and the length of 10.0cm mixing tube afforded the best results with good sensitivity and reproducibility.

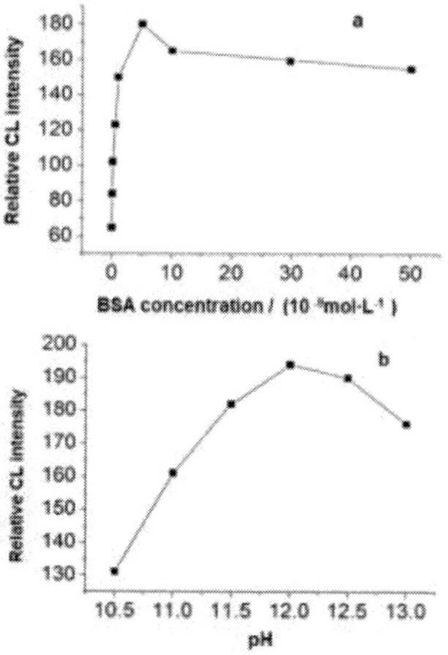

Fig.1 Effect of BSA concentration (a) and pH (b) on the CL intensity

Standard Curve and Detection Limit. Under the selected conditions described above, the calibration graph of CL decrement (ΔI, relative unit) versus the logarithm of diacerein concentration was linear in the range of 5.0 ~ 7.0×103 pg·mL-1 with a detection limit of 1.0 pg·mL-1 (3σ). The regression equation was ΔI = 38.0 lgC − 27.6 with a correlation coefficient of 0.9968 (n=6). The relative standard deviation was 2.2 % for 50 pg·mL-1 diacerein in 8 replicate measurements. At flow rate of 2.0mL·min−1, both sampling and washing could be accomplished in 30 s. So the sample measurement frequency was calculated about 120 samples·h−1. Repeated injection of standards and samples leads to easily reproducible data. It indicated that FI-CL was a non-time consuming strategy for the quantification of diacerein.

Interference. In order to assess the possible analytical application of the proposed method to dosage forms, the effect of foreign substances was studied by analyzing a standard solution of 50 pg·mL-1 diacerein to which increasing amounts of interfering species and potential interfering substances were added. Assuming interference at a level of ±5%, tolerable ratios of foreign species with respect to 5.0 × 102 pg·mL-1 diacerein were over 1.0 × 104 times for the excipients commonly existed in the capsules, such as glucose, sucrose, starch, methanol, ethanol and fructose; 2.0 × 103 for Na+ , K+, Ba2+, Mg2+, Ca2+, NH4+, F-, Cl-, Br-, I-, SO42-, NO3-, CO32-; 1.0 ×103 for Zn2+, Ni2+, Mn2+, Cd2+, Ag+, Cu2+, Hg2+; 1.0×102 for the tryptophan, tyrosine, dopamine and peroxidase. In fact, the effect of above species and some other foreign substances in saliva and serum can be eliminated by calibrating the blank prior to diacerein determination. Thus the influence on the determination of diacerein at picogram levels can be neglected under the experimental conditions.

Analysis of Real Sample. Following the procedure described previously, the proposed CL method was applied to the determination of diacerein in commercially available Artrodar (diacerein capsules, labeled 50 mg per capsule). The results were listed in Table 1, and each result was the average of six determinations and agreed well with the labeled values. It can be seen that the recoveries were from 97.8% to 101.2% and RSDs were less than 3.0% (n = 6).

Determination of diacerein in human saliva and serum samples. To validate the expanded applicability of the method for the determination of diacerein in the complex biological matrix, the spiked human saliva and serum samples prepared in the experimental section were quantified, and the results were listed in Table 2 and, Table 3, respectively. Each result was the average of five determinations. It was clear that the recoveries ranged from 96.4% to 104.0% and RSDs were less than 4.0% (n = 6). A good agreement between the added and found concentration values was achieved. The results proved that the proposed method could be applied to the determination of diacerein in biofluids.

Table 1. Results for the determination of diacerein in Artrodar capsules

Sample	Proposed method [a]($mg\ cap^{-1}$)	Added ($pg\ mL^{-1}$)	Recovered [a] ($pg\ mL^{-1}$)	Average recovery (%)
1	49.39±1.3	10.00	10.12±2.8	101.2
2	50.02±1.8	20.00	19.57±1.9	97.8
3	49.65±2.6	30.00	30.08±2.3	100.3
4	49.50±2.9	20.00	19.92±2.6	99.6
5	49.18±2.1	40.00	40.19±2.4	100.5
6	49.89±1.9	60.00	58.93±2.1	98.2

Table 2. Results for the determination of diacerein in spiked saliva

Sample	Added($pg\ mL^{-1}$)	Found($pg\ mL^{-1}$)	RSD(%)	Recovery(%)
1	0/20.0	50.1/70.2	3.1/2.5	100.5
2	0/50.0	50.5/101.8	3.6/3.2	102.6
3	0/100.0	49.6/150.4	2.3/1.9	100.8
4	0/200.0	49.4/242.2	2.8/2.6	96.4
5	0/300.0	50.2/349.6	3.4/2.8	99.8
6	0/500.0	49.8/558.5	2.5/1.7	101.7

Table 3. Results for the determination of diacerein in spiked serum

Sample	Added($pg\ mL^{-1}$)	Found($pg\ mL^{-1}$)	RSD(%)	Recovery(%)
1	0/10.00	20.8/31.5	2.6/3.5	107.0
2	0/20.00	20.1/40.9	2.1/3.6	104.0
3	0/50.00	20.6/71.2	3.1/2.8	101.2
4	0/100.00	19.8/118.5	2.9/2.2	98.7
5	0/200.00	19.2/220.8	3.2/3.7	100.8
6	0/300.00	19.5/317.4	2.4/3.0	99.3

4 Conclusions

A simple, sensitive and rapid flow injection CL method of luminol-BSA reaction in alkaline medium for the determination of diacerein was developed for the first time. Even with a simple set-up, diacerein can be sensitively detected with a 1.0 pg·mL-1 detection limit (3σ). Moreover, the proposed procedure was successfully applied to the determination of diacerein in pharmaceutical formulation, human saliva, and serum samples without interference from its potential impurities. Satisfactory performance of the assay of diacerein in pharmaceutical preparations and biofluids has proved the ability of this method for analyzing complex biological matrix with separation-free, indicating its potential value in pharmaceutical and clinic researches.

Acknowledgments

This work was supported by the Undergraduates' Innovative Training Program (No. 1647, 2015DXS1-22), the Ph.D. Startup Foundation of Xi'an Medical University of China (No.2012DOC12), the Scientific Research Program Funded by Shaanxi Provincial Education Department (No. 14JK1620) and the discipline construction of Xi'an Medical University.

References

1. Tamura, T. Shirai, T. Kosaka, N. Ohmori, K. Takafumi, N. Eur. J. Pharmacol, **448** (2002)
2. Pelletier, J.P. Yaron, M. Haraoui, B. Cohen, P. Nahir, M.A. Choquette, D. Wigler, I. Rosner, I.A.; Beauleu, A.D. Arthritis Rheum, **43** (2000)
3. Cui-e He, Y.M. Wang, Journal of clinical research. **28** (2011)
4. S Singh, A Jain, S K Mishra, et al. Osteoarthr Cartilage, **20** (2012)
5. J.-M. Lin, Yamada, M. Trends Anal, Chem, **22** (2003)
6. Giannellini, V. Salvatore, F. Bartolucci, G. Coran, S.A. Bambagiotti-Alberti, M. J. Pharm, Biomed. Anal. **39** (2005)
7. S Dhaneshwar, V Patel, D Patil. Bioorg, Med. Chem. Lett, **23** (2013)
8. H.C. Yao, X.F. Yang, H. Li, J. Chin, Chem.Soc, **54** (2007)
9. S Bais, I Singhvi, A Chandewar, M Shirsat, Asian J. Pharm. Ana, **3** (2013)
10. K. R. Sreejith, Dr. K. Premalatha, International J. Research in Pharm, Biomed. Sci. **2** (2011)
11. Y. Zhang, S.Y. Gao, Advanced Material Research, **884-885** (2014)

Heparin Coating Technology in Cardiopulmonary Bypass

Kun Zhang[1] ,Rui Xue[2,*] and Shuwei Shen[1]

[1]Department of Anesthesiology, Jingzhou Central Hospital, The Second Clinical Medical College, Yangtze University, Jingzhou, Hubei Province,434020,China.
[2]Department of Anesthesiology, Renmin Hospital, Hubei University of Medicine, Shiyan, Hubei, China.
Corresponding Email: 7885909@qq.com

Abstract. Heparin coating technology has become increasingly mature, which is conducive to the material of blood, biocompatibility proved very important, therefore, heparin coating technique has become a development cycle is very useful for in vitro selection of patients. On the other hand, the overall cost of medical treatment will increase because of the use of heparin coated materials and devices, so the clinical use of heparin coated materials is limited. In this paper, the principle of heparin coating technology, heparin coating technology should be applied in the norms of the external circulation, and related heparin coating technology development status analysis.

Keywords. Cardiopulmonary bypass , heparin coating technology, principle.

1 Introduction

The researches have proved that there is the heparin-like molecular structure (heparin sulfate) in a microvascular endothelial cell; when the solidification state of blood changes, such endothelial cells will militate [1-3]. Therefore, it may after all be accepted as a reasonable method to modify the abiotic surface with heparin substances to make it to have high blood compatibility. Heparin solidification, which has been widely studied on the blood contact surface of CPB equipment or for heparin coating, has been currently regarded as the most promising method to improve the biocompatibility.

2 Principle of heparin coating technology

As shown in the past researches, the heparin coating surface is formed by enhancing the catalysis of antithrombin-III and reducing thrombus. Although the favorable side of the technology has not been proven yet, the latest researches show that the advantage of heparin coating technology is the selective attachment of plasma proteins, which can rapidly form a pseudomembrane of blood compatibility on the surface of material and, meanwhile, can prevent albuminous degeneration from enhancing the activity of attachment protein and blood cell [4]. It is also found through related researches that the plasma proteins, such as blood coagulation factor XII, fibrinogen and transparent binding protein, on the surface of artificial material can enhance the anticoagulant activity.

3 Principles of heparin coating technology for cpb equipment

Universal surface principle: its representative method is to combine polyethyleneimine (PEI) with the surfaces of all sorts of materials by means of sedimentation or ionic bond so that all the material surfaces have affinity with the heparin degraded with nitrous acid.

Universal coating principle: it is to process heparin with binding agent so that its physico-chemical property can be changed; then, the heparin will have high affinity with the surfaces of all sorts of materials and its anticoagulation property can be kept simultaneously. Duraflo Ⅱ heparin is namely an ideal universal coating applicable to heparin coating of CPB equipment. After it is diluted to the concentrated needed with organic solvent, 1, 1, 2trichloro-1, 1, 2fluoro-ethane, it is injected into the human's CPB loop for circulation, is emptied and is dried for standby. The coating will not cracks or expand under stress and the residual solvent can be easily drained.

4 Heparin coating technology in cpb

The heparin covalent bond coating technology can be divided into three types: ionic bond combination; covalent bond combination and fixed end point. Its molecular formula is shown in Figure1.

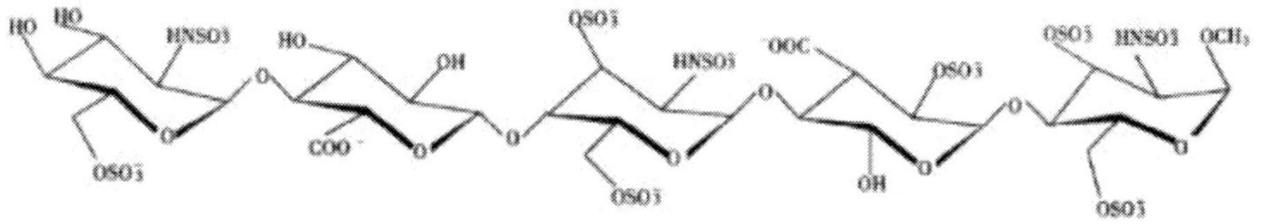

Fig.1. Schematic diagram for molecular structure of Heparin.

The heparin with its ionic bond combination carries a lot of negative electricity. The positive ion inducer is used to lead positive ions to the surface of synthetic material and attracts heparin with static electricity. The benzylamine-heparin with benzylamine as its inducer has been widely used in clinic but is not applicable to CPB because it is very easily soluble in normal saline [5-7]. TDMAC-heparin (hydrochloric acid 3-12 alkyl methylamine - heparin) is a very practical heparin coating material synthesized by Grode and the shunting pipes with TDMAC-heparin are available now. The combination between some positive ion genes, such as N-N dimethylamide vinyl, and polymer substrate can also help the heparin attachment. After the complexation of poly ammonia 1 and heparin, it can adhere to polyurethane and glass surface.

The mode of covalent bond combination is to fix heparin by using the functions of –CNS radical, -NH2 radical, -OH radical and other special active groups on the surface of high polymer material and the –OH radical of heparin molecule [8]. As heparin has plenty of functional groups which are combined with the surface of material excessively firmly, resulting in the loss of activity in heparin molecule, the expected anticoagulation effect usually cannot be reached.

The mode of fixed end point is firstly to prepare a binding ion complex on the surface of biomaterial; then, activate the heparin through a controlled nitrous acid degradation step so that the heparin molecule can form the final terminal free aldehyde group. This heparin and the glutamic acid dialdehyde can have a chemical reduction reaction through the ammoniation reducing action so as to fix heparin in the mode of covalent bond on the surface of material containing amino [9,10]. It's worth noting that part of such heparin molecules will degrade to lower the catalytic activity in solution. However, because of its very proper binding site between such heparin molecules and the surface of biomaterial, this mode can truly generate the anticoagulation coating materials.

5 Conclusion

One's whole body needs to be heparinized in the CPB operation. The heparin modification to CPB lines can make them to have better biological compatibility. This coating method, which can lead to relatively secure heparin binding within 8 hours allowed by an CPB operation, is applicable to clinics. To sum up, the heparin binding technology, which is a newly developed technology in recent years, can improve CPB lines and the biological compatibility of blood and can reduce the inflammatory reaction incurred by contacting the surface of a foreign matter.

References

1. Gravlee GP.Heparin-coated cardiopulmonary bypass circuits.Journal of cardiothoracic and vascular anesthesia. (1994) Apr; 8 (2):213-22.
2. Stammers AH , Christensen KA , Lynch J , Zavadil DP .Quantitative evaluation of heparin-coated versus non-heparin-coated bypass circuits during cardiopulmonary bypass.The journal of extra-corporeal technology 1999 Sep; 31 (3):135-41.
3. Ovrum E , Tangen G , Oystese R, Ringdal MA , Istad R .Comparison of two heparin-coated extracorporeal circuits with reduced systemic anticoagulation in routine coronary artery bypass operations.The Journal of thoracic and cardiovascular surgery 2001 Feb; 121 (2):324-30.
4. Van Poucke S, Stevens K, Wetzels R, Kicken C, Verhezen P, Theunissen M, Kuiper G, van Oerle R, Henskens Y, Lancé M.Early platelet recovery following cardiac surgery with cardiopulmonary bypass.Platelets. 2016 May 10:1-7.
5. Olsson A, Alfredsson J, Hkansson E, Svedjeholm R, Berglund J, Berg S.Protamine reduces whole blood platelet aggregation after cardiopulmonary bypass.Scand Cardiovasc J. 2016;50(1):58-63.
6. Kehara H, Takano T, Ohashi N, Terasaki T, Amano J.Platelet function during cardiopulmonary bypass using multiple electrode aggregometry: comparison of centrifugal and roller pumps.Artif Organs. 2014 Nov;38(11):924-30.
7. Ortmann E, Klein AA, Sharples LD, Walsh R, Jenkins DP, Luddington RJ, Besser MW.Point-of-care assessment of hypothermia and protamine-induced platelet dysfunction with multiple electrode aggregometry (Multiplate) in patients undergoing cardiopulmonary bypass.Anesth Analg. 2013 Mar;116(3):533-40.
8. Rubens FD.Cardiopulmonary bypass technology transfer: musings of a cardiac surgeon.Journal of biomaterials science. Polymer edition 2002 ; 13 (4):485-99.
9. Chryssos A , Stroup SJ , Pifer MM , Tawil M,Conrad CG .Management of Heparin-Resistant Patients with Benefits? Maximizing Biocompatibility in Cardiopulmonary Bypass: Combining ATryn Recombinant Antithrombin III and Carmeda Heparin-Bonded Perfusion Circuits: A Case Series.The journal of extra-corporeal technology2015 Mar; 47 (1):44-7.
10. Zhang R, Huang Y, Zhang M, Liu J.Bivalirudin utilization in rats undergoing cardiopulmonary bypass: preventing the increase of antiheparin/platelet factor 4 antibody in perioperative period.Clin Appl Thromb Hemost. 2014 Mar;20(2):196-8.

Clinical Application of Ophthalmic Implant Materials and Biocompatibility

Feng Ke and Ying Wang*

Department of Ophthalmology Renmin Hospital, Hubei University of Medicine, Shiyan, Hubei, China
Corresponding E-mail: hubeiwenbo@163.com

Abstract. A biomedical material, which is a material to diagnose, treat and repair a living body or to replace its damaged tissue and organ, or to enhance its functions, is one of effective tools that human uses to fight against diseases.

Keywords. Ophthalmology, implant materials, biocompatibility.

1 Introduction

Eye health care and the treatment of eye diseases are always the focuses concerned in the medical field, especially after the improvement of medical level in recent years. Ophthalmic materials are indispensable parts in the treatment and heath care of eye diseases. The treatment of many parts, such as the treatment of organic injury in the ocular region, needs the implantation of artificial substitutes, such as artificial eyeball and artificial tear duct. Besides, some patients with corneal diseases can restore the eyesight only by implanting an artificial cornea in their eyes; it is required to implant intraocular lens for the treatment of cataract. An ideal ophthalmic material needs to not only have stable chemical property but also have good biocompatibility so that there won't be any rejection reaction and irritation after implantation and thus patients won't feel any discomfort. Of course, materials for different purposes should have corresponding characteristics to ensure patients' comfort and safety.

2 Clinical application of ophthalmic implant materials

2.1 Intraocular lens

Intraocular lens are mostly used in the cataract surgery to replace a patient's own turbid lens removed and are the precision optical components implanted into human eyes. Based on materials, Intraocular lens can be divided into hard materials such as polymethylacrylic acid (PMMA) which is the polymethyl methacrylate that we often mentioned, soft materials including silica gel and hydrogel etc., and some acrylic vinegar Intraocular lens derived from PMMA [1-3].

The performance of Intraocular lens as an eye implant which attracts the greatest attention of people is its biocompatibility, which includes the cytotoxicity, blood compatibility and histocompatibility of materials [4,5]. The surface design of biomaterial has a good control action on the interface of a living body and transplantation material. Meanwhile, the chemical structure and surface features of chemical material also have great impacts on the biocompatibility, mainly the free energy on material surface, hydrophilic-hydrophobic balance, electric charge, chemical structure and functional group, relative molecular mass, scalability of polymer, morphological structure and surface roughness etc.

2.2 Artificial cornea

At present stage, some achievements have been made in the research on tissue engineering cornea. However, there are still some shortcomings in its strength and toughness and it is transparent to a certain extent, so there is certain difference between the curative effect of tissue engineering cornea and that of living cornea. For this reason, it is the current key direction of research on tissue engineering cornea to solve such issues [6].

The ideal support materials of tissue engineering cornea have the following characteristics: (1) it has good biocompatibility; (2) with good tenacity, the tissue engineering cornea has the performance similar to that of human cornea and can form a three-dimensional structure; (3) the degradation rate of support materials can match with the degradation rate of tissue and organ formed by implanted corneal cells; after offering templates to the corneal tissue, the support can be completely degraded and absorbed, or become a integration part with new tissues; it can be molded based on defects in characteristics and technologies of corneal tissue to form a curvature, so it can well perfect the morphological repair [7]. Two types of common corneal support materials are briefly described below:

(1) Fibrous Protein

The corneal epithelial cells and fibrous protein are highly compatible, so the limbal stem cells cloned on fibrous protein grow very well. The fibrous protein-based stroma supports the growth and differentiation of corneal epithelial cells. Fibrous protein can also be used in other tissues and organs, such as skin and arm muscle. Other researches show that the skin epithelial cell and arm muscle epithelial cell can grow on fibrin gel and can be transplanted successfully [8].

(2) Amniotic Membrane

With good biocompatibility, the amniotic membrane, which is a semitransparent tissue, can provide basilar membrane and extracellular matrix, meanwhile, help to enhance the epithelial differentiation and strengthen epithelial connectivity. In addition, it also has the advantages such as anti-inflammation, anti-fibrosis and antiangiogenesis. Compared with the collagen lamellar, the amniotic membrane and corneal stroma lamella with the epithelium removed have better performance and are very helpful to the growth of limbal stem cells [9]. The decellularized amniotic membrane can greatly eliminate the cells and soluble constituents in tissue, can reserve the stroma and reticulate structure and has good cytocompatibility and histocompatibility, so it can serve as tissue defect material and corneal support material for tissue engineering.

3 Biocompatibility of ophthalmic implant materials

3.1 Biocompatibility of intraocular lens

The biocompatibility of intraocular lens can be divided into the biocompatibility of uvea and that of phacocyst. There are a lot of factors influencing biocompatibility, mainly including the material characteristics and shapes of intraocular lens. Besides, there are also some preoperative risk factors, such as glaucoma, diabetes mellitus and uveitis. Scientific and proper selection of intraocular lens can effectively enhance the postoperative clinical effect. It is true that the application of intraocular **lens** also has some complications, mainly including the following types: (1) inflammatory reactions, such as intractable uveitis, infectious endophthalmitis and toxic symptom complex of intraocular lens etc. (2) turbid refractive mesenchyme (3) Intraocular lens-induced mechanical complications (4) complications induced in intraocular lens (4) corneal complications such as corneal decompensation, edema and diastasis etc. (5) corneal complications etc. (6) Iritis complications such as iris abrasion and corrosion, iris prolapsed and peripheral anterior synechia of iris etc. [10,11].

3.2 Biocompatibility of artificial lens

Among several types of biomaterials of artificial cornea, the biocompatibility of collagen and chitosan are slightly poorer. The integration between materials and host is poor; the degradation rate fails to match with the generation rate of new tissues; the materials have no surface specificity. However, fibrous protein, amniotic membrane, PGA and polylactide

PGA can serve as tissue defect materials and corneal support materials for tissue engineering and have good biocompatibility [12-14].

In these years, the optic materials in the center of artificial cornea include silica gel and polymethylmethacrylate, both of which have the common characteristics such as long-term stability, high toughness, clearness, transparency and low toxicity. However, their obvious shortcoming is that they don't support the growth of epithelial cells, go against the diffusion of nutrient substances and cannot be degraded. Hydrogel can swell in water, is elastic, transparent, highly flexible and highly elastic, can help the diffusion of nutrient substances and is very helpful to the growth of epithelial cells after treatment [15]. The artificial cornea has good and stable optical performance but its biocompatibility is poorer than that of tissue engineering cornea. New composite materials, such as collagen and hydrogel, formed by organically combining the two types of materials will probably become the new composite biological support materials.

4 Conclusion

Ophthalmic implant materials have been widely used in the treatment of all sorts of ophthalmic diseases; artificial implants made from different biomaterials are more and more popular in the treatment of ophthalmic diseases. Thus, properly selecting an implant with good biocompatibility can attain a satisfactory clinical effect.

References

1. Ehrich D, Stein A, Wilhelm F, Duncker GI. Hydrophobic AcrySof lenses or heparin-modified PMMA lenses in combined vitrectomy surgery -- results of a four-year study.Klin Monbl Augenheilkd. 2002 Nov;219(11):806-9. German.
2. Schmidbauer JM, Apple DJ, Peng Q, Arthur SN, Vargas LG.Cloudiness of a PMMA intraocular lens. "Snowflake" degeneration.Ophthalmologe. 2002 Apr;99(4):306-7. German. No abstract available.
3. Ram J, Kaushik S, Brar GS, Gupta A.Neodymium:YAG capsulotomy rates following phacoemulsification with implantation of PMMA, silicone, and acrylic intraocular lenses.Ophthalmic Surg Lasers. 2001 Sep-Oct;32(5):375-82.
4. Tomlins PJ, Sivaraj RR, Rauz S, Denniston AK, Murray PI.Long-term biocompatibility and visual outcomes of a hydrophilic acrylic intraocular lens in patients with uveitis.J Cataract Refract Surg. 2014 Apr;40(4):618-25.
5. Kao EC, McCanna DJ, Jones LW.Utilization of in vitro methods to determine the biocompatibility of intraocular lens materials.Toxicol In Vitro. 2011 Dec;25(8):1906-11.
6. Wang HY, Wei RH, Zhao SZ.Evaluation of corneal cell growth on tissue engineering materials as artificial cornea scaffolds.Int J Ophthalmol. 2013 Dec 18;6(6):873-8.
7. Liu XY, Chen J, Zhou Q, Wu J, Zhang XL, Wang L, Qin XY..In vitro tissue engineering of lamellar cornea using human amniotic epithelial cells and rabbitcornea stroma.Int J Ophthalmol. 2013 Aug 18;6(4):425-9.
8. Kimura N, Mizokami A, Oonuma T, Sasano H, Nagura H.Immunocytochemical localization of androgen receptor with polyclonal antibody in paraffin-embedded human tissues.J Histochem Cytochem. 1993 May;41(5):671-8.
9. Deihim T, Yazdanpanah G, Niknejad H.Different Light Transmittance of Placental and Reflected Regions of Human Amniotic MembraneThat Could Be Crucial for Corneal Tissue Engineering.Cornea. 2016 Jul;35(7):997-1003.
10. Xu SC, Gauthier AC, Liu J.The Application of a Contact Lens Sensor in Detecting 24-Hour Intraocular Pressure-Related Patterns.J Ophthalmol. 2016;2016:4727423.
11. Tang XJ, Yao K, Xu W, Shentu XC, Chen PQ. Study of clinical application of aberration-free aspheric intraocular lens.Zhonghua Yan Ke Za Zhi. 2010 Apr;46(4):328-31.

12. Agrawal A, Pratap VB.Amniotic membrane transplantation (AMT) without the use of sutures/fibrin glue.Nepal J Ophthalmol. 2015 Jul;7(14):173-7.
13. Toker E, Eraslan M.Recurrence After Primary Pterygium Excision: Amniotic Membrane Transplantation with FibrinGlue Versus Conjunctival Autograft with Fibrin Glue.Curr Eye Res. 2016;41(1):1-8.
14. Nguyen LP, Wang Z, Molina J, Tellez A, Chemoriya T.Complications of fibrin glue in pterygium surgery with amniotic membrane transplant.Eye Sci. 2012 Mar;27(1):19-24.
15. Builles N, Bechetoille N, Justin V, André V, Burillon C, Damour O.Variations in the characteristics of keratocytes in culture in relation to their location in humancornea.Biomed Mater Eng. 2008;18(1 Suppl):S87-98.

Experiment on the Influence Factors of Steam Distillation Rate of Crude Oil in Porous Media

Guoqing Tian[1], Wenzhuang Xu[1], Hangjuan Huang[1], Zhiti Xie[1], Junzhou Huo[1] and Yongsheng Li[2]

[1]The sixth gas production plant, Changqing Oilfield Branch, Petro China, Shanxi, China
[2]Chongqing Guotian Technology Co. Ltd. Chongqing, China

Abstract. To explore the influence of complexity of reservoir properties in porous media and the diversity of operating conditions on the steam distillation rate of crude oil in the process of heavy oil exploitation with steam injection, steam distillation simulation devices are used to study steam distillation rate of crude oil in porous media. Then steam distillation ratio is obtained under the condition of different core permeability, oil saturation, steam temperatures, system pressure, steam injection rates and steam distillation rates with different viscosities of crude oil. The results show that the steam distillation rate of crude oil in porous media depends mainly on the nature of the crude oil itself, for temperature and pressure are the key factors compared with the pore structure, the initial oil saturation and steam injection rate. The experimental results help estimate the amount of crude oil and the required steam in the reservoir in the steam drive process, aiming to facilitate the optimization design and operation of steam drive.

Keywords. Steam distillation rate, steam drive, influencing factors, porous media, experiment, heavy oil.

1 Introduction

The distillation of crude oil in porous media is one of the major mechanisms in viscous oil recovery driven by steam, which is determined by distillation characteristics of crude oil at high temperature [1, 2]. Willman, et al, believed that the recovery of steam distillation mechanism was up to 5%~19% when the temperature reached 270 ℃ [3]. Johnson pointed that the recovery range of steam distillation was 54.7 - 94.0% by volume of crude oil [4]. Farouq Ali held the view that the total steam distillation accounted for 5% to 10% of the steam drive recovery [5]. K.C. Hong estimated that for the reservoir with the crude oil viscosity of 5mPa·.s, the contribution of steam distillation to the steam drive recovery was 18% [6]. Several reports on site test items of steam drive indicate that the remaining oil saturation of the steam drive region was less than 10%, the contribution of the steam distillation to the recovery was estimated at around 10% [7-11].

Although the steam distillation mechanism is considered as an important mechanism of steam drive recovery, there are few researches on the intensified distillation in the porous medium reservoirs, letting alone the factors affecting the steam distillation rates. The intensified distillation theory and technology in the chemical process could facilitate the development of oil filed development to a degree. Obviously, different from the distillation device in the plant, it is necessary to establish the reservoir distillation theory for the porous medium reservoir based on the chemical distillation theory, thus forming the new oil recovery technology.

2 Experimental device

The simulated experimental device of steam distillation is as shown in Figure 1. The main body of the device is a distillation chamber, outside the distillation chamber is low carbon steel and the middle section is corrosion-resistant stainless steel. The length of the distillation chamber is 600mm and the diameter which is 178mm can be placed vertically and horizontally. The steam distillation chamber is divided into upper and lower parts. As for the upper part of the distillation chamber, there are baffle plate, separator and baffles; the baffle plate system is used to prevent the upward drive of oil and only the condensate can go through the buffer system, and then the separator system. In the lower part, there are no baffle plate, separator and baffles, which are used for core sand and oil filling. The steam is injected from the lower part of the distillation chamber and then the steam distillation is conducted by the core sand and crude oil. The products pass through the baffle plate and baffle system and then through the valve, condenser and back-pressure regulator to enter the gas-liquid separation device. To get a uniform temperature distribution, there are 10 electric boiling plates on the external part of the distillation chamber surrounded by the thermal insulation materials, and the temperature can be controlled automatically.

The flow rate of the steam generator ranges from 1ml/min ~ 180ml/min and can be controlled and adjusted automatically. The middle part between the steam generator and steam chamber is wrapped by the electric heater and insulating jacket to prevent heat loss. The steam temperature of the pipeline 3~5℃higher than that of the steam chamber is to prevent condensation of steam in the pipeline.

The fluid collection system consists of a condenser, back pressure regulators, gas-liquid separator, dry ice room and wet type gas meter.

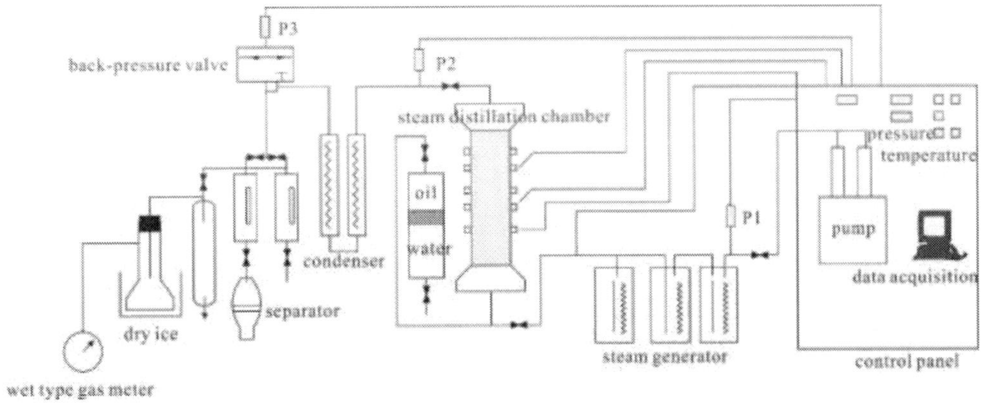

Fig. 1. Simulated device of steam injection distillation.

3 Experimental procedures

Put the oil sample in the steam distillation chamber or fill in the porous medium first and then load the crude oil. To carry out the latter method, it is necessary to determine the porosity and permeability, and the other procedures are consistent.

Clean the steam distillation chamber with cyclohexane (or xylene), inject acetone (or ethanol), and then dry it with hot nitrogen to determine the porosity and permeability.

Fill in the upper part of the distillation chamber with nitrogen after being vacuumized and pump the bottom of chamber to saturate oil and water. Then close the top and bottom valves of the chamber.

Heat the distillation chamber, pipelines and the steam generator according to the pre-set temperature, and supply water to the steam generator at a constant rate. Control the outlet pressure by the back-pressure regulator (the back pressure is slightly lower than the saturation pressure-0.02MPa under the experiment temperature). Open the top valve of the distillation chamber to release the steam and open the bottom valve, then the steam goes into the steam chamber, so the experiment of steam distillation begins.

Collect the produced fluids (oil and water) regularly and terminate the experiment when the content of oil is less than 0.4%. Calculate the amount of residue by the mass balance.

4 Experimental results

4.1 The influence of permeability on the steam distillation of crude oil

The crude oil is reserved in the porous media and it is heterogeneous in the longitudinal direction and horizontal plane. As the pore structure and connectivity of the rock may pose certain influence on the steam distillation rate of the crude oil [12], the influence of permeability on the steam distillation rate of crude oil is respectively measured at three levels at four different temperature points, and the experimental results are as shown in Table 1.

Table 1 indicates that under the constant experimental temperature and pressure, the steam distillation rate of the same oil slightly increases with the rise of permeability with no significant difference found. The maximum difference between the three measured permeability is 2.0% at the same temperature. The reason is that on the one hand, the steam distillation of steam leads to the significant decline in the interfacial tension of oil – water and oil - rock, so that some of the semi-enclosed oil is eroded, increasing the chance of drive, as shown in Figure 2; but on the other hand, the core used in the experiment is artificial. Although there are differences in permeability and porosity, the connectivity is good and there is almost no stagnant pore

DOI: 10.1051/matecconf/201710004021

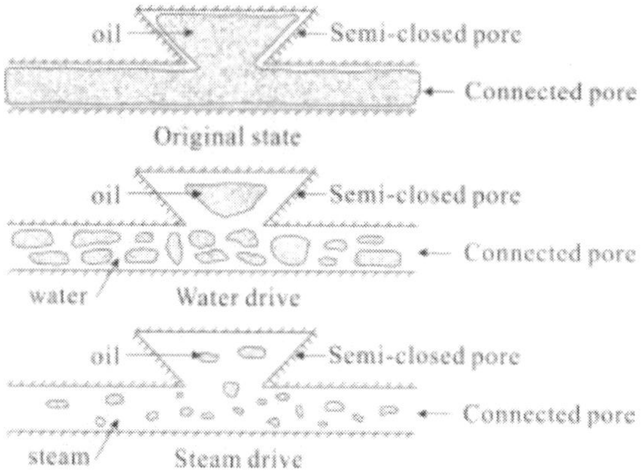

Fig. 2. The erosion effect of steam drive in water-wet system.

Table 1. The steam distillation rate of crude oil under different permeability.

NO	Permeability μm²	Temperature ℃	Pressure MPa	Vw/Voi cm³/cm³	Distill Rate %
1	0.783			1.712	11.1
2	1.56	150	0.442	1.688	12.2
3	3.15			1.715	12.4
4	0.783			2.012	13.6
5	1.56	200	1.53	1.948	14.4
6	3.15			2.021	15.3
7	0.783			3.318	18.7
8	1.56	250	3.95	3.292	19.6
9	3.15			2.324	20.7
10	0.783			3.423	23.8
11	1.56	300	8.55	3.398	24.5
12	3.15			3.421	25.6

4.2 The influence of oil saturation on the steam distillation rate of crude oil

The fluids in the reservoir typically include three phases-oil, gas and water. For the heavy oil reservoirs, there are only two phases-oil and water, because there is no gas. In the steam injection production of heavy oil, the injected steam should be used to heat the reservoir rock, crude oil and water, so the water saturation of the reservoir may influence the distillation mechanism of steam [13]. This experiment is to determine the distillation rate of crude oil under different oil saturation in the reservoir. In addition to different oil saturation of the core in each experiment, other experimental conditions are the same.

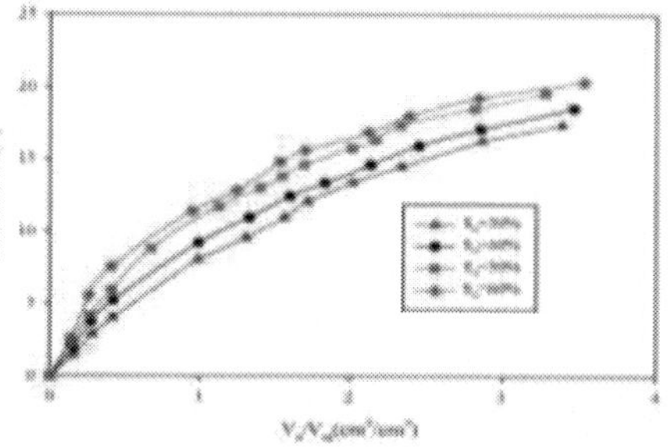

Fig. 3. The steam distillation rate of crude oil with different oil saturation ratios.

Figure 3 shows the steam distillation curve of the actual core (the permeability is 1.56μm²) with different saturation ratios when the crude oil is at 250℃. The results indicate that with the increase of injected steam, the steam distillation rate of the crude oil continues to rise, when the volume of the accumulated steam water is 3.5 times of that of the original oil (V_m/V_{oi}=3.5), the steam distillation rate of crude oil is up to about 20%. The distillation rate of crude oil in the core increases with the rise of oil saturation, the reason is that high water content in the core inhibits the vaporization of the light components of crude oil. In the core with high oil content, the small moisture content is conductive to the vaporization of light components of crude oil.

4.3 The influence of steam temperature on the steam distillation rate of crude oil

One of the important mechanisms of steam injection production of heavy oil is the effect of viscosity reduction by heating. For this mechanism, the temperature of the injected steam does not accord with the rule-the higher, the better. Therefore, the injected steam should make the formation oil become Newtonian fluid to meet the Darcy flow condition [14]. For another important mechanism-the steam distillation, the higher temperature and dryness means better steam distillation effect.

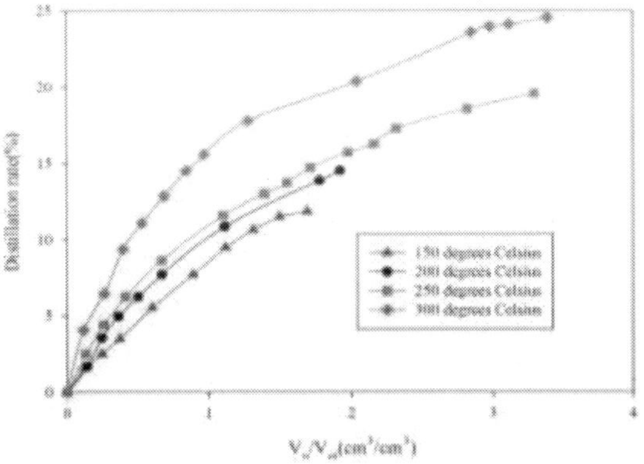

Fig. 4. The steam distillation rate of crude oil at different temperatures.

Figure 4 shows the steam distillation curves of crude oil in experimental core (the permeability is 1.56μm²) at different temperatures. It can be seen from the figure that under saturated conditions, when the temperature increases from 150 ℃ to 200 ℃, 250 ℃ and 300 ℃, the steam distillation rate of crude oil respectively rises by 2.6%, 7.8% and 12.7%. This indicates that the increase in steam temperature may gradually raise the distillation rate. For under saturated conditions, the rise in temperature may lead to the increase in enthalpy of wet saturated steam.

4.4 The influence of system pressure on the steam distillation rate of crude oil

The experiment to explore the influence of system pressure on the steam distillation rate of crude oil is to measure the changes of steam distillation rate under different system pressure when the steam is heated to 300 ℃. The saturated steam pressure is 8.583MPa when it is heated to 300 ℃, so the steam whose system pressure is less than 8.583MPa is called superheated system.

Figure 5 shows the curves of distillation characteristics of crude oil in the rock core when the pressure is respectively 1.0MPa, 2.0MPa, 4.0MPa, 6.0MPa and 8.55MPa at the temperature of 300 ℃. The experimental results indicate that at the same temperature of 300 ℃, the distillation rate of crude oil is respectively 39.4%, 36.8%, 32.4%, 27.4% and 24.5%. Compared with that under the saturated condition (8.55MPa), with the system pressure, the distillation rate respectively rises by 2.9%, 7.9%, 11.3% and 14.9%. This is because at the same temperature, with the decrease in the system pressure, the wet saturated steam is gradually converted into superheated steam whose enthalpy value is greater than that of dry saturated steam (enthalpy with 100% dryness) and far outweighs that of the wet saturated steam.

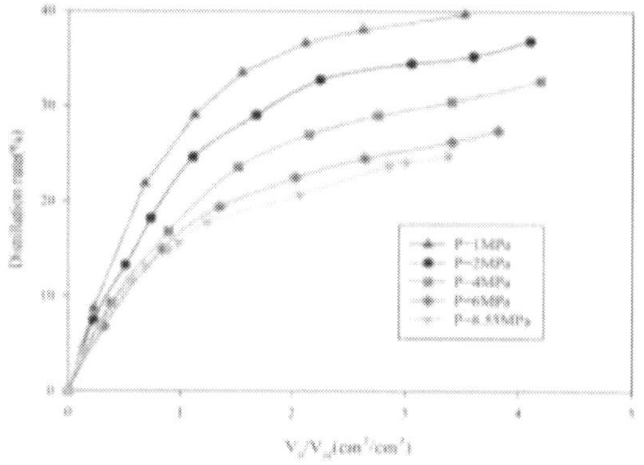

Fig. 5. The steam distillation rate of crude oil under different system pressure.

4.5 The influence of injection rate on the steam distillation rate of crude oil

In the actual operation process of the steam injection, there is an optimal steam injection rate. If conditions permit, the increase in injection rate is conducive to improving the steam dryness at the bottom and the effect of steam distillation [12]. However, with the same dryness, the influence of steam injection rate on the steam distillation rate of crude oil is not clear. This experiment is to measure the steam distillation rate of crude oil under different injection rates.

The experiment is carried out in the rock core with the temperature of 250 ℃, the system pressure of 3.95MPa, and the permeability of 1.602μm^2. In this experiment, four different injection rates are adopted, 2.0ml / min, 5.0ml / min, 10.0ml / min, and 15.0ml / min, and the results are as shown in figure 6. The maximum difference between distillation rates with these four injection rates is only 0.8%, which means that the steam injection rate has little effect on the distillation rate of crude oil.

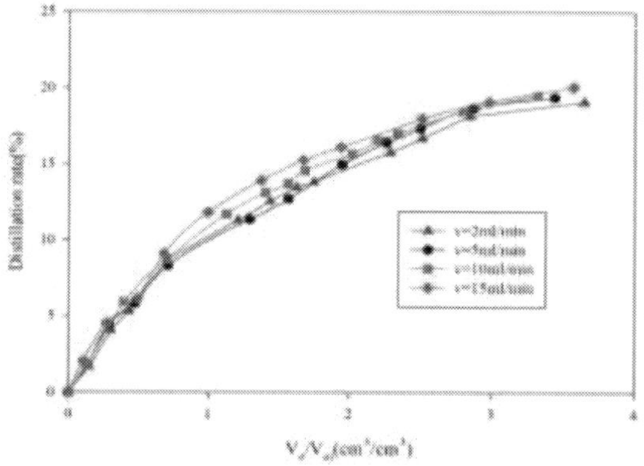

Figure 6. Steam distillation rates of crude oil with different steam injection rates.

4.6 The influence of viscosity on the steam distillation rate of crude oil

For the crude oil with different properties, the major difference is reflected in the viscosity [15]. This experiment is to explore the viscosity of representative heavy oil, aiming to analyze the relationship between the viscosity and steam distillation rate.

The experimental results are as shown in Figure 7. The steam distillation rate decreases with the increase of viscosity. For example, when the temperature is 250 ℃, the steam distillation rate of crude oil with small viscosity oil (<2500mPa.s) is more than 10%, while for the one with great viscosity (>5000mPa.s), the distillation rate is less than 10%. However, it is not reliable to find out how the distillation rate changes with the viscosity of crude oil because distillation rate is not only affected by viscosity, which is also related with other relative chemical properties of crude oil.

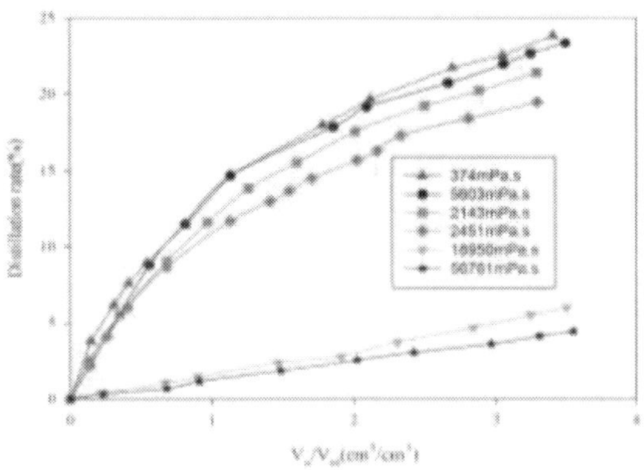

Fig. 7. The steam distillation rate of crude oil when the viscosities are different.

5 Conclusion

• The natures of crude oil are the inherent factors affecting its steam distillation, while temperature and pressure are the key external factors.
• Under the same conditions of steam injection, the distillation rate decreases with the increase of system pressure.
• Under the saturated condition, the steam distillation rate slightly increases with the rise of steam temperature. The superheated steam could greatly improve the steam distillation rate of crude oil, which is particularly evident for the heavy oil.
• The permeability, initial oil saturation and injection rate in the porous medium have subtle effects on the steam distillation rate. There is a slight increase in the steam distillation rate with the rise of the permeability and initial oil saturation, while the influence of the injection rate is almost negligible.

References

Online references will be linked to their original source, only if possible. To enable this linking extra care should be taken when preparing reference lists.

1. RICHARDSON W C, BELADI M K, WU C H. Steam Distillation Studies for the Kern River Field. SPE Reservoir Evaluation & Engineering, 2000, 3(1): 13 - 22.
2. DUERKSEN J H, HSUEH L. Steam Distillation of Crude Oils. Society of Petroleum Engineers Journal, 1983, 23(2): 265 - 71.
3. B.T.WILLMAN, V.V.VALLEROY, W.RUNBERG G, et al. Laboratory studies of oil recovery by steam injection. Journal of Petroleum Technology, 1961, 13(7): 681-90.
4. JOHNSON F S, WALKER E G, BAYAZEED A F. Oil vaporization during steamflooding . Journal of Petroleum Technology, 1971, 23(4): 731–42.
5. ALI S M F, K.D.KIMBER, V.R.PUTTAGUNTA. New Scaling Criteria And Their Relative Merits For Steam Recovery Experiments. Journal of Canadian Petroleum Technology, 1988, 27(4): 67-72.
6. K.C.HONG, J.W.AULT. Effects of Noncondensable Gas Injection on Oil Recovery by Steamflooding . Journal of Petroleum Technology, 1984, 36(12): 2160 - 70.
7. MURER A S, MCCLENNEN K L, LARSON D C, et al. Steam Injection Project in Heavy-Oil Diatomite . SPE Reservoir Evaluation & Engineering, 2000, 3(1): 2-12.
8. BYCHKOV A, BOERRIGTER P M, HEEL A P V. Steam Injection Into Fractured Carbonates - The Physical Recovery Mechanisms Analyzed and Upscaled. Abu Dhabi International Petroleum Exhibition and Conference. Abu Dhabi, UAE. 2008.
9. RIVEROS G L V, BARRIOS H. Steam Injection Experiences in Heavy and Extra-Heavy Oil Fields, Venezuela . SPE Heavy Oil Conference and Exhibition. Kuwait City, Kuwait. 2011.
10. TANG G-Q, INOUYE A, LOWRY D, et al. Recovery Mechanism of Steam Injection in Heavy Oil Carbonate Reservoir . SPE Western North American Region Meeting. Anchorage, Alaska, USA. 2011.
11. DENNEY D. Steamflooding To Enhance Recovery of a Waterflooded Light-Oil Reservoir. Journal of Petroleum Technology, 2012, 64(1): 64-6.
12. A.SHAH A, P.FISHWICK R, A.LEEKE G, et al. Experimental Optimization of Catalytic Process In Situ for Heavy-Oil and Bitumen Upgrading. Journal of Canadian Petroleum Technology, 2011, 50(11-12): 33-47.

13. DEHGHANI K, EHRLICH R. Evaluation of the Steam-Injection Process in Light-Oil Reservoirs. SPE Reservoir Evaluation & Engineering, 2001, 4(5): 395 - 405.
14. A.MOLLAEI, B.MAINI. Steam Flooding of Naturally Fractured Reservoirs: Basic Concepts and Recovery Mechanisms. Society of Petroleum Engineers Journal, 2010, 49(1): 65 - 70.
15. A.YAZDANI, MAINI B B. Measurements and Modelling of Phase Behaviour and Viscosity of a Heavy Oil/Butane System . Journal of Canadian Petroleum Technology, 2010, 49(2): 9-14.

Clinical Observation after Multifocal Intraocular Lens Implantation

Ying Wang and Feng Ke*

Department of Ophthalmology, Renmin Hospital, Hubei University of Medicine, Shiyan,Hubei, China
Corresponding E-mail: hubeiwenbo@163.com

Abstract. Objective: To evaluate the binocular vision function, validity and safety of implanted multifocal intraocular lens eye. Method: ultrasonic emulsified combined ReSTOR multifocal intraocular lens implantation surgeries were conducted for 25 eyes of 21 patients, including 17 patients with monocular surgeries and 8 patients with binocular surgeries. Then, patients' post-operation distant and near visions, far and near visual functions and stereoscopic visual function were observed; the off-glasses rate, poor vision and satisfaction [1] were invested through questionnaires. The post-operation follow-up lasted for 3 to 18 months. Result: the distant and near visions of naked eyes of patients implanted with ReSTOR intraocular lens are respectively 0.84±0.20 and 0.62±0.27 for 3 months after surgeries; all patients have good near and far stereoscopic visual functions; the satisfaction about moderate-distance vision is 84.6% (21/25); the satisfaction about near vision is 93%; the near-vision off-glasses rate is 88% (25/22). Conclusion: multifocal intraocular lens can provide good distant and near visions, far, near and stereoscopic visual functions for patients; the patients implanted such lens in both their eyes can recover the stereoscopic visual function and their vision satisfaction is higher than that of patients implanted such lens in one of their eyes.

Keywords. Multifocal intraocular lens, implantation, post-operation effect.

1 Experimental material and method

1.1 Material

A whole cluster of 21 patients with cataract and 25 eyes from January 2013 and December 2014 in the hospital were selected; the ages of patients were between 25 and 75 years old (average age: 58±14). Inclusion criteria: the patients' best corrected visual acuity is within 4.7; corneal astigmatism ≦1.5D; pre-operation vision: index~0.4; nuclear hardness: Class II-IV. Patients with other lesions in their eyes should be excluded. Preoperative examination: vision, corrected visual acuity (CVA), intraocular pressure, A ultrasound, B ultrasound, corneal endothelial cell counting, corneal curvature, corneal topography, IOL-master, OCT, ocular anterior segment photographing, and lacrimal passage irrigation [2]. Materials and devices: Infiniti Vision system ultrasonic emulsifying machine and AcrysofReSTOR SN6A0D1 multifocal intraocular lens. All patients included had no severe physical illness so there's no contraindication in surgeries.

1.2 Method

In order to reduce errors in experimental results caused by human factors during operations, all operations were conducted by one doctor. Compound tropicamide was used for sufficient mydriasis; then, Benoxil eye drops were used in the eyes with operation at intervals for three times in total; the eye drops were used once every ten minutes. The eyes with operation were disinfected paved with cloth. The operative field needed to be fully exposed. In the axial direction of high curvature of cornea, a one-off incision scalpel was used to cut a main transparent corneal incision with a width of 2.2mm and a length of about 2.0mm from the corneal limbus; then, viscoelastic agent was injected into the anterior chamber; an auxiliary corneal incision was made; the lens anterior capsule was continuously and annularly tore in the middle till the opening diameter was about 5.5mm; then, hydro dissection was conducted; the lens nucleus piece ultrasonic emulsified aspiration was made with the Infiniti Vision system ultrasonic emulsifying machine; later, the rest cortex was gettered with I/A; the viscoelastic agent was injected into the capsular bag; then, the intraocular lens was implanted and its position is adjusted so that it could be in the center; next, the viscoelastic agent was completely gettered and the incision was closed. The TobraDex eye ointment was applied and the eye with operation was covered [3].

2 Experimental result and discussion

2.1 Result

The vision for 3m after operation: the patients' average distant naked vision is 0.84±0.20 and their average near naked vision is 0.62±0.27. Their postoperative corrected distance visual acuity spherical equivalents are -0.50～+0.25, which are approximate to the pre-conceived refractive state of 0.25～+0.25D. The specific vision distribution of patients after three months is shown in Table 1.

Table 1. Vision Distribution of Patients after Three Months.

Total number of eyes	Far naked vision distribution			Near naked vision distribution		
	<05	05~0.9	≧1.0	≦0.33	0.4~0.9	≧1.0
25	1	17	7	3	18	4

All patients have good far and near stereoacuities of naked eyes as well as the near stereoacuity under the corrected near visual acuity.

Postoperative off-glasses rate for 3m after operation. According to the investigation of patients' degree of dependence on glasses through questionnaires, 22 patients (88%) totally take off their glasses.

As for postoperative long-distance vision, patients' satisfaction is 95%; as for postoperative moderate-distance vision in doing housework, vegetable washing, cutting and cooking, and walking upstairs and downstairs, the satisfaction is 84.6% (21/25); as for short-distance vision in book and newspaper reading, using mobile phone and reading labels in the supermarket, the satisfaction is 93%. This result is consistent with that found in researches and investigations at home and abroad; namely, after implantation of ReSTORMIOL, the patients' satisfaction about the moderate-distance vision is lower than the satisfactions about the far and near visions of naked eyes.

Postoperative poor vision symptoms and questionnaire survey on patients' glare for 3m after operation: most patients suffered halation in early days after operation but this didn't influence their life and work. 2 patients suffered the night glare and 1 patient felt shadows when watching the mobile phone; but all such symptoms disappeared 2 months after operation. The stereoscopic visual functions of patients with one of their eyes implanted with lens were recovered but such patients' vision satisfaction is slightly poorer than that of patients with both their eyes implanted with lens; there's no increase in the poor vision symptom.

2.2 Discussion

Although patients can completely have their far vision recovered after single-focus intraocular lens (intraocularlens, SIOL) implantation operation after the traditional cataract operation, the monofocal IOL fails to provide sufficient near vision; for this reason, patients usually need to wear low-degree spectacles for hypermetropia to watch near things and read books. The methods to solve this problem include: monovision, adjustable IOL and multifocal IOL [4,5,6]. Monovision is to implant IOL in the dominant eye for far vision and implant IOL in the non-dominant eye for near and moderate vision. However, not all people can accept mono vision due to the loss of perception of depth. The clinical application of multifocal intraocular lens (MIOL) helps patients with cataract to obtain rather satisfactory far, near and moderate visions. Such lens implanted in the eyes can provide two focuses for near and far visions so the patient's near vision ability is greatly enhanced, compared with the monofocal IOL [7]. According to the optical imaging principle, the MIOLs mainly include DMIO, RMIO and mixed type of DMIO and RMIO [8].

After cataract operation, a patient can get the regulating function similar to that of a human eye. The research in the direction of intraocular lens is always to get rid of the limitation of single-point focusing of intraocular lens, which is also a project researched for years [9, 10, 11]. At present, the application of MIOL is the best way to solve the loss of accommodation power after cataract operation [12, 13, 14]. Through its special design, the MIOL greatly improves the postoperative vision quality of patients with cataract and effectively enhances their off-glasses rate.

The application of MIOL largely improves patients' functional vision and greatly enhances their living quality, so it provides an important direction for the current ophthalmic medical technology [15, 16, 17]. However, we should also know that superficial vision is not always directly related to the patient's daily functional vision. As MIOL cannot provide the best focus at all sorts of functional distances and MIOL is expensive and has high financial conditions of patients, almost all patients pay more attention to the actual visual quality in daily living activities. Therefore, it is a very meaningful research direction for patients' postoperative visual quality to expand samples in researches, combine objective and subjective tests, and organically combine the researches on clinical and functional tests.

References

1. Hida WT, Motta AF, Nakano CT, Tzelikis PF.Inadvertent implantation of a reversed-optic Tecnis ZM900 multifocal intraocular lens: case report.Arq Bras Oftalmol. 2012 Nov-Dec;75(6):427-9.
2. Dexl AK, Zaluski S, Rasp M, Grabner G.Visual performance after bilateral implantation of a new diffractive aspheric multifocal intraocularlens with a 3.5 D addition.Eur J Ophthalmol. 2014 Jan-Feb;24(1):35-43.
3. Peng C, Zhao J, Ma L, Qu B, Sun Q, Zhang J.Optical performance after bilateral implantation of apodized aspheric diffractive multifocalintraocular lenses with +3.00-D addition power.Acta Ophthalmol. 2012 Dec;90(8):e586-93.
4. Khoramnia R, Auffarth GU, Rabsilber TM, Holzer MP.Implantation of a multifocal toric intraocular lens with a surface-embedded near segment after repeated LASIK treatments.J Cataract Refract Surg. 2012 Nov;38(11):2049-52.
5. Rabsilber TM, Kretz FT, Holzer MP, Fitting A, Sanchez MJ, Auffarth GU.Bilateral implantation of toric multifocal additive intraocular lenses in pseudophakic eyes.J Cataract Refract Surg. 2012 Aug;38(8):1495-8.
6. Agresta B, Knorz MC, Kohnen T, Donatti C, Jackson D.Distance and near visual acuity improvement after implantation of multifocal intraocular lenses in cataract patients with presbyopia: a systematic review.J Refract Surg. 2012 Jun;28(6):426-35.

7. Lin HT, Chen WR, Ding ZF, Chen W, Wu CR.Clinical evaluation of two multifocal intraocular lens implantation patterns.Int J Ophthalmol. 2012;5(1):76-83.

8. Gayton JL, Mackool RJ, Ernest PH, Seabolt RA, Dumont S.Implantation of multifocal intraocular lenses using a magnification strategy in cataractous eyes with age-related macular degeneration.J Cataract Refract Surg. 2012 Mar;38(3):415-8.

9. Solomon R, Barsam A, Voldman A, Holladay J, Bhogal M, Perry HD, Donnenfeld ED.Argon laser iridoplasty to improve visual function following multifocal intraocular lens implantation.J Refract Surg. 2012 Apr;28(4):281-3.

10. Schmickler S.Postoperative changes in refraction and anterior chamber depth in different multifocalintraocular lenses.J Refract Surg. 2008 Mar;24(3):306-8.

11. Lee ES, Lee SY, Jeong SY, Moon YS, Chin HS, Cho SJ, Oh JH.Effect of postoperative refractive error on visual acuity and patient satisfaction after implantationof the Array multifocal intraocular lens.J Cataract Refract Surg. 2005 Oct;31(10):1960-5.

12. Mesci C, Erbil H, Ozdoker L, Karakurt Y, Bilge AD.Visual acuity and contrast sensitivity function after accommodative and multifocal intraocular lens implantation.Eur J Ophthalmol. 2010 Jan-Feb;20(1):90-100.

13. Woodward MA, Randleman JB, Stulting RD.Dissatisfaction after multifocal intraocular lens implantation.J Cataract Refract Surg. 2009 Jun;35(6):992-7.

14. Kuo IC, Reviglio VE.Wavefront-guided refractive surgery after multifocal lens implantation.Curr Opin Ophthalmol. 2009 Jul;20(4):255-9.

15. McNeely RN, Pazo E, Millar Z, Richoz O, Nesbit A, Moore TC, Moore JE.Threshold limit of postoperative astigmatism for patient satisfaction after refractive lens exchange and multifocal intraocular lens implantation.J Cataract Refract Surg. 2016 Aug;42(8):1126-34.

16. Baig R, A Chaudhry T, Kukreja S, Shakil S, Ahmad K.Patients' satisfaction and spectacle independence after cataract surgery with multifocal intraocular lens implantation in a tertiary care hospital.J Pak Med Assoc. 2016 Jun;66(6):745-7.

17. Chen X, Zhao M, Shi Y, Yang L, Lu Y, Huang Z.Visual outcomes and optical quality after implantation of a diffractive multifocal toric intraocularlens.Indian J Ophthalmol. 2016 Apr;64(4):285-91.

A Model of Trusted Measurement Model

Zhili Ma[1], Zhihao Wang[2], Liang Dai[1] and Xiaoqin Zhu[1]

[1]Electric Power Research Institute, State Grid Gansu Electric Power Corporation, Lanzhou 730050, China

[2]Global Energy Interconnection Research Institute, Beijing 102209, China

Abstract. A model of Trusted Measurement supporting behavior measurement based on trusted connection architecture (TCA) with three entities and three levels is proposed, and a frame to illustrate the model is given. The model synthesizes three trusted measurement dimensions including trusted identity, trusted status and trusted behavior, satisfies the essential requirements of trusted measurement, and unified the TCA with three entities and three levels.

Keywords. Trusted measurement, trusted identity, trusted status, trusted behaviour

1 Introduction

Trusted measurement is a complicated security mechanism, which involves a wide range of factors while the measurement methods are varied. Many scholars have conducted extensive research on it. In the paper [1-6], a famous formal trusted calculation model is proposed for the first time, and this model can accurately describe the trust, but because of the complexity which makes it difficult to be widely used. In the paper [5-17], the PTM model is supported by the European IST FP6, reflects that trusted derivation and trusted evolution are strict punitive, and well reflects the trust degree changes dynamically with the change of time and context. With the good computing performance and strong scalability, yet the model can't adapt to different application scenarios. In the paper [18-20], Hassan et al proposed another trusted measurement model, which is based on vector and establishes a mathematical model with some uncertainty. Because of the introduction of trust, history, time and so on, this model makes significant progress when comparing with other models. It has dynamic adaptability, has certain defenses against malicious behavior, but lacks of corresponding risk assessment mechanism. In the paper [21-24], Sun et al proposed a trust model based on entropy. On the basis of the fact that the nature of the uncertainty of trust relationship can be displayed according to the entropy in the information theory, the model can reflect the global trust degree by the trust chain transfer and update the trust dynamically. But Sun didn't give a specific mathematical model explicitly while the definition of the behavior was ambiguous. This paper presents a unified reliability model, which synthesizes three trusted measurement dimensions including trusted identity, trusted status and trusted behavior and extends the trusted measurement mechanism at present. The model bases on identity measurement and status measurement, regards the behavior measurement as the core, and carries out the comprehensive measurement for the status and identity of access requesters by the feedback of behavior.

2 The measurement model supporting behavior measurement

The measurement model supporting behavior measurement describes as Quintuple, defined as $M = (SID, IPM, BM)$

(1) ID is the result of identity measurement for access requester, the formal description is as follows, including the identity of sender and receiver.

(2) IM is the measurement of the platform configuration, the formal description is as follows:

$IPM = \{M\_BIOS, M\_OSLoader, M\_OS, M\_APP1, ..., M\_APPn,\}$, which represent the integral measurement of BIOS, OS Loader, OS Kernal and Applications

(3) BM is the result of behavior measurement for access requester, the formal description is as follows:

$BM = (SD, Data, EV)$, in which SD indicates the identity of sender and receiver, Data is the content of the behavior, EV is the environment of the behavior. The identity measurement model

In trusted network, the identity measurement focuses on the authentication of access requesters and the verification of authorization data, and the identity authentication comprises non anonymous authentication and anonymity authentication. Then non anonymous and anonymous authentication algorithm are described separately as follows.

2.1 Non anonymous authentication

2.1.1 Platform authentication

(1) The establishment of parameters: suppose (Ga, Gb) is a pair of bilinear group, $G1$ and $G2$ are both cyclic group, the given bilinear map is $e: Ga \times Ga \rightarrow Gb$, ga and gb are the generators of Ga and Gb respectively, the given hash function is $H\{0,1\}^* \rightarrow Zp^*$ without collision, then select $s \in Zp^*$ as private key of system and $Pks = ga^S$ as public key, finally the public parameter of system is $(G1, G2, ga, gb, p, Pks, H)$.

(2) The key generation: the trusted cyber requesters select $x \in RZ_P^*$ where x is the private key of requesters, then calculate $IDPK = ga^X$ which is regarded as public key of identity for requesters, send ID of requesters and $IDPK$ to the managers of trusted network policy. After that, the managers register users according to ID and $IDPK$ $IPK = ga^X$, meanwhile, issue a certificate including ID and $IDPK$ to the requesters, then managers calculate $I = IDPK^S$.

(3) Authentication: At first the trusted cyber requesters send $IDPK$ to the managers of certified trusted network policy, then the managers select $r \in RZ_P^*$ to the requesters who calculate $T = ga^{\frac{1}{x+r}}$ and sent it to managers back. Finally, the managers determine whether the requesters are legal by checking whether $e(T, Iga^r) = e(ga, ga)$ is OK.

(4) Since the legitimacy of identity of requesters has been judged, the requesters and managers will consult with a common session key K. Then, the managers encrypt K and the public key of TMP and sent them to users. The requesters encrypt users and authorization data by using K to send to server. The managers can complete the identity authentication of the requesters by verifying their authorization data. Finally the requesters and the managers can carry on the data transmission through the session key.

2.1.2 Security analysis

The security of the algorithm can be proved through that constructs an attacker A to interact with the IPK owner C, and then pretends to be C. Construct a function F which is the attacker of K-CCA hard problem, that is, this function knows the set of data as $\{h1, h2......hk \in Z_p, g, g^x, g^{h1x}, g^{h2x} ......g^{hkx}\}$. The function F treats A as its own call subroutine, and simulates the honest IPK owners to produce the corresponding A.

2.2 Anonymous authentication

2.2.1 Platform authentication

(1) The establishment of parameters: suppose Ga, Gb is large prime number multiplication group of P order, g is the generator of G_a, the bilinear pairing is $e(G_a, G_a) \to G_b$, and $e(R, g^r PK) = I$, then the public parameters are $(G_a, G_b, g, P, e(R, g^r PK) = I)$.

(2) The key generation: the trusted cyber requesters select $x \in {}_R Z_p^*$ and compute $PK = g^x$, make PK as public key of user platform, and request certificates from the managers of certified trusted network policy, which include platform identity and public key.

(3) Authentication: At first the trusted cyber requesters send PK to the managers of certified trusted network policy, then the managers select $r_1, r_2 \in {}_R Z_p^*$ to the requesters who compute $R = g^{\frac{1}{r_2 + r_1 x}}$ and sent it to managers back. Finally, the managers determine whether the platform is legal by checking whether $e(R, g^r (PK)^{r_2}) = I$ is OK.

2.2.2 Security analysis

Authentication security depends on the security of platform identity authentication, so the following will analyze the security of the platform identity authentication process. The security of the algorithm can be proved through that constructs an attacker A to interact with the PK owner C, and then pretends to be C. Construct a function F which is the attacker of K-CCA hard problem, that is, this function knows the set of data as $\{h_1, h_2......h_k \in Z_p, l_1, l_2......l_k \in Z_p ? \neg g, g^x, g^{\frac{1}{h_1 + l_1 x}}, g^{\frac{1}{h_2 + l_2 x}} ......g^{\frac{1}{h_k + l_k x}}\}$. The function F treats A as its own call subroutine, and simulates the honest PK owners to produce the corresponding A.

Therefore in the stage of the identity authentication, firstly we should verify the legitimacy of PI, and then the trusted cyber policy manager will generate a session key K, use the PK of requesters to encrypt session key K, send the encrypted K to requesters. The requesters encrypt the user identity UI and authorization data AUD by using the session key K to managers, who will verify whether the authorization data is legal, and if so, the authentication will be passed.

3 The status measurement model

Current research on the platform status measurement including integrity measurement for TCG, based on this, this paper proposed the network connection method to measure the status of the access requesters. When an access requester successfully has passed identity measurement, he will go on the status measurement. The status measurement is conducted firstly according to security policy for trusted network management strategy, then according to the users' authorization data for real time status measurement. The status measurement process is described in detail as follows.

3.1 The status measurement for basic trusted configuration

Suppose $TPCR = \{TPCR[1], TPCR[2], ...TPCR[n]\}$ as the PCR of basic trusted configuration elements of requesters, the network service providers keep the requesters' desired configuration indicated as $PCR = \{PCR[1], PCR[2], ...PCR[n]\}$, the measure result function is defined as $MRbt$, whose description is as follows:

$$MC = \sum_{i=1}^{n} TPCR[i] \oplus PCR[i]$$

Given a MC and the threshold MC_0, if $MC > MC_0$, the measure result of requesters' basic trusted configuration conform to the security policy on trusted network server end.

3.2 The real time status measurement

The most important thing of real time measurement for network requesters is to confirm whether the network request process is malicious. Here we refer to malicious code measurement method to confirm whether the network request process is malicious. The most typical method to determine malicious code is static and the method has been widely used in all kinds of antivirus software, however, as the most serious defects, this method can't judge variant or unknown malicious code accurately.

The real time status of network requesters is the real time status of N processes of recorded access requesters, because of some shared characteristics of malicious processes such as unauthorized access of Trojan, virus self-replication and tampering with files, Worm network attacks, etc. Real time status measurement is mainly to measure the malicious process of the trusted network requesters, so the definition of the malicious degree of the network requesters' processes can confirm whether the real time status is credible. Suppose BQ indicates the malicious degree of processes, $BQ = \{bq[1], bq[2]......bq[m]\}$ indicates the malicious degree of each process, the trusted network service providers set the set of permissions.

$$MR = \sqrt{\lambda_1 bq^2[1] + ... + \lambda_m bq^2[m]}$$

Given a threshold Mt, if $MR < Mt$, the measure result of requesters' basic trusted configuration conform to the security policy on trusted network server end.

4 The behavior measurement model

The description is represented by the following quintuple $M = (SID, IPM, BM)$. This paper will introduce the relationship between identity, status and behavior, and the feedback effect of behavior on identity and state.

Suppose in the time of ΔT, if the behavior measurement of network requesters is trusted, the trusted network policy manager can improve the permission for user's access and modify the user's authorization data. The results of behavior measurement can change the user's identity and status information. Behavior feedback on identity information is mainly refers to the access permissions and authorization data information in the network access control mechanism, and the behavior is defined as feedback behavior. Besides, the behavior of requesters can be categorized as trusted and non-trusted behavior (Including threat and malicious behavior).

The behavior has a feedback effect on the identity and state of network request, such as in a given period of time ΔT, if network policy devices are trusted for the measurement on identity, status, behavior of requesters, then the trusted network policy manager can improve the permission for user's access and modify the user's authorization data. Whether it is trusted can be judged by the behavior feedback function F, which is defined as follows:

$$F(SID, IPM, BM) = True(SID, IPM, BM)$$

5 Conclusion

In view of the current state that trusted measurement mechanism lack of systematic measurement model and are based on static integrity, a multidimensional unified reliability model named UTM is proposed in this paper. And the model established unified measure relationship in trusted network through integrating trusted measure elements such as identity, status, behavior, etc. The model is a combination of static and dynamic measurement and an integration of multi-dimensional elements. With characteristics such as fine granularity, dynamic, behavioral measurement and so on, the model can provide the basis to develop a more fine-grained security policy. The model is based on the TCA with three entities and three levels, and the measurement method is easy to be integrated.

References

1. ITU-T Recommendation X.509, ISO/IEC 9594-8. Information Technology. Open systems interconnection. The Directory: Public-key and Attribute Certificate Frameworks. Draft ITU-T Rec. X.509, May, 2001: 22-44
2. Marsh, S. P., Formalizing Trust as a computational Concept. Doctoral Dissertion, University of Stirling, 1994: 7-33
3. Almenarez F, Matin A, Diaz D, Sanchez J. Developing a model for trust management in pervasive devices. In: Bob Werner, ed. Proc of the 3rd IEEE Int'l Workshop on Pervasive Computing and Communication Security. Washington, IEEE, 2006: 245-250
4. Jameel H, Hung LX, Kalim U, Asjjad A, Lee SY, Lee YK. A trust model for ubiquitous systems based on vectors of trust values. In Proc of the 7th IEEE Int'1 Symp. on M ultimedia. Washington: IEEE Computer Society Press, 2005: 674-679
5. Sun Y, Yu W, Han Z, Liu KJR. Trust modeling and evaluation in ad hoc networks. In: Proc of the Global Telecommunications Conf, Globecom 2005. Washington: IEEE Computer Society Press, 2005: 1-10
6. Li DY, Meng HJ, Shi XM. Membership clouds and membership clouds generator. Journal of Computer Research and Development, 1995, 32, 6: 15-20
7. Sailer, X Zhang, T Jaeger, L vail Doom, Design and implementation of TCG-based integrity measureme architecture. In Proceedings of USENIX Security Symposium. Lake Tahoe, California, USA: ACM Press, Aug. 2004: 223-238
8. Haldar, D Chandra, M Franz, Semantic remote attestation a virtual machine directed approach to trusted computing. In Proc of the Third virtual Machine Research a Technology Symposium. San Jose, CA, USA: USENIX, 2004: 29-41
9. Seshadri M, Luk E, Shi A, *etal.* Pioneer: verifying integrity and guaranteeing execution of code on legacy platforms. In the Proc ACM SIGOPS Operating Systems Review, SOSP'05, Brighton: ACM Press, 2005: 1-16
10. Barka, E.; Ravi Sandhu. Role-based delegation model/hierarchical roles (RBDM1). Computer Security Applications Conference, 2004. 20th Annual Digital Object Identifier: 10.1109/CSAC.2004.31 Publication Year: 2004: 396-404
11. Gang L X, Bei G. Unified Trusted Measurement Model of Trusted Network[C]// Industrial Control and Electronics Engineering (ICICEE), 2012 International Conference on. IEEE, 2012:1082-1084.
12. Yang L, Shi Y, Tang J. A trusted measurement scheme for wireless sensor networks[C]// Communication Software and Networks (ICCSN), 2015 IEEE International Conference on. IEEE, 2015.

13. Gong B, Zhang J, Xiaolie Y E, et al. A Trusted Measurement Scheme Suitable for the Clients in the Trusted Network[J]. Wireless Communication Over Zigbee for Automotive Inclination Measurement China Communications, 2014, 11(4):143-153.
14. Ning Z H, Shen C X, Zhao Y, et al. Trusted measurement model based on multitenant behaviors.[J]. Scientific World Journal, 2013, 2014:384967-384967.
15. Chunlei W, Dongxia W, Yiqi D. Towards a Unified Framework for Network Survivability Measurement[C]// Proceedings of the 2010 Second International Conference on Networks Security, Wireless Communications and Trusted Computing - Volume 01. IEEE Computer Society, 2010:129-134.
16. Xiao Y, Cao J, Lai X, et al. Trusted network connect system based on tri-element peer authentication: US, US8191113[P]. 2012.
17. Brickell E F. Method of using signatures for measurement in a trusted computing environment: US, US8631507[P]. 2014.
18. Wang X, Liao G, Wang F, et al. Study of Rapeseed Information Website Trusted Measurement Based on Fuzzy Comprehensive Evaluation[J]. Chinese Agricultural Science Bulletin, 2015.
19. Xin S Y, Zhao Y, Liao J H, et al. Dynamic trusted measurement model of operating system kernel[J]. Journal of Computer Applications, 2012, 32(4):953-945.
20. Kanstrén T, Evesti A. Security Metrics, Secure Elements, and Operational Measurement Trust in Cloud Environments[M]// Security and Trust Management. Springer International Publishing, 2015.
21. Tang Y, Zhang Y P, Chen H Y. Trusted Component Measurement Model Based on Specific Areas[J]. Computer & Modernization, 2014.
22. Challener D C, Cromer D, Locker H, et al. Local verification of trusted display based on remote server verification: US, US8205248[P]. 2012.
23. Horman N R T, Paris E L. Network access control for trusted platforms: US, US8832811[P]. 2014.
24. Ning Z H, Shen C X, Zhao Y, et al. Trusted measurement model based on multitenant behaviors.[J]. Scientific World Journal, 2013, 2014:384967-384967.

Numerical simulation of the fluid dynamics within the tissue engineering scaffolds micro-tubes

Jingzhi Jiang [1], Yuyao Jia [1], Haiting Cui [1], Peiying Peng [1] and Qinggang Liu [1]

[1] Mechanical engineering Department, Hebei University of Science and Technology,Hebei, China

Abstract. The core of tissue engineering is the fabrication of a complex three-dimensional space with cells and biomaterials. In the development of porous scaffolds in vitro, no matter the seed cells ran into the scaffold, or its excretion of waste discharged scaffolds, they both need the nutrient solution to be brought into or taken out. Therefore, the flow of the nutrient solution, cells and metabolic waste for the profitable living in vitro culture plays a significant role. the paper constructed bone scaffold models of different geometric parameters, analog cells, and nutrient solution flow conditions in the scaffolds using the FLUENT software. By making a contrastive analysis with the simulated result, the internal organizational design parameters of scaffold for tissue engineering is optimized and a certain amount of data and a theoretical basis for the internal bone scaffold structure design is provided.
***Keywords*:** Tissue engineering; Scaffold models; Operating parameters; Velocity distribution

1 Introduction

From the source map of blood supply for bone, the outer of the Haversian canal has multi-layer bony plates those are centered around the center, arranging in concentric circles, which constitute bone unit with Haversian canal, known as the Harvard system. Harvard system connects with the Volkmann canals compose mainly channel for transport nutrients, which provides a good space and access for the seed cell proliferation, scaffolds in growth as well as the exchange of nutrients and metabolites discharge.

Based on the design and development of artificial bone, in view of the ratio of the actual bony structure, the model of bony scaffold was established and the fluid dynamics of cell and nutrient solution within the scaffolds are stimulated.

2 Model establishing

Nutrient solution and bone cells are transported to bone scaffolds by Volkmann canals. The fluid flow in the bone scaffold is a mixture of nutrient solution and bone cells. The density of nutrient solution used is $1100kg/m^3$, viscosity is 0.0018Pa.s. The density of bone cell used is $1200kg/m^3$ and volume fraction is 0.02 which ratio of the mixture is small. The average velocity of the nutrient solution flow is 0.19m/s. Based on above, the density of the mixture of nutrient solution and bone cell is determined as $1100kg/m^3$, the viscosity is 0.0018Pa·s, and the initial flow rate is set to 0.19m/s[1-3]. Because of the flow rate of the nutrient solution and the cell is small, and its viscosity is large, according to the Reynolds number formula, the Reynolds number is calculated as 244, which is much less than 2000. That means the flow is an incompressible laminar flow. The laminar flow model calculation is adopted[4].

3 Simulation results and analysis

3.1. The effect of Haversian canal length

Fig.1 is the flow velocity and pressure distribution of the mixture of the nutrient solution and cells in the scaffold micro-tube. The intersection angle of Harvard's and Volkmann canal is 45°. The diameter of Harvard's canal is 0.6mm and the lengths are respectively 1.5mm, 2mm and 3mm. The figures of velocity distribution show that the flow rate decreases obviously when the mixture solution flow from the first layer of Haversian canal to the Volkmann canals. In the second layer of Haversian canal, the flow rate of the mixture solution in the central three Haversian canals increased, but lower than the rate first layer of Haversian canal. With the increasing of the Haversian canal length, the fluid flow rates of each tube will be decreased.

 DOI: 10.1051/matecconf/201710004024

When the fluid flows to the Volkmann canals, the flow is influenced by the local resistance generated the bifurcation canal, which makes the rate of the Volkmann canals reduce. In the central three Haversian canals, fluids from two Volkmann canals converge there, flow increases and results in the flow rate among increases.

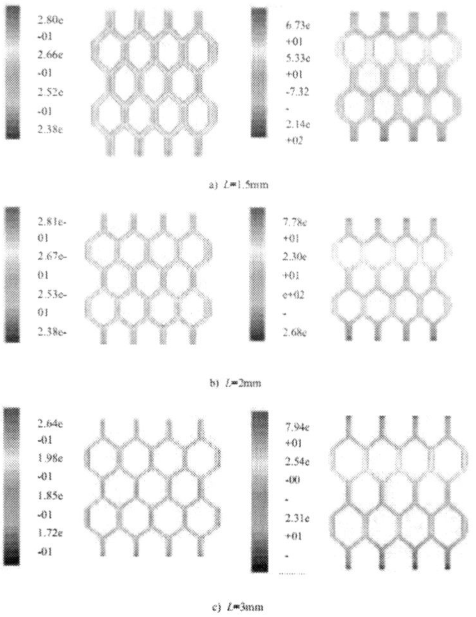

Fig. 1. Velocity and pressure distribution in different lengths of Harvard canal

3.2 The effect of Haversian canal diameter

Fig 2 is the flow velocity and pressure distribution of the mixture of the nutrient solution and cells within the scaffold. The intersection angle of Harvard's and Volkmann canal is 45°.The length of the Haversian tube is 1.5mm, and the diameter is respectively 0.4mm, 0.6mm and 0.8mm. Comparing to the velocity

distribution of Figure 1 a)and Figure 2 a), b), when Harvard's canals diameter is 0.4mm and 0.6mm, the velocity of the Volkmann canal and Haversian canal are greatly different which shows an adverse impact on the deposition of cell in

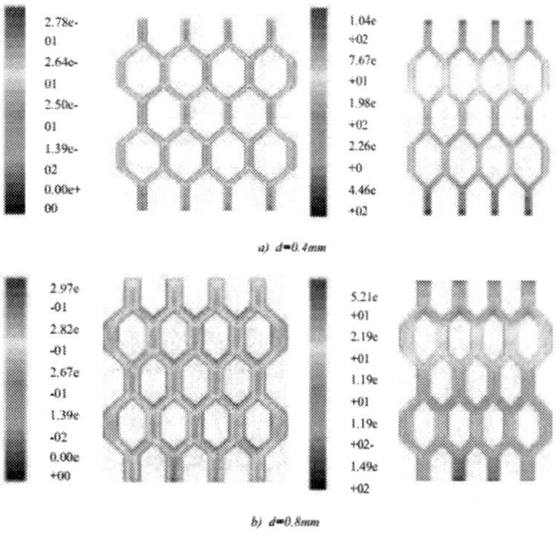

Fig. 2. Velocity and pressure distribution in different diameter of Harvard canal

micro-tubes. In the bone scaffold of Haversian canal with the diameter of 0.8mm, fluid flows within each micro-tubes are uniform and the average flow rate reaches around 0.2m/s, which is beneficial to the transport of cells and nutrient solution.

When Haversian canal diameter decrease, the entrance resistance increase, the loss of local energy increases, the velocity of the fluid in the Volkmann canal is reduced quickly. Therefore, the fluid of the first layer in the Volkmann canal has a larger flow rate, when the fluid in the Volkmann canals confluence into Haversian canal, due to Haversian canal's diameter is bigger than the Volkmann canal, the flow rate will increase after the fluids confluence, but the increase is not obvious. The situation in other layers is similar to the first layer. Within the range of this simulation, Haversian canal diameter is larger, the fluid flow rate of in each micro-tube bone scaffold is relatively uniform, and the average flow will reach to about 0.2m/s.

3.3 The effect of the angle between Haversian and the Volkmann canal

Fig2 b)and Fig3 a), b) (the left is velocity distribution) are the flow velocity and pressure distribution of the

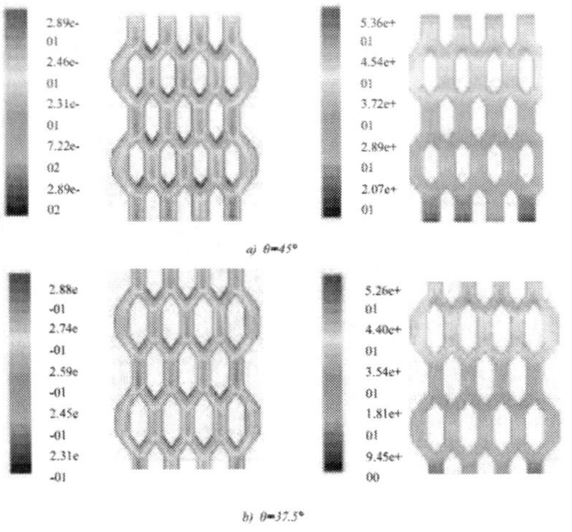

Fig. 3. Velocity and pressure distribution in different angle between Harvard's tube and Volkmann tube

mixture of nutrient solution and cells in the scaffold. The length of the Haversian canal is 1.5mm, the diameter is 0.8mm and the angle between Harvard's and the Volkmann canal are respectively 30°, 37.5° and 45°[5-6]. When the angle is 30°, the upper of the Volkmann canal fluid flow rate is greater, which is 2.17e-01 m/s, but the flow rate in the tube is not uniform.

With the angles become bigger, the flow rate of fluid in the Volkmann canals is gradually reduced and the fluid flow within the tubes is uniform. When the angles are 37.5 °and 45 °, the flow rate of the upper of the Volkmann canal are 1.93e-01m/s and 1.88e-01m/ s respectively; meanwhile, with the angle increase, the fluid rate of each micro-tube in bone scaffold are more uniform.

When the angle is large, most of the fluid in the Haversian canal is shunt by the Volkmann canal. When the fluid flows into the bottom of Harvard canal, the volume and velocity of fluid decrease. Whether the fluid flow from the Haversian canal into the Volkmann canals or on the contrary, the flow of fluid are uniform, but the uniform flow area is smaller than the area of 45°[7].

4 Conclusion

With the Haversian canal length decreasing, the fluid flow rate of each of the microtube are slightly reduced, the flow rate in each the mirotube is uniform, which is conducive to the delivery of nutrient solution and cells. With Haversian canal diameter increasing, the fluid flow rate of in the each pipeline bone scaffold is relatively uniform, and the average flow will reach to about 0.2m/s. With The angle between the Harvard's tube and the Volkmann canal increasing, it is easier to transport the cells and the nutrient solution and more suitable for tissue engineering scaffolds. Within the range of this simulation, the optimized structure parameter of bone scaffold is the length of Haversian canal of 0.15mm, Haversian canal diameter of 0.6mm, the angle between the Harvard's tube and the Volkmann canal of 45°.

References

1. C. PHILIPPE, D. STEPHANE, PAURELIEN, et al. Chapter 28-Blood Rheology, Blood Flow and Human Health. Nutrition and Enhanced Sports Performance, 2013: 283-293
2. Qin T J. Hemorheological Beijing: People's Medical Publishing House, 1999
3. Philippe Connes, Stephane Dufour, Aurelien Pichon, et al. Nutrition and Enhanced Sports Performance. Amstardam:Elsevier, 2013:283-293
4. Liu H L,Hoff B H,Anthonsen T. Chemo-enzymatic synthesis of the antidepressant duloxetine and its enantiomer . Chirality., 2000, 12 (1):26-29
5. Emerson dos Reis, Leonardo Goldstein Jr. Fluid dynamics of horizontal air-water slug flows through a dividing T-junction. International Journal of Multiphase, 2013,50:58-70
6. Tao Z L. Biological Fluid Mechanics [M] Beijing: Science Press, 1984
7. Huang W X, Chen W M Engineering Fluid Mechanics Beijing: Chemical Industry Press, 2001

Model-based Adjustment of Droplet Characteristic for 3D Electronic Printing

Na Lin[1a], Shikai Jing[2b], Hanning Chen[3c] and Wenxin Liu[4d]

[1]Beijing Shenzhou Aerospace Software Technology Co. Ltd., Beijing, China
[2]School of Mechanical Engineering, Beijing Institute of Technology, Beijing, China
[3] School of Computer Science and Software, Tianjin Polytechnic University, Tianjin, China
[4] Department of Electrical and Computer Engineering, Lehigh University, Bethlehem, USA.
[a]linn@htrdc.com, [b]jingshikai@126.com, [c]perfect_chn@hotmail.com, [d]wliu@lehigh.edu

Abstract. The major challenge in 3D electronic printing is the print resolution and accuracy. In this paper, a typical mode - lumped element modeling method (LEM) - is adopted to simulate the droplet jetting characteristic. This modeling method can quickly get the droplet velocity and volume with a high accuracy. Experimental results show that LEM has a simpler structure with the sufficient simulation and prediction accuracy.

Keywords. Model-based adjustment, Droplet characteristic, 3D electronic printing

1 Introduction

3D electronic printing has been widely applied on manufacturing the low-cost electronic products, such as flexible electronics, radio frequency identification (RFID) tags, solar cells, organic light emitting diode (OLED) flexible packaging electronic devices, such as medical equipment [1].A classical implementation for 3D electronic printing is to use drop-on-demand (DoD) piezoelectric (PZT) inkjet printhead to deposits conductive, dielectric, and non-conductive ink onto various substrates [2].

A typical DoD piezoelectric inkjet printhead for 3D electronic printing comprises a number of ink channels that are arranged in parallel [3]. A piezoelectric actuator covers all ink channels, which can generate pressure oscillations inside the ink channel to push the ink droplet out of the nozzle. Nowadays the developments of DoD printing are moving towards higher productivity and quality. Therefore, the adjustable small droplet sizes fired at high jetting frequencies must be implemented.

The printing quality of DoD piezoelectric inkjet printhead is reflected in many respects, such as the jetting direction, the droplet velocity, the droplet volume and so on. In practice applications, to meet the various requirements of 3D electronic printing and get a higher inkjet performance, the droplet characteristics of the ink have to be tightly controlled. For improving the droplet jetting performance, several operational issues, such as the residual vibrations and cross talk, should be taken a care.

To solve such problem, researchers have proposed many methods. One of most typical methods is using numerical techniques to model the DoD piezoelectric inkjet printhead [4], which focuses on the liquid filament volition of droplet formation process with known pressure input boundary conditions after solving the nonlinear Navier-Stokes equations. This method provides directly visualized information about the acoustic pressure wave travelling inside the channel [5].Another typical method is using analytical modeling method with a higher simulation speed. This method can describe the ink channel dynamics, although the accuracy of analytical models is lower than numerical models. Based on several assumptions and simplifications, analytical models can provide a simple and timesaving way to model-based control droplet generator with a sufficient accuracy [6].

In this work, we adopted a typical analytical model - lumped element modeling method - to simulate the droplet formation process. The advantage of this method is its simpler structure with the sufficient simulation accuracy. After setting a series of desired droplet characteristics, a grid search method is applied to determine the parameters of printhead driving waveform.

The rest of this paper is organized as follows: In Section 2, a review of original lumped element modeling droplet generator is present. The simulation results of droplet volume and velocity decide the most suitable model in Section 3. The prediction process of droplet characteristics is built in Section 4. Finally, concluding remarks are collected in Section 5.

2 Lumped Element Modeling

A classical LEM is given to simulate jetting characteristics of PZT printhead, as shown in Figure 1. The equivalent circuit model is constructed by the energy storage elements and ideal dissipative terminal. In this electroacoustic system, pressure and voltage are independent variables, while current and volumetric flow rate are dependent variables. Model structure shows that the energy converts from electrical energy to mechanical energy, then to fluidic/acoustic energy, and finally to kinetic energy, as described in Figure 1(b). The droplet generator structure can be characterized by equivalent acoustic mass (representing stored kinetic energy) and acoustic compliance (representing stored potential energy), in which the corresponding equivalent circuit models are supported by various fluid mechanisms, as represented in Fig. 1(a) and 1(c). Furthermore, the piezo-ceramic model is constructed based on the electro-fluidic/acoustic theory [7]. The neck model is built on the velocity profile function [8]. The nozzle model is constructed according to the end correction for an open tube theory [9].

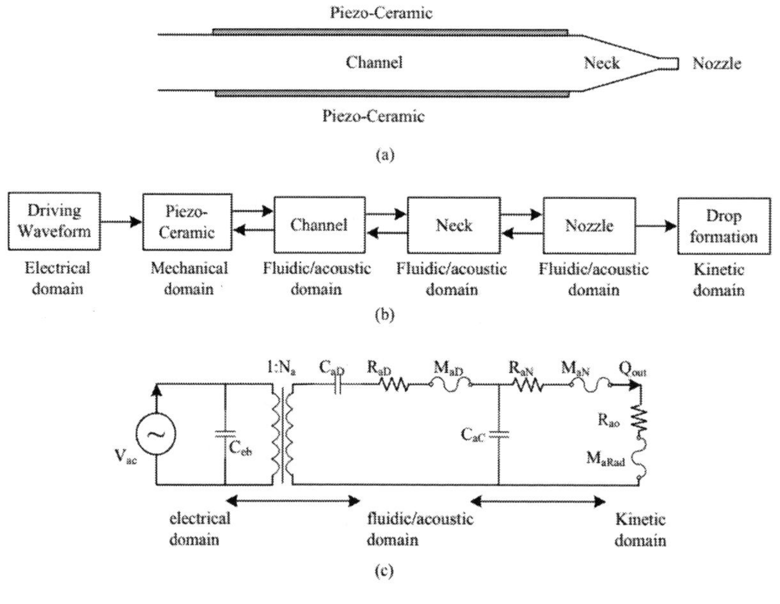

Fig.1. Schematic overview of lumped-element modeling for PZT printhead.

(a) Droplet generator structure (b) Model structure (c) Equivalent circuit model

As the equivalent circuit model shown in Figure 1, excitation voltage V_{ac} is applied to piezoelectric ceramic to create mechanical deformation. Coupling coefficient ϕ_a representsa conversion from mechanical domain to acoustic domain. C_{eb} is the blocked electrical capacitance of the piezoelectric material. In the acoustic domain, C_{aD} and C_{aC} represent the acoustic compliance of piezo-ceramic and channel. R_{aD},R_{aN}, and R_{aO} are the acoustic resistance due to structural damping, neck tapering, and fluid flowing out of nozzle, respectively. M_{aD}, M_{aN}, and M_{aRad} represent the acoustic mass of piezoceramic, neck, and nozzle in proper order, respectively.

The dimensions of droplet generator and physical properties of nano-silver ink provided by the LEED-PV Corporationare listed in Table 1, and calculation formulas of LEM modelparameters are provided in Eq. 1 -9.

Table 1. Droplet generator parameters

	Parameter	Unit	Value
Printhead	Channel volume, V_0	*pL*	166000
	Channel length, L_0	*um*	7200
	Nozzle radius, a_0	*um*	13.5
	Neck length, L	*um*	45
	Piezoceramic thickness, h	*um*	25
	Free electrical capacitance, C_{eF}	*nF*	2.8
Ink	Free electrical capacitance, σ	*N/m*	$45*10^{-3}$
	Viscosity,u	*Kg/(m*s)*	15
	Density, ρ_0	*Kg/m³*	$1.05*10^3$
	Sonic speed, C_0	*m/s*	1400

$$R_{aN} = (2+\frac{a_0}{r})\frac{\sqrt{2\mu\rho_0\omega}}{\pi r^2} \tag{1}$$

$$M_{aN} = \frac{\rho_0(t+\Delta t)}{\pi r^2} \tag{2}$$

$$\Delta t = 0.85r(1-0.7\frac{r}{a_0})+0.85r \tag{3}$$

$$C_{aD} = \frac{\pi r^6(1-v^2)}{16Eh^3} \tag{4}$$

$$M_{aD} = 4\rho_0 L/3\pi a_0^2 \tag{5}$$

$$R_{aD} = 2\xi\sqrt{(M_{aD}+M_{aRad})/C_{aD}} \tag{6}$$

$$C_{eB} = C_{eF}(1-\kappa^2) \tag{7}$$

$$R_{aO} = \frac{\rho_0 c_0}{\pi r^2}[1-\frac{2J_1(2kr)}{2kr}] \tag{8}$$

$$M_{aRad} = 8\rho_0/3\pi^2 a_0 \tag{9}$$

where u is viscosity, w is wave frequency, and a_0/r is gradients ratio, V_0 is volume of the cavity, E is the elastic modulus, v is Poisson's ratio, and h is the thickness, ξ is the experimentally determined damping factor [8], J_1 is the Bessel function of the first kind, and $k = w/c_0$, κ^2 is electroacoustic oupling factor.

3 Simulation

For industry applications in printable electronics fabrication, the printhead must work at a certain status to meet many restrictive conditions. Among them, the most important problem is choosing the appropriate combinations of the driving waveform parameters for the used conductive inks. However, an exhausting manual selective process inevitably wastes a lot of time. Therefore, the computer-aided methods are urgently needed for searching these appropriate combinations efficiently and robustly.

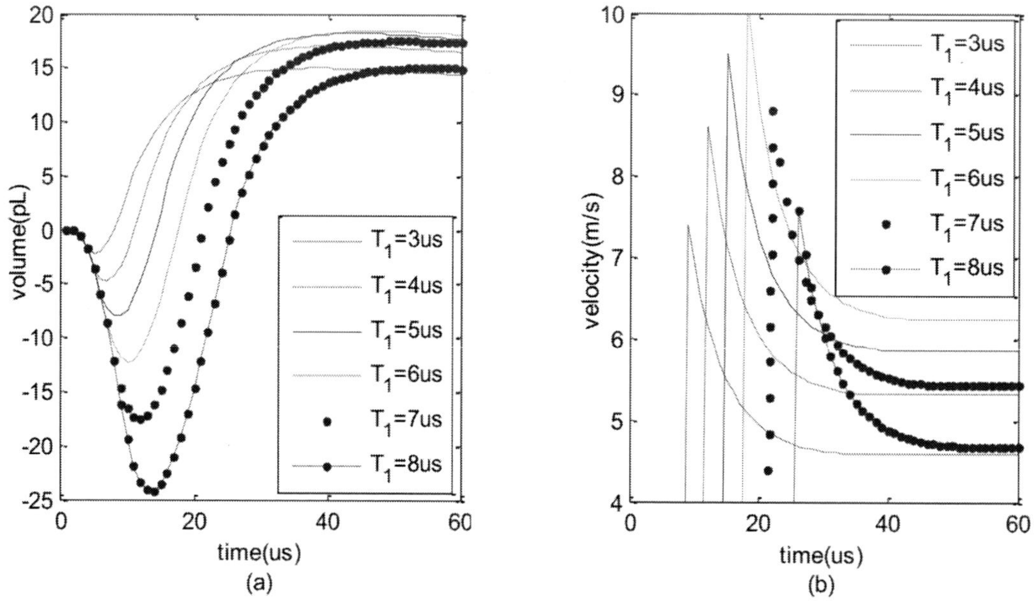

Fig.2. Predictive droplet volume and velocity with different with various dwell time T_2=2*T_1

(a) Simulation curves of outflow change (b) Simulation curves of droplet average velocity

As mentioned above, the LEM model is chosen to simulate jetting characteristics for the nano-silver ink. Here, an illustrative example is given to display the effect of predicting the typical combinations of waveform parameters with different dwell times (T_1 = 3~8us, T_2 = 2*T_1) and the same amplitude (12V). The properties of nano-silver ink provided by the LEED-PV Corporation have been listed in Table 1. In Figure 2, the predicted outflow and average velocity curves with various dwell times are depicted. We can observe from Fig. 2(a) that, before the droplets rush out of the nozzle, the depth of the inhaled meniscus is proportional to dwell time. From Figure 2(b), the time of the droplet appearing is postponed with the increase of dwell time. That is, both two simulation phenomena consist with the theoretical analysis results in [10].

Table 2. Predictive droplet volume (*pL*) and velocity (*m/s*)

T_2 \ T_1	7	7.5	8.5	9	9.5	10	10.5	11	11.5	12	12.5	13

4	V	11.9	12.7	14.3	14.6	14.9	15.1	15	14.	14.	Defaul	Defaul	Defaul
	S	3.95	4.17	4.78	4.91	5.00	4.89	4.7	4.5	4.3	Defaul	Defaul	Defaul
4.	V	11.7	12.5	15	15.5	15.9	16.2	16	15.	15.	14.7	13.9	Defaul
5	S	4.21	4.51	5.16	5.32	5.48	5.60	5.4	5.2	5.0	4.82	4.66	Defaul
5	V	Defaul	Defaul	15.1	15.7	16.3	16.7	16.	16.	16.	15.7	15	14.2
	S	Defaul	Defaul	5.31	5.53	5.72	5.87	6.0	5.8	5.6	5.40	5.25	5.09
5.	V	Defaul	Defaul	15.1	15.9	16.5	16.8	16.	17.	16.	16.5	16	15.4
5	S	Defaul	Defaul	5.24	5.50	5.72	5.93	6.0	6.2	6.0	5.71	5.51	5.37
6	V	Defaul	Defaul	Defaul	Defaul	15.8	16.3	16.	16.	16.	16.5	15.8	14.95
	S	Defaul	Defaul	Defaul	Defaul	4.95	5.15	5.3	5.4	5.6	5.44	5.19	4.94
6.	V	Defaul	Defaul	Defaul	Defaul	Defaul	Defaul	15.	15.	16	15.5	15.0	14.3
5	S	Defaul	Defaul	Defaul	Defaul	Defaul	Defaul	4.7	4.8	4.9	5.03	4.90	4.69

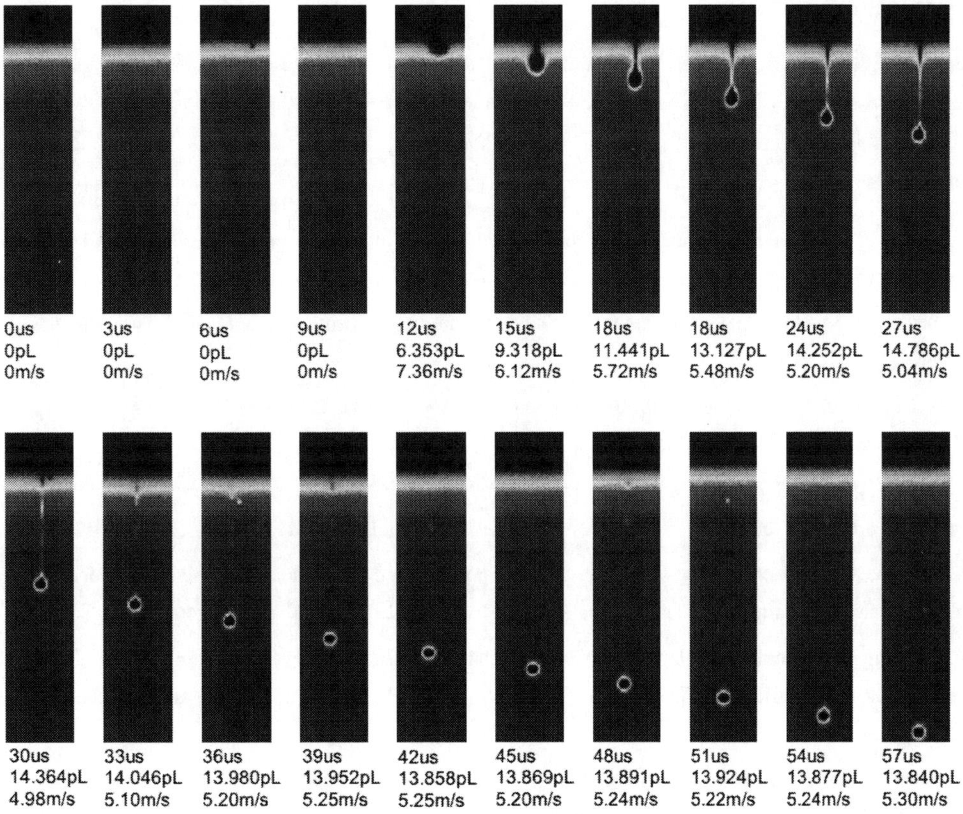

Fig. 3. A sequence of pictures of droplet falling from nozzle

4 Prediction

In this experiment, an illustrative example under desired conditions is presented to demonstrate the feasibility and effectiveness of the proposed method for predicting jetting characteristics. The dimensions of droplet generatorand properties of the nano-silver ink are listed in Table 1. In order to reach a high printing resolution and good quality of droplet impact with the substrate, the desired conditionsinvolve that the droplet volume is smaller than 14pL and the droplet velocity is larger than 5 m/s.

Based on the prediction procedure, the predictive valuesof drop volume/velocity with various dwell times T_1 and T_2 are listed in Table 2. According to the desired conditions, two combinations of parameters are found out: T_1 = 5us/T_2 = 13us, and T_1 = 5us/T_2 = 7.5~8 us. After a small range search around predictive values in the actual test, the combination of T_1 =4.9 us/T_2 = 7.7 us is finally chosen. The dynamic effect of jetting characteristics driven by this combination is shown in Figure 2. From this figure, the jetting characteristics satisfy the specified desired conditions quite well. Meanwhile, almost no satellite droplet emerges after jetting the main droplet.

5 Conclusion

The lumped element modeling based on linear model of piezoelectric ceramic has been successfully used to simulate piezo-electric droplet generator for 3D electronic printing. In this work, the droplet volume/velocity curves simulated can carry sufficient information to distinguish nanosilver droplet formation process. LEM model is also successful in predicting jetting characteristics under desired conditions.

Acknowledgements

This research is partially supported by National Natural Science Foundation of China (Grand Nos. 51575158, and 61305082), and a self-planned task of CAS Key Laboratory of Networked Control System (Grand No. WLHKZ2015001).

References

1. J. Hwang, A. Wan, and A. Kahn, Energetics of metal-organic inter faces: new experiments and assessment of the field, MaterialsScience and Engineering R: Reports, **64**, 1-2 (2009), p. 1-31.
2. M. Singh, H. M. Haverinen, P. Dhagat, and G. E. Jabbour, Inkjet printing-process and its applications, Advanced Materials, **22**, 6 (2010), p. 673-685.
3. J.-C. Liou and F.-G. Tseng, Multi-dimensional data registration CMOS/MEMS integrated inkjet printhead, Microelectronic Engineering, **88**, 6 (2011), p. 888-901.
4. J. Liu, S.-H. Tan, Y. F. Yap, M. Y. Ng, and N.-T. Nguyen, Numerical and experimental investigations of the formation process of ferro fluid droplets, Microfluidics and Nanofluidics, **11**, 2 (2011), p. 177-187.
5. F. Sarrazin, K. Loubi`ere, L. Prat, C. Gourdon, T. Bonometti, and J. Magnaudet, Experimental and numerical study of droplet shydrodynamics in MicroChannel, AIChE Journal, **52**, 12 (2006), p. 4061-4070.
6. X. Q. Xing,D. L. Butler, S. H. Ng, Z.Wang, S. Danyluk, and C. Yang, Simulation of droplet formation and coalescence using lattice Boltzmann-based single-phase model, Journal of Colloid and Interface Science, **311**, 2 (2007), p. 609-618.
7. S. Prasad, S. Horowitz, Q. Gallas, B. Sankar, L. Cattafesta, and M. Sheplak, Two-port electroacoustic model of an axisymmetric piezoelectric composite plate, in Proceedings of the 43rd AIAA/ASME/ASCE/AHS Structures, Structural Dynamics, and Materials Conference, AIAA Paper 2002-1365, AIAA, Denver, Colo, USA, April 2002.
8. D. T. Blackstock, in: *Fundamentals of PhysicalAcoustics*, John Wiley & Sons, New York, NY, USA (2000).
9. F. M. White, in: *Fluid Mechanics*, McGraw-Hill, New York, NY, USA (1979).
10. H. Wijshoff, The dynamics of the piezo inkjet printheadoperation, Physics Reports, 491, 4-5 (2010), pp. 77-177.

Intelligent Image Segment for Material Composition Detection

Xiaodan Liang[1a], Na Lin[2b], Hanning Chen[1c] and Wenxin Liu[3d]

[1]Tianjin Polytechnic University, Tianjin, 300387, China

[2]Beijing Shenzhou Aerospace Software Technology Co. Ltd., Beijing, China

[3] Department of Electrical and Computer Engineering, Lehigh University, Bethlehem, USA.

[a]lxdtjpu@163.com, [b]linn@htrdc.com, [c]perfect_chn@hotmail.com, [d]wliu@lehigh.edu

Abstract. In the process of material composition detection, the image analysis is an inevitable problem. Multilevel thresholding based OTSU method is one of the most popular image segmentation techniques. How, with the increase of the number of thresholds, the computing time increases exponentially. To overcome this problem, this paper proposed an artificial bee colony algorithm with a two-level topology. This improved artificial bee colony algorithm can quickly find out the suitable thresholds and nearly no trap into local optimal. The test results confirm it good performance.

Keywords. Image segment, artificial bee colony, material composition detection

1 Introduction

The target of image segmentation is to extract meaningful objects from the image. Segmentation is having major importance and elementary place in image processing for interpretation of any image. It is useful in discrimination an objective from some objectives or background.

In recent years, researchers proposed many methods for image segmentation. Among the existing segmentation techniques, multilevel threshold is a simple but effective tool to isolate objects of interest from the background, which required multiple threshold values to accomplish segmentation [1]. Several such methods have originally been developed for bi-level threshold and later extended to multilevel threshold described in [2]. However, all of these methods have a common problem that the computational complexity rises exponentially when extended to multilevel threshold due to the exhaustive search employed, which limits the multilevel threshold applications.

Recently, to address this concerned issue, several swarm intelligence (SI) algorithms as the powerful optimization tools have been introduced to the field of image segmentation owing to their predominant abilities of coping with complex nonlinear optimizations [3]. Among them, artificial bee colony algorithm (ABC) is one of the most popular members of the SI family, which simulates the social foraging behavior of a honeybee swarm [4]. Due to its simple arithmetic and good robustness, the ABC algorithm has been widely used in solving many numerical optimizations and engineering optimization problems. However, when solving complex multimodal problems, ABC algorithm suffers from the following drawbacks: (1) It is easily trapped into a local minimums in the early generations, which leads to low population diversity in successive generations. (2) With the dimension increasing, the information exchange of each individual is limited in a random dimension, resulting in a slow convergence rate. (3) Due to the random selection of the neighbor bee and dimensions, food sources with higher fitness are not utilized, which influences the ability of global search.

In this paper, we propose an improved artificial bee colony algorithm with a two-level topology controlling the information exchanging. The experimental results confirm it's a good performance.

The rest of the paper is organized as follows. In Section 2 the proposed hierarchical artificial bee colony is given. Section 3 presents the experimental studies of the improved artificial bee colony algorithm and the other algorithms with descriptions of the involved benchmark functions, experimental settings, and experimental results. And its application to image segmentation has been presented in Section 4. Finally, Section 5 outlines the conclusion.

2 Improved Artificial Bee Colony Algorithm

2.1 Canonical artificial bee colony algorithm

The artificial bee colony (ABC) algorithm, proposed by Karaboga in 2005 [4] and further developed by Basturk and Akay et al. [5, 6] for real-parameter optimization, which simulates the intelligent foraging behavior of a honey bee swarm, is one of the most recently introduced swarm-based optimization techniques.

The entire bee-colony contains three groups of bees: employed bees, onlookers and scouts. Employed bees explore the specific food sources; meanwhile pass the food information to onlooker bees. The number of employed bees is equal to that of food sources, in other words, each food source owns only one employed bee. Then onlooker bees choose good food sources based on the received information, and then further exploit the food near their selected food source. The food source with higher quality would have a larger opportunity to be selected by onlookers. There is a control parameter called "*limit*" in the canonical ABC algorithm. If a food source is not improved anymore when *limit* is exceeded, it is assumed to be abandoned by its employed bee and the employed bee associated with that food source becomes a scout to search for a new food source randomly.

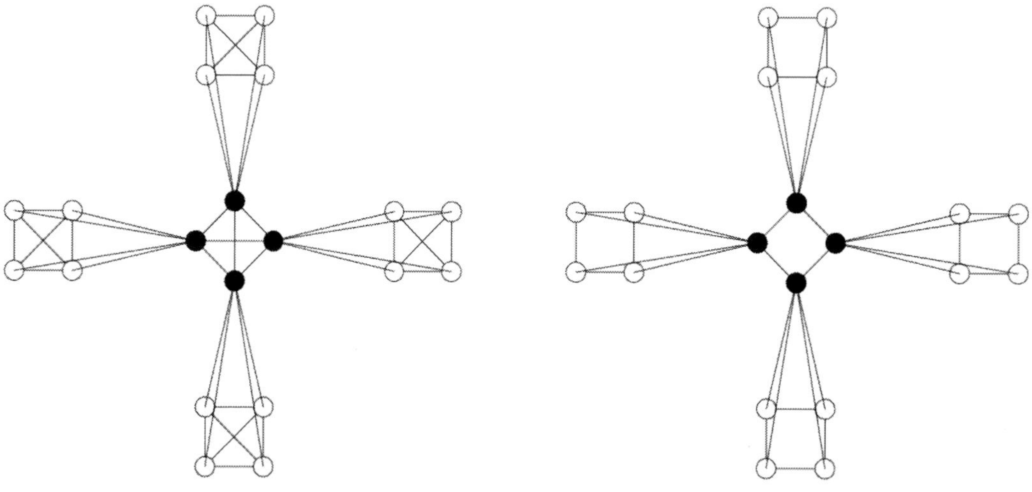

(a) (b)

Fig. 1. Two-level interaction topologies

(a) Hierarchical star topological structure (b) Hierarchical ring topological structure

2.2 Hierarchical Artificial Bee Colony Algorithm with a Two-level Topology (HABC)

In our model, the cooperation occurred in two levels, i.e., species level (interaction between species) and individual level(interaction within species). Two hierarchical interaction topologies have been employed in this paper to realize this two-levelcooperative mechanism. In the first topology (namely the 2-level star topology, shown as in Figure 1(a)), each individualis influenced by the performance of its own species and all the other species in the ecosystem. In the second topology(namely the 2-level ring topology, shown as in Figure 1(b)), each individual is influenced only by *n* closest neighbours from itsown species and other *n* species from the ecosystem.

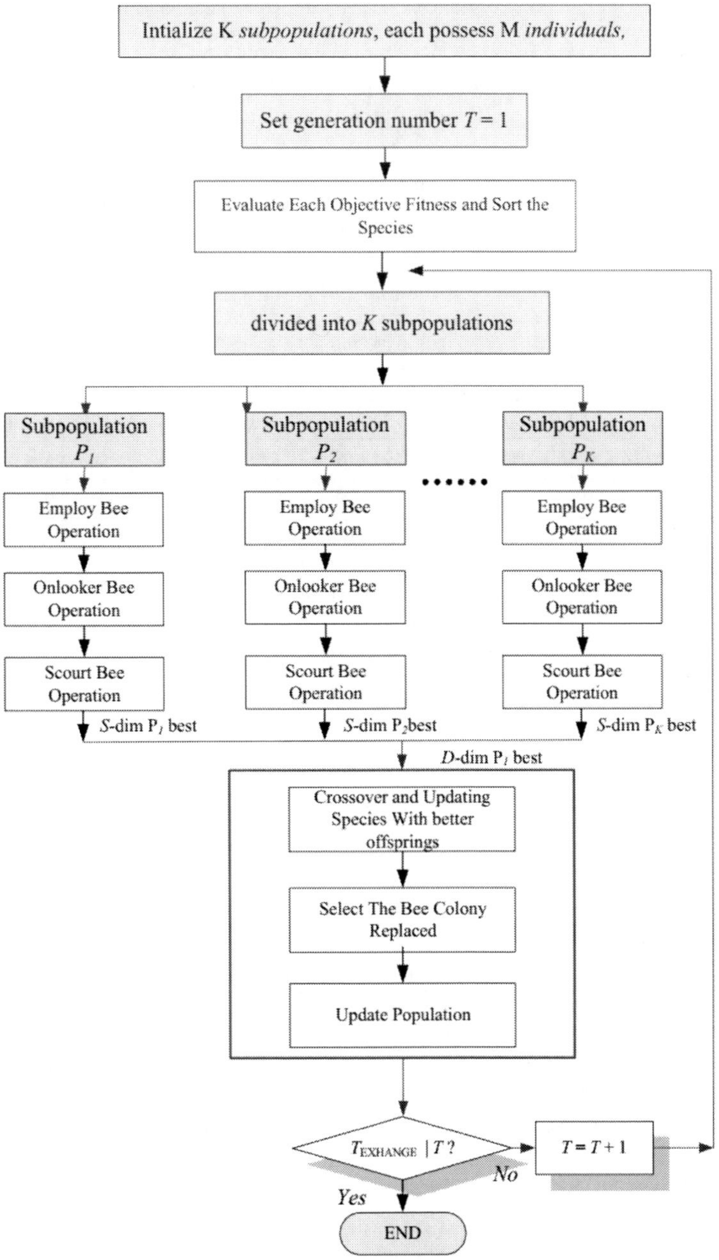

Fig.2. The flowchart of the proposed artificial bee colony algorithm

The main steps of the proposed algorithm summarized as follows: According the constraints and dimensions of the search problems, stochastically generate *M* species, and each species owns *N* individuals with *D* dimensions. According to the selected topological structure, determine the neighbor individuals of each individual in its own species and the neighbor species of its own species for later information communication. The flow of the proposed artificial bee colony algorithm is shown in Fig. 2.

3 Experimental study

In this experiment, we compared four variant versions of Hierarchical Artificial Bee Colony Algorithm with a Two-level Topology with the canonical ABC algorithm.

Table 1 shows the basic statistical results (i.e., the mean and standard deviations of the function values found in 30 runs) of the 30*D* benchmark test functions – griewank, rastrigin, rosenbrock and ackley. Obviously, the canonical ABC algorithm demonstrates the worse performance than other four algorithms. The ABC's variants outperform the canonical ABC algorithm. ABC algorithm with the star colony and the star swarm is comparatively good among these algorithms. The results of the ABC algorithm with the ring colony and the star swarm are also competitive. The two variants of ABC algorithm with the ring colony and the ring swarm and ABC algorithm with the star colony and the ring swarm have nearly the same results. This is attributed to that the topological structure of the upper level and lower level both with starscan improve the convergence rate.

Table 1. Results for allmulti-colony algorithms (ring) on benchmark functions

Function	Criteria D=30	ABC_ original	colony= ring swarm=star;	colony= star swarm=star;	colony= star swarm=ring	col= ring swarm=ring
griewank [-600,600]	Mean	3.5273e-002	2.0072e-002	1.6706e-002	1.8036e-002	2.9307e-002
	Best	2.0460e-002	1.2149e-002	9.9191e-004	4.6484e-003	2.1519e-009
	Worst	1.1344e-001	1.9317e-001	7.5833e-002	1.1030e-001	1.8270e-002
	Std	2.9923e-002	3.0132e-002	6.1813e-003	5.3601e-003	2.0087e-002
rastrigin [-15,15]	Mean	2.3766e+000	1.3091e+001	1.4505e+001	1.5822e+000	7.0624e+001
	Best	7.3355e-001	2.3292e-001	2.1183e+000	3.9955e-001	4.3973e-001
	Worst	8.7333e+000	6.9749e+000	2.4977e+001	7.1948e+000	1.5009e+001
	Std	2.2283e+000	2.3365e+000	6.1069e+000	2.1939e+000	3.8349e+000
rosenbrock [-15,15]	Mean	5.5504e+001	8.0374e+001	2.3365e+001	1.2928e+001	4.5391e+001
	Best	1.0025e+001	1.5002e+001	5.1353e+000	1.8746e+000	1.0252e+001
	Worst	7.5467e+001	1.7651e+002	3.4417e+001	1.9675e+001	6.7366e+001
	Std	2.0495e+001	1.1712e+002	1.1957e+001	1.0906e+001	2.5445e+001
ackley [-32.768,32.768]	Mean	4.6581e-002	3.1423e-001	1.4241e-002	9.0066e-001	3.0116e+000
	Best	5.3002e-003	5.8978e-002	2.0973e-003	4.3395e-002	9.3130e-001
	Worst	9.2545e-002	4.2501e-001	8.3362e-002	3.2369e+000	9.6034e+000
	Std	4.2248e-002	1.7765e-001	3.4167e-002	8.1111e-001	2.1700e+000

4 Multilevel threshold for image segmentation

The Otsu multi-threshold entropy measure [7] proposed by Otsu has been popularly employed in determining whether the optimal thresholdmethod can provide image segmentation with satisfactory results. Here, it is used as the objective function for theinvolved algorithms and its process can be described as follows:

Let the gray levels of a given image range over [*0, L-1*] and *h(i)* denote the occurrence of gray-level *i*.

Let

$$N = \sum_{i=0}^{L-1} h(i),\, P(i) = h(i) / N \text{ for } 0 \le i \le L-1 \tag{1}$$

$$Maximize\ f(t) = w_0 w_1 (u_0 - u_1)^2 \tag{2}$$

$$\sigma(t) = w_0 \times \sigma_0^2 + w_1 \times \sigma_1^2 + \ldots w_N \times \sigma_N^2 \tag{3}$$

$$\delta_0 = \sum_{i=0}^{t-1}(i-u_0)^2 \times P_i / w_0 \quad \delta_1 = \sum_{i=t}^{L-1}(i-u_1)^2 \times P_i / w_1 \quad \delta = \sum_{i=0}^{L-1}(i-u)^2 \times P_i / w$$

$$w = \sum_{i=0}^{L-1} P_i \quad u = \sum_{i=0}^{L=1} i \times P_i / w$$

Expanding this logic to multi-level threshold

$$f_{12}(t) = w_0 \times \delta_0^{\ 2} + w_1 \times \delta_1^{\ 2} + w_2 \times \delta_2^{\ 2} + \ldots\ldots + w_1 \times \delta_N^{\ 2}$$

where N is the number of thresholds. The above functionis used as the objective function to be optimized (minimized).

The experimental evaluations for segmentation performance by HABC are carried out on a wide variety of image datasets. These datasets involve a set of popular tested images used in previous studies, namely avion.ppm, house.ppm, and lena.ppm [8]. Table 2 presents the fitness function values, mean computation time, and corresponding optimal thresholds (with M-1 = 2, 3, 4) obtained by Otsu. Due to that when M-1>4 the consumption of CPU time is too long to bear, the correlative values are not listed in our experiment. It is noteworthy that the term CPU time is also an important issue in the real-time applications.

Table 2. Objective values and thresholds by the Otsu method.

Image	M-1=2		M-1=3		M-1=4	
	Objective values	Optimal thresholds	Objective values	Optimal thresholds	Objective values	Optimal thresholds
avion	3.4445E4	108,167	3.4583E4	92,146,192	3.4543E4	83,130,174,206
house	2.8486E4	106,178	2.8623E4	86,138,182	2.8576E4	72,116,155,184
lena	1.7784E4	95,156,	1.7764E4	812,129,173	1.7834E4	76,111,143,183
Mean CPU time	1.4427		61.421		2644.543	

As can be seen form Table 3, the proposed algorithm with the ring colony and the ring swarm generally performs close to the Otsu method in term of fitness value when M-1=2,3,4, whereas the performance of the improved ABC on time complexity is significantly superior to its counterpart Otsu. In other words, the proposed algorithm consumes less computing times than the traditional one. This is mainly due to the fact that the crossover-based social learning strategy used in the proposed algorithm to have faster convergence speed. Furthermore, the HABC-based algorithm achieves the best achievements among the population-based methods in most cases. This can be explained that the HABCwith two-level ring topology has a high convergence rate.

5 Conclusion

In this paper, we propose a novel hierarchical artificial bee colony algorithm with two-level topology to improve theperformance of solving complex problem. The concept and main idea is extending single artificial bee colony algorithm to hierarchical and cooperative mode by combining the multi-population cooperative co-evolution approach. The proposed algorithm is applied in solving image segmentation. Experimental results show that the proposed algorithm has a good performance on image segment for material composition detection.

Acknowledgements

This research is partially supported by National Natural Science Foundation of China (Grand Nos. 51575158, and 61305082), and a self-planned task of CAS Key Laboratory of Networked Control System (Grand No. WLHKZ2015001).

References

1. J. Kittler, J. Illingworth, Minimum error thresholding. Pattern Recognition, **19** (1986), p. 41–47.
2. T. Pun, Entropy thresholding: A new approach. Computer Vision Graphics Image Processing, **16** (1981), p. 210–239.
3. Chander, A. Chatterjee, P. Siarry, A new social and momentum component adaptive PSO algorithm for image segmentation, Expert Systems with Applications, **38** (2011): p. 4998-5004.
4. D. Karaboga, *An idea based on honey bee swarm for numerical optimization,* Technical Report-TR06, Erciyes University, Engineering Faculty, Computer Engineering Department, (2005).
5. D. Karaboga, B. Basturk, On the performance of Artificial Bee Colony (ABC), Applied Soft Computing, **8**, 1 (2008), pp. 687–697.
6. D. Karaboga, B. Basturk, Artificial Bee Colony (ABC) optimization algorithm for solving constrained optimization problems, in: P. Melin, et al. (Eds.), IFSA 2007, LNAI 4529 (2007), p. 789–798.
7. N. Otsu, A threshold selection method from gray-level histograms, IEEE Transactions on System, Man, and Cybernetics, **9** (1979), p. 62–66.
8. Information on http://decsai.ugr.es/cvg/dbimagenes/

Effect of Pt Concentration on the Grain Growth of TiO2sol-gel Films

Y.J. Jia[1a], Na. An[2b], H.N. Chen[3c] and Yi Chen[4d]

[1]School of Mechanical Engineering, Tianjin Polytechnic University, Tianjin 300387, China
[2]Siemens Mechanical Drive Systems Co., Ltd, Tianjin 300400, China
[3] School of Computer Science and Software Engineering, Tianjin Polytechnic University, Tianjin 300387, China
[4]Department of mechanical engineering, Inha University, 253 Yonghyun Dong, Nam Gu, Incheon, 402-751, South Korea
[a]jiayanjun@tjpu.edu.cn, [b]na.an@siemens.com, [c]chenhanning@tjpu.edu.cn, [d]chenyi6970@hotmail.com

Abstract. Pt metallic has been supported on TiO2 surface using different methods, Here, Pt doped TiO2 (Pt-TiO2) sol gel thin film were successfully produced by reducing chloroplatinc acid (H2PtCl6). The structures of prepared composites were investigated using X-ray diffraction (XRD). The physical morphologies of the composites were examined using transmission electron microscope (TEM). The grain size of Pt-TiO2 thin film after annealing was also measured by atomic force microscope (AFM) images.
Keywords. TiO2, Pt Annealing, Gran size

1 Introduction

Recently, extensive research for titanium dioxide (TiO2) has been carried out in efforts to develop variety of application fields including capacitors for photovoltaic cells [1], sensors [2], antireflection films [3], white pigments [4], optical coatings [5], and high density dynamic random access memory devices [6]. TiO2 have three different crystal phases, such as anatase, rutile, and brookite [7]. Rutile is the most common natural form of TiO2. Brookite phase is crystallized in the orthorhombic system and exhibits no photocatalytic activity. Anatase TiO2 has excellent photocatalytic activity, physical and chemical stabilities and antimicrobial activity [8-11].

Earlier studies have revealed that the photocatalytic activity of TiO2 can be improved significantly by doping with noble metals such as Pt, Au, Ag, etc [12, 13]. It is well known that platinization of TiO2 often shows a high photocatalytic activity. The doping of platinum on TiO2 can form the Schottky diode barrier among the metalsand the electronic potential barrier at the metal-semiconductor heterojunction, and the platinized TiO2 traps the photogenerated electrons efficiently [14]. The doping of Pt on TiO2 surface has been widely reported to improve photocatalytic performance for the split of water and the degradation of different harmful compounds [15].

In this paper, we report characteristic properties of TiO2 and Pt composite including UV-visible spectra, transmission electron microscope (TEM) images, x-ray diffraction (XRD) patterns and atomic force microscope (AFM) images.

2 Experimental details

Titanium isopropoxide was used as a sol-gel precursor. The metal alkoxide was mixed with ethanol. The resulting solution became a milk color, and white precipitation was observed. Hydrochloric acid was slowly added to the solution with vigorous stirring until the solution become transparent. Various amounts of chloroplatinc acid (H2PtCl6) (8wt. % in water) was added to the TiO2 sol-gel solution. Sodium borohydride (NaBH4) was dispersed in ethanol and slowly added into the H2PtCl6/TiO2 solution to reduce [PtCl6]2- to Pt0. The aged sol was spin coated onto the glass and silicon substrateby spin-coated at 2000 rpm and heated at 200 °C for 5 min.. UV-visible spectra were obtained with diode array spectrophotometer (Hewlett-Packard 8452 A).The sol-gel was diluted to ethanol and drop to the copper TEM grid. Using this copper grid sample, TEM images were obtained with CM200 (Phylips) microscope. The TiO2 thin films coated onto silicon wafer were annealed in agas environment tube furnace (EM Tech).The annealing process started from room temperature to 750°C in steps of 5 °C/min and then maintained the temperature for 1 h. The films naturally cooled down to room temperature. The TiO2 thin films doped with Pt were subject to XRD analysis with X'Pert MPD Pro diffractometer (Phylips). The surface of the film also analyzed with NanoScope 3D (Veeco) microscope.

3 Results and discussion

UV-visible spectra of bare TiO2 and Pt-TiO2 prepared with different pt concentration are shown in Figure 1. The single broad intense absorption peak of Pt-TiO2 between 300 – 350 nm can be attributed to the charge-transfer from the valence band to the conduction band. The UV-vis absorption peaks also showed a red shift due to the modification of TiO2 with Pt [16].

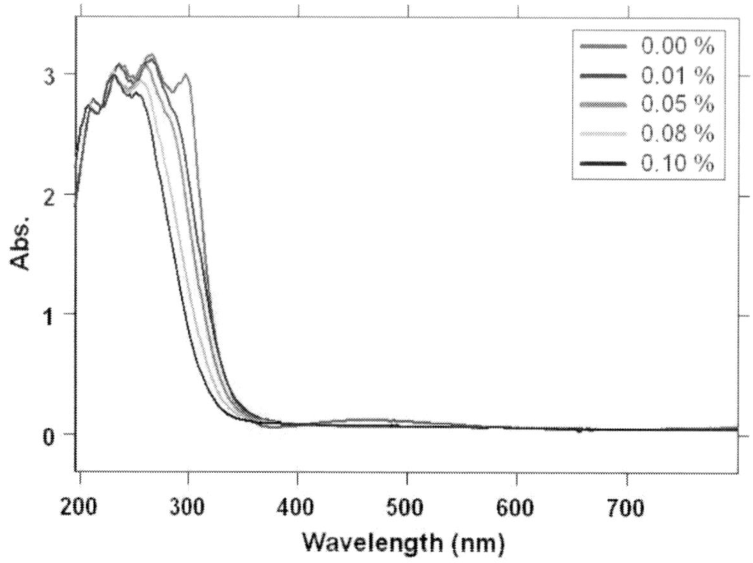

Fig. 1. UV-Visible spectra of bare TiO2 and Pt-TiO2 sol-gel thin film after annealing.

The UV-visible spectra show the possible uniform distribution of the Pt-nanoparticles throughout the sol-gel films. To obtain the particle size and distribution, the TEM was employed. Figure 2 (a) shows the TEM image for the 0.1 % of Pt-TiO2film with scale bar size of 100. The aggregated area was further focused and obtained the image with scale bar size of 20 nm as shown in Figure 2 (b). Pt-nanoparticles do not have big aggregates, distribute relatively uniformly, and distribute particle sizes between 4–8 nm.

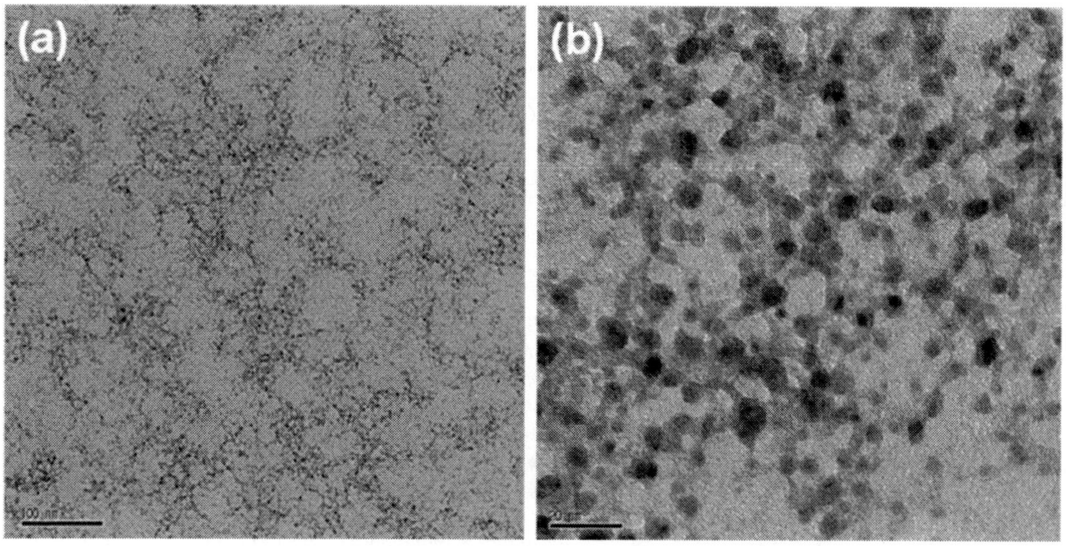

Fig.2. TEM image of TiO_2-Pt with 0.1% Pt.

Figure 3 (a) shows XRD patterns for Pt-TiO2films prepared by sol-gel method with various amount of Pt. All these samples show typical TiO2diffraction peaks in form of anatase. It shows almost coincide with a pure TiO2 diffraction peaks and shows no diffraction peaks due to the doped Pt. It can be assumed that the amount of Pt doped particle was very low, which resulted in non-appearance of the Pt crystalline peaks [17,18]. However, diffraction intensity increased with the increase of Pt concentration. The crystalline size of the Pt-TiO2 samples after annealing were estimated from line broadening using Scherrer equation based on (101) peak of anatase TiO2 [19].

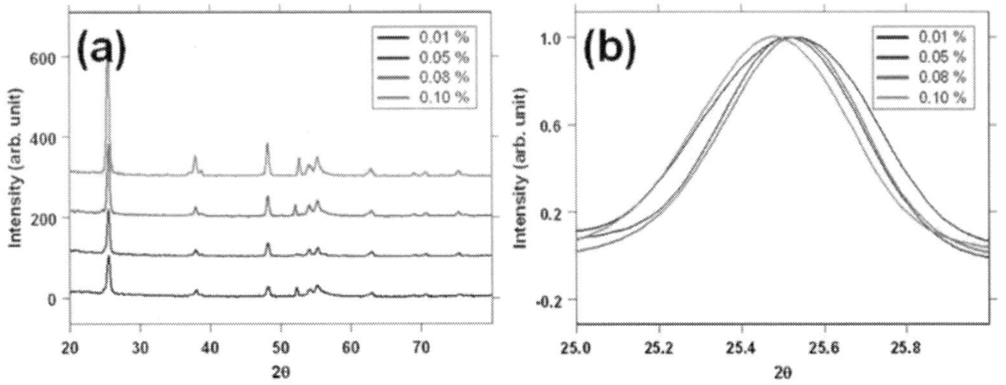

Fig. 3. XRD pattern of TiO_2-Pt thin film depends on Pt concentration.

$$D = 0.9\lambda / \beta \cos\theta$$

where λ is the X-ray wavelength of Copper Kα radiation, θ is Bragg's angle and β is the pure full width of the diffraction line at half of the maximum intensity (Figure 3(b)). The calculated grain sizes are 15 nm (0.01%), 20 nm (0.05%), 35 nm (0.08%) and 40 nm (0.1%), separately. The result refers to that doping Pt also retard the grain growth of TiO2 thin film.

AFM images of the Pt-TiO2 surface were used to monitor the grain size of the sol-gel thin film depends on Pt concentration. 1 μm × 1 μm scans revealed it is clear that the grain size of Pt-TiO2 gradually increased with Pt loading.

4 Conclusions

Metallic doping is one factor for TiO2anatase phase grain growth. The intensity of the main anatase diffraction peak increased with increasing Pt concentration. The grain sizes estimated from anatase XRD major diffraction peaks were 15, 20, 35 and 40 nm for the sample with Pt concentration of 0.01%, 0.05%, 0.085 and 0.1%, respectively. The measured grain sizes using AFM images also increased depends on Pt concentration. The governing factor to determine the grain size of the TiO2 thin films is Pt doping.

References

1. H. W. Wang, C.F. Ting, M.K. Hung, C.H. Chiou, Y.L. Liu, Z. Liu, K.R. Ratinac, S.P. Ringer, Nanotechnology, **20** (2009), pp. 055601.
2. J. Bai, B. Zhou, Chem. Rev. **114** (2014), pp. 10131.
3. V. Zoulalian, S. Monge, S. Zrcher, M. Textor, J.J. Robin, S. Tosatti, J. Phys. Chem. B, **110** (2006), pp. 25603.
4. N.B. Chaure, A.K. Ray, R. Capan, Semicond. Sci. Technol. **20** (2005), pp. 788.
5. H. Yaghoubi, N. Taghavinia, E.K. Alamdari, Surf. Coating Tech. **25** (2010) pp. 1562.
6. Q. Li, K. Ali, S. Lulia, P. Christos, H. Xu, P. Themistoklis, Scientific Reports, **4** (2014), pp. 4522.
7. W. Zhang, S. Chen, S. Yu, Y. Yin, J. Crystal Growth, **308** (2007), pp. 122.
8. T. Ohno, K. Sarukawa, M. Matsumura, New J. Chem. **26** (2002), pp. 1167.
9. X.H. Yang, Z. Li, C. Sun, H.G. Yang, C. Li, Chem. Mater. **13** (2011), pp. 3486.
10. G. Fu, P.S. Vary, C. Lin, J. Phys. Chem. B, **109** (2005), pp. 8889.
11. H. Zhang, G. Chen, Environ. Sci. Technol. **43** (2009), pp. 2905.
12. D. Morris, Y. Dou, J. Rebane, C.E.J. Mitchell, R.G. Egdell, Phys. Rev. B, **61** (2000), pp. 13445.
13. M. Enachi, M. Guix, T. Braniste, V. Postolache, V. Ciobanu, V. Ursaki, O.G. Schmidt, I. Tiginyanu, Surf. Eng. Appl. Electrochem. **51** (2015), pp. 3.

14. W.Y. Park, G.H. Kim, J.Y. Seok, K.M. Kim, S.J. Song, M.H. Lee, C.S. Hwang, Nanotechnology, **21** (2010), pp. 195201.
15. D. Eder, M. Motta, A.H. Windle, Nanotechnology, **20** (2009) 055602.
16. H. Tada, F. Suzuki, S. Yoneda, S. Ito and H. Kobayashi, Phys. Chem. Chem. Phys., **3** (2001), pp. 1376.
17. L. Ravichandran, K. Selvam, B. Krishnakumar, M.. Swaminathan, J. Hazardous Mater, **167** (2009), pp. 763-769.
18. H. Wang, Z. Wu, Y. Liu, Y. Wang, Chemosphere, **74** (2008), pp. 773-778.
19. M. Beghi, P. Chiurlo, L. Costa, M. Palladino, M.F. Pirini, J. Non-cryst. Solids, **145** (1992), pp. 175.

Preparing Methods and Its Influencing Factors about Nanoparticles Based on Dendritic Polymer

Jianwei Zhang[1*] and Jeff Li[2]

[1]Department of Environmental Science, Qingdao University, Qingdao 266071, China
[2]Department of Earth Sciences, University of waterloo, waterloo N2L 3G1, Canada
Corresponding Email: dragonzjw@126.com

Abstract. Based on the properties, structure and application of dendritic polymer, this paper analysed the methods of the preparation of nanoparticles using dendritic polymer, detailed preparation process, technical parameters and application effect about a single metal nanoparticles, bimetallic nanoparticles, sulfide and halide nanoparticles. The influencing factors of the preparation about nanoparticles were discussed, including the molecular algebra, the molar ratio of the metal ions to the dendritic polymer, and so on.

Keywords. Nanoparticles, preparing methods, dendritic polymer, influencing factors

1 Introduction

Dendritic polymer is used as a template to prepare nanoparticle [1] application range is more and more wide in recent years [2]. At present, based on dendritic polymer, it has been successfully prepared a variety of metal nanoparticles, oxide nanoparticles and halide nanoparticles, such as Au, Ag, Pt, Cu, AgBr, Au-Ag and ZnO etc. The prepared nanoparticle has the advantages of good dispersion, high stability, large surface area and high activity, and it is widely used as catalysts for industrial production.

The dendritic polymer are highly branched, which is a large molecule of single disperse with dendritic three-dimensional structure. It is obtained by the method of convergent or divergent based on the branch element (dendron).

Polyamide (PAMAM) is the development of more mature dendritic polymers [3]. According to the different of reproduction, divided into different Algebra (Gn) PAMAM dendritic polymers. OCH_3 terminal is half generation PAMAM dendritic polymers, and NH_2 terminal is whole generation PAMAM dendritic polymers. Then the specific algebra can be determined by the number of times that the branching unit is repeated.

Dendritic polymers have the following structural characteristics:

(1) Precise molecular structure.

(2) The high geometry symmetry.

(3) A large number of functional groups, and the surface modification and modification can be carried out.

(4) The intermolecular in the cavity.

(5) Relative molecular weight control.

(6) The molecule itself is a nanometer size.

Based on the above characteristics, dendrimers have many advantages:

(1) The unique chemical composition and structure of the tree form, so that it can get a very good "reproduction", to ensure that the product of the single dispersion.

(2) There are a large number of cavities inside the dendritic polymer, which can fully wrap the nanoparticles to prevent the aggregation of the nanoparticles and improve their stability.

(3) The nanoparticles are mainly restricted by steric hindrance, the surface of which is not passive, and still keep high activity.

(4) In the catalytic reaction, the special structure of the dendritic polymer can be used as a nano grade filter to select the substrate.

2 Preparation of nanoparticles

The cavity of the dendritic polymer provides space for the formation of nanoparticles, so that the product is stable and controllable in size. Under different reaction conditions, such as the water phase, the organic solvent, the reducing agent and the electromagnetic wave radiation, the preparation of the nano particles by the dendritic polymer is more and more mature. In the process of preparing nano materials, based on the preparation of single metal nanoparticles, bimetallic nanoparticles, halide nanoparticles, sulfide and oxide nanoparticles were synthesized.

(1) Single metal nanoparticles

According to the position of metal nanoparticles and dendritic polymer, it can be divided into two types: the molecular template and the outer template. Molecular template is within the metal ions and dendritic polymer active functional group, so a single molecule formed nanoparticles in the cavity. The outer molecular template is the active functional group of the metal ions and the outer surface of dendritic polymer, which is reduced to form a group of nano particles. In the preparation of metal nanoparticles, this two methods are very practical. Synthesis of Cu nano clusters with PAMAM dendritic polymer as molecular template, the synthesis of nanoparticles is simple, easy to operate, and the controllability of product characteristics. Thereafter, Pt, Au, Ag, Pd, Ni, Rh and other metal nanoparticles were successfully prepared.

When using methanol as solvent, the PAMAM G5.5 dendritic polymer as template, the metal Pt nanoparticles can be obtained by $NaBH_4$ reduction. When the ratio of the amount of G5.5 and PAMAM Pt^{2+} is changed, the two different synthesis methods of the molecular template and the outer template can be realized. The particle size of nanoparticles prepared by the template method is smaller than that of the outer template method. Knecht et al. considered that the reduction reaction is a self catalytic process.

Dendritic polymer as a template to heat 70 min at 100℃, the metal Ag nanoparticles with high stability and small particle size can be synthesized in one step, and the particle size of the product increases with the increase of temperature.

In the aqueous phase, the metal Ag nanoparticles were prepared by UV irradiation with the dendritic polymer as the template. Moreover, the size of the nanoparticles was increased with the increase of the reaction time, until the Ag ion was completely reduced. So that the appropriate reaction time is very important.

Pd metal nanoparticles were Prepared of using G6.0PAMAM dendritic polymer as templates, and to G4.0 PAMAM dendritic polymer as template can get Rh and Pt nanoparticles [4].

As can be seen, the direct reduction method for the synthesis of metal nanoparticles has two main steps using dendritic polymer as template. The first step is the complexation reaction of metal ions and dendritic polymer. The second step is generates metal element with complex metal ions by chemical means, then to form nanoparticles.

(2) Bimetallic nanoparticles

Bimetallic nanoparticles are different from single metal nanoparticles. It is the synergistic effect of the two kinds of metal, so that it can greatly improve the catalytic range and catalytic efficiency. The advantage about the preparation of bimetallic nanoparticles by dendritic polymer is very prominent.

There are three methods for the preparation of bimetallic nanoparticles: synchronous complexing method, sequential complex method, partial substitution method. Among which the former two types are the main methods.

Synchronous complexing method is a method that two metal ions mixed with the dendritic polymer at the same time to prepare bimetallic nanoparticles. Au-Ag bimetallic nanoparticles can be obtained in the aqueous phase, and the test proves that the bimetallic nanoparticles are not a simple mixture of single metal particle or core shell structure, but mixed together in the form of alloy. When near room temperature, the prepared Pt-Au bimetallic nanoparticles have higher activity for the catalytic oxidation of CO. Based on the PAMAM dendritic polymer templates, when $N_2H_4 \cdot H_2O$ as a reducing agent, it can get small size and uniform spherical Ag-Cu bimetallic nanoparticles; but with $NaBH_4$ as a reducing agent, get a long rod shaped Ag-Cu bimetallic nanoparticles [5].

It is found that the Bimetallic nanoparticles prepared by the sequential complex method are mostly the product of core-shell structure, but the Bimetallic nanoparticles synthesized by the synchronous complexing method are mostly in the form of alloy.

(3) Preparation of halide and sulfide nanoparticles

The nature of the halide and sulphide nanoparticles was influenced not only by the particle size, but also influenced by the particle size distribution, particle shape and stable performance. Therefore, it is important to prepare the shape and particle size uniformity, stable performance of halide and sulfide nanoparticles. With the further development of dendritic polymer in the preparation of nano materials, the halide and sulfide nanoparticles with special applications in the optical field can also be prepared.

When the AgBr nanoparticles were prepared, the ratio of the amount of Ag and PAMAM dendritic polymer was changed, and the nanoparticles with different particle sizes were obtained. Their catalytic ability gradually increased with decreases of the size about the nanoparticles. The AgI nanoparticles under the same preparation conditions, to G4.5-$COOCH_3$ template of the average particle size is 6nm, and PAMAM G5.0-NH_2 as template from particle diameter is 9nm. This is mainly due to the metal ion dendritic macromolecular template complexation reactions in different parts, the aggregation degree of different [6].

The stable CdS nanoparticles can be obtained by the template of dendritic polymer, and the particle size of CdS nanoparticles increases with the increase of molecular algebra [7].

The reaction process about synthesis halide or sulfide nanoparticles with dendritic polymer: first is the complexation of dendritic polymer with dendritic polymer template, then halide ions or sulfur ions with metal ions combine to form halide or sulfide nano clusters, so it is wrapped in the dendritic polymer to obtain nanoparticles.

(4) Preparation of oxide nanoparticles

PAMAM dendritic polymer were used as templates to prepare ZnO nanoparticles with high photocatalytic activity. Compared with other methods, the method has the advantages of low cost, simple operation, controllable particle size and so on.

3 Influence factors about the size of nanoparticle

The synthesis of nanoparticles with dendritic polymer is used in industrial production. The catalytic efficiency is closely related to the specific surface area of the catalyst. In general, the greater the specific surface area, the higher the catalytic efficiency. And the size of the specific surface area is related to the size of the nanoparticles, the smaller the size, the larger the specific surface area. So its size is very important to prepare nano particles. Influence factors about the size of nanoparticles include molecular algebra, metal ions and the molar ratio of dendrimers, the concentration of the inner and outer templates, the means of reducing etc.

(1) Molecular algebra

According to the different of molecular algebra, the absorbance of Au nano clusters shows that with the increase of molecular algebra, the absorbance increases, and the larger of the absorbance, the smaller of the nanoparticles.

(2) Molar ratio of metal ions to dendritic polymer

Dendritic polymer have a certain amount of capacity to react with metal ions. When the metal ion is too long, the metal atoms will be reduced to a large number of atoms together, the size will become larger. In addition, too much metal ions may also be combined with the external functional groups of dendrimers, so that the branches of large molecules act as an outer template. And when the metal ion is less, only a small amount of metal ions are reduced to form a small size of the nanoparticles.

When the other conditions were the same, the size of the nanoparticles increased with the increase of the molar ratio of the metal ions to the dendritic polymer.

4 Conclusion

The preparation of nanoparticles based on dendritic polymer is simple, easy to operate, and the product is controllable, which ensures the excellent performance in catalysis. From single metal nanoparticles to bimetallic nanoparticles, and then to metal compound nanoparticles, it is also possible to prepare more superior multi metal nanoparticles on this basis.

But due to the active characteristics of nanoparticles, it will have a variety of interference factors in the preparation process, so nanoparticles morphology control still need further study, the control standard of various reaction conditions need further exploration.

References

1. Percec V., Ahn C. H. and Ungar G. Controlling polymer shape through the self-assembly of dendritic side-groups: Nature Vol. 391 (1998), p. 161-164
2. Bhadra D., Bhadra S., and Jain S. A pegylated dendritic nanoparticulate carrier of fluorouracil: Int. J. Pharm. Vol. 257 (2003), p. 111-124
3. Rosen B. M., Lligadas G., and Hahn C. Synthesis of dendrimers through divergent iterative thio-bromo"Click" chemistry: J. Poly. Sci. PartA: Poly. Chem. Vol. 47 (2009), p. 3931-3939
4. Carino E. V., Knecht M. R., and Crooks R. M. Quantitative analysis of the stability of Pd dendrimer- encapsulated nanoparticles: Langmuir Vol. 25 (2009), p. 10279-10284
5. Li G. P. and Luo Y. J. Preparation and characterization of dendrimer-templated Ag-Cu bimetallic nanoclusters: Inorg. Chem. Vol. 47 (2008), p. 360-364
6. Weir M. G., Knecht M. R. and Frenkel A. I. Structural analysis of Pd Au dendrimer-encapsulated bimetallic nanoparticles: Langmuir Vol. 26 (2010), p. 1137-1146
7. Gu Y. L., Xie H., and Gao J. X. AFM characterization of dendrimer-stabilized platinum nanoparticles: Langmuir Vol. 21 (2005), p. 3122-3131

Status and Prospect of Application about the New Materials Graphene

Jianwei Zhang[1*], Jeff Li[2]

[1]Department of Environmental Science, Qingdao University, Qingdao 266071, China
[2]Department of Earth Sciences, University of waterloo, waterloo N2L 3G1, Canada
Corresponding email: dragonzjw@126.com

Abstract. In this paper, the related properties of graphene were described as the new materials. Through the comparison, the advantages and disadvantages about the preparation methods of graphene were analyzed. According to the present situation of its application, application areas, existing products, etc. of the graphene materials, that Graphene materials are widely used in the future. And the possible environmental pollution risk in the process of development and application are put forward.
Keywords. Graphene, application status, preparation method, application prospect, new material.

1 Introduction

Graphene was a new material that is found in the first form of a single atom. In 2004, it was successfully separated from the graphite and confirmed that it could exist alone. Graphene is a nano material, that it has the thinnest, maximum strength and strongest conductive thermal conductivity [1], also known as the "wonder material", "black gold", "king of the new materials". Since the discovery of graphene materials, technical personnel in various countries to make full use of their excellent performance, expected to be applied in many industrial areas [2], but so far there is still not a large-scale industrial production and application.

Graphene is separated from the graphite material. It is a two-dimensional crystal consisting of carbon atoms with only one layer of atomic thickness (two-dimensional carbon material). Its thickness is about 1/300 million mm [3].

2 Characteristics

2.1 Physical and chemical properties

Graphene has perfect two-dimensional crystal structure. It is the hexagonal lattice surrounded by six carbon atoms. They are bonded with each other to form a honeycomb between carbon atoms by strong conjugated bond to SP2 hybrid method. This unique stable structure makes graphene with excellent mechanical strength and structural rigidity [4]. Each carbon atom of graphene has a p orbital electron, which can move freely in the crystal, and moving speed up to 1/300 times the speed of light, so that graphene has excellent electrical conductivity [5].

2.2 Performance analysis

DOI: 10.1051/matecconf/201710004029

The mechanical strength of graphene can be made into various kinds of graphite carbon materials, such as coiled carbon nanotubes or globular material, and pencil graphite is equivalent to countless graphene stacked together. The layered graphene includes monolayer graphene, double-layer graphene (two layers of hexagonal honeycomb structure stacking), few-layer graphene (3-10 layer hexagonal honeycomb structure stacking), multi-layer graphene (thickness more than 10 layers below 10 nm) and a variety of materials.

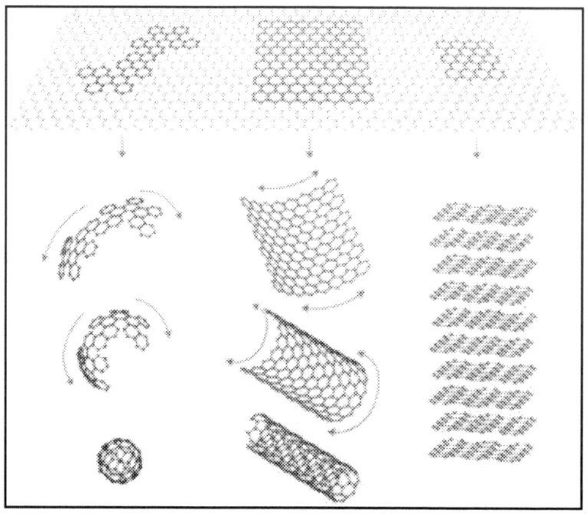

Fig. 1. Structural relationships of spherical, tubular, and tabular carbon materials [4]

The conductivity of graphene is closely related to the electronic band structure. Because there is only one layer of atoms, the motion of electrons is restricted to a plane, which is much faster than the speed of motion of electrons in a general conductor.

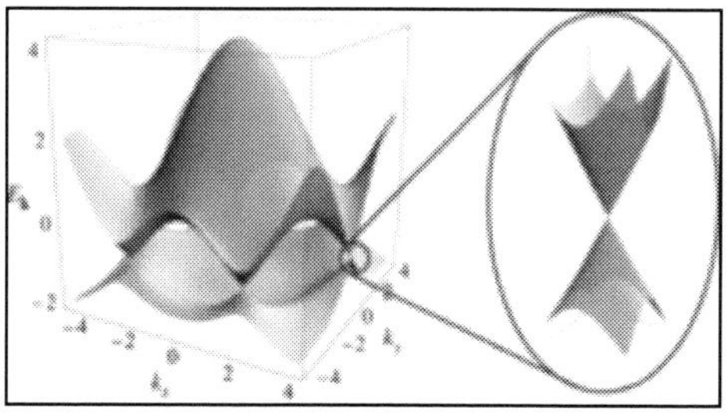

Fig. 2. Schematic diagram of graphene electronic energy band structure [5]

2.3 Possible application areas

According to the strength characteristics of graphene, graphene is the thinnest material, but also the strength and toughness of materials, good elasticity, than steel high fracture strength of 200 times. It can provide a thin, light, strong and tough new material for industrial manufacturing industries, such as automobiles, aircraft and satellites.

Graphene resistivity is only about $10^{-6}\Omega\cdot cm$, is the smallest resistivity material. Due to its low resistivity, electron migration is extremely fast, so the future can replace silicon, to develop the thinner and the faster conduction velocity that new generation of electronic components or ultra micro transistor. It will make the speed of the computer processor running several hundred times. In the future, it can be used to produce super computer.

Graphene is almost completely transparent, and very dense, even the smallest gas atoms (helium atoms) cannot penetrate it. Therefore, it is very suitable as raw materials for transparent electronic products, such as transparent touch control display, light emitting plate, etc. And in combination with its flexibility, it can be used to produce curved screens.

3 Application status

3.1 Preparation method

The production and preparation of graphene materials are the most important issues in the development and utilization. Initially, the two sides of the graphite sheet is stuck on a special adhesive tape, tear tape, can be divided into two pieces of graphite, and constantly operate, finally, it is only a layer of carbon atoms of graphene. So far, the commonly used methods of preparation are mechanical stripping, chemical stripping, chemical synthesis, catalytic generation (chemical vapor deposition, epitaxial growth method), graphite intercalation method, high temperature quenching method, etc. [6]

Mechanical stripping method is the most simple and direct method, the use of friction and relative motion between the object and the graphene, mechanical stripping to prepare graphene. The method has higher quality, the majority of graphene has a complete crystal structure, but the film layer is small, the production efficiency is low, and the cost is high.

Chemical stripping method is the strong oxidizing substances reaction between graphene and strong acid to produced the graphite oxide, after ultrasonic dispersion made precursor graphene oxide, increasing the spacing of the layers, then to separated it through physical methods. Add reducing agent to remove the oxide oxygen gene of graphene surface, then it can obtain graphene by chemical reduction and thermal reduction method. The method is simple in operation and high in yield, but the product quality is low.

Catalytic generation method includes chemical vapor deposition method, epitaxial growth method and so on. The chemical vapor deposition method is to put the transition metal film in the hydrocarbon gas, transition metals as catalysts, in the container to the hydrocarbon pyrolysis, and deposited on the substrate to form graphene film. Epitaxial growth method using growth matrix of atomic structure, under ultra high vacuum in the high temperature environment, silicon atoms sublimed from material, residual C atoms self remodeling group. Namely high temperature carbon dissolved in the metal phase, forming interstitial impurities and low temperature carbon segregation at the surface, in a metal substrate surface grown monolayer or double-layer graphene. This method can obtain high quality graphene, but the equipment requirements are higher.

Graphite intercalation method using natural flake graphite as raw material, between the layers and insert some atoms and molecules of non carbon, to form a new layered compounds, increase the spacing between the layers of the graphene, so as to achieve the stripping.

High temperature quenching method is to heat up to 1000 degrees, quickly into the mass fraction of 1% of ammonium bicarbonate solution, high temperature quenching, 5min cooling to room temperature, then it can obtaine the graphene material.

3.2 Application status of graphene

In 2013, the researchers of University of California at Los Angeles developed a micro super capacitor based on graphene, it is 1000 times the rate of charge faster than ordinary batteries.

China also has a unique advantage in the application of graphene. At present, researchers have successfully made the first 15 inch single layer graphene, and successfully applied to the resistance touch screen graphene transparent electrode, made a 7 inch graphene touch screen. The flexible composite thin film with high conductivity and high thermal conductivity is developed, which is of thin thickness, high thermal conductivity and high tensile strength. Researchers from Chinese University of Science and Technology for the first time found that graphene can be used as a good proton conducting membrane. The world's first batch of graphene phone has been released in China, opened a new era of graphene industry applications.

3.3 Major problems

The main problem of the current application of graphene is how to increase the production of graphene materials, and reduce production costs [7]. Generally the method for preparing graphene with high yield, such as chemical vapor deposition method, output of graphene on edges tend to be disrupted into five membered or seven membered ring, does not guarantee complete alignment, so that in the interior of the graphene exist structural weakness, easy to fracture and fracture has been extended, to reduce the intensity of the graphene.

4 Application prospects

Graphene application prospects are very broad, many countries and regions have established graphene technology research and development center, to promote the industrialization and commercialization of graphene.

The European Commission will take the graphene as the future flagship technology project, set up special research and development plans. The British government has also invested in the establishment of the national graphene research institute, to promote the material can be from the laboratory into the production line and the market in the future. Spain Graphenano company plans to develop graphene electric vehicles. South Korean researchers successfully synthesized a wafer level high quality multilayer graphene, making it closer to the commercial application, and achieved the cross of graphene from potential materials to the actual application.

Following the implementation of the integration strategy about "graphene global mergers and acquisitions" in China, Beijing-Tianjin-Hebei Industrial Development of graphene alliance was established in 2015. In the future it will be set the graphene industry cluster of research and development, production, inspection, financial services and others.

5 Conclusions

New nano materials graphene with the high strength, conductivity and other excellent properties to get people's attention, although the current application is subject to material production and its costs, but it is broad prospects for future applications. And we should pay attention to the risk of environmental pollution when we are committed to graphene application technology research and development.

References

1. K. S. Novoselov, V. I. Fal'ko and L. Colombo. A roadmap for grapheme: Nature Vol. 490 (2012), p. 192-200
2. Z. G. Song, VI. Artyukhov, and BI. Yakobson. Pseudo Hall-Petch Strength Reduction in Polycrystalline Graphene: Nano Lett. Vol. 13 (2013), p. 1829-1833
3. Yang L., Liu D. and Chen A. Y. The New Star Nanomaterials-Grapheme: G.Z. Chem. Ind. Technol. Vol. 43 (2015), p. 9-11
4. Huang H. D., Ni Z. F. and Huang G. D. Preparation of Graphene Oxide and Property of Its Composites: J. S.X. Univ. (Nat. Sci. Ed.) Vol. 38 (2015), p. 675-680
5. A. Schinwald, F. A. Murphy and A. Jones. Graphene-Based Nanoplatelets: A New Risk to the Respiratory System as a Consequence of Their Unusual Aerodynamic Properties: ACS Nano Vol. 6 (2011), p. 736-746
6. Xiao S. J., Yu S. W. and Tan X.Y. Properties Characterization of Graphene and Its Preparation: Plas. Sci. Technol. Vol. 43 (2015), p. 30-34
7. Zhou H., Piao M. X. and Li Q. Preparation and properties of graphene oxide nanosheeets/epoxy composite: Acta Mater. Comp. Sin. Vol. 32(2015), p. 1309-1315

The Enrichment and Transfer of Heavy Metals for Two Ferns in Pb-Zn Tailing

Jiajie Mai [a], Qihe Yang [b], Yingfei Zhang[c], Xianlu Zeng [d], Yanmei Zhong [e] and Deliang Liu [f]

School of Life Science, Jiaying University, Meizhou 514015, China
[a]neusee@163.com, [b*]yangqh@jyu.edu.cn, [c]fei314@foxmial.com, [d]xyz@jyu.edu.cn, [e]zym@jyu.edu.cn, [f]ldl@jyu.edu.cn

Abstract. The enrichment and transfer of 8 heavy metals of *Equisetum ramosissimum* and *Pteris vittata* growing naturally close to edge of the sewage pool in Bencun Pb-Zn Tailing, Eastern Guangdong were investigated. The results indicated that the pollution of Cd, Pb, Hg, Zn was very severe in this tailing, followed by Cu and Mn. The potential ecological risk of heavy metals was assessed to be very strong based on soil background values of Guangdong Province and at high risk according to criteria of the second grade State Soil Environmental Quality Standard, and Cd, Hg, Pb were the main factors leading to potential ecological risk. The content of 8 heavy metals in the two ferns did not reach critical content of hyperaccumulator, so neither of them was typical hyperaccumulator, but both had a certain tolerance to these heavy metal pollution. Underground parts of *Pteris vittata* had an enrichment coefficient above 1 and that of *Equisetum ramosissimum* had a value near 1, therefore the two ferns could be utilized as potential enrichment plants. The two ferns have strong adaptability to the tailing habitat and can be used as pioneers in ecological restoration of Pb-Zn tailings.
Keywords. *Equisetum ramosissimum*, *Pteris vittata*, heavy metal phytoremediation

1 Introduction

The enrichment and transfer of 8 heavy metals of *Equisetum ramosissimum* and *Pteris vittata* growing naturally close to edge of the sewage pool in Bencun Pb-Zn Tailing, Eastern Guangdong were investigated. The results indicated that the pollution of Cd, Pb, Hg, Zn was very severe in this tailing, followed by Cu and Mn. The potential ecological risk of heavy metals was assessed to be very strong based on soil background values of Guangdong Province and at high risk according to criteria of the second grade State Soil Environmental Quality Standard [1], and Cd, Hg, Pb were the main factors leading to potential ecological risk. The content of 8 heavy metals in the two ferns did not reach critical content of hyperaccumulator, so neither of them was typical hyperaccumulator, but both had a certain tolerance to these heavy metal pollution. Underground parts of *Pteris vittata* had an enrichment coefficient above 1 and that of *Equisetum ramosissimum* had a value near 1, therefore the two ferns could be utilized as potential enrichment plants. The two ferns have strong adaptability to the tailing habitat and can be used as pioneers in ecological restoration of Pb-Zn tailings.Soil resource is one of the main sources of livelihood and the basis of terrestrial ecosystem and agricultural development. It has important significance to maintain the species diversity and determine the yield and quality of agriculture. The heavy metal pollution in soil is a kind of irreversible pollution process and also a severe threat to human, which directly leads to crop production and brings serious damage to human health through a negative feedback effect. In mine tailings, such as Pb-Zn tailings, their components and residual flotation reagent bring about serious contamination to ecological environment. The acidic water produced by sulfide in tailings caused the leaching of heavy metals to contaminate the soil and water environment in tailing area.

Bencun Pb-Zn Mine Tailing (24o22'54"E, 116o13'12"N), with a slope of 15o ~20o (Figure 1) , is one of mine tailings with the most largest volume of industrial solid waste in Meizhou City, East Guangdong, South China. The mine started operation in 1970 and abandoned in 1980. The mineral waste residue had been directly discharged and piled up in top of this slope without any purification measures till over a decade ago, the contaminants from waste residue leached by rainwater which flows into pools and farmlands down along this slope and then resulted in severe ecological degradation [2]. Meizhou city enjoys a warm and humid subtropical monsoon climate with abundant sunshine, plenty and concentrated rainfall; the average temperature ranges from 20.6~21.4oC, average annual sunshine hours 1714.6~2010.5 h, average rainy days is about 150 d, average annual rainfall is 1483.4~1798.4 mm, with 75% of which being concentrated in months from April to September, and a frost-free period is 309 d.

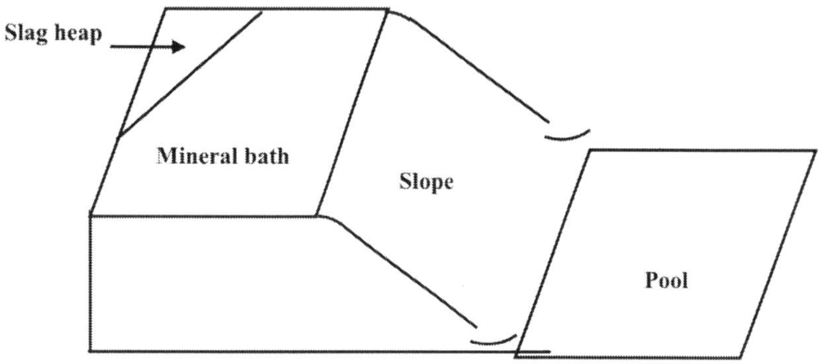

Fig.1. Schematic diagram of Bingcun Pb-Zn tailings area

Leaching of heavy metals flows into the water and farmland in the mine area and results in serious pollution. Due to the complexity of the mechanism of heavy metal toxicity to plants and their stability in the soil, the control and treatment of heavy metal pollution has been a hot research task for decades [3-4]. At present, heavy metal pollution in soil and water can be remediated by one or more of the following technologies: isolation, immobilization, toxicity reduction, physical separation or extraction [4-5]. But physical and chemical methods always require a great amount of engineering and investment, and can only repair small contamination area application. Phytoremediation is an effective polluted soil restoration technology developed since the mid19th century, namely heavy metal hyperaccumulator plants are screened, which grown in the polluted soil to absorb and remove heavy metals when they are harvested from the field [6-7]. It is relatively less cost and applicable for remediation of large area of heavy metal polluted soil especially in economic less-developed areas.

Equisetum ramosissimum and *Pteris* vittata are two dominant plants in Bencun Pb-Zn Tailing. On the edge of slope near the pool, the two ferns can grow strong almost as vigorously as in areas far from tailings, but other plants are rarely distributed (Figure 1). In this study, the characteristics of heavy metal absorption and enrichment of *Equisetum ramosissimum* and *Pteris vittata* growing close to the edge of a sewage pool in the Pb-Zn Mining Tailing were investigated through analysis on heavy metal content in mine soil around the roots, the aboveground and underground parts of the two ferns in order to provide theoretical basis for heavy metal pollution remediation and ecological restoration in Pb-Zn tailing area.

2 Materials and methods

2.1 Collection of soil and plant samples

Samples of the ferns and soil (0-10 cm below soil surface) close to edge of the sewage pool were collected in winter of 2014 at interval points in Bencun Pb-Zn Tailing, sieved (with 2 cm holes) and packed on the spot and taken back laboratory for further procedures of analysis.

2.2 Pretreatment of plant samples

Plant samples were divided into two parts: aboveground parts and underground parts, rinsed 3 times with tap water, and then cleaned with ultrapure water to avoid interference from heavy metal ions in soil. The samples were dried for 5 min at 105°C and at 80°C for 8 h, and broken in an organization crusher, then ground into fine powder in mortar. Powder samples were preserved for 2 h in an oven at 105oC and used in subsequent experiments described below after being sieved with 200 mesh sieve.

2.3 Pretreatment of soil samples

Soil samples were firstly air-dried, from which the stones and animal and plant residues were removed, and milled with agate rods and sieved through 2 mm nylon sieve, then ground into fine power in agate mortar and followed by sieved with 100 mesh nylon sieve (0.149 mm), finally preserved for subsequent experiments as described below.

2.4 Determination of soil pH value

10 g soil samples after pretreatment were taken into a 50 ml beaker, added 25 ml boiled distilled water, stirred vigorously for 1~2 min with a glass rod and let stand for 30 min, then measured by pH meter.

2.5 Digestion of plant and soil samples

Digestion of plant and soil samples was basically carried out as described byYang et al.(2012), but made some subtle changes. Briefly, 0.5 g of plant and soil powder was placed into a 50 ml ptfe high-pressure digestion tanks (KH-100 ml, Shanghai Yingdi Instrument Equipment Co., LTD.), and wet with 4 drops of double distilled water, then transferred to a ventilation cabinet for an overnight cold digestion after slowly adding 5 ml HNO_3 to remove organic matter after soaking for 0.5 h. Followed by addition of 2 ml H_2O_2, the digestion tanks were taken into the supporting steel liners. After being covered inner lids and firmly tightened stainless steel sheathes, the tanks with plant samples were transferred to a constant oven and heated for 3 h after the temperature rose to 120°C, but those with soil samples were transferred to oven and heated for 9 h at 160°C. When all the tanks naturally dropped to room temperature (25~30 °C) in the oven, they were placed on electric heating plates to dispel HNO_3 in the digestion liquid after their lids were carefully opened. Digestion liquid was transferred to 50 ml calibration test tubes for constant volume with 0.5% HNO_3 solution which was then shook, let stand, and used in subsequent tests. For each batch of soil samples, double distilled water was replaced samples in 2 copies of reagent blank test.

2.6 Determination the heavy metals of plant samples and soil samples

Heavy metal (Cr, Mn, Ni, Cu, Zn, Cd, Hg and Pb) contents in digested solution of plant and soil samples were determined in parallel using X Series 2 inductively coupled plasma mass spectrometry (ICP-MS, Thermo Fisher).

2.7 Assessment of heavy metal enrichment capability of ferns and soil heavy metal pollution

Heavy metal enrichment capability of plants was evaluated by indicators of TF (transfer factor) and EC (enrichment coefficient). TF of heavy metals for plant is the ratio of metal content in above-ground parts to that in underground parts. EC of heavy metals for plant is the ratio of metal content in organs or parts of plant to that in the soil where the plant grows, which reflects the plants' capability to enrich heave metals from its soil or water environment. The larger the enrichment coefficient of a metal is, the easier the element will be enriched. Single factor pollution index and Nemerow pollution index were applied to assess heavy metal pollution, and potential ecological risk index was calculated according to the method proposed by Hakanson [8]. The potential ecological risk level was assessed according to criteria of the second or third grade Environmental Quality Standard in China for Soils (GB 15618-1995) based on soil background values of Guangdong Province [9-11].

The calculation formula of single factor pollution index is $P_i=C_i/S_i$. P_i refers to the single factor pollution index of pollutants in soil; C_i refers to the measured content of pollutants in soil; S_i to the evaluation standard of pollutant. The formula of Nemerow synthetic pollution index is $P=\{[(C_i/S_i)^2_{max}+(C_i/S_i)^2_{ave}]/2\}^{1/2}$. $(C_i/S_i)_{Max}$ refers to the maximum of pollution index in soil; $(C_i/S_i)_{Ave}$ to the average pollution index in soil.

The formula of potential ecological risk index is: $E_r^i= T_r^i\times C_f^i=T_r^i\times C_i/S_i$; $RI=\sum E_r^i$. E_r^i refers to the single heavy metal potential ecological risk index; T^i_r refers to heavy metal toxicity coefficient, RI to the potential ecological risk index of a variety of various metals. The heavy pollution level and potential ecological risk of heavy metal pollution were determined according to the classification standard of heavy metal pollution level [12-13].

2.8 Data analysis

All data were proceeded and analyzed with Excel2003 and SPSS17, and multiple comparison of means was carried out by Duncan's（$P<0.05$）multiple range tests.

3 Results

3.1 Soil pH

The pH value of soil was 7.78, which revealed a weak alkaline. When evaluating the pollution level of heavy metal in soil, the numerical criteria should refer to the range of pH value above 7.5 in the second grade SSEQS (State Soil Environmental Quality Standard) in China or greater than 6.5 in the third grade SSEQS. Guo et al. found that in Guangdong Province, the spatial distribution pattern of soil pH in the Province had less change (mainly acidic), except that in Pearl River Delta and part of Qingyuan and Shaoguan (weak alkaline) City in the past over 30 years and industrialization and mining increased the soil pH in some areas[14].

3.2 Heavy metal content of soil

The data showed that the heavy metal content in soil was in descending order of Pb, Mn, Zn, Cu, Cd, Cr, Ni, Hg. Pb content in soil reached the maximum (13332.94 mg.kg-1), but Hg content was the minimum (only 4.12 mg.kg-1). The contents of Pb, Zn, Cd and Hg were far above the third grade SSEQS, and were 26.67, 10.81, 39.62, and 2.75 times more than the third grade SSEQS, respectively. Cu content was above the second grade SSEQS, but Cr and Ni content did not exceed national standard. In addition to Cr, the content of other heavy metals exceeded the background value in soil of Guangdong Province, that of Cd was even 707.50 times more than the background value, it is reported that the medium standard of Mn content in soil was 170 ~ 1200 mg kg-1, so the Mn content in the region exceeded the medium standard; the Ni content didn't exceed the national standard, but also 1.32 times more than the soil background value in Guangdong province (Table 1).

Table 1. The content of heavy metals in soil.

Heavy metal	Cr	Mn	Ni	Cu	Zn	Cd	Hg	Pb
Content(mg kg^{-1})	20.32	12575.67	19.05	272.92	5404.75	39.62	4.12	13332.94

3.3 Evaluation on the single factor and synthetic pollution index of heavy metals

According to the second grade SSEQS (but Mn content has not been set in the SSEQS), the single factor pollution index of Cr and Ni belonged to the clean degree, other five metals all had reached the pollution level. The decreasing order of single factor pollution index was Cd, Pb, Zn, Hg and Cu, the single factor pollution index of Cd, Pb, Zn, Hg, Mn and Cu reached the heavy level, but that of Ni reached the light degree (Table 2). Both the synthetic pollution index of heavy metals calculated on the basis of the second SSEQS and the soil background value in Guangdong Province reached the heavy pollution level.

Table 2. The index and grade of soil heavy metal pollution.

Heavy metal	P_{Cr}	P_{Mn}	P_{Ni}	P_{Cu}	P_{Zn}	P_{Cd}	P_{Hg}	P_{Pb}	$P_{综}$
S_1	0.08C	NA	0.32C	1.36L	18.02H	66.03H	4.12H	38.09H	48.45H
S_2	0.40C	45.07H	1.32L	16.05H	114.27H	707.50H	52.82H	370.36H	513.46H

Note: S means security, W means alert; C means clean; L means light pollution; M means moderate pollution; H means heavy pollution. S_1 was calculated based on the second grade State Soil Environmental Quality Standard, S_2 was calculated based on the soil background value in Guangdong Province, the second grade State Soil Environmental Quality Standard, NA means the data cannot be calculated because Mn content has not been set in the State Soil Environmental Quality Standard, the same as below.

3.4 Evaluation of potential ecological risk of heavy metals in soil index

Based on the second grade SSEQS, the descending order of ecological risk index was Cd, Pb, Hg, Zn, Cu, Ni, Cr, the potential ecological risk index for Cd was the highest and reached very strong level, those of Pb and Hg both reached strong degree, but those of Zn, Cu, Ni and Cr was lower and reached minor level, and even that of Cr was below 1 and had hardly any hazard. Based on the soil background value in Guangdong Province, the decreasing order was Cd, Hg, Pb, Zn, Cu, Ni and Cr, the potential ecological risk index of Cd, Hg and Pb reached extremely high degree, those of Zn and Cu reached high degree, but those of Ni and Cr was relatively low and reached slight degree (Table 3).

Table 3. The index and grade of potential ecological risks of soil heavy metal pollution.

Heavy metal	E_r^{Cr}	E_r^{Mn}	E_r^{Ni}	E_r^{Cu}	E_r^{Zn}	E_r^{Cd}	E_r^{Hg}	E_r^{Pb}	RI
S_1	0.16A	NA	1.59A	6.82A	18.02A	1981.00E	164.80D	190.47D	2362.86D
S_2	0.80 A	NA	6.61A	80.27C	114.27C	21225.00E	2112.82E	1851.80E	25391.57D

Note: A. Slight; B. Moderate; C. Strong; D. Very strong; E.Strmely strong

3.5 Heavy metal content and transfer coefficients of two ferns

Enrichment pattern of heavy metals including Cr, Mn, Ni, Cu, Zn, Cd, Hg and Pb in *Equisetum ramosissimum* was different from that in *Pteris vittata*. For both the ferns, heavy metal contents of the aboveground parts were below that of underground parts, namely the heavy metals were mainly concentrated in the underground parts. For *Equisetum ramosissimum*, the decreasing order of heavy metal content in aboveground parts was Zn, Mn, Pb, Cu, Ni, Cd, Cr, Hg, and that in underground parts was Zn, Mn, Pb, Cu, Cd, Ni, Cr, Hg. For *Pteris vittata*, the decreasing order of heavy metal content was Zn, Mn, Pb, Cu, Ni, Cr, Cd, Hg in aboveground parts, while that was Zn, Mn, Pb, Cu, Ni, Cd, Cr, Hg in underground parts, i.e, besides that Cd and Cr were in reverse order, the other 6 metals are in the same order.

Except that the transfer coefficient of Hg in *Equisetum ramosissimum* was above 1, all the other transfer coefficients of heavy metals in both *Equisetum ramosissimum* and *Pteris vittata* were below 1. As for the same heavy metal, the transfer coefficients of *Equisetum ramosissimum* were above those of *Pteris vittata* (Table 4).

Table 4. Heavy metal content (mg·kg^{-1})of above and under ground parts and transfer coefficients.

Species	*Equisetum ramosissimum*			*Pteris vittata*		
Part	Above-ground	underground	TF	Above-ground	underground	TF
Cr	1.82	4.63	0.39	1.40	6.10	0.23
Mn	427.16	1487.33	0.29	56.78	1233.80	0.05
Ni	9.81	12.26	0.80	4.06	20.35	0.20
Cu	52.70	110.24	0.48	9.00	92.18	0.10
Zn	819.41	2734.91	0.30	176.55	1460.52	0.12
Cd	6.88	18.12	0.38	0.88	11.76	0.07
Hg	0.17	0.11	1.55	0.11	0.29	0.38
Pb	300.37	697.97	0.43	36.23	783.08	0.05

3.6 Enrichment coefficient of above and underground parts for two ferns

Table 5. Heavy metal enrichment coefficients of above and under round parts for two ferns.

Species	*Equisetum ramosissimum*		*Pteris vittata*	
Part	Above-ground parts	Underground parts	Above-ground parts	Underground parts
Cr	0.09	0.25	0.07	0.30
Mn	0.03	0.13	0.004	0.10
Ni	0.51	0.73	0.21	1.07
Cu	0.19	0.35	0.03	0.34
Zn	0.15	0.33	0.03	0.27
Cd	0.17	0.26	0.02	0.30
Hg	0.04	0.03	0.03	0.07
Pb	0.02	0.05	0.0027	0.06

Except that the enrichment coefficient of Ni in *Pteris vittata* was above 1, the enrichment coefficients of all metals in both ferns were below 1. In *Equisetum ramosissimum*, the descending order of 8 heavy metals in the aboveground parts were Ni, Cu, Cd, Zn, Cr, Hg, Mn, Pb, and that in underground parts were Ni, Cu, Zn, Cd, Cr, Mn, Pb and Hg. In *Pteris vittata*, the order of enrichment coefficients in aboveground parts were Ni>Cr>Cu=Zn>Cr>Pb, and that in underground parts were Ni>Cu>Cr=Cd> Zn>Mn>Hg>Pb. The ECs of both ferns were low, and that in aboveground parts were below in the underground parts (Table 5).

4 Discussion

4.1 Evaluation of heavy metal content and pollution in soil

Soils in most regions of Guangdong Province is acid[10], but the soil in Bencun Pb-Zn Tailing was weak alkaline, which may be due to large volume $CaCO_3$ application during the mine construction and operation in recent years. The soil heavy metal pollution in this tailing is undoubtedly a composite pollution, the content of Pb, Zn, Cu, Cd and Hg in the soil was all above SSEQS and the soil background value in Guangdong Province, the Mn content was 45.07 times above the soil background value in Guangdong Province and also above Mn moderate standard. Liu et al. found the content of Pb, Zn, Cd and Cu in soils on a ladder slope in Bencun Pb-Zn Tailing[6], the results were shown that the content of 4 heavy metals were below that in soils close to edge of the sewage pool in this study, which suggested that the soil pollution of heavy metals close to pool, including Pb, Zn, Cd, Cu, was more serious than other areas in the tailing, because long term rain erosion caused accumulation of heavy metals in pools surrounding the tailing. Based on the analysis of soil heavy metal pollution index and pollution levels, it is not difficult to find that the main factors in the soil heavy metal pollution were Cd, Pb, Zn and Hg which had reached heavy pollution level, so the accumulation of these 4 heavy metals was key causes for the soil pollution. Soil RI of heavy metals is high, indicating the soil has high potential ecological risk of heavy metal pollution. In Bencun Pb-Zn Tailing, Cd, Hg and Pb were the main factors of the potential ecological risk, followed by Cu, Zn, so the soil in this tailing does not meet the requirements for growing crops. These results were consistent with studies made by Chu & Luo [15] that in Qixiashan Pb-Zn Tailing, the main pollution elements were Pb, As, Cd and Zn, as well as a mild Cu pollution in some soils. In Changhua Pb-Zn Tailing, the heavy metal pollution, including Cd, Cu, Pb and Zn, reached very high pollution levels [16]. As described by Li et al. [17], in Shuikoushan Pb-Zn Tailings, Cd pollution was the most serious, and there also was a serious pollution of Cd, Pb, Hg and mild compound pollution of Zn and As. Therefore, in Pb-Zn tailings, there usually exists a serious composite pollution of Cd, Pb, Zn and Cu, which also occurred in the present study. In addition, Hg pollution cannot be ignored in Bencun Pb-Zn Tailing.

4.2 The characteristics of heavy metal absorption and enrichment of two ferns

The definition of heavy metal hyperaccumulator widely used at present is the reference values of heavy metals in plants proposed by Baker and Brooks [18], the critical content of Zn is 10000 $mg.kg^{-1}$, Cd 100 $mg.kg^{-1}$, Hg 10 $mg.kg^{-1}$, Pb, Cu, Ni, Co 1000 $mg.kg^{-1}$, and both EC (enrichment coefficient) and TC (transport coefficient) of these plant should be above 1.

In the present study, Pb or Zn content of underground parts in *Equisetum ramosissimum* and *Pteris vittata* are very high and above the normal content range in plants (0.1~4.17 $mg.kg^{-1}$ for Pb, 1.0~160 $mg.kg^{-1}$ for Zn)[7], but the content of two ferns were far below the critical content standard of hyperaccumulators. When the heavy metal pollution in soil is heavy, the plant growth and crop yield may decrease. It was reported that when the Zn concentration reached 140~310 $mg.kg^{-1}$ in soil [3], it could inhibit the growth and development of barley, rye grass, clover, buckwheat and other plants, but *Equisetum ramosissimum* and *Pteris vittata* could grow in Bencun Pb-Zn Tailing, and the appearance and growth of these individuals distributing in the tailing were almost the same as those distributing in the non-polluted areas. But the heavy metal content and TC were below the critical standards of hyperaccumulator, for example, the Pb content of above ground parts in *Pteris vittata* was only 36.23 $mg.kg^{-1}$, the TC was only 0.05, so the two ferns were not typical hyperaccumulators.

Cd content in plants is usually below 3.00 $mg.kg^{-1}$, but that in those growing in Cd-enriched soil may be above 20.00 $mg.kg^{-1}$, even if growing in Cd-contaminated soils, it is rarely above 100.00 $mg.kg^{-1}$[5]. In this study, the Cd content of *Equisetum ramosissimum* and *Pteris vittata* was above 3.00 $mg.kg^{-1}$, suggesting the two ferns had relatively strong Cd tolerance, even though Cd content did not exceed the threshold for hyperaccumulator.

Mn content in plants is usually in the range of 1.0~700.0 $mg.kg^{-1}$[7], but that of *Equisetum ramosissimum* and *Pteris vittata* in this study was above the normal value range. The EC was too low and below 1. So the two ferns are likely to have strong adaptability to high Mn-enriched soil and can be used as a pioneer in ecological restoration of Pb-Zn tailings.

Cu content of *Equisetum ramosissimum* was above that of general plants whose Cu content is usually 0.4~45.8 $mg.kg^{-1}$[7] and mainly accumulated in underground parts. TC and EC were both below 1, but it has been reported in other researches that EC of *Equisetum ramosissimum* was above 1, indicating *Equisetum ramosissimum* has some Cu enrichment ability and can be regarded as a phytoremediation plant in Cu-polluted soils[19].

The normal Ni content for most plants is 0.05~5.00 $mg.kg^{-1}$. When Ni content in the dry matter of plants reaches 0.01~0.15 $mg.kg^{-1}$, leaves appear on urea toxic symptoms and had tip necrosis [20]. In this study, except in the above ground parts of *Pteris vittata*, Ni content of other parts in two ferns was above 5.00 $mg.kg^{-1}$, so the two ferns can be regarded as ideal Ni pollution tolerant plants.

Hg content of *Equisetum ramosissimum* and *Pteris vittata* was far above that reported in other plants (0.010~0.050 $mg.kg^{-1}$) [21], which confirmed the two ferns were subject to a certain level of Hg pollution, but they were still not typical Hg hyperaccumulators because Hg content of the two ferns was below the critical content value (10 $mg.kg^{-1}$) and EC was below 0.1. Han (2012) carried out a cultivation experiment of *Pteris vittata* in 10 $mg{\cdot}kg^{-1}$ Hg solution and found its TC and EC were 0.38~0.71 and 0.01~0.06 respectively [22], these results were similar to the results in this study. The Cr content in plants are usually 0~8.4 $mg{\cdot}kg^{-1}$ [7], that of these two ferns were also within the range.

In conclusion, the soil in Bencun Pb-Zn Tailing was severely polluted by heavy metals, especially Cd, Cu and Mn. The potential ecological risk of heavy metals was assessed to be very strong based on soil background values of Guangdong Province and at high risk according to criteria of the second grade SSEQS, and Cd, Hg, Pb were the main factors leading to potential ecological risk. The content of 8 heavy metals in *Equisetum ramosissimum* and *Pteris vittata* did not reach critical content of hyperaccumulator, so neither of them was typical hyperaccumulator, but both had a certain tolerance to the combined pollution of Pb, Zn, Mn, Cu, Cd, Hg and Ni. The underground parts of two ferns had a relatively strong enrichment, the EC of Ni in underground parts of *Equisetum ramosissimum* was near 1, and that in *Pteris vittata* was over 1, so the two ferns could be utilized as Ni potential enrichment plants. The transfer ability of 8 heavy metals by the two ferns was not strong, except for Hg by *Equisetum ramosissimum*, the TC of other metals by *Equisetum ramosissimum* and that of the 8 heavy metals by *Pteris vittata* were below 1. The heavy metal pollution in ground soil could increase after the ferns withered, but the heavy metals could be fixed in the soil and could not unlimitedly migrate and diffuse, water heavy metal pollution was then reduced. The two ferns can grow vigorously in the tailing indicating they had strong adaptability to the tailing habitat, so they can be used as pioneers in ecological restoration of Pb-Zn tailings.

Acknowledgements

This work was financially supported by Key Project of Science and Technology Innovation of Colleges and Universities in Guangdong Province (cxzd1132); Guangdong Province Natural Science Foundation Project (815140150100003) and Special Funds of Talent Introduction of Colleges and Universities in Guangdong Province.

References

1. Chinese Environmental Protection Administration. State soil environmental quality standard. Beijing: China Standard Press (1997).
2. Q.H.Yang, H.N. Liu, Q.H. Li, J.Q. Li and C. Q. Tu. Study on the microbial acitivity of rhizosphere soil for three dominant plants in Pb-Zn wasteland in Eastern Guandong Province. Chinese Agricultural Science Bulletin, Vol. 28(2012),p.56-64.
3. J.H. Yang. Soil Pollution and Control. Science Press, Beijing (1995).
4. C.A. Evanko and D.A. Dzombak: Remediation of metals-contaminated soils and groundwater, in:Technology Evaluation Report, TE 97-0-1, Ground-water Remediation Technologies Center, edited by P.A. Pittsburgh (1997).
5. S.R. Tang. Phytoremediation of contaminated soils. Science Press, Beijing (2006).
6. H.N.Liu, Q.H.Yang, H.S.Yang, J.Q.Li and D.L.Liu. Characteristics of heavy metals Pb-Zn mine absorption and accumulation of for three dominant plants in tailings eastern Guangdong. Guihaia,Vol.32(2012), p.743-749.
7. C.Lu, Y.H.Wang and J.W.Yang. Soil heavy metal pollution and dominant plants selection in Pb-Zn mining areas of Guangxi. Chinese Journal of Soil Science, Vol.41(2010),p. 1471-1474.
8. L.Hakanson. An ecology risk index for squatic pollution control: A sedimentological approach.Water Research, Vol.14 (1980), p. 975-1001.
9. S.D.Wu. The atlas of soil environmental background value in the People's Republic of China. China Environmental Science Press (1994).
10. L.F. Xu and T.H. Liu. The zonal differentiation of soil environmental background values and critical contents in Guangdong. Journal of South China Agricultural University, Vol.17(1996), p. 58-62.
11. S.W.Chai, Y.M.Wen, Y.Zhang and J.F.Zhao. Application of Index of Geoaccumulation (Igeo) to Pollution Evaluation of Heavy Metals in Soil. Journal of Tongji University (Natural Science), Vol. 34 (2006), p. 1657-1661.
12. Q. Li, Y.F. Han, D.S. Xu and Y. Gao. Enrichment characteristics and contaminant assessment of heavy metals in three food crops around the gangue dumps. Journal of safety and Environment, Vol.14(2014),p.321-326.
13. Y.Q.Luo, Y.P.Chen, L.Tao, Y.Q.Li and X.M.Wang. Investigation and evaluation on heavy metals pollution in farmland soil in Lanzhou City. Journal of Gansu Agricultural University, Vol.45(2011),p. 98-104.
14. Z.X.Guo, J.Wang，M.Chai, Z.P.Chen, Z.S.Zhan, W.P.Zheng and X.G.Wei. Spatiotemporal variation of soil Ph in Guangdong Province of China in past 30 years. Chinese Journal of Applied Ecology, Vol.22(2011), p.425-430.
15. B.B. Chu and L.Q. Luo. Evaluation of heavy metal pollution in soils from Nanjing Qixiashan Lead-Zinc Mines. Rock and Mineral Analysis,Vol.29(2010),p. 5-8,13.
16. S.F.Lu. Soil heavy metal elements content characteristics and pollution evaluation in Changhua lead-zinc mine tailings wasteland of Hainan. Hainan Normal University, Hainan, China (2012).
17. G. Li, F.P. Tong and Z.H. Liu. Analysis on pollution of heavy metal in Shuikoushan Pb-Zn mining and smelting area in Hengyang. Journal of Central South University of Forestry and Technology, Vol.32(2012),p.105-110.
18. A.J.M. Baker and R.R.Brooks. Terrestrial higher plants which hyperaccumulate metallic elements–a review of their distribution, ecology and phytochemistry. Biorecovery, Vol.1(1989), p. 81–126.
19. Y.Li and L.Chu. The uptake and accumulation of Cu in *Hippochaete ramosissimum*. Acta Ecologica Sinica, Vol.28 (2008), p.1565-1573.
20. X.X.Long and X.E. Yang. Nickel nutrition of plants. Chinese Journal of Soil Science, Vol. 31 (2000), p.39-42.
21. D.M. Zheng, Q.C. Wang and Z.S. Zhang. Transfer of mercury in soil-plant system of Wuli river, Huludao City. China Enviromental Science, Vol.27(2007),p. 676-680.
22. S.H. Han. A Comparative study on Removal Performance of Hg and Cd in Agriculture Soils for Several Plants. Donghua University, Shanghai (2012).

Fast Tomography Imaging System for Material Surface Based on Doublet-cylinder-lens in Intelligent OCT

Shijun Xu [1,a] and Xiaoling Ren [2,3,b]

[1]School of Science, Xi'an Technological University, Xi'an 710021, China
[2]School of Computer, Xi'an Polytechnic University, Xi'an 710048, China
[3]Department of Materials Science, University of Delaware, Newark, DE 19716, USA
[a]xushijun000@sina.com, [b]renxiaoling2@163.com

Abstract. A fast lateral line-scan mode in OCT was investigated. The elimination of spherical aberration, the beam mode and the focus of doublet-cylinder-lens, were theoretically analyzed. The parameters and signal characteristics of the confocus microscopy system with doublet-cylinder-lens were analyzed and simulated. The relation between the transverse resolution and the numerical aperture and the depth of focus (DOF) was discussed in the OCT. Our results show that the spherical aberration and the sine aberration are effectively eliminated in the doublet-cylinder-lens with K9-ZF2 glass. The beam is uniformly focused on the arc-sagitta-plane of cylinder lens, and the parallel interference pattern with axial symmetry is in favor of improving transverse resolution. The transverse resolution depends upon the numerical aperture and the DOF. The scanning velocity of doublet-cylinder-lens confocus microscopy system is 103 times faster than the traditional spot-scan, while the transverse resolution is 15μm and the constraints of aperture 50.8 mm. This advanced system will be valuable in real-time imaging to material surface.
Keywords. Material surface testing, optical coherence tomography (OCT), confocus microscopy, line-scan, doublet-cylinder-lens

1 Introduction

For optical coherence tomography system (OCT) on testing material surface, main restriction or shortcoming focus on the imaging velocity of the scanning process, which complexity inclued two parts called axial scanning and transverse scanning. Thus, many solutions to solve this problem has been explored. Specifically, one way was to enhance the velocity of axial scanning by application of different scanning devices [1]. Although the velocity could be boosted, the demand for high speed, stability and accuracy may cause very high investment. Another way was proposed in 2000 by Bourquin etc, which could achieve the parallel OCT of video rates (Parallel OCT) [2]. But this full-range detecting technology scanning has a very small range, so it need the intelligent CMOS image sensor, the lack of communication mode makes it vulnerable to interferences made by different $1/f$ noise sources, as well as lack of narrowband heterodyne demodulation technology makes interferences from noises relying on bandwidth intensified.

Line focus is proposed in 1999 by Suekuni etc. researching about spectrum OCT scanning mode. This way can get optical path difference directly through the Fourier transform and spectral density function [3-5]. This line scanning mode is similar with scanning mode, mainly has expansion interms of the transverse scan and reduced the cycle of transverse scan process. The use of slit in the expanded way of light decreased the flux while restricting the shape of the beam, thus reduced the efficiency of light source use, therefore, and caused less depth imaging and low SNR.

In our study, a rectangular aperture is used behind the expanded beam, and we obtain line focus with a cylinder lens. To reducing the scanning time, samples are linearly scanned through the cylinder lens of focal microscopy system. This line focusing model could transfer horizontal point scanning into parallel scanning mode, simplified the scanning process, and the shortened the scanning time. In the mean time, it can also retain the advantages of point scanning, which can effectively restrain the stray light with the confocal scanning. The time for getting 3D chromatography image could be shortened by three orders of magnitude.

2 Drawbacks of OCT with Point-focus-scan Mode

In the line scanning of OCT with point-focus-scan mode, the light beam going through the beam splitter is focused to a point via focusing lens. Light focus point will stay or suspend on each scan position of the sample. Finished the axial scanning, it will be moved to the next position by the line scanning device, and then going on the axial scanning. Firstly, one keeps Y direction invariable and conducts point scanning along the X direction. Secondly, the light focus point is moved to the next row of Y direction. Finally, the above point scanning along the X direction is repeated. Obviously, the repeating scanning along the X & Y point by point, which is "static-accelerate -slow-static" mode, will exhaust much time [1-2].

3 Line Scanning Structure and Characteristic

In the general OCT system, after expanding the source light beam we set a rectangular or square aperture in front of the expanding beam, and put a confocus microscopy system with double-cylinder-lens near the sample. If we use the linear CCD or CMOS detector, which match with the double-cylinder-lens, a rapid transverse line scanning should be realized.

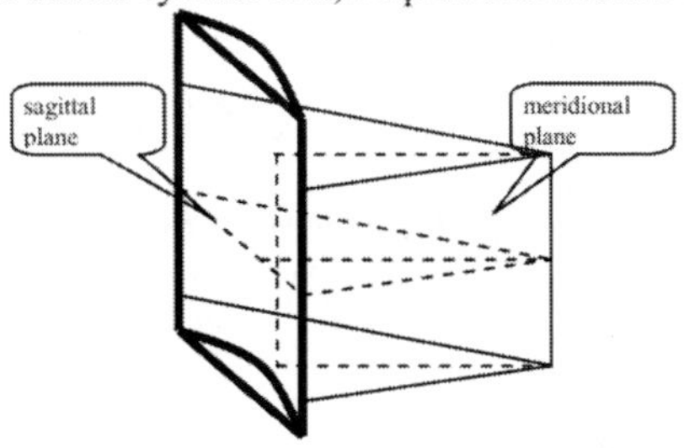

Fig.1. Imaging mode of cylinder lens.

The Beam Mode and the Focus of Doublet-cylinder-lens. As shown in the figure1, to show the three-dimensional light beam form, we plot two mutually perpendicular planes, one plane parallels to the beam axis and contains the interface normals (meridional plane), the other plane also parallels to the beam axis and is perpendicular to the meridional plane, that is to say a sagittal plane. The characteristics of the cylindrical lens are as follows: the beam focus on the sagittal plane with the lens, and the light parallelling the meridional plane will not fracture. Therefore, these characteristics not only realize the line focusing but also can ensure the same strength on every point of the focusing line in theory.

The focal length of the cylindrical lens can be calculated according to the formula of the thin lens

$$f=r_1r_2/(n-1)(r_2-r_1) \tag{1}$$

where n represents refractive index of single lens, r_1 and r_2 represent two spherical radiuses respectively. As the flat convex cylinder lens, Eq.(1) can be simplified as follows

$$f=r_1/(n-1) \tag{2}$$

Common Spherical Aberration of the Cylindrical Lens. Generally, spherical aberration appears in imaging process of the optical systems. A monochromatic light beam from a point on the axis goes through an optical system, if the beam with different apertures focuses on different positions of the axis, there will be a dispersion light spot near the image point, which is named spherical aberration of the imaging system. Therefore, spherical aberration is the optical aberration from an imaging system to axial object points with monochromatic and wide (in diameter) beam.

The junior spherical aberration to cylindrical lens is

$$\delta L' = -\frac{h^4}{2n_1u_1'^2}\left\{\frac{n+2}{n}\phi\rho^2 \cdot \frac{2n+1}{n-1}\phi^2\rho + \frac{n^2}{(n-1)^2}\phi^3\right\} \tag{3}$$

If the lens are exposed to the air, then $n_1=1$, $h=l\times u$ in which l is object distance, u is light aperture angle, h is beam height content in normal section of the object space. Where u_1' represents aperture angle of the image space, φ is light focal power of lens, ρ is lens curvature of the first surface. According to Eq.(3), when $\mathrm{d}\delta L'/\mathrm{d}\rho=0$, we obtain the minimum of spherical aberration as follows

$$\delta L_0' = \frac{n(4n-1)}{8(n-1)^2(n+2)}h^2\varphi \tag{4}$$

As per Eq.(4), the spherical aberration can not be eliminated with single thin lens.

Corresponding to thin lens group, the junior spherical aberration [6] is

$$\delta L' = -\frac{h^4}{2n_1u'^2}[A_1(n_1,\varphi_1,\rho_1,\sigma_1)+A_2(n_2,\varphi_2,\rho_2,\sigma_2)] = -\frac{h^4}{2n_1u_1'^2}\left\{\frac{n+2}{n}\varphi\rho^2 \cdot \frac{2n+1}{n-1}\varphi^2\rho + \frac{n^2}{(n-1)^2}\varphi^3\right\} \tag{5}$$

Elimination of Spherical Aberration of Doublet-cylinder-lens. For Spherical aberration correction, there are three methods: (1) Use diaphragm to cover part of the light through the edge of lens, thus to reduce the aperture angle, though simple, this method can narrow the optical diameter of the lens and reduce the intensity of illumination on image plane; (2) Via integration of two curved surface of the lens will become appropriate aspheric surface, which can eliminate the Spherical aberration, but the treating of non-spherical surfaces are very difficult; (3) applications of double glued structures or double separation and reunion structures can eliminate the Spherical aberration, so in order to eliminate the Spherical aberration, the doublet-cylinder-lens has a lot of potential.

However, no matter what kind of optical system we use, we can only eliminate the Spherical aberration caused by paraxial rays, but not the Spherical aberration caused by light of all aperture angles, so still remains the spherical aberration. Application of he doublet-cylinder-lens could solve this problem efficiently, specific explanation s are as follow.

Because plus lens often generate positive spherical aberration while concave lens often generate minus spherical aberration, the doublet-cylinder-lens system can be considered as optical system which was made up by plus lens and concave lens, which have three refraction surfaces. In generally, one makes the lens curvature ρ_1 of glued surface as variable, and chooses the glasses pair as K9+ZF2. Relevant parameters are as follows: K9 glass n_D=1.51637, v=64.1, n_F=1.52196, n_c=1.51389; ZF2 glass n_D=1.67268, v=32.2, n_F=1.68747, n_c=1.66662. Where v is Abbe number. Substituting all the parameters into Eq.(5) and setting $\delta L'$=0, we get

$$\delta L_0' = -\left(19090\rho_1^2 + 871.33\rho_1 + 9.943\right) = 0 \tag{6}$$

The discriminant of Eq.(6) is less than zero (the discriminant is very small), so that the solution ρ_1 can be considered as the solution of the elimination of spherical aberration. According to calculation ρ_1=-0.02282/mm, r_2=-43.82mm, we get the sine aberration OSC=0.00049. The OSC is very small, thus it indicates that the spherical aberration and the sine aberration could be corrected in the same time.

Using the junior spherical aberration Eq. (6) of doublet-cylinder-lens and Matlab software, we get three sorts of curve about spherical aberration respectively (in figure 2). Figure 2(a) is the spherical aberration curve with Full Correction method, the curve demonstrates that other rays all have certain spherical aberration, except for the paraxial rays and the edge rays. Figure 2(b) is the spherical aberration curve with Over Correction method. Figure 2(c) is the spherical aberration curve with Under Correction method.

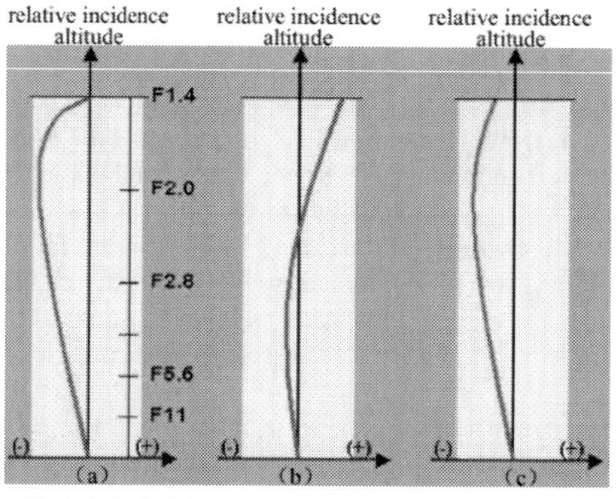

Fig.2. Spherical aberration curve of doublet-cylinder-lens.

Focal Length of Doublet-cylinder-lens. Considering the correction of spherical aberration and sine sent in transverse scan of the samples, doublet-cylinder-lens was used. Its focal length can be thought of the focal length as two thin lens joint together, whose focal length of the negative lens is

$$f_2=r_1r_2/(n-1)(r_2-r_1) \tag{7}$$

Whereas the focal length of the doublet-cylinder-lens is

$$f_{double}=r_1r_2/[r_1(n_2-1)-r_2(n_1+n_2)] \tag{8}$$

where n_1 is the refractive index of the plus lens, while n_2 is the refractive index of the negative lens.

Confocus Microcopy with Doublet-cylinder-lens. In the scanning process of using confocus microcopy system with doublet-cylinder-lens, the stray light was greatly eliminated, the surrounding fluorescence generated background and stray light interference signals was restrained, so, this system has the gateway role[4,7]. The light outside of the objective lens focal plane was blocked, the signal-to-noise ratio of the probe got greatly improved, so the imaging surface has a very high contrast and clarity, and also can the system provide a high resolution.

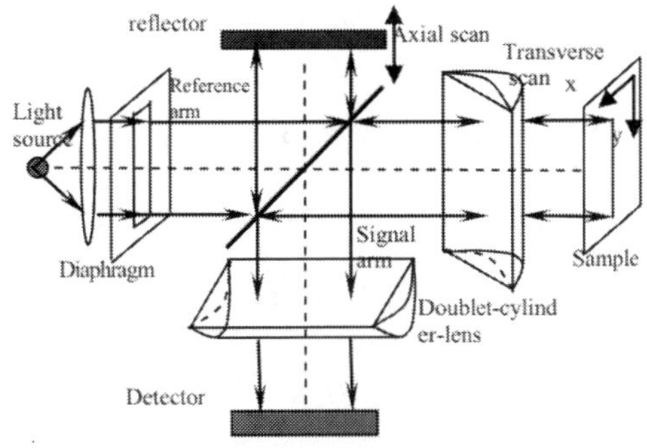

Fig. 3. OCT system with doublet-cylinder-lens.

Signal Characteristics. Fig.3 is the sketch of the OCT system with doublet-cylinder-lens. The light vector complex amplitude of the light from the light source can be expressed as

$$E=A_0\exp(\mathrm{i}\varphi) \tag{9}$$

After beam splitter, the light was divided into two beams, the first reflected by the beam splitter, then through the reflector, light vector of reference ray is

$$E_r=2^{(-1/2)}A_0\exp[\mathrm{i}(\varphi+kl_r)] \tag{10}$$

where k is the wave vector. The second beam goes through the splitter, then it is focused by the doublet-cylinder-lens and projected to the surface and inner of the sample. The optical path differences caused by different focal points vary with different space position. So that the light vector distribution from the sample arm is

$$E_s=2^{(-1/2)}A_0'\exp[\mathrm{i}(\varphi+kl_s)] \tag{11}$$

where $l_s=l_s(y)$ is the optical length of the sample arm, A_0' is the amplitude of the signal light modulated by the sample. After line focused by the cylinder lens, signal light reflected by samples and reference light reflected back by reflection lens were received by the detector. The overall interference light vector is

$$E=E_r+E_s \tag{12}$$

Assume $I_0=A^2/4$, $I'=A'^2/4$ (A and A' are considerated with loss of light intensity caused by the components like the doublet-cylinder-lens), then the stacked light intensity by two waves is

$$I=\langle EE^*\rangle=I_0+I'+2\sqrt{I_0I'}\cos k[l_r-l_s(y)] \tag{13}$$

Let $\Delta=l_r-l_s(y)$, if $I_0=I'$, we obtain

$$I=2I_0(1+\cos k\Delta) \tag{14}$$

According to the type above, changes in the level of interference is still relevant with optical path difference. Because of the waken coherent of the light sources, the only reflected light or scattering light from the sample with a particular depth can coherent with the reference beam. The reflected lights from different depth of the samples have different phase delay, along with the increase of delay, interference signals fell sharply. The interference signals contain information of the sample, Therefore, after filtration and amplification we can get images of the samples.

In addition, the confocus microscopy system can increase the resolution of the system. Because the signal strength from outside of the focus is much weaker than within, and can greatly reduce the unwanted flickering noise and the noise from the scattering light.

System Signal Simulation. In the simulation to the above optical system, we adopted the ZEMAX from the INFOTEK corporate with monochromatic light (the wavelength is 1310 nm).

 DOI: 10.1051/matecconf/201710004031

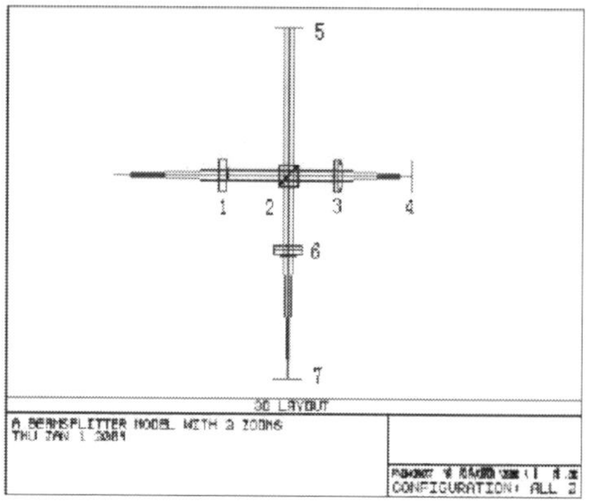

Fig.4. Simulation of the OCT system.

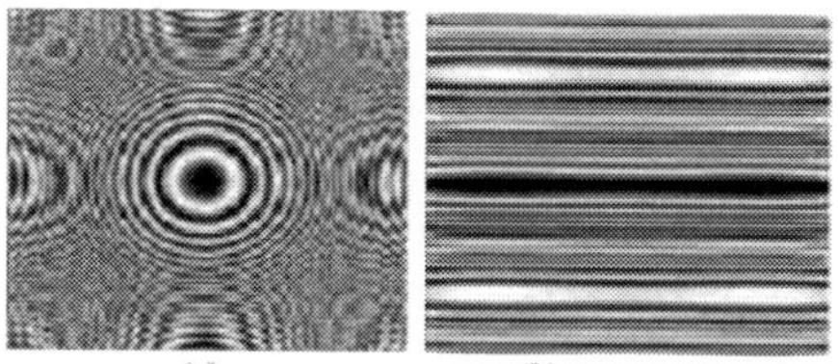

(a) (b)

Fig.5. Two kinds of interference fringe.

The point lamp or light source are expanded to a collimator beam with a single lens. As shown in figure 4 of 3D shape of the main light through the center, component 1 is expanding lens, component 2 is the beam spliter, 3 and 6 are the doublet-cylinder-lens, 4 and 5 the plane mirror, 7 the imaging surface or detector, the light path contains plane mirror 5 is the reference arm, the light path contains the doublet-cylinder-lens 3 is the sample arm. In designing the length for the reference arm, we have already compensated the light path difference caused by cylinder mirror, thus the light paths on two arms is equal.

Simulation experiments are divided into two kinds of cases. One is thin round beam interference patterns on Michelson Interferometer, the simulation results showed in figure 5(a). The other simulation results are as shown in figure 5(b), with the doublet-cylinder-lens, as the same specification in the light path of the sample arm, as in front of the detecting surface, and as the compensate for the optical path difference caused by them.

In figure 5(a) the interference patterns are concentric circles, while in figure 5(b) they are parallel interference plan stripes, indicates that under the limit of beam, the cylindrical lens are the same in terms of focusing light, namely on the line of focus, every points have the same optical path difference. This kind of axial symmetry parallel interference patterns are beneficial to enhance the Transverse resolution of the line scanning OCT system.

4 Resolution Ratio and Focal Depth of Doublet-cylinder-lens Microscopy System.

Theoretical Formula of the Resolution Ratio and Focal Depth. The adoption of doublet-cylinder-lens line scanning confocus system has the same axial resolution with point scan of OCT system, and they are all dependent on the coherent length of the light sources [8]

$$\Delta z_{FWHM} = \frac{2\ln 2}{\pi}\cdot\frac{\lambda^2}{\Delta\lambda_{FWHM}} \tag{15}$$

where λ is the centre wavelength, $\Delta\lambda_{FWHM}$ is the spectrum width of the light source.

The horizontal resolution Δx do not simply depend on the numerical aperture(NA) of the beam, it is also relevant with the distances between image sensitive units on the detector. The horizontal resolution of this doublet-cylinder-lens confocus microcopy system can also be shown as Eq. (16)

$$\Delta x = \frac{4\lambda}{\pi}\cdot\frac{f}{d} = \frac{4\lambda r_1 r_2}{\pi d[r_1(n_2-1)-r_2(n_1+n_2)]} \tag{16}$$

where d is the caliber of the parallel beams of the arm (and a square aperture caliber consistent).

If the calculated transverse resolution Δx is greater than the distance of all probe light-sensing units. The transverse resolution of the line OCT scanning system is the value of Δx; If the calculated transverse resolution Δx is smaller than the distance, the transverse resolution of the line OCT scanning system is the size of the light-sensing units. Obviously, the transverse resolution of the line OCT scanning system can not be higher than the size of the linear array CCD light-sensing units.

For the line scanning OCT system, Its focal depth (DOF) was decided by two times beam parameters of the confocus Gaussian beams, using Eq.(16), we have

$$b=2Z_R=(\pi/2)(\Delta x^2/\lambda) \quad (17)$$

Relationship between Transverse Resolution & NA & Focal Depth. When increasing the transverse resolution, the gathering depth will decrease, which is similar with the nature and optical microscope. When using low numerical aperture, we can get great converged depth, but at this time the converge radius of the converging beams are bigger, so the transverse resolution will be lower; And when the system using high numerical aperture, the beam waist radius of converge beams will be lower at this time, which will increase the transverse resolution, but decrease the Confocal parameters, thus to decrease the converging depth.

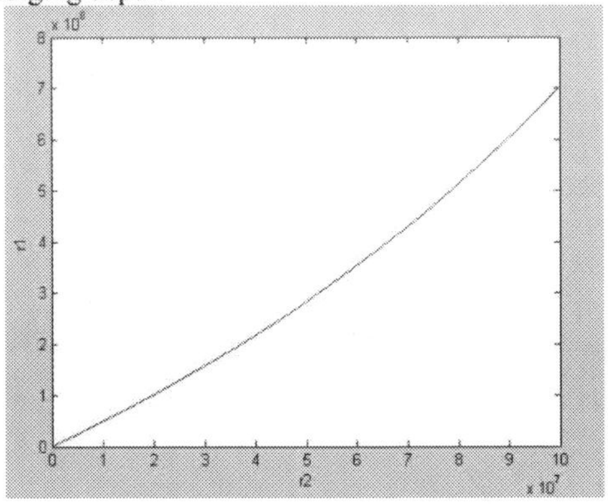

Fig.6. Relationship of r_1 & r_2 in doublet-cylinder-lens.

In order to improve the scanning speed, the wider width of the line scanning is the better. But in fact, the line width is not unlimited of increase. We should make optimization for those parameters. (1) On the scanning line, energy reduce of single point will cause the whole detecting ability to reduce; (2) To increase the line width, it have to increase beam diameter after beam expanding, this will also increase the numerical aperture, and decrease the focal depth; (3) The lager the line width, the more stray light on every detecting point, These stray light will interact with backscatter light from other points, which will reduce the signal-to-noise ratio(NSR), and affect the quality of scanning.

The parameters designed for this work are as follows: choosing the light center wavelength 1310nm, the caliber of the parallel beams with the sample arm for d 50.8mm. For glasses pairs, taking *K9+ ZF2* (above parameters have be given). As shown in figure 6, when the transverse resolution is 15μm, using Matlab software, we obtained the relationship between r_1 and r_2. When the axial resolution is 15μm, we know the bandwidth $\Delta\lambda_{FWHM}$ is 50.5nm according to Eq. (15).

5 Summary

The doublet-cylinder-lens not only inherited advantages of the traditional OCT system as its depth as point scanning and independent transverse scan resolution but can also Make up faults for traditional point scanning OCT technology, as to change the point scanning into line scanning, and the transverse scan resolution is also increased.

The main advantage of using confocus scanning mode is the flitting of the out focusing background. By the spherical aberration analysis for the cylindrical lens, and by test and verify via K9+ZF2 glasses, it's clear that this method can effectively eliminate the spherical aberration and the sine aberration. After analysis about the relationship between r_1 and r_2, we design the specific parameters for the doublet-cylinder-lens system. But because the line scan will distract energy of the detecting light, its detecting depth and signal-to-noise ratio will decrease with the increase of the line length, so, detectors with high sensitivity of detecting is essential.

Under the aperture 50.8mm×50.8mm, the horizontal resolution is equal to 15μm. In comparison with the point scanning mode under the same conditions, the new scanning speed at least achieves 3387(50.8mm/15μm) times than that of single point scanning mode, theoretically, the scanning speed can be improved by three orders of magnitude. Thus the imaging velocity was significantly promoted. In terms of living body imaging, the high-speed-scan mode can effectively reduce the influence from the physiological activities to the living tissue. Therefore, this research has higher value of applying to the real-time imaging on material surface.

Acknowledgements

This work was sponsored by the Scientific Research Fund from Shaanxi Province Education Committee (No.06JK283) and supported in part by a grant from the National Natural Science Foundation of China (No.F051205).

DOI: 10.1051/matecconf/201710004031

References

1. X.M. Liu, M.J. Cobb, X.D. Li: Optic Letters.2004, 29(1), P. 80-82.
2. S. Bourquin, V. Monterosso, P. Seitz, et al.: Optics Letters, 2006,25(2), P. 102-104.
3. M. Suekuni, Y. Teramura: J. Opt A: Pure Apple Phys,2005,38, P. 2519-2535.
4. D. Huang, E.A. Swanson, et al: Science,1991,254, P. 1178-1181
5. J.G. Wu, M. Conry, C.H. Gu: Optics Letters, 2015,40(9), P. 1265-1267.
6. G. Li, W. Shao, K.J. Wu, et al.: Proceedings of SPIE, 2012, 5916, P. 536-542.
7. P.H. Tomlins, R.K. Wang: J.Phys D: Appl Phys, 2005,38, P. 2519-2535.
8. A.F. Fercher, W. Drexler, C.K. Hitzenberger: Rep Prog Phys, 2013, 76, P. 239-303.

Protective Material of the Security Mechanism Friction Surface on the Ship Lift

Han Wang

Sch Power & Mech Engn, Wuhan Univ, Wuhan 430072, Hubei Province, Peoples R China
Corresponding Email: 2530707676@qq.com

Abstract. The security mechanism of the ship lift has the needs of increasing the friction and being anti-corrosive. This article tried an antirust grease used on wire rope with high friction coefficient (Alice IRIS-400M grease) and rayed it with ultraviolet. We find that ultraviolet rays can effectively accelerate the speed of solidification. It needs to ray at least 12h before the friction coefficient of the surface can meet the requirements (more than 0.1).

1 Introduction

With the vigorous development of China's water conservancy project, ship lift is increasingly being used, the screw and the nut column friction surface of security mechanism give self-locking condition to ensure the safety operation of ship lift. The self-locking conditions of safety mechanism is: screw friction surface's frictional angle must be greater than the helix angle (γ=5.64°), which means the equivalent coefficient of friction fv > arctanγ=0.1. Considering the need of practical application, the safety mechanism is in an open environment. It is necessary to stop them rusting.

Before leaving the factory, people coated the spiral surface of safety mechanism with antirust oil. In the test, a set of security mechanism could not be self-lock. While after draining the water, with antirust oil thoroughly removed, mechanism self-locked successfully. Thus, the actual situation of spiral surface has a great influence on self-lock. If we daub the general oil, it is difficult to achieve both the requirements of antirust and self-locking. In this article, we want to explore a suitable protective material, without changing the structure of the safety mechanism. The material can not only ensure the equivalent coefficient of friction >0.1, but also be antirust and convenient for construction.

2 Testing program

2.1 Sample preparation

Considering the actual materials of the screw and nut column of security mechanism on ship lift (screw's is 42CrMo4, nut column is material GS-25, CrNiMo4), we made friction specimens using the similar material 40Cr. And we daub the protective material on it to do the test of rust and friction. .Making Ra3.2 to ensure the surface roughness of the sample is similar to that of the reality. The direction of the surface working is the same with that of sliding in the test. The size of our slip samples: 30×20×20mm; the stationary samples: 160×75×8mm. Shown as Figure1.

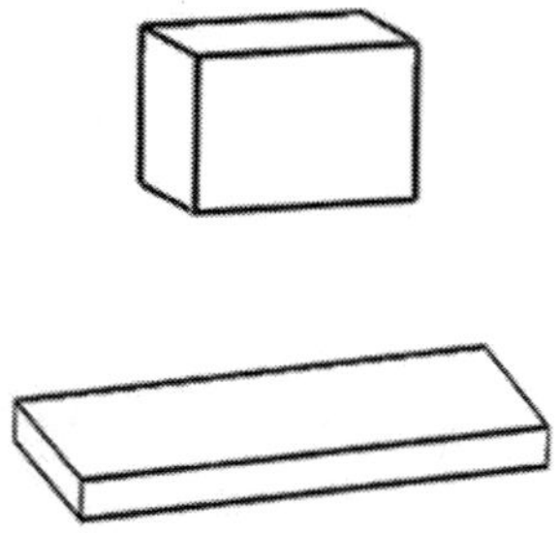

Fig. 1. Slip sample and stationary sample.

DOI: 10.1051/matecconf/201710004032

2.2 Protective material selection

The wire rope used in coal mine has the same need: increasing the friction, and antirust. In order to solve this problem, people developed an antirust grease used on wire rope with high friction coefficient, as the literature [1-5] introduced. It can protect the wire rope from rust, and the friction coefficient between wire rope and the friction pad (whose material is polyurethane) can reach 0.25, which is able to meet the requirements of lifting. But the environment is the friction between steel and the friction pat, which is different from the friction between steel itself like that on ship lift. So it is still necessary to have some experiment to exam whether the protective material is feasible.

We choose the Alice IRIS-400M grease produced by Alice Technology (Yangzhou) Co .Ltd, which is "black oil" in liquid state at normal temperature. The protective material is daubed on the surface of the sample, and then gradually solidified. In experiment (January 2015), we found it need 20 days before the friction coefficient of protective material to go over 0.1, which is not acceptable in actual construction. So, it is necessary to explore a new method to accelerate the consolidation.

In the experimental observation, we found that the weather affected the velocity of concreting, so we think that ultraviolet radiation may be useful to accelerate the concreting, which can shorten time. We used OSRAM UV bulb (OSRAM ULTRA-VITALUX), 300W. Light the bulb facing the sample, and the ultraviolet radiation eradiate the protective coating materials directly. Keep the distance: 15~20cm, the temperature at 60~80℃.We tested how coating thickness and the ultraviolet irradiation time affected the frictional behavior. It is shown as Figure 2.

Fig. 2. Ray the samples with ultraviolet.

2.3 Frictional behaviour test

Common used friction testing machine in china cannot meet our needs because of the special working load and environmental condition requirements. So we developed a static friction testing machine with digital display. Shown as Figure 3.

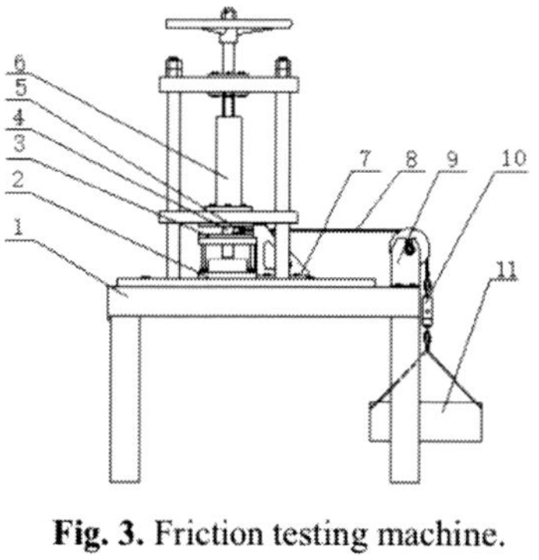

Fig. 3. Friction testing machine.

When it is working, a vertical load FN is applied to the test sample by means of screw pressure, and the horizontal direction load FT is applied by adding a heavy load on the tray. In the sensor, pressure and pulling force were measured. The horizontal force increases gradually, through the observation of the tension we can see whether the sample is sliding. Shown as Figure 4. A glass cover is arranged on the upper part of the friction testing machine. We can change the environment temperature and humidity by using heater and humidifier.

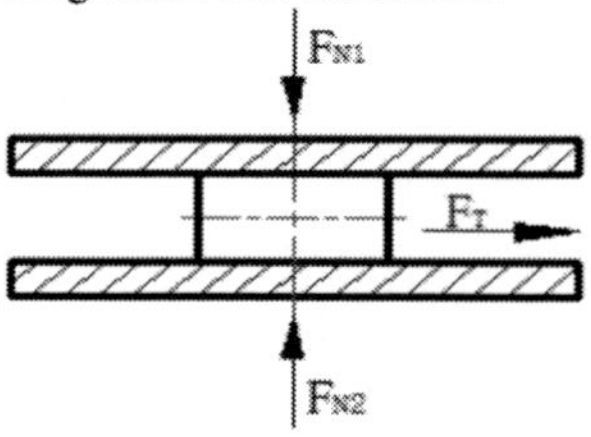

Fig. 4. Force diagram.

Coefficient of friction between materials

$$f = \frac{F_T}{2F_N} \tag{1}$$

Equivalent friction coefficient of screw nut column friction pair

$$f_v = \frac{f}{\cos\alpha} = \frac{F_T}{2F_N \cos\alpha} \tag{2}$$

$\alpha = 20^o$ is the angle of tooth of nut column mechanism
The self-locking condition of screw nut column friction pair is:

$$\gamma \le \arctan f_v = \arctan(\frac{F_T}{2F_N \cos\alpha}) \tag{3}$$

γ=5.64° is the helix angle of friction pair of screw nut

2.4 Test in different thickness and time

We did test of friction coefficient of no protective material and shell motor oil, and friction coefficient of different thickness and different irradiation time. The results are shown as the Table 1 and Table 2.

Table 1. Friction coefficient of no protective material and shell motor oil.

State of friction surface	Temp [℃]	relative humidity	f_v
Clean, no protective material	13.2	85%	0.126
Daub shell motor oil	13.2	85%	0.048

2.4.1 Thickness at 30~50μm, ultraviolet light at 12h

Protective coating thickness is controlled at 30~50μm, the time of using ultraviolet light is 12 hours, test the friction coefficient of the sample surface (ambient temperature 13 degrees, relative humidity 83%). When the pressure is 298Kgf, pull force is 61Kgf the sample started sliding. After sliding a distance of about 0.5mm the sample become static. When the coating thickness is 30~50μm, ultraviolet irradiation for 12 hours, in the case of micro sliding, the equivalent friction coefficient of the protective materials f_v=0.109.

Repeatability: after scratching the surface repeated friction test (static) showed there is little change.

2.4.2 Thickness at 30~50μm, ultraviolet light at 24h

Protective coating thickness is controlled at 30~50μm, the time of using ultraviolet light is 24 hours, test the friction coefficient of the sample surface (ambient temperature 13.6 degrees, relative humidity 61%). When the pressure is 309Kgf, pull force is 70Kgf the sample started sliding. After sliding a distance of about 0.3mm the sample become static. When the coating thickness is 30~50μm, ultraviolet irradiation for 24 hours, in the case of micro sliding, the equivalent friction coefficient of the protective materials f_v=0.120.

Repeatability: after scratching the surface repeated friction test (static) showed there is little change.

2.4.3 Thickness at 50~70μm, ultraviolet light at 12h

Protective coating thickness is controlled at 30~50μm, the time of using ultraviolet light is 12 hours, test the friction coefficient of the sample surface (ambient temperature 13.2 degrees, relative humidity 85%). When the pressure is 300Kgf, pull force is 58Kgf the sample started sliding. After sliding a distance of about 1mm the sample become static. When the coating thickness is 50~70μm, ultraviolet irradiation for 12 hours, in the case of micro sliding, the equivalent friction coefficient of the protective materials f_v=0.103.

Repeatability: after scratching the surface repeated friction test (static) showed there is little change.

2.4.4 Thickness at 50~70μm, ultraviolet light at 24h

Protective coating thickness is controlled at 50~70μm, the time of using ultraviolet light is 24 hours, test the friction coefficient of the sample surface (ambient temperature 13.6 degrees, relative humidity 61%). When the pressure is 301Kgf, pull force is 64Kgf the sample started sliding. After sliding a distance of about 0.5mm the sample become static. When the coating thickness is 50~70μm, ultraviolet irradiation for 24 hours, in the case of micro sliding, the equivalent friction coefficient of the protective materials f_v=0.113.

Repeatability: after scratching the surface repeated friction test (static) showed there is little change.

Those results are shown at Table 2 together.

Table 2. Friction coefficient of different thicknessn and different irradiation time.

Thickness	Irration time[h]	Force [Kgf]	Pressure [Kgf]	f_v
30~50μm	12	61	298	0.108
30~50μm	24	70	309	0.121
50~70μm	12	58	300	0.104
50~70μm	24	64	301	0.113

2.5 Test in different Loading Conditions

2.5.1 Loading Conditions is 2.3MPa

The contact area between slip sample and stationary sample is 600mm2.When the pressure force is 140Kgf, the corresponding pressure is 2.3MPa. When the pressure is 140Kgf and the pull force is 37Kgf, the sample started sliding. After sliding a distance of about 0.5mm the sample become static. When the Loading Conditions is 2.3MPa, the equivalent friction coefficient of the protective materials fv=0.143.

Repeatability: after scratching the surface repeated friction test (static) showed there is little change.

2.5.2 Loading Conditions is 3.1MPa

When the pressure force is 190Kgf, the corresponding pressure is 3.1MPa. When the pressure is 190Kgf and the pull force is 44Kgf, the sample started sliding. After sliding a distance of about 0.5mm the sample become static. When the Loading Conditions is 3.1MPa, the equivalent friction coefficient of the protective materials f_v=0.119.

Repeatability: after scratching the surface repeated friction test (static) showed there is little change.

2.5.3 Loading Conditions is 4.8MPa

When the pressure force is 293Kgf, the corresponding pressure is 2.3MPa. When the pressure is 293Kgf and the pull force is 54Kgf, the sample started sliding. After sliding a distance of about 0.5mm the sample become static. When increase the pull force to 66Kgf, the sample started sliding. After sliding a distance of about 1.1mm the sample become static. When the Loading Conditions is 4.8MPa, the equivalent friction coefficient of the protective materials f_v=0.119.

Repeatability: after scratching the surface repeated friction test (static) showed there is little change.

2.5.4 Loading Conditions is 9.7MPa

When the pressure force is 596Kgf, the corresponding pressure is 9.7MPa. When the pressure is 596Kgf and the pull force is 103Kgf, the sample started sliding. After sliding a distance of about 1mm the sample become static. When increase the pull force to 122Kgf, the sample started sliding. After sliding a distance of about 1mm the sample become static. When the Loading Conditions is 9.7MPa, the equivalent friction coefficient of the protective materials fv=0.110. Repeatability: after scratching the surface repeated friction test (static) showed there is little change.

Those results are shown at Table 3 together.

Table 3. f_v in different Loading Conditions.

Loading Conditions	Temperature [℃]	relative humidity	f_v
2.3 MPa	20.4℃	78%	0.143
3.1 MPa	20.4℃	78%	0.119

4.8 MPa	20.4℃	78%	0.119
9.7 MPa	20.4℃	78%	0.110

2.6 Salt spray test

The resistance to corrosion can obviously increase of the service life of metal work [6]. Salt spray test can effectively simulate the corrosion of the workpiece under actual conditions [7,8]. One sample was not coated with protective material and the other was coated with protective material.we put them into salt spray test, to compare the performance of protective materials. The thickness of protective material is 50~70μm, and the time of using ultraviolet light is 24 hours. Set temperature at 35 degrees Celsius, humidity at 95%-98%, mass fraction of the saline water is 5%. After 3h, 5h, 10h, 24h, have observation of corrosion.

(1)After 3-hours test, the samples without coating protective materials had began to be corroded, with time going by, the corrosion has become more and more serious;

(2)After 24 hours in the salt spray test, the surface coated with protective material has serious corrosion, and the other coated with a protective material is not corroded. After removing of the protective material, no corrosion on the surface of the substrate found.

3 Analysis

We can see from the test results above, all of them can meet the needs. Because screw friction surface's frictional angle must be greater than the helix angle (γ=5.64°), which means the equivalent coefficient of friction fv > arctanγ=0.1. Antirust grease are developed to protect the metal meterial better [9,10]. It works well n different thickness and different load condtions. So we think the antirust grease works well both in protect the material and increase friction.

4 Conclusions

1.The thickness of the protective material coating has little effect on the friction coefficient of the friction pair.

2.Ultraviolet radiation is conducive to the solidification of protective materials.

3.We did a twenty-four hour salt spray test.(35℃, humidity 95%-98%, percentage of salt water 5%)： IRIS-400M protective material has good corrosion resistance, which can meet the needs. Put the sample in the laboratory for one hundred days (From February 8th to May 20th), the surface kept in good condition. It shows that the protective material has good ability of oxidation resistance in the air.

4.IRIS-400M protective material can meet the needs of the security mechanism friction surface on the ship lift, and has great application prospect.

References

1. S.X. Duan, R.Z. Cheng, Z.D. Wang, Coal Mine Mechanical and Electrical, **4** (1989).
2. S.X. Duan, R.Z. Cheng, Coal Science and Technology, **5** (1991).
3. F.J. Liu, Metalwork, **6** (2004)).
4. S.Z. Tang, Metalwork, **2** (2008).
5. X.Y. Li, Zhongzhou Coal, **1** (2002).
6. H. Morris, Ironmaking & Steelmaking, **15** (1988) .
7. S.L. Esfahani, Z. Ranjbar, S. Rastegar, British Corrosion Journal, **51** (2016).
8. K. Raeissi, M.R. Toroghinejad, Isij International, **48** (2008).
9. L.J. Yu, Synthetic Lubricants, **32** (2005).
10. X.J. Yuan, Lubricating Oil, **22** (2007).

Graphene and Carbon Nanotubes Synergistically Improved the Thermal Conductivity of Phenolic Resin

Han Wang

Sch Power & Mech Engn Department, Wuhan University, Wuhan 430072, Hubei Province, Peoples R China
Corresponding Email: 2530707676@qq.com

Abstract. People discover the synergistic effect of graphene and carbon nanotubes on heat conduction in graphene carbon nanotubes / epoxy resin hybrid composites. In this article we added them into the phenolic resin and test the thermal conductivity. We found the thermal conductivity was increased by 6.5% in the phenolic resin by adding 0.45wt% graphene and 0.15wt% single wall carbon nanotubes (maintain the mass ratio 3:1). So if graphene and carbon nanotubes are added in proportion, thermal conductivity of phenolic resin will be improved significantly which is better than only carbon nanotubes or graphene.

1 Introduction

Graphene and carbon nanotubes are good conductors for electricity and heat because of their special structure. Now there are many useful way to produce Graphene [1] and carbon nanotubes [3, 4]. They are also very suitable to produce resin composite materials. In particular, the structure of carbon nanotubes is similar to polymers, whose main component is carbon. Carbon nanotubes and polymer composite materials will form a very completely binding interface which will have excellent performance [2]. Carbon nanotubes are easy to agglomerate [5], we can use ultrasonic to solve the problem [6]. Recently, people discover the synergistic effect of graphene and carbon nanotubes on heat conduction in graphene carbon nanotubes / epoxy resin hybrid composites, which has provided a new idea for the study of high thermal conductivity polymer composites. Some researchers found that the graphene carbon nanotube / epoxy hybrid composites have higher thermal conductivity than the same content of graphene composites and carbon nanotube composites. In the case of no change in the interface, the thermal conductivity of graphene and carbon nanotubes can significantly improve the thermal conductivity of epoxy resin composites. So, we are curious about whether it will has the same effect if we add them into the phenolic resin? It needs experiments to verify.

Because the acting force between adjacent graphite layers is weak, it is extremely easy to separate it and make many slices. Many scientists found graphene has good thermal conductivity and added it into some materials to make composite materials, which can improve its performance. At present, there are three ways to produce graphene. First is called chemical stripping: the first step is the preparation of graphite oxide, and then use the thermal expansion reduction to obtain graphene. The second is synthesis method using micro molecular or macro molecule as presoma. Which has high quality and high purity. The third method is called catalytic growth method.

The thermal conductivity of carbon nanotubes bear on its special structure and size. The thermal conductivity of can reach that of the graphene base. Which has the best performance of the heat conducting in nature. Carbon nanotubes has very high aspect ratio, which make it be very perfect for reinforced composite material. The structure of carbon nanotubes is similar to polymers, whose main component is carbon. Carbon nanotubes and polymer composite materials will form a very completely binding interface which will have excellent performance. Recently, carbon nanotube nanocomposites have attracted wide attention, most of them are carbon nanotube polymer composites, also there are few about the metal and nanotube reinforced composites with ceramic matrix.

2 Testing program

2.1 Preparation of materials

The materials we used in the experiment are layered oxidized graphene, which can be dispersed in liquid. In order to improve the dispersion of layered graphene, the following experiments are made:

Weigh 0.2g layered graphene, 20ml ethanol in a beaker. Add graphene into ethanol, and use ultrasonic cleaning apparatus oscillating for 1 hour, aiming to make them uniformly dispersed.

The turbid liquid of the oxidized graphene is poured into the filtering device, and a vacuum is drawn to accelerate the separation of the ethanol. Then put the filter paper into the heat treatment furnace, adjust to 180℃. Half an hour after the filter paper being completely, take it out from the heat treatment furnace. Scrape the graphene on the filter paper and pour it back to the small glass bottle. The results showed that this method can improve the dispersion of graphene.

The dispersion of carbon nanotubes in the application process is an important step in the modification of the composite materials and the fabrication of nano devices. The dispersion of carbon nanotubes directly affect the performance of materials, therefore, how to effectively disperse carbon nanotubes, has become an important step in the practical application. Mechanical dispersion method is a simple physical method, which mainly depends on mechanical energy like external shear force or impact force in the dispersion of carbon nanotubes. Such as ball-milling or mixing. Ball-milling reduces carbon nanotube clusters but also has some destroy to the structure. Some people try to use the low speed ball mill the multi-wall carbon nanotubes, he found that multi-wall carbon nanotube open the ports and the length becomes shorter. After further milling for 20h, carbon nanotubes have obvious agglomeration. The energy generated by the collision between ceramic balls and carbon nanotubes breaks the carbon nanotubes. It is said that by controlling the grinding time, a large number of appropriate length of carbon nanotubes can be produced. We adopt the method of ball milling.

2.2 Laboratory apparatus

For sintering mixed sample we use the spark plasma sintering system in Japan Sumitomo Coal Mining Corporation production (SPS), just like Figure 1. Control the temperature at 200 degrees Celsius, and the pressure under the condition of 15MPA, the diameter of graphite mold is 15mm, sintering for ten minutes. To test the thermal diffusivity we use the laser thermal conductivity meter LFA457. It is shown in Figure 2.

Fig. 1. SPS system.

Fig. 2. Laser thermal conductivity meter.

2.3 Preparation of composite materials

The experimental scheme is as follows:

(1) Weigh 4 portion of powder (phenolic resin) with scales for 7g each, grind them into uniform powder, adding to four beakers.

(2) Weigh 0.042g carbon nanotubes (amount to 0.6wt% of phenolic resin) into the No. 2 beaker. Weigh 0.042g graphene (amount to 0.6wt% of phenolic resin) into the No. 3 beaker. Weigh 0.0315g graphene and 0.0105g carbon nanotubes (mass ratio of graphene and carbon nanotubes is 3:1) into the No.4 beaker. The beakers were respectively mixed on mechanical grinding for an hour until evenly mixed, forming powder of uniform thickness.

(3) We used the spark plasma sintering system, and the diameter of graphite mold is 15mm (the apparatus for testing thermal diffusivity requires the sample's diameter is 12~17mm, and the thickness is 2-3mm). Keep the temperature at 200 degrees Celsius, the pressure under the condition of 15MPa, and sinter the sample. About ten minutes later we got the sintered samples.

2.4 Test of density (ρ)

Polish the surface of samples with sandpaper of 200, making it a smooth surface, and measure the density by the way of drainage. We use the way of drainage to measure the volume of the sample, and measure the quality. We know that

$$\rho = m / V \tag{1}$$

We got the density of them just like Table 1.

Table 1. Four groups of density.

	Blank	CNTs	G	CNTs/G
Volume[cm3]	0.3	0.5	0.3	0.5
Quality[g]	0.4215	0.6890	0.4165	0.6965
density [g/cm3]	1.405	1.378	1.388	1.393

2.5 Calculation of specific heat capacity(C)

We know that:

$$Q = C \times M \times \Delta t \tag{2}$$

And because our experiment is carried out at room temperature, the temperature is constant, the whole process of absorption of heat and heat release is balanced.

So we can have:

$$C1 \times m1 = C2 \times m2 + C3 \times m3 \tag{3}$$

We can know the specific heat capacity:

(1) Blank group: 1650J/ (kg*K).
(2) CNTs group: 1648.5J/ (kg*K).
(3) G group: 1649.1J/ (kg*K).
(4) CNTs/G group: 1648.95J/ (kg*K).

2.6 Calculation of thermal diffusivity (α)

After testing the thermal diffusion coefficient we use the laser thermal conductivity meter LFA457. We can have the thermal diffusion coefficient shown as Table 2.

Table 2. Thermal diffusivity (α) of four groups

Property	Blank	CNT	G	CNT/G
Thickness[mm]	1.666	2.633	1.500	2.820
Diameter[mm]	15.10	15.00	15.00	15.10
α3	0.180	0.183	0.178	0.185
α2	0.168	0.181	0.179	0.187
α3	0.171	0.182	0.175	0.187
average value	0.173	0.182	0.177	0.186

We can use:

$$\lambda(T)=\alpha(T)\times Cp(T)\times\rho(T). \quad (4)$$

Then the thermal conductivity is:

(1)Blank group: 0.401 W/ (m*K).
(2)CNTs group: 0.413 W/ (m*K).
(3)G group: 0.405 W/ (m*K).
(4)CNTs/G group: 0.427 W/ (m*K).

3 Analysis

3.1 Synergistic heat conduction theory

At present, there are more and more research on graphene and carbon nanotube composites. However, the mechanism of the graphene and carbon nanotubes co thermal conduction is not clear enough. There are many kinds of statements.

3.1.1 Aninterconnected networks is produced

Some people say that the causes synergistic thermal conductivity of graphene and carbon nanotubes are one-dimensional carbon nanotubes (CNTs) and two-dimensional graphene (GNPs) cooperate with each other to form a crosslinked network, the network greatly reduces the thermal resistance, and they have the Percolation Effects [7]. Other people also support this view through the establishment of models and the use of statistical mechanics [8].

3.1.2 Increase of the contact area

Yang believes that the contact between carbon nanotubes is zero dimensional, and the contact between graphene and carbon nanotubes is one-dimensional [9]. The increase of the contact area produced a percolation effects, so it will produce the effect of synergistic heat conduction.

3.1.3 Graphene and carbon nanotubes stimulate the percolation effect together.

Chu established a theoretical model to describe the synergistic thermal conductivity of graphene and carbon nanotubes using the effective medium theory [10]. Through the analysis, they believe that the synergistic effect of graphene and carbon nanotubes may be due to the percolation effect of their co excitation.

3.2 Analysis of the data

Seeing the thermal conductivity from the experiment, we found the addition of graphene will increase the thermal conductivity, adding carbon nanotubes alone will make the thermal conductivity has a small increase, while adding graphene and carbon nanotubes can greatly improve the thermal conductivity. We can clearly verify the synergistic thermal conductivity of graphene and carbon nanotubes. With all kinds of conjecture mentioned above, they are all proposed reasonably but all have disadvantages, if we want to know what the real reason is, we need more experiments, observing the structure from the micro scale, and the establishment of a more practical models.

4 Conclusions

The thermal conductivity of the sample was improved by 2.9% in the phenolic resin by adding the single wall carbon nanotubes and the thermal conductivity of the sample was improved by 1.0% in the phenolic resin by adding. The thermal conductivity was increased by 6.5% in the phenolic resin by adding 0.45wt% graphene and 0.15wt% single wall carbon nanotubes (maintain the mass ratio 3:1). From above, we can see that if graphene and carbon nanotubes are added in proportion, thermal conductivity of phenolic resin will be improved significantly. And the effect is better than only adding carbon nanotubes or graphene.

References

1. A A Balandin: Nat Mater, **10**(8): 569-81.(2011)
2. Z D Han, A Fina: Prog Polym Sci, **36**(7): 914-44.
3. J R Potts, D R Dreyer, C W Bielawski, et al, Polymer, **52**(1): 5-25(2011)
4. T Ramanathan, A A Abdala, S Stankovich, et al,Nat Nanotechnol, **3**(6): 327-31. (2008)
5. A Yu, P Ramesh, X Sun, et al, Adv Mater, **20**(24): 4740-4 (2008)

6. S Y Yang, C C M Ma, C C Teng, et al, Carbon, **48**(3): 592-603. (2010)
7. Y W Yi, X S Zeng, Materials Review,**20**:81-3.(2012)
8. R P Luo, East China University Of Science.(2013)
9. Y C Jia, K J Yu, K Qian, et al, Fiber Reinforced Plastics/Composites,**2**:69-73.(2013)
10. K Chu, C C Jia, W S Li. Appl Phys Lett, **101**(12): 211903. (2012)

Mechanical properties and experimental researches of new CSIPs sandwich panels

Wenfeng Du, Xue Yang, Liming Zhu* and Yun Sun

Institute of Steel and Spatial Structures in School of Civil Engineering and Architecture, Henan University, Kaifeng 475004 China

*Corresponding author: dwf@henu.edu.cn

Abstract. The advantages of glass fiber reinforced composites (FRP) and SIPs (structural insulated panels) are combined, and a new type of sandwich panel called composite structural insulated panels (CSIPs) is proposed. Through the adhesive bonding, CSIPs are made of FRP as face sheets and expanded polyethylene foam (EPS) as a core. To master the mechanical characteristics of CSIPs, firstly, adopting the large deflection theory of Reissener in this paper derived the calculation formula of displacement and the stability critical load of CSIPs. Then, ANSYS software was used to carry on the analysis of finite element simulation. Finally, a testing piece of CSIP with length 1000mm and breath 1000mm was made and a test was done. The results show that the theoretical analysis results, finite element simulation results and test results are basically coincide. So the calculating formula of deformation and bearing capacity of CSIPs are correct. And CSIPs have the outstanding advantages of light weight and high strength.

1 Introduction

The earliest sandwich construction originated from the corrugated board structure invented by the British in 1856. It adds a corrugated paper core between the two layers of paper to make the soft paper to obtain a certain stiffness[1].In 1935, a forest products laboratory (FPL) in the United States proposed sandwich panel structures, namely SIPs (Structural Insulated Panels) ,which is composed of two layers of the Oriented Strand Board (OSB) as face sheets and a layer of thermal insulation laminated composite panel as a core[2-3].Currently SIPs have been extensively used all over the world, however, the SIPs have the disadvantages of poor fire resistance, easy occurrence of pest and warping deformation.

In recent years, application of FRP (Fiber Reinforced Plastic Composites) in civil engineering structure developed rapidly. The FRP is comprised of enhanced fiber material and matrix material through unique moulding process with the advantages of light weight, high strength, corrosion resistant, the ability of customize, durability, energy conservation and environmental protection and so on. Based on the traditional SIPs, a new kind of sandwich panel structure, namely CSIPs (Composite Structural Insulated Panels), was proposed by this paper using the FRP instead of the OSB. CSIPs have inherited sandwich structural characteristic and production process of SIPs, at the same time, effectively overcome the disadvantages of poor fire resistance, easy occurrence of pest and warping deformation. Moreover, because the strength of FRP is much higher than wood, CSIPs can improve mechanical properties of SIPs greatly.

CSIPs are made of low-cost orthotropic thermoplastic glass/polypropylene (glass-PP) laminate as face sheets and expanded polyethylene (EPS) foam as a core through a hot-melt thermoplastic spray adhesive bonding together. Glass-PP laminates provide high strength-to-weight ratio, excellent impact resistance, and high durability. EPS, on the other hand, are marked by light weight, thermal insulation, good fire resistance and excellent impact properties. All these features lead finally to superior performance of the panels. Therefore, CSIPs have broad application prospects in the wall and roof structures [4-6].

In order to provide a good foundation for the application of CSIPs, the mechanistic properties of CSIPs were studied by means of theoretical analysis, finite element simulation and experimental study.

2 Theoretical analysis

2.1. Basic assumptions

In this paper, the calculation formulas of internal force and displacement in sandwich panels are derived from Reissner theory, which takes into account the basic high-order partial differential equation for the sandwich plate with large deflection bending. The following assumptions are advanced on CSIPs [7-8]:

(1) For the thickness of the surface layer is very small relatively comparing to the whole sandwich panel, a thin film was used to simulate face sheets. That is assuming that the surface layer is in the state of film stress;

(2) Due to the core is soft, it can be ignored that force components of sandwich panel in parallel the XY plane, this is assuming that $\sigma_x=\sigma_y=\sigma_{xy}=0$;

(3) It is assumed that the strain in the core and the surface layer along the thickness direction is zero ($\varepsilon_z=0$);

(4) The stress σ_z along the thickness direction is very small;

(5) The transverse shear effect was considered and the straight line segment along the direction of thickness keeps a straight line after the deformation even though it is not perpendicular to the middle plane.

2.2. The displacement and stress components of the sandwich panel

The displacements of any point in the sandwich panel are as following:

The upper face sheet

$$u_1 = u + \frac{1}{2}c\varphi_x \quad v_1 = v + \frac{1}{2}c\varphi_y \quad w_1 = w \tag{1}$$

The core

$$u_2 = u + z\varphi_x \quad v_2 = v + z\varphi_y \quad w_2 = w \tag{2}$$

The lower face sheet

$$u_3 = u - \frac{1}{2}c\varphi_x \quad v_3 = v - \frac{1}{2}c\varphi_y \quad w_3 = w \tag{3}$$

In the above equations, u_i, v_i, w_i are the displacement components of the upper, the lower face sheet and the core. $\varphi_x(x,y)$ and $\varphi_y(x,y)$ are rotation angles in the XY and YZ plane.

The geometric equations of sandwich panel are as following:

$$\begin{aligned}
\varepsilon_{xi} &= \frac{\partial u_i}{\partial x} + \frac{1}{2}\left(\frac{\partial w_i}{\partial x}\right)^2 \\
\varepsilon_{yi} &= \frac{\partial v_i}{\partial y} + \frac{1}{2}\left(\frac{\partial w_i}{\partial y}\right)^2 \\
\gamma_{xyi} &= \frac{\partial v_i}{\partial x} + \frac{\partial u_i}{\partial y} + \frac{\partial w_i}{\partial x}\frac{\partial w_i}{\partial y} \\
\varepsilon_{zi} &= \gamma_{xzi} = \gamma_{yzi} = 0 \qquad (i=1,3) \\
\gamma_{xz2} &= \frac{\partial u_2}{\partial z} + \frac{\partial w_2}{\partial x} \\
\gamma_{yz2} &= \frac{\partial v_2}{\partial z} + \frac{\partial w_2}{\partial y} \\
\varepsilon_{x2} &= \varepsilon_{y2} = \varepsilon_{z2} = \gamma_{xy2} = 0
\end{aligned} \tag{4}$$

In the above equations ε_{xi}, ε_{yi}, ε_{zi}, γ_{xyi}, γ_{yzi} and γ_{xzi} (i=1,2,3) are the strain components of the upper, the lower face sheet and the core.

The strain components of the upper face sheet

$$\begin{aligned}
\sigma_{x1} &= \sigma_{x0} + \frac{E_f c}{2\left(1-v_f^2\right)}\left(\frac{\partial \varphi_x}{\partial x} + v_f\frac{\partial \varphi_y}{\partial y}\right) \\
\sigma_{y1} &= \sigma_{y0} + \frac{E_f c}{2\left(1-v_f^2\right)}\left(\frac{\partial \varphi_y}{\partial y} + v_f\frac{\partial \varphi_x}{\partial x}\right) \\
\tau_{xy1} &= \tau_{xy0} + \frac{Gc}{2}\left(\frac{\partial \varphi_y}{\partial x} + \frac{\partial \varphi_x}{\partial y}\right)
\end{aligned} \tag{5}$$

The strain components of the lower face sheet

$$\sigma_{x3} = \sigma_{x0} - \frac{E_f c}{2(1-v_f^2)}\left(\frac{\partial \varphi_x}{\partial x} + v_f \frac{\partial \varphi_y}{\partial y}\right)$$

$$\sigma_{y3} = \sigma_{y0} - \frac{E_f c}{2(1-v_f^2)}\left(\frac{\partial \varphi_y}{\partial y} + v_f \frac{\partial \varphi_x}{\partial x}\right) \quad (6)$$

$$\tau_{xy3} = \tau_{xy0} - \frac{Gc}{2}\left(\frac{\partial \varphi_y}{\partial x} + \frac{\partial \varphi_x}{\partial y}\right)$$

The strain components of the core

$$\tau_{xz2} = G_{core}\left(\varphi_x + \frac{\partial w}{\partial x}\right)$$

$$\tau_{yz2} = G_{core}\left(\varphi_y + \frac{\partial w}{\partial y}\right) \quad (7)$$

Where ε_{xi}, ε_{yi}, ε_{zi}, ε_{xyi}, ε_{yzi} and ε_{xzi} (i=1,2,3) are the strain components of the upper, the lower face sheet and the core. E_f and v_f are elastic moduli and poisson's ratios. G and G_{core} are the shearing modulus of the face sheet core.

$$\sigma_{x0} = \frac{E_f}{(1-v_f^2)}\left[\frac{\partial u}{\partial x} + v_f \frac{\partial v}{\partial y} + \frac{1}{2}\left(\frac{\partial w}{\partial x}\right)^2 + \frac{v_f}{2}\left(\frac{\partial w}{\partial y}\right)^2\right]$$

$$\sigma_{y0} = \frac{E_f}{(1-v_f^2)}\left[\frac{\partial v}{\partial y} + v_f \frac{\partial u}{\partial x} + \frac{1}{2}\left(\frac{\partial w}{\partial y}\right)^2 + \frac{v_f}{2}\left(\frac{\partial w}{\partial x}\right)^2\right] \quad (8)$$

$$\tau_{xy0} = G\left(\frac{\partial v}{\partial x} + \frac{\partial u}{\partial y} + \frac{\partial w}{\partial x}\frac{\partial w}{\partial y}\right)$$

Where σ_{x0}, σ_{y0} and τ_{xy0} are the strains of intermediate surface in sandwich panel.

2.3. General bending moment and shear in sandwich panel

Because the transverse equilibrium equations are the same as those of the single layer panel, the bending moment, the torque and the shear force in sandwich panel can be obtained:

$$M_x = -D\left(\frac{\partial \varphi_x}{\partial x} + v_f \frac{\partial \varphi_y}{\partial y}\right)$$

$$M_y = -D\left(\frac{\partial \varphi_y}{\partial y} + v_f \frac{\partial \varphi_x}{\partial x}\right) \quad (9)$$

$$M_{xy} = \frac{D}{2}(1-v_f)\left(\frac{\partial \varphi_x}{\partial y} + \frac{\partial \varphi_y}{\partial x}\right)$$

$$Q_x = G_{core} c\left(\varphi_x + \frac{\partial w}{\partial x}\right)$$

$$Q_y = G_{core} c\left(\varphi_y + \frac{\partial w}{\partial y}\right) \quad (10)$$

Where D is the flexural rigidity of the sandwich panel.

2.4. The fundamental equations and boundary conditions

According to the elastic strain energy formula, the total potential energy of the sandwich panel can be obtained:

DOI: 10.1051/matecconf/201710004034

$$U=\frac{t}{E}\iint_{s}\begin{bmatrix}\left(\sigma_{x0}+\sigma_{y0}\right)^{2}+\\ 2(1+v_{f})(\tau_{xy0}^{2}-\sigma_{x0}\sigma_{y0})\end{bmatrix}dxdy$$
$$+\frac{D}{2}\iint_{s}\begin{bmatrix}\left(\frac{\partial\varphi_{x}}{\partial x}\right)^{2}+\left(\frac{\partial\varphi_{y}}{\partial y}\right)^{2}+\\ 2v_{f}\frac{\partial\varphi_{x}}{\partial x}\frac{\partial\varphi_{y}}{\partial y}+\frac{1-v_{f}}{2}\left(\frac{\partial\varphi_{y}}{\partial x}+\frac{\partial\varphi_{x}}{\partial y}\right)^{2}\end{bmatrix}dxdy \quad (11)$$
$$+\frac{G_{core}c}{2}\iint_{s_2}\begin{bmatrix}\left(\varphi_{x}+\frac{\partial w}{\partial x}\right)^{2}\\ +\left(\varphi_{y}+\frac{\partial w}{\partial y}\right)^{2}\end{bmatrix}dxdy-\iint_{s}qwdxdy$$

So according to potential energy principle:

$$\delta U=0 \quad (12)$$

$$\frac{D}{G_{core}c}\left(\frac{\partial^{2}\varphi_{x}}{\partial x^{2}}+\frac{1-v_{f}}{2}\frac{\partial^{2}\varphi_{y}}{\partial y^{2}}+\frac{1+v_{f}}{2}\frac{\partial^{2}\varphi_{y}}{\partial x\partial y}\right)$$
$$-(\frac{\partial w}{\partial x}+\varphi_{x})=0$$
$$\frac{D}{G_{core}c}\left(\frac{\partial^{2}\varphi_{y}}{\partial y^{2}}+\frac{1-v_{f}}{2}\frac{\partial^{2}\varphi_{y}}{\partial x^{2}}+\frac{1+v_{f}}{2}\frac{\partial^{2}\varphi_{x}}{\partial x\partial y}\right) \quad (13)$$
$$-(\frac{\partial w}{\partial y}+\varphi_{y})=0$$
$$G_{core}c(\frac{\partial\varphi_{x}}{\partial x}+\frac{\partial\varphi_{y}}{\partial y}+\nabla^{2}w)$$
$$=-q-2t\left(\sigma_{x0}\frac{\partial^{2}w}{\partial x^{2}}+\sigma_{y0}\frac{\partial^{2}w}{\partial y^{2}}+2\tau_{xy0}\frac{\partial^{2}w}{\partial x\partial y}\right)$$

The boundary conditions

$$when \quad x=0,a$$
$$\sigma_{x0}=0 \quad or \quad \delta u=0;$$
$$\tau_{xy0}=0 \quad or \quad \delta v=0; \quad (14a)$$
$$G_{core}c\left(\varphi_{x}+\frac{\partial w}{\partial x}\right)+2t\left(\sigma_{x0}\frac{\partial w}{\partial x}+\tau_{xy0}\frac{\partial w}{\partial y}\right)=0$$
$$or \quad \delta w=0;$$
$$D\left(\frac{\partial\varphi_{x}}{\partial x}+v_{f}\frac{\partial\varphi_{y}}{\partial y}\right)=0 \quad or \quad \delta\varphi_{x}=0;$$
$$\frac{D\left(1-v_{f}\right)}{2}\left(\frac{\partial\varphi_{y}}{\partial x}+\frac{\partial\varphi_{x}}{\partial y}\right)=0 \quad or \quad \delta\varphi_{y}=0;$$
$$when \quad y=\pm\frac{b}{2}$$

$$\begin{aligned}
&\tau_{xy0} = 0 \quad or \quad \delta u = 0; \\
&\sigma_{y0} = 0 \quad or \quad \delta v = 0; \\
&G_{core}c\left(\varphi_y + \frac{\partial w}{\partial y}\right) + 2t\left(\frac{\partial w}{\partial x} + \sigma_{y0}\frac{\partial w}{\partial y}\right) = 0 \\
&or \quad \delta w = 0; \\
&D\left(\frac{\partial \varphi_y}{\partial y} + v_f\frac{\partial \varphi_x}{\partial x}\right) = 0 \quad or \quad \delta\varphi_y = 0; \\
&\frac{D(1-v_f)}{2}\left(\frac{\partial \varphi_y}{\partial x} + \frac{\partial \varphi_x}{\partial y}\right) = 0 \quad or \quad \delta\varphi_x = 0
\end{aligned} \tag{14b}$$

Then according to Hu Haichang's methods which cope with the small deflection equations sandwich panel, φx, φy can be express using the new function ω, f [9]:

$$\begin{aligned}
\varphi_x &= \frac{\partial \omega}{\partial x} + \frac{\partial f}{\partial y} \\
\varphi_y &= \frac{\partial \omega}{\partial y} - \frac{\partial f}{\partial x}
\end{aligned} \tag{15}$$

We can get:

$$\begin{aligned}
&\frac{\partial}{\partial x}\left(\frac{D}{G_{core}c}\nabla^2\omega - \omega - w\right) + \frac{\partial}{\partial y}\left[\frac{D(1-v_f)}{2G_{core}c}\nabla^2 f - f\right] = 0 \\
&\frac{\partial}{\partial y}\left(\frac{D}{G_{core}c}\nabla^2\omega - \omega - w\right) - \frac{\partial}{\partial x}\left[\frac{D(1-v_f)}{2G_{core}c}\nabla^2 f - f\right] = 0
\end{aligned} \tag{16}$$

ω, f can be regarded as two separate parameters, so that we can obtain:

$$\begin{aligned}
&\frac{D(1-v_f)}{2G_{core}c}\nabla^2 f - f + i\left(\frac{D}{G_{core}c}\nabla^2\omega - \omega - w\right) \\
&= F(x+iy)
\end{aligned} \tag{17}$$

The corresponding homogeneous equations are:

$$\begin{aligned}
&\frac{D(1-v_f)}{2G_{core}c}\nabla^2 f - f = 0 \\
&\frac{D}{G_{core}c}\nabla^2\omega - \omega - w = 0
\end{aligned} \tag{18}$$

The complete solution is

$$\begin{aligned}
&w = \frac{D}{G_{core}c}\nabla^2\omega - \omega \\
&D\nabla^4\omega = -q - 2t \times \left(\frac{D}{G_{core}c}\nabla^2\omega - \omega\right) \\
&\times\left(\sigma_{x0}\frac{\partial^2}{\partial x^2} + \sigma_{y0}\frac{\partial^2}{\partial y^2} + 2\tau_{xy0}\frac{\partial^2}{\partial x\partial y}\right)
\end{aligned} \tag{19}$$

2.5. The single triangle progression method

The exact solution for the bending of the thin rectangular panel under a uniformly distributed load with four edges simply supported is obtained by the single triangle progression method. The vertical displacement can be expressed:

$$w(x,y) = \sum_{m=1}^{\infty} Y_m(y)\sin\frac{m\pi}{a}x \tag{20}$$

Where Y_m is an arbitrary function of y, and m is any positive integer.

Y_m is selected which can be satisfied by the formula (20) and the boundary conditions, then:

$$\nabla^4 w = \frac{q}{D} \tag{21}$$

$$\frac{q}{D} = \frac{2}{a}\sum_{m=1}^{\infty}\left[\int_0^a \frac{q}{D}\sin\frac{m\pi x}{a}dx\right]\sin\frac{m\pi}{a}x$$

$$Y_m^{(4)}(y) - 2\left(\frac{m\pi}{a}\right)^2 Y_m^{(2)}(y) + \left(\frac{m\pi}{a}\right)^4 Y_m(y) \tag{22}$$

$$= \frac{2}{aD}\sum_{m=1}^{\infty}\int_0^a q\sin\frac{m\pi x}{a}dx$$

Whatever x takes on an arbitrary value, the formula (23) can be satisfied, we have

$$Y_m = A_m ch\frac{m\pi y}{a} + B_m\frac{m\pi y}{a}sh\frac{m\pi y}{a}$$

$$+C_m sh\frac{m\pi y}{a} + D_m\frac{m\pi y}{a}ch\frac{m\pi y}{a} + f_m(y) \tag{23}$$

$$w = \sum_{m=1}^{\infty}\begin{pmatrix} A_m ch\frac{m\pi y}{a} + B_m\frac{m\pi y}{a}sh\frac{m\pi y}{a} \\ +C_m sh\frac{m\pi y}{a} + \\ D_m\frac{m\pi y}{a}ch\frac{m\pi y}{a} + f_m(y)\end{pmatrix}\sin\frac{m\pi}{a}x$$

Where $f_{m(y)}$ is any of particular solutions and A_m, B_m, C_m, D_m are constants.

$$f_m(y) = \frac{2qa^4}{\pi^5 Dm^5}(1-\cos m\pi) \tag{24}$$

Because the panel is symmetry related to the x direction, the formula (24) retains the even terms of y. Thus, C_m=0, D_m=0. By substituting the formula (23) into the formula (19), the final deflection formula is obtained:

$$w = \sum_{m=1,3,5\cdots}^{\infty}\begin{pmatrix}\frac{2qa^4}{\pi^5 Dm^5}(1-\cos m\pi) \\ +A_m ch\frac{m\pi y}{a} \\ +B_m\frac{m\pi y}{a}sh\frac{m\pi y}{a}\end{pmatrix}\sin\frac{m\pi x}{a} \tag{25}$$

$$w = \frac{5qa^4 b}{384D} + \frac{qa^2 b}{8U} \tag{26}$$

Rectangular sandwich panel of equal surface layers is analyzed with four simply supported edges under uniformly transverse load q. The size of the panel is $a\times b$. The load on sandwich panel can be obtained:

$$P = qab \tag{27}$$

When the material and thickness of sandwich panel are same, the deflection calculation formula at mid-span can be derived under uniform load:

$$w = \frac{k_b Pa^3}{D} + \frac{k_s Pa}{U} \tag{28}$$

The first term on the right-hand side of Eq.(29) is the deflection due to bending, and the second term is the deflection due to shear. k_b and k_s are the bending and shear deflection coefficients, respectively. The values of both k_b and k_s depend on the loading and boundary conditions. The bending and shear deflection coefficients under different loading and boundary conditions can be obtained. k_b and k_s are 5/384 and 1/8, respectively.

According to the theory of combination beam, formula for translation of axis and ASTM C-393, the formula of moment of inertia per unit width sandwich panel can be obtained[13-17]:

$$I = \frac{(t_1^3 + t_2^3)}{12} + \frac{t_1\times(t_c+t_2)^2 + t_2\times(t_c+t_1)^2}{4} \tag{29}$$

Where e t1, t2 are surface layer thicknesses and tc is core thickness.

 DOI: 10.1051/matecconf/201710004034

By substitution of the size of CSIPs as shown in Fig.1. into the formula (30),we can obtain the flexural and shear rigidities:

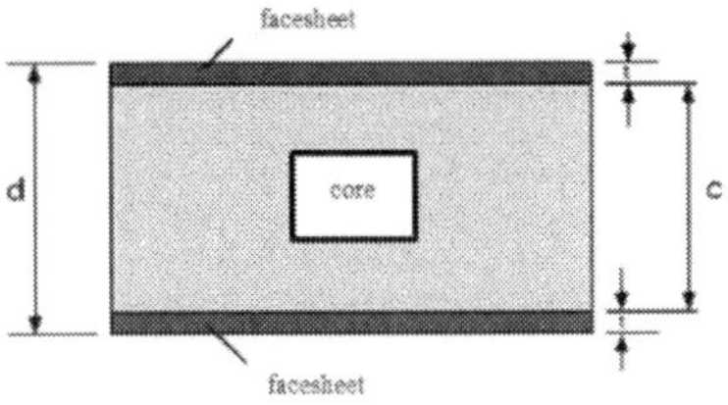

Fig. 1. Schematic figure for the CSIPs

$$t_c = c$$
$$t_1 = t_2 = t = \frac{d-c}{2} \quad (30)$$
$$I = \frac{\left(d^3 - c^3\right)b}{12}$$

$$D = E_f I\left(1 - v_{xy}^2\right) = \frac{E_f\left(d^3 - c^3\right)b}{12}\left(1 - v_{xy}^2\right) \quad (31)$$

Where d_{eff} should be used to instead of d in the formula (31) and (32) when determining the flexural and shear rigidities[9]:

$$d_{eff} = t\left(1 + \frac{c}{d}\right) + 0.3\frac{c^2}{d} \quad (32)$$
$$c_{eff} = d_{eff} - 2t$$

Using the same way, the shear rigidity can be obtained[8-9]:

$$U = \frac{G(d+c)^2 b}{4c} \quad (33)$$
$$\Delta = \frac{k_b P L^3}{D} + \frac{k_S P L}{U}$$

Where b and L are the width and length of the panel, and Δ is the displacement of the panel.

According to ASTM C-393, ultimate flexural bearing capacity of CSIPs is determined from Eqs.(34) [6-9]:

$$q_{cr} = \frac{[\sigma]}{\frac{L^2}{8ct} + \frac{L^2}{8c^2} + \frac{E_f t}{2cG_{core}}} - \frac{G_m}{bL} \quad (34)$$

3 Analysis of an example

CSIPs are made up the glass fiber reinforced composites (FRP) and expanded polyethylene (EPS) foam. The thickness is 103mm with a surface layer thickness of 1.5mm and a core thickness of 100mm. Material properties of CSIPs are listed in Table 1, 2.

3.1. Theoretical calculations

By substitution of each parameter of CSIPs into the deducing procedure, the deformation and bearing capacity of CSIPs are obtained as following:

Table 1. Material properties of FRP

Project	Numerical
Thickness of face sheet (t)	1.5mm
Density (ρ_f)	980kg/m^3
Glass fiber content	70%
Modulus of elasticity along length(E_x)	15,169MPa
Modulus of elasticity along width (E_y)	15,169MPa
Modulus of elasticity along depth(E_z)	1050MPa
Poisson's ratio (v_{xy}, v_{yz} , v_{xz})	0.11,0.22,0.22
Shear modulus (G_{xy}, G_{yz}, G_{xz})	1800MPa,1800MPa,750MPa
Tensile strength	690MPa
Compassion strength	317MPa

Table 2. Material properties of EPS

Project	Numerical
Thickness of core (t)	100mm
Density (ρ_f)	16kg/m^3
Modulus of elasticity (E_c)	1.2-1.5MPa
Poisson's ratio (v)	0.25
Shear modulus (G)	1.9-2.2MPa
Tensile strength	0.11-0.14MPa
Compassion strength	0.07-0.1MPa

$$d_{eff} = t\left(1+\frac{c}{d}\right)+0.3\frac{c^2}{d}$$

$$=1.5\times\left(1+\frac{100}{103}\right)+0.3\times\frac{100^2}{103}$$

$$=32.0825mm$$

$$c_{eff} = d_{eff} - 2t = 32.0825 - 3$$

$$=29.0825mm$$

$$D = E_f I = \frac{E_f\left(d^3-c^3\right)b}{12}\left(1-v_{xy}^2\right)$$

$$=\frac{15169\times(32.0825^3-29.0825^3)\times1000}{12}\times(1-0.11^2)$$

$$=1.0520\times10^{10}N\cdot mm^2$$

$$U=\frac{G_{core}\left(d+c\right)^2 b}{4c}=\frac{2.2\times\left(32.0825+29.0825\right)^2\times1000}{4\times29.0825}$$

$$=7.0752\times10^4 N$$

$$\Delta=\frac{k_b PL^3}{D}+\frac{k_S PL}{U}$$

$$=\frac{5\times1000\times1000^3}{1.0520\times10^{10}\times384}+\frac{1000\times1000}{7.0752\times10^4\times8}$$

$$=3.0045\text{mm}$$

$$q_{cr}=\frac{[\sigma]}{\dfrac{a^2}{8ct}+\dfrac{a^2}{8c^2}+\dfrac{E_f t}{2cG_{core}}}-\frac{G_m}{ab}$$

$$= \frac{690}{\frac{1000^2}{8\times29.0825\times1.5}+\frac{1000^2}{8\times29.0825^2}+\frac{15169\times1.5}{2\times29.0825\times2.2}} -82$$

$$=122.3821\,KN/m^2$$

3.2. Finite element simulation

The universal finite element program-ANSYS12.0 was employed to establish model and analysis using 8-node three-dimensional entity unit element (SILOD45). Due to bond together, glass fiber enhanced composite board and polystyrene foam has no relative movement model in sandwich panel. In this paper, the panel is simplified as ideal elastic material.

u_x, u_y, u_z for all nodes in the edge of the sandwich panel carry on the restraint. In the static analysis, the uniform load of 1kN/m^2 on the sandwich panel is applied in the model. The finite element analysis model is shown in Fig.2.

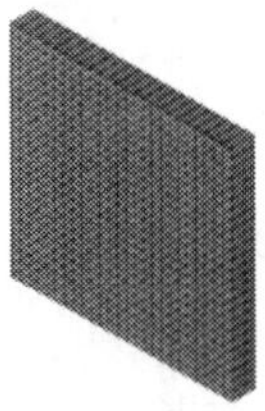

Fig. 2. Finite element model of CSIPs

Fig. 3. Displacement of C

Fig. 4. The first bucking mode of CSIPs panel

The results of the finite element static analysis are shown in Fig.3. The maximum displacement value is located in the middle of the span, which is 3.1024mm. The results are in good agreement with those of the theoretical analysis. Eigenvalue buckling analysis was carried out on the model to obtain the buckling mode and the characteristic value of the coefficient, as shown in Fig.4. The first eigenvalue coefficient of Sandwich panel is 21.117, which is slightly smaller. It is caused mainly by the low shear strength of the core panel.

3.2. Model tests

The three same specimens were designed and made whose material are made up the glass fiber reinforced composites(FRP), expanded polyethylene (EPS) foam and a hot-melt thermoplastic spray adhesive. The glass fiber reinforced composite panel and polyethylene foam panel are cut into the length 1000mm and breath 1000mm panel using cutting tools. Applying evenly thermoplastic spray adhesive to the FRP, then placing it on the polyethylene foam board and bringing pressure to bear on it. The specimen after curing was shown in Fig.5.

DOI: 10.1051/matecconf/201710004034

In loading process of experiments, firstly, the wires of the pressure sensor on the self- balancing anti force frame are connected with the equipment and the computer, and opening the test software. Then placing the specimen on the support and simulating the whole process from loading to failure. The calculation results of displacement and the critical load of the specimen are automatically recorded by the sensor. Experimental loading conditions are consistent with the boundary condition of the finite element model.

Fig. 5. The testing specimen and the loading equipment **Fig. 6.** Failure pattern of the specimen

With the increase of load, specimens made a slight sound. We cannot find significant change with the naked eye observation. The displacement of the mid-span increased with the increase of load; in loading process of experiments, specimens made "crack" sound and bended down into arc after observation; at a later stage of loading, obvious tearing sound can be heard and panels on the surface warped. Now it was found that core board and the edges of the upper layer appeared slight debonding phenomenon; with the further increase of load, the core and the lower layer bonding edge also appeared slight debonding phenomenon, the failure area extended constantly to the center and bearing capacity suddenly dropped, specimen failure is shown in Fig.6.

4 Conclusion

(1) A new type of sandwich panel, namely CSIPs (Composition Structural Insulated Panels), was proposed in this paper. CSIPs are made of low-cost orthotropic thermoplastic glass/polypropylene (glass-PP) laminate as face sheets and expanded polyethylene (EPS) foam as a core and are bonded together through a hot-melt thermoplastic spray adhesive.
(2) CSIPs have the advantages of light weight, high strength, better corrosion-resistant, high durability, energy conservation and environmental protection, low maintenance cost. It is also provided with features of convenient transportation and construction, especially the production methods of industrialization and assembled on site can greatly shorten the building cycle.
(3) Due to the relatively low elastic modulus of the FRP in the composite sandwich panel, the stiffness of the structure is small. Therefore, it is necessary to use the large deflection theory to analyze.
(4) Theoretical calculation, finite element simulation and experimental results are consistent. It is showed that the calculating formula of deformation and bearing capacity of CSIPs are correct.

References

1. J.F. Wu, F.H. Xu, D. Gao. Vibration isolation performance of precompression multi-layer calcium plastic corrugated board, *Packaging Engineering* **19**,27(2015).
2. S.G. Wang, L.Zhang, D.S. Du,et al. Full scale shaking table tests on frame structure with new SIP filling wallboard, *Journal of vibration and shock* **18**,100(2015).
3. S.G. Wang, T. Wu, W.Q. Liu, et al. The basic properties test of new SIP wall panel, *New Materials* **40**, 72(2013).
4. U. Nasim, W.F Du. New thin shells made of composite structural insulated panels, *Journal of Reinforced Plastics and Composites* **33**, 1954(2014).
5. A. Vaidya, N. Uddin, U. Vaidya. Design and analysis of composite structural insulated panels (CSIPs) for exterior wall applications, Journal of Composite Construction **28**, 1587(2010).
6. M. Mousa, N. Uddin. Structural behavior and modeling of full-scale composite structural insulated wall panels, *Journal of Engineering Structures* **41**, 320(2012).
7. M. Mousa, N. Uddin. Debonding of composites structural insulated sandwich panels, *Journal of Reinforced Plastics and Composites* **29**, 3380(2010).
8. W.F. Du, N. Uddin. Behavior and connection of new CSIPs thin shell structures// *Proceeding of 15 National Conference on Advances in Modern Structures* Kaifeng, China, 483 (2015).
9. X.Y. Wang, F.B. Yang, J.C. Zeng. The design principle and its application of sandwich structural composites. *Beijing: Chemical industry press,* 2(2007).

DOI: 10.1051/matecconf/201710004035

Relationship on the Relational Benefit Effect of Oil Products on Customer Satisfaction

Muxue Jing [1,*] and Xilai Jin [2]

[1]School of Management, Wuhan University of Technology, Wuhan, China, 430074
[2]School of Engineering, University of Florida, Gainesville, FL, US
Corresponding author: jmuxue@126.com

Abstract. In order to find out the relationship of relational benefit effect of oil products on customer satisfaction, the local fractional algorithm is proposed in this paper for data analysis. It is investigated the adjustment effect of alternative brand competitiveness and customer characteristics to this mechanism. The results show that the proposed algorithm can thus improve overall system performance substantially.

1 Introduction

With the acceleration of economic globalization pace, the environment that enterprises confront has changed enormously, that enterprises participate in the fierce international competition only by its own resources had been far from enough, It is necessary to achieve the goal of responding to the market fast with the help of the resources of other enterprises. So, supplier, manufacturer, retailer etc, must break the traditional boundary line to change from the traditional buying and selling relation into new-type partnership relation, However, some domestic enterprises have implemented supply chain management, and has made certain effect [1], but because of the lack of coordination and cooperation among enterprises and the lack the long-term plan and the whole optimization of supply chain, its application level and range still await to improve further. Rob's paper [2] regarding the optimization of the whole supply chain as the goal, beginning with the supply partnership relation interests, study the interest mechanism of the supply chain partnership relation. Its main content consists of the forming mechanisms, security mechanisms and allocation mechanisms of supply chain partnership relation interests [3].

The forming mechanism of supply chain partnership relation interests explain the reasons of the forming of the supply chain partnership relation interests to improve enterprise's the understanding to the supply chain partnership relation and its interests [3-4]; the security mechanisms of the supply chain partnership relation interests provide direct for enterprises to establish supply chain partnership relation, to keep and improve the operation efficiency of the whole supply chain; the allocation mechanism of the supply chain partnership relation have decisive actions on the foundation and successful operation of supply chain partnership relation ,because it is all to obtain some economic benefits for any enterprise participating in supply chain partnership relation, if the synergy interests emerged in supply chain partnership relation cannot be allocated fairly and reasonably among the partners, supply chain partnership relation can't be for a long time, and its interests are not known where to begin either. The study of supply chain partnership relation interest mechanism can offer some guide for the domestic enterprises to establish successful supply chain partnership relation to participate in the international competition [5].

2 Basic Framework

To replace the transaction-oriented marketing as relationship-oriented is the most representative of marketing theoretical breakthrough and marketing practices during the past two decades. Different from traditional transaction marketing, relationship marketing focuses on the long-term interests of enterprises, by maintaining and strengthening the long-term relationships with customers to achieve enterprises' marketing objectives. Brand relationship is an important manifestation of the customer relationship, and seeking relationship stability and durability is the point of brand relationship management. With language expression of brand relationship management, brand relationship management should focus on the management of whole life cycle of brand relationship, because the long-term relationship-oriented marketing concept plays a key role for enterprises to achieve a competitive advantage status. From the perspective of the complete life cycle of brand relationship, brand relationship management should include the brand relationship to establish, maintain and renew. The management objects of the three stages are new customers, old customers, and relationship fracture customers [6-7].

At present, both from the enterprise management practice and from the academic research level, brand management focus on relationship establishment and maintain, the attention to the relationship fracture customers is far less than the first two stages. In fact, with the increased competition of customer resources and customer development and maintenance cost of ascension, to re-develop the relationship of fracture customers should be the new focus on brand relationship management. Under the condition of buyer's market, due to the selective enhancement of consumers, customer resources become more competitive. The tentacles of enterprise brand relationship management have to extend to relationship revival of fracture customers. The premise of effective brand relationship revival strategy is correctly grasp of what factors drive the customer-brand relationship revival behavior and how these factors work. However, the academic study on brand relationship revival has just begun. Driving factors of the brand relationship revival, especially the thinking of mechanism of action is still very insufficient. The biggest problem of the only scattered results is the lack of a clear theoretical basis and analysis framework of the system.

Aiming at the shortcomings of the theory research of the brand relationship revival and confusion of management practice, it is puts forward three issues to be addressed. Firstly, what factors drive customers to renew the fracture brand relationship? Secondly, how these factors act on the customer's intention to renew the brand relationship? Thirdly, which factors could influence the driving effects that the brand relationship revival had on relationship revival intention? Around the three questions, it is proposed the concept of customer perceived relationship revival value, and use critical incident technology, focus interview and other qualitative research methods developed the constituent elements of perceived relationship revival value. Guided by the theory of value-driven behavior of social exchange, based on the theory of rational behavior and technology acceptance theory, it is developed theoretical framework of perceived relationship revival value-emotion of relationship revival-brand relationship revival intention. And using 856 questionnaires customer data restaurants empirical test reveals the mechanism of customer perceived relationship revival value works on customer relationship revival intention by acting on customer perceived emotional of relationship revival, instead of brand competitiveness and customer characteristics and investigates. It is also investigated the adjustment effect of alternative brand competitiveness and customer characteristics to this mechanism.

3 Algorithm

The customer behavior intention of brand relationship revival has characteristics of situation dependencies and peculiarity dependencies. Consumer decision process is a process of complicated mentality. On the one hand, for the same customer, even though he tends to rebuild the relationship because of recovery value and relationship benefit, competing brands could impact the customer's decision. The more competitive the alternative brands are, the less possible the customer is to rebuild the relationship. On the other hand, Customer purchase behavior affected by its own characteristics. It is shows that the level of confidence and demand for diversity, involvement and other customers individual traits play a regulating effect on the relationship between customer relationship revival emotion and behavioral intentions in relation adjourned, which means that positive emotions may not be able to produce revival intention. The relationship revived is a prepaid trust to the service failure companies. In other words, it is an uncertain decision. Generally speaking, the customers with lower confidence have higher level of perceived risk than the general level, and they are likely to avoid the risk when faced with risk decisions. The customers with high level of demand diversity tend to make the decision by comparing the different brands, and they have low degree of stickiness, once there are ideal alternative brand, they are not willing to rebuild the relationship. Similarly, customers with high purchase involvement have features that they may shop around, make the decisions discreetly, assess brand fully, then it will not contribute to the management of relationship revival. The equation of basic function is as equation (1) as follows:

$$\partial_j(C_{ijkl}\partial_k u_l + e_{kij}\partial_k\varphi) - \rho\ddot{u}_i = 0 \tag{1}$$

Under the linear relationship, basic equation is shown in equation (2):

$$\partial_j(e_{ijkl}\partial_k u_l - \eta_{kij}\partial_k\varphi) = 0 \tag{2}$$

The linear differential equation can be expressed into the following simplified forms:

$$L(\nabla,\omega)f(x,\omega) = 0,$$

$$L(\nabla,\omega) = T(\nabla) + \omega^2\rho\mathrm{J} \tag{3}$$

In which,

$$T(\nabla) = \begin{Vmatrix} T_{ik}(\nabla) & t_i(\nabla) \\ t_k^T(\nabla) & -\tau(\nabla) \end{Vmatrix}, \quad \mathrm{J} = \begin{Vmatrix} \delta_{ik} & 0 \\ 0 & 0 \end{Vmatrix},$$

$$f(x,\omega) = \begin{Vmatrix} u_k(x,\omega) \\ \varphi(x,\omega) \end{Vmatrix} \tag{4}$$

$$T_{ik}(\nabla) = \partial_j C_{ijkl}\partial_l, \quad t_i(\nabla) = \partial_j e_{ijk}\partial_k, \quad \tau(\nabla) = \partial_i\eta_{ik}\partial_k$$

Consider an infinite situation, we have the equation (5) in the following:

$$L^0 = \begin{Vmatrix} C^0_{ijkl} & e^0_{kij} \\ e^{0T}_{ikl} & -\eta^0_{ik} \end{Vmatrix} \tag{5}$$

The Experiment and Analysis

The theoretical framework can lay a foundation for the related research on brand relationship revival, and provide guidance for service enterprises brand relationship management practice.

Table 1 shows the common model-fit indices, recommended values and results of the test of structural model fitness. As shown in Table 1, comparison of all fit indices with their corresponding recommended values, the evidence of a good model fit was revealed. Fig. 1 presents the structural model results for the model.

The relationship benefit is focusing on the customer psychology and behavior characteristics in the latter part of brand life cycle enrich the theory research results of brand relationship from a dynamic perspective. How to revive the relationship, win back lost customers, extended the length of every customer relationship life cycle, to enhance the contribution of the customer relationship full-life cycle to service provider, is a new focal point in brand management and customer relationship management to service company. In-depth analysis of customer value demands and psychological-behavioral mechanisms in this stage, not only is beneficial to profound grasp of the brand relationship full-life cycle, but also enriched research results of the brand relationship from a dynamic perspective.

Table 1. The result of measurement model.

Fit indices	*Recommended value*	*Result*
X^2/df	*<3.00*	*2.73*
GFI	*>0.90*	*0.94*
RMSEA	*<0.08*	*0.05*

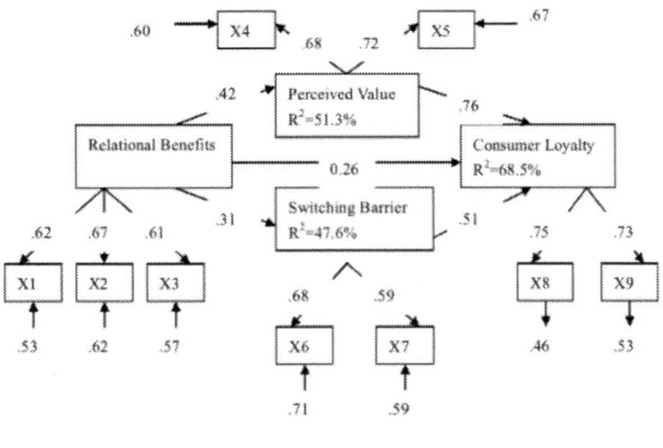

Fig. 1. Results of structure equation model.

References

1. Babak A, Kristian K, Martin M et al. Exploiting symmetries for scaling loopy belief propagation and relational training. Machine learning, 92(2013):91-132.
2. Babak A, Kristian K, Sriraam N et al., Lifted Online Training of Relational Models with Stochastic Gradient Methods. Machine learning and knowledge discovery in databases, 5(2012):585-600.
3. Chingarande A, Mzumara M, Karambakuwa RT. Does MERSOSUR possess comparative advantage? Journal of Economics, 5(2014): 341-349.
4. Eric T. G. W, Jeffrey C. F. T, Varun G et al., Examining the Relational Benefits of Improved interfirm information processing capability in Buyer-Supplier Dyads. MIS quarterly, 37(2013):149-173.
5. Gilles B, Cédric F, Benjamin G et al. Probabilistic Relational Verification for Cryptographic Implementations. ACM SIGPLAN Notices: A Monthly Publication of the Special Interest Group on Programming Languages, 49(2014): 193-205.

6. Haixia G, Dawei L. Relationship of trustworthiness and relational benefit in electronic catalog markets. Electronic markets: The international journal on networked business, 24(2014): 67-75.Benrun Huang, Li Xia, Zhengguo Wu, Weiping Zhou. Control Strategy of A Dynamic Voltage Restorer (DVR) Based on Line Voltage Compensation. Energy Procedia, (2011), pp. 13-26.
7. Wei Dai, Mostafa Bassiouni. An improved task assignment scheme for Hadoop running in the clouds. Journal of Cloud Computing, (2013), pp. 21-33.

DOI: 10.1051/matecconf/201710004036

Relationship on the Price Sensitivity and Actual Market Acceptance Degree of Metallic Materials

Muxue Jing [1,*] and Xilai Jin [2]

[1]School of Management, Wuhan University of Technology, Wuhan, China, 430074
[2]School of Engineering, University of Florida, Gainesville, FL, US
* Corresponding author: jmuxue@126.com

Abstract. In order to find out the relationship between the price sensitivity and actual market acceptance degree of metallic materials, the database ensemble learning model is proposed in this paper. Due to the variety and class imbalance of customers, a database marketing model based on supervised clustering and ensemble learning is used for the model. The results show that the database ensemble learning model can thus improve the calculation accuracy and time-efficiency substantially.

1 Introduction

A key problem of database marketing is to correctly locate target clients. In the example of Baesens etc. [1] the accurate target client locating can add 500,000 Euro extra earnings for every extra percentage. Knott [2] etc point out that for a retail service bank, 0.7% extra percentage correct target client locating can help to improve 20% revenues for each customer. Judging from the perspective of data mining, the target client locating can be beckoned as classification problem, namely to predict the percentage for customers buying or to buy product based on their consumption characteristics. However, the diversity and unbalanced category of consumer groups restrict traditional classification predication technique. To begin with, among consumer groups, the number of target consumers is far lower the number of non-target clients, in other word, a class imbalance problem. The traditional classification predication technique aims to minimize the risk and it will be hard to effectively deal with class imbalance [3-4]; what's more, consumer groups are extensive and diverse, so single classification predication model cannot accurately reflect various consumption model as well as characteristics and over-fitting problem for learning model may appear [5].

In order to solve this problem, 2e consider targeting customers in database marketing as a classification and prediction problem in data mining, i.e., to predict whether a customer would purchase a product or the probability of purchasing based on his/her characteristics. Due to the variety and class imbalance of customers, a database marketing model based on supervised clustering and ensemble learning is proposed. The empirical study indicates that the proposed approach is able to improve the performance of database marketing.

Even though database marketing research has achieved important progress, an existing problem is to effectively optimize the difference and individual performance among various learning models so as to improve the accuracy of database marketing. The main idea of the learning model generated in existing ensemble studying research is to repeatedly and randomly take samples in sample space or feature space which yet cannot guarantee the big difference among various learning models. At the same time, the training subset created by random sampling cannot accurately depict the diversity and difference among consumer groups and as a result the accuracy of establishing learning model based on training subset will decrease. In order to solve above problems, this paper comes up with ensemble learning model based on supervised clustering which firstly adopts K-Means to gather the non-target consumers and divide them into several consumer subgroups with big differences. Then it combines subgroups with few respondent consumers so as to have several training subsets to train artificial neural network sample learning model in several subgroups to carry out integration and overcome the data imbalance problem meanwhile improve the learning performance of every single learner.

2 Basic Framework

The ensemble learning based on supervised clustering can be divided into three stages of data pre-processing, supervised clustering and ensemble learning.

The first stage is data pre-processing. The dimension property of consumer is different (for example, if the property of income is over 1000, the property of age is dozens). In clustering analysis dimension discrepancy will cause a big influence for those properties with big differences while for those with small differences, the influence can be ignored. Therefore, before carrying out clustering analysis, we have to carry out data pre-processing so as to make the dimension property consistent. This paper adopts min-max normalized methods to conduct linear conversion on data and the formula shows as follows:

$$\bar{x}_{ij} = \frac{x_{ij} - \min_j}{\max_j - \min_j}, (i = 1, 2, \cdots, n; j = 1, 2, \cdots, m) \quad (1)$$

In this formula, n means the table cardinality and m refers to data attribute number (not including class label), x_{ij} refers to the original value of i record's j property, $\bar{x}_{ij}$ refers to the value of i record's j property after standardized implementation, $\min_j$ is the minimum value of j property while $\max_j$ is the maximum value of j property . After standardizing formula (1) we can make the value range of all properties become $[0,1]$ so as to avoid the influence caused by different dimension properties.

The second stage is supervised clustering. Firstly calculate the proportion of majority class and minority class, set K is equal to the proportion in the *K*-Means algorithm and then we carry out clustering analysis on the majority class and cluster the majority class into K aggregates of data. Based on this, we recombine each aggregate of data with minority class to form K sub set sample with relevant balance.

Suppose the training sample set of N consumer information is $G = \{y_i, \mathbf{x}_i\}_{i=1}^{N}$. The individual information property of each consumer (such as age, sex and family members) shows in the form of $\mathbf{x}_i \in R^p$, among which p refers to the property number of customers. Suppose G_{pos} and G_{neg} ($G_{pos} \cup G_{neg} = G$) respectively refers to the customer set who response and do not response to marketing activities and in general. In order to explore the consumer subgroup with different consumption model and characteristics and at the same time overcome the problem of class imbalance we set the K value in *K*-Means algorithm is $|G_{neg}|/|G_{pos}|$ and adopts this value to carry out clustering analysis on the majority class (which does not affect consumers). Based on this, we combine every aggregate of data and minority class (which can affect consumers) to form K subgroups which are relevantly balanced. The algorithm flow chart for supervised clustering is shown as follows:

The third stage is to establish basic learning model and carry out ensemble learning. BP neural network has good learning ability and fault tolerant ability in classification predication and this paper adopts it as the basic learning algorithm.

In supervised clustering stage, we train the BP neural network of every sample set $subG^k, (k = 1, 2, \cdots, K)$ so as to have K basic learning model $Learner^1, Learner^2, \cdots, Learner^K$. Different basic learning models have different accuracy for different data samples and how to choose the ensemble learning method for basic learning model is a key issue. Comparing from choosing the optimal model among all basic learning models, a better approach is to choose different learning modes for different data samples (commonly known as dynamic integration)[27]. This paper takes weighted voting based on sample neighborhood learning accuracy as the integration method which belongs to dynamic integration.

Concerning the unknown data record $\mathbf{x}$, the first step is to find out the close sample NN_x^k of $subG^k$ based on KNN algorithm so as to get the predict output of $Learner^1, Learner^2, \cdots, Learner^K$ and $\mathbf{x}$ as well as its close sample. P_x^k is the predict percentage on $\mathbf{x}$ carried out by basic learning model $Learner^k$ while φ_x^k is the accuracy of $Learner^k$ in NN_x^k. When integrate the predict output of K basic models we can finally get the integrated forecasting output FP_x and the formula shows as follows:

$$FP_x = \sum_{k=1}^{K} P_x^K \frac{\varphi_x^K}{\sum_{k=1}^{K} \varphi_x^K} \quad (2)$$

$$(k = 1, 2, ..., K)$$

The output of ensemble learning in whole testing set is $FP = (FP_1, FP_2, \ldots FP_x)$.

Based on the above three stages, the overall flow chart for database marketing model based on supervised clustering and ensemble learning shows in figure 1.

DOI: 10.1051/matecconf/201710004036

3 Discussion and Conclusion

This paper adopts the data of predict contest of COIL (Computational Intelligence and Learning) in 2000 as the empirical research. This data includes the sample data of 9822 European families buying the car insurance. What's more, this paper divides this data set into training set and testing set, among which training set includes 5822 data which is used to establish the ensemble learning model put forward in this paper while another 4000 data is used to assess the effect of this model. Among the 5822 training data set, there are only 348 minor class sets, occupying 5.97% of the training set. Among the 4000 testing data, there are 238 minor class sets, occupying 5.95%. Therefore, we can conclude that the data set has obvious class imbalance problem which can be beckoned as the data source of this paper.

Singe ANN approach has the slowest lift speed which demonstrates that it has poor ability to deal with the imbalance class problem in database marketing while ANN based on SMOTE, ANN based on FN and GA/ANN approaches can create better effect. The model put forward in this paper has obvious advantages from 5%-35% depth and is worse than GA/ANN from 40%-50% depth. In reality, there is a dazzling array of customers in database marketing and therefore we pay more attention to the performance in smaller depth. In a summary, the above result indicates that the model in this paper can have higher hit rate in smaller depths which can be effectively affect the database marketing.

The ensemble learning method based on supervised clustering can have higher hit rate with the depth from 5%-35% comparing to other four methods especially from the depth 5%-25%, the hit rate is 32.00%, 23.25%, 18.67%, 15.75 and 14.30% which have obvious prediction advantages over other approaches. To conclude, the ensemble learning method based on supervised clustering is able to have higher Hit Rate with smaller depth which can improve the database marketing efficiency.

At present, both from the enterprise management practice and from the academic research level, brand management focus on relationship establishment and maintain, the attention to the relationship fracture customers is far less than the first two stages. In fact, with the increased competition of customer resources and customer development and maintenance cost of ascension, to re-develop the relationship of fracture customers should be the new focus on brand relationship management. Under the condition of buyer's market, due to the selective enhancement of consumers, customer resources become more competitive. The tentacles of enterprise brand relationship management have to extend to relationship revival of fracture customers. The premise of effective brand relationship revival strategy is correctly grasp of what factors drive the customer-brand relationship revival behavior and how these factors work. However, the academic study on brand relationship revival has just begun. Driving factors of the brand relationship revival, especially the thinking of mechanism of action is still very insufficient. The biggest problem of the only scattered results is the lack of a clear theoretical basis and analysis framework of the system.

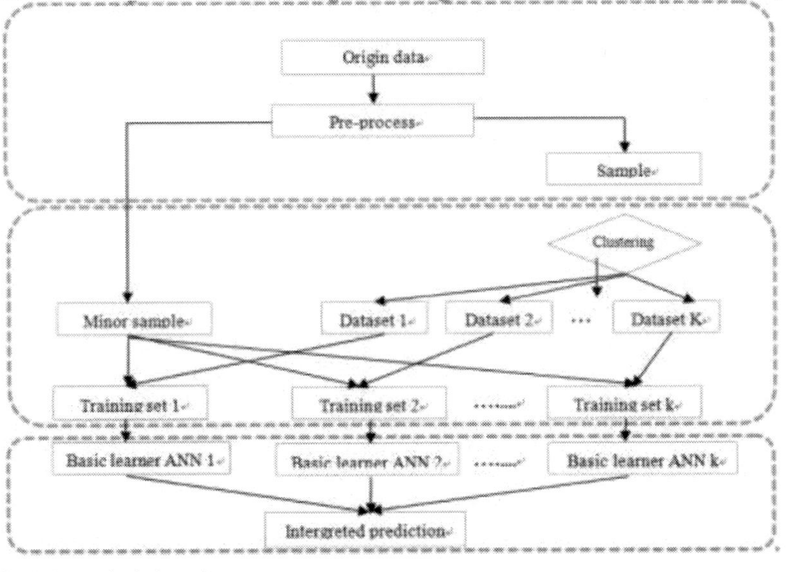

Fig. 1. Supervised clustering process.

This model believes that inside the data there are consumer aggregate of data with similar characteristics, so we adopt K-Means to carry out supervised clustering to conduct integration and classification with artificial neural network based on this. After carrying out experimental analysis and comparison among several learning methods, we can verify that the model in this paper is able to effectively deal with the imbalance class problem in database marketing and improve the accuracy. The following researches can be carried out from two aspects: firstly, extract the data characteristics to improve performance; secondly, carry out optimization as well as integration with other algorithms.

References

1. Gilles B, Cédric F, Benjamin G et al. Probabilistic Relational Verification for Cryptographic Implementations. ACM SIGPLAN Notices: A Monthly Publication of the Special Interest Group on Programming Languages, 49(2014): 193-205.
2. Wei Dai, Mostafa Bassiouni. An improved task assignment scheme for Hadoop running in the clouds. Journal of Cloud Computing, (2013), pp. 21-33.Babak A, Kristian K, Martin M et al. Exploiting symmetries for scaling loopy belief propagation and relational training. Machine learning, 92(2013):91-132.
3. Babak A, Kristian K, Sriraam N et al., Lifted Online Training of Relational Models with Stochastic Gradient Methods. Machine learning and knowledge discovery in databases, 5(2012):585-600.
4. Chingarande A, Mzumara M, Karambakuwa RT. Does MERSOSUR possess comparative advantage? Journal of Economics, 5(2014): 341-349.
5. Eric T. G. W, Jeffrey C. F. T, Varun G et al., Examining the Relational Benefits of Improved interfirm information processing capability in Buyer-Supplier Dyads. MIS quarterly, 37(2013):149-173.

Research on the Optimization Coupling for Enhancing Polymer Optical Fiber Communication

Zhengqin Dong [1,*] and Yun Lu [2]

[1] School of Computer Science and Technology, Wuhan University of Technology, Wuhan, China, 430074
[2] School of Naval Architecture and Ocean Engineering, Huazhong University of Science and Technology, Wuhan, China, 430074
[*] Corresponding author: 15926424120@163.com

Abstract. Aiming at improving the performance for optical fiber communication, the various factors to enhance the optical fiber bandwidths are discussed in detail. A novel design with cost effective coupling is explored. The proposed package with the integrated signal conditioners will make the transceiver packages simple and compact. This signal conditioner also helps in effective optical coupling without the need of any additional light. This connector less package design, universal to any light source proposes to enhance the data rate and is widely manufacturing at an ease of installation and low cost.

1 Introduction

In the rising data rate market, short distance communication requires higher bandwidths and low cost components. One of the potential areas that demands high data rate is premise networks like broadband access network and home automation. POFs are being considered as a practical solution for this kind of application. Their major advantages include a very flexible structure, lower connection costs and immunity to electromagnetic interference. The large core structure of POF is advantageous in installation while it allows a large number of modes to propagate through the core.

The presence of multimode limits the bandwidth of POF due to intermodal dispersion. The bandwidth enhancement studies have got foremost interest to keep POFs attractive in the growing data rate market. The three major component's characteristics, say transmitter, fiber and receiver, decide the bandwidth capacity of a multimode fiber link. Since these networks are implemented in premises and local area networks, overall cost of the components is also of major concern. Even though LED sources are cost effective transmitters, the requirement for higher bandwidths made them less attractive for POF communications [1-2]. The high power, high data rate, narrow spectral width and smaller spot size compared to LEDs raised low cost 850 nm VCSEL as an alternative solution for such applications. Connector less packages and fiber plug in modules also helped to maintain POF system installation cost lower. There are new generations of POF, which offer a higher bandwidth are manufactured by engineering the core index profile. A graded index (GI) POF can transmit four times higher data rates than a step index POF. But they are offered at higher price than normal POFs.

Another alternative to increase the bandwidth of POF is by external coupling mechanisms, which selectively excites either the fundamental modes lying at the center of the core or higher order modes lying towards the cladding. One of the widely used methods to achieve this condition is attaching a mode conditioning single mode fibers with the multimode fiber. But this will have the additional process of splicing or connecting the SMF to POF [3].

2 The Model Bandwidth

The bandwidth of a fiber is decided by the aggregate performance of all the modes in the fiber or in other words the modal bandwidth depends on the number of modes excited in the MMF. The optical launch conditions have a greater effect on deciding the number of modes excited. The lower order modes are centered on the core, while a higher order mode finds their paths towards the cladding. Over filled and under filled launches are the two general terms associated with the light propagation through fiber. In the former case the light launched in to the fiber over fills the core of the fiber and hence excites the maximum number of modes possible. In the latter case, the light launched fills only a certain portion of the fiber and excites limited number of modes. Restricting the modal excitation by means of an undefiled launch condition is known as restricted mode launch (RML) [4]. Figure 2 shows the example of overfilled and under filled launches.

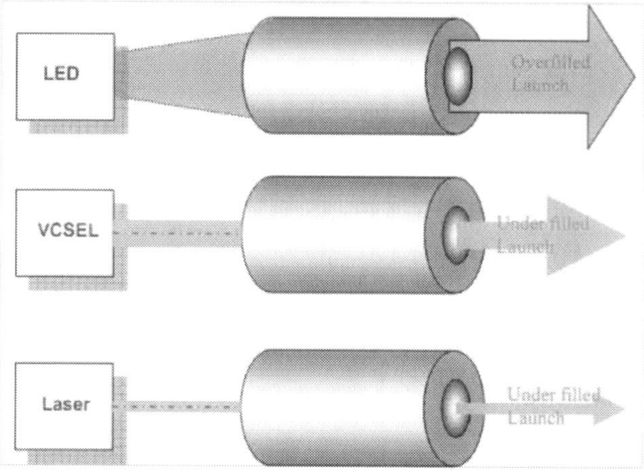

Fig. 1. The example of overfilled and under filled launches.

The RML can be either center launch or offset launch. In the center launch, the optical power from a single mode fiber or laser is coupled to the center of the multimode fiber. This excites the lower order modes, fundamental mode, centered at the core. Alternatively in the offset launch, the power is launched to an area offset from the center of the fiber. In this method the power is coupled to the higher order modes having relatively low modal dispersion.

Both these methods have proven very effective in increasing the bandwidth of multimode fibers by controlling the number of modes excited. But the fibers having discontinuities in their refractive index profile near the center badly affects the performance under center launch scheme. The variation in refractive indices offers different speeds to light and ends up in the phenomena known as Differential Mode Delay (DMD), which considerably reduces the effective bandwidth. In an offset launch, the launch position can be selected to avoid this defective region.

The equilibrium mode distribution (EMD) and steady state mode distribution are the other two concepts related to light launch conditions. EMD is the mode distribution obtained after going through a system, where the output mode distribution is unaffected by the input mode distribution; independent of launch conditions. This situation is seldom met [5].

Steady state distribution is reached when the mode distribution is no longer changes along the fiber. This depends on launching conditions. The maximum steady state is reached in an over filled launch condition. In the absence of strong mode coupling, restricted launch condition also creates a steady state mode distribution. Once the steady state is reached, the transfer of energy between the modes will stop.

Mode stripping is another method used to reduce the number of higher order modes propagating in a fiber and thereby improving the bandwidth further. They are of two types: core mode filter and cladding mode stripper. In the core mode filter the fiber is wound over a mandrel to reduce the higher order modes in the core. During this process the higher order modes leaks to the cladding and is being lost. This results in an exit numerical aperture less than the material NA of the fiber. A fiber that has reached modal equilibrium, along with the reduced NA, is said to exhibit long-launch conditions. Cladding mode stripper is required in short distance applications since the higher order modes leaked to the cladding couples back to the detector at the receiver side. It is necessary to remove this mode using a cladding mode stripper. The cladding mode stripper is nothing but a higher refractive index material coated on the fiber cladding. Here the cladding buffer is removed initially and coated with a higher index material. Fig. 2 shows the equilibrium steady state mode propagation.

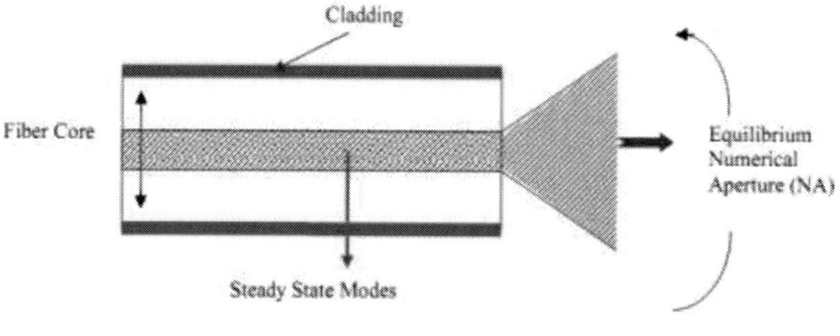

Fig. 2. Equilibrium steady state mode propagation.

3 Experiment Results and Discussion

To verify the effects of light source parameters on modal distribution, we set up an experiment to launch the light into POF having a core diameter 980μm. In the initial experiment, light from a 1310 nm laser diode is launched to the POF using a single mode fiber of core diameter 9 μm and the near field patterns are captured by a beam analyzer. The second set up replaced the single mode fiber with a multimode fiber of 62.5 μm to repeat the experiment.

The design of RML coupling requires the information on encircled flux (EF) of the light source used. The encircled flux is the percentage of the power within a given radius when light is launched by a transmitter into a multimode fiber. The modal distribution of the fiber depends on the numerical aperture and spot size at the launch end of the fiber. If the spot size is small and its NA is less than the fiber core, the optical power will be concentrated at the center of the fiber core. Figure 3 shows the experimental setup for optimizing the launch conditions and figure 4 (a) shows the near field output pattern of POF, when light launched with MMF and (b) shows the launched with an SMF.

The bigger spot size and larger NA, leads to higher order mode excitation. Hence in any optics design for mode control it should take care of a matching spot size and NA of the source and fiber. A steady state equilibrium mode distribution state can be realized by an identical far and near field patterns. The radiation patterns are compressed in the case of RML conditions and enlarged in the case of OFL conditions.

Keeping in mind the necessity of a cost effective compact system, we explored the possibility of integrating the signal conditioning elements inside the package. In the proposed design, at the transmitter side the molded plastic package consists of a restricted mode launcher and at the receiver side it involves a mode stripper. Restricted mode launcher, which helps to keeps the encircled flux of the source to remain the same irrespective of the light sources used. The design of this element is optimized to generate the same effect of launching as with a single mode fiber of NA 0.1. The signal conditioner at the transmitter side provides center launch with undefiled launching conditions. This helps to excite only the lowest order modes and hence reduces the modal dispersion. The concept of mode stripper is implemented in such a way that it not only helps to radiate the higher order modes excited but also couples the light effectively to the detector. The dimensions of the signal conditioners are optimized based on the optical simulations with a VCSEL source of divergence half angle 160, SI POF having a core diameter of 980μm and a detector radius of 50 μm.

Based on the theoretical calculations, POF of 980 μm dimensions will support thousands of modes in over filled condition. The design of RML is done to reduce the number of modes excited to a minimum by concentrating the light only to excite the fundamental mode. There by reducing the modal dispersion and increasing the modal bandwidth. The mode stripper at the receiver side is designed to eliminate higher order modes and have efficient coupling to the detector. The mode stripper design is such a way that it will eliminate minimum 10% higher order modes. So with the help of both RML and Mode stripper, the data rate will be enhanced considerably. The fabrication of this integrated components are proposed to do with plastic molded packages. The transmitter and receiver plastic packages can be fabricated with optimized signal conditioner structure cavities. These cavities are then filled with optically transparent epoxy of suitable index of refraction. Either ultra violet curing or thermal curing can be used for reinforcing the structure. It is better to use an epoxy having Tg > 95 C having a shrinkage rate less than 2% of the initial volume.

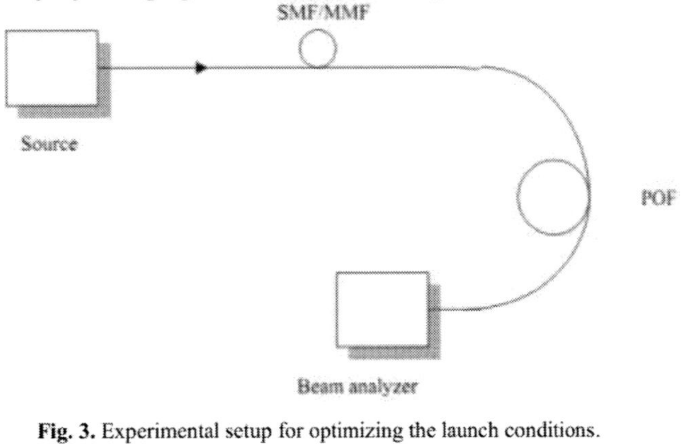

Fig. 3. Experimental setup for optimizing the launch conditions.

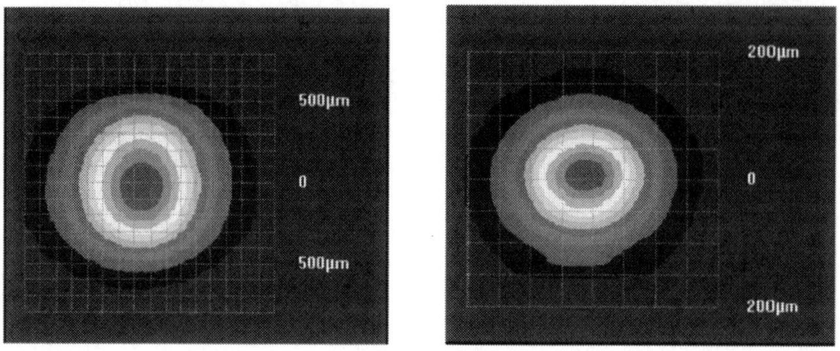

Fig. 4. (a) Near field output pattern of POF when light launched with MMF(b) launched with an SMF.

References

1. Wu He. Examining students' online interaction in a live video streaming environment using data mining and text mining. Computers in Human Behavior, (2013), pp. 291-304.
2. Fu Xiao, Cheng Fan. Data mining in building automation system for improving building operational performance. Energy & Buildings, (2014), pp. 75-83.
3. Benrun Huang, Li Xia, Zhengguo Wu, Weiping Zhou. Control Strategy of A Dynamic Voltage Restorer (DVR) Based on Line Voltage Compensation. Energy Procedia, (2011), pp. 13-26.
4. Wei Dai, Mostafa Bassiouni. An improved task assignment scheme for Hadoop running in the clouds. Journal of Cloud Computing, (2013), pp. 21-33.
5. Wei Kuang Lai, Yi-Uan Chen, Tin-Yu Wu, Mohammad S. Obaidat. Towards a framework for large-scale multimedia data storage and processing on Hadoop platform. The Journal of Supercomputing, (2014), pp. 681-696.

Effect on Communication System of Dispersive Multiplexing in Optical Fiber

Zhengqin Dong [1,*] and Yun Lu [2]

[1] School of Computer Science and Technology, Wuhan University of Technology, Wuhan, China, 430074
[2] School of Naval Architecture and Ocean Engineering, Huazhong University of Science and Technology, Wuhan, China, 430074
[*] Corresponding author: 15926424120@163.com

Abstract. In order to research on the transmission effect on communication system of dispersive multiplexing in optical fiber, the concept is applied to data transmission over multimode optical fiber, and the result is an optical multiplexing that can increase the capacity of such fiber in the engineering. Experiment result shows that the proposed method can thus improve the overall system performance substantially.

1 Introduction

From the standpoint of information theory, multimode optical fiber (MMF) has more information capacity than single-mode optical fiber (SMF). However, this assertion becomes meaningless when one tries to send a high data rate down any sizable length of multimode fiber, owing to the effects of modal dispersion, which severely limit the achievable data rates in current MMF transmission systems. The information theory viewpoint relies on the notion that the additional spatial modes in the transmission medium provide additional degrees of freedom and, therefore, additional capacity.

In practice, however, there is no simple way to independently address and detect the individual modes of the fiber (which number in the hundreds), and this, coupled with the presence of mode-mixing during propagation (often present to some degree), makes it nearly impossible to exploit the additional capacity in any straightforward manner. Because of their cost advantages, systems employing MMF are finding increased use in certain data networking applications, and MMF promises to replace copper as the transmission medium of choice for future high-speed local area networks, perhaps eventually reaching the desktop. Any practical technique that could effectively use the spatial degrees of freedom in the fiber to improve transmission bandwidth could potentially find extensive application in these future systems.

We describe a transmission scheme that uses the effects of modal dispersion to enhance the transmission capacity of MMF. The method does not require orthogonal coupling to the individual modes of the fiber and can tolerate some degree of mode-mixing. The technique is based on recently developed ideas in the field of wireless communications. Using information theory, Foschini and Gams have shown that a rich multipath scattering environment pro-vides a dramatic capacity enhancement for wireless systems employing multiple antennas at both the transmitter and receiver. Under conditions of sufficiently strong scattering, the information capacity of a multiple transmit receive system scales linearly with the number of transmitters, even as the total transmit power is held constant. This extraordinary result occurs, in a fundamental sense, because the system is able to use the spatial degrees of freedom inherent in the scattering environment to improve the capacity. Experimental work in this area has resulted in multiple antenna systems with spectral efficiencies as high as 40 bits, a value unobtainable by conventional transmission methods (7, 8). We demonstrate here that these basic concepts can be applied to optical transmission in multimode fiber, in a system using multiple lasers and multiple detectors. Although there are important differences between the optical and wireless versions of this technique, we show that for the optical system, the most important property, the linear scaling of the capacity with the number of transmitters, is preserved.

2 The Model and Framework

The transmission system (Fig. 1) consists of N lasers coupled together onto a single multimode fiber and received at the end of this fiber link by M detectors. N independent bit streams are first modulated onto identical radio frequency (RF) carriers, and these electrical signals modulate the N lasers. The RF modulation format is arbitrary, the data clocks in each arm are synchronized, and the symbol rate of the individual bit streams is constrained to operate within the bandwidth of the fiber. At the receiver, the optical power is split evenly between the M detectors, which convert the optical signals to electrical signals through direct detection. Each detector receives power from all N transmitting lasers. These signals are demodulated by coherent microwave detection, and the resulting M baseband signals (which contain a mix of all of the data) are sent to a signal processing circuit that recovers the original N data streams. No wavelength division multiplexing (WDM) is used in the combining or splitting of the optical signals, i.e., the wavelengths of the transmitting lasers do not have to be precisely specified.

The system works because of a process referred to here as "modal-coupling diversity." At the input to the system, each laser excites a slightly different modal power distribution within the fiber. There will be a tendency toward this type of diversity regardless of the actual technique used to couple the N beams together. At the output of the fiber, further modal-coupling diversity causes each detector to receive power from a slightly different distribution of modes (even when the coupling optics is designed to split the total optical power equally).

One can define an M X N channel matrix that characterizes the system: Each element of this matrix is a complex number that relates the (electrical) magnitude and phase of the received RF carrier in a given detector to that transmitted from a given laser. These matrix elements are determined by the impulse response of the channel. As a result of modal dispersion, a carrier injected into the front end of the system will appear as a series of randomly phased sine waves at the receiver. Under conditions where sufficient modal dispersion exists in the fiber (such that the impulse response length which is at least the RF cannier period), the elements of the transmission matrix can be modeled as uncorrelated Gaussian random variables. This is a consequence of the large number of modes, the dispersion properties of the fiber, and the modal coupling diversity and is analogous to the Rayleigh fading model used for multipath propagation in wireless systems. Measurements of actual MMF transmission have confirmed this property (Fig. 2). For a 2-km length of 62.5-pm fiber, independent Rayleigh fading statistics are realized for RF carriers greater than -500 MHz. It is this similarity between the optical and wireless regimes that suggests the applicability of multiple antenna theories to MMF transmission.

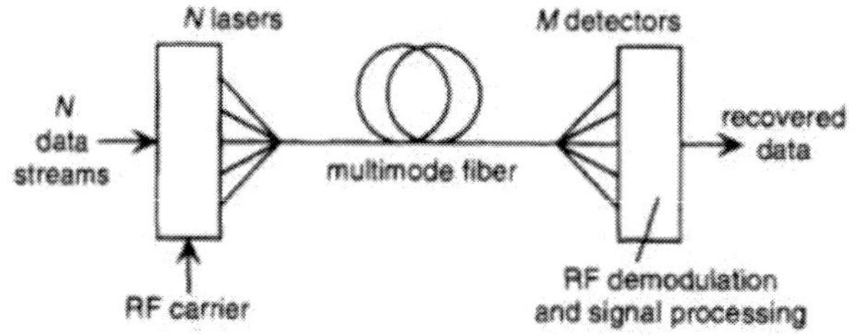

Fig. 1. Diagram of the proposed optical transmission system.

The transmitter and receiver "diversity" is essential for the system shown in Fig. 1 to work. In order to recover the transmitted information, the channel matrix must be known (channel estimation can be accomplished with a series of training bits preceding the data stream) and it must be invertible. A single-mode fiber, for example, will necessarily produce a noninvertible channel matrix. For a MMF, if the elements of the matrix are partially correlated (because of insufficient diversity or insufficient modal dispersion), the matrix will be poorly conditioned, leading to difficulties in achieving error-free data recovery. These issues can be quantified by using information theory to derive the capacity of the transmission system. The formalism used to derive capacity for multiple-antenna wireless systems can be applied here, but only after a modification of the basic assumptions underlying the details of the transmission. The most important difference between the optical and wireless regimes occurs at the receiver. In the wireless system, the total received power increases as the number of receiving antennas increases. In the optical system, a fixed amount of optical power anises at the end of the fiber, and this power is split between the M detectors, reducing the average received signal-to-noise ratio (SNR). The power splitting at the receiver can be compensated at the transmitter, with the per-channel SNR preserved by setting N = IM (an equal number of lasers and detectors). This assumes lossless coupling of the N lasers to the fiber, a requirement that is physically allowed for multimode fiber. This requirement is in contrast to the wireless system, where the total transmitted power is fixed independent of N.

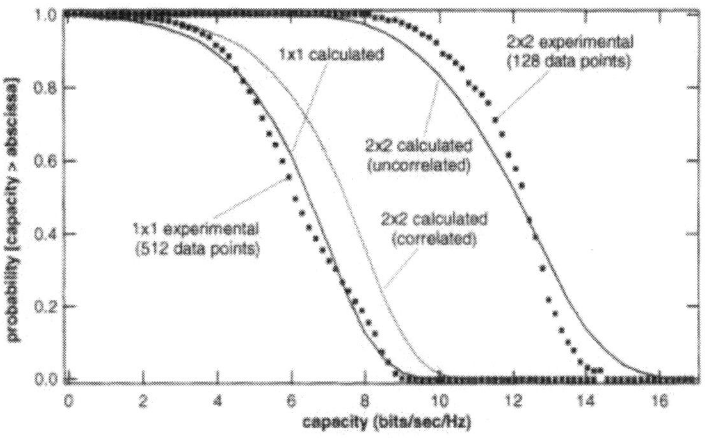

Fig. 2. Squares Cumulative distribution functions (CDFs) of capacity.

3 Experiment Results and Discussion

The linear scaling of capacity with N is significant, as it implies that one can multiplex independent signals onto "virtual" spatial channels that are resolvable through the dispersive properties of the modes. The depraving of the RF carrier at high frequencies enables this to happen; however, this depraving also introduces a notable power penalty relative to baseband transmission to 20 dB for 1-km MMF transmission). This must be overcome to realize true capacity enhancement in the link. Because capacity degrades logarithmically (not linearly) with lost power (- 1 bit s-' Hz1 for each 3 dB of power), this is achieved with a relatively modest number of transmitters. For p =21 and 9 dB, capacity enhancements are realized for N > 3 and N > 5, respectively, assuming an 18-dB power penalty. We have built a 2 X 2 experimental demonstration of this technique, operating at a relatively low bit rate (50 Mbits per channel), but clearly demonstrating the feasibility of this type of multiplexing. For simplicity, a conventional 1 X 2 splitter is used to couple the two lasers into the fiber, introducing 3 dB of optical power loss which, though tolerable in this initial experiment, will need to be eliminated for higher bit-rate systems. Two independent data streams are binary phase shift key (BPSK) modulated onto identical RF carriers (-1 GHz), and these signals are used to directly modulate the drive current of the lasers. At the receiver, the detectors convert the received optical signals to RF electrical signals and these are quadrature phase shift key (QPSK) demodulated to recover the baseband signals. QPSK demodulation is required to recover the complete phase information of the received signal. The recovered baseband data are fed to an analog processing circuit, described below, which recovers the transmitted bit streams.

The constellation diagrams display the phase space of the received data and serve to illustrate the basic properties of the transmission. For back-to-back transmission (top plots), there is negligible modal dispersion, and we observe a three-point constellation in detector 1. The outer two points correspond to the data pairs (1,l) and (- 1,- l), while the center point consists of two overlapping data pairs, {(I,- l), (- 1,1)}. This is indicative of two interfering (and unresolvable) BPSK signals and results from an attempt to send two independent signals over a single channel. Although we do observe a slight splitting of the center point in detector 2 (attributable to residual "power-splitting" diversity in the MMF couplers), the channel matrix for the top plots is poorly conditioned (condition number >20), and data recovery would require a very high SNR. In the bottom plots, 1 km of MMF has been inserted into the system. The received constellations now consist of four clearly resolved points in each detector, and the constellation seen by each is different, demonstrating the diversity present in the system.

The condition number of the channel matrix in this example is 4.5. Application of the signal-processing algorithms results in recovery of the transmitted data (Fig. 3B). The signal-processing steps required for data recovery consist of estimating the channel matrix and determining, then applying, a set of complex weights to cancel the interfering signals. The optical channel was observed to be relatively stable, suggesting that the channel can be estimated, inverted, and tracked "off-line" (at rates much slower than the bit rate). Even so, the degree of signal processing required to apply the complex weights at gigabit per second data rates is daunting and led us to consider an analog (rather than digital) approach. We have constructed a real-time analog processing circuit, using 500-MHz bandwidth analog multipliers.

The processor recovers a single dimension in phase space (sufficient for BPSK) for one channel. The processor bandwidth in the initial implementation was limited to 50 MHz to simplify circuit layout. Channel estimation is accomplished by viewing the eye diagram in real-time on an oscilloscope and adjusting the weights (four potentiometers on the circuit board) until the interfering channel is eliminated. In the experiments, each of the transmitted channels is recovered in sequence (two of these circuits would be required in a complete system). These curves show a small (- 1.2 dB) optical power penalty associated with the presence of the interfering channel. Although not shown on the plot, BERs below 10^{-9} have been measured.

DOI: 10.1051/matecconf/201710004038

References

1. Wei Dai, Mostafa Bassiouni. An improved task assignment scheme for Hadoop running in the clouds. Journal of Cloud Computing, (2013), pp. 21-33.
2. Wei Kuang Lai, Yi-Uan Chen, Tin-Yu Wu, Mohammad S. Obaidat. Towards a framework for large-scale multimedia data storage and processing on Hadoop platform. The Journal of Supercomputing, (2014), pp. 681-696.
3. Zhijian Chen, Wenhai Luo, Dan Wu, Xiang Huang, Jian He, Yuanhuan Zheng, Di Wu. Exploiting application-level similarity to improve SSD cache performance in Hadoop. The Journal of Supercomputing, (2014), pp. 703-715.
4. Benrun Huang, Li Xia, Zhengguo Wu, Weiping Zhou. Control Strategy of A Dynamic Voltage Restorer (DVR) Based on Line Voltage Compensation. Energy Procedia, (2011), pp. 13-26.
5. Wei Dai, Mostafa Bassiouni. An improved task assignment scheme for Hadoop running in the clouds. Journal of Cloud Computing, (2013), pp. 21-33.
6. Wei Kuang Lai, Yi-Uan Chen, Tin-Yu Wu, Mohammad S. Obaidat. Towards a framework for large-scale multimedia data storage and processing on Hadoop platform. The Journal of Supercomputing, (2014), pp. 681-696.

Relationship between Oxidation of Metal Surface Layer and Air Quality in Wuhan

Yiwen Wang [1,*] and Venda Garilli [2]

[1] School of Resources and Environment Engineering, Wuhan University of Technology, Wuhan, China, 430074
[2] School of Engineering, National University of Singapore, Singapore
[*] Corresponding author: 15926424120@163.com

Abstract. In order to find out the relationship between the oxidation of surface layer and air quality, the correlation analysis method and elemental analysis is used in this paper. Based on the analysis of the database of RAPIDC Corrosion Project, dose-response functions for the tested metals were derived through non-linear regression method. The established dose-response functions for carbon steel and zinc are applicable to use in Wuhan and the results show that the proposed method can thus improve overall system performance substantially.

1 Introduction

Wuhan City used to be known as a region of heavy air pollution and acid rain in China. Many efforts have been invested in the control of air pollution and acid rain since the 1980s. It is found these two nuisances have negative impact on both ecological environment and structural materials, endangering some cultural heritages in this region. In recent years, the local government has taken many actions such as "Clean Energy", "Clean Sky Action" and so on, to reduce SO_2 emission and improve the quality of atmospheric environment in Wuhan. As a result, the situation has ameliorated gradually in urban areas. Thus corrosion of materials in this new mitigating environment becomes an interesting subject once again. Our research which was one of the subprograms for corrosion problems of an international project, Regional Air Pollution in Developing Countries (RAPIDC), was targeted at the impacts of air pollution and acid rain on corrosion of three metals—carbon steel, zinc and copper- at an urban site Wuchang and a rural site Qingshan in Wuhan. It was accomplished by cooperation with Swedish Corrosion Institute (SCI) and Swedish Environmental Institute (SEI), in 2002.

Standard specimens of carbon steel, zinc, and copper were exposed at an unsheltered position on racks. At the same time, atmospheric pollution data were gathered bi-monthly with passive samplers for SO_2, NO_2, HNO_3, O_3 and particle matter. Data of precipitation, and precipitation pH, relative humidity and temperature were obtained from nearby automatic air quality monitoring stations. This methodology was established by the cooperation of SCI and SEI, and it had been successfully applied to exposure experiments in Europe and Northern America [1].

2 Result and Discussion

A variety of materials are subject to atmospheric corrosion, such as metals, limestone, paint coatings, etc. In this work, we concentrated only on carbon steel, zinc, and copper, for which the atmospheric corrosion rates of first-year exposure were respectively (782.5, 9.1, 24.2) g m^{-2} a^{-1} at GYQ and (492.0, 11.6, 17.9) g m^{-2} a^{-1} at TSP. Clearly, the more serious air pollution in the urban area gave rise to the higher corrosion rates of carbon steel and copper at GYQ, but for zinc, because of its high sensitivity to humidity and precipitation acidity, its corrosion rate at GYQ was a little lower than that at TSP where the environment was characterized by high humidity and precipitation acidity. This abnormal phenomenon was also observed in Cuba by Mendoza, et al [2].

Compared with corresponding results previously reported [3], it can be seen that the annual corrosion rates for carbon steel, zinc and copper in the year from 1987 to 1988 at GYQ were as much higher as 2.8, 4.0 and 3.2 times those in the year from 2002 to 2003, respectively. The decrease in corrosion rates could be, to a large extent, ascribed to the improved situations of SO2 and acid rain pollution. However, the reduction trend of corrosion rate of carbon steel in the same period was not as obvious as that of SO_2 concentration, which may partly reflect the combined effects of multi-pollutant situation on atmospheric corrosion of materials.

In the RAPIDC Corrosion Project, exposure tests were carried out at 16 sites in an extended geographical zone that covered 8 countries in Asia (mainly South Asia) and Southern Africa (Fig. 1), which involved rural, urban and

 DOI: 10.1051/matecconf/201710004039

industrial atmospheres. The nearly synchronous tests made the comparison among RAPIDC sites possible and meaningful. The atmospheric environment characteristics of RAPIDC sites were shown in Fig. 1. In general, the pollution situation of SO2, particle matter and acid rain (H+) at both urban and rural site of Wuhan was more serious than that at most other RAPIDC sites. Accordingly, the atmospheric corrosion rates of carbon steel, zinc and copper at the two sites in Wuhan were higher than those at most of other RAPIDC sites. These comparisons confirm us the importance and necessity to further abate air pollution in Wuhan. Grey system theory was first proposed by Deng in 1982 [4], and it has been successfully applied to many fields including atmospheric corrosion. Grey system theory can be applied to any uncertain system in which information does not suffice to support a relational analysis. Therefore, this method is a helpful tool to find out intrinsic regularity existing in atmospheric corrosion process whose mechanism remains relatively unclear due to the complexity of the process itself and the variability of atmospheric environment. We used the grey system theory to analyze the data in this study. Here, the atmospheric corrosion rate of a tested metal was taken as the reference sequence, and the environmental parameters as compared sequences. The grey relational grades between tested metal and environmental parameters were calculated according to the principle of grey relational analysis.

A factor with an order is not larger than 4 was regarded as the main factor that affected the atmospheric corrosion of the tested metal. We derived the following conclusions from grey relational analysis:

1) The main factors that affected atmospheric corrosion of carbon steel and copper were almost identical, which included SO2, Particle and H+, and herein SO2 was especially crucial.

2) For atmospheric corrosion of zinc, the primary factors were H+, relative humidity, precipitation amount and SO2 in order, which could well explain the abnormally high corrosion rate of zinc at Qingshan, Wuhan;

3) In general, the two important indicators of acid environment, SO_2 and H+, were common primary factors that affected atmospheric corrosion of all tested metals, which demonstrated again the strong destructive effect of acid precipitation on metals. Principal component analysis (PCA) is a pattern recognition technique to find out similarities among the observations, and also to ascertain which variables are related to each other. It is noted that all variables are treated simultaneously in an n-dimensional space by PCA, whereas in a simple statistical study only two variables are considered at the same time, usually by means of a two-dimensional diagram [5].With PCA method, a loading plot can be made for all variables, and shown as an abstract projection in the k-dimensional space. In this work, k=2.

It can be seen that Component 1 had higher loadings on variables such as Zn, H (H+), RH , and TOW (time of wetness, which is defined as a period of time when the metal surface is sufficiently wetted for corrosion to occur [6]); whereas Component 2 had higher loadings on variables such as Fe, Cu and S. Therefore, Component 1 was mostly correlated with zinc corrosion rate, and Component 2 with corrosion rates of carbon steel and copper. Also, in a loading plot, variables in close vicinity of each other are usually, but not always, correlated. Accordingly, S is likely related to Fe, and by less extent related to Cu, while H is likely related to Zn. All of above basically agree with the grey relational analysis. In addition, the variables TOW and RH were also closely related which was consistent with the definition of TOW and reflected the reality and reliability of the PCA processing.

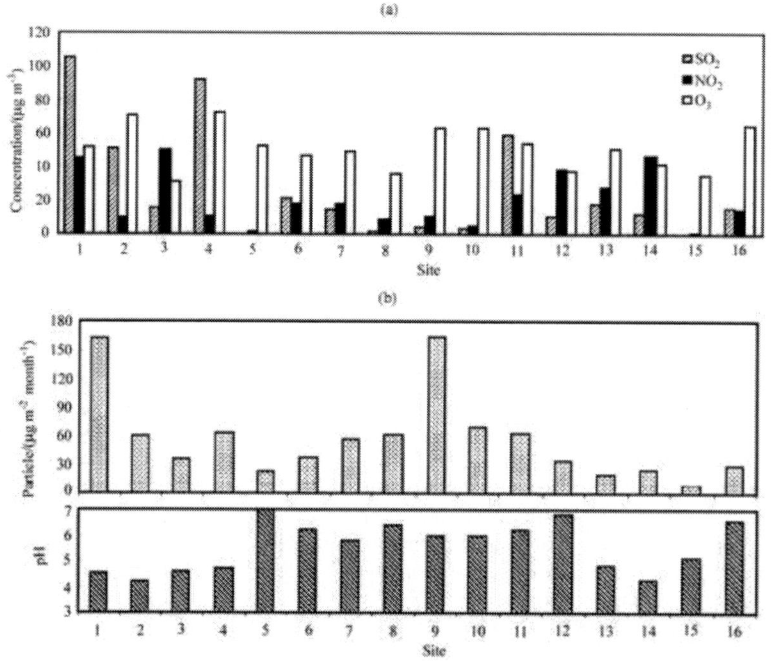

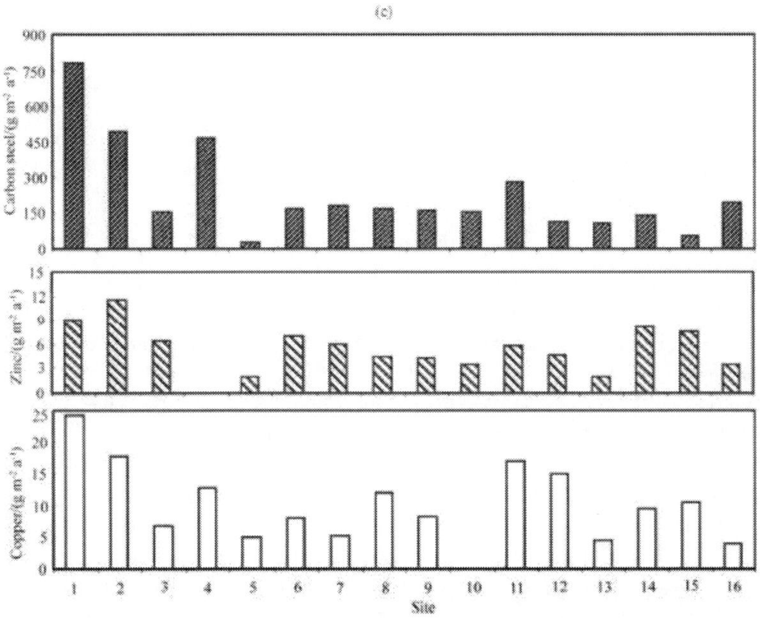

1: Wuchang, Wuhan, China 2: Qingshan, Wuhan, China 3: Hong Kong, China; 4: Kitwe, Zambia
5: Magoye, Zambia; 6: Ho Chi Minh, Vietnam 7: Hanoi, Vietnam 8: Mytho, Vietnam
9: Bhubaneswaru, India 10: Bhubaneswarr, India 11: Phrapradaeng, Thailand 12: Bangkok, Thailand
13: Johannesburg, South Africa; 14: Kuala Lumpur, Malaysia 15: Tanah Rata, Malaysia 16: Harare, Zimbabwe

Fig. 1. Comparison of (a) main gaseous pollutants; (b) particles and precipitation pollution; (c) tested metals at sites in RAPIDC corrosion project.

The obtained dose-response functions (DRF) confirmed the dominant effect of dry deposition of SO_2 on atmospheric corrosion of the three metals, which agreed well with the outcomes of grey relational analysis and principal component analysis.

In the case of zinc, the effect of wet deposition of H+ was sometimes also as important as that of dry deposition of SO2, especially in the places with acid rain of high frequency and large amount. This could be clearly demonstrated by comparison of the experimental data of GYQ with that of TSP in Wuhan. So, in the following part, we focused on only the corrosion of carbon steel and zinc. The established dose-response functions except the one for copper had a relatively high R2, which means they agreed well with the experimental data. However, there were still some uncertainties on whether they could be applied to areas other than the RAPIDC sites. So, we performed a preliminary verification of the DRF for carbon steel using some publicly available data. The calculated carbon steel corrosion results are shown in Fig. 2, in comparison with the reported data. The DRF prediction turned out to be very accurate, especially at sites such as Wuhan and Shanghai. The unavailability of reliable data about zinc and copper corrosion frustrated our attempt to verify the other two dose-response functions respectively for zinc and copper. However, it can also be deduced by some certainty that they have a comparable prediction power to that for carbon steel, based on the fact that SO2 and H+ were common crucial factors influencing corrosion of these three metals. Thus, we applied the three functions to further research of corresponding metal corrosion in Wuhan.

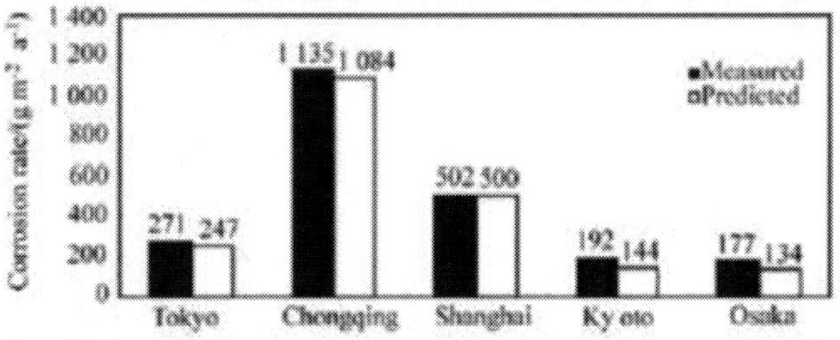

Fig. 2. The carbon steel corrosion rates calculated by the dose response function for outdoor carbon steel compared with corresponding reported data.

3 Conclusion

The combined results of the grey relational analysis and principal component analysis suggest that the two typical agents of acidic environment, SO_2 and H+ were the common factors that contribute most to the atmospheric corrosion of carbon steel, zinc and copper, which demonstrates again the strong destructive effect of acid deposition on metals. Based on the analysis of the database of RAPIDC Corrosion Project, dose-response functions for the tested metals were derived through non-linear regression method. The established dose-response functions for carbon steel and zinc are applicable to use in Wuhan, but the function for copper needs further modifying considering the relatively low R^2 value.

References

1. Wu He. Examining students' online interaction in a live video streaming environment using data mining and text mining. Computers in Human Behavior, (2013), pp. 291-304.
2. Fu Xiao, Cheng Fan. Data mining in building automation system for improving building operational performance. Energy & Buildings, (2014), pp. 75-83.
3. Benrun Huang, Li Xia, Zhengguo Wu, Weiping Zhou. Control Strategy of A Dynamic Voltage Restorer (DVR) Based on Line Voltage Compensation. Energy Procedia, (2011), pp. 13-26.
4. Wei Dai, Mostafa Bassiouni. An improved task assignment scheme for Hadoop running in the clouds. Journal of Cloud Computing, (2013), pp. 21-33.
5. Y. Mishing, in: Diffusion Processes in Advanced Technological Materials, edtied by D. Gupta Noyes Publications/William Andrew Publising, Norwich, NY (2004), in press.
6. G. Henkelman, G.Johannesson and H. Jónsson, in: Theoretical Methods in Condencsed Phase Chemistry, edited by S.D. Schwartz, volume 5 of Progress in Theoretical Chemistry and Physics, chapter, 10, Kluwer Academic Publishers (2000).

Research on Corrosion Effects on Materials in Air Pollution Condition

Yiwen Wang [1,*] and Venda Garilli [2]

[1] School of Resources and Environment Engineering, Wuhan University of Technology, Wuhan, China, 430074
[2] School of Engineering, National University of Singapore, Singapore
[*] Corresponding author: 15926424120@163.com

Abstract. In order to find the corrosion effects on materials in air pollution condition and to investigate the corrosion behavior of structural metals, weight change and mass loss were studied. The results obtained are discussed and their plausible interpretation is attempted. The results for the modern glass samples indicate that the increasing trend of the total deposited mass of particles per surface unit and haze, demonstrates a cumulative non-linear phenomenon with an asymptotic level of saturation.

1 Introduction

The damage and effects caused by air pollution on humans and the environment are gradually recognized. Air pollution includes outdoor air pollution and indoor air pollution. Outdoor air pollutants include gaseous pollutants, particle pollutants etc., and indoor air pollution is caused by excessive harmful materials in indoor air, which will affect human health. The harmful materials include formaldehyde, benzene, ammonia, radioactive radon etc. In the past20/30years, haze weather has been rapidly increasing in the eastern areas of our country, and at the same time, the visibility has been getting worse significantly. Haze pollution is becoming one kind of the main air pollution in cities. And ordinary people and scientific community have paid more and more attention to it. However, it still lacks systematic and in-deep research on health risks caused by air pollution, related economic loss, indoor air pollution characteristics and pollutants.

Reports point out: The average PM10concentrations of25cities in Yangtze River Delta Region during 2009-20015 were 52.8-112.1μg/m^3, and among them, the average PM10 concentrations of Nanjing (107.3μg/ m^3), Xuzhou (112.1μg/ m^3), Yancheng (107.8μg/ m^3) and Hangzhou (108.2μg/ m^3) were higher than the other cities of YRDR. The PM 10 concentrations in the winter were the highest, about110.5μg/ m^3, followed by spring (95.9μg/ m^3) and summer (85.5μg/ m^3), with the lowest in autumn (78.2μg/ m^3). The potential health risks of respiratory system disease caused by atmospheric particulates in Xuzhou, Yancheng, Hangzhou and Nanjing were 2.11, 2.04, 2.04 and 2.02 times respectively as much as in Zhoushan. The average concentrations of SO_2, $NO_2$and PM10in the urban areas of Ningbo were 1.13, 1.25and 1.41times as much as those in suburban areas. The health risks of urban residents were higher than the residents in the suburbs. The urban residents had the health risks caused by NO_2 as high as the11times the acceptable level. The concentrations of PM10, NO_2 and SO_2 (63.6μg/ m^3, 45.3μg/ $m^3$and 27.5μg/ m^3respectively) during haze pollution period in2009inShanghai was higher than those (63.6μg/ m^3、 45.3μg/ m^3 and 27.5μg/ m^3 respectively) in non-haze pollution period. And the concentrations of PM10, NO_2 and SO_2 were highest in winter, followed by spring and summer, and the lowest in winter. The outpatients of two large pediatric hospitals investigated have more outpatients in winter than other seasons. The number of outpatients in non-haze pollution period is larger than haze pollution period, which shows the hysteresis effect of haze pollution on local residents' health.

2 Experiment

The MULTI-ASSESS project was divided into seven work packages (WPs). The University of Athens has participated in WP1, WP2 and WP4. The main experimental observations were obtained in the frame of WP4, which took place during 2013-2014, and are presented here. More specifically in this work package, specimens of structural metals, glass, natural and artificial stone materials were exposed to different environmental conditions. Some of these materials were selected because they have been used in the construction of monuments, while the rest of them are often used in the construction of modern buildings. For the purpose of this project, a station was installed on the building of the Ministry of Health and Social Solidarity of Wuhan City. This building is located in central Wuhan next to the "ring road", at a crossroads. One facade of the building is facing south, whilst the other faces to the east. Some of the specimens were exposed on the roof, while the rest of them were exposed at specific height levels on the east facade of the building. Due to the needs of the experiment two kinds of racks were installed at this building. A wooden rack (the main rack) was installed on the roof, facing south, while a group of three aluminum racks were installed on the ground floor, the 1st, 2nd, 4th, 5th and 7th floor (height levels of 3m, 6m, 9m, 15m, 18m and 21m from the ground, respectively).

The wooden rack consisted of an inclined plane, for the specimens' display under unsheltered conditions, and an aluminum box with open bottom, for the exposure of specimens under sheltered conditions. The display of the aluminum racks was under sheltered conditions and they were distributed on the east facade of the building. The exposed structural metal samples included zinc (99.99%), copper (Cu 99%, P 0.015 to 0.04%), unalloyed carbon steel (C<0.2%, P<0.07%, S<0.05%, Cu<0.07%) and bronze (Cu 81%, Sn 5.8%, Pb 6.7%, Zn 4.5%, Ni 1.6% and trace elements). Three samples of each material were exposed at an unsheltered situation on the inclined plane of the main rack for one year. All the samples had the same dimensions (100×150×(1−2)mm3). Two samples of zinc, copper and steel were exposed at each height level, mounted on the aluminium racks for one year. All the samples had the same dimensions, but different from the samples exposed on the main rack (100×50×(1−2)mm3). Two different types of glass were exposed, medieval glass (SiO_2 48%, K_2O 25.5%, CaO 15%, MgO 3%, Al_2O_3 1.5%, P_2O_5 4%, Na_2O 3%, in mass percentage) and modern glass (SiO_2 71.7%, Na_2O 13.1%, CaO 9.6%, MgO 4.1%, K_2O 0.3%, Al_2O_3 0.7%, Fe_2O_3 0.1%, in mass percentage). Six samples of medieval glass (10×10×2mm3), mounted three on each of two glass-fibre reinforced plastic plates, were exposed in the aluminum box, while four samples, mounted two on each of two of the same kind of plastic plates, were exposed at the specified levels. The duration of the exposure was 6 months for half of the specimens and one year for the other half. Modern glass samples were exposed on the main rack. Eight samples, mounted on a wooden plate, were placed on the inclined plane and another group of eight samples, mounted on an aluminum plate, was placed into the aluminum box. All specimens had the same dimensions (100×100×2mm3). Every three months a sample of each group was with-drawn. During the experiment it was decided to expose the last two specimens of each group for a longer time period.

The natural stone materials included Carrara marble (Italy), Baumberger sandstone (Germany), Gotland sandstone (Sweden), Saaremaa dolomite (Estonia) and three types of Portland limestone (England) with smooth, medium and rough surfaces. In parenthesis the country of origin of each material is depicted. Figure 1 shows the physical properties of the natural stone materials. Reference concrete was used as artificial stone material, prepared in laboratory conditions, using Portland cement CEM I 32.5N, fine-middle aggregates, Water/Cement (W/C) ratio 0.7 and having: apparent density: 2294 kgm−3, open porosity: 16.8%, water absorption: 5.7%, saturation coefficient: 0.8.n. All the samples had the same dimensions (50×50×(8±2) mm3). Three samples of each material, medium type for Portland limestone, were exposed as sheltered and unsheltered on the main rack, mounted on special carousels.

Half of the samples were withdrawn after one year of exposure, while the rest of the samples remained exposed for two years. In addition, two samples of each type of Portland and concrete were exposed at the specific height levels, mentioned above.

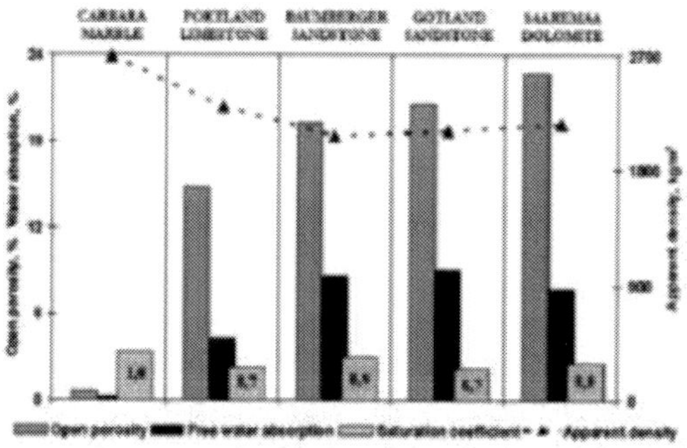

Fig. 1. Physical properties of stone materials.

3 Results and Discussion

In order to investigate the corrosion behavior of structural metals, weight change and mass loss were studied. Figure 2 shows the measurements obtained for the weight change of the exposed specimens at the specific height levels. It should be kept in mind that the weight change is defined as the difference between the mass of an exposed sample and its initial mass before the exposure. Obviously the weight change depends on the particular environmental conditions. In this context, under sheltered conditions, where there is no run-off, weight change is always positive, because of the uptake of material from the environment.

Under unsheltered conditions, except of the uptake process, the run-off of dissolved metal or patina components leads to a diminution of environmental compounds and weight change can be positive or negative. According to the results obtained, the unalloyed carbon steel is the most sensitive metal, among the exposed ones, to weight change, while copper is the most durable. The highest value of weight change for the steel was noticed at the level of 18 m, while the lowest at the level of 6m. For zinc and copper specimens values of weight change were almost constant with the height. The experimental results of the mass loss, for the same specimens, are presented in Fig. 3. Keeping in mind that the mass loss is defined as mass difference between the original metal and the remaining metal after removing the corrosion layer, we reach to the conclusion that this parameter is directly correlated to the loss of metal due to run-off and compounds within the corrosion layer. The unalloyed carbon steel arises to be the most sensitive metal, among the exposed ones, to the mass loss, while copper is the most durable. For all the metals, the lowest values were noticed at the level of 6m with an increase with the height.

The same parameters were also estimated for the specimens exposed as unsheltered on the main rack. The results for the weight change can be seen in Fig. 4. Unalloyed carbon steel was the material with the greatest value of weight change, while bronze was the only material with negative value of this parameter. The latter result can be explained by considering the influence of run-off on the specimens exposed under unsheltered conditions, which cause diminution of environmental compounds. Figure 5, illustrates the results for the mass loss of the same specimens as above. These results confirm that the unalloyed carbon steel is the most sensitive material to the corrosion. Zinc mass loss values were almost 70% greater than the same values of copper and bronze. Sulphate dominates among the water soluble anions, while nitrate concentrations are also high. Chloride concentrations are lower than nitrate. The water soluble anion with the lowest concentration detected is fluoride. Concentrations of nitrate, chloride and fluoride decrease with height against sulphate concentration, which increasing up to the level of 18m.

The most critical parameter for the specification of the weathering of medieval glass 15 specimens is the leaching depth of the constituents and especially K and Ca (d(K), d(Ca)) because these elements present the most significant deterioration. After the appropriate preparation of the samples these parameters were measured using scanning electron microscopy (SEM) in combination with energy dispersive x-ray microanalysis (EDX). Mean values of the leaching depths have 20 been calculated for the specimens exposed for 6 and 12 months. According to these results, K is the element with the highest leaching depth.

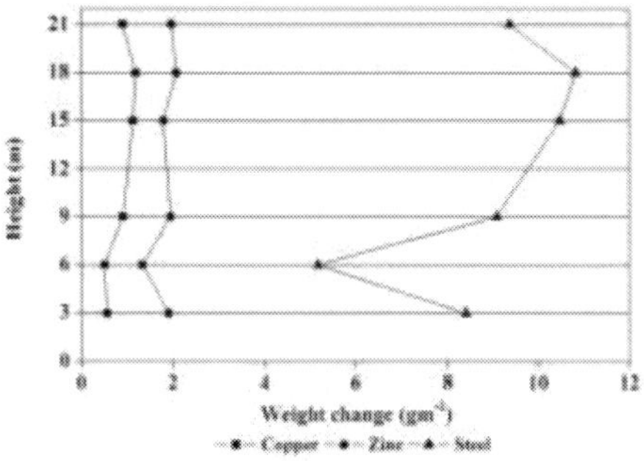

Fig. 2. Weight change of the specimens exposed

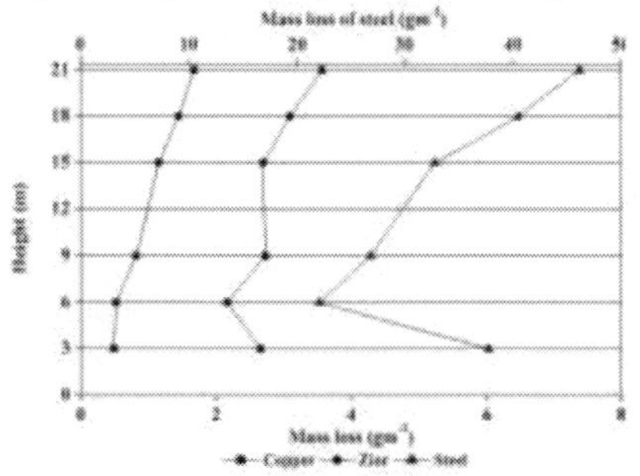

Fig. 3. Mass loss of the specimens exposed

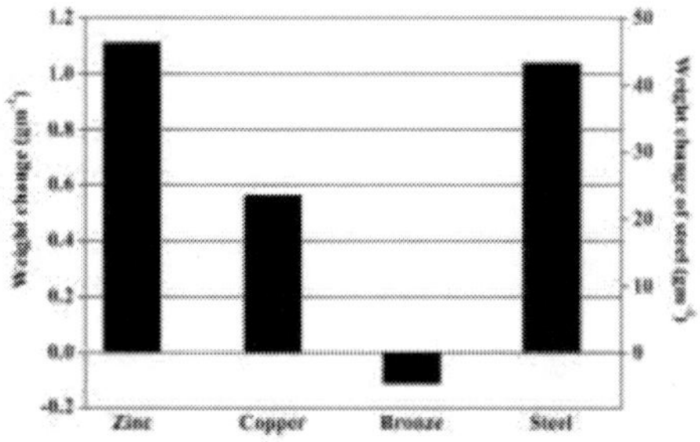

Fig. 4. Weight change of the specimens exposed for one year on the main rack (unsheltered).

 DOI: 10.1051/matecconf/201710004040

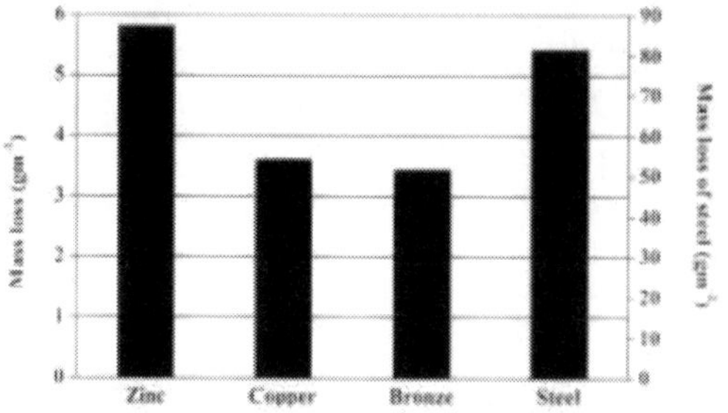

Fig. 5. Mass loss of the specimens exposed for one year on the main rack (unsheltered)

References

1. Wu He. Examining students' online interaction in a live video streaming environment using data mining and text mining. Computers in Human Behavior, (2013), pp. 291-304.
2. Fu Xiao, Cheng Fan. Data mining in building automation system for improving building operational performance. Energy & Buildings, (2014), pp. 75-83.
3. Benrun Huang, Li Xia, Zhengguo Wu, Weiping Zhou. Control Strategy of A Dynamic Voltage Restorer (DVR) Based on Line Voltage Compensation. Energy Procedia, (2011), pp. 13-26.
4. Wei Dai, Mostafa Bassiouni. An improved task assignment scheme for Hadoop running in the clouds. Journal of Cloud Computing, (2013), pp. 21-33.
5. Y. Mishing, in: Diffusion Processes in Advanced Technological Materials, edtied by D. Gupta Noyes Publications/William Andrew Publising, Norwich, NY (2004), in press.

Relationship between Oxidation of Metal Surface Layer and Air Quality in Wuhan

Mengmeng Hou [1,*] and West Gabllian [2]

[1] School of Management, Wuhan University of Technology, Wuhan, China, 430074
[2] School of Engineering, University of Florida, Gainesville, FL, US
[*] Corresponding author: 2426935187@qq.com

Abstract. In order to find out the relationship of steel futures price with different manganese content, a local fractional algorithm is proposed in this paper. We use three analytical tools - price discovery analysis, arbitrage analysis and a case study approach to assess the magnitude of price divergence risk and determine the hedging effectiveness of steel billet futures. The results show that the proposed method can thus improve overall system performance substantially.

1 Introduction

Steel is one of the most important alloys in the world. It is used to make a variety of products that are used in agriculture, construction, healthcare, industry and transportation. There are two types of steel products – flat products and long products. Flat products include plates, hot-rolled strips and sheets, and cold-rolled strips and sheets. These products are used in automotive, heavy machinery, pipes, and tubes, construction, packaging, and appliances. Long products include billets, blooms, re-bars, wire rods, rails and drawn wire. Long products are used in construction, mechanical engineering, and energy industries. Steel products are generally produced from iron ore or steel scrap – recycled steel. Global steel production has increased by leaps and bounds in the last few decades, from 568 million tons in 1980 to 1548 million tons in 20122. Few decades back, major steel producers were located in First World countries in North America and Europe. However, in the last 20 years, the focus of steel production has shifted to developing countries such as China, India, Brazil and Turkey. Today, steel has emerged as one of the most widely traded commodities in the world, with several countries active in the steel trade, either as net steel importer or exporter. The global steel industry has historically been a highly cyclical industry. It is strongly linked to expectations of future economic conditions. As a result, steel prices are generally very volatile. High price volatility affects the economic decisions of steel manufacturers, merchants and users in several industries, particularly in automotive and construction industries. Unlike other metals and alloys such as copper, lead and gold, steel is a highly differentiated product. There are several varieties of steel that fall under the flat and long products categories [1-2]. These varieties differ in chemical composition, dimensions and other parameters. Due to such high variety of steel products, managing steel price risk is a difficult proposition, as each variety has a different market and thus pricing benchmark. But given the sheer size of the industry and the importance of the alloy for business, it was inevitable that a hedging tool be developed to effectively manage steel price risk. Hence, in the last decade, the steel industry witnessed the introduction of several futures contract, to assist the process of steel price risk management. Amidst all the contracts, the one that gained most traction amongst steel users was the steel billet futures contract launched by the London Metal Exchange.

2 LME Steel Billet Futures Contract

The London Metal Exchange (LME) is the leading global exchange for trading of metals and alloys. In 2008, the LME introduced futures contract for steel billet that could be delivered in two regions – the Mediterranean and the Far East. Steel billet is a semi-finished steel product that is used for making other long steel products such as bars, rods and pipes. In 2010, the two contracts were merged into a single contract. The specifications of the contract are provided in Appendix A. The contract served the following purposes [3]:

Hedge price volatility: The primary objective of the steel billet futures was to help producers, users and merchants in hedging price volatility of steel products. Market of last resort: The users of the contract could use the London Metal Exchange as a market of last resort, and thus sell or buy steel billet whenever it was not possible for them to do so in the physical steel billet market. Benchmark pricing: As mentioned earlier, steel is a highly differentiated product, with each variety having its own unique market and pricing mechanism. Hence, it is difficult to obtain a benchmark steel price, given the high variety. The futures price would serve as a benchmark for all components in the steel supply chain. For example, scrap traders, who deliver scrap for production of steel billet, as well as wire rod manufacturers could use steel billet futures prices to price their products [4].

Long term fixed sales price: The steel billet futures was also introduced to enable manufacturers to offer long term fixed sales price for their steel products and thus lock in profit margins. Initially the steel billet futures contract was divided into the Mediterranean and Far East contracts. Volumes in the first few years were not significantly high, as steel users were skeptical of the hedging and benchmarking function of the futures contract [5]. Thus, volumes in the first two years of the contract were not high. However, since the Mediterranean and Far East contracts were merged into a single steel billet futures contract in 2010, traded volumes increased, reflecting the growing popularity of the LME steel billet futures contract. During this period, the number of participants in the steel billet futures market increased significantly and major banks such as the Deutsche Bank used the contract significantly. However, in the last couple of years traded volume has decreased significantly. The declining trend in traded volumes is shown in the Figure 1. The data for Figure 1 excluded traded volume on July 31st, 2012 where the number of contracts traded was 660. This observation was in stark contrast to the significantly lower volumes observed on days before and after this date. Thus, to show the clear declining trend, this observation has been excluded in Figure 1.

Tracking physical steel billet markets: The price of the futures contract of any commodity should be closely interlinked with the price of the physical commodity in different regions. If the futures and physical market prices diverge, then it implies inefficiency in pricing of the commodity in either market. In Figure 2, the price of LME 3-month steel billet futures has been shown against the price of Black Sea Export FOB billet and East Asia Import Billet CFR.

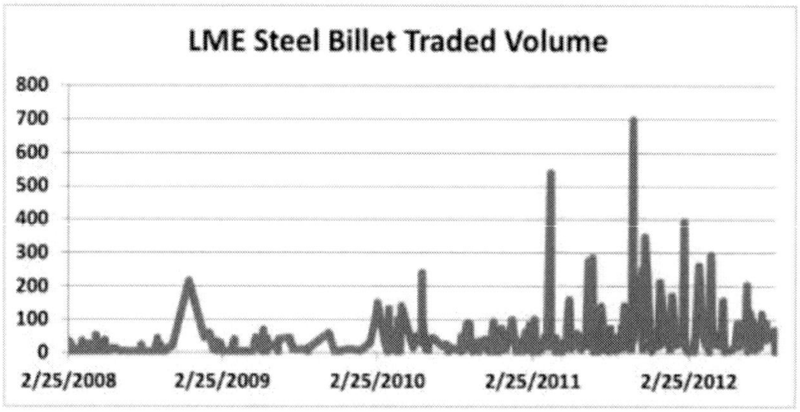

Fig. 1. The traded volume of the LME steel billet futures contract

These two billet benchmark prices are important as the Black Sea and East Asia regions are the largest trading hubs of steel billet in the world. Please note that the terms FOB and CFR, along with export and import billet are not important in our discussion. There is difference in the absolute values of all three prices, due to the difference in steel billet prices in different regions. We notice that the gap between the LME steel billet price graph and the other two billet prices graphs is insignificant in the first few years of the futures contract. This indicates that the steel billet futures price was an excellent benchmark of global steel billet, since most of the steel billet trade is centered in the Black Sea and East Asia regions. However, the gap increases in the last two years of the contract. This indicates that in the last two years, the steel billet futures price has not closely tracked the price of steel billet in major physical markets. Thus, the benchmarking ability of steel futures contracts has declined in the last two years.

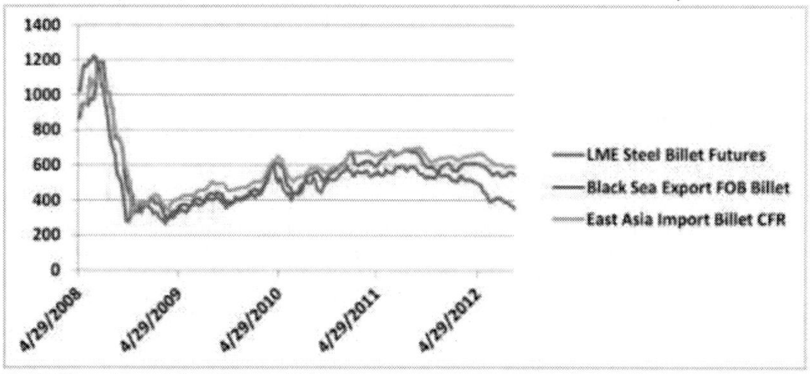

Fig. 2. The prices of the LME steel billet futures against the prices of spot billet in important physical markets in East Asia and the Black Sea region.

3 The Algorithm

The price discovery analysis entails analysis of the price discovery process in the futures market and the correlation between price changes in the futures and cash markets. The analysis of price discovery will help us determine which of the two markets is dominant over the other in terms of flow of information. Perfect and complete flow of information between the two markets ensures that prices in both markets always move in synchronization. Analysis of price discovery also presents us with information of elasticity of supply of arbitrage services, a measure that will be defined in the next section. The elasticity of supply of arbitrage services, along with price discovery analysis helps make inferences about the price divergence risk in using the futures contract for hedging. The local fractional algorithm can be expressed as following:

$$f^{(\alpha)}(x0) = \left.\frac{df(x)}{dx^{\alpha}}\right|_{x=x_0} = \lim_{\delta x\to 0}\frac{\Delta^{\alpha}(f(x)-f(x_0))}{(x-x_0)^{\alpha}} \tag{1}$$

for $0 < a \le 1$ where

$$\Delta^{\alpha}(f(x)-f(x_0)) \cong \Gamma(1+\alpha)\lim_{x\to\infty}\Delta(f(x)-f(x_0)) \tag{2}$$

And local fractional integral of $f(x)$ defined by Eq.3.

$${}_aI_b^{(\alpha)}f(t) = \frac{1}{\Gamma(1+\alpha)}\int_a^b f(t)(dt)^{\alpha} = \frac{1}{\Gamma(1+\alpha)}\lim_{\Delta t\to 0}\sum_{j=0}^{j=N-1} f(t_j)(\Delta t_j)^{\alpha} \tag{3}$$

With $\Delta t_j = t_{j+1} - t_j$ and $\Delta t = \max\{\Delta t_1, \Delta t_2, \cdots, \Delta t_j, \cdots\}$, where for $j = 1, 2, \cdots, N-1$, $[t_j, t_{j+1}]$ is a partition of the interval $[a,b]$ and $t_0 = a$, $t_N = b$.

If $f(x)$ is defined on the real line $-\infty < x < \infty$, its local fractional Hilbert transform, denoted by $f_x^{H,\alpha}(x)$ is defined by

$$H_{\alpha}\{f(t)\} = \hat{f}_H^{\alpha}(x) = \frac{1}{\Gamma(1+\alpha)}\oint_R \frac{f(t)}{(t-x)^{\alpha}}(dt)^{\alpha} \tag{4}$$

Where x is real and the integral is treated as a Canchy principal value, that is,

$$\frac{1}{\Gamma(1+\alpha)}\oint_R \frac{f(t)}{(t-x)^{\alpha}}(dt)^{\alpha} = \lim_{\varepsilon\to 0}\left[\frac{1}{\Gamma(1+\alpha)}\int_{-\infty}^{x-\varepsilon}\frac{f(t)}{(t-x)^{\alpha}}(dt)^{\alpha} + \frac{1}{\Gamma(1+\alpha)}\int_{x+\varepsilon}^{\infty}\frac{f(t)}{(t-x)^{\alpha}}(dt)^{\alpha}\right] \tag{5}$$

To obtain the inverse local fractional Hilbert transform, write again Eq. (4) as

$$\hat{f}_H^{\alpha}(x) = \frac{1}{\Gamma(1+\alpha)}\int_{-\infty}^{\infty}\frac{f(t)}{(t-x)^{\alpha}}(dt)^{\alpha} = \frac{1}{\Gamma(1+\alpha)}\int_{-\infty}^{\infty} f(t)g(x-t)(dt)^{\alpha} = f(x) * g(x), \tag{6}$$

The elasticity of supply of arbitrage services is a measure of the rate of convergence of cash and futures prices. Garbade and Silber postulate that if elasticity of supply of arbitrage services is zero then the futures market is a poor substitute for the cash market position. Prices in both markets follow uncoupled random walks and the futures markets' risk transfer and price discovery functions are eliminated. The futures contract is thus an ineffective hedging tool for the hedger as price divergence risk is high. On the other hand, if elasticity of arbitrage services is infinite, then futures contract is a perfect substitute for the cash market position. In this case, prices are discovered simultaneously in both markets and the futures contract serves as a perfect hedging tool since price divergence risk is low. For non-zero and non-infinite values of elasticity, the two markets follow an intertwined random walk, with one market dominant over the other in terms of price determination. In this case, the futures contract can be used to hedge but the hedge will not be perfect due to price divergence risk.

References

1. Benrun Huang, Li Xia, Zhengguo Wu, Weiping Zhou. Control Strategy of A Dynamic Voltage Restorer (DVR) Based on Line Voltage Compensation. Energy Procedia, (2011), pp. 13-26.
2. Wei Dai, Mostafa Bassiouni. An improved task assignment scheme for Hadoop running in the clouds. Journal of Cloud Computing, (2013), pp. 21-33.
3. Wei Kuang Lai, Yi-Uan Chen, Tin-Yu Wu, Mohammad S. Obaidat. Towards a framework for large-scale multimedia data storage and processing on Hadoop platform. The Journal of Supercomputing, (2014), pp. 681-696.
4. Zhijian Chen, Wenhai Luo, Dan Wu, Xiang Huang, Jian He, Yuanhuan Zheng, Di Wu. Exploiting application-level similarity to improve SSD cache performance in Hadoop. The Journal of Supercomputing, (2014), pp. 703-715.
5. Y. Mishing, in: Diffusion Processes in Advanced Technological Materials, edtied by D. Gupta Noyes Publications/William Andrew Publising, Norwich, NY (2004), in press.

Research on the Price Features of Oil Stochastic Model Based on the Continuous Jump Model

Mengmeng Hou [1,*] and West Gabllian [2]

[1] School of Management, Wuhan University of Technology, Wuhan, China, 430074
[2] School of Engineering, University of Florida, Gainesville, FL, US
[*] Corresponding author: 2426935187@qq.com

Abstract. Aiming at calculating the price changes under the price features of oil stochastic model, the continuous jump model is proposed in this paper for data processing. The procedure is flexible, may be used with market prices of any oil contingent claim with closed form pricing solution, and easily deals with missing data problems. The results show that the accuracy can thus be improved overall the proposed system substantially.

1 Introduction

This paper develops a parsimonious three-factor model of the term structure of oil futures prices that can be easily estimated from available futures price data. In addition, it proposes a new simple spreadsheet implementation procedure. The procedure is flexible, may be used with market prices of any oil contingent claim with closed form pricing solution, and easily deals with missing data problems. The approach is implemented using daily prices of all futures contracts traded at the New York Mercantile Exchange between 1991 and 2001. In-sample and out-of-sample tests indicate that the model fits the data extremely well. Though the paper concentrates on oil, the approach can be used for any other commodity with well-developed futures markets [1-2].

Enlightening as they may be, these stochastic models have been adopted rather slowly by practitioners. One possible reason for this could be that even though two-factor models behave reasonably well most of the time, for some market conditions they behave poorly, making daily estimations somewhat unreliable[3-5]. Also, most parameter estimation procedures proposed in the literature are rather involved and require extensive data aggregation, which translates into substantial information loss. The Kalman filter approach, however, has relatively high implementation requirements. One of the main issues is the handling of missing observations in the data, which is particularly important when using financial instruments, like futures, that only trade for specific maturities. In typical implementations of this procedure contracts with close, but different maturities are grouped together to complete the data sets used in the estimation. The consequence of these groupings is that a substantial amount of data is discarded with its associated information loss. This problem is particularly acute when new contracts are introduced in the market and there are no complete time series data for a given maturity. For example, oil markets currently trade futures contracts for over thirty-five different maturities while only a few years ago they only traded futures contracts for less than twenty different maturities [6].

Current implementations of the Kalman filter methodology for this type of models require linearity in the pricing expressions, which restricts its applications to linear payoffs such as futures. Nonlinear Kalman filter estimations do exist in the literature, but are much harder to apply. Even though in the current implementation of our model we only use futures prices in the estimation, it would be easy to include contracts with nonlinear payoffs such as options.

2 The Algorithm and Data Analysis

This modification considerably simplifies the estimation procedure since we only have to estimate seven parameters. In what follows we will be using this modified model in all the calculations. The basic equation for the routing algorithm is shown below [7-8]:

$$\overline{x_i} = \frac{x_i - b_i}{a_i - b_i} \tag{1}$$

Experimental data to determine, we can also experience the value of the formula (2) the decision.

$$n = \log_2 m \tag{2}$$

Hidden node output is calculated as follows:

$$h_j = f\left(\sum_{i=1}^{m} w_{ij}x_i - \theta_j\right) \tag{3}$$

The output of the output node is calculated as follows:

$$f\left(\sum_{i=1}^{m} w_{ij}x_i - \theta_j\right) = f\left(f\left(\theta_j\right)\right) \tag{4}$$

Where in θ is an output node threshold.
Put Equation (3) into Equation (4), then we can get the S-type function:

$$f\left(\sum_{i=1}^{m} w_{ij}x_i - \theta_j\right) = f\left(f\left(\theta_j\right)\right) \tag{5}$$

In the structure of GA algorithm, we can get the optimization equation as the following equation (6):

$$h_j = \exp\left(-\frac{\|X - C_j\|}{2b_j^2}\right), \quad j = 1,2,\ldots,m \tag{6}$$

The output of the network is given as:

$$y_m(k) = wh = w_1h_1 + w_2h_2 + \ldots + w_m h_m \tag{7}$$

Assuming the ideal output is y (k), the performance index function is:

$$E(k) = \frac{1}{2}\left(y(k) - y_m(k)\right)^2 \tag{8}$$

The equation of basic function is as equation (9) as follows:

$$\partial_j(C_{ijkl}\partial_k u_l + e_{kij}\partial_k \varphi) - \rho \ddot{u}_i = 0 \tag{9}$$

Under the linear relationship, basic equation is shown in equation (2):

$$\partial_j(e_{ijkl}\partial_k u_l - \eta_{kij}\partial_k \varphi) = 0 \tag{10}$$

The linear differential equation can be expressed into the following simplified forms:

$$L(\nabla,\omega)f(x,\omega) = 0,$$

$$L(\nabla,\omega) = T(\nabla) + \omega^2 \rho \mathrm{J} \tag{11}$$

In which,

$$T(\nabla) = \begin{Vmatrix} T_{ik}(\nabla) & t_i(\nabla) \\ t_k^T(\nabla) & -\tau(\nabla) \end{Vmatrix}, \quad \mathrm{J} = \begin{Vmatrix} \delta_{ik} & 0 \\ 0 & 0 \end{Vmatrix},$$

$$f(x,\omega) = \begin{Vmatrix} u_k(x,\omega) \\ \varphi(x,\omega) \end{Vmatrix} \tag{12}$$

$$T_{ik}(\nabla) = \partial_j C_{ijkl}\partial_l, \quad t_i(\nabla) = \partial_j e_{ijk}\partial_k, \quad \tau(\nabla) = \partial_i \eta_{ik}\partial_k$$

Consider an infinite situation, we have the equation (5) in the following:

$$L^0 = \begin{Vmatrix} C_{ijkl}^0 & e_{kij}^0 \\ e_{ikl}^{0T} & -\eta_{ik}^0 \end{Vmatrix} \tag{13}$$

Consider the propagation, instead the equation (13) with the following form:

$$C(\mathrm{x}) = C^0 + C^1(\mathrm{x}), \quad e(\mathrm{x}) = e^0 + e^1(\mathrm{x}), \quad \eta(\mathrm{x}) = \eta^0 + \eta^1(\mathrm{x}), \quad \rho(\mathrm{x}) = \rho_0 + \rho_1(\mathrm{x}) \tag{14}$$

Then we have equation (15):

$$C^1 = C - C^0, \quad e^1 = e - e^0,$$

$$\eta^1 = \eta - \eta^0, \quad \rho_1 = \rho - \rho_0 \tag{15}$$

The proposed procedure provides results reasonably close to those of the more formal Kalman filter estimation. For example, applying both estimation procedures to the weekly oil futures prices traded at NYMEX between 1990 and 1995, we find very similar mean square errors (MSE). Figure 1 shows that the estimated values for the state variables are almost indistinguishable between the two procedures. Figure 2 shows that the percentage differences between the estimated values, using both procedures, for the six and twelve-month futures prices are very small and generally smaller than 0.5%.

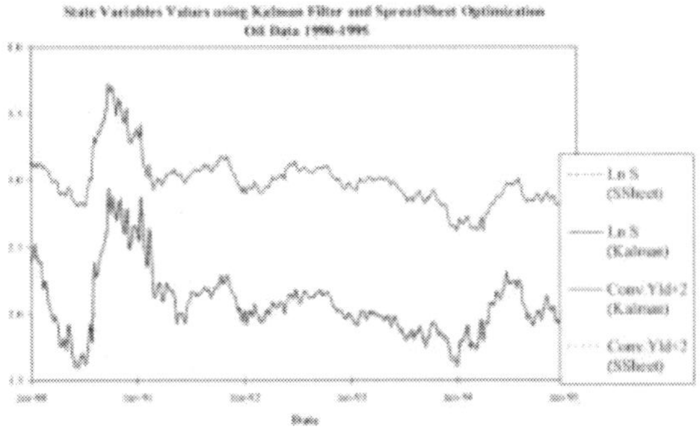

Fig. 1. State variables values of the two-factor model in Schwartz.

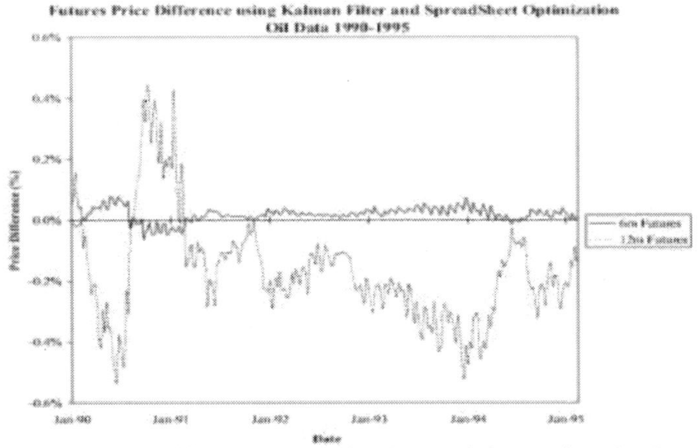

Fig. 2. Percentage differences in 6 month and 12 month futures price estimation.

3 Conclusion

To make our out-of-sample analysis we use Data-set 2 and analyze the performance of the model over the last three years (1999 to 2001). Our initial model calibration, for transactions in the first quarter of 1999, is done with Data-set 1. The model is then recalibrated on a quarterly basis (on the last day of March, June, September and December of each year) and parameter values are not updated for the following three months. This out-of-sample analysis is done using only past information, available at the time of the calibration and represents an upper bound for the errors. More frequent calibrations could reduce model errors, since more recent prices would be included.

Results for the three-year out-of-sample data are now presented. We start by noting that the model is very flexible in that it can deal with very different shapes of the term structure of futures prices. To illustrate this point we select arbitrary dates in the sample period in which futures oil prices exhibit either backwardation or contango. The consequence of these groupings is that a substantial amount of data is discarded with its associated information loss. This problem is particularly acute when new contracts are introduced in the market and there are no complete time series data for a given maturity. In this article we develop a parsimonious three-factor model of the term structure of futures oil prices, which fits the data extremely well. In addition, and very importantly, we propose an implementation procedure that significantly simplifies the estimation methods proposed in the literature. These factors make the proposed approach ideally suited for practical applications in the valuation and hedging of real and financial oil-contingent claims. The method can be used for other commodities as well; for example, we have also implemented the model using copper futures price data with similar results.

References

1. Wei Dai, Mostafa Bassiouni. An improved task assignment scheme for Hadoop running in the clouds. Journal of Cloud Computing, (2013), pp. 21-33.
2. Wei Kuang Lai, Yi-Uan Chen, Tin-Yu Wu, Mohammad S. Obaidat. Towards a framework for large-scale multimedia data storage and processing on Hadoop platform. The Journal of Supercomputing, (2014), pp. 681-696.

3. Zhijian Chen, Wenhai Luo, Dan Wu, Xiang Huang, Jian He, Yuanhuan Zheng, Di Wu. Exploiting application-level similarity to improve SSD cache performance in Hadoop. The Journal of Supercomputing, (2014), pp. 703-715.
4. Y. Mishing, in: Diffusion Processes in Advanced Technological Materials, edtied by D. Gupta Noyes Publications/William Andrew Publising, Norwich, NY (2004), in press.
5. Zhijian Chen, Wenhai Luo, Dan Wu, Xiang Huang, Jian He, Yuanhuan Zheng, Di Wu. Exploiting application-level similarity to improve SSD cache performance in Hadoop. The Journal of Supercomputing, (2014), pp. 703-715.
6. Y. Mishing, in: Diffusion Processes in Advanced Technological Materials, edtied by D. Gupta Noyes Publications/William Andrew Publising, Norwich, NY (2004), in press.
7. Xie Qun, Zhang Yubin, Zhang Jibiao, etc. Cluster analysis and comprehensive evaluation on summer coastal marine environment quality of Leizhou Peninsul. Marine Environmental Science ,33 (2014): 543-549.
8. Zhao Dongling, Feng Yanruo, Pan Zhengyun. Research on clustering analysis based Algorithm K- grid, Journal of Technology, 30(2014):175-178.
9. Gong Liyan, Meng Xianjun. Quality evaluation of apple processing based on Principal Component Analysis and Cluster Analysis, Agricultural Engineering, 30 (2014): 276-285.

Exploration of Application Oriented Talent Training Mode of Material Science in Colleges and Universities

Fang Wei [1,*]

[1] Wuhan University of Technology, Wuhan, China, 430074

[*] Corresponding author: 3033523@qq.com

Abstract. Material science has achieved more and more attention because of its singular response to the electromagnetic wave. A new application oriented talent training mode of material science in colleges and universities is studied in this paper. The new model of cultivating talents is explored, and the practical teaching system of two step education curriculum platforms is used for constructing the model. The investigation result shows that an increased learning for the student whom were more active on the platform independent on their previous performances.

1 Introduction

The rapid popularization of higher education plays an important role in providing talents for the reunification and national rejuvenation, which yet make talents cultivation, deviate from the market demand. On the one hand, the society is desperate for excellent talents while on the other hand graduation students are unable to find a job. The fundamental reason for this phenomenon is that the advanced talent training structure is unbalanced which is keen on cultivating academic talents while ignoring the cultivation of applied talents. The expansion of students' recruitment brings along with development for various universities which yet ask higher requirement [1-2]. To innovate the talent cultivation mechanism for universities is the most direct and clear requirement stipulated in Decision for higher education reform in 3rd Plenary Session of 18th CPC Central Committee which also points out the direction for development. The State Council Standing Committee meeting held by Premier of the State Council explicitly proposed to guide a series of universities to transfer to pay attention to the applied talents cultivation. Under such circumstance, the necessary path for common colleges and universities to achieve continuous development is to innovate the core competence to cultivate high quality applied talents [3-4].

General colleges and universities have to stick to the education philosophy of applied undergraduate education as well as the school operation thoughts of market-orientation, highlight employment, paying attention to quality, strengthening characteristics to come up with the talents cultivation targets of refining professions, strengthening practice and highlight characteristics [5-6].

Solid professional knowledge is the basis to cultivate applied talents and schools should obey the whole correlation of knowledge structure, and make general knowledge as guarantee to systematically grasp the professional knowledge of science and lay a foundation. The brand of school, faculty quality, subject foundation, and management experience and culture background can be powerful guarantee for students to consolidate their professional knowledge.

Paying attention to the practical ability of university students aims to acclimate the development trend of higher education and cultivate the internal quality of applied talents. The cultivation of practical ability is in accordance with the recognition regulation of university students. Education, as a recognition process should demonstrate its subject position-practice.

School characteristics are special connotations accumulated with a long-term operation. Only when sticking to the characteristics, general colleges and standards can form a brand and being competent in fierce competitive environment. Taking full advantage of the special characteristics of schools and the social demand, setting multidisciplinary specialty as well as urgently wanting specialty, teaching students according to their aptitude with classified guidance, and determining the cultivation direction based on individual personality can help to improve students' employment ability [7-8].

2 Design principles and content

Our a-priori goal was to create an active and fluid environment, wherein learners could engage with the curriculum on multiple cognitive dimensions. We implemented a points-badges-leaderboard framework, which awarded points to the users for completing tasks, and a weekly badge for exceeding a threshold number of points. The students were aware that neither the points nor the badges would have direct in fluency on their grades, but just as the conventional theoretical exercises Student-Researcher was a part of the course material in preparation for the oral exam. General education course platform is divided into general courses for whole school and general course for certain majors. General courses for whole school are public elementary courses opened for all majors while general courses of certain majors are opened for some special majors.

Adhering to the quality education philosophy, strengthening public elementary courses and basic courses, scientifically construct the knowledge structure for students and strengthening the basic knowledge as well as basic quality cultivation are necessary methods. The teaching quality assessment system can be adopted to carry out the system for teachers to take up a job through competition which can maximize the excellent teacher resources with a high sense of responsibility and good teaching effect. Besides, schools have to consolidate the basic knowledge.

Professional education course platform include professional basic course, specialized course, elective course and practice. The setting of professional basic course and specialized course should be set according to the professional features and sticks to the principle of improving students' ability. The setting for elective course should adhere to the principle of developing students' personalities based on expanding the knowledge structure. The practice should adhere to the continuous system for four years, including cognitive practice, specialized investigation, professional internships and employment practices.

Colleges and universities have to construct a stepladder-type teaching system from cultivating basic courses and specialized courses to cultivate specialized talents from the perspectives of knowledge structure, ability quality, and adaptation. Based on the target of cultivating specialized talents, basic courses should be carried out in the freshmen year while specialized courses can be carried out after stepladder year.

Divide the majors into superior majors and common majors to carry out distribute training. Superior majors are brands of school which have created a stable and good pattern in terms of recruiting, training, and employment. Common majors can take advantage of superior majors to positively develop interdisciplinary courses with independent settings. It takes employment as the direction, focusing on reforming the teaching plan arrangement and launching cross-disciplinary special courses, allowing interdisciplinary education, paying attention to the comprehensive integration of knowledge to break the major barriers.

3 Results and Discussion

Adhere to implement the teaching approach based on internal class teaching and out of class activities so as to realize open practice teaching, encourage students to study independently, fully mobilize the enthusiasm and initiative of students. What's more, schools can develop expanded and flexible cultivation models such as academic competitions, community Club, school and enterprise Alliance, industry research, scientific research innovation and students can be granted with quality grades and the main contents as well as approaches show as follows:

Encourage teachers and students to actively participate in influential professional skills contest to cultivate students' competitiveness, develop their expertise. Through participation in academic competitions, their learning, practical and innovative ability can be improved. Figure 1 shows the cultivation model for applied talents.

Actively promote the development of students ' professional associations, and guide students to develop professional and innovative activities which can develop their abundant skills, team spirit as well as leadership skills and practice innovation skills a variety of Club activities, develop students ' professional skills, teamwork and leadership skills.

The alliance can help to train applied talents, who can take the initiative to adapt the changing social needs. The working experience via alliance can help students learn the production model, and running philosophy of enterprises, which can also help them establish cooperation relation with enterprises and is beneficial for their employment after graduation.

Organize students to design questionnaire, survey, data analysis and write articles to help them fully understand their professions too scientifically and correctly plan for the future. Schools should encourage students to independently carry out research activities or actively participate in teachers ' scientific research project, regularly organize students to apply for research report and support them on excellent projects. What's more, schools have to regularly hold exchanges meetings in terms of graduate entrance exam, scientific research innovation and research project application. As a result, students can innovate their thinking ability after referring to literature, improve their practical ability while participating in experiment which are beneficial for employment.

Launch series rules and regulations to standardize the school teaching and management. Students are encouraged to participate in various skill training classes, specialty classes, professional skill competition to carry out innovation and entrepreneurship. Besides, capitals should be appropriated to encourage and award excellent students and tutors.

Build a dynamic teaching team dominated by full-time teachers supplemented by part-time teachers. Besides, these teachers have proper age, education background and title structure. Moreover, teachers can be assigned to cooperative enterprises to be trained, professional skill training can be carried out and enterprise managers with abundant working experience can be recruited so as to optimize the faculty structure.

Construct a multifaceted, three-dimensional and connected information platform for teaching so as to realize the modernization of teaching management means. While taking advantage of comprehensive resources such as education funds, time, information, and teaching condition the regulated management for teaching management activities also provides guarantees. Establish three-level teaching quality supervision organization made up of school, dean's office and college. In each term, supervised teacher will randomly evaluate the teaching quality, which can be also assessed based on student evaluation and feedback so as to assess the teaching quality.

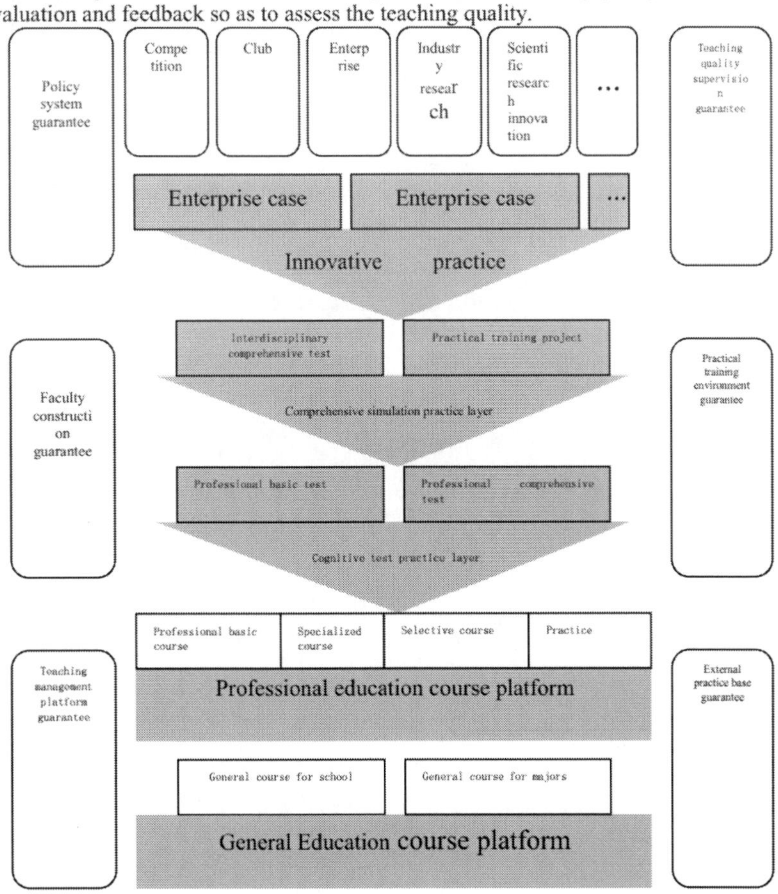

Fig. 1. The cultivation model for applied talents.

Increase education funds; actively promote the hardware environment construction for experiment and training. In spare time, laboratories are open and experiments can be carried out on weekends or in the evening. At the same time, special lectures can be conducted to stimulate students' enthusiasm. The excellent experimental training environment provides good conditions for teachers' teaching and scientific research as well as innovative practice for students. Establishing talent cultivation mechanism with enterprises to innovate internship model, provide scientific service, deliver talents, combine production, teaching and research, provide professional training can greatly improve students' practical, innovative and entrepreneurship capacity.

References

1. B H Li, L Zhang, S L Wang, F Tao, J W Cao, X D Jiang, X Song, X D Chai, Cloud manufacturing: a new service-oriented networked manufacturing model, Computer Integrated Manufacturing Systems, vol.16, (2010) , pp.1-16.
2. D. McDonagha, A. Brusebergb, C. Haslamc. Visual product evaluation: exploring users' emotional relationships with products. Applied Ergonomics, 33, (2002), p.p.231-240.
3. L Zhang, Y. L. Luo, F Tao, L Ren, H Guo, Key technologies for the construction of manufacturing cloud, Computer Integrated Manufacturing Systems, vol.16, (2010), pp.2510-2520.
4. J. D. Ullman, NP-complete scheduling problems, Journal of Computer and System Sciences, vol.10, (2005), pp.384-393.
5. Benrun Huang, Li Xia, Zhengguo Wu, Weiping Zhou. Control Strategy of A Dynamic Voltage Restorer (DVR) Based on Line Voltage Compensation. Energy Procedia, (2011), pp. 13-26.

6. Wei Dai, Mostafa Bassiouni. An improved task assignment scheme for Hadoop running in the clouds. Journal of Cloud Computing, (2013), pp. 21-33.
7. Wei Kuang Lai, Yi-Uan Chen, Tin-Yu Wu, Mohammad S. Obaidat. Towards a framework for large-scale multimedia data storage and processing on Hadoop platform. The Journal of Supercomputing, (2014), pp. 681-696.
8. Zhijian Chen, Wenhai Luo, Dan Wu, Xiang Huang, Jian He, Yuanhuan Zheng, Di Wu. Exploiting application-level similarity to improve SSD cache performance in Hadoop. The Journal of Supercomputing, (2014), pp. 703-715.

An Empirical Study on the Reform of Talent Cultivation Mode of Material Science in University Based on PBL

Fang Wei [1,*]

[1] Wuhan University of Technology, Wuhan, China, 430074

* Corresponding author: 3033523@qq.com

Abstract. The purpose of this study is to research on the reform of talent cultivation mode of material science in university. The construction of knowledge system and curriculum system is the important part in the program of cultivating talents and the course construction projects vary in different institutions with different trade backgrounds and talent cultivation plans. The Problem-Based Learning (PBL) is used for constructing a specific model for the material science major and the empirical study is done in the experiment part. The result shows that the PBL model can improve the efficiency and effectiveness in the reform process of talent cultivation mode of material science in university.

1 Introduction

At the quick change time, organizational learning and learning organization are considered as the best methods for improving the organization learning ability, accommodating the existent environment of the quick change, but the implement is difficult, the long-term success cannot be obtained. For the reason, it is for the education people obtained have not given them the real learning, thus forming the barrier of the organizational learning, this problem should be solved at the procedure of education. At the same time, after the popularization of high education in China, the talent cultivation model cannot adapt the change situation; every university faces the pressure of talent cultivation and occupation. So under the view of high education, guided by the theory of organizational learning and learning organization, using systematical idea, integrated theory and practice studies, the dissertation has the importance on studying the talent cultivation model in learning university.

The talent cultivation model is composed of four factors, i.e. target, procedure, system and evaluation of cultivation. Through analysis of distance measuring model for talent cultivation, it can be found that there have differences between the actual result and expectation of talent cultivation in university, which is disadvantageous for the realization of talent cultivation model. PBL is problem-Based Learning short, usually translated as "problem-based learning." In material education, PBL concept is often used as a teaching method, teaching mode or teaching philosophy to use. "Classic" PBL refers to material students by solving ill-structured problems material cases to learn a learning strategy material curriculum. PBL has the advantage of changing the traditional teaching mode passive learning to active learning, group members mutual division of labor, through various channels to find their own solution to the problem of information, students learn the necessary knowledge in the process of solving problems, helping to train material students independent the ability to learn and lifelong learning, communication skills, teamwork, spirit of humane care and practical application of the ability to solve difficult material problem [1].

In recent years, with the advancement of education reform, education and more attention from the perspective of culture clinicians were practicability knowledge education of material students. Education has changed from "cramming" the traditional teaching to the "issue-based" PBL teaching methods [2-5]. PBL this new teaching model has been affirmed in the world of material education, and gradually become the new direction of material education mode reform. Therefore, the problem how to apply this model to our national material practice teaching effectively, improve the exist problems in the implementation of PBL teaching is worthy of thinking for every material educators.

The current principal contradiction in our material education is no longer a shortage of material personnel quantity, but the quality of the material personnel to improve. Whether it is from our social, economic, scientific and technological development and higher requirements for the proposed health, or from the enhanced competitiveness of international cooperation and urgent material needs, all we need to improve the quality of material education, training material talents. As a country in the world of higher material education more popular, with the development of foreign exchange, China's material education gradually internationalized [6].

Educational reform in developed countries focus on the reform of teaching methods, and reforms in developing countries is mainly reform of teaching content. Currently, the United States as a typical representative of the North American material education has a leading position in the world, after 200 years of rapid development and several major material education reform, has been completed and an effective mechanism and outstanding material education of material personnel education model is rest of the world in material education reference point [7].

2 The Framework of PBL

With the social development and technological progress, the material model from the biomaterial model into a bio-psycho-social material model, marking a healthy material science, has entered a new period of development, promote the transformation of medicine all fields to accelerate prevention, updated diagnosis and treatment technology, so that the whole of material science fundamentally changed the face of occurrence. Each branch of medicine continue to expand and lead to mutually cross between disciplines continue to rise to a large number of interdisciplinary and multidisciplinary material curriculum Division is getting smaller, new technologies are emerging, material education is facing new challenges.

PBL is problem-based learning. PBL material education and the traditional model of material education are taught based learning (Lecture-Based Learning, LBL) that is very different. PBL in the late 1960s produced in Canada McMaster University School of Medicine, in the late 1980s in North America gained rapid development. 1991, 70 percent of US material schools have adopted varying degrees PBL mode of teaching. Since the 1990s, parts of school of medicine in European have begun to experiment PBL curriculum. University of Hong Kong in 1997 and started teaching PBL, PBL teaching currently account for 60% of all material school education and Harvard Material School have all been substituted LBL application of PBL teaching.

The basic equation for the routing algorithm is shown below:

$$\overline{x_i} = \frac{x_i - b_i}{a_i - b_i} \quad (1)$$

Experimental data to determine, we can also experience the value of the formula (2) the decision.

$$n = \log_2 m \quad (2)$$

Hidden node output is calculated as follows:

$$h_j = f\left(\sum_{i=1}^{m} w_{ij} x_i - \theta_j\right) \quad (3)$$

The output of the output node is calculated as follows:

$$f\left(\sum_{i=1}^{m} w_{ij} x_i - \theta_j\right) = f\left(f\left(\theta_j\right)\right) \quad (4)$$

Where in θ is an output node threshold.

Put Equation (3) into Equation (4), then we can get the S-type function:

$$f\left(\sum_{i=1}^{m} w_{ij} x_i - \theta_j\right) = f\left(f\left(\theta_j\right)\right) \quad (5)$$

In the structure of GA algorithm, we can get the optimization equation as the following equation (6):

$$h_j = \exp\left(-\frac{\left\|X - C_j\right\|}{2b_j^2}\right), \quad j = 1, 2, \ldots, m \quad (6)$$

3 The Experiment and Data Analysis

PBL pedagogy emphasize problem-based learning, students of all learning content around the "problem" This spindle start. It is mainly through "OK Threads -> analysis of the problem -> gathering information -> self-discussed -> Establishment Solutions -> summary evaluation and feedback" and other steps will introduce students to the problem situation. Pay attention to inspire and induce students to form a master self-learning ability and ability to solve problems in the whole process of material teaching. Based on the completion of each student while addressing the issue of teaching, and also completed the reproduction of knowledge, deepen the knowledge and understanding of the consolidation. Therefore, PBL teaching method has its unique advantages and effects in material teaching, mainly as follows:

(1) Help to stimulate students' interest in learning, improve learning efficiency

In PBL teaching mode, teachers will introduce students to the problem situation, students become problem-solving real "masters." By the classroom portion of the initiative to the students, not only to provide an opportunity for students

DOI: 10.1051/matecconf/201710004044

to active learning, promote student autonomy, actively speak, to fully express their views, but also to develop independent learning ability to obtain knowledge and discovery issues, pole greatly improves the students' interest in learning.

(2) Help improve the overall quality of students

Based on material teaching and training by PBL teaching method designed to enable students not only have much to gain in terms of material expertise, and through active participation and problem solving can cultivate students' consciousness, innovation, competition and team spirit, in line with institutions of higher learning to implement concepts and requirements of quality education.

(3)Help build a theoretical study and practice cohesive platform

Problem-solving process, you need to curricular and extra-curricular, internal and external, direct and indirect experience closely together, gathering information through various channels, the expansion of knowledge to solve problems. Such training with a problem to solve the problem in training in learning, learning theory and practice can make a good combination of cognitive; improve the ability of students to apply theory to solve practical problems.

(4) Contribute to teachers to improve their professional quality and achieve a total growth of teachers and students

Figure 1 shows the comparison between effectiveness and sustainability of teaching methods in inside and outside, women and children and figure 2 shows the students' appraisal about credibility and sustainability of skill test.

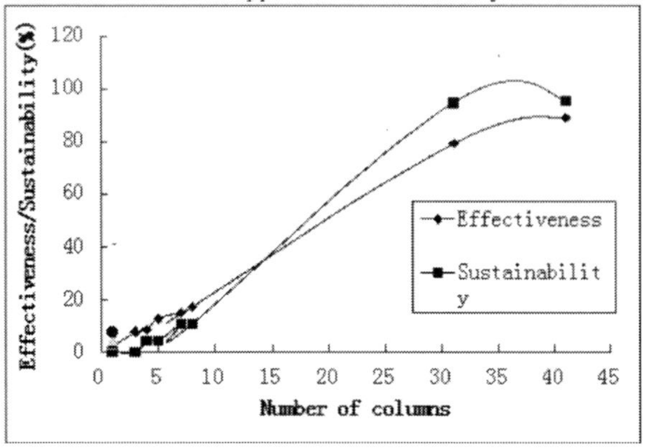

Fig. 1. The comparison between effectiveness and sustainability of teaching methods.

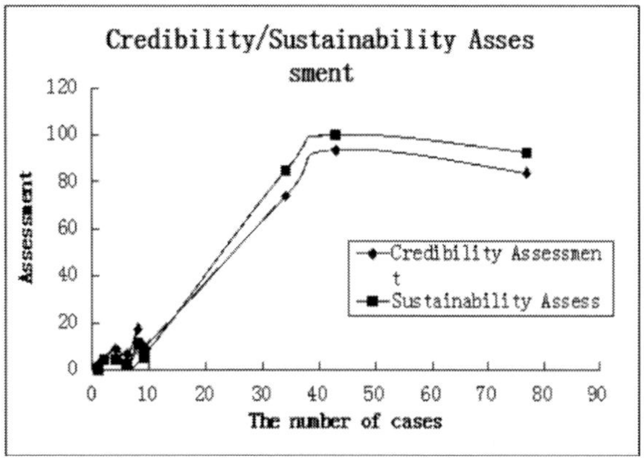

Fig. 2. The students' appraisal about credibility and sustainability of skill test.

In PBL teaching methods, problem-driven, integrated practice will not only help students to information sources and information tools to understand and use, and help students information analytical thinking and inquiring mind, to enable students to think independently, initiative exploration, self-discovery opportunities

Practice runs found because of differences in English reading ability, the experimental group students to master English to retrieve most difficult to retrieve than the Chinese, but to master the Chinese improve the ability to retrieve them also helped Further, since the practice time is short, in less time in evidence-based material education, students grasp the depth and difficulty enough, but less than nothing if the student to be familiar with the basics of evidence-based material education, such as the concept of authenticity background step evaluation of the implementation, but also basic to teaching results of the study.

References

1. Jing Liangyou. Impact of heterogeneous cooperative learning groups on the students with learning difficulties in English studying. Teaching and performance management: theory version, (2013): 72-74.
2. Xie Qun, Zhang Yubin, Zhang Jibiao, etc. Cluster analysis and comprehensive evaluation on summer coastal marine environment quality of Leizhou Peninsul. Marine Environmental Science ,33 (2014): 543-549.
3. Zhao Dongling, Feng Yanruo, Pan Zhengyun. Research on clustering analysis based Algorithm K- grid, Journal of Technology, 30(2014):175-178.
4. Gong Liyan, Meng Xianjun. Quality evaluation of apple processing based on Principal Component Analysis and Cluster Analysis, Agricultural Engineering, 30 (2014): 276-285.
5. Zouyao, Qi Guinian, Liu Tingting, etc. Ingredients and cluster analysis on the principal componentr of Sichuan ottawa heap of black tea quality, Food Industry Technology, (2014): 304-307.
6. Wu He. Examining students' online interaction in a live video streaming environment using data mining and text mining. Computers in Human Behavior, (2013), pp. 291-304.
7. Wei Kuang Lai, Yi-Uan Chen, Tin-Yu Wu, Mohammad S. Obaidat. Towards a framework for large-scale multimedia data storage and processing on Hadoop platform. The Journal of Supercomputing, (2014), pp. 681-696.

Research on Communication Resource Management System based on Metamaterials

Zhou Rong [1,*]

[1] School of Information Engineering, Wuhan University of Technology, Wuhan, China, 430074

* Corresponding author: 125850438@qq.com

Abstract. Electromagnetic Metamaterial (EMM), which cannot be found in nature and possesses the novel electromagnetic properties of negative refraction, reversals of both Doppler shift and Cherenkov radiation, is a kind of artificial composed material. In order to solve the problem of long delay and high packet loss rate, the paper proposed a communication resource management system based on Metamaterials. By using the electromagnetic simulation method, the triple band and tunable transmission/reflection characteristics are investigated. Experiment result shows that the proposed method can improve the performance in the communication resource management.

1 Introduction

Electromagnetic Metamaterial (EMM), which cannot be found in nature and possesses the novel electromagnetic properties, e.g. negative refraction, reversals of both Doppler shift and Cherenkov radiation, is a kind of artificial composed material. It has very important potential applications in the physical areas of including electromagnetics, optics, and materials, and also the engineering areas of such as wireless communications and electronic engineering. Recently, it has excited highly research interests in the world, since the first experimental realization of such EMM in2000. The realization methods of such EMM developed in recent years include the transmission-line structure, metallic resonance structure, and full dielectric structure. Also, the Photonic Crystals can also lead to some novel electromagnetic properties as negative refraction, and therefore it can be considered as a kind of EMM. Among the above mentioned design methods, the metallic resonance structure has huge research merit due to the advantages such as simple structure, easy fabrication method, and flexibility to construct 3D realization. However, for the metallic resonance EMM, the major drawback is that the operating bandwidth is fixed and very narrow.

Software engineering data refers to the data generated in the whole life cycle of software, including source codes, specification documents, and bug reports and so on. In most situations, software engineering data is the only information source for developers. With the rapid enlargement of software scale, to acquire manually the relevant information of software development and maintenance is becoming more and more difficult. Data mining technology can help to discover useful information from software engineering data automatically, which thus speeds up the process of software development.

Wu's [1] paper investigates some data mining problems in software engineering data. Identifier names play an important role in software maintenance. A good method name explicitly indicates the function of the method to the developers, while bad names mislead them. However, except for developers' individual experiences and some naming rules, the developers lack effective tools to guide them to name methods. We consider the problem of automatically choosing proper action names, or verbs, for methods as a classification problem. We build an automated method name based on text categorization. Experiments show that our approach can achieve up to70%naming accuracy for Java methods.

In Fu's [2] paper, process neural networks is introduced into time series data mining to deeply studying on clustering, classification and prediction problems, combining with wavelet multi-resolution analysis and related technologies. A time series clustering method is proposed base on wavelet and improved self-organization process neural network for time series clustering problem. Original time series data is decomposed by wavelet. Under the principle of reserving clustering characteristics, the signal is reconstructed. And then reconstructed signal which has been fitted into time-varying functions is used as the input of process neural network. Self-organization PNN is trained by improved competition algorithm [3]. Making use of time-varying input characteristic of PNN, the timing signal characteristics processed by wavelet is considered adequately in clustering analysis. Network extracts implicit process mode characteristics of input function to self-organize. The improved competition learning algorithm is given. Clustering result of UCI datasets shows that the proposed approach can be applied to timing clustering effectively.

2 The Framework of Dynamic Voltage Restorer

In contrast to the rapid development of software industry, especially open source projects, we lack effective and efficient plagiarism detection tools to protect them from piracy. A less self-disciplined developer, who committed software plagiarism, may elude punishment by seriously disguising the original open source projects. Existing Program Dependence Graph (PDG) based approaches can identify a few kinds of intentional disguises. However, they are rather slow and can be defeated by PDG affecting disguises. We develop a Type Dependence Graph (PDG) based plagiarism detection tool called TPLAG. TPLAG reduces the time complexity of graph based plagiarism detection from exponential to polynomial, and can see through several disguises that confuse existing tools.

Since hundreds of bugs are reported every day for large software projects such as Mozilla, bug triage is obviously a heavy burden for developers. Automated triage can relieve developers from such boring work. Existing work treats bug triage as a text categorization problem [4-5]. We use the Latent Dirichlet Allocation (LDA) model to reduce the dimension of documents, and then classify the bug reports to implement bug triage. Software maintenance highly depends on the information implied in bug reports. A bug report with little useful information is of poor quality and very likely to be ignored. We propose a topic model based approach to analyze bug report and quantify the information in bug reports, i.e., bug report quality. Understanding WHAT has gone wrong and WHEN did it happen with an open source software project can provide high-level and useful information for software maintenance. We try to acquire such information from the bug repository of the project. The LDA model is applied to answer the WHAT question, and an event detection tool is developed to answer the WHEN question. Experiments validate the proposed methods in Chen's [6] paper. Figure 2 shows the basic procedure of data mining technology.

Figure 1 shows the composition of data mining technology.

Fig. 1. The composition of data mining technology

Time series problem involves several fields, such as economy, meteorology, water conservancy, forestry etc. At present, time series data mining has been the focus of data mining, which has strong theoretical and practical significance. Because of some complex characteristics of time series data, for example, time variation, high dimension, noise jamming and volatility, time series data mining is always one of the difficulties in data mining research. Process neural network (PNN) is a development of traditional neural network in the time domain.

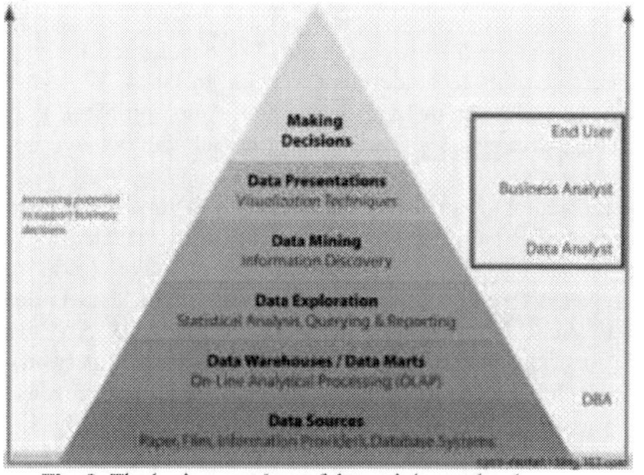

Fig. 2. The basic procedure of data mining technology

3 The Algorithm

A less self-disciplined developer, who committed software plagiarism, may elude punishment by seriously disguising the original open source projects.

The basic model for online error identification as follows:

$$TSP(t)=\begin{cases} TSP_1 & 0\le t\le \Delta t \\ TSP_2 & \Delta t\le t\le 2\Delta t \\ \cdots & \\ TSP_n & (n-1)\Delta t\le t\le n\Delta t \end{cases} \quad (1)$$

$$\min d(T)=\sum_{k=1}^{n}\sum_{i=1}^{N}\sum_{j=1}^{N}c_{i,j}(k\square\Delta t) \quad (2)$$

$$s.t \qquad \Delta t=\frac{T}{n},\frac{\Delta c_{i,j}}{\Delta t}=0 \quad (3)$$

We may get the calculating method for the main index in the following equation (4)-(5):

$$M_{ij}=\exp\left(-\frac{\left\|x_i-x_j\right\|^2}{\sigma^2}\right) \quad (4)$$

$$L=\begin{bmatrix} L_1 & & \\ & \ddots & \\ & & L_k \end{bmatrix} \quad (5)$$

Their matching eigenvectors matrix is shown in the following equation (6):

$$H=\left[h_1,h_2,...,h_k\right]=A^{1/2}E \quad (6)$$

So, we can get:

$$U_{ij}=\frac{H_{ij}}{\sqrt{\sum_{t=1}^{k}H_{it}^2}},i=1,...,n,j=1,...,k \quad (7)$$

$$P=I-A^{-1/2}MA^{-1/2} \quad (8)$$

According to the equation (6), the calculating formula can be obtained in equation (7)-(10).

$$g\,(x,\omega)=\frac{1}{(2\pi)^3}\int g\,(k,\omega)\exp(-ik\,x)dk \quad (9)$$

$$g\,(k,\omega)=\begin{Vmatrix} G_{ik}(k,\omega) & \gamma_i(k,\omega) \\ \gamma_k^T(k,\omega) & g(k,\omega) \end{Vmatrix} \quad (10)$$

$$G_{ik}=(\Lambda_{ik}+\frac{1}{\lambda}h_ih_k^T)^{-1},\quad g=-(\lambda+h_i^T\Lambda_{ij}^{-1}h_j)^{-1},\quad \gamma_i=\frac{1}{\lambda}h_k^TG_{ki}, \quad (11)$$

4 Results and Discussion

Process neural network (PNN) is a development of traditional neural network in the time domain. Process input of PNN relax synchronization instantaneous limit on inputs in traditional neural network models, and is more general artificial neural network model. It has its own unique advantages in dealing with time related problems. Figure 3 shows the result for communication resource management system on the data optimization.

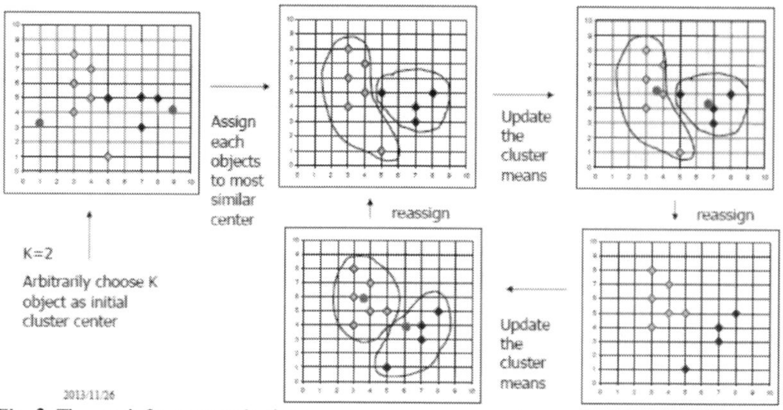

Fig. 3. The result for communication resource management system on the data optimization

This algorithm is defined as Valued K-means Genetic Algorithm. Based on the actual needs of the field of telecommunications knowledge discovery, the Data Mining Grid is an effective means of settlement to the problem that the efficiency of data mining algorithms will continue to fall when the data scale is increasing in geometric patterns. The improved competition learning algorithm is given. Clustering result of UCI datasets shows that the proposed approach can be applied to timing clustering effectively.

References

1. Wu He. Examining students' online interaction in a live video streaming environment using data mining and text mining. Computers in Human Behavior, (2013), pp. 291-304.
2. Fu Xiao, Cheng Fan. Data mining in building automation system for improving building operational performance. Energy & Buildings, (2014), pp. 75-83.
3. Benrun Huang, Li Xia, Zhengguo Wu, Weiping Zhou. Control Strategy of A Dynamic Voltage Restorer (DVR) Based on Line Voltage Compensation. Energy Procedia, (2011), pp. 13-26.
4. Wei Dai, Mostafa Bassiouni. An improved task assignment scheme for Hadoop running in the clouds. Journal of Cloud Computing, (2013), pp. 21-33.
5. Wei Kuang Lai, Yi-Uan Chen, Tin-Yu Wu, Mohammad S. Obaidat. Towards a framework for large-scale multimedia data storage and processing on Hadoop platform. The Journal of Supercomputing, (2014), pp. 681-696.
6. Zhijian Chen, Wenhai Luo, Dan Wu, Xiang Huang, Jian He, Yuanhuan Zheng, Di Wu. Exploiting application-level similarity to improve SSD cache performance in Hadoop. The Journal of Supercomputing, (2014), pp. 703-715.

Application and research on business intelligence in audit business

Lei Che [1,2,a], Xiaoping Yang [1,b*] and Fan Jiang [3,c]

[1] Renmin University of China, China
[2] Beijing Information Science & Technology University, China
[3]University of Pittsburgh, America
[a]chelei@bistu.edu.cn, [b]yang@ruc.edu.cn, [c]ramily_jf@outlook.com

Abstract. With audit data requirements, taking for example the audit data of construction industry, the audit data integration requirements are analyzed and application architecture of business intelligence technology in the audit business is put forward. The modeling process and the populating process of data warehouse are detailed, and the scheme of data pre-processing is set out. The report display methods are proposed, such as proportion chart, tendency chart, index plots and etc. The application model and value of business intelligence technology in audit business are analyzed, which is based on data mining technology such as outlier detection. This research has been applied to National Twelfth Five-Year Plan Science and Technology Support Program and has achieved better analysis and prediction of audit.
Keywords: Business intelligence, Audit business, Data warehouse, Multidimensional data analysis.

1 Introduction

With the development of information technology, audit data is becoming increasingly large and complex, and the original data management pattern has formed large and small islands of data. Audit trails have been overshadowed by the growing complexity of information systems and mass business data, so that variety storage form of data sources need to be analyzed by auditors, which is very difficult to be realized with the traditional data query and statistical analysis methods. Beyond that, many frauds have also noted the importance to maintain the balance of some financial ratios, the role of audit gradually weakening due to artificial control important ratio. With the development of the computer, the auditor can use computer to write audit plan, audit analysis, and carry out information systems audit. But the traditional data analysis method cannot deal with the huge database system, outdated technology tools is bound to affect the breadth and depth of audit. As far as application research of computers concerned, with the advent of big data, business intelligence and other analysis-oriented and decision-oriented techniques have emerged, which will undoubtedly not only provide new ideas and methods for modern audit, but also provide a new way for the development of auditing information. Only if business intelligence technology is applied to audit business, a structured data warehouse and multidimensional data set being built, auditors can organize data and analysis data from different dimensions, even exploit valuable experience and rules, to provide decision support for the audit work1.

2 Business intelligence technology

2.1 Basic concepts

Business intelligence refers to using data warehouse and data mining techniques to store and manage the customer data, and using a variety of data statistical analysis tools to analyse customer data and provide all kinds of analysis report, such as customer evaluation, service evaluation, marketing evaluation, the future market demand and so on, which provide decision making information for the enterprise23. Business intelligence technology includes data warehouse technology, multidimensional data analysis technology and data mining technology.

In computing, a data warehouse (DW, DWH), or an enterprise data warehouse (EDW), is a system used for reporting and data analysis4. Integrating data from one or more disparate sources creates a central repository of data, a data warehouse (DW). Data warehouses store current and historical data and are used for creating trending reports for senior management reporting such as annual and quarterly comparisons5. Multidimensional analysis is a data analysis process that groups data into two or more categories: data dimensions and measurements. Data mining (sometimes called data or knowledge discovery) is the process of analysing data from different perspectives and summarizing it into useful information that can be used to increase revenue, cuts costs, or both6. Data mining software is one of a number of analytical tools for analysing data. It allows users to analyse data from many different dimensions or angles, categorize it, and summarize the relationships identified. Technically, data mining is the process of finding correlations or patterns among dozens of fields in large relational databases.

2.2 Software selection

There are a lot of BI products. However, according to the size, market awareness and development momentum, the products provided by SAP, Oracle, IBM, Microsoft, SAS, the five large multinational BI provider, are dominating the market, and are also the pacesetter for BI. As one of the major platform providers in the industry, Microsoft has provided an integrated suite of business intelligence products in the SQL Server, including related data warehouse, data analysis, data mining, data integration, development, and management tools. The main three tools for BI architecture are: SQL Server Integration Services and SQL Server Analysis Services and SQL Server Reporting Services. At the same time, Microsoft's BI products relate closely to other products, and provide a complete solution from the data store files to the deployment server display, in which some commonly used Office software are also. Therefore, this study selects Microsoft's BI products.

3 Data warehouse model design

3.1 Audit data integration requirements

There are more than 20 industries in audit business. And there are financial data, business data and etc., which is related to audit business.

In audit business, auditors can understand cash flow condition, operation capacity, debt paying ability through analysing enterprise financial situation, so that auditors can evaluate lasting management ability of enterprises and find doubtful points of audit. Audit financial data mining and analysis are the decision-making tools to provide correct information for auditors. Therefore, auditors can understand the enter-prise profit and debt paying ability, so that they can discover the financial problems of enterprises and take corresponding measures in time. There are some typical case of the computer audit work in big data environment, such as quality analysis of credit assets, financing platform risk analysis, two sets of mort-gage analysis, capital abnormal flow analysis and so on. Decision support scheme based on audit business data can not only provide important decision support for the enterprise, but also brings the convenience for the auditors. Audit business has strong demand for decision support solutions, because auditors need to do a lot of analysis work on financial statements. The construction of data warehouse is the key step to realize decision support.

3.2 Business intelligence technology in the application of audit services architecture

As Figure 1 shows, we can see that the application architecture of business intelligence technology in audit business includes Data Sources, ODS, ETL, DW, DM, OLAP server, Data Analysis, Data Mining and so on.

The data of Data Sources come from the data of industries involved by audit business, such as financial data, enterprise data and so on. These data format can be structured, unstructured, or semi-structured. Taking for example the audit data of the construction industry, this paper illustrates the building of data warehouse model, data analysis based on data warehouse and data mining methods. The public data come from the official website, whose data format is an EXCEL file. The industry data includes cash flow table, profit table, asset and liability table and financial indicator table. In order to make up for the data synchronization different between the business system and Data Warehouse, Operational Data Store (ODS) is inserted between the Data source and Data warehouse. ETL is the tool to extract data, transfer data and load data from Data source to ODS and from ODS to DW.

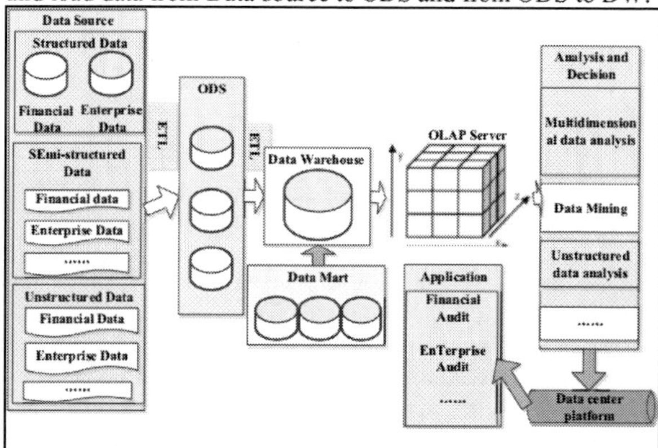

Fig. 1. Application architecture of business intelligence technology in audit business

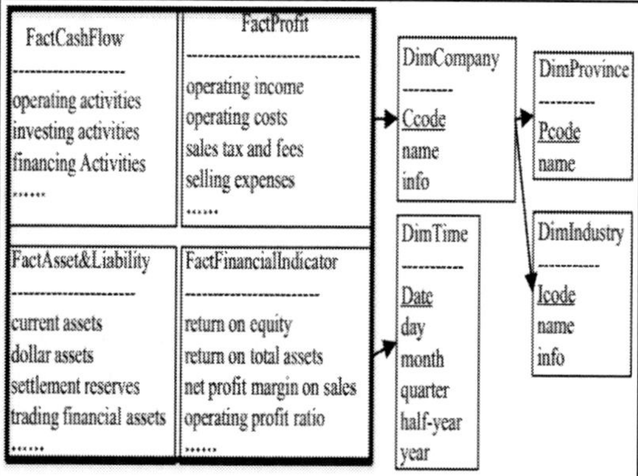

Fig. 2. Multidimensional data model

4 Data warehouse construction

4.1 Multidimensional data model construction

Data warehouse includes construction determining topics, dimensions tables, fact tables and populating the data into data warehouse or data market.

Determining the topic is to determine the topic of data analysis or reports showing. In specific audit process of the construction industry, the following results should be gotten based on existing industry data:

(1) Enterprises of every industry, the amount of the major issues money of the industry, the proportion of assets of the enterprise in its industry

(2) The trend diagram comparing the asset-liability ratio with the average of the enterprise;

(3) The information about the enterprise profits, cash flow indicators, operating income and etc.

In view of the financial audit data gathered from the construction industry, the topics which need to be studied and analyzed include the following 4 categories: "financial indicators of the construction industry", "the balance sheets of the construction industry" and "the cash flow of the construction".

Dimension refers to the different analysis angles. Taking for example the topic of construction industry, we should study financial indicators through time dimensions, enterprise dimensions, industry dimensions, province dimensions and etc.

This model of data warehouse is based on the Fact Groups Model, in which dimensions Star Models or Snowflakes Models connect dimensions Fact Tables through the same dimension tables such as time dimension, enterprise dimension and so on. From Figure 2 we can see the local conceptual model of the data warehouse according the collected data, among which there are 4 dimension tables and 4 fact tables. The dimension tables are DimTime, DimProvince, DimCompany and DimIndustry. The fact tables are FactCashFlow, FactProfit, FactAsset&Liability and FactFinancialIndicator.

When the dimension table is used to achieve dimension, the dimension level need to be represented, by the following three methods: by the different fields of dimension table, by association between the tables and by the associated fields. In those representations, the representation based on the fields is the simplest, and has higher query efficiency, because the queries are single-table queries. Representing the dimension levels by different tables can extract many common dimension tables, improve the maintainability of dimension tables, which has low query efficiency because the query requires multiple correlation between tables. The third method can represent uneven and irregular dimensions well. The "DIMTime" table is the application of the first method.

4.2 Populating the data warehouse

Populating data warehouse refers to data extracting, data transformation and data loading, namely the ETL (Extract - Transform – Load). SSIS is Microsoft SQL Server Integration Services, which is an integration solution generating high performance data, and the ETL platform provided by Microsoft. The ETL process will be discussed based on the collected data of the construction industry. The data source is from Excel. Every indicator corresponds to an excel file, at the same time, every sheet in every excel file corresponds to an enterprise. ETL steps are as follows:

(1) Accessing Excel file, and deleting meaningless data by the data cleansing work. Normalizing every fields' name of Excel files according to the corresponding mapping relationship, and designing corresponding relationship between the excel files and the database.

(2) Importing all sheet of all excel files into the local temporary tables by SSIS graphical operation. That's to say, if there are 4 indicator tables and 50 enterprises in every indicator table, there are 200 local temporary tables would be imported data into.

(3) Emerging the same indicator information of all enterprises into one permanent indicator tables using Procedure of the database.

4.3 Data preprocessing

Multidimensional data analysis needs to analyse data with different comprehensive degrees, therefore the details data should be integrated. Data integration is a very time-consuming operation. In order to shorten query response time, OLAP uses precomputed technology, by which the whole or part data will be calculated and stored in advance. During the financial analysis of the audit business in the, the auditor should not come to a conclusion not just by the information of cash flow, but should combine with assets & liabilities and profit. Only in such way, the audit can make more comprehensive and correct evaluation on business activities. Therefore, the following data models are the example ones which should be precomputed.

Data Model 1: $\frac{FactCashFlow.[operating\ activities\ net\ cash\ flows]}{FactAsset\&Liability.[enterprises\ equity\ ]}$

Function Information: Representing the ability of share capital generating cash and the ability of the largest corporate dividend distributions.

Data Model 2: $\frac{FactCashFlow.[operating\ activities\ net\ cash\ flows]}{FactAsset\&Liability.[total\ assets]}$

Function Information: Representing the ability of the total assets generating cash.

Data Model 3: $\frac{FactCashFlow.[operating\ activities\ net\ cash\ flows]}{FactAsset\&Liability.[current\ liabilities]}$

Function Information: Representing the ability of cash acquired in business activities repaying short-term debt.

Data Model 4: $\frac{FactCashFlow.[operating\ activities\ net\ cash\ flows]}{FactAsset\&Liability.[full\ liabilities]}$

Function Information: Representing the ability of cash acquired in business activities repaying full debt.

Data Model 5: $\frac{FactCashFlow.[operating\ activities\ net\ cash\ flows]}{FactProfit.[net\ profit]}$

Function Information: Representing how much cash will support when enterprises achieve the profit of one yuan book and the superior and inferior quality of corporate profits to some extent.

Data Model 6: $\frac{FactCashFlow.[cash\ received\ from\ sales\ of\ goods\ or\ rendering\ of\ Services]}{FactProfit.[major\ business\ income]}$

Function Information: Representing the corporate sales recovering cash situation and the quality of corporate sales.

5 Showing report

Once the data warehouse is built, the user can access and query the information and data by data analysis tools, real-time query and reporting tools, and the user can see variety of reports of data analysis results, as shown in Figure 3. We can make the visual solutions according to business requirements. There are 3 type reports designed for the current commands: proportion charts, trend charts and indicator charts. The pie charts are suitable for showing the proportion charts, the line charts are suitable for showing the trend charts, and the bar charts are suitable for showing the indicator charts. In SQL Server BI solutions, the SSRS can be used for showing reports to the users.

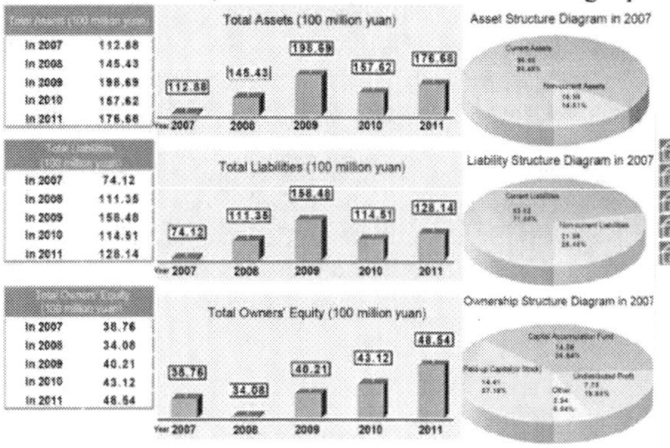

Fig. 3. Multidimensional reports of one corporate assets & liabilities information

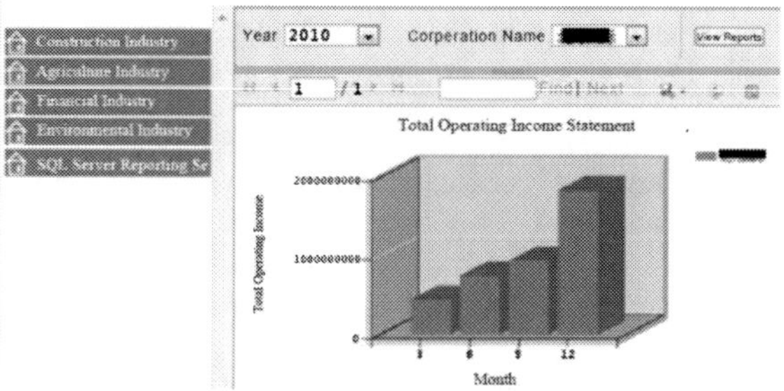

Fig. 4. Reports showing the total operating income

If the report shown in the Figure 4 need to be created, the corresponding procedure in SQL Server and the reporting service should be created and executed or deployed. The procedure is created as follows:

```
--Procedure: Showing the total operating income
CREATE PROCEDURE TOI_Pro
@Year int, @CompanyName nvarchar(255)
AS
  SELECT SUM(FactProfit.opratingincome) AS TotalOpratingIncome, Max(DimCompany.name) AS Company,
        Max(DimProvince.name) AS Province,YEAR(FactProfit.Date) AS Year,MONTH(FactProfit.Date) AS Month
  FROM FactProfit JOIN DimCompany ON FactProfit. Ccode = DimCompany. Ccode JOIN DimProvince ON
DimCompany. Pcode = DimProvince. Pcode
  WHERE YEAR(FactProfit.Date)=@Year and DimCompany.name=@ CompanyName
  GROUP BY YEAR(FactProfit.Date), DimCompany.name, MONTH(FactProfit.Date)
```

The parameters of the report model are showed as follows:

Connection Character String: *Data source=localhost;Initial Catalog=AuditDemo*

DataSet1: *TotalOperatingIncome*
- DataSourceName:*AuditDB*
- Query Type:*Procedure*
- Procedure Name: *TOI_Pro*

DataSet2:*YearList*
- DataSourceName:*AuditDB*
- Query Type:*Text*
- Query: *SELECT DISTINCT YEAR(Date) AS Y FROM DimProfit ORDER BY Y DESC*

DataSet3:*CompanyList*
- DataSourceName:*AuditDB*
- Query Type:*Text*
- Query: *SELECT DISTINCT name FROM DimCompany*

6 Data mining

Data Mining refers to gaining the knowledge in which people are interested from the data of large databases. These information is implicit, previously unknown and potentially useful. Data mining technology is a static computer data analysis method, which obtain implication modes in the data internally by analysis of the static data, such as classification 8, Clustering, Association, etc. And these modes can help auditors to analyse vast amounts of electronic data of the audited entities, and find the audit trail to determine audit doubts. Data Mining techniques and methods would enhance the ability of auditors to analyse audit data, and increase efficiency and quality of the audit work. Therefore, Data Mining technology will get more applications and development in the field of audit. The Data Mining models SQL Server supporting are: Decision Tree, Bayesian Classification, Association Rules, Clustering Analysis and Sequential Cluster and so on.

Some problems in Computer Audit can be solved based on these Data Mining models 910.

(1) Based on clustering technology, auditors can carry out data grouping operation, and then master the rules among data of each group and differences among data of different groups, so that auditors will find the characteristics of the checked problem.

(2) Based on the classification and prediction technology, auditors can find the inner link between input data and output data from the massive amounts data of audited units, then use the found rules to analyse the historical data. Finally, audits can compare the predicted value and actual value, and lock the auditing emphasis more quickly and easier, so as to find doubtful points of audit more correctly.

(3) Based on association analysis, auditors can extract frequent pattern of data, so as to sum up the two events or items rules hidden in the data. But these rules can only be precondition of the next judgement.

(4) Based on outlier detection technology, auditors can mine the isolated points from the data, so as to improve the probability of finding abnormal data, find possible regulation violating phenomenon.

(5) The audit corpus of every industry audit can be built based on Text Mining technology, which is used for processing of the field of every industry auditing language, management information, thereby supporting the optimization of auditing methods.

7 Conclusions

With the rapid development of economy and computer technology, the operation, management and accounting mode of audited units are experiencing outstanding changes, more and more dependent on efficient, complicated and huge information system. In the face of huge amounts of data, it is necessary for auditors to introduce BI technology into auditor process. It's the new requirements, which help auditors to carry out auditor work in big data environment, and provide auditors auditor idea and audit key identification. This research has been applied to National Twelfth Five-Year Plan Science and Technology Support Program ("The research and development and application demonstration of audit simulation and simulation platform to national audit immune system") and has achieved better analysis and prediction of audit.

Acknowledgments

This work was financially supported by these projects as following: Beijing Higher Education Young Elite Teacher Project (NO. YETP1503), The Scientific Research Project of Beijing Educational Committee (NO. KM201511232016 and SM201511232004), Teaching Reform Project of Beijing Information Science & technology University (NO. 2015JGYB30), National Natural Science Foundation (NO. 61272513).

References

1. Isabel Pedrosa, Carlos J. Costa, Computer Assisted Audit Tools and Techniques in Real World: CAATT's Application and Approaches in Context, Computer Information Systems and Industrial Management Applications, 4(2012)
2. Elbashir M.Z., Collier P.A., Sutton S.G., The role of organizational absorptive capacity in strategic use of business intelligence to suppport integrated management control systems, The Accounting Review, **86**, 1(2011)
3. Chaudhuri S., Dayal U., Narasayya V., An overview of business intelligence technology. communications of the ACM, **54**(2011)
4. Jindal R., Taneja S., Comparative study of data warehouse design approaches: a survey. International of Database Management Systems, **4**,1(2012)
5. QU X. , Multi-granularity Temporal and Spatial Approximate Aggregate Query on RFID Data Warehouse, Computer Science, **39**, 6(2012)
6. Han Jiawei, Micheline Kamber, Jian Pei, DATA MINING: Concepts and Techniques, Third Edition.Beijing: Machinery Industry Press, **84**, 92-99(2012)

7. R. Uday Kiran, P. Krishna Reddy, Improved approaches to mine rare association rules in transactional databases. Proceedings of the Fourth SIGMOD PhD Workshop on Innovative Database Research, (2010)

8. Gyorgy J. Simon, Vipin Kumar, Peter W. Li, *A simple statistical model and association rule filtering for classification, Proceedings of the 17th ACM SIGKDD international conference on Knowledge discovery and data mining*, (2011)

9. Lalit Wadhwa, Dr. Virender pal, Forensic Accounting and Fraud Examination in India, International Journal of Applied Engineering Research, **11**(2012)

10. Roger S. Debreceny, Glen L. Gray, Data mining journal entries for fraud detection: An exploratory study, International Journal of Accounting Information Systems, **11** (2011)

Research on the water environment capacity of Qingyi River (Xuchang Section) with GIS technology

Li Wang[1], Luji Yu[2], Chunxiao Liu[2], Nan Sun[3] and Minquan Feng[1,*]

[1]State Key Laboratory Base of Eco-Hydraulic Engineering in Arid Area, Xi'an University of Technology, Xi'an, Shanxi, China
[2]College of Water Conservancy & Environmental Engineering, Zhengzhou University, Zhengzhou, Henan, China
[3]Faculty of Chemistry Water Science, University Duisburg-Essen, Essen, Germany
Corresponding Email: mqfeng@xaut.edu.cn

Abstract. Water environment capacity calculation is the foundation of basin environment management. Due to lack of basic materials and data, the water environment capacity in small basin was not massively researched with appropriate calculating method. This paper mentioned a water capacity calculating method suitable for environment management. The method was based on the study of Xuchang Section of Qingyi River and described with details as follows: Xuchang Section was divided into four control units with GIS technology. The river pollution loads of non-point source pollutants from farmland runoff, rural life, livestock and poultry were calculated with the in-site and statistical data of pollution resource. Meanwhile the calculated river pollution loads of non-point / point source pollutants were statistically analyzed on the basis of control units. Then a water quality module was tested and verified, in which the predicted value tallied with the measured value. The parameter of this water quality module corresponds to the in-site data within relative error ±14%. This module was used to estimate and calculate water environment capacity. With this module the available water environment capacity of each control unit and pollutant reduction amount can be earned through deducting the river pollutant load of point pollutant. The results showed that the utilized method in this paper can satisfy the requirement for the calculating accuracy of small basin water environment capacity.
Keywords. Water environment capacity, geographic information system(GIS), non-point source pollutant (NSP), small basin

1 Introduction

Water environment capacity is a nature resource which exists objectively and also defines as a maximal permissible quantity of pollutants. It satisfies the requirement for particular water quality in a given water body with a target flowrate [1]. The utilization of water environment capacity has already become the scientific basis of China's water quality management and is also the theoretical basis of environmental policy such as capacity control, emission trading and permit system of pollution discharged [2]. China led on to many studies on water environment along with stricter and stricter control system of total emission [3-11]. Generally, most of the researches aim on rivers of big basins and important lake reservoirs, which can be deeply studied with professional software, since many monitoring sections of aquatic environmental quality and hydrological stations exist in those areas and a large amount of basic materials and data can be supplied. Small basins lack in hydrological materials and basic data to research on water environment capacity, therefore there was no suitable method to calculate water environment capacity of small basins. The calculation of water environment capacity in small basins should be more frequently considered to serve the development of the social economy, with the refinement of water environment protection. Recently the total pollutant control focuses mainly on point source pollutant, in contrast the non-point source pollutant has not been valued. The total pollutant control of point and non-point source pollutant used in many years is just starting in China [12]. Due to lack of data, the data of non-point source pollutant cannot be utilized to calculate water environment capacity. Therefore Li, Jake [13] preliminarily proposed a total pollutant control model with considering the effect of non-point source pollutants, and it was practically verified in Xi'an Section of Weihe River. Zhang, Likun [14] monitored the Ashi small river basin mainly with area pollution sources and calculated their environment capacities. Ji, Dingyu [15] researched on a small basin, Huangla Stream of the hilly area in Sichuan, and calculated river pollution load of non-pollutant resources according to rural life pollution, livestock pollution, chemical fertilizer and pesticide pollution. Another difficulty calculating water environment capacity is to determine the designed flowrate [16], so it can only be roughly calculated with the statistics data of water pollutant discharge amount.

This paper focuses on calculating water environment capacity of small basin by GIS technology [17], basing on study of Qingyi River (Xuchang Section). COD values from each control unit and available environment capacity of ammoniacal nitrogen were earned with considering the effect of non-point pollution resource from each aquatic and terrestrial control unit. This method enhanced the accuracy of water environment capacity and supplied environment protecting department the scientific basis to manage and control the total pollution load in small basin.

2 Material and Methods

Study Area. Xuchang Section of Qingyi River as the research target, belongs to the Huai River basin and is the largest and significant tributary of Shaying River (see Figure 1.) Xuchang Section of Qingyi River with length 79km has basin area 1585 km^2. The average annual precipitation adds up to 671.1–736.0 mm. Qingyi River is regarded as the main water body in the function of flood protection, flood drainage, waste drainage and landscape water. It is mainly influenced by industrial wastewater and urban wastewater.

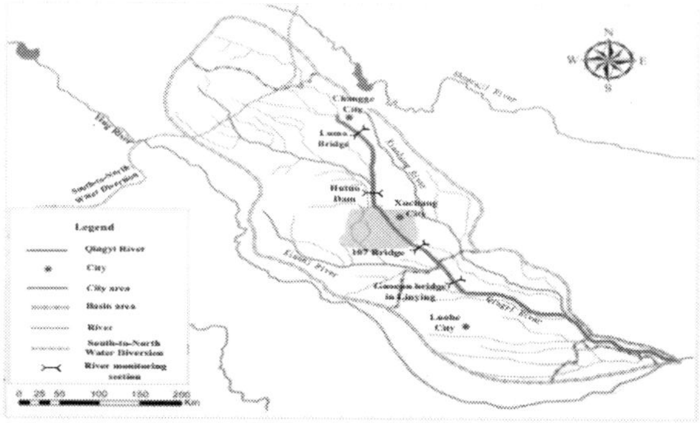

Fig. 1. Schematic diagram of scope of the study.

Research method. *1) Water quality model:* The water quality of Qingyi River can be simulated with one – dimensional model. To describe the steady decay law with one – dimensional model, assumes that water body keeps also the steady status under pollution and local hydrological conditions(the pollutant concentration in water body doesn't change with time.), moreover the effect of longitudinal diffusion is also ignored. The differential equation formulates:

$$u\frac{\partial C}{\partial x}=-KC. \tag{1}$$

Under the initial condition $C=C_0$, equation 1 is converted:

$$C_X=C_0\bullet e^{-kx/u}. \tag{2}$$

$$C_0=\frac{(QC_1+qC_2)}{(Q+q)}. \tag{3}$$

C_X : Calculated pollutant concentration in a certain section [mg/L]; C_0 : Pollutant concentration in initial section [mg/L]; K : Pollutant degradation coefficient [d^{-1}]; X : River length in a certain section [km]; u : average flowrate in a certain section [m/s]; C_1 : Pollutant background concentration in river [mg/L]; C_2 : Pollutant concentration of discharged wastewater [mg/L]; Q : River flowrate [m^3/s]; q : Discharged wastewater flow [m^3/s]

2) Calculating model of water environment capacity: Water environment capacity is calculated with one-dimensional model by trial-and-error method, which utilizes a dynamic water quality model. The used "River Water Environment Capacity Analysis System" (6th Edition) was developed by Li, Yunsheng (Department of Pollutant Prevention of ministry of Environmental Protection). The rate of small river sections, in which the calculated pollutant concentration reaches a standard value, is more than 60%, so that it can ensure the accuracy of water environment capacity calculation.

3 Results

Aquatic and terrestrial range of control unit. Determining aquatic and terrestrial range of control unit is the basis to calculate water environment capacity[2]. According to water environment functions and water quality monitoring points, DEM was analyzed by hydrologic analysis tool (ArcGIS) to select river networks and to divide river basin, then to calculate the closed catchment area. The terrestrial catchment range of control unit is finally determined through the adjustment of river data by GPS (see table 1).

Table 1: Water area and land area of control units

control units	Control units Starting——Terminals	Length (km)	Catchment Area(km$^{2)}$	Water quality in functional area
Qing-1	Xuchang——Lumaqiao	21.3	193.2	Ⅳ
Qing-2	Lumaqiao——Hutuozha	7.7	358.5	Ⅳ
Qing-3	Hutuozha——Dongchengqu	12.9	65.5	Ⅳ
Qing-4	Dongchengqu——Linyinggaoqiao	11.6	486.2	Ⅳ

Pollutant load in various pollution emission sites. The data of calculating water environment capacity of surface water comes from general national pollution resource investigation in 2007, in addition the new data of 2010 contributed also to the calculation. The data of COD and NH_3-N discharge were statistically arranged according to industry, agriculture and life activities (see Figure.2 and 3).

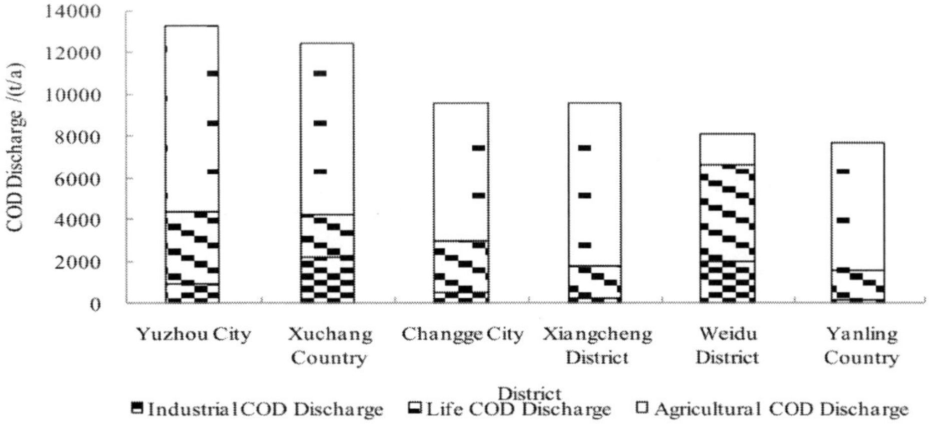

Fig. 2. Consist of COD discharge in county and district of XuChang.

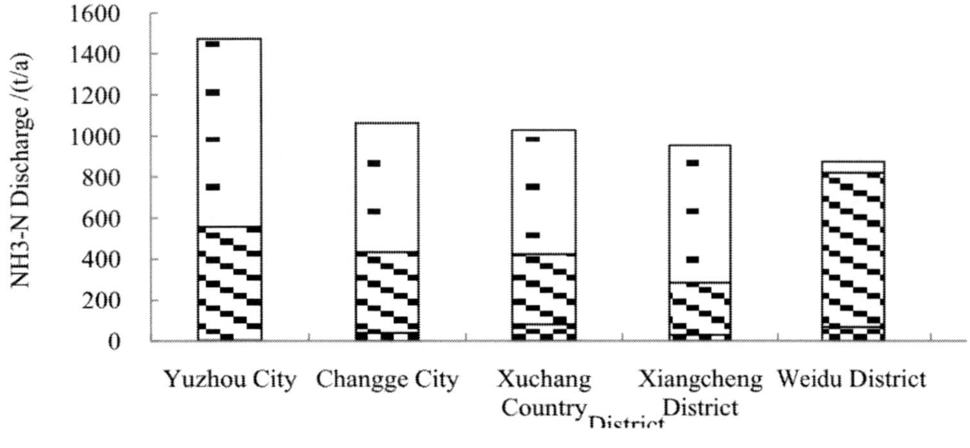

Fig. 3. Consist of NH_3-N discharge in county and district of XuChang.

Point and non-point source pollutant load. The investigation of point and non-point source pollutant depends on connection of pollution source, pollution discharge outlets and contaminated water body. The pollution discharge outlets are navigated by GPS in order to determine pollutant discharge and the main pollution discharge outlets, which influence the water quality in functional regions. River load of point and non-point source pollutants is statistically analyzed at the point of control units (see table 2).Table 3 shows the pollutant loads calculated separately according to point and non-point source pollutant.

Table 2. Loads of point pollutant sources in control units.

Control	Wastewater Load	COD Load	NH_3-N Load

units	(10 thousand t/a)	(t/a)	(t/a)
Qing 1	1700.9	1994.5	262.2
Qing 2	439.2	176.2	14.2
Qing 3	6816.6	7231.9	525.7
Qing 4	6.8	10.5	0.4
Total	8963.5	9413.1	802.5

Table 3. Loads of non-point pollutant sources in control units.

Control units	Agriculture (kg/a)		Rural life (kg/a)		Livestock (kg/a)		Total (kg/a)	
	COD	NH_3-N	COD	NH_3-N	COD	NH_3-N	COD	NH_3-N
Qing 1	/	550	84050	8050	154370	12210	238420	20810
Qing 2	/	60	12160	1210	162450	11700	174610	12970
Qing 3	/	110	0	0	74110	2650	74110	2760
Qing 4	/	60	12160	1210	40610	2880	52770	4150
Total		780	108370	10470	431540	29440	539910	40690

Determination and verification of water quality model parameter. COD and NH_3-N degradation coefficient with 0.151/d and 0.181/d are the calculated coefficient resulting from the measurement between upper section in Maluqiao and lower section in Hutuozha. The monitoring data in November, December, January and February, 2009 were used to determine the initial concentration of upper boundary in model validation units. The hydrological parameter in low-water period are entered into the model to estimate the concentration of lower boundary with one-dimensional water quality model. From figure 4 and 5 the predicated value is corresponding with measured value (relative error ±14%).It can reflect the changes of practical value. Therefore the chosen model parameter are suitable for the state of Qingyi River.

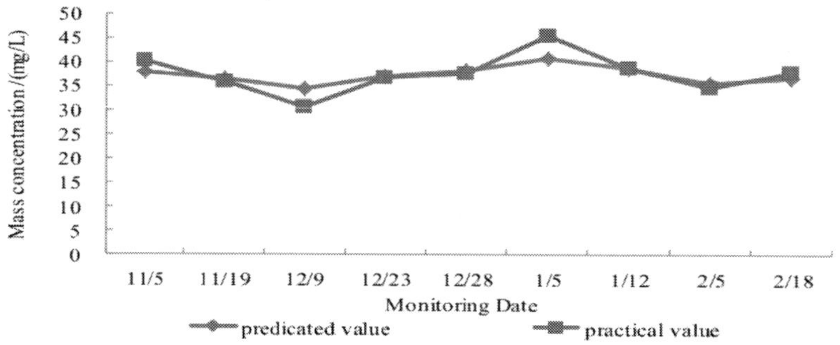

Fig. 4. Comparison of COD predicated and practicaldata at HuTuoZHA.

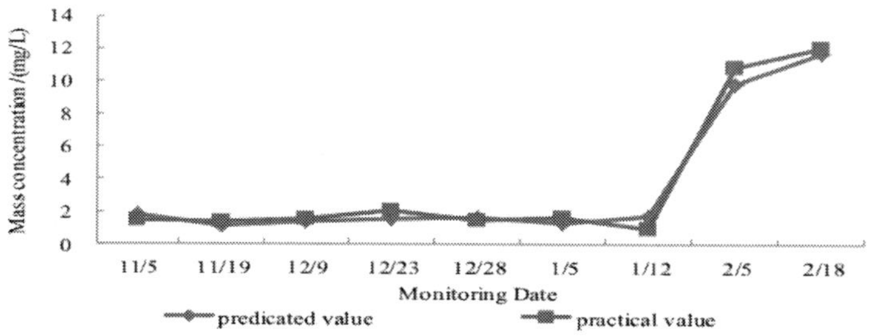

Fig. 5. Comparison of NH_3-N predicated and practical data at hutuozha.

Water environment capacity calculating. The water quality model was at first verified through deducting non-point source pollutant load of each control units. Water environment capacity was calculated under the design conditions of driest months of recent 10 decade. The calculated available water environment capacities were shown in table 4. Meanwhile the results were earned with considering of 1.2 m^3/s biological water transfer. The water environment capacity in Qingyi River was also calculated, ignoring the biological water transfer.

Table 4. Water environmental capacity calculation results.

River	Control units	District	Available water quality in the driest month of 10 years (t/a)		Non-ecologic water transfer (t/a)		Max. admitted pollution load (t/a)		Present Pollution load (t/a)		Reduced amount (t/a)	
			COD	NH_3-N	COD	NH_3-N	COD	NH_3-N	COD	NH_3-N	COD	NH_3-N
Qingyi	Qing1	Changge	427.58	33.19	427.58	33.19	567.27	44.03	2631.98	426.95	2064.71	382.92
	Qing2	Xuchang	5.39	2.03	5.39	2.03	6.32	2.38	205.64	17.50	199.32	15.12
	Qing3	Weidu	2566.89	193.24	2112.89	105.24	3371.43	264.14	8980.51	1175.55	5609.08	911.41
	Qing4	Weidu	1388.23	72.85	1272.23	66.85	2191.94	121.42	15.00	0.87	-2176.94	-120.55
Total			4388.09	301.31	3818.10	207.31	6136.96	431.97	11833.13	1620.87	5696.17	1188.90

4 Discussion

According to the monitoring-data (2011) of processing profile in Qingyi River, COD rises but NH_3-N decreases in Xuchang Section of Qingyi River. In 2009 the calculated concentrations of all sections in Qingyi River are higher than the standard value. COD and NH_3-N concentrations gradually decrease according to the following sequence: high-water period, normal water period, low-water period. The concentrations in high-water and normal water period are lower than the average value. COD and NH_3-N concentrations reach a maximum in high-water period, since the pollutants are diluted with high water load.

COD concentration rises gradually from upper reaches to lower reaches and the high COD load occurs in the middle and lower reaches of Qingyi River to lead to a heavy pollution in those regions. In contrast NH_3-N concentration is higher in upper reaches than in lower reaches. This concentration change illustrates that the NH_3-N load is higher in upper reaches.

The total carbon emission amounts to 60956.32t including centralized carbon emissions. The 64.85 % of the total emission comes from agriculture, which takes up a large slice. Life activities is the main COD emission source in Weidu District and it is the largest part with 57.09 % of the total emission in this District. In contrast COD emission in the other five boroughs accounts for about 66 %-82 % and comes mostly from agriculture. However in Xuchang Country industrial COD emission takes up the largest part and reaches 2123.06 t/a. The total NH_3-N emission is 6189.04 t/a in the six boroughs (including centralized emission), in which agriculture is the largest emission source with 54.95 %. In Weidu District life activities is also the major COD emission source, taking up 86.04 % of the total emission. In other five boroughs agricultural NH_3-N emission with 54 % - 71 % occupies still the first place. At the point of the view of industrial NH_3-N emission, Xuchang Country and Weidu District are considered particularly noteworthy, with the emission amount of 79.34 t/a and 66.66 t/a.

The calculated result of water environment capacity indicates that COD and NH_3-N emission at the other three control units doesn't reach the standard except at the fourth control unit. Therefore the emission at the other three control units should be reduced. Among these three control units the reduction maximum arises at the third control unit and amounts to 5609.08 t/a and 911.41 t/a. Its emission maximum is attributed to the location of the third control unit, in which a large amount of domestic waste water and industrial waste water emit.

5 Conclusion

The river basin distribution graph was drawn with ArcGIS9.0 hydrological analysis. The river data were adjusted by onsite GPS location and were added in the graph. The finally gathered information were used to earn the catchment area of control units. The contour line graph was also used as an auxiliary tool for analysis.

The pollutant data came from point source online monitoring data and national pollution investigation data. The point source pollution was investigated according pollution resource, pollution discharge and the corresponding aquatic and terrestrial relationship of control units. The pollution drain outlet was located by GPS to determine the main pollution drain outlet which influences the river quality and its pollution discharge. The point Pollution discharge was statistically analyzed by control units. Non-point source pollutant includes rural life pollutant, livestock pollutant and farmland runoff pollutant. To get the discharge load of the rural life pollutant, pollutant load was calculated with recommended value of pollutant coefficient by national water environment capacity verification manual. The earned discharge load was distributed into the rivers according to the distribution principle: the emitted pollutant load is distributed at the point of view of resident sites and basin area of control units.

The representative river basin was select to verify the water quality model with water quality monitoring data in driest water period. The measured value had a relative low error with the predicted value. The verified water quality model was used to calculate the available water environment capacity of control units through the removal of non-point source pollutant load. Conclusion showed: the emission differences of aquatic and terrestrial control units were used to earn a practical calculated value with GIS technology, considering with the effect of non-point pollutant. Meanwhile the pollutant load was separately calculated to make the environment more scientific.

Acknowledgment

This work was supported by Major Science &Technology Program for Water Pollution Control and Treatment (2015ZX07204-002-005).

References

1. China's environmental sarft. 2004. The technical review key point of the surface water environmental capacity for approval.
2. China's environmental sarft. 2003. The technical review key point of the surface water environmental capacity for approval.
3. Y. Zhou, X.D. Zhou and M.Q. Feng: Journal of Xi'an University of Technology Vol. 27 (2011), p. 7-11.
4. G. Zhou, K. Lei, G. Fu and G.J. Mao: Journal of Hydraulic Engineering Vol. 45 (2014), p. 227-234+242.
5. Y. Zhang, Y. Liu and X.G. Li: Technology of Soil and Water Conservation. 3 (2015), p. 23-25.
6. H. Ma. 2006. Study of Water Environmental Capacity of Songhua River along Harbin City.Harbin:Harbin institute of thchnolology.
7. J.G. Lu, X. Zhong, Y. Liu and H. Wang: Water Resources and Power Vol. 33 (2015), p. 51-54.
8. S.T. Cheng, Y.C. Qian and H.J. Zhang: Acta Scientiae Circumstantiae Vol. 33 (2013), p. 2848-2855
9. S.F. Ge. 2012. Study on water environmental capacity of Liaohe River in Jilinprovince. Changchun:Jilin University.
10. Z.L. Huang, Y.L. LI, J.X. LI and Y.C. Chen: Journal of Hydraulic Engineering. 3 (2004), p. 7-14.
11. Y.Y. Liang. 2004. Research on the Water Environmental Capacity in Bengbu Urban Section of Huai River.Hefei: Hefei University of technology.
12. L.M. Zhang, Y. Liu, W.H. Sun and B. Bian: Journal of Lake Sciences Vol. 21 (2009) , p. 502-508.
13. J.K. Li, J.J. Zhou, H.E. Li, M. Cai and L. Zhu: Journal of Xi' an University of Technology Vol. 28 (2012), p. 269-277.
14. L.K. Zhang. 2014. Study on Water Pollutants Total Amount Control Technique in Ashi River Basin Based On Non-point Source Pollution Control.Beijing: Chinese Research Academy of Environmental Sciences.
15. D.Y. Ji, Q.G. Wang, H.Y. She, D.R. Wang and X. Zhang: Journal of Water Resources & Water Engineering Vol. 22 (2011), p. 81-84.
16. Y.X. Zhao, M. Xu and Y. Chen: Environmental Pollution & Control Vol. 37 (2015), p. 69-75.
17. C.X. Liu, L. Wang and P. Zhang: Journal of Anhui Agricultural Sciences Vol. 40 (2012), p. 2531-2533.

Studies on Construction Technical System of Ecological Irrigation District in the Lower Yellow River Basin

Shuang Zhang[1,2] and Guibin Pang[3*]

[1]Han-dun Irrigation Management Bureau of Binzhou city, Binzhou, China;
[2]School of civil engineering, Shandong University, Jinan, China;
[3]School of Resources and Environment, University of Jinan, Jinan, China;
Corresponding Email: stu_panggb@ujn.edu.cn

Abstract. Under the background of global water crisis, the construction of ecological irrigation district has become the inevitable trend of the development of irrigation district and new rural construction. In this paper, Handun Irrigation District in the lower Yellow River Basin, as the research object, through analyzing the present operation situation and the existing problems, the technical system of ecological irrigation district in the lower Yellow River Basin are constructed, including water and sediment regulation, saline-alkali land management, efficient use of water resources, water environment protection, irrigation management, water culture and landscape. Finally, a series of the feasibility measures are proposed, including engineering measures, management measures and ecological measures, which has an important reference value for construction of ecological irrigation district in the lower Yellow River Basin.
Keywords. Yellow River irrigation district, ecological irrigation district, constriction, measures

1 Introduction

The Irrigation District in the lower Yellow River Basin is one of the largest continuous gravity irrigation area, there are 104 Yellow River irrigation districts, involving more than 70 cities, covering irrigation area of 1.9 million hm^2, which play a very important role of agricultural development, urban and rural water supply and ecological environment in the lower Yellow River Basin. In recent years, the construction of irrigation district has always been to engineering benefit maximization as the goal, the regional ecological and environmental quality are ignored, combined with the global climate extremes, irrigation district water environment deterioration, soil quality degradation, farmland non-point source pollution, decline of biodiversity, ecosystem destruction and so on, a series of outstanding problems which restrict the development of the irrigation district sustainable future constantly emerging, this requires irrigation district must be from a single production functions to convert to ensure the safety of food security, water supplyment, water-saving society construction and water ecological environment protection, rural landscape construction and so on. Water ecological civilization should be as a guide to the construction of the new stage in the irrigation district, highlighting the efficient utilization of water resources and ecological environment protection as well as the surrounding supporting system are the main content of ecological irrigation district construction [1, 2]. The construction of ecological irrigation district become a necessary guarantee of our country food security and water security.

2 Material and methods

Handun Irrigation District was founded in 1958, located in the hinterland of the Yellow River delta. Scope of irrigation area in Zhanhua, Bincheng and Lijin county, 13 townships, 504 natural villages. The east of irrigation district is Chuguan river and Maxin river, the west of irrigation district is Qintai ditch and Qinkou river, the south is Yellow River, and the bohai sea in the north. There are 0.15 million hm^2 total area, 0.08 million hm^2 arable land, 0.064 million hm^2 designed irrigation area, which belongs to large irrigation district of country.

At present, the irrigation engineering mainly include the following sections. Firstly, the main diversion canal, 28.9 km length, has the entire whole section lining seepage, slope, the design water diversion capacity reach to 60 m^3/s. Maxin, Chaohe, Jianghe and Guotu, four main branch canal reach a length of 147 km, 89 lateral canals, reach a length of 396 km. Secondly, there are Yellow River gate and intercept sand gate, the intercept sand gate located in 1.8 km before the Yellow River gate, also three controlling gates located in Gujia, Niuwang and Wuqi Management stations, as well as a hub named Jiangji At the end of the canal, and more than 463 buildings. Thirdly, the main drainage channels in the irrigation district, including Tuhaihe, Chaohe, Maxinhe, Qinkouhe river and other 38 drainage channels upper the branch. Fourthly, there are several water storage projects, including Dongjiao, Longqihu, Qintai, Maojiawa reservoirs and other more than 40 water storage pits, the total storage capacity reach to 130 million m^3.

3 Analysis of the existing problems

In recent years, the coming water from Yellow River Basin decreased year by year, it is becoming more and more difficulty to maintain the design water supply reliability in the irrigation district, the contradiction between supply and demand is prominent [3]. Irrigation water use efficiency is lower than 0.5, and the contradiction between upstream and downstream, industry and agriculture, production and living water in the irrigation district are prominent. In addition, unreasonable water storage and water irrigation in the area, causing the rise of ground water level, soil secondary salinization caused by the problem is still serious, and unreasonable drainage way led to the water ecological environment further deterioration, intensifying the problems of irrigation water shortage [4].

Due to the excessive use of pesticide and fertilizers in irrigation agriculture, the residue with flow into rivers and lakes and formed the agricultural non-point source pollution [5]. At present, the return water of farmland irrigation lead to environmental pollution, the problem such as regional water environmental problems appear constantly. There are a lot of waste and sewage in the irrigation district discharged arbitrarily, at the same time, irrigation return water carrying large amount of nitrogen and phosphorus into surface and groundwater, causing the exceeding standard of nitrogen and phosphorus in surface water and groundwater generally. Agricultural non-point source pollution has become the main pollution sources of channels in irrigation district.

Excess exploitation of water in upstream leading to the downstream flow decreasing, and the Yellow River slows down, so the river channel of the hydrological environment under the natural conditions are destroyed, common damage situation including zero flow of lower reachs of the river, sediment deposition, narrow flow section and so on, which will seriously affected the downstream and the surrounding ecological environment of agricultural production. In addition, in order to reduce leakage and loss in the process of water, it is commonly to lay the engineering such as concrete and masonry on main channel, however, the ecological functions of ditches are weakened, the soil and biological separated from drainage, river canal lost self-purification ability, which lead to the deterioration of ecological environment directly [6].

Due to the influence of traditional planned economy system, it is difficult to establish the mechanism of water price adjustment based on market supply and demand and cost changes, thus lead to the serious deviate from the cost of water price and water supply. At the same time, water rate collection and water metering is not standard, it is failed to play the function of water rate economic leverage. Since 1982, Handun irrigation district has been carrying out paid for the water supply, water measurement method for the irrigation district management unit to the county, the water price according to the charge to the county, due to water consumption under the lateral canal can not measurement, the county water fee payable by mu evenly spread to the villagers, then the financial departments collected by degrees and assigned to the irrigation district. However, the lower water price and non-standard water management lead to the lack of enough money to projects supporting the construction and maintenance by the water administrative department directly.

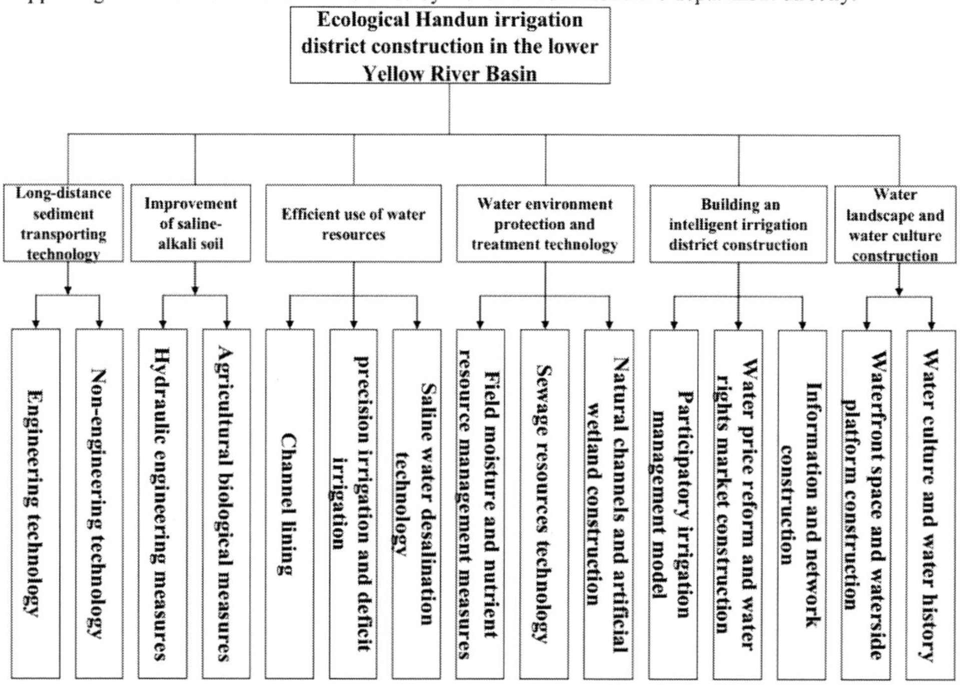

Fig. 1. Technical system framework of ecological Handun irrigation district construction

4 Results

As is shown in the figure 1, Firstly, long-distance sediment transporting technology in the lower Yellow River Basin irrigation district should be accepted. Sediment deposition is one of the major problems in the lower Yellow River Basin irrigation district, how to improve the ability of channel sediment is the key technology of distributing and deploying sediment over a long distance, including the block sand sluice engineering in Handun canal head, cross section shape optimization, channel lining engineering technology such as longitudinal slope and roughness set, as well as water diversion of joint scheduling of water and sediment discharge conditions, non-engineering technology such as water-saving and sediment reduction.

Secondly, it is important to improve comprehensive management of saline-alkali soil in the Yellow River irrigation district. It is suggested that in order to build an improved and optimized integrated technical system of salinization soil resources in the Yellow River irrigation district, we should accord to the principles of adaptation to local conditions and comprehensive prevention, as well as water conservancy engineering measures combine with agricultural biological engineering measures. Two measureas are applied, including hydraulic engineering measures, such as perfecting the irrigation and drainage facilities, controlling salt migration, Subsurface pipe draining off water and salt, decreasing groundwater level, flatting framland, building platform field, also agricultural biological measures, such as water-saving agriculture, increasing organic fertilizer, salt tolerance plant and breeding salt-tolerant varieties, and so on.

Thirdly, Several methods are proposed to improve the utilization efficiency of water resources in the Yellow River irrigation district. Researching new materials, new technology, new standards of channel lining, integrating application of precision irrigation control technology and new theory of the deficit irrigation, building efficient water-saving regulation technology and promotion mode, developing saline water desalination technology in shallow groundwater buried area, proposing multi-water resources joint configuration theory and regulation decision-making method, which embody sustainable utilization of water resources, social and economic development and ecological environment friendly[7].

Fourthly, Water environment protection and comprehensive treatment technology need to be adopted in the Yellow River irrigation district. First of all, field moisture and nutrient resource management measures such as factors influencing the pollutant migration should be analyzed, agricultural non-point source pollution control of the sewage sludge management technology system are established. Then, it is necessary to explore sewage resources technology and method, which is suitable for small enterprises, village life, small-scale livestock and poultry farms. Finally, it would like to study on non-point source pollutants from farmland to nearby water ditch wetlands intercept and removal mechanism in the process of migration, and establish channels of natural and artificial wetland construction technical programme, it is "reducing the pollution - source controlling - intercepting - conducting - repairing", thus water environment of irrigation district are improved.

The fifth, it would like to building an intelligent Yellow River irrigation district. it is necessary to Explore irrigation water management mode accord with the practice of irrigation district, dynamic, scientific and reasonable, proposing the participatory irrigation management model of "administrative organization - water users association - farmers", improving water price forming mechanism, establishing agricultural water subsidies and reward system, building the water rights market, promoting efficient use of water resources. It is important to etablish irrigation system of water monitoring and dispatching system, also irrigation district environment monitoring system and information network should be created, as well as the ecological irrigation system for information management and other advanced technology need to be researched.

In the end, it need to accelerate the Water landscape and water culture construction of Yellow River irrigation district. First of all, according to the goal of integrated multiple functions in irrigation district, it is necessary to research four waterfront sapce modes including main town canal, river, lake, wetland and four kind of ecological waterscape types including paddle type, enjoy water type, cognitive and recreation and so on. Then, full investigation and digging of water conservancy and water culture, religious belief, classic water conservancy scientific literature and poetry, stone inscription, historical records and legend, idiom, proverb, construction and other water culture resources, exploring the irrigation area water culture building framework, principle and its reasonable form.

5 Discussion

It is necessary to adopt the concrete measures of ecological irrigation district construction. The first is engineering measures, including main canal lining, check gate and intake sluice maintenance of main canal, building new management and canal bank road, upgrading substation and power distribution unit. In order to control main canal seepage, on the basis of the vegetative retaining wall brick with concrete to protect inner slope angle, at the same time, increasing water line and pure soil protective layer. The second is management measures, in order to enhance irrigation district management level, the main measures are rely to advanced technology. First of all, we need to establish a set of computer automatic monitoring system, the system include measuring water system and water control system, at the same time, it is necessary to equip monitoring machine to implement the collaborative scheduling in the irrigation district management office, multiple management not only can greatly save manpower, but also can improve efficiency and comprehensive benefits in terms of supply and demand of water. The third is ecological measures, it mainly include establishing three-dimensional botanical retaining body and wetland ecological area. Cultivating plant near the surface and planting grass in the water, aquatic plants can stop the sediment brought up by wind and hydraulic for a long time, then send accumulating sediment back in the water, which can provide nutrients required for aquatic plants and improve the water self-purification ability, also adjust the structure of the ecosystem circulation. Through planting aquatic plants such as algae, duckweed to establish wetland ecosystem, not only can improve the nitrogen and phosphorus pollutants absorption ability, but also can beautify the landscape of irrigation district [8].

6 Conclusions

Compared with the traditional irrigation, ecological irrigation district has the more complex and more comprehensive function, it should implement efficient utilization of soil and water resources, pay more attention to the water ecological environment protection and biodiversity maintenance, rational allocation of resources, maintaining the sustainable development of the irrigation district. In this paper, Handun Irrigation District in the lower Yellow River Basin, as the research object, Through analyzing the present operation situation and the existing problems, the technical system of ecological irrigation district in the lower Yellow River Basin are constructed, after that a series of the feasibility measures are proposed, including engineering measures, management measures and ecological measures, which has an important reference value for ecological irrigation district construction in the lower Yellow River Basin.

Acknowledgements

This work was financially supported by Shandong Natural Science Foundation(NO. ZR2014EEQ020) and Shandong Water Conservancy Science and Technology Practice Project(NO. SDSLKY201502).

References

1. P. L. Yang, Y. K. Li, X. H. Zeng, J. H. Yang, Q. Guo, S. M. Ren and Y. P. Su: China Water Resources. Vol. (14)(2009), p. 32-35
2. B. J. Gu, C. Wang and P. F. Wang: China Rural Water And Hydropower. Vol. (12) (2005), p.7-9
3. J. M. Goncalves, L. S. Pereira, S. X. Fang and B. Dong: Agricultural Water Management. Vol. 94(1-3)(2007), p. 93-108
4. Z. Z. Feng, X. K. Wang and Z. W. Feng: Agricultural Water Management. Vol. 71(2)(2005), p. 131-143
5. J. C. Valenzuela: Agricultural Water Management. Vol. 96(2)(2009), p. 188-194
6. B. J. Gu: China Rural Water and Hydropower. Vol. (10)(2006), p. 4-6
7. D. Xu and S. H. Gong: Journal of Hydraulic Engineering. Vol. 38(7)(2007), p. 806-811
8. B. J. Gu, C. Wang and P. F. Wang: Journal of Hydraulic Engineering. Vol. 37(2)(2006), p. 178-183

DOI: 10.1051/matecconf/201710005004

Relationship between Venture Capital and Enterprises' Performance of GEM Listed Companies

Jianguo Wang[1,*], Jun Wu[1] and Guomin Ji[1]

[1]School of Management, Harbin Institute of Technology, Harbin, China
Corresponding Email: 490782424@qq.com

Abstract. With the Growth Enterprise Market (GEM) opening, China's capital market system has been gradually improved, the total amount of venture capital investment has emerged explosive growth. This article focused on relationship between venture capital and enterprises' performance of GEM listed companies in China. The empirical results found that venture capital involvement will enhance enterprises' performance, a high proportion of venture capital holdings is conducive to enhancing enterprises' performance, and the high reputation of venture capital institution is able to enhance enterprises' performance.

1 Introduction

In the current GEM market, the percent of venture-backed companies is more than 70%, which indicated that the venture capital playing an increasingly important role in the market, and also a growing number of venture capital involved in companies listed on the GEM indicates venture capital is playing an increasingly important role on high-tech enterprises. Based on the study of venture capital and corporate performance from domestic and foreign, the article selected GEM companies to conduct empirical research.

Nahata [1] empirically studied the venture enterprises, and he found that the involvement of venture capital is helpful to improve the performance of enterprises. Especially, the reputation of venture capital institutions has a notable impact on the performance of the enterprise. In Krishnan's [2] study, the market share of the IPO enterprises invested by the venture capital institution is used to measure the reputation of the venture capital institution.

Cuifeng W. [3] examines the investment preference and pattern of venture capital firms, from the perspective of the investment preference, when the geographic distance between the venture capital firm and the entrepreneurial firm is shorter, or when the entrepreneurial firm has financial strength, the probability of the venture capital firm's investment will be higher. Xiangyu G. et al. [4] found that the UNDPRC of company invested by venture capital firms with foreign capital and mixed type background is lower than that of the state owned company, and the stock return rate is higher when the company fist IPO. Wei Z. [5] investigated the relationship between venture capital and the performance of the strategic emerging industries listed companies in China. The empirical result shows that: Venture capital has significant inhibited effect on the strategic emerging industries listed companies. Hochberg's further research shows that the venture capital with high-shareholding and high-reputation can significantly improve the performance, while number of venture capital cannot. Bing L. [6] studied on the relationship between slack resources and firm performance in entrepreneurial firms with the sample of China GEM listed companies from the perspective of managerial discretion theory, and specifically focuses on how venture capital affects this relationship for post-IPO entrepreneurial firms.

2 Instructions on sample data collection

The article collected the explanatory variables that can reflect whether the venture capital is involved, the proportion of venture capital investment, reputation and combined venture capital investment. Venture capital intervention is defined as VC, when there is a venture capital investment institution in the top ten shareholders of the enterprise, then VC = 1, otherwise VC = 0, the proportion of venture capital is defined as Shr, the proportion of venture capital investment is the sum of the proportion of the venture capital institutions in the top ten shareholders of the enterprise, reputation of venture capital was defined by Fame, in this paper, the Zero2IPO of the venture capital institutions ranked was chosen for judgment, when the venture capital investment institutions ranked in the top fifty, it means the venture capital investment institutions has high reputation, then Fame = 1, otherwise Fame = 0, joint investment is defined as Comb, when the current ten major shareholders of the companies have two or more venture capital investment institutions, Comb = 1, otherwise Comb = 0.

And also the article collected the explained variable that can reflect the enterprise's performance. Performance of enterprises is reflected in four dimensions., they are the profitability of the enterprise, the operating ability, the development ability and the innovation ability which can be measured by the net assets income rate (ROE), total assets turnover rate (TOA), operating income growth rate (Growth) and R&D revenue ratio (Ratio).

In addition, in order to be able to correctly study on relationship between venture capital and corporate performance, the article selected scale of sample firms (lnScale), equity concentration (Top1), financial leverage (Lev) and cash flow of operation (Cfo) as control variables.

The article chose data of Shenzhen GEM listed companies from 2009 to 2013. According to the relevant provisions of the securities law of PRC: shares held by other shareholders of the company at the time of the IPO since the date listed on the 12 months shall not be transferred, so within 12 months after the IPO companies won't have venture capital exit, therefore this article selected empirical data of Shenzhen GEM listed companies in the year of 2009-2013, and the research eliminated companies that lack data.

3 Descriptive analysis of the model designed and statistical analysis of variables

In this article, Gompers' research method was used to analyse the relationship between venture capital and its characteristics and company's performance. The empirical study was took by two steps, in the first part, the influence of venture capital on firm performance was studied, and the model 1 is as follows.

$$ROE=C+\beta_1 VC+\beta_2 lnScale+\beta_3 Top1+\beta_4 Lev+\beta_5 Cfo+\varepsilon \quad (1)$$

In the second part, the influence of characteristics of venture capital and the proportion of venture capital was studied, and the model 2 is as follows.

$$ROE=C+\beta_1 State+\beta_2 Fame+\beta_3 Shr+\beta_4 Comb+\beta_5 lnScale+\beta_6 Top1+\beta_7 Lev+\beta_8 Cof+\varepsilon \quad (2)$$

C is intercept, β_1—β_5 is coefficient, ε is residual, the rest are interpreted variables, explanatory variables, and control variables.

SPSS and Stata statistical software were used for empirical research. Descriptive statistics of the sample population can be seen from table 1. It mainly analyses the maximum, minimum, mean and standard deviation of the venture capital intervention and the enterprise performance, table 1 shows that Venture Capital Holdings accounts for 72% of the Sample Firms, which means that the risk of investment in China's GEM Listing Companies play an important role.

Table 1. Descriptive Analysis of Sample.

	N	Min	Max	Mean	SD
ROE	326	.0031	.2817	.1252	.04214
TOA	326	.1316	1.8568	.5674	.24565
Growth	326	-1.0642	.5634	.1505	.21322
Ratio	326	.0028	3.3800	.0729	.19298
VC	326	.0000	1.0000	.7209	.44927
lnScale	326	19.4910	22.4975	20.6127	.506208
Top1	326	.0877	.6887	.33938	.127021
Lev	326	.0110	.5255	.15255	.105756
Cfo	326	-3.708E8	6.2386E8	1.541E7	7.726E7

Table 2. Descriptive statistics on the classification of venture capital investment.

	N		Min	Max	Mean	SD
ROE	VC	235	.0031	.2817	.129403	.0440708
	non VC	91	.0491	.2164	.127263	.0368133
TOA	VC	235	.1316	1.8568	.558272	.2322009
	non VC	91	.1791	1.7395	.530803	.2774084
Growth	VC	235	-1.0642	.5634	.191316	.2229609
	non VC	91	-.7855	.5495	.148495	.1868492
Ratio	VC	235	.0030	3.3800	.105940	.2248443
	non VC	91	.0160	.3990	.065433	.0505320
lnScale	VC	235	19.4910	22.4975	20.620855	.5071598
	non VC	91	19.5314	21.7827	20.591565	.5059252
Top1	VC	235	.0891	.6887	.345106	.1245387

	non VC	91	.0877	.6148	.324582	.1327767
Lev	VC	235	.0110	.4471	.151817	.1010311
	non VC	91	.0126	.5255	.154438	.1174643
Cfo	VC	235	-3.7086E8	6.2386E8	1.606858E7	8.1381779E7
	non VC	91	-2.3034E8	2.1214E8	1.370103E7	6.5829932E7

As can be seen from table 2, venture capital invested enterprises' return on equity (ROE) and total asset turnover (TOA) mean are higher than the non-venture capital firms, the difference between them is not too much. Venture capital invested enterprises' average operating income growth rate (Growth) and R&D income ratio (Ratio) were higher than the values of the non-venture investment enterprises, respectively, higher than 5% and 4%, by comparing the venture and non-venture invested enterprise, the involvement of venture capital on enterprise performance has certain influence. In terms of scale of the enterprise (lnScale), venture investment enterprises average scale is higher than non-venture investment enterprises, which is consistent with the theory, because compared to non-venture enterprises, the venture investment enterprises can get financing from venture capital, and the scale of investment is relatively large. In terms of assets and liabilities rate (Lev), compared with non-venture investment enterprises, the venture capital invested enterprises' assets and liabilities rate levels were lower, the reason is that the GEM listed companies are generally high and new technology enterprises, due to their uncertainty development, it is very hard for them to obtain funds through traditional banks and other financial institutions, therefore, the assets and liabilities rate is lower. Cash flow of operation (Cfo) of the venture investment enterprises is higher, this is because the venture capital means a kind of rights and interests, venture capital investors can not only provide funds for the enterprises, but also provide their own professional knowledge and investment experience, which are helpful to improve management level of enterprises they invested, the cash flows from operating activities is the main source of cash, it can represent enterprise's operation and management level.

Table 3. Correlation analysis of venture investment intervention.

	VC	lnScale	Top1	Lev	Cfo
VC	1	.026	.073	-.011	.014
lnScale	.026	1	-.014	.173	.142
Top1	.073	-.014	1	.043	.073
Lev	-.011	.173	.043	1	-.323
Cfo	.014	.142	.073	-.323	1

Table 3 shows that the partial correlation coefficient between the independent variables and control variables absolute most are less than 0.3, and the partial correlation coefficient that less than 0.2 accounted for 90%, which indicates that there is no significant correlation between venture capital and other control variables in the model designed, and regression analysis can be carried out.

Table 4 shows the correlation analysis of characteristic variables of venture capital, there is no significant correlation between variables and control variables.

Table 4. Correlation analysis of characteristic variables of venture capital.

	Shr	Comb	Fame	State	lnScale	Top1	Lev	Cfo
Shr	1	.374	.030	-.154	.139	.043	-.034	.188
Comb	.374	1	.207	-.036	.080	.011	.035	.012
Fame	.030	.207	1	.182	.035	-.114	-.052	-.070
lnScale	.139	.080	.035	007	1	.008	.179	.131
Top1	.043	.011	-.114	-.124	.008	1	.096	.089
Lev	-.034	.035	-.052	.013	.179	.096	1	-.203
Cfo	.188	012	-.070	-.071	.131	.089	-.203	1

4 Regression analysis on relationship between venture capital and enterprises' performance

4.1 Regression of venture capital intervention on enterprises' performance

Regression of venture capital intervention on enterprises' performance was shown in table 5, the involvement of venture capital has an impact on corporate performance. The R2 were 0.355, 0.445, 0.148, 0.324, the adjusted R2 were 0.327, 0.412, 0.112, 0.309, the model fit was better and the Sig. value has passed the significant test.

According to table 5, venture capital intervention has positive effect on firms' performance, the t values of venture capital on ROE, growth and Ratio were 2.321, 196.2, 1.951 in 10% level, it means that the three variables all reached the significant level, on the other hand, the t value of regression analysis of the venture capital on TOA has not reached the significant level, but they were positively correlated. Through the analysis, it can be concluded that venture capital has a positive effect on firms' performance, the enterprises with venture capital always perform better than those without venture capital.

Table 5. Regression of venture capital intervention on firms' performance.

	ROE	TOA	Growth	Ratio
VC	0.003**	0.032	0.001*	1.315*
	(2.321)	(1.112)	(1.962)	(1.951)
lnSca.	-0.006	-0.072**	0.029	-1.558*
	(-1.273)	(-2.773)	(1.178)	(-1.948)
Top1	0.006	0.165*	0.064	-15.167*
	(0.396)	(1.939)	(0.679)	(-1.984)
Lev	0.007*	0.807***	0.028**	-16.592*
	(1.961)	(6.166)	(2.428)	(-1.949)
Cfo	0.000***	-0.000	0.000	0.000
	(4.923)	(-1.467)	(0.384)	(0.442)
Inter.	0.239***	1.903	-0.469*	46.048
	(5.106)	(0.419)	(-1.938)	(1.024)
r2	0.355	0.445	0.148	0.324
r2_a	0.327	0.412	0.112	0.309
F	5.495	10.874	2.485	4.565
Sig.	.000	.000	.087	.017

4.2 Regression of characteristics of venture capital

Regression of characteristics of venture capital shows in table 6, the venture capital investment shareholding ratio (Shr) has a positive effect on the performance of the enterprise. The venture capital investment is an equity investment, in order to realize the value of capital, it will use its own experience and expertise to provide added services to the invested companies, which can promote the improvement of enterprise performance. The fame of venture capital institutions has positive effect on firms' performance, the impact of the fame of venture capital institutions on ROE and Growth is significant at the 10% level, and the impact on R&D is significant at the 5% level, the fame of venture capital institutions also has a positive impact on the TOA, but it is not as significant as the other indicators. Effect of combine investment on firms' performance is negative, the main reason can be explained that venture capital institutions want to take a free ride from each other.

Table 6. Regression of characteristics of venture capital.

	ROE	TOA	Growth	Ratio
Shr	0.015**	0.003*	0.122*	0.005
	(2.957)	(1.931)	(1.949)	(0.290)

DOI: 10.1051/matecconf/201710005004

Comb	-0.012*	-0.003	-0.040	-0.007
	(-1.925)	(-0.100)	(-1.278)	(-1.116)
Fame	0.007*	0.033	0.008*	0.017**
	(1.995)	(1.358)	(1.941)	(2.427)
lnSca.	-0.003	-0.038	0.064**	-0.001
	(-0.557)	(-1.306)	(2.232)	(-0.162)
Top1	0.020	0.268**	0.123	-0.036
	(0.886)	(2.352)	(1.071)	(-1.574)
Lev	0.042	0.676***	0.061	-0.136***
	(1.441)	(4.538)	(0.406)	(-4.586)
Cfo	0.000***	-0.000	0.000	0.000
	(4.923)	(-1.467)	(1.315)	(0.658)
Inter.	0.185	1.157*	-	0.118
	(1.626)	(1.966)	1.193**	(1.005)
			(-2.021)	
r2	0.148	0.160	0.062	0.166
r2_a	0.117	0.130	0.029	0.136
F	4.826	5.298	1.853	5.551
Sig.	.000	.000	.0232	.0394

5 Conclusion

In this article, a data analysis for the venture capital investment and performance of the GEM companies is mainly conducted, by regression analysis, the following conclusions can be obtained.

5.1 Intervention of venture capital investment contributes to the improvement of enterprises' performance

Venture capital investment produces positive effect on firms' performance, indicating that as an equity investment the venture capital investment not only provides capital for the enterprises, but also increases their own value using their professional knowledge and experience, so that they can achieve their own asset appreciation.

5.2 Proportion of venture capital investment has impact on enterprises' performance

As the shareholders of the enterprises, venture capital investors concerned all aspects of the invested enterprises in order to realize their high return target. The good performance of the enterprise is the basis for the shareholders to obtain high returns. Therefore, the higher proportion of venture capital holdings of enterprise shares, the enterprise more likely to get operating guidance, so that it is possible for the venture capital to achieve their own investment income.

5.3 Fame of venture capital institutions has a positive effect on enterprises' performance

In order to establish a good reputation to make the venture capital later refinanced smoothly, they will make full use of their professional knowledge and experience to provide consulting guidance for the invested enterprises. And also, the venture capital with higher reputation is recognized by the market and has a certain market value, and they can get better performance.

5.4 Combined venture capital has a negative influence on the invested enterprises' performance

The main reason may be that although the combined investment of each venture capital institution has rich investment experience and resources, but each of them want to take a free rider, and they didn't give full play to their initiative, each of them expected to gain profit from other venture capital institutions.

DOI: 10.1051/matecconf/201710005004

References

1. Nahata R. J. FINANC. ECON. **90**, 2(2008)
2. Krishnan, C. N. V. , Ivanov, V. I. , Masulis, R. W. & Singh, A. K. J. F. Q. A. **46**, 5 (2011)
3. Cuifeng W. Shinong W. Wei L. N. K. Biz Rev. **17**, 5(2014)
4. Xiangyu G. Yuhua L. F & A. 1 (2014)
5. Wei Z. Jun W. Ind. Eco. Rep. 2 (2015)
6. Bing L. Chaoliang L. Zhengping F. Nan Kai Biz Rev. **19**, 1 (2016)

Incentive Mechanisms for Tacit Knowledge-Sharing in Master-Apprentice Pattern Based on The Principal-Agent Theory

Ruihua Wang[1,2,*] and Yongbo Lv[1]

[1]School of Traffic and Transportation of Beijing Jiaotong University, Beijing, China
[2]Management Engineering Department of Zhengzhou University, Zhengzhou, China
Corresponding Email: wangqiqi0305@163.com

Abstract. Continual knowledge sharing is the key to improve the competitive ability, operation ability and innovative ability of the organization. Through the comparison of the game theory, knowledge market theory and the principal-agent theory, the principal-agent theory is more suitable for the research on the incentive mechanism of tacit knowledge sharing in master-apprentice Pattern. During the process of tacit knowledge sharing in master-apprentice Pattern, different types of tacit knowledge determine the different design of incentive mechanisms.When the master and the apprentice share the inexpressible tacit knowledge, the organization does not need to design any incentive mechanism for promoting the master's knowledge sharing.When master and apprentice share the expressible tacit knowledge, the organization needs to design different incentive mechanisms for the master and the apprentice respectively. Moreover the organization needs to take into account the different master-apprentice models in different departments. So the organization needs to design different incentive mechanisms for different departments in order to furtherance the tacit knowledge sharing in master-apprentice pattern.

1 Introduction

In the era of knowledge economy, the competition among organizations has evolved into the competition of knowledge products or knowledge services [1]. But the knowledge within the organization will not appear in the air. Knowledge of the organization comes from the individual based on the practical knowledge theory. Therefore, knowledge within the organization is mainly derived from the mutual influence of individual knowledge in the organization [2].Lawson et al.(2009) defined the interaction process within the organization as the knowledge sharing process, namely, the process of exchanging knowledge and creating knowledge [3]. The acquisition and accumulation of knowledge in the organization can not be separated from the knowledge sharing among individuals in the organization.Knowledge sharing within the organization is more effective, the more likely that the organization can survive in the fierce competition and keep higher productivity [4].So knowledge sharing is the key to the successful implementation of knowledge management. Knowledge sharing can not only improve the application ability and innovation ability in order to enhance competitiveness of the organization, but also can accelerate the development of new products through reducing the production cost of the organization [5].Thus how to promote knowledge sharing in the organization becomes the key to enhance the competitive ability, the operation ability and the innovation ability of the organization.

2 State of the art

2.1 Incentive mechanism

Kimbal(2012)considered that information technology can promote knowledge sharing in the organization to a certain extent [6].But the effect of knowledge sharing is not significant. The reason is that information technology can create more convenient conditions for knowledge sharing among individuals. But it can not really improve the individual motivation of knowledge sharing. Thus knowledge sharing is still not easy to occur.Knowledge is a symbol of the individual rights or organization rights. Knowledge is embedded in the individual's thinking. In the organization which is full of a wide competition, in the rational behavior, employees with knowledge reject knowledge sharing in order to maintain their competitive advantage in the organization [7].But the success of the organization is based on the individual knowledge in the process of knowledge sharing.Employees must be willing to share knowledge. Individual knowledge can be transformed into organizational knowledge and intellectual capital of the organization. Accordingly, in order to improve the competitive advantage of the organization, the organization is bound to promote the knowledge sharing in the organization through the designing of incentive mechanisms.Based on the analysis of people's motives and needs, incentive mechanisms emphasizing the rational behavior of people, can enable employees to generate the motivation of knowledge sharing,and ultimately realize the knowledge sharing between individuals in the organization.

Compared with information technology, why incentive mechanisms are more capable of promoting knowledge sharing within the organization? The main reason is the existence of information asymmetry in the process of knowledge sharing].Managers can not directly monitor the level of individual share effort in the process of knowledge sharing. Just because of the difficulty of monitoring the degree of individual share effort, it is to raise the speculative opportunity of the individual. Therefore it is needed to design the incentive mechanism from the perspective of organization to heighten the degree of individual share effort in order to solve the problems caused by information asymmetry.

As a form of knowledge sharing in organizations, knowledge sharing in master-apprentice pattern is the process of the tacit knowledge transferring from the master to the apprentice. The master generally refers to the senior staff with the long work experience in the organization. And the apprentice generally refers to the junior staff with a lack of work experience in the organization [8]. At present, the knowledge sharing in master-apprentice pattern is still the main form of knowledge transferring, especially tacit knowledge transferring in most organizations in China.Compared with the explicit knowledge, the tacit knowledge is more invisible [9].The employee needs to pay greater efforts in the process of tacit knowledge sharing than that in the process of explicit knowledge sharing. Therefore, organizations need to design special incentive mechanism for tacit knowledge sharing in master-apprentice pattern.

2.2 Research methods of incentive mechanisms

At present, the research on the incentive mechanism for knowledge sharing in the organization mainly focuses on three methods. These are the game theory, knowledge market theory and the principal-agent theory.

Although the game theory is of great significance in understanding the strategic relationship between the knowledge provider and the knowledge receiver in the organization. But the research hypothesis of the game theory, which is that each participating subject has the same sharing cost and sharing benefit, is not in conformity with the actual situation of the organization. Because there are different individuals in the organization, and different individuals with different private knowledge have different sharing costs and benefits in the process of knowledge sharing. For example, in the process of tacit knowledge sharing in master-apprentice pattern, Whether the sharing cost or the sharing benefit, that of the master with the long work experience is obviously much higher than that of the apprentice with a lack of work experience.

Despite knowledge market theory is helpful to understand which knowledge of the organization can be used in the knowledge transaction, and the interests of the two parties in the process of knowledge transaction.But the premise of the knowledge market theory is that the knowledge is a kind of tangible goods, and the process of knowledge transaction can be directly monitored by the organization,which is clearly not consistent with the characteristics of information asymmetry in the organization. In other words, the organization is not directly monitoring the knowledge sharing behavior of the individual.

The principal-agent theory is generally used to analyze the information asymmetry between the two economic entities in Economics [10].Although the organization manager as the principal is difficult to directly monitor the degree of individual effort as the agent.But the degree of individual effort can be affected by the incentive mechanism in the organization.Also, according to incentive mechanism in the organization, the employee as the agent can maximize his benefits through the choice of different degree of individual effort [11].Therefore, in order to research the incentive mechanism for knowledge sharing in master-apprentice pattern under the condition of asymmetric information,compared with the game theory and the knowledge market theory, the principal-agent theory is more suitable because of the explicit consideration of the non-observable behavior of the agent to solve the incentive problem.

In addition, the knowledge sharing in master-apprentice pattern is more practical knowledge sharing.On the basis of the division of the explicit knowledge and the tacit knowledge of Nonaka (1994) [12],Blair(2002) divided the tacit knowledge into two categories. One is the inexpressible tacit knowledge; the other is the expressible knowledge [13]. It is obvious that there are the inexpressible tacit knowledge and the expressible tacit knowledge during the process of knowledge sharing in master-apprentice pattern.But, during these two kinds of tacit knowledge sharing process, individuals pay a different level of shared effort due to the different tacit knowledge types. To this end, the organization needs to design different incentive mechanisms according to different types of tacit knowledge sharing in master-apprentice pattern. So, on the basis of existing research, this paper will discuss the incentive mechanism for tacit knowledge sharing in master-apprentice pattern by taking into account the different types of tacit knowledge based on the principal-agent theory.

3 The principal-agent model

There are three main bodies during the process of tacit knowledge sharing in master-apprentice pattern, the organization, the master as the knowledge provider and the apprentice as the knowledge recipient. Based on the principal-agent theory, the organization is the principal. The master and the apprentice are agents.The task of the principal is to design a reasonable incentive mechanism according to the observed information, and to encourage agents to choose the beneficial behavior strategy for the principal.

3.1 Output of tacit knowledge sharing in master-apprentice pattern

The process of tacit knowledge sharing in master-apprentice pattern is actually the process of knowledge transferring from the master to the apprentice. Obviously, the efforts of the master or the apprentice to choose the cooperative behavior ,such as knowledge sharing, is much higher than that to choose the non-cooperative behavior, such as knowledge hoarding or knowledge rejecting. So we define the degree of shared effort paid by the master in the choice of knowledge sharing strategy as mh and define that paid by the master in the choice of knowledge hoarding strategy as ml.Likewise, we define the degree of shared effort paid by the apprentice in the choice of knowledge sharing strategy as ah, and define that paid by the apprentice in the choice of knowledge rejecting strategy as al. It is obviously that mh>ml and ah>al.

Moreover, for the organization, only when the master and the apprentice choose the cooperative behavior, the tacit knowledge of the master can be translated into the intellectual capital of the organization by virtue of the absorption and reuse of the apprentice.Therefore, from the perspective of the output of tacit knowledge sharing in master-apprentice pattern, we assume that the total output of tacit knowledge sharing in master-apprentice pattern is a linear function of the master's knowledge sharing output and the apprentice's knowledge sharing output. Furthermore the master's knowledge sharing output is a linear function of the master's knowledge sharing efforts. And the apprentice's knowledge sharing output is a linear function of the apprentice's knowledge sharing efforts. The total output of tacit knowledge sharing in master-apprentice pattern can be expressed as $\prod(m,a)=\prod(m)+\prod(a)$

When the master pays a higher degree of sharing effort, that is "mh", the master's knowledge sharing output has two results, one is the higher output of knowledge sharing, that is $\prod_h(mh)$, and the other is the lower output of knowledge sharing, that is $\prod_l(mh)$. That is the same with the apprentice's efforts.

3.2 Cost of tacit knowledge sharing in master-apprentice pattern

In the process of tacit knowledge sharing in master-apprentice pattern, whether the master or the apprentice needs to pay a certain cost of knowledge sharing. C(m) is the knowledge sharing cost of the master. C(a) is the knowledge sharing cost of the apprentice. The master or the apprentice will obtain different sharing costs in different levels of sharing efforts. Obviously, the share cost in the high degree of sharing effort must be greater than that in the low level of shared effort.That is C(mh)>C(ml) and C(ah)>C(al).

3.3 Optimal reward incentive

The principal needs to provide the agent with a certain reward in order to encourage the agent to pay a higher degree of sharing effort in the process of knowledge sharing. The reward of the principal should reflect the difference of the agent's reward in consideration of the different degree of sharing efforts.In the process of tacit knowledge sharing in master-apprentice pattern, the reward of the master is $R(m)$. And the reward of the apprentice is $R(a)$. However because the organization as the principal is unable to monitor directly the agent's sharing efforts. The organization provides the agent with the corresponding rewards according to the sharing output.So the reward of the agent is the function of the sharing output. That is $R(\prod(m,a))=R(\prod(m))+R(\prod(a))$

$R(\prod(m))$ is the reward of the master. $R(\prod(a))$ is the reward of the apprentice. Because the reward of the agent is the function of the sharing output, the reward of the high sharing output is higher than that of the low sharing output.That is $R_h(\prod_h(m))>R_l(\prod_l(m))$, $R_h(\prod_h(a))>R_l(\prod_l(a))$.

3.4 Expected utility

Assuming that both the principal and the agent are risk neutral, the expected utility of the organization, the master and the apprentice is equal to the respective expected return. The expected utility of the organization is
$\prod(m,a)$—R(m)—R(a)= $\prod(m)+\prod(a)$—R(m)—R(a)

The expected utility of the master is R(m)—C(m).

The expected utility of the apprentice is R(a)—C(a).

We assume that when the master or the apprentice chooses the cooperative behavior and pays a higher degree of sharing efforts, P is the possibility that the master or the apprentice creates the higher knowledge sharing output. And (1-P) is the possibility that the master or the apprentice creates the lower knowledge sharing output. When the master or the apprentice chooses the non-cooperative behavior and pays a lower degree of sharing efforts, Q is the possibility that the master or the apprentice creates the higher knowledge sharing output. And (1-Q) is the possibility that the master or the apprentice creates the lower knowledge sharing output. It is obviously that P≥Q. In addition, the organization as the principal expects the master and the apprentice to choose the cooperative behavior in the process of knowledge sharing in master-apprentice pattern. Only when the master and the apprentice choose the cooperative behavior, the organization can achieve the maximum expected utility.That is

$$\text{Max}\ \begin{matrix} P\times[\prod_h(mh)+\prod_h(ah)-R_h(\prod_h(m))-R_h(\prod_h(a))] \\ +(1-p)[\prod_l(mh)+\prod_l(ah)-R_l(\prod_l(m))-R_l(\prod_l(a))] \end{matrix}$$

3.5 The basic incentive model

Based on the principal-agent theory, the maximum expected utility of the organization as the principal is constrained by the individual rationality constraint and the incentive compatibility constraint of the agent. The individual rational constraint is that the maximum expected utility obtained by the agent from the acceptance of the incentive contract can not be less than that from the non-acceptance of incentive contract.The maximum expected utility obtained by the agent from the non-acceptance of the incentive contract is general represented by $\overline{U}$ [14].The incentive compatibility constraint is that when the agent chooses the cooperative behavior in accordance with the way the principle is expected to,the best efforts of the agent is the satisfactory efforts for the principal, and is to maximize the benefits of the agent. The realization of the maximum benefit of the principal can be realized through the maximum utility of the agent.The incentive mechanism of the principal should ensure that the benefits obtained by the employee from the cooperative behavior are greater than that of the non-cooperative behavior [15].To this end, the basic incentive model of the organization can be expressed as follows.

$$\text{Max } P\times[\Pi_h(mh)+\Pi_h(ah)-R_h(\Pi_h(m)) - R_h(\Pi_h(a))]+(1-p)[\Pi_l(mh)+\Pi_l(ah) - R_l(\Pi_l(m))-R_l(\Pi_l(a))] \quad \text{S.T.}$$

$$P\times R_h(\Pi_h(m))+(1-p)R_l(\Pi_l(m)) - C(mh)\ge\overline{U_m} \tag{1}$$

$$P\times R_h(\Pi_h(a))+(1-p)R_l(\Pi_l(a)) - C(ah)\ge\overline{U_a} \tag{2}$$

$$P\times R_h(\Pi_h(m))+(1-p)\times R_l(\Pi_l(m)) - C(mh)\ge Q\times R_h(\Pi_h(m))+ (1-Q)\times R_l(\Pi_l(m))-C(ml) \tag{3}$$

$$P\times R_h(\Pi_h(a))+(1-p)\times R_l(\Pi_l(a)) - C(ah)\ge Q\times R_h(\Pi_h(a))+ (1-Q)\times R_l(\Pi_l(a))-C(al) \tag{4}$$

Formula (1) and (2) is called the individual rationality constraint of the master and the apprentice (IR). Formula (3) and (4) is called the incentive compatibility constraint of the master and the apprentice (IC). In addition, taking into account the tacit knowledge sharing in master-apprentice pattern, the types of tacit knowledge are ordered by the inexpressible tacit knowledge and the expressible tacit knowledge.

4 Incentive mechanisms

4.1 The inexpressible tacit knowledge sharing in master-apprentice pattern

The main feature inexpressible tacit knowledge is the highest degree of obscure.The knowledge owner is not even aware of the existence of this kind of tacit knowledge in the working process, so it is difficult to use the system language to express it.For this type of tacit knowledge, regardless of whether the master will pay a high degree of sharing efforts or low level of sharing efforts, the organization can only monitor a lower sharing output. The probability of producing a high sharing output is almost zero.That is P=0 and Q=0. So when the master provides the tacit knowledge which is difficult to express, the incentive model can be expressed as the incentive mechanism for the master.That is

$$\text{Max } \Pi_l(mh)-R_l(\Pi_l(m))$$

$$R_l(\prod_l(m)) - C(mh) \geq \overline{U_m} \tag{5}$$

$$R_l(\prod_l(m)) - C(mh) \geq R_l(\prod_l(m)) - C(ml) \tag{6}$$

We can see from formula (6) that when the master is supported by the same incentive mechanism, the sharing cost in the high degree of sharing effort is not more than that in the low level of shared effort. That is C(*mh*)≤C(*ml*), which is inconsistent with C(*mh*)>C(*ml*) as mentioned above. So we can draw the following conclusions.

When the master shares the inexpressible tacit knowledge, the organization does not need to design any incentive mechanism for the master.The reason is that when the master shares the inexpressible tacit knowledge, regardless of the master to pay a high or low level of sharing efforts, the shared output of the organization is always little, and even zero.The incentive mechanism is designed in the light of the sharing output, rather than the degree of sharing efforts. So in the case of the little sharing output or zero sharing output, the organization will not spend a certain amount of costs to motivate the master to share tacit knowledge. Moreover in the case of lack of organizational reward incentive, the master will give up inexpressible tacit knowledge sharing in the face of higher sharing costs naturally.

4.2 The expressible tacit knowledge sharing in master-apprentice pattern

The expressible tacit knowledge is between the inexpressible tacit knowledge and the explicit knowledge.Although the expressible tacit knowledge has certain obscure too, but can be translated into explicit knowledge through hard work and shared with other people. When the expressible tacit knowledge is shared between the master and the apprentice, there is the higher uncertainty between the degree of sharing efforts and the sharing output. For example, the master or the apprentice with the high degree of sharing efforts may produce a high sharing output or a low sharing output. And the master or the apprentice with the low degree of sharing efforts may produce a high sharing output or a low sharing output. Therefore the values of P and Q are varied, and are not determined. From the foregoing, we can see that 0<P<1 and 0<Q<1. In the process of expressible tacit knowledge sharing in master-apprentice pattern, the sharing output is not only related to the sharing output of the master, but also related to the sharing output of the apprentice. Therefore, the incentive model for expressible tacit knowledge sharing in master-apprentice pattern can be expressed as follows.

$$\text{Max} \begin{array}{l} P\times[\prod_h(mh)+\prod_h(ah)-R_h(\prod_h(m))-R_h(\prod_h(a))] \\ +(1-p)[\prod_l(mh)+\prod_l(ah)-R_l(\prod_l(m))-R_l(\prod_l(a))] \end{array}$$

$$P\times R_h(\prod_h(m))+(1-p)R_l(\prod_l(m))-C(mh)\geq \overline{U_m} \tag{7}$$

$$P\times R_h(\prod_h(a))+(1-p)R_l(\prod_l(a))-C(ah)\geq \overline{U_a} \tag{8}$$

$$\begin{array}{l} P\times R_h(\prod_h(m))+(1-p)\times R_l(\prod_l(m))-C(mh) \\ \geq Q\times R_h(\prod_h(m))+(1-Q)\times R_l(\prod_l(m))-C(ml) \end{array} \tag{9}$$

$$\begin{array}{l} P\times R_h(\prod_h(a))+(1-p)\times R_l(\prod_l(a))-C(ah) \\ \geq Q\times R_h(\prod_h(a))+(1-Q)\times R_l(\prod_l(a))-C(al) \end{array} \tag{10}$$

In consideration of the objective function, the organization always expects to obtain the maximum expected profit by paying the less reward for employees. Therefore, in order to meet the optimal of the objective function, individual rationality constraints and incentive compatibility constraints of the master and the apprentice should be binding. In other words, when formula (7), formula (8), formula (9) and formula (10) is the equation, the expected profit of the organization is biggest.

Firstly we analyze the constraints of the master. Formula (7) and formula (9) is transformed into the equation as follows.

$$P\times R_h(\prod_h(m))+(1-p)R_l(\prod_l(m))-C(mh)=\overline{U_m} \quad (11)$$

$$P\times R_h(\prod_h(m))+(1-p)\times R_l(\prod_l(m))-C(mh) \geq Q\times R_h(\prod_h(m))+(1-Q)\times R_l(\prod_l(m))-C(ml) \quad (12)$$

The solutions of equation (11) and equation (12) are as follows.

$$R_h(\prod_h(m))=\overline{U_m}+\frac{(1-Q)\times C(mh)-(1-P)\times C(ml))}{P-Q} \quad (13)$$

$$R_l(\prod_l(m))=\overline{U_m}+\frac{P\times C(ml)-Q\times C(mh)}{P-Q} \quad (14)$$

The solutions of equation (13) minus equation (14) are as follows.

$$R_h(\prod_h(m))-R_l(\prod_l(m))=\frac{C(mh)-C(ml)}{P-Q} \quad (15)$$

From equation (13) , it is seen that the reward of the master is not only related to the sharing cost, but also related to the values of P and Q under different shared outputs.As the foregoing, because of C(mh)>C(ml) and P>Q, we can infer $R_h(\prod_h(m))>R_l(\prod_l(m))$ from equation (15), which shows that the more sharing cost of the master, the more sharing output can be created, and the organization will provide more rewards to make up the sharing cost of the master.

In addition, it can also be seen from the formula (13) and type (14) that with the increasing value of P, the possibility of creating the high sharing output by paying a high degree of the master's sharing efforts is increasing. When the value of P is close to one, the master can create a high sharing output by paying a high degree of sharing efforts only.So we can assume that Q=0, and P=1 and Q=0 are taken in equation (13) and equation (14) respectively.

$$R_h(\prod_h(m)=\overline{U_m}+C(mh) \quad (16)$$

$$R_l(\prod_l(m)=\overline{U_m}+C(ml) \quad (17)$$

It is seen from equation (16) and equation (17) that there is the positive correlation between the master's rewards provided by the organization and the master's sharing costs. In other words, the higher the master's sharing costs, the greater rewards provided by the organization.It is obviously consistent with the idea of more pay for more work in China and is also most advantageous to the organization.

With the increasing value of Q, the possibility of creating the higher sharing output by paying a lower degree of the master's sharing efforts is increasing. When the value of Q is close to one, the master can create a high sharing output by paying a low degree of sharing efforts only. As time goes on, the master will give up the sharing efforts, which will lead to the lower value of P. So we can assume that Q=1, and P=0 and Q=1 are taken in equation (13) and equation (14) respectively.

$$R_h(\prod_h(m)=\overline{U_m}+C(ml) \quad (18)$$

$$R_l(\prod_l(m)=\overline{U_m}+C(mh) \quad (19)$$

It is seen from equation (18) and equation (19) that there is the negative correlation between the master's rewards provided by the organization and the master's sharing costs. In other words, the higher the master's sharing costs, the lesser rewards provided by the organization. And the lower the master's sharing costs, the greater rewards provided by the organization. So the master will give up the high sharing costs in order to gain the high rewards provided by the organization, which is clearly consistent with the idea of gain without doing any work in China and is unfavourable to the organization.

Analysis of the apprentice's constraints is the same with the master,so we will not repeat .

It can be seen from.the analysis of master's constraints and the apprentice's constraints that the rewards of the master and the apprentice provided by the organization are related to the sharing costs of agents, but also to the value of P and Q. Only when P>Q,incentive mechanisms of the organization are consistent with the idea of more pay for more work in China and are also most advantageous to the organization.In other words, incentive mechanisms of the organization should allow agents to deem that the higher sharing output can be created only by expending a higher degree of sharing efforts, and then the organization will provide higher rewards for the agent's higher output.

5 Conclusion

It can be found through the aforementioned that in the process of tacit knowledge sharing in master-apprentice pattern, any organization may not have one general incentive mechanism, which can be applicable to all processes of knowledge sharing in master-apprentice pattern.

One reason is that tacit knowledge in master-apprentice pattern can be divided into the inexpressible tacit knowledge and the expressible tacit knowledge. The organization needs design different incentive mechanisms according to the different types of tacit knowledge in master-apprentice pattern. For example, the organization does not need to design any incentive mechanism for the inexpressible tacit knowledge sharing in master-apprentice pattern.But for the expressible tacit knowledge sharing in master-apprentice pattern, incentive mechanisms of the organization need to be divided into incentive mechanisms for the master and incentive mechanisms for the apprentice. Furthermore the design of incentive mechanisms is related to the sharing cost of the master and the apprentice and the value of P and Q. Only when P>Q,incentive mechanisms of the organization are most advantageous to the organization.

The other reason is that there are also different types of tacit knowledge in different departments within the organization . For example, although there is the expressible tacit knowledge in R & D department, marketing department, financial department and so on. But there are different sharing costs and the different estimated value of P and Q in different departments. Because incentive mechanisms are not only related to the sharing cost, but also the value of P and Q. So in order to design incentive mechanisms for tacit knowledge sharing in master-apprentice pattern within different departments, first of all, the organization should estimate the sharing costs of different departments, and the value of P and Q value, and then design corresponding incentive mechanism for different departments. Therefore how to estimate reasonably the sharing cost, and the value of P and Q value, is the key to the design of incentive mechanisms for tacit knowledge sharing in master-apprentice pattern.

Since the organization does not expend the greater rewards for the master and the apprentice with the lower sharing costs. So how to reduce the sharing cost of the master and the apprentice is one of the keys to design incentive mechanisms of the organization.For the master, the focus of the organization is on the mining and re-organizing of the master's tacit knowledge, and on making the master's tacit knowledge transformed into explicit knowledge, which is contributing to the knowledge absorption of the apprentice.Moreover the organization should pay attention to the training of the master's skills in imparting tacit knowledge. Even if it is necessary, the organization should set up the special training department, whose aim is how to impart tacit knowledge of the master. Therefore, owing to the master having had the imparting tacit knowledge during the process of tacit knowledge sharing in master-apprentice pattern, sharing costs of the master and the apprentice are greatly reduced.

For the apprentice, the organization can improve the apprentice's ability to learn by educating and training in terms of learning methodology for the apprentice. For example, the organization can provide lectures with the different content and demonstration about different positions for the apprentice in order to reduce the apprentice's sharing cost.

As for the estimated value P and Q, the organization should take advantage of a large amount of information about sharing outputs, sharing costs and rewards and so on in the past of tacit knowledge sharing in master-apprentice pattern with the help of information technology. How to use the past information to estimate accurately sharing costs and the value of P and Q is another problem that needs to be studied.

Acknowledgment

This work was supported by a grant from Key scientific research projects of Higher Education of Henan province (NO. 16A630033).

References

1. Nassuora, A.B. Knowledge sharing in institutions of higher learning. International Journal of Economics and Management Sciences, 1,3,29-36 (2011).
2. Kaiji L.*Principles and applications of knowledge management*.Beijing: Tsinghua University press,(2010) .
3. Lawson, B., Petersen, K., Cousins, P., Handfield, R.Knowledge sharing in Inter-organizational product development teams: the effect of formal and informal socialization mechanisms, Journal of Product Innovation Management, 26,2,156-172 (2009).
4. van den Hooff, B. ,de Ridder, J.A. Knowledge sharing in context: the influence of organizational commitment, communication climate, and CMC use on knowledge sharing, Journal of Knowledge Management, 8,6,117-130 (2004).
5. Blomkvist, K.Knowledge management in MNCs: the importance of subsidiary transfer performance. Journal of Knowledge Management, 16,6,904-918 (2012).
6. Sheng Wang & Raymond A.Knowledge sharing: A review and directions for future research. Human Resource Management Review,2,115-131(2010).
7. Kimbal F & Simon K. Effects of Employee Governance and Operational Control on Psychological Ownership and Perceived Justice. New Zealand Journal of Psychology, 41,3,13-20 (2012).
8. Ning Nan. A principal-agent model for incentive design in knowledge sharing. Journal of Knowledge Management. 12,3,101-113 (2008).
9. Xi S, Nan L & Xinduo F. Evaluation on the effectiveness of knowledge sharing and knowledge transferring in master-apprentice pattern. Information Studies: Theory and Application, 36,7,76-80 (2013).
10. Xianyin L & Wanming C. Evolutionary Game of Knowledge Sharing in Organization. Statistics and Decision,12,170-173 (2008).
11. Mas-Colell, A., Whinston, M.D. & Green, J.R.. *Microeconomic Theory*. New York, Oxford University Press (1995).
12. Nonaka, I. A dynamic theory of organizational knowledge creation. Organization Science,.5 ,1, 14-37 (1994).
13. Blair, D.C.Knowledge management: hype, hope or help?. Journal of the American Society for Information Science and Technology.53,12,19-28 (2002).
14. Keying G,Wei G & Xiaohua S..Research of Motivate Mode on Knowledge Sharing in Knowledge-based Organization. Journal of Information. 9,42-44 (2008).
15. Zhizhu L, Changqing D & Guanzhong G. Research on incentive mechanism of enterprise knowledge sharing based on the principal agent theory. Science and Technology Management Research. 10,209-211(2008).

The Governance Mechanism of Knowledge Sharing Hostility within E-business Enterprise

Liang Zhao[1], Shuhui Fan[1] and Ruihua Wang[1]

[1]Management Engineering Department, Zhengzhou University, Zhengzhou, China
Corresponding Email: fsh0209@163.com

Abstract. Organizational knowledge sharing is the key to improve the core competitiveness and production efficiency. How to govern the behaviour of knowledge sharing and promote knowledge sharing has become the focus of organizational knowledge governance. Under the exploration of the relationship between knowledge governance, knowledge sharing hostility and knowledge sharing within the Internet organization and empirical analysis, we draw the following conclusions: formal and informal knowledge governance have a negative and weak impact on the reasons of knowledge hoarding and rejecting and the attitude toward the failure knowledge; The reasons of knowledge hoarding and rejecting have a significant and negative impact on knowledge sharing, but passive attitudes towards failure knowledge have a positive influence on knowledge sharing; The reasons of knowledge rejecting and passive attitudes towards failure knowledge exist partial mediating effect between formal and informal knowledge governance and knowledge sharing, but the reasons of the knowledge hoarding does not exist.

Keywords. Knowledge governance, knowledge sharing hostility, knowledge sharing, mediating effect.

1 Introduction

Organizations can integrate internal and external knowledge by knowledge sharing within the organization, which can help enterprises to find the leakage of the original knowledge system, help companies to get out of the governance blind area, form a new knowledge system and produce more innovative ideas [1]. Knowledge, as a kind of intangible assets, has two attributes of "individual property" and "public goods". In the process of knowledge sharing, there are two opposing rights: proprietary right of knowledge and knowledge sharing [2]. It is the mutual restriction of these two kinds of rights that make the social dilemma when knowledge sharing is carried out. From the perspective of enterprise, knowledge sharing is a kind of collective behaviour, and employees should select organization's common interests instead of personal interests [3]. Similarly, some people contend that knowledge is intangible private property. When the organization takes the heterogeneous knowledge as an important indicator to evaluate the value of staffs, the staffs will repel knowledge sharing in order to preserve their own competitive advantage, characterized by knowledge hoarding, rejection behaviour and attitude towards the errors of knowledge transformation, etc. Snejina (2003) called this kind of behaviour of not willing to share knowledge as knowledge sharing hostility. Studies show that the behaviour of knowledge sharing hostility is common in the organization. Enterprises will take the corresponding knowledge governance mechanism to break the phenomenon of knowledge sharing hostility in order to enhance the overall core competitive advantage.

E-commerce is a kind of business activity that using information technology to optimize of enterprise business activities continuously, and e-commerce enterprises are those organizations who engaged in e-commerce activities. E-commerce enterprises are high-tech industries and knowledge intensive enterprises, whose operations cannot be separated from the knowledge sharing between employees. Luo (2010) pointed out that the core elements of the high-tech industry are the staff education, experience and professional quality and so on, exclusively and fuzzily [4]. It is these characteristics that cause the enterprise's dilemma of knowledge sharing. In previous researches on knowledge sharing, few domestic scholars have studied the hostility of knowledge sharing, while the research on the governance mechanism of knowledge sharing hostility in the e-commerce enterprise is more scarce. This paper studies the governance mechanism of knowledge sharing hostility in the e-commerce enterprise based on the social dilemma theory and the theory of knowledge governance mechanism (KGM).

2 Social dilemma theory

2.1 Social dilemma

In the knowledge-intensive enterprises, the heterogeneous knowledge is an important source of core competitiveness of enterprises, and the heterogeneous knowledge grasped by employees is a manifestation of their competitive advantage. Heng-Li Yang (2008) found that employees are not willing to share their knowledge in order to maintain their position in the organization, but the organization is bound to taking certain measures to promote knowledge sharing from the perspective of overall interest, which causes governance dilemma between the individual and organizational, namely "social dilemma" [5]. Wang (2015) pointed out that the main problem of knowledge governance lies in the knowledge sharing hostility among employees [6]. There are three main aspects of knowledge sharing behaviour: knowledge hoarding, knowledge rejecting and negative attitude towards mistakes. Knowledge Hoarding refers to the behaviour of employees unwilling to share their knowledge with others; knowledge rejecting refers to the behaviour of employees unwilling to accept others' knowledge in knowledge sharing; false knowledge refers to the mistakes staff made in the work or what they have learned from the failure knowledge, and their attitude towards mistakes will be different because of different management system within the organization and personal different subjective feelings. Diverse personal interest of employees, differentiation of educational background and knowledge structure, different jobs will affect knowledge sharing behaviour the organization. The greater the knowledge distance is between employees, the more difficult they acquire knowledge from each other, the less likely it is to occur knowledge sharing behaviour [7].

2.1.1 Reasons of knowledge hoarding and knowledge sharing

There are two main reasons for the knowledge hoarding: individual reasons and the organizational environment. From personal point of view, employees are afraid of losing their competitive advantages after knowledge sharing, who want to retain their absolute control right of knowledge and reserve their own personal competitive advantage [8]; in addition, the staff think they can attain nothing in knowledge sharing. If they spend their time and energy on their work rather than on knowledge sharing, their capacity will be further improved and more conducive to their promotion [8]. In terms of organizational environment, when sharing knowledge, employees are in a hostile environment, which means that colleagues regard their knowledge with suspicion. If so, knowledge sharing and cannot achieve the desired results and will produce exactly the opposite result [8]. Therefore, employees do not want to share their knowledge in such an environment.

Reasons of knowledge hoarding in e-business enterprise: One is related to the organizational atmosphere. Unfriendly and unfamiliar colleagues tend to choose knowledge hoarding because they don't have enough trust [9]. Second, they worry that their knowledge is false. They will mislead colleagues if sharing false knowledge.

Thus, we propose:

H1: Reasons of knowledge hoarding have a negative effect on knowledge sharing.

2.1.2 Reasons of knowledge rejecting and knowledge sharing

Reasons of knowledge rejecting also has two aspects: individual and organization. From a personal point of view, first, knowledge receivers may not trust the quality of others' knowledge and they have a fear that external knowledge will bring them trouble. Another reason may be personal and professional pride means that some argue that they have grasped enough knowledge so there is no need to accept others' knowledge. From the organizational standpoint, corporations refuse to learn from competitors, lack of trust in other enterprise and lack of a sense of belonging for external knowledge [10].

In addition to the reasons mentioned above, another important reason of knowledge rejecting in e-business enterprise is that demanding of knowledge areas is different. The knowledge provided by the other party is not their own needs, then knowledge sharing the two sides will repel each other the knowledge passed. If the knowledge that the opposite party offers is not what they want, both of the knowledge receiver and knowledge provider will reject knowledge sharing.

Therefore, we hypothesize that:

H2: Reasons of knowledge rejecting have a negative effect on knowledge sharing.

2.1.3 Passive attitudes toward false knowledge and knowledge sharing

When it comes to mistakes, most people always hide them instead of sharing. Their own psychological perception is one of the reasons. They are afraid of that others look at them strangely. Another explanation is that they worry their leaders blame them or deduct wages and bonus etc. In organizations characterized by strong knowledge sharing hostility, making mistakes has always been banned [11]. Managers do not attach enough importance to encouraging staff to considerate mistakes and give feedback, but just focus on "who is to blame", which makes employees more afraid to tell colleagues their own mistakes, which is not conducive to knowledge sharing. As a high-tech industry, the work in e-commerce enterprise is demanding, and management measures are also relatively harsh, which causes employees to understand mistakes biased to a certain extent, let alone share false knowledge. This leads us to hypothesize that:

H3: Passive attitudes toward false knowledge have a negative effect on knowledge sharing.

2.2 Theory of knowledge governance

2.2.1 Formal knowledge governance and informal knowledge governance

Grandori (1997) was the first person to put forward the concept of knowledge governance, he believes that knowledge governance is to use the organization system to promote intellectual activity and management knowledge exchange, transfer and sharing Inside and outside the organization [12]. Based on structure view, process view and system view, Wang (2013) studied the model and evolution process of organizational knowledge governance [13]. Li (2007) studied the dimensions of knowledge governance from the perspective of transaction cost economics [14]. From the cognitive perspective, Ren (2007) divided knowledge governance into formal and informal knowledge governance [15]. Referring to Ren's classification of knowledge governance, this paper divides knowledge governance into formal and informal knowledge governance. Ren (2007), meanwhile, pointed out that knowledge governance is that organization utilizes the formal system design to influence the informal activities within the organization so as to choose, create, share and use knowledge optimally [15]. Xiang and Cao (2014) found that formal knowledge governance has a guiding role on informal knowledge governance through the empirical research [16]. Zhao and Liu (2015), found that both formal knowledge governance and informal knowledge governance have a positive effect on the creativity of the organization through empirical research and formal knowledge governance is in a dominant position [1]. Thus:

H4: Formal knowledge governance has a positive effect on informal knowledge governance.

2.2.2 Knowledge governance and reasons of knowledge sharing hostility

Formal knowledge governance means that organizations use organizational design, material incentives to govern the behaviour of knowledge sharing.

Based on the theory of social exchange, employees pay more attention to the value gained in the knowledge sharing activities. So, material motivation, as a means of formal knowledge governance, can alleviate the knowledge sharing hostility among employees, and effectively promote the knowledge sharing behaviour. Jin (2011) pointed out that incentives can make employees feel that organization encourages knowledge sharing, and when they get material rewards because of knowledge sharing, their willingness of knowledge sharing will be more intense, hence material incentives have a positive effect on individual subjective intention of knowledge sharing [17]. Zhao and Liu (2015) pointed out that the intellectual rewards can not only motivate employees to share knowledge, but also stimulate them to acquire new knowledge, improve the internal staff interact with external environment actively, access to external knowledge [1]. Husted Kenneth et al. (2012) pointed out that the transaction-based mechanisms can effectively weaken employees' negative attitude towards the mistakes they have made, and organizational incentives can promote the employees to participate in the knowledge sharing actively [3]. The organization design should include the design of performance evaluation system, and involve the contribution to the efficiency of others, which helps to reduce the occurrence of knowledge hoarding behaviours [18]. Wang (2015) pointed out that the development of education and training system in the organization design, help new employees to understand the distribution of knowledge within the organization, and reduce the suspicion of other people's knowledge, to promote their participation in knowledge sharing [6]. The organization design based on team training system can weaken the negative attitude of the employees to the false knowledge, so as to promote the knowledge sharing [3].

Therefor the following hypotheses are proposed:

H5a: Formal knowledge governance has a negative effect on the reasons of knowledge hoarding.

H5b: Formal knowledge governance has a negative effect on the reasons of knowledge hoarding.

H5c: Formal knowledge governance has a negative effect on the negative attitude towards false knowledge.

H6a: The reasons of knowledge hoarding have a mediating effect between formal knowledge governance and knowledge sharing.

H6b: The reasons of knowledge rejecting have a mediating effect between formal knowledge governance and knowledge sharing.

H6c: The negative attitude towards false knowledge has a mediating effect between formal knowledge governance and knowledge sharing.

Informal knowledge governance is to use organizational culture and trust interfere knowledge sharing behaviour among employees.

Organizational culture can help employees to realize that individuals and organizations are a whole and can be integrated into the organization as soon as possible, and the culture of unity can enhance the trust among employees [19], which can diminish knowledge hoarding and knowledge rejecting and promote knowledge sharing. Organizational culture, based on trust, reduce staffs' doubts of other people's knowledge to a certain extent [6], and promote them to accept the knowledge shared by colleagues with pleasure. Wang et al. (2013) found that if employees feel that they are in a friendly, fair atmosphere in which knowledge sharing is encouraged and supported, they will be more actively involved in the knowledge sharing [9]. Zhang and Zhu (2012) pointed out that colleagues' attitudes towards knowledge sharing will directly affect the willingness of knowledge sharing of both sides Good interpersonal relationships among colleagues will promote knowledge sharing among employees [20]. Whether managers pay attention to the exchange of the false knowledge between employees, will directly affect the attitude to the wrong knowledge is positive [8]. Managers' creating a relaxed atmosphere of the organization will help employees to weaken the negative attitude to the mistakes they have made, and promote knowledge sharing.

On the basis of these arguments, the following hypotheses are advanced:

H7a: Informal knowledge governance has a negative effect on the reasons of knowledge hoarding.

H7b: Informal knowledge governance has a negative effect on the reasons of knowledge hoarding.

H7c: Informal knowledge governance has a negative effect on the negative attitude towards false knowledge.

H8a: The reasons of knowledge hoarding have a mediating effect between Informal knowledge governance and knowledge sharing.

H8b: The reasons of knowledge rejecting have a mediating effect between Informal knowledge governance and knowledge sharing.

H8c: The negative attitude towards false knowledge has a mediating effect between Informal knowledge governance and knowledge sharing.

According to the above assumptions, the following theoretical model is proposed. FKG stands for formal knowledge governance; IFKG represents informal knowledge governance; RKH means reasons of knowledge hoarding; RKJ denotes reasons of knowledge rejecting; AFK signifies negative attitude towards false knowledge; KS expresses knowledge sharing:

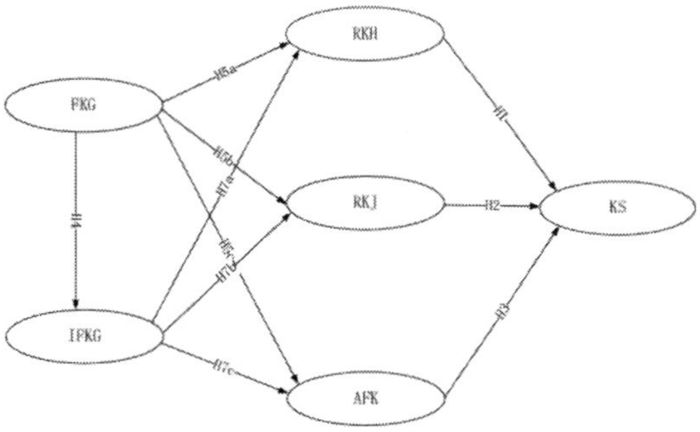

Fig. 1. Research model.

3 Research design and hypothesis testing

3.1 Data source

In this study, we use the method of questionnaire survey to collect data, and research the e-commerce enterprises in Henan province. We first conduct research for 10 individuals in each enterprise before the formal questionnaire survey, and then we amend the questionnaire according to their recommendations and research needs. The research questionnaire was sent to 200 members, with 185 respondents were returned, of which 177 questionnaires were valid. Basic statistics are shown in Table 1. EB means educational background; WL stands for work life:

Table 1. Basic statistics.

title	category	number	percentage
sex	male	61	34.5%
	female	116	65.5%
age	Under 25	80	45.2%
	26-35	94	53.1%
	36-45	3	1.7%
EB	Junior College	65	36.7%
	Undergraduate	101	57.1%
	Master's degree and above	11	6.2%

WL	Less than a year	88	49.7%
	1-3 years	66	37.3%
	3-6 years	20	11.3%
	7 years and above	3	1.7%

3.2 Measurement of variables

In this study, we measured 5 latent variables: formal knowledge governance, informal knowledge governance, reasons of knowledge hoarding, reasons of knowledge rejecting, negative attitude towards false knowledge. Measurement of knowledge governance draws Cao's (2012) measurement scale of knowledge governance and items of reasons of knowledge hoarding, reasons of knowledge rejecting, negative attitude towards false knowledge refer to Kenneth Husted et al. measurement to the knowledge sharing hostility. All items were measured by using a five-point Likert scale (1 = strongly disagree, 5= strongly agree).

(1) Reliability Analysis

In this paper, we test the internal consistency of all variables by using Cronbach alpha value in reliability analysis. We used SPSS to analyse relevant data of questionnaire survey. The Cronbach 's alpha coefficient was 0.795 and the standard of the Cronbach s' alpha coefficient was 0.797, both of which were larger than the critical value of 0.7, which shows that the statistical results of the questionnaire have a good reliability.

(2) Validity Analysis

Validity analysis is based on the reliability analysis of the data obtained from the questionnaire to ensure the stability and consistency of the research results. The study of measurement indicators can reflect the degree of the problem characteristics to be measured. We performed the factor analysis of the data. The KMO value was 0.859 and the significant level of Bartlett ball type test is 0.000. According to the above research, the data of this study has good reliability and validity, meeting the requirements of research.

3.3 Empirical methods and results

The main variables for statistical analysis and correlation analysis descriptive results are shown in Table 2. M means Mean value and SD stands for standard deviation. As can be seen from the data in Table 2, the correlation between variables is well.

Table 2. The main variables for statistical analysis and correlation analysis.

	M	SD	1	2	3	4	5	6
FKG	18.25	3.453	1	0.762**	-0.174*	-0.242**	0.325**	0.467**
IFKG	22.19	4.401	0.762**	1	-0.215**	-0.263**	0.233**	0.502**
RKH	13.31	3.966	-0.174*	-0.215*	1	0.679**	-0.003	-0.191*
RKJ	5.96	2.082	-0.242**	-0.263**	0.679**	1	0.030	-0.299**
AFK	10.45	1.898	0.325**	0.233**	-0.003	0.030	1	-0.251**
KS	15.06	2.497	0.467**	0.502**	-0.191*	-0.299**	-0.251**	1

notes: N=177; Significant level: **Express P<0.01; *Express P<0.05

(1) The empirical study of social dilemma theory

We used the method of linear regression analysis to verify the impact of reasons for knowledge sharing hostility on knowledge sharing. The results are shown in Table 3.

Model 1 tests the relationship between reasons of knowledge hoarding and knowledge sharing. The results of analysis indicate that the reasons of knowledge hoarding have a significant, negative and weaken effect on knowledge sharing (β=-0.136, P=0.004), so H1 is supported by the data.

Model 2 tests the relationship between reasons of knowledge rejecting and knowledge sharing. The results of analysis indicate that the reasons of knowledge hoarding have a significant, negative and weaken effect on knowledge sharing (β=-0.401, P=0.000), so H2 has been verified.

Model 3 tests the relationship between negative attitude towards false knowledge and knowledge sharing. Contrary to our hypothesis, the negative attitude towards false knowledge has a positive effect on knowledge sharing (β=0.333, P=0.001), so H3 has not been verified.

Table 3. Regression analysis of reasons of knowledge sharing hostility and knowledge sharing.

variable	KS		
	Model 1	Model 2	Model 3

sex	-0.526	-0.586	-0.388
age	0.261	0.148	-0.063
EB	-0.085	-0.008	-0.037
WL	0.568*	0.632*	0.660*
OP	0.182	0.230	0.135
RKH	-0.136**		
RKJ		-0.401**	
AFK			0.333**
R^2	0.093	0.155	0.111
Adjust R^2	0.061	0.125	0.079
F value	2.922	5.193	3.526

notes：N=177；Significant level：**Express P<0.01；* Express P<0.05

(2) The empirical study on knowledge governance theory

Table 4 presents the results of the regression analysis of knowledge governance theory.

Model 1 tests the relationship between formal knowledge governance and informal knowledge governance, the results show that formal knowledge governance has a significant positive effect on informal knowledge governance (β=0.971, P=0.000). H4 is supported by the data.

The regression results of model 2 and model 4 suggest that the formal knowledge governance has a significant and negative effect on reasons of knowledge hoarding and rejecting (β=-0.213, P=0.014; β=-0.153, P=0.001) which indicates that H5a and H5b have been verified. The regression results of model 6 show that formal knowledge governance has a significant and positive effect on attitude towards false knowledge, assuming that H5c has not been verified.

The regression results of model 3 and model 5 suggest that the informal knowledge governance has a significant and negative effect on reasons of knowledge hoarding and rejecting (β=-0.197, P=0.003; β=-0.125, P=0.000) which indicates that H7a and H7b have been verified. The regression results of model 6 show that informal knowledge governance has a significant and positive effect on attitude towards false knowledge, assuming that H7c has not been verified.

Table 4. Regression analysis of theory of knowledge governance.

variable	IFKG	RKH		RKJ		AFK	
	Model 1	Model2	Model 3	Model 4	Model 5	Model 6	Model 7
sex	0.058	-1.191	-1.178	-0.559	-0.549	-0.479	-0.492
age	0.729	-0.350	-0.210	-0.384	-0.300	0.318	0.266
EB	0.396	-0.120	-0.035	0.120	0.181	0.208	0.139
WL	-0.380	0.420	0.134	0.252	0.197	-0.266	-0.207
OP	0.008	0.446*	0.445*	0.281*	0.278*	0.087	0.098
FKG	0.971**	-0.213*		-0.153**		0.181**	
IFKG			-0.197**		-0.125**		0.096**
R^2	0.593	0.069	0.083	0.113	0.119	0.140	0.082
Adjust R^2	0.579	0.037	0.051	0.082	0.088	0.110	0.050
F value	41.311	2.116	2.564	3.623	3.830	4.625	2.546

notes：N=177；Significant level：**Express P<0.01；*Express P<0.05

(3) The empirical study on the mediating effect of reasons of knowledge sharing hostility

Table 5 shows the mediating effect of reasons of knowledge sharing hostility.

the results of table4 show that formal knowledge governance has a significant positive effect on informal knowledge governance; formal and informal knowledge governance have a significant impact on the reasons of knowledge hoarding and rejecting and negative attitude towards false knowledge; the test of mediating effect table 5 model 3 and model 4, while adding formal and informal knowledge governance (independent variables) and reasons of knowledge hoarding, reasons of knowledge rejecting and negative attitude towards false knowledge (mediator) to predict knowledge sharing (dependent variable). Reasons of knowledge rejecting and negative attitude towards false knowledge have a significant effect knowledge sharing. Moreover, the effect of formal and informal knowledge governance is significant and coefficients are reduced, which indicates that the reasons of knowledge rejecting and passive attitudes towards failure knowledge exist partial mediating effect between formal and informal knowledge government and knowledge sharing. On the contrary, the reasons of rejecting have no mediating effect between formal and informal knowledge governance and knowledge sharing.

Table 5. Regression analysis the mediating effect of reasons of knowledge sharing hostility.

variable	KS		KS	
	Model 1	Model 2	Model 3	Model 4
sex	-0.339	-0.36	-0.406	-0.386
age	0.247	0.053	0.070	-0.094
EB	0.041	-0.09	0.042	-0.063
WL	0.463	0.583	0.590*	0.687**
OP	0.082	0.087	0.148	0.138
FKG	0.327**		0.248**	
IFKG		0.281**		0.227**
RKH			0.018	0.032
RKJ			-0.319**	-0.315**
AFK			0.189*	0.221*
R^2	0.251	0.291	0.361	0.358
Adjust R^2	0.224	0.266	0.279	0.324
F value	9.477	11.647	8.558	10.366

notes：N=177；Significant level：**Express P<0.01；*Express P<0.05

4 Discussions and limitations

4.1 Discussions

This paper discusses governance mechanism of knowledge sharing hostility in the e-commerce enterprise. The research results are as follows:

(1) The effect of reasons of knowledge sharing hostility on knowledge sharing

In e-commerce enterprises, the analysis results in Table 3 show that the reasons of knowledge hoarding and rejecting have a significant and negative effect on knowledge sharing, but negative attitude towards false knowledge has a significant and positive effect on knowledge sharing. When analysing three factors' influence on knowledge sharing, we found that the working life has a significant positive effect on knowledge sharing(β=0.568，P=0.031；β=0.632，P=0.013；β=0.660，P=0.018). Thus, those employees who are experienced and senior are more willing to share knowledge. Therefore, during the knowledge governance, organization can establish an interdependent system to enable older employees to help new employees, and to find the reasons for knowledge sharing hostility among new employees so as to promote knowledge sharing through psychological counselling for them. In addition, the three model's goodness of fittest is not good (R2=0.093; R2=0.155; R2=0.111). One possible explanation is that there exist other factors in knowledge sharing, such as relationships among employees within organization etc.

(2) The effect of formal and informal knowledge governance on reasons knowledge sharing hostility

Through the analysis of the data, it verifies that formal knowledge governance has a significant positive effect on informal knowledge governance; formal and informal knowledge governance have a significant and passive impact on reasons of knowledge hoarding and rejecting. Contrary to our hypothesis, the effect of knowledge governance on the negative attitude towards the false knowledge is positive. This may be because the company's managers pursue the perfect and corporation's governance mechanism is too harsh, so employees fear to be punished because of making mistakes. Therefore, managers should fully consider the feelings of employees when making knowledge governance system, and they should also encourage employees to share the experience learning from the mistakes. In our study, we found that operating post has a significant positive effect on reasons of knowledge hoarding and knowledge rejecting from the analysis in Table 4. One possible explanation of the reasons may be that knowledge fields or background span is relatively large among different jobs, and employees may be not familiar with each other, which promote the behaviour of knowledge sharing hostility. Huang (2009) pointed out that the trust mechanism can enhance transactive memory and promote knowledge sharing. Therefore, organization can establish transactive memory system in knowledge governance mechanism in which employees can strengthen the exchange of knowledge and know more about what kind of knowledge colleagues grasp [21]. At work, if you do not know a piece of knowledge, you can ask your colleagues who are master of this filed to solve problems.

(3) The mediating effect of reasons of knowledge sharing hostility

Through the analysis of the data, the reasons of knowledge rejecting and the negative attitude towards the false knowledge have partial mediating effect between formal and informal knowledge governance and knowledge sharing. However, R2 value of 0.361 indicates that the goodness of fitness is not well. The possible reason is the limited number of our questionnaire, and the sample size is not big enough, so this result may be change after increasing sample size, and reasons of knowledge sharing hostility may exist partial or total mediating effect between formal and informal knowledge governance and knowledge sharing.

4.2 Limitations

(1) We didn't distinguish the same department or cross functional departments when studying the governance mechanism's impact on knowledge sharing hostility, only taking into account behaviours knowledge sharing in hostile in the same organization. While knowledge sharing is a dynamic process, the causes of the knowledge sharing between the same department and the cross sector may be different, and the further studies can be carried out in this area.

(2) The thought of Internet + and The Belt and Road make the development environment of e-commerce enterprises more uncertain. In this paper, we do not take into account the internal external environment for the organization, but any governance of an enterprise cannot be separated from the environment of the organization.

(3) In the research of knowledge sharing behaviour, social capital is also a more important point. In this paper, we didn't consider social capital, so the further research should examine the effect of social capital on reasons of knowledge sharing hostility.

Acknowledgement

This research was conducted with the support of key scientific research projects of institution of higher learning in Henan Province in 2016.

References

1. Zhao, X. and Liu, X.M., *The promoting mechanism of social knowledge governance on organizational creativity*, Science of Science and Management of S. & T., Vol. 36 No. 6, pp. 94-97, 2015.
2. Li,J., Modern Finance & Economics, *Research on the governance mechanism of knowledge sharing in large scale collaboration*, Vol. 30 No. 9, pp.69-73,(2010)
3. Kenneth H., Snejina M., Dana B.M. and Torben P., *Knowledge-sharing hostility and governance mechanisms: an empirical test*, Journal of Knowledge Management, Vol. 10 No. 5, pp.754-773, 2012.
4. Luo, M., Zhang, S.Y. and Liu, Y.J., *The research on new and high technology enterprise' s knowledge governance performance*,Science Research Management, Vol. 31 No. 3, pp.1-9, 2010.
5. Heng-Li Yang , Ted C.T. Wu. *Knowledge sharing in an organization*, Technological Forecasting & Social Change, No.75, pp.1128-1156,2008.
6. Wang, R.H. and Lv, Y.B., *Research on the governance mechanisms of mentoring knowledge sharing hostility*, Research on Library Science, No.21, pp. 96-101, 2015.
7. Chen, T., Wang, T.N. and Zhu, Z.M., *The distance of knowledge, environmental turbulence and the relationship of knowledge sharing in organization: an empirical research which contains moderating effect*, Studies in Science of Science, Vol. 31 No. 10, pp. 1532-1539,2013.
8. Snejina, M., Kenneth, H., *Knowledge-Sharing Hostility in Russian Firms*, California Management Review, Vol. 45 No.3, pp. 59-75, 2003.
9. Wang, S.H., Xu, B. and Peng, J.S., *The impact of organizational climate perception on employees innovation behavior: Based on the medium role of knowledge sharing intention*, Science Research Management, Vol.34 No.5, pp. 130-135, 2013.
10. Diaz-Diaz, N., Aguiar-Diaz, I. and De Saa-Perez, P., *the effect of technological knowledge assets on performance: the innovation in Spanish firms*, Research Policy, Vol.37 No.9, pp. 1515-1529, 2008.
11. Husted, K. and Michailova, S., *Diagnosing and fighting knowledge sharing hostility*, Organizational Dynamics, Vol.31 No.1, pp. 60-73,2002.
12. Grandori, A.,*Governance structures, coordination mechanisms and cognitive models*, Journal of Management and Governance, Vol.1 No.1, pp. 29-42, 1997.

13. Wang, Y., *Research on knowledge governance pattern of organization and its evolutionary process*, Forum on Science and Technology in China, No.3, pp. 102-108, 2013.
14. Li, Y., *New development of governance: latitude analysis of knowledge management*, Jiangxi Social Science, pp. 173-176,2007.
15. Ren, Z.A., *Go beyond knowledge management: Conceptions, framework and application of knowledge governance theory*, Science Research Management, Vol.28 No.1, pp. 20-26,2007.
16. Cao, Y. and Xiang, Y., *Study on the relationship among knowledge governance，knowledge sharing and employee innovation based on the mediating of social capital and the moderating of absorptive capacity in enterprises*, Studies in Science of Science, Vol.32 No.1, pp. 92-102, 2014.
17. Jin, H., Yang, Z. and Feng, F., *Rewards, perceived ownership of knowledge and organizational knowledge sharing incentive mechanism*, Studies in Science of Science, Vol.29 No.7, pp. 1036-1055, 2011.
18. Zhang, C.Z., Jiang, X.R. and Xu, H.B., *Research on the influence of organizational design on knowledge sharing*, Science & Technology Progress and Policy, Vol.30 No.3, pp. 128-133, 2013.
19. Jin, H., Yang, Z., Huang, Y.T. and Wu, J., E*xploring how organizational incentives and organizational culture affect knowledge sharing based on modified SIT*, Studies in Science of Science, Vol.31 No.11, pp. 1697-1707, 2013.
20. Zhang, X.D. and Zhu, M., *The impact of incentive，attitude toward colleagues，and personal culture on the knowledge sharing*, Science Research Management, Vol.33 No.10, pp. 97-105, 2012.
21. Chi-Cheng Huang. *Knowledge sharing and group cohesiveness on performance: An empirical study of technology R&D teams in Taiwan*, Technovation , No.29, pp.786-797, 2009.

Product Image to Build the Core Competitiveness of Enterprises

Yan Shi [1,*] and Laura Hjaltaso[1]

[1] Chengdu Vocational & Technical College of Industry, Chengdu, China
Corresponding Email: 510266833@qq.com

Abstract. As a new specialty research field, Product Image is being paid more attention. This article introduced the concept of Product Image and Core-competition, and illustrated how to conduct Core-competition of enterprise from two aspects of the foundation of Product Image and the communication of Product Image.

1 The concept of product image

Product identity (PI, product image, more translated into product identity) and chivalrous is refers to the product's appearance, including product geometric shape, color, material and man - machine interface, brand logo graphics, in addition to product packaging, display, advertising, logo, construction, marketing, and service and products related to the subject of the external factors [1].In a broad sense, the product image also includes the product of the enterprise concept, spirit, vision, culture and brand concept, passed to the consumer and the credibility of the establishment of the concept of content.

Product image and the overall image of the enterprise is closely linked, you can imagine, the merits of the product image will have a profound impact on the consumer's psychology. It can stimulate or inhibit the purchase desire of consumers, and then influence the establishment of brand image and enterprise image. Visible, product image design is also an effective means to reflect and maintain the brand image.

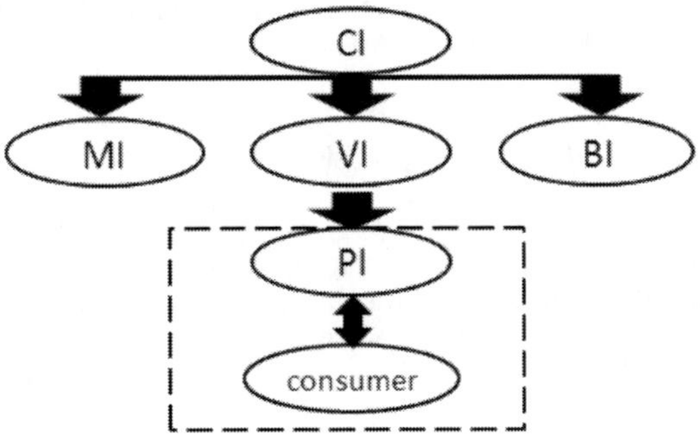

Fig. 1. The relationship of CI and PI.

As shown in Figure 1, the corporate image includes three parts, respectively is the concept of identification (MI), visual identification (VI), behavior recognition (BI) product image (PI) is a part of the corporate image (CI), but is the most important part, because it is the primary medium for communication between the enterprises and consumers. The recognition of the manufacturing enterprises and enterprises of all business activities are carried out around the product, an enterprise wants to continue to grow and develop, the advent of new products constantly bestselling, let enterprise products by consumers, continue to tap the potential of consumer, consolidate the existing consumer, establish a brand image, let enterprise image rooted in the minds of consumers, and product is the corporate image of the best physical carriers. Therefore, the image of the product naturally becomes the core part of corporate image [2].

2 The concept of core competence

Enterprise core competence (Core-Competition) has three basic characteristics: first, it has obvious competitive advantage, the two is to have the potential to expand the application, the three is difficult to imitate competitors. According to the core competence of enterprises, the core competence can be divided into three categories: core products, core technology and core competence. The three are closely related, the product comes from the technology, and the technology comes from the ability. Specific performance for the following aspects:

2.1 Innovative technology

Whether the enterprise has the innovation technology often has the decisive function to its development. Technological innovation, it is required to achieve the functionality of the product, uniqueness, as well as beyond the industry's average level of cutting-edge. This advantage of the technology, the enterprise will bring more than ordinary enterprise customer attention and market wide degree.

2.2 Talents with innovative ability

Even in the information age, the emergence of a variety of intelligent devices greatly reduced the requirements of human resources, but the ability to innovate is still rare in this era of wealth. Because of innovative technology, and ultimately have to be able to create talent to complete the development and design. Therefore, in an enterprise, innovation talent is always an enterprise can lead the industry trend of the most important factors; it is the necessary condition for the enterprise to build the core competitiveness.

2.3 Excellent corporate culture

Compared to the enterprise culture, also belong to the category of abstract consciousness, and some factors of production, the value of corporate culture is often is difficult to be judged, in spite of this, in the modern enterprise system, corporate culture status is is widely recognized and respected. This is because the cultural connotation of a business, the impact of the management of the enterprise, the level of the construction of personnel, such as the more specific aspects. At present, whether a company has excellent culture, is no longer the problem of internal staff attention, more and more consumers in the choice of products, will take into account the culture of a business. This is because a good cultural connotation of the enterprise, it will be in the social responsibility, quality and safety and other aspects of access to consumer trust, which is an important soft power of enterprise construction.

2.4 Brand influence

Brand is the product of market competition; more and more enterprises attach importance to brand strategy to create. In today's highly convergence of goods, consumers have been difficult to judge from the value of the level to determine what kind of product is to meet their needs; the use of value has become a lower level of demand. The brand is an important symbol of a company's products, which is different from other enterprise products. It is also a symbol of corporate culture, value and characteristic. In modern society, brand influence means that the degree of accumulation of wealth, with a wide range of influence, good reputation of the brand has a vital role in the development of enterprises. Brand building is a long accumulation of the road, but the destruction of brand is an overnight thing, so to build brand influence and need to enterprise long-term adherence [2].

Product image in a sense is the vision of the core competitiveness of enterprises, product image shaping helps to enhance the core competitiveness of enterprises. For example, Switzerland's Swatch watch, it is through the product image of the re design of a distinctive fashion watch culture, enhance the value and quality of products, and thus enhance the core competitiveness of Swatch company. In this case, the image of the watch product has become the core competitiveness of Swatch company.

3 The establishment of product image

Product image design is to design products as the core around people's demand on product, limits for individual participants demand, and changing people's way of life, enhance an image of the quality of life and level of creative activities. Today, people's sense of all kinds of products has been from the internal function and simple geometric shape change design; rise to the inside and outside of the unity of the image of the design. Product of the open, operation, touch, form and other image significance has become the focus of today's product design. For example, the pen holding and writing the image of the induction, the computer keyboard operation image induction, electric iron ironing image induction, etc., are entirely dependent on the external image to display.

The establishment of a good brand image depends on the product of both innovation and a continuation of the image of the design. This can be to understand the product image design: manufacturing enterprises toward the market many kinds of products because of its excellent industrial design innovation system of rigorous planning and gradually advancing, in the minds of the consumer market and establish the unified style and distinctive features of product image. For example, the famous ZIPPO lighter, its product design has long been beyond the shackles of the material function, and then evolved into a decorative culture. People buy not only can spark the function, more important is the visual image of products to buy it.

According to AIDA theory, people buy products when the decision-making process is generally divided into four steps: Attention - Interest - Desire - Action. From the above process, we can see that the first step "attention" is particularly important, because it directly led to the production of several steps behind [3].for the product, in order to get success must be able to attract the attention of the dog dog. Such as automobile rear-view mirror with flashing steering lamp, makes a chic ringing phone. All of these can become to attract the focus of consumer attention; all these attract the attention of consumers and make them interested in place are part of the image of the product. Consumers to buy the product decision-making process, in fact, is the product image from the cognitive to the identity, until the process of purchase, the successful product image design is a decisive factor in the purchase decision of consumers.

4 Product image transmission

4.1 Enterprise publicity

Enterprise reality and its characteristics are not perceived outside, can not form a corporate image. The formation of the corporate image must be based on the fixed communication channels and means of communication. As mentioned earlier, the spread of corporate image information points for direct communication and indirect communication, direct communication is certainly important, but in terms of the modern enterprise, the mass media to the audience a wide range, information propagation speed quick characteristics become the main channel for the dissemination of corporate image. The mass media can be divided into three forms, the print media, electronic media and outdoor media.

（1） print media

Print media include newspapers, magazines, books, and corporate information and print advertising, etc.. Among them, the newspaper's information dissemination speed is the quickest, the widest range, the greatest impact. The magazine and book information capacity is big, the reading cycle is long, but is affected by the publication cycle, the information dissemination speed is slow. Enterprise printing materials and printing ads targeted at a stronger, faster transmission of information, but the spread of smaller, it is difficult to have a greater impact. Enterprises can choose appropriate printing media according to their own needs [4].

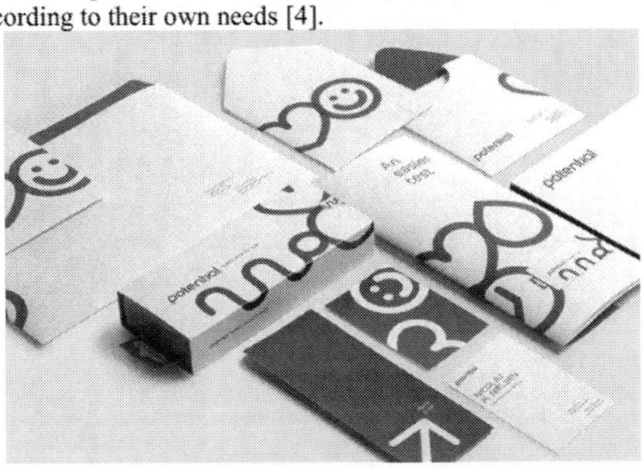

Fig. 2. The enterprise printing media.

（2） electronic media

Electronic media including TV, radio, film, video, DVD and Internet etc.. One of the most important mass media television, radio and the Internet, the rapid development of A new force suddenly rises. TV is the first choice for enterprises to transmit information. Its characteristics are vivid and acoustic painting, spread fast, wide coverage. Broadcast also has the characteristics of fast spread, wide coverage, has been the main form of mass communication, but with the popularity of television, its influence has been shrinking. But the broadcast media has the characteristics of low cost and fast, and it is still an important channel for information dissemination. Internet as the modern information society, the revolutionary mark, along with the rapid growth of the popularity and users of the home computer, the importance of encroaching upon the television broadcasting may also become the main channel for the dissemination of popular electronic [5].

Fig. 3. The enterprise electronic media.

(3) outdoor media

Outdoor media include outdoor public places such as buildings, traffic signs, traffic tools, outdoor electronic screen release advertisement information of various communication media. Outdoor media generally have magnificent and huge volume characteristics; can produce a strong visual impact. Enterprises through outdoor advertising media to promote their own, you can use the characteristics of long duration of outdoor advertising; the information will be repeatedly transmitted to the audience, so that it leaves a deep impression. But the outdoor media coverage is narrow, the information update is slow, so it can only be used as a supplementary means for corporate image communication, and other mass media, the use of organic unity, to complete the task of image communication.

Fig. 4. The enterprise outdoor media.

4.2 The image of the product itself

Everything in our daily life products are formed through form, color, material, function of visual image information to people that a specific information, people through the reading of the information to achieve the purpose of understanding and using products. From the perspective of semiotics, the external shape of the product is a series of visual communication symbols, in the production and life has become a specific form of expression of a certain meaning. Product modelling in line with the basic nature of the general symbol, in the structure can be divided into material carrier and emotional carrier. It is based on user's stimulation, stimulate their experience and their past life experiences or behavior associated with a Lenovo, so that consumers have a strong interest in the product, and then form a general impression of the use of the products, this impression had decided whether to continue to buy the same brand of goods, if any products comply with the people's living habits, convenient for people to use and satisfy the function of the product and psychological needs, so that the product will be favored by people, and has been generally recognized, in the course of time, will form a brand, the brand image will be deeply rooted in people's heart, which will not only bring huge the economic benefits to the enterprise, will become a kind of intangible assets, the value of intangible assets is.

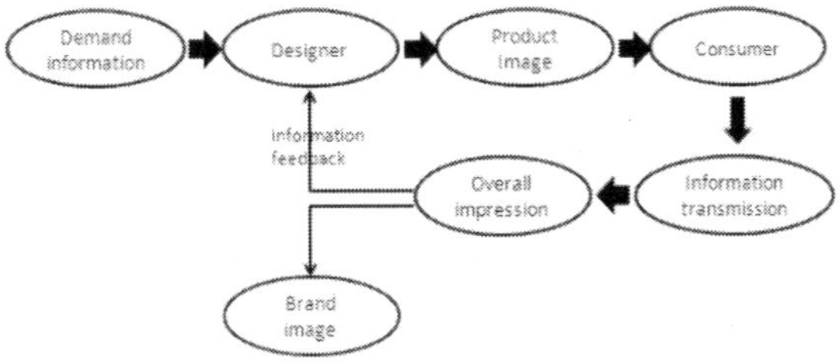

Fig. 5. The sketch map of product image communication.

5 The non - material development trend of product image

5.1 Product image is the formation of the spiritual wealth of the factors

Coca-Cola has succeeded in transforming a carbonated beverage product into a spirit of culture, promoting the spirit of the United States and the American culture, from tangible to intangible assets. Not only that, we from the world famous enterprises such as Microsoft, IBM, IT, Intel, apple...... Car industry's Mercedes Benz, BMW, Audi, general motors, TOYOTA......; them every day, every minute of every in the left and right and the control of our way of life and the way of thinking, the impact is already is more than stay in through the use of products to obtain the level of functional requirements, more is a kind of spiritual needs. Cannot be imagined, if one day these companies disappeared overnight, we do not know what life is! Can also constitute our present society?

Product image can be produced by the value of material value, but, more importantly, it can produce the spiritual value of the non - material side. In the process of sustainable development, green design, green economy, and the process of non-material globalization, it is a unique way to stimulate enterprises to consider the concept of image more seriously than before. Globalization forces the domestic market and the domestic economy to move towards global integration at a slow but steady pace. In separated by vast oceans of the global enterprise, globalization has given impetus to the unity of people's experience, wisdom, faith and feelings, these people do not know each other, they come from different countries, speak different languages, believe in different cultures. But these people are gathered together, competing in a land of endless challenges, and they are going to meet these challenges. "Who are we?" the question has never been as urgent as it is now. In fact, they come from the image; they have a business activity, as if they were in charge of their own destiny.

5.2 The image of the product as a factor of intangible assets

Product image has become an effective means to establish and maintain the credibility of products. On the one hand, product image as a tangible material function part, with form and function, to meet the basic requirements of people's material; on the other hand, product image as the invisible spirit of part, influence people's attitude towards life and value orientation.

In the foreseeable future, the establishment of product image reputation and value will be an important task of image economy; enterprises will have to use more energy to manage these intangible assets. Through the product image to establish in the minds of the consumer loyalty to the product, so that enterprises continue to create a lasting economic value.

As a result of the image, the strength of the product image presents a unique feature - image everywhere, nowhere to be seen, in our table, wardrobe, bedroom bed, window, and even the toilet....... We live, work, entertainment and so on, everywhere is the role of the force, it is no longer short or fleeting, long-term results, strength is almost endless, change the mode of life of individuals and businesses.

In the fierce competition in the modern society, between the enterprises technical difference is almost the same, the output of each enterprise products can be public recognition is key to the success of the enterprise, the relationship between products and consumers are most closely. Therefore, product image is crucial. From a certain meaning, the endorsement of the product image is the corporate image.

References

1. Wang Zhaoxia, Liang Hao, Tang Lin. Image design - the key to modern product design. Journal of Tianjin University of Science and Technology, 2004:.22-27
2. Liu Yun, who can continue innovation and product image design industry.IT.2003:35-38 decoration.
3. Zhang Linghao, Liu Gang. Visual design of product image. Nanjing: Southeast University press,.2005:46-55
4. Tao Yi. On the brand communication strategy of modern enterprises. Shandong economy, 2005.7
5. Yu Xianglong, Li Juan.VI design. Nanning: Guangxi Fine Arts Publishing House,.2003.6

Top-Level Design and Practical Applications of Wisdom Campus in Chinese University——A Case of Wisdom Beihang

Qingshan Xu[1,2,*], Jianhua Zhang[3], Lihua Yang[1] and Liu Zhang[1]

[1] College of Public Administration, Beihang University, Beijing, China
[2] Office Automation Technology Center of Gansu, Gansu, China
[3] Network Information Center, Beihang University, Beijing, China
Corresponding Email: xuqs@buaa.edu.cn

Abstract. Wisdom education is the direction of educational informatization development. Wisdom campus is the advanced stage of the development of educational informatization. The construction of wisdom campus is a complicated systematic project, requiring not only strengthening top-level design from the macro level, but also considering the specific operational implementation from the micro level. On the background and connotation of wisdom campus development, this paper reviews the domestic research literature, and analyses the top-level design of wisdom campus. The organization structure is the key; the business process is the main line; the institution standard system is the focus; the campus culture is the essence; the platform construction is the core; the public service is the target; the operation and maintenance is the support; the information security is the basis; the supervision and assessment is the guarantee. At the same time, combined with the practical application of wisdom campus construction of Beihang University, the One-card service, integrated services, One-stop service, wireless network, unified identity authentication, data center, and iHome comprehensive application platform provide a reference case for the top-level design and construction of wisdom campus.

1 Introduction

With the deepening of the continuous development of information technology and educational reform, the educational informatization in our country has experienced changes from traditional campus to the stage of digital campus, and then moves towards a new stage of construction of the wisdom campus. Educational informatization has been gradually showing characteristics of digital, network, platforms, intelligence, open, and social networking. Wisdom campus has become the theme and the trend of the current development of educational informatization. The informatization of university is the priority among priorities of practice and development in national educational informatization, being an important indicator of a country's education level, and furthermore an important manifestation of a country's comprehensive competitiveness.[1] How to improve the informatization and intelligence level in the education field, especially making the top-level design and practical application of the wisdom campus, and how to develop with wisdom medical service, transportation, community and city at the same pace are real problems and great challenges in the development of wisdom education.

There are few dissertations and core journal papers on the study of wisdom campus in China. As of December 2015, there are 123 papers on wisdom campus searched in the CNKI, among which 2 papers are doctoral dissertations, 60 papers are master's dissertations, and 61 papers are core journal in the Chinese Social Sciences Citation Index (CSSCI), as shown in Table.1. Overall, it shows a trend of rapid growth year by year.

Table.1 Annual statistics of wisdom campus paper.

Name	2010	2011	2012	2013	2014	2015	Sum
doctoral and master's dissertations	0	0	4	10	25	23	62

CSSCI journals	1	3	14	6	16	21	61
total	1	3	18	16	41	44	123

These papers can be divided into engineering technology category, education and management category, educational technology category and others category according to subject classification. According to this classification, there are 60 papers (49%) in the engineering technology category, 14 papers (11%) in the education and management category, 17 papers (14%) in the educational technology category, and 32 papers (26%) in others category.(see Figure 1) These analysis show that we have paid more attention to the technical attributes of wisdom campus in a long time, but not enough to the educational attributes and the top-level design. So the biggest obstacle is the top-level design and management mechanism, rather than the technology in the construction to promote the process of wisdom campus. Thus, we should put greater emphasis on researches about the design and mechanism of wisdom campus.

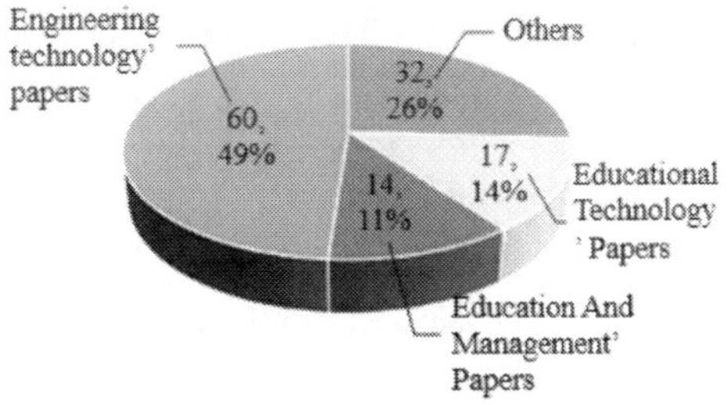

Fig. 1. Statistics in subject classification of wisdom campus.

2 The background of wisdom campus development

In 1970s, Massachusetts Institute of Technology firstly proposed the "Electronic Campus"(E-campus) program. In 1990, Kenneth C. Green, a professor of Claremont University, firstly proposed the concept of "Computing Campus", and presided over the Campus Computing Project(CCP) research projects, opening the pace of educational informatization construction in the world and providing a lot of digital campus experience for other countries [2]. Since the implementation of CCP, in the year from 1996 to 2010, the United States government has promulgated and implemented four national "Educational Technology Program". These programs adopts a joint model that the government, schools and enterprises work together. It enables every student to use the computer, every classroom to access the Internet, and every youth to receive higher education which is made through Internet. It achieves the connection of "Person, Machine, Road, Net" from the primary school to university. It completely changed the way, means and process of education and learning in university education in the United States. And then the U.S. educational informatization has always been in the leading position in the world. The U.S. has also esTablelished the world's most influential institution EDUCAUSE to lead the development direction of education informatization [3]. Information technology in American education has experienced a transformation process from the infrastructure construction to system application, to practical effect, and then to adjust the structure. It has always followed the strategic policy of "Information Technology to Promote Education Reform and Development". The development process of American educational informatization has been gradually developing from the electronic campus to digital campus, and then to the wisdom campus [4].

The Britain, Australia, New Zealand, and Japan, have also formulated and implemented a series of educational informatization promotion program [5]. For example, Britain has released the Super JANET and 2010-2012 Educational Development Strategy. Australia has the Building the Education Revolution(BER). New Zealand has the Information and Communication Technologies Professional Development(ICTPD) and An E-Learning A ction Plan for School(2006 -2010). Japan has e-Japan, u-Japan and i-Japan. In the European Union, the Comenius Program is implemented for the basic education, the Socrates Program for the higher education, the Leonardo Da Vinci Programme for vocational education. etc [6]. These programs and strategies provide a typical reference and experience for other countries regarding the transformation from digital campus to wisdom campus.

Educational informatization of university in China has been influenced by foreign factors, and also has experienced from the electronic campus to digital campus and then to the wisdom campus. The China Education Technology Network(CERNET) was officially opened in 1995, which not only achieved the interconnection between province level, but also laid the foundation for the esTablelishment of electronic campus. In 2004, the New CETNET(CERNET 2) using IPv6 was built [7]. Chinese has made strategy of The Education Promotion Motion Plan(2003-2007). On the one hand, it could accelerate the development of education information technology infrastructure construction and construction of information resource. On the other hand, it could promote the construction of digital campus in universities, using change traditional teaching methods and models to improve the information level of education. In 2006, the Chinese Ministry of Education set up the office of educational information, which was responsible for the top-level design and unified management of various types of education information construction. On July 29th 2010, Ministry of Education issued The Reform and Development Plan Outline of National Long-Term Education(2010-

2020), which has included education information into the national information development strategy [8]. These plans have accelerated the construction of digital campus, improved the educational information system, and promoted the modernization of educational content, means and methods. Zhejiang University was the first one which put forward the wisdom campus construction among domestic universities. Its goal is green and energy conservation, peace and coordination, scientific decision-making, and convenient service [9]. Nanjing University of Posts and Telecommunications, Tongji University, Tsinghua University, Beijing University and Beihang University have put forward the construction of wisdom campus.

3 The connotation of wisdom campus

So far, there is no accurate and scientific definition of "wisdom campus". Experts and scholars in different fields have different understanding of it. From the perspective of education, the principle of wisdom campus is to provide personalized service to students and teachers. Through the system, we can be fully aware of the physical environment, identify the learners' individual characteristics and learning situation, and use information technologies to provide an open learning environment, a comforTablele living environment and rich campus culture [10]. These are the starting point and destination of the wisdom environment, wisdom teaching and wisdom learning, which eventually aims at developing the wisdom of the students [11]. From the perspective of information technology, benefited from the development of new technology, wisdom campus is an open, interactive and collaborative integrated service platform featuring informationized and intelligent, with the technology of internet, cloud computing and internet of things, providing the intelligent service with numbers on the internet [12,13]. In this way, the majority of teachers and students are able to acquire comprehensive teaching information, effectively share data resources, making the teaching activities and common life more convenient [14,15].

To sum up, wisdom campus is a broad concept. It generally describes a system combining "technology, education, management, service and ideas".(see Figure 2) Wisdom campus is based on infrastructures of educational informatization, collecting and integrating all kinds of educational information resources with the use of cloud computing, network of things, big data, virtual technology, sensor technology and mobile communication technology of a new generation of information technology. It focuses on wisdom service to diversified customers, including students, teachers, administrative staffs, alumnus, parents, and the public. Guided by teaching, learning, researching, using, management and other business needs, we build a digital, networking, integrated, collaborative, amalgamation, comprehensive, personalized and intellectualization resources platform. We hold the beliefs of innovation, coordination, green, open and sharing, deeply combine technology, education, management, system and culture, to esTablelish a new integrated educational model capable of providing wisdom learning, teaching, scientific research, applications, social interacts, management, decision-making, services and life. It will be a new wisdom environment in the campus, which is convenient for learning, teaching, working and social life.

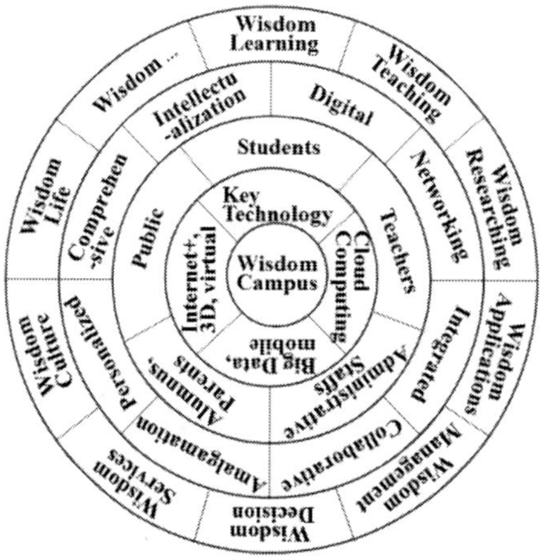

Fig. 2. Logical relation of the wisdom campus connotation.

4 The Top-Level Design of Wisdom Campus

As an engineering concept, the top-level design, with the system theory method, is to make an overall plan on a task or project, making considering of all the facts and elements. Thus, we can focus on the advantages of resources, to achieve the goal in an efficient way [16]. Top-level framework of the U.S. E-government Federal Enterprise Architecture(FEA) consists of five parts [17]: strategic business planning reference model(BRM), data standard reference model(DRM), technical framework reference model(TRM), service reference model (SRM), performance evaluation reference model(PRM).

Construction of the wisdom campus is not a simple information-engineering project, but a complex engineering system, combining information technology of the new generation with education, management, system and culture, which needs to strengthen top-level design from the overall situation. Wang Yukai, an information expert, believes that the top-level design is the specification of overall planning, a means of implementation of the overall plan [18]. The top-level design of wisdom campus comprehensively considers all aspects of environmental factors in technology, business, management, service, mechanism and culture as a whole, in order to make comprehensive and systematic planning and design. There should be coordination, matching, integration and support, between the various environment elements. The design sticks to the principle of "unified planning, intensive construction, unified deployment, resource sharing, unified implementation, and result oriented". It is a designing process combining the method of "from above to below" and the operational implementation of "from below to above", featuring top-level decision-making, operational, unified specification and overall coordination. Specifically, the overall structure of the top-level design of the wisdom campus includes nine systems that are organizational structure, business processes, institution standard, campus culture, platform construction, public service, operation and maintenance, information security, supervision and assessment.(see Figure 3)

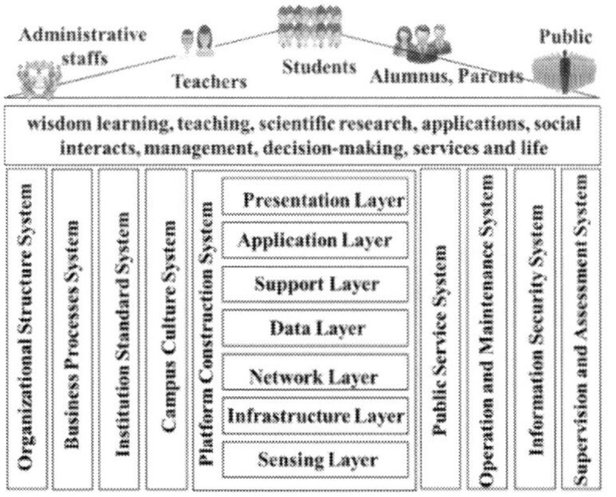

Fig. 3. Top-level design framework of the wisdom campus.

4.1 Organizational structure system is the key

Organizational structure system is the organizational structure, functional responsibilities and coordination mechanisms to promote the construction of wisdom campus, including systems of leadership decision-making, implementation and expert consultation. Leadership decision-making system is on the leadership and decision-making level, in charge of making decisions and coordination management of major issues. The implementation system is on the implementation level, responsible for the construction of wisdom campus construction and implementation of the specific work. Expert advisory system is on the auxiliary decision-making level, responsible for providing supplementary decision-making and professional support. The system of wisdom campus in Beihang University is made of the information management committee, executive team and the expert advisory committee composition(see Figure 4), which forms two levels of information management mechanism. The information management committee by the president, executive vice president and the leaders of each department is mainly responsible for the overall planning of information and the top-level design of the wisdom campus. It is in charge of the information technology management office (often set in the network information center), mainly responsible for the coordination and management work. The information implementation group is staffed by directors of the network information center and other departments, working on works about information and the building, implementation and application of the wisdom campus. Information expert advisory committee by the professor, senior engineers and external experts, provides auxiliary advices and solid support for the planning, construction and implementation of information issues.

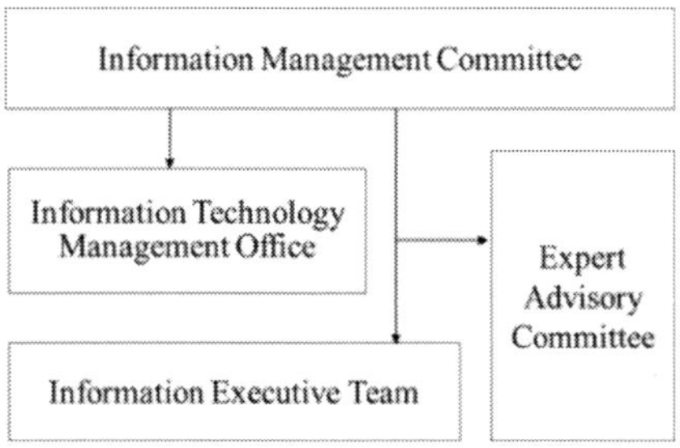

Fig. 4. The organizational structure system of wisdom campus in Beihang University.

4.2 Business process system is the main line

The process of business contains procedures of cooperation and undertaking between different departments with diversified business. It is transforming the known issue into events that is valuable to the customers.[19] The business process system is to simplify, refine, re-simplify, and re-define the business according to actual needs of each department, each job and each character. By constructing a wisdom business processing system, unified, scientific, standard and formal, satisfying needs and corresponding to reality, we can finally realize the reinventing of business process and reusing of information. The business process system of Beihang University is based on basic business needs of the university. It draws lessons from life cycle management of business process management (BPM) in enterprise resource planning (ERP) to define different kinds of business, set up models, make simulations, and launch the project, and then followed by monitoring, analyzing and optimizing it. Setting up examples based on BPMN2.0, the system ensures every unit of business and event to be quantified and standard, thus, it enables every part of the process to be optimized, instead of only part of the process.

4.3 Institution standard system is the key point

Institution standard system includes standards, systems and mechanisms of techniques, management and services. From the technical level, we need to make sure that the information collection, data processing, data exchange and data analyses are qualified in standards, and that the data structure and the data interface are suiTablele for each other. Therefore, it can satisfy the need of data exchange and lay a solid foundation for data integration and resource sharing. From the management level, we need to esTablelish regulations and systems of the project planning and design, engineering implementation management, information resource management, system processing management, information security management, performance assessment, in order to improve the intelligent level of management. From the service level, according to the actual business needs of teaching, learning, researching, using and management, we esTablelish a people-oriented wisdom service concept in order to upgrade the level of intelligence services. In terms of policy standards, the wisdom campus of Beihang University formulated management systems of informatization project, methods of operation and maintenance management, network and information security measures. We have assured the technical standards of the data structure, business architecture, technical architecture, and application architecture, compiled the technical manual and operation guidelines of different systems, as well as Chief Information Officer (CIO) system, personnel training system, performance appraisal system and so on.

4.4 Campus culture system is the essence

Campus culture is created in the long-term education practice process of a school, reflecting the school values, thinking methods and behavior norms, embodying the common belief and spiritual pursuit of all teachers and students. It is a spiritual force to hold all parts of the school together [20]. In the age of information technology rising up, human wisdom and environment interact with each other. Under the circumstances like that, wisdom culture inherited the traditional spirit of the campus and formed innovation culture with characteristics of network culture, reflecting advanced campus spirit and cohesion of "positive energy" of the era. As a subset of the construction of campus culture, the construction of wisdom campus is a like tool for the development of campus culture, and it is the only way to school education and culture [21]. Campus culture system mainly includes four aspects that are material culture, system culture, behavior culture and spiritual culture. The wisdom campus of Beihang University formed unique wisdom culture, such as the small "i" culture, the culture of the wisdom website, and the new media culture.

4.5 Platform construction system is the core

The platform construction system includes perception layer, infrastructure layer, network layer, data layer, support layer, application layer and presentation layer. (1)Sensing layer is to use sensors, RFID (Radio Frequency Identification), two-dimensional code, 3D printing, camera, etc., for the realization of the overall perception between "human - machine - materials". (2)Infrastructure layer includes server storage devices, network security equipment, terminal equipment, cloud platform equipment, computer rooms and teaching and office spaces, etc. (3)Composed of the campus wired network, wireless network, mobile network, Network layer refers to private cloud, public cloud, mixed cloud, and wisdom cloud. (4)Data layer includes all kinds of business daTablease, data warehouse, resource daTablease, information resource catalog, data backup and remote storage, etc. (5)The support layer includes components of shoring of foundations, service support, public support, and a unified support platform. (6)Application layer refers to all kinds of business application software system serving in teaching, scientific research, management, service and life. (7)The presentation layer is to provide personalized information service through the website, students or teachers' home page, micro-blog, WeChat and Weibo for teachers, students and the public. Based on the cloud platform and to meet all kinds of business needs, as in teaching, learning, researching, using and management, the wisdom campus of the Beihang University built corresponding software system platforms, such as systems for teaching, scientific research, finance, personnel, assets, logistics, office, school, enrollment, employment and iHome integrated application platform.

4.6 Public service system is the goal

Public service is an important measure to improve the overall teaching, scientific research, management and service level of colleges and universities, and it is also an important way to improve the comprehensive strength and core competitiveness of colleges and universities [22]. The public service system of the campus includes education resource services such as the library, online education, scientific research equipment, infrastructure, students' service, sports, talent recruiting, culture and information security. Constructing the public service system of the wisdom campus can effectively promote the education reform and the transformation of college management and service mode. Specifically, it is to transform services from face-to-face to online service, from the single service to the diversified, from the passive service to active, from facing the campus to the outside, from the school service to the social service. The public service system of wisdom Beihang University adopted the exploration of "1+1" mode and "The Belt and Road" mode, to unite the related colleges and universities. Also, we use wisdom campus platform to provide wisdom public services to the school and the society. Among the services are the books, scientific research, online education, information technology, talent, culture, etc.

4.7 The system of operation and maintenance is the support

Operation and maintenance system includes organizing personnel, processing, systems, supporting platform and management object, etc [23]. The organizational structure and personnel of operation and maintenance is to have a dedicated organizations and specialized team in accordance with the different professionals and different levels of security personnel. Operation and maintenance process is to develop a standardized workflow, to ensure that the operation and maintenance carried out in order and the operation problem to be solved. The system of operation and maintenance requires a series of management system, to ensure that the operation and maintenance is efficient and coordinated. The operation and maintenance platform refers to various software platforms that are applied in the work. The objects in the management consist construction of infrastructure, business application systems, daTablease, information resources and staff of operation and maintenance. The wisdom campus of Beihang University esTablelished the unified management system for the operation and maintenance personnel, technology, resources, processes and institutions, by way of centralized monitoring, collaborative linkage, responsible classification and standard service. It provides 24/7 services, providing a safe and reliable environment for the planning, construction, implementation and management of the wisdom campus.

4.8 Information security system is the foundation

Information security system consists of physical security, network security, data security, application security, content security, user security and safety management, etc [24]. Physical security includes safety equipment, medium security, backup and disaster recovery, electricity safety; network security including access security, risk assessment, anti-intrusion detection, anti- virus attacks, security rank protection, etc. Data security includes daTablease security, data encryption, digital authentication, digital signature, data audit and so on. Application security includes security monitoring and warning, emergency response and security policy planning, etc. Content security contains information filtering, privacy protection, classified information management; user security consists user authentication, access permissions. Safety management includes the safety management system, safety organizations and responsibilities, security emergency measures, etc. The wisdom campus of the Beihang University strengthens organization, management, technology and applications from multiple levels, with the use of the firewall, anti-virus, intrusion detection, unified authentication, network auditing, data encryption, access control, virtual private network (VPN) and other key technology to guarantee the information security.

4.9 The system of supervision and assessment is for guarantee

The evaluation index of information supervision in universities is an important tool to measure the construction, application and development of the wisdom campus. It has the characteristics of skopos, targeted, systematic, scientific, operational and dynamic [25]. The system of supervision and evaluation includes development strategy status, organization management and guarantee, infrastructure construction, information talent construction, information resources construction, information technology application, information security and so on. Combining with the actual characteristics of the university, the wisdom campus of Beihang University developed detailed assessment indicators that consist the top-level design, organization and management, infrastructure, information resource, data center, web application, application for the system, information security, information management, process management, talent construction and system construction. With information openness, resource sharing, online services, online evaluation, network supervision as means of information, supervision and evaluation, it enables us to achieve wisdom supervision and assessment of the campus.

5 Practical applications of wisdom campus

5.1 One-card service

One-card service is one of the most important applications of Wisdom Beihang, and it is the combined application of integration, networking, amalgamation and intelligentization. The background of Wisdom Beihang One-card system is a complex system composed of hardware system, software system, daTablease, data exchange platform, terminal, etc. It is the integration of cards, licenses, Tableles, bills and tickets (e.g. Student ID card, library card, dining card, bath card, electric card, water card, access card, attendance card, commuter card, medical card, payment card, consumer card, etc.). In addition, the multi-application of one card in certification, management, service, consumption and finance is completed so that "One card in hand, No worry in campus" is achieved. The application of One-card is throughout the whole life cycle and every aspect required by teachers and students from entering to leaving the school. By sharing various types of data to solve the information isolated island problem, using various types of resources to avoid repeated construction, integrating various tapes of business demands to improve the quality and efficiency of teaching, optimizing various types of process standards to improve management and service levels, the One-card provides teachers and students with convenient, efficient and satisfactory Intelligent service in study, work and life.

5.2 Integrated service

Integrated service is one of the most unique services in Wisdom Beihang. This service can be in the form of integrating virtual online service hall and Entity service hall. It has effectively helped teachers and students solve "shuttle run problem" and "circle run problem" during studying, working and living. Moreover, the quality of school integrated management and public service have been enhanced. Wisdom Beihang esTablelished service hall of teachers and students, logistics service hall and financial settlement center. Meanwhile, it also appointed various functional departments moving into service hall. These service halls provide services including the school seal, status management, graduate, abroad, employment, household, scientific research, personnel, accommodation, logistics, tuition and financial. The online service hall, on the other hand, provides integrated online services. A seamless transition between online and offline services is the combined embodiment of intelligent service and integrated Service of Wisdom Beihang.

5.3 One-stop service

Portal website is an important platform for universities to do information releasing, image promoting, brand displaying and service supplying. It is also an important "window" and "face" for first-class universities. By using advanced design concepts, sophisticated technology and a complete website management system, Wisdom Beihang has constructed a one-stop service website group management system based on cloud platform. This one-stop platform polymerizes various types of resources in the way of Web2.0, supports website group data ESB(Enterprise Service Bus) and distributed deployment. Meanwhile, this background system of one-stop platform mainly adopts CMS(Content Management System). One-stop service of Wisdom Beihang has not only realized portal site's intensive construction of various faculties and departments by unified standards, unified planning, unified framework, unified release, unified management, and unified maintenance, but it has also solved the problem of coordinating standards, capital, personnel, management, maintenance and safety in order to esTablelish a new management and service mode of website group with the characteristics of one-net and one-stop.

5.4 Wireless network

Wireless network of Wisdom Beihang construction mainly uses WiFi(Wireless Fidelity) and mobile internet technology. Combined with the wired network, it has achieved full coverage on teaching building, office buildings, laboratory buildings, dormitories, libraries, restaurants and other places in two campus (Beihang Xueyuan Road Campus and Shahe Campus).Wireless network realizes following functions: in the aspect of scale, it achieves the WiFi coverage of every corner of the campus; in terms of speed, it achieves the high speed and high quality transmission; in the view of the conformity, it achieves interconnection, intercommunication and sharing between wireless network and wired network; in the perspective of application, it enables wireless network to carry all kinds of business application system; in the aspect of using, it supports mobile PDA, Tablelet PCs, smart phones and other types of mobile terminal equipment. Wireless Network of Wisdom Beihang has broken through the boundaries of space, and provides teachers and students with a convenient, simple, personalized, safe, reliable, sTablele, efficient and lightweight mobile intelligent services.

5.5 Uniform identity authentication system (UIAS)

Based on the unified coding rules, unified technical standards and unified process specifications, Wisdom Beihang UIAS mainly uses SSO (Single Sign-On) and LDAP (Lightweight Directory Access Protocol), and its interface supports JAVA, C, C#, PHP and several other languages. By using unique user name and password to log on different business application system, the system provides users with the only electronic identity authentication from the interface layer, data layer application layer, and other layers, realizing the unified identity authentication and management based on 4A specification (Authentication, Account, Authorization, Audit). Furthermore, Wisdom Beihang UIAS, unified portal site and unified public data exchange platform have been put together to implement SSO and DS(Data Synchronization) in order to achieve the true meaning of the unified identity authentication.

5.6 Data center

Data center is the most importance in Wisdom Beihang, and it's also one of the important contents of Beihang University 13th Five-Year informatization planning. Wisdom Beihang uses cloud computing and virtualization technology to set up a virtual data center architecture. Taking advantage of this architecture, it can realize the hardware resources, network resources, and data storage resources in comprehensive integration. In addition, it can optimize the computing pool, storage pool, network pool, security pool, application pool, and resource pool in an all-sided way. Furthermore, the data center has implemented full virtualization, a high degree of integration and optimization of all kinds of resources sharing. The Wisdom Beihang Data Center, which consists of the storage infrastructure, data acquisition system, data exchange platform, intelligent analysis system, intelligent decision-making system, etc., processes, prepares, analyzes and mines deeply on multi-sources data (e.g. teaching, scientific research, experiment, archives, books, management, service, talent, international cooperation, and other multi-source data). Eventually, "speaking with data", "deciding by data", "managing by data" and "innovating by data" has been achieved, and data is used as a strategic resource for Wisdom Beihang to provide safe and reliable support.

5.7 iHome integrated application platform

Wisdom Beihang has constructed a comprehensive application platform "iHome" with the concept of "Internet+". "iHome+ Teaching" integrates online courses center, course record system, cloud video center, student educational system, teaching evaluation system and others to achieve the function of online education. "iHome+ Management" integrates student management system, personnel management system, financial management system, asset management system, online office system and others to achieve the coordination management functions on arranging human, financial, material, and other things. "iHome+ Service" integrates enrollment and employment system, household registration security system, medical treatment system, online payment system, logistics service system and scientific research reporting system to achieve online service functions. "iHome+ Community Spaces" integrates personal homepage of teachers and students, communicating space, BBS, academic forum, club activities to accomplish online community and online communication. iHome of Wisdom Beihang shows the personalization features of "Internet+ Education", and it reflects the important features of wisdom campus including digitalization, networking, integration, collaboration, amalgamation, integration, personalization, intellectualization etc.

6 Conclusion

Based on the new information technology, such as the cloud computing, Internet of things, big data and mobile Internet, and learning from the experience and outcomes of the construction of digital and wisdom campus in domestic and the world, the top-level design of it can be strengthed from multi-levels, including technique, business, management, standard, institution and security. At the same time, we are ought to make constant innovations in building many wisdom platforms of the colleges and universities, featuring digital, networking, integrated, collaborative, amalgamation, comprehensive, personalized and intellectualization, in order to make learning, teaching, scientific researching, applications, management, decision-making, services, and normal life more intelligent. All those above are the final goal of the construction of wisdom campus of colleges and universities, and the irresistible trend of educational informatization.

References

1. Y. L. Jiang, L. Chen. Enlightenment and Research Comparison between American and Chinese on University Education Informatization. J. E-Education Research, **118**, 234 (2012)
2. L. Li, Z. Li. Research on the Development and Driving Factors of Digital Campus in American Colleges and Universities. J. CEI, **17**, 24 (2009):
3. Q. L. Yu, X. D. Zheng. The Reality and Challenges of Eduational Informatizationin America's Higher Education. J. Open Education Research, **110**, 16 (2010)
4. J. H. Sui, Y. Wang, D. R. Yang. Analysis of the Development Trend of University Education Informatization at Home and Abroad. J. CEI, **5**, 24 (2008)
5. B. Q. Liu, S. H. Sun. Conception, Design and Implementation of Digital Campus. *University of Science Technology China Press* (2014)
6. Y. B. Hu, X. Li, Y. J. Zhao. State of European Union ICT in Education in K-12 Schools. J. CET, **123**, 349 (2016)

7. Q. T. Hu, K. Zheng, N. H. Lin. Development and Transformation of Educational Information. J. CET, **35**, 324 (2014)
8. M. Y. Gu. Study and Interpretation on State Planning Out Line for Medium and Long-Term Education Reform and Development (2010 -2020). J. Higher Education, **5**, 31 (2010)
9. D. M. Lu. Not Just A Concept on Smart Campus. J. CEN, **29**, 12 (2011)
10. H. R. Huang. J. B. Zhang, Y. B. Hu. Smart Campus: The Developing Trends of Digital Campus. J. OER, **12**, 18 (2012)
11. Z. T. Zhu, B. He. Wisdom Education: New Realm of Education Information. J. E-Education Research, **5**, 236 (2012)
12. X. M. Yang, S. Q. Yu. The Architecture and Key Support Technologies of Smart Education. J. CET, **77**, 336 (2015)
13. D. H.Yan, X. M. Chen. The Application of the Internet of Things in Digital Campus. J. MET, **125**, 21 (2011)
14. J. Shen. Y. X. Huang. A Preliminary Exploration on Smart Campus and its Construction. J. Fujian Institute of Education, **122**, 12 (2011)
15. Y. Bi. Smart Campus Need Resources and Applications. J. CEN, **31**, 13(2011)
16. J. He. Study on the Top-level Design and Implementation Measures of Smart City. J. Urban Development Studies, **72**, 20 (2013)
17. Z. Y. Lan, C. X. Sun. American Public Policy in Action. *Renmin University of China Press* (2007)
18. Y. K. Wang. Accurate Understanding of the Top-Level Design of Reform. J. LS, **20**, 36 (2012)
19. Y. R. Wang. Process revolution 2.0: Let the Strategic Management of the Landing Process. *Peking University press* (2011)
20. F. Qiao, L.Y. Fang. On the Construction of Campus Culture in Digital Campus. J. Chongqing Jiaotong University, **100**, 7 (2007)
21. Z. C. Song. Educational Culture Theory. J. ER, **4**, 33 (2012)
22. Y. J. Sun. Exploratory Thinking on the Universities Campus Public Service. J. HUE, **31**, 10 (2015)
23. Information Technology Service Center of ShangXi. Implementation and Service of E-government Public Platform. J. E-Government, **25**, 12 (2013)
24. Z. G. Zhang, Y. Tian. Exploration and Practice of the Digital Campus Information Security System. J. ETM, **114**, 29 (2012)
25. S. R. Pu, J. Q. Liu. A Probe into the Design and Estimation of Grade -3 Evaluation Index System of Informatization in Universities. J. SNU, **850**, 33 (2010)

The Integrated Design for Micro - environment Monitoring System of Showcase in Museum

Chan Dong[1], Yang Zhang[2], Jingjie Li[1] and Hui Zhang[2,*]

[1]Department of Instrument Intelligent Manufacturing Technology, Hefei University of Technology, Hefei, China
[2]Department of Instrument Science and electronic engineering, Hefei University of Technology, Hefei, China
Corresponding Email: zhanghui@hfut.edu.cn

Abstract. In order to improve the current environment quality of cultural relics in museum and make them preserved for a long time, the paper proposes the integrated design for micro - environment monitoring system of showcase in museum, the system mainly monitors the micro-environment of heritage, such as ultraviolet, light, formaldehyde, CO_2, TVOC, PM2.5, choosing cotex-M3 ARM microcontroller STM32F103ZET6 as the control core. Based on the concept of integration, the system integrates the single air monitoring instruments. The design of circuit mainly includes the process of digital power supply and analog power supply, the acquisition and processing of 5 analog signals from sensors, and the design of reserved interface. In the aspect of interaction, the serial port lcd module was uesd for the display and control, which can get rid of the control of PC and achieve the functions of environmental monitoring, environmental warning, environmental assessment, historical data query. The integrated design for Micro-environment of Showcase in Museum achieves a monitoring platform successfully which is easy for users' operation and access to display information conveniently . The advantages of the system are strong portability,low cost and short development cycle.
Keywords.STM32, micro-environmental monitoring, museum showcase, human-computer interaction.

1 Introduction

The 12th Five-Year plan for the development of the National Museum of Antiquities put forward that establishment of the support system of museum quality monitoring [1], environmental assessment and regulatory should be considered as a major research content. In recent years, the related industries are committed to the research of the micro-environmental monitoring of showcase [2]. As a symbol of national cultural soft power, the protection of cultural relics has always been a top priority in museum management, the air quality of the showcase is vital to the preservation of exhibits.

At present, the parameters monitored by the monitoring terminal on the market are relatively simple, which cannot achieve the all-round monitor. So the comprehensive micro-environmental monitoring system was designed for a showcase. This system integrates a plurality of various single-parameter air monitor, the monitoring of the four major polluting gases can be integrated, also include light and ultraviolet, the major four polluting gases include formaldehyde, CO_2, PM2.5 and TVOC, which have bad effects on the exhibits [3]. Meanwhile, the system comes with a good interactive interface, which is convenient for the users to get the information of the current micro-environment and deal with it in time.

2 Overview of museum micro- environment monitoring system

The museum micro-environment monitoring system which is used to monitor the environment of cultural relics in the showcase. The purpose is to preserve the cultural relics in a safe and applicable condition of exhibition, in order to maintain the overall appearance of the showcase; the whole system is hid in the bottom of the showcase.

To comprehensively improve the level of environmental monitoring, control of showcase and storehouse. Technical Research Institute of Machinery Industry Instrument and Meter Industrial Technology Research Institute held a meeting which is about the standard for protective equipment of cultural relics. Including 28 standards of protection and safety requirements of cultural relics prevention equipment [4], evaluation methods, performance monitoring, intelligence terminal, multifunctional mobile showcase platform. For monitoring terminal. Online monitoring content should include temperature and humidity, formaldehyde, particulate matter (PM10 and PM2.5), light and ultraviolet light, carbon dioxide, volatile substances [5], etc. and the lowest performance index of the product is established.

According to this standard, The integrated design of six kinds of parameters was realized, which is including ultraviolet, light and illumination, formaldehyde, PM2.5, TVOC and CO_2, considering the different adaptability of different cultural relics to the environment, the system can set the threshold, different cultural relics set different monitoring intervals.

In the design of the structure, the system adopts modular design, each module will be treated appropriately, in order to obtain higher precision. At the same time, the system comes with interactive interface, which have brought great convenience to the museum staff and visitors, the interactive interface focus on art design, strive to achieve consistent aesthetics with cultural relics.

3 Hardware design of monitoring system

3.1 Composition of monitoring system

Showcase micro-environment monitoring system block diagram was shown in Figure 1.

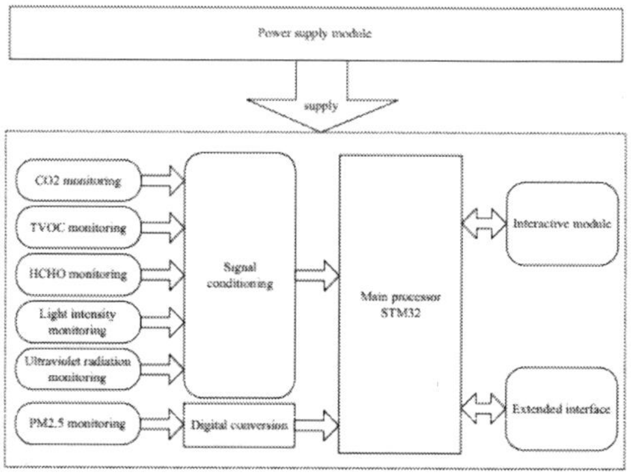

Fig. 1. Micro - environment monitoring system block diagram of showcase.

The system is based on the STM32F103ZET6 ARM cortex-M3 micro controller. The chip is rich in resources and fully meets the design requirements. That included power module, 5-way AD signal acquisition, signal conditioning, 1-way RS485 communication, interactive module, and reserved expansion interface.

The main power module provides the main processor, interactive modules, each sensor with their own required voltage.

AD signal conditioning mainly for the precision of the measurement requirements, a reasonable choice of internal AD or external AD for data collection [6].

Interactive module is the mainly based on 7-inch serial LCD and the main controller through RS232 communication to achieve a variety of interactive interface.

Taking into account higher air quality requirements of showcase micro environment, the resolution and precision of the air parameters must meet the requirements.Therefore, the system of the power module, the hardware design of the data acquisition part, the sensor output signal processing as well as the PCB has done the corresponding processing in order to obtain the higher measurement precision.

3.2 Power supply module

Stable and reliable power supply is an important part of system, at the same time; communication acquisition module requirements are relatively high not only on the system power supply stability but also anti-jamming performance. The system power supply block diagram was shown in Figure 2.

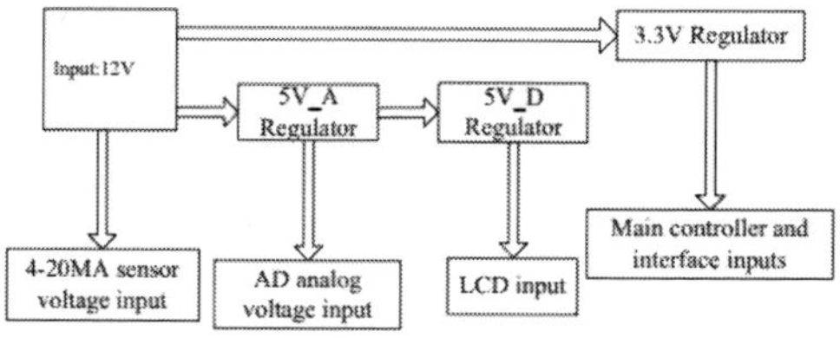

Fig. 2. Micro - environment monitoring system power management block diagram of showcase.

DOI: 10.1051/matecconf/201710005009

Taking the influence of power circuit on signal transmission into account, the power isolation was done in the part of mutual interference, the 12V DC power provide input for the light sensor, UV sensor. 12V will be converted to 5V_A through the three-terminal regulator LM2576. While the use of isolated DC-DC voltage module will convert 12V to 5V_D, supply power for the LCD. And then through the corresponding filter for external AD power supply. the 5V_D converted to DC 3.3V voltage through ASM1117-3.3 [7], provides voltage input for the main controller, interface, and so on.

In addition, the analog power and digital power through the 0 Ω resistor to connected GND and 3.3V respectively. Loop current can be suppressed, thereby suppressing the high-frequency noise of the digital ground [8]. When the PCB is covered with copper, separate the AD part from the rest that can reduce the interference of the circuit to AD.

3.3 Data acquisition module

The quality of the system was affected directly by Data collection accuracy. Therefore, In order to improve the accuracy of the monitoring system, the appropriate treatment was used for different types of sensors, also can meet the requirements for the indicators of the monitoring terminal, which was promulgated in the "cultural relics protection equipment standard assembly (a)" in 2016.

In the signal processing, light illumination sensor and ultraviolet sensor, which have a larger output current, the standard provides the measurement accuracy of the two sensor, respectively 10% and 5%, so the internal AD was choosed to meet the measurement requirements directly.

According to T / WWXT0015-2015 standard [9],the measurement range and accuracy of CO_2 monitoring terminal were {2000ppm, ± (60ppm ± 3%full range)}, the model of CO_2 sensor is COZIR-A2000 in the system, the range of measurement is 2000ppm, so that meet the standard requirements fully. The output voltage of CO_2 sensor was related to the input voltage, when the input voltage is 3.4V, the output is 3.4V, CO_2 measurement accuracy is higher, so external AD was used for data collection, but the circuit voltage divider was need at first, followed by AD processing.

The model of formaldehyde sensor is EC803, the measurement accuracy and measurement range are able to meet the T / WWXT0015-2015 standards, the input voltage is 5V and the output voltage is 0-2V, it is the analog voltage output, formaldehyde detection required the higher resolution, so external AD was used for data processing, TVOC and other sensors are also selected in accordance with standard requirements.

External AD hardware design of circuit was shown in figure 3, the (1) is external AD signal acquisition, and the (2) is condition circuit design.

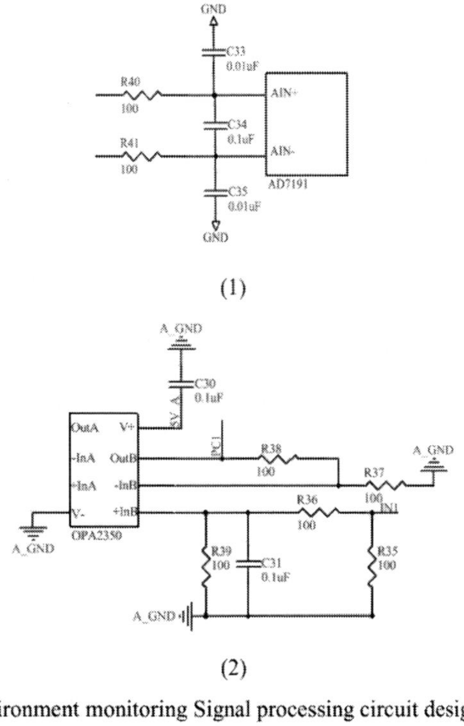

Fig. 3. Micro - environment monitoring Signal processing circuit design of showcase.

Since the input voltage of CO_2 is 3.3V, it should be divided. In order to ensure the stability of the voltage, the voltage follower circuit is added after the separated voltage, and finally the AD processing, the accuracy of the Measurement is ± (50ppm ± 2%full range), fully meet the requirements.

3.4 Interface design

STM32F103VET6 master chip has very rich peripheral resources, 13 communication interfaces, clocked frequency up to 72Mhz. PM2.5 sensor uses serial to RS485 communication, and the rest are using ordinary interface input, a serial port to RS232 to communicate with the LCD screen, the interface information shown in Figure 4.

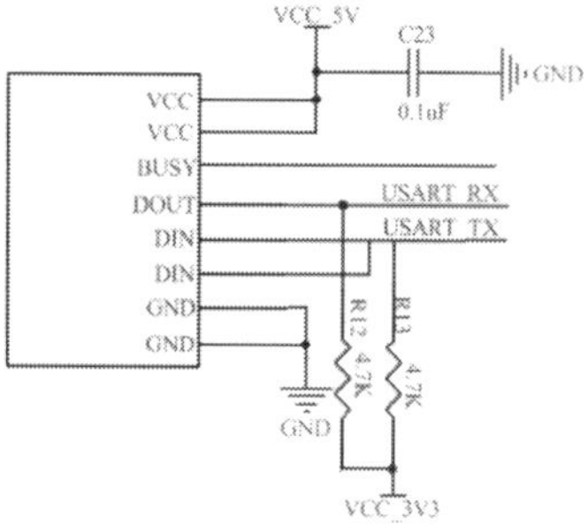

Fig. 4. Micro - environment monitoring LCD interface circuit design of showcase.

Take the expansion of the monitoring system to consideration. set aside a serial port for hardware and software debugging in the remaining resources, a RS485 communication interface is connected to the control, A USART interface is used for wireless communication.

4 Software design of monitoring system

Software design is based on hardware design in the system, including analog-digital conversion, real-time acquisition of 5-way sensor analog signal, a RS485 signal acquisition, processing, display, save and so on. The software design logic block diagram shown in Figure 5.

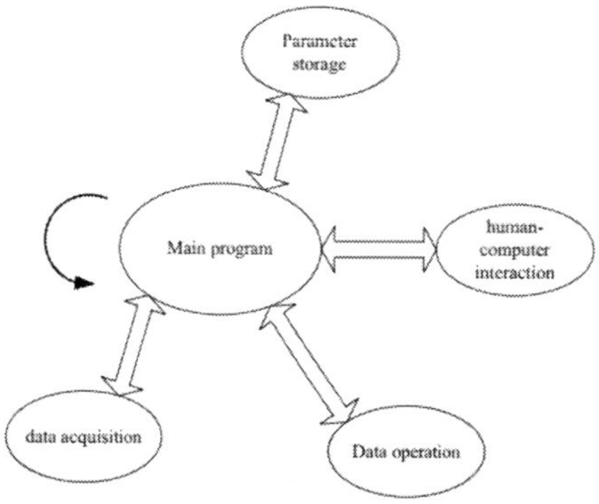

Fig. 5. Micro - environment monitoring system software design framework of showcase.

When the main program running, the timing full-period sampling of the system is opened, once enter a sample cycle, each sensor module to start data acquisition, the data of the collection will be operated, which includes the average, remove the gross error, the unit conversion, etc. The converted data into the human-computer interface for real-time display and update the data according to the set period.

Software Design of Interactive Interface.Flow chart of interactive software design shown in Figure 6. First, the whole initialization mainly includes the system equipment, the serial port, the clock, each variable and so on, second, go to the main interface of monitoring, all operations were responsed by users to user interface, the serial port will return the appropriate instructions, and then enter the communication subroutine and human-computer interaction subroutine respectively, the communication subprogram mainly completes the analysis to the serial port instruction, finally, sub-menu interface conversion operation was completed by human-computer interaction subroutine.

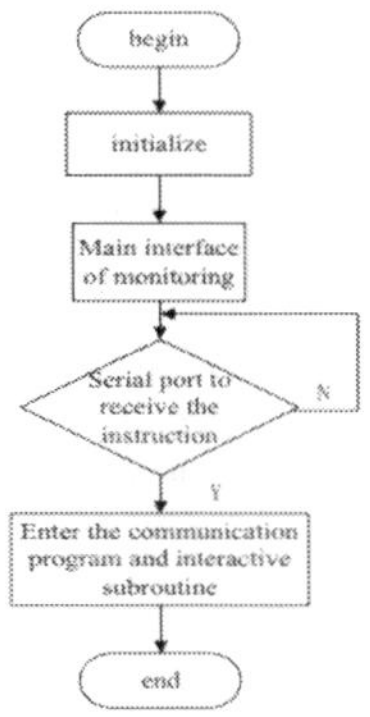

Fig. 6. Flow chart of interactive software design.

At the same time, the system can also be implemented the sensor state settings, threshold settings and other information on the parameters of storage, on the other hand, the host computer can send a time calibration command to calibrate the serial port screen real time.

5 Realization of interactive interface

At present, 80% of the micro-environmental monitoring terminal can not be separated from the PC-side control [10],that users can not access the current environment or related information directly.

Human computer interaction is gradually developing towards the direction of user experience. Users communicate with the system through interactive interface and a series of operations can be performed. A good human-computer interaction interface design can enhance the user's understanding of the system.

5.1 On-line emulation debugging

The user has defined the relevant items of the human-computer interaction, the simulation and online testing of human-computer interaction system can be completed without the support of the hardware, which is better than other display tools, using the serial port screen simulation software VisualTFT on the PC, virtual serial port screen. In the keil development environment, the virtual serial port contact with host computer through the virtual serial port, keil program debug simulation. The serial port screen will realize the function that corresponding communication record, also can monitor the host to the screen issued by the instruction , facilitate the process of analysis and debugging later.

Thus, the serial port screen is applied to the human-computer interaction system. It is a relatively independent unit, which can be synchronized with the development of instrument hardware, and will greatly shorten the project cycle, saving the project cost.

5.2 Realization of Interactive Interface

The main interface of environmental monitoring includes four secondary interfaces, there are environmental monitoring, environmental assessment, environmental warning, and historical data. Environmental monitoring to achieve the description of cultural relics information in current showcase, the display of monitoring data, the setting of working mode, as well as threshold settings.Set password-protected at the mode setting and threshold setting, non-staff members can not enter. Environmental assessment realizes the curve description of four kinds of detection gas. Environmental warning realizes the abnormal alarm mainly. Historical data realizes the current showcase address query and upload data query.

Take a second-level interface as an example, which is the environment warning communication display as shown in Figure 7. The second-level interface contains three three-level interfaces, including the early warning description, the time of warning, the type of warning. At the same time, the early warning description refers to the monitoring terminal mainly, it can tell the users which is abnormal at present. The warning time mainly refers to the time period that is viewed. The early warning type is classified according to the different abnormal conditions, mainly including the primary warning, secondary warning and high-risk warning.

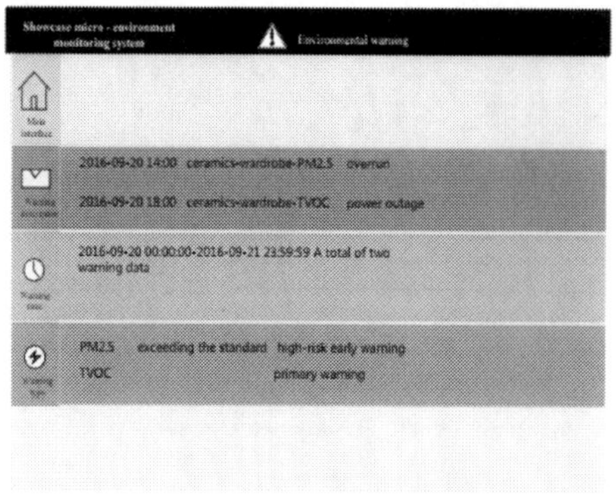

Fig. 7. Function diagram of Human computer interaction.

6 Conclusion

The micro environmental monitoring platform of showcase this paper has introduced, implant the concept of the integration into this design, and making the function of the products diversification. The products features can be combined freely, and the wide variety of type can meet the various needs of users. At the same time, for the visitors, the interactive interface design not only enhance the understanding of museum products, but also more conducive to museum staff to grasp the current environmental information timely, and then make effective management in time. The function of serial LCD online test also greatly shorten the development cycle and development costs were reduced.

From the perspective of product scalability, some functions will be increased, such as the environmental control, voice interaction, SMS alarm and so on. The monitoring system will has good development prospects.

Acknowledgement

This research was finacially supported by Key Project of Transformation and Industrialization of Scientific and Technological Achievements in Anhui Province.

References

1. Laiming Wu, The seventh China Museum and related products and Technology Expo. **6**(2016) 88-96.
2. Laiming Wu, Promote Cultural Relic Prevention and Protection Ability by Developing Heritage Protection Equipment. **11**(2015) 1134-1145.
3. Bo Zhang, Open heritage monitoring and early warning and protection decision-making management of the new model. **10** (2015) 266-276.
4. Joan Cahill, Nick Mc Donald, Human computer interaction methods for electronic flight bag envisionment and design. **9**(2015)113-123.
5. Shengming WU. Smart home interactive terminal design and implementation. **14**(2013).
6. Salaheldin Mohamed Ibrahim Edam, Key Technologies of Wireless Ad-hoc and Sensor Networks for Optimized Transmission Management in Disaster Situation. International Journal of Rock Mechanics & Mining Sciences. **3**(2014) 59-67.
7. Hui Tian, Design and Research of Data Center Micro - Environment Monitoring System Based on Zigbee. **5**(2016) 207-214.
8. Xiaoting Li, Development of Serial LCD Based on Embedded System. **4**(2014) 446-486.
9. Cultural relic protection equipment industry and application of collaborative work platform, *Compilation of Standards for the Protection of Cultural Relics.* (Cultural Relics Press, 2016) .
10. Gabor Aranyi, Paul Schaik, Modeling user experience with news websites. **4**(2015) 59-66.

Colorimetric Temperature Measurement Method Considering Influence of Ambient Temperature

Bin Zheng[1,*] and Guoqing Ding[1]

[1]Instrument Science and technology Department, Shanghai Jiao Tong University, Shanghai, China
Corresponding Email: zhengbin22@sjtu.edu.cn

Abstract. Colorimetric temperature measurement has important applications in industrial production. However, it's usually applied to high temperature measurement because it lacks accuracy in medium-low temperature measurement. Aimed at improving the low accuracy of traditional colorimetric temperature measurement method in medium-low temperature measurement, this paper pays attention to the influence of ambient temperature on colorimetric temperature measurement method, especially focusing on the influence of ambient temperature on the measured target and the non-refrigerated infrared detector, and improves the traditional method by considering the influence of ambient temperature. Experimental results show that in the experimental temperature range, the relative error of the method proposed in this paper is less than 1.2%, while the relative error of traditional method is less than 4.0%. So this method can significantly improve the precision of temperature measurement especially in medium-low temperature measurement.

1 Introduction

Temperature measurement has important applications in industrial production.Temperature measurement can be divided into contact and non-contact measurement. In non-contact measurement, infrared temperature measurement is one of the most important methods. In the application of infrared temperature measurement, there are four methods, including radiation measurement, monochromatic measurement, colorimetric measurement and multi spectrum measurement. Among them, colorimetric temperature measurement method has all the advantages of radiation measurement and monochromatic measurement. What's more, colorimetric temperature measurement can work without knowing the emissivity of the measured target, which avoids the difficulty of emissivity measurement of the target [1,2].

Colorimetric temperature measurement method is usually applied to high temperature measurement(over 600 degree) [3,4]. However, it lacks accuracy in medium-low temperature measurement because the influence of ambient temperature should also be considered. This paper discusses the basic principle of colorimetric temperature measurement, then analyses the influence of ambient temperature on the target temperature and infrared detector, and how traditional colorimetric measurement method of calibration and measurement method is improved considering the influence of ambient temperature. Experiments show that compared to traditional method, this method obtains higher accuracy especially in a medium-low temperature measurement.

2 The mathematical model of the traditional colorimetric temperature measurement method

In order to reduce the influence of the emissivity of measured target, colorimetric temperature measurement method is put forward. It's based on the measurement of two given wavelength λ_1 and λ_2, and then calculates the radiation power ratio under the two wavelength. In this way, the temperature of the object can be calculated, while the measured temperature is called colorimetric temperature [5].

Colorimetric temperature measurement system has two channels, the output signals are corresponding to the radiation signals of the two channels, which can be expressed as:

$$V_1=K_1\varepsilon_1\lambda_1^{-5}e^{-C_2/\lambda_1 T},\ V_2=K_2\varepsilon_2\lambda_2^{-5}e^{-C_2/\lambda_2 T} \tag{1}$$

In Eq. 1, K_1 and K_2 respectively represent the characteristic constants of the two channels.
The ratio of the outputs of the two channel is:

$$R= V_1/ V_2= K_1\varepsilon_1\lambda_1^{-5}\ e^{-C_2/\lambda_1 T} /(K_2\varepsilon_2\lambda_2^{-5}\ e^{-C_2/\lambda_2 T}) \tag{2}$$

From Eq. 2 it can be known that if the ratio R can be calculated, the temperature of the object T can be calculated too. When using this method to measure the temperature of blackbody T_S, the ratio of the outputs of two channels is:

$$R= V_1/ V_2= K_1\varepsilon_1\lambda_1^{-5}e^{-C_2/\lambda_1 T_S}/(K_2\varepsilon_2\lambda_2^{-5}e^{-C_2/\lambda_2 T_S}) \quad (3)$$

According to the definition of colorimetric temperature, the relationship between colorimetric temperature and real temperature of the measured target can be expressed as:

$$1/T-1/T_S=\lambda_1\lambda_2(\ln\varepsilon_2- \ln\varepsilon_1)/[C_2(\lambda_2-\lambda_1)] \quad (4)$$

3 Measurement considering ambient temperature

There are two kinds of methods to solve temperature drift in the infrared temperature measurement, one method is to change the circuits of the instrument, reducing the influence of ambient temperature on the system. Another method is software compensation, which studies the relationship between the output signal and measured temperature. However, no matter which method is chosen, the system still lacks accuracy especially in medium-low temperature. Therefore we need to further study the relationship between ambient temperature and the measured target temperature. The effect of ambient temperature can be divided into the temperature of external environment, the temperature of infrared detector, the temperature of infrared lens, the temperature of circuit board. Among them, ambient temperature and the temperature of infrared detector are the most important.

Other articles mainly consider the influence of ambient temperature on the target temperature. In fact, in industrial measurement, in order to reduce cost, non-refrigerated infrared detector is commonly used. In this situation, the influence of ambient temperature on non-refrigerating infrared detector should also be considered. In this paper, from these two aspects, the effect of ambient temperature on colorimetric temperature measurement method is analysed in detail.

2.1 The influence of ambient temperature on the target temperature measurement

Firstly, establish the mathematical model between target temperature and ambient temperature.

The environment around the measured target has great influence to colorimetric temperature measurement. If measured target temperature is T_1, ambient temperature is T_2, the radiation emitted per unit area is $A\varepsilon\sigma T_1^4$, the radiation absorbed per unit area is $A\varepsilon\sigma T_2^4$, then the net radiation emitted by objects Q is:

$$Q= A\varepsilon\sigma T_1^4- A\alpha\sigma T_2^4 \quad (5)$$

In Eq. 5, A: unit area; ε: emissivity of the measured target; α: absorptivity of the measured target.
If $\varepsilon=\alpha$, then Q is:

$$Q= A\varepsilon\sigma(T_1^4- T_2^4) \quad (6)$$

Table 1 shows one example of energy error of a traditional colorimetric temperature measurement device(wavelength ranges from 9-12 μm) [6].

Table 1. The varying relationship of the energy error according to the target temperature and the ambient temperature %.

target temperature	ambient temperature T_2/K						
T_1/K	270	280	290	300	310	320	330
500	3.64	4.35	5.12	5.96	6.86	7.82	8.83
600	2.23	2.66	3.15	3.68	4.25	4.86	5.52
700	1.53	1.84	2.17	2.54	2.94	3.37	3.84
800	1.14	1.37	1.62	1.90	2.20	2.53	2.88
900	0.93	1.12	1.33	1.55	1.80	2.07	2.35
1000	0.75	0.90	1.07	1.25	1.45	1.66	1.90

2.2 The influence of ambient temperature on the non-refrigerated infrared detector

2.2.1 Voltage-current characteristic of InGaAs PIN photodiode

Since the price of non-refrigerating infrared detector is much lower than that of the refrigerated infrared detector, non-refrigerated infrared detector is widely used in industrial measurement, so the influence of ambient temperature on infrared detector should be analysed. Here takes one of the most frequently used infrared detector InGaAs PIN photodiode for example.

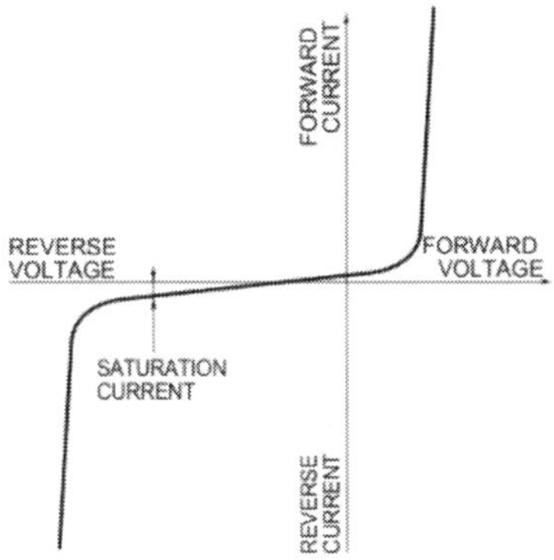

(a) In darkness

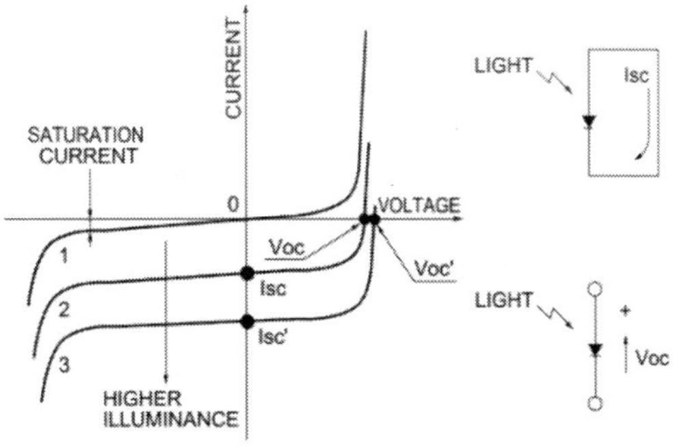

(b) Illuminated

Fig. 1. Voltage-current characteristics.

Figure 1 shows how light can be measured by measuring the photoelectric current. InGaAs PIN photodiode works as follows. When a voltage is applied to an InGaAs PIN photodiode in darkness, a voltage-current characteristic like that shown in Figure 1(a) is obtained. When incident light enters the InGaAs PIN photodiode, the voltage-current characteristic curve becomes as shown in 2 of Figure 1(b). When much stronger incident light enters, the curve shifts as shown in 3 of Figure 1(b). Here, when both terminals of the photodiode are left open, a forward voltage V_{op} appears, and when both terminals are shorted, a current of I_{sc} flows in the reverse direction.

2.2.2 Equivalent circuit of InGaAs PIN photodiode

The equivalent circuit of an InGaAs PIN photodiode is shown in Figure 2. From this equivalent circuit, the short circuit current I_{SC} is given by Eq. 7. The limit of I_{SC} linearity is determined by the 2nd and 3rd terms of this formula.

$$I_{SC}=I_L-I_S[\exp(q*I_{SC}*R_S/kT)-1]-I_{SC}*R_S/R_{sh} \tag{7}$$

In Eq. 7, q: Electron charge; k: Boltzmann constant; T: Absolute temperature of element; I_S: Saturation current of photodiode in the reverse direction.

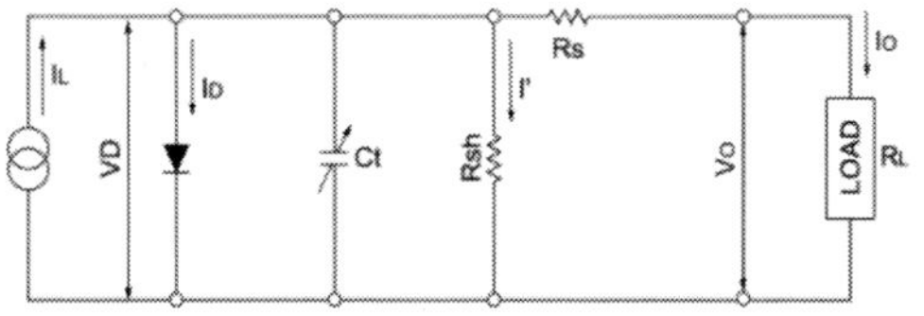

Fig. 2. Equivalent circuit of InGaAs PIN photodiode.

In Figure 2, I_L: Current generated by incident light(Proportional to the quantity of light); I_D: Diode current; C_t: Terminal capacitance; R_{sh}: Shunt resistance; R_S: Series resistance; I': Shunt resistance current; V_D: Voltage across the diode; I_O: Output current; V_O: Output voltage.

2.2.3 Temperature characteristics of InGaAs PIN photodiode

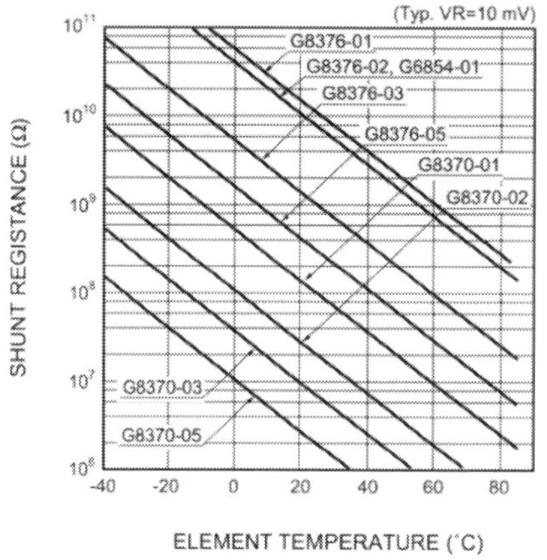

Fig. 3. The impact of element temperature on shunt resistance.

However, spectral response characteristics vary with detector temperature. Temperature characteristics of shunt resistance R_{sh} are shown in Figure 3. Thus, the influence of detector temperature can be analysed.

3 Experiment

3.1. The building of the experiment platform of the colorimetric temperature measurement

The device needed for experiment mainly consists of blackbody furnace, 24V switching power supply and computer. The emissivity of the blackbody furnace HFY-202B is 0.99 and is not an ideal model of 1. Its accuracy is 1 degree. The computer is used to receive and calculate the experimental data. The environment temperature detector is embedded in the measurement system. In addition, to weaken the influence of other environmental factors as much as possible, experimental environment should be dry and windless.

3.2 Experimental procedure

3.2.1 Calibration experiment of the influence of ambient temperature on infrared detector

Cover the lens of the temperature measurement system, that is to say, there is no input of infrared radiation. In such situation, put the temperature measurement system input thermostat, record the output of the system in different ambient temperatures in Figure 4

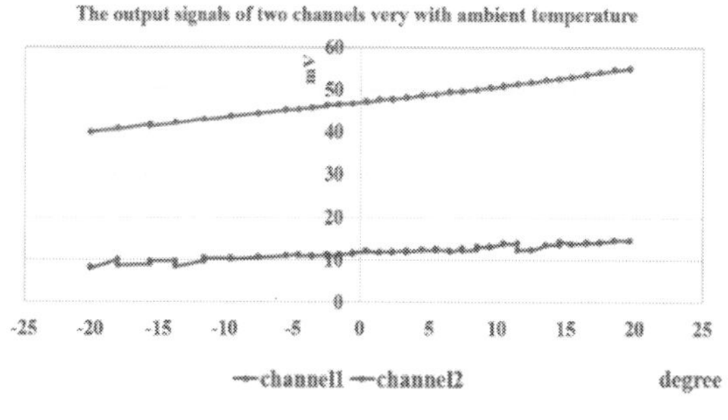

Fig. 4. The output signals of two channels very with ambient temperature.

From Figure 4, it can be concluded that there is linear correlation between the output signal and the temperature of infrared detector.

3.2.2 Integrated influence of ambient temperature on temperature measurement system

Since this system cannot be measured in thermostat, that is to say, we cannot individually measure the influence of ambient temperature on target temperature. This experiment measures the system with the influence of ambient temperature both on the measured target and the infrared detector.

Measure the accuracy of the method proposed in this article and traditional method in 10 different temperatures of object being measured. The results are shown in Table 2.

Table 2. Results of proposed method and traditional method.

Temperature/ degree	Measured temperature of proposed method/ degree	Relative error of proposed method/ %	Measured temperature of traditional method/ degree	Relative error of traditional method/ %
400	395.5	1.13%	415.7	3.93%
420	415.9	0.98%	430.6	2.52%
440	436.0	0.91%	451.1	2.52%
460	455.7	0.93%	469.4	2.04%
480	476.1	0.81%	487.9	1.65%
500	497.0	0.60%	507.8	1.56%
600	603.4	0.57%	605.4	0.90%
700	703.9	0.56%	705.9	0.84%
800	803.5	0.44%	805.5	0.69%
900	902.9	0.32%	904.2	0.47%

3.3 Result analysis

Table 2 shows the results of the two methods in 10 different temperatures. Usually, temperature more than 600 degree is defined high temperature while temperature lower than 600 degree is called medium-low temperature. In high temperature measurement, the precision of the two methods are relatively high, that is to say, ambient temperature has little effect on high temperature measurement. However, in medium-low temperature measurement, take 400-500 degree for example, for the method proposed in the article, the absolute error is within 4 degree and the relative error is within 1.2%, which still has high accuracy. For the traditional method, the absolute error is within 6 degree and the relative error is within 4%, which lacks accuracy. That is to say, if the influence of ambient temperature discussed in this paper is compensated, the accuracy of colorimetric temperature measurement method can be significantly improved especially in medium-low temperature measurement.

In addition, it can be concluded that with the increase of temperature of object being measured, the influence of the ambient temperature becomes smaller and smaller, which also explains why traditional colorimetric temperature measurement method is suitable for high temperature measurement.

From the above analysis, it can be concluded that the traditional colorimetric temperature measurement method lacks precision for medium-low temperature object measurement. However, the method proposed in the article takes into consideration of the influence of ambient temperature on the measured target and the infrared detector, and compensate the influence of ambient temperature. This method is not only suitable for high temperature target, but also can significantly improve accuracy for medium-low temperature target.

4 Summary

Compared to other methods of infrared temperature measurement, traditional colorimetric method can eliminate the effects of emissivity, but ignores the influence of ambient temperature, so it lacks accuracy especially in medium-low temperature measurement. This paper thoroughly analysis the influence of ambient temperature on the measured target and the infrared detector and compensate the errors brought by ambient temperature. Experiments show that this method can significantly improve accuracy for medium-low temperature target, which will be of great significance in practical applications.

References

1. D. Shi, Xinyang Teachers College, *Chinese Journal of Scientific Instrument*, **26**(12): 1238-1243 (2005)
2. Y. Si, Doctoral dissertation, *North University of China* (2013)
3. J.H. Zhu, Doctoral dissertation, *North University of China* (2011)
4. C. Romero, X. Li, S. Keyvan, *Applied Thermal Engineering, 25*(5–6), 659-676 (2005)
5. P. Sun, Doctoral dissertation, *Jilin University* (2007)
6. Y. Min, C.W. Peng, Y.H. Yan, *Journal of Hunan University*, **31**(5): 110-112 (2004)

Collaborative Innovation Research on High-tech Industry in the Center Delta

Jingdong Yan[1],Li Li[2,*] and Xiangmiao Liu[3]

[1]Managment School, Wuhan University of Technology, Wuhan, China
[2]Managment School, Wuhan University of Technology, Wuhan, China
[3]Managment School, Wuhan University of Technology, Wuhan, China
Corresponding Email: 1554103284@qq.com

Abstract. As a strong point in the Middle Rises Strategy, the Center Delta is of great significance to high-tech industry development. While the collaborative innovation is an effective way to promote the coordinated development of regional economy. This article selected electronic and communication equipment manufacturing industry of 3 provinces in the Center Delta as samples, built evaluation index system of collaborative innovation of high-tech industry, and put the relevant data into the system coordination degree model, in order to get the synergy of industrial innovation system and innovation environment system of 3 provinces. The empirical results shown that the industrial innovation system and innovation environment system coordination degree of Hubei province was the highest, while Jiangxi province was the lowest. Based on the empirical result analysis of the above-mentioned systems, this article put forward suggestions to promote the development of the high-tech industry collaborative innovation in the Center Delta.

Keywords. Collaborative innovation, innovation system, innovation environment.

1 Instruction

The Middle Reaches of the Yangtze River, which named "the Center Delta", is the major high-tech industry, advanced manufacturing, modern services and regional scientific and technological innovation bases in China. The Center Delta consist of Hubei wuhan city circle, ring of Changsha-Zhuzhou-Xiangtan urban agglomeration in Hunan province, Jiangxi PoYang lake ecological economic zone.And they are important middle urban agglomerations to carry out the Middle Rises Strategy as well [1]. The Center Delta in the narrow sense refers to the Hubei Wuhan city circle, ring of Changsha-Zhuzhou-Xiangtan urban agglomeration in Hunan province, Jiangxi Poyang Lake ecological economic zone, which contains seven central city(Hubei Wuhan, Changsha, Zhuzhou and Xiangtan in Hunan province, Jiangxi Nanchang, jingdezhen and Yingtan)and urban community formed by 19 cities, municipals or sections. The Center Delta in the broad sense refers to three provinces (Hubei, Hunan and Jiangxi), radiating driven by three major cities (group). Since the development of high-tech industries of the three provinces are mainly concentrated in the three major urban agglomerations (circle), and statistics calibers are given priority to the provinces, we chose to extend the area of research to the whole provinces in where the Center Delta located.

On September 23th, 2009, the State Council promoted in the Opinions about Vigorously Promoting Central Region Rise Strategy that the state should strengthen modern equipment manufacturing and high-tech industry base, rely on the leading enterprise, master the core technology as the breakthrough point, and give full play to the leading role of Wuhan, Changzhutan regional comprehensive national high-tech industry base and professional radiation national high-tech industry base; in order to form a new industrial clusters with core competitiveness and ultimately would achieve the transformation and upgrading of traditional industries under high-tech conditions [2]. The opinions had strategic importance in putting surrounding provinces on an innovative and developing path which was leading by the Center Delta city group and supported by high-new technology. As an important driving force to promote sustainable development of social economy,high-tech industry with the characteristics of human-financial investment, sophisticated requirements and environment and social sustainable development,undertakes an important role in promoting the regional economy of the Center Delta.

However, as a regional economies, there exists an obvious gap among this region and the Yangtze River Delta and the Pearl Rive Delta, which specific embodiment in serious segmented regional market, isolated development, unclear division of regional industrial labor, low conversion rate of scientific research, unreasonable resources configuration of resources that have great advantages,etc [3].In addition, the gap of high- new technology industry development degree is also obvious among the 3 provinces.All of these factors have weakened the competitiveness of the region, hindering the coordinated development of regional economy in the Center Delta. Thus, it is of great research significance for the Center Delta to find methods of achieving complementary advantages, optimizing the allocation of high-new technology industries resources, promoting the competitiveness and fusion of science&technology and productivity, forming an overall advantage,taking strong pulling effect on the prosperity of the central region's economy situation, and becoming the engine driving the rise of central China.

2. Related concepts

Center Peter A. Gloor, a researcher in MIT Sloan, first gave definition to collaborative innovation: self-motivated staffs with a common vision constituted a group, and aimed to achieve shared goals by exchanging ideas and information through network. Researches based on collaborative innovation began to rise In 1990s [4]. In 1992, the State Economic and Trade Commission, Ministry of Education and Chinese Academy of Sciences jointly organized and implemented the "Joint Development of Research Projects", which had given rise to large quantities of relevant studies. In 2011, collaborative innovation had been put forward. In order to establish an innovative country; universities, research institutes and enterprises were encouraged to expand the depth of cooperation; and innovative elements of vitality (talent, capital, information, technology) should be released.

For the essence of collaborative innovation connotation, domestic scholars had different views. This article tended to adopt Zhitai Zhou's opinion (2013), that the collaborative innovation was an activity about the circulation and integration of inter-organizational and inter-regional human, financial, material and information [5]. It was launched by enterprises, government, universities, research institutes, intermediary service organizations and users and aimed at achieving technological innovation. Collaborative innovation promoted the complementarity of resources and advantages, cooperated to carry out technical innovation and achievement transformation activities through policy guidance and mechanism arrangement, the integration of enterprises, universities and research institutes as well as innovative resources.

3 Collaborative innovation model

In this paper, we have studied the triangular cooperative innovation of high-tech industry and have built industry theory model of collaborative innovation. System coordination degree model have been used to evaluate the synergy degree of high-tech industry in the Center Delta on the basis of system coordination degree model established by Qingsong Meng, Wenxiu Han (2000) [6].

The order parameter in a economic system development process is ei=(e11, e12,...,e1n), $n \geq 2, \beta 1i \leq e1i \leq \alpha 1i, i \in [1, n]$. e11,e12,...,e1n are the indexes to describe the economic system operation mechanism and operation condition.When the indexes are positive,the bigger e11, e12, ..., e1n are,the higher the order degree is,otherwise the opposite.When the indexes are negative,the bigger e11, e12, ..., e1n are,the lower the order degree is,otherwise the opposite.Therefore, the following equation is defined as a system component of the order parameter of the degree of order:

$$u_1(e_{1i}) = \begin{cases} \frac{e_{1i}-\beta_{1i}}{a_{1i}-\beta_{1i}}, & e_{1i} \text{ are positive indexes} \\ \frac{a_{1i}-e_{1i}}{a_{1i}-\beta_{1i}}, & e_{1i} \text{ are positive indexes} \end{cases} \tag{1}$$

$u_1(e_{1i}) \in [0,1]$, the smaller $u_1(e_{1i})$ is, the smaller contribution has been made to the industry innovation subsystem degree of order.

Subsystem degree of order can be measured by index degree of order. The contribution order parameter e_1 made to industry innovation system can be achieved through the integration of $u_1(e_{li})$. Geometric method and weighted sum method are commonly used in practice, that is:

$$u_1(e_{li}) = \sqrt[n]{\prod_{j=1}^{n} u_1(e_{ij})}$$

$$\text{or: } u_1(e_{li}) = \sum_{j=1}^{n} \omega_j u_1(e_{li}), \omega_j \geq 0, \sum_{j=1}^{n} \omega = 1$$

This paper adopted weighted sum method, and environment subsystem degree of order can be calculated as $u_2(e_{2i})$.

The synergy degree of high-tech industries collaborative innovation system can be identified according to the order degree of innovation subsystem. Collaborative innovation of the system refers to the degree of coordination of the various subsystems in the innovation process,and was decided by degree of order collective of each subsystem. The improvement of individual system degree of order can not lead to the improvement of coordination degree of the whole system. The improvement of system coordination degree is decided by ordered degree of the whole system.

System order degree of innovation System order parameter is $u_1^0(e_1)$ at the given initial point.So for t1 in the development of system evolution, if the System order degree of innovation System order parameter is $u_1^1(e_1)$, the order degree of innovation environment system is $u_2^1(e_2)$, and $u_1^1(e_1) \geq u_1^0(e_1)$, $u_2^1(e_2) \geq u_2^0(e_2)$ are both satisfied,it can be said that high-tech industry collaborative innovation system is developing collaboratively in the period from t0 to t1. The following is a model for computing system coordination degree:

$$c = \lambda \sqrt{|u_1^1(e_1) - u_1^0(e_1)| \cdot |u_2^1(e_2) - u_2^0(e_2)|}$$

$$\lambda = \begin{cases} 1, & u_1^1(e_1)-u_1^0(e_1)\geq 0 \text{ and } u_2^1(e_2)-u_2^0(e_2)\geq 0 \\ -1, & \text{other} \end{cases}$$

As is shown in the system coordination degree model: $c\in[-1,1]$, the smaller c is,the lower the degree of coordinated development system is, otherwise the opposite.

If one of the subsystems' the increasing amount of order degree is larger, while the other subsystem has smaller increasing amount , then the synergy degree of the whole system is at a lower level, which means the system is not in a state of good coordination or not harmonious.

4 Empirical analysis

4.1Collaborative innovation evaluation index system

In this paper, we have studied the collaborative innovation between industrial innovation system and innovation environment system, then have established an index system based on the characteristics of elements and systems.

Industrial innovation system indicated innovation capacity and performance of the industry. Ten indexes were chosen in this study, including S&T Personnel, Scientists and Engineers, Full-time Equivalent of R&D Personnel, New Fixed Assets, Funding for S&T Activities, Intramural Expenditure on R&D, Patent Applications, Output Value of New Products, Export Sales Revenue of New Products, GDP at Current Prices [7].

(1)S&T Personnel refers to person who are engaged in the activities of science and technology directly or indirectly , as well as people providing service for science and technology activities. This index reflects human input of industry innovation system.(2)The Number of Scientists and Engineers refers to the number of people with medium and senior technical titles and those who have received college or higher education but don't have medium or senior technical titles. This index reflects the R&D investment of the industry innovation system and creativity of science and technology. (3) Full-time Equivalent of R&D Personnel refers to the number of person-years of full-time personnel accounting of R&D project research personnel, management personnel and direct service personnel. The index reflects the actual workload of R&D investment of industrial innovation system. (4)New Fixed Asset refers to the increasing value of fixed assets in the reporting year. The index reflects cumulative capacity of industrial innovation system infrastructure.(5)Funding for S&T Activities reflects the science and technology investment situation of industry innovation system.(6)Intramural Expenditure on R&D refers to internal spending for research activities. This index reflects the investment of research and development funds of industrial innovation system.(7)Patent Applications refers to the number of patents that a particular industry applied, reflects the industrial R&D ability.(8)Output Value of New Products refers to new products adopted new technology, new design; or new products which performance and functions are significantly improved, a significant increase in product output and use function, reflects the innovation output of the system.(9)Export Sales Revenue of New Products refers to the new products export value proportion of the total export value. The index reflects the new products' market competitiveness of industrial innovation system [8].(10)GDP at Current Prices refers to industrial production output value of the current period. The index reflects the production capacity of industrial innovation system.

Innovative Environment System indicates the industry's innovation environment. Three indexes were chosen in the study, including the Number of Research Institutions, Microelectronic Control Equipment and the Government Funds in the investment.

(1)The Number of Research Institutions refers to the institutions engaged in scientific research and technological development activities, and had science and technology activities expenditure. The index reflects creating, accessing and the spreading situation of knowledge and technology of the innovative environmental system.

(2)Original value of Equipment Controlled by Micro-electronics represents the industry informatization level, informatization of production equipment is conducive to improve production efficiency and reduce production costs.

(3)Government Funds in the Investment reflects the government's efforts to support the industrial development.

4. 2 Data sources and analysis

This article collected 5 years'(2005, 2008, 2009, 2010, 2011) electronic and communication equipment manufacturing industry data from"China Statistical Yearbook on High Technology Industry" as shown in the table 1 and table 2. The difference between the evaluation of observation units in the subsystems resulted in significant differences between the observed values. So the raw data should be standardized before empirical analysis.

After standardizing the data, we put them into formula(1), computed degree of order of industry innovation system and innovation environment syste , at the same time, analyzed the correlation of each index system, the results are shown in table 3 and table 4.

CRITIC objective weighting method was used to calculate the weight of each component of the industry innovation system and the innovation environment in the Center Delta as shown in table 5 and table 6.

Put the degree of order and weighted value of the components in the industry innovation system into formula(3), degree of order of industry innovation system could be concluded, degree of order of innovation environment system could be calculated in the same way. Then put the degree of order data of the two systems into formula(4), the coordination degree of system could be obtained. The results are shown in table 7.

Table 1. The raw data of Industry Innovation System Index in the Center Delta.

Region	Year	S&T Personnel (person)	New Fixed Asset (100million yuan)	Funding for S&T Activities (10000 yuan)	Scientists And Engineers (person)	Full-time Equivalent Of R&D Personnel (man-year)	Intramural Expenditure on R&D (10000 yuan)	Patent Applications (person)	Output Value of New Products (10000 yuan)	export sales revenue of new products (10000 yuan)	GDP at current price (100 million yuan)
Hubei	2007	12,139	35.61	6652	2568	8765	62963	12	1258699	553879	447.5
	2009	17,558	59.36	8569	2986	8965	103056	89	1889872	586972	847.5
	2011	19,574	113.49	9158	3522	7790	183112	120	3584687	669875	1039.5
	2013	21,497	160.93	16998	4622	9532	166053	269	3886245	887754	1312.0
	2015	22,142	229.47	17223	5890	13258	372111	390	4002587	10258796	1721.0
Hunan	2007	6,908	30.39	3560	1768	3458	18603	1	985443	200154	219.3
	2009	11,490	50.46	4225	2213	3698	46367	3	1344126	324877	522.8
	2011	12,910	81.60	4889	2610	1998	76954	25	1523954	387514	648.8
	2013	15,776	124.40	5336	3325	2598	84600	66	2103987	455536	930.6
	2015	21,407	153.06	7325	4476	3120	128016	127	3158976	514783	1544.9
Jiangxi	2007	10,987	46.54	3882	650	5769	42928	2	688521	784423	234.0
	2009	17,267	160.18	4153	710	6582	52963	29	1325897	254648	586.0
	2011	18,571	215.18	4523	779	6132	77507	37	1569872	354742	755.7
	2013	21,810	356.27	7960	890	5887	101994	96	2225987	365872	1037.5
	2015	23,999	329.79	8763	936	6117	133261	189	3004894	420457	1418.6

Table 2. The raw data of Environmental Innovation System Index in the Center Delta.

Region	Years	Scientific &technical	Original value of Equipment	Government funds

		Institutions (unit)	Controlled by Micro-electronics (10000 yuan)	(10000 yuan)
Hubei	2007	25	100, 234	1, 586
	2009	31	112, 350	3, 562
	2011	38	128, 756	5, 523
	2013	42	135, 692	8, 895
	2015	50	142, 388	10, 256
Hunan	2007	7	3, 215	982
	2009	11	3, 865	1, 235
	2011	16	4, 230	1, 890
	2013	19	5, 663	2, 360
	2015	21	6, 349	3, 356
Jiangxi	2007	6	1, 963	1, 095
	2009	8	2, 536	2, 089
	2011	13	3, 025	2, 698
	2013	14	3, 560	3, 469
	2015	16	3, 881	4, 499

Table 3. Index correlation coefficient of Industry innovation system in the Center Delta.

Hubei Province								
1.000	0.023	0.919	0.995	0.712	0.824	0.895	0.919	0.925
0.023	1.000	0.072	0.034	0.596	0.205	0.071	0.354	0.365
0.919	0.072	1.000	0.889	0.607	0.849	0.977	0.783	0.812
0.995	0.034	0.889	1.000	0.729	0.774	0.880	0.936	0.940
0.712	0.596	0.607	0.729	1.000	0.298	0.662	0.482	0.483
0.824	0.205	0.849	0.774	0.298	1.000	0.725	0.744	0.754
0.895	0.071	0.977	0.880	0.662	0.725	1.000	0.775	0.808
0.919	0.354	0.783	0.936	0.482	0.744	0.775	1.000	0.998
0.925	0.365	0.812	0.940	0.483	0.754	0.808	0.998	1.000
0.887	0.094	0.775	0.858	0.558	0.907	0.661	0.766	0.757
				Hunan Province				
1.000	0.646	0.904	0.955	0.798	0.638	0.942	0.379	0.181
0.646	1.000	0.528	0.660	0.066	0.426	0.686	0.195	0.187
0.904	0.528	1.000	0.918	0.771	0.250	0.978	0.130	0.034
0.955	0.660	0.918	1.000	0.763	0.506	0.941	0.345	0.113
0.798	0.066	0.771	0.763	1.000	0.487	0.692	0.683	0.266
0.638	0.426	0.250	0.506	0.487	1.000	0.359	0.705	0.380
0.942	0.686	0.978	0.941	0.692	0.359	1.000	0.093	0.041
0.379	0.195	0.130	0.345	0.683	0.705	0.093	1.000	0.198
0.181	0.187	0.034	0.113	0.266	0.380	0.041	0.198	1.000
0.393	0.784	0.526	0.566	0.047	0.132	0.596	0.448	0.646
				Jiangxi Province				
1.000	0.962	0.941	0.988	0.059	0.950	0.909	0.825	0.968
0.962	1.000	0.956	0.939	0.049	0.962	0.838	0.729	0.987
0.941	0.956	1.000	0.942	0.327	0.925	0.894	0.860	0.987
0.988	0.939	0.942	1.000	0.156	0.963	0.962	0.890	0.963
0.059	0.049	0.327	0.156	1.000	0.103	0.342	0.543	0.195
0.950	0.962	0.925	0.963	0.103	1.000	0.924	0.805	0.967
0.909	0.838	0.894	0.962	0.342	0.924	1.000	0.963	0.899
0.825	0.729	0.860	0.890	0.543	0.805	0.963	1.000	0.823
0.968	0.987	0.987	0.963	0.195	0.967	0.899	0.823	1.000
0.986	0.948	0.956	0.999	0.184	0.970	0.962	0.892	0.974

Table 4. Index correlation coefficient of Environmental Innovation System in the Center Delta.

	Hubei Province			Hunan Province			Jiangxi Province	
1.000	0.785	0.603	1.000	0.272	0.711	1.000	0.978	0.325
0.785	1.000	0.582	0.272	1.000	0.271	0.978	1.000	0.319

0.603	0.582	1.000	0.711	0.271	1.000	0.325	0.319	1.000

Table 5. The weight of each component of Industry innovation system in the Center Delta.

Region	W1	W2	W3	W4	W5	W6	W7	W8	W9	W10
Hubei Provinc	0.0567	0.2747	0.0717	0.0592	0.1124	0.0969	0.0782	0.0841	0.0859	0.0797
Hunan Provinc	0.0647	0.0856	0.0696	0.0684	0.0974	0.1119	0.6358	0.1768	0.1587	0.1285
Jiangxi Provinc	0.0564	0.0638	0.0688	0.0557	0.4041	0.0625	0.0758	0.1007	0.0628	0.0552

Table 6. The weight of each component of Innovation Environmental System in the Center Delta.

Hubei Province			Hunan Province			Jiangxi Province		
W1	W2	W3	W1	W2	W3	W1	W2	W3
0.2854	0.3057	0.4201	0.3408	0.4369	0.2235	0.2528	0.2551	0.4921

Table 7. Degree of coordination of High-tech industry collaborative innovation in the Center Delta.

Year	Hubei Province			Hunan Province			Jiangxi Province		
	Industrial innovation system degree of order	Innovation environme nt system degree of order	Degree of coordinati on	Industrial innovation system degree of order	Innovation environmen t system degree of order	Degree of coordinati on	Industrial innovation system degree of order	Innovatio n environm ent system	Degree of coordinati on
2007	0.0202	0.0245		0.3520	0.5175		0.0775	0.5125	
2009	0.4911	0.3031	0.3623	0.2070	0.3946	0.1335	0.0433	0.000 0	-0.1324
2011	0.3393	0.2579	0.2729	0.5689	0.2367	-0.2468	0.4919	0.1645	-0.3798
2013	0.7150	0.4676	0.5549	0.6523	0.4651	0.3148	0.4452	0.1793	-0.3500
2015	0.7433	0.6480	0.8399	0.6586	0.6698	0.3286	0.6011	0.4506	0.4208

4.3 Result analysis

From2009, the ordered degree of Industry innovation system calculation in 3 provinces has been always increasing. Hubei province reached the highest ordered degree of Industry innovation system in 2015, the Funding for S&T Activities and the Patent Applications number were significantly higher than the other years according to the raw data. Since 2009, ordered degree of industry innovation system of Jiangxi province has improved obviously [9]. S&T Activities number, New Fixed Assets and Scientists and Engineers number have been increasing. It could be seen that the input of innovation system has put great impact on the ordered degree of industry innovation system.

The calculation of the ordered degree of Innovation environment system shown the following information: In addition to 2011, ordered degree of innovation environment has been gradually improving. The Degree of order of Innovation environment system from 2009 to 2013 in Hunan province has been obviously increasing, while for the past two years, the government funds was much higher than the other years. As for Jiangxi province, the government funds in 2011 was much higher. It could be seen that government plays an important role in the innovation environmental system.

The Synergy degree calculation results of Industry innovation system and Innovation environment system indicated that from 2007 to 2015, Collaborative degree has increased compared to last year. All regions have attached great importance to developing Industry innovation as well as improving Innovation environment, which has contributed to the coordinated development of industry innovation system and innovation environment system.

5 Conclusions

The synergy of industry innovation system and innovation environment system has positive effects on the rapid and benign development of high-tech industries. By means of collaborative innovation, product performance has been improved constantly, the yield of new product has increased, and industry innovation and innovation environment have promoted mutually, which enhance the competitiveness of high-tech industries [10].

5.1 Cultivation and introduction of talents

Through the empirical study of coordination degree of the industry innovation system and innovation environment system, the number of scientists and engineers in the industry innovation system has been found having important effects on the synergy degree. Therefore, talent factor, not only the cultivation of the talent but also the flow of talent, should be paid great attention to. In recent years, Hubei province and Hunan province have cultivated and attracted a large number of talents through their respective strengths. Relatively speaking, Jiangxi province is weaker on education and scientific research, and lacks in cultivating talents as well. What's more, Jiangxi province started later in introducing talents. From an overall perspective, market orientation should be adopted, in order to speed up the talent flow and promote the communications among scientific research personnel through the guidance of profit-driven or policies.

5.2 Increase the science and technology funds

The mount of funds raised for science and technology activities andgovernment funds have a great impact on the system synergy. Thus the government should pay more attention to play a important role in industrial innovation system and innovation environment system in the development of collaborative innovation. Economic methodsshould be taken to provide direct financial support and indirect incentive and supporting policies for the development of high-tech industry. Such as tax relief, establishing a financial system that contains a diversified venture capital system , etc. These related policies not only guide enterprises to increase R & D investment, but also broaden the financing channels of high-tech enterprises so that it can provide powerful fund guarantee for new high-tech enterprises and drive them to develop collectively and form a large scale.

5.3 Encourage technological innovation

The number of patent applications has a certain influence on the order degree of industrial innovation system and the synergy degree of the whole system [11]. Therefore, the government should formulate and implement the corresponding policies which are conducive to the collaborative innovation development of high-tech industry, create a good atmosphere of innovation environment, for example, establishing and perfecting the legal protection system, improving the patent protection system , in order to safeguard the legitimate rights and interests of enterprises and scientific research personnel earnestly and then advance the begin development of high and new technology industry.

References

1. S.L. Zhang, D. Hu, J.C. Liu. The analysis of industrial synergy innovation in Hi-tech industry among Jing-Jin-Ji region. Journal of Hebei University of Technology, (2012)

2. Y.Q. Xu, X.Q. Xin. Research on high-tech industry cluster innovation ability evaluation index system—based on Collaborative Innovation Theory.Industrial & Science Tribune, 18:48-50,(2014)

3. China vertical and horizontal planning research group in 12th Five-Year,"12th Five-Year" planning outline, 3:32-33,(2010)

4. Opinions of the State Council on Implementing the Strategy of the Rise of Central China,(2012)

5. B.L. Liu, Y.C. Ma. Study on Coordinative Development between Hi-tech Industry Cluster and Innovation-oriented City .Science & Technology Progress and Policy , 15:71-73,(2010)

6. S.H. Hu, R.Y. Mu. Situation analysis and strategic countermeasures of Wuhan east lake national innovation demonstration zone.Wuhan press, 1: 54-55,(2011)

7. J.Y. Zhang. High-tech industry cluster innovation collaborative mechanism. Contemporary Economics, 10:140-141, (2011)

8. J. Yu. Collaborative innovation model research on high-tech industry cluster. Research on Financial and Economic Issues, 12:41-44, (2008)

9. M. Shapiro. Innovation: A Blueprint for Surviving and Thriving in an Age of Change. McGraw-Hill, 102-111,(2002)

10. J.Y. Zhang. Collaborative Innovation research of high-tech industry cluster in Wuxi. Science & Technology Information, 05:273,(2013)

11. B. Tucker. Driving Growth through Innovation. Berrett-Koehler Publishers, Inc. San Francisco, (2002)

The Mental Health Condition of Manufacturing Front-line Workers: The Interrelationship of Personal Resources, Professional Tasks and Mental Health

Qian Zhang[1], Zhanwu Zhang[2] and Mingbo Sun[1,*]

[1]School of Management Engineering, Zhengzhou University, Zhengzhou, China
[2]Hong Fu Jin Precision Electronics (Zhengzhou) Co. Ltd, Zhengzhou, China
Corresponding Email: sunmingbo@zzu.edu.cn

Abstract. Manufacturing front-line workers were more likely to experience mental health problems. Personal resources and professional tasks were the major factors of workers' mental health. Therefore, this study was to explore the interrelationship of these three key factors. A questionnaire including the revised Occupational Stress Inventory (OSI-R) and the Symptom Checklist (SCL-90) covered 480 manufacturing front-line workers to measure their personal resources, professional tasks and mental health. Results showed that among manufacturing front-line workers, the status of mental health and professional tasks were below the national average level, and the personal resources were relatively deficient as well. Correlation analysis indicated a negative relation between the indicators of mental health and professional tasks (except responsibility), while personal resources and mental health were significantly positive correlation. These findings suggested that personal resources and professional tasks were highly related to mental health in manufacturing front-line workers.
Keywords. Mental health, Personal resources, Professional tasks, Front-line workers.

1 Introduction

With the economic globalization, China has become the "global factory". The total number of manufacturing workers has increased from 32.1 million in 2005 to 52.4 million in 2014 [1]. At present, most companies only focus on material welfare. However, a higher level of mental welfare hasn't been widely considered. There are increasing number of mental health problems in recent years. It becomes the significant factor to restrain the development of companies [2].

Domestic and international studies have already investigated the factors of workers' mental health [2-5]. The literature has suggested that the relevant factors of workers' mental health mainly come from society, profession and workers themselves. First, the common perceptions of mental health care are still outdated [3]. There is widespread prejudice against psychological counseling and treatment. Besides, workers are unwilling to acknowledge their mental health problems, at least until the severity of the symptoms makes it clear to colleagues or family. Consequently, workers fail to seek professional help in early stage and quickly benefit from the early treatment. In addition, several studies have provided that high-intense professional tasks are related to poor mental health outcomes [4]. Meanwhile, Xu et al. have demonstrated that the mental health condition is positively correlated with negative personal resources [5]. Moreover, coupled with the lack of necessary social support and the ways to decompress, these workers can easily suffer from mental health problems when they engage in high-intensity, monotonous tasks [2]. Based on past studies, personal resources and professional tasks have contributed to new challenges for mental health management. The influence of single factor on workers' mental health has been mainly discussed. However, there are few studies exploring the interrelationship of these factors, especially for manufacturing front-line workers.

Therefore, this study further explores the interrelationship of personal resources, professional tasks and mental health among manufacturing front-line workers using the revised Occupational Stress Inventory [6] (OSI-R) and the Symptom Checklist [7] (SCL-90). By understanding the relationship, companies may be able to devise interventions or strategies to mitigate the negative effects. The study also investigates the applicability of the Conservation of Resource Theory [8] (COR) in the eastern countries like China, which is originally established in the western countries.

2 Literature review

2.1 Mental health

Workers' mental health problems have strong negative impact on the companies performance. H.B.English has conceptualized mental health, which is a level of psychological well-being, or an absence of mental health problems[9]. It is the "psychological state of someone who is functioning at a satisfactory level of emotional and behavioral adjustment". From the perspective of positive psychology or holism, mental health may include an individual's ability to enjoy life, and create a balance between life activities and efforts to achieve psychological resilience [9]. In the past few decades, studies have focused on the condition and relevant factors of workers' mental health. Paek et al. found that front-line workers with high psychological capital were more engaged with their work and more likely to display job satisfaction and organizational commitment [10]. Loi et al. thought coworker support reduced foreign workers' mental health problems [11]. Huang et al. implied changes in Chinese migrant workers' mental health by using the SCL-90. And results described that their mental health level was decreasing over time [12]. In addition, Zhong et al. revealed that labor workers' mental health was related to their own psychological adjustment mode and the pressure caused by working environment, working intensity, working conditions and management style. And the SCL-90 was used to measure workers' mental health [13]. These studies have covered the present condition of workers' mental health and its relevant factors, but the functioning mechanism is still lack of study. Therefore, this study further explores the condition of mental health on manufacturing front-line workers using the SCL-90, which has already been wildly used in China.

2.2 Professional tasks and mental health

Professional tasks are a set of the working conditions, including role overload, role insufficiency, role ambiguity, role boundary, responsibility, work environment and so on. They have significant impact on work stress, and can be called stressors [14]. A large number of studies have confirmed that professional tasks of workers vary among different companies. Zhang et al. found that professional tasks had significant impact on mental health of the electronic manufacturing front-line workers by using the OSI-R to measure professional tasks [14]. Sharma et al. revealed that work-family conflict had a direct negative relationship with the mental health of female nurses, signifying that the lifestyle of female nurses was that their work and family roles interfered with each other [15]. Ganster et al. stated that there was consistent evidence supporting a link between work stressors and primary Allostatic Load indicators for self-reported affective outcomes such as anxiety, job attitudes, and job-related tension [4]. Chen et al. studied the condition of the work stress among Chinese offshore oil workers using the OSI-R, and the result showed professional tasks were the key factor impacting mental health [16]. In addition, He et al. revealed that manufacturing front-line workers often suffered from the mechanical work with high speed and repetition for a long time [3]. Therefore, in this study, the professional tasks are considered as the major factor of workers' mental health, and the Occupational Research Questionnaire which is the sub-scale of the OSI-R is used to measure workers' mental health because of its good reliability and validity.

2.3 Personal resources and mental health

Personal resources refer to the ability of workers to cope with professional tasks. Hobfoll's Conservation of Resource Theory (COR) bridges the gap between environmental and cognitive aspects by introducing behavioral perspectives to analyze the situation [8]. According to the COR model, people suffer from mental health problems due to: (1) the threat of the loss of resources, (2) the actual loss of resources, or (3) a lack of resources gain following the investment of resources[17]. People strive to either acquire or maintain resources that have value for them. The COR model describes how the professional tasks cause mental health problems evidenced by depression, phobic and anxiety due to the loss of personal resources in the process, and it's suitable for this study to use the COR model to explore the interrelationship of personal resources, professional tasks and mental health. The study of college students has appeared that social support mediated the relationship between emotional intelligence and mental health [18]. In the study focusing on medical workers' health, different kinds of coping resources were considered to have different impacts on mental health, and the coping resources were measured by the OSI-R[19]. Above all, studies have suggested that personal resources change the way that external environment impacts mental health. Examination of the relationship between personal resources and mental health may shed light on how individuals respond in certain situations and provide us with some information on how to intervene. Therefore, the personal resources are taken as another significant factor of workers' mental health in this study, and the Personal Resources Questionnaire which is the sub-scale of the OSI-R is also used to measure workers' personal resources because of its comprehensive content.

3 Materials and methods

3.1 Participants

The front-line workers were investigated from a typical electrical manufacturing company in Zhengzhou by random sampling, which was a leading company of global electrical manufacturing industry with 200,000 workers. The investigation was conducted through asking front-line workers to answer the questionnaires distributed to them. Prior to the data collection informed consent was obtained from the workers.

From March 2016 to May 2016, a total of 480 questionnaires were delivered to front-line workers, and 392 qualified questionnaires remained for analysis during the data cleaning process in this study. Males comprised 54.85% of the total sample. The majority of participants were the only child (82.91% of the sample). 68.11% of participants ranged in age from 20 to 29. And 62.76% were married at the time the study took place. Only 16.33% of participants were college graduates. Additionally, workers with countryside origin comprised 73.47% of the total sample. In summary, The structure of sample was similar to that of company manpower.

3.2 Measures

Data was collected by using the OSI-R and the SCL-90 which were proved valid and reliable in past studies. These items were all rated on a 5-point Likert scale. The OSI-R included 140 items, and was divided into 3 sub-scales: Occupational Research Questionnaire (ORQ), Personal Resources Questionnaire (PRQ) and Personal Strain Questionnaire (PSQ) [6]. The ORQ and PRQ were selected to measure workers' professional tasks and personal resources respectively. Besides, mental health was measured by using SCL-90 with possible answers ranging from 1=All of the time to 5=None of the time [7]. After pretest, Cronbach alphas for three scales were 0.87, 0.90 and 0.88. Results showed that these scales were suitable for manufacturing front-line workers. In addition, questionnaire also included a brief report form that collected information on demographic characteristics.

4 Results

T-test and Pearson correlation analysis were conducted in this study using SPSS 17.0.

4.1 Description of the variables

A simple linear transformation was proposed to standardize the collection data of the OPQ and PRQ.The conversion formula was presented below. x stands for the actual score of item; n stands for the total number of items, and S is the standard score [14].

$$S = \frac{\sum x}{n \times 5} \times 10 \tag{1}$$

After standardized treatment, the descriptive statistics of the variables were shown in table 1.

Table 1 showed that the average scores of all indicators was significantly different from the national norm except interpersonal sensitivity. In professional tasks, the scores of role overload, role insufficiency and role boundary were higher than the national norm. However, the scores of role ambiguity, responsibility and work environment were below the national norm. All scores of indicators in personal resources were below the national norm. The GSI (M=1.62, SD=0.55) of SCL-90 was significantly different from the national norm, and the scores of indicators were much higher than the national norm except interpersonal sensitivity. The results indicated that among manufacturing front-line workers, the condition of mental health and professional tasks were lower than the national average level, and the personal resources were relatively deficient as well.

4.2 Correlation analysis

The results of Pearson correlation analysis were shown in table 2, indicating that professional tasks and personal resources were both correlated strongly with mental health. It stated a positive relationship between the indicators of mental health problems and professional tasks (except responsibility), while personal resources and mental health problems were strongly negative correlation. Besides, the correlation coefficients of role overload, social support, role insufficiency and role boundary were obviously higher than other indicators.

Table 1. Descriptive statistics of the variables.

Factor		Mean	SD	Norm	t
professional tasks	Role overload	6.42	1.06	5.79±1.16	11.75^{***}
	Role insufficiency	6.59	0.88	6.05±1.38	12.28^{***}
	Role ambiguity	5.42	1.17	5.64±2.11	-3.79^{***}
	Role boundary	5.29	1.12	4.96±1.02	5.83^{***}
	Responsibility	4.82	1.29	4.93±1.27	-1.64^{*}
	Work environment	4.58	1.74	5.18±1.47	-6.85^{***}
personal resources	Recreation	4.85	1.21	5.48±1.09	-10.45^{***}

	Self-care	5.31	1.17	5.90±1.15	-10.07***
	Social support	6.71	1.49	7.32±1.30	-8.13***
	Rational coping	6.63	1.32	7.13±1.19	-7.59***
mental heath	Somatization	1.55	0.62	1.37±0.48	5.71***
	Obsessive compulsive	1.84	0.62	1.62±0.58	6.96***
	Interpersonal sensitivity	1.69	0.60	1.65±0.51	1.33
	Depression	1.64	0.63	1.50±0.59	4.51***
	Anxiety	1.56	0.58	1.39±0.43	5.62***
	Hostility	1.64	0.65	1.48±0.56	4.71***
	Phobic anxiety	1.43	0.52	1.23±0.41	7.75***
	Paranoid ideation	1.58	0.58	1.43±0.57	4.97***
	Psychoticism	1.53	0.56	1.29±0.42	8.33***
	GSI	1.62	0.55	1.44±0.43	6.47***

Note: *: P<0.05, **: P<0.01, ***: P<0.001.

Table 2. Correlations on personal resources, professional tasks and mental health.

	Somatization	Obsessive compulsive	Interpersonal sensitivity	Depression	Anxiety	Hostility	Phobic anxiety	Paranoid ideation	Psychoticism	GSI
Role overload	0.536**	0.521**	0.489**	0.513**	0.509**	0.528**	0.511**	0.482**	0.413**	0.523**
Role	0.315**	0.323**	0.273**	0.327**	0.352**	0.311**	0.309**	0.318**	0.362**	0.336**
Role ambiguity	0.119*	0.153**	0.164*	0.094*	0.138**	0.135*	0.106	0.122*	0.101*	0.143*
Role boundary	0.322**	0.339**	0.275**	0.328**	0.306**	0.206**	0.314**	0.268**	0.341**	0.312**
Responsibility	0.075	0.082	0.109*	0.097	0.069	0.073	0.084	0.089	0.067	0.092
Work	0.138*	0.113*	0.231**	0.184*	0.122*	0.147*	0.155*	0.137	0.121*	0.181*
Recreation	-0.287**	-0.319**	-0.298**	-0.311**	-0.323**	-0.316**	- **	-0.308**	-0.281**	-0.291**
Self-care	-0.211**	-0.267**	-0.172**	-0.237**	-0.196**	-0.305**	- **	-0.231**	-0.216**	-0.215**
Social support	-0.412**	-0.489**	-0.504**	-0.478**	-0.459**	-0.448**	- **	-0.484**	-0.466**	-0.483**
Rational coping	-0.206**	-0.249**	-0.301**	-0.212**	-0.244**	-0.254**	- **	-0.227**	-0.253**	-0.237**

Note: *: P<0.05, **: P<0.01, ***: P<0.001.

5 Discussion

To study the mental health condition of the manufacturing front-line workers, the interrelationship of personal resources, professional tasks and mental health was explored in this study.

The questionnaire about professional tasks of 392 front-line workers indicated that the average scores of role overload, role insufficiency and role boundary were higher than the national norm. On the contrary, the scores of role ambiguity, responsibility and work environment were lower than the national norm. Results implied that manufacturing front-line workers were lack of responsibility for heavy tasks. However, the duty of job was defined clearly and work environment was tolerable. Essentially, many manufacturing companies were labor intensive companies in China. The consecutive work for front-line workers exceeded 10 hours a day. Companies usually implemented shift work operation and asked workers to work overtime on weekends or holidays. The manufacturing front-line workers engaged in high-intensity, monotonous tasks day by day, which led to an increase in the number of mental health problems in recent years.

In this study, the scores of all indicators in personal resources were below the national norm. It implied that workers lacked the necessary social support and ways to relieve stress when they engaged in the high-intensity tasks. The indicators included in this study depicted the individual conditions of the COR model. The high-intensity tasks caused an actual loss of resources due to the multiple demands of work, and the fear of resource loss created mental health problems. The theory of conservation of resources advocated that personal resources were exploited in the process of relieving stress. Thus, it was imperative for managers to take workers' mental welfare into frame policies to promote workers' physical and mental health, so as to achieve high quality job.

The GSI of SCL-90 in this study was significantly different from the national norm, and the score of its indicators were much higher than the national norm except interpersonal sensitivity. It implied that the condition of mental health among manufacturing front-line workers was poor. And it's below the national average level. Combined with the present survival status of manufacturing workers, high-intensity tasks and harsh management style can be considered as the major factors of the mental health problems. Because manufacturing front-line workers with heavy tasks had fewer opportunities for human interaction. Moreover, in order to maximize profits, companies regarded cheaper labor as a tremendous competitive advantage. Thus, they usually formulated draconian management system and treated workers as machines. Under the special working condition and inflexible management system, workers endured excessive work stress and lacked the way to release their emotion. These factors will inevitably led to mental imbalance.

Correlation analysis indicated a negative relationship between the indicators of mental health and professional tasks (except responsibility), while personal resources and mental health were significantly positive correlation. It implied that the mental health problems were severer in population with heavier tasks and less personal resources. In professional tasks, the correlation coefficients of role overload, role insufficiency and role boundary were obviously higher than other factors. In addition, social support was the key indicator of the correlation between personal resources and mental health. Because manufacturing front-line workers were usually away from home and lacked of proper care from relatives and friends. Meanwhile, their experience of solving severe mental health problems was insufficient. So that, the support from organization, colleagues and friends had a considerable effect on workers' mental health. According to the results, managers can improve workers' mental health by increasing mental welfare. For instance, increasing entertainment facilities in the factory, organizing mental health-care activities, implementing job-rotation regularly and improving the psychological help system. Moreover, recruiting local workers could also improve their personal resources, which was the major factor of workers' mental health.

6 Conclusion

The study has explored interrelationship of personal resources, professional tasks and mental health through the analysis of the questionnaire. Several tentative conclusions may be drawn: (1) The mental health problems are severer among manufacturing front-line workers with heavier tasks and less personal resources. (2) The indicators of personal resources are below the national norm. In professional tasks, the scores of role overload, role insufficiency and role boundary are higher than the national norm. However, the scores of role ambiguity, responsibility and work environment are lower than the national norm. The average scores of SCL-90 are much higher than the national norm except interpersonal sensitivity. (3) Personal resources and mental health are significantly positive correlation, while the indicators of professional tasks (except responsibility) have a significantly negative impact on mental health. So, it is an important way to ensure the front-line workers' mental health by improving workers' mental welfare, like enhancing personal resources and reducing professional tasks.

References

1. NBS. *China Statistical Yearbook* (Chin Stats Press, Beijing, 2014)
2. Z.G. Lin, F.M. Fan, J. Wu, Chin. J. Health. Psy, **20**, 1345 (2012)
3. J.L. He, R. Zhao, Talent Exploitation, **7**, 19 (2010)
4. D.C. Ganster, C.C. Rosen, J. Mgmt, **39**, 1085 (2013)
5. M.J. Xu, X.N. Huang, Z.Y. Feng, J. Environ & Occup. Med, **33**, 134 (2016)
6. S.H. Osipow, PAR-Psy. Assess. Res. Inc, (1998)
7. L.R. Derogatis, R.S. Lipman, K. Rickels, Behav. Sci, **19**, 1 (1974)
8. S.E. Hobfoll, Appl. Psy, **50**, 337 (2001)
9. H.B. English, A.C. English, **115**, 957 (1958)
10. S. Paek, M. Schuckert, T.T. Kim, Int. J. Hosp. Mgmt, **50**, 9 (2015)
11. R. Loi, O.K.Y. Ao, A.J. Xu, Int. J. Hosp. Mgmt, **36**, 23 (2014)
12. S.L. Huang, W.J. Hou, M. Zhang, Acta Psy. Sinica, **47**, 466 (2015)
13. Q.Y. Zhong, K. Li, G.Q. Li, Chin. J. Health. Psy, **12**, 1815 (2013)
14. Z.W. Zhang, B. Liu, B.F. Li, Ind. Eng. Mgmt, **18**, 141 (2013)
15. J. Sharma, R.L. Dhar, A. Tyagi, Appl. Nurs. Res, **30**, 268 (2016)
16. W. Chen, T. Wong, T. Yu, Work. Stress, **17**, 287 (2003)
17. L.M. Little, D.L. Nelson, M.J. Quade, J. Vocat. Behav, **79**, 724 (2011)
18. M. Zeidner, G. Matthews, Pers. Indiv. Diff, **99**, 196 (2016)
19. W. Zhu, Q.S. Jiang, Z.M. Wang, J. Zhengzhou. Univ. (Med. Sci), **43**, 468 (2008)

A Gas Scheduling Optimization Model for Steel Enterprises

Honghai Niu*, Xin Geng and Jun Chen

Research and Development Center of NR Electric Co., Ltd., Nanjing 211102, China
Corresponding Email: niuhh@nrec.com

Abstract. Regarding the scheduling problems of steel enterprises, this research designs the gas scheduling optimization model according to the rules and priorities. Considering different features and the process changes of the gas unit in the process of actual production, the calculation model of process state and gas consumption soft measurement together with the rules of scheduling optimization is proposed to provide the dispatchers with real-time gas using status of each process, then help them to timely schedule and reduce the gas volume fluctuations. In the meantime, operation forewarning and alarm functions are provided to avoid the abnormal situation in the scheduling, which has brought about very good application effect in the actual scheduling and ensures the safety of the gas pipe network system and the production stability.

Keywords. Steel enterprises, gas scheduling optimization, process state calculation, soft measurement, operation forewarning.

1 Introduction

Steel industry is one of the energy-intensive industries which shares 10% total energy consumption of the national economy and the energy cost of steel enterprises takes about a third of the production cost [1]. High energy consumption not only increases the cost of steel products but brings about much pollution and emission. In practice, due to the changes of factors such as the production and maintenance schedule, gas production and consumption keep changing. Gas system exhibits imbalance in a short time which results in low gas utilization rate. The aforementioned problems need to be solved by optimizing the gas system scheduling.

Recently, researches about the energy optimization model of steel enterprises have been carried out extensively. Ref. [2] described the relationship between the various units of gas pipeline network as the form of matrix then modeled and solved the problem with partial least squares regression method. Ref. [3-5] proposed the gas supply and demand forecast model of steel enterprises and realized the gas tank safety control and integrated power generation benefit maximization.

Although the above algorithm is theoretically reasonable and the simulation results are satisfactory, the instantaneity and effectiveness of are not enough for the gas scheduling. In this research, according to the characteristics of gas pipeline network system and the actual scheduling process in steel enterprises, the generation and consumption of gas are scientifically and effectively estimated and the by-product gas system is modeled and analyzed. Then the gas scheduling optimization model is proposed according to the rules and priorities. The proposed model is of great significant in reducing the gas loss and balance of the whole gas system.

2 Problem description

The by-product gas system of steel enterprises is mainly composed of blast furnace gas and converter gas system. The gas system is illustrated in Figure 1:

The basic data of energy control system of steel enterprises is relatively consummate, telephone scheduling has been replaced by centralized scheduling. However, balance scheduling operation is still dependent on the experience of scheduling staff. Gas scheduling is based on the expertise of scheduling staff which is highly influenced by human intervention besides the following problems:

(1) When the process system state changes, the gas consumption variation is very larger. The gas will exhibit large fluctuation and loss if the scheduling center is not informed timely.

(2) At present, the measured value of some gas meter different from the actual value. For examples: the gas consumption power plant boiler is not accurate, the gas generation meter of blast furnace is not accurate in small scale range.

(3) When there is a large fluctuation of gas, the extra blast furnace gas is usually released to maintain the pipe network pressure. For examples: re-blowing of blast furnace after maintenance will cause large amount of gas, maintenance of rolling accident will decrease the gas consumption sharply.

(4) There are no operation forewarning and alarm functions in existing gas scheduling. For example: when there is gas release while the power plant boilers capable to consume the gas and the boiler is running normally, the boiler load should be raised to reduce gas release.

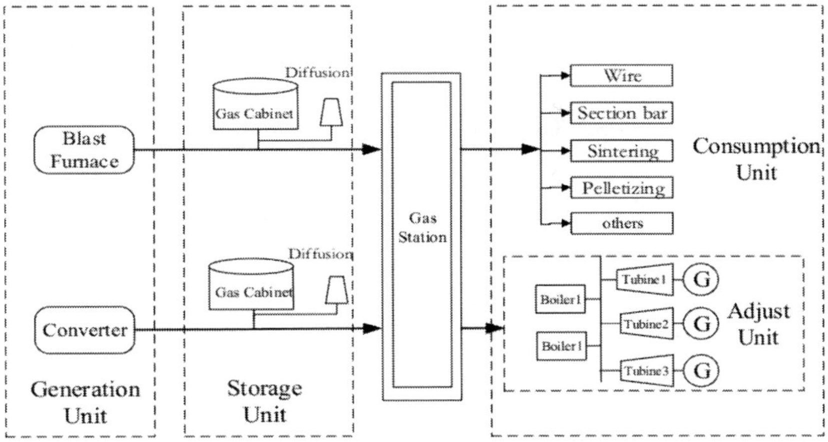

Fig. 1. The gas system diagram.

3 Gas scheduling optimization design

According to the features of steel enterprises' gas pipe network, the gas scheduling optimization is proposed based on the energy control system, including:

(1) Process state calculation

By the calculation of process states including blast furnace, cowper stove, steel rolling furnace, sintering, thermal power plant, etc., the switching state of the main gas units are judged so that the scheduling staff can see each process state in real time.

(2) Gas soft measurement

When the process states change, the gas consumption of the current states can be estimated according to the features of the gas system. Then the fluctuation quantity of gas consumption caused by process states change can be calculated.

(3) Building up the gas scheduling optimization model based on the rules

According to the effective scheduling rules, the fluctuating range and the consumption amount of gas are determined. Then the gas can be scheduled according to the priorities.

(4) Forewarning and alarm of scheduling rules

According to the scheduling guidelines, forewarning and alarm will be given against violation of scheduling rules to correct the unreasonable operation of scheduling staff.

The optimized scheduling process is illustrated in Figure 2:

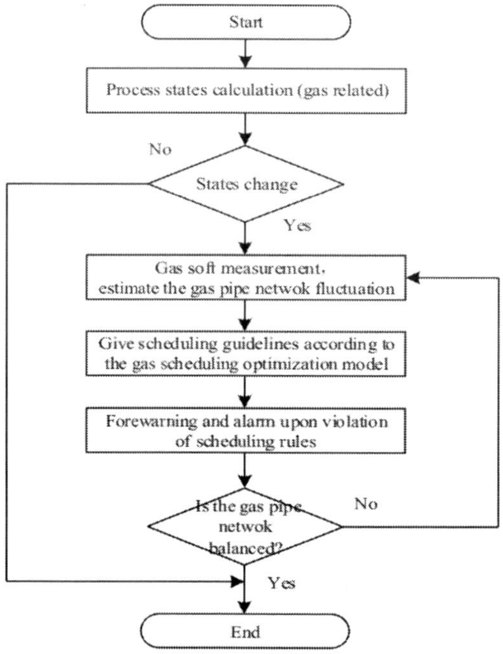

Fig. 2. The gas scheduling optimization diagram.

3.1 Process states calculation

The processes related to the gas production and consumption, including sintering, blast furnace, converter, steel rolling furnace, etc., are calculated. When the states change, the scheduling staff will be informed by warning so that the real-time process states can be taken into account for gas scheduling optimization. Take the sintering process as an example, the state logic is shown in Figure 3:

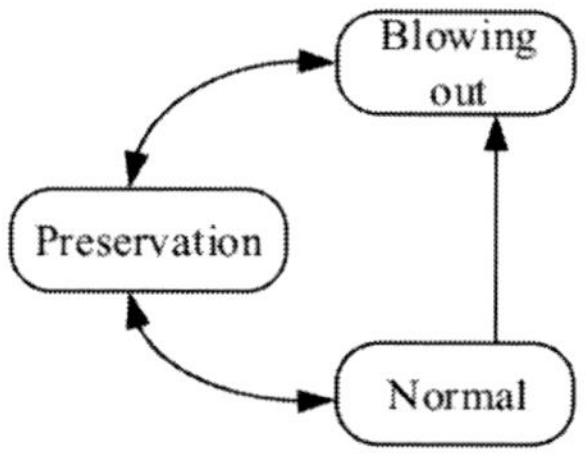

Fig. 3. Sintering state logic.

The sintering ignition furnace have three operation states including ignition, preservation, and blowing out. The dynamic model of gas consumption is shown below:

$$g(t)=\begin{cases} Q_n, & t \in \text{normal production} \\ Q_p, & t \in \text{preservation} \\ 0, & t \in \text{blowing out} \end{cases} \tag{1}$$

where: Q_n is the gas consumption of the sintering furnace in normal state, Nm^3/h; Q_p is the gas consumption of the sintering furnace in preservation state, Nm^3/h.

The state calculation logic of the sintering ignition furnace is described below:

(1) Blowing out: the gas consumption of sintering furnace is below the setting value (such as 100 m3/h) or the average firing temperature of sintering furnace is below the setting value (such as 400 °C), and the temperature rate is 0 m/min.

(2) Preservation: the average ignition temperature of ignition furnace is above the setting value (such as 400 °C), and the temperature rate is 0 m/min.

(3) Ignition: the temperature rate of sintering furnace is above the setting value and the average ignition temperature of ignition furnace is above the setting value (such as 900 °C).

3.2 Gas soft measurement

The measurement accuracy of gas meter is bad considering small range measurement and the impurities and soft measurement should be carried out to calculate and correct the gas production. At the same time, when the process states change, gas production and the gas consumption of power plant boiler can be calculated according to the correlations between the processes' data, so that the scheduling staff can be aware of the pipeline network fluctuating quantity. When the deviation between estimated and actual value is intolerant, warning will be given to remind the scheduling staff to optimize the scheduling of gas consumption. Take the gas consumption of power plant boiler as an example, the algorithm model of gas soft measurement is introduced as below.

Regarding the gas consumption calculation of single boiler, the boiler heat absorption is first calculated according to the main steam enthalpy rise and gas flow, then the boiler heat release can be obtained from the calorific value of gas. Finally, the gas consumption per hour can be calculated based on the heat efficiency of boiler:

$$V_G = \frac{(H_{zzq} - H_{gs}) * Q_{zzq} * 1000}{\eta H_m} \quad (2)$$

where: V_G is the gas consumption of boiler (m^3/h); H_{zzq} is the main steam enthalpy of gas boiler (KJ/kg); H_{gs} is the water feed enthalpy (KJ/kg); Q_{zzq} is the main stream flow (t/h); η is the heat efficiency of gas boiler; H_m is the calorific value of gas (KJ/m^3) which can be obtained from the calorific value analyzer.

3.3 Function description of scheduling based on rules

In this research, gas scheduling algorithm is designed based on the rules, which includes:

3.3.1 Scheduling priority division

In order to further utilize the gas resource and satisfy the needs of actual production, each unit of the gas system is simplified and the units are divided into three categories according to the applications as shown in Table 1:

Table 1. Categories of the users.

Categories	Contents
Type 1	Users must be ensured in main production.
Type 2	Users with small correlation between main production.

Type 3	Buffer and adjust the user.

Set the scheduling priority for each type of user. According to the priority ranking, gas supply of some users are suggested to be cut off upon gas shortage. The definition of priority can be given by the user. The priorities are represented by integers, the greater value means that the user is less important and the adjustment is faster, which will be adjusted to relieve the gas pipeline pressure when the gas is not balanced.

3.3.2 Scheduling rules description

The type 3 users are first adjusted upon gas fluctuation, type 1 and 2 users will be adjusted when the adjustment range of type 3 is beyond. Specific scheduling principles depend on the fluctuation quality, unit of action, state of the consumption user, and the adjustment ability of the unit user, the purpose of adjustment is the balance between production and consumption of gas. Negative fluctuation quality means gas oversupply and the consumption of type 3 users should be increased to balance the gas supply. If the fluctuation quality is in the adjustment range of a certain type of users, the users of high priority value will be adjusted. When the fluctuation quality is beyond the adjustment range of a certain type of users the adjustment will minimum the number of adjusted users. For gas oversupply, the gas consumption of type 3 users will be adjusted to the maximum value and the extra gas will be released. For gas undersupply, the gas consumption of type 3 users will be adjusted to the minimum value and some users will be cut off according to the priorities.

3.3.3 Scheduling function process

The gas medium of the new district includes blast furnace gas (BFG) and Linz-Donawitz process gas (LDG). The specific algorithm process depends on the gas medium as illustrated in Figure 4:

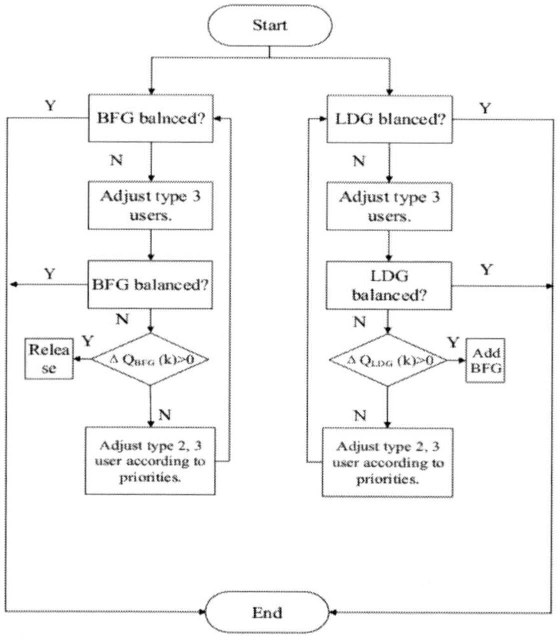

Fig. 4. Scheduling process diagram.

The residual fluctuation quality $\Delta Q_{BFG}(k)$, $\Delta Q_{LDG}(k)$ represent the residual fluctuation beyond the adjustment range of blast furnace and converter after the adjustment of three types of users.

3.4 Scheduling forewarning and alarm functions

Based on the scheduling results of model calculation in the previous section, forewarning and alarm will be given against violation of scheduling rules to correct the unreasonable operation of scheduling staff.

3.4.1 Scheduling forewarning

During the gas scheduling, the scheduling staff should timely response to the changes of pipeline network system. The forewarning functions are designed to remind about some significant operations and help the scheduling staff to make actions in advance. For examples:

When there is a blast furnace plan damping down, it is suggested to maintain every gasometer (blast furnace, converter) at a high counter, the unsatisfied counters will result in forewarnings.

(1) When the blast furnace gas production sharply decreases, or the variation rate beyond the set value, forewarning will be given.

(2) At the beginning of the converter blowing (or the blowing step 1), it is suggested to recover the converter gas and check the counters of the gasometer so that to improve the converter gas recovery rate.

3.4.2 Scheduling alarm

Violation of scheduling rules will be displayed and prompted on screen in the form of alarm, for example:

(1) Power plant boiler has the ability to consume and operates normally while there is gas release.

(2) There is gas release while the gas tank have space of absorption.

(3) The CO and O2 concentration of converter gas are in safe range while the gas is not recovered when possible.

4 Application effect

Regarding the specific gas scheduling problems of steel enterprises, the aforementioned model is applied to the following two practical production cases: case 1 represents a certain normal production case, case 2 is based on case 1 while the profile production line is shut down due to unplanned maintenance. The gas optimization results are shown in Table 2.

Table 2. Gas optimization of different production cases.

Process	Case 1			Case 2		
	Production (t/h)	Blast furnace gas (m3/h)	Converter gas (m3/h)	Production (t/h)	Blast furnace gas (m3/h)	Converter gas (m3/h)
Sintering	298	15830	0	298	15830	0
Puddling	272	405782	0	272	405782	0
Steelify	235	4255	7270	235	4255	7270
Bar	152	64865	0	152	64865	0
Profiles	65	10980	0	0	0	0
Power plant		192905	15745	0	203885	15745
Release		0	0	0	0	0

Table 2 indicates that the energy supply of each process has been timely redistributed by the scheduling optimization model when the production case changes. There is no gas release after the optimization which confirms the validity of the proposed rule-based gas scheduling optimization model. In addition, the forewarning and alarm functions improve the fault treating ability of scheduling staff.

5 Conclusions

The proposed gas scheduling optimization functions take the different characteristics and technical requirements of various gas units of steel enterprises in practical production process in account. The rule-based gas scheduling optimization model has been established. Algorithm models of Process state calculation and gas consumption soft measurement have been proposed to deal with specific scheduling problems. Based on the rules of scheduling optimization, the scheduling staff are timely informed about the gas consumption state of various processes then schedule gas in time and reduce fluctuation quality. The model can effectively avoid the abnormal situations by forewarning and improve the safety of the gas tank and the stability of production, which has obtained well application effect in practice. This model is of great theoretical and practical significance in reducing gas release and realizing the balance of the whole gas system.

References

1. ZHANG qi, CAI Jiu-ju, PANG Xing-lu, et al. Optimal distribution of by-product gases in iron and steel complex. Journal of Northeast University, 2011, 32(1): 98-101.
2. ZHANG qi, CAI Jiu-ju, DU Tao. Optimum utilization of gas system in incorporated business enterprises of iron and steel. Energy for Metallurgical Industry, 2005, 24(5): 9-16.
3. Su C L. Probabilistic load-flow computation using point estimate method. IEEE Transactions on Power Systems, 2005,20(4):1843-1851.
4. Coello C A C, Pulido G T, Lechuga M S. Handling multiple objectives with particle swarm optimization . IEEE Transactions on Evolutionary Computation, 2004,8(3):256-279.
5. Kim J H, Yi H S, Han C H. Plant-wide optimal byproduct gas distribution and holder level control in the iron and steel making process, Korean Journal of Chemical Engineering, 2003, 20(3): 429-435

Discussion on Manufacturing and Services Integration Based on E-Commerce Platform

Jinfa Li[1] and Guangcheng Liu[1]

[1]School of Management Engineering, Zhengzhou University, Zhengzhou 450001, China
Email: lijinfa@zzu.edu.cn, 272572784@qq.com.

Abstract. The integration of manufacturing and services is the key to adapt to the inevitable requirement of the revolution of the development of a new round of mobile Internet industry. They are also the key and trend of the next period to enhance the quality and efficiency of Chinese's manufacturing industry to release potential services development fully, and cultivate new economic growth point.Meanwhile it makes it more effective than before.The enterprises can publicize their own values as well as collect and deal with consumer feedback of the product experience more effectively.And then,it can grasp the market trends and adapt to developping world more quickly.
Keywords. Manufacture, services, industry integration, mobile internet.

1 Introduction

With the rapid development of modern information technology, manufacturing and services integration have become the mainstream trend of modern industrial development, And it also the main driving force of promoting global industrial upgrading. The services and the manufacturing show a trend interactionsy and mbiotic interdependence.And then it constantly give birth to new industries and new format, so as to promote economic transformation from the industrial structure to services economy [1] plays an increasingly more important role.And the fierce competition in the E-economic platform made the competition of original price and quality of goods be the current competition of services [2].The services includes price, quality, and user experience and so on [3]. The academic papers and related institutions concerned more about the question of the integration of manufacturing and service industries national [4]or regional [5] as well as in the fusion path [6]. Now this paper will explore the process of how they interact with each other in the process of integration manufacturing and services.

2 The Problems and Current Situation Before The Integration Of Manufacturing And Services

2.1Manufacturers have low market-perceived capabilities

Manufacturers need spend serval and even more experie-ncing product research, design, selection of materials be-fore production.In the era of morden market,information changes fast, consumer's demand is diverse and volatile. If the manufacturer's perceive abil- ity and foresight abi- lity are very strong, it may be able to adapt to market ch-anges, or it will be difficult to survive.

2.2 Distributor are more sensitive to the market

Distributor can know the evaluation of product featu- res, style in service and quality directly when consu-mption.Especially in the era of e-commerce, in the e-commerce platform, the customer care know the cu- stomers' satisfaction level of product's feature,style and material when they talk each other.Then distrib-utor can summarize the good's advantages and disa-dvantages.After market customer service will know about the feeling about the customer experience m-ore intuitively when they communicate with consum- ers.Therefore, the distributor will conclude the demand of the good basing on the information ,then they find the target product.By this way,they can improve their own serivices constantly.In the ear of information today,it not only accelerates the spreed of information, but also acce-lerates the competition.It is more difficult for manufactu- rers to survive especially smaller.Only distributors must

2.3 Cooperating with each other is difficult

Distributor and manufacturers 'focusing point is different. For the same to capture high profits, distributors focus the point of improving service levels. But manufacturers are more care about their own products and whether the concept of the design product is recognized or not by society. Distributors and manufacturers concerned about the different groups, The species of manufacturers' product is singal, so manufacturers focuse more about a single product and its corresponding group is single too, But the number of the group is larger. Distributors' product species is more than manufacturers'.so the target groups smaller, and its personality and characteristics are more obvious. In the ear of the information, the group of pursuiting fast fashion is restless, and the younger, the more serious.

Normally, the scale of distributors is small, so most distributors are unable to publicize his goods on the e-commerce platform On a large scale. In order to increase the number of customer, distributors will choose more types of products. But manufacturers don't have so many species especially for small enterprises. In order to allow people to accept their products, manufacturers must publicize their own product ideas so that it spreads in the crowd of target group. Therefore, when the product can't meet the needs of the market in time, the manufacturer cannot get the consumers' feedback of product form distributor, the distributor cannot yet get support from the manufacturer about product optimization.so they cannot cooperate with each other too.

3 The positive meaning of industrial integration

3.1 Improve the control ability and the service level in terms of consumption

Information transmission between enterprises and consumers is the most direct and rapid through e-commerce platform. Perceptions of consumer truly reflects the consumer experience, the direct dialogue improve the directness and effectives of information transmission. According to the consumer experience, enterprises can optimize the product in order to make every customer satisfied. Accurate positioning of the market, fully display the brand culture, spread the value of enterprises.

3.2 Reduce inventory costs and operational risk

Inventory risk is the problem of traditional enterprise is inevitable. From the early Lining to XTEP and PEAK, they were all malaise for years or even bankrupty because of inventory problem. While e-commerce enterprises can base on the platform of large data monitoring product popularity and purchase quantity and more accurately predict product sales quantity. But traditional business is difficult to achieve it. The first brand of clothing electronic commerce Inman, sold over one million pieces of clothing with the inventory less than 10% at end of the year which below the average inventory level of 70% inventory of traditional women brand.

4 The diagram of services and manufactures integration model

4.1Industrial convergence model diagram

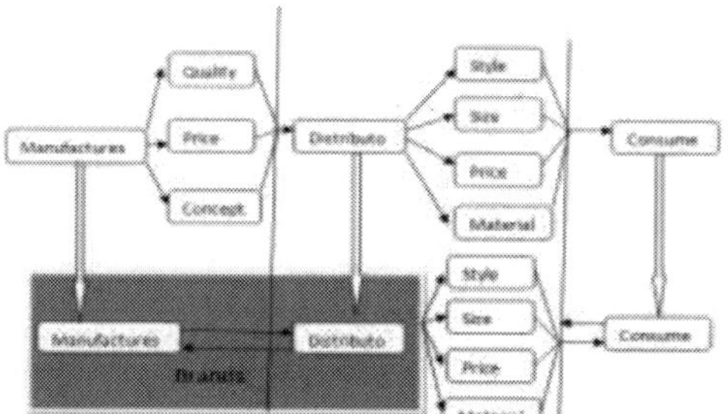

Fig. 1. Industrial convergence model diagram.

Manufacturers and service providers, consumer was the state of the upstream and downstream before the integration of industries. The information from the consumer to the service provider and then to the manufacturer should go through two stages, the transmission effect decreasing step by step. After the industry convergence， manufacturers and service merge into one, and its ternary interaction before fusion become binary interaction. From two levels of information transmission to one level and greatly increase the transmission speed of information. The change between manufacturer and service provider of information transfer is: from the external transfer into internal transfer; from passive transfer into active transfer, this improves the transmission speed and efficiency. Consumer's information feedback is adopted, and it improves the interactive enthusiasm of consumers. The manufacturer and service provider improve their company's performance due to the utilization of information. After the integration of the industry, manufacturers and service providers are not only simple sale goods, but more delivery the value of the brand from selling goods to do the brand, from short-term shop to long-term brand building.

4.2 Industry convergence case: Inman [7]

4.2.1 Business background

2005, Fang Jianhua joined Alibaba integrity, the initial contact with electronic commerce. 2007 registered Inman brand, began to test the domestic market. 2007 began to develop the mobile Internet, Fang Jianhua divides the company's business, Inman officially settled in Taobao, turnover of several million, profit for the loss. The same year to stay Taobao Mall (the predecessor of the Tmall). 2009, 2010, Inman loss of the previous year to reach more than two million in the electricity business. 2010 Inman for the first time to participate in the double eleven, the day sales record of 6.8 million, annual trading volume of 70 million, the basic line of profit and line flat. 2011, with 17.87 million sales ranked Tmall women's wear brand. The annual turnover of 140 million. 2012, with a single day sales of 70 million refresh brand history, order trading volume of 330 thousand. 2012, annual turnover of more than 300 million. 2013, Inman day sales exceeded 120 million, becoming the first full-line sales of women's brand. 2013 annual turnover exceeded 590 million. 960 million in 2014. 2015 annual turnover is more than 1.1 billion mark.

4.2.2 Industrial integration performance

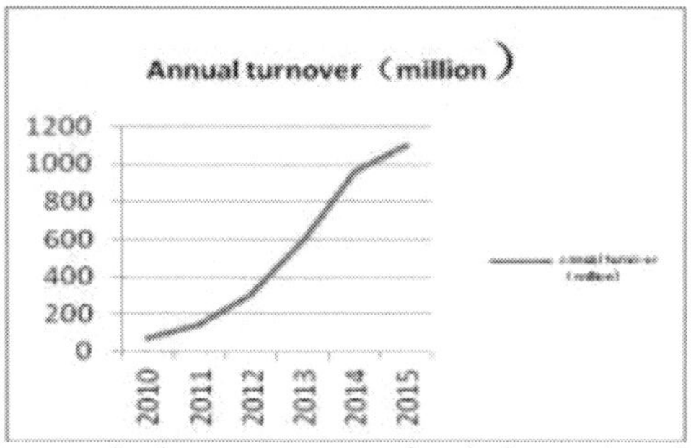

Fig. 2. Inman calendar year turnover after the integration of industry

5 Conclusion

Consumers continue to highlight the personality, small and beautiful things continue to be sought after. Mass pro duction of traditional enterprises is becoming increasing ly difficult to adapt the fast fashion consumer environment.

The development of the whole industry chain has become a new profit growth mode, with the manufacturing and service integration continues to increase, New technologies, new formats, new management models are emerging, to promote more refined industrial division of labor, the value chain of the industry value chain has become more and more, Products from the development, production to marketing, distribution, maintenance, and so on, the formation of the value chain process began to decompose, integration and reconstruction, Products in the manufacturing process only a small part of the time, most of the time in R & D, procurement, storage, operation, sales, aftersales service, etc, The industry chain more c-oncentrated in producer service sectors, the efficiency , valueadded space, profit margins of these services on the whole industry chain are great influence, thus becoming the industry-wide value-added, the main source of profit, but also the whole industry control decision facto- rs.With ecommerce from the sales side to the designside penetration, reduce production costs, optimize product features will be the main function of e-commerce.

Enhance interaction with consumers, the development of fan economy. Fans of the economy are now a new marketing model, the company through the use of brand promotion, to win fans to achieve economic efficiency. Fans of the economy from the identification of the prod-uct, is the product of the concept of affirmation, fans of the economy continues to improve the quality of the product, but also the continuous improvement of user experience. This need to continue to increase the depth of industrial integration.

References

1. Wang Xiaohong. Six Trends Of The Convergence Of Manufacturing And Service.New Chongqing.2014.08

2. Xing Xiangjun.Study On The Influencing Factors Of E-commerce Service Industry In China. Commercial Times.2014.07
3. Lin Nan To Promote Industrial Upgrading By The Integration Of Services And Manufacturing Industry. Macroscopic economy.2010.08.
4. Li Jianzhong Grasp the trend of sthe integration of manufacturing industry and services .Zhengjiang Economy.2015.
5. Li Bo An Study on Interactive of Producer Services and Manufacturing Industry: A Case Study of Dalian 2011.05
6. The Integration of Manufacturing and Service Based on Independent Innovation: Mechanism and Path. Journal of Nanjing University.2012.
7. The Interview Of Wang Jiangzhong.2013.09

Construction and Empirical Analysis of Innovation Index System of Zhengzhou-Luoyang-Xinxiang Self-dependent Innovation Demonstration Area

Meng Zhan

School of Management Engineering, Zhengzhou University, Zhengzhou,P.R. China
Corresponding Email: mengzhan217@163.com

Abstract. As the main cities of the Central Plains Urban Agglomeration, Zhengzhou, Luoyang, Xinxiang has established evaluation index system which is based on innovative environment,innovation inputs and innovation outputs.Determining of the one-level index weight by Analytic Hierarchy Process (AHP), and using the entropy weight method to determine the two-level index weight. By analyzing the results, obtained the following conclusions:Zhengzhou has a distinct advantage in the innovation environment, innovation inputs and innovation outputs,but there is a downward trend in innovation environment and innovation input.;Luoyang is in a stable stage basically,and with great advantages in terms of innovation output. Xinxiang at a disadvantage in the three regions,but its better environment for innovation,innovation inputs continued to increase,innovation output tends to increase.

Keywords. Zhengzhou-Luoyang-Xinxiang self-dependent innovation demonstration area, innovation index, index system.

1 Introduction

In April 2016, the State Council approved that Zhengzhou-Luoyang-Xinxiang national high tech Zone to build the national independent innovation demonstration zone, and to carry out scientific and technological system and mechanism innovation actively, and exploration in some aspects as follows: scientific research projects and funds management, equity incentive, the combination of technology and finance, the use of intellectual property and protection, personnel training and introduction, scientific and technological achievements, technological evaluation. As of July 2016, the State Council officially approved the 17 National Innovation Demonstration Zone. National Independent Innovation Demonstration Zone play an important lead, radiation and drive role in further improve the mechanism of scientific and technological innovation system, accelerate the development of strategic emerging industries, promote innovation-driven development, accelerate the transformation of economic development mode and so on. Furthermore, National Independent Innovation Demonstration Zone has become a great supporting strength of China's economic development, industrial transformation and upgrading, and innovation-driven development, under the condition of Chinese New Normal. After several years of development, some national independent innovation demonstration zone, such as Zhongguancun, Zhangjiang, Donghu, has become the main driving force of regional economic development, and the model of China's National Innovation Demonstration Zone. The gross value of industrial output of Zhongguancun is 410 billion 410 million yuan, accounts for 51.1% of the city. The number of patent applications of Zhongguancun enterprises reached 27966, accounts for 32.9% of the city. Silicon Valley initiated a science, technology, production of three-dimensional integrated development model, with small and medium-sized enterprises as the core. It has created a global scientific and technological innovation, transformation of scientific and technological achievements and the development of high-tech industry model, becoming the world's most dynamic economic region. Tsukuba, Bangalore, Hsinchu Taiwan, China, and other world famous high-tech Industrial Development Zone has provided a good experience for the good development of China's National innovation demonstration zone. Considering the difference of regional economic development, we put forward to the development model of national independent innovation demonstration zone, which government-led, cluster development of electronic information, advanced manufacturing, new materials and other strategic emerging high-tech industry, the industrial-academic-research collaborative development.

Creativity is the soul of a nation's progress.With the implementation of the strategy of innovation-driven development, China's major innovation index has reached the leading level of the world. But the self-innovation capability still faces a severe test.Similar to the function of high-tech park, National Innovation Demonstration Zone is mainly to accelerate the development of strategic emerging industries, promote the strategy of innovation-driven, accelerate the transformation of the mode of economic development.In our country, the national innovation demonstration zone has the functions of strengthening financing, innovation incentive, risk aversion and achievement transformation as well.

In the case of the United States, foreign scholars N. Link Albert and T. Scott John study the impact of Science and Technology Park on the university. They believe that once the establishment of science and Technology Park, the University and other resources near to the science and technology parks will have a huge impact, may make university turn to applied research from basic research directly [1] .Francis C.C. Koha, Winston T.H. Koh and Feichin Ted Tschanga proposed framework includes growth mechanism, technical ability and global role and market integration, analyzed the Silicon Valley, Hsinchu, Cambridge Science and Technology Park, Singapore Science Park, they believe that technological capability, especially R&D which is the primary driving force to promote the development of science and technology parks [2] . Through the study of the Greek Science Park, Yiannis L Bakourosa, Dimitri, C Mardasb anad Nikos and C Varsakelisc found that the enterprise and universities in Science and Technology Park are generally built with informal contact, and synergy only appear in business transactions and social interaction, while in scientific research is not [3] . The domestic scholar Sha Dechun make comparative study of Silicon Valley index and evaluation index system of national high-tech Zone, and conclude that the two have significant differences in the design idea, index content, system structure, making subject and the index function [4] .Taking Silicon Valley and Hsinchu Science Park as an example,Lin Lijian,Teng Tangwei studied the difference and convergence of the fusion of first-class Science Park. [5] Hu Shuhua through the construction of "four-three structure" model, take Wuhan East Lake national innovation demonstration zone as an example to analyze the input and output, the main body of innovation, innovation content and innovation output, and puts forward three measures of the independent innovation project [6] .According to the analysis of complex network structure, Gong Yuhuan propose that the innovation capability of Zhongguancun industrial cluster is highly dependent on a small number of large enterprises [7] .Zhou Hongyu make comparative study to the innovation capability of Zhongguancun, Donghu and Zhangjiang, by construct the evaluation model of innovation capability of National Independent Innovation Demonstration Zone, which is regard innovative input, innovative talents, innovation output, innovation body and innovation environment as the as the indicator information [8] .

2 Contruction of index system

The index theory originated from the price index, and later was widely used in the field of economic statistics, such as CPI, PPI, import and export price index, etc. With the development of economy and society, innovation has become one of the main indicators of national economic development and comprehensive national strength. Innovation index is a comprehensive index to reflect the degree of change of science and technology innovation, and it is used to reflect the relative number of technological innovation and technological innovation ability [9] .The typical innovation index includes: the innovation alliance score card, the global innovation index(GII), the global competitiveness index(GCI) and the Silicon Valley index. Among them, the innovation alliance score card is used to analyze the innovation advantages and disadvantages of EU countries. It includes three one-level indexes such as Innovation vitality, innovation activity and innovation output, and 29 two-level indexes such as human resources, business investment, economic performance, etc. On the basis of comprehensive analysis, furthermore, it divides the EU member states into leaders, followers, general and low innovation countries. In China, the relatively authoritative indices are as follows: national innovation index, national high tech Zone Innovation Index, China Innovation Index, and some indexes that reveal the National Innovation Demonstration Zone such as Zhongguancun index, Zhang Jiang innovation index, Hangzhou innovation index, etc. Overall, from the aspects of scientific objectivity and feasibility principle the evaluation index system establishes tree evaluation index system, and uses a large number of statistical indicators of science and technology, economic indicators, according to the research emphases. Zhengzhou-Luoyang-Xinxiang National Innovation Demonstration Zone studied in this paper is composed of Zhengzhou high-tech Industrial Development Zone, Luoyang high-tech Industrial Development Zone, Xinxiang high-tech Industrial Development Zone. It is a National Innovation Demonstration Zone that takes the city group as the basic unit. While the South of Jiangsu national independent innovation demonstration zone is composed of eight High-tech Industrial Development Zone and Suzhou Industrial Park, is China's first National Independent Innovation Demonstration Zone that tack the city group as the unit. Therefore, when establishing the evaluation index system of Zhengzhou-Luoyang-Xinxiang the National Independent Innovation Demonstration Zone, this paper refers to the relevant research on the innovation index system of Independent Innovation Demonstration Zone in South of Jiangsu by Zhou Yong, Ge Lufei and others [10] .

Table 1. Index and weight of all levels.

one-level index	two-level index	weight
Innovation environment (0.1104)	Number of students per million include graduate and above(10 000 persons)	0.2237
	Number of legal units in industrial agglomeration area(individual)	0.1980
	Per capita GDP(10 thousand yuan)	0.2084
	Number of R & D institutions(individual)	0.1488
	Number of high-tech enterprises identified in the year(individual)	0.2212
Innovation investment (0.3230)	R&D funds accounted for the proportion of GDP(%)	0.2142
	activity personnel of R&D(persons)	0.2110
	the proportion of local financial expenditure on science for total financial expenditure(%)	0.3290
	R&D expenditure(100 million yuan)	0.2459
Innovation output (0.5666)	Invention patent authorization amount(pieces)	0.2355
	Number of published scientific papers(pieces)	0.0725
	sales revenue of new product(100 million yuan)	0.0218
	labor productivity in above-scale enterprise(10 thousand yuan/person)	0.2262
	Ratio of patent license and patent application(%)	0.2108
	the proportion of high-tech industrial output value for above-scale output valu	0.2333

2.1 The index system of innovation index

According to the research above, we can establish an index system of innovation index, which is consisted of 3 one-level indexes such as Innovation environment, innovation investment and innovation output, and 15 two-level indexes such as the number of undergraduates per million, Per Capita GDP, the proportion of total social research funding for GDP, invention patents authorization amount, and the proportion of high-tech industrial output value for above-scale output value etc.See Table 1.

2.2 Determination of the one-level index weight

Due to the one-level target only contains the three aspects content of innovation environment, innovation input and innovation output, so we use Subjective Weight-assign Method when determine the weight of the first-level target. The popular subjective weighting approach includes analytic hierarchy process (AHP), expert investigation method (Delphi method), TACTIC, least square method and two coefficient method, etc.AHP is the most commonly used methods in practical application, and it is able to simplify and layering the complex issue. Thus to use the analytic hierarchy process to determine the one-level index weight.

Design the questionnaire that intercomparsion between innovation environment, innovation investment and innovation output, and ask experts in related fields to make a comparative judgment. And then, organize relevant results according to the Saaty scale table. Finally obtain a comparison matrix A of one-level index, as shown in Table 2. In the third rows and second columns of comparison matrix of the one-level index, there is a number 5, which is express that innovation investment has a significant effect on the overall innovation, compared with the innovation environment.

Table 2. comparison matrix of one-level index.

	innovation environment A_1	innovation investment A_2	innovation output A_3
innovation environment A_1	1	1/5	1/3
innovation investment A_2	5	1	1/3
innovation output A_3	3	3	1

Solve the maximum eigenvalue of the matrix by using MATLAB

$$[x, y] = eig(A) \tag{1}$$

$$\lambda_{max} = \max(abs(sum(y))) \tag{2}$$

Then solve the normalized feature vector that corresponding to λ_{max}

$$\omega = x(:,1) / sum(x(:,1)) \tag{3}$$

According to the λ_{max}=3.2948 solved by Formula (2) and (3), the normalized feature vector that corresponding to λ_{max} is $\omega = (0.1104, 0.3230, 0.5666)^T$.

During the construction of the comparison matrix A, we do not consider the consistency of the matrix, and therefore, according to the consistency checking procedure proposed by Saaty [11], the consistency check of the comparison matrix is carried out. The procedures are as follows.

(1)Calculating the consistency index: C.I.; C.I.= $(\lambda_{max} - n)/(n-1)$=0.1474;

(2)Calculating the average random consistency index R.I, and find the R.I.=0.58 by looking for the past data.

(3)Calculate the consistency ratio: C.R.: C.R.=C.I./R.I.=0.254. When C.R.<0.3, the consistency of the matrix is acceptable, otherwise it is necessary to revise the matrix. Therefore, It can be considered that the consistency is betterIt can be learned that the weight of the three one-level index are:0.1104、0.3230、0.5666.

2.3 Determination of the two-level index weight

Given the artificial subjective factors are not easy to determine and the two-level index contains too many indicators, thus to use the objective weighting method to determine the one-level index weight. Using historical data to study the relationship between indicators, objective weighting method excluding the impact of objective human factors. At present, the main methods of objective weighting are entropy weight method, least square method, variation coefficient method, multi objective programming method, principal component analysis method and CRITIC method. Because of the superiority of entropy weight method in determining the weight of each evaluation index of multi-decision making problem, and its maturity of usage, we using entropy weight method to determine the weight of the two-level indicators

The procedures are as follows.

$$r_{ij} = [x_{ij} - \min_{(xi)}]/[\max_{(xi)} - \min_{(xi)}] \tag{4}$$

According to the definition of information entropy in information theory, calculate the information entropy of every two-level index.

$$E_j = -(\ln n)^{-1}\sum_{i=1}^{n} p_{ij} \ln p_{ij} \tag{5}$$

For the $p_{ij} = r_{ij} / \sum_{i=1}^{n} r_{ij}$ in formula (5),if p_{ij} =0,then $\lim_{i \to n} p_{ij} \ln p_{ij} = 0$.

Confirming the weight of each two-level index:

$$\omega_j = E_j / \sum_{j=1}^{t} E_j \tag{6}$$

In the formula above is the number of two-level index for each one-level index. This model is based on the two-level index of innovation environment such as the number of undergraduates per million, the number of legal units in industrial agglomeration area ,Per Capita GDP, number of R & D institutions, the number of high-tech enterprises identified in the year as an example. T value is 5. According to the analysis above, we can obtain the weight of each two-level index. See Table 1.

3 Empirical analysis

3.1 Calculation method and data collection

After the analysis above, we can solve the innovation index that correspond to one-level index and two-level index in each region. Calculating the one-level index fraction by weighted methods, the calculation formula is as follows:

$$Y_{ij} = \sum_{i=1}^{t}\sum_{j=1}^{3} r_{ij} \cdot \omega_i \tag{7}$$

In the formula (7), i is the number of two-level index, t is the number of two-level index that correspond to one-level index, j represents years, ω_i is the weights that correspond to two-level index.

Based on the formula (7), we can calculate the total innovation index of a region by using the one-level index's weights that concluded in 3.2. The calculation formula is as follows:

$$Y_t = \sum_{t=1}^{3} Y_{ij} \cdot \omega_a \tag{8}$$

In this formula, t represents the three regions, ω_a is one-level index weight.

As used herein, the data from the "Henan Statistical Yearbook --2013,2014,2015", Zhengzhou, Luoyang and Xinxiang 2012 - 2014 National Economic and Social Development Statistics Bulletin and open government website, the part of the data collected in the Ministry of Science and Technology torch High-tech industry development Center Web site. Due to differences in statistics, some indicators used in the three cities is the data used to replace of cities, such as per capita GDP, the whole society to R & D expenditure as a percentage of GDP, etc. And when identify the indicators of high-tech enterprises, we used the statistics results of Zhengzhou, Luoyang, Xinxiang three national high-tech zones, which are lie in the list of passed high-tech enterprise identify that yearly batch released by Henan province.

First of all, according to the formula 4, Standardizing the data of the 15 two-level index of the three regions of Zhengzhou, Luoyang and Xinxiang. The standardization eliminates the difference of measurement units and numerical value in the evaluation of multi index. Limited to space, this paper uses the range of processing methods, and obtain a standard matrix. See Table 3.

Table 3. Standard matrix.

	Zhengzhou			Luoyang			Xinxiang		
	2012	2013	2014	2012	2013	2014	2012	2013	2014
1	0.87	0.92	1.00	0.00	0.07	0.11	0.46	0.59	0.71
2	0.96	0.90	1.00	0.00	0.26	0.27	0.34	0.49	0.59
3	0.75	0.89	1.00	0.38	0.43	0.47	0.00	0.06	0.11
4	0.84	0.93	1.00	0.00	0.02	0.05	0.17	0.22	0.20
5	1.00	0.91	0.00	0.64	0.45	0.27	0.27	0.27	0.09
6	0.00	0.13	0.15	0.35	0.20	0.24	0.92	0.97	1.00
7	0.65	0.85	1.00	0.15	0.20	0.22	0.00	0.04	0.07
8	1.00	0.93	0.38	0.81	0.83	0.86	0.00	0.15	0.22
9	0.69	0.87	1.00	0.24	0.25	0.29	0.00	0.04	0.09
10	0.53	0.60	0.88	0.63	0.48	1.00	0.00	0.04	0.36
11	1.00	0.92	1.00	0.01	0.01	0.00	0.07	0.04	0.08
12	0.12	0.84	1.00	0.00	0.03	0.02	0.00	0.01	0.02
13	0.20	0.45	0.99	1.00	0.61	0.72	0.00	0.04	0.46
14	0.00	0.00	0.13	0.71	0.51	1.00	0.14	0.10	0.83
15	0.57	0.79	1.00	0.02	0.04	0.00	0.63	0.72	0.52

And then, according to the above analysis, calculate the three regions of the total innovation index and the horizontal and vertical analysis of the innovation index.

3.2 Overall analysis of innovation index

Based on the formula and the method of 3, we can get the total index of innovation in Zhengzhou, Luoyang and Xinxiang, as well as the growth rate in 2012 as a benchmark. The details are shown in Table 4.

Table 4. Total innovation index and growth rate of 2012-2014.

region/year	2012 innovation index	2013 innovation index	2014 innovation index	2013 growth rate	2014 growth rate
Zhengzhou	0.5171	0.6248	0.7288	0.1076	0.1040
Luoyang	0.4650	0.5532	0.6695	0.0882	0.1163
Xinxiang	0.1950	0.4549	0.6185	0.2599	0.1636

From Table 4, it is not difficult to see that Zhengzhou has been ahead of Luoyang and Xinxiang in the overall innovation index. Zhengzhou lead and promote economic restructuring and upgrading support innovation and development in Henan, and is of great significance to speed up the rise of the Central Plains, Henan revitalization and enriching Province. From the angle of growth rate, Zhengzhou is in a relatively stable position. This is because Zhengzhou has been concerned about innovation, and the policy of committed to leading innovation in Henan development. And Luoyang is showing a larger increase, its 2014 growth was even exceeds Zhengzhou. Although the growth rate of Xinxiang has slowed down, the growth rate is still large, ahead of Zhengzhou and Luoyang.

As a whole, Zhengzhou's innovation ability is the strongest, Luoyang is in the chase phase, Xinxiang is weak, and its growth rate has slowed down.

3.3 Index analysis of one-level index

Table 5 is the index of innovation environment, innovation investment and innovation output of the three cities Zhengzhou, Luoyang and Xinxiang during 2012-2014. From the point of view of innovation environment in Table 5, we can see that Zhengzhou has been in a leading position, Xinxiang is in a period of rapid growth, while Luoyang is in the volatility. From the point of view of innovation investment, there are fluctuations in Zhengzhou but in absolute advantage, Xinxiang, Luoyang are showing growth potential, while the larger growth rate in Xinxiang. From the point of view of innovation output, Luoyang and Zhengzhou are in a period of rapid growth. This is mainly due to the development of the information industry and the transformation of traditional enterprises. In Henan Province, the second batch of high-tech enterprises to be identified in the list of 62 enterprises in Zhengzhou, the transformation of traditional industries in the field of 26. And Luoyang has to be identified 22 high-tech enterprises, the transformation of traditional industries in the field of 13. To a certain extent, it shows that the transformation and upgrading of traditional enterprises have a greater promotion effect on the output of high -tech enterprises.

Table 5. The index of innovation environment, innovation investment and innovation output in 2012-2014.

	region/year	2012	2013	2014
Innovation environment	Zhengzhou	0.8877	0.9093	0.7788
	Luoyang	0.2192	0.2609	0.2422
	Xinxiang	0.2560	0.3333	0.3479
Innovation investment	Zhengzhou	0.6360	0.7264	0.6115
	Luoyang	0.4349	0.4222	0.4536
	Xinxiang	0.1961	0.2766	0.3214
Innovation output	Zhengzhou	0.3776	0.5119	0.7867
	Luoyang	0.5305	0.3686	0.6099
	Xinxiang	0.1469	0.1683	0.1222

4 Conclusion and discussion

Based on the above analysis, we can draw the following conclusions. Zhengzhou has a clear advantage in terms of innovation environment, innovation investment and innovation output, but the innovation environment and innovation input has a declining trend. Luoyang is basically in a stable stage, and has a greater advantage in terms of innovation output, as a result of the increase of traditional industrial transformation and upgrading. Xinxiang are basically at a disadvantage in the three regions, but the innovation environment for good performance. Its innovation investment continued to increase, and the innovation output showing a growing trend as well. It has a greater potential in the green environmental protection household electrical appliances, biological new medicine and other industries.

In this paper, the statistical data of Zhengzhou, Luoyang and Xinxiang are used to analyze the innovation ability of the three places. We have reached the index of innovation environment, innovation investment and innovation output, which is in line with the actual development. As the main park of Zhengzhou-Luoyang-Xinxiang National Innovation Demonstration Zone, its innovative ability is related to the innovation and development of the whole demonstration area. The analysis of this paper can be used as policy guidance. In the future we can establish a system suitable for the overall innovation index evaluation, based on the innovation index evaluation system of Independent Innovation Demonstration Zone in other countries and the data of Zhengzhou-Luoyang-Xinxiang. This is the direction of the future need to work hard as well.

References

1. Albert N.Link,John T. Scott.U.S. science parks: the diffusion of an innovation and its effects on the academic missions of universities.International Journal of Industrial Organization, 2003,21(9):1323-1356.
2. Francis C.C. Koha,Winston T.H. Koh,Feichin Ted Tschanga. An analytical framework for science parks and technology districts with an application to Singapore.Journal of Business Venturing.2005,20(2):271-239.
3. Yiannis L Bakourosa, Dimitri C Mardasb,Nikos C Varsakelisc.Science park, a high tech fantasy: an analysis of the science parks of Greece.Technovation.2002,22(2):123-128.
4. Dechun S. The Comparative Study on the Index of Silicon Valley and the Evaluation Index System of China's National Hi-tech Zones.Forum on Science and Technology in China,2012,12(12):74-81.
5. Lijian L,Tangwei T. The Differentiation，Convergence and Its Implications of the Park'Integration into the City in the World-class Science Parks—Case Study of Silicon Valley and Hsinchu Science — based Industrial Park.Science and Technology Management Research,2014,8:33-38.
6. Shuhua H,Song W,Hengjin D.Study on the Development of National Independent Innovation Demonstration Park Based on Four-Three Structure Model——The Analysis of Wuhan East Lake National Independent Innovation Demonstration Park.Science & Technology Progess and Policy,2011,28(9):29-32.
7. Yuhuan G,Linhua P,Qingwei M. An Analysis of Zhongguancun Industrial Cluster＇s Innovation Ability Based on Complex Network Structure.Science of Science and Management of S.&T.,2009(5):56-60.
8. Hongyu Z. A Comparative Study on Innovation Capability of National Independent Innovation Demonstration Zone-Taking Zhongguancun, East Lake , Zhang Jiang as an Example.Science & Technology Progess and Policy,2015,32(22):34-39.
9. Hugo H,Nordine E S. Innovation Union Scoreboard 2014.Belgium:European Commission,2005:1-11.
10. Yong Z,Hufei G.Analysis on Innovation Index and Industrial Development of Southern Jiangsu National Innovation Model Zone.Southeast University Press,2015(12),153-162.
11. Alexander J, Saaty T L. The forward and backward processes of conflict analysis. Behavioral Science,1997,22(2):87-98.

Research on Tacit Knowledge Acquisition Based on Gray Theory and RBF Neural Network

Xiaomin Guo, Zhiwei Zhou, Yamin Zhao, Mengmeng Zhang, Zhikang Lian, Lin Wei and Jianhua Zhang

School of Management Engineering, Zhengzhou University, Zhengzhou 450001 China
Corresponding Email: 530378118@qq.com

Abstract. In the era of knowledge economy, Knowledge has become the main means of production and the core source for value creation. However, especially tacit knowledge acquisition, knowledge acquisition is bottleneck in all kinds of knowledge system, cutting down the benefit of knowledge sharing, application and innovation. In view of this, in this paper, the main ideas and deficiencies of the existing methods of knowledge acquisition are generalized. And then, the basic principles, characteristics and the integrated advantage of gray theory and RBF neural networks are analyzed. On this basis, three kinds of models are designed and discussed, which is able to obtain tacit rule set by using gray theory integrated with RBF neural network. Simultaneously, an empirical analysis is carried out to analyze application results of tacit knowledge acquisition, which includes five kinds of models using independent and integrated strategies.

1 Literature Review and the starting point

1.1 Knowledge Acquisition connotation

The academic world on the "knowledge acquisition" of the understanding there are multiple versions, representative are: Breuker suggested that knowledge acquisition is to obtain useful knowledge from a specific source to solve the problem of knowledge or experience, it is the foundation for people to understand and transform the the world [1];Cassiman and Veugelers argue that knowledge acquisition is the fundamental activity that underpins the organization to achieve knowledge innovation [2];In China, Wu Quan-yuan and Liu Jiang-ning pointed out that knowledge acquisition is the process of extracting knowledge from external knowledge sources into the computer system [3].In view of the foregoing, we believe that knowledge acquisition refers to the analysis and extraction of problem-solving knowledge from knowledge sources, formats and codes them by appropriate knowledge representation methods then stored in the system knowledge base.

1.2 The Main Methods of Acquiring Knowledge and Its limits

Tacit knowledge and explicit knowledge as knowledge of the two basic forms, for its different characteristics of the commonly used knowledge acquisition method is different.There are two common methods of explicit knowledge acquisition: distributed search and data mining.The former is a search strategy, which can be used to query and retrieve knowledge of RDB, special documents, Internet sites, Web pages, and so on, and can gain explicit knowledge in related fields [4,5]. However, the distributed search algorithm is generally high complexity, low search efficiency.Data mining is to extract credible, effective and understandable new knowledge from a large amount of data. However, the knowledge gained by this method does not necessarily have good Interpretability.

Manual knowledge acquisition,semi-automatic knowledge acquisition and automatic knowledge acquisition are the main methods of tacit knowledge acquisition [6,7].Manual knowledge acquisition through face-to-face communication with domain experts and experts, which is very inefficient.Compared with manual knowledge acquisition,Semi-automatic knowledge acquisition improves the efficiency of knowledge acquisition to a certain extent. However, relatively frequent I/O operations still restrict the efficiency of knowledge acquisition, and the production rules of manual access are often redundant or contradictory.The way of automatic knowledge acquisition is high efficiency, high degree of intelligence is the future direction of development; its main drawback is that algorithms maturity is not high, systematic training time-consuming data demand.

1.3 Starting point

Explicit knowledge is easy to code and transferable effectively. However, in the "knowledge iceberg" model, explicit knowledge is only part of the knowledge iceberg floating on the water surface. Tacit knowledge hidden beneath the water surface accounts for knowledge Iceberg about 90%.On the other hand, the famous SECI model shows that tacit knowledge plays an important role in ensuring and promoting the application, sharing and innovation of knowledge[8].Tacit knowledge includes many types, such as informal experience and skills, difficult to express skills, recessive rule sets of data systems.Tacit knowledge in terms of the proportion of knowledge or the role of knowledge are far greater than the explicit knowledge. Therefore, how to do tacit knowledge acquisition well is the focus of knowledge management research.

1.4 Model selection

The gray RBF neural network model, which combines the RBF neural network model with the gray model, not only has the characteristics of small sample data modeling for gray system, but also has the advantages of neural network model with adaptive ability to nonlinear and inexact laws ; The model reduces the demand for raw data and the training time consumption of automatic knowledge acquisition to a certain extent.

2 Grag RBF neural network model

2.1.The GM (1,1) model

Gray GM (1,1) modeling process is as follows[9,10]: the original data sequence:

$$X^{(0)} = \{x^{(0)}(1), x^{(0)}(2), \ldots, x^{(0)}(n)\}.$$

And a new sequence is obtained by accumulating it once

$$X^{(1)} = \{x^{(1)}(1), x^{(1)}(2), \ldots, x^{(1)}(n)\},$$

In this formula,

$$X^{(1)}(i) = \left\{\sum_{j=1}^{i} x^{(0)}(j) \middle| i = 1,2,\ldots,n\right\}.$$

Order

$$D(i+1) = -\frac{1}{2}\left[x^{(1)}(i) + x^{(1)}(i+1)\right] \tag{1}$$

create the matrices **B** and $\mathbf{Y_n}$:

$$B = \begin{bmatrix} D(2) & 1 \\ D(3) & 1 \\ \cdots & \cdots \\ D(n) & 1 \end{bmatrix}, Y_n = \begin{bmatrix} x^{(0)}(2) \\ x^{(0)}(3) \\ \cdots \\ x^{(0)}(n) \end{bmatrix} \tag{2}$$

Let the series satisfy the first - order univariate ordinary differential equation:

$$\frac{dx^{(1)}}{dt} + ax^{(1)} = b. \tag{3}$$

Among them, a is the development coefficient, b is the gray effect, both a and b are non-0 constant. The least squares method is used to solve a and b respectively, then:

$$p = \begin{bmatrix} a \\ b \end{bmatrix} = (B^T B)^{-1} B^T Y_n \tag{4}$$

Thus, the cumulative series prediction model can be obtained:

$$\hat{x}^{(1)}(i+1) = \left[x^{(0)}(i) - \frac{b}{a}\right]e^{-ai} + \frac{b}{a}, (i = 0,1,2,\ldots) \tag{5}$$

The predicted values for the series is:

$$\hat{x}^{(0)}(i+1) = \hat{x}^{(1)}(i+1) - x^{(1)}(i) \tag{6}$$

This is a function model based on time response and is the core formula of GM (1,1) model.

2.2 RBF neural network

RBF neural network, or radial basis function neural network, is a three-layer forward network. RBF neural network is superior to ordinary feed-forward network [11], because the convert of RBFcan transform a nonlinearly separable sample point in one space into a linearly separable sample point in another space.RBF network structure similar to the BP network.The first layer is the input layer, which is composed of the source nodes. The second layer is the hidden layer and the number of nodes depends on the requirements of described questions . The third layer is the output layer, which responds to the input mode. In this network, the transformation from the input space to the hidden layer space is nonlinear, and the transformation from the hidden layer space to the output layer space is linear; the transform function of the hidden layer unit is a radial basis function, which is a nonnegative nonlinear function of the locally distributed radial symmetric attenuation of the central point.

RBF network has a good nonlinear function approximation ability.It does not require the sample has a specific distribution and only uses the neural network itself reflects the non-linear change of things which directly estimate the non-linear change characteristics of things; At the same time, because it affects the network output with a few connection weights in the local area of the input space, it converges faster than the traditional neuron model, and is not easy to fall into the local minimum point and has better robustness.

2.3 Gray RBF Neural Network Combined Model

Gray GM (1,1) and RBF neural networks achieve integration in three way by series, parallel and embedded[12-14]. Among them, tandem Gray RBF neural network as shown in figure 1 , the gain results of the rules knowledge of the multiple Gray GM (1,1) model as input of RBF network model, and then use RBF network 's non-linear mapping function to find out the Gray GM (1,1) model weight, so that the output get better accuracy of the rules knowledge output.

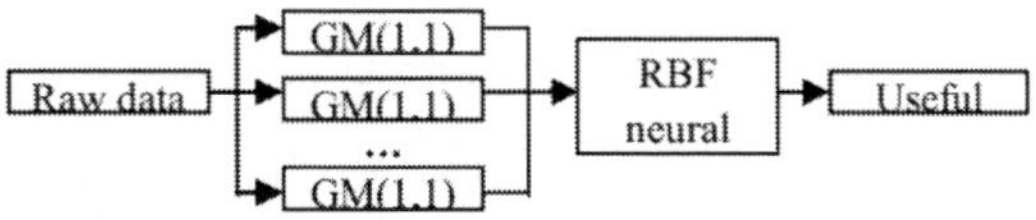

Fig.1 .Tandem Gray RBF neural network.

For initial datas, The parallel gray RBF neural network model use gray GM (1,1) model and the RBF neural network model to modeling of problem.. Then, the weighted average of the two models' output is obtained.

The embedded gray RBF neural network uses the gray GM (1,1) model to gray the sample data before inputting the data. the gray output of the RBF neural network is reduced, that is, the gray GM (1,1) model is used for the RBF output. And whitened to obtain a final output.

3 Empirical Analysis

3.1 Examples Background and Data Acquisition

Stock prices are affected by many factors, with a high degree of complexity and nonlinearity. Stock opening and closing prices, the highest and lowest prices as the two extreme performance of stock prices, they and the stock of the actual price between the implied association rules. Changes in stock volume and turnover can reflect the level of stock activity. One of the most important indicators is the closing price, which is the benchmark for investors to make money and lose money. Due to the opacity of the market, investors can not get all the factors affecting the closing price.

In view of this, assuming that the government policy, the company's basic business conditions remain unchanged, we select Yutong Bus (600066) February 7, 2014 to March 18, 2014 the stock historical data as the original data 1) model, RBF neural network and three gray RBF neural network models were used to analyze and acquire the rule knowledge contained in the stock closing price data set. Then, Is applied to the closing price forecast for the stock, which is the forecast value of the closing price of the stock from March 19, 2014 to March 26, 2014; and finally, the actual value in the segment (see Table 1 Dashed line below) to test the accuracy of the predicted value, and then determine the level of validity of the acquired hidden rule knowledge.

3.2 Implicit Rule Knowledge Obtainment of Five Different Models

3.2.1 Through the GM (1,1) model to obtain hidden rules

The closing price of the part above the dotted line in Table 1 is taken as the original data, and the recessive rule is obtained by GM (1,1) model. First of all, the original data is accumulated to obtain the accumulated sequence $X^{(1)}$:

$$X^{(1)} = (18.01,37.41,56.7,75.63,95.08,114.67,134.45,153.61,172.44,190.86,\\ 209.32,226.84,243.93,261.12,278.71,296.81,315.5,333.09,350.48,\\ 367.81,385.47,402.56,419.65,436.4,453.2,469.61,486.41,503.14)$$

On this basis, the matrix B and Y are established and P is obtained.

The simulation sequence of $X^{(1)}$ is further calculated and the simulated sequence of s is generated by accumulation. The accuracy of the residual test $p^{(0)}$ is:

$$p^{(0)} = (1-\bar{\varepsilon})\times 100\% = 97.2\% \tag{7}$$

$$B=\begin{bmatrix} D(2) & 1 \\ D(3) & 1 \\ \cdots & \cdots \\ D(n) & 1 \end{bmatrix}=\begin{bmatrix} -27.71 & 1 \\ -47.055 & 1 \\ \cdots & \cdots \\ -494.775 & 1 \end{bmatrix} \tag{8}$$

$$Y_n=\begin{bmatrix} x^{(0)}(2) \\ x^{(0)}(3) \\ \cdots \\ x^{(0)}(n) \end{bmatrix}=\begin{bmatrix} 19.4 \\ 19.29 \\ \cdots \\ 16.73 \end{bmatrix} \tag{9}$$

$$p=\begin{bmatrix} a \\ b \end{bmatrix}=(B^T B)^{-1}B^T Y_n=\begin{bmatrix} 0.0066 \\ 19.7283 \end{bmatrix} \tag{10}$$

3.2.2 Obtain hidden rules through RBF neural network

RBF neural network is calculated by using MATLAB software, RBF neural network is trained by newrb function, the simulation function is sim. The opening price, the highest price, the lowest price, the transaction volume and the transaction number of the part above the dotted line in Table 1 are taken as the output layer, and the number of input layer nodes is R = 5. The closing price as the output layer, the output layer node number S =1. The newrb function has the function of automatically increasing the number of nodes in the hidden layer. Therefore, the hidden layer nodes are automatically adjusted by the program and are no longer set. All the sample data are normalized to remove the influence of different units on the RBF neural network model.
In this paper, the adopted normalization method is

$$\bar{x}(k)=\frac{2(x(k)-x_{\min})}{x_{\max}-x_{\min}}-1 \tag{11}$$

in which $k=1,2,\ldots,34$ represents the value of each of the different influencing factors, and $x_{\min}$ and $x_{\max}$ represent the minimum and maximum values of each influence factor, respectively.The normalized sample data is selected as the training sample, and the spread value is tested as the test data under the dotted line in Table 1 (numbers 29 to 31). Select the spread value from 37 to 45, and draw the simulated data and the actual data of different spread values , see Table 2.As can be seen from the table, when the spread value 43, the closest to the true value. The final RBF neural network is created as: net = newrb (P, T, 0.005, 43).

3.2.3 Obtain recessive rules through tandem gray RBF neural network

GM (1,1) model was used to calculate the opening price, the highest price, the lowest price, the trading volume, the transaction number and the closing price of Yutong Company (600066), and the corresponding simulation sequence was obtained. As the posterior error ratio of turnover and the number of transactions are $C_{\text{volume}}=0.8526$ and $C_{\text{turnover}}=0.7842$ respectively, the accuracy of which can not meet the requirements. Therefore, only the opening price, highest price and the lowest simulation sequence of Yutong Bus (600066) are selected as input layer and the nodes number of input layer R is 3.The closing price as the output layer, the nodes number of output layer S is1.Use the sample data of Nos. 1 to 28 in Table 1 as the training samples and the sample data of Nos. 29 and 30 as the test data to test the spread value. Debugging found that it is the closest to the true value when the spread value of 4. The final RBF network is: net = newrb (P, T, 0.005,4). The results of the implicit rule application are shown in Table 3.

Table 1 Yutong Bus (600066) stock data.

numb	Date	Opening quotation	highest	lowest	Volume	Turnover number	Closing
1	2014/2/7	17.78	18.08	17.7	22845	2802	18.01
2	2014/2/10	18.24	19.81	18.24	239398	17491	19.4
3	2014/2/11	19.38	20.17	19.18	195687	12882	19.29
4	2014/2/12	19.43	19.5	18.83	104345	11025	18.93
5	2014/2/13	19.15	19.77	19	180840	15579	19.45
6	2014/2/14	19.22	19.8	19.01	92542	8034	19.59

7	2014/2/17	19.87	19.95	19.41	82475	9618	19.78
8	2014/2/18	19.67	19.75	19.03	104130	10903	19.16
9	2014/2/19	19.2	19.38	18.8	99349	9690	18.83
10	2014/2/20	18.92	18.98	18.29	90619	10047	18.42
11	2014/2/21	18.4	18.6	18.26	54283	6321	18.46
12	2014/2/24	18.35	18.36	17.5	145797	14773	17.52
13	2014/2/25	17.52	17.66	16.47	209073	16751	17.09
14	2014/2/26	17.1	17.42	16.88	106920	9067	17.19
15	2014/2/27	17.23	17.9	17.06	180925	10830	17.59
16	2014/2/28	17.43	18.4	17.21	167603	11277	18.1
17	2014/3/3	18.11	19.2	18.11	199555	13515	18.69
18	2014/3/4	18.28	18.3	17.5	218370	17132	17.59
19	2014/3/5	17.75	17.85	17.35	119701	9734	17.39
20	2014/3/6	17.4	17.44	17.08	52471	5361	17.33
21	2014/3/7	17.4	17.82	17.35	68688	5642	17.66
22	2014/3/10	17.49	17.49	16.99	78765	6695	17.09
23	2014/3/11	17.09	17.15	16.85	65032	8971	17.09
24	2014/3/12	17.05	17.05	16.5	70396	6838	16.75
25	2014/3/13	16.66	16.98	16.66	35826	3584	16.8
26	2014/3/14	16.7	16.8	16.38	72118	6725	16.41
27	2014/3/17	16.61	16.83	16.4	82789	6623	16.8
28	2014/3/18	16.85	16.88	16.63	87685	8244	16.73
29	2014/3/19	16.72	16.72	15.52	257000	19285	15.98
30	2014/3/20	15.96	15.96	15.1	201594	18219	15.22
31	2014/3/21	15.17	15.8	14.86	172199	11498	15.57
32	2014/3/24	15.61	15.71	15.25	119907	9226	15.47
33	2014/3/25	15.72	16.2	15.68	137274	10072	15.86
34	2014/3/26	15.91	16.47	15.91	155977	14163	16.39

Table 2. The simulated data and the actual data of different spread values.

Spread values	Output value of Sample 29	actual value	error	Output value of Sample 30	actual value	error	Output value of Sample 31	actual value	error
37	-0.6688		0.0018	-0.806		-0.194	-0.8255		-0.0205
38	-0.6704		0.0034	-0.8081		-0.1919	-0.8265		-0.0195
39	-0.6718		0.0048	-0.8099		-0.1901	-0.8275		-0.0185
40	-0.6731		0.0061	-0.8117		-0.1883	-0.8283		-0.0177
41	-0.6743	-0.667	0.0073	-0.8133	-1	-0.1867	-0.8291	-0.846	-0.0169
42	-0.6755		0.0085	-0.8148		-0.1852	-0.8299		-0.0161
43	-0.6765		0.0095	-0.8161		-0.1839	-0.8305		-0.0155
44	-0.6775		0.0105	-0.8174		-0.1826	-0.8312		-0.0148
45	-0.6784		0.0114	-0.8186		-0.1814	-0.8318		-0.0142

3.2.4 Parallel gray RBF neural network model

Using the parallel gray RBF neural network model, the system output of the GM (1, 1) model and the RBF neural network based on the weight coefficient combination for the hidden rule acquisition is:

$$f_c = \frac{Var(e_2)}{Var(e_1)+Var(e_2)} f_1 + \frac{Var(e_1)}{Var(e_1)+Var(e_2)} f_2 = 0.088 f_1 + 0.912 f_2 \tag{12}$$

Correspondingly,The results of the implicit rule application are shown in Table 3.

3.2.5 Embedded gray RBF neural network model

Add up the sample data of the opening, the highest price, lowest price, volume, turnover and the closing price, and then use the previously described method to normalize the cumulative data. The normalized results of sample numbers 1 to 28 in Table 1 are selected as the training samples, and the sample data of numbers 29 to 31 are used as the test data to test the spread value. Debugging found that when the spread value of 9 when the closest to the true value. The final RBF neural network is: net = newrb (P, T, 0.005,9), which is the closest to the true value when the spread value is 9.

3.2.6 Single input layer embedded gray RBF neural network model

The embedded gray RBF(5) neural network model is adopted, but the sample data only adopts the closing price. Add up the sample data of the closing price, and then use the previously described method to normalize the cumulative data. The sample data to do the cumulative accumulation of the closing date, then the cumulative data using the methods described above normalized. The normalized results of sample numbers 1 to 28 in Table 1 are selected as the training samples, and the sample data of numbers 29 to 31 are used as the test data to test the spread value. Debugging found that when the spread value of 50 when the closest to the true value. The final RBF neural network is: net = newrb (P, T, 0.005, 50), which is the closest to the true value when spread is 50.

3.3.Results analysis

Table 3 describes the comparison of the output results from the five model implicit rule sets described above. It can be seen that the parallel RBF neural network model has better results than GM (1,1) model and RBF neural network model in the extraction and application of implicit rules of stock closing prices. And, in the implicit rule extraction and application of the stock closing price it also has more advantages than the RBF neural network model and the embedded RBF neural network model; On the other hand, the embedded RBF neural network model increases the influence factor of the closing price of single day after the completion of the ashing process, and the result of the application of its recessive rule set has a big error, while the single input layer RBF neural network model has been less influenced, the application result of the recessive rule set of it is less.

Table 3. Model rule extraction value.

Numb		29	30	31	32	33	34	Mean
actual data		15.98	15.22	15.57	15.47	15.86	16.39	15.748
GM(1,1)Model	Extract value	16.366	16.259	16.152	16.046	15.941	15.836	16.100
	Absolute error	0.386	1.039	0.582	0.576	0.081	-0.554	0.352
	Relative error/%	2.359	6.390	3.603	3.590	0.508	-3.498	2.159
RBF neural network model	Extract value	15.66	15.776	16.165	16.122	16.91	17.013	16.274
	Absolute error	-0.32	0.556	0.595	0.652	1.05	0.623	0.526
	Relative error/%	-2.043	3.524	3.681	4.044	6.209	3.662	3.180
tandem gray RBF neural network	Extract value	15.969	15.723	15.76	15.798	16.001	16.16	15.902
	Absolute error	-0.011	0.503	0.19	0.328	0.141	-0.23	0.154
	Relative error/%	-0.069	3.199	1.206	2.076	0.881	-1.423	0.978
Parallel gray RBF neural network model	Extract value	16.013	15.719	15.667	15.648	16.189	16.529	15.961
	Absolute error	0.033	0.499	0.097	0.178	0.329	0.139	0.213
	Relative error/%	0.206	3.175	0.619	1.138	2.032	0.841	1.335

Embedded gray RBF neural network model	Extract value	23.034	20.622	17.538	15.321	16.027	17.751	18.382
	Absolute error	7.054	5.402	1.968	-0.149	0.167	1.361	2.634
	Relative error/%	30.624	26.195	11.221	-0.973	1.042	7.667	12.630
Single input layer embedded gray RBF neural network model	Extract value	15.846	15.058	15.376	15.24	15.59	16.074	15.531
	Absolute error	-0.134	-0.162	-0.194	-0.23	-0.27	-0.316	-0.218
	Relative error/%	-0.85	-1.08	-1.26	-1.51	-1.73	-1.97	-1.40

4 Conclusion

Automatic knowledge acquisition can effectively save the cost of knowledge acquisition, reduce the time consumption, improve the access efficiency, and become the improvement direction of KM knowledge acquisition "bottleneck". Gray RBF neural network model combines the characteristics of gray system "weakening data randomness, better revealing the data state" and RBF neural network has higher self-learning and self-adaptability; Different tacit knowledge acquisition environments provide flexible support. Knowledge explicitization is the basic activity of SECI helix, which is helpful to deepen knowledge cognition and benefit knowledge sharing, application and innovation benefit. Follow-up research direction.

Acknowledgement

This work is supported by the Excellent Young Teacher Development Fund Project of Zhengzhou University (Foundation No. 2015SKYQ15).

References

1. Joost Breuker. A Cognitive Science Perspective on Knowledge Acquisition.International Journal of Human-Computer Studies,2013,71(2):177-183.
2. Bruno Cassiman, Reinhilde Veugelers. In Search of Complementarity in Innovation Strategy: Internal R&D and External Knowledge Acquisition.Management Science,2006,52(1):68-82.
3. Wu Quan-yuan, Liu Jiang-ning.Artificial Intelligence and Expert System. Beijing: China Railway Publishing House, 1995.
4. WU Bao-gui, DING Zhen-guo.Research on Distributed Search Engine Based on Map / Reduce .New Technology of Library and Information Service, 2007 (8): 52-55.
5. JIANG Jian-hong.Research on the Technology of Distributed Distributed Search Engines. Science Technology and Engineering, 2007,7 (10): 2418-2424.
6. Kaj U. Koskinena, Pekka Pihlantob,Hannu Vanharanta.Tacit knowledge acquisition and sharing in a project work context.International Journal of Project Management,2003,21(4):281–290.
7. Zhang Jianhua, E-government knowledge management. Beijing: Science Press, 2010.
8. Ikujiro Nonaka, Ryoko Toyama, Noboru Konno. SECI, Ba and Leadership: a Unified Model of Dynamic Knowledge Creation. Long Range Planning, 2000, 33(1):5-34.
9. LIU Si-feng, DENG Ju-long.Application of GM (1,1) model .Systems Engineering -Theory & Practice, 2000 (05): 121-124.
10. Yang Hualong, Liu Jinxia, Zheng Bin. Improvement and application of grey prediction GM (1,1) model. mathematics practice and understanding, 2011,41 (23): 39-46.
11. CHANG Yu-qing, ZHANG Hong-yan, WANG Shu. Injection Speed Control Based on RBF Neural Network Inverse System. Journal of Northeastern University (Natural Science), 2013, 34 (2): 174-177.
12. YUAN Jing-ling, ZHONG Luo, LI Xiao-yan.Study and development of gray neural network.Journal of Wuhan University of Technology, 2009,31 (3): 91-93.
13. Xing X. Gray neural network improved algorithm and its application. Huazhong University of Science and Technology Master's thesis, 2011.
14. ZHANG Yi, YANG Jianguo.Thermal error modeling of machine tools based on gray neural network .Journal of Shanghai Jiaotong University, 2011,45 (11): 1581-1586.

A Price Index Model for Road Freight Transportation and Its Empirical analysis in China

Zhishuo Liu*, Kuan Zhao, Jingmiao Ma and Chunfang Wang

School of Traffic and Transportation, Beijing Jiaotong University, Beijing ,China

Abstract. The aim of price index for road freight transportation (RFT) is to reflect the changes of price in the road transport market. Firstly, a price index model for RFT based on the sample data from Alibaba logistics platform is built. This model is a three levels index system including total index, classification index and individual index and the Laspeyres method is applied to calculate these indices. Finally, an empirical analysis of the price index for RFT market in Zhejiang Province is performed. In order to demonstrate the correctness and validity of the exponential model, a comparative analysis with port throughput and PMI index is carried out.
Keywords. Logistics, road freight transportation, price indices, Laspeyres method.

1 Introduction

The road transport industry is often low running and "market failure" condition often appears because the road transport market price, cost, supply and demand information are not transparent and asymmetric. At the same time, transport enterprises have insufficient grasp of the market. The dynamic indices and general indices are so less that they can't meet the needs of industry analysis and decision making. As a result, the information used to reflect the climate condition of the transport market is inadequate and lack of quantitative analysis. So a quantitative indicators to evaluate prices and the condition of the transport market are needed urgently. The price index for RFT can comprehensively reflect the transport price, costs, climate information of shipping business. They can also reflect the situation of the road transport market, forecast industry trends and provide strong evidence for government and corporate decision. Firstly, a price index model for RFT based on the data from Alibaba logistics platform was built. Then the road transport market in Zhejiang Province was selected to make an empirical analysis, and the results show that this model can reflect the market situation perfectly.

2 The research status of road transport price index

Scholars have studied the Price Index for RFT in research methods. Hanchao Guo et al. [1] made a price index system using weighted average index method and geometry average method based on Beijing freight market by gathering point time price information. Gu Jingyan et al. [2] analyzed the market for RFT in Yiwu as an example using the Fisher index method. Bin Liu et al. [3] constructed a four levels index system for the transport market in Ningbo by the arithmetic mean method. Guangcong Yan et al. [4] analyzed a transport market of a city which belongs to a pilot project launched by the government. They used the weighted average method for index to work out the indices after gathering data from the sample enterprises and transport routes.

As for the application status of price index methods, there are some price indices for freight transport at home and abroad. At the aspect of shipping price indices, Baltic Exchange issued the Baltic dry freight index(BFI) [5] for the worldwide dry bulk shipping market in 1985.The index system revealed the supply and demand for dry bulk shipping market and related markets such as shipbuilding market. In 1998, China's export Container Freight Index (CCFI) [6] issued by the Shanghai Shipping Exchange for existent several liner for container transport market. They used the Laspeyres [7] method to calculate the indices so as to reflect comprehensive indicators of China's export container transportation market prices. In terms of price indices for RFT, in July 2006, China Federation of Logistics & Purchasing (CFLP) and Best Express jointly issued a national general cargo freight index [8] based on the data of medium-sized and small enterprises. It was used to reflect the changes of supply and demand of road transport market. In June 2015, CLFP and a Logistics enterprise called Linan jointly issued Highway Price index (HPI) [9]. The index was based on the dynamic transaction information which came from logistics platform of road represented by the Linan logistics network. Although the data is true and reliable, it belonged to the single enterprise and was not comprehensive. In 2012, Zhejiang Provincial Communication Department and local government jointly released of price index for RFT for Yiwu [10]. It was used to reflect Yiwu area and the Zhejiang Province of road transport market based on different routes which start from Yiwu and ended in other cities. In June 2015, Ministry of Transport of the People's Republic of China (MOT) released of China Freight Price Index (CFI) [11]. The data came from a pilot work results which belonged to a project carried out by MOT. Although the data covered a wide range, the authenticity, reliability and sustainability of this data needed to be tested.

It can be seen from the above analysis that methods of constructing a price index for road become more and more mature. As for the construction of the index system, the reliability of data representation and the data sources are deficiency. So this paper build a three levels index system including total index, classification index and individual index. This indices divide the index system from multiple dimensions according to the practical situation. In this way, the difficulty of data collection can be reduced and the project will be launched easily.

3 The price index model for RFT

3.1 Selecting the method of price index of RFT

Laspeyres index method and Paasche index method are widely used in practice because they are simple, practical. Although the results of the two methods can indicate the degree of the change in indices, there are some differences between them in terms of the economic significance. As for price of freight transport index, Laspeyres index method can indicate the change of price and the impact on total transportation cost based on the transport volume during the base period. And Paasche index method is based on the transport volume during the report period. Therefore, both of them have real economic significance. Paasche index method based on the report is more realistic compared to Laspeyres index method. In order to reflect the impact to the change degree of price in different periods because of pure price change and the size of the calculated values, the Laspeyres index method is applied to calculate the indices. The calculation formula is as follows:

$$I_t = \frac{\Sigma q_0 p_t}{\Sigma q_0 p_0} \times I_0 \tag{1}$$

p_0-the transport price during the base period.p_t-the transport price during the report period.q_0-transport volume during the base periodI_0-the transport price during the base periodI_t-the transport price during the report period

3.2 Designing the transport price index system

Considered the actual situation and data acquisition, the transport price index system was built including total index, classification index and individual index as shown in Figure 1.Total index refers to index for RFT. Individual index refers to the route of transport. Classification index refers to the index which divided by departure place, mileage, the way of organizing goods, the type of goods, speed, calculating period and the choice of base period.

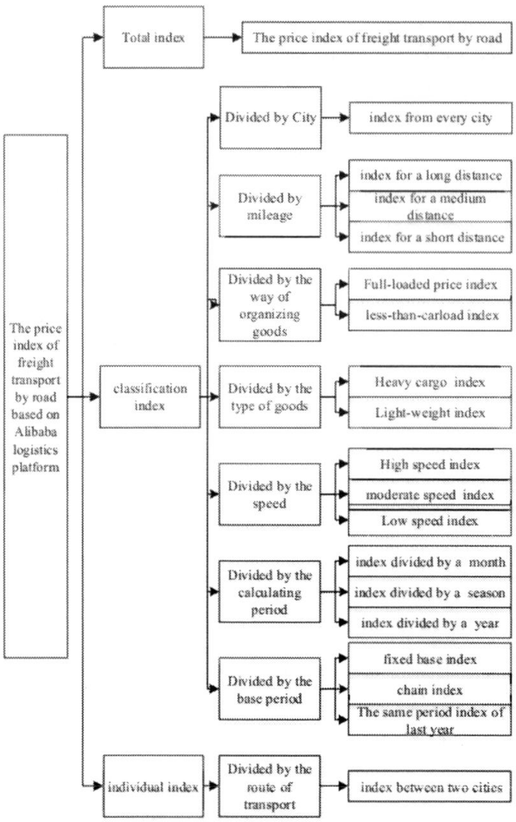

Fig. 1. The transport price index system.

3.3 Calculating the transport price index

According to the road transport price index system, the formulas of calculating indices at various levels are as follows.

3.3.1 The calculation formula of individual index

The individual index is divided by the routes of transport when the transport price and volume from logistics enterprises are taken into consideration. Supposed that the amount of logistics enterprises is M on each route, the formula is as follows.

$$I_1^t = \frac{\sum_{i=1}^{i=M} p_1(i)q_0(i)}{\sum_{i=1}^{i=M} p_0(i)q_0(i)} \times I_0^t \tag{2}$$

As in (2), I_1^t represents the transport price index on the t line during report period and I_0^t represents the index during the base period. $p_0(i)$ represents the transport price from the I logistics enterprise during report period and $p_1(i)$ represents the price during the base period. $q_0(i)$ represents the transport volume from the I logistics enterprise during the base period.

3.3.2 The calculation formula of classification index

Firstly, the index is divided by different departure city and an example of the transport price index from t city is taken. Supposed that the amount of routes departing from k city is N and there are M logistics enterprises on each route, the formula is as follows.

$$I_1^k = \frac{\sum_{j=1}^{j=N}\sum_{i=1}^{i=M} p_1(ij)q_0(ij)}{\sum_{j=1}^{j=N}\sum_{i=1}^{i=M} p_0(ij)q_0(ij)} \times I_0^k \quad (3)$$

As in (3), I_1^k represents the transport price index for the k city during the report period and I_0^k represents the index during the base period. $p_0(ij)$ represents the transport price from the I logistics enterprise on the route j during the base period and $p_1(ij)$ represents the price during the report period. $q_0(ij)$ represents the transport volume from the I logistics enterprise on the route j during the base period.

Then the index is divided by transport mileage, including long-distance index, medium distance index and short distance index. Supposed that there are K departures with N routes and each route has M logistics enterprises, the formula is as follows.

$$I_1^l = \frac{\sum_{k=1}^{k=K}\sum_{j=1}^{j=N}\sum_{i=1}^{i=M} p_1(ijk)q_0(ijk)}{\sum_{k=1}^{k=K}\sum_{j=1}^{j=N}\sum_{i=1}^{i=M} p_1(ijk)q_0(ijk)} \times I_0^l \quad (4)$$

As in (4), I_1^l represents the transport price index for the l kinds of mileage during the report period and I_0^l represents the index during the base period. $p_0(ijk)$ represents the transport price for the i logistics enterprise on route j departing from city k during the base period. $p_1(ijk)$ represents the price during the report period.

Next, the index is divided by types of organizing transport, including full-loaded index and less-than-carload index. Supposed that there are k departure city with N routes and each route has M logistics enterprise, the formula is as follows.

$$I_1^r = \frac{\sum_{k=1}^{k=K}\sum_{j=1}^{j=N}\sum_{i=1}^{i=M} p_1(ijk)q_0(ijk)}{\sum_{k=1}^{k=K}\sum_{j=1}^{j=N}\sum_{i=1}^{i=M} p_1(ijk)q_0(ijk)} \times I_0^r \quad (5)$$

As in (5), I_1^r represents the transport price index by the r type of organizing transport during the report period and I_0^r represents the index during the base period. The meaning of other variables are the same as mentioned above.

Next, the index is divided by the type of goods including heavy cargo index and light-weight index. The formula is as follows.

$$I_1^s = \frac{\sum_{k=1}^{k=K}\sum_{j=1}^{j=N}\sum_{i=1}^{i=M} p_1(ijk)q_0(ijk)}{\sum_{k=1}^{k=K}\sum_{j=1}^{j=N}\sum_{i=1}^{i=M} p_1(ijk)q_0(ijk)} \times I_0^s \quad (6)$$

As in (6), I_1^s represents the transport price index for the s type of goods during the report period and I_0^s represents the index during the base period. The meaning of other variables are the same as mentioned above.

Finally, the index is divided by transport speed including the high speed index, moderate speed index and low speed index. The formula is as follows.

$$I_1^u = \frac{\sum_{k=1}^{k=K}\sum_{j=1}^{j=N}\sum_{i=1}^{i=M} p_1(ijk)q_0(ijk)}{\sum_{k=1}^{k=K}\sum_{j=1}^{j=N}\sum_{i=1}^{i=M} p_1(ijk)q_0(ijk)} \times I_0^u \quad (7)$$

As in (7), I_1^u represents the transport price index at the u kind of speed during the report period and I_0^u represents the index at the u kind of speed during the base period. The meaning of other variables are the same as mentioned above.

3.3.3 The calculation formula of complex index

The complex index consist of different classification indices. For example, the transport price index of full-loaded for a city involves two classification indices: the index divided by city and the index divided by the type of goods.

3.3.4 The calculation formula of total index

Although there are kinds of formula for total index, they originate from Laspeyres method, the formula to calculate the total index is designed according to the city, the mileage, the types of organizing transport, the types of goods and transport speed. Supposed that the amount of cities is K and the city k has r kinds of types of organizing transport. At the same time, there are s kinds of goods in each types of organizing transport. Then there are u kinds of speed in s kinds of goods. There are N routes in the u kinds of speed when the amount of logistics enterprise is M in the route n. So the formula is as follows.

$$I_1 = \frac{\sum_{k=1}^{K}\sum_{r=1}^{R}\sum_{s=1}^{S}\sum_{u=1}^{U}\sum_{j=1}^{N}\sum_{i=1}^{M} p_1(ijusrk)q_0(ijusrk)}{\sum_{k=1}^{K}\sum_{r=1}^{R}\sum_{s=1}^{S}\sum_{u=1}^{U}\sum_{j=1}^{N}\sum_{i=1}^{M} p_0(ijusrk)q_0(ijusrk)} \times I_0 \qquad (8)$$

As in (8), I_1 represents the total transport price index during the report period and I_0 represents the index during the base period.

Generally, each transport price index value during the base period is set to 100.

4 Price index for RFT complied instance

4.1 Selecting RFT samples

The first step to compile the transport price index is to determine the sample. There are some principles should be followed such as typical principle, the relevant principles and information accessible principles. So the sample routes must occupy numerous market share. At the same time, it can indicate the change degree of the total transport price through its change. According to the principles mentioned above, the sample routes is selected which start from Zhejiang province and end in other cities based on the Alibaba logistics planform.

4.2 Data source and processing

The transport price data and transport volume data have been got from the Alibaba logistics platform using data capture technology. These raw data include the name of the 11 departure cities and destination cities, the name of logistics enterprises, weight of goods, transport mileage, transport time and speed.

Meanwhile, there are some missing values and outliers needed to be processed at first. First of all, the method of using data capture technology to get the data may lead to miss some information such as route and the weight of freight. So a step to simulate the missing data for the route is necessary according the history data and a nearby transport route. Next, there are outliers in raw data. An average value of the data can be used to replace the outlier.

4.3 Statistics and analysis

The data is input and the index is calculated according to the data collected. The excel table consists of date, provinces, departure, destination, logistics enterprises, the price of heavy and light-weight goods, weight, volume, the full-loaded price, the number of vehicles, transport time, mode of transport and the type of goods.

4.3.1 The total index for RFT

In this paper, the fixed base index is taken as an example. The index curve are drawn from September 2015 to August 2016 according the road transport price index model mentioned above. This indices include total index, classification index and individual index, as shown in Figure 2 to Figure 4.

In order to analyze the fluctuation of the total index for RFT, the correlation between the port cargo throughput and the total index for RFT in Zhejiang is examined by using bivariate similarity theory. The results of the analysis are shown in Table 1.

Table 1. Correlation test.

Correlation coefficient				
			Throu-ghput	Total index
Spearman's rho	Throughput	CC	1.000	.721
		Sig	.	.000

		N	41	41
	Total index	CC	.721	1.000
		Sig	.000	.
		N	41	41
The correlation was significant at the 0.01 level of confidence.				

From the correlation result, it can be seen that the original hypothesis that the two variables are uncorrelated should be denied according to the Confidence Sig=0<0.05. And the correlation coefficient is equal to 0.721. From the result of above analysis, the port cargo throughout has an effect on the price index for RFT in Zhejiang and this result can gave us an evidence to compare the two variables. At the same time, the total index for RFT is compared with PMI index which can reflect the trend of microeconomic development. In this way, the reason for the index fluctuation will be found. The results are shown inFigure 2.

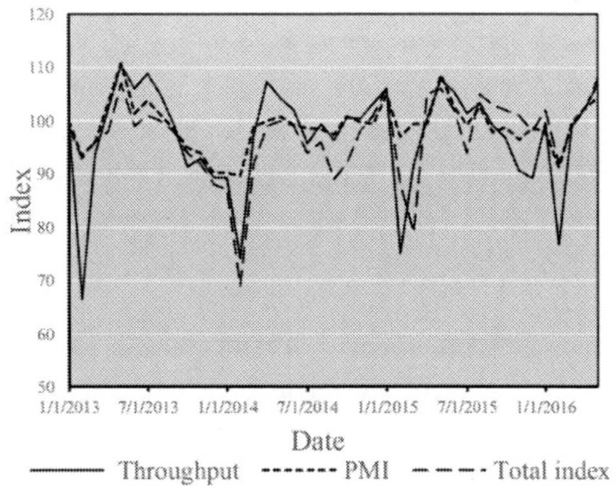

Fig.2. The total price index for RFT in Zhejiang.

From Figure 2, the changing trends of the three line graphs are basically the same. When compared to the port throughput in Zhejiang, the total index for RFT is declining form April 2013 to January 2014. And the freight volume for input and output is decreasing in Zhejiang province, which means the decline of shipping volume will have a negative effect on the market for RFT. From July 2014 to January 2015, both of them are rising. This result means that if the shipping volume become larger, the transport price should be higher and the market for RFT should be boom. When compared to PMI index, the total index also shows a slight downward trend but not obvious. It was significantly decreased until April 2014, which is caused by the lag of linkage response. From July 2014 to January 2015, both of them are rising. During this period, the microeconomic has developed well. There are export-oriented economy and strong export in Zhejiang province. So, the demand for RFT is larger and the market is more active.

4.3.2 The classification index for RFT

In this paper, the freight volume from Zhoushan port is taken as an example. The relationship between port throughput in Zhoushan and the individual index is analyzed. The results are shown in Figure 3.

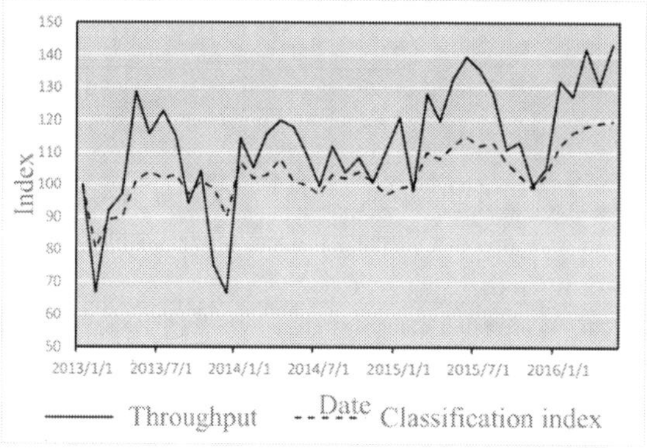

Fig.3. The classification price index divided by city.

From Figure 3, it can be seen that the overall trend of the two curves is upward from January 2013 to May 2016. And they fluctuate in the same way. For example, from January 2015 to July 2015, the cargo throughput of Zhoushan port increased, which indicated the demand of shipping freight was bigger. When using the road way and water way to import and export, the road freight traffic volume rises. Then the road freight market is more active and its trading volume is larger. Then it can be seen that the trend of the freight index is consistent with the freight throughput.

4.3.3 The individual index for RFT

In this paper, the Ningbo-Beijing highway freight line is taken as an example. The date of October 2013 is set as the base month and the index is compared to the National Road Freight Price Index (CFI) from October 2013 to May 2016. The results are shown in Figure 4.

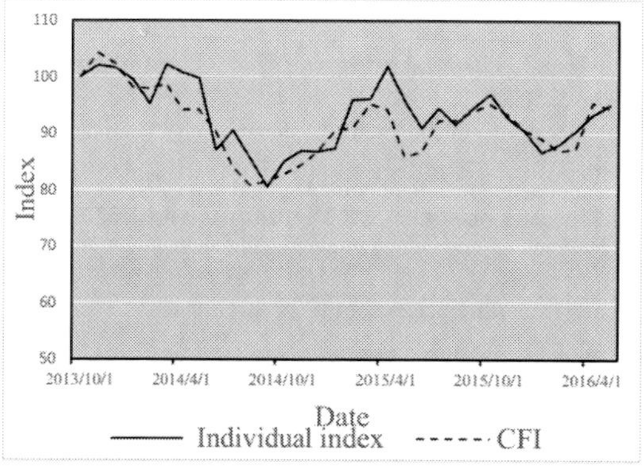

Fig.4. The individual price index divided by route.

From Figure 4, it can be seen that the change trend of the branch line index is consistent with the overall change trend of the CFI index. This indicates that both of the two indexes can reflect the road freight market prosperity of the Ningbo - Beijing line.

5 Conclusion and discussion

In this paper, the background, existing problems and research significance of Chinese road transport are described firstly. Then some typical domestic and foreign price indices are summarized. Next, the Price Index model of Road Freight Transport based on the Alibaba logistics platform is built. At the same time, an empirical analysis is performed by using the Laspeyres index method after designing the way for collecting data. Finally, the corresponding index curves are drawn according to the indices. By analyzing and comparing the results to PMI index and port throughput, it turns out that the index can well reflect the demand and trading situations of transport market.

This paper puts the transport price index research into practice, and it's also an improvement of transport index system. At the same time, it will contribute to the investment and management decisions-making of freight industry subjects; it's conductive to optimize resource allocation of transport market and improve the freight industry training mechanism; and it could help the industry supervision authorities to monitor the operations of market and enhance their service capability.

Acknowledgement

The research is supported by National Science and Technology Support Program Grant No.2014BAH24F02.

References

1. H.C. Guo, L.M. Sun, BJ.Fif.S.*S*.C, pp.7 (2009).
2. J.Y. Gu, W.J. Fan, TST, **09**:154-158(2013).
3. B. Liu, G.Y. Jiang, J.D. Chen, NBU.J, **03**(2015).
4. G.C. Yan, H.D. Nie, G.Q. Shi, R&M, **05**:62-64(2015).
5. X.R. Cai, CA.Tra.Sci, **04**(2012).
6. L.J. Chen, S.Q. Yu, Chi.Wat, **02**:12-13(2004).
7. G.X. Xv, *China Statistics Press*, 11-12(2004).
8. Y.B Ma, Int.Tra, **09**:69-72(2007).
9. J.Y. Li, C.Sto&Tra, **08**:72-73(2015) .
10. W.G. Zhao, H. Wang, Y. Zhong, TST, **02** (2012).
11. J. Qi, Tran.Man.Wor, **07**:50-51(2015).

Analysis on Causes and Countermeasures of Bullwhip Effect

Jianhua Dai[1,*], Shibiao Li[1], and Shengbo Peng[1]

[1] School of Economics and Management, Communication University of China, Beijing, China

Abstract. Bullwhip effect is an inevitable phenomenon in supply chain management, because of its objective existence. This phenomenon is very common and harmful to make the operating costs of enterprises double and become one of the main concerns of many enterprises. In this paper, the causes of the bullwhip effect are explored through the methods of literature research and investigated consultation to weaken the bullwhip effect. This paper analyzes the key countermeasures with Wal-Mart successful logistics management case. And according to the reason of bullwhip effect, a mathematical programming model of maximizing the efficiency of supply chain is established, which provides a way to solve the negative effect of bullwhip effect and has certain reference value.

1 Introduction

Bullwhip Effect refers to a kind of distortion occurring in the process of transmitting order information upstream, which is a bigger fluctuation in upstream order quantity caused by the fluctuation of downstream demands.[1] This is a common phenomenon in supply chain.[2, 3] The existence of bullwhip effect makes it difficult for enterprises to grasp market demands, causing an overstock and reducing the operational efficiency for the whole supply chain.[3-5] The best way to solve bullwhip effect is to reduce knots of supply chain as far as possible, thus to greatly ensure accuracy of information. Using efficient supply chain management system can reduce bullwhip effect and realize real-time response, directly reducing operating costs of enterprises.[6] The factors causing bullwhip effect include the following several aspects which are demand forecast amendment, fluctuations in prices, order quantity decision, shortages game, inventory imbalances, lead time, etc.

2 Bullwhip effect in supply chain

2.1 What is the "bullwhip effect"

In the early 1990s, P&G discovered a wonderful phenomenon, the United States of infant birth rate and baby diaper usage has remained stable, but Procter & Gamble distributors to their factory orders volatility is more pronounced, and Procter & Gamble The volatility of orders placed with their suppliers is greater. The degree of order and inventory levels moving up the supply chain is always increasing - from customers to retailers to distributors to suppliers to suppliers and even suppliers. In addition to Procter & Gamble, Hewlett-Packard found slight fluctuations in its sales of its retail stores, And HP's printer division to its integrated circuit sector orders fluctuate even greater. All this shows that the demand for information in the transmission process has been seriously distorted, as shown in Figure 1. In addition to Procter & Gamble, Hewlett-Packard found slight fluctuations in its sales of its retail stores, And HP's printer division to its integrated circuit sector orders fluctuate even greater. All this shows that the demand for information in the transmission process has been seriously distorted, as shown in Figure 1.

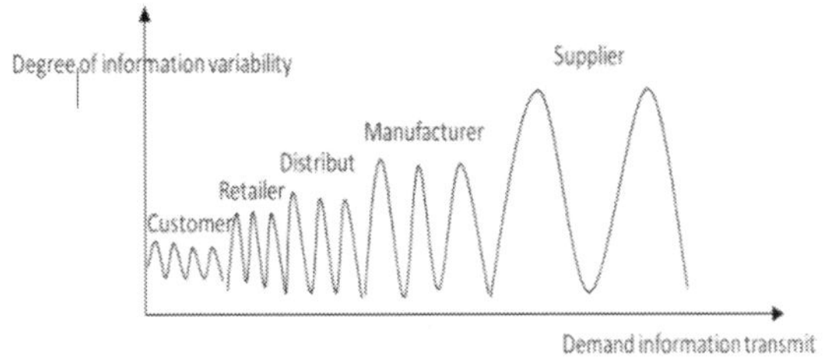

Fig.1. Schematic diagram of bullwhip effect.

From the above chart we can see that the slight changes in customer demand back to the upstream business order fluctuations are enormous, which is the bullwhip effect appeared in the supply chain phenomenon, and the more the level of the supply chain, the more obvious of bullwhip effect. Because of this phenomenon, upstream suppliers often need to maintain more inventory levels than downstream suppliers.

In summary, the bullwhip effect is a typical complex phenomenon caused by the uncertainty of the supply chain is the distortion of demand information in the supply chain to convey a description of the image. When the information flow from the final client to the original supplier side, can not effectively and accurately achieve information sharing, making the information distorted and level-by-step amplification, resulting in a great demand for information fluctuations, such information distorted like a rejection of the bullwhip, the root of a small change, it will lead to huge changes in the tail. Therefore, some people will link "bullwhip effect" and "butterfly effect".

2.2 The harm caused by bullwhip effect

Bullwhip effect exists in the supply chain objectively with high risk, it brings the consequences of self-evident. Mainly in the following aspects:

Inventory backlog. Distortion of demand information, making the company's inventory and the actual demand for a conflict, results in a large backlog of inventory.

Increased operating costs. Demand information volatility, enterprises to avoid the risk of stock, to expand production capacity to meet the demand for orders, which undoubtedly greatly increased the total cost of the supply chain.

Reduced customer service levels. Amplify the distortion of the demand information makes it difficult for the nodes of the real needs of the market to make accurate forecasts and the right decision, can not really meet customers' needs.

In summary, the bullwhip effect makes the supply chain operation disorders, the work of the supply chain has had a significant negative impact, we are very necessary to further analysis of the causes of bullwhip effect and propose solutions.

2.3 The Cause of Bullwhip Effect

Professor Stein, came from the Massachusetts Institute of Technology, did a famous experiment - the beer experiment. In the experiment, there are four students representing consumers, producers, wholesalers and retailers, respectively, to form a simple supply chain. The requirements of the experiment are: any upstream and downstream enterprises can not exchange any business information, Only downstream enterprises are allowed to transfer orders to upstream enterprises, consumers can only order to the retailer. The result of the experiment shows that the asymmetric information of each node and the maximization of its own interests make the demand information distorted in the supply chain, which is the bullwhip effect. There are several reasons for this phenomenon:

2.3.1 Demand forecast information single and demand information distortion

As shown in the beer experiment, each node is always based on its direct downstream demand information as the basis for their demand forecast, which leads to the future demand forecast is not accurate. And in order to avoid out-of-stock risk, companies will always increase part of the inventory, resulting in the demand inflated, bullwhip effect followed.

2.3.2 Supply chain fluctuations in the price of each node enterprises

Retailers and distributors in the face of price volatility, discounts or promotions, natural disasters, etc., tend to take the practice of increasing inventories, so that orders are far greater than the actual demand. For consumers, they may change purchases during periods of price volatility, but this does not reflect consumers' long-term real needs, since during this period they postpone or advance some of their needs. For example, the annual holiday, businesses will be promotional, consumers will postpone some of the pre-holiday demand, will also be part of the demand ahead of time, in order to focus on holiday spending, so that changes in demand will be very large, causing Bullwhip effect.

2.3.3 Bulk order and lead time

In the supply chain, each enterprise will order to its upstream. Normally, after receiving only one order, the seller will not immediately order the goods from his supplier, they will consider the inventory and freight costs in one cycle or aggregate to a certain number of re-orders. In order to reduce the frequency and reduce the cost of ordering, vendors tend to increase the amount of orders.

Because frequent orders will increase the workload and cost of suppliers, suppliers often require vendors to achieve a cycle or a certain number of orders after the order. To prepare for the occasion, vendors tend to increase the order quantity, thus leading to the bullwhip effect.

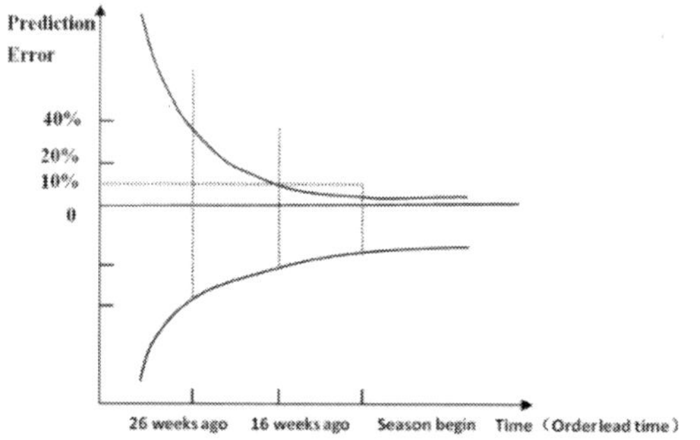

Fig.2. The Relationship Between Order Lead Time and Prediction Error.

Because the order information needs to go through these links: Customers - Retailers - Wholesalers - Manufacturers - Suppliers for its information processing. The process will inevitably lead to delays in information and logistics. So the order quantity of information is not timely amended, for the upstream business, to consider the delay phase of the demand, will naturally increase the safety stock. Therefore, the companies in the expected inventory are included in the lead time, the longer the lead time, small changes in inventory changes caused by the greater. As shown in Figure 2, with the increase in the lead time of the order (from right to left), the prediction error gradually increases, and will affect the size of the inventory, the entire supply chain nodes are very unfavorable.

2.3.4 The complexity of the supply chain itself and multi-level

Supply chain is composed of many suppliers, manufacturers, vendors and customers of a complex network of chain system, the demand information pass up from the end of the network chain to another end. Under the dual effects of external market environment interference and conflict of objectives among internal stakeholders, the demand information presents the complexity characteristics of time and space evolution each other, such as uncertainty, delay, nonlinearity, dynamicity, time interleaving, transfer structure and so on.

In addition, the supply chain has a multi-level, at all levels of enterprises will be set up safety stock, the accumulation of multi-storey system will increase the amount of inventory products in the supply chain, as shown in Figure 1. This, in turn, has slowed down the response to market volatility and led to the bullwhip effect when it comes to raising the level of customer service, especially the level of stock availability and improving safety stock.

2.3.5 Poor information exchange and information game

In the beer experiment, any upstream and downstream enterprises can not exchange any business information, companies can not grasp the real needs of the downstream and upstream supply capacity, but to store their own goods. At the same time, if the supply chain can not achieve the availability of stock exchange and transfer allocation, can only have their own high inventory, which will undoubtedly lead to bullwhip effect.

Supply chain nodes Enterprises are interdependent, interest-related partnerships. However, under incomplete information, as the rational participants, in order to maximize their respective interests, each node enterprise will compete and compete with the demand, price, inventory, product development and other information. The results of the game is one of the causes for the bullwhip effect.

2.3.6 The product in short supply

Another mechanism that causes the bullwhip effect is a shortage of products. For example, during the market introduction period of the best-selling new products, the manufacturer will implement a sales quota system, and each retailer can get only a fraction of the order quantity from the manufacturer. Aware of the quota behavior of suppliers, even if retailers know that they do not need and can not sell additional products, will continue to expand its order quantity, known as a false order. Manufacturers receive more orders than their forecast demand, will further invest in expanding production. Unfortunately, when the real demand appears, the supply may be far more than demand, false orders will be revoked by retailers, suppliers can only sell a very low price of the remaining products.

3 The countermeasures and case analysis of weakening bullwhip effect

For the obstacles which lead to the bullwhip effect, we need began from overcoming these causes to start. Therefore, the basic countermeasures to weaken the bullwhip effect are: to achieve information sharing, to avoid the multi-demand forecast; implementation of outsourcing services to shorten the lead time; reduce bulk orders; reduce price promotions, avoidance of variability; circumvent short game situations; Inventory responsibility and so on. This paper focuses on three strategies to effectively weaken the bullwhip effect, the first two for the establishment of supply chain strategic alliance and application of Internet of things, combined with examples of Wal-Mart's logistics management analysis. The third one is the supply chain efficiency maximization model, which combines various factors of bullwhip effect, and provides a valuable method to solve the practical problem of bullwhip effect.

We must make clear that the bullwhip effect can not be completely eliminated from the supply chain, but if the method properly, realistic, you can greatly weaken the bullwhip effect on the supply chain management of the negative impact.

3.1. The establishment of supply chain strategic alliance

For the demand forecast is not accurate and the information communication is not smooth, we can adopt the strategy of establishing the supply chain strategic alliance. The traditional upstream supply chain enterprises can obtain the demand information only from the downstream enterprises. The information flows are single and not flexible enough. If the upstream and downstream nodes enterprise information is symmetry and interoperability, the formation of a more stable supply chain strategic alliance, then forecast the accuracy of demand will be greatly improved. In this way, the structure of the supply chain will be as follows. See Figure 3 below.

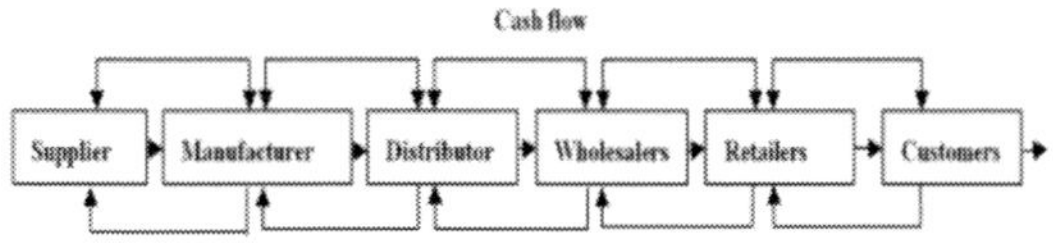

Fig.3. Supply chain strategic alliance.

In this regard, Wal-Mart and Procter & Gamble's production and marketing strategic alliance to fully reflect the advantages of this logistics model and a strong foundation. P & G is the world's largest manufacturer of daily necessities; Wal-Mart is the world's largest commercial retail enterprises. The cooperation between them is not smooth sailing, also experienced the "cold war", Procter & Gamble is always trying to control the Wal-Mart's sales prices and sales conditions to its priducts, and Wal-Mart to be outdone, threatened to end Procter & Gamble product sales, or left it the worst shelf.

However, the two sides soon realized the benefits of in-depth cooperation. Procter & Gamble chose to cooperate with Wal-Mart to establish a strategic alliance of production and marketing partnership. As the main organization for the implementation of the cooperation, Procter & Gamble and Wal-Mart formed a special cooperation team of about 70 people, including personnel from the financial, circulation, production and other functional departments, the team was sent to Walmart to implement collaborative management. Wal-Mart began in 1989 to implement the supply chain management for Procter & Gamble, the specific form of the two sides are through EDI (electronic data interchange) and satellite communications to achieve accurate understanding of networking information. With the help of information systems, Procter & Gamble in addition to quickly know Wal-Mart logistics center diaper inventory, but also to keep up-to-date diaper stores in Wal-Mart sales, inventory, price and other data, so accurate information base allows Procter & Gamble formulate development of market demand in line with the production and research and development plans timely, but also on the Wal-Mart's stock for replenishment, to prevent excessive inventory and out of stock status. At the same time, Wal-Mart on the basis by gotting information from Procter & Gamble through EDI, can make decisions timely of decision-making shelves and purchase of goods. Wal-Mart gives the logistics center or warehouse management right to the Procter & Gamble. In this way, Wal-Mart not only do not engage in specific logistics activities, and because the two sides do not have on each transaction conditions (such as distribution, price issues, etc.) to negotiate, greatly reducing the entire business process time of goods from the purchase, storage, sorting, replenishment to sales of.

Wal-Mart and Procter & Gamble's production and marketing strategic alliance, to reduce unnecessary inventory, save the cost of circulation, to achieve the information sharing and elimination of information game, making the two companies get a win-win situation, have a absolute effective control of the bullwhip effect.

3.2 Application of Internet of things

Internet of things is a network which through the radio frequency identification, infrared sensors, global positioning systems, laser scanners and other information sensing equipment, according to the agreement, to connect any object with the Internet and exchange of information and communication to achieve the intelligent objects Identification, location, tracking, monitoring and management.

Internet of things is one of the products of logistics network. The so-called logistics network is simply the logistics of information, network management, to achieve logistics information's real-time cross-regional transmission. In the logistics network system, the weakening of the advantages of the bullwhip effect are:

1. Logistics nodes implement information management in the general. Each node's information is the base of logistics system information in the logistics process, through the analysis and mining of logistics information, to maximize the use of effective information on logistics activities management.

2. The whole system has unlimited openness. The entire system is built on the Internet, can quickly exchange data, with unlimited openness and expand capacity.

3. Information flow in the entire logistics process to guide and integrate the role. Information flow throughout the entire logistics activities in the logistics process has played a pre-forecast, after the feedback role, greatly reducing the error, making a significant decline in the pause time.

4. The system has obvious scale advantage. In the logistics network, the system to form a network structure, large-scale joint operations to reduce the overall operating cost of the system.

Here, have to mention Wal-Mart's logistics network construction. The world's largest retailer Wal-Mart is the world's first group to achieve 24-hour computer logistics network monitoring of the enterprise. In the 1970s, Wal-Mart established a logistics management information system to handle system reports, which effectively speeded up operations. In the early 1980s, Wal-Mart and Hughes co-operated with the launch of logistics communication satellites, logistics communication satellite makes the global networking, Wal-Mart therefore had a jump development. In 1983 when Wal-Mart uses a POS machine, full name Point Of Sale, is the starting point of sales data system. After the establishment of the EDI, the electronic data interchange system, paperless operation, all the information all in the computer operation. In 1986, when it established a QR, known as the rapid response mechanism, quickly pull the market demand. In 2004, Wal-Mart asked its top 100 suppliers to use radio frequency identification technology (RFID) when sending pallets and crates to its distribution centers.

RFID technology is based on computer technology and communication technology of highly automated information and data collection of integrated technology. Is the application of the main technology of things. RFID has many advantages: it is not restricted to the line of sight. Video identification card has a strong ability to read and write, while difficult to copy and intelligence is very high. RFID does not require human intervention to complete the identification work, while moving up to 500km per hour to identify the moving object, you can also identify multiple RF tags at the same time.

With these advanced information technology of the Internet of Things, Wal-Mart effectively weakened the information can not be shared and lagging behind the possibility of lead to bullwhip effect, making the supply chain more rapid and smooth flow of information from upstream to downstream transmission more accurate , So that even more powerful enterprises, has made leading and long-term development.

3.3 Establish a supply chain efficiency model

In order to reduce the impact of bullwhip effect, we should establish a supply chain efficiency model, the bullwhip effect makes the efficiency of the supply chain greatly reduced, The establishment of this model makes the indicators adjusted to maximize the efficiency of the supply chain, that is, the objective function of the efficiency model to determine the maximum efficiency of the supply chain as a whole, you can weaken the bullwhip effect to a certain extent.

The objective function is established

$$\text{Max } Z = f\,(x_1, x_2, x_3, x_4, x_5, x_6) \tag{1}$$

Among them, $x_1, x_2, x_3, x_4, x_5, x_6$ respectively on behalf of the demand forecast correction, order batch decision-making, price fluctuations, shortage of information game, lead time and supply chain structure, Corresponding to the above analysis of the bullwhip effect of the causes. F represents the parameters of each node in the supply chain, such as firm size, efficiency, and the number of workers, for the multi-factor matrix.

Constraints established:

Suppose there are five main nodes in a supply chain, namely raw material suppliers, producers, distributors, third-party logistics enterprises, and retailers. According to their respective infrastructure conditions, operating conditions, financial strength, operational efficiency, technical conditions and other major factors, it forms the formation of each node enterprise conditional constraint equation and the final formation of the objective function and constraint equations.

Objective function:

$$\max Z=c_1x_1+c_2x_2+\cdots\cdots c_6x_6 \tag{2}$$

$$\begin{cases} \sum_{i=1}^{6} Ax_i \le b \\ A=(P_1,P_2,P_3,P_4,P_5) \\ b=(b_1,b_2,b_3,b_4,b_5)^T \end{cases} \tag{3}$$

In the constraint equation, A denotes the condition factor of each node enterprise, which is expressed by a matrix. B denotes the upper limit of the comprehensive limiting factors of each node enterprise.

The knowledge of operational research is used to solve the efficiency model function to determine the optimal efficiency of the supply chain under the influence of the variable factors, and the conditions that each node enterprise should have in order to weaken the negative effect of the bullwhip effect and quantify the harm caused by the bullwhip effect, to bring a positive effect for the supply chain management. You are free to use colour illustrations for the online version of the proceedings but any print version will be printed in black and white unless special arrangements have been made with the conference organiser. Please check with the conference organiser whether or not this is the case. If the print version will be black and white only, you should check your figure captions carefully and remove any reference to colour in the illustration and text. In addition, some colour figures will degrade or suffer loss of information when converted to black and white, and this should be taken into account when preparing them.

4 Concluding remarks

Economic informationization and globalization have brought competition among supply chains for enterprise commercial competition. The survival and development of enterprises largely depends on whether the supply chain is running smoothly. While the bullwhip effect has brought great harm to the supply chain. Based on the analysis of the causes of the bullwhip effect and the case of Wal-Mart, this paper focuses on the countermeasure of weakening the bullwhip effect, and establishes a mathematical model to provide some reference for enterprises to reduce the harm caused by bullwhip effect.

Acknowledgements

This paper was supported by the Research Tasks of Communication University of China (No. 3132014XNG1458, CUC16A17), the Program for Young Scholars of Beijing (No. YETP0633), and the National Natural Science Foundation of China (No. 71203202).

References

1. Chen Rongqiu, Ma Shihua: Production and operations management (Third Edition), Beijing: Higher Education Press, June 2011.
2. Jesse Sheffer: Flex - MIT's Supply Chain Management Essence. Shanghai Sanlian Bookstore .2008.11.
3. Zhong Heng: Analysis of Causes and Weakening Strategies of Bullwhip Effect in Supply Chains, Chengdu University of Information Technology, Chengdu, China .
4. Ouyang Xiaoya: Comprehensive Control of Bullwhip Effect Based on Logistics Network System, Jiangxi University of Science and Technology, Science and Technology Information Development and Economy, 2005.8.
5. Zhang Jingfang: Study on the Case of Bullwhip Effect in Supply Chain and Its Solution, Lanzhou Jiaotong University, Logistics Technology, 2007.
6. Li Yanqing, Qin Qiuli: Effect of Internet of Things on Reducing Bullwhip Effect Based on RFID Technology, Beijing Jiaotong University, 2011.

Exploring the Mediating Effect of E-social Capital Between Community Members Interaction and Consumer Engagement

Bingsheng Yan[1], Lihua Li [2] and Hongtao Sun [3,*]

[1,2,3]Staff Development Institute of China National Tobacco Corporation, 7# Road Xinyuan, Zhengzhou, Henan, China
Corresponding Email: 14877884@qq.com

Abstract. This article explored the effect of instrumental interaction and relational interaction on consumer engagement (community engagement and brand engagement) among community members. The mediating effect of E-social capital was investigated as well. The research results showed that: both instrumental interaction and interpersonal interaction promote the formation of E-social capital (online trust and online reciprocity); online trust plays a partial mediating role between community interaction (instrumental interaction, relational interaction) and community engagement, but the influence of online reciprocity on community engagement is not significant; community engagement leads to brand engagement. The findings enrich the theories of brand community and consumer engagement and contribute to the virtual community management.

1 Introduction

Interaction among community members is an important factor to form a positive community relationship. The users engaging in online community activity are related with the frequent interaction between members. This interaction reflects the social, Continuous interaction can improve member status, trust and influence in the community that it is shown by the increase of social capital. According to Hsiao & Chiou (2012) and Bartolini & Sarracino (2014), social capital would influence brand loyalty, subjective well-being and other psychological attitude or behaviour [1,9] .Since it is so, whether it can be deemed as a positive role in consumer community relationship and behavior such as engagement? From the perspective of the influence path, what role does social capital play between the members' interaction and engagement in a online brand community?

Based on the theory of social capital, we explore the effect of the interaction among community members on engagement. This article combines the theory of social capital and brand to study how E-social capital influence consumer engagement. There are meanings for management understanding how to manage the online brand community and to guide the behavior of the community members.

2 Theoretical background

2.1 Community interaction

Online interaction among brand community members is a kind of wide spread phenomenon. Lots of famous corporations have set up their own online brand community in order to maintain long-term contact with their current or potential consumers [5,23] . In an online brand community, there are complicated interaction related to consumers, brand, products and firms. The interaction among consumers is especially frequent and important. It is critical for the improvement of brand relationship. Hudson et al. (2015) found that the emotional attachment of consumers would be generated through the interaction of social media, then affect the quality of brand relationship and promote the spread of the brand word of mouth [10].

Network interactive includes instrumental interaction and relational interaction. According to social exchange theory, the purpose of community members' interaction is to get some interest and their loyalty to community to a large extent depends on whether the community meet their needs [2,6,7] .Chen et al. (2013) regards social interests of community members as emotional support, friendship, relationship, social identity and self-expression, and social value community members perceived have an important impact to community attitudes [3]. Zhou et al. (2013) suggests that viewing posts will make the visitors obtain information value and social value in the online brand community and ultimately affect their participation [24]. Instrumental interaction and relational interaction just right satisfy both values. The instrumental interaction can be accurate exchange or common exchange and members' interactive motivation is more obvious. Relational interaction, by contrast, can be called fuzzy returns, which are uncertain or not immediately available and may need to repeat.

2.2 E-social capital

Coleman (1990) pointed out that social capital is composed of social organizations and provides convenience for the individual within the organization and the individuals with social capital will be conducive to realize their goals [4]. Nahapiet and Ghoshal (1998) defined social capital as "the sum of the actual and potential resources embedded in personal and organizational relationship network" that is different from other capital and owned by the group [16]. Although the individual can acquire resources by organization, no member can enjoy social capital alone. They suggested that social capital should be divided into three parts, namely, the relational dimension, structural dimension and the cognitive dimension.Lin (2001) also regarded social capital as a kind of resource embedded in the social network that exists in social relations and social networks rather than individuals [13]. According to the above definitions, social capital is embedded in a particular social network and studying social capital would be meaningless without social network.

Scholars have explored the influence mechanism of social capital in addition to its concept, structure and properties. Several scholars explored the antecedents of social capital from the perspective of social interaction because interaction and communication are the most active links to display human sociality [20,22].The influence of social capital on economy and society is very extensive.

With the popularity of Internet technology, scholars found that online interaction is very important antecedent to form social capital as same as offline. There are a lot of heterogeneous members and they can mutually contribute novel information under the network environment. This kind of information is one of values the members obtained when they got the social capital of network [11] . For a online community, the increase of network social capital members owned is beneficial to the innovation and sharing of knowledge, which can be used to accelerate the development of new products, to strengthen the relationship with my partners, and increase consumer brand loyalty and so on by firms [18]. It shows that the social capital in network have a positive impact on corporate performance like offline social capital. These findings suggest that the network social capital has become one of important resources of individuals, businesses, social, and even national development when the network space becomes more and more popular in people's life.

Sicilia and Palazon (2008) held that offline brand community and online brand communities are similar in nature [19]. Mathwick et al. (2008) using "voluntary law, reciprocal norms and trust to measure the social capital of virtual community, namely, E-social capital can be differentiated from the social capital of offline community [15]. We adopt the measurement method of Mathwick et al. (2008) for E - social capital because our object of study is online brand community.

2.3 Consumer engagement

Community engagement will promote members more willing to help others and achieve own goal and eventually get many benefits. Brodie et al. (2013) believed that the most important is the interaction between members of brand community [2]; Second, engagement is a extensive, dynamic and interactive process; Third, it is a multi-dimensional variable that includs cognitive, emotional, and behavioral dimensions; Finally, consumer engagement plays an important role in members' relationship exchange. Vivek (2009) suggested that consumer engagement refers to the degree of connection and participation between consumers and firm's products or activity, such as consumer' enjoyment for Ikea experience, i-phone, etc [21].

We refer to the definition of Vivek (2009) about consumer engagement and define community engagement as the degree of connection and participation between the members of the community. Brand engagement is defined as the degree of connection and attention between consumer and a specific brand. Community engagement and brand engagement are divided into three dimensions: enthusiasm, participation and social contact [21].

3 Hypotheses

3.1 Community interaction and community engagement

Online interactions between community members not only include words, pictures and other information exchanges, but also include the participation of online activities, net friend relationship maintenance, making new friends and so forth, such as "thumb up" on this kind of new form of interaction. Owing to the different purposes of community members interaction, this paper divides community interaction into instrumental interaction and relational interaction. Community interaction manifests as social contact, participating in activity and dedicating community. These purpose of interaction are to get some information related to their own interests, ask for help, make new friends and strengthen interpersonal relationship. Community members will generate a series of psychologies or behaviors that is beneficial to

 DOI: 10.1051/matecconf/201710005019

community, such as participating in positive network word-of-mouth if the interactions of different purposes in the community can bring the happy social experience or solve problems [4].

Interaction is the main behavior of online community members and social contact, participation activity and so on can be regarded as different forms of interaction that they have very close connection between them. For example, Lin (2001) found that the interaction in social networking platform has a positive effect on social participation, which is the principal form of community engagement [13]. When firm unites their websites with social media platforms, the communication between firm and consumers can significantly improve consumer satisfaction so that they are reluctant to leave the social networking platform that this condition would further strengthen the relationship between the user and the community at the psychological level. Therefore, based on the above realistic observation and theoretical analysis, this study proposed the following hypotheses:

H1a: Instrumental interaction among members has a positive effect on community engagement;

H1b: Relational interaction among members has a positive effect on community engagement.

3.2 Community interaction and E-social capital

Some scholars have proved the importance of community interaction on social capital formation. Studies have shown that the online interaction quality of members would affect the social presence and ultimately affect the formation of the E-social capital [13].The interaction effect of microblog contributes to improve social contact, trust, reciprocal norms and shared vision of social capital. Counterparties in bidding auction sites positively utilize members interaction to enhance the score of social capital dimensions and strengthen the thought to stay online trading [22].Above all, it can be found that online community interaction affect E - social capital.

Frequent interaction for some important information or opinions of both sides can easily establish mutual trust. Positive and continuous interaction is beneficial to enhance each other's presence and trust from each other in uncertain environment. So, this paper puts forward the following assumptions:

H2a: The instrumental interaction of members has positive influence online trust.

H2b: The relational interaction of members has positive influence online trust.

Another dimension of E-social capital, reciprocity, is very similar to social exchange theory. The underlying mechanism of exchange is reciprocity. People participate in brand community is that they can get useful information, establish related relation with other participants and get return through reciprocal behavior in the community [22]. This relationship will not last if only pay no return. This reciprocal relationship is formed through long-term interaction [15]. Members access to information through instrumental interactive and will also be willing to provide to other members understand their known information under the influence of the reciprocal norms. A member' reciprocal intention would not be very strong if he(she) has never participated in any interaction. Thus, we put forward the following assumptions:

H2c: The instrumental interaction of members has positive influence on online reciprocity;

H2d: The relational interaction of members has positive influence on online reciprocity.

3.3 The influence of E-social capital on community engagement

The reinforcement of trust among members, expectant improvement for others and continuous reciprocal behavior are all the performance that members have raised the level of E - social capital, which is beneficial to reinforce members' relationship with community and brand. Mathwick et al. (2008) suggested that E - social capital exerted positive influence on information value and social value and ultimately affected the online brand community commitment [15]. Especially when community members perceive exclusive shared information, their participation will be significantly increase. Shared vision, trust and identity have significant effect on members participating in community affairs. According to the findings of Mathwick et al. (2008), when information values of members are satisfied the relationship between members and community will be affected. Therefore, as the level of trust among community members raise online interaction frequency will be more and higher and will develop into a more intimate relationship among members. According to the above analysis, trust can promote the development of the relationship between members of the community [15]. So, the following hypothesis is provided:

H3a: Online trust will significantly affect the community engagement of members.

With respect to the study of reciprocity in brand community, scholors empirically investigated the antecedents and consequences of new members to join the online community [3]. The results show that members' similarity and reciprocity will affect the will of new members to join and reciprocity positively affects the member satisfaction and community participation intention. The reciprocity can promote community commitment. Lee, et al. (2011) held that the altruism motivation of community members promotes community engagement [12]. Therefore, the online reciprocity of members will affect community engagement. Reciprocity enables the members in the community to get what they want, such as understanding, a sense of belonging and value identity or entertainment, etc. Along with the continuous increase of community members and reinforcement of participation willingness, the connection among members will be more frequent and close their relationship will be endured and deepen. Therefore, there is the following hypothesis:

H3b: Online reciprocity would positively affect the community engagement of members.

3.4 The mediating effect of E-social capital on community interaction and engagement

With the popularity network interaction, scholars gradually transfer research focus from the real offline to online. Existing studies related to social capital have evaluated the relationship quality of community members established by interaction and how these relationships affect members' activities [2].

Study shows that interaction can promote interpersonal trust [14,16].Which would effectively promote members to actively participate in social activities that enhance the relation between the members and the community. Community members hope to reciprocate others through participating in community activity. Reciprocity is an important foundation for the efficient users operation of online brand community, strengthen the interdependence among members, form a more intimate and friendly relationship and reinforce the community sense of belonging of members.As community members to interact, their contribution will increase to the community and gradually form the social capital (e.g. trust, reciprocity). Community members will contribute their behavioral, cognitive and emotional aspects to the community and are willing to spend more time in community immerse in the community.

Combining with the hypothesis H1a,H1b, H1c, H1d and H2a, H2b, this paper puts forward the following assumptions:

H4a:Online trust plays a mediating role between the instrumental interaction and community engagement;

H4b: Online trust plays a mediating role between the relational interaction and community engagement;

H4c: Online reciprocity plays a mediating role between the instrumental interaction and community engagement;

H4d: Online reciprocity plays a mediating role between the relational interaction and community engagement.

3.5 Community engagement and brand engagement

Brodie et al. (2013) suggested that activity participation and social presence are merely the antecedent of consumer engagement, its consequences may include commitment, trust, self-brand links, brand attachment and brand loyalty, etc. Among them, loyalty and commitment are most prominent in online brand community [2]. Vivek etal. (2012) established a theoretical model of consumer engagement that activity participation and joining community are as antecedents of consumer engagement and value, loyalty, trust, commitment, word-of-mouth and community involvement etc. are all the consequence variables [21]. Scholors not only ulteriorly pointed out the three dimensions (cognitive, emotional, and behavioral) of consumer engagement, but also proved that the consumer engagement and brand engagement will appear simultaneously in online community [8,17].

Community and brand belong to two different research category. However, they have certain connections. The brand is the core element of a community and community is built for a brand. Members' behavioral intention or emotions related to community will transfer to the brand when there is high similarity between the community and the brand in terms of users' characteristics, values, management philosophy. The deeper the interaction of the community members, the closer their links with the brand will be, and will pay more attention and passion to the brand and brand engagement appears. Then, there is the following assumption:

H5: The community engagement of members has a significant and positive effect on brand engagement.

Based on the mediating role of E-social capital, the members' interaction of online brand community will affect community engagement, which and then affects brand engagement (Figure. 1):

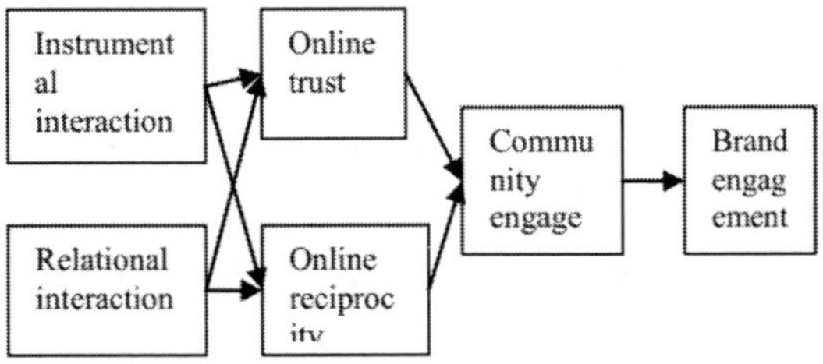

Fig.1. Study model.

4 Method

4.1 Measurement

The test items of main variables in the model are adapted from maturity scale. Community interaction includes instrumental interaction and relational interaction, which each contains four test item [16].E-social capital consists of online trust and reciprocity that the former is measured by four items and the latter includes five test items [3].Community engagement and brand engagement all include enthusiasm, participation and social contact and community engagement and brand engagement use ten items to measure respectively [21].

In order to accurately explore the relationship among main variables and eliminate the influence of irrelevant factors, we investigates the control variables in terms of community engagement and brand engagement. The control variables of community engagement include gender, time for joining community, community size, community identity and community type; the control variables of brand engagement include brand trust brand loyalty and brand satisfaction.

4.2 Data Collection

After questionnaires design was completed, we carried out questionnaire survey through a online platform. 1062 effective questionnaires were recovered after two weeks. Then we eliminated 57 ineffective questionnaires and eventually get 1005effective questionnaires (effective questionnaire rate: 94.63%).

5 Empirical analysis

5.1 Reliability and validity

We used SPSS 17.0, AMOS17.0 and Smart PLS 2.0 to analyze the data. Internal consistency reliability and composite reliability were adopted to test reliability. Apart from Cronbach's α of instrumental interaction is 0.701, Cronbach's a of brand loyalty is 0.789, Cronbach's α of other latent variables were greater than 0.8.When we deleted arbitrary item the new Cronbach's α of all latent variables are less than the original Cronbach's α coefficient, which suggests that the questionnaire and variables have high internal consistency. The composite reliability of all latent variables are greater than 0.8, which shows that the composite reliability of all latent variables are reasonable.

In this paper, the process of questionnaire design is very strict and the content validity can be guaranteed. Next step we test the structure validity of each variable that include convergent validity and discriminant validity. Convergent validity is usually tested by confirmatory factor analysis and every item factor loading are above 0.7.Fit indexes of model are respectively: $\chi 2$ (1005) = 1093.725, $\chi 2$ /df = 4.971 (< 5), RMSEA = 0.063 (< 0.08), NNFI = 0.927 (0.9) >, CFI = 0.937 (0.9) >, IFI = 0.937 (> 0.9).At the same time, the value of average variance extract from the same latent variables are between 0.542- 0.815, which are greater than 0.5. Therefore, we can determine all latent variables in the model have high convergent validity.

In addition, the item loading of each latent variable should be greater than the loading in other latent variable. The cross loadings of all latent variables conform to this standard that suggest that all test items of latent variables in the model are with high discriminant validity.

In conclusion, based on above analysis, the reliability and validity are reasonable and the further test is permitted.

5.2 The test of goodness of fit

Goodness of fit is generally used to test the fit of model.According to the formula; we can calculate the value of goodness of fit is 0.615, which suggests that the fit degree among latent variables in the model is very good.

5.3 Hypothesis test

We established two models to test the mediating role of E-social capital. Model 1 directly tests the effect of community interaction (instrumental interaction and relational interaction) on the community engagement. Model 2 tests the mediating role of E-social capital (online trust and reciprocity) between community interaction and community engagement. According to the results of software PLS test, the path coefficients of instrumental interaction and relational interaction to the community engagement are 0.205 and 0.205 (H1a, H2a: t-value is 4.768 and 4.768, respectively, are greater than 1.96, $p < 0.001$) in model 1 that suggests that there is significant correlation between community interaction and community engagement and community engagement has significant and positive influence on brand engagement (H5: β= 0.667, t = 0.667, $p < 0.001$). Therefore H5 is supported.

In Model 2, E-social capital is applied as an mediator variable to add in. First of all, Model 1 has proved that community interaction (instrumental interaction and relational interaction) significantly affects community engagement in the absence of E-social capital as a mediator; Then the effect of instrumental interaction on online trust is tested(H2a: β= 0.156, t = 5.795 > 1.96, $P < 0.05$) and the effect of online trust on community engagement is significant as well(H3a: β= 0.189, t = 6.014 > 1.96, $P < 0.01$) that show H1a and H2a are supported. When the mediator online trust is added in the path coefficient of from instrumental interaction to the community engagement is smaller than before ($\Delta\beta$= 0.049, Δt = 0.049, $P < 0.001$) but the effect between the variables is still significant (H1a: β= 0.156, t = 0.156, $P < 0.001$), which shows that online trust plays partial mediating effect between instrumental interaction and community engagement; Adopting same testing method proves that online trust plays mediating effect between relational interaction and community engagement; However, when we examined the effect between instrumental interaction and online reciprocity the effect is not significant between online reciprocity and community engagement (H3b: β= 0.052, t = 0.052 < 1.96, $p > 0.05$) that suggests there is no mediating effect between community interaction and community engagement.

At the same time, we can find from the Model 2 that adding time (β = 0.042, t = 0.042, $p < 0.05$) and community identity (β = 0.348, t = 0.348, $p < 0.001$) are reasonable for the controlling effect of the community engagement in the control variables of community engagement and the controlling effect of gender, BBS size and BBS type on community engagement are not significant ($t < 1.96$, $p > 1.96$).Among the control variables of brand engagement, the controlling effect of brand trust and brand satisfaction and brand loyalty on community engagement are not significant ($t < 1.96$, $p > 1.96$).

To sum up, except the mediating effect of E-social capital between community interaction online reciprocity is not supported all other hypothesis are supported under controlling several variables.

6 Conclusions and discussions

Based on above empirical analysis, we found conclusions as follows:

First, the instrumental interaction and relational interaction of members significantly and positively influence online trust and online reciprocity in online brand community. Instrumental interaction and relational interaction are two types of online interaction in online brand community. Interaction would promote the formation of social capital [20] that is consistent with the results of our study.

Second, online trust plays mediating effect. After long time interactions among members, trust to other members will promote commitment to others. As a result, online trust play a role of bridge among members and lead to the sense of belonging and community engagement. At the same time, the empirical analysis results show that the online reciprocity does not have significant influence on community engagement and the mediating effect between member interaction and community engagement is also not significant.

Third, the community engagement of members significantly affects brand engagement. This suggests that the relationship between the members and the community can promote the improvement of brand relationship that is consistant with Gulati et al. (1995) [12]. The in-depth interaction among community members will be able to bring them social capital, which leads to the trust and reciprocal behaviors for the community and produces strong sense of belonging to the community and community engagement.

Managements should actively create community atmosphere and encourage members to participate in interaction, properly guide to community members, such as organizing various kinds of activities to attract the different needs of members. Marketers should strengthen the guidance of the community and improve the members' trust to the community. Marketers should apply with the certain measurements to ensure the usefulness and authenticity of information and the normalization of behaviors, such as ban on advertising, spreading untrue comments, timely handling conflicts among members. In addition, the community managers should strengthen the connection between the members and the community, encourage community members to participate in community activities and make them immersed in community. And then spread brand recognition and brand culture through community to make the members have strong brand consciousness. As such they not only can improve relationship between the members and the community, but also can enlarge the brand appeal and influence, strengthen the brand value and personality, eventually achieve the goal of improving consumer - brand relationship.

References

1. Bartolini S, Sarracino F. Happy for How Long? How Social Capital and Economic Growth Relate to Happiness over Time. Ecological Economics, 2014(108): 242-256.
2. Brodie R J,Ilic A, Juric B. Consumer Engagement in a Virtual Brand Community:An Exploratory Analysis. Journal of Business Research, 2013, 66(1): 105-114.
3. Chen G L, Yang S C, Tang S M. Sense of Virtual Community and Knowledge Contribution in a Virtual Community: Motivation and Experience. Internet Res, 2013, 23(1): 4-26.
4. Coleman J S. Foundations of Social Theory. Cambridge, MA: Harvard University Press, 1990.
5. Gabarro J J. The Development of Trust, Influence, and Expectations. Interpersonal Behaviors:Communication and Understanding in Relationships. Englewood Cliffs. 1978. NJ: Prentice Hall,1978: 290-303.
6. Granovetter M S. Economic Action and Social Structure: The Problem of Embeddedness. American Journal of Sociology, 1985, 91(3): 481-510.
7. Gulati R. Does Familiarity Breed Trust? The Implications of Repeated Ties for Contractual Choice in Alliances. Academy of Management Journal, 1995, 38(1): 85-112.
8. Halpern D. Moral Values, Social Trust and Inequality—Can Values Explain Crime?. British Journal Criminology, 2001, 41(2): 236-251.
9. Hsiao C C, Chiou J S. The Effects of a Player, s Network Centrality on Resource Accessibility, Game Enjoyment, and Continuance Intention:A Study on Online Gaming Communities. ElectronicCommerce Research and Applications, 2012, 11(1): 75-84.
10. Hudson S, Roth M S, Madden T J, Hudson R. The Effects of Social Media on Emotions, Brand Relationship Quality, and Word of Mouth: An Empirical Study of Music Festival Attendees. Tourism Management, 2015(47): 68-76.

11. Kobayashi,Tetsuro. Bridging Social Capital in Online Communities: Heterogeneity and Social Tolerance of Online Game Players in Japan.Human Communication Research,2010,36(4):546-569.
12. Lee R.Social Capital and Business and Management:Setting a Research Agenda. International Journal of Management Reviews, 2009, 11(3): 247-273.
13. Lin N. Social Capital. Cambridge: Cambridge University Press, 2001.
14. Loury G. A Dynamic Theory of Racial Income Differences. Women, Minorities, and Employment Discrimination, 1977(153): 86-153.
15. Mathwick A,Young R. Affording to Exchange: Social Capital and Online Information Sharing.Cyber Psychology, Behavior & Social Networking,2013,16(8):588-592.
16. Nahapiet J, Ghoshal S. Social Capital, Intellectual Capital, and the Organizational Advantage.Academy of Management Review, 1998, 23(2): 242-266.
17. Putnam R, Leonardi R, Naetti R. Making Democracy Work: Civic Traditions in Modern Italy.Princeton: Princeton University Press, 1993.
18. Sheng M,Hartono R. An Exploratory Study of Knowledge Creation and Sharing in Online Community :A Social Capital Perspective. Total Quality Management & Business Excellence.2015,26(1/2) :93-107.
19. Sicilia M, Palazon M. Brand Communities on the Internet: A Case Study of Coca-cola's Spanish Virtual Community. Corporate Communications: An International Journal, 2008, 13(3): 255-270.
20. Tsai W, Ghoshal S. Social Capital and Value Creation: The Role of Intrafirm Networks. Academy of Management Journal, 1998, 41(4): 464-476.
21. Vivek S D. A Scale of Consumer Engagement. Tuscaloosa: The University of Alabama, 2009:1-233.
22. Wang J C, Chiang M J. Social Interaction and Continuance Intention in Online Auctions: A Social Capital Perspective. Decision Support Systems, 2009, 4(47): 466-476.
23. Zaglia M E. Brand Communities Embedded in Social Networks. Journal of Business Research, 2013,66(2), 216-223.
24. Zhou Z M, Wu J P, Zhang Q Y, Xu S. Transforming Visitors into Members in Online BrandCommunities: Evidence from China. Journal of Business Research, 2013, 66(12): 2438-2443.

The Study on the Key Elements in Strategic HRM: from Strategy to Organizational Performance

Yunlan He

China State Key Laboratory of Coal Resources and Safety Mining, China University of Mining & Technology, Beijing,100083, China
Corresponding Email: 151023631@qq.com

Abstract. Ever since the end of last century, strategic HRM scholars have devoted a great and increasing deal of efforts to the exploration and examination of contributions of HRM practices/systems into organizational performance. In spite of the lack of a conclusive message about the HRM-performance relationship, the current literature has documented vast of the evidence, which worth reviewing and may generate inspirations for future research. Thus, this research reviews the literature and teases out the key elements in strategic HRM research. These elements can be further categorized into several sub-groups: namely, HRM systems, business strategy, external and internal organizational contextual factors. Specifically, HRM systems include high control work systems, high commitment work systems, high involvement work systems, high performance work systems, high reliability work systems, and customer services-oriented work systems. The strength of the relationships between these HRM systems and performance will be varied by the changes of business strategies and organizational contextual factors. This study reviews the five types of fits between the key elements in strategic HRM literature. Moreover, it has been argued and proved that the fits between the key elements will results in better organizational performance. This article can generate not only theoretical implications for scholar, who attempt to deepen our understanding of the mechanism through which HRM systems improve organizational performance, but practical implications for managers, who will be able to identify the key factors impacting their effectiveness in managing people in real world.

1 Introduction

As cross-over study of strategic management and human resource management (HRM), strategic HRM means a series of planed and strategic human resources deployment and management behaviours in order to realize goal of enterprises[1]. As a whole, the research in this field involves various factors or variables. This paper has two purposes, the first one is summarizing and teasing main factors and variables involving in research of strategic HRM; and the second one is summarizing these variables how to interact and influence performance of organization. Main factors or variables involving in research of strategic HRM can be divided into HRM system, organizational strategy, organizational performance, and internal and external environment of organization. These factors interact in different ways, and forms match (contingency) status, which realizes optimization of organizational performance.

2 HRM system

The style of strategic HRM has comparatively independent human resource policy and practice, and human resource bundle or human resource system, which can be categorized into 6 kinds of systems. Control mode of HRM system. proposed that control mode of HRM system was a management system to reduce direct labour costs and improve efficiency by means of improving employee's compliance level for policy and regulation, and connecting employee's compensation with labour output together[2].

High commitment HRM system was making employee to learn the goal of organization and try to realize it by means of creating a series of conditions[3]. These management styles included formal work team, employee involvement team and cross-team job rotation, etc.[4,5] . Huselid (1995) thought that high performance work system could improve knowledge, technology and ability of existing employees and potential employees of organization[6]. In empirical aspects, researchers also proved that high performance work system related to variables of organizational performance, including turnover rate of employees, financial performance, and first market performance, etc. [6]. Since occupational safety caused a great loss to enterprises and suffering to individual, it was necessary for enterprise to guarantee occupational safety, Zacharatos et al. (2005) put forward occupational safety-oriented high performance work system[7].

The function of customer service-oriented work system is helping employees to provide service for customers by means of creating a kind of service-oriented organization climate. Schneider et al. (1998) found that service-oriented organization climate and subsequent evaluation results of service quality from customers depended on some basic problems[8]. Liao and Chuang (2004) further proposed that customer service-oriented high performance work system included employee involvement, training and incentive mechanism based on evaluation[9].

3 Organizational strategy

In the field of strategic HRM, people mainly pay much attention on influence of strategy to HRM and influence of match between strategy and HRM to organizational performance [10-12]. Generally, the competitive strategy proposed by Schuler based on strategic classification of Porter (1985) was widely used. In addition, there were strategic classification of Wiles and Snow (1984) and strategic framework of Wright and Snell (1991).

Miles and Snow (1984) divided organizations into 4 types: defending type, aggressive type, analytical type and reaction type. Wright and Snell (1991) thought organizations could be categorized into growth type, stable type and shrinking type. Although classification of competitive strategy owes to Schuler, one point must be explained, organization strategy classification which is most widely used in present strategic HRM, is proposed by Schuler and Jackson (1987a,1987b)[13,14] on the basis of competitive strategy framework of Porter, and they described behavior characteristics of needed employees in each strategy. In total, there are 3 competitive strategies, including innovation strategy, quality commission strategy and cost saving strategy.

4 Organizational performance

In research of strategic HRM, organizational performance can promote HRM exactly. Meanwhile, people have strong interest on what kind of organizational performance and how organizational performance can be influenced by HRM. The first batch of models of human resource system atically described involving in organizational performance appeared in 1980s, which were proposed by Beer et al. (1985) and Fombrun et al. (1984) respectively[15,16]. Models of Beer, etc, also were called Harvard Models. They though that HRM policy influenced effect of human resource firstly and further produced some long-term effects and promoted input-output rate. These long-term effects included individual welfare of employees, organizational efficiency and social welfare. Fombrun et al. (1984) put forward a so-called individual circling model of human resource[16]. The content of human resource involving in organizational performance relationship was 4 kinds of key HRM policies and practices, including selection, remuneration, evaluation and training, which promoted both individual performance and organizational performance.

5 Internal and external environment of organization

As BudhwarandAryee (2008) had pointed out that the most important contextual factors of organization was influence of economic, technology, social and political factors to human resource which should be mainly considered in research of strategic HRM[17]. Thus it can be seen the significance of organization and environmental factors in strategic HRM. Jackson and Schuler (1995) proposed it systematically in research of strategic HRM[18]. They unified organizational variables and environment variables to contextual variables of organization. These variables were divided into internal contextual variables and external contextual variables of organization.

5.1 Internal contextual variables of organization

At present, there are 6 forms of ownership in China, including state-owned enterprises, private enterprises, collectively-owned enterprises, joint venture companies and foreign-owned companies [19]. Furthermore, township enterprises exist, although township enterprises belong to collectively-owned enterprises, their corporate structure and governance situation are greatly different from other collectively-owned enterprises'.

According to institutionalism opinion, a series of management activities adopted in forming stage of organization would be internalized in organization, and had deep influence to future development of organization. In China, age of organization related to forms of ownership. For example, age of state-owned enterprises would be older than age of joint venture enterprises or foreign-owned enterprises[20]. In a number of empirical studies, size of organization would influence style and effectiveness of HRM significantly, especially in growing and transforming stage of organization. Since the size of organization was larger, the structure would be unwieldy comparatively, organizational inertia would be greater, and further hindered the transformation process of organization[20]. This factor was proposed by, and they suggested that internal technical level of organization should be considered in research of strategic HRM. Meanwhile, internal technical level should be a determinant factor in choosing pattern of HRM. In different life cycle stages, such as initial period, growth stage, mature stage and decline stage, organization needs different support of HRM effect. Therefore, life cycle of organization was an important influencing factor in HRM[18].

5.2 External contextual variables of organization

The technological change here means changes in macro level. In Hongkong companies they surveyed, Fields et al (2000) found that pressure caused by technological change significantly influenced core strategic position of HRM department[21].

Industry means a group of production or profit organization sets with same characteristics. Jackson and Schuler (1995) indicated that difference of industry would influence HRM of enterprises[18]. Boxall and Purcell (2008) divided industry change into 3 phases: confusion of early establishment, mature and regenerative confusion[22]. An enterprise will pay more attention on cultivation of sustainable competitive advantages when it's competitive pressure is higher. In Hongkong companies they surveyed, Fields et al (2000) found that competitive pressure would greatly influence enterprise to adopt what kind of human resource strategy[21]. The influence of legal factors to HRM mainly reflected in utilization of laws and regulations, such as Labour Law, Environmental Protection Law, and Tax Law, etc. [18]. Political factors mainly reflect in political stability and political risks. The political stability can bring confidence to enterprises for long-range program.

Many theoretical and empirical studies proved that labour union would influence HRM. For example, in China, the existence of labour union can't significantly promote enterprise to adopt innovative strategic[23]. On the other hand, western scholars Ramaswamy and Schiphorst (2000) found that the existence of labour union would suppress organization on the contrary, especially when management staffs of organization tried innovative management activities[24].

In aspect of geographical location, the management mode of HRM was more refined in relatively developed areas, but in underdeveloped areas, the management mode of HRM trended to traditional personnel management mode [21]. The most popular measurement framework of national culture was proposed by Hofstede (1980)[25]. Wallach (1983)classified company culture into bureaucrat type, innovative type and supporting type. In field of strategic HRM, Jackson and Schuler (1995) pointed out very few people directly studied influence of culture to HRM, while most people compared difference of HRM between two countries or areas, and thus thought culture would influence mode of HRM[18].

6 Matching: road to improvement of organizational performance

Main research purpose of strategic HRM is verifying the positive influence of HRM system to corporate performance, and the level of influence even is influenced by organizational strategy, internal and external environment of organization. In functions of this system, match of each factor has decisive function.

Chenevert and Tremblay (2009)classified match in strategic HRM into 5 types: environment match, longitudinal match, internal match of organization, internal match of human resource system and internal activity match in sub-function of HRM[26]. Environment match emphasizes match between HRM policy and external environment of organization. According to this view, HRM system should adapt characteristics of social, cultural and legal factors. Longitudinal match emphasizes connection between human resource and organizational strategy, considers HRM system should be adjusted timely for supporting realization on strategic target of organization. Internal match of organization means match between HRM system and other internal systems of organization, such as technological level, structure, production system and organizational culture. Internal match of human resource system means match of different functions in HRM system, for example, training and development functions should match with salary management. Internal activity match in sub-function of HRM means match of policy, activity and concrete actionin each function of HRM, for example, the internal policies, activities and concrete actions of human resource recruitment and selection should match. Only when these factors match in different ways, strategic HRM policy and practice will produce significant and positive influence to organizational performance.

7 Conclusions

This paper teases present important factors involving in research of strategic HRM from China and the West, reviews main factors in previous research of strategic HRM, and classified these factors into HRM system, organizational strategy, internal and external environment of organization. The relationship strength between HRM system and organizational performance will change with changes of organizational strategy and internal and external environment of organization. The match level of each factor in strategic HRM will improve influence of HRM system to organizational performance. Through the study, on the one hand, in management research, researchers can further explore important variables that are not involved, and deepen research on relation between variables. On the other hand, in management practice, administrator should pay more attention on influence of these factors in whole management process, attach importance influence caused by changes of these factors in management decision and execution, reasonably enlarge positive influence, and try to avoid negative influence of these factors, for improving organizational performance in maximum limitation.

Acknowledgement

This research is supported by Youth Fund of Natural Science Foundation of China (41602168), Postdoctoral Fund of China (2016M591286), and Research Project for Oil and Gas Centre of China Geological Survey (12120115102101).

References

1. P. M.Wright, G. C.McMahan, B.McCormick and W. S.Sherman. Strategy, Core Competence, and HR Involvement as Determinants of HR Effectiveness and Refinery Performance, Human Resource Management, **37**,1 (1998)
2. J. B.Arthur. 'The Link between Human Resource Systems on Manufacturing Performance', Academy of Management Journal, 37, (1994)
3. E. M.Whitener. Do "high commitment human resource practices affect employee commitment? A cross-level analysis using hierarchical linear modeling. Journal of Management, 27 (2001)
4. J. P.MacDuffie. Human resource bundles and manufacturing performance: Organizational logic and flexible production systems in the world auto industry. Industrial andLabor Relations Review, 48 (1995)
5. P.Osterman. How common is workplace transformation and who adopts it? Industrialand Labor Relations Review, 47 (1994)
6. M. A.Huselid. The impact of human resource management practices on turnover,productivity, and corporate financial performance. Academy of Management Journal, 38 (1995)
7. A. Zacharatos, J.Barling and R. D. Iverson. High-performance work systems andoccupational safety. Journal of Applied Psychology, 90 (2005)
8. B.Schneider, S. S.White, M. C.Paul. Linking service climate and customer perceptions of service quality: Test of a causal model. Journal of Applied Psychology, 83 (1998)
9. H.Liao, & A.Chuang. A multilevel investigation of factors influencing employee service performance and customer outcomes. Academy of Management Journal, 47 (2004)
10. R. E.Miles and C. C..SnowDesigning Strategic Human Resource System. Organizational Dynamics, (1984)
11. M. E.Porter. Competitive Strategy: Techniques for Analyzing Industrial and Competitors. New York: Free Press (1980)
12. P. M.Wright and S. A.Snell. Toward an integrative view of strategic human resourcemanagement. Human Resource Management Review, 1 (1991)
13. R. S.Schuler and S. E.Jackson.. Organizational Strategy and Organization Level as Determinants of Human Resource Management Practices. Human Resource Planning, **10**,3 (1987a)
14. R. S.Schuler and S. E.Jackson. Linking Competitive Strategies with Human Resource Management Practices. The Academy of Management Executive, **1**,3 (1987b)
15. B.M.Beer, P. R.Spector, D.Lawrence, Quinn Mills, and R. E.Walton. Human Resource Management: A General Manager's Perspective, New York: Free Press. (1985)
16. C. Fombrun and M. A. Devanna. Strategic human resource management. New York: John Wile (1984)
17. P.Budhwar and S. Aryee. An Introduction to Strategic Human Resource Management, In The Aston Centre for Human Resources (2008), Strategic Human Resource Management, Building Research-Based Practice, London: CIPD, (2008)
18. S. E.Jackson and R. S.Schuler. Understanding Human Resource Management in the Context of Organizations and Their Environment, in Rosenzweig, M. R. and Porter, L. W. (eds), Annual Review of Psychology. Palo Alto, CA: Annual Reviews, 46 (1995)
19. F. L.Cooke. HRM, Work and Employment in China. New York: Routledge (2005)
20. Z. D.Ding and S.Akhtar. The Organizational Choice of Human Resource Management Practices: A Study of Chinese Enterprise in three Cities in the PRC. The International Journal of Human Resource Management Journal, **12**,6 (2001)
21. D. Fields, A.Chan and S. Akhtar. Organizational context and human resource management strategy: a structural equation analysis of Hong Kong firms. The International Journal of Human Resource Management, **11**,2 (2000)
22. P. Boxall and J. Purcell. Strategy and Human Resource Management, New York: Palgrave Macmillan. (2008)
23. I.Ng and D.Maki. Trade Union Influence on Human Resource Management Practices. Industrial Relations, 33 (1994)

24. E.A.Ramaswamy and F. B.Schiphorst. International Journal of Human Resource Management. **11**,4 (2000)
25. G.Hofstede. Culture's Consequences: International Differences in Work-Related Values. Berverly Hills, CA: Sage. (1980)
26. D.Chenevert and M.Tremblay. Fits in Strategic Human Resource Management and Methodological Challenge: Empirical Evidence of Influence of Empowerment and Compensation Practices on Human Resource Performance in Canadian Firms. The International Journal of Human Resource Management, **20**,4 (2009)

Market Analysis of Soundproof and Noise Reduction Plate

Heyu Chang[1,*] and Yaxing Wang[2]

[1]Management Science and Engineering Department, Zhengzhou University, Henan Province, China
[2]Management Science and Engineering Department, Zhengzhou College, Henan Province, China
Corresponding Email: 1406987621@qq.com

Abstract. Since the reform and opening up, China has a booming economy development, transportation industry as an important part of the economic processes has made a considerable progress, and continue to promote the rapid economic development. At the end of 2012, China's highway mileage has reached 4.238 million km and highway mileage ranks first in the world. Transportation in promoting rapid economic development, but also brings the traffic noise which has a tremendous influence on residents' daily life, and this influence has gradually deepening and widening, in this situation the production of noise reduction panel gradually emerged. In this paper, a noise barriers manufacturer' business model in Beijing Daxing District and analyze existing data will be analyzed, it will give much guidance and reference for a new noise reduction factory. The study has shown that plant will target all levels of government departments in charge of roads and the property of management department of residence as our customer groups, increasing technological innovation and improving product quality to establish a more extensive business relationships, and gradually formed noise barriers market competitiveness.

1 Introduction

The establishment of socialist market economy gradually relaxed the access threshold for enterprises, so small and medium enterprises have sprung up. By the end of 2014, the number of registered small and medium enterprises have reached 42000 thousand, which hold 99 percent amount all registered enterprises in China, and its contribution rate of GDP, tax and employment reached 58.5%, 52.2% and 80% respectively [1]. There is no doubt that an economic boom and high-speed growth depend on the manufacturing industry because the real economy is the drive for social development. The state-owned economic system reform of last century in 90s has gradually turned social economic development from the quota system dominated by the major state-owned enterprises to the market competition system dominated by the small and medium enterprises [2]. And the state has encouraged development of small and medium enterprises in terms of preferential tax and opened up a Easy Access for entrepreneurs [3].

A good creative project is the basic premise of success, but not business results. Only through effective action the brain ideas can be transformed into practical productive forces [4]. In this paper, based on the Potter five forces industry analysis, the opportunity and challenge of noise reduction board factory machine will be discussed, and the feasibility of a series of problems of production, price, marketing and finance will be analyzed. Through analyzing the development expectations and possible problems of the noise reduction plate in Beijing Senseshixing Environmental Engineering Department, some countermeasure will be offered for future entrepreneurs.

2 Marketing analysis of soundproof and noise reduction plate

2.1 The definition of soundproof and noise reduction plate

Soundproof and noise reduction plate is also called noise barrier. Its operating principle is reducing the propagation strength of noise through altering the propagation path of sound artificially, with the aim of soundproof and noise reduction. Noise pollution has become the fourth pollution following solid waste pollution, water pollution, air pollution, and it has huge impact on people's daily lives [5, 6].

According to the survey data from the related organizations of environment ministry in 2014 people's degree of satisfaction of life noise is only 50%, just above their degree of satisfaction of air pollution, ranking second from bottom [7].

With the rapid construction of national highways, railways and people's higher and higher requirement for the quality of living environment, the rapid development phase of the noise barrier industry has begun [8]. The soundproof principle of noise barrier is that acoustic waves are very easy to be a reflected back to sound source when the wavelength of acoustic waves of the noise is smaller than the surface size of the barrier, and thus a sound shadow region where it cannot be affected by the noise is formed behind the noise barrier. It is shown in Figure 1.

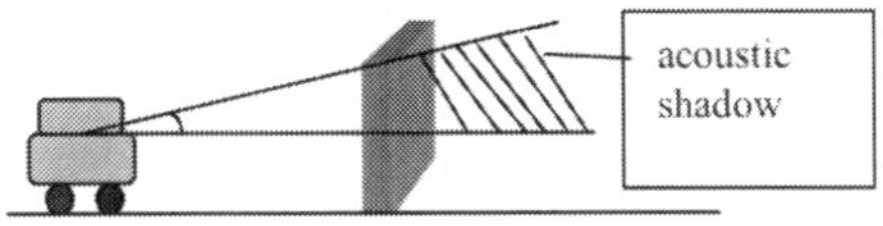

Fig. 1. Acoustic shadow of noise barrier.

2.2 The market development of noise barrier in China

Since the policy of China's reform and opening-up was carried out, transportation industry has gotten rapid development. At present highways has covered more than 90% of the medium-sized city in our country, and total highway mileage ranks first in the world [9]. With the unceasing construction of highways, railways and other transportation facilities, the need for building noise barrier along the road will also be increased accordingly.

So far, the Chinese manufacturers of sound barriers are around 5000 by conservative estimates, and products range from the transparent soundproof PC plate, bridge and railway noise barriers to airport fence noise barriers. There are many manufacturing enterprises in this industry [10]. Based on the different production process and soundproof effects, noise barriers of most manufacturers are priced at 50-400 RMB per square meter. It is shown in Table 1.

Table 1. The price of all kinds of noise barrier(Unit: RMB).

Core Materials	Textures	Acrylic	Iron or zinc plating	PC soild sheet
PC board		50	100	150
Glass wool		100	150	200
Rock wool		150	200	300

The price of noise barrier is mainly determined by its internal core material and its external material. The price of noise barrier is the cheapest one if its core material is PC plate and its external material is acrylic, and the market price has maintained at the level of 50 RMB per square meter. The price of noise barrier is about 150 RMB per square meter if its core material is glass wool and its plate is iron plate or galvanized iron plate. The price of noise barrier is the most expensive one if its core material is rock wool and external material is PC plate, and its market price is roughly 300 RMB per square meter.

2.3 Consumption analysis of noise barrier

The construction of national railways has leaped forward since 1995. By the end of 2013 China's highway traffic mileage has reached 104468 kilometers, ranking first in the world [11]. The increase of highway mileage adds the possibility to build highways beside cities and towns. it is bound to have certain impact on the residents' daily life along the highway [12]. And it has become the inevitable choice to build noise barrier along the highway to minimize the impact of noise on the residents. Figure 2 reflects market demand growth on China's noise barrier market from 2005 and 2012.

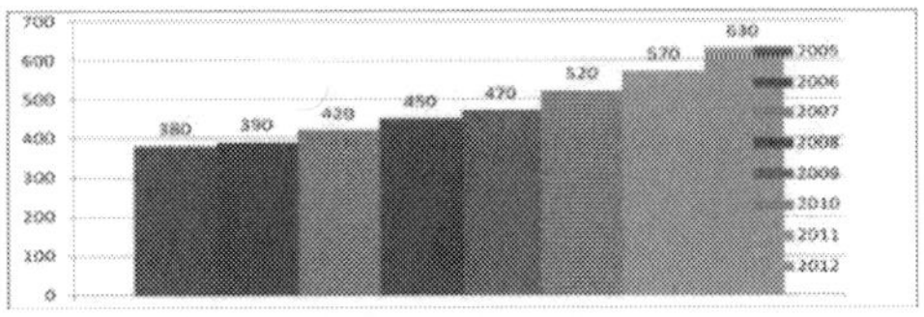

Fig. 2. The market demand of noise barrier from 2005 to 2010.

Demand for noise barrier is in steady trend of rising, from 3.8 million tons in 2005 (a ton equivalent to 100 square meters) to 6.3 million tons in 2012, and the market demand growth rate reached 65.7%. Looking from the data, the market demand for noise barrier is very big, and the growth of noise barrier has stable momentum. It indicates that investment risk in this field is not particularly big, and it is a relatively stable equipment manufacturing industry that is conducive to the development of the national environmental protection.

3 External environment analysis of noise barrier market

3.1 Customer analysis

The main group of consumption of noise barrier is government management departments of highway, railway and airport construction at all levels, whose requirement for the quality of noise barrier is relatively high. They are not particularly sensitive to the product price, so more business profits can be gotten. Secondary customers are all kinds of living community built along important arteries. Noise barrier construction bid determines the success or failure of the project, but market share of this part is very big [13]. Scale effect can be obtained by using the strategy of gaining high turnover through meager profit. The rest consumers some service sectors determined by the operating characteristics, such as KTV. Consumer groups of this part pursue the maximization of the soundproof effect, so their bargaining power is low. However, because of the unbalanced development of those places providing leisure service, their market demand is small.

3.2 Supplier analysis

The manufacturing technology of noise barrier produced by the environmental protection department of Center INT in Beijing is high, and it involves a good deal of product material. Figure 3 is a simple picture of the noise barrier produced by this company.

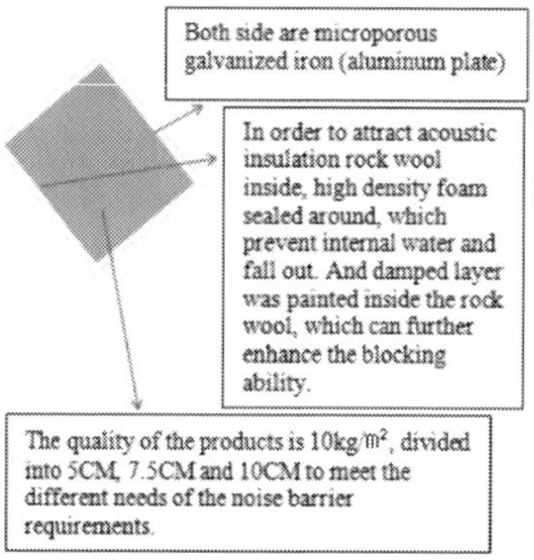

Fig. 3. Simple engineering drawings of noise barrier.

The quality of the noise barrier produced by Center is high. And many suppliers are involved. But the basic raw materials are ordinary industrial products, and there are many suppliers.

3.3 Competitor analysis

Competition among noise barriers belongs to differentiation competition of products of the same industry [14]. Along the railways and airports only noise barrier can be used to obtain efficient soundproof and noise reduction effect. According to the product type the competitors' products can be divided into three types. Specific conditions are shown in Table 2.

Table 2. Three main noise barrier product in sound insulation market (Unit:RMB/㎡).

Type of sound insulation products	Product structure features	Price	Advantages and disadvantages

Product one	The mainly external material is acrylic and the core material is PC board.	30-100	Advantages: low cost, low engineering cost, Suitable for village sound insulation construction along the way. Disadvantage: poor noise insulation effect.
Product two	The mainly external material is galvanized sheet and the core material is fiber cotton board.	100-200	Advantages: sound insulation effect is good and its price is moderate. Disadvantages: waste fiber cotton board can easily cause pollution.
Product three	The mainly external material is senior endurance plate and the core material is rock wool.	Around 300	Advantages: sound insulation effect is the best, and the product is beautifu. Disadvantage: higher prices

The new-established factory is going to produce product type 2 based on the manufacturing technology of Center. The quality of this product is at above the average level in the industry, and it has the characteristics of airproofing, waterproofing, fireproofing, strong surface corrosion resistance, good soundproof effect and moderate price. Thus it can be used to participate in the bidding of government construction projects and serve the program of community noise control. And it can maximize the management efficiency of the company.

3.4 Substitute analysis

Noise barrier is the most effective way to solve the problem of traffic noise along the highway, railway and airport. It is difficult to find a substitute. Substitutes with the similar soundproof effect are landscape greening woods along the road such as lawns and trees and concrete block walls. Noise barrier is by far the most effective way to control traffic noise. It not only has good soundproof effect and also less construction difficulty. And the later maintenance is easy. It has the appeal of the landscape. Threat from the substitutes is small.

3.5 Potential entrants

With the continuous development of national road construction, development of noise barrier becomes faster and faster. Where there is demand there will be a market. Keen and rational economic men will grasp the opportunity, and become the potential entrants of noise barrier industry. Currently potential entrants include landscaping company and individuals or companies who have operating experience of the noise barrier industry and certain funds.

Opportunities of the industry at present are concluded through the analysis of competition of noise barrier in noise barrier industry: (1) The construction of national infrastructure is striding forward, and noise barrier market will usher in a golden period of development. (2) Noise barrier industry is a capital-intensive as well as technology-intensive industry. The use value of the products can be increased through improving the manufacturing technology and making the production process finer. Space of improvement and reform of noise barrier is large. And it also gives new business opportunities of development. (3) The bargaining power of suppliers and customers is not high, therefore it also ensures timely realization of corporate profits. (4) Substitutes to replace the effect of noise barrier is not obvious. They are unable to completely replace the soundproof and noise reduction effect of noise barrier. Noise barrier can keep the product uniqueness for quite a period of time. (5) The threat of potential entrants is small. Due to the restriction factors of obtaining soundproof effect and practical operation difficulty. It is difficult for the landscaping companies and the all alike to enter the field of noise barrier. Threat mainly comes from competition within the industry. There are more than 5000 enterprises producing soundproof facilities on domestic market. Although the quality of them is uneven, they all can impose a certain impact and influence on fixed market.

4 Marketing and sales

4.1 Market orientation

4.1.1 The siting of factories

Noise barrier factories are manufacturing factories. Their geographical location should be sited in small and medium-sized industrial parks far away from business districts and residential areas. There transportation is convenient, and all kinds of industries cluster in the parks, and complete infrastructure is established. Besides they are far away from the residential and business district, they will not impose much impact on the residents nearby.

4.2.2 Customer orientation

The consumption subject of noise barrier is government administrative departments at all levels of highways, railways and airports. It also includes the medium-to-high grade residential areas near traffic roads and a small number of service sectors needing acoustic environment. For the relevant government departments, companies should use the high quality of soundproof equipment and a price with competitive advantage to win the market.

For buildings near the highways, railways and airports, opinions of related management department about project of building noise barrier should be actively understood. Right of construction of the project should be strived to obtain. For service sectors needing soundproof equipment, customers' bargaining power is low. Although the market is not very big, considerable benefits can still be gotten by selling products at a premium. Therefore this part of market should also be won.

4.2 Marketing strategy

4.2.1 Product strategy

Most noise barrier has fixed production specifications. Researchers suggest that the object's geometric shape can effectively weaken the noise effect, so noise barrier produced by companies will adopt laced resonant devices to improve soundproof effect. This not only enhances the functionality of noise barrier, and also makes the appearance of the product more ornamental. At the same time a layer of light blue zinc powder is plated on the outside of the soundproof plate so that the overall landscape effect of it will be improved, and this makes the product look not so stiff.

4.2.2 Pricing strategy

According to different consumption subject different product pricing strategy is chosen. The method of public bidding is adopted for government projects. Product price is relatively transparent. The profit of the products are kept at a stable and reasonable level, so while normal profit is maintained, it is better to offer a rational price after the production cost of the product is deducted. Under special circumstances unprofitable ways can be adopted to develop a long-term cooperation relationship with the government. For community soundproof plate construction and service sector needing soundproof measures, the product can be sold at a higher premium to improve the level of corporate profits.

4.2.3 Channel strategy

Soundproof products are different from other products or services, and it belongs to the special construction materials. Consumers involved are relatively fixed, and they are primarily the government, the residential property management and service sector needing soundproof effect. So way of marketing is single. The method of direct selling is suitable for the characteristics of the product itself.

4.2.4 Promotion strategy

The price of soundproof product is determined by comprehensive factors such as cost of the product, equipment losses and labor. Generally speaking the price is relatively fixed. Only large-scale construction orders are possible to have the corresponding discount.

5 Conclusion

5.1 The market prospect of noise barrier is vast

On the one hand, with the rapid and stable development of economy, Chinese Road, rail and air transport industry will usher in the golden period of road construction, which means the road that need to install the insulation section increase, and this phenomenon objectively promotes the development of sound insulation screen industry. On the other hand, with the increasingly high demands on the life quality, noise pollution has become an important issue. Many NPC representatives proposed a bill aimed at noise pollution at the annual meeting of the Chinese People's Political Consultative Conference from 2012 to 2014, and the demand of one community near the road on installing sound insulation screen increases too.

5.2 The independent development can be established based on the existing resource

In the early stage, the company can smoothly get through challenges through imitating product characteristic of Beijing Senseshixing. However, the company need to rely on its ability of independent development of production to keep growing. Therefore, they need to establish their own image in the industry, continue to strengthen the innovation ability on the basis of reduced cost and exploit their own business and customer.

5.3 Perfecting marketing system and maximizing marketing resource

A problem of maximizing corporate profits the sound insulation screen obstacle is the product sales, the company's ability to solve this problem determine whether this company can develop rapidly or not. Therefore, the company should actively develop new customers and new resources based the advanced information network, such as register their own website on the Internet and put related information products on the Internet, only do this can make customers intuitively understand products and contact us.

References

1. Siyu Lin, Ma Zhong. "Restless City: Noise Pollution Issue", Ecological Economic, June 2014.
2. Weiwei Liu. "Current Situation of Noise Pollution and Controlling Research in Dalian", Green Technology, April 2014.
3. Wei Meng, Fuchen He. "Current Situation of Ambient Noise in Fushun City and Its Control Countermeasure", Liaoning's Urban Environmental Science, April 2006.
4. Lirong Zhang, Zhijun Wen, Dongchuan Wang. "Measure Research of
5. Jun Xiao, "Low Carbon Energy Saving and Decreasing Bibration Insulation of New Material Makes the Care More Quiet", Bird's Eye View of World Environments, July 2012.
6. Li Ma, "High Performance Highway Acoustic Barriers", Building Materials Information, July 2013.
7. Decai Liang, "Innovation Research of Sounds Insulation and Noise Reduction in Traffic Artery", Resource and Environmental Technology Information, June 2008.
8. Long Long, "Discuss of Upper Noise Barrier and Every Products Performance", North Traffic, 2012.
9. Expert group of Noise Barriers Information Portal, "Noise Barriers Technology and Material Selection Manual" Beijing: China Machine Press Files, August 2011.
10. Yunfen Wang. "Road Traffic Environmental Engineering". Beijing: People's Communications Press Files, 2001.
11. Fuqing Dan. "Highway Construction and Environmental Protection Research", Harbin:University of The Northeast Forestry Press Files, 2010.
12. Wentang Ren, "Traffic Noise and Control", Beijing: People's Communications Press Files, 1984.
13. Fvkuda T, Ozawa S. "Propagation of Compression Ware in a Long Tunnel With Slab Tracks", Quarterly Report of RTRI, 2005, 46(3): pp.188-193.
14. Howe Ms, Iida T. "Rapid Calculation of the Compression Wave Generated by a Train Entering a Tunnel With a Vented Hood", Journal of Sound and Vi-bration, 2006, 297(1): pp. 267-292.

The Construction of Early-warning Index System in Regional Circular Economy of China

Yanqing Kang [1,*] and Rui Zhang [2]

[1] College of Administrative Engineering, Zhengzhou University, Zhengzhou, 450001, China
[2] College of Management and Economics, Tianjin University, Tianjin, 300072, China
Corresponding E-mail: kang4006@126.com

Abstract. In this paper, the model of early warning system for circular economy development system was studied, the early-warning system for circular economy provides a quantitative model to support the pollution control. Specifically, this paper built an early warning index system of regional circular economy, and put forward the measure model for the development of circular economy. In addition, this paper constructed prediction model of the circulation economy development. Finally, by the proposed early warning model, this paper analysed the circular economy of Henan province and targeted policy suggestions were put forward for the construction of circular economy of Henan province.

1 Introduction

Circular economy is a thought of sustainable development and thought the ecology rule to realize the recycling of resources. It is taken the resources as highly efficient and recycling materials, with low consumption, high efficiency and low emissions as a basic feature of economic growth mode. In addition, circular economy pursues the maximum of using energy and recourses and the minimum of discharge of wastes, doing like this is in order to build the chronic circular development.

At present, the theory and operating mechanism of circular economy system have become increasingly attention and the quantitative researches are mainly through the construction of index system to measure the circular economy development level, Such as Zhang [1], Yang [2], Ouyang [3], Zhao [4], Xiang [5], mostly considered "3R" principle as the starting point, through investigating the circulation economy connotation and development model to building can measure the index system of circular economy for the corresponding research. Specific to regional research, Zhong [6], Cao [7], Huang and Ping [8] discussed the development phase and mode of regional circular economy to build the corresponding index system and give a comprehensive evaluation of the development level of regional circular economy in the paper.

Resent some years, through building scientific index system then according to the index system of regional circular economy to predict the future development trend gradually becomes important, literature such as Tong [9], Qin [10] and Qian [11] has studied the construction of early warning system for regional circular economy, but the methods are different. On the basis of these articles, this paper will evaluate the development level of regional circular economy and gives corresponding judgment and warning about future development through the establishment of the corresponding early warning index system of Henan province.

The rest of this article is presented as follows. Firstly, this paper conducts the construction of regional index system and introduces the weight given by the method in section 2. Secondly, section 3 introduces and processes the data. Thirdly, this paper builds the early warning system through the index system and analyses the development level of circular economy of Henan province in section 4. Finally, section 5 summarizes the conclusions and puts forward suggestions.

2 The index system and weight method

2.1 The conduction of index system

This paper discusses the necessity of constructing an index system of regional sustainable development and basic principles while constructing the system. Index system must be able to compare the construction of a comprehensive response to a development level of regional circular economy, but also need to consider on index selection of independent and representative principle, this article refers to various research literature on the basis of fully, constructs in the following evaluation index system of regional circular economy in Table 1.

Table 1. The index system of circular economy.

Index layer		characters of index
Per capita GDP (yuan/person)	X1	positive
The third industry of GDP (%)	X2	positive
Disposable income of urban residents	X3	positive
Per capita net income of farmers	X4	positive
High technology production value of industrial output value (%)	X5	positive
R&D share of GDP(%)	X6	positive
Environmental pollution control investment of GDP (%)	X7	positive
The standard discharge rate of industrial wastewater(%)	X8	positive
Industrial dust removal rate （%）	X9	positive
The standard discharge rate of industrial SO_2 emissions (%)	X10	positive
The standard discharge rate of industrial soot emissions (%)	X11	positive
The repetitive use rate of industrial water (%)	X12	positive
The treatment rate of urban wastewater (%)	X13	positive
The comprehensive utilization rate of industrial solid waste (%)	X14	positive
"Three wastes" comprehensive utilization rate of gross industrial output value (%)	X15	positive
Elasticity coefficient of energy consumption	X16	negative
Fertilizer usage per hectare (kg)	X17	negative
Pesticides usage per hectare (kg)	X18	negative
Energy consumption (tce/ten thousand yuan)	X19	negative
Power consumption (KWH/ten thousand yuan)	X20	negative
Energy consumption per unit value added of industry (tce/ten thousand yuan)	X21	negative
water consumption (m^3 / ten thousand yuan)	X22	negative
SO_2 emissions intensity (m^3 / ten thousand yuan)	X23	negative
COD emissions intensity (kg/ten thousand yuan)	X24	negative
Urban residents engel's coefficient (%)	X25	negative
Built-up area green coverage rate (%)	X26	positive
Per capita green area of park (m^2)	X27	positive
Hazard-free treatment rate of urban living garbage (%)	X28	positive
Spending on education career of GDP (%)	X29	positive
Registered urban unemployment rate (%)	X30	negative
Per capita area of urban road (m^2)	X31	positive

2.2 The weight method

Considering weight coefficient of index belong to different hierarchy should measure the overall index under the degree of influencing on other indicators. The thought of variance drive exactly conforms to the basic ideas, empowerments of the original information directly from the indexes provide the amount of information, and thus to determine the weight coefficient of each index. The specific method is introduced as follows.

Suppose there are m evaluation objects x_1, x_2, ..., x_m, corresponding weight coefficient of w_1, w_2, ..., and w_m. Use x_{i1}, x_{i2}, ..., X_{in} express the ith value of each evaluation index, $X=\{x_{i,j}\}_{m*n}$. Comprehensive evaluation function is expressed as:

$$y_i = w_1 x_{i,1} + w_2 x_{i,2} + \ldots + w_m x_{i,m} \tag{1}$$

Here

$$y=(y_1, y_2, \dots, y_n)^T, w=(w_1, w_2, \dots, w_m)^T \quad (2)$$

When the original data are standardized, $\bar{y} = 0$ ($\bar{y}$ is the average of y), there are

$$ns^2 = y^T y = w^T X^T Xw = w^T Hw \quad (3)$$

$$s^2 = \frac{1}{n}\sum_{i=1}^{n}(y_i - \bar{y})^2 = \frac{y^T y}{n} - \bar{y}^2$$

Where $H = X^T X$.

The maximum value of Equation (3) will be able to get the biggest information differences between various indicators, so as to realize different drivers. Through calculating the largest eigenvalue of H and the corresponding eigenvectors and normalizing all the characteristics of the object vector, it can get the corresponding weighting system structure (w), then take it to Equation (1) to get the ith a comprehensive evaluation value (y_i) of the object.

2.3 The division precinct

This paper shows the classification standard of early warning system for circular economy in Table 2.

Table 2. Classification standard of early warning system.

Range	(0.8,1]	(0.6,0.8]	(0.4，0.6]	(0.2,0.4]	[0,0.2]
Warning degree	non police	light police	intermediate police	heavy police	giant police

3 The data

The data of this paper are from Henan statistical yearbook, Henan statistical bulletin, China statistical yearbook of science and technology, China statistical yearbook, China's environmental statistics yearbook. Besides, the data are conducted by dimension.

4 Early warning system for circular economy of Henan province

The rest of this article will study the construction of early warning system of regional circular economy, taking Henan province as an example. This paper will make the comprehensive analysis about the development stage of circular economy in Henan, then this paper will study the development level of circular economy in Henan using the index system, and then do early-warning analysis of Henan circular economic sustainability.

4.1 The evaluation of Henan circular economy

According to the approach of variance drive, given the corresponding weights, the comprehensive calculation of comprehensive evaluation of Henan circular economy presented in Table 3.

Table 3 shows that circulation economy of Henan maintain has a stable development from 2005-2013. The development degree of circular economy gradual increases from 0.1990 to 0.9013, the level of economic development rises from 0.0840 to 0.1672, the level of green development increases from 0.0839 to 0.4826, the level of social development rises from 0.0311 to 0.2248. These show that the development of Henan's circular economy is growing steadily.

Table 3. Comprehensive evaluation of circular economy in Henan.

	2005	2006	2007	2008	2009	2010	2011	2012	2013
Level of economic development	0.0840	0.1142	0.1285	0.0802	0.0903	0.0842	0.1477	0.1586	0.1672
Level of green development	0.0839	0.1654	0.2351	0.3465	0.3945	0.4363	0.4533	0.4725	0.4826
Level of social development	0.0311	0.0567	0.1276	0.0967	0.1273	0.1755	0.1991	0.2136	0.2248
Development degree of circular economy	0.1990	0.3363	0.4912	0.5234	0.6121	0.6961	0.8001	0.8542	0.9063

Considering the green index is the core of development level of the circular economy, this paper also carries on the further subdivided to green development index study, the results are shown in Figure 1.

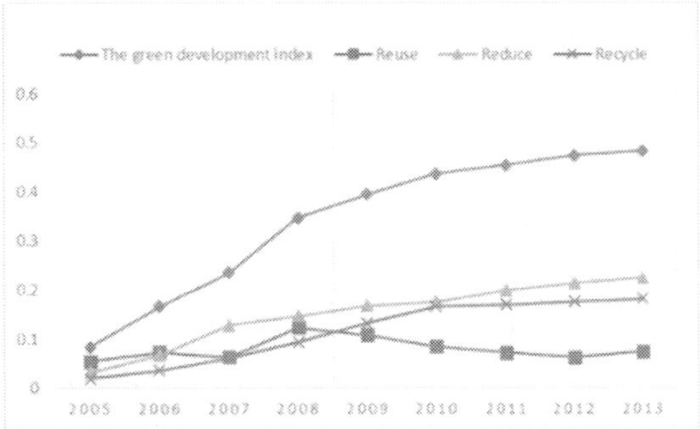

Fig. 1. Trend chart of green development index of Henan province.

Figure 1 shows that recycling and reduction index of Henan province belong to a steady rise trend in 2005-2013, but the reusing index presents a rising trend first, then showed a gradual decline trend after 2008 years. These show that there is no sustainable road in the resource utilization of Henan province.

4.2 The construction of early warning system for Henan's circular economy

Before undergoing early warning system for the construction of circular economy, this paper first determines the target value of various indicators and the determination of the target value need to consider the data authority, accessibility, as far as possible in accordance with the principle of circular economy. In this paper, the target values of various indicators use the value in the Henan's twelfth five-year plan for national economic and social development and the Henan's five-year plan of environmental protection planning as early warning target. If there are no specific targets in this papers, this paper will consider a certain index that does not affect the validity of the evaluation results. Specific results see Table 4.

Table 4. The target value of early warning index.

Index	Expre-ssion	Characte-rs of the index	Weight	Target value
Per capita GDP (yuan/person)	X1	positive	0.0284	38000
The third industry of GDP (%)	X2	positive	0.0409	46.8
Disposable income of urban residents	X3	positive	0.0295	24460
Per capita net income of farmers	X4	positive	0.0277	8460
High technology production value of industrial output value (%)	X5	positive	0.0265	7
R&D share of GDP(%)	X6	positive	0.0305	1.6
Environmental pollution control investment of GDP (%)	X7	positive	0.030	1.39
The standard discharge rate of industrial wastewater(%)	X8	negative	0.0282	95.3
Industrial dust removal rate(%)	X9	negative	0.0373	92.35
The standard discharge rate of industrial SO_2 emissions (%)	X10	negative	0.0362	99.74
The standard discharge rate of industrial soot emissions (%)	X11	negative	0.0331	97.3
The repetitive use rate of industrial water (%)	X12	negative	0.0284	93.5
The treatment rate of urban wastewater (%)	X13	positive	0.035	85
The comprehensive utilization rate of industrial solid waste (%)	X14	positive	0.0507	72
"Three wastes" comprehensive utilization rate of gross industrial output value (%)	X19	negative	0.0279	0.937
Elasticity coefficient of energy consumption	X20	negative	0.0301	721.53
Fertilizer usage per hectare (kg)	X21	negative	0.0293	1.837
Pesticides usage per hectare (kg)	X22	negative	0.0366	1.1455
Energy consumption (tce/ten thousand yuan)	X23	negative	0.04	3.17
Power consumption (KWH/ten thousand yuan)	X24	negative	0.0338	3.4

Energy consumption per unit value added of industry (tce/ten thousand yuan)	X25	negative	0.0355	36.3
water consumption (m^3 / ten thousand yuan)	X26	positive	0.036	45
SO_2 emissions intensity (m^3 / ten thousand yuan)	X27	positive	0.0285	12
COD emissions intensity (kg/ten thousand yuan)	X28	positive	0.0245	80
Urban residents engel's coefficient (%)	X29	positive	0.0363	4
Built-up area green coverage rate (%)	X30	negative	0.0296	3.38
Per capita green area of park (m^2)	X31	positive	0.0284	13.8

We conduct the standardization of every index. Then use the calculation of early warning index of the development of Henan's circular economy to make a comprehensive index. The warning degree classification of early-warning of Henan's circular economy are shown in Table 5.

Table 5. The early warning of Henan's circular economy from 2005-2013.

Year	2005	2006	2007	2008	2009
Warning degree	0.0625	0.1724	0.3357	0.4534	0.5025
Year	2010	2011	2012	2013	
Warning degree	0.6134	0.7241	0.7725	0.8316	

It can be seen from Table 5 that the warning degree of early warning system for Henan's circular economy in 2005 and 2006 are giant police, heavy police in 2007, intermediate police in 2008 and 2009, and light police from 2010 to 2013. The development level of Henan's circular economy shows a better trend in the whole. Considering that Henan pays attention to circular economy, takes sustainable development road, and doesn't increase outputs are at the expense of resources and environment. Hence, it is imperial to attach importance to environmental protection in developing recycling economy and promoting new-type development road.

5 Conclusion

This paper built an early-warning index system of regional circular economy, and put forward the measure model for the development of circular economy. By the proposed early-warning model, this paper analyzed the circular economy of Henan province. The development level of Henan's circular economy shows a better trend in the whole. Specifically, the warning degree of early warning system for Henan's circular economy are giant police in 2005 and 2006, heavy police in 2007, intermediate police in 2008 and 2009, and light police from 2010 to 2013. These mostly are due to Henan pays attention to circular economy, takes sustainable development road, and doesn't develop economy at the expense of environment, which is the important content of the viewpoint for the scientific development.

Based on the analysis of early warning system of Henan, early warning of the development of circular economy in Henan province over the next ten years will enter a police precinct, economic production mode will be in line with the principle of circular economy that is resource conservation, less pollution, high degree of resource reuse. But there still exist many problems, such as the efficiency of resource utilization is a little low, pollution is still in a high precinct, pollution control funds still need to add. If Henan does not adjust and control timely, it may exacerbate the early-warning degree.

Henan province is the largest part of the central economic zone provinces, is also a big economic province and major agricultural province, is the leading provinces in the Midwest, but at the same time is also a big resource-consumption province. Therefore, Rapid economic development is important, but from the point of view of sustainable development, simultaneously must ensure that the emissions shall not exceed the maximum environmental capacity. Resource utilization must be in a planned way, blindly predatory use of the resource must cause the alert degree of Henan's circular economy rise in the future.

References

1. Zhang B. Research and empirical evaluation of index system of circular economy. Chinese population, resources and environment, 2005 (3):22-25.
2. Yang HF. The construction of evaluation index system of circular economy. Management science and technology, 2005 (9) :123-128.
3. Ouyang LW. Research on index system of circular economy development. Environmental science and management, 2006 (6):117-120.
4. Zhao B. The design of evaluation index system of circular economy development. Economic system reform, 2007 (1):173-176.

5. Xiang LS. Analysis of evaluation index system of circular economy. Chinese population, resources and environment, 2007 (2) : 76-78.
6. Zhong TY. Assessment of regional circular economy: method, index system and empirical research. resources science, 2006 (3) : 154-161.
7. Cao XL. Index system, Evaluation method and empirical study of regional circular economy. Journal of chongqing university (social science), 2008 (3) : 30-35.
8. Cao HP. Dynamic evaluation of regional circular economy development base on MFA with AHP, resources science, 2009, 31 (2) : 278-287.
9. Tong JP. Early warning system for regional circular economy. System engineering, 2008 (9) : 104-110.
10. Qin LB. The application of sustainability assessment of comprehensive ecological suitability in regional innovation system. Systems engineering theory and practice,2011(5):87-93.
11. Qian WY, Dang RG. The application of early warning model of development of circular economy system. System engineering theory and practice, 2013 (7) :56-63.

The optimization of Three-echelon Agricultural Supply Chain under Crop Revenue Insurance

Juncao Wu[1,*] and Qiyuan Shi[2]

[1] Management Engineering Department, Zhengzhou University, Zhengzhou, China
Corresponding Email:9936133182@qq.com

Abstract. As the agricultural supply chain of is vulnerability due to the dual effects of natural disaster and marketable risk, the paper deals with the optimization of agricultural supply chain by introducing the crop revenue insurance for the first time. Firstly, this paper established the profit model of supply chain without any insurance. Then the crop revenue insurance is introduced into the supply chain, and the impact of supply chain is discussed and profit optimization of the supply chain with yield insurance and crop revenue insurance is analyzed as well. Finally, a numerical example is used to verify the results. The study shows that the supply chain system can be optimized through crop revenue insurance based on the principle of sharing insurance cost and claim, which can make up profit loss of natural risk and market price risk.

1 Introduction

China is the large agricultural nation since the ancient times. Agriculture is the foundation of our country's economy. In recent years, however, agricultural planting industry are facing the more complex risks that include the irresistible natural risk and the market price risk. In addition, the independently small farmers unit is the main organization mode of China's agricultural planting, and the mode is not only small in scale, but also low efficiency and more difficult to avoid the fluctuation of risks. Duing the risks, the supply chain is unstable, and even rupture. Therefore improving the supply chain's ability to resist risk and reduce the external risk such as natural risk and marketable risk is play a vital significance for the supply chain stable.

Since the agricultural risk events can bring about huge impact on the normal operation of supply chain and the supply chain risk management has become a hotspot in the research of the supply chain. YJ. Zhou et al. [1] firstly, reviewed risk identification and management process of the research results at home and abroad. Then, points out that strengthen the recognition of risk in supply chain and reinforce the quantitative risk assessment research, moreover, tighten up the management of risks' feedback mechanism in the dynamic environment. B. Schuster [2] pointed out that risks can be modified by buying insurance, and be reduced by signing contract terms. Moreover, he also proposed that use derivatives to reduce market risk, such as futures, options and forward. B. Dan et al. [3] designed a new risk compensation mechanism that the bad weather will not break the coordination state of the supply chain and motivate producers input the optimal level of agricultural materials. To a certain extent, the risk compensation mechanism can reduce the adverse weather's influence and improve the stability of the supply chain. Ling et al. [4] demonstrated that share the risks of output and demand in the contract can not only increase the profits of all members, at the same time also make mutual coordination and promote development of supply chain. Yang and Li [5], put forward agricultural insurance can be divided into yield insurance and revenue insurance, and agricultural insurance can be used as a way to transfer risks and improving the capacity of natural disaster risk management in our country.

In [6-8], the authors summarized the experience of foreign crop revenue insurance. Cao et al. [9] demonstrated the yield insurance can realize further optimize agricultural products supply chain under the natural risk. In [10-14], the authors considered the supply and demand uncertainty of the agricultural product that introduce options contracts to increase the agricultural products supply chain flexible to avoid risk. Above all, the importance of supply chain risk management has become increasingly obvious. Experts and scholars used a variety of methods and studied the supply chain risks from different angles to take further steps to transfer risks. Though the foreign scholars in the agricultural products supply chain has carried out a lot of research of the risk management, but the study of China's agricultural products is relatively poor, especially use the agricultural insurance to transfer the agricultural supply chain risk research are even less.

Using the agricultural insurance to transfer the risks is the important measure to reduce supply chain profit loss. Most of the studies focused on the production of agricultural insurance and summarized the experience of foreign crop revenue insurance, but the research that in the quantitative analysis of the impact of supply chain profit under the crop revenue insurance is still in a blank state. With the principle of sharing insurance cost and claim, the innovation of this paper is that introduced crop revenue insurance in three-level agricultural supply chain for the first time, and demonstrated that the crop revenue insurance can realize the further optimization of the supply chain, and the crop revenue insurance can also inspire dealers and distributors to increase quantity. In addition, this paper can also provide a basis gist for the supply chain risk management decision makers.

2 Problem descriptions

There is a single cycle model of three-level supply chain that composing a producer M, a distributor D and a retailer R The retailer that faced with the market demand is a random variable x and its density function and distribution function are $f(x)$ and $F(x)$. Before the start of production cycle, firstly, the retailer makes the revenue sharing contract with the distributor, and then determines the quantity q_r. Secondly the distributor makes the revenue sharing contract with the supplier, and decides its wholesale quantity q_d. Finally, according to the quantity, the supplier determines its investment yield A. The supplier's production yield is a times more than q into production, $A=(1+a)q, 0<a<0.5$.

At the beginning of production, according to principle of sharing cost L and claim H, members of the supply chain decision to place the crop revenue insurance, and the insurance company will pay out in claim after the end of the production. In the selling season, the producer supplies agricultural products T, moreover the retailer expect sales S and retail price p.

Using β to describe the different damage degree of agricultural products under the different natural risk., but the supplier's actual output affected by natural risk and the available supply to the distributor is T, $T=(1+a)(1-\beta)q$, the retailer expects sales S

$$S=\int_0^T xf(x)dx+\int_T^{+\infty} Tf(x)dx = T-\int_0^T F(x)dx \tag{1}$$

The following assumptions are assumed to hold for this model.

(1) All decision makers are risk neutral and perfectly rational, in addition, this paper don't consider stock losses and the rest of the products of the salvage value is zero.

(2) If the supply chain's makers decide to place insurance, they will place the insurance for all production.

(3) $q_r=q_d=q$.

(4) Agricultural products market is a completely competitive market, and the retail price p is decided by the market price.

The subscript M, D *and* R respectively represent the supplier, distributor and retailer. The farm production cost c_0. c_1, c_2 and c_3 respectively represent the supplier, distributor, retailer marginal cost. The retailer and distributor income share coefficient are λ_1, λ_2. And the expected profit π.

3 The three-echelon supply chain profit model without insurance

In this situation, revenue sharing contract are used to coordinate the supply chain. In the model, the distributor is at a low wholesale prices w_1 to buy agricultural products from the supplier, and promises to give the supplier 1-λ_1 proportion of earnings in return. Then the retailer take a lower wholesale at w_2 to gain the production from the distributor, and promises to give the distributor 1- λ_2 proportion of earnings in return. So the profit model of overall profit, retailer, distributor and supplier can be as below.

$$\pi_{I1} = \text{pS-}(c_1+c_2+c_3)(1+\alpha)(1-\beta)q - c_0(1+\alpha)q \tag{2}$$

$$\pi_{R1} = \lambda_2\text{pS}-(c_3+\text{w}_2)(1+\alpha)(1-\beta)q \tag{3}$$

$$\begin{aligned}\pi_{D1} &= \lambda_1\left[\text{w}_2(1+\alpha)(1-\beta)q+(1-\lambda_2)p\text{S}\right]\\ &-(c_2+\text{w}_1)(1+\alpha)(1-\beta)q\end{aligned} \tag{4}$$

$$\begin{aligned}\pi_{M1} &= \text{w}_1(1+\alpha)(1-\beta)q - c_0(1+\alpha)q\\ &+(1-\lambda_1)\left[\text{w}_2(1+\alpha)(1-\beta)q+(1-\lambda_2)p\text{S}\right]\\ &-c_1(1+\alpha)(1-\beta)q\end{aligned} \tag{5}$$

Lin et al. [15] demonstrated that only the revenue-sharing contract parameters meet the conditions that $\text{w}_1 = \lambda_1 B - \lambda_1 c_3 - c_2$, $\text{w}_2 = \lambda_2 B\text{-}c_3$, $\pi_{R1} = \varphi_1\pi_{I1}$, $\pi_{D1} = \varphi_2\pi_{I1}$, $\pi_{M1} = (1-\varphi_1-\varphi_2)\pi_{I1}$ The supply chain can achieve coordination, $\varphi_1 = \lambda_2$, $\varphi_2 = \lambda_1(1-\lambda_2)$, $c = c_1+c_2+c_3, B = \left[cA + c_0(1+\alpha)\right]/A$, $A = (1+\alpha)(1-\beta)$.

4 The three-echelon supply chain profit model with insurance

In order to reduce the effects of natural risk and marketable risk of supply chain profit, Decision makers can make a good use of the crop revenue insurance to transfer risks at external, and then, achieve the goal of both natural and marketable risk management. With the contract of sharing cost and claim, they appointed the sharing proportion. The retailer, distributor and supplier's sharing proportion respectively are φ_1, φ_2, $1\text{-}\varphi_1\text{-}\varphi_2$ .Finally the supplier represents the supply chain to sign a insurance contract with the insurance company.

Reference to American ARP-HPE[14], the agricultural crop revenue insurance's claim rules as follows.

(1) Guaranteed income per unit area = the average yield of history * forecast price * security level. This article assumes that security level is 70%.

(2)The supply chain participants should pay insurance premium of each production cycle and $K\omega_S P_S/t$. The security level is K. ω_S on behalf of the insurance premium rate. The security price P_S.

(3) According to the actual income of harvest, the insurance company will compensate the part that under the guaranteed income which specified in the insurance contract. The claim expression is $N_S\left[(1+\alpha)qKp_S\text{-p}(1+\alpha)(1-\beta)q\right]V$ .the coefficient of claim N_s . Difference growth period of the highest standard of compensation V .

$$N_S = \begin{cases} 1 & \text{p}(1-\beta) < p_S K \\ 0 & \text{p}(1-\beta) \geq p_S K \end{cases}$$

$$V = \begin{cases} 1 & \beta < 0.9 \\ V(i) & \beta \geq 0.9 \end{cases}$$

Adding the crop revenue insurance to the supply chain, so the profit model of overall profit, retailer, distributor and supplier can be as below.

$$\pi_{I2} = \text{pS-}(c_1 + c_2 + c_3)(1+\alpha)(1-\beta)q - c_0(1+\alpha)q + E \tag{6}$$

$$\pi_{R2} = \lambda_2 \text{pS} - (c_3 + \text{w}_2)(1+\alpha)(1-\beta)q + \varphi_1 E \tag{7}$$

$$\begin{aligned}\pi_{D2} &= \lambda_1\left[\text{w}_2(1+\alpha)(1-\beta)q + (1-\lambda_2)pS\right] \\ &\text{-}(c_2 + \text{w}_1)(1+\alpha)(1-\beta)q + \varphi_2 E\end{aligned} \tag{8}$$

$$\begin{aligned}\pi_{M2} &= \text{w}_1(1+\alpha)(1-\beta)q - c_0(1+\alpha)q \\ &+ (1-\lambda_1)\left[\text{w}_2(1+\alpha)(1-\beta)q + (1-\lambda_2)pS\right] \\ &- c_1(1+\alpha)(1-\beta)q + (1-\varphi_1-\varphi_2)E\end{aligned} \tag{9}$$

$$E = \text{-}H + L, H = (1+\alpha)qKp_S\omega_S$$
$$L = N_S\left[(1+\alpha)qKp_S \text{ - p}(1+\alpha)(1-\beta)q\right]V$$

5 The analysis of the crop revenue insurance on the optimization of supply chain profit

Comparing with the model of the third and fourth sections, and get the following propositions.

Proposition1. The supply chain decision makers place the crop revenue insurance, and when β increases to the condition that the actual revenue less than the guaranteed income, the profit of total supply chain, retailer, distributor and supplier will be advanced with the growth of β

proof.

$$\frac{\partial \pi_{I1}}{\partial \beta} = p\frac{\partial S}{\partial \beta} + c(1+\alpha)q = (1+\alpha)q\left[p(F(T)-1)+c\right]$$

$0 \le F(T) \le 1, -1 \le F(T) - 1 \le 0, p >> c, \dfrac{\partial \pi_I}{\partial \beta} < 0$ With the increase of β, the profit of total supply chain without insurance will reduce.

$$\begin{aligned}\frac{\partial \pi_{I2}}{\partial \beta} &= (1+\alpha)q\left[p(F(T)-1)+c\right] + N_S p(1+\alpha)qV \\ &= (1+\alpha)qp\left[F(T) - 1 + c + N_S V\right]\end{aligned}$$

$$\text{p}(1-\beta) \ge p_S K, N_S = 0, \frac{\partial \pi_{I2}}{\partial \beta} = \frac{\partial \pi_{I1}}{\partial \beta}$$

$$\text{p}(1-\beta) < p_S K, N_S = 1, \frac{\partial \pi_{I2}}{\partial \beta} < 0 < \frac{\partial \pi_{I1}}{\partial \beta}.$$

Proving by the same methods, and gained that the profit of retailer, distributor and supplier would be advanced with the growth of β

Proposition2. The supply chain decision makers placed the crop revenue insurance, and when marketable risk lead to lower price to the condition that the actual revenue less than the guaranteed income, the profit of total supply chain, retailer, distributor, supplier will be advanced with the decrease of retail price p

proof. $\frac{\partial \pi_{I1}}{\partial p} = S$, $\frac{\partial \pi_{I2}}{\partial p} = S - N_S(1+\alpha)(1-\beta)qV$,*S*>*0*, $\frac{\partial \pi_I}{\partial p} > 0$. With the decrease of *p*, the total profits of supply chain without insurance is reduce.

$$p(1-\beta) \geq p_S K, N_S = 0,\ 0 < \frac{\partial \pi_{I2}}{\partial p} = \frac{\partial \pi_{I1}}{\partial p}.$$

$$p(1-\beta) < p_S K,\ N_S = 1, \frac{\partial \pi_{I2}}{\partial p} < 0 < \frac{\partial \pi_{I1}}{\partial p}.$$

Proving by the same methods, and gained that the profit of retailer, distributor,supplier would be advanced with the decrease of *p*.

6 Numerical study

In this section, the paper first use an example of Soybean production in Heilongjiang province, and describe the experimental parameters used in the numerical solutions. Then analyze these results and compare and contrast the performance of the channel without insurance and with the crop revenue insurance.

Assuming that the demand is a normal distribution function $x=\sigma$, $\sigma \sim N(200,8)$ $\alpha=0.03$, $p=7$yuan/kg,$\sigma=200$t,$c_0=1.5$yuan/kg,$c_1=0.15$yuan/kg,$c_2=0.15$yuan/kg,$c_3=0.15$yuan/kg. The average yield is 2000 kilograms per mu, $P_S=7$ yuan/kg, $K=70\%$, $\omega_S=6\%$ per mu.

According to the model in the third and fourth quarters, the paper consider the influence of natural risk on the profit of the supply chain. And then, the channel profit assessment of the insurance is implemented by "with-and-without comparison method".

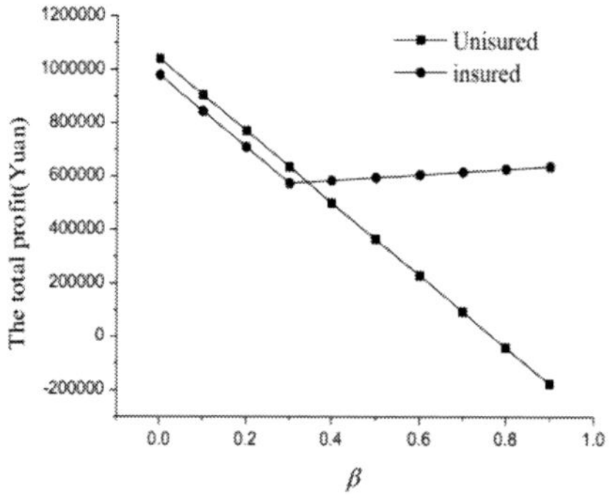

Fig. 1. Natural risk impact on the overall profit.

The Figure1 shows that the profit of the participants in the supply chain that without insurance, with the increase of β the profit will reduce. $\beta > 0.8$ the profit of supply chain participants are negative, and the natural risk threat to the supply chain's profit. But the supply chain place the crop revenue insurance, although has premiums, through share premium and claim ,and the insurance's characteristics of low cost and high claims, its impact on the supply chain profit is very small. On the contrary with the increase of β, the gap between the income and actual income increase, and members of the supply chain profit will increase slightly, so under the effect of the natural risk, the crop revenue insurance can optimize the supply chain profit.

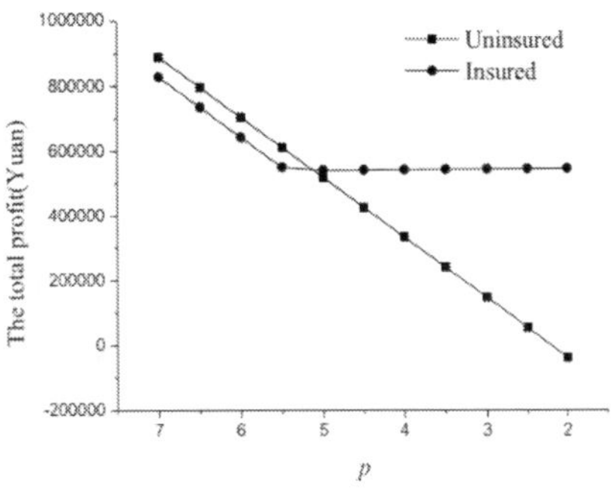

Fig. 2. Marketable risk impact on the overall profit.

The Figure 2 shows that the profit of the participants in the supply chain without insurance, with the marketable risk lead to lower price ,the profit of supply chain fail sharply ,and the marketable risk threat to the supply chain's profit. But the supply chain place the crop revenue insurance, with the increase of the marketable risk lead to lower price, members of the supply chain profit will increase slightly, so under the influence of the marketable risk, the crop revenue insurance can optimize the supply chain profit.

7 Conclusion

According to the principle of sharing cost and claim, this article introduces the crop revenue insurance into the three-level agricultural supply chain, and discuss the impact on supply chain profit. In the demonstration part, first of all discuss the situation that the supply chain do not have insurance ,the revenue sharing contract was used to optimize the supply chain cooperation between members ,and make the supply chain in the coordination state that the overall profit achieve the optimal. Then, based on the revenue sharing profit model, and join the crop revenue insurance to the model and draw the following conclusion. The crop revenue insurance is able to transfer part of the risk to outside, and the claims of the crop revenue insurance can make up the loss profit that caused by the natural risk and marketable risk. Therefore the supply chain on the basis of the coordination of revenue sharing contract is able to realize the further optimization.

References

1. Y.J. Zhou, G.H. Qiu, Z.R. Wang. It summarized research progress of supply chain risk management and analysis. Syst Eng **24** (2006), 1-7.
2. D. Barnes - Schuster, Y. Bassok, Anupindi r. Coordination and Flexibility in Supply Contracts with the Options. M&SOM **4** (2002),171-207.
3. B. Dan, H.Y. Fu, G.Y. Xu. etc. The weather affect the output of the agricultural products supply chain coordination under risk aversion. J Syst Eng **29** (2014).
4. L.Y. Ling, X.L. Guo. Based on the random output and random demand of the agricultural products supply chain risk-sharing contract. Chinese Journal of Management Science **21** (2013), 50-57.
5. X. Yang, Y. Li. China's agricultural natural disaster risk management research, and theory of the development of agricultural insurance. At zhongnan university of economics and law journal (2010), 34 - 37.
6. J. Cheng. Economic crop insurance for farmers revenue impact study. Journal of Liaoning administration college (2015), 41-43.
7. Y.G. Xia, Y.H. Liu. U.S. federal crop insurance products system, running mechanism and enlightenment. Issues in Agricultural Economy (2014), 101-109.

8. X. Zhang, L.Y. Wang. The crops in China the necessity and feasible scheme of income insurance pilot. Agricultural outlook (2015), 18 to 24.
9. W.J. Cao, Y.Y. Shi. Based on the option and insurance of the agricultural products supply chain coordination optimization study. IE&M **21** (2016), 8-16.
10. F. Chen, M. Parlar. The Value of a put option to the risk - averse newsvendor. IIE TRANS **39** (2007), 481-500.
11. Y. Zhao, X. Meng, S. Wang, et al. The Coordination of Supply Chains with Bidirectional Option Contracts. EJOR **229** (2013), 375-381.
12. L. Yang, X.X. Cui, Z.Y. Zhang. Demand depend on the effort level of agricultural products supply chain option contracts optimization. Mathematics in Practice and Theory (2014), 44-55.
13. L. Liu, C.X. Teng. Supply chain coordination based on option contract research. Operations Research and Management Science **20** (2011), 48-57.
14. G.H. Sun, L. Xu. Secondary agricultural products supply chain under stochastic demand and supply an option contract coordination study. Journal of Industrial Engineering and Engineering Management **28** (2014), 201-210.
15. L. Lin, S.P. Yang, Dan bin. Revenue sharing contract under the fresh agricultural products three-stage supply chain coordination. J Syst Eng **25** (2010).

Study on Customer Experience Dimension Based on Online Shopping reviews of MI's Products

Lin Wang [1,2,*], Huijie Tang [1,2], Dong Liu [1,2] and Caiyan Xing [1,2]

[1] School of Business and Management, North eastern University, Shenyang 110819, China
[2] School of Management, North eastern University at Qinhuangdao, Qinhuangdao 066004, China
Corresponding Email: wanglin05vip@163.com

Abstract. This study has selected sorts of comments about MI's products from its customers' as research objects, then we used the theory of Rooting to analyze those objects, and explored the Customer Experience Dimensions' type, characteristics and the way to impact others. The results come as follows: First of all, the Customers Experience Dimensions are consisted of nine major categories defined as Value Perception Experience, Sensory Experience, Quality Assurance Experience, Parts Experience, Purchase Experience, Service Experience, Brand Comparison Experience, Compensating Experience, Brand Reliance Experience respectively. Secondly, we put these nine major categories above all to compare the conclusions with the previous research, then we analyzed and integrated them, and finally obtained seven major categories containing Purchasing Website, Product Rendering, Delivery Experience, Transaction Convenience, Safety and Security, Brand Reliance, Marketing Communications respectively. In addition, we have conducted a further investigation to draw three types about the Customer Experience named Function Experience, Purchasing and Service Experience, and Brand Experience. Last but not the least, this paper has definitely provided the basis theory of how does the Customer Experience affect the Brand Loyalty for future research, as well as guided a definite orientation.

1 Introduction

With the creation of MI and the rapid development of the recent six years, a lot of small brands have begun to rise in the field of mobile digital industry, follow the various models of MI. So how to further improve the loyalty of MI users becomes increasingly important. According to the comprehensive analysis of documents and the data collected by questionnaire, we find the elaboration about the customer experience dimension is not perfect. Therefore, the specific research on the customer experience dimensions has an important guiding significance for the enterprise.

2 Theoretical review

Although the customer experience is developing, there is still not a uniform definition about it until now. Considering that the process of consumption involves experience, Lofman divided the customer experience into 6 dimensions: thought, emotion, feeling, scene, action and evaluation [1]. Pine II and James H•Gilmore divided customer experience into Education, Entertainment, Escape and Aesthetic [2]. Josko Brakus divided customer experience into 5 dimensions from cognitive theory. They're Emotion, Intelligence, feeling, social communication and body [3].

Besides the overseas scholars have studied the customer experience from various angles, domestic scholars also put forward some different views.

Chen Jian Xun thought customer experience should be divided by the level of warm when consumers were shopping. From negative experience to height experience, there are No experience, Low-degree experience and Moderate experience [4]. Song Wei divided B2C customer experience into 3 dimensions: Brand, Emotion and Function[5]. The specific topic of each dimension is shown in Table 1.

Table 1. The Dimensions of customer experience.

Dimension	Element	Measurement
Brand Experience	Brand Building	The experience of brand names, brand identity, brand reputation and brand awareness
	Marketing Communication	The experience of sales and advertisement
Emotion Experience	Relax	Feel relaxed and alleviate the pressure and trouble when experience
	Pleasure	Feel pleased or excited and bring surprise when experience
Function Experience	Website Experience	The experience of page design of website, product information, site speed and stability, he convenience of browse and purchase, security mechanism
	Product Experience	The experience of the price, quality, package and kinds of product
	Service Experience	The experience of pre-sale service, after-sale service, logistic service, paying and personalized service

3 Research method

3.1 Sample selection

The sample in this research comes from the comments put forward by online purchasers of MI. In the process of sample's data collection, we mainly adopt the following methods. First of all, formal researching channel is needed, which means we collect assessments of MI users through formal channels. Secondly, the comments we collect must be both comprehensive and focused. Thirdly, it is significant to ensure the representativeness of sample as well as the randomness of comments. Lastly, we neither add any personal subjective factors nor modify any original users' comments in the process of entire collection.

3.2 Theoretical sampling

When using the above methods to do our collection, we have got 1742 messages of comments on purchasing. Moreover, in the stage of data organization, We have got 870 pieces of effective comments, which is the basis of our following analysis.

3.3 Grounded theory research method

Our research is on account of the criteria and the process of grounded theory, research tool is selected from ROST Content Mining, a content mining software system, designed by professor Shen Yang. As shown in Figure.1.

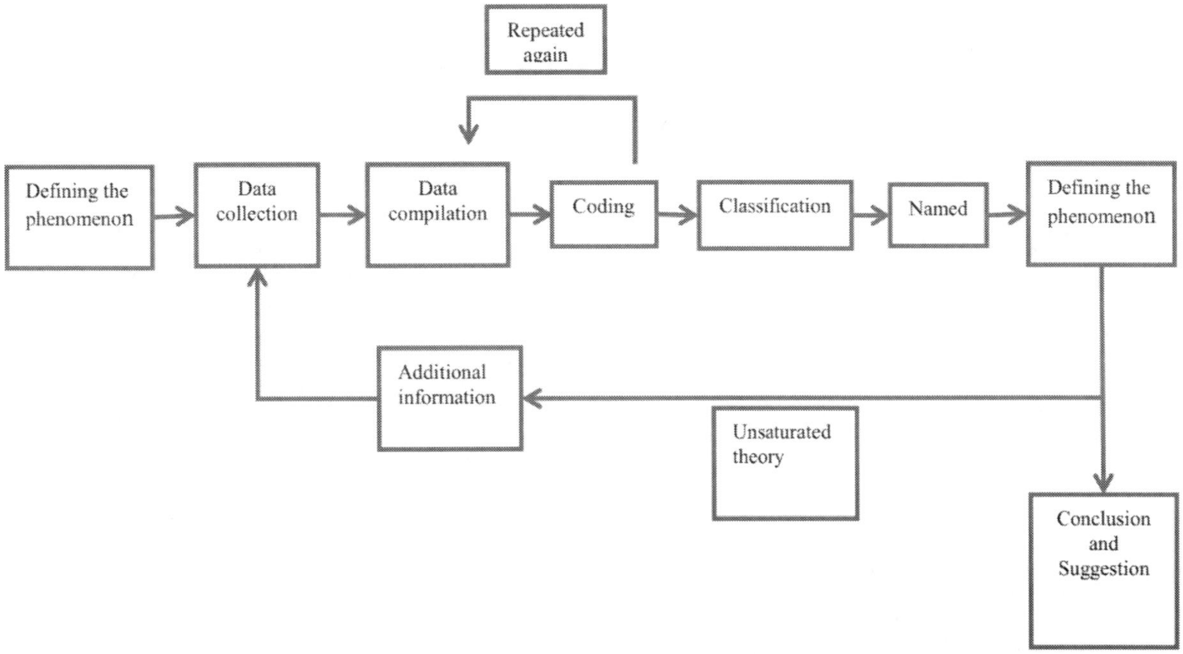

Fig. 1. The standard and process of Grounded theory.

4 Research result

4.1 Open code

In this section we will encode the original statement openly. Then, we form 25 categories, as shown in Table 2.

Table 2. Preliminary coding categories table.

ID	Category	Annotation (article)	percentage	Typical source statements (original concept)
N01	Satisfaction	113	12.99%	Put it down ; I am quite satisfied
N02	Customer service experience	96	11.03%	Is patient ; It's why eat cat customer service
N03	Relation between clues	82	9.43%	Help colleagues to buy ; For my father bought it
N04	Buying experience	65	7.47%	Rob times didn't get the 4G version of the move ; It's hard to take
N05	Logistics experience	58	6.67%	Express fast ; The delivery speed of the slowest ever
N06	Comparing experiences	48	5.52%	Wins on price is not high, the system works a little better than Huawei ; The strongest signal is not red rice
N07	Experience product attributes	47	5.40%	Feels good ; Not press
N08	Brand perception	42	4.83%	Millet mobile phone is good ; Shop big guests
N09	Accessories gift experiences	30	3.45%	Nothing was sent, and sent a shell with foil, at least ; Buy back even the headphone cable is not open
N10	Payment experience	26	2.99%	The address to fill a fault, do not give change ; Took a couple of things, why should a single pay the freight
N11	Return experience	25	2.87%	Please return they refused to apply ; Buying the wrong type of refunds but it was too difficult

N12	Experience of product defects	24	2.76%	Bought 4 M 4, double 11 split ; The back of the phone has received scratches
N13	Product feature experience	24	2.76%	Serious sound, no video;Sound good
N14	Value for money experience	24	2.76%	Value for money ; Price / performance king
N15	Brand loyalty-aware	23	2.64%	Buy millet things, has become a habit ; From M 1 to M 4
N16	Exchange experiences	21	2.41%	Order Shun Feng Express, so why go after replacement post ; Replacement of the speed is very fast
N17	Quality of Experience	21	2.41%	Cargo is true; control, sooner or later, nobody would buy it
N18	Rice injury experience	17	1.95%	Loyalty rice noodles have a problem, customer service has been not to answer ; For the millet consumed all
N19	Packaging experience	17	1.95%	Packaged very carefully and tightly ;The product in a Fedex box, not securely fastened
N20	Accessories experience	16	1.84%	Bad cable ; Cable cannot be recharged
N21	Purchase channels experience	15	1.72%	It's hard to get the official website ; Flagship store quality is absolutely reliable!
N22	Graphic describes the experience	14	1.61%	Prettier than the picture; I don't feel so nice in the picture
N23	Risk Perception experience	10	1.15%	Broken screen cannot buy the official website of insurance ; Don't see broken screen when buying insurance option, there is no hint to buy!
N24	Promoting sales experience.	7	0.80%	Forget to use coupon, unhappy ;A $20 coupon when you buy useless
N25	Turn powder experience	5	0.57%	The first transfer from Apple powder bought millet mobile phone ; You don't have to buy Apple, good affordable price

4.2 Spindle coding

Based on the original statement and part of the document definition above, we have these categories categorized into 9 categories and defined them, as shown in Table 3.

Table 3. Table of 9 main categories.

Fundamental category	Correspondence category	Meaning of category
Value perception experience	Product function experience	The use of hardware and software products, such as the use of feelings
	Cost performance experience	The comparison of the performance of the product or service, the price matching degree
	Promotion experience	On the way of brand sales, profits, limited coupons to attract customers when purchasing experience
Perceptual experience	Comparison of picture and text description	Perception of the actual situation of the product or service and the picture and text comparison
	Product attribute experience	The use of product configuration, parameters, process and other aspects of the experience
	Packaging experience	In the receipt of goods, the packaging of the product, the exquisite degree of perception
Quality assurance experience	Quality experience	The quality of the product to buy the environment, genuine feeling
	Risk perception experience	The product accident insurance purchase, return shipping fee dispute feeling
Accessories experience	Experience of giving away	In the purchase of the main products, the feelings of the business as a gift to other parts of the behaviour

Buying experience	Accessories experience	In addition to the purchase of the main products, the use of free accessories for the feeling
	Purchasing channel comparative experience	The comparison of the different channels to buy the brand products
	Panic buying experience	Feelings of the brand by letting users buy the way to allocate existing products
Service experience	Customer service experience	Customer service recovery rate, processing efficiency, service attitude and logistics speed, attitude
	Logistics experience	On the brand's courier service attitude, as well as the delivery speed and other aspects of the experience
	Exchange experience	After receiving the goods are not satisfied, to apply for a replacement, replacement of a series of process experience
	Return experience	After receiving is not satisfied with the product, for the application to return, return and a series of processes, such as the feeling
	Payment experience	In the purchase order, payment of postage, fill in the address, and other aspects of the experience
Brand comparative experience	Perceived brand awareness	Users of the brand has a certain understanding or awareness of the use of products
	Comparative experience	Through different brands, the same brand of different series or the same series of different models of the use of product experience comparison
	Powder transfer experience	From other brands of loyal users to the brand loyalty user experience
Compensation experience	Rice injury experience	As a brand loyal user, because there is no expected service or product experience feelings
	Product defect experience	Experience in product defects, defects, functional / design defects, etc.
Brand dependent experience	Satisfying experience	A comprehensive evaluation of the satisfaction of product identification, complaints, complaints and so on.
	Relationship experience	Feelings of people who have a certain social relationship with buyers to buy the brand's products
	Brand loyalty perception	As a brand loyal users in the purchase process of a series of feelings

4.3 Determination of the dimensions of customer experience

This paper compares previous results with 9 experience categories in Table 3, analysis and integrates them, we get customer shopping experience. They are functional experience, buying and service experience and brand experience. Specific such as Table 4.

Table 4. Determination of the dimensions of customer experience.

Customer experience dimensions	Main category	Subcategory
Feature experience	Buy website	Web site design
		Classified display
	Product rendering	Product quality
		Product price
		Product information
Purchase and service experience	Delivery experience	Graphic description
		Packaged experiences
		Accessories gift experiences
	Transaction convenience	Buying experience
		Ordering experience
		Customer service experience
	Safety and security	Logistics experience
		Return and exchange experience
		Accident insurance
Brand experience	Brands rely on	Relationship experiences
		Brand experience

Marketing communications	Promotions
	Advertisement

4.4 Theoretical saturation test

Usually, in qualitative research credible degree is an important measure standard used to evaluate the quality [6]. At the same time, the validity is also used to evaluate the quality [6]. At the same time, the validity is also used to evaluate a measure of the quality of qualitative research, which can reflect the online scene clues, etc. We will use the Raw Material Inspection and Inspection and Feedback Method to verify the effectiveness of the model and the results of the study.

4.4.1 The raw material inspection

Concepts and relations are in return for the model review. That "concept" must be derived from the original data [7]. All of the selections of data in this paper are from MI's setting of formal channels for evaluation of the original users' information. It is effectively to ensure the information from the user's comments in the reliability of the dimensions of customer experience.

4.4.2 The expert evaluation method

Successively in the whole process of this research has four experts in the fields of Marketing and E-commerce, under their suggestions, eventually, we got the effective customer experience dimensions.

To sum up, using the Original Material Test, Expert Evaluation Method and the Research Results have good consistency and good validity. Therefore, this research is effective, credible and reliable.

5 Discussion and conclusions

In this study, we selected two dimensions of customer experience, which contain functional experience and service experience, into the purchase and service experience. In addition, this study can also develop the research, delivered by Song Wei, about dividing class on the dimensions of customer experience.

5.1 Theoretical significance

First of all, this study used the method of grounded theory to get customer experience dimensions: functional experience, buying and service experience, brand experience. These dimensions are constituted by seven main categories, which respectively are purchasing website, product, delivery and transaction convenience, security assurance, dependence on branding, marketing communications.

Secondly, the findings revealed a phenomenon in the period named as "experience economy" that product experience is a good way to understand a brand, what is more, a well customer experience is powerful for the brand loyalty. Therefore, according to the customer experience, we proposed the following recommendations derived from two aspects.

5.1.1 Recommendation for MI's products

In addition to designing a more simple, comfortable MI's products purchased website with the products exhibition and the category clearly, the most important thing that MI should focus on is continuing produce the offerings with good quality, even making users to scream.

5.1.2 Recommendation for service experience

The company should promise that especially unblocking the process in snapping up, promising the snapping up is fair, and if the snapping up is not success, the company should make comfort measures, like sent comfort SMS. In terms of the customer's remedies, it is allowed that according to the times of customers' buying, the waiting time, the failure times and so on, to category different customer types with varying degrees of comfort measures.

5.2 Limitations and future research directions

There are also some limitations in this study, specific description comes as follows. Firstly, the size of sample in this study is limited, the validity of the results is limited and cannot exclude the influence of the external factors. Secondly, the dimensions of customer experience have been obtained through this research and these results can be seen as a guiding role to enhance the customer's brand loyalty for MI, but how does the customer experience influence the customer's brand loyalty are not well-known.

Acknowledgements

This research is supported by funding of China's central university basic scientific research expenses Project No. N152304008.

This research is supported by funding of Northeastern University at Qinhuangdao in China,which Humanities and Social Sciences Key Project No. XNR201601.

References

1. A. Samuel, O. W. Fred, Q. Zhou, etc. A1 Transformation Leadership, Innovative Behavior, and Task Performance: Test of Mediation and Moderation Processes, **25**, 1 (2012)
2. Joseph Pine II, James II. Gilmore. The experience economy. translated by XIAYELIANG(Beijing mechanical industry publishing house, 2008)
3. Joskol Brakes, A theory of consumer experiences (Columbia University, 2001)
4. CHENJIANXUN, Multiple levels of customer experience and prolong the life cycle of strategic choices (Statistics and decision making, 2005)
5. SONGWEI, B2CE-commerce customer experience impact on customer loyalty research (JINAN:SHANDONG University, 2012)
6. J.X. Wu, X.T Huang, Credit structure (Psychology Newspaper, 2012)
7. D. Trafimow, Considering Quantitative and Qualitative Issues Together (Qualitative Research in Psychology, 2014)

The Impact of Brand Awareness and Customer Experience on the Brand Loyalty of MI

Yang Zhao [1,2],Tong Tong[1,2], Guanchu Li [1,2] , Shuang Ma [1,2] and Lin Wang [1,2]

[1] School of Business and Management, North eastern University, Shenyang 110819, China
[2] School of Management, north eastern University at Qinhuangdao, Qinhuangdao 066004, China.
Corresponding Email: wanglin05vip@163.com

Abstract. This study discusses the impact of brand awareness and customer experience on brand loyalty of MI. This study randomly selected the people who have purchased the MI's product to complete the questionnaire. The variables of this questionnaire included population statistics variables, the brand awareness of consumers, the experience of consumers when purchasing, and the customer's loyalty of its brand. By using the SPSS to analyze the date from reliability analysis, validity analysis, correlation analysis and descriptive statistical analysis of each variable. As the results shown, brand awareness can positively predict brand loyalty. Service experience and emotional experience can positively affect brand loyalty. There existed significantly positive correlation among brand awareness, service eexperience, emotional eexperience and brand loyalty. Finally, summarizing the empirical results, and bringing up some limitation and expanding direction of this research.

1 Introduction

With the development of MI, the company has been gaining sorts of loyalty customers. But in the last two years, more competitors began to replicate the marketing strategies of MI, they gained the market share from MI. As to MI, further enhancing customer loyalty became quite noticeable. In this context, considering most users of MI are attracted by its brand profile, and MI's product has huge sales online. So the customer experience seems to be particularly important. This paper made some researches on whether customer experience and brand awareness can promote Customer Loyalty. According to the research of MI's development paths and strategies in recent years, combining the development backgrounds of the current domestic market for smart product, two problems were discussed in this paper. First, what is the relationship between brand awareness of MI and customer experience? Second, how the brand awareness and customer experience influence brand loyalty. Therefore, discussing the relationship among Brand Recognition, Customer Experience and Brand Loyalty is a crucial question which must be faced in today's business. It will be helpful to draw sustainable development strategic policy for MI.

2 Theoretical review and Hypothesis

2.1 Brand recognition

Due to the high profile of MI, when investigating the cognition of MI's users about brand , using the views of Keller. It is dividing brand awareness into brand profile and brand image. These two dimensions basically cover all the content of brand awareness. Brand profile represents the depth and breadth of brand awareness. Brand image includes the different content of brand awareness [1]. Therefore, when designing the questionnaire, this study investigated from brand profile and brand image.

2.2 Customer experience

Up to now, there is no uniform definition of customer experience. Different scholars have different views. By analyzing the previous classification of customer experience, Song Wei combined his study on the characteristic of B2C industry, and divided customer experience into three dimensions: brand, emotion and function [2]. Consumers buy MI's product online, it is B2C customer experience. So this research used the view of Song, divided customer experience into functional experience, brand experience and purchase and service experience.

2.3 Brand loyalty

After studying the different views of scholars, Yang Yongpan proposed that attitude and behaviour loyalty make up the brand loyalty. Attitude loyalty includes attachment psychology and proposition of attitude. Behaviour loyalty involves buying more and maintaining relation [3]. In conclusion, this study selected attitude and behaviour to design the questionnaire.

2.4 Research hypothesis

The way MI's branding is different from the traditional enterprise. The traditional path is to hit brand recognition first. Then build the brand. Finally, increase customer loyalty. But for the Internet enterprise, product is brand. MI's branding focused on loyalty first, marketing itself by word of mouth. After strengthened to some extent, they began to get more brand recognition. It is the initial way of MI. With the high brand profile of MI, most consumers have the awareness of brand, then they decided to buy MI's product. Focusing on whether this way can form brand loyalty, this study raised two issues.

Question one: What is the relationship between brand awareness and customer experience of MI?

Question two: How much impact that MI brand awareness and customer experience of MI caused on brand loyalty?

According to the literature review, this paper established a theoretical model preliminarily, as shown in Figure1.

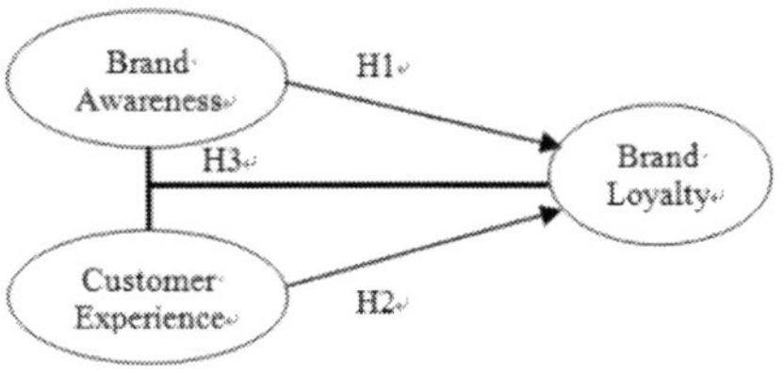

Fig. 1. Prediction of Causal relationship among Four variables.

By the above model, this paper makes the following assumptions, as shown in Table 1:

Table 1. Hypothesis of the Research.

	hypothesis
H1	Brand awareness can positively predict brand loyalty.
H2	Customer experience has a positive predictive effect on brand loyalty.
H3	Brand awareness, customer experience and brand loyalty are significantly correlated.

3 Research variables

This study used empirical research method. The questionnaire's design focused on the study of brand awareness, brand loyalty and purchase experience, variables included the population statistics variables, brand awareness of consumer to MI, purchase experience(involving service experience and emotional experience), and consumer's brand loyalty of MI.

Population statistics variables included sex, occupation, level of education, age, personality traits and so on, analyzing whether individual character affects brand awareness, purchasing experience or the degree of brand loyalty. We can know the consumer's brand loyalty of MI through an investigation of MI's profile and brand image in the minds of its users. By investigating consumer's functional experience, service and purchase experience to know the their satisfaction when they are shopping. From two aspects, attitude loyalty and behaviour loyalty, to investigate the brand loyalty.

The Research of Impact of Brand awareness and Purchase experience on the MI's Brand loyalty through specialized online survey platform to answer with multiple-choice questions. Questionnaire holds 23 answer questions and 5 multiple-choice questions, covering a wide range , and its number of questions is moderate. This study began in May 2015 and finished the distribution of questionnaires to users who have bought MI's product. 157 questionnaires were returned, including 11 invalid questionnaires, surplus effective questionnaires were 146. Eventually the survey results were analyzed by SPSS to draw the relevant conclusions, and to prove hypothesis.

4 Research results

4.1 Descriptive analysis

4.1.1 Independent sample T test

In order to confirm the rationality of the 33 questions in the questionnaire, this paper carried out the independent sample T test. According to the F value and sig value of A1~A33 can judge the 33 questions have reached a significant level. These date proved that 33 questions can be left to continue studying.

4.1.2 Reliability analysis

This study adopts reliability which Cronbach's α-coefficient evaluation used, it is generally believed that if the scale α-coefficient greater than 0.7, it is reliable. the sub scale reliability of this research is greater than 0.7, the overall scale α-value is 0.968, which shows that scale has the ideal reliability.

4.1.3 Validity analysis

The difference between the observation values of the scale of validity analysis is used to reflect the real difference of the measured characteristics of things. According to Kaiser's (1974) points, if the KMO value is greater than 0.5, it is suitable for factor analysis. With varimax rotation was used for factor analysis, the analysis results showed KMO value is 0.934, the significant probability value is 0.000 < 0.05, Bartlett test is remarkable. It indicates that is suitable for factor analysis.

The results of factor analysis showed that the characteristics of the 4 principal components are greater than 1. According Kaiser's guidelines, extract four common factors, after the ingredients are rotated, get the Table 2:

Table 2. Results of Rotated component matrix[a] Analysis.

Item	Rotational component matrix[a]			
	1	2	3	4
A1	0.777			
A7	0.753			
A2	0.727			
A9	0.691			
A5	0.665			
A11	0.657			
A14	0.653			
A8	0.649			
A12	0.643			
A6	0.639			
A13	0.622			
A10	0.596			
A19	0.593			
A31		0.834		
A32		0.778		
A33		0.733		
A26		0.731		
A28		0.683		
A22		0.64		
A23		0.636		
A29		0.613		
A27		0.592		
A30		0.592		
A24		0.573		
A25		0.573		
A21			0.71	
A18			0.694	
A20			0.638	
A17			0.638	
A16			0.619	
A3				0.786
A4				0.741
A15				0.545

The result of the above table is the composition matrix obtained after sixteen iterations, Items were arranged according to the factors of the load, and the four factors were extracted. Rename these four factors. For the first factor named brand awareness, including item for A1, A2, A5, A6, A7, A8, A9, A10, A11, A12, A13, A14, A19. The second factors named service experience, including item for A16, A17, A18, A20, A21. The third factors named brand loyalty, including item for A22, A23, A24, A25, A26, A27, A28, A30, A31, A32, A33. The fourth factor is named emotional experience, including item for A3, A4, A15. Because the factor loading of the variables on each common factor level are more than 0.500, indicating that the latent variables can reflect the index variables.

The results of factor analysis showed that the scale had good construct validity, and it was suitable for further analysis.

4.1.4 Correlation analysis

The following information is the results of correlation analysis between the four variables of brand awareness, service experience, emotional experience and brand loyalty, as shown in Table 3:

Table 3. Results of correlation analysis.

	mean value	standard deviation	1	2	3	4
1 brand awareness	64.0342	15.08432	--	--	--	
2 service experience	21.7945	6.09456	**.734****	--	--	
3 emotional experience	11.863	4.01141	**.548****	**.571****	--	
4 brand loyalty	48.3356	15.54583	**.736****	**.710****	**.607****	--

From the table we can see that the relevance between brand awareness and other variables, they are all between 0.5 to 0.8, said the four variables showed a significant correlation, and positive correlation. Hypothesis 3 have been proved. The degree of brand awareness and brand loyalty is the highest which up to 0.736.

Table 4. Descriptive statistics among variables （n=146）.

	M	SD	1	2	3	4	5	6	7	8	9	10
1.Brand awareness	64.0342	15.08432	1									
2.Service experience	21.7945	6.09456	**.734****	1								
3.Emotional experience	11.863	4.01141	**.548****	**.571****	1							
4.Brand loyalty	48.3356	15.54583	**.736****	**.710****	**.607****	1						
5.Sex	1.4	0.491	.050	.025	.035	.024	1					
6.Age	2.16	0.494	.144	**.238****	**.199***	**.208***	-.118	1				
7.Level of Education	2.86	0.563	-.047	-.155	.089	-.100	.108	-.141	1			
8.Monthly income	1.84	1.202	.101	.129	.169*	.102	-.139	**.379****	-.085	1		
9.Native place	1.53	0.807	.080	.106	-.020	-.069	.112	-.054	-.044	.015	1	
10.Personality traits	2	0.621	-.055	-.020	-.003	.001	0.000	-.090	-.059	.083	.014	1

4.2 Empirical analysis

4.2.1 Descriptive statistical analysis of each variable

The average, the standard deviation and correlation of each variable are presented in Table 4. As can be seen from the table, brand awareness, service experience, emotional experience and brand loyalty are in significantly positive correlation. It can also be seen that population statistics variables and brand loyalty existed in varying degree of negative correlation. Age and service experience are positively correlated, that means the older feels better about Mi service experience of MI. Age and brand loyalty is also significantly positive correlation, the older the consumers are, the stronger their loyalty is. Considering the impact of population statistics variables on the independent variables and dependent variables, the research need to controll population statistics variables in follow-up studies.

4.2.2 Regression analysis of the four variables

In order to investigate the impact of brand awareness, service experience and emotional experience on brand loyalty, we conducted a hierarchical regression analysis of brand loyalty. Take the population statistics variables as the first-order, brand awareness, service experience, emotional experience as the second-order. According to the views of Aiken and West, when high-order interactions are significant, there is no need to consider whether low-order interactions are significant , as shown in Table 5. Therefore, according to the results of the second stage, we can see that brand awareness, service experience and emotional experience have significant positive predictive effects on brand loyalty. Hypothesis 1 and 2 are verified.

Table 5. Hierarchical regression analysis of brand loyalty.

variable		brand loyalty	
		First order	**Second order**
First order	Sex	0.067	0.009
	Age	0.189*	0.039
	Level of education	-0.08	-0.058
	Monthly income	0.033	-0.033
	Native place	-0.071	0.129*
	Personality traits	0.012	0.035
Second order	Brand awareness		**0.418****
	Service experience		**0.279****
	Emotional experience		**0.219****
	R^2	0.057	0.658
	F	1.405	29.061
	ΔR^2	0.016	0.635

According to the actual situation, when the customer's awareness of a brand is 0, and never experience its products and service, then his loyalty must be 0. It means the constant is equal to 0, so this paper selected the standard factor. From the above table, The normalized coefficient of the dependent variable Y (brand loyalty) to the independent variable X1 (brand cognition) is 0.418, to the independent variable X2 (service experience) is 0.279, to the independent variable X3 (emotional experience) is 0.219. Therefore, the regression model in this study is established.

$$Y = 0.418X1 + 0.279X2 + 0.219X3 \quad (1)$$

Brabd loyalty = 0.418 * Brand awareness + 0.279 *Service experience + 0.219*Emotional experience (2)

5 Discussion and conclusions

5.1 Theoretical significance

This paper adopted the method of Questionnaire to explore the causal relationship among brand awareness, customer experience and brand loyalty. And it brought up three hypothesis. Through an empirical analysis of the survey data, we got four factors, according to the factor analysis changed original three variables into four variables, including the change of the customer experience is divided into the service experience and emotional experience. Because the customer experience included Service experience and emotional experience, it would not affect the hypothesis which is put forward. So in this paper, we draw the following conclusions about the three hypothesis.

The hypothesis that "Brand awareness can positively predict brand loyalty" has been proved in this paper. Through the hierarchical regression analysis, we found that the higher the degree of users' awareness of MI, the more can promote the formation of the brand loyalty. The regression model shows that when the user's awareness of MI degree increased by 10%, the user's Brand Loyalty of MI will be increased by 4.18%. Because the Brand Awareness is a foundation about forming the Brand Loyalty, the more those users know about MI, the more they can feel the value of the brand. So it is necessary to do the positive popularity of Brand Awareness.

The hypothesis that "Service experience and emotional experience can positively affect brand loyalty" has been proved in this paper. The good customer experience can promote the form of brand loyalty. The customer experience includes service experience and emotional experience. When the customer buy a MI's product with service experience and emotional experience respectively increased by 10%, then the customer of MI's brand loyalty will be increased by 2.79% and 2.79% respectively. Combined with the actual situation, we can see that if MI pays more attention to customer experience, as its CEO Lei Jun said, the company will do its best to in every contact point to provide best customer experience.

The hypothesis that "Brand awareness, customer experience and brand loyalty are significantly correlated" has been verified. Through the related analysis, we found that the brand awareness, service experience, emotional experience and brand loyalty between these four variables both two have great relevance. The correlation of brand awareness and service experience is 0.734. The correlation of brand awareness and emotional experience is 0.548. The correlation of

DOI: 10.1051/matecconf/201710005025

brand awareness and brand loyalty is 0.736. The correlation of service experience and emotional experience is 0.571. The correlation of service experience and brand loyalty is 0.710. The correlation of emotional experience and brand loyalty 0.607.

5.2 Limitations and future research directions

This research also has certain limitation, specific description comes as follows:

To begin with, this sample that this study selected cannot be ruled out potential invalid samples' influence on the results of the study, thus, the results of this study are expected to be promoted with further verification. In addition to the sample size, the research ought to select wider dimensions and longer time spanning to cover in the future.

Then, although this study used the method of empirical study and concludes the relationship among brand awareness, customer experience, and brand loyalty, it has not gained accurate analysis of the data, and it also has not variable interactions between specific proportions. Thus, it is necessary to adopt Root Research to further clear the causal relations between the variables in the future.

Finally, the index system of brand loyalty still needs to be further perfected.

Acknowledgements

This research is supported by funding of China's central university basic scientific research expenses Project No. N152304008.

This research is supported by funding of Northeastern University at Qinhuangdao in China, which Humanities and Social Sciences Key Project No. XNR201601.

References

1. K.L. Keller., Conceptualizing, measuring and managing customer-based brand equity (Journal of Marketing, 1993)
2. W Song. Research on the impact of B2C e-commerce customer experience on customer loyalty (Jinan: Shan Dong University, 2012)
3. Y.P. Yang, The study on the factors affecting online shopping consumer loyalty (Chengdu: Southwest University of Finance and Economics, 2013)

Financial Development, Government R&D Subsidies and Green Innovation

Shuying Wang[1], Bobo Li[1] and Shuijuan Zhang [1]

[1]School of Management Engineering, Zhengzhou University, Zhengzhou 450001, China
Email: wangshuying@zzu.edu.cn,1796149399@qq.com,1563536373@qq.com

Abstract. In the low carbon economic era, green innovation has become an important basis for an enterprise to obtain and maintain a sustained competitive advantage.In paper, explores the path of green innovation using the panel data of 30 regions in China from 2008 to 2014.It finds that financial development has heterogeneity and different dimensions of financial development have different effects on green innovation of enterprises. That is, financial development structure has a negative impact on enterprise green innovation, and financial development efficiency has a positive impact on green innovation. Howere, the impact between financial development scale and green innovation is not significant. In addition, the study also finds that government R&D subsidies has no significantly effect on green innovation. Finally, this paper aims to provide the government and enterprises some theoretical and practical guidance to carry out policy-making of green innovation.

Keywords. Green innovation, financial development structure, financial development efficiency, financial development scale, government R&D subsidies.

1 Introduction

With the increasingly serious environmental problems, green sustainable development has become a common choice of countries around the world. In this context, green innovation has caused a great concern as an important way to alleviate environmental problems. From the national level, green innovation can produce positive externalities through the spillover of knowledge and technology, which is helpful to improve the efficiency of resource utilization and speed up the transformation of the national economic development mode, becoming the inevitable choice for the country to obtain and maintain the core competitive advantage. From the enterprise level, green innovation has a certain impact on the innovation activities of the whole enterprise (Aghion, 2012) [1],which helps to promote enterprises in production, operation and management changes and has an important impct on enhancing enterprise competition advantages and performance. In terms of the development of green innovation theory and practice in China , the research of green innovation in our country is still in the initial stage. Therefore, it is necessary to learn and utilize the green innovation system, management, technology and service in developed countries to provide important support for green innovation of Chinese enterprises.

Financial development and government R&D subsidies have important effects on the innovation activities of enterprises, which provide dynamic support for enterprise innovation through market stimulation and government regulation. Financial development, as a kind of market pull, can effectively integrate investment information, mobilize savings, improve resource allocation efficiency, improve enterprise supervision and timely disperse risk, which plays an important role in innovation activities (Levine, 1997) [2]. Relative to financial development, as a kind of government regulation and control, government R&D subsidies can mitigate the possible "market failure" (Hall, 2000) [3] and plays a crucial role in the innovation activities of enterprises.

The vast majority of past research has examined the effects among financial development, the government R&D subsidies on the general technology innovation, and provided substantial material and theoretical reference for the research of this paper, but there are still some limitations. First, most of the past studies ignored the heterogeneity of financial development. Thus, this article divids financial development into financial development scales, financial development structure and financial development efficiency, and respectively explores the three dimensions' impact on green innovation. Second, the existing research about the effects of government R&D subsidies on green innovation has not yet reached a consensus. Third, few studies integrate financial development, government R&D subsidies, and green innovation into the same research framework to systematically analyze the role of variables.

Based on the above analysis, using panel data of 30 regions in China, this article constructs the theoretical framework to further explore the impact of financial development and government R&D subsidies on green innovation, hoping to further expand green innovation theory and provide useful reference for enterprises to carry out green innovation activities.

2 Theory and hypotheses

2.1 Financial development and green innovation

Green innovation is a high-input process with a high degree of uncertainty and risk. In contrast to other activities, companies are often faced with financial constraints when undertaking green innovations (Hall & Lerner, 2010) [4].The financial system takes on the foremost financial activities and is an important means for the effective allocation of resources in the process of enterprise innovation. On the one hand, financial development can mitigate the financial constraints and improve the efficiency of resource allocation. Enterprises need to invest lots of R&D funds to green innovation. Companies can only seek help through external financing when a shortage of funds within the enterprise. Financial intermediaries and financial markets through various types of financing mechanisms to integrate social idle funds to meet the financing needs of green innovation. A well-developed financial development system not only provides sufficient external financing for corporate innovation to mitigate financial constraints (Ben & Goaied, 2008) [5], but also helps investors to obtain information on enterprise investment and financing decision-making, reduce information asymmetry, improve the allocation of financial resources efficiency (Habia, 2008) [6].On the other hand, financial development can reduce the uncertainty and risk of green innovation investment. Due to the high risk of green innovation, many enterprises and individuals tend to put capital into low risk areas, which hinders the development of green innovation. According to Levine (1997) [2] , financial development has the function of dispersing risk, which can reduce the uncertainty and risk in the process of green innovation to a certain extent.

Sun and Wang (2013) [7] studied the relationship between financial development and technological innovation, showing that financial development has a catalytic role in technological innovation.Bo et al (2013) [8]suggested that financial development scale, financial development efficiency and financial development structure are important factors impacting technological innovation capacity. Li et al (2013) [9] believed that the financial development scale has a significant positive effect on R&D investment. He et al (2011) [10] considerded that the strengthening of financial market development, expansion of direct financing channels and the scale is more conducive to enterprise independent innovation. Zhang (2012) [11] found that financial development scale and efficiency has significant R&D innovation effect. Based on the above analysis, the following hypotheses are proposed:

H1: There is a positive correlation between financial development scale and green innovation.

H2: There is a positive correlation between financial development efficiency and green innovation.

The financial structure depicts the relative importance of the stock market in comparison with the banking intermediaries. According to the current development, China's financial development structure is basically a bank-led financial structure (He & Chang,2010 [12]Sun&Wang,2013 [7]).Roger(2011) [13] and Ferreira (2012) [14] argue that banks would tend to deny loan support for high-risk, high-yield innovative projects in order to achieve soundness in their operations, which would make it difficult for banks to obtain credit for these projects. Therefore, it can be found that the financial system dominated by banks is not conducive to the innovation activities. Sun and Zhu (2008) [15],He and Chang(2010) [12]also believed that bank-led financial structure seriously restrictes the independent innovation of enterprises. Based on the above analysis, the hypothese is proposed:

H3: There is a negative correlation between financial development structure and green innovation.

2.2 Government R&D subsidies and green innovation

Green innovation of enterprises may face the high risk of R&D investment and external spillover. If the government does not carry out R&D subsidies, the enthusiasm of enterprises R&D investment can be dampened. From the point of risk control and R&D costs, government R&D subsidies can stimulate the enthusiasm of enterprises to innovate and expand their R&D investment (Yager & Schmidt, 1997) [16].In addition, government R&D subsidies has significant leverage on R&D innovation of enterprises, which can largely compensate for the "market failure" of green innovation. The existing literatures discussed the impact of government R&D subsidies on innovation of enterprise. Spencer and Brander (1985) [17] found that government R&D subsidies has similar utility as export subsidies, which can stimulate innovation activities and affect the equilibrium behavior of R&D competition. Hewitt and Roper (2010) [18]conducted a study of Ireland and Northern Ireland data for 1994-2002 and found that government subsidies has apositive impact on the proportion of enterprise R&D activities and product improvement innovation activities. Bai (2011) [19] conduted an empirical analysis, regarding that government R&D subsidies significantly improves the efficiency of R&D innovation. Ding and Zhu (2016) [20]analyzed the impact of government R&D subsidies on the performance of technological innovation. The results shown that government R&D subsidies has positively influenced the performance of technological innovation in a high-tech industry. In sum, government R&D subsidies can help to compensate for the lack of investment in innovation, to solve the problem of knowledge spillovers, and to reduce the risk of R&D, thus stimulating the enterprise green innovation. Based on the above analysis, the hypothese is proposed:

H4: There is a positive correlation between government R&D subsidies and green innovation.

3 Research design

3.1 Variable selection

3.1.1 Dependent variable

At present, the measurement of green innovation (gi) has not yet reached a consensus in China. Li et al(2013) [21] adopted the new product unit energy consumption, that is, the ratio of energy consumption to new product output. Wang and Guo (2015) [22] used the ratio of new product sales revenue to energy consumption. Taking into account the availability of data, this paper uses the indicator given by Wang and Guo (2015) [22] to measure green innovation.

3.1.2 Explanatory variables

The explanatory variables of this paper are financial development scale (fd1), financial development structure (fd2), financial development efficiency (fd3) and government R&D subsidies (s).According to the research of Bai et al. (2013) [8] , Chang and Ju (2015)[23], the financial development scale is measured by the ratio of the sum of financial institutions' deposits and loans balance to GDP in each region; the financial development structure is measured by the ratio of stock's total market value to financial institutions' loans balance in each region; the financial development efficiency is measured by the ratio of financial institutions' deposits balance to loans balance in each region. According to the research of Wang (2010) [24] , government R&D subsidies (s) is measured by the funds from the government to raise funds for enterprise science and technology.

3.1.3 Control variables

The control variables are human capital (hum) and economic development (ed).Human capital (hum) is the main body of enterprise innovation activities and reflects the enterprise's digestion and absorption abilities of new knowledge. This article uses the average schooling years of labor force to measure human capital (Barro,2001) [25].The formula is: the average schooling years of labor force = $6a_1+9a_2+12a_3+16a_4$,which a_1, a_2, a_3, a_4 respectively represent the proportion of primary school, junior high school, high school ,college education and above residents to the population aged 6 and over in the provinces. The economic development can provide a strong support for enterprises' innovation. This article uses per capita GDP to measure the economic development in in each region.

3.2 Sample data and sources

Considering the availability and validity of the data,this paper chooses the industrial enterprises above designated size in 30 provinces of China as the research object, and takes the panel data from 2008 to 2014 as the sample(due to the lack of data in the Tibet Autonomous Region is serious, this article removes the Tibet Autonomous Region). The data are from China Statistical Yearbook, China Statistical Yearbook of Science and Technology, China Statistical Yearbook on Population and Employment, Statistical Yearbook of Science and Technology Activities of Chinese Industrial Enterprises and China Financial Yearbook.The data is analyzed using Stata14.0.

3.3 Estimation Method

Based on the above analysis, in order to better investigate financial development, government R&D subsidies on the impact of green innovation, this paper establishs the panel model:

$$\begin{aligned} gi_{it} &= \alpha_0 + \alpha_1 fd1_{it} + \alpha_2 fd2_{it} \\ &+ \alpha_3 fd3_{it} + \alpha_4 s_{it} + \alpha_5 hum_{it} \\ &+ \alpha_6 ed_{it} + \varepsilon_{it} \end{aligned} \quad (1)$$

gi_{it}, $fd1_{it}$, $fd2_{it}$, $fd3_{it}$, s_{it}, hum_{it}, ed_{it} respectively represent green innovation ,financial development scale, financial development structure, financial development efficiency, government R&D subsidies, human capital and economic development of region i in the period t. ε_{it} represents the error term of region i in the period t. In order to eliminate the impact of heteroskedasticity, government R&D subsidies, human capital and economic development are all natural logarithm.

4 Empirical snalysis

4.1 Descriptive statistical analysis

Descriptive statistics are shown in Table 1. Means, standard deviations, maximum and minimum values are summarized for all variables. As shown in Table 1, firstly, in terms of the average 21.65 and the standard deviation 53.74,the paper finds that green innovation has great regional differences. Secondly, from the level of financial development, the financial development scale, the financial development structure and the financial development efficiency are obvious regional differences. It implies that China 's financial development is unbalanced. The mean value of the financial development structure is 0.43, which shows that China's financial structure is a banking-oriented financial structure in general (Sun & Wang (2013)[7],He & Chang(2010)[12]).

Table 1. Descriptive statistics.

Variables	Mean	Std. dev.	Min	Max
gi	21.65	53.74	0.67	466.03
fd1	2.92	2.19	0.90	28.35
fd2	0.43	0.62	0.02	4.56
fd3	1.77	3.68	0.16	47.34
s	10.75	1.31	5.09	12.87
hum	9.16	5.97	5.77	90.93
ed	9.07	0.23	8.66	11.42

4.2 Panel data regression analysis

4.2.1 Unit root test

Before the panel regression, this paper carries out the stationarity test to examine the smoothness of the data for the purpose of preventing false regression or pseudo-regression. In this paper, ADF-Fisher test method is used. The test results show that the p-values of the ADF-Fisher test are less than 0.01, rejecting the null hypothesis containing unit roots, indicating that each sequence is stationary and eliminates the existence of spurious or pseudo-regressions.

4.2.2 Panel model selection

Before the regression, it is necessary to test the form of the model to avoid the deviation caused by improper model setting in the regression process. In this paper, the Hausman test is used to examine the model. The Hausman test has a p-value of 0.006,which indicates that the null hypothesis is rejected at the 1% level, so the fixed effect model is chosen.

4.2.3 Regression results and analysis

Based on the above analysis, this paper selected the fixed effect model to regress the model, the regression results are shown in Table 2.From the regression results in Table 2, the regression coefficient of financial development efficiency is 4.84, and through the 1% significance test. It shows that the financial development efficiency is positively related to the green innovation of enterprises, that is, the financial development efficiency has significantly promoted the enterprise green innovation activities. This conclusion is similar to the finding of Ran et al (2013) [26] and H2 is verified. The regression coefficient of the financial development structure is -49.26,and the 1% significance test is adopted. It shows that Chinese.

Table 2. Panel data regression results.

gi	Coef.	Std. Err.	t	P>|t|
fd1	-0.13	1.19	-0.11	0.91
fd2	-49.26	10.25	-4.81	0.00***
fd3	4.84	1.19	4.07	0.00***
s	-3.91	4.83	-0.81	0.42
hum	0.57	0.95	0.60	0.55
ed	-15.24	25.90	-0.59	0.56

Note: ***, **, * represent the 1%, 5%, 10% confidence level was significantly.

Banking-led financial structure has a significant hindrance to green innovation and is not conducive to the development of green innovation. H3 has been supported. This finding is consistent with the conclusions of Kang and Zhang (2008) [27], He and Chang (2010) [12],Sun and Wang (2013) [7].

From the regression results, the regression coefficient of financial development scale is negative, but not significant (Coef=-0.13,p=0.91), which means that H1 has not passed the test. The possible causes of this result are the following. First, it may be because the scale of financial development indicators have yet to be improved or data sources are inadequate, resulting in the regression results do not support the hypothesis. Second, in recent years, as the state has issued a series of policies to encourage the financial system reform, China's financial development scale has been developed. Howere,due to capital gains and hedging effect, financial resources in different regions and industries can not be effectively configured, thus making financial development scale can not play an effective role. According to the study by Shi and Wang (2010) [28], Qian et al. (2010) [29]; Lu and Li(2016) [30],the impact of financial development on innovation has a certain threshold effect. The scale of financial development in many parts of China has not yet crossed the higher threshold. Therefore, financial development scale can not to play an effective role in promoting green innovation, and even inhibit corporate green innovation.

As shown in Table 2, it implies that the regression coefficient of government R&D subsidies is negative but not significant (Coef =-3.91, p=0.42).It is assumed that H4 has not been validated. Analysis of its causes, may have the following points. First, it may be due to government R&D subsidies has a "crowding out" on enterprise R&D investment. When the government R&D subsidy rate is high, it will have crowding-out effect on enterprise R&D investment, leading to negative impact. Second, moral hazard may occur because of imperfect information disclosure mechanism and capital supervision mechanism. In this case, enterprises may investe government subsidies in activities that are not related to green innovation. Third, the impact of government R&D subsidies may also be affected by the policy environment and the nature of the enterprise. Enterprise which has close ties with the government and strong policy background would be more likely to achieve the purpose of rent-seeking, while, those which really need government R&D subsidies for green innovation may not get the government's support. Just as Liu et al. (2012) [31]found, the poorer the regional policy environment and the weaker the anti-corruption, the less the effect of government R&D subsidies.

5 Conclusions

Based on panel data of 2008-2014, this paper studies the impact of financial development scale, financial development structure, financial development efficiency and government R&D subsidies on the green innovation of enterprises in 30 provinces of China. The following conclusions are drawn and the corresponding suggestions are given .

First, the different dimensions of financial development have different effects on green innovation. There is a negative correlation between financial development structure and green innovation, which shows that bank-oriented financial development structure is not conducive to the green innovation activities of enterprises.Financial development structure will be affected by historical conditions, legal environment, culture, politics and other factors. It is hard to say that the financial market-based structure is better or bank intermediaries based financial structure is more advantage. But in terms of promoting enterprise innovation, the direct financing structure dominated by financial markets is more effective. Therefore, in the structure of financing model, China should actively develop financial markets, increase the proportion of direct financing, and build a financial market system that coordinates the development of direct financing and indirect financing.

On the current development situation, China's financial development scale has yet to be expanded. All regions should speed up the development of multi-level capital market, encourage the development and construction of small and medium-sized bank financial institutions, investment financial institutions and financing intermediary services .Encourage private capital to enter the banking industry and build a property rights coordination, mixed all and effective competition in the financial service system. At the same time, the government should regulate the development of Internet finance and build a mainstream format and the new format coordinated development of the financial system to weaken the capital gains and risk. And the government is committed to the effective allocation of financial resources in different regions and different industries.

Financial development efficiency has a significant positive effect on the green innovation of enterprises. Therefore, goverment should pay attention to the continuous improvement of financial development efficiency. Simultaneously, enterprises should actively respond to the call for of the "13th Five-Year" financial system to deepen reform, assimilating the green development concept into the enterprise development process. And continuously improve green innovation capacity, making enterprises more competitive advantage.

Second, in view of the government R&D subsidies for green innovation is not significant, this paper puts forward three proposals. Firstly, the government should establish an effective supervision and control mechanism and do it well in market research to reduce the asymmetry of information, implementing government R&D subsidies. Secondly, the government should strengthen the anti-corruption education to the relevant government staff and enterprises, construct the healthy policy environment, standardize the subsidy procedure and make the government R&D subsidies more standardized and transparent. Thirdly, enterprises should rationally plan their own R&D investment so as to reduce the negative effects from government R&D subsidies "crowding out".

References

1. Acemoglu D, Aghion P, Bursztyn L, et al. The environment and directed technical change. The American Economic Review, 2012, 102(1): 131-166.
2. Levine R. Financial development and economic growth: views and agenda. Journal of economic literature, 1997, 35(2): 688-726.
3. Hall B, Van Reenen J. How effective are fiscal incentives for R&D? A review of the evidence. Research Policy, 2000, 29(4): 449-469.
4. Hall H, Lerner J. The financing of R&D and innovation. Handbook of the Economics of Innovation, 2010, 1: 609-639.
5. Ben S, Goaied M. The determinants of commercial bank interest margin and profitability: evidence from Tunisia. Frontiers in Finance and Economics, 2008, 5(1): 106-130.
6. Habib A. Corporate transparency, financial development and the allocation of capital: empirical evidence. Abacus, 2008, 44(1): 1-21.

7. Sun Wuqin, Wang Wei. Chinese Financial Development and Promotion of Technological Innovation. Management World, 2013 (6): 172-173.
8. Bai Ling, Jiang Lei, Zhan Benfu. Finance Development System, Capacity and Transformation of Technological Innovation Output - An Empirical Study from Provincial Dynamic Panel Data.Comment on the Sankei, 2013 (1): 15-25.
9. Miao Miao, Xiao Hongjun, Zhao Shuang.Research on the Relationship among Financial Development, Technological Innovation and Economic Growth - Based on Panel Data of Provinces and Cities in China.Chinese Journal of Management Science, 2015 (2): 162-169.
10. Guo Guohua, Liu Lintao, Chang Xinxin. Research on the Relationship between Financial Structure and Independent Innovation in China. Technology Economics, 2011 (3): 19-24.
11. Zhang Zhiqiang.Financial Development, R&D and Regional Technology Deepening . Economic Review, 2012 (3): 82-92.
12. He Guohua, Chang Xinxin.The Characteristics of China's Financial Structure and Its Restrictions on Enterprise Independent Innovation. Wuhan Finance, 2010 (2): 8-11.
13. Roger S, Vlček J. Macroeconomic costs of higher bank capital and liquidity requirements. IMF Working Papers, 2011: 1-51.
14. Ferreira A, Matos P. Universal banks and corporate control: Evidence from the global syndicated loan market. Review of Financial Studies, 2012, 25(9): 2703-2744.
15. Sun Wuqin, Zhu Shunlin.The Efficiency of Financial Development to Promote Technological Innovation: An Analysis Based on Malmuquist Index. Statistical Research, 2008 (3): 46-50.
16. Yager L, Schmidt R. The Advanced Technology Program. American Enterprise Institute, 1997.
17. Brander A, Spencer J. Export subsidies and international market share rivalry. Journal of international Economics, 1985, 18(1): 83-100.
18. Hewitt-Dundas N, Roper S. Output additionality of public support for innovation: evidence for Irish manufacturing plants. European Planning Studies, 2010, 18(1): 107-122.
19. Bai Junhong, Li Jing.Government R & D Funding and Enterprise Technology Innovation: An Empirical Analysis Based on Efficiency.Financial Research, 2011 (6): 181-193.
20. Ding Kai, Zhu Shunlin.Study on the Impact of Government R & D Subsidy on Technological Innovation Performance - Based on China's High-tech Industry Perspective. Science and Technology, 2016 (1): 37-41.
21. Li Wanhong, Bi Kexin, Sun Bing.The Impact of Environmental Regulation Intensity on Green Technology Innovation in Pollution-intensive Industries: An Empirical Test Based on Panel Data from 2003 to 2010 .Research and Development Management, 2013 (6):72-81.
22. Wang Fengzheng, Guo Xiaochuan.Effects of Environmental Regulation Intensity on Green Technology Innovation in Resource-based Industries: An Empirical Test Based on Panel Data from 2003 to 2011.China Population, Resources and Environment, 2015 (1): 143-146.
23. Tao Changqi, Ju Zexia.The Empirical Analysis on the Relationship among Financial Development, Environmental Regulation and Technology Innovation - Based on the Panel Threshold Regression Model.Journal of Jiangxi Normal University (Natural Science Edition), 2015 (1): 27-33.
24. Wang Jun. An Empirical Study on the Impact of R & D Subsidies on R & D Input and Innovation Output of Enterprises .Research in Science, 2010 (9): 1368-1374.
25. Barro J, Lee W. International data on educational attainment: updates and implications. oxford Economic papers, 2001, 53(3): 541-563.

26. Ran Guanghe, Xu Kun, Lu Zhaoyang.Financial Development, the Impact of FDI on Regional Innovation Capacity. Scientific Management, 2013 (7): 45-52.
27. Kang Zhiyong, Zhang Jie.Study on the Influence of Chinese Financial Structure on Independent Innovation Capability. Statistics and Decision, 2008 (19): 130-133.
28. SHI Wenming, WANG Yu-huai.The Threshold Effect of Financial Development on Technological Progress: An Empirical Study Based on Provincial Panel Data.Shanxi University of Finance and Economics, 2010 (9): 38-45.
29. Qian Shuitu, Zhou Yongtao. Analysis on the Threshold Model of Financial Development, FDI and China's Independent Innovation Capability. South China Finance, 2010 (10): 24-29.
30. Lu Yingfei, Li Zheng. The Threshold Effect of Financial Development on Innovation Efficiency. Journal of Northeast Normal University (Philosophy and Social Sciences), 2016 (5): 75-79.
31. Liu Hong, Xiao Meifeng, TANG Qing-quan. The Incentive and Crowding-out Effects of R & D Subsidy on Enterprise R & D Expenditure: An Empirical Analysis Based on Chinese Listed Companies Data.Economic Management, 2012 (4): 19-28.

The Relationship between Institutional Environment and Enterprise's Technology Innovation Performance——The Visual Angle Based on MOA Theoretical Model

Shuying Wang[1] and Xingye Xiao[1]

[1]School of Management Engineering, Zhengzhou University, Zhengzhou 450001, China
Email: wangshuying@zzu.edu.cn, 408640203@qq.com

Abstract. This paper discusses the relationship between institutional environment and technological innovation performance from the perspective of MOA, and analyzes the moderating role of external social capital and top management team heterogeneity in the relationship between institutional environment and technological innovation performance. It is found that good institutional environment can improve the performance of technological innovation significantly; social capital as an informal, non-mandatory system to regulate individual behavior and promote the company's technological innovation performance; senior management team heterogeneity and technological innovation significantly Negative correlation. Finally, it discusses the management revelation of the conclusion and gives some policy suggestions.

Keywords. Institutional environment, technological innovation performance, executive team heterogeneity, corporate social capital.

1 Introduction

Technological innovation is the root of the enterprise to obtain and maintain the core competitive advantage. However, with the intensification of market competition, the increase of economic environment uncertainty and the rapid development of knowledge economy, China's enterprise technology innovation is faced with unprecedented challenges and enormous pressure. From the practical level, the total R&D investment in China's total industrial added value accounted for only 1.9%, with the United States, Germany, South Korea, 28 countries and Japan and other advanced developed countries a far cry, indicating that China's R&D Low, a serious shortage of technological innovation capacity [1]. From a theoretical point of view, technological innovation is the enterprise in order to seek profit maximization, to reduce the external suppression of technology caused by self-defense behavior taken [2]. The existing literature on the enterprise network capabilities, financial development, internal and external control and other aspects of the impact of technological innovation of enterprises [3]. However, there are few literatures on the internal influence of external environment on the performance of technological innovation, especially from the perspective of institutional environment to study the relationship between technological innovation. Specifically, how does the institutional environment affect the technological innovation of enterprises is still a "black box" that has not been opened yet, and whether the improvement of regional institutional environment can enhance the performance of technological innovation? Does the heterogeneity of internal competencies (externalities) and external opportunities (corporate social capital) of firms have an innovative heterogeneity in response to regional institutional environments? The answers to these questions have important theoretical value and practical significance to promote the development of technological innovation of listed companies and to solve the effective integration of resources [4].

Based on the MOA theory, this paper integrates the regional institutional environment, corporate social capital and executive team heterogeneity into a unified analytical framework. By analyzing the impact of institutional environment on the technological innovation performance of enterprises, (Heterogeneity of senior management team) and the external environment (social capital) to try to provide the theoretical interpretation and empirical answer for the above-mentioned problems. Firstly, based on the concept of resource integration, the innovation performance of enterprises is regarded as the common result of power, opportunity and ability. The integration of theoretical model and the effect of technological innovation performance on enterprise's technological innovation performance is discussed in this paper. Factors of the empirical analysis. Second, in the study of "high-level echelon theory", empirical research on corporate executives and technological innovation capacity is rare. This paper uses listed companies as research samples, and increases the diversity of the literature. Thirdly, the enterprise scale, the enterprise age and the net profit are taken as the control variables, and the impact of the institutional environment on the enterprise's technological innovation performance is studied, which enriches the enterprise innovation literature research.

2 Theory and hypotheses

MOA theory points out that only having the motive, opportunity, and ability elements at the same time is a necessary condition for producing certain results. As the institutional environment on the enterprise R&D investment have a direct impact on enterprises to enhance technological innovation performance of the "motives" [5]; as the executive team of innovation-oriented, innovative decision-making and innovative resource integration irreplaceable role, thus affecting innovation performance Social capital is the key "opportunity" for technological innovation of enterprises because of the important support for the grasp of technological innovation opportunities by corporate social capital. Based on the above analysis, this paper proposes the following hypotheses on the basis of the MOA theory: institutional environment, executive team heterogeneity and social capital.

2.1 Institutional environment and technological innovation performance of enterprises

The innovation literature generally argues that when faced with increasing external uncertainties, firms tend to pursue more advanced and competitive innovation strategies to protect their market position [6]. Xu, Wenjun and Feng Tao (2016) pointed out that the improvement of the institutional environment can significantly increase the total amount of capital allocation and the duration of allocation of technological innovation projects to promote technological innovation [7]; Liu Jibing, Wang Qi and Ma Huanyu 2015) pointed out that the regional policy environment and market environment for enterprise innovation ability has a significant positive impact, and thus promote the technological innovation performance [8]. Liao Kairong and Chen Shuangying (2011) take the national private enterprise survey data in 2006 as a sample to analyze the impact of external institutional environment on the R&D investment of private enterprises [9]. The research shows that a good external institutional environment will promote R&D investment of private enterprises, (2011) study shows that the institutional context deeply affects the enterprise's R&D expenditure behavior [10]; Rainder (2012) found that the competitive market environment in the (2016) analyzed the impact of institutional environment on the performance of technological innovation in 266 enterprises in two traditional industrial clusters in Zhejiang province [11]. The results show that the efficiency of technological innovation is higher and the innovation efficiency is faster. Found that there is a significant role between the two [12]. Therefore, based on the above analysis, this paper proposes the following assumptions:

H1: Good institutional environment is promoting enterprise technological innovation.

2.2 Top management team heterogeneity and enterprise technology innovation performance

Team heterogeneity refers to the differences in the demographic characteristics of team members as well as important cognitive concepts, values, and experience [13]. Specific to the heterogeneity of executive team, including the level of education, entrepreneurial experience and other easy to observe the external heterogeneity and cognition, values, preferences, attitudes, entrepreneurial commitment and other deep internal heterogeneity [14]. The echelon theory points out that the different cognitive bases, values, insights, and processes that influence the strategic choice and organizational performance of the organization's members [15]. As the leader of strategic decision-making, the top management team influences their behavior by their cognition and values, and maximizes the role of senior management team in strategic choice and development direction. However, with the extension of executive tenure, corporate strategy will be the lack of change, easy to fall into the rigid corporate governance model, which will not conducive to enterprise management. In other words, with the growth of executive tenure, senior management team easy to produce too much self-confidence, resulting in management thinking and behavior may be sticky, resulting in decreased technological innovation output. Based on this, this paper, senior management team heterogeneity of the term of senior management team heterogeneity on the impact of technological innovation performance.

O.Relly & Flatt (1989) argue that, in the enterprise life cycle, enterprise innovation and senior management team heterogeneity is negatively correlated [16]; Ma Fuping and Guo Xiaochuan (2010) pointed out that executive team heterogeneity of technological innovation 2006 to 2010 China's small and medium-sized enterprises listed companies as a sample of the empirical study of entrepreneurial enterprise executive team heterogeneity of the direct impact of entrepreneurship-oriented research [17], the study has a positive impact on the performance of China's A-share listed companies in manufacturing research shows that the heterogeneity of executive team R&D investment from the negative phase of [18, 19]; the same time, Chen Rui (2005) take the data samples of listed companies in Fujian Province and Shandong Province as an example, the results show that the higher the team heterogeneity, the greater the investment in technological innovation [20]; Bai Jingkun And Li Sha-sha (2015) show that there is a significant negative correlation between executive team heterogeneity and innovation performance [21]; Lei Hui and Liu Peng (2013) take the listed companies of China's SME sector as the research object, This indicates that executive team heterogeneity is significantly negatively correlated with technological innovation [22]. Therefore, based on the above analysis, this paper proposes the following assumptions:

H2: There is a negative correlation between executive team heterogeneity and enterprise technological innovation performance.

2.3 Social capital and enterprise technology innovation performance

Putnam (2014) argues that social capital refers to trust, norms, and networks that can improve social efficiency through coordinated action, and thus social capital is embodied in mutual trust, civic awareness, and mutual benefit. It has created a good interactive environment for enterprise technological innovation and economic and talent exchange [23]. Social capital includes social trust, social organization, social participation and so on. As an institutional framework, social capital has created a policy and institutional environment foundation for enterprise technological innovation, which provides an opportunity for the improvement of technological innovation performance [24, 25].

On the one hand, a good external social capital, to promote internal mechanisms to play a role. In other words, external social capital through the enterprise's innovation power, innovation process and innovation results in three aspects to influence the enterprise's innovation ability. Cheng Cong and Xie Hongming (2012) pointed out that social capital is one of the key factors to determine whether enterprises can obtain the performance of technological innovation. On the other hand, social capital can promote technological innovation by using external resources and knowledge integration. Zhang Honghui (2014) pointed out that in the R&D activities, social capital can promote the sharing of knowledge, and thus has a fundamental decision on technical activities [26]. Capello & Faggian (2005) argues that the exchange of R&D activities and knowledge is synchronized, and the network of individual interactions of social capital increases the trust between people and strengthens communication between individuals. Type "link ties. Such weak ties are a useful complement to the success of R&D activities for strong ties such as 'family'. Therefore, based on the above analysis, this paper proposes the following hypothesis:

H3: Social capital is promoting enterprise technology innovation performance.

3 Research design

3.1 Sample selection and data sources

The paper chooses the data of A-share listed companies in 61 high-tech industries (communication equipment, software development, semiconductors and electrical equipment) in Shanghai and Shenzhen in 2009-2015 as the samples, and makes a comparative study on the relationship between institutional environment and technological innovation performance:(1) Try to choose more frequent R&D activities of high-tech industry listed companies; (2) To exclude listed companies that have been or are being used by ST, ST, SST, S*ST and S; (3) Elimination of missing data of listed companies, the final sample of 61 listed companies. Due to limited space, not a detailed list of the specific list of sample companies. The original data of this article is listed company's annual report and listed company information disclosed on the website of Shanghai Stock Exchange and Shenzhen Stock Exchange.

3.2 Variable measurement

Dependent variable: enterprise technology innovation performance. Most of the variables in the literature measure technological innovation activities are divided into two categories: one is the input, that is, investment in R&D, R&D personnel input; the second is output, that is, the number of patents, new products, new Product sales. In view of the reliability and availability of data, this paper draws lessons from Liu Yun-guo and Liu Wen (2007), because of the low number of new products and the comparability of data in China's annual reports, and the sales income index is easy to be "manipulated by earnings" [27].The ratio of R&D expenditure to the total assets at the end of the period is taken as the proxy variable of the enterprise's technological innovation performance [28].

Independent variables: institutional environment, executive team heterogeneity, and social capital. The Institutional Environment Indicator (Evm) of the paper is also based on the China Business Environment Index for 2013 published by Wang Xiaolu, Yu Jingwen and Fan Gang (2013) [29]. In this report, the operating environment index has the data of seven years from 2009 to 2015, according to the data of the seven-year average, the average higher than the national average of the sample into the "institutional environment better" group; Into the "institutional environment is poor" group, the system environment is better assigned a value of 1, or 0. Institutional environment indicator is a comprehensive index, including the level of marketization, the level of government governance, the level of the legal system and so on. Executive team heterogeneity, senior management team performance for the rich cognitive, management, interest and experience in the accumulation process. The heterogeneity of tenure has two sides to the enterprise's technological innovation performance, which provides a good opportunity window for observing the heterogeneity of the executive team. Therefore, this paper draws on Hu Wangbin, Zhang Yuli and Yang Jun (2014) to study the heterogeneity of senior management team by the tenure heterogeneity (the ratio of the standard deviation of the senior management age and the average value). Social capital, from the relevant research literature on the "social capital" of the definition and measurement methods, this paper from the number of regional social organizations, the natural logarithm of the measurement of regional social development, social capital data from the "China City Commercial Credit Environmental Index official website" and China Statistical Yearbook for the years 2009-2015.

Control variables: In order to control the impact of other factors on the research model, this paper select the enterprise size, corporate net profit, corporate age as a control variable.

Table 1. Variable definitions and variable design.

The type of the variable	The variable name	Variable symbol	Variable definition
Explained variable	Technological Innovation Performance	RD	R&D investment / total assets
Explanatory variables	Executive Team Heterogeneity	MH	Executive Team age standard deviation / age mean
	Social capital	SC	The natural logarithm of social organization
	Institutional environment	INDEX	System environment is good 1, otherwise 0
Control variables	Enterprise size	SIZE	The natural logarithm of the number of employees in an enterprise
	Business age	AGE	Based on 2009, there are 7 dummy variables
	Net profit	P	The natural logarithm of the net profit of the enterprise at the end of the year

3.3 Descriptive statistical analysis

Table 2 shows the descriptive statistics of the main variables. From the sample company's technical innovation indicators (exp) point of view,the mean value is 0.03 and the standard deviation is 0.05, which indicates that the technological innovation activity of listed companies in our country is weak, and there are big differences among the companies; The mean value of executive team heterogeneity (mh) is 0.17 and the standard deviation is 0.04, which indicates that there is a big difference in executive team; The maximum of social capital (sc) is 11.18, the minimum is 7.77, and the standard deviation is 0.68, which shows that there is significant difference in the social capital of each region; The maximum value of the system index is 1.00, the minimum value is 0.00 and the standard deviation is 0.48, which indicates that the development of institutional environment is not balanced in our country at present.

Table 2. Variable descriptive statistics.

	RD	MH	SC	INDEX
Mean	0.03	0.17	9.82	0.34
Median	0.02	0.17	9.91	0.00
Maximum	0.47	0.29	11.18	1.00
Minimum	0.00	0.04	7.77	0.00
Std. Dev.	0.05	0.04	0.68	0.48
Skewness	4.97	0.04	-0.60	0.66
Kurtosis	34.58	2.83	3.06	1.43
Jarque-Bera	19503.95	0.62	25.54	74.45
Probability	0.00	0.74	0.00	0.00
Sum	14.09	73.05	4192.30	147.00
Sum Sq. Dev.	0.91	0.72	198.85	96.39

3.4 Research design

Based on the research questions and the research variables, this paper constructs an empirical analysis model of institutional environment and technological innovation performance.

$EXP_{it} = \beta_0 + \beta_1 MH_{it} + \beta_2 SC_{it} + \beta_3 Index_{it} + \sum_{i=7}^{n} \beta_i CON_{it} + \varepsilon_{it}$ EXP_{it} indicates the Technological Innovation Performance of Enterprise i in t year; MH_{it} indicates the Executive Team Heterogeneity of Enterprise i in t year; SC_{it} indicates the Social capital of Enterprise i in t year; $Index_{it}$ indicates the Institutional environment of Enterprise i in t year; CON_{it} indicates the Control variable set; ε_{it} indicates the Random error term.

4 Result and analyze

4.1 Regression analysis

First of all, this data belongs to the panel data, reducing the time series of multi-collinearity problem, so the first to determine whether the time series is stationary, and the existence of "pseudo-regression" phenomenon. A unit root test using Eviews 7.2 yields the following results (Table 3):

Table 3. Unit Root Test of Variables.

Method	Statistic	Prob.**
Im,Pesaran and Shin W-stat	-11.44	0.00
ADF-Fisher Chi-square	1277.39	0.00
PP-Fisher Chi-square	1669.25	0.00

As can be seen from Table 3, the selected data is stable. Following the use of cointegration test, test whether there is a "pseudo-regression" equation, as shown in Table 4:

Table 4. Co-integration test of variables.

Variable	Coefficient	Std. Erro	t-Statistic	Prob.
RESID?(-1)	-1.34	0.09	-13.90	0.00
D(RESID?(-1))	0.16	0.06	2.38	0.01

As can be seen from Table 4, at 5% significant level to reject the null hypothesis through the KOA test. Thus, there is a cointegration relationship between institutional environment, social capital, and executive team heterogeneity in panel data on the performance of technological innovation.

Secondly, the F-test and the T-test are used to test the fixed effect and the random effect. The test results reject the null hypothesis, indicating that the fixed effect and the random effect are the significant influencing factors in the model. Thirdly, according to the Hausman test, the model is set, and the result shows that the fixed effect should be used in the static panel data model. Limited by space, the specific regression analysis did not give a report.

4.2 Analysis of regression results

According to the fixed-effect regression results, it can be seen that the index coefficient of the system is 0.01 and is significant at the level of 5%, the results show that the institutional environment is positively related to the performance of technological innovation. In other words, the more perfect the institutional environment, the easier the technological innovation and the higher the innovation performance. The coefficient of external social capital (sc) is 0.01, and is significant at the level of 5%, the results show that there is a positive correlation between external social capital and technological innovation performance. In other words, the more abundant external social capital is, the easier it is for technological innovation and the higher the technological innovation performance. The coefficient of heterogeneity (mh) for the top management team is -0.17 and is significant at the 5% level, the results show that there is a negative correlation between executive team heterogeneity and enterprise technological innovation performance. In other words, with the growth of senior management, senior management team is easy to produce too much self-confidence, resulting in management thinking and behavior may be sticky, resulting in decreased technological innovation output, technological innovation performance of listed companies decreased.

Table 5. Fixed effect regression results.

Variable	Coefficient	Std. Error	t-Statistic	Prob.
C	-0.07	0.04	-1.74	0.08
MH	-0.17	0.06	-3.11	0.00
SC	0.01	0.00	3.20	0.00
INDEX	0.02	0.01	3.39	0.00
AGE	-0.00	0.00	-4.32	0.00
SIZE	0.01	0.00	2.06	0.04
P	0.00	0.00	0.24	0.81

5 Discussion and conclusion

Today's domestic and foreign economic situation is so grim, in addition to some large or state-owned enterprises, most of the enterprise development is difficult, its survival and growth faces enormous challenges, and only continue to carry out technological innovation activities, access to and build the core competitiveness of enterprises is imperative. Based on the data of A-share listed companies in Shanghai and Shenzhen stock markets, this paper constructs the relationship model between corporate social capital, institutional environment, executive team heterogeneity and technological innovation performance. Through empirical research, the following conclusions are drawn: First, corporate social capital, institutional environment, technological innovation performance of enterprises has a significant impact, but there are differences in the impact factor. The influence degree is the institutional environment, social capital from high to low respectively; Second, the improvement of China's regional institutional environment is more conducive to improving the level of technological innovation performance of private listed companies; social capital can have a positive stimulating effect on corporate R & D investment, corporate social capital is conducive to the improvement of technological innovation performance ; Thirdly, in the listed companies, the heterogeneity of senior management team has negative effect on the performance of technological innovation. The higher the senior team is, the longer the term is, the more unfavorable for enterprise technological innovation.

In the process of enterprise development, often due to the lack of effective support of the organizational structure of the inter-enterprise technical interruption caused by enterprises to improve the level of technological innovation, by adjusting the senior management of gender, age, education, professional background, functional background and

Education background and functional background heterogeneity, choose more young, highly educated, science and engineering background, engaged in output-type functional background of senior management staff to makes the team heterogeneity level is low. For the listed companies, through external social capital can obtain more technological innovation resources, enterprises can effectively strengthen the external network through organizational change, in order to better learn new knowledge, new experience, and effectively promote the enterprises to achieve technological innovation.

Based on the above analysis, this paper proposes the following policy recommendations:

First, optimize the regional policy and institutional environment, expand diversified government support channels. The government should broaden the channels of support, in addition to giving certain tax incentives and support funds, but also to build a better social innovation environment. First of all, through the improvement of laws and regulations, and earnestly protect the fruits of enterprise innovation, from the source to prevent infringement, theft of patented technology and other acts; Secondly, the government should promptly give innovative industries a certain degree of government procurement tilt indicators; Finally, continue through the financial, tax and other channels to reduce business costs, thereby reducing R & D risk, encourage enterprises to achieve profit growth through innovation.

Second, promote regional social capital development level, strengthen the innovation and interaction between enterprises. The social relationship-oriented cultural background makes Chinese enterprises in the intricate social network, which makes our enterprises have greater development space by raising the level of corporate social capital to improve the financing efficiency. Enterprises should pay attention to the construction and consolidation of the credit network with customers and suppliers, and actively strengthen the exchange of senior management between enterprises, as well as business and government and financial institutions, communication and other cooperation between the enterprises to reduce Transaction costs, reduce corporate finance risk, improve the efficiency of corporate finance.

Third, the scientific establishment of senior management team. Senior management team is younger, a strong sense of innovation, education is generally high, which is conducive to the development of enterprise management decision-making is very important. The recruitment of talent and promotion of personnel, the appropriate join the young, highly educated and has extensive experience in the talent to the senior management team to help optimize the team structure, reduce the heterogeneity of senior management team, and the need to gradually upgrade the R&D department Employees to the management, so that they have the technology on the basis of strengthening the ability of enterprise management, to strengthen the senior management team to do a good job of human resources planning.

References

1. Xu Lingling. Institutional environment, ownership structure and technological innovation .Soft Science, 2015 (12): 22-26.
2. Ancona G,Caldwell F. Demography and design:predictors of new product team performance.Organization Science,1989,3(3):321-341.
3. ZHEN Hong-Xian, ZHANG Xian-Zhi, CHI Gui-Tai.The Impact of Institutional Environment and Ultimate Control on Corporate Performance: An Intermediary Effect Test Based on Agency Cost.Financial Research 2015 (12): 162-177.
4. Williams Y. demography and diversity in organizations: A review of 40 years of research. Research in organizational behavior, 1998, 20 (3): 77-140.
5. Blumberg M, Pringle D. The missing opportunity in organizational Research: some implications for a theory of work performance. Academy of Management Review, 1982, 7 (4): 560-569.

6. Freel S. Perceived environmental uncertainty and innovation in small firms. Small Business Economics, 2005, 25 (1): 49-64.
7. Xu Hao, Wen Jun, Feng Tao.Institutional Environment, Financial Development and Technological Innovation.Shanxi University of Finance and Economics, 2016 (6): 44-53.
8. Liu Jibing, Wang Qi, Ma Huanyu. Institutional environment on the strategic emerging industry innovation ability. Science and technology progress and countermeasures 2015 (23): 54-61.
9. Liao Kaorong, Chen Shuangying.Empirical Research on the Impact of Institutional Environment on R & D Investment of Private Enterprises.Science Research, 2011 (9): 1342-1348.
10. Zhao Huifang, Zhou Yanlei, Zhang Xuan. The Influence of Institutional Environment on Enterprise R & D Investment. Friends of Accounting, 2011 (31): 31-35.
11. Simons T, Smith K A. Makinguse Of Difference: diversity, debate, and decision comprehensiveness intopmanagementteams. Academic Journal of Management, 1999, 42 (6): 662-673.
12. Zhou Jianhua.Network Relation Embedding, Regional Institutional Environment and Technological Innovation.Journal of Tianjin University of Finance and Economics (in Chinese), 2016 (5): 50-62.
13. Finkelstein S. Top-Management-Team tenure and organizational outcomes: the moderating role of managerial discretion. Administrative Science Quarterly, 1990, 35 (3): 484-503.
14. Allan J D. The effects of reduction in trout density on the invertebrate community of a mountain stream. Ecology, 1982, 63 (5): 1444-1455.
15. ZHANG Biwu, SHI Jintao.Characteristics of Board of Directors, Executive Compensation and Pay Sensitivity: An Empirical Analysis of Chinese Listed Companies .Management Science, 2005 (4): 32-39.
16. SUN Yin-feng, LI Xiao-qing. Heterogeneity, Financial Innovation and Financial Performance of Commercial Banks: Evidence from Listed Banks in Shanghai and Shenzhen.Financial Theory and Practice, 2014 (6): 44-50.
17. Lei Hui, Liu Peng. SME executive team characteristics on the impact of technological innovation - based on the nature of ownership perspective. Journal of Zhongnan University of Economics and Law, 2013 (4): 30-52.
18. ZHU Jin-wei, PENG Jin-jin, LIU Jing.Study on the Influence of Top Management Team Characteristics on the Input of Technological Innovation in Enterprises - The Regulation Effect of Incentives, Science in Policy, 2014 (8): 17-33.
19. Chen Rui, Jing Runtian. The Effect of Team Heterogeneity on Team Members' Innovation Performance. Technology Economics, 2012 (12): 13-21.
20. ZHENG Xiao-Dan, HUANG Yi. Analysis on the Effect of Top Management Team Heterogeneity on the Input of Technology Innovation in Enterprises - An Empirical Analysis Based on Fujian Province and Shandong Province. Journal of Hunan University of Commerce, 2015 (6): 51-57 .
21. BAI Jing-kun, LI Sha-sha. Research on Heterogeneity, Strategic Decision Quality and Innovation Performance of Top Management Teams. Journal of Northeast University of Finance and Economics, 2015 (5): 3-9.
22. XIE Feng-hua, YAO Xian-guo, GU Jia-jun.Empirical Research on the Relationship between Top Management Team Heterogeneity and Firm's Technological Innovation Performance.journal of Scientific Management, 2008,29 (6): 65-73.
23. Putnam M. Bridging network divides: building capacity to support aging with disability populations through research .Disability & Health Journal, 2014,7 (7): S51-S59.

24. Zhang Ying. The Impact of Social Capital on Corporate Debt Maturity Structure: Evidence from Listed Companies in China. Journal of Zhejiang Finance and Economics University, 2016 (5): 53-61.
25. Cheng Cong, Xie Hongming, Chen Ying, et al.Study on the Relationship among Network, Internal and External Social Capital and Technological Innovation. Scientific Research Management, 2013 (11): 1-8.
26. Robert A. Ape, o Alessandra Fag, Ian. Cullective learning and relational capital in local innovation processes .regional studies, 2005, 39 (1): 75-87.
27. LIU Yun-guo, LIU Wen.The Term of Office and R & D Expenditure of Chinese Listed Companies. Management World, 2007 (1): 128-136.
28. Wang Xiaolu, Yu Jingwen, Fan Gang.Analysis of the Indexes of Business Environment in China by Province in 2013 (abstract). Journal of National School of Administration, 2013 (4): 24-34.
29. HOU Wang-bin, ZHANG Yu-li, YANG Jun.Homogeneity or Heterogeneity: The Role of Entrepreneurial Orientation in Regulating the Relationship between Technology Entrepreneurial Teams and New Firm Performance. Management World, 2014 (6): 92-109.

Risk Assessment of Engineering Project Financing Based on PPP Model

Qiuli Ma and Mingbo Sun

School of Management Engineering, Zhengzhou University, Zhengzhou 450001, China

Abstract. At present, the project financing channel is single, and the urban facilities are in short supply, and the risk assessment and prevention mechanism of financing should be further improved to reduce the risk of project financing. In view of this, the fuzzy comprehensive evaluation model of project financing risk which combined the method of fuzzy comprehensive evaluation and analytic hierarchy process is established. The scientificalness and effectiveness of the model are verified by the example of the world port project in Luohe city, and it provides basis and reference for engineering project financing based on PPP mode.

1 Research background

The problems faced by the development of city infrastructure in our country: The amount of investment of the city infrastructure construction is not sufficient. The capital is lack, and the investment unit is too single, and the financing method is too little in the urban infrastructure projects. Meanwhile, the city project investment and financing mechanism is not perfect, and the system is not complete. Furthermore, after the completion of the urban infrastructure projects, the lack of effective asset management mechanism, there is no proper operating institutions, resulting in a large number of funds invested did not make an inventory, and the construction of urban infrastructure has been far from meeting the great demand of the public.

From the above problems, the implementation of urban infrastructure projects in the process there is still a big problem. Government investment advantages are: quick, relatively large. Investment disadvantage is: the project economic efficiency is poor, the cost is more, and the loss is more serious. Now the problem is that the private sector has accumulated a large amount of money, and the government sector funds are obviously short, so a public-private partnership (PPP) came into being. PPP clever use of the advantages of the government departments and private enterprises respectively, and established the relationship of mutual cooperation, it introduced high efficiency and strong support of government departments and the management of advanced high technology, capital of private enterprises on the can provide policy and strong communication abilities, both sides reached the project than alone the obtained benefits.

In order to provide a reference for the risk identification of highway project under PPP financing mode, Cao Cong [1] constructed the risk identification system through two dimensions of the five steps, and emphasized the identification of key risk factors. Liu Yanwu [2] analyzed the causes of the financial risk of the PPP project, and found out the way to solve the risk. The difficulty and risk of Beijing Metro Line 4 risk sharing share basic program was analyzed according to the manifestation of the risks in each stage. Obtaining the results of risk sharing and quantitative analysis, and it provided theoretical support and practical guidance for the future of China's rail transit construction project PPP mode decision scientific and reasonable risk sharing [3]. Ai Bing and Lu Xiaocheng [4] analyzed the main PPP project risk, policy risk, exchange rate risk, technology risk, financial risk, operational risk, risk prevention measures of government project procurement PPP financing mode is mainly the government to speed up the change of role, strengthen the qualification certification and optimize the risk structure. Kong Qingli [5] studied the characteristics of PPP model, the particularity of risk allocation and the risk management methods of urban infrastructure projects, which had reference value for the promotion of PPP projects. Wu Aman [6] carried out a detailed analysis on the financing risk of the project through its structure, related elements, forms and typical case. It summarized the measures of financing risk prevention two and beneficial help to avoid the risk of project financing.

In this paper, the fuzzy comprehensive evaluation model of project financing risk which combined the method of fuzzy comprehensive evaluation and analytic hierarchy process is established. The risk of the project financing is assessed by the example of the world port project in Luohe city, and it provides basis and reference for engineering project financing based on PPP mode.

2 Construction of risk assessment model of PPP model project financing

(1) the weight of each risk factor is calculated:

The weight of each index can be calculated by the following formula:

$$\omega_i = \frac{\sqrt[n]{\prod_{j=1}^{n} a_{ij}}}{\sum_{k=1}^{n}\sqrt[n]{\prod_{j=1}^{n} a_{ij}}} \qquad (i=1,2,3,\ldots,n) \tag{1}$$

ω is normalized:

$$\omega = \frac{\omega_i}{\sum_{i=1}^{n}\omega_i} \tag{2}$$

The first-level indexes B1 ~ B8 can be calculated by this formula which relative to the total weight of the project target

A weight vector: ω =(ω_1 ω_2 .... ω_n)^T, similarly the second indicators relative to a target weight value can be got, In this way, the risk factors and the overall objective of the two level indicators are obtained.:

$$W_{1(3)} = \sum_{j=1}^{8} p_{ij}^{(3)}\omega_j^{(2)} \tag{3}$$

(2) The consistency of the judgment matrix is tested

Consistency check need to do when there is only one principle in an engineering project. The proportion of the lump sum ratio according to the following formula is calculated: CR=CI/RI. When the index value CR<0.1, it can be considered that the judgment matrix of the engineering project is consistent with the consistency test. Where, CI= (λ_{max} -1) / (n-1) that is the consistency of the indicators, and λ_{max} is the so-called matrix to determine the greatest characteristics of the root, after the solution of the value calculated by the following formula can be calculated:

$$\lambda_{max} = \frac{1}{n}\sum_{j=1}^{n}(AW)_i / \omega_i = \frac{1}{n}\sum_{i=1}^{n}(\sum_{j=1}^{n} a_{ij}\omega_j) / \omega_i \quad (4)$$

The $(A\omega)_i$ index value of the above formula can be expressed as the i component index of $A\omega$. The CI and RI are the average consistency index. In the following table 1, the random mean consistency index of the positive and negative matrix of the 1 to 15 order is given.

Table 1. Average random consistency index.

Order number n	1	2	3	4	5	6	7	8
RI	0.00	0.00	0.52	0.89	1.12	1.26	1.36	1.41
Order number n	9	10	11	12	13	14	15	
RI	1.46	1.49	1.52	1.54	1.58	1.58	1.59	

Using the fuzzy comprehensive evaluation method to evaluate the project

①At first, the sets of factors, weights and evaluation are set up.

The 26 categories of 8 small class as a factor set of projects based on the PPP mode of project financing risks are selected: U={U1, U2... , U26}. According to the importance of project financing risk, the weight of each risk is calculated, and the risk of the project can be divided into three levels, namely the evaluation set V: "high risk, medium risk and high risk".

②The fuzzy comprehensive evaluation matrix of the individual factors is established in the project

The the degree of membership that the i in the last step is relative to the j in the assessment set V= "risk is greater, the risk of medium, small risk" is r_{ij} . After r_{ij} is determined, the value can be given by expert scoring method according to the degree of membership of the factors. When the degree of membership is greater, the value is closer to 1, whereas the more close to 0.

③Fuzzy comprehensive evaluation matrix of multi factors in the project is built

After the normalization of B, the fuzzy comprehensive evaluation set B=$(B_1\ B_2\ \ B_3)$ is obtained. And set B elements in the fuzzy evaluation have been quantified, but at this time elements in the evaluation set V are not quantified. The financing risk assessment under the PPP mode is still unable to be got a quantitative measure. At this point, the V begins to be quantified, and quantitative values are found in table 2:

Table 2. Standard table of quantitative risk.

Grade	Greater risk	Medium risk	Less risk
assignment	100--86	85-66	<65

The M value is made as the risk assessment results. According to the resulting matrix, M=$\lfloor B,C^T \rfloor$, which C1=85, C2=75, C3=65, and finally the size of the risk is determined according to the M value.

3 Empirical analysis

Luohe Huading world port project is a city greening leisure theme park with the entertainment function responsed by the government, and the budget is 800 million yuan. In order to solve the government pressure, the project investment of about 500 million yuan was invited public bidding by PPP financing mode. According to the project risk identification, the risk is divided into 8 first-level indexes and 26 second-level indexes.

3.1 Determination of weight of AHP

According to the evaluation index system in construction of PPP mode, the target layer is set to the financing risk of project PPP financing mode. The first layer is the 8 first-level indexes with political, financial, market, force majeure risks, and second layer is the 26 second-level indexes with policy and legal changes, price risk.

In this paper, the data survey mainly through the expert investigation and questionnaire survey method to determine the weight of the factors, the results of the scoring table 3:

Table 3. Elements of the scoring table

First-level indexes	Fraction	Second-level indexes	Fraction
Political risk B_1	6	Policy risk C1	5
		Law change risk C2	6
Market risk B_2	9	Price risk C3	8
		Demand risk C4	7
		Competition risk C5	7
Financial risk B_3	6	Inflation risk C6	4
		Interest rate risk C7	4
Non class risk B4	4	Weather risk C8	3
		Earthquake and other risks C9	2
Technology risk B5	7	Design technical risk C10	5
		Construction technology risk C11	6
		Other reasons for technical risk C12	4
Construction risk B6	9	Land relocation and compensation risk C13	4
		Timely completion risk C14	8
		Quality risk C15	8
		Cost expenditure risk C16	7
		Safety accident risk C17	6
		Private investment default C18	5
		Contract unit breach of contract C19	6
Operational risk B7	7	Supply risk C20	5
		Equipment maintenance risk C21	4
		Operating unit management C22	5
		Project unit breach of contract C23	3
Environmental class riskB8	3	Pollution fines C24	4
		Environmental costs are too high C25	3
		To correct the error of investment is too large C26	3

After the hierarchical structure of the project is constructed, the relationship between the elements of the relationship is generally clear. Then the analytic hierarchy process is used to calculate the relative weight of each factor and combined with the level of various factors to conduct a comprehensive assessment, the judgment matrix is as follows:

$$A=\begin{pmatrix} 1 & 6/9 & 1 & 6/4 & 6/7 & 6/9 & 6/7 & 6/3 \\ 9/6 & 1 & 9/6 & 9/4 & 9/7 & 1 & 9/7 & 9/3 \\ 1 & 6/9 & 1 & 6/4 & 6/7 & 6/9 & 6/7 & 6/3 \\ 4/6 & 4/9 & 4/6 & 1 & 4/7 & 4/9 & 4/7 & 4/3 \\ 7/6 & 7/9 & 7/6 & 7/4 & 1 & 7/9 & 1 & 7/3 \\ 9/6 & 1 & 9/6 & 9/4 & 9/7 & 1 & 9/7 & 7/3 \\ 7/6 & 7/9 & 7/6 & 7/4 & 1 & 7/9 & 1 & 7/3 \\ 3/6 & 3/9 & 3/6 & 3/4 & 3/7 & 3/9 & 3/7 & 1 \end{pmatrix}$$

$$B_1=\begin{pmatrix} 1 & 5/6 \\ 6/5 & 1 \end{pmatrix}$$

$$B_2=\begin{pmatrix} 1 & 8/7 & 8/7 \\ 7/8 & 1 & 1 \\ 7/8 & 1 & 1 \end{pmatrix}$$

$$B_3=\begin{pmatrix} 1 & 1 \\ 1 & 1 \end{pmatrix} \quad B_4=\begin{pmatrix} 1 & 3/2 \\ 2/3 & 1 \end{pmatrix}$$

$$B_5=\begin{pmatrix} 1 & 5/6 & 5/4 \\ 6/5 & 1 & 6/4 \\ 4/5 & 4/6 & 1 \end{pmatrix}$$

$$B_6=\begin{pmatrix} 1 & 4/8 & 4/8 & 4/7 & 4/6 & 4/5 & 4/6 \\ 8/4 & 1 & 1 & 8/7 & 8/6 & 8/5 & 8/6 \\ 8/4 & 1 & 1 & 8/7 & 8/6 & 8/5 & 8/6 \\ 7/4 & 7/8 & 7/8 & 1 & 7/6 & 7/5 & 7/6 \\ 6/4 & 6/8 & 6/8 & 6/7 & 1 & 6/5 & 1 \\ 5/4 & 5/8 & 5/8 & 5/7 & 5/6 & 1 & 5/6 \\ 6/4 & 6/8 & 6/8 & 6/7 & 1 & 6/5 & 1 \end{pmatrix}$$

$$B_7=\begin{pmatrix} 1 & 5/4 & 1 & 5/3 \\ 4/5 & 1 & 4/5 & 4/3 \\ 1 & 5/4 & 1 & 5/3 \\ 3/5 & 3/4 & 3/5 & 1 \end{pmatrix} \quad B_8=\begin{pmatrix} 1 & 4/3 & 4/3 \\ 3/4 & 1 & 1 \\ 3/4 & 1 & 1 \end{pmatrix}$$

The weights are calculated by the formula mentioned in the former part in this paper, the formula (3) is used to calculate the maximum eigenvalue of the matrix $\lambda_{max}=8.5834$. And $CI=\frac{\lambda_{max}-n}{n-1}$, then $CI=\frac{8.5834-8}{8-7}=0.08332$.

According to Table 2.1, the average value of random consistency index can be checked RI=1.41.Because $CR=\frac{CI}{RI}=\frac{0.08332}{1.41}=0.059<0.10$, the matrix is fit to the consistency, and the weight of each factor of the first level index is: $\omega=(0.118\ 0.176\ 0.118\ 0.078\ 0.137\ 0.176\ 0.137\ 0.059)$; Similarly, the maximum characteristic value of the second-level indexes can be obtained: $\lambda_{max1}=8.0544$, $\lambda_{max2}=8.7895$, $\lambda_{max3}=8.0852$, $\lambda_{max4}=8.0771$,

$\lambda_{max5}=8.1250$, $\lambda_{max6}=8.0073$, $\lambda_{max7}=8.0981$, $\lambda_{max8}=8.0155$,CI means $CI_1=0.07551$, $CI_2=0.08731$, $CI_3=0.08013$, $CI_4=0.07892$, $CI_5=0.08455$, $CI_6=0.07954$, $CI_7=0.07642$, $CI_8=0.07883$. After the calculation can be obtained CR are less than 0.100, so all the judgment matrix to meet the consistency test. Second-level indexes of the weight of the elements are:

$\omega_1=(0.455\ 0.545)$ $\omega_2=(0.364\ 0.318\ 0.318)$ $\omega_3=(0.500\ 0.500)$ $\omega_4=(0.600\ 0.400)$ $\omega_5=(0.333\ 0.400\ 0.267)$
$\omega_6=(0.091\ 0.182\ 0.182\ 0.159\ 0.136\ 0.114\ 0.136)$ $\omega_7=(0.294\ 0.235\ 0.294\ 0.176)$ $\omega_8=(0.400\ 0.300\ 0.300)$

According to the above calculation, in the project risk of PPP financing model, the market risk and proportion of construction risk is the largest, followed by technical risk, financial risk, political risk. In the study, in the project financing risk process of PPP mode, we should focus on taking into account the large proportion of the risk, and the only way to effectively prevent the risk of the project.

3.2 Fuzzy comprehensive evaluation of project financing risk

The 26 risk factors mentioned in this paper is different from the impact of the whole project. According to the different factors, the corresponding weight is given, and the weights of 26 second-level indexes risk are got by the analytic hierarchy process:

A={0.455,0.545,0.364,0.318,0.318,0.500,0.500,0.600,0.400,0.333,0.400,0.267,0.091,0.182,0.182,0.159,0.136,0.114，0.136,0.294,0.235,0.294,0.176,0.400，0.300，0.300}.

And then through finding the project related experts, the expert scoring method is used for these risk factors scoring, scoring table as shown in table 4.

When the first-level indexes are got matrix, the basic formula $B=\omega\bullet R$ can be calculated by fuzzy comprehensive evaluation:

$$B_1=\omega_1\cdot R_1=(0.378\ \ 0.346\ \ 0.423)$$

$$B_2=\omega_2\cdot R_2=(0.343\ \ 0.319\ \ 0.338)\ B_3=\omega_3\bullet R_3=(0.242\ \ 0.199\ \ 0.514)$$

$$B_4=\omega_4\bullet R_4=(0.314\ \ 0.259\ \ 0.427)$$

$$B_5=\omega_5\bullet R_5=(0.399\ \ 0.410\ \ 0.190)$$

$$B_6=\omega_6\bullet R_6=(0.368\ \ 0.269\ \ 0.363)$$

$$B_7=\omega_7\bullet R_7=(0.249\ \ 0.284\ \ 0.467)$$

$$B_8=\omega_8\bullet R_8=(0.359\ \ 0.337\ \ 0.305)$$

Table 4. Effect of PPP mode on the project financing risk factors scoring table.

First-level indexes	weight	Second-level indexes	weight	Comment set		
				Greater risk	Medium risk	Less risk
Political risk B_1	0.118	Policy risk C1	0.455	0.322	0.375	0.303
		Law change risk C2	0.545	0.425	0.321	0.524
Market risk B_2	0.176	Price risk C3	0.364	0.442	0.239	0.319
		Demand risk C4	0.318	0.247	0.318	0.435

		Competition risk C5	0.318	0.325	0.412	0.263
Financial risk B_3	0.118	Inflation risk C6	0.500	0.157	0.213	0.630
		Interest rate risk C7	0.500	0.327	0.275	0.398
Non class risk B4	0.078	Weather risk C8	0.600	0.456	0.249	0.295
		Earthquake and other risks C9	0.400	0.102	0.273	0.625
Technology risk B5	0.137	Design technical risk C10	0.333	0.319	0.438	0.243
		Construction technology risk C11	0.400	0.468	0.373	0.159
		Other reasons for technical risk C12	0.267	0.397	0.432	0.171
Construction risk B6	0.176	Land relocation and compensation risk C13	0.091	0.631	0.233	0.136
		Timely completion risk C14	0.182	0.732	0.138	0.130
		Quality risk C15	0.182	0.375	0.239	0.386
		Cost expenditure risk C16	0.159	0.237	0.368	0.394
		Safety accident risk C17	0.136	0.118	0.299	0.583
		Private investment default C18	0.114	0.216	0.273	0.511
		Contract unit breach of contract C19	0.136	0.227	0.361	0.412
Operational risk B7	0.137	Supply risk C20	0.294	0.312	0.218	0.470
		Equipment maintenance risk C21	0.235	0.421	0.312	0.267
		Operating unit management C22	0.294	0.128	0.275	0.597
		Project unit breach of contract C23	0.176	0.116	0.372	0.512
Environmental class risk B8	0.059	Pollution fines C24	0.400	0.413	0.279	0.308
		Environmental costs are too high C25	0.300	0.366	0.432	0.202
		To correct the error of investment is too large C26	0.300	0.279	0.318	0.403

That is: R= $(B_1\ \ B_2\ \ B_3\ \ B_4\ \ B_5\ \ B_6\ \ B_7\ \ B_8)^T$. It is calculated that: $B = W \bullet R = (0.401\ \ 0.362\ \ 0.375)$

According to the theoretical part of this paper, the M value is the result of risk assessment, and then:

$$M=(B,C^T)=(0.401\ \ 0.362\ \ 0.375)(85\ \ 75\ \ 65)^T=78.24$$

Finally, the size of the risk is 78.24 according to the M value, so the risk of the project is medium.

4 Conclusions

(1) For the assessment and prevention of risk financing project under PPP mode, firsty, the risks of project construction in PPP model is identified. Secondly, the PPP mode of project financing risk index system is established, and the AHP method is used to confirm the weight of each risk factor and clear the proportion, which is conducive to the participation in the project the main risk preventive measures in a timely manner.

(2) The scoring results were calculated and analyzed by using the fuzzy comprehensive evaluation method of expert survey. The risk factors of PPP project financing mode is divided into 8 categories and 26 items, so as to the comprehensive weight of project financing risk factors under the PPP model is obtained. In order to more accurately express the data of the fuzzy parameters of the fuzzy comprehensive evaluation, which can make quantification persuasive, and using fuzzy comprehensive evaluation method has rigorous mathematical logic, scientific and practical.

(3) In this paper, the above comprehensive research method is applied to the Luohe PPP project financing risk management, and the PPP mode of our country under the application of the general engineering project has played an important guiding significance.

References

1. C. CAO. A Study on Identification of the Risk Factors of Highway Projects with PPP (Public-Private Partneiship) Financing Mode.Jouranl of Chongqing Three Gorges University, 3: 75-78(2013).
2. Y.W. LIU. Management Research of the Financial Risks of Project Financing Based on PPP Model. Value Engineering, 2015(12):16-18.
3. Y.L. YIN, X.L. YIN. Study on Risk Share in PPP Construction Mode of Beijing Metro Line 4. Railway Transport and Economy, 10:6-11(2013).
4. B. AI, X.C. LU..Model comparison and risk prevention of PPP financing of government project procurement. Seeker, 4:19-21(2008).
5. Q.L. KONG. Risk Management of Urban Infrastructure Construction Based on PPP Model. Construction Engineering Technology and Design,10(2015).
6. A. M. WU. Risk Analysis and Prevention of Engineering Project Financing. Architectural knowledge: Academic Journal, 3:249-249 (2013).

Weight Calculation for Cases Generated by Tacit Knowledge Explicit Based on RS-FAHP

Yue Cao [1,*], Zengmao Guo [1,2] and Jianhua Zhang [1]

[1] Management Engineering Department, Zhengzhou University, Zhengzhou, China
[2] Henan Electronic Commerce Association, Zhengzhou, China
Corresponding Email: 1217200588@qq.com

Abstract. In the knowledge economy, it becomes the core competence of persons, groups and organizations to effectively organize and manage tacit knowledge, affecting their sustainable development. Case explicit for tacit knowledge is an effective way to improve their clarity, improve management efficiency. it determines the validity of the case view to calculate legitimately the weights for the case aspects or attributes, and further affect the application benefit of the explicit knowledge. The case view affected seriously by the subjective, obtaining via traditional direct weighting method, and the objectivity of the result is not strong. On the other hand, the objective weights configuration is not only ignored the expert knowledge, but also lead to the acceptance barriers for the body of knowledge to accept the result. Therefore, in this paper, relying on rough set (RS) theory, the integrating algorithm of two objective weight configuration is analyzed Systematically, based on conditional entropy and property dependence. Simultaneously, Fuzzy Analytic Hierarchy Process (AHP) is studied to take into account the operational experience and knowledge of experts in the field. And then, case attribute RS-FAHP comprehensive weight placement algorithms is designed, based on the integration of subjective and objective thinking. The work mentioned above can improve and perfect the traditional configuration of weights, and support to apply and manage the tacit knowledge explicit cases effectively.

1 Introduction

In the era of knowledge economy, knowledge evolved into the core of production, become the main source of value creation [1, 2].In the main body of knowledge structure, tacit knowledge not only accounts for most of it, but also directly determines the realization of knowledge application and innovation; However, due to its low level of clarity, the organization and management efficiency of tacit knowledge in the domain of practice is relatively low. This restricts the implementation of knowledge management (Knowledge Management, KM) performance. Artificial intelligence based on case-based Reasoning (Case-based Reasoning, CBR) system is a case of tacit knowledge explicit management of self-learning technique, it is based on historical problems and its solutions to assist users to make decisions or solve existing similar problems. So here, the author through the CBR technology implementation of KM tacit knowledge explicit organization and management, to improve efficiency and effectiveness of KM implementation.

In the traditional "4R" cycle of CBR model (Retrieve-Reuse-Revise-Retain), the retrieve is a core activity, its speed and accuracy directly decides the actual level of CBR system. Case retrieval is based on matching calculation, which requires a reasonable and efficient case-based weight vector computing mechanism. There are many traditional weight calculation method, such as direct weighting method, analytic hierarchy process (AHP) and Delphi method, these are based on the experience of the main methods, the results are greatly influenced by subjective factors, objectivity is relatively weak, persuasion is not high. Rough Sets (RS) theory is applied to analyze and deal with incomplete and uncertain data sets, so as to discover the potential law (knowledge) [3].Weight allocation method based on RS theory, based on the existing data set, it is not subject to subjective factors and has strong objectivity. On the other hand, domain experts, experts in the experience is precious knowledge resources, it is a valuable to determine the case weight vector. Therefore, on the basis of the above-mentioned objective method, we introduce Fuzzy Analytic Hierarchy Process (FAHP), which is a qualitative and quantitative system analysis method, and further through subjective experience preference factor, seek to establish subject-object fusion mechanism of empowerment, to promote the organization, application and management efficiency of the case of tacit knowledge externalization.

2 Complete objective weight allocation based on RS theory

There are two ways to calculate the weight based on RS theory:A method based on attribute dependence and a method based on conditional entropy (attribute information).The former is to measure the importance of attribute by the influence of the classification of subsets in domain of discourse, the calculation process is intuitive and easy to understand, but the result is relatively rough; The latter measures the importance of attributes from the perspective of the amount of information contained in the condition attributes, and has a finer degree of differentiation, but the process is less intuitive relatively [4-6].

2.1 Based on the dependence of weights allocation

Different knowledge on the theory of domain classification ability is different, the weight difference leads to its important degree. Different knowledge on the domain of the classification ability is different, leading to its important degree difference namely weight difference.

The known knowledge expression system $KS=(U,A,V,f)$,among them, U is nonempty finite domains of knowledge expression system, A is attribute set of knowledge expression system, and $A=C\bigcup D$; f is assignment function, determines the attribute value of each object, namely $f(A\to V)$.Decision attribute (set) D completely depends on the attribute sets(C) that composed of multiple condition attributes($a_1,a_2,\ldots,a_n$);However, its dependence on each condition attribute is different, so it can calculate the importance of each condition attribute. Its basic rule is: According to the knowledge representation system, two-dimensional decision table is established in order to eliminate each condition attributes, calculate out before and after decision attribute (set) of the dependence of condition attribute set. The greater the dependency change, the more important the conditional attribute is relative to the decision attribute(set), the higher the weight; On the contrary, the lower.

Conversely, the solving steps are as follows:

(1) Collecting the original data of the evaluation object, discretizing the continuity data and establishing the two-dimensional decision table.

(2) Calculate D's C domain $pos_C(D)$, on the basis of the calculation D for the condition attribute set C's dependence:

$$\gamma_C(D)=\frac{|pos_C(D)|}{|U|}$$

(3) The new positive domain $pos_{C-a_i}(D)$ is calculated after the conditional attribute a_i is removed, on this basis, the dependence on the new conditional attribute set ($C-a_i$) is calculated:

$$\gamma_{C-a_i}(D)=\frac{|pos_{C-a_i}(D)|}{|U|}$$

(4) Calculate the importance of attribute a_i to D:

$\sigma(a_i)=\gamma_{C-a_i}(D)-\gamma_C(D)$, and then perform step (3) again, until the condition attribute is overwritten.

(5) Normalization for the whole $\sigma(a_i)$,namely, the weight of condition attributes a_i :

$$\omega_1(a_i)=\frac{\sigma(a_i)}{\sum_{i=1}^{n}\sigma(a_i)}$$

$$\omega_1(a_i)=\frac{\sigma(a_i)}{\sum_{i=1}^{n}\sigma(a_i)}$$

2.2 Weight assignment based on conditional entropy

Entropy is a measure of uncertainty or disorder state, the entropy can be used to judge the degree of discretization of some aspect. The greater the degree of discretization, the greater the effect on the evaluation of case view, ie, the higher the weight [7].Different condition attributes $a\in C$ for decision attribute (set) D provide different information (negative entropy) to decision attributes (sets).When a conditional attribute is added to the conditional attribute set, the amount of information provided by conditional attribute set to decision attribute (set) has changed, showing the reduction degree of uncertainty; The greater the amount of information provided, the greater the uncertainty reduction, indicating that the more important condition attributes.

The idea of weighting based on information view is as follows: according to the knowledge representation system, the two-dimensional decision table about evaluation object is established, and the conditional entropy of C and D is calculated [8].Then, the condition attributes are removed in turn and the conditional entropy is recalculated. Finally, according to the degree of change of conditional entropy, we obtain the importance degree of each condition attribute relative to decision attribute (set).

A knowledge representation system $KS=(U,A,V,f)$,among them $A=C\cup D$, $C=\{a_1,a_2,\ldots,a_n\}$ is conditional attribute set, D is decision attribute (set), the information weight calculation procedure is as follows:

(1) Collecting the original data of the evaluation objects, discretizing the continuity data and establishing the two-dimensional decision table.

(2) The conditional entropy of D with respect to C is calculated $H(D|C)$;

(3) Remove the conditional attribute a_i in turn, calculating the conditional entropy $H(D|C-a_i)$ of D with respect to $C-a_i$.

(4) Calculate the importance of condition attribute a_i to decision attribute D: $Sig(a_i,C,D)=H(D|C-a_i)-H(D|C)$,and perform step (3) again, until the condition attribute is overwritten.

(5) Normalization for the whole $Sig(a_i,C,D)$,namely, the weight of condition attributes a_i:

$$\omega_2(a_i)=\frac{Sig(a_i,C,D)}{\sum_{i=1}^{n}Sig(a_i,C,D)}$$

2.3 Information concept weight allocation

Through the foregoing two methods to calculate the weight of the objectivity of the results are good, but if the result will be the only two methods simply count up and normalized processing, the phenomenon of "large number eat decimals" may occur, not rigorous. Taking into account the characteristics of both, this paper introduces the coordination factor λ to take into account the advantages of two, and in practice according to the application of demand characteristics of flexible adjustment in order to calculate the benefits and understandable results to find a balance.

Thus, based on the above-mentioned idea and the results of the two methods, the conditional attribute of a_i of the complete RS weight is:

$$\omega_{RS}(a_i)=\lambda\omega_1(a_i)+(1-\lambda)\omega_2(a_i)$$

3 Based on FAHP of case of quantificating attribute subjective weight configuration

Analytic Hierarchy Process (AHP) is a combination of qualitative and quantitative system analysis method, is suitable for the complex structure, much decision criteria and decision problem is not easy to quantify. AHP quantitative characterization of the inference process, to avoid the problem of decision-making complex and more programs caused by the logic of decision-makers reasoning errors a certain extent [9-11];However, this method also have shortcomings as following:(1) It requires decision-makers to determine the value of the project scale, but in practice decision-makers subjective judgments are ambiguous, the determination of the value scale is difficult to eliminate due to the impact of subjective judgments [12, 13].(2) When the order of the AHP judgment matrix is high, it is difficult to check its consistency;When the consistency of the judgment matrix can't be met,need to repeat adjustment and testing, exhausting process, lower efficiency [14].

The feasible way to deal with the aforementioned shortcomings is:combining the AHP method with the fuzzy theory, it fully integrates the fuzziness of the decision-maker's thinking, makes the judgment matrix more reasonable and easier to pass the consistency test, that is fuzzy analytic hierarchy process（FAHP）[15, 16].It constructs a fuzzy judgment matrix based on fuzzy numbers (such as triangular fuzzy number, trapezoidal fuzzy number,type of L-R fuzzy number, etc.).The triangular fuzzy numbers represent the fuzzy thinking of the decision maker in the form of triples (l,m,u), which is more practical and the method is simple and easy to operate.Among them, l , m , u is the the lower bound, median and upper bound of the fuzzy number respectively.Under normal circumstances,meet $l<m<u$;when

$l=m=u$,the fuzzy number evolves to an exact value;Besides, $(u-l)/(l+m+u)$ the larger the value, the more fuzzy the fuzzy number.

Based on Fuzzy Theory,the fuzzy scale takes values within $[0,1]$,the rules of the values are shown in Table 1.

Table 1. Fuzzy scale and its meaning.

Scale value C_{ij}	meaning
0.1	Object j is more absolutely important than object i
0.2	Object j is more strongly important than object i
0.3	Object j is more obviously important than object i
0.4	Object j is slightly more important than object i
0.5	Object j and I are equally important
0.6	Object i is slightly more important than object j
0.7	Object i is more obviously important than object j
0.8	Object i is more strongly important than object j
0.9	Object i is more absolutely important than object j

Definition 1:If all elements meet $0 \le r_{ij} \le 1$ of square matrix $R=(r_{ij})_{n\times n}$,it is the fuzzy matrix.

Definition 2:If all elements meet $r_{ij}+r_{ji}=1$ of fuzzy matrix $R=(r_{ij})_{n\times n}$,it is complementary.

Definition 3:If all elements meet $r_{ij}=r_{ik}-r_{jk}+0.5$ of fuzzy complementary phalanx $R=(r_{ij})_{n\times n}$,it is consistent array.

Assuming that there are k experts to evaluate the importance of the n properties to be judged by comparing the two judgments, FAHP steps are as follows:

(1) Comparison of two pairs of properties to be evaluated,a fuzzy complementary judgment matrix $R=(r_{ij})_{n\times n}$ is established based on triangular fuzzy numbers.

$$R=\begin{bmatrix} (l_{11},m_{11},u_{11}) & (l_{12},m_{12},u_{12}) & \cdots & (l_{1n},m_{1n},u_{1n}) \\ (l_{21},m_{21},u_{21}) & (l_{22},m_{22},u_{11}) & \cdots & (l_{2n},m_{2n},u_{2n}) \\ \vdots & \vdots & \vdots & \vdots \\ (l_{n1},m_{n1},u_{n1}) & (l_{n2},m_{n2},u_{n2}) & \cdots & (l_{nn},m_{nn},u_{nn}) \end{bmatrix}$$

In $R=(r_{ij})_{n\times n}$, $l_{ii}=m_{ii}=u_{ii}=0.5$, $l_{ij}+u_{ji}=1$, $u_{ij}+l_{ji}=1$.

(2) On the basis of,calculating probability matrix $B=(b_{ij})_{n\times n}$ and fuzzy evaluation matrix $S=(s_{ij})_{n\times n}$ [14]:

$$b_{ij}=\frac{l_{ij}+4m_{ij}+u_{ij}}{6},$$

$$s_{ij}=1-\frac{u_{ij}-l_{ij}}{2(l_{ij}+m_{ij}+u_{ij})}.$$

Among them,the s_{ij} smaller the value of S , the larger the fuzzy interval, the fuzziness of the fuzzy number is stronger and the confidence level is lower.

(3) The aforementioned matrix is multiplied $t_{ij}=b_{ij}s_{ij}$,to adjust it to obtain complementary and fuzzy judgment matrix $R^{'}=(r_{ij}^{'})_{n\times n}$.Among them:

$$r_{ij}^{'}=\frac{1}{2}(1+t_{ij}-t_{ji})$$

(4)Adjust $R^{'}=(r_{ij}^{'})_{n\times n}$ for the consistency matrix $R^{''}=(r_{ij}^{''})_{n\times n}$,among them:

$$r_{ij}^{''}=\frac{r_i^{'}-r_j^{'}}{2(n-1)}+0.5\,,r_i^{'}=\sum_{a=1}^{n}r_{ia}^{'}\,.$$

(5)Weighting average fuzzy judgment matrix for each evaluation expert and to get the comprehensive matrix $\tilde{R}=(\tilde{r}_{ij})_{n\times n}$,

$$\tilde{r}_{ij}=\frac{1}{k}\sum_{p=1}^{k}r_{ij}^{''}(p)$$

Among them: $r_{ij}^{''}(p)$ represents the first evaluation expert's in $R^{''}$ corresponding evaluation elements.

(6)The final weight vector is obtained by the following equation $\omega_{FAHP}=(\omega_{FAHP}^{1},\ \omega_{FAHP}^{2},\cdots,\omega_{FAHP}^{n})$:

$$\omega_{FAHP}^{i}=\frac{\sum_{j=1}^{n}\tilde{r}_{ij}+\frac{n}{2}-1}{n(n-1)}$$

The above algorithm is based on the fuzzy consistent matrix expansion process, which avoids the process of consistency and continuation of the traditional AHP method ,to ensure the efficiency of solving.

4 Case attributes RS-FAHP comprehensive weight configuration

Based on the implicit knowledge explicit case model, the condition attribute set and decision attribute are established;Importing tacit knowledge in the explicit case database through discretizing of case data, and the knowledge expression system $KS=(U,A,V,f)$ is established.Thereinto, $A=C\bigcup D$, $C=\{a_1,a_2,\ldots,a_n\}$ is conditional attribute set, D is decision attribute.

Using ω_i ($i=1,2,\cdots,n$) to express RS-FAHP comprehensive weight,Obviously ω_{RS}^{i}, ω_{FAHP}^{i} should meet the following conditions:

$$\sum_{i=1}^{n}\omega_{RS}^{i}=\sum_{i=1}^{n}\omega_{FAHP}^{i}=\sum_{i=1}^{n}\omega_i=1,$$

$0\le\omega_{RS}^{i}\le1\,,0\le\omega_{FAHP}^{i}\le1\,,0\le\omega_i\le1\,.$

The fusion factor ε (meet $0\le\varepsilon\le1$) was introduced,establish the optimal model:

$$\min\{\sum_{i=1}^{n}[\varepsilon(\frac{1}{2}(\omega_i-\omega_{RS}^{i})^2)+(1-\varepsilon)(\frac{1}{2}(\omega_i-\omega_{FAHP}^{i})^2)]\}$$

To solve the above equation, Lagrange function:

$$L(\omega_i,\lambda)=\sum_{i=1}^{n}\left[\varepsilon(\frac{1}{2}(\omega_i-\omega_{RS}^{i})^2+(1-\varepsilon)(\frac{1}{2}(\omega_i-\omega_{FAHP}^{i})^2)\right]+\lambda(\sum_{i=1}^{n}\omega_i-1)$$

Find the partial derivatives of ω_i , λ respectively,and the other

$$\frac{\partial L}{\partial\omega_i}=0\ ,\frac{\partial L}{\partial\lambda}=0\ ,$$

$$\begin{cases}\varepsilon(\omega_i-\omega_{RS}^{i})+(1-\varepsilon)(\omega_i-\omega_{FAHP}^{i})=0\\ \sum_{i=1}^{n}\omega_i-1=0\end{cases}$$

Furthermore: $\omega_i=\varepsilon\omega_{RS}^{i}+(1-\varepsilon)\omega_{FAHP}^{i}$ .It can be seen that the fusion factor contains the subjective experience preference of the evaluation subject, the smaller the value is, the stronger the subjective preference is.

5 Example analysis

Assuming that an explicit case model with implicit knowledge:the condition section is composed of four aspects, and the decision section is composed of one aspect,respectively for $C=\{a_1,a_2,a_3,a_4\}$ and $D=\{d\}$.The case subspace in this mode is 10,after case data were imported and discretized, the knowledge expression system $KS=(U,A,V,f)$ was established,as shown in table 2.

Table 2. Knowledge representation system based on explicit case subspace of tacit knowledge.

U	a_1	a_2	a_3	a_4	d
x_1	1	1	0	0	1
x_2	1	0	0	0	1
x_3	0	0	0	0	0
x_4	1	1	0	1	0
x_5	1	2	0	1	2
x_6	2	2	0	2	2
x_7	2	2	2	2	2
x_8	1	2	4	1	3
x_9	2	2	4	2	3
x_{10}	2	1	0	0	3

In order to establish the case view of the case model, the calculation of the case is based on the RS-FAHP method. Firstly,weight algorithm based on degree of dependence $\omega_1(a_i)$ (among $i=1,2,3,4$),calculated as follows:

$$U/d=\{\{x_1,x_2\},\{x_3,x_4\},\{x_5,x_6,x_7\},\{x_8,x_9,x_{10}\}\}\ U/C=\{\{x_1\},\{x_2\},\{x_3\},\{x_4\},\{x_5\},\{x_6\},\{x_7\},\{x_8\},\{x_9\},\{x_{10}\}\}$$

$$U/\{C-a_1\}=\{\{x_1,x_{10}\},\{x_2,x_3\},\{x_4\},\{x_5\},\{x_6\},\{x_7\},\{x_8\},\{x_9\}\}\ U/\{C-a_2\}=\{\{x_1,x_2\},\{x_3\},\{x_4,x_5\},\{x_6\},\{x_7\},\{x_8\},\{x_9\},\{x_{10}\}\}$$

$$U/\{C-a_3\}=\{\{x_1\},\{x_2\},\{x_3\},\{x_4\},\{x_5,x_8\},\{x_6,x_7,x_9\},\{x_{10}\}\}\ U/\{C-a_4\}=\{\{x_1,x_4\},\{x_2\},\{x_3\},\{x_5\},\{x_6\},\{x_7\},\{x_8\},\{x_9\},\{x_{10}\}\}$$

The results are as follows:

$$\gamma_C(d)=1,\gamma_{C-a_1}(d)=0.7,$$
$$\gamma_{C-a_2}(d)=0.8,\gamma_{C-a_3}(d)=0.5,$$
$$\gamma_{C-a_4}(d)=0.8.$$

Thus, the importance of each attribute can be obtained as:

$$\sigma(a_1)=0.3,\sigma(a_2)=0.2,$$
$$\sigma(a_3)=0.5,\sigma(a_4)=0.2.$$

The above results are normalized, and the weight of each attribute based on dependency is as follows:

$$\omega_1(a_1)=0.2500,\omega_1(a_2)=0.1667,$$
$$\omega_1(a_3)=0.4166,\omega_1(a_4)=0.1667.$$

Secondly,Weight algorithm based on conditional entropy $\omega_2(a_i)$.

$$H(d|C)=0,\qquad H(d|C-a_1)=0.4,$$
$$H(d|C-a_2)=0.2,\qquad H(d|C-a_3)=0.4755,$$
$$H(d|C-a_4)=0.2$$

According to $Sig(a_i,C,D)$,further calculation of the importance of each attribute is:

$$Sig(a_1,C,D)=0.4,\quad Sig(a_2,C,D)=0.2,$$
$$Sig(a_3,C,D)=0.4755,Sig(a_4,C,D)=0.2$$

The above results are normalized, and the weight of each attribute based on conditional entropy is as follows:

$$\omega_2(a_1)=0.3136,\omega_2(a_2)=0.1568,$$
$$\omega_2(a_3)=0.3728,\omega_2(a_4)=0.1568.$$

Based on the above results, considering the high degree of differentiation based on conditional entropy , taking into coordination factor $\lambda=0.45$,according to 1.3 Section information concept weight allocation,the complete objective weight of each attribute based on RS theory is:

$$\omega_{RS}(a_1)=0.2850,\omega_{RS}(a_2)=0.1613,$$

$\omega_{RS}(a_3)=0.3924, \omega_{RS}(a_4)=0.1613$.

Then,based on the aforementioned FAHP algorithm to solve the subjective weight of each attribute of the case.In this case, the importance of each case attribute based on triangular fuzzy number is given by three experts ($k=3$)in the field,the result is as follows:

$$R_1=\begin{bmatrix}(0.5,0.5,0.5) & (0.5,0.6,0.7) & (0.3,0.4,0.5) & (0.5,0.6,0.7)\\(0.3,0.4,0.5) & (0.5,0.5,0.5) & (0.2,0.3,0.4) & (0.45,0.5,0.55)\\(0.5,0.6,0.7) & (0.6,0.7,0.8) & (0.5,0.5,0.5) & (0.65,0.7,0.75)\\(0.3,0.4,0.5) & (0.45,0.5,0.55) & (0.25,0.3,0.35) & (0.5,0.5,0.5)\end{bmatrix}$$

$$R_2=\begin{bmatrix}(0.5,0.5,0.5) & (0.55,0.6,0.65) & (0.35,0.4,0.45) & (0.5,0.6,0.7)\\(0.35,0.4,0.45) & (0.5,0.5,0.5) & (0.25,0.3,0.35) & (0.45,0.5,0.55)\\(0.55,0.6,0.65) & (0.65,0.7,0.75) & (0.5,0.5,0.5) & (0.65,0.7,0.75)\\(0.3,0.4,0.5) & (0.45,0.5,0.55) & (0.25,0.3,0.35) & (0.5,0.5,0.5)\end{bmatrix}$$

$$R_3=\begin{bmatrix}(0.5,0.5,0.5) & (0.65,0.7,0.75) & (0.35,0.4,0.45) & (0.5,0.6,0.7)\\(0.25,0.3,0.35) & (0.5,0.5,0.5) & (0.25,0.3,0.35) & (0.4,0.5,0.6)\\(0.55,0.6,0.65) & (0.65,0.7,0.75) & (0.5,0.5,0.5) & (0.65,0.7,0.75)\\(0.3,0.4,0.5) & (0.4,0.5,0.6) & (0.25,0.3,0.35) & (0.5,0.5,0.5)\end{bmatrix}$$

On the basis of this, the probability matrix and the fuzzy judgment matrix are calculated as follows:

$$B_1=B_2=\begin{bmatrix}0.5 & 0.6 & 0.4 & 0.6\\0.4 & 0.5 & 0.3 & 0.5\\0.6 & 0.7 & 0.5 & 0.7\\0.4 & 0.5 & 0.3 & 0.5\end{bmatrix}$$

$$B_3=\begin{bmatrix}0.5 & 0.7 & 0.4 & 0.6\\0.3 & 0.5 & 0.3 & 0.5\\0.6 & 0.7 & 0.5 & 0.7\\0.4 & 0.5 & 0.3 & 0.5\end{bmatrix}$$

$$S_1=\begin{bmatrix}1 & 0.944 & 0.917 & 0.944\\0.917 & 1 & 0.889 & 0.967\\0.944 & 0.952 & 1 & 0.976\\0.917 & 0.967 & 0.944 & 1\end{bmatrix}$$

$$S_2=\begin{bmatrix}1 & 0.972 & 0.958 & 0.944\\0.958 & 1 & 0.944 & 0.967\\0.972 & 0.976 & 1 & 0.976\\0.917 & 0.967 & 0.944 & 1\end{bmatrix}$$

$$S_3=\begin{bmatrix}1 & 0.976 & 0.958 & 0.944\\0.944 & 1 & 0.944 & 0.933\\0.972 & 0.976 & 1 & 0.976\\0.917 & 0.933 & 0.944 & 1\end{bmatrix}$$

The corresponding probability matrix is multiplied by the fuzzy evaluation matrix, and the computation of the complementary fuzzy judgment matrix $R^{'}=(r_{ij}^{'})_{n\times n}$ is as follows:

$$R_1^{'}=\begin{bmatrix}0.5 & 0.5998 & 0.4002 & 0.5998\\0.4002 & 0.5 & 0.3002 & 0.5\\0.5998 & 0.6998 & 0.5 & 0.7\\0.4002 & 0.5 & 0.3 & 0.5\end{bmatrix}$$

$$R_2^{'}=\begin{bmatrix} 0.5 & 0.6 & 0.4 & 0.5998 \\ 0.4 & 0.5 & 0.3 & 0.5 \\ 0.6 & 0.7 & 0.5 & 0.7 \\ 0.4002 & 0.5 & 0.3 & 0.5 \end{bmatrix}$$

$$R_3^{'}=\begin{bmatrix} 0.5 & 0.7 & 0.4 & 0.5998 \\ 0.3 & 0.5 & 0.3 & 0.5 \\ 0.6 & 0.7 & 0.5 & 0.7 \\ 0.4002 & 0.5 & 0.3 & 0.5 \end{bmatrix}$$

Further, the fuzzy complementary matrix $R^{'}=(r_{ij}^{'})_{n\times n}$ is adjusted to be fuzzy consistency matrix $R^{''}=(r_{ij}^{''})_{n\times n}$:

$$R_1^{''}=\begin{bmatrix} 0.5 & 0.4334 & 0.5666 & 0.4334 \\ 0.5666 & 0.5 & 0.6332 & 0.5 \\ 0.4334 & 0.3668 & 0.5 & 0.3668 \\ 0.5666 & 0.5 & 0.6332 & 0.5 \end{bmatrix}$$

$$R_2^{''}=\begin{bmatrix} 0.5 & 0.4334 & 0.5667 & 0.4334 \\ 0.5666 & 0.5 & 0.6333 & 0.5 \\ 0.4333 & 0.3667 & 0.5 & 0.3667 \\ 0.5666 & 0.5 & 0.6333 & 0.5 \end{bmatrix}$$

$$R_3^{''}=\begin{bmatrix} 0.5 & 0.4067 & 0.5567 & 0.4234 \\ 0.5933 & 0.5 & 0.65 & 0.5167 \\ 0.4433 & 0.35 & 0.5 & 0.3667 \\ 0.5766 & 0.4833 & 0.6333 & 0.5 \end{bmatrix}$$

On the basis of this, the matrix formula $\tilde{R}=(\tilde{r}_{ij})_{n\times n}$ is obtained based on the fuzzy consistent complementary judgment matrix:

$$\tilde{R}=\begin{bmatrix} 0.5 & 0.4235 & 0.5666 & 0.4334 \\ 0.5755 & 0.5 & 0.6388 & 0.5056 \\ 0.4367 & 0.3612 & 0.5 & 0.3667 \\ 0.5699 & 0.4944 & 0.6333 & 0.5 \end{bmatrix}$$

At this point, through the calculation method of the final weight vector $\omega=(\omega_1,\ \omega_2,\cdots,\omega_n)$, the weight vector of the algorithm based on FAHP algorithm is:

$$\omega=(0.2751,0.1859,0.3524,0,1866)$$

Finally,in the "outstanding objectivity, taking into account the subjective experience" principle, take the fusion factor $\varepsilon=0.65$,according to the algorithm shown in Section 3,the results are as follows:

$$\omega=(0.2751,0.1859,0.3524,0.1866).$$

6 Conclusion

In the era of knowledge economy, knowledge management, especially for the effective management of tacit knowledge is to enhance the core competence of the organization, to achieve an important foundation for sustainable development.In this paper, the idea of CBR system is introduced, which is based on the external of tacit knowledge, the RS system is based on the two kinds of weight calculation method, which is based on the attribute dependence and conditional entropy,at the same time, by taking into account the valuable experience of experts in the field of FAHP, the RS-FAHP weight configuration of the subjective and objective fusion of the explicit case view of tacit knowledge is realized.Compared with the traditional subjective experience empowerment method, this method is more rigorous data base and reasoning process, strong objectivity, good interpretability; compared with pure objective method, this method is to fully mix the subjective experience of experts in the field, not only realize the flexible complement to objectively, but also improve the acceptance level of empowerment results.

Acknowledgement

This work is supported by the Excellent Young Teacher Development Fund Project of Zhengzhou University (Foundation No. 2015SKYQ15).

References

1. Shah Bano, John Taylor. Universities and the knowledge-based economy: perceptions from a developing country. Higher Education Research & Development,**34,**242-255(2015).
2. Ahrweiler Petra1, Pyka Andreas, Gilbert Nigel. A new model for university-industry links in knowledge-based economies. Journal of Product Innovation Management, **8**,218-235(2011).
3. Zdzisław Pawlak. Rough Sets.International Journal of Computer and Information Sciences, **11**,341- 356(1982).
4. Zhang Jianhua. Multi-Case Derivational Adaptation Mechanism of KML-Case for Knowledge Management.JournalofSystems& Management, **23**,845-853(2014).
5. Zhu canhong,Chen nenghua.Improvement of Method of Determining Entropy Weight of Conditional Information in Rough Set.Statistics & Decision,**8**,154-156(2011).
6. Bao XinZhong,Liu cheng.A New Method of Ascertaining Attribute Weight Based on Rough Sets Theory.Chinese Journal of Management,**6**,729-736(2009).
7. Zhang Jianhua.*Electronic commerce knowledge management*. Beijing: Chinese Science Publishing,(2010).
8. Zhai JunChuang,Che WeiWei.A Spam Filtering Algorithm Based on Conditional Entropy. JiSuanJi Yu XianDaiHua,**2**,129-132(2014).
9. Li ChunHao,Sun YongHe,Jia YanHui.Analytic Hierarchy Process Based on Variable Weights.Systems Engineering-Theory &Practice, **30**,723-731(2010).
10. Liu Jun,Li XiaoNan.Handover Algorithm for WLAN/Cellular Networks with Analytic Hierarchy Process.Journal on Communications,**34**,65-72(2013).
11. Cai Yan,Wu Min,Zhou JinNi.AHP-based Fuzzy Multiple Criteria Optimization Methods for Storage Location Choice.Journal of Hunan University(Natural Sciences),**40**,103-108(2013).
12. Thomas L. Saaty. Modeling unstructured decision problems-the theory of analytical hierarchies. Mathematics and Computers in Simulation,**47**,147-158(1978).
13. Wang HuaJi,Zong HuaFu,Guan Xin.Method of Determining Weights Subjective Evaluation Indexes for Car Handing and stability Based on Fuzzy Analytic Hierarchy process.Journal of Mechanical Engineering,**47**,83-90(2011).
14. Cheng Kai.An Evaluation Study on Application Condition of Residential PCa Technology on Fuzzy Analytic Hierarchy Process.HangZhou:ZhengJang University,(2012).
15. Lei Lei.Evaluation of Critical Success Factors of Industry Based on FAHP and Fuzzy Delphi Method.Statistics and Decision Making,15,44-47(2014).
16. Tan BaiYing.The Research of Scale System and Ranking Method of FAHP.NanNing:GuangXi University,(2005).

Study on the Coupling Development between Urbanization and Ecosystem-- The Comparative Analysis Based on Guizhou, Yunnan, Hunan and Zhejiang Province

Lingyun Luo[1,*],Chenggang Li[1],Jue Hu[2],Bing Yang[1] and Kang Pan[1]

[1] Faculty of Finance, Guizhou University of Finance and Economics, Guiyang, China
[2] College of Textiles, Donghua University, Shanghai, China
Corresponding Email: lly0089@163.com

Abstract. Urbanization and ecological environment coordinated development is the development direction of contemporary urbanization. In this paper, the coupling degree model is used to analyze the coupling degree and coupling coordination degree of Guizhou, Yunnan, Hunan and Zhejiang Province in 2014. The analysis results showed that, four provinces are highly coupled in the coupling degree. Guizhou province is in the low degree of disorder and the lag of urbanization for the coupling coordination degree; Yunnan province is also in a low degree of disorder and urbanization hysteresis; Zhejiang province is in a good coordination, while its urbanization is slightly lagging behind; Hunan province is in a low degree of imbalance, and there is a coordinated development of urbanization and ecosystem. In this paper, the research results help four provinces above to better realize the coordinated development of urbanization and ecological environment.

1 Introduction

Urbanization refers to the historical transformation process from the primary industry-based agricultural society to the second and third industry-oriented modern urbanization of one country or region during the development of economy. Nowadays China is vigorously promoting the development of urbanization. As domestic demand is the fundamental driving force of China's economic development, and urbanization helps in expanding the domestic demand. Therefore, urbanization can be a strong engine to keep the sustained and healthy development of the economy. At present, China's urbanization rate of resident population is 53.7%, and the urbanization rate of household registration population is only about 36%, which is 80% lower than that of the developed countries. Guizhou Province as a western province, the urbanization rate is far behind the national average urbanization value. The urbanization rate in western China is 44.8%, while the urbanization rate in Guizhou is only 42%. In addition,the distribution of urbanization is unbalanced. As a result of this phenomenon, the urbanization potential of Guizhou is quite huge. So urbanization is regarded as the main strategy for Guizhou's development [1, 2]. From the past experience, the development of urbanization will inevitably bring about the destruction of ecological environment. Over the past few years, China has only focused on the develop of economy, while do not pay any attention to the protection of the ecological environment. Thus it leads to enormous environmental pollution. In recent years, our country decided to take the road of sustainable development, which means, when we focus on the development of economy, we should pay attention to the protection of the ecological environment at the same time. In this paper, the coupling model is used to analyze the relationship between urbanization and ecological environment in order to promote the coordinated development of urbanization and ecological environment.

2 Literature review

Domestic and foreign scholars have carried out a series of researches on the coordinated development of urbanization and ecological environment. By the panel regression empirical study, Grossman (1995)found that the relationship between the ecological environment and the economic development was inverted U-shaped. In the early stage, because of the economic development, people ignored the protection of the environment. After a certain stage, people began to pay attention to environmental protection and governance, and then put forward the famous environmental Kuznets Curve [3]. Haughton G et al. (1994) considered that the coupling relationship between urbanization and ecological environment is the sum of all kinds of non-linear relations between the elements of urbanization and the elements of ecological environment [4]. Domestic experts and scholars also carried on a certain research on the coupling model and the urbanization.Yan-ling Xiao (2013) used coupling model to calculate the coupling degree of urbanization and ecological environment coordinated development in Guizhou Province from 2001 to 2010, and obtained the conclusion that the coupling degree of urbanization and ecological environment coordinated development in Guizhou is increasing, In 2010, the coupling degree has reached 0.9545 [5]. Neng-sheng Luo et al. (2014) calculated the coupling degree of urbanization and environmental coordinated development in three cities of Changsha, Zhuzhou and Xiangtan. It was concluded that the coupling degree of Changsha was higher, from 0.293 in 2001 to 0.758 in 2011. However, Zhuzhou and Xiangtan City's coupling degree is low, they destroyed the ecological environment when they were developing [6]. Pei-chao Dai et al. (2004) calculated the coupling degree of Xuzhou with the model of urbanization and ecological environment coupling. It was concluded that Xuzhou entered a state of high coupling from 2001 onwards, and urbanization and ecological environment coordinated development [7]. Bo Li (2015) researched coupled spatiotemporal differences of 29 prefecture level cities of 5 central provinces by using a coupled model of the city, and concluded that the level of urbanization and the ecological environment quality of the coupling development degree in space showed obvious local spatial autocorrelation clustering pattern. The overall trend shows that the in the degree of southeast is higher than that in the northwest [8].

Based on the above literatures, scholars at home and abroad use the coupled model to study the coupling degree of urbanization and ecological environment in different places, but there is no comparative research between the coupling degree of Guizhou Province and other provinces. This paper selects the eastern province of Zhejiang, the central province of Hunan, Guizhou province and Yunnan Province located in the Yunnan Guizhou Plateau. The author collects the data of urbanization and ecological environment, and uses the coupling degree model to calculate the coupling degree of urbanization and ecological environment, and makes a comparative analysis of the four provinces.

3 Indicators selection and data sources

The data of this paper is divided into two systems, which are urbanization system and ecological environment system. Urbanization system includes population urbanization subsystem, spatial urbanization subsystem, Economic Urbanization subsystem and social urbanization subsystem. Among them, the population urbanization subsystem includes the proportion of urban population, the proportion of non-agricultural population and percentage of employment in the Tertiary Industry. System space towns including urban area (square kilometers), city population density (per km2), and per capita City Road area (square meters). The sub-system of urbanization includes the per capita disposable income of urban residents (yuan), the proportion of added value of the tertiary industry to GDP and the per capita revenue (yuan). The social urbanization sub-system includes the total retail sales per capita (ten thousand yuan), the number of buses owned by ten thousand people (vehicles), the proportion of the number of Internet users, and the proportion of college students. Ecological environment system includes ecological environment level subsystem, ecological environment pressure subsystem and ecological environment protection subsystem. Among them, the ecological environment subsystem includes water supply per capita (ton/person), city gas penetration rate, per capita green area (m2/person), built-up area green coverage. The pressure of ecological environment subsystem including industrial wastewater emissions per capita (ton/person), per capita amount of industrial solid wastes (ton/person). The ecological environment protection subsystem includes investment in per capita industrial pollution control (yuan), investment in per capita wastewater treatment project (yuan), and investment in per capita pollution control project (yuan).

The data used in this paper are taken from the National Bureau of statistics, as well as the statistical yearbook of the provinces in 2015.

4 Empirical analysis

4.1 Original data standardization

The original data has positive and negative differences. For example, ecological environment, the index of the ecological environment of pressure subsystems for the reverse index, the higher the index is indicate that the worse the environment is. The positive indicator is in accordance with the formula: $X=(x-x_{max})\div(x_{max}-x_{min})$. Inverse index is according to the formula: $X=(x_{max}-x)\div(x_{max}-x_{min})$. The raw data are processed according to the formula. x_{max} is the maximum value of the data in the index. x_{min} is the minimum value of the data. Urbanization indicators X is defined as X_i (i=1,2,3...13). Ecological environment indicators X is defined as X_j (j=1,2,3... .11).

4.2 Coefficient of variation and weight of each index

In this paper, we use the coefficient of variation method to determine the index weight. Variation coefficient method can be better to reflect the possession of weight among all the indicators, the formula is as follows:

$$V_i=\sigma_i/\overline{X_i} \qquad V_j=\sigma_j/\overline{X_j} \tag{1}$$

The weights of each index are obtained from the subsystem of urbanization and ecological environment, and the formula is as follows:

$$\alpha_i = V_i \div \sum_{i=1}^{13} V_i \quad (2)$$

Note: i i is for the urbanization of 13 subsystems V_i value.

$$\beta_j = V_j \div \sum_{j=0}^{11} V_j \quad (3)$$

Note: j is for the ecological environment of the 11 subsystem V_j value. α is the weight of each indicator of the urbanization subsystem, and β is the weight of each index of the eco-environment subsystem.

4.3 Coupling

Calculate the score of the urbanization system using the following formula.

$$F = \sum_{i=1}^{13} \alpha_i X_i , \sum_{i=1}^{13} \alpha_i = 1 \quad (4)$$

α is every index in the urbanization subsystem. X_i is the data in the process of Urbanization.
Calculation of ecological environment subsystem score using the following formula.

$$G = \sum_{j=1}^{11} \beta_j X_j \ \sum_{j=1}^{11} \beta_j = 1 \quad (5)$$

β is the weight of each index in the eco-environment subsystem. X_j is the value of the ecological index after treatment.

Calculate the urbanization subsystem score F and the eco-environment subsystem score G of the four provinces, substitute the coupling degree formula and couple the two systems. Couple the two systems and calculate the coupling degree.
Coupling degree formula:

$$C = 2 \times \frac{\sqrt{(F \times G)}}{F + G} \quad (6)$$

F value, G value and the coupling degree of the four provincial capitals are shown in Table 1.

Table 1. Coupling degree.

	Guizhou	Yunnan	Zhejiang	Hunan
F	0.1928235	0.1277376	0.802272	0.322798
G	0.2763830	0.2646786	0.850411	0.33653
coupling degree	0.9840148	0.9371344	0.999576	0.999783

From Table 1, we can see the coupling degree above four provinces are relatively high, which are belong to the highly coupling state . The coupling of Zhejiang and Hunan are close to 1. The coupling degree of Guizhou and Yunnan are slightly behind, which are higher than 0.9. The data indicate that the coupling degree between the two systems of the four provinces are in high state, and the urbanization subsystem and the ecological environment subsystem are in the mutual influence.

4.4 Coupling coordination degree

Coupling degree can only explain the interaction between the two systems, and the level of urbanization and the coordinated development of the ecological environment can not be explained. So the coupling coordination degree model is introduced to explain this question.

Formula of Coupling Coordination Degree is as the following:

$$D = \sqrt{C \times T}, T = \alpha F + \beta G \quad (7)$$

In the formula, α and β are the contribution coefficient for the two systems in the development of the city. Here we think that the urbanization subsystem and ecological environment subsystem of urban development is equally important. We set value $\alpha = \beta$=0.5. D is the coupling coordination degree. C is theCoupling degree. The coupling coordination degree is shown in Table 2

Table 2. Coupling coordination degree.

	Guizhou	Yunnan	Zhejiang	Hunan
Coupling coordination degree	0.480472	0.428805	0.908841	0.574102

The coupling coordination degree level and the division standard can be seen in Table 3

Table 3. The coupling coordination degree level and the division standard.

D	Degree of coordination	F>G	F<G
0.8 < D≤1	Good coordination	The ecological environment lag	Urbanization lag
0.6 < D≤0.8	Moderate coordination	The ecological environment lag	Urbanization lag
0.4 < D≤0.6	Low coordination	The ecological environment lag	Urbanization lag
0.2 < D≤0.4	Moderate disorder	The ecological environment lag	Urbanization lag
0 < D≤0.2	Severe disorder	The ecological environment lag	Urbanization lag

From Table 3 we can find out that Guizhou province is in a level of low imbalance and urbanization hysteresis. The governments of Guizhou province care both the development of economy and the environmental protection, but the coordination degree between the two systems is not high. The *F* value of urbanization subsystem is 0.1928, which is much less than that of Zhejiang (0.8023), and 0.3228 of Hunan Province. Therefore, there is still room for the improvement of the urbanization of Guizhou province. The Urbanization subsystem and ecological environment subsystem has a large gap in Guizhou province, which means Guizhou province not only focused on the development of urbanization but also paid attention to the protection of the ecological environment, and the government takes the road of sustainable development. Yunnan province is in the level of low coordination and urbanization hysteresis, and the urbanization level in Yunnan is much lower than the ecological environment subsystem. This can be due to its main industry (tourism industry). Yunnan Province has paid great attention to the protection of the environment, and because the development of tourism industry is quite good, the villages have some functions of cities and towns, thus limiting the urbanization of Yunnan Province to a certain extent. Zhejiang Province, as the eastern economic strong province of China, is in the level of high coordination and Zhejiang's urbanization index is far ahead of the other two western provinces and Hunan Province. And because the economy is well developed, the investment in ecological environment is also increased, therefore the ecological environment subsystem value in Zhejiang province is also in high level. Urbanization drives the development of economy, while economic development is also promoting the construction of ecological environment. Hunan is in the level of low imbalance, and its urbanization is almost keep pace with the ecological environment. As a province with a certain economic base in central China, Hunan Province has developed rapidly and has a relatively high urbanization system. Hunan Province also invest a certain amount for the protection of ecological environment. But there are multiple economic indicators are relatively left behind, and its per capita investment funds in environmental protection is less than that of the two western provinces, so there are a lot of room for the improvement of urbanization and ecological environment protection investment in Hunan province.

5 Conclusions and recommendations

In this paper, the coupling degree model is used to analyze the coupling development of urbanization and ecological environment in Guizhou Province, Yunnan Province, Hunan Province and Zhejiang province. The results of the comparative analysis show that the coupling degree of Guizhou province is highly coupled, which is as high as 0.9840. Among these four provinces, the coupling degree of Guizhou province occupies the third position, which is only 0.4805, and the gap between Hunan province and Zhejiang province is not very large. The index of urbanization subsystems lower than the index of ecosystem in Guizhou province. The coupling degree and coupling coordination degree of Zhejiang province are both high, and they are highly coupled and in good coordination. The coupling degree of Hunan province is highly coupled, and the coupling coordination degree is in the level of low coordination. The coupling degree of Yunnan province is highly coupled, and the coupling coordination degree is inthe level of low coordination. According to the research results of this paper, here we present the following policy recommendations:

Firstly, Guizhou provincial government should vigorously promote urbanization and strengthen the construction of ecological environment. Under the trend of the national vigorously promoting urbanization, Guizhou province should grasp the opportunity to promote urbanization and enjoy the benefits of economic development. At the same time, the government should strengthen the construction of ecological environment, promote the construction of urban green space, increase the investment of environmental protection, and vigorously promote the tourism industry and the development of large data industry, in order to promote the harmonious development of urbanization and ecological environment.

Secondly, the tourism industry of Yunnan province helps the development of small towns and rural economy. In the future, Yunnan province should continue to put ecological environment protection as the first priority, and continue to promote the construction of tourist attractions around the urbanization. Zhejiang province should vigorously develop green economy, complete economic transformation, and further promote the development of urbanization and ecological environment. As a population outflow province, Hunan province should formulate policies to promote the older generation of migrant workers to settle in towns and enjoy the welfare of social development. In the construction of ecological balance, Hunan province is a developed province in the steel making industry and chemical industry. In the past, the ecological environment of Hunan province has been polluted seriously. Therefore, Hunan province should pay attention to the experience, and push on the coordinated development of ecological balance and urbanization.

References

1. People's Publishing House. National new urbanization planning, People's Publishing House, (2014).
2. Z.G. Sun, The people's government of Guizhou province in 2016, "the government work report". Guizhou daily, (2016).
3. G.M. Grossman, A.B. Krueger, Economic growth and the environment. Q J. Econ, **110**(2):353–377, (1995).
4. G. Haughton, C. Hunter, Sustainable cities, London: Jessica Kingsley Publishers Ltd, (1994).
5. Y.L. Xiao, Study on the coupling development model of urbanization and ecological environment in Guizhou. Guiyang: Guizhou University of Finance and economics, (2013).
6. N.S. Luo, J.J. Li, Fu, F.Z. Luo, The coupling relationship between urbanization and ecological environment - A case study of Changsha, Zhuzhou, Xiangtan city group. Hunan forum, **27**, (2014).
7. P.C. Dai, Z.P. Shen, Y.L. Zhao et al, Changes in coupling relationship between urbanization and ecological environment: a case of Xuzhou City, Journal of Jiangsu Normal University (Natural Science Edition), **32**, (2014).
8. B. Li, J.X. Zhang, The regional differences and temporal and spatial evolution on urbanization and ecological environment in the central plains economic region, Regional Research and Development, **34**, (2015).

 DOI: 10.1051/matecconf/201710005031

The Technological Input-output Efficiency of High-technology Enterprises in China Based on the DEA method

Jing Lu[1,*], Guozhu Jia[1], Hengshan Zong[2] and Wei He[1]

[1]School of Economics and Management, Beihang University, Beijing 100191, China
[2]China Aerospace Academy of Systems Science and Engineering, Beijing 100191 P.R. China
Corresponding Email: lujing@buaa.edu.cn

Abstract. High-technology enterprises play the leader role in the regional economic progress, but nowadays technological resources have many problems such as division, separation and dispersing. It's an urgent problem to be solved that how to evaluate input-output productivity of technological resources towards industries and regions scientifically and efficiently. Firstly the article analyzes the input-output efficiency status of technological resources at home and abroad. Then high-technology enterprises in 29 provinces, as the subject of evaluation, are analyzed for their operation efficiency by DEA. The article states the suggestions like balanced development among regions and reasonable structure of input and output by comparing the differences among regions. And the article estimates the technological resources configuration efficiency of 5 leader industries and explains the developmental characteristics and direction among industries. And on these bases the article conducts hypothesis testing for inner elements which might have an influence on the operational efficiency by Tobit model. Finally, the paper proposes many suggestions to the benefits of promoting the productivity of high-technology enterprise according to the comprehensive analysis.

1 Introduction

In modern society, the science and technology plays an increasing role, and its "primary productive forces" status has attracted the attention of all countries. At the same time, the leading role of science and technology to the business, region and even country has been deeply understood by policy makers and resource managers. The effect of leading role is amplified to several times or even hundreds of times through market economy, so increasing investment in science and technology resources has gradually become the main policy of many countries. Meanwhile, science and technology resources has risen to a strategic level [1-3].

Although in recent years there is great progress on science and technology output achievements in China, the scientific and technological achievement of Chinese high-tech enterprises still has much room for improvement compared with the United States and other developed countries. For example, there is imitation and learning stages in the field of aerospace, aircraft technology and Internet applications. The existence of this phenomenon can not only be attributed to the status of national economy development and current nation policy. But the enterprises should also reflect on themselves.

To improve science and technology output with the recent limited resources, it is necessary to analyze input-output efficiency of science and technology, which can help to find the key factors affecting the efficiency. Only in that way, the targeted rationalization proposals can be obtained. Considering that technology investment itself is on a macro level, and there have been many scholars analyzed the macro technology input-output before [4-5]. In highly developed market economy, enterprises have become the main body of carrying technology. High-tech enterprises often have the core intellectual property, along with research and experimental team, and are able to grasp the direction of economic and technology development, which makes them take the lead in the enterprises.

In order to conducting systematic research and analysis, this paper elects the appropriate data from "China Statistical Yearbook on High Technology Industry" and "China Statistical Yearbook of Science and Technology". With selecting a number of important affecting factors of input and output, the same period high-tech companies' operating efficiency in Chinese different regions and different sectors are calculated and analyzed by DEA method, and the high-tech industry input-output efficiency level of various regions are compared and analyzed. Through using the DEA method, the efficiency values can be analyzed by selecting a number of possible factors affecting technology efficiency. To obtain a more objective and comprehensive analysis of the factors impacting the science and technology input-output efficiency, Tobit regression model method has been used to analyze the efficiency values. From the view of government and high-tech enterprises, the paper puts forward some reasonable opinions and suggestions with the results of research and comparison of the different results.

2 Research status of input-output efficiency of science and technology resources

Previous scholars have conducted a series of studies of science and technology resources allocation efficiency with qualitative research and achieved fruitful results. OECD believed that it is necessary to integrate the various input factors and regard them as an organic whole to improve the regional input-output efficiency [6]. Maria applied participating evaluation method to the regional input-output efficiency evaluation, and put the results of evaluation into practice [7]. Through researching and analyzing, Song Yu found that the inefficient and ineffective phenomenon exited in the technology market, which would result in incomplete market environment. Instead of like Adam Smith's theory in which the social resources allocation are adjusted by the market, enterprises and government forces should develop rationalized policies to promote rational distribution and allocation of resources [8]. Ye Rufei and others used the method of institutional economics. They thought information asymmetry which caused the low efficiency of science and they provided the method of improving the efficiency at the same time [9-11]. Wei Shuyan analyzed the experience of foreign countries, and provided that resources sharing strategy should be raised to the strategic perspective in our country and be formulated with corresponding regulations and policies so that the resources sharing could work smoothly [12]. Some scholars also provided suggestion for science and technology resources sharing policy. Through establishing the methods of analysis model and field research, they proposed that in order to improve the overall science and technology resources sharing, it was necessary to increase the sharing of resources flow in a region and improve the usefulness of the market [13-16]. Ge Huili studied varieties of regions and industries, and pointed out the role of government in science and technology resource input-output efficiency. She thought that situation of China was in a weakened economy and market environment and there are many problems in current scientific research system, resources must be controlled by the government. It was possible to improve the control of resources by supervision and evaluation and then improve the efficiency of science and technology input and output [17].

With many advanced analysis methods, scholars around the world also have studied the quantitative research of science and technology resources allocation efficiency for a long time. From the aspect of optimization methods, Li and other groups constructed optimization model through studying the open system of analysis and establishing a function indicators with the numbers of researchers and funds of science and technology in different regions [18]. Li Jianhua and others considered that in the optimization method, only when human and financial resources were in effective restraint state, it could achieve the most matching optimization status [14,15]. With the regression analysis methods, Li Shizhu and others did the quantitative analysis to the various elements and summarized the factors affecting the results of science and technology resources allocation efficiency [19]. In terms of principal component analysis, Li Dongmei showed that the constituent elements of science and technology resources would have an impact on technology efficiency by using the principal component analysis to different regions of China, but the effect of other factors were reflected on the human and financial elements. Thus, it was possible to analyze the main elements of human and financial resources, and grasp the overall aspects [20,21]. As well as the data envelopment analysis (following referred to "DEA") method, there were many articles based on the domestic research for the DEA method itself. Because of that, it was able to have a deep understanding of this method and have more reasonable application by using this method [22]. Wu Hecheng measured the different regions' input-output efficiency by DEA method [16]. Science and technology capabilities of different regions were evaluated and analyzed by Sun Baofeng and others by using DEA method[23]. The science and technology efficiency of China in all regions in recent years were analyzed and evaluated by He Defang and others, and they proposed that it was supposed to increase the size of the efficiency [24]. In regional analysis, Niu Shuhai analyzed the connection with technology resource development regional differences and economic growth by region analysis, which did the basic work for reducing regional disparities [25]. Kerssens van Drogelen's(1999) R&D performance measurement system was mainly: contingency factors influencing R&D performance, measurement methods and techniques, metrics, time frequency, matching method and selection principles [26]. From an empirical point of view, Jooh Lee &Eunsup Shim studied the high-tech industry innovation and efficiency in the United States and Japan, and researched the relationship among investment in technology, market share and corporate long-term performance [27]. Romijn [28] and Neel-ankavil [29] studied the high-tech industry innovation efficiency using multiple regression method and Granger causality method.

Overall, existing researches mainly have following disadvantages: firstly, foreign scholars' research of science and technology input-output efficiency aimed at developed countries, while those empirical results of the study did not necessarily meet the circumstance in China; secondly, the study of the existing literatures on the science and technology input-output efficiency mainly focused on the regional and industry, which basically belong to the meso or macro level, but few scholars studied science and technology input-output efficiency from the microscopic perspective. Thus, they ignored the innovation subjects--enterprises' output efficiency of science and technology investment; thirdly, a part of studies of enterprises science and technology input-output efficiency mainly focused on the aspects such as influencing factors and evaluation system construction etc., but few scholars used quantitative methods to evaluate the input and output efficiency for the business from the empirical point of view. Most researches of enterprises technology input-output efficiency were superficial and did not analyze the basic factors which fundamentally affected science and technology input-output efficiency.

Therefore, by studying the high-tech enterprises in China with DEA method, this paper comprehensively evaluates and analyzes input-output efficiency. This paper broadens the ideas of science and technology input-output efficiency of domestic research and enriches the existing research, which has strong innovation value and theoretical significance. At the same time, the study identifies the deep reasons why high-tech enterprises technology input-output efficiency in China is low, which is an exploratory analysis and has strong practical significance.

3 Evaluation model of input-output efficiency of science and technology investment in Chinese enterprises

3.1 Selecting evaluation index

According to the general principles of system design and the relative effectiveness based on DEA, in order to make sure that input and output indicators are real effective, there are some principles that should be followed when selecting indexes: scientific principle, feasible principle, independent principle, simplified principle, comparable principle and goal-oriented principle. The design for input and output indicators should have the same input-output status of enterprises technology investment and be consistent with the scientific meaning of the respective areas. In addition, when the indexes include the whole information, the indicators should be relatively simplified. The comparable indicator is an important prerequisite to compare and evaluate multiple decision units. According to Index selecting principles and characteristics of DEA, human and financial resources input-output status can be integrated as the appropriate indicators. While there are many factors impacting the enterprises technology input and output, the actual operation should not consider too many factors to avoid lots of DMU's value becoming 1. Dividing the indexes which can replace or complement others into the same type makes sure that there is no correlation between indexes.

Based on feasibility and ease of the operation principle, there are two main selection of science and technology input indicators, including science and technology human and financial resources elements. First, science and technology human resources is the element with the strongest initiative and variability; Second, science and technology financial resources element is the basis and premise for scientific and technological activities. Related indicators reflecting the R&D activities are added to the technology of human and financial resources in the article, which also reflects principles of representation and principles of diversity in index system design.

High-tech enterprises are knowledge production-oriented enterprises, and the patent as a direct manifestation of knowledge is to be included in the output indicators; New product is also coagulation of technology and knowledge, which can be used as indicators to measure the output of science and technology resources; Finally, to evaluate disposition of technology resources from the view of overall development of high-tech business, it is necessary to have a total index so the added value of enterprises can display the effect of technology resources investment properly.

Referring index selected principles, Chinese basic national conditions and the availability and timeliness of the data, the index system is initially identified as follows:

Science and technology input indicators are: R&D Equivalent to full-time equivalent staff(X1), New Product Development Expenditure(X2), R&D personnel amount(X3) and R&D expenditure(X4).

Science and technology output indicators are: Sales of new products(Y1) and the amount of valid invention patents(Y2).

Each indicator can be briefly represented in the following table:

Table 1. Science and technology input-output indexes.

input indicators	output indicators
R&D Equivalent to full-time equivalent staff	Sales of new products
New Product Development Expenditure	valid invention patents
R&D personnel amount	
R&D expenditure	

Each indicator is explained as below:

R&D equivalent to full-time equivalent staff is the summary of R&D full-time staff and part-time staff work according to the actual working hours.

New product development expenditure is the funding of new product research.

R&D personnel amount refers to the total number of R&D institutions in enterprises.

R&D expenditure means the cost of internal R&D activities of enterprises in reporting period.

Sales of new products is new products' sales revenue in the prime operating business income and in other operating income.

The amount of valid invention patents refers to the number of patents which one enterprise has in an investigation period.

3.2 Selecting sata

The data of R&D in equivalent full-time equivalent personnel, new product development expenditure, R&D personnel amount, R&D expenditure, sales of new products and the effective number of patent in this study are from the "China Statistical Yearbook on High Technology Industry 2013", "international Statistical Yearbook 2013" and the "China Science and Technology statistical Yearbook 2005-2013".

High-tech enterprises input and output data in 29 provinces of mainland China from 2000 to 2012 are selected as a sample in this paper. Based on the general requirements of using DEA model to measure efficiency, there are 29 samples selected, which are more than twice as the number of input and output data which meet the model requirements. At the same time the industry's DMUs are also evaluated to obtain comprehensive efficiency evaluation.

3.3 Research methods

Efficiency evaluation method used in this paper is DEA method. The basic idea of DEA method is to define different kinds of objects as one decision-making unit(referred to as the "DMU"). For each DMU, selecting the same input and output indicators and creating the appropriate models can help to calculate and analyze these input and output indicators. Calculating the relative indicators for each DMU and adjusting parameters can make sure the maximum efficiency is 1, which means that efficiency value is fully effective while other DMUs not being 1 means that they are not completely effective. This method analyzes the causes and gaps and provide decision-making information for researchers [22,30]. This method can achieve accurate calculation of the multiple decision units of a plurality of input and output indexes, and the result is compared with efficiency values, which is more objective and accurate without the interference of external human factors.

4 Example for science and technology input-output efficiency evaluation

4.1 Case study

According to the raw data of the medium-sized high-tech enterprises investment production in 29 provinces from 2000 to 2012, it is possible to obtain the technology efficiency of the provinces and the industry as well as pure technology efficiency and scale efficiency by using software DEAP 2.1 and BCC model. 2012 concrete results are shown in Table 2.

From Table 2 and Table 3, there are only 4 regions having the whole science and technology input-output efficiency in 2012 high-tech enterprises. That is to say, only 13.8% region's input and output efficiency are both scale effective and technology effective. And 86.2% of the area which needs to improve science and technology inputs and outputs are non-relatively effective. What's more, the enterprises in China have been increasing science and technology investment in general. They just don't make full use of those investment to get the maximum output.

Table 2. High-tech enterprises input-output efficiency in 2012.

DMU	TE	PTE	SE	RTS
Beijing	1.000	1.000	1.000	-
Tianjin	1.000	1.000	1.000	-
Hebei	0.629	0.653	0.962	drs
Shanxi	0.523	0.649	0.805	drs
Inner Mongolia	0.645	0.985	0.655	irs
Liaoning	0.427	0.442	0.966	drs
Jilin	0.772	0.815	0.948	drs
Heilongjiang	0.412	0.713	0.578	drs
Shanghai	0.514	0.560	0.918	drs
Jiangsu	0.647	1.000	0.647	drs
zhejiang	0.610	0.610	1.000	-
Anhui	0.514	0.536	0.960	drs

Fujian	0.634	0.843	0.752	drs
Jiangxi	0.540	0.543	0.994	irs
Shandong	0.470	0.723	0.650	drs
Henan	0.396	0.401	0.988	drs
Hubei	0.511	0.591	0.864	drs
Hunan	0.773	1.000	0.773	drs
Guangdong	1.000	1.000	1.000	-
Guangxi	0.615	0.622	0.988	irs
Hainan	0.320	0.332	0.964	drs
Chongqing	0.685	0.696	0.984	drs
Sichuan	0.475	0.483	0.983	drs
Guizhou	0.682	1.000	0.682	drs
Yunnan	0.876	0.900	0.973	irs
shanxi	0.384	0.611	0.629	drs
Gansu	0.349	0.355	0.982	irs
Ningxia	0.365	0.459	0.795	irs
Xinjiang	1.000	1.000	1.000	-

Table 3. 2012 high technology businesses input-output efficiency in various areas.

DMU	TE		PTE		SE	
	amount	proportion	amount	proportion	amount	proportion
effective	4	13.8%	7	24.1%	5	17.2%
Non-effective	25	86.2%	22	75.9%	24	82.8%
	CRS	5	IRS	6	DRS	18

From the statistic results, the level of efficiency value has not been necessarily connected with the location in Chinese eastern, central or western area. For example, cities in eastern China such as Beijing, Tianjin and others have fully effective input-output efficiency, while in the other cities in the east such as Anhui and Shandong, the technology efficiency are 0.514 and 0.470. Central and western regions in China also have the similar phenomenon. Not as the same as common understanding, when technology investment income and high-tech enterprises income are overall lower than other areas, its technology efficiency would be relatively higher, such as high-tech enterprises in Xinjiang in 2012. The science and technology input-output efficiency value is 1 in Xinjiang's enterprises, which is perfectly valid. Also in the west area like Yunnan in 2012, the high-tech enterprise technology input and output of TE, PTE and SE are 0.876, 0.900 and 0.973, which are relatively at a high level. Science and technology input-output efficiency is a comprehensive evaluation index, which refers not only the amount of the science and technology output, but also considers the ratio of input and output. So there are many factors which can influence the indicator like science and technology input, R&D staff full-time equivalents in an area and new product R&D expenditures, and it is also related to the local scientific research departments, business administration, even the tax administration's comprehensive management level. Only when the investment of technology, scale of local enterprises, employees and relative policies can be matched legitimately, the science and technology input and output efficiency in the area can reach to a higher level.

From the view of returns to scale, few areas are at constant returns to scale. From Table 4, the number of regions around 2010, which were at constant returns to scale, are relatively more than other years. The reason why DEA is so invalid is that uncoordinated inputs and outputs are at the current scale situation, which means that the input is inadequate or excessive at the same output level. Therefore, 24 regions should adjust their input or output to improve the efficiency of inputs and outputs in 2005.

Table 4. 2000--2011 region input-output efficiency statistics of high-tech enterprises.

Year	DMU	TE		PTE		SE		Scale changes		
		amount	proportion	amount	proportion	amount	proportion			
2011	Effective	6	20.7%	8	27.6%	5	17.2%	CRS	IRS	DRS
	Non-effective	23	79.3%	21	72.4%	24	82.8%	6	4	19
2010	Effective	7	24.1%	11	37.9%	7	24.1%	CRS	IRS	DRS
	Non-effective	22	75.9%	18	62.1%	22	75.9%	7	6	16
2009	Effective	6	20.7%	11	37.9%	7	24.1%	CRS	IRS	DRS
	Non-effective	23	79.3%	18	62.1%	22	75.9%	7	6	16
2008	Effective	3	10.3%	8	27.6%	3	10.3%	CRS	IRS	DRS
	Non-effective	26	89.7%	21	72.4%	26	89.7%	4	11	15
2005	Effective	3	10.3%	8	27.6%	3	10.3%	CRS	IRS	DRS
	Non-effective	26	89.7%	21	72.4%	26	89.7%	3	1	24
2000	Effective	4	13.8%	12	41.4%	4	13.8%	CRS	IRS	DRS
	Non-effective	25	86.2%	17	58.6%	25	86.2%	4	4	21

Table 5. High-tech and medium-sized enterprise in high-tech industry technology input-output evaluation index.

Input index	Output index
R&D Equivalent to full-time equivalent staff	New product sale revenues
R&D internal expenses	

Table 6. High-tech and medium-sized enterprises in high-tech industries input-output efficiency in 2012.

DMU	TE	PTE	SE	RTS
Pharmaceutical	0.81	0.827	0.979	drs
Aerospace vehicles and equipment	0.313	0.353	0.887	irs
Electronics and communications equipment	1.000	1.000	1.000	-
Computer and office equipment	1.000	1.000	1.000	-
Medical equipment and instruments	0.728	1.000	0.728	irs

Science and technology input-output efficiency trend during these seven years shows that technology input and output efficiency in Beijing always stayed in 1 during this period. It indicates that the input-output efficiency in recent years has been effective. As for Tianjin, technology efficiency has been always fully effective except in 2008. The region which Technology efficiency was less than or equal to 0.6 has been further analyzed, and the amount was 16, 21, 22, 14, 14, 18 and 14. In the period, there has been a large proportion of all the regions which technology efficiency value was less than or equal to 0.6, which indicates that the relative efficiency was low in mang areas. Therefore, these areas where efficient value is relatively low can improve the efficiency of input and output in the region through organizational learning, studying from Beijing and Tianjin, adjusting the allocation of resources, developing suitable policies and encouraging personnel training.

From the above data processing and analysis, the conclusions can be drawn as follows: (1) the output efficiency in technology investment of the high-tech enterprises in China is low; (2) From the perspective of returns to scale, a large part of the region are in increasing or decreasing phase to the return of scale, and they can achieve efficient scale through increasing investment and efficiency and adjusting the direction and amount of output; (3)Chinese medium-sized high-tech enterprises science and technology input and output efficiency values are distributed in the lower range, and the efficiency has not been improved significantly in the seven-year study period.

In order to have more understanding of input and output efficiency in Chinese medium-sized high-tech enterprises, the paper has further analyzed and researched different industries.

Select the following five industries as analysis units: 1.Pharmaceutical Manufacturing; 2. Aerospace vehicles and equipment manufacturing; 3.Electronics and communications equipment manufacturing; 4.Computer and office equipment manufacturing; 5.Medical equipment and instruments. To make the efficiency analysis more intuitive and more representative, R&D Equivalent to full-time equivalent staff(X1) and R&D internal expenses(X2) are selected as input indicators, while new product sales(Y) is selected as output indicator.

Calculating the medium-sized high-tech enterprises input-output data in five high-tech industries in 2012 according to the selected samples by using DEAP2.1 software, CCR and BCC models, it is easy to get technology efficiency in various industries(TE), pure technology efficiency(PTE) and scale efficiency(SE). 2012 concrete results are shown in Table 6.

From Table 6, enterprises technology input and output are overall effective in five high-tech industries elected in 2012, namely the two of indexes (TE, PTE and SE) were both 1, accounted 40% for selected DMUs. Three industries' pure technology efficiency were 1, accounted for 60%. Although the amount of elected DMUs was few, from the results of the analysis, overall science and technology input and output indicators of the high-tech industry were relatively balanced, and they were all at a high level. There was only one industry which the overall efficiency was less than 0.6, accounted for 20% of all the DMUs. According to its own characteristics of the DEA method, it can be considered that science and technology efficiency of the high-tech industry is more average in China.

In these five industries, only the efficiency values of aviation, the spacecraft and the equipment manufacturing industry are low, and the overall efficiency value has been only 0.313. From further analysis, the electronics and communications equipment manufacturing, the medical manufacturing, the computer and office equipment manufacturing, the medical equipment and the instruments which have the high values of efficiency are in a high degree of market. Close connection with the market will lead management system and the system to be market-oriented. The role of market in resource allocation is very obvious, especially with efficiency value calculated by the sales of new product being higher. The aerospace industry is mostly state-owned enterprises. With rigid institutions, not being sensitive for market and even the existence of political tasks, there are some reasons that make low efficiency in the industry. Through the field research, many phenomena and factors are consistent with analysis. For example, by investigating some certain aerospace institutes, we find that large equipment's utilization is low, which is because they don't have access to use the equipment. But because of the funding management system, these enterprises have to buy some large equipment. In response to this phenomenon, high-tech enterprises of the aerospace industry are proposed to meet the market integration as soon as possible and make full use of the advantages to break the traditional system limitation, and then they finally form a Chinese characteristics route of high efficiency. Government also should introduce policies from the macro management level to promote traditional state-owned enterprises breaking the rigid system and to meet the market integration as soon as possible. At the same time, it is important to use advanced technology to improve the industry technology input-output efficiency.

In addition, the aviation and spacecraft and equipment manufacturing and pharmaceutical manufacturing also expose input and resources wasted phenomena by analysis of DEA method. On the one hand, those industries should reduce unnecessary investment; on the other hand, what is more important is to adjust the industrial scale, optimize the industrial results and improve resource control system. Only in this way, they can increase the technology input-output efficiency, expand the status and role of the industry in the entire field, and finally achieve promoting the development of the social economy with high technology.

4.2 Analysis outside influence of input-output

After calculating the regional high-tech enterprises technology input and output efficiency in 2000, 2005, 2008-2012 by DEA method, and using Tobit regression model, we identify outside efficiency affecting factors except the efficiency of input items and output items to provide more scientific and rational policy recommendations which can improve the science and technology input and output efficiency.

In order to understand the efficiency affected factors and degree of these factors, there are two-stage method derived from DEA. This method firstly analyzes the efficiency values of DMU by DEA, and then the method uses regression analysis with taking the efficiency values as the dependent variable and selecting possible factors as independent variables.

Standard Tobit model is as follows:

$$y^* = \beta x_i + \varepsilon$$
$$y_i = y^*(if\ \ y^* > 0)$$
$$y_i = 0(if\ \ y^* \le 0)$$

Here ε~N(0, σ2); β is the regression parameter vector; x_i is explanatory variable vector; y^* is the potential for variable vector; y_i is the observed dependent variable. When $y^* > 0$, observed dependent variable y is equal to y^*, but when $y^* < 0$, y is equal to 0.

Tobit model can analyze more deep-seated reasons related to the efficiency values. This article selects regions business prime operating revenues(a hundred million yuan, F1), high-tech enterprise taxes(a hundred million yuan, F2), delivery value of export(a hundred million yuan, F3), technical reform costs(a hundred million yuan, F4) and new fixed assets(ten thousand yuan, F5) as the environmental variables of science and technology input-output efficiency, and the paper makes the assumption that eastern region technology input-output efficiency is higher than in the west.

The Tobit model is built as follows:

$TE_i = \beta 1F_{1i} + \beta 2F_{2i} + \beta 3F_{3i} + \beta 4F_{4i} + \beta 5F_{5i} + \beta 6Region + \varepsilon$ The dependent variable TE is a enterprises technology input-output efficiency value calculated by DEA; *Region* means the location that every region belongs to (eastern=1, western=0); ε is the error term in regression model; $\beta 1$-$\beta 6$ are the regression coefficients of independent variables; i means region(i=1,2,…,29).

This article selects the indexes of factors impacting the high-tech enterprises technology investment value output efficiency of 2012 as sample data in regression analysis, which descriptive statistics are shown in Table 7 by using Eviews 6.0.

Table 7. Tobit results of Influencing Factors of high- tech enterprises science and technology input-output efficiency in 2012.

	Coefficient	Std. Error	z-Statistic	Prob.
enterprise prime operating revenue (F1)	8.05E-05	4.89E-05	1.647877	0.0994
enterprise taxes (F2)	-0.000613	0.000576	-1.063875	0.2874
delivery value of export (F3)	-7.25E-05	5.96E-05	-1.21606	0.224
Technical reform cost(F4)	-8.70E-07	3.30E-07	-2.633749	0.0084
new fixed assets (F5)	-6.42E-05	0.000191	-0.336845	0.7362
C	0.565513	0.025377	22.28407	0

The regression results show that: enterprise prime operating revenue and high-tech enterprises technology input-output efficiency has the significantly positive correlation; technology reform costs and high-tech enterprises technology input-output efficiency has the significantly negative correlation; corporate taxes, delivery value of export and high-tech enterprises science and technology input-output efficiency have shown no presence of significantly negative correlation, while new fixed assets has no significant correlation with high-tech enterprises science and technology input-output efficiency.

The significantly positive correlation between high-tech enterprises science and technology input-output efficiency and enterprise prime operating revenue indicates that the high-tech enterprises science and technology input-output efficiency, the whole region economies of scale and the economic trend have a relatively close relationship. When the overall economic situation is good, input-output efficiency value is relatively high, whereas when the overall economic situation is weak, the input-output efficiency value is lower than when economic situation being good. It can be also analyzed if the local government policies on economic management is reasonable and effective from the view of enterprise prime operating revenue. These policies affect the input-output efficiency of high-tech enterprises in the same way. The amount of technology market trade contracts and regional GDP show a positive correlation with high-tech enterprises science and technology input-output efficiency.

The technical reform costs and high-tech enterprise science and technology input-output efficiency have a negative correlation. In the tradition, it is common that the reform is bound to the increase of output value. But the analysis has indicated that the cost of investment in technical reform is redundancy. In a nationwide analysis, the reason for such a result is that the region high-tech enterprises fund is wasted and idled. However, from the analysis above, it has shown that high-tech enterprises in technical reform fund still have room for improvement, where the conversion of investment in research funding and technical reform fund have existed inefficiency. Because of the phenomenon, the use of research fund must be strictly controlled by local governments to make sure that the fund is fully used and used in the right way.

Meanwhile, high-tech enterprises should also control the inside use of the research fund. Only in that way, it is able to ensure the input-output efficiency and maintain the competitiveness of enterprises.

Delivery value of export has non-significantly negative correlation with input-output efficiency, which indicates that the number of exported goods transaction does not play a significantly positive role in boosting the overall business conditions of enterprises. High-technology is the core competence of high-tech enterprises. The high-technology has higher requirement on the core intellectual property rights. The analysis results describe that most parts of China should not only just focus on the delivery value of export, but also should build their own technology and industry to boost economic growth.

In addition, enterprises taxes and the location of the high-tech enterprises have a negative correlation with science and technology input-output efficiency, which indicates that Chinese tax policy could seriously affect the high-tech enterprises technology input-output efficiency. The negative relationship also shows that the force of imposing taxes of high technology enterprises has greatly affected the input-output efficiency of enterprises. With the economic development, the state can provide more preferential tax policies for high-tech enterprises. Location of enterprises and the input-output efficiency have the negative correlation, which can be explained as the results of the analysis described above. With the Chinese economic development and the continuous increase of science and technology resources investment, the eastern region has been changed from the status of lack funds to ample funds status. But the idle and waste of funds will take place with the change. In this situation, putting too much fund in the east has no benefit to improve the efficiency. However, the investment of resources in the western region is still short, and there are still some western regions which actually have certain technology development capacity, such as Xinjiang province whose efficiency value is 1.

Overall, government should focus on building a whole region economic policy, strictly controlling the use of research funding in high-tech enterprises, increasing investment to where needed and avoiding waste and idle. Meanwhile, it is necessary to improve the level of corporate governance to make all Chinese high-tech enterprises holding science and technology input-output efficiency in a high level.

5 Summary

5.1 Studying results

In this paper, in order to analyze Chinese high-tech enterprises, we select high-tech enterprises of 29 regions in 2000, 2005, 2008-2012 as samples, using descriptive statistical analysis, DEA method and Tobit regression analysis to analyze and evaluate the current situation of science and technology input-output in China. We research enterprises science and technology input-output efficiency in the targeted area and industry. And we find out the other factors affecting enterprises science and technology input-output efficiency, and make recommendations to rationalize the allocation of resources. The main conclusions are as follows:

(1)High-tech enterprise science and technology input-output efficiency is generally low by region. By using DEA, the evaluation of the high-tech enterprise science and technology input-output efficiency of various provinces shows that the number of provinces which achieves science and technology effective are less than a half of them, and the high-tech enterprises science and technology input-output efficiency are low on the whole. We can't find the significant relationship among the input-output efficiency, technology investment and province location, and there is also the overall effective province in western regions while some provinces in the east are still having low efficiency. A large part of the regions are in increasing or decreasing returns to scale phase, which need to increase investment or adjust the direction and amount of output to improve efficiency and achieve the efficient scale.

(2)Tobit regression analysis shows that the enterprise main operating revenue and high-tech enterprises science and technology input-output efficiency have a significantly positive correlation. The technology reform costs and the high-tech enterprises science and technology input-output efficiency have significantly negative correlation. Meanwhile, there are the negative correlation among the enterprises taxes, delivery value of export and high-tech enterprises science and technology input-output efficiency.

5.2 Policy recommendations

According to the research and analysis above, there are some policy recommendations to improve the Chinese high-tech enterprises science and technology input-output efficiency.

Allocate science and technology resources rationally and improve the input-output efficiency.

When each region has continuously increased investment in the science and technology resources, they also need to focus on the science and technology resources input-output efficiency. Making full and effective use of science and technology resources and improving the allocation efficiency to achieve higher levels of output are important. Otherwise, serious waste of human and material resources could take place. Throughout all the provinces whose DEA are effective, all of them have shown the importance of science and technology resources input-output rational structure. That is, only when the ratio of science and technology research fund and the amount of research personnel are proper and the structure of them are reasonable. It is possible to achieve maximum output with fixed science and technology resources investment, or put minimum of resources under a given output. The area might takes advantage of a variety of investment resources and achieve high input-output efficiency of science and technology resources. In recent years, although the Chinese high-tech enterprises science and technology input has been growing, the output efficiency still has not been increased as the same proportion. The proportion of input in science and technology has not reached the optimal allocation ratio, and the redundancy and inadequate of individual technology resources makes the formation of a barrel effect, which resulted in a waste of the rest of resources. We need to regard the intensification of resource utilization as the core concept, improve the operation of enterprises resources investment and management capabilities, allocate the research fund rationally, strengthen scientific equipment procurement plan, and strive to improve the quality of research Specialist Staff to avoid input redundancy or shortage.

Develop all regions and strengthen the support to central and western enterprises.

Thorough comparing various science and technology resources input, it has shown that the amount of science and technology input is relatively few in central and western regions. It also indicates a downward trend in recent years, but the results of DEA has indicated that there is no significant difference of the high-tech industry science and technology resources input-output efficiency between the west and the east. Tobit regression analysis has shown that the increasing science and technology investment in the west has benefits for increasing the overall efficiency. But increasing science and technology investment in the eastern region does not have the obvious benefit to the overall efficiency compared with the western region, because of the sufficient resources and exceeding the overall enterprises configuration. In addition, government should increase science and technology input in the western region from the perspective of coordinated regional development, and enhance scientific and technological competitiveness in these areas and drive the overall economic development by high-tech enterprises. At the same time, the development of the overall economic situation will have the react on high-tech enterprises science and technology input-output efficiency as well.

(3)Government can implement balanced development policy and improve the necessary industry and region in high-tech enterprises.

Regional technology market turnover contract amount and area GDP have a positive correlation with high-tech enterprises science and technology input-output efficiency, which indicates that high-tech enterprises science and technology input-output efficiency, the whole economies of scale and the economic situation have a relatively close relationship. Development of high-tech enterprises depends on the overall economic situation of the entire region, so the government must establish a regional economic policy which begins with the guide of government, and it is supported by national policy. Besides, the policy should regard the high-tech enterprises as the vanguard and we must focus on the overall economic development. Only in this way, the development and improvement of high-tech enterprises will be more durable and the Chinese technology industry can achieve long stable development.

What's more, the government should establish benign tax policy, reward science and technology output and make some supporting policies to create the environment where the attitude of scientific research is positive and healthy. Besides, the government should try to maintain the vitality of the national high-tech enterprises and let the enterprises lead the regional economy healthy and stable.

Acknowledgement

The authors would like to thank the reviewers and the editor for their insightful comments, which helped improve the quality of the paper significantly. The research is supported by the National Natural Science Foundation of China under Grant No.71372007, the Ph.D. Program Foundation of Ministry of Education of China (Grant 20111102110025).

References

1. J.Z. Zhou, Science and technology resources, 107-113(1999).
2. P. Shi, Research on the relationship between science and technology resource allocation and institutional arrangement, (2001).
3. Q. Liu, Journal of Luoyang Normal University, **19**,2, 41-53(2000).
4. B. Sun, J.H. Li, Social Science Front, **5**, 36-39(2001).
5. P. Shi, Y. Li, China Soft Science, **12**, 55-57(2000).
6. OECD, "National Innovation System", Paris(1997).
7. A. Maria A, Decentralization and Evaluaion, **4**, 276-315(2000).
8. Y. Song, Quantitative and Technical Economics, **10**, 29-31(1999).
9. R.F. Ye, X.R. Chen, X.B. Yu. Research and Development Management, **5**, 113-118(2004).
10. J.J. Mei, S.Z. Li, Journal of Beijing Insititute of Machinery Industry, **2**, 59-64(2002).
11. Z.J. Yang, Science and Technology Management, **2**, 213-216(2007).
12. S.Y. Wei, Innovation and Technological Progress, 95-97(2005).
13. C.Q. Wu, Principle of regional economics, 89-90(2008).
14. J.H. Li, S.J. Zhou, B.F. Sun, Scientific Management Research, **6**, 72-76(200l).
15. Y. Ma, Y.L.Gao, Journal of Northeast Normal University, **3**, 24-28(2002).
16. H.C. Wu, C.Y. Zheng, Scientific Management Research, **10**, 93-96(2003).
17. H.L. Ge, Technology Management Research, **24**, 14-16(2010).
18. X.Q. Li, K.F. Xie, J.Z. Li, Technology Economy, **11**, 48-50(1999).
19. S.Z. Li, D.M. Li, W.X. Tang, Scientific Management Research, **2**, 60-63 (2003).
20. D.M. Li, S.Z. Li, W.X. Tang, Journal of Beijing Insititute of Machinery Industry, **1**, 50-55(2003).
21. S.H. Wei, G.S. Wu, Studies in Science of Science, **4**, 467-473(2005).
22. Q.L. Wei, Science Bulletin, **17**, 1793-1808(2000).
23. B.F. Sun, J.H. Li, Y.S. Yang, Application of Statistics and Management, **2**, 52-58 (2004).
24. D.F. He, Information Journal, **6**, 740-748(2006).
25. S.H. Niu, F.J. Jin, Y. Liu, "The difference of regional science and technology resource allocation", Resources Science, vol.1, pp.61-68(2004).
26. K. Drogelen, J. Bilderbeek, R&D management, **29**, 1, 35-46(1999).
27. J. Lee, E. Shim, The Journal of High Technology Management Research, **6**, 179-191(1995).
28. H. Romijn, A. Mike, Regional Studies, **36**, 1, 81-86(2002).

 DOI: 10.1051/matecconf/201710005031

29. J.P. Neelankavil, V.T. Alaganar, Journal of Business Research, **6**,493-502(2003).
30. V. Korhonen, R. Tainio, J. Wallenius, European Journal of Operational Research, **130**, 121-132(2001).

Does Financial Development Necessarily Lead to Economic Growth? Evidence from China's Cities, 2007-2014

Shiyong Zhao

School of Business, Macau University of Science and Technology, Macau, China
Corresponding author: syzhao@must.edu.mo.

Abstract. Using data from 286 Chinese cities over the period 2007-2014, this paper investigates the impact of financial development on economic growth. Our results from traditional cross-sectional regressions, first-differenced GMM and system GMM regressions all show that financial development does not have any significant positive effects on economic growth, while some indicators of financial development show significant negative effects on growth. Our results are consistent with many existing studies that a state-ruled banking sector, such as that of China, hinders economic growth because of the distoring nature of the government. To examine the sensitivity of our results, different sets of control variables sets are experimented with. Our results are shown to be robust. Our finding shows that to let the financial sector play a more efficient and effective role in promoting real economic growth, China has to further reform its financial system.

1 Introduction

Joseph Schumpeter argued in his seminal work The Theory of Economic Development (1911) [1] that the services provided by financial intermediaries are essential for technological innovation and economic development. They mobilize savings, evaluate projects, manage risk, monitor mangers, and facilitate transactions. Earlier empirical works by Goldsmith (1969) [2] and McKinnon (1973) [3] illustrate the close ties between financial and economic development for a few countries. Numerous influential economists, however, believe that finance is a relatively unimportant factor in economic development. Notably, Robinson (1952) [4] argues that financial development simply follows economic development. Lucas (1988) [5] terms the relationship between financial and economic development 'over-stressed.' In this paper we study whether higher levels of financial development are positively associated with economic growth using data on 286 Chinese cities from 2007 through 2014.

To examine whether Schumpeter (or Lucas) was right, we must define 'financial development' empirically. We construct four indicators of financial development that are designed to measure the services provided by financial intermediaries. First, we compute the traditional measure of financial depth, which equals the overall size of the formal financial intermediary system, i.e. the ratio of loans to GDP. Second, we measure the ability of financial intermediaries to mobilize capital, i.e., the ratio of deposits to GDP. Third, the ratio of households savings to GDP to measure the ability of financial intermediaries to attract household savings. Fourth, share of fixed asset investment financed by bank loans relative to state budgetary appropriation. It is believed that bank loans have more role than government appropriation in disciplinary the recipients. Although each financial indicator has shortcomings (sometimes due to data limitations), using this array of indicators provides a richer picture of financial development than if we used only a single measure.

China has been experiencing rapid economic growth and remarkable expansion of financial sector since 1978, when China started its economic transformation. China's real GDP has been growing at an annual average rate of 9.65 percent over 1978-2015 (NBS, 2016) [6]. Over the same period, the total loans outstanding in its financial institutions divided by GDP has increased from 51 percent to 147 percent (NBS, 2010, 2016)[6, 7]. As the largest emerging market and the second largest economy with rapid growth and dramatic transformation, China presents us with an intriguing case for study to test Schumpeter's view posed in 1911. Does finance play an important role in China's recent growth? This paper investigates this question by using a rich and most updated set of city-level data.

Many cross-country studies have shown that finance has an important impact on growth (to be reviewed in next section). China has been viewed as a counterexample to the common insight of the finance-growth literature, because on the one hand China has achieved fast economic growth for more than three decades, and on the other, China's financial sector has been controlled by the state very rigidly. For example, private investors are still not allowed to open a bank in China even today. Boyreau-Debray (2003) [8] finds that financial intermediation has a negative impact on local economic growth in China over 1990-1999 because the banking sector was mainly supporting the loss-making state-owned enterprises in the 1990s. Using provincial level data over 1986-2002, Hasan et al. (2009) [9] also finds that the financial sector has a negative impact on China's economic growth. Chen (2006) [10], however, finds that China's financial development has a positive effect on economic growth by using provincial-level data over 1985-1999. Cheng and Degryse (2007) [11] show that the development of banking services contributes positively to economic growth over 1995-2003.

Our finding in this paper suggests that financial development does not show a positive effect on economic growth over 2008-2013. This result is consist with some existing literature conclusions, which states that state-controlled banking sector impedes economic growth because banks tend to lend money mostly to inefficient large SOEs due to various reasons (to be explained later), although millions of small and medium-sized private firms have an urgent need for money. While deposits did climb as a share of GDP, there was little expansion of credit to the private sector, as many state banks remained in existence, and their habits of directing credit died hard (World Bank, 2005) [12]. Our finding contrasts sharply with the significant, robust relationship between the level of financial development and both the current and future rate of economic growth (e.g., King and Levein, 1993) [13].

2 Literature review

It is undoubted that financial market is essential to a country's economic growth and prosperity. Financial intermediaries are indispensable for the smooth functioning of financial markets. It may be in this sense that Miller (1988) [14] said, 'that financial markets contribute to economic growth is a proposition almost too obvious for serious discussion.' Other influential economists, however, believe that finance is a relatively unimportant factor in economic development. Lucas (1988) [5], for example, as mentioned above, said 'the importance of financial matters is very badly over-stressed in popular and even much professional discussion.' Apparently, there is no consensus among economists on the role of finance in promotion growth at least theoretically. It seems that this dispute should be solved empirically.

Early cross-country studies suggest a positive correlation between financial development and economic growth. King and Levine (1993) [13] present cross-country evidence consistent with the view that the financial system can promote economic growth, using data on 80 countries over 1960-1989 period. Compared with earlier work by Goldsmith (1969) [2] who arrives at the same conclusion, King and Levine (1993) [13] add more control variables affecting growth to the regression model, such as trade, education and political stability. Levine and Zervos (1998) [15] show that stock market liquidity and banking development can predict economic growth. We should remind ourselves that correlation does not prove causation. It is possible that a more developed financial sector is correlated with various other growth-enhancing factors, and it is other factors that promote growth. Technically, it is not easy to disentangle the effect of financial development from other factors. As Robinson (1952) [4] contends, financial development simply follows economic growth.

Using a country's legal origin as a valid instrumental variable, Levine (1999) [16] finds that financial development has a significant positive effect on economic growth. Levine et al. (2000) [17] confirms this conclusion by using the generalized moments of method (GMM) for dynamic panel data on a panel of 71 countries over 1960-1995. From the perspective of econometric techniques, dynamic panel models allow the use of instrumental variables for all the explanatory variables so that more precise estimates could be obtained than cross-sectional data. This is why more recent studies on the relationship between finance and growth use dynamic panel models. Rioja and Valey (2004) [18] explore the impact of financial development on sources of growth in different groups of countries with a panel dataset covering 74 countries. Beck et al. (2000) [19] uses GMM estimators for dynamic panel data and find that financial development has a significant and positive effect on total factor productivity (TFP) growth. Moreover, Rousseau and Wachtel (2002) [20] and Beck and Levine (2004) [21] find that some exogenous components of bank and stock market development have a large impact on economic growth. This result is also echoed by studies with time-series data, such as Xu (2000) [22], Christopoulos and Tsionas (2004) [23], and Bekaert et al. (2005) [24].

Apart from cross-country studies, there are also many researchers who focus on specific countries. This is advantageous because country-specific factors could be avoided in regression analysis. Jayaratne and Strahan (1996) [25] provide evidence that financial markets can directly affect economic growth by studying the relaxation of bank branch restrictions in the United States. They argue that the observed changes in growth are the result of changes in the banking system, and improvements in the quality of bank lending appear to be responsible for faster growth. Dehejia and Lleras-Muney (2003) [26] examine the effect of state-level banking regulation on financial development and economic growth in the U.S. from 1900 to 1940. In particular, they document that not all forms of financial development have a positive effect on economic growth, for example, indiscriminate lending can negatively impact economic growth. Japan's Meiji Restoration started from 1868, which successfully transformed Japan into an industrial power. Rousseau (1999) [27] studies this period from 1880 to 1913 and offers evidence that financial factors played a key role in leading to Japan's rise. Examining Italy's experience, Guiso et al. (2004) [28] suggest that local financial development is an important determinant of the economic success of an area even in an environment where there are no frictions to capital movements.

Besides the above studies based on country-level data, there are also a lot of studies using more micro-level data, such as industry- or firm-level data. Rajan and Zingales (1988) [29] find that financial development facilitates economic growth, and the rationale is that financial development reduces the costs of external finance to firms. They find that industrial sectors that are relatively more in need of external finance develop disproportionately faster in countries with more-developed financial markets. Kumar et al. (1999) [30] show that size of firms in industries dependent on external finance is larger in countries with better financial markets, which means that the growing of firms relies on the support of financial markets. Wurgler (2000) [31] shows that financial markets 'appear to improve the allocation of capital,' thus conducive to economic growth. Cetorelli and Gambera (2001) [32] investigate the role played by the market structure of the banking sector on the dynamics of capital accumulation. This paper provides evidence that bank concentration promotes the growth of those industrial sectors that are more in need of external finance by facilitating credit access to younger firms. Claessens and Laeven (2003) [33] find evidence consistent with better property rights leading to higher growth through improved asset allocation. Quantitatively, the growth effect is as large as that of improved access to financing due to greater financial development. Using a unique firm-level survey database covering 54 countries, Beck et al. (2008) [34] investigate the effect of financial, legal, and corruption problems on firms' growth rates. They show that financial and institutional development weakens the constraining effects of financial, legal and corruption obstacles.

It seems that an overwhelming weight of the evidence from this body of research is that Schumpeter was right, i.e., financial development is good for economic growth, though, a consensus has not yet to emerge. Guariglia and Poncet (2008) [35] use data for 30 Chinese provinces over the period 1989-2003 and find that traditionally used indicators of financial development is negatively associated with growth its sources. According to them, this is because of financial distortions in China, which represent an impediment to growth. Hasan et al. (2009) [9] also argue that bank loans that are predominantly non-performing loans to SOEs are hardly likely to be growth inducing. From the literature, we conjecture that financial development is conducive to growth only when the financial system is efficient and the financial market is not heavily distorted.

3 Variables and data

We will use city-level data to examine the relationship between financial development and economic growth in recent China. Before that we must construct a number of indicators to measure financial development. And we also list some control variables and put them into different conditioning sets.

3.1 Variables

The dependent variable in our regression model is growth, which is measured by the annual growth rate of real GDP. The independent variables consist of variables indicating financial development and a conditioning information set controlling for other factors affecting economic growth.

We construct four indicators to measure financial development.

Loans, which is the ratio of total loans of national banking system to GDP. This ratio measures financial deepening.

Deposits, which is the ratio of total deposits of national banking system to GDP. This ratio measures the overall size of financial intermediaries and their capacity to mobilize capital.

Savings, which is the ratio of total household savings deposited in the financial institutions to GDP. This ratio serves as a proxy for the degree of China's financial development in mobilizing household savings.

Loan to appropriation, which is the share of fixed asset investment financed by domestic loans relative to the share financed by state fiscal appropriation. Fixed asset investment, which forms the capital, may be financed by difference sources, such as domestic loans, government appropriation, foreign investment, and self-raised funds. Among these various sources, loans are considered more efficient than state appropriation in terms of capital allocation. Following the literature (i.e., Liu and Li, 2001 [36]; Chen, 2006 [10]; Zhang et al., 2012 [37]), we use this ratio to measure the substitution of more market- and profit-oriented financial transactions for state fiscal appropriation in order to allocate capital more efficiently.

We put control variables into different conditioning information sets. The following control variables are chosen partly because they are conventionally used in the finance-growth literature, and partly because they are available in the data. We divide these control variables into four different conditioning information sets mainly to further test the sensitivity of the empirical results. The sets are defined as follows (and these variables are taken logarithms before entering into regression).

Small set. The logarithm of GDP in pervious year (Initial GDP) to capture the convergence effect, the logarithm of 'students enrollment of regular institutions of higher education per 10,000 persons' (Human capital) to control for the effect of human capital on growth.

Medium set. The small set plus the share of 'persons employed in private enterprises and self-employed individuals in urban areas' in 'persons employed in various units at year end' (Private) as a proxy for the progress of economic reforms, and the Consumer Price Index (CPI) to control for inflation.

Large set. The medium set plus the ratio of foreign direct investment to GDP (FDI) to measure the degree of openness of the local economy, and government expenditure to GDP (Government) to control for the government size of each city and its role.

Full set. The large set plus the ratio of business volume of postal and telecommunication services to GDP (Postal & Telecom) to indicate the role of information communication, and the density of roads (Infrastructure) as a proxy for local infrastructure condition. The density of roads is measured by the ratio of area of city paved roads to total land area of that city.

3.2 Data

Data in our sample are city-level. China has altogether 658 cities at three levels. Level 1: Four municipalities which are directly under the central government. They are Beijing, Shanghai, Tianjin and Chongqing. These 4 municipalities together with the other 22 provinces and 5 autonomous regions constitute the 31 provincial-level regions of mainland China. Level 2: 286 prefecture-level cities which are under the leadership of the provinces and autonomous regions. Level 3: 368 county-level cities which are under the leadership of the prefecture-level cities.

Our sample consists of 4 municipalities and 282 prefecture-level cities (we delete 4 prefecture-level cities from all the 286 because they lack complete data; one of them was just established in 2012).

Our dataset covers 286 Chinese cities over 2008-2013. All the data are from China City Statistical Yearbook and China Statistical Yearbook for various years.

4 Finance and growth: cross-sectional analyses

We first conduct a cross-sectional analysis focusing on the 'initial values' of financial indicators and control variables in the conditioning information sets of 2008 with the average value of the dependent variable over 2008-2013. That is, there is only one observation in each city. The basic cross-sectional model is as follows:

$$Growth_i = \alpha + \beta Finance_i + \left[Control\ Variables\ in\ the\ Set\right]_i' \gamma + \varepsilon_i \quad (1)$$

where i=1, 2, …, 286 represent the 286 cities in our sample. The dependent variable is the average real GDP growth rate over 2008-1023, Finance takes values of each of the four financial indicators defined in Section 3.1, and control variables in different sets are also defined in Section 3.1.

The initial value regression can alleviate the critical weakness of the 'contemporaneous' regressions where dependent and independent variables are all averaged over the same period, notwithstanding, this analysis doe not address the issue of causality. There is a pitfall in the contemporaneous regressions, i.e., it is possible that a common shock to the dependent and independent variables occur in the same period (rather than the causation between them) that drives the empirical results. In other words, contemporaneous regressions may disregard the potential endogenous determinations of the dependent and independent variables.

Table 1 summarizes the OLS regression results with control variables in different sets. The results indicate a negative relationship between financial development and economic growth. We only report the coefficients on financial indicators to save space. Loans basically do not show any significant impact on growth, while the other three financial indicators all demonstrate significant negative effect on growth. In a word, our cross-sectional analyses do not show any positive role of financial development in economic growth.

Table 1. Finance and growth: OLS estimations (initial value regressions).

Conditioning information set		Loans	Deposits	Savings	Loan_to_Appro.
Small set	Coefficient	0.148	-0.492	-0.435	-0.334*
	Standard error	0.3663	0.3604	0.4739	0.1790
	R-squared	0.103	0.110	0.107	0.117
	Observation	274	274	274	274
Medium set	Coefficient	0.087	-0.435	-0.366	-0.249
	Standard error	0.3545	0.3416	0.4537	0.1521
	R-squared	0.176	0.181	0.179	0.183
	Observation	273	273	273	273
Large set	Coefficient	-0.194	-0.704*	-0.685	-0.391**
	Standard error	0.2953	0.3832	0.4232	0.1681
	R-squared	0.200	0.210	0.208	0.214
	Observation	265	265	265	265
Full set	Coefficient	-0.085	-0.575	-0.520	-0.349**
	Standard error	0.3078	0.4366	0.4853	0.1661

	R-squared	0.206	0.212	0.210	0.217
	Observation	264	264	264	264

***, **Significance levels at 1% and 5% respectively.

5 Finance and growth: dynamic panel analyses

5.1 Model specification and methodology

To examine the effect of financial development on GDP growth, we construct our basic regression model as follows:

$$Growth_{it} = \alpha + \beta Finance_{it} + \left[Conditioning\ Information\ Set\right]_{it}{}'\gamma + \mu_i + \lambda_t + \varepsilon_{it} \quad (2)$$

where the subscript i stands for city i (i=1, ..., 286) and t stands for year t (t=2008, ..., 2013). μ_i are unobserved city-specific effects, λ_t are time fixed effects, and ε_{it} is an idiosyncratic error term.

Table 2. Finance and growth: first-differenced GMM estimation.

Conditioning information set		Loans	Deposits	Savings	Loan_to_Appro.
Small set	Coefficient	-4.950	-3.049*	-2.801*	-1.699**
	Standard error	3.0238	1.8078	1.6030	0.7442
	Sargan test (p-value)	0.605	0.365	0.392	0.650
	Instruments	14	14	14	14
	Observations	1588	1588	1586	1615
Medium set	Coefficient	-3.331	-2.312	-2.617	-1.993***
	Standard error	3.1666	2.0885	1.8856	0.7253
	Sargan test (p-value)	0.225	0.153	0.172	0.454
	Instruments	17	17	17	17
	Observations	1566	1566	1565	1593
Large	Coefficient	-8.702**	-9.309*	-8.431***	-2.021***
	Standard error	4.0807	4.7401	3.1242	0.4752
	Sargan test (p-value)	0.522	0.345	0.518	0.826
	Instruments	21	21	21	21
	Observations	1494	1494	1494	1511
Full	Coefficient	-6.643	-2.623	0.585	-2.020***
	Standard error	4.4252	4.5618	4.2678	0.3660
	Sargan test (p-value)	0.153	0.227	0.295	0.228
	Instruments	25	25	25	25
	Observations	1478	1478	1478	1495

Notes: The test statistics and standard errors are asymptotically robust to heteroscedasticity.
***, **, and * stand for significant level at 1%, 5%, and 10% respectively.

We use generalized method of moments (GMM) estimators to examine the effect of financial development on economic growth based on our sample dataset. There are several advantages of using GMM estimators for dynamic panel data (DPD) models. Firstly, we can control for both the city-specific effects and time effects. Secondly, we can choose appropriate lags of the independent variables as instruments to address the possible endogeneity in the regressions. As to the relation between financial development and economic growth, the joint determination between them may cause a simultaneous bias which would produce inconsistent estimators. Meanwhile, variables in the conditioning information sets may also suffer from the endogeneity problem. By using GMM estimators, we can solve these econometric problems with lagged observations of the independent variables as instruments. In this way we are capable of examining reliably the effect of exogenous components of financial development on economic growth. To save space, we omit the description of the econometric details of the GMM regressors.

Table 3. Finance and growth: system GMM estimations (full set)

Regressors	model 1	model 2	model 3	model 4

	(collapsed)	(collapsed)	(collapsed)	(collapsed)
Loans	-2.146 (1.8646)			
Deposits		-2.126 (3.7707)		
Savings			-5.335 (3.6973)	
Loans to Appro				-2.057*** (0.4028)
Initial GDP	-0.183 (0.3142)	-0.244 (0.2897)	-0.168 (0.3125)	-0.344 (0.3220)
Human capital	-0.333 (1.0909)	0.096 (1.3941)	-0.439 (0.8653)	-0.417 (1.1815)
Private share	0.958 (1.0989)	0.981 (1.2927)	0.312 (1.1591)	0.947 (1.0985)
CPI	0.140 (0.3507)	0.024 (0.3969)	0.020 (0.3328)	-1.132 (0.3031)
FDI	0.577 (0.8492)	0.815 (0.8171)	0.585 (0.7983)	0.602 (0.6857)
Government	-8.294* (4.4644)	-10.538** (4.3793)	-6.333 (5.7600)	-8.681* (4.6916)
Postal & Telecom	-0.021 (1.0395)	0.495 (1.0255)	0.629 (1.1645)	0.219 (1.0610)
Infrastructure	-0.3432 (2.0860)	-1.544 (2.2108)	-0.523 (1.6755)	-0.762 (2.1514)
Dummy yr2009	3.005 (2.4339)	2.595 (2.6965)	2.026 (2.2450)	0.4158 (1.9432)
Dummy yr2010	6.162*** (1.9070)	6.326*** (1.8547)	5.450*** (2.1014)	4.386*** (1.5854)
Dummy yr2011	2.456* (1.2511)	3.152*** (1.1970)	2.005 (1.5926)	1.908 (1.3007)
Dummy yr2012	2.622 (1.8921)	3.275* (1.8505)	1.903 (2.2233)	0.869 (1.7179)
Dummy yr2013	2.393 (2.3417)	2.826 (2.6401)	1.677 (2.6113)	0.642 (2.1280)
Dummy yr2014	10.855 (11.1488)	11.826 (12.9924)	4.040 (11.7567)	8.2945 (10.8745)
Lagged growth rate	0.481*** (0.0719)	0.517*** (0.0813)	0.475*** (0.0791)	0.436*** (0.0657)
Constant	-27.095 (48.6725)	-9.342 (56.5285)	-2.108 (42.9100)	12.225 (42.3333)
Observations	1788	1788	1788	1807
Instruments	26	26	26	26
Sargan test (p-value)	0.990	0.920	0.997	0.988
Arellano-Bond test for AR(2) (p-value)	0.742	0.904	0.868	0.605

Notes: The standard errors are in parentheses. The test statistics and standard errors are asymptotically robust to heteroscedasticity. ***, ** and * stand for significance levels at 1%, 5% and 10% respectively.

5.2 Regression results

Table 2 presents the regression results based on differenced GMM estimators. The results show that financial developmnet has no significant positive effects on economic growth: some financial indictors show a statistically negative impact on growth, for example, the variable Loan_to_Appro has negative effects on growth in all four models with four different sets of control variables, and all the four financial development indicators present significant negative effects on growth in the large set. The Sargan test of over-identifying restrictions, as well as the test for AR(2), is satisfactory. In a word, Table 2 does not offer any evidence of positive effect of financail development on growth.

Table 3 reports the system GMM estimates of Eq. (2) with the full set. We conduct two different regressions for each financial indictors by collapsing and un-collapsing the instruments in order to guarantee the credibility of system GMM estimators. Table 3 reports only the results from the collapsed regressions. Regression results in Table 3 are basically consistent with that of Table 2. Only Loan_to_Appro shows a significent negative effect on growth, and the other three financial indicators do not show any significant effects on economic growth. All the regressors pass the second serial correlation test and the Sargan test. All the p-values of the Sargan test comfortably satisfy the conventional significance levels with an average value of 0.9738.

6 Conclusion

This paper examines the relationship between financial development and economic growth in China. We use data from 286 Chinese cities over the period 2008-2013. We employ first-difference and system GMM estimators for dynamic panel data. We do not find any significant positive effect of financial development on growth over this period. Most of the financial indicators do not have any significant effect on economic growth, while some of the indictors eve have significant negative effect on growth in the cross-sectional analyses. Our findings are consistent with many existing studies on China that suggest that financial development hinds economic growth due to the distorting nature of state-ruled (even monopolized) banking sector, but run contrary to most cross-country studies on the relationship between financial development and economic growth. Thus, our findings suggest that the scarce capital had not flowed to the healthy and efficient firms who urgently need money, but was diverted to those inefficient large SOEs. The capital allocation mechanism of China's financial system has not been wholly market-oriented yet, which implies that there is still a long way to go for China's financial system reform.

Theoretically, we have a few remarks. Few would be inclined to deny that there is a rough parallel between economic and financial development. As real income and wealth increase, so do the size and complexity of the financial superstructure. Yet this is a loose relationship. It is hard to argue that a given volume or composition of financial assets is a sufficient condition for the development of real sectors of the economy - or even a necessary condition, given that rapid growth has sometimes taken place during periods of deliberate financial repression. We do admit, however, that financial innovation has at times sparked off virtuous circles of growth in particular sectors and regions, such as the successful microcredit movement in Bangladesh. As Greenwood and Jovanovic (1990) [38] said, if building a functional formal sector of financial intermediaries is arduous and costly, the evolution of financial structure and real economic development may well be mutually determined, with causation flowing in both directions.

References

1. J.A. Schumpeter, *The Theory of Economic Development*. Cambridge, MA: Harvard University Press (1911).
2. R.W. Goldsmith, *Financial Structure and Development*. New Heaven: Yale University Press (1969).
3. R.I. McKinnon, *Money and Capital in Economic Development*. Washington DC: Brookings Institutions (1973) .
4. J. Robinson J., *The Rate of Interest and Other Essays*. London: Macmillan (1952) .
5. R.E. Lucas Jr., Journal of Monetary Economics **22**, pp. 3-42 (1988).
6. NBS (National Bureau of Statistics of China), *China Statistical Yearbook 2016,* Beijing: China Statistics Press (2016).
7. NBS (National Bureau of Statistisc of China), *China China Compendium of Statistics 1949-2008*, Beijing: China Statistics Press (2010).
8. G. Boyreau-Debray, The World Bank, Policy Research Working Paper Series 3027 (2003).
9. I. Hasan, P. Wachtel, M. Zhou, Journal of Banking & Finance **33**, pp. 157-170 (2009).
10. H. Chen, China Economic Review **17**, pp. 347-362 (2006).
11. X. Cheng, H. Degryse, Bank of Finland Institute for Economies in Transition, BOFIT Discussion Papers 22/2007 (2007).
12. World Bank, *Economic Growth in the 1990s: Learning from a Decade of Reform*. Washington, DC: World Bank (2005).
13. R.G. King, R. Levine, Quarterly Journal of Economics **108**, pp. 717-138 (1993).
14. M.H. Miller, Journal of Applied Corporate Finance **11**, pp. 8-15 (1998).
15. R. Levine, S. Zervos, American Economic Review **88**, pp. 537-558 (1998) .
16. R. Levine, Journal of Financial Intermediation **8**, 8-35 (1999) .
17. R. Levine, N. Loayza, T. Beck, Journal of Monetary Economics **46**, pp. 31-77 (2000) .
18. F. Rioja, N. Valey, Economic Inquiry **42**, pp. 127-140 (2004).
19. T. Beck, R. Levine, N. Loayza, Journal of Financial Economics **58**, pp. 261-300 (2000).
20. P.L. Rousseau, P. Wachtel, Journal of International Money and Finance **21**, pp. 777-793 (2002) .
21. T. Beck, R. Levine, 2004, Journal of Banking & Finance **28**, pp. 423-442 (2004).
22. Z. Xu, Economic Inquiry **38**, pp. 331-344 (2000).

23. D.K. Christopoulos, E. G. Tsionas, Journal of Development Economics **73**, pp. 55-74 (2004).
24. Bekaert B.G., C.R. Harvey and C. Lundblad, 2005, 'Does financial liberalization spur growth?' Journal of Financial Economics, Vol. 77, pp. 3-55.
25. J. Jayaratne, P. Strahan, Quarterly Journal of Economics **111**, pp. 639-670 (1996) .
26. R. Dehejia, A. Lleras-Muney, National Bureau of Economic Research, Working Paper No. **9551** (2003).
27. P.L. Rousseau, Japan and the World Economy **11**, pp. 185-198 (1999).
28. L. Guiso, P. Sapienza, L. Zingales, Quarterly Journal of Economics **119**, pp. 929-969 (2004).
29. R. Rajan, L. Zingales, American Economic Review **88**, pp. 559-586 (1998).
30. K. Kumar, R. Raghuram, L. Zingales, NBER Working Paper No. w.**7208**. Available at SSRN: http://ssrn.com/abstract=227556 (1999).
31. J. Wurgler, Journal of Financial Economics **58**, pp. 187-214 (2000) .
32. N. Cetorelli, M. Gambera, Journal of Finance **56**, pp. 617-648 (2001).
33. S. Claessens, L. Laeven, Journal of Finance **58**, pp. 2401-2436 (2003).
34. T. Beck, A. Demirguc-Kunt, L. Laeven, R. Levine, Journal of Money, Credit and Banking **40**, pp. 1379-1405 (2008).
35. A. Guariglia, S. Poncet, Journal of Comparative Economics **36**, pp.633-657 (2008) .
36. T. Liu, K. Li, Journal of Asian Economics **12**, pp. 245-262 (2001).
37. J. Zhang, S. Wang, Journal of Comparative Economics **40**, pp. 393-412 (2012).
38. J. Greenwood, B. Jovanovic, J. Poli. Econ. **98**, pp. 1076-1107 (1990).

Study on Co-branding Strategy- Dongfeng Yueda Kia as Example

Jinrong Yu [1,2*] , Ying Song [1], Yin Lu [1,2] and Dong Qi [1]

[1]Nanchang Hongkong University, Nanchang, Jiangxi 330063, China
[2]Jiangxi University of Finance and Economics, Nanchang, Jiangxi 330013, China
Corresponding E-mail: tecomyjr@126.com

Abstract. Owing to the fact that the science and technology is developed increasingly, the product life cycle trend to shorten. Under this fierce competition, enterprises have to launch new product quickly in order to survive. Now, a growing number of enterprises release the new product based on existing brands, which has been successful. The status quo of joint brand strategy, combined with the co-branding of Dongfeng Yueda Kia, has been analyzed by regression analysis method. By doing empirical validation through developing the theory model, the result concluded that product compatibility and brand compatibility have a noticeable impact on co-branding evaluation. At the same time, the attribute of product has also some adjusting impact. That is, when enterprises develop a joint brand strategy, the attribute of product should be concerned. The paper put forward recommendations and solutions about how to improve a joint brand strategy from enterprises choosing partners, consumer's perception to co-branding.

1 Introduction

As the product homogeneity, enterprises have to launch differentiated product which could adapt to the changing market demand in order to gain competitiveness for sustainable development. Brand has been key strategy resource of enterprises development. The continuous improvement of science and technology has been getting shorter the product life cycle and enterprises face increasingly fierce competition. At present, through releasing new products from existing brands, more and more enterprises have succeeded. But a growing number of enterprises come to use co-branding strategy as the brand extension strategy may dilute the parent brand's assets. Co-branding strategy is growing up in the business for it could overcome faults of brand extension.

Dongfeng Yueda Kia (DYK) is familiar to many, which often list in consumer's Plan B. In spite of this, it is ignored by consumers in their purchase. DYK want to enhance the market share through combining with IMAX(Wanda Cinema) to improve the brand awareness. By analyzing the case, the paper study four dimensions, that is: product compatibility, brand compatibility, product attribute, co-branding evaluation, and their impact to the co-branding strategy, which can provide some academic bases for enterprises choosing brand partner and promotion.

2 Theory background and hypothesis

2.1 Theory background

The theory of co-branding comes from brand extension. From the beginning, it focuses on entity. Most of scholars define co-branding in terms of brand extension. In his study, Riezebos(2003) suggests that different corporations or enterprises combine two separate brands into a unified name and then release a new product jointly.

Gammoh & Voss(2006) note that when new brands enter the consumer market and combine with the high profile brands, it will gain more better market effect than new brands do alone [1]. Park(1999) concluded that when two noted brand are combined and extended to other product area, this extension product has stronger market response than do the product whose brand is directly extended. He also note that co-branding product which is combined by two highly complementary. brand could make consumers form more positive attitude [2].Rao(2002) applied the signal theory and found that when consumers are not sure to the quality of co-branding product, it could raise perceived quality of the new product that the brand combines with famous brand product with good image [3]. Paul (2008) applied long-memory models and concluded that consumer's trust to co-brands will play an important role in brand alliance and the trust could prompt consumers to make a positive assessment [4]. Washburn (2004) suggests that brand equity of co-brands affects consumer's attitude to brand alliance considerably, and make consumers try the product, which will further deeply the affect [5].

Form the economics, JiNan Xu(2005)notes that after enter the market, two jointed brands gain more profit than two brand do respectively, which has great economic significance [6] . ChangHui Ning(2006) studies the key factors impacting brand joint in theory. The result shows that brand alliance has a very important strategy operating value [7]. Yang Liu (2008)focus on consumer's evaluation factors on the attitude of co-branding . His viewpoint is that consumer's attitude to brand alliance as well as product compatibility and brand compatibility has a great affect on evaluation [8]. ZhongCheng Guan (2006) discusses perception of consumer has close relation with their attitude to brand alliance. He points that each co-brands has its appropriate position respectively in consumer's origin memory. When consumer could aware that some kind of position of two co-brands are highly complementary, then it will reinforce perceived performance, enhancing consumer's perceived effect to new co-brands [9]. ZhiGang Fan (2009) holds that consumer's origin attitude and familiarity to co-brands has impact on their evaluation and matching degree between co-brands is crucial [10]. From the consumer learning theory, GongGuang Fan (2008) studies empirically that brand relationship quality has regulatory effect on brand alliance compatibility and evaluation. The result demonstrates that the better brand relationship is, the more obvious their path is [11].

Overseas scholars study brand alliance earlier and have had a lot of achievement compared to domestic ones, who have limit studies, single perspective and most focus on existing summary about the theory of brand alliance, lacking of adaption. Through studying the impact on co-branding evaluation from product compatibility, brand compatibility and product attribute, the paper emphasizes product attribute and whether it can adjust product compatibility and brand compatibility. The conclusion will be drawn by empirical, which could provide theoretical basis to enterprises

2.2. Status of co-branding matching

There have various ways to co-branding either in physics or in forms. In the existing market and studies, two brands combine is common. So the paper just discusses this case.

The type of co-branding could determine its effect on the result. The co-branding is to be classified as follows based on how the size of value will be created [11].

(1) The type of contact & cognition: the two cooperating brands take advantage of resources each other to complement, enlarging the popularity in the new target market, which increases the combined brands' recognition.

(2) The type of value endorsement has higher requirements to enterprises. It wants the cooperating brand to improve the established value, leading to the less choosing partners. The two brands could increase popularity through co-branding respectively; enlarging product's marketing and gaining much spillover effect.

(3) The type of element association is common in the market. A product that contains a lot of good material element is taken as one of element combined with another high quality brand, a basic law of this type. Together with increasing value each other, the partners could also share publicity expenses. Manufacturers and suppliers are willing to use it to deliver product quality information to consumers.

(4) The type of complementary capability is the highest level, which not only combines a few elements but allows two complementary partners to put into their most influential core technology to produce product or service. In this type, cooperating time is longest and then has the highest level of participation and value.

3 Research hypothesis

According to theoretical model, as well as the research basis, raise the hypothesis of co-branding evaluation of product attribute and brand attribute based on questionnaires an empirical.

Park and other scholars, through the experiment, pair chocolate "Ferroero", weight-loss good spirulina and ice-cream "Haagen-dazs". "Ferroero" is delicious but high-calorie. Spirulina is low calorie but poor taste and "Haagen-dazs" is the same to "Ferroero". The former combination is major brand allying the modifier which has high level of compatibility and low level of popularity but the later is low level of compatibility and high level of popularity. The result shows that the former is more appealing as the new product has both ideal features, that is, delicious and low calorie.

When combined by the complement of the product function, how could DYK pair to appeal to consumers and impulse the purchasing desire?

So, put the hypothesis as follows:

H1: product compatibility has a significant positive correlation to co-branding

Domestic scholar YuTian Huang defines it as brand compatibility. That is, when two brands are combined, consumers may feel the degree to which the typical products can further meet their special demand. If consumers believe there is high level of compatibility between two brands, then there will be more cooperation and complementary effect of their professional ability and technical support, which can not only perform the advantages moderately, leading to the new co-branding product more practical and perfect, but enjoy convenient and value in one shopping. They usually don't consider the unsuitable between the brands, but heighten degree to identity for new product shortly. By contrast, considering low level of compatibility, consumers will suspect why two products are combined and even develop the bad evaluation.

Put forward to the hypothesis accordingly:

H2: brand compatibility has a significant positive correlation to co-branding evaluation

There are usually three ways to classify products: (1) by use: (2) by tangible or not; (3) by endurance, which is simple and understandable but rough to classify products through attribute themselves. In our study, products will be classified functional ones and enjoyable ones according to attribute. Mittal suggests products could be classified two. One emphasizes performance or function, such as, razor and micro oven, and the other self-expression or pleasure, such as, cosmetics and beauty services. However, these attributes are not absolute. That is, there are not simple functional products or enjoyable ones, but have two attributes more or less.

Some study theory, while there are not too much ones deeply on product attributes impact on co-branding evaluation, could provide the base. When considering the impact of combined compatibility on its evaluation, Simonin and other study vehicles. In some ways, cars have not been mere transportation, but the symbol of status and are enjoyable products now. The finding concludes that to the combined compatibility, co-branding plays a dominant role. The domestic scholar Fang Wu takes milk as study subject. To consumers, milk is more for basic nutritional need and is the functional products. The conclusion is that the compatibility of products has more significant to co-branding evaluation.Thus, the paper concludes that the different dimension of the combined compatibility has different impact on co-branding and has certain adjustable effect on product attribute. Put forward to the hypothesis accordingly:H3: when functional attribute surpasses enjoyable, product compatibility has more impact on co-branding evaluation.H4: when enjoyable attribute surpasses functional, brand compatibility has more impact on co-branding evaluation. The paper emphasizes combined compatibility, which product compatibility and brand compatibility have direct effect, and product attribute has adjustable effect.

4 Methods

4.1 Samples and data

The research takes the co-branding of DYK as subject and Chinese individual consumers as object. Given co-branding is a new marketing pattern, and young people has a strong ability to accept new knowledge, they are suitable research objects. The samples are fresh graduates, which has planned to purchase car in five years. Form Feb. 2014 to June 2014, the survey distributed 180 questionnaires and collected 176, and the collected rate is 97.8%, which have 167 valid ones, the rate is 92.8%.

4.2 Questionnaire design and survey

The measurement of co-brand evaluation is designed by combined compatibility and co-brand evaluation. The questionnaire refers to other scholars' researches (Aaker(1991) co-brand evaluation; Park, Milberg, Lawson(1990), Simonin(1998) product compatibility evaluation; Varadrajan(1985), Keller(1990) brand compatibility evaluation; Aaker, Keller(1990)co-brand evaluation) and modify these questionnaires combining our research. All measurements quesinsert five scales, including "very agree", "agree", "general", "disagree", "strongly disagree" five options, and give each level corresponding score, successively define of 5 points,, 4 points, 3 points, 2 points, 1 point, the respondents can answer according to their own consumer experience.This study analyzed reliability measurement scale by SPSS20.0, analysis results shown in Table 1. The results show that the total scale reliability was 0.862, indicating that this empirical study of scale has high internal consistency, meet the requirements of research. Brand evaluation, product matching, four dimensions Cronbach α values brand matching , co-branding evaluate were 0.740,0.885,0.868,0.872, are larger than the standard value of 0.7, indicating that the four dimensions has internal consistency.

Table 1. Reliability range: Cronbach.

Measurement table	Cronbach	total	judgment
total reliability analysis	0.862	14	effective
brand evaluation	0.740	5	Effective
product matching	0.885	2	Effective
Brand Matching	0.863	3	effective
Co-Branding evaluation	0.872	4	effective

4.3 Questionnaire analysis of reliability and validity

We use internal consistency coefficient (Cronbach'α) test variables reliability, variables validity include content validity and construct validity. Since the measurement scale used in this paper are from abroad representative literature, through a number of certification and inspection to ensure that scale used herein has good content validity. This study analyzed reliability measurement scale by SPSS20.0, analysis results shown in Table 1. The results show that the total scale reliability was 0.862, indicating that this empirical study of scale has high internal consistency, meet the requirements of research. Brand evaluation, product matching, four dimensions Cronbach α values brand matching , co-branding evaluate were 0.740,0.885,0.868,0.872, are larger than the standard value of 0.7, indicating that the four dimensions has internal consistency.

4.4 Sample description

The total questionnaires survey is 180, returned 176 questionnaires, 167 valid questionnaires, the effective rate of 92.8%, the analysis results are shown in Table 2.

Table 2. Sample characteristics describes: education, gender, income.

Sample characteristics	total number of sample	(%)
sale	70	42%
female	97	58%
Government / institution personnel	1	0.5%
Employees	33	19.8%
Financial	6	3.5%
others	26	15.6%
0-1000 元 yuan	47	28.1%
1000-3000 yuan	57	34.1%
3000-5000 yuan	62	37.1%
More than 5000 yuan	1	0.07%

4.5 Handling test

The study asked two key indicators as a measure of matching, in this paper, select "Shampoo + Hair Dryer, and" watch + phone" as a product mix high product matching, and "Shampoo + computer" and "watch + TV", as a matching low product portfolio. This article will use the independent sample T test (Independent Samples Test) method to test products matching level if it cause the difference between the average index, and compare its size.

Can be seen from Table 3, respondents in the higher co-branding products matching case, matching the product index score is higher than in the case of products matching low scores, and the difference is remarkable. Therefore, this study can be considered that the product matching handling achieve desired effect.

In this study, the two asked matching item as a brand metrics, In this paper select "KIA + Michelin" , as a high matching product portfolio, and "KIA + Wanda" as a low matching product portfolio. This article will use the independent sample T test (Independent Samples Test) method for testing products matching the level which causing the difference between the average index, and compare its size, the test results shown in Table 4.

As can be seen from Table 4, respondents under the brand co-branded high matching case, matching the brand index score is higher than in the case of the brand matching low scores, and the difference is remarkable. Therefore, this study can be considered the brand match manipulation to achieve the desired effect.

Table 3. Items matching data manipulation.

product mix		mean	F value	P value
Shampoo + Hair dryer	high match	6.03	75.84	0.000***
Shampoo + computer	matching low	3.81		

Table 4. Brand matching data manipulation

Portfolio		mean	F value	P value
KIA + Michelin	high match	4.35	9.359	0.008***
KIA + Wanda	low match	2.73		

4.6 Multiple regression analysis

In this study, the product matches, the brand as independent variables, co-branding evaluation as the dependent variable, product attributes as the moderator, then begin to multiple regression.

H1 Inspection:

In this paper, conducted a meta regression analysis to independent variable (product matching) and the dependent variable (Brand Joint Evaluation) using SPSS19.0 , the results shown in Table 5.

The regression analysis of Table 5, to all brands, when enterprises branding, product matching significantly influence consumer about brand joint evaluation ($P < 0.05$). That is, the higher product matching between the co-branded, the higher consume brand joint revaluation. Thus, assuming the establishment HI.

H2 Inspection:

Perform simple regression analysis to independent variables (brand match) and the dependent variable (brand joint evaluation) by SPSS19.0, the results are shown in Table 6.

From Table 6 Regression analysis shows that, to be analyzed for all brands, when corporate brand association, brand matching significantly affect consumer brand joint evaluation ($P < 0.05$), that is to say the higher the brand match between co-branding , the higher consumer brand joint evaluation. Thus, assuming the establishment H2.

H3 Inspection

Dongfeng Yueda Kia is the leading brand of the two portfolio, product functionality that is product matching attributes, enjoy attributes is brand matching, perform binary regression model analysis the relation between two matching joint with co-branding evaluation, the results shown in Table 7.

Table 7 Regression analysis shows that the product matching coefficient is 0.513, more than brand matching coefficient 0.226, indicating when the KIA and the Michelin tire co-branding, product functional properties beyond the product enjoyment property, and adjusted R^2 is 0.821, explanation that the model was well fitted, table 4 ANOVA F statistic is 57.356, the P values is less than the significance level ($P < 0.05$). Thus, assuming the establishment of H3.

H4 Inspection:

Table 8 regression analysis can be seen, when compare with products matching coefficient and brand matching coefficient, partial regression coefficient of T test between two variables is less than the probability P significance level (0.05), the brand matching coefficient is greater than the product matching coefficient, description brand matching influence larger than the product matching. In addition, Table 6 variance analysis shows that, F statistic is 40.665 less than probability P, namely two independent variables have significant linear relationship to the dependent variable brand joint evaluations . Therefore, assuming the establishment of H4.

Table 5. Items matching regression analysis.

Model		Non-standardized coefficients		standard coefficient	T value	Sig.	R2
		regression coefficient	Standard error	trial version			
1	(constant)	1.931	0.064		0.000	1.000	0.861
	(Products match)	0.565	0.064	0.565	8.789	0.000***	

Table 6. Brand Match Regression Analysis.

model		Non-standardized coefficients		standard coefficients Beta	T value	Sig.	R2
		regression coefficients	standard error				
1	(constant)	1.417	0.065		0.000	0.000	0.733
	1Match brand 1	0.541	0.065	0.541	8.258	0.000***	

Table 7. KIA with Michelin tires co-branding

Model		Non-standardized coefficients		standard coefficient	T value	Sig.	R2	adjusted R2
		regression coefficient	Standard error	Beta				
1	(constant)	0.048	0.060		12	0.000	0.854	0.821
	Products match	0.513	0.066	0.513	7.790	0.000***		
	Brands match	0.226	0.066	0.226	3.436	0.001**		

Table 8. KIA with Wanda Cinema co-branding.

Model		Non-standardized coefficients		standard coefficient	T value	Sig.	R2	adjusted R2
		regression coefficient	Standard error	Beta				
1	(constant)	0.032	0.064		1.65	0.000	0.745	0.685
	Products match	0.380	0.123	0.380	3.095	0.002**		
	Brands match	0.516	0.123	0.516	7.962	0.000***		

Table 9. Summary of findings.

Hypothesis	research	results
H1	products matching has a significant positive correlation to the brand joint evaluation	established
H2	brand matching has a significant positive correlation to brand joint evaluation	established
H3	when co-branding functional attributes beyond the enjoyment property, the greater impact on the product matching to brand joint evaluation	established
H4	when enjoyment of property of co-branding products exceed the functional properties, the greater impact on brand matching to brand joint evaluations	established

5 Discussion

Through the H3, H4 binary regression analysis, when KIA with Michelin tires joint, the product functionality attributes played a leading role, while union with Wanda Cinema, product enjoying attributes played a dominant role. We found that in the two sets of experiments, the product attributes play a regulation role in co-branding and joint matching evaluation.

6 Conclusions

In this paper, empirical studies show that joint matching has two-dimensional structure, the higher joint matching, the more positive consumer brand joint evaluations. Joint matching is mainly reflected in the product matching and brand matching two levels, where the product matching is mainly reflected product function has a certain complementarity is that whether total product utility of consumers is improved; and brand matching is reflected in whether brand image, brand position and brand personality are the same of co-branded. When companies choose to cooperate brand, not only consider whether product level bring a better functional value, but also consider whether two sides of brand level coordinated combination.

DOI: 10.1051/matecconf/201710005033

Conflict of interest

The author confirms that this article content has no conflict of interest.

Acknowledgements

This work was financially supported by project of national social science funds: (14BJY075), Zhejiang natural science funds: (LY13G030034),Nanchang social science funds Jj201506.

References

1. Gammoh B. S, Voss K. E, Chakraborty G. Consumer Evaluation of Brand Alliance Signals.Psychology & Marketing, 2006, 23(6): 26-42.
2. Park C. W, Jun S. Y, Shocker A. D. Composite Branding Alliance: An Investigation of Extension and Feedback Effect. Journal of Marketing Research, 1999, 33(4): 53-66.
3. Rao A.R, R.W. Ruekert. Signaling Unobservable Product Quality through a Brand Ally.Journal of Marketing Research, 2002, 36(2): 258-268.
4. Paul Benjamin Lowry, Anthony Vance, Greg Moody. Explaining and Predicting the Impact of Branding Alliance and Web Site Quality on Initial Consumer Trust of E-Commerce Web Sites.Journal of Management Information Systems, 2008, 24(4): 199-224.
5. Wash burn, BD Till, R Priluck. Brand Alliance and Customer-based Brand-equity Effects.Psychology and Marketing, 2004, 16(22): 48-50.
6. Xu Jinan. Co-branding. Jiangxi University of Finance and Economics, 2005, 12 (4): 38-41.
7. Ning changhui. Antecedents and Spillover effects analysis Co-branding .Finance and Trade Economics, 2006, 12 (8): 87, 90.
8. Liu Yang, Qiu Ling, Chen Jingyu. SR model test and expand based on brand joint evaluation consumer attitudes. Statistics and Decision, 2008, 8 (5): 89-92.
9. Guan Zhongcheng, Cheng Gang. Brand Joint strategy study. Chongqing University journal, 2006, 6 (12): 38-43.
10. Fan Zhigang, Wu Xiaobo. Thinking about the corporate co-branding strategy .Commercial Research, 2009, 18 (4): 21-24.
11. Fan Gongguang, Wu Fang. Moderating Factors of brand matching and co-branding Evaluation - Based on Empirical analysis of brand relationship quality . technical and economic, 2008, 27 (10): 108-109.

Research in the Agglomeration Level of Textile Industry in Yangtze River Delta Region of China

Hao Zhou * and Jiaona Wan

School of Economics, Wuhan Textile University, Wuhan, China
Corresponding Email: zhouhaoli@163.com

Abstract. This paper uses the EG index, which adjust accordingly to the textile industry data of Yangtze River Delta region of China to measure the level of textile industry agglomeration in Yangtze River Delta region of China from 2003 to 2014. The results show that the agglomeration level of textile industry in Yangtze River Delta region of China has experienced four stages: rapid growth, steady development, rapid decline and slow recovery. Through the analysis of the causes of the changes of agglomeration level namely EG index in each stage, this paper has found that the textile industry in Yangtze River Delta has several problems in industrial structure and enterprise development. Therefore, the corresponding recommendations were presented to promote the healthy development of the textile industry in Yangtze River Delta region.

1 Introduction

Yangtze River Delta has always been a highly developed area of China's textile industry, with its unique source of raw materials, low transport costs, low labor costs and other natural advantages, attracting a lot of capital invests into the textile industry. In its development process, the textile industry of Yangtze River Delta gradually formed industrial agglomeration. The idea of industrial agglomeration firstly appeared in Marshall's "external economic theory" and "internal economic theory" of "Principles of Economics ". Porter believes that industrial agglomeration is a form of spatial organization that creates competitive advantages in terms of efficiency, benefit and flexibility. It is generally accepted that industrial agglomeration refers to the agglomeration in a particular space of related firms in a particular industry and its associated supporting structures [1].

At present, the industrial agglomeration measurement methods used by domestic and foreign scholars mainly include industry concentration degree, location entropy, Herfindahl index, space Gini coefficient and EG index. Xiao Lei and Botang Han(2011) used the EG index to measure the industrial agglomeration level of Chinese manufacturing industry, and Get 2003-2007 industries' agglomeration level of Chinese manufacturing industry [2]. Chaoping Sun, Huiming Liu and Yong Wu (2015) used the two-dimensional namely economy and scale location entropy to measure the agglomeration level of textile industry in 26 regions of China in 2011 and 2013. It is found that China's textile industry agglomeration level is not complete a pattern that high in east and low in west [3]. Jing Hu, Luping Chen and Mi Zhou (2015) used the Herfindahl index and the spatial Gini coefficient to measure the agglomeration level of tourism industry in Hubei Province from 2005 to 2012, and compared the changes of tourism industrial agglomeration level before and after the opening of high-speed railway. On this basis, they analyzed the differential impact of high-speed railway on tourism industry and relevant industries [4].

In industrial agglomeration research field, it is clear that foreign scholars started earlier and has formed a mature theoretical system. Since the 90s, domestic scholars have also been studying industrial agglomeration and combining with Chinese national conditions to output a large number of empirical research results. It is in this context, this paper uses the EG index which is suitable for measuring the agglomeration level of the textile industry to measure the textile industry agglomeration level in Yangtze River Delta region of China in 2003-2014, which stands on the sustainable development of the textile industry in Yangtze River Delta region of China and making reasonable suggestions. In addition, the scope of Yangtze River Delta region of China in this paper includes Shanghai, Jiangsu and Zhejiang provinces.

2 Measurement of agglomeration level in textile industry

2.1The measurement method of industrial agglomeration level

At present, the industrial agglomeration measurement methods used by domestic and foreign scholars mainly include industry concentration degree, location entropy, Herfindahl index, space Gini coefficient and EG index. Considering the applicable conditions of the various measurement methods, this paper finally choose EG index to measure the agglomeration level of textile industry [5].

2.1.1 Ellison - Glaeser index

Ellison - Glaeser index referred to as "EG index", Ellison and Glaeser proposed the EG index which improved largely based on the space Gini coefficient G and Herfindahl index H to overcome the drawback that a large enterprise in a region may lead to a inflated G index in space Gini coefficient; It also avoids the uncertainty of existence of the positive correlation between Herfindahl index and industrial agglomeration level. The EG index is calculated as:

$$\gamma = \frac{G - (1 - \sum_i x_i^2)H}{(1 - \sum_i x_i^2)(1 - H)} \tag{1}$$

Among them, $G = \sum_i (s_i - x_i)^2$, $H = \sum z_j$, s_i is the ratio of employment of one industry in i region to employment of the industry in the whole nation; x_i is the ratio of employment in i region to total employment in the whole nation; z_j is the market share of one firm within one industry in j region [6].

2.1.2 The adjustment of Ellison - Glaeser index

Taking into account the availability of data, the Yangtze River Delta textile industry employment data is not available, so when calculating the space Gini coefficient namely G uses the enterprises' number of Textile Industry in Yangtze River Delta region and the whole nation to replace it. The adjusted G is:

$$G' = \sum_i (n_i - x_i)^2 \tag{2}$$

Where n_i is the ratio of the enterprises' number of one industry in i region to the enterprises' number of the industry in the whole nation.

In addition, Herfindahl index H refers to the sum of square of market share of all enterprises within an industry in a region. Obviously, the market share of all enterprises within the textile industry in the Yangtze River Delta is not available, so the adjusted H is approximately calculated as the following formula:

$$H' = \sum_j \frac{1}{m_j} s_j^2 \tag{3}$$

Where m_j is the enterprises' number within one industry in j region; S_j is the ratio of the total assets of one industry in j region to the total assets of the industry in the whole nation [7].

The adjusted EG index is:

$$\gamma' = \frac{G' - (1 - \sum_i x_i^2)H'}{(1 - \sum_i x_i^2)(1 - H')} \tag{4}$$

2.2 Data selection and measurement results

Based on the adjusted EG index, the data selected in this paper is the number of textile enterprises, the total assets of the textile industry and the number of employees in the textile industry from 2003 to 2014 in Shanghai, Jiangsu, Zhejiang and the whole nation. Where the data of the number of textile enterprises and the total assets of the textile industry from 2003 to 2014 in Shanghai, Jiangsu, Zhejiang and the whole nation is derived from the State Council Development Research Center Information Network Database, while the data of the number of employees in the textile industry from 2003 to 2014 in Shanghai, Jiangsu, Zhejiang and the whole nation is derived from the National Bureau of Statistics and the Local (Shanghai, Jiangsu, Zhejiang)Bureau of Statistics. The measurement results of the textile industry agglomeration level of Yangtze River Delta region are shown in Table 1 and Figure 1(In order to more intuitively observe the changes of textile industry agglomeration level in 2003-2014 in Yangtze River Delta region, the Figure 1 is specially made for reference).

Table 1 and Figure 1show that the textile industry agglomeration level -EG index in Yangtze River Delta region was in a rapid growth stage from 2003 to 2005, with growth rates of 7.1% and 24.2% respectively. From 2005 to 2010, the agglomeration level of the textile industry -EG index in Yangtze River Delta region was in a steady development stage .In addition to 2007 compared with 2006 increased by 7.8%, the EG index rose less than 3% in the rest of the years. Furthermore, the EG index was always between 0.8 and 0.9 in this stage. In 2010-2011, the agglomeration level of the textile industry -EG index in Yangtze River Delta region was in a sharp decline stage and it decreased by 32.9%

DOI: 10.1051/matecconf/201710005034

while still greater than 0.05. In 2011-2014, the agglomeration level of the textile industry -EG index in Yangtze River Delta region was in a slow recovery stage and increased as the time went on, with the growth rates of 2.1%, 1.2% and 4.6% respectively.

Table 1. 2003-2014 Textile industry agglomeration level-EG index in the Yangtze River Delta.

year	H′	G′	EG index
2003	2.78774E-05	0.06168604	0.061986165
2004	2.59456E-05	0.066074332	0.066399493
2005	2.05806E-05	0.082023463	0.082444926
2006	1.894E-05	0.08031534	0.080736289
2007	1.69384E-05	0.086544721	0.087024817
2008	1.63908E-05	0.084887202	0.085393906
2009	1.4579E-05	0.085023886	0.085551865
2010	1.42133E-05	0.087422193	0.087971553
2011	2.18174E-05	0.058698918	0.059056311
2012	2.22723E-05	0.059907022	0.060270608
2013	2.02819E-05	0.060595597	0.060970614
2014	1.8255E-05	0.06339555	0.063788203

Source: the State Council Development Research Center Information Network Database, the National Bureau of Statistics and the Local (Shanghai, Jiangsu, Zhejiang) Bureau of Statistics.

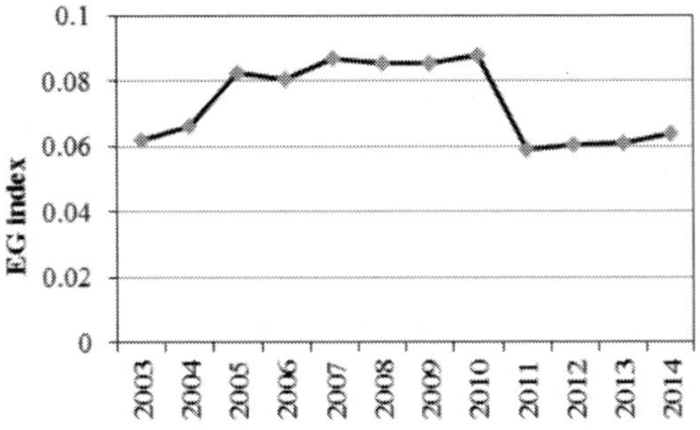

Fig. 1. 2003-2014 Textile industry agglomeration level-EG index in the Yangtze River Delta.

3 Analysis of the changes of textile industry agglomeration level in different stages

The Yangtze River Delta region is always the more developed areas of Chinese textile industry. It is generally acknowledged that if industrial agglomeration level>0.05, the industry is considered to be highly agglomerated in the region. The data shows that the textile industry agglomeration level-EG index in Yangtze River Delta were all greater than 0.05 in 2003-2014, which also squares with the current situation of the textile industry in Yangtze River Delta region. According to the measurement results of industrial agglomeration, the change of textile industry agglomeration level in Chinese Yangtze River Delta can be divided into four stages.

3.1 Rapid growth stage

From 2003 to 2005, the textile industry in Yangtze River Delta region was in the agglomeration efficiency increasing stage, in another word, the higher the agglomeration level of textile industry in the region, the more output a single textile enterprises will obtain from a certain proportion of the capital input. This is due to the textile enterprises in

Yangtze River Delta region had a cost advantage at the stage: sources of abundant and inexpensive raw materials, less transportation costs formed by perfect transport infrastructure and its inherent geographical advantage and lower labor costs. While, the cost advantage led to the profit level of textile industry in the Yangtze River Delta was higher than industrial average profit level, which attracted a lot of capital into the textile industry, expanded the industrial scale, and in return the agglomeration level of textile industry -EG index was increasing.

3.2 Steady development stage

In 2005-2010, the agglomeration level of textile industry in Yangtze River Delta region had been maintained at a high level of 0.08-0.09. At this moment the textile industry in Yangtze River Delta region was in the agglomeration efficiency constant stage and the market was saturated, and the capital input couldn't bring ultra - proportional income. Capital flowed in and out but had been maintained at a equalizing level and the scale of the industry was also nearly unchanged, so the agglomeration level of textile industry maintained at around 0.08during this period.

3.3 Sharp decline stage

Causing the agglomeration level of textile industry in Yangtze River Delta region dropped significantly in 2011, mainly due to a large number of textile enterprises closed down. The reasons are as follows: on the one hand, the decline of international competitiveness of textile enterprises. The majority of textile enterprises did not pay attention to innovation and were committed to producing some low value-added low-end textiles; On the other hand, the increase in labor costs. The labor costs of some Southeast Asian countries (Such as India, Laos, Cambodia and etc.) is cheaper than our country, China has gradually lost its labor cost advantage; In addition, the land resources in Yangtze River Delta region is increasingly scarce, which makes the cost of fixed assets is also increasing. These unfavorable factors make the profits of the textile enterprises in the region greatly shrunk, which led to many small and medium textile enterprises can't make ends meet and one after another closed down.

3.4 Slow recovery stage

In 2011-2014, the agglomeration level of textile industry -EG index in Yangtze River Delta region was slowly increased year by year, and had been maintained at around 0.06. After the closure of the tide in 2011, capital greatly flowed out of the textile industry, and the market was again in the agglomeration efficiency increasing stage. Furthermore, the textile industry in Yangtze River Delta region actively participated in technological innovation, industrial upgrading, improving the industrial chain and enhancing the international competitiveness in this period.

4 Conclusions

By using the EG index to measure the agglomeration level of textile industry in Yangtze River Delta region of China from 2003 to 2014, this paper has found that the agglomeration level of textile industry in Yangtze River Delta region of China has experienced four stages: rapid growth stage, steady development stage, sharp decline stage and slow recovery stage. In this paper, the following suggestions are proposed:

Firstly, since the agglomeration level of textile industry in Yangtze River Delta region has been at a high level, namely the market has been saturated and the cost (such as land resource costs and labor costs) increased year by year, the textile industry in Yangtze River Delta region can be considered to transfer to central and western regions whose agglomeration level of textile industry is lower.

Secondly, for the textile enterprises in the Yangtze River Delta region, it is imminent to improve the enterprise's innovation consciousness, complete the upgrading of machinery and equipment and improve the core competitiveness of enterprises.

Finally, for the textile industry in the Yangtze River Delta region, with the increasing international competitiveness, it should continue to adhere to in-depth promotion of transformation and upgrading and constantly improve the industrial chain to maintain steady growth of the textile industry.

Acknowledgement

The research was supported by Hubei Provincial Department of Education Science Research Major Project under Grant No. 14zd022.

References

1. Wei Zou, Modern economy,3(2007).
2. Xiao Lou, Botang Han, Science and Technology Progress Countermeasure, 8(2012).
3. Chaoping Sun, Huiming Liu, Yong Wu, Industrial Technology & Economy, 12(2015).

4. Jing Hu, Luping Chen, Mi Zhou, Journal of Chongqing traffic University, 5(2015).
5. Jian Hu, Chunshi Dong, Statistics and Information Forum, 1(2013).
6. Ellison G, Glaeser E. JPE,**105**,5(1997).
7. Hongjiao Yang, Linyan Sun, Anbo Wu, China Industrial Economy, 4(2008).

DOI: 10.1051/matecconf/201710005035

Management Practice of Supply Chain Quality Management in Service-oriented Manufacturing Industry

Ping Gu[1], Ruyi Song[2,*] and Xuyun Chen [3]

[1,2] College of Economics and Management, Jiangsu University of Science and Technology, Jiangsu, China
[3] Bachelor Business of Administration, University of New Brunswick, Canada
Corresponding Email: songruyi2544@163.com

Abstract. Supply chain quality management (SCQM) in service-oriented manufacturing industries is needed in delivering value to customers in all process of supply chain management based on total quality management(TQM).In the paper, we identify the latest themes through reviewing prior quality management and supply chain management(SCM) literature. In particular, we find manufacturing firms transform from providing products towards providing services, which means they need to go through fundamental changes especially in supply chains. We use a case study of Heilan Home to illustrate the SCQM themes and their ways in industrial practice. Based on our research, the case study, the experience of working with this firm, we propose a SCQM performance evaluation framework, as well as four strategies for other industries to improve customer satisfaction and added value. In this way can promote the transformation and upgrading of the manufacturing industry in the perspective of quality.

1 Introduction

The trend of transformation towards servitization has taken an important role in the development of global economy. Service industry in developed countries accounted for more than 70% of GDP, which is the source of economic growth. At the central economic work conference of China, leaders proposed that the primary task of promoting the structural reform of the supply side is resolving excess capacity actively and steadily in 2016.This means the manufacturing industry of China will transform from excess capacity to better supply. Service upgrade of manufacturing industry is the only way to develop Chinese economy currently. Enterprises pay more attention to the quality focused on providing better products and services for customers rather than quantity only. It will carry out the manufacturing industry better development and transformation by building "customer-centered" innovation collaboration value chain platform. This depends on the supply chain innovation and quality control. More and more scholars and enterprises try to integrate the supply chain management and quality management , expand the scope of application of quality management and endow it with the new connotation. The effective use of quality management can significantly affect the overall performance of the supply chain and the enterprise performance [1].Previous researches on quality management focus on the internal quality and external service quality, however, the core of supply chain quality management is centered on quality and win-win cooperation instead of cost and competition among enterprises on supply chain [2].

2 Literature review

2.1 Service-oriented manufacturing industry

With the development of economic globalization and informatization, the adjustment of industrial structure has become an important way for our national economy to transform. The boundary between manufacturing and service industry is constantly blurred. The differences between manufacturing and service industry in terms of product characteristics, functional distinction and organizational features have been weakened, then, the trend of integration of manufacturing and service industry is increased, then the development of modern manufacturing services has become the key to ensure the efficiency development of national economy [3]. At first, the manufacturing enterprise is the core of service-oriented manufacturing industry, then, which realize servitization by service industry as a tool. It is a strategic choice for manufacturing enterprises in their development process. Through "smile curve", firms can choose the high value added process of the left side-R&D and service, right side-marketing and brand, spun off its low value added process-manufacturing in the middle to improve the competitiveness and efficiency of enterprises. Vandermerwe S [4]gave many real-life examples, the authors assess the main motives driving corporations to servitization, and point out that its cumulative effects are changing the competitive dynamics in which managers will have to operate. The special challenge for top managers is how to blend services into the overall strategies of the company. Zahir Ahamed [5] examined the needs of servitization from customer perspectives, particularly the IT industry of emerging market 'Bangladesh'. The survey results showed that the current suppliers cannot satisfy the customer needs at this moment, because customers are not happy anymore with the IT goods only; they also require solutions, knowledge and reliability as well. Christian Schnürmacher [6] pointed out in contrast to the product development the service development approach is highly dependent on the project manager. There is no service development process defined and most of the employees and project managers are not used to work in processes. Thus it was important to train the fundamentals and benefits of a process based development to the people from the service development first before introducing the new method to ensure a successful implementation of an integrated product and service development process. Yong S K [7] contented that service elements are added and enhanced to make manufacturing companies more innovative and grow revenue. The servitization process could take many different routes reflecting various properties of the manufacturing firm and its business contexts. King Fengguan [8] described the basic purpose of servitization is to promote the combination of existing service and new service, the combination of goods and services and the commercialization of services. Furthermore, White [9] focused on the relationship between service and environment, defined the servitization as the transformation from the goods provider to service provider for the manufacturers.

2.2 Supply chain quality management

Kuei and Madu [10]defined supply chain quality management (SCQM) with three simple equations where each equation represents the letters that make up SCQM. The definition is as follows:

-SC=a production–distribution network;

-Q=meeting market demands correctly, and achieving customer satisfaction rapidly and profitably;

-M=enabling conditions and enhancing trust for supply chain quality.

In SCQM, Pu Guoli [11]discussed the relationship between supply chain members and customers, the important problem of supply chain is providing satisfied products and services for customers through joint effort of all members in supply chain, the concept of process has also been the focus instead of products. In essence, SCQM is not the study of supply chain, but the further understanding of QM from the point of view of supply chain. Kuei and Madu [12] note that the focus of quality-based paradigm has shifted from thetraditional company-centered setting to complete supply chain systems. Tan et al. [13] discussed the commitment involved to integrate channel suppliers, manufacturers, and customers in order to achieve both long-term growth. Collaborative relationships are also topics reviewed by Choi and Hartley [14], Forzaand Filippini[15] Obtaining customer satisfaction (caused construct) requires greater attention to factors which concern downstream relations with customers such as the involvement in quality improvement programs(causing construct=TQM link with customers) Waterson et al.[16], Most successful practices for quality improvement: TQM, Team-based groups, Manufacturing Cells and Integrated computer-based technologies. Most common practices: SC partnering, TQM, JIT, team-based working and integrated computer-based technology. Mangiameli and Roethlein[17], Multi-directional quality awareness and communication between channel partners (case study) can be a competitive advantage .Fynes andVoss[18] Strong buyer-supplier relationships will improve design quality.

In the past, QM pracitices pay much attention to inter-enterprise, so SCQM can transform the external problems of supply chain into the internal process, which include the process of expanding supplies at upstream such as negotiation, selection and improvement of suppliers' performance and the process of service at downstream such as customer service and after-sales service etc. Logistics flow, business flow, capital flow, information flow in supply chain, each flow can run correctly and effectively, then firms can obtain excellent performance.

2.3 Characteristics of supply chain quality management under the servitization

Service demand is the precondition of servitization [19], which is a transformation for specific enterprises to meet development needs. Under the trend of service-oriented manufacturing, enterprises integrate customer satisfaction into SCQM, deliver the behavior and perception of customers into the entire supply chain process, that is collaborative creation and acquisition of value through customer participation. Thus, supply chain quality must further improve the value on supply chain by customer participation and experience. "Customer" includes not only the ultimate objects accepting service, but also the member of the inner supply chain one by one.

In this paper, we will focus on SCQM in service-oriented manufacturing industry, which is to improve the overall quality on the supply chain through the attention to customer satisfaction and customer demand. In this model, the supply chain will have a multidimensional trend, deliver products and service. The path of information flow and logistics flow will connect with service demand and priorities of different customer groups. The relationship between the customer and the supplies is a kind of value added relationship, which makes the structure of the supply chain and operation rules have new features.

Logistics management is the core of the traditional supply chain, and the service flow is introduced in the service network organization, which is invisible and lack of material carrier. In fact, it is a kind of ability to transfer.

The traditional manufacturing supply chain raw materials and parts logistics are coupled by product structures and related needs; there is specific products and services existing in service network, which is connected with a new mechanism.

In service network, supply chain is customer demand-oriented; furthermore, customer participation and customer experience is more important, the core enterprises need provide direct and indirect service, service configuration and service platform.

Under the trend of economic globalization and the development of technology information, customer consumption has already shifted from the previous consumption based on the product to the service based experiential consumption. Service oriented manufacturing is the necessary means to enhance the manufacturing industry in China [20]. The transformation of manufacturing service makes the core business of the enterprise changed, which affects the entire supply chain of enterprises. The core enterprises will affect its own production and operation mode to realize the servitization, and have a significant impact on the entire supply chain members [21].

By focusing on the customer, the customer needs to be found, and the demand is delivered to the entire supply chain network. The effect of customer satisfaction on enterprise's operation ability is significantly higher than that of the enterprise's profitability. The impact of input- servitization on the traditional supply chain is mainly reflected in the procurement and production processes. Enterprises shift business focus to R&D and procurement management, that makes the production outsourcing and vendor management inventory (VMI). Manufacturing supply chain structure also changes.Input-servitization for the core enterprises means increasing investment in R&D. Output-servitization means that the enterprises transform products providing to service based on products providing. This change will affect the entire distribution and after-sales service model, downstream of the supply chain will be significantly reduced. Enterprises will face directly to terminal customers as "service-distributor" based products; another situation is that the downstream distributors and retailers will be replaced by service partners.

3 Research framework

The research on SCQM is established on comprehensive systematic and objective basis, according to the different development status of the industry, characteristics and situation of this enterprise-- Heilan Home, this paper mainly includes the following:

a)research on supply chain quality management practice.

b)research on the strategy of supply chain quality management based on customer satisfaction.

The main contents of this case as shown in the following Figure1.

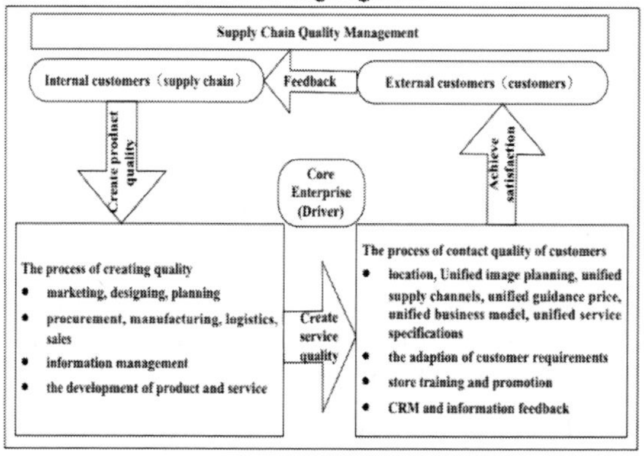

Fig. 1. Main contents.

3.1 The selection of research method

This paper takes a single case study method, and based on archival materials, interviews, field investigation to Heilan Home(Heilan) that is a typical and representative enterprise. It is heuristic for others. So the method of selecting single case study is reasonable. This company attaches to the construction of internal management system, adhering to the "market-oriented, customer satisfaction ", articulates the marketing positioning, makes market segmentation, identifies target customer groups, implements marketing research, designs unique men's clothing integrated into the corporate culture with its brand and resources integration capability. Heilan company has established and consolidated win-win relationship with customers. The managers of Heilan are actively involved in product and service quality by promoting and highlighting the importance of customer requirements. This concept is emphasized in staff training, daily meetings, weekly management meetings. Furthermore, Heilan has its employees in quality-solving teams and customer site visits to solve those quality issues. In particular, Heilan has its unique ways to avoid and solve quality issues. This processes include (1) product design, (2) suppliers selection,(3) production stage, (4) inspection before storage,(5) patrolling , inspection, secret enquiry. Heilan is a ideal case of this research for its quality-oriented culture, and its communication and partnership activities are strategically integrated with their customers. In this interview with their senior executives and department managers we communicate about performance index, management system and process activities, continuous improvement programs, customer demands as well as the quality project quality plan.

3.2 Data sources and data processing

In this paper, the data collection carried out through field research, interviews, which is helpful for understanding the theoretical basis, research process and the background of this case. We collected the information by means of Heilan Group official website, journal articles, interviews and investigation including: (1) group reports, (2) journals and speeches of senior managers ,(3) five-year strategic planning,(4) department development planning,(5) company management process, system files and so on. The members of the project group had in-depth experience-sharing and discussion with senior managers including R&D, brand, quality, marketing department to guarantee the integrity and efficiency of information.

4 A case study of management practice

4.1 Joint operations

Heilan takes the ways of payment after the completion of the sale, the return of unsalable goods and procurement of secondary purchases to get the interests together between the brands and manufacturers. The relationship of Heilan and other suppliers is "returnable joint venture", which means suppliers are no longer OEM manufacturers. They have to improve dynamic pin ratio and profit margin, understand market trends, cooperate closely with Heilan, produce readily marketable goods. That would help improve dynamic pin ratio and the efficiency of stores. It is the benefit-sharing mechanisms that maintain Heilan with suppliers cooperate closely. Heilan has formulated a series of rules and regulations on the supply chain management to ensure effective operation including the selection and evaluation of supplier, supplier production process tracking and management, supplier sample design selection and modification, product purchasing and quality control, supplier accounting, supplier elimination system, logistics and so on.

Heilan develops the new products jointly with suppliers, and makes full use of suppliers' design resources. The design center of Heilan adhere to market-oriented, reference the domestic and international fashion design principles, focus on the 20-45 year old Chinese man as customers, pay attention to their clothing habits and taste, analyze and summarize the sales data, plan and design products. Designers complete the product development proposal, fully communicate with the design department of suppliers. Suppliers' designers conduct the overall proofing. Heilan's designers select and modify the old proofing, at last, determining the next season product styles. Under the joint development mode, Heilan not only masters the dominate the product design, but also give full play to suppliers' subjective initiative, and make full use of the suppliers' design resources.

Heilan signs procurement contracts of returning unsalable goods with suppliers, which is beneficial for company to make full use of the channel resources to control the product quality, sales channels, brand advantage. the stores provide high quality, cost-effective products and services. There is little left in two sale quarter. In order to take advantage of the company's resources, Heilan's stores would place more marketable commodity to reduce the occupancy of unsalable goods. At the same time, this contract would help suppliers more responsible to design marketable products and keep the quality. This kind of procurement contracts agree to return unqualified products and still return unsalable goods after a certain period.

Suppliers and Heilan become the organic combination, assume different responsibilities, share the interests, develop with each other. In this purchasing mode, suppliers provide high quality products, safeguard brand reputation and pay more attention commodity management, supply chain management and sales management. Only provide marketable products can suppliers have more interests, otherwise, both would have a loss.

4.2 Union and control

In sales link, Heilan has taken the mode of the ownership and operation-right separated from franchisee. Franchisee need bear the rent, renovation costs, utilities, staff salaries and other expenses of its store, but not participate in the management of the store. Heilan investigates the store location, assesses the future revenue, decides the distribution, manages the store, controlls the whole market. There are more than 3000 stores with unified image planning, unified supply channels, unified guidance price, unified business model, unified service specifications. This helps the cooperation and control. Sales settlements take the trust-surrogate mode. The franchisee does not bear the inventory unsalable risk, according to the contract to settle business income.

- Store expansion

The location is very important for the stores, and passenger flow is the key elements. The company requires the store must located in the main business district, which is in charge by the market department. Heilan take the screening mode, according to the business development plan to decide the store expansion goal, through the level of "city-commercial distinct-store", find the qualified store, at the same time, actively develop the local private owners with strength to become franchisee.

- Operation management

Heilan has established a series of standardized management system about the daily store operation, delivery, settlement, inventory management system, customer communication and service managementmethods and so on. All stores must be strictly enforced. The staff salary(including bonuses) of all stores is decided by the company, all the expenses are charge by franchisee .

- Sales information management of franchisee

The pos system of store is connected with DRP system of Heilan. All after-sales data is gathered and analyzed by Heilan. In addition to electronic sales list, franchisee need provide written consignment list every month, after checked by Heilan, that would confirm sales revenue according the consignment list.

- "three-inspection"

In order to realize the target of standardized stores, Heilan has set up the inspection department to conduct regular or irregular inspections of all stores to check the implementation of the rules and regulations of service. The company would analyze the issue of feedback of inspection department, request the business manager to process and track, which is helpful to realize customer satisfaction.

4.3 Focus on management

As the model of service-oriented manufacturing industry, Heilan has taken the asset-light strategy, kept R&D and sales, peel off the production and processing. The company is committed to build a community of interests, on the other hand, enhances operational efficiency of stores, which would strengthen the core competitiveness. In the management mechanism, it implements standardized management, that is "management institutionalization, process standardization, supervision tracking, assessment digitization". In the work process, in addition to R&D center and logistics center, there is development, quality control, deployment, brand management and other departments, around the new business model and the trend of servitization, it gives full play to the advantages of "headquarters economy", dedicated to provide the whole process for the upstream and downstream of industry chain. At the same time, Heilan has set supplier requirements, with the advantages of scale to reduce procurement costs. Otherwise, the establishment of the store manager system and patrol clerk team, help to enhance the store management level. In Heilan's SAP project, the implementation system involves all aspects of retail management, product planning, logistics and warehousing, financial management, personnel management, data analysis and other, which support the headquarter ,retail stores and suppliers to share information and coordinate.

Heilan has taken two methods to establish current formats in the process of transformation, finally developed as a new type of retail industry "service-oriented manufacturing industry" . One is the reconstruction of the business process of supply chain based on TQM, another is the establishment of the most advanced IT technology to support the construction of the logistics system.

Heilan has taken the flat centralized distribution model of "supplier-headquarter-store" differing from the most clothing companies' mode of "regional warehouse storage and sales agent". It established the strategic cooperative relationship with the domestic logistic enterprises. Thousands of goods from more than 300 suppliers are transported to headquarters, after sorted, which then distributed to nationwide stores. This could let the company orders quickly and accurately delivered to the stores, furthermore, grasp the distribution of goods resources. Headquarter distribution center allocates reasonably and manages scientifically, not only to ensure efficient distribution but also control the cost of storage, labor and freight. Intensive distribution mode requires enterprises to have a certain storage capacity. Heilan group invested more than 500 million to build intelligent automatic tridimensional storage system, covers 10 million square meters, including 2 tridimensional warehouse, 3 delivery hall, a distribution center. This is the largest and the most technologically advanced automatic garment logistics park with the most advanced SAP information management system. Regardless any one item, from query, positioning, library and logistic, this process just needs 30 seconds to achieve "zero error", which completed from manpower to brainpower. With SAP, Heilan can grasp all the information of warehousing , distribution, sales and other aspects, realize the whole process tracking, conduct the overall control of the market.

4.4 SQCM performance index

According to the SQCM practice of Heilan, the following (Table 1) is the SCM process performance index. Through setting the reasonable performance indicators, it is helpful to implement the process.

4.5 Lessons and suggestions

Through the analysis of Heilan, taking different methods to analyze different scale enterprises, the level of transformation and reform are not same. From the perspective of servitization , according to the process of its organizational reform, considering the necessity and importance of transformation and if discovers and uses the key factor of the success elements of current business fields as shown in the following Figure2. Firstly, it is the customer satisfaction, which depends on the sophisticated analysis by the e-commerce, big data platform for customers, understands the customer demand and consumption. Secondly, it is business model innovation. Heilan's business model is developed from SPA(Speciality retailer of Private label Apparel) ,the characteristics of this model is integrated the research, production, logistics, sales by IT in order to realize the efficiency and the speed of commodity supply plan. Heilan stripped the lowest part of the value chain to gain competitive advantage. Thirdly, service configuration and construction of service industry platform is very important, service is apperceived by customer, how to transfer customer perception into whole supply chain, that relies on the platform. Last but not least, forming the co-creative service means full participation, including all stakeholders, it is also the core of supply chain quality management, through this kind of joint participation and creativity to determine the content and value of service.

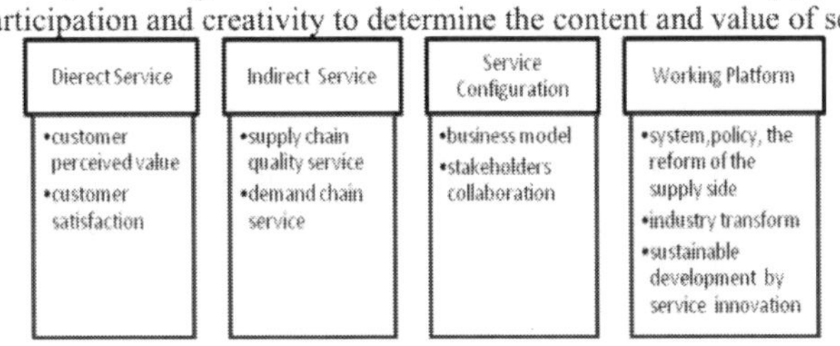

Fig. 2. The points of SCQM practice.

5 Conclusion

The focus of this paper is supply chain quality management in service-oriented manufacturing industry. On one hand, it provides better theoretical and empirical support for service-oriented manufacturing enterprises to achieve total quality management and improve customer satisfaction, on the other hand, lays the path foundation for some enterprises to transform itself from a traditional manufacturing to a service-oriented company. As this paper demonstrates, it is easy to see how future research in SCQM that seeks to integrate quality and process improvement strategies across the entire supply chain will benefit the stakeholders, the concept of TQM should be carried out in the whole process of business management, the trend of servitization has entered into an era of providing more added value into service process.

Table 1. Performance evaluation index.

Key Process	Stakeholders						Require	Process Requirement	Indicator
	Shareholders	Customers	Employees	Vendor	Partner	Community			
R&D management	•	•	•	•		•	※&◎	Meet market demand	Dynamic pin ratio
									sales
									Storage attainment rate
									Product quality
Supplier management	•	•	•	•			※★◆▲	Select the products of meeting brand culture	Prenatal sample confirmation rate
									Documentary efficiency
									Delivery quality
Logistics management	•	•	•	•	•		※★◆▲&◎	Improve goods accurate and targeted	Storage time rate
									Time delivery rate
									Delivery accuracy
									Inventory accuracy
									Sales completion rate
									Store turnover

Key Process	Stakeholders						Require	Process Requirement	Indicator
	Shareholders	Customers	Employees	Vendor	Partner	Community			
									Delivery success rate Store demand satisfaction rate
Store management		•		•	•	•	※◆◎ ⋆	Achieve business objectives with unified store image, model, service standards	Sales completion Personnel management The third superintendence Store development success rate
IT management	•	•	•		•		※★◆ ▲◎	Meet the accuracy , integrity, reliability, timeliness, safety and ease using of information and knowledge management	Technology capability Business ability Communication and coordination Annual comprehensive performance

★——provide unique products and services,
◆——provide the effective solution with the best value,
▲——improve market share and share value,
&——import advanced technology and tools,
◎——enhance corporate citizenship and social responsibility,
⋆——provide high quality products .

References

1. Robinson, Carol J., and M. K. Malhotra. "Defining the concept of supply chain quality management and its relevance to academic and industrial practice."International Journal of Production Economics 96.3(2005):315-337.
2. Pu Guoli, Su Qin. "The integration of supply chain and quality management". Industrial Engineering Journal, 13.6(2010):114-120. (in Chinese)
3. Yao Xiaoyuan. The New model of integrated process between manufacturing and service industry-the servitization of manufacturing.Journal of Shanghai Normal University(Philosophy&Social Sciences Edition), 2014:60-71. (in Chinese)
4. Vandermerwe, Sandra, and J. Rada. "Servitization of business: Adding value by adding services." European Management Journal 6.4(1988):314-324.
5. Ahamed, Zahir, A. Kamoshida, and T. Inohara. Servitization of Business: An Exploratory Case Study of Customer Perspective. The 8th International Conference on Knowledge Management in Organizations. 2014:3-12.
6. Christian Schnürmacher, Haygazun Hayka, and Rainer Stark. "Providing Product-Service-Systems - The Long Way from a Product OEM towards an Original Solution Provider (OSP) ☆." Procedia Cirp 30(2015):233-238.
7. Yong, Se Kim, et al. "Product-Service Business Concept Design: Real-world Case of a Small Furniture Manufacturing Firm ☆." Procedia Cirp30(2015):257-262.
8. KING Fengguan. "Customer management". Seoul,2000. (in Korea)
9. White, Allen L., et al. "Servicizing: The Quiet Transition to Extended Product Responsibility." Tellus Institute (1999).

10. Chu - Hua Kuei, C. N. Madu, and C. Lin. "The relationship between supply chain quality management practices and organizational performance."International Journal of Quality & Reliability Management volume 18.18(2001):864-872.
11. PU Guoli, Su Qin, LIU Qiang. "A Review of Supply Chain Quality Management: A New Disciplin Direction". Science of Science and management of S.&T, 32.10(2011):70-79. (in Chinese)
12. Chu - Hua Kuei, et al. "Developing supply chain strategies based on the survey of supply chain quality and technology management."International Journal of Quality & Reliability Management19.7(2002):889-901.
13. Tan, Keah Choon, et al. "Supply chain management: an empirical study of its impact on performance."International Journal of Operations & Production Management 19.19(1999):1034-1052.
14. Choi, Thomas Y., and J. L. Hartley. "An exploration of supplier selection practices across the supply chain."Journal of Operations Management14.4(1996):333-343.
15. Forza, Cipriano, and R. Filippini. "TQM impact on quality conformance and customer satisfaction: A causal model."International Journal of Production Economics 55.1(1998):1-20.
16. Waterson P E, Clegg C W, Bolden R, et al. The use and effectiveness of modern manufacturingpractices: A survey of UK industry. International Journal of Production Research, 1999, 37(10):2271-2292.
17. Mangiameli, Paul, and C. J. Roethlein. "An examination of quality performance at different levels in a connected supply chain: a preliminary case study." Integrated Manufacturing Systems volume 12.12(2001):126-133.
18. Voss, Chris, and B. Fynes. "The moderating effect of buyer-supplier relationships on quality practices and performance." International Journal of Operations & Production Management volume 22.22(2002):589-613.
19. CAI Sanfa , WANG Qingyu, HUANG Zhiming. "The Discussion on The Core Process of Business Servitization". Economics Forum, 6(2013):61-65. (in Chinese)
20. LIN Win-jin , JIANG Zhi-bin , LI Na."A Survey on the Research of Service-Oriented Manufacturing". Industrial Engineering and Management, 14.6(2009):1-6. (in Chinese)
21. LI Hong-lei, YU Yang."Manufacturing Servitization Oriented Supply Chain Re-engineering". Value Engineering, 7(2013):17-18. (in Chinese)

 DOI: 10.1051/matecconf/201710005036

Research on Supply Chain Coordination of Fresh Agricultural Products under Agricultural Insurance

Pei Zhang[1,*], Zhifei Chen[2] and Wujun Cao[2]

[1] Department of Electrical and Information Engineering, Zhengzhou University of Light Industry, Henan, China
[2] Department of Management Engineering, Zhengzhou University, Henan, China

Abstract. Based on the fact that the current fresh agricultural products are susceptible to natural risks and the coordination of supply chain is poor, This paper constructs the supply chain profit model under the two models of natural risk and agricultural insurance, Firstly, studying the coordination function of the supply chain system under Two-part Tariff; Then discussing the setting and claiming mechanism of agricultural insurance, compares the influence of agricultural insurance on supply chain profit and supply chain coordination; Finally, giving an example to validate the model results and give decision - making opinions. Research shows that the supply chain of fresh agricultural products can coordinated under Two-part Tariff, but the supply chain cooperation is poor in the natural risk , need to further stabilize and optimize the supply chain; When the risk factor is less than the non-participation insurance coefficient, not to participate in agricultural insurance is conducive to maintaining the coordination of the supply chain system; When the risk coefficient exceeds the non-participation insurance coefficient, the introduction of agricultural insurance can not only effectively manage the natural risks, but also help to improve the coordination of the supply chain system.

1 Introduction

Fresh agricultural products are closely related to the lives of the residents of vegetables, fruits, meat, aquatic products, milk, etc, occupy an important position in the daily life, but it has a characteristics about perishable, long production cycle, vulnerable impacted of the external environmental [1] .By the products' own characteristics and the impact of the external environment, fresh agricultural products supply chain is more unstable than the average product supply chain in the coordination. Specific performance for the uncontrollable natural risk are the uncertain supply of products, the unbalanced distribution of profits among supply chain members, the low coordination of the supply chain system, eventually leading to damage to the interests of consumers and members [2, 3].

China Federation of Logistics and Purchase defining the supply chain coordination as a application process management methods in the enterprise decision-making and planning, which produce a coordinated production, marketing, market effective management mechanism, thereby enhancing the overall operational efficiency of enterprises. While the supply chain's coordination can lower inventory levels, enhance information sharing, improve communication between members to maintain strategic co-operation between the partners, etc [4].

Supply chain contract as an effective business cooperation model, not only makes the overall coordination level of the supply chain increase, but also makes the supply chain members to achieve "win-win", has been widely recognized theorists and the business community [5] Therefore, this paper uses the contract to coordinate the supply chain and achieve the initial stability of the supply chain system.

A large number of scholars have shown that the revenue sharing contract, the option contract and the two charge contract can coordinate the supply chain system, but the Two-part Tariff has obvious advantages in the cooperation of the stable supply chain members [6]. So this paper uses Two-part Tariff to coordinate the supply of fresh agricultural products. Literature [7] made a detailed introduction about Two-part Tariff and its application, literature [8] describes the Two-part Tariff on the industrial supply chain coordination role. However, few scholars at home and abroad to study the Two-part Tariff's coordination role on the fresh agricultural supply chain system, for example, Wang Yu [9] used Two-part Tariff to coordinate the secondary supply chain of fresh agricultural products, but not to the coordination of three-level supply chain, so this paper uses Two-part Tariff to achieve the coordination of supply chain and the initial stability of member cooperation.

Under the influence of natural risks, the loss of fresh agricultural products in China is serious, Due to natural disasters caused by agricultural disasters reached more than 30 billion in Xinjiang each year[10], 134 million mu of vegetable crops affected, 83 million acres of disaster, 7.3 million mu of crops, but the insured farmers less than 10% in HeBei Province from November 2015 to January 2016[11], which result the supply chain members of the profit declined and the cooperation collapsed, supply chain coordination declines. Therefore, how to reduce the degree of natural damage to protect the profits of stakeholders to become another problem at all levels of the supply chain members. The agricultural insurance is designed for accidental factors such as natural disasters, after the occurrence of natural disasters, the manufacturer by paying low insurance premiums to obtain high claims to cover losses. Some scholars have verified that agricultural insurance can not only effectively manage natural risks, but also increase the profits of producers. Such as Yao Shoufu [12] quantitative analysis the economic effects about the development of China's agricultural insurance by the 1982-2014 insurance data, which shows that the agricultural insurance not only increase the farmers benefit, but also stable product production. Li Yaqi, etc. [13] selected 1985-2014 agricultural insurance premiums, compensation and agricultural output data to do a Co-integration test, which concluded that agricultural insurance premiums and compensation has a weak positive effect on agricultural output. However, there are few literatures study the effect of agricultural insurance on agricultural product supply chain, such as Cao Wujun [14] based on the option contracts and agricultural insurance to achieve the supply chain coordination and optimization in the secondary supply chain of farm produce under natural risk, but not to the three-level supply chain.

Based on previous studies, this paper studies the supply chain system of fresh agricultural products, which is composed of single producer, distributor and retailer. The paper discusses the coordination function of the Two-part Tariff to the supply chain system under the natural risk; and then introduces agricultural insurance into the fresh agricultural products three supply chain profit model; analyzes the impact of agricultural insurance on the quantity and profit of supply chain system; finally, an example is given to validate the model results and give decision - making opinions.

2 Problem description and assumptions

This paper studies the three-level supply chain system of fresh agricultural products, which consists of a single supplier(S)-distributor (D)-retailer(R). The retailer has an order opportunity before the sales season. The relevant assumptions are as follows:

Hypothesis 1: Supply chain members are fresh and rational risk, only to discuss a production cycle's activities, regardless of transport losses, inventory costs and management costs;

Hypothesis 2: Natural risks occur only in the process of agricultural production, which produces a threat to production, does not affect product quality, the supplier bears the loss.

Hypothesis 3: t_1 is the time interval at which the supplier delivers the product to the distributor, t_2 is the time interval at which the distributor will ship the product to the retailer, t is the time interval at which a retailer's product reaches the consumer, $t_1 + t_2 + t \le T_0$, T_0 is the longest product life cycle, so $t \in (0, T_0 - t_1 - t_2)$,assumed $T = T_0 - t_1 - t_2$.

Hypothesis 4: Freshness of fresh agricultural products decreased with the increase of transportation time of supply chain members and was affected by the degree of fresh effort of supply chain members. Based on the freshness iteration formula of fresh agricultural products in Ref. [15], it is assumed that the freshness factor of fresh agricultural products of retailers at the appropriate temperature is $\theta_{(t)} = 1 - \eta * (\frac{(1-k_{s1}*\tau_s)*t_1^2 + (1-k_{d1}*\tau_d)*t_2^2 + (1-k_{r1}*\tau_r)*t^2}{T_0^2})$, η is the agricultural products' maximum value of the natural attenuation. $\tau_i \in (0,1)$ is the freshness level of member i ($i = 1$ indicates the Supplier, $i = 2$ indicates the Distributor, $i = 3$ indicates the Retailers), k_{i1} is the freshness of the degree of preservation of the sensitivity coefficient.

Hypothesis 5: For the supply chain, only concerned about whether consumers buy and the capacity of the entire market, Reference [16] in the consumer utility function, set it to $U_{(t)} = U_0 - \alpha * p + \beta * \theta_{(t)}$, U_0 is the initial cognitive value of the product to the consumer, α、β are sensitive coefficient about consumers of fresh agricultural products price and freshness, $\alpha + \beta = 1$. Only then $U_{(t)} > 0$, consumers will choose to buy products, or give

up the purchase, assumed δ is the market size measure, at any time t, the retailer's market demand is $D(t) = \delta p(U_{(t)} > 0)$,so $D_{(t)} = \delta * P * (U_0 - \alpha * P + \beta * \theta_{(t)} > 0 = \delta * (1 - \alpha * P + \beta * \theta(t))$ In a sales cycle T within the supply chain system, the total demand is $Q = \int_0^T \delta * (1 - \alpha * p + \beta * \theta_{(t)}) dt$, assumed $M_1 = \beta * T * (1 - \eta * \frac{t_1^2 + t_2^2 + \frac{1}{3} * T^2}{T_0^2})$, $A_r = \beta * T * \eta * \frac{k_{r1} * T^2}{3 * T_0^2}$, $A_d = \beta * T * \eta * \frac{k_{d1} * t_2^2}{T_0^2}$, $A_s = \beta * T * \eta * \frac{k_{s1} * t_1^2}{T_0^2}$, so the actual demand of the supply chain system is $Q = \delta * ((1 - \alpha * p) * T + M)$, assumed $M = M_1 + A_s * \tau_s + A_d * \tau_d + A_r * \tau_r$.

Hypothesis 6: When the supply chain members take fresh, there must have the cost of fresh inputs and the degree of freshness preservation is positively related to the cost of preservation, Ref. [17] is used to express the cost function and combined with the article required, assumed the fresh-keeping cost function of member i in case j (j =1 indicates natural risk, j =2 indicates agricultural insurance) is $c_{ij} = \frac{1}{2} * k_{i2} * \tau_{ij}^2$, $k_{i2} > 0$ is the effect of preservation effort on the cost of preservation, the formula shows that the higher the degree of preservation efforts to preserve the higher the cost. τ_{ij} is the level of freshness effort of member i under mode j.

Hypothesis 7: Assume that the natural risk coefficient is μ, Q is the order quantity of the retailer, the actual production of the supplier is $\frac{Q}{1-\mu}$.

Hypothesis8: The supplier estimates the natural risk factor for the product is $\overline{\mu}$, Δ is the non-participation insurance coefficient. When $\overline{\mu} \le \Delta$, the risk is small, the supplier does not buy insurance, at this point, the supply chain system's optimal order quantity and the degree of preservation efforts to change small, which treated as invariant. When $\overline{\mu} > \Delta$, the risk is bigger, the supplier purchases agricultural insurance, so the insurance decision-making coefficient value is $k_{\mu} = \begin{cases} 0, \overline{\mu} \le \Delta \\ 1, \overline{\mu} > \Delta \end{cases}$.

Other related symbols: Q is the retailer's order quantity, p is the product per price, c_i is the unit production cost of the member i, assume $C = c_1 + c_2 + c_3$, π_1 and π_2 are the overall profit function of the supply chain under the natural risk and agricultural insurance, π_{i1}、π_{i2} are the profit function of the supply chain member $i\,(i=r, i=d, i=s)$ based on the two-part Tariff under the natural risk and the agricultural insurance. w_{2j}, w_{1j} represent the wholesale price of the distributor to the retailer and the manufacturer to the distributor in the natural risk and the agricultural insurance.

3 Supply chain coordination of fresh agricultural products

3.1 Research on the coordination of the supply chain system under natural risk

Researching a fresh agricultural products supply chain, which are composed of supplier, distributor and retailers. From the literature [18] shows that non-contract under decentralized decision-making supply chain's efficiency is low, there is a double marginal effect between the supply chain members, while the two-part Tariff can not only eliminate the double marginal effects of the supply chain, but also stabilize the cooperative relationship of members. Therefore, this chapter uses two-part Tariff to coordinate the supply chain of fresh agricultural products. The decision-making process is retailers according to their objective function and the reaction function of supplier and distributor to determine the channel fee paid to the distributor f_{r1}, the selling price of the product p_1, the degree of preservation effort τ_{r1}, then distributor according to their objective function and the reaction function of supplier and retailer to determine the channel fee paid to the supplier f_{r1}, the selling price of the product w_{21}, the degree of preservation effort τ_{d1}, finally, the supplier according to the reaction function of distributor and retailer determine the selling price of the product w_{11}, the degree of preservation effort τ_{s1}.

The expected profit function of the retailer is:

$$\max \pi_{r1}(p_1) = (p_1 - c_3 - w_{21})Q_1 - c_{r1} - f_{r1}$$

$$s1\begin{cases}(w_{21})\in\arg\max\pi_{d1}=(w_{21}-c_2-w_{11})Q_1-c_{d1}+f_{r1}-f_{d1}\\(w_{11})\in\arg\max\pi_{s1}=(w_{11}-\frac{1}{1-\mu})Q_1-c_{s1}+f_{d1}\end{cases}$$

The total profit of supply chain system is

$$\pi_1=(p_1-C-\frac{\mu}{1-\mu}*c_1)Q_1-c_{r1}-c_{d1}-c_{s1} \tag{1}$$

Two-part Tariff can achieve a necessary and sufficient coordination of supply chain conditions:

Proposition 1 If the retailer, distributor, supplier's profit is an affine function of the total profit of the supply chain system, the supply chain is said to be coordinated under the two-part Tariff.

Proof: For the profits of the supply chain system, the second-order Hessian matrix of π_1 in formula (1) is:

$$H=\begin{bmatrix}\frac{\partial^2\pi_1}{\partial p_1^2},\frac{\partial^2\pi_1}{\partial p_1\partial\tau_{i1}}\\\frac{\partial^2\pi_1}{\partial\tau_{i1}\partial p_1},\frac{\partial^2\pi_1}{\partial{\tau_{i1}}^2}\end{bmatrix}=\begin{bmatrix}-2\alpha T,\delta A_i\\\delta A_i,-k_{i2}\end{bmatrix}$$

As a result of the matrix is $-2*\alpha*T<0$, when $2*\alpha*T*k_{i2}-\delta^2*A_i^2>0$, Order Hessian matrix is negative, at this point π_1 is a concave function about p_1 and τ_{i1}, indicating that there is a unique p_1 and τ_{i1} to maximize the profit of the supply chain system. Many parameters are solved from the first-order optimality condition

$$\frac{\partial\pi_1}{\partial p_1}=0,\frac{\partial\pi_1}{\partial\tau_{r1}}=0\ ,\frac{\partial\pi_1}{\partial\tau_{d1}}=0,\ \frac{\partial\pi_1}{\partial\tau_{s1}}=0.$$

The optimal retail price of the supply chain system is:

$$p_1=\frac{T+M_1+\alpha*T*(C+\frac{\mu}{1-\mu}*c_1)+A_r*\tau_{r1}+A_d*\tau_{d1}+A_s*\tau_{s1}}{2\alpha T}$$

The optimal degree of freshness preservation for retailers is:

$$\tau_{r1}=\frac{\delta*A_r(T+M_1-\alpha*T*(C+\frac{\mu}{1-\mu}*c_1))}{2*\alpha*T*k_{r2}-\delta(A_r^2+\frac{A_d^2*k_{r2}}{k_{d2}}+\frac{A_s^2*k_{r2}}{k_{s2}})}$$

The optimal dilution effort of the distributor is:

$$\tau_{d1}=\frac{\delta*A_d(T+M_1-\alpha*T*(C+\frac{\mu}{1-\mu}*c_1))}{2*\alpha*T*k_{d2}-\delta(A_d^2+\frac{A_r^2*k_{d2}}{k_{r2}}+\frac{A_s^2*k_{d2}}{k_{s2}})}$$

The optimal degree of freshness preservation of the supplier is:

$$\tau_{s1} = \frac{\delta * A_s(T + M_1 - \alpha * T * (C + \frac{\mu}{1-\mu} * c_1))}{2 * \alpha * T * k_{s2} - \delta(A_s^2 + \frac{A_r^2 * k_{s2}}{k_{r2}} + \frac{A_d^2 * k_{s2}}{k_{d2}})}$$

At the same time when the members of the supply chain coordination at all levels, the wholesale price of distributors is $w_{21} = \frac{1}{1-\mu} c_1 + c_2$, the supplier's wholesale price is

$$w_{11} = C + \frac{1}{1-\mu} c_1 - \frac{T + M_1 + A_s * \tau_{s1} + A_d * \tau_{d1}}{\alpha T}$$

At this point the optimal order quantity of the supply chain system is

$$Q_1 = \frac{\delta(T + M_1 - \alpha * T * (C + \frac{\mu}{1-\mu} * c_1) + A_r * \tau_{r1} + A_d * \tau_{d1} + A_s * \tau_{s1})}{2}$$

The maximum profit function of the supply chain system is

$$\pi_1 = (p_1 - C - \frac{\mu}{1-\mu} * c_1) * Q_1 - \frac{1}{2} k_{r2} * \tau_{r1}^2 - \frac{1}{2} k_{d2} * \tau_{d1}^2 - \frac{1}{2} k_{s2} \tau_{s1}^2$$

So the retailer's profit function is $\pi_{r1} = \pi_1 - f_{r1}$

The distributor's profit function is $\pi_{d1} = 2\pi_1 + f_{r1} - f_{d1}$

The supplier's profit function is $\pi_{s1} = -2\pi_1 + f_{d1}$

That is to say, the profits of retailers, distributors and supplier are the affine functions of supply chain coordination profit under the Two-part Tariff, therefore, the optimal quantity Q_1 for the retailer is the optimal demand of the system and the supply chain system allocates the profit by adjusting the contract parameter f_{r1} and f_{d1}, the supply chain in Two-part Tariff is coordinated.

It is worth noting that $\tau_{ij} \in (0,1)$, $2 * \alpha * T * k_{i2} - \delta^2 * A_i^2 > 0$, so this article is discussed under the condition of $T + M_1 - \alpha T(C + \frac{\mu}{1-\mu} * c_1) \geq 0$, it also means $C \leq \frac{T + M_1}{\alpha T} - \frac{\mu}{1-\mu} * c_1$, when $\tau_{ij} \geq 1$, freshness efforts can only take 1, when $0 < \tau_{ij} < 1$, τ_{ij} takes the optimal solution.

Corollary 1: The fixed channel fee changes do not affect the supply chain cooperation in certain range, which was only the revenue game between supply chain members.

Prove: According to the profit function of retailers, distributors, suppliers, assume the channel fee charged by the manufacturer to the distributor in general is f_{r1}, the channel fee charged by the distributor to the retailer is f_{d1}, f_{r1} and f_{d1} satisfies the requirement that the profits of all members of the supply chain under the contract are greater than the profits without contract, if the manufacturer to the distributor fees charged to increase the $f_{r1}+\Delta f$, f_{d1} unchanged, the retailer's profits decline, the distributor's profits rise, the supplier's profits remain unchanged, supply chain partnerships remain unchanged. It can be seen, fixed channel fee is the supply chain members of the game between the gains.

Corollary 2: With the increase of natural risk coefficient μ, supply chain system profits down, supply chain system is not coordinated.

Prove: In the other conditions remain unchanged, with the increase of natural risk coefficient μ, product prices down, preservation efforts to decline, the optimal order quantity of the supply chain system decreases, supply chain's overall profit decline, supply chain system is not coordinated.

Based on the analysis of supply chain model of fresh agricultural products under the Two-part Tariff, the following conclusions are drawn:

Conclusion 1: The Two-part Tariff can not only coordinate the supply chain of fresh agricultural products, but also to stabilize the cooperation between supply chain members.

From proposition 1 and corollary 1, Two-part Tariff can effectively eliminate the double marginal effect of supply chain, so that supply chain members achieve a win-win situation and achieve the coordination of the supply chain system. At the same time in the Two-part Tariff under the coordination of the fresh agricultural supply chain, downstream members of the upstream members of the order to pay a certain fixed channel fee, and then lower the wholesale price of access to upstream goods, making members of their respective profits will be greater than the profits without contracts, in the actual transaction in a fixed channel to do the profit protection, the transaction between members is not easy to interrupt; and the lower wholesale price will encourage members to produce or order more products, which is not only has theoretically meaningful but also has practical significance in the cooperative relation of the stable supply chain members.

Conclusion 2: Under the natural risk, the cooperation relationship of the supply chain of fresh agricultural products coordinated is low by Two-part Tariff.

Once the natural risk coefficient increases from the corollary2, supply and demand at all levels of the supply chain change, supply chain coordination declines. Therefore, introduced agricultural insurance in the coordination of fresh agricultural supply chain as a risk management tool, through the low payment of insurance premiums to obtain high claims to make up for losses in order to achieve a coordinated supply chain system at all levels of members of the further stable cooperation.

3.2 Research on the coordination of the supply chain system under agricultural insurance

Suppose that producers engaged in the production of fresh produce are able to circumvent the risks by purchasing product yield insurance. According to the "Insurance Practice of Planting Industry of China People's Property Insurance Co" insurance and claims rules, the policy of premium payment and claims for fresh agricultural products yield insurance is as follows:

Insurance Premium: In reference to the literature [19] under the incentive of government subsidies, the purchase of insurance for each production cycle when the producer only pay unit insurance $20\%*w*c_B$ (Per kilogram), among them, w is the premium rate, c_B is the unit yield insurance amount. Assume that the fresh yield of agricultural product yield insurance is $\Delta Q=(1-\Delta)*Q$, Δ is the non-participation insurance coefficient, which is part of the supplier supply Q, then the supplier to the insurance company's insurance costs $0.2*w*c_B*\Delta Q$.

As the natural risk factor is μ, the actual production of the supplier is $\dfrac{Q}{1-\mu}$.

Claims: The insurance company according to the actual yield of the supplier and the insurance production of the difference between the compensation, the specific rules are as follows:

When $\mu \le \Delta$, the insurance company does not compensate.

When $\mu > \Delta$, the insurance companies according to the actual output of agricultural production and insurance production of the difference between compensation, by the agricultural insurance claims rules can be obtained that the insurance company claims H for the supplier of the expression is: $H = k_0 * (\Delta Q - (1-\mu) * Q) c_B = k_0 * (\mu - \Delta) * Q * c_B$

Among them k_0 is the claim coefficient, $k_0 = \begin{cases} 0, \bar{\mu} \le \Delta \\ 1, \bar{\mu} > \Delta \end{cases}$, The actual insurance company to the manufacturer's claim amount is: $H - 0.2 * w * c_B * \Delta Q = (k_0 * (\mu - \Delta) - 0.2 * w * (1-\Delta)) * Q * c_B$

Assumed that $B = (k_0 * (\mu - \Delta) - 0.2 * w * (1-\Delta)) * c_B$ is insurance factor, the actual insurance company to the supplier claims amount $B * Q$. It is shows that Agricultural insurance claims is lower than the producer's normal production income, which is also a strategy of the insurance company for incentives to purchase insurance in the event of disaster can take positive measures to save the loss.

Under the above agricultural insurance rules, according to the supply chain coordination function of the Two-part Tariff, agricultural insurance is applied to fresh agricultural products supply chain, the decision-making process is as described in above3. Establishing agricultural insurance under the supply chain decision-making model, the overall profit function of the supply chain system is:

$$\pi_2 = (p_2 - C - \frac{\mu}{1-\mu} * c_1 + k_\mu * B) * Q_2 - c_{r2} - c_{d2} - c_{s2}$$

The expected profit function of retailer is:

$$\max \pi_{r2}(p_2) = (p_2 - c_3 - w_{22}) * Q_2 - c_{r2} - f_{r2}$$

$$s.t. \begin{cases} (w_{22}) \in \arg\max \pi_{d2} = (w_{22} - c_2 - w_{12}) * Q_2 - c_{d2} + f_{r2} - f_{d2} \\ (w_{12}) \in \arg\max \pi_{s2} = (w_{12} - \frac{1}{1-\mu} c_1 + k_\mu * B) * Q_2 - c_{s2} + f_{d2} \end{cases}$$

The same decision-making process as Proposition 1 is used to obtain the parameters of the supply chain in the case of purchasing agricultural insurance, as shown in Proposition 2

Proposition 2: The contract parameters of the supply chain system coordination under the agricultural insurance are as follows:

The optimal selling price of agricultural products is:

$$p_2 = \frac{T + M_1 + a * T * (C + \frac{\mu}{1-\mu} * c_1 - k_\mu * B) + A_r * \tau_{r2} + A_d * \tau_{d2} + A_s * \tau_{s2}}{2aT}$$

The optimal degree of freshness retention for retailers is:

 DOI: 10.1051/matecconf/201710005036

$$\tau_{r2}=\frac{\delta * A_r(T+M_1-\alpha * T*(C+\frac{\mu}{1-\mu}*c_1-k_\mu * B))}{2*\alpha * T * k_{r2}-\delta(A_r^2+\frac{A_d^2*k_{r2}}{k_{d2}}+\frac{A_s^2*k_{r2}}{k_{s2}})}$$

The optimal dilution effort of the distributor is:

$$\tau_{d2}=\frac{\delta * A_d(T+M_1-\alpha * T*(C+\frac{\mu}{1-\mu}*c_1-k_\mu * B))}{2*\alpha * T * k_{d2}-\delta(A_d^2+\frac{A_r^2*k_{d2}}{k_{r2}}+\frac{A_s^2*k_{d2}}{k_{s2}})}$$

The optimal degree of freshness of the supplier is:

$$\tau_{s2}=\frac{\delta * A_s(T+M_1-\alpha * T*(C+\frac{\mu}{1-\mu}*c_1-k_\mu * B))}{2*\alpha * T * k_{s2}-\delta(A_s^2+\frac{A_r^2*k_{s2}}{k_{r2}}+\frac{A_d^2*k_{s2}}{k_{d2}})}$$

The supplier's wholesale price is:

$$w_{12}=\frac{1}{1-\mu}c_1+c_2+k_\mu * B$$

The wholesale price of the distributor is:

$$w_{22}=C+\frac{1}{1-\mu}*c_1-\frac{T+M_1+A_s*\tau_{s2}+A_d*\tau_{d2}}{\alpha T}-k_\mu * B$$

At this point the optimal order quantity of the supply chain system is:

$$Q_2=\frac{\delta(T+M_1-\alpha * T*(C+\frac{\mu}{1-\mu}*c_1-k_\mu * B)+A_r*\tau_{r2}+A_d*\tau_{d2}+A_s*\tau_{s2})}{2}$$

The maximum profit function of the supply chain system is:

$$\pi_2=(p_2-C-\frac{\mu}{1-\mu}*c_1+k_\mu * B)*Q_2-\frac{1}{2}k_{r2}*\tau_{r2}^2-\frac{1}{2}k_{d2}*\tau_{d2}^2-\frac{1}{2}k_{s2}\tau_{s2}^2$$

The retailer's profit function is: $\pi_{r2}=\pi_2-f_{r2}$

The profit function of the distributor is: $\pi_{d2}=2\pi_2+f_{r2}-f_{d2}$

The profit function of the supplier is: $\pi_{s2}=-2\pi_2+f_{d2}$

It is worth noting that $\tau_{i2}\in(0,1)$, including 0 and 1, when $\tau_{i2}\geq 1$, freshness efforts can only take 1, when $0<\tau_{i2}<1$, τ_{i2} takes the optimal solution.

Corollary 3: When the supplier estimates the natural risk factor $\overline{\mu} \leq \Delta$, if the actual natural risk factor $\mu \leq \Delta$, the optimal decision of the supply chain is not changed; if the actual natural risk factor $\mu > \Delta$, the entire supply chain profits decline.

Prove: When the supplier estimates the natural risk factor $\overline{\mu} \leq \Delta$, the supplier is not insured, if the actual natural risk factor $\mu \leq \Delta$, from hypothesis 7, at this point the optimal retail price of agricultural insurance, the optimal degree of preservation efforts, the optimal demand, supply chain members at all levels of profit are the same as natural risk. That is to say when the supplier estimates the natural risk factor $\overline{\mu} \leq \Delta$ and the actual natural risk factor $\mu \leq \Delta$, which makes the optimal decision of the supply chain is not changed. If $\mu > \Delta$, the supplier's actual supply and demand for a serious inconsistency, which eventually leading to insufficient supply of suppliers, the entire supply chain profits decline.

Corollary 4: When the supplier estimates the natural risk factor $\overline{\mu} > \Delta$, if the actual natural risk factor $\mu \leq \Delta$, the supply chain overall profit decline, which is not conducive to supply chain coordination; if the actual natural risk factor $\mu > \Delta$, agricultural insurance can not only effectively manage natural risks, but also increase the supply chain as a whole and members of the income at all levels, help to strengthen the coordination of the supply chain.

Proof: When the supplier estimates the natural risk factor $\overline{\mu} > \Delta$, if the actual natural risk factor $\mu \leq \Delta$,the supplier is insured, suppliers pay more insurance costs, making the cost of the supply chain rise, the profits decline, which is not conducive to supply chain coordination. If the actual natural risk factor $\mu > \Delta$, take any number $\overline{\mu} \in (\Delta,1)$, due to the introduction of agricultural insurance claims amount in $\overline{\mu}$, so $\tau_{i2} > \tau_{i1}$, $p_2 > p_1$, $w_{21} > w_{22}$, $w_{11} > w_{12}$, thus making $Q_2 > Q_1$, $\pi_2 > \pi_1$ .Which means the supply chain system's profit under agricultural insurance is than the profit under natural risk, supply chain members at all levels of profits also become larger, the agricultural insurance can not only manage the natural risk effectively, but also can increase the profit for the whole supply chain and the members at all levels, which is helpful to strengthen the coordination of the supply chain system.

Corollary 5: With the natural risk factor μ increases, the decline of the profit of the supply chain system under the natural risk is larger than that of the agricultural insurance under the supply chain. Supply chain coordination under the agricultural insurance increased compared to the supply chain under natural risks,

Prove: In the agricultural insurance mode, with the natural risk factor μ increases, because the manufacturer has been the amount of agricultural insurance claims,

the rate of decline in the supply chain profits compared to the natural risk of supply chain profits decline is slower. Therefore, agricultural insurance under the supply chain coordination has increased under the natural risk of supply chain system coordination.

Based on the analysis of supply chain coordination model of fresh agricultural products under agricultural insurance and Two-part Tariff, the following conclusions are drawn:

Conclusion 3: The agricultural insurance is introduced into the fresh agricultural product supply chain under Two-part Tariff, supply chain coordination can be realized again by setting contract parameters;

Conclusion 4: The introduction of agricultural insurance not only can effectively manage the natural risks, but also increase the supply chain as a whole and members of the income levels, making the supply chain coordination increased.

From the inference (3) (4) (5) shows, agricultural insurance to make up for the lack of the supply chain system profits declined, part of the natural risk is transferred to insurance institutions outside the supply chain, reduce the risk of natural risks on the supply chain system, increase the supply chain as a whole and members of all levels of income, making the supply chain coordination increased.

4 Case study

In order to more clearly reflect the two charges and agricultural insurance system on the supply chain coordination role, the results of the above model are validated by onion. Suppose a region of the onion supply chain is composed of a supplier, a large fresh wholesaler, a retailer (such as a supermarket), the market size of onion in this area is roughly measured $\delta = 10000$ (kg). Other variables in the model: $c_3 = 0.3$ RMB/kg, $c_2 = 0.5$ RMB/kg, $c_1 = 1.2$ RMB/kg, $t_1 = 3$ day, $t_2 = 2$ day, $T = 10$ day, $\eta = \alpha = \beta = 0.5$, $k_{r1} = 0.1$, $k_{d1} = 0.2$, $k_{s1} = 0.3$, $k_{r2} = 50$, $k_{d2} = 80$, $k_{s2} = 100$.

According to the insurance company claims, the onion claim parameter is $w = 4\%$, $c_B = 0.5$ RMB/kg. Insurance claims are calculated on the basis of natural disasters occur in the maturity of crops.

4.1 Research on the coordination of supply chain

Substituting the above variables into the models in Sections 3.1 and 3.2, when the risk factor is determined, the optimal parameters and profit of the supply chain system under natural risks and agricultural insurance are obtained, the conclusion is: Under certain conditions, whether the agricultural insurance, supply chain members achieved a win-win fee system are in the Two-part Tariff, the supply chain system achieved effective coordination.

4.2 Supply chain coordination and optimization analysis

According to the relevant models given in Sections 3.1 and 3.2, $0 < \tau_{ij} \le 1$ $(i = r, d, s)$ .Using Matlab to calculate the sensitivity coefficient for the supply chain coordination of fresh agricultural product under Two-part Tariff and agricultural insurance.

(1) Under the agricultural insurance, with the natural risk coefficient μ increases, agricultural insurance under the supply chain orders and profits are down, but the decline is not the same. As shown in Figure 1 below, when the natural risk coefficient μ in 0.1 step size gradually close to the non-participation insurance coefficient $\Delta = 0.2$, the order quantity of the supply chain rises slowly with about 260Kg for the step, the profits of the supply chain system gradually increased to about 190 yuan for the step. When the natural risk coefficient μ in 0.1 step away from the non-participation insurance coefficient $\Delta = 0.2$, the order quantity of the supply chain rises slowly with about 400Kg for the step, the profits of the supply chain system gradually increased to about 220 yuan for the step. Which means with the increase of natural risk coefficient μ, agricultural insurance supply chain orders and profits are down, the actual life of the risk factor is too large, the supplier in order to avoid risks, generally do not produce such products.

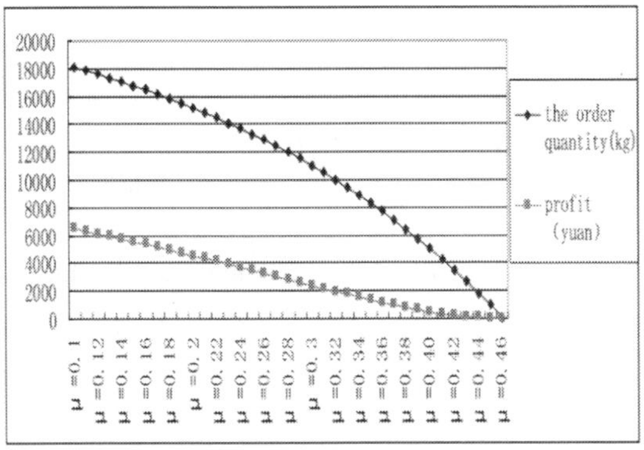

Fig. 1. Optimal order quantity and profit graph of supply chain system under agricultural insurance.

(2) When the coefficient of natural risk is less than the non-participation insurance coefficient ($\mu \leq \Delta$), the order quantity and profit of supply chain under natural disasters are larger than those of agricultural insurance. As shown in Figure 2 and Figure 3, when the non-participation insurance coefficient $\Delta = 0.1$, the disaster loss coefficient gradually increases to 0.1 in 0.01 steps, the order quantity and profit of the supply chain under the natural disaster are all bigger than the order quantity and profit under the agricultural insurance. But with the increase of the natural risk coefficient μ, in natural risk, the order quantity of the supply chain rises slowly with about 300Kg for the step, the profits of the supply chain system gradually increased to about 300 yuan for the step. In agricultural insurance, the order quantity of the supply chain rises slowly with about 200Kg for the step, and gradually close to the natural risk of supply chain under the order quantity, the profits of the supply chain system gradually increased to about 150 yuan for the step, and gradually close to the natural risk of supply chain profits. That is to say, when the natural risk coefficient is less than the non-participation insurance coefficient, due to the supplier to pay insurance costs, and insurance companies do not give claims, which making agricultural insurance under the cost of supply chain systems rise. The order quantity and profit of supply chain under agricultural insurance are all less than the order quantity and profit of supply chain under natural disaster. At this point the supplier is not suitable for insurance.

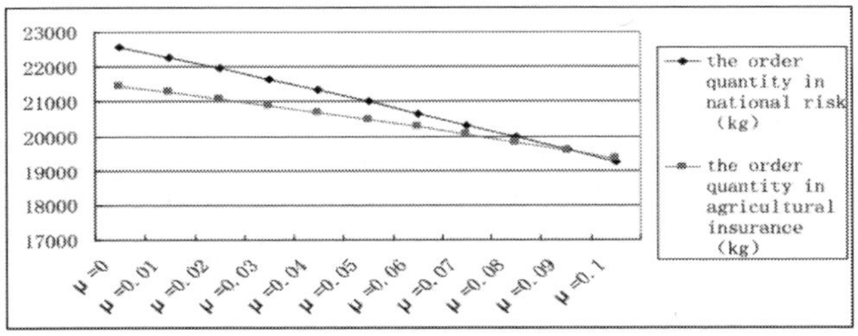

Fig. 2. $\mu < \Delta$ The Optimal Order Quantity of Supply Chain Under Natural Risks and Agricultural Insurance.

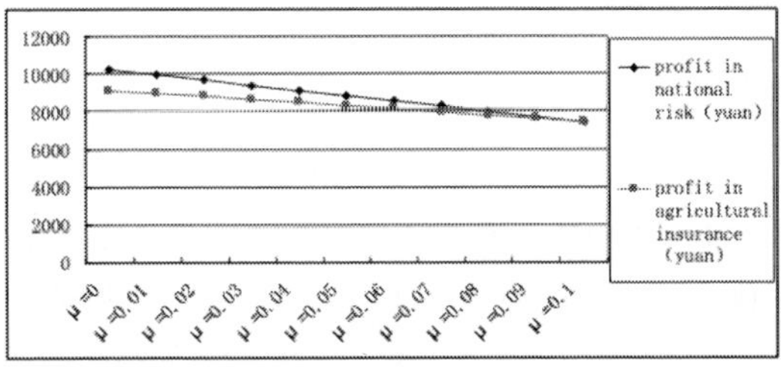

Fig. 3. $\mu < \Delta$ The Profit of Supply Chain Under Natural Risks and Agricultural Insurance.

(3)When the coefficient of natural risk is greater than the non-participation insurance coefficient ($\mu > \Delta$), the order quantity and profit of the supply chain under the agricultural insurance are all larger than the order quantity and profit under the natural risk. As shown in Figure 4 and Figure 5, when the non-participation insurance coefficient $\Delta = 0.1$, the natural risk coefficient increases gradually from 0.1 by 0.01 for the steps, the order quantity and profit of the supply chain under the agricultural insurance are all larger than the order quantity and profit under the natural risk. But with the increase of the natural risk coefficient μ , in natural risk, the order quantity of the supply chain rises slowly with about 400Kg for the step, the profits of the supply chain system gradually increased to about 300 yuan for the step; in agricultural insurance, the order quantity of the supply chain rises slowly with about 300Kg for the step, the profits of the supply chain system gradually increased to about 200 yuan for the step. That is to say, when the natural risk coefficient exceeds the non-participation insurance coefficient, the insurance company will reduce the cost of the supply chain system under the agricultural insurance, and the order quantity and profit of the supply chain under the agricultural insurance are all larger than the order quantity and profit under the natural risk. At this point the supplier chooses to purchase agricultural insurance.

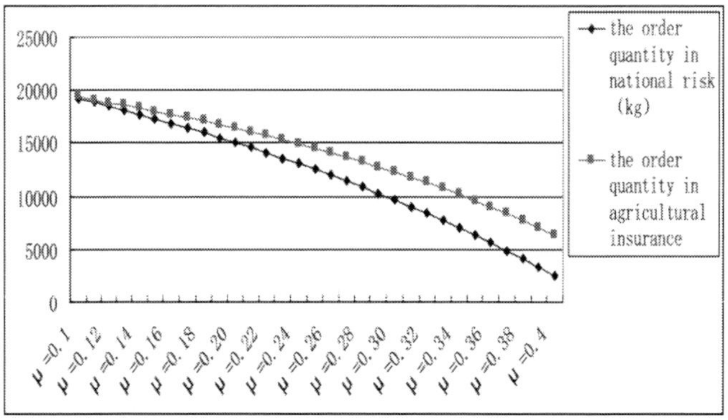

Fig. 4. $\mu \geq \Delta$ The Optimal Order Quantity of Supply Chain Under Natural Risks and Agricultural Insurance.

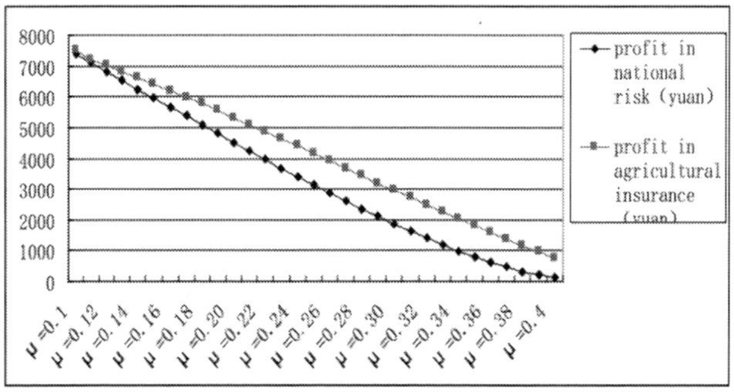

Fig. 5. $\mu \geq \Delta$ The Profit of Supply Chain Under Natural Risks and Agricultural Insurance

4.3 Application strategy of fresh agricultural products

Through an example analysis, with the increase of natural risk coefficient, the agricultural supply chain of agricultural insurance not only avoids natural risks, but also slows down the supply chain as a whole and the interests of its members, and makes the supply chain coordination increase .Here are three fresh agricultural products supply chain application strategy:

(1)Using Two-part Tariff to achieve supply chain coordination and supply chain members of the stable cooperation, first set the parameters to satisfy either Proposition 1 or Proposition 2, then the two sides to negotiate a fixed channel fee, and finally to achieve the overall coordination of the supply chain and the stability of cooperation between members.

(2)In the supply chain operation process to join the agricultural insurance in order to reduce the natural risk damage of supply chain system profit. Part of the damage is transferred to the insurance company. At the same time, due to the addition of agricultural insurance, the supply chain members are more stable in the cooperative relationship.

(3)In practice, agricultural insurance not only makes the supply chain as a whole and the members of the natural risk increases, the interest decline slowed down, but also promote the information construction of the fresh agricultural product supply chain, improving the degree of information sharing, making the supply chain members of information symmetry, and ultimately to ensure the overall coordination of the supply chain.

5 Conclusion

This paper discusses optimal supply chain decision and supply chain coordination of fresh agricultural products, combined with the actual situation, using fresh iteration relation of fresh agricultural products, comprehensive use of inventory theory, utility theory, cooperative game, for a fresh agricultural products three supply chain, which consist suppliers, distributors, and retailers. The paper focuses on the coordination of Two-part Tariff under the agricultural insurance system to the supply chain system, research shows: ① the Two-part Tariff contract can not only coordinate the supply chain of fresh agricultural products, but also stable supply chain members of the cooperative relations, but under natural risks, the coordination of fresh agricultural products supply chain system is low. ②Agricultural insurance can not only effectively manage natural risks, but also increase the supply chain system benefits, and further strengthen the coordination of the supply chain system. At the same time, this paper demonstrates that whether there is Two-part Tariff under the agricultural insurance system to coordinate the supply chain and supply chain cooperation relationship changes by an example based on the model, which has theoretical guidance and practical significance for deeper research on supply chain coordination and supply chain cooperation. Because this article only single insurance for suppliers, the practice will involve the supply chain to other members of the various insurance. Therefore, it is the next research direction to study the coordination of multi-member and multi-member agricultural insurance supply chain.

References

1. Xiong Feng, Peng Jian, Jin Peng, Zhang Xiangyang, et al.Study on the Stability of Supply Chain Relationship in Fresh Agricultural Products - From the Perspective of Subsidy Mode of Cold Chain Facilities. Chinese Journal of Management Science. 2015.23(8):102-111
2. WU Zhonghe, Chen Hong, Zhao Qian, et al. Time-constrained Fresh Agricultural Supply Chain Emergency Coordination Contract. Journal of Systems Management,2014,23(1):49-56+61
3. SUN Yuling, SHI Kuiran, ZHANG Lin, Ordering Decision of Fresh Agricultural Products for Retailers with Loss Avoidance under the Constraint of Inventory Capability. Systems Engineering - Theory & Practice,2013,33(12):3020-3027
4. Qu Han Yu. Supply chain members of the relationship between the matching and stability of the study - based on business needs perspective. Jiangsu University of Science and Technology.2014
5. Liu Qiu, Zhang Chengke, and Leng Bin. Supply chain contract stability and its optimization in option game perspective.Chinese Journal of Management Science. 2016.3(24):71-79
6. WU Zhonghe, Chen Hong, Liang Cui-lian. Association model of emergent events in supply chain of fresh agricultural products under time constraint. Chinese Journal of Management Science. .2015.23(6):126-134
7. Cui Xiao-chen. Monopoly industries Two-part Tariff pricing method. Journal of Business Economics.2015 （6）:127-129

8. Xu Hao, Li Jiachuan. Research on Coordination Mechanism of Dual Channel Supply Chain in Cost and Demand Disturbance. Prediction.2014,33(4)：70-75
9. Wang Yu. Based on Two-part Tariff contracts of fresh agricultural products online retailers and logistics business co-game. Logistics Technology. 2015.34:136-139
10. Ma Wenhai, Li Jingyue. Study on Development Path of Agricultural Catastrophe Reinsurance in Xinjiang. Agricultural Economy. 2016(2):77-78
11. Zhao Hongmei. Since the beginning of winter, 134 million mu of facilities in our province affected vegetables, but farmers insured rate of less than 10%. Hebei Daily.2016.1.7（006)
12. Yao Shoufu, Ma Liyao, Liang Xiaofeng. Thoughts on Accelerating the Development of Agricultural Insurance - An Empirical Study Based on the Broker Effect of Agricultural Insurance in China. West China Economic Management Forum. 2016.27(1):33-38+70
13. LI Ya-qi, Han Xing-yong. Research on the Interactive Relationship between Agricultural Insurance and Agricultural Output Based on VAR Model. Journal of Anhui Agricultural Sciences.2016.44(1):292-297
14. Cao Wujun, Shi Yangyang. Research on Coordination and Optimization of Agricultural Product Supply Chain Based on Option and Insurance. Industrial Engineering and Management.2016.21(1):8-16
15. LI Ying-tang, Qiao Zhong. Research on Collaborative Decision Method for Cost Control of Fresh Keeping in Fruits and Vegetables Supply Chain. Science and Economy.2014.(4):35-38
16. Akcy Y,Natarajan H P,Xu S H.Joint dynamic pricing of multiple perishable products under consumer choice.Management Science,2010,56(8):1345-1361
17. Cai X Q ,Chen J, Xiao Y B,et. Optimization and coordination of fresh product supply chains with freshness-keeping effort. Production and Operations Management, 2010 19(3): 261-278.
18. Lin L, Lei Xiaoyan, Dan Bin. Three-echelon Supply Chain Coordination of Fresh Agricultural Products Based on Linear Time-Varying Demand. Industrial Engineering.2011.14(4):46-51
19. The People's Property Insurance Company of China. Agricultural Insurance Underwriting Claims [EB/OL].(2012-2-13)[2015-4-20]http://www.piccnet. com.cn

Research on Fiscal Expenditure and Tourism Development of Huangshan City: Based on Grey Relational Analysis

Wenhao Wu[1] and Huibin Zhan [1]

[1]Huangshan University 39 Xihai Rd. 245041 Huangshan, Anhui, China

Abstract. Using grey relational analysis, the impacts of various fiscal expenditure items on local tourism development were detected by employing relevant statistical data from Huangshan City during 2008- 2013. Top five fiscal expenditure items influencing local tourism development, including general public services, education, public security, social security and employment, as well as urban and rural community affairs, were revealed in this analysis. According to the results, expenditure efficiency and its important domains are able to be recognized and verified under current economic and social circumstances. As a result, in order to optimize the expenditure structure, we suggest that financial strengthen, existing economic and geography resource, as well as the requirements of tourism development for tourism demand in the new stage should be combined.

Keywords. Fiscal expenditure, tourism development, grey relational analysis, Huangshan City.

1 Introduction

Chinese tourism is in transition stage with constant development of self-service, pluralism, popularization and normalization. New features of tourism consumption and demand have emerged in various aspects, causing changes in public needs of tourism. Therefore, government functions should be transformed with adjustment in structure of government expenditure, thus improving the level and quality of tourism in new development context.

Basis of government fiscal function in tourism development is reflected in public goods, externality, intra-generational and inter-generational factors, as well as market failure such as monopoly [1]. In China, financial support for tourism is also closely related to development goals of tourism which should meet the requirements of the rapidly developed economy, society and culture. Since the 1980s, most of the local governments of Chinese provinces, autonomous regions and municipalities have introduced financial support policies in succession to promote tourism development based on the development goals of local tourism industry [2].

Urban tourism attracts comprehensive attentions in recent years. Wang and Pei (2014) established an index system and a grey comprehensive evaluation model to analyze the sustainability of urban tourism [3]. Zhou & Dong (2014) use similar approach to investigate tourism development of Xinjiang. More specifically [4], He, Chen & Zhai (2012) examined the correlation between forest tourism and three industrial sectors in Eastern Liaoning from 2000 to 2009 [5], and Liu, Gao & Xue (2013) evaluated tourism festival activities impact on water environment in DongPing Forest Park [6].

Further, some scholars exhibit more interests on the relationship between fiscal expenditure and tourism development. Ishikawa and Fukushige (2009) examined the impact of fiscal expenditure and the number of tourists on per capita taxable income in Japanese remote islands, and find that the former exhibit positive significant relationship while the latter not [7]. Avramescu & Baldan (2011) use degree of taxation, interest rate and exchange rate to analyzes the impact of Romania`s main fiscal and monetary measures on the development of the tourism sector, results showing that negative relationship existed and other financial solutions can boost Romanian tourism [8]. Feliziani & Monni (2013) analyzed the role of fiscal policies in the tourist sector, and find that governance is one of the major causes of weakness [9].

Based on previous researches, our analysis focus on the following questions: what kind of financial supports does tourism development need? How to promote and adapt to transformation and structural adjustment of tourism industry at this stage through financial expenditure? The domains and tourist effects are different for the influence of different items of fiscal expenditure. It is very important to understand the different impacts brought by different components of tourism expenditure, which is a critical step for further improvement of efficiency of fiscal expenditure.

2 Research methods, index selection and data sources

2.1 Research methods

As a complex and integrated system, tourism system has limited statistic data with greater fluctuation and less typical distribution laws. Thus it is difficult to define all factors of tourism system and mechanism interacted among them. In this context, we regard tourism system as a grey system. To be specific, Grey system theory was first proposed by Chinese scholar Deng Julong in 1982, and had been widely used in social, economic and ecological domains. In the study, grey correlation analysis in gray system theory was employed to quantify the correlation between fiscal expenditure and tourism, revealing the relationship between expenditure and tourism development.

2.2 Index selection

2.2.1 Selection of related indexes for fiscal expenditure

The main indexes selected to measure fiscal expenditure in Huangshan City include General public services (X1), Public security (X2), Education (X3), Science and technology (X4), Culture, sports and media (X5), Social security and employment (X6), Health care (X7), Environmental protection (X8), Urban and rural community affairs (X9), Agriculture, forestry and water affairs (X10),Transportation (X11), as well as Industrial, commercial and financial affairs (X12).

2.2.2 Selection of related indexes for tourism development

There were four first-grade indexes and five second-grade indexes selected to measure tourism development in Huangshan City (See Table 1). The indexes reflect current situations of tourism development in Huangshan City from different aspects. Tourism-related industries were defined based on classification criteria[1] of the World Tourism Organization [10]. Meanwhile, the availability of data was taken into account.

Table 1. Indexes for tourism development in Huangshan City.

First-grade indexes	Second-grade indexes
Tourist income (Y_1)	Total tourism income (Y11)
	Domestic tourism income (Y_{12})
	Foreign exchange earnings from tourism (Y_{13})

[1] According to the classification criteria of the World Tourism Organization, tourism can be divided into total-tourism sector and partial-participation tourism sector.

Tourist person-time (Y_2)	Domestic tourist person-time (Y_{21})
	Foreign tourist person-time (Y_{22})
Tourist person-day (Y_3)	Domestic tourist person-day (Y_{31})
	Foreign tourist person-day (Y_{32})
Tourism-related industries (Y_4)	Transportation, warehousing and postal service (Y_{41})
	Wholesale and retail (Y42)
	Accommodation and catering (Y43)
	Financial industry (Y_{44})
	Leasing and business service (Y_{45})
	Culture, Sports and Entertainment (Y_{46})
	Public Management and Social Organization (Y_{47})
	Residents service and other services (Y_{48})

2.3 Data sources

The data used in this study were extracted from the Statistical Yearbook of Huangshan City from 2008 to 2013 (See Table 2 and 3).

3 Gray relational analysis

In grey relational theory, determinate factors are selected among multiple factors in the system for comparison. The basic concepts include: the relevance can be judged through the level of similarity between comparison sequence curve and reference sequence curve; the strength and weakness of evaluation objects are analyzed though grey relational degree. Relational degree indicates the correlation among things and factors, which quantitatively describes relative change of things or factors. The main relationship among these factors can be revealed using grey relational analysis with certain methods, exploring important factors affecting target values, thus clarifying the correlation among these factors.

Table 2. Indexes of fiscal expenditure in Huangshan City (Unit: ten thousand RMB).

	2008	2009	2010	2011	2012	2013
General public service	77473	87700	117308	141696	175065	189750
Public security	24642	39868	45655	48570	59728	57861
Education	58586	66834	82230	106105	129427	128797
Science and technology	7491	9745	14074	21190	28881	31778
Culture, sports and media	9739	13982	21508	29854	41349	41004
Social security and employment	80157	105394	82583	120365	137634	150135
Health care	29359	50524	60542	82761	96424	102273
Environmental protection	14201	18266	26998	33407	87447	84162
Urban and rural community affairs	24533	21453	25261	37046	71183	140851
Agriculture, forestry, water	46055	71413	86427	122671	145258	175459
Transportation	15079	23828	18774	32780	83207	88507
Industrial, commercial, financial affairs	38594	31215	45206	23553	67916	71255

Table 3. Indexes of tourism development in Huangshan City (Unit: ten thousand RMB).

	2008	2009	2010	2011	2012	2013
Total tourism income (ten thousand RMB)	1409000	1681500	2021400	2510200	3029801	3145400
Domestic tourism income (ten thousand RMB)	1259300	1535537	1822700	2260103	2723730	2843576
Foreign exchange earnings from tourism	22010	24401	30100	38500	48211	48739
Domestic tourists (person-time)	17202700	20348106	24396899	29230300	34809900	35720000
Domestic tourists (person-day)	21719190	27149715	33278404	39934085	47585133	48829239

Foreign tourists (person-time)	810405	877094	1050301	1313609	1602699	1605919
Foreign tourist (person-day)	1160400	1337633	1673961	2776107	3386505	2772152
Transportation, warehousing and postal service	156000	164200	170400	188630	207500	226961
Wholesale and retail	146000	165322	188732	234973	268402	296961
Accommodation and catering	83000	93900	105515	127747	146360	160886
Financial industry	67000	73136	87747	104447	123544	148068
Leasing and business service	27000	33900	38910	48292	58243	83600
Residents service and other services	82000	88560	90668	108802	132000	136014
Culture, sports and entertainment	42000	42420	47688	58121	65727	76978
Public Management and Social Organization	158000	170939	178109	203100	232144	247000

3.1 Definition and description of factors in correlation analysis

In correlation analysis of an abstract system, data sequences reflecting behavior characteristics of the system should be accurately selected—this process is called "selection of mapping variable reflecting system behavior". Then, relevant factors affecting main behavior of the system should also be confirmed. Generally, dependent variables constitute reference sequence Y_i, while independent variables constitute comparison sequence X_j.

3.2 Calculation of correlation coefficients

Based on given data of comparison and reference sequences, correlation coefficients and relational degree of comparison and reference sequences could be calculated by using the following methods and procedures. Then, the influence degree of each comparison sequence was analyzed. Analytical methods and procedures were described as follows.

(1) Initialization of raw data

Dimension or unit of each factor in raw data might be different. Therefore, the data were difficult to compare directly. Instead, they should be converted into comparable data series. There were lots of dimensionless methods for raw data. Based on researches of related scholars, the following formula was adopted:

$$\chi_i'(\kappa) = \chi_i(\kappa) / \chi_i \, .$$

(2) Calculation of absolute difference between reference sequence and comparison sequence

$$\Delta_i(\kappa) = \left|\chi_0'(\kappa) - \chi_i'(\kappa)\right|$$

(3) Calculation of correlation coefficient

$$\xi_i(\kappa) = \frac{\min_i \min_\kappa \Delta_i(\kappa) + \rho \max_i \max_\kappa \Delta_i(\kappa)}{\Delta_i(\kappa) + \rho \max_i \max_\kappa \Delta_i(\kappa)},$$

where ρ was resolution ratio, taking 0.5; $\min_i \min_\kappa \Delta_i(\kappa)$ denotes secondary minimum difference; $\max_i \max_\kappa \Delta_i(\kappa)$ denotes secondary maximum difference.

(4) Calculation of relational degree

$$\gamma_i = \frac{1}{n}\sum_{\kappa=1}^{n} \xi_i(\kappa)$$

3.3 Strength analysis

In correlation analysis, correlation matrix could be constructed when there were more than one reference sequences and comparison sequences. Then, the strengths and weaknesses of elements could be analyzed through correlation matrix. For example, Y_1, Y_2,…… Y_s was the characteristic behavior sequence of the system, while X_1, X_2,…… X_m was relational behavior sequence; Y_i and X_j had the same length; γ_{ij} (i = 1,2, ... s; j = 1,2, ... m) was the grey relational degree of Y_i and X_j, thus constituting the correlation matrix R: $\begin{pmatrix} \gamma_{11} & \gamma_{12}\cdots & \gamma_{1m} \\ \vdots & \ddots & \vdots \\ \gamma_{s1} & \gamma_{s2}\cdots & \gamma_{sm} \end{pmatrix}$.

Elements of the i-th line were grey relational degrees between characteristic behavior sequence of the system Y_i (i = 1, 2, ... s) and relational factor sequence X_j (j = 1,2, ... m). If l and $j \in \{1, 2, \ldots\ldots m\}$, and $\gamma_{il} \geq \gamma_{ij}$, then it indicated that system factor X_i was superior to X_j.

3.4 Dimensionless method

Using the first-year value of each index to divide values of other years, then dimension could be eliminated (See Table 4).

Table 4. Dimensionless values of indexes.

	2008	2009	2010	2011	2012	2013
X_1	1.0000	1.1320	1.5142	1.8290	2.2597	2.4492
X_2	1.0000	1.6179	1.8527	1.9710	2.4238	2.3481
X_3	1.0000	1.1408	1.4036	1.8111	2.2092	2.1984
X_4	1.0000	1.3009	1.8788	2.8287	3.8554	4.2422
X_5	1.0000	1.4357	2.2084	3.0654	4.2457	4.2103
X_6	1.0000	1.3148	1.0303	1.5016	1.7171	1.8730
X_7	1.0000	1.7209	2.0621	2.8189	3.2843	3.4835
X_8	1.0000	1.2862	1.9011	2.3524	6.1578	5.9265
X_9	1.0000	0.8745	1.0297	1.5100	2.9015	5.7413
X_{10}	1.0000	1.5506	1.8766	2.6636	3.1540	3.8098
X_{11}	1.0000	1.5802	1.2450	2.1739	5.5181	5.8696
X_{12}	1.0000	0.8088	1.1713	0.6103	1.7598	1.8463
Y_{11}	1.0000	1.1934	1.4346	1.7815	2.1503	2.2324
Y_{12}	1.0000	1.2194	1.4474	1.7947	2.1629	2.2581
Y_{13}	1.0000	1.1086	1.3676	1.7492	2.1904	2.2144
Y_{21}	1.0000	1.1828	1.4182	1.6992	2.0235	2.0764
Y_{22}	1.0000	1.2500	1.5322	1.8387	2.1909	2.2482
Y_{31}	1.0000	1.0823	1.2960	1.6209	1.9777	1.9816
Y_{32}	1.0000	1.1527	1.4426	2.3924	2.9184	2.3890
Y_{41}	1.0000	1.0526	1.0923	1.2092	1.3301	1.4549
Y_{42}	1.0000	1.1323	1.2927	1.6094	1.8384	2.0340
Y_{43}	1.0000	1.1313	1.2713	1.5391	1.7634	1.9384
Y_{44}	1.0000	1.0916	1.3097	1.5589	1.8439	2.2100
Y_{45}	1.0000	1.2556	1.4411	1.7886	2.1571	3.0963
Y_{46}	1.0000	1.0800	1.1057	1.3269	1.6098	1.6587
Y_{47}	1.0000	1.0100	1.1354	1.3838	1.5643	1.8328
Y_{48}	1.0000	1.0819	1.1273	1.2848	1.4693	1.5633

3.5 Results analysis and discussion

Calculating relational degree between various factors of fiscal expenditure and tourism development, then correlation matrix R could be established as follows.

	X_1	X_2	X_3	X_4	X_5	X_6	X_7	X_8	X_9	X_{10}	X_{11}	X_{12}
Y_{11}	0.8635	0.7274	0.9304	0.6862	0.6462	0.7229	0.6150	0.7301	0.7865	0.6722	0.7324	0.7159
Y_{12}	0.8712	0.7408	0.9194	0.6911	0.6502	0.7213	0.6195	0.7334	0.7843	0.6772	0.7342	0.7083
Y_{13}	0.8577	0.7106	0.9439	0.6728	0.6356	0.7183	0.6049	0.7205	0.7990	0.6605	0.7309	0.7314
Y_{21}	0.7980	0.6713	0.8628	0.6749	0.6365	0.7678	0.6009	0.7222	0.7863	0.6577	0.7254	0.7506
Y_{22}	0.8908	0.7634	0.8975	0.7029	0.6591	0.7136	0.6301	0.7419	0.7765	0.6897	0.7334	0.6940
Y_{31}	0.7437	0.6266	0.7880	0.6521	0.6182	0.8015	0.5803	0.7058	0.8023	0.6346	0.7232	0.7971
Y_{32}	0.7708	0.6887	0.7366	0.7424	0.6909	0.6125	0.6969	0.7674	0.8023	0.7629	0.7521	0.6463
Y_{41}	0.5884	0.5068	0.5947	0.6066	0.5796	0.7485	0.5293	0.6705	0.7803	0.5793	0.6861	0.7785
Y_{42}	0.7452	0.6240	0.7922	0.6542	0.6201	0.8244	0.5803	0.7074	0.7956	0.6342	0.7245	0.8050
Y_{43}	0.7160	0.5995	0.7516	0.6473	0.6142	0.8770	0.5721	0.7021	0.7986	0.6257	0.7210	0.8375
Y_{44}	0.7502	0.6442	0.8158	0.6531	0.6198	0.8011	0.5823	0.7056	0.8029	0.6346	0.7207	0.7888
Y_{45}	0.8102	0.6694	0.8168	0.7169	0.6760	0.6842	0.6665	0.7458	0.7938	0.7113	0.7469	0.6620
Y_{46}	0.6286	0.5384	0.6409	0.6202	0.5915	0.8303	0.5444	0.6805	0.7946	0.5956	0.6972	0.8224
Y_{47}	0.6269	0.5487	0.6457	0.6192	0.5911	0.8496	0.5460	0.6809	0.8013	0.5962	0.6993	0.8529
Y_{48}	0.6156	0.5255	0.6259	0.6165	0.5881	0.7835	0.5393	0.6781	0.7853	0.5904	0.6950	0.8001

According to relational degree rankings between fiscal expenditure and tourism development, there was slight difference in the impact degrees of various items of local fiscal expenditure on tourism revenues, tourist person-time, tourist person-day and tourism-related industries in Huangshan City. The top five items of local fiscal expenditure showing the largest impacts were General public services, Education, Public security, Social security and employment, as well as Urban and rural community affairs; the last three items of local fiscal expenditure influencing tourism development were Culture, sports and media, Science and technology, as well as Environmental protection.

The item of local fiscal expenditure showing the largest impacts on tourism of Huangshan City was general public services. Public management and services of tourism, supported by expenditure of general public services, can achieve effective communication between different subjects in tourism market. Meanwhile, market failures of information incompletion and asymmetry can be improved, promoting effective definition and protection of property rights; a well-performed institutional environment can be formed, thereby reducing transaction costs in tourism market, with improved efficiency of tourism economy.

Local education expenditure was mainly invested in local educational administration, primary and secondary school education, as well as primary and secondary vocational education. The impact degree on tourism development of Huangshan City was not as high as general public services, indicating that education, especially for primary and secondary school education and vocational education, plays an important role in local tourism development. The reason may be that local expenditure on education provides sufficient human capital. Thus, tourism knowledge and skills with generality, professionalism and innovation can be cultivated for tourism enterprises and related departments [11]. Meanwhile, credit capital stock of tourism has been increased to promote development of tourism [12].

Public security was the third item of fiscal expenditure greatly influencing tourism development of Huangshan. It indicates that local security environment, traffic safety and other public security of tourism are also important factors affecting tourism development.

Social security and employment expenditure positioned fourth for the impacts on tourism development of Huangshan City. In the correlation between tourist person-time and development of tourism-related industries, social security and employment expenditure ranked in the first place. Various local expenditure and subsidies on social security, public service employment, as well as employment and vocational training promote accumulation of tourism employment, venture capital investment, as well as human capital of individuals and enterprises in Huangshan.

Expenditure on urban and rural community affairs plays an important role in promoting the development of urban and rural community planning, public infrastructure and public welfare establishments. In addition, it encourages urban and rural environmental renovation, governance of community public affairs, as well as the construction of new and beautiful countryside. Local traditions and culture can be inherited with renaissance in Huangshan, improving living spaces, public cultural space and living environment. Attraction of urban and rural tourism is enhanced, and development of local tourism accelerated thereafter.

Traffic is an essential bridge of the tourism and the infrastructural industries for tourism development. However, correlation between transportation expenditure and tourism development of Huangshan was in the sixth place. Huangshan is one of the earliest cities with tourism development. In early stage, transportation expenses were larger with more important role in tourism. However, when tourism has developed to a certain stage with continuous improvement of transportation system, marginal contribution ratio of local transportation expenditure to tourism began to showing declining trend. In addition, improvement of traffic shortened the distance between Huangshan and the Yangtze River Delta as well as other surrounding cities. The radiation Yangtze River Delta cities and other surrounding cities had promoted the development of Huangshan Tourism. However, traffic improvement also generated substitution effects of tourism demand between Huangshan and these cities. Therefore, changes of modern traffic environment had generated complex effects on tourism development.

4 Conclusions and recommendations

Above all, results show that the top five items of local fiscal expenditure significantly promote local tourism development are General public services, Education, Public security, Social security and employment, as well as Urban and rural community affairs. According to the results, expenditure efficiency and focuses of local government could be recognized and varied under current economic and social environment. Thus, the structure of fiscal expenditure can be further optimized, providing scientificity and rationality of decision-making in fiscal expenditure [13].

As a traditional famous tourist city, Huangshan also faces pressure and confusion of industrial upgrading and public services, as well as public administration innovation, along with changing external environment. Firstly, the combination of culture and tourism, as well as environment and tourism is not perfect enough. Secondly, new changes in traffic environment, such as high-speed rail, have brought changes of regional advantages. This has changed substitutional and complementarity relations of tourism resources and products between Huangshan and surrounding cities. Thirdly, the proportion of self-catering individual travelers has become keep increasing, thus enhancing pressures for tourism public service and public management innovation.

In the analysis of expenditure growth, Musgrave and Rostow emphasized changes of expenditure structure resulted from public need changes in different stages of economic development. Similarly, the structure of local fiscal expenditure should adapt to changes of public needs in different stages of tourism development. Government should combine its financial advantage, existing economic and resource conditions of Huangshan, as well as requirements of tourism development for tourism public demand in new stage. Meanwhile, scientificity and rationality of decision-making in fiscal expenditure should be further improved for tourism development. Government should promote industrial transformation and upgrading with industrial development goals in new stage of tourism, thus achieving fast tourism development of Huangshan and building a leading tourism city in China.

Acknowledgements

The work was supported by Anhui Province Philosophy and Social Science Planning Project (AHSKY2015D34), grant from Huangs han University Research Foundation (2014xhwh007).

References

1. M. Thea Sinclair, Mike Stabler (Compilers), Song Haiyan, Shen Shujie (Translators), 2004. *Economics of Tourism*. Beijing: Higher Education Press, pp. 145-148.
2. Wang Zhidong, 2005. Empirical Research on Policy Support of Chinese Local Government to Promote Tourism Development. *Dongyue Tribune*, Vol 5, pp. 69-76.
3. Zheng-Xin Wang & Lingling Pei, 2014. A systems thinking-based grey model for sustainability evaluation of urban tourism. Kybernetes, Vol. 43 Iss 3/4 pp. 462 – 479.
4. Zhou, L., Dong Q, M, 2014. Research on tourism development of Xinjiang based on grey relational analysis and gravity model, Grey Systems: Theory and Application, Vol. 4 Iss 3.
5. He D, Chen K & Zhai Y. 2012. Study on Forest Tourism and County Economy by Grey Correlation Analysis——Taking six counties Eastern Liaoning as a case. Advanced Materials Research Vols. 347-353, pp 874-878.
6. LIU S, GAO J, & XUE D. 2013. The evaluation of tourism festival activities impact on water environment in DongPing Forest Park based on the grey relational recognition model. *Advanced Materials Research* Vols. 726-731, pp 1391-1395
7. Ishikawa, N & Fukushige, M 2009. Impacts of tourism and fiscal expenditure on remote islands in Japan: a panel data analysis. Applied Economics, Vol 41, 921–928.
8. Avramescu, T,C., Baldan, C,F. 2011. Some considerations on the government involvement in tourism development through fiscal and monetary measures. Anale. Seria Ştiinţe Economice. Timişoara
9. Feliziani V., Monni S. 2013. Towards new tourism policies: The role of fiscal policy in increasing competitiveness. Tourism Review International, Vol. 16, pp. 227–237.
10. Chris Coope, John Fletcher, et al. (Compilers), Zhang Lili, Cai Liping, et al. (Translators) 2004. *Tourism: Principles and Practice (Second Edition)*, Higher Education Press, pp. 11.
11. Hu Qinglong, 2009. Relationship between Human Capital Accumulation and Economic Growth of Huangshan Tourism—An Analytical Framework Based on Neoclassical Theory. *Journal of Huangshan University*, Vol 1, pp. 53-56.
12. Liu Changsheng, Jane Yufeng, Yin Hua,2009. Tourism Credit, Human Capital Development and Tourism Industry. *Tourism Tribune*, Vol 11, pp. 13-20.
13. Paul F.J. Eagles, 2014. Fiscal implications of moving to tourism finance for parks: Ontario Provincial Parks, Managing Leisure, 19:1, 1-17.

The Mechanism of Knowledge Associated Integration of Interdisciplinary Research Team Based on Absorption Ability

Ziqi Jin[1], Bowen Zhong[2,*], Xiaohong Wang[3] and Hua Jiang[3]

[1] The School of Mechanical and Electric Engineering, Soochow University, P.R.China
[2] The School of Mechanical and Electric Engineering & Collaborative Innovation Center of Suzhou Nano Science and Technology, P.R.China
[3] School of Management, Harbin Institute of Technology, P.R.China
Corresponding author: zhbw@suda.edu.cn

Abstract. Interdisciplinary research team (IDRT) to solve complex, major multidisciplinary problem, is an important organization form. Members thought fusion and multidisciplinary knowledge integration is the key of implementing interdisciplinary team sustainable competitive advantage. According to the theory of knowledge absorptive capacity, this paper has analyzed knowledge associated integration mechanism of interdisciplinary team. First, the paper analyses the characteristics of absorbing ability cognitive mediation, subject link together mechanism, dimension conversion mechanism and reciprocal decision mechanism. Second, this paper builds an absorption evolution model of knowledge associated integration of interdisciplinary research team. Finally, it simulates and analyzes the evolution characteristics of interdisciplinary team knowledge network, and proposes management strategy.

1 Introduction

Big science times, scientific and technological development tends to complex multidimensional characteristics. Solution of major problem of science and technology gradually change to mutual cooperation way. With collaborative innovation and use different knowledge resources, interdisciplinary research team becomes an important organizational form of knowledge production. Knowledge integration and thinking between the different disciplines is the key to achieve interdisciplinary team unique competitive advantage. Study on mechanism of knowledge associated integration put forward the management strategy, which has the important research meaning and value to promote team knowledge innovation elements effective integration and fusion depth and ultimately to serve the society and regional economic development.

Knowledge production activities in interdisciplinary research team can not be separated off member knowledge absorption. Knowledge absorption stimulate creative thinking is the foundation of knowledge innovation. Thinking associated with the knowledge integration is the effect of knowledge absorption. The action relationship and mechanism of the absorption capacity and knowledge associated integration is one of the hot research topics in recent years. But so far, to absorb the outside world knowledge, the thinking associated way of members with heterogeneity knowledge of the interdisciplinary team and the knowledge integration mechanism between different members are relatively shortage. Deep analysis knowledge associated integration mechanism of interdisciplinary research team based on the knowledge absorptive capacity theory provides some suggestions for interdisciplinary team management and development.

2 Connotation of knowledge associated integration of interdisciplinary research team

Interdisciplinary research team as a link to complexity science mission and technical problem, gather different disciplines knowledge resources advantages based on group cooperation with openness organization, which has the nature of discipline knowledge integration and fusion of cognitive thinking to comes true scientific research and knowledge innovation target [1]. Through the comprehensive perspective integration of different disciplines and fusion heterogeneity of thinking, interdisciplinary team makes use of interdepartmental and the interaction of the subject creative thinking to solve complex scientific problems involving multidisciplinary field.

Knowledge integration and thinking associated is the key to effective integration and fusion depth of interdisciplinary research team elements of innovation.

Interdisciplinary knowledge associated integration means the member with heterogeneity discipline knowledge on the basis of problem solving cognitive psychological mechanism screening and using the advantage of resources from their own internal and external knowledge element actively carry out interdisciplinary communication, learning and cooperative innovation of issues field. By knowledge selecting, links, and convergence points, then formed interdisciplinary teams new knowledge needed and realize the value of knowledge, which promotes individual cognition of interdisciplinary team member sustainable development [2]. There are two main clues in interdisciplinary knowledge associated integration, subject knowledge integration and creative thinking association. Knowledge integration accompanied knowledge acquisition, knowledge application and Knowledge transfer, which reflect interdisciplinary knowledge innovation activities creative, unpredictability and uncertainty. Thinking associated elaborates that individual cognitive development follows general psychological processes of problem solving. It contains analysis and characterization and generation and selection the solution scheme, implement and evaluate solution scheme. Subject knowledge integration and thinking associated are two interdependent and mutual promotion spiral sequence. The reciprocal action of multiple members thinking and disciplinary knowledge formation of knowledge network system, which finally resolves the problem achieve knowledge innovation. The members of interdisciplinary team keep absorption ability(including access, transformation and utilization) face to external knowledge, promotion knowledge open flow and high efficiency transfer, stimulation creativity thinking associated and subject knowledge integration, which improves the innovation and development ability of interdisciplinary team in the process. It is the key of the intermediate links to success of the interdisciplinary research.

3 Absorption mediation characteristic and evolution mechanism of interdisciplinary research team knowledge associated integration

Knowledge absorptive capacity referred to as absorption capacity. It put forward by Cohen and Levinthal in 1990. Knowledge absorptive capacity is defined as identifies new information on the outside and digest and application ability of an organization. It is a function of organization knowledge, which is crucial for the organizational innovation ability. The fact of science and technology development tells us that interdisciplinary team as a kind of social organizational system and social activities always attached to certain social environment. The individual knowledge absorptive capacity is regarded as team core starting point and cognitive mediation of knowledge production activities to guide the interdisciplinary research team knowledge associated integration and affects the multidisciplinary knowledge collaborative innovation. Characteristics of collaboration, series feedback and triadic theory are precondition of promotion interdisciplinary research team knowledge associated integration and stimulation creative thinking.

3.1. The collaborative features and subject link together mechanism of IDRT knowledge associated integration

Thinking associated and knowledge integration with absorption new knowledge is a microscopic behaviour, but also is the collaborative innovation between team members. Members are the basis of interdisciplinary knowledge innovation. Absorptive capacity of members is the foundation of the team absorptive capacity, but also is the innovation source of interdisciplinary team knowledge associated integration. It significantly embodies the highest level on organizational learning and member communication, promotion the access of unique information resource acquisition, transmission and application. It is the fundamental guarantee of interdisciplinary team knowledge innovation. Member knowledge absorptive capacity based on the absorptive capacity of team flexible uses and gives full expression to information and resources advantage on organize internal and external. On individual level, the individual of members through social networks set the resource network that can absorb the knowledge and technology. Available knowledge resources is assembled in the individual mind associated cognition then materialized knowledge and innovative behaviour. On team level, team absorptive ability based on the member absorptive capacity could realize the long term development and performance growth of the team organization. Therefore, knowledge absorption ability collaborative features make team and personal knowledge associated integration has the connected effect supplement each other. Personal effective exploration, transformation and utilization of knowledge promote directly individual and organization knowledge absorptive capacity growth and improvement of performance. In conclusion, knowledge absorptive capacity on team and members with "dual belong to one" attributes has subject link together mechanism. It then is able to realize that individual knowledge absorption- individual knowledge innovation team knowledge creation in spiral interactions on individual cognitive thinking and team knowledge innovation is spiral growth. It improves the efficiency of team knowledge collaborative innovation.

3.2 The continuous feedback characteristics and dimension conversion mechanism of IDRT knowledge associated integration

Absorbing ability closely related to organizational learning is a dynamic knowledge conversion process [3].Absorbing ability not only depends on the prior knowledge but also enhance knowledge associated integration degree through the study of continuous feedback.

This paper takes example by Zahra and George dimensions of absorptive capacity, with cognitive psychology problem solving process perspective and the dynamic theory of knowledge innovation, and divides members of interdisciplinary team knowledge absorptive capacity into three continuous feedbacks dimensional of exploration, transformation and utilization [4]. On the one hand, knowledge absorptive capacity is the foundation of knowledge innovation, which provides knowledge coupling material and extension space for problem solving. Three dimensions are relatively independent and complement each other, move in circles and spiral, and change the members of the individual knowledge quantity and nature of knowledge to improve the interdisciplinary research innovation performance and innovation ability. On the other hand, these three kinds of ability is a process of the fuzzy knowledge discovery to become clear knowledge innovation. Cohesion exploration ability and utilization ability, transformation ability embodies subject knowledge integration and thinking associated with knowledge-based association rules. Exploring ability is the potential knowledge absorptive capacity, providing the breadth and depth of knowledge absorption of innovative individual. It reflects results expectations of absorbing knowledge. Use is real knowledge absorption ability reflects actual results to absorb the knowledge. It plays a direct role in promoting to individual knowledge innovation behaviour. The continuous feedback three dimensions characteristics of exploration, transformation and utilization knowledge ability, step by step excitation knowledge integration and thinking association, expand knowledge system space. Relative to the external knowledge, it also highlighted the explicit knowledge and tacit knowledge members of the innovation are equal importance. Individual cognitive mechanisms of assimilation and comply with stimulates interaction between knowledge and thinking. With transformation and fusion, knowledge grows up and be shared to other members absorption and recycling.

3.3 The ternary interaction characteristics and reciprocal decision mechanism of IDRT knowledge associated integration

Interdisciplinary research is a collective activity. The social cognitive ternary interaction theory tells us behaviour is combined action result of the cognition and environment. Knowledge absorption is a kind of people's cognitive activities. Knowledge integration and thinking associated is concrete behaviour caused by this kind of cognitive activity. This process constantly affected by the member environment.

The ternary interaction theory proposed by social cognitive scientist Bandura points out three reciprocal actions of environment, cognitive and behaviour, behaviour control by the cognitive and affected by the environment at the same time. Visibility, environment, main cognitive and behaviour are relatively independent as well as interdependent. The relationship of affects and restricts between them similarly external environment change individual cognition, individual cognitive change individual behaviour. Chini indicates human behaviour lead by rules of organization situation and social resources. The knowledge transferred by individual behaviour matches the receiving element situation needs [5]. Therefore, the reciprocal action on factors of knowledge, personality and environmental will affect member knowledge associated integration by influencing the members to absorb knowledge. In turn, knowledge associated integration feedback to promote adjustment various factors in the environment further influence member knowledge absorption. The ternary interaction theory has revealed the member knowledge absorptive capacity and knowledge associated integration and team setting has the mutual decision relationship. Interdisciplinary research team knowledge associated integration absorption mediation characteristic as shown in the Figure1.

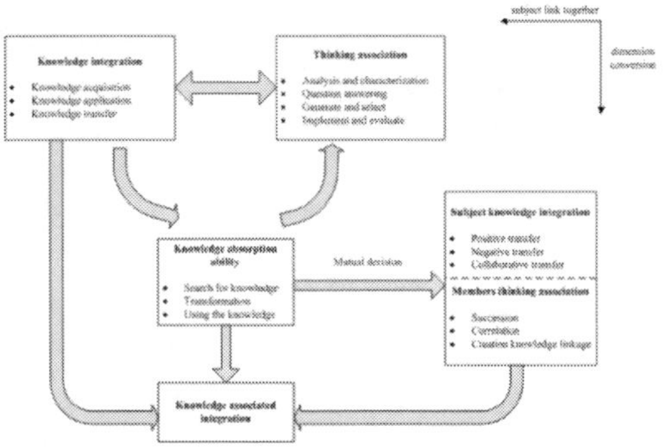

Fig.1. The absorption mediation characteristic of interdisciplinary research team knowledge associated integration.

4 The evolution model and analysis of absorption mediation characteristic of IDRT knowledge associated integration

Interdisciplinary knowledge associated integration contains two main threads subject knowledge integration and members thinking association. Knowledge absorption as a cognitive mediation plays connection function of macro knowledge and micro thinking. Knowledge integration including knowledge acquisition, knowledge application and knowledge transfer gives expression to creative, uncertainty and unpredictability on interdisciplinary knowledge innovation activities.

Thinking associated elaborates individual cognitive development following general psychological process of problem solving, including the analysis and characterization of problem, trouble shooting, generation and selection the solution scheme, implement and evaluate solution scheme, with reflection cross discipline thinking starting, respectively, jumping, activation. Subject knowledge integration can be divided into positive transfer, negative transfer and collaborative transfer three types. Thinking associated includes the succession, correlation and creation knowledge linkage. Deep analysis of IDRT knowledge associated integration process is favour of improving the efficiency of team innovation.

4.1 The evolution model of absorption mediation characteristic of IDRT knowledge associated integration

4.1.1 Interdisciplinary research team knowledge system

Let $N=\{n_1,n_2,n_3,\cdots,n_N\}$ denote a set of N knowledge points. At time t, if there is a directed edge from knowledge node j to node i, at time t, then expression node i can absorb knowledge discovered by agent j, $j\in N-\{i\}$. Therefore, all the directed edges $e_{i,j}$ make up a set of edges $E(t)$, $E(t)=\{e_{i,j}\mid i,j=1,2,\cdots,N\}$. The collection on N and E, $G(t)=(N,E(t))$, represents the topology of collaborative knowledge innovation network under dynamic absorption. Knowledge absorption prompts knowledge point of individual cognitive distance smaller link and fusion. It then forms knowledge network and comes into being obvious influence on the network structure. Members knowledge absorptive capacity, subject knowledge distance, probability of knowledge integration affected by knowledge, probability of knowledge integration affected by the personality, it common effect innovation efficiency. Knowledge associated integration can use the following Eq. (1).

$$S = S_K + S_T = f_{(K,E,C,t)} + g_{(T,U,C,t)} \tag{1}$$

S is knowledge system of interdisciplinary research team. $S_K$is the interdisciplinary team discipline knowledge integration system. $f_{(K,E,C,t)}$ is the knowledge integration system. S_T is interdisciplinary research team member thinking association system. $g_{(T,U,C,t)}$ is the member thinking association system function. E is subject knowledge cross weight coefficient. $0 \le E_{(i,j)} \le 1$, Its value decided by subject function distance between the knowledge i and j. U is the knowledge absorptive capacity function. In normal conditions, knowledge i can't completely absorbed j. So the node i for the absorption coefficient of knowledge is random in the scope of [0, 1]. Knowledge absorptive capacity by causing high level knowledge overflow between members in the knowledge transfer process produces a dynamic effect on team innovation activities. The knowledge stock of members is continuously increasing along with the knowledge exchange and sharing activities. The higher knowledge updates speed, the greater knowledge depreciation. Therefore, set the maximum difference between absorption learning and knowledge depreciation isCU. Knowledge absorptive capacity function obeys uniform distribution between $[0, CU]$. K、T is a function of time t. C is the system state variables and is a constant.

4.1.2 Interdisciplinary research team discipline knowledge integration system

$$f_{(K,E,C,t)} = K_{i(t)} = E_{(i,j)} \cdot T_{i(t-1)} \cdot P_{K\cdot j} + T_{i(t)} \tag{2}$$

Discipline knowledge integration system includes two parts. At time t, the $T_{i(t)}$ denotes the thinking increment of knowledge node i and absorption knowledge stock from its neighborhood node j, until time t-1,$E_{(i,j)} \cdot T_{i(t-1)} \cdot P_{K\cdot j}$. $T_{i(t-1)}$ is the incremental thinking of nodes i, at time$t-1$. $E_{(i,j)}$ is subject knowledge cross weight coefficient. It has diminishing exponential relationship of subject function distance $d_{(i,j)}$. The smaller subject function distance, the closer the cognitive, the higher the efficiency of knowledge diffusion. The greater subject function distance, the lower the efficiency of knowledge diffusion, the slower the growth rate of knowledge. Considering the distant subjects' innovation is less likely, there is no cognition resonance and cooperative relationship. Subject distance $d_{(i,j)} \neq \infty$ is beyond the scope of consideration. The impact factor of subject knowledge distance lessens, along with cultural gap decrease. When the culture gap is very minimal, the practical significance of the integration innovation with comprehensive knowledge different subjects in interdisciplinary research is lose away. $P_{K\cdot j}$ is for the node j the affected by the knowledge probability in Knowledge integration. Knowledge is mainly divided into explicit knowledge and tacit knowledge. In comparison with the theory and method, cutting-edge technology knowledge show more tacit knowledge. It can't make oneself master skill in a short period of time. The probability of knowledge effectively integrate is small. The wider is prior knowledge scope and difference, the easier is knowledge absorption. Knowledge accumulation and structure, knowledge expressiveness increase the chance to acquire knowledge. Knowledge heterogeneity and complexity increased the difficulty of communication and exchanges between scientific research personnel, which reduce the knowledge associated integration efficiency.

4.1.3 Interdisciplinary research team members thinking associated systems

$$g_{(T,U,C,t)} = T_{i(t)} = T_{i(t-1)} \cdot P_{T\cdot j} + U_i \cdot K_{i(t-1)} \quad (3)$$

Member thinking associated systems describes implicit cognitive growth process of personal mental enlightenment. It consists of two parts. The existing association thinking on node i, at time t-1, is thinking associated incremental, $T_{i(t-1)} \cdot P_{T\cdot j}$. And the stock of knowledge associated is $U_i \cdot K_{i(t-1)}$ from the absorption of other nodes at time t-1. $P_{T\cdot j}$ is the probability of node i personality effect on node i absorption thinking j. Learning engagement of members of IDRT for the innovation practice is the power to explore knowledge. Everyone involved in the cooperation has some mental freedom and feeling of control is favor of cooperation innovation [6]. The rigor and complexity and occasionality of interdisciplinary research activities decide to study difficulty.

So the individual of the members has the good psychological quality and personality traits directly affect thinking associated integration efficiency. The courage to face difficulties, open innovation thinking, tolerance of trial and error, creative thinking, individual learning, outcome expectation, self assessment of the results and behavior, self efficacy and so on. Knowledge absorptive capacity function is U_i. Due to knowledge aging effect, knowledge absorptive capacity function obeys uniform distribution among $[0, CU]$.

4.2 The simulation analysis of IDRT knowledge associated integration evolution model

There are six nodes in an interdisciplinary research team. Each initial increment of the nodes, $T_{i(t)}$, randomly generated within the $[0,1]$ in the system. Under normal conditions, interdisciplinary is easier to produce major knowledge innovation. Subject knowledge cross weight coefficient is greater than knowledge absorptive capacity coefficient. $E_{(i,j)} > U_i$. In order to see each knowledge node quantity transformation, this paper combining with the coefficient values range set $E_{(i,j)} = 0.05$, $U_i = 0.01$. Regardless of the network nodes no path connected and knowledge type impact on the interdisciplinary innovation, it conducts the evolution simulation of knowledge associated integration mechanism. The simulation calculation realized by Matlab programme. In order to reduce the simulation result influence of the initial random data, all simulation models simulate 40 times. The results are shown in Figure2.

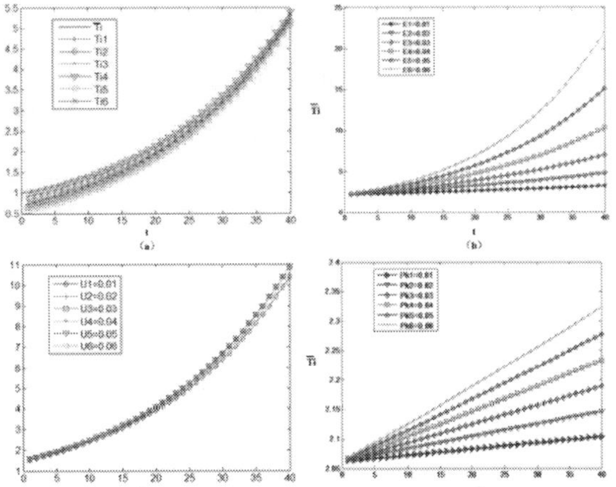

Fig.2. The simulation of IDRT knowledge associated integration.

4.2.1 The evolvement analysis of the network overall knowledge level ($\overline{T_{(t)}}$)

The network overall knowledge level is measuring knowledge created by knowledge network over a period of time. It is average value of all the knowledge nodes in the network, Eq. (4).

$$\overline{T_{(t)}} = \frac{1}{N}\sum_{i=1}^{n} T_{i(t)} \tag{4}$$

The evolvement simulation of interdisciplinary team knowledge associated integration network see Figure2（a）. The network overall knowledge level and each node incremental overall shows a trend of increasing over time. The growth rate of overall levels of knowledge reflects knowledge transfer speed. First, the network overall knowledge level gradually increase. After reached an extreme, it increases rapidly. This shows that a period of stage of discipline understanding and knowledge exchange between members in early interdisciplinary team building, because of discipline culture, logic and language disacquaintance [7-8]. The knowledge collaboration degree increases rapidly among members, when effective fusion of the acquisition knowledge and prior knowledge. Thinking association and knowledge integration happen frequently.

Knowledge is growing fast. Despite the initial values of different knowledge level, but growth trend is the same.

4.2.2 Analysis of the influence of discipline knowledge cross weight coefficient

Subject knowledge distance has nothing in common with each other within the interdisciplinary team. The smaller the subject knowledge distance, the greater the knowledge cross weight coefficient [9]. The greater knowledge innovation opportunities, the faster network overall knowledge promotion. Parameter is the same as set previously except discipline knowledge cross weight coefficient. The simulation result is Figure2（b）. Upgrade discipline knowledge cross weight coefficient, innovation efficiency increases accordingly. This shows that the greater the chances of knowledge cross fusion with donor and receptor disciplines in the process of interdisciplinary research, the more prone to collaborative innovation and the faster the overall level of knowledge network improvement.

4.2.3 Analysis of the influence of knowledge absorption ability

Parameter is the same as set previously except knowledge absorption ability. The simulation result is Figure2（c）. Team node has different knowledge absorptive capacity. The stronger absorption capacity, the faster network overall knowledge level promotion. Upgrade knowledge absorptive capacity, the overall level of knowledge network is a growing trend. Each ability level shows the characteristics of obvious differences on the whole intellectual growth. In the initial stage of the team, the effect was not significant with knowledge absorptive capacity on the whole knowledge growth. But when the team development to a certain stage the effect was more and more manifest with knowledge absorptive capacity on the whole knowledge growth. The stronger knowledge absorptive capacity, innovation ability, the overall knowledge enhancement is fast. Knowledge absorptive capacity is weak [10]. So the knowledge innovation growth is slower. The network overall knowledge level is significantly lower than the knowledge network the stronger knowledge absorptive capacity.

4.2.4 Analysis of the influence of probability of knowledge integration affected by knowledge

Other assumptions invariability, conditioning probability of knowledge integration affected by knowledge, analysis of the influence of the probability variation to the evolutionary character of the network overall knowledge level, the simulation is shown in Figure2（d）. The probability of knowledge integration affected by knowledge is greater. The knowledge variation is more significant. That is to say the collaboration efficiency of a single node and other nodes depends on the knowledge flow type. The type of single, simple and explicit knowledge easy to transfer to make full use of knowledge, which affected by the type of knowledge small have no obvious effect on the network overall knowledge level. when there is heterogeneity discipline knowledge members to join, bring the flood of resource types of discipline theory, method and technique, the more knowledge type affected by the knowledge impact probability is bigger exerts significant influence on the network overall knowledge level.

5 Conclusion

Interdisciplinary team members' knowledge absorption ability, keep the absorption of external knowledge, including access, transformation and utilization, to promote knowledge open flow and high efficiency transfer [11]. It stimulates creative thinking and knowledge integration and is the work foundation of collaborative between team members. The knowledge absorptive capacity theory is introduced in interdisciplinary team knowledge associated integration mechanism. It reveals interdisciplinary team knowledge associated integration is precondition of giving full play to the team multidisciplinary knowledge advantages. First, the characteristic of cognitive mediation of absorptive capacity is analyzed in the interdisciplinary team knowledge associated integration process, subject link together mechanism, dimension conversion mechanism and reciprocal decision mechanism. Second, the evolution model of interdisciplinary team knowledge associated integration is constructed and the simulation analysis of the four evolutionary trend of interdisciplinary team knowledge network. In conclusion, interdisciplinary team members have the good psychological quality and personality characteristics, the courage to face difficulties, open innovation thinking, tolerance of trial and error, creative thinking, individual learning, outcome expectation, self assessment of the results and behavior, self efficacy and so on, directly affect the thinking associated integration efficiency. With the team culture and common goal as the glue of interdisciplinary knowledge resources, interdisciplinary collaborative innovation practice can promote interdisciplinary team innovation performance in the system design to form under the dual role of interaction between team members (internal cause) and the external environment condition (external cause) to control the dynamic knowledge correlation integration process. It is in favor of team management work in the interdisciplinary.

Acknowledgement

This paper was supported by the National Natural Science Foundation of China (No.71473057) , the Jiangsu province Natural Science Foundation (No. BK20140345), the Natural Science Foundation of Young Teachers of Soochow University(No.SDY2015A06), the Jiangsu province Colleges and Universities Natural Science Foundation (No.14KJB460025), the National Science Foundation for Post-doctoral Scientists of China (No. 2014M551651) and the Jiangsu province Natural Science Foundation for Post-doctoral Scientists (No. 1401073C).

References

1. D. M. Amidon. Research Technology anagement, 39(1993).
2. Wang Liying, Chen Jin. Thailand: The 4thInternational Conference on Management Innovation& Technology,(2008).
3. Enyu Zhuang, Guanrong Chen, Gang Feng. Physica A: Statistical Mechanics and its Applications, 390(2011).
4. Zhong Bowen, Jin Ziqi, Wang Xiaohong, Jiang Hua. Key Engineering Materials, (2014).
5. Yuxian Liu, Ismael Rafols, Ronald Rousseau. Journal of Documentation, 68(2012).
6. Rafols, I. and Meyer, M. Scientometrics, 82(2010).
7. Hamilton john, Tee singwhat. TQM Journal, 1(2016).
8. Fontaine, Philippe. Journal of the history of the behavioral sciences, 1(2015).
9. Schaffer Scott P, Chen Xiaojun, Zhu Xiumei. Journal of engineering education, 1(2012).
10. Haythornthwaite, Caroline. Journal of the American society for information science and technology, 8(2006).
11. Full Robert J, Dudley Robert, Koehl M. A. R. Integrative and comparative biology, 5(2015).

 DOI: 10.1051/matecconf/201710005039

Analysis and Countermeasures of PPP Project Implementation Difficulties

Yifeng Xu [1] and Wei Chang [2]

[1] School of management and engineering, Zhengzhou University, China
[2] China Machinery TDI International Engineering Co.,Ltd, China

Abstract. Since the beginning of 1995, the model of PPP project construction was introduced in China. It has been widely used in many fields such as energy, municipal, and so on. It has accumulated a lot of experience in improving the investment system structure and guiding the social capital into the infrastructure field. But the mode of PPP project construction in the implementing process in different environment often encounters a lot of problems. Based on a large amount of analysis on the existing literature, this paper resorts out the PPP project model combining the problems occurred in the implementation process of PPP project, and summarizes corresponding countermeasures aiming at the problems in order to provide help for PPP project implementation in the future.
Keywords. Public Private Partnership, Problem, Strategy.

1 Introduction

PPP originates in western countries. In 80s of last century, in the overall privatization ideas, the British government began to advocate PPP mode. The British government is a pioneer advocate of PPP, and it is also the promoters of PPP [1]. PPP is the abbreviation of three English words which include private, partnership, and public. PPP itself is a concept whose meaning is covering a wide range. Because of the different ideologies, all countries have different interpretations of PPP. The international mainstream translation is capital cooperation between the government and society [2,3].

The essence of PPP is that the government provides public services to the public. On the surface, PPP's origins began with the government's lack of funds in public spending. Many governments choose to develop PPP when they are under great pressure to provide infrastructure. It is conducive to reduce the short-term public expenditure pressures, and to enhance the efficiency and quality of public services to a certain extent. In the final analysis, the definition of PPP is actually a sentence which refers that the government makes full use of the money, technology, talents and management experience of the social capital in order to improve the quality and efficiency of public service supply through sign a contract.

The main role of the government in the PPP model will be providers of public services rather than direct producers. But there is a lack of perfect legal system in the PPP construction mode, although the relevant laws can be referred to, but cannot fully cover a variety of problems which may be occurred in the PPP construction projects [4]. In addition, because of some government officials lacking the necessary professional knowledge, skills and scientific assessment methods, or the using of the old management method management, it is easy to appear mistakes in decision-making. At the same time, ineffective government regulation will affect the implementation of the PPP construction model. Based on this, this paper analyzes the current implementing situation and the problems occurring of the PPP construction mode, puts forward the corresponding countermeasures, and aims at improving the implementing effect of PPP construction mode [5].

The contents of this paper are as follows. The second section describes the characteristics of PPP construction mode. The third section analyzes the difficulties in the implementation process of the current PPP construction mode. The fourth section summarizes the corresponding countermeasures to solve the problem. Finally, the conclusion discusses the research content of this paper and puts forward the prospect.

2. Main characteristics of PPP

2.1 Partnership

Partnership is the first major character of PPP, and all the PPP projects which are successfully implemented are built on the partnership [6]. It can be said that the partnership is one of the most basic and critical relationship in PPP, and there will be no PPP if there is no partnership.

Compared to other relationships, a significant unique point in the partnership between the private sector and the government's public sector in PPP is consistent with the project objectives. The reason why the public sector and private sector cooperate and form partnerships is that they have the same goal. The goal is to achieve the most products or services with the least resources in a specific project. Private sector is to achieve the pursuit of their interests according to this goal, while the public sector is to achieve the pursuit of public welfare and interests according to this goal.

2.2 Benefit sharing

Benefit sharing is the second character of PPP. What needs to be clear here is that the public sector in PPP and private sector do not share the profits, but need to control the probable high profits in the private sector. That's to say that the private sector is not allowed to form an excess profit in the implementation process of the project. The main reason is that any PPP project is a public welfare project, and do not aim to maximize profits.

It is worth to note that benefit share here is not only means sharing the social outcomes of PPP, but also means making the private sector, private enterprises or institutions which are as the participants to obtain relatively peaceful, stable investment returns [7].

2.3 Risk sharing

The third feature of PPP is risk sharing. Partnership not only means benefit sharing, but also means risk sharing. In PPP, which is compatible with the market economy rules, the benefit and the risk are also corresponding. Risk sharing is another foundation of partnership besides the benefit sharing. If there is no risk sharing, it is impossible to form the partnership.

In PPP, the character of risk sharing between the public sector and the private sector is a significant sign to differ other forms of transaction between the public sector and the private sector [8].

If each of the risks can be assumed by partners who can deal with the risks well, there will be no doubt that the cost can be minimized throughout the infrastructure project.

3 Analysis of difficulties in the implementation of PPP project in China

3.1 The partnership is not solid enough

Partnership in China refers to that a natural person, legal person and other organizations establish in the territory of China in accordance with the Law of the People's Republic of China on Partnerships. It is a profit organization in which there are two or more than two partners make the partnership agreement, joint contribute and adventure, share the benefit and risk to operate the common cause.

The partnership in the PPP mode is more special. It is reflected in the following respect. One of the partners is the government or designated by the government. Under the background of public ownership as the foundation of economy in our country, the government has a deeper meaning. This leads the partnership's implementation of the legal principle, four principles, based on the mutual trust to be a problem. On the one hand, government's trust in social capital, especially private capital is not high. A lot of requirements are not high in the PPP project which may only cost two or three billions but the governments often require enterprises which have billions of dollars of its own funds to participate in the project. On the other hand, the universal acceptance of private capital for the government's sincerity, the corresponding policy continuity is low. The cycle of PPP project is generally longer, and the uncertainty is very high. All of these lead to that the partnerships are hard to build in the specific PPP project.

3.2 It is hard to achieve the benefit sharing

The public sector and the private sector in PPP projects are not to share profits. And the government also needs to control the high profits of the private sector. That is not allowing private sector form the excess profit in the process of the implementation of the project. In essence, it produces an incompatible contradiction with the capital of the profit driven. In our country, the development time of private capital is relatively short, foundation and strength is not strong, and it is in the growth and cultivation period. It requires high returns on profits in internal logic, which is the reflecting of scarcity of capital.

While in the PPP project, the government is often in a strong position. Pricing power and the price adjustment of public services is determined by the government in general. When the market has not yet formed, due to the aircraft effect which means the two sides jointly strive to cultivate the market makes the benefit sharing exist in a short period of time. However, once the market is formed, the inconsistency of the goal between the two sides often lead the balance of benefit sharing soon to break, leading to part company each going his own way. It can be more fully reflected in the case of the Quanzhou Ci tong bridge in which private capital hope that the government does not build other competitive projects in order to get the stable high income, while the government wants to build other competitive projects in order to achieve better social services, which leads to the difficulty of the operation of PPP project. Working together can be realized but sharing the good things cannot be realized, benefit sharing becomes a kind of ideal to some extent.

3.3 Risk sharing evolving into undertaking risk by oneself

In PPP, it is generally required that the public sector assume associated risks in which they have their own advantages as far as possible and let the other party to bear the risk as little as possible to minimize the cost of the entire project. But due to the specific to a single PPP project, the barriers between departments highlights. Implementation of land and the planning, matching water, electricity, roads and so on should be the responsibility of the government to assume and defuse the risk, but be assumed by private party as an inescapable duty. Risk sharing evolves into undertaking risk by oneself. For an example, after the completion of a PPP sewage treatment plant, it is failed to put into operation for many years just because the construction of the pipe network needed is not in place.

Especially after the increasing of the PPP project, even though the government's ability to share the risk and coordination capacity increase, its increase range is far behind the increase rate of PPP project. For a single PPP project, the sense and the ability of risk sharing of government is weakened. It pushes partners in PPP project to bear the corresponding risks in order to ensure their own interests have so that the cost of the project increases.

For the government, especially for the government department, it is the promotion of a number of PPP projects. The specific progress of a single PPP project does not affect the government offer the public service. The awareness to assume the risk forwardly in subjective is not strong. In extreme circumstances, the risk of the project may be increased unconsciously. And for private investors, a PPP project may determine whether they can survive. Their risk prediction and risk appetite determine that they have to take more risks. Sometimes, to get some projects, they have to agree with all the things.

4 Corresponding countermeasures

In the strict sense, the division of the analyzing difficulties in the third part is not very scientific. The main reasons are as the following. First, the distinction boundary between the difficulties is not obvious. Second, the perplexing relationship exists among each other, and it can't be simply separated. Using the classification and description method is only in order to analyze the problem relatively incisively. Therefore, in the discussion of the corresponding countermeasures, it is not suitable to use the method of point-to-point. It is necessary to consider the relationship between countermeasures and the various difficulties, solving mechanism, solving degree and so on. The following is a brief discussion on the corresponding countermeasures.

4.1 Cultivate the spirit of contract combining with the characteristics of Chinese market economy

From the perspective of the connotation of contract spirit, the agreement between it and the value system the structure of social ethics of Chinese society are in a high degree. They generally recognize the honest and trustworthy, encourage the hard-working rich, and abandon the gains without working. In civil activities, the freedom and equality are highly guaranteed. These are the favorable conditions for fostering the contract spirit in China. It should be noted that family oriented economies are shrinking with the substantial increase in social productivity. However, the change of traditional concept requires a lot of time. In the country which cries up the rule by man and pecking order for thousands years, implementing the concept of rule of law which uses the all men are born equal as the core with force is not easy to success.

It should be said that the China market economy is still largely rely on family oriented. Individual industrial and commercial households, small processing plants, and the small family business are still the mainstream of our country. In economic activities, in addition to obey the laws and regulations of the state, most of these small and micro enterprises also obey the constraints of Chinese social ethics and values. This can also be referred to as the contract spirit of Chinese style. With the accumulation and subliming of this spirit coupled with the further development of the economy, the complete idea of strictly abide by contracts and keep one's word will gradually internalize into our social ethics structure, which makes the value base of the implementation of PPP project to be stable.

4.2 Understand the significance of PPP project correctly, eliminate the prejudice and make partner be the real partner

The substance of the PPP project is the government to provide public services to the public. PPP originated from the government's lack of money, but it is definitely not to solve the problem of government having no money. The government makes full use of the money, technology, talents and management experience of the social capital in order to improve the quality and efficiency of public service supply with the way of making the contract. It not only solves the problem of money, but also solves the problem of the long-term mechanism. In fact, the real source power which influences the development of PPP is the power of public management reform. Increase and optimize the supply and management efficiency of public goods and services through changing the institutional mechanisms, reducing market access, breaking the monopoly and introducing competition.

This requires that the PPP project participants eliminate the bias between each other, especially the bias of government departments on private capital. They should carefully examine, select partners from the perspective of increasing and optimizing the supply and management efficiency of the government's public goods and services, rather than hold the wrong ideas and prejudices that PPP is borrowing money in a disguised form, digesting the crisis brought by local government debt to response to demand of the higher level of leadership, and finally make private capital in the PPP project be in a true partnership position.

4.3 Encourage, foster private capital to further expand the scale

After thirty years of reform and opening up, China's private capital has been developed by leaps and bounds from scratch and from small to large, and greatly improves and enriches the market economy. According to the relevant data of the National Bureau of statistics, the total amount of private capital has exceeded the state capital. However, the single body of private capital is very small, and it cannot be compared with the state-owned capital. For the PPP project, the initial funding requirements limit the involvement of private capital to a certain extent. Therefore, encourage and foster private capital to further expand the scale, so that it can adapt to the rigid requirements of the PPP project. At this stage, the government can try to make the pension fund, social security fund enter the PPP project conditional, replenish the short board of scale shortage of private capital.

5 Conclusion

Above all, PPP model has been applied in many fields in our country. As a new mode of operation, PPP reflects the more tenacious vitality. On the one hand, China's public service infrastructure is still not perfect, it needs a huge investment. But the government's financial resources are limited and need funds to enter. On the other hand, the total amount of private capital is gradually increasing, but the investment channel is narrow. The generalized M2 currency is high, and it is necessary to open up new areas of investment. Under the common effect of the two factors, the PPP project mode of our country will be further extended and used in the next period of time.

Based on the concept, connotation and essence of PPP project, this paper makes a brief analysis of the difficulties in the implementation of PPP project. In comparison with the implementation environment of the PPP project in other countries, think qualitatively that the lack of contract spirit affects the implementation of the PPP project, and the inherent prejudice of private capital affects the acceptance and absorption capacity of PPP project to private capital. Besides, the partnership which is loose and lack of trust will directly affect the operation of the PPP project, and may lead to the failure of the project.

Based on the policy perspective, on the macro, PPP project needs more developed market economy, and needs the good faith as the core of the general contract spirit. Besides, it needs the government who has the courage to break the barriers and bent over. What' s more, it needs social public to be understanding and tolerant. On the micro level, it needs to change bias of the no business is not evil and believe and understand each other. It needs reasonable profit distribution and appropriate risk sharing. And it needs third party coordination platform which is objective and fair and can make the dialogue equal.

The operation cycle of PPP project is very long. There is no project which has experienced the whole process so far in our country. The experience of operation, maintenance and transfer and existing problems need to be further considered. In addition, another important role in the PPP project is that the user's behavioral characteristics and its impact on the PPP project have not yet been described and need further investigation and study.

References

1. Tang L Y, Shen Q, Cheng E W L. A review of studies on Public - Private Partnership projects in the construction industry. International Journal of Project Management, 2010, 28(7):683-694.
2. Forrer J, Kee J E, Newcomer K E, et al. Public - Private Partnership and the Public Accountability Question. Public Administration Review, 2010, 70(3):475 - 484.
3. Hwang B G, Zhao X, Gay M J S. Public private partnership projects in Singapore: Factors, critical risks and preferred risk allocation from the perspective of contractors. International Journal of Project Management, 2013, 31(3):424-433.
4. Chan A P C, Yeung J F Y, Yu C C P, et al. Empirical Study of Risk Assessment and Allocation of Public-Private

Partnership Projects in China. Journal of Management in Engineering, 2011, 27(3):136-148.
5. Yongjian Ke, ShouQing Wang, Albert P. C. Chan. Risk Misallocation in Public-Private Partnership Projects in China. International Public Management Journal, 2013, 16(3):438-460.
6. Cheung E, Chan A P C. Revitalizing Historic Buildings through a Partnership Scheme: Innovative Form of Social Public - Private Partnership. Journal of Urban Planning & Development, 2014, 140(1):94-95.
7. Lee C H, Yu Y H. Service delivery comparisons on household connections in Taiwan's sewer public - private-partnership (PPP) projects. International Journal of Project Management, 2011, 29(8):1033-1043.
8. Bank W. Building Integrated Markets within the East African Community: EAC Opportunities in Public-Private Partnership Approaches to the Region's Infrastructure Needs. World Bank Publications, 2014.

Study on the Coordination of Supply Chain Based on Carbon Emissions Trading Considering the Retailers' Competition

Daoping Wang[1] , Xiaoyan Li[1,*] and Kun Ding[1]

[1] Donlinks School of Economics and Management, University of Science and Technology Beijing, Beijing 10083, China
Corresponding author: lixiaoyan.24@163.com

Abstract. This paper studies the coordination of supply chain in the context of carbon emissions trading mechanism, which considering the competition between retailers. Centralized and decentralized supply chain models were constructed to discuss the price of product, to avoid the losses of profit from the decentralized decision-making, the revenue-sharing contract was introduced to coordinate the supply chain. Research shows that the carbon emissions trading reduce emissions effectively, but the higher price of carbon emissions trading cut down the total profit of supply chain; The competition between retailers upgrades the supply chain members' profit; Coordination was achieved by introducing the revenue-sharing contract. Finally, numerical example was given to illustrate the validity of the revenue-sharing contract, and the sensitivity analysis of parameters such as the price of the emissions trading and the retailers' competition were presented.

1 Introduction

The emissions of greenhouse-gases is deemed as the main causes of global warming, with people gradually in-depth understanding of environmental protection and sustainable development, the government and the corporate attaches great importance to the carbon emissions. To achieve the target of emission reduction and implement the growth of carbon economic, the government need to establish the market mechanism gradually, encourage enterprises to reduce carbon emissions. the carbon emission trading is widely promoted as a kind of effective mechanism to control carbon emissions. The manufacturers are starting to pay attention to their carbon emissions to achieve low carbon development. At present, the research on the carbon trading mechanism has got some achievements, however, many studies consider one retailer most of the time, but in actual the retailers show competition for the market. Therefore, study the coordination of supply chain under the carbon trading mechanism and considering retailers' competition has more theoretical value.

2 Literature review

Carbon trading refers to the department considering the economic growth and the energy structure comprehensive to determine the enterprise' free carbon quotas, under this background, the enterprise adjust the production plan to mitigation and makes carbon emissions trading to maximize their benefit 1. In recent years, domestic and foreign scholars take research on carbon trading mechanism, Such as Zakeri analysis the reduction mechanism of carbon trading and carbon tax' influence on the supply chain, studies have shown that the carbon trading shows better reduction effect 2. Du put forward a new type of supply chain in the case of considering carbon trading, study the two sides game process based on the newsboy model 3, on the basis, he see the quotas of the government' carbon emission as a variable parameter, research the carbon emission's impact on the profit of the supplier 4. Abdallah studies how to maximize the enterprises' interests with the economic and environment influence that under the carbon trading mechanism 5. Li solves the logistics distribution routing problem under the carbon emission trading mechanism, introducing a calculation method that consider the vehicle load and speed, make some interesting observation about how to control cost and reduce carbon emissions 6. Ma assuming that a local government has launched a regional carbon trading market in the purpose of maximizing the profit of whole system, the result shows some valuable managerial insights on the decision making of the company and the government 7.

Researchers have further study on supply chain which considering the competition between retailers. Bernstein on the basis of competitive relationship between multiple retailers, studies the decentralized supply chain when faced with uncertain demand equilibrium strategies 8. Yao in the perspective of retailers provide competitive services, study the operation strategy of supply chain which is composed of one manufacturer and two competitive retailers 9. Fan presents a channel pricing strategy which is combated of the quantity discount pricing and slotting fees, on the basis of considering the retailers' competition 10. Gao in the case of two competitive retailers face random demand, to decide the product price, quantity and the responsive to price 11. Cao study the pricing and coordination problems of closed-loop supply chain under stochastic demand, where the two retailers face the different market demand and the product recycled affected by competition price 12.

The literature on carbon trading mechanism has achieved fruitful results, but there are little literature considering the effects of competition between retailers at the same time. Therefore, this article researches the coordination of supply chain that considering retailers' competition under carbon trading mechanism, the optimal decision of the manufacturer and retailers in the mode of centralized and decentralized are concluded, and by introducing the revenue-sharing contract to coordinate the supply chain.

3 Model establishment and analysis

3.1 Description of problem and hypothesis

This paper studies the supply chain that composed of one manufacturer and two competitive retailers. In carbon trading mechanism, the government gives carbon emission quota of E to manufacturer for free, the manufacturer response the government mechanism to reduce emission, and the manufacturer has the carbon trading with price of p_c to maximize its profit. e_H and e_L are the carbon emission of manufacturer that before and after the reduction respectively, where $\lambda(0\leq\lambda<1)$ is the purification rate, and the cost of reduction is $c_L=\mathcal{E}(\lambda e_H)^2/2$, $\mathcal{E}(\mathcal{E}>0)$ is the parameter for reduction investment. The unit producing cost of product is c. The manufacturer sales the products to two competitive retailers with wholesale price w, the sale price of two retailers are p_1, p_2, which are the decision variables for retailers. The market demand of two competitive retailers are expressed as: $q_1=\varphi_1-p_1+\beta p_2$, $q_2=\varphi_2-p_2+\beta p_1$, where φ_1, φ_2 are the market demand size of two retailers respectively, $\beta(0\leq\beta\leq1)$ is the degree of competition between retailers.

3.2 Decision of supply chain based on the carbon emissions trading considering the retailers' competition

The centralized and decentralized decisions were made which based on the carbon emissions trading considering the retailers' competition, and the reverence-sharing contract was introduced to coordinate the supply chain.

3.2.1 Centralized Decision-making

The centralized decision-making is a kind of ideal system structure, the decision target of the supply chain members is to maximize the total profit of supply chain. The profit of supply chain under centralized decision-making is:

$$\Pi_{Sc}=(p_{1c}-c)q_{1c}+(p_{2c}-c)q_{2c} - p_c\left(e_L(q_{1c}+q_{2c})-E\right)-\frac{1}{2}\varepsilon(\lambda e_H)^2 \tag{1}$$

By formula (1), take partial derivative of p1c and p2c is: $\frac{\partial\Pi_{Sc}}{\partial p_{1c}}=\varphi_1+(1-\beta)(c+p_ce_L)-2p_{1c}+2\beta p_{2c}$, $\frac{\partial\Pi_{Sc}}{\partial p_{2c}}=\varphi_2+(1-\beta)(c+p_ce_L)-2p_{2c}+2\beta p_{1c}$, the second order derivative is $\frac{\partial^2\Pi_{Sc}}{\partial p_{1c}^2}=-2,\frac{\partial^2\Pi_{Sc}}{\partial p_{2c}^2}=-2,\frac{\partial^2\Pi_{Sc}}{\partial p_{2c}\partial p_{1c}}=\frac{\partial^2\Pi_{Sc}}{\partial p_{1c}\partial p_{2c}}=2\beta$, considering that 0≤β≤1, so the Hessen matrix of ΠSc(p1c, p2c) is negative, therefore the optimal solution is exist, let $\frac{\partial\Pi_{Sc}}{\partial p_{1c}}=0$ and $\frac{\partial\Pi_{Sc}}{\partial p_{2c}}=0$, the sales price of two retailers in centralized decision-making is:

$$p_{1c}^*=\frac{(1-\beta^2)(c+p_ce_L)+\beta\varphi_2+\varphi_1}{2(1-\beta^2)} \tag{2}$$

$$p_{2c}^*=\frac{(1-\beta^2)(c+p_ce_L)+\beta\varphi_1+\varphi_2}{2(1-\beta^2)} \tag{3}$$

Then take p_{1c}^* and p_{2c}^* into the demand function of market, generate the optimal order quantity of two retailers is:

$$q_{1c}^*=\frac{\varphi_1+(\beta-1)(c+p_ce_L)}{2} \tag{4}$$

$$q_{2c}^*=\frac{\varphi_2+(\beta-1)(c+p_ce_L)}{2} \tag{5}$$

Take p_{1c}^*, p_{2c}^*, q_{1c}^*, q_{2c}^* into formula (1), the total profit of the supply chain under centralized decision-making is:

$$\Pi_{Sc}^*=\frac{\left(\beta\varphi_2+\varphi_1-(1-\beta^2)(c+p_ce_L)\right)\left(\varphi_1-(1-\beta)(c+p_ce_L)\right)}{4(1-\beta^2)} +\frac{\left(\beta\varphi_1+\varphi_2-(1-\beta^2)(c+p_ce_L)\right)\left(\varphi_2-(1-\beta)(c+p_ce_L)\right)}{4(1-\beta^2)} +p_cE-\frac{1}{2}\varepsilon(\lambda e_H)^2 \tag{6}$$

3.2.2 Decentralized Decision-making

In this section assume that the manufacturer and retailers have a Stackelberg game, the manufacturer is in the dominant position, and the retailers in follow status, and Nash game was carried on between two competitive retailers. In decentralized decision-making model, the manufacturer determines the wholesale price firstly to maximize its own profit, then the two retailers decide the sales price to maximize their own profit respectively. The profit of the manufacturer and the two retailers are expressed as:

$$\Pi_{Md}=(w_d-c)(q_{1d}+q_{2d})-p_c\left(e_L(q_{1d}+q_{2d})-E\right) -\frac{1}{2}\varepsilon(\lambda e_H)^2 \tag{7}$$

$$\Pi_{R1d}=(p_{1d}-w_d)q_{1d} \tag{8}$$

$$\Pi_{R2d}=(p_{2d}-w_d)q_{2d} \tag{9}$$

Reverse induction method is used to solve the problem. By formula (8) and (9), take partial derivative of sales price is: $\frac{\partial\Pi_{R1d}}{\partial p_{1d}}=\varphi_1-2p_{1d}+\beta p_{2d}+w_d$, $\frac{\partial\Pi_{R2d}}{\partial p_{2d}}=\varphi_2-2p_{2d}+\beta p_{1d}+w_d$, considering $\frac{\partial^2 \Pi_{R1d}}{\partial p_{1d}^2}=-2$, $\frac{\partial^2 \Pi_{R2d}}{\partial p_{2d}^2}=-2$,therefore the optimal solution is exist. Let $\frac{\partial \Pi_{R1d}}{\partial p_{1d}}=0, \frac{\partial \Pi_{R2d}}{\partial p_{2d}}=0$, deduce $p_{1d}=\frac{2\varphi_1+\beta\varphi_2+(2+\beta)w_d}{4-\beta^2}, p_{2d}=\frac{2\varphi_2+\beta\varphi_1+(2+\beta)w_d}{4-\beta^2}$ Take p_{1d}, p_{2d} into formula (7), let $\frac{\partial \Pi_{Md}}{\partial w_d}=0$, obtain the optimal wholesale price of manufacturer under decentralized decision-making is:

$$w_d^*=\frac{\varphi_1+\varphi_2+2(1-\beta)(c+p_c e_L)}{4(1-\beta)} \tag{10}$$

Take wd* into p1d, p2d , the optimal sales price of retailers under decentralized decision-making is:

$$p_{1d}^*=\frac{4(1-\beta)(2\varphi_1+\beta\varphi_2)+(2+\beta)(\varphi_1+\varphi_2+2(1-\beta)(c+p_c e_L))}{4(1-\beta)(4-\beta^2)} \tag{11}$$

$$p_{2d}^*=\frac{4(1-\beta)(2\varphi_2+\beta\varphi_1)+(2+\beta)(\varphi_1+\varphi_2+2(1-\beta)(c+p_c e_L))}{4(1-\beta)(4-\beta^2)} \tag{12}$$

Then take (11) and (12) into the demand function of market, generate the optimal order quantity of two retailers is:

$$q_{1d}^*=\frac{(6-\beta)\varphi_1+(3\beta-2)\varphi_2-2(2+\beta)(1-\beta)(c+p_c e_L)}{4(4-\beta^2)} \tag{13}$$

$$q_{2d}^*=\frac{(6-\beta)\varphi_2+(3\beta-2)\varphi_1-2(2+\beta)(1-\beta)(c+p_c e_L)}{4(4-\beta^2)} \tag{14}$$

Take p_{1d}^*, p_{2d}^*, q_{1d}^*, q_{2d}^* into formula (7), (8) and (9), the profit of manufacture and the two retailers that under the decentralized decision-making is:

$$\Pi_{Md}^*=\frac{(\varphi_1+\varphi_2-2(1-\beta)(c+p_c e_L))^2}{8(1-\beta)(2-\beta)}+p_c E-\frac{1}{2}\varepsilon(\lambda e_H)^2 \tag{15}$$

$$\Pi_{R1d}^*=\left(\frac{(6-\beta)\varphi_1+(3\beta-2)\varphi_2+2(\beta^2+\beta-2)(c+p_c e_L)}{4(4-\beta^2)}\right)^2 \tag{16}$$

$$\Pi_{R2d}^*=\left(\frac{(6-\beta)\varphi_2+(3\beta-2)\varphi_1+2(\beta^2+\beta-2)(c+p_c e_L)}{4(4-\beta^2)}\right)^2 \tag{17}$$

3.2.3The Coordination of Supply Chain with Revenue-sharing Contract

Here we introduce the revenue-sharing contract to coordinate the decentralized decision-making. The revenue-sharing contract (w_1,w_2,η_1,η_2) means: manufacture sales products to two retailers with wholesale price w_1 and w_2 respectively, after the two retailers sell products, they give the manufacture $(1-\eta_1)p_1q_1$ and $(1-\eta_2)p_2q_2$ sales revenue separately, where $0<\eta_1,\eta_2<1$. The profit of manufacture and two retailers that with the coordination of revenue-sharing contract is:

$$\begin{aligned}\Pi_M=&(w_1-c)q_1+(w_2-c)q_2-p_c(e_L(q_1+q_2)-E)\\&-\frac{1}{2}\varepsilon(\lambda e_H)^2+(1-\eta_1)p_1q_1+(1-\eta_2)p_2q_2\end{aligned} \tag{18}$$

$$\Pi_{R1}=(\eta_1 p_1-w_1)q_1 \tag{19}$$

$$\Pi_{R2}=(\eta_2 p_2-w_2)q_2 \tag{20}$$

Reverse induction method is used to solve the problem. By formula (19) and (20), take partial derivative of sales price with the retailers' profit is:

$\frac{\partial \Pi_{R1}}{\partial p_1} = \eta_1(\varphi_1 - 2p_1 + \beta p_2) + w_1, \frac{\partial \Pi_{R2}}{\partial p_2} = \eta_2(\varphi_2 - 2p_2 + \beta p_1) + w_2$, the two order derivative is $\frac{\partial^2 \Pi_{R1}}{\partial p_1^2} = -2\eta_1, \frac{\partial^2 \Pi_{R2}}{\partial p_2^2} = -2\eta_2$, considering that $0 \leq \eta_1, \eta_2 \leq 1$, therefore there is optimal solution, let $\frac{\partial \Pi_{R1}}{\partial p_1} = 0, \frac{\partial \Pi_{R2}}{\partial p_2} = 0$, obtain the optimal sales price:

$$p_1^* = \frac{\eta_1(2\varphi_1 + \beta\varphi_2) + (2+\beta)w_1}{\eta_1(4-\beta^2)} \tag{21}$$

$$p_2^* = \frac{\eta_2(2\varphi_2 + \beta\varphi_1) + (2+\beta)w_2}{\eta_2(4-\beta^2)} \tag{22}$$

When the optimal decision behavior of the decentralized decision-making is equal to the centralized decision-making, the supply chain achieves coordination, in this model means $p_1^* = p_{1c}^*$, $p_2^* = p_{2c}^*$, we obtained:

$$w_1^* = \frac{3\eta_1\beta^2\varphi_1 + \beta\eta_1\varphi_2(2-\beta^2) + \eta_1(4-\beta^2)(1-\beta^2)(c+p_c e_L)}{2(1-\beta^2)(2+\beta)} \tag{23}$$

$$w_2^* = \frac{3\eta_2\beta^2\varphi_2 + \beta\eta_2\varphi_1(2-\beta^2) + \eta_2(4-\beta^2)(1-\beta^2)(c+p_c e_L)}{2(1-\beta^2)(2+\beta)} \tag{24}$$

Then take w_1^*, w_2^*, p_1^* and p_2^* into formula (18) ~ (20), the profit of manufacture and retailers that after introducing of the reverence-sharing contract is:

$$\begin{aligned}\Pi_M^* = &\left(\frac{3\eta_1\beta^2\varphi_1 + \beta\eta_1(2-\beta^2)\varphi_2 + (1-\eta_1)(2+\beta)(\beta\varphi_2+\varphi_1)}{2(1-\beta^2)(2+\beta)} + \frac{(\eta_1 - 1 - \beta\eta_1)(c+p_c e_L)}{2}\right) \\ &* \left(\frac{\varphi_1 + (\beta-1)(c+p_c e_L)}{2}\right) \\ &+ \left(\frac{3\eta_2\beta^2\varphi_2 + \beta\eta_2(2-\beta^2)\varphi_1 + (1-\eta_2)(2+\beta)(\beta\varphi_1+\varphi_2)}{2(1-\beta^2)(2+\beta)} + \frac{(\eta_2 - 1 - \beta\eta_2)(c+p_c e_L)}{2}\right) \\ &* \left(\frac{\varphi_2 + (\beta-1)(c+p_c e_L)}{2}\right) + p_c E - \frac{1}{2}\varepsilon(\lambda e_H)^2\end{aligned} \tag{25}$$

$$\begin{aligned}\Pi_{R1}^* = &\left(\frac{\eta_1(6+\beta-\beta^2)(1+\beta)(c+p_c e_L) + \eta_1(3\beta+2)\varphi_1 + \eta_1\beta^2\varphi_2}{2(1+\beta)(2+\beta)}\right) \\ &* \left(\frac{\varphi_1 + (\beta-1)(c+p_c e_L)}{2}\right)\end{aligned} \tag{26}$$

$$\begin{aligned}\Pi_{R2}^* = &\left(\frac{\eta_2(6+\beta-\beta^2)(1+\beta)(c+p_c e_L) + \eta_2(3\beta+2)\varphi_2 + \eta_2\beta^2\varphi_1}{2(1+\beta)(2+\beta)}\right) \\ &* \left(\frac{\varphi_2 + (\beta-1)(c+p_c e_L)}{2}\right)\end{aligned} \tag{27}$$

4 Numerical example

For more in-depth research, the model is analyzed by using Matlab software, to investigate the influence of the revenue-sharing contract on the profit of the supply chain. The parameters in the model are assigned as follows: E=1000, φ_1=1000, φ_2=800, η_1=0.6, η_2=0.5, β=0.4, c=100, e_H=10, ε=1000, γ=0.5, λ=0.4, m=3, p_c=50.

The result of Centralized decision-making, Decentralized decision-making and decision-making with the revenue-sharing contract follows in table1.

Table 1. The profit of supply chain members in different decision-making model.

Decision-making model	Centralized decision-making	Decentralized decision-making	The decision-making with revenue-sharing

The sales price of retailer1	985.7	1197.9	985.7
The sales price of retailer2	914.3	1114.6	914.3
The profit of manufacturer	—	268875.0	272915.6
The profit of retailer1	—	61462.7	91344.8
The profit of retailer2	—	27087.7	44311.1
The total profit of supply chain	408571.4	357425.3	408571.4

Table1 shows after introducing of the revenue-sharing contract, the manufacturer and the retailers' profit are higer than the decentralized decision-making that without the contract. The manufacturer' profit relatively increased 1.5%, the profit of the two retailers grow 48.6% and 63.6% respectivily. At the same time, the total profit of supply chain rises 14.3%, that is equal to the profit of the supply chain under the centralized decision-making, that means the revenue-sharing contract coordinate the supply chain effectively.

Table 2 shows the influence of carbon emissions trading' price on the total carbon emissions, the profit of supply chain members and the total profit of supply chain.

Table 2. The influence of carbon emissions trading price on the total carbon emissions and profit in decentralized decision-making.

The price of carbon emissions trading	Total carbon	The profit of manufacture	The profit of retailer1	The profit of retailer2	The total profit of
10	3015.0	338675.0	85800.2	43925.2	468400.3
20	2880.0	319200.0	79336.1	39336.1	437872.2
30	2745.0	301075.0	73125.2	35000.2	409200.3
40	2610.0	284300.0	67167.4	30917.4	382384.7
50	2475.0	268875.0	61462.7	27087.7	357425.3

Table 2 shows: (1) With the risen in the price of trading, the total carbon emissions has been controlled effectively, the carbon trading mechanism plays positive role in the process of low carbon economy; (2) With the risen in the price of trading, the profit of manufacturer, retailers and the overall supply chain are both on decline, so the government needs to control the price reasonably, neither too low nor too high. The low carbon trading price can't stimulate the manufacture to reduce the carbon emissions, on the contrary, the high carbon trading price will aggravate the burden on supply chain members.

Table 3. The influence of retailers' competing degree on the profit that after coordination.

The degree of retailers'	The profit of manufacturer	The profit of retailer1	The profit of retailer2	The total profit of supply chain
0.1	122814.2	61430.8	24300.5	208545.5
0.2	157311.2	69669.1	29436.4	256416.7
0.3	205210.2	79338.9	35868.3	320417.6
0.4	272915.6	91344.8	44311.1	408571.4

0.5	372133.3	107200.0	56000.0	535333.3

The Table3 analyzes the influence of the retailers' competing degree on the profit of supply chain members and the total profit of supply chain after introducing of the contract. Table3 shows: with the increase of retailers' competition, the profit of manufacturer, two retailers and the total profit of the supply chain were both promoted, it means the existence of the competition of retailers making the retailers have motivation to promote for more profit, and the effective market competition also increases the profit of manufacturer.

5 Conclusion

Under the background of carbon trading mechanism, this paper studies the coordination problem of supply chain. By building the centralized and decentralized decision-making model of supply chain, the optimal pricing strategy of manufacturer and retailers were obtained, and in order to solve the problem of profit losses which resulting from the decentralized dicision-making, the revenue-sharing contract was introduced to coordinate the supply chain. Research shows that: (1)With the revenue-sharing contract, the supply chain members' profit and the pross profit of supply chain both run up to the centralized decision-making level; (2)The implementation of the carbon trading mechanism reduces carbon emissions in the process of production effectively; (3)With the risen in the price of carbon trading, the profit of the manufacturer, the retailers and the overall supply is on the decline, so the government need to control the price of carbon trading; (4)With the increase of retailers' competition ,the profit of manufacturer, two retailers and the total profit of the supply chain were both promoted.

But we have not considered the competition between manufacturers, it will be the main content of further research.

References

1. Baldwin R. Regulation lite: The rise of emissions trading. Regulation & Governance, 2015, 2(2): 193-215.
2. Zakeri A, Dehghanian F, Fahimnia B, et al. Carbon pricing versus emissions trading: A supply chain planning perspective. International Journal of Production Economics, 2014, 164: 197-205.
3. Du S, Ma F, Fu Z, et al. Game-theoretic analysis for an emission-dependent supply chain in a 'cap-and-trade' system. Annals of Operations Research, 2015, 228(1): 1-15.
4. Du S F, Zhu L L, Liang L, et al. Emission-dependent supply chain and environment-policy-making in the 'cap-and-trade' system. Energy Policy, 2013, 57: 61-67.
5. Abdallah T, Farhat A, Diabat A, et al. Green supply chains with carbon trading and environmental sourcing: Formulation and life cycle assessment. Applied Mathematical Modeling, 2012, 36(9): 4271-4285.
6. Li J, Zhang J H. Study on the effect of carbon emission trading mechanism on logistics distribution routing decisions. Systems Engineering-Theory & Practice, 2014, 34(7): 1779-1787.
7. Ma Q Z, Song H Q, Chen G Y. Research on product and production decision-making under the supply chain environment considering the carbon-trading mechanism. Chinese Journal of Management Science, 2014, 22(8) : 37-46.
8. Bernstein F, Federgruen A. Decentralized supply chains with competing retailers under demand uncertainty. Management Science, 2010, 51(1) :18-29.
9. Yao S J, Chen J H. After-sales service capacity of operation strategy considering retailers competition. Engineering Management, 2016, (1): 88-95.
10. Fan X J, Chen H M. Channel coordination strategy mixed by quantity discount and slotting allowances under the encironment of retailer competition. Engineering Management, 2016, (1): 155-160.

11. Gao X M, Liu X Z, Zuo X L. Competitive retailers' pricing strategy under stochastic demand. Operations Research and Management Science, 2016, (1): 133-144.
12. Cao X G, Wen H. Pricing and coordination of closed-loop supply chain considering competition between retailers under stochastic demand. Operations Research and Management Science, 2015, (1): 34-39.

Analyzing The Value Drivers and Path Selection of Grid Corporations From the View of The Long-term Value

Ruobing Shuai[1]and Yuanbin Jiang[2]

[1] Guangdong Power Grid Corporation Foshan Power Supply Bureau, Foshan 528000, China

[2] State Grid Chengdu Power Supply Company, Chengdu 610000, China

Corresponding author: jybthebest@163.com

Abstract. Based on the business characteristics and EVA, the paper develops an econometric model for the long-term value of grid corporations in China, and analyzes the value drivers and path selection correspondingly. The study shows that, the growing long-term value of grid corporations derives from the creation of future value, and the fundamental factor is an increase in investment. Grid corporations should target at investment management and capital optimization, realizing the goal of long-term value.
KeyWords. Economic Value Added, Long-term Value, Value Drivers, Path Selection.

1. Instruction

Enterprise Value is a measure of the comprehensive enterprises competitiveness, and obviously continuous value creation reflects enterprise's success. Therefore, achieving the maximum of enterprise value becomes the ultimate objective for the enterprise management.

Like other peers, grid corporations in China(State Grid Corporation of China and China Southern Power Grid Corporation) invests in, constructs and operates power grids, providing safer, cleaner, and more economic and sustainable power supply. Meanwhile, as the state-owned enterprises(SOEs), they are supervised by the State-owned Assets Supervision and Administration Commission of the State Council(SASAC) and evaluated with Economic Value Added(EVA) which targets at the value created in excess of the required return of the shareholders rather than simply the expansion of scale and profit. Hence, it is essential to develop a econometric model for the long-term value of grid corporations, and analyzes the value drivers and path selection correspondingly.

2. The econometric model for the long-term value and its value drivers

2.1 The econometric model

For the long-term financial performance and value maximization, the management should concentrate on two dominant elements of value creation: revenue growth and accomplishing a return on invested capital (ROIC) beyond the weighted average cost of capital (WACC). Obviously, grid corporations are no exception. Moreover, SASAC has introduced the EVA system to evaluate their performance, which stresses a firm's economic profit, or the value created in excess of the required return of shareholders. The formula is:

$$EVA = (r - c) \cdot TC$$

where:

r= a return on invested capital

c= weighted average cost of capital

TC= the economic capital employed

Based on the existing EVA system and going-concern assumption, we introduce the EVA discount model to measure the long-term value of grid corporations, which is:

$$EV = A_0^* + \sum_{t=o}^{\infty} \frac{EVA_t}{(1+c)^t}$$

where:

EV= the long-term enterprise value

A_0^*= the opening balance of the net asset value

c= weighted average cost of capital

t= the duration of business

Given the certain opening balance of the net asset value, EVA each period in the future and discount rate are two key factors affecting the long-term enterprise value. As for the discount period, in theory, it should be indefinite. However, the longer the discount period is, the lower the accuracy of forecast is. According to the expected make-value management theory, it is reasonable to set the five-years forecast, and the management usually apply such period as the time interval in planning.

The long-term enterprise value is targeting at the discount EVA deriving from both the current and future capital, making up the shortcoming of period indicators fundamentally. It largely decreases the possibility of manipulation and unduly high growth in scale and profit at the cost of sustainable development in the long term.

2.2 The value drivers

Based on the econometric model, there are three key components of grid corporations' enterprise value: the opening balance of the net asset value, EVA and WACC in each period. Given the certain the opening balance of the net asset value, we mainly focus on EVA and capital charge in each period.

There are two components of EVA: net operating profit after taxes (NOPAT) and capital charge(WACC x capital).

$$EVA= (r-c)\cdot TC= r\cdot TC-c\cdot TC$$

$$= NOPAT- c\cdot TC$$

where :

r= a return on invested capital

c= weighted average cost of capital

TC= the economic capital employed

NOPAT= net operating profit after taxes

Here is the EVA index decomposition of grid corporations:

$$EVA= NOPAT- c\cdot TC$$

≈ *(Electricity Sales Volume x Average Electricity Sales Price + Other Revenue)- (Electricity Purchase Volume x Average Electricity Purchase Price + Transmission & Distribution Expenses + Gain/Loss + Taxes) – (Current Assets + Fixed Assets + Other Assets) x WACC*

Given the characteristics of gird corporations, we can give a snapshot of factors affecting its EVA, as the Figure 1 shows:

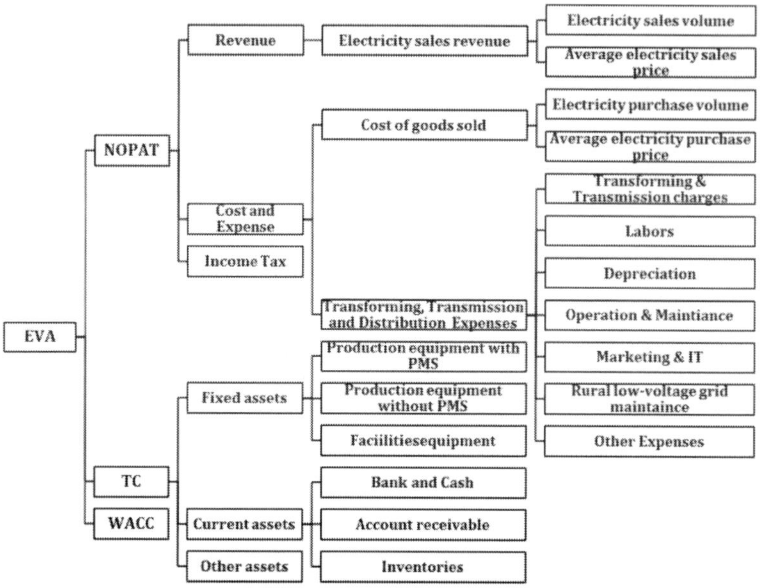

Fig.1. Factors influencing EVA of grid corporations

2.2.1 Factors influencing net profit after tax (NOPAT)

Obviously, NOPAT is the key element of grid corporation's EVA, consisting of revenue, cost and income tax. These three mainly reflect the grid corporation's financial appearance during the period of time [1].

(1) Revenue

For grid corporations, the electricity sales revenue is the dominant resource of the revenue, around 90% of the total. It is determined by electricity sales volume and electricity sales price. Geographically, the electricity sales revenue is divided into provincial electricity sales revenue and interprovincial electricity sales revenue, which are determined by electricity sales price and volumes respectively:

Electricity Sales Revenue= Electricity Sales Volume x Average Electricity Sales Price

= Provincial Electricity Sales Volume x Average Provincial Electricity Sales Price + Interprovincial Electricity Sales Volume x Average Interprovincial Electricity Sales Price

The average electricity sales price is determined by electricity sales structure (such as large industry usage, commercial usage and resident usage) and consumer-differentiated pricing correspondingly.

(2) Cost and Expense

Cost is necessary for the revenue generating and it is not reasonable to blindly cut the cost and expense in cost of safety and sustainable development, especially for grid corporations balancing the profit-pursing and social-responsibility. The management should target at enhancing the gross margin and thus an increase in NOPAT.

In general, there are two key elements of cost and expense of grid corporations: Cost of goods sold and Transforming, Transmission and Distribution expense.

1) Cost of goods sold

Cost of goods sold is up to 75% of the total cost and expense of grid corporations, and it is determined by electricity purchase volume and electricity purchase price. Notably, the management must take not only the electricity demand but also the line loss which is the unavoidable loss during the transmission and distribution. The electricity purchase price is determined by electricity purchase structure and supplier-differentiated pricing correspondingly [2].

$$Cost\ of\ goods\ sold = Electricity\ Purchase\ Volume \times Average\ Electricity\ Purchase\ Price$$

$$= \sum (Electricity\ Purchase\ Volume_i \times Electricity\ Purchase\ Price_i)$$

2) Transforming, Transmission and Distribution Expense

For grid corporations, transforming, transmission and distribution expense mainly include transforming & transmission charge, labors, depreciation, operation & maintenance, marketing & IT, rural low-voltage grid maintenance, etc.

Transforming &transmission charge consists of transmission network use of system (TNUoS) charge and ancillary service charge.

Labors. Like other SOEs, grid corporation's labor costs consist of salary, social insurance, welfare expenses, education and training funds, union funds, housing provident fund (including incremental subsidies), non-monetary benefits, termination benefits and other labor costs.

Depreciation. Due to a tremendous amount of fixed assets, grid corporations have to shoulder the heavy burden of depreciation, a large proportion of cost and expense.

Operation & Maintenance is necessary to keep on the go for grid corporations, including materials, fixing, etc. Most fixed assets are equipment, which means vast investment, long investment cycle and thus heavy operation & maintenance cost.

Marketing & IT. Enterprise resource planning (ERP) system is widely applied in grid corporations, and marketing informatization is one of the essential parts. Such investments help improve energy efficiency, power supply reliability and consumer satisfaction. Rural low-voltage grid maintenance is to guarantee the reliably power supply in rural areas, mainly including rural line loss, remuneration for rural electric management and rural grid operation fees.

Other expenses is the necessary expense for excluding items above [3].

(3) Income Tax

Income tax payable and income tax rate determine the income tax. For grid corporations, the income tax rate was 25%. So, the income tax expense analysis is mainly focused on the income tax payable. There are mainly two aspects to decrease income tax payable and thus reducing the tax burden:

1) Tax planning. The existing enterprise income tax law is more flexible in deduction, expanding the space of tax planning, such as salary restructure, proper choice of accounting policy, transfer pricing, etc.

2) Utilization of Tax preference. Grid corporation can adjust to the change of macropolicy while decision making, especially for investment in technology innovation, environment protection and agriculture support. These fields are strongly encouraged and more likely to utilize corresponding tax preference from the government.

2.2.2 Factors influencing capital employed (TC)

Considering grid corporations, there are two aspects of capital employed: create enterprise value through assets employed; increase the use efficiency of assets and thus enterprise value. These two aspects mainly show in occupancy of fixed assets, current assets, and other assets.

(1) Occupancy of fixed assets

Grid corporations are typically asset-intensive enterprise, and fixed assets make up two thirds of total assets. Meanwhile, a vast of fixed assets costs heavy, and their depreciation and obsolescence incur great capital occupancy, which largely weaken value-creating ability.

(2) Occupancy of current assets

Current assets mainly include bank and cash, account receivable and inventory. Holding enough current assets can strong the liquidity, but excessive current assets, especially cash and cash equivalents, are regarded as a waste of opportunities and better return, and thus decreasing the enterprise value. Therefore, it is necessary to set proper cash balance.

(3) Occupancy of other assets

Other assets are assets excluding fixed assets and current assets, such as intangible assets, which are essential parts of asset management.

2.2.3 Factors influencing weighted average cost of capital (WACC)

WACC is not only the key element of EVA, but also the discount rate in enterprise value calculation, affecting the current operation value directly and the growth value indirectly. It is determined by the capital structure during the financing process. Here is the formula:

$$WACC = \frac{D}{D+E} \cdot K_d + \frac{E}{D+E} \cdot K_e \cdot (1-t)$$

where:

D= total debt

E= total equity

K_d= cost of debt

K_e= cost of equity

t= income tax rate

Notably, grid corporations in China are completely SOEs and their capital result from government capital and some external debt financing. According to the evaluation requirement from SASAC, the WACC is 4.1% for both equity capital and debt capital, stressing the safety of capital structure rather than economical efficiency [4].

3. The path selection of long-term value improvement

3.1 Investment management based on value development

Due to the value composition of grid corporation, its enterprise value consists of current operation value and growth value in the future, and the latter decides the long-term value. For gird corporation, the emphasis is the investment, and cherry-picking the projects with better return. The increases in return of incremental capital will largely enhance EVA in the future.

3.1.1 The scope of investment

(1) Fixed-asset investment includes medium and large construction, expansion, technical innovation, informatization, and other fixed-asset investments.

(2) Equity investment includes establish new companies, mergers and acquisitions (M&A), and other equity investments.

Achieving better return is the topic never out of topic in investment management. Under EVA, principles below must be followed: firstly, the investment project must make a contribution towards EVA positively during the life cycle, and the corresponding EVA target and rejection standard must be re-established in case of oversize scale and low efficiency; secondly, such project must conform to the company goals, highlight the main operation, and keep ahead in competitive business; lastly, such project must secure the compliance of the law and policy [5].

3.1.2 Investment analysis

While investment decision-making, an investment is considered valuable when it generates more NOPAT than cost of capital. For grid corporations, investment driving consists of investment in various business segments and the expansion in certain area, and the basis of decision-making is positive EVA. It means the investment can bring excess return, and thus creating value. Based on growth-oriented investment behaviors, both EVA return rate, which is return on invested capital (ROIC) less WACC, and net investment growth rate should be considered. Here is the 2d scatter plot of these two:

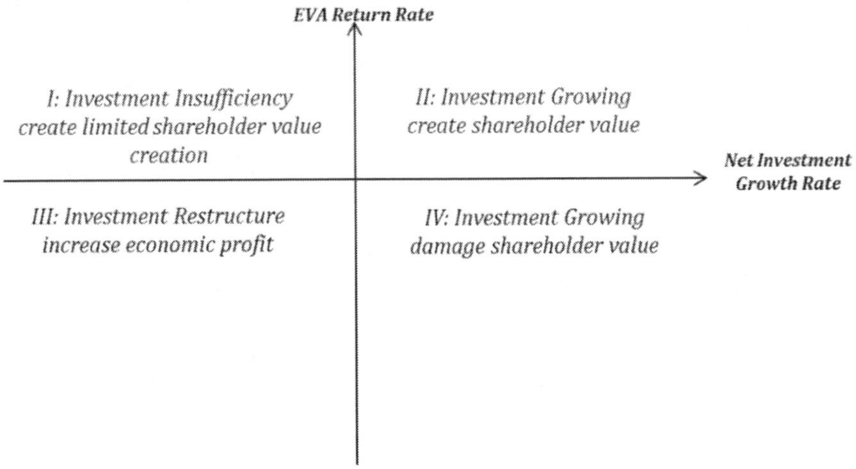

Fig.2. Investment Behavior Analysis

3.1.3 Capital optimization based on EVA

Introducing EVA could make the management keep eyes on capital optimization because it stresses the cost of capital employed. The management could adjust the capital structure through more flexible financing method and tools.

Based the analysis of value drivers and its resources, capital optimization should be considered in reducing capital occupancy and decreasing the cost of capital:

(1) Accelerate the capital flows

From the view of the whole company, as a result of different life cycle of each business segment, the volume of capital needed vary according to the growth rate of market and market share. The capital can be better utilized through adjusting the capital distribution in each business segment. During the redistribution, the grid corporation can shift the capital into the high value-added business from the low value-added businesses, improving the profitability of whole company. Such flow path can be clearly shown with BCG Matrix [6].

1) Question Marks mean businesses with low market share in a high growth market. It is regarded as a starting pointing with great potential but uncertainty as well. The best path is becoming the stars and finally cash cow after fast growing. If fail, it will enter the quadrant of dogs.

2) Stars mean business with high market share in a high growth market. It is the "superstar" because of both continuous growing and great profitability. As the strategic segment of enterprise objective, it makes large contribution to the whole enterprise value for both the current and future, and needs to increase the investment for expanding business and keeping competition advantage.

3) Cash cows mean businesses with high market share in a low growth market. It is regarded as mature business with high profit margin but less potential. It holds vast of surplus capital and does not need large-scale market. Therefore, it is the shield of other businesses, especially the stars.

4) Dogs mean businesses with low market share in a low growth market. It is considered to be a declining period with low margin or even deficit, and can hardly create value for the enterprise.

Before introducing EVA, grid corporations make efforts to launch projects as many as possible due to adverse selection. It causes enormous waste of resource and capital idle. In terms of EVA, any capital idle incurs cost, and blind investment will largely weaken the financial performance. In particular, applied the BCG Matrix into the business of grid corporations, the traditional grid network business, especially large-industry customer market, is the typical cash cows with stable return but slow growing (or even declining). On one hand, the stable return results from the price scissors between buying and selling, determined by the government. On the other hand, the market is growing slowly, or even declining, due to challenge from overproduction and corresponding supply-side reform. Therefore, to enhance EVA, investing smart grid has become the inevitable tendency. Smart grid is a fully automatic power transmission network with a variety of operational and energy measures, controlling each connection and guaranteeing the information and power flow in both direction. It mainly consists of user interaction, intelligent device, internet of things, big data, self-prediction & self-healing & self-adjustment, which can largely strong the EVA and power supply reliability in the future. Currently, such area is still the Question Marks, which means great potential but less profit and market share, and cost heavily, even reduce the EVA. But, based on the long-term value, smart grid is a chance to win the future and can be continuously "milked" by our cash cows.

(2) Decrease the capital cost rate

Decreasing the capital cost is another effective way to optimize the capital. Given the certain capital employed, lower capital cost rate can decrease the total capital cost and thus enhancing EVA. Emphasizing the existence of cost of equity capital, EVA makes the management consider the balance between the cost of equity capital and the cost of debt equity (which is relatively tax-saving), and then make the right decision in financing. It is worth thinking about whether the financing will increase the EVA, to avoid blindly financing expansion and capital idle from it.

4. Conclusion

Based on the business characteristics and EVA, the paper develops an econometric model for the long-term value of grid corporations in China, and analyzes the value drivers and path selection correspondingly. The study concludes that:

4.1 The value drivers of grid corporations

We have analyzed the value drivers of grid corporations in the future growth value and current operation value. Given the certain current operation value, we conclude that the root cause of future growth value is an increase in investment. Some factors happened currently will influence the value in the future. For example, asset overhaul not only affect current operating cost, but also affect the cost of overhaul in the future. Therefore, the decision-making of these factors should reflect the value maximization in the whole life cycle.

4.2 The path selection of long-term value improvement

The analysis of value drivers helps the management clarify the mechanism of value creation, rationalize the distribution of resources and eventually enhance the enterprise value. Based on the value drivers, the study has established the path of achieving the long-term value enhancement: investment management based on value development and capital optimization based on EVA.

References

1. Chen Zhibin, 2006. Analyzing Model of Value-Creation Factors from the View of Free Cash Flow Management [J]. Accounting Research 4, 58-62
2. Zhao Zhenquan, Geng Yuxin, 2005. The Application of Value Added Method to Corporate Evaluation [J]. Contemporary Economic Research 4, 43-47
3. Pan fei, Wen Donghua, 2006. Recent empirical study on managerial accounting and future research direction of China: based on value-based management [J]. Accounting Research 2, 81-86
4. Ittner, C., Larcker, D., 2001. Assessing empirical research in managerial value based management perspective [J]. Journal of Accounting and Economics 32, 349-410
5. Lloyd, W., DAVIS, L., 2007. Building Long-Term Value- Objective financial analysis focuses clients on business revenue growth [J]. Journal of Accountancy 204.5, 56-62
6. Federal Energy Regulatory Commission Assessment of Demand Response & Advanced Metering. United States Federal Energy Regulatory Commission. United States Federal Energy Regulatory Commission.

A Study of Female Board Directors Correlation With the Board Decision - Making Efficiency-- Based on Automobile Manufacturing Industry Listed Companies' Empirical Evidence

Dandan Dong[1,*] and Aimin Li[2]

[1]Engineering Management Department, Zhengzhou University, Zhengzhou , China
[2]Engineering Management Department, Zhengzhou University, Zhengzhou , China
Corresponding author: Dong Dan-dan 1392525437@qq.com

Abstract. This paper studies the correlation between the female board directors and Board decision-making behavior, from a gender perspective to our board ,using multivariate regression method,choosing the number of women directors, the proportion of female directors and women directors average educational background as a measure of the Board female directors feature variable index,and the frequency of board meetings as a measure of the Board decision-making efficiency, and China's automobile manufacturing industry listed companiesdata between 2006 and 2014 as research samples. This paper find that the number of women directors, the proportion of female directors and women directors average educational background are significantly negatively correlated with the Board decision-making efficiency.
Keywords. Female directors, board of directors, board decision - making efficiency.

1 Introduction

The board of directors plays a dominant role in the corporate governance. The efficiency of the corporate governance is closely related to the function of the board. while the rationality of the board structure directly determines the effective functioning of the board of directors. As the economic is developing dramatically, more and more women participate in the work, and gradually enter into the company's management positions.

At present, although domestic academia has gradually begun to study the relevance between board gender diversity and corporate governance, the literatures which directly studied the relation between female directors and board decision - making efficiency in Chinese companies are very little. Therefore, this paper carries on the empirical analysis to the relationship between the female director and the board decision-making efficiency in automobile manufacturing industry listed companies .

2 Literature review

Some foreign or chinese scholars have started to study the relationship between board and enterprise performance from the gender perspective, but the conclusions of the studies are not very consistent, a large part of the studies found that female director has a positive impact on board behavior, and others showed that the impact is negative or uncertain .

Triana, Maria del Carmen [1]et al. (2014) showed that at least three women was the first condition for promoting gender equality. Schwab, Andreas and Werbel, James D. et al. [2](2016) studied the relationship between managing gender diversity and firm performance ,they found that the introduction of female managers has a positive impact on firm performance. In addition, Boulouta, Ioanna [3](2013) explored the link between gender diversification and business performance follows the inverted U shape based on panel data collected from 151 German public companies from the period 2000-2005,based on the critical mass theory. While,Abdullah, Shamsul N., Ismail, Ku Nor Izah Ku and Nachum, Lilac [4] (2016) found that gender diversification sometimes positively affects board behavior and creates value for some firms, However, it sometimes has negative impact on board of directors conduct, and take a loss to the company.

Yan Yonghai [5]inspected the relationship between board of directors gender diversification and financial performance of the board, and found that gender diversity and corporate financial performance have a positive correlation. While, Cheng Rui et al. [6]studied the risk aversion and board assurances of female director, and found the higher the proportion of female directors and directors in the company, the more conservative the decisions made by the board of directors. .But, ChangMengmeng [7] found the gender of that CEO orchairman don not have a significant impact on company's financing decisions .

3 Research designs

3.1 Study samples

According to the classification standard of the industry in 2012, this paper selects listed companies in Shanghai and Shenzhen A-share automobile manufacturing industry from 2006 to 2014 as research samples to study the correlation between female directors and board decision-making efficiency.

In this paper, the original data were screened according to the following principles: (1) remove ST companies; (2) remove companies which have incomplete data . After the above-mentioned treatment, we get the sample observations of 42 listed companies. Data from Guotai'an database, the data are calculated using Minitab for analysis.

3.2 The selection and definition of variables

(1) Interpreted variable. In this paper, the Board of Directors annual meeting frequency (BMF) is choosed to measure the efficiency of board decisions, the higher the frequency of the annual meeting of the Board, indicating that the lower the efficiency of board decision-making, the contrary is higher.

(2) Explanatory variables. this paper mainly used these three variables , the number of female directors (FD), female directors Percentage (FADPER), the average level of education of female directors (EDU).As according to related researches, this paper divides the academic degree of female directors into five categories, including secondary, tertiary, undergraduate, master and doctor, the corresponding scores are 1 point , 2 points, 3 points, 4 points, and 5 points, the total score of all the female directors in the sample companies is added together and then is divided by the average number of female directors to get average value which is the average level of education of the company's female directors.

(3) Control variables. In order to maximize avoiding other factors effect on the research results, the size of board of directors (BS), the proportion of independent directors (IDR), size of company (SIZE),equity checks and balances (SHS) and asset-liability ratio (DAR) were choosed to control the company's size and capital structure drawing on existing researches in this paper.

Explained variables, explanatory variables and control variables are defined and described in Table 1.

Table 1. Variable definitions and descriptions.

Project	**Variables Name**	**Definition and Metering**
Explained variable	BMF	Number of meetings held by the Board each year
Explanatory variables	FD	The number of women in the board of directors, and if companies have no female director,it will be zero
	FADPER	The ratio of the number of female directors to the size of the board of directors
	EDU	Female director's grade scores are added together to arrive at the sum divided by the number of female directors and get the average value
Control variables	BS	Total number of directors
	IDR	The ratio of the total number of directors to the board size
	SIZE	The natural log of the total assets of the company
	SHS	The ratio of the number of shares held by the top ten shareholders to the total share capital at the end of the year
	DAR	The ratio of total liabilities to total assets

3.3 Model design

From the above theoretical analysis, this paper uses multiple linear regression model as follows:

$$BMF=\beta_0+\beta_1 FD+\beta_2 BS+\beta_3 IDR+\beta_4 SIZE+\beta_5 SHS+\beta_6 DAR+\varepsilon \quad (1)$$

$$BMF=\beta_0+\beta_1 FDI+\beta_2 BS+\beta_3 IDR+\beta_4 SIZE+\beta_5 SHS+\beta_6 DAR+\varepsilon \quad (2)$$

$$BMF=\beta_0+\beta_1 EDU+\beta_2 BS+\beta_3 IDR+\beta_4 SIZE+\beta_5 SHS+\beta_6 DAR+\varepsilon \quad (3)$$

Where β_0 is the constant term of the model, β (i = 1, 2, ... 6) is the coefficient of the explanatory variable and the control variable, and ε is the random error term.

4 Empirical analysis

4.1Descriptive statistical analysis of variables

Table 2. Descriptive statistics of variables.

	Number of valid data	**Number of missing data**	**Minimum**	**Maximum**	**Mean**	**Standard deviation**	**variance**
FD	378	0	0.000	4.000	0.794	0.958	0.918
FADPER	378	0	0.000	0.444	0.079	0.096	0.009
EDU	378	0	0.000	5.000	0.601	1.351	1.824
BS	378	0	5.000	22.000	10.048	2.463	6.067
IDR	378	0	0.200	0.600	0.356	0.054	0.003

SIZE	378	0	18.939	26.751	22.097	1.339	1.792
SHS	378	0	0.189	0.880	0.545	0.153	0.024
DAR	378	0	0.000	1.226	0.572	0.161	0.026
BMF	378	0	3.000	36.000	8.484	3.442	11.847

From the descriptive statistics of the variables in Table 2, it can be found that the average frequency of annual board meetings in sample enterprises is 8.484,The average number of female directors is 0.794,the minimum proportion of female directors is 0, the average is 7.9%, the overall proportion of female directors is very low. The average educational background of female directors is 0.000, the highest is 5.000. While the average is 0.601,it has not reached the level of college, meaning that the average overall educational background of female directors in Chinese automobile manufacturing industry is low and not dominant

Table 3. Annual distribution of female directors.

FD	**2006**		**2007**		**2008**		**2009**		**2010**	
Number	Frequecy	Percentage	Frequency	Percentage	Frequency	Percentage	Frequency	Percentage	Frequency	Percentage
0	18	42.86	22	52.38	20	47.62	21	50.00	22	52.38
1	18	42.86	12	28.57	12	28.57	13	30.95	13	30.95
2	5	11.90	7	16.67	9	21.43	6	14.29	4	9.52
3	1	2.38	1	2.38	1	2.38	0	0	3	7.14
4	0	0	0	0	0	0	2	4.76	0	0
Total	42	100	42	100	42	100	42	100	42	100

FD	**2011**		**2012**		**2013**		**2014**	
Number	Frequency	Percentage	Frequency	Percentage	Frequency	Percentage	Frequency	Percentage
0	23	54.76	21	50.00	21	50.00	19	45.24
1	11	26.19	10	23.81	9	21.43	11	26.19
2	4	9.52	6	14.29	10	23.81	10	23.81
3	3	7.14	3	7.14	1	2.38	2	4.76
4	1	2.38	2	4.76	1	2.38	0	0
Total	42	100	42	100	42	100	42	100

Table 4. Correlation statistical analysis.

	FD	FDI	EDU	BS	IDR	SIZE	SHS	DAR	BMF
FD	1.000 ------								
FDI	0.960*** (0.000)	1.000 ------							
EDU	0.366*** (0.000)	0.411**(0.000)	1.000 ------						
BS	0.167*** (0.001)	-0.021 (0.685)	-0.089* (0.084)	1.000 ------					
IDR	-0.052 (0.318)	-0.033 (0.526)	0.013 (0.800)	-0.174** *(0.001)	1.000 ------				
SIZE	-0.199** * (0.000)	-0.234** *(0.000)	-0.071(0. 167)	0.325** * (0.000)	-0.226** *(0.000)	1.000 ------			
SHS	-0.279** * (0.000)	-0.298** (0.000)	0.079 (0.127)	0.031(0. 552)	-0.088*(0.088)	0.348*** (0.000)	1.000 ------		
DAR	-0.031 (0.554)	-0.028 (0.594)	-0.065 (0.209)	0.099*(0 .055)	-0.010 (0.850)	-0.047 (0.364)	-0.154** (0.003)	1.000 ------	

BMF	0.157*** (0.002)	0.082 (0.114)	0.059(0.250)	0.280*** (0.000)	-0.025 (0.633)	0.113** (0.028)	0.036 (0.487)	0.010 (0.840)	1.000 ------

Notes. *** indicates a significant correlation at 0.01 level (both sides), ** indicates significant correlation at 0.05 level (both sides), and * indicates significant correlation at 0.1 level (both sides)

Table 5. Summary of models.

Model	R	R square	Adjust the R square	Standard estimation error	Change the statistics		
					R-side change	F change	Sig.change
1	0.313	0.098	0.084	3.564	0.098	6.714	0.000
2	0.303	0.092	0.077	3.584	0.092	6.238	0.000
3	0.293	0.086	0.071	3.476	0.086	5.816	0.000

Table 6. Regression analysis.

Variable	Model 1		Model 2		Model 3	
	B	Sig.	B	Sig.	B	Sig.
Constant	0.027	0.994	0.453	0.900	2.071	0.552
FD	0.534	0.006				
FDI			4.268	0.025		
EDU					0.211	0.099
BS	0.335	0.000	0.379	0.000	0.393	0.000
ID	2.764	0.398	2.686	0.413	2.043	0.533
SIZE	0.149	0.316	0.121	0.415	0.084	0.566
SHS	1.025	1.406	0.930	0.454	0.076	0.950
DAR	-0.289	0.793	-0.412	0.708	-0.495	0.656

As can be seen from Table 3, there are about 50% of listed companies have no female directors every year. In addition, during this period also find most of the listed companies have one female director, the proportion is23%or more.From 2013 to 2014, the proportion of listed companies with twofemale directors is on the rise. From 2009 onward, with four female directorsof listed companies appear. However, its proportion is less than 10% .

4.2 Correlation analysis of variables

From the data in Table 4, it can be found that the number of female directors is significantly positively related to the frequency of board meeting, but the proportion of female directors is not significant with it. And the correlation coefficients of the three explanatory variablesare mostly less than 0.5, so there is little likelihood of multiple collinearities between them, willnot affect the multiple linear regression band.

4.3 Regression analysis

From the model summary and regression coefficients in Table 5, it can be found the R-level coefficients are 0.098, 0.092 and 0.086, all the significant results are 0.000. Since the R-level coefficients are positive, it can be obtained there is no linear relationship between the explanatory variables and interpreted variables, and the significance level is0.000,which was far less than 0.001, indicating a high level of significance. All of the three models had good goodness of fit, proving the three explanatory variables could explain the interpreted variables well, and the coefficient of predictive variables reachsignificance level.

As can be seen from Table 6, there is a positive correlation between the number of female directors, the proportion of female directors, the average educational background of female directors and frequency of board meetings, which are significant at the levels of 0.01, 0.05 and 0.1 respectively. This shows that female directors are negatively correlated with board decision making efficiency.

5 Robustness test

In order to test the rationality and robustness of the model, the stability test is carried out.

(A)In the descriptive statistical analysis we find that there was a big gap in the companies asset-liability ratio, so this article removes the companies whose asset liability rate is more than 100% , then take regression analysis,the result is in agreement with the above analysis.

(B)In order to ensure that this screening does not affect the results , this article joins this part of the companies in regression analysis, the result is consistent with the above analysis.

(C)As with combination of the two cases, removing the companies whose asset liability rate was more than 100% and adding the companies by ST, then take regression analysis, the result is consistent with the above analysis.

6 Conclusions

This paper studies the influence of female directors on board decision - making efficiency in 42 listed companies in Shanghai and Shenzhen A-share automobile manufacturing industry from 2006 to 2014.The results shows that the number of female directors, the proportion of female directors and the average educational background of female directors are negatively related to board decision - making efficiency. Although, according to the individual cognitive limitations theory, the participation of female directors can help the company to make a comprehensive scientific decision-making.This paper gets the conclusion that the female director was negatively correlated with board decision - making efficiency. First of all, this may be related to the samples selected in this paper. Only took 42 companies in the listed companies in the automobile manufacturing industry as the sample, the scope of the sample is not big enough .Second, the study results would be affected by the survey interval, the selected study interval is not long enough.Finally, in practice, the board decision - making efficiency may also be affected by other factors,for example, the higher the average educational background of female directors, the more likely it will be the phenomenon of contention, which will bring some impact to the research result.

References

1. Triana, Mariadel Carmen, Miller, Toyah L., Trzebiatowski, Tiffany M..TheDouble-EdgedNature of Board Gender Diversity: Diversity, Firm Performance, and the Powerof Women Directors as Predictors of Strategic Change.J.Organization Science,**25**,2(2014).
2. Schwab, Andreas,Werbel, James D., Hofmann, Heike.Managerial Gender Diversity and Firm Performance: An Integration of Different Theoretical Perspectives.J.Group & Organization Management,**41**,1(2016).
3. Boulouta, Ioanna.Hidden Connections: The Link Between Board Gender Diversity and Corporate Social Performance.J.Journal Of Business Ethics,**11**,3(2013).
4. Abdullah, Shamsul N., Ismail, Ku Nor Izah Ku and Nachum ,Lilac.Does Having Women on Boards Create Value? The Impact of Societal Perceptions and Corporate Governance in Emerging MarketsJ.Strategic Management
5. Yan Yonghai, Gender Diversification of Board of Directors and Corporate Financial Performance - Empirical Evidence Based on Small and Medium Board Listed Companies .J. Taiyuan: Friends of Accounting, **18**(2014).
6. Cheng Rui, Ni Hengwang, Li Changqing .Study on the Risk Evasion of Women Directors and the Guaranty Behavior of the Board of Directors - Evidence from the Shanghai Stock Market .J. East China Economic Management(2016).

 DOI: 10.1051/matecconf/201710005042

7. Chang Mengmeng.CFO gender, financial stability and corporate finance decision-making relationship analysis .J. Business Economics Research(2015).

The Study on Activation Strategy of Time-honored Brand

Jiankang Mu

Business School, Sias International University, Henan, China

Corresponding author: author@e-mail.org

Abstract. Time-honored brands have profound cultural heritage, a large part of glorious time-honored brands has gradually disappeared from the market. Brand activation is the core of management strategic of brand equity as well as the fundamental requirement of time-honored enterprise recovery. This paper constructs the time-honored brand purchase model and finds that consumers nostalgic psychology for the consumer's perception and buying behavior has a positive effect.

1 Instuduction

Time-honored brand refers to the brand management has a long history, with generations of products, production techniques or services have a distinct Chinese traditional culture. They were founded before 1956, achieved wide recognition of society, have a good reputation.

With the development of economic globalization, more and more foreign brands poured into China. Because of the lack of brand consciousness and core competitiveness, China's time-honored brand enterprises encountered hard-hit and difficult to continue development. Part of the old brand market share has fallen sharply, even withdraw from the market. China's time-honored brand enterprises face enormous survival crisis.

2 literature review

2.1 Time-honored brand

Brand is a name, term, mark, symbol or design, or their mutual combination, it is a product or service used to identify a consumer or a group of consumer. It makes different with competitors' products or services. Brand is a kind of intangible assets, brand owners can get competitive advantage and huge economic benefits from it. According to the needs of article research, we choose Time-honored brand as the object of this research. To be specific, these Time-honored brands must have the following typical characteristics: 1)With long historical period; 2)Distinctive national character; 3)With the reputation of social recognition; 4) An ongoing philosophy of culture; 5)With the traditional process of high quality.

2.2 Consumers nostalgic

Nostalgia is a kind of preferences (for people, places or things), it's a kind of good feelings. These things are common in younger. Brand activation based on consumers' nostalgia is a common strategy in western countries when solve the problem of aging brands. They use nostalgic psychology of consumers and stimulate consumers nostalgia, eventually led to purchases.

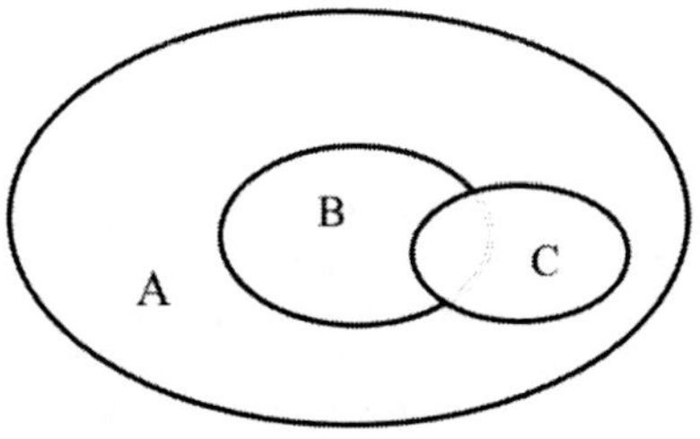

Fig.1. Time-honored brand relationship diagram.

A:Time-honored brand B: Time-honored brand of China C:Time-honored brand of Chinese

3 Theoretical framework

3.1. Theoretical model

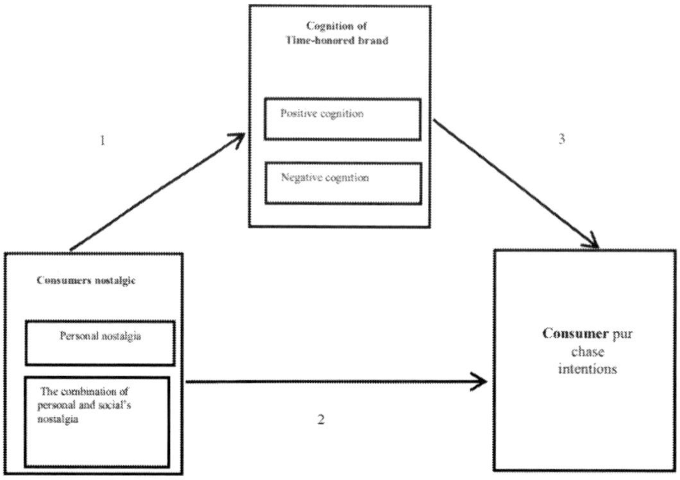

Fig.2. Time-honored brand purchase behaviour.

3.2 Research Model

Place the figure as close as possible after the point where it is first referenced in the text. If there is a large number of figures and tables it might be necessary to place some before their text citation. If a figure or table is too large to fit into one column, it can be centred across both columns at the top or the bottom of the page.

Based on two groups of consumers score, we judge the relationship between consumer nostalgia and consumer-awareness .We put forward these assumptions as follows:

H1a: "Consumer nostalgia" to "Positive Consumer Cognition of Time-honored brand" has a positive effect.

H1b: "The combination of personal and social's nostalgia" to "Positive Consumer Cognition of Time-honored brand" has a positive effect.

H1c: "Personal nostalgia" to "Negative Consumer cognition of Time-honored brand" has a positive effect.

H1d: "The combination of personal and social's nostalgia" to "Negative Consumer cognition of Time-honored brand" has a positive effect.

Due to the nostalgia on consumers purchase intentions of consumers to be able to make a difference has been confirmed by numerous studies, we put forward these assumptions as follows:

H2a: "Personal nostalgia" to "Consumer purchase intentions" has a positive effect.

H2b: "The combination of personal and social's nostalgia" to "Consumer purchase intentions" has a positive effect.

Consumer cognition of Time-honored brand can has an impact on Consumer purchase intentions. So we put forward these assumptions as follows:

H3a: "Positive Consumer Cognition of Time-honored brand" to "Consumer purchase intentions" has a positive effect.

H3b: "Negative Consumer cognition of Time-honored brand" to "Consumer purchase intentions" has a positive effect.

4 Empirical study

4.1 Case selection

Here is an example of the Pechoin. Founded in 1931, the company is a research, development, production, sales and service as one of the professional of cosmetics. With 80 years of history, Pechoin is one of China's first cosmetics brands, and its products are not only affordable, but also well received by consumers, such as Vaseline cream, body lotion and others. As domestic cosmetics pioneer, Pechoin is different from other domestic cosmetics manufacturers in the product category, product packaging, and brand image. The corporation has a higher distinction degree because of grasping the herbal essence series primarily for Chinese female characteristics and preferences.

4.2 Questionnaire design and data collection

Table 1. Descriptive statistical analysis.

Basic situation	Classification	Number	Effective percentage
Age	Under 20	1	1.1
	21-30	59	64.8
	31-40	22	24.2
	41-50	5	5.0
	51-60	3	3.3
	Over 60	1	1.1
Qualifications	Under Junior High School	4	4.4
	Senior high school, Technical secondary school	30	33
	Junior college, Undergraduate course	50	55
	Over postgraduate	7	7.6
Monthly income	Under 2000	11	12
	2000-5000	50	55
	5000-10000	28	30.8
	Over 10000	2	2.2

As can be shown from the diagram, respondents were concentrated in the 21-40 years of age, besides less than 20 years old and over the age of 40 respondents were only 11%. It has a larger relationship with the paper selected case. Women, who age from 21 to 40 years old, are the main consumers of cosmetics such products. Therefore, the samples of age at this stage is a relatively regular phenomenon. In another, the level of education of these respondents are mainly at a high school, secondary school, college or university. This phenomenon in general is also very compatible with current consumers. The monthly income of these people is concentrated in 2000 to 10000, which reaches 85.8% percent of the sample. And this conclusion is more in line with the actual situation of the general consumers.

5 Data analysis

5.1 Reliability analysis

Reliability refers to credibility of the test. Reliability is a credibility indicator of evaluating respondents filling in the questionnaire data.

Table 2. Reliability statistics.

Cronbach's Alpha	number of item
.788	13

Alpha coefficient in 0.65-0.7 as the minimum acceptab-le values, 0.7-0.8 is very good, 0.8-0.9 is extremely good. From the table above, in this paper, scale reliability is 0.788, the Scale reliability in this paper is very high.

5.2 Correlation analysis

According to last figure, the correlation coefficient of personal nostalgia to positive consumer cognition of Time-honored brand, error is less than 0.01,within 99% confidence interval correlation, the correlation coefficient is 0.831, it is a significant positive correlation, we can preliminary validate the existence of H1a.

The combination of personal and social's nostalgia to positive consumer cognition of Time-honored brand, within 95% confidence interval correlation, the correlation coefficient is -0.293.So there is a negative correlation relationship between the combination of personal and social's nostalgia to positive consumer cognition of Time-honored brand. What's more, personal nostalgia to negative consumer cognition of Time-honored brand, error is 0.0137, the correlation coefficient is negative, so there is a negative correlation relations-hip between them. The correlation coefficient of personal nostalgia to positive consumer cognition of Time-honored brand is 0.809, error is less than 0.01.

Table 3. Correlation analysis.

		Positive	Negative
Personal nostalgia	Pearson correlation	0.831**	-0.20*
	Significant	0.000	0.0137
	N	91	91
The combination of personal and social's nostalgia	Pearson correlation	-0.293*	0.809**
	Significant	0.005	0.000
	N	91	91

Personal nostalgia to positive consumer cognition of Time-honored brand has a positive effect, we can preliminary validate the existence of the hypothes H1d.

According to last figure, perason correlation is 0.375, within 99% confidence interval correlation. So positive consumer cognition of Time-honored brand to consumer purchase intentions has a significant positive correlation. We can preliminary validate the existence of the hypothes H3a.

The correlation coefficient of negative consumer cognition of Time-honored brand to consumer purchase intentionsis -0.293, within 95% confidence interval correlation. So there is a negative correlation relationship between them.

Table 4. Correlation analysis (Purchase intentions).

		Purchase intentions
Personal nostalgia	Pearson correlation	0.316**
	Significant	0.040
	N	91
The combination of personal and social's nostalgia.	Pearson correlation	-0.26*
	Significant	0.05
	N	91

According to last table, the correlation coefficient of personal nostalgia to consumer purchase intentions is 0.316, within 95% confidence interval correlation. So personal nostalgia to consumer purchase intentions is a significant positive correlation. We can preliminary validate the existence of the hypothes H2a. The correlation coefficient of the combination of personal and social's nostalgia to consumer purchase intentionsis -0.26, there is a negative correlation relationship.

Table5. Correlation analysis.

		Purchase intentions
Positive cognition	Pearson correlation	0.375**
	Significant	0.001
	N	91
Negative cognition	Pearson correlation	-0.293*
	Significant	0.04
	N	91

6 Recommendation

6.1The countermeasures and suggestions to activate the brand of pechoin

Opinions vary on this point. Because of the products, services, history, culture and many other aspects are different, the brand activation strategy is also different. Combined with the actual situation of the brand development, starting from the research direction of this article, the process of Pechoin's brand activation should pay more attention to purchase intentions.

6.2 Make use of the nostalgic psychology contributing to purchase behavior

The Pechoin was founded in 1931, is one of the earliest cosmetics brand in China, it has more than eighty years old so far, And in the 1960s, create a monopoly in the market. So we can learn that the Pechoin was very popular to consumers at the time. Therefore, for the Pechoin's situation at present, this paper argues that we can make use of the nostalgic psychology contributing to purchase behavior of consumers. First of all, for the former customers, Pechoin can make the products they were familiar with, such as pearl cream, glycerin, cheap and fine old products, to attract the customers to buy them again. Second, for the customers who never buy this brand, due to the elders and nostalgic psychology drives around, they will produce high evaluation and expectations of Pechoin. In the process of marketing should be paid attention to the counter display, advertising campaign, product packaging, Marketing staff and other ways to Transfer memories of the past. For example, make the boutique brand productionas same as the past, play old songs in store and in addition to show the products, they can add some old items for the consumer to create a nostalgic shopping environment. Finally, in view of the consumers have never heard of the Pechoin, should help them know the brand from the old identity of the Pechoin, let consumers know about it as much as possible of the current information, one of the most important part is hidden behind the brand of history and legend. To arise the attention of consumers, and translate this into buying power.

6.3 Distinguishing standards for each segmentation of the market marketing plan

When making marketing plan should be targeted. And according to different age, different degree of consumer's nostalgia emotions is also different, the same plan for all consumers would be unwise to marketing. In simple terms, aim at young consumers, they prefer to purchase cosmetics on the Internet, but the Pechoin in the online sales model is not mature. The design of the web pages need to improve, the display of store if couldn't arise consumer's nostalgic psychology, the Pechoin will the same as other general brands. So, increase the element of nostalgia in web designment to attract the attention of consumers is very important.

6.4 Stimulate consumers to accept the culture and emotion recognition of the time-honored

Largely, the time-honored brand is different from general brands .For one thing, the old brand has a rich cultural heritage .For another, it gained a good reputation in the process of long-term development, such as Pechoin ,which has more than 80 years of history. In the 1930s, Pechoin's skin care products sold well and became the first cosmetic choice of the celebrities. Ruan lingyu, Zhou Xuan and other international stars led the youth of an era, with the unique scent of Pechoin. Even the Song sisters, wandering in Shanghai that year, also respected the products of Pechoin among the United Kingdom, Germany, France and other diplomatic envoys to China. To some extent, they felt that the brand contained the Oriental fashion. Only mentioned in the advertisement, or marked on the product packaging are not enough. To make sure a consumer to remember this story, we must strengthen the impression of this information in the consumer's mind. Here are some ways: First of all, the corporation can sponsor the related TV series, which appear the actors using its product shots. In order to attract consumers ' attention, the company can also cooperate in the development and sale of actual products, launching consistent with the historic porcelain-cans cosmetics. If the marketing budget allows, the sponsorship of such television or movies can be increased in the promotion of the brand. Secondly, it can directly add some elements which represent the Republic of China era to the ad plot design. The plot may contains a certain story so that the ad content is more plentiful and different from the general advertising of cosmetics ,which only emphasize product efficiency .

References

1. Keller, K.L.. Managing Brands for the Long Run: Brand Reinforcement and Revitalization Strategies. California Management Review(2004).
2. Brown, S., R.V. Kozinets, and J.F. Sherry. Teaching Old Brands New Tricks: Retro Branding and the Revival of Brand Meaning. Journal of Marketing(2003).
3. Torelli, Carlos. Iconic brands and iconic categories: Consequences for the use of the cultural knowledge in judgments about foreign brands. Working Paper, University of Illinois(2009).
4. Holbrook, Morris B, and Robert M Schindler. Echoes of the dear departed part: some work in progress on nostalgia. Advances in Consumer Reshearch(1991).
5. Reisenwitz, Timothy H. Rajesh lyer, and Bob culter. Nostalgis advertising and the influence of nostalgia proneness. The Markrting Management Journal(2004).

Study on Behavior Value Analysis and Decision Methodology of Grid Corporations in China

Yuanbin Jiang[1] and Ruobing Shuai[2]

[1]State Grid Chengdu Power Supply Company, Chengdu 610000, China

[2]Guangdong Power Grid Corporation Foshan Power Supply Bureau, Foshan 528000, China

Corresponding author: jybthebest@sina.com

Abstract. Based on grid corporations, business environment and value characteristics in China, the article analyzes its behavior value factors, and then divides its behavior into three categories: the behavior only affecting the current Economic Value Added(EVA), the behavior affecting both the current and future EVA, and the behavior only affecting future EVA. Finally, the article studies such corporations, behavior value decision making based upon analyses and classifications above.

KeyWords. Value-Based Management(VBM), Economic Value Added(EVA), Behavior Value Factor Behavior, Value Decision Methodology.

1 Instruction

Early this century, the idea of Value-based Management(VBM) and EVA was introduced into China from the United States. In 2010, the State-owned Assets Supervision and Administration Commission of the State Council(SASAC) has begun to evaluate the performance of all state-owned enterprises(SOEs) with EVA system, and opened a new era different from the tradition simply focusing on the enterprise scale and profit. In China, with the development of SOEs, introducing EVA is to help SOEs accomplish strategic objectives, improve management mode, and solve problems economically and efficiently.

As significant parts of SOEs, grid corporations construct and operate power grids as the core business, to provide safer, cleaner, and more economic and sustainable power supply. These corporations target at creating sustainable growth in EVA under VBM, in planning, forecasting, evaluation and other enterprise behaviors. Therefore, it is necessary to establish a set of behavior value analysis system for grid corporations in China.

2 Behavior value factors analysis of grid corporations

Behavioral value factors analysis refers to factors affecting value inflow and outflow within the enterprise (value added and impairment), thus assessing various enterprise behavior qualitatively and quantitatively. According to the influence on EVA, we can divide enterprise behavior into three categories: the behavior only affecting the current Economic Value Added(EVA), the behavior affecting both the current and future EVA, and the behavior only affecting future EVA, as the Figure shown below:

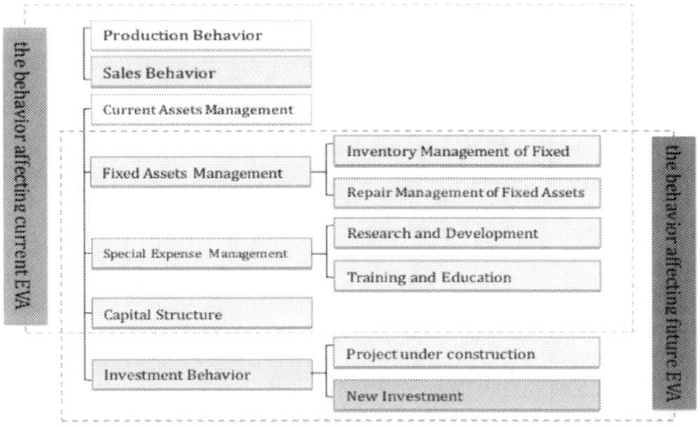

Fig.1. Enterprise behavior classification based on long-term value.

3 Behavior value decision method of grid corporations

Grid corporations should establish a set of behavior control system targeting at sustainable EVA growth. The system can clarify the value contribution of the management and the staff respectively to the enterprise, check whether the decision-making deviates from the VBM target, and guide the direction of corporate behavior. Corresponding with the behavior value factors, the behavior value decision method of grid corporations consist of three aspects: the behavior value decision method affecting the current EVA only, both the current and future EVA, and the future EVA only.

3.1 The behavior value decision method affecting the current EVA only

Specifically, it is made up of decision methods of income management, cost management, current assets management.

The behavior value decision method of income management. The business process of electricity purchasing (Electricity Sales Income) reflects the company's operating scale and industry status, determined by electricity sales, electricity sales structure, the average unit price of selling. After the decomposition of revenue drivers, we can find associated indicators supporting that. The management can analyze the existing revenue through these indicators under VBM.

3.2 The behavior value decision method of cost management

3.2.1 Electricity purchasing cost

The business processes of purchasing and selling electricity determine the gross profit level jointly. In terms of electricity purchasing cost analysis, we keep eyes on the quantity and unit price of purchasing electricity from different power suppliers. Doing such analysis is to clarify the impact of electricity purchasing structure and unit price on the electricity purchasing cost. After the decomposition of cost drivers, we can find associated indicators supporting that. The management can analyze the existing cost through these indicators under VBM [1].

3.2.2 Electricity transmission and distribution cost

Transmission and distribution cost is the transportation and power grid enterprises to provide electrical energy costs that occurred in the transmission and distribution sectors. It includes the cost of fixed assets and grid operating costs.

3.2.3 Other costs

Other costs of grid corporations are costs that not directly related to the volume of business, including office expenses, travel, utilities, and vehicle usage fees. These costs result in an outflow of economic benefits and an reduction in EVA index directly. However, these costs are necessary to maintain the regular operation of company, naturally helping the current inflow of economic benefits. Therefore, we cannot conclude that it is of no value, and the company should carefully control such costs in the manner of quota management.

3.3 The behavior value decision method of current assets management

Current assets include cash and cash equivalents, inventory and accounts receivable, etc., belonging to the scope of the overall enterprise value stream. The quota management system is a good option for current assets, analyzing its specific behavior under VBM. It can determine the amount of current assets held by the best, strengthen the turnover, and improve the utilization efficiency [2].

3.3.1 Cash and cash equivalents management

Companies hold cash and cash equivalents for three needs: transactional demand, cautious demand and speculative demand. Under the premise of guaranteeing regular operation, stronger cash management can make full use of surplus cash to capture potential investment opportunities.

Obviously, the following question is how to determine the best cash holding. In theory, there are three modes: cost analysis model, inventory model and random model.

1) Cost analysis model suggests that the cost of holding cash includes opportunity cost, management cost and shortage cost. The opportunity cost is directly proportional to the amount of cash holdings. There is no significant relationship between the management cost and cash holdings (given that certain cash holdings are relatively unchanged). The shortage cost is inversely proportional to the amount of cash holdings. Therefore, there is optimal cash holding, minimizing the sum of the three kinds of cost.

2) Inventory model suggests that holding cash incurs opportunity cost (positive with holding cash) and transaction cost (negative with holding cash). The minimization of these two is the optimal amount of cash holding for companies.

3) Random model suggests that the cash demand is difficult to predict, the management determine the lower and upper of cash holding based on experience. When the cash demand exceeds the upper limit, the management converts securities into cash to make up for the insufficiency; when the cash demand is lower than the lower limit, the management investment securities for better return.

3.3.2 Inventory management

Inventory is a necessary part to maintain the regular operation of grid corporations, generally including spare parts for repairing, low-value consumptions and so on.

Under VBM, inventory management can be analyzed in three aspects: the claim management system for existing inventories, the utilization efficiency of inventories in need, and the reserve quota.

1) The claim management system should be used for existing inventories. The management ought to claim inventories in details, and distinguish the inventories really needed from others, dealing with unnecessary ones in time.

2) In terms of the utilization efficiency of inventories needed, the management targets at the turnover ratio, checking whether the liquidity and the fund occupied is sensible. Along with the actual usage, the quantities of inventories need to buy can be decided at reasonable levels.

3) Given the maintenance of normal operation, the reserve quota is the economic order quantity. With the change of order quantity, ordering cost and storage cost shift. Economic order quantity is the quantity with the lowest sum of storage cost and ordering cost in a certain period [3].

$$Total\ cost = Ordering\ cost + Storage\ cost$$

The economic order quantity is the quantity that makes the total of these two kinds of cost minimum, as the formula expressed:

$$TC = \frac{Q}{2} \times C + \frac{A}{Q} \times F$$

TC: the total cost of inventories

Q: the economic order quantity

C: the unit inventory storage cost

A: the total inventories in demand

F: the inventory ordering cost per time

3.3.3 Account Receivable Management

In terms of consumer classification differentiated pricing, the power supply can be divided into large industrial electricity, general industrial electricity, commercial electricity, agricultural electricity and residential electricity. Among them, the residential electricity users pay in advance, and it does not incur accounts receivable. While other categories of electricity users, there are sales on credit resulting in accounts receivable. It weakens the current assets turnover, thus impacting on EVA.

Based on accounts aging, the management should implement the responsibility system for corresponding accounts receivable recovery departments, and evaluate such departments in quota control and the reclaim speed. The quota control of accounts receivable can accelerate capital returns and reduce current assets occupancy. Meanwhile, enhancing the recovery rate can improve accounts receivable turnover. Both of them make a contribution to EVA increase directly. According to the control of quota and turnover, the budget of accounts receivable can be reasonable made for next period.

3.4 The behavior value decision method affecting both the current and future EVA

3.4.1 Analysis and decision methods

For the analysis and decision method of behaviors affecting both the current and future EVA, there are basically two methods: the quantitative evaluation method (formula method) for quantifiable project; the expert evaluation method for un-quantifiable project.

1) The quantitative evaluation method (formula method)

Quantitative evaluation method is to calculate the economic benefits generated and costs and expenses incurred for specific enterprise behavior in each accounting period, thus estimating its impact on EVA for different periods and discounting it with reasonable rate. If the present value of EVA is greater than the cost, such spending is valuable. On the contrary, it damages the enterprise value. This method is suitable for the management of existing fixed assets and its overhaul, which is quantifiable. The formula for calculating the present value of EVA is:

$$V(X) = \sum_{i=1}^{n} \{[(R - C_1) - TC_1 \times WACC_1] / (1+r)^n\}$$

V (X): the contribution value of project X
R: expected growth in sales resulting from project X
C_1 *: the cost of the project X incurred this period*

r : the discount rate of EVA

n : the expected effective years

Expert evaluation method is suitable for un-quantifiable behavior, such as R&D expense. Expert evaluation method, based on qualitative analysis, uses scoring and other techniques to make a comprehensive evaluation of the results with mathematical statistic characteristics. Such method focuses on experts' experience and opinion determining the weight of each index, and obtains satisfying results with constant feedback and modification. Specific steps are as follows [4]:

The first step is choosing 9~15 experts with both rich practical experience and solid theoretical foundation.

The second step is to determine n indicators, and then ask selected experts to give independent weight values of each indicator under certain rules.

The third step is to select the results, calculate the mean and standard deviations of each index weight, and check whether to re-determine the weight.

The fourth step is to repeat the third step until the difference between index weight and its average does not exceed the pre-specified criteria, which means the experts' opinions are basically converging. The average index weight with constant modified is what the management wants.

3.4.2 Existing fixed assets control method - claim system

Up to 85% of total assets, the fixed asset management plays a significant role in the operation of grid corporations. For existing fixed assets, the claim system is a sensible choice, and it is a certain way to identify their real value and clarify the responsibility of corresponding departments. Specifically, each department claim necessary fixed assets in the classification of production equipment with power production management system, production equipment without power production management system, and non-production equipment. Comparing the difference between existing assets and claim assets, each department can determine the capital occupancy and utilization efficiency, and then the un-claim capital with proper depreciation rate is the measured impact on EVA.

3.4.3 The behavior value decision method of capital structure

Capital structure determines the capital cost, directly impacting on the EVA. Optimal design of capital structure should balance development and efficiency, comprehensively considering VBM factors. Considering the debt structure, liquidity, capital cost and financial leverage, it can be found the existing problems of grid corporations in capital structure [5].

3.5 The behavior value decision method affecting the future EVA only

In terms of issues affecting the future growth value, the grid corporations target at fixed assets investment (including project management under construction) in four stages: feasibility study, construction, operation and economic benefit evaluation.

3.5.1 Behavior value analysis of feasibility study

By studying the investment behavior of grid corporations, we notice that the value consideration in project establishment is not comprehensive enough, without showing value contribution from certain project. In the existing feasibility study report, the management simply focuses on the process input of construction, while ignoring the feasibility of setting standards and the analysis of the value contribution.

1) The project increasing the economic return

For projects with better economic return, the expected increasing electricity sales and corresponding revenue are quantifiable. Based upon the EVA formula, the expected EVA can be calculated with the capital occupied of fixed assets. According to the expected effective years, the sum of discounted EVA for each period is the quantifiable value of fixed asset invested. As the formula below:

$$V(X)=\sum_{i=1}^{n}\{[(R-C_1-C_2\cdots C_m)-TC_1\times WACC_1]/(1+r)^n\}$$

V (X): the contribution value of project X;

R: the expected sales revenue;

$C_1 \cdots C_m$: *the cost of the project X incurred in each period;*

r: the discount rate of EVA;

n: the expected effective years

2) The project controlling the risk

The expected return is measured and calculated by the method of combining the EVA calculation with the probability distribution. Due to the uncertainties in reduction of risk and loss of fixed assets investment, the probability distribution of the expected EVA occurrence needs to be recalculated for precise prediction. Based on the expected utility theory, the expected EVA value is discounted as random variable X, and the probability value, i=1, 2, 3…n, is set according to the expected operation condition of the investment project... N, the expected value of the random variable (EVA) is calculated based on the expected EVA value of the probability determined:

$$U(X)=\sum_{i=0}^{n}[u(XN)]=P_1u(x_1)+P_2u(x_2)+P_3u(x_3)+\cdots+P_nu(x_n)$$

U(X): the expected contribution value of project X

3) The project improving the management level

It is difficult to measure the value of project improving the management level mathematically. The reasons are: Firstly, they cannot directly make contribution to the growth in economic benefits. In the actual production process, the improvement on management level from such project can hardly be quantified by the economic benefits, and companies have less desire to collect and process the relevant information.

3.5.2 Behavior value analysis of construction

1) Duration management

Extension of time will inevitably lead to additional costs, and the assets under construction do not produce value. It delays the output of future EVA, negative with EVA. It is necessary to set a clear requirement of duration during the feasibility study, and strongly follow the schedule under construction.

2) Cost control

Cost and expense is the incremental items of capital invested, negative with EVA. These aspects should be considered: pre-project cost calculation, cost carry-over monthly and project settlement. As the main control factor, cost management plays a significant role throughout the power grid project.

3) Quality control

There is a close interdependence of construction cost and quality level. The quality cost consists of control cost and failure cost. Control cost belongs to the quality assurance fee, proportional to the level of quality. While failure cost is simply loss, negative with the level quality. The quality goal should be reasonable determined with the standard of project contracts.

3.5.3 Behavior value analysis of operation

When project goes into operation, it will create value and the value-creating ability reflects the scale structure of capital invested and its sustainable development. Based on the classification of investment projects, and the uncontrollably macro factors, the management should target at the enhancement of profitability, risk control ability, and asset management capability, through streamline, life-cycling and intensification.

3.5.4 Behavior value analysis of evaluation

Although there is already an project evaluation system, it is not comprehensive enough because the feasibility study has not closely attached to the value target indicators. In terms of the evaluation, together with stages analyzed above, the management should pay more attention to the project's impact on the future growth value of grid corporations, and the difference between the expected contribution value and the actual one.

4 Conclusion

The enterprise value-creation results from the enterprise behavior based on specific markets and resources background. Generally, the enterprise behavior consist of production, sales, cost management, asset management, capital cost management and so on. In terms of different value drivers, the enterprise behavior, or business activities, are reclassified. This paper considers that, due to the different impact on the current and future EVA , the enterprise behavior can be divided into three categories: behaviors only affecting the current EVA, behaviors affecting both the current and future EVA, and behaviors only affecting the future EVA. The enterprise should establish a set of behavior control system targeting at sustainable EVA growth. The system can clarify the value contribution of the management and the staff respectively to the enterprise, check whether the decision-making deviates from the VBM target, and guide the direction of corporate behavior. Based on this, this paper has built a behavior vale analysis and decision methodology for gird corporations, including ones affecting the current, the future and both periods EVA.

References

1. Gang Wang. The Application of EVA Model in Grid Company.Friends of Accounting, 2008(5):106-107. (In Chinese)
2. SASAC. The Guidance on the improvement of Central SOEs' Value Management based on EVA. (In Chinese)
3. Martin J D, William Petty J. Value Based Management: The Corporate Response to the Shareholder Revolution. OUP Catalogue, 2001.
4. Zhenjin Zhao, Yuxin Di. The Application of EVA in the Investment Evaluation of Chinese companies. Contemporary Economic Research, 2005(4):43-47 (In Chinese)
5. Xin Xiao. The Research on EVA and the improvement. Northwest University, 2010. (In Chinese)

The Distribution Strategies of Convenience Stores Chain in China from Japan 7-11

Ting Huang [1] and Chunxiong Liu [1]

[1]Department of Management ang Engineering, Zhengzhou University, Zhengzhou 450001, China

Abstract. With the rapid development of China's supermarket chain industry, logistics distribution technology has become more and more important. Regarded as the core technology of chain operation, it is the basis of success in chain operation. At present, logistics distribution system's backward has hindered the development of China's chain enterprises. Based on the success of Japan 7-11, this paper makes a research on centralized shop strategy, diversified information management and the high efficiency of logistics distribution strategy, and seeks to optimize profits of logistics distribution from the perspective of enterprise strategy. Besides, this paper analyses the advantages and disadvantages of the mode of chain enterprises and gives advice on promoting chain enterprises.
Keywords. Logistics distribution, chain operation, logistics cost.

1 Introduction

As an important form of enterprise organization, chain enterprises play an important role in the development of China's economy. The rise of chain operation is in China's coastal city in the last century 80s and 90s. After 20 years' development, chain operation has made great progress and has developed into a retail business model in the world with great vitality and development prospects. However, there are still a series of problems compared with other countries in the world. For example, it does not have a high level in convenience store chain, and its scale is limited. Besides its business characteristics are not clear, and market positioning and supermarket have in common. The lack of features and service merchandise is also a problem. Logistics distribution lags behind and development in different areas is unbalanced. Japan 7-Eleven convenience store chain, as the world's largest chain enterprises, its experience in the logistics distribution of chain enterprises worth learning for China. Japan 7-11 convenience store chain has obtained great achievement today, it cannot do without the effective operation of the brand, unique management style, and market strategy based on region, comprehensive management information system, diversified logistics distribution are the key to success.

2 The overview of Japan 7-11 convenience store chain

7-11 brand was of the South Company of America, and became a Japanese company in 2005. Founded in 1927 in Texas, the name of the 7-11 came from the year of 1946, and its business hour was from 7 am to 11 pm. Then it was introduced by the Japanese retail manager Ito Yokado to Japan in 1974, and changed to 24 hours a day of business from the beginning of 1975. Until now, its shops are permeated with China, the United States, Japan, Singapore, Malaysia, Philippines, Thailand and other countries and regions [1].

Japan 7-11 convenience store's business area is basically in about 100 square meters. Its varieties are about 3000 varieties and they are all popular goods. Among them, food accounts for about 75%, magazines, daily necessities is about 25%. The headquarter recommends 80 new varieties monthly to its franchisee, so that the variety of merchandise business is often replaced to adapt to market changes as well as gives customers a sense of freshness.The convenience store business not only satisfy daily necessities, but also can help the region take charge of electricity rates, gas rates, water rates, fast broadcast rates, postage rates and so on. It is a real convenience store which attends to people.

3 The analysis of Japan 7-11 convenience store distribution strategy

Japan 7-11's success is mainly due to the design and management of its supply chain.Japan 7-11's business purpose is to provide them with products when the customers need. From a strategic point of view, one of the main objectives of the company is to seek the micro balance between supply and demand through the regional, seasonal and daily schedule. Then this paper will analyses the 7-11 convenience store distribution strategy.

3.1 Market strategy closed to the region

One of the main location strategy of 7-11 company is to follow a high density centralized stores, and enter a any new market. It is about to establish 50-60 intensive shops supported by the distribution center [2].

7-11 convenience store distribution system develops from a single wholesaler stage to the intensive development of the distribution system, then to common distribution system. It delivered by the manufacturer to the distribution center, and the distribution center through the milk run route to store delivery. Common distribution system saves transportation costs and reduce the cost of the store to receive goods. The shop can save less inventory through frequent replenishment.

Therefore, the implementation of the distribution strategy can get the following results.

3.1.1 Saving advertising costs for each store

The main advertising media of 7-11 is the television advertisement. If we consider the radiation range of every convenience store as a small area, when the area has many convenience stores who need for publicity through television advertising, which can also achieve the same effect by television advertisement. The advertising cost of one store or a few stores are the same, but to every store is different. Therefore, the more stores there is, the less advertising costs each store has.

3.1.2 Conducive to improve the efficiency of the 7-11 chain of the department of service

Japan 7-11 convenience store chain has the position of manager, that is OFC (Fiber Contr01 Open). Each area manager who is responsible for guiding the number of stores is about 7 or 8. This figure, based on the work of the area manager, has not changed from the past to the present. It is estimated that the future will not change. Now, we assume a situation, if the number of stores in the area is reduced and scattered, so the time cost for the administrator in the guidance process is relatively high, and then to the store instruction frequency will be reduced, and the time to the store guidance and communication will be reduced, which will reduce the quality of guide service. When stores are concentrated, the time cost for the manager to every district is relatively small, then the effect will be good. It will increase guidance frequency for the store to enhance the quality of guidance.

3.1.3 Brand aggregation effect

We can imagine, walking in the streets of Japan, we will be able to see a 7-11 convenience store signs every distance, which will produce a kind of visual perception [3]. The more signs we see, the more favorable to the company's popularity to improve, which will attain the effect of brand image multiplication. At the same time, it can also avoid a temporary closure of the accident or temporary shortage caused by the lack of sales opportunities, and to affect customer satisfaction or company reputation.

3.2 The implementation of a comprehensive Information system management strategy

7-11 convenience store has developed a comprehensive information system. It can be said that the success of the 7-11 benefits mostly to its comprehensive information system. This system is installed in each stores of 7-11, so you can keep in touch with the headquarter, suppliers and 7-11 distribution centers. The system is processing the day's sales data before 11 o'clock every night , so that the second day's morning will be the analysis of the business data.

The use of information systems of 7-11 can better maintain the balance of supply and demand of the store. Stores can adjust the goods on the shelves based on a day's consumption patterns. Information is delivered directly to the supplier and distribution center when the store orders. The supplier received the order to start the stock to complete the production. There is no necessary for people at the scene when the 7-11 store deliver goods. Using scanning equipment to collect goods, so the entire system is based on trust.

The using of multimedia to carry out the distribution of goods is also a major feature of 7-11. Through the satellite communication system, we can send multimedia advertising, weather forecast and cloud charts, and even electronic products to the store. In this way, we can achieve the multimedia information, product sales and distribution of multimedia. In the 7-11 weather forecast, it actively explores the sales of food and weather correlation. The information system will collect weather reports five times every day, so that the local store can get a reference. Based on the weather forecast, the convenience store can master the weather conditions in advance. According to the weather conditions, we can select the number of distribution of goods and optimize the distribution route. The information system also collects the age information of customers to better grasp the customers' demand. Data is generally evaluated by the company, the area, and the store.

3.3 High efficient logistics distribution

Logistics distribution is the last link between producers and consumers in the circulation of goods. The speed of logistics and distribution efficiency directly affect the degree of customer satisfaction. The important reason for the rapid development of 7-11 is that there is a strong rear logistics support system. In addition, its logistics system can match the common distribution model. Regional centralized shop strategy for the use of common distribution model, that is, the establishment of joint distribution centers in each region to achieve high frequency, varieties, and small unit distribution effect. Acted as a common distribution center fully reflect the sales of goods in transition and inventory information. The reducing of transportation costs, 7-11 gradually grasps the dominant power of the whole industry chain.

There are two main characteristics of 7-11 in the logistics management.

3.3.1 Implementing the aggregated distribution, and establishing a joint distribution center

Summary distribution is a unique distribution system of 7-11 company. Japan 7-11 company's circulation is implemented by the manufacturer designated wholesalers purchase of the special wholesale system. The same product has different manufacturers, so special wholesalers are also different. So, if we want to complete many manufacturers of goods here, it requires the distribution channels and distribution of different delivery vehicles, and then increase the delivery vehicles. In order to solve this problem, 7-11 company adopted a summary distribution method. The distribution area of a commodity which was responsible by various wholesalers in the past is distributed to each wholesaler. A wholesaler is responsible for all stores and distribution area, thus increasing the number of distribution to each store, reducing the number of receiving stores, and raising the efficiency of transportation. Owing to the narrow distribution area, it also can shorten the delivery distance and time distribution.

3.3.2 Quick response to customer needs

In Japan 7 - Eleven chain convenience stores, when the customers went to the cashier checkout, the cashiers will scan the bar code firstly, and offer the customers a quote. After paying, the cashier will firstly input the amount paid by the customers. Then according to their estimate of age and gender, they press buttons above the POS which blue is for men and pink is for women. There are numbers 12, 19, 29, 39, 50 according to the corresponding digital open POS cash register and change. After checkout the data is passed through the network to the headquarters of the mainframe computers. So OFC and store headquarters can get instant updated data information and data analysis of customers' demand. Through the analysis of the data, OFC can grasp the trend of every hour and every customer's demand, and the price and inventory of every customer. Through the integration of these areas, products and time data, the speed of the headquarters mastering the customer demand is greatly improved. It can help to forecast the sales trend timely every day, to adjust the price, tally and distribution activities.

4 The distribution of chain supermarket enterprises in our country

The chain supermarket in our country has become the most dynamic industry in the commercial field. Over the past few years, large and medium-sized chain supermarket sales volume increased year by year. Sales growth was significantly higher than the total retail sales of social commodities and the growth of traditional department stores [4].

Despite the development of China's supermarket chain, in general, China's supermarket chain is still in the primary stage. Whether from its share of the total retail sales of social goods, or from its development stage and management level, there are still a lot of mistakes and the lack of management level .

4.1 High distribution costs

In order to achieve high profit and high service level, it is necessary to reduce the distribution cost and increase the profit of the enterprise. Therefore, the company need to consider from the aspects of the purchase, the distribution of vehicles, the design of distribution routes, and so on.

4.2 The low degree of information

The application of computer information technology is not high in Logistics distribution center, and many important decisions have been done still in a state of artificial. The problems which were found in the distribution process and feedback are not timely, which leads that he treatment is not timely, and the efficiency of the work is reduced. In addition, the application of the computers is limited to the management of daily affairs. Most of the centers have not been fully established, which affects not only the quality of the distribution but also the speed of distribution.

4.3 Low service level

The function is not completed, which further leads to the backwardness of the service level. Distribution is the organic combination of "distribution" and "delivery". It is a modern logistics activity, and it also emphasizes the function of coordination and integration. But from the current situation of chain supermarket distribution in our country, there are still a lot of problems. On the one hand, although some distributions do inventory, goods, distribution, but it can only achieve the function of storage, and can not be door-to-door due to the lack of necessary transport vehicles, resulting in the seperation between "coordination" and "send". On the other hand, most of the goods are just to be sent, rarely considering the distribution, distribution processing and other functions.

Based on the above three points, we can draw a conclusion that if China's local supermarket chain enterprises want to keep their own position, the reform is urgent to achieve the expansion of the scale and market share of the upgrade.

5 Japan 7 - eleven chain convenience store distribution strategy can be used for reference in china's chain supermarket

As the world's largest chain enterprise, 7-Eleven can be said to achieve a big success today. In addition to the effective use of the brand, the agglomeration effect, as well as its unique way of management, it mainly lies in the effective logistics strategy, and this point is just the "bottleneck" in Chinese supermarket chain industry development. We can learn from the distribution strategy of 7-Eleven in the success of some successful experiences, combining with the characteristics of China's chain enterprises, so that we can establish a chain of retail mode in line with the development of China's national conditions [5].

After comprehensive analysis, we can improve from the following aspects.

5.1 Improving the efficiency of logistics distribution of chain enterprises in our country

From the successful experience of Japan 7- Eleven convenience store , the high efficiency of logistics and distribution, the building of joint distribution center and the rapid response of customer demands are the key to its success. Starting with the details, for example, the analysis of consumer data, optimal design of logistics distribution route, we need to set the improvement of the distribution efficiency as the top priority of the work.

5.2 Improving the construction and management level of the chain enterprise's information system in our country

The degree of information is not high and most of them are artificial, which is a common problem of many enterprises in our country. It seriously affected the rapid and healthy development of enterprises. Japan 7-Eleven company has made good use of advanced information technology, and established a comprehensive enterprise internal multi-faceted information management system to achieve the information sharing between the stores. If the chain enterprises in our country want to get rid of the shackles of the existing development, we should change the idea firstly, fully utilize advanced technology to establish a complete information system, and strengthen the monitoring, analysis and improvement of each links. At the same time, like the 7-Eleven, we ought to put the climate change, the changes in the market environment, the age of the customer base, customer's purchase preferences and other factors into the system to achieve stable old customers and constantly open up new customers.

5.3 Changing the existing business strategy to improve service level

Retail convenience stores are different from ordinary supermarkets. Convenience stores have longer service time as well as a small store, which are closer to life. But looking at the development of chain enterprises in our country, influenced by Chinese traditional concepts, the development still can not get rid of the family style business model and ideas. At the same time, the phenomenon of high cost of distribution, lack of funds and lack of core competitiveness still exist. Japan 7-Eleven convenience stores take a series of effective measures to reduce the cost, open stores and improve market share and distribution efficiency. Therefore the chain enterprises in our country should change the business strategy, pay attention to the purchasing ability, brand effect, traffic condition and the improvement of the logistics cost.

5.4 Improving the quality control of the distribution of goods

7-11 convenience is famous for operating a wide variety of goods, and the following chart gives the proportion of its sales of goods.

Table 1. The proportion of 7-Eleven store sales of different types of goods.

Type of commodity sales	Percentage (%)

Processed foods (snacks, beverages)	31.64
Snack foods (rice, lettuce, fried food, etc.)	30.25
Day sales of food (milk, yogurt, etc.)	12.75
Non food (magazine, daily necessities, game software, etc.)	25.36
Total	100

Data from the above chart can be seen that 7-Eleven convenience stores operate a wide variety of goods. Enterprises implement the logistics and distribution management in accordance with the different characteristics of different products, and thus ensure the freshness of products. The new management method is to establish distribution system according to different temperatures. 7-Eleven's temperature management of goods is mainly divided into four forms, as shown in Table 2 below.

Table 2. Types of temperature management of goods.

Type of temperature management	**The main application of goods**
Freezing type (20 degrees below zero)	Normal Initials followed by last name
Micro cold type (5 degree)	Normal Abbreviated
Normal temperature type	Italic
Warm type (20 degrees)	Bold

Different products have different temperature managements according to the characteristics, then it can meet the customers' requirements of product freshness and commodity quality. For local chain enterprises in our country, being strict with quality and temperature division to ensure product freshness and taste, then we can seize market share in the fierce market competition to win the magic weapon.

The Japan 7-Eleven convenience store chain becomes the world's largest chain enterprise due to its effective logistics strategy, comprehensive information system management strategy and efficient logistics distribution mode. According to the characteristics of the development of local chain enterprises in our country, we should draw the effective operation and distribution strategy of 7-Eleven convenience store. We can make some improvement from the distribution cost, service level, information construction and other aspects and we will be able to build a successful chain of supermarket industry with Chinese characteristics.

References

1. Baidu Encyclopedia.7-11 convenience store. Search engine.
2. Du Haiqing. Close to the region: Japanese 7-11 market. Value of China, (03/19/2006).
3. Cheng Yanfei. Japan 7-11 convenience store market expansion strategy of. information and computer, 64-65 (2010).
4. Xiao Chun, Wen Jinjun. The development of retail industry in China under the joint distribution mode - Japan 7-11 convenience store logistics innovation inspiration. Logistics technology, volume twenty-ninth, 29(11), 27-29 (2006).
5. Liu Jianmin.7-11 convenience store logistics operation of China's Chinese fast food logistics enlightenment. Guangdong Logistics technology, 25-27 (2008).

The Measurement of Urbanization Level Based on Entity Space: A Case Study of JingJinJi Region, China

Fenghua Wen[1,2,*] and Lirong Ren [2]

[1] School of National Fiscal Development, Beijing,China
[2] School of Government, Central University of Finance and Economics, Beijing,China
Corresponding author: wen_fh@163.com

Abstract. On the basis of analyzing the pros and cons of different urbanization level measurement methods, this paper puts forward urbanization level measurement method based on entity space and then uses entity space data to describe the urbanization rate of various districts and counties of JingJinJi Region in 2005 and 2009. According to the analysis of time and space differences on such basis, JingJinJi may be divided from high to low order according to the urbanization rate of the entity space: central urban area, urban development area and suburb, which respectively bear different functions. According to the comparison of entity space urbanization rate and population urbanization rate in the aspects of sequence, primacy ratio, zipf index, etc., we find the non-intensive land use status quo of urban development area of JingJinJi and the feature that the distribution of entity space urbanization scale is much uniform than distribution of population urbanization scale. Accordingly, the paper puts forward the following suggestions : (1) The space construction should depend upon function division; (2) The urban development area should pay attention to the construction of the supporting facilities and meanwhile enhance land use rate; (3) The urban construction should be enhanced in small and middle regions and counties surrounding Beijing and Tianjin.

1 Introduction

Karl Heinrich Marx firstly mentioned "rural urbanization" [1] when discussing urban and rural division and urban development in the *Critique of Political Economy* in 1858. In the following 1867, the Spanish Engineer Serda definitely proposed the conception of city urbanization in *Basic Principle of City Urbanization* and the conception became popular as world urbanization accelerated. Chinese scholars extended the conception as "urbanization" in consideration of Chinese actual conditions [2]. In regards to domestic definition of urbanization, the principles and methods of different subjects are different. Population urbanization refers to the population gathering course towards city and is the core of the urbanization [3]; land urbanization is the carrier of the urbanization and is the enhancement of urban space extension and land use efficiency enhancement in essence [4]; economic urbanization is the power of the urbanization and is the transition of economic structure in essence; social urbanization is the inevitable result of the urbanization and is population, land and economy urbanization and the change course of people's production form, life form, behavior habits and values [5]; ecological urbanization is the development form of the urbanization and is the external extension of new-type urbanization under sustainable development view[6]. To sum up, urbanization [7] is comprehensive urbanization of population scale, space extension, economic form, industrial structure, infrastructure construction and resource and environment support, etc.

The urbanization level measurement is one of important contents of urbanization research. Different subjects have different definitions about urbanization level. Accordingly, the measurement methods of the urbanization level show diversified. The urbanization level measured based on traditional population, land, economy, society and ecology has respective advantages and disadvantages. What's popularly used in the educational circles is the ratio of the urban population to the total population, namely population urbanization. The indicator is simple, clear, and easy for statistics and disposal and is of great practicability and is the mainstream method [8] to measure the urbanization level at present. The urban population here is also divided into registered population and permanent resident population. Different statistics methods may lead to different results. In addition, the change of the administrative division and population mobility also leads to the result difference [9]. Therefore, the educational circles also propose to use land measurement method and indicators such as economy, society, and ecology to measure the urbanization level. Land urbanization level refers to the ratio of urbanization area to the total area, namely urbanization level = (area of finished region/total area) x100%. The measurement method is simple and high pragmatic and can reflect the urbanization level in the angle of land use. The measurement result comparatively conforms to the facts. However, it may be influenced by population density, construction speed and quality of the urban area as well as the geological technology, especially monitoring and statistics of remote sensing technology. The data acquisition is difficult. In general, compound measurement method is adapted to measure economic, social and ecological urbanization level. For instance, the indicators of economic urbanization include ratio of the tertiary industry to GDP, total annual fiscal expenditure and per capita disposable income, etc. [6] When multiple indicators are adopted, the proportion of each indicator in the constitution of urbanization level is different. Therefore, it is necessary to firstly adopt clustering analysis and analytical hierarchy process (AHP), etc. in statistics to analyse the proportion of each indicator and then obtain the urbanization level after weighting. Since the diversity of the selected indicators, the result obtained with this method is relatively comprehensive and calculation method is rational. Hence, it can comprehensively reflect the urbanization level of certain region or certain city and the conclusion is scientific. However, due to the difference of the research area and research property, selected indicators, methods to obtain data, calculation methods and ratios of various indicators, the urbanization level of the same region may be different and even differ greatly and the vertical comparison of the same region in different years and the horizontal comparison of different regions in the same year are meaningless [10]. Since the retrievable conditions of various indicator data of various regions are different, it is hard to form comparatively general indicator system and measurement method.

Entity is geographic feature and geographic phenomenon in real life and mainly consists of point entity, line entity, surface entity and body entity. Space refers to regular or irregular Euclid space that the length and angle can be measured. Multiple space relationships such as topology, order, distance and orientation exist among entities. What's called entity space is the space structure of entity facilities. Urban entity refers to various urban facilities (for instance, house, building, road and pipeline) and urban landscape mainly consisting of non-farming land and non-farming economic activities. The distribution scope and space structure of these urban facilities and urban landscape are the urban entity space [11]. Here propose urbanization level measurement method of JingJinJi (short for Beijing, Tianjin and Hebei province) Region to calculate the entity space urbanization level and compare with traditional population urbanization level. The research scope is 207 districts and counties inside JingJinJi Region, including 18 districts and counties of Beijing, 18 districts and counties of Tianjin, and 171 districts and counties of 11 prefecture-level cities of Hebei Province [12].

2 Method and Data

The measurement method of entity space urbanization rate in JingJinJi Region: "entity space urbanization rate = entity space area/total area of administrative region*100%". Wherein, the area data of administrative region of districts and counties of JingJinJi Region in 2005 and 2009 were from Chinese County (Urban) Social Economic Statistics Yearbook 2006/2010. The area data of administrative region of Beijing and Tianjin were respectively from *Beijing Statistics Yearbook 2006/2010* and *Tianjin Statistics Yearbook 2006/2010*.

Meanwhile, through using ENVI software to handle MCD12Q1 remote sending image of JingJinJi region in 2000 and 2010, we can obtain vector data of pattern spot of constructed area of JingJinJi region. The handling steps are as below:

(1) Projection transformation: transform data into WGS84 coordinate system.

(2) Transform into vector data: resample the raster data and refer to materials "ENVI analysis on MCD12Q1 data of MODIS product" with resolution rate of 1000m. According to vegetation classification standard of University of Maryland, if dn is 13, it is urban area and constructed area. Extract the remote sensing image data that dn is 13 and it is the area of urban constructed region.

(3) Area calculation of residence zone

After extracting residing area image from MODIS image, transform raster data of residing area into vector data and calculate the area of the residing region. Since the resolution rate of the data is 1000m, the area of the constructed region should be at least $1km^2$. Secondly, the author considers that the area of the constructed region where the urbanized population is distributed is not smaller than $1km^2$.

The pattern spot data of 2000 and 2010 JingJinJi Region obtained through processing are as below:

Fig.1. Area of Entity Space of JingJinJi Region in 2005 (Left) and 2009 (Right).

3 Result

According to above urbanization rate calculation formula of the entity space, extract the area data of entity space and area data of administrative region of various districts and counties of JingJinJi Region in 2005 and 2009, correct the original data, analyze and process and obtain the urbanization level of the entity space of the districts and counties of JingJinJi Region in 2005 and 2009.

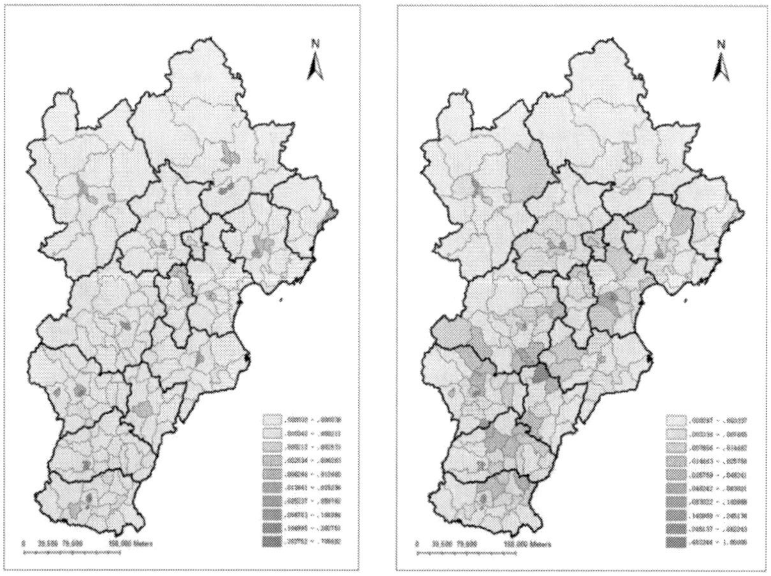

Fig.2. Entity Space Urbanization Rate of Districts and Counties of JingJinJi Region in 2005 (Left) and 2009 (Right).

Among various districts and counties of JingJinJi Region in 2005, the urbanization level of the entity space of the districts and countries was much higher, namely Dongcheng District, Xicheng District, Chongwen District, Xuanwu District (Present Dongcheng District and Xicheng District) of Beijing, Heping District, Hedong District, Hexi District, Nankai District, Hebei District and Hongqiao District of Tianjin, Changan District of Shijiazhuang, Lunan District and Lubei District of Tangshan, Haigang District, Shanhaiguan District and Beidaihe District of Qinghuangdao, Hanshan

DOI: 10.1051/matecconf/201710005046

District and Congtai District of Handon, Qiaodong District and Qiaoxi District of Xingtai, Xinshi District, Beishi District and Nanshi District of Baoding, Qiaodong District, Qiaoxi District, Xuanhua District and Xiahuayuan District of Zhangjiakou, Shuangqiao District and Shuangluan District of Chengde, Xinhua District and Yunhe District of Cangzhou, Anci District and Guangyang District of Langfang, and the urbanization level of entity space of other districts and counties was very low.

In 2009, the urbanization level of the entity space of the districts and countries was much higher, namely Dongcheng District, Xicheng District, Chongwen District, Xuanwu District of Beijing, Heping District, Hedong District, Hexi District, Nankai District, Hebei District and Hongqiao District of Tianjin, Changan District, Qiaodong District, Qiaoxi District, Xinhua District and Jingxingkuang District of Shijiazhuang, Lunan District, Lubei District, Guye District and Kaiping District of Tangshan, Haigang District, Shanhaiguan District and Beidaihe District of Qinghuangdao, Hanshan District, Congtai District, Fuxing District and Fengfengkuang District of Handon, Qiaodong District and Qiaoxi District of Xingtai, Xinshi District, Beishi District and Nanshi District of Baoding, Qiaodong District, Qiaoxi District, Xuanhua District and Xiahuayuan District of Zhangjiakou, Shuangqiao District, Shuangluan District, Yingshouyingzikuang District of Chengde, Xinhua District and Yunhe District of Cangzhou, Anci District and Guangyang District of Langfang, Taocheng District of Hengshui and the urbanization level of entity space of other districts and counties was very low.

4 Discussion

In the spatial and temporal difference analysis on the urbanization rate of entity space of various districts and countries of JingJinJi, we should firstly analyze according to the time change law, then analyze the space region difference and finally compare the entity space urbanization rate and the population urbanization rate; find out difference in aspects of urbanization order, primacy ratio and zipf indicator and research the distribution features of the entity space urbanization rate.

4.1 Time change law of entity space urbanization rate

According to the calculation, the overall entity space urbanization rates of JingJinJi Region in 2005 and 2009 were respectively 0.766% and 0.782%. It was on the rise but enhanced slowly. The detailed analysis on districts and counties is as below.

(1)The overall change trend of the entity space urbanization rate is stable;

(2)The development of economic and technological development zone has obviously enhanced the entity space urbanization rate;

(3)The change of administrative region is the main reason for the drop of the urbanization rate of some districts and counties.

4.2 The regional difference of the entity space urbanization rate is obvious

According to the entity space urbanization level of JingJinJi Region, in consideration of the development features of various districts and counties, and following the high-to-low order of entity space urbanization level, all districts and counties of JingJinJi Region are divided into three teams: central urban area, urban development area and suburb.

Table 1. District Standard of Entity Space Urbanization Rates.

District Type	Entity Space Urbanization Rate Section
Central Urban Area	[10%,1)
Urban Development Area	[1%,10%)
Suburb	(0,1%)

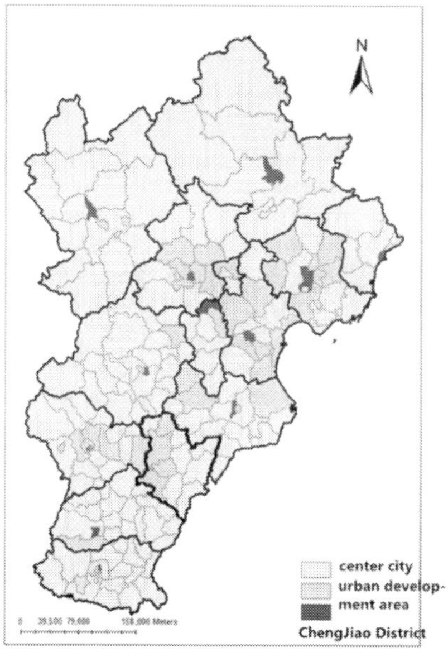

Fig.3. Entity Space Urbanization Rate District Figure of JingJinJi in 2009.

We can found that:

(1)Central urban area has perfect facilities and bears urban core functions.

(2)Urban development area bears urban extension function.

(3)Suburb bears ecological conservation function.

4.3 Comparison of entity space and population urbanization rate

During the urbanization course, the population change and space extension are two kinds of expressions [13]. As Chinese urbanization schedule accelerates, the virtual high population urbanization, disordered urban space construction and even out-of-control problem will become gradually extruded [14]. The paradox phenomenon of urban shrinking appears [15]. According to the comparison of entity space urbanization rate and traditional population urbanization rate in the aspects of urbanization order, primacy ratio and zipf indicator, we find the population growth and entity space extension law during the urbanization course in JingJinJi Region.

(1)The population of three teams is inconsistent with the entity space urbanization order.

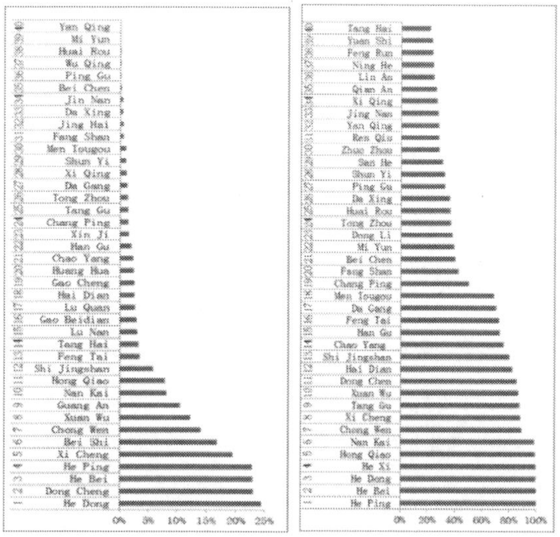

Fig.4. Population (Left) and Entity Space (Right) Urbanization Order (The First 40) of the Districts and Counties of JingJinJi Region in 2009.

(2)The population and entity space urbanization rate of central urban area and suburb is consistent.

Fig.5. Comparison of Population Urbanization Rate and Entity Space Urbanization Rate of Representative Districts and Counties.

(3)The land use of urban development area is not intensive.The population and entity space urbanization rate in urban development area are not consistent and shows two different types: (1) the entity space urbanization rate is faster than population urbanization rate; (2) the entity space urbanization rate is slower than the population urbanization rate. Now two situations are classified and discussed as below.

(4)The primacy ratio of population urbanization is higher than the primary rate of entity space.

The common index that balances the distribution of the regional city scale is urban primacy ratio and was proposed by American scholar Jefferson at earliest in 1939: "the ratio of the population of the largest city to the population of the second largest city of one country should be the primacy ratio". In the 1980s, the scholars Yuemin Ning and Chongmin Yan introduced the primacy ratio conception into China [16] and later the scholars deepened the conception of the primacy ratio and proposed to reflect the scale of urban other functions in one region with economic primacy ratio, industrial primacy ratio, technological primacy ratio, talent primacy ratio and cultural primacy ratio [17]. Yixing Zhou improved the primacy ratio index and proposed much effective and rigorous four-city index and eleven-city index [18], namely: four-city index(1) and eleven-city index(2),Pn refers to the population scale of the nth city.

$$S=P1/(P2+P3+P4) \tag{1}$$

$$S=2P1/(P2+P3+\ldots\ldots+P11) \tag{2}$$

Introduce the conception of the primacy ratio into urbanization level and calculate the primacy ratio of the population and entity space urbanization of JingJinJi Region, listed in the following table. The two-city index of the primacy ratio of the population urbanization of the districts and counties of JingJinJi Region is close to 1, the four-city index is close to 1/3 and the eleven-city index is close to 2/11, meaning that the population urbanization level of JingJinJi Region almost shows "multi-core" status with the second-order districts and counties and also conforming to the fact that the population urbanization rates in the central urban area of Beijing and Tianjin are not very different. However, the two-city index of the primacy ratio of the entity space urbanization is 1.36 and the four-city index and eleven-city index are all higher than the primacy ratio of the population urbanization. The "multi-core" status of the entity space urbanization level is no longer obvious. It reflects that according to the integration degree of the districts and counties, the integration of the population urbanization of the central urban area has tended to maturity, while there is still development space for the integration of the entity space urbanization.

Table 2. Primacy Ratio of Urbanization of JingJinJi Region.

Index	Primacy Ratio of Population Urbanization	Primacy Ratio of Entity Space Urbanization
Two-city Index	1.00032	1.36018
Four-city Index	0.33392	0.47296
Eleven-city Index	0.23545	0.41486

(5)The zipf index of the entity space urbanization is higher than that of the population urbanization.

Order—the most general scale rule is zipf rule, namely(3). After being changed for logarithm form, it is(4) . Wherein, Pi is the population of the ith city; Ri is the city order; q is Zipf index [19]. Through researching the size of the Zipf index, it may reflect the urban scale and form. The larger the q value is, the much uniform the city distribution is. If q=+∞, the city distribution is equal-size and uniform. The ideal city system according to Zipf should satisfy q=1[20].

$$Pi=P1*Ri^{-q} \quad (3)$$

$$LnPi=LnP1-qLnRi \quad (4)$$

Introduce Zipf rule into urbanization rate and use Pi to express the urbanization rate of the ith city. Adopt formula(5) and regression analysis to respectively calculate the Zipf index q of the population urbanization rate and the entity space urbanization rate in 2005 and work out that the Zipf index of the population urbanization is 0.5051 and the Zipf index of the entity space urbanization is 1.8076. The zipf index of the entity space urbanization is obviously higher than that of the population urbanization. It means that the scale distribution of the entity space urbanization is much uniform than that of the population urbanization.

$$LnPi=LnP1-qLnRi \quad (5)$$

5 Conclusion

On the basis of researching the measurement method of the traditional population, land, economy, society and ecology urbanization level, the paper proposes urbanization level measurement method based on entity space. Calculate the entity space urbanization rates of the districts and counties of JingJinJi Region in 2005 and 2009 according to the formula "urbanization level = area of entity space/area of administrative region*100%" and analyze the time and spatial difference to work out that the change of the administrative division and the development of the economic and technological development zone are the main factors to influence the change of the entity space urbanization rates of the districts and counties of JingJinJi Region during 2005-2009. Divide the districts and counties of JingJinJi Region into three teams according to the high-to-low order of the entity space urbanization rate: central urban area, urban development area and suburb, and find that the three areas respectively bear different city functions. Compare the entity space urbanization rate and the population urbanization rate of the districts and counties of the three teams in the aspects of order, primacy ratio and zipf, etc. and find the features: in the urban development area of JingJinJi Region, the land use is not intensive; the primacy ratio of the population urbanization is higher than that of the entity space urbanization rate; the entity space urbanization rate scale distribution is much uniform. According to above findings, the paper proposes following advices about the integration development of Beijing-Tianijn-Hebei Region: (1) the space construction should be decided according to the function division. The city division according to the entity space urbanization rate and the function division according to state policy are basically consistent. The central area bears core functions; the development area bears external extension function; the suburb bears ecological conservation function. (2) The urban development area should pay attention to the construction of supporting facilities as well as enhancing the land use rate. Change the extensive land use form of the urban development area, improve infrastructure and meanwhile absorb population through corresponding supporting industries and let population urbanization and entity space urbanization consistent. (3) Enhance city construction and enhance the entity space urbanization rate in small and middle districts and counties surrounding Beijing and Tianjin.

Acknowledgements

Supported by the Project of The National Natural Science Foundation "Urban Hierarchical System Evolution and Mechanism Study Based on Population Migration Network: Take Beijing-Tianjin-Hebei Region as Example" (No.: 41301178) and the Project of School of National Fiscal Development of Central University of Finance and Economics "Research on Local Governmental Behaviors during Chinese Urbanization Strategy Course" (No: 024050314002/004).

References

1. LIU Jiehong,On Urbanization.Journal of Northwest A&F University(Social Science Edition), 2009,9(4): 58-62.
2. LIU Yaolin,LI Jiwei,HOU Heping, LIU Yan-fang, Study on urbanization rate of urban-rural construction land and its influencing factors: A case study of Hubei Province. Geographical Research,2014,33(1):132-142.
3. GU Chaolin,WU Liya,Productions of the Research of China's Urbanization, Urban Problems,2008(12):2-12.
4. WANG Yang,WANG Shaojian, QIN Jing,Spatial evaluation of land urbanization level and process in Chinese cities,Geographical Research,2014（12）：2229-2238.
5. XU Xueqiang, ZHOU Yixing,NING Yue-min, Urban Geography .Bei Jing:Higher Education Press,2010.
6. CHEN Mingxing.Research progress and scientific issues in the field of urbanization. Geographical Research,2015（4）:614-630.
7. Lu Dadao. The research content framework of urbanization field in Geography. Scientia Geographica Sinica, 2013, 33(8): 897-901.
8. CHEN Mingxing, LU Dadao, ZHANG Hua.Comprehensive Evaluation and the Driving Factors of China's Urbanization. Acta Geographica Sinica,2009(4):387-398.
9. WU Jie,QUAN Shaowei.Discussion of China's population urbanization rate statistics and calculation method.World Research ,2013,(7):44-47.
10. HENG Wen-sheng,WANG Xiao-fang,LI Cheng-gu,the Spatial Disparties of Regional Comprehensive Urbanization level of Vice Province City in China from 1997,Economic Geography, 2007(2).256-260.
11. HU Li. Definition of Urban Proper Scope Based on ETM density of construction land—A Case Study of Dong Guan City.Guang Zhou:South China Normal University.2010.
12. YANG Dongfeng,LONG Ying，YANG Wenshi,SUN Hui,Losing Population with Expanding Space:Paradox of Urban Shrinkage in China. Modern Urban Research.2015(9):20-25.
13. LIU Fawei，XU Hengzhou,WANG Shu,Analysis on Spatial- temporal Coupling Coordinate Degree among Population，Land and Economy Urbanization: Based on China Provincial Panel Data,Urban Development Studies, 2014（8）：7-11.
14. YIN Hongling,XU Teng,The Mismatch between Population Urbanization and Land Urbanization in China,Urban Planning Forum, 2013(2)：10—15.
15. LONG Ying, WU Kang,WANG Jianghao. Shrinkage Cities in China, Modern Urban Research, 2015(9):14-19.
16. YAN Zhongmin,NING Yuemin,Changes of the urban population development of China//Research on population, Shang Hai:East China Normal University Press,1981：20—37.
17. LU Xuefa,SHEN Huifang.The Status Quo and Countermeasure Research of the first degree of Hang Zhou city Zhejiang Statistics，2008(6)：32—34.
18. ZHOU Yixing，YANG Qi，A Review on the Urban-size Hierarchy of China and the Territorial Types of the Hierarchy on Provincial Level. Acta Geographica Sinica, 1988，(2)：97 — 111.
19. Zipf,G.K. Human Behavior and the principle of Least Effort. Addison—Wesley, Reading, MA.
20. CHEN Kaiming，ZHUANG Yanjie, Spatial Econometric Analysis of The Rank-size Rule for Urban System:A Case of Prefectural-level cities in China's Middle Area, Scientia Geographica Sinica, 2012.8（8）:905-912.

Comprehensive Transportation Logistics Network Level Layout Based on Principal Component Factor and Cluster Analysis

Jingrong Zhang

School of Management Engineering, Zhengzhou University, Zhengzhou Henan, China
Corresponding author: 13676995513@139.com

Abstract. Comprehensive Transportation Logistics Network (CTLN) acts as a crucial prop and fundamental carrier for regional economic and social development. Firstly, an index system for evaluating the development of regional Comprehensive Transportation Logistics (CTL) nodes is established; then regional CTLN nodes are ranked according to their importance by the method of Principal Component Analysis(PCA), and main factors affecting the development of regional CTL nodes are analyzed by applying factor analysis, and regional CTL nodes are classified according to their feature similarities by applying cluster analysis; and then level structure of constructing regional CTLN is proposed. Finally, combined with geographic locations of different nodes, level layout model of CTLN of the whole region is obtained. Taking Henan province this region as an instance, level layout model of hub-and-spoke CTLN taking Zhengzhou at its core is proposed after analysis, providing a reference basis for constructing CTLN in whole province scientifically and reasonably.

1 Introduction

Transportation logistics remains a vital precursor and basis for the development of modern logistics, also a key link and crucial component for the development of regional logistics, as well as acts as the significant driving force for developing regional manufacturing industry and commercial circulation business. And transportation logistics network is an important prop and fundamental carrier for the development of regional economy and society and for joint development of logistics industry and manufacturing industry. With constant development and integration of modern logistics and transportation, as well as with deepening supply-side structural reform in transportation logistics field, transportation logistics is striding to an advanced stage of CTL where various modes of transportation are balanced, cooperated according to their functions and integrated organically, meanwhile transportation logistics network during construction is approaching to an advanced stage of seamless, interconnected and reasonably arranged CTLN.

In recent years, studies on transportation logistics network have been conducted by many scholars from various respects, and many achievements have been gained. For instance, Shi Jin et al. [1] have discussed integrated construction of cross-border subregional logistics transportation, Chen Wenqiang et al. [2] have probed into reasonable pricing of toll roads in logistics transportation network by applying game theory approach; He Qichao et al. [3-5] have inquired into influences to regional logistics network planning from the views of carbon tax constraint and transportation cost; Han Shilian et al. [6-8] have studied route planning, scheduling and optimizing issues concerning transportation logistics network; Zhang Shixiang et al. [9-11] have studied the construction and cases of hub-and-spoke logistics network respectively.

However, throughout existing study achievements, documents concerning transportation logistics network layout from the respect of comprehensive transportation are not retrieved. Given this, the present paper proposes nodes ranking of regional CTL according to their importance by the method of PCA, then main factors affecting the development of regional CTL nodes are analyzed by the method of factor analysis, and regional CTL nodes are classified according to their similarities by the method of cluster analysis. Thereafter, based on network nodes importance ranking, influencing factors analysis and similarity classification, level structure of regional CTLN is proposed, then combined with geographical location of various nodes, level layout model of CTLN of the whole region is obtained, providing a reference basis for constructing regional CTLN system scientifically and reasonably.

2 CTL evaluation index system

To measure CTL development level in different nodes, and in line with index selecting principles of interpretability, accessibility, comparability, comprehensiveness and completeness, an evaluation index system is established after consulting abundant documents, see Table 1.

Table 1. CTL evaluation index system.

Primary index	**Secondary index**
Node Comprehensive Development Level A1	Total Output Value B1
	Per Capita Total Output Value B2
	Fixed Asset Investment B3
CTL Market Size A2	Total Society Freight Volume B4
	Freight Turnover B5
	Total Retail Sales of Social Consumption B6
	Total Export Import Volume B7
	Added Value of Industries above a Designated Scale B8
	Business Volume of Postal and Telecommunication Services B9
	Volume of Telecommunication Services B10
CTL Facility Level A3	Length of Highways on Operation B11
	Railway Lines Transit Condition B12
	Inland Navigation Condition B13
	Air Cargo Loading and Unloading Capacity B14
	National Highway Freight Hub Condition
	National Railway Logistics Base Condition

Wherein, node comprehensive development level (A1): reflects economic and social foundation of CTL development in different nodes, including total output value (B1), per capita total output value (B2) and fixed asset investment (B3).

CTL market size (A2): reflects CTL development level and size in different nodes, including total society freight volume (B4), turnover volume of freight transport (B5), total retail sales of social consumption, total export import volume (B7), added value of industries above a designated scale (B8), business volume of postal and telecommunication services (B9) and volume of telecommunication services (B10).

CTL facility level (A3): reflects material foundation and facility condition of CTL development in various nodes, including length of highways on operation (B11), railway lines transit condition (B12), inland navigation condition (B13), air cargo loading and unloading capacity (B14), national highway freight hub condition (B15) and national railway logistics base condition (B16).

3 CTL nodes ranking according to their importance based on PCA

Though indexes being screened, a certain degree of correlation and information overlapping still exist. While by PCA [12], original indexes can be replaced by fewer comprehensive principal component indexes through dimensionality reduction, where new indexes have reserved most information of original ones and new indexes are independent and uncorrelated with each other, getting issues simplified. Therefore, PCA can be applied in comparing and ranking CTL development level in different nodes to determine different importance degree of various nodes in CTLN.

Supposing, CTL evaluation sample matrix : $X^* = (x^*_{ij})_{m\times n}$, $i = 1,2,\cdots,m$, $j = 1,2,\cdots,n$ is obtained with n indexes of m nodes. Firstly, indexes are standardized through formula (1), then standard evaluation matrix is obtained: $X = (x_{ij})_{m\times n}$

$$x_{ij} = (x^*_{ij} - \overline{x_j^*}) / s_j^* \tag{1}$$

Wherein, $\overline{x_j^*}$ and s_j^* presents sample mean of the jth index and sample standard deviation respectively.

Secondly, calculate correlative coefficient matrix $R_{n\times n}$ among indexes and its eigenvalues $\lambda_1 \geq ... \geq \lambda_n \geq 0$ as well as regularized eigenvector e_j;

Then, principal component is obtained

$$Y_j = (y_{ij}) = Xe_j \tag{2}$$

Wherein, variance contribution rate of the jth principal component is $\alpha_j = \lambda_j / n$. When accumulated variance contribution rate $\alpha = \sum_{j=1}^{k} \alpha_j$ reaches a certain value (generally ≥85%), it can be regarded that the first k principal components Y_1 - Y_k have embodied information of original indexes comprehensively with fewer indexes.

Then, taking variance contribution rate α_j of the first k components as weight, comprehensive evaluation function is obtained through linear weighing sum

$$z_i = \sum_{q=1}^{k} \alpha_q y_{iq}, \quad q = 1,2,...,k \leq n \tag{3}$$

A higher value of z_i indicates higher comprehensive strength and competitiveness, namely a higher importance degree of this city as logistics node among the whole regional CTLN; while a lower value of z_i indicates lower comprehensive strength and competitiveness.

4 Influencing factor analysis of ctl development based on factor analysis

Factor analysis [13] applies a few abstract common factors to express data structure of original indexes through dimensionality reduction and simplification. Compared with PCA, factor analysis is more explanatory, therefore, it is usually conducted on the basis of PCA to study main influencing factors of index variables. Thus factor analysis model further inspecting main factors influencing the development of regional CTL nodes is obtained:

$$Y_j = \mu_j + A_j F_k + \varepsilon_j, \ j = 1,2,\cdots,n, \ k \leq n \tag{4}$$

Wherein, μ_j is mean value, A_j is factor loading matrix, F_k is common factor, and ε_j is specific factor.

To make common factors more explanatory, factor matrix is usually rotated. What the present paper applies is the commonest varimax orthogonal rotation.

5 CTL nodes similarity analysis based on cluster analysis

Cluster analysis [14] gradually classifies individuals or groups according to feature similarities of their index variables to obtain classification system that can reflect close or distant relationship among individuals or groups. The first k principal components are adopted to analyze and classify CTL nodes according to their feature similarities, which greatly simplifies the present issue.

Firstly, the first k principal components comprise a new matrix $Y^* = (y^*_{iq})_{m\times k}$ and distances among new sample data of various nodes are calculated according to the formula (5). What the present paper applies is the commonest Euclidean distance. A smaller distance coefficient indicates a higher feature similarity level of CTL development among nodes, namely, a higher similarity level among CTL nodes; While a lower one indicates a lower feature similarity level.

$$d_{iq} = \left[\sum_{q=1}^{k} (y^*_{iq} - y^*_{jq})^2 \right]^{1/2}, \ q = 1,2,\cdots,k \leq n \tag{5}$$

6 CTLN level layout

In the present paper, different nodes in CTLN are compared and ranked according to their comprehensive competitiveness and importance by the method of PCA; main factors affecting the development of different nodes in CTLN are analyzed by the method of factor analysis; and different nodes in CTLN are compared and classified according to their similarity degree and similar features by the method of cluster analysis.

Based on the above-mentioned analysis to importance ranking, influencing factor analysis and similarity classification of network nodes, corresponding level and category of various nodes in CTLN can be proposed, therefore network level structure of regional CTLN can be determined. Then combined with geographical locations of different nodes, level layout model of CTLN in the whole region can be determined.

7 Case analysis: henan province CTLN level layout

Henan province is a vital comprehensive transportation hub in China, thus there is the old saying "who gains the Central Plains gains the world" since ancient times. In national strategic planning, one important position of Henan is "national important comprehensive transportation hub and modern logistics center" [15]. At present, economic and social development in this province is increasingly accelerated, and the level of opening up rises gradually. The implementation of "One Belt One Road" strategy, the construction of Central Plains economic zone, as well as gradually deepening implementation of Henan free trade zone, Zhengzhou airport comprehensive experimental area and other national strategies all have put forward new requirements to the construction of CTLN in Henan province.

7.1 Related data

Latest data of this study (2015), are mainly from two ways:

(1) From statistical material. The following data can be obtained from Henan Province Statistic Yearbook and statistical bulletin of each city respectively: total output value (B1), per capita total output value (B2), fixed asset investment (B3) in node comprehensive development level (A1); total society freight volume (B4), turnover volume of freight transport (B5), total retail sales of social consumption (B6), total export import volume(B7), added value of industries above a designated scale (B8), business volume of postal and telecommunication services (B9) and volume of telecommunication services (B10) in CTL market size (A2); as well as length of highways on operation (B11), inland navigation condition (B13), air cargo loading and unloading capacity (B14) in CTL facility level (A3).

(2) Based on expert scoring. The three qualitative indexes: railway lines transit condition (B12), national highway freight hub condition (B15) and national railway logistics base condition (B16) can be determined by expert scoring on highway and railway infrastructure construction according to specialty statistic materials of each city and "National Highway Transportation Hub Layout Planning" from former Ministry of Transportation and "Railway Logistics Base Layout Planning & 2015-2017 Construction Planning" from China Railway Corporation. In expert scoring, interval values from 0-10 is adopted, for instance, Zhengzhou is crossed by "米"shaped main line railway, therefore its score is 10; city has no railway, then the score is 0; similarly the scores of other cities can be obtained by the same method.

7.2 CTL nodes ranking according to their importance

Firstly, data to be analyzed is standardized, and then corresponding eigenvalue and variance contribution rate of each principal component are obtained by applying principal component operation in SPSS, see Table 2.

Table 2. Corresponding eigenvalue and variance contribution rate of each principal component.

Principal compone nt	Eigenvalu e	Variance contributio n rate α_j	Cumulative variance contribution rate α
Y1	9.910	61.939	61.939
Y2	2.903	18.143	80.081
Y3	1.004	6.276	86.357
Y4	0.601	3.758	90.115
Y5	0.529	3.307	93.423
Y6	0.408	2.551	95.974
Y7	0.221	1.382	97.356
Y8	0.181	1.134	98.490
Y9	0.128	0.802	99.292
Y10	0.067	0.419	99.711
Y11	0.028	0.176	99.887
Y12	0.009	0.058	99.945
Y13~Y16	≈0	≈0	100

To reduce information loss as much as possible and to ensure the following operation close to initial condition to the greatest extent, the first 6 principal components are selected, then the accumulated variance contribution rate is $\alpha = 95.97\%$, namely the 6 principal components have reflected the information of original indexes with the precision of 95.97%. Then obtain

$$z = 0.619Y1 + 0.181Y2 + 0.063Y3 + 0.038Y4 + 0.033Y5 + 0.026Y6 \quad (6)$$

Table 3 presents nodes comprehensive ranking according to their importance in Henan CTLN. Since centralized standardized processing method is adopted, there exist plus z_i values and minus z_i values.

Table 3. CTL nodes ranking according to their importance.

City	Zhengzhou	Luoyang	Nanyang	Zhoukou	Xinyang	Xinxiang	Shangqiu	Anyang	Jiaozuo
Score	6.86	1.74	1.66	0.62	0.61	0.22	0.18	-0.25	-0.4
Rank	1	2	3	4	5	6	7	8	9
City	Xuchang	Zhumadian	Luohe	Pingdingshan	Kaifeng	Puyang	Sanmenxia	Hebi	Jiyuan
Score	-0.7	-0.71	-0.85	-0.89	-1.2	-1.42	-1.43	-1.9	-2.13
Rank	10	11	12	13	14	15	16	17	18

According to Table 3, CTL nodes according to their importance in this whole province can be sorted into 3 levels as shown in Table 4. Wherein, the score of Zhenghou is far ahead, belonging to the first level; Luoyang, Nanyang, Zhoukou, Xinyang, Xinxiang, Shangqiu as CTL nodes with general importance fall into the second level (z>0); An'yang, Jiaozuo, Xuchang, Zhumadian, Luohe, Pingdingshan, Kaifeng, Puyang, Sanmenxia, Hebi and Jiyuan as CTL nodes with lower importance fall into the third level (z<0).

Table 4. CTL nodes importance level.

Level	CTL node
The first level	Zhenghou
The second level	Luoyang, Nanyang, Zhoukou, Xinyang, Xinxiang, Shangqiu
The third level	An'yang, Jiaozuo, Xuchang, Zhumadian, Luohe, Pingdingshan, Kaifeng, Puyang, Sanmenxia, Hebi and Jiyuan

7.3 Main Influencing factors of CTL nodes development

Since common factors usually realize their explanation to original indexes through conceptual abstraction, it is inadvisable to adopt too many common factors, and generally accumulated variance contribution rate is $\alpha \geq 80\%$. On the basis of the above-mentioned PCA, factor loading matrix is orthogonally rotated by maximum variance, and then factor loading and common factor variance after orthogonal rotation are obtained as shown in Table 5. To make analysis more convenient, the first 3 common factors are extracted, and the factor variance contribution rate reaches 92%, satisfying analysis requirements well [12,13].

In Table 5, factor loading matrix has provided correlation coefficient of each index on the first 3 common factors. A greater correlation coefficient indicates greater importance and influence of this index on certain common factor, while communality describes the contribution of all common factors make to the total variance of index variables. Meanwhile, communalities of most variables in Table 5 are high after extracting common factors, implying that most information is reserved while original variables are rotated to factor variables.

Table 5. Factor loading matrix after orthogonal rotation.

Index	**B1**	**B3**	**B6**	**B7**	**B8**	**B9**	**B10**	**B11**
Factor 1	**0.944**	**0.953**	**0.951**	**0.887**	**0.914**	**0.971**	**0.955**	-0.002
Factor 2	-	-0.219	-0.278	0.299	-0.025	-0.114	-0.203	**-0.945**
Factor 3	0.180	0.097	0.062	0.098	0.217	-0.001	0.144	0.247
Communality	**0.974**	**0.966**	**0.985**	**0.885**	**0.883**	**0.956**	**0.974**	**0.953**
Index	**B12**	**B14**	**B15**	**B4**	**B5**	**B2**	**B13**	**B16**
Factor 1	0.132	0.233	0.127	0.427	0.059	0.763	-0.002	0.749
Factor 2	-	**-0.907**	**-0.932**	-0.323	-0.182	-0.383	-0.945	-0.451
Factor 3	0.139	0.271	-0.029	**0.868**	**0.956**	0.247	0.247	0.264
Communality	**0.962**	**0.950**	**0.886**	**0.877**	**0.950**	**0.790**	**0.953**	**0.834**

Note：KMO=0.776>0.5，Sig.=0.000<0.05。

From Table 5, it can be known that there are 3 principal factors (namely common factors) affecting development comprehensive competitiveness of regional CTL nodes. The first common factor is determined by total output value (B1), fixed asset investment (B3), total retail sales of social consumption (B6), total export import volume (B7), added value of industries above a designated scale (B8), business volume of postal and telecommunication services (B9), volume of telecommunication services (B10), which can be defined as economic development factors. The second common factor is determined by length of highways on operation (B11), railway lines transit condition (B12), air cargo loading and unloading capacity (B14), which can be defined as comprehensive transportation infrastructure factors. The third common factor is determined by national highway freight hub condition (B15) and freight turnover (B5), which can be defined as CTL supply and demand factors, and these two indexes not only reflect the supply and demand in CTL development, but directly present CTL development.

7.4 CTL nodes similarity analysis

New sample matrix is formed by the obtained 6 principal component indexes, and then cluster analysis is conducted by applying Ward cluster approach according to Euclidean distance in SPSS. When Euclidean distance is 20, cities in Henan province as CTL nodes can be divided into 3 categories based on Euclidean distance similarity, as shown in Figure 1.

Wherein, the first category only includes Zhengzhou; the second category includes Zhoukou, Xinyang, Luohe, Shangqiu, Xinxiang, Nanyang and Luoyang; the third category includes Jiaozuo, Jiyuan, Sanmenxia, An'yang, Pingdingshan, Zhumadian, Kaifeng, Puyang, Xuchang and Hebi.

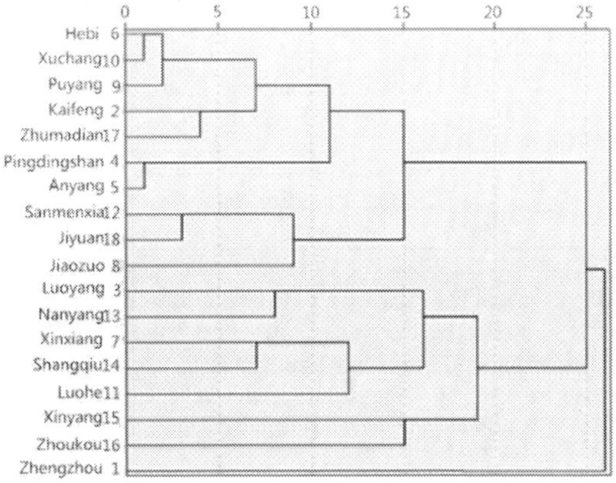

Fig.1. CTL nodes cluster figure.

Table 6. CTL nodes similarity category.

Category	CTL node
The first category	Zhengzhou
The second category	Luoyang, Nanyang, Xinxiang, Shangqiu, Luohe, Xinyang, Zhoukou
The third category	Jiaozuo, Jiyuan, Sanmenxia, An'yang, Pingdingshan, Zhumadian, Kaifeng, Puyang, Xuchang, Hebi

(1) Zhengzhou, the provincial capital of Henan province, is the political, economic and cultural center of this province, as well as the comprehensive transportation and communication hub equipped with national highway, railway, aviation and communication, and also is one of the 21 national logistics node city. Many industrial and commercial enterprises cluster in Zhengzhou, therefore there are large number and size of transportation logistics enterprises. At present, Zhengzhou is being constructed as a national central city and international logistics center, whose strategic position in future will be more prominent. Therefore, only Zhengzhou falls into the first category.

(2) Luoyang, Nanyang, Xinxiang, Shangqiu, Luohe, Xinyang and Zhoukou possess favorable geographical location and transportation condition and are crossed by "┼" shaped national main line railway and high-grade highway, belonging to national highway transportation junctions planned by former Ministry of Transportation. And all of these cities enjoy fine economic and social development and have solid industrial base and dynamic social consumption. Except Luohe, these cities possess large administrative areas; Xinyang, Zhoukou and Luohe also have water transportation. Therefore, these cities are sorted into the same category.

(3) Though Jiaozuo, Jiyuan, Sanmenxia, An'yang, Pingdingshan, Zhumadian, Kaifeng, Puyang, Xuchang and Hebi also are crossed by national main line railway and high-grade highway, yet the shape generally is "—"; and they are far away from Zhengzhou, except Kaifeng and Xuchang, some of them locate in the border of two provinces; these cities have not so large area and population. Therefore location and transportation advantages of these cities are faded, and transportation logistics radiating capacity is lowered. But these cities have their own features in economic and industrial development. Thus these cities are sorted into the same category.

7.5 Henan CTLN level layout

Based on the above mentioned analysis, it can be seen that: there are many similarities in analysis results in nodes importance ranking (Table 4) and nodes similarity clustering (Table 6). Except the divergence in recognition of "Luohe" in the second level (the second category), other levels (categories) are the same. The difference in Table 4 and Table 6 is caused by different study emphasis placed by the two methods. PCA lays its emphasis on the study of comprehensive competitiveness and importance of each node in CTL development, while cluster analysis places its study emphasis on feature similarity among nodes in CTL development.

Then Henan CTL level structure is obtained as shown in Table 7. Wherein, Zhengzhou is far ahead others as the first level node of CTL; Luoyang, Nanyang, Zhoukou, Xinyang, Xinxiang and Shangqiu possess obvious advantages in regional transportation, economic and social development as the second level nodes of CTL; Though Luohe is similar to Luoyang and other five cities to certain extent, its node importance in CTL is low, thus it together with other cities falls into the third level nodes.

Table 7. CTLN level structure.

Node	CTL node
The first level node	Zhengzhou
The second level node	Luoyang, Nanyang, Zhoukou, Xinyang, Xinxiang, Shangqiu
The third level node	An'yang, Jiaozuo, Xuchang, Zhumadian, Luohe, Pingdingshan, Kaifeng, Puyang, Sanmenxia, Hebi, Jiyuan

Further more, combined with geographical locations of different cities in Henan, hub-and-spoke CTLN layout can be established with Zhengzhou as its axis city, Luoyang, Nanyang, Zhoukou, Xinyang, Xinxiang and Shangqiu as its radiating cities and other cities supporting nodes as shown in Figure 2. Modern CTLN with Henan characteristics where various modes of transportation are closely connected, efficiently linked up and functions are complemented, supporting the rapid economic and social development in Henan province.

Fig.2. Henan CTLN level layout diagram.

8. Conclusion and Suggestion

(1) The primary factor influencing CTL development is regional economic development level, namely, transportation logistics serves for regional economic development and also is benefited from regional economic development.

(2) Departments of transportation logistics should fully consider various development features of CTL in different areas, and issue corresponding guiding measures to promote joint development and overall uplift of CTL in the whole region.

(3) While constructing CTLN in the whole province, layout and implementation should be conducted in accordance with the three logistics nodes levels: national level, provincial level and regional level. Wherein, Zhengzhou can be positioned as national hub; Luoyang, Nanyang, Zhoukou, Xinyang, Xinxiang and Shangqiu the provincial hubs, other rest cities the regional hubs. Zhengzhou functions as leader, core and growth pole in CTLN in the whole province.

References

1. J. Shi, Y.J. Liu, Econ Geogr. **32**, 98(2012).
2. W.Q. Chen, Y.N. Liu, Log Technol. **33,** 101(2013).
3. Q.C. He, W.P. Luo, L.H. Li, J. RY Sci Eng. **11**, 131(2014).
4. W. Wei, Y.H. Cao, W.D. Cao, Sci Geogr Sin, **29**, 485(2009).
5. X.L. Wang, S.Y. Wang, Pro Geogr, **31**, 1583(2012).
6. S.L. Han, X.W. Liu, Ops Res Manage Sci, **23**, 55(2014).
7. G. Zheng, Q. Li, Bull Sci Technol, **29**, 109(2014).
8. S.G. Song, A.P. Li, L.Y. Xu, J. TJU(nat sci), **38**, 113(2010).
9. S.X. Zhang, J.Z. Huo, Chin J. Manage, **2(S2)**, 194(2005).
10. W. Ding, L. Zhang, J. Li, Chn Soft Sci Mag, **25,** 160(2010).
11. W.B. Li, Y.S. Zhang, Econ Geogr, **31**, 1335(2011).
12. H.M. Lin, Z.F. Du, Stat Res, **30**, 25(2013).
13. J. Yu, W.B. Lu, J. CDU (Nat sci), **22**, 37(2003).
14. X.R. Li, J SD Educ Inst,**14,** 23(2007).
15. National Development and Reform Commission, *Central Plains Economic Zone Planning*(2012).

Strategy for Developing Cold Chain Logistics in Zhengzhou Air Harbor Comprehensive Economic Experimentation Zone Based on PEST Analysis

Jing Feng [1] and Chunxiong Liu [1]

[1]School of Management Engineering, Zhengzhou University, Zhengzhou 450001, China

Abstract. Regional economic cooperation has become a common economic phenomenon in various parts of the world under the background of economic globalization. Logistics, information flow, capital flow and other circulation problems continually emerged. Logistics activities is becoming more important as a sign of economic development. There are less researchers in regional cold chain logistics in China, which is incompatible with the development of cold chain logistics proposed in 12th five-year plan. If the theoretical research lacked, it will restrict the development of cold chain logistics. In this paper, the research is based on the actual situation of Zhengzhou, and it uses the PEST process to analyze the environment and politics. It accordingly proposed the corresponding strategy and suggestions.

Keywords. Zhengzhou air harbour, cold chain logistics, PEST analysis, development strategy.

1 Introduction

With the popularity of the Internet and the increasing of people's living standard, their needs have transformed the traditional single spot trading into the fast variety type of trading now. Aviation logistics started late but developed fast in China, the technology is not mature, and the formation of system is not standard [1]. To compare with developed countries, there is an obvious gap. Cold-chain logistics has high demand which need low temperature environment to keep the supply chain system in technology in logistics industry. As a big agricultural province of Henan, the circulation of flour, rice, vegetables, fruits and other agricultural products all require technical support of cold chain logistics. At present, cold-chain logistics industry has been developing well, but there are fewer researches for a particular region. Therefore, it is necessary to carry out research for this article.

2 Development status of Zhengzhou air harbor comprehensive economic experimentation zone

"Fast" has become a major feature of the current market competition, and there is rapid and sustained development that air transport has advantages with fast, comfortable safe and so on. In recent years, as an inland province with large population, if Henan can seize the opportunities for the development of aviation, they are looking forward to aviation accelerate the development of airport industry. Zhengzhou air harbor comprehensive economic experimentation zone is the first domestic airport economic experimentation area which was approved by the State Council. It situated in the hinterland of the Central Plains, and it developed with Xinzheng International Airport as the center in Zhengzhou city. Besides, Zhengzhou has multiple cross traffic lines combined with North, South, East and West. In 2015, Zhengzhou air harbor has opened the whole cargo shipping international lines as many as 30, covered the trip to Europe, America and Southeast Asia and other regions [2]. In addition, the government steady increased the investment to Zhengzhou air harbor comprehensive economic experimentation zone logistics industry with the national strategy to boost the development of the central region, and it attracted the domestic and foreign well-known logistics company entered air harbor Logistics Park, such as USP, FedEx, Damco, SF, etc. There is fast development pattern in logistics industry in Zhengzhou. Therefore, it is necessary to make a scientific research in the development of cold chain logistics in Zhengzhou air harbor, to construct its theoretical basis, to guide for a scientific development of the cold chain logistics industry. And it is also the urgent problem to be solved in the current development of cold chain logistics in Zhengzhou air harbor comprehensive economic experimentation zone.

3 PEST analysis of cold chain logistics nodes in Zhengzhou air harbor comprehensive economic experimentation zone

3.1 Politics

At the beginning of the national government approved the construction of the airport in 2013, the government issued 81 policies included the financial, port construction, finance, industry development, personnel protection and other aspects, it gave full support and protection in developing logistics. In order to carry out the ideology which construct a large hub, cultivate large industry and shaping the metropolis, six provinces in Central China has opened 14 international cargo routes, and 13 of which are in Zhengzhou at present [3]. In addition, there are comprehensive bonded zone, aviation logistics area and land and air transportation collection center in Zhengzhou air harbor, and focusing on the development of the aviation logistics, bonded processing, exhibition trade and other industries.

It actively create logistics environment well in the process of becoming international logistics hub, formed Zhengzhou as the center of the Central Plains logistics city group. Then this group will radiate to the national logistics system, and become a mutual accommodation logistics hub from western region to eastern. The government put forward building strongly competitiveness of international air cargo hub in Development plan of Zhengzhou air harbor comprehensive economic experimentation zone. It point out three aspects to strengthen the construction of the logistics system. First, to promote Zhengzhou airport freight transportation system function, second, to improve the function of land transportation, third, to develop multimodal transport. At the same time, Zhengzhou air harbor have cooperation with relevant logistics industry to promote the construction of logistics infrastructure, transportation, information services, technology research and other functions to enhance the service level in logistics, and promote faster and better development.

Henan province has a good foundation of agriculture and food processing industry. Agricultural development ranked third in China, the domestic well-known food companies Sinian, Sanquan Pangge, Yunhe and other brands of frozen food became one of the ten major elements of the city of Zhengzhou. The government has made a lot of planning for the logistics development of cold chain products on the basis of these good development foundation. In December 2015, Zhengzhou Yongzhixing air cold chain logistics base project officially started construction in exhibition logistics area of Zhengzhou air harbor comprehensive economic experimentation zone after Zhengzhou Xianmeilai project in 2014. The project is a important project which is planning 8 million yuan invested in our province, it covers an area of 257 acres and a totally more than 23 square meters construction area. The project as one of the integrated project included air cargo transport, land transport, inter city regional distribution, import and export goods, food storage, bonded logistics, customs supervision, container transportation. It will provide world-class standards for the central region if the project was completed. And it can allocate resources optimally, further accelerate the construction of international freight distribution center, accelerate the formation of the aviation logistics industry cluster, greatly enhance the core competitiveness of the regional aviation logistics industry.

Conclusions as a result, Zhengzhou air harbor has good environment and policy support to develop cold chain logistics. On the one hand, the promoted freight hub and extensive geographical location provide good environment for the development of cold chain logistics; on the other hand, the government gave financial support and fully industrial development policy for the construction of cold chain logistics, they provide a good opportunity to develop cold chain logistics.

3.2 Economy

The economic development situation often reflects a social level in a region, the logistics industry as an important part of the third industry services, which have a significant impact on the development of society and economy.the author chose three aspects of the economic level of Zhengzhou city: Per capita GDP of Zhengzhou, GDP of there industry of Zhengzhou and import and export value of Zhengzhou.

3.2.1 Per capita GDP and GDP of there industry of Zhengzhou

Table 1 can be seen that per capita GDP reached 77217 yuan in Zhengzhou city in 2015, increase 7.9 percent. According to statistical results of Zhengzhou communique, the primary industrial added value is 15 billion and 100 million yuan, increase 3 percent; the second industrial added value is 362 billion and 550 million yuan, increase 9.4 percent; the third industrial added value is 353 billion and 870 million yuan, up 11.4 percent. GDP increased more stable, and the third industry developed faster than the first and second industry, it provided a good opportunity to develop cold chain logistics.

Table 1. Per capita GDP and Rate of Zhengzhou: 2011-2015.

(100 million yuan)

Year	Per capita GDP	Rate（%）
2011	56086	9.7
2012	62046	9.9
2013	68070	7.9
2014	73056	7.5
2015	77217	7.9

3.2.2 Import and export value of Zhengzhou

Table 2 shows that changing situation of imports and exports have an increasing trend year by year in Zhengzhou city in recent years. Frozen industry, food processing industry and electronic technology have well developed with the international routes decreasingly opened in recent years, and they can find a good breakthrough for the development of cold chain logistics.

Table 2. Import and export value and rate of Zhengzhou: 2011-2015.
(100 million dollars)

Year	Import and export value	Rate (%)	Export value	Rate (%)
2011	160	210.1	94.1	182.9
2012	358.3	124	202.6	110.3
2013	427.5	19.3	250.66	23.6
2014	464.3	8.6	266.57	6.4
2015	570.3	22.9	312.5	17.2

3.3 Society

The factors of social environment are mainly composed of population scale, cultural background, personal values, consumption and income levels and social mobility.

The development direction for the Zhengzhou air harbor is positioned as international logistics hub, and the logistics park is the main carrier of international logistics function, so its development analysis mainly conclude the population, income and expenditure, social mobility and other aspects:

The first aspect: Statistically, as the core of the Central city, there is 9.03 million resident population in Zhengzhou City by the end of 2012, population movement reached 3 million 230 thousand people, floating population density ranks second after Guangzhou in China, which provides adequate labor for Zhengzhou. Accompanied by the rapid development of Internet development of China, this environment can provide the rapid expansion of the logistics industry, and give an opportunity to the development and operation to cold chain logistics.

The second aspect: members of society income and spending level directly respond the regional economic development level and the living conditions, and there is the interaction of payments and economic development, only making a balance investment of living level and spending level, can it have a good return in income; only satisfying consumers demands, can it have a better residents economic construction [4]. This section measures the social level by the analysis of residents income and expenditure data of Zhengzhou city in nearly five years.

Table 3. Income and expenditure data and rate of Zhengzhou residents: 2011-2015.

(Yuan)

Year	Per capita disposable income	Rate (%)	Consumer expenditure	Rate (%)
2011	21612	14.4	14500	13.4
2012	24246	12.2	16610	14.6
2013	26615	9.8	18672	12.4
2014	29095	9.3	20122	7.8
2015	31099	8.7	21692	7.8

From statistical data of table 3 in recent five years, it can see that residents per capita disposable income and expenditure increased year by year in Zhengzhou city, and put two digit incremental trend, which indicates that the living standards of residents is increasing year by year. It can predict the next few years, payments situation in Zhengzhou city will continue to show growth trend, and it will promote market economy as a well condition, at the same time, it will promote the development of Logistics Park of Zhengzhou air harbor comprehensive economic experimentation zone.

3.4 Technology

The government issued the latest policy named The construction system mechanism innovation demonstration zone overall program Zhengzhou air harbor economic comprehensive experimentation zone of Henan Provincial People's Government Office of CPC Henan Provincial Committee General Office, which involves four innovative aspects: financing, trade, logistics and regulation. In terms of logistics, it put forward the logistics facilitation with global vision to implement logistics strategic layout, and to make the construction of connecting the world's logistics network, innovation of operation management mode and the safeguard mechanism, to establish and improve the modern logistics service system, to improve the supply chain management and logistics service level.

It is necessary to have technical support to operate cold chain logistics,but to obtain information is also the key. Earlier, in March 2014, Zhengzhou has become one of the pilot cities, which is the first batch of 10 smart city cloud computing platform construction. It will provide technical support for the development of cold chain logistics in air harbor once the project has been completed.

4 Development strategy of Zhengzhou air harbor comprehensive economic experimentation zone

4.1 Change the industrial structure and promote technological innovation

The main service object of cold chain logistics is agriculture, fisheries, animal husbandry and frozen industry, so to develop cold chain logistics, firstly, the managers need to change the traditional agriculture mode and the introduction of new technologies to improve the first industrial output; secondly, to accelerate the research on mechanical equipment of frozen products with high technology, and introducing base construction, packaging technology, advanced equipment and sustainable green development. Besides, there are always food safety problems in the cold chain logistics industry, so how to implement the whole process with closure and low temperature operation. It is the most important work at present.

4.2 Improve the regional logistics network and actively develop multimodal transport

Zhengzhou air harbor has a good environment to develop transportation, so the managers should tighten up on supporting the land transportation and logistics network, and build an integrated transport network system among with Zhengzhou air harbor, logistics park node and land transportation. The managers should have the positive attitude to develop multimodal transport to achieve efficient activities; At the same time, they should broaden the transportation routes, and cooperate on cold chain logistics with the main cities in Yangtze River Delta and Bohai. Hope expanding incidence of Zhengzhou air harbor comprehensive economic experimentation zone.

4.3 Stimulate cold chain logistics demands by sophisticated industry

Henan province is the birthplace and leader of frozen food. It has become the largest frozen food production and processing base of China since 1992. Now there are more than 300 different sizes of frozen enterprises. Zhengzhou Sanquan Foods Co. Ltd., Zhengzhou Sinian Food Co. Ltd., Henan Sijipangge Co. Ltd. ranked in the top ten in 2014 of China frozen food company. Therefore, the managers should vigorously promote the retail industry and frozen wholesalers, sophisticated enterprises lead small and medium-sized enterprises. Then it can accelerate the integration between the industries and exert clustering effect to enhance the comprehensive competitiveness of cold chain logistics.

4.4 Regulate the system, rationally create a healthy environment

Cold chain logistics development can not be separated from the laws and regulations, which is the inevitable requirement of the healthy development of the cold chain logistics [5]. First of all, the managers should set a set of operation standard for Zhengzhou air harbor comprehensive economic experimentation zone. This rules should contain strict control and supervision from producing to selling, and satisfying development needs of the cold chain logistics. Secondly, reasonably construct equipment of cold chain logistics. The target is that we can achieve the development model with an optimization system in the cold chain logistics. Finally, draw lessons from foreign in cold chain logistics operation mode. The managers should take more international practice to realize the status which is the continuous operation and mutual connection of cold chain logistics system.

4.5 Strengthen information platform construction, train logistics professionals

It often appears chain rupture phenomenon in china, it is not only the uncompleted facilities and premature technology, but also blocked information and lack professionals. Therefore, the manager can establish cold chain quality information and responsibility system by using Zhengzhou smart city temporal information cloud. Besides, the managers should train professionals in charge of all aspects of supervision, included producing, transporting, distributing and selling. And they can ensure that all aspects of the whole system have clear division of labor and clear duties.

5 Conclusions

Logistics activities support the regional economy development, effective logistics activities can originate from the industry development and the government support. The research can make the conclusion from the PEST strategy analysis method: firstly, Zhengzhou city has good policy support at present to develop cold chain logistics, the government regards constructing Zhengzhou air harbor comprehensive economic experimentation zone as the primary goal at present, especially since the State Council approved the Zhengzhou air harbor comprehensive economic experimentation zone development plan. Secondly, the economic development of Zhengzhou city is relatively stable through the relationship between the city of Zhengzhou the per capita GDP and economy and logistic analysis, but the managers need to pay attention to technology reform and innovation in first industry to develop logistics industry. Furthermore, people's consumption level is steadily rising trend through the analysis of the social environment. Finally, it still need improve and innovate technology of cold chain logistics in Zhengzhou city.

References

1. Li Xiao, Jinzhou new area scale of cold chain logistics park and function research. Dalian Maritime University, 2011
2. Tian Zhenzhong, Strategy for Developing Aviation Logistics in Zhengzhou Air Harbor Comprehensive Economic Experimentation Zone Based on SWOT Analysis. Logistics Technology, **21**,77-79(2014).
3. The general office of the general office of the CPC Henan Provincial Committee and the general office of the people's Government of Henan Province on the comprehensive demonstration system for the construction of the comprehensive economic experimentation area of Zhengzhou Airport. People's Government of Henan Province.
4. Li Ke, Research on the development strategy of Urban Agglomeration Logistics Industry in Central Plains Based on Grey Theory and coordination theory. Zhengzhou University, 2010
5. He Xiaolin, Sunli. Bottleneck analysis and Discussion on the development of cold chain logistics in China. Logistics technology and Application, 48-50(2014).

The Evolution Process on Information Technology Outsourcing Relationship

Weihua Duan[1]

[1]School of Business Administration, Jilin University of Finance and Economics, Changchun 130117, China
Corresponding author: duanweihua@jlufe.edu.cn

Abstract. Information technology outsourcing relationship is one of the key issues to IT outsourcing success. To explore how to manage and promote IT outsourcing relationship, it is necessary to understand its evolution process. Firstly, the types of IT outsourcing based on relationship quality and IT outsourcing project level will be analyzed; Secondly, two evolution process models of IT outsourcing relationship are proposed based on relationship quality and IT outsourcing project level, and the IT outsourcing relationship evolution process is indicated; Finally, an IT outsourcing relationship evolution process model is developed, and the development process of IT outsourcing relationship from low to high under the internal and external power is explained.

1 Introduction

In [1], Qi had pointed out that information technology (IT) outsourcing has become the most important way of enterprise information construction. Many scholars form domestic and foreign have carried on the research on the IT outsourcing, and have achieved many research results, but the research on the IT outsourcing relationship is relatively little [2], and the related research is not systematic and deep. Goles research shows that there is an important link between the successes of IT outsourcing and IT outsourcing relationship [3-6]. Understanding the evolution process of IT outsourcing relationship can effectively guide enterprises to manage the IT outsourcing relationship, so as to improve the success rate of IT outsourcing. This paper analyses the evolution process of IT outsourcing relationship from the following two dimensions: Relationship quality between IT outsourcing subject and its objective IT outsourcing project.

2 Type partition of IT outsourcing relationship

IT outsourcing relationship is a dynamic cooperation process, which is built on the customer outsourcings its IT project to vender by contract. IT outsourcing relationship emphasizes three points: (1) two partners of the relationship are vender and customer; (2) IT outsourcing project is the foundation of the relationship; (3) relationship is a dynamic cooperation process. Thus, the subject and object of the outsourcing relationship are the essential elements of the relationship, that is, the two partners and the outsourcing project are necessary elements for the relationship. This paper bases on the relationship quality of partners and the level of IT outsourcing project to divide the IT outsourcing relationship.

2.1 Relationship quality between IT outsourcing partners

Relationship quality is an important concept in relationship marketing theory. Relationship quality is the degree of trust and commitment between customers and service providers [7], which is partners' common recognition to the degree of relationship satisfaction [8]. In the field of IT outsourcing, the relationship quality is considered as "the degree of the expected results of the relationship between vender and customer" [9], which is an important indicator to measure the relationship between IT outsourcing relationship.

In this paper, according to the relationship quality, IT outsourcing is divided into market relationship, cooperative relationship and partnership. The characteristics of relationships are shown in Table 1.

Table 1. Contrast of market relationship and partnership.

Characteristics	**Partnership**	**Market Relation-ship**
communication	Important, Frequent	Unimportant, Not frequent
Trust	High	Low
Cooperation	High	Low
Uncertainty	High	Low
Relationship Time	Long	Short
Control Mode	Trust, Right and Price	Price
Conversion cost	High	Low
Interests and Risk	Sharing	Clearly defined in contract

2.2 The impact of IT outsourcing project on the customer's strategic

IT outsourcing project is a bridge between the two sides of the IT outsourcing. The impact of IT outsourcing project on the customer's strategic directly affect the outsourcing relationship[10-11].IT outsourcing project is different level [12-14], the higher the level of IT outsourcing project, and the greater the degree of impact on customer. In this paper, using the method of Kevin, according to the degree of the impact of the strategy, we divide the project into three levels: High, Medium and Low, such as Table 2.

Table 2. IT outsourcing project level.

Project level	**Impact to strat-egy**	**Main content**
Low	Low	Equipment management, System management
Medium	Medium	Business process, Value chain integration
High	High	Strategic development and sup-port

3 Analysis on the evolution process of IT outsourcing relationship

3.1 Analysis on the evolution process of IT outsourcing relationship quality

The evolution of relationship quality is the development process of relationship quality from low to high. Arnold and Joshi believes that both vender and customer can establish a high relationship through the repeated flexibility, information exchange behavior and the positive cycle of solidarity, which is advantageous to reduce the opportunism tendency and promotes mutual trust [15].

In the book "beyond the partnership: innovation strategy and lean supply ", cording to the development process of things, Lamming divided enterprises relationship into five stages: (1) the traditional relationship stage; (2) free competition stage (3) cooperative relationship stage; (4) partnership stage; (5) strategic alliance stage. From above all, we can see that the relationship between enterprises is a continuous evolution process from market relationship, cooperative relationship to partnership. Thus, the evolution of the relationship between enterprises is a gradual process, the relationship between IT vendor and customer is no exception, and the evolution of relationship quality between the two sides is shown in Figure 1.

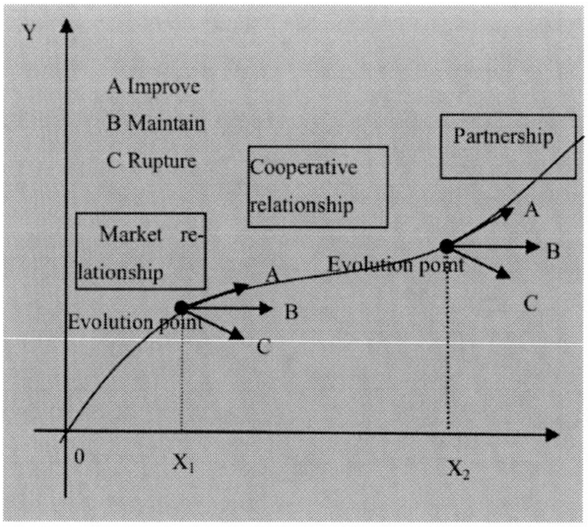

Fig.1. The evolution process model of outsourcing relationship quality.

In Figure 1, X indicates the Time, Y represents the evolution process of the IT outsourcing relationship quality. As X goes on, Y is a continuous evolution process. Based on the quantitative change, relationship quality has qualitative change in X1 and X2, X1 and X2 are called evolution points. After evolution points, the IT outsourcing relationship quality will be split in three directions:

(1)Evolutionary path A shows that the degree of trust between vender and customer achieves quantitative improvement and development by communication and cooperation.

(2)Evolutionary path B shows that vender and customer maintain their trust level by communication and cooperation.

(3)Evolutionary path C shows that the vender and customer find that the other side is not worthy of trust, trust is destroyed, the relationship ruptures.

3.2 Analysis on the evolution process of IT outsourcing project

IT customer investigates the IT vendor through the implementation process of each IT outsourcing project. In the start-up phase of implementation process of the IT outsourcing project, the IT customer pays attention to the service capability of IT vendor; in the implementation process of IT outsourcing project, the IT customer will track and evaluate the implementation of IT project, so as to further understand the comprehensive ability of the IT vendor; With the implementation of IT outsourcing project, the IT customer gradually learn and absorb the IT technology, management skills, IT knowledge and experience; At the end of implementation process of the IT outsourcing project, the IT customer will have a more comprehensive understanding and evaluation to the overall capacity of the IT vendor. In fact, with the need of strategic adjustment and process innovation, the IT customer will produce multiple needs of IT outsourcing. That is to say, the development of IT outsourcing project' level is a multi cycle and dynamic evolution process, as shown in Figure 2.

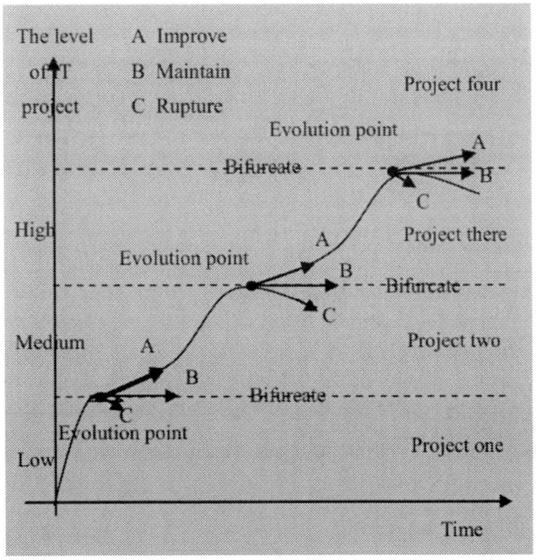

Fig.2. The evolution process model of IT outsourcing project level.

Implementation process of each IT outsourcing project is the process of IT customer to understand the IT vendor's ability. The completion point of this IT project may be the beginning of the next IT project, and this point is called evolution. After evolution points, the level of IT outsourcing project will be split in three directions:

(1)Evolutionary path A shows that the cooperation of this IT outsourcing project is very successful, the customer is very satisfied to service ability of the vender and will consider a higher level of IT outsourcing project, so that the level of IT outsourcing project gets improvement.

(2)Evolutionary path B shows that the cooperation of this IT outsourcing project is basically successful, the customer is basically satisfied to service ability of the vender, but the customer is not very sure that the service ability of higher level IT project, it will maintains the cooperation of this IT project and not consider other cooperation.

(3)Evolutionary path C shows that the customer is not satisfied to service ability of the vender, so it eliminates the vendor and results in a broken relationship.

4 The evolution process model of IT outsourcing relationship

IT outsourcing relationship is the relationship of two organizations between vendor and customer; its evolution process is bound to be influenced by various powers from internal and external environment of the organizations. IT outsourcing relationship is constantly evolving under the influence of the external and internal powers of the organizations. According to the principle of synergetic, when powers from the government, market, society and technology are enhanced, exchange trend of product and service and competitive position will be changed.

The IT customer feels the pressure of enterprise information and understands the potential value of using information technology to enhance the enterprise performance, thus resulting in a more intense need of IT outsourcing, so the IT customer will further strengthen cooperation with the IT vendor and further consider outsourcing the next higher level of IT outsourcing project. At the same time, the IT vendor feels the IT need of IT customer and knows customer's ability requirements of IT outsourcing service, and then continue to improve the relationship and the service ability of high-level IT outsourcing project. The synergy between the customer and vendor forms the internal driving force of the evolution of the IT outsourcing relationship. The IT customer, vendor and other subjects from environment mutually promotes the evolution of IT outsourcing relationship.

Under the pressures of external and internal powers, the IT outsourcing relationship is becoming more and more closely, and the level of IT outsourcing projects is higher and higher. Under the mutual pressures IT outsourcing relationship is evolved from junior to senior, from simple to complex.

In the whole system of IT outsourcing relationship, all subsystems are in a common environment, and each subsystem is in a small environment. IT outsourcing relationship is evolution adaptively in the small environment, and this adaptive capacity is a kind of intelligent expression. In the evolution process of IT outsourcing relationship, the subjects constantly adjust their own behaviors to adapt to the needs of the environment, and continue to accumulate experience to ensure that they do better next time. In turn, the behaviors of a large number of subjects of IT outsourcing relationship in the environment are constantly changing the environment; coupled with the change of environment itself, the dynamic environment affects subjects of IT outsourcing relationship as a form of "constraints". After repeated cycles, the subject and environment of IT outsourcing relationship are in a process that is long-term, mutual influence, mutual interaction, and mutual evolution.

In summary, the evolution of IT outsourcing relationship is a dynamic feedback process in which subjects to external environmental stimulation is from perception to response. The evolution of relationship quality and IT outsourcing project has certain periodicity. In the process of the IT outsourcing relationship evolution, the relationship quality is periodic development along Low - Medium - High degree of trust, the level of IT outsourcing projects is periodic development along Maintenance - Improve - Bifurcate - Maintain, as shown in Figure 3.

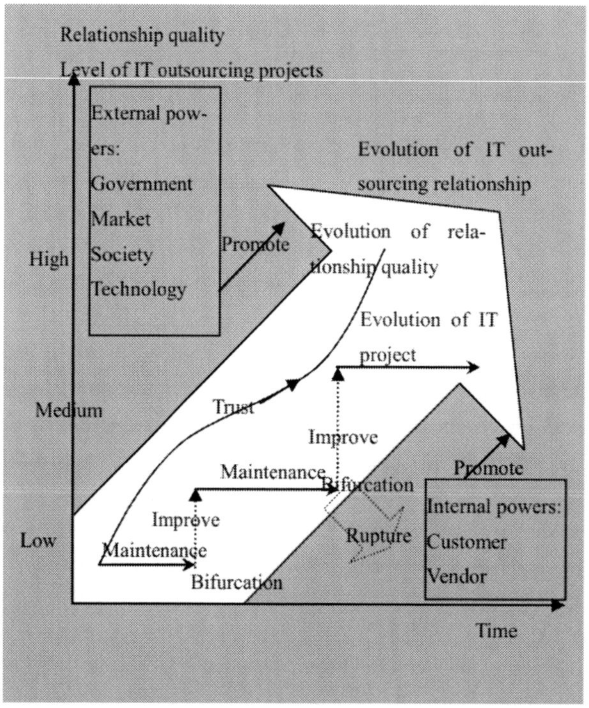

Fig.3. The evolution process model of IT outsourcing relationship.

5 Conclusions

In order to discuss how to manage the IT outsourcing relationship, this paper deeply analyzed the evolution process of IT outsourcing relationship. Firstly it analyzed the types of IT outsourcing based on relationship quality and IT outsourcing project level; Secondly two evolution process models of IT outsourcing relationship was proposed based on relationship quality and IT outsourcing project level, and they indicated the IT outsourcing relationship evolution process; Then an IT outsourcing relationship evolution process model was developed, and it explained the development process of IT outsourcing relationship from low to high under the internal and external power.

This research can help the IT customer to understand the evolution process of outsourcing relationship between IT customer and vender, so as to effectively guide the IT customer to manage the IT outsourcing relationship and improve the success rate of IT outsourcing.

Acknowledgment

This work was supported by the Social Science Foundation of Jilin province education office (Grant No. 2016068).

References

1. Xiaoyun Qi. An empirical study on the macro growth process of enterprise information system in China.Changchun: management college of Jilin university, 2008.
2. Kern T. The Gestalt of an Information Technology Outsourcing Relationship: An Exploratory Analysis. Atlanta Georgia 18thInternational Conference on Information Systems, 1997, December: 15-17.
3. Goles T. The Impact of the Client-Vendor Relationship on Information Systems Outsourcing Success. Bauer College of Business. Houston: University of Houston University Park, 2001.
4. Grover V, Cheon M J, Teng J T C. The Effect of Service Quality and Partnership on the Outsourcing of Information Systems Functions. Journal of Management Information Systems, 1996, 12(4):89-116.
5. Lee J N, Kim Y G. Effect of Partnership Quality on IS Outsourcing Success: Conceptual Framework and Empirical Validation. Journal of Management Information Systems. 1999, 15(4):29-61.
6. Qi C, Patrick Y K. Does interpersonal trust also matter? Exploring the role of trust in successful IT outsourcing. The Proceedings 0f the 40th Hawaii imitational conference on system sciences, 2007:1530-1605.
7. Ravald A, Grobnroo s C. The Value Concept and Relationship Marketing. European Journal of Marketing, 1996, 30 (2): 19-30.
8. Yi Qin. Study on IT outsourcing relationship quality. Journal of management, 2006, 3(6):669-672.
9. Alborz S, Scheepers R, Seddon P B. Impact of Configuration on IT Outsourcing Relationships. Paper presented at the Tenth Americas Conference on Information System, New York, 2004.
10. Nam K, Rajagopalan S, Rao H R, Chaudhury. A two-level investigation of information system outsourcing. Communications of ACM 7, 1996, 39(7):36-44.
11. Kern T, Willcocks L P. The Relationship Advantage: Information Technologies, Sourcing, and Management. Oxford: Oxford University Press, 2001.
12. Kevin J D. IT outsourcing relationships: An Exploratory Study of Inter-organizational Control Mechanisms. Cambridge: Harvard University Graduate School of Business Administration, 1996.
13. Anyi Tian. Study on the dynamic mechanism of enterprise information.Beijing: Chinese Academy of Social Sciences, 2008.
14. Jianbin Sun, Yingying Li. Comparative analysis of information technology outsourcing model. Commercial era, 2008(31):85-86.
15. Joshi A W, Amold S J. The Impact of Buyer Dependence on Buyer Opportunism in Buyer Supplier Relationships: The Moderating Role of Relational Norms. Psychology & Marketing, 1997, 14(18):823-845.

The Effect Analysis of Perceived - expected Organizational Support Consistency on Employee Performance

Hongwei Wu[1,*] and Yahui Wang[2]

[1] School of Economics, Huaibei Normal University, Huaibei 235000, China
[2] School of Management, Shanghai University, Baoshan 200444, China
Corresponding author: whw121588@163.com

Abstract. The relationship between job performance and organizational support has attracted attention of the scholars in recent years, but empirical research on the mechanism of perceived - expected organizational support consistency on employee performance is not much. Using Polynomial regression and the response surface analysis method, this paper discusses effect of the consistency between expected organizational support and perceived organizational support on employee job performance. The data analysis results show that when the expected organizational supports exactly the same consistency as perceived organizational support, the higher the perceived organizational support, the higher job performance is. When expected organizational support is higher than perceived, organizational support and job performance are in negative correlation. Then the paper makes several management suggestions from the working environment, executive support, tools support, emotional support and so on, in order to improve the effect of human resource management.

1 Forward

Globalization, informationization and servitization have made enterprises in an ever-changing environment. How to motivate employees' potential and make them work independently is the key to the development of enterprises, and it has an immeasurable promotion effect on improving enterprise competitiveness. Eisenberger (1986) [1] believed that organizational support is the employee's subjective perception of the organization's evaluation of the work results the and welfare, which can meet the emotional needs of employees to a certain extent, employees will increase the dedication and efforts after perceiving organizational recognition and awards. Borman (1993) [2] put forward that employee personality traits, responsibility, superior power distance, demographic characteristics such as age, gender, educational level, and length of service also have a certain effect on the perceived organizational support. Borman (1993) initially put forward job performance, and further subdivided into task performance and peripheral performance, in which task performance refers to the series of behavior under mandatory rules of the organization, peripheral performance refers to the employee's initiative services for the organization that do not even require income in return. By studying the performance of US Air Force employees, Motowidlo & Van Scotter (1994) [3] found that task performance and peripheral performance act on the overall job performance independently of each other, and the roles of peripheral performance and task performance were separately analyzed from a practical point of view.

With regard to the relationship between the perceived organizational support and job performance and by using empirical data, Eisenberger(1986) [1] demonstrated that the perceived organizational support can reduce staff absence rate and promote job performance. Based on the above research and in the context of globalization, we have to think how modern enterprises, employee expectations and perceived organizational support have specific impact on the job performance? What is the specific internal mechanism? In this paper, response surface analysis was used to establish a polynomial regression equation, and the regression equation coefficient was used to reflect the basic characteristics of the three-dimensional graph. The non-linear response surface graph among the three variables of the equation were described using the statistical test results, the practical effect of the response surface's basic characteristics on the consistency level is expounded, and then the polynomial regression equation and its coefficient calculation process and the consistency empirical study are combined organically.

2 Research Design

2.1 Operational definition and measurement of variables

Indirect measures are mainly used by researchers on the consistency effect to assess the existing situation of independent employees and organizations without interference, and to evaluate the consistency by means of polynomial regression and other methods. This study focuses on the influencing mechanism of expected organizational support - perceived organizational support consistency on the individual job performance, therefore, we mainly conducted indirect measurement by individual direction to assess the consistency degree of expected organizational support - perceived organizational support.

2.1.1 Perceived organizational support

The data are measured with reference to the POS scale (Eisenberge, 1986) [1] and McMillin's instrumental organizational support concept, a new two-dimensional scale was formed through consolidation, the instrumental and emotional organizational support were selected for the real measurement. Instrumental organizational support mainly consists of three items, such as "the unit is always trying to provide me with high-quality working environment and supporting facilities", "the unit is always trying to provide me with job-related knowledge and staff support", etc; and seven items are specifically covered in emotional organizational support, such as "the organization values my contribution", "the organization respects my values and set goals", etc.

2.1.2 Expected organizational support

The data are measured with reference to the Perceived POS scale in item (1), instrumental and emotional organizational support were selected for the real measurement. Instrumental organizational support mainly consists of three items, such as "I expect the unit to provide me with high-quality working environment and supporting facilities", "I expect the unit to provide me with job-related knowledge and staff support", etc; and seven items are specifically covered in emotional organizational support, such as "I expect the organization to value my contribution", "I expect the organization to respect my values and set goals", etc.

2.1.3 Job performance.

Task performance scale (3 items), such as "I can accurately complete my work objectives", "the quality of my work has been kept at a high level", etc; peripheral performance scale (4 items), such as "I am willing to stay in my department", "I am willing to do my best to work", etc.

Likert 7-point scale is the main measurement tool used in this paper. It consists of 7 degrees, and 1-7 means 7 different degrees from "totally disagree" to "fully agree".

2.1.4 Control variable

Some scholars have found that non-research variables such as age and gender have a deviation influence on employee job performance, and these four variables are also set as the control variables in job performance associated study. Therefore, this study will also control these variables to avoid errors.

2.2 Questionnaire collection and sample characteristics

The research object includes 24 enterprises and the geographical distribution of enterprises is in Shanghai, Sichuan, Guangdong. A total of 341 questionnaires were distributed and 238 were collected. The questionnaires with obvious similarities and some gaps were removed during data processing, finally obtaining a total of 152 credible samples.

From the sample descriptive statistics, men accounted for 52.8%, women accounted for 47.2%; degree of education: middle school or below (2.2%), high school and college (11.9%), junior college (25.1%), college (53.2%), graduate and above (7.6%); employees sources include Shanghai (46.7%), Sichuan (20.8%) and Guangdong (32.5%); traditional manufacturing industries (5.3%), real estate (6.4%), financial sector (35.5%), IT sector (8.2%), service sector (10.8%), public institutions (9.6%). Employee's years of service: 1 year (excluding) accounted for 3.7%, 1 year (inclusive) or above and less than 5 years accounted for 9.9%; 5 years (inclusive) or above and less than 10 years accounted for 35.9%; 10 years (inclusive) or above and less than 20 years accounted for 27.7%; 20 years (inclusive) or above and less than 30 years accounted for 16.5%; above 30 years accounted for 7.3%.

2.3 Validity evaluation of data structure

The experimental data are less likely to produce statistical and systematic errors. The data processing is randomly reorganized by using the index Parceling, and then the reliability and validity of each concept are tested. SPSS19.5 and Lisrel8.0 were used in the test, and CR (composite reliability), Cronbach's Alpha coefficient and SMC (square multiple correlation) were mainly tested in reliability test.

The experiment results show that the CR value of each variable is more than 0.5, Cronbach's Alpha coefficients all exceed 0.7 (as shown in Table 1), and the SMC value of each item is 0.42, greater than the critical value. All of the items have passed reliability test and the above results showed that the overall reliability of the questionnaires was high.

Table 1. Descriptive statistics and reliability and validity test of major variables

	Mean	S.D.	Cronbach coefficient	CR	AVE	1	2	3	4	5
Perceived emotional POS	4.11	.89	.862	.791	.546	(.768)				
Perceived instrumental POS	3.60	.76	.783	.823	.507	.329**	(.719)			
Perceived emotional POS	3.24	.92	.825	.817	.512	.341**	.398**	(.717)		
Perceived instrumental POS	3.34	.96	.893	.743	.534	.262**	.473**	.398**	(.743)	
Job performance	3.63	.97	.867	.757	.565	-.206*	-.314**	.244**	-.314**	(.741)

The validity of the questionnaires was checked by convergent and discriminate, respectively. Convergent uses the CFA tool to test whether the different observed variables can be used to evaluate the same construct (latent variable). Experiments show that the AVE values of each variable are more than 0.5 (the critical criterion). The variables reflect the absolute fitness index of covariance matrix generated by model against the overall model, the relative fitness index meets the specified requirements, which proves that the convergent of the data meets the specified level.

Discriminate tests whether there is significant difference between variables by comparing the AVE square root value of compare variable and the absolute value of the correlation coefficient between relevant variables. We calculated the AVE square root value of each variable and put it into the descriptive statistics and the diagonal line of correlation coefficient matrix table for a comparison, the results are shown in Table 1 (see Table 1). The AVE square root values of all variables are greater than the absolute value of the row-related and column-related coefficients, indicating that there is no discriminate problem of the data.

2.4 Response surface analysis

2.4.1 Theoretical hypothesis and response surface features

The polynomial regression and response surface analysis are used in this study. The polynomial regression method is used to construct the model, and the perceived organizational support and the expected organizational support are centralized by the center of the scale (value of 4). The polynomial regression method extends the constancy effect model of expected organizational support-perceived organizational support from the traditional two-dimensional space (plane) to three-dimensional space (three-dimensional). Job performance is taken as the dependent variable, expected organizational support and perceived organizational support are taken as independent variables for equation regression, the model formula is as follows:

$$Z=b_0+b_1x-b_2y-b_3x^2+b_4xy—b_5y^2+e \quad (1)$$

Where z denotes job performance, x denotes employee's perceived organizational support, and y denotes employee's expected organizational support. The formula allows for more detailed consideration of the mechanism of subtle relative differences between the two independent variables (x and y) on the primary dependent variable z. It is easy to know by the structure of the equation that there are three high-order regression terms (including three terms of the third order or higher) in the polynomial equation, which are the second squared term of perceived organizational support, the interaction term of expected organizational support and perceived organizational support, and the quadratic term of expected organizational support.

It is assumed that the level of job performance increases with the consistency degree increase in expected organizational support and perceived organizational support, and that the consistency or complete matching of expected organizational support and perceived organizational support at the highest numerical value marks the job performance to a maximum for the employee. The basic assumption is mainly expressed in the three-dimensional graph as follows: (1) If the consistency of the maximal values of the expected organizational support and perceived organizational support makes the job performance to the maximum, the ridge of the response surface image is projected on the XY plane as a Y = X straight line. In other words, the first principal axis of the response surface coincides with Y = X (p_{10} =

0, p_{11}= 1); (2) If the consistency effect between expected organizational support and perceived organizational support promotes the improvement of job performance, the response surface graph should be concave; (3) If the positive impact of the two dependent variables (perceived organizational support and expected organizational support) on the major dependent variable (job performance) is statistically significant, then the slope of the response surface image along the second principal axis should be 0, but the curvature is not significantly 0 (Edwards, 2007) [4].

Based on the above consistency theory and the geometric characteristics of corresponding response surface image, it is proved that the verification of the consistency effect of the two independent variables by using polynomial regression equation and response surface analysis shall start from the following three aspects. First, verify that the response surface is centered on the original coordinate axes (X and Y), and verify that the inflection point (extreme value) of the response surface is at the origin and whether the two principal axes coincide with Y = X and Y = -X. By checking that the slope of the first principal axis and the second principal axis is not significantly 1 (-1), it is possible to judge that the response surface rotates significantly along the principal axis, by verifying that the values of $-p_{10}/(1+p_{11})$ and $-p_{20}/(1+p_{21})$ are significantly not 0, it is possible to judge that the response surface shifts along Y = X or Y = -X. Second, the curvature and slope of the response surface along the two principal axes are calculated. If the curvature is not significantly 0, the groove of the response surface (if curvature <0, the transversal is concave) or the ridge (if curvature> 0, the transversal is convex) is not smooth; if the slope is not significantly 0, it indicates that there is a slope at the intersection between the response surface along the transversal of the two principal axes and Y-axis on the XY plane (or the intersection between response surface and the Y-axis); finally, if the two principal axes of the response surface do not coincide Y = X or Y = -X, it is also necessary to further calculate the curvature and slope of the response surface along the Y = X and Y = -X to analyze the three-dimensional shape of the response surface. If the curvature or slope of the response surface image along the Y = X plane is not significantly 0, it means that the independent variables have different effects on the dependent variables at different levels; the slope and curvature of response surface along Y = -X are visual indicators of the effect intensity of different consistency degree on the dependent variables.

2.4.2 Test results and analysis

In this paper, the SPSS 19.5 software is used to analyze and test the coefficients of the polynomial regression equation, and then the three-dimensional graphs of the response surface are drawn using the professional graphics software Origin 8.5. Figure 1 and Table 2 show the three-dimensional graphs of response surface, estimated values of the key indexes and data significance test results.

Table 2. Response surface data analysis (N=141)

Estimated parameters	Model analysis results	
	Estimated value	Standard deviation
X_0	-2.66	4.034
Y_0	0.04	0.065
P_{10}	3.29	0.018
P_{11}	1.21	0.027
Slope (the first principal axis)	1.07	0.051
Curvature (the first principal axis)	0.20**	0.049
P_{20}	-2.15	1.541
P_{21}	-0.82	0.023
Slope (the second principal axis)	0.35	0.051
Curvature (the second principal axis)	0.07	0.075
Slope (Y=X)	0.20*	0.012
Curvature (Y=X)	0.08	0.028
Slope (Y=−X)	0.40*	0.003
Curvature (Y=−X)	0.16*	0.047

Note: *$p<0.05$; **$p<0.01$; ***$p<0.001$, two-tailed test

According to the analysis of the response surface image and the results of the polynomial calculation, the following conclusions are drawn: First, the consistency between expected organizational support and perceived organizational support and employee job performance (Figure 1). The visualization of the image shows that the response surface is concave, and the inflection point coordinates (-2.66, 0.04) are calculated by the polynomial equation, and the value of p10/ $(1+p_{11})$ is 1.49, the value of $-p_{20}/(1+p_{21})$ is 11.94, both are significantly not 0, we can determine that there is no obvious translation of the response surface image along Y = X and Y = -X. The transversal curve of the response surface along the first principal axis is a concave curve (curvature= 0.20, n.s.), and the slope is greater than zero at the origin of the projection (slope = 0.35, probably because the estimated standard error is not significant), indicating that the expected organizational support and perceived organizational support completely deviate from the impact on the job performance, appearing an inverted "U" shape. That is to say, with the increase of the expected organizational support, job performance shows a trend of increasing first and then decreasing, along the second principal axis is a straight line with almost positive slope (slope= 0.35, $p < 0.01$; curvature= 0.07, n.s.), indicating that when the expected organizational support deviates from perceived organizational support, with the increase of perceived organizational support, job performance also increases. The slope of the response surface along the Y = X transversal is a straight slope (slope=0.2, $p < 0.01$; curvature= 0.08, n.s.), indicating that the higher the level of the consistency between the expected organizational support and perceived organizational support, the higher the level of job performance. While the transversal along Y = -X is concave (curvature = 0.16, $p < 0.01$), and the slope at the origin of the projection is 0.4 ($p < 0.05$), indicating that when the expected organizational support and perceived organizational support is consistent, the individual's job performance level is higher than when the two are inconsistent. At the same time, when the expected organizational support deviates from perceived organizational support, with the increase of deviation degree, job performance level is declining faster and faster.

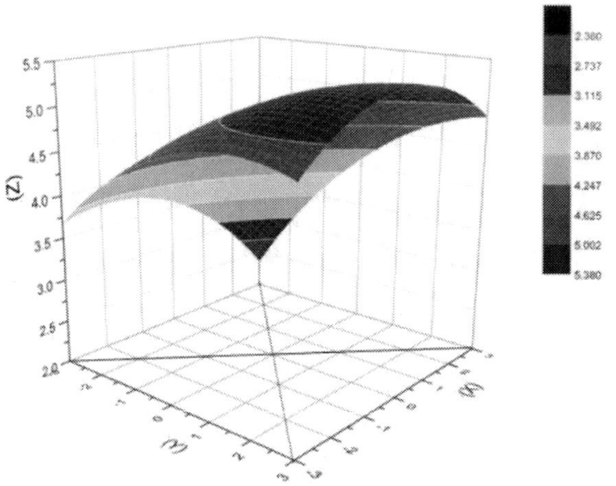

Fig. 1. Perceived - expected organizational support and employee job performance.

Note: (Z): Job performance; (Y): Perceived organizational support; (X): Expected organizational support

3 Research Implications

3.1 Theoretical contribution of the research results

Polynomial regression equation and response surface analysis can be used to obtain more information that can not be obtained by conventional methods, which can be expressed in the following three aspects: First, whether there is smooth and complete consistency between the two variables is verified exactly. The experiments show that binomial regression equation and response surface analysis method can objectively show the significant difference in the consistency effect of the independent variables at different levels (image display). Second, the main effect of single independent variable, the influence of consistency effect between two variables and the main effect of single independent variables on the consistency effect of two variables are investigated. Third, under the condition of incomplete consistency, the dependent variable values and changes with the independent variables are proved.

Edwards & Cable (2009) [5] , Edwards & Parry(1993) [6] showed that there is very little perfect consistency effect. Under the premise of not completely matching, polynomial regression equation and response surface analysis have a great practical value to potential information. For example, the results of response surface analysis in this paper show that in some cases the consistency of expected organizational support and perceived organizational support is higher than that of the job performance at a high level of inconsistency, because the consistency effect of the dependent variable is weaker than the perceived outcome of a single dependent variable (organizational support and expected organizational support). If we use the traditional single method to analyze, simply pursuing the consistency degree between the expected organizational support and perceived organizational support, regardless of the level of a single variable, it may cause wrong guidance of the practical situation. Along with the in-depth research on consistency from domestic management and other aspects, the combination method and architecture of polynomial regression and response surface analysis will bring new thoughts to related domestic research, and are of great significance both in theoretical construction and practical guidance.

3.2 Enlightenment to practice

Organizational support shall be brought into day-to-day organizational affairs to improve business-to-staff attention and support in order to reduce staff pressure, improve job performance, bring value-added business and thereby enhancing competitiveness of the industry.

3.2.1 Optimize working environment, improve management methods

First, companies should start from improving the working environment of employees, such as adjusting the workplace light, cleaning regularly and keeping the air fresh so as to create a work space with a sense of belonging as "home", and striving to reduce the pressure or obstruction caused by environmental discomfort on the staff. Moreover, the quality of interpersonal relationship in working environment is also closely related to employee performance, employees with a good interpersonal environment will have higher job satisfaction and more organizational commitment, virtually increasing employee's ability to adapt to and adjust organizational pressure. The communication between the horizontal (the same level) and the vertical (different levels) levels needs to be strengthened, and the employees are more likely to play their own creativity in an open and harmonious atmosphere, thus effectively improving the individual job performance.

3.2.2 Learn from foreign management experience, implement EAP program

Increase the intensity of executive support. Senior leadership is the embodiment of business for the grassroots employees, their support for employees will directly affect the employee's perceived organizational support intensity, and will effectively promote organizational commitment. Employees should be provided with good upper and lower levels of support whenever and wherever they need to improve their perceived organizational support. Improve the relationship between employees and direct supervisors, the trust relationship between employees and direct supervisors is one of the main factors affecting organizational support. Through communication and information transparency, we can reduce the distrust between employees and supervisors, managers can even achieve "govern by non-interference", the key lies in the establishment of mutually supportive relationship. This relationship will help to positively encourage employees' positive attitude and initiatives. On the contrary, if the direct leadership ignores the daily work of employees, do not give necessary emotional support to employees, it will lead to employees' negative slack, even job-hopping.

Provide tools support initiatively. Tools support mainly refers to the professional and technical training support and good external environment, etc. Employees from an enterprise that provides staff with initiative support have the highest perceived organizational support. In the long-term engagement in a particular industry and job, due to technical constraints, employees will have a strong sense of dependence to their current occupation, while they will have a large perception gap between their living skills in other industries and the actual experience, and a certain gap between reality perception, even questioning on their own future career planning, self-worth or the overall quality and ability. So managers need to strive more instrumental support for employees. Individuality shall be considered in the widespread use of initiative tools, different employees have different needs for vocational and technical education and training. For example, an employee in long-term engagement in accounting can take extra specialized knowledge training such as management expertise. In addition, the staff's gender, education, length of service, marital status and other factors will also affect the tools support. In short, tools support can achieve the best results only after developing an organizational support plan for each employee.

Emphasize emotional support. Emotional support emphasizes concerns of employees' emotional appeal and strong support. The organization needs to pay great attention to employee's well-being perception and related welfare measures. From the daily work and management point of view, the organization should always listen to the voice from the grassroots level, only the first-line employees understand the organization's operating conditions best. In addition, the organization can conduct regular psychological education and counseling for staff, once the staff have doubts on their own career planning, or career development encounters "ceiling", the organization needs to actively alleviate the pressure through psychological counseling to make them adapt to changes in the external environment as soon as possible. Furthermore, the organization's assessment of employee performance should be timely and effective, public and transparent feedback shall be made to the staff, providing effective solutions to problems encountered at work, practically concerning and safeguarding the rights and interests of employees, and effectively improving employees' perceived organizational support.

Acknowledgement

We are grateful to the referees for their constructive comments. We also appreciate the support of The 2016 Key Program in the Youth Elite Support Plan in Universities of Anhui Province (gxyqZD2016103).

References

1. R. Eisenberger, R. Huntington, S. Hutchison, J. Appl. Psychol. 71, 500-507 (1986).
2. W. Borman,.Human Performance, 6, 12-21 (1993).

3. S. Motowidlo, J. Scotter, J. Appl. Psychol. 4, 475-480 (1994).
4. J. Edwards, Perspectives on organizational fit , 4, 361-372 (2007).
5. J. Edwards., D. Cable, J. Appl. Psychol. 94, 654-677 (2009).
6. J. Edwards, M. Parry, Acad. Manage. J. **36**, 1577-1613 (1993).

Thoughts on The Marketing Mode of Le TV & Super TV

Aimin Li[1] and Jiajia Han[1]

[1] Management Engineering Department, Zhengzhou University, Henan province, China

Abstract. Motivated by the nation's policy of Tri-networks integration, the maturity of technology platform, and the user's demand, more and more enterprises start to launch the research of smart TV. In order to offer the reference to other smart TV enterprisers, the status of the smart TV industry is presented in this thesis. The mode of operation and profitable of Le TV·super TV, is also discussed , as well as the underlying problems during its development.
Keywords. Smart TV, LeTV, CP2C.

1 Introduction

20 years ago, the combination of computers and the Internet, so that only the original calculation, office tools, computers have become almost omnipotent, Google, Tencent and other strong rise; 5 years ago, mobile phone and Internet combination, so that the original communication tools are mobile phones have colorful, from Apple to dominate the world. Past experience tells us that once the combination of traditional products and the Internet, it is possible to change the fate of this product, at the same time will spawn a number of great enterprises. Today, cross-border convergence of television and the Internet should be a major trend in the future.

Market research firm NPD Display Search research data show that in 2011 China's smart TV market total of 400 million units in 2012 16.75 million units in 2013 reached 26 million units, all the way up. At the same time, Orville Consulting released the first quarter of 2013 TV brand sales data show that the smart TV market penetration has reached 37.2%, compared with 19% in 2012 increased nearly doubled, 2015 smart TV market penetration will rise To 90%. CCID Consulting also predicted that the next three years, China's smart TV production scale annual compound growth rate is expected to reach 65.1%, which means that smart TV will soon become the protagonist of the TV market. Can be expected, the blowout of the industry will bring to home entertainment a huge change. In view of this, the music TV • super TV turned out worth pondering.

2 Le, Le TV • super TV

The so-called smart TV, is the same as smart phones, with a full open platform, equipped with the operating system, you can install and uninstall the software, games and third-party service providers to provide programs, through such programs on the color TV function To expand, and can be wired, wireless network to achieve the Internet surfing such a general term for color TV [1]. Simple point of view, the arrival of smart TV, conform to the TV network, high-definition, intelligent trend, but also is expected to break the traditional TV getting lower and lower the situation.

November 2004, Le established, announced in 2007 earnings, August 12, 2010 listed on the GEM in China, the video industry is the world's first IPO company, is currently the only Chinese A-share listed video company. When other video sites are still a substantial loss of time, it has continued to profit; when other video sites for copyright hit, it took the film and television distribution rights to sit to bid; when other video sites still standing in the radio and television door Outside the wandering, it has become a television manufacturer [2]. Le, Le TV, music video industry as the three major business lines is the main business as music, the introduction of Le TV • super TV is completed the whole industry chain layout.

Le relies on its own "content + platform + terminal + application" video marketing ecological chain to redefine the traditional TV, 416 engineering staff, after 1253 days to build Le TV • Super TV has 107 patents, 54 copyright, R & D personnel are no longer limited to the TV itself, but focus on creating a complete eco-system based on the big screen - the super configuration + experience LeTV UI system + for large screen Letv Store + China's largest film library + super Cloud Video Platform + ...

3 Smart TV market trends

3.1 Smart TV industry policy

In February 2011, the Ministry of Industry issued a "digital television and digital home industry," second Five-Year "plan", the Ministry of Industry and Information Technology, the Ministry of Industry and Information Technology, , That will be in the "second Five-Year" period to vigorously promote the digital TV industry sales growth, and is expected to 2015, digital television and digital home-based audio-visual industry sales output to achieve "double than 2010" On October 28, the State Administration of Radio Film and Television General Office issued a notice issued by the Office of the "Internet TV license plate operation and management requirements," the notice to allow video content directly or through the set-top box into the TV set Screen, but only if the license must cooperate with, and the content must accept control [3]. So far this smart TV legalization, but the right to operate only 7: CNTV (China Network Television), China, BesTV, South Media, Hunan TV, CRI (China Radio International) and CNR (Central People's Broadcasting Station). January 12, 2012, Le and one of the seven license side of the CNTV reached a strategic cooperation agreement, which for music TV as a platform strategy to open a cut.

From this we can see, the State Administration of Radio on the attitude of smart TV has evolved from a simple regulatory development to encourage the operation, which is the smart TV industry for the first time in years of development. In the triple play environment, based on an open software platform for smart TV will become an important carrier of triple play, play the important task of home multimedia information platform.

3.2 Smart TV will become the next hot market

What's the hottest keywords on the Internet in 2013? Not "mergers and acquisitions", not "market", but "cross-border." "Cross-border" in the traditional hardware industry is no stranger to the industry's frequent integration, so that hardware vendors have horizontal cross-border. If the horizontal cross-border hardware is the industry's normal, then the Internet cross-border hardware, in addition to Google outside the world before this is also no two [4]. The Internet industry have to enter the smart TV, for Internet companies is undoubtedly the second collective "collision" (the first time is a smart phone), which is sufficient to prove that the smart TV is a trend in the future. In the real advance of triple play, the implementation of broadband strategy in China and the SARFT issued under the premise of Internet TV license, smart TV is expected to get rapid development.

At present people's viewing habits are moving from the PC to the TV-screen transfer, the traditional TV industry is also facing an important stage of transformation to the living room entertainment, smart TV will gradually become a family entertainment center. The first listed TV business revenue of more than 10 billion US dollars in a research report that Samsung, the modern family to the TV as the living room core, which companies can get control of television business status, to get into the network family business pass, And the gold content of this pass is no less than the PC Internet and mobile Internet.

Data show that in 2012 the average smart TV activation rate of 27.5%, smart TV active users is about 10%, although this figure is slightly lower, but the overall application environment has been formed. August 17, 2013, the State Council issued a "broadband China" strategy, called for the end of 2015, the basic realization of urban fiber to the home floor, rural broadband into the rural villages, urban families up to 20Mbps. Looking to the future, smart TV is expected to become the smart phone after another hot application.

Figures and tables, as originals of good quality and well contrasted, are to be in their final form, ready for reproduction, pasted in the appropriate place in the text. Try to ensure that the size of the text in your figures is approximately the same size as the main text (10 point). Try to ensure that lines are no thinner than 0.25 point.

4 Mode of Le TV operation

4.1 based on Le unparalleled resources

Le adhere to the "genuine + paid" profit model, and achieved good results. Internet content playback, the Le TV • Super TV will rely on music sources rich source of film, Le now has more than 90,000 sets of TV series, more than 50,000 film network copyright, monopoly over the past three years, 60% To achieve more than 95% of film and television coverage. The number of video sites than the other more than 10 exclusive two times the sum of more than the sum of film and television copyright. In addition there are a wealth of homemade, animation, fashion, documentary and other content, and the industry's most comprehensive sports events, covering La Liga, Serie A, Ligue 1, Super League, AFC Champions League, CBA and other top events. According to Le Watch their own description, they hold the copyright of the film and television assets in 2007 is about 1.87 million yuan, but by 2011 has been as high as 855 million yuan, an increase of 450 times in 4 years, which undoubtedly provides Le TV Strong competition.

4.2 Le TV to create the public marketing model

Le TV as the pioneering use of "CP2C" mode (public marketing), so that as the TV TV • TV direct access to the whole process. The so-called "CP2C" (customer planning to customer), is initiated by the consumer products, as well as some of the flexible requirements of the DIY to the manufacturers, and manufacturers in the next single from the time, you can give the whole production schedule and Product tracking. It has two meanings: one is to allow consumers to put forward their own needs, to achieve targeted production; the second is to change the past in accordance with the batch production, and by the order to drive production [5]. Obviously, "CP2C" model can not only achieve the product's individuality, but also greatly shorten the product update cycle.

Le CEO Jia Yue Ting said that with the social media and online video era, the existing marketing model is out of date. R & D, dissemination, sales, after-sale and operation, recycling to product design closed-loop every link can be the whole process of direct access to the user, and the user can be deeply involved in the whole process of each of the "CP2C" mode, the realization of product design. A link, truly "tens of millions of people discontent, millions of people involved, tens of millions of R & D, tens of millions of people to use, millions of people spread." In the design and research and development, "CP2C" allows users to "custom DIY", and truly create a suitable interest and demand for products; sales, Le TV with "custom DIY + continuous sales" instead of "hungry marketing" , That is, start the open pre-sale mode, the purchase and payment channels will be long-term open, consumers do not need to make an appointment and buy; while "CP2C" mode in the dissemination of more efficient, music TV will give full play to the value of social communication, in addition to micro- And other widely used means of social communication, will also play Le the first professional long video site advantage, play the Le the ecological fusion effect.

4.3 The profit model of Le TV

With the traditional home appliance manufacturers to do the largest scale to maximize the cost-sharing, that is, with the scale for low-cost completely different, Le the road is to use low-cost in exchange for the scale. Le the purpose is to first take a low-cost hardware circle massive users, and then by charging content service fees to earn real money, and ultimately create a traditional TV manufacturers do not have the "content + platform + terminal + application" of the ecosystem. Its core profit model is to use music TV • super TV to open up its integration with the network port, smart TV APP application development, while expanding the audience, then through its additional services and advertising revenue.

"Technology + copyright" two-wheel drive, "science and technology + entertainment" dual gene, Le the "content + platform + terminal + application" vertical industry chain integration equivalent to multiplication effect, leading the upstream and downstream industry chain performance doubled release. Content is the basis for the application of value-added services, intelligent TV terminal is to provide users with a good experience, and ultimately build cloud video open platform. Traditional TV manufacturers can only rely on hardware sales of this single business model, because they have no real control of content and platform operations, can only be a device manufacturer. The music can open up with its cloud video platform and massive copyright content, providing rich content and applications to change the profitability of traditional television models and the establishment of new business model, a single hardware sales profit upgrade to "content revenue + hardware Income + terminal advertising + application into "the four sources of income. Super TV's four-fold profit model, making music TV • Super TV no longer rely on hardware profits, so super TV pricing more flexible.

As of September 2013, the average daily average UV (Unique Visitor) 23 million, peak 32 million, average daily PV (Page View) 110 million, average daily VV1.2 (Video View) billion peak 160 million. If you can convert this level of video users into super TV users, to pay 490 yuan per household annual fee calculation, music will be earned as a bowl full of surplus. Not only that, China has 230 million TV users, if you can win 500 million end-users, according to the annual fee of 490 yuan revenue, Le an annual increase of 2.5 billion. If coupled with advertising revenue, music, when the credit is even more expensive.

5 Thinking on the Le TV • super TV

5.1 Industry competition

Music TV as the first day of opening an appointment on the approximation of 20,000 people, higher than the number of works machines sold nearly 100 times; almost the same time, including Changhong, a number of traditional household electrical appliance enterprises have also introduced 60-inch new products, the price of Of the low-music TV. However, Haier, Changhong, Samsung and other companies to introduce low-cost products is actually a competitive strategy, hoping to suppress a single product to suppress super TV, because Le, the traditional TV companies have a full range of products, through other models Complementary form to maintain the super TV siege. On the other hand, with the Le TV • super TV frontal confrontation, not only the traditional TV manufacturers. September 3, 2013, the video site leader Fantasy together with the TV giant TCL jointly launched the smart TV "TV +", massive video content, announced the "free life", which is equal to completely cut off the music charges, September 5 millet In Beijing National Convention Center held a new conference, introduced millet TV ......

5.2 Profit model to be market test

Le the 2012 earnings report, the annual operating income of 1.167 billion yuan, while Le only 75 million paid subscribers, high-definition video services revenue of only 1.5 billion. Obviously, Le the pay conversion rate on the

Internet is still very low, therefore, Le TV UI smart TV platform, how much money from the user that circle, it is necessary to put a big question mark. Apple's revenue for the second quarter of 2013, for example, from the Mac revenue of $ 5.447 billion; from iPone's revenue of 22.95 billion US dollars; from the iPad's revenue of 8.476 billion US dollars; and from iTunes / software / services Revenue was only 4.114 billion US dollars. Even if it is to open up the whole industry chain of Apple, its main revenue is still from the hardware products, rather than content services [6].

On the other hand, the charging model does not seem to be "national conditions", at least the current level of acceptance is not very high - too many free online resources, not all movies and television programs have to see high-definition. For the future of TV renewals (Le TV • super TV service fee of 490 yuan / year), can achieve the "must" point is still unknown.

5.3 Logistics short board

As a result of its self-built platform for the sale of music video TV • super TV, the implementation of the channel to the strategy, not with third-party electronic business and channels to cooperate, which means Le the need for a strong logistics system backing. July 3, 2013, as the TV TV • super TV officially on sale, ten thousand TV sold out in 49 minutes, but the problem ensues, in the sale of two weeks, still ordered the first batch of super TV consumption The customer service phone is always busy, even if fortunate enough to get through the customer service phone, they can not be promised in the customer service within the time to get back ... ... music promise to order the user's spot "Arrived at the regional transit center, the core city (North Canton wide) orders the next day, two cities (regional transit centers covered by the city) orders every other day," but the reality and commitment vary widely.

References

1. Xu Jian. Smart TV technology and industry development status and trends. Cable TV technology, 2012, 3: 93-96.
2. Chen Xinyan. Super music super TV. China Small and Medium Enterprises, 2013,7: 26-29.
3. Wang Chuanzhen. Triple play under the background of the Internet TV business development. Radio and Television Information, 2012,10: 47-49
4. Ma Zhenmin, Li Xin, Wang Ziju. Cross-border new "as" wild. Successful marketing, 2013,9: 37-38.
5. Zhang Nan. Comments: Le "public marketing" really can subvert it? .http://tech.163.com/13/0508/00/8UAIBTB9000915BD.html, 2013-05-08.
6. Zheng Yuanchun, Cui Jing. Lonely Le. Chinese Economy and Informationization, 2013,20: 73-75.

Research and Some Operation Suggestions on the Relationship Between Project and Sub-Project in R & D Project

Wei Tang[1,2]

[1] Hohai University, Jiangsu, China
[2]Nanjing xiaozhuang University, Jiangsu, China

Abstract. With the increasing competition in the market and the increasing complexity of the products, more and more enterprises are trying to use the project management to complete the relatively complex tasks. How to manage the hierarchical project effectively, which is related to the effective implementation of the strategic objectives of the enterprise. Taking a high-tech company as an example, this paper analyzes the common problems of project management, and puts forward the concrete improvement ideas. This management optimization idea has a wide range of reference for other enterprises.

1 Introduction

With the aggravation of the market competition, product complexity is increasing. Project management approach is commonly used in more and more product development. All activities will be considered as a project management. If the size of the product development task is very complex, the first action is setting up the corresponding portfolio management, and the second is setting up the project group in the project portfolio, the third is setting up the project in the project group. Project management is not only to help enterprises to solve complex research and development issues, but also to increase the difficulty of the management of enterprises, and put forward new challenges.

How to regulate the behavior of enterprise management and the staff by hierarchical project management and portfolio management? Practice and theory circles are working hard to clarify the relationship between project activities. All this is to promote the realization of the enterprise's overall strategic objectives.

They have different standards for different companies to divide the project and sub-project. When we can take the project and sub-project as a whole, it follows the general rules of project management. This paper focuses on the relationship between the project and its sub projects, through a high-tech company project management problem, to seek the improvement of project management.

2 Basic theories on project sub-project management

2.1 The level of the project and the intrinsic relationship

According to the different strategic objectives, current industry divides different levels to manage, such as: project portfolio, project group, single project, sub project, etc.

PMBOK (Project Management Body Of Knowledge) proposed: above the level of project, there are Portfolio, program and so on; the project portfolio is composed of a series of programs, program is composed of a number of projects. Refer to the above rules, below the level of project, we can set up sub project; a project can be composed of several sub projects, and then launch a series of activities around the WBS decomposition of the various specific goals.

The relationship between program, project, sub-project, for different levels of management, may be WBS (Work Breakdown Structure) in different decomposition level or the relationship between scope and activities. Therefore, the division is the need for project hierarchical management. We firstly defined the meaning of Project, so as to define the corresponding Program and Sub project, and then define the boundaries of the scope and activities in the project naturally.

In addition, in the PMI (Management Institute Project) of the WBS standard practice, it focuses on how to establish WBS and the degree of WBS decomposition. Through the decomposition of the standard, it can be seen to what extent is appropriate, according to the management of the underlying content (Manageable) to the management. The main content from the complexity, cost, time, resources and risks should be considered. This should also be taken into account with the relationship between WBS and organizational structure. These considerations of management can also help different levels of management to consider the depth of project management.

2.2 The project group and the management of two level project

Project management is a hierarchical management, the project group, project, sub-project can be seen as the three layer management structure:

Project management is a relatively new concept. As an extension of the concept of the project, its meaning is not yet finally unified. Different scholars, organizations have different understanding of it. PMI proposed project is a group of interrelated activities and need to coordinate the management of the project based on the link of project. The characteristic of the project group is composed of a number of projects, which are composed of a number of simultaneous or overlapping parts of the project. These items have a certain logical relationship, or they have no logical relationship, but they have similar characteristics. The project group is managed as a whole, but each project operates independently .In general, group project will not interfere in the internal management of a project, but the delay of a project may affect another project.

The project is the foundation of the project group and the project combination. The whole project is a project main body, can have a number of sub projects, in accordance with the standards of the project management norms.

Sub-project is the minimum management unit of project management. In order to facilitate the overall project management, according to the contents of the work and the different participants, according to the professional group we can set up many sub-projects. Sub-project operation is relatively independent; the parent project can also interfere with the operation of the sub-project.

Project group, project, sub-project can be transformed into each other according to the management of the different particles. If the project size is larger, the decomposition level is more.

3 The problems existing in the project management

3.1 Process and specification

When sub-project is set up, the project management does not know about it; when the project makes some adjustment, sub-project can't get the notice in time. The decision making process of project adjustment is lack of process and specification. Sub-project needs to serve several projects, but the specification and process is not enough to support project information exchange.

3.2 Resource management

Due to the project resources can't be shared, project resources aren't unified budget, resulting in repeated investment and low efficiency. Sub-project from multiple management, it is difficult to carry out resources of coordination.
It is not easy to reach the goal through the trading system of resources.

3.3 Assessment and incentives

Sub-project award should be determined by the parent project to assign, but IT system can't support. Subprojects working hours can't get data automatically. It will spend a lot of manpower and time to obtain relatively accurate data. So its process is often simplified.

4 The improvement measures of two-level project management

4.1 Issue some regulations to strengthen the project management

In the light of the exposure of project management problems of a high-tech company, The company project management office introduce series of related systems to promote and project management rules based on a large amount of basic research, combined with the practice experience. These rules are as follows:

4.1.1 "Father-son relationship management norms in R&D project"

The specification in the field of project management carries out the provisions, including these areas such as decision-making, change, monitoring, and knot items, In the field of resource management, the project and sub-project 's budget, resource transactions, the actual workload has also been clearly defined. In the field of assessment, some related issues are also explained with the project performance evaluation, project award and cohesion award.

4.1.2 "R&D project establishment management procedures"

This specification relates to the requirements of the project of two-level -project. The process of the project is clearly defined, and the requirement of input and output of the project is put forward.

4.1.3 "R&D project management"

The specification involves the management of parent-child project management requirements; product development team may set up two level projects in accordance with the product development process and the characteristics of the project, and to develop a two level plan. First level plan is for the product development project, product manager is for the person in charge; the second level is for the professional group of sub projects, the professional group representative is the person in charge of the corresponding sub project.

4.1.4 "General requirements for product development work"

In the definition of typical project life cycle, the relationship between project and sub-project is related to the project. Product development should adopt the project management mode; all activities of the product development are treated as a project to manage. In order to facilitate the overall project management, the professional group project can be further divided into a number of project management according to the content of the work and personnel involved in the establishment of different projects, according to the professional group. For example, after the completion of system design, they will be divided into hardware, software sub project and structure sub project etc.

4.2 Transform the IT system to support the two-level-project management

In addition to publishing the corresponding standard, training and publicizing, the company also invested considerable power in improving the current project management IT supporting system; its aim is to standard project management activities using the electronic process. All the management activities are all based on E-business. The system supports the subsequent resource allocation, assessment incentives and so on.

4.2.1 Set project

In the project management aspect, adjusting project process, the project management system is electronic transformed to support the two-level -project. The requirements of two-level-project management are as follows:

Father and son project should be set up in the project management system after the formal entry into force; the project should be associated with the project's parent project. Before the end of the parent project it does not allow the following new project establishment.

The specific timing of sub project can be flexible according to the characteristics of the sub project: project proposal started within two weeks after the completion of the father project; for test equipment, it is recommended to start to collect them after reviewing the parent project system design.

According to the parent project target, the parent project manager decompose the sub project work scope and objectives, and assign the sub- project manager to prepare the R$D task book. Sub projects directly follow the parent project follow-up phase information (including the stage name, the stage start and end time). The examination and approval of the project shall include the parent project manager.

4.2.2 Decision-making

In the decision management aspect, according to the parent and son project management decision-making process, The company carried out the E-system to support parent-son project decision-making. Parent-son project decision-making requirements are as follows:

Sub-project has no need to set up a separate of conventional decision points, but together with the conventional decision-making. However, according to the actual situation of the sub project may be held in accordance with the actual situation of the interim decision, the provisional decision team members should include the parent project manager.

4.2.3 Adjust

In the project change aspect, according to the two level project change process, the project team carried out electronic transformation to support the change of two-level project management. The request for a change in a son project is as follows:

Parent project change: no matter what is the impact to the sub-project, the change should be notified to the sub project manager. If there is an impact on the sub project, the change of child project should take effect under the approval of the change of the parent project.

Sub- project changes: no matter what is the influence on the parent project, the change of sub-project should be approved by the parent project manager, if the project has an impact on the parent, should wait for the approval by the parent project. After the approval, the sub-project may be changed.

4.2.4 Monitoring

In the project supervision and management aspect, according to the two-level -project monitoring process, the project management team carried out the E-system to support the two-level-project monitoring. And project monitoring requirements are as follows: the project manager should report the progress regularly to the parent project manager project, including project weekly report, monthly report, milestone report and so on.

4.2.5 Project closing

During the application of knot, according to the two level project's knot procedure, The company carried out the electronic transformation to support the two level project knot. The requirements of the parent and child items are as follows: only when the parent project is under the knot, the parent has the ability to get the knot.

4.2.6 Assessment and motivation

In the assessment and motivation aspect, according to the two-level-project process, The company carried out electronic system to support two level project incentives. Bonus is needed to reserve for sub-project separately, the team member of sub-project declare the bonus according to milestone, but the details and results of bonus distribution of sub-project members should be transparent to the sub-project manager.

When the project milestone bonus amount is calculated, the sum of the manpower budget of two-lever-project should be adopted.

5 The conclusion

There are many similarities between two-level-project and project group management. In some conditions, the two level project and project group can be transformed into each other. In the project management, the company needs to define the project group, the parent project, sub project for they have different particle size. The project should have a clear classification of the particle size standards, and carry out the necessary project portfolio management and hierarchical management.

The project portfolio management for R & D project is of great strategic significance, such as some large, comprehensive or multi-unit cooperative project, Company executives or senior supervisor is responsible for the execution. In the management mode, it is mainly to establish the mode of the project portfolio management office, which mainly has the overall planning and coordination management of the resources. Project portfolio management office should submit relevant definition of the organizational structure, role definition, document templates and other related documents.

This paper puts forward some suggestions for making specification, process management and standardizes the process in E tools, managers need to utilize the project management theory to make some management innovations and optimize the project management activities. Only we continuously promote the project operating level of the enterprise, we can improve the market competitiveness of enterprises.

References

1. Fricke SE, Shenhar AJ. Managing multiple engineering projects in a manufacturing support environment. HE Transactions on Engineering Management 2000; 47(2)
2. Jennifer Bates, PLANNING & SCHEDULING -DEVELOPING ACTIVITY LOGIC. AACE International, Inc. AACE International Recommended Practices, March 26, 2004
3. Gregory t. Hogan, Effective Work Breakdown Structures, Mechanical industry press , 2005-07
4. Shelly A. Brotherton, Robert T. Fried, Eric S. Norman, Applying the Work Breakdown Structure to the Project Management Lifecycle, PMI Global Congress Proceedings©,2008
5. WAN Li ,CHEN Xing-pan, LIU Qing-hua, Project management oriented to product-development process in PDM .COMPUTER INTEGRATED MANUFACTURING SYSTEMS
6. GUI Wei—min, YANG Nai—ding, JIANG Ji—jiao, Enterprise Strategic Project Management, CHINA SOFT SCIENCE, 2004(5)
7. Li Zhe, Zhou Huadong, Li Yan , Experience and Inspiration from Foreign Professional S&T Program Management Agency: An Example of DLR-PMA. Forum on Science and Technology in China, 2015 (1)

8. Paul O. Roberts. Supply Chain Management: New Directions for Developing Economies. World Bank Report. 2001.

Current Situation and Developing Countermeasures of Postgraduate Innovation Center in China

Can Li[1*], Sujuan Liu[2],Yuzhou Chen[3] and Hongxiang Cao[1]

[1]Administrative Engineering College, Zhengzhou University, Zhengzhou, Henan, China
[2]School of History, Zhengzhou University, Zhengzhou, Henan, China
[3]School of Law, Zhengzhou University, Zhengzhou, Henan, China
Corresponding author: 365083792@qq.com

Abstract. Innovation is the driving force to promote the development of our nation. Postgraduate education, as an important approach to cultivate high-level talents, is a major part of the national innovation system. The establishment of the Postgraduate Innovation Center has an important significance for the cultivation of innovative talents in colleges and universities. This paper expounds the connotation and characteristics of Postgraduate Innovation Center and analyzes its current situation, operation mode, function orientation, institution settings and the innovative products. The paper also points out the problems in the current situation of the Postgraduate Innovation Center and finds the factors which restrict the development of the center. Finally, the paper puts forward some corresponding countermeasures.

Keywords. Postgraduate, innovation center, situation, countermeasures.

1 Introduction

Postgraduate construction is not only an important indicator to evaluate the development of the research universities, but also the main force to promote the development of society. Since 2003, for the implementation of the Postgraduate Education Innovation Project, Ministry of Education successively approved national universities set up Postgraduate Innovation (Entrepreneurship) Center, (hereinafter referred to as the Postgraduate Innovation Center or PIC) in order to promote graduate education, improve the social adaptation ability. After ten years of development, each of the PIC has an achievement. As the accelerating social development, however, the PICs should try the best to satisfy some new requirements, which make the PICs improve continuously, to enhance the comprehensive competitiveness of graduate students.

2 Current situation of the construction of PICs in China

In 2003 the Ministry of Education began to deploy the postgraduate education innovation project, and in 2005, the Ministry of Education issued a document putting forward opinions for implementation of graduate education innovation project, strengthen the research of the innovation ability of graduate student training, further improve the quality of postgraduate education. Since the establishment, the PICs have carried out many kinds of special activities, which not only enrich the postgraduates' extracurricular life, but also improve the comprehensive ability of graduate students.

According to the statistics of 19 Postgraduate Innovation Centers[1], there are four types of the activities that PICs held, which are graduate innovation project, research fund, innovation and entrepreneurial salon innovation contest, some other characteristic activities. All universities have stimulated the graduate student innovative projects, to have a different degree of innovation awareness of the graduate students to be supported, which have a strong catalytic effect on the graduate students' innovative consciousness. Innovation salon gives inspiration to the majority of graduate students through the theme speech of teachers both inside and outside school and some brilliant students, and provides graduate students opportunities to communicate with successful entrepreneurs. Many universities have established venture funding scheme, the school provides basic conditions and start-up funding for postgraduates, on one hand to stimulate postgraduate innovation potential, on the other hand also injected fresh blood for the society. Graduate innovation competitions of many types, is to increase the exchanges of postgraduate inside school and intercollegiate through the means of competition. Universities also hold different characteristics activities according to their own characteristics, not only enrich the graduate school life, but also provide special exchange platform for graduate students [1].

According to the content, the activities held by the Postgraduate Innovation Centers could be divided into two categories, basic activities and enhanced activities. The basic activities are mainly to meet the basic needs of the research and exchange of research students. The so-called innovation is not river without source, nor tree without root, only by inheriting first would we have the innovation, basic activities provide many chances for graduate students, training teaching mainly focuses on the scientific research ability, and basic activities can provide more thinking opportunity for postgraduate students to learn to use. The enhanced activities are to select from the majority of graduate students have a sense of innovation and innovation, and in the event will be fully demonstrated innovative consciousness, and to become a typical incentive to other graduate students. The enhanced activities are mainly launched in form of competition and funded project. Although only part of the graduate students could involve in, they can set up a typical, fanning out from point to area of effect [2].

3 Difficulties in the construction of the PIC in China

3.1 The inside factor of the difficulties in the construction of the PIC

There are many kinds of activities carried out by the Postgraduate Innovation Centers, it has greatly enriched the life of the graduate students, but after graduation, the students still face the pressure from work and life. In the face of reality, a lot of students do not know how to transform intelligent into productivity, in another word, do not know how to learn for use [3]. A lot of requirements is proposed by the pressure of reality for graduate students, for example, how to adapt to the work as soon as possible, how to use professional knowledge to solve practical problems, how to solve practical problems by cross disciplinary. These issues will become the obstacles to the development of graduate students, but also will become the PICs' problems [4].

The Postgraduate Innovation Center aims to help the graduate students to improve the level of academic research, to enhance the innovation ability, to improve the comprehensive qualities and to enrich the campus cultural life. But for now, the main activities held by the PICs focused on the basic activities, many activities have to attract graduate students to participate in as much as possible, while the graduates maybe not very interested in. On the other hand, the

[1] This paper references the Postgraduate Innovation Center, Henan province postgraduate innovation center, Jilin University postgraduate innovation center, Dalian Technology University of postgraduate innovation center, Dongbei Technology University postgraduate science Service Center, Northwest Agriculture and Forestry University postgraduate innovation center, Huazhong Agricultural University postgraduate innovation engineering, Beijing Forestry University postgraduate innovation center, Hehai University postgraduate innovation center, Chongqing University innovation practice base, Northwestern Polytechnic University postgraduate innovation experiment center, Nanchang University postgraduate innovation center, Huazhong Science and Technology University postgraduate innovation and entrepreneurship practice base, Liaoning Technical University postgraduate innovation center, Shenyang Medical University Postgraduate Innovation Center, Northeast Forestry University postgraduate of science and technology innovation center, Hunan Postgraduate innovation and entrepreneurship center, Zhejiang University postgraduate center for innovation and entrepreneurship, Beijing Jiaotong University postgraduate center for innovation and innovation center, Nanjing Technology University postgraduate innovation center.

enhanced activities focused on innovative competition, which the main purpose is to achieve better goals in the national competition. Although postgraduates can also get exercise, the utilitarian purpose is contrary to the original intention of the establishment of the competition and so the PIC [5].

3.2 The outside factor of the difficulties in the construction of PIC

The Postgraduate Innovation Centers have two main forms, one is relying on the establishment of the university, affiliated the postgraduate department, while the other is establishing a cooperative relationship with enterprises. Affiliated the postgraduate departments in charge of the form, PICs are easy to be affected by the administrative, even though still managed by the postgraduates and freely develop, it does not prevent the school will judge the goals of the PIC linking to the graduate department, in a certain extent by results of quantitative evaluation of the PIC. While the mode of cooperation with enterprises, are more likely to exhibit the utilitarian nature, these PICs indeed increases the opportunity for graduate students to practice, but too much to cater to some enterprises is easy to limit of graduate students innovation ability and the ability to explore [6].

Innovation based on market demand and social development is of great significance, but it is far more important to lead the development of the market and the society. In recent years, there goes an innovation slogan, called encourage people to start their own businesses and to make innovations. There is still a big gap between the developed countries and China in innovation. The transformation from made in China into made by China still need to continue to advance. In addition, undergraduate and graduate groups have the phenomenon of weak entrepreneurial ability, adapting to the situation in long cycle. The graduate students have not been exerted fully in competition of social work.

4 Analysis on the causes of the difficulties in the PIC construction in China

4.1 Heritage

The development of Postgraduate Innovation Center dilemma from many reasons, as any innovation is not achieved overnight, so is the development of the PICs. The primary factor for the building of PICs is heritage. In order to innovate, a team should be built up first, which could cooperate and inherit. Some innovations may need a long incubation period, even the researches would make some progress with the help of the Postgraduate Innovation Center, it maybe still a problem for team disbanded. Especially in the interdisciplinary field of innovation research, it is more common to lead to the whole innovation plan stagnation if the core innovation researchers left.

4.2 Cross discipline

The inadequate development of cross discipline is another important factor restricting the development of Postgraduate Innovation Centers. Innovation must be based on the heritage and to make a breakthrough. However, it is not easy to make a breakthrough in a single subject or field, and it may take a long time to develop. And innovation in a single subject or field maybe restrict in the practical application. It is always not easy to solve the complex practical problems with the research of one single subject or field independently. Interdisciplinary research tend to be more conducive to solve practical problems, cross disciplinary exchanges are also more likely to make a breakthrough innovation. At present, more and more attention has been paid to the Interdisciplinary researches, but the PIC is lack of providing the platform for the exchange of researches crossing fields [7]. However, professors tend to be more focused research in a particular field or discipline, not many disciplines involved, while the graduate students on the other hand, have certain understanding of their own subject, have not form a fixed pattern of thinking in their majors and have a strong accepting ability. However, most of the PICs do not have sufficient capacity to provide the necessary research equipment and funds for the cross discipline [8].

4.3 Research period

Research period of innovation plan and innovative project is long, while the funding cannot meet the need of the research. Although each Postgraduate Innovation Center has funded the independent innovation project of graduate students, the funding is usually not enough.[2] Generally speaking, the research ability of graduate students is weaker than that of the professors, so the postgraduate is not supported like the professor when doing research. But currently the PIC is difficult to give enough support to the neither innovation programs nor innovative projects. And due to the time limit of graduate students, graduate innovation projects are often limited to one year, while is not sufficient for some of the graduate innovation project, which could only afford to some stage results. Fixed period and limited research support have been limited to the results of the graduate innovation project.

5 Breakthrough in the predicament of the PIC construction in China

From the perspective of the establishment of Postgraduate Innovation Center, it is a platform for graduate students to plan independently, manage independently, develop freely, innovate independently, but also the link between the school and postgraduate. The construction of PIC requires the lead of the supervisor, but more freedom to develop. The supervisor arranges the PIC basic activities, in addition to the PIC need brainstorm and hold more enhanced activities. It could regard the basic activities as to inherit, while the enhanced activities can be considered to be the focus of breakthrough and innovation. Graduate students not only need to learn the existing knowledge to absorb the existing successful experience, but also need independent innovation, free development. The first job of PIC is to mobilize the enthusiasm of the postgraduate as far as possible, especially to hold the characteristic activities, which according to the different situation of the schools, to carry out rich and colorful activities, to improve the graduate student's enthusiasm for innovation through the characteristic activities.

From the perspective of the basic situation of the Postgraduate Innovation Center, it mainly provides the development platform for the graduate students. As mentioned above, the innovative advantages of graduate students are in cross disciplines, the PIC should provide platforms for different professional graduate students to increase the impact of thinking as much as possible. And PIC should do whatever it could to meet the basic conditions for the development of interdisciplinary research. The graduate students completed the professional development in their training units to complete which does not require additional resources, but each training unit typically do not have the conditions for the development of interdisciplinary research. As a platform for the exchange of graduate students, The PIC should do it best to provide the opportunity to cross disciplinary exchanges for the graduate student. It is necessary to set up laboratory or communicating room for a further research of cross disciplinary, but should not waste the resources of the university.

From the perspective of the needs of Postgraduate Innovation Center development, the PIC should be the leader of the innovation of the graduate students. PIC not only use the innovation program as a means to encourage innovation and innovation awareness, but also need to try other ways to stimulate the innovation willingness of graduate students. It is not right to make innovation formalized, even if there is no innovation project graduate students should also develop research habits and thinking habits. The PIC should try to do a good job in guiding the postgraduates, through the establishment of a typical way, to further stimulate the innovative awareness and innovative desire.

6 Suggestions on the development of the PIC in China

Since 2005, the domestic Postgraduate Innovation Center has achieved remarkable results, but also basically developed to a bottleneck. We must strive to break through this bottleneck in order to make the PIC to develop continually, but also in order to make the PIC to provide better service for the China Postgraduate Innovation Program.

[2] Some of the PICs annually funded dozens of key projects, while the general project does not provide funding, others fund a large number of Graduate Innovation Center projects, which funding for each project is not much.

DOI: 10.1051/matecconf/201710005053

According to the analysis above, the root of innovation lies in inheritance, the Postgraduate Innovation Centers should not only strengthen the development of basic activities, but also should improve the practical ability and basic working ability of graduate students. What's more, for the development of PIC itself, PIC needs to optimize the enhanced activities and increase the communication between graduate students in different majors. And the PIC should also attract more graduate students to participate in innovative activities, and strive to create a complete and continuous innovation team. The specific suggestions for the development of the PIC are as follows,

6.1 Improve the basic abilities of the postgraduate

As the graduate student's basic abilities restrict the innovation ability, the Postgraduate Innovation Center, as the leader of the innovation of the graduate student, should undertake the responsibility of improving the basic abilities of the graduate students. Entrepreneurship and innovation not only requests the students have the agile thinking but also need to have solid basic abilities. Therefore, the PIC should try its best to improve the students' practical ability, to avoid unrealistic expectations for graduate students.

6.2 Constructing innovation teams

As the Postgraduate Innovation Center is an autonomous organization of postgraduate, it is easier for PIC to unite the graduate student from different schools and build an innovation team. After building the graduate innovation team, the PIC should try to maintain a relatively stable team system, and make great efforts to perfect the echelon construction of the innovation team, in order to create famous brand innovation team and fan out from point to area, to leading role. And the PIC could combine with the specific circumstances of the university or the local, and build a targeted innovation team, of which the innovative results could lay particular emphasis on local specific circumstances and practicality. What's more, the improvement of the echelon construction of innovative teams can overcome the problem of a longer period of innovation research. Hence some of the innovation projects with a long period of research will not be terminated because of some members leaving [9].

6.3 Hold the enhanced activities

To improve innovation consciousness of graduate students mainly depends on holding the highly quality enhanced activities. Each Postgraduate Innovation Center has characteristic activities, which combine the local characteristics or the university. Compared to traditional activities like salons, seminars, characteristic activities could excite graduate students innovation consciousness and enthusiasm greater. And enhanced activities are easy to select innovative talent, which is conducive to the construction of innovative team.

6.4 The interdisciplinary innovation

The interdisciplinary innovation can bring into play the advantages of the Postgraduate Innovation Center. But interdisciplinary innovation is also more likely to encounter obstacles especially the space, equipment. The PIC should give full play to the network resources and data resources, and provide space for interdisciplinary innovation research as much as possible, and apply for idle equipment to school, try to make school existing equipment to maximize the use of. If the existing equipment cannot meet the conditions, the PIC could also apply for the purchase of equipments, which the equipment application should be based on a complete innovation team and a forward-looking innovation program in order to avoid waste.

Conclusion

The analysis of the article about the Postgraduate Innovation Center cannot cover the entire problem the Postgraduate Innovation Center faces. We still wish this article would take more attentions on the construction of PIC in China. The development of postgraduates is the some kinds of the future of China. So the PIC should shoulder the mission of improving the innovation ability of graduate students. And the main problems for the development of PIC are heritage, cross discipline and research period. And the suggestions that mentioned above could make a great help to the construction of the PIC.

References

1. Z.Tian, P.Jiao, Chin. High. Educ. Res, 8, 41-42(2005)
2. J.Zhang, J.Xue, Ac. Deg. Grad. Educ. 1, 25(2005)
3. J. Xue, J.Zhang, Chin. High. Educ. Res, 6, 36-38 (2004)
4. Z. Dong, Q. He, H. Zhang, Ac. Deg. Grad. Educ. 2, 1-5(2013)
5. H. Zhu, W. Li, Z. Zuo, J. High. Educ. 2, 74-82(2011)
6. Y. Liu, J. N.E. Norm. Un(Phil. Soc. Sci.), 1,163-166(2014)
7. J. Qiu, J. Xue, F. Cui, Chin. High. Educ. Res, 3, 23-25(2005)
8. X. Song, H. Li, W.Zhou, J. Grad Educ. 6, 29-32(2015)
9. Z.Fang, Z.Gong, X.Zheng, B.Feng, Chin. High. Educ. 9, 15-16(2005)

Study on Village Cadres' Working Enthusiasm on Voluntary Provision of Village-level Public Goods: Take "One Case, One Meeting" System as an Example

Xiuyi Li[1, 2],Weiping Liu[1,*] and Xiaoyan Xing [3]

[1]School of Economics, Fujian Agriculture and Forestry University, Fuzhou, Fujian, China
[2]School of Public Affairs, Fujian Jiangxia University, Fuzhou, Fujian, China
[3]Departments of Management, Fujian Chuanzheng Communications College, Fuzhou, Fujian, China
Corresponding author: lwp789326@126.com

Abstract. Because of the collaborative dilemma in the " One Case, One Meeting " system, the villagers cannot agree on village-level public goods supply. After the financial rewards and subsidy innovation in 2008, many villages were witnessed to break through the collaborative dilemma. Village cadres has played a key role on the provision of village-level public goods in the "One Case, One Meeting" system.which factors transform village cadres' behavior from "idleness" to "entrepreneur"?Based on " One Case, One Meeting" cases in Fujian Province, by means of quantitative analysis ,the paper obtained a results that national performance assessment, election and villagers' meeting have significant impact on village cadres' work enthusiasm .

1 Introduction

1.1 Background

China's rural public goods supply mechanism has undergone tremendous changes since 2000: the supply of basic public goods and services in rural areas was up to government finance; whereas the supply of village-public goods such as small village irrigation, village road and bridges, sanitation facilities, public facilities used inclusively by villagers was up to "one case, one meeting" system, which means it is the villagers who discuss, decide, finance the village-level public goods,and it is the village cadres who construct the village-level public goods .

1.2 Issues raised

Statistics from the Ministry of Finance showed that the number of villages built public projects through the "One Case, One Meeting" mechanism accumulated to 2008 was less than 10% [1].The government innovated the mechanism by providing financial awards to match investment from villagers, accounting for 50% of villagers' investment, [2]. Good results were achieved by this innovation: data from the Ministry of Finance showed by 2012, the number increased from 14% in 2008 to 37.3% [1].Data from Department of Finance of Fujian province, the number of villages received financial awards was 7443,accounting for 51.56% only in the year 2011.

More and more villages were witnessed to have built public goods after financial rewards and subsidy being provide d to the construction.The motivation of villagers can be attributed to the financial rewards.: Most scholars attributed the reduction of villagers cooperation cost and improvement of villagers' participatory enthusiasm to the successful implementation of financial awards, neglected or not gave due attention to roles played by village cadres. In fact, village cadres pay a huge effort to integration of the villagers' will, funds collection, project construction and maintenance [3,4]. Some scholars noticed village cadres' key role in the development of "one case, one meeting" project, what factors motivate them to put on efforts to the supply of rural public goods? The existing research gave no further discourse [5,6].This essay will take village construction of public goods in Fujian Province as an example, in order to identify factors contributing to villager cadres' working enthusiasm, and what can do to improve their enthusiasm.

2 Text theoretical analysis of influential factors of village cadres' behavior

Village main cadres (including village chief and village secretary) are elected by villagers, but employed and assessed by town government, so their behavior is also subject to the dual constraints from demands of both villagers and government. Oi Jean and others noted as "dual representative of the interests of the government and villagers," village cadres strike a balance between the will of the government, the requirements of villagers and their interests [7].This paper holds that financial awards encouraged villagers to participate actively in "one case, one meeting" program, also asked the village cadres to actively contribute to public welfare projects, government motivated village cadres by adjusting the village cadres performance evaluation indicators, putting more importance to village public construction. That is to say, the dual pressures from both villagers and government have substantial influence on village cadres' behavior. L. J. Li and others noted that their behavior was impacted by wages and job generation method [8]; L. X. Zhang noted that gender of village cadres, and village scale also affect their behavior [9].So this paper argues that:Town government influence village cadres' behavior through performance appraisal, villages influence village cadres' behavior through the villagers' democratic.

3 Variables and the model

3.1. Data sources

From January 2013 to March 2013, the author had conducted a questionnaire survey on village cadres from 203 villages in the prefecture-level city in Fujian. Survey data show that only 9.6% of the villages had carried out village-level public construction before 2008, and from 2009 to March 2013 we witnessed the ratio was 70.86%, with 256 public projects constructed in total. Village cadres played a key role in the process, with 58.58% and 89.45% being initiated and implemented by the village committee respectively.

3.2 Variables description

The dependent variable is the estimated ratio of time village cadres' spending on the "one case, one meeting" issues versus total working time . Based on data of the ratio in 2012, the author assumed that:the higher the ratio, the better. Main independent variables are as follows:

(1)Influences from government. Firstly, the proportion of score allocated to public welfare project in the performance appraisal system.village cadres will invest more time on the village-level public goods when the town government lay more emphasis on this issue We assumed that: more emphasis on construction of village-level public and higher wages prompt village cadres to put more time in public issues.

(2)Influences from villagers. Firstly, because of village cadres' being elected by villagers, they respond positively to villagers' requirements.Secondly,the villagers meeting times.We assumed that: more meeting times prompt village cadres to put more time in public issues.

Control variables are as follows:

(1)The village characteristic variables.the village numbers, distance from village to town.

(2) The village cadre's personal characteristics. Village cadres age, education level, gender will also affect the ratio. We assign "1" for male, "0" for female. Female cadres may want to spend more time on family compared to male, so gender may have influence on time spent on public issues.

(3) How much village migrant workers are paid also has impact on the ratio. Being village cadres means giving up the wage obtained as migrant workers. So we assumed that: the more village migrant workers earn, the less the ratio.

3.3. The model

The model is to test the compound impact of government and villagers on the dependent variable:

$$\text{Tratio} = c_0 + \beta_1 \text{score} + \beta_2 \text{ wage} + \beta_3 \text{ meeting} + \beta_4 \text{ title} + \beta_5 \text{ controls} + \mu \quad (1)$$

Variable Tratio refers to time percentage of village cadres' spending on rural public goods,which is a continuous variable.Variable score refers to the proportion of score allocated to public goods project in the performance appraisal system,a continuous variable too.The aforementioned two variables are measured by percentage.Variable wage denotes village cadres' position wage.Variable meeting refers to meeting times of villager representative in a year, title is a zero, one binary variable,with 0 coded for village secretary,1coded for village director.The controls denotes control variables as follows: villagers group ,village cadres' age(measured by year),village cadres' education (measured by year),gender(coded as a zero, one binary variable,with 1 coded for males), and migrant wage (village migrant annual income measured by ten thousand yuan),distance (distance from village to town)

4 Empirical results and analysis

Based on main variables' statistical features, using STAT12 statistical software to process data, taking OLS estimation method, a one-time regression. the regression results in Table 1 show that the model simulation is better, goodness of fit is 0.3272; and at 1% significance level. Specific analysis of the results is as follows:

Table 1. Regression results.

Variable	coefficient	standard deviation	t	p
Score	2.676	0.175	14.54	0.000***
Wage	0.133	1.172	0.11	0.909
Meeting	2.168	0.386	5.63	0.000***
Title	3.191	1.021	3.13	0.018**
Villages group	0.078	0.039	1.98	0.058*
Migrant wage	-0.002	0.001	-2.59	0.051*
Distance	0.055	0.132	0.42	0.675
Education	-0.058	0.236	-0.26	0.794
Age	-0.062	0.118	0.52	0.602
Gender	0.141	2.018	0.007	0.944
C	0.431	5.843	0.01	0.995
F(11, 287) = 13.27		Prob > F = 0.0000		
R-squared = 0.3272		Adj R-squared = 0.3118		

From results listed in Table 1, we can conclude that:

(1) Influence from government .Firstly, the regression shows the influence of performance appraisal on village cadres' working time spent on public welfare,if government putting more importance to village-level public goods in the performance appraisal system, can significantly affect village cadres' working time ratio in the "One Case,One Meeting",significant at the 1% statistical level; secondly, influence of salary on village cadres' working time ratio is positive, but not significant, which means the role of salary in motivating village cadres is not as important as imagined. Many village cadres surveyed claimed that they pay more attention to the right to collective resource allocation, contact brought to their business, and the corresponding fame status, social security.

(2) Influence from villagers. Firstly, the meeting number of villagers has a positive impact on time spent on "one case, one meeting" by village cadres and significant at the 1% statistical level.As a formal way to express the will of the villagers,meeting plays an important role in coordinating the interests of different villagers and strengthening the villagers' cooperation,furthermore exerting pressure on the village cadres to work enthusiastically.Secondly, the election procedure contributes to enthusiasm for work, and positively affects the village director' enthusiasm more than the village secretary', significant at the 5% statistical level.

(3) Influence from control factors. Firstly, the village migrant workers' wages have a negative influence on village cadres' working time, but significant only at the 10% statistical level. Maybe village cadres are deprived of other working opportunities, the more village migrant workers' wages earn, the more opportunity cost to village cadres, the less time they put in "one case, one meeting". Secondly, the number of group has positive effect on village cadres' working time, and significant at the 10% statistical level.

5 Conclusions and policy recommendations

Empirical results show village cadres, are "double agent of government and villagers ",whose time spent on "one case, one meeting" issues is significantly affected by requirement of government and villagers. The impact of wages is not significant, perhaps non-wage returns brought by the position is more valued than a specific salary.

Based on these conclusions, in order to motivate village cadres spent more time on "one case, one meeting" system, several recommendations are raised to further motivate village cadres.

Firstly, More importance should be laid on the rural public goods supply in village cadres' performance appraisal system.improve village cadres' performance evaluation mechanism to make "village public welfare projects" account for prominent share; determine village cadres in reference to migrant workers' wages, make sure it is equal to or moderately above migrant workers' wages, basic wage should be correlated to village scale; while combine wage and non-wage motivations, we should notice that hidden motivation, intrinsic motivation, and prestige incentives also play due role in motivating village cadres.

Secondly, fully employ role of villagers' meeting, make sure meeting be held regularly, to supervise village cadres' behavior; plus, perfect the direct election mechanism, make the election of village cadres reflects villagers' demands.

Last, strengthen cooperation between government and villagers, common requirements of the two party can promote the "double agent" to devote more energy to "one case, one meeting" issues.

Acknowledgements

This paper is one of the periodic accomplishments of the National Social Science Fund Project, namely, "government - Village relations and crack of dilemma of collective action in the supply of village -level public goods (15BGL178).

References

1. Liu, H. and Yan, H. (2008), "The problems and solutions of 'One Case One Meeting' system aboutthe village-level public goods", Truth Seeking, Vol. 8, pp. 89-92.
2. Li Xiuyi, Liu Weiping;The Solution to the Cooperation Dilemma in rural after the Complementation of Financial Reward——a Case Study about 39 Villages in Fujian,Journal of Agro-Forestry Economics and Management,vol. 14 (2015), p.91-100.(In Chinese).
3. Zhou Mi, Zhang Guangsheng: The Operation Mechanism of "One Case, One Meeting" System and its Applicability, Issues in Agricultural Economy, 2010 ,No.2, p.38-43.(In Chinese).
4. Tu Shengwei: Community,Enterprise,Cooperative Organization and Rural Public Goods Supply,Economic Science Press,China,2011:22-24.(In Chinese).
5. Liu, H. and Yan, H. (2008), "The problems and solutions of 'One Case One Meeting' system about the village-level public goods", Truth Seeking, Vol. 8, pp. 89-92.
6. Chenjie,Liu Weiping,Yu Liyan: Study on Performance and evaluation of "One Case, One Meeting"financial subsidy system——take Fujian Province as an example,Fujian Tribune (The Humanities & Social Sciences Bimonthly), 2013,No.9, p.135-138.(In Chinese).
7. JC. Oi , S. Rozelle:Elections and Power: The Locus of Decision-making in Chinese Villages,The China Quarterly, 2000,No. 162, p.513-523.
8. L.J. Li:Elections and Popular Resistance in Rural China, China Information, vol. 15 (2001) No.2, p.2-15.
9. Zhang Linxiu,Luo Renfu,Liu Chengfang,Scott Rozelle.An Analysis on the Determinants of Public Good Investment in China's Rural Communities, Economic Research Journal, 2005,No.11, p.76-83. (In Chinese).

Zhengzhou Distribution Research and Countermeasures

Wujun Cao* and Lulu Wan

Management and engineering Department, Zhengzhou University, Henan province, China

Abstract. With the acceleration of urbanization process in China, the city has become an increasingly important part of the logistics. This paper analyzes the current development in Zhengzhou logistics and distribution, and use the SWOT analysis model to analyze the Zhengzhou logistics and distribution strengths, weaknesses, opportunities and challenges presented in Zhengzhou distribution tertiary network architecture, and finally summarized the Zhengzhou logistics countermeasures and suggestions distribution.
Keywords. Zhengzhou; Urban logistics; Distribution

1 Introduction

With the increasing number of urban population and vehicles, the status of urban logistics becomes more and more important. City logistics is the logistics for the city, it serves the needs of urban development, refers to the physical flow of goods within the city, the city and the external area of goods distribution and urban waste clean-up process, and there are different models,systems and existence form, and other forms of logistics have some differences. Urban logistics is part of the logistics activities of the whole society, it is an important part of the modern logistics service system of the city, and it is the key to supply the goods or goods or services to the customers in time. However, with the increase of urban population, Urban logistics and distribution has become increasingly difficult, urban logistics distribution problems need to be resolved.

Urban logistics has many characteristics such as large amount of information, frequent logistics activities, short transportation distance, many logistics nodes, small quantity of delivery, high variety, high frequency, many factors of urban planning and control, and uneven distribution of logistics facilities, especially the current situation of logistics in our country, urban logistics started late, faster than the other services of form a complete set of urbanization process. However, with the development of urban economy, new demands and challenges have been put forward in China's urban logistics. As China has long attached great importance to the production process and neglected the circulation of the logistics, appeared in the circulation attaches great importance to the flow phenomenon of light logistics, resulting in China's logistics, especially urban logistics lack of coordination, and now the city logistics has become a huge obstacle to improve the social and economic benefits, which restricts the development of regional economy.There are some problems in China's urban logistics:

•urban transportation infrastructure is backward, the ability of transport and freight volume is limited, a serious impact on the efficiency of logistics. At the same time, because of the existence of the roundabout logistics and the backward, cross flow, the circulation link has been increased, and the cost of logistics has been increased;

•due to the lack of unified and effective management of urban logistics, repeated transport, one-way traffic and other traffic congestion caused the phenomenon of heavy traffic, and also has a huge impact on society;

•The continuous process of logistics is divided into different parts, and these belong to different departments, seriously affecting the city logistics rationalization, efficiency , etc.

This paper will take Zhengzhou as an example, analyzes the domestic and foreign research status of city logistics, through the analysis of the status of city logistics distribution in Zhengzhou, using the SWOT analysis to clear the advantages, opportunities, threats and challenges of city logistics distribution in Zhengzhou, and puts forward the corresponding countermeasures.

2 Status and Characteristics of Logistics in Zhengzhou

2.1 The total economy of Zhengzhou has increased year by year, and the new requirement of logistics capability has been put forward.

Between 2005 and 2015, GDP and the growth rate of Zhengzhou. Specific as shown in Table 1. Although Zhengzhou GDP has increased year by year, the growth rate has decreased year by year, it is related to the economic situation, but it is also related to the high cost of social logistics industry.

 DOI: 10.1051/matecconf/201710005055

Table 1. Zhengzhou GDP and growth rate

Year	GDP(Billion)	Growth rate(%)
2008	3004	12.2
2009	3300	12
2010	4040	13
2011	4912	13.2
2012	5547	12
2013	6202	12.1
2014	6800	9.3
2015	7315	10.1

Source: Henan statistical yearbook

2.2 Types of distribution of goods and freight volume

City logistics distribution is given priority to with consumer products related to people's daily life, according to the statistics of Zhengzhou Bureau of statistics, social consumer goods can be divided into food, drink, smoke wine, clothing, shoes and hats, knitwear, gold and silver jewelry, class, furniture, etc. At this stage, Zhengzhou has not yet built a sound statistical system of urban logistics and distribution of cargo volume, statistical methods and different caliber, can not get the freight volume data of the logistics industry in Zhengzhou . But commodity circulation and logistics has a close link, the data is easy to get on sales of goods, we use the total amount of social consumer goods data to estimate the distribution of Zhengzhou , the results of the data, as shown in Table 2.

Table 2. Zhengzhou distribution of all kinds of goods information in 2015

Product category	Commodity value (billion)	The relationship between the volume factor and commodity value(million per ton)	Commodity city logistics distribution freight volume(Million tons)	Accounted for the proportion of urban logistics freight (%)
Food, beverage, tobacco and alcohol category	159.6	1.63	97.91	2.363
Clothing, shoes and hats, knitwear category	142.6	2.75	51.85	1.251
Gold and silver jewelry category	39.8	46.62	0.85	0.021
Commodity category	62.8	1.76	35.68	0.861
Hardware, electrical equipment category	14.4	1.17	12.31	0.297
Furniture	11.4	1.57	7.26	0.175
Petroleum and related products	136.4	0.54	252.59	6.095
Automotive	702.6	2.85	246.53	5.948
Coal and products	10.1	0.09	112.22	2.708
Accommodation, catering and related services	461.6	0.14	3297.14	79.557
other	32.1	1.07	30.00	0.724

Source: bureau of Zhengzhou city

2.3 Compared with the surrounding cities, some of the logistics industry's poor development in Zhengzhou

Although in recent years, Zhengzhou logistics volume index has grown steadily, but compared to cities like Wuhan,Jinan,Xi 'an, there is a larger gap. Table 3 for the contrast of total freight volume between Zhengzhou and its peripheral cities in 2015.

Table 3. The contrast of total freight volume between Zhengzhou and its peripheral cities in 2015

	Unit	The rank of Zhengzhou	Zhengzhou	Xi 'an	Wuhan	Jinan
Freight volume	Million tons	4	29367	50119	44529	36617
Highway	Million tons	2	26202	49243	25023	17570
Railway	Million tons	3	3101	858	9010	19043
Civil aviation	Million tons	3	4	17.9	9.8	3.8
Freight turnover	Billion tons per kilometer	3	685.3	647.2	2556	1696.9
Highway	Billion tons per kilometer	2	467.5	416.9	468.4	305.8
Railway	Billion tons per kilometer	4	217.2	229.3	1198.2	1389.1
Civil aviation	Billion tons per kilometer	4	0.6	1	1.5	2

Source: Henan, Shanxi, Hubei, Shandong Statistical Yearbook

2.4 Zhengzhou logistics is mainly based on the industrial logistics

Total social logistics in Henan province to 95538.48 billion in 2015.The total amount of industrial logistics occupies 82.1% of the total amount of social logistics in Henan province(The Total Quantity and Composition of Social Logistics in Henan province in Figure 1). We can see from the chart,solve the problem of industrial logistics, build a smooth flow of regional logistics network, promote the coordinated development of regional logistics and city logistics are particularly important.

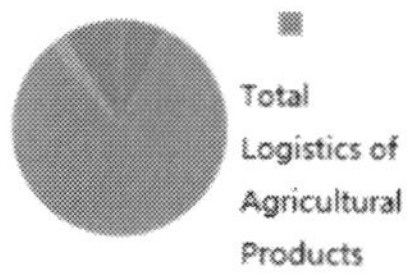

Figure 1. The Total Quantity and Composition of Social Logistics in Henan province

Source: Henan Statistical Yearbook

2.5 The logistics industry divisions, outstanding characteristics, adjust measures to local conditions

Zhengzhou has 1011 logistics enterprises (as of 2014), the completion of operating income of 725 billion, business tax 13.6 billion yuan, sub-sectors of view the number of units in the top three is the road transport, handling and storage industry reached 425, 344, 178, these three industries accounted for 93.7% of the total number of logistics enterprises. Zhengzhou Railway Bureau in the transport industry outcrop status and the opening of the train between Zhengzhou and European, making the rail transport industry ahead of other industries.

Points in the counties, due to the unique to Erqi district railway administration support, thus Erqi district ranked first; Zhengdong New District as warehousing development is rapid, thus Zhengdong New District is in the front rank; Economic and technological development zone is to drive the development of logistics industry in Zhengzhou in recent years, Zhengzhou international land port, logistics park areas, comprehensive bonded zones and a series of industry construction growing vigorously, made economic and technological development zone logistics development prospect great.

3 The SWOT Analysis of Zhengzhou Logistics Distribution

SWOT analysis model, also known as the situation analysis method. SWOT analysis represents the strengths, weaknesses, opportunities, and threats of an organization. This method is actually a comprehensive summary of all aspects of internal and external conditions, and analyzes the advantages and disadvantages of the organization, through the analysis of opportunities and threats, can help enterprise resources and action focused on their own projects and the place that has more opportunities.

Through the SWOT analysis we can see that the advantages of Zhengzhou logistics distribution:logistics market demand for large, commercial circulation for large. But there are obvious deficiencies, such as Industrial policy support lack of intensity, the extensive development pattern of logistics distribution, lack of competitiveness, etc. At the same time, the external opportunities are greater, national policy support, foreign capital entry, the construction of the airport port, Zhengzhou logistics distribution have played a catalytic role.In the next few years , according to the development of e-commerce, they can take the SO strategy, develop the electric business logistics and agricultural logistics, strengthen the logistics infrastructure construction, increase investment in logistics park, etc. Henan has more logistics enterprises, but they lack competitiveness, through the WO strategy to accelerate the development of logistics enterprise groups, speed up the logistics enterprise information construction, thereby improving the Zhengzhou distribution status.

Table 4. SWOT analysis on the competitiveness of Zhengzhou logistics development

	Strength: 1.The unique geographical traffic advantages 2.Good logistics infrastructure and park conditions 3.Huge logistics market demand space 4.Active trade circulation 5.Convenient customs clearance environment	Weakness: 1.Imperfect industrial policy system 2.the extensive development pattern of logistics distribution, lack of competitiveness 3.Logistics education lag, talent shortage
Opportunity: 1.Governments at all levels attach more and more importance to provide macro-policy support 2.Strong industrial base 3.Social economic sustained high growth 4.Opportunities for the construction of the Airport economic comprehensive experimental area 5.Infiltration of foreign logistics enterprises	SO strategy(take advantage of opportunities) 1.Further promote the development of e-commerce, strengthen the logistics industry support 2.Strengthening the connection and coordination of logistics infrastructure 3.Accelerate the construction of key parks	WO strategy(take advantage of opportunities to overcome weaknesses) 1.Seize the opportunity and improve the competitive power of logistics enterprises 2.Accelerate the development of logistics enterprise groups 3.Accelerate the informationization process of the logistics industry, change the extensive development mode
Threat: 1.Unbalanced regional development 2.Increased competitive pressures from international and domestic parties	ST strategy (use advantage to avoid threats) 1.Adjusted to local, stress the characteristic 2. The implementation of "bringing in", "sending out"	WT strategy(Minimizing weaknesses and avoid threats) 1.Accelerate the training of logistics person 2.Study advanced areas of advanced experience, enhance the comprehensive competitiveness

After the above analysis, we can see that although Zhengzhou logistics has the advantages of the policy, but there are also the following questions:

- The concept of modern logistics is lagging behind.
- Service quality and management level is relatively lacking.
- There is a lack of logistics professionals, standardized training and certification
- Scale is not high.

4 The construction of urban logistics distribution system

According to the above analysis, combining with some of Zhengzhou logistics policy, constructing the Zhengzhou logistics distribution system, proposing Zhengzhou to support the logistics park, the public distribution center and the end of the joint distribution, to build a three level city logistics distribution network system. Specifically to speed up the urban area of Zhengzhou and its surrounding logistics park construction, Zhengdong New District, High-Tech Zone, South Central and North Central as the base point, building four large-scale distribution center, based on common distribution in urban areas, take "logistics park - distribution center - the distribution point" of the three city distribution system.As shown in Figure 2, it is Zhengzhou logistics distribution system of the three level architecture.

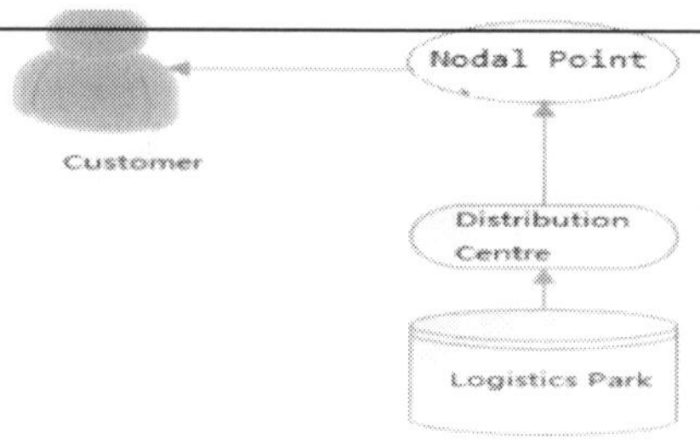

Figure 2. Zhengzhou logistics distribution system of the three level architecture

4.1 Logistics Park

Logistics Park is a place where logistics operations are concentrated in the area, centralizes layout of a variety of logistics facilities and different types of logistics enterprises in the space, is also a certain scale and a variety of service functions of the logistics enterprise assembly point. Zhengzhou currently has five provincial logistics parks, while Henan Bonded Logistics Center and Zhengzhou International Airport Logistics Park enter China's first demonstration logistics park. The two logistics parks have grown up in the Henan province, and have a good replication, promotion and construction of the logistics park construction and operation mode in Henan province, which has obvious promotion effect on improving the overall operation level of our logistics park.

4.2 Distribution Center

Distribution center is from the hands of suppliers to accept a variety of large quantities of goods, the flip, sorting, storage, circulation processing and intelligence processing and other operations, and then in accordance with the needs of many orders to prepare the goods in order to satisfactory water distribution services central flow-level distribution facilities. At present, Zhengzhou is building a city distribution center and regional distribution center. Urban distribution center is a symbol of urban modernization, because the scope of urban distribution are generally in the range of the experience of motor transport, so this distribution center can be directly delivered to the final user, and the use of motor transport for distribution. Short distance, strong ability can respond to engage in more varieties, less bulk, multiple user distribution. Regional distribution center is taking into account the distribution of the surrounding cities of Zhengzhou and other cities in the country.

4.3 Nodal Point

The distribution point is to undertake the distribution center, a link between customers. At present most express distribution nodal point distributes in Zhengzhou city each district, office buildings and other crowded areas. Under normal circumstances, when people have a demand, the nodal point will quickly meet people's needs and requirements.

At present, Zhengzhou has a lot of nodal point, but its security measures and the daily management of most do not meet the specification, this is also the need to further regulate the place. Distribution point can be said to be an important part of the whole logistics chain, is the most direct impact on customer satisfaction, because the distribution point is a direct contact with the customer's front.

Mature logistics distribution system have a greater role in optimizing the whole process, promoting the progress of the industry, using good social resources. The establishment of a good city logistics distribution system in Zhengzhou not only has a positive role in promoting Zhengzhou city logistics distribution, and has a great effect on the development of the urban logistics industry in other cities around the city.

5 Countermeasures and Suggestions

After the above analysis, it show the Zhengzhou city distribution of the status and problems, using SWOT analysis is to let us know the advantages and disadvantages of Zhengzhou logistics distribution, as well as measures to be taken in the future. This paper believes that Zhengzhou logistics can be strengthened from the following aspects of the construction, to create a suitable for Zhengzhou distribution system.

• Support the common distribution model. Common distribution has already been implemented in Japan, Germany, Singapore and other countries, and formed standardized system. The essence is to improve the efficiency through the integration of scale. Common distribution can not only reduce the owner's operating costs, but also to save resources for the community.

Construction of urban distribution transportation channel, dedicated parking spaces. Daily necessities distribution, food distribution, distribution of other social materials need to occupy a public location, especially now the distribution frequency is more and more high, there are more and more distribution categories. Therefore it needs special distribution channel, special parking space, etc. At present Beijing, Shenzhen and other cities are set for the city's distribution of a dedicated transport channel. As a developing city, Zhengzhou needs to set up a special distribution channel for urban distribution and improve the efficiency of urban distribution.

Guide the logistics enterprises to improve the level of information technology. Although many people are engaged in urban transport and distribution of personnel in Zhengzhou,, but there is no reasonable use of good social resources and advanced technology development. Establishing a public logistics platform and using internet of things, RFID and positioning and other advanced technologies can significantly improve the quality and level of service and save part of the cost for enterprises, etc.

Improve the urban distribution management system and give full play to the role of industry organizations. Distribution of the city needs a large number of enterprises to participate in.In Henan province, a large number of logistics and distribution enterprises concentrated in Zhengzhou. However, there is no industry organization in which to play a good role. So it is urgent to have a reasonable organization of the industry organization.

6 Conclusion

This paper first analyzes the status of Zhengzhou logistics present situation, proposing the existing problems, using the SWOT analysis method to make us more directly see the strengths weaknesses opportunities and challenges of city logistics distribution, putting forward Zhengzhou logistics distribution system architecture. In the end, it gives four countermeasures and suggestions to the urban logistics distribution in Zhengzhou. But there are some shortcomings, such as some data is not found, although this paper refers to the data using methods of some other papers, but it did not verify the rationality of these data.

References

1. Ying Li,Zhaoqiang Liu.Research on the present situation and distribution system of Guangzhou City.*Highways & Automotive Applications*.J.E 2,82-87(2015).
2. Guang Hua,Na Dong.The bottleneck and countermeasure of urban distribution in China.*China Logistics & Purchasing*.J.E 6,74-75(2014).
3. Lieying Luo,Zhongcheng Zheng.Study on the problems and Countermeasures of the development of city distribution logistics in Wenzhou.*Business*.J.E 4,240-241(2015).
4. Yuchao. Hu.Study on Establishing Urban Green Logistics and Distribution System.*Logistics Technology*.J.E 31, 56-59(2012).
5. Guoqing Dong.Analysis on the Advantages and Disadvantages of the Third Party Logistics Enterprise Common Distribution Mode.*Modern Business Trade Industry*.J.E 20, D (2012).
6. Hongpeng Wang.Construction strategy and policy measures of urban logistics distribution system in Xiamen.*China Logistics & Purchasing*.J.E 5, D (2010).
7. Hongpeng Wang.Con.struction strategy and policy measures of urban logistics distribution system in Xiamen.*Comprehensive Transportation*.J.E 5, D (2010).
8. Xiuming Niu.On the development of urban common distribution.*Logistics Technology*.J.E 6, D (2013).

Sports dance artistic expression culture analysis

Zegang Chen

Academy of music Department,HeiHe University, Heilongjiang, China

Abstract. At present, the sports dance has entered every stage of the people's life, has become the public's favorite sport. Sports dance has been well developed. This article mainly uses the literature material law to carry on the detailed analysis to the sports dance constitution, elaborated in detail the sports dance artistic expression. The composition of sports dance elements; sports dance is a form of dance art show; sports dance through the dance art can be divided into three aspects, namely, form, music, shape of the expressive force. In this paper, the study will be more in-depth excavation of the cultural connotation of sports dance, and promote the development of sports dance can be more comprehensive. In 20s of last century, Chinese Sports Dance Association officially joined the International Sports Dance Association, which also makes our country's sports dance and international exchange more frequent. However, due to China's sports dance sports dance learning time is not long, while learning is influenced by Chinese traditional culture, the sports dance movements are too conservative, there is a very large gap and international enthusiasm, bold and unrestrained, the pursuit of individual sports dance in the dance style, music and performance hand. Sports dance originated from abroad, it is produced in the daily life of people in foreign countries. China's domestic sports dance players in learning dance at the same time, the production and the connotation of dance is not very understanding, therefore, it is difficult to better reflect the emotional expression of sports dance. Although the sports dance is a kind of similar to the competitive projects, but it is also a kind of dance culture, and to constitute a force from the dance art show a detailed study, detailed mining playing officer of sports dance performance further, reducing China's sports dance and international sports dance gap.

1 Sports dance is acollection of all kinds of elements

1.1 Sorts elements

As the name suggests, the sports dance must contain the sports element. The origin of sports dance can be traced back to the earliest period of ancient Greece, ancient Greece was very popular in competitive games, adding some characteristics of the dance sport in the combat process in competitive games in. Platon, the great philosopher of ancient Greece, also commented on the dance, which is considered to be a kind of dance that not only shows the fighting force, but also shows the unique elegance of the dance. There is a kind of antagonism in competitive sports, which is the most essential embodiment of human life. Just as people are in the process of participating in a game to get better results of the value is the same.

Sports dance and other sports as we know, need to be very confrontation in the fierce competition, the hope can through confrontation show better level, and strive to win the final victory of force. However, the sports dance and other athletic sports is not exactly the same, the sports dance dancers need to express what they want to express feelings through dance form, it is a pure physical performance, achieve competitive sports in the course of the performance, and in this process also allows the audience was a kind of art aesthetic experience. The sports dance is not only a kind of body language skills through the dance show dance, it is a kind of sport, is through the process of sports dance performances and sports to show dancers for the art and aesthetic pursuit.

1.2 At elements

When it comes to art, people will think of dance and music. Art is an aesthetic value, is a kind of perceptual cognition, form of art is the most fundamental of perfect life, in the aesthetic process can feel a sense of self existence, to reach a new level of life. For sports dance, the art of sports dance is mainly through the dance movements of emotional expression, through the movement of the body to allow viewers to experience the feelings of the dancers want to express. Different art also has different classification, mainly reflected in the aesthetic aspects, aesthetic aspects. It can be said that art is the most direct way to reflect the human aesthetic consciousness.

Although the sports dance alone to become independent of the arts have certain difficulties, but in sports dance incorporates a lot of aesthetic consciousness, aesthetic value is very high, the sports dance people can get a kind of aesthetic and spiritual enjoyment of the arts. In addition, the sports dance is not separated from the music, it needs to dance through the rhythm of the music, dance special costumes also make the charm of sports dance more prominent. Therefore, the art of sports dance is very obvious, it is not replaced by other factors.

1.3 Cltural elements

People in the process of production and labor continue to create a lot of social life reflects the dance form. Influence of all nationalities of the world dance by different regions and different ethnic societies, historical origin and the production of labor, customs, living environment and other cultural factors, the accumulation of different dance movements, forms, forming a different style of regional, national dance art.

The sports dance is divided into two categories, each including five kinds, national characteristics and style of each species and the birthplace of the living habits, aesthetic concepts, self, self related standard dance originated in the folk dance, introduced into the European Court, is an agency European aristocracy way, with European culture Latin dance is the body; in combination with African culture, Indian culture and southern European culture, western "individualism and creative" cultural characteristics but also deeply affects the sports dance development process, the formation of many characteristics of modern sports dance, and the formation of social, material and spiritual charm of the special one good dancer, not only have the skill of dance technique, but also have profound cultural connotations and aesthetic taste, in order to match with the transformation to meet the concert dance Dance movements of environmental and cultural artistic conception.

2 Sorts dance is a form of dance art

2.1 The connotation of dance culture

Dance culture is produced by the social practice of human survival, production and so on. German musician Curt Sachi once praised: "dance is the mother of art". The dance culture summed up and concentrated, which is in the form of dance movements, their attitude and other performers as the media, taking human movement changes in amplitude, intensity and angle on the performance of people's emotion and aesthetic pursuit of the spirit of the times and performing arts. Dance is an art form to express the inner feelings of the people, it is easy to lyrical and expressive, not known to directly reflect the character and plot, mainly focus on emotional aspects.

2.2 Dance Aesthetics is an important aspect of the study of sports dance culture.

Among the many kinds of art, the relationship between dance and sports is probably the most closely related. Dance is an important form of direct expression of emotion. "In a lot of art, dance art is the real human body art". At the same time, it is also the most closely related to sports. The American dance critic John Martin once said: "the dance is usually the most difficult of all art, but it is used for medium material body movement reflects the distance on the surrounding life experience than any other art closer." As the art of human body art is not dependent on any other means to rely solely on their own body movement to form the complete art. It can be said, "dance is a kind of artistic expression, which is based on the principle of aesthetics and body movement as the medium."

2.2.1 Dnce beauty: the beauty of movement

The essence of dance and Dance Aesthetics lies in its dynamic nature. As a human form of the "moving" art, the beauty of the dance is the body of a variety of "shape" of the movement 3ll]. The movement of the body is in the thoughts and feelings under the control of rhythm and sense of purpose and organization movement. The dance of "moving" as the breath of life, exercise more than; breath stop, dance and fresh vitality will not disappear. As the author of "art aesthetics" said: "the image of the dance is the flow of image...... The flow here is not only to emphasize the dancer's dynamic and kinetic energy, but also to emphasize the beauty of the dance movements of fluency and coherence." 4ll] is precisely this kind of dynamic image to give the person with can be nearly as far as the mystery, show the irresistible charm of dance.

2.2.2 Dnce beauty: the beauty of love

The essence of dance beauty is not only reflected in the body is light, elegant, very pretty and charming, more people originated from the inner feelings. Just because of the emotion, only then gives the dance the fresh activity and the time character. Any dancer, can not be in the performance of the person's own emotional experience. Duncan was the dancer is divided into three kinds, the first kind of pure physical dance, attention is elegant and coordinated action, but no personality; second can control the limbs, subjecting it to a plot rhythm and rich third souls to dance to appeal, the body becomes clear and transparent, so that the audience can feel the fluctuation of the 5ll] of his soul. Just like third dancers that can make people realize the beauty of the dance, to understand why the human ancestors dance as sacred. As many contemporary great dancer dancers should have a consensus: the "heart" to dance through the heart but inspired by dance, really exudes the beauty of the dance. Here said the heart, that is, from the inside out to show the true feelings.

2.2.3 Dnce beauty: magic beauty

Dancers, through their body in time and space in the form of different "force", the body's body is limited, but the expression is infinite. The strength of the dancers is very complicated, because people's feelings are colorful. Usually, we cannot figure out what is the force movement, how to behave, but can feel the "unreal", can feel its emotions: or gentle, or passionate. Dance in this "unreal" is difficult to say, the language is always weak in the face of the dance. Wigman, one of the founders of the German modern dance, once said that if you can use language clearly, there is no need to use dance to show. The beauty of dance, as with other arts, is often covered up, so this kind of aesthetic object has a certain fuzziness. It is this true or false, let the dance art bring people a unreal world, can let a person on the wings of imagination, creativity and imagination. Therefore, the real dancers, but also to the Dance Camp caused such a kind of aesthetic field, enjoy the illusory beauty, so that the viewer feel the real life more strongly than the real beauty of the dance.

3 Atistic expression is an important way to show the beauty of sports.

3.1 Te connotation of artistic expression

"Artistic expression" refers to a unified representation of the inner spirit and external action, is accomplished through two aspects of facial expressions and body language performance, the audience can enjoy the visual interpretation of the style of music performers and emotions, such as joy, joy, passion, sexy, romantic and form. Artistic expression is often the key to winning the competition. The artistic expression of the sports dance is an important factor in the evaluation of the competition value. The artistic expression of sports dance is the ability of the athletes in the game will be their own internal emotional experience in the rules within the scope of the allowed by the body movements, facial expressions, special skills, etc.. A good sports dance players will also have good artistic expression, he (she) will be action fusion dance music harmony into one's own feelings and understanding of dance and the most incisive display to the audience and the judges and audience identity, so as to obtain the applause, eventually get good athletic performance.

3.2 Sorts dance artistic expression of the external form of expressionPhysical expression of

3.2.1Sports dance

Form expression is a prerequisite for the performance of sports dance. Sports dance shows the charm of artistic charm, the players through the trunk and limbs of the parts of the action as well as to show the beauty of the form, action, shape, expression. Sports dance form "to express love, xingshenjianbei". Dance form is artistic expression, because men and women can form the physiological characteristics and emotional characteristics of their own unique directly and fully reveal, reflect their own unique artistic beauty: Men's muscular body exudes a vigorous masculine and steady beauty; women's graceful body full of curves sense of exudes femininity, blooming beauty. A yin and Yang form in sharp contrast will produce a strong form of expression.

3.2.2 Sports dance music performance

Music is an important part of sports dance, is the soul of sports dance. Because the ten kinds of action, the sports dance style is not the same, so the players according to the different dance, different styles of movement with suitable music through music performance show dance, the dance sports more artistic expression and appeal. To make the music really play the role of the soul, to different styles of sport dance music with a specific type of dance and movement organically, so as to make the sports dance music performance most incisive play. In the sports dance in different genres of music can also create a different artistic expression, and to effectively control the show and dance.

3.2.3 Sports dance form expression

According to the characteristics of different kinds of sports dance and dance style, set different styling. Dance movement modelling should be consistent with the music, music players and characteristics, in order to achieve the best effect of dance. Sports dance in each other for both men and women are static and dynamic modeling gives sports dance artistic expression. Rich and colorful dance style can be divided into dynamic modeling and static modeling: dynamic modeling including male and female athletes walking and jumping, twisting, swinging, lifting, reflexive and tilt, the other performance is instantaneous; static modeling refers to the action of more than two small pause music rhythm section above all start balance and static movement, two other causes without stop effect in time and space, the combination of static and dynamic, firmness and flexibility, against each other.

3.3 The relationship between artistic expression and sports dance skills

Sports dance skills acquisition and consolidation must go through repeated practice, and practice is the only way to strengthen all the skills of learning, learning any dance skills are not "at will" or "to understand", dance is a cruel art, not through hard work, can not become a true dancer. Performance is the art of players in the game will show or emotional experience to show the ability, through technical action simply, appeal show artistic expression is in the works of art and the feeling of dance. In the dance teaching, rehearsal, creation or performance, often listen to a word is: this player is very good sense of dance here, the feeling is the stage, it can be said that the artistic expression of the players. The sports dance makers need not only acquired a good dance skills, have a good sense of self that artistic expression should control and regulate their body movement with the artistic expression of the state, the dance gradually skilled, so as to cause the audience to feel good and sympathetic, dancers to the stage to shine, bright.

Conclusion

Sports dance is a combination of health and beauty of a sport, you can show the health of modern people to give people the enjoyment of beauty. Sports dance is an important form of dance art, music and form can be a very good display. Sports dance has a long history and rich cultural heritage. Chinese sports dance in order to have a breakthrough in the international arena, we must deeply analyze the elements of sports dance composition, tap the cultural connotation of sports dance, deepen the understanding and understanding of the artistic expression of sports dance. Study on essence of sports dance artistic expression is to fully demonstrate the sports dance artistic heritage, deep promote the sports dance embodies the "humanism", to promote the healthy development of sports dance in China to enhance the competitive level of sports dance.

References

1. Macbeth.Kristina.Invitation to the Dnace. Dnace Mgaazine,2004.
2. George Gollin.The Physics of Dance. Journal of Women s Health,2001.
3. IDSF International Dancesport Federation Statutes. Journal of Women s Health,2001.

 DOI: 10.1051/matecconf/201710005057

Role of CEO Age in Determining Executive-employee Pay Gap in Chinese Listed Manufacturing Companies: A Perspective of Risk Aversion

Changzheng Zhang [1*], Qian Guo [1] and Lili Hu [1]

[1]Xi'an University of Technology, School of Economics & Management, Xi'an China

[*]Corresponding author: zcz7901@163.com, 383531334@qq.com,251158817@qq.com

Abstract. The paper investigates the effect of CEO age on executive-employee pay gap (EEPG), a newly focused dimension of executive compensation arrangements by the literature, by choosing the panel data consisting of 3495 firm-years in Chinese listed manufacturing companies during 2010-2014 as the research sample. Empirical analysis based on multiple regression analysis adopting the method of OLS by applying SPSS19.0 makes a new finding, i.e., there is a positive relationship between CEO age and EEPG, which holds robust with the change of the measures. Further investigation on the determination mechanism of CEO age on EEPG show that older CEOs intend to set relatively lower EEPG out of their risk aversion needs.

1 Introduction

A key question in leadership, corporate governance and strategy management research is whether chief executive officer (CEO) matters. A growing literature in the above fields points to the central role of CEOs' individual characteristics in explaining corporate outcome variables such as investment, mergers, or financial decisions, given their role as the top decision-makers in the firm and their data availability due to their positions [1]. The spectrum of CEO traits ranges from age, gender, tenure, risk-aversion, education, and childhood experiences to behavioral biases such as loss aversion, confirmation bias, or overconfidence[2,3]. Among the mentioned CEO traits, CEO age is receiving more and more attention from the scholars, since CEO age may be systematically related to risk preferences both due to the changing incentives over the CEO's career and due to the changing CEO characteristics, both physiological and psychological, as the CEO gets older[4,5].

Effects of CEO age on the choices of organizational policies have been investigated to good degree. Though Serfling (2014) [5] cites anecdotal evidence that portrays younger CEOs as more risk-seeking, Chevalier and Ellison (1999)[6] suggests younger CEOs display risk-averse behavior because they face greater risk of being fired following poor performance. Bertrand and Marianne (2003) argue that physiological changes may lead older CEOs to prefer 'the quiet life' [7], which could pull them away from corporate policies or events that require expending high levels of energy. DavidsonIII et al. (2007) [8] find that firms in which CEOs are nearing retirement age have large discretionary accruals in the year prior to turnover. Nulla(2014)[9] indicates that there is a relationship between CEO compensation and CEO age using accounting performance as a benchmark. Matthew A. Serfling (2014)[5] documents a negative relation between CEO age and stock return volatility, which is consistent with the prediction that risk-taking behaviour decreases as CEOs become older; Brandon(2015)[10] confirm that CEOs promoted during their 40s negatively influence firm value, while CEOs in older age brackets show a positive abnormal return on firm value

Yet, although CEO age and executive-employee pay gap (EEPG) are readily observable, there is surprisingly little evidence on how a CEO's age affects the decision of EEPG in the CEO's corporate. EEPG is a critical organizational policy, which would determine the behavior choices of the top executives, the employees and thus firm performance to a large degree. CEOs with different ages would naturally have the willingness to manipulate EEPG according their own favorites in order to accomplish their own objectives, which may not be the same the organizational objectives. The paper tries to investigate the relationship between CEO age and EEPG by taking the Chinese listed manufacturing companies over 2010-2014 as the sample.

The new finding of the study is that CEO age has positive effect on EEPG in Chinese listed manufacturing companies, which means older CEOs have stronger intention to enlarge EEPG out of their risk aversion, while younger

CEOs have an intention to control the unlimited growth of EEPG out of their firm-served motivations and career concerns.

The next parts of this paper are arranged as follows: section II provides the literature and hypothesis, section III describes the research method and research sample, section IV provides the empirical results and discussion, and the final section is the conclusions.

2 Literature and hypothesis

Incentives compensation, as a frequently used incentives in business practice, has always been a hot topic in theorists, especially its antecedents and real effect on firm performance. EEPG, as one of such main incentives pay methods, is one of the critical issues of corporate governance in Chinese listed companies, and its wild growth is receiving more and more attention from researchers all across the world. Based on the extant literature, beyond all question, EEPG is a true sense of double-edged sword which can potentially and essentially improve or harm firm performance. It means that the function mechanisms of EEPG's effect on firm performance are complicated and it is rather difficult to understand accurately.

EEPG may have both positive effects proposed by the scholars holding tournament theory and negative effects proposed by the scholars holding behavioral theory on firm performance at the same time. However, each effect has its preconditions for getting a dominant position compared with another effect. In other words, different contextual factors will adjust the relative strength of the two EEPG's opposite effects on firm performance. On one side, when EEPG is not too large and the cooperation among employees and top manages is of high performance consequences, the positive effect derived from interpersonal competition and financial incentives due to higher EEPG will beat the negative effects; On the other side, under some other conditions, for example, the relationship among top manages and front-line employees is poor and EEPG is already too large, the negative effects derived from the counterproductive behavior and bad emotions due to higher EEPG will beat the positive effects. Consequently, for scholars, it is critical to find out such contextual antecedents and further investigate its effect on EEPG.

CEOs are responsible for maximizing their companies' value. Though such core function applies to all employees in a firm, the CEO ultimately is responsible for this central task of business, and reports directly to the board of the directors of [11]. Because CEOs bear this ultimate accountability, they are commonly given rather broad authority within their companies and are charged with the overarching leadership, strategy, and direction of their firm running. Therefore, CEOs would certainly be also charged with the setting of EEPG. Different characteristics of CEOs would have essential effect on their choices on different level of EEPG.

Age has been demonstrated to affect personal risk taking in general and more specifically corporate risk taking. For one thing, many studies investigate the role of investors' age on portfolio choices. For example, Ackert et al. (2002) find investors age is closely related to a lower percentage of their risky assets in equities [12], as compared to bonds, consistent with increasing risk aversion with age. Similarly, Bodie and Crane (1997) indicate investors age is significantly and negatively related to percentage of risky assets held in equity [13], while Samanez-Larkin et al. (2010) find unique effects of investors age on risk taking in financial decision making [14]. For another thing, some other studies study risk taking by managers in a corporate setting. For example, Vroom and Pahl (1971) find older managers are more risk averse by adopting a questionnaire administered to managers [15]. Wiersema and Bantel (1992) argue that the firms most likely to undergo changes in strategy generally have top management teams with lower average age[16]. Therefore, we hold the view that older CEOs have higher risk preferences. From the perspective of risk-aversion, CEO age has significant association to the setting of EEPG.

First, older CEOs intend to empower employees with relatively more limited discretion in order to control the possible risks for unexpected consequences potentially derived from the discretionary behavior of the widely distributed employees both regionally and emotionally. Employee discretion would not only enlarge employees' bargaining power with the executives on compensation issues, but enhance employees' potential capability in determining firm performance. In this way, higher employee discretion would necessarily be correlated to higher employees' wage. Since older CEOs want to limit employee discretion, then employee compensation would be limited to a lower degree. Here, given the level of executive compensation, lower employee pay would lead to larger EEPG.

Second, older CEOs intend to delegate executives with relatively higher discretion in order to minimize the possible operation risks due to CEOs' limited energy, knowledge and expertise when they run the firms completely depending on themselves. Executives with higher discretion would be motivated to work hard toward improving firm performance, and the success possibility of firm running would be improved to a better degree than that under the condition of lower executive discretion. Obviously, higher executive discretion needs the match of higher executive compensation. Simultaneously, in order to minimize the possible uncontrollable risk due to higher executive discretion, a relatively higher pay than the theoretically reasonable pay would be a good choice, which can not only motivate executives work harder, but can improve the default cost of executives when they misuse their discretion. Of course, older CEOs have better experience and capability in monitoring and evaluating their executives. In this case, under the condition of older CEOs, executive pay would be set at higher level out of the reason of risk aversion. Here, given the level of employees' compensation, higher executive pay would lead to larger EEPG.

According the discussion above, the paper proposes the following hypothesis.

H1: CEO age has positive effect on EEPG.

3. Method

3.1. Measures

CEO age (CAGE). CAGE is measured by the number of the years from CEO's birth year to the sample year. We do decimals to round up and round down numbers.

EEPG is measured by logarithm of difference between the average of executive compensation and the average of employees' compensation [17].

Control variables. Following the popular literature in investigating the antecedents of EEPG, Seven control variables are chosen. (1) EAST is coded as 1, when the sample firm is located in the eastern China; otherwise, EAST is coded as 0. (2) LNTA is the logarithm of total assets of the sample firm. (3) ROA is the return on assets which measures firm performance. (4) RID represents the ratio of the number of independent directors to board size. (5)CSH is coded as 1 when CEO holds certain share of the listed company; otherwise, CSH is coded as o. (6)CDUA is coded as 1 when CEO also takes the position of the Chairman of the board ; otherwise, CDUA is 0; (7) TMTN represents the logarithm of the size of the top management teams.

3.2 Sample and data

Taking all the listed manufacturing companies which exist during 2010-2014 in Shanghai and Shenzhen stock exchange as the sample framework, the paper selects the final research sample according to the following requirements: (1) The firms should go public before 2009; (2) The firms should not be ever punished publicly by the government authorities for any reasons; (3) The firms should not be marked with ST, PT, SST, or *ST, etc.; (4) The firms should disclose all the data that the paper needs according to our measure design; (5) The firms should run relatively normally and smoothly, i.e., EEPG should be negative. According to such rules, 709 firms are chosen, and a panel sample consisting of 3495 firm-years are designed. The descriptive statistics analysis and correlation analysis results are shown in Table.1.

Table1. Descriptive statistics and correlation coefficients

	Mean	Standard Deviation	1	2	3	4	5	6	7	8	9
1. EEPG	11.990	.962	1								
2. CAGE	52.510	6.929	.162**	1							
3. EAST	.560	.496	.176**	.070**	1						
4. LNTA	9.570	.514	.348**	.110**	-.030	1					
5. ROA	.600	2.934	.184**	.079**	.045**	.099**	1				
6. RID	.370	.056	.066**	-.012	-.042*	-.004	-.027	1			
7. CSH	.350	.478	.242**	.131**	.128**	.082**	.071**	.009	1		
8. CDUA	.210	.405	.064**	-.046**	.022	-.079**	.024	.015	.170**	1	
9. TMTN	17.520	3.915	.070**	.037*	-.102**	.425**	-.007	-.089**	-.030	-.160**	1

N=3495
** represents significance level of 0.05
*** represents significance level of 0.01

4 Empirical test

4.1 Main results

In order to test H1, we design Model 1 based on OLS which takes EEPG as the dependent variable, CAGE as the predictor, and EAST, LNTA, ROA, RID, CSH, CDUA and TMTN as the control variables.

$$EEPG_{it} = \alpha + \beta_0 CAGE_{it} + \beta_1 EAST_{it} + \beta_2 LNTA_{it} + \beta_3 ROA_{it} + \beta_4 RID_{it} + \beta_5 CSH_{it} + \beta_6 CDUA_{it} + \beta_7 TMTN_{it} + \varepsilon_{it} \quad (1)$$

The paper executes the regression analysis on Model 1 by adopting the panel data with 3495 firm-years. Results are shown in Table 2.

According to Table 2, the standardized regression coefficient of CAGE is 0.087 (T=5.754, P<0.0001). That means H1 is confirmed. Our data supports that CEO age has positive effect on EEPG. Specifically speaking, one unit change of CAGE would lead to 0.087 unit of change of EEPG toward the same direction.

With regard to the control variables, it is worthy of pointing out that: (1) Compared with the firms in western China, the firms in eastern China would generally have higher EEPG; (2) Compared the smaller firms, larger firms would favor much higher EEPG; (3) The more higher firm performance is, the higher EEPG; (4) The ratio of independent directors favor higher EEPG, instead of constraining the enlargement of EEPG, which is not confirming to our normal expectation; (5) When CEOs hold the share, EEPG would get larger, which may rely on the higher power of CEOs in determining EEPG at their own interests; (6) When CEO duality is true, EEPG is higher; (7) There is a negative relationship between the size of top management teams and EEPG.

Basically, the conclusions on the control variables are rather in agreement with the extant literature, except for the results of the ratio of independent directors. There is a good need to investigate the relationship between the ratio of independent directors and EEPG in the future study. Besides, the fact that most of the variables show similar effects with those in the other literature show that the validity of the data and methods in this study. From this perspective, the main research finding of this paper, i.e. the positive relationship between CEO age and EEPG, which is found for the first time as far as we know, can be regarded as believable.

4.2 Robustness test

In order to test the robustness of the empirical results, the paper uses REEP as the alternative measure of EEPG, which is calculated by the logarithm of the ratio of the average of executive compensation to the average of employees' compensation, and adopts LNCAGE as the alternative measure of CEO age, which is calculated by the logarithm of AGE. Then Model 2 is given.

$$REEP_{it} = \alpha + \beta_0 LNCAGE_{it} + \beta_1 EAST_{it} + \beta_2 LNTA_{it} + \beta_3 ROA_{it} + \beta_4 RID_{it} + \beta_5 CSH_{it} + \beta_6 CDUA_{it} + \beta_7 TMTN_{it} + \varepsilon_{it} \quad (2)$$

Considering the potential time lag effect of CEO age and other control variables on EEPG, the paper chooses EEPG in the year of "t+1" as the dependent variable, while chooses the data of CEO age and all the other control variables in the year of "t" as the independent variables(t=2010, 2011, 2012, 2013). Then Model 3 can be built.

Table 2. Regression results of Model 1

Model		Unstandardized Coefficient		Standardized Coefficient	T	Sig.
		B	Standardized Error	β		
Model 1	Constant	4.669	.303		15.412	.000
	EAST	.293	.029	.151	9.964	.000
	LNTA	.635	.031	.339	20.292	.000
	ROA	.041	.005	.126	8.336	.000
	RID	1.244	.257	.072	4.834	.000
	CSH	.329	.031	.164	10.581	.000
	CDUA	.127	.036	.054	3.493	.000
	TMTN	-.010	.004	-.042	-2.487	.013
	CAGE	***.012***	***.002***	***.087***	***5.754***	***.000***
	R	.476	R2	.227	Adjusted R2	.225

	F	127.758	Model Sig.	.000	N	3495

Note: Dependent variable is EEPG.

Regression results of Model 2 with the data of the whole sample are shown in Table 3, while the regression results of Model 3 are not provided in detail due to the limitation of paper length. Both results show that H1 still holds. That is to say, CEO age is still positively related with EEPG. The result does not change with the change of variable measures, and it still does not change when we consider the time lag effect.

Table 3. Regression results of Model 2

Table 3. Regression results of Model 2Model		Unstandardized Coefficient		Standardized Coefficient	T	Sig.
		B	Standardized Error	β		
Model 2	Constant	-3.377	.356		-9.491	.000
	EAST	.119	.021	.091	5.701	.000
	LNTA	.329	.022	.260	14.764	.000
	ROA	.014	.004	.065	4.092	.000
	RID	.610	.183	.053	3.330	.001
	CSH	.241	.022	.178	10.914	.000
	CDUA	.087	.026	.054	3.364	.001
	TMTN	-.014	.003	-.085	-4.779	.000
	CAGE	***.354***	***.078***	***.073***	***4.538***	***.000***
	R	.375	R2	.141	Adjusted R^2	.139
	F	71.478	Model Sig.	.000	N	3495

Note: Dependent variable is REEP.

4.3 Discussion

Our theoretical analysis and empirical data show that CEO age is positively related to EEPG in Chinese listed manufacturing companies. The fact indicates that older CEOs favor higher EEPG. Though the main reasons of such a result have been explained to good degree from the risk aversion perspective in section 2, there still are some other alternative perspectives.

For example, a possible perspective may rely on the CEO's affiliation needs to his or her closest workmates, i.e., the top executives. Older CEOs would favor interpersonal relationship of better quality than younger CEOs would, and especially it is true in China. The fulfillment of CEO's affiliation needs mainly comes from the good relationship between CEOs and their executives, because they work together with the highest frequency and the longest time. In this case, older CEOs intend to provide their executives with higher compensation than the reasonable level it should be. Here, EEPG gets larger and larger with the implementation of meeting CEO's affiliation needs.

Another possible perspective may rely on the various degrees of familiarity of older CEOs with their top executives and the employees. Besides the possible differences in personal feelings, older CEOs would have much higher recognition on top executives' capability and they would think the top executives make higher contributions to firm performance than the front-line employees do. Even being out of firm-served motivations, older CEOs would enlarge EEPG since they know the capability and performance of top executives much better than they know those of the employees.

5 Conclusions

This paper investigates the role of CEO age in determining EEPG from the perspective of CEO's risk aversion, which has not been explored till today by the scholars. Data from the Chinese listed manufacturing companies over 2010-2014 prove that CEO age can improve the enlargement of EEPG out of the purpose of lowering the operation risks. It is a new finding of great significance since it does break the traditional view on the differences between older CEOs and younger CEOs in China, which thinks the former would favor more equal distribution than the later do. The fact is that older CEOs would enlarge EEPG at their own discretion. When facing the situation of letting older CEOs set their firm's EEPG, the board and the shareholders should pay more attention to the possible unreasonable enlargement of EEPG, since the older CEOs are more self-served and risk-averse; while when the newer CEOs are setting the EEPG, higher discretion should endowed to them since they are more firm-served with great passion.

In the future, more empirical studies should be done in the following aspects: (1) Samples should go beyond the limitation of manufacturing companies; (2) Methods should adopt more strict empirical models which consider the possible endogenous problems; (3) The measure of EEPG should be refined to a better degree according to the newest accounting practices.

Acknowledgements

This research was supported by Projects of the National Social Science Foundation of China under the Grant "15BGL109".

References

1. Carpenter, Mason A.,. The Handbook of Research on Top Management Teams (Edward Elgar Publishing, Inc., Northampton, MA). (2011)
2. Musteena, Martina, Vincent L. Barker IIIa and Virginia L. Baeten, CEO attributes associated with attitude toward change: The direct and moderating effects of CEO tenure, Journal of Business Research 59, 604–612. (2006)
3. Malmendier, U., & Tate, G. Who makes acquisitions? CEO overconfidence and the market's reaction. Journal of Financial Economics, 89, 20–43 (2008)
4. Elsaid, Eahab and Ursel, Nancy D. Age, CEO succession, and risk taking. Accounting and FinanceResearch,1(2).76-86. http://scholar.uwindsor.ca/odettepub/13 (2012)
5. Matthew A. Serfling. CEO age and the riskiness of corporate policies. Journal of Corporate Finance, Volume 25, Pages 251–273 (April 2014)
6. Chevalier, J., Ellison, G.. Career concerns of mutual fund managers. Quarterly Journal of Economics 114, 389-432 (1999)
7. Bertrand, Marianne, and Antoinette Schoar, Managing with style: The effect of managers on firm policies, Quarterly Journal of Economics 118, 1169-1208 (2003)
8. Wallace N. DavidsonIII, Biao Xie, Weihong Xu, Yixi Ning. The influence of executive age, career horizon and incentives on pre-turnover earnings management, Journal of Management & Governance. Volume 11, Issue 1, pp 45–60 (March 2007)
9. Nulla, Yusuf Mohammed, The Effect of CEO Age on CEO Compensation Using Accounting Performance as a Benchmark: An Empirical Study on NYSE Index Companies. Journal of Management Policies and Practices, March 2014, Vol. 2, No. 1, pp. 119-133 (June 8, 2014)
10. D'Ewart, Brandon H., "The Effect of CEO Gender, Age, and Salary On Firm Value". CMC Senior Theses. Paper1059.http://scholarship.claremont.edu/cmc_theses/1059 (2015)
11. Lin, Tom CW. "CEOs and Presidents." UC Davis Law Review 47: 1351. Web. 9 (Nov.2014)
12. Ackert, L., Church, B., & Englis, B. The asset allocation decision and investor heterogeneity: A puzzle? Journal of Economic Behavior and Organization, 47 (4), 423-433 (2002)
13. Bodie, Z., & Crane, D. Personal investing advice, theory, and evidence. Financial Analysts Journal, 53 (6),13-23 (1997)
14. Samanez-Larkin, G., Kuhnen, C., Yoo, D., & Knutson, B. Variability in nucleus accumbents activity mediates age-related sub-optimal financial risk taking. Journal of Neuroscience, 30, 1426-1434 . (2010)
15. Vroom, V., & Pahl, B. Relationship between age and risk taking among managers. Journal of Applied Psychology, 55 (5), 399-405 (1971)
16. Wiersema, M., & Bantel, K. Top management team demography and corporate strategic change. Academy of Management Journal, 35 (1), 91-121. http://dx.doi.org/10.2307/256474 (1992)
17. Taekjin Shin. Explaining Pay Disparities between Top Executives and Non-executive Employees: A Relative Bargaining Power Approach, Social Forces Volume 92, Number 4, pp. 1339-1372 (June 2014)

The Advantages of BIM Application in EPC Mode

Hongbo Qin*

School of Management Engineering, Zhengzhou University, Henan Vocational College of Judicial Police Officers, Zhengzhou 450001, China

*Corresponding author: 121425031@qq.com

Abstract. Applying BIM in the model of EPC General contracting, general contractor, designer, constructor, Supervisor and other professional sub contractors can share resource which can achieve the project information, standardization, visualization, collaborative and ultimately make the project quality, schedule, cost, safety and other objectives to achieve maximum.

Key words. EPC model BIM apply Advantages

1The mode and characteristics of epc general contract

1.1 EPCMode

Design, procurement and construction (EPC) / Turnkey: design, procurement and construction contract refers to the general contractor enterprises in accordance with the contract, to undertake the project design, procurement, construction, commissioning and service work, quality, safety, schedule and cost of the Contractor shall be responsible for the overall project. Turnkey general contracting is an extension of the general contracting business and the responsibility of the design procurement construction. It is a project that meets the use function and the conditions of use.

Design build (D-B): Design - construction contract refers to the general contractor enterprises in accordance with the contract, to undertake the project design and construction, and the quality, safety, schedule and cost of the Contractor shall be responsible for the overall project. According to the different size and type of the project and the owner's requirement, the general contracting of the project can also adopt the design, procurement and construction (E-P), procurement and construction contracting (P-C, etc.).

1.2 The characteristics of EPC

1.2.1 To give full play to the role of the market mechanism. Not only the owner of the project as an investment project first, but also the architect, the contractor, the construction division are from this priority. In order to participate in the competition effectively, the whole project is divided into several relatively independent work packages. Responsible for the design, manufacture or supply of materials and components for each work package, and responsible for the construction and installation of all the work packages. This dual coordination feedback, relying on the project related parties are able to follow the common recognition of the control procedures, national norms and technical standards.

1.2.2 Consulting engineers to provide a complete design of the professional, but the design stage only to the initial design or expand the depth of the preliminary design, the detailed design of the construction plans, and the latter is completed by the contractor. In particular, some of the more independent sub sub project construction plans, and sometimes also known as the two design, by the professional subcontractor to complete, but need to be approved by the architect.

1.2.3 architect known as the market situation, they will not only design as its architectural works, and implement the cost control in the whole design process, must have overall awareness of cost control, cost control is particularly important in this stage. Set up design is the key to influence the success or failure of the owner's investment.

2 The application of bim technology

BIM is a kind of technology, that is, how to use information technology to carry out the construction of technical activities. At the same time BIM is a kind of thinking, how to analyze the whole process of thinking, how to use the analysis of the problem and solve the problem of thinking.

BIM is the abbreviation of building information model. Building information model integrated geometric model information, functional requirements and performance of all components, the use of all the information the whole life cycle of a building model covers separate buildings within the project, including the process of information, such as the construction schedule, construction process, maintenance management and property management etc..

BIM technology is a platform (integrated professional information model, field model, mechanical model and temporary model, schedule, budget, drawings and other business data; (three) the direction of technology, production, business), five core (integrated model, construction simulation, schedule control, cost control, quality tracking and management, ten application: basic application; pipeline management, construction simulation, schedule control, volume and price inquiry, material extraction, animation disclosure, quality tracking, drawings management, collision detection. As an important tool of information management, work efficiency can be greatly improved.

3 The advantages of general contracting using bim under the epc mode

Design, procurement, construction and integration in EPC mode and facilitate the application of BIM, it can realize the design, procurement and construction of collaborative work, shorten the construction period, improve management efficiency, to achieve integration, construction project information management.

3.1 The Advantages Of the design side

The designer can make large scale, time, complex design content and work easier, using various features of the BIM model can be analyzed and simulated and optimized performance of various design schemes of each stage (such as sunshine, wind, thermal environment, landscape visibility, noise, energy consumption, emergency treatment cost, etc.), the best performance determination of building design standards, the overall project is fundamentally improved. The designer has built a warm water and related information by using BIM software to program 360 all-round display; issued construction plans; according to the characteristics of EPC project issued by the engineering quantity statistics and engineering cost, provide visual and standardized data.

BIM design means compared with the traditional means of design, multi person collaborative design is also one of the highlights. The advantages are as follows: 1, improve the design efficiency, communication and non professionals, practical design results directly; 2, improve the drawing quality and efficiency, quickly and accurately find the drawings of German "riddled bruise"; 3, simulate the construction, optimization of the procedure; 4, convenient construction phase of the exchange and communication; 5, facilitate the operation and maintenance phase of the work; 6, To integrate design and cost control.

3.2 The Advantages of construction side

In the construction design support can radically reduce the construction cost, shorten the construction period, improve the quality, and the stages of design and construction of the complete building information can be reserved for all owners in the construction operation and maintenance and lay a solid foundation, fundamentally changed the traditional management information lost caused by serious situation information. BIM construction schedule simulation and other functions can enhance the owner's control of the entire project. After the completion of the building, with the complete information of the BIM can provide for the owners of the operation, maintenance and management. Construction is a highly dynamic and complex process, the current construction project management is often used to represent the progress of the network plan, due to professional, low degree of visualization, can not clearly describe the progress of construction and a variety of complex relationships, it is difficult to image expression Dynamic Change Process of Engineering Construction. The integration of spatial information, time and cost information into a visual 5D (3D + Time + Cost) model can be intuitively and accurately reflected throughout the construction process and virtual Image progress. 5D construction simulation technology can be in the construction process of the project to develop a reasonable planning, accurate grasp of the progress of construction, and optimize the use of construction resources and scientific site layout, the entire project construction schedule, resources and quality of unified management and control to shorten the construction period, Reduce costs, improve quality. In addition, with the help of 5D model, construction companies in the project bidding will be competitive advantage, BIM can let the owners intuitive understanding of the bidding unit of the tender construction of the main control methods, construction arrangements are balanced, the overall plan is reasonable, Units of construction experience and strength to make effective assessment.

3.3 Design and Construction Methods in EPC Mode

In the EPC general contracting mode, the design side and construction side as a union, the design side through the BIM application fundamentally improve the performance of the building and less change, improve production management efficiency, it can reduce construction costs to improve the construction Stage of the profit rate, from the construction phase to obtain the income outside the design fee; the construction side can be in the design side with the support from the fundamental lower construction costs, the construction stage of the profit margin. In other words, in the EPC mode design and construction as a whole can achieve mutual benefit and win-win benefits can promote the application of BIM to provide a powerful driving force. This model is also suitable for China's current construction project development of the overall requirements.

3.4 Calculation of Engineering Quantity with BIM Technology

BIM design methods in the amount of statistical engineering compared to the original CAD drawings of equipment and materials have a greater statistical advantage. After the drawing is finished, the engineering quantity statistics can be carried out by the function of the software itself, and the relevant price information can be obtained, so that the engineering cost can be obtained, and the working efficiency of the cost budget can be greatly improved compared with the CAD design. But in this regard is not perfect function, such as electrical professional cable length of the statistics, there is no way to select the line, and the actual amount of a gap. Need to wait for further improvement of the software to solve these problems. It can also provide more accurate and intuitive data and views for final accounts.

3.5 The advantages of the purchaser

Using BIM technology, the integration of design, procurement, and construction is beneficial to the purchaser in the management of the project, to coordinate with the design and construction, to unify and standardize the model, to construct the physical characteristics and functions of the project Which is based on building information and provides decision-making for each stage of the construction project. The purchaser can know the material type and quantity more accurately through the model and better understand the structure of the material in the collision test.

In the EPC general contracting mode, BIM technology is used to integrate various professional contents into the BIM model. The total package, design, construction, supervision and professional subcontractors share resources and communicate with each other in a timely manner. A very good social benefits. EPC mode BIM application of a useful exploration, will be able to optimize the allocation of labor, material equipment approach, cost control, program management, improve the efficiency of management, but also speed up the construction of the project speed, So that the various BIM participants in different environments can benefit, has great economic value and potential, but also for the green, environmental protection, low-carbon construction provides a reliable data information support.

Acknowledgments

EPC project general contracting is to achieve "project function" as the ultimate goal which includes planning, design, procurement, construction, completion and acceptance, trial operation, operation and other stages of the implementation of a number of stages or the whole process。And BIM technology can play the role of information, visualization, standardization in the whole process of EPC project management.

References

1. Li Songdong. EPC general contracting model design management research. construction economy, 2012 (7): 68-70.
2. Wang Shuangjun. Combined with the research on schedule management mode of. housing industry EPC engineering BIM technology, 2016 (7): 46-60.
3. Liu Huan, Liu Shangyang. BIM technology applied to the total contract cost management advantage points Analysis of. building economy, 2013 (6): 31-34.
4. Tang Xiaoling, Tian dawn. Building information model (BIM) technology diffusion and Application Research Study. building economy, 2013 (6): 98-100.
5. Liu Wenping. BIM EPC in Guo REDCOLLAR in highway engineering application mode of. construction economy, 2014 (9): 31-34.

Study on the Withdrawal Way of Silk Road Fund under the Belt and Road Initiative Based on Game Theory Model

Yingying Qu and Shuying Wang

Management Engineering Department, Zhengzhou University, Zhengzhou, China

Abstract. The Silk Road Fund is an important financial support of The Belt and Road Initiative,it has a strict investment operation process to realize the capital appreciation,so the withdrawal is crucial.This paper models the utility game and find the game mechanism between Silk Road Fund and target enterprise.And then we put forward some referential suggestions such as choosing suitable withdraw opportunity, perfecting talent team and information transmission mechanism,strengthening cooperation with local government,making effective use of gambling agreement and actively exploring new exit mode.

1 Introduction

Silk Road Fund mainly bases on equity investment,through diversified investment and financing methods such as equity,debt,and loans,focusing on commitment to the Belt and Road Initiative to provide investment and financing support of economic and trade cooperation and bilateral multilateral interoperability between the countries within the Belt and Road Initiative Frame,then it can promote China and the countries and regions along the Belt and Road Initiative to achieve common development and prosperity.

2 The importance of the Silk Road Fund withdrawal mechanism

Silk Road Fund has a strict investment operation process, “financial, investment, management, retirement”the four links complement each other,any problem on any part will affect the normal operation of the Silk Road Fund.Silk Road Fund’s ultimate goal is to achieve capital appreciation,rather than to always hold shares of enterprises. Silk Road Fund’s profit does not rely on the control of enterprise management,but through participation in project management to gain long-term capital earnings,it is a capital cycle and value-added process.The exit of the fund can also provide the basis for judging the success of the target project.And the investment return rate of the Silk Fund can be used to measure the investment efficiency of the target project.In order to maximize the benefits,the fund manager must choose the appropriate exit mode according to the target enterprise’s operating conditions and their own situation.

3 The classification of Silk Road Fund withdrawal way

At present,Private Equity fund exit methods are mainly IPO,target corporate repurchase,external transfer,liquida-Tion and so on.

(1)the IPO.IPO way can obtain the largest value and the highest degree of return, the investment rate of return can be substantially more than expected,it can bring maximum economic benefits to the Silk Fund and the target companies.For the Silk Road Fund’s commercial projects,it tends to use the IPO withdrawal way.

(2)the target corporate repurchase.This approach is simple,and its time is short, the fund can quickly and completely exit,it’s more suitable for small investment projects.

(3)external equity transfer.Through the equity transfer the Silk Fund and the target companies can seek capital appreciation.This approach is simple,low cost, good capital mobility,high flexibility and low risk.But for the Silk Road Fund is a special PE funds to service the Belt and Road Initiative,there may be local government participation in the situation.

(4)bankruptcy liquidation.Bankruptcy liquidation means the failure of fund investment.Since bankruptcy liquidation is the sign of investment failure,the game analysis of the follow-up exit mode does not include bankruptcy liquidation.

4 The impact factors of Silk Road Fund withdrawal way

Silk Road Fund is a long-term development investment fund to achieve the Belt and Road Initiative,it is also a financial innovation under the Belt and Road Initiative. The Belt and Road Initiative advocates "all-dimensional,multi-level" interconnection, Silk Road Fund withdrawal way will inevitably be affected by various factors.

First of all,from a macro point of view,the withdrawal of the Silk Road Fund is mainly affected by three aspects:(1)business cycle.Silk Road Fund is an open investment platform,the countries and regions within the Belt and Road Initiative are likely to benefit.(2)capital market conditions.The capital market of each country and region along the route is characterized by its own characteristics.The higher the perfection of the capital market system,the greater the advantage of the equity transfer. (3)laws and regulations.The laws and regulations of the countries and regions within the Belt and Road Initiative to a certain extent impact the Silk Road Fund exit,strict laws and regulations can increase the difficulty of the Fund to withdraw.

Secondly,from the microscopic point of view,the withdrawal of the Silk Road Fund is mainly affected by two aspects:(1)the Fund's own preferences.Return on investment (ROI) is a factor that fund managers often need to consider.(2)the status of the target enterprise itself.The industrial characteristics of the target enterprise,the development stage,the financial status and team changes and other factors also affect the withdrawal of the Silk Road Fund.

		Silk Road Fund	
		α_1	α_2
target enterprise	β_1	[I-F+M-(1-π)N，(1-π)N-C+rI]	[π(I-F)+M，(1-π)(I-F-C)+rI]
	β_2	[I-(1-π)N，(1-π)N]	[πI，(1-π)I]

Figure 1 The Utility Game Matrix of Target Enterprise and Silk Fund

5 The game theory model of Silk Road Fund withdrawal way

As a important financial support of the Belt and Road Initiative,Silk Road Fund's investment success or failure is related to the implementation of the strategy.On the choice of withdraw methods,the purpose of the Silk Road Fund and the target enterprise is not the same.The game theory is mainly to study the conflict of interest and consistency of the main body how to make decisions,as well as the interaction and impact of these decisions.Therefore,the withdrawal of the Silk Road Fund can be seen as a game between the Fund and the target enterprise.Since IPO and stock transfer are the most common and the most profitable two exit methods,this paper uses the two methods as the basic assumptions for game analysis.

5.1 Basic assumptions

(1)Suppose the man-made Silk Road Fund and target enterprises.
(2)Assume that The Silk Road Fund has two exit strategies:non-convertible securities (denoted as strategy α_1) and converted securities (denoted as strategy α_2); The target firm has two kinds of exit strategies:IPO (denoted as strategy β_1) and equity transfer (denoted as strategy β_2).
(3)Assuming under the IPO exit mode,the total cost of the target enterprise to be listed is F,the Silk Road fund as a national strategy to support the financial innovation, due to the particular nature of their own,in order to assist the successful listing of the target business will also have some labor costs,time costs and opportunity costs,then it can be denoted by C(C<F);Equity transfer's cost of the two sides is very little,then it can be ignored.
(4)Assume that the expected total return after the withdrawal of the Fund is I.And under the IPO,Silk Road Fund can get a reputation earning denoted by R, assume R is a linear function of I,then R=rI(r>0);As the equity transfer is not open to market transactions,therefore the equity transfer's reputation earning can be ignored.
(5)Assume that before the exit,the Silk Road fund's share is 1-π,the target enterprise's share is π.
(6)Assume that the revenue of the target firm under the control right is M;when the Silk Road Fund does not implement the conversion of securities,the gain is (1-π)N(N<I-F-C).

5.2 Game utility matrix and model analysis

According to the above hypothesis,the utility combination matrix of the exit strategy choice between the Silk Fund and the target enterprise is established.As shown in Figure 1.

As can be seen from Figure 1,(1) from the perspective of the target enterprise analysis:①under the $\alpha_1\beta_1$ combination strategy,the gain is I-F+M-(1-π)N,under the $\alpha_1\beta_2$ combination strategy,the gain is I-(1-π)N.If I-F+M-(1-π)N>I-(1-π)N,that is when M>F,the target company will select the IPO withdraw way;If M<F,it indicates that the listing fee is too high,the target companies tend to choose the stock transfer withdraw way.②under the $\alpha_2\beta_1$ combination strategy,the gain is π(I-F)+M,under the $\alpha_2\beta_2$ combination strategy,the gain is πI.If π(I-F)+M>πI,that is when M>πF,The target enterprise choose IPO,on the contrary it choose stock transfer.(2)From the perspective of the Silk Road Fund analysis:①under the

$\alpha_1\beta_1$ combination strategy,the gain is $(1-\pi)N-C+rI$,under the $\alpha_1\beta_2$ combination strategy,the gain is $(1-\pi)N$.If $(1-\pi)N-C+rI>(1-\pi)N$,that is when $rI>C$,the reputation earning is larger than the additional costs,then the Silk Road Fund can choose IPO withdraw;on the contrary it choose stock transfer.②under the $\alpha_2\beta_1$ combination strategy,the gain is $(1-\pi)(I-F-C)+ rI$,under the $\alpha_2\beta_2$ combination strategy,the gain is $(1-\pi)I$.If $(1-\pi)(I-F-C) +rI>(1-\pi)I$,that is when $rI>(1-\pi)(F+C)$,the Silk Road Fund can choose IPO withdraw;on the contrary it choose stock transfer.(3) For the whole market,the sum of the two sides of the game to be maximize is the best strategy.①under the $\beta_1\alpha_1$ combination strategy,the sum of the two sides of the game is $I-F+M-(1-\pi)N+(1-\pi)N-C+rI=I-F-C+M+rI$, denoted as U_1; under the $\beta_1\alpha_2$ combination strategy,the sum of the two sides of the game is $\pi(I-F)+M+(1-\pi)(I-F-C)+rI=I- F-(1-\pi) C+M+rI$,denoted as U_2.For $1-\pi<1$,then $U_2> U_1$, It indicates that through the IPO way out,the Silk Road Fund should choose to convert securities, and now the total utility is U2.②under the $\beta_2\alpha_1$ combination strategy,the sum of the two sides of the game is $I-(1-\pi)N+(1-\pi)N=I$,denoted as U_3;under the $\beta_2\alpha_2$ combination strategy,the sum of the two sides of the game is$\pi I+(1-\pi)I=I$,denoted as U_4;then we can see that $U_3=U_4$,It indicates that through the stock transfer way out,it is just the redistribution of utility between the Silk Road fund and the target enterprise,and now both parties should make an integrated decision based on the financial situation or other performance indicators to arrive at an optimal exit plan.

Through the above analysis, we can see that reputation and listing fees is great impact factors when the fund managers to make decision.In theory,if the Silk Road Fund withdraw from the IPO,the total utility of the conversion of securities is greater than the total utility of not converting securities.The result may be caused by the social effect of the IPO exit and capital realization effect.If the Silk Road Fund to exit through the equity transfer,whether converse securities or not,the total effect is the same,and now,the financial situation or other performance indicators are also factors that fund managers should take into account when make the decision.

6 Summary and recommendations

The successful withdrawal of the Silk Road Fund is the end of the current round of capital cycle, but also the beginning of the next round of capital circulation. In view of how the Silk Road Fund could improve its exit mechanism, this article put forward some reference suggestions from the following several aspects.(1) the appropriate exit timing is very important. Under the IPO withdraw way after the target business growth period is the ideal high-speed exit time. Under the equity transfer withdraw way, the general selection is the maturity of the target enterprise to exit.(2) strengthen their own management, improve the personnel,perfec the information transfer mechanism. As the important financial innovation support of the Belt and Road Initiative, the Silk Road Fund should break the original economic and management model, attract interdisciplinary top-level talents and rich practical experience professional talents, establish a more sound, efficient comprehensive information access and delivery mechanism.(3) strengthen cooperation with the national and regional government within The Belt and Road Initiative. Silk Road fund's investment of large-scale projects usually has a certain social benefits to the countries and regions within The Belt and Road Initiative. This a mutually beneficial and win-win cooperation model. (4) make the effective use of the gambling agreement. Gambling agreements can be risk neutral and can reduce losses in the event of a capital appreciation failure at the end of the round.(5) actively seek other new withdraw ways. Silk Road Fund's political, economic and cultural environment is more complex and changeable, with the continuous development of the Silk Road Fund, we can follow the traditional withdrawal of the road while moving forward, and at the same time we can continue to explore more new withdraw ways which is suitable for the Silk Road Fund itself, meets the requirements of the new financial era, adapt to international capital market demand. This is in line with The Belt and Road Initiative strategy's requirements.

References

1. YANG Li-Hua,ZHOU Li-Ping,WENG Dong-Ling.Foundation of Silk Road,PPP and Construction of The Belt and Road Initiative-Based on Game Theory.Asia-Pacific Economic Review,2016(2):25-26.
2. Wei Lei.Silk Road Fund boost The Belt and Road Initiative interoperability. International Business Finance and Accounting,2015 (4):7-8.
3. Yang Hua.YOZMA government to guide the operation mode of the fund and the inspiration to China's Silk Road Fund.International Vision,2015 (11):52-53.
4. Hu Yanwei.Study on the he advantages and exit mechanism of PE institutions to participate in overseas mergers and acquisitions.Business Review,2015 (2):108.
5. ZHANG Tie-shan,SHI Kai,HE Qian.Gaming Choice of Venture Capital Exit Pattern.Accounting communication,2009(2):134-136.

Risk Evaluation in Auto Spare Parts Transport Based on the AHP Method

Rong Zeng[1] and Chang Xu[2]

[1]Wuhan Huaxia University of Technology, Wuhan, China
[2]Wuhan University of Technology, Wuhan, China

Abstract By way of example we analyze various risks exist in auto spare parts transportation, building engineering logistics risk evaluation index system of auto spare parts transportation, and the introduction of AHP, be more rational auto spare parts transportation risk assessment methods, strengthen risk management auto spare parts auto spare parts transportation to provide a reference for other engineering logistics auto spare parts transportation.

1 Introduction

Auto spare parts primarily relates to transport objects of transport, handling, packaging and storage, these relatively independent but closely linked aspects constitute the complete logistics process. In particular, the process object transportation risk is an objective reality, and has a dynamic variability, diversity, but also by subjective factors, so the analysis of engineering logistics risk is very important. [1].

Currently there are many research concerning logistics risk[2-3], but not a specific risk assessment system of auto spare parts transportation. Taking a highway transport precast beams Tianjin, for example, the establishment of a transport risk factor assessment model, a comprehensive evaluation of the risk presented for the segment prefabricated beams risk assessment system and risk control measures, to hedge against the risk of auto spare parts transportation auto spare parts services.

2 AHP

Analytic Hierarchy Process (Analytic Hierarchy Process, AHP) is home Satty American operations research in the early 1970s to make his decision analysis method for quantitative and qualitative. It policymakers thought process complex decision-making system will be modeled quantify. Application of this method, the makers of complex problems down into a number of factors and a number of levels, between the various factors pair wise comparisons and calculations, get different weights, provide the basis for selecting the best solution[4-5]. AHP method can be divided into four steps: First, analyze the relationship between the various factors in the system, create a system of hierarchical structure; then the same level of importance of various elements on a layer corresponding guidelines pair wise comparison, construct pair wise comparison judgment matrix; and then by the judgment matrix calculation is to compare elements of the guidelines should be relatively heavy weight; and finally calculate the levels of the elements of the overall target system combining weights, and sort.

3 The establishment of auto spare parts transportation risk assessment model

 DOI: 10.1051/matecconf/201710005060

3.1 Construction of evaluation index hierarchy Project

logistics auto spare parts transportation is a very complex process, involving many areas of technology, management, science, etc., is also related to the relevant factors affecting the transport business, management section, etc., for the selected indicators are more in the actual process should remove duplicate factors avoid larger deviations from the actual situation. Tianjin highway below a certain precast transport for example.

3.2 Judgment matrixt

After the establishment of the hierarchy analysis model, the relative importance of each factor on each score given by experts. Two more things to make an accurate comprehensive judgment, is assessed by expert opinion, referring to *SUTTY* nine scale method, in proportion to the scale defined by the extent of importance assigned to construct the index levels of judgment matrix. To avoid errors of judgment, but also to construct judgment matrix consistency test[6-7].

Judgment matrix calculation utilizing the relative weight W_i and each index coefficient matrix characteristic root, as follows:

(1) calculate the matrix product of all elements in each row: $M_i = \prod_{j=1}^{n} a_{ij}$, where i =1,2, ..., n, n is the order of the matrix;

(2) M_i be opened the n-th power: $\overline{W}_i = \sqrt[n]{M_i}$;

(3) $\left(\overline{W}_1, \overline{W}_2, ..., \overline{W}_n\right)^T$ will be normalized: $w_i = \frac{\overline{w}_i}{\sum_{j=1}^{n} \overline{w}_j}$;

(4) Calculate the characteristic roots: $\lambda_{\max} = \sum_{i=1}^{n} \frac{(BW)_i}{nW_i}$.

Calculate the characteristic roots level by a single sort consistency index: $C.I = \frac{\lambda_{\max} - n}{n-1}$;

Random Consistency Index is then calculated: $C.R = \frac{C.I}{R.I}$; if $C.R \le 0.1$ that is acceptable consistency of judgment matrix, wherein the average random consistency index RI is derived using a random sampling method, the value shown in Table 1.

Table 1. Average random consistency index R.I

dimension	1	2	3	4	5	6	7	8
R.I	0.00	0.00	0.58	0.96	1.12	1.24	1.32	1.41

Criterion level factors environment, operation, management, auto spare parts vehicle to the target layer U of total auto spare parts transpotation risk judgment matrix in Table 2-6.

Table 2. Judgement matrix of U-U_i

U	U_1 U_2 U_3 U_4	W_0	**index**	
U_1	1 1/3 1/5 1/3 3	0.077	$\lambda_{\max}$=4.104	C.I=0.035
U_2	1 1/3 2	0.238		
U_3	5 3 1 3	0.517	R.I=0.96	C.R=0.036<0.1
U_4	3 1/2 1/3 1	0.168		

Programmer layer factor U_{1j} of U_1 judgment matrix in Table 3.

Table 3. Judgement matrix of U_1-U_{1j}

U_1	U_{11} U_{12} U_{13}	W_1	index
U_{11}	1 1/3	0.106	λ_{max} =3.04 C.I=0.02
U_{12}	1/5	0.258	
U_{13}	3 1	0.636	R.I=0.58
	1/3		C.R=0.03<0.1
	5 3		
	1		

Table 4. Judgement matrix of U_2-U_{2j}

U_2	U_{21}	U_{22}	W_i	*index*
U_{21}	1	1/3	0.25	λ_{max} =2 C.I=0
U_{22}	3	1	0.75	
				R.I=0 C.R=0<0.1

Table 5. Judgement matrix of U_3-U_{3j}

U_3	U_{31}	U_{32}	U_{33}	*Wi*	*index*
U_{31}	1	1/3	1/5	0.106	λ_{max} =3.04
U_{32}	3	1	1/3	0.258	
U_{33}	5	3	1	0.636	C.I=0.02
					R.I=0.58
					C.R=0.03<0.1

Table 6. Judgement matrix of U_4-U_{4j}

U_4	U_{41} U_{42} U_{42}	W_i	*index*
U_{41}	1 1/2 1/6	0.090	λ_{max} =3.013 C.I=0.007
U_{42}	2 1	0.322	
U_{43}	1/4	0.588	R.I=0.58
	6 4		C.R=0.011<0.1
	1		

When λ_{max} = 2, $C.R$ = 0 is exactly the same. *CR* larger the value, the worse the consistency of judgment matrix, the matrix due to the judgment *CR* value is less than 0.1, it is considered that they have a satisfactory consistency.

It should be noted that in practice, the hierarchical model of the number of layers of factors and factors of pairwise comparisons between two values are not static, but according to the specific circumstances by experienced, expert judgment and strong decision.

4 Fuzzy comprehensive evaluation

4.1 Identify factors set and evaluation set

As the auto spare parts complex reality, even the experts can not give an accurate assessment can only be a vague assessment. Linguistic variables can be approximated to describe the complex phenomenon not clearly defined characteristics. Such as "risk" as a factor set "high risk, high risk, moderate risk, general risk, low risk, no risk" can be set as one of its judgment. On precast beam transported example, the judge set V = {a very high risk, high risk, general risk, low risk, low risk}, and quantify.

Table 7. Quantified risk grade

Risk level	Very high risk	High risk	General risk	Low risk	Very low risk
Indicator Value	(8,10)	(6,8)	(4,6)	(2,4)	(2,0)

4.2 Determine the fuzzy relation matrix and membership

Fuzzy Comprehensive Evaluation of Multi-level system is to start step by step to make a more comprehensive evaluation from the lowest level up, get up to the highest level of target comprehensive evaluation results of the original problem. Here departure from the program level indicators, the first layer of each factor fuzzy comprehensive evaluation criteria, and then the target layer fuzzy comprehensive evaluation factors.

(1) The second level fuzzy comprehensive evaluation factor. Order membership $R_{ij}=\{r_{ij1},r_{ij2},...r_{ijn}\}$, then the second level of single factor evaluation matrix:

$$R_i = \begin{bmatrix} r_{i11} & r_{i12} & \cdots & r_{i1n} \\ r_{i21} & r_{i22} & \cdots & r_{i2n} \\ & & & \\ r_{im1} & r_{im2} & \cdots & r_{imn} \end{bmatrix}$$ factor to obtain a second level fuzzy comprehensive evaluation level is

$B_i'=W_i \circ R_i=[b_{i1}, \ b_{i2}, \ \cdots, \ b_{i3}]$, where"$\circ$"is fuzzy Operators can select the appropriate algorithm according to different situations. Assessment by the first group of 10 people on the second level of each factor single factor evaluation factors, such as the factors set U_1 = {environment, weather, geology} three factors of risk evaluation, assessment team voting results are shown in Table 8.

Table 8. U_{11} risk evaluation by expert

	Very high risk	High risk	General risk	Low risk	Very low risk
U_{11}envir-onment	0	0	0	3	7
U_{12}Weather	0	0	0	5	5
U_{13}geology	0	0	0	5	5

4.3 Calculate the weight of each layer combining weights

The right combination of computing the weight of each layer are shown in Table 9, and consistency checking.

Table 9. The total sort Computation sheet of U

U	***U₁*** ***0.077***	***U₂*** ***0.238***	***U₃*** ***0.517***	***U₄*** ***0.168***	***weight***
U_{11}	0.106				0.008
U_{12}	0.258				0.020
U_{13}	0.636				0.049
U_{21}		0.25			0.060
U_{22}		0.75			0.179

U_{31}	0.106		0.055
U_{32}	0.258		0.133
U_{33}	0.636		0.329
U_{41}		0.090	0.015
U_{42}		0.322	0.054
U_{43}		0.588	0.099

$C.I=\sum_{i=1}^{n} W_i^B (C.I)^T$ =(0.077,0.238,0.517,0.168)(0.02,0,0.02,0.007)T=0.0131

$R.I=\sum_{i=1}^{n} W_i^B (R.I)^T$ =(0.077,0.238,0.517,0.168)(0.58,0,0.58,0.58)=0.4420

$C.R=\sum_{i=1}^{n} W_i^B (C.R)^T$ =0.03<0.1,indicating that meet the conformance requirements.

4.4 Ealuation

As can be seen from Table 7 that U_2 *and* U_3 manage operational impact on auto spare parts transportation safety prefabricated beams relatively large, and relatively small impact on the environment. U_{33} level in the second stage security personnel awareness maximum impact on transport. In the process of precast auto spare parts transportation to ensure safe and smooth it can reach the destination, the risk must be controlled from the management and operation of two aspects, in particular, to strengthen the security personnel education.

5 Conclusion

Step article using AHP to Tianjin, a highway auto spare parts transportation precast beams, for example, to build a hierarchy of risk, combined with expert opinion to determine the various levels of judgment matrix, right by the formula to calculate the factors of weight, and consistency inspection, forming a total weight sort table, laid the foundation for further evaluation of the risk of auto spare parts transportation. However, because data collection is limited, this risk hierarchy needs to be further improved.

Acknowledgments

The authors thank the reviewers who gave a through and careful reading to the original manuscript. Their comments are greatly appreciated and have help to improve the quality of this paper. This work is supported in part by the Scientific Research Plan of Hubei Province (No.B2016408).

References

1. Wang, Ailing. Fuzzy risk assessment of construction project contractors based on entropy-weight coefficient method. Electric Technology and Civil Engineering (ICETCE), 2011 International Conference on IEEE, (2011).
2. Markowski A S, Mannan M S. Fuzzy risk matrix Journal of Hazardous Materials, (2008).
3. Li, Chaoyang, C. Ye, and Y. Shen. Risk Assessment of Subway Pit Construction Based on Fuzzy Comprehensive Evaluation. Chinese Journal of Underground Space & Engineering 10 (2014).
4. Erkut, Erhan, and A. Ingolfsson. Transport Risk Models for Hazardous Materials: Revisited. Social Science Electronic Publishing
5. ERKUT Erhan, and INGOLFSSON Armann. Transport risk models for hazardous materials: revisited. Operations Research Letters (2005).

6. Verma, Manish. Railauto spare parts transpotationation of dangerous goods: A conditional exposure approach to minimize transport risk. Transportation Research Part C Emerging Technologies 19 (2011).
7. Chapman, Rebekah L., et al. The impact of school connectedness on violent behavior, transport risk-taking behavior, and associated injuries in adolescence. Journal of School Psychology 49 (2011).
8. Paulo Sérgio, Lucas Da Silva, E. D. A. Vania, and R. Maria Eunice. Assessing outcome in interhospital infant transport: the transport risk index of physiologic stability score at admission. American Journal of Perinatology 29(2012)

Protection of Reservoir Water Quality Combining with Rocky Desertification Present Situation

Weirui Chu*

College of Architecture and Civil Engineering, Nanyang Institute of Technology
*Corresponding author: fbdb.happy@163.com

Abstract. Middle Route Project of S-N water diversion will be completed this year, and by that moment, the dream of a "stream of clear spring sent to Jingjin" will be come true. The water source of Middle Route Project of S-N water diversion-Danjiangkou water reservoir's more than half area is in Xichuan town's Nanyang City. To investingate and protect the environment situation around water source, we have gone deep into many mountains, soil sampling, stone sampling and make chemical laboratory analysis after plenty of preparation work in early stage. We come up with feasible plan about Xichuan area's rocky desertification, to plant plants with higher survival life in the barren mountain with advanced ways, which can achieve water conservancy on one hand, and on the other hand, to prevent from water and soil loss so as to protect water quality of Dan Jiangkou reservoir.

1 Current situation and forming reasons of xichuan's stony desertification

1.1 Basic situation of Xichuan's stony desertification

Stony desertification is the abbreviation of rocky desertification, and is the extreme form of karst area's land deterioration. As the stone desertification area's water storage capacity is low, which can easily lead to drought, water exhaustion, woods decline and soil poor, stone desertification is also called "land cancer".

Xichuan's stony desertification spreads around Danjiangkou and drainage basins such as Dan Jiang river, Guan river, Qi river and Tao river,etc, 14 towns which covers about 0.96 million banbo. The annual soil erosion modulus is 41tons/hectare. The mountain rock of Xichuan rocky desertification is mainly limestone, and the soil is mainly ballast soil. As this area is washed by rain water all the year around, the bed rock exposed, gravel buildup, soil thin and woods rare, it is called vividly "black stony mountain" by local people.Our country has no uniform standards for division of stony desertification's degree yet at present, so considering factors such as objective basis, value basis and theory basis, etc, we achieve the following distinguishing standards[1] in table (1),according to Xiong Kangning and Liping's <<Remote sensing of Kasite's stony desertification-GIS classis research—GIS>>. Following is pictures of all kinds of

Fig.1 mild rocky desertification

rocky desertification taken during Xi Chuan investigation in fig.(1).,fig.(2).,fig.(3).

1.2 Formation cause of stony desertification

1.2.1 Natural Factor

Xichuan belongs to the Southwest of Nanyang basin, and has many mountains ups and downs, peaks rising one after another, ravines crossbar, rock bared, bad water binding capacity, low vegetation coverage and serious water and soil loss, and it belongs to classic mountainous town of limestone mainly.

Fig.2 moderate rocky desertification

Fig.3.serve desertification

Normally, limestone's main chemical composition is as following in table(2):We collected rocks in 22 places in seven towns in Xichuan, and the average content of CaO in the collected rocks is above 60%, which proves that the place we are investigating is carbonate rocky desertification beyond all question.

Table1. Index table of impure carbonate rock Karst area's stony desertification degree division

Strength grade	Bedrock exposure/%	Regolith/ %	Gradient/ °	Vegetatin +bedrock /cm	Average soil thickness/cm	Agriculture use value
No obvious rocky desertification	<40	>60	<22	>70	>20	Suitable for water conservancy
Potential rocky desertification	>40	<60	>22	50~70	<20	Suitable for forestry and grassland farming
Mild rocky desertification	>60	<30	>25	35~50	<15	Critical suitable for forestry and grassland farming
Moderate rocky desertification	>70	<20	>30	20~35	<10	Difficult to use land

Table2. Chemical composition content of limestone(%)

Name of rock	SiO_2	CaO	Fe_2O_3	Al_2O_3	MgO	MnO	SO_3	P_2O_5
limestone	1.01	56.27	0.27	0.27	0.057	0.0065	0.009	Trace amount

Natural factors is the basic condition of rocky desertification's forming. The abundant carbonate in karst area has the features of easily leached and slow soil forming, which is the material basis of forming rocky desertification. High mountain, steep slope and concentrated rain prove erosion power and dissolution conditions for stony desertification.

Strong karst process promotes the forming and development of stony desertification [2]: the quicker corrosion speed not only corrode all of mother rock's soluble constituent, but also takes away most of the insoluble substance and reduce carbonate's soil-building ability.

Stony desertification is often spreaded over pure limestone area [3], and this is because the soluble matter substance not easily to be become in pure limestone is easily to lose, only some soil after weathering. Besides, under present climate conditions, it takes about 0.5 million years to form 1m thickness soil, while as in lack of soil and plants in

Stony desertification area, the karstification is weakened relatively, then the time of forming 1m thickness soil will be much more longer.

1.2.2 Hunan factors

The human factor is the main reason for the formation of rocky desertification land.

In the 1950s due to the construction of the Danjiangkou Reservoir, 543000 acres of land in Xichuan inundated(among this 285000 acres is farmland), relocation of 202,000 immigrants (including 127000 million after placement by the county). As mostly farmland were flooded, the after-by immigrants had to steep land clearing, deforestation, cultivation, grazing in order to maintain their livelihood , which leads to the destruction of vegetation, serious soil erosion and the mountain rocky desertification being intensified.

When we had investigation in Biaochi Town, Jinhe, the elders in the local place told us that there were a large number of virgin forest and some trees had to be encircled by two people in the forties and fifties of last century.In those years of refining iron and steel, a large number of trees were felled for ironmaking, and the whole mountain was almost hackled.People's environmental awareness was very weak at that time, and reforestation projects was not carried out immediately.The mountain has been washed by rainwater through the year, which gives rise to the seriouc rocky desertification problem at present.Damage of rocky desertificationSerious soil erosion, frequent natural disasters Xichuan rocky desertification region's forest ecosystems has been severely degraded, forest soil and water conservation capacity has been weakened and limestone bedrock was corroded and brushed by rain and surface water, and the water and soil loss is severe. With the intensifying of water and soil loss, it results in the loss of much of the land surface and a serious decline in productivity. In addition, due to the rocky desertification land vegetation coverage is very low[4], soil's ability to conserve water is poor, which intensify degree of drought in winter and spring.Water and soil loss leads to land desertification, rocky desertification seriously affects the ability of soil's capacity of water and soil conservation, and this vicious cycle has become one of the major factors in the induction of natural disasters in the region. "when there's rain there's flood, no rain namely drought " is the main feature of rocky desertification area[5].

2 Affect water quality of Danjiangkou Reservoir water source

Danjiangkou Reservoir watershed involves Danjiangkou City of Hubei Province, Yun County, Yunxi, Zhangwan etc and Xichuan, Dengzhou of Nanyang including all 16 towns in Xichuan.

According to the tasks indicators Nanyang Municipal Environmental Monitoring Station have been completed within the monitoring in 2013[6], now most of the rivers within Xichuan territory belong to II Grade water quality, some individual rivers belong to Grade III water quality.Namely, now most of the water quality of the river water diversion is in full compliance with the requirements. However, the soil test results show that most of the soil samples we got are acidic soils. The pH value of Large Shiqiao's three sampling locations[7] are as following table (3).

 DOI: 10.1051/matecconf/201710005061

Table3 Average ph value of soil sample in Yangying town DaShiqiao

Sampling site	Dashiqiao Yangying adjacent foothills	Dashiqiao Yangying adjacent mountainside	Dashiqiao Yangying hilltop neighborhood
pH value	6.5	6.5	6.4

Elevation of these three borrow points respectively H = 176m, H = 204m, H = 281m. Getting ten soil samples within a certain range of each earth point to ensure that the measured results eliminate certain chance. To finish the Danjiangkou Reservoir water supply, its water level needs to reach an altitude of at least 170m, so all of the areas within Yangyin town adjacent mountain have huge threat to water quality of Danjiangkou Reservoir potentially.

Xichuan Danjiangkou reservoir has drainage area of 2616 square kilometers, and takes 56.9% of the total basin area. Rocky desertification within Xichuan caused serious soil erosion, a lot of soil and sand flow into Dan Jiangkou Reservoir, which on one hand leads to servoir water turbidity and nitrogen exceeded; on the other hand, it causes reservoir sedimentation. Statistics show that for 20 years from 1960 to 1979, the total sediment of Danjiangkou reservoir area is

897.155 million cubic meters, with an average annual deposition of 44.86 million cubic meters, which has seriously affected the service life of the reservoir.

2.1 Affect people's production and life, economic development and social stability

As we have learned from the County Land Resources Bureau, Xichuan's arable land is not rich originally, and nearly 14 acres of Xichuan's arable land was submerged after the Danjiangkou reservoir was built and completing water diversion, which make the original not rich arable land become stretched.

This town losses 2.2 million tons of soil due to rocky desertification, water and soil loss on one hand leads to deficiency of large amounts of nitrogen, phosphorus and potassium nutrients, one the other hand exacerbates land barren, declines arable land and conflicts between people and land in the reservoir become more prominent. Currently, the county's average per person farming is less than 0.8 acres, of which more than 10,000 people less than 0.3 acres of arable land per capita; on the other hand, rocky desertification's water source is dried up during winter and spring, which not only affect agricultural irrigation water, but also cause difficulty in drinking water for people and animals. Up to now there are seven impoverished townships of the county, 60 poor villages, 85,000 poor people, of which 2.3 million can't even solve the basic food and clothing problem, and this have severely affected social stability.

3. Governance program of stony desertification

3.1 Existing prevention measures of Xichuan

First, intensify propaganda, close hillsides to facilitate afforestation. Take measures of closing hillsides to facilitate afforestation for areas of lighter degree. Slight rocky desertification can adopt moderate grazing.

Second, take moderate reforestation activities according to the existing land conditions of the mountains. For middle-degree rocky desertification areas, Xichuan will take large-scale afforestation activities from March to April each year, and most of them are arborvitae trees.

Third, explore the rescue plan of rocky desertification land actively. There's no complete programs for severe rocky desertification area at present yet. Policy, technology, finance, specific management modes , etc. are all constraints.

3.2 Put forward prevention measures for rocky desertification aiming at investigation conditions

3.2.1 Prevention measures of mild stony desertification issues

To offer the problem of slightly rocky desertification appropriate policy guidance, then its development can be relatively easily to be controlled First of all, the local forestry sector can continue to adhere to the basic policy of critical grazing farmland eliminate grazing. Large-scale farmers cab be guided around town to develop aquaculture, and carried out large-scale farming. Administrations can carry out matchmaking between multiple households, so that a certain number of farmers to unite and establish cooperatives, introduce advanced management and technology, and make use of Xichuan's geographical advantages to create a group of local characteristics aquaculture industry.

Second, local forestry sector can carry our management innovation on the basis of the original policy. Slight rocky desertification mountain areas does not mean can't conserve any water source, but need a certain growing conditions. The famous Fussen Pharmacy in Henan Province has an annual output of nearly 400 million yuan , and Chinese medicine raw materials has very large gap.Growing herbs requires special mountain climate, Xichuan county government is totally able to make policy concessions, reach certain agreement with Fussen Pharmacy, first try to grow moderately in a small range of Chinese herbal medicines, and then allow Fussen medicine for large-scale cultivation of needed medicines after obtaining some results in accordance with relevant laws and regulations, which is threefold thing. In this way, rocky desertification can be effectively prevented from developing into moderate desertification and severe rocky desertification, can create a considerable part of the profits for the company, can promote local rural labor force.

In the course of the investigation, we take back a lot of soil samples in the local rocky desertification mountain and make analysis. After laboratory analysis by Soil and Fertilizer Station of Nanyang City, the organic matter, nitrogen, and phosphorus contained in the soil are very considerable.As long as the climate is suitable , planting ordinary herbsis not a problem in general.For example, golden cypress, berberine, magnolia and plum are all very suitable for growing[9].

Xichuan's Slight rocky desertification area is 827000 hectares, which accounts for more than one-third of the total area of rocky desertification.If we make good use of all these woodlands, its economic profits and environmental effects can not be underestimated.

3.2.2 Solutions of moderate and severe rocky desertification problems

Some of the severe rocky desertification areas are directly high mountains, some places are like slopes, and some is fundamentally barren stone ground. The problem we need to solve is how to make these rocky desertification areas own soil, and use what ways to conserve the soil. If we manage to solve these two problems, the problem of rocky desertification half will be settled half.

3.2.2.1 Governance model should be innovated

To change the situation of low survival rate of planted trees , it is necessary to abandon this backward cultivation methods, and forest planting can adopt the method of responsible woods for planting. Xichuan has many river network and the groundwater can completely be utilized, and they can directly drill a vertical shaft in the side which can solve water problem. Our country has made great achievements in the governance of rocky desertification, and gave birth to a large number of forestry companies living on dealing with stony desertification exclusively in the meantime of making achievements. carry They carry out desert tourism economy in the original rocky desertification area and has very good effect. Forestry can make contact with these companies achieving results, emulate the governance policies during desertification control, give appropriate concessions and jointly govern the rocky desertification problem relying on those companies' superior technology, personnel and financial strength.

3.2.2.2 Specific measures of dealing with moderate and severe stony desertification

In order to completely solve the rocky desertification problem, we must not let the bare bedrock exposed. In the research process, we saw a lot of arable land has been hoarding in one place.These earthworks was about 3-5 meters above the ground, with an average size of two football fields. Grass mats woved by sraw are covering around all sides of the soil from bottom to the side to prevent from loss of rail soil. There are many earthworks like this piled in the total Xichuan, and by people's maintance after transforming these soil to a specific location, we can not only prevent the soil from losing, but also keep it a certain fertility. In the future, these soil will be to moved to relatively barren land area, which is the famous Xichuan fertilizing project.

Then why need to dig these soil? Now Danjiangkou

Reservoir water level is at an altitude of 142 meters, if store water, its water level will reach at least 176 meters, therefore, Xichuan has made immigration of nearly 200000 people. The digged and transferred soil is immigrants'arable land. Once the farmland is flooded by the reservoir, the water is bound to absorb soil nutrients, cause eutrophication of water bodies, especially exceeded nitrogen which will seriously affect water quality and safety and cause water turbidity; on the other hand, after so many soil isflooded, it will definitely cause a lot of sediment deposition in the reservoir area.

County government and relevant department can actively communicate with companies having successful treatment experience in stony desertification, invite professional forestry technician, soil and water conservation specialists, planting technicians , soil laboratory staff to form an expert team to the county for more deep, comprehensive and detailed research; these research can be used in the method of combining government and the relevant agriculatural forestry companies who are interested.Thus,on one hand, the government can reduce some of the financial burden;on the other hand, related companies can get first-hand information on investigation and research, and the government and departments can both make the first judgment according to related research report.

The local government departments can provide a range of land, soil and other preferential policies when the agriculture and forestry companies take advantage of rocky desertification issue to do scale cultivation. Administrations can sign dedicated rocky desertification land use agreement with the agriculture and forestry company, and the agreement can clearly define that the agriculture and forestry company have to be engaged in economic activities related to rocky desertification treatment and both sides should implement according to the terms of the agreement.

In the first two years when the land was originally approved, the agriculture and forestry companies can quickly perform "experimental field" according to survey results. "Experimental field" can be carrid out in layered manner.

The first step is to find ways to solve the problems of soil covering the bedrock on the rock desertification mountain.It is obviously not pratical to cover every place in the stony desertification area with uniform soil.The feasible solution is not to requir covering every inch of rocky desertification bedrock with soil, but use some engineering measures - to establish a proper platform for accumulation of soil, so that the soil has a stable "foothold".In order to achieve better water and soil conservation, the tilt angle of the "starting point" must be different from the hillside.If the vertical distance between the two "foothold"keeps at about 10 meters, then the effect will be very good.

The second step is to solve the problem of selecting what plants to grow [9]. Grow plants of different species that can produce economis of scale in different zone, which in the present situation is one of the more viable option. We measured its soil pH is normal, not acidic [8]; the local climate in winter and spring is still very dry, so Xichuan's mountain area is not suitable for growing tee tree.

According to the agriculture and forestry staff, the output of pepper, walnuts, pomegranate is very commercioganic, but this method did not continue.

We adopt advanced management mode, and the introduction of companies can be a good solution to this problem. After research we found that the late economic effects of these crops do not show up, and the main reason is still a lack of water resources.After discussion, the final solution is that the agriculture and forestry companies build a large reservoir in the peak; then establish a comprehensive crop water irrigation system. In this regard, Israel has the world's leading advanced drip irrigation technology, and we can actively refer to their mature technical experience [10].

All in all, government support needs policies and companies rely on policies for financial investment; the local people can gain some income from this and are willing to take participation in, which can be said to be the secret of solving stony desertification problems.

4 Conclusion

Our country has made great achievements in the treatment of land rocky desertification, and the gradually emerging desert tourism and "comprehensive economic profit in desert" is just our successful experience in prevention and control of desertification.From these experiences, we can actively absorb and utilize in the treatment of rocky desertification.In summary, according to the successful treatment experience in land rocky desertification,we should do well in the following points for dealing with rocky desertification:rocky desertification can cause great impact on the water quality of Danjiangkou Reservoir, and it is also a very negative factor for the water level guarantee of the Danjiangkou Reservoir.To completely solve the problem,there must be in need of government's policy support,society's positive participation, raise awareness while dealing with rocky desertification and strengthening research;scientific vertification,deep analysis;improving effectiveness of governance relying on science;improving the monitoring system and implementation of dynamic monitoring. Ensure the treatment to be carried out in stages and step by step, and only in this way, can the wish of a steady stream of clear spring to be conveyed to Beijing and Tianjin come true.

References

1. Li Lihan, Zhang Nanlu, Sun Daquan, Yang Qun. Road Engineering Materials. Beijing: China Communications Press, 2010
2. Wang Shangyan. Walking into the Triassic Marine of Guizhou. Beijing: Geological Publishing House, 2007
3. Li Dinglong, Li Hongdong. Civil Engineering Geology. Beijing: Science Press, 2009
4. Liu Tuo. Karst Rocky Desertification in China: current situation, causes and control. Beijing: China Forestry Publishing House, 2009
5. National Standard of People's Republic of China. Surface Water Environmental Quality Standard (GB3838-2002). Beijing: China Environmental Science Press .2002
6. Zhang Shaoqian, Jia Donghong, Li Li. Nanyang City
7. Environmental Quality Summary. Nanyang Environmental Monitoring Station .2013
8. The Soil Test Report. Nanyang City Fertilizer
9. Monitoring Laboratory Centre .2014
10. Wei Yunxun. Chinese Medicine Planting New Technology. Beijing People's Publishing House, 2009
11. China's 21 Century Agenda Forestry Action Plan Environment. 2010
12. Liutuo, Zhang Kebin, Linqiong. China Desertification Control Strategy.Beijing: China Forestry Press, 2006

Study on the Present Situation and Cause of Formation of Rocky Desertification in Xichuan County

Sensen Li*, Hongwei Du and Weirui Chu

School of Civil Engineering, Nan Yang insititute of Technology, NanYang, China

*Corresponding author: 402894195@qq.com.org

Abstract. As the eco-environmental security guarantors for water source, Xichuan and its hydro-ecological condition play important roles in Middle Route Water Transfers Project from South to North. Via analyzing the soil, water quality, rock and vegetation et al., we studied the current extend of rocky desertification in Xichuan. Considering the local situation, we analyzed the human factors and natural factors that contribute to the rocky desertification in Xichuan, which are helpful in reducing rocky desertification and worthy of promoting.
Key Word. Rocky Desertification, Present Situation, Cause Analysis.

1 Foreword

Xichuan County, Nanyang City, is adjacent to Danjiangkou reservoir bank. It is the water source ecological environment security area for the water diversion project in the middle line of South-to-North Water Diversion Project, and the last barrier to purify the inflow of water. Perfect soil and water conservation at the local place has a decisive role for ensuring the water quality. Due to the influence of various factors, there is a big proportional of rocky desertification area in Xichuan County, and rocky desertification area causes very great potential risk to the water quality in reservoir area.

The rocky desertification in Xichuan County of Nanyang City not only affects the water quality of the water source in the middle line of South-to-North Water Diversion Project, but also degrades the fertility of local soil and affects the local agricultural and forestry economy. The combination of the control of rocky desertification with the improvement of the living conditions of farmers is beneficial to the local social and economic development.

2 The Status of Rocky Desertification in Xichuan County

Xichuan County has stretching mountains, various rivers and steep slopes and deep valleys throughout the territory. It has total area of 2820 square kilometers, most of which is located within Danjiangkou reservoir basin. The hilly area covers 1815 square kilometers (2.723 million mu), accounting for 64.9%; gryada and flat land cover 588 square kilometers (882 thousand mu), accounting for 21%; the water area covers 395 square kilometers (592 thousand mu), accounting for 14.1%. The total population of the county is 612834 people (in year 2012) [1].

Figure 1. Rocky Desertification Status in Jinhe Town, Xichuan County

Xichuan County now has rocky desertification area of 2 million mu, accounting for nearly 50% of the county's total area, of which the severe rocky desertification area is 484 thousand mu, the moderate rocky desertification area is 689 thousand mu, the shallow rocky desertification area is 827 thousand mu, and the average annual soil erosion modulus is

41 tons / ha [2-3]. The rocky desertification areas are distributed in the surrounding area of Danjiang reservoir, Danjiang River, Taohe River, Guanhe River, Qihe River and other river basins, including 12 towns and townships, such as: Jinhe, Shangji, Madeng, Dashiqiao, Laocheng, Jingguan, Siwan, Xihuang, Maotang, Taohe, Cangfang. The rocky desertification mountain rock in Xichuan County is most limestone and the soil is mostly tyrosinance subgroup ballast soil. Due to long eroding of rainfall, the base rock is exposed, and the gravel is aggregated, so the soil is barren and the vegetation is sparse.

In order to better grasp the first-hand information about the status of rocky desertification in Xichuan County, we conducted a field research in Xichuan County, and sampled the local soil, water, rock, vegetation, etc. for analysis.

Table 1. Soil Nutrient Test Report of Nanyang Soil and Fertilizer Monitoring and Analysis Center

Sampling site	#	Total N (%)	Total P (%)	Total Potassium (mg/kg)	Rapidly Available phosphorus (mg/kg)	Slowly available potassium (mg/kg)	Rapidly available potassium (mg/kg)	Organic matter (%)	pH value (pH)	Humidity (%)
Xichuan Wangjialing (peak)	1	0.464	--		17.6	624	208	2.75	6.2	--
Xichuan Wangjialing Mountain	2	0.171	--	--	16.3	672	256	1.35	6.2-	--
Xichuan Wangjialing Mountain (mountain foot)	3	0.198	--	--	17.2	396	196-	1.03	-6.1-	--
Shangji Yanshan Mountain(peak soil)	4	0.243	--	--	35.8	824	248--	1.24	6.3--	--
Shangji Yanshan Mountain (hillside soil)	5	0.220	--	--	33.2	788	316--	0.97	6.2--	--
Shangji Yanshan Mountain (mountain foot soil)	6	0.805	--	--	28.0--	532--	140--	2.04	6.2--	--
Jinhe Biaochi Village (peak)	7	0.366	--	--	41.1--	656--	368--	2.27	6.4--	--
Jinhe Biaochi Village (in mountain)	8	0.879	--	--	31.9--	672--	224--	3.37	6.5--	--
Jinhe Biaochi Village (Mountain Foot)	9	0.293			39.8	240	392	0.78	6.4	
Madengxi Valley (Peak)	10	0.592	--	--	39.5	696	408	4.18	6.3	--
Madengxi Valley (hillside)	11	0.293	--	--	18.6	1456	204	3.13	6.3	--
Madengxi Valley (base station)	12	0.226	--	--	17.7	1528	200	1.64	6.3	--
Taohe Baishagang (Peak)	13	0.141	--	--	17.2	2084	280	1.44	6.5	--
Taohe Baishagang (in mountain)	14	0.211	--	--	15.8	1716	268	2.40	6.5	--
Taohe Baishagang (mountain foot)	15	0.144	--	--	17.2	2072	232	0.52	6.4	--
Yangying Village (peak soil)	16	0.199	--	--	17.2	1380	220	2.75	6.5	--
Yangying Village (in mountain)	17	0275			18.1	1984	256	3.61	6.5	
Yangying Village (mountain foot)	18	0.107			17.6	1620	108	1.44	6.4	

Xianghua Yangying Village	19	0.113	--	--	9.9	2064	240	0.52	6.3	--
Taohe Hujia	20	0.110	--	--	17.2	1612	148	0.76	6.4	--
Shengwan Jiawan	21	0.193	--	--	18.1	528	112	0.15	6.7	--
Shengwan Wamiao	22	0.107	--	--	19.4	320	160	0.69	6.8	--

2.1 Soil Analysis

We used random sampling method in the 7 sever rocky desertification towns and townships, to collect 22 soil samples, and detected the nitrogen, phosphorus, potassium, organic matter, pH and moisture content in the samples. The inspection report is shown in Table 1.

According to this table, we can see that the sampled soil (7, 8, 9) in Jinhe Town showed weak acidity; although the weak acidity of the soil does not have great effect on water quality, but it poses a potential threat to the water quality in the long term, so we should take appropriate measures to control.

Affected by depth of water storage in the middle line of South-to-North Water Diversion Project, part of the arable land will be flooded; in order to avoid the impact of originally applied fertilizers in the part of arable land to the water quality, Xichuan County transferred the fertilized soil in these area to the barren highland area to achieve a multiplier effect. We can see from Table 1, in the soil transferred area, the soil elements in the sampled soil in Jinhe Town has very little impact to the water quality and can meet the requirements of the water supply quality.

2.2 Water Quality Analysis

Table 2. Nanyang City Surface Water Testing Schedule Unit: mg/L

River / Item	Qihe River Xichuan Gaowan	Danjiang River Xichuan Zijingguan	Danjiang River Xichuan Shijiawan	Laoguanhe River Xichuan Zhangying
PH Value	8.10	8.07	7.96	7.88
Dissolved oxygen	7.44	7.46	7.35	7.27
Potassium permanganate	2.60	3.03	2.81	2.89
Biochemical requirement	2.34	2.79	2.38	2.53
Ammonia nitrogen	0.21	0.42	0.25	0.24
Petroleum type	0.01	0.04	0.02	0.01
Volatile phenol	0.0016	0.0016	0.0016	0.0014
Mercury	0.00002	0.00002	0.00002	0.00002
Lead	0.005	0.005	0.005	0.005
Chemical oxygen demand	12.62	14.18	12.99	13.28
Total nitrogen	1.87	3.53	3.35	1.67
Total phosphorus	0.04	0.19	0.06	0.05
Copper	0.0005	0.0005	0.0005	0.0005
Zinc	0.025	0.025	0.025	0.025
Fluoride	0.18	0.025	0.50	0.19
Selenium	0.00025	0.40	0.00025	0.00025

Arsenic	0.0005	0.00025	0.00025	0.00025
Cadmium	0.005	0.00025	0.0005	0.0005
Hexavalent chromium	0.002	0.003	0.010	0.006
Cyanide	0.002	0.004	0.002	0.002
Anionic surface-active agent	0.025	0.025	0.025	0.025
Sulfide	0.01	0.01	0.01	0.01
Water quality category	II	III	II	II

According to the water quality summary (as shown in Table 2) of main rivers in "Nanyang City Environmental Quality Summary (2013 year)", on the basis of the "Surface Water Environmental Quality Standards", it can be judged that the water quality of main rivers in Xichuan County can basically meet the requirements of class II and class III water quality, and meet the requirements of current water supply.

2.3 Rock Analysis

By random sampling rocks in Jinhe Town, Xichuan County, for analysis(as shown in Table 3), the content of calcium oxide in the rock reached 60%, so we can determine the rock in Biaochi Village in Jinhe Town is limestone.

2.4 Vegetation Analysis

Jinhe Town belongs to moderate rocky desertification area, with less vegetation in the rocky area. Through field investigation, the vegetation in the rocky desertification area mainly includes: pines, cypresses, apricots, tung trees, walnut trees and cherry trees, etc.

Table 3. Chemical Composition Table of Rock

Rock name	Silicon oxide	Calcium oxide	Ferric oxide	Alumina	Magnesium oxide	Manganese oxide	Sulfur trioxide	Anhydrous phosphoric acid
Limestone	1.01	56.27	0.27	0.27	0.057	0.0065	0.009	
Granite	69.62	1.81	2.60	15.69	0.022	0.022	0.14	0.02
Quartz	98.43	0.21	1.23	0.09		0.006	0.21	0

3. Formation Mechanism of Rocky Desertification

The aggravation of soil and water loss is the main factor that caused the rocky desertification, and the enlargement of the area of rocky desertification has further aggravated the soil erosion. The factors influencing the rocky desertification include natural causes and human causes.

3.1 Natural Factors

Xichuan County belongs to limestone mountainous county, where the lime soil can be easily leached, and the geographical condition is poor; the bedrock is exposed due to the perennial rainfall erosion and the surface is filled by gravel; because rocky desertification is the result of erosion and soil loss, they remain relevant. Therefore, the rocky desertification development trend is influenced and controlled by rock, soil properties, vegetation, terrain slope, rainfall duration, rainfall intensity and other factors[4].

3.1.1 Soil and Terrain Analysis

The rock in Xichuan County is mainly the carbonate rock series, and it has strong resistance against wind erosion, and the soil forming process is slow, which is the objective background condition and the basic reason for the shallow soil and easy rocky desertification in Xichuan County. The surface of Xichuan mountainous area is rugged and broken, there are many mountains and the slopes are steep. The hilly area of the county covers 1815 square kilometers (2.723 million mu), accounting for 64.9% of the total area of the county; the surface structure with many mountains and steep slopes is not conducive to the preservation of water and soil resources. The surface structure with many mountains and steep slopes aggravated the loss of water, soil and fertilizer, and under the impact of some human activities, vast karst mountains become rocky desertification.

The lime soil can be easily leached, and the geographical condition is poor; the bedrock is exposed due to the perennial rainfall erosion and the surface is filled by gravel; because rocky desertification is the result of erosion and soil loss, they remain relevant[5]. Therefore, the rocky desertification development trend is influenced and controlled [4] by rock, soil properties, terrain slope, and other terrain factors.

3.1.2 Effect of Precipitation and Vegetation Rate

Xichuan has rich rainfall, and the annual rainfall is between 391.3 to 1423.7 mm; the average annual rainfall is 804.3 mm; the initial flood time in this territory is very early and the final flood time is very late; the flood season is from June to September, during which the rainfall accounts for 58% of the total annual rainfall; the highest percentage is up to 79.8% in certain years; the geographical distribution of rainfall is not uniform, with the feature of more rainfall in northwest and less rainfall in southeast. The average rainfall is between 731.2 to 809.4 mm; the average annual rainfall in the mountainous area is basically between 900 mm and 1300 mm; the heavy rain occurs mainly in spring (accounting about 40%) and summer (accounting over 55%). The spring and early summer are the sowing seasons for large area of arable slope; the crops (corn, rape, green manure, etc.) are at the seedling stage, and the loose slope soil is not well covered, so the rainstorms in spring and early summer exacerbate the rocky desertification development.

The vegetation can slow down the velocity of runoff on slope surface and absorb the flowing rain and prevent erosion of rain to the surface. The more vegetation coverage, the more slowly the soil and water will lose. The effect of vegetation on soil erosion usually depends of the forest coverage and canopy density; the higher the forest coverage rate is, the greater the canopy density will be, and the slighter the soil erosion problem will be. If a large range of vegetation is destroyed, it will reduce the forest coverage, and the soil erosion problem will be more serious, and this will result in the some rocky desertification.

3.2 Human Factors

Xichuan County has a big population and backward economic development, and the population and limited land per capita form a contradiction; the irrational use of land promoted the development of rocky desertification. Especially in areas with low forest coverage and large population and steep slope terrain, the rocky desertification is very serious.

3.2.1 Population Factor

Xichuan County has total population of 746 thousand, of which agricultural population is 658 thousand, accounting for 88.2% of the county's total population. In recent decades, the county had very high birth rate and low mortality rate resulted in an obvious rising trend of population. The size of population density directly determines the demand for resources, so the region with high population density cannot guarantee enough land resources. In order to meet the needs of survival, the expansion of cultivated area increased the destruction of vegetation. With the continuous weakening of the ability of self rehabilitation of the environment, the rocky desertification has become more and more serious.

Rapid population growth, high population density and low population quality caused Xichuan fall into a vicious cycle: population increase - over reclamation - the degeneration of soil due to erosion - rock desertification - economic poverty.

3.2.2 Economic Level

As a state-level poverty-stricken county, the backward regional economy in Xichuan County has seriously restricted the development of agriculture and the improvement of living standards. The predatory operation, deforestation, excessive reclamation, turf damage, digging roots, straw burning and other phenomena frequently took place in that area, and the farming methods are very backward.

Agricultural activities for excessive reclamation are carried out in the absence of cultivation conditions or protective measures. The cultivation of steep slope and slope planting in Xichuan County is a serious problem. The annual tillage soil accelerates the loss of soil resources and increases the rocky desertification area.

The ecosystem in Xichuan has very poor self protection ability. Along with the continuous development of infrastructure construction, the demand for timbers continued to increase. In order to obtain economic benefits, some people deforested in a wide range regardless of the consequences. In addition, the traffic conditions in some areas of Xichuan are quite underdeveloped, and the people are living at a very low standard; most of the people use natural plants as fuel in their lives. These factors result in a wide range of deforestation, so the original vegetation was severely damaged and the carrying capacity of the local environment was decreased.

In the 1950s, the construction of Danjiangkou reservoir caused 543000 mu of land to be flooded in Xichuan (including 285000 mu of arable land) and the relocation of 202000 residents; as the arable lands of these relocated residents were flooded, they planted crops on the steep slopes and deforested to make a living. Their behaviors damaged the vegetation and caused serious soil erosion and the rocky desertification in mountainous areas.

4. Concluding remarks

In order to protect the canal ecological environment, enhance the ability of local soil conservation, the ability of water conservation ability and reservoir water purification capacity, reduce non-point source pollution, improve local residents' living environment and increase their income, improve their life quality, explore the reasonable and efficient exploitation mode of ridge and slope region with rocky desertification to realize the regional ecological environment protection and the coordinated development of regional economy, it is necessary to select representative ridge and slope area within the narrow basin near the reservoir bank in Xichuan County and carry out special research on rocky desertification control.

References

1. Yong Hao, Guo Zhansheng, Yang Zhaoxing, Luo Zhifa. "Study on the Current Status and Control Measures of Rocky Desertification in Danjiangkou Reservoir Area - Taking Xichuan County, Henan Province as an Example". Hebei Agricultural Science, 2012:75-77
2. Yang Zhaoxing, Feng Dexian, Guo Zhansheng. "Rocky Desertification and Control Measures in Water Source Area in the Middle Line of South-to-North Water Diversion Project". Regional Research and Development, 2013:104-107
3. Xu Yan, Wang Ruipeng, Ren Bupan; Canal Water Source Area and Main Canal Ecological Corridor Construction in the Middle Line of South-to-North Water Diversion Project". Journal of Xinyang Normal University (Natural Science Edition), 2014:235-238
4. Su Weici. The Current Status, Reasons and Optimized Control Model of Rocky Desertification in Karst Mountainous Area in Southwest China. Journal of Soil and Water Conservation, 2002:29-32
5. Li Anding, Zhu Xiaoke, Yu Lifei, Wu Shizhang. Current Status of Karst Rocky Desertification in Luodian County and Comprehensive Prevention and Control Measures. China Soil and Water Conservation, 2007:56-61

Fragmented Production in East Asia: What are Their Implications for the Sino-U.S. Trade

Feiting Lin*

Department of economics, Minjiang University, China
*Corresponding author: feitinglin@163.com

Abstract. Fragmented production means organization of production in which different stages of production are divided among different suppliers that are located in different countries. East Asia is one of the most important manufacturing centers in the world, China imports components from other East Asian economies for processing and then exports to the United States. The volume of export from China to the United States increase in traditional international trade statistics. Therefore, the solution of Sino-U.S. trade imbalance should pay attention to the relationship between the United States, China, and other East Asian economies.
Key words. Fragmented Production; East Asia; Sino-U.S. Trade; Other East Asian Economies; Replace Trade Imbalance;

1 Introduction

Since the 1990s, with the international division of labor deepened and refined, fragmented production has emerged. Fragmented production means organization of production in which different stages of production are divided among different suppliers that are located in different countries. Now products traded between firms in different countries are components instead of final products. Most of the final products have been sold to outside the region in which fragmented production happens. East Asia is one of the most important manufacturing centers in the world, and the final products are mainly exported to Europe and the USA. Therefore, the study of fragmentation focus on East Asian countries, this kind of fragmentation is an important part of contemporary globalization. This kind of new international division of labor has some similar interpretation, such as vertical specialization、Intra-product specialization、international division of production, localization, outsourcing, etc. However, compared with other explanations, fragmented production focuses on the division of labor among different countries, mainly used for the analysis of East Asian production. As fragmented production has grown, so too has the potential for gross flows of trade to mislead.

2 The Current Situation of Sino-U.S. Trade Imbalance

Since the 1990s, Sino-U.S. trade is booming. According to U.S. Department of Commerce Bureau of Economic Analysis, the United States and China bilateral goods import and export volume topped 600.82 billion dollars in 2015, increased 1.3%. The United States exported to China 116.70 billion dollars in goods, down 6.1%, imported 481.12 billion dollars, up 3.2%, Sino-U.S. trade imbalance up to 367.42 billion dollars in goods, increased 6.6%. Sino-U.S. trade imbalance accounting for 48.18% of the total U.S. trade deficit in goods, which is the largest source of deficit countries in the world.

2.1 Trade surplus in goods coexists with the trade deficit in services in Sino-U.S. trade imbalance.

The United States is the largest exporter countries in services, also is the largest services surplus countries. In Sino-U.S. trade imbalance, Trade surplus in goods coexists with trade deficit in services. As shown in Table 1, the United States and China bilateral import and export in services is increasing year by year[ii], and the volume of U. S. export to China in services is also increasing rapidly. The United States has stable earnings from Sino-U.S. service trade. The Sino-U.S. trade deficit in services up to 333.36 billion dollars in 2015, increased 9.24%. From the view of proportion, The Sino-U.S. trade deficit in services accounting for 0.19% of the Sino-U.S. trade surplus in goods in 2006, but accounting for 9.07% in 2015, the proportion increased year by year. Compared with the Sino-U.S. trade volume in goods, the volume of trade in services is relatively small, which can't reverse the overall Sino-U.S. trade surplus. In Sino-U.S. trade imbalance, trade surplus in goods coexists with trade deficit in services, and the volume of surplus and deficit both are increasing year by year. Sino-U.S. trade surplus in goods in the main reason of Sino-U.S. trade imbalance.

Table 1 Balance on goods and services in Sino-U.S. trade Units: Billion Dollars

YEAR	2006	2007	2008	2009	2010	2011	2012	2013	2014	2015
Exports of services	10.57	13.13	15.85	17.06	22.5	28.44	33.04	37.5	44.50	48.44
Imports of services	10.14	11.8	10.92	9.56	10.61	11.78	13.04	13.9	14.0	15.11
Balance on goods	-234.4	-258.7	-268.2	-227.2	-273.1	-295.2	-314.9	-318.8	-345.0	-367.4
Balance on services	0.43	1.34	4.92	7.5	11.89	16.65	20.0	23.62	30.52	33.34

Source: U.S. Department of Commerce, Bureau of Economic Analysis
http://www.bea.gov/international/factsheet/factsheet.cfm?Area=650
*Due to the impact of American subprime mortgage crisis and international financial crisis, the service volume U. S. import from China decreased in 2008 and 2009.

2.2 The Sino-U.S. trade surplus in goods transferred commodity structure from labor-intensive products to capital and technology-intensive products.

Analyzed from the commodity description of Sino-U.S. trade, there is a big changed between trade imbalance in goods. According to Standard International Trade Classification (SITC) of the United Nations, classify the exports and imports to ten categories reflected raw materials、semi-finished products and finished products. As shown in Table 2, the top-3 categories of Sino-U.S. trade surplus in goods comes from categories 6、7、8, which concentrated in the field of industrial finished products and semi-finished products. The category of Sino-U.S. trade deficit in goods mainly come from category 2, which concentrated in the field of raw materials. Then analyzed from the commodity description of Sino-U.S. trade imbalance, SITC8 including furniture、clothing、footwear and small goods, is Labor intensive products. The proportion of SITC6 in Sino-U.S. trade imbalance in goods is decreased year by year, accounting for 73.5% in 1996, which is the main commodity source of Sino-U.S. trade imbalance in goods. down to 59.5% in 2001, 42.8% in 2006, 39.8% in 2012, 38.6% in 2015. SITC7 including electronic storage, radio transmitter, recorder and digital accessories, excavator and so on. The proportion of SITC7 in Sino-U.S. trade imbalance in goods is increased year by year, accounting for 21.3% in 1996, 29.7% in 2001, 46% in 2006, up to 57.6 in 2012, 53.7% in 2015, which is the main commodity source of Sino-U.S. trade imbalance in goods. The Sino-U.S. trade surplus in goods are transferred commodity structure from labor-intensive products to capital and technology-intensive products. Machinery and transport equipment substituted for Miscellaneous manufactured articles, which has been the main commodity source of Sino-U.S. trade imbalance in goods. Source: U.S. Census Bureau. http://censtats.census.gov/cgi-bin/sitc/sitcCty.pl

Table2 The commodity description of Sino-U.S. trade imbalance Units: Billion Dollars

SITC	Dec.	1996	2001	2006	2012	2015
0	Food and live animals	1.2	-6.3	-21.6	4.8	20.2
1	Beverages and tobacco	-0.1	-0.3	0.4	1.5	2.4
2	Crude materials, inedible, except fuels	15.1	25.5	122.4	334.8	221.4
3	Mineral fuels, lubricants and related materials	-4.5	-2.9	-9.6	22.2	17.6
4	Animal and vegetable oils, fats and waxes	1.1	0.1	0.5	2.9	-0.1
5	Chemicals and related products, n.e.s.	6.9	1.5	0.0	-9.2	-20.5
6	Manufactured goods classified chiefly by material	-37.7	-97.0	-325.0	-401.4	-483.1
7	Machinery and transport equipment	-84.2	-246.6	-1076.0	-1814.3	-1964.9
8	Miscellaneous manufactured articles	-290.3	-494.2	-1002.1	-1255.2	-1411.5
9	Commodities and transactions not classified elsewhere in the SITC	-2.7	-10.2	-29.9	-36.6	-38.5
	total	-395.2	-830.5	-2341.0	-3150.5	-3656.9

3 The impact of fragmented production in East Asia on the Sino-U.S. trade imbalance

Fragmented production theory focus on production network in East Asia, emphasize the disassembly production process according to factor endowment in the East Asia. products traded in different countries are components instead of final products. Now products traded between firms in different countries are components instead of final products. most of the final products are sold to Europe and the USA. Therefore, this paper focus on components and trade flows to analyze the impact of fragmented production on Sino-U.S. trade imbalance.

3.1 Analyzed from commodity structure, China undertake terminal processing and assembly process in the East Asia fragmented production network.

According to the second part of this paper, the analysis of the current situation of Sino-U.S. trade imbalance, can be concluded that Sino-U.S. trade surplus of goods has transferred from labor-intensive products to capital and technology-intensive products. According to factor endowment theory, The United States is capital and technology abundant, China is labor abundant, so Sino-U.S. trade should be exported labor-intensive products from China, exported capital and technology-intensive products from the United States, and Sino US trade surplus in goods should focus on labor-intensive products, while the actual Sino-U.S. trade surplus in goods is contrary to the factor endowments of the two countries.

Analyzed from the perspective of fragmented production in East Asia, East Asian economies according to the advantages of natural resources to disassembled the commodity process, Japan, South Korea, and other East Asian economies have the advantage in producing machinery and transport equipment, China with the advantage of labor resources to participated in the fragmentation, such as assembly processing.

Take the hard disk drive production for example, four biggest hard disk drive manufacturers are Seagate and WD from United States、Toshiba from Japan、Samsung from South Korea, most of the components used by them are from East Asia. As shown in Figure 1, substrate suppliers all come from Japan, such as Furukawa Electric and Kobe Steel supply aluminum substrate, Hoya supply glass substrate. Then the biggest disc's supplier is Showa Denko from Japan Ishida, and Seagate and WD. The electric motor market monopolized by Nidec from Japan. The manufacturers of Magnetic read-write head also come from Japan, such as TDK、Showa Denko、Fuji Electric and so on.

Because of the labor cost in Japan is higher, so the components of hard disk drive suppliers choose other place in Asian area to organize production to reduce the costs through the division of labor in fragmented production. Then the hard disk drive manufacturers use the same way to organize production. Therefore, with the fragmented production establishing in East Asia, China has become the capital and technology intensive products export country because of its low labor cost in processing.

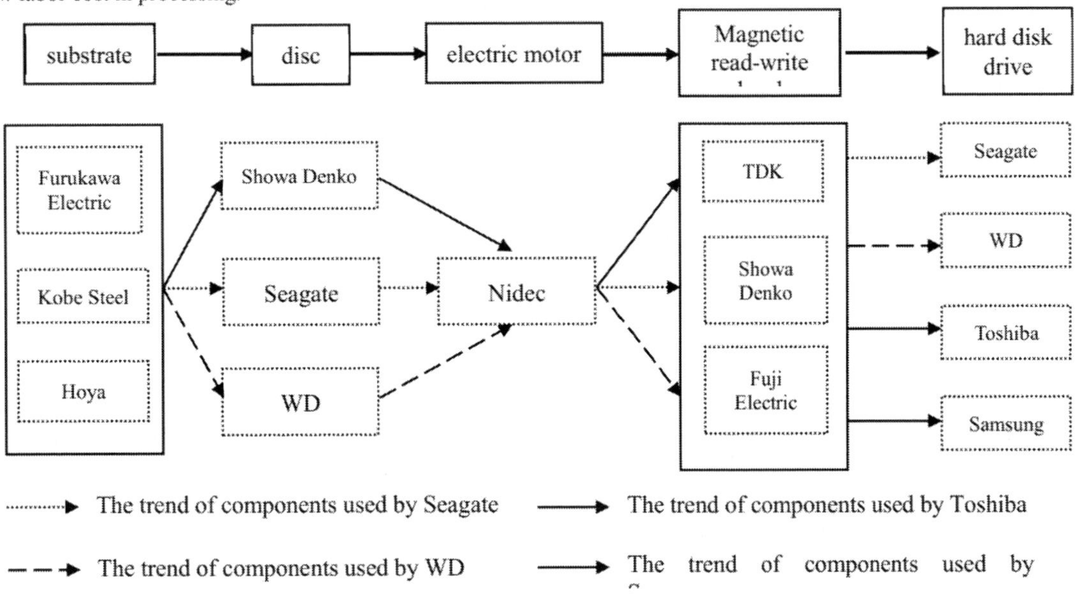

Figure 1:

The trend of components used by hard disk drive

Source: organize and summarize by author.

*Due to the Diaoyu Islands issue, the trade between China and Japan decreased in 2012, China reduces imports from Japan.

3.2 Analyzed from the flow of trade, the trend changes between China's imports from East Asia and the United States exports to China are the same

Fragmented production theory emphasized that most of the final products are sold outside the region. Analyzed from the flow of trade, if the trend changes between China's imports from East Asia and the United States exports to China are the same, it's can be considered that China undertakes assembly processing of Capital and technology intensive products to substitute for other East Asian economies to export to the United states.

Take SITC7 (Machinery and transport equipment) for example, as shown in figure 2, the changes between China export to American and China import from Japan, South Korea and ASEAN move in the same direction[iii]. Because of the impact of the international financial crisis in 2009, China reduces the imports from Japan, South Korea and ASEAN, then China exports to the U.S. also decreased. Therefore, in the traditional international trade statistics method, they just count the final product export and import, the assembly products from East Asia in China is all count for China export to American, ignoring the triangle trade relationship between the United States, East Asia and China, resulting in the exaggerated volume between China and American, twisted the Sino-U.S. trade imbalance.

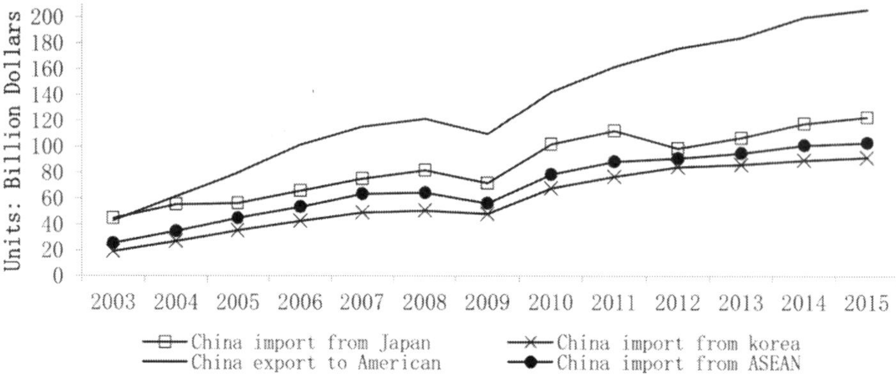

Figure 2: The volume of SITC7 that China export to American and China import from Japan、Korea and ASEAN

Source: United Nations Commodity Trade Statistics Database. http://comtrade.un.org/db/dqQuickQuery.aspx
*The statistics of other East Asia's economies in here only include: Japan, Korea, Taiwan, ASEAN.

3.3 Fragmented production in East Asia exaggerated the Sino-U.S. trade imbalance

East Asia is one of the most important manufacturing centers in the world, if take countries in East Asian as a whole, the United States is continued to be in deficit. According to the U.S. Department of Commerce statistics, China replace Japan as the largest source of trade deficit in 2000, is the second largest trading partner、 the third biggest export market and the biggest source of imports with the United States. According to Chinese customs statistics, the trade deficit between China and other East Asian economies are coexisted with Sino-U.S. trade surplus. As shown in figure 3, With the increase of Sino-U.S. trade imbalance, the trade deficit between China and other East Asia's economics[iv] is also increased, the trend of change in volume are nearly the same. Therefore, China participate in the fragmented production in East Asia, replaced the other economies in East Asian to export to American, exaggerated the Sino-U.S. trade imbalance.

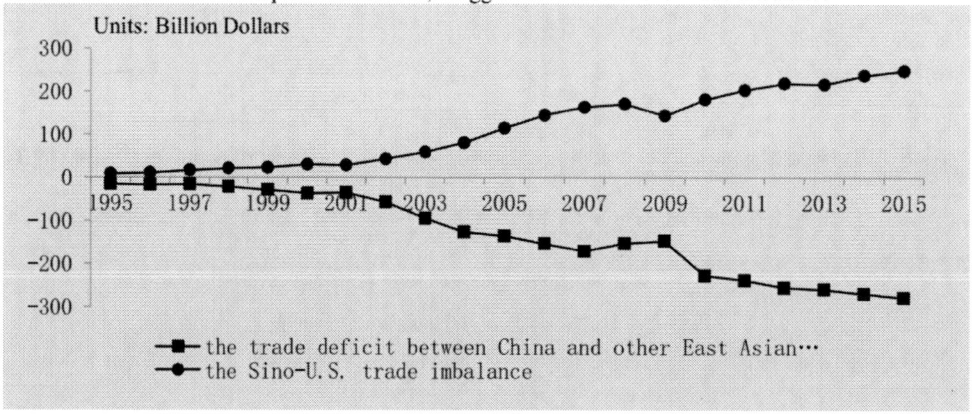

Figure 3: The Sino-U.S. trade imbalance and the trade deficit between China and other East Asian economies

Source: Statistical Yearbook of China

4 Conclusion

With fragmented production in East Asia, ASEAN and other East Asian economies transfer the labor-intensive part of production to Chinese, by using the low labor costs in China to process. Those economies in East Asian was also the mainly source of imports from the United States, so when they transferred the part of fragmented production to China, they also transferred their trade surplus with the United States to China with the traditional international trade statistics by using final product for accounting standard. Because of China participate in the fragmented production in East Asia by using labor endowment, at a low value-added chain, profit is very few, while the United States import from other East Asian economies due to lower labor costs in China is cheaper than before, so American consumers are getting more benefit. Fragmented production in East Asia ignored the triangle trade relationship between the United States, East Asia and China, resulting in the exaggerated volume between China and the United States, twisted the Sino-U.S. trade imbalance. The Sino-US trade imbalance does not reflect the gains, so the development of bilateral economic and trade relations should pay attention to the relationship between the United States, China and other East Asian economies instead of the volume of Sino-U.S. trade imbalance, asked China to open the market, to reform RMB exchange rate, to introduce trade protection can't fundamentally solve the problem of Sino-U.S. trade imbalance.

4.1 Correctly deal with the impact of exchange rate on Sino-US trade imbalance, and steadily promote the reform of the RMB exchange rate

The fundamental cause of the Sino-U.S. trade imbalance is China substitute for other East Asian economies to export to American in the fragmented production network in East Asia, whether appreciation or devaluation of the RMB has no effect to solve the Sino-U.S. trade imbalance. Correctly deal with the impact of exchange rate on Sino-US trade imbalance, China government should insist the independent exchange rate policy, to plan, step by step to steadily promote the reform of the RMB exchange rate.

4.2 Taking technological innovation as the goal, by participating in the process of East Asia fragmented production to upgrade the domestic industrial structure

Other East Asian economies transfer labor-intensive part of capital and technology intensive products to China not only because of labor endowment, but also because China enterprises lags behind other countries in R&D, can only in assembly process in the industrial-chain division. Therefore, for the purpose of technological innovation, China involved in fragmented production in East Asia actively, should learn the advanced experience from other East Asian economies to upgrade domestic industrial structure at the same time, to promote economic growth turn to high value-added capital and technology intensive production.

4.3 Implement the "going out" overseas investment strategy actively, to achieve product diversification and country of origin diversification

The advantage of China's participation in fragmented production is the endowment of labor factors in East Asia. With economic growth, per capita income grows, the cost of labor also grows, China enterprises should implement the "going out" overseas investment strategy actively, master technique from other developed countries through fragmented production, then transfer labor-intensive part to other developing economies to relief trade friction, which to realize the sustainable growth of China's economy.

Acknowledgements

This research was supported by the Education Department of Fujian Province (project number: JAS160422) and Minjiang University (project number: YSY13008).

References

1. [1] Jianjiang Liu, Xizhen Yang. Study on the trade interests of Sino US trade imbalance from the perspective of intra product specialization. international trade issue. 8(2011)68-80.
2. [2] Liwei An, Yeqing Ma. Journal of international fragmentation of production and the Sino US trade imbalance. Nanjing University. 3(2012)31-39.
3. [3] Jihong Lin, Jing Meng. The influence of international vertical specialization on Sino US trade surplus. world economy research. 11 (2012)38-43.
4. [4] Koopman, R., Z. Wang & S.J. wei, How Much of Chinese Exports is Really Made in China? Assessing Domestic Value-Added When Processing Trade is Pervasive[EB/OL]. NBER Working Paper 14109.
5. [5] Athukorala, P. & N. Yamashita, Global Production Sharing and Sino-US Trade Relations. China & World Economy. 3(2009)39-56.
6. [6] Fung, K.C. and L.J. Lau. The China-United States Bilateral Trade Balances: How Big is it Really?. Pacific Economic Review. 3 (1998)33-47.
7. [7] Bart Los, Erik Dietzenbacher, Robert Stehrer, Marcel Timmer and Gaaitzen de Vries. Trade Performance in Internationally Fragmented Production Networks: Concepts and Measures. WIOD Working Paper 11.
8. [8] Blyde J, Molina D. Logistic infrastructure and the international location of fragmented production. Journal of International Economics. 95(2015)319-332.

9. [9] Mattoo A, Wang Zhi, Wei S Jin. Measuring trade in value added when production is fragmented across countries: An overview. Trade in Value Added. 25(2013).
10. [10] Bai Y, Liu Y. Causes and Countermeasures for Sino-US Trade Imbalance: From the Perspective of International Trade. Journal of Chinese Economics. 2(2014)1.

Research on the application of VRP theory in logistics transportation

Ruiqi Zhu and Yunkai Zhai

School of Management Engineering, zhengzhou University

Abstract. Logistics transportation problem is a hot issue in today's research, with the diversification of transport vehicles, the complexity of the transportation network, the choice of logistics transportation route is a NP problem in combinatorial optimization. The paper discusses the relevant knowledge of VRP and physical distribution routing, such as the relationship between them, how to describe the distribution problem with VRP, how to build up mathematical model and so on and meanwhile, points out the direction of research.

1 Introduction

Nowadays, with the integration of the logistics industry, information technology and the trend of globalization, the whole logistics distribution occupies more and more important status in the system.Transportation system is one of the most important subsystems in the distribution system, which accounts for about 50% of the total logistics cost, so it is the first to reduce the cost of logistics distribution.Among them, whether the transportation route directly affects the distribution speed, cost and benefit, especially the determination of distribution line is a complicated system engineering.Select appropriate vehicle routing, can speed up the response to customer demand, improve service quality, enhance customer satisfaction to the logistics link, reduce service provider operating costs.Optimizing the transportation logistics and reducing the transportation cost is one of the effective ways for enterprises to improve the competitiveness of enterprises, especially in the logistics distribution enterprises.Therefore, since Danting and Ramser put forward the Vehicle Routing Problem in 1959, VRP has become a research hotspot in logistics field in recent years .

2 Logistics distribution and VRP

The study of VRP generally has the following assumptions:
- the material distributed can be mixed;
- the location of the individual users and requirements are known;
- From the distribution center to the transportation distance between individual users is known;
- distribution center has enough resources for distribution, and has enough transport capacity.

VRP program clearly stipulates that the number of vehicles should be dispatched to meet the constraints, models and the specific driving route of each vehicle. The implementation of VRP transport scheme can ensure on time and quality of the day to complete the transportation task, and can make the total travel at least. Visible in the logistics and transportation issues, the introduction of VRP is very necessary.

3 Mathematical model of VRP in logistics distribution

3.1 logistics distribution in the VRP description

A distribution center offers logistics services to a certain geographical scope (customer demand),amount of goods required each of demand points are small (less than the capacity of the vehicle), and each demand point distribution center as well as the needs of the distance between points is known. If the amount of goods distribution center can meet the needs of the customers, and the distribution center can cooperate with transportation companies through its own or rent or the way enough capacity available for deployment, a load weight per delivery trucks exceeds its rated load, and the total running distance of each vehicle has a certain limit.

In order to improve the utilization rate of vehicles, how to arrange the vehicle routing and vehicle scheduling can meet the distribution task, and make the total mileage of the vehicle to be the shortest. That is to say, in order to complete the transport task, the distribution center should send a plurality of cars, all delivery routes for several major routes (loop); each delivery trucks starting from the distribution center, along a route covering several users (loop) delivery, and then return to the distribution center. At this time, including our problem to be solved: what customers should be assigned to a loop (that is, what the customer goods should be arranged in the same car); bypass order on each route of customers.

3.2 Mathematical model of VRP in logistics distribution

A typical VRP model can be used to express the loop, you can also put the time, distance, cost into transportation costs , the basic principles are the same. Here we use the loop representation.

basic conditions. Existing n vehicles parked in a common source point (that is, the logistics center) O, it needs to provide goods to m customers. Assuming the customer is a1, a2,... , am, and the source and customer locations are known.

model target. Determine the required number of vehicles N, and assign these vehicles to a loop, including the routing and scheduling within the loop, so that the total transport distance S minimum.

constraint conditions.

- $N \leqslant n$;
- Use Q_{ij} to express the first i car in its sub circuit corresponding to the requirements of the first j customer. Due to the previous assumptions, the number of customer demand Q_{ij} smaller, so the customer demand is not likely to exceed the weight of the car. Therefore, each customer can only accept a car at the same time service. A sub circuit corresponding to a car, let the number of customers of each loop is 1, i is the vehicle corresponding to the number, c_i said the i collection by the path corresponding to the car, c_{ij} said the order of the corresponding sub circuit for j customers. If $l_i=0$, then said the car did not participate in the service. From this we can know :

$$\sum_{i=1}^{n} l_i = m$$

- After the vehicle is completed, the task is to return to the source a_0.
- The vehicle can not exceed the maximum load W_i and maximum running length limit of I_i.

mathematical model. Assume that $d_{i\,(j-1)\,j}$ between the j-1 car the corresponding line in the 1 customer and j customer distance; $d_{i\,(LI)\,(0)}$ between the l_i client and said the source i car in the distance of the route of the corresponding v_0. The mathematical model can be established as follows:

$$\min s = \sum_{i=1}^{N}[\sum_{j=1}^{l_i} d_{i(j-1)j} + d_{i(l_i)(0)} \cdot sign(l_i)] \quad (1)$$

s.t. $$\sum_{j=1}^{l_i} Q_{ij} \le w_i \quad (2)$$

$$\sum_{j=1}^{l_i} d_{i(j-1)j} + d_{i(l_i)(0)} \cdot sign(l_i) \le L_i \quad (3)$$

$$0 \le l_i \le m \quad (4)$$

$$\sum_{i=1}^{N} l_i = m \quad (5)$$

$$c_i = \{c_{ij} \mid c_{ij} \in \{v_1, v_2, \cdots, v_n\}, j = 1,2, \cdots, l_i\} \quad (6)$$

$$c_i \cap c_j = \Phi, \forall i \ne j \quad (7)$$

$$sign(l_i) = \begin{cases} 1, l_i \ge 1 \\ 0, 其他 \end{cases} \quad (8)$$

$$N \le n \quad (9)$$

(1)the objective function, that is, the total running distance of the vehicle in the completion of the distribution task is the shortest.

(2)the vehicle capacity constraints, that is, the total demand for each sub circuit customer does not exceed the maximum load of vehicles, to ensure that the needs of all customers can not exceed the vehicle's total weight of the vehicle itself.

(3) the operating length of the vehicle in each sub circuit is not more than the maximum mileage of the vehicle.

(4) the number of customers per vehicle is not more than the total number of customers.

(5)the total number of customers involved in the operation of the vehicle, the total number of customers and the actual number of customers are equal, that is, to ensure that each customer service.

(6) the corresponding customer set for each vehicle service.

(7) each customer can only be served by a car at the same time.

(8) the i car is involved in the service.

4 Logistics distribution in the VRP algorithm

4.1 Basic algorithm of Hamilton map

Traveling salesman problem with the theory of graph theory, that is, in a complete graph of a weighting, to find a minimum weight of the Hamilton circle, said the circle as the optimal circle. At present, there is no effective algorithm to solve the problem of minimum Hamilton circle, only to find a better algorithm to solve the problem. One possible way is to first ask for a Hamilton turn C, and then modify the C appropriately to get another Hamilton loop with the smaller weight. After a series of changes in the series, and finally get a circle, can not be improved by this method. This last lap is almost certainly not optimal, but there are reasons to think that it is often relatively good. In order to get a higher precision, this program can be repeated several times, each time from different circles to start. It can be said that the use of Hamilton to solve the smallest circle to find a relatively simple type of TSP or VRP.

4.2 Other classical algorithms

Solving the vehicle routing problem is very much, basically can be divided into 2 categories: accurate algorithm and heuristic algorithm. The exact algorithms for solving VRP problems include the branch and bound method, the cut plane method, the dynamic programming method, the network flow algorithm, etc.. Heuristic algorithm mainly has saving method, scanning method, two stage algorithm, tabu search algorithm, genetic algorithm, duplicate matching, etc..

Branch and bound method. The basic idea of branch and bound method to solve the VRP problem, with no integer constraints to the VRP problem (B) optimal solution as the starting point, if the solution is an integer solution, then the solution is the original VRP problem (A) of the optimal solution, otherwise B non integer solutions of adjacent integer additional conditions the formation of 2 branches, namely, 2 sub problems are solved. If the above 2 sub problems are solved, the optimal solution is the integer solution, then stop the branch of the sub problem, or continue to solve the problem. This method can only be applied to solve the small VRP problem. Kolen at alhas used this method to solve the problem of vehicle traveling with time windows, and found that when the nodes are expanded to 12, the computer has a memory problem.

Dynamic programming method. The basic idea of dynamic programming method to solve the VRP problem, the VRP problem is considered as a decision problem of phase n, and then transformed into decision problems in solving N with the recursive relation of single stage, so as to simplify the calculation process. The optimal solution of VRP can be obtained by this method, but it is only suitable for small scale optimization problems.

Two stage heuristic algorithm. Christofides, Mingozzi, Toth in 1979 proposed two stage heuristic algorithm, in order to improve the construction algorithm to solve the problem. The first phase of the algorithm commonly used heuristic algorithm to obtain a feasible solution, second stage method, Insert method、2- exchange、2- swap、2-OPT、3- OPT and other technical improvements, the adjustment in the solution remain feasible, trying to close to the optimal target, each step produces another feasible the solution to replace the original solution, the objective function is improved, continue to improve the objective function so far. Two stage method in the process of solving, often using interactive optimization technology, the person's subjective initiative to join the VRP problem solving process.

Tabu search algorithm.The idea of tabu search is proposed by Glover in 1986. It is a kind of global optimization algorithm. It introduces a tabu list, local records have searched the advantages of using tabu table information in a search or not selectively search these points, in order to jump out of the local advantages, so as to realize the global optimization. In 1994, Garcia et al firstly applied the tabu search algorithm to the VRPTW problem. Tabu search algorithm is used to solve the VRP problem, although it can ensure the exploration of different effective search methods, but it is not easy to implement in practice because it involves complex neighborhood transformation and solving strategies.

Simulated annealing method. Simulated annealing is proposed by Metropolis in 1953, and it is a kind of random search algorithm based on the principle of thermodynamics. The simulated annealing algorithm for VRP problem is studied in 1993, and the simulated annealing method is proposed to solve the Osman problem. When the VRP problem is solved by simulated annealing, the energy of the atoms in the physical annealing is equal to the distribution of the optimal node, and the atomic vibration simulation is a random search in the search space of the line. The search process is completed by the execution of the pairwise exchange of the distribution, the initial solution for the distribution of the total cost of the distribution. In order to find the optimal route, the distribution of the distribution is improved by using arbitrary pair wise exchange in the existing allocation. Simulated annealing algorithm is suitable for large-scale demand fixed or random demand of the VRP problem. In theory, the global optimal solution can be obtained by this method, but in practical application, due to the limitation of the computation time, the optimal solution of the VRP problem can only be given. In order to make the approximate solution obtained by the simulated annealing algorithm more accurate, the simulated annealing method is repeated, and the optimal solution is selected as the final approximate optimal solution.

Genetic algorithm. Genetic algorithm is a kind of improved algorithm for local search. When genetic algorithm is used to solve the VRP problem, it is mainly used to search for the common character between the biological evolutionary generation and the optimal route, and to search the optimal solution of the distribution route in the global scope. The same expenses in case of the same level distribution node, are considered to be involved in the next step of iteration, and by subsequent steps gradually eliminated, and ultimately determine the optimal delivery route; different expenses in case of the same level distribution node, "the fittest survive", to get the right distribution by iteration branch to spend less, but can not determine the optimal qualifications, must participate in the subsequent elimination step, until the end of the iteration, the optimal distribution route. Genetic algorithm has good robustness, flexibility and versatility, and is especially suitable for solving large scale VRP problems. But because of some defects still exist in the algorithm itself, such as premature convergence, poor local search ability, therefore, can not guarantee the maximum probability of convergence to the global optimal solution. Zhang Liping et al in 2002 by introducing a novel crossover operator to construct a kind of improved genetic algorithm. This algorithm gets rid of the diversity of population, there is no common traditional genetic algorithm premature convergence problem, optimization can effectively obtain the solution of VRP. Lang Maoxiang in 2006 through the construction of partheno genetic algorithm, effectively improved the traditional genetic algorithm for complex VRP problem of low search efficiency and easy to fall into the defects of "premature convergence".

Ant colony algorithm. Ant colony algorithm is a kind of intelligent optimization algorithm based on swarm intelligence proposed by Dorigo. The basic principle of the algorithm is to simulate the behavior of ants in search of food: ants find food, they can always find a path from food to the nest between the optimal path. This is because ants in the search path, will be released on the path of a special pheromone. When they come across a road that has not yet been passed, they randomly pick a path and release the pheromone that is related to the length of the path. The longer the path, the lower the release of the hormone concentration. When the later came to the intersection of the ants, the choice of higher probability of hormone concentration path will be relatively large. This forms a positive feedback, the optimal path of the hormone concentration is more and more large, and the other path on the hormone concentration will be decreased with the passage of time, and ultimately the entire ant colony will find the optimal path. The algorithm has a good effect in solving the VRP problem, but it is easy to fall into the local optimum problem if the calculation time is long. Huang Lan et al. In 2002, the specific application of ant algorithm in the traveling salesman problem is presented, and the problem is solved and the local optimization is solved by the opt 3- method and the crossover strategy. Hao Jin and others in 2002 for the basic ant colony algorithm, the design of a novel random perturbation ant colony algorithm, and its application to solve complex TSP problems. Wu Jianjun and others in 2004 for the logistics distribution routing problem, proposed a hybrid ant colony algorithm to overcome the classical ant colony algorithm time complexity is too difficult.

Particle swarm optimization algorithm. PSO algorithm is an evolutionary computation technique based on swarm intelligence, which was proposed by Kennedy and Eberhart in 1995. The algorithm is to solve the optimization problem by simulating the foraging behavior of birds. Firstly, in the solution space, the birds are randomly initialized, each bird is called "particle", these "particles" move in the solution space, and find the optimal solution after several iterations. In each iteration, the particle updates itself by tracking the 2 "extremes". The first one is the optimal solution of the particle itself ("Pbest"); the other extreme is the optimal solution ("Obest"), which is found in the whole particle swarm. Find these 2 extreme, each particle according to their own flight speed, determine their own direction and flight distance. PSO algorithm to solve the VRP problem with fast convergence speed, dependent on the experience of small parameters, simple, easy to use and so on. But like other evolutionary optimization algorithms, it is easy to fall into the local optimum. Huang Lan et al. By introducing the concept of exchange and exchange order in 2003, a special particle swarm optimization algorithm is constructed to solve the traveling salesman problem. Xiao Jianmei designed a new real number encoding scheme in 2005, the vehicle routing problem is transformed into a quasi continuous optimization problem, and the penalty function method to deal with constraints, so as to solve the particle swarm algorithm is applied to the vehicle routing problem.

5 Comparison of various basic algorithms

Exact algorithm is based on strict mathematical means of the algorithm, in the case can be solved, the solution is usually better than the artificial intelligence algorithm. However, after the introduction of strict mathematical methods, it can not avoid the exponential explosion problem, so that the class algorithm can only solve the determination of small and medium scale VRP. Specific to each algorithm, they have their scope of application and characteristics. Given the lower bound and the branch and bound algorithm is related to the point of access from the perspective of the establishment, therefore is suitable for symmetric VRP, also applies to non symmetric VRP; three subscript vehicle flow equations in the model introduces effectively represents the time window variables, thus the utility model is applicable to general assignment problem (GAP) and with time window TSP (TSPTW); two subscript vehicle flow equations by SYM equation and the expansion of TSP, the removal of the number of vehicles on behalf of the subscript, the form is more compact, with fewer constraints, so only applies to symmetric CVRP and DVRP, but also for the two star to ask VRP loose constraints. K is thus the method is applicable to CVRP and DVRP; the feasibility conditions of tabu search are also available, so the scope of application and the two stage method: repeat the same matching algorithm in the model considering the time constraints and capacity constraints, so it is suitable for strong constraint VRP.

Deterministic VRP is the basis of the research of non deterministic VRP. At present, our country is still in the initial stage of complex VRP research. But in real life the most common or uncertain VRP, most of the uncertainty of VRP is the simplification of uncertainty VRP. The existing SVRP results have been able to solve the small scale of SVRP, but can not be competent for the larger scale of SVRP. A practical way to deal with large scale SVRP is still the artificial intelligence method, but to get a general efficient algorithm is difficult, so we should study and different algorithms with different neighborhood feature problems with neighborhood features. For FVRP, the fuzzy possibility theory should be combined with the research.

References

1. Pan Lijun, Fu Zhuo, Liu Ximei. The open vehicle routing problem with working time and time windows. *computer engineering*. 2014 (04)
2. Zhu Yongli, Li Kai. Research and application of multi objective model of B2C e-commerce logistics distribution vehicle routing optimization problem. *Chinese business world (the second half)*. 2015 (12)
3. Yu Lei. The problem and reason of the distribution of Y in the Anhui branch. *express company logistics engineering and management*. 2013 (05)
4. Wang Jun, Guo Tingting. Research on Logistics Distribution Routing Based on improved ant colony algorithm. *value engineering*. 2013 (02)
5. Liu point, Mao Chao. E-commerce distribution center location problem and optimization. *system engineering*. 2015 (10)

Analysis of Collaborative Relationship among Regional Medical Service Collaboration Based on Game Theory

Yan Qiao [1] and Yunkai Zhai [1,2,*]

[1]School of Management Engineering, Zhengzhou University, Zhengzhou, China
[2]The First Affiliated Hospital, Zhengzhou University, Zhengzhou, China
*Corresponding author: zhaiyunkai@zzu.edu.cn

Abstract. Regional medical service collaboration is an effective way to solve the problem of uneven distribution of medical resources in China. This paper analyzes the cooperative relationship between the main bodies of the regional medical service from the point of view of game theory and uses the "prisoner's dilemma" model in game theory to analyze the cooperative game relationship between medical institutions in the region. Then it obtains the Nash equilibrium condition, and puts forward the strategy to maximize the efficiency of medical institutions according to the game results.
Key words. medical service collaboration, "prisoner's dilemma" model, cooperative relation

1 Introduction

Optimizing the allocation of medical and health resources has been the focus and difficulty of the medical and health industry in China. In recent years, the emergence of Regional Medical Association (hereinafter referred to as the Medical Union will help to promote the quality of medical resources and improve the efficiency of resource utilization in the region [1]. And in the hospital, all levels of the hospital are supposed to cooperate with each other in order to truly play the role of the Medical Association. With the promotion of health care reform, many places actively explore the new model to improve the health care. The general office of the State Council also promulgated "the guiding opinions of the general office of the State Council on promoting the construction of the grading system of diagnosis and treatment" in 2005 to encourage the establishment of a grading system of diagnosis and treatment, and Hierarchical diagnosis and treatment is exactly the specific practice of regional medical collaborative services.

Service collaboration is a service model of multi agent cooperative interaction in modern service industry. As a special social service, medical service is the most important one to solve the relationship among different subjects, that is, the relationship between medical institutions in the region.

Constructing the collaborative service system of Medical Association under the background of new medical reform is of great significance for the establishment of an orderly and effective new medical and health service system [2]. The existing literature analysis the regional medical service collaboration model, the existing problems and platform construction extensively. Du Xin and et al. [3] divided the regional medical collaboration into 3 kinds of cooperative service mode, that is, longitudinal medical collaborative service model among multi-level medical institutions, remote medical collaboration model among different medical institutions and horizontal medical collaboration model among large medical institutions. He believes that regional medical collaboration is an effective way to solve the exchange and sharing of medical resources among medical institutions and believes it is significant to optimize the allocation of medical resources, reduce the cost, and promote the new mode of regional medical collaborative service. Wang Shu [4] established the model of regional cooperative medical system according to the principle of CO students and identified the key order parameters. Through the analysis of the force on the regional cooperative healthcare system, he analyzed the coordination mechanism of regional medical institutions, and gives suggestions and measures to strengthen the allocation of resources to the top-level design and system planning, the establishment of regional coordination mechanism, a series of adjustment and integration of medical resources. Jiang Xianfei and et al. [5] based on the practice of regional collaborative medical service in West China Hospital of Sichuan University. He pointed out that in the western region of regional medical cooperation in promoting the lack of power, regional coordination of medical policy lag, lack of standardized medical information system platform construction standards and other issues, and put forward the corresponding improvement measures. Yu Bo [6] from the Pudong New Area Shanghai regional medical center combined the area of 11 community health service centers. He practically analyzed on the medical and health collaboration network of business closely and believed that it could solve the problem of regional medical collaboration when the hospitals achieve the same level of health care, information technology support and a reasonable cost management. Zhou Leming and et al. [7] pointed out that the factors that affect regional collaboration services are not only technical reasons but also management reasons. Sun Zhonghai[8] believed that information synergy is the most important in regional medical service collaboration, and the most important is to establish the standard of interoperability[9].

The American health care system is based on market economy and managed care or management and its regional medical service collaboration efficiency is very high. As the most important cost control mechanism, payment system and managed care led to the medical service market competition is more intense. American medical institutions began to develop collectivize and large-scale. A way of horizontal integration is that many of the poor management of the hospital were purchased, mergers, the direction of the development of the group across the region. The other way is vertical integration, that is, different levels of medical institutions to form a large-scale urban hospital as the core, community hospitals, rehabilitation hospitals, registered home care institutions as one of the hospital group. American hospital group achieve large-scale medical equipment, information, management and service sharing, management functions such as personnel, logistics, and all the implementation of social, substantial increase in service efficiency [10].

The construction of regional medical collaboration system has important theoretical and practical significance to reduce the cost of health care, improve the efficiency of medical treatment and optimize the allocation of medical resources. However, a great deal of literature has focused on the analysis of regional medical coordination model, problems, platform construction and other macro level, and they lack of relevant theoretical research how to obtain the maximum benefit and how to analyze the main body of the system from the perspective of game theory. This paper based on this starting point, through analysis of the game relationship between the main regional service cooperative medical systems, and draw conclusion on how to make full use of the game system between the main interests of regional medical institutions.

2 Analysis of the competition and cooperation among medical institutions

According to the synergy theory, if the regional medical institutions form collaborative services, the whole region will be the quality of medical services to improve the quality of production. It cannot be achieved for a single medical institution and it plays an important role in improving the social cooperative medical service system. The goal of collaborative service implementation has not only exceeded the level of the participating parties, but also a higher social level [11].

Hospitals will compete with each other in China. In the process of cooperation of regional medical institutions, the relationship between medical institutions is a kind of cooperative competition, not a purely competitive competition any more. It is mainly manifested in the following aspects. First of all, for the superior medical institutions, too many patients mean to occupy more medical resources. There are not so many superior hospitals and their accommodation for patients is limited, while the superior hospitals' high quality medical resources should be given to severe patients. It will lead to a decline in the quality of hospital service if mild patients crowed high-quality medical resources. For the primary medical institutions, there are many primary hospitals and they have high accommodation. It can improve the benefit of the hospital with patients increasing and it can improve the level of primary medical services. Therefore, superior hospital in the region can be guided by the government to allow mild patients to diagnose in the primary hospital or to use telemedicine to make high-quality medical resources benefit the primary hospitals. Their sharing resources and complementary advantages like these can overcome the problems that difficult and expensive to see a doctor in China for a long time. It also makes the original competition in the medical institutions to achieve a win-win situation. Therefore, it embodies a kind of cooperative competition in the process of service collaboration of medical institutions in the region.

3 Analysis of the cooperative relationship based on game theory

3.1 The basic nature of the game model among medical institutions

There are a lot of medical institutions in the medical association, and medical service collaboration can occur between two or more medical institutions. In order to facilitate the model, this paper simplifies the relationship between medical institutions and medical institutions to a superior hospital and a primary hospital. The whole collaborative service is a dynamic process, and it is also a process of continuous selection. In the whole process, there exist two kinds of state——cooperation and betrayal.

According to the basic theory of game theory, there are characteristics in the process of cooperative game between the superior hospital and the basic level hospital as follows:

(1) Hospital has the right to choose independently, without government intervention. That is hospitals can trust each other, to take cooperative action; but also can take the betrayal.

(2) Every hospital is rational. The starting point of the hospital selection is to obtain the maximum benefit with the minimum investment. In the process of the game, when the expected benefits of cooperation are greater than the betrayal of the hospital, the hospital will choose to cooperate, and vice versa, will choose to betray.

(3) Repeated game. From the first cooperative game to the termination, the whole process of the game is a dynamic process, which is the result of many times.

3.2 Game model construction

This paper uses the classic prisoner's dilemma model in game theory to analyze. The prisoner's dilemma model was proposed by Tucker Albert. It mainly describes the contradiction between individual rationality and collective rationality, namely, the behavior of individual interests as the starting point of a lot of times and cannot realize the maximum benefit of the group. And it also explains the inherent contradictions of individual rationality, namely, the behavior of individual interests as the starting point is not necessarily guaranteed to achieve the maximum benefit of the individual [12].

In order to build the model expediently, this paper makes the assumptions as follows:

Hypothesis 1, Participant input. On the assumption that the input of each cooperative need in the game process is expressed by I(Including manpower, material resources, financial resources, information, technology and other resources in the input, in the form of a unified currency translation). Let IH represent the investment of the superior hospital and let IL represent the investment of the basic level hospital.

Hypothesis 2, Cooperation benefit. It assumes that if the superior hospital and the primary hospital trust each other and cooperate with each other, the output benefit is related to the investment of each other. And it is a kind of linear output, that is to say, the size of the cooperative benefit is positively correlated with the cooperative effect coefficient k ($k>1$).

Hypothesis 3, Betrayal punishment. It assumes that the superior hospital and the basic level hospital, one side to take the trust behavior, one side to take the betrayal. As a result, who takes the trust of one party's income is 0, and who betrayals can get the trust of the investment, but it will get the corresponding punishment b (Including the credibility of the hospital and other non-quantitative indicators in the form of currency conversion).

Hypothesis 4, Trust incentive. The cooperation between the superior hospital and the basic level hospital makes the investment cost of the cooperation be affected by the negative feedback. Let 1-μ(0<μ<1) represent it. The more times of the number of collaboration and cooperation are, the lower of the cost of the corresponding cooperation is.

Hypothesis 5, Collaborative process. The whole process is dynamic. It assumes that before the superior hospital and the primary hospital collaborate for n times, the hospitals have been mutual trust and cooperated all the time. As the result, the hospitals will continue to maintain the choice of collaborative. Once a hospital chooses betrayal for the n stage, then the superior hospital and the basic level of hospital cooperation will no longer be carried out [13].

Let p represent the probability of superior hospital to take cooperative action, and its probability of choosing betrayal is 1-p. Let q represent the probability of cooperation behavior of the primary hospital, and its probability of choosing betrayal is 1-q. According to the above assumptions, this paper constructs the payment matrix on the time node of the cooperation between the superior and the basic level hospitals in the n times. Table 1 is the collaborative service payment matrix of superior hospital and primary hospital.

Strategic combination 1. According to the assumptions, when the superior hospital and the primary hospitals are selected to cooperate, the payment matrix of superior hospital is the benefits of superior hospital excepting its investment cost.

$$U_{1H} = I_H kpq - I_H(1-\mu)^{n-1}pq = I_H pq[k-(1-\mu)^{n-1}] \tag{1}$$

$$U_{1L} = I_L pq[k-(1-\mu)^{n-1}] \tag{2}$$

Table 1. collaborative service payment matrix of superior hospital and primary hospital

The probability of superior hospital	**The probability of primary hospital**	
	cooperation（q）	betrayal（1-q）
cooperation（p）	U_{1H}, U_{1L}	U_{2H}, U_{2L}
betrayal（1-p）	U_{3H}, U_{3L}	U_{4H}, U_{4L}

Strategic combination 2. When the superior hospital selects cooperation, and the primary hospital select betrayal, the superior hospital will not have the benefits, but it will pay the cost; the primary hospital will get the cost of superior hospital, but it will be punished out of betrayal.

$$U_{2H} = -I_H p(1-q)(1-\mu)^{n-1} \tag{3}$$

$$U_{2L} = I_H p(1-q)(1-\mu)^{n-1} - b \tag{4}$$

Strategic combination 3. When the superior hospital selects betrayal, and primary hospital select cooperation, its situation is the same as strategic combination 2.

$$U_{3H} = I_L(1-p)q(1-\mu)^{n-1} - b \tag{5}$$

$$U_{3L} = -I_L(1-p)q(1-\mu)^{n-1} \tag{6}$$

Strategic combination 4. When the superior hospital and the primary hospital are both select to betray, the superior hospital and the primary hospital will not have benefits, and they will be punished out of betrayal.

$$U_{4H} = -b \tag{7}$$

$$U_{4L} = -b \tag{8}$$

3.3 Analyzing game process

This paper uses the prisoner's dilemma game model. In order to make the model be able to get the solution, and make the superior hospital and the primary hospital get a win-win strategy s*, that is the superior hospital and primary hospital both selecting cooperation, the model must be made to achieve the Nash equilibrium condition. According to the Nash equilibrium theory, for all the participants, when the other participants choose the strategic combination of s* given in the given strategy, the choice of s* in the corresponding strategy to obtain the payment will be not less than the choice of other strategies [14]. That is the following conditions:

From the point of view of the superior hospital, when the primary hospital selects to cooperate, the payment of superior hospital to select cooperation should not less than the payment of selecting betrayal. And when the primary hospital selects betrayal, the payment of superior hospital to select cooperation should not less than the payment of selecting betrayal as well.

$$\begin{cases} U_{1H} \ge U_{3H} & p = 1, q = 1 \\ U_{2H} \ge U_{4H} & p = 0, q = 1 \end{cases} \tag{9}$$

Put the payment result into it.

$$\Rightarrow \begin{cases} I_H pq[k - (1-\mu)^{n-1}] \ge I_L(1-p)q(1-\mu)^{n-1} - b \\ \qquad p = 1, q = 1 \\ - I_H p(1-q)(1-\mu)^{n-1} \ge -b \quad p = 0, q = 1 \end{cases} \Rightarrow \begin{cases} I_H[k - (1-\mu)^{n-1}] \ge I_L(1-\mu)^{n-1} - b \\ - I_H(1-\mu)^{n-1} \ge -b \end{cases}$$

$$\Rightarrow \begin{cases} I_H k \ge (I_H + I_L)(1-\mu)^{n-1} - b \\ I_H(1-\mu)^{n-1} \le b \end{cases} \tag{10}$$

From the point of view of the primary hospital, similarly, when the superior hospital selects to cooperate, the payment of primary hospital to select cooperation should not less than the payment of selecting betrayal. And when the superior hospital selects betrayal, the payment of primary hospital to select cooperation should not less than the payment of selecting betrayal as well.

$$\begin{cases} U_{1L} \ge U_{2L} & p = 1, q = 1 \\ U_{3L} \ge U_{4L} & p = 1, q = 0 \end{cases} \tag{11}$$

$$\Rightarrow \begin{cases} I_L pq[k - (1-\mu)^{n-1}] \ge I_H p(1-q)(1-\mu)^{n-1} - b \\ \qquad p = 1, q = 1 \\ - I_L(1-p)q(1-\mu)^{n-1} \ge -b \quad p = 1, q = 0 \end{cases} \Rightarrow \begin{cases} I_L k \ge (I_H + I_L)(1-\mu)^{n-1} - b \\ I_L(1-\mu)^{n-1} \le b \end{cases} \tag{12}$$

It can achieve the conditions of Nash equilibrium of superior hospitals and primary hospitals to combine formula (10) and formula (12):

$$\Rightarrow \begin{cases} I_L k,\ I_H k \ge (I_H + I_L)(1-\mu)^{n-1} - b \\ I_L(1-\mu)^{n-1},\ I_H(1-\mu)^{n-1} \le b \end{cases} \tag{13}$$
$$(k > 1, 0 < \mu < 1, n = 1,2,3\ldots)$$

It assumes that the investment of superior hospital is more than primary hospital. Let $I_H=I_L+\alpha, \alpha>0$ represent the investment of superior hospital.

$$\begin{cases} b \ge (2I_L + \alpha)(1-\mu)^{n-1} - I_L k \\ b \ge (I_L + \alpha)(1-\mu)^{n-1} \end{cases} \tag{14}$$

Let

$$\begin{cases} B_1(n) = (2I_L + \alpha)(1-\mu)^{n-1} - I_L k \\ B_2(n) = (I_L + \alpha)(1-\mu)^{n-1} \end{cases} \tag{15}$$

$$\begin{aligned} B(n) &= B_2(n) - B_1(n) \\ &= I_L k - I_L(1-\mu)^{n-1} \\ &= [k-(1-\mu)^{n-1}]I_L > 0 \end{aligned} \tag{16}$$

$$\Rightarrow b \geq I_H(1-\mu)^{n-1} \tag{17}$$

From the above, the condition of Nash equilibrium of superior hospital and primary hospital to collaboration is formula 17. That is to say, there must be enough punishment to constrain their selection in the initial cooperation. Even in the first time to cooperate, the punishment of betrayal should be more than the investment of superior hospital. When cooperation is carried out step by step, the punishment will be quickly reduced. Even when the number of cooperation is enough for a long time, it will also make the game to be Nash equilibrium without punishment of betray.

3.4 Analyzing the result of game

When the game can reach the Nash equilibrium, and superior hospital and primary hospital take the choice of cooperation all the time before the cooperation for the n times, the total benefit for the n times is:

$$\begin{aligned} W(n) &= U_{1H} + U_{1L} \\ &= I_H pq[k-(1-\mu)^{n-1}] + I_L pq[k-(1-\mu)^{n-1}] \\ & \qquad p = 1, q = 1 \\ &= (I_H + I_L)[k-(1-\mu)^{n-1}] \end{aligned} \tag{18}$$

the total benefit is:

$$\begin{aligned} W &= \sum_{n=1}^{n} W(n) \\ &= (I_H + I_L)[nk - \frac{1-\mu-(1-\mu)^n}{\mu}] \end{aligned} \tag{19}$$

It can obtain the following results according to the formula 18 and 19:

(1) When k and$\mu$are invariant, if let the superior hospital and the primary hospital to achieve more income, it will need to increase the I_H and I_L, and they should keep cooperating. It can be seen from the formula 18, with the increase of n, that is, the increase in the number of cooperation, the single cooperation benefit will increase. That means it will enhance the trust of both parties with each cooperative selection. And it will reduce the possibility of betrayal in the next collaborative selection. This is conducive to format a long-term stable cooperative form of cooperation.

(2) When I_H, I_L andμ are invariant, if they want to get more benefit, it needs to increase the value of k. That is to increase the cooperation effects.

(3) When I_H, I_L and k are invariant, if they want to get more benefit, it needs to increase the value of μ, and let the value of μ close to 1. That is to reduce more the cost of each collaborative selection than ever.

3.5 Analyzing the conclusion of game

According to Nash equilibrium theory, if s* is not Nash equilibrium, then there is at least one participant will deviate from the strategy that the agreement s* provided. And it makes the agreement s* cannot be implemented. Therefore, if the agreement can be implemented automatically, it is necessary to make the agreement be Nash equilibrium [14]. Therefore, first of all, it should make the hospital consciously reach the intention of cooperation, and it should have enough betrayal punishment to constrain. If it can conform a long-term stable cooperation and win-win situation between hospitals or not, it to a large extent depends on the position and attitude to cooperation between the two sides of hospitals [15]. This punishment can be signed by the hospital agreement between the two parties to bear the additional loss of betrayal. After the game can reach the Nash equilibrium, it will be beneficial to both sides and achieve the goal of win-win.

According to the analysis of 3.4, this paper believes it should be taken to improve the synergy between medical institutions as the following aspects:

(1) The hospital itself needs to increase the investment of collaborative cooperation. Initially, it can be guided by the government. When the hospitals get the benefits of cooperation and identity the cooperation model, they will increase the cooperative investment consciously.

(2) Improving the effect coefficient of collaborative service cooperation among hospitals. The greater the effect of hospital collaboration is, the more willing to continue to cooperate with the hospitals. And hospitals are increasingly willing to continue cooperation. It can improve the effect of hospital collaboration through optimizing cooperation mode, strengthening hospital contact, upgrading hospital hardware facilities and so on.

(3) Increase the range of cost reduction to multi-collaborative. When the hospital formed a stable form of cooperation, the investment cost of each cooperative selection will be reduced exponentially. For example, it can achieve the superior hospital and primary hospital cooperative cooperation through the way of remote consultation. In the initial collaboration, they need to pay a lot of hardware costs, labor costs, learning costs, etc. When cooperation is getting more and more mature, the cost of cooperation will be reduced quickly. In order to improve the scope of the cost reduction, the cooperation mode can be realized through regulating the cooperation mode and strengthening the information exchange.

4 Conclusion

This paper analyzes the synergy relationship between medical service collaboration based on the theory of game theory and research methods. And put forward some corresponding suggestions according to the model. It aims to stimulate the enthusiasm of each hospital in the system, and to maximize the benefits of cooperation among the subjects in the system. And then it promotes the Medical Union to realize the service collaboration through the classified diagnosis and treatment; realizes the reasonable allocation of medical resources effectively; makes all-level hospitals function reasonably, distinctly, practically and efficiently.

References

1. SUN Tao, ZHANG Wei, GE Si-ao, et al. A framework of evaluation index system of medical alliance stability for the regional integrated care organizations. Chinese Hospital Management. **36**,5-7(2016)
2. LIANG Hong, et al. Policy value and practical significance of the synergistic service of medical association in China. Chinese Journal of Health Policy. **12**,1-5(2013)
3. DU Xin, WANG Yi-lin, Li Gang-rong. Research on the regional collaborative health care model and the key technology of collaborative platform construction. China Digital Medicine. **9**,38-40(2014)
4. Wang Shu, Wang Hengshan, Wang Yunguang. Research on synergetic mechanism and countermeasure of regional cooperative healthcare system for medical resources optimization deployment. Science & Technology Progress and Policy. **20**,38-42(2010)
5. JIANG Xian-fei, TANG Yu, ZHONG Yan, XIE Juan. Practice and thinking about regional medical collaborative service in western region of China. Chin J Evid-based Med. **14**,1286-1288(2014)
6. YU Bo, WANG Wei, et al. Practice of construction of medical and health care network based on the medical cooperation pattern under the trend of graded medical treatment system. Chinese Hospital Management. **12**,7-9(2015)
7. Zhou Leming, Yu Zhongxin, et al. The study of influencing factors of the regional medical collaborate service based on the platform. Chinese Journal of Health Informatics and Management. **01**,65-69(2014)
8. SUN Zhonghai, SUN Wei, WANG Jiwei. New model of collaborative regional medical service. Chinese Health Quality Management. **04**,15-18(2010)
9. LIU Wen-sheng. Regional information collaboration go forward with heavy burden. China Hospital CEO, **24**,48-51(2015)

10. Goldman,d.p. Managed care as a public cost-containment mechanism[J]. RAND journal of Economics,**26**(2),277-295
11. Huxham, Chris. 1993. Pursuing Collaborative Advantage. Journal of Operational Research Society. **44**(6),599-611
12. HUO Ying. Research on the efficiency of service industry's support for the strategic emerging industries: evidence from Chinese northeast Provinces' electronic and communication equipment manufacturing industries. Scientific Decision Making. **7**,25-40(2013)
13. ZHANG Weiying. *Game theory and information economics*,364-380(Shanghai people's Publishing House, 2004)
14. LUO Yunfeng. *Game Theory*,28-29(Tsinghua University Press, 2007)
15. XU Jian-zhong, ZHAO Wei-feng, WANG Li-jing. Analysis of the relationship among agents of collaborative innovation system in equipment manufacturing industry based on Game Theory. China Soft Science. **07**,161-171(2014)

Research on the development strategy of Pingdingshan iron and steel industry cluster with Wugang Company as the core

Shuai Wang*and Hongxiang Cao

School of Management Engineer, Zhengzhou University
*Corresponding author: 1024333549@qq.com

Abstract. Arouse the great potential of the industrial clusters will play a significant role in promoting the development of local WuGang company. The article takes the Pingdingshan City as the research object, who is a typical resource-based city and develops on the basis of the local resources of coal and iron ore, the core enterprise WuGang Company is a large state-owned enterprise based on the wide and thick steel plate production. Many problems also exit in the cluster innovation, company network construction, company cooperation. The article is from the perspective of steel industry cluster and cluster innovation. Firstly, the article introduces the correlation theory about the steel industry cluster and the basic development information of Pingdingshan steel industry cluster, and promotes the research purpose and meaning.The article will give the analysis to the problems from the Pingdingshan steel industry cluster and develop the reform proposals and development strategy according to the actual situation.

1 Background

1.1 Introduction to the concept of iron and steel industry cluster

Cluster is a small and medium-sized enterprise and institutions of a specific industry in a large number of clusters gathered in a certain geographical range and the formation of a stable, sustainable competitive advantage. There are many types of industrial clusters, the iron and steel industry cluster is a form of resource-based industrial cluster the classification, is relying on the iron ore resources, through the development and utilization of resources, production and processing as a link, regional resource-based industrial clusters and the large and medium-sized state-owned steel enterprises as the core form. The iron and steel industry cluster mainly relies on the iron ore this non renewable resources, the industrial structure and the supply chain has led to a lack of diversity and complementarity, limits among organizations of the steel industry chain of cooperation, the cluster characteristics of its network is difficult to fully show.

1.2 Introduction to the basic situation of Pingdingshan iron and steel industry cluster

Pingdingshan city is located in central Henan Province, has a strong industrial foundation and rich mineral resources, iron ore resources in Wugang city under the jurisdiction of the rich, as one of China's ten largest iron ore district, Wugang iron and steel company core enterprise clusters in Pingdingshan in Wugang province. Currently, cluster enterprise network there are 8 mining enterprises, 2 iron enterprises, 2 steel companies, 2 enterprises, 23 steel plate processing enterprises, 12 steel, more than and 140 steel materials production enterprises and sales enterprises, the formation of a from mining, beneficiation, ironmaking, steelmaking, rolling steel, steel processing to complete the production materials of iron and steel industry chain. Pingdingshan iron and steel industry cluster compared to other local industrial clusters, strong basic, the total amount of more development speed, well-known brands are more prominent.

2 Development status of Pingdingshan iron and steel industry cluster

2.1 development course and current situation of Pingdingshan iron and steel industry cluster

2.1.1 The development course of core enterprise Wugang company

The formation of iron and steel industry in Pingdingshan is to start from the core business of Wugang company, Wugang company was founded in 1970, formerly the national subordinate enterprises, as the beginning of national defense project construction. In June 1986, the first phase of Wugang steelmaking project was completed and put into operation, which ended the history of China's iron and steel industry. In 2000, the Wugang municipal government support, set up with Steel Corp and other enterprises, thus gradually established a relatively complete industrial chain of iron and steel, Pingdingshan iron and steel industry of the early embryo. Since then, the rapid rise of Wugang Company. At present, the total assets of over ten billion yuan, with more than 2 employees, operate income of over 20 billion yuan of the actual production and operation capacity.

2.1.2 General situation of Pingdingshan iron and steel industry cluster and its development status

There are about more than and 210 enterprises in the cluster enterprise network. Cluster of coal enterprises and steel enterprises. At present, there are four main enterprises in the cluster, which runs through the production chain of the whole industrial cluster.

Henan iron and Steel Group Wuyang Mining Co., Ltd., is a large state-owned iron ore production enterprises in Henan province. Their reserve of 3.23 tons of ore reserves, Wugang is an indispensable and important raw material base for the company. An iron and Steel Group, Wuyang iron and Steel Co., Ltd. has its own iron ore and the surrounding large and small iron ore for the production of blast furnace to provide adequate supply of raw materials. What the Pingdingshan Coal Group as the iron and steel industry of coal resources supply enterprises is the largest enterprise in the cluster, but also the domestic ultra million tons of large coal production bases, coal production capacity of 70 million tons, production and sales in the forefront of the country. Pingdingshan Yao Meng Power Engineering Co., Ltd. is one of the backbone power plants in Central China power grid, in order to provide power energy.

At present, Wuyang special steel rolling mineral company mining and processing, in the same sintering and pelletizing and Steel Corp and Wuyang pig iron smelting Steel Corp formed a complete industrial chain, between the upper and lower process complementary advantages, reduce the circulation, reduce logistics costs, achieve mutual benefit and win-win three party. Downstream machinery manufacturing industry formed a steel plate, composite plate, mechanical processing, equipment manufacturing, the industry chain.

2.2 Organizational structure of Pingdingshan iron and steel industry cluster

Single core cluster is also called the wheel type cluster on a single core network only one core enterprise and other enterprises around the core enterprise around, provide spare parts production and a production process or other production and service life of the. Pingdingshan iron and steel industry with the main characteristics of the single core cluster, is the development of small and medium-sized enterprises in Wugang's large state-owned enterprises such as the core to drive around the whole cluster in Wugang company's value chain and the production process as the main line of division of labor, the cluster followed by the raw material supplier (Wuyang mineral company, and Steel Corp) - pig iron producers (with Steel Corp) - steel enterprises (Wugang) - steel scrap processing enterprises (with Steel Corp) - Sales trader (Wugang tengwu Co.) in series. Wugang brand has a strong influence in the international wide steel plate market. Wugang Company is one of the five production plants with the influence of the world by virtue of its technological advantages, and it has a significant market. With these advantages, Wugang, the company holds most of the resources of the enterprise network, but also determines the direction of development of other enterprises. Wugang company with raw material suppliers and waste processing enterprises, etc. have maintained a high degree of vertical cooperation, but also the existence of long-term trading and commitment.

3 The main problems existing in the development of Pingdingshan iron and steel industry cluster

3.1 affected by the macro factors, the company suffered serious losses

By the impact of the global economic slowdown and the negative impact of China's economic growth, automobile, shipbuilding, machinery manufacturing and other downstream steel industry slowdown, weakens domestic demand for steel. China's steel industry average sales profit margin has dropped to 0.47%, large and medium-sized enterprises expanded to 32.5%, while China's steel industry's huge production capacity, further exacerbated the disorderly competition in the steel industry. Wugang Company is a professional manufacturer of wide heavy plate, although the production technical level is at the leading position in the same industry, but due to the homogenization of competition to high-end products spread, makes the profit level of iron and steel enterprises continue to decline. The competition strategy of Wugang variety and market for the other medium and heavy plate enterprises are facing the unprecedented challenge and impact.

3.2 Innovation capacity of the cluster is generally insufficient

At present, although the formation of the opponent to imitate the unique competitive advantage of Pingdingshan iron and steel industry core competitiveness in the wide and heavy plate production, but the lack of innovation ability and professional talents are weak still exist in other areas of production of Wugang steel company, cluster enterprises in the overall innovation capability is weak.

First of all, from the development of iron and steel industry clusters in Pingdingshan, to the traditional industry and mining of mineral resources industry, the product is still in the stage of the supply of raw materials or semi-finished products, it is difficult to form a long-term competitive advantage, reproducible. Secondly, the Pingdingshan iron and steel industry cluster has not formed the perfect innovation mechanism and the innovation network, the cluster sustainable development ability is worrying. In terms of the cluster enterprise, the removal of large state-owned enterprises such as Wugang, other companies generally lack of R & D institutions and R & D team building, talent education and training system is still not perfect. Most of the enterprises in the cluster are low in the proportion of product innovation. Again, the lack of enhancement of cluster innovation capability of Pingdingshan iron and steel industry are limited to the professional talents, different from the other in the more developed areas of iron and steel industry clusters, has a strong research team supported by universities in the province ranked relatively in the metallurgical field. This is based on the small town developed in the industrial cluster of enterprises in the individual, family oriented enterprises, low starting point, small scale, in attracting professional talent, there are obvious shortcomings.

3.3 The lack of intermediary organizations

Pingdingshan iron and steel industry cluster enterprise network within the form of intermediary service agencies specialized in cluster within the enterprise services. At present, the development of cluster only focus on the development of the core enterprise itself, the lack of targeted financial intermediaries, recruitment agencies, productivity service center, the assessment of advisory bodies and related legal advisory bodies, etc.. Wugang City, there are only two consulting companies and enterprises in the cluster and the demand is not matching; number within the cluster of legal advisory bodies, but for business in Wugang city is not strong; only through the recruitment network announced the recruitment information, recruitment information for no complex. In addition, due to the lack of financial intermediaries in the cluster, many small and medium-sized enterprises in the bank to obtain funds through bank loans to support the case, mainly adopts private funds to raise funds to expand production and operation.

3.4 Mining waste discharge is serious in the cluster

Pingdingshan city is a typical resource-based city, mainly depends on iron ore resources and coal resources, but because of the non renewable resources, the development of the cluster also has an important impact. At present, the northern city of Pingdingshan, in the West due to excessive exploitation of coal resources has been formed in subsidence area contiguous, mining difficult again. As the development and utilization of coal technology behind, the recovery rate of mining and mining waste two development and utilization of technology is still scarce, resulting in a large number of waste of resources available. With the exploitation of resources, the collapse accidents and natural disasters have destroyed the production and life in the cluster.

4 The development strategy of Pingdingshan iron and steel industry cluster

In this paper, through the support of relevant industrial cluster innovation theory, and analysis of the status quo and problems faced by Pingdingshan iron and steel industry cluster, the development strategy of Pingdingshan iron and steel industry cluster is proposed the following suggestions.

4.1 To guide the implementation of professional management strategy of Wugang company

Wugang company should rely on its own technical advantages and advantages in the production of extra thick steel plate, to take specialized business strategy.

First of all, the market development should be fully grasped . For Wugang companies, the first job is the market and the contract. To contract information, market development, export increment, volume and price balance and incentive assessment is the most important. Secondly, it should establish the concept of the customer as the center. Make clear the demand of terminal users, through appropriate to extend the service chain and product chain, to provide better products; to clear business direction always start production "around the customers, enterprises to form effective linkage and quick reaction ability of problem solving. Again, the Wugang company should adhere to the "new special" road. To adhere to carry forward the Wugang Company in the brand, high-end product line, product research and development capabilities and the advantages of high-end market. Only sustained development of new products, committed to the development of high-end market to meet the needs of the domestic and international market for heavy steel plate.

4.2 To establish the cluster innovation based on Knowledge Network

4.2.1 Encourage innovation and interaction among enterprises in cluster

First, the Wugang Company as a local large-scale state-owned enterprise should act as a leading role to promote inter enterprise cooperation model. Other steel production enterprises within the cluster enterprise network, such as iron and Steel Group, etc., should be more interactive with Wugang companies to carry out product innovation, improve their production technology and scientific research level. At the same time when the other small and medium enterprises actively brokered area of technical innovation alliance, to carry out extensive cooperation in the field of innovation. Secondly, until the cluster enterprise network development and maturity within the cluster can be set up a research fund, funded by the various enterprises, specializing in the production and service innovation. This kind of cooperation is actually a kind of simplified technology alliance, which is helpful to the formation of the unique competitiveness of knowledge sharing, human resources and technological advantages.

4.2.2 Seek cooperation opportunities from external agencies

In the era of knowledge economy, enterprises should actively look for external joint venture and technical partners to improve the efficiency of cluster technology, and inject fresh vitality into cluster innovation. Enterprises within the cluster to ensure long-term stable contact with external agencies, the formation of a timely knowledge of the flow. At present, enterprises in the cluster are mostly solved by external experts to solve technical problems, and lack of long-term and stability. Wugang companies and other enterprises should establish long-term cooperation with universities in the relevant areas of expertise, such cooperation, including cooperation in the field of scientific research and talent output.

4.3 To encourage the sharing of resources among enterprises within the cluster

4.3.1 Use industry website to promote the sharing of information resources between enterprises

It should draw on the successful experience of Zhejiang's industrial clusters using network technology to promote the development of industry, to encourage the steel industry cluster to establish their own professional website. Conditions within the cluster enterprises should be established in Wugang as a benchmark for the industry's Web site, the industry will share the information in this industry. In addition to publishing the cooperation information and market dynamic information of each industry, it should also provide various policies and regulations, as well as scientific research content.

4.3.2 Organization of industry associations to expand the sharing of knowledge resources between enterprises

In Zhejiang, Shenzhen these industrial clusters to develop more mature areas of the association is worth our reference. Association of interests because of strong consistency and high degree of specialization often has a good ability to act, to achieve a good effect of information communication between enterprises in the same industry, develop the market in learning and training, the development of industry practices such as more likely to agree. Pingdingshan iron and steel industry cluster should draw on this form of organization, with Wugang Company as the core to establish the industry association of enterprise network, and support its development activities.

4.4 Perfect the external environment of cluster development

4.4.1 To ensure the competitive order of enterprises in the cluster

Wugang municipal government and other public management departments should take on the important task of ensuring the competitive order of the iron and steel industry cluster, strengthen its supervisory role. First, the government departments should strengthen market supervision, fight against shoddy work, shoddy, deceive customers, fake and shoddy products, the protection of intellectual property rights and patent technology. Second, in terms of specific work, technical supervision departments and industry associations should cooperate with each other, to develop industry technical standards, through industry standards to guide enterprises to upgrade technology. Third, the local government for the iron and steel industry cluster in mining enterprises, coal mining enterprises, steel enterprises, lower emissions disorderly exploitation costs arising from environmental pollution problems timely monitoring, and resolutely investigate and deal with illegal enterprises for fines, eliminate arbitrary row misplacing problem is.

4.4.2 Strengthen the functions of intermediary service agencies in the inter enterprise communication

Pingdingshan municipal government and the Wugang municipal government should support the establishment of intermediary institutions, at different levels to provide support services for the enterprise technology innovation. The enterprises in the cluster should form a close interest relationship with the intermediary organizations, and reduce the barriers of knowledge flow. Intermediary service agencies should be responsible for helping enterprises to provide cooperative information, the right choice of partners; the company released new technologies, new products, and new investment opportunities in the role of information.

4.2.3 Create a good business environment within the cluster.

The relationship between enterprises cannot be separated from the informal non-governmental exchanges, these activities can be a good way to cultivate corporate social responsibility, promote the exchange and sharing of knowledge. Actively carry out a variety of cultural exchange activities, to effectively promote the cluster within the enterprise to establish a harmonious and trusted business environment. The Wugang company should have such basic superior natural environment in the city, focusing on the construction of service facilities, has formed its own cultural environment, attract more investors and social capital. Various cultural undertakings should be developed in the region, to create a more harmonious social living space.

References

1. Propris, Systemic Flexibility L.D., Fragmentation and Cluster Governance . European Planning Studies Production, 2001, Vol 9
2. Meng Yue, Chi Wencheng, industrial cluster technology learning mechanism research, economic management publishing house, 2009
3. Liu Yan, empirical study on the change of China's strategic emerging industry agglomeration degree, Shanghai economic research, 2013 .02
4. Liu Junying, the development of industrial agglomeration in the transformation of resource based cities -- Taking Pingdingshan city as an example, the people's forum, 2014.19
5. Wang Shuying, the evolution of industrial clusters and regional economic development research, Guangming Daily Press, 2009
6. Peng Peng, Xiao Rui, study on the transformation of resource based cities -- Taking Pingdingshan as an example, Journal of Xianning University, April 2010
7. Meng Tao, industrial cluster organization under the network perspective, China Social Science Press, 2009
8. Zhao Zhonghua, Research on the network structure and performance of innovative industrial clusters, Harbin Institute of Technology press, 2009
9. Xiong Xiaohui, "the iron and steel industry is on the verge of comprehensive loss", China Business News, 2011

A Three-dimensional Analysis Model and Application about Energy Conservation& Emissions Reduction Technology System and Synergistic effect of an Energy Intensive Product

Jiliang Zheng and Qingqing Gu

Faculty of Management & Economics, Kunming University of Science & Technology, Kunming China

Abstract. It is an important project to improve the synergistic effect of energy conservation &emissions reduction (ECER) through building technical system now. Energy intensive products are the key to promote ECER strategy in China. It is necessary to study the relationship between the construction of technology system and synergistic effect of ECER. Thus, this paper built a three-dimensional model to analyze ECER technology system and synergistic effect in production of a product by the analysis method of value flow and Euclidean space coordinate system. With an energy intensive product as example, this paper used the model to analyze the ECER synergistic effect of technology system in each stage of production. The results show that the three-dimensional analysis model has a certain value of theoretical and practical application for further studying the construction of the technology system and synergistic effect of ECER of energy intensive products.

1 Introduction

There were some researches on the relationship between technology system and synergistic effect of energy conservation &emissions reduction (ECER) in manufacturing industry so far. The representative research has the following. Martin(2015) selected 22 ECER measures in the iron and steel industry to evaluate and compare the emission reduction potential and cost as well as synergistic effect of each measure. Mao Xianqiang(2012) proposed a method to evaluate the synergistic control effect of technical reduction measures on sulfur, nitrogen and carbon in iron and steel industry, and examined the synergistic control effect of different measures from multi angle. Xing Fangfang(2013) analyzed the characteristics of synergistic control technology about sintering flue gas multi pollutant in iron and steel industry in China and abroad, and made a comparative analysis on the technical economy and emission reduction effect. Zhu Guilin (2010) proved that the technology innovation of comprehensive utilization of steel slag had improved significantly the ECER effect of the iron and steel industry based on 2005-2010 data. Han Yijie(2011) found that technological innovation is the endogenous mechanism of ECER and has a positive effect on ECER performance through researching on energy efficiency and ECER potential of iron and steel industry. Zeng Ping (2013) got a conclusion that ECER technology innovation of enterprises has a significant stimulating role, and the enterprises of advanced energy-conservation or reaching to discharge standards have higher new product output rate through research and empirical analysis on 348 manufacturing enterprises in Guangdong Pearl River Delta.

International research on this area is less. Most of the related research was done by Chinese scholars to discuss the issue of ECER in Chinese enterprises or industries in international journals. Representative papers have the following. Li Li (2011) researched on ECER policies in China's electric power industry. Zhang Zhaoguo (2013) studied the influence factors of ECER in China's listed companies. Yan Gao (2011) evaluated the effect of energy conservation and pollutant emission reduction of the coal mine enterprise. Pimenteira (2004) researched on energy conservation and carbon emission reduction by recycling in Brazil. It is seen that the research on the relationship between technology system and synergistic effect of ECER is rare from micro perspective of an energy intensive product so far. Therefore, in this paper, the theory and application of ECER of an energy intensive product was studied by using two dimensional or three dimensional coordinate system analysis and value flow analysis methods on the aspect of energy intensive product. This paper tries to constructs a 3D model, and to demonstrate the rationality of the model through the application on a case. It will provide a micro level of theoretical guidance and practical analysis method to promote ECER progress of energy intensive products, and so as to enrich the related theory system.

2 A three-dimensional analysis models about ecer technology system and synergistic effect of an energy intensive product

2.1 The ECER technology system and the concept of synergistic effect of an energy intensive product

The ECER technology system of an energy intensive product refers to the integration of energy conservation technology, emissions reduction technology, waste recycling technology, and the scale production of by-products in production. The technical synergistic effect of an energy intensive product refers to the integrated ECER effect generated through the integrated application of a technology system in the production of products. The ECER synergistic effect of a product may be measured through value flow analysis. From perspective of circular economy of "resource value" concept, value flow describes the value change of material flow in the circular motion by means of accounting theory and material flow route (Lin Yanyu, 2013). The purpose of technology system construction of an energy intensive product is to promote the most efficient use of resources and maximize the recycling of waste products in the production and operation of energy intensive products. It requires each product to minimize the resources consumption and waste emission to obtain the maximum comprehensive benefits from the input of resources, resource recycling to the output of the resources.

In the whole process of production, ECER synergies of technologies may be divided into two situations as below. First, a single technology may improve energy conservation or emissions reduction effects simultaneously of a product, which is called as the single technical synergistic effect here. Second, the integration of technologies improves a product's ECER effect, which is called as the inter-technology synergy effect. The integrative synergistic effects of both single technology and inter-technology come to the total synergistic effects of a technology system.

2.2 Analysis models of synergistic effect on a single technology and inter-technology

In the two-dimensional Euclidean space coordinate system, the ordinate represents energy conservation effect, abscissa represents emission reduction effect. Each technology position in coordinate system shows the comprehensive effect of a technology, which was referred to ECER synergistic effect (Mao Xianqiang, 2011), as shown in Figure 1. The synergistic effect of inter-technology refers to meet the requirements of ECER targets through the space distribution of a number of technologies.

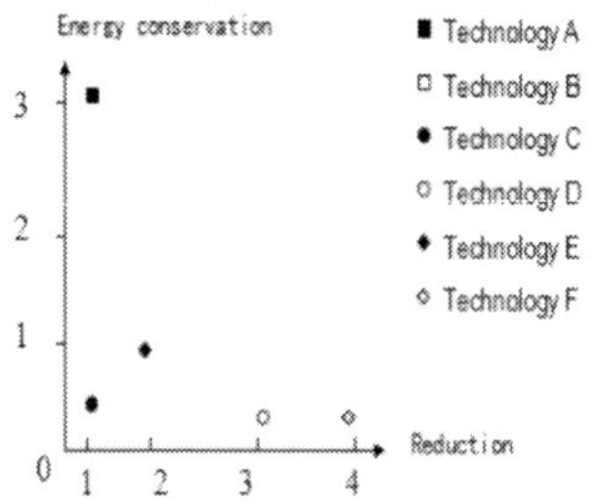

Figure 1. ECER synergistic effect of a single technology

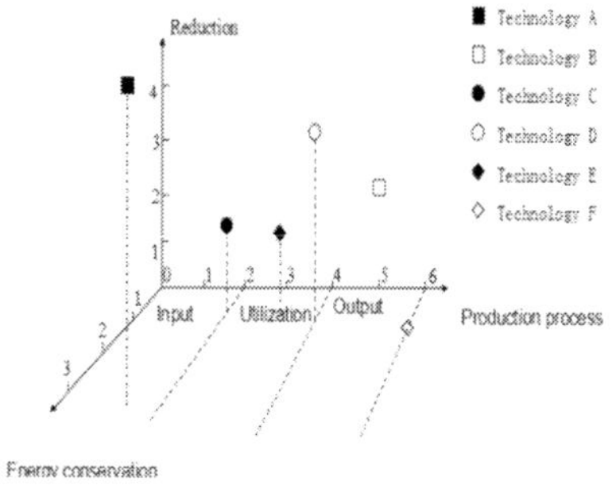

Figure 2. Three-dimensional map of ECER synergy effects of production technology system

2.3 A three dimensional analysis model on synergistic effect of a product's technology system

A product production is a process of integration applications of a number of technologies. Production process may be divided into three stages generally: resource input, resource utilization and resource output stage. Each stage involves the application of ECER technologies. The integration of all technologies forms a product's technology system. Therefore, the three-dimensional Euclidean space coordinate system may be used to express the three-dimensional relationship between the production process and ECER effect, as shown in figure 2.

2.4 Reference standard for collaborative evaluation of technical system

According to the cooperative type of single technology and inter-technology, the collaborative evaluation reference standard of a technical system is designed as shown in Table 1.

Table 1. Reference standard for technical collaborative evaluation

NO.	Judging index	Reference standard for judging result
1	More than a type of synergy, whether there is a technical synergy is OK.	Good synergy
2	A type of synergy + inter-technology synergy	General synergy
3	A type of synergy and no inter-technology synergy	Poor synergy

2.5 A value flow analysis model on ECER synergistic effect in a product's technology system

It is suitable to use value flow method to evaluate and sum the value flow produced by ECER technology system to obtain the comprehensive effects or synergistic effects of a product as shown in Figure 3. The value flows in Figure 3 indicate the economic effect of ECER after applying a certain technology, including reducing the consumption of energy and raw materials, or reducing the disposal cost of waste and pollution. The sum of the value added in the whole production is the comprehensive effect of ECER, which is produced by the synergy in technology system. If a synergistic effect may be formed between technologies, it will achieve the optimization of ECER effects and even the maximization. Figure 3 expresses not only the application scope of ECER technology, and also considerate waste utilization and optimization in product production.

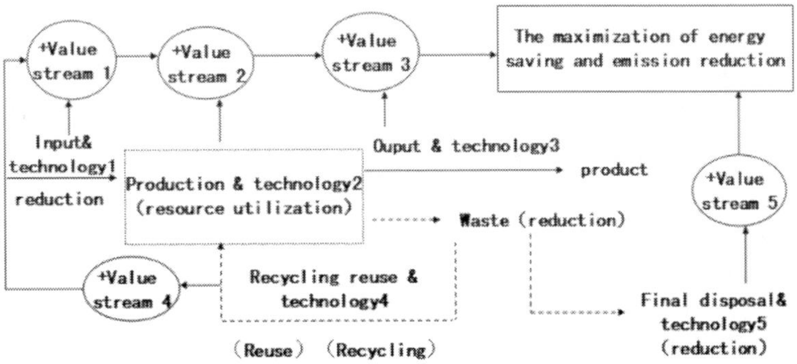

Figure 3. A value flow analysis model of ECER synergistic effects in a product's technology system

3 Case study of an energy intensive product

3.1 Product introduction

Take a PVC product of a chemical firm as an example to analyze the ECER effect produced by the construction of technology system. Polyvinyl chloride (PVC) is a kind of widely used plastic. Owing to energy intensive production and serious pollution, the calcium carbide process is one to eliminate in the world. However, due to the situation of lack oil and gas, as well as maturity technology and less investment, it provides a greater market space for the calcium carbide acetylene process with coal as raw material. This firm has taken a series of technical measures on the PVC product to improve cleaner production, energy conservation and emission reduction in order to ensure the normal operation of the firm.

3.2 Process flow analysis

Using local lignite coal, the firm produces steam for the use of caustic soda and PVC production, waste heat is used to generate electricity which is called cogeneration. It produces PVC resin by using chlorine and hydrogen by-products from production of caustic soda as well as calcium carbide raw material. The calcium carbide slag produced in the process of polyvinyl chloride production is calcined and decomposed into cement clinker, mixed with the fly ash produced from thermoelectric boiler to produce cement. There is no discharge in the whole productive process. Resources and by-products in the productive process are reused efficiently to achieve the maximum utilization of resources.

The ECER system for this product includes Waste hydrochloric acid analytical technology (technology1), Waste sodium hypochlorite recycling technology (technology2), Variable pressure adsorption technology (technology3), Biochemical treatment technology on mother solution water (technology4), Acetylene recovery from slurry waste water (technology5) and Desulfurization technology of calcium carbide slag slurry (technology6). Among them, the technologies in the resource input stage is Tech. 1 and Tech. 2; the technologies in the resource utilization is Tech. 3 and Tech. 4; the technologies in the resource output stage is Tech. 5 and Tech. 6, as shown in Figure 4.

3.3 The effects and value flow analysis of ECER technology

3.3.1 Effects and value flow analysis on water saving technologies

(1)Waste sodium hypochlorite recycling technology (technology2)

ECER effect. The technology is used to remove sulfur and phosphorus impurities in the process of acetylene purification and is more conducive to reduce the difficulty of sewage treatment, and the cost and amount of sewage. The environmental protection cost has been reduced so.

Value flow analysis. Reusing waste sodium hypochlorite decreases greatly water production. Daily water of 1000 tons is saved based on calculation of 130 thousand tons of PVC.

(2)Biochemical treatment technology on mother solution water (technology4)

Mother liquid water produced from PVC device is wastewater, which may be cycled for cooling transmission equipment, polymerization production and water production. It will save about 800 thousand tons water each year.

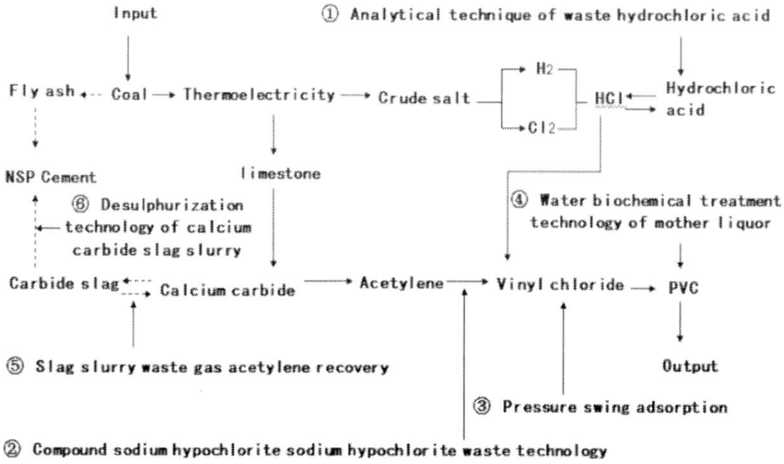

Figure 4. PVC production process and ECER technology system

3.3.2 Effect and value flow analysis on waste gas recovery technology

(1)Variable pressure adsorption technology (technology3)

ECER effect. Using the technology to recover vinyl chloride gas, acetylene gas and hydrogen in the tail gases, make the waste gas in effective recycling.

Value flow analysis. The hydrogen recycling may reduce power consumption in water electrolysis. 40,000 degrees electricity will be saved based on 8000 hours a year.

(2)Acetylene recovery from slurry waste water (technology5)

ECER effect. The acetylene gas dissolved in the waste water solution of calcium carbide slag slurry after reaction is recovered to cut the investment on the power equipment, and decrease other production medium for clean gas.

Value flow analysis. 3,000 tons calcium carbide consumption per year will be cut, and the original consumption of calcium carbide of 1.56 tons / per ton PVC is lowered to less than 1.45 ton / per ton PVC. In terms of economic efficiency, the annual 9 million Yuan will be cut, and 9,000 cubic meters of water may be saved.

3.3.3 Effect and value flow analysis on solid waste comprehensive treatment technology

(1)Desulfurization technology of calcium carbide slag slurry (technology6)

ECER effect. Carbide slag may be used for desulfurization in a heat and power plant, which and yellow phosphorus slag in the enterprise are reused for cement production, increasing greatly the comprehensive utilization amount of solid waste.

Value flow analysis. By the desulfurization technology of carbide slag slurry, the effect is obvious on the energy conservation, emission reduction, and resource recycling.

3.3.4 Effect and value flow analysis on recycling and utilization of industrial by-products

(1)Waste hydrochloric acid analytical technology (technology1)

ECER effect. It may prevent the production of industrial waste acid caused by the excessive proportion of hydrogen chloride in the production of vinyl chloride, and increase the economic value of recycling hydrogen chloride.

Value flow analysis. The annual reduction of sodium chloride is 6,000 tons, produces 2,100,000 Yuan economic benefits and the annual 12,3 000 tons of water can be saved.

3.4 Synergistic effect analysis on the ECER technology system

The technology system of the above six technologies brought the comprehensive effect of ECER, including water saving, electricity conservation, reducing the amount of raw materials. By using the method of value flow analysis, the cost of water saving and electricity conservation may be obtained, and the economic benefits of these six technologies were obtained as shown in Table 2. From Table 2, we see that the ECER effect of technical system may bring about the economic value of up to 4.117 million Yuan. Among them, the largest contribution is the technology 6, the economic value of lime conservation is up to 3 million Yuan. The value flow of water saving is 61.4 thousand Yuan, involving four technologies, the largest contribution technology is technology 4, which saves water 800,000 tons per year, and the economic value is 37,900 Yuan.

Through technical cooperation between 2 and 4, it reduced sewage discharge and using water, and improves the utilization rate of circulating water. Through technology 5 and 3, it realized recycling of waste gas, and ECER. Through technology 6 and 1, the reuse of wastes and by-products had been realized, and the additional economic value was created.

Technology 5 and 1 each had the dual synergistic effects of ECER or water saving and emission reduction. Technology 6 had only the energy-conservation and emission reduction synergies. The technology 3 had only power-conservation and emission reduction synergies, and had no synergistic effect with other technologies. According to table 2, the ECER collaborative evaluation table of the technical system of the energy intensive product was obtained as shown in table 3.

From table 3 we see that, in the resource input stage, the synergistic effect of technology 1 is good, while the technology 2 is general. In the stage of resource utilization, the synergistic effect of technology 3 is poor, while the technology 4 is general. In the resource output stage, the synergistic effect of technology 5 is good, while the technology 6 is poor. Therefore, we obtain the result that the front control of technology system of the energy intensive product has a good synergistic effect, while the process control and the end control measures have general synergistic effect.

Table 2. Value flow analysis on ECER effect of the technology system

NO.	Technology contribution	Technology Source	Annual requirement (tons)	Standard coal coefficient	Standard coal （ton）	Cost （Yuan）
1	Water saving	Technology 6	123,000	0.0857 Kg/t	11.7	5,800
		Technology 1	365,000		34.6	17,300
		Technology 2	800,000		75.8	37,900
		Technology 4	9,000		0.8	400
Sum			1,297,000		122.9	61,400
2	Power conservation	Technology 3	40,000 degrees	0.329 Kg/ kW·h	13.16	6,600
3	calcium carbide	Technology 4	3,000 tons	——	——	9,000,000
5	lime	Technology 5	100,000 tons	——	——	30,000,000
6	NaCl	Technology 6	6,000 tons	——	——	2,100,000
Sum						41,168,000

Table 3. Collaborative evaluation on ECER synergy of the technical system

NO.	Collaborative types of the single technology	Technology collaboration	Judge result	Production process
Technology1	ECER	Yes	Good synergy	Resource input stage
	Water saving and emission reduction	Yes		
Technology2	Water saving and emission reduction	Yes	General synergy	
Technology3	Power conservation and emission reduction	No	Poor synergy	Resource utilization stage
Technology4	Water saving and emission reduction	Yes	General synergy	
Technology5	Water saving and emission reduction	Yes	Good synergy	Resource output stage
	ECER	Yes		
Technology6	ECER	No	Poor synergy	

Based on the conclusion that the end of pipe control measures had not synergistic effect, while the front control and process control measures had good synergistic effect (Mao Xianqiang, 2011), the technology2,3,4 and 6 remained to be improved in the production process of energy intensive products. To analysis further, technology 3 had not synergistic effect with other technology and its economic value was only 400 RMB, so it is necessary to seek for the replace technology. Technology 6 had poor synergistic effect, and has brought about economic value of 30,000,000 Yuan, while it is at the end of the production chain, so it may be limited to improve more. Water saving and emission reduction effect of technology 4 and technology 2 was remarkable. Technology 4 and technology 2 had good technical collaboration effect, and had produced inter-technology effect, so the two technologies were limited to improve further. Technology 6 and technology 3 have not had a synergistic effect with other technologies yet, it may have greater potential to improve the synergistic effect.

4 Conclusions

The ECER progress of an energy intensive enterprise depends on the ECER effect of the energy intensive products. ECER technology system construction of an energy intensive product is the main path and endogenous mechanism for enterprises to achieve ECER targets. Among them, the construction and implementation of the technical system in the production is the key factor. Technology system in production includes ECER technologies in three stages: resource input, resource utilization and product output. The synergistic effects of ECER technology system include two aspects: single technology and inter-technology, which may be summarized and calculated by the value flow analysis on each technology. This thesis tries to construct a 3D analysis model of ECER technology system and synergistic effects of an energy intensive product, and link the productive process, the position and effect of ECER technology. It will provide a useful method to analyze space arrangement, effectiveness regulation and coordinating control of ECER technology system. Through a case study for the application on an energy intensive product, we had taken in-depth analysis on the ECER system construction and the ECER effects achieved and the shortcomings in production, and some suggestions were given. The application results show that the three-dimensional analysis model will have a good theoretical and practical application value for in-depth study on the relationship between ECER technology system and ECER synergistic effect of an energy intensive product. The next step is to study further on improving the ECER technology system of an energy intensive product to optimize the ECER synergistic effects.

References

1. Martin, Chen Wenying. Analysis on the synergy effect of emission reduction technology in China's iron and steel industry. China Environmental Science, **35 (1)**: 298-303(2015)(in Chinese)
2. Mao Xianqiang, Zeng Eucalyptus. Evaluation on the coordinative control effect of sulfur, nitrogen and carbon of technical emission reduction measures in the steel industry. Environmental science, **32** (5):1253-1260(2012)
3. Xing Fangfang, Jiang Qi. Analysis on the cooperative control technology of multiple pollutants in sintering flue gas in iron and steel industry. Environmental engineering, (4): 75-78(2014)
4. Zhang Shuling, Zhu Guilin. Technological innovation, circular economy and ECER about comprehensive utilization of iron or steel slugs. The proceeding of the eighth annual meeting of China Iron and steel, 169-176(2011)
5. Han Yijie, Liu Xiuli. Eenergy efficiency and ECER potential of iron and steel industry in every regions of China based on super efficiency DEA model. System science and mathematics, **31** (3):287-298(2011)
6. Zeng Ping, Deng Tengzhi, Wu Xiaojie. ECER and technological innovation: from the enterprise experience in Guangdong Pearl River Delta region. Economic system reform, (1):108-111 (2013)
7. Li Li, Jianhui Wang. ECER policies for the electric power industry in China. Energy Policy, **39**: 3669–3679(2011).
8. Zhaoguo Zhang, Xiaocui Jin. An empirical study on the institutional factors of energy conservation and emissions reduction: Evidence from listed companies in China. Energy Policy, **57**: 36–42(2013).
9. Gao Yan, Zhang Ling. Performance evaluation of coal enterprises energy conservation and reduction of pollutant emissions base on GRD-TOPSIS. Energy Procedia, (5): 535–539(2011).
10. C.A.P. Pimenteira, A.S. Pereira. Energy conservation and CO_2 emission reductions due to recycling in Brazil. Waste Management, (24): 889–897(2004).
11. Yan Linyu. Comprehensive analysis on material flow and value flow of Chinese iron resource. Chinese Geology University, Master Thesis, 2013
12. Mao Xianqiang, Zeng Eucalyptus. Study on the evaluation of the cooperative control effect of technical emission reduction measures. China Population, Resources and Environment, **21** (12): 1-7(2011)

A Security Architecture Research Based on Roles

Linbo Tao[1,*], Jianjing Shen[1], Bo Liu[2] and Zhenyu Zhou[1]

[1]School of Arts and sciences, Information Engineering University, Zhengzhou 450001, China
[2]Shenyang NanChang High School, Shenyang 110002, China
[*]Corresponding author: taotlb@126.com

Abstract. Security has always been the top issue against the cloud computing, scientific security architecture is the foundation of other security technologies. Comparing existed cloud computing security architectures and traditional security architectures, there are some common ground and new problems. In cloud computing environment, different users need different security requirements, so if we can assign them reasonable authority, there will be great efficiency improvement for data security and system efficiency. Role is an authority division and management method, it gains very good effect for its classification to user and data operations. Inspired by this, a security cloud computing architecture based on role has been designed, the security of the architecture has been evaluated at last.

1 Introduction

Cloud computing is a type of service that service vendors provide different types of services by a cluster of servers through networking, such as online software services, hardware rental, data storage, calculation etc [1]. These resources or services are supplied as the way of water or electricity which only to be paid according to the amount you actually used [2].

The biggest advantage of cloud computing is the elasticity of supply and low cost, the biggest obstacle is data security[3]. The security issues include the risk of privacy leakage and data abuse led by remote data storage, data tampering, lacking of censorship on the service provider, imperfect verification mechanism on user login check. For the data actually controlled by service providers, users have no rights to develop specifications on data management and security measures. So the cloud services providers and academia have designed a variety of security cloud computing infrastructure to maximize the security. These architectures should consider the practicality and scalability besides security.

Role is a collection of certain number of privileges. It refers to a collection including resources accessing and appropriate operating permission to complete a task. As an agent layer between user and authority, role is expressed as the relationship between authority and users. All authorities should be given to roles rather than directly to a user or group.

So a security cloud computing architecture based on roles is supposed on the following. The architecture will compares existing cloud computing security architectures first, then analyzes characters of roles, including authority division, security requirements division, the security of the architecture will be evaluated at last.

2 Security Model and Architecture

Cloud computing is the integration results of distributed computing, parallel computing, utility computing, network storage, virtualization, load balancing, hot standby redundancy and other traditional computer and network technologies[4]. It has a hierarchical structure, each level involving a wide range of properties and security issues.

Santos et al has proposed and designed a trusted cloud computing platform, TCCP, including a series of trusted nodes (N), trust coordinator (TC), untrusted cloud manager (CM) and external trust entity (ETE) etc. Wherein TC is maintained by a specific external trust entity (ETE) [7]. TCCP secures guest virtual machine by providing a closed box operating environment, it also allows users to test and verify the security.

The advantage of this architecture is to fundamentally solve the security issues of cloud computing architecture, the disadvantage is the excessive dependence on hardware which is contradictory with compatibility and cheapness of cloud computing. Another security architecture is based on isolation which can achieve data and operation isolation from hardware and software, it isolates multi-tenants data efficiently to a certain extent and has certain maneuverability.

Currently Cisco, VMware and other software, hardware providers have put forward their own isolation scheme, the core idea is to achieve logical space and level division through appropriate technologies which can isolate services provided to users. But this scenario is still high dependence on hardware and high cost.

Security as a service based SOA draws lessons from concept of SOA, it takes and packages security as services, different user can choose different services according to their needs. The classical SOA security architectures are IBM , EasySaaS security architectures.

3 Security Architecture Based on Roles

Sharing property of cloud infrastructure always give users the illusion that data is easier to lose when they are stored in the cloud. In fact, the overall technical architecture of cloud computing provides centralized management by security experts can achieve more security objectives[8] than the decentralized management of individual and non-professional management .

It is different from traditional security architecture who emphasizes border protection. The cloud computing divides logic isolation instead of physical border protection when different users applying their services[9]. But there are similarities between the two architectures on corresponding levels. Fig.1 has shows the comparison between cloud security architecture and traditional security architecture.

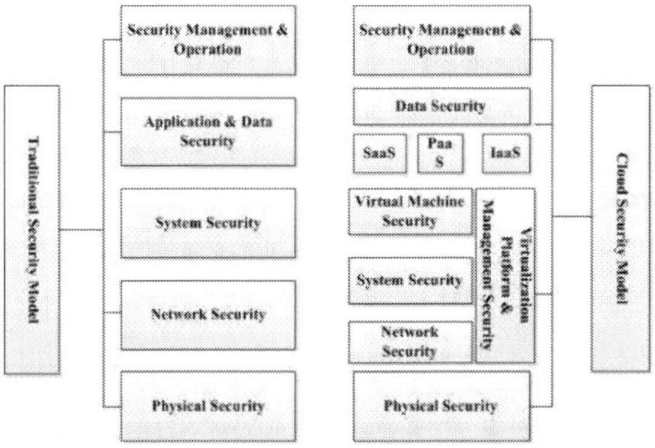

Fig.1. Comparison between cloud security and traditional security

The results show that cloud computing has introduced virtualization technology who change the service model, but not overturn the traditional security model. Cloud computing security and traditional security have the same security objectives, system resource type, basic security technologies. The cloud computing has its own specific security issues, including virtualization security and security issues about cloud computing sub-service model[10]. Generally speaking, cloud security is the inheritance and development of traditional security in a cloud computing environment. They are substantially similar in the level of security classification.

In cloud computing environment, cloud computing need to incorporate virtualization security precautions, due to the introduction of virtualization technology. In the basic level, sophisticated traditional security technologies can still provide security [11].

3.1 Security Architecture Design

Trusted security architecture and isolation architecture have good security, but poor extension. SOA architecture is flexible, but lack of clear secure margin [12], its level of responsibility is not clear enough.

In order to find a secure cloud computing management structure with both flexibility and security, user management and rights assignment should be the key. The concept of role has many good applies in rights management. Such as roles management in database are much successful in categorizing users, unifying licenses and unifying management. The concept of group in windows operating system is a role management too. Roles represent a class of users with same rights. Certain role has a set of certain rights.

The advantage role management is that designers can focus more attention on division of authority without concerning for someone specific. When the members of role changed, the content of role can still remain unchanged. In the cloud computing environment, the main bodies include cloud service providers, users, distributed hardware and software applications. The cloud service providers are responsible for providing hardware and software services to users and ensuring their data security. Users pay for their actual amount of consumption and storage.This mode allows data separated from the control of their owners, so its the actual reason why users would not save their data in the cloud.

In order to help users actually control their data, we must think more on rights assignment. If rights of cloud service providers and users are clear, they will get their own role, users will trust cloud computing more for their actual control of their data[13].

Here we divide the roles according to storage, data management, data ownership and cloud service providers. Combining with the cloud computing security architectures above and comparing with the traditional security models, a cloud computing security architecture based on roles has been proposed. as figure 2

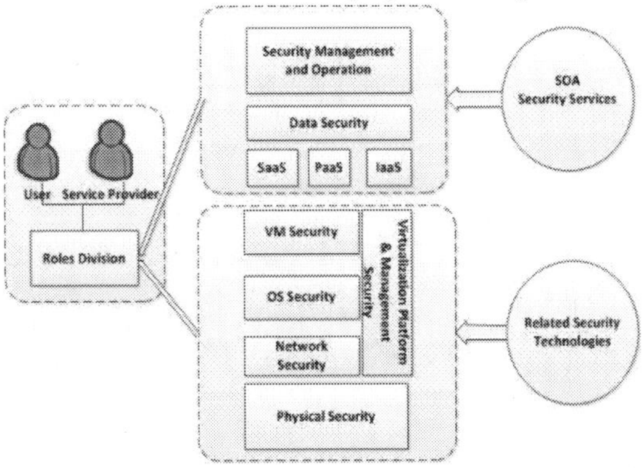

Fig.2. Security Cloud computing architecture based on roles

From the perspective of roles, cloud service providers are only the role that data kept in them, they are not the data owners and managers. The data owners are the only manager of their data who have all rights to their data management. Other users must be authorized by the user if they want to access his data. Any other users can not grasp the actual content of his data without his authorization, so data owner is the only administrator of his data.

3.2. Roles Division and Rights Assignment

Here roles are divided into four types including service provider, data administrator, inspector and user. The relationship among the four types are service providers build cloud environments and provide services, accept service requests, save data, recycle resources, user apply services and manage his own data, he is only the administrator of his own data, but users to others.

Let any user is u_i, his corresponding data set is D_i, rights set is F, $F=\{f_1,f_2,\cdots f_n\}\rightarrow\{\text{copy},\text{delete},\cdots\}$, each f_i corresponds a operating authority, such as moving, delete, encryption, copying etc. $f_i=0$ represents illegal operation, $f_i=1$ represents legal operation. The four types of roles above are the subset of F, let them as F_p,F_a,F_i,F_u.

Let user u_i and u_j get the F_a role to their own datasets D_i,D_j after they apply cloud services successfully. To each $f_a\in F_a$ there is

$$u_i(D_i)\xrightarrow{f_a}true \quad (1)$$

$$u_j(D_j)\xrightarrow{f_a}true \quad (2)$$

$$u_i(D_j)\xrightarrow{f_a}false \quad (3)$$

$$u_j(D_i)\xrightarrow{f_a}false \quad (4)$$

That is only the user itself has the F_a role to his data, other users do not have the role. To the service provider, role p_i and inspector role s_i, they do not have the F_a role, so they don't have rights to operate users' data, that is $F_p\cap F_a\cap F_i=\phi$.

Because of the openness of the cloud, the user can not only operate their own data, they need to share other users' data or share their data with others. But the sharing degree depends on the authority degree they get or give. The user get authority having the role F_u, they can get part or all rights of administrator, this depends on the authority degree by the administrator, so F_u is the subset of F_a. To any data d_i, $d_i\in D_i$ of user u_i, the operation authority relationships is shown as table1.
the relationship of roles and authorities

Table 1. Roles and Authority

authorities / roles	F_p	F_a	F_i	F_u
Service provider	1	0	0	0
administrator	0	1	0	1
inspector	0	0	1	0
user	0	e	0	1

From table1, we can see the roles have no operating authority overlapping except administrator and user, this approach is more conducive to ensure the security of data. For data those role F_u has rights to operate, the role F_a certainly have rights to operate them. In turn, the data operated by role F_a do not mean they can be operated by the role F_u unless they get the authority by role F_a. So the role value of F_u is e, $e \in \{0,1\}$, when the role get the operating authority $e=1$,else $e=0$.

In order to ensure the authenticity and legality of the operation among administrators, authorized users and service providers, a trusted third party as an inspector is needed to monitor the operation of the other three roles. The content of monitoring is the authenticity check and operation rights test. The role of inspector has no operating authority to data, but they has the rights to limit some unusual or dangerous operation operated by administrator or user, then give user a warning or terminate the dangerous operation.

The authenticity check is judged by the integrity and the user's operation records. If the user's identification information is incomplete, no logical association with the identification information or the reservation information is incorrect. When the unauthorized operation or dangerous operating application appearing, the role F_i will strictly control the user's service request and give a risk warning to the corresponding role F_a timely. Each broken access trying of roles F_p, F_a, and F_u will be recorded so that to facilitate judging the safety level of this operation trying by tracking their operating locus. The relationship among the four roles are shown as figure 3.

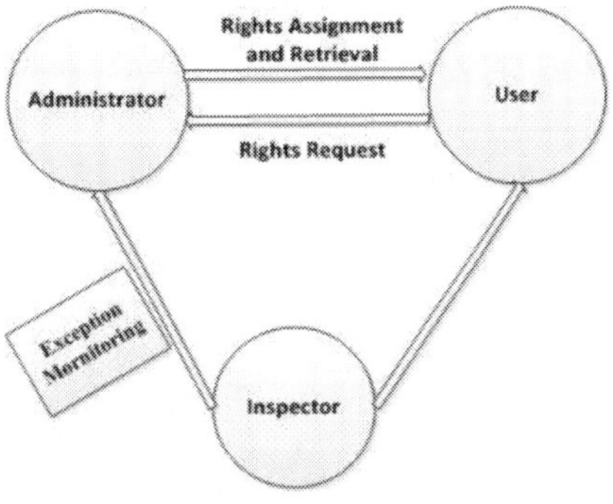

Fig.3. Relationship between roles and authority

4 Security analysis

Considering the complexity and cost of evaluation, the mathematical analysis is undoubtedly the most convenient and efficient methods[14]. The following is the security analysis procedure of cloud computing security architecture based role division.

Comparing to the previous architectures, the security architecture based role division has more clearer levels and responsibilities [15]. So we compare them from the angle of having roles division and without roles division. The architecture with roles division has the four roles F_p, F_a, F_i, F_u, and F_u is the subset of F_a. We can regard sets F_p, F_a, F_i as three vector spaces. Respectively take any sequence $A=\alpha_1\alpha_2\cdots\alpha_n$,

$B=\beta_1\beta_2\cdots\beta_n$, $C=\gamma_1\gamma_2\cdots\gamma_n$ from the three vector spaces, there are $\vec{A}\bullet\vec{B}=0$, $\vec{A}\bullet\vec{C}=0$, $\vec{B}\bullet\vec{C}=0$.

This shows the three spaces are orthogonal, so the vectors correspond to a series of authorities, the orthogonal vectors means the their vector spaces have no crossover authorities. But to any two operating series $\Gamma=\delta_1\delta_2\cdots\delta_m,(\delta_i\in F)$, $X=\chi_1\chi_2\cdots\chi_m,(\chi_j\in F)$, in the architecture without roles division, there are not always $\Gamma\bullet X=0$, so there are the danger to users that they may face operation without authority.

Through the analysis above, two objects u_i, u_j both have the role F_a, they will be closed to their data operation, so the Cartesian product constituted by different users and their data are orthogonal, that is $(F_a\times D_i)\perp(F_a\times D_j)$,the unauthorized probability is $P(f_i\times d_j)=0$.

In the architecture without roles division, any data $d_i, d_i\in D_i$, its owner is u_i, but d_i is stored in the remote place, so the service provider p_i becomes to the actual administrator of d_i, if p_i is untrusted, the dangerous probability to d_i is $P(d_i)\geq\frac{1}{2}$, because the architecture is lack of role F_i, user u_i can not get unauthorized operation of d_i or unauthorized operation can not be terminated timely.

If p_i is trusted, but without roles division, p_i is the actual administrator of d_i, so there are probability that p_i authorize the operating authority of d_i to other users. So the worst unauthorized operating probability of d_i is $P(d_i)=\frac{1}{2}P(u_j)$, $P(u_j)\leq\frac{1}{2}$, $P(u_j)$ is the probability of p_i authorize other users to operate d_i without the permission of u_i. The probability under the condition with roles division will be $P^{'}(d_i)\leq\frac{|D_i|}{\left|\sum_{i=0}^{n}D_i\right|}$, and $P^{'}(d_i)$ will be smaller along with increasing of data scale.

From the analysis above we can see architectures without roles division would likely to be confusion on authorization between data owners and those sharing the data or the condition that attackers get the authority of service providers. The attackers can do illegal operation and further damage penetration after they illegally obtain the service providers authority. But under the architecture with roles division, the attackers can only get one role, so they can't do more harm to others data. It shows that the architecture with roles division can bring higher security.

5 Conclusion

Existing cloud computing architecture and security model have been summarized and compared to the traditional security structures in this thesis. A cloud computing security architecture based on roles division has been proposed, the architecture combines the common characteristics with other cloud computing security architectures, divides their security levels more clearly, against the different authorities of different roles and their cooperation with each other, the data security and efficiency can work better.

References

1. Ashutosh Kumar Singh, Dr. Ramapati Mishra, Fuzail Ahmad, et al. A Review of Cloud Computing Open Architecture and Its Security Issues. International Journal of Scientific & Technology Research, Vol.6, No.1, 65-67 (2012)
2. Fei Hu, Meikang Qiu, Jiayin Li, et al. A Review of Cloud Computing:Design Challenges in Architecture and Security. Journal of Computing and Information Technology, Vol.1, No.19, 25-55 (2011)
3. Chen Ke-you, The Research On Data Security And Privacy Issues In Hybrid Cloud Computer. NanChang, JiangXi Normal University (2013)
4. Chi Li-ying. Research on Security Architecture of Cloud Computing and Its Key Technologies. Computer Development & Applications,Vol.6, No25, 20-22 (2012)
5. Zhang tao. The Cloud Computing Security Architecture Research Status Analysis Domestic and Foreign. Safety Broadcasting & Monitoring, No.11, 123-127 (2011)
6. JerichoForum. Cloud Cube Model:Selecting Cloud Formations for Secure Collaboration. Version1.0. www.JerichoForum.org (2009)
7. Lin Chuang, Su WenBo, Meng Kun,Liu Qu, Liu WeiDong. Cloud Computing Security:Architecture, Mechanism and Modeling. Chinese Journal of Computers,Vol.9, No.36, 1765-1781 (2013)
8. Sanjaya Dahal. Security Architecture for Cloud Computing Platform. Stockholm,Sweden:Master of Science Thesis (2012)
9. Muthu Ramachandran. Component-Based Development for Cloud Computing Architectures. Computer Communications and Networks,No.10, 91-114 (2011)
10. Masayuki Okuhara, Tetsuo Shiozaki, Takuya Suzuki. Security Architecture for Cloud Computing.Fujitsu Sci. Tech. J,Vol.4, No.46, 397-402 (2010)
11. Vic (J.R.) Winkler. Securing the Cloud:Cloud Computer Security Techniques and Tactics. Edition 1, Publisher: Syngress (2011)
12. Chen Chi,Yu Jing. Cloud computing security system. Edition 1, Science Press (2014)
13. Zhang Wei, Dong Qunfeng. Design and implementation of cloud security comprehensive analysis system. Computer Engineering and Applications, Vol.19, No.50, 89-94 (2014)
14. Dan Gonzales, Zeremy Caplan, Evan Saltzman, et al. Cloud-trust-a Security Assessment Model for Infrastructure as a Service(IaaS) Clouds. .IEEE Transactions on Cloud Computing. PP(99), 1-14 (2015)
15. Jinag Zheng-wei, Zhao Wen-rui, Liu Yu. Model for Cloud Computing Security Assessment Based on Classified Protection. Computer Science,Vol.8, No.40, 151-156 (2013)

Zhengdong New District Smart Government Application System Construction and Its Development Strategy Research

Bing Li *, Xianghua Xing, Hua Xuan and Xueyuan Wang

School of Management Engineering, Zhengzhou University, Zhengzhou, China
*Corresponding author: lbing@zzu.edu.cn

Abstract. Smart government is the advanced stage of e-government development, whose essence is the fusion of Internet, Internet of things, cloud computing, big data and other high and new technology and e-government, to achieve the efficient management of all kinds of information resources and make the government services more transparent, scientific and wisdom. Based on the introduce, basic principle and construct meaning, application system construction, main problems and developing countermeasures and other aspects of smart government to make a comprehensive analysis in the construction of Zhengdong new district smart government, then advise the government departments to speed up the transformation of government management functions and social service mode to integrate the business processes, and promote the construction of service-oriented and smart-oriented government actively.

Keywords. Smart government, application system, problem analysis, development countermeasures.

1 Introduce

As the Internet, the Internet of things（IOT）,the service-oriented architecture (SOA),cloud computing, big data and other high and new technology applied widely, in order to speed up the transformation of the mode of government management and service and improve government scientific decision-making ability as soon as possible, and build smart and service-oriented government quickly, the Chinese government has issued "National new urbanization plan (2014-2020)"and "Guiding Opinions on Promoting the Healthy Development of the smart city", clearly put forward to accelerate the construction of the smart city.

Smart government affairs is the important constituent parts in the construction process of the smart urban and development foundation, provide important support for other areas of intelligence applications, and is also the advanced stage of e-government, whose essence is the fusion of Internet, IOT, SOA, cloud computing, big data and other high and new technology and e-government [1], to achieve the efficient management of all kinds of information resources and has become an important way to change the functions of management and social services, and integrate business process. Smart government can provide convenient and efficient service for the general public [2], also can provide scientific decision-making basis for government departments to accelerate the construction of smart and service-oriented government.

Smart government mainly has perception, collaborative, intelligence and service-oriented characteristics. Among which, the perception refers to the changes in the data sources, collaborative refers to the use of high-tech to break the boundaries between government departments and change collaboration efficiency, intelligence is the change in the use of data, servitization refers to the transformation of government departments and management of service delivery. The perception, collaborative and intelligence is the basis of servitization, the servitization is the ultimate goal of perception, collaborative and intelligence.

2 Basic principle and construct meaning

2.1 Basic principle

Based on "Top-level design of smart city of Zhengdong new district" and "Eastern wisdom Thirteen Five Development Plan", combined with the building requirements of smart government of national and provincial government, identify the major purposes of building smart government at Zhengdong new district is in order to improve the management of government departments and public service capabilities, accelerate the transformation of the city' upgrading, and further enhance the convenience and comfort of people's lives, the overall design of Zhengdong new district smart government should follow the following principles:

(1) Overall planning and intensive construction. Smart government is the important constituent parts and development foundation in the construct process of the smart city, its planning and construction need to be a good cohesion with the overall urban planning wisdom [3]. In the specific planning and design, need to focus on the overall situation, implement the national and provincial smart government's framework relevant standards, with smart eastern district's the actual construction of "3+ 1+1", to avoid the relevant departments' fragmentation, dispersion and duplication of investment in construction and other phenomena.

(2) Integrate resources and share information. Based on public information resources' sharing, integration and business collaboration, implement disclosure the government information and government business collaboration system actively, to achieve real-time exchange of information interconnection and government business systems, set up covering the entire area of the cloud data center of smart government, benefit maximize information resources.

(3) Uniform standards to ensure safety. Depth study the policy documents' information and standards of national and provincial government how to construct smart government, focusing on the regulatory system, security system and operation and maintenance management system's construction of smart government[3], realize the unity of the command, the coordination and the maintenance within the region and smart government become the leading and demonstrating building in Zhengzhou.

(4) Moderate prospective and leapfrog development. The rapid development of high-tech, government institutional restructuring mechanism deepening, the construction of smart government should be based on the current situation, but also with a modest-looking, to meet the needs of government departments managed services model and high-tech development to ensure sustainable and leaping development of smart government.

2.2 Construct meaning

Implementing the new smart city of Xi Jinping General Secretary's important speech, reflecting the people-oriented and service-oriented smart governing philosophy, the main significance of constrict Zhengdong new district smart government as following:

(1) Solve the "information island" problem effectively. By using high-tech applications and constructing standard system, security system and operation management system of smart government, can help resolve the "islands of information" during the administrative examination [4], and can help achieve the information resources' integrating and sharing.

(2) Reduce costs and optimize resources. With sharing and integrating basic information resource, introducing and using a series of advanced high-tech tools and services means to accelerate the construction of government cloud data center, to achieve effective interconnection of various information data, can not only save equipment and labor costs, and it can also integrate and optimize resources effectively.

(3) Improve the efficiency of government departments effectively. The government cloud data center can address interoperability of data resources between Zhengdong new district offices, can promote the integration and sharing and can also promote the good synergy between the services, can help government departments to achieve "one-stop" office functions, in order to promote the government's work efficiency.

(4) Effectively assist scientific decision-making. The smart government can help leaders understand the dynamics of city management, economic, social development and other information, can provide early warning and predictive analysis and other technical support for the leaderships' scientific decision-making, to enhance the region's management and the overall effectiveness of services business.

3 Smart government application system

Within the scope of public administration, all kinds of complex factors, it is necessary to consider the amount and quality of data and efficiency and economy of data analysis, but also think about new synthetic methods and analytical data, in order to enables data to realize its own value. Based on this consideration, when construct the overall framework of Zhengdong new districts mart government applications system should be based on the smart government unified standards system, security system and operation management system as the basis, take the construction of the cloud computing center that provides hardware and software environment in "Eastern wisdom" project planning as a support, Mainly includes the following content:

3.1 Cloud data center of smart government

Smart government cloud data center is an important part of smart government's construction process, is also the bottom support of the city data, is also highly resource information sharing breakthrough, can provide foundation for information sharing and integration between government departments and collaboration between services business, therefore we must accelerate the use of high-tech to establish smart government cloud data centers including population database, corporate libraries, library and other basic spatial information database[5]. Among them, the population information repository can help government manage the basic information about population; Corporate library is a quality inspection organization code based on indexes, take business registration and tax registration as supplemented index forming the enterprise underlying database, to achieve information sharing and dynamic updates of corporate units; Spatial geographic information database based on the electronic map, mainly including basic and thematic data, such as administrative divisions, urban ecological environment, scientific research institutes.

3.2 "Five list one single network" information system

From the "Five list one single network" to start optimize and improve government service management mechanism, which is a basic requirement to construct service-oriented and smart government, is also an important measure to improve public service satisfaction and speed up building a harmonious society. "Five list one single network" refers to the list of responsibilities of government, the list of administrative examination and approval matters, the list of industrial clustering enterprise investment project management, the list of administrative fees and charges, the list of government investment funds, the establishment of the region's unified government service information network [6] . "Five list one single network" smart government system mainly includes office project approval platform, project approval service platform, integrated query platform for project approval, project approval statistical analysis platform.

3.3 The city's comprehensive management platform of smart eastern district

"The city's comprehensive management platform of smart eastern district" takes the scientific intelligent city management as the goal, on the basis of urban grid management application system, the construction of the city's comprehensive management platform is in order to promote urban management components to solve things standardization and participation of the whole idea implemented. On the basis of the existing The city's comprehensive management platform of smart eastern district's works and extend other units of video access points, to improve collaboration and coordination between the services and the lateral sectors and establish and improve urban management information base, and design and build the system by using disposal cases, GIS maps, statistical analysis, performance evaluation , system management, intelligent collection terminals and other standard modules [7].

3.4 Economic operation information platform of smart eastern district

"Economic operation information platform of smart eastern district" takes the tax-related information private network as the carrier, based on the software platform to set up a "government-led, sector assistance and social participation" as the main content of socialization comprehensive management tax running system [8]. The system is mainly based on the digital three-dimensional geographic information data and mobile private network, can full use of information technology's advantage, and set up information sharing platform for tax management, strengthen the relevant government departments of tax-related information resources' integration, in order to fully grasp the scale, distribution, composition and changes of tax sources, plug the loopholes in collection to prevent the loss of tax revenue, mainly including sources of revenue ledger management, tax source analysis, coordination department, tax co-grid, building economic and other sources module [9].

4 Main problems

Zhengdong new district smart government can provide an effective way for the government comprehensive upgrade of public administration and public services, can help improve the government's governance capabilities, also is a self-flagellation for the government. But the building of smart government is an complex system engineering, there is a lot of uncertainty related to the Internet, IOT, SOA and other high-tech, so Zhengdong new district government authorities should make sufficient attention to smart government.

4.1 Information resources' quality and safety should be improved and guaranteed

Government information resource database has a lot of personal information, the quality and safety of information resources is the prerequisites of smart government healthy development. Professionals of internet, IOT, SOA, cloud computing, big data and other high-tech aspects are relatively scarce, resulting in the deviation of information resource quality and increase risk of safety [10]. Information security issues of Zhengdong new district smart government is mainly reflected on information security technology, information security management, information security laws and regulations, information security awareness [11].

4.2 Responsibilities inventory systems needs to be deepened

With the reform and establishment of laws and regulations and the adjustment of institutions and functions, the published list needs dynamic updates in a timely manner. Public service matters that are closely related to mass production and life is incomplete, should according to the laws and regulations conduct further sort to approve, in order to reduce handling process, link and time limits and improve the standard.

4.3 Government affairs service system needs to be improved

The city's administrative examination and approval service platform is incomplete relatively, still need to strengthen top-level design, and actively implemented. The existing functions of the position and reform requires of offices service center can't match exactly, matters entered service hall are also inconsistent completely, and many convenient service centers' hardware facilities, services function, construction and other mechanisms have to be further standardized.

4.4 Smart government management system needs to be strengthened

The building of Zhengdong new district smart government platform lacks strong leadership, planning and operation and maintenance team, the authorities lack effective monitoring and support in the planning, management and implementation of the project, when coordinate the constructing work of Zhengdong new district smart city, the department and daily management is always inadequate.

5 Developing countermeasures

5.1 Improve the data quality and information security system

(1) Establish information security system. Clear scientific information security legal protection philosophy, strengthen information security laws and regulations, strengthen information security ahead of legal efficiency, based on four levels that physical layer security, internet layer security, security system layer and application layer security, then establish security risk assessment mechanism and comprehensive security audit system of smart government.

(2) Strengthen personnel security system. Set up Internet, IOT, big data and other high-tech programs in Longzihu university campus, to train high-level, highly skilled person for the wisdom of government. Explore service policy of talents, simplify the process of introduction of talents, make full use of scientific research institutions and other platforms to absorb professionals of the smart government affairs.

(3) Increase capital investment. Strengthen the integrated management of special funds in the construct of informational government departments, to ensure the construction of information technology applications of various departments. Accelerate to innovate investment and financing and operating model in the construct of smart government, help Zhengdong new district smart government form a good situation comprehensively.

(4) Improve the system of accountability. Based on the collection and classification, statistical analysis, application management and other links of data, to formulate behavior standards and secrecy system, and formulate identification and punishment system for the abuse, private and generic of data [12].

5.2 Deepening the Reform of the "Five list one single network"

(1) Speed up administrative responsibilities matters into network. Formulate inspect together, judge and estimate, evaluate performance and other supporting mechanisms actively, for the operation of all levels of inventory for dynamic monitoring, tracking judged, performance evaluation. Start to handle case and enforce law online and trace the whole actively, implement publicity system of the enforcement of administrative law, explore administrative expropriation, compulsory administrative matters, and other types of administrative powers and responsibilities into network, realize all administrative powers and responsibilities matters can be running in the network.

(2) Speed up administrative responsibilities matters star-management. Strengthen the analysis of accountability matters, and combine with counseling, complaints and suggestions of the public, to simplify the optimization of processes online. Use big data means to sort out administrative powers and responsibilities matters, to achieve star-management of administrative accountability matters as soon as possible.

(3) Deepen the sharing of information resource on administrative service network. Continue to promote the "Internet + government services", built and apply data sources service platform of Zhengdong new district smart government, and improve the resident population information database, corporate database and geospatial information repository database and other basic database.

(4) Continue to deepen the preparation of inventory systems. Clear the running processes and responsibilities of every job of each matter's all aspects, make measures on how to control risk of integrity, to promote the implementation of administrative accountability matters of personal responsibility, and to promote the lists' implementation.

5.3 Improvement smart government affairs service system

Specification the service functions, operation mechanism, personnel management and other respects of the two governments that district and township (office), unify the name of all government service center as soon as possible, so as to make a good match between the institutional functions and tasks. Accelerate the construction of convenient service centers covering the whole township (office), to provide basic security services for three linkage of Zhengdong new district smart government to provide basic security services three linkages.

5.4 Strengthen smart government affairs management mechanism

Clear construct leading group mechanism of smart government to promote the comprehensive building work of smart government. Establish an independent operation team to participate in projects all the time as soon as possible, to enhance the achievements of the operation and maintenance work in the constructing of smart government, to ensure the healthy and sustainable development of the smart government. Establish professional institutions to strengthen the daily management work of smart government, coordinate the construction work of Zhengdong new district smart government with Zhengzhou city and other relevant departments, clear the responsibility, mission and assess goals of relevant departments.

6 Conclusions

In summary, smart government is an important component and developing foundation of smart city, is also an more advanced stage of e-government, has become an important way for government departments to change management functions and integrate business processes and change the way of social services in the information age. Smart government let government departments have a more detailed and scientific management of the society, also have great significance in achieving a comprehensive well-off society and promoting the modernization of social governance and building a harmonious society.

Acknowledgements

This research is supported by the Ministry of Education of Humanities and Social Science Project (Grant no. 15YJC630148).

References

1. Zhang XH, Deng XF. The Overall Architecture Design Ideas and Implementation Path of Nanjing Smart Government. E-Government, 2013, (9): 103-109.
2. Xu XL, Zhu GW. Smart Government: The Life Path of Electronic Information Society Governance. Journal of Dialectics of Nature, 2012, (34):95-100.
3. Li CM. Construction of Smart Government in the Eyes of Modern Management. Socialism Studies, 2014, (4):81-88.
4. Gu SZ, Yang JW, Liu JR. Current Problems and Measures in the Construction of Smart City in Our Country. China Soft Science, 2013, (1): 6-12.
5. Napitupulu, D. The critical success factors study for e-Government implementation. Int. J. Comput, 2014, 89 (16), 23–32.
6. Po-Ling Sun, Cheng-Yuan Ku. An implementation framework for E-Government 2.0.Telematics and Informatics, 2015, 32(3):504-520.
7. Hui Zhang, Jianying Xiao. Diffusion of e-government: A literature review and directions for future directions [J]. Government Information Quarterly, 2014, 31(4):631-636.

DOI: 10.1051/matecconf/201710005069

8. Song G, Meng QG. Government2.0: Government Innovation in the Eyes of Innovation 2.0. E-Government,2012, (2-3):53-61.
9. Agus Prahono, Elidjen. Evaluating the Role e-Government on Public Administration Reform: Case of Official City Government Websites in Indonesia. ScienceDirect, 2015, 59(3): 27-33.
10. Mohammad Sarrayrih, B. Sriram. Major challenges in developing a successful e-government: A review on the Sultanate of Oman[J]. Journal of King Saud University- Computer and Information Sciences, 2015, 27(2):230-235.
11. Li Y. Informational Security Strengthen and Impact Analysis in the Construction of Smart City. Library and Information Service. 2012, 56(6):20-24.
12. Li DR, Yao Y, Shao ZF. Smart City in Big Dat. Geomatics and Information Science of Wuhan University. 2014, 39(6):631-640.

An Efficiency of Resource Allocation Research Based on DEA Within 30 Ethnic Minority Autonomous Prefectures

Jianzhong Zhang[1] and Qiusheng Lu[2,*]

[1] Economic and Management Department, Hubei University for Nationalities, Enshi, China
[2] Economic and Management Department Hubei University for Nationalities, Enshi, China
[*] Corresponding author:qiusheng1413@163.com

Abstract. The 12th Five-Year is an important strategic opportunity period for the economic and social development in ethnic minority regions of China. The development strategy of ethnic minority areas should adhere to Scientific Outlook on Development as a guide, suit the measure to local condition, constantly optimize and improve the efficiency of resource allocation then further achieve economic and social scientific development with the national policy support for ethnic minorities. In this paper, the evaluation index system of input and output efficiency of resource allocation in Minority Autonomous Prefecture is established, and the DEA model is the main analysis tool for comparing China's resources allocation efficiency in 30 Ethnic Minority Autonomous Prefectures, and proposes some policy recommendations on optimization of resources allocation.
Keywords. DEA Model; Ethnic Minority Autonomous Prefecture; Resource Allocation.

1 Introduction

China is a multi-national country composed of 56 ethnic groups. To guarantee the construction and promotion of economic and social development in Ethnic Minority Areas, the Party and the Government design national regional autonomy system according to "Big Mixing-up, Small Inhabit" nation distribution characteristic and the historical evolution of Chinese nation. Among them, the ethnic minority autonomous prefecture is situated between the autonomy region and autonomy county level and unit, but also is a part of our local government. At present, there are 30 ethnic autonomous prefectures in Yunnan, Xinjiang and other nine provinces (see table 1).

Table 1. The Basic Situation of China's Ethnic Autonomous Prefecture.

Serial Number	Autonomous Prefecture	Affiliation	Establishment Time	State Capital
DMU_1	Korean Autonomous Prefecture of Yanbian	Jilin Province	1952-9-3	Yanji City
DMU_2	Tibetan Autonomous Prefecture of Ganan	Gansu Province	1953-10-1	Hezuo City
DMU_3	Hui Autonomous Prefecture of Linxia	Gansu Province	1956-11-19	Linxia City
DMU_4	Tibetan Autonomous Prefecture of Yushu	Qinghai Province	1951-12-25	Jiegu Town
DMU_5	Tibetan Autonomous Prefecture of Hainan	Qinghai Province	1953-12-6	Qiabuqia Town
DMU_6	Tibetan Autonomous Prefecture of Huangnan	Qinghai Province	1953-12-22	Longwu Town
DMU_7	Tibetan Autonomous Prefecture of Haibei	Qinghai Province	1953-12-31	Xihai Town
DMU_8	Tibetan Autonomous Prefecture of Golog	Qinghai Province	1954-1-1	Dawu Town
DMU_9	Haixi Mongolian Tibetan Autonomous Prefecture	Qinghai Province	1954-1-25	Delingha City
DMU_{10}	Bayinguoleng Mongolian Autonomous Prefecture	Xinjiang Province	1954-6-23	Korla City

DMU_{11}	Mongolian Autonomous Prefecture of Bortala	Xinjiang Province	1954-7-13	Bole City
DMU_{12}	Kirgiz Nationality Autonomous Prefecture	Xinjiang Province	1954-7-14	Artux City
DMU_{13}	Hui Autonomous Prefecture of Changji	Xinjiang Province	1954-7-15	Changji City
DMU_{14}	Yili Kazak Autonomous Prefecture	Xinjiang Province	1954-11-27	Yining City
DMU_{15}	Xiangxi Tujia and Miao Autonomous Prefecture	Hunan Province	1952-9-3	Jishou City
DMU_{16}	Enshi Tujia and Miao Autonomous Prefecture	Hubei Province	1983-12-1	Enshi City
DMU_{17}	Qiandongnan Miao and Dong Autonomous Prefecture	Guizhou Province	1956-7-23	Kaili City
DMU_{18}	Qianxinan Buyei and Miao Autonomous Prefecture	Guizhou Province	1982-5-1	Xingyi City
DMU_{19}	Qiannan Buyei and Miao Autonomous Prefecture	Guizhou Province	1956-8-8	Douyun City
DMU_{20}	Xishuangbanna Dai Autonomous Prefecture	Yunnan Province	1953-1-24	Jinghong City
DMU_{21}	Dai-Jingpo Autonomous Prefecture of Dehong	Yunnan Province	1953-7-24	Luxi Town
DMU_{22}	Lisu Autonomous Prefecture of Nujiang	Yunnan Province	1954-8-23	Lushui Town
DMU_{23}	Bai Autonomous Prefecture of Dali	Yunnan Province	1956-11-22	Dali City
DMU_{24}	Tibetan Autonomous Prefecture of Diqin	Yunnan Province	1957-9-13	Zhongdian County
DMU_{25}	Honghe Hani and Yi Autonomous Prefecture	Yunnan Province	1957-11-18	Mengzi County
DMU_{26}	Wenshan Zhuang and Miao Autonomous Prefecture	Yunnan Province	1958-4-1	Wenshan County
DMU_{27}	Yi Autonomous Prefecture of Chuxiong	Yunnan Province	1958-4-15	Chuxiong City
DMU_{28}	Tibetan Autonomous Prefecture of Ganze	Sichuan Province	1950-11-24	Kangding County
DMU_{29}	Aba Tibetan and Qiang Autonomous Prefecture	Sichuan Province	1953-1-1	Barkam County
DMU_{30}	Yi Autonomous Prefecture of Liangshan	Sichuan Province	1952-10-1	Xichang City

Decision of the Central Committee on Some Major Issues Concerning Comprehensively Deepening the Reform was adopted at the Third Plenary Session of the 18th Central Committee of the Communist Party of China in which "improving the general transfer payment growth mechanism, focusing on increasing the growth of the transfer payment to minority areas and poor areas, accelerating the improvement of modern market system" have been presented . China's natural resource bases are uneven in minority areas, the development degrees are different, economic and social development of ethnic minority regions are out of balance. This paper applies DEA model for quantitative comparison and analysis the resource allocation in ethnic autonomous prefectures in China in order to optimize the resources allocation, set some efficiency of differentiated support policy recommendations for the relevant government departments.

2 An analysis model of resource allocation efficiency in ethnic minority autonomous prefecture

Under specific product conditions, groups of various production factors inputs possible producing maximum output is the core of the economic production theory. The production possibility boundary which is described by production function in the production theory model is generally called production frontier. The gap between actual production status and production boundary reflects the efficiency level of the production, that is, the technical efficiency, which also is known as resource allocation efficiency, describes a degree of resource allocation for production factors. The study on measure of resource allocation efficiency began in 1957 by economist Farrell

Michaell, who oriented to production frontier ideas, developed and proposed the parametric method and non parametric method for research of the production frontier. Parameter method needs to construct a specific production function, which estimates the function parameters in production frontier, completes constructing the frontier function parameters; The non parametric method does not require prior assumptions and estimate parameters to structure the concrete function form, but observe a large number of actual data according to a certain production efficiency standard to find the relative effective point in production frontier.Non parametric effect of production frontier theory-Data Envelopment Analysis (abbreviated as DEA) is put forward by U.S economist A.Charnes et al to appraise relative validity of decision making unit in 1978,which also is an econometrics method.

2.1 C^2R Model and Comprehensive Efficiency Measure of Data Envelopment Analysis

Use DEA to determine the comprehensive efficiency of resources allocation, the general use of C^2R model was determined and described as follows:

There are n decision making units (DMU), each of which has a m DMU type of "input X", as well as S type of "output Y". Record the input variable of the first j DMU as $X_j = (x_{1j}, x_{2j}, \ldots, x_{mj})^T$,j =1, 2,... n; the output variable of first j DMU is recorded as $Y_j=(y_{1j}, y_{2j} \ldots y_{nj})^T$, j =1, 2,... n.Using (X_j, Y_j) represents the first j decision making unit DMU_j, and its data structure is shown in table 2.

Table 2. N Decision Making Unit DMU Input and Output Data Structure.

v	DMU_1	DMU_2	…	DMU_n	u	DMU_1	DMU_2	…	DMUn
v_1	x_{11}	x_{12}	…	x_{1n}	u_1	y_{11}	y_{12}	…	y_{1n}
v_2	x_{21}	x_{22}	…	x_{2n}	u_2	y_{21}	y_{22}	…	y_{2n}
…	…	…	…	…	…	…	…	…	…
v_m	x_{m1}	x_{m2}	…	x_{mn}	u_s	Y_{s1}	Y_{s2}	…	Y_{sn}

Setting $v=(v_1,v_2 \ldots v_n)^T$ represents m input indicator corresponding to the weight coeffective vector, Setting $u=(u_1,u_2 \ldots u_s)^T$ represents s output indicator corresponding to the weight coeffective vector, so each decision making unit DMU_j has corresponding efficiency evaluation index:

$$h_j = \frac{u^T Y_j}{v^T X_j} \quad j=1,2,\ldots,n \tag{1}$$

Proper choice of the weight coeffective of v and u, makes them meet $h_j \leq 1$, j =1,2, …., n. Evaluate the efficiency of j_0 decision making unit DMU_{j0} (X_0, Y_0).For any DMU_j, the production possibility set is :

$$T=\left\{(X,Y)\middle| \textstyle\sum_{j=1}^n X_j \lambda_j \leq X, \sum_{j=1}^n Y_j \lambda_j \geqq Y, \lambda_j \geqq 0, j=1,2,\ldots,n\right\}$$

Construct the following maximization model($\bar{p}$)):

$$(\bar{p})\begin{cases} \max h_o = \frac{u^T Y_o}{v^T x_o} = V_{\bar{p}} \\ \text{s.t.} \\ h_j = \frac{u^T Y_j}{v^T x_j} \leq 1, j = 1,2\ldots,n \\ v \geqq 0, u \geqq 0 \end{cases} \tag{2}$$

let $t=1/(v^T X_0)$,w = tv,μ= tu, using Charnes-Cooper transformation, $\bar{p}$ can be described as an equivalent linear programming problem(P_1) (that is, C^2R model):

$$(P_1)\begin{cases} \max \mu^T Y_0 = V_{p1} \\ \text{s.t.} \\ w^T x_j - u^T Y_j \geqq 0, j = 1,2,\ldots,n \\ w^T x_0 = 1 \\ w \geqq 0,\ \mu \geqq 0 \end{cases} \tag{3}$$

Given the following C^2R model DEA effective definition:

If Optimal solution of linear programming problem(P_1)w_0 and u_0 meet $V_{p1}=u_0^T Y_0=1$,the DMU_{j0} is called weak DEA efficiency; if the linear planning problems (P_1) has an optimal solution of w_0 and u_0 meet the $V_{p1}= u_0^T Y_0=1, w_0>0, u_0>0$, DMU_{j0} is called DEA efficiency

The Non Aqi Mead infinitesimal ε is introduced into C^2R model which is described as ($P_{1\varepsilon}$):

$$(P_{1\varepsilon})\begin{cases} \max \mu^T Y_0 = V_{p1\varepsilon} \\ \text{s.t.} \\ w^T x_j - \mu^T Y_j \geqq 0, j = 1,2,\ldots,n, \\ w^T x_0 = 1 \\ w^t \geqq \varepsilon e^{-T}, \mu^T \geqq \varepsilon e^{+T} \end{cases} \tag{4}$$

where, $e^{-T} = (1,1,\ldots,1) \in E_m, e^{+T} = (1,1,\ldots,1) \in E_s$, ε is Non Aqi Mead infinitesimal.
Introducing variable s^- And s^+, pairing C^2R model ($D_{1\varepsilon}$) is:

$$(D_{1\varepsilon}) = \begin{cases} \min[\theta - \varepsilon(e^{-T}s^- + e^{+T}s^+)] = V_{D1\varepsilon} \\ s.t. \\ \sum_{j=1}^{n} X_j \lambda_j - s^- = \theta X_0 \\ \sum_{j=1}^{n} X_j \lambda_j - s^+ = X_0 \\ \lambda_j, j = 1,2,\ldots,n \\ s^- \geqq 0, s^+ \geqq 0 \end{cases} \quad (5)$$

Suppose the optimal solution of C^2R model ($D_{1\varepsilon}$) are λ^*, s^{-*}, s^{+*} θ^*, when$\theta^* = 1$, and $s^{-*} = 0$, $s^{+*} = 1$, the DMU_{j0} is DEA effective; when $\theta^* < 1$ or $s^{-*} = 0$, $s^{+*} \neq 1$, the DMU_{j0} is non DEA efficiency. According to **C^2R** model operation results $\theta(0 \leq \theta \leq 1)$ is comprehensive efficiency of decision making unit.

2.2Data Envelopment Analysis BC2 Model and Scale Efficiency and Technical Efficiency Measurement

The scale efficiency and technical efficiency of resource allocation were determined by DEA, and the BC^2 model was used.

There is a hypothesis in prior C^2R model ($D_{1\varepsilon}$): C^2R model ($D_{1\varepsilon}$) is scale invariant, the hypothesis is very strict, and difficult to meet this condition in practice. So, adding a convexity assumptions in the C^2R model ($D_{1\varepsilon}$):$\sum_{j=1}^{n} \lambda_j = 1$,then production possibility set becomes as follows:

$T_{BC^2} = \{(X,Y) \mid \sum_{j=1}^{n} X_j \lambda_j \leq X, \sum_{j=1}^{n} Y_j \lambda_j \geqq Y, \sum_{j=1}^{n} \lambda_j = 1, \lambda_j \geqq 0, j = 1,2,\ldots,n\}$

After removing cone condition in C^2R model, make DEA efficiency evaluation which is the input of the DEA model— the BC^2 model (P_2) , bases on the production possibility set T_{BC^2}:

$$(P_2) \begin{cases} \max\left(\mu^t Y_0 + u_0\right) = V_{P_2} \\ s.t. \\ w^T X_j - \mu^T Y_j - u_0 \geqq 0, j = 1,2,\ldots n \\ w^T x_0 = 1 \\ w \geqq 0, \mu \geqq 0 \end{cases} \quad (6)$$

And dual problem DEA model -- BC^2 model (D_2)

$$(D_2) \begin{cases} \min \theta = V_{D_2} \\ s.t. \\ \sum_{j=1}^{n} X_j \lambda_j \leq \theta X_0 \\ \sum_{j=1}^{n} Y_j \lambda_j \geqq Y_0 \\ \sum_{j=1}^{n} \lambda_j = 1 \\ \lambda_j \geqq 0, j = 1,2,\ldots,n \end{cases} \quad (7)$$

Introduce Non Aqi Mead infinitesimal ε in DEA model—BC^2 model,get BC^2 model($P_{2\varepsilon}$):

$$(P_{2\varepsilon}) \begin{cases} \max\left(\mu^T Y_0 + \mu_0\right) = V_{P_{2\varepsilon}} \\ s.t. \\ w^T X_j - \mu^T Y_j - \mu_0 \geqq 0, j = 1,2,\ldots,n \\ w^T X_0 = 1 \\ w^T \geqq \varepsilon e^{-T}, \mu^T \geqq \varepsilon e^{+T} \end{cases} \quad (8)$$

Where ,$e^{-T} = (1,1,\ldots,1) \in E_m, e^{+T} = (1,1,\ldots,1) \in E_s$,ε is the Non Aqi Mead infinitesimal.
when introduction of variable s^- and s^+, dual BC^2 model ($D_{2\varepsilon}$) is :

$$(D_{2\varepsilon}) \begin{cases} \min[\theta - \varepsilon(e^{-T}s^- + e^{+T}s^+)] = V_{D_{2\varepsilon}} \\ s.t. \\ \sum_{j=1}^{n} \lambda_j X_j + s^- = \theta X_0 \\ \sum_{j=1}^{n} \lambda_j Y_j - s^+ = Y_0 \\ \sum_{j=1}^{n} \lambda_j = 1, \lambda_j \geqq 0, j = 1,2,\ldots,n \\ s^- \geqq 0, s^+ \geqq 0 \end{cases} \quad (9)$$

In which, $e^{-T} = (1,1,\ldots,1) \in E_m, e^{+T} = (1,1,\ldots,1) \in E_s$,ε is the Non Aqi Mead infinitesimal.

Suppose the optimal solution of C^2R model ($D_{2\varepsilon}$) are λ^*, s^{-*}, s^{+*} θ^*, when$\theta^* = 1$, and $s^{-*}=0$, $s^{+*}=1$, the DMU_{j0} is DEA effective; when $\theta^* < 1$ or $s^{-*} = 0$, $s^{+*} \neq 1$, the DMU_{j0} for non DEA effective.

In short, the previous C^2R model represents the comprehensive effect of the evaluated object, while the BC^2 model represents the technical efficiency of the evaluated object, that is, the BC^2 model operating results $\theta(0\leq\theta\leq1)$ is to be evaluated technical efficiency.

3 Evaluation indicators system of resource allocation efficiency in ethnic minority autonomous prefecture

Efficiency refers to the relationship between input and output or cost and benefit, and its essence is the general name of effective resources allocation, input versa output capability. the efficiency of evaluation object ,according to different purposes of study ,can be divided into technical efficiency, scale efficiency and comprehensive efficiency. Among them, the technical efficiency reflects the ability of the maximum output, while the scale efficiency reflects the suitability of the investment scale. Using C^2R model to determine the comprehensive efficiency and utilization BC^2 models to determine the technical efficiency, can get scale efficiency = comprehensive efficiency/ technical efficiency, and then comprehensive measure evaluated object scale efficiency, technical efficiency and overall efficiency as well as efficiency comparison in similar evaluated object.

3.1 Input Indicator

Resource allocation efficiency of the input indicators mainly include three aspects of human, financial and material resources, capital investment and human capital which constitute the basic elements of economic development, especially the contradiction between economic development and demand for talent becomes an important factor restricting in minority areas. Generally selecting the number of people with a certain degree of education as a measure of indicator, this paper replaces permanent resident population as this indicator. China has increased the financial support to the minority regions, such as high speed growth of fixed assets investment and the scale of fiscal expenditure. Therefore, selection of permanent resident population, financial expenditure, fixed assets investment as Input Indicators to measure the efficiency of resources allocation.

3.2 Output Indicator

Local fiscal revenue is an indicator of regional comprehensive strength, which shows the financial resources of local government as well as a very important role in measuring the efficiency of resource allocation in a region. Average capita fiscal revenue is an important indicator to measure economic development and economic strength for a country or a region. GDP reflect the new creation value, and it is the final outcome of the national economic activities. Urban disposable income and rural per capita net income indicate free disposable income of urban residents and the development of rural economy, respectively. Therefore, in this paper, selection of local fiscal revenue, GDP, per capita income, urban disposable income ,rural per capita net income as output indicators for measure of resource allocation efficiency. The concrete index system is shown in Table 3

Table 3. Evaluation Index System of Resource Allocation Efficiency in Minority Autonomous Prefecture.

Primary Index	Secondary Index
Input	Permanent Resident Population
	Plant Asset Investment
	Per Capita Fiscal Expenditure
Output	Local Fiscal Revenue
	GDP
	Per Capita FiscalRevenue
	Urban Disposable Income
	Rural Per Capita Net Income

4 Determination of resource allocation efficiency based on the c2r model and bc2 model inminority autonomous prefecture

Import input indicators and output indicators data for 30 Minority Autonomous Prefecture into DEAP2.1 software, resource allocation efficiency results based on C^2R and BC^2 model are showed in table 4.

Table 4 shows that the technical efficiency of the 30 minority autonomous prefecture is 0.918, the average technical efficiency is 0.949. From its resource allocation efficiency score, it's known that a total of 18 ethnic autonomous prefectures are DEA effective, which represent resource allocation efficiency optimized when input fixed.

Among them there are 12 Minority Autonomous Prefecture of resource allocation with non DEA effective, DMU14, DMU15 pure technical efficiency value is 1.000, which indicate that the 2 Mnority Autonomous Prefecture with non DEA effective because of low scale efficiency, requiring reasonable control scale of investment to improve the efficiency of resource allocation; the comprehensive technical efficiency and pure technical efficiency of DMU_2, DMU_3, DMU_4, DMU_5, DMU_{16}, DMU_{17}, DMU_{19}, DMU_{21}, DMU_{28}, DMU_{29} do not equal to 1. 000, indicating that the non DEA effective results from not only inappropriate scale of inputs but also lower output level. According to the 10 Minority Autonomous Prefecture, the improvement measurements are put forward by introducing slack variables into DEA model. Analysis for Resource allocation efficiency in 30 Minority Autonomous Prefecture, non DEA effective accounts for 40%, which shows that there is a great potentiality in resource allocation in the Minority Autonomous Prefecture.

Table 4. Efficiency of Input-Output in China's Minority Autonomous Prefectures.

Decision Making Module	Technical Effective	Pure Technical Efficiency	Scale Efficiency	Appraisal Result	Scale Efficiency State
DMU_1	1.000	1.000	1.000	DEA Effective	Invariable
DMU_2	0.593	0.678	0.874	NonDEA Effective	Decreasing
DMU_3	0.603	0.620	0.974	Non DEA Effective	Decreasing
DMU_4	0.663	0.881	0.752	Non DEA Effective	Decreasing
DMU_5	0.779	0.817	0.954	Non DEA Effective	Decreasing
DMU_6	1.000	1.000	1.000	DEA Effective	Invariable
DMU_7	1.000	1.000	1.000	DEA Effective	Invariable
DMU_8	1.000	1.000	1.000	DEA Effective	Invariable
DMU_9	1.000	1.000	1.000	DEA Effective	Invariable
DMU_{10}	1.000	1.000	1.000	DEA Effective	Invariable
DMU_{11}	1.000	1.000	1.000	DEA Effective	Invariable
DMU_{12}	1.000	1.000	1.000	DEA Effective	Invariable
DMU_{13}	1.000	1.000	1.000	DEA Effective	Invariable
DMU_{14}	0.943	1.000	0.943	Non DEA Effective	Decreasing
DMU_{15}	0.988	1.000	0.988	Non DEA Effective	Increasing
DMU_{16}	0.778	0.780	0.997	Non DEA Effective	Increasing
DMU_{17}	0.813	0.900	0.904	Non DEA Effective	Decreasing
DMU_{18}	1.000	1.000	1.000	DEA Effective	Invariable
DMU_{19}	0.899	0.942	0.955	Non DEA Effective	Decreasing
DMU_{20}	1.000	1.000	1.000	DEA Effective	Invariable
DMU_{21}	0.973	0.978	0.995	Non DEA Effective	Increasing
DMU_{22}	1.000	1.000	1.000	DEA Effective	Invariable
DMU_{23}	1.000	1.000	1.000	DEA Effective	Invariable
DMU_{24}	1.000	1.000	1.000	DEA Effective	Invariable
DMU_{25}	1.000	1.000	1.000	DEA Effective	Invariable
DMU_{26}	1.000	1.000	1.000	DEA Effective	Invariable
DMU_{27}	1.000	1.000	1.000	DEA Effective	Invariable
DMU_{28}	0.623	0.902	0.691	Non DEA Effective	Decreasing
DMU_{29}	0.818	0.979	0.836	Non DEA Effective	Decreasing
DMU_{30}	1.000	1.000	1.000	DEA Effective	Invariable

Average Value	0.918	0.949	0.964		

In 30 Minority Autonomous Prefecture, 18 0f which are scale efficiency invariable , 9 of which keep decreasing state,3 of which keep increasing state, and the proportion are 60%, 30% and 10% respectively.

Those show that the scale efficiency of the 30 minority autonomous prefecture keep the same or increasing state. The reason is that in recent years, China government increase input in Minority Autonomous Prefecture. For the Minority Autonomous Prefecture of decreasing scale efficiency, it should be appropriately controlled to increase output efficiency. For the Minority Autonomous Prefecture with increasing scale efficiency , it should be appropriate to expand the scale of investment in order to improve the efficiency of input-output.

5 Policy suggestion

From the above analysis, this paper proposes some suggestions to optimize Minority Autonomous Prefecture resource allocation efficiency.

5.1 Adhere to the Principles of Classified Guidance

The reasons for non DEA effective in Ethnic Minority Autonomous Prefecture resource allocation are different, therefore, we should first find out the factors ineffective of resource allocation, then optimize resources allocation measures in the Minority Autonomous Prefecture to improve the efficiency of resource allocation. In view of the low efficiency of resource allocation, the investment scale of fixed assets should be controlled, guide direction of investment, optimize the structure of fiscal expenditure and establish the regional economic and social development to match the personnel training plan, strengthen the construction of talent team according to the actual need; for resource allocation problems resulting from pure technical efficiency, not only need reasonable adjustments of fixed assets, fiscal expenditure scale and structure, but also pay attention to strengthen the building of local resources, maintain a certain economic growth rate, improve urban disposable income and rural per capita income to improve the efficiency of resource allocation.

5.2 Implement of Differentiated National Support Policies

China's natural resource bases in ethnic minority regions are with traits of uneven, different levels of economic and social development, weak self accumulation capacity, especially require funding support. But the "GSP" financial support without considering the actual needs of local, the pertinence is not strong. Therefore, it should differentiate national funds support policy, establish national regions transfer payment mechanism according to the input redundancy and output shortfall analysis results; determine funding support level according to actual demands. All those can effectively improve the efficiency of resource allocation in minority regions.

5.3 Multiple Supports for Economic and Social Development in Minority Areas

The factors that affect the efficiency of resource allocation in ethnic areas from many aspects, such as insufficient investment scale, irrational industrial structure, uneven educational level and the difficulty of talent introduction, which affect their economic and social development. Resource allocation in minority areas needs national policy support, financial support, education and personnel support, but in every minority area of specific performance is not identical from only unilateral financial support to a full range of support. Therefore, when the state and the government take specific support steps for the minority regions, it is necessary to consider natural resource endowment, the degree of market development, economic and social development level, self accumulated ability factors. The implementation of multiple support programs, abandoning "foot painful medicine foot" and "headache medicine head".

Acknowledgement

This research was financially supported by the State Ethnic Affairs Commission Foundation (NO15064) and Hubei University for Nationalities Teaching Research Project (NO2015JY027).

References

1. Chen MK, "Two Forms of An Equity Index for Resource Allocation to Minority Groups,"A Journal Of Medical Care Organization, Provision And Financing ,1976,13 (3),pp.228-32
2. Chen N, et al, "Equity of Health Resources Allocation in Minority Regions of Sichuan Province,"Journal of Sichuan University. Medical Science Edition, 2016, 47 (2), pp.238-43

3. Jiang Shuxia, et al, "An Evaluation of Efficiency and Productivity Change Based on DEA for Fujian Science and Technology Enterprises," Statistical Research, 2008, 25 (12), pp.30-35
4. Ye Rui, Yang Jianfei, Chang Gao, "The Measures and Decomposition for the Efficiency of China's Provincial High Technology Industry,"Quantitative Economic and Technical Economic Research, 2012 (7), pp.3-17
5. Hao Qian,"Minority Regional Industrial Structure Optimization Based on Financial Return Support ,"Guizhou National Studies, 2015, 36(172) ,pp.124-127
6. Pan Bao,"The Dilemma of Natural Resources in the Economic Development of China's Minority Nationalities , "Guizhou Ethnic Studies, 2014,35(156) ,pp.94-97
7. Liu Guiqing,"A Research on the Coordinated Development of Population, Resources, Environment and Economy in Ethnic Minority Areas, "Guizhou National Research, 2013,34 (151) ,pp.79-82
8. Lu QS,Gan SD, Zhang JZ, "Developing Comprehensive Abilities of Accounting Undergraduates through Curricular Reconstruction ," ICESD2015,Nanjing,Vol .1,pp278-282,December 2015
9. Zhang JZ, "EnterpriseInternal Control's Fuzzy Comprehensive Evaluation and Application," CME 2014, Shanghai,Vol.2, pp1795-1803, May 2014.
10. Zhang YT, Wang T, Li, XD, "Statistical Analysis Model of Money Laundering Risk Based on Behavior Detection Technology, " ITME2014,Hongkong,Vol.1,PP174-179,April 2014.

Equilibrium contract strategies under asymmetric product substitutability

Baixun Li and Yaru Nan

School of Business Administration, Guangdong University of Finance and Economics, Guangzhou 510320, China
*Corresponding author: libaixun2002qq@163.com

Abstract. This paper investigates how asymmetric product substitutability affect the equilibrium contract choices as well as the performances of the supply chain players. We consider a market with two competing supply chains, including two competing manufacturers and two common retailers. Two frequently used contracts: wholesale-price and quantity-discount are considered. We find that under whether the wholesale-price or quantity-discount contract can be chosen by manufacturers would depend on the degree of product substitutability.
Keywords: game; product substitutability; wholesale-price contract; quantity-discount

1 Introduction

It is a common phenomenon in reality that a manufacturer sells through many competing retailers, whereas the retailers also sell the products provided by the other manufacturer competing with him. In other words, the manufacturers sell through common retailers, and the retailers carry both products. We define this phenomenon as 'overlapping' or 'cross sales' by referring to Trivedi[1] (1998) and Wu and Mallik[2](2010). A typical example is the consumer goods market, where multiple retailers sell products of multiple competing brands in one area. For instance, the retailers Wal-Mart and Carrefour sell products of multiple competing brands (e.g., PHILIPS, SONY, SIEMENS, TCL, and so on) in the same area. One objective of this paper is to investigate the behaviors that all parties will behave in two competitive supply chains with multiple manufacturers and multiple retailers, and to study how the degree of product substitutability affects the contract choices and the performances of the players. Our results can be used to explain the reason why various contract types have been adopted in real supply chains and, especially, to explain why the sophisticated contracts have not completely eliminated the Pareto inferior wholesale-price contract in the real world.

The setting of this paper relates to the work on chain-to-chain competition and contract. Cachon and Kök [3](2010) study the problem of contract choice based on the many-to-one supply chain. Zhao and Shi [4](2011) consider two competing supply chains, which multiple upstream suppliers produce complementary products and sell to a single buyer. Govindan et al.[5](2013) gives a complete overview of the literature that addresses coordination by contracts. Zhang et al.[6](2014) explore the issue of supply chain coordination by considering trade credit. Chakraborty et al.[7] (2015) study a supply chain comprising two competing manufacturers who sell their products through a common retailer. Ye et al. [8](2016) quantify the efficiency of decentralized competitive reverse supply chains (RSCs) with quality-dependent price-only contracts.

The remainder of this paper is organized as follows. Section 2 presents our model setting. Section 3 establishes and explores the sub-game solution for WW, QQ, WQ and QW case. Section 4 establishes the equilibrium contract structure. Finally, we summarize the paper, and suggest research extensions in Section 5.

2 Model setting

In this paper, we consider a market with two competing supply chains (SC_1 and SC_2), including two competing manufacturers (M_1 and M_2) and two common retailers (R_1 and R_2). The manufacturers act as the Stackelberg leader, and have more channel power than the retailers. In stage one, each of the two manufacturers simultaneously and independently chooses a contract type and designs their individual contract to maximize their individual profit. In stage two, after observing the contracts set by the two manufacturers, the retailers will choose their own purchase volumes simultaneously and independently.

In addition, each manufacturer will provide either *wholesale-price contract* (denoted by W) or *quantity-discount* contract (denoted by Q), representing the "simple" contract and the "sophisticated" contract, respectively. Consequently, in the two supply chains, there will be four possible contract combinations: WW, WQ, QW and QQ, where WW denotes that both manufacturers are restricted to choose wholesale price contracts; WQ means that manufacturer 1 (M_1)

is restricted to choose the wholesale-price contract while manufacturer 2 (M_2) is restricted to choose the quantity-discount contract. QW and QQ are similar.

For expositional simplicity, we elect to work with the inverse demand function described as follow.

$$p_{ij}=1-\left(d_{ij}+d_{i,3-j}\right)-\theta_i\left(d_{3-i,j}+d_{3-i,3-j}\right),\qquad i=1,2;j=1,2. \tag{1}$$

As reflected in (1), the smaller the degree of product substitutability, the more differentiable of the two products, therefore the less potential price competition. In particular, θ_i=0 implies that the products are independent of each other, while θ_i=1 implies the perfect substitution between products

3 Model solution

In this section, we can derive closed form equilibrium solutions for various quantities such as prices and profits for different contract structures, including the WW, QQ, WQ, and QW case. In the QQ case, the retailer j's problem is

$$\max_{d_{ij}^{QQ},d_{3-i,j}^{QQ}}\pi_{R_j}^{QQ}=\sum_{i=1}^{2}\left[\left(p_{ij}^{QQ}-w_i^{QQ}\right)d_{ij}^{QQ}+v_i^{QQ}\left(d_{ij}^{QQ}\right)^2/2\right] \tag{2}$$

where the superscript QQ denotes the quantity discount contracts offered by both manufacturers.

To derive the retailers' best response functions, we apply the first order conditions to their profit functions based on manufacturers' and competitors' decisions, thus getting,

$$\partial\pi_{R_j}^{QQ}/\partial d_{ij}^{QQ}=1+\left(v_i^{QQ}-2\right)d_{ij}^{QQ}-d_{i,3-j}^{QQ}-\left(\theta_i+\theta_{3-i}\right)d_{3-i,j}^{QQ}-\theta_i d_{3-i,3-j}^{QQ}-w_i^{QQ}=0 \tag{3}$$

$$\partial\pi_{R_j}^{QQ}/\partial d_{3-i,j}^{QQ}=1+\left(v_{3-i}^{QQ}-2\right)d_{3-i,3-j}^{QQ}-d_{3-i,j}^{QQ}-\left(\theta_i+\theta_{3-i}\right)d_{i,3-j}^{QQ}-\theta_{3-i}d_{i,j}^{QQ}-w_{3-i,3-j}^{QQ}=0 \tag{4}$$

then solving for the conditional Nash equilibrium values for $d_{i,j}^{QQ}$ as a function of w_i^{QQ} and v_i^{QQ}.

$$d_{ij}^{*QQ}\left(w_i^{QQ},v_i^{QQ}\right)=\frac{d_{3-i,3-j}^{QQ}\left[\theta_{3-i}+\theta_i\left(v_{3-i}^{QQ}-1\right)\right]+d_{i,3-j}^{QQ}\left(\theta_{3-i}^2+\theta_i\theta_{3-i}+v_{3-i}^{QQ}-2\right)+\left(\theta_i+\theta_{3-i}\right)w_{3-i}^{QQ}+\left(v_{3-i}^{QQ}-2\right)w_i^{QQ}+2-\theta_i-\theta_{3-i}-v_{3-i}^{QQ}}{4-\theta_i^2-2\theta_i\theta_{3-i}-\theta_{3-i}^2-2v_{3-i}^{QQ}+v_i^{QQ}\left(v_{3-i}^{QQ}-2\right)} \tag{5}$$

We now turn to the manufacturers' decisions. The manufacturers' reaction functions can be derived from the following first-order conditions of the manufacturers' profit maximization problems.

$$\max_{w_i^{QQ},v_i^{QQ}}\pi_{M_i}^{QQ}=\sum_{j=1}^{2}\left[w_i^{QQ}d_{ij}^{*QQ}\left(w_i^{QQ},v_i^{QQ}\right)-v_i^{QQ}\left(d_{ij}^{*QQ}\left(w_i^{QQ},v_i^{QQ}\right)\right)^2/2\right] \tag{6}$$

By differentiating Equation (6) first partially with respect to w_i^{QQ}, we can determine Nash equilibrium wholesale prices.

$$w_i^{*QQ}\left(w_{3-i}^{QQ},v_i^{QQ}\right)=\frac{\left(2\theta_i^2+5\theta_i\theta_{3-i}+2\theta_{3-i}^2+3v_{3-i}^{QQ}-9\right)\left[\left(2\theta_i+\theta_{3-i}\right)\left(1-w_{3-i}^{QQ}\right)+v_{3-i}^{QQ}-3\right]}{\left(v_{3-i}^{QQ}-3\right)\left[4\left(\theta_i^2+\theta_{3-i}^2\right)+10\theta_i\theta_{3-i}+6v_{3-i}^{QQ}-v_i^{QQ}\left(v_{3-i}^{QQ}-3\right)-18\right]} \tag{7}$$

Substituting $w_i^{*QQ}\left(w_{3-i}^{QQ},v_i^{QQ}\right)$ to Equations (6), and differentiating partially with respect to v_i^{QQ}, we get

$$\frac{\partial\pi_{M_i}^{QQ}}{\partial v_i^{QQ}}=\frac{\left[\left(2\theta_i+\theta_{3-i}\right)\left(1-w_{3-i}^{QQ}\right)+v_{3-i}^{QQ}-3\right]^2}{\left[4\left(\theta_i^2+\theta_{3-i}^2\right)+10\theta_i\theta_{3-i}+6v_{3-i}^{QQ}-v_i^{QQ}\left(v_{3-i}^{QQ}-3\right)-18\right]^2}>0 \tag{8}$$

It can be found that given the retailers' best response functions and optimal wholesale price, the manufacturer's profit function increases with quantity-discount coefficient. Therefore, each manufacturer's best response is to choose upper bound of quantity discount coefficient, and let $v_i^{QQ}=2-\theta_i-\theta_{3-i}$. Substituting $v_i^{QQ}=2-\theta_i-\theta_{3-i}$ into Equation (5) and (7) respectively will yield the equilibrium decisions of the two retailers and manufacturers. Thus, we can obtain the equilibrium values of profit for manufacturers and retailers under the QQ cases, which are only a function of the degree of product substitutability. By similar ways, we can obtain the equilibrium values of profit for manufacturers and retailers under the WW\WQ\QW cases.

4 Contract strategies under asymmetric product substitutability

In reality, if two products belong to different class brand type, the degree of product substitutability from product 1 to product 2 is the different with the latter to former, that is, the degree of product substitutability is asymmetric. In this section, we will consider the problem of contract choice game between two overlapping supply chains given that the degree of product substitutability is asymmetric. For notational convenience, we show the definitions of several functions as follows.

(a) $f_a(\theta_1,\theta_2)=\pi_{M_1}^{*WW}(\theta_1,\theta_2)-\pi_{M_1}^{*QW}(\theta_1,\theta_2)$, (b) $f_b(\theta_1,\theta_2)=\pi_{M_1}^{*WQ}(\theta_1,\theta_2)-\pi_{M_1}^{*QQ}(\theta_1,\theta_2)$,

(c) $f_c(\theta_1,\theta_2)=\pi_{M_2}^{*WW}(\theta_1,\theta_2)-\pi_{M_2}^{*WQ}(\theta_1,\theta_2)$, (d) $f_d(\theta_1,\theta_2)=\pi_{M_2}^{*QW}(\theta_1,\theta_2)-\pi_{M_2}^{*QQ}(\theta_1,\theta_2)$,

(e) $f_e(\theta_1,\theta_2)=\pi_{M_1}^{*WW}(\theta_1,\theta_2)-\pi_{M_1}^{*QQ}(\theta_1,\theta_2)$.

And define,

$A_1 = \{(\theta_1,\theta_2) | f_a < 0, f_b < 0, f_c < 0, f_d < 0, f_e < 0, and, 0 < \theta_1 < 1, 0 < \theta_2 < 1\}$;

$A_2 = \{(\theta_1,\theta_2) | f_a < 0, f_b < 0, f_c < 0, f_d < 0, f_e > 0, and, 0 < \theta_1 < 1, 0 < \theta_2 < 1\}$;

$A_3 = \{(\theta_1,\theta_2) | f_a < 0, f_b > 0, f_c < 0, f_d > 0, f_e > 0, and, 0 < \theta_1 < 1, 0 < \theta_2 < 1\}$;

$A_4 = \{(\theta_1,\theta_2) | f_a > 0, f_b > 0, f_c > 0, f_d > 0, f_e > 0, and, 0 < \theta_1 < 1, 0 < \theta_2 < 1\}$.

It is obvious to have $\bigcup_{n=1}^{n=4} A_n = \{(\theta_1,\theta_2) | 0 < \theta_1 < 1, and, 0 < \theta_2 < 1\}$. Figure 1 describes graphically the manufacturers' Nash equilibrium of contracts choice game in the (θ_1, θ_2) plane. The outcomes can be classified into four distinct regions by five critical curves. (a) f_a=0, (b) f_b=0, (c) f_c=0, (d) f_d=0, and (e) f_e=0.

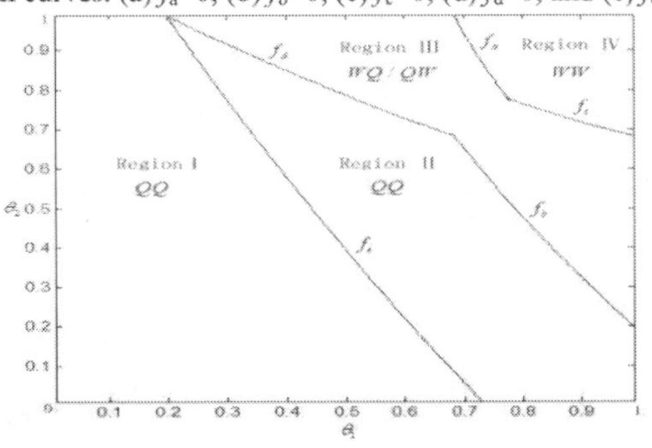

Figure 1 Sketches of four equilibrium type regions of manufacturers

Region Ⅰ. This region is defined by the set A_1, and suggests that on the one hand, no matter the manufacturer chooses quantity-discount contract or wholesale-price contract, it is optimal for the other manufacturer to choose quantity-discount contract, therefore, quantity-discount contract is the dominant strategy for both manufacturers in this scenario. On the other hand, both manufacturers are better off with the pure quantity-discount contract than with the pure wholesale-price contract. Therefore, in this region, the pure quantity-discount contract is not only the unique equilibrium, but also the best option for both manufacturers.

Region Ⅱ. This region is defined by the set A_2, and suggests that on the one hand, similar to Region Ⅰ, the pure quantity-discount contract is the unique equilibrium. However, both manufacturers are better off with the pure wholesale-price contract than with the pure quantity-discount contract. Hence, in this region, although the manufacturers would choose quantity-discount contract if they are not restricted in their contract choice, they could be better off if they have been restricted to offer only wholesale-price contract, then, the problem of choosing an optimal contract structure is a classical prisoners' dilemma game.

Region Ⅲ. This region is defined by the set A_3, and suggests that as the degree of product substitutability increases, if the manufacturer chooses quantity-discount contract, the other manufacturer can obtain much more profits by choosing wholesale-price contract, in contrast, if the manufacturer chooses wholesale-price contract, the other manufacturer can obtain much more profits by choosing quantity-discount contract. As a result, counter to the intuition, the mixed contract structure WQ/QW case is Nash equilibrium between two manufacturers.

Region Ⅳ. This region is defined by the set A_4, and suggests that when the degree of product substitutability increases farther, especially, both θ_1 and θ_2 approach one, on the one hand, no matter the manufacturer chooses quantity-discount contract or wholesale-price contract, it is optimal for the other manufacturer to choose wholesale-price contract, that is, wholesale-price contract is the dominant strategy for both manufacturers in this scenario. On the other hand, both manufacturers are better off with the pure wholesale-price contract than with the pure quantity-discount contract. Therefore, in this region, the pure wholesale-price contract is not only the unique equilibrium, but also the best option for both manufacturers.

In summary, Figure 4 describes the general trend that in the Manufacturer Stackelberg game, as the degree of product substitutability increase, the manufacturers will gradually switch from quantity-discount contract to wholesale-price contract. Our results are consistent with the fact that when the degree of product substitutability (degree of competition) is low, especially, the manufacturers function monopolistically, the relative power (the channel bargaining power) of retailers is small, the manufacturers will extract concessions from retailers, and are better off with the pure quantity-discount contract structure (i.e., in Region Ⅰ and Ⅱ). On the other hand, the manufacturers may tend to drop retail prices to capture more market demand by adopting quantity-discount contract. However, as the degree of product substitutability increases, in other words, the two manufacturers compete more and more severely, the manufacturers can prevent the retail price from dropping too much by adopting the pure wholesale-price contracts, owing to the fact that in the Manufacturer Stackelberg game, the retail price is always higher under wholesale-price contract structure than within quantity-discount contract structure.

These results indicate that, to prevent fierce price competition, it is important for the manufacturer to invest product differentiation by considering important non-price factors such as the provision of product information, faster check-out, or after-sales service. That is to say, the manufacturers can benefit from the use of sophisticated contracts (e.g., quantity-discount contract) with high brand loyalty (e.g., low product substitutability) by important non-price factors.

5 Conclusions

In this paper, we consider a market with two competing supply chains, where the manufacturers produce competing products, and distribute through two common retailers. We investigate the impact of asymmetric product substitutability on the equilibrium contract structure and the supply chain players' performances. We achieve this by deriving and comparing the equilibria that arise under WW, QQ, WQ and QW case respectively. Finally, we would like to mention some potentially interesting directions for future research. First, we model competition as a single stage (static) game in this paper. It is interesting but challenging to study an infinitely repeated (dynamic) game. Second, we have assumed that information is common knowledge. Relaxing this assumption would lead to the study of games with asymmetric information, and would generate insights on the impact of information on the contract choice of the supply chain members. Yet another extension would be to consider this problem under demand uncertainty. We believe that the ideas and models presented in this paper lay the motivational ground for future research in these directions.

Acknowledgement

This research was supported in part by the National Natural Science Foundation of China under Grants nos. 71401042, 71472143, the Youth Foundation of Ministry of Education in China, Humanities and Social Science Research Projects under Grant nos. 13YJC630068. the Guangdong Natural Science Foundation under Grants nos. 2014A030313606.

References

1. M. Trivedi, Manage Sci, **44(7)** , 896-909(1998).
2. C. Q.,Wu, S., Mallik, Int J Prod Econ, **126(5)**, 158-167(2010).
3. G., Cachon, A., Kök, Manage Sci, **56(3)**, 571–589(2010).
4. X., Zhao, C., Shi, Int J Prod Econ, **134(2)**, 434-446(2011).
5. K., Govindan, M.N., Popiuc, A., Diabat,.J Cle Prod, **47**,319-334(2013).
6. Q.H., Zhang, M., Dong, J.W., Luo, et al. Int J Prod Econ, **153**, 352-360(2014).
7. T., Chakraborty, S.S., Chauhan, N., Vidyarthi, Int J Prod Econ, **166**,103-118(2015).
8. S.,Ye, Z.J.,Ma, Y., Dai, Transport Res E-Log, **89**, 86-107(2016).

City, Urban Planning and the Creation of Urban Culture
——Taking the Ancient City of Xi'an as an Example

Yiqing Qiao [1, 2,*]

[1] Xi'an Academy of Fine Arts Department of environment teacher, Shaanxi Xi'an 710065, China

[2] Chang'an University college of Architecture urban planning and design doctor, Shaanxi Xi'an 710064, China

*Corresponding Email:Email:echo_q117@sina.com

Abstract. City is the concentrated reflection of city culture. Urban planning as a synthetic discipline which that both social science and natural science, it plays an important role in the urban development and layout. Fully understand the city's cultural implication, helps us went up from nature know and grasp the city, city to city planning and construction of theory research and practice has important guiding significance.

This paper argumentation the relationship between the city, urban planning and city culture, believes culture is the city context. Have taken xi 'an as an example, discussion the creation city culture in the process of planning and design from two aspects of the traditional Chinese urban culture inheritance and the creation of the contemporary city culture, then expounded the positive significance of creating urban culture, and putted forward reasonable scientific method of urban planning and urban design means to have a positive role in contemporary urban development.

Key words. city, city culture, urban planning, the Xi'an city.

1 Introduction

The creation of urban culture mainly relying on the city and urban planning in common sense.

City, is a kind of the main human living form and living space. Urban planning, it is human to maintain public life in the development of the city space order and make space for future development will, is reasonable and the creation of effective, fair and orderly living space environment. City's condition and environment construction place foil urban culture.

Different age of urban planning discipline in order to adapt to the needs of social development. The development and change of the concept of urban planning have connected closely with social mode of production, life style and level of science and technology, culture and art features .

The present urban construction rapid rise in China, the city seems to be towards a unified, stereotypes, there is no difference between the image of the city development, in this paper, through the demonstration cities and the relations between urban planning and urban culture, urban culture is the connotation of the city. The essay discusses from two aspects of inheritance and creation in the process of planning and design for the creation of the urban culture .

2 Methodology

The city is a kind of "living environment" that it has two aspects about human settlements and the environment. First, the city as a living form; secondly, people involved in the activities of the city, includes live、work、 transportation and recreation etc.

2.1. City and urban culture

The city as a way of life, there is little difference between urban and rural areas. But people inhabited the city can form a variety of social relations and expand the scope of exchanges, thus creating more wealth, but also to obtain different spiritual and rural life, life is also greatly enrich the content.

People were in the city with a variety of lively activities, if the urban physical environment is the space vector of people's lives, is the characterization; so, urban living culture is reflected in the concept of urban form, is the deep structure. Exterior combines complementary jointly build a city of this special living environment.

Different cities have different cultural backgrounds city, in essence were to some extent, represent the local as well as a wider range of better life. Leave a person's life, the city can at best be called the ruins of civilization, or historical sites. Associated with the nature of the city between the people living on this, it is crucial to our understanding of urban culture.

2.2 Urban and urban planning

Urban planning is to achieve a certain period of economic and social development goals of the city, the city determined the nature, size and direction of development, rational use of urban land, coordinating comprehensive and specific arrangements for the deployment of urban space layout and the construction done. Urban planning is the fundamental basis for the construction of the city and the management of the city, on the basis of urban space to ensure effective allocation of resources and the rational use of the land, the city is one of the important means of achieving economic and social development goals.Implementation and development of urban planning, the following aspects influence

2.2.1 rationalist view of the ancient Chinese city planning

Urban construction thought in ancient China can be divided into two categories:

First, is the influence of Confucianism in the ritual system planning ideology targeted.

Second, is the pursuit by Taoist and Buddhist influence of "natural" planning ideas.

"Heaven" is one of the main characteristics of traditional Chinese philosophy emphasizes a holistic view of Heaven's way of thinking as well as the integration of social harmony based on the spirit of humanism and respect for nature. Under the influence of this idea, the ancient Chinese city planning to create a structure rigorous, primary and secondary orderly, harmonious unity of highly rational beauty.

2.2.2 modern Western urban planning concept of idealism

15 to the 16th century Renaissance to break the shackles of spiritual people, human history has brought a great ideological emancipation movement. Ideology is reflected in urban planning, highlighting the authority of the "people" as the main place of "God", "ideal city" thinking germination, planning and construction of more to reflect the requirements of the public life.

From the early 17th century Scamozzi practice, Plato "Utopia", Thomas Moore • "Utopia," Campanella "Sun City", Robert Owen 1852 Experimental "New Concord Village" to Howard (Ebenezer Howard) in the "garden city", which reached the peak of the evolution of urban planning concepts.

2.2.3 Functionalism and technology of modern post-Marxist plan view of the city's

A large number of 19th century science and technology breakthroughs and inventions, for the city's construction, transportation, communication provides a great convenience, the spatial structure of the city began a historic change brings hope for people from the functional improvement and urban planning . Le Corbusier, represented by reformers to "new construction movement" to penetrate into the building in the field of urban planning. New ideas to change the college style to the plan as a building designed to expand the practice to abandon the formalistic approach, focusing on functional and reasonable layout, emphasizing the concept of functional zoning and planning work to think is a technical work. Think everything is planning techniques and physical environmental determinism.

Figure 1. Xi 'an city wall

Under the guidance of this concept appeared in a number of well-known planning, such as the 1944 Greater London Plan, General Plan 1935 Moscow, 1948 Copenhagen planning. In public buildings and urban road construction has made enough proud of the results of a series of new ideas to adapt to urban production and living needs from the function, the function and order as a solution to urban development in the various contradictions raised by this period and the primary means of problem, then adapted to the needs of urban development, modernization and scientific urban planning has made a landmark contribution.

3 City planning and urban culture technology embodied - A Case Study in Xi'an

The complexity and diversity of the socio-economic structure of the city functions, be regarded as a single technology makes the process of urban planning, has been far can not meet the needs of modern society. Therefore, the focus on technology, but also pay attention to the planning of urban culture to create. The saying goes "side soil support people," the Earth is only one culture can be diverse, urban and regional cultural diversity is as essential urban culture brand.

In the modern city culture shaping, we both splendid cultural tradition to be proud of, but also on the creation of modern culture is full of confidence. We need to adopt means of urban planning, urban design methods, from the perspective of urban culture to examine the city's involvement in the construction and creation, not only to maintain the traditions of their own culture, and endowed with its modern life, and thus to participate in the global civilization.

Xi'an, for example, as a historical and cultural city with culture as the carrier of urban planning and urban culture.

3.1 Reflect the city's character

Figure2.Xi'anhistoricalheritage

Characteristics and other characteristics of the city's art is essentially different, it has a physical nature, multi-level nature of the integrated arts, is materialized art form, is the emergence and development of certain natural environment and human activities in the process. Xi'an city lies in the protection of historical relics protection, protection of Xi'an personality.

Zhou Fenggao City Fanggong Qin, Han Chang'an City and Chang'an City of the four sites is a major part of the historical and cultural city of Xi'an, on its meaning far beyond the other remains, they are either from outside or from the visual experience, as reflected in material and spiritual culture, reflects the characteristics of Xi'an, the focus should be protected.

First, it reflects the process of urban development, is a living history of urban development. From week Fenggao city Fanggong Qin, Han Chang'an City, the Ming and Qing dynasties until today, the city of Xi'an, Xi'an, the city and the rise and fall of this series evolving process, to understand the history of the city of Xi'an, and Xi'an in China's economy since ancient times status and role of the enormous political aspects, and for the future development of Xi'an foundation. Secondly, it Xi'an urban spatial development put forward new requirements for the future. Starting from urban development, urban space layout reasonable, properly handle the relationship between site protection and development of urban space, forming a historical and cultural characteristics of urban spatial structure of Xi'an city's future planning and management major issue. Third, the four sites is not only reflected from the surface of the city personality, and it bears the Qin and Han civilization, Tang Treasures, a profound impact on Shaanxi earth, forming a distinctive cultural practices. Therefore, they are important for the continuation of the city's historical context.

3.2 Promote urban development

Many cities have their own historical relics from the surface, which remains on the development of the city play a certain impediment. In essence, the history of the city has a role in promoting heritage, is the driving force of urban development. First, a large number of relics constitute a rich tourism resources, but also the most powerful way to show the historical and cultural city, and brought great economic benefits. Tourism foreign exchange earnings, $ 194.1 million in 1998; $ 225.25 million in 1999; $ 270 million in 2000. Which in 2000 received a total annual foreign tourists 15,670,000 passengers. Second, in order to protect the main sites of green space system, not only effectively protect the historic sites, beautify the urban environment, but also increase the cultural connotation of the city, highlighting the city's personality, to promote improvement of the ecological environment in Xi'an, an increase of city vitality. Third, the historical and cultural city, and deep cultural connotation deep-rooted cultural tradition of urban development plays an active role in promoting. Xi'an is the ancient tradition of outstanding cultural hub, colleges and universities gathered here today, the strength of science and technology for urban development has provided intelligence and technological support.

3.3 Protection of Xi'an city planning and urban relations

Xi'an as a historical and cultural city, its main functions are: the inheritance and development of historical and cultural traditions at the same time, to meet the new needs of the city's growing capabilities. This must take into account the population carrying capabilities, living functions, land features, traffic function, landscape features and other features coordinated development. Must deal with coordination between the city's main functions and other functions of the city, in highlighting the city's main function taking into account other functions, so that the two complement each other and promote each other for common development.

Mainly related to the following areas

1) clear functional layout, reasonable arrangement city, ensuring land sites

2) city protection and exploitation complementary

3) handle the relationship between the protection of historical and cultural city and between urban innovation

4) in order to focus on green heritage site, building, and the historic city green space system to adapt.

4 Conclusion

Therefore, This study demonstrated that to avoid the thousands of the city has the same image of city in the urbanization process, with the creation of the urban planning and the culture of the city by xi'an as example. It is important to note that the way is urban planning and the performance is culture and needs planner to do that：

1) planning activities, systematic and dynamic

2) planning of humanism

3) sustainable planning

4) have regional characteristics of urban culture

5) both inheritance and meet the needs of contemporary urban culture

Urban planning technology and the urban culture, urban culture connotation and development of urban construction, planning, culture and development between supplement each other, rely on each other.

City is always accompanied by the development of social economy and continuous development, the development is the main melody of urban evolution. With the rapid development of urbanization process, In addition to historical and cultural cities likes xi 'an, but also has the nature of a certain amount of regular general cities. This analysis has concentrated on xi'an that historical and cultural city. The development of historical and cultural city is not achieved overnight, but it is a long-term project, development and planning protection work needs the joint efforts of several generations to achieve. So the key to the development of modern cities is to deal with the relationship between historical and cultural cities and general city development, handle properly the relationship between urban planning and urban development, deal with the modern city development reflect the relationship with their own culture. In the future , only to properly handle these relations can conform to the urban sustainable development, adapt to the needs of social progress, promote the health development of the harmonious urban.

References

1. Lewis Mumford. The City in History, its origins, its transformations, and its prospects. London. Secker & Warburg,1963.
2. Lewis Mumford. The Culture of Cities .New York .Harcourt， Brace and Company,1938.
3. Lewis Mumford. Technics and Civilization .New York .Harcourt， Brace and Company,1934.
4. Wu Liangyong. Urban studies - to meet the coming of the new century (1986-1995), Beijing China building industry publishing

DOI: 10.1051/matecconf/201710005072

5. Wu Liangyong. The generalized architecture. Beijing's tsinghua university press, 1989.
6. bao of the bank's chief editor: "the new concept of urban planning method", the commercial press, 1993.
7. albala c: talking about ancient Chinese religious influence on city planning, city planning transactions, no.4, 1991.
8. f. gibbard waiting, the town design, YaoYi cheng, China building industry press, 1983
9. Qiao yiqqing, Shallow of historical heritage and xi 'an city protection northwest art, 2010

Consumer Credit Card Use Intention and Influence Factors Analysis

Yantao Wang

School of Economics and Law, University of Science and Technology Liaoning, Liaoning, China
Email:wangyantao1968@163.com

Abstract. Credit card as a kind of advanced means of payment and the new way of consumer credit in the world widely, but in our country is faced with some problems, such as their per capita spending is low, open the card number is small, motionless bank card is a lot. In this paper, using the Logit model to analyze the Anshan city in Liaoning province consumer credit card usage. And combined with empirical analysis to provide banking management Suggestions and marketing countermeasures, including targeted to provide different kinds of credit card.
Keywords. Credit card, willingness to use, factors affected.

1 Introduction

The characteristics of the credit card also said borrow write down card, it is the cardholder can get certain without guarantees and the mortgage credit, the cardholder can brush calorie of consumption within the credit; The cardholder need not be paid in full within this grace period interest. The cardholder to pay for the corresponding interest under the premise of can pay a certain percentage of the first overdraft; Cardholder recycled within credit funds; Cardholder can also directly in an ATM cash withdrawal, but for the withdrawal of the cardholder does not directly given a grace period to enjoy, calculation the corresponding interest from the date of withdrawal [1] .

Credit card is the earliest prototype is derived from the 1880 s Britain, money cannot default for a long time, no credit. In March 1950, Mr McNamara to cooperate with friends schneider founded in New York like Club (big Club), the first credit card appeared [2] . In 1952, franklin national bank becomes the first release the credit card of bank, since then reveal the prelude of bank-issued credit CARDS [3] .in China, however, was born in June 1985, credit card issued by bank of China zhuhai branch of the first credit card, the silver card. The use of credit card in our country faces a late start, lack of social inspection, payment system is not sound and so on characteristics, as an important means of new global payment in urgent need of the development and improvement. Credit card as a product of modern financial innovation, which is the advanced means of payment and settlement tool, is also a modern economic development, scientific and technological progress and an important symbol of social civilization. Some predict that the future of the world's largest credit card market in Asia, Asia's largest market in China.

2 Literature review

This study from the perspective of the consumer consumers intend to the analysis of the affecting factors and the use of credit CARDS. Due to consumers' willingness to subjective psychological factors directly affect the use, and most researchers will intend to model references of social psychology to study consumer behavior theory. Its research has experienced three stages: rational behavior theory, planned behavior theory, the technology acceptance model. Fishbein and Ajzen first put forward the theory of rational behavior in 1975, provide a method to predict consumer behavior. The theory of the two factors as assumptions: (1) everyone is rational, can act according to their wishes, (2) everyone will consider when to take actions to contain the behavior of the implicit meaning. Conclusion: the subjective norms and behavior attitude determines a person's behavior intention. When the subjective norms or behavior attitude tend to be more positive behavior intention is higher, lower instead. Ajzen improved on the basis of the theory of rational behavior in 1985, considering the consumers sometimes can't control their own behavior, and to relax the assumptions for individual cannot control their own behavior under the condition of the relationship between intention and behavior, puts forward the theory of planned behavior [4]. the theory not only think the actual behavior of the individual subject to the will of individual, but also by the resources required for behavior, opportunity, the ability to control behavior three factors mutual influence.

Davis, 1989, on the basis of the theory of rational behavior of technology acceptance model is put forward and emphasized personal attitude and subjective norm interaction affect individual apiration, will ultimately affect individual behavior, subjective perceived usefulness and subjective attitude include ease of use [5] .for a deeper understanding of the external factors internal to the user attitude and intention.

Through the literature, the individual behavior is influenced by individual apiration, personal willingness and directly affected by subjective norms, subjective perceived usefulness, subjective usability three aspects factors. By the technology acceptance model, assuming that the influencing factors of consumer credit card use intention; Through literature research, the present scholars involved the study of consumer desire, use of credit CARDS is relatively small, most of existing research adopted descriptive statistics analysis method to deal with statistical results, For example: Weian Wang, HongGuo Fan and Jianguo Zhang [6] Yi Yan, Fan Zhangl [7] found in the literature research, scientific measure of the lack of empirical research, thus the innovation of this research lies in the Logit model is adopted to statistical results of quantitative analysis.

3 Analysis of consumer credit card use intention

3.1 Personal basic information statistical analysis

The respondents, aged 20 and the following accounted for as follows: 5.2%;As follows: 30.1% of the total 20 to 30 years old; As follows: 11.9% of the total 30 to 40 years old; As follows: 28.5% of the total 40 to 50 years old; As follows: 14.9% of the total 50 to 60 years old; Due to the bank for a credit card application conditions is 60 years old the following, thus for respondents aged 60 and older people to be outside the scope of research. After actual survey found, however, the current social population age structure, the credit card application age limit at 60 apparent irrationality. Bank staff believe that limit credit card application conditions to 60 years old is out of consideration for bank risk prevention, because there are quite a part of the population aged over 60 will soon face retirement problem, its income persistence and stability may be affected, so bank of risk aversion is wrong people aged 60 and above open a credit card. The scholar thinks, along with the progress of an aging population, the population aged 60 and above accounted for the proportion of rising year by year, which also contains a number of retired people have higher retirement pension, social insurance, commercial insurance safeguard mechanism, they have a stable source of income, have good credit level, for those who are willing to accept to open or continue to use a credit card, if the bank refuses to provide relevant services for them, and social discrimination of discontent is inevitably produced, lose the social fairness. For the Banks and social aspects, the two sides, however, the reform will take time, Banks should strengthen risk prevention emergency measures, and gradually ease restrictions on age in the next few years, to provide high age group targeted credit CARDS accepted system, fill in the blanks.

3.2 Consumer credit card use intention situation analysis

Table1. Consumer credit use will statistics.

Use will	Frequency number(people)	The percentage （%）
Willingness to	169	54.7%
Dont want to	140	45.3%
Combined meter	309	100.0%

Table 2. Consumer card use dependent on statistics.

Start a number （a）	Frequency number （people）	The percentage （%）
no	128	41.4%
a piece of	85	27.5%
two pieces of	62	20.1%
three or more	34	11.0%
combined meter	309	100.0%

Overall, 41.4% of the potential clients never use credit card payment method, but also can't rule out the remaining open card no sleep card customers; By starting the share of the credit card number can be seen that people hold two or more credit CARDS accounted for 31.1%, far more than the proportion of 20.8% hold a credit card number.Card number is in the majority of consumers have card sleeping probability is small, to a certain extent reflects their credit card use for dependence is stronger, after visiting survey, how much to open credit card consumers most people are not willing to spend more time and energy management of credit card just due to provide credit is low, forced to hold more credit CARDS.

4 Consumer credit card use will affect the factor analysis

4.1 Model method

As set of questionnaires were explained variables for binary classification is "yes" and "no", so using binary choice model of logistic curve model (Logit model).

Used to model the dependent variable Y has only two values: 0 (" not willing to use ", 1 "willing to use". Many factors affect the dependent variable Y in the independent variable Xi (I = 1, 2, 3,..., n) said. Consumers are willing to use probability for (Y = 1 P | Xi) , reluctant to use the probability of the P (Y = 0 | Xi) , i.e. P = 0 | Xi) (Y = 1 - P | Xi) (Y = 1.Will consumers use will happen than expressed with odds, namely: the odds = P | Xi) / P ((Y = 1, Y = 0 | Xi) odds take logarithm in willing to use value, its value satisfy the linear function of the independent variable Xi, namely the log odds = alpha + beta Xi, these odds, Xi, alpha, beta, are based on vector group, alpha is a probability density function of the intercept vector, beta is a regression coefficient vector. Will log odds = log P (Y = 1 | Xi) / 1 - P = 1 | Xi) (Y = alpha + beta Xi inverse solution: P = e (Y = 1 | Xi) alpha + beta x/alpha 1 + e + x function expressions can be seen by the function is increasing function, the model is a nonlinear regression model.[9] as for alpha, beta, solving, is due to the discrete function, maximum likelihood method is adopted to establish the likelihood function is used to estimate parameters. Make Y = (y1, y2,..., yn) for n observations of a random variable, the likelihood function expression is: L (Y) = P (∏ Yi = 1) Yi x P (1 - Yi Yi = 0) .According to the basic methods to solve the likelihood function, to take the logarithmic likelihood function, after finishing equation equation: L (Y) = ∑ yiln (P = 1) / P ((a Yi Yi = 0)) + ∑ lnP (Yi = 0) , the likelihood equation for partial derivative makes it equal to zero, simultaneous equations:

$$\partial l(\alpha,\beta \mid Y)\ /\ \partial\alpha=\sum y_i-\sum e^{\alpha+\beta x}{}_{/1}+e^{\alpha+\beta x}=0 \quad (1)$$

$$\partial l(\alpha,\beta \mid Y)\ /\ \partial\beta=\sum x_i y_i-\sum x_i e^{\alpha+\beta x}{}_{/1}+e^{\alpha+\beta x}=0 \quad (2)$$

Algorithm using iterative method and parameter alpha, beta.The probability density function:

$$P\{Y=1 \mid X\}=e\alpha+\beta x/1+e\alpha+\beta x \quad (3)$$

$$P\{Y=0 \mid X\}=1-P\{Y=1 \mid X\}=1/1+e\alpha+\beta x \quad (4)$$

Thus, as long as the known vector set of values that xi and can find out the probability value, judgment, under a certain factors the probability of consumers are willing to use a credit card size, through the probability value judgment whether the factors for the significant influence factors, sum up significant weights and calculate the significant influence of affecting factors, for coefficient and probability of solving adopt spss19.0 iterative calculation software.

4.2 Variable selection

Logit model qualitative variable quantified, you should take the use of general quantitative value is 0 or 1.Because in the design of the questionnaire is interpreted as variables: have you opened a credit card: "is interpreted as" willing to use a credit card "" value set to" 1 ";"Not" interpreted as "do not want to use a credit card" value set to "0".Such variables in statistics called virtual by D (Dummy), in the model used to describe the qualitative variables.

: D = 0 (reluctant to use)
1 (to use)

According to the matrix form of the model are defined as: Y = X beta + u Y for the observed value in column of 1 s and 0 s, X as explained variable observation matrix. Selects the Logit analysis to choose between two possibilities of qualitative data, calculating index the characteristics and problems of the size of the incidence of they.

To explain the selection of variables in the following segment: personal basic information, personal attitude, banking services, security concerns, risk concerns factors five categories, including personal basic information factors include age, education, occupation, income, monthly consumption amount, six factors, risk appetite explanatory variables will therefore pose matrix Xi (I = 1, 2, 3,..., 10).Under an explanation variable assignment is a detailed description:

For the choice of respondents age, "20 years old and under" values for "zero";"20 to 30 years old" value is 1;"30 to 40 years old" value is 2;"40-50" values for 3;"50 to 60 years old" value of 4.For education, occupation, income, monthly consumption amount, risk preference the assignment of the basic information of the other people all use same method;

Attitude to respondents personal factors in the questionnaire puts forward two problems, perceived usefulness and perceived ease of use of judgment, respondents have one topic to choose "no" can be interpreted as the investigators personal attitude towards credit card is not optimistic, the reason is that there may be difficulties of perceived usefulness and perceived ease of use, the values for "zero"; When both of select "yes", explained by the respondents personal attitude towards credit card is optimistic, values of "1".

Respondents were satisfied with the banking services is embodied in three issues, respondents in one of the topic to choose "no" is considered respondents to provide support for Banks is not satisfied with the service of credit card, there may be a few charge merchants, ATM or POS terminal manufacturers products production and research and development innovation, timely remind not, preferential card not significant dissatisfaction factors, such as the values for "zero"; Instead of three questions at the same time, select "yes", think respondents to existing services provided by the bank are very satisfied with the bank without improvement, values of "1".

Respondents security concerns about the existence of credit card use become the factors influencing the use of set up three questions, including a topic to choose "yes", assign a value to "0", the explanation for the personnel security concerns as he is not willing to accept credit card use. For example, worry about personal information leakage, poor record of personal credit, hacker attacks, and the influence of factors such as money laundering safety concerns. For such customer concerns, the bank shall be by strengthening management of personal information protection, network security technology to prevent operation, solve the customer safety concerns. On the other hand, the respondents choose "no", completely assignment for "1", explained by the respondents to bank secrecy existing commitments and technology more trust, safety factor will not become the main concerns whether he is willing to use a credit card.

Table3. Variables describe.

classification	variable		The assignment
The base	age educationdegree monthly income monthly consumption		(0，20] =0 (20，30] =1 (30，40] =2 (40，50] =3 (50，60] =4 (60，70] =5 High school of following=0 undergradute master's Dr=1 undergraduate course=2 master=3 doctor=4 education=0 finance=1 manufacturing =2 construction =3 accommodation and catering=4 Wholesale and retail=5 traffic、keep grain and postal service=6 realty industry=7 Hygienism and work=8 culture、physical culture and amusement share=9 Service maintenance and the others=10 (0，4000] =0 (4000，6000] =1 (6000，8000] =2 (8000,10000] =3 (10000,+∞)=4 (0,2000]=0 (2000,4000]=1 (4000,6000]=2 (6000,8000]=3 (8000,10000]=4 (10000,+∞)=5
Characteristic variable	The demands services Safety worries		true，true=1； the others=0 true，true，true=1； the others=0 false，false=1； the others=0 false，false=1； the others=0

Obtained from the above results, the variable of occupation, income, risk preferences, customer requirements, the bank service the five variables by more than 5% level of significance test, fully demonstrates the five factors of consumers' willingness to credit card use significant impact [8] .

The results showed that (1) career will affects the use of credit CARDS. Industry in general, consumers are easier to open a credit card for its stable bank, for example: the state-owned enterprises, institutions, self-employed as a legal person, civil servants, etc. Professional in a certain extent can represent the stability of the work, income, is also the symbol of the identity and status. Therefore, professional a decent consumers are willing to open credit.(2) income for the use of credit card will have a positive correlation. The people with higher income not only satisfied with the cost of daily necessities, at the same time have a higher share in the cost of luxury goods. High income people generally have more normative financial habits, so the consumption is more inclined to lending consumption, good at using the time value of money, financial management, more willing to accept to use credit card (3) the degree of risk appetite will have positive correlation with credit card use. These groups have a bold personality of trying new things, encouraging them to always be society try to pioneer and guide in the new products. Therefore, to be willing to accept new things and risk preferences of investors are more likely to promote the use of credit card (4) customer demand for the use of credit card will have positive correlation. Customer requirements can be classified as a customer for credit card using the perceived usefulness and perceived ease of use of the two factors, combination between subjective evaluation to the credit card directly affect consumers.(5) banking services has positive correlation to the use of credit CARDS [9] .

Therefore, to attract more customers to open a credit card should strengthen the consumption of no ATM and POS terminal places to provide product support; Strengthening consumer warning, emergency report the loss of STH, overdue reminder, telephone counseling, such as after-sales service, avoid the bank warned that generates customer overdue interest and not in time existence personal credit record to disgust and give up the use of credit card; With several known, often used for consumption of the enterprise start co-branded credit card, such as set up shopping mall shopping preferential activities, promote people brush calorie of consumption [10] .

5 Conclusion

Personally, credit card consumption credit function by using the time value of money is the best means to improve the service efficiency of funds; In terms of society, the widespread use of credit card can be rapidly expand the personal credit system database covers the number, so as to improve the personal credit system, strengthen the social credit culture. At the same time, the establishment of personal credit database is the important of the big data era, relying on the large data analysis technology for user preferences and behavior information; In terms of enterprises, the use of credit card e-commerce payment processes are greatly improved, especially in the field of B2B, via credit card fast consumption method to optimize enterprise capital collection management, increasing the service efficiency of funds, the use of credit CARDS has successfully promoted the P2P era; In terms of nations, the credit card payment facilities can improve the propensity to consume, extensive use can effectively expand domestic demand, boosting economic growth.

References

1. Ajzen,I. Attitude, personality, &behavioral. Milton Keynes: Open University Press,1989.330-334
2. Ajzen,I. From intentions to actions: A theory of planned behavior. In J. ,Kuhland J. Beekmann(Eds.),Action controls: from cognition to behavior,New York:Springer Veflag,1985:11-39. Ajzen,I. From intentions to actions:A theory of planned behavior. Heidelberg:Springer,1985.231-245
3. Davis,F. D. Perceived usefulness, perceived ease of use, and user acceptance of information technology. MIS Quarterlg 1989b,319—340.
4. Davis,F.D.Technology acceptance model for empirically testing new end—user information system: theory and results,Ph.D.dissertation,MIT Sloan School of Management,Cambridge,MA.1986.246-267.
5. Davis,F.D.User acceptance of information technology: System Characteristics, User perceptions and behavioral impacts.International Journal of Man·Machine Studies,1993,38:475-487.
6. Weian Wang HongGuoFan : hangzhou institute of university of finance and trade, zhang jianguo: ningbo branch of agricultural bank of China. The present situation and the development of credit card - the investigation and analysis on the condition of credit card use. Journal of zhejiang finance, 1998, 8:22.
7. Yi Yan, sail. Problem analysis - credit card market in China in shijiazhuang city residents' credit card usage, for example. Journal of hebei university school of economics, hebei province, baoding, hebei finance, 2009, 10:19-22.
8. Zhang Jiahong. And influencing factors of consumer credit card use intention research. Zhejiang university, master's degree thesis. 2008.347-358.
9. Liang Yuzhou. Based on the a-share market data Logstic regression empirical research on enterprise credit rating. Master degree theses of master of university of science and technology of China. 2010.332-341.
10. Zhu Xingliang, Joyce ge, good chrysanthemum. Based on the Probit model empirical analysis of the affecting factors of consumer credit card payments. Journal of Harbin institute of technology, shenzhen graduate school. Consumer economy. 2013, 4, 48-51.

Discussion on the Relevant Factors of General Surgery Incision Infection and Prevention Methods

Baotao Jin[1*], Changsheng Lv[1] and Hu Kai[1]

[1]Department of General Surgery, Jingzhou Second Renmin Hospital, Jingzhou, Hubei Province,434000,China
Corresponding E-mail:270882285@qq.com

Abstract. There are many reasons that can lead to incision infection of general surgical patients. The main reasons include weight, age, body albumin level, surgical time, observation ward, etc. This paper analyzes the clinic data of patients with incision infection after general surgery based on clinic practice and study on the reasons that have impact on general surgical incision infection and gives relevant prevention countermeasures.

Keywords. General surgery department, incision infection, prevention countermeasures.

1 Introduction

The department of general surgery, which is a clinical department with surgery as its main treatment means, has significant curative effect in treating liver, intestines, stomach, vascular disease, biliary tract, thyroid and tumor etc. Wound infection is the factor which affects the curative effect of the department of general surgery most seriously. Besides, postoperative wound infection is one of the most common clinical complications in the department of general surgery. According to the incomplete data statistics, its incidence rate is 9.45%～14.9%, so it has become the most important factor influencing a patient's life safety [1]. Furthermore, there are numerous factors clinically causing a patient's post-operative wound infection. In order to analyze the factors causing a patient's wound infection.

2 Data

Case selection: 679 patients with surgical treatment at the department of general surgery in our hospital from January 2009 to December 2014, including 412 male patients and 267 female patients with their ages from 6 to 76 years old.

Diagnostic criteria: refer to the *Diagnostic Criteria for Nosocomial Infection (Trial)* issued by the Ministry of Health.

3 Methods

The retrospective investigation method was used. The investigation mainly included the pre-operative factors: patients' ages, combination of underlying diseases (anemia and diabetes mellitus), operation timing, season of operation, duration of operation, type of operation, suture by medical staff and post-operative hospitalization duration etc. In this study, the SPSS statistical software (18.0 version) was used to analyze and process data. Besides, as for the clinical data obeying normal distribution in this study, two independent samples t were used for inspection; as for the enumeration data, chi-square was used for inspection. The result of study was $P<0.005$, so the difference is of statistical significance.

4 Result

The infection factors of operative incision of the department of general surgery are related to ages, time of operation, property of surgery and type of incision to some degree, as shown in Table 1.

Table 1. Analysis on predisposing factors of operative incision of the department of general surgery.

	Number of operations	Number of infected cases	Infection rate (%)	$\chi 2$ value	P
Age				4.624	< 0.05
<30 years old	156	5	3.21		
30~60 years old	342	8	2.34		
>60 years old	181	10	5.52		
Time of operation				17.628	<0.01
<120min	463	6	0.88		
≥120min	216	17	7.87		
Type of incision				14.629	<0.01
Type I	346	5	1.45		
Type II	243	10	4.12		
Type III	90	8	8.89		

5 Discussion

Operative wound infection is one of the main infections among nosocomial infections. About 2%~5% of patients receiving abdominal external aseptic operation and 20% of patients receiving intra-abdominal operation will suffer wound infection. As indicated in related data, about 500,000 patients suffer the infection of surgical site in America every year. Besides, compared with the patients without infection, those with infections are affected in their death danger, hospitalization time and cost and they also probably would require hospitalization again. In addition, the infection rate increases with the degree of seriousness of a disease, because a large operation will increase the time of operation and the invasion of bacteria. Then, a patient's skin and mucosa are cut up during an operation, so their organism lacks the defense against bacteria, raising the infection rate. One of the most common infections is the infection of surgical incision by the department of general surgery, which concerns a patient and all links during an operation [2-4].

5.1 Factors causing infection of surgical incision by the department of general surgery

It can be found through the above-mentioned study that the infection of operative wound by the department of general surgery is related to the following factors: ① a patient' s age: the degradation of intracorporeal organs of the aged and lowered immunity will lower a patient' s wound healing ability and thus result in wound infection. Due to the not mature immune system, a minor patient has a poor resistance against the trauma of wound. Therefore, more clinical attention shall be paid to the old or young patients so as to avoid the deterioration of their wounds. ② Is a patient obese: an obese patient' s high content of internal fact will result in his/her weakened immunity, which will easily cause wound infection. ③ Does a patient have the medical history of diabetes: the granulation tissue formed in a diabetic patient' s body prolongs the duration of wound healing and provides a favorable condition for the bacterial infection of wound. For such patients, insulin treatment is required before an operation to alleviate a patient' s symptom of diabetes and keep his/her glucostasis [5-7]. ④ Type of operation: a type II incision is a hazardous factor leading to infection mainly due to the not strict operation of the surgical process. During a surgery of the department of general surgery, the operating room shall strictly follow the requirement of aseptic technique, strengthen the awareness of aseptic technique and thus reduce the possibility of wound infection of a patient. ⑤ Time of operation: the duration of operation is largely related to a patient' s wound infection. A longer duration of operation requires a longer exposure of a patient' s wound, so the possibility of infection will increase accordingly. Clinically, it is required to conduct corresponding clinical inspections before an operation, determine the therapeutic schedule, accelerate the process of operation, shorten the duration of operation, reduce the a patient' s parts exposed and guarantee strict sterile operation.

5.2 Strategies for preventing infection of surgical incision by the department of general surgery

An incision of surgery of the department of general surgery has a high risk of infection and there are abundant complicated causes causing an infection, so targeted protective measures are required during the perioperative period to guarantee the smooth post-operative recovery of a patient and lower the impact of incision infection. (1) As for an old patient, it is required to assess his/her overall physical condition firstly, strictly execute aseptic techniques during an operation and actively treat the complication; (2) As for a patient with a high body mass index, it is required to cut the subcutaneous tissue with one cut, avoid forcibly rubbing the fat layer, stop bleeding thoroughly, reserve no dead space during suture and flush the incision with stroke-physiological saline solution before suture; as for a patient with a thick fat layer, layered suture can be adopted and relaxation suture can also be conducted in a special condition. (3) The nutritional support must be in place. Albumin can be conveyed to a patient when necessary so as to increase his/her resistance. (4) Strengthen the technical operation level of medical staff and enhance the teamwork so as to guarantee the quality of operation and reduced duration of operation simultaneously. (5) Use an electrotome for cutting and coagulation in a scientific and effective way; avoid the excessive increase in the output frequency of an electrotome; in an operation involving fat and fascia tissue, impose certain tensile force to separate tissue and minimize the thermal energy contact between an electrotome and tissue so as to reduce the inactivation and liquefaction probabilities of tissue [8, 9]. Provide separate wards; separate patients with wound infection from those with aseptic operation; strengthen the management over ware environment; control the visitors; respectively arrange the family visit time and centralized dressing change time.

The existence of underlying diseases lowers the body immunity and resistance and increases the incision infection rate, such as merged diabetes mellitus. The hyperglycemia facilitate the bacterial growth in tissue and diabetic patients suffer the metabolic disorder at varying degrees, which is further aggregated by the stress response of an operation, so the bactericidal ability of their cells in their body is weakened to cause the poor tissue healing ability and thus to cause easy infection[10]. For this reason, ideally control the blood glucose and urine glucose with insulin during the perioperative period. Timely adjust the dosage of insulin after operation according to the metabolic disorder of blood glucose and the level of sensitivity for insulin. In addition, improve a patient's nutrition, preventively use antibiotics and select a proper anesthetic method and medicine to minimize the occurrence of complication and thus ensure that a patient can safely live through the period of operation.

6 Conclusion

All in all, the incision infection during the surgery of the department of general surgery is caused by multiple factors, so we should take measures from many aspects to reduce the incision infection, such as shortening the length of incision, strengthen aseptic technique of surgery and using antibiotics.

References

1. Ventral Hernia Outcome Collaborative, Mitchell TO1, Holihan JL2, Askenasy EP3, Greenberg JA4, Keith JN5, Martindale RG6, Roth JS7, Liang MK.Do risk calculators accurately predict surgical site occurrences?J Surg Res. 2016 Jun 1;203(1):56-63.

2. Golub AV1, Kozlov RS2, Pleshkov VG1, Moskalev AP3, Alibegov RA4, Chelombit'ko MA5.Surgical Site Infections after Open Appendectomy and Effectiveness of Complex Approach to TheirPrevention.Khirurgiia (Mosk). 2016;(6):68-76.
3. Zhou ZY1, Liu YK1, Chen HL2, LiF.Prevention of Surgical Site Infection After Ankle Surgery Using Vacuum-Assisted Closure Therapy in High-Risk Patients With Diabetes.J Foot Ankle Surg. 2016 Jan-Feb;55(1):129-31.
4. Ngai IM1, Van Arsdale A, Govindappagari S, Judge NE, Neto NK, Bernstein J, Bernstein PS, Garry DJ.Skin for Prevention of Surgical Site Infection After Cesarean Delivery: A Randomized Controlled Trial.Obstet Gynecol. 2015 Dec;126(6):1251-7.
5. Kiernan M1. Prevention of surgical site infection: compliance is key.Br J Nurs. 2015 Sep 24-Oct 7;24(17):856.
6. Nodzo SR1.Commentary to "CDC Will Soon Issue Guidelines for the Prevention of Surgical Site Infection".Am J Orthop (Belle Mead NJ). 2015 Sep;44(9):E360.
7. Matsumoto M1, Shigemura K, Yamamichi F, Nakano Y, Tanaka K, Arakawa S, Fujisawa M.Prophylactic antibiotic for prevention of surgical site infection in urological laparoscopic surgeries.Kobe J Med Sci. 2012 May 24;57(4):E137-44.
8. Beckmann A1, Doebler Khttps://www.ncbi.nlm.nih.gov/pubmed/?term=Graf K[Author]&cauthor=true&cauthor_uid=21277786.Sternal surgical site infection prevention - is there any room for improvement?Eur J Cardiothorac Surg. 2011 Aug;40(2):347-51.
9. Matsumoto M1, Shigemura K, Yamamichi F, Nakano Y, Tanaka K, Arakawa S, Fujisawa M.Prophylactic antibiotic for prevention of surgical site infection in urological laparoscopic surgeries.Kobe J Med Sci. 2012 May 24;57(4):E137-44.
10. Weiser MC1, Moucha CS1.The Current State of Screening and Decolonization for the Prevention of Staphylococcus aureus Surgical SiteInfection After Total Hip and Knee Arthroplasty.J Bone Joint Surg Am. 2015 Sep 2;97(17):1449-58.

A Comparative Study on New Media Marketing in Domestic and Foreign Public Libraries

Yan Xu

Library of Beihua University, NO.15, Jilin Road, Jilin City, Jilin Province, China
Email:494660880@qq.com

Abstract. Currently, new media has become the popular platform for brand marketing and services promoting of public libraries. Using the method of online survey, literature analysis and comparative analysis, this paper chooses 4 foreign public libraries and 4 domestic public libraries as the research samples and compares the differences in new media marketing between them from new media marketing tools, marketing content and marketing strategies, then makes some proposals and measures to perfect domestic public libraries' new media marketing, so as to provide references to improve the libraries' social awareness and benefits.

Keywords. Public libraries, new media marketing, comparative analysis, library marketing.

1 Introduction

With the rapid development of Internet technology, new media, such as smart phones, tablet PCs, social networks, digital TV and so on, has become the main platform for users to obtain and share information. According to eMarketer data show that in 2016 the Chinese Internet media access time for the first time than the traditional media, accounting for 50.6% of the daily media time [1]. The new media with its massive user base, rapid information exchange and other characteristics, as each industry marketing and product promotion of the best channel, but also opens up a new way for public libraries to increase social awareness and the promotion of information and knowledge service.

At present, the new media marketing related research has become the focus of attention of scholars at home and abroad. Foreign scholars E.Vassila analyzes the role of Twitter as a new social media, in promoting library service marketing and improving the user's attention [2]. D.S.Negi and other research on how to use mobile devices for library readers to provide more convenient information services [3]. The domestic scholar Zhang Ling takes 13 libraries as an example, analyzes the problems existing in the promotion of resources and services, and puts forward the strategy of promoting the application of the library with the new media technology [4]. From the existing research results, most of the new media in the library service promotion and brand marketing in the application mode and strategy. However, the comparative study of new media marketing for public libraries at home and abroad is relatively small. Therefore, this paper selects several public libraries at home and abroad as an example, from the new media marketing tool type, the main contents and strategies to compare the differences between them, so as to provide reference for the development of China's library.

2 New media marketing of public library

2.1 New media and new media marketing

The new media is a new media form with respect to traditional media, domestic and foreign scholars generally believe that the new media is based on network technology, mobile communication technology, digital technology, relying on the computer, intelligent mobile phone, digital TV and other terminal equipment, to achieve information exchange and sharing between users of media forms, including portals, search engines, social media the application of digital TV, mobile, etc.. New media marketing, just as its name implies, is the use of new media platforms to carry out marketing activities. Baidu encyclopedia is defined as: the realization of enterprise soft penetration business strategy in the new form of media, so that consumers recognize some concepts, ideas and analysis of ideas by means of the media usually expression and dissemination of public opinion, so as to achieve the enterprise brand promotion, product sales to [5]. At present, the mainstream of the new media marketing methods for WeChat, micro-blog, client marketing, foreign mainly for social media, online video marketing.

2.2 The significance of Public Libraries in developing new media marketing.

The concept of library marketing first appeared in 1990s, is to enhance the library brand image, promotion of information service and to gain public recognition, so as to maximize the social benefits as the fundamental purpose, to carry out a series of targeted activities for existing and potential users. At present, with the emergence of all kinds of new media, people tend to get more intuitive and convenient digital resources and network information. The function and status of the public library are threatened, and the service promotion and brand publicity should be strengthened while strengthening the digital and mobile information services. New media has brought new opportunities for the public library marketing. First, the new media users group public opinion advantage helps public library awareness; second, the public library can make full use of new media dissemination speed, the variety of the content features to improve the efficiency and quality of information service; thirdly, the characteristics of new media can strengthen the interaction between the public library and user communication, so as to provide accurate personalized service.

3 Domestic and foreign public library new media marketing comparative analysis.

3.1 Sample selection and research methods

The main content of the public library marketing is nothing more than two points: brand marketing and service marketing, and foreign public library service marketing in the subdivision level is more abundant. First of all, in terms of brand marketing, foreign public libraries flexible use of social media and other types of websites to attract the attention of the majority of users. For example, many public libraries will use Pinterest and YouTube pictures, video sharing websites brochures and video; the New York public library by Twitter and user real-time communication, the topic of organization and a variety of interactive games, increase site traffic. Secondly, in the service marketing, including the news announcement, theme activities; library resources, academic literature, video and other resources promotion services; training courses, online education, business counseling and other public service. For example, the Australian National Library by Facebook post pictures, video or news release important news, activities in the form of links to resources, release all kinds of theme speech or report by the official podcast website provides online library [6]; digital resources or scientific literature to readers; Boston Public Library through Linkedin website cooperation, provide employment guidance for users. In addition, the foreign public library has also carried out the payment of the project marketing, such as the British National Library Home page to the user sales of literary and artistic exhibition tickets.

3.2 The new media marketing of Public Library

The main content of the new media marketing of domestic public library also involves the brand marketing and service marketing. In the aspect of brand marketing is mainly embodied in the creation of library image and brand culture, for example, the Hangzhou library "Wenlan lecture series of culture, integration of resources to WeChat, mobile client, digital TV and other new media platform, users can free access to the Hangzhou area, has greatly enhanced the scope of the audience[7]. The library of Hangzhou library social media the current marketing situation and Strategy Research -- take" Wenlan auditorium "as an example; Shanghai library with the Oriental TV TV platform to carry out, urban culture, access to a wide range of social benefits. In the new media service marketing, the domestic library is usually in WeChat, micro-blog, the mobile client provides event announcements push, readers information query service, reference service, exquisite articles and books and other information to the library service promotion. For example, the Shenzhen port mobile phone clients to integrate the digital literature resources of Shenzhen library, Shenzhen University city library, provides a great convenience for the readers to obtain literature resources[8]; Hangzhou Library in the "offline" lectures, art exhibitions, micro theater and other special activities, and online publicity and reports by WeChat national public number; the WeChat platform not only push the fragmentation of exquisite articles, announcements, also provides a classic audio-visual and poetry reading service.

3.2.1 Comparative analysis of new media marketing tools

The main platform tools of the new media marketing of the new media of the four sample libraries in foreign countries are shown in table one.

Table 1. New media marketing tool of Public Library in foreign countries.

Library name	New media marketing tools	Features new media marketing tools
New York Public Library	Facebook, Twitter, YouTube, Blog, podcast, Google+, NYPL Mobile（Mobile Library）,Official website, etc.	Facebook, Twitteretc. Social media
Burlington Public Library	BPL Mobile（Mobile Library）,Facebook, YouTube, Official website	BPL Mobile （Mobile Library ）
National Library of the United Kingdom	Tripadvisor, Facebook, twitter, YouTube, Google+, pinterest, Official website	Tripadvisor platform
Australian National Library	Twitter, Facebook, Youtube, Blog, Flickr, Sound Cloud, Podcasts, Official website	Social media, video sites
Boston Public Library	Twitter, Facebook, linkedin, Flickr, Pinterest, YouTube, Instagram, Boston Public Library Community Social Media page	Boston Public Library Community Social Media page

Can be seen from the table, often use foreign public library marketing new media tools with social networking, video sharing websites, digital broadcasting, photo sharing website, micro blog, mobile client, the official website of the digital library etc.. Among them, Facebook, Twitter and Google+ belong to social media, their huge user group, interactivity and immediacy, become the mainstream channels of library marketing; Pinterest, Flickr for image share of new media, the library can show the style, activity or celebrity pictures through such platforms; Youtube, Viemo is a video sharing site playing videos, activities or lecture video provides the best way to public library; SoundCloud and Podcasts in the form of digital radio podcast website provides new channel for reports, news stories push library. In addition, foreign public libraries will be through cooperation with other industry website publicity, such as the British National Library and Tripadvisor (tourism website) to establish links. Visible, foreign public library marketing a large number of new media tools, a wide range of media content presented in the form of diversification, both text class, as well as pictures, audio and video.

3.2.2 The main tools of new media marketing in the domestic public libraries

Table 2. New media marketing tool of Public Library in China.

Library name	New media marketing tools	Major new media marketing tools
Shanghai library	WAP portal, mobile client APP, WeChat, micro-blog, digital TV, digital library official website	Micro-blog marketing (pictured above)
National Library of China	Digital library official website, micro-blog, WeChat, RSS subscription, SMS, mobile client (Palm figure APP)	WeChat, mobile library
Shenzhen Library	Digital Library's official website, news network, WeChat, micro-blog, mobile phone client	WeChat, news network
Hangzhou Library	Digital library official website, digital TV, WeChat, micro-blog, mobile phone client	Micro-blog, mobile client

Can be seen from the table, the main domestic new media marketing of public library with the use of mobile phone library, mobile client application (APP), WeChat and micro-blog and other social media, digital library, digital TV website. Among them, the public library and public official WeChat micro-blog account is currently the most popular marketing platform, can be used to release the news, recommended reading, event notifications, and interact with the reader to provide reference service; SMS and WAP portal is the earlier emergence of new media platform, through information for the users to send notifications the book to the library, the basic message book, has been rarely used; mobile client application can satisfy the user access to library consultation, whenever and wherever possible bibliographic retrieval and reading needs, public libraries often open up special plate in the client, building brand display activities; fusion with digital TV, very well with the television media this propaganda the channel, expand the scope of the audience. From the whole, the new media marketing channels of China's public libraries gradually widened, social media and digital media utilization increased.

3.3 Comparative analysis of new media marketing content

3.3.1 The main contents of new media marketing in foreign public libraries

The main content of the public library marketing is nothing more than two points: brand marketing and service marketing, and foreign public library service marketing in the subdivision level is more abundant. First of all, in terms of brand marketing, foreign public libraries flexible use of social media and other types of websites to attract the attention of the majority of users. For example, many public libraries will use Pinterest and YouTube pictures, video sharing websites brochures and video; the New York public library by Twitter and user real-time communication, the topic of organization and a variety of interactive games, increase site traffic. Secondly, in the service marketing, including the news announcement, theme activities; library resources, academic literature, video and other resources promotion services; training courses, online education, business counseling and other public service. For example, the Australian National Library by Facebook post pictures, video or news release important news, activities in the form of links to resources, release all kinds of theme speech or report by the podcast; library official website to provide online digital resources or scientific literature to readers; through Boston Public Library and Linkedin site cooperation, provide employment guidance for users. In addition, the foreign public library has also carried out the payment of the project marketing, such as the British National Library Home page to the user sales of literary and artistic exhibition tickets.

3.3.2 The main contents of the new media marketing in the domestic public libraries

The main content of the new media marketing of domestic public library also involves the brand marketing and service marketing. In the aspect of brand marketing is mainly embodied in the creation of library image and brand culture, for example, the Hangzhou library "Wenlan lecture series of culture, integration of resources to WeChat, mobile client, digital TV and other new media platforms, Hangzhou area users can free access to the library, greatly enhanced the scope of the audience by Shanghai library; Oriental TV, TV platform to carry out urban culture, access to a wide range of social benefits. In the new media service marketing, the domestic library is usually in WeChat, micro-blog, the mobile client provides event announcements push, readers information query service, reference service, exquisite articles and books and other information to the library service promotion. For example, the Shenzhen port mobile phone clients to integrate the digital literature resources of Shenzhen library, Shenzhen University city library, provides a great convenience for the readers to obtain literature resources; Hangzhou Library in the "line" art exhibitions, lectures, and other characteristics of the micro theatre activities, and online publicity and reports by WeChat public No.; the National Library of WeChat platform not only push the fragmentation of exquisite articles, Activities Notice also provides a classic audio-visual and poetry reading service.

3.4 Comparative analysis of new media marketing strategy

3.4.1 New media marketing strategy of Foreign Public Libraries

New media marketing strategies of foreign public libraries are abundant, mainly including mobile marketing, celebrity marketing, marketing to the game, users participate in the construction of marketing. Among them, the activities of marketing is to take action as the carrier, to guide readers to participate in interactive, enhance the user experience at the same time pass the brand value. For example, make full use of micro Facebook and Twitter activities of Australian public libraries, users carry out self sent push, photo sharing, knowledge competition and diversified activities, arouse their enthusiasm, thereby increasing the library visits. Taking advantage of marketing is the use of celebrity fans effects of celebrity, invite them to offer lectures, interviews, or invite shooting star library theme pictorial, video or photo sharing to social networking sites, to attract the attention of fans. Game marketing is loved by the majority of users, with entertainment driven propaganda. Such as the New York public library development of library online interactive fun games, game winners can enjoy activities such as library tickets, discount stores and other privileges [9]. User participation in the construction of marketing is to allow users to participate in the construction of library resources, strengthen the user experience. For example, the library's collection of Boston Public Library is connected with social media, allowing users to tag, comment, and share a list of books [10].

3.4.2 New media marketing strategy of domestic public libraries

Domestic public libraries are constantly absorbing foreign experience, enrich the new media marketing strategy. At present, there are mainly activities interactive marketing, hot events or topic marketing, micro film marketing, etc.. Among them, the activities of marketing and foreign marketing is similar, interesting activities will enhance the user experience. For example, micro-blog, WeChat and other Shanghai library to carry out knowledge quiz, book prize, recommended online entertainment exhibition activities, and readers one-on-one interaction, providing personalized service precision. Hot event or topic marketing is the public library planning, the use of social influence of the popular events or topic of public opinion, to attract public attention, to achieve the purpose of publicity. Hangzhou library, for example, has become a hot topic for micro-blog, becoming a hot topic in the world. It has been named "the warmest Library in history". It has successfully promoted the idea of library service [11]. Micro film marketing is a relatively new mode, the real story about through the lens of the library, library culture and publicity services, to attract readers to pay more attention to and accept the library service. For example, the national library has held a "library story readily shoot" creative video contest "met the most beautiful you", released to all kinds of new media platform for promotion [12].

4 Domestic and foreign public library new media marketing differences and enlightenment

4.1 New media marketing differences between domestic and foreign public libraries

By comparing the new media marketing of the domestic and foreign public library analysis, found that the differences are the following: first of all, from the point of view of marketing tools, the use of foreign public libraries of new media types, a large number of social networking sites, including pictures and video websites, digital broadcast platform, mobile phone client, job sites, travel sites etc.. The new media marketing tool of domestic public library mainly include social media, mobile phone APP, library's official website, digital tv. Although the number and type of growth trend, but compared to foreign countries are still less, but also in the content of the more text, pictures, audio, video based platform for less. Secondly, from the marketing point of view, the contents of service marketing in foreign public libraries more abundant, includes not only the basic reading service, information resources, activities and push notifications, skills training, employment guidance and other public services. In contrast, the domestic public library service marketing is mainly the basic information resources push, news and activities, and so on, the more scarce. Finally, from the point of view of marketing strategy, the domestic and foreign public libraries have different characteristics. Foreign attention to the user experience, such as through the game, the user to participate in the construction of resources and other forms to attract their attention, and can work with the field of cooperation to broaden the channels of brand promotion. Domestic is good at using the power of public opinion to achieve the effect of publicity, with interesting things to promote, but compared to foreign countries, new media marketing channels and strategies are still to be widened.

4.2 The enlightenment to the new media marketing of Public Library in China

In this paper, through the comparative analysis of new media marketing of the domestic and foreign public library, summed up the differences, and learn from foreign advanced marketing mode, brings the following enlightenment to our public library: a new media marketing culture building full participation, strengthening the librarian's new media skills, to mobilize the enthusiasm of all staff participation service promotion. Two, open and comprehensive utilization of various new media marketing platform, not limited to the mobile phone client, portals, social media, and can also strengthen the tourism website, job site, watercress reviews website cooperation, broaden the scope of marketing. Three, to explore new media marketing content rich, expand service areas, not only to meet the basic public access to information, knowledge of the demand, but also for user education, employment, entrepreneurship, leisure to provide reference and guidance, for example, to carry out online training courses, push occupation information, entrepreneurial knowledge etc.. Four, combined with a variety of marketing strategies, should actively absorb foreign game marketing, user participation marketing, celebrity marketing, combined with excellent domestic marketing mode, according to different marketing object, platform and content users loved the way. Five, pay attention to the user experience, to attract users to participate, give full play to the function of the user from the media to build the library resources and services online, to mobilize the enthusiasm of the user, and through their interaction in social circles, the "viral marketing" strategy, promotion of Library brand.

5 The conclusion

This paper selected foreign New York public library, the Australian National Library, the British Library, Boston Public Library, and Shanghai library, Shenzhen library, Hangzhou National Library and the National Library as a sample, using the network investigation, document analysis, comparative analysis methods from the new media marketing tools, content and strategy of the three aspects of comparative analysis at home and abroad the public library and the similarities and differences of new media marketing. At last, according to the current situation of domestic public library's new media utilization, the corresponding countermeasures and suggestions are put forward. The research of this paper has a certain positive significance to broaden the new media marketing channels, innovative brand marketing strategy and service marketing content, and improve the social awareness and social benefits of the library. But this research also has certain limitation, for example, the sample size is less, and the universality is not guaranteed. The future research will expand the sample scope, and further discuss the new media marketing mode of domestic and foreign public library.

References

1. eMarketer. 2016 Chinese Internet media access time for the first time more than traditional media:[2016-07-20]. Vassilakaki E, Garoufallou E.The impact of Twitter on libraries: a critical review of the literature. The electronic library, 2015,33(4):795 - 809.
2. Vassilakaki E, Garoufallou E.The impact of Twitter on libraries: a critical review of the literature. The electronic library, 2015,33(4):795 - 809.
3. Negi D S.Using mobile technologies in libraries and information centers. Library hi tech news, 2014, 31(5):14-16.
4. Zhang Ling. Library publicity and promotion strategy under the new media technology environment. Library and information science, 2015,59 (01): 78-81.
5. Baidu encyclopedia. New media marketing.[2016-07-20].http://dwz.cn/3SKksk.
6. Lu Zhenbo, Li Xiaodong, Cheng Yanxia. A case study on social media marketing of foreign libraries and Its Enlightenment. Journal of university library, 2014 (04): 76-82.
7. Li Huihui, Sun Yanxiang. The Hangzhou library social media marketing present situation and Strategy Research -- take "Wenlan auditorium" as an example of. Communication of science and technology, 2015 (21): 20-21+34.
8. Ma new media service. library and information service of public library, 2014, 58 (6): 70-74.
9. Li Jinbo. New York public library social media marketing practice. Library forum, 2016 (04): 120-125.
10. Yin Yuehuan. Public Library of. media marketing method, Northeast Normal University, 2014
11. Yang Mei. Hangzhou micro-blog Promotion Strategy Research Library. Library forum, 2012,32 (05): 157-160.
12. Zuo Bin Ningbo. Social media library marketing strategy under the situation of the implementation of. Library research, 2015 (16): 16-20.16-20.

Exploration on the Ways of Thinking for Government by Law in the Governance of the Grassroots Society

Xuzhi Fan

Department of Humanities and Social Science, Hubei Polytechnic University, Huangshi, Hubei, China
Email: 1302928893@qq.com

Abstract. Social governance of the grassroots is the foundation for maintaining social stability and building a harmonious society. The obvious problems of weakened power, abnormal petitioning, hidden danger of social security, weakness of virtual social control, and vacancy in management of the special group in the vast grassroots area did exist. Faced with the problems, the rapid development of urbanization and the relative lack of public service, the public's strengthened consciousness of rights safeguarding and not sound interests safeguard mechanism, heavy tasks on social governance of the grassroots and not enough resources, and the imbalance between the public's improved expectation and the ability of the cadres in grass-root level, the level of grass-roots social governance needs to be improved by employing the "five-in-one" thinking mode of governance by law.

Keywords. Grass-root, social governance, thinking mode of governance by law, mass disturbance.

1 Introduction

Although "social management" and "social governance" are not quite different from each other literally, they are different in the aspects of behavioral agents, operation modes and purposes of implementation, etc. The transformation from top-down "management" to the combination of top-down "governance" employed by the government and top-down "governance" of the society is the great reform in idea of governing a country [1]. In such profound reform, ability of cadres at the grass-roots level shall be improved so as to cultivate their work of "deepening reform, promoting development, solving contradiction, and maintaining stability" by employing the thinking mode of governance by law and the method of governance by law [2].

2 The main problems exist in social governance of grassroots level

Firstly, it is about the existence of weakened power in grassroots level. The grass-roots regime is situated in the bottom of the pyramid-structured state regime. Determined by its historical uniqueness, its dependence on the grass-roots government in managing everything well cannot be fulfilled. Power of the grassroots social organizations is often weak and dispersed, and power of the rural community is more limited, which results in the fact that a lot of social affairs in the countryside are unable to be dealt with. Thus, there are many blank points and weak points. Election would be affected by clan power in some area.

Secondly, the tendency of abnormal petition remains high in level. Petition is the public's solution to the problems relating to their own vital interests, and dealing with the petitions well is also the responsibility of the grass-roots party organizations. In recent years, proportion of collective petition for grass-roots society, repeated petition, petition bypassing the immediate leadership and abnormal petition is relatively high, and such cases would lead to the abnormal petitions of trouble-making petition, direct petition to the provincial capital and Beijing and even mass incidents if they are not fairly dealt with. As in 2015, collective petition in city A is about 318 batches and 6816 person-time, which rose 8.9% and 38.2% respectively on year-on–year basis, and they accounted for 19.4% and 76.3% respectively of the total petition that year; the number of repeated petition is 646 batches and 4247 person-time, which accounted for 39.3% and 47.6% respectively of the total petition that year [3].

Thirdly, there are large quantities of hidden danger in the problems of social security, while social disputes are complex and diverse. Along with the social openness and further enhanced mobility, many new situations exist in security management and the total volume of social criminals in the grassroots in our country in recent years shows a rising tendency. Burglary, purse-cutting in public places, and loot etc. seriously influence the public's personal property security; the illegal criminal activities, such as problems on rental houses which is linked to security in life, occur frequently; economic crimes involving the mass, such as illegal fund raising, financial crime, telecom fraud, financial fraud, and pyramid schemes etc., shows a tendency of extension.

Fourthly, the control over virtual society is still weak. The development of "Internet +" is both a blessing and a curse at the same time. Along with the gradually strengthened management and control system of urban network, the crime, committed by employing "high-tech + network", is extending to the grassroots, especially the vast rural areas. Crime presents a dynamic, organized, and intelligent trend [4, 5]. For instance, illegal and criminal activities, such as fraud, gambling and pornography etc. committed through network, even form the elusive criminal organizations that are based on the network platform and crime of the new form and new means is being repeated brought out in grassroots society.

Fifthly, there is much vacancy in the control over special groups. In promoting economic development, the middle-aged, which accounts for more than 50% of the population in rural area, work out as migrant workers and the left-behind children reaches more than 30% of the total population. Because they lack parents' affection, care, education and guidance, plus the induction of some adverse factors on the minor in the society, the issue of "problem child" raises severe challenges to grassroots social governance.

3 Reason-analysis on the causes to the of the dilemma of social governance in the grassroots

1.The contradiction between the rapid development of urbanization and relative shortage of public services

The quickened progress of urbanization provides economic and social development with vitality. At the same time, the rapidly expanded urban population leads to intensive allocation of resources. Moreover, the obvious lag of public service and the lack of public facilities results in increased conflicts in all sorts of interest. Thus, contradiction would easily happen in large number in reconstruction of old city, land requisition and demolishment, enterprise restructuring, medical disputes, labor disputes, job placement, law and litigation involved, and food and drug safety etc.

2.Contradiction between enhanced right-protection awareness of the public and not sound safeguard mechanism on interests

Channels for the public's interest appeal are not smooth, and the mechanism with smooth communication for contradiction resolving covering the mechanism of troubleshoots and early warning is not sound. In addition, some departments pay insufficient attention to the masses and they'd shuffle. Moreover, problem solving stays in form, and someone would even show the mood of resistance. Part of the grassroots cadres are still used to administrative commands or take compulsory measures. According to survey, 38% of the cadres in city A think that the major obstacles to grassroots work is that the method is not suitable, and it is believed that 46% of the cadres lack service consciousness and service ability. What's more, some people would have instinct panic in the work of grassroots land requisition and demolishment, as well as petition and safeguard stability, and the masses' satisfaction in the village and town over the cadres is less than 70%.

3.The contradiction between heavy tasks and not enough resources in grassroots social governance

Grassroots communities (villages) undertake the heavy task of social governance, but they have not yet reached the goal of "power goes along with responsibility and the passing of costs system accompanies duty", which leads to the phenomenon of "3 Nos" in part of grassroots departments—"no measures in their work, no power owned by the coordinating departments, and no resources to serve the public". Especially in rural areas, the public's enthusiasm in participating villagers' autonomy, social undertakings and social governance would be reduced. Instead, about 70% or 80% of the villagers thought that they can only gain the sense of belonging by participating the clan-related activities, such as being written on the pedigree of their clan, building ancestral temple, and amending their family tree. Some villagers are attracted by superstition power in clan and they'd choose the latter as to leadership of the party organizations at the grass-roots level and clan power. Cadres and the masses at the grass-roots level are deeply affected by the ideas of power over legislation, love over legislation, representing legislation with customs, being ashamed of litigation and the inability of legislation to the majority.

4.The contradiction between the public's expectation and ability of grassroots cadres

The grassroots masses not only focus on social development dynamic, but also hope to have a platform where they can make comments and express their viewpoints. Some cadres at the grass-roots level have no broad vision and no liberated thoughts, and they'd work by depending on their own feelings, experience and authority work. Thus, they'd do less work on demonstration, guidance service, and more work on certain relationship. Moreover, they'd depend on law less and lack the concept of legal thinking. Considering the questionnaire survey in rural area and grassroots that is aimed at the party branch secretary of the villages, 53% of the respondents in city A think that the most prominent problem lies in the not strong consciousness of dedication; 27.7% of the respondents think that their motivation of being in office is seeking for gains of themselves and their clans; 29.7% of the respondents believe that they want to be spokesperson of the interests owned by the small collectives.

4 Exploration on the ways of thinking for government by law in the governance of the grass-root society

Based on the above-mentioned problems and difficulties, it is discussed from the macroscopic angle of legislative thinking mode that whether most party members and cadres at the grass-roots level should treat legislation and the way to obey the law as an important value orientation in planning work, judging the right and the wrong and dealing with problems so as to deal with problems according to the provisions and procedure of law. Legal thinking has certain directivity, and it has the characteristics of openness. Legal thinking should include the five basic connotations of systematicness, fairness, constructiveness, contractuality and proportionality, and it plays a important role of guidance, support and security in promoting the ability of social governance in grassroots.

1. Realize the certainty of the grassroots social governance by employing regular thinking mode. To build the grassroots society with the rule of law where impersonalization prevails, people at the grassroots level shall obey the same rules. As people at the grassroots level has strengthened consciousness of rights protection and relatively weak consciousness of rules, grassroots cadres shall think about the universality of problems that they are faced with. If relevant work are done at the expense of the authority of law, or problems are dealt with for a short time by attaching importance to specialty, and the present and individual interest is satisfied by sacrificing the excessively high social cost, then it will not only fails the grassroots people's expectations, but also lead to new social problems. If things continue this way, it will seriously interfere with the process of our modernization.

2. Realize the harmony of the grassroots social governance by employing fair thinking mode.

Fair thinking refers to the thinking mode which is employed by all the social members who maintain their own interest and harms not the interest of others, our country and the collective in the legal framework. Public power refers to the truth that "law is infeasible if it is unauthorized" and civic power refers to the fact that "law is feasible if it is not forbidden". Practice shows that the grassroots attaches great importance to the equity of resources and public service. If cadres at the grassroots level abuse their power to damage fairness and justice, it will greatly reduce the grassroots' trust on organizations of public power. To maintain the dignity of the rule of law, the bad thoughts of attaching importance to power and management while thinking lightly of rights and service should be full eradicated so as to realize the justice of procedure and entity.

3. Achieve the sustainability of grassroots social governance by using constructive thinking.

Constructive thinking is generally a kind of thinking mode that starts from the general problems so as to find the common factors, and then it is ascended and solidified into the system that can be implemented and generally obeyed. Thus, the public can make judgment that can be expected by their own behavior. The grassroots cadres shall set up orientation of the public will, respect the public's pioneering spirit, find problems from the public opinion, take analysis and settlement of the public will into consideration, and nurture people's belief toward the rule of law.

4. Achieve the honesty of grass social governance by using contractual thinking. Contractual thinking refers to the communication in life and production with contract (system) as the link and honesty as the basis. Chinese society in the past, especially the grass-roots society, is relatively closed, and relationship among people is called the "society of acquaintance". Moreover, the main body of social interaction is also the network of acquaintance. The current China is going through the profound transformation period from the "society of acquaintance" to the "society of strangers" [4]. The times of contract's gradual growth and formation is also the transformation period from "identity" to "contract". Cadres at the grass-roots level shall deepen their understanding on the trend of social development and grasp the direction and trend of legal construction scientifically so as to build up trust and undergo communication with contract and system as the link, and build up the harmony of grassroots society by the establishment of contract, following the contract, and safeguarding the contract.

5. Achieve the uniformity of grass social governance by using balanced thinking. Things in the world are complicated and they'd relate to each other. The legal issues shown by the contradictions that are interweaved and interplayed contain more complicated economic and social problems. It is required that the issues should be viewed from two, three and even more angles. If it is analyzed and dealt with from the angle of law and technique, it is difficult to be settled and it may also lead to chain reaction. If it is considered more thoroughly and the scheme is designed more carefully, the negative impact can be minimized and win-win effect between the pursuit of personal value and collective value even multi-win can be attained. Thus, the unification between development and harmony, rights protection and stability safeguarding, and formal justice and substantive justice can be achieved. In the process of concrete practice, law should be abided and subjective initiative should be exerted so as to reasonably employ the philosophy of law enforcement. That is, comprehensive factors should be taken into consideration in evaluating the advantages and disadvantages, right or wrong, and the pros and cons of things.

Acknowledgements

1.Connotative Development and Research for Public Cultural Service of Grassroots Community supported by open fund of public culture for key study base of the humanities and social science in Hubei province in 2015 Item No.: 2015 kyb06

2. Analysis on the Response and Disposal of Sudden Mass Disturbance supported by Scientific Research Projects of Hubei Polytechnic University in 2009

References

1. Wei Wanbin, The New Success in the Concept of Governing a Country. Journal of History of the Party in Beijing. Forum Vol.1(2014),p.23
2. Walking along the Path of Socialism with Chinese Characteristics and Striving to Build A Well-off Society in an All-round Way—Report on the 18th National Party Congress of the Communist Party of China.

3. Information on http://www.xj.xinhuanet.com
4. Sun Xixin, Zhang Zhaoli, Don't Make Clan "Obstacle" in Maintaining Harmonious Stability in Rural Area. Journal of Life of Party Branch (Shandong). Forum Vol.7 (2010),p.29.
5. Zhang Yan, Law and Morality of Strangers' Society. Journal of China Book Review. Forum Vol.3 (2012),p.45.

Environmental Regulation, Industrial Heterogeneity and Value Compensation

Lisha Chen[1,2,*],and Changqing Ding[1]

[1]Business School of Hohai University, Nanjing 211100, China
[2]School of Business Administration of Nanjing Audit University, Nanjing 211815, China
Corresponding Email: chenlisha006@163.com

Abstract. This article divided environmental regulation tools into market-oriented environmental regulation, environmental administrative control and environmental pollution regulation, divided enterprises into pollution intensive and clean production enterprises. Take Chinese manufacturing listed companies as research object, through panel data analysis, this article measured short-term and long-term value compensation effect and the possible existence of the threshold effect and cumulative effect of different environmental regulation tools for different types of micro enterprises to inspect Porter hypothesis and to inspect the effect of Chinese environmental regulation.
Keywords. Environmental regulation, heterogeneity, value compensation.

1 Introduction

Deterioration of living environment forced governments to take more efforts on environmental regulation. Under the pressure of environmental regulation, the corresponding coping behavior would be taken by enterprises to reduce the impact of environmental regulation. The coping behaviors may be end-of-pipe control, technology innovation, change of operation direction, etc. All the coping behavior world undoubtedly lead to energy conservation and emissions reduction, and will eventually be reflected in the enterprise value changes. So it is ponderable to link environmental regulation and enterprise value to search the influence of environmental regulation on enterprise, to observe whether value compensation effect do exist, and whether industry heterogeneity affect the influence.

2 Literature review and research hypothesis

2.1 The influence of environmental regulation on corporate value

Economics essence of environmental regulation is internalization of negative externalities of environmental pollution as enterprise's cost. To achieve more profits, micro enterprises, inevitably faced with the tradeoffs between cost and enterprise performance. To solve this problem, different enterprises make different decisions. From strategic perspective, enterprises would choose different environment strategic and make different environmental management decisions. From investment perspective, on the one hand, environmental regulation can make the enterprise to strengthen technology research to improve energy efficiency and reduce energy consumption and pollution, on the other, environmental regulation may force companies to invest in other areas to avoid environmental regulation to reduce business risk. No matter what kinds of decisions are made, the goal of rational corporate is to improve competitiveness and finally increase the enterprise value.

Hypothesis 1a: environmental regulation affects enterprise value.

Enterprise value includes short-term and long-term value. The impact of technological innovation on enterprise value has cumulative effect and threshold effect. Technology innovation present as corporate spending in short term, that is why the feedback of enterprise technology given raised by environmental regulation to enterprise value of the short-term performance is negative or of no significant role. In the long run, green innovation accumulation bring about promotion of product image, as well as the green innovation especially the predictive green innovation can bring about reformation of products structure and industry structure which will infuse new competitive advantage into enterprises. Therefore, there is a positive relation between environmental regulation and the enterprise long-term value.

Hypothesis 1b: the relationship between environmental regulation and short-term enterprise value is negative or of no significant role.

Hypothesis 1c: there is a positive relation between environmental regulation and enterprise long-term value

2.2 The influence of different environmental regulation to the enterprise value is different

Environmental regulation is a complex policy tool system. Different types of environmental regulation have different operation mechanisms. And the influence mechanism on enterprise strategy formulation and implementation so as to influence value-added effect must be different. According to Ling Li and Feng Tao (2012) [1] this paper divided environmental regulation into administrative-controlling environmental regulation and market-oriented environmental regulation.

Administrative-controlling environmental regulation mainly includes emission standards, product technical standard and penalty which can be divided into two parts of legislation (environmental administrative control) and law enforcement (environmental pollution regulation). Environmental administrative control is to directly affect the choice of the polluters to make beneficial to environmental protection by laws, regulations, policies and systems legislated. So it is a kind of beforehand supervision, which can give clear direction to enterprise to establish long-term friendly environment strategies. Environmental pollution control is a kind of after-event supervision. When companies choose to discharge illegally, they need play game with government. When government improve the pollution control, the likelihood of be punished for illegal discharge is increased, which would encourage enterprises to select technology investments or change direction of enterprise management for less pollution as much as possible. Therefore the latter influences short-term decisions of enterprises, and can't guide enterprises to take the initiative to establish friendly environmental strategy. Market-oriented environmental regulation mainly includes the pollution tax, tax deduction, the clean development mechanism (CDMs). The primary of market-oriented environmental regulation tools in China is emission charges [2]. Emission charges increase the cost of enterprises and reduce their profits, but the technical level is a significant variable to adjust the function of enterprise cost and the profit. Market-oriented environmental regulation can give companies a clear direction of cleaner production and encourage enterprises to pursuit long-term strategy.

Hypothesis 2: different types of environmental regulation have different effects on corporate value.

2.3 The industry heterogeneity in the impact of environmental regulation on enterprise value

From degree of contamination enterprises can be divided into pollution intensive enterprise and clean production-oriented enterprises. The influence of environmental regulation on the two types of enterprises is different. The higher level of pollution enterprises do, the more environmental regulation would impose on them, and environmental compliance costs are relatively higher. On the one hand, there are more hinders for enterprises to invest technology innovation, on the other hand, only by investing more in technology to improve efficiency pollution intensive enterprises can make up for the high environmental costs. Thus it can effectively motivate pollution intensive enterprise to take technical innovation. And for the industry of low level of pollution, effect of environmental regulation on industry production action is not obvious, so the influence of environmental regulation on its production and operation is not obvious.

Hypothesis 3: Impact of Environmental regulation on enterprise value exist industry heterogeneity.

3 Empirical design

3.1 Variable definitions and data sources

Dependent variables: enterprise value (V) was divided into short-term and long-term value. The short-term value measured in this article is based on accounting indicators, using return on assets (ROA) to illustrate; long-term value is the expected value of the enterprise, use enterprise Tobin's q value to illustrate which represent the ratio of the enterprise's market value and replacement cost.

Independent variables: Environmental regulation variable was divided into market-oriented environmental regulation (PEG), environmental administrative control (GOV) and environmental pollution regulation (CAS). Refer to the practice of Shu Li and WeiGuo Weng(2014) [3],GOV were expressed by the number of regional administrative rules and regulations promulgated; CAS in various areas was measured by environmental administrative penalty cases accepted; PEG was measured by the ratio of regional emission charges and the gross industrial output value.

Control variables: control variables in this paper includes property rights attributes (State), Industry (Industry), the joining together of two position (Dual), company size (Size), share ratio of the largest shareholder (First), equity Balance degree (i.e., the second - fifth big shareholder's stake, Balance), the proportion of management shareholding (Manager), financial leverage (Lev), enterprise Growth ability (Growth), Age (Age).

Sample selection and data sources: Because this article is mainly to investigate the influence of environmental regulation on industrial enterprises, so agriculture, mining, and all companies in service industries were eliminated from the samples. Companies of ST, PT, SST * and those missing variable samples were also eliminated. Given the availability of raw data of relevant indexes, the study period this paper chose is 2001-2013.Among them, the number of regional administrative rules and regulations promulgated and the data of emission charges are from China's Environmental Yearbook; Enterprise asset data are from CSMAR database, annual reports of listed companies and other kinds of data are calculated from China Statistical Yearbook.

3.2 Research methods and models

To validate compensation effect of environmental regulation on value of industrial enterprises, regression models were built as follows:

$$ROA = \alpha_0 + \beta_1 X + \beta_2 State + \beta_3 Industry + \beta_4 Dual + \beta_5 Size + \beta_6 First + \beta_7 Balance + \beta_8 Manager + \beta_9 Lev + \beta_{10} Growth + \beta_{11} Age + \varepsilon \tag{1}$$

$$TobinQ = \alpha_0 + \beta_1 X + \beta_2 State + \beta_3 Industry + \beta_4 Dual + \beta_5 Size + \beta_6 First + \beta_7 Balance + \beta_8 Manager + \beta_9 Lev + \beta_{10} Growth + \beta_{11} Age + \varepsilon \tag{2}$$

In model (1), (2), X represents the intensity of GOV, CAS and PEG.

4 Empirical analysis and results discussion

Due to the hysteresis, we designed the environmental regulation variable lag one period when we do the statistical analysis to inspect the effect of different environmental regulation tools on both short-term and long-term value of listed companies.

4.1 Model estimate result

By using Stata14.0 the panel data of Chinese listed companies was deal with, the results are shown in Table 1.

Table1. Estimate Results of the Whole Samples.

Variable	ROA			Tobin Q		
Old	0.00260***	0.00364***	0.00320***	0.0860***	0.0967***	0.0923***
	(4.89)	(7.67)	(6.85)	(8.98)	(11.28)	(10.95)
Dual	0.00763*	0.00736*	0.00724*	0.116	0.0938	0.110
	(1.82)	(1.76)	(1.72)	(1.52)	(1.24)	(1.45)
First	0.00000942	0.000359**	0.000349**	-0.0203***	-0.00864***	-0.00942***
	(0.07)	(2.41)	(2.32)	(-8.49)	(-3.21)	(-3.48)
Balance	0.000134	0.000435***	0.000446***	-0.0153***	-0.00515*	-0.00575*
	(0.87)	(2.64)	(2.69)	(-5.53)	(-1.74)	(-1.92)
Size	0.00321*	-0.00396*	-0.00353	-0.459***	-0.690***	-0.689***
	(1.77)	(-1.73)	(-1.53)	(-13.95)	(-16.74)	(-16.54)
Lev	-0.190***	-0.186***	-0.187***	-1.328***	-1.207***	-1.181***
	(-24.49)	(-23.97)	(-23.94)	(-9.41)	(-8.62)	(-8.38)
Growth	0.0413***	0.0428***	0.0424***	0.270***	0.323***	0.323***
	(13.22)	(13.68)	(13.50)	(4.75)	(5.71)	(5.70)
Gov_{t-1}	0.000938***			0.0190***		
	(5.59)			(6.24)		
PEG_{t-1}		5.357***			77.88***	
		(3.44)			(2.77)	
CAS_{t-1}			-9.69e-08			-0.0000100*
			(-0.34)			(-1.95)
C	0.00537	0.144***	0.147***	12.26***	16.48***	16.66***
	(0.14)	(3.22)	(3.26)	(18.06)	(20.46)	(20.46)
Adi.R^2	0.1718	0.1780	0.1771	0.1284	0.1172	0.2995
F	3.53	3.61	3.56	6.19	5.68	5.56
Model	FE	FE	FE	FE	FE	FE
Sample Size	358	358	358	358	358	358

Note: * indicates significant at the 0.1 level, * * significant at the 0.05 level, * * * significant at the 0.01 level

It can be seen from table 1 that environmental administrative control and market-oriented environmental regulation has a significant positive effect on short- and long-term value. The effect of environmental pollution regulation on short- and long-term corporate value is negative. This result account for the fact that punishment of national and local environmental protection department for illegal discharge did not force enterprises to carry out technical innovation to reduce pollution emissions. Overall, environmental regulation tools affect the enterprise value, so the hypothesis 1a was verified. The influence of various environmental regulation tools on short- and long-term value is different. The impact of environmental administrative control and market-oriented environmental regulation on enterprise long-term value were greater than on short-term effects. Although not as hypothesis 1b predicted, the model proved that the effect of technology innovation caused by environmental regulation have a cumulative effect and the threshold effect on enterprise value: in the short term, technology innovation is expenditure to enterprises, its short-term impact is not big; In the long run, technological innovation can bring reformation of product and industry structure which infuse new competitive advantage for enterprises. So there is a positive relationship between environmental regulation and the enterprise long-term value. Therefore, hypothesis 1c was verified.

4.2 Industry classification regression statistics

4.2.1 The model results of environmental regulation and enterprise short-term value

According to the division in Environmental Information Disclosure Guide of Listed Company (draft) announced by the Ministry of Environmental Protection on September 14, 2010, and Industry Classification Management Catalog of the Environmental Protection Verification of Listed Companies, coal, petrochemical, mining, paper making, fermentation, textile, pharmaceutical, chemical, leather, thermal power, cement, iron and steel, electrolytic aluminum, building materials, metallurgy, brewing 16 sectors were included in heavy pollution industry. The rest were classified as clean production enterprises. Statistical results are as follows:

Table 2. Classified Statistic Results of the Effect of Environmental Regulation on Short-term Value.

Variable	Heavy Pollution Industry			Clean Production		
Old	0.00255***	0.00355***	0.00311***	0.00318***	0.00411***	0.00358***
	(3.32)	(5.11)	(4.56)	(4.57)	(6.65)	(5.92)
Dual	0.00678	0.00535	0.00540	0.00511	0.00549	0.00520
	(1.09)	(0.86)	(0.86)	(0.95)	(1.03)	(0.98)
First	-0.000187	0.000169	0.000140	0.000196	0.000609***	0.000641***
	(-1.02)	(0.81)	(0.67)	(1.06)	(2.97)	(3.12)
Balance	0.00000497	0.000292	0.000292	0.000315	0.000728***	0.000760***
	(0.02)	(1.28)	(1.26)	(1.49)	(3.21)	(3.33)
Size	-0.00556**	-0.0131***	-0.0125***	0.0126***	0.00470*	0.00472
	(-2.05)	(-3.81)	(-3.60)	(5.51)	(1.65)	(1.65)
Lev	-0.198***	-0.194***	-0.195***	-0.175***	-0.169***	-0.171***
	(-18.16)	(-17.84)	(-17.75)	(-16.32)	(-15.88)	(-16.03)
Growth	0.0315***	0.0329***	0.0330***	0.0595***	0.0611***	0.0599***
	(7.70)	(8.02)	(7.98)	(12.38)	(12.78)	(12.53)
Gov_{t-1}	0.000879***			0.000915***		
	(3.56)			(4.16)		
PEG_{t-1}		4.886**			7.749***	
		(2.44)			(3.12)	
CAS_{t-1}			-0.000000267			0.000000135
			(-0.66)			(0.36)
C	0.214***	0.361***	0.362***	-0.215***	-0.0773	-0.0635
	(3.89)	(5.39)	(5.33)	(-4.45)	(-1.37)	(-1.13)
Adi.R^2	0.1612	0.1707	0.1700	0.2315	0.3659	0.3710
F	3.25	3.34	3.3	4.06	4.06	4.04
Model	FE	FE	FE	FE	FE	FE
Sample Size	202	202	202	156		153

Note: * indicates significant at the 0.1 level, * * significant at the 0.05 level, * * * significant at the 0.01 level

To both clean production enterprises and pollution intensive enterprises, the influence of environmental administrative controls on short- and long-term value, although small, but is all significantly positive. Emission charges as market-oriented environmental regulation promoted enterprise short-term value. The influence of the environmental administrative controls on pollution intensive enterprise is less than on clean production-oriented enterprises. As the same, the influence of emission charges on pollution intensive enterprise is less than on clean production-oriented enterprises. The effects of environmental pollution regulation on short-term value of pollution intensive enterprise is negative, of clean production enterprise is positive. Both for pollution intensive enterprise and clean production-oriented enterprises, the impacts of environmental pollution regulation are small, and not significant.

4.2.2 The model estimate results of environmental regulation and enterprise long-term value

Table 3. Classified Statistic Results of the Effect of Environmental Regulation on Long-term Value.

Variable	Heavy Pollution Industry			Clean Production		
Old	0.0874^{***}	0.101^{***}	0.0955^{***}	0.0824^{***}	0.0899^{***}	0.0856^{***}
	(6.85)	(8.77)	(8.47)	(5.59)	(6.87)	(6.61)
Dual	0.204^{*}	0.149	0.161	0.0143	0.0153	0.0347
	(1.96)	(1.45)	(1.56)	(0.13)	(0.14)	(0.31)
First	-0.0207^{***}	-0.00889^{**}	-0.00985^{***}	-0.0210^{***}	-0.00975^{**}	-0.0105^{**}
	(-6.75)	(-2.56)	(-2.84)	(-5.36)	(-2.25)	(-2.38)
Balance	-0.0171^{***}	-0.00714^{*}	-0.00759^{**}	-0.0136^{***}	-0.00294	-0.00369
	(-4.82)	(-1.88)	(-1.99)	(-3.02)	(-0.61)	(-0.76)
Size	-0.446^{***}	-0.694^{***}	-0.686^{***}	-0.475^{***}	-0.678^{***}	-0.684^{***}
	(-9.84)	(-12.16)	(-11.95)	(-9.81)	(-11.21)	(-11.17)
Lev	-1.736^{***}	-1.620^{***}	-1.603^{***}	-0.686^{***}	-0.561^{**}	-0.535^{**}
	(-9.54)	(-8.97)	(-8.84)	(-3.02)	(-2.49)	(-2.35)
Growth	0.162^{**}	0.217^{***}	0.217^{***}	0.468^{***}	0.508^{***}	0.508^{***}
	(2.37)	(3.18)	(3.18)	(4.59)	(5.01)	(4.97)
Gov_{t-1}	0.0211^{***}			0.0161^{***}		
	(5.11)			(3.45)		
PEG_{t-1}		72.58^{**}			103.8^{**}	
		(2.19)			(1.97)	
CAS_{t-1}			-0.00000822			-0.0000119
			(-1.23)			(-1.48)
C	12.09^{***}	16.80^{***}	16.85^{***}	12.42^{***}	15.88^{***}	16.27^{***}
	(13.14)	(15.12)	(15.01)	(12.14)	(13.32)	(13.55)
Adi.R^2	0.1519	0.3620	0.3640	0.0857	0.2175	0.2120
F	6.05	5.60	5.51	6.26	5.65	5.51
Model	FE	FE	FE	FE	FE	FE
Sample Size	202	202	202	156		153

Note: * indicates significant at the 0.1 level, * * significant at the 0.05 level, * * * significant at the 0.01 level

It can be seen from the chart that to all enterprises the environment administrative controls have significantly positive on long-term. Promoting effect of emission charges on long-term value of enterprises is most obvious. Environmental pollution regulation has negative effects on two types of companies, although numeric is small and not significant, but it is enough to declare the current environmental administrative penalty does not play the role of negative reinforcement to force enterprises to implement innovation.

5 Policy enlightenment

The government should implement targeted environmental regulation measures based on industry classification. For pollution intensive enterprises, through rules and regulations, illegal punishment, emission charges, new energy and new material subsidies and other measures not only lead them to speed up the process of shifting from end-of-pipe treatment to clean production, but also get revenue from the process, so that to encourage enterprises to respond to the government gradually stricter and sophisticated environmental protection policy actively. For clean production-oriented enterprises, it is reasonable to improve the industry standard and strength of environmental regulation to push production process optimization through the government's external constraint, but at the same time pay attention not to only improve environmental regulation standards, be careful that too much environmental regulation may beyond the withstand scope of enterprises.

Acknowledgement

A Project Funded by the Priority Academic Program Development of Jiangsu Higher Education Institutions and National Natural Science Foundation of China(Grant No. 71472090)

References

1. Ling Li, Feng Tao. Selection of Optimal Environmental Regulation Intensity for Chinese Manufacturing Industry. China Industrial Economics,2012,5:70-82.
2. Qian Zhang. Impact of Market based Incentive Environmental Regulations on Different Types of Technology Innovation and Regional Heterogeneity.Forward Position or Economics,2015,3(2):36-48.

3. Shu Li, WeiGuo Weng. Local environmental regulation and total factor productivity growth in China. Journal of Financial Research, 2014, 40 (2) : 19 – 29.

Research on the Mechanism of Cross Organizational Knowledge Sharing in BIM Competitive Environment

Wei Zhong[1], Liping Li[1], Jin Xu[1] and Xiaoyu Ma[1]

[1]School of Management, Tianjin University of Technology, Tianjin 300384, China

Abstract. Co-opetition under the environment of BIM implementation process because the level is not high resulting in the project BIM application value of incomplete knowledge sharing among organizations, thus establishing good knowledge can effectively solve this problem and achieve the overall benefit and benefit sharing mechanism in the project organization. This paper defines five competing modes according to the competing degree of organization between different BIM applications, including imperfect competition, competition, competition and cooperation, full cooperation and cooperation, and put forward the conceptual model and related assumptions. Analysis of the effect of path and effect of project determined in BIM application mode, the concurrence of knowledge sharing, efficiency and overall efficiency of the project within the organization through the survey and empirical results, and according to the proposed contract, the distribution of benefits and work three kinds of knowledge sharing mechanism implementation path.

1 Introduction

The application of BIM in construction projects is deepening day by day. However, due to the multidisciplinary integration of BIM technology, the application value of BIM needs to be realized by multi-party cooperation. Therefore, BIM application process inevitably occurs. Inter - organizational knowledge sharing.

Claudin [1] proposed four configuration mechanisms for knowledge sharing among organizations. Song Chao [2] His research on the influencing factors of knowledge sharing and knowledge protection among organizations and systematically proposed knowledge sharing and protection strategies and mechanisms. Soekijad and Andriessen [3] put forward the requirement of benign knowledge sharing in competitive alliances.Liao Nuo[4] model of the relationship between the theory of competition and cooperation, knowledge sharing and innovation performance. Shen Nan [5] on the impact of different competitive cooperation environment in the inter-organizational knowledge sharing game and knowledge sharing unit cost relationship. Yang Wei [6] based on the perspective of competition and cooperation, proposed the existence of cooperation and competition is an important mechanism to solve the problem of border dilemma. Yuan Li [7] obtained the knowledge acquisition and interaction of the competing type to the participant by studying the factors that influence the knowledge acquisition in the competitive environment, thus highlighting the importance of competition and cooperation to the knowledge management. These scholars have put forward their own viewpoints and innovative achievements on knowledge sharing in different environments, but they have not studied the application of BIM technology to the cross-organizational knowledge sharing mechanism in competing environment.

Based on the analysis of BIM technology application points according to the interest-related attributes, five competing modes are defined according to the degree of competition and cooperation among BIM application implementation organizations and the knowledge sharing , Organizational benefit and overall benefit of the project, and constructs a theoretical model of BIM application of inter-organizational competition-cooperation model, knowledge sharing and knowledge sharing efficiency, and uses empirical questionnaires and SEM structural equation model to do empirical research, To develop an inter- organizational knowledge sharing mechanism for BIM projects.

2 Research hypothesis

BIM technology throughout the project, as a fusion of collaborative tools much professional knowledge, the need to participate in the organization from different angles and professional direction, collaboration and management across organizations in different depth, in order to better realize the goal of the BIM project application[8]-[10]. BIM technology is a comprehensive application of multi-disciplinary expertise, technical software and professional experience, so this paper uses the way of BIM application to express the knowledge transfer process among organizations.

The BIM implementation process of the project needs to be coordinated by the multi-party organizations. The nature of the organization determines the direction of BIM implementation and the difference of the benefits distribution after BIM implementation, which forms the BIM application environment [11]. According to the competing theory proposed by Brandburger and Nalebuff, the competition and cooperation relations among enterprises are divided into three types: low cooperation low competition, high cooperation high competition, low cooperation high competition and high competition from the relative strength of competition and cooperation. Cooperation and low competition "[12,13]. Based on Luo's four types of competing relations, Wang Liang divided the competing relations into five types: partnership, cooperation, moderation, struggle and isolation [14,15]. In this paper, five competing modes are defined according to the degree of inter-organization competition in the implementation of BIM, including imperfect competition, competition, cooperation, incomplete cooperation and cooperation. Therefore, according to the application of BIM to realize the path (Figure 1), the BIM application based on five kinds of model were classified and made the impact of knowledge sharing across organizations in different degrees of competing project in the hypothesis.

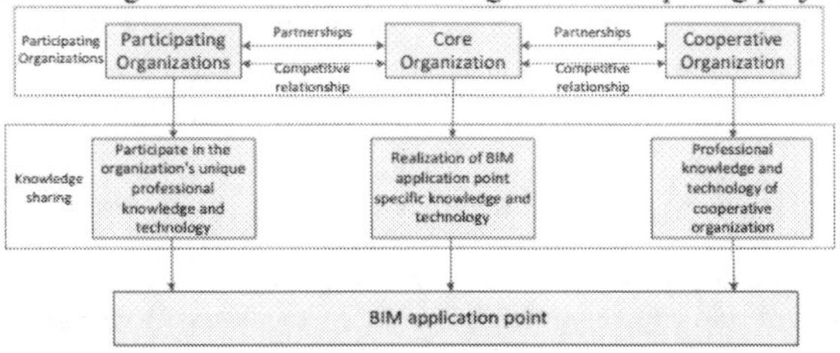

Fig. 1. BIM application point to achieve the path.

2.1 Competitive mode

Competition refers to the competition and cooperation mode exists only competition between organizations, the main features of the application for organizing special core competence, the realization of the BIM application process and no other organizations in the implementation of the results has a positive impact on the maximization of the interests of the organization, the organization that cooperate with other organizations to complete this application implementation there is a risk of knowledge, leakage, so the organization competition environment is difficult to carry out knowledge sharing [16].Accordingly, this paper proposes the following assumptions:

H1 $_{(-)}$.Competitive model has negative impact on cross-organizational knowledge sharing in engineering projects with BIM technology.

2.2 Imperfect competition model

Refers to the presence of imperfect competition between organizations in competition based, a kind of competition cooperation model as its main feature, the application for the organization strong core competence, but the need for external organizations to cooperate to complete and achieve multi benefit amount. When the unique professional knowledge of a BIM application belong to the group, after completion of the BIM application, to participate in Inter Organizational cooperation and organization will be less, because the protection of its own core knowledge and influence the level of knowledge sharing [17]. Accordingly, this paper proposes the following assumptions:

H2 $_{(-)}$.The imperfect competition model has a negative impact on cross-organizational knowledge sharing in engineering projects with BIM technology

2.3 Co-opetition model

Co-opetition between the organization and the organization refers to both the existence of competition is a mode of cooperation, the specific performance of the application is two or more than two organizations have, and each organization believes that through cooperation with other organizations to enable the application to obtain the maximum lifting at minimum cost. Competition and cooperation mode due to the organization to obtain benefits through cooperation, but with the passage of time, competition and cooperation will occur because of the competition level, the existence of such behavior increases with time and the influence of knowledge sharing [18].Accordingly, this paper proposes the following assumptions:

H3 $_{(+)}$.Competitive cooperation model has a positive impact on cross-organizational knowledge sharing in engineering projects with BIM technology

2.4 Incomplete cooperation model

Incomplete cooperation refers to the existence of cooperation between the organization and the organization, supplemented by a competitive mode of cooperation, the main features of the organization to participate in the hope that through cooperation to enhance their ability to enhance their own. Under the imperfect cooperation mode, the tendency of cooperation is greater than the competition will promote the knowledge sharing between organizations.Accordingly, this paper proposes the following assumptions:

$H4_{(+)}$.Incomplete cooperation model has a positive impact on cross-organizational knowledge sharing in engineering projects with BIM technology

2.5 Cooperation model

Cooperation refers to the application points can only be completed through two or more than two organizations, and achieve the purpose of multi benefit. The application of BIM model to achieve the path of cooperation exists in the organization of two or more than two cooperation organization, this process, by cooperation and cooperation organization A B, cooperative organization and participate in organizing and participating organizations and participating organizations cooperate to complete. Cooperation model, the willingness to participate in the organization of strong, high level of knowledge sharing. Accordingly, this paper proposes the following assumptions:

$H5_{(+)}$.Collaborative models have a positive impact on cross-organizational knowledge sharing in engineering projects with BIM technology

2.6 Organizational performance, the overall effectiveness of the project

Different competing models can bring about different knowledge sharing benefits, while knowledge sharing benefit is divided into organization benefit and project overall benefit in the Inter Organizational cooperation. It is precisely because of the existence of this kind of knowledge sharing performance, which will affect the behavior of each organization in the knowledge sharing, so it will be the factors that affect the overall efficiency of the organization and the overall benefit of the project. Accordingly, this paper proposes the following assumptions:

$H6_{(+)}$.Knowledge sharing has a positive effect on the intra-organizational benefits of engineering projects with BIM technology.

$H7_{(+)}$.Knowledge sharing has a positive impact on the overall benefit of the engineering project in the application of BIM technology.

$H8_{(+)}$.There is a positive correlation between the overall benefit of the project and the benefit within the organization.

Based on the above assumptions, the theoretical model of the relationship between the organization of BIM application and the benefit of knowledge sharing and knowledge sharing is constructed (Figure 2).

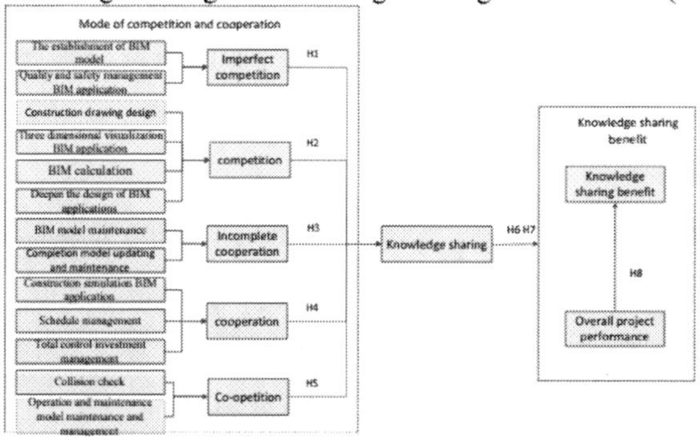

Fig. 2. BIM application of inter-organizational cooperation model, knowledge sharing and knowledge sharing effectiveness of the theoretical model.

3 Rmpirical research

3.1 Questionnaire

This research mainly with reference to domestic and foreign mature measurement table has the relevant variables, sharing and knowledge sharing within the organization efficiency, the preliminary design of the project overall benefit from the five competing measure of knowledge, the organization mainly discusses the application of BIM technology in the project of different degree of competition and cooperation of knowledge sharing and knowledge sharing benefits the impact and the corresponding mechanism, the questionnaire for the object with the organization nature, organization scale, BIM application ability as control variables in the study of influence factors. Survey questionnaire using the form of Likert five component to measure variables, variable range from 1 to 5 points, 1 of the total does not meet, 5 said in full compliance, followed by. The questionnaire issued a total of 300 questionnaires, 198 valid questionnaires were recovered, the recovery rate was 66%.

3.2 Reliability and validity test

In this paper, SPSS 22.0 software for questionnaire survey data reliability and validity of the empirical analysis of variables for the reliability and validity test. The calculation of all the factors of Cronbach 'alpha value between 0.7 to 0.9, which indicates that the scale has high reliability, good internal consistency and validity; found from the test results, all factor KMO values are greater than 0.6, and each item of the load factor is greater than 0.6 and less than 0.8, which indicates that the scale has good structural validity, optimal model with a good degree.

4 Empirical result analysis

4.1 Theoretical model checking

In this paper, we use the SEM21.0 structural equation model to estimate the path coefficient and its fitting index, and the test parameters are $\chi^2 = 23.835$, $df = 12$, $\chi^2/df = 1.743$, between the ideal reference value 1~3, the model matches well;$CFI = 0.986$, $IFI = 0.972$, $TLI = 0.934$, both were larger than 0.9, which showed good fit for the model;$RMSEA = 0.066 < 0.08$, all indexes reach the acceptable range, which indicates that the theoretical model constructed in this paper is reasonable.

4.2 Hypothesis test

The standard path coefficient model is obtained by the method of structural equation model (Figure 3), of which 1) competition on knowledge sharing effect standard coefficient is - 0.23, $P < 0.01$, assuming H1 support;2)imperfect competition on knowledge sharing effect of standard coefficient is -0.19, $P < 0.01$, H2 supported hypothesis 3); the competition and cooperation of knowledge sharing standard coefficient of 0.09, $P < 0.01$, assuming H3 is supported; 4) do not fully cooperate on knowledge sharing standard coefficient of 0.19, $P < 0.01$, assuming H4 is supported; 5) cooperation on knowledge sharing standard coefficient of 0.26, $P < 0.01$, H5 hypothesis is supported; 6) knowledge sharing within the organization efficiency standard coefficient is 0.09 , $P < 0.01$, assuming H6 is supported; 7) knowledge sharing standard coefficient on the overall efficiency of the project is 0.39, $P < 0.01$, H7 hypothesis is supported; 8) project The overall effectiveness of the standard coefficient of efficiency within the organization is 0.35, $P < 0.01$, assuming that H8 is supported. As a whole, the degree of competition and cooperation directly affects knowledge sharing, which indirectly affects the efficiency of the organization and the overall benefits of the project.

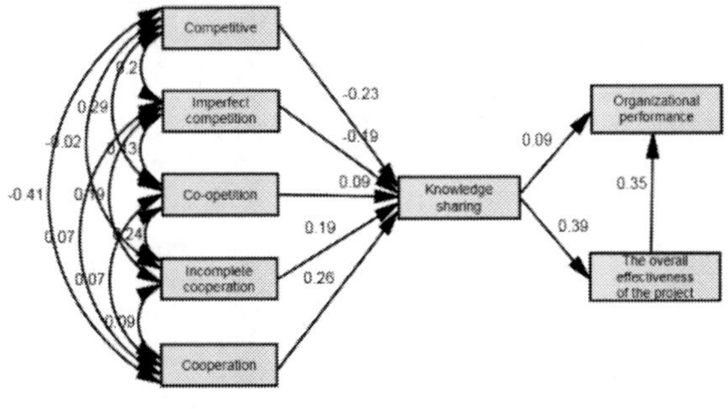

Fig. 3. Standard parameter estimation model.

5 Knowledge sharing mechanism

The organization according to the results of the study and application of BIM technology combined with the project of cooperation and competition of the actual situation, put forward the contract mechanism, interest distribution mechanism and knotworking cooperation mechanism of three kinds of cross organizational knowledge. The empirical results suggest that the greater the degree of competition in the knowledge sharing level is low, the greater the degree of cooperation of knowledge sharing the higher the level of knowledge sharing level directly affect the overall benefit and project organization, this paper puts forward the knowledge sharing mechanism to improve the organization and participation degree, knowledge sharing motivation and determine the application of BIM model set up (Figure 4). BIM competing environment of inter organization knowledge sharing mechanism path is as follows: firstly, according to the different competing model for contract mechanism, competition mode, imperfect competition mode and competition mode of cooperation, focusing on formulating the participation constraint scope of the organization work related to the terms of the contract, increase participation organization cooperation degree; according to incomplete cooperation mode, cooperation mode, focus on make the protection involved in the organization of professional knowledge related terms, due to the degree of cooperation too closely to lead to knowledge leaks; in the premise of good knowledge sharing environment, establish a reasonable profit distribution mechanism, in order to improve the knowledge sharing motivation; finally, according to the work mechanism, breaking the boundaries of knowledge of BIM application, to achieve the overall benefits of project.

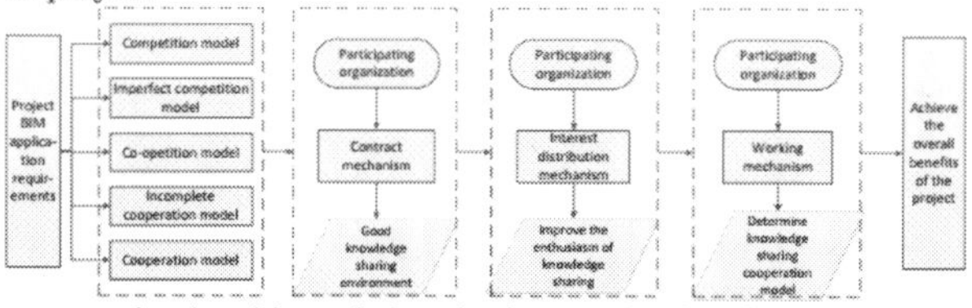

Fig. 4. the role of cross organizational knowledge sharing mechanism in BIM competitive environment.

6 Conclusion

By constructing the structural equation model, this paper studies the influence of the relationship between the organization of the engineering project and the application of BIM technology on the knowledge sharing and the benefit of knowledge sharing, according to the empirical results of the organization with the application of BIM technology in engineering project of the competition and cooperation of the actual situation, put forward the contract mechanism, interest distribution mechanism and node (knotworking) mechanism of three kinds of knowledge sharing mechanisms. 1) the contract mechanism, in the organization of knowledge sharing, should be in advance of the norms of the organization and the scope of responsibility and knowledge sharing, create a good knowledge sharing environment, in order to improve the enthusiasm of participating in organizational knowledge sharing; 2) the interest distribution mechanism, set up a good benefit distribution system, in order to encourage the participation of the organization of knowledge sharing, manufacturing knowledge sharing virtuous circle; 3) knotworking cooperation mechanism, can quickly break the boundaries of knowledge, improve the efficiency of knowledge sharing. The contract mechanism establishes a good knowledge sharing environment, the interest distribution mechanism improves the participation of the organization knowledge sharing initiative, the working mechanism provides a knowledge sharing mode of work.

References

1. Claudia L,Paul C.Managing inter-organizational knowledge sharing. Journal of Strategic Information Systems ,2016, (25) : 4-14.
2. Song Chao. Research on the mechanism of knowledge sharing and protection in the Inter Organizational cooperative enterprise [D].JINLIN: Jilin University.2015.
3. Soekijad M,Andriessen E. Conditions for knowledge sharing in competitive alliances. European Manage-ment Journal,2003, (21) : 578-587.
4. Niao Nuo,Li Xiaoyan. Research on the influence of horizontal cooperation strategy on information sharing and innovation performance in the supply chain . Scientific and technological progress and Countermeasures,2016,33 (13) : 27-33.
5. ShenNan,Li Wanhong. Research on knowledge sharing game of green innovation system in manufacturing industry . Study and exploration,2015,25 (9) : 117-121.
6. YangWei,JiangXu. Knowledge acquisition and knowledge leakage in strategic alliances: Based on the perspective of Co-opetition . Research and development management,2016,33 (3) : 1-11.
7. Yuan Li,Yi Liu,Heng Liu. Co-opetition, distributor's entrepreneurial orientation and manufacturer's knowledge acquisition: Evidence from China. Journal of Operations Management,2011, (29) : 128 - 142.
8. Man Qingpeng,Sun Lianying. Analysis of influencing factors on the implementation of Inter Organizational BIM collaborative application . Journal of Engineering Management,2016,30 (3) : 26-30.

9. A. Volkov,P. Chelyshkov,D. Lysenko.Information Management in the Application of BIM in Construction. The Roles and Functions of the Participants of the Construction Process.Procedia Engineering,2016,（153）：828–832.
10. Mehmet Yalcinkaya,Vishal Singh.Patterns and trends in Building Information Modeling (BIM) research: A Latent Semantic Analysis. Automation in Construction,2015,（59）：68–80.
11. Xu Jiefeng,Lei Xinghui. China's engineering contracting enterprise BIM supply chain partnership based on the investigation and classification . China Civil Engineering Journal,2015,48（6）：122-128.
12. Brandenburger A M,B J Nalebuff .Coo-petition.New York：Doubleday,1996：35-46.
13. Augustine A,Nancy G,Susan C.Competition,Cooperation, and the search for economic rents:a syncretic model.Academic of Management Review,1997,（22）：110-141.
14. Wang Liang,Liu Yi. A study on the relationship of cooperation relationship among enterprises, knowledge sharing and innovation performance in the process of business process outsourcing . Scientific and technological progress and Countermeasures,2013,30（7）：84-89.
15. LUO Y. Toward coopetition within a multinational enter-prise : aperspective from foreign subsidiaries.Journal of Word Business,2005,（40）：71-90.
16. Ricarda B,Sascha Kraus.Innovation in knowledge-intensive industries: The double-edged sword of co-opetition.Journal of Business Research ,2013,（66）：2060－2070.
17. Murat Akpinar,Zsuzsanna Vincze. The dynamics of coopetition: A stakeholder view of the German automotive industry. Industrial Marketing Management ,2016,（57）：53－63.
18. Jao-Hong Cheng.Inter-organizational relationships and knowledge sharing in green supply chains—Moderating by relational benefits and guanxi.Transportation Research Part E ,2011,（47）：837－849.

Risk Management of Assets Dependency Based on Copulas Function

Lei Cheng and Xiaofang Chen

School of Management, Wuhan University of Technology, Wuhan, China
E-mail: chenglei_cl@126.com, cxf2010611@163.com

Abstract. As the two important form of financial market, the risk of financial securities, such as stocks and bonds, has been a hot topic in the financial field; at the same time, under the influence of many factors of financial assets, the correlation between portfolio returns causes more research. This paper presents Copula-SV-t model that it uses SV-t model to measure the edge distribution, and uses the Copula-t method to obtain the high-dimensional joint distribution. It not only solves the actual deviation with using the ARCH family model to calculate the portfolio risk, but also solves the problem to overestimate the risk with using extreme value theory to study financial risk. Through the empirical research, the conclusion shows that the model describes better assets and tail characteristics of assets, and is more in line with the reality of the market. Furthermore, Empirical evidence also shows that if the portfolio is relatively large degree of correlation, the ability to disperse portfolio risk is relatively weakness.

Keywords. Portfolio, Correlation analysis, Copulas function, Risk management.

1 Introduction

Economic globalization and financial integration greatly enhanced the interdependence between the global economy and financial markets; The synergy of prices in financial markets made local fluctuations of any areas have an influence on other financial markets, triggered the financial crisis. During the Asian financial crisis in 1997, Hengsheng index in Hong Kong dropped from the top of the nearly 16000 point to less than 8000 points. And during the global financial crisis in 2007, the subprime crisis originated in America spread to the global market. A-share index fell from 6200 to 2000, at the same time, Hengsheng index also fell from 30000 to 12000, indicating a strong following up. So the study of the dependent features and risk factors among the assets structure is an on-going process in the financial risk analysis.

For foreign references [1, 2, 4], Frees, Valdez and Embrechts studied simulation and estimation of individual asset dependency structure between among time series, without studying multiple assets dependency structure. Nelsen introduced the Copulas function into the finance, and discussed the limitation of using the linear correlation index to measure the dependency in the financial markets. Furthermore he suggested to estimate the joint distribution with Copulas theory among the random variables. Domestic study of Copulas connection theory started late [9, 10, 11]. Zhang Yaoting (2002) theoretically discussed the feasibility of the application of Copulas function in the financial industry, and pointed out that using Copulas technology to analyze the correlation structure between the variables can be more reliable between the variables. Wei YanHua and Zhang Shiying (2004) applied the Copulas to GARCH model to measure the autocorrelation structure of financial time series.

The above research either concerns with a single portfolio risk, or the static study of portfolio. In order to study the relationship of portfolios, these research has been neither conducted from the perspective of dynamic effect, nor considered assets as additive and independent. This paper will introduce Copulas function to study dependency structure of multiple assets, to measure the correlation between each other, and to verify the correlation with the empirical analysis. Therefore, not only the effective analysis of assets dependency will be achieved, but also the purpose of forecast and control of the financial risk will be attained.

2 The tail correlation and copulas function

Financial market is full of risk, but Markowitz creatively put forward that diversified investment can reduce risk level under a certain range. In general, investors will choose assets of small correlation to reduce investment risk. So correlation analysis of assets portfolio has great influence on the returns, which needs further interpretation.

Copulas function does not contain any information about marginal distribution, but contains relevant information between marginal distribution. This correlation is often non-linear. Generally we use Kendallτor Spearmanρrank correlation coefficient to measure the correlation of marginal distribution. Because correlation coefficient has a strong robustness, and is strongly resistant to the impact of abnormal data. Therefore it has been used to measure the correlation. T is defined as the difference of two random variables between the same direction and the reverse direction.

Definition: if $(x_1,y_1),(x_2,y_2)$ is the vector of independent and identically distributed, $x_1,x_2 \in x, y_1,y_2 \in y$,

$$\tau = p\{(x_2-x_1)(y_2-y_1) > 0\} - p\{(x_2-x_1)(y_2-y_1) < 0\} \tag{1}$$

then τ measures the consistency of the variability of x and y, which proves

$$\tau = 2p\{(x_2-x_1)(y_2-y_1) > 0\} - 1, \quad \tau \in [-1,1] \tag{2}$$

According to research and deduction of Kendall's rank correlation coefficient, we can get the following conclusion: τ changes reflect the extent of consistency of the variability of x and y.

Table 1 shows two Copulas functions with only one parameter, and if the estimated value of τ is calculated, the whole nature of Copulas function is sure to know.

Table 1. The Copula function parameters and the correlation coefficient.

Type	Copula distribution	$\theta - Range$	$\tau - Parameter$
Gumbel	$\exp\{-[(-\ln u)^{1/\theta}+(-\ln v)^{1/\theta}]^{\theta}\}$	$[1,\infty)$	$\theta/(\theta+2)$
Clayton	$(u^{-\theta}+v^{-\theta}-1)^{-1/\theta}$	$[-1,0)\cup(0,\infty)$	$\theta/(\theta\text{-}2)$

3 Single asset yield model (garch –t)

When we use Copulas technology to build the model, we can separate correlation structure between marginal distribution and random variable. This will help us to analysis and understand a lot of financial problems; At the same time, for conditional heteroscedasticity and tailing characteristics of the financial time series, the following research will analyze financial time series model of single variable distribution---- building of marginal function.

Traditional fluctuation model believes that exogenous variables, such as the exchange rate or interest rate, can cause yield heteroscedasticity (volatility). But conditions mean and variance may change by time, leading to the change of the variance. So Engel (1982) [2] proposed conditional heteroscedasticity (ARCH) model to describe the time series with the conditions of the second order moment in nature, and describe fluctuations of time-varying and aggregation through the change of conditional heteroscedasticity. In the ARCH model, Engel points out that the mean ε_t interference conditions do not change by time. But in the ARCH model, variance (h_t) depends on the time series of past observations. So according to the description of the ARCH model, under the condition of known information, y_t obey normal distribution where ε_t=0, h_t . Thus the ARCH model can be used to describe the condition of variable marginal distribution.

Because in practical application the ARCH model takes a lot of order, leading to deficiency in calculation, Bollerslev (1986) proposed the Generalized regressive conditional heteroscedasticity (GARCH) model, which is an important extension of the ARCH model. The model has faster lagged convergence than the ARCH model, and the similar structure with the ARMA model.

GARCH (p, q) - T model is defined as follows

$$\begin{cases} r_t \left| I_{t-1} \sim \mathrm{t}_v \right. \\ h_t = \partial_0 + \sum_{i=1}^{q} \partial_i \varepsilon_{t-i}^2 + \sum_{i=1}^{p} \beta_i h_{t-i} \\ r_t = h_t^{1/2} v_t \end{cases} \quad (3)$$

Where, r_t is the series of return, $p \geq 0, q > 0; \partial_0 > 0$,

$\partial_i \geq 0, i = 1, \ldots, q; \beta_i \geq 0, i = 1, \ldots, p$。IF p=0, GARCH process becomes ARCH (q) process. So, you can consider ARCH process as a special case of GARCH process.

4 Multi-asset portfolio model and parameter estimation

4.1 Construction of multi-asset portfolio model

Supposing, portfolio have n kind of financial assets, $i(i = 1, 2, \ldots, n)$, and build model directly by using GARCH - t model, according to the latest m phase of the historical data

N for return on assets, $(R_{i,t-1;} \ldots, R_{i,t-\mathrm{m}})$, the rate for return time series is $r_t = (r_{1t}, r_{2t}, \ldots, r_{nt})$, if the prior information set before t moment shows

$I_{t-1} = \{r_{t-1}, h_{t-1}, r_{t-2}, h_{t-2}, \ldots\} = \bigcup_{i=1}^{n} I_{i,t-1}$, $I_{i,t-1} = \{r_{i,t-1}, h_{i,t-1}, r_{i,t-2}, h_{i,t-2}, \ldots\}$, h_{it} is conditional volatility for the information set ($I_{i,t-1}$) of singular asset(r_{it}), $C\left(\cdot \middle| I_{t-1}\right)$ shows Copulas function for n ,

$F_i\left(r_{i,t} \middle| I_{i,t-1}\right)$ is the conditional distribution of i-th component .

Many empirical studies have shown that asset return sequence obeyed the GARCH (1, 1) model. Bollerslev (1987) point out that t-student distribution can reflect more characteristics for financial time series than the normal distribution can do, such as the peak of time series and thick tail. Here we also assume that r_{it} satisfy the model. After filteried the fluctuation of time-varying about a single series, we propose a time-varying model ---Copulas- GARCH (1, 1) - t (shorthand for Copulas- GARCH - t). The term used to describe the time-varying dependence structure of n assets, and the model is as follow:

$$\begin{cases} r_{i,t} = u + \varepsilon_{i,t}; \\ \varepsilon_{i,t} = \sqrt{h_{i,t}} \cdot v_{i,t} \ ; \\ h_{i,t} = \alpha_0 + \alpha_1 \varepsilon_{i,t-1}^2 + \beta_1 h_{i,t-1} \\ F\left(v_t \middle| I_{t-1}\right) = C_{\Sigma}\left[F_1\left(v_{1,t} \middle| I_{1,t-1}\right), F_2\left(v_{2,t} \middle| I_{2,t-1}\right), \ldots, F_n\left(v_{n,t} \middle| I_{n,t-1}\right) \middle| I_{t-1}\right] \end{cases} \tag{4}$$

where, r_{it} is the series for return, u is the average return rate for singular asset, $p \geq 0, q > 0; \partial_i \geq 0, \beta_i \geq 0$; $F\left(v_t \middle| I_{t-1}\right)$ is the joint distribution portfolio on the moment of T, C_{Σ} is Copula function. α_0, α_1, β_1, v and Σ are parameters to be estimated.

4. 2 Parameter estimation of Copulas - GARCH function

Parameter estimation of function model includes parameter estimation and non-parameter estimation. Parameter estimation method generally adopts maximum likelihood estimation (MLE) and the torque estimation. The maximum likelihood estimation is the most commonly and most effective method of Copulas function model. Parameters α_0, α_1, β_1 to be estimated in the model for single asset are obtained through the Eview software, and the parameters of the portfolio will be obtained by maximum likelihood estimation.

If we set random variables $x_1, \ldots, x_N$ in the Copula function as $C(.,\ldots,.)$, density distribution is $c(.,\ldots,.)$,the marginal distribution $F_n(.,.)$,the edge density function $f_n(.,.)$,n=1 , ... , N , then the parameters $\theta = (\theta_1, \ldots, \theta_N; \theta_c)$ of the logarithm likelihood function is:

$$\ln L(x_1, \ldots x_N; \theta) = \sum_{t=1}^{T} \left[\ln c(F_1(x_1; \theta_1), \ldots, F_N(x_N; \theta_N), \theta_C)) + \sum_{n=1}^{N} \ln f_n(x_{n,t}, \theta_n) \right] \tag{5}$$

We can directly use the maximum likelihood estimation for parameter estimation. But from the logarithmic likelihood function we can find that the marginal distribution density and the density of Copulas likelihood function is separate. Therefore, we can first estimate the edge parameter, then the Copulas function parameter. The estimation of parameter is divided into two phases.

$$\text{The first phase:} \quad \hat{\theta}_n = \arg\max \sum_{t=1}^{T} \ln f_n(x_{n,t}, \theta_n) \text{ ,n=1,...,N} \tag{6}$$

$$\text{The second phase:} \quad \hat{\Sigma}_c = \arg\max \sum_{t=1}^{T} \ln c(F_1(x_{1,t}; \hat{\theta}_1), \ldots, F_N(x_N; \hat{\theta}_N), \theta_C) \tag{7}$$

First estimate the parameter in the marginal distribution $\theta_n, n = 1,..., N$, and then put the estimated value $\hat{\theta}_n$ into Copula function of different kinds, thus estimate the parameter of Copula function $\hat{\Sigma}_c$.

4.3 The empirical analysis

Here we only analyze two stock market index of daily closing price: Baoli Real Estate and Wanke Real Estate; the data range from July 31, 2006 to May 9, 2013. It is important to note that because there may be one day when two stocks are not trading at the same time, in order to accurately describe the correlation of the two stocks, we consider data observed as the data for that day when only one stock has dealings (t stands for trading day). After pretreatment, valid data of observation samples reach 1372, namely T = 1372. Secondly, the daily closing price is selected as poor stock index, logarithmic LPR, defined as LPR = log (pr) - log (pr (1)).

(1) The figure of the return rate of the two stocks, in which bp stands for Baoli Real Estate's daily yield rate, wp stands for Wanke Real Estate closing price.

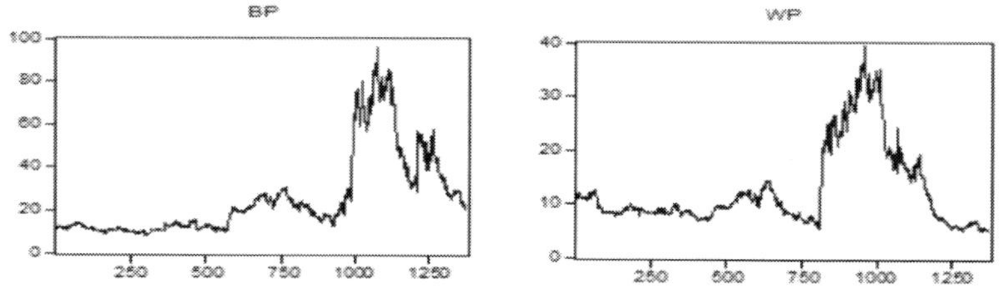

Fig. 1. The daily yield rate of the two stocks.

(2) Descriptive statistics of the two stocks

Table 2. Descriptive statistics of the two stocks.

	Mean	Median	Maxi	Min	Std. Dev	Skewness	Kurtosis	J-Bera	Prob
P	25.715	19.055	96.23	8.52	19.52	1.605	4.654	745.244	0
P	12.632	9.3	39.5	4.93	7.60	1.578	4.511	699.837	0

From the table, we can draw the following analysis. During the sample period, the mean of bp stands at 25.7, standard deviation 19.5 with 1.6 of skewness, and 4.7 kurtosis. Compared with the normal distribution, bp has the tendency to slide right with rush thick tail phenomenon. JB normality test also confirmed this, with JB=745.2, P=0, and refused null hypothesis. These show that under the confidence level ∂=0.05, Baoli Real Estate significantly differs from those of normal distribution. Similarly, Wanke Real Estate also has the tendency to slide right with rush peak phenomenon and significantly differs from those of normal distribution.

(3) Parameter estimation

Table 3. Two property parameter estimation.

	u	∂_0	α_1	β_1	v	Σ_c
lbp	1.000002 (2525.385)	0.0000134 (3.499094)	0.004938 (2.740976)	0.983879 (239.6179)	3.872199 (12.86543)	$\begin{bmatrix} 1,0.078442 \\ 0.078442,1 \end{bmatrix}$
lwp	0.999609	0.001052	0.103511	-0.00604	3.433089	

	(1946.871)	(9.165047)	(2.179955)	(-0.18711)	(12.11436)	

From the above table, we know that α_1 and β_1 in logarithm return rate (LBP) is respectively equal to 0.004938 and 0.983879 in the conditional variance equation, and they are significantly different. It shows that the return series has significant volatility cluster. Secondly, in Baoli Real Estate's LBP, the sum of ARCH and GARCH coefficients equals to 1, so the GARCH (1,1) process is stable, that is, the effects of fluctuations in the past on the future is gradually fading. Similarly, we can get logarithm yield values of the parameters for Wanke Real Estate.

(4) Equation residual test

Autocorrelation	Partial Correlation		AC	PAC	Q-Stat	Prob
		1	-0.000	-0.000	0.0003	0.987
		2	0.004	0.004	0.0221	0.989
		3	-0.002	-0.002	0.0266	0.999
		4	-0.030	-0.030	1.2946	0.862
		5	-0.024	-0.024	2.0710	0.839
		6	-0.002	-0.002	2.0791	0.912
		7	-0.030	-0.030	3.3004	0.856
		8	-0.009	-0.010	3.4130	0.906
		9	0.001	-0.000	3.4140	0.946
		10	-0.020	-0.021	3.9806	0.948

Fig. 2. Baoli Real Estate autocorrelation.

The residual test of equation shows Q statistic of the logarithm return rate on Baoli Real Estate, at k=10, p=0.948. It rejects the original hypothesis, namely the residual sequence is independent from each other, which meets the requirements.

Autocorrelation	Partial Correlation		AC	PAC	Q-Stat	Prob
		1	-0.001	-0.001	0.0006	0.980
		2	-0.031	-0.031	1.3540	0.508
		3	0.011	0.011	1.5218	0.677
		4	0.008	0.007	1.6148	0.806
		5	-0.010	-0.010	1.7591	0.881
		6	0.006	0.006	1.8056	0.937
		7	-0.019	-0.020	2.2974	0.942
		8	0.016	0.017	2.6506	0.954
		9	0.001	-0.000	2.6516	0.977
		10	-0.009	-0.007	2.7562	0.987

Fig. 3. Wanke Real Estate autocorrelation.

The residual test of equation shows Q statistic of the logarithm return rate on Wanke Real Estate ,at k=10, p=0.987. It rejects the original hypothesis, namely the residual sequence is independent from each other, which meets the requirements.

5 Conclusion

Based on the conditional heteroscedasticity GARCH-t model for portfolio, this thesis attempts to simulate spike and thick tail characteristics, and describes the characteristics of time-varying and state independence of the distribution on financial assets, and solves the problem proposed in the beginning; By using Copulas function, the asset correlation is reflected, and the portfolio selection problem is solved in the risk management. And from an empirical analysis of the Baoli Real Estate and Wanke Real Estate, the following conclusion can be drawn:

Conclusion 1: There are significant GARCH effects in two stocks, namely the conditional heteroscedasticity. Positive risk premium of Baoli is higher than that of Wanke, which reflects the Baoli stock investors are more risk-averse than Wanke investors.

Conclusion 2: There is obvious leverage in two stocks, which the effect of bad news on the stock market will cause more volatility than that of the good news. For example, in 2008, after increasing the stamp duty, there was a sharp decline in stock market. There was two-way spillover effect between the two stocks, which indicates that a stock fluctuations can cause the fluctuation of the other stock.

References

1. Engle,R.E. Dynamic Conditional Correlation:A Simple Class of Multivariate Generalized Autoregressive Conditional Heteroskedasticity Models. Journal of Business&Economic Statistics, 2002,7:339-350.
2. J.C.Rodriguez. Measuring Financial Contagion: A Copula Approach. Journal of Empirical Finance, 2003,14: 401-423.
3. Embrechts P, Hoeing A, Juri A. Using copula to bound the value at Risk for functions of dependent risks . Finance and Stochastics, 2003, 7:145- 167.
4. Fan J·, Jiang, J·Zhang, etal, Time-Dependent Diffusion Models for Term Structure Dynamics, Statistica Sinica, 2003, 13: 965~992.
5. Chen M, Chen G. A nonparametric test of conditional autoregressive heteroscedasticity for threshold autoregressive models . The Canadian Journal of Statistics, 2008,4:649 – 666.
6. R B Nelsen·,An Introduction to Copulas , Springer, New York,2006:51-155.
7. ZHANG Xuegong,XUE Zhichao,etal. Analysis of financial market based on Copula-SV-t Pair-Correlation, journal of Finance and Economics, 2016.04:71-76 (In Chinese).
8. YI Wen-de.Conditional dependence modelsof assets based on and its applications of portfolios Copula functions.Systems Engineering-Theory and Practice, 2011, 6:1004-1013 (In Chinese).
9. WEN X ,WEI Y,HUANG D.Measuring contagion between energy market and stock market during financial crisis:a copula approach. Energy economics,2012,34(5):1435-1446(In Chinese).
10. Kunlapath S,Tatevik Z,etal.Interdependence of oil pricess and stock market indices:a copula approach.Energy Economics, 2014,44(4):331-339.
11. YAO Den-bao,LIU Xiao-xing,7HANG Xu.Dynamic Correlation Structure Between Market Liquidity and Market Expectation:An Analysis Based on ARMA-GJR-LARCH-Copula Model, Chinese Journal of Management Science, 2016.2 (4):1-10(In Chinese).
12. YU Wen-hua.Study on Extreme Value Dependence Measurement of Shanghai and Shenzhen Composite Index—Based on the Four Types of Time-Varying Copula Model,Journal of southwest jiaotong university2013,7(4):90-96(In Chinese).

Study on the Coupling Coordination Degree between Metropolitan Economic System and Water Environmental System - Taking Beijing as an Example

Ming Chen[1,2,]* and Changxin Xu[1]

[1]Business School, Hohai University, Nanjing, China
[2]Gulf Coast Research & Education Center, University of Florida, Wimauma, The United States
Corresponding Email: chenming.1008@163.com

Abstract. The generation of metropolis is the inevitable outcome of the development of urbanization to a certain stage. The economy and society of metropolis are in the rapid development, but this process brings great pressure to the ecological environment, especially the water resources environment at the same time. In this paper, the relationship between metropolitan economic system and water environmental system is deeply studied, and the concept of "coupling" is introduced. Based on the framework of "pressure-state-response" (PSR), 12 detailed indexes, such as total population and water consumption per ten thousand Yuan GDP, were selected to construct four subsystems. The coordination degree measurement model has been used to calculate the degree of coupling between the economic system and water environmental system. On this basis, the following conclusions are drawn through the example of Beijing city: (1) In the process of metropolitan development, the economic system and the water environmental system are interrelated and there exist a complex coupling mechanism. (2) With the adjustment of economic structure and the progress of technology, the coupling index between the metropolitan economic system and the water environmental system has been increasing, and this process shows an upward trend

Keywords. Metropolis, economic system, water environmental system, coupling.

1 Introduction

In recent years, the number and the population of large cities continues to increase because of China's rapid economic development. As of the end of 2014, China has 16 mega-cities, in the forefront of the world. Urbanization was initially led by agglomeration development in order to achieve economies of scale. When the scale of urban development to a certain stage, it began to spread out from the center of the city, radiated surrounding areas and re-stage, eventually formed city or metropolitan area. As China's opening up efforts to expand the internal structure of the economy continue to optimize the transition, China's economic development situation is good. The economic and social situation of Beijing, Shanghai and other cities is highly developed. However, with the rapid development of the economy, the extensive economic industry still occupies a considerable proportion of the energy consumption, which causes the corresponding water environment system of the region to bear too much pressure. The water pollution of these cities is serious, and the water quality self-recovery ability is poor [1].

Beijing is the capital of China, at the same time it is a typical representative of China's metropolis. Beijing is in the north of China and is located in the northern part of the North China Plain. And Beijing is the largest city in northern China, it has jurisdiction over 18 counties, with a total area of 16410.54km^2 and total population of 2151.6 million. Beijing's winter is cold, summer is hot and humid. It is a typical temperate continental climate, with an average annual rainfall of about 600mm, precipitation distribution is extremely uneven, 80% concentrated in 6,7,8 three months. Beijing's economic and social development is relatively advanced, but the corresponding water resources environment is relatively scarce, it belongs to the typical water shortage city. The economic development brought greater pressure to the water supply system. It discharges a large amount of waste water, causing great pollution to water resources. However the corresponding water environment system in Beijing due to limited supply capacity, can not meet the Beijing daily Production and life requirements of the necessary water supply requirements. Its corresponding water resources carrying capacity is relatively limited, their own digestion, their ability to digest sewage is relatively inadequate.

2 Research methods

2.1 Construction of index system

Based on the three principles of scientific, representative and feasible, 12 indicators were selected according to the "Pressure-State-Response (PSR)" framework to construct four subsystems. These four subsystems include social population subsystem, economic development subsystem, water environment supply subsystem and water environment carrying subsystem. Among them, pressure indicators reflect a range of factors caused by human activity to the system, status indicators show the current status of the system, and the response indicators reflect human's positive efforts to further improve the system. Three types of indicators have differences and relations, jointly push forward the development of the system (see Figure.1)

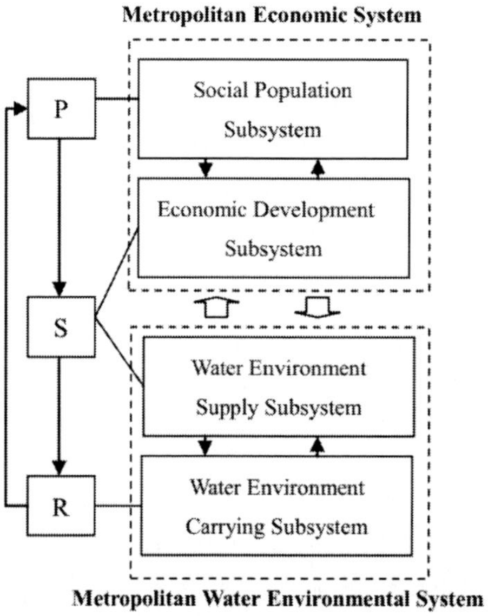

Fig.1. PSR running mechanism and coupling relationship diagram.

2.2 Data Sources

This paper takes Beijing as one of the most representative metropolises in China. All the data are obtained from "Beijing Statistical Yearbook 1997-2013" and relevant water resources bulletins.

2.3 Data Processing

2.3.1 Data preprocessing

Since units of the index in evaluation system are different, we must first carry out data preprocessing in order to compare between different indicators and determine the weights, i.e. for each data dimensionless normalized. This paper uses the maximum difference normalization method for further data processing. There are two types of evaluation, efficiency index and cost -based index. Efficiency index refers that the attribute values show a positive correlation with coordination degree of economy and environment, the greater its value, the better. While the cost-based indicators, by contrast, refers to the negative correlation between its attribute values and coordination degree of economy and environment [2-3].

efficiency index $$X'_{ij}=\frac{X_{ij}-\min(X_j)}{\max(X_j)-\min(X_j)} \tag{1}$$

cost -based index $$X'_{ij}=\frac{\max(X_j)-X_{ij}}{\max(X_j)-\min(X_j)} \tag{2}$$

In this formula, X_{ij} means the original value of index j in year i, X'_{ij} is the normalized value, $\max(X_j)$ and $\min(X_j)$ refer to the maximum and minimum value of index j.

2.3.2 Mean square difference method to determine the index weight

In this paper, an objective mean square difference method is used to determine the index weights. And in order to eliminate the impact of different indicators means, the concept of variation coefficient is introduced.

$$CV_j=\frac{\sigma_j}{\overline{X}_j} \tag{3}$$

$$W_j=\frac{CV_j}{\sum_{j=1}^{n}CV_j} \tag{4}$$

In this formula, CV_j represents a variation coefficient of index j, σ_j says the average variance of index j (i.e., standard deviation), and $\overline{X}_j$, W_j respectively refer to their mean value and weight value.

2.3.3 Metropolitan economy and metropolitan water environment comprehensive index calculation

Multiply the standardized values of each index by the corresponding weights and come to the comprehensive index of metropolitan economy f(x) and the comprehensive index of metropolitan water environment g (y).They respectively refer to the Metropolitan comprehensive development of economic systems and water environmental systems.

$$f(x)=\sum_{i=1}^{m}X'_{ij}W_x \tag{5}$$

$$g(y)=\sum_{i=1}^{n}Y'_{ij}W_y \tag{6}$$

2.3.4 Coordination degree analysis

Coordinate degree is a quantitative indicator used to reflect the degree of harmony between systems in the process of development. Comprehensive evaluation index is a measure of the overall function and benefits of the parent system. But in some cases there will be two parent system, such as A and B, whose coordinate degrees are the same while comprehensive evaluation indexes are different from each other. In this case, even if they have the same coordination degree, it is also difficult to show that their coordinated development situations are consistent.

$$C=\left\{\frac{f(x)\times g(y)}{\left[\frac{f(x)+g(y)}{2}\right]^{2}}\right\}^{k} \tag{7}$$

$$T=\alpha f(x)+\beta g(y) \tag{8}$$

$$D=\sqrt{C\times T} \tag{9}$$

In this formula, C is coordination degree, K means the adjustment coefficient (the k=2),T is the comprehensive evaluation index of urban economy and the environment system, α and β coefficients to be determined. In this paper the development of metropolitan economy and Metropolitan water environmental protection are equally important, so take α = β = 0.5. D refers to the coordinated development coefficient [4-7].

3 Results and analysis

According to the above steps , the data processing results can be obtained in Table 1 .f(x) means comprehensive index of economic system, g(y) means comprehensive index of water environmental system and D means coordinated development coefficient.

Table 1. The results of the coordinated development of economic and water environmental systems in Beijing (1997-2013).

Year	f(x)	g(y)	D
1997	0.1177	0.2268	0.0072
1998	0.1196	0.2185	0.0071
1999	0.1625	0.2331	0.0120
2000	0.1966	0.3357	0.0239
2001	0.1861	0.4104	0.0277
2002	0.2379	0.5014	0.0504
2003	0.2168	0.5069	0.0444
2004	0.3532	0.5891	0.1102
2005	0.3797	0.6968	0.1501
2006	0.4336	0.7053	0.1859
2007	0.6539	0.7400	0.3649
2008	0.6603	0.7370	0.3685
2009	0.7057	0.7607	0.4231
2010	0.6723	0.7610	0.3937
2011	0.8801	0.7607	0.5894
2012	0.9094	0.7684	0.6266
2013	0.9829	0.7657	0.7011

DOI: 10.1051/matecconf/201710005080

3.1 The comprehensive index analysis of economic system in Beijing

From the above results, we can see that the economic performance of Beijing in 1997 to 2013 shows a stable growth trend (see Figure 2). The economic system composite index increased from 0.1177 in 1997 to the highest Value of 0.9829 in 2013.

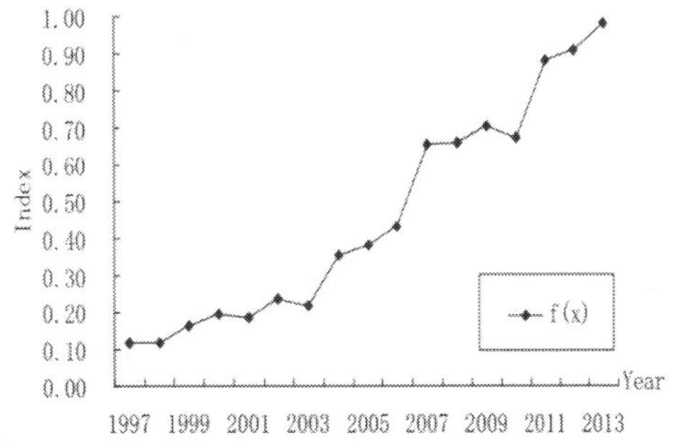

Fig.2. The comprehensive index trend chart of economic system in Beijing (1997-2013).

3.2 The comprehensive index analysis of water environmental system in Beijing

The data of the water environment system from 1997 to 2013 in Beijing were processed, and the water environmental system comprehensive index also showed a simple overall upward trend (see Figure 3). The lowest value of water environment index appeared in 1998, the index value was 0.2185, while the highest value appeared in 2012, the index value was 0.7684.

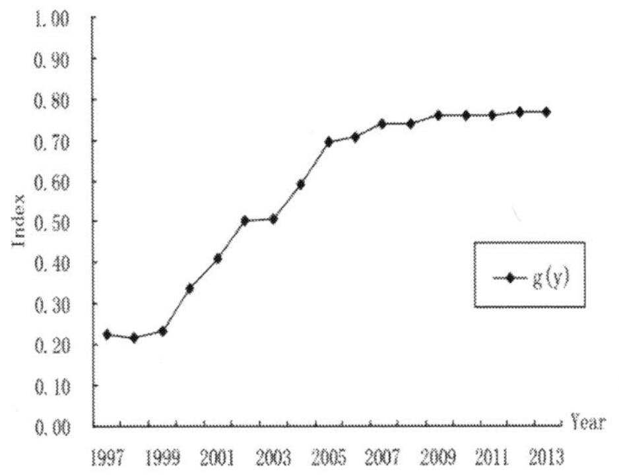

Fig.3. The comprehensive index trend chart of water environmental system in Beijing (1997-2013).

3.3 Coordination degree analysis between economic system and water environmental system in Beijing

According to the results of data processing in Table 2, the trend of the coordinated development coefficient of economic and water environmental systems in Beijing is plotted (see Figure 4). It can be seen from the figure that during 1997 to 2013, the relationship between Beijing's socio-economic and its protection of the coordinated development of water environment was good. The degree of coordination between them showed an overall upward trend, and the degree of coupling continuous optimization. The coefficient of coordinated development of economic system and water environmental system in Beijing increased from the lowest value of 0.0071 in 1998 to the highest value of 0.7011 in 2013.

DOI: 10.1051/matecconf/201710005080

On the basis of summarizing the existing research results, this paper divides the coupling states into 5 different stages from the serious uncoordinated stage to the high quality coordinated stage (see Table 2).

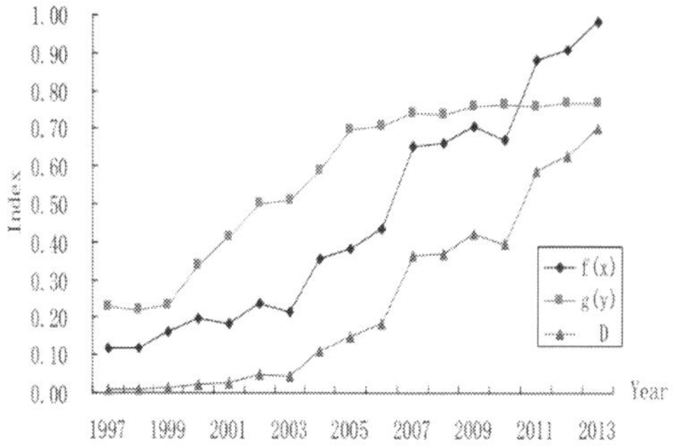

Fig.4. The coordinated development coefficient of economic system and water environmental system in Beijing (1997-2013).

Table 2. The classification standard of coordinated development coefficient.

Coordinated development coefficient（D）	Standard
0≤D≤0.1	Serious uncoordinated stage
0.1<D≤0.3	Uncoordinated stage
0.3<D≤0.5	Basic coordinated stage
0.5<D≤0.7	Good coordinated stage
0.7<D≤1	High quality coordinated stage

According to the classification criteria, the coordinated development coefficient of economic system and water environmental system in Beijing from 1997 to 2013 is divided into the following stages [8]:

(1)Serious uncoordinated stage (1997-2003)

In the past, the development of water conservancy in Beijing has been neglected due to the high attention paid to economic growth and urbanization by the state and the Beijing municipal government. Serious water pollution happened caused by the weak sewage treatment ability of Beijing. So the coordinated development coefficient of economic system and water environmental system in Beijing was so poor in a very long period of time, it was in the serious uncoordinated stage.

(2) Uncoordinated stage (2004-2006)

In 2004, with the success of China's accession to the WTO and the accelerated adjustment of the national economic structure, Beijing's economic development also ushered in a new opportunity to optimize the economic structure. It eliminated high water consumption and high pollution industries. At the same time Beijing increased investment in water pollution control, its water pollution treatment capacity also increased to a certain extent. However, due to the lack of investment in water resources, sewage discharge still increased. The coordinated development coefficient of economic system and water environmental system in Beijing was in the uncoordinated stage

(3) Basic coordinated stage (2007-2010)

After 2007, with the Beijing Olympic Games approaching, the environmental protection of Beijing was imminent. The investment in the field of water environmental protection increased significantly, and the adjustment of industrial structure was also further accelerated. Beijing has introduced a large number of advanced sewage treatment technology, the daily sewage treatment capacity and sewage treatment rate improved significantly. During this period, the coordinated development coefficient of economic system and water environmental system in Beijing was in the basic coordinated stage.

(4) Good coordinated stage (2011-2012)

From 2011 to 2012, the construction of energy-saving economy in Beijing was increasingly perfect. During this period, Beijing's per capita GDP continued to increase, investment in water supply increased significantly, while the industrial water consumption and domestic water consumption remained basically unchanged. The degree of coordination between the Beijing economic system and the water environmental system was further improved, and it was in a good coordination stage.

(5) High quality coordinated stage (2013)

In 2013, the scientific concept of development has been subjected to more attention by all walks of life. Beijing's economic level of urbanization, per capita GDP and many other indicators reached the maximum in recent years. At the same time, Beijing's water pollution treatment capacity continued to improve, industrial water, domestic water demand was stable. The coordination degree between the Beijing economic system and the water environmental system reached the optimal state and was in the stage of high quality coordination.

4 Conclusions

(1) In the process of metropolitan development, the economic system and the water environmental system are interrelated and there exist a complex coupling mechanism.

The economic system and the water environment system are two basic systems which constitute the metropolitan system, and there exist a complex coupling mechanisms. Metropolitan economic system consists of two subsystems: population society subsystem and economic development subsystem. Metropolitan water environment system consists of water environment supply subsystem and water environment carrying subsystem. There is a "pressure-state-response (P-S-R)" relationship among the four subsystems. The operation of metropolitan economic system transports a large number of industrial and domestic wastewater to the water environment system, and also has brought greater pressure to it. At the same time, the metropolitan water environment system provides the necessary water resources for the further development of the economic system, and provides the water environment feedback to the metropolitan economic system [9].

(2) With the adjustment of economic structure and the progress of technology, the coupling index between the metropolitan economic system and the water environmental system has been increasing, and this process shows an upward trend

China's economic and social development presents a typical dual structure. Metropolis, as a priority area for economic and social development in China, enjoys all kinds of policy dividends and makes full use of all kinds of resources, so it develops rapidly. But it the early time, due to the Government's one-sided emphasis on economic construction, environment especially the water environment suffered serious pollution. The coordinated development coefficient of economic system and water environmental system in was in the uncoordinated stage. With the optimization and adjustment of China's economic structure, the whole society has increased investment in water pollution control. Metropolis enhanced the sewage treatment technology, and reduced water consumption. In this period, the coordination degree between the metropolitan economic system and the water environmental system is gradually increasing, and a rising trend appears.

Acknowledgments

This paper is supported by ①National Natural Science Foundation of China (No: 51279058 & No: 41001377); ②Graduate students' Program of Scientific Research and Innovation in Jiangsu Province (No: KYZZ15_0158); ③The Joined PhD student program funded by China Scholarship Council (No: 201506710057).

References

1. C.Z.Wang and Y.Liu. A Review on the Harmony of Economy and Environment Research, China Popu. Res. Envir. **12(3):**32-36 (2002).
2. S. Asci, T. Borisova and J. J. VanSickle, Role of economics in developing fertilizer best management practices, Ag. Water Manage. **12(3):**251-261 (2015).
3. D. Blandford, J.B. Braden and J.S. Shortle, Economics of Natural Resources and Environment in Agriculture, Encyc. Ag. Food Syst. **122:** 18-34 (2014).
4. X.Q.Zhang and P.J.Shi. The Assessment of Urban Ecosystem Health Based on PSR Model--A Case Study of Lanzhou City, J. Rid Land Res. Envir. **24(3):**77-82 (2010).
5. C.H.Zhu and X.T.Zhan. Study on the Environmental Kuznets Curve, Ecol. Econ. **250(3):**48-53 (2012).
6. M.Chen and C.X.Xu. Study of Urban Economic and Environmental System Coordinated Development Measure, Inter. J. App. Envir. Sci. 9(4): 1603-1613(2014).
7. K.Feng, C.F.Wu and Y.Liu. Spatial Difference of Urban Land Intensive Use in Zhejiang Province--A Case Study Based on The Principal Component Analysis and PSR Model, China Soft Sci. **2:**102-108 (2007).
8. J.M.Spiegel. Developing Ecosystem Health Indicators in Centro Habana: A Community-based Approach, Ecos. Health, **7(1):**15-26 (2001).
9. M.Chen. Study on Relationship Between Changes of Industrial Structure and Economic Growth in Jiangsu Coast Areas, Econ. Res. Guide, **132(22):**168-169 (2011).

A Discussion about American Mathematics Education from an Administrative Perspective

Tao Wang[1] and Miaomiao Yang[2]

[1]Base Depart, Zhengzhou University of Science & Technology
[2]Zhengzhou Institute of Technology, Zhengzhou, China

Abstract. In this paper we reflect on the roles and responsibilities of academic leaders in encouraging faculty in mathematics departments to value contributions to mathematics teaching and learning. We discuss how academic leaders can and should use their perspective, position and influence to encourage productive dialogue between practitioners of mathematics and mathematics education; to use assessment of student learning as an opportunity to further this dialogue; and to value and reward work in mathematics teaching and learning in the hiring, evaluation, tenure, promotion, and merit processes.
Keywords. Assessment, mathematics education, scholarship of teaching and learning.

1 Introduction

Faculty at American colleges and universities are responsible for teaching, research, and service, or what the American Association of University Professors describes as "student-centered work," "disciplinary- or professional-centered work ," and "community-centered work" [1]. Academic leaders in today's colleges and universities, especially deans, are responsible for supporting this tri-partite work of the faculty with the overall goal of promoting excellence in their institutions and advancing institutional mission. As such, an academic leader must have a future-oriented perspective; take a wide, cross-campus view; and prioritize the support and nurturing of activities that contribute to the institution's broad and strategic goals .

Here we discuss how academic leaders can use their perspective, position, and influence to encourage and value the work of mathematics faculty in mathematics education. We regard the definition of mathematics education and the associated contributions to be intentionally broad and to encompass work in pedagogy, curricula, and outreach, as well as research in mathematics education. We focus primarily on approaches to supporting the professional-centered work of the mathematician in mathematics education, as this connects more closely with our backgrounds and experiences. We also provide examples to illustrate what academic leaders actually do and what results they achieve. An important note: for convenience, we will often refer to the academic leader as the dean. In reality, the academic leader could be a department chair, a program director, a division head, an associate dean, an associate or vice president for academic affairs, a provost, or even a president. What matters here is that the individual is a respected leader, has some financial discretion and some influence, and possesses a viewpoint that can encompass both local issues and global issues.

We consider the role of academic leaders in facilitating productive interactions and discuss their contributions to this area. We also address the increasing emphasis on assessment of student learning in higher education and the relationship between assessment and disciplinary-centered work. Finally we discuss contributions to the teaching and learning of mathematics in the context of a faculty member's professional and career development, with particular attention to how these contributions may be evaluated. Throughout the paper, we highlight the important role of academic leaders in supporting and valuing all forms of contributions to mathematics education.

2 Mathematics education: as facilitation of productive interactions

Our fundamental conviction is that practitioners in mathematics and mathematics education have much to learn from one another; in fact, these practitioners could and should be extended to include in-service teachers, education faculty, psychology faculty, and policy makers. We acknowledge the difficulties of past interactions as summarized. But we agree with Ralston and others that cooperation is essential for real progress to be made in K-12 and, therefore, post-secondary mathematics education in America. A significant challenge in this context is that research mathematicians have not always demonstrated an understanding of, or appreciation for, the nature of work in mathematics education. Ball and Forzani addressed this challenge and noted: One impediment is that solving educational problems is not thought to demand special expertise. Despite persistent problems of quality, equity, and scale, many Americans seem to believe that work in education requires common sense more than it does the sort of disciplined knowledge and skill that enable work in other fields. Few people would think they could treat a cancer patient, design a safer automobile, or repair a bridge, for these obviously require special skill and expertise. In mathematics departments, this misunderstanding frequently results in an undervaluing of the work performed by mathematics faculty whose focus has shifted to mathematics education. As Mc Callum diplomatically put it, "Collaborative efforts between mathematicians and mathematics educators are sometimes hampered by a general lack of mutual respect between the two fields". Hyman Bass articulated the challenges further, arguing that there are two common myths regarding research mathematicians becoming involved in mathematics education. Mathematicians promulgate the first myth, sharing "... a common belief ... that attention to education is a kind of pasturage for mathematicians in scientific decline." Educators are responsible for the second myth, with doubts about "... the relevance of contributions made by research mathematicians, whose experience and knowledge is so remote from the concerns and realities of school mathematics education". Bass acknowledged that mathematics and mathematics education are not the same, but that "productive interactions" between these fields can exist. Mathematics departments should bear the primary responsibility for supporting these collaborations and "productive interactions." Sometimes, however, they need help in initiating or sustaining these efforts. The silo-like nature of today's higher education makes this a challenge. The prevalent organizational structure in US higher education is one of departments within colleges or schools. This somewhat vertical structure supports discipline-based teaching and research, but isolates departments and inhibits collaboration. As a result, faculty may not see opportunities to work with colleagues in other departments or in other areas of the institution.

3 Academic leader working on collaborations

Our experience is that an academic leader who resides outside the department and who has an understanding of the need for cooperation across different areas of the university can bring faculty together around projects that may lead to deeper collaborations. We have seen this at several different institutions. In each of the following examples, an academic leader identified encouraged, or brought together faculty members from across the institution to work on a common project. In several cases, the results went beyond the immediate project to include the deepening of interdisciplinary understandings, the implementation of curricular change, or the advancement of new research partnerships.

(1) Education faculty wanted to develop a graduate degree in education for in-service high school mathematics teachers. Because such a degree needed to have a significant amount of mathematics content (both as a good practice and as a requirement for advanced certification), the input of mathematics faculty was vital. The resulting collaboration led to the development of a joint graduate degree [2]. In a different institution, faculty in the mathematics department sought out a collaboration with faculty in education to design a subject-based master's degree in mathematics education for in- or pre-service teachers. The resulting discussions brought mathematics and education faculty together in new ways, leading to other joint projects that included grant proposals and curriculum development [3]. With support from a federal

grant, mathematics and education faculty at a 4-year institution and a community college met over the period of a year to compare syllabi for first- and second-year mathematics courses, discuss student success concerns, and review transfer policies. Mathematics and mathematics education faculty served together on a state-wide committee charged with evaluating mathematics placement testing and the role of developmental mathematics in public higher education in that state [4]. A mathematics department chair initiated a collaboration with education faculty and K-12 teachers, supported by National Science Foundation funding, to vertically bridge the school curriculum to research-level mathematics. The innovative partnership benefitted graduate students and faculty at the institution, as well as teachers and students in local K-12 schools, and provided a model for other institutions.

How can deans help? The nature of their role is that they take a cross-institutional perspective. This perspective gives them insights into connections and opportunities, as well as synergies, across a college or university. With this perspective, deans bring together groups of faculty to initiate new programs or build out areas of potential strength. In this context, collaborations that connect mathematics faculty and mathematics education faculty, or that support mathematics faculty with research interests in teaching and learning, should be connected to specific programs (such as the graduate education programs mentioned earlier) or focused to support assessment of student learning (more on that below), student retention goals, or revisions of academic support services. Bennett's chapter in this volume describes work with education faculty who had concerns about a required course for a mathematics education program. The request to replace this course with a "mathematically rigorous capstone course for secondary mathematics teachers that would make explicit connections between college and high school mathematics" [5] gave Bennett the opportunity to not only think carefully about what these students should learn, but to begin a research project in the scholarship of teaching and learning.

Sultan and Artzt presented an example of how faculty from two different departments, with two different worldviews, can collaborate. What began as an initial discussion about the preparation of secondary mathematics teachers led to National Science Foundation funding for a project to recruit high school seniors into a mathematics teacher preparation program. As they noted, "If there is one thing that we have learned it is that collaboration is a complex process. We have to be willing to learn from each other, we have to respect each other, and we have to be willing to change". Holm's chapter in this volume echoes this sentiment. She wrote, "I believe strongly that the mathematical sciences community must maintain the bridges between researchers in mathematics education and practitioners of mathematics education, particularly at the post-secondary level. Moving forward we need to improve our communication and collaboration." [6].

4 How to build productive interactions

The call for assessment of student learning in higher education has grown across the United States over the last decade. The use of business terminology in this context, such as "return on investment" and "value-added," is increasingly common at the administrative level and this usage suggests that non-academic models and approaches are being imposed on higher education .Colleges and universities are under tremendous pressure to not only articulate student learning outcomes in general education and in the major, but to assess progress toward these learning outcomes and to demonstrate that this progress is a direct result of the educational experience. These pressures come from federal and state agencies including the United States Department of Education, the media and the public. At the departmental level, regular program reviews may include an expectation for assessment of the curriculum, of student learning, and of other areas such as tutoring and other academic support. While the immediate pressure for assessment and departmental reviews may be external, or from the higher administration, there are potential benefits to a department that engages seriously in this work. The guide-lines prepared by the Mathematical Association of America's Committee on Departmental Reviews frames a self-study as an opportunity for renewal, suggesting that, "...it provides an opportunity for members of a department to move forward together with a shared understanding, a shared set of goals , and a shared commitment".

5 The role of faculty in the assessment of student learning

Faculty may rightly argue about the emphasis on assessment in today's higher education world, raising questions about time and effort along with the concern that "the entire premise of 'assessment to improve instruction'—especially if it is offered by outsiders—is that there is something wrong with instruction to begin with" [7]. But when faculty have significant roles in assessment, the results can be meaningful. Discussions about assessment should become conversations about student learning, which then should become transformative conversations about curricula and teaching. As Steen noted in his introduction to Gold et al., "Assessment not only places value on things, but also identifies the things we value". The value of assessment itself is described by Pat Hutchings of the Carnegie Foundation for the Advancement of Teaching, as follows: "... the real promise of assessment—and the area in which faculty involvement matters first and most—lies precisely in the questions that faculty, both individually and collectively, must ask about their students' learning in their regular instructional work: what purposes and goals are most important, whether those goals are met, and how to do better".

Deans should use assessment expectations (or mandates) as opportunities to bring faculty together to have discussions to identify what is valued, to ask questions about students learning within the context of their discipline, and to think deeply about course and curriculum goals .The assessment discussions also present an opportunity for deans to demonstrate to faculty involved in mathematics education that their work is valued. This group not only focused on the student learning outcomes in designing the course, but thought care-fully about the pedagogy and the learning environment. Catepillán's chapter in this volume outlined a different challenge: how to use her background in ethno mathematics to develop a mathematics course that also met the student learning outcomes to qualify as a diversity course. Deans play a role in supporting these discussions and this work in multiple ways.

The most fundamental should be the encouragement, if not the expectation, for broad faculty involvement and leadership in the development and assessment of student learning outcomes. In the context of mathematics, this assessment work presents a very natural opportunity for mathematicians with interests in and experience with mathematics education to take meaningful and significant roles. Deans should provide financial support to begin and sustain these activities. This can take several forms beyond offering stipends or course releases for assessment work; the results often last well beyond the initial activity. We have seen faculty who receive funding to attend conferences and workshops on the assessment of student learning and the scholarship of teaching and learning return energized and excited. This enthusiasm can be infectious, leading to real change in teaching and learning at the institution. In some cases this engagement revitalizes a faculty member's research interests, leading to new work in the scholarship of teaching and learning.

Deans should also lead conversations about recognition and rewards for faculty work in assessment (see below for a broader discussion of this). This can take many forms: providing guidance to hiring committees as they write position descriptions; supporting graduate program directors in the incorporation of assessment into the preparation of graduate students (see Chap. 6 in this volume by Lai et al. for more information about professional development for mathematics graduate teaching assistants); working with faculty to incorporate student self-assessment (in the form of an e-portfolio, for example) into the curriculum ;and leading campus-wide discussions about assessment as scholarship and as part of the departmental and institutional reward system . These discussions may fi t naturally within a larger campus discussion about scholarship .In 1990, Boyer called for a rethinking of how higher education institutions prioritize the "activities of the professoriate". He argued for a new, shared vision of scholarship , one that defines "the work of the faculty in ways that reflect more realistically the full range of academic and civic mandates ". These discussions are as "vital" now (in Boyer's terminology) as they were more than 25 years ago.

6 Faculty professional and career development

Many faculty (ourselves included) tend to view faculty careers in a traditional progression: first finish graduate school, then perhaps have a post-doctoral position or other short-term experience, then move to a tenure -track position as an assistant professor. From there, the steps are tenure, promotion to associate professor, and then promotion to full professor. (And as we can attest, a move to the administration may be part of a faculty member's academic journey.) Of course there are many versions of this progression, and alternate career paths could include an appointment as a full-time lecturer or instructor (perhaps a " professor of the practice ," as the teaching-intensive positions are commonly called), working in industry or in government before taking a position as a faculty member, or laboring as an adjunct at several different institutions. Regardless of the path, an individual faculty member's professional development should never stop. Department chairs and deans should not only recognize this, but should work actively with faculty at all stages of their career to help them reflect on their own professional development and to set short- and long-term goals. Bremser's chapter in this volume provides a compelling reflection of how faculty may naturally change focus and direction as their careers develop. Educated as a research mathematician, Bremser described how she "...began to direct more intellectual energy toward educational issues, with the explicit goal of finding constructive ways to get involved". This led to participation in a workshop on social justice, which in turn led to scholarly engagement in a number of new and unexpected ways.

7 Conclusion

The program grew out of an identified need to improve the mathematical and pedagogical content knowledge of local middle school mathematics teachers. The approach meant doing something new: designing an interactive and in-depth experience of mathematics for these teachers using the Math Circle model. As the authors noted, the impact of this work went beyond the audience of middle school teachers. All three authors benefitted from the experience and saw positive impacts on their careers. It also benefitted the department, since the authors brought their reflections about the program back to the department through discussions at department meetings.

References

1. AAC&U. (2016). Trends in learning outcomes assessment: Key findings from a survey among administrators at AAC&U member institutions. Resource document. Association of American Colleges and Universities. Retrieved April 15, 2016.
2. AAUP. (n.d.). What do faculty do? Resource document. Retrieved March 30, 2016.
3. Ball, D., & Forzani, F. (2007). What makes education research "educational"? Educational Researcher, 36 , 529–540.
4. Bass, H. (2005). Mathematics, mathematicians, and mathematics education. Bulletin of the American Mathematical Society, 42 (4), 417–430.
5. Boyer, E. (1990). Scholarship reconsidered: Priorities of the professoriate. San Francisco, CA: Carnegie Foundation for the Advancement of Teaching. Jossey-Bass.
6. Gold, B., Keith, S., & Marion, W. (Eds.). (1999). Assessment practices in undergraduate mathematics. Washington, DC: The Mathematical Association of America.
7. Hutchings, P. (2010). Opening doors to faculty involvement in assessment. National Institute for Learning Outcomes Assessment. Occasional Paper #4. Retrieved March 30, 2016.

Hot Charging and Transporting to Continuous Discussion on Casting Slab at Tanggang

Jun Dai

Mechanical and electrical engineering Department, TangShan college,Tang Shan 063000,Hebei, China

Abstract. This paper introduces technological design and equipment of hot delivery and hot charge in Tanggang and Steel Co and its notable economic benefit.At the same ,the difficulty and problems in realizing hot delivery and charge are point out.

1 Foreword

Hot charging and hot delivery process of continuous casting billet has the advantages of saving energy, improving the production capacity of heating furnace, reducing the burning loss of billet, decreasing the rolling fault, improving the rate of finished products, speeding up the logistics movement, reducing billet storage warehouse and liquidity occupation etc. It is widely used in many iron and steel enterprises [1].The basic conditions for the realization of hot delivery and hot charging of continuous casting billet is [2]:

1) Continuous casting process of high temperature defect-free billet production technology;
2) The production capacity of steel rolling can match the production capacity of steel making;
3) The process of steel making, continous casting and steel rolling is stabe and efficient;
4) The shape,size and temperature of the billet can meet the rolling product varieties and specifications;
5) Set up a throughout production management and quality assurance system which can realize process integration.

The continuous casting billet hot delivery form that Tanggang originally used is hot delivery and hot charging rolling.First,we need putting the high temperature cleaned billet into the soaking pit offline ,then dispatching billet from the soaking pit according to the rolling plan and placing them into the heat furnace (charging temperature is about 400℃-700℃) to heat,and then rolling.In this way, it cost much time for the hot delivery and has a great loss of the slab temperature, it also can't make full use of the heating furnace production capacity. In view of this situation, technical personnel redesigned a new equipment to realize the continuous casting billet hot delivery and hot charging rolling. It can directly put the high temperature cleaned billet into the heat furnace (charging temperature is about 400℃-700℃) to heat according to the order of billet, and then rolling. It removed the soaking pit buffer storage between the steel making——continuous casting and steel rolling, only retained the heating furnace as the production rhythm buffer means of regulation [3]. Compared with the original hot delivery and hot charging rolling, direct hot charging rolling improved the degree of continuous and the energy saving effect of the steel making——continuous casting and steel rolling, truly realizing the continuous casting billet hot delivery and hot charging.

2 Hot delivery and hot charging process layout of the continuous casting billet

Tanggang bar plant can match with the second Tanggang steel making plant of on its third stage continuous casting process. It is very suitable for hot delivery and hot charging process of the continuous casting billet. From the layout in Figure 1, we can see that the 6# continuous casting machine of the second Tanggang steel making plant and the heating furnace steel loading station of Tanggang bar plant are propped in the same bay, getting a good natural conditions for heat delivery. In order to realize hot delivery and hot charging of the continuous casting billet,we set up a special roller tablel for continuous casting billet hot delivery between the terminal of the conticaster ejection roller table and the inlet of the heating furnace in the design process, so that the conticaster can integrate with the rolling system, forming a new production line. The new production line retained the whole original canticaster system and the heating furnace steel charging system, making full use of the function of the billet transfer car of the conticaster and the roller table of the heating furnace emtrance. Not only can it save the project investment, but also it make the whole production line more flexible. Under normal circumstances, we can use the heat delivery roller table for hot charging, under special circumstances, we can use the original system to product cold billet and conduct cold charging.

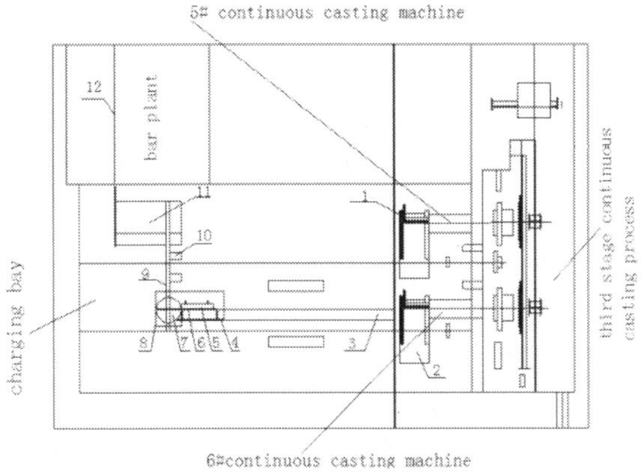

Fig.1. Process plan.

1— Slab transfer car 2—Step by step turn cooling bed
3—Hot roller 4—Weighing roller 5—chain elevator
6—The transition table 7—rotary disk 8—dead plate
9—The heating furnace roller 10—With steel table
11—heating furnace 12—main rolling line

To guarantee the continuous casting billet hot charging temperature, the time to send heat is critical. The shorter the time is , the smaller the casting billet's temperature drop. The temperature of the casting billet when it arrives the hot delivery track is about 850℃, and the temperature to ensure hot charging is 700℃, the delivery time should be less than 10 minutes[3]. The distance from the third phase of continuous casting to bar factory is 140m. The slab is delivered from hot delivery roller bed 3 to 7, raised 5 meters up by chain hoist 4 to transition roller bed 3 and delivered to turntable 4. The casting blank is rotated 90°and put into the hot stove [4]. In order to finish this process within 10 minutes, we adopted the following technical measures in our design:

1) Instead of using single roll to drive the hot roller, we divide the rollers into 12 groups. Each group uses drive. To achieve stable operation and reduce noise, the speed is set to 1.5m/s. In this way it takes only 1minute and a half which save the investment in equipment and guarantee the slab temperature.

2) Chain hoist installs two hold claw on the chain which is shown in Figure 2. It raises the billet 5m, shipments the casting to the transition roller and increases the speed to 0.25m/s.

3) We don't use the normal gear transmission in the turntable, because the turntable needs frequent start-braking. Although the gear transmission is smooth, the turntable only rotates 90° each time. The big gear bears uneven force on the circumference of the circle and it is easily damaged in the mesh. The processing manufacturing of big gear is difficult and costs high. Thus we designed a new organization which is shown in Figure 3. The turntable' center is used as shaft, and the wheel is installed under the turntable. The wheel is driven by the motor and operates on the circular orbit. The turntable is designed to be able to turn positive and negative 90° whose rotation angle is controlled by the proximity switch.. This design can solve the problem of cable winding, which has simpler structure compared with gear transmission structure. It also has the advantage of less investment, easy maintenance and performs well in practice.

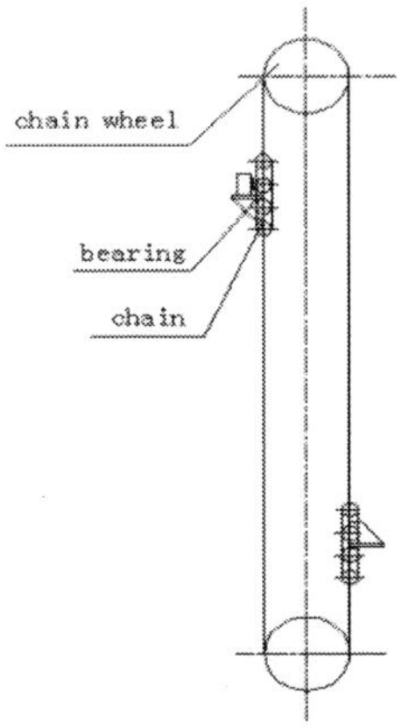

Fig.2. Schematic diagram of chain hoist.

3 Other problems

Because the whole hot charging line control is a very complicated system. Each production link is closely related to another. The disorder of a target value will influence the global process [5-6]. To ensure the realization of the hot charging process, the following problems should be concerned.

1) Steelmaking, continuous casting and steel rolling workshop should be scheduled together, thus the maintenance of two workshop can be carried out at the same time.

2) Mill roll change should better be done between mill's minor repair time. In general, the change of roll and groove should also be arranged in the time when the continuous casting billet is not offering heat or in the gap of continuous casting.

3) The management of steelmaking, continuous casting and steel rolling workshop should be strengthened. Operation should be standardized to guarantee the stability of production.

4) Set in the heating furnace or before the heating furnace of steel rolling the billet insulation and caching facilities in order to improve the rate of hot charging.

When estimating the casting billet hot charging rate, in addition to consider the stable relation of production per hour between continuous casting and rolling, we should also consider the working systems of two workshops. There are many factors influencing the production. If productions of two workshops match well, the hot charging scheme is reasonable, the enterprise's management level is high, the continuous casting billet hot charging rate can reach 80% - 90%.

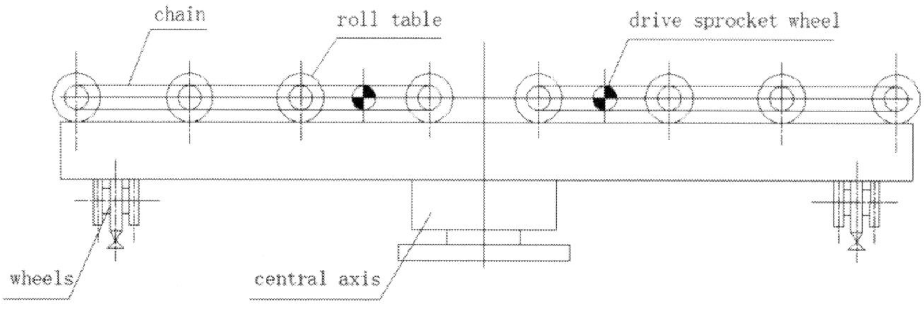

Fig.3. Turntable structure diagram.

4 Conclusion

It usually takes 2~3 hours from the casting blank of Tang Gang 3 period continuous casting being casted to being poured out to the furnace, and the degree decreases a lot. By adopting this set of equipment, it only takes 4~5 minutes from the third phase of continuous casting to bar feeding platform, and the temperature when the billet is in the furnace can reach 700℃ above, or even higher. Assuming the hot charging temperature is 700℃, the hot charging rate is 90%, then each ton of steel can save 8kg heavy oil. Therefore, the benefit of it is obvious and the efficiency is greatly improved.

Production practive proves that the equipment has simple structure, less investment, reliable and stable operation which realizes the process of hot charging.

References

1. Liu Ruining, Li-wu zhang, Chen Gongwei. Continuous casting billet hot delivery practice hot charging process. Iron and steel.2000(08).
2. Ji wei. Casting billet hot delivery directly hot charging. Tianjin metallurgical.2001(05).
3. YiPeiLing. Small wire production technology of casting billet hot delivery and hot charging. Heavy Machinery Science and Technology.2001(Z1).
4. Fan Jing Biao, Fande Zhong, YANG Shu-sen. Billet hot charging of practice. Baotou Iron and Steel Technology.2004(05).
5. Luo Hui. Guangzhou Iron and Steel Plant to achieve billet hot charge of the plant to discuss. Metallurgical Collections. 2001(04).
6. Tsai H T.The Characters and Improves of Cold Rolled Strip Defect. Steelmaking Conference Proceedings . 1990.

Application of Fuzzy Borda Combination Evaluation in Subjective and Objective Evaluation Methods in Comprehensive Evaluation of Population Development

Jiajia Yu[1], Hongqi Ma[2] and Yuqi Han[3,*]

[1]Construction Management and Real estate Department, Chongqing JianZhu College, Chongqing, China
[2]Agricultural Economics Department, China Agricultural University, Beijing, China
[3]Construction Management and Real estate Department, Chongqing JianZhu College, Chongqing, China
Corresponding author: 43222240@qq.com

Abstract. This paper constructs the comprehensive index evaluation of population development from the three aspects of human development, population and economic society and population, resources and environment, according to the movement course and changing trend of population development at home and abroad, and the guiding ideology of national population development strategy system.The index system consists of 3 first level indexes, 8 Secondary indexes,22 tertiary level indexes and 42 fourth level indexes.And on the basis of commonly used comprehensive evaluation methods. The comprehensive evaluation of the population development in China in 2013 was made by using the combined evaluation method, which effectively overcomes the shortcomings of the objective evaluation method and the subjective evaluation method, and obtains a more credible comprehensive evaluation value.

1 Introduction

China is the world's most populous country. The US Bureau of Population and Human Services published the 2010 WORLD POPULATION in July 2010 DATA SHEET "shows that in the foreseeable 2050 China will be 1.437 billion population to become second only to India, the world's second most populous country [1]. Population has always been China to consider all the problems of the starting point and end point, the population problem is always a key factor restricting the development of sound and fast. As the most populous developing country in the world, by the 1950s to encourage childbearing and 70 years after the birth control of the population policy of the two major impact in less than 30 years, the population reproduction type from the "high birth, low death, high Growth "to" low birth, low death, low growth. " The change of the type of population reproduction in western developed countries is the natural result of economic and social development and modernization; the main body of population development as modernization is the realization of the modernization of economy and society. Modernization process. In the past 30 years of reform and opening up, industrialization,marketization,urbanization,informatization and economic globalization have promoted China's rapid economic and social development. In the past 30 years, China's industrialization and marketization, urbanization, informationization and economic globalization have promoted China's rapid economic and social progress. In the transitional period, the population development lags behind the economic and social development and faces an unprecedented complex situation. Population quality, the transfer of rural surplus labor force, the problem of population mobility, the problem of young labor shortage, population aging, population poverty, social security, gender imbalance, the one-child problem, population development and economic society, resource and environment And many other problems appear. The national population development strategy put forward: Adhere to the people-oriented, give priority to investment in the comprehensive development of human beings, realize the transition from the big population to the powerful country of human capital and realize the coordination and sustainable development of the population and the economic and social resources and environment. The 11th Five-Year Plan for Population Development and the 2020 Plan clearly defined the guiding ideology of population development: "With Deng Xiaoping Theory and the important thinking of the 'Three Represents' as the guide, comprehensively implement the scientific concept of development, adhere to the requirements of building a harmonious socialist society, People-oriented, promote institutional innovation, give priority to investment in the overall development of the people. Stable low fertility, improve the quality of population, improve population structure, guide the population reasonable distribution, guarantee population security .Promotional population to the human capital power transformation, promote population and economy, Social, resource, environmental coordination and sustainable development. "The implementation of population development strategies and plans requires a scientifically constructed population development monitoring system to ensure the accuracy and effectiveness of the underlying statistical sources. It is worth noting that population development includes not only the total population and the growth of the scale, more importantly, the quality of the population, the natural environment of population activities and the improvement of the social environment. The monitoring of population development includes not only the monitoring of total population, but also the monitoring of relevant factors such as population health quality, cultural quality, population structure, population movement, economic activities of population, living environment and so on. Therefore, population development monitoring is an ambitious and complex system engineering. It not only needs to integrate the statistical path of demographic development data source, but also needs to construct a scientific and reasonable comprehensive evaluation system and monitoring model. Through the computer network system, The population monitoring objectives, to ensure that the authority of the decision-making of the population of the scientific and accurate; At the same time for the academic community in-depth study of China's population problems, revealing China's population quantity, quality, structure, mobility, and population economic activities, People and the natural environment and other inherent laws to provide a real and reliable data source; also for enterprises to predict the market, to provide decision-making reference.

2 Index system

Based on the study of population development both at home and abroad, this paper aims to promote the construction of the evaluation system of population development from the theoretical and practical perspectives. According to population size, population quality, population and economic society and population, resource and environment, four comprehensive evaluation methods of population development are constructed. The four comprehensive evaluation methods of principal component analysis, factor analysis, entropy and expert scoring are used respectively. Finally, the composite evaluation method was applied to evaluate the four comprehensive evaluation results, and the comprehensive evaluation results were obtained.

2.1 Comprehensive evaluation index system of China 's population development

From the connotation of population development, this paper makes an investigation on the indicators of population development in various functional departments (statistics, public security, development and reform, population and family planning, labor and social security, civil affairs, education, health, resources, transportation and other sectors) A total of more than 200, according to the basic principles of the index system to build the screening indicators, to build a comprehensive population evaluation index system, including the population of its own development index system, population and economic and social indicators system, population and resources and environmental indicators system three sub- system [4].

The comprehensive evaluation system of population development includes three indicators: population self-development, population and economic society [5], population and resource and environment index system, and secondary index system 8 (population size, population structure, population quality, population distribution, population and population). Economy, population and society, population and resources, population and environment); 22 index system of level 3; 42 index system of 4 levels in Table1.

Table 1. Population evaluation index system.

Level indicators (3)	two indicators (8)	three indicators (22)	four indicators (42)
Population self - development indicator system	Population size	Population total fertility rate (person)	Population total fertility rate (person)
		Natural population growth rate (‰)	Natural population growth rate (‰)
		Birth rate (‰)	Birth rate (‰)
		Mortality (‰)	Mortality (‰)
	Population Quality	Average life expectancy (years)	Average life expectancy (years)
		Public health expenditure per unit GDP (%)	Public health expenditure per unit GDP (%)
		Expenditure on Educational Expenditure per Unit of GDP (%)	Expenditure on Educational Expenditure per Unit of GDP (%)
		Literacy rate (%)	Adult literacy rate
		Average years of schooling (years)	Average years of schooling (years)
		The number of scientific and technical personnel per 10,000	population (person) the average number of scientific and

 DOI: 10.1051/matecconf/201710005083

			technological personnel per million population
		Number of Ordinary Institutions of Higher Education	Number of Ordinary Institutions of Higher Education
	Population structure	Child support ratio (%)	Child support ratio (%)
		Old-age dependency ratio (%)	Old-age dependency ratio (%)
		Gender structure (female = 100)	Gender structure (female = 100)
	Population distribution	Population density (person / sq km)	Population density (person / sq km)
		The net migration rate (‰) of the population	The net migration rate (‰) of the population
		Non-agricultural population (%)	Non-agricultural population (%)
Population and Economic and Social Indicators System	Population and Economy	GDP Population Carrying Capacity	GDP Population Carrying Capacity
		Population distribution Population density (person / sq km)	Population distribution Population density (person / sq km)
		Labor productivity (yuan / person / year)	Labor productivity (yuan / person / year)
		income disparity	Gini coefficient
			Per capita income ratio between urban and rural areas (rural = 1)
		Population employment	employment population carrying capacity
			employment rate(%)
			Female employment ratio (%)
			Tertiary industry employment ratio (%)
	Population and Social	Social Security	Old-age Insurance Rate (%)
			Medical Insurance Rate (%)
			Unemployment insurance rate (%)
			Low coverage (%)
		Infrastructural	Internet penetration rate (%)
			Road network density (km / 100 km2)
		Living conditions	per capita housing area (square meters / person)
Population and resources and environment index system	Population and resources	natural resources population carrying capacity	water resources population carrying capacity
			Population carrying capacity of forest resources
			Energy production population carrying capacityEnergy production population carrying capacity
		Population carrying capacity of urban built - up area	Land population carrying capacity
			rural arable land population carrying capacity
	Population and Environment	Per capita public green area (square meters / person)	Per capita public green area (square meters / person)
		Environmental Pollution	Unit Area Industrial Wastewater Discharge (ton / km2)
			Sulfur dioxide emissions per unit area (tons / km2)

2.2 Multi - level combination evaluation model

This paper attempts to construct a multi-level combination evaluation model:

The first step is to use representative principal component analysis, factor analysis and entropy method as the objective weighting method, using the expert scoring method as the subjective weighting method, respectively, to make a comprehensive evaluation of population development, and lists the different evaluation .The results of the evaluation of the method were compared for comparison. The second step, according to the results of the comprehensive evaluation of portfolio evaluation, and finally give the comprehensive development of population evaluation value [2].

Combined evaluation method is a comprehensive evaluation method in which a variety of evaluation methods can not draw a convincing and unanimous conclusion, and then use a variety of evaluation methods to conduct a comprehensive evaluation of the same evaluation object or the same index system, and then according to the ranking of each comprehensive evaluation. The results of a combination sorts an evaluation method. The methods of sorting the various combinations are: average method, Borda method, Copeland method, fuzzy Borda method. The combination of fuzzy Borda method and the Borda method and Copeland method only consider the difference of the sorting order of the evaluation results, which is based on the difference of the scores and the ranking difference of each method. Therefore, fuzzy Borda combination evaluation method is relatively more scientific. The final combination results of this project are based on the fuzzy Borda combination evaluation method. The combination of fuzzy Borda method is: [6]

1.calculate the membership

Membership refers to the evaluation of a certain object in the evaluation object in the "superior" membership, assuming membership for U_{ij} ,then

$$U_{ij}=\frac{z_{ij}-\min\{z_{ij}\}}{\max\{z_{ij}\}-\min\{z_{ij}\}},i=1,2,...,p \tag{1}$$

z_{ij} is the comprehensive evaluation value of the i-th evaluation scheme under the condition of the j-th evaluation method.

2.Fuzzy frequency calculation

Fuzzy frequency is mainly used to reflect the difference of comprehensive evaluation value. According to membership degree can calculate the fuzzy frequency[2]

$$f_{ih}=\sum_{R=1}^{m}\sigma_{ih}U_{ij} \tag{2}$$

$\sigma_{ih}=\{^1_0$ (When the first i kind of evaluation scheme in a certain evaluation method ranked h, the value is 1, otherwise 0).Then the fuzzy frequency is

$$p_{ih}=\frac{f_{ih}}{\sum f_{ih}} \tag{3}$$

3.The ranking order of the evaluation object which is derived from the comprehensive evaluation is converted into the scoring form

$$S_h=\frac{(n-h)(n-h+1)}{2} \tag{4}$$

S_h represents the score,and h represents the ranking order.

4.Calculate the fuzzy Borda values.The formula is

$$B_i=\sum P_{ih}S_{ih} \tag{5}$$

B_i is Fuzzy Borda number.

3.Solving

In order to obtain the final comprehensive evaluation result of population development, and to make full use of the advantages of various evaluation methods to reduce the errors caused by its shortcomings, this paper adopts the fuzzy Borda combination evaluation method on the basis of the four kinds of comprehensive evaluation and sorting results.

Based on the data of the 2013 index data, the population development of China is evaluated. Firstly, comparing the differences of the four comprehensive evaluation methods, principal component analysis, factor analysis, entropy and expert scoring methods, in principle, the degree of difference of all kinds of comprehensive evaluation methods can not be too large, otherwise, the combination evaluation significance. In this paper, spearmen rank correlation coefficient to test a variety of comprehensive evaluation results of the degree of association, spearmen rank correlation coefficient is: [7]

$$r_{ps} = 1 - \frac{6\sum_{t=1}^{n}(P_t - S_t)^2}{n(n^2 - 1)} \quad (6)$$

r_{pt} is the relationship between the p th comprehensive evaluation method and the t th comprehensive evaluation method. P_t, S_t respectively show the rank number of the t th evaluation object in the comprehensive evaluation method of the p type and the S type. n represents the number of objects to be evaluated. Accordingly, the spearmen rank correlation coefficient matrices of four comprehensive evaluation methods are shown in Table 2.

Table 2. Spearmen 's Rank Correlation Coefficient Matrix of Four Comprehensive Evaluation Methods.

	Principal component analysis	factor analysis	entropy weighting	expert scoring
Principal component analysis	1	0.774	0.818	0.782
factor analysis	0.774	1	0.942	0.894
entropy weighting	0.818	0.942	1	0.921
expert scoring	0.782	0.894	0.921	1

According to the spearmen correlation coefficient matrix of the four comprehensive evaluation methods in Table 2, the four comprehensive evaluation methods all have a high degree of association, and are suitable for combinatorial analysis.

In this paper, fuzzy Borda combination evaluation method is used to do comprehensive evaluation on four comprehensive evaluation methods of principal component analysis, factor analysis, entropy and expert scoring, and the fuzzy Borda combination evaluation value and ranking are obtained in 2013 in Table3.

Table 3. Comprehensive Evaluation and Ranking of Population Development Based on Fuzzy Borda Combination Evaluation [3,8]

	Principal component analysis		factor analysis		entropy weighting		expert scoring		Fuzzy Borda Combination Evaluation	
Serial number	Area	Integrated value	Area	Integrated value	Area	Integrated value	Area	Integrated value	Area	Integrated value
1	Beijing	0.16	Beijing	1.43	Beijing	0.72	Beijing	0.69	Beijing	435
2	Shanghai	0.16	Shanghai	1.22	Shanghai	0.69	Shanghai	0.63	Shanghai	406.0
3	Jilin	0.11	Tianjin	0.73	Tianjin	0.57	Tianjin	0.55	Tianjin	370.3
4	Tianjin	0.01	Guangdong	0.45	Guangdong	0.52	Liaoning	0.50	Guangdong	319.6
5	Hei longjiang	0.01	Liao ning	0.36	Zhejiang	0.51	Jilin	0.48	Jilin	304.6
6	Guangdong	0.08	Jiang su	0.35	Liaoning	0.50	Guangdong	0.48	Liaoning	302.9
7	Jiang su	0.07	Zhejiang	0.25	Jiang su	0.49	Zhejiang	0.47	Jiang su	276.7
8	Liaoning	0.07	Shandong	0.15	Fu jian	0.48	Jiang su	0.46	Zhejiang	267.4
9	Fu jian	0.06	Fu jian	0.12	Jilin	0.44	Fu jian	0.45	Heilongjiang	245.6
10	Zhejiang	0.05	Jilin	0.07	Heilongjiang	0.43	Heilongjiang	0.43	Fu jian	236.3
11	Hubei	0.05	Hainan	-0.06	Hainan	0.43	Hubei	0.42	Shandong	172.1
12	Neimenggu	0.04	Hubei	-0.07	Hubei	0.42	Hainan	0.42	Hainan	157.1
13	Shaanxi	0.03	Hebei	-0.08	Shandong	0.42	Shandong	0.41	Hubei	148.6

14	Shandong	0.02	Heilongjiang	-0.08	Chongqing	0.42	Neimenggu	0.41	Neimenggu	138.2
15	Sichuan	-0.002	Chongqing	-0.09	Neimenggu	0.42	Chongqing	0.40	Chongqing	118.6
16	Chongqing	-0.01	Xinjiang	-0.12	Sichuan	0.41	Jiangxi	0.40	Shaanxi	105.6
17	Hunan	-0.02	Neimenggu	-0.16	Jiangxi	0.40	Shaanxi	0.39	Sichuan	87.5
18	Hebei	-0.02	Shaanxi	-0.18	Shanxi	0.39	Shanxi	0.39	Hebei	86.6
19	Gansu	-0.03	Ningxia	-0.19	Xinjiang	0.39	Sichuan	0.39	Henan	71.1
20	Anhui	-0.03	Henan	-0.20	Hebei	0.38	Hebei	0.39	Jiangxi	63.3
21	Guizhou	-0.05	Shanxi	-0.22	Shaanxi	0.38	Hunan	0.38	Shanxi	58.4
22	Shanxi	-0.06	Jiangxi	-0.23	Hunan	0.38	Henan	0.38	Hunan	57.1
23	Henan	-0.06	Hunan	-0.24	Ningxia	0.37	Ningxia	0.38	Xinjiang	54.6
24	Xinjiang	-0.08	Sichuan	-0.28	Henan	0.37	Xinjiang	0.36	Gansu	39.3
25	Guangxi	-0.09	Guangxi	-0.30	Qinghai	0.36	Qinghai	0.35	Ningxia	34.3
26	Hainan	-0.09	Anhui	-0.30	Anhui	0.35	Anhui	0.35	Anhui	28.8
27	Qinghai	-0.09	Qinghai	-0.42	Guangxi	0.33	Guangxi	0.35	Guizhou	25.9
28	Jiangxi	-0.11	Gansu	-0.57	Gansu	0.31	Gansu	0.30	Guangxi	11.5
29	Ningxia	-0.13	Yunnan	-0.59	Guizhou	0.30	Yunnan	0.29	Qinghai	10.4
30	Yunnan	-0.16	Guizhou	-0.64	Yunnan	0.30	Guizhou	0.29	Yunnan	0.5

Table 3 lists the four basic evaluation methods and the combination evaluation methods together for comparison. Taking Chongqing as an example, the comprehensive values of the four comprehensive evaluation methods are -0.01, -0.09, 0.42, 0.40, respectively, in the principal component analysis method, factor analysis method, entropy method and expert scoring method. The membership degree is 0.53, 0.34, 0.35 and 1.57 respectively, the fuzzy frequencies are 0.34, 0.22, 0.23 and 0.20, respectively, and the rank scores are 105, 120, 136,120, respectively, rank ranked 16,15,14,15, finally concluded that the fuzzy fuzzy Borda value of 118.6, ranked 15 ranking. It can be seen that the fuzzy Borda combination method can better reflect the results of the first four evaluation methods. From the blurred Borda integrated value, it can be seen that the comprehensive evaluation value of population development in the eastern provinces is higher than that in the central and western regions, and the future population development in the central and western regions is the focus of China's population strategy work, and its improvement can quickly improve the population development as a whole Level.

4.Conclusing

On the basis of reviewing the domestic and foreign papers on the comprehensive evaluation of population development, according to China's guiding ideology of population development strategy which is "stick to people-oriented,System innovation, give priority to investment in the comprehensive development of human beings, stabilize the low fertility level, improve the quality of the population, improve the population structure, guide the population reasonable distribution, guarantee the population security; realize the population change to the human capital power",this paper constructs the comprehensive evaluation index system of population development in China.They are the first index system of 3,the second index system of 8, the third index system of 22;thr fourth index system of 42. The principal component analysis method, the factor analysis method, the Entropy Weight Method and Expert Score Method have been selected in this paper,which includes both the objective analysis method and subjective analysis method.In order to overcome the inherent shortcomings of subjective analysis and objective analysis, the fuzzy Borda combination method is used to evaluate the four comprehensive evaluation methods, and a relatively reliable evaluation result is obtained.The results show that the comprehensive evaluation value of population development in the eastern provinces is higher than that in the central and western regions. In the evaluation of population self-development, the top ten are the eastern regions with the exception of Jilin, Heilongjiang and Shanxi. And the latter ten are mostly in the central and western regions. In the evaluation rankings of population and economic and social development, the eastern region is in the front and the central and western regions are backward; the central and western regions rank relatively higher in the evaluation of population and resources and environment.

Acknowledgement

This paper is supported by Chongqing Municipal Education Commission of Science and Technology Research Program (NO.KJ1404105).

References

1. UNDP.Development of the Human Race 2010. UNDP (2010).
2. Ma Hongqi,The Index System and the Comprehensive Evaluation of China's Population Development.South China Population, **27**, **3**, 3-11(2009).
3. Chen Zhongchang, Model for Monitoring, Measuring and Evaluating Population Development and Its Application for 31 Chinese Provinces, Chinese Journal of Population Science,**5**, 72-79 (2007).
4. Chen Youhua, The Research on the Index Assess System of Population Modernization. Chinese Journal of Population Science,**3**,60-66(2003).
5. Hong Yingye,Scientific Development Perspective of Population and Resource Environment Coordinately Develop Research,Advances in Social Sciences,**1**,31-36(2012).
6. Yang Yongjie, Exact algorithms for weighted and unweighted Borda manipulation problems, Theoretical Computer Science,**3**, (2016).
7. Liu Man, The Correctional Model of Population Development Equation, Modeling and Numerical Simulation of Material Science,**3**,139-141 (2013).
8. Nation Bureau of Statistics of China. China Statistical Yearbook 2014, China Statistics Press,(2014).

The Intervention Effect Assessment on Social Support Condition of the First Settlers in Dan jiangkou Reservoir Area, China

Jinfeng Sun[1], Baoqing Shi[2], Jianxiang Shi[3], Xinpei Yue[1], Liang Sun[1] and Qingfeng Tian[1*]

[1]Department of Social Medicine and Health Service Management, College of Public Health, Zhengzhou University, 450001 Zhengzhou, PR China;
[2] College of Public Health, Zhengzhou University, 450001 Zhengzhou, PR China;
[3]Cancer Autoimmunity Research Laboratory, Department of Biological Sciences, University of Texas, El Paso, TX 799 68, USA.
Corresponding Email: tianqf@zzu.edu.cn

Abstract. Objective: To assess the effect of psychological intervention on social support condition of the first settlers in Dan jiangkou reservoir area. **Methods:** Using the Social Support Rating Scale (SSRS) to measure the social support condition of the first batch of immigrants before and after the intervention, and then compare it with the immigrants who were not intervened. **Results:** Compared with the immigrants who were not intervened, the immigrants who received intervention have a higher score on the availability of social support ($P<0.05$).**Conclusion:** Psychological intervention can improve the social support condition for immigrants, especially in enhancing the availability of social support.
Keywords. Dan jiangkou, immigrant, social support, intervention

1 Introduction

Social support is an interpersonal system, taking the individual as the core. It is composed of several aspects of supporting behaviors from both the individuals themselves and others supporters around [1]. Previous prospective study found that, it increased the risk of death among people who suffered from the poor quality or the lacking of social relations. Both human and animals experiments showed that social isolation was a risk factor of death[2]. Epidemiological studies have shown that social support is associated with lower morbidity and mortality, and benefits the cardiovascular, neuroendocrine, and immune system[3]. Since Dan jiangkou reservoir served as the source of the diverted water, the life of residents living in this area was totally disturbed. About 11 townships, 185 administrative villages, and 1312 village groups were flooded directly because of this project. Nearly 15,000 m^2 land was flooded and 2,583,800 m^2 houses were destroyed, affecting up to 10.73 million people[4]. The immigration for of South-to-North Water Diversion Project was mandatory, and residents were forced to move to new places. Unwillingness of leaving the former residence along with the unaccustomedness to the resettlement leaved psychological trauma among these immigrants. Changes in natural and social environment, especially the loss of farming and social relationship, made this situation even worse.[5]. Previous studies show that the mental status of these immigrants were negatively affected, and it's difficult for them to eliminate negative emotions.[6]. This study assessed the intervention of quo-immigrants' social support and explored the effectiveness of this measures, which may provide useful information and advice to the policy-makers about psychological comfort for immigrants.

2 Material and Methods

Research Subjects. Subjects were the first batch of immigrants who moved out Xichuan county, Dan jiangkou reservoir in 2010. The study design was approved by the ethical committee of Zhengzhou University. Written informed consent was obtained from all participants. Inclusion criteria: Participant must be aged 15 years and over. Exclusion criteria: people who have deaf, mental underneath, mental patients and severe cardiovascular, and cerebrovascular diseases were excluded.

Intervention programs. The immigrants were divided into intervention group and control group. The control group had no treatment, while the intervention group was treated with the following steps:

1. Carry out mental health education

Health education pamphlets and relevant materials were distributed; radio, blackboard, and propaganda showcase, etc. were used to transmit mental health knowledge; thematic seminar and lectures were given for mental health education among immigrant schoolchildren

2. Establish mental health records and psychological warning information network

The immigrant mental health records and psychological warning information network were established.

3. Provide psychological counseling

Face to face psychological counseling was carried out. High-risk groups were screened according to the results of the psychological investigation, then psychological counseling was provided.

4. Provide health service

We configured health services for immigrants, improving immigration physical health.

5. According to the intervention to immigration subjects, the immigrants' health guidelines which included health knowledge, common sense psychology and so on were drafted to make the resettlement progress smoothly.

Questionnaire survey. The Social Support Rating Scale (SSRS) was used to evaluate the situation of social support. The scale was established by Xiao, Shuiyuan [7]. It has 10 items and is divided into three aspects: subjective support, objective support and support utilization. Subjective support is subjective experience or emotional support, which mainly refers to the individual's emotional experience and satisfaction to be respected, supported and understood in society; objective support refers to the objective, actual or perceived support, including direct material assistance and social networks; utilization of support refers to the individual on the utilization of social support.

The scoring method of Social Support Rating Scale is as follows [8]: ① Items 1 ~ 4,8 ~ 10: Selecting items 1,2,3,4 will get 1,2,3,4 points respectively. ②Item five: the total scores are divided into A, B, C, D 4 parts, each count from none to full support will get 1 to 4 points separately. ③ Articles 6 and 7, if the answer is "no source", the score is 0 points, whose answer is "following sources", he will get the score according to the amount he choose. ④ Total score is the sum of 10 items; objective support score is the sum of item 2, 6, 7; subjective support score is the sum of item 1, 3, 4, 5; the utilization of supporting score is the sum of 8,9,10. The higher the score you get, the better your social support have.

Statistical Analysis. Before inputting the data, we conducted quality audits on each questionnaire and removed the incomplete questionnaires. The Software of Epidata3.0 was used to establish a database. Double entry was conducted to avoid input error. SPSS17.0 statistical software was used to conduct chi-square test, two-sample t-test and other statistical analysis for the data. Significance level $\alpha = 0.05$.

3 Results

Demographic Characteristics. We issued 1420 questionnaires, and retrieved 1372 questionnaires, the effective rate was 96.6%. The intervention group has 522 males and 445 females; the mean age is 46.74 ± 15.85 years old. Control groups have 204 males and 201 females; the mean age is 44.75 ± 14.51 years old.

Comparison of the score on social support between intervention and control groups before and after the intervention. Before the intervention, there was no statistically significant differences between the two groups (P>0.05) in social support, subjective support, objective support and the utilization of social support. After the intervention, the intervention group had a higher score on the utilization of social support than the control group (P<0.05); no statistically significant difference was observed on the other three aspects (P>0.05). The results are shown in Table 1.

Table 1. Comparison of social support between intervention and control groups before and after intervention

Group	Before intervention				After intervention				t value			
	Total score	Subjective support	Objective support	Utilization	Total score	Subjective support	Objective support	Utilization	Total score	Subjective support	Objective support	Utilization
Intervention group	37.95±8.68	22.63±6.13	8.36±2.75	6.92±2.43	38.62±8.46	22.68±5.36	8.60±3.17	7.73±2.68	1.231	0.200	0.969	4.904#
Control group	38.84±7.10	23.59±4.95	8.60±2.69	6.66±1.96	37.80±7.79	22.93±4.95	8.09±2.75	6.77±2.11	-1.228	-1.186	-1.640	0.488
t value	-1.088	-1.676	-0.784	1.139	1.376	-1.242	1.763	5.420#	/	/	/	/

Note: # means P<0.05

Hierarchical comparison of social support between immigrants before and after intervention. After the intervention, the scores on utilization of social support are statistically higher than immigrants before the intervention both in male, female, 15~34 and 35~54 age groups, primary group, junior high school group, unmarried group, married group and farmer (P<0.05), Comparing the two groups, the differences are not statistically significant (P>0.05) in the scores of social support, subjective support and objective support. The results are shown in Table 2.

Table 2 Hierarchical comparison of social support between immigrants before and after intervention

Project Group	Before intervention				After intervention				t value			
	Total score	Subjective suppor	Objective suppo	Utilization	Total score	Subjective suppor	Objective suppo	Utilization	Total scor	Subjective suppor	Objective suppo	Utilization

		t	rt			t	rt		e	t	rt	
Gender												
male	38.73±8.51	23.24±5.98	8.43±2.70	7.02±2.58	39.16±8.73	23.04±5.48	8.45±3.18	7.68±2.72	-0.552	0.379	-0.077	-2.810#
female	36.73±8.82	21.66±6.27	8.25±2.85	6.75±2.18	38.23±8.21	22.07±5.26	8.52±3.23	7.63±2.68	-1.740	-0.713	-0.864	-0.404#
Age												
15～	39.53±8.65	23.94±6.03	8.51±2.77	7.03±2.61	41.43±8.94	23.99±5.22	9.36±3.63	8.08±2.58	-1.494	-0.064	-1.847	-2.829#
35～	37.56±8.89	21.92±6.18	8.47±2.89	7.06±2.35	38.33±8.26	22.20±5.47	8.41±3.07	7.69±2.70	-0.893	-0.488	0.196	-2.455#
55～	37.47±8.30	22.57±6.21	8.20±2.47	6.75±2.38	37.56±8.36	22.04±5.29	8.26±3.18	7.29±2.71	-0.087	0.742	-0.162	-1.634
Education												
Primary and below	35.72±8.26	21.16±6.12	8.01±2.56	6.55±2.28	36.84±8.63	21.22±5.59	8.27±3.05	7.33±2.65	-1.288	-0.097	-0.925	-3.071#
Junior high school	39.07±8.77	23.53±6.10	8.49±2.83	6.95±2.52	39.72±8.28	23.32±5.21	8.68±2.29	7.74±2.73	-0.774	0.389	-0.634	-3.080#
High school and above	42.13±7.69	24.70±5.12	9.28±2.98	8.15±2.09	40.97±7.57	24.14±4.36	8.44±3.37	8.40±2.61	0.771	0.621	1.314	-0.522
Marriage												
Unmarried	39.15±9.06	23.33±6.75	8.74±2.57	7.07±2.12	37.81±9.08	21.37±5.48	7.94±3.91	8.53±2.41	0.641	1.455	0.980	-2.696#
Married	38.52±8.29	23.14±5.80	8.47±2.75	6.85±2.43	39.31±8.18	23.07±5.25	8.73±3.04	7.52±2.71	-1.313	0.179	-1.213	-3.542#
Others	30.19±8.58	16.58±5.88	6.71±2.53	6.90±2.53	33.51±9.01	18.96±5.15	6.73±3.28	7.88±2.83	-1.620	-1.886	-0.028	-1.550
Profession												
Farmer	37.98±8.63	22.64±6.13	8.40±2.77	6.89±2.42	38.99±8.54	22.64±5.48	8.59±3.18	7.75±2.74	-1.561	0.000	-0.851	-4.450#
Non-farmer	37.80±8.98	22.57±6.19	8.16±2.69	7.05±2.52	37.90±8.31	22.35±5.17	8.20±3.26	7.38±2.58	-0.076	0.267	-0.089	-0.867

Note: # means $P<0.05$

4 Discussion

Differences between the two investigations among control group are not statistically significant ($P>0.05$in social support, subjective support, objective support and the utilization of social support. It shows that with the psychological intervention, their social support situation has not been improved, and the immigrants had the same social support with the resettlement. Comparing with the score before the intervention in intervention group, immigrants after the intervention have higher scores in the use of social support and the differences are statistically significant ($P<0.05$). After the intervention, the intervention group get a higher scores on the utilization of social support than the control group, and the differences are statistically significant ($P<0.05$). It shows that the implementation of psychological intervention can improve the utilization of social support, and exclude the adaption of resettlement and other factors. Interventions' improvement of social support utilization is still valid. Comparing the scores before and after the intervention in intervention group, the differences are not statistically significant ($P>0.05$) in social support, subjective support, objective support and the utilization of social support. Considering after the relocation, immigrants lived collectively in the form of immigrant village. Immigrants have very little contact with local residents in daily life, their objective sources of support are all from immigrants. After the remove, the external environment break the immigrants' original production and lifestyle, disrupting its natural kinship, the long-regional relations and the social support systems formed of other social networks, social support occurred tremendous changes. With these changes, objective source of support reduced, causing the reducing of subjective support immigrants feel. By implementing psychological interventions in the key populations, immigrants' utilization of social support improved to a certain extent, but the new social network is still not established, objective and subjective support did not change significantly.

Comparing the scores before and after the intervention in intervention group, the differences are not statistically significant (P>0.05) in social support, subjective support, objective support and the utilization of social support. Considering after the relocation, immigrants lived collectively in the form of immigrant village. Immigrants have very little contact with local residents in daily life, and their objective sources of support are all from immigrants. After the remove, the external environment broke the immigrants' original production and lifestyle, disrupted their natural kinship, the long-regional relations and the social support systems formed of other social networks which indicated the available social support occurred tremendous changes. With these changes, objective source of support reduced, and subsequently caused the decrease of subjective support in immigrants' feelings. By implementing psychological interventions in the key populations, immigrants' utilization of social support improved to a certain extent, but the new social network is still not established, objective and subjective support did not change significantly.

After the intervention, the scores on utilization of social support are higher than immigrants before the intervention both in male and female group, suggesting the intervention effect in both genders is close. After the intervention, the scores on utilization of social support are higher than immigrants before the intervention both in 15～34 and 35～54 age groups. Before and after the intervention, the differences are not statistically significant (P> 0.05) in every dimensions in 55～ age group. It shows that effect of intervention program in people over 55 years is not as efficient as that in the other age groups, considering the mobility of the older was limited. Since their social adaptability was reduced, they are more difficult to establish new social support network. After the intervention, the scores on utilization of social support are higher than immigrants before the intervention both in primary group and junior high school group. Before and after the intervention, the differences are not statistically significant (P>0.05) in every dimensions in high school age group. This shows that psychological intervention has an obvious effect on immigrants who have the lower level of education, probably because the immigrants of higher educational level accept more information. They learned their current standard of living and the gap between them and other rural areas through a variety of channels, which causing psychological negative emotions that affect the intervention. After the intervention, the scores on utilization of social support are higher than immigrants before the intervention both in unmarried and married group, and the differences are statistically significant (P<0.05). Before and after the intervention, the differences are not statistically significant (P>0.05) in every dimension in other marital status group. Interventions have less effect on immigrants who were divorced, widowed and other conditions, considering the quality of marriage has a certain impact on the intervention effect. After the intervention, the scores on utilization of social support are higher than immigrants before the intervention in farmer group, and the differences are statistically significant (P <0.05). Before and after the intervention, the differences are not statistically significant (P>0.05) in every dimension in non-farmer group. Indicating that interventions help farmers adapt to new farming environment gradually, they are getting used to resettlement life.

5 Conclusion

In conclusion, we can get the following conclusions from our study:

1 Psychological intervention in our study has an effect on improving the social support condition, especially on enhancing the availability of social support.

2 Intervention programs have a significant effect on immigrants 55 years old or younger, lower education level, and farmers.

References

1. Ou, Y.:*Social support on the mental health of college students.* Youth Studies, Vol.**03**(2003), p. 29-33+38.
2. House, J.S., K.R. Landis, and D.Umberson: *Social relationships and health.* Science (New York, N.Y.), Vol. **241**(1988), p. 540-5.
3. Uchino, B.N.: Social support and health: A review of physiological processes potentially underlying links to disease outcomes. Journal of Behavioral Medicine, Vol.29(2006), p. 377-387.
4. Xue, X.: he thinking of Danjiangkou reservoir resettlement work in Henan. China Water Resources, Vol. 10(2009), p. 40-41.
5. Chengbin Liu, X.F.: Degree of Satisfaction on Migration of the Three-Gorges Migrants: Transition and Causes. Population Research, Vol. 01(2007), p. 76-85.
6. Zhang, Q.: The social support network of migrants from Three Gorges Reservoi. Social, Vol: 01(2000), p. 30-31.
7. Xiao, S.: he theoretical basis and research applications of the "Social Support Rating Scale," Journal of Clinical Psychiatry, Vol. 02(1994), p. 98-100.
8. Huang Wanqi, Z.D., Luo Tongyong: Characteristics of social support and coping style of four pop-ulations. Chinese Journal of Clinical Research, Vol. 42(2006), p. 37-39.

Talking about Problems and Countermeasures of Small-sized Apartments Design

Wenye Gong[1, a], Tianwei Bai[2, b] and Donglin He[3, c]

[1]North China University of Water resources and Electric Power, Zhengzhou, 450046, China

[2]Jilin University, Changchun, 130012, China

[2]University of Illinois at Chicago, Chicago, Illinois, USA

[3]Kaifeng University, Kaifeng, 475000, China

[a]13663851982@126.com, [b]btw@126.com, [c]kfuhdl@yahoo.com.cn

Abstract. In order to promote the healthy development for small apartments, the author made a full investigation to Zhengzhou of China, analyzed the existing problems of small-sized houses, and proposed improvement suggestions from traffic space, family-unit design and plan-combination. These measures will help to create a suitable living environment for users.

Keywords. Small-sized apartment, traffic space, house type, plan combination

1 Introduction

With the rapid development of the national economy, the improvement of the consumption and the prosperity of the market, the real estate sales are full of competition. Due to less investment, low down-payment and high rent returns, the small-sized apartment is nearly the favourite house for young people and most investors. What about the quality of the small apartment? Take Zhengzhou City of Henan province as an example, there are obvious shortcomings and defects in the daily use according to spot investigations. For instance, high density will lead to traffic congestion, monotonous house-type will cause the use-inconvenience, and the space of over saturation will give a non-communication environment to residents. How to do a small-apartment design? How to achieve "small but complete" and "dense but comfortable"? This paper mainly discusses the improvement measures of small apartments from the aspects of traffic space, house-type design and plan combination, which to create a suitable habitat environment and promote sound development of small-sized apartments.

2 Present situation of small-sized apartments

In general, small-sized apartments are given characteristics as surrounding by convenient traffic, locating in the downtown of the city and having a perfect life support. The advantage of low price can satisfy people's ideal of urban life. Today, in the fast-paced city environment of high housing prices, small apartments play an important role in solving the housing needs of low-income families and saving land resources.

In Zhengzhou, for example, with a recovering housing market since 2015, sales volume of small-sized apartments have accounted for about 60% of the total. Small dwelling house can be divided into two kinds according to the size of the area. One kind is absolute small-house with 20~30m^2, some having hall and some having not, such as Changsheng Square, Longzihu Micro-era and Yingzuo Square. The other kind is relative small-house of 40~50m^2, which can be designed into "one hall+ one room+ one toilet" or "one hall+ two rooms+ one toilet", and characterized by compact

 DOI: 10.1051/matecconf/201710005085

housing space. For example, Meijing City, Qihuihao Home and International City, which living-room area is not more than 20m^2 and the bedroom is not more than 10m^2.

3 Problems in small-sized apartments

High density leads to traffic congestion. With the rising price of land, the construction of high-rise buildings is undoubtedly the best choice for estate businessmen, so that existing small apartments are mostly high-rise buildings relying on elevators to achieve vertical transportation. Zhengzhou as an example, in International City community, the first-stage project has 12 floors, the second-stage is 18 and the third is built to 26 stories, which accommodate thousands of families. Similarly, Yinzuo Square plan a new concept of small-family, composed of two blocks of 30 stories business-living buildings, which 1-4 floors are shops and 5-30 floors for dwelling houses, all composed of 36-59m^2 small-sized apartments,[1] so the living density is high. Residents in small-apartments are mostly young people with frequent activities, which hundreds of people go in and go out in the rush hours. On average, each elevator will carry 100-200 households every day, so the waiting time and crowded degree can be imagined. If the power-outage occurs, people must be in an awkward situation of "can't go up and can't get down".

Duplicate design of dwelling unit leads to the use of inconvenience. At present, small-sized apartments in Zhengzhou are mostly one-room type like standard-room of the hotel, there is no natural ventilation and only one side has natural light. Common unit-types in the market are shown in figure.1. The merger of living-room and bedroom led to having no daily reception space, balcony or corridor is usually occupied for cooking space, and the lack of storage space generally exists.[2] Monotonous apartment design cause a long and narrow "gray space". The bedroom is too bright and the living room is too dark, which is not conformed to people's living habits. As shown in figure.2, 30m^2-40m^2-50m^2-60m^2 are all simply designed into one room, which unreasonable design is obvious.

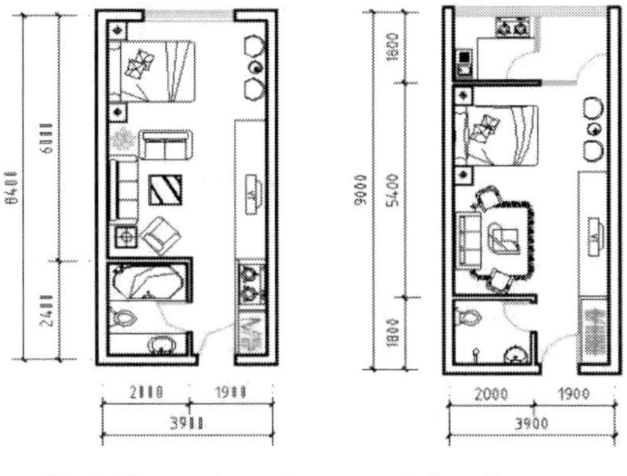

Fig.1. The one-room house type of small apartment

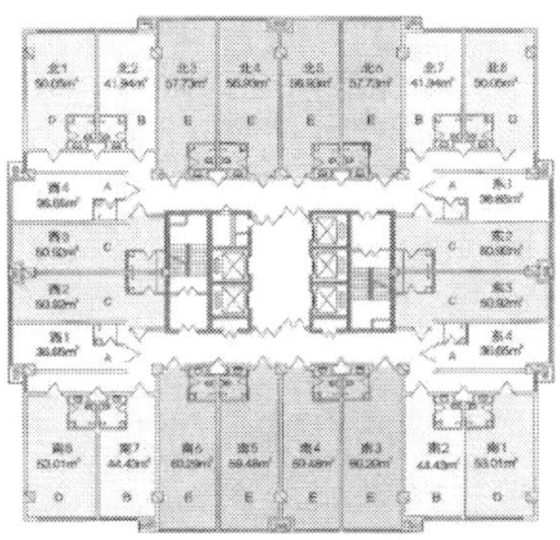

Fig.2. The plan of Yinzuo Square

Too compact plan-combination causes the loss of communication. In recent years, with the increasing floors of high-rise buildings, the communication between the neighbourhoods has been greatly reduced. Real estate developers often require an airtight design in order to seek maximum benefits. Apart from long-narrow aisles and cramped elevator lobbies, no space can be used to carry out public interaction. Buildings in this design ideal will inevitably lead to the deterioration of the living environment and indifference of neighbourhood relations. At the same time, each layer accommodated a large number of households, which has resulted a complicated and large- interference environment. The residents are not familiar with each other, which will lead to the occurrence of many unsafe incidents in a way. Moreover, hundreds of household-doors in building are decorated as a unified style without any sign, easy to make people lost direction.

4 Discussion on the design strategy of small-sized apartments

Increase the number of lifts and the width of corridors to ease traffic. Buildings of small-sized apartment mainly locate in three areas: one is downtown of bustling cities, the other is close to the city's main traffic lines, and the third is working areas of young men.[3] In these regions, the rent is higher and the demand for small apartments is larger. Therefore, the developers want to build houses as much as possible. In high-density buildings, if designers increase the width of corridors and the number of elevators appropriately, traffic pressure will be reduced. For example, if the aisle be widened from 2.4m to 2.7m, and service-number of each elevator be reduced from 5-6 to 3-4 households, (figue.3) traffic efficiency will be greatly improved. At the same time, the partition service of lifts can reduce the waiting and running time to relieve the anxiety.

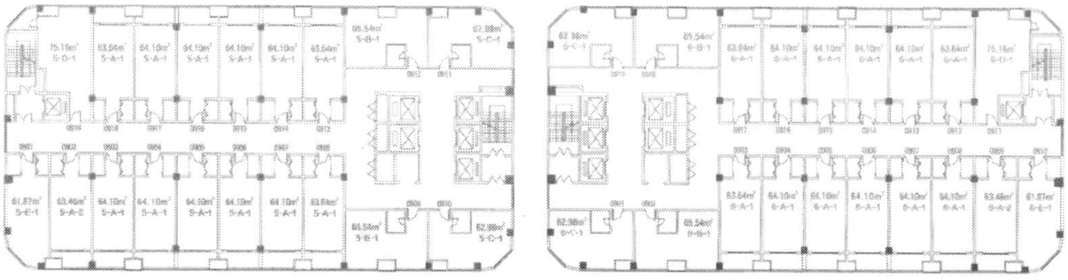

Fig. 3. Building5 and building6 of Changsheng Square of Zhengzhou

Try the diversified design of house-type to meet the use requirements. As stated earlier, the monotonous design of small houses directly leads to unreasonable rooms and poor living function. If designers increase face-width of each household appropriately in the premise of land-saving, and change vertical flat into horizontal flat, (figure 4) many problems can be avoided.[4] As shown in figure 5, if the span is widened to 5.4m and the surface is recessed into a small piece, the bedroom and living room can be intersected. In this way, designers can not only effectively control the area, but also open more windows to achieve natural lighting for living room, bedroom and kitchen. Another example is shown in figure 5, two households sharing 1.8m-width kitchen and toilet space, all spaces such as reception, rest and cooking play their proper roles, [5] which can improve the living quality greatly.

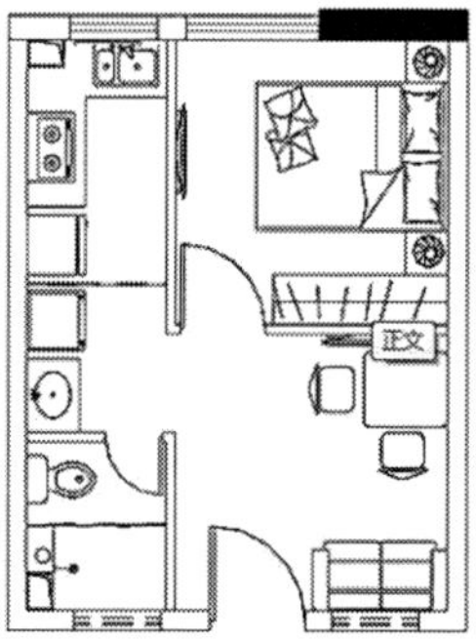

Fig.4. The horizontal flat

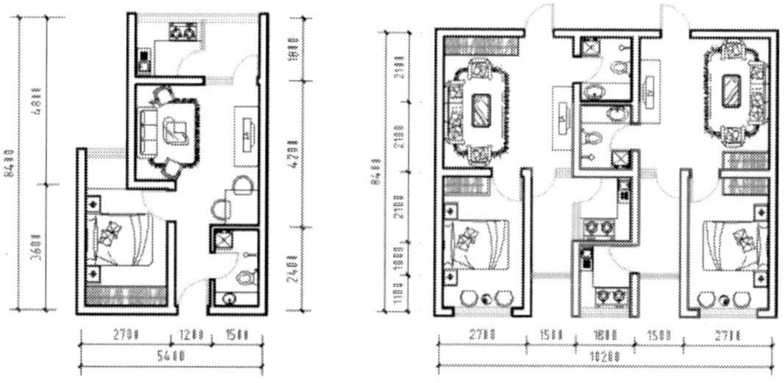

Fig.5. The combination design of horizontal flat

Try the diversified design of plan-combination to promote neighbourhood communication. House is the place where people live every day, so it must be humanized. It should meet people's life needs in physiology and psychology. In the view of physiology, although the area is limited, small apartments should have complete function and natural lighting. From the angle of psychology, in addition to privacy requirements, small-sized house should consider people's communication needs. As shown in figure 6, some concave-processing of building-plane can achieve the nature lighting and ventilation for small-sized apartments. [6] At the same time, designers set aside some wide concave corridors or verandas in the plan nod, which can be properly set up seats or green plants to form a small hanging garden [7] to provide conditions for the communication between the neighbours. These measures will help to improve the neighbourhood relationship in high density environment, enhance people's familiarity and establish a sense of responsibility. Of course, the establishment of public space is a common thing of their neighbouring residents, the increasing of shared area will aggravate the economic burden, so it is necessary to conduct a thorough discussion and extensive publicity in the planning and design stages.

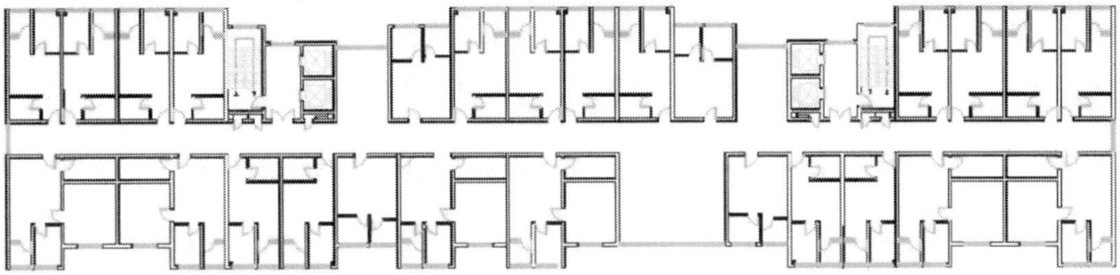

Fig. 6. The plan-combination of small apartments with concave corridors

5 Conclusions

With the flow and aggregation of urban population, the demand for small houses will increase, and the market share will be high more and more. The optimized design of traffic space, family-unit and plan-combination can improve the quality of small apartments. In addition, perfect infrastructure and high-level property management will also play an important role in elevating the living environment for residents.

References

1. Information on http://www. newhouse.zz.fang.com
2. Y. Tian: Creative Design of the Space of Small Apartment (China Building Industry Press, Beijing 2010).
3. L. Hong and J.T. Ying: Research on Small-sized Apartments in Modern Buildings. Jiangxi Building Material. Vol. 13 (2014), p. 24
4. S.F. Lu, L. Zhang and J. Zhou: Introduction to design plan of small family houses. Housing Science, Vol. 12 (2013), p.19-22
5. Y. Jiang: The sustainable development of small and medium sized unit. Architectural Journal. Vol.11 (2007), p.19-22
6. H. Chen, YH. Lv and Q. Ji: Middle and small family housing flexibility and variable designing. Chinese & overseas architecture. Vol.4(2007), p.13-15
7. J. Choi and T. Cho: Comparing Perception Concerning the Importance of Apartment Complex Components between Consumers and Housing Providers. Journal of Asian Architecture and Building Engineering. Vol.1(2014), p.109-116

Experiment Study on Elastic Indicator of Thermal Shock Ceramic Materials——Implementation of Students' Innovative Research Project of Shandong University of Science and Technology

Yanxia Wang[1,*], Tiantian Sun[1], Yongli Bao[1], Xiaomin Ma[1] and Zhendong Liu[2]

[1]School of Transportation and Vehicle Engineering, Shandong University of Technology, Zibo 255049, china
[2]School of Transportation Engineering, University of Southampton, United Kingdom
Corresponding Email: wangyx200@163.com

Abstract. In order to improve the quality of undergraduate education and combine theory and practice, Shandong University of science and technology organized innovative research activities project for undergraduates. Combined with the characteristics of engineering mechanics course, teachers of engineering mechanics teaching and research section guided students to take an active part in scientific research and innovation practice teaching, which has obtained a good teaching effect. This paper introduces the concrete implement process of the college students' innovative scientific research project "Experiment Study on Elastic Indicator of Thermal Shock Ceramic Materials", which measures elastic indicator of ceramics using the ultrasonic method. This paper studies elastic indicator change rule of the mullite ceramic samples under different factors such as temperature difference, thermal shock times and so on. Studies have shown that in the condition of air-cooling, with the increase of thermal shock temperature difference and thermal shock times, the elastic modulus value, shear modulus and Poisson's ratio are in a falling trend. The project implementation have proved that implement undergraduate innovation research projects could effectively arouse students' learning enthusiasm, cultivate students' scientific research innovation and analytical abilities to solve practical scientific research problems.
Keywords. Ceramic materials, elastic modulus, shear modulus, poisson's ratio, thermal shock.

1 Introduction

In order to strengthen students' innovation consciousness and practical ability, Shandong University of Technology launched a college student innovation research project since 2009, and by 2015 it has been successfully held for consecutive 7 sessions. Through the implementation of this project, leading undergraduate student to participate innovation activities, not only could help the students to understand the theoretical knowledge and improve their practical skills, but also could let them access to the most advanced knowledge of the subject and increase their interesting in learning. It will be very helpful for the students' further research training. For this purpose, Department of Engineering Mechanics with their own characteristics applied a program: Thermal shock resistance of ceramic materials-Determination of elastic parameters and it has been successfully completed [1, 2].

Mullite ceramics, with a unique combination of properties including high temperature resistance, oxidation resistance, corrosion resistance, low creep rate and other excellent characteristics, is widely used in metallurgy, energy, electricity, chemical industry, aerospace and so on. In the use process, mullite ceramic structure is induced thermal stress due to heat shock, and led to crack, flake, fracture and other thermal shock damages [3-6]. These damages will directly affect the service life of the structure, so the study of thermal shock resistance on ceramic materials is great significance to the practical application. At present, the research of thermal shock resistance on ceramic material, is mostly concentrating on the determination of critical temperature difference [7], thermal shock cycles [8], residual strength [9] and the occurrence and extension of crack [10-12]. However, few papers had a systemic research in the influence of temperature difference, thermal shock cycles and cooling medium on the materials in the process of thermal shock. This paper focused on the changing rule of the mechanical properties indexes on the mullite materials in the thermal shock process, and then cultivates students the research ideas and research methods of fatigue failure.

To measure the sample's elastic index, an ultrasonic method is introduced which calculates the material's elastic performance indicators such as Young's modulus, shear modulus and Poisson's ratio by determining the density of materials, speed of propagation of transverse waves and longitudinal waves in the material. This method has a lot advantages, like no pasting of strain gauge, low cost, short cycle and multipoint measurement. It will be helpful for fostering students' independent thinking and hands-on ability.

2 Experiments Method

Thermal Shock Experiments. The specimen made by high temperature solid method, the size of the test piece for $10\times20\times100$ mm. Table 1 shows the experiment design.

Table 1. Thermal shock experiment.

Temperature difference [°C]	Air cooling cycles
400	60
600	60
800	60

Ultrasonic Experiments. Elastic index is measured by ultrasonic method under the condition of air-cooling. Wave velocity measurement system is shown in figure 1.

Principle of operation: firstly, the pulse generator sends signals (through wire 1 and wire2) to two different transmitting probes which will send transverse waves and longitudinal waves respectively and at the same time the signals are send to a digital oscilloscope (through wire 6). Secondly, the ultrasound goes into the sample through coupling agent and reflects multiple times. Thirdly, each echo signal that reaches probe will be received and transferred to amplifier (through wire 2 and wire 3) and then send back to digital oscilloscope (through wire 4 and wire 5). At last, current sent pulse signal and multiple echo signal is showed on numerical oscilloscope and once the time difference is calculated the propagation time will be obtained [13,14].

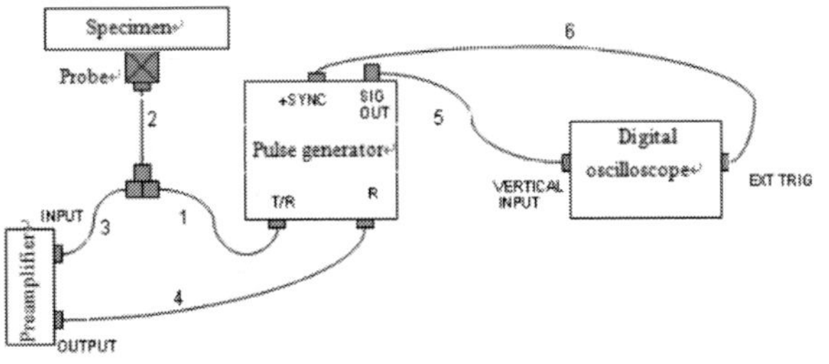

Fig.1. Ultrasonic wave velocity measurement schematic diagram.

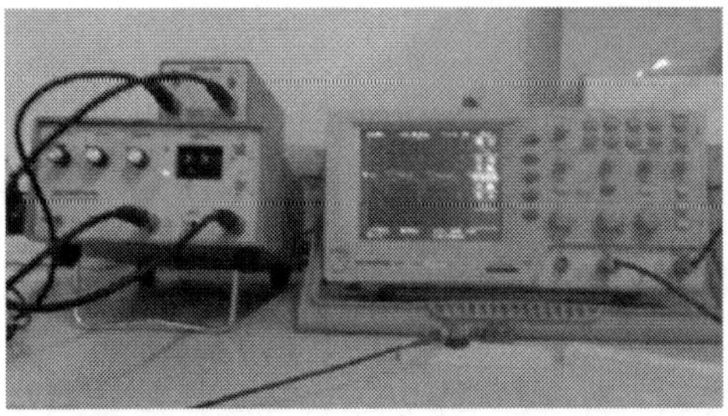

Fig.2. Working scheme of testing instruments.

In short, the working process is that pulse signal generated by pulse generator is conversed into high-frequency mechanical energy and then transferred to specimens through coupling agent, propagate, reflect, refract, attenuate to death and finally the propagation time is shown on oscilloscope. Wave velocity measurement device consists of OLMPUS-ultrasonic pulse generator, preamplifier, probe and digital oscilloscope. Figure 2 shows the instruments.

3 Research on Elastic Index

Research Principle. The elastic modulus E, shear modulus G and Poisson's ratio μ of materials are calculated according to the following formulas:

$$E = \rho c_T^2 \left[3 - \frac{1}{\left(c_L / c_T\right)^2 - 1} \right] \quad (1)$$

$$\mu = \frac{1}{2} \frac{c_L^2 - 2c_T^2}{c_L^2 - c_T^2} \quad (2)$$

$$G = \rho c_T^2 \quad (3)$$

Where C_L is the velocity of longitudinal waves (m/s), C_T is the velocity of shear waves (m/s), ρ is density.

Research Results. The curve of elasticity modulus E changing with thermal shock cycles are showed in figure 3. It shows that elasticity modulus gradually decrease with the increment of thermal shock cycle times, and when the temperature difference of thermal shock is small, the change of elasticity modulus is small. But the change will be bigger when the temperature difference grows to 600°C and 800°C showed in the figure. The decline of elastic modulus indicates that the shape changing capacity of the material has become weak, so the thermal shock has caused much more damages to the sample's elasticity modulus.

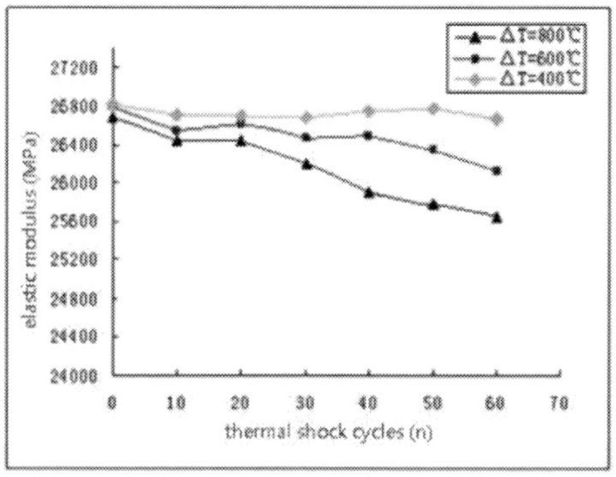

Fig.3. Elastic modulus changing with the thermal shock cycles times.

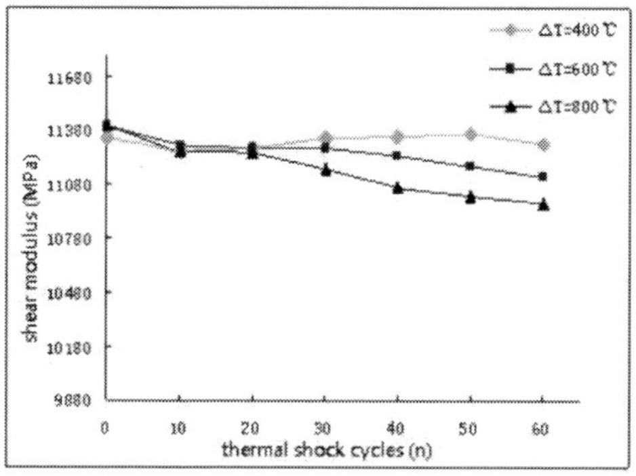

Fig.4. Shear modulus changing with the thermal shock cycles times.

Shear modulus G also changes with thermal shock cycles times, see figure 4. When the temperature difference of thermal shock is 400°C, shear modulus will change like a wave. But when the temperature difference rises to 600°C or 800°C, Shear modulus will go down with the growth of thermal shock cycles times.

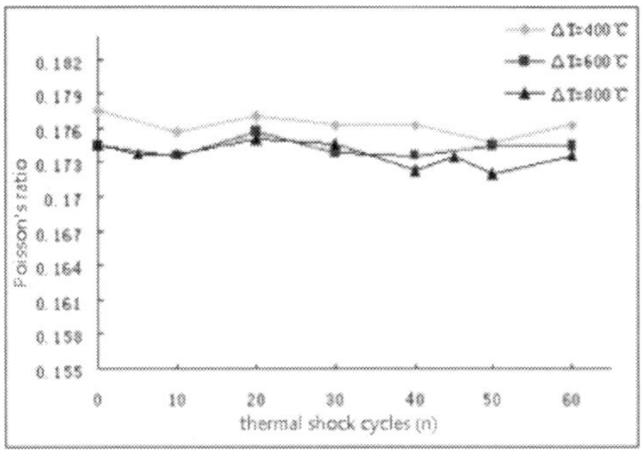

Fig. 5. Poisson's ratio changing with the thermal shock cycles times

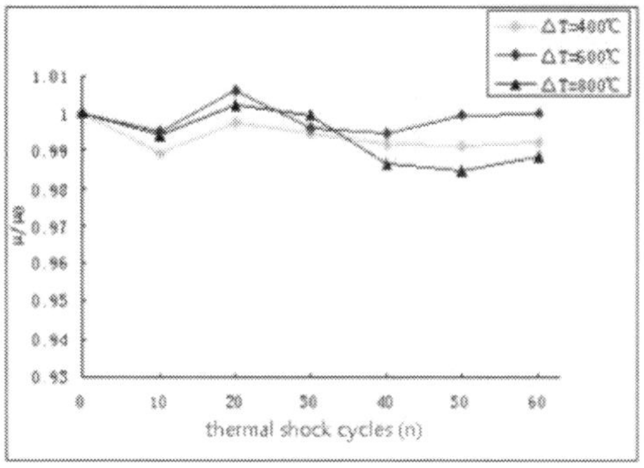

Fig. 6. Normalized processing of Poisson's ratio data

Figure 5 is the curve of Poisson's ratio variation under thermal fatigue. As showed the Poisson's ratio change is not obvious with increasing thermal shock cycle, the changing curves also look like a wave. Figure 6 is the Poisson's ratio variation under 60 times thermal shock. It tells that the bigger thermal shock temperature difference is, the bigger fluctuation Poisson's ratio will have. That means the Poisson's ratio of 400°C and 600°C will be smaller than the 800°C. And when the temperature difference is 800°C and thermal shock cycle is 60, Poisson's ratio decreased by 1.17%.

Research Conclusion. Through experiment, the changing curves of the elasticity indexes under different temperature differences and different thermal shock cycles are measured and the extents of damages of thermal shock mechanical properties of the mullite ceramics were obtained.

In the thermal fatigue experiment, the value of the elastic modulus E, shear modulus G and Poisson's ratio μ were reducing gradually with increasing of the thermal fatigue cycles, so that the damage degree of the samples increased.

4 Summary

Department of Engineering Mechanics at Shandong University of Technology has always put raising student's innovation ability and practical problems solving skills at the first place. Considering the current Engineering Mechanics experiment course contain a little knowledge of optics and acoustics, the department launched the project: the ultrasonic method for measuring the elastic properties of materials. The results show that by connecting practice teaching with theory teaching could expand students' knowledge, enhance students' learning interest and enthusiasm, enable students to learn theoretical knowledge well and master basic experiment skills, and cultivate and enhance their innovative capacity, which lays a good foundation for the subsequent teaching activities.

Acknowledgements

This work was financially supported by the excellent course in Shandong Province "Engineering Mechanics", a grant from National High Technology Research and Development Program of China (863 Program) (2009AA063202) and the Shandong Province Natural Science Fund (ZR2009FQ023, ZR2011EL017, ZR2013EEQ008).

References

1. G.G. Zhao, P.G. Rao and M. Lu: China Ceramics, Vol. 42 (2006) No.9, p.13.
2. B. Chen, P.D. Ding and C. Cheng: Nai Huo Cai Liao, Vol. 38 (2004) No.4, p.234.
3. H.W. Liu: *Material Mechanic* (Higher Education Press, China 2011).
4. Z.L. Xu: *Elasticity Mechanic* (Higher Education Press, China 2010).

5. Z.Y. He, Y.F. Zhao: *Basical Theory of Sonic* (Defense Industry Press, China 1981).
6. Y.W. Shi: *Ultrasound Detection* (Mechanical Industry Press, China 2005).
7. F. Osterstock: Mater. Sci. Eng Vol. A168 (1993), p.41.
8. K.T. Faber, M.D. Huang and A.G. Evans: Journal of the American Ceramic Society, Vol. 64 (1981) No.5, p.296.
9. J.K. Li, J. Zhou and X. Liu: Journal of Ceramics, Vol. 31 (2010) No.1, p.101.
10. J. Lu, Z.X. Zheng and Z.H. Jin: Material Composite Science, Vol. 18 (2001) No.4, p.76.
11. B. Chen, P.D. Ding and C. Cheng: Journal of the Chinese Ceramic Society, Vol. 32 (2004) No.6, p.718.
12. T. Andersson, D.J. Rondiffe: Journal of the American Ceramic Society, Vol. 79 (2005) No.6, p.1509.
13. J. Lu, Z.X. Zheng and Z.H. Ling: PTCA, Vol. 39 (2003) No.1, p.14.
14. J.W. Li: *NDT handbook* (China Machine Press, China 2012).

Research on the Relationship between Economic Growth and Environmental Technologies in China

Haibo Zhang

School of Management and Economics, Beijing institute of technology, Beijing, China

Abstract. Based on the time series data from 1990 to 2012,with the expanded IPAT model, the economic-environment coefficient (GT) was constructed to measure the impact of economic growth on the environmental technologies. The results showed that there was a diminishing trend of the impact in the period of 1990 -1998 except the year of 1996, meaning the coordinated development between economic growth and environment technologies. In order to further analyze the measures and methods of energy saving and emission reduction, DEA model was using for the calculation of input redundancy and output deficit value. The results revealed that the coal consumption keeps Pareto optimal all the time. The emission of pollution decreased and the employment rate increased gradually. The input redundancy decreased to efficiency in the second and third industries.

Keywords. Environmental technologies, economic growth; IPAT model; DEA model

1 Introduction

Since reform and opening up, with rapid economic development, the GDP ratio of China to the whole world rises to 15% from 5% [1]. Along with rapid economic growth, ecological environment is continuing deterioration and energy consumption is too excessive. China became the country owing the largest energy consumption, exceeding America in 2010. The ratio of Chinese energy consumption to the one of the whole world was one fifth, meaning that energy consumption was significant to the rapid economic development. With the way of extensive economic development, it is difficult for China to realize the goal of constructing one country which is resource saving and environment friendly. Therefore research on the intrinsic relationship between economic growth and energy environment is helpful to safeguard energy security and rapid economic growth, and promote the coordinated development among energy, economy and environment.

2 Literature review

Research on the relationship between economic growth and environment has been concerned by scholars in recent years. According to the empirical data, Grossman and Krueger (1991) [2] proposed the hypothesis of inverted U type relationship between the pollution and income for the first time. The hypothesis was that in the early stages of economic development, per capita income was negatively related to environmental quality, and positively when economy beyond the critical value. Bruce (2004), Dinda (2004) [3, 4] and other scholars had done theoretical and empirical studies on EKC. Based on Chinese input-output table, Garbaccio,R.F (1999) [5] analyzed the effect on energy consumption reduction from technology progress and structural change. The results showed that sectional technology progress reduced the energy output ratio, that structural change increased energy consumption, and that the import of energy intensive products lowered the energy intensity from 1987 to 1992. Hannesson,R. (2002) [6] studied the relationship between the growth rate of energy utilization and the growth rate of economy, by dividing 1950-1997 into 5 time periods which were low oil price period, the first energy crisis, high oil price period, 1974-1986 and oil prices falling back period. The results showed that after the first energy crisis, their relationship became weak. But it became strong again after 1986. The energy utilization rate is lower than the GDP growth rate in rich countries, with the contrary situation in poor countries.

With Environmental Sustainability Index (ESI), Hyun-Hoon Lee (2011, 2012) et al [7, 8] done regression analysis, concluding 2 pollution indexes which were air pollution and water pollution, and 9 kinds of efficiency index, to research the relationship between environmental stability and income. The results showed that GDP per capital had positive relationship with environmental stability index and pollution index, and negative relationship with eco-efficiency index. Through nonlinear STR model, Jinwen Zhao and Jitao Fan (2007) [9] studied the intrinsic structure relations between energy consumption and economic growth in China, and found that the effect characteristics of economic growth on energy consumption were nonlinear, non-symmetry and phase. By provincial panel data during 1991 and 2006, Shuai Shao (2008) [10] analyzed the correlation and transmission mechanism between energy development and economic growth in the western region, finding that energy development had negative correlation with economic growth. Based on decoupling theory model, Chongmei Wang (2010) [11] found that economic growth and energy consumption were relative decoupling after 2000 in China, considering historical data from 1990 to 2007 as sample. According to the index of energy consumption per unit output, Yanxu Wei et al. (2011) [12] compared two stages which were one from 1953 to 1977 and one from 1978 to 2009, to study the relationship between energy consumption and economic growth with the help of IPAT equation. The conclusions were that in the former 30 years, the per capita energy consumption and per capita GDP were synchronous growth, meaning economic growth dependence on energy consumption. But in the later 30 years, along with the growth of per capita GDP, per capita energy consumption increased volatility, and more depended on technological progress, with less dependence on energy consumption. Adopting the modified model, Peng Li (2013) [13] empirically test the interdependence between energy consumption and economic growth, selecting the panel data from 1995 to 2008. The result was that economic growth was the one-way Grainger reason to energy consumption.

Based on the above research results, according to the actual situation of China and data, this paper studied the impact of economic growth on the environment, with the IPAT and expanded model. And energy economic efficiency change over the years, the input redundancy of the output and output deficit situation were analyzed by DEA model to reflect the actual efficiency of economy growth and implementation effect of energy-saving emission reduction policies. Those empirical analysis which were effect of environment on economy growth and environment efficiency, were used to provide a theoretical basis for energy saving and emission reduction policies.

3 Assessment on the impact of economic growth on energy environment

3.1 Establishment of model

70's last century, Ehrlich and Holden [14,15] proposed IPAT model to measure impact of economic activities on environment, to find out the main driving force of human activities to environment. All the factors are classified into three categories in the model, which are population, affluence and technology. The model can be written as

$$I = P \times A \times T \tag{1}$$

Where I is environmental impact index, meaning impact to represent energy consumption, or waste emission and so on, P being population to be represented by the total population at the end of the year, A being affluence, to be represented by per capita GDP, that is $A = GDP / P$, T being technology, to be represented by the environmental impact of per unit GDP, that is T=I/GDP [16], G means GDP, then the formula (1) can be transformed into

$$I = G \times T \quad (2)$$

Assuming that the GDP of base year is G_0, with an average annual growth rate of g, the technical level of the base year denoted as T_0, with an average annual growth rate of t, then the G_n, T_n and I_n can be written as

$$G_n = G_0 \times (1+g)^n \quad (3)$$

$$T_n = T_0 \times (1-t)^n \quad (4)$$

$$I_n = G_n \times T_n \quad (5)$$

The nth environmental impact can be obtained by bring (3) and (4) into (5).

$$I_n = G_0 \times T_0 \times (1+g)^n \times (1-t)^n = G_0 \times T_0 \times (1+g-t-gt)^n \quad (6)$$

In the formula (6),

(1) if $g-t > gt$, then the environmental impact increases year by year.

(2) if $g-t = gt$, then the environmental impact stays stable.

(3) if $g-t < gt$, then the environmental impact decreases year by year.

Economic environment coefficient GT is defined as index to measure the environmental impact.

$$GT = g-t-gt \quad (7)$$

If GT>0, then the environmental impact is larger than last year. If GT=0, then the environmental impact is the same with last year. If GT<0, then the environmental impact is smaller than last year.

3.2 The empirical analysis

In the paper, energy consumption is selected as the substitution variable to environmental impact I, g being GDP added value. In order to eliminate the inflation factor, data are parity modified based on the price of 1980 as base one. The falling rate of energy consumption is represented by t. All data come from China Statistical Yearbook 1991-2012. The GT in every year is shown in Figure 1, by formula (7).

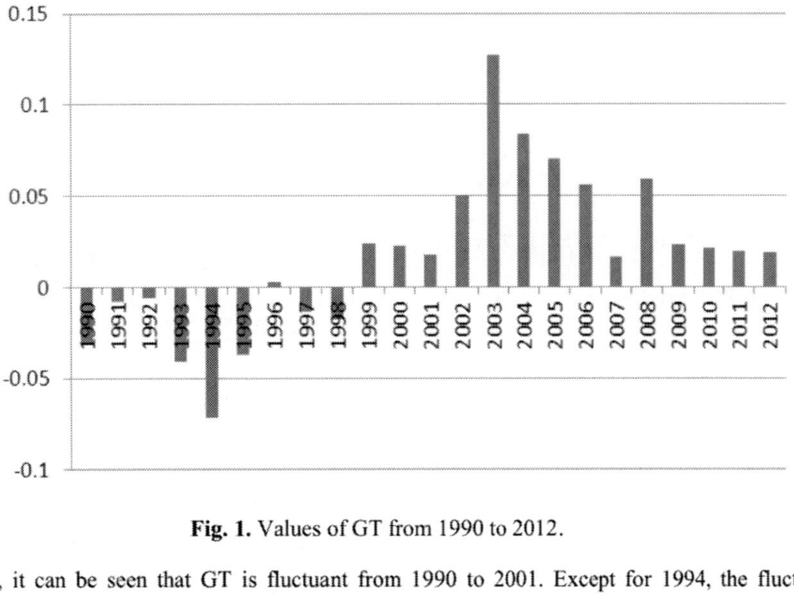

Fig. 1. Values of GT from 1990 to 2012.

From Figure 1, it can be seen that GT is fluctuant from 1990 to 2001. Except for 1994, the fluctuation values maintained in the vicinity of -0.05-0.05, with the good overall trend. The GT values were negative from 1990 to 1998, but a few years, meaning that the impact effect of economic growth on the environment was declining, and that economy and environment were coordinated development. The values turned to be positive after 1998, but relatively small, meaning that economic growth led to the deterioration of the environment, but the deterioration degree was small. Since 2001, the damage degree increased, and to the maximum 0.127 in 2003. In the following years, the GT values decreased. The minimum 0.0163 appeared in 2007. After a small rise in 2008, the values stayed relatively stable fluctuation.

In 1998, China suffered severe flooding. At the same time, the international economic situation also had great changes, including that severe financial crisis took place in Southeast Asia, that the government changed moderately tight economic policies, into expansion ones of domestic demand, to stimulate economic growth, and that 70% of billions of treasury bonds was used in infrastructure construction with the positive fiscal polies. So GT turned into positive since 1998, and be the maximum in 2003. Environment deterioration prompted government to increase the energy-saving emission reduction efforts and to increase investment in science and technology. The effect was remarkable, so GT began to decline after 2003. While the international financial crisis in 2008 led the government to increase national expansion of infrastructure construction, therefor GT rebound because of reduced protection on the environment.

Based on the growth rate in every year, the average growth rate of GDP calculated is 0.1012 from 1990-2011, energy consumption decline rate 0.0889, and total GT 0.01681. The results showed that due to the national energy-saving emission reduction policies, economic growth can basically maintain coordinated development with environment. Then on the basis, DEA model can be applied to evaluate the redundant and deficit to further analyze the measure and methods of energy-saving emission reduction.

4 DEA evaluation of economic growth and energy environment

4.1 DEA model

In 1978, Data envelopment analysis (DEA) was proposed by A.Charnes, W.W.Cooper and E.Rhodes to evaluate the relative effectiveness among departments [17]. Its essence is a kind of method to measure the relative efficiency of decision-making units, with the same types of inputs and outputs, to decide whether they are located in the production frontier of production possibility set [18].

Assuming that there are n DMU_j in DEA model, which own the m kinds of inputs X_j and s kinds of outputs Y_j, then input index vector $X_j = x_{1j}, x_{2j}, \cdots, x_{mj}{}^T > 0$ and output vector $Y_j = y_{1j}, y_{2j}, \cdots, y_{sj}{}^T > 0$ represent the input and output of DMU_j, $j = 1, 2, \cdots, n$. Then the validity of scale and technology of DMU_{j_0} can be circulated by mathematical programming as

$$\min\left[\theta - \varepsilon(\hat{e}^T s_r^- + e^T s_p^+)\right]$$

$$\left(D^T_{C^2R}\right)\begin{cases}\sum_{j=1}^{n} X_j\lambda_j + S^- = \theta X_0 \\ \sum_{j=1}^{n} Y_j\lambda_j - S^+ = Y_0 \\ \lambda_j \geq 0, j = 1, 2, \cdots, n \\ S^- \geq 0, S^+ \geq 0\end{cases} \tag{8}$$

θ in formula(8) represents relative efficiency of $DMUj_0$, also called efficiency evaluation index. $\varepsilon = 10^{-6}$. λ_j is the combination ratio of the i DMU, with the restructured effective DMU combination compared to DMU_{j_0}. S^- and S^+ are slack variables. S^- is vector of m dimensions, $S^- = (s_1^-, s_2^-, \cdots, s_r^-)^T$, while S^+ is vector of s dimensions, $S^+ = (s_1^+, s_2^+, \cdots, s_p^+)^T$.

$k=\sum_{j=1}^{n}\lambda_j$. When $k=1$, the validity of scale of DMU_{j_0} is invariant. When $k<1$, the validity of scale of DMU_{j_0} is increasing. When $k>1$, the validity of scale of DMU_{j_0} is decreasing.

If and only if $\theta=1$ and $S^-=S^+=0$, DMU_{j_0} can be called DEA effective, reaching the Pareto optimal state. On the contrary, DMU_{j_0} can be called DEA weak effective or DEA non-effective. If DMU is not DEA effective, then the original input vector and output vector can be adjusted to make it DEA effective [19].

The point after adjusted is the projection of decision-making unit in the production frontier. Projection formula can be written as

$$\begin{cases}\hat{X}_0=\theta^0X_0-S^-\\ \hat{Y}_0=Y_0+S^+\end{cases} \tag{9}$$

4.2 The empirical analysis

According to the study results from Boqiang Lin [20, 21], Shaozhou Qi [22], Qiang He [23] and Angang Hu [24], in the selection of indicators, sulfur dioxide emission, coal consumption and population index are selected as input indicators, while the added value of the second industry, the added value of the third industry and GDP as output indicators [25]. The data come from China Statistical Yearbook, and the sample period is 1990-2012. The results calculated are shown in Tab.2.

Table 2. The results of DEA model.

DMU	Energy economic efficiency value	Input			Output		
		Sulfur dioxide	Population	Coal consumption	GDP	added value of the second industry	added value of the third industry
1990	0.314	250	26209	0	0	574	2207
1991	0.331	295	26791	0	0	445	2137
1992	0.37	331	28715	0	0	204	1771
1993	0.414	267	20273	0	0	0	1840
1994	0.466	36	1355	0	465	0	2234
1995	0.507	56	0	0	1097	0	1935
1996	0.536	20	0	0	1459	0	2116
1997	0.628	0	266	0	1497	0	2122
1998	0.719	270	6015	0	812	0	1560
1999	0.768	77	6778	0	597	0	1174
2000	0.761	184	6486	0	750	0	973
2001	0.812	117	5643	0	260	0	407
2002	0.901	139	0	0	287	42	305
2003	0.862	149	0	0	1698	99	0
2004	0.841	162	0	0	2610	169	0
2005	0.845	371	0	0	4030	0	326
2006	0.876	334	0	0	1530	0	426
2007	0.894	187	0	0	3318	41	0
2008	0.901	175	0	0	1248	86	0
2009	0.923	164	0	0	1101	75	0
2010	0.945	152	0	0	1098	69	0
2011	0.951	148	0	0	1065	61	0
2012	0.968	132	0	0	1048	73	0

As can be seen from Tab.2, energy input-output efficiency increased year by year, showing that the energy technology was in improving efficiency. Since 2000, the efficiency had been maintained at above 0.9. The decline in 2003 was due to the energy waste and one-sided pursuit of economic interests. Compared with the GT in Fig.1, GT in 2003 reached the highest level since 1990, showing that the environment indeed was destroyed. Through adjustment in several years, especially for the eleventh five-year plan proposed, the efficiency value was obviously increased and reaching Pareto optimality in 2008.

From the input perspective, the utilization efficiency of coal consumption was higher than the others in China. However there was no a decreasing trend of the sulfur dioxide produced by coal. The next step for China is to do more efforts on the treatment of waste gas, by more intensive rectification, to promote the coordinated development of economy and environment. From the output perspective, with the open market, cost could be reduced the most degree, to increase profit and effectively use the input. But due to the non-coordinated development between environment and economy, it needed great efforts to clean up the environment. As for output index, the second industry was always located in output efficiency. Owing the perfection of market economy and economic crisis diffusion in 2008, output redundancy of the third industry was on the decline, to produce fully effective in recent years.

As to input of population, there were two stages of population redundancy.

One was the period from 1990 to 1994, which was a crucial transition for China from planned economy to market economy. Under the planned economy, the enthusiasm was not high, with not guaranteed production efficiency, but also great waste of personnel was caused. The reason was that creativity was severely inhibited because of guiding plan. In that condition, the employees of state-owned enterprises formed a fixed and limited mode of thinking, because production and sales were planned by state-owned enterprises just following national mandatory, with marketing of high security from national plan. In the change to the market, state-owned enterprises were changed into company with state-owned shares as the leading, instead of contract system from 1994. With the changed policies, the population redundancy immediately dropped to one fifteen of the original, with remarkable results, in 1994.

The other one was period from 1997-2001, which was Asian financial crisis. Owing that the time of the crisis was June 1997, the employment had been impacted less. The following years were far-reaching ones by financial crisis. With effective regulatory policies of Chinese government, the population redundant was back to effective in 2002.

5 Conclusion

Based on the former research results, considering Chinese situation and data obtained, IPAT model was introduced to this paper, with economic environment coefficient (GT) as indicator to measure environmental impact. The results showed that the environment impact of economic growth showed decreasing trend from 1990 to 1998 except for a few years, with coordinated development of economy and environment. After 1998, economic growth led to the deterioration of the environment, with small degree. Since 2001, the deterioration increased, reaching the maximum value in 2003. The deterioration of environment improved after a small peak in 2008. In order to further analyze the measures and methods of energy saving and emission reduction, DEA model was applied to evaluate the redundant and deficit. The DEA results showed that from the input, the coal consumption was always in Pareto efficient stage, with gradually decreased sulfur dioxide. The employment rate increased and was Pareto efficiency from 2002 to 2009. From the output, output gradually become more effective, along with the second industry in effective and the third industry with declined redundancy. Overall, energy-saving and emission reduction policies are moderate intensity, with reasonable implementation. The results achieved initial effective. Economic growth can basically maintain the stable and coordinated development with environment.

DOI: 10.1051/matecconf/201710005087

References

1. Research group of the State Council Development Research Center, Yutai Zhang, Shijin Liu, et al. Study on country account of carbon dioxide emissions: governance framework to climate change and economic growth of green [J]. Economy Research, 2011 (2).(In Chinese)
2. Grossman G M, Krueger A B. Environmental impacts of a north american free trade agreement [J]. National Bureau of Economic Research Working Paper Series, 1991, No. 3914.
3. Dinda S. Environmental Kuznets Curve hypothesis: a survey [J]. Ecological Economics, 2004, 49(4):431-455.
4. Yandle B, Bhattarai M, and Vijayaraghavan M. Environmental Kuznets curves: a review of findings, methods, and policy implications [J]. Research study, 2004, 2:1-16.
5. Garbaccio R F, Ho M S, Jorgenson D W. Why has the energy-output ratio fallen in China? [J]. The Energy Journal, 1999:63-91.
6. Hannesson R. Energy use and GDP growth, 1950–97[J]. OPEC review, 2002, 26(3):215-233.
7. Lee H, Chung R K, Koo C M. The relationship between economic growth and environmental sustainability [J]. Research in Economics, 2011, 4(65):353-364.
8. Zhan S F, Zhang X C, Ma C, et al. Dynamic modelling for ecological and economic sustainability in a rapid urbanizing region [J]. Procedia Environmental Sciences, 2012, 13(0):242-251.
9. Jinwen Zhao, Jitao Fan. Empirical Research on inherent relationship between economic growth and energy consumption [J]. Economy Research, 2007 (08): 31-42. (In Chinese)
10. Shuai Shao, ZhongYing Qi. The western region's energy development and economic growth -- Empirical analysis based on "resource curse" hypothesis [J]. Economy Research, 2008 (4): 147-160. (In Chinese)
11. Chongmei Wang. decoupling analysis on Chinese economic growth and energy consumption[J]. China Population, Resources and Environment, 2010,20 (003): 35-37. (In Chinese)
12. Yanxu Wei, Gennian Sun, Jing Li. Based on technological progress Energy consumption and economic growth in China: compare the two 30 years [J]. Resources science, 2011, 33 (7): 1338-1345. (In Chinese)
13. Peng Li. Economic growth and energy consumption in China -- An Empirical Analysis Based on dynamic panel data [J]. Economic management, 2013 (1): 1-10. (In Chinese)
14. Ehrlich P, Holden J. One-dimensional economy [J]. The Bulletin of the Atomic Scientists, 1972,28(5):16-27.
15. Ehrlich P, Holden J. Impact of population growth [J]. Science, 1971,171:1212-1217.
16. Yu Dai, Yiwen Liu. An Empirical Study on city cluster in Chang Zhu Tan of economic growth and energy consumption Based on The IPAT model [J]. Mathematics economics, 2009, v.26 (02): 65-71. (In Chinese)
17. Quanling Wei. The DEA method of evaluation the relative effectiveness: operations research new field [M]. Renmin University of China press, 1988 (In Chinese)
18. Jun Chen, Jinhua Cheng. Efficiency evaluation of Chinese non-renewable energy production: the empirical research based on data envelopment analysis method [J]. Economic review, 2007 (05): 65-71. (In Chinese)
19. Lixin Xu, Wang Jin Shi. Based on two stage of DEA China's real estate Corporation business analysis [J]. Journal of University of Electronic Science and Technology (SOCIAL SCIENCE EDITION), 2013 (1). (In Chinese)
20. Boqiang Lin, Weixian Wei, Pidong Li. China long-term coal demand: effect and policy choice [J]. economic research, 2007,2:48-58. (In Chinese)
21. Boqiang Lin. The econometric analysis on Chinese energy demand [J]. Statistical research, 2001 (10): 34-39. (In Chinese)
22. Shaozhou Qi, Bo Yun, Kai Li. Research on convergence and mechanism analysis on Chinese economic growth and energy consumption [J]. Economy research 2009 (04): 56-64. (In Chinese)

23. Qiang He, Guangming Lv. Analysis on the effect on ecological environmental based on IPAT model -- a case study of Beijing [J]. Journal of Central University of Finance and Economics based on, 2008 (12): 83-88. (In Chinese)
24. Angang Hu, Jinghai,Zheng Yuning Gao et al. Ranked the provincial technical efficiency considering environmental factors (1999 - 2005) [J]. Economics (Ji Kan), 2008 (3): 933-960. (In Chinese)
25. Dingtao Zhao, Huiping Dong. Analysis on technical efficiency and factors of regional tourism industry-- empirical research Based on two stage DEA-PCR model [J]. Journal of University of Electronic Science and Technology (SOCIAL SCIENCE EDITION), 2012 (6): 51-56. (In Chinese)

AUTHOR INDEX

AUTHOR INDEX

AUTHOR INDEX

AUTHOR INDEX

AUTHOR INDEX

AUTHOR INDEX